DOE/TIC -27601
(DE84005177)

Atmospheric Science and Power Production

Darryl Randerson, Editor

Weather Service Nuclear Support Office
National Oceanic and Atmospheric Administration
United States Department of Commerce

Prepared for
United States Department of Energy
Office of Energy Research
Office of Health and Environmental Research

To John with best regards,
Darryl Randerson
22 october 1984

1984

Published by
Technical Information Center
Office of Scientific and Technical Information
United States Department of Energy

ABOUT THE TECHNICAL INFORMATION CENTER

The Office of Scientific and Technical Information, Technical Information Center in Oak Ridge, Tennessee, has been the national center for scientific and technical information for the Department of Energy (DOE) and its predecessor agencies since 1946. In developing and managing DOE's technical information program, the Center places under bibliographic control not only DOE-originated information but also worldwide literature on scientific and technical advances in the energy field and announces the source and availability of this information. Whereas the literature of science is emphasized, coverage is extended to DOE programmatic, socioeconomic, environmental, legislative/regulatory, energy analysis, and policy-related areas. To accomplish this mission, the Center builds and maintains computerized energy-information data bases and disseminates this information via computerized retrieval systems and announcement publications such as abstracting journals, bibliographies, and update journals. Direct access to the Center's most comprehensive data base, the Energy Data Base, is available to the public through commercial on-line bibliographic retrieval systems. The Energy Data Base and many of the Center's energy-related data bases are available to DOE offices and contractors and to other government agencies via DOE/RECON, the Department's on-line information retrieval system. The Center has developed and maintains systems to record and communicate energy-related research-in-progress information, to maintain a register of DOE public communications publications, to track research report deliverables from DOE contractors, and to test and make available DOE-funded computer software programs with scientific and management applications. The Center also maintains a full-scale publishing capability to serve special publication needs of the Department. To effectively manage DOE's technical information resources, the Center's program is one of continual development and evaluation of new information products, systems, and technologies.

ABOUT BOOKS PUBLISHED BY
THE TECHNICAL INFORMATION CENTER

Books published by the Center include undergraduate, graduate, and workshop textbooks; handbooks and manuals; reference books; monographs; specialized reviews; and DOE-sponsored symposium, workshop, and short-course proceedings.

Library of Congress Cataloging in Publication Data
Main entry under title:

Atmospheric science and power production.

"DOE/TIC-27601."
Includes bibliographical references and index.
1. Air quality—Mathematical models. 2. Electric power-plants—Environmental aspects. 3. Atmospheric chemistry. 4. Atmospheric diffusion. I. Randerson, Darryl. II. United States. Dept. of Energy. Technical Information Center.
TD883.1.A85 1984 628.5′32 82-18197
ISBN 0-87079-126-5

Available as DE84005177 (DOE/TIC-27601) for $29.50 from

National Technical Information Service
U. S. Department of Commerce
Springfield, Virginia 22161

Printed in the United States of America

July 1984

FOREWORD

This book is the third volume in a series spanning nearly three decades. The first, *Meteorology and Atomic Energy,* was prepared for the U. S. Atomic Energy Commission (AEC) by the U. S. Department of Commerce (U. S. Weather Bureau) and was published in 1955. It soon became the primary reference work for preparing estimates of the atmospheric concentrations resulting from postulated routine and accidental releases from AEC site operations and the new atomic energy industry. With considerable prescience the Foreword in that first volume predicted that the information developed by the meteorology programs of the AEC would find additional use in application to nonnuclear issues involving not only energy but also pollution from a variety of other commercial and industrial activities. And so it did.

The next volume in the series, *Meteorology and Atomic Energy—1968,* was again prepared for the AEC by the U. S. Department of Commerce (Air Resources Laboratory). This volume also proved to be very useful in AEC site operations, in the nuclear energy industry, and in a variety of other air pollution-producing situations. As in the case of the 1955 version, the meteorological research program of the AEC furnished much of the substance. The utility of *Meteorology and Atomic Energy—1968* is amply attested to by citation to it throughout the air pollution literature and by its status as among the best sellers of the DOE Technical Information Center's publications.

Atmospheric Science and Power Production was prepared for the U. S. Department of Energy by the U. S. Department of Commerce (U. S. National Weather Service) continuing this long-time relationship. The trend established in the first two volumes is continued; that is, the information in the volume, although tailored to the interests of the energy community, will be invaluable in many other contexts involving the health of man and his environment. This is particularly true of the first three sections dealing with the fundamentals of atmospheric transport, dispersion, chemistry, and removal processes.

We are pleased to be able to present this work to the public and wish to compliment those involved in its preparation for their dedication to this noteworthy undertaking.

Alvin W. Trivelpiece
Director, Office of Energy Research
U. S. Department of Energy

PREFACE

Atmospheric Science and Power Production is the third in a series of scientific publications sponsored by the U. S. Atomic Energy Commission and the two later organizations, the U. S. Energy Research and Development Administration, and the U. S. Department of Energy. The first book, *Meteorology and Atomic Energy,* was published in 1955; the second, in 1968. The present volume is designed to update and to expand upon many of the important concepts presented previously. However, the present edition draws heavily on recent contributions made by atmospheric science to the analysis of air quality and on results originating from research conducted and completed in the 1970s. Special emphasis is placed on how atmospheric science can contribute to solving problems relating to the fate of combustion products released into the atmosphere.

Since the publication of *Meteorology and Atomic Energy—1968,* a new direction and emphasis have been placed on atmospheric science and its role in assessing local and regional air quality. In 1969, Congress mandated a specific direction by enacting the National Environmental Policy Act, which declared a national environmental policy. New emphasis was defined in the Amendments to the Clean Air Act of 1963. Of particular importance to those working in air-quality analysis, control, or regulation were the Amendments of 1970 and 1977. For example, the Amendments of 1977 established (1) air-quality goals in the form of national ambient air-quality standards, (2) emissions standards for new and existing stationary sources, and (3) the concept of the prevention of significant deterioration of air quality.

When we attempt to relate air-quality goals to the specific limitations on emissions, it becomes abundantly clear that a reliable method of estimating temporal and spatial changes in atmospheric processes is needed. A useful mechanism for simulating these processes is through mathematical models based on sound physical theories. Such air-quality models can be designed to account for the transport, dispersion, deposition, and transformation of effluents released into the atmosphere.

The framework of this book is built around the concept of air-quality modeling. Fundamentals are addressed first to equip the reader with basic background information and to focus on available meteorological instrumentation and to emphasize the importance of data management procedures. Atmospheric physics and field experiments are described in detail to provide an overview of atmospheric boundary layer processes, of how air flows around obstacles, and of the mechanism of plume rise. Atmospheric chemistry and removal processes are also detailed to provide fundamental knowledge on how gases and particulate matter can be transformed while in the atmosphere and how they can be removed from the atmosphere. The book closes with a review of how air-quality models are being applied to solve a wide variety of problems.

Awareness for a needed update to *Meteorology and Atomic Energy—1968,* was anticipated by Dave Slade and Harry Moses in 1973. Through their conceptual inspiration, the original framework for the book was laid out, potential authors were contacted, and some preliminary writing was begun. In November 1975, I became editor of this volume. After reviewing the scope of the book, I added new chapters (such as Chapters 18 through 20) and visited all the authors. Most of the authors began conscientious writing in 1977. However, each author had to accommodate prior commitments before he could allot the required time to this work. Some authors started writing at an early date, while others were unable to meet their commitment until later. Therefore a wide variety of completion dates are associated with the 20 chapters. Chapter 1 was prepared last so that it would provide a general overview of the contents of the book. Essentially all chapters were completed by late 1980, and most of the galley proofs were reviewed in late 1981.

During the final preparation of Chapter 6, the author, Dr. S. K. Kao, died. Dr. Kao was an internationally known and respected scientist who drew together the concepts of fluid dynamics and air-quality modeling. We are fortunate to have his last major work available in this document. Chapter 6 reflects the depth to which Dr. Kao viewed problems pertaining to atmospheric science. Unfortunately, Dr. Kao did not live to review the galley proofs of his chapter. I reviewed the galleys and believe the chapter to be in the form he would have desired.

As was true with its predecessors, this book can be used as a textbook, a handbook, and a guide. University professors and students can use this volume as a textbook. Professional workers in the various disciplines related to power production and to air-quality analysis will find this volume useful as a handbook. The reader requiring a general knowledge of the factors relating to the power production industry can use the book as an introductory guide to the concepts and terminology used by atmospheric scientists, by health physicists, and by those responsible for regulating air quality.

Darryl Randerson, Editor

Las Vegas, Nevada
September 1981

ACKNOWLEDGMENTS

Preparation of this book was supported by the Office of Health and Environmental Research, U. S. Department of Energy. D. H. Slade and H. Moses laid out the original framework for the book and provided the initial thrust from concept to development.

Many individuals contributed directly or indirectly to this book. The authors include K. H. Abel, S. Barr, G. A. Briggs, W. E. Clements, H. L. Crutcher, R. R. Draxler, P. H. Gudiksen, S. R. Hanna, D. M. Hardy, J. W. Healy, R. P. Hosker, Jr., S. K. Kao, J. B. Knox, F. M. Luther, C. J. Mason, H. Moses, M. M. Pendergast, L. A. Rancitelli, D. Randerson, N. I. Sax, G. A. Sehmel, W. G. N. Slinn, E. G. Walther, J. J. Walton, J. A. Young, R. L. Zalkan, and K. F. Zeller. These authors not only lavished much time on the preparation of their chapters but also served as reviewers for other chapters.

Reviewers of various chapters and sections include D. Ballentine, B. Chapman, F. E. Courtney, T. V. Crawford, W. Culkowski, M. H. Dickerson, A. N. Dingle, H. W. Ellsaesser, F. Gifford, J. Halitsky, B. B. Hicks, A. H. Huber, J. C. Hunt, B. A. Hutchison, R. B. Jurgens, K. M. Kothari, P. Long, M. C. McCracken, W. C. Malm, W. E. Marlatt, R. N. Meroney, P. Nickola, V. E. Quinn, K. S. Rao, R. C. Rhodes, R. Thompson, B. Turner, V. D. Urban, I. Van der Hoven, A. Venkatram, J. Weil, and D. J. Wilson.

The Technical Information Center, Office of Scientific and Technical Information, U. S. Department of Energy, was responsible for the editing, composition, proofreading, artwork, design, and layout of the book. Editorial support was provided by R. N. Bean, M. C. Fox, A. M. Goulden, J. L. Littlepage, J. Roberts, P. B. Rose, and J. S. Smith.

Darryl Randerson, Editor

CONTENTS

POWER PRODUCTION AND THE ATMOSPHERE

1

Darryl Randerson, Weather Service Nuclear Support Office

1-1 INTRODUCTION

In 1978 there were approximately 920 coal-fired power plants in the United States. Of this total, there were roughly 150 plants with generating capacities of 1000 MW or more. These power plants, as well as the smaller ones, release voluminous amounts of effluents to the atmosphere. For example, each day a typical 1000-MW coal-fired power plant emits approximately 10^7 kg of CO_2, 10^5 kg of SO_2, and 10^6 kg of fly ash.† Not included in these figures are the thousands of kilograms of CO and NO_x generated daily. Additional by-products include toxic metals and the 10^6 kg of lime sludge (e.g., calcium sulfate or gypsum) generated while scrubbing the stack gases so that 90% of the SO_2 is removed after combustion of high-sulfur coal. Beckmann (1978a) notes that a 1000-MW coal-fired power plant produces 5×10^4 kg of solid waste per hour. Radioactive material is also released from this power plant because coal contains small quantities of ^{238}U, ^{235}U, ^{232}Th, and their radioactive daughter products (Goldstein, Sun, and Gonzales, 1971). On the basis of the calculations of McBride et al. (1978), approximately 65 g of uranium and 125 g of thorium are released daily from a 1000-MW coal-fired power plant. By contrast, Cohen (1977) notes that the total volume of radioactive wastes produced annually by a 1000-MW nuclear reactor is approximately 2 m^3. The spent fuel comprising this volume is primarily uranium so that 2×10^4 kg of wastes are produced annually or roughly 2 kg/h. Oil-fired power plants also release effluents into the atmosphere. For example, a 1000-MW oil-fired plant burning 2.5% sulfur oil can release 10^5 kg of SO_2 per day and 10^3 kg of particulate matter per day plus trace amounts of radioactive species (Lave and Freeburg, 1973; and Enger and Högström, 1979).

To keep a 1000-MW coal-fired plant operating, approximately 8×10^6 kg of coal are consumed daily, enough to fill 100 railcars. A 1000-MW nuclear plant can be refueled once a year by a few truckloads of fuel rods. A 1000-MW oil-fired plant can consume 5×10^6 kg of oil per day.

The thrust of the preceding paragraphs is intended only to direct attention to the fact that any type of power generation can create potentially hazardous by-products that are, in part, released into the atmosphere. Even the development of an alternative energy resource (see Chap. 18) such as a solar power plant will result in additional effluents being released to the atmosphere. Beckmann (1978b) notes that a 1000-MW solar power plant would require 2×10^9 kg of concrete, 7×10^7 kg of glass, 5×10^7 kg of steel, 3×10^7 kg of aluminum, and 7×10^6 kg of silver, plus other materials. Energy is required to produce these materials. For example, the energy content of coal is approximately 7.2×10^6 g-cal/kg (see Table 1.1), and it takes 10^8 kg of coal to produce 5×10^7 kg of steel. Consequently nearly 10^9 kW·h of energy is required to produce 5×10^7 kg of steel. Furthermore, a high-energy solar power plant is, in the parlance of the day, land intensive. Beckmann (1977) estimates that approximately 130 km^2 of land would be required for a 1000-MW solar facility. Solar energy is also a restricted resource because the sun will not provide us with more than 1 kW/m^2 at sea level (Beckmann, 1977; Randerson, 1981).

TABLE 1.1 Approximate Energy Content of Some Commonly Used Fuels†

Fuels	10^6 g-cal/kg	kW·h/kg
Wood	2.8	3.3
Denatured alcohol	6.4	7.5
Coal	7.2	8.5
Fuel oil	10.8	12.6
Natural gas	11.0	12.9
Gasoline	11.5	13.4
^{235}U fission	2.0×10^{13}	2.2×10^7

†Available only for energy conversion processes that are 100% efficient.

We need not highlight the constraints of a few energy technologies. All energy production techniques have some proven or potential environmental, health, and safety problems (Chap. 18 and Department of Energy, 1979) that will require careful management by the producer, whether private or public. Part of the management responsibility is

Randerson: Weather Service Nuclear Support Office, National Weather Service, National Oceanic and Atmospheric Administration, Las Vegas, Nevada.

†Fly ash produced and collected at coal-fired power plants was considered the sixth most abundant mineral in the United States with 6×10^{10} kg being produced in 1977 (*Chemical and Engineering News*, 1978). See Chap. 10 for additional information.

to assess and to assure the air quality in the vicinity of energy production facilities. More recently, the degradation of air quality at distances of hundreds to thousands of kilometers from power plants has become an important issue with regard to the acid precipitation phenomena and visibility impairment. In fact, with the Clean Air Act Amendments of 1977, Congress gave the Environmental Protection Agency (EPA), the Federal Land Managers, and the states the responsibility to protect visibility as well as the air quality of wilderness areas and of the national parks. Details of the Clean Air Act and its amendments are described in Chap. 19.

1-2 WORK, ENERGY, AND POWER

The word "work" is commonly used to identify any form of activity requiring the exertion of muscular or mental effort. However, in physics, the word has a very precise technical meaning—it is used to represent the work done on a body by some external agent that exerts a force on the body and thereby produces motion. A differential of work is expressed as $dW = F \cos \theta \, dx$, where the increment of work done (dW) by force, F, is defined as the product of the displacement increment, dx, and the component of the force ($F \cos \theta$) in the direction, x, of the displacement. Angle θ is that formed by vectors representing the direction of the force and the direction of motion. If the applied force and the resulting motion are parallel, the work done equals the force times the distance of the displacement. Work is a scalar quantity with dimensions ML^2/T^2. In the cgs system, the unit of work is the dyne-centimeter. A dyne is defined as the force acting on a 1-gram mass that produces an acceleration of 1 cm/s^2 or 1 dyne = 1 g-cm/s^2. In SI units, work is measured in joules. For basic definitions see Table 1.2, and for an alphabetical listing of conversion factors for energy and power units see Tables 1.3 and 1.4, respectively.

Energy is considered to be the capacity to do work. Energy can appear in the form of motion, or the potential for motion, or in a form related to heat. For a body of

TABLE 1.2 Definitions of Selected Energy and Power Units

Units	Definition
Energy	
Btu (British Thermal Unit)	The quantity of heat required to raise 1 lb of pure water 1° F from 63° F to 64° F
g-cal (gram-calorie)	The quantity of heat required to raise 1 g of pure water 1°C from 14.5°C to 15.5°C
ft-lb (foot-pound)	The work done when a force of a 1-lb weight is applied over a distance of 1 ft
J (joule)	The work done when a force of 1 newton moves a distance of 1 m. A newton is equal to 1 kg·m/s^2
Wh (watt-hour)	The energy supplied by 1 W of power in 1 h
Power	
W (watt)	Rate of work of 1 J/s
hp (horsepower)	Rate of work when a force of 550 lb is raised a distance of 1 ft in a time of 1 s

TABLE 1.3 Conversion Factors for Selected Units of Energy

Multiply	By	To obtain
Btu	251.98	g-cal
	777.97	ft-lb
	1054.80	J
	2.93×10^{-4}	kW·h
	0.252	kg-cal
	1.055×10^{10}	ergs
ft-lb	1.285×10^{-3}	Btu
	3.240×10^{-1}	g-cal
	1.356	J
	3.766×10^{-7}	kW·h
	3.240×10^{-4}	kg-cal
	1.356×10^{7}	ergs
g-cal	3.968×10^{-3}	Btu
	3.087	ft-lb
	4.186	J
	1.163×10^{-6}	kW·h
	10^{-3}	kg-cal
	4.186×10^{7}	ergs
J	1.020×10^{-1}	Btu
	7.376×10^{-1}	ft-lb
	2.389×10^{-1}	g-cal
	2.389×10^{-4}	kg-cal
	2.778×10^{-7}	kW·h
	10^{7}	ergs
kWh	3.413×10^{3}	Btu
	2.655×10^{6}	ft-lb
	8.60×10^{5}	g-cal
	3.60×10^{6}	J
	8.60×10^{2}	kg-cal
	3.60×10^{13}	ergs

TABLE 1.4 Conversion Factors for Selected Units of Power and Energy Flux

Multiply	By	To obtain
Btu/min	1.758×10^{-2}	kW
	12.97	ft-lb/s
	1.758×10^{8}	erg/s
	4.2	g-cal/s
	17.58	W (J/s)
kW	56.896	Btu/min
	4.425×10^{4}	ft-lb/min
	2.389×10^{2}	g-cal/s
	10^{11}	erg/s
	10^{3}	W (J/s)
hp	42.42	Btu/min
	0.745	kW
	3.3×10^{4}	ft-lb/min
	178.13	g-cal/s
	7.457×10^{9}	erg/s
kW/m^2	317.27	Btu/hr/ft^2

mass, m, and speed, V, its kinetic energy is $\frac{1}{2}$ mV^2, whereas its energy of position in a gravitational field is called potential energy (mgz), where g is the acceleration due to gravity, and z is the height above a horizontal reference level. The internal energy of composition is called chemical energy and the internal energy of motion is heat. Some other forms of energy are radiant, electrical, and nuclear energy. Work and energy have the same dimensions $[ML^2/T^2]$.

In the preceding paragraphs, one important factor has been omitted—the element of time. In many cases, it is necessary to account for the rate at which work is done so that the concept of power is used. Power, Φ, is defined as the work, dW, done in time period dt or $\Phi = dW/dt$. Therefore the dimensions of power are ML^2/T^3. The watt is the international unit for power. Note that if we integrate power with respect to time, the result is work (or energy). The work done by the local electric utility in our homes is equal to the watts used multiplied by the amount of time required to consume the wattage. We thereby end up with energy units such as kilowatt-hours (kW•h). A kW•h is the work done in 1 h by a process working at the constant rate of 1 kW. Such a unit creates a common misconception that there is something inherently electrical about a watt. It is true that electrical power is usually expressed in watts, but the power consumption of a common light bulb could just as well be expressed in horsepower or calories per second, or an automobile engine could be rated in kilowatts. Some common conversion factors for power units are listed in Table 1.4.

The energy content of different fuels varies over a wide range of values. As can be seen in Table 1.1, the available energy content ranges from just over 1 kW•h/kg of chemical energy for wood to over 10^7 kW•h/kg of nuclear energy for ^{235}U. These amounts of energy are only available if the energy-generating facility operates at 100% efficiency. For example, the most efficient coal-fired power plants and nonbreeder reactors operate at 30 to 40% efficiency.

Amounts of energy involved with a variety of other processes are tabulated in Table 1.5. The listed data range from the extremely small quantities of energy carried by a single photon of visible light to the daily use of energy by all human beings. The table also permits you to compare the energy values of different processes.

1-3 ENERGY PRODUCTION AND CONSUMPTION†

In 1977 the total consumption of energy on a global scale was approximately 270 Q. One Q, or quad, is the equivalent of 10^{15} Btu or 2.93×10^{11} kW•h of energy. Of this energy consumption, countries with centrally planned economies used 32%; the United States, 29%; western Europe, 18%; Japan, 5%; and the rest of the world, 16%. Petroleum accounted for 45% of the global energy; coal, 32%; natural gas, 20%; hydroelectric and geothermal sources, 2%; and nuclear power plants, 1%. For the past 20 yr, the global demand for energy has been increasing at a rate of roughly 8 to 10 Q/yr. As with the global demand, energy consumption in the United States has been increasing steadily (Fig. 1.1). Indeed, Fig. 1.1 shows that energy consumption has been increasing steadily since the depression years, and it has doubled in the past 25 yr. In addition, Fig. 1.1 reveals that, in the United States, energy consumption exceeded local production after 1952.

Of the 77 Q consumed in the United States in 1977, approximately 45% was generated by the combustion of petroleum products; 32% by coal; 19% by natural gas; and the remaining 4% by nuclear, hydropower, and other sources. In terms of the distribution of the 77 Q, Fig. 1.2 illustrates that 30% was used by electric utilities, 26% by transportation, 25% by industry, and 19% by residential and commercial users. Each of these processes acts as a source for gaseous and particulate material that is released to the atmosphere.

In 1979 the total energy consumed in the United States increased only slightly over the 1977 and 1978 levels. This slow increase is believed to result from voluntary conservation efforts associated with rapidly rising fuel costs, a slowdown in economic growth, and unrest in the world. These factors helped keep gasoline consumption in the United States at approximately 4×10^5 barrels/day, roughly 5% below 1978 levels. In addition, the consumption of other petroleum products (home heating oil, diesel fuel, jet fuel, and residual oil) remained at 11.4×10^6 barrels/day, near the 1978 levels. Annual natural gas consumption in 1979 remained about the same as in recent years at 19.5×10^{12} ft^3. However, coal consumption grew by more than 50×10^6 tons. This increase resulted largely from increased coal use by electric utilities. Approximately 2×10^6 tons are used daily.

There is little doubt that there are abundant coal resources in the United States relative to petroleum resources. However, because of government intervention (Guccione, 1977; Reynolds, 1977) and economic factors (Mead, 1977), petroleum products are used to generate 45% of our energy. Consequently the United States energy system relies most on the least plentiful of its domestic resources and least on the most plentiful. But, setting aside economics and politics, there will continue to be a high demand for energy worldwide. Private and public power companies will continue to try to meet this demand. However, many problems must be faced by the power industry. Although Nagel (1978) notes that the primary problem facing utility management is uncertainty, general problem areas can be categorized as (1) power needs, (2) facility siting, (3) regulatory delays, (4) environmental constraints, (5) primary fuel supply, and (6) financing. Atmospheric science enters in most of these problem areas. In the United States the principal information requirement comes from the variety of air pollution analyses, forecasts, and scenarios required by various provisions of the National

†Data used in this section were obtained from the *Statistical Abstracts of the United States* (U. S. Department of Commerce, 1979).

TABLE 1.5 Other Approximate Energy Values

	g-cal	kW·h
Energy in one photon of visible light	10^{-21}	10^{-27}
Gravitational energy on 1 kg of mass 1 km above sea level	2.3	2.7×10^{-6}
Energy required to melt 1.0 kg of ice at 0°C	8.0×10^{1}	9.3×10^{-5}
Energy required to evaporate 1.0 kg of H_2O	6.0×10^{2}	7.0×10^{-4}
Maximum solar radiation on 1 m² in 1 h at sea level†	8.6×10^{5}	1.0
Solar energy on 1 m² normal to rays in 1 h at top of atmosphere	1.2×10^{6}	1.4
Maximum observed daily solar radiation/m² at Las Vegas, Nev., in June†	7.7×10^{6}	9.1
Chemical energy released in burning 1 kg of wood	2.8×10^{6}	3.3
Chemical energy released in burning 1 kg of coal	7.2×10^{6}	8.5
Chemical energy released in burning 1 m³ of natural gas	8.9×10^{6}	10.3
Chemical energy released in burning 1 kg of fuel oil	1.0×10^{7}	11.6
Nuclear energy released by the fission of 1 kg of ^{235}U	2.0×10^{13}	2.2×10^{7}
U. S. daily energy consumption (1977)	5.3×10^{16}	6.2×10^{10}
World daily energy consumption (1977)	1.8×10^{17}	2.2×10^{11}

†For a horizontal surface. Values 10% greater may be observed at higher elevations in desert regions.

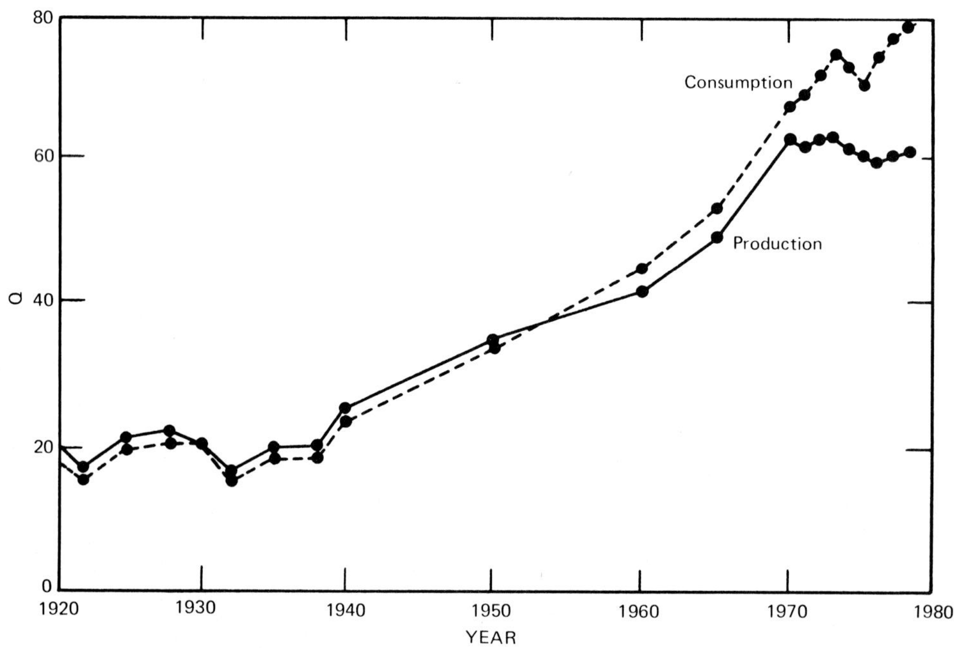

Fig. 1.1 Total energy consumption vs. production in the United States since 1920.

Environmental Policy Act, the Clean Air Act and its amendments, and by various environmental laws in each state. In addition, information pertaining to potential effluent concentration patterns is needed for facility siting and design and for many other industrial and social planning† purposes.

1-4 DATA

Accurate documentation of meteorological and air-quality conditions is required to provide the researcher and the practitioner with information needed to understand, to model, and to regulate air quality. The success of any model development and evaluation program depends heavily upon the quality of the data collected. In turn, the quality of the monitoring data depends upon the quality assurance activities specified for the project. According to Crutcher (1970 and Chap. 4), quality is a peculiar and essential character that can be defined and quantified. Quality control is an essential and important part of any project that collects and uses data (information). A procedure for maintaining the quality of a program consists of (1) quality control, (2) quality assurance, (3) design and reliability, and (4) management. The details of these concepts are described in Chap. 4. Additional information on quality assurance and data management has been given by the Environmental Protection Agency (1976a), by Jurgens and Rhodes (1976), and by others. On a related topic, Hunter (1980) has addressed the concept of quality control at the national level.

†For view opposing social planning, see Hayek (1977).

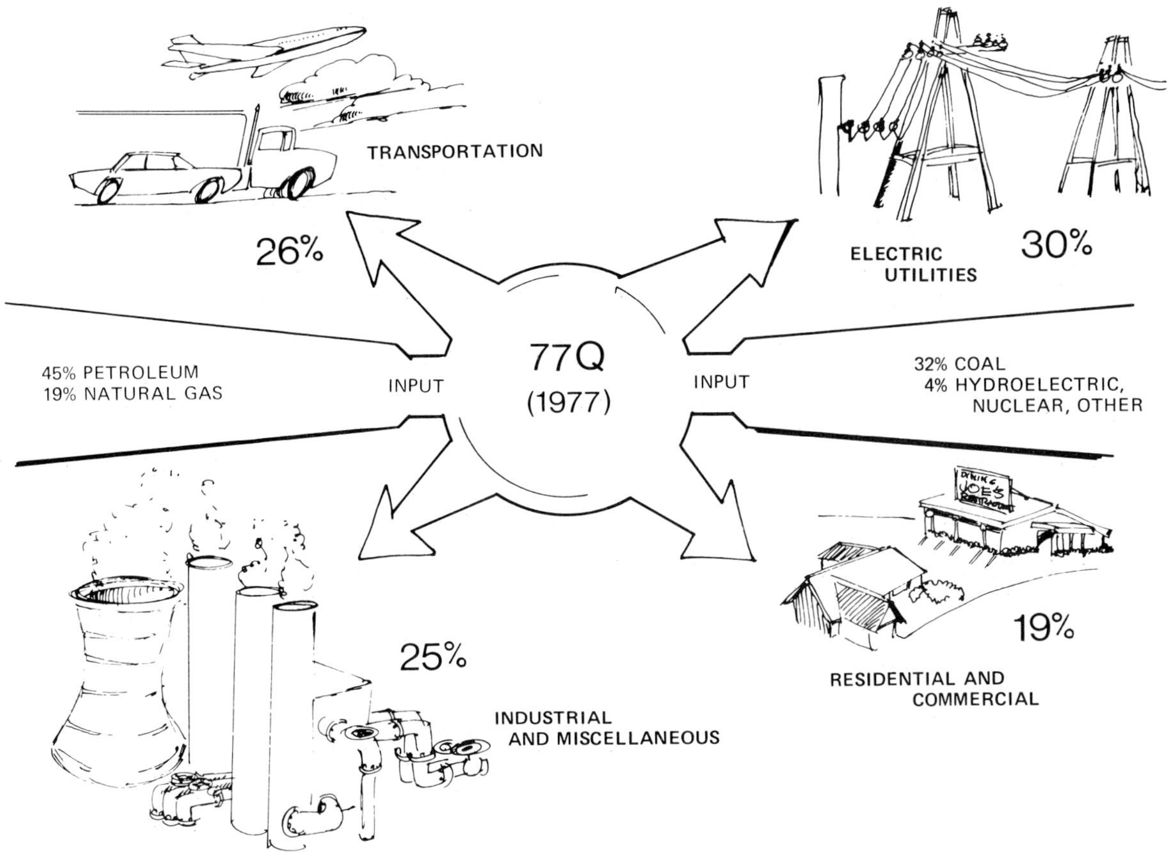

Fig. 1.2 Energy uses in the United States for 1977.

Atmospheric processes are stochastic and cover a wide spectrum of wavelengths. Instruments can detect these processes but only with a certain precision (see Chap. 3). In general, measurements of atmospheric processes represent averages in time. Consequently an important problem in collecting data is to select the instrument and/or instruments that will resolve the phenomena to be simulated and studied. In air-quality modeling studies, the meteorological data problem usually focuses on two important aspects: (1) meteorological instrumentation and data collection requirements needed to characterize the airborne dispersion of effluents and (2) representativeness of the meteorological data in time and space. Traditional meteorological instruments continue to meet many needs, but more use is being made of remote sensing techniques. Standard meteorological instruments and remote sensing instruments are described in detail in Chap. 3. Additional information on meteorological instruments was brought together during an EPA-sponsored workshop held in early 1980 (Strimaitis, Hoffnagle, and Bass, 1980, and Hoffnagle et al., 1981). One of the important products of this workshop was a list of recommendations on instrument accuracy, resolution, sampling time, and response time. These recommendations are summarized in Table 1.6 (for additional information see Chap. 3.). In Table 1.6, wind velocity is separated into its horizontal (u, v) and vertical components (w) and its horizontal (θ) and vertical (ϕ) directional components. The vertical temperature gradient from 2 to 10 m is given by ΔT, and T is the ambient air temperature. The depth of the convectively mixed boundary layer is represented by z_i and that of the mechanically mixed layer by z_m (see Chap. 5).

Representativeness is a term that is used frequently by meteorologists to describe a desirable attribute of data collected. However, no term in meteorology is more poorly defined. For example, the term does not appear in the *Glossary of Meteorology* (1959) published by the American Meteorological Society. An early attempt to define representativeness was made by Petterssen (1940, pp. 2 and 3) who suggested that it was ". . . the same as saying that the observations should not be influenced locally." Petterssen and Middleton (1963) warned the neophyte meteorologists to beware of unrepresentative data. Representativeness was alluded to and discussed in "Meteorological Observations and Instrumentation" (Teweles and Giraytys, 1970) in the contributions by Teweles (p. 4), Noble (p. 16), and Bellamy (p. 74). Noble drew attention to "the degree to which small-scale fluctuations appear under the alias of large-scale" fluctuations. He also noted that for surface observations, "consideration should be given to reporting an additional wind which represents a time mean corresponding to the needs of a network with the given horizontal spacing." Noble emphasized that the time-aliasing problem can be eliminated by selecting sensors having time constants compatible with the reading interval. In this way, individual readings would represent averages over the reading interval rather than point values within the continuum of values.

TABLE 1.6 Response Characteristics Required of Meteorological Instruments Used to Measure Atmospheric Characteristics Needed to Characterize Dispersion from Point Sources [After Hoffnagle et al., On-Site Meteorological Instrumentation Requirements to Characterize Diffusion from Point Sources: A Workshop, 15–17 January 1980, Raleigh, N. C., *Bull. Am. Meteorol. Soc.*, 62: 255-261 (1981)]

Measurement†	Height (AGL)	Accuracy†	Resolution	Averaging time	Response time
u, v, w	10 m	0.2 m/s		1–5 s	Distance constant ⩽5 m
θ	10 m	5° A		1–5 s	Distance constant ⩽5 m
		1° R			Damping ratio ⩾0.4
ϕ	10 m	5° A		1–5 s	Distance constant ⩽5 m
		1° R			Damping ratio ⩾0.4
ΔT	2–10 m	0.1°C	0.02°C	30 s	1 min
z_i	50–200 m	50 m	10 m		10 s
z_m	0–200 m	10–50 m	10 m		10 s
T	2 m	0.5°C	0.1°C	30 s	1 min

†Where accuracies are within the given values and where A is the absolute accuracy and R is relative accuracy, u and v are the horizontal wind components, w is the vertical wind component, θ is the horizontal wind direction, ϕ is the elevation angle of the wind, ΔT is a temperature difference, z_i is the depth of the convective boundary layer, z_m is the depth of the mechanically mixed boundary layer, and T is temperature. Regarding the use of ΔT, Pendergast and Crawford (1974) and others have noted the problems in using ΔT over depth Δz to determine atmospheric stability.

Instead of integrating a number of observations to produce a datum representative of a large area, Zemel and Lyons (1976) note there is a need to specify the area (and volume) over which a single observation may be usefully applied.

Analysis of macroscale data does not require observations that are a detailed sensing of a very small space and time domain. Petterssen and Middleton (1963) suggest that a more useful datum would be one that "has been so modified as to eliminate or reduce to a minimum its contamination with additive disturbing noise as well as bona fide atmospheric fluctuations of a smaller scale than observable with a given measurement network." To a meteorologist analyzing macroscale data, such a datum would constitute a representative observation.

In considering representativeness from the aspect of the scales of atmospheric processes, a degree of commonality can be reached because of the nature of the science of meteorology. Meteorology is fundamentally an observational science. Observational data have in all cases a certain statistical character. Isolated, irrelevant effects cannot be completely eliminated, nor is it desirable to do so, because the isolated, rare effect is often the one of paramount concern. The rare event is frequently the one society is not prepared to deal with and for which society requires forewarning.

For air-quality management activities, there are stringent requirements for representative weather data. A major problem in obtaining representative data is that traditionally weather data have been sampled regularly in time but erratically in space. Spatial irregularity has always been burdensome to the synoptic meteorologist and probably even more so to the air pollution meteorologist. The detailed weather information necessary to evaluate properly the environmental impact of a proposed source is almost never available at the time critical decisions are made as to the location, size, and operating characteristics of the source. A frantic search for representative weather data usually ensues.

The need for representative meteorological data is uniformly recognized. The measures to characterize the representativeness of a data set are not widely agreed upon. Neither are procedures to establish values that characterize adequate representativeness. As Hayes (1979) discusses with respect to choosing performance standards, it seems appropriate to resort to the pragmatic–historic approach to arrive at the needed decision. Objective measures are applied and the results documented wherever possible. Redundancy of data should not be avoided.

1-5 ATMOSPHERIC PROCESSES

As effluents are released into the atmosphere, they are immediately acted upon by atmospheric processes. Figure 1.3 represents a conceptual view of some of these processes. If the emitted material is warmer than the ambient air, the effluent will rise buoyantly to an equilibrium level. At this level, the effluent tends to elongate and acquire a plume-like shape in the downwind direction. Wind can transport effluent material great distances. Turbulence mixes the air and thereby disperses the effluent vertically and horizontally. Precipitation scavenging and dry deposition (including impaction and soil and foliar absorption) eventually remove most effluent material from the atmosphere. In the presence of solar radiation, water vapor, and other atmospheric gases, a myriad of potential chemical reactions are possible, transforming the effluent into other substances. Additional material can be added to the atmosphere by the process of resuspension. If the source happens to be a nuclear facility, the radioactive nature of

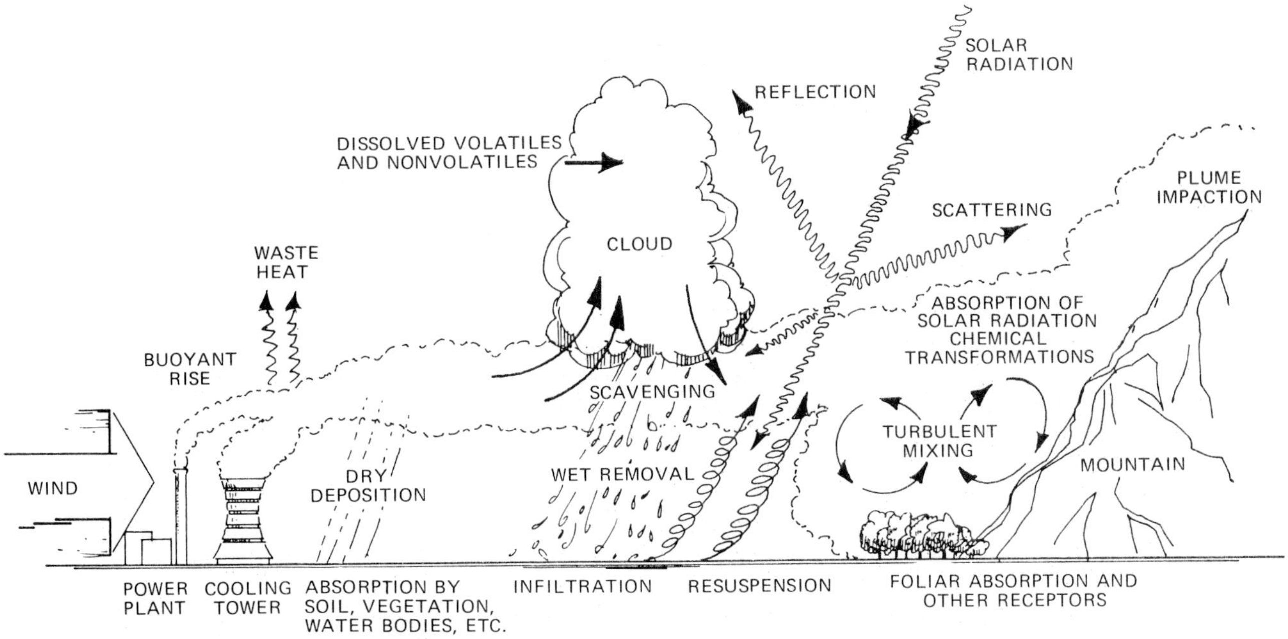

Fig. 1.3 Conceptual view of atmospheric processes that can act upon effluents released into the atmosphere.

the material must also be considered. Data are needed to document all these processes and a variety of instruments are required to collect the data.

This book is designed to provide information on how atmospheric science can be applied to solve problems arising within the meteorological domain of power production. Both the theoretical and practical aspects of the applications are addressed. For example, following a review of the fundamentals of meteorology in Chap. 2, the operational characteristics of a variety of meteorological instruments are described in Chap. 3. Data management principles are addressed by Chap. 4. The fundamental physics of the atmospheric boundary layer is reviewed in Chap. 5, while Chap. 6 addresses the theoretical concepts of atmospheric transport and diffusion. The effects of obstacles on the atmospheric flow field are reviewed in Chap. 7. Plume rise and buoyancy processes are described in Chap. 8. The results of field measurements of transport and diffusion processes are summarized in Chap. 9. Atmospheric diffusion, a process that depends on the state of atmospheric turbulence at any location and time, is addressed in different ways in Chaps. 2, 5, 6, 9, 13, and 14. The atmospheric chemistry and transformation processes of inorganic emissions from power generation are discussed in Chap. 10. Precipitation scavenging, dry deposition, and resuspension are described in Chaps. 11 and 12. The atmospheric effects of energy production are reviewed in Chap. 15. Radiological dosage principles are described in Chap. 16 and some of the health effects of various compounds are outlined by Chap. 17. Potential contributions of atmospheric science to alternative energy resources are addressed in Chap. 18. Environmental impact statements and the concept of atmospheric visibility are described in Chaps. 19 and 20, respectively.

1-5.1 Plume Rise

Stack height and the buoyant rise of effluents are important in reducing ground-level concentrations. The importance of these two parameters has been known for nearly 65 yr (e.g., Wells, 1917). Bosanquet and Pearson (1936) demonstrated that, under certain atmospheric conditions, the maximum ground-level concentration depends on the inverse square of stack height. Later, the stack height in their formulation was replaced by the "effective stack height," which represents the sum of the actual stack height and the height of rise of the plume above the top of the stack. In Chap. 8, Briggs demonstrates that the effective stack height, h_e, is very important for determining the maximum ground-level concentration from most stacks. For example, plume rise can increase h_e by values ranging from 2 to 10 times the stack height. Consequently, since the maximum ground-level concentration is inversely proportional to h_e^2 (see Chap. 8, Eq. 8.103), plume rise can reduce the maximum ground-level concentration by a factor of as much as 100. However, atmospheric stability, turbulence, vertical wind shear, and moisture conditions complicate the computation of h_e and can affect plume rise. Procedures for incorporating these atmospheric conditions into the computation of h_e are described in Chap. 8.

1-5.2 Transport

The atmospheric transport of anthropogenic and natural effluents has been widely studied theoretically and experimentally (see Chap. 9). Emphasis has been placed on steady-state conditions over uniform terrain on scales of 0.1 to 10 km. However, it has been generally recognized that pollution can be transported great distances by the wind. Early documentation of large-scale transport was provided by both Smith (1950) and Wexler (1950) in their descriptions of the transcontinental migration of a large smoke pall generated by a great forest fire in western Canada. Plummer (1912) provided documentation of the occurrence of large smoke palls within the United States since 1706. Machta (1958), Junge (1963), and others reported on the global transport of radioactive debris released from nuclear tests in the atmosphere. More recently, Hirst, Stedman, and Hurst (1967) detected airborne spore concentrations approximately 700 km from their source in England. Prospero (1968) demonstrated that dust from Africa can traverse the tropical Atlantic Ocean, and Millian and Chang (1977) detected SO_2 400 km from its source. Hall et al. (1973) reported on a large-scale pollution episode over the central United States in which an effluent pall was observed to advance as much as 1100 km from the source region. Based on changes in visibility, the pall affected an area of the order of 10^6 km^2.

Long smoke plumes and pollution palls have been observed from manned and unmanned spacecraft (Randerson, 1968, 1977; Randerson, Garcia, and Whitehead, 1971). Both man-made and naturally induced episodes have been detected. For example, during the Skylab-4 mission, an effluent pall was observed over the Los Angeles basin and the Gulf of Santa Catalina. The entire pall covered a geographical area of approximately 13 000 km^2. Parmenter (1977) documented GOES (Geostationary Operational Environmental Satellite) imagery of smoke plumes from forest fires in Canada. The plumes were quite distinct and showed a marked downstream spreading in a very unstable air mass. The plumes covered an area of the order of 10^5 km^2. On the other hand, a spreading plume generated by the eruption cloud emanating from the Sakurazima volcano was estimated to be 255 km long and to cover an area of 20 000 km^2 on Jan. 8, 1974 (Randerson, 1977). The NOAA-5 satellite pictured a large dust veil from a volcanic eruption on the island of Hawaii (NOAA, 1978) in September 1977; however, no quantitative information is available to document the areal extent of the eruption cloud. During the 1980s, detailed documentation will be available of the Mount St. Helen's volcanic eruption of May 18, 1980.

The transport of combustion products over long distances has become an important issue in the analysis and control of air quality. Of particular concern is the transformation of gaseous emissions into particulate matter and the consequent removal of the particles hundreds to thousands of kilometers from the source. For example, the SO_2 released from coal-fired power plants can be oxidized to sulfur compounds (sulfates).† However, this transformation is a slow process (of the order of 1%/h, see Sec. 1-5.1 and Chap. 10). Consequently, far from the source, transformed SO_2 can be deposited at ground level as sulfate by wet or dry removal mechanisms. This process in the sulfur budget has been invoked to explain the acidification of precipitation, the reduction in visibility, and the deposit of sulfur compounds on buildings and vegetation, all occurring far from sources. The need for quantitative information to document and to study the atmospheric sulfur budget has led to the implementation of major field programs. Some examples of these projects are the OCED Long-Range Transport of Air Pollutants (LRTAP) study in Europe (Ottar, 1976), the Canadian regional study (Whelpdale, 1978), the Electric Power Research Institute (EPRI) Sulfate Regional Experiment (SURE) in the United States (Perhac, 1978), the Midwest Interstate Sulfur Transformation and Transport (MISTT) study (Wilson, 1978), the Multi-State Atmospheric Power Production Pollution Study (MAP3S) which addressed a large spectrum of effluents (MacCracken, 1978), and EPRI's Plume Model Validation Project (Bowne and Hilst, 1981). Projects of these types are providing the scientific community with much needed data for resolving the important questions relating to the transport, transformation, and removal of effluent species from the atmosphere.

1-5.3 Diffusion

As an atmospheric process, diffusion represents the spreading of gaseous and/or particulate matter caused by turbulent motions in the atmosphere. Consequently what is actually being described is turbulent diffusion as opposed to molecular diffusion, which is trivial in the atmosphere relative to horizontal transport and vertical mixing. Dispersion represents the sum of the processes of transport and diffusion. An effective mechanism for dispersing effluents released into the atmosphere is a change in wind direction and speed with height (i.e., vertical wind shear). In other words, effluents released at two different levels in the presence of large directional shear will disperse in different directions.

When effluents are released into the atmosphere, their concentration patterns are controlled to a large extent by turbulent diffusion. This process depends on how turbulent the atmosphere is at a given time and place. But atmospheric turbulence is difficult and expensive to measure. For this reason, much research has been directed toward relating atmospheric turbulence to routine meteorological measurements (e.g., see Chap. 9). Emphasis has been placed on describing boundary-layer turbulence in terms of meteorological values averaged over time periods of 30 to 60 min. In particular, many attempts have been made to link boundary-layer turbulence to the vertical gradients of temperature and wind, to the average temperature, and to the horizontal wind. Theoretical relationships between these average quantities and turbulence have been established (e.g., see Haugen, 1973; Panofsky, 1974; and Chaps. 6, 9, and 13). However, the relation between these quantities and diffusion is not well understood. Therefore it has been necessary to develop empirical connections between the meteorological quantities and atmospheric diffusion. These empirically based procedures are known as turbulence-typing schemes. They have been reviewed in detail by Gifford (1961, 1976) and by others.

†Hitchcock (1976) has noted that some sulfate concentrations can be caused by biogenic sources.

Perhaps the most widely used turbulence-typing scheme is the one developed by Pasquill (1961) and modified by Gifford (1961) so that the resulting curves are frequently referred to as the Pasquill–Gifford curves. For the purpose of turbulence typing, Pasquill (1961) viewed turbulence near the ground as having a mechanical and a thermal component. The mechanical component was conceived as being created by frictional wind shear and the thermal component as arising from vertical heat flux. The relative importance of these two components determines the turbulence type. For example, type A is associated with a small mechanical but a large thermal component, whereas type D is all mechanical. Type A corresponds to very unstable conditions, type F to very stable conditions, and type D to neutral conditions (see Chaps. 6, 9, and 13 for details). Gifford (1961) converted the plume spread data into families of curves of the standard deviations of the plume concentration distribution, σ_y and σ_z, where σ_y is for the crosswind spread and σ_z is for the vertical spread.† It is important to emphasize that the Pasquill–Gifford curves were developed from diffusion data collected for distances of up to only 0.8 km. A nonbuoyant tracer gas was used and was released near the ground. Consequently estimates of plume spread based on Pasquill's categories are intended to be applied only in specific situations: ground-level emissions, flow over uniform terrain, wind speeds greater than 2 m/s, and nonbuoyant plumes.

There are many flow fields that can be classified as exceptional. Such flow regimes lie outside the limits of Pasquill's basic system because turbulence is not generated solely by the mechanical and thermal components identified by Pasquill. Examples of such flows are very light winds (<2 m/s), channeling effects, flow separations around obstacles, flows over cities and flows with buoyancy. Attempts have been made to modify the PG curves to account for exceptional flows. Indeed, the turbulence categories have been extended in attempts to account for diffusion over cities, diffusion over water, diffusion with buoyancy, diffusion along coastlines, and diffusion at great distances (>10 km). Results from many of these field programs are described in Chap. 9. Gifford (1976) notes that Pasquill concluded that if good-quality emissions and meteorological data are available, estimates of atmospheric concentrations based on typing methods may be accurate to within 20% for long-term averages but may exceed a factor of 2 for short-term values. This conclusion has also been emphasized by the American Meteorological Society (1978).

Questions have been raised about the representativeness of the PG curves. For example, at the Specialist Conference on the EPA Modeling Guideline (Roberts, 1977), some attendees emphasized that significant discrepancies can occur between PG categories and those derived from tower ΔT data. In addition, the PG curves for σ_z were questioned with the suggestion that, for tall stack emissions, stability class A (and to some extent, class B) curves may result in large overestimates of short-term, ground-level concentrations. Hovind, Edelstein, and Sutherland (1979) reported on the concern of determining σ_y and σ_z over complex terrain. Rather than rely upon a single PG stability class for both horizontal and vertical dispersion estimates, perhaps separate values for each of these components would be more meaningful.

In an effort to move away from the discrete nature of the typing methods and to allow for a continuum of stability, Irwin (1979) has proposed a different method for making practical estimates of σ_y and σ_z. The technique has been reviewed by a panel of experts (Randerson, 1979). Irwin's technique builds upon the recommendations of a workshop on stability classification schemes and sigma curves (Hanna et al., 1977). The technique makes use of the convective velocity scale w_* (see Chap. 5), the mixed layer depth, z_i, and the formula (see Chap. 9)

$$\frac{\sigma_y}{\sigma_v(t)} = f\left(\frac{t}{t_i}\right)$$

to estimate σ_y. In this expression, σ_v is the standard deviation of the horizontal crosswind component of the wind, t_i is the Lagrangian time scale, a parameter that is proportional to the convective time scale ($t_* = z_i/w_*$). The functional form of t/t_i has been discussed by Draxler (1976 and Chap. 9), by Pasquill (1976), and by Irwin (1981).

General formulas for determining σ_y and σ_z from observed plumes have been proposed by Gifford (1959, 1980). According to Gifford, the appropriate formulas can be obtained by integrating the steady-state Gaussian plume formula (see Chaps. 6 and 13) in either the y- or z-direction, expressing the result in terms of a constant edge concentration, and collecting and eliminating all the resulting constants. This process leads to the formulas

$$\sigma_y^2 = \frac{y_E^2}{\ln (e y_m^2/\sigma_y^2)}$$

and

$$\sigma_z^2 = \frac{z_E^2}{\ln (e z_m^2/\sigma_z^2)}$$

where $e = 2.718$, and y_E and z_E represent half the width and depth of the visible plume in the y- and z-directions, respectively. The subscript m is for the maximum values of y_E and z_E at the plume's widest point. The advantage of using these two formulas is that the forms of σ_y and σ_z are not predetermined; they are functions of time. These two transcendental equations can be solved either graphically or numerically. Gifford (1980) has used the same two equations to evaluate σ_y and σ_z from photographs of smoke plumes.

1-5.4 Source Effects

Buildings, stacks, and cooling towers obstruct the air flow and consequently can contribute to changing the rate

†The relationship between σ_y and σ_θ, the standard deviation of the horizontal wind direction, has been addressed by Doran, Horst, and Nickola (1978), by Draxler in Chap. 9, and by others.

of transport and diffusion of effluents released from power plants. For this and other reasons, some utility companies have built tall stacks to help assure that their effluents are released high above the ground and are thereby not influenced by surface structures. However, for any stack, the aerodynamic effects of the stack itself may become important in dispersing the effluent plume. The lower pressure in the wake of a stack tends to draw the plume downward behind the stack. This downwash can be effectively prevented by keeping the escape velocity of the effluents, w_e, at a value greater than the wind speed at stack height, U_s. In Chap. 7, Hosker notes that downwash will generally not occur if $w_e/U_s > 1.5$. For values less than 1.5, Briggs (1973) has suggested that the plume can downwash an amount, h_d, given by

$$h_d = 2\left(\frac{w_e}{U_s} - 1.5\right)D$$

where D is the diameter of the stack.

From wind-tunnel studies and field experiments, our understanding of how air flows around obstacles is improving. These studies demonstrate that, on the windward face of a cubic obstacle, there is a stagnation point approximately two-thirds of the way up the face of the cube. Below this point, air flows downward and then away from the obstacle so that a circulation eddy is established on the windward side of the obstacle. Vortices can be generated at the windward and leeward edges of the cube as air flows over and separates from (and sometimes reattaches to) the roof and sides, and a recirculating wake cavity zone forms on the downwind side of the obstacle. A turbulent wake, perhaps influenced by trailing vortices, can exist sometimes to large distances downwind. These features and their influence on the transport and diffusion of effluents are described in detail in Chap. 7. In addition, the author develops empirical formulas for the cavity length. Techniques to estimate the effects of these aerodynamic phenomena on plumes emitted upwind, close to, or downwind of buildings are summarized in Chap. 7. In particular, methods are given for predicting the stack height necessary to avoid troublesome concentrations near or because of buildings. Empirical formulas for the extent of the roof and of wake recirculating cavities are given; these regions are found to depend on the particular building geometry. Methods for estimating ground-level, wake-centerline concentrations are discussed and compared to data.

1-5.5 Atmospheric Effects

Power production as well as various industrial processes result in the release of not only gases and particles into the atmosphere, but also moisture and heat. The observed and potential effects of this moisture and heat source are described in Chap. 15. In general, most of the effects on the atmosphere and to vegetation and structures are most apparent near the source. Examples of these local effects are (1) reduction in insolation; (2) increased incidence of fog; (3) increased deposition from cooling tower plumes; (4) concentration of local vorticity near large power parks; (5) changes in cloudiness, precipitation, temperature, and in the concentration of atmospheric trace constituents; and (6) interaction with plumes from other sources. For details, see Chap. 15.

According to Hanna and Gifford (1975), the most pronounced atmospheric effects from power plants are likely to be produced by power parks in which many power plants would be sited. The total power output by these parks is likely to be from 10 000 to 50 000 MW. Hanna and Gifford show that waste energy from such a concentration of power plants is approximately equal to the atmospheric dissipation of energy by such geophysical phenomena as thunderstorms, volcanoes, and large brush fires.

Reductions of insolation by industrial effluents have been documented by Sheppard (1958), Randerson (1970), Peterson, Flowers, and Rudisill (1978), and others. These studies show that the greatest attenuation of solar radiation by effluents is in the ultraviolet where depletions of 10 to 30% can occur. Bridgman (1978) has documented depletions of over 50% for wavelengths less than 0.53 μm. The attenuation of insolation by cooling tower plumes has been addressed by Junod et al. (1974) and by others.

As noted in Chap. 15, the deposition of fungicides and salts contained in cooling tower plumes have been recently studied in more detail. For example, Davis (1979) has summarized drift deposition data for the cooling towers at Chalk Point, and Talbot (1979) has studied the relationship of the dosage of saline aerosols to vegetation.

Clouds produced by industrial processes have been studied by Auer (1976) and others. He measured many quantities, including the vertical velocity in a cumulus cloud spawned by an oil refinery. Updraft speeds of 6 m/s were measured. Hobbs, Stith, and Radke (1980) have measured the concentrations of cloud condensation nuclei (CCN) in the plumes from coal-fired electric power plants. They note that the in-plume CCN concentrations are generally two to five times greater than in ambient air. They also found CCN production by gas-to-particle conversion to be of the order of 10^{13} to 10^{15} CCN/(h•mole) of SO_2. More important, they reported that, after travel times of about 1 h, most of the in-plume CCN had been produced by gas-to-particle conversion rather than being emitted directly from the stack (see Sec. 1-6 and Chap. 10).

Otts (1976) has documented two cases of the enhancement of snowfall by the water vapor plumes from the cooling towers of the 2900-MW John E. Amos plant in West Virginia. His report documents "plumes" of heavy snowfall (depths up to 15 cm) downwind of the cooling towers. An important point is that little or no snow fell anywhere else. The heavy snowfall occurred at distances of 5 to 40 km from the power plant. Similar observations have been documented by Parungo, Allee, and Weickmann (1978); and Koenig (1980) has used LANDSAT imagery to document snow streaks from power plant plumes.

1-5.6 Dry and Wet Removal†

Essentially all effluents released into the atmosphere are removed by the natural cleansing processes referred to as precipitation scavenging and dry deposition (Slinn, 1977a). Precipitation scavenging is the removal of any material from the atmosphere to the surface of the earth by various types of precipitation (Slinn, 1977b). It proceeds through the incorporation of effluent material into the precipitation and the subsequent deposition of the effluent onto the ground, vegetation, and/or structures. Obviously this process can significantly influence the chemical composition of the precipitation (see Sec. 1-8 and Chap. 10). Dry deposition proceeds without the aid of precipitation and denotes the direct collection of gaseous and particulate species on land or water surfaces.

Knowledge of the deposition rates of gases and particles onto ground-level surfaces is essential in determining the residence time of atmospheric compounds, the distances they can be transported, and their airborne concentrations. For example, without removal processes, global sulfur concentrations in the atmosphere could increase at the rate of about 70 $\mu g \cdot m^{-3} \cdot yr^{-1}$ (Hales, 1978). By comparing this value with the current United States annual standard (40 $\mu g \cdot m^{-3} \cdot yr^{-1}$), it is easy to comprehend the importance of removal processes in relation to air quality.

Methods used for measuring the dry deposition of pollutant species has been reviewed and critiqued by Hicks, Wesley, and Durham (1980). In general, there are three methods in use: (1) estimates of accumulation or mass balance method, (2) flux monitoring, and (3) flux parameterization. None of these methods has proven to be a panacea for dry deposition measurements.

An abundant amount of data is available pertaining to the wet and dry removal of gaseous SO_2 and sulfates. Quantitative information on the removal of other gaseous and particulate species is limited. Consequently the deposition of sulfur will be emphasized in the summary that follows. The other important removal process, atmospheric chemical transformation, is summarized in Sec. 1-6 and is described in more detail in Chaps. 10 and 11.

Dry deposition velocities‡ for SO_2 are of the order of 0.1 to 1.0 cm/s, depending on atmospheric stability (e.g., see Wesley et al., 1977). European studies (Smith and Jeffrey, 1975; Barnes, 1977; Garland, 1978; and others) demonstrate that SO_2 is removed by dry deposition with an average velocity of 0.8 cm/s. This value is probably correct to within a factor of 2 (Garland, 1978) when used to provide estimates of annual deposition over flat, open country dominated by grass. According to Sehmel (1980), deposition velocities for SO_2 to a variety of surfaces range from 0.1 to 10.0 cm/s. The largest values seem to occur over cities, but results are still uncertain. Small deposition velocities are for SO_2 on snow. They appear to range from approximately 0.1 cm/s to 0.5 cm/s (Barrie and Walmsley, 1978).

Hales (1978) notes that sulfur budget studies suggest that 50 to 90% of the atmospheric sulfur burden is removed by wet processes. This estimate implies that 10 to 50% of the sulfur may be removed by dry deposition. The variability in these numbers points to the difficulties involved with measuring the removal of sulfur from the atmosphere. Furthermore, effluent material does not always simply fall out of the sky like raindrops. Some material may remain nearly suspended in the air and actually flow around an exposed surface. Moreover, gases such as SO_2 can reach natural (or man-made) surfaces by absorption so that gases can dissolve into a water body and into moisture in the soil or on vegetation. The rate of removal of gases (and particulate matter) from the atmosphere may depend on whether or not the contacted surface is wet or dry, hot or cold, rough or smooth, and on many other factors.

An important problem in defining the removal of effluents from the atmosphere is to determine the residence times for these effluents. There appears to be a whole spectrum of atmospheric residence times (e.g., see Table 1.7) ranging from a few hours or less for reactive gases and heavy particles, to about 10^7 yr for helium. Henmi and Reiter (1978) calculated SO_2 residence times of 20 to 25 h over the eastern United States in winter and 30 to 35 h in summer. Higher rates were computed for the west-central United States. The average residence time for mid-latitudes for SO_2 is about 1 d (*Atmospheric Environment*, 1978). Residence times for sulfate are estimated to be approximately 100 h (e.g., Schwartz, 1979) with a range varying from about 2 to 7 d (Hitchcock, Spiller, and Wilson, 1980).

1-6 BASIC CHEMICAL TRANSFORMATIONS

Chapter 10 emphasizes that energy production yields a wide variety of by-products. Many of these products are released directly or indirectly into the atmosphere. The major constituents released are oxides of sulfur (SO_x), oxides of nitrogen (NO_x), carbon monoxide (CO),§ hydrocarbons (HC), and particulate matter (Part.). Figure 1.4 portrays the sources and the amounts of these effluents released annually in the United States. In Fig. 1.4, fuel combustion represents those effluents released into the atmosphere from stationary sources, such as power plants. To compare the global anthropogenic emissions with natural emission, refer to Table 1.6. This table demonstrates that on a global scale natural emissions are generally much larger than anthropogenic emissions, except for SO_2.

From the numbers tabulated in Fig. 1.4, a little calculation demonstrates that just over 80% of the total

†For details, see Chaps. 11 and 12, as well as the review of modeling procedures by Hosker (1980) and the review paper by Slinn et al., 1978.

‡For more information, see the tables in Chap. 12 or (1980) and McMahon and Denison (1979).

§Carbon dioxide (CO_2) is not normally considered a pollutant and will be addressed in Sec. 1-6.3 and in Chap. 10.

TABLE 1.7 Estimates of the Global Scale Production of Various Trace Gases Released into the Atmosphere from Natural and Man-Made Sources Annually. Based on the Table Constructed by Rasmussen, Taheri, and Kabel (1975) with Updates

Gas	Some sources	Estimated emission × 10^9 kg/yr: Anthropogenic	Natural	Estimated background, ppbv†	Estimated residence time
SO_2	Combustion of coal and oil, volcanoes	210[1]	295	0.03–0.1[2]	1 day[3]
H_2S	Biological decay, volcanoes, chemical processes, sewage treatment	3	40[1]	0.05[4]	10–20 h[5]
CH_4	Anaerobic decomposition of organic matter, coal mining, industrial losses	200[6]	1000[7]	1400[6]	3–10 yr[7,8]
CO	Oxidation of CH_4, combustion processes	640[9]	1500[10]	50–120[11]	0.1–0.2 yr[7,8]
NO_X	Microbiological processes, industrial processes, lightning	55	310	0.1[12] (NO_2)	100 s[13] (NO) 4 d (NO_2)
NH_3	Biological processes, volatilization of NH_3 in nitrogenous fertilizers	4	170	0.2–6.0[14]	7 d[14]
N_2O	Denitrifying bacteria, fertilizers, combustion	1[15]	150[10]	325[16]	4–70 yr[14]

†To convert ppbv to $\mu g/m^3$, multiply ppbv by (0.045 × Molecular Weight) at 0°C and 760 mm of pressure. To convert ppbv to $\mu g/m^3$, multiply ppbv by (0.041 × Molecular Weight) at 25°C and 760 mm of pressure.

(1) Cullis and Hirschler, 1980.
(2) Sze and Ko, 1980.
(3) *Atmospheric Environment*, 1978.
(4) Slatt et al., 1978.
(5) Bolin, 1979.
(6) Ehhalt, 1974.
(7) Peters and Jouvanis, 1979.
(8) Chameides, 1978.
(9) Seiler, 1974.
(10) Approximation using data in Chap. 10.
(11) Rasmussen, Khalil, and Dullage, 1981.
(12) Noxon, 1975.
(13) Liu, 1978.
(14) Stewart, Hameed, and Pinto, 1978.
(15) Conrad and Seiler, 1980.
(16) Rasmussen and Pierotti, 1978.

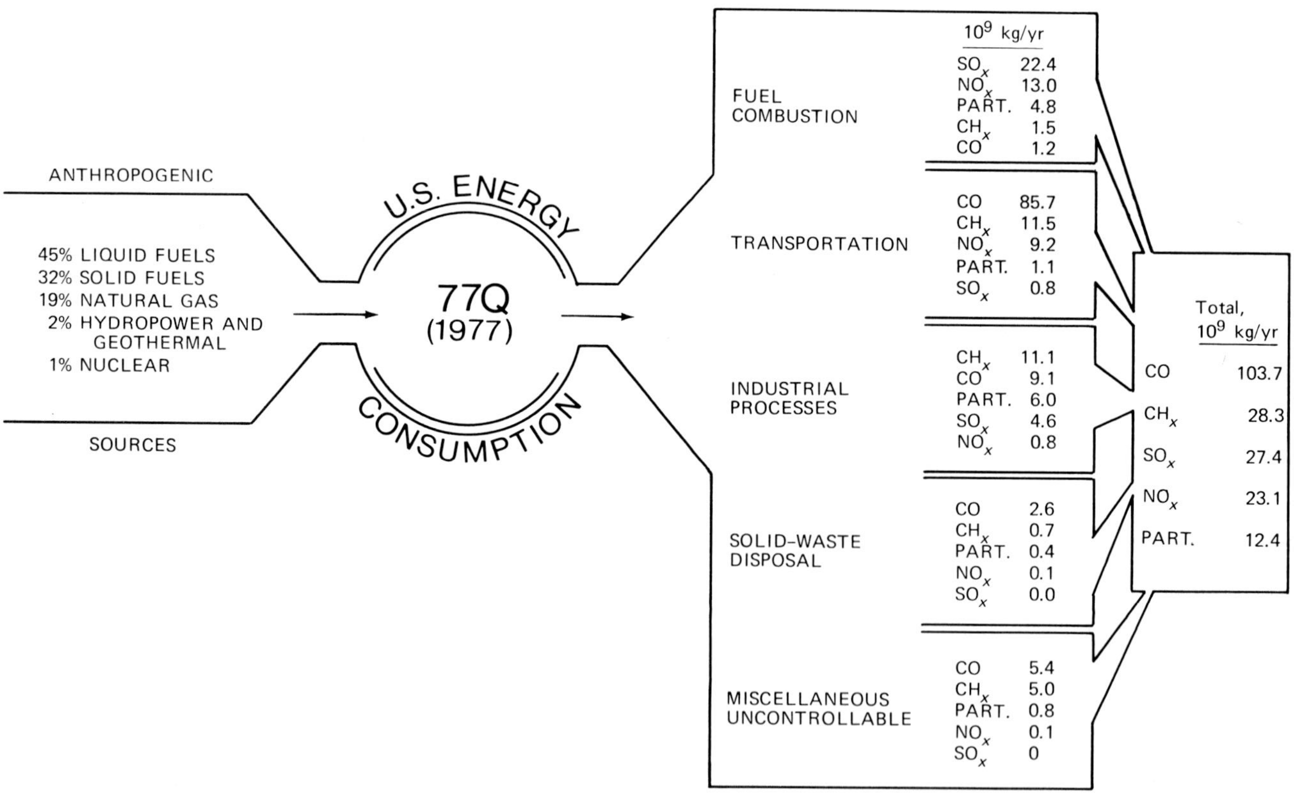

Fig. 1.4 Anthropogenic effluents released into the atmosphere in the United States in 1977.

annual release of SO_x in the United States can be attributed to the fuel combustion cycle. This cycle also accounts for 55% of the NO_x and for nearly 40% of the particulate matter released into the atmosphere. The primary source of CO is the transportation cycle which accounts for a little over 80% of the CO released annually in the United States. Transportation and industrial processes contribute about 40% each to the total release of hydrocarbons. Industrial processes also account for nearly 50% of the particulate matter released into the atmosphere.

From a different aspect, consider the following observation. In 1977 approximately 218×10^6 people lived in the United States. Consequently, of the total annual SO_x production, nearly 350 g of SO_x were produced per person per day; 290 g of NO_x per person per day, and 1300 g of CO per person per day. Using Fig. 1.4 and Table 1.7, simple calculations suggest that of the 210×10^9 kg of man-made SO_2 released annually over the globe, approximately 13% can be attributed to anthropogenic sources in the United States. In a similar manner for NO_x, the United States contribution to the global burden appears to be roughly 40% and for CO, approximately 16%. By casting the effluents in this context, we must not forget the many benefits much of the world derives from the vast production of energy in the United States. For example, some note that countries such as Sweden and Switzerland consume less than half as much energy per capita as the United States and have a high standard of living. These nations derive much of their energy from hydroelectric sources. Of course, the United States could live on half its energy consumption, but at the sacrifice of energy-intensive industries such as agriculture, mining, smelting, petrochemical and paper production, and a host of other industries. Consequently much of our food and work-saving machines would have to be imported. A major reduction in the production of fertilizers and in the consumption of energy by the agricultural sector would result in a reduction of food stocks that the United States can export. This reduction in the global food supply would have to be produced by other nations or an increase in starvation could be anticipated.

1-6.1 Sulfur Compounds

A key issue in the future of energy conversion is the economic and environmental costs of increased coal usage as a replacement for petroleum. Environmental shortcomings of coal are related to mining, to transporting, and to releasing combustion products such as SO_2, NO_x, CO_2, and fly ash into the atmosphere. Coal combustion results in the production of sulfur which is present in coal as organic sulfur, sulfate, and sulfide. The sulfates become ash and the organic sulfur and sulfides oxidize to SO_2 and SO_3.

Based on molecular weights (M.W.), twice as much SO_2 (M.W. = 64) can be produced by weight as the sulfur (M.W. = 32) content of the fuel. Coal typically contains 1 to 3% sulfur after cleansing. Therefore relatively large amounts of SO_2 can be produced when great quantities of coal are burned. As an estimate, the amount of SO_2 produced in kilograms is equal to two times the sulfur content of the coal times the amount of coal burned in kilograms. For example, if 2% sulfur coal is burned by a coal-fired power plant that consumes nearly 10^7 kg of coal per 1000-MW day, then 4.0×10^5 kg of SO_2 will be emitted daily [SO_2 emitted = $(2.0)(0.02)(10^7$ kg$)$]. Fuel oil also contains sulfur, usually 1 to 2% by weight. The amount of SO_2, in kilograms, produced through the combustion of fuel oil can be estimated as two times the sulfur content times the total number of liters burned. For example, a typical 1000-MW oil-fired power plant can consume nearly 10^7 L of 2% sulfur oil daily so that 4.0×10^5 kg of SO_2 will be produced.

Sulfur dioxide is a colorless gas possessing a penetrating and characteristic odor. It is a weak absorber of radiant energy between 0.29 and 0.40 μm, with an absorption peak near 0.37 μm. The weight of a gram-molecular volume of the gas is 64 g, which makes it twice as heavy as air. Sulfur dioxide is very soluble in water, with about 4000 volumes of the gas dissolving in 100 volumes of water at 20°C. Consequently one of the most important chemical reactions of SO_2 is its union with water to form sulfurous acid (H_2SO_3) which is unstable and exists only in aqueous solution. Another important reaction for SO_2 is that it oxidizes easily to SO_3, which in the presence of water forms sulfuric acid (H_2SO_4).

Sulfur is so widespread, it is sometimes difficult to distinguish between polluted and unpolluted areas. For example, Bolin (1979) notes that the amount of sulfur deposited on the ground annually is often less than 1% of that in the soil. Large amounts of sulfur are also released into the atmosphere with nearly 500×10^9 g of SO_2 (250 g of S) being released annually (Cullis and Hirschler, 1980). Of this total amount of sulfur, Cullis and Hirschler estimate that, in 1976, nearly 104×10^9 g S/yr were from man-made sources and 147×10^9 g S/yr originated from biogenic or natural sources. Consequently, on a global scale, anthropogenic sources account for approximately 40% of the sulfur burden and natural sources account for 60%. However, in the Northern Hemisphere, the man-made contribution is greater than natural sources (98×10^9 g S/yr vs. 76×10^9 g S/yr). On the basis of the data tabulated by Cullis and Hirschler, the release of SO_2 from man-made sources has been increasing at nearly 5×10^9 g/yr since 1960. If this rate of increase is maintained, mankind will release as much SO_2 into the atmosphere as natural sources in 17 yr.

Most of the anthropogenic sulfur is produced in the eastern United States and in northwestern Europe. In fact, Galloway and Whelpdale (1980) estimate that anthropogenic sources exceed natural ones by a factor of 10 over eastern North America. Husar et al. (1978) report an emission rate of SO_2 of 20 to 30 $g \cdot m^{-2} \cdot yr^{-1}$ for the states in the Ohio River Valley. Ground-level concentrations of SO_2 range from background values of 0.03 to 0.10 ppb in remote areas to values of 1 to 3 ppm in heavily polluted urban environments.

When SO_2 is released into the atmosphere, oxidation begins, forming sulfates (e.g., see Möller, 1980). Experimental results show that SO_2 is oxidized homogeneously to a variety of sulfoxy radicals and compounds by reaction with OH and HO_2 radicals (see Chap. 10 and Sze and Ko, 1980). The two most important reactions for the gas-phase conversion of SO_2 to sulfate (SO_4^{2-}) appear to be (Calvert et al., 1978; Graedel, 1976; Payne, Stief, and Davis, 1973; and others)

$$OH + SO_2 + M \rightarrow OHSO_2 + M$$

$$HO_2 + SO_2 \rightarrow SO_3 + OH$$

where the fate of $OHSO_2$ is not known and the second reaction may play a role in the removal of SO_2.

The sulfur trioxide (SO_3) is rapidly converted to sulfuric acid via

$$SO_3 + H_2O \rightarrow H_2SO_4$$

Calvert et al. (1978) estimate that SO_2 is oxidized to $OHSO_2$ by OH at about 0.4% h^{-1} in clean air and at 2.7% h^{-1} in dirty air.

In general, then, SO_2 can be oxidized to SO_3 and converted to H_2SO_4 aerosol, or it may form sulfite ions that are then oxidized to sulfate. Subsequent to the oxidation, H_2SO_4 may interact with other materials to form sulfate compounds (for more details, see Chap. 10). The final fate of these compounds apparently is removal through precipitation scavenging (acid rain), and soil and surface water absorption (see Chaps. 10 to 12). Möller (1980) concludes that about 50% of emitted SO_2 is removed by dry deposition, mainly as SO_2, and that 50% is removed from the atmosphere as SO_4^{2-} by precipitation.

During the 1970s, the European meteorological community was quite active in documenting and studying the flow of atmospheric pollution across national boundaries. Perhaps the most famous study is the one the Swedish government did on atmospheric sulfur for the United Nations Conference on the Human Environment (Royal Ministry for Foreign Affairs, 1971). More recently, a comprehensive insight into the long-range transport of atmospheric pollutants in Europe was presented in the Final Report of the Organization for Economic Cooperation and Development (OECD, 1977). This report has been summarized by Barnes (1977), Ottar (1978), and others. The report shows that the average SO_2 concentrations for rural sites range from near 20 $\mu g/m^3$ (7 ppb) near the major sources in Germany, the United Kingdom, and the Netherlands to near 2.0 $\mu g/m^3$ (0.7 ppb) in remote areas. The annual mean aerosol sulfate concentration ranges from 10 $\mu g/m^3$ near major sources to 0.5 $\mu g/m^3$ in remote rural areas. Sulfate found in precipitation shows a geographical distribution similar to that of the air concentrations with annual mean values ranging from 5 mg SO_4^{2-}/L near major sources to less than 1 mg SO_4^{2-}/L in the far north of Scandinavia. The data show that the weighted mean pH of precipitation varies from near 4.1 in Germany to near 6.2 in western France. The average sulfate background in air originating over the Atlantic Ocean was found to be 0.8 mg SO_4^{2-}/L. It is speculated that the source of this sulfate is North America. However, Nyberg (1977) argues that North Atlantic air probably has a natural sulfur content of 0.1 mg/L or less. He also suggests that wet scavenging of sulfur in central Europe does depend on emissions in North America to a small extent (<10%).

Within the United States, the largest sulfate values occur in the northeastern states where geometric means for urban areas range from 10 to 25 $\mu g/m^3$. Daily averages in urban areas range from 5 to 50 $\mu g/m^3$; however, there are episodes when concentrations of 50 to 100 $\mu g/m^3$ can occur (*Airborne Particles*, 1979, and Boldt et al., 1980). Such concentrations are much larger than the 3 to 4 $\mu g/m^3$ measured in the western United States. In fact, Hoffer, Kliwer, and Moyer (1979) measured a minimum 24-h sulfate concentration of 0.09 $\mu g/m^3$ in the desert southwest, while Whitby (1978) has reported 0.04 $\mu g/m^3$ over remote ocean areas. For more information, see Chaps. 10 and 11.

In southeastern Canada, Whelpdale (1978) found sulfate concentrations of 10 to 15 $\mu g/m^3$; however, values of over 20 $\mu g/m^3$ have been observed. During episodes of high sulfate concentrations, 24-h average values of 40 to 50 $\mu g/m^3$ were observed along the northern shores of Lakes Ontario and Erie. He also noted that over western and northern Canada, daily sulfate concentrations are typically less than 5 $\mu g/m^3$. In extreme southern Ontario, the sulfate content of precipitation was typically 2 to 7 mg SO_4^{2-}/L and pH values were usually less than 4.5.

The rate of oxidation of SO_2 to sulfate occurs in competition with deposition so that documentation of this reaction rate is important in studies of the sulfur budget. In general, the rate of conversion of SO_2 to SO_4^{2-} in power plant plumes is of the order of 0.5 to 5.0% h^{-1} (Forrest and Newman, 1977). Rates as high as 5.7% h^{-1} have been observed in plumes by Hegg and Hobbs (1980) for travel times of 10 to 162 min. Results of field experiments tend to show that SO_2 conversion rates increase with plume age for travel times of roughly 500 min. (Meagher et al., 1978). In addition, there appears to be a diurnal variation in the conversion rate. Values of 0.1 to 0.5% h^{-1} seem typical at night, whereas values of 1 to 5% h^{-1} are typical of midafternoon. Theoretical considerations of smog kinetics point to daytime conversion rates as large as 10 to 20% h^{-1} over urban areas (Miller, 1978). These theoretical rates have been confirmed by Forrest, Schwartz, and Newman (1979) who observed conversion rates as large as 25% in urban plumes. The higher values tend to be associated with increased travel time, high insolation, high temperatures, and, in some cases, high relative humidity.

1-6.2 Nitrogen Compounds

By volume, approximately 78% of the troposphere is composed of molecular nitrogen that is chemically a rather inert gas. An additional 21% of the tropospheric composition is molecular oxygen. Nitrogen and oxygen can combine to form several stable oxides (see Chap. 10). Two oxides identified as important "pollutants" are nitric oxide (NO) and nitrogen dioxide (NO_2). These two gases are normally listed together as oxides of nitrogen (NO_x). Nitric oxide is a colorless, odorless gas having approximately the same density as air. On the other hand, NO_2 is 1.6 times more dense than air. At low concentrations NO_2 has a yellowish-orange color, whereas in high concentrations it takes on a reddish-brown color. Nitrogen dioxide also has a very pungent odor and a high oxidation rate and is extremely corrosive (for details see Chap. 17). The primary anthropogenic source of NO_x is the combustion of fossil fuels for transportation and industrial processes (~55×10^9 kg/yr). Specifically, NO is produced during high-temperature combustion. Rapid cooling of the gases in the combustion chamber prevents NO from returning to N_2 and O_2. Upon release into the atmosphere, NO quickly oxidizes to NO_2 through the reaction

$$NO + O_3 \rightarrow NO_2 + O_2$$

In the atmosphere, this conversion gives NO a lifetime of about 100 s and most of the NO_2 is photolyzed back to NO (Liu, 1978). However, some NO_2 combines with OH to give nitric acid (HNO_3) which can be scavenged by precipitation. The NO_x plays an important role in the chemistry of the troposphere because it involves the production of ozone (see Chap. 10) and because changes in the NO_x (and CO) abundance can result in changes in the abundance of OH radicals (see Sec. 1-6.4 and Chap. 10). The background concentration of NO_2 appears to be of the order of 0.1 ppb (Noxon, 1975).

Different emission factors for NO_x result when utilities burn various fuels. For example, the combustion of coal yields 10 g NO_x/kg of coal burned or, on an energy equivalent basis, 1.3 g NO_x/kW·h. Fuel oil produces approximately 12.5 g NO_x/L of oil burned or 1.1 g NO_x/kW·h. Natural gas combustion yields nearly 6.2 g NO_x/m^3 of gas burned or 0.6 g NO_x/kW·h. The energy equivalent values suggest that in terms of the production of NO_x, natural gas combustion yields roughly half the NO_x generated by burning coal.

On the basis of the data collected by EPA in the National Emissions Data System (Environmental Protection Agency, 1976b), the greatest anthropogenic emission rates of NO_x are found along the Michigan–Ohio–Indiana border, in northeastern Illinois and in Pennsylvania. Emission rates of NO_x of 4 $g/m^2 \cdot yr$ occur in these areas.

Natural sources of NO_x are bacterial action in the soil, lightning, and forest fires. The estimated annual global NO_x emission from biologic sources is roughly 300×10^9 kg/yr, from lightning 10×10^9 kg/yr, and from forest fires 10^9 kg/yr for a total natural source of approximately 310×10^9 kg/yr. However, this estimate is subject to change. For example, Chameides, Liu, and Cicerone (1977) estimated that lightning produced 30 to 40×10^9 kg N/yr, accounting for roughly 10% of the total source of NO_x for the atmosphere. More recently, Dawson (1980) and Hill, Rinker, and Wilson (1980) suggested that 3 to 13×10^9 kg N/yr is perhaps a more realistic estimate.

Another atmospheric nitrogen compound, nitrous oxide (N_2O), is believed to be created by the decomposition of nitrogen compounds in the soil by anaerobic bacterial action. This gas plays an important role in the chemistry of ozone in the atmosphere (see Chap. 10). Near the ground, its concentration ranges from about 0.25 to 0.60 ppm, and it is 1.5 times more dense than air. Recently the influence of N_2O derived from fertilizers has been examined in terms of potential effects on the global ozone cycle (Crutzen, 1974; Liu, Cicerone, and Donahue, 1977; Conrad and Seiler, 1980; and others). However, there is very little known about the production rates and sources of N_2O. For example, estimates of the oceanic production of N_2O ranges from 6 to 250×10^9 kg/yr (Hahn and Junge, 1977; Cohen and Gordon, 1979). Anthropogenic releases of N_2O into the atmosphere of 1.2×10^9 kg/yr have been estimated by Conrad and Seiler (1980) and this value, rounded to 1.0, appears in Table 1.6. Additional information is given in Chap. 10.

Perhaps the most important consequence of the production of NO and its release into the atmosphere, especially over urban areas, is the role played by NO in producing what is called photochemical smog (Haagen-Smit, 1952). The primary ingredients contributing to this condition are sunlight and emissions of both NO and a wide variety of hydrocarbons. All these constituents interact in the atmosphere, producing a host of secondary products such as ozone (O_3) and a variety of sulfates and nitrates.† In the atmosphere, the principal set of reactions that lead to the production of O_3 and involve NO, NO_2, and sunlight ($h\nu$) are (see Chap. 10)

$$NO_2 + h\nu \rightarrow NO + O \quad (1)$$

$$O + O_2 \rightarrow O_3 \quad (2)$$

$$NO + O_3 \rightarrow NO_2 + O_2 \quad (3)$$

The first reaction demonstrates the role played by solar radiation in the formation of O_3, and the third reaction shows that the destruction of O_3 by NO is a direct process. In addition to the above reactions, it is generally believed that hydrocarbon emissions also affect O_3 concentrations because peroxy radicals (RO_2) that result from hydrocarbons can provide a means for oxidizing NO and NO_2 without the loss of O_3 through

$$NO + RO_2 \rightarrow NO_2 + RO$$

†Singh, Ludwig, and Johnson (1978) emphasize that the primary reservoir for O_3 is the stratosphere and that this region is the major source for tropospheric O_3. In general, minimum ground-level concentrations of O_3 in remote areas range from 10 to 30 ppb (1-h averages) in the winter to maximum values of 40 to 80 ppb (or more) in the spring.

The oxidation of NO to NO_2 is only one of a large number of atmospheric reactions that involve free radicals. For example, the chains leading to the above reactions can be initiated by the hydroxyl radical (see Sec. 1-6.4 and Chap. 10) to produce sulfates, nitrates, and carbonyls.

Trijonis et al. (1978) note that since NO is the principal oxide of nitrogen emitted by combustion, O_3 can be removed effectively near strong NO sources and can be formed by oxidized products elsewhere. In fact, Cleveland and McRae (1978), Trijonis et al. (1978), Karl (1978), and Cleveland and Graedel (1979) found very high levels of O_3 at distances of 45 to 60 km from major NO sources. Dramatic urban–rural differences were reported by Grennfelt (1980) who found gaseous nitrate concentrations three times greater over a rural site than over the urban source located 40 km north. The reason for this difference is reported to be due to the absorption of the gaseous nitrates by particles over the urban area.

Another important consequence of releasing NO and NO_2 to the atmosphere is their absorption of electromagnetic energy. Nitric oxide absorbs effectively in narrow bands for wavelengths between 0.15 and 0.23 μm. Since NO does not absorb visible light, it is a colorless gas. On the other hand, NO_2 absorbs radiant energy over the entire visible spectrum (e.g., see Chap. 20, Fig. 20.12). The absorption is strongest at the short wavelengths and is a maximum near 0.4 μm. Consequently the presence of NO_2 tends to reduce the brightness and contrast of distant objects, causing the horizon sky and white objects to appear pale yellow to reddish-brown. The extent to which NO_2 can color the horizon sky and reduce visibility depends on its atmospheric concentration, on the viewing distance, and on the ambient aerosol concentration (see Chap. 20). The presence of other photochemical aerosols or particulates suppresses the coloration effect of NO_x but increases the visibility reduction effect.

One of the important reaction products of NO_x in the atmosphere is particulate nitrate. However, the conversion rate of NO_x to nitrate is not well documented. Preliminary results from field programs (e.g., Hegg and Hobbs, 1980) suggest rates of 0.1 to 0.5% h^{-1}. The data also point to an increase in the conversion rate as a function of travel time and time of day. For example, during midday summertime conditions, conversion of NO_2 to HNO_3 may occur very early (3 to 5 h) in the chemical life of a power plant plume. The consequences of this condition could result in HNO_3 being the major acid species present near the source.

Studies of NO_x reactions suggest that nitrates initially account for only a small fraction of the NO_x that is transformed to other products under photochemically reactive conditions. However, under less reactive conditions (i.e., during long-range transport), particulate nitrate can make up a significant fraction of the oxidized nitrogen in the lower troposphere. Some examples of gaseous nitrates identified in the atmosphere include nitric acid (HNO_3), peroxyacetyl nitrate (PAN), and peroxypropionyl nitrate (PPN). But even though we know these products exist, our understanding of the distribution and formation mechanisms is too limited to cope with any potential need to control them.

Nitrogen dioxide can enhance the oxidation of other gases. For example, Cofer, Schryer, and Rogowski (1980) demonstrated that the oxidation of SO_2 onto carbon particles was markedly enhanced by the presence of a trace (10 ppm) amount of NO_2. On the basis of laboratory experiments, Cox (1974), Meyer (1977), and others suggest that NO_2 and NO are essential in the formation of H_2SO_4 from SO_2 (see Chap. 10 for details).

1-6.3 Carbon Compounds

The principal carbon-containing gases in the atmosphere are carbon dioxide (CO_2), methane (CH_4), and carbon monoxide (CO). Carbon dioxide is 1.5 times more dense than air. However, CO is slightly less dense (97%) than air and CH_4 has roughly half (55%) the density of air. The average global concentration of CH_4 is about 1.4 ppm (Ehhalt, 1974). By 1980 the average global concentration of CO_2 is expected to be almost 340 ppm. Liu (1978) reported that carbon monoxide has an average tropospheric concentration of about 0.20 ppm in the Northern Hemisphere, a value that is about four times larger than that for the Southern Hemisphere (0.05 ppm). This hemispheric asymmetry may indicate that anthropogenic processes are a primary source of CO in the Northern Hemisphere.

Available data show that CO_2 has been increasing steadily since accurate measurements began in the early 1900s (e.g., see Junge, 1963, p. 28). Measurements made between 1870 and 1900 point to a mean value of about 290 ppm with concentrations ranging from near 265 ppm to 310 ppm (Machta, 1979). On the basis of these data, it seems that an average concentration between 280 and 300 ppm was likely before significant amounts of fossil fuels were burned. So, it would appear that during the last 100 yr, the CO_2 content of the atmosphere has increased by 15 to 20%. No one has successfully related this increase to climatic change, though there is an abundance of theories proclaiming various climatic consequences, the most popular suggesting a 2 to 4°C warming of the global mean temperature if the CO_2 concentration is doubled. However, some recent calculations (e.g., Newell and Dopplick, 1979; Idso, 1980; and "Letters" in *Science*, 1980) suggest that the CO_2 problem may have been overdramatized by an order of magnitude. Clearly, more will be written on this topic in the future.

Imbedded in the long-term, increasing trend in the concentration of CO_2 are seasonal and diurnal variations. A decrease in CO_2 is observed during the warm season when plants are growing and consuming CO_2. An increase in CO_2 occurs during the winter when many plants are dormant. This seasonal variation ranges from approximately 5 to 15 ppm (e.g., Woodwell, 1978). The smaller seasonal variations appear to occur in the Southern Hemisphere, especially over the polar latitudes. Large seasonal variations occur in the high northern latitudes. It is speculated that these large variations stem from the seasonal alternation between ice cover and vegetation. Diurnal variations in CO_2

in the lower troposphere show a peak just before sunrise and a minimum in midafternoon. Large values occur at night when darkness depresses the consumption of CO_2 by photosynthesis and when sources of CO_2, such as the decomposition and respiration of organic matter continue.

The total carbon reservoir for the earth–atmosphere–oceans system appears to be of the order of 10^{19} g of carbon (for details, see Chap. 10). Roughly 75% of the carbon is in the oceans, 20% is in the form of fossil carbon, 3% is in the land biosphere, and nearly 2% is in the atmosphere, primarily as CO_2. The primary sources for atmospheric CO_2 appear to be the oceans (slightly $<10^{17}$ g/yr) and the land biosphere (10^{17} g/yr). However, if the net fluxes of carbon are considered, the ocean actually appears to absorb more CO_2 than it releases and the oxidation of carbon in the land biosphere seems to be balanced by photosynthetic processes. Consequently analyses of available data suggest that the combustion of fossil fuels is the main source (5×10^{15} g/yr) of atmospheric CO_2 (Baes et al., 1977). For this reason some suggest that a thorough analysis be made of the expanding use of fossil fuels. The climatic issue arises because CO_2 absorbs electromagnetic energy effectively in the infrared, especially between 13 and 17 μm.

Methane and CO are less abundant than CO_2, but they are also important members of the carbon group. Methane is a colorless, odorless, flammable gas that is a product of the decomposition of organic matter. It is the chief component of natural gas. Methane appears to be uniformly present in the troposphere with an average concentration of about 1.4 ppm. It has a residence time of 3 to 10 yr (Peters and Jouvanis, 1979; Chameides, 1978). At least 80% of the CH_4 is produced by the anaerobic decomposition of organic matter (e.g., in swamps, marshes, etc.). The remaining 20% appears to come from natural gas wells, seepage through the soil, and seepage from the oceans (e.g., Brooks, 1979). There are about 4×10^{15} g of CH_4 in the atmosphere. As with CO_2, CH_4 also absorbs infrared radiation but in a narrow band at 7.7 μm. The major sink for CH_4 appears to be through oxidation by free OH radicals, via

$$CH_4 + OH \rightarrow CH_3 + H_2O$$

Of course, the effectiveness of this reaction depends on the abundance of OH (see Sec. 1-5.4 and Chap. 10).

Carbon monoxide is a colorless, odorless gas that is quite toxic (see Chap. 17) at high concentrations (30 to 50 ppm for 8 h). Approximately 6.4×10^{14} g CO/yr are released to the atmosphere by anthropogenic sources, and about 15×10^{14} g CO/yr are released by natural processes. The primary sink for CO appears to be oxidation by OH, via

$$CO + OH \rightarrow CO_2 + H$$

and by soil uptake (Ingersoll, Inman, and Fisher, 1974). The photochemical lifetime of CO in the troposphere appears to range from approximately 30 days (Chameides, 1978) to about 60 days (Peters and Jouvanis, 1979).

1-6.4 Hydrogen Compounds

Obviously the most abundant member of this group is water vapor, which is present in the atmosphere as a consequence of evaporation from the oceans and other water bodies. Not only does water vapor support life on our planet but it is also a significant absorber and emitter of long-wave radiation. In fact, between 5 and 8 μm, there is nearly total absorption by water vapor. It is estimated that the absorption of long-wave radiation by water vapor, CO_2, and a few other gases is responsible for raising the mean temperature of the earth by approximately 30°C (Cole, 1975). According to Robinson (1977), the average residence time of a water molecule in the atmosphere is about 10 d.

In clean tropospheric air, OH and HO_2 appear to be very important radicals (Levy, 1971; Warneck, 1974; and others). These two radicals are important because they are generally inert towards N_2, O_2, Ar, CO_2, and H_2O, but they react easily with trace gases like CO, NO_2, and SO_2. Water vapor is an essential precursor for the formation of free OH radicals through the reaction of $O(^1D)$ atoms with water vapor, via

$$O(^1D) + H_2O \rightarrow 2(OH)$$

where the (^1D) represents an electronic state of the oxygen atom. The production of the metastable oxygen atoms appears to occur via

$$O_3 + h\nu \rightarrow O(^1D)$$

for wavelengths less than 0.31 μm. Thus the rate of production of OH depends on the rate of photodissociation of O_3 for the production of $O(^1D)$. The OH radical is an effective sink for the majority of the natural and anthropogenic trace gases released to the atmosphere. In fact, observations indicate that SO_2 can be oxidized by OH under very dilute concentrations of SO_2. Such a reaction can yield various sulfur-type acids at great distances from the source (>100 km). Similar reactions can occur with NO_x gases, yielding nitric acid (e.g., see Rodhe, Crutzen, and Vanderpol, 1981). The formation of these sulfate and nitrate aerosols can result in visibility degradation and acidic precipitation far from the source. The details of these reactions are described in Chap. 10. Measurements of OH radicals in the atmosphere yield a number density of the order of 10^6 cm^{-3} up to an altitude of 6 km, with the highest concentrations over the equator (Sze and Ko, 1980). Concentrations above the atmospheric boundary layer appear to range from 0.28 to 0.70 ppm (Davis, Heaps, and McGee, 1976). According to Crutzen and Fishman (1977), the greatest OH concentrations prevail in summer at low altitudes between 40°N and 40°S with maximum values over the equator. In addition, OH concentrations in the Southern Hemisphere are calculated to be three times greater than those in the Northern Hemisphere. This difference may be accounted for by the oxidation of OH by the more abundant concentration of CO in the Northern Hemisphere. In other words, the primary sink for OH is CO. Chameides (1978) estimates that the total production of OH molecules near the ground is 1.5×10^5 molecules/$cm^3 \cdot s$.

Another important hydrogen compound is the hydroperoxyl radical (HO_2). This radical is formed by the reaction of CO and N_2 with OH. Experimental data suggest that the concentration of NO_x may be important in determining the concentration of HO_2 and OH radicals. In fact, the reaction

$$NO + HO_2 \rightarrow NO_2 + OH$$

appears to be a major source of hydroxyl in urban atmospheres. This reaction can be followed by $NO_2 + OH \rightarrow HNO_3$ creating nitric acid. For details, see Chap. 10.

Hydrogen peroxide (H_2O_2) is also found over urban and rural areas. As a gas, H_2O_2 can act as a sink for hydroperoxyl radicals and as a source for hydroxyl radicals when photolyzed. In rainwater, H_2O_2 may play an important role in the oxidation of SO_2 to H_2SO_4 (Penkett et al., 1979). Crutzen and Fishman (1977) emphasize that knowledge of the gaseous concentration of H_2O_2 is important for modeling OH concentrations. Measurements of H_2O_2 by Kok et al. (1978) in the vicinity of Los Angeles revealed concentrations of 10 to 30 ppb. For a rural air shed, Kelley, Stedman, and Kok (1979) observed concentrations of 0.3 to 3.0 ppb. In the vicinity of Claremont, Calif., Kok (1980) has measured H_2O_2 concentrations of 1 to 160 ppb in rainwater.

1-6.5 Organic Compounds

Little is known about the distribution of organic material in the atmosphere. However, it is generally recognized that a vast number of organic chemicals are released into the atmosphere daily (e.g., see Brice and Derwent, 1978). Duce (1978) estimates that the global anthropogenic emissions of nonmethane hydrocarbons are approximately 8×10^{10} kg/yr. Global anthropogenic emissions of gaseous organics appear to be an order of magnitude less than natural emissions. Organic material from natural and from man-made sources are dispersed worldwide (National Academy of Sciences, 1978; and Atlas and Giam, 1981). The potential health effects of these complex substances are poorly known and the sources and sinks are not well known. Data for documenting anthropogenic and naturally generated organics are sparse and require the use of special sampling techniques (e.g., see Lillian and Singh, 1974; and Sawyer, 1978). Solutions to procedures for measuring organic chemicals in the atmosphere are paramount to proving answers to the many questions pertaining to the interrelationships between organic compounds, the environment, and health. Some progress has been made in sampling atmospheric organics. For example, Singh et al. (1979) have identified the distribution and the sources and sinks of some select halocarbons and hydrocarbons. Measurements of chlorinated hydrocarbons such as polychlorobiphenyls (PCBs) and total DDT have been reported by Harvey and Steinhauser (1974), by Hartung and Klinger (1970), and by others. Lovelock, Maggs, and Rasmussen (1972) suggested that organic sulfides such as dimethyl sulfide may constitute the bulk of the sulfide emitted from biological decay processes. Atlas and Giam (1981) documented the concentrations of nine organic compounds in a remote area (Enewetak Atoll), and they also measured the concentration of selected organic compounds in precipitation. The largest concentration in precipitation was for di(2-ethylhexyl) phthalate with a concentration range from 5.3 to 213 ng/L of precipitation. Singh et al. (1981) measured the concentrations of many organic chemicals in large urban areas. They also calculated the average daily-dose levels for exposures to halomethanes (including fluorocarbons), haloethenes, chloroethylenes, chloroaromatics, aromatic hydrocarbons, and secondary organics. The documentation and study of the distribution, transport, and transformation of organic compounds in the atmosphere is a relatively new and growing science with new information appearing in the literature frequently.

1-7 RADIATION†

Radiation occurs in the power production industry because radioactive by-products are released into the atmosphere by many types of power plants (Sec. 1-1). Mining and milling operations associated with uranium production are also a source of radiation. However, these sources of radiation should be evaluated with respect to the contribution to humans from natural sources. For example, proposed methods of conserving energy in homes may lead to increased exposure to radiation inside houses and buildings (e.g., *Natural Radiation Environment III*, 1980). The main issue revolves around the potential increase in health effects resulting from energy conservation practices for homes and buildings. Conservation efforts tend to reduce the exchange of outside and inside air and thereby permit radon gas (and other material) concentrations to increase inside dwellings. Radon is a radioactive gas that emits alpha particles, but the primary concern is with the short-lived, alpha-emitting daughters of radon that can irradiate the bronchial epithelium after being deposited in the respiratory tract. Much more will be said about this issue in the future.

In general, there are two types of radiation; namely, ionizing and nonionizing. Nonionizing radiation is in the low energy region of the electromagnetic spectrum and includes radio and TV waves; microwaves; and infrared and visible light. By contrast, ionizing radiation has two fundamental sources. Some ionizing radiation originates in the high energy region of the electromagnetic spectrum from X rays and cosmic rays. Another source of ionizing

†See Chap. 16 for details.

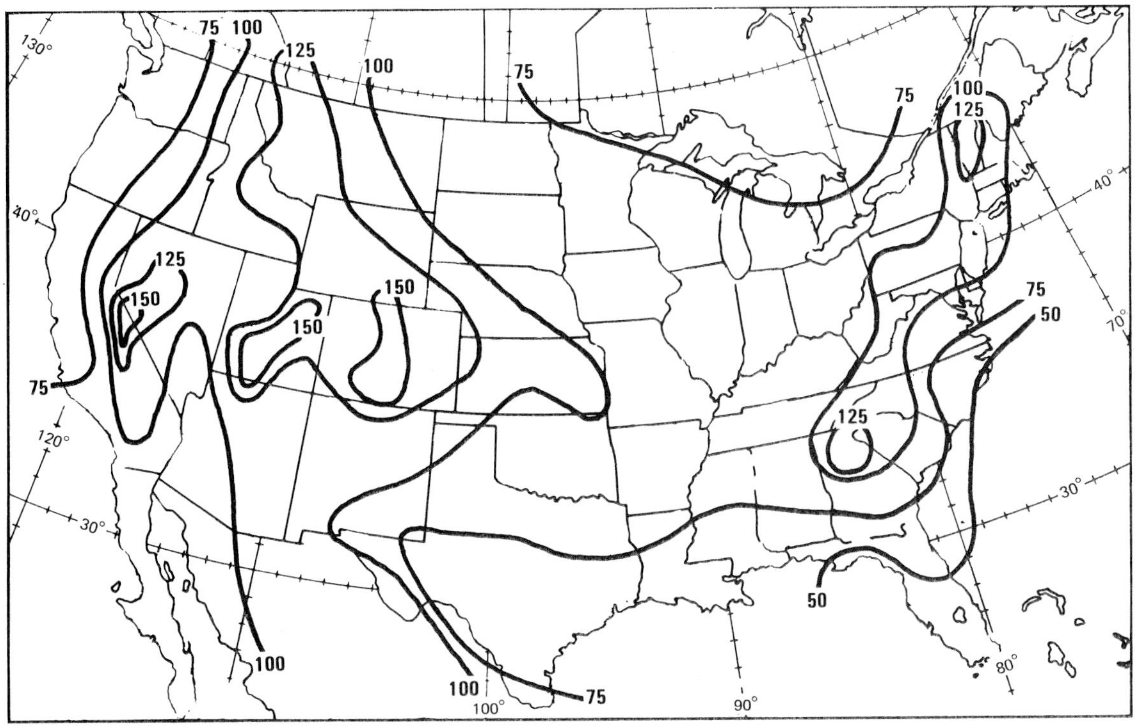

Fig. 1.5 Generalized distribution of the natural background whole-body dose rate to people in the United States from terrestrial and cosmic radiation. Based on data collected by C. L. Lindeken et al., Geographical Variations in Environmental Radiation Background in the United States, in *Proceedings of the Second International Symposium on the Natural Radiation Environment*, Vol. I, Aug. 7–11, 1972, Houston, Tex., NTIS.

radiation results from radioactivity in which the nuclei of radioactive elements disintegrate spontaneously, emitting radiation in the process.

Radiation from radioactive materials is of three types: alpha, beta, and gamma rays. Alpha particles are massive helium nuclei that are absorbed in a few inches of air or by a sheet of paper. Beta particles are electrons that can be absorbed by about 1 m of air. Gamma radiation is highly penetrating, but lightly ionizing per unit of path length when compared to alpha radiation.

A quantity of radiation is not usually measured directly. Instead, it is measured by the ionization produced when radiation passes through a medium. Several units are used to represent a measured quantity of radiation. The quantity may refer to charge, energy, or dose. Those units with the widest acceptance are the roentgen (R), rad, and rem. The roentgen is a unit of exposure to radiation or a measure of the ionization in air due to gamma (or X) radiation. The rad is the basic unit of absorbed dose of ionizing radiation, whereas the rem is a unit of biological dose of radiation and is also the unit of dose equivalent (see Chap. 16, Sec. 16-3.2.3 for details).

Radiation is present everywhere in the environment. It originates from natural and man-made sources. Radiation from natural sources is the major source of radiation exposure to people. Most of this radiation comes from radioactive material in the ground and from cosmic sources. Figure 1.5 portrays the generalized distribution of the natural background dose rate to people in the United States from terrestrial and cosmic radiation. Figure 1.5 demonstrates that, by simply traveling from the coastal areas of the Gulf of Mexico to Colorado, dose rates from this component of the natural background radiation can double or even triple. But these data are only for the United States. The natural background radiation varies widely throughout the world. For example, the monazite sand area of Kerala, India, exhibits a terrestrial background ranging from 130 mrem/yr to 2800 mrem/yr and Guarapari, Brazil, has dose rates of 17 000 to 24 000 mrem/yr (Lapp, 1979). The cosmic component of the background radiation can increase dramatically with elevation. This increase is illustrated in Fig. 1.6 which is based on observations and calculations made by O'Brien (1972).

According to the National Council on Radiation Protection and Measurements (NCRP) (1975), whole-body dose rates from natural sources range from 65 to 125 mrem/yr in the United States. The average dose rates from natural background sources are 28 mrem/yr for cosmic radiation, 26 mrem/yr for terrestrial radiation, and 27 mrem/yr for radionuclides in the body (NCRP, 1975), for a rounded total of 80 mrem/yr. Additional exposures from building materials† and from radioactive material in the atmosphere yields an average background dose rate of approximately 100 mrem/yr in the United States.

†Lapp (1979) reports dose rate levels as high as 525 mrem/yr in Grand Central Station in New York City.

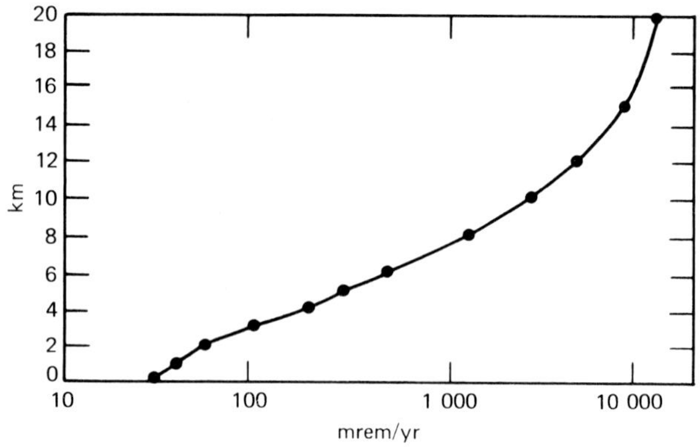

Fig. 1.6 Cosmic radiation dosimetric profile for the lower atmosphere at 55°N latitude. Based on observations and calculations reported by K. O'Brien, The Cosmic Ray Field at Ground Level, in *Proceedings of the Second International Symposium on the Natural Radiation Environment*, Vol. I, Aug. 7–11, 1972, Houston, Tex., NTIS.

Man-made radiation comes primarily from medical equipment, with X-ray equipment delivering by far the greatest dose rate—nearly 100 mrem/yr to the average person. Fallout from atmospheric nuclear tests gives about 5 mrem/yr, whereas color TVs give about 1 mrem/yr. Nuclear power plants account for about 0.1 to 1.0 mrem/yr to people living nearby. McBride et al. (1978) calculated a maximum dose rate of 2 to maybe 5 mrem/yr on the grounds of a 1000-MW nuclear power plant and about 2 mrem/yr near a 1000-MW coal-fired power plant. Beckmann (1976) estimates that the average citizen of the United States is exposed to nearly 250 mrem/yr of radiation from all sources.

In an effort to protect public health, the National Council on Radiation Protection and Measurements and the International Commission on Radiation Protection (ICRP) have recommended limits for occupational exposures and for exposures to populations. Both the NCRP (1971) and ICRP (1977) suggest an occupational limit of 5000 mrem/yr for whole-body radiation to any individual. For exposures to populations, these two groups also recommend dose limits of 500 mrem per person in any one year in addition to exposures from the natural background and from medical equipment. The NCRP (1971) also recommends a limitation for the control of possible genetic effects. An annual average of no more than 170 mrem per person from all sources is suggested for the population of the United States. This value is in addition to that from natural background and medical equipment.

The procedures for calculating external and internal dose rates are described in detail in Chap. 16. In addition, the methods for calculating dose rates from ingestion are discussed.

1-8 AEROSOLS

An aerosol is a colloidal system consisting of small particles, either solid or liquid, in a gaseous suspension (Dennis, 1976). If there is no need to differentiate between the droplets and particles of an aerosol, the collective term "particulate matter" can be used. Particles are aggregations of either solid or liquid matter that are larger than molecules. Consequently the smallest particles have diameters of the order of 10^{-3} μm. Maximum diameters are near 10^2 μm. The size distribution spectrum for atmospheric aerosols as a function of surface area is described in Chap. 11, Fig. 11.27. The figure shows that the size (mass) distribution of particles is quite variable and depends strongly on the type of source (and on the distance from the source). For additional information on particle-size distribution, see the book by Cadle (1975), the report on aerosol size distributions over northern Europe (Cress, 1980), and the paper by Lodge et al. (1981).

There are many sources of atmospheric aerosols. They are the product of combustion, volcanic eruptions, and chemical reactions in the atmosphere. Additional sources are the resuspension of particles from the ground through the action of the wind and from cultivation of land. The oceans are large producers of aerosol particles, accounting for nearly half the natural source of airborne particles (Robinson, 1977). Blanchard and Woodcock (1957) demonstrated that almost all sea aerosol particles >1 μm in diameter are formed by the bursting of small air bubbles in waves. In addition, they showed that large quantities of particles with diameters of about 0.1 μm are formed in sea spray.

In the United States, approximately 1.2×10^9 kg of aerosol were produced by humans in 1977 (U. S. Department of Commerce, 1979). Globally, the annual production of aerosols by anthropogenic sources is of the order of 10^{11} kg, whereas natural sources are estimated to be of the order of 10^{12} kg (Robinson, 1977). Consequently 90% of all aerosols are derived from natural sources. An important natural source of aerosols is volcanic eruptions. For example, Stith, Hobbs, and Radke (1978) reported that the St. Augustine volcano in Alaska produced 4×10^5 kg of particles per second during violent eruptions. During less-active events, 40 to 300 kg/s were generated. On the basis of their calculations, Stith, Hobbs, and Radke estimated that St. Augustine emitted 2.5×10^8 kg of particles to the atmosphere from 1976 to 1977. Global estimates of the production of particulate matter by volcanoes range from 10^9 to 10^{11} kg/yr (*Science News*, 1978).

Aerosols are an important atmospheric constituent because of the important roles they play. They are important in cloud microphysics, where they act as condensation nuclei. Atmospheric visibility is determined by the mass and size distribution of aerosols. For example, visibility reductions of up to 50% have been attributed to urban aerosols (Komp and Auer, 1978). Particulate matter is a potential health hazard because particles can be inhaled.

Most of the sulfur released to the atmosphere ends up as sulfate aerosols having diameters of 0.1 to 1.0 μm. On the basis of a review of available sulfate data, Whitby (1978) concluded that the average diameter of sulfate aerosols was nearly 0.5 μm and he suggested that 95% of the atmospheric sulfate (and ammonium) consists of particles with diameters of 0.07 to 2.0 μm. Mamane and Pueschel (1980) have confirmed this conclusion. They found that 80 to 90% of the sulfate particles in one power plant plume had diameters slightly less than 0.5 μm. The large-size particles were mostly fly ash. Based on measurements from many field programs, Whitby (1978) concluded that marine aerosols contain 30 to 60% sulfate, whereas continental aerosols contain 15 to 30% sulfate. At normal humidities, aerosol concentrations of 20 to 30% sulfate are enough to make fine particles hydroscopic and ensure that they behave as optically ideal droplets. For additional information, see Chaps. 11 and 12.

Aerosol particle concentrations decrease markedly with elevation for the first 4 to 5 km above sea level. Below the 5-km level, decreases of a factor of 5/km can be deduced from published data for Aitken nuclei (e.g., Junge, 1963; and Wallace and Hobbs, 1977). Ground-level measurements of Aitken nuclei in large urban areas yield concentrations of 10^4 particles/cm^3.

On the basis of data collected by the National Air Sampling Network (NASN) in the United States, the spatial distribution of particulate matter can be found. The NASN concentrations are expressed in mass/unit volume, and particles >50 μm are not detected (Corn, 1968). In pristine areas of the western United States, 24-hr average suspended particulate concentrations are of the order of 1 to 10 μg/m^3. In built-up areas, average daily concentrations of 100 μg/m^3 are common, whereas in large cities and in areas where soil can be resuspended easily, average daily concentration can range from 100 to 1000 μg/m^3.

The primary hazard to health by particles comes from our ability to inhale them. Particles with diameters less than 50 μm can be inhaled (Lidwell, 1964) with normal breathing, but the more important health effects occur for smaller particles. Particles with diameters of 5 to 15 μm can enter upper respiratory areas and are especially important to persons with allergies. Particles with diameters less than 5 μm are considered respirable so that they can enter the lung and cause infection. For additional information, see Chap. 17.

Chemical reactions between gases and particles can occur in the atmosphere (see Sec. 1-6 and Chap. 10). Available evidence demonstrates that such reactions occur on the surface of particles and at rates proportional to the available surface. Size-distribution charts such as the one in Chap. 11 (Fig. 11.27) show that much of the surface area is on particles in the size range from 0.1 to 1 μm in diameter. Therefore many of the gas–particle interactions take place in this size range. By contrast, particle-to-particle interactions tend to occur at rates proportional to particle number concentrations. Junge (1963), Whitby (1978), and others have shown that the greatest number of particles are associated with diameters between 10^{-2} and 10^{-1} μm so that particle-to-particle interactions are concentrated within this size range.

1-9 ACIDIC PRECIPITATION†

Upon entering the atmosphere, water vapor can condense into the liquid phase as droplets. These droplets absorb atmospheric gases, forming acidic or alkaline solutions. The consequences of this phenomenon have led to the study of what is popularly referred to as "acid rain." A common example of the formation of acidic precipitation is the absorption of CO_2 by water droplets, forming carbonic acid (H_2CO_3). The initial pH of this type solution is approximately 5.5 to 5.6. However, there are many other constituents in the atmosphere that can combine with water to form acids (and bases). For example, soil particles and sea spray act as sources of positive ions (cations) such as Na^+, K^+, Ca^{2+}, and Mg^{2+}. Decaying organic matter is an important natural source of the ionic form of ammonia (NH_4^+). Atmospheric gases from natural and man-made sources can be oxidized and hydrolyzed into acidic solutions in droplets. For example, SO_2 can be oxidized to yield sulfate ions and sulfuric-type acids whereas nitrogen oxides eventually end up as nitrates in precipitation. All these reactions tend to increase the acidity of precipitation. The net result is acidified precipitation with a pH ranging from near 2 to near 7 (neutral). Precipitation can be alkaline too. Allan and Jonasson (1978) found pH values of snow in Winnipeg and Ottawa, Canada, to range from 7 to 11. They speculated that the alkaline snow was related to Ca and Mg being released into the atmosphere from a limestone roasting plant.

It has been widely publicized that the acidity of precipitation is (1) increasing, (2) caused by coal-fired power plants, and (3) causing harm to the environment (e.g., see Likens et al., 1979; and Chap. 10, Sec. 10-7.4.3). One area where acid rain is thought to be increasing is in the state of New York, but the data tend to yield conflicting results. For example, Fig. 1.7 shows the 14-yr variation in mean monthly pH at two stations located in New York. These two stations were selected for their length of record and because similar variations appeared in the other nine records, although the one for Upton, N. Y., (1965 to 1973) showed variations confined between pH values of 4.0 and 5.0. The data were collected by the United States Geological Survey (Albany, N. Y.), were plotted by the Electric Power Research Institute (EPRI), and made available by Perhac (1980). The charts in Fig. 1.7 are self-explanatory and demonstrate that (1) acid rain does occur, (2) there is considerable variation in pH from one location to another, (3) the pH can change markedly from one month to the next, and (4) the most acidic monthly mean value of pH is near 4.0 and the least acidic events have values near 7.0. In addition, the curves do not suggest a trend toward more acidic precipitation; in fact, Perhac (1980) noted that statistical analysis of the data used in Fig. 1.7 yields a 95% confidence of no change in 14 yr. A similar result was reported by Likens et al. (1977). They analyzed the hydrogen ion content of precipitation col-

†See Chaps. 10 and 11 for more details.

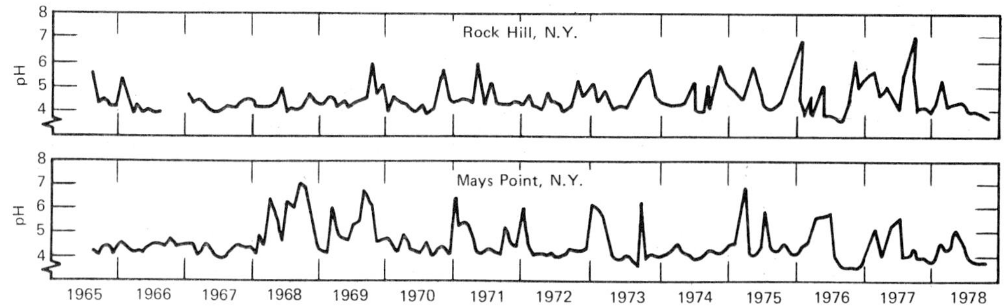

Fig. 1.7 The 14-yr variation in the mean monthly pH at two sites in New York.

lected at the Hubbard Brook Experimental Forest located near West Thornton, N. H. The result of this analysis showed that there was a slight upward trend in the annually weighted average concentrations of hydrogen ion for 1964 to 1965 and for 1970 to 1971. The 1970 to 1971 trend was followed by a downward trend until 1973 to 1974. For the 1964 to 1974 period, no trend in the hydrogen ion content was found to be statistically significant. In addition, the authors noted that the concentrations for SO_4^{2-} and for NH_4^+ did vary from year to year but there were no significant statistical trends for the 10-yr period.

In retrospect, the data seem to give conflicting results on whether or not the acidic nature of precipitation is increasing. The data must be used with caution. Conflicting data may be masking the importance of any power plant contribution to the acidity of precipitation. Moreover, Patrick, Binetti, and Halterman (1981) emphasize that the acidification of lakes can occur from both natural and man-made causes. It is important to separate these two components to understand any anthropogenic contribution. Furthermore, air pollution control equipment may be removing alkaline particles that once served as buffers in power plant plumes.

1-10 VISIBILITY

Visibility is the degree of clearness of the atmosphere. Visibility is a concept and visual range is a physical variable used as a measure of visibility. Visual range is the greatest distance toward the horizon that a prominent black object can be identified visually with the naked eye if viewed horizontally through a homogeneous atmosphere. Visibility is not only influenced by the clarity of the air between the observer and a given object, but also by the sun angle, by the interplay of light and shadow, and by the colors of the object and of its background. Atmospheric particulates and gases can scatter and absorb light (McCartney, 1976); however, the important effects on visibility appear to be caused by the scattering of light (Charlson, 1969) from particles with diameters of 0.1 to 1.0 μm and by the absorption of light (Roessler and Faxvog, 1981) by sulfates and nitrates and by gases, such as NO_2. Relative humidity plays an important role in determining the size distribution of particles and on their ability to scatter light. In fact, for relative humidities above 70%, the visibility decreases rapidly (Neuberger, 1957 and Chap. 20, Fig. 20.11). For example, Wallace and Hobbs (1977) note that the amount of light scattered by sea-salt particles increases by about a factor of 3 as the relative humidity increases from 60 to 80%.

Contrast determines visibility. Contrast, of both color and brightness, is reduced by the scattering of light into the line of sight to a black object. If B_b is the brightness of the background and $B_b + dB_b$ is the brightness of an object (B_o) seen against the background, the brightness contrast, C, is defined as $C = dB_b/B_b$. In the absence of color contrast, an object is visible as long as $B_b - B_o/B_b \geqslant E$, where E is the brightness contrast threshold of the eye. For an average eye under daylight conditions, E is taken as 0.02, but it can vary. Using Beer's Law (irradiance decreases monotonically with increasing path length through an atmospheric layer), it can be shown (e.g., Tricker, 1970) that $C = -e^{\sigma r}$, where σ is the extinction coefficient of the air and r is the distance to the target. Consequently the visibility or visual range, V, is obtained when the brightness contrast is equal to the threshold value E, or, $C = -E = -e^{-\sigma r}$ so that $V = (1/\sigma) \ln (1/E)$. If $E = 0.02$, then the visual range is simply $V = 3.912/\sigma$ or approximately $4/\sigma$. For a particle-free atmosphere at sea level, σ is about 1.5×10^{-2} km^{-1} so that V is nearly 270 km. For the U. S. Standard Atmosphere (1962), σ is about 0.16 km^{-1} so that V = 25 km.

Sulfates, nitrates, and other materials can degrade visibility. According to Robinson (1977), if the concentration of sulfate particles is 10 μg/m^3, if their density is 1.67, and if they all have a radius of 0.3 μm, then V is approximately 65 km at sea level. However, for concentrations as high as 80 μg/m^3, the maximum visibility would decrease to about 8 km for the same particle density and radius. Robinson notes that the size distribution of the particles is very important in determining V. According to Trijonis and Yuan (1978a), sulfate is the single major contributor to haze in the northeastern United States. They reported that over this area sulfates appear to account for approximately 50% of the total extinction. In the southwestern United States, Trijonis and Yuan (1978b) noted that sulfates account for 32 to 53% of the extinction of light and that nitrates account for 23 to 37%. They estimated an extinction coefficient of 0.04 ($\times$10 km)$^{-1}$/μg/m^3 for sulfates in the southwestern states; slightly smaller than the model calculations of Waggoner et al. (1976) and others. However, their estimated extinction coefficient for sulfates in the northeastern states was similar to those computed by Waggoner et al. (1976). Investigation of the details of visibility degradation from sulfates, nitrates, and other materials is just beginning so that more information will be available in the future.

Visibility has become an important factor in the development of air pollution control strategies. With the Clean Air Act Amendments of 1977 (Sec. 169A), Congress gave the Environmental Protection Agency (EPA), the states, and the Federal Land Managers the responsibility to protect air quality (including visibility) in the national parks and in wilderness areas that have been designated Class I areas.† Proposed regulations for visibility have been recommended by EPA (*Federal Register,* 1980a). For more information, see Chap. 19. For a discussion of visibility and air pollution, see Chap. 20.

1-11 AIR-QUALITY MODELS

In the United States, atmospheric pollution is controlled through a complex array of programs established under the Clean Air Act (see Chap. 19 for details). This act establishes air-quality goals in the form of National Ambient Air Quality Standards (NAAQS). The ambient standards are numerical representations of the act's objectives that air pollution not endanger the public health or welfare.‡ On the basis of these objectives, two standards were promulgated to assure that neither the public health nor the public welfare is threatened. Public health is protected through the primary standards and public welfare by the secondary standards (see Chap. 19). Responsibility for achieving these standards rests primarily with the states. The basic tool for achieving these standards is provided to the states through the State Implementation Plan (SIP). According to the Clean Air Act, the states are required to devise SIPs that designate specific emission limitations for sources to ensure that ambient levels are not exceeded. When attempting to relate air-quality goals to the specific limitations on emissions, it becomes abundantly clear that a reliable method of simulating atmospheric processes is needed. A useful mechanism for simulating these processes is a mathematical model. Basically, such a model is a relationship between the amount of pollution released into the atmosphere by the source and the corresponding incremental contribution the source makes to ambient concentrations. Indeed, this representation is so important that it is commonly referred to as an air-quality model. Such models are essential tools in the management of air quality. To date, they appear to be the only objective procedure available to predict the results of proposed actions. Chapters 13 and 14 describe the fundamental concepts on which air-quality models are established. The Environmental Protection Agency has issued a document on guidelines for air-quality models (EPA, 1978). Turner (1979); Drake, McNaughton, and Huang (1979); Budiansky (1980), and others have reviewed air-quality models and procedures. Some long-range (10^2 to 10^3 km) transport models have been reviewed by Eliassen (1980).

The use of models for air-quality management has been formalized by the Clean Air Act Amendments (CAAA) of 1977. For example, these amendments established the concept of the Prevention of Significant Deterioration (PSD) to be applied before the construction of all new or modified major emitting facilities. Projected emissions from these facilities should be modeled to determine if these sources (in combination with other sources subject to PSD increments) would cause the PSD increments for SO_2 and total suspended particulates (TSP) to be exceeded. Since a PSD review is conducted before construction, the focus of the permit process is on air-quality modeling. Other requirements for air-quality models in the CAAA can be found in provisions for air-quality maintenance plans, in new source performance standards, and in the protection of visibility in pristine areas. Consequently, in many cases, results from models can be a deciding factor in whether or not the construction of a new facility can proceed. Recognizing the burden of responsibility that the CAAA provisions place on models, Congress sought to ensure that the best possible models be used. For this reason, the CAAA requires EPA to hold a national modeling conference at 3-year intervals. Such a conference was scheduled to be held in 1981 (*Federal Register,* 1980b). The first modeling conference was held by EPA in December 1977 in Chicago, Ill.

At the five public meetings preceding the first national modeling conference, as well as at the conference, numerous recommendations were made to EPA. For example, many participants emphasized the need for consistency between air-quality models used in the regulatory process. In addition, the participants noted the need for performance measures and performance standards. Such standards would provide a means of objectively evaluating models and of judging between models.

The drive toward consistency began as a result of disputes between modelers whose predictions were at some degree of variance with each other. In particular, when one model prediction is below the appropriate ambient air-quality standard and the other is above it, a good deal of legal and regulatory energy may be expended. The proposed solution to this problem is to standardize the application of models. Responding to this request, EPA developed a draft modeling guideline document and subjected it to wide technical review. Included in this review was the convening of a specialist's conference to review the concept and the draft. A subgroup discussing validation and calibration of models at this conference recommended that

†A Class I area is one where any additional degradation of air quality is limited to specified increments of pollution (see Chap. 19, Table 4).

‡There are investigators (e.g., Tabershaw, I. R., 1976; and Ellsaesser, 1979, 1981) who argue that the NAAQSs are invalid because there is no believable evidence of health effects at such low concentrations. The American Medical Association (1981), through its Council on Scientific Affairs, has noted that in the United States there is no definitive epidemiologic evidence of an overall increase in the incidence of cancer related to high concentrations of pollutants in the environment.

models meet certain standards of performance if they are to be endorsed in the *Guideline* (EPA, 1978) for general use. This recommendation was made because air-quality simulation models were changing rapidly as a result of much development work in progress and because much of the necessary data was not in place.

Since the formal promulgation of the *Guideline* in April 1978, EPA has responded with a number of studies designed to identify various aspects of the performance of air-quality simulation models. But establishing model performance standards is not done easily (Hayes, 1979). The criteria used to select models for recommendation in the *Guideline* were mostly oriented around convenience. Where there were a number of models competing for selection in the *Guideline,* factors such as model availability and familiarity to the regulatory community weighed very heavily. Although these are not insignificant concerns, it seems more logical that the basis of model selection should at least include some formal consideration of how well model outputs compare with the phenomena it simulates, i.e., a comparison with data. Furthermore, there seems to be a general consensus that such models should have a valid scientific and physical basis.

In September 1980, under a cooperative agreement with EPA (American Meteorological Society, 1979), the American Meteorological Society convened a workshop to address performance measures and standards. The workshop participants made specific recommendations on statistical procedures to be used to evaluate model performance, but they decided it was not appropriate to propose and establish specific numerical standards for model performance (Fox, 1981). The participants judged that there were insufficient cases of rigorous model evaluations to enable a standard to be suggested.

It is important to emphasize that models represent approximations to real conditions and are therefore subject to error. One error results from the fact that atmospheric motions can only be measured to within a certain tolerance (see Chaps. 2 and 3). This tolerance could be expressed in terms of confidence limits on the simulations. Also, only average conditions can be measured and, therefore, predicted. Furthermore, emission data are usually incomplete and/or inaccurate and thereby contribute to the overall model error. Consequently simulations by air-quality models can be no more accurate than the input data. For this reason, models should not be used as the sole determinant in a pass/fail context for making decisions concerning air-quality impacts. Models must not replace rational and informed decisión making. Decision makers must become more sensitive to the fact that model predictions are reliable within certain confidence limits or range of values. Modelers must be more diligent in reporting these limits.

1-12 RISK ANALYSIS AND POWER PRODUCTION

For every method of producing power, there are certain inherent dangers that relate to occupational and public health effects. A method of accounting for these dangers is known as risk analysis. In power production, risk manifests itself in the form of accidents and disease, both of which can result in injury or death. Risk is part of the social cost of power production and includes such other consequences as land abuse, depletion of resources, and air and water pollution.

During the 1970s, there were many scholarly studies of the risks of power production. One of the earliest was by Hull (1971) who analyzed the environmental risks from nuclear and fossil-fuel power plants. He corroborated the suggestion by Eisenbud and Petrow (1964) that the radioactivity released from coal-fired power plants was more significant than that from pressurized-water reactors (PWR). However, he also found that the radioactivity from coal-fired plants was less significant than from boiling-water reactors (BWR). In a similar study, Lave and Freeburg (1973) analyzed the health effects of electricity generation from coal, oil, and nuclear fuels. They concluded that both nuclear and coal-fired plants discharge radioactivity into the environment in amounts that have little effect on background radiation levels. In the most conservative comparison made, they found that a PWR appears to offer 1.8×10^3 times less health risk than a corresponding coal-fired plant, and a BWR, with a 30-min holdup of stack gases, offered 24 times less health risk than the coal-fired plant. A similar result has been reported by McBride et al. (1978).

Lave and Freeburg (1973) emphasized that a proper evaluation of risk should include the entire fuel cycle rather than just the power-generation phase. Using this concept, Inhaber (1978, 1979) has completed an analysis of risk. He compared the risks from conventional power sources (e.g., coal, oil, nuclear, and natural gas) and nonconventional sources (e.g., solar, methanol, and ocean thermal gradients). In his study, Inhaber identifies seven sources of risk: (1) fuel production, (2) component fabrication, (3) plant construction, (4) transportation, (5) plant operation and maintenance, (6) waste disposal, and (7) public health. He totals all these factors in an evaluation of the total risk (occupational and public health effects) for nine different power production facilities. He concluded that risks associated with nonconventional power sources can be as great or even greater than those of conventional power sources. Power production based on natural gas seems to offer the least total risk, with nuclear power second.

The public health component in the risk of power production has been evaluated by the Office of Technology Assessment. According to an investigation conducted at the Brookhaven National Laboratory in 1978 and reported in *The Direct Use of Coal* (Office of Technology Assessment, 1979), emissions from coal combustion are responsible for approximately 2.5% (48 000 deaths) of the annual mortality in the United States. Projections are that increased use of coal could raise the mortality rate to nearly 56 000 people/yr by 1990. In a similar study at the request of the U. S. Senate Committee on Public Works, the National Academy of Sciences, in cooperation with the National

Academy of Engineers and the National Research Council, related coal combustion to public health. This report (U. S. Senate, 1975) estimated that each year the sulfur emissions from a modern, remote-sited, 1000-MW coal-fired plant would cause 20 premature deaths, 40 000 cases of chronic respiratory disease, 400 000 person-days of aggravated heart-lung disease symptoms, 80 000 asthma attacks, and 10 000 cases of children's respiratory disease.

Studies such as these are often challenged by many in the public health community. In fact, there is widespread skepticism that any air pollutants at current levels of concentration are killing people in large numbers (e.g., Lipfert, 1980). For example, as long as people die and effluents such as SO_2, NO_x, CO, and particulates are released into the atmosphere (as they would be in the absence of humans), statistical analyses will imply deaths due to the released effluents. However, such reports represent the first comprehensive attempts to formulate a quantitative methodology for assessing the health effects of power production. For additional information on the potential health effects of power production, see Chap. 17. A detailed and extensively referenced report on the health effects of the products of combustion is being prepared by the Office of Technology Assessment as an appendix to *The Direct Use of Coal.*

1-13 ROLE FOR THE METEOROLOGIST IN THE POWER PRODUCTION INDUSTRY

Utility companies draw on the science of meteorology in four phases: (1) site selection, (2) technical services, (3) routine operations, and (4) during or after accidental releases of hazardous materials. The relative importance of these four factors may vary from one utility to another. Scientific support for these various needs originates in research and development programs.

Site selection is a complicated decision and depends not only on the local climatology but also on such factors as the availability of land, water, labor, and transportation. In recent years, the meteorologist's role in site selection has been broadened by laws passed by Congress in an effort to control the release of man-made effluents into the atmosphere, as well as to land and water surfaces. Consequently an important responsibility of the meteorologist in the power production industry is to determine the air-quality impact of proposed power plants. This determination is made through field measurements and mathematical models so as to specify emission control requirements and to determine acceptable sites for the power plant.

The meteorologist may be called upon during the design phase of the power plant to provide engineers with information on severe or extreme weather conditions that might occur at the selected site. Power-generating equipment, cooling towers, and buildings must be designed to withstand severe weather events. Furthermore, complete meteorological data sets are needed to provide proper information for the efficient operation of heat-dissipation systems.

Technical services pertaining to atmospheric sciences are also provided by the meteorologist. Technical guidance is needed by utilities if they convert from one fuel to another. A meteorologist usually prepares the environmental impact statement for, say, the conversion from oil to coal. The meteorologist may also be asked to testify before state or federal agencies responsible for monitoring and regulating the release of combustion products from power plants. Utilities often ask meteorologists to represent utility views at public hearings pertaining to air-quality regulations and air-quality modeling techniques. Consequently a meteorologist who plans to work for a utility must become familiar with and keep up to date on air-quality regulations and changing legislation that might affect the operation of the company.

Energy policy issues are becoming important in planning for various fuel strategies, especially for those relating to air-quality offsets. In other words, based on air-quality regulations, where should the company burn natural gas rather than, say, coal? Where can coal be used and still permit the company to remain in compliance with air-quality regulations? Because of uncertainties in the cost and availability of fuels, the meteorologist may be called upon to advise a utility on the environmental consequences of using different fuel strategies.

Monitoring air quality levels at the power plant is necessary to ensure conformity with Federal Air Quality Standards. Decisions concerning the number and locations of on-site air quality and meteorological stations must depend on the judgment of a competent meteorologist who can work effectively with power engineers and health and safety personnel.

Meteorological information is also used by utilities during routine operations (e.g., see Suchman, Auvine, and Hinton, 1979). Advance warning of extreme changes in weather conditions are beneficial to effective plant operation. Prior knowledge of unexpected peak demands due to weather are helpful in the smooth, efficient operation of complex energy networks. In addition, advance warning of heavy ice or wind loading on power lines, or intense lightning discharges are also beneficial to operational considerations. Prediction of the occurrence of prolonged periods of air stagnation are important because they are conducive to the local accumulation of combustion products in the atmosphere. The transport of effluents over great distances is now important in the consideration of visibility degradation. In the event of accidental release of hazardous materials to the atmosphere, prediction of the likely path and concentration of the effluent will be essential to decisions on control, treatment, or evacuation.

In the domain of alternative energy sources, the meteorologist may be called upon to address the potential environmental impacts that could result from their implementation (see Chap. 18). For example, tapping the natural fluxes of energy flow in the atmospheric boundary layer for wind or solar energy involves interference with geophysical budgets whose responses must be evaluated before alternative energy sources can move ahead to large-scale use. These evaluations may include climatological summaries for use in determining wind speed and solar radiation flux densities available in a given area so that the practicality of using these energy sources can be assessed objectively. For example, would tapping ocean currents for power production alter climate? Or, would large-scale development of a geothermal energy source significantly increase regional scale (Randerson, 1976) concentrations of H_2S, NH_4, CH_4, and water vapor? Chapter 18 emphasizes that these and other pertinent questions must be addressed if alternative energy sources are to be judged with the same environmental and economic standards imposed on conventional energy sources.

The preceding paragraphs are not intended to be a complete review of the possible duties of a meteorologist working in the power production industry. Rather, the intent was to emphasize the broad responsibilities a meteorologist can have. In fact, the topics reviewed in this chapter and described in detail in the following chapters emphasize that a meteorologist needs a broad background in atmospheric science to work effectively and authoritatively in the power production industry. The intent of this book is to provide the fundamental knowledge needed to accomplish these and other duties effectively.

REFERENCES

Airborne Particles, 1979, Subcommittee on Airborne Particles, Committee on Medical and Biological Effects of Environmental Pollutants, National Research Council, National Academy of Sciences, University Park Press, Baltimore, Md.

Allan, R. J., and I. R. Jonasson, 1978, Alkaline Snowfalls in Ottawa and Winnipeg, Canada, *Atmos. Environ.,* 12: 1169-1173.

American Medical Association, 1981, Carcinogen Regulation, *J. Am. Med. Assoc.,* 246: 253-256.

American Meteorological Society, 1979, Air-Quality Modeling, *Bull. Am. Meteorol. Soc.,* 60: 1224.

——, 1978, Accuracy of Diffusion Models, *Bull. Am. Meteorol. Soc.,* 59: 1025-1026.

Atlas, E., and C. S. Giam, 1981, Global Transport of Organic Pollutants: Ambient Concentrations in the Remote Marine Atmosphere, *Science,* 211: 163-165.

Atmospheric Environment, 1978, Sulfur in the Atmosphere, 12: 18.

Auer, A. H., 1976, Observations of an Industrial Cumulus, *J. Appl. Meteorol.,* 15: 406-413.

Baes, C. F., H. E. Goeller, J. S. Olsen, and R. M. Rotty, 1977, Carbon Dioxide and Climate: The Uncontrolled Experiment, *American Scientist,* 65: 310-327.

Barnes, R. A., 1977, Sulphur Deposit Account, *Nature,* 268: 92-93.

Barrie, L. A., and J. L. Walmsley, 1978, A Study of Sulphur Dioxide Deposition Velocities to Snow in Northeastern Canada, *Atmos. Environ.,* 12: 2321-2332.

Beckmann, P., 1978a, Coal Waste Management, *Access Energy,* 6(1): 2-3.

——, 1978b, Is Solar Energy Without Wastes?, *Access Energy,* 6(3): 3.

——, 1977, Solar Power, Not Solar Piddling, *Access Energy,* 4(5): 4.

——, 1976, *The Health Hazards of Not Going Nuclear,* The Golem Press, Boulder, Colo.

Blanchard, D. C., and A. H. Woodcock, 1957, Bubble for Motion and Modification in the Sea and Its Meteorological Significance, *Tellus,* 9: 145-158.

Boldt, K. R., C. P. Chang, E. J. Kaplin, J. M. Stansfield, B. R. Wuebber, 1980, Impact of a Primary Sulfate Emission Source on Air Quality, Report EPA-600/2-80-109, Research Triangle Park, N. C.

Bolin, B., 1979, On the Role of the Atmosphere in Biogeochemical Cycles, *Quart. J. Roy. Meteorol. Soc.,* 105: 25-42.

Bosanquet, C. H., and J. L. Pearson, 1936, The Spread of Smoke and Gases from Chimneys, *Trans. Faraday Soc.,* 32: 1249-1264.

Bowne, N. E., and G. R. Hilst, 1981, Overview of the EPRI Plume Validation Model Validation Project, in *Proceedings of the Fifth Symposium on Turbulence, Diffusion, and Air Pollution,* Mar. 9–13, 1981, pp. 1-2, Atlanta, Ga., American Meteorological Society, Boston, Mass.

Brice, K. A., and R. G. Derwent, 1978, Emissions Inventory for Hydrocarbons in the United Kingdom, *Atmos. Environ.,* 12: 2045-2054.

Bridgman, H. A., 1978, Direct Visible Spectra and Aerosol Optical Depths at Urban and Rural Locations During the Summer of 1975 at Milwaukee, *Sol. Energy,* 21: 139-148.

Briggs, G. A., 1973, Diffusion Estimation for Small Emissions, in *Atmospheric Turbulence and Diffusion Laboratory 1973 Annual Report,* NOAA/ATDL Publication No. ATDL-106, Oak Ridge, Tenn., pp. 739-747.

Brooks, J. M., 1979, Deep Methane Maxima in the Northwest Caribbean Sea: Probable Seepage Along the Jamaica Ridge, *Science,* 206: 1069-1071.

Budiansky, S., 1980, Dispersion Modeling, *Environ. Sci. and Tech.,* 14: 370-374.

Cadle, R. D., 1975, *The Measurement of Airborne Particles,* Wiley-Interscience, New York.

Calvert, J. G., F. Su, J. W. Bottenheim, and O. P. Strausz, 1978, Mechanism of the Homogeneous Oxidation of Sulfur Dioxide in the Troposphere, *Atmos. Environ.,* 12: 197-226.

Chameides, W. L., 1978, The Effects of Increased CO and NO_X upon Tropospheric OH and CH_4, in *Man's Impact on the Troposphere,* Lectures in Tropospheric Chemistry, J. S. Levine and D. R. Schryer (Eds.), NASA-RP-1022, pp. 81-108, Science and Technical Information Office.

——, S. C. Liu, and R. J. Cicerone, 1977, Possible Variations in Atmospheric Methane, *J. Geophys. Res.,* 82: 1795-1798.

Charlson, R. J., 1969, Atmospheric Visibility Related to Aerosol Mass Concentration, *Environ. Sci. Technol.,* 3: 913-918.

Chemical and Engineering News, 1978, Power Plant Ash Disposal: A Growing Problem, *Chem. Eng. News,* 56(45): 26.

Cleveland, W. S., and T. E. Graedel, 1979, Photochemical Air Pollution in the Northeastern United States, *Science,* 204: 1273-1278.

——, and J. E. McRae, 1978, Weekday–Weekend Ozone Concentrations in the Northeast United States, *Environ. Sci. Technol.,* 12: 558-563.

Cofer, W. R., D. R. Schryer, and R. S. Rogowski, 1980, The Enhanced Oxidation of SO_2 by NO_2 on Carbon Particulates, *Atmos. Environ.,* 14: 571-575.

Cohen, B. L., 1977, The Disposal of Radioactive Wastes from Fission Reactors, *Sci. Am.,* 236: 21-31.

Cohen, Y., and L. I. Gordon, 1979, Nitrous Oxide Production in the Ocean, *J. Geophy. Res.,* 84: 347-353.

Cole, F. W., 1975, *Introduction to Meteorology,* John Wiley & Sons, Inc., New York (p. 171).

Conrad, R., and W. Seiler, 1980, Field Measurements of the Loss of Fertilizer Nitrogen into the Atmosphere as Nitrous Oxide, *Atmos. Environ.,* 14: 555-558.

Corn, M., 1968, Nonviable Particles in Air, *Air Pollution,* Vol. 1, A. C. Stern (Ed.), Academic Press, New York.

Cox, R. A., 1974, Particle Formation from Homogeneous Reactions of Sulphur Dioxide and Nitrogen Dioxide, *Tellus,* 26: 235-240.

Cress, T. S., 1980, Airborne Measurement of Aerosol Size Distributions over Northern Europe, AFGL-TR-80-0178, Environmental Research Papers, No. 702, Air Force Geophysics Laboratory, Hanscon Air Force Base, Massachusetts.

Crutcher, H. L., 1970, Centralized Quality Control and Evaluation Programs, *Meteorol. Monogr.,* 11: 137-140.

Crutzen, P. J., 1974, Estimate of the Possible Variations in Total Ozone Due to Natural Causes and Human Activities, *Ambio,* 3: 201-210.

——, and J. Fishman, 1977, Average Concentrations of OH in the Troposphere and the Budgets of CH_4, CO, H_2, and CH_3Cl_3, *Geophys. Res. Lett.,* 4: 321-324.

Cullis, C. F., and M. M. Hirschler, 1980, Atmospheric Sulfur: Natural and Man-Made Sources, *Atmos. Environ.,* 14: 1263-1278.

Davis, D. D., M. Heaps, and T. McGee, 1976, Direct Measurements of Natural Tropospheric Levels of OH via an Aircraft-Borne Turnable Dye Laser, *Geophys. Res. Lett.,* 3: 331-333.

Davis, E. A., 1979, *Environmental Assessment of Chalk Point Cooling Tower Drift and Vapor Emissions,* JHU-PPSP-CPCTP-28 by the Johns Hopkins University, Applied Physics Laboratory, Johns Hopkins Road, Laurel, Md.

Dawson, G. A., 1980, Nitrogen Fixation by Lightning, *J. Atmos. Sci.,* 37: 174-178.

Dennis, R., 1976, Introduction: Objectives, Organization, and Applications, in *Handbook on Aerosols,* R. Dennis (Ed.), Chap. 1, pp. 1-8, ERDA Report TID-26608, NTIS.

Department of Energy, 1979, *Environmental Readiness of Emerging Energy Technologies,* Summary Report, January 1979, Report DOE/ERD-0022, NTIS.

Doran, J. C., T. W. Horst, and P. W. Nickola, 1978, Variations in Measured Values of Lateral Diffusion Parameters, *J. Appl. Meteorol.,* 17: 825-831.

Drake, R. L., D. J. McNaughton, and C. Huang, 1979, *Available Air Quality Models,* Appendix D of Mathematical Models for Atmospheric Pollutants, Report EPRI-EA-1331, Project 805, Electric Power Research Institute, Palo Alto, Calif.

Draxler, R. R., 1976, Determination of Atmospheric Diffusion Parameters, *Atmos. Environ.,* 10: 99-105.

Duce, R. A., 1978, Speculations on the Budget of Particulate and Vapor Phase Nonmethane Organic Carbon on the Global Troposphere, *Pure Appl. Geophys.,* 116: 244-274.

Ehhalt, D. H., 1974, The Atmospheric Cycle of Methane, *Tellus,* 26: 58-70.

Eisenbud, M., and H. G. Petrow, 1964, Radioactivity in the Atmospheric Effluents of Power Plants that Use Fossil Fuels, *Science,* 144: 288-289.

Eliassen, A., 1980, A Review of Long-Range Transport Modeling, *J. Appl. Meteorol.,* 19: 231-240.

Ellsaesser, H. W., 1979, Air Pollution: A Different View, *Water, Air, Soil Pollut.,* 11: 115-127.

——, 1981, Statement Prepared for Presentation at the Hearings on Consolidation Coal Company's Petition to Amend the Primary Ambient Air Quality Standards for Sulfur Dioxide, *The Science of the Total Environment,* 17: (in press).

Enger, L., and U. Högström, 1979, Dispersion and Wet Deposition of Sulfur from a Power Plant Plume, *Atmos. Environ.,* 13: 797-810.

Environmental Protection Agency, 1978, *Guideline on Air-Quality Models,* Report EPA-450/2-78-027, Research Triangle Park, N. C.

——, 1976a, *Quality Assurance Handbook for Air Pollution Measurement Systems,* Vol. I, *Principles,* Report EPA-600/9-76-005, Research Triangle Park, N. C.

——, 1976b, *1973 National Emissions Report,* Report EPA-450/2-76-007, May 1976, Technical Publications Branch, Research Triangle Park, N. C.

Federal Register, 1980a, Visibility Protection for Federal Class I Areas, 45: 34762-34779.

——, 1980b, *Guidelines on Air Quality Models,* 45(61): 20157-20158.

Forrest, J., and L. Newman, 1977, Further Studies on the Oxidation of Sulfur Dioxide in Coal-Fired Power Plant Plumes, *Atmos. Environ.,* 11: 465-474.

——, S. E. Schwartz, and L. Newman, 1979, Conversion of Sulfur Dioxide to Sulfate During the DaVinci Flights, *Atmos. Environ.,* 13: 157-167.

Fox, D. G., 1981, Judging Air-Quality Model Performance, *Bull. Am. Meteorol. Soc.,* 62: 599-609.

Galloway, J. N., and D. M. Whelpdale, 1980, An Atmospheric Sulfur Budget for Eastern North America, *Atmos. Environ.,* 14: 409-418.

Garland, J. A., 1978, Dry and Wet Removal of Sulphur from the Atmosphere, *Atmos. Environ.,* 12: 349-362.

Gifford, F. A., 1980, Smoke as a Quantitative Atmospheric Diffusion Tracer, *Atmos. Environ.,* 14: 1119-1121.

——, 1961, Use of Routine Meteorological Observations for Estimating Atmospheric Dispersion, *Nucl. Safety,* 2(4): 47-57.

——, 1976, Turbulent Diffusion-Typing Schemes: A Review, *Nucl. Safety,* 17(1): 68-86.

——, 1959, Smoke Plumes as Quantitative Air Pollution Indices, *Int. J. Air Pollut.,* 2: 42-50.

Glossary of Meteorology, 1959, American Meteorological Society, Boston, Mass.

Goldstein, N. P., K. H. Sun, and J. L. Gonzales, 1971, *Trans. Am. Nucl. Soc.,* 14: 66.

Graedel, T. E., 1976, Sulfur Dioxide, Sulfate Aerosol, and Urban Ozone, *Geophys. Res. Lett.,* 3: 181-184.

Grennfelt, P., 1980, Investigation of Gaseous Nitrates in an Urban and a Rural Area, *Atmos. Environ.,* 14: 311-316.

Guccione, E., 1977, Why Coal Won't Be America's Energy Salvation, *Reason,* 9: 16-20.

Haagen-Smit, A. J., 1952, Chemistry and Physiology of Los Angeles Smog, *Ind. Eng. Chem.,* 44: 1342-1346.

Hahn, J., and C. Junge, 1977, Atmospheric Nitrous Oxide: A Critical Review, *Z. Naturforsch.,* 32a: 190-214.

Hales, J. M., 1978, Wet Removal of Sulfur Compounds from the Atmosphere, *Atmos. Environ.,* 12: 389-399.

Hall, F. F., Jr., C. E. Duchon, L. G. Lee, and R. R. Haugen, 1973, Long-Range Transport of Air Pollution: A Case Study, August 1970, *Mon. Weather Rev.,* 101: 404-411.

Hanna, S. R., G. A. Briggs, J. Deardorff, B. A. Egan, F. A. Gifford, and F. Pasquill, 1977, AMS Workshop on Stability Classification Schemes and Sigma Curves: Summary of Recommendations, *Bull. Am. Meteorol. Soc.,* 58: 1305-1309.

——, and F. A. Gifford, 1975, Meteorological Effects of Energy Dissipation at Large Power Parks, *Bull. Am. Meteorol. Soc.,* 56: 1069-1076.

Hartung, R., and G. W. Klinger, 1970, Concentration of DDT by Sedimented Polluting Oils, *Environ. Sci. Technol.,* 4: 407-410.

Harvey, G. R., and W. G. Steinhauser, 1974, Atmospheric Transport of Polychlorobiphenyl to the North Atlantic, *Atmos. Environ.,* 8: 777-782.

Haugen, D. A. (Ed.), 1973, *Workshop on Micrometeorology,* American Meteorological Society, Boston, Mass.

Hayek, F. A., 1977, Planning Our Way to Serfdom, *Reason,* 8: 26-30.

Hayes, S. R., 1979, Performance Measures and Standards for Air-Quality Simulation Models, Report EPA-450/4-79-032, Research Triangle Park, N. C.

Hegg, D. A., and P. V. Hobbs, 1980, Measurements of Gas-to-Particle Conversion in the Plumes from Five Coal-Fired Electric Power Plants, *Atmos. Environ.,* 14: 99-116.

Henmi, T., and E. R. Reiter, 1978, Regional Residence Time of Sulfur Dioxide over the Eastern United States, *Atmos. Environ.,* 12: 1489-1495.

Hicks, B. B., M. L. Wesley, and J. L. Durham, 1980, Critique of Methods to Measure Dry Deposition, Report EPA-600/9-80-050, September 1980, Environmental Protection Agency, Washington, D. C.

Hill, R. D., R. G. Rinker, and H. D. Wilson, 1980, Atmospheric Nitrogen Fixation by Lightning, *J. Atmos. Sci.,* 37: 179-192.

Hirst, J. M., O. J. Stedman, and G. W. Hurst, 1967, Long-Distance Spore Transport: Vertical Sections of Spore Clouds over the Sea, *J. Gen. Microbiol.*, 48: 357-377.

Hitchcock, D. R., 1976, Atmospheric Sulfates from Biological Sources, *J. Air Pollut. Control Assoc.*, 26: 210-215.

——, L. L. Spiller, and W. E. Wilson, 1980, Sulfuric Acid Aerosols and HCl Release in Coastal Atmospheres: Evidence of Rapid Formation of Sulfuric Acid Particulates, *Atmos. Environ.*, 14: 165-182.

Hobbs, P. V., J. L. Stith, and L. F. Radke, 1980, Cloud-Active Nuclei from Coal-Fired Electric Power Plants and Their Interactions with Clouds, *J. Appl. Meteorol.*, 19: 439-451.

Hoffer, T., J. Kliwer, and J. Moyer, 1979, Sulfate Concentrations in the Southwestern Desert of the United States, *Atmos. Environ.*, 13: 619-627.

Hoffnagle, G. F., M. E. Smith, T. V. Crawford, and T. J. Lockhart, 1981, On-Site Meteorological Instrumentation Requirements to Characterize Diffusion from Point Sources: A Workshop, 15–17 January 1980, Raleigh, N. C., *Bull. Am. Meteorol. Soc.*, 62: 255-261.

Hosker, R. P., Jr., 1980, Practical Application of Air Pollution Deposition Models: Current Status, Data Requirements, and Research Needs, in *Proceedings of the International Conference on Air Pollutants and Their Effects on the Terrestrial Ecosystem*, Banff, Alberta, Canada, May 10–17, 1980, S. V. Krupa and A. H. Legg (Eds.), John Wiley & Sons, New York.

Hovind, E. L., M. W. Edelstein, and V. C. Sutherland, 1979, Workshop on Atmospheric Dispersion Models in Complex Terrain, Report EPA-600/9-79-041, Environmental Protection Agency, Research Triangle Park, N. C.

Hull, A. P., 1971, Radiation in Perspective: Some Comparisons of the Environmental Risks from Nuclear- and Fossil-Fueled Power Plants, *Nucl. Safety*, 12(1): 185-196.

Hunter, J. S., 1980, The National System of Scientific Measurement, *Science*, 210: 869-874.

Husar, R. B., W. E. Wilson, M. C. MacCracken, and R. M. Perhac, 1978, Transport Processes and Ecological Effects, *Energy Environment III*, Report on the Third National Conference on the Interagency R&D Program, Report EPA-600/9-78-022, pp. 75-94, Environmental Protection Agency, Research and Development Technical Information Staff, Cincinnati, Ohio.

Idso, S. B., 1980, The Climatological Significance of a Doubling of Earth's Atmospheric Carbon Dioxide Concentration, *Science*, 207: 1462-1463.

Ingersoll, R. B., R. E. Inman, and W. R. Fisher, 1974, Soil's Potential as a Sink for Atmospheric Carbon Monoxide, *Tellus*, 26: 151-159.

Inhaber, H., 1978, *Risk of Energy Production*, Atomic Energy Control Board, Report AECB-1119/REV-1, Ottawa, Canada.

——, 1979, Risk with Energy from Conventional and Nonconventional Sources, *Science*, 203: 718-723.

International Commission on Radiological Protection, 1977, ICPR Publication No. 26, Recommendations of the International Commission on Radiological Protection, *Annals of the ICPR*, 1: 3.

Irwin, J., 1981, Estimating Plume Dispersion for Elevated Point Sources, *J. Appl. Meteorol.* (in press).

——, 1979, Estimating Plume Dispersion: A Recommended Generalized Scheme, in *Proceedings of the Fourth Symposium on Turbulence, Diffusion, and Air Pollution*, Jan. 15–18, 1979, Reno, Nev., pp. 62-69, American Meteorological Society, Boston, Mass.

Junge, C. E., 1963, *Atmospheric Chemistry and Radioactivity*, Academic Press, New York.

Junod, A., J. Hopkirk, D. Schmeiter, and D. Haschke, 1974, Meteorological Influences of Atmospheric Cooling Systems as Projected in Switzerland, in *Cooling Tower Environment—1974*, ERDA Symposium Series, College Park, Md., Mar. 4–6, 1974, Steven R. Hanna and Jerry Pell (Coords.), CONF-740302, NTIS.

Jurgens, R. B., and R. C. Rhodes, 1976, Quality Assurance and Data Validation for the Regional Air Monitoring System of the St. Louis Regional Air Pollution Study, in *Proceedings of the Conference on Environmental Modeling and Simulation*, Cincinnati, Ohio, Apr. 20–22, 1976, Report EPA-600/9-76-016, pp. 730-735, Research Triangle Park, N. C.

Karl, T. R., 1978, Ozone Transport in the St. Louis Area, *Atmos. Environ.*, 12: 1421-1431.

Kelly, T. J., D. H. Stedman, and G. L. Kok, 1979, Measurements of H_2O_2 and HNO_3 in Rural Air, *Geophys. Res. Lett.*, 6: 375-378.

Koenig, L. R., 1980, Anomalous Snowfall Caused by Natural-Draft Cooling Towers, *A Rand Note*, Report RAND/N-1479-DOE, Rand Corporation, Santa Monica, Calif.

Kok, G. L., 1980, Measurements of Hydrogen Peroxide in Rainwater, *Atmos. Environ.*, 14: 653-656.

——, T. Holler, M. Lopez, H. Nachtrieb, and M. Yuan, 1978, Chemiluminescent Method of the Determination of Hydrogen Peroxide in the Ambient Atmosphere, *Environ. Sci. Technol.*, 12: 1072-1076.

Komp, M. J., and A. H. Auer, Jr., 1978, Visibility Reduction and Accompanying Aerosol Evolution Downwind of St. Louis, *J. Appl. Meteorol.*, 17: 1357-1367.

Lapp, R. E., 1979, *The Radiation Controversy*, Reddy Communications, Inc., Greenwich, Conn.

Lave, L. B., and L. C. Freeburg, 1973, Health Effects of Electricity Generation from Coal, Oil, and Nuclear Fuel, *Nucl. Safety*, 14(5): 409-428.

Levy, H., 1971, Normal Atmosphere: Large Radicals and Formaldehyde Concentrations Predicted, *Science*, 173: 141-143.

Lidwell, O. M., 1964, Microbiology of the Atmosphere and Airborne Infection, in *Medical Climatology*, S. Licht (Ed.), pp. 131-158, Waverly Press, Baltimore, Md.

Likens, G. E., F. H. Bormann, R. S. Pierce, J. S. Eaton, and N. M. Johnson, 1977, *Biogeochemistry of a Forested Ecosystem*, Springer-Verlag, New York.

——, R. F. Wright, J. N. Galloway, and T. J. Butler, 1979, Acid Rain, *Sci. Am.*, 241: 43-51.

Lillian, D., and Singh, H. B., 1974, Absolute Determination of Atmospheric Halocarbons by Gas Phase Coulometry, *Anal. Chem.*, 46: 1060-1063.

Lindeken, C. L., K. R. Peterson, D. E. Jones, and R. E. McMillen, 1972, Geographical Variations in Environmental Radiation Background in the United States, in *Proceedings of the Second International Symposium on the Natural Radiation Environment*, Vol. I, Aug. 7–11, Houston, Tex., NTIS.

Lipfert, F. W., 1980, Sulfur Oxides, Particles, and Human Mortality: Synopsis of Statistical Correlations, *J. Air Pollut. Control Assoc.*, 30: 366-371.

Liu, S. C., 1978, Possible Nonurban Environmental Effects Due to Carbon Monoxide and Nitrogen Oxide Emissions, in *Man's Impact on the Troposphere*, Lectures in Tropospheric Chemistry, J. S. Levine, and D. R. Schryer (Eds.), Report NASA-RP-1022, pp. 65-79, Science and Technical Information Office, Washington, D. C.

——, R. J. Cicerone, and T. M. Donahue, 1977, Sources and Sinks of Atmospheric N_2O and the Possible Ozone Reduction Due to Industrial Fixed Nitrogen Fertilizers, *Tellus*, 29: 251-263.

Lodge, J. P., A. P. Waggoner, D. T. Klodt, and C. N. Crain, 1981, Non-Health Effects of Airborne Particulate Matter, *Atmos. Environ.*, 15: 431-482.

Lovelock, J. E., R. J. Maggs, and R. A. Rasmussen, 1972, Atmospheric Dimethyl Sulfide and the Natural Sulfur Cycle, *Nature*, 237: 452-453.

MacCracken, M. C., 1978, MAP3S: An Investigation of Atmospheric Energy Related Pollutants in the Northeastern United States, *Atmos. Environ.*, 12: 649-659.

McBride, J. P., R. E. Moore, J. P. Witherspoon, and R. E. Blanco, 1978, Radiological Impact of Airborne Effluents of Coal and Nuclear Plants, *Science*, 202: 1045-1050.

McCartney, E. J., 1976, *Optics of the Atmosphere: Scattering by Molecules and Particles,* John Wiley & Sons, New York.

McMahon, T. A., and P. J. Denison, 1979, Empirical Atmospheric Deposition Parameters: A Survey, *Atmos. Environ.,* 13: 571-585.

Machta, L., 1979, Atmospheric Measurements of Carbon Dioxide, in *Workshop on the Global Effects of Carbon Dioxide from Fossil Fuels,* Miami Beach, Fla., Mar. 7–11, 1977, W. P. Elliott, and L. Machta (Eds.), USDOE Report CONF-770385, pp. 44-50, NTIS.

——, 1958, Discussion on Meteorological Factors and Fallout Distribution. Environmental Contamination from Weapons Tests, USAEC Report HASL-42, pp. 310-325, NTIS.

Mamane, Y., and R. F. Pueschel, 1980, Formation of Sulfate Particles in the Plume of the Four Corners Power Plant, *J. Appl. Meteorol.,* 19: 779-790.

Mead, W. J., 1977, An Economic Appraisal of President Carter's Energy Program, *Science,* 197: 340-345.

Meagher, J. F., L. Stockburger, E. M. Bailey, and O. Huff, 1978, The Oxidation of Sulfur Dioxide to Sulfate Aerosols in the Plume of a Coal-Fired Power Plant, *Atmos. Environ.,* 12: 2197-2203.

Meyer, B., 1977, *Sulfur, Energy, and Environment,* Elsevier, New York.

Miller, D. F., 1978, Precursor Effects on SO_2 Oxidation, *Atmos. Environ.,* 12: 273-280.

Millian, M. M., and Y. S. Chang, 1977, Detection of a Plume 400 km from the Source, *Atmos. Environ.,* 11: 939-944.

Möller, D., 1980, Kinetic Model of Atmospheric SO_2 Oxidation Based on Published Data, *Atmos. Environ.,* 14: 1067-1076.

Nagel, T. J., 1978, Operating a Major Public Utility Today, *Science,* 201: 985-993.

National Academy of Sciences, 1978, *The Tropospheric Transport of Pollutants and Other Substances to the Oceans,* Chap. 6, pp. 146-168, *Halogenated Hydrocarbons,* Washington, D. C.

National Council on Radiation Protection and Measurement, 1971, Basic Radiation Protection Criteria, *NCRP Publication No. 39,* Washington, D. C.

——, 1975, Natural Background Radiation in the United States, *NCRP Publication No. 45,* Washington, D. C.

Natural Radiation Environment III, USDOE Report CONF-780422(Vol.2), 1980, T. F. Gesell and W. M. Lowder (Eds.), pp. 1211-1445, NTIS.

Neuberger, H., 1957, *Introduction to Physical Meteorology,* The Pennsylvania State University, University Park, Pa.

Newell, R. E., and T. G. Dopplick, 1979, Questions Concerning the Possible Influence of Anthropogenic CO_2 at Atmospheric Temperature, *J. Appl. Meteorol.,* 18: 822-825.

NOAA, 1978, The View from Space, *Nat. Oceanic Atmos. Adm., NOAA Tech. Rep.,* 8: 30-31.

Noxon, J. F., 1975, Nitrogen Dioxide in the Stratosphere and Troposphere Measured by Ground-Based Absorption Spectroscopy, *Science,* 139: 547-549.

Nyberg, A., 1977, On Airborne Transport of Sulphur over the North Atlantic, *Quart. J. Roy. Meteorol. Soc.,* 103: 607-616.

O'Brien, K., 1972, The Cosmic Ray Field at Ground Level, in *Proceedings of the Second International Symposium on the Natural Radiation Environment,* Vol. I, August 7–11, 1972, Houston, Tex., J. A. S. Adams, W. M. Lowder, and T. F. Gesell (Eds.), ERDA Report CONF-720805-P1, pp. 15-54, NTIS.

OECD Programme on Long-Range Transport of Air Pollutants, Measurements and Findings, 1977, Director of Information, Organization for Economic Cooperation and Development, 2 rue André-Pascal, 75775, Paris, Cedex, 16, France.

Office of Technology Assessment, 1979, *The Direct Use of Coal,* GPO, Washington, D. C.

Ottar, B., 1978, An Assessment of the OECD Study on Long-Range Transport of Air Pollutants (LRTAP), *Atmos. Environ.,* 12: 445-454.

——, 1976, Monitoring Long-Range Transport of Air Pollutants: The OECD Study, *Ambio,* 6: 203-206.

Otts, R. E., 1976, Locally Heavy Snow Downwind from Cooling Towers, *NOAA Technical Memo, NWS-ER-62,* Scientific Services Division, National Weather Service, Eastern Region.

Panofsky, H., 1974, The Atmospheric Boundary Layer Below 150 Meters, *Annu. Rev. Fluid Mech.,* 6: 147.

Parmenter, F. C., 1977, Convective Cloud Plumes Mark Canadian Fire Sites, *Weather,* 32: 424-427.

Parungo, F. P., P. A. Allee, and H. K. Weickmann, 1978, Snowfall Induced by a Power Plant Plume, *Geophys. Res. Lett.,* 5: 515-517.

Pasquill, F., 1976, Atmospheric Dispersion Parameters in Gaussian Plume Modeling, Part II, Report EPA-600/4-76-030b, U. S. Environmental Protection Agency, Research Triangle Park, N. C.

——, 1961, The Estimation of the Dispersion of Windborne Material, *Meteorol. Mag.,* 90: 33-49.

Patrick, R., V. P. Binetti, and S. G. Halterman, 1981, Acid Lakes from Natural and Anthropogenic Causes, *Science,* 211: 446-448.

Payne, W. A., L. J. Stief, and D. D. Davis, 1973, A Kinetics Study of the HO_2 with SO_2 and NO, *J. Am. Chem. Soc.,* 95: 7614-7619.

Pendergast, M. M., and T. V. Crawford, 1974, Actual Standard Deviations of Vertical and Horizontal Wind Direction Compared to Estimates from Other Measurements, in *Proceedings of the Symposium on Atmospheric Diffusion and Air Pollution,* Sept. 9–13, 1974, pp. 1-6, Santa Barbara, Calif., American Meteorological Society, Boston, Mass.

Penkett, S. A., B. M. R. Jones, K. A. Brice, and A. E. J. Eggleton, 1979, The Importance of Atmospheric Ozone and Hydrogen Peroxides in Cloud and Rainwater, *Atmos. Environ.,* 13: 123-137.

Perhac, R. M., personal communication, August 1980.

——, 1978, Sulfate Regional Experiment in the Northeastern United States: The SURE Program, *Atmos. Environ.,* 12: 641-647.

Peters, L. K., and A. A. Jouvanis, 1979, Numerical Simulation of the Transport and Chemistry of CH_4 and CO in the Troposphere, *Atmos. Environ.,* 13: 1443-1462.

Peterson, J. T., E. C. Flowers, and J. H. Rudisill, 1978, Urban–Rural Solar Radiation and Atmospheric Turbidity Measurements in the Los Angeles Basin, *J. Appl. Meteorol.,* 17: 1595-1609.

Petterssen, S., 1940, *Weather Analysis and Forecasting,* McGraw-Hill, Inc., New York.

——, and D. Middleton, 1963, On the Representative Observations, *Tellus,* 15: 387-405.

Plummer, F. G., 1912, Forest Fires, *U. S. Dep. Agric., For. Serv. Bull.,* 117: 15-22.

Prospero, J. M., 1968, Atmospheric Dust Studies on Barbados, *Bull. Am. Meteorol. Soc.,* 49: 645-652.

Randerson, D., 1981, Solar Radiation as a Sole Source of Energy for Photovoltaics in Las Vegas, Nevada, for July and December, *NOAA Technical Memorandum,* NWS-WR-161, National Weather Service, Western Region, Salt Lake City, Utah.

——, 1979, Review Panel on Sigma Computations, *Bull. Am. Meteorol. Soc.,* 60: 682-683.

——, 1977, Quantitative Analysis of Atmospheric Pollution Phenomena, *Skylab Explores the Earth,* Report NASA-SP-80, Chap. 14, L. B. J. Space Center Scientific and Technical Information Office, National Air and Space Administration, Washington, D. C.

——, 1976, Overview of Regional Scale Numerical Models, *Bull. Am. Meteorol. Soc.,* 57: 797-804.

——, J. G. Garcia, and V. S. Whitehead, 1971, Photogrammetric and Photometric Investigations of a Smoke Plume Viewed from Space, *J. Appl. Meteorol.,* 10: 1122-1131.

——, 1970, A Comparison of the Spectral Distribution of Solar Radiation in a Polluted and a Clear Air Mass, *J. Air Pollut. Control Assoc.,* 20: 546-548.

——, 1968, A Study of Air Pollution Sources as Viewed by Earth Satellite, *J. Air Pollut. Control Assoc.,* 18: 249-253.

Rasmussen, K. H., M. Taheri, and R. L. Kabel, 1975, Global Emissions and Natural Processes for Removal of Gaseous Pollutants, *Water, Air, Soil Pollut.*, 4: 33-64.

Rasmussen, R. A., M. A. K. Khalil, and R. W. Dalluge, 1981, Atmospheric Trace Gases in Antarctica, *Science*, 211: 285-287.

——, and D. Pierotti, 1978, Interlaboratory Calibration of Atmospheric Nitrous Oxide Measurements, *Geophys. Res. Lett.*, 5: 353-355.

Reynolds, A., 1977, Energy Crisis: Made in Washington, *Reason*, 8: 16-20.

Roberts, J. J., 1977, Report to the U. S. Environmental Protection Agency of the Specialists' Conference on the EPA Modeling Guidelines, Under Interagency Agreement No. EPA-IAG-D7-0013 Between ERDA and EPA, Argonne National Laboratory, Chicago, Ill.

Robinson, G. D., 1977, Effluents of Energy Production: Particulates, in *Energy and Climate*, pp. 61-71, National Academy of Sciences, Washington, D. C.

Rodhe, H., P. Crutzen, and A. Vanderpol, 1981, Formation of Sulfuric and Nitric Acid in the Atmosphere During Long-Range Transport, *Tellus*, 33: 132-141.

Roessler, D. M., and F. R. Faxvog, 1981, Visibility in Absorbing Aerosols, *Atmos. Environ.*, 15: 151-155.

Royal Ministry for Foreign Affairs, Royal Ministry of Agriculture, 1971, *Air Pollution Across National Boundaries: The Impact on the Environment of Sulfur in Air and Precipitation*, P. A. Norstedt and Söners Förleg AB, Stockholm.

Sawyer, L. D., 1978, Quantitation of Polychlorinated Biphenyl Residues by Electron Capture Gas-Liquid Chromatography: Collective Study, *J. Assoc. Off. Anal. Chem.*, 61: 282-291.

Schwartz, S. E., 1979, Residence Times in Reservoirs Under Non-Steady-State Conditions: Application to Atmospheric SO_2 and Aerosol Sulfate, *Tellus*, 31: 530-547.

Science, 1980, Letters, 210: 6-8.

Science News, 1978, Measuring Volcanic Particle Production, *Sci. News*, 114: 200.

Sehmel, G. A., 1980, Particle and Gas Deposition: A Review, *Atmos. Environ.*, 14: 983-1011.

Seiler, W., 1974, The Cycle of Atmospheric CO, *Tellus*, 26: 116-135.

Sheppard, P. A., 1958, The Effects of Pollution on Radiation in the Atmosphere, *Intern. J. Air Pollut.*, 1: 31.

Singh, H. B., F. L. Ludwig, and W. B. Johnson, 1978, Tropospheric Ozone: Concentrations and Variabilities in Clean Remote Atmospheres, *Atmos. Environ.*, 12: 2185-2196.

——, L. J. Salas, H. Shigeishi, A. J. Smith, and E. Scribner, 1979, Atmospheric Distributions, Sources, and Sinks of Selected Halocarbons: Hydrocarbons, SF_6, and N_2O, Final Report to the U. S. Environmental Protection Agency, Report EPA-600/3-79-107, Project 4487, SRI International, Menlo Park, Calif.

——, L. J. Salas, A. J. Smith, and H. Shigeishi, 1981, Measurements of Some Potentially Hazardous Organic Chemicals in Urban Environments, *Atmos. Environ.*, 15: 601-612.

Slatt, B. J., D. F. S. Natusch, J. M. Prospero, and D. L. Savoie, 1978, Hydrogen Sulfide in the Atmosphere of the Northern Equatorial Atlantic Ocean and Its Relation to the Global Sulfur Cycle, *Atmos. Environ.*, 12: 981-991.

Slinn, W. G. N., 1977a, Some Approximations for the Wet and Dry Removal of Particles and Gases from the Atmosphere, *Water, Air, Soil Pollut.*, 7: 513-543.

——, 1977b, Proposed Terminology for Precipitation Scavenging, in *Precipitation Scavenging (1974)*, ERDA Symposium Series, Champagne, Ill., October 14–18, 1974, R. E. Semonin and R. W. Beadle (Coords.), pp. 813-818, CONF-741003, NTIS.

——, L. Hasse, B. B. Hicks, A. W. Hogan, D. Lal, P. S. Liss, K. O. Munnich, G. A. Sehmel, and O. Vittori, 1978, Some Aspects of the Transfer of Atmospheric Trace Constituents Past the Air–Sea Interface, *Atmos. Environ.*, 12: 2055-2087.

Smith, C. D., 1950, The Widespread Smoke Layer from Canadian Forest Fires During Late September 1950, *Mon. Weather Rev.*, 78: 180-184.

Smith, F. B., and G. H. Jeffrey, 1975, Airborne Transport of SO_2 from U. K., *Atmos. Environ.*, 9: 643-659.

Stewart, R. W., S. Hameed, and J. Pinto, 1978, The Natural and Perturbed Troposphere, *Man's Impact on the Troposphere*, Lectures in Tropospheric Chemistry, J. S. Levine and D. R. Schryer (Eds.), NASA Report NASA-RP-1022, pp. 27-64, Science and Technical Information Office, Washington, D. C.

Stith, J. L., P. V. Hobbs, and L. F. Radke, 1978, Airborne Particle and Gas Measurements in the Emissions of Six Volcanoes, *J. Geophys. Res.*, 83(C8): 4009-4017.

Strimaitis, D., G. Hoffnagle, and A. Bass, 1980, On-Site Meteorological Instrumentation Requirements to Characterize Diffusion from Point Sources—A Workshop, Contract No. 68-02-3282, Environmental Research and Technology, Concord, Mass.

Suchman, D., B. A. Auvine, and B. H. Hinton, 1979, Some Economic Effects of Private Meteorological Forecasting, *Bull. Am. Meteorol. Soc.*, 60: 1148-1156.

Sze, N. D., and M. K. W. Ko, 1980, Photochemistry of COS, CS_2, CH_3SCH_3, and H_2S: Implications for the Atmospheric Sulfur Cycle, *Atmos. Environ.*, 14: 1223-1239.

Tabershaw, I. R., 1976, Oxides of Sulfur, *J. Occup. Med.*, 18: 360-361.

Talbot, J. J., 1979, A Review of Potential Biological Impacts of Cooling Tower Salt Drift, *Atmos. Environ.*, 13: 395-406.

Teweles, S., and J. Giraytys (Eds.), 1970, Meteorological Observations and Instrumentation, *Meteorological Monographs*, 11, American Meteorological Society, Boston, Mass.

Tricker, R. A. R., 1970, *Introduction to Meteorological Optics*, American Elsevier, New York.

Trijonis, J., T. Peng, G. McRae, and L. Lees, 1978, Oxidant and Precursor Trends in the Metropolitan Los Angeles Region, *Atmos. Environ.*, 12: 1413-1420.

——, and K. Yuan, 1978a, Visibility in the Northeast, Report EPA-600/3-78-075, August 1978, Research Triangle Park, N. C.

——, 1978b, Visibility in the Southwest, Report EPA-600/3-78-039, April 1978, Research Triangle Park, N. C.

Turner, D. B., 1979, Atmospheric Dispersion Modeling: A Critical Review, *J. Air Pollut. Control Assoc.*, 29: 502-519.

U. S. Department of Commerce, 1979, *Statistical Abstract of the United States*, 100th ed., GPO, Washington, D. C.

U. S. Senate, 1975, *Air Quality and Stationary Source Emission Control*, Committee on Public Works, Serial No. 94-4.

U. S. Standard Atmosphere, 1962, GPO, Washington, D. C.

Waggoner, A. P., A. J. Vanderpol, R. J. Charlson, S. Larsen, L. Granat, and C. Tragardh, 1976, Sulfate-Light Scattering Ratio as an Index of the Role of Sulfur in Tropospheric Optics, *Nature*, 261: 120-122.

Wallace, J. M., and P. V. Hobbs, 1977, *Atmospheric Science*, Academic Press, Inc., London.

Warneck, P., 1974, On the Role of OH and HO_2 Radicals in the Atmosphere, *Tellus*, 26: 39-46.

Wells, A. E., 1917, Results of Recent Investigations of the Smelter Smoke Problem, *Ind. Eng. Chem.*, 9: 640-646.

Wesley, M. L., B. B. Hicks, W. P. Dannevik, S. Frisella, and R. B. Husar, 1977, An Eddy-Correlation Measurement of Particulate Deposition from the Atmosphere, *Atmos. Environ.*, 11: 561-563.

Wexler, H., 1950, The Great Smoke Pall of September 24–30, 1950, *Weatherwise,* 3: 129-134, 142.

Whelpdale, D. M., 1978, Large-Scale Atmospheric Sulfur Studies in Canada, *Atmos. Environ.,* 12: 661-670.

Whitby, K. T., 1978, The Physical Characteristics of Sulfur Aerosols, *Atmos. Environ.,* 12: 135-159.

Wilson, W. E., 1978, Sulfates in the Atmosphere: A Progress Report on Project MISTT, *Atmos. Environ.,* 12: 537-547.

Woodwell, G. M., 1978, The Carbon Dioxide Question, *Sci. Am.,* 238: 34-43.

Zemel, Z., and J. Lyons, 1976, An Objective Method for Assessing Representativeness of a Station Network Measuring Minimum Temperature near the Ground, *Boundary-Layer Meteorol.,* 10: 3-14.

Chapter 1 covers research prior to June 1981.

METEOROLOGICAL FUNDAMENTALS

Malcolm M. Pendergast, Savannah River Laboratory

2

2-1 INTRODUCTION

Meteorological elements influencing the transport and diffusion of atmospheric pollutants are known to contain a wide variety of wave phenomena of greatly differing time and space scales. Historically, these phenomena have been classified according to their horizontal dimensions into microscale, mesoscale, and macroscale. The microscale contains dynamical processes having space scales of the order of 2 km and smaller and time scales of the order of minutes. Macroscale processes have space scales of 2000 km and larger and time scales of the order of days. The intermediate, or mesoscale, lies between these two categories.

The various atmospheric processes are discussed according to their contribution to each of the three atmospheric scales to provide the reader with a basic understanding of the atmospheric processes influencing the transport and diffusion of atmospheric pollutants. The discussions in this and subsequent chapters are related to several basic laws of chemistry and physics. A number of these laws are reviewed at the beginning of this chapter.

2-2 BASIC LAWS AND DEFINITIONS

Pressure, p, is defined as force (F) per unit area (A). Pressure forces at any point in a parcel of air are directed uniformly in all directions. On the recommendation of Bjerknes and Sandstrom in 1906, the *bar* unit is used for expressing a pressure of 1 million dynes/cm^2 (1 dyne = 1 g·cm/s^2). Notice that the bar system has the units of pressure. Atmospheric pressures are rather small; so the unit of a millibar (mb) is used in practice (1 mb = 10^{-3} bars). In addition to the millibar, the following relations are used: 1 in. Hg = 33.86 mb; 1 atm = 1013.25 mb; and 1 mb = 100 Pa.

Temperature is a property often confused with *heat*. To understand the concept of temperature, we must distinguish between sensible heat and the heat content of a body. The concept of hotness, coldness, or the absence of either is based on the ability of our sensory organs to detect this difference. Consequently the property that is detected is the sensible heat or cold of an object. This degree of hotness or coldness is commonly referred to as the temperature. However, the kinetic theory of gases states that temperature is directly proportional to the *mean* translation kinetic energy of the molecules. On the other hand, heat is a form of energy that can be transferred by virtue of a difference in temperature. Heat energy can be transferred from one place to another by the processes of (1) conduction transfer by contact, (2) convection transfer by mass transfer, and (3) radiation transfer without the involvement of a physical substance. In another context, heat can be defined as the *total* translational kinetic energy of the molecules.

Temperature Scales. In measuring temperature, three different scales are in general use—Celsius, Fahrenheit, and the Kelvin, or absolute system. The Celsius scale is used throughout most of the world. For pure water, this system sets 100°C as the steam point and 0°C as the ice point at sea level. Similarly, the Fahrenheit system sets 212°F as the steam point and 32°F as the ice point. In thermodynamics the Kelvin, or absolute system, is used. In this system, 373°K is the steam point, 273°K is the ice point, and 0°K is absolute zero—the hypothetical temperature at which all molecular motion ceases. When we convert from one system to the other, the following expressions are useful: °C = 5/9 (°F − 32); °F = 9/5 (°C + 32); and °F = °C + 273.16, where the 273.16 in the last expression is usually rounded to 273.

Density, ρ, is defined as the mass of a substance per unit volume (ρ = M/V). The densities of solids and liquids vary slightly with changes in pressure and temperature, whereas the densities of gases vary greatly with changes in pressure and temperature. Consequently, when reporting the densities of gases, the corresponding pressure and temperature should be specified. For example, at 0°C and 1013.25 mb, the density of dry air is 1.293 × 10^{-3} g/cm^3, that of carbon dioxide is 1.977 × 10^{-3} g/cm^3, and that of oxygen is 1.429 × 10^{-3} g/cm^3.

In meteorology we frequently find it convenient to express the volume as a *specific volume* α. The specific volume is inversely equal to the density so that

$$\alpha = \frac{1}{\rho} = \frac{V}{M}, \text{ or } \rho\alpha = 1$$

Pendergast: Savannah River Laboratory, Aiken, South Carolina.

Boyle's Law. In 1662 Robert Boyle described how a volume of gas is related to the pressure it exerts. From experiments Boyle found that at constant temperature the volume (V) of any gaseous mixture (such as dry air) varies inversely as the pressure, or mathematically,

$$\frac{p_0}{p_1} = \frac{V_1}{V_0}$$

A more convenient form for the equation in terms of specific volume is

$$p_0\alpha_0 = p_1\alpha_1 = \text{constant}$$

which states that the product of the pressure and the volume of an enclosed gas, at constant temperature, remains constant. In a practical sense, Boyle's law tells us that by halving the pressure on a fixed mass of gas we will double the volume or by doubling the pressure we will half the volume occupied by the gas.

Charles's Law. Another relationship between the variables of state was discovered independently by Jacques Charles in 1787 and by Joseph Gay-Lussac in 1802. On the basis of laboratory experiments, these two researchers found that at constant pressure a fixed mass of gas will undergo a volumetric change in direct proportion to the change in temperature. Mathematically this result is expressed as

$$\frac{T_1}{T_0} = \frac{V_1}{V_0}$$

or, in terms of specific volume, $\alpha_0/T_0 = \alpha_1/T_1$ = constant, the constant being different from the one in Boyle's law.

Real gases follow Boyle's and Charles's laws only approximately. For this reason it is useful to introduce the concept of an ideal or perfect gas which is defined as a gas that follows these two laws exactly.

The Equation of State. Boyle's and Charles's laws can be combined to formulate a relationship containing the variables of state (p, α, and T). This relation is referred to as the equation of state. To develop this equation, let the specific volume of the ideal gas at temperature T and pressure p be written as $\alpha(p,T)$. Then, according to Charles's law,

$$\frac{\alpha(T,p_0)}{T} = \frac{\alpha(T_0,p_0)}{T_0}$$

and writing Boyle's law so that $\alpha(T,p_0)$ is the common state,

$$p\alpha(T,p) = p_0\alpha(T,p_0)$$

then by eliminating $\alpha(T,p_0)$, we obtain

$$p\alpha(T,p) = \frac{p_0 T}{T_0}\alpha(T_0,p_0)$$

where the quantity $[p_0\alpha(T_0,p_0)]/T_0$ is a constant for any given gas.

This constant has the dimensions $[L^2/T^2\theta]$ and is specified as R, the *specific gas constant.* Consequently the equation of state becomes

$$p\alpha = RT = \frac{pV}{m}$$

where the temperature is in degrees Kelvin. The specific gas constant can be defined for any given mass through the use of Avogadro's law, which states that at equal temperatures and pressures a gram molecular weight (mole) of any two gases will occupy the same volume (22.4×10^3 cm^3). For standard conditions of p = 1013.25 mb and T = 273.16°K,

$$mR = \frac{pV}{T} = 8.3144 \times 10^7 \ \frac{g \cdot cm^2}{s^2 \cdot mol \cdot {}^\circ K}$$

where the constant is the *universal gas constant* ($R^* = mR$) for all gases. The mean gram molecular weight for dry air is 28.966 g/mol so that the specific gas constant $R = R^*/\overline{m}$ for dry air is 2.8704×10^6 cm^2/(s^2·°K).

The Hydrostatic Equation. The variables of state have been related to each other. In meteorology we are concerned with how these variables change with altitude. For example, to relate pressure and height (z), assume the atmosphere is in static equilibrium so that the net upward pressure force is balanced by the force of gravity. Consider these forces to be acting within a vertical column of air having a cross-sectional area A. Furthermore, let the density of a small section, Δz, be ρ with the pressure of the bottom slice p and that of the top slice $p - \Delta p$.

The net pressure gradient force acting on the volume is $-\Delta p\,A$, with the negative sign indicating that the force is acting upward. The force of gravity acting on the volume is $g\rho A\,\Delta z$. Then, from our assumption, the net vertical force on the volume must be zero so that

$$-\Delta p\,A = g\rho A\,\Delta z$$

or

$$\frac{\Delta p}{\Delta z} = -\rho g$$

This expression is the hydrostatic equation which tells us the vertical pressure gradient force is exactly balanced by the force of gravity. This equation is one of the best approximations in meteorology and can be applied to all meteorological phenomena except those in which large vertical accelerations may exist. However, in this form the hydrostatic equation is not very useful to us because density is not a directly measured quantity. To alleviate this shortcoming, substitute for ρ from the equation of state so that

$$\frac{1}{p}\frac{\partial p}{\partial z} = -\frac{g}{RT_v} \qquad (2.1)$$

where T_v is the *virtual temperature* which must be used to account for any moisture in the atmosphere. The virtual temperature is the temperature of dry air having the pressure and density of a given sample of moist air. The virtual temperature is always greater than the temperature and is given by

$$T_v = T + 0.61rT$$

where r is the mixing ratio which is the ratio of the mass of water vapor to the mass of dry air with which the water vapor is mixed. As an approximation

$$T_v \cong T + \frac{r}{6}$$

where r is expressed in grams per kilogram.

Equation 2.1 is quite remarkable since it tells us that when T_v is large (warm) the vertical rate of change in pressure is small; however, if T_v is small (cool), then the vertical rate of change in pressure is large.

Integration of the hydrostatic equation as expressed above between levels z_1 and z_2 yields

$$z_2 - z_1 = \Delta z = K\overline{T}^* \ln \frac{p_1}{p_2}$$

where $\overline{T}^*$ is the mean virtual temperature through thickness Δz and $K = 29.2898$ m/°K. This equation is the *hypsometric equation* and permits us to compute the thickness between any two known pressure surfaces.

First Law of Thermodynamics. The first law of thermodynamics is based on empirical evidence and is a statement of the law of conservation of energy. The law states that an increase in the internal energy, dU, of a gaseous parcel can occur through the external addition of heat, dH, to the parcel, by performing work, dW, on the parcel, or by a combination of both. This process is expressed mathematically, for a unit mass system, as

$$dU = dW + dH$$

A nonviscous fluid can do work through the pressure force available along the surface of a parcel. The pressure exerted on a hypothetical parcel is applied equally in all directions so that the change in volume is given by $A\,dn = dV$, where A is the parcel's surface area and dn is the distance over which the compression (expansion) occurs. For unit mass, we may write $A\,dn = d\alpha$. The work, dW, done on the parcel is given by the product of the force, F, acting on the parcel surface, times the distance, dn. However, from the definition of pressure, p, we can write $F = pA$ so that $dW = pA\,dn = -p\,d\alpha$.

Experimental evidence tells us that internal energy per unit mass for an ideal gas varies only with its absolute temperature; so for a constant volume process, we have $Du = c_v\,dT$ and, for a constant pressure process, $dU = c_p\,dT$. In the first expression, c_v is called the specific heat capacity at constant volume and, in the second expression, c_p is for the constant pressure process. For dry air, $c_p = [1.005\ J/(g\cdot°C)]$ and $c_v = [0.712\ J/(g\cdot°C)]$ so that $c_p > c_v$; this means it takes more heat to raise the temperature 1° in a constant pressure process than in a constant volume process because part of the pressure force is used in doing work (expanding the parcel) whereas the rest is used to raise the temperature. For a unit mass, the first law of thermodynamics can be written in a more explicit form as

$$dH = c_v\,dT + p\,d\alpha \tag{2.2}$$

for a constant pressure process.

The differential form of the equation of state is $\alpha\,dp + p\,d\alpha = R\,dT$ so that through substitution and use of the relation $c_p - c_v = R$, Eq. 2.2 becomes

$$dH = c_p\,dT - \alpha\,dp \tag{2.3}$$

This form of the first law of thermodynamics is more useful than Eq. 2.2 because the increments dT and dp are easily measured. Now, heat can also be added to a parcel of air by such processes as the turbulent transfer of heat, condensation of water vapor, and radiation. However, Deardorff (1972) has shown that up to periods of greater than 1 d, these heating processes are of secondary importance. Consequently, if we assume that for short time periods (<1 d) no heat is added to or taken away from a parcel of dry air ($dH = 0$), then Eq. 2.1 becomes $c_v\,dT = \alpha\,dp$. A process involving no heat exchange between the parcel and its environment is called an *adiabatic process.* The temperature lapse rate for such a process is uniquely defined by differentiating Eq. 2.3 with respect to height (z) so that $c_p\,dT/dz = -g$. Consequently, after proper conversion of units [$c_p = 10^7\ cm^2/(s^2\cdot°C)$; $g = 980\ cm/s^2$], $dT/dz = -9.8$°C/km which is referred to as the *dry adiabatic lapse rate* and represents the theoretical change in temperature (±9.8°C/km) of a parcel of dry air as it rises (or falls) through the atmosphere.

Potential Temperature. With Eq. 2.3 by assuming an adiabatic process and substituting for α from the equation of state, we obtain

$$c_p\frac{dT}{T} = R\frac{dp}{p}$$

Integrating this expression from an initial state (p_0, T_0) to a final state (p,T) yields *Poisson's adiabatic equation*

$$\frac{T}{T_0} = \left(\frac{p}{p_0}\right)^k$$

where $k = R/c_p = 0.286$. This equation relates adiabatic temperature changes, experienced by an air parcel undergoing vertical displacement, to the field through which it moves. If the initial pressure is 1000 mb and the initial temperature is identified as θ, then

$$\theta = T\left(\frac{1000}{p}\right)^k$$

where θ is the potential temperature. This expression tells us that the potential temperature is the temperature a sample of gas would have if it were compressed (or expanded) adiabatically from a given state (p,T) to a pressure of 1000 mb. It should be obvious that θ is a characteristic property of an air parcel and that it is invariant during an adiabatic process. In other words, θ is a conservative property. Theoretically, then, θ can be used to identify a parcel of air to facilitate its identification from day to day because empirical and theoretical evidence tells us that dry atmospheric motions are approximately adiabatic.

Scales of Atmospheric Motion. Scientists have historically studied atmospheric processes from two points of view. To predict the evolution of weather patterns in the atmosphere, meteorologists have devoted much attention to the understanding of large-scale dynamics having space scales greater than 1000 km and time scales of the order of a week. In trying to explain the detailed behavior of the immediate surface environment in which man lives, they have developed the science of micrometeorology, which deals with processes having space scales on the order of several meters and time scales on the order of a minute. Many important atmospheric phenomena have scales that are intermediate between these two scales, such as thunderstorms, fronts, and heat islands.

Our knowledge of the intermediate-scale processes has not developed to the extent that it has for processes on the large and small scales. The reason is not a lack of interest but rather the difficulty in obtaining useful observational data concerning them. However, the recent surge of interest in the environmental effects of power production and in weather associated with severe storms has led to a rapid advancement of our understanding of the intermediate meteorological processes.

Orlanski (1975) summarized a number of different scale classifications based on space and time scales and concluded the horizontal scale of motion to be the better quantity for classifying atmospheric phenomena. In this chapter, atmospheric motions are assumed to fall into one of three scales: macro, meso, and micro, named after Greek words equivalent to large, intermediate, and small. Furthermore, macroscale motions are assumed to involve space scales exceeding 2000 km; microscale motions involve space scales less than 2 km; and mesoscale motions contain the remaining scales.

2-3 MACROSCALE SYSTEMS

Earth's atmosphere is a gaseous environment extending approximately 200 km away from the earth's surface. Nitrogen and oxygen account for about 99% of the volume of the atmosphere. There are lesser amounts of water vapor, argon, ozone, carbon dioxide, dust, and several smaller constituents. The portion of the atmosphere which is affected by man's activity and conversely which affects man's activities can be considered to be within the lowest 15 km. Within this zone, called the troposphere, the atmosphere can be considered to be made up of dry air and water vapor. The physical and dynamic characteristics of this fluid are what we realize as characteristics of weather—wind, rain, heat, and cold. The variability of the weather can be attested to by just about everyone. Much of the variability is attributed to large masses of air that drift over the surface of the earth. In this section we will discuss the cause–effect relationship between the large-scale systems and the temperature patterns, wind patterns, and dispersion of atmospheric pollutants.

2-3.1 Energy Budget: Radiation

All energy on the earth is derived directly or indirectly from a single star called the sun. This energy is created through a complex nuclear reaction that converts hydrogen to helium. In the process, mass is converted to energy in accordance with Einstein's famous relation, $E = mc^2$. Through this conversion, it is estimated that the sun radiates approximately 2.3×10^{28} J of energy every minute. By the time this radiant energy, E, travels the mean distance to the earth ($r = 1.5 \times 10^{13}$ cm), the amount of energy, S_0, falling on a unit area on a spherical shell of radius, r, in 1 min is

$$S_0 = \frac{E}{4\pi r^2}$$

which amounts to 8.372 J/(cm^2·min). This value is referred to as the *solar constant* and is believed to be accurate to within a few percent.

The amount of radiant energy incident at the top of our atmosphere being known, the next task is to determine how much of this incoming solar radiation (insolation) arrives at the surface of the earth. The most important factors affecting the energy received at a point on earth at any particular time is the distance to the sun, the 24-h rotation of the earth about its own axis, the 12-month revolution around the sun along the plane of the ecliptic, and the inclination of the axis of rotation to the plane of the ecliptic. These factors affect the angle, h, at which the solar radiation strikes a particular location on the earth's surface. Figure 2.1 shows how the insolation at the earth's surface, S, is a function of the incident angle, h, and the solar constant S_0:

$$S = S_0 \sin(h)$$

Practically speaking, this incident angle is a function of time of day, season of the year, and location and slope of the surface. Measurements at the ground indicate that about one-half of the possible radiation from the sun is received at ground level. This depletion is due to reflection, absorption, and scattering of insolation in the atmosphere.

The earth–atmosphere system is in a steady-state condition with respect to the radiant energy it receives from the sun. Proof of this is that the average temperature of the atmosphere does not vary significantly from year to year. Through measurements and theoretical considerations, scientists have been able to determine the relative

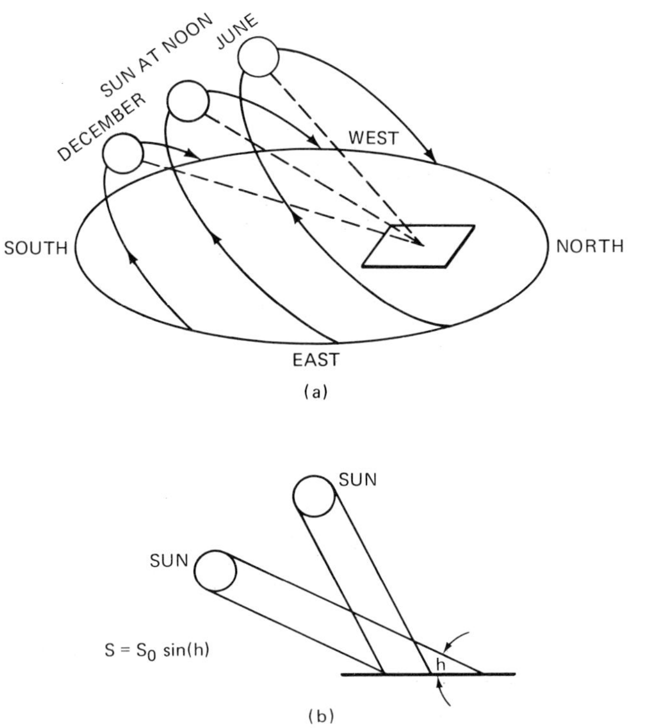

Fig. 2.1 (a) Seasonal variation of the length of day as shown by the length of the arc of the apparent path of the sun. (b) The relative seasonal change in the angle that the noon sun makes with a horizontal surface. (From H. A. Byers, *General Meteorology*, 2nd ed., McGraw-Hill Book Co., New York, 1944.)

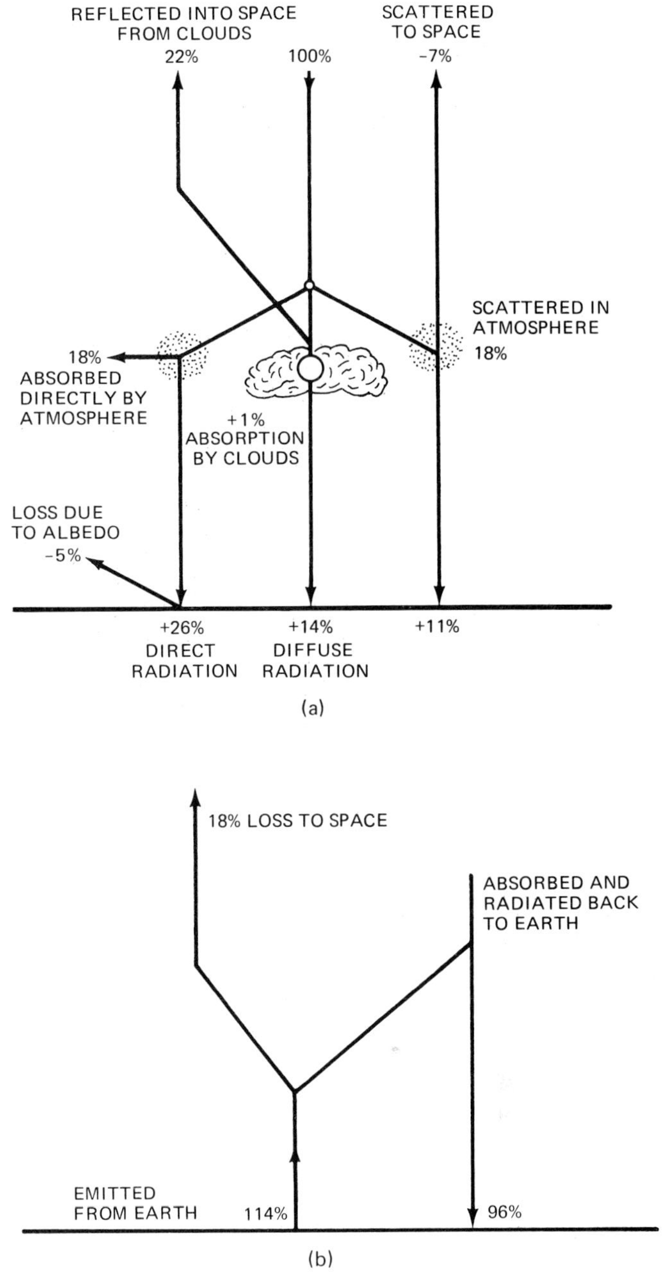

Fig. 2.2 Schematic representation of the radiation budget in the earth–atmosphere–sun system: (a) Shortwave radiation; (b) Long-wave radiation. (From H. Flohn, *Climate and Weather*, McGraw-Hill Book Co., New York, 1969.)

importance of various factors affecting the radiation budget needed to preserve the heat balance of the atmosphere. Figure 2.2 shows schematic diagrams of the radiation budget for shortwave and longwave radiation.

Reflection. All solids and liquids reflect sunlight. The percentage of the sunlight that is reflected is called the *albedo.* The albedo varies significantly from surface to surface ranging from 2% for dark soils to about 90% for cloud tops (see Fig. 2.3). The average albedo for the earth's surface is about 10%.

The reflection of insolation is determined by the amount of cloud cover and by the albedo of the earth's surface. On the average, 22% of insolation is reflected from cloud tops and 5% is reflected from the ground. At any one time, however, the amount of reflected radiation is quite variable. The reflection from the surface can be treated relatively simply if the albedo and the sun angle are known. The treatment of the more significant reflection from clouds, however, is much more difficult. The location of clouds is determined by the macroscale weather systems. The dynamics of these weather systems is in turn related to the energy budget at the earth's surface. The energy budget in turn is significantly affected by cloud amount. This complicated nonlinear interaction, much like a dog chasing his own tail, is one reason that improvements in weather forecasting remain so elusive.

Absorption. Since absorption of shortwave radiation by gases is selective, more energy is absorbed in some wavelengths than in others. Water vapor is responsible for most of the absorption of insolation, although oxygen, carbon dioxide, ozone, suspended particles, and cloud droplets play a minor part in insolation. On the average, 18% of the insolation is absorbed by the atmosphere. As will be shown later, long-wave radiation is much more effectively absorbed by the atmosphere.

Scattering. When light passes through a medium containing particles (including molecules) with diameters less than the wavelength of the light, a portion of the light is scattered in all directions. This type of scattering, called Rayleigh scattering, is inversely proportional to the fourth power of the wavelength of the light. Sunlight is white light made up of a whole spectrum of wavelengths of various

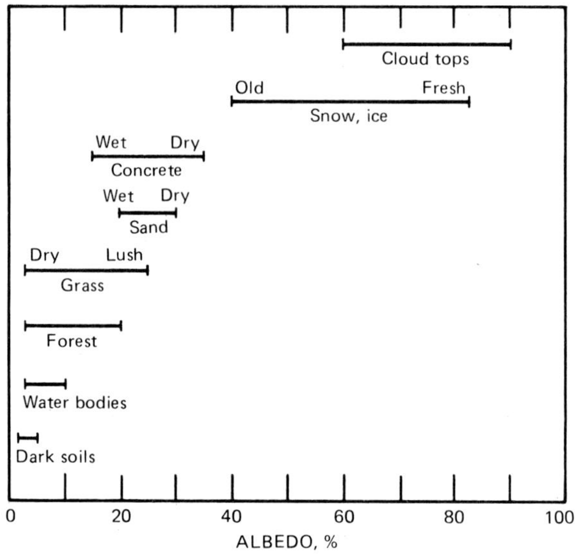

Fig. 2.3 The albedo (percent reflectivity in visible spectrum only) for various surfaces. (From D. H. Slade, *Meteorology and Atomic Energy—1968*, USAEC Report TID-24190, 1968.)

colors. The sky appears blue because of the intense scattering of the short wavelength (blue) which illuminates surrounding particles. Without scattering, the sky would be black and the sun would appear as a bright white sphere in the sky casting strong shadows (exactly as we observed from television broadcasts from the moon).

The efficiency of the scattering process is dependent on the length of the path the incident light travels and the number of scattering particles. An example of this can be observed at sunrise and at sunset when the sun is low in the sky. The sky and clouds near the horizon at these times have a reddish appearance caused by the extremely efficient scattering of all the short wavelengths away from the incident sunlight. The result is that much of the blue light is removed and most of the remaining light appears in blends of red and orange. The great variability in the color of sunsets and sunrises attests to the variability of concentrations of the scattering particles.

Clouds are also responsible for the scattering of sunlight. The size of the obscuring particles within a cloud are much larger than are the wavelengths of the incident light, and this produces a different type of scattering than that created by small particles in the atmosphere. Because the light incident to the cloud is scattered equally for all wavelengths, the clouds appear to be white. On the average, the scattering of light by clouds accounts for slightly more energy being scattered to the ground than results from scattering by the atmosphere. The energy at the surface due to scattering by clouds represents approximately 14% of the total incident radiation; that due to scattering from the atmosphere represents about 11% of the total incident radiation.

Scattering accounts for the depletion of a significant amount of solar insolation. Scattering also presents a mechanism by which energy can be more uniformly, and perhaps more efficiently, distributed across the surface of the earth. On the average the amount of diffuse, or scattered, radiation received at the surface of the earth is equal to the amount of direct radiation. In certain instances, such as in polar regions when the sun is on the horizon, scattering and reflection from clouds and snow may account for as much as 90% of the total solar radiation received at the surface.

2-3.2 Energy Budget: Heat

Approximately 71% of the total incoming solar radiation is available for heating the atmosphere. Approximately 52% of the radiation is received at the earth's surface and about 19% is absorbed directly by the upper atmosphere. This absorbed energy is converted to heat, and most atmospheric phenomena derive their energy from this heat. The study of the heat budget of the earth–atmosphere system is one of the central problems of meteorology.

There are significant differences between land and water in terms of specific heat capacity, albedo, and the amount of water available for evaporation. The most important difference between land and water with respect to the surface heat budget is the fact that solid ground depends on transfer of heat by molecular interaction called conduction, whereas, in exchange of heat in fluids, such as water and air, the heat transfer is largely the result of motions of the fluid called convection.

Land and water surfaces with identical amounts of solar energy falling upon them do not acquire similar temperatures. In bodies of water, motions of the water in the form of waves, drifts, currents, and tides tend to distribute absorbed solar energy throughout a large layer (mass) with the consequence that surface heating and cooling are relatively slow. Obviously, no similar vertical mixing can occur in land areas so that only a relatively thin layer becomes heated. The rate of temperature change is inversely proportional to the mass and heat capacity of the medium. Since heat capacity and the mass of affected matter are both smaller for land areas, a land surface responds more quickly than a water surface to the heat supplied by radiation and reaches higher temperatures. Conversely, land surfaces cool more rapidly. Continental climates, therefore, are characterized by a large daily and seasonal range in temperature, whereas maritime climates show much smaller ranges.

Figure 2.4 shows the mean air temperatures at the surface of the earth for January and July for the Northern Hemisphere. Evident is the marked contrast between surface temperatures over land and water. The contrast is much stronger in January than in July. A synopsis follows on the effects of this observed surface temperature pattern (i.e., heat sources) on the atmosphere in terms of the four mechanisms of transfer of sensible heat to the atmosphere: radiation, convection, conduction, and water-vapor evaporization and condensation.

Radiation. All bodies having a temperature above 0°K radiate energy. The wavelength of the maximum radiation is given by Wien's displacement law, $\chi_{max} = 2897/T$, where

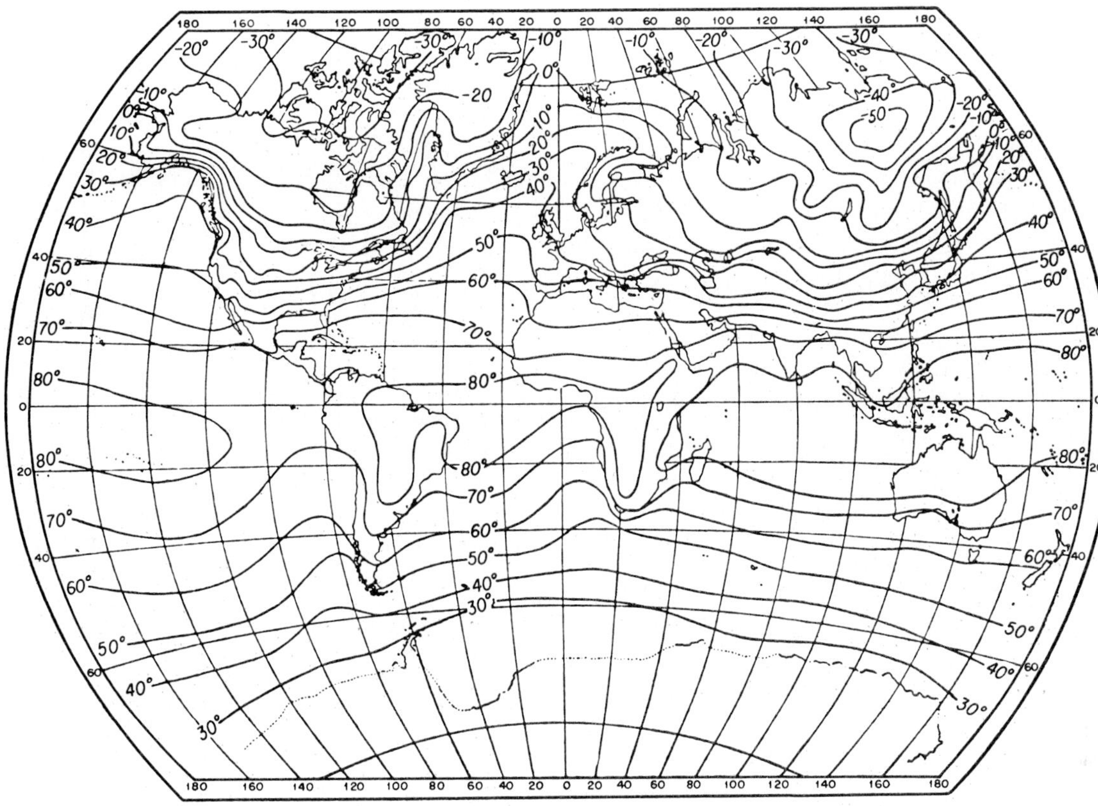

JANUARY
(a)

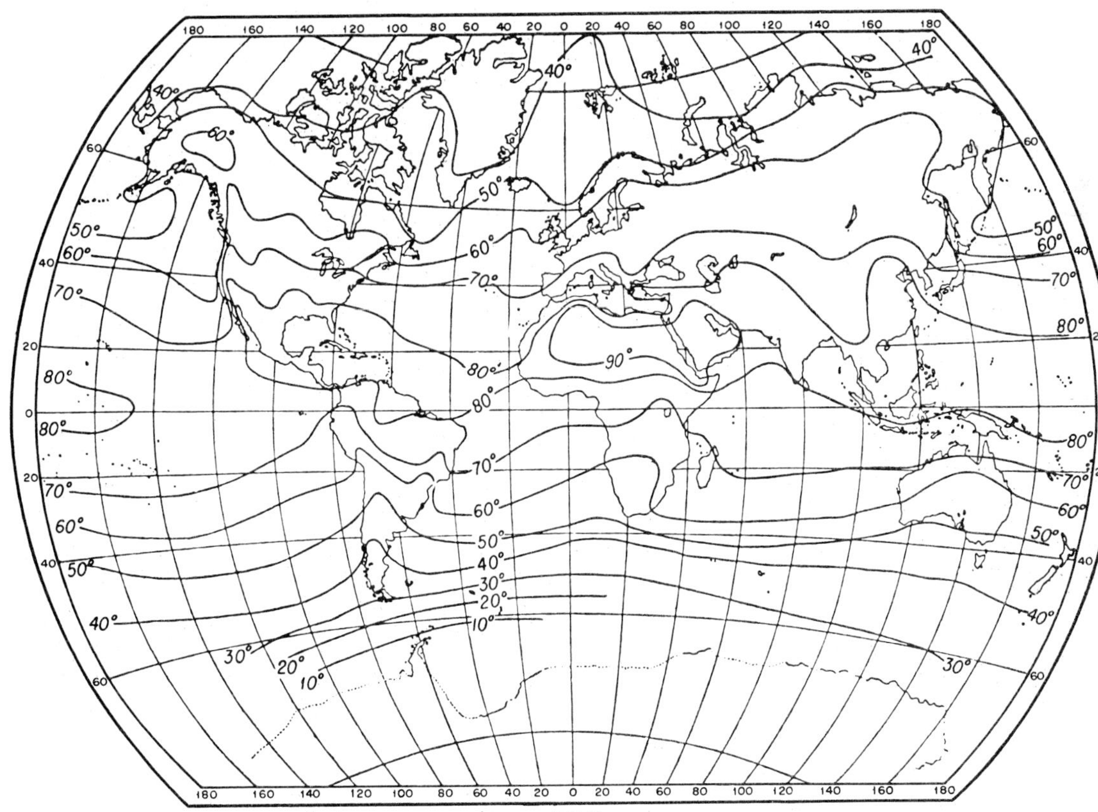

JULY
(b)

Fig. 2.4 Mean air temperature at the surface for (a) January and (b) July. (From H. A. Byers, *General Meteorology*, 2nd ed., McGraw-Hill Book Co., New York, 1944.)

λ_{max} is wavelength and T is the absolute temperature. The temperature of the earth is about 300°K and the effective temperature of the sun is about 6000°K. Thus the wavelength of terrestrial radiation is about 20 times longer than solar radiation. The atmosphere is approximately 95% opaque to terrestrial radiation. Consequently the atmosphere receives most of its heat indirectly from the sun but directly from the ground. [This process, responsible for heating the atmosphere, has been called the "greenhouse" effect. Scientists believed that the glass of the greenhouse was nearly transparent to shortwave radiation and opaque to long-wave radiation just as was the atmosphere. Recently Businger (1963) has shown this explanation to be incorrect. He suggests that the air inside the greenhouse is heated owing to a lack of exchange of the outside air.]

The amount of energy radiated from the surface of earth, F, is dependent on the absolute temperature given by the Stefan–Boltzman law

$$F = \sigma T^4$$

where T is temperature in °K and σ = [0.342 J/ (cm^2·min·°K^4)]. For example, if the temperature over a land mass is 260°K and the temperature over the ocean is 293°K, the amount of energy radiated from the land surface would be about 62% of the amount of energy radiated from the water surface at the same latitude. This difference in the radiated energy greatly affects the air temperature as demonstrated, for example, by the difference in the January air temperature over Iceland (affected by the Gulf Stream) relative to the temperature over Siberia.

Convection. The transfer of heat by convection is a complex mechanism and will be discussed in later sections. For purposes of this discussion, let it suffice to say that convection is the process whereby rising and falling air currents transfer heat from a surface that is warmer than the air above it. In meteorology, convection is used to denote vertical transfer only. Another term, advection, is used to describe the horizontal transfer of a substance (heat) in the atmosphere.

Conduction. Conduction is an important mechanism for heat transfer in the ground but is fairly unimportant in air and water. The conductivity or rate of heat transfer of air is about 2.5 × 10^{-4} J/(s·cm·°C). The conductivity of water and soil is about 20 and 100 times larger, respectively. As a result, heat transfer by conduction from land or water surfaces would affect only the lowest few meters of the atmosphere during the course of a day. Heating by conduction is primarily a daytime and a summertime process. It follows that cooling by conduction is primarily a nighttime and wintertime process.

Evaporation–Condensation. Because the oceans occupy approximately 70% of the earth's surface, much energy is directed toward evaporation of seawater. In fact, of the total solar energy absorbed at the surface during the course of a year, an estimated 50% is consumed in evaporating seawater (Trewartha, 1954). The latent (stored) heat used in evaporation is released when the water vapor is converted back to liquid through condensation, at which time this heat is given off to the surrounding air. The condensation forms clouds in the atmosphere. In this way, heat from the surface is effectively transmitted to higher altitudes. Naturally this mechanism is most important in the areas where clouds are continually being formed, such as the intertropical convergence zone.

2-3.3 Average Temperature Structure of the Atmosphere

The most important processes that govern the temperature of the earth and its atmosphere have been described. The present understanding of these processes has been brought about by a number of theoretical studies supported by radiation and temperature observations. A basic knowledge of the physical causes for temperature change can help in the analysis of observed temperature data. Hence the temperature distribution in the atmosphere should not be considered as merely a tabulation of average values but as the integration of the effects of several physical processes. Its study permits an easy means to evaluate and place in perspective the terms in the energy budget relationships.

2-3.3.1 Horizontal Variation of Temperature. The seasonal average patterns of surface temperature presented in Fig. 2.4 show a familiar pattern, the inverse relationship between temperature and latitude. The colder temperatures at the poles occur because the surface energy budget is small or negative in the polar regions. The temperature pattern shows the average location of large-scale weather systems called cold and warm air masses. Not so obvious, but very important, is the relationship between the magnitude of the horizontal temperature gradient and the wind. Temperature differences lead to pressure differences which lead to the motion of the air.

2-3.3.2 Vertical Variation of Temperature. Another major feature of temperature distribution within the atmosphere is the relationship between temperature and elevation. The general decrease of temperature with increasing elevation is an indication that the earth's surface is the primary source of heat energy for heating the atmosphere. The nature of the vertical temperature distribution (temperature profile) is well-known to meteorologists; it is not so well-known to nonmeteorologists. The importance of the vertical temperature profile is because of its relationship with atmospheric stability. Atmospheric stability is an important consideration in almost all meteorological processes including atmospheric diffusion.

Mean Temperature Profile. Figure 2.5 shows the mean temperature profile between the surface and 150-km level, referred to as the standard atmosphere. The profile is characterized by two minimums and one maximum that serve as boundaries between layers of the atmosphere having similar vertical temperature gradients. The temperature minimum at 15 km defines the height of the tropopause, which is the upper limit for clouds and weather systems contained within the troposphere. The temperature

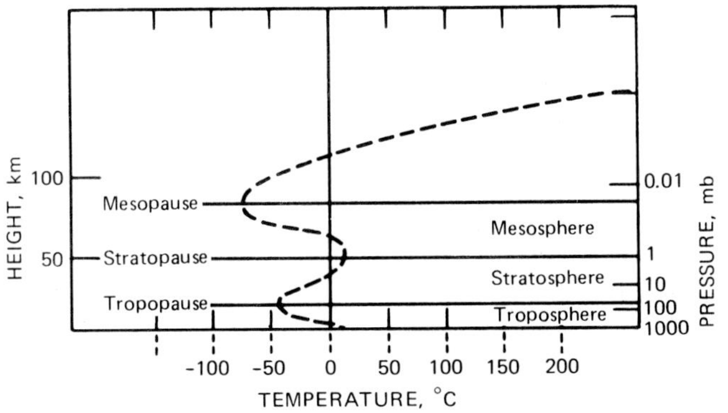

Fig. 2.5 Mean temperature distribution in the atmosphere.

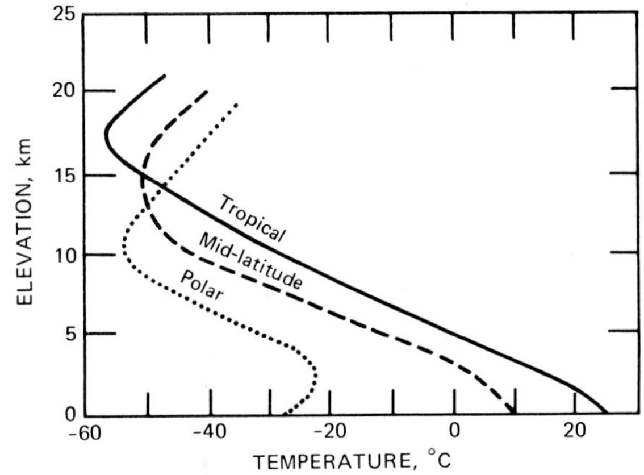

Fig. 2.6 Annual mean temperature profiles over a tropical, mid-latitude and polar region.

maximum at 50 km, caused by the synthesis of ozone from oxygen involving absorption of shortwave radiation from the sun, is called the stratopause. The stratosphere extends between the tropopause at 15 km and the stratopause at 50 km. The second temperature minimum occurring at an elevation of 80 km is referred to as the mesopause; in turn, the area between the mesopause and the stratopause is the mesosphere.

The troposphere is characterized by an average temperature gradient of −6.5°C/km (the negative sign indicates a decrease in temperature with increasing elevation). At the tropopause and extending several kilometers into the stratosphere, the gradient of temperature is approximately zero. Within the stratosphere the temperature gradient is about +6.5°C/km (the positive sign indicates an increase in temperature with increasing height). In common meteorological terminology, a negative gradient is referred to as a lapse condition, a zero gradient as an isothermal condition, and a positive gradient as an inversion condition. The average temperature distribution within the standard atmosphere contains lapse conditions between the surface and 10 km, isothermal conditions between 10 and 15 km, and inversion conditions between 15 and 20 km.

Figure 2.6 also shows mean temperature profiles between the surface and 20 km for tropical, mid-latitude, and polar locations. The height of the tropopause over the tropics is approximately 16 km, lowering to approximately 10 km over the polar regions. The lapse rate of temperature in the lowest 3 km shows a marked change from the tropics to the polar regions. Over the polar regions the colder surface temperatures cause the formation of a surface-based temperature inversion, whereas in the warm tropics the inversion is not observed. The temperature profile between the tropopause and 3 km is similar for all locations.

The mean temperature profile over any of the three locations varies with the season. For example, a mid-latitude location during the summer months has a mean temperature profile similar to a tropical location. During the winter months the major features of the mean profile look more like the polar-temperature profile (i.e., a lower tropopause and an isothermal or inversion layer near the surface).

Figure 2.7 shows a cross section of temperature within the lowest 30 km, extending from the North Pole to the South Pole for the period December through February. The left side of the figure (North Pole to Equator) represents a distribution of temperature for winter, the right side (Equator to South Pole) represents a distribution of temperature for summer. The information in this figure is similar to although more detailed than the information derived from the temperature profiles. Note that the largest changes between the winter and summer cross sections occur within the lowest 3 km. Several features of the temperature distribution will be shown in the next section in an explanation of wind.

Adiabatic Lapse Rate. The hypsometric equation (see Sec. 2-2) shows that the pressure exerted by the atmosphere decreases logarithmically with height. Note the atmospheric pressure on the vertical coordinate in Fig. 2.5. If a parcel† of dry air is carried upward in the atmosphere, the pressure of that parcel will decrease in response to the surrounding pressure. Accompanying this decrease in pressure will be an increase in the volume and a decrease in the temperature of the parcel as predicted by Boyle's and Charles's laws. If the parcel is moved rapidly, the process can be assumed to be adiabatic. The rate of change of temperature with height as determined from this adiabatic process is called the adiabatic lapse rate, the term lapse rate referring to the rate at which the temperature decreases with increasing height, 9.8°C/km.

A careful distinction must be made between the temperature profile and the adiabatic lapse rate. The temperature profile depicts the actual temperature of the air as we go upward through the atmosphere, and the adiabatic lapse rate depicts the temperature that a parcel would attain if it was lifted mechanically through the atmosphere. Under certain atmospheric conditions, the environmental lapse rate and the adiabatic lapse rate will be the same. More often than not, however, the environmental lapse rate is different from the adiabatic lapse rate. The degree to which the environmental lapse rate and the adiabatic lapse rate differ is a measure of the thermal or

†An air parcel is defined as a volume of air, ~3 x 10^4 cm^3, having uniform characteristics throughout.

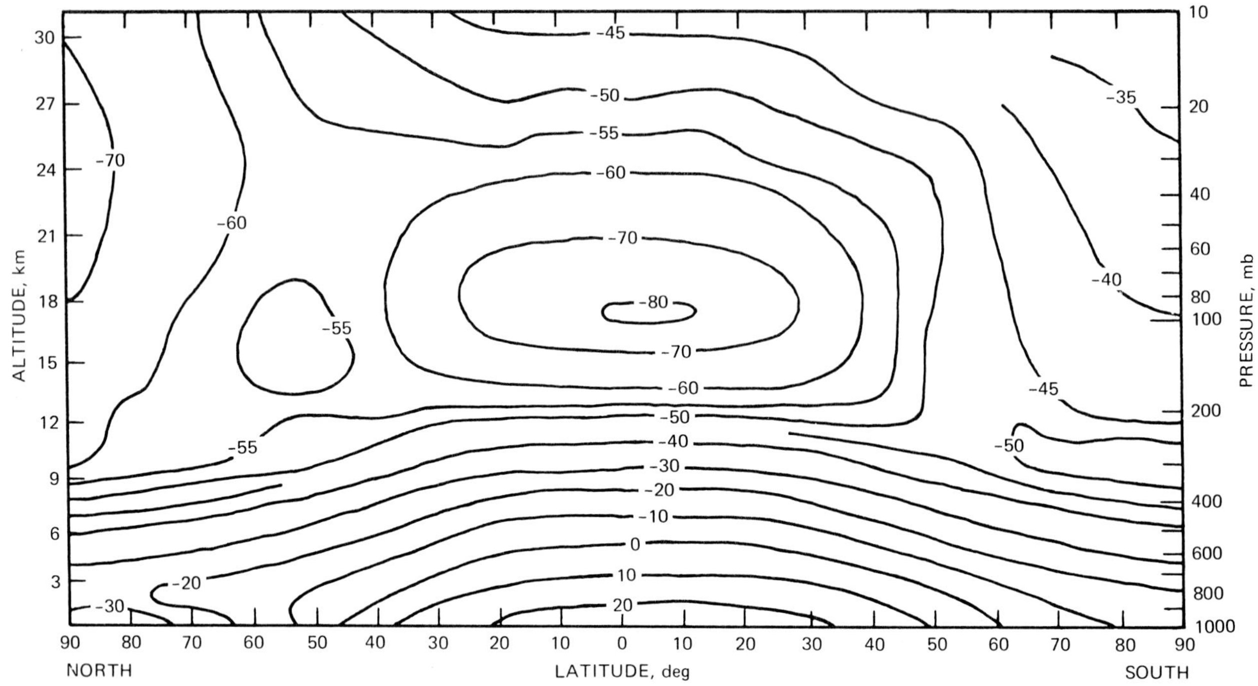

Fig. 2.7 Temperature pattern is given for a north–south cross section of the atmosphere for the 3-month period December–February. The isotherms, or contour lines of equal temperature, indicate the temperature in degrees Celsius. The coldest air is over the Equator and near the winter pole. In the latter region the temperature sometimes reaches −85°C at certain longitudes. [From R. G. Newell, The Global Circulation of Atmospheric Pollutants, *Scientific American*, 224(1): (1971).]

static stability (often simply called stability) of the atmosphere.

Stability. Figure 2.8 compares temperature profiles associated with different stability conditions. The heavy solid line is the normal lapse rate. The dashed line is the temperature that a parcel originally at the surface at a temperature of 30°C would have if it were lifted upward and cooled at the adiabatic lapse rate. You can see that, after being lifted approximately 1 km, the parcel would have a temperature approximately 3.5°C colder than the temperature of the surrounding air. If the force is removed, the parcel would descend because of its greater density relative to that of the warmer surrounding. The parcel would continue to fall until it achieved equilibrium with the environment, a condition referred to as stable.

The temperature profile indicated with a dotted line in Fig. 2.8 is an example of a temperature inversion reaching 35°C at a height of 1 km. A parcel cooling at the adiabatic lapse rate will attain a temperature about 13°C cooler than the environmental temperature at 1 km. The buoyant restoring force for this case is much greater than for the normal case and represents a condition that suppresses turbulent mixing.

The thin solid line in Fig. 2.8 shows an environmental lapse rate greater than the adiabatic lapse rate. For this case, a parcel lifted at the adiabatic lapse rate would attain a temperature warmer than the environmental temperature. When the lifting force is removed, the parcel, being less dense than the surrounding environment, would tend to continue its upward motion. Such a case is referred to as an unstable condition.

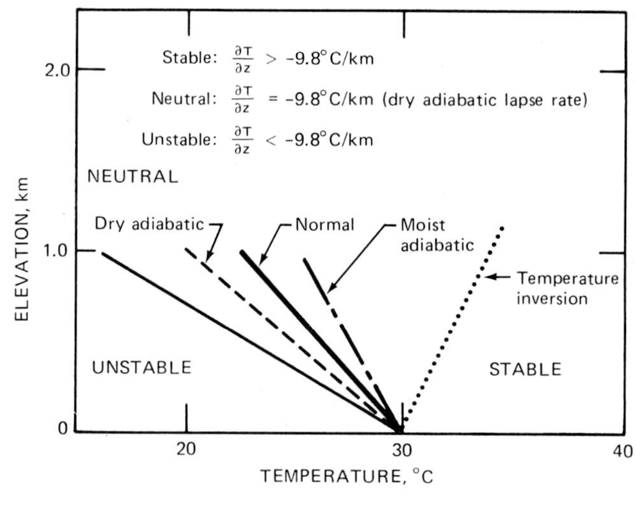

Fig. 2.8 Comparison of several different temperature profiles with the dry adiabatic lapse rate and the moist adiabatic lapse rate.

The atmosphere is not dry, although it acts as a dry atmosphere as long as no condensation of water vapor occurs. As soon as condensation of water vapor occurs, however, the dry adiabatic lapse rate is no longer appropriate. As the water vapor condenses, it gives off the latent heat of vaporization; thus the rate at which saturated air cools as it expands adiabatically is smaller than the rate at which unsaturated (or dry) air cools adiabatically because part of the cooling is canceled by the latent heat released (see Fig. 2.8).

2-3.4 Macroscale Wind Structure

2-3.4.1 Basic Concepts. Wind occurs as a result of pressure gradients caused by the unequal distribution of temperature in the atmosphere. For the simplest explanation, it is necessary to consider only the surface temperatures. The area of earth that lies between latitudes of 30°N and 30°S gains by net radiation a total of 2.6×10^5 (g•J)/s averaged for the year. Between 30° latitude and the poles, there is a net loss of heat at about the same rate. If there were no circulation to carry heat from equatorial to polar regions, an unbearable situation would soon develop with bitter cold at the high latitudes and intense heat near the equator. Such a circumstance does not occur because of the significant transfer and exchange of heat from the equator to the poles and return.

To describe the mechanisms responsible for the production of the large-scale transport of air and energy, we must define the major forces acting on any parcel of air. These are an apparent force called the Coriolis effect and forces due to gravity, pressure gradient, and friction.

Coriolis Effect. Newton's first law of motion states that a body in motion will continue in motion with a constant speed in a straight line unless directed from its path by another force. Although this law very nearly applies to a parcel of air traveling in the atmosphere, to an observer on earth the motion is curved since the coordinate system (in this case the lines of latitude and longitude) rotates. The apparent force that would have caused this motion is referred to as the Coriolis effect, f, given by $f = 2V\Omega \sin \Phi$, where V is wind speed, ω is the angular rotation of earth, and Φ is latitude (see Fig. 2.9). In the Northern Hemisphere the Coriolis force will always tend to move a parcel to the right of the motion; in the Southern Hemisphere, to the left of the motion. If the parcel is not moving or is located at the equator, the Coriolis effect will be zero since it is directly related to wind speed.

Gravity. The constant downward force of attraction exerted by the mass of the earth on each parcel represents the force of gravity, F = mg, where m is mass of the parcel and g is acceleration due to gravity, 980 cm/s^2. Additional gravitational forces are exerted on the parcel owing to the moon and the sun. These small forces are generally ignored in meteorology, although they produce atmospheric tides analogous to ocean tides. Gravity holds the atmosphere to the earth.

Pressure Gradient. Air pressure is a measure of the force exerted by the mass of air over a unit area. Air pressure decreases with increasing height above the ground. The hydrostatic equation (see Sec. 2-2) shows that the colder (and therefore the denser) the atmosphere, the faster will be the rate of upward decrease in pressure. The warmer (and therefore the lighter) the atmosphere, the slower will be the rate of upward decrease in pressure. This principal is shown in Fig. 2.10, where the left side of the diagram depicts the height of isobars (surfaces of constant pressure)

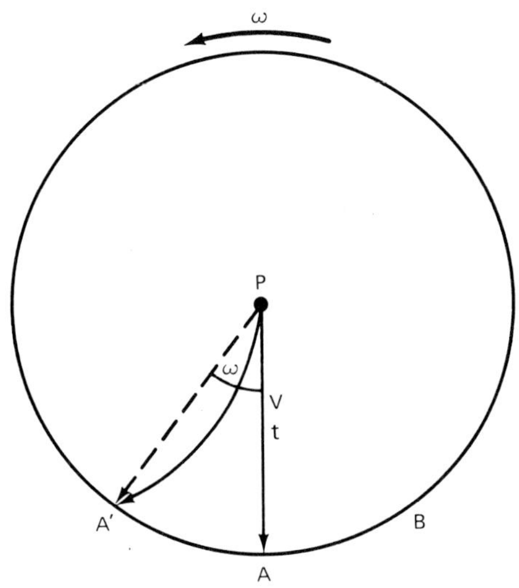

Fig. 2.9 The Coriolis force. If we consider a particle of air moving horizontally away from the North Pole (P) at a velocity v in the direction of point A, which it should reach in time t, it will soon become affected by the earth's rotation ω. Since the earth's surface is rotating with counterclockwise angular velocity ω, the point A will have moved to B in the time t, and point A′ will have moved to A. Thus, relative to the earth's surface, the particle will appear to have followed path PA′ represented by the solid line, i.e., to have undergone a continuous deflection of its movement to the right. (From H. Flohn, *Climate and Weather*, McGraw-Hill Book Co., New York, 1969.)

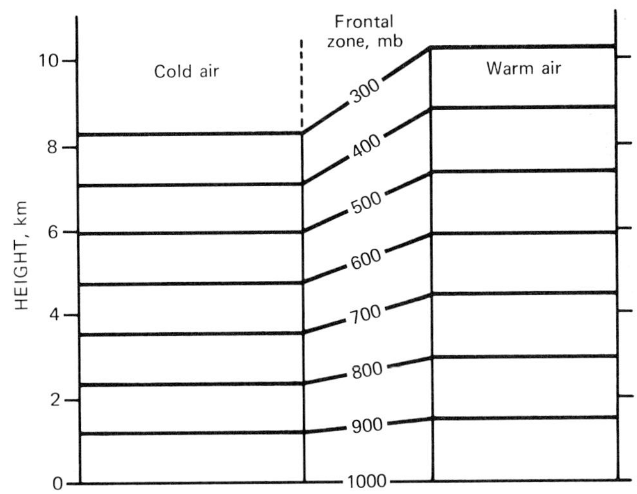

Fig. 2.10 The differential reduction of pressure with altitude in cold and warm air. This feature is explained theoretically by the use of the hypsometric equation which shows that the thickness of the layer between two pressures is directly proportional to the temperature of the layer.

between the surface and 10 km over a cold region. On the right is shown the height of the same isobars over a warm region.

A cold air mass lying next to a warm air mass, as shown in Fig. 2.10, produces a horizontal pressure gradient from warm air toward cold air at higher levels. The pressure at 6 km above the surface in the warm air is 600 mb, and the pressure in the cold air is 500 mb. This horizontal pressure gradient causes an air parcel at these higher levels to move toward lower pressure. The stronger the gradient, the stronger the force exerted on the parcel. Since the pressure gradient at the surface in our example is zero, no surface wind will be induced by this configuration. If the earth were not rotating, the induced pressure gradient at the higher levels would cause tropical air to flow straight north to the pole. This motion would tend to decrease the pressure over the tropics since upper-level air has moved out of the tropics. By contrast, the pressure at the poles increases; this increase in pressure also occurs at the surface, and thus a north–south pressure gradient would be established at the surface. The result would be a return flow near the surface from the poles to the tropics. In this way there would be established a complete circulation involving one cell, such that we would have north winds at the surface (in meteorological terminology wind direction is from whence it comes) all over the globe and southerly winds at higher levels with upward motion over the equatorial latitudes and downward motion over the polar latitudes. By inducing the differential heating rates at the surface over continent and water masses and introducing energy transport caused by the release of latent heat of condensation, an appreciably different flow pattern would result from this idealized north–south pattern.

The rotation of the earth changes the circulation pattern drastically. Once the parcel of air responds to the pressure gradient force and starts its motion, it causes the Coriolis effect to act in a direction perpendicular to the movement of air. For equilibrium to be achieved, then, the deflecting force caused by the rotation of the earth must necessarily be equal and opposite to the force of the pressure gradient. For such conditions, the motion of the parcel would be perpendicular to the pressure gradient force and also perpendicular to the Coriolis force as depicted in Fig. 2.11. The resulting motion would be along the isobars. Such motion frequently occurs within the atmosphere at elevations of the order of 1 km above the ground and is called the *geostrophic wind.*

$$U_g = \frac{1}{\rho f}\frac{\partial p}{\partial n}$$

where U_g is the geostrophic wind speed directed parallel to the isobars, f is the Coriolis parameter, ρ is the air density, and $\partial p/\partial n$ is the pressure gradient measured perpendicular to the isobars. Thus the effect of rotation on the flow is to add an easterly component to northerly winds and a westerly component to southerly winds.

Friction. Below about 1 km, the retarding effects of friction produce a drag on the air which reduces the speed of the parcel. However, the magnitude of the Coriolis effect is proportional to wind speed so that friction reduces the Coriolis effect. When this occurs, the pressure gradient force exceeds the Coriolis effect and the wind blows across isobars rather than parallel to the isobars as is the case in a geostrophic wind. (See Fig. 2.11.) The angle between the surface wind and the isobars depends on atmospheric stability and, to a lesser degree, on the roughness of the

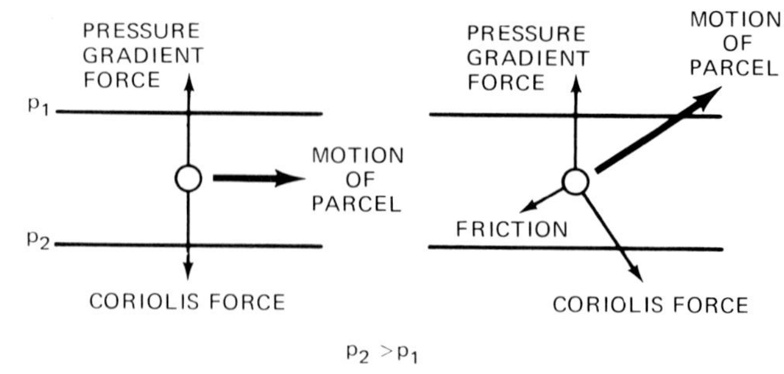

Fig. 2.11 A vector diagram to show forces acting on an air parcel in the Northern Hemisphere. (From H. Flohn, *Climate and Weather*, McGraw-Hill Book Co., New York, 1969.)

surface. This angle is approximately 10° during unstable conditions and increases to about 35° during very stable conditions (Arya, 1975).

2-3.4.2 General Circulation. The simple single-cell circulation described previously (proposed by Hadley in 1735) is not the process by which the tropical excess of temperature is transported poleward. One theory assumed that this movement required standing meridional circulation cells in the vertical plane, as shown at the left edge of Fig. 2.12. Three such cells were envisaged: a trade-wind (or Hadley) cell, a mid-latitude cell, and a polar cell. Observations failed to confirm the existence of the two northern cells, and an alternate method of transport was sought. In the late 1940's meteorologists revived a theory proposed by Jeffreys (1926) who showed that the required transport could be achieved horizontally by means of stirring by traveling large-scale circulation patterns—the familiar highs and lows of the daily weather map. It is now generally conceded that this is the method that maintains poleward transport in the mid- and upper-latitude regions. The trade-wind cell remains a necessity for heat transfer in lower latitudes.

Wind observations indicate that the prevailing motion of the greatest volume of the earth's atmosphere is from west to east. Figure 2.13 shows the average *zonal* or east–west component of the winds over the earth in summer and winter. Lines of equal wind speed, called *isotachs,* encircle zones having higher wind speeds in one direction. The shaded areas represent east winds, and the unshaded areas represent west winds.

The westerlies become more pronounced and steadier with increasing elevation and reach a sharp maximum of speed between 10 to 12 km above the surface. The north–south wind speed component is generally an order of magnitude or more slower than the east–west component except in the lower 2 km of both the polar and tropical regions.

There are two major exceptions to the westerly winds. The first is the shallow layer of east winds in the polar regions called the polar easterlies. They are generated in the polar high-pressure systems and flow toward the subpolar low-pressure cells. The other exception to the westerly winds is found in the tropics. These are the trade winds,

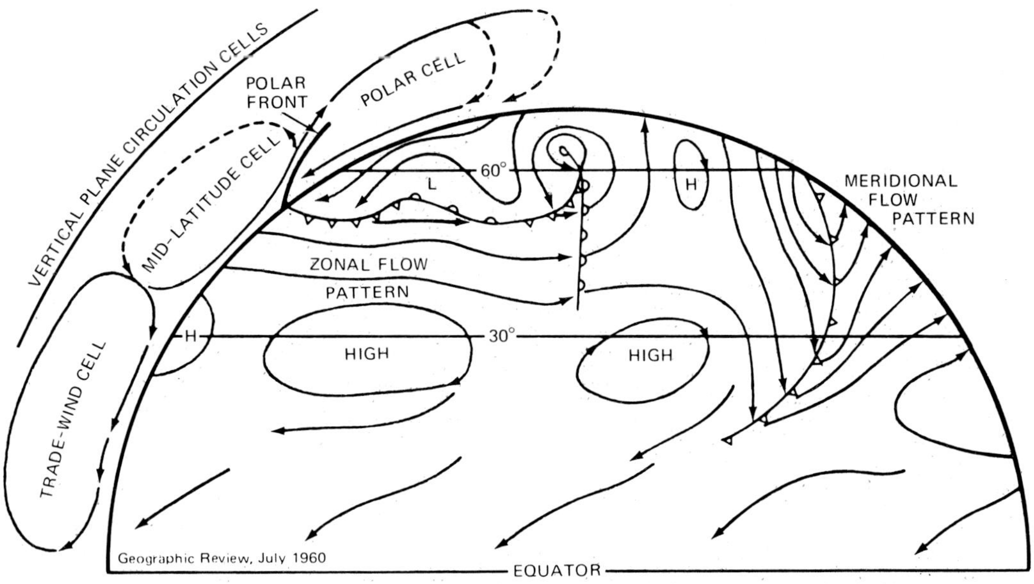

Fig. 2.12 The discredited meridional circulation model. The left edge of the sketch shows the three meridional-plane circulation cells once thought necessary to convey the tropical excess of heat, moisture, and angular momentum toward the polar sink. It is now thought that only the trade wind (or Hadley) cell actually exists. North of the subtropics the traveling westerly waves, cyclones, and anticyclones bring about the required transport by means of predominantly horizontal stirring, especially in areas of strong meridional flow (right). There may be a minor polar front cell just north of the subtropical jet, but this is denied by most authorities. [After F. K. Hare, The Westerlies, *Geophysical Review*, 50(1960): 345-367 (1960).]

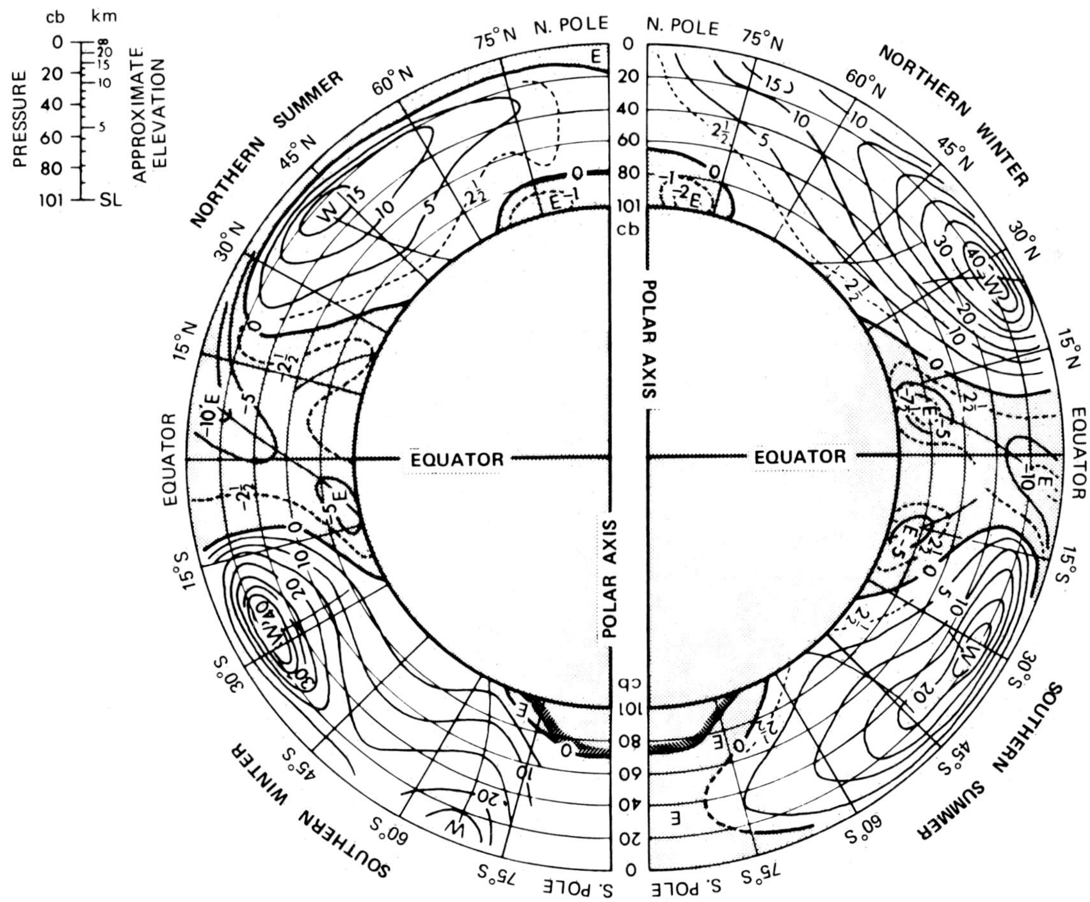

Fig. 2.13 Mean zonal winds over the earth in summer and winter. Isotachs are in meters per second. Regions of easterly winds are hatched. (From H. Flohn, *Climate and Weather*, McGraw-Hill Book Co., New York, 1969.)

which are much stronger and extend to greater heights than the polar easterlies and which are caused by the meridional circulation of the Hadley cell.

The observed distribution of wind in Fig. 2.13 is reasonably well explained in terms of the mean temperature distribution as shown in Fig. 2.7. The relationship between the temperature and wind distribution is best explained through an analysis of the geostrophic wind at two levels. The east–west, u_g, and the north–south, v_g, components of the geostrophic wind evaluated at a constant pressure surface are given by

$$u_g = -\frac{g}{f}\frac{\partial h}{\partial y}$$

and

$$v_g = \frac{g}{f}\frac{\partial h}{\partial x}$$

where h is the height of the pressure surface and x and y are coordinates in the east–west and north–south directions, respectively. The differences between the components at two altitudes (1) and (2) are given by

$$u_{g_2} - u_{g_1} = -\frac{g}{f}\frac{\partial(\Delta h)}{\partial y}$$

and

$$v_{g_2} - v_{g_1} = \frac{g}{f}\frac{\partial(\Delta h)}{\partial x}$$

where Δh is the thickness between the two pressure surfaces p_1 and p_2. The hypsometric equation (Sec. 2-2) shows that the thickness, Δh, is proportional to the average virtual temperature of the layer $\overline{T}_v$ and that

$$u_{g_2} - u_{g_1} = -R \ln\left(\frac{p_2}{p_1}\right)\frac{\partial \overline{T}_v}{\partial y} = u_{th}$$

and

$$v_{g_2} - v_{g_1} = R \ln\left(\frac{p_2}{p_1}\right)\frac{\partial \overline{T}_v}{\partial x} = v_{th}$$

(Note: Meteorologists often approximate T with the air temperature, T, which seldom results in an error greater than 1%.)

The preceding expressions show that the change in wind speed with height is related to the horizontal temperature gradient. Hence the vector difference in the geostrophic wind over a layer is called the *thermal wind.* The above-mentioned equations define u_{th} and v_{th}, the components of the thermal wind.

Between the surface and about 12 km (see Fig. 2.7), the equator–polar temperature gradient ($\partial T/\partial y$) is negative and above 12 km it is positive. Thus u_{th} is positive below 12 km, and u_g increases with height. Above 12 km, u_{th} is negative; so u_g decreases with height. The temperature gradient is largest in the mid-latitudes corresponding to the wind speed maximum at an elevation of about 10 km. Since the east–west temperature gradient ($\partial T/\partial x$) on the average is much less than the north–south temperature gradient ($\partial T/\partial y$), v_{th} is correspondingly smaller. (See, for example, the distribution of the mean surface temperature in Fig. 2.4.)

In summary, the important points to remember are that the motion of air is the result of a combination of three forces: (1) the pressure gradient force caused by differential heating of the earth's surface, (2) the Coriolis effect caused by the rotation of earth, and (3) the retarding effects of friction. Moreover, the greatest pressure gradient generally occurs in the mid-latitudes so that the general motion of the upper atmosphere responds to this pressure force and develops westerly winds aloft. However, since the motion is not exactly toward the east, there is a transfer of heat between the poles and the tropics. In terms of wind speed, this flow is a very small percentage of the average east–west, or zonal, wind component. This poleward flow in the upper troposphere induces an equatorial flow at the surface levels.

Geographical features, such as land and water, have a very important influence on the wind patterns. This very complicated interaction of various forces and motions produces the surface weather features we observe in our daily weather, which is by no means constant and in some cases is difficult to predict.

2-3.5 Weather Systems

The method of weather analysis practiced by meteorologists is called the synoptic method. (Weather observations made at the same time at many locations plotted on a map or chart show a synopsis or picture of the weather conditions over an area. A satellite photograph is another form of synoptic weather map.) This method is based on the study of a series of synoptic weather-depiction charts that reveal the structure of the all-important macroscale weather systems, such as low-pressure centers and high-pressure centers. In the early days of meteorology, the synoptic maps were based on few and scattered observations. As the observation network increased in density, the synoptic maps could be constructed over larger areas until at the present time we can construct synoptic weather maps for the entire globe. For many years these synoptic maps were limited to observations made at the earth's surface, and it was not until relatively recent years (the 1940s) that upper-air data could be used regularly. Although numerical weather prediction is based almost exclusively on information obtained from the upper-level observation network, the general public is still more familiar with the surface weather maps. Justifiably so, since the surface weather map represents an integration of the entire atmosphere down to the level where most people reside.

To better understand the variations in the surface weather map, we must start with an understanding of the more simplified upper-air maps. Surface weather maps are excessively complicated, with a number of low-pressure centers, called *cyclones,* and high-pressure centers, called *anticyclones.* The pattern revealed by charts representing the upper troposphere at pressure levels near the height of the maximum wind band is less complicated, at least in its appearance. Figure 2.14 reproduces a portion of the upper-level wind pattern at a height equivalent to a pressure of 500 mb for four typical cases. Since to a large extent the two hemispheres behave independently, it is convenient to restrict attention to the Northern Hemisphere only. The arrows indicate the location of the band of strong winds and point in the direction of the motion. This band of strong winds has been shown previously to be the result of a sharp contrast in temperature between the equator and the polar regions. The axis of this band of strong winds is called the *jet stream.* The jet stream is defined as a narrow band of strong winds (⩾25 m/s) in the westerlies. It may extend around the hemisphere, although frequently it is discontinuous and varies greatly from day to day.

Probably the most important characteristic of the jet stream is its meandering property. Instead of existing in one smooth circle around the pole, the band of winds inevitably exhibits a waviness with more or less pronounced displacements toward the pole (in the so-called *ridges* where the pressure is relatively high for the latitude) and toward the equator (in so-called *troughs* where pressure is correspondingly low). Frequently a system of waves retains its shape for an extended period (1 to 2 weeks), but the whole pattern drifts slowly (5 to 15 m/s) toward the east. By measuring the distances in the west-to-east direction between successive troughs or ridges, a wave length can be defined. It was Rossby (1939) who developed this mode of thinking about planetary waves (Rossby waves, or long waves) and made important contributions to the theoretical explanations of the existence and behavior of the large-scale features.

The simple diagrams in Fig. 2.14 show that the Rossby waves vary greatly in amplitude from one time period to another and may actually break and throw off detached closed circulations on either the polar or the tropical side. The significance of these cutoff circulations is that the westerly, upper-level winds are frequently divided into two parts, one being displaced northward and the other southward, with the middle zone being left as a region of little or no westerly wind. Since the upper-level winds are largely responsible for the migration of weather systems across the globe, these cutoff circulations block the normally westward progression of systems. In other words, stagnation of the upper-level wind pattern produces a corresponding stagnation in the progression of weather systems at the surface. These stagnating systems can produce the long spells of cold weather in certain situations or the unseasonably warm weather in other situations.

The motion of weather systems (which are essentially located in the lower part of the troposphere) are greatly influenced or steered by the upper-level wind patterns. A system located directly under the jet stream will move considerably faster than a system not so located. A glance at Fig. 2.14 shows that systems over the polar or equatorial regions are frequently beyond the influence of these steering winds. Thus large bodies of air remain stationary over homogeneous surfaces; soon the air takes on the temperature, moisture, and stability values characteristic of the surface. Such bodies of air are referred to as *air masses* and are named according to their source regions.

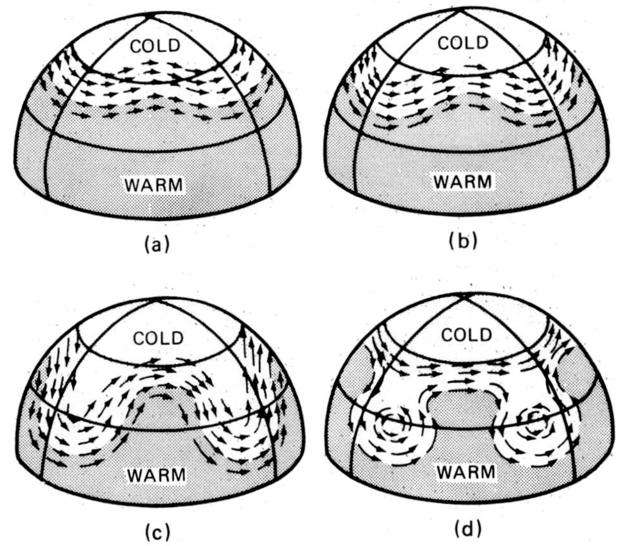

Fig. 2.14 Upper-level wind pattern at approximately 500 mb for four typical cases. The areas indicate the location of a band of strong winds the center of which is called the jet stream. The diagram shows a sequence of disturbances (called the Index Cycle) in these upper-air waves. The amplitude of the waves increases until both waves become unstable and break, pools of cool air being left in low latitudes with warmer air at higher latitudes. These cutoff circulations give rise to unusual weather in the respective latitudes. (From H. Flohn, *Climate and Weather,* McGraw-Hill Book Co., New York, 1969.)

When the jet-stream pattern shifts to that shown in Fig. 2.14c, previously stationary air masses can be caused to move from their source regions. As the air mass moves out of its source region and over different types of surfaces, its properties are modified until gradually the air mass loses its identity as a recognizable system. For example, a cold, continental air mass moving out to sea takes on maritime characteristics in its lower levels. While this modification to the air mass is occurring because of its movement, the upper-level circulation pattern may change again, and conditions over the source region may become conducive to the formation of a new air mass.

Almost everyone interested in the causes of inclement weather is aware that the boundary between two air masses of different densities (temperatures) is associated with rainy weather. This boundary is called a *front.* A cold front occurs when cold air displaces warm air, and a warm front occurs when warm air replaces cold air. The location of the front differentiating the polar and tropical air masses approximately corresponds to the position of the jet stream. The reason? Both are caused by the same polar–equatorial temperature gradient.

A typical winter surface weather map for North America (with an upper-level map for comparison) is shown in Fig. 2.15. The polar front, which corresponds approximately to the jet-stream position at the upper levels, meanders irregularly from the Atlantic to the Pacific Ocean with the cold air generally located north of the front. On the basis of hundreds of surface observations, an analysis has been made of the pressure field, using lines of constant pressure, called *isobars,* to delineate the various air masses and pressure systems. The heavy solid lines are used to indicate the position of important sections of the polar front (not shown in Fig. 2.15 is a weak portion of the polar front extending from west of Hudson Bay through Minnesota and Arkansas and ending south of Cuba). Precipitation areas are indicated with hatched lines. The variation of the weather parameters across the Northern Hemisphere should be noted. The changes of these same parameters with time become apparent from an examination of a sequence of weather maps.

The upper-air maps are constructed for a fixed pressure surface (in this example, 500 mb), and, instead of isobars, lines of constant height above sea level (called *isohypses*) are drawn. These lines are called contours because of their similarity to the contours on topographic maps. The contours delineate highs and lows just as the surface isobars do. At this altitude (≅6 km) the closed patterns are much less common than they are at the surface. The wind blows along contour lines with the lowest contour height to the left. Considerably less variation in the contour patterns is evidenced on these high-level charts. The uniformity is due to two factors: the absence of surface effects (a true physical cause) and the greater spacing between upper-air stations, which may result in a simplicity that more complete data would show to be unwarranted.

2-3.6 Macroscale Motions and Dispersion

For many applications the macroscale circulation patterns, as formerly described, are responsible for the transport of pollution, and smaller scale motions are responsible for the diffusion of the pollutant. Transport and diffusion are shown schematically in Fig. 2.16. The left side of Fig. 2.16a depicts a typical pollution cloud in an atmosphere containing many small atmospheric motions

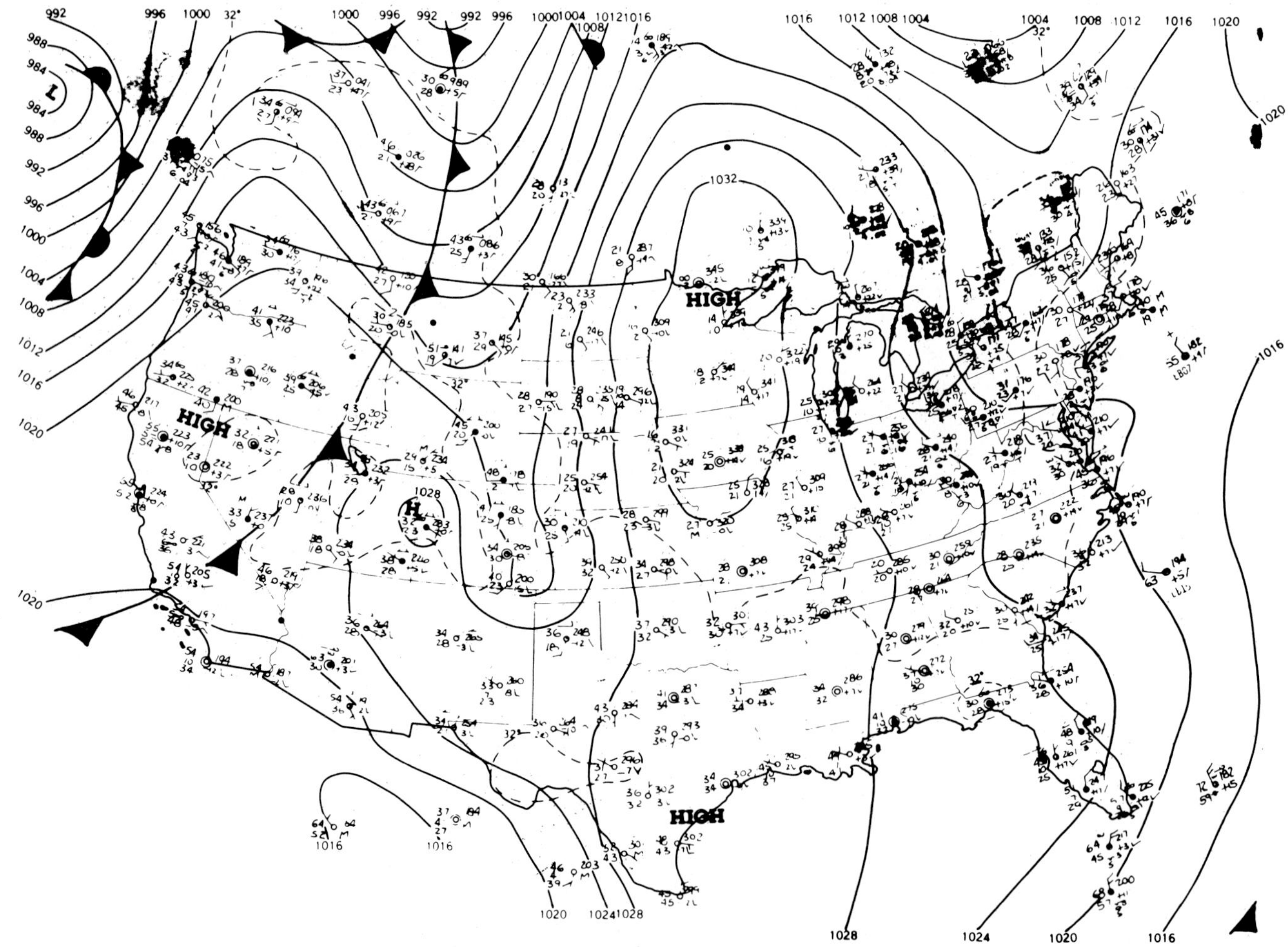

(a)

Fig. 2.15 (See facing page for legend.)

called *eddies*. Since the size of the eddies is much smaller than the size of the cloud, these eddies tend to slowly diffuse the cloud and make it grow larger in time as evidenced by the right side of Fig. 2.16a. When the size of the cloud is much smaller than the eddies, as shown in Fig. 2.16b, the motions tend to transport the cloud rather than diffuse it. When the cloud and eddies are nearly the same size, the cloud grows rapidly.

A reexamination of the surface weather map shown in Fig. 2.15 shows vast areas of the United States over which the pressure pattern is quite uniform. For example, in eastern South Dakota the isobars are almost parallel, indicating that the wind direction is nearly constant. Thus an atmospheric release made in this area would tend to flow with the macroscale wind patterns.

Pollutants are introduced to the atmosphere in different ways. Sources of pollution can be at the surface or elevated and can be classified as either area, line, or point sources. Pollution from a power-production facility is normally released through an elevated stack. The duration of the release ranges from an instantaneous release (from an explosion or a broken pipe) to a continuous release (from an unavoidable release during routine operations). Since the size of the pollutant cloud is very small, the ultimate location of the pollution is determined by the macroscale wind pattern, and the growth is controlled by smaller scale motions. At times the path of the pollution will be straight and quite easy to predict; at other times the path will be curved and difficult to determine. In either case the downwind concentration of the pollution is a function of the rate at which the plume or the cloud will diffuse. At times this diffusion rate will be dependent on the magnitude of the small-scale eddies. As the pollutant cloud approaches macroscale dimensions, the macroscale wind field also contributes to the diffusion processes.

A measure of the effect of the synoptic weather patterns on the motion of a particular pollutant release is most easily exemplified for the case in which an instanta-

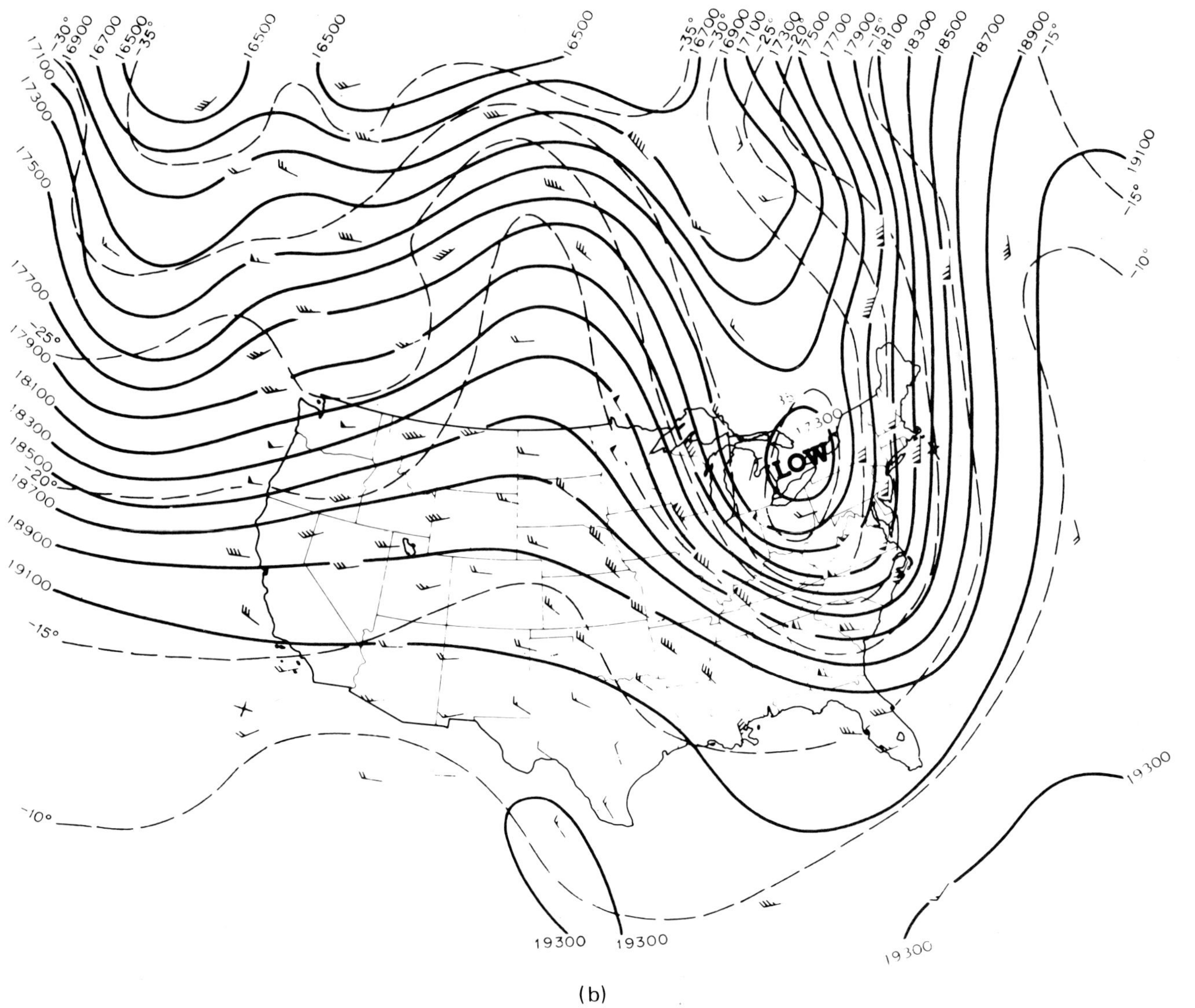

(b)

Fig. 2.15 Typical winter weather maps for North America (a) at the surface and (b) at 500 mb.

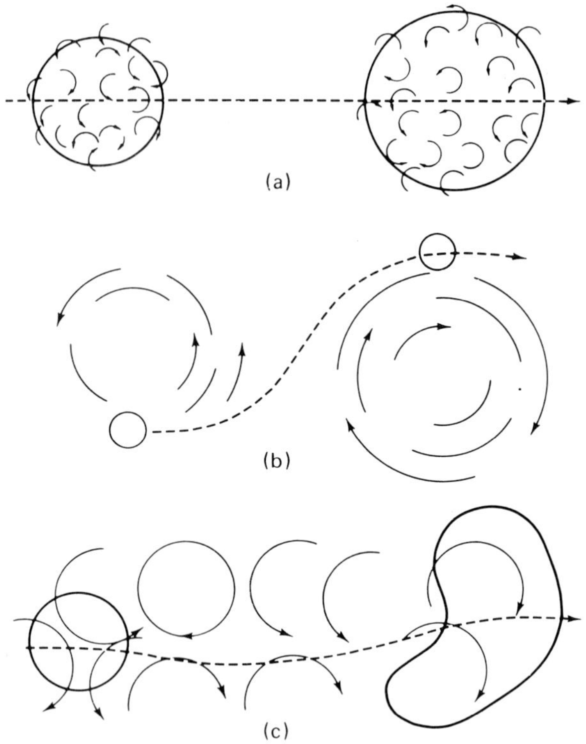

Fig. 2.16 Idealized dispersion patterns. (a) A large cloud in a uniform field of small eddies, (b) a small cloud in a uniform field of large eddies, (c) a cloud in a field of eddies of the same size as the cloud. (From D. H. Slade, *Meteorology and Atomic Energy—1968*, USAEC Report TID-24190, 1968.)

neous release (puff) is assumed. A number of observational studies have been made of the role of the large-scale weather patterns on the transport of puffs. Several have been made in association with the testing of nuclear reactor propelled rocket engines (Crawford, 1968; Randerson, 1972). Figure 2.17 shows an actual cloud tracked by airplanes at an altitude of 3 km for which the concentration of the radioactive elements was used to define the dimensions of the cloud. Note that 6 h from time of release until approximately 24 h, the cloud size increased with some distortion in shape. After 24 h from the time of release, the cloud grew at a faster rate, and the distortion became more pronounced, primarily because the large-scale motions tended to tear the cloud apart (see Fig. 2.16c) until, at a time 64 h after release, the cloud became quite diffuse.

Weather maps analyzed at 6-h intervals for the appropriate level can be used to reconstruct the path or trajectory of the cloud. This trajectory is estimated by the addition of the 6-h wind vectors originating at the point of release. Figure 2.18 shows four such trajectories calculated from wind observations at four different altitudes. The large dashed line indicating the trajectory at the 3-km level agrees very well with the resulting cloud path. Note the vast difference in the position of the end point of the trajectories after a 3-d travel time. This variation of the position of the end point is solely due to variation in the large-scale weather patterns at each constant height level. Furthermore, these four trajectories are representative of variation of 3-d trajectories having initial times differing by about 6 h. These examples point out that, to forecast the effect of a release from a power plant over long distances, we must take into account the changing synoptic weather patterns.

The preceding case study shows that a trajectory can be estimated by using observed data to simulate or reproduce the path of the release. In an emergency a forecast of the trajectory must be made, and the present state-of-the-art of synoptic forecasting can provide forecasts of these large-scale wind patterns for up to 48 h. Furthermore, twice-daily prediction of air trajectories originating at a number of locations is currently available from the National Weather Service [Forecast Office Facsimile Network (circuit) chart numbers F042C and F107C].

Machta (1958) suggests that the synoptic circulations and global circulations would tend to distribute an effluent uniformly in the Northern Hemisphere after 14 d. The rate at which an effluent would be spread would be dependent on the height at which the pollutant is injected into the atmosphere as well as on the synoptic patterns. Machta's calculations indicate that pollutants emitted in the Northern Hemisphere tend to stay in the Northern Hemisphere with little transport across the equator.

An interesting verification of this hypothesis is provided by observations of Freon in the atmosphere (more specifically, the term Freon applies to several chlorofluorocarbons, namely, CCl_2F_2 and CCl_3F, used in aerosol dispensers). The Freon is unusually stable chemically and only slightly soluble in water; therefore it persists and accumulates in the atmosphere. The interest lies in its potential use as an "inert" tracer for the study of mass-transfer process in the atmosphere. Figure 2.19 shows data collected during an ocean voyage in 1971 to 1972 (Lovelock, Maggs, and Wade, 1975) showing the average concentration of Freon as a function of latitude. The solid circles indicate observed concentrations in units of 10^{-12} by volume, and the solid line represents the numerical predictions by Machta (1958). For his model, Machta assumed that the Freon was released between latitudes 35°N to 55°N.

In summary, the role of the macroscale motions in the dispersion of pollutants with respect to the power industry is primarily that of transport of the pollutant. (For calculations involving continents or the entire globe, macroscale motions are treated as diffusion.) The variability of the synoptic motions in both space and time causes much difficulty in the determination of the location of a particular plume or cloud. If the location cannot be determined accurately, it makes little difference whether the diffusion rate is correctly assessed. Thus, for transport of effluents for more than several hours, the effects of synoptic transport must be included.

2-4 MESOSCALE SYSTEMS

Macroscale motions were assumed to be made up of motions for which the horizontal scale exceeded 2000 km.

Fig. 2.17 The effluent trajectory and diffusion pattern measured from the NATS aircraft between the altitudes of 2.9 to 3.2 km MSL. (From T. V. Crawford, *The Long-Range Diffusion of the Effluent Cloud from Phoebus-IV Reactor Test of February 23, 1967*, University of California, Livermore, Lawrence Radiation Laboratory, April 1968.)

This section will consider a smaller scale motion called the mesoscale, including motions having horizontal dimensions between 2 and 2000 km. To understand mesoscale motions, we must examine in more detail a number of the mechanisms already discussed with respect to macroscale motions. During this examination we must recognize the interplay between the different scales of motion. A number of mesoscale features can exist only during specific macroscale patterns (i.e., only during certain weather conditions). For example, for a land–sea breeze to occur, the winds should be light and sky conditions should be clear.

2-4.1 Local Energy Balance

The rate at which the surface receives and absorbs solar energy determines the temperature distribution of the surface and, ultimately, the characteristics of the air over the surface. In the macroscale only two surface types were considered, land and water. For many mesoscale motions the subtle differences between surface types need to be considered. An important point to reemphasize is that all atmospheric motions are related to the gradients of density (and hence temperature and pressure) in both horizontal and vertical directions. Therefore we need to consider only those features of the surface energy balance which significantly affect the density gradients.

2-4.1.1 Surface Effects. A number of factors which are significant with respect to the production and maintenance of horizontal density gradients on the mesoscale are the albedo, heat capacity, emissivity of the surface, and the rate of evaporation from the surface. In some cases these factors

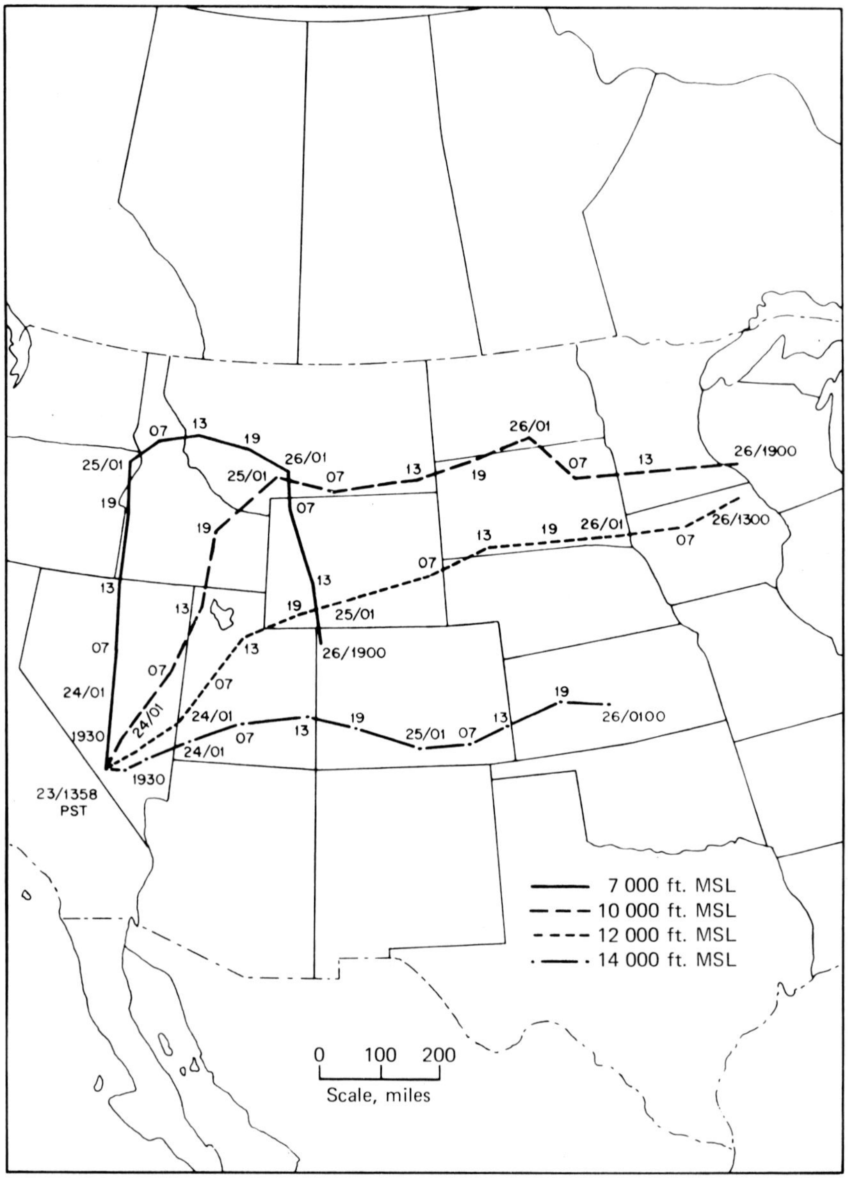

Fig. 2.18 Long-range meteorological trajectories. (From T. V. Crawford, *The Long-Range Diffusion of the Effluent Cloud from Phoebus-IV Reactor Test of February 23, 1967*, University of California, Livermore, Lawrence Radiation Laboratory, April 1968.)

may have a compensating effect, whereas in other cases one factor may predominate. Therefore, for an understanding of mesoscale motions, the importance of all these factors must be evaluated.

Albedo. Table 2.1 gives representative values of albedo for several different surface types. Also shown is the variation of albedo with the elevation angle of the sun for several surfaces. Note, for cloudy days, the clouds mask the direct rays of the sun. In these cases the sky can replace the sun as the effective source of radiation. Man's activity can drastically change the albedo of an area and hence can affect the heat balance. For example, clearing of forested areas for the construction of a city can alter the albedo by a factor of 2.

Heat Capacity. Heat capacity (called specific heat when expressed per unit mass) is a measure of the amount of energy required to increase the temperature of a surface. For example, Table 2.1 shows that the heat capacity for water is 4.2 J/g·°C), which means it requires one unit of heat (defined as a calorie) added to 1 g of pure water to raise the temperature to 1°C. Subjecting different surfaces with different heat capacities to the same flux of radiation would thus produce different surface temperatures. Variations of heat capacity from one surface to another can have a great influence on the ultimate production of horizontal temperature (density) gradients near the ground.

Infrared Emissivities. The last column in Table 2.1 lists the infrared emissivities (100.0 minus the infrared albedo) which relates the ability of a particular surface to radiate energy. An emissivity of 100 indicates that all of the sunlight absorbed at the surface is reemitted. An emissivity of 80 indicates that the surface will radiate infrared energy

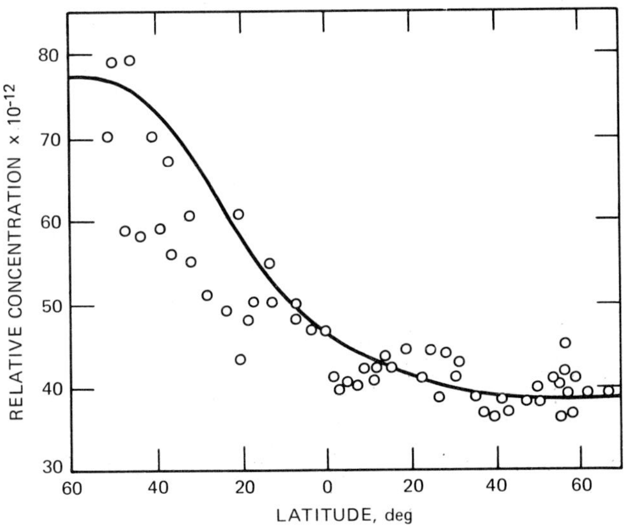

Fig. 2.19 Distribution of CCl_3F in and over the North and South Atlantic Ocean measured by Lovelock and Maggs; (○), theoretical calculation obtained by Machta. (From L. Machta, in *Proceedings of the United Nations Conference on the Peaceful Uses of Atomic Energy*, Geneva, Vol. 18, p. 519, New York, 1958.)

with an efficiency equal to 80% of that of a black body at the same temperature. Thus 20% less energy would be available for absorption by the overlying atmosphere.

Evaporation and Evapotranspiration. Evaporation is a cooling process. The energy required for a liquid located on a surface to change from a liquid to a gas, or to evaporate, is absorbed from the surface. This heat is called *latent heat* or stored heat. Analogously, when a gas is caused to convert back into a liquid under the process called condensation, this latent heat of evaporation is given off to the surroundings. The effects of latent heat and evaporation were discussed in section 2.3.2 with respect to the evaporation and condensation of water droplets in clouds.

Differential evaporation from surfaces can play an important role in the energy balance. The combined effects of the transpiration of water by plants and the evaporation of transpired water is called evapotranspiration. For most plants the surface is saturated so that the effects of evaporation are directly related to the temperature of the surface (in this case, water), the gradient of water vapor above the surface, and the amount of air flowing over the evaporating surface (i.e., wind speed). Decreasing any of these three terms decreases the rate of evaporation and transpiration. When a large portion of the surface is covered by plants or water surfaces, the amount of energy used in evaporation becomes extremely significant in the overall energy budget. Figure 2.20 shows maps of the distribution of the annual values of the radiation balance of the earth's surface and the annual loss of heat in evaporation. Note the annual loss of heat due to evaporation is about the same magnitude as the net radiation balance. The highest loss of heat due to evaporation occurs over the oceans between latitudes 20°N to 30°N and 10°S to 30°S which have a high income of solar radiation and relatively constant wind patterns. The loss of heat in evaporation decreases noticeably near the equator owing to the lower wind speeds and a reduction in the vertical gradient of vapor pressure. The influence of temperature on evaporation is amply demonstrated by the influence of ocean currents. Warm currents, such as the Gulf Stream off the east coast of the United States, cause a marked increase in evaporation rate. Evaporation rate over the Gulf Stream is about four times as large as the rate over the cold water of the California current on the west coast of the United States.

2-4.1.2 Man-Made Heat Sources. People are using an ever increasing amount of fuel. Most fuel is converted to energy, a significant amount of which is wasted. Most of the wasted energy is in the form of heat that is discharged into the atmosphere. For the United States the amount of wasted heat is estimated to be 1.7×10^6 MW. In areas of heavy industrialization, normally centered around cities, this man-made heat can represent a significant amount of the total energy in the local energy budget. For example, the heat input to the atmosphere due to man-made sources over New York City during the summer is 70 W/m^2, or about 10% of the natural heat supply, 600 W/m^2. During the winter the heat produced over the city is 60 W/m^2 which is about 200% of the natural heat supply, 30 W/m^2. Influence of this man-made heat is already felt in the form of increases in temperatures over cities and increases in rainfall downwind of the cities (Lowry, 1967). Other consequences of man-made heat are difficult to assess for they depend on the area over which the energy processes occur, the weather conditions, and the many feedback mechanisms acting in the atmosphere (Hanna and Swisher, 1971). Large power parks continuously releasing 100,000 MW of heat from an area between 5 to 100 km^2 could cause whirlwinds or possible tornadoes (Hanna and Gifford, 1976).

2-4.1.3 Effect of Changing Makeup of the Atmosphere on the Radiation Balance of the Ground Surface. In the section on macroscale radiation balance (Sec. 2-3.1), the effect of particulates and cloud cover were discussed as they affect the energy budget for the global circulation pattern. A similar but much smaller effect is important with respect to mesoscale motions. The most important interferences to the energy budget are produced by air pollution (either gas or particulates), clouds, and water vapor.

Air Pollution. The effects of air pollution on the radiation balance are to absorb, reflect, and scatter incoming solar radiation and thus to prevent this radiation from reaching the ground. The energy absorbed by the pollution cloud then can be radiated in the infrared wave length.

Investigations of solar radiation within a large city have indicated that the incoming solar radiation can be decreased by amounts of 20 to 30% of the radiation received on a day with little pollution (Randerson, 1970). In addition, the polluted air mass can increase the downward infrared flux of radiation by as much as 7%. The total effect of the pollution on the temperature near the surface is to reduce the daytime temperature owing to the reduction in the

TABLE 2.1 Albedo, Heat Capacity, and Infrared Emissivity of Several Surface Types† ‡

Surface	Albedo, %	Heat capacity, joule/(g·°C)	Emissivity, %
New snow	75–95	0.50	82–99.5
Old snow	40–70	0.50	89
Ice (−5°C)	30–40	0.50	96
Water	6–21	1.00	92–96
Tall grass	10–20		90
Dry sand	35–45	0.26	90
Wet sand	20–40	0.34	95
Pine forest	5–15	(0.5–0.65)	90
City, central business district	13	(0.55)	92
City, suburban	16	(0.52)	98

Percent Albedo for Varying Elevation Angles, Surfaces, and Sky Cover §

Surface	10	20	37	52	60
Bare ground					
Morning	0.14	0.10	0.11	0.13	0.13
Afternoon	0.41	0.26	0.18	0.14	0.12
Peas					
Morning	0.30	0.17	0.16	0.16	0.12
Afternoon	0.70	0.34	0.23	0.18	0.13

Cloudiness	10	22	37	52	60
Overcast					
Morning	0.07	0.07	0.09	0.11	0.12
Afternoon	0.39	0.25	0.15	0.12	0.10
Intermittent cloudiness					
Morning	0.19	0.15	0.13	0.14	0.12
Afternoon	0.42	0.27	0.16	0.14	0.12
Clear					
Morning	0.24	0.18	0.15	0.14	0.11
Afternoon	0.44	0.28	0.18	0.15	0.11

†Albedo represents average over different surfaces: bare ground, peas, green beans, red cabbage, and potatoes.

‡William D. Sellers, *Physical Climatology,* University of Chicago Press, Chicago, 1965.

§Lawrence C. Nkemdirim, A Note on the Albedo of Surfaces, *J. Appl. Meteorol.,* 11(5): 867-874, 1972.

amount of sunlight and to increase the nighttime temperature owing to reradiation in the infrared region at night.

Clouds and Humidity. As mentioned earlier, the latent heat of evaporation can play an important role in the surface energy budget. The amount of evaporation or evapotranspiration taking place is governed by the amount of moisture in the air above the evaporating surface. For example, if the air is saturated, no evaporation will take place no matter what the temperature of the evaporating surface. Moreover, cloud cover influences mesoscale motions, such as the land–sea breeze and the mountain-valley breeze. During cloudy periods the temperature over the land does not go through the extreme ranges that occur during clear periods. This reduces the daytime and nighttime land–sea temperature contrast. Cloudy skies have a similar effect on the mountain–valley regions. In addition, on cloudy days most insolation at the surface is diffuse (no shadows) rather than direct; therefore heating proceeds almost equally on the hillside and the valley floor.

2-4.1.4 Mesoscale Circulations Caused by Variations in Energy Balance. The three most common circulation patterns caused by local changes in the energy balance are the sea breeze, the mountain–valley breeze, and the urban–rural heat island. Other important mesoscale patterns exist, such as the snow breeze, but they are not as common. Figure 2.21a shows a diagram of the basic cellular circulation pattern established by two adjacent surfaces at different temperatures. The depth of the circulation, h, as well as the strength of the circulation, is a function of the temperature difference of the two surfaces. The arrows

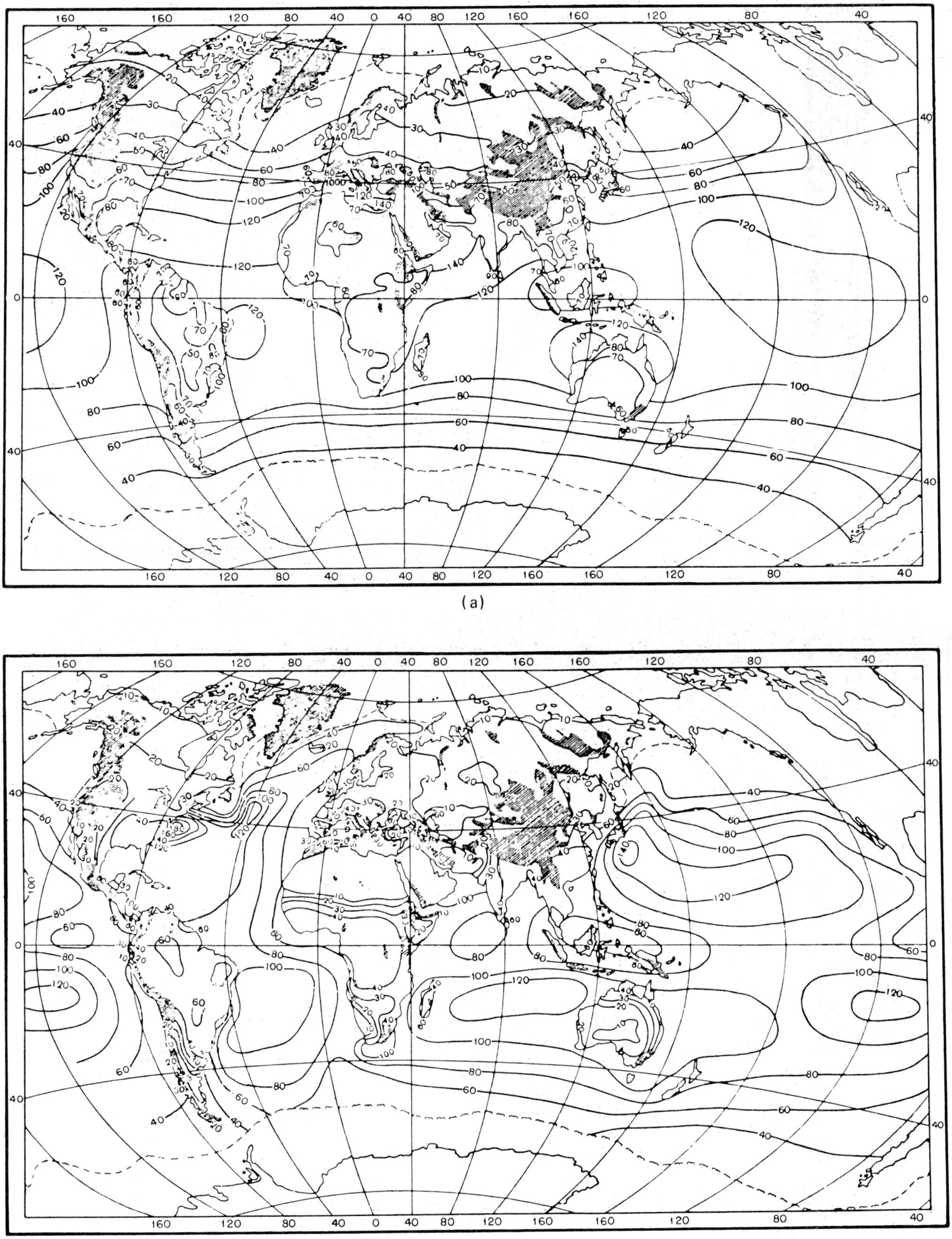

(a)

(b)

Fig. 2.20 Annual net radiation and the loss of heat in evaporation for the world. [Based on M. I. Budyko et al., *The Heat Balance of the Surface of the Earth, Soviet Geography, Review and Translation,* 3(5): 3-16 (1962).]

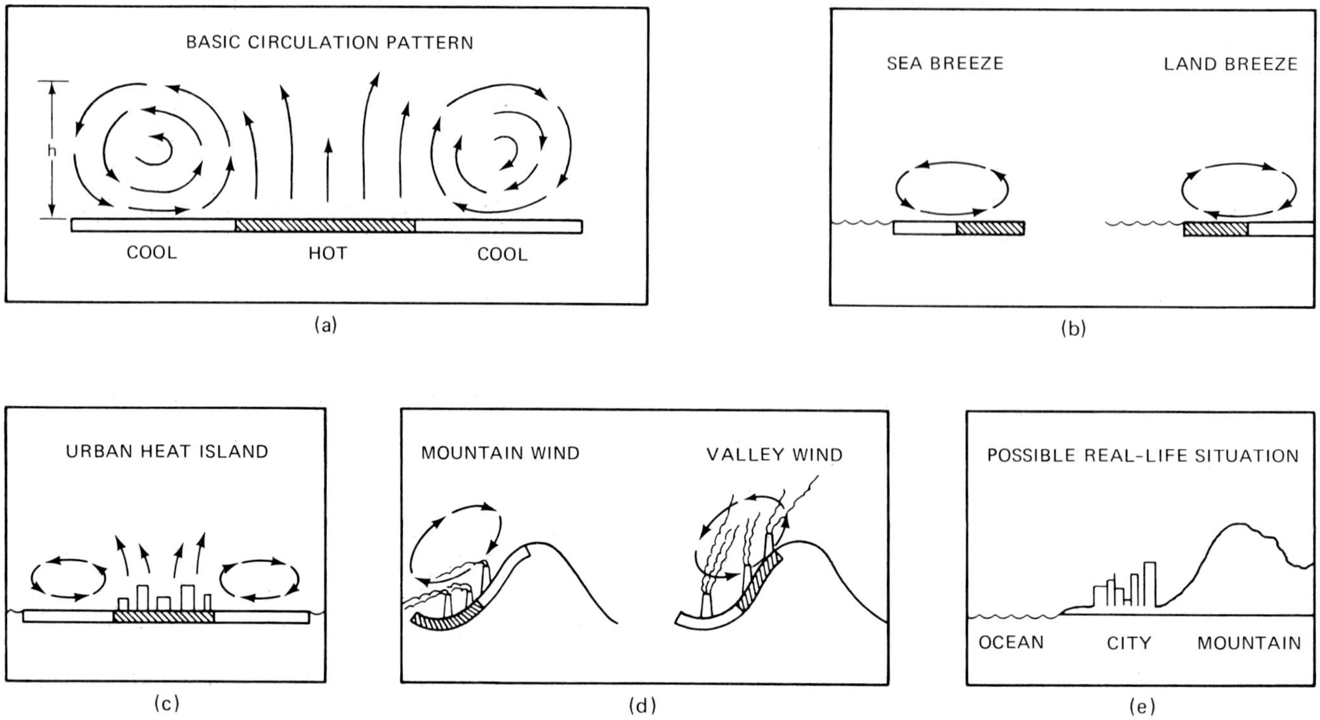

Fig. 2.21 (a) Schematic diagram of cellular circulation induced by two surfaces at different temperatures and (b) to (e) analogous circulations observed in nature.

point in the direction of the flow, and the thick arrows indicate highest wind speeds. Also shown in parts (b) to (d) of Fig. 2.21 are simple diagrams of specific circulation patterns associated with the three mesoscale systems. Notice all are essentially the same as the basic diagram.

Frequently, highly industrialized areas are located near population centers. More often than not, the highly populated areas are along seacoasts and near mountains, so that it is not unusual for power industry to be located where effects from all three of these mesoscale systems can be felt as in Fig. 2.21e.

Although the end result of the circulation pattern is similar for all three cases, the density gradients are established in different ways. The land–sea temperature difference related to the sea-breeze circulation is due in part to the differing heat capacities and albedos of water and land but is due primarily to the much larger mixing depth over water than over land. (This process was discussed in Sec. 2-3.2.)

The urban–rural heat island circulation is directly related to differences in specific heat capacities, albedo, and latent heat of evaporation. As mentioned several times previously, it is not the values of these particular elements but the contrast between the rural and urban areas that produces the necessary density gradients.

The mountain–valley breeze circulation could be generated over areas having identical surface conditions with respect to specific heat capacity, albedo, and emissivity. Exposure to incident radiation is the principal mechanism for developing the temperature gradient. In effect, this exposure alters the albedo by changing the incident angle that the sun makes to the various surfaces so that, when the sun is low in the sky, the side of the hill facing the sun would receive more radiation than the valley floor. During the late evening, the air nearest the ground on the shady side of the hill is cooled first (because of its proximity to the surface). At the top of the shady side of the hill, this air would be cooler than the air at the same elevation over the valley floor (which is further from the surface and thus does not cool very quickly). Gravitational flow begins, and a circulation pattern is established. Conversely, during the afternoon as the surface is heated, the air at the top of the sunny side of the hill would be warmed faster than the air at the same elevation over the shady valley. This rising warm air eventually develops an up-valley flow regime, although it is not as strong as the nighttime circulation. Naturally a mountain–valley location with spatial variations in emissivity, heat capacity, and albedo (caused by inequal distribution of snow cover, for example) could increase the density gradients and produce stronger circulations.

We must remember that, in all the above-mentioned mesoscale circulations, macroscale weather conditions can be significant. However, when macroscale wind speeds are low, mesoscale features tend to dominate and hence are more readily observed.

2-4.2 Moist Convection and Severe Storms

The mesoscale systems with the most dramatic and immediate effect on people are the severe storms consisting of an individual convective cloud or a group of convective clouds. Severe storms can be defined as any storm producing strong winds, hail, or precipitation capable of damaging property or people either on the ground or in the air. The impetus for extensive research on severe storms is to improve predictions on their occurrence and probable extent of damage. The following discussion of severe storms is directed toward understanding their effect on power production. For additional information, the reader is referred to a meteorological text on the subject.

2-4.2.1 Convective Clouds. On clear days with relatively light winds, a superadiabatic lapse rate is usually found in the lowest few hundred meters above the surface (see Fig. 2.22a). Intense turbulent motions exist within this

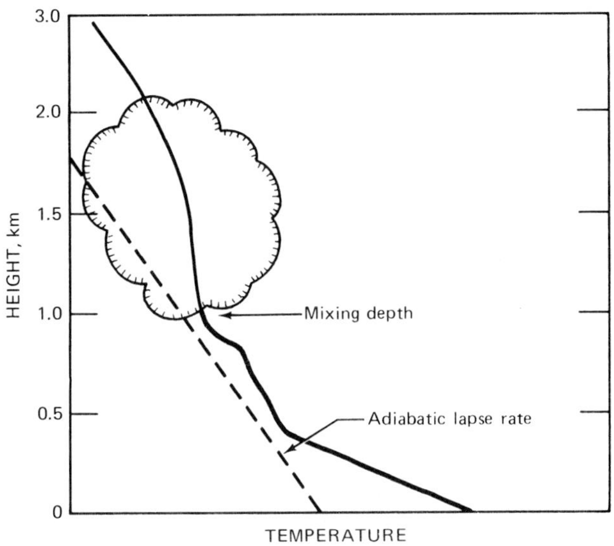

(a)

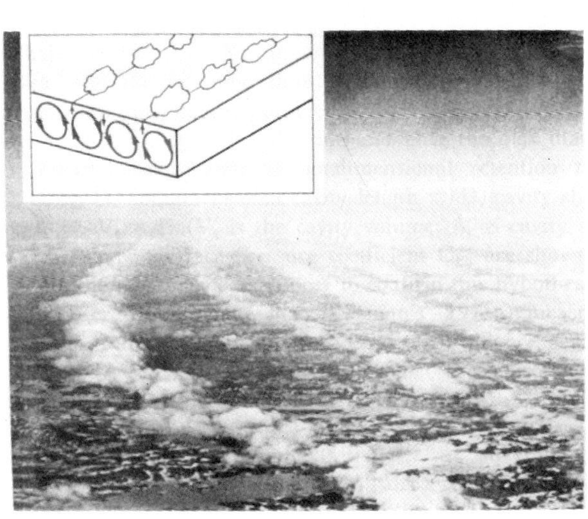

(b)

Fig 2.22 (a) Temperature lapse rate during a warm afternoon corresponding to cumulus cloud development, (b) cloud streets observed on a warm afternoon. [H. Tennekes, The Atmospheric Boundary Layer, *Physics Today*, 27(1): 152-163 (1974).]

unstable layer, and these motions guarantee rapid lateral and vertical mixing between an emitted pollutant and the clean air. Above a few hundred meters, the lapse rate is usually less unstable, and the chaotic low-level turbulent flow shows some organization into discrete columns of rising and falling air convection. The ascending columns may be topped by convective clouds if the moisture supply is sufficient and the rising air cools until the condensation temperature is reached. The height of the base of the clouds indicates the extent of a mixed layer. In the absence of convective clouds, the mixed layer can be defined from temperature profiles. The procedure for determining mixing depths from temperature profiles will be discussed in a later section.

Frequently clouds are seen to line up along the mean (macroscale) winds, forming rows of clouds called streets. Generally, the height of these clouds is less than 2 km. Figure 2.22b is a photograph of such clouds, and the insert shows the associated wind pattern. The cloud streets are caused by buoyancy and wind shear. Cloud streets and their accompanying flow patterns have an appreciable influence on a pollution plume. Of particular interest is the case in which a pollution plume is forced to the ground by descending air. LeMone (1973) has pointed out that cloud streets do not generally occur unless the mean wind speed is approximately 10 m/s. These strong winds would tend to dilute pollutants so that, over an extended period, the effect of a cloud-street circulation would not be detrimental to the overall air quality of a point some distance downwind from the source. This hypothetical receptor downwind from this source could, however, receive high concentrations over short periods of time. Stationary cloud streets with respect to a source could cause a significant reduction in the overall air quality at a downwind receptor for extended periods. Current understanding of these features is rather limited, and their persistence over a site is not well tabulated.

2-4.2.2 Thunderstorms. The small cumulus clouds previously discussed can develop into a cumulonimbus cloud or thunderstorm. A thunderstorm is defined as a cumulonimbus cloud with lightning and thunder. A severe thunderstorm is one in which any or all the following occur: surface winds exceeding 13 m/s, rainfall exceeding 2.5 cm/h, and hail reaching the ground. Figure 2.23 shows the various stages in the evolution of a thunderstorm as well as the predominant flow patterns within and in the vicinity of a thunderstorm. Note that the mature thunderstorm is considered a mesoscale feature with horizontal distances of 5 to 7 km and vertical distances up to 10 km.

The physics behind the formation of a thunderstorm is reasonably well understood. In fact, a skilled meteorologist can predict the time and the general area in which certain types of thunderstorms will develop. This prediction is based primarily on atmospheric stability, vertical motion, and availability of moisture. The stability and moisture in advance of cold fronts are normally adequate to produce thunderstorms. The severity of the thunderstorms is well correlated with the magnitude of the temperature lapse

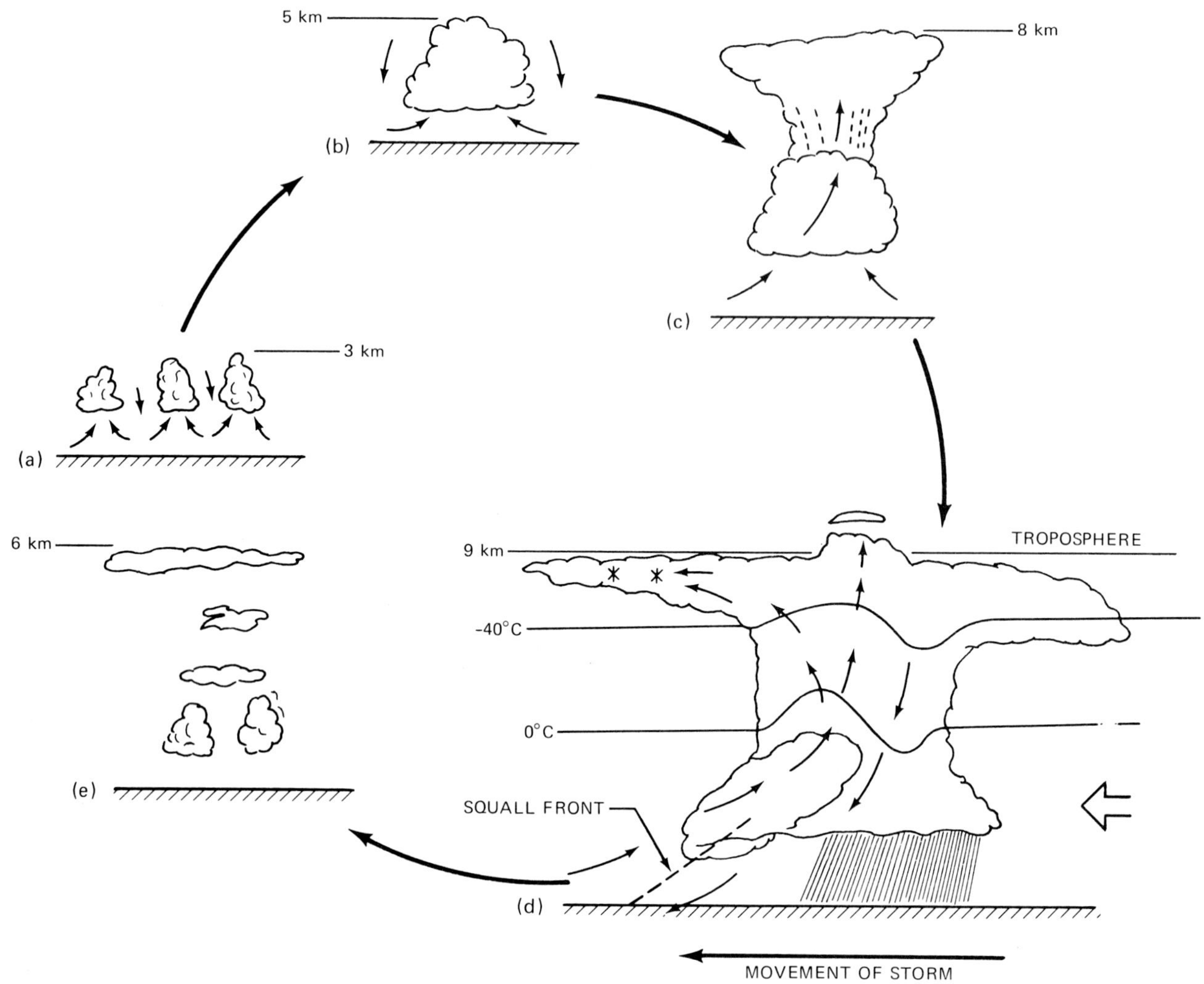

Fig. 2.23 Various stages in the life of a thunderstorm showing approximate height of the cloud and general air flow.

rate. The more unstable the lapse rate, the greater the potential for a severe thunderstorm.

Frequently the temperature distribution within an entire air mass is such that the addition of small amounts of heat at the surface is sufficient to cause a thunderstorm to develop. In these situations, the probability for the development of thunderstorms is highest at the time of maximum surface temperature. This type of thunderstorm is called an air-mass thunderstorm and occurs more or less randomly over an area in the afternoon. Frequently thunderstorms generated in this way dissipate in approximately 40 min. Parts (d) and (e) of Fig. 2.23 show that the rain showers and downward rushing cold air developed by the thunderstorm are responsible for its destruction. At other times, when the thunderstorm circulation system is large, a persistent structure may evolve. The cloud mass drifts with the macroscale flow. This motion, coupled with the rising warm air and descending cold air, can produce a density discontinuity, called a gust front, between the advancing rain-chilled air moving outward from the storm and the warm air ahead (Fig. 2.23d). Peak wind gusts of about twice the speed of movement of the cloud are common (Goff, 1976). The warm air ahead of the front rises above the cold air and is swept up into the cloud mass to maintain the whole system. The almost vertical currents in the center of the thunderstorm frequently have a diameter of 100 m and a vertical speed of up to 30 m/s. Lightning discharges can occur between the negatively charged base of the thunderstorm cloud and the positively charged top. In addition, the negative charge at the cloud base can induce a positive charge at the ground (particularly on tall structures such as towers and trees) such that a lightning discharge occurs between cloud and ground. Generally, only one in three lightning discharges strikes the ground. The life of a single thunderstorm is only 30 to 60 min, but, amalgamated into huge organized systems, they persist for hours. The potential damage from severe thunderstorms is important in the design and operation of power-production facilities. Transmission-line towers and tall smoke stacks are particularly vulnerable to strong winds and lightning strikes. The frequency of occurrence of thunderstorms throughout the United States is shown in Fig. 2.24a.

2-4.2.3 Tornadoes. A tornado is a violently rotating column observable as a funnel cloud suspended from a cumulonimbus cloud or thunderstorm. The diameter of

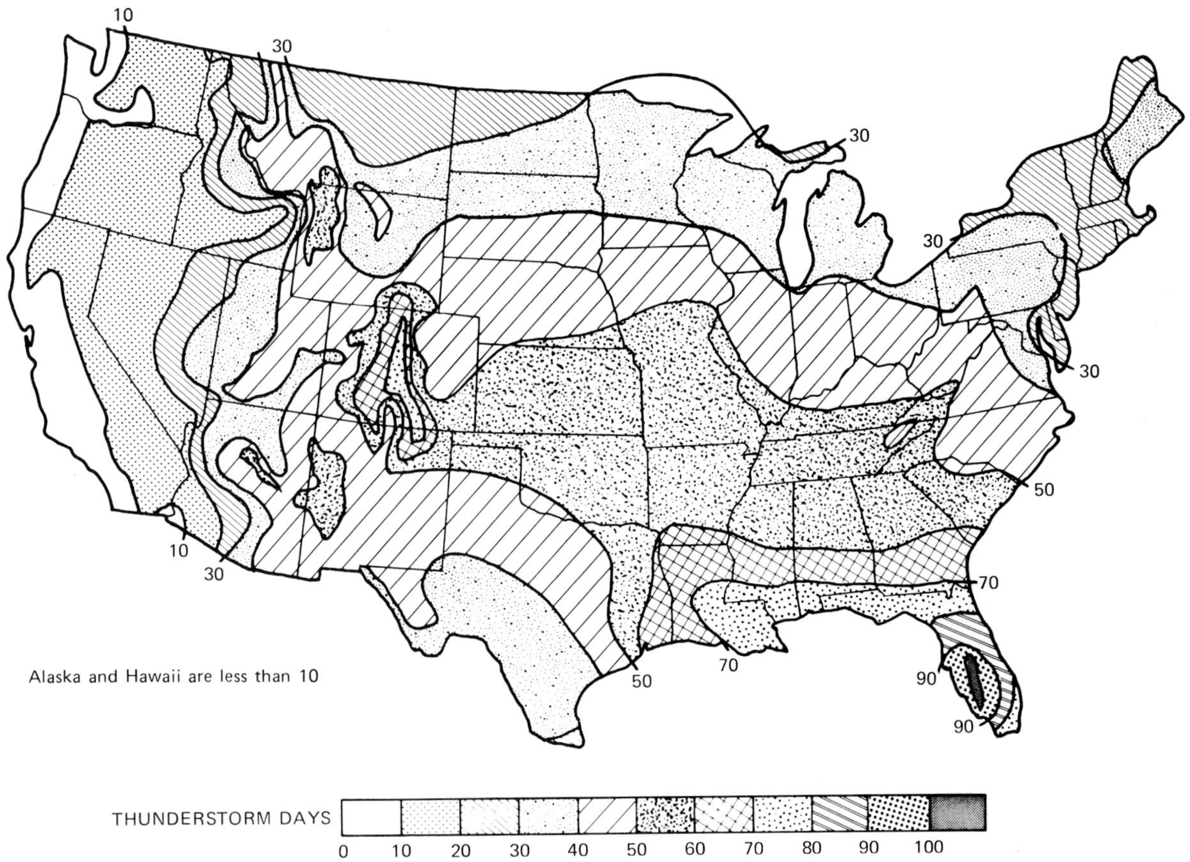

Fig. 2.24 Thunderstorms in the United States. (From National Oceanic and Atmospheric Administration paper 75009, 1976.)

tornadoes generally ranges from several hundred meters to several kilometers, and the vertical dimensions generally are about 1 km. Cumulonimbus clouds bearing tornadoes occur during periods having extreme thermal instability as well as large amounts of moisture in the lower levels (>10 g of water vapor per kilogram of dry air).

A statistical analysis of tornadoes reported over a 30-yr period was the basis for calculating the tornado strike probability (Fig. 2.25) and design criteria (Fig. 2.26) for the continental United States (Markee, Beckerley, and Sanders, 1974).

It is estimated that if a tornado were to cause structural damage to a containment building the debris or pollutants would probably be dispersed into the parent cloud and immediately mixed throughout owing to the strong vertical motions. Surrounding the tornado, however, are peripheral winds with speeds of up to 50 m/s. These strong winds may cause structural damage that might cause a pollutant to be released without a direct hit by the tornado vortex itself. In instances of this kind, the wind pattern surrounding the tornado could cause the pollutant to be dispersed without being drawn into the parent thunderstorms.

An awareness of mesoscale features associated with convective storms, as well as other mesoscale systems discussed previously, is most useful in the design of power-production facilities. On occasion, however, circulations induced by these mesoscale systems might be helpful to make judgments concerning effective transport of pollutants by these systems.

2-4.3 Temperature

Temperature distribution is one of the principal variables to consider in order to assess the significance of many mesoscale systems. The horizontal temperature gradient largely determines the magnitude of the heat island, land–sea breeze, and mountain–valley circulations. The vertical temperature gradient (i.e., atmospheric stability) is an important indicator used to predict the magnitude of convective clouds. Similarly the vertical temperature profile of the lower atmosphere is important because the turbulent structure governing the rate of growth of pollution palls is intimately related to stability.

When assessing the environmental effects of power-production facilities at distances in excess of 80 km (i.e., travel times on the order of 10 h), the temporal variation of mesoscale systems must be considered (such as the day–night wind shift associated with land–sea breezes). Analogously, the day–night or diurnal variation in the vertical temperature profile has a very significant effect on pollutant concentrations. Furthermore, at mesoscale distances the influence of the mixing depth must also be considered.

Diurnal Temperature Pattern. Figure 2.27 portrays the daily cycle of the vertical temperature pattern constructed

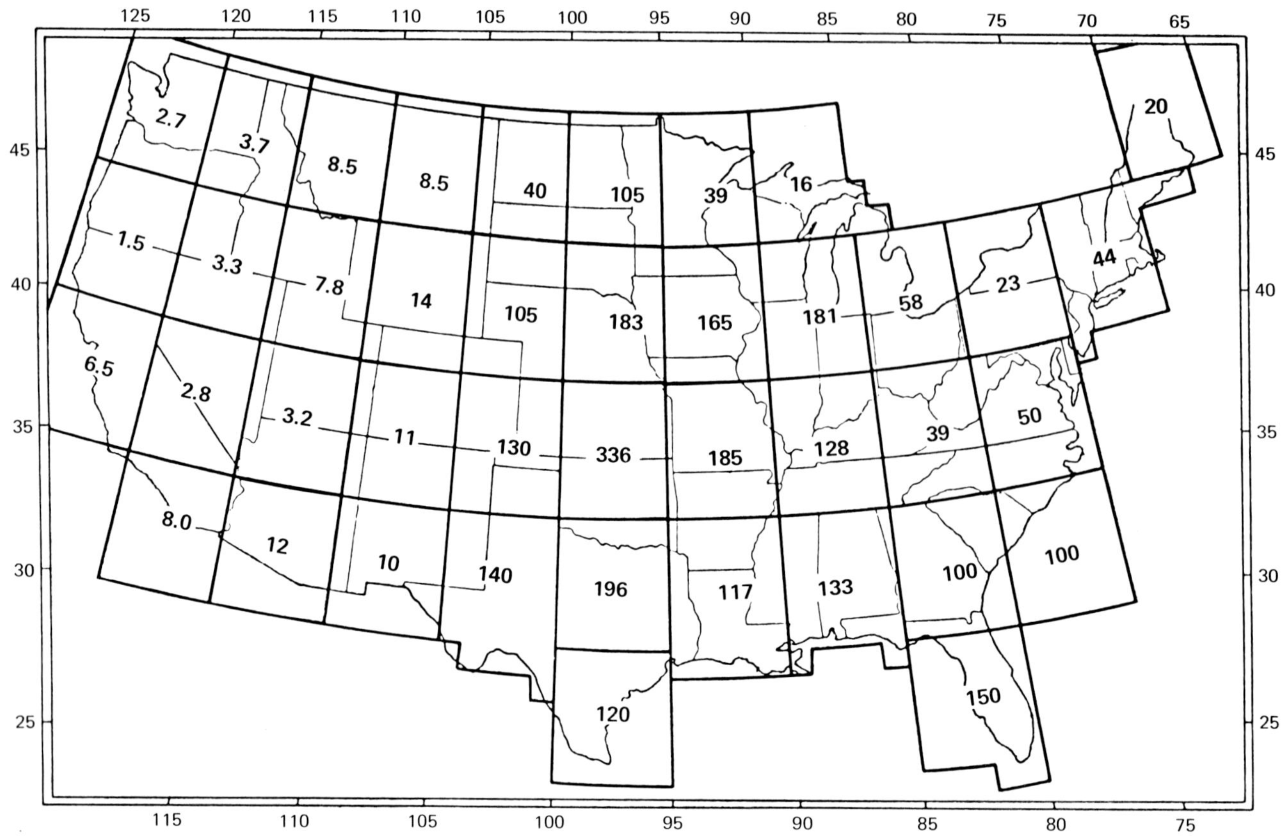

Fig. 2.25 Tornado strike probability within 5-degree squares in the contiguous United States (units are 10^{-5} probability per year). (From E. H. Markee, Jr., J. G. Beckerley, and K. E. Sanders, *Technical Basis for Interim Regional Tornado Criteria*, USAEC Report WASH-1300, 1974.)

from data taken during the late summer and early fall in the eastern United States. At 0600 local time, a surface inversion caused by heat loss due to radiation and conduction by the ground is the predominant feature of the temperature profile. By 0800 local time, the heating effect of the sun destroys the lowest part of the inversion, and, by 1000 local time, no trace of the nocturnal inversion exists. By this time an unstable layer exists nearest the ground (superadiabatic lapse rate), and the entire layer of air near the ground is heated. The return of the nocturnal inversion is evident at 1800, and the cooling of the entire layer progresses through the night. The presence of low-level cloud cover and pollution would reduce the intensity of the inversion since clouds or pollution absorb the energy leaving the ground and radiate it back to the ground. A higher temperature is maintained during these circumstances than that realized during dry, cloud-free conditions. In addition, strong winds caused by a pressure gradient give rise to mechanical turbulence. The resulting mixing caused by the mechanical turbulence keeps the layer well mixed and even eliminates the inversion completely.

Mixing Depths. The height to which the atmosphere is uniformly mixed is called the mixing depth. Section 2-4.2.1 introduced the concept of a mixing depth and indicated that a useful method for determining the height of the mixed depth is to observe the height of the base of clouds. A post facto estimate of the mixing depth can be determined from the observed morning temperature profile and the observed afternoon maximum surface temperature. The height of the mixing depth is defined by the intersection of two temperature profiles: the observed temperature profile and a profile constructed from the observed maximum surface temperature and the adiabatic lapse rate. This simple approach, shown in Fig. 2.28, can be used to determine the mixing depth at any time, provided the initial temperature profile and the surface temperature are known. Errors in the mixing-depth estimate would be greatest during periods when the temperature structure changes by methods other than turbulent mixing, e.g., by advection or radiation.

The method has been tested extensively, using visual observations of clouds, analyses of data collected with acoustic sounders, and comparisons with accelerometer and temperature measurements made through the well-mixed layer. In many cases all methods agree, but several studies have shown that this method overestimates the overall height by about 50% (Pendergast, 1974; Schubert, 1976).

The height of the mixing depth is useful as a simple assessment of the pollution potential of an area. For example, if pollutants are well-mixed within this layer, the concentration at the ground level would be the same as that at every level within the layer. Therefore, for a constant source of pollutant, the concentration would be inversely related to mixing depth. The lowest concentration would

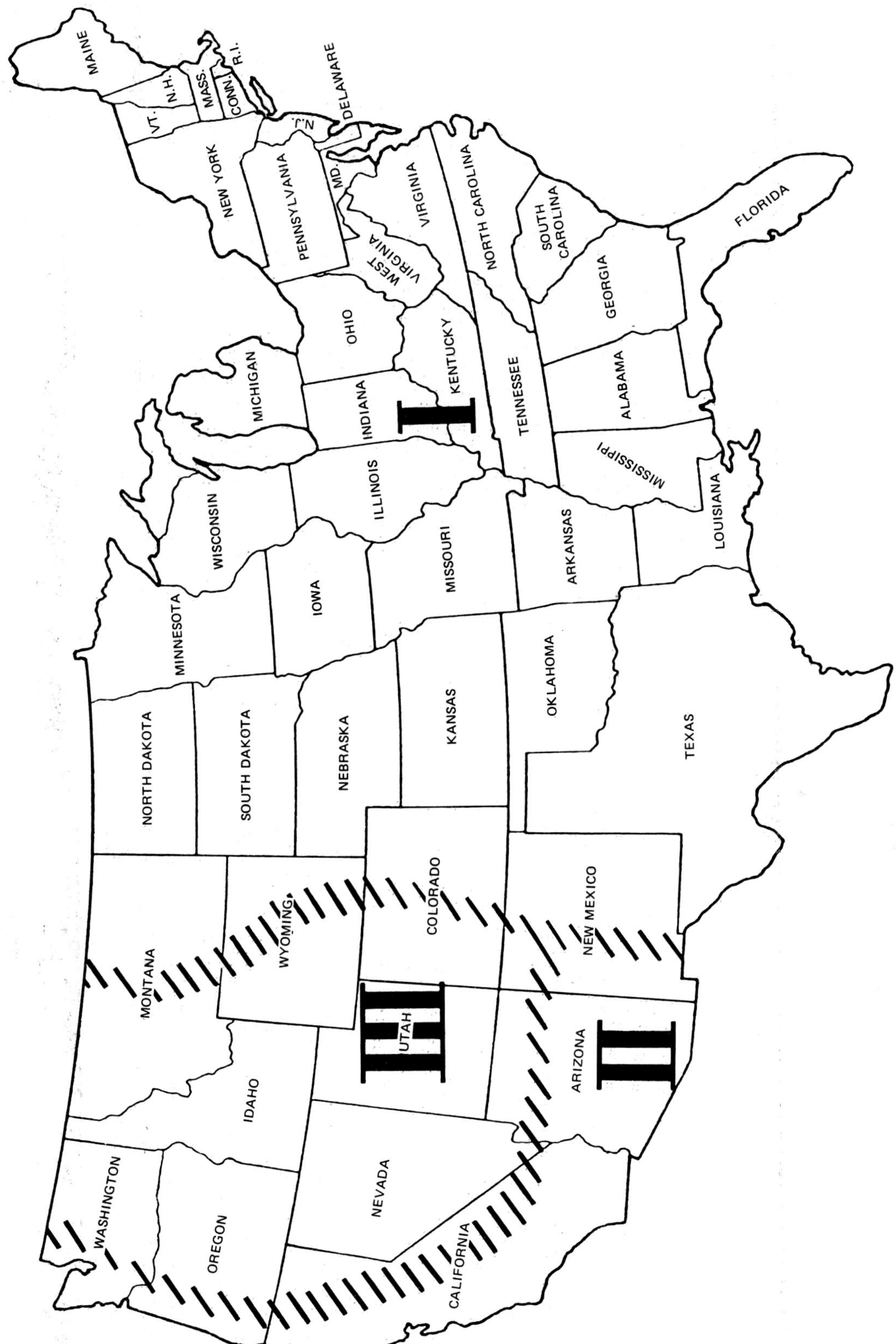

Fig. 2.26 Wind speeds and pressure drop for the design basis tornado for the three tornado intensity regions of the continental United States. (From E. H. Markee, Jr., J. G. Beckerley, and K. E. Sanders, *Technical Basis for Interim Regional Tornado Criteria*, USAEC Report WASH-1300, 1974.)

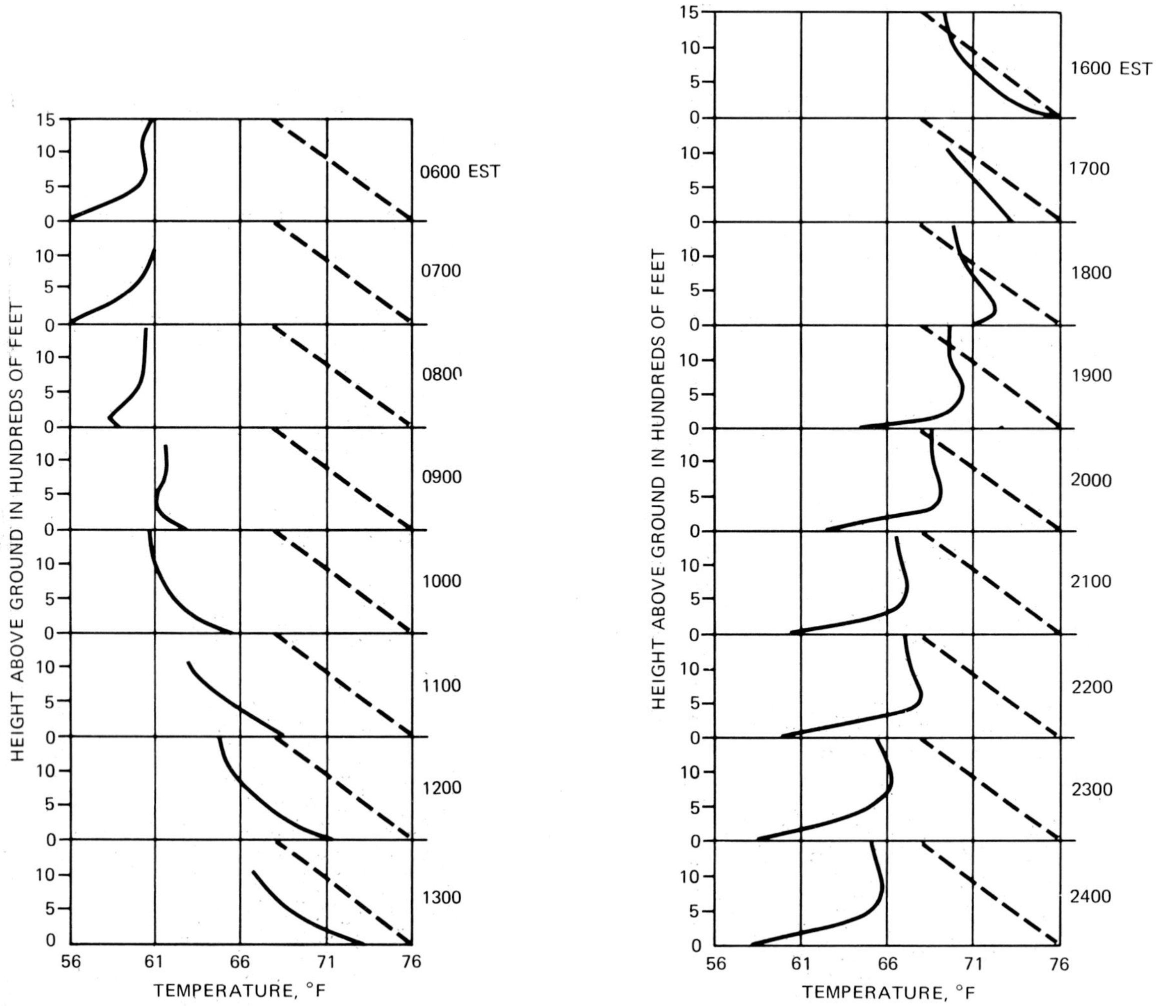

Fig. 2.27 The average diurnal variation of the vertical temperature structure at the Oak Ridge National Laboratory during the period September to October 1950. The data were obtained from captive-balloon temperature soundings; — — —, represents the adiabatic lapse rate. (From D. H. Slade, *Meteorology and Atomic Energy—1968*, USAEC Report TID-24190, 1968.)

occur at the time of maximum mixing depth. Seasonal changes in maximum mixing depths, therefore, could be related to seasonal changes in expected pollutant concentration.

Figure 2.29 shows maps of maximum mixing depth for the United States during January and July. The variation of this maximum mixing depth gives information helpful in deciding times when air quality can degrade. We must be aware that to assess the air quality of an area many factors other than mixing depth must be considered.

The day-to-day variation in mixing depth may be important during power-production operation—particularly if downwind air concentrations approach hazardous levels. The height of temperature inversions is well correlated with mixing depths. Furthermore, changes in the mixing depth can be related to the formation, duration, and destruction of temperature inversions. Although there are a great number of different temperature inversions observed over various locations, most can be traced to one of four types of inversions: radiation, subsidence, turbulence, and frontal.

The mechanism responsible for the radiation inversion is radiational heat loss from the surface levels and cooling by conduction. The height of the radiation inversion is restricted to the lowest several hundred meters and persists during the nighttime until being destroyed by surface heating.

The subsidence inversion is caused by large-scale downward motion (normally associated with an anticyclone). The best example is the tropical trade-wind inversion. This type of inversion is always elevated with a normal base height approximately 1 km above the surface. The sub-

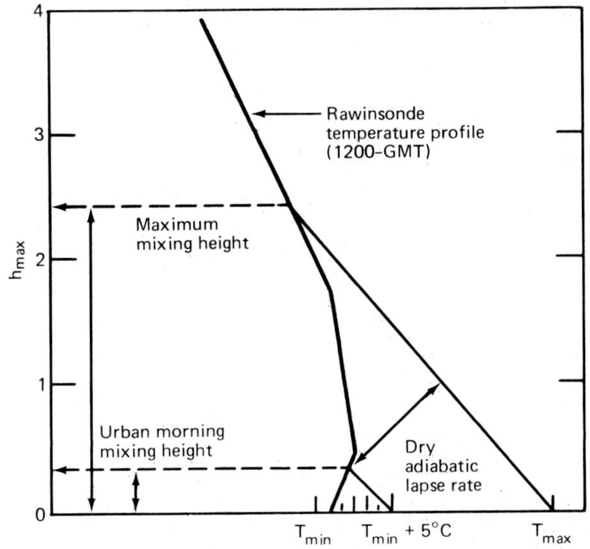

Fig. 2.28 An example of a technique used to obtain a post facto estimate of the mixing depth by using the observed morning temperature profile and the observed afternoon maximum surface temperature.

sidence inversion can prevail for weeks or months depending on its location.

A turbulence inversion is one in which the original temperature profile has been altered by vigorous vertical mixing that develops an adiabatic lapse rate throughout the layer in which the mixing is taking place. The temperature inversion occurs at the top of this mixed layer and is also an elevated inversion. The vertical extent of the mixing determines the base height of the inversion. Normally the turbulence inversion persists during the afternoon hours and is altered by the formation of a radiation inversion at night.

A frontal inversion causes a sharp reduction of temperature when cold air from the front moves into the lower layers. The frontal inversion is formed at the interface between the cold air in the lower levels and the warm air in the upper levels. The frontal inversion is an elevated inversion and can be located at almost any height from several meters above the ground up to the middle troposphere. It can persist over a site for several days, depending on the speed of movement of the air mass.

2-4.4 Wind Structure

As discussed in Sec. 2-2.4, the wind speed and direction are determined primarily by the interaction of three forces: horizontal pressure gradient force, Coriolis force, and friction. In the upper layers of the atmosphere, the pressure gradient and Coriolis forces are the dominant forces driving the airflow. Near the surface, however, drag exerted on the atmosphere by the surface must be considered. This force is analogous to that arising when a viscous fluid flows over a fixed plane surface, such as syrup over a tabletop. The fluid has a speed of zero at the surface, but the speed increases at increasing distances from the surface. If the unaffected fluid some distance from the plane is moving in response to a given balance of forces, then the flow near the surface can be explained by the introduction of an additional force due to the property of the fluid called *molecular viscosity*. (This viscosity would be characteristic of the syrup and the tabletop.) In the atmosphere the role played by a molecular viscosity near the surface of the earth is small in comparison with an apparent viscosity due to turbulent eddies that transfer momentum or any other property from regions rich in that property to regions deficient in it. The nature of these fluctuations and the methods by which they are measured are discussed in Chaps. 5 and 6.

2-4.4.1 Wind Profiles. The friction caused by air passing over a rough surface tends to reduce the speed of the wind near the ground through a mechanism called eddy viscosity. At ever increasing distances from the ground the effect of this surface drag is reduced. The level at which the effects of friction are no longer important is called the top of the atmospheric boundary layer (ABL); at this level the wind is nearly geostrophic. The frictional force also changes the vector balance of forces so that the air within the ABL moves at some angle to the geostrophic wind. In general, the airflow at the earth's surface is directed to the left (in the Northern Hemisphere) of the geostrophic flow at an angle that varies from about 10° during the day (unstable) over a smooth surface to about 50° at night (stable) over rough terrain (Arya, 1975). Above the surface, the observed wind usually turns slowly in a clockwise manner (with increasing distance from the surface) so that the angle between the wind and the geostrophic wind gradually decreases with height until, at the top of the ABL, the observed wind is parallel and equal in magnitude to the geostrophic wind. The theoretical development of this observed feature will be discussed in Chap. 5, under the term Ekman spiral. Above the top of the ABL, the characteristics of the wind, such as average speed, direction, and turbulence, are divorced from surface influences and reflect the forces acting in the free air.

Since both wind speed and direction vary with height near the surface, a vector diagram is frequently used to illustrate this combined change. Figure 2.30 shows average wind vectors constructed from data obtained during a number of summer nights and days at a location in central Nebraska. During the day the vector difference between the lowest level wind (16 m) and the upper-level wind (1000 m) is smaller than it is during the night.

2-4.4.2 Stability Effects on Wind Profiles. Since eddy viscosity is not a permanent characteristic of the fluid but depends on the amount of turbulent mixing present, it can be associated not only with mechanical features, such as roughness of the surface, but also with other factors. A major factor to be considered is the presence of buoyant forces associated with the occurrence of heating or cooling at the surface. During the day, when the ground is heated by the sun, turbulent mixing is enhanced, and the influence of the surface may be detected to a comparatively great height. At night, however, the stable stratification of the air

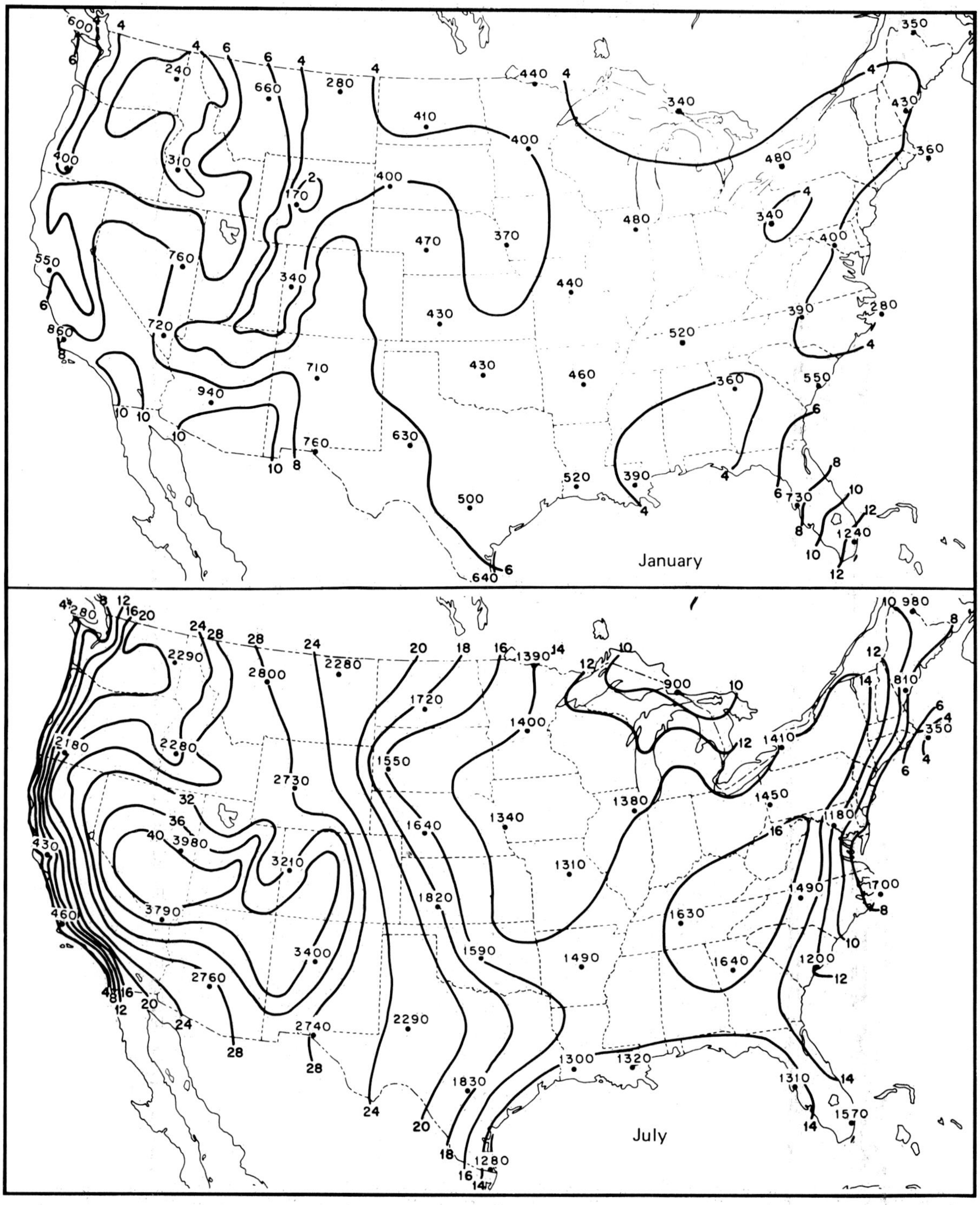

Fig. 2.29 The mean maximum depth in meters for January and July. These data were computed from atmospheric temperature soundings obtained at 45 points in the United States, [From G. C. Holzworth, Estimates of Mean Mixing Depths in the Contiguous United States, *Mon. Weather Rev.*, 92(5): 235-247 (1974).]

limits mixing and produces a situation in which the surface effects are limited to the very lowest layers. Above the ABL the average wind speed may be expected to be higher at night than during the day. Below this level, in the region affected by the nocturnal temperature inversion, highest wind speeds occur during the day when the momentum of the faster moving air aloft is transferred downward toward the surface. This effect is shown in Fig. 2.30 for the daytime case. During the night the wind-speed maximum occurs near the top of the ABL. This phenomenon, known as the nocturnal jet, has been the object of much study in recent years. It is thought to be due largely to the decoupling of the upper layers of the ABL from the surface and the setting up of inertial oscillations. A more detailed discussion of the nocturnal jet with respect to turbulence in the planetary boundary layer will be discussed in Chap. 5.

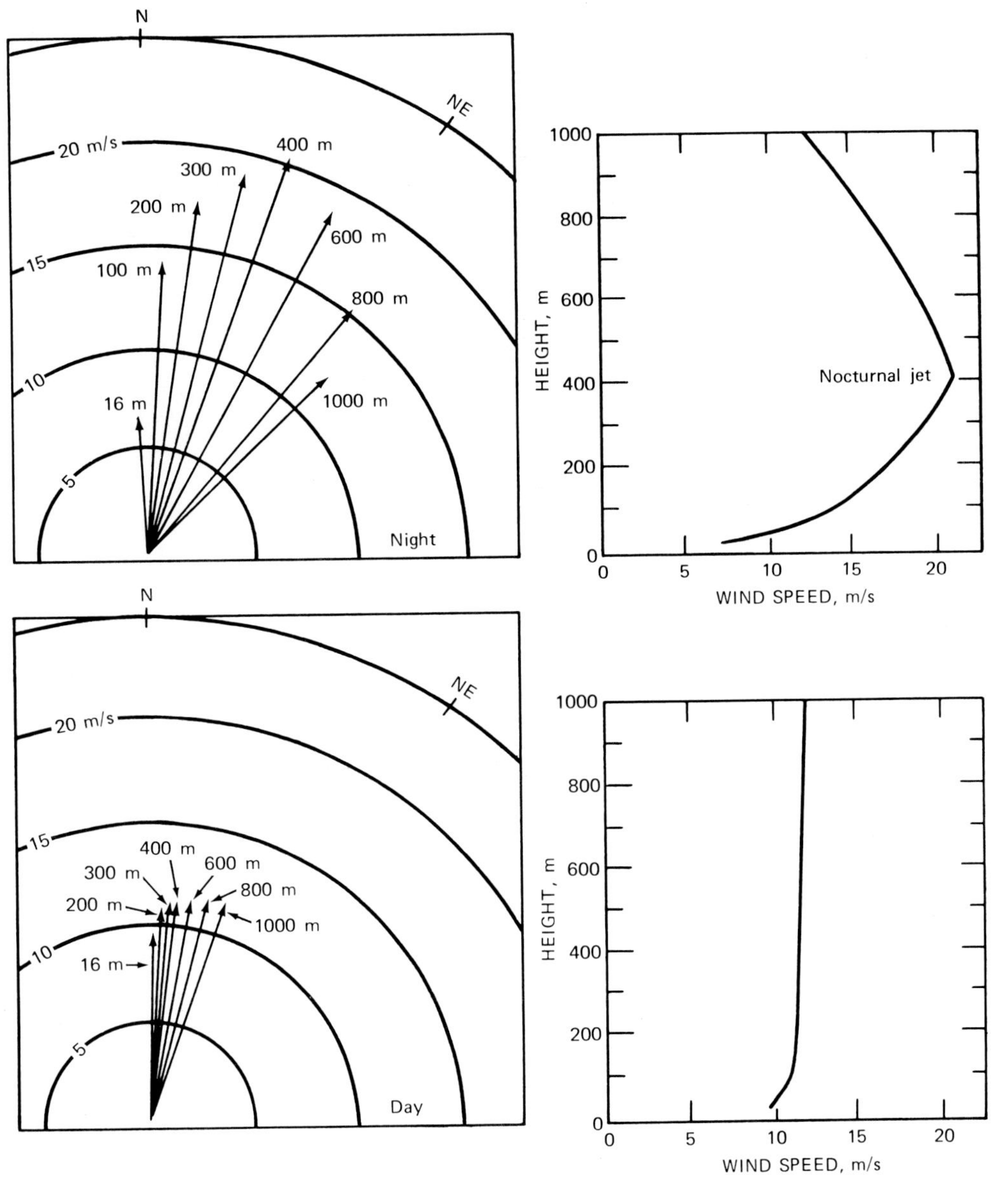

Fig. 2.30 Average vectors constructed from night and day wind observations obtained during the Great Plains Turbulence Field Program in central Nebraska. Also shown are profiles of the scalar wind speed as a function of height. (From H. Lettan and B. Davidson, *Exploring the Atmosphere's First Mile*, Vols. I and II, Pergamon Press, New York, 1957.)

2-4.4.3 Effect of Topography on Wind Patterns. A large class of stationary atmospheric circulation patterns is caused by the passage of air over topographic or man-made obstacles. Vertical and horizontal wind patterns display rotary or oscillatory motion as air is forced to flow over or around obstacles. These motions may be associated with clouds or even precipitation if the air is sufficiently moist. Mountain ranges produce widespread uplifting of air moving over the barrier. In the state of Washington, the Olympic and Cascade mountains furnish a dramatic example of large-scale uplifting because they are situated normal to the predominantly westerly flow of moist air. Figure 2.31 shows the average annual precipitation over the state of Washington in relation to the major mountain ranges.

The great variety of topographic obstacles and the great diversity of weather situations create an unlimited range of local circulation patterns. Under appropriate meteorological conditions, the disruption of an airstream by even a small hill can extend many kilometers vertically or horizontally downstream from the point of generation.

Parts (a) to (c) of Fig. 2.32 show a cross section of topographic areas perpendicular to the general wind flow. Over hills and valleys several patterns are possible depending upon stability. Heating of the earth's surface tends to induce local circulation patterns that have large vertical components; radiational cooling, however, results in air motions that are primarily horizontal or parallel to the slope of the land. The pattern showing separation of the flow into a vortex in the lee of an obstacle is more likely to occur when the ABL is unstable.

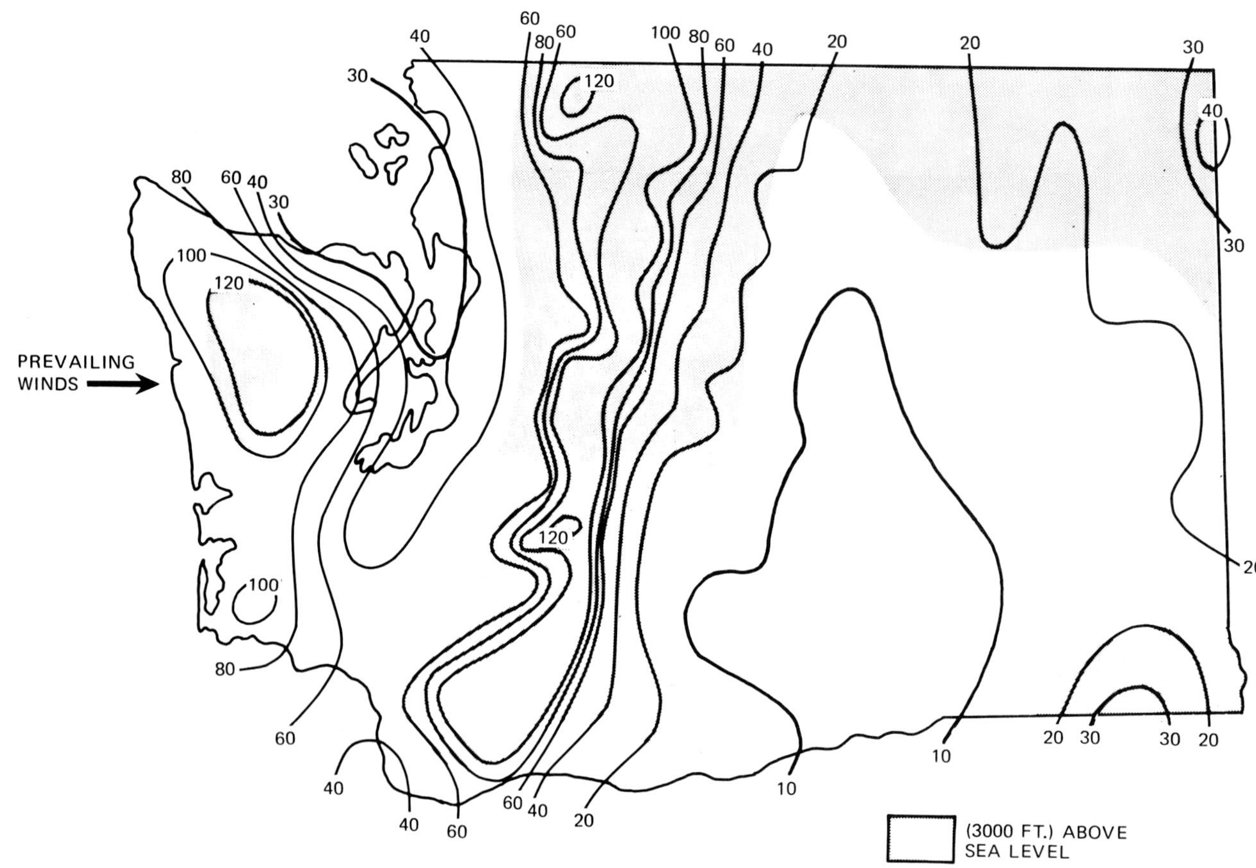

Fig. 2.31 Annual rainfall (in inches) for the state of Washington. Note large amounts of rainfall on mountain tops caused by the induced upward motions. Small rainfall amounts to lee of mountain produced by subsiding drier air. (From *Meteorology and Atomic Energy—1968,* USAEC Report TID-24190, 1968.)

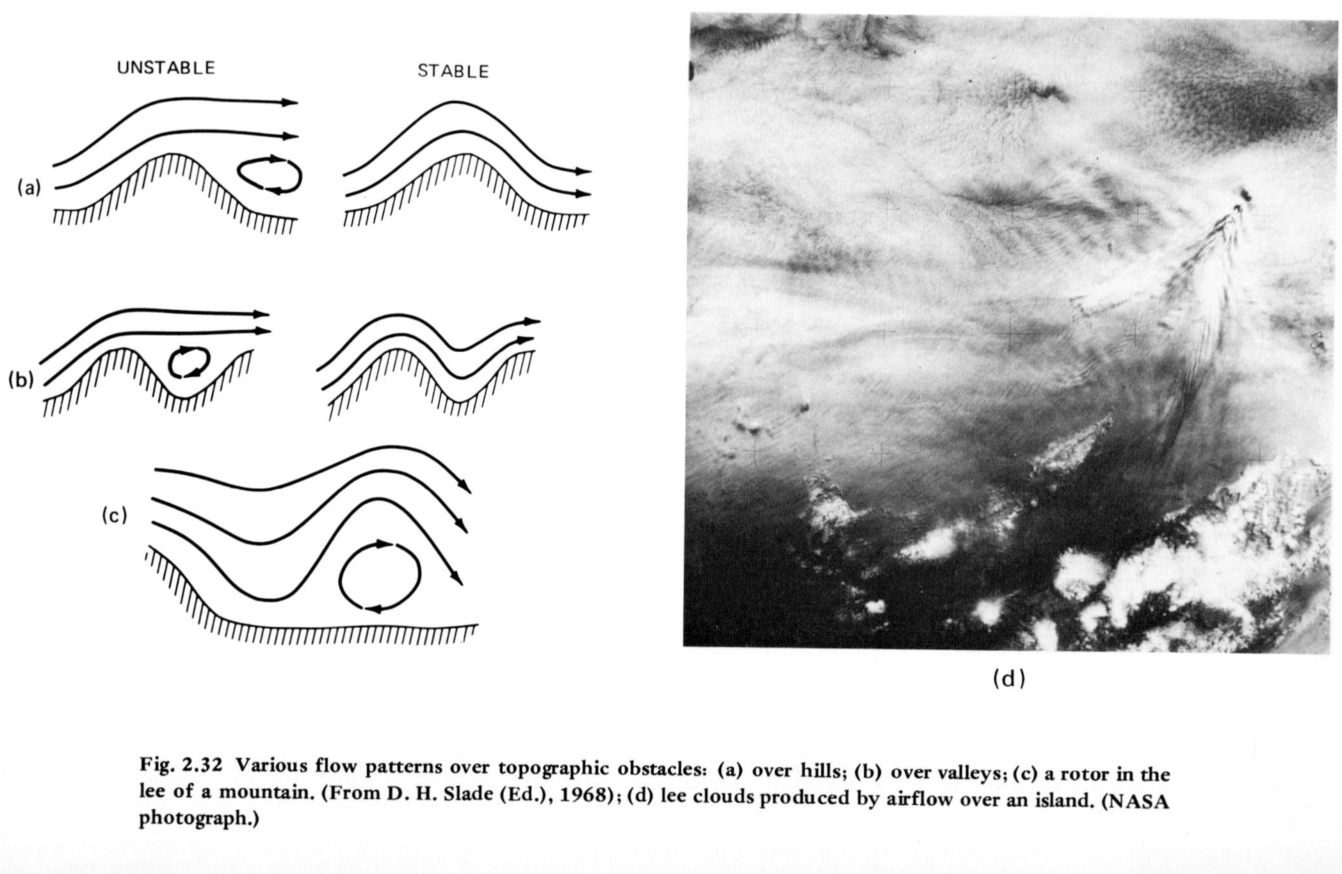

Fig. 2.32 Various flow patterns over topographic obstacles: (a) over hills; (b) over valleys; (c) a rotor in the lee of a mountain. (From D. H. Slade (Ed.), 1968); (d) lee clouds produced by airflow over an island. (NASA photograph.)

The pattern shown in Fig. 2.32c is found when stable air lifted over a barrier begins to oscillate about some equilibrium value and gives rise to lee-side wave clouds, as shown in Fig. 2.32d. When the amplitudes of these waves are large, the rotary motion causes violent turbulence and rapid diffusion.

Many different flow patterns are possible when air flows at some angle to a valley. One such configuration is shown in Fig. 2.33a. The thermally produced up-or-down valley flow may enhance or subtract from the component of the wind in the valley caused by channeling. Another common occurrence of channeling is shown in Fig. 2.33b, where high-speed local winds are produced by rather gentle flow through a pass in a mountain barrier. The Santa Anna wind of southern California and the Columbia winds are examples of these phenomena in the United States.

Wind flows around obstacles greatly complicate the calculation of pollutant concentration (Start, Dickson, and Wendell, 1975). A detailed experimental program or a careful theoretical examination of the various factors that influence the wind is necessary to determine the significance of such obstacles on wind and diffusion conditions.

2-4.4.4 Wind Power. For centuries man has attempted to harness the power of the wind. Until recently the most common use of wind power in the United States has been for pumping water with the old-fashioned wind mills that dot the countryside of rural America. Now, wind-powered electrical generators with power ratings of 100 kW and higher, e.g., NASA's Plumbrook unit, are a reality (Ruthoff and Sirocky, 1974). These generators can provide supplemental, or in some cases the total, power requirements of an area if sufficient wind is available.

Estimates of wind-power potential in the United States (Thomas, 1945; Reed, 1974) have been compiled on the basis of the cube of the mean wind speed observed at various sites across the country. The kinetic energy of the wind is $E = \frac{1}{2}mV^2 = \frac{1}{2}\rho AV^3$, where V is the wind speed and m is the mass of a volume of air having density ρ, cross sectional area A, and speed V; power is energy per unit time.

Although the average output of a wind-powered generator is proportional to the average wind speed, it also depends on the variance of the wind speed. Justus et al. (1976) have computed wind-power potential for the United States on the basis of the frequency distribution of wind speeds and realistic relationships between power output and wind speed. Figure 2.34 shows annual capacity factors, ratio of average power output to rated power output, for a 100-kW generator installed at a height of 61 m.

We can conclude that wind-powered generators with operating characteristics near those of the NASA Plumbrook units can be expected to operate over substantial areas of the country with capacity factors of better than 50%.

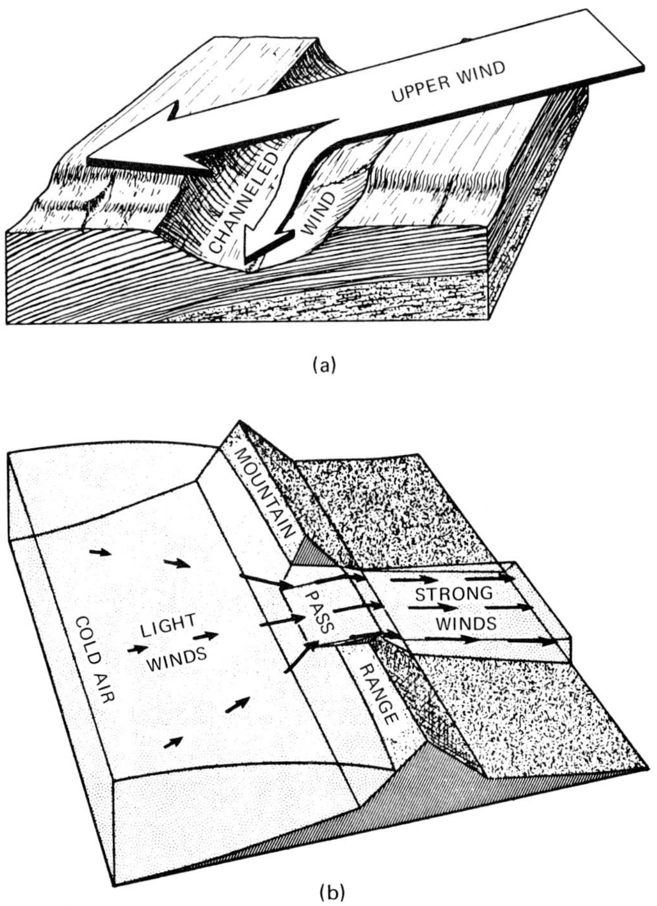

(a)

(b)

Fig. 2.33 Distortions of the wind flow by topographic obstacles. (a) Channeling of the wind by a valley, (b) the effect of a mountain pass on the wind flow.

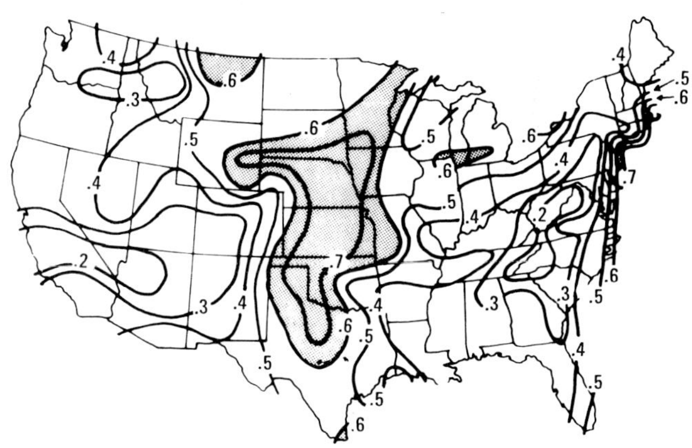

Fig. 2.34 Contours of annual capacity factor at a height of 61 m above flat terrain for a 100-kW generator with V_0 = 3.6 m/s and V_1 = 8.0 m/s.

2-4.5 Meteorological Observations on the Mesoscale

The data requirements for the analysis of various mesoscale phenomena in the ABL could normally be satisfied by wind and temperature measurements within the lowest 1 km of the atmosphere and over an area with horizontal dimensions in hundreds of kilometers. Furthermore, owing to the diurnal alternation and rapid changes in the possible structure of the flows, these data should be available at intervals of about 1 h. With these minimal requirements then, it is easy to see that a macroscale observation network is not an adequate source of data. Usually the surface network of weather stations is spaced far too sparsely. The vertical temperature and wind profiles have sufficient resolution but do not provide the required horizontal and temporal resolution.

A great number of pertinent investigations of mesoscale features can be made with macroscale data alone. For example, a study of sea-breeze circulations can be made through a critical examination of the surface wind data at a location near the coast. However, the horizontal and vertical extent of the sea breeze will have to be estimated. These estimates are improved by studying available temperature profiles and results of other sea-breeze investigations using a meteorological network more appropriate to the scale of the phenomena investigated. Some examples of these techniques follow.

Tall Towers. A number of television and radio towers have been instrumented for meteorological purposes. Since these represent towers of opportunity, the expense is usually minimal, although the maintenance of instruments on a tall tower can be a difficult problem. Most data collected on tall towers in the United States are for the lowest 300 m of the ABL. There are several 500-m towers in use. Tower data can be used to obtain insight into more subtle variations in the ABL, as well as providing information about the rate of change and persistence of mesoscale features.

Tetroons. The tetroon is a constant-level balloon which is followed by a radar-tracking technique similar to that used in tracking the conventional rawinsonde. Since the tetroon is designed to flow with the mean wind at any level (constant density surface), it is useful for the investigation of effects of various mesoscale or topographic features on the horizontal wind. Distance traveled is limited to the radars' line of site so that distances on the order of a few hundred kilometers and travel times up to 24 hr are approximate limiting values. Tetroon measurements are usually confined to short-term surveys. Simple tetroons are released in a series, or a number of tetroons can be released at the same time, depending on the objective of the experiment.

Mesoscale Tower Networks. An increasing number of mesoscale networks, consisting of meteorological towers, are used to study the wind and temperature pattern of a site. The network on the National Reactor Test Site at Idaho Falls, for example, is used to study the effects of topography on wind patterns in the ABL in support of reactor testing operations. When the need (or location of the site) is such that this detailed wind information is required, the expenditures are justified. The networks provide real-time support in the event of an accidental release in which rapid response is necessary. Tower networks provide similar data as tetroon flights with the added advantage of being able to collect data continuously rather than for just a short-term experiment.

Remote Sensing. With the new technology in the aerospace and satellite industry, the use of remote sensing devices is increasing our capability for examining the structure of the lower atmosphere. Such remote sensing devices as acoustic sounders and surface based and airborne optical radars have provided much data concerning the structure of the ABL.

Aerial and Surface Transects. During many past experimental programs, helicopters and airplanes have been used to obtain data on meteorological parameters and pollutant distribution in both the horizontal and vertical directions. Surface transects by automobiles have been used extensively to collect data to study the effect of urban–rural temperature patterns (Lowry, 1967). Short-term and selective surveys of this nature can be very helpful for obtaining detailed information of mesoscale systems. Naturally the success of a transect depends on whether it was carried out at the right time and place.

Sources of Data. The real value of the various observational techniques used for the mesoscale features is the application of results of the observed sites to other sites. For the most part, the extensive mesoscale networks are site specific. These networks were required due to the inability of meteorologists to adapt knowledge obtained in other experiments for that particular site. For this reason, results from one site may not be directly applicable to another. They can, however, provide information on the type of instrumentation needed to delineate specific meteorological motions on the mesoscale. See Chap. 8 where a number of the more extensive observational programs are listed.

2-4.6 Role of Mesoscale Motions in the Dispersion of Atmospheric Pollutants

Some of the experimental programs established to investigate mesoscale motions are designed to answer such engineering questions as the effect of wind loading on buildings. Other programs have been established to investigate the effect of mesoscale motions on the transport and diffusion of pollutants. As previously mentioned, many sites have predominate mesoscale motions that are important with respect to diffusion of pollutants. Most important, however, is the need for determining where a cloud or plume will go once it is released. For this reason, the impetus has been on diagnosing characteristics of the flow field rather than on examination of the diffusive capabilities representative of the particular mesoscale feature. Figure 2.35 summarizes significant features of the various mesoscale circulation patterns and shows their effect on airborne pollution plumes. At times the flow patterns corresponding to mesoscale features are beneficial (i.e., through downward mixing or change in wind direction) and lead to improved air quality. Frequently these patterns add to the degradation of the overall air quality. The problem for the meteorologists in the power-production industry is to determine the position and time at which this degradation will appear.

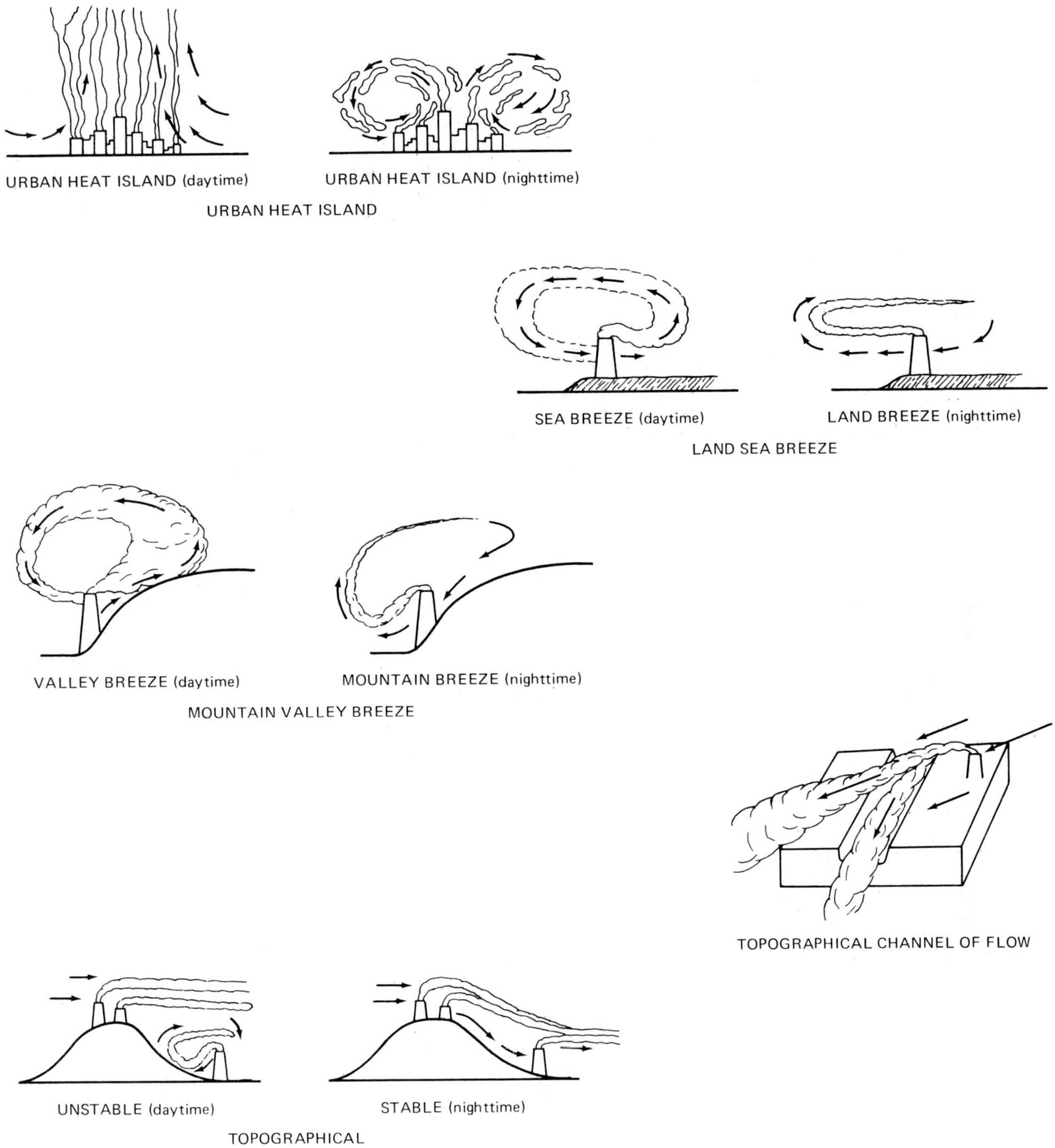

Fig. 2.35 Effect of mesoscale motions on typical smoke plumes. (Synoptic scale flow removed from three upper figures.)

2-5 MICROSCALE SYSTEMS

Microscale motions are those atmospheric motions occurring on time scales between seconds and minutes and over space scales less than 2 km. Most human activities are restricted to the lowest several hundred meters of the atmosphere. Of prime importance, however, are the motions within the lowest 30 m.

Numerous meteorological experiments have been conducted in the last 50 yr to observe meteorological motions in this lowest surface layer (see Chaps. 5 and 9). An impetus for such experimental programs is the importance of these motions on the well-being of people, but undoubtedly many programs can be attributed to the relative ease with which data can be collected at these low elevations. These experiments have produced a general understanding of the nature of atmospheric motions within the lowest 30 m. The focus of the following section is to describe the measurable characteristics of microscale motions that are responsible for the diffusion of atmospheric pollutants.

2-5.1 Energy on Microscale

Two types of energy are considered with respect to microscale motions. The first involves radiant and heat energy relating to physical processes in the manner described with respect to mesoscale and macroscale motions. The heat and radiation budget affects air motions on the microscale. The second type of energy is turbulent kinetic energy, which is a characteristic of the flow, i.e., the wind, and determines the rate of growth and deformation of a cloud or plume released into it.

2-5.1.1 Energy Balance on Microscale. Items discussed in the energy budget relationships on the mesoscale, e.g., as heat capacity, emissivity, albedo, and radiation from pollutants, could at times influence the energy budget and hence the flow characteristics near a receptor. Receptors may be either buildings, vegetation, or animal life. Such subtle differences as the amount of shade or variations in evaporation rates may provide the impetus for the development of a very small but significant perturbation. This perturbation to the circulation would be particularly important when macroscale and mesoscale winds are calm.

On the microscale, individual differences from one plant to another are significant so that more-detailed parameters of physical characteristics must be made. In many locations, such as under the canopy of trees or in the canyons of streets, the location and orientation of various surfaces are complex. For example, in a pine forest, the percentage of the total area covered by vegetation varies greatly depending on the height aboveground. The amount of radiation received at the ground, for example, is related to an integration of leaf areas as a function of height above the surface, reflection and radiation of incident light being taken into account. These complicating influences result in much difficulty in assessing the rate of exchange of a pollutant to the biotic plant as well as the rate of exchange of heat energy from the surface up through the canopy. The importance for the correct consideration of these fluxes lies in the fact that sources and sinks of energy are located within the canopy. Only by knowing the proper fluxes can the effect of the overall system be evaluated.

An important effect in the consideration of the sink for pollution in the plant canopy is called the boundary-layer resistance. As pollutants are diffused through the canopy and come very near the leaf surface, the boundary-layer flow of air near the leaf (on the order of 1 mm in depth) tends to keep the pollutant from being deposited on the leaf. The pollutant particle has to overcome this boundary resistance before it can intercept the plant surface. Analogously, a pollutant has to overcome a similar boundary-layer flow along the surface of a building before it can be deposited on, or interact with, the building surface.

2-5.1.2 Turbulent Kinetic Energy. At a given point of observation, the interaction of all weather systems on all scales results in a three-dimensional wind that varies continuously in time. The magnitude of the statistical variance of the wind is directly related to the kinetic energy available for the diffusion of pollutants.

Turbulent kinetic energy (TKE) is dependent on three factors: (1) mechanical effects of objects protruding into the airstream, e.g., the roughness of the earth's surface, (2) the vertical wind shear, and (3) the vertical temperature structure of the atmosphere. The amount of TKE contained within each eddy size determines the rate of growth of a pollution plume or cloud.

The budget of TKE is usually studied through applications of statistical techniques rather than the dynamic methods that characterize macroscale and mesoscale motions. The statistical technique most used to study TKE is the distribution of variance with frequency through use of spectrum analysis.

The distributions of the variance of horizontal wind velocity (σ_v^2) with frequency and period (i.e., reciprocal of frequency) for two time series are shown in Fig. 2.36.

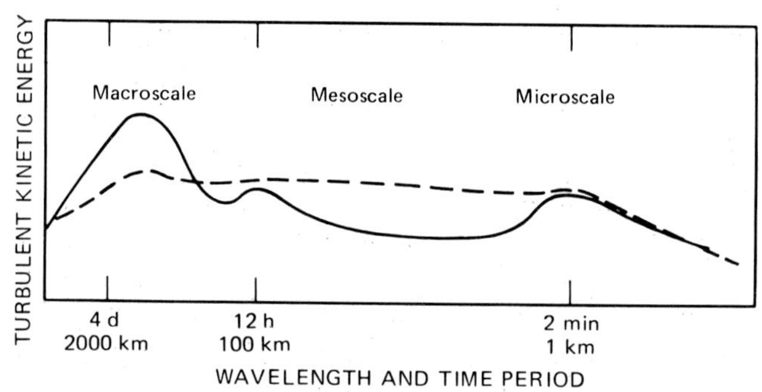

Fig. 2.36 Spectra of wind speed near the ground for two different synoptic conditions. The spectrum indicated by a solid line shows the existence of the "mesoscale gap"; the dashed line represents a spectrum for a period during which no mesoscale gap exists.

These spectra are referred to as power or energy spectra because they have units of kinetic energy per unit mass (m^2/s^2). Note, energy is merely the variance of the wind velocity; the peaks occur at frequencies having the greatest contribution to the overall variance.

Peaks in the spectra correspond to production of TKE. For example, the peak near a period of 4 d (see Fig. 2.36) represents the macroscale peak. The 4-d period corresponds to the average travel time of weather systems over a point in the mid-latitudes. The peak near 1 min is the microscale peak. The portion of the spectrum to the right of the microscale peak is called the inertial subrange (Kolmogarov, 1941). Theoretically, energy in the inertial subrange is neither created nor destroyed but transfers to higher and higher frequencies (i.e., smaller and smaller scales) until it is dissipated by friction into heat. The spectra in Fig. 2.36 show the distribution of TKE for two different meteorological conditions. The spectrum for series 1, indicated by a solid line, shows a large macroscale peak and a small microscale peak. The reduced energy (i.e., absence of a peak) in the mesoscale region is referred to as the *mesoscale gap*.

The dashed line in Fig. 2.36 represents an energy spectrum for series 2 during atmospheric conditions for which the mesoscale gap did not exist. This implies that energy is produced in this range by mesoscale motions (thunderstorms, land–sea breeze, etc.) or that energy produced on a larger scale is broken down into this scale. Consider a pollutant cloud released during periods with meteorological conditions corresponding to the two spectra in Fig. 2.36. Initially the two clouds would grow at the same rate since the cloud is most affected by eddies the same size as the cloud (see Fig. 2.16) and the two series have similar magnitudes for energy up to a scale of 1 km. As the clouds grow to mesoscale sizes, the cloud released during series 1 would grow about twice as fast as the cloud released during series 2 since about twice as much energy is available at these frequencies. As the clouds continue to grow to larger and larger sizes, their growth rates will be determined by the amount of energy at that scale. Differences in growth rates of the two clouds are proportional to differences in the energy spectra.

It is not always clear where the line should be drawn between mean transport and diffusion of a puff or plume. The scale of averaging and the nature of the phenomena being studied must enter into the decision. The energy spectra afford a means by which the energy available for diffusion can be categorized according to the scale of motion dominating the transport and diffusion processes. Most successful attempts to relate spectra and diffusion theoretically have been concerned with eddies having dimensions less than 1 km or periods less than 2 min, i.e., the inertial subrange region, which will be discussed in Chap. 6.

2-5.2 Sources of Turbulence

The amount of turbulence of the wind varies both in time and space. Changes in production rates are primarily responsible for these variations. Turbulence in the microscale range can be assumed to be produced two ways, mechanically and by convection. Much of the knowledge about these two mechanisms was obtained through spectral analyses of meteorological data. Brief discussions of the characteristics of each process follow. Energy spectra are used to show relative magnitudes of each process for different conditions.

Mechanical Turbulence. Mechanical turbulence is produced by the frictional forces acting on air passing over surface irregularities. The larger the irregularities, the greater the mechanical turbulence. The magnitude of the turbulent eddies is a function of surface roughness, elevation aboveground, wind speed, and, to a lesser extent, stability. The effects of surface roughness on flow in the ABL are handled analytically through use of the roughness length, z_0, which ranges between a few millimeters over smooth water to several meters over urban areas (see Fig. 2.37). The influence of surface roughness, wind speed, elevation, and stability on the production of mechanical turbulence is demonstrated by spectra of the horizontal and vertical wind direction in Fig. 2.38.† The area under the spectra curve of the horizontal wind direction, θ, is the variance of θ, denoted by σ_θ^2, and the area under the spectra curve of the vertical wind direction, ϕ, is the variance of ϕ, denoted by σ_ϕ^2.

Peaks in the spectra of the horizontal and vertical wind direction at a frequency near 0.3 Hz (1000 cycles/h, corresponding to a period of 3.6 s) correspond to mechanical turbulence. These spectra show that mechanical turbulence is (1) not a function of stability, (2) directly related to surface roughness, (3) directly related to wind speed, and (4) inversely related to elevation.

Convection. Convection is defined as predominantly *vertical* atmospheric motions resulting in vertical transport and mixing of atmospheric properties. Convective eddies or turbulence arise from hydrostatic instability as the result of surface heating. These eddies are larger and occur at a lower frequency than eddies produced by mechanical turbulence. Spectra of the horizontal wind direction show peaks at a frequency of approximately 0.003 Hz (10 cycles/h). Figure 2.38a shows spectra of *horizontal* wind direction during stable and unstable conditions. Convective turbulence, unlike mechanical turbulence, is strongly related to stability. The large spectral peak appearing during unstable conditions is almost gone during stable conditions. The low-frequency peak appearing during stable conditions is not caused by convection. This peak occurs during nighttime and cannot be explained simply although it is established by observational evidence. The low-frequency oscillations contributing to the peak are probably due to slow-moving systems which are embedded in the stable air and which have horizontal dimensions ranging from hundreds to a few thousand meters. These systems are of considerable importance in predicting dispersion at night since they may lead to a much greater degree of horizontal diffusion than might normally be expected during very stable conditions.

Figure 2.38b shows that the spectral peak of the horizontal wind direction owing to convection is not a strong function of surface roughness. However, the convective peak for the vertical wind direction (not shown) is directly related to surface roughness. The spectra of vertical wind direction [parts (c) and (d) of Fig. 2.38] show the effects of wind speed and elevation on the convective turbulence. Note that convective turbulence, unlike mechanical turbulence, is indirectly related to wind speed. Turbulent eddies in the vertical direction are generally smaller than horizontal eddies because their growth is limited by the surface and elevated inversions. This causes the spectral peak of the vertical wind direction to occur at a higher frequency [about 0.03 Hz (100 cycles/h)] than the peak of the horizontal wind direction. The closer the observations are taken to the ground, the more severely the energy at lower frequencies is limited. Thus the convective

†Note: These spectra were used rather than spectra of the components of the wind because there was good correlation between σ_θ and σ_ϕ and the horizontal and vertical spread of pollution plumes.

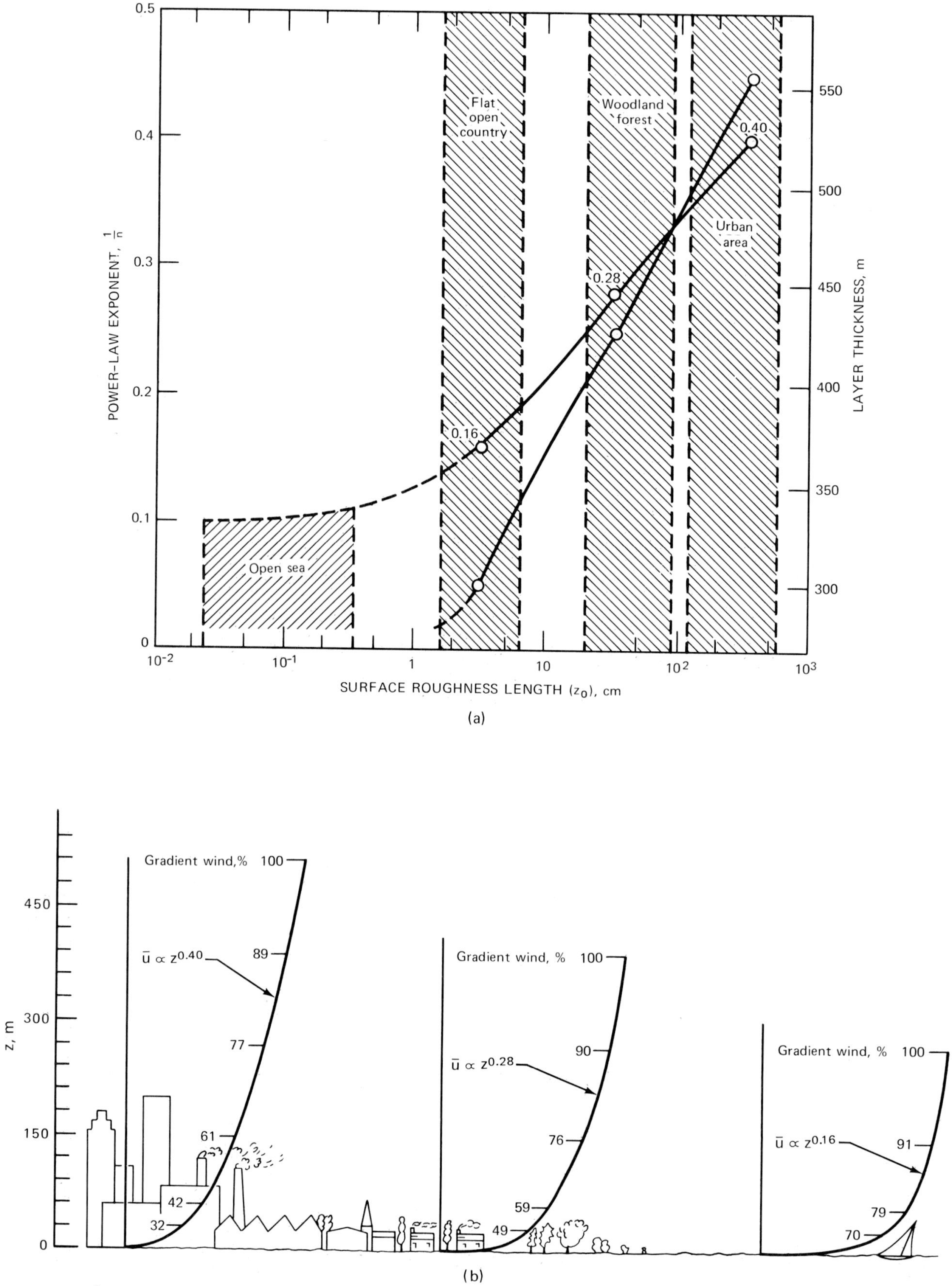

Fig. 2.37 (a) The exponent in the power law and the height of the boundary layer as functions of the roughness length and the type of terrain. (b) Empirical power laws over different terrain. (From A. G. Davenport, 1965, The Relationship of Wind Structure to Wind Loading, in *Wind Effects on Buildings and Structures*, Symposium No. 16, pp. 54-102, National Physical Laboratory, Her Majesty's Stationery Office, London, 1965.)

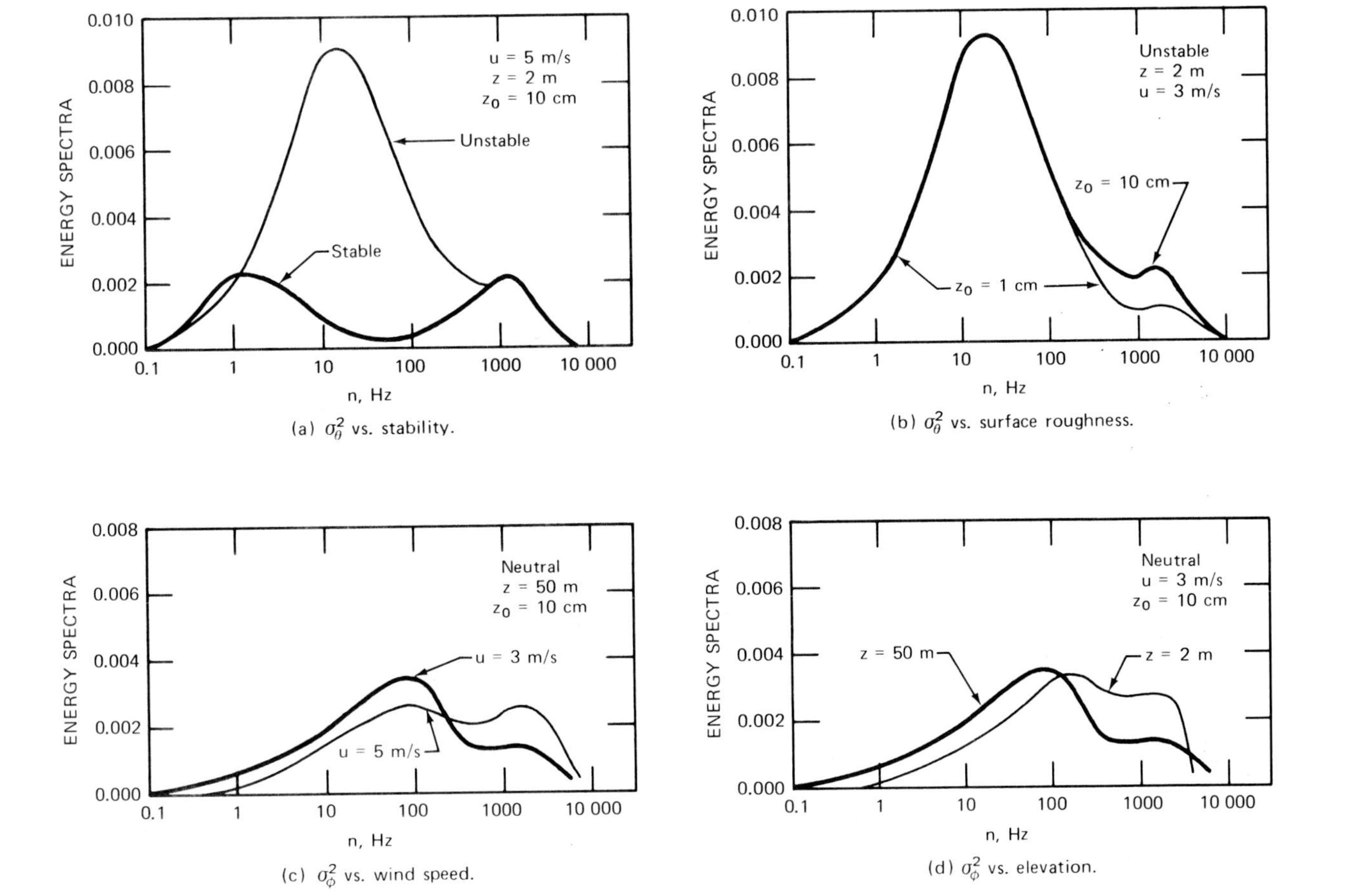

Fig. 2.38 Spectra of the horizontal and vertical wind direction fluctuations showing the effects of stability, surface roughness, wind speed, and elevation. (Spectra constructed by author using theoretical and observed spectra summarized in J. L. Lumley and H. A. Panofsky, 1964.)

peak shifts to lower frequencies with elevation (Fig. 2.38d). Theoretical relationships for mechanical and convective turbulence will be given in Chap. 6.

2-5.3 Turbulence Parameters

Since the equations governing turbulent flow are very complicated, it is tempting to treat the diffusive nature of turbulence by means of a properly chosen effective diffusivity. In doing so, the idea of trying to understand the turbulence itself is partially discarded. If we use an effective diffusivity, we tend to treat turbulence as a property of the fluid rather than as a property of the flow. Conceptually, this is a very dangerous approach; however, it often makes the mathematics tractable. The problem of turbulent diffusion is approached by two alternative methods: the statistical theory and the gradient transport theory. Neither method can be categorically eliminated from consideration since each has areas of utility that do not overlap the other. Statistical diffusion theories consider motion following fluid particles and thus can be described as Lagrangian. On the other hand, diffusion at a fixed point in the atmosphere, according to the gradient transport theory, is proportional to the local concentration gradient. Consequently it could be said that this theory is Eulerian in that it considers properties of the fluid motion relative to a spatially fixed coordinate system. Both approaches attempt to reproduce the observed diffusion of atmospheric releases. Both approaches rely on empirical relationships between the dispersive powers of the atmosphere and the various parameters of turbulence, which are discussed in Chap. 6.

In Sec. 2-5.2, we saw that the rate of growth of a pollutant cloud or plume is determined by the spectrum of atmospheric turbulence. State-of-the-art methods for estimating diffusion rates do not use the full energy spectra owing to limitations of theory, computational resources, and measurements of appropriate meteorological variables. As a result, present practices are to characterize diffusion with measurable, theoretically tractable, and easily applied parameters.

Most parameters used to estimate the diffusion rates of the atmosphere in the horizontal and vertical directions involve $\bar{u}$, σ_θ, σ_ϕ, $\partial T/\partial z$, cloud cover, and solar radiation (Gifford, 1976). Generally these parameters characterize diffusion rates in one direction better than the other. Furthermore, some parameters are acceptable only within limited ranges of stability conditions. The most acceptable parameters to represent horizontal and vertical diffusion rates are the turbulent intensities, σ_θ and σ_ϕ, respectively. Diffusion rates can be determined through such relationships as those developed by Draxler (1976) and Pasquill

(1976), provided estimates of u, σ_θ, and σ_ϕ are available at the release height and represent a 1-h sample period.

2-5.3.1 Turbulence Intensity. If, in a Eulerian coordinate system, u is the component of the wind along the x axis (usually assumed to coincide with the mean vector wind direction), v is the component in the y direction, and w is the component along the z axis (assumed to be vertical), then the actual wind components at any instant can be thought of as being composed of mean values and deviations from the mean:

$$u = \bar{u} + u'$$

$$v = \bar{v} + v'$$

$$w = \bar{w} + w'$$

The fluctuation quantities, u′, v′, and w′ can be represented as time series that pass through positive and negative values. These series can be subjected to variance, spectra, and autocorrelation analyses.

The intensity of the turbulence along each axis is defined by

$$i_x = \frac{\sigma_u}{\bar{u}}$$

$$i_y = \frac{\sigma_v}{\bar{u}} \simeq \sigma_\theta$$

$$i_z = \frac{\sigma_w}{\bar{u}} \simeq \sigma_\phi$$

where σ is the standard deviation of the velocity distribution. These statistics of turbulence bear a direct relation to the diffusing power of the atmosphere.

Statistics of wind fluctuations which have found wide use in diffusion estimation techniques are the standard deviations of the horizontal wind direction angle, σ_θ, and the standard deviation of vertical wind direction, σ_ϕ. These parameters can be measured directly from the output of a bidirectional wind vane and are approximately equal to the intensity of turbulence in the cross-wind direction (i_y) and the vertical direction (i_z). The numerical values of σ_θ and σ_ϕ vary depending on elevation and meteorological conditions. Their values also depend on sampling and averaging times that are inherent characteristics of a sample of data. This is not unexpected in view of the earlier discussion of sources of microscale turbulence which showed that the value of σ_θ or σ_ϕ is determined by turbulence that varies in size and magnitude according to elevation, surface roughness, and meteorological conditions.

Relationship with Height and Stability. The horizontal and vertical diffusion rates of a puff or plume correlate best with u, σ_θ, and σ_ϕ observed at the height of release. Frequently, meteorological measurements are available at an elevation other than the release height. For this situation an estimate of u, σ_θ, and σ_ϕ at release height is highly desirable especially when large changes in u, σ_θ, and σ_ϕ with height are expected (e.g., during very stable conditions). A relationship commonly used for this purpose is called the power-law profile,

$$x_1 = x_2 \left(\frac{z_1}{z_2}\right)^{P_x}$$

where x_1 is the value of variable x at height z_1 and x_2 is the value at height z_2. The exponent, P_x, is an appropriate exponent obtained experimentally or theoretically. Table 2.2 gives values of exponents for several stability conditions for three variables, u, σ_θ, and σ_ϕ. These exponents were determined by the method of least squares from 15-min averaged data applicable below 300 m. For a positive exponent the variable increases with height; for a negative exponent the variable decreases with height. Note that wind speed increases and σ_θ decreases with height for all situations. During unstable conditions, σ_ϕ increases with height, whereas, in stable conditions, σ_ϕ decreases with height. The exponents are largest for moderately stable conditions. For example, between 10 and 100 m, u would increase by a factor of 2.6, and σ_θ would be reduced to 30% of its value at 10 m.

TABLE 2.2 Power-Law Exponents Used to Describe the Vertical Variation of Turbulence†

Stability	Variables u‡ (p_u)	σ_θ (p_θ)	σ_ϕ (p_ϕ)
Extremely unstable	0.08	−0.06	0.02
Moderately unstable	0.10	−0.15	0.04
Slightly unstable	0.11	−0.17	0.01
Near neutral	0.18	−0.23	−0.14
Slightly stable	0.31	−0.38	−0.31
Moderately stable	0.42	−0.53	−0.49

†The power law is in the form, $X_1 = X_2 (Z_1/Z_2)^{P_x}$. The exponents are averages over a 1-yr period and apply between 10 to 304 m. (From M. M. Pendergast, 1977.)

‡The variable u is fitted to a power law with exponent p_u.

2-5.3.2 Relationship with Sampling. The values of σ_θ and σ_ϕ are strongly dependent on sampling and averaging times. It will be shown that, for the range of averaging times used in meteorological studies (15 min to 1 h), the value of σ_θ can vary by a factor of 2 or more. A brief discussion of terms follows with data showing the effect of sampling rates and sampling period on the magnitudes of σ_θ and σ_ϕ.

The wind-direction trace shown in Fig. 2.39 could represent a measurement by a wind vane at a particular point, in which case it would be called a Eulerian time series. It could, however, represent the path of a particle following the wind, in which case it would represent a Lagrangian series.

Considering the curve as a Eulerian time series produced by a wind vane, we can define the sampling and averaging

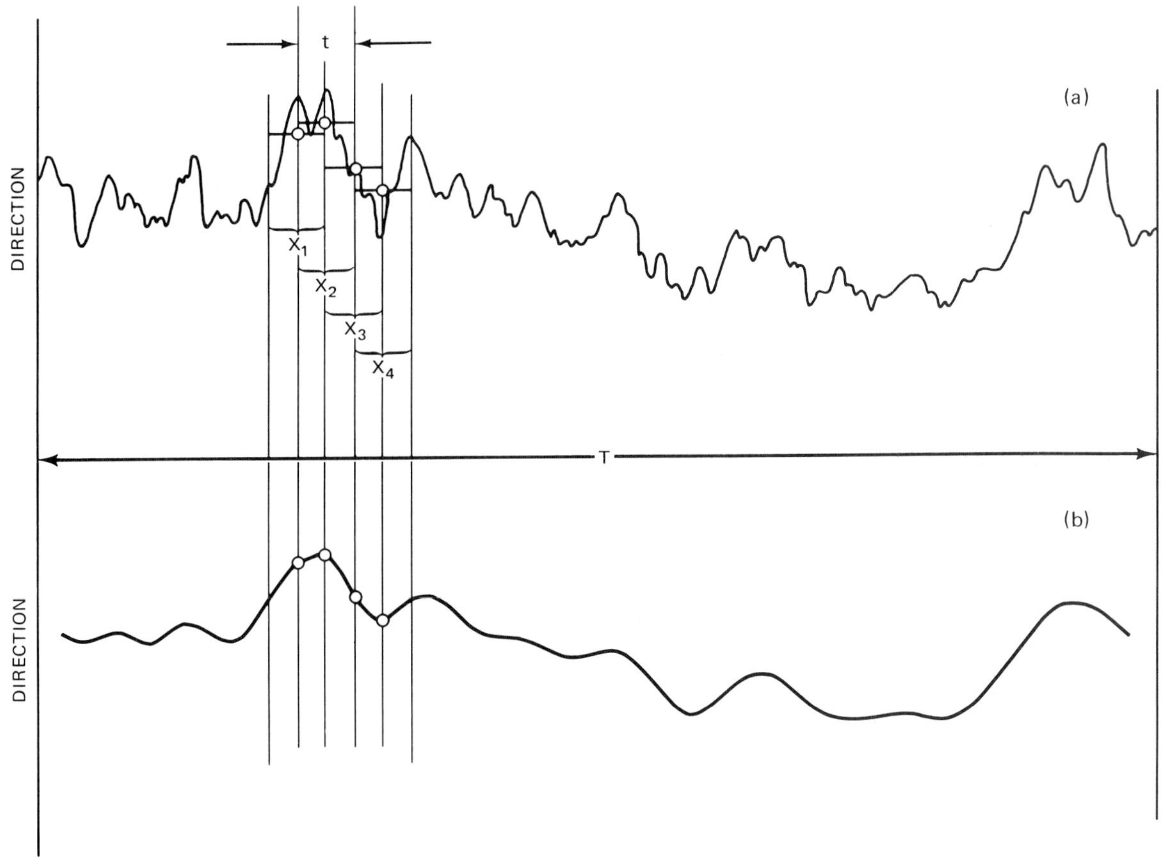

Fig. 2.39 An instantaneous wind-direction trace (a) and the time-averaged trace derived therefrom (b). The time interval t represents the averaging time, and T represents the total length of record, or the sampling time. [From D. H. Slade, 1968.]

times as follows: the sampling time, T, is that interval during which the data are measured and is represented by the total period of the record; the averaging time, t, is some time interval over which the continuously varying trace may be represented by a constant (average) value. If the illustration represents the trace taken directly from a wind-direction recorder, the averaging time will be that resulting from the inertia of the wind vane and recorder system and their resulting inability to respond faithfully to the very highest frequency fluctuations impressed upon them. This averaging time is usually in the range of a few seconds or less for the high-response instruments used in diffusion work. For some applications it may be desirable to eliminate high-frequency fluctuations that are caused by instrument errors. Smoothing can be done by constructing a digital series from the analog series. The digital series might be composed of the average values of the trace over successive overlapping equal intervals of time, t, where t is the averaging time. Connecting these average values produces the time-averaged, or time-smoothed, trace shown in Fig. 2.39b. The application of the averaging time has the effect of removing the fluctuations of periods shorter than t.

The standard deviation of wind-direction fluctuations, whether horizontal or vertical, normally increases with increasing sampling time up to some point. Further, the standard deviation decreases with increasing averaging time. The increase of standard deviation with increasing sampling time follows because more of the long-period fluctuations are sensed during the longer sample periods.

On the other hand, the increase in averaging time damps out the contribution of the shorter period fluctuations to the total standard deviation. The application of a sampling and averaging process to a time series therefore constitutes a filter, or window, through which only a portion of the total energy in the spectrum of wind fluctuations can be observed. The values of σ_θ and σ_ϕ can be related to sampling time in a manner analogous to the power-law profile

$$\sigma_{\theta_1} = \sigma_{\theta_2}\left[\frac{T_1}{T_2}\right]^{b_\theta}$$

where σ_{θ_1} and σ_{θ_2} are obtained over sample periods, T_1 and T_2, respectively.

The value of b_θ and b_ϕ varies with the stability, wind speed, and height above the surface. Table 2.3 contains estimates of b_θ and b_ϕ applicable at a height of 10 m with sampling times between 2.5 min and 3 h. Note that σ_ϕ is only a function of sampling time for very stable conditions whereas σ_θ is a function of sampling period for neutral and very stable conditions.

TABLE 2.3 Variation of σ_θ and σ_ϕ with Sampling Period and Stability†

	b_θ	b_ϕ
Unstable	0.00	0.00
Near neutral	0.22	0.00
Very stable	0.71	0.28

†Value of b_θ and b_ϕ in the relationship $\sigma_x(T_1) = \sigma_x(T_2)\ (T_1/T_2)^{b_x}$ at a height of 10 m. (From M. M. Pendergast, 1977.)

2-5.4 Role of Small-Scale Motions in the Dispersion of Atmospheric Pollutants

From a knowledge of the values of the accompanying meteorological variables, many formulations have been advanced for estimating concentration within a continuously emitting plume. We will discuss these formulas or models in greater detail in Chap. 13. These models have found considerable application in estimating the downstream concentration distribution from a variety of effluent sources, including continuous point sources, instantaneous point sources, line sources, and area sources. Such techniques have been particularly useful when applied over the range of distances that are usually of interest in nuclear reactor safety analyses, i.e., from a few hundred meters to about 10 km. Over these ranges of distances, the mesoscale and macroscale influences are assumed to be minimal. The primary mechanism for diffusion is the microscale turbulence. The following sections list several dispersion models and present examples of diffusion through an observational study of smoke plumes and meteorological parameters.

2-5.4.1 Dispersion Models. Knowledge of the configuration of the source of an effluent is a necessary starting point for diffusion estimates. For purposes of description, it will be assumed in the following discussion that the effluent is composed of a gas or of particles which faithfully follow the motions of the atmosphere. This implies that the effluent is neither negatively nor positively buoyant if it is a gas and that it has a negligible settling velocity if it is composed of discrete particles.

The Instantaneous Point Source. The instantaneous point source is the conventional approximation to the type of release associated with a very short venting of material to the atmosphere, i.e., an explosion. The term "instantaneous point" is, of course, fiction since even a very small explosion will have finite time and space dimensions. The puff, once formed, moves away from the source with a speed and direction determined by the wind at the moment of release. The mean speed and direction of the puff may be expected to change from the original values during its travel as the wind pattern, in which it is imbedded, changes with time. As the puff moves, it will expand about its center owing to the action of turbulent fluctuations.

In the real atmosphere, turbulence is the composite result of all the motion systems discussed earlier. The turbulent elements are not regular circles. A puff of smoke released into the atmosphere will always grow larger since there almost always will be eddies smaller than or of the same size as the puff. The amount of material in a puff from an instantaneous point source is usually given as the total quantity of material released (curies, grams, etc.), and the downwind distribution of the material after it has been dispersed is frequently expressed in terms of exposure, or the time integral of the concentration as the puff passes the point of observation.

The Continuous Point Source. Diffusion meteorologists have paid more attention to the continuous point source than to any other type of release. It is a good approximation to releases from stacks, from small fissures in structures containing pollutants, and from small spills of volatile materials. The approximation of the true source configuration by a point becomes increasingly better as the distance at which the concentration is observed grows increasingly large compared with the dimensions of the source.

A continuous plume may be considered to be made up of an infinite number of puffs released sequentially with a vanishingly small time interval between puffs. Initially each puff moves with the wind direction at the moment of release. The quantity of material released is usually expressed in terms of a release rate (curies per second, kilograms per hour, etc.), and the downwind material distribution is usually expressed in terms of average concentration (curies per cubic meter, micrograms per cubic meter, etc.) over the period of release. Linear dimensions of the plume perpendicular to the plume axis are often given in terms of the standard deviation of the concentration distribution since the average cross-plume distribution is usually close to a normal curve with boundaries at infinity.

The left side of Fig. 2.40 denotes, schematically, the boundaries of a smoke plume as it would appear at any given instant (such as in a snapshot) or as it might appear having been averaged over 10 min or 2 h (as in a time exposure).

The right side of the figure represents the distributions of the concentration for each of the three plume types. The snapshot shows an irregularly varying plume, which gradually expands with distance from the source. The corresponding concentration distribution is irregular and quite peaked. The 10-min average shows a smooth and regular boundary that can be considered as the envelope containing all the short-period fluctuations of the plume. The average concentration distribution perpendicular to the axis of such a plume has been found to resemble a normal (Gaussian) distribution about the centerline, as shown on the right side of the figure. The 10-min cross-plume average concentration distribution is broader and smoother than in the instantaneous case but has a lower peak value. The 2-h plume is wider yet with an even flatter average concentration pattern.

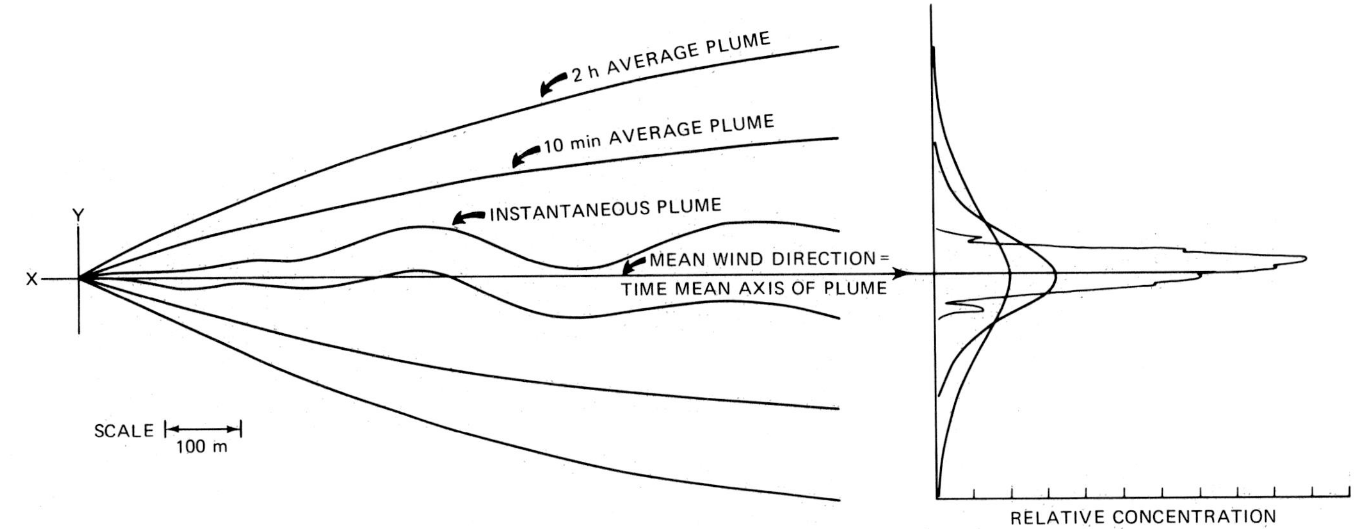

Fig. 2.40 The diagram on the left represents the approximate outlines of a smoke plume observed instantaneously and of plumes averaged over 10 min and 2 h. The diagram on the right shows the corresponding cross-plume distribution patterns. [From D. H. Slade, 1968.]

The time-averaged plume from a continuously maintained point source expands both laterally and vertically with downwind distance from the source. At the same time, the centerline concentration within the plume will decrease continuously with distance. The rate at which the centerline concentration decreases varies with the magnitude of the turbulent fluctuations.

Although continuous sources located at the earth's surface may be postulated, the elevated source represented by a continuously emitting chimney is a more common occurrence. The surface concentration pattern from an elevated source is typically like that shown in Fig. 2.41. The concentration is essentially zero near the base of the chimney, rises rapidly to a peak value at some downstream distance, and decreases regularly thereafter. The concentration is always higher along the projection of the plume axis on the ground than on either side.

Line and Area Sources. Line and area sources are extensions of the continuous point source in one and two directions, respectively. A busy highway or a line of closely spaced vents are examples of line sources. The continuous area source, such as from a city, is as yet of minor importance in the power industry. Although increased attention will be given to area source models as power parks are developed.

Limits of a Vertical Plume Spread by Stable Layer. The preceding dispersion models normally provide best estimates of concentration within 1 km from the source. However, the rate of vertical growth that is observed over such distances cannot be maintained indefinitely. The lower boundary of the plume is, of course, fixed by the surface of the earth. The top of the plume is bounded by a stable layer aloft. The daytime temperature profiles show an average trend toward greater stability with height. On a clear summer day, the superadiabatic lapse rate will rarely extend more than a few hundred meters above the surface. Above this unstable layer there may be a smooth transition to a more stable structure, or there may be a sharp change to stable conditions. These changes may arise because of subsidence or more subtle large-scale meteorological processes. As a result of the stabilization aloft, a distinct lid is placed on the further vertical growth of a plume, and any future decrease in concentration must be due to horizontal spreading. This condition is frequently referred to as "trapping," and the lid is in the vertical and along-wind directions.

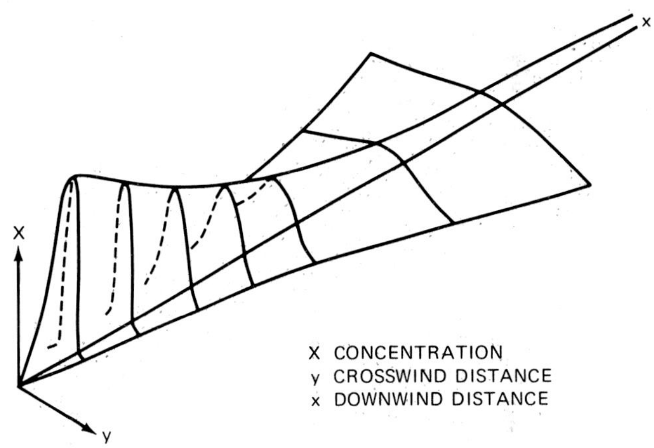

Fig. 2.41 The surface concentration pattern downwind from an elevated source. [From D. H. Slade, 1968.]

2-5.4.2 Smoke Plume Patterns. It was noted earlier that the width of a time-averaged plume increases with increasing sampling time. If a plume were emitted continuously over a period of many weeks or months, it would move many times during this period through all the directions of the compass. The concentration distribution therefore would not fall into some finite sector but, rather, would be represented by isopleths about the source. Such a distribution shows the average concentration decreasing in all directions from the source but most slowly in the directions of greatest wind frequency.

Although the characteristics of horizontal spreading of time-averaged smoke plumes are difficult to observe, the vertical appearance of a real plume makes no demands on the observer; it offers considerable information on the thermal and dynamic state of the lower atmosphere. Some hypothetical patterns of smoke spreading are shown in Fig. 2.42. At the left of the figures are vertical profiles of temperature, wind speed, σ_θ, and σ_ϕ to help in the understanding of the meteorological variability causing the observed plume spreading.

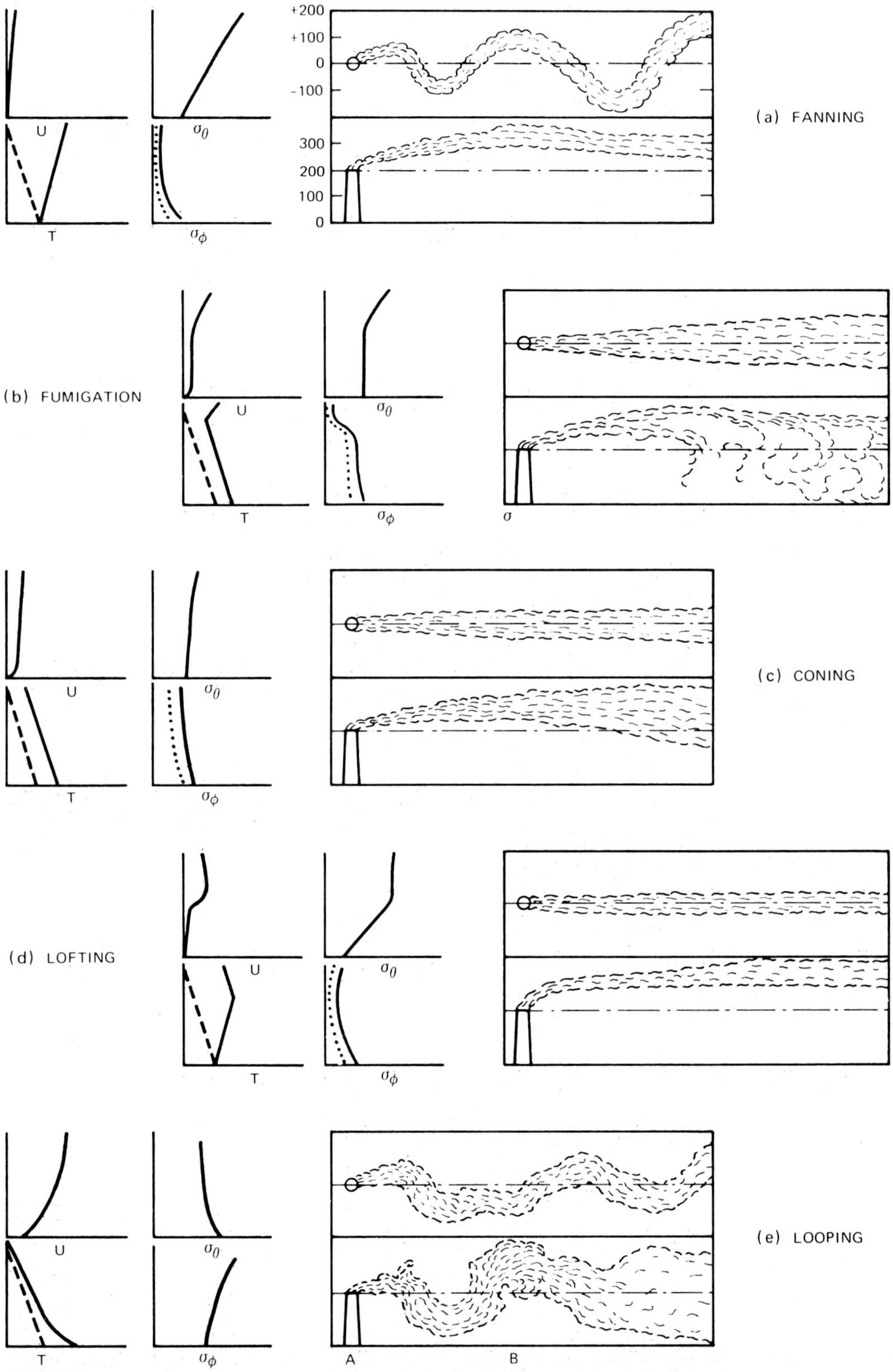

Fig. 2.42 Instantaneous position of pollution plumes released during periods of varying conditions.

REFERENCES

Arya, S. P. S., 1975, Geostrophic Drag and Heat Relations for the Atmospheric Boundary Layer, *Q. J. R. Meteorol. Soc.*, 101: 147-161.

Budyko, M. I., N. A. Yefimova, L. I. Zubenok, and L. A. Strokina, 1962, The Heat Balance of the Surface of the Earth, *Soviet Geogr.*, 3(5): 3-16.

Businger, J., 1963, The Glass House (Greenhouse) Climate, in *Physics of Plant Environment*, W. R. Van Wijk (Ed.), North Holland Publishing Co., Amsterdam.

Byers, H. A., 1944, *General Meteorology*, 2nd ed., McGraw-Hill Book Co., New York.

Crawford, T. V., 1968, *The Long Range Diffusion of the Effluent Cloud from the Phoebus 1B EP-IV Reactor Test of February 23, 1967*, USAEC Report UCRL-50418, University of California, Livermore, Lawrence Radiation Laboratory, April 1968.

Davenport, A. G., 1965, The Relationship of Wind Structure to Wind Loading, in *Wind Effects on Buildings and Structures*, Symposium No. 16, pp. 54-102, National Physical Laboratory, Her Majesty's Stationery Office, London.

Deardorff, J., 1972, Numerical Investigations of Neutral and Unstable Planetary Boundary Layers, *J. Atmos. Sci.*, 29: 91-115.

Draxler, R. R., 1976, Determination of Atmospheric Diffusion Parameters, *Atmos. Environ.*, 10: 99-105.

Flohn, H., 1969, *Climate and Weather*, McGraw-Hill Book Co., New York.

Gifford, F. A., 1976, Turbulent Diffusion-Typing Scheme: A Review, *Nucl. Safety*, 17(1): 68-86.

Goff, R. C., 1976, Vertical Structure of Thunderstorm Outflows, *Mon. Weather Rev.*, 104(11): 1429-1440.

Hanna, S. R., and F. A. Gifford, 1975, Meteorological Effects of Energy Dissipation at Large Power Parks, *Bull. Am. Meteorol. Soc.*, 56(10): 1069-1076.

——, and S. D. Swisher, 1971, Meteorological Effects of the Heat and Moisture Produced by Man, *Nucl. Safety*, 12(2): 114-122.

Hare, F. K., 1960, The Westerlies, *Geophys. Rev.*, 50: 345-367.

Holzworth, G. C., 1974, Estimates of Mean Mixing Depths in the Contiguous United States, *Mon. Weather Rev.*, 92(5): 235-247.

Jeffreys, H., 1926, On the Dynamics of Geostrophic Winds, *Q. J. R. Meteorol. Soc.*, 52: 85-104.

Justus, C. G., W. R. Hargraves, and Ali Yalcin, 1976, Nationwide Assessment of Potential Output from Wind-Powered Generators, *J. Appl. Meteorol.*, 15(7): 673-678.

Kolmogorov, A. N., 1941, The Local Structure of Turbulence in Incompressible Viscous Fluid for Very Large Reynolds Numbers, *Dokl. Akad. Nauk, SSSR*, 30: 301.

Lee, Richard, 1978, *Forest Microclimatology*, Columbia University Press, New York.

LeMone, M. A., 1973, The Structure and Dynamics of Horizontal Roll Vortices in the Planetary Boundary Layer, Ph.D. Thesis, Department of Atmospheric Science, University of Washington, Seattle, Wash.

Lettan, H., and B. Davidson, 1957, *Exploring the Atmospheres First Mile*, Vols. I and II, Pergamon Press, New York.

Lovelock, J. E., R. J. Maggs, and R. J. Wade, 1973, Halogenated Hydrocarbons in and over the Atlantic, *Nature*, 241: 194-196.

Lowry, W. P., 1967, The Climate of Cities, *Sci. Am.*, 217: 15-23.

Lumley, J. L., and H. A. Panofsky, 1964, *The Structure of Atmospheric Turbulence*, Interscience Publishers, John Wiley & Sons, Inc., New York.

Machta, L., 1958, Global Scale Dispersion by the Atmosphere, in *Proceedings of the Second United Nations Conference on the Peaceful Uses of Atomic Energy, Geneva*, Vol. 18, p. 519, New York, 1958.

Markee, E. H., Jr., J. G. Beckerley, and K. E. Sanders, 1974, *Technical Basis for Interim Regional Tornado Criteria*, USAEC Report WASH-1300.

Newell, R. E., 1971, The Global Circulation of Atmospheric Pollutants, *Sci. Am.*, 224(1): 32-42.

National Oceanic and Atmospheric Administration, Paper 75009, 1976, *Thunderstorms*.

Oke, T. R., 1979, *Review of Urban Climatology*, World Meteorological Organization Technical Note No. 169, WMO No. 539.

Orlanski, I. O., 1975, Rational Subdivision of Scales for Atmospheric Processes, *Bull. Am. Meteorol. Soc.*, 56(5): 527-530.

Pasquill, F., 1976, *Atmospheric Dispersion Parameters in Gaussian Plume Modeling*, Part II, EPA Report EPA-60014-76-0306.

Pendergast, M. M., 1974, *A Study of the Effect of the Urban Mesoclimate on Local and Regional Pollution Potential in SE Texas*, Ph.D. Thesis, Texas A&M University, College Station, Tex.

——, 1977, Estimating Diffusion Coefficients from Meteorological Data, submitted to *Journal of Applied Meteorology*. Also SRL Report DP-MS-76-64, Savannah River Laboratory, E.I. Du Pont de Nemours & Co., Aiken, S.C.

Randerson, D., 1972, Temporal Changes in the Horizontal Diffusion Parameters of a Single Nuclear Debris Cloud, *J. Appl. Meteorol.*, 11(4): 670-673.

——, 1970, Spectral Distribution of Solar Radiation in a Polluted and a Clear Air Mass, *Air Pollut. Control Assoc.*, 20(8): 546-548.

Reed, J. W., 1974, Wind Power Climatology, *Weatherwise*, 27: 237-242.

Rossby, C. G., 1939, Relations Between Variation in the Intensity of the Zonal Circulation of the Atmosphere and the Displacements of the Semipermanent Centers of Action, *J. Mar. Res.*, 2(1): 38-54.

Ruthoff, R. L., and P. J. Sirocky, 1974, *Preliminary Design of a 100-kW Wind Turbine Generator*, Report NASA-TM-X-71585, NTIS.

Schubert, J. F., 1976, A Climatology of the Mixed Layer Using Acoustic Methods, in *Third Symposium of Meteorological Instrumentation*, Feb. 10—13, 1975, Washington, D. C.

Sellers, William D., 1965, *Physical Climatology*, University of Chicago Press.

Slade, D. H. (Ed.), 1968, *Meteorology and Atomic Energy*, USAEC Report TID-24190, NTIS.

Start, G. E., C. R. Dickson, and L. L. Wendell, 1975, Diffusion in a Canyon Within Rough Mountainous Terrain, *J. Appl. Meteorol.*, 14: 333-346.

Tennekes, H., 1974, The Atmospheric Boundary Layer, *Phys. Today*, 27(1): 152-163.

Thomas, P. H., 1945, *Electric Power from the Wind*, Federal Power Commission.

Trewartha, Glenn T., 1954, *An Introduction to Climate*, McGraw-Hill Book Co., New York.

Chapter 2 covers research prior to January 1979.

METEOROLOGICAL INSTRUMENTATION 3

Conrad J. Mason, University of Michigan
Harry Moses, U. S. Department of Energy

3-1 APPLICATION OF METEOROLOGICAL MEASUREMENTS

Important uses of meteorological instrumentation include the evaluation of potential sites for power generating stations, the preparations of environmental impact statements and reports, and for site monitoring to ensure conformity with Federal Air Quality Standards during the operation of the plant. The primary areas in which meteorological information is required are: (1) the dispersion of contaminants emitted; (2) the dissipation of heat; (3) the structural design of the equipment to withstand severe weather conditions; (4) the design of natural draft cooling towers for efficient operation; (5) the design of solar-power and wind-power systems; and (6) the development of improved forecasting techniques to aid power load dispatching.

Most of the power-producing facilities release effluents and excess heat to the atmosphere. Consequently the problems dealing with effluent dispersion are of essential importance. The meteorological variables that must be monitored to determine dispersion rates are turbulent intensity and wind speed and direction. Turbulent intensity can be measured directly from a wind vane that is free to pivot in both the horizontal and vertical directions. However, turbulent intensity is usually related to atmospheric stability as determined from a measure of the temperature difference between two levels.

Both the fossil-fuel and fission-fuel technologies convert only about one third of the energy released to electricity. The other two thirds is dissipated as waste heat by natural or forced-draft cooling towers, by once-through cooling in rivers, lakes, or the ocean, by artificial ponds or spray canals, or by combinations of these techniques. These discharge methods are strongly affected by temperature, humidity, and rainfall. For cooling towers, gustiness is also of concern. We must, therefore, take meteorological measurements to provide the needed information for the efficient operation of heat-dissipation systems.

In the initial design of any power-generating equipment, we must decide upon the design limits for the structures. Tornadoes present a particularly crucial problem for nuclear reactors, since the uncontrolled release of radioactive material cannot be permitted. Statistical information must be available for determining the probabilities of such severe wind conditions. For this purpose, the measurement of wind speed is essential.

During the operation of any power-generating system, the load dispatcher must know at least 6 to 24 h or more in advance what the load will be. Although the load is determined primarily by temperature, the meteorological variables of wind, relative humidity, and solar radiation also enter into forecast models.

As the preceding considerations show, many meteorological variables are important for the operation and design of power-generating sources, as well as for input to models that allow us to forecast both the conditions under which operations will occur and their long-term safety. In short, meteorological measurements are of extreme importance to the power industries.

The types of meteorological systems required for the power industry vary with the nature of the power-generating technology, with the geographical and meteorological features of the site, and with the quantity and character of potential receptors. At some locations, such as the Savannah River Laboratory at Aiken, SC, just about every sensor to be discussed in this chapter has been used. In contrast, the observational requirements of many small facilities have been met by one or a few commercially available systems of the simplest variety, i.e., systems providing only relatively crude records of wind, temperature, and precipitation.

The number of locations on a site at which meteorological measurements are necessary depends on the complexity of the terrain and the estimated distances to which environmental effects would cause concern. For example, the characterization of wind-flow patterns in a hill–valley complex that materially influences these patterns may require many wind-measuring stations. The same is true near large bodies of water over which sea-breeze circulations may form, in large forested areas, and in cities. Decisions concerning the number and locations of on-site meteorological stations must depend on the judgment of a competent meteorologist working in close association with power engineers and with health and safety personnel.

Mason: Department of Atmospheric and Oceanic Science, University of Michigan, Ann Arbor, Michigan.
Moses: Division of Biomedical and Environmental Research, U. S. Department of Energy, Washington, D. C.

The operational modes of the facility should be kept in mind when the meteorological equipment for the initial site survey is selected, since it may be possible to fulfill the preoperational and operational requirements with the same instrument set. However, no attempt will be made here, and, indeed, no attempt can be made to specify the types and quantities of meteorological instruments and equipment required in particular situations. Instead, a broad survey of appropriate instrumentation is presented. In the case of wind instruments, the characteristics of the sensors are discussed in some detail, since a correct choice will depend in some measure on these features.

Because many meteorological surveys involve considerable detail and, therefore, extensive processing of the data, it is important to consider at the outset the form of recording that is most desirable for each individual problem. In some instances a large number of measurements are made routinely, especially at the larger installations, and the handling of data becomes a serious problem unless automatic techniques are adopted. Therefore a section dealing with this subject is included at the end of this chapter.

3-2 PLATFORMS FOR METEOROLOGICAL MEASUREMENTS

3-2.1 Surface-Based Platforms

When elevated measurements of wind, temperature, or relative humidity are made, a platform is needed for mounting the meteorological sensors. Such platforms can take many forms, e.g., poles, towers of various cross-sectional sizes and of different types of lattice construction, and even buildings, although the latter form is seldom recommended.

Meteorological instruments can be mounted on wood poles such as the conventional utility pole. These poles have a diameter of 30 cm or more near the ground and taper slightly upward. Wood utility poles are quite suitable for mounting anemometers to a height as great as 25 m above the ground. Such poles are often unguyed but are sunk about 2 m into the ground. Steel step bolts can be used for climbing.

A wide variety of towers is commercially available. These towers are mass-produced, are comparatively inexpensive, and are adequate for conducting a survey of several years' duration. Some are hinged, others have ladder rungs, and still others are telescoping to provide access to the instruments.

Towers of intermediate size, i.e., from about 25 to 60 m, can be furnished either with or without guying. An unguyed tower, however, costs approximately two to three times as much as one with guying. In deciding upon a tower, we must be careful to determine the stresses to which it may be subjected. For example, some towers will withstand stresses from winds of about 30 m/s; others will withstand stresses from 70-m/s winds. The probability of high winds must always be considered in selecting a suitable platform.

Very high towers, i.e., those exceeding 100 m, are used only rarely and for special purposes, e.g., the 300-m tower operated by the National Oceanic and Atmospheric Administration (NOAA) and the National Center for Atmospheric Research in Boulder, CO. It is interesting to note that, as of mid-1977, 352 television or radio towers 300 m or taller have been built in the United States. There exist 281 towers 1000 to 1499 ft tall, 60 towers 1500 to 1599 ft tall, and 11 towers 2000 ft tall or taller. Although the expenses and complications of placing meteorological sensors on these tall towers are great, there are instances that justify such an effort. A current listing of the heights and locations of towers, as well as other tall structures that may obstruct air navigation, can be obtained from the Federal Aviation Administration, Washington, D. C. The *Television Fact Book,* an annual publication prepared by Television Digest, Inc., Washington, D. C., also lists the heights and locations of television towers.

Side arms or booms extending from towers are generally about 3 m in length. The direction in which the booms extend should be selected with a knowledge of the least-frequent wind direction to minimize the tower wind-shadow effect (Moses and Daubek, 1961). The errors in measurement of wind speed and direction for sensors mounted on openwork towers and on stacks are discussed by Gill, Olsson, and Suda (1966) and by Angell and Bernstein (1976). At low speeds the tower effect can produce a substantial reduction in the measured wind. Anemometers are generally mounted some distance above the booms, and thermal elements are mounted below the booms. Humidity elements can be mounted on the booms.

Any high tower causes a deformation of the equipotential surfaces of the earth's electric field, especially if the tower is located in an unobstructed area. Because a high tower or even one about 15 m in height may invite lightning strokes, it is essential to take necessary precautions to ground the tower suitably and avoid lightning damage to the meteorological instrumentation. Many groups using a meteorological tower for the first time fail to take the necessary precautions. The resulting damage produced by direct strikes upon the tower and also by voltage surges arising from lightning strikes on power lines some distance away necessitates costly equipment repair or replacement.

3-2.2 Atmospheric Measurements with Balloons

Most meteorological observations necessary for the energy industry have been made from sensors mounted on short or tall masts or towers. Occasionally, when measurements above tower heights are necessary, balloons have been used either as the sensor or as a platform for sensors. Since balloon techniques require the services of at least one observer and frequently more than one, their use in a long-term observation program results in considerable cost.

3-2.2.1 Free Balloons. The first of three primary types of balloon systems used for obtaining meteorological information is the buoyant free balloon. In this system the horizontal motion of the rising balloon is used to infer the wind-speed profile. The balloon can carry pressure, temperature, and humidity sensors that transmit readings by radio to a ground station. A complete outline of these various techniques as practiced by the National Weather Service and most other national meteorological services can be found in the *National Weather Service Manual of Winds-Aloft Observations,* Federal Meteorological Handbook No. 5, and *Manual of Radiosonde Observations,* Federal Meteorological Handbook No. 3; these publications are available from the Superintendent of Documents, U. S. Government Printing Office, Washington, D. C.

3-2.2.2 Tethered Balloons. In the second type of balloon system, the balloon is tethered to a ground station, and the tether carries signals from sensors carried by the balloon. This system is sometimes called a wiresonde. A multiconductor nylon-coated wire with braided fiberglass covering is often used as the tether and as the means for transmitting the signals. The data can be recorded automatically by a self-balancing bridge or recorder, or manually by an observer using a balancing bridge or reading the indicators, or by other suitable electronic equipment. Atmospheric soundings using tethered systems have been made at many locations in the United States and abroad, such as those at the Oak Ridge National Laboratory (Meyers, 1952), at the Idaho National Engineering Laboratory (Humphrey, 1951), in England by Thompson (1962), and in Australia by Tapp and Lile (1978).

With the wiresonde, a balloon or blimp with a volume of several tens of cubic meters is usually used. Commercially available blimps with a volume of 17 m^3 have been found to be very satisfactory. Helium is generally used as the inflating gas in this country, although hydrogen is also suitable if appropriate safety precautions are observed during inflation. Measurements from tethered blimps at heights of 1500 m have been reported (Smith, 1961). Specially designed balloons with fins have desirable aerodynamic properties that provide greater stability and permit observations to somewhat greater heights than do conventional blimps.

The tethered-balloon technique is cumbersome and sometimes hazardous. During high winds the blimp is extremely hard to handle. Special precautions must be taken if there is danger of lightning or the buildup of static electricity. For this reason soundings are avoided beneath or near convective clouds. Ordinarily the equipment must be well grounded, and the blimp on its return to the surface must be grounded before it is touched.

3-2.2.3 Constant-Density-Level Balloons. The third type of balloon used in meteorological investigations has been designed to float at a constant density level for low-level studies (below 2 to 3 km). Such a balloon is produced by filling a sack made of an inelastic film with a lighter-than-air gas mixture that at the equilibrium altitude completely inflates the envelope to a positive superpressure. Changes in the temperature of the superpressured gas within the balloon will not result in an appreciable balloon-volume change. Thus the buoyancy force will remain relatively constant, and the balloon will respond only to ambient density changes and to the forces exerted by the three-dimensional wind field. However, the balloons used in the troposphere to date have not been capable of maintaining their height in precipitation.

Constant-density-level balloons can take a variety of shapes, but the tetrahedron has been the most commonly used shape. The balloon shown in Fig. 3.1 has a nominal volume of about 1 m^3 and is made of Du Pont Mylar with a thickness of 0.05 mm. Such balloons are called tetroons—a combination of the words tetrahedron and balloon. The tetroons are inflated to a superpressure of several tens of millibars above ambient pressure and at the floating level may have a superpressure of slightly under 100 mb. Mylar has a high stress-to-strain ratio; thus its volume changes little during its rise, i.e., a few percent for 100-mb superpressure (Fig. 3.2). The tetroon floating level can be predicted from information concerning tetroon volume as a function of superpressure, the total weight of the system, and the variation of atmospheric density as a function of height. The floating level can be varied by adding weight to or subtracting it from the system. As with the wiresonde blimps, helium is ordinarily used as the inflating gas in the United States, but hydrogen is also suitable.

Constant density-level balloons can be tracked by three different methods. The first method consists of the double-theodolite technique with a baseline exceeding 1 km. This technique suffers from the usual difficulties inherent in visual tracking, such as loss of the target in clouds or in the sun and limitations in the length of the run. The second method involves tracking with radar. In this case the balloon material can be metalized to serve as a passive reflector, or auxiliary reflectors can be used; the latter have been more satisfactory. A difficulty associated with tracking by reflection is the interference due to ground clutter when the elevation angle is low. The third and most successful technique uses a 403-MHz transponder attached to the tetroon. Both the X-band (3 cm) and the S-band (10 cm) radars can be used: the radar signal triggers the transponder to transmit its 403-MHz signal. The radar circuitry is modified to receive this signal, which is then used to provide data on the range, azimuth, and elevation. The radar-transponder system can be used to track constant-level balloons to distances of well over 100 km, even when the balloon floats at 1000 m or less.

3-2.3 Satellite Data-Collection Systems

Since the mid-sixties, satellites have been used as relays to transmit data from ground-based sensors to receivers some distance away. The government-operated Geostationary Operational Environmental Satellites (GOES) I and II are both capable of handling data from Data Collection Platforms (DCPs). The commercial firm COMSAT—

Fig. 3.1 A superpressurized constant-level tetrahedron balloon (tetroon).

GENERAL and the Canadian firm TELESAT also provide facilities for DCPs. In addition, the three geostationary satellites launched in connection with the Global Atmospheric Research Program (GARP) by Japan, Russia, and the European Community each have DCP facilities.

In the GOES system, the information from the DCP is transmitted to the satellite at UHF frequencies (~400 MHz) but is retransmitted at about 1700 MHz to the Command and Data Acquisition (CDA) Station at Wallops Island, VA, and from there by land lines to the NOAA World Weather Building, Marlow Heights, MD. As of mid-1979, there were well over 900 data-collection platforms associated with the GOES satellite which provided data for over 30 different users. A data collection system using meteor trails as reflectors (Project SNOTEL) became operational early in 1979 for the Department of Agriculture. The National Aeronautics and Space Administration (NASA) polar-orbiting satellites Landsat I and II and Nimbus 6 have also been used on an experimental basis as DCP relays; the CDA for these is at the NASA Goddard Space Flight Center in Greenbelt, MD.

The DCPs are used to sense many different types of variables. In the hydrological field, they transmit data on water temperature, stream flow, and water levels; a particularly useful application is the capability to warn of impending floods. The DCPs are also used to measure seismic activity and to provide meteorological data on temperature, humidity, and precipitation. The largest users of DCPs are the National Weather Service, the U. S. Geological Survey, the Army Corps of Engineers, and the Department of Agriculture.

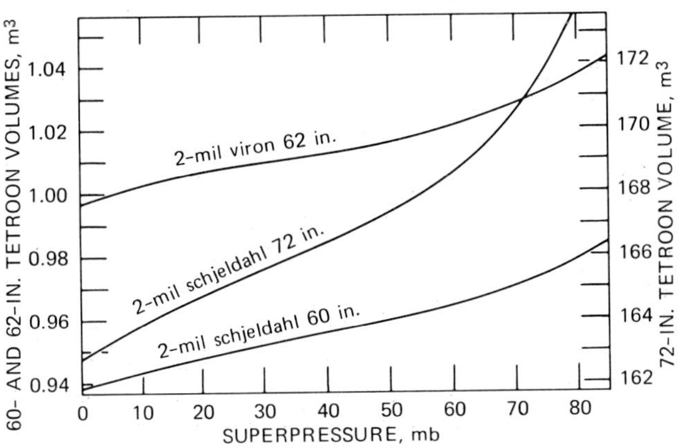

Fig. 3.2 Volume as a function of superpressure for three different tetroons.

There are many potential uses for DCPs in the power-generation industry. For example, for ocean thermal-energy converters, offshore oil drilling or offshore power generation, buoy-mounted monitors to measure meteorological and oceanographic variables would be desirable. In complex terrain, such as the Rocky Mountains or forested areas where access is difficult, a network of DCPs to measure meteorological and air-quality variables can be practicable. The individual data-collection platform can be powered by four 6-V dry cells that will last for about 6 months; the platform may also be powered by solar cells. The cost for the DCP unit is around \$3000 to \$5000 in 1979 dollars.

3-3 WIND-SPEED SENSING

3-3.1 Rotation Anemometers

Rotation anemometers are the most commonly used instruments for the measurement of wind speed. They consist of two subassemblies: the sensor (rotor) and the transducer (signal generator). The angular velocity of the sensor is usually directly proportional to the component of the wind speed (over the useful speed range of the instrument) in the direction to which the sensor responds. Large deviations from linearity may occur near the threshold speed.

Because of the rotary motion of the sensor, many different types of transducers are used: alternating-current (a-c) generators, direct-current (d-c) generators, light-beam choppers, magnetically actuated switches, and simple make–break mechanical contactors are used. Generally, the transducer selection is governed by how responsive to changes in wind speed the instrument is to be and what type of readout device is used.

3-3.1.1 Cup Anemometer. The sensor of a cup anemometer consists of a number of hemispherical or conical cups mounted symmetrically about a vertical axis of rotation; the open ends of the cups are parallel to a vertical plane passing through the rotation axis. Several types of cup anemometers are shown in Fig. 3.3.

Cup anemometers measure the horizontal component of the wind vector. Their response is almost independent of the magnitude of the vertical component as long as the wind vector itself is within ±10° of the horizontal. Beyond this, response usually drops off, but, except in unusual situations, sustained winds at these larger angles are rarely encountered.

During steady rotation of a cup anemometer, the linear speed of the wind is about two to three times as great as the linear speed of a point on the center of a cup. When Robinson (1850) wrote about the cup anemometer, he indicated that this ratio was equal to 3; later work by Brazier (1914), Patterson (1926), and others, however, indicates that this ratio varies between 2 and 3, depending upon such factors as the arm length, cup diameter, moment of inertia of the cups, ratio of cup diameter to arm length, and density of material from which the sensor is made.

Brazier (1914) describes the relation between wind speed and speed of cup center as $U = a + b\ U_c + c\ U_c^2$. In this equation, a, b, and c are constants, U is the speed of the steady wind,† and U_c is the linear speed of the cup centers. Brazier found that when the ratio d/D is made approximately 0.5, where d is the diameter of the cup and D is the diameter of the circle described by the cup centers, the constant c goes to zero and the relation is linear. Subsequent work by Spilhaus (1934) indicates that the ratio d/D is also a function of the cup size for linearity.

Originally, four cups were used. Patterson showed that a wheel of three cups is preferable because of a more uniformly applied torque around the entire revolution. Furthermore, a three-cup wheel made of the same materials produces a greater torque per unit weight and is more responsive to changes in wind speed, since the moment of inertia is smaller.

It might be expected that the maximum torque produced by a single cup occurs when the wind blows directly into the concave side of the cup. Instead, Patterson (1926) and later Brevoort and Joyner (1935) show that the maximum occurs when the concave side of the cup makes an angle of about 45° with respect to the wind direction. Figure 3.4 shows the relation between the angle of attack and the normal force coefficient for hemispherical and conical cups.

High-quality instruments that meet Weather Service and Federal Aviation Agency specifications are durable and accurate and will give years of reliable service with little maintenance. Such instruments usually use a cup wheel 38 to 45 cm in diameter from the center of rotation to the center of the cups and use conical beaded cups about 10 cm in diameter. Above speeds of 1 m/s, the rate of rotation of

†In practice, the "steady" wind speed referred to here is that observed in a well-maintained wind tunnel where most wind-sensor calibrations are performed. This steady speed fluctuates little about the mean speed, and the mean speed itself varies little over successive time intervals. The wind-tunnel type of steady flow is that represented by U in this and following equations.

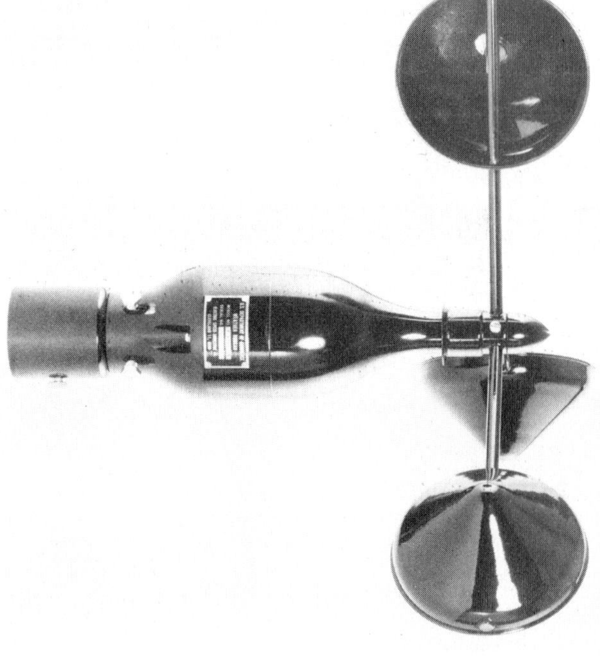

(a)

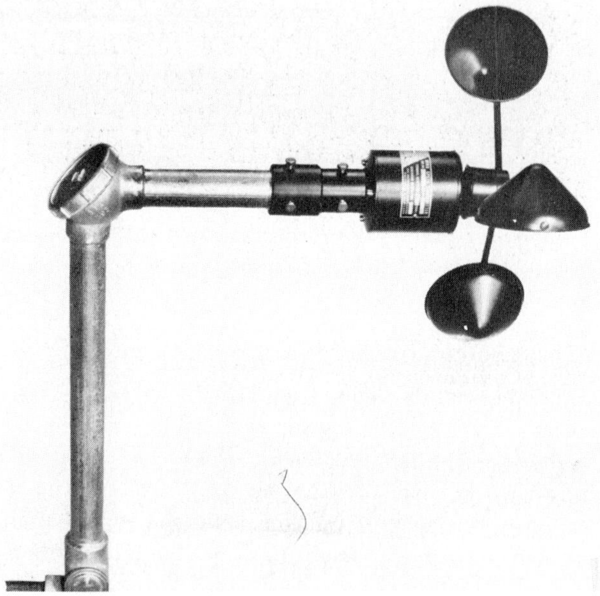

(b)

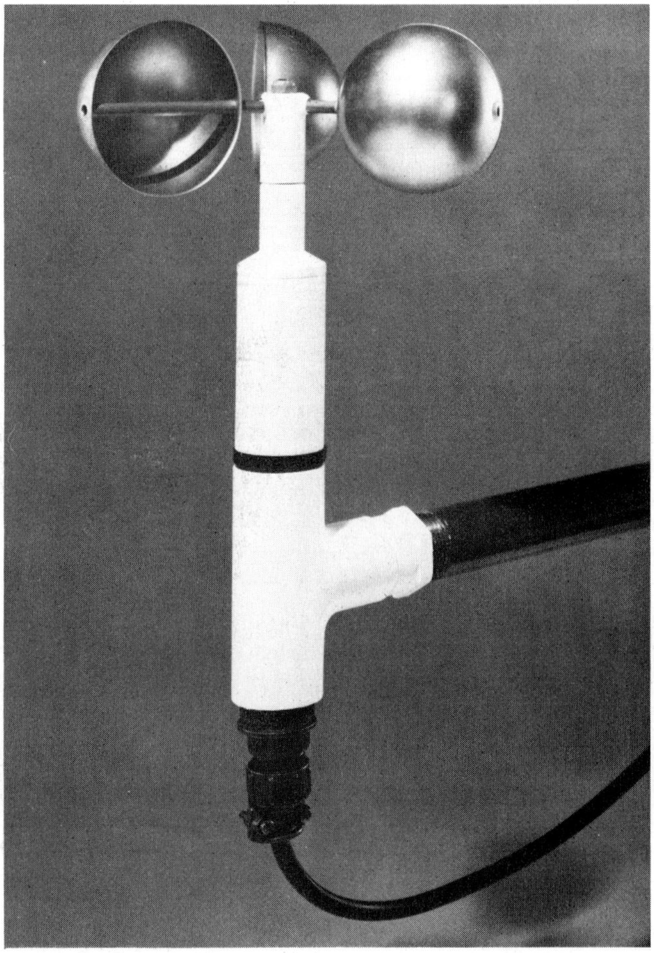

(c)

(d)

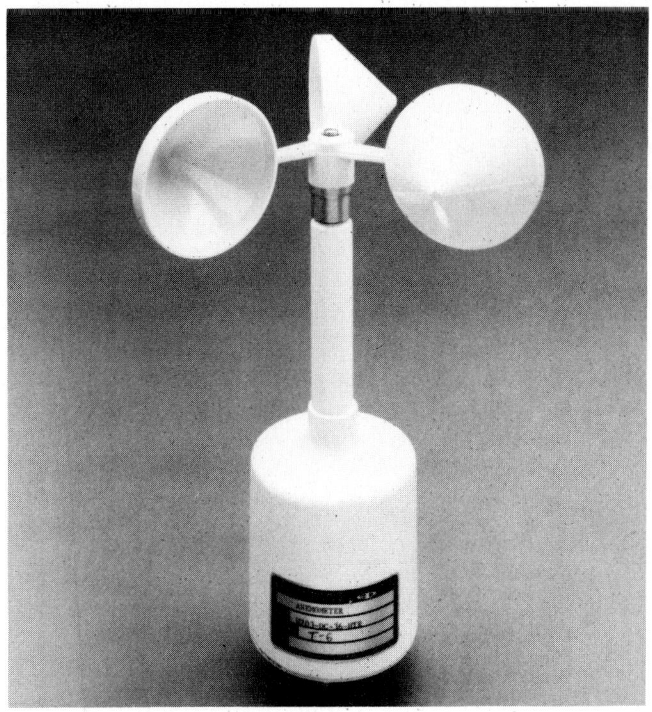

(f)

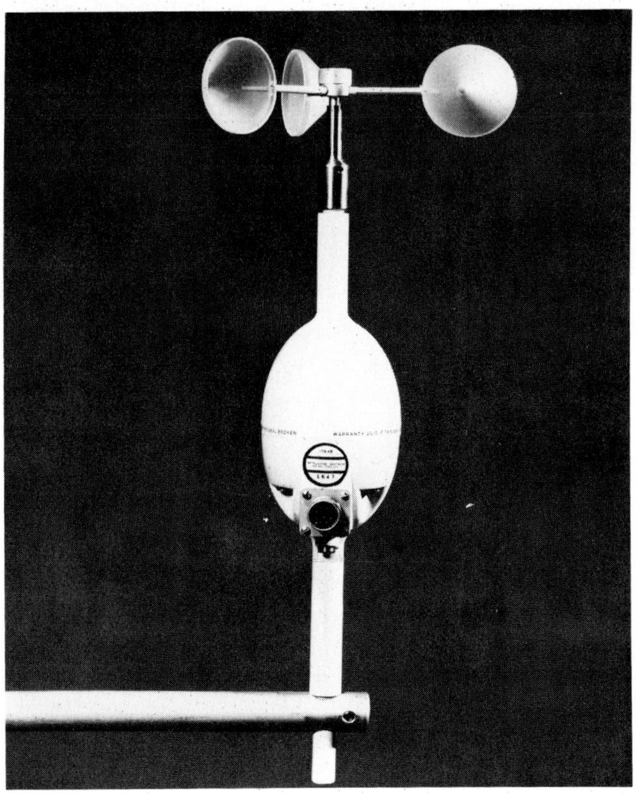

(e)

Fig. 3.3 Cup anemometers. (a) Electric Speed Indicator Co., Inc., (b) Belfort Instrument Co., (c) R. M. Young Co., (d) Sierra-Misco Inc., (e) Teledyne Geotech, (f) WeatherMeasure Corp.

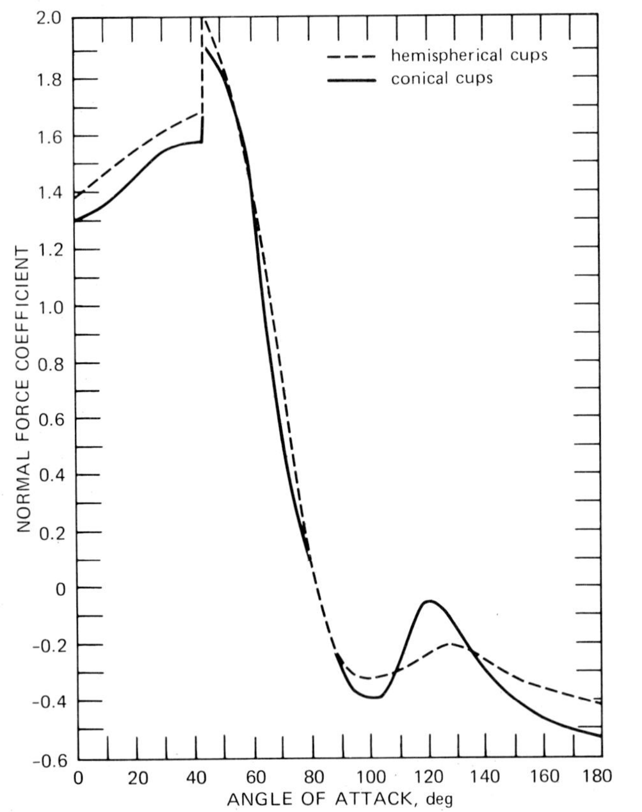

Fig. 3.4 The normal force coefficient as a function of angle of attack for hemispherical and conical anemometer cups at a Reynolds number of 8×10^4. (From M. J. Brevoort and U. T. Joyner, Technical Report 513, National Advisory Committee for Aeronautics, Lander Memorial Aeronautical Laboratory, 1935.)

the cups is essentially linear with wind speeds up to about 30 m/s. An anemometer of this type is shown in Fig. 3.3.

For accurate wind measurements at low speeds and for more responsiveness, small three-cup anemometers have been developed. These have a cup wheel diameter of about 15 cm and a cup diameter of 5 cm. The cups are made of either plastic or very thin aluminum. The moment of inertia of such an anemometer is considerably smaller than that of the standard Weather Service or Federal Aviation Agency type; to keep the moment of inertia small, a light-beam chopper is sometimes used as a transducer. In addition, to minimize friction, miniature ball bearings support the shaft assembly. Several lightweight cup anemometers are shown in Fig. 3.3.

3-3.1.2 Propeller Anemometer. The propeller anemometer utilizes a propeller or windmill as a speed sensor; the sensor rotates on a pivoted shaft that is turned into the wind by a vane. Several such anemometers are pictured in Fig. 3.5.

Several current versions of these anemometers use propellers based on a modified helicoid. The helicoid was invented by Dines (1887) but saw little use until 1940, when the Aerovane (Fig. 3.6) came into production. The helicoid is a primary sensor: its rate of rotation depends on only the wind speed and its pitch. Gill (1973) points out that it is not necessary to calibrate the sensor in the wind tunnel by comparison with any other sensor. He also shows that over a fluid density range greater than 10^4 (from water at 1.00 g cm^{-3} to rarefied air at 5×10^{-5} g cm^{-3}), the same calibration factor prevails. Of course, wind tunnel testing is necessary to determine deviations from linearity at low wind speeds when the transducer load and bearing friction represent a significant fraction of the developed torque. Above this point, however, the response is linear.

For the most part, the response of a helicoid to off-axis winds follows a cosine law, i.e., $U_i = U \cos \alpha$, where U_i is the indicated wind speed, U the steady speed, and α the angle between the axis of rotation and the wind vector—the angle of attack. In practice, the modifications to the pure helicoid which are required for a durable propeller (thicker hub and web, rounded edges and corners) sometimes affect the cosine-response characteristic. Figure 3.7 compares the response of a particular helicoid anemometer to the cosine curve. Note that the largest deviations of approximately 10% occur at about ±50° and again at about ±130°. As Gill (1973) notes, attempts to improve the cosine-response characteristic of this unit by altering blade number and contour have not produced any marked improvement.

A propeller anemometer for measuring the vertical component of the wind is shown in Fig. 3.8. The sensor is mounted on a fixed vertical shaft. Because the propeller can reverse its direction, the output signal from a suitable transducer (d-c generator) not only is proportional to the magnitude of the vertical component of the wind flow, but at the same time a polarity reversal indicates whether this component is directed upward or downward.

3-3.1.3 The uvw Anemometer. As mentioned previously, the response of a helicoid to off-axis winds is very nearly a true cosine curve. Consequently the helicoid can be used as a device to measure wind vector components. For example, a pair of helicoid anemometers 90° apart in a horizontal plane form an anemometer array to measure the horizontal wind vector (both magnitude and direction). The addition of a third sensor at right angles to the first two extends the unit's capabilities to measure the total wind vector (Gill, 1975). Such a device, called a uvw anemometer, is shown in Fig. 3.9. The supporting structure for the three anemometers interferes with the smooth flow of wind over the sensors, particularly when the wind is coming from the back. In this case, even more marked deviations from the cosine response are produced owing to the induced turbulence. A careful calibration of the entire unit permits the magnitude of these errors to be known and, further, allows corrections to be made to the indicated response of each unit to compensate. Horst (1973) and McBean (1972) describe two such correction schemes.

The wind vector is reconstructed from the corrected indicated response of each sensor. For any extensive data set, computing facilities are required owing to the large number of computations involved. An on-site microprocessor or real-time access to a minicomputer is perhaps best,

(a) (b) (c)

Fig. 3.5 Propeller anemometers. (R. M. Young Co.)

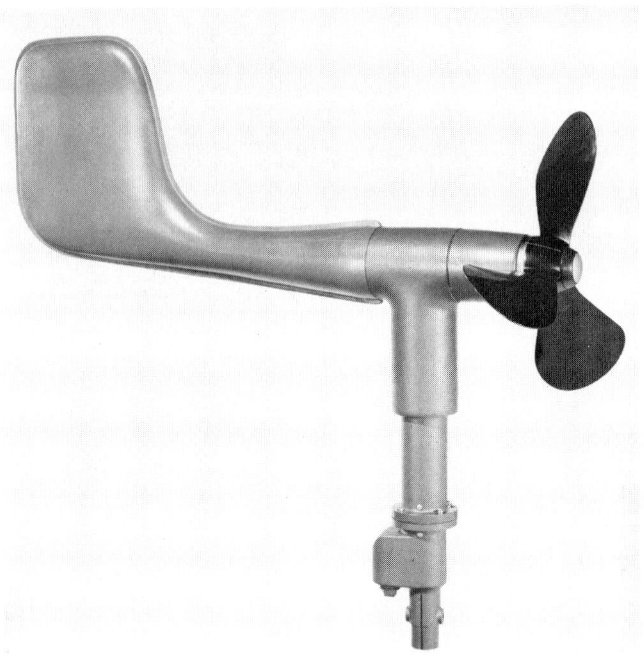

Fig. 3.6 An Aerovane wind speed and direction sensor. The propeller is a modified helicoid. (Belfort Instrument Co.)

Fig. 3.8 Vertical wind-speed sensors. (R. M. Young Co.)

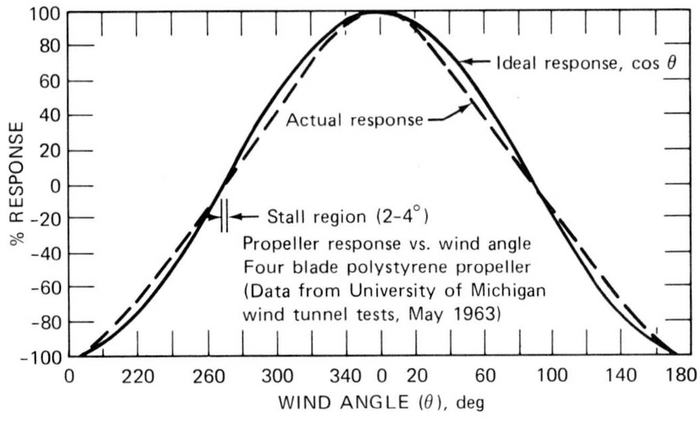

Fig. 3.7 The response of a helicoid propeller to angle of attack. The actual propeller response is compared to an ideal cosine-response curve. [From G. C. Gill, The Helicoid Anemometer, *Atmosphere*, 11: 145-155 (1973).]

Fig. 3.9 A uvw anemometer to measure the three wind-velocity components simultaneously. (Courtesy, R. M. Young Co.)

since immediate displays of wind speed, azimuth, and elevation are then possible.

Finally, note that the three sensors of a uvw anemometer are spatially separated; for the unit pictured (Fig. 3.9), the straight-line hub-to-hub distance is 60 cm. Consequently the three velocity components are not measured at the same point in space. But to the extent that the wind field can be regarded as homogeneous over the dimensions of the anemometer, the interpretation of the output as the prevailing wind vector is valid.

3-3.1.4 Transducers. Many of the rotation anemometers previously described are coupled to small permanent-magnet direct-current (d-c) generators that are connected electrically by two-conductor cables to indicating meters of the D'Arsonval type. The output of such a generator is linearly proportional to the turning rate of the cup wheel or propeller. This transducer has low brush and bearing friction and needs routine servicing two or three times per year. Brush and commutator maintenance is usually a minor problem; however, the friction is great enough to cause the starting speed to be in the range of 1 to 2 m/s. If the D'Arsonval galvanometer drives the inked pen or stylus of a chart recorder, a continuous trace of the varying wind speed can be obtained.

Some manufacturers use alternating-current (a-c) generators instead of d-c generators. Because a-c generators do not have brushes or commutators, maintenance of these items is eliminated and friction is reduced. Two-, four-, six-, or eight-pole permanent-magnet rotors are used, the larger number giving a correspondingly larger number of a-c pulses per revolution of the sensor. A frequency counter can be used to count these pulses, or they must be rectified for suitable recording.

Another class of transducers utilizes the rotation of the anemometer to activate switches or relays. There are three general types of transducers in this category: mechanical contacts similar to telephone relays, light-beam interrupters, and magnet-actuated switches. In the mechanical-contact type, a rotating shaft is coupled through one or more gears to an eccentric cam that opens and closes a mechanical switch after the passage of a predetermined amount of wind (e.g., a fraction of a kilometer); in turn, the switch operates a suitable counter or recorder. An event recorder (Fig. 3.10) with suitable circuitry can be used to provide contacts for each 10-, 100-, and 1000-switch closures and simplify data extraction.

Rotation anemometers often use the interrupted light beam-photocell technique, i.e., a photo-chopper, to reduce the starting speed and to increase responsiveness. Here, electrical impulses are generated each time a perforation in the shaft or a slotted disk attached to it rotates into a position that allows a light on one side of the shaft or disk to shine on a photocell on the other side. The shaft or disk can be drilled with one or many such holes. When only a few electrical impulses are generated for each rotation of the shaft, counters or recorders are usually used for totaling. When many pulses originate with each revolution of the shaft, i.e., six or more, it is convenient to feed them to a frequency-counting circuit that produces an output voltage directly proportional to the speed of the rotation of the sensor. Recording this voltage provides a continuous record of the wind speed. Power must be supplied to the anemometer's light source and provisions made for its easy replacement. However, starting speeds in the range 0.1 to 0.2 m/s are attainable with this technique for helicoid anemometers.

Fig. 3.10 An event recorder. (Courtesy, Esterline–Angus Instrument Co.)

3-3.2 Hot-Wire Anemometers

The speed of air moving over a fine heated wire can be measured by the cooling effects it produces. The hot wire can range from 0.003 to 0.01 cm in diameter and from 0.1 to 10 cm in length. Various conductors may be used, e.g., platinum, Nichrome, or tungsten. The wire is usually heated to temperatures of from several hundred to well over 1000°C to make the response independent of ambient-air temperature. The basic equations describing the theory of the hot-wire anemometer stem from the work of King (1914). Three modes of operation are possible: constant voltage, constant current, and constant power (i.e., temperature). For example, an approximate equation relating electric current and wind speed at a constant wire temperature is

$$i^2 = i_0^2 + K(U)^{1/2} \qquad (3.1)$$

where U = steady wind speed
i = electric current in the wire at wind-speed U
i_0 = electric current in the wire when U = 0
K = an experimentally determined constant

In this mode, the power to the circuit is regulated electronically to maintain a Kelvin bridge in balance as the wind speed varies. As the preceding equation shows, the response of a hot-wire anemometer is manifestly nonlinear; consequently its sensitivity is a function of the wind speed. However, by an adroit choice of the operating point, maximum sensitivity can be achieved over the range of wind speeds important to the particular experiment. Hot-wire anemometers can respond rapidly to fluctuations in wind speed; commercial units are available with a frequency range of up to 50 to 60 kHz. With suitable electronic circuitry, the turbulence spectra can be obtained automatically. Thus important variables of wind speed, such as the root-mean-square velocity, the correlation coefficient between two anemometers placed at different positions, or even lag correlations can be derived.

3-3.3 Thermocouple Anemometers

Another technique for measuring wind speed uses three thermojunctions in a T-formation, as shown in Fig. 3.11.

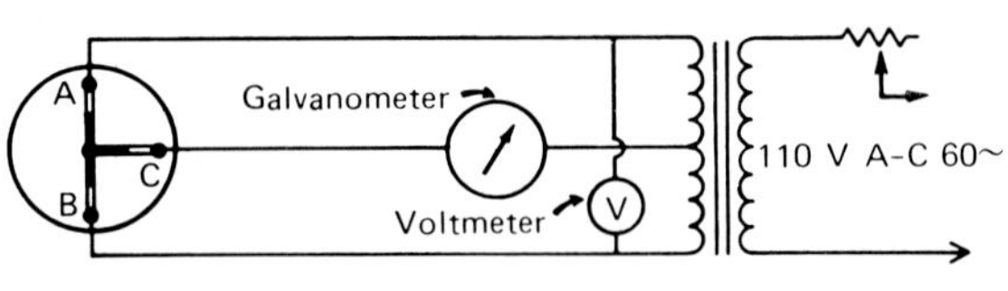

Fig. 3.11 Schematic drawing of a heated thermocouple anemometer. Thermojunctions are labeled A, B, and C.

The thermocouple formed by junctions A and B faces into the wind; junction C is a temperature-compensating junction. The thermojunctions use fine wires of nickel—chromium (chromel) and copper—nickel (constantan) alloys. The thermocouple is heated by alternating current provided by a transformer; thermojunction C is unheated. The millivoltmeter, V, measures only direct current because the transformer is center-tapped, and the resistances of the heated junctions are equal. An anemometer of this type has been used by Cramer, Gill, and Record (1957).

3-3.4 Pressure-Plate Anemometers

The force exerted upon a plate oriented perpendicularly to the wind direction is proportional to the square of the wind speed. Such pressure-plate anemometers have been used for detailed studies of the wind (Sherlock and Stout, 1931). Although such instruments offer the potential of high-frequency response and conceptual simplicity, they suffer from the difficulties of a nonlinear drag coefficient and the necessity of orientation into the wind.

If a rotation anemometer is prevented from rotating, it develops a torque proportional to the square of the wind speed. A measurement of this torque therefore provides a measure of the wind speed. An instrument based on this principle is called a bridled-cup anemometer. Because of its insensitivity and inaccuracy at low speeds, as well as its distorted recording of gust speeds, it is rarely used.

3-3.5 Pressure-Tube Anemometers

The response of the pressure-tube anemometer is based on the Bernoulli principle. Static and dynamic pressure holes lead to a float manometer that measures the pressure difference between them and transforms this difference into a linear displacement recordable on a chart wound about a clock-driven drum. Although the pressure difference is proportional to the square of the wind speed, the float manometer is designed to provide a linear relation between pen displacement and wind speed. Standard electrical transducers are also available to provide wind-speed readings based on the static and dynamic pressure difference. The pressure-tube anemometer must be oriented into the wind. A more familiar version of this sensor, the pitot tube, is often used as the reference to measure speeds in wind tunnels.

3-3.6 Anemometer Distance Constant and Dynamic Response Characteristics

Owing to the moment of inertia and friction, a cup or propeller anemometer has appreciable lag, i.e., peaks and lulls in the wind speed are attenuated for frequencies greater than one cycle per several seconds. Schubauer and Adams (1954) have shown that the lag characteristics of this anemometer can be expressed by a length called the distance constant. If an instantaneous change in wind speed occurs, this constant represents the length of the wind column that must pass the anemometer before it indicates $(1 - 1/e)$, or 63% of the change. For example, an anemometer can be placed in a 10-m/s wind, but its rotational rate is artificially retarded to correspond to a 9-m/s reading. When the retarding force is removed, the distance constant represents the distance the wind travels before the anemometer indicates a value of 9.63 m/s, i.e., 63% of the change. For all anemometers, the distance constant is independent of the wind speed. Values of the distance constant range from about 0.7 to 8.0 m for commercially available anemometers (Table 3.1).

The dynamic response characteristics of a cup or propeller anemometer, neglecting bearing friction, can be represented by the differential equation for a first-order system

$$\tau \frac{dU_i}{dt} + U_i = F(t) \qquad (3.2)$$

where τ = the time constant [i.e., the time for the system to respond to $(1 - 1/e)$, or 63% of a step change]

TABLE 3.1 Anemometer Distance Constants

Manufacturer and instrument	Model	Distance constant, m
Belfort Instrument Co.		
3 cup	Type M	7.6
Electric Speed Indicator Co.		
3 cup	Type F-420C	7.9
R. M. Young Co.		
3 cup	Model 12101	2.7
4-blade, 9-in. propeller	Model 27105	0.94
2-blade, 9-in. propeller	Model 27106	0.73
Propeller vane	Model 05105	3.3
Sierra–Misco Inc.		
3 cup	No. 1005	2.4
Teledyne Geotech		
3 cup (polycarbonate)	Series 1500	1.5
3 cup (stainless steel)	Series 1500	2.4

U_i = the indicated wind speed
t = time
F(t) = a time-dependent forcing function

For a step-forcing function, assume that an anemometer is held fixed and is suddenly released in a steady wind of U m/s. Its response can be described by

$$\tau \frac{dU_i}{dt} = U - U_i \tag{3.3}$$

The solution of Eq. 3.3 can be written

$$U_i = U\left[1 - \exp\left(-\frac{t}{\tau}\right)\right] \tag{3.4}$$

Let $L = U\tau$ and $x = Ut$; then

$$U_i = U\left[1 - \exp\left(-\frac{x}{L}\right)\right] \tag{3.5}$$

We can easily see that L is the distance constant or the length of travel of the wind before the anemometer attains 63% of a step change. Experiments by Schubauer and Adams (1954) and by MacCready and Jex (1964) show that the distance constant is independent of wind speed for cup and propeller anemometers over the usual range of atmospheric speeds. Since $L = U\tau$, τ must be inversely proportional to wind speed. Schubauer and Adams give the relation

$$\tau = \frac{Ik^2}{\rho Uc} \tag{3.6}$$

where ρ = the density of the air
I = the moment of inertia of the anemometer
c = an effective drag coefficient appearing in the formula for the torque T exerted by an anemometer as shown by the relation $T = cU^2Ar/2$, where A is the blade or cup area and r is an effective radius
k = the anemometer factor U_c/U, where U_c is the linear speed of the center of the cup

Thus

$$L = U\frac{Ik^2}{c}\frac{1}{\rho}\frac{1}{U} = \frac{K}{\rho} \tag{3.7}$$

or

$$L\rho = K = \text{constant} \tag{3.8}$$

where K depends on only the mechanical properties of the anemometer. The distance constant is therefore inversely proportional to the air density and directly proportional to the moment of inertia of the sensor.

If the forcing function is taken as

$$F(t) = U + A\sin(bt) \tag{3.9}$$

where U = steady wind speed
A = amplitude of an imposed sinusoidal variation
f = frequency in hertz
$b = 2\pi f$

then Eq. 3.2 can be transformed to

$$\tau\frac{dU_i}{dt} + U_i = U + A\sin bt \tag{3.10}$$

Neglecting transient effects, we can write a particular solution of Eq. 3.10 as

$$U_i = U + \frac{A}{(1 + b^2\tau^2)^{1/2}}\sin(bt - \beta) \tag{3.11}$$

where β is a phase-lag angle and is equal to arc tan(bt). According to Eq. 3.11, the anemometer reading will show an amplitude reduced by a factor $(1 + b^2\tau^2)^{-1/2}$ and will lag a sinusoidally varying wind by an angle β.

Since $b = 2\pi U/\lambda$, where λ is the gust wavelength and $L = U\tau$, Eq. 3.11 can be written

$$U_i = U + \frac{A}{[1 + (2\pi L/\lambda)^2]^{1/2}}\sin(bt - \beta) \tag{3.12}$$

Figure 3.12 shows the percentage of attenuation for a sinusoidally varying wind as a function of λ/L.

It should be emphasized that the preceding analysis is only approximate, since equations for both the propeller and cup anemometers are much more complex than indicated (see MacCready, 1966, and Corcoran and Esau, 1964).

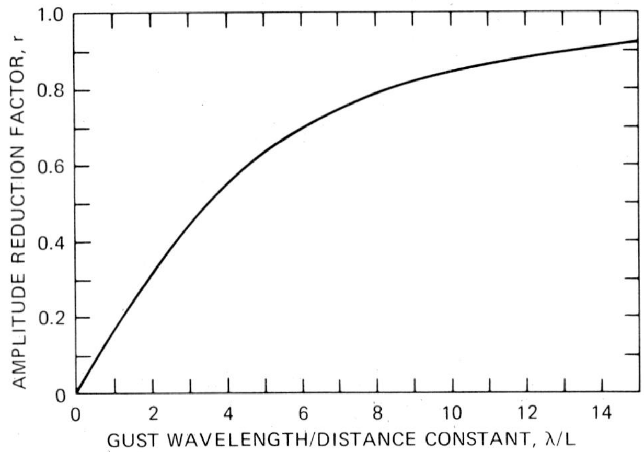

Fig. 3.12 Cup anemometer response to a sinusoidally varying wind speed as a function of λ/L. The amplitude reduction factor $r = [1 + (2\pi L/\lambda)^2]^{-1/2}$ from Eq. 3.12. (From G. B. Schubauer and G. H. Adams, *Lag of Anemometers,* Bureau of Standards, Washington, D. C., 1954, unpublished.)

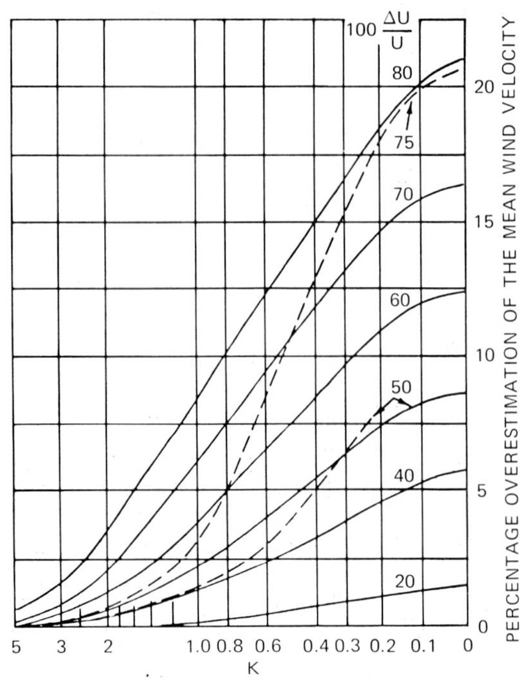

Fig. 3.13 The overestimation of the Sheppard anemometer (– – –) compared to Schrenk's (1929) results (——) for an anemometer with hemispherical cups for various values of K (see Eq. 3.13). The amplitude of the speed fluctuations as a percentage of the mean speed (100 ΔU/U) is indicated. [From E. L. Deacon, The Over-Estimation Error of Cup Anemometers in Fluctuation Winds, *J. Sci. Instrum.,* 28: 231-234 (1951).]

3-3.7 Errors in Wind-Speed Measurements

There are many sources of error in wind-speed measurements. Improper instrument siting is responsible for some of these. For example, an anemometer mounted on a short mast located on the roof of a building will often indicate higher than true wind speeds. Even if an anemometer is properly mounted on a boom from a tower, the wind-shadow effect (Moses and Daubek, 1961) will often cause the indicated wind speed to be other than the true speed. A number of errors are inherent in the instrument itself. These errors arise from its response characteristics, misalignment of the rotational axis, nonlinear response, and atmospheric density variations.

In Sec. 3-3.6, it was shown that the instantaneous anemometer reading lags behind the true reading for both a step-function change and a sinusoidally fluctuating wind. In addition, under the latter condition, the indicated amplitude is attenuated. Because some anemometers speed up faster than they decelerate, the indicated mean wind speeds may be slightly higher than the true speeds. Schrenk (1929), Deacon (1951), and Schubauer and Adams (1954) discuss this error. Figure 3.13 shows the errors that may be expected in a fluctuating wind. The K-values are plotted on the abscissa, where K is defined as

$$K = \frac{A\rho\, TUa^2\, r^2}{4I} \qquad (3.13)$$

where A = a constant
ρ = air density
T = period of fluctuations
U = steady wind speed
a = cup radius
r = arm radius
I = moment of inertia of the cup rotor

Deacon used the value of 2.7 for the quantity A. For an air density of 1.2×10^{-3} g/cm^3, U expressed in meters per second, and T in seconds, Deacon finds that K = 0.031 UT for a Sheppard cup anemometer. The dashed lines in Fig. 3.13 are the curves for a Sheppard anemometer, a type frequently used in this country.

Ower (1949) made an analysis for a propeller anemometer to determine the expected errors from wind fluctuations of about 1 Hz. He assumed that $U_i = U(1 + A \sin bt)$ and derived the relation

$$\bar{U}_i = U\left(1 + \frac{A^2}{2}\right) \qquad (3.14)$$

where $\bar{U}_i$ is the indicated mean wind speed, U is the true steady wind speed, and A is an amplitude parameter. Note that the percentage error is 100 $A^2/2$. MacCready (1966) indicates that this error is ordinarily quite small, about 2.0% or less.

The effects of turbulence on the measurement of horizontal mean wind speeds were studied by MacCready (1966). He noted that various horizontal speed sensors respond differently to vertical fluctuations of the real wind. Part (a) of Fig. 3.14 shows the vertical-angle effect for several representative mechanical sensors. Since the true horizontal wind speed is V cos ϕ, where V is the total three-dimensional wind speed and ϕ is the angular departure of the total wind direction from the horizontal, it is seen that the conventional cup anemometer can give unrepresentative readings. As discussed in Sec. 3-3.1.2, the response of a helicoid to off-axis winds is nearly a cosine curve; this

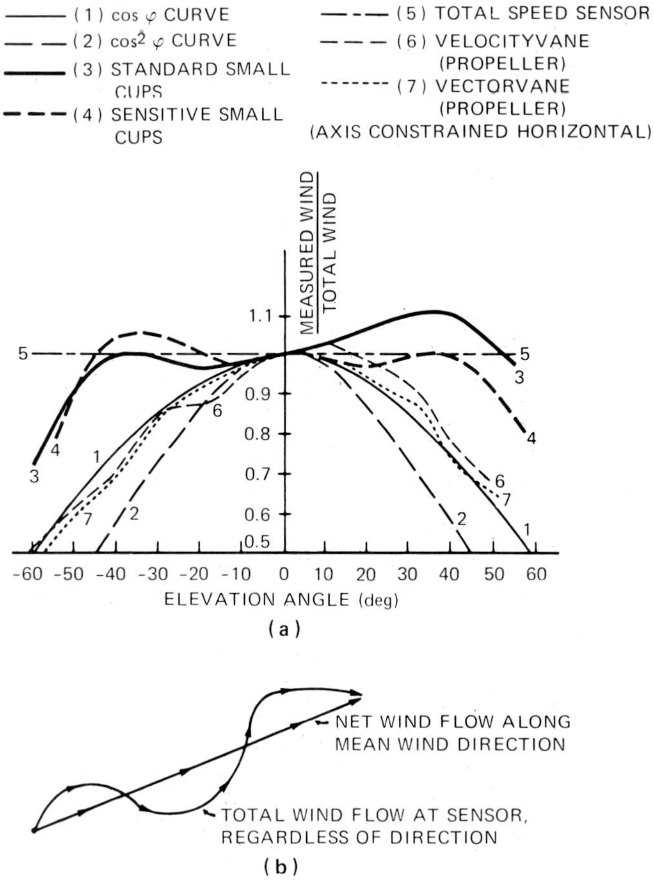

Fig. 3.14 (a) The response of different wind-speed sensors as a function of vertical wind-direction angle. (b) An example of a data-processing error, i.e., the error in the measurement of the net wind flow in the direction of the mean wind with a sensor that measures the total wind flow. [From P. B. MacCready, Jr., Mean Wind Speed Measurements in Turbulence, *J. Appl. Meteorol.*, 5: 219-225 (1966).]

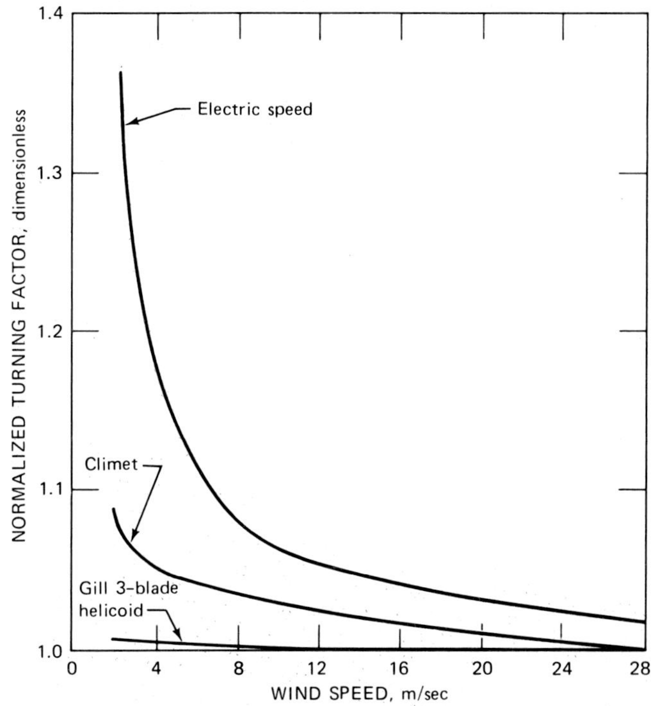

Fig. 3.15 Normalized turning factors for different anemometers. [From H. W. Baynton, Errors in Wind Run Estimates from Rotational Anemometers, *Bull. Am. Meteorol. Soc.*, 57: 1127-1130(1966).]

type of response is shown by the VectorVane and the VelocityVane. From data collected by Swanson and Cramer (1965), MacCready calculated the error between the true and indicated horizontal wind speeds; he found the error to be about 6% for cup anemometers and negligibly small for a cosine propeller anemometer.

MacCready further discussed the error due to the inability of the vane-mounted propeller anemometer to follow faithfully the horizontal fluctuations of the wind. Again using the Swanson and Cramer data, he found that this error amounted to about 1.0%. A third type of error discussed by MacCready was that arising from along-wind turbulent fluctuations. A brief evaluation indicated that this error could become increasingly important with typical sensors as the height of measurement approached ground level.

Recently, Baynton (1976) investigated the response of both cup and propeller anemometers. He found that true helicoid anemometers obey the response equation $U = bR$, while other types obey the equation $U = a + bR$, where U is the wind speed, a and b are empirically determined coefficients, and R is the rate of rotation. Figure 3.15 shows the normalized turning factors (U/R) for various anemometers; it is obvious that each turn of some of these sensors does not correspond to a constant wind run. Baynton concludes, however, that errors introduced by ignoring this finding are generally small, except when ratios of wind speeds are being determined, as in near-surface wind profile measurements. Also, he notes that many manufacturers provide an electrical analog output signal that automatically corrects for this zero offset.

In addition to the errors introduced by the instrument, the manner in which the wind data are analyzed may introduce additional errors. For example, a totalizing anemometer, i.e., one in which the wind is integrated over a changing horizontal direction, will indicate a wind run that differs from the net run in the mean wind direction, as shown in (b) of Fig. 3.14. This error can be circumvented by recording the wind components. In one case, the magnitude of this error was found to be 4.5% (MacCready, 1966). Thus, even with good response, a cup anemometer in a strongly fluctuating wind can produce a 10% difference between the indicated speed and the horizontal component of the mean vector speed.

For the usual variation of pressure and temperature, the changes in density are generally too small to affect materially the calibration of cup or propeller anemometers. Nevertheless, we must remember that calibration does depend upon density and that density has a second-order effect on the calibration.

3-4 WIND-DIRECTION SENSING

3-4.1 Wind Vanes

Wind vanes can take many different forms, from the very ornate, as seen on some farmhouses and older public buildings, to the relatively simple counterbalanced flat plate. For meteorological purposes, flat-plate vanes are easy to construct, are simple and durable, and provide reliable data. For increased torque at low wind speeds, some splayed vanes, i.e., two rectangular plates joined at their forward vertical edges and spread out at an angle of up to 20° at their back edges, have been used. Sanuki (1950) has shown that overall performance of the splayed vane is not superior to that of a flat-plate vane. Some manufacturers use a vertical airfoil as a vane. Gill (1959) indicates that, for certain angles of attack, up to 15% more torque is produced by an airfoil vane compared to a flat-plate vane of the same size, but, because of the added mass and, therefore, a larger moment of inertia for the airfoil, the increased torque is offset by poorer dynamic performance. Various wind vanes are shown in Fig. 3.16. Suitable sheet-metal stock—i.e., stainless steel or aluminum—or expanded polystyrene are both suitable materials for vane construction. The metal vanes are more rugged but exhibit poorer dynamic response compared to the lighter, more fragile plastic-finned vanes.

*3-4.1.1 **Bivanes.*** The horizontal (azimuth) and vertical (elevation) components of the wind direction can be measured with a bivane. The bivane can take either of two forms. The first arrangement consists of two flat plates perpendicular to each other, counterbalanced, and mounted on a gimbal so that the unit can rotate both horizontally and vertically. The second type consists of a counterbalanced annular fin, either an airfoil or a flat metal sheet, with a similar gimbal arrangement. Various bivanes are shown in Fig. 3.17. In lieu of a counterweight, a propeller anemometer can be mounted on a bivane; the output from such an instrument can be used to determine all three components of the total wind vector.

3-4.2 Transducers

A direct mechanical linkage of the wind vane to a direction-indicating device is rarely used as a transducer because of its high moment of inertia and the inflexibility imposed on the location of the vane. Generally, electrical transducers are used instead.

The simplest of these transducers is a commutator system in which a rotating contactor or brush is connected to the vane shaft. This rotating brush makes contact with a series of conducting sectors (four or eight) and, with the appropriate circuitry, wind direction can be indicated (four or eight points). An improved commutator system couples the vane to two brushes fixed together as a unit and spaced about 22.5° apart. These brushes contact one or two of eight 45° conducting sectors. Each of the sectors corresponds to 45° of wind direction, i.e., one point of an eight-point compass. When both brushes make contact in one sector, the direction corresponds to one of the eight compass points, e.g., north, northeast, east. When the brushes contact two adjacent sectors simultaneously, the direction corresponds to an intermediate direction, such as north-northeast, east-northeast, or east-southeast. By this means, direction to 16 points is obtained. This system can be used to energize light bulbs for visual readout or to operate electromagnetic relays that drive recorder pens. Similar systems can be used for dividing the compass into any other number of sectors.

A frequently used transducer is a 360°-potentiometer whose sliding contact is coupled to the vane shaft. For a known voltage applied across the potentiometer winding, any angular position of the vane will correspond to a particular voltage on the sliding contact; this output voltage can be applied to a voltage-sensing readout device. The chief advantage of this system is its low moment of inertia, especially if microtorque potentiometers are used. One disadvantage of the potentiometer system is the excessive wear produced as the contact moves along the wires; however, present-day conductive-plastic potentiometers are rated for 2×10^6 revolutions, which gives them a lifetime of 2 to 3 yr under normal conditions. Another drawback to this arrangement is the discontinuity that occurs when the contact moves across the gap in the potentiometer. When the wind direction oscillates about a position corresponding to the gap, the output voltage fluctuates between zero and full value. If a recording voltmeter is used (as is frequently the case), the pen repeatedly swings from one edge of the chart to the other and causes record "painting." This difficulty can be circumvented by using two potentiometers whose sliding contacts are 180° out of phase and driven by a common gear affixed to the vane shaft. In this case, the recording voltmeter uses a 540°-wide chart scale and is connected to a voltage discriminator that switches inputs from one sliding contact to the other. When the pen reaches either the upper or lower limit of the chart record, the voltage discriminator switches to the other contact, causing the pen to move in the appropriate direction a distance corresponding to 360° of vane rotation. Since a 540° chart is used, the pen in its new position, i.e., after the 360° change, is positioned about one third of the distance from one end of the scale.

Perhaps the most trouble-free technique for recording the angular position of a vane is one that utilizes a synchro-motor system. The vane is mechanically coupled to the shaft of a synchro transmitter, which is electrically connected to a synchro receiver; the latter can be situated 100 m or more from the vane location. The receiver shaft rotates in synchronism with the shaft of the transmitter and can operate the pointer of a wind-direction-indicating dial or drive a pen of a strip-chart recorder. However, the synchro-motor rotors both have high moments of inertia which adversely affect the dynamic response of any vane connected to them. Consequently these transducers cannot be attached to fast-response sensors.

3-4.3 Dynamics of the Wind Vane

The behavior of a wind vane can be described by the differential equation for a second-order system

$$\frac{d^2\theta}{dt^2} + 2\omega_n\zeta\frac{d\theta}{dt} + \omega_n^2\theta = F(t) \quad (3.15)$$

where θ = angular displacement of the vane with respect to a fixed wind direction
ω_n = natural or undamped angular frequency of the system
ζ = damping ratio
t = time
F(t) = a time-dependent forcing function

On the basis of the results of Corcoran (1962) and Garbell (1957), this equation can be expressed as

$$\frac{d^2\theta}{dt^2} + \frac{k\rho Ar^2 U}{2I}\frac{d\theta}{dt} + \frac{k\rho ArU^2}{2I}\theta = F(t) \quad (3.16)$$

where k = an aerodynamic factor that depends on the geometry of the vane
ρ = density of the air
A = area of the vane
r = distance from the vane pivot to the vane centroid, i.e., arm length
U = steady wind speed
I = moment of inertia of the vane system

From Eqs. 3.15 and 3.16 we see that the natural frequency of the vane is

$$\omega_n = U\left(\frac{k\rho Ar}{2I}\right)^{1/2} \quad (3.17)$$

and that the damping ratio is

$$\zeta = \frac{r}{2}\left(\frac{k\rho Ar}{2I}\right)^{1/2} \quad (3.18)$$

Note that the damping ratio ζ is independent of the wind speed. From Eq. 3.17 we have

$$2\pi f_n = U\left(\frac{k\rho Ar}{2I}\right)^{1/2} \quad (3.19)$$

where f_n is the natural vane frequency. If we introduce the natural wavelength λ_n defined by $f_n\lambda_n = U$, we have

$$2\pi f_n\lambda_n = 2\pi U = \lambda_n U\left(\frac{k\rho Ar}{2I}\right)^{1/2} \quad (3.20)$$

or

$$\lambda_n = 2\pi\left(\frac{2I}{k\rho Ar}\right)^{1/2} \quad (3.21)$$

Thus λ_n is also independent of the wind speed and is a useful parameter. Table 3.2 shows values of λ_n and ζ for some frequently used wind vanes.

The dynamic characteristics of a wind vane are ordinarily determined by observing its behavior in a wind tunnel. A method frequently used is to displace the vane an angle θ_0 from the mean wind direction and then to release it. The resulting oscillations are recorded on a fast-response chart recorder as an output voltage proportional to the vane displacement. This procedure allows us to determine the parameters of Eq. 3.15, ζ and ω_n, as described below.

If we take F(t) = 0 in Eq. 3.15, i.e., no forcing function is present, the equation will have three characteristic solutions depending on whether the vane is overdamped ($\zeta > 1$), critically damped ($\zeta = 1$), or underdamped ($\zeta < 1$). Assuming boundary conditions of $\theta = \theta_0$ and $d\theta/dt = 0$ at t = 0, we have for $\zeta > 1$

$$\theta = \theta_0 e^{-\zeta\omega_n t}\left[\cosh \omega_n(\zeta^2 - 1)^{1/2} t + \frac{\zeta}{(\zeta^2 - 1)^{1/2}} \sinh \omega_n(\zeta^2 - 1)^{1/2} t\right] \quad (3.22)$$

For $\zeta = 1$, we have

$$\theta = \theta_0 e^{-\omega_n t}(1 + \omega_n t) \quad (3.23)$$

and for $\zeta < 1$, we have

$$\theta = \theta_0 \frac{1}{(1 - \zeta^2)^{1/2}} e^{-\omega_n\zeta t} \cos\left[\omega_n(1 - \zeta^2)^{1/2} t - \beta\right] \quad (3.24)$$

where

$$\beta = \arctan \frac{\zeta}{(1 - \zeta^2)^{1/2}} \quad (3.25)$$

It is customary to let

$$\omega = \omega_n(1 - \zeta^2)^{1/2} \quad (3.26)$$

when ω is the observed angular frequency. Then by substitution we have

$$\theta = \theta_0 \frac{1}{(1 - \zeta^2)^{1/2}} \exp\left[\frac{-\omega\zeta}{(1 - \zeta^2)^{1/2}} t\right] \cos(\omega t - \beta) \quad (3.27)$$

All wind vanes have a damping ratio $\zeta < 1$; in fact, it is virtually impossible to construct a practical vane with a damping ratio $\zeta > 0.7$, owing to the lack of a strong atmospheric viscous damping force. Therefore Eq. 3.27 is the solution of interest; as expected, for the case of underdamped motion, it describes an oscillatory motion with a period of $2\pi/\omega$.

Figure 3.18 shows a typical record of a vane test. If the ratio of successive peak amplitudes is measured, the damping ratio ζ can be determined; and, if the angular

(a)

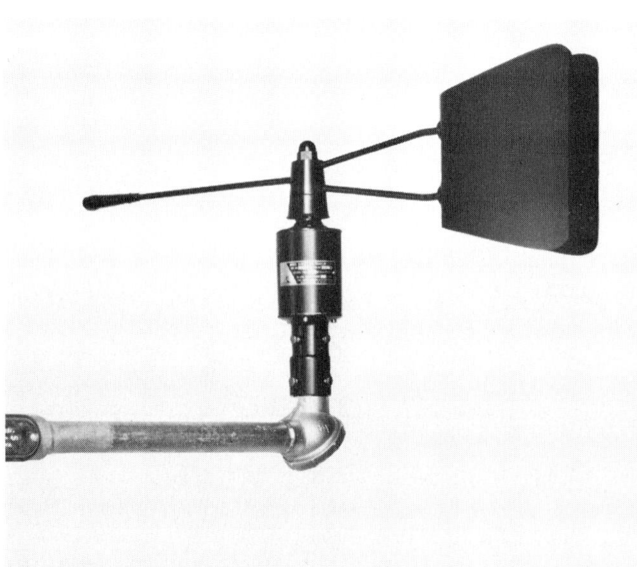

(b)

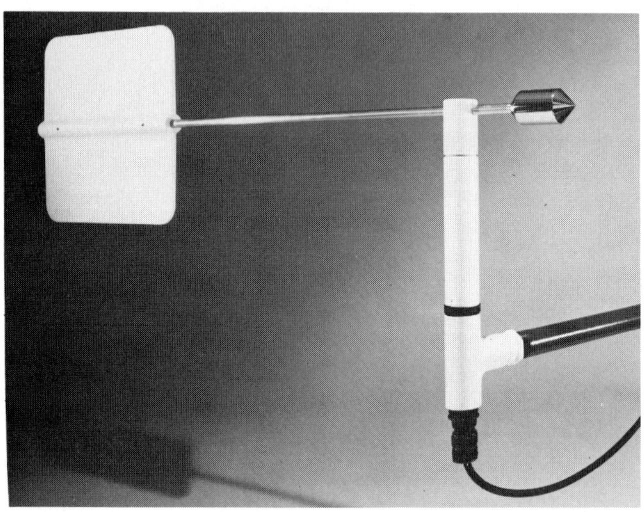

(c)

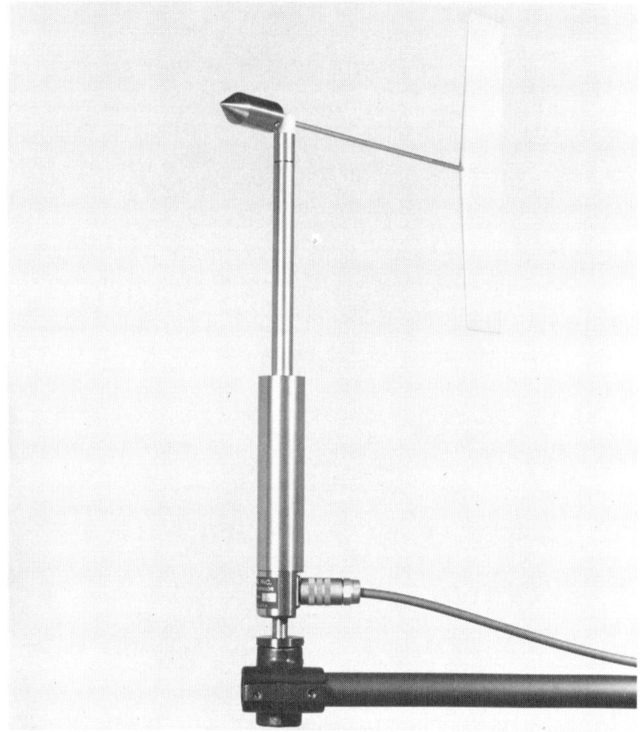

(d)

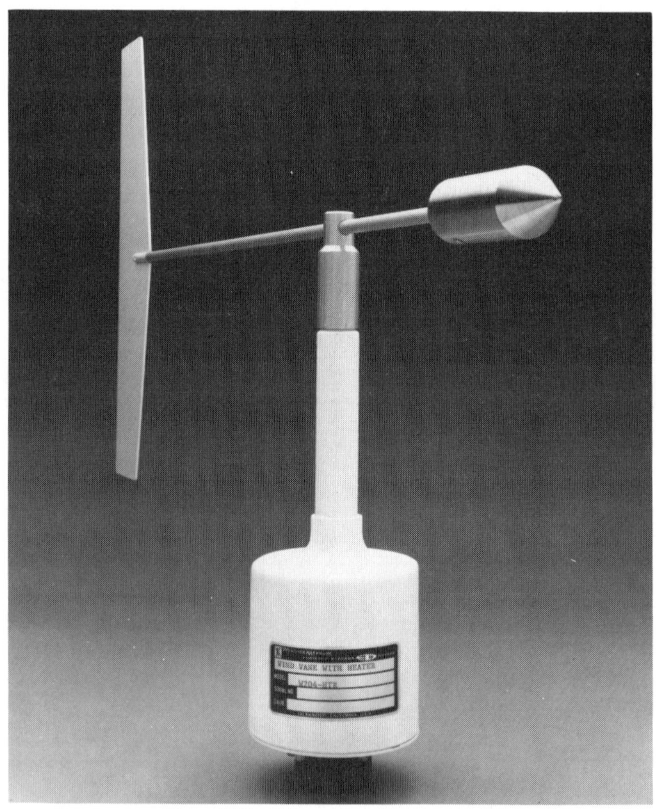

(e)

Fig. 3.16 (See facing page for legend.)

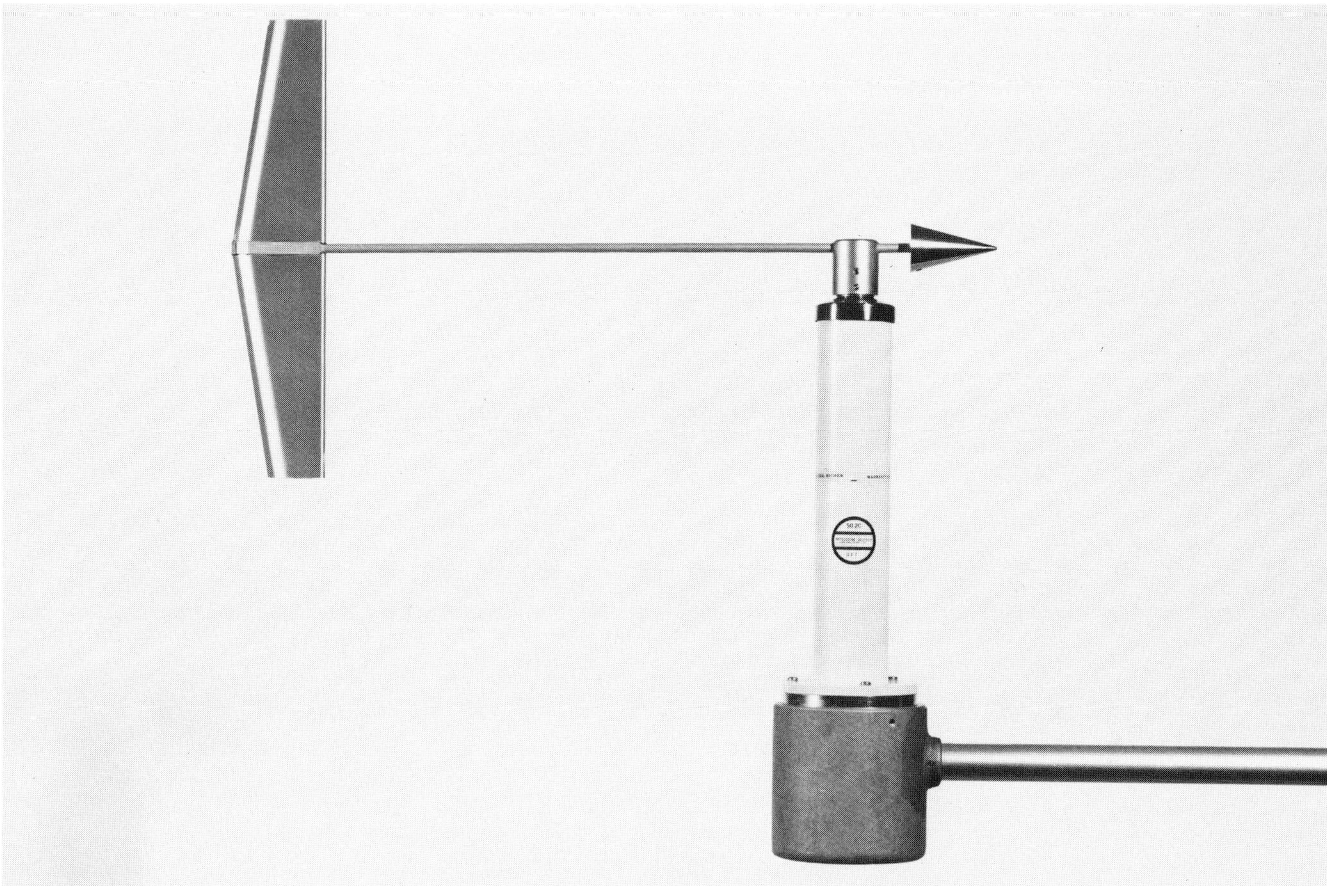

(f)

Fig. 3.16 Wind vanes. (a) Electric Speed Indicator Co., Inc., (b) Belfort Instrument Co., (c) R. M. Young Co., (d) Met One, Inc., (e) WeatherMeasure Corp., (f) Teledyne Geotech.

(a) (b)

Fig. 3.17 Bidirectional (bivane) wind direction sensors. (a) Met One, Inc., (b) R. M. Young Co.

TABLE 3.2 Wind Vane Parameters

Manufacturer and instrument	Model	Damping ratio (ζ)	Natural wavelength (λ_n), m
Belfort Instrument Co.			
Splayed vane	Type M	0.20	9.5
Electric Speed Indicator Co.			
Splayed vane	Type F-420C	0.14	17.7
Met One, Inc.			
Vane	Model 020	0.4–0.6	
R. M. Young Co.			
Microvane	Model 12301	0.44	4.8
Bivane	Model 17002	0.53	4.1
Propeller vane	Model 05101	0.30	
Propeller vane	Model 35001	0.47	5.2
Teledyne Geotech			
Quick One	Series 1500	0.60	
Quick Two	Series 1500	0.40	
Standard	Series 1500	0.20	

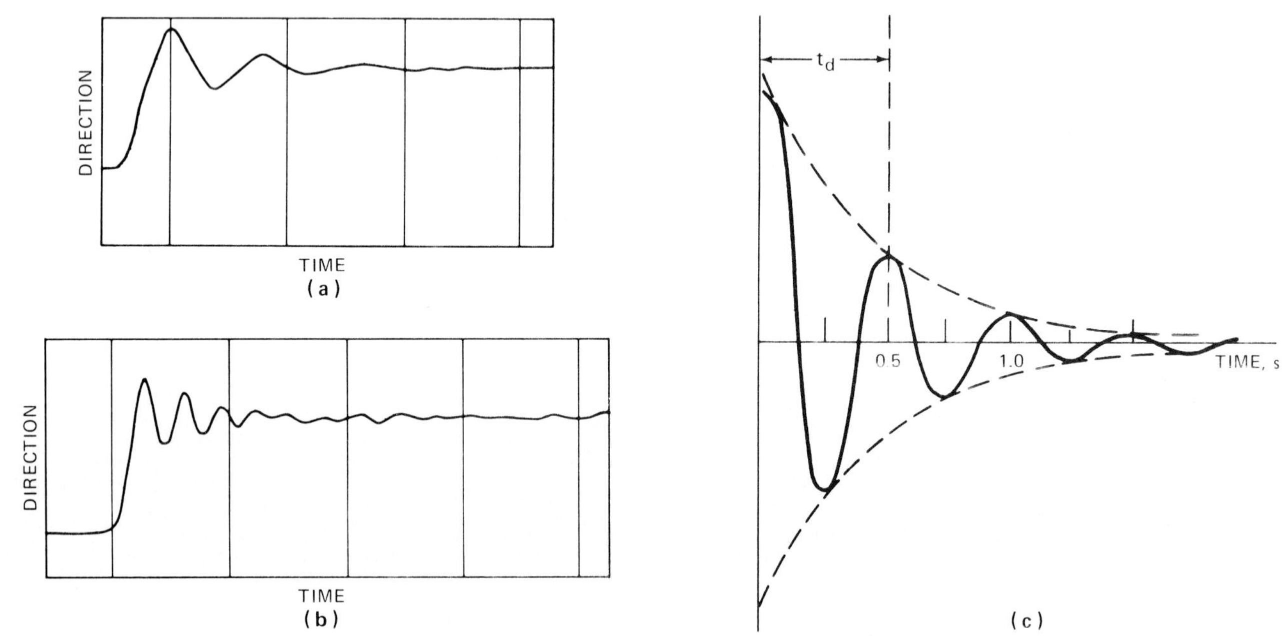

Fig. 3.18 Response of a wind vane to a step-function direction change. Time markers are at 1-s intervals. (a) Displacement, 45°; wind speed, 4.5 m/s. (b) Displacement, 45°; wind speed, 9.0 m/s. (Courtesy of Beckman & Whitley, Inc.) (c) Response of a wind vane initially held off-axis and released in a steady wind; the period of oscillation is t_d and, in this case, is 0.5 s.

frequency ω is measured, the natural wavelength λ_n can be found. In practice, it is easiest to determine the ratio of the amplitude of the first overshoot on the opposite side (θ_{opp}) to the amplitude of the initial displacement (θ_0), especially for highly responsive vanes whose motion is quickly damped. Since $t = (\beta + \pi)/\omega$ at the first overshoot on the opposite side, Eq. 3.27 can be used to show that, for this case

$$\frac{\theta_{opp}}{\theta_0} = \exp\left[\frac{-\pi\zeta}{(1-\zeta^2)^{1/2}}\right]$$

from which

$$\zeta = \frac{\ln(\theta_0/\theta_{opp})}{[\pi^2 + \ln^2(\theta_0/\theta_{opp})]^{1/2}} \tag{3.28}$$

in agreement with MacCready and Jex (1964). The angular frequency ω is determined by a direct measurement of the oscillatory period as recorded, i.e., $\omega = 2\pi/t_d$ where t_d is the time interval between successive maxima or one half the time interval between the initial displacement and the first overshoot on the opposite side. Then the natural wavelength can be found by combining Eqs. 3.20 and 3.26, namely,

$$\lambda_n = 2\pi \frac{U}{\omega_n} = 2\pi \frac{U(1-\zeta^2)^{1/2}}{\omega} \tag{3.29}$$

When the second-order system of Eq. 3.15 has a sinusoidal forcing function impressed on it, i.e.,

$$\frac{d^2\theta}{\omega_n^2\, dt^2} + \frac{2\zeta}{\omega_n}\frac{d\theta}{dt} + \theta = A \sin bt$$

the general solution can be expressed as the sum of the complementary and particular solutions. Thus

$$\begin{aligned}\theta &= C_1 \exp[-\omega_n\zeta - \omega_n(\zeta^2-1)^{1/2}t] \\ &\quad + C_2 \exp[-\omega_n\zeta + \omega_n(\zeta^2-1)^{1/2}t] \\ &\quad + \frac{A \sin(bt-\beta)}{\{4\zeta^2(b^2/\omega_n^2) + [1-(b^2/\omega_n^2)]^2\}^{1/2}}\end{aligned} \tag{3.30}$$

where

$$\beta = \arctan\left(\frac{2\zeta\,\omega_n\, b}{\omega_n^2 - b^2}\right) \tag{3.31}$$

When $\zeta < 1$, the contribution of the first two terms will become negligibly small in a short time, and the motion can be described by

$$\theta = \frac{A \sin(bt-\beta)}{\{4\zeta^2(b^2/\omega_n^2) + [1-(b^2/\omega_n^2)]^2\}^{1/2}} \tag{3.32}$$

The denominator of Eq. 3.32 is equal to 1/G, where G is referred to as the gain. Figure 3.19 shows the relation between the gain and the ratio of the forcing angular frequency, b, to ω_n, the natural angular frequency. When $b = \omega_n$, for ζ less than 0.5, amplification occurs, since $G = (2\zeta)^{-1} > 1$, as can be seen from Eq. 3.30. Figure 3.19 indicates that, for $\zeta = 0.6$ and $b/\omega_n < 0.8$, "faithful" response is to be expected in the sense that a gain, G, not very different from unity prevails over a wide range of frequency ratios.

The angle β is the phase lag between the forcing and response functions. This relation as a function of b/ω_n for a range of damping factors is shown in Fig. 3.20. Even for a damping factor of $\zeta = 0.6$, the phase lag is appreciable.

The foregoing analysis is based on a sinusoidally varying wind direction of a given single frequency. Real wind-direction fluctuations can be regarded as the superposition of many sine waves—each with a different frequency—acting simultaneously. Thus we might expect errors in the instantaneous indicated direction because the gain, G, as well as the phase angle, β, will be different for each frequency. Where the wind has frequency components comparable to or higher than the natural frequency of the vane, these will be attenuated as indicated in Fig. 3.19.

The preceding discussion of the dynamics of a wind vane applies to a vane without a propeller. When a sensor with a propeller is considered, such as an anemometer bivane, the analysis becomes considerably more complicated. Analyses of the dynamics of the vane with a propeller are given by MacCready and Jex (1964) and by Sanuki, Kimura, and Baba (1960).

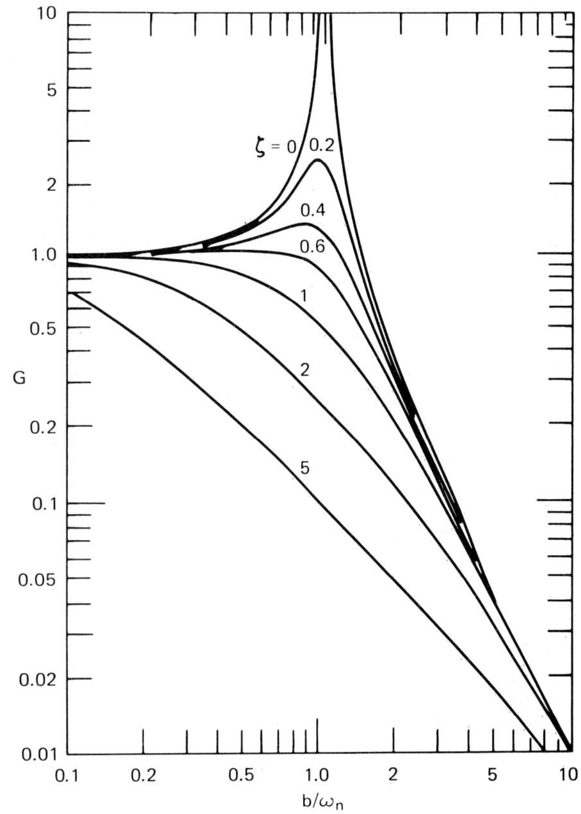

Fig. 3.19 Second-order gain characteristics. (From J. D. Trimmer, *Response of Physical Systems*, John Wiley & Sons, Inc., New York, 1950.)

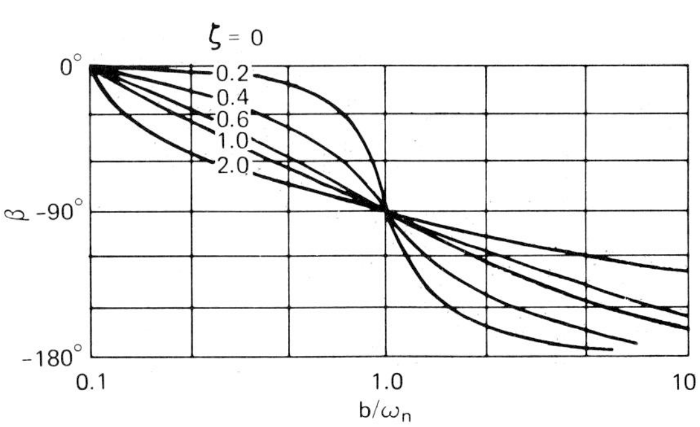

Fig. 3.20 Second-order phase characteristics (From J. D. Trimmer, *Response of Physical Systems*, John Wiley & Sons, Inc., New York, 1950.)

3-4.4 Standard-Deviation Measurements

The standard deviation of the mean wind direction σ_θ can be related to the dispersive capabilities of the atmosphere and, in particular, to the dispersion coefficients σ_y and σ_z which characterize a plume-concentration distribution. Therefore, if the dispersive capabilities of a given site are to be evaluated using the above relation or if data from different sites are to be compared, we must be able to interpret properly the wind-vane records in view of the dynamic response characteristics of the particular vane used to acquire the data.

The standard deviation of the wind-direction distribution (σ_θ or σ_ϕ) over time periods approximately equal to the duration of the release period has been used in many experiments as an index of diffusion. These quantities can be computed from high-speed analog traces from wind-direction recorders or from other analog or digital data records. With data from a vane or bivane having a damping factor appreciably less than 0.4, the resonant oscillations must be removed mathematically by averaging the data over a 10- or 20-s interval and using these averages for the computations. The most appropriate interval will depend on the characteristics of the vane in use.

A simple electronic system, popularly called a sigma meter, presents a continuous real-time record of the standard deviations of wind direction without recourse to an intermediate collection of raw wind-direction data. This instrument has been discussed by Jones and Pasquill (1959) and has been used in diffusion research studies (e.g., MacCready, Smith, and Wolf, 1961). Using the nomenclature and notation of Jones and Pasquill, we can write

$$\sigma_{T,t}^2 = \sigma_\infty^2 \int_0^\infty F(n) \left[1 - \frac{\sin^2 \pi nT}{(\pi nT)^2}\right] \frac{\sin^2 \pi nt}{(\pi nt)^2} dn \qquad (3.33)$$

where $F(n)$ = fraction of variance (σ_∞^2) occurring between n and n + dn

T = time interval of interest

t = incremental time of each sample ($t \ll T$)

$\sigma_{T,t}^2$ = the variance accounted for by the function F(n) during the period T, as sampled from periods t

If the signal is a continuous one, as is the output signal from the wind vane, $t \to 0$, and Eq. 3.33 becomes

$$\sigma_T^2 = \sigma_\infty^2 \int_0^\infty F(n) \left[1 - \frac{\sin^2 \pi nT}{(\pi nT)^2}\right] dn \qquad (3.34)$$

Equation 3.34 shows that, with an electrical filter whose characteristics approach $1 - [\sin^2 \pi nT/(\pi nT)^2]$ through which the vane signal is passed, the variance of the signal can be read directly.

The sigma meter is a high-pass resistance–capacitance filter that passes all frequencies above a set limit and attenuates frequencies below this limit in accordance with Eq. 3.34. The alternating current (across the capacitors) can be read directly by a simple voltmeter or, as is more often the case, rectified and filtered to produce a d-c voltage suitable for recording. The sigma meter itself is available from several manufacturers and can be readily adapted to most telemetered systems and to direct readout on charts or digital displays.

3-4.5 Distance Constant of a Vane

The distance constant for an anemometer was defined as the distance over which the wind travels corresponding to a (1 − 1/e) or ~63% change in indicated response to a step function. A similar quantity can be defined for the wind vane. But we must emphasize that the concept of a wind-vane distance constant is fundamentally very different from that for the anemometer. In the latter case, the distance constant appears as a natural parameter in the first-order differential equation characterizing the response of the anemometer. Because a second-order equation is required to characterize vane response, two parameters are required to specify it completely. As we have seen, the damping ratio and the natural wavelength—both of which are independent of the wind speed—are the two natural parameters that appear in the vane-response equation. Any other introduced quantity ultimately can be written as a function of these two.

To show that, for a given wind vane, the fraction of the change from the initial to final position during the first oscillation is a function of wind distance rather than speed, we can rewrite Eq. 3.24

$$\frac{\theta}{\theta_0} = \frac{1}{(1-\zeta^2)^{1/2}} e^{-\omega_n \zeta x/U} \cos\left[\omega_n (1-\zeta^2)^{1/2} \frac{x}{U} - \beta\right] \qquad (3.35)$$

since $t = x/U$, where x is the distance of wind travel in time t. From Eqs. 3.17 and 3.21, we see that

$$\omega_n = 2\pi U/\lambda_n \qquad (3.36)$$

Hence, when we substitute for ω_n in Eq. 3.35, the wind-speed U cancels out, and we have

$$\frac{\theta}{\theta_0} = \frac{1}{(1-\zeta^2)^{1/2}} e^{-2\pi\zeta x/\lambda_n} \cos\left[\frac{2\pi}{\lambda n}(1-\zeta^2)^{1/2} x - \beta\right] \qquad (3.37)$$

Since x is the only variable on the right-hand side of the equation, the distance the wind must travel for the vane to adjust to any fraction of its initial displacement is independent of wind speed. Therefore it is possible to define a distance constant corresponding to the distance the wind must travel for the vane to reach (1 − 1/e) of its initial displacement. As is evident, Eq. 3.37—the defining equation for the distance constant when $\theta/\theta_0 = (1 - 1/e)$—is a transcendental equation. For this reason, no simple expression for the vane distance constant can be written, unlike the analogous quantities for temperature sensors or anemometers. As a result, the distance constant concept is of limited use in characterizing vane performance. Normally, the damping ratio, ζ, and the natural wavelength, λ_n, are the two parameters chosen to specify vane response. Note that these two parameters appear in Eq. 3.37 above.

3-4.6 Matching of Speed and Direction Sensors

For a propeller vane or anemometer bivane, the error in measuring the wind vector can be minimized by matching the characteristics of the speed and direction sensors. This matching is done by making the phase lags of these two sensors equal. Because of phase lag, the vane or anemometer provides a reading corresponding to the wind some distance downstream. For large eddies, the relation between phase angle and frequency is approximately linear.

For an anemometer, since $\beta = \tan^{-1}$ (bt),

$$\beta \approx bt \qquad \text{for small bt} \tag{3.38}$$

and, similarly, for a vane when $b/\omega_n \ll 1$

$$\beta \approx \frac{2\zeta b}{\omega_n} \tag{3.39}$$

since $\tan \beta = 2\zeta(b/\omega_n)/[1 - (b/\omega_n)^2]$. If these two phase angles are set equal to each other, then

$$bt = \frac{2\zeta b}{\omega_n} \tag{3.40}$$

or

$$t = \frac{2\zeta}{\omega_n} \tag{3.41}$$

In terms of the anemometer distance constant,

$$L = U\tau = \frac{2U\zeta}{\omega_n} = \frac{2U\zeta}{2\pi f_n} = \frac{\lambda_n \zeta}{\pi} \tag{3.42}$$

since $U = \lambda_n f_n$ and $\omega_n = 2\pi f_n$.

The speed lag refers to the propeller, and the direction lag refers to the vane centroid. If there is an appreciable distance between the two, an additional correction equal to the distance, d, between the vane and the propeller should be applied, i.e.,

$$L = \frac{\lambda_n \zeta}{\pi} + d \tag{3.43}$$

MacCready and Jex (1964) indicated that this technique will provide a phase mismatch of less than 1° for eddies larger than five times the natural wavelength and 10° for eddies of the order of twice the natural wavelength.

3-5 TEMPERATURE MEASUREMENTS

At meteorological installations established by the power industry, two types of temperature measurements are commonly made. The first and most frequent measures the temperature of the air at about 1.5 m aboveground; this value is frequently referred to as the surface-air temperature. The second type measures the temperature of the air at various elevations within about the first 100 m to determine the thermodynamic stability of the atmosphere; often the temperature differences between levels are measured directly. Both types of measurements are also used to establish general climatologies.

A third type of temperature measurement sometimes made is that of soil temperature. A vertical soil-temperature profile can be constructed from soil-temperature measurements at various subsurface levels.

3-5.1 Exposure of Temperature-Sensing Equipment

When any object used as a thermal sensor is exposed in the atmosphere, it may reach an equilibrium temperature quite different from that of the surrounding air. The temperature it reaches is the result of (1) the radiation it absorbs from the sun, sky, earth, and surrounding objects, (2) the radiation it emits (which, in turn, depends upon its own temperature and emissivity), (3) the heat transfer between it and the ambient air by convection and conduction, and (4) the heat conduction from surrounding materials, such as electrical lead wires or supporting mounts.

Usually the thermal sensor is placed in a well-ventilated enclosure to minimize unwanted radiation effects. At installations where only air temperatures are measured, the enclosure is an instrument shelter such as that shown in Fig. 3.21. The shelter is normally mounted on an open framework to position the thermometer about 1.5 m above the surface. Typical dimensions for the enclosure itself are 73 cm long, 52 cm deep, and 83 cm high. The shelter is painted white to enhance its reflectivity to incident radiation. It usually has louvers on all four sides, a slatted

Fig. 3.21 Standard instrument shelter. Minimum (upper) and maximum (lower) thermometers and a thermograph are usually positioned as shown.

bottom, and a double roof, with a 5.0- to 7.5-cm air space between the roof panels. On one side of the shelter is a door, hinged at the bottom. The shelter is positioned so that the door faces north. With this orientation, solar radiation does not impinge on the instruments when observations are made. The shelter normally houses mercury-in-glass thermometers for measuring ambient and wet-bulb temperatures, but it can also hold a hygrothermograph and maximum and minimum thermometers. The shelter should be sited over a grass surface in an unobstructed area away from surrounding structures. It should never be placed over hard gravel or cement surfaces. Either a hand-driven or a motor-driven fan can be used to improve the ventilation before making the observation. However, in the latter case, precautions must be taken to ensure that the heat from the motor does not influence the air passing over the thermometers.

The variation of temperature with height is a measure of the stability of the atmosphere. To measure the vertical temperature structure, thermal sensors must be placed on platforms to determine the temperatures at various levels aboveground. A simple platform is a boom, 2 or 3 m in length, extending from a structure such as an open-lattice tower. (The use of rooftops or hills as platforms is inappropriate since the air temperature near such objects can be quite different from that at the same height some distance away.) Resistance thermometers or thermocouples are the most common sensors used to measure the variation of temperature with height. These sensors are normally enclosed in a radiation–ventilation shield (see Fig. 3.22) to minimize radiation effects and to allow them to come into equilibrium with the ambient air. The shield is often made of thin sheet metal, such as aluminum, and is in the form of two concentric cylinders. The outer surfaces of these cylinders are made highly reflective to minimize the heating effects of incident radiation (Fuchs and Tanner, 1965). The shield incorporates a baffle at the bottom end to block out reflected or emitted radiation from objects in the field-of-view. An umbrella-like shield is sometimes provided to protect against precipitation and, to a lesser extent, against radiation. The shield is normally mounted at the end of a boom; it may be naturally—or artificially—ventilated. In the latter case, an aspiration blower is mounted at the other end of the boom to draw in air through both the inner cylinder and the surrounding annular space. Recently the vacuum-bottle type of radiation–ventilation shield has received increasing use. This device is a double-walled glass shield with silvered interior surfaces; the dead air space between the two walls of the shield is evacuated and sealed. Tanner (1963) discusses other types of radiation shields and materials, as well as the problems related to measurement of air temperatures and temperature profiles.

Aspiration of artificially ventilated shields can be obtained by either individual blowers or a centrally located fan. On the meteorological towers at Brookhaven National Laboratory and at Battelle–Northwest Laboratories, centrally located blower systems provide the necessary ventilation, whereas at Argonne National Laboratory individual blowers are used. Individual blowers are economical, will operate for 3 or 4 yr with minimum maintenance, and are readily available commercially. The aspiration rate is set so that the airflow over the thermal sensor is about 5 m/s. At typical atmospheric wind speeds, dynamic heating is not a problem.

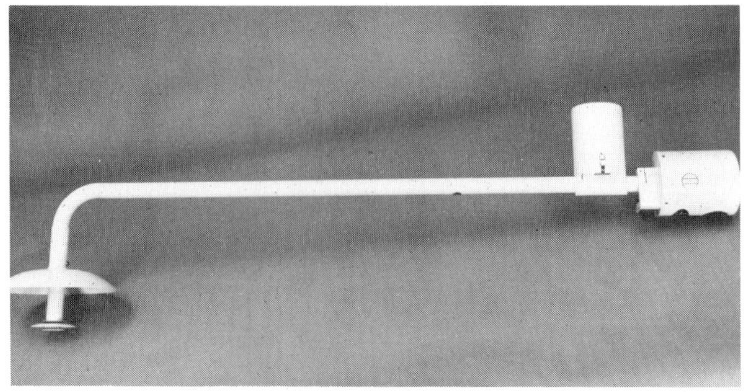

(a)

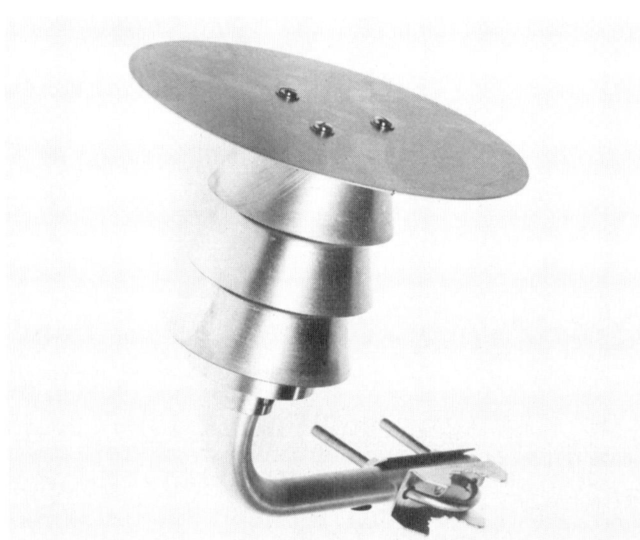

(b)

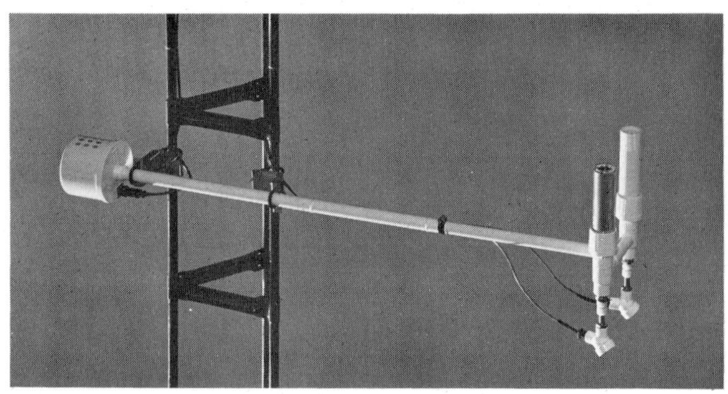

(c)

Fig. 3.22 Ventilated radiation shields for temperature and dew-point sensors. (a) Teledyne Geotech, (b) Texas Instruments, Inc., (c) R. M. Young Co.

3-5.1.1 Soil-Temperature Measurements. The World Meteorological Organization (Blanc, 1958) recommends that soil-temperature measurements be made at the following levels below the ground surface: 10, 20, 50, 100, 150, and 300 cm. When soil-temperature sensors are installed, the soil structure is usually altered. The soil above and around the sensor itself should be disturbed as little as possible so that such thermal properties as volumetric heat capacity, conductivity, and moisture content are not changed. Alteration of the soil profile can be minimized by the following procedure. With an auger, dig a hole about 1 m in diameter to a depth slightly below the lowest level of soil-temperature measurement; now drill a horizontal hole along a radius of the pit through the wall. This last hole should be about 1 m in length with a diameter of about the same size as the protective shield of the soil thermometer. Drill similar horizontal holes at each level at which soil temperatures are to be measured. The soil thermometers should fit snugly into these horizontal holes. Finally, when the large pit is refilled, make an effort to restore each soil layer to its original position and compaction. For a continuous recording of soil temperatures, resistance thermometers or thermocouples are most frequently used. These sensors should be enclosed in metallic shields welded to lead-covered cables to prevent moisture from entering the system.

3-5.2 Types of Thermal Sensors

In general, any property of a substance that changes with temperature can be used for measuring temperature. In this section, sensors based on the properties of thermal expansion, electrical resistance, and thermoelectricity will be discussed. Thermometers based on such properties as a change in the vapor pressure of a liquid with temperature or the dependence of the velocity of sound in air on temperature are used infrequently in meteorological installations and will not be considered.

3-5.2.1 Thermal-Expansion Sensors. There are three commonly used thermometers that depend upon thermal expansion: the liquid-in-glass thermometer, the bimetal thermometer, and the Bourdon tube.

3-5.2.1.1 Liquid-in-Glass Thermometers. The liquid-in-glass thermometer consists of a bulb containing a liquid, e.g., mercury or alcohol, at one end of a capillary tube. The length to which the capillary tube is filled by the liquid is a measure of the temperature of the bulb. Scale readings can be etched on the stem itself or on the mounting plate. Although the volumetric expansion of such organic liquids as ethyl alcohol is several times greater than that of mercury, mercury is preferred because thermometers using it have greater inherent precision and better repeatability. Organic liquids must be used for very low temperatures, since mercury freezes at $-38.9^\circ C$, but ethyl alcohol, for example, does not freeze until the temperature is well below $-100^\circ C$.

Although liquid-in-glass thermometers are simple to use and require little maintenance, they are subject to a number of problems, e.g., separation of the liquid in the capillary, distillation of the liquid, hysteresis, and secular changes. When the thermometer is exposed to high temperatures and then quickly placed in a cold environment, a calibration shift may occur, i.e., the zero point may change; after a few days the readings will return to normal. However, over a period of a year, a calibration change of the order of $0.1^\circ C$ is not uncommon. Therefore, if accuracy of this magnitude is required, annual calibration checks are desirable. A calibration check can be made by immersing the thermometer in a well-stirred mixture of distilled water and ice made from distilled water.

Some liquid-in-glass thermometers are calibrated as partial-immersion thermometers, i.e., they read correctly only when the bulb and a part of the stem are immersed. Only total-immersion thermometers are suitable for meteorological measurements.

One special type of liquid-in-glass thermometer is the maximum thermometer (Fig. 3.23). A maximum thermometer can be used to measure the highest temperature that occurred over a selected time interval, such as 24 h. This sensor is similar to an ordinary mercury-in-glass thermometer except for a constriction near the bulb. When the temperature rises, the liquid column remains intact; when the temperature falls, the column breaks at the constriction. Because the liquid above the constriction cannot return to the bulb, the maximum reading is retained. The ordinary fever thermometer is an example of a maximum thermometer.

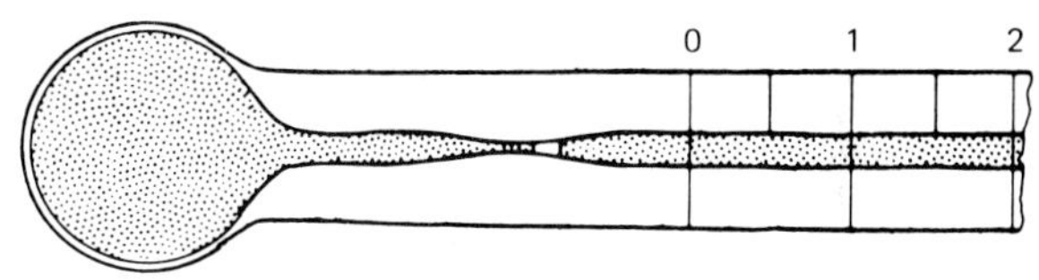

Fig. 3.23 A section of a maximum thermometer showing the bulb and the constriction in the capillary. A mounted maximum thermometer is shown in Fig. 3.21.

It is also possible to obtain minimum temperatures with a liquid-in-glass thermometer. In this case, the capillary tube contains a clear organic liquid and movable index (Fig. 3.24). When the temperature falls, the surface tension of the contracting liquid column at the end of the capillary causes the index to move toward the bulb. When the temperature rises, however, the liquid flows around the index. Thus the position of the end of the index farthest from the bulb provides a measurement of the minimum

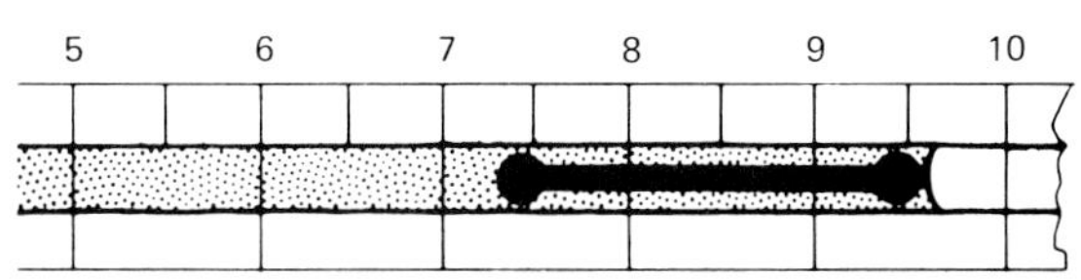

Fig. 3.24 A section of the capillary of a minimum thermometer showing the sliding index that marks the lowest temperature attained. A mounted minimum thermometer is shown in Fig. 3.21.

temperature. For reliable results, the minimum thermometer must be mounted at an angle of about 5° from the horizontal with the bulb down. Minimum and maximum thermometers mounted in a weather shelter are shown in Fig. 3.21.

3-5.2.1.2 Bimetallic Thermometers. If two dissimilar metals that have differing coefficients of thermal expansion, such as invar and brass or invar and steel, are welded together to form a single strip, this assembly will bend with a change in temperature. Consequently, if one end of the unit is fixed in position, the deflection of the other end can be used to measure ambient temperature. The welded strips can take the form of a simple strip usually about 0.1 cm thick, less than 2.0 cm wide and several centimeters long; alternately, the strip may be longer and wound into a coil or helix. The sensitivity of the strip varies inversely as its thickness, directly as the square of its total length, and directly as the difference in the expansion coefficients of the two metals.

The bimetallic sensor can withstand rough handling. As a result it is frequently used in such portable temperature-measuring instruments as the thermograph. In this last unit, a mechanical linkage relays the position changes of the sensor's free end to a pen; the pen, in turn, makes an inked trace on a chart which is mounted on a clock-driven rotating drum. A continuous record of temperature vs. time for the period corresponding to one drum revolution is obtained.

3-5.2.1.3 Bourdon-Tube Thermometers. The Bourdon tube is a thermal sensor consisting of a metallic enclosure filled with an organic liquid. The tube may be 5 cm or more in length and usually has an elliptical cross section. In operation the Bourdon tube is similar to the bimetal thermometer, in that temperature changes cause the tube to bend. One end of the tube is clamped, and the other end is attached to a pen; as before, the pen provides a temperature–time record on a clock-driven, rotating drum. The Bourdon tube is commonly used as the temperature sensor in commercial thermographs (Fig. 3.25). Although this apparatus is subject to some sources of error, such as a calibration change owing to polymerization of the organic liquid, inability to withstand rough handling, or exposure to temperatures above its design range, it has nevertheless proved to be a successful design. Such sensors in thermographs can operate for many years with only routine maintenance.

3-5.2.2 Resistance-Change Sensors. There are two types of sensors that detect temperature changes by a change in the resistance of an electrically conducting medium. The first sensor uses a metal as the conducting medium; the second, a semiconductor compound of metallic oxides.

3-5.2.2.1 Metal-Resistance Thermometers. The resistance of such metals as platinum, nickel, copper, silver, or iron increases with a rise in temperature. This property can be used to determine the temperature of the element and, in turn, the temperature of the medium in which it is immersed. The relation between resistivity and temperature can be expressed as

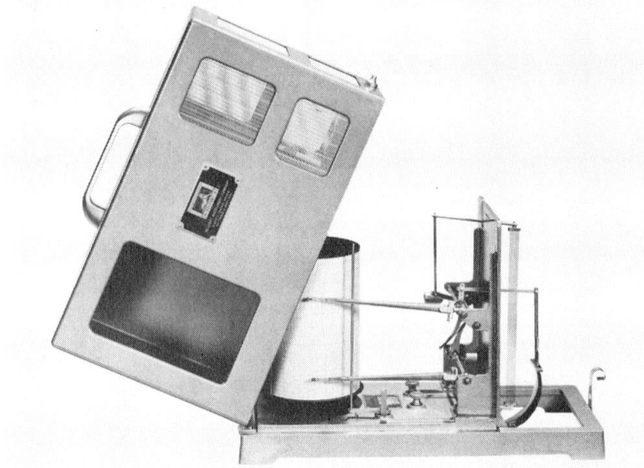

Fig. 3.25 A hygrothermograph. (Belfort Instrument Co.)

$$R = R_0 [1 + a(T - T_0) + b(T - T_0)^2] \qquad (3.44)$$

where R is the resistance at temperature T, R_0 is the resistance at the reference temperature T_0, and a and b are calibration constants. Both copper and platinum show a fairly linear relation between temperature and resistance. Therefore the coefficient b is quite small, and these two metals are frequently used.

Resistance thermometers can take several different physical forms. Often the sensor is a coil with enough wire to have a resistance of about 100 ohms at room temperature. For example, if copper is used as a 100-ohm element, a 1.0°C change in temperature will provide an approximately 0.4-ohm change in resistance. Other resistance thermometers of various sizes and resistance values are commercially available.

A number of precautions must be observed if resistance thermometers are used. Because the sensitivity of the thermometer is low, contact resistances between electrical connections must be kept small. Changes in the temperature of the lead wire are usually compensated for by making the lead wire part of the bridge. Finally, the current passing through the wires must be kept small, so that self-heating effects are minimized.

A resistance thermometer is generally used with a self-balancing Wheatstone bridge and a strip-chart recorder. With ordinary precautions, all these instruments will give years of reliable service, although calibration checks every 6 months to a year are necessary.

3-5.2.2.2 Thermistor Thermometers. The second type of resistance thermometer is the thermistor; this is made from a mixture of metallic oxides compressed together and heat-treated. Thermistors are available as rods, flakes, beads, or wafers of various sizes. This sensor shows a change of resistance with temperature that is one or two orders of magnitude higher than that of metallic resistance thermometers and in the opposite direction, i.e., as the temperature increases, its resistance decreases. The relation between resistance and temperature for the thermistor can be expressed as

$$R = a \exp\left(\frac{b}{T}\right) \tag{3.45}$$

or in a differential form

$$\frac{1}{R}\frac{dR}{dT} = -\frac{b}{T^2} \tag{3.46}$$

In spite of the advantage of a large resistance change with temperature, these sensors have two deficiencies. First, the temperature-vs.-resistance relation is extremely nonlinear, as Eqs. 3.45 and 3.46 show; a sensor with such a response curve does not exhibit constant sensitivity to identical temperature changes at different points within its measurement range. Second, thermistor stability is not as long-lived as that of metallic resistors. As a result, these sensors are satisfactory for short-term, specialized studies where calibrations are frequently checked; however, attempts to use them for long-term climatological measurements may lead to measurement errors.

The problems associated with the nonlinear response characteristic of a thermistor can be circumvented by the use of a composite device. This device consists of two precision thermistors encapsulated as a single three-terminal component which, in conjunction with two external resistors, exhibits a linear change of resistance with temperature. Alternately, the device can be used to produce an output voltage that varies linearly with temperature. These units are available commercially.

3-5.2.3 Thermoelectric Sensors. When two dissimilar metals, such as copper and constantan (approximately 60% copper and 40% nickel with small amounts of manganese, iron, and carbon), are brought together as shown schematically in Fig. 3.26 and the two junctions are at different temperatures, a current will flow. For a given pair of metals, the relation between the potential difference ΔE and the temperature difference $(T_A - T_B)$ between the two junctions can be expressed as

$$\Delta E = c_1(T_A - T_B) + c_2(T_A - T_B)^2 \tag{3.47}$$

where T is temperature, c_1 and c_2 are coefficients, and the subscripts A and B refer to the junctions A and B. For copper—constantan at room temperature, about 40 μV/°C is generated. Wire size determines the response time of the junction to a change in temperature. Thermopiles (several thermojunctions in series) are often used to increase the sensitivity, i.e., the potential difference per degree of temperature difference.

The output of a thermocouple is often measured by a self-balancing recording potentiometer. Except for precision work, a cold junction is not necessary; instead, a compensating cold-junction spool is used in the recording potentiometer. This spool consists of a reference thermojunction embedded in a resistance coil designed to compensate for changes in ambient temperature.

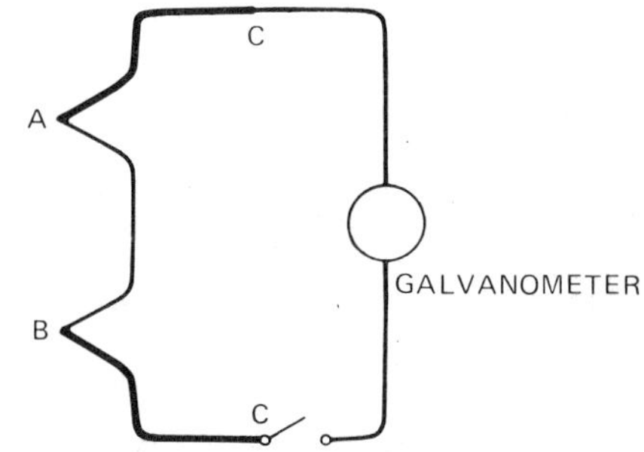

Fig. 3.26 A thermocouple. In operation, the junctions A and B are at different temperatures, one of which is a known reference temperature. If neither metal is copper, the junctions C should be at the same temperature.

The installation of a temperature-measuring system using thermocouples has many pitfalls, and assistance should be sought from those who have had previous experience in this area. For example, it is important to guard against induction currents from nearby ac sources: thermojunctions can act as rectifiers and produce dc potentials that are added to those generated by the temperature difference between the junctions. Errors of as much as 2°C may be introduced in this manner, and only proper electrical shielding can eliminate them. It is also important to minimize the number of connecting junctions to avoid spurious contact potentials. Finally, during damp weather, rain, or snow, wet thermojunctions may produce spurious voltages arising from galvanic action.

3-5.3 Dynamic Response Characteristics of Thermometers

Thermal sensors are generally considered first-order systems because their response can be characterized by the first-order linear differential equation

$$\tau \frac{dT}{dt} + T = f(t) \tag{3.48}$$

where τ = time constant
T = indicated temperature
t = time
f(t) = time-dependent forcing function

The dynamic response of such a thermal sensor to a change in temperature is analogous to that of the cup or propeller anemometer to a change in wind speed.

When a thermal element at an initial temperature T_0 is suddenly immersed in an environment of temperature T_e, the response is described by the differential equation

$$\tau \frac{dT}{dt} + T = T_e \tag{3.49}$$

whose solution is

$$\frac{T - T_e}{T_0 - T_e} = \exp\left(-\frac{t}{\tau}\right) \tag{3.50}$$

At time $t = \tau$, approximately 63% of the change has already taken place.

The time constant, τ, is a function of the product of air density and wind speed raised to a power n. This relation can be expressed as

$$\tau = \frac{K}{(\rho U)^n} \tag{3.51}$$

where ρ is the air density and U is the steady wind speed. The exponent n has a value of about ½. At very high wind speeds, the exponent may approach the value of 1 (Peterson and Womack, 1937). Moreover, Middleton and Spilhaus (1953) point out that for still air the time constant for rising temperatures is slightly larger than for falling temperatures, but, even with a little ventilation, this difference becomes small.

The time constant for a mercury-in-glass thermometer with a spherical bulb of about 1-cm diameter is 56 s for a ventilation rate of 5 m/s (Great Britain, Meteorological Office, 1956). In addition to wind speed, the time constant depends on other factors, such as the conductivity of the glass and mercury or other organic material, the specific heat and mass of the bulb and the liquid, and the geometric shape and roughness of the bulb. The latter two factors influence the turbulence and consequently the heat transfer from the sensor to the surrounding atmosphere.

When the temperature changes at a constant rate, the temperature indicated by the sensor will lag behind the actual temperature. If we assume that the temperature of the environment is changing at a rate of α degrees per second, i.e.,

$$T_e = T_i + \alpha t \tag{3.52}$$

where T_i is the initial temperature of the environment, then the differential equation describing this condition can be written

$$\tau \frac{dT}{dt} + T = T_i + \alpha t \tag{3.53}$$

The solution of this equation is

$$T - T_e = -\alpha\tau + (\alpha\tau)\, e^{-t/\tau} \tag{3.54}$$

Thus for a period of time that is large compared to τ, the thermometer will lag behind the actual temperature by a difference equal to $\alpha\tau$ degrees.

Suppose the environment undergoes a sinusoidal temperature fluctuation described by

$$f(t) = T_i + A \sin\left(\frac{2\pi}{P}\right) t \tag{3.55}$$

where A is the amplitude of the fluctuation and P is its period. The differential equation for the response of the sensor is

$$\tau \frac{dT}{dt} + T = T_i + A \sin\left(\frac{2\pi}{P}\right) t \tag{3.56}$$

and its solution is

$$T = T_i + C\, e^{-T/\tau} + \frac{A}{[1 + (2\pi\tau/P)^2]^{1/2}} \sin\left(\frac{2\pi}{P} t - \beta\right) \tag{3.57}$$

where T_i is the initial temperature, C is a constant, and β is the phase angle, given by

$$\beta = \arctan \frac{2\pi\tau}{P} \tag{3.58}$$

Table 3.3 shows the relative attenuation of sinusoidal temperature fluctuations for various ratios of fluctuation period to time constant. The phase shift is also given.

Middleton and Spilhaus (1953) recommend that the thermal sensor have a lag greater than 30 s for ordinary uses. At a power-station site, the 30-s time constant is quite acceptable. Special studies, such as those to determine heat flux in which the covariance of vertical wind and temperature fluctuations must be determined, require fast response sensors with time constants of 1 s or less.

3-5.4 Calibration and Testing of Thermometers

Equipment for the immersion testing and calibration of liquid-in-glass and electrical thermometers is available commercially. A thermometer calibrated by the National Bureau of Standards for use as a secondary standard is essential for carrying out thermometer calibrations.

The calibration procedure compares the sensor to be calibrated with the secondary standard. The comparison is effected by immersing both sensors in a large vessel (15 to 20 L) filled with a nonflammable liquid of low volatility, such as one of the silicone oils. The temperature of the bath is varied, and the indicated temperatures are read at selected points in the range of interest. A thermostatically controlled electric immersion heater is usually suitable for heating the bath to about 45°C, and either dry ice or cooling coils containing a circulating brine solution can be used for cooling the bath to subzero temperatures. Ordinarily this range of temperatures is satisfactory for all meteorological conditions normally encountered. Care must be taken to ensure that the bath temperature is at equilibrium during the comparison of the two thermometers. A sparkless stirring motor is used to provide uniform temperatures throughout the vessel; its stirrer is enclosed in a metal case with suitable openings to allow liquid flow. If temperature comparisons are made for temperatures 20°C above or below room temperature, the vessel should be thermally insulated to reduce heat transfer through the walls. Water can be used as the bath liquid for temperature calibrations above 0°C.

TABLE 3.3 Thermometer Response Characteristics†

Ratio of fluctuation period to lag coefficient (P/τ)	0.2	0.4	0.6	0.8	1.0	2.0	4.0	6.0	10.0
Ratio of response to time amplitude	0.04	0.06	0.09	0.13	0.16	0.31	0.54	0.69	0.85
Phase lag (β), deg	88	86	85	83	81	72	58	46	30

†Great Britain, Meteorological Office, Instruments for Surface Observations, in *Handbook of Meteorological Instruments, Part I*, p. 92, Her Majesty's Stationery Office, London, England, 1956.

There are a number of pitfalls in the calibration of thermocouples or resistance thermometers. For example, if the temperature of the bath differs from that of the room, a sufficient length of the lead wire should be immersed in the bath to prevent heat conduction through the wire and a consequent erroneous temperature indication. If a thermopile not enclosed in a protective cover is to be calibrated, the electrical conductivity of the liquid must be checked, especially if a bath liquid such as ethyl alcohol or water is used. In these two cases, dirt from the dry ice as well as condensed water on the dry ice itself may dissolve in the bath liquid, raise the conductivity, and cause a serious calibration error. The lead wires of a thermojunction must be shielded to avoid inductive ac pickup, and if a metallic vessel is used, it must be electrically grounded to minimize this source of error.

Thermographs using bimetals or Bourdon tubes are best calibrated in an air chamber. The reference thermometer must be placed as close as possible to the thermal element.

3-6 SOLAR AND TERRESTRIAL RADIATION MEASUREMENTS

Solar radiation measurements are required for a variety of reasons: (1) for climatological and agricultural purposes, (2) for the design of solar collectors and energy-storage facilities, and (3) for forecasting the availability of solar energy. In addition, the intensity of solar radiation or of net radiation flux is closely related to the stability of the atmosphere and provides a measure of its diffusive capacity. Net radiation flux measurements can also be used to indicate cloudiness day and night. Finally, measurements of solar radiation are desirable in air pollution studies since it is solar energy that drives the photochemical reactions in the atmosphere which produce ozone and other contaminants.

3-6.1 Solar Radiation Instruments

There are many types of instruments for measuring solar radiation. The most common is the pyranometer, which measures the total amount of both diffuse and direct solar radiation incident on a horizontal surface. In some instances we must measure the total radiation on a tilted surface, e.g., solar heating and cooling applications in which the collectors are located on a tilted roof; a tilted pyranometer can be used for these measurements. The pyrheliometer measures the solar radiation incident on a surface perpendicular to the solar beam or at normal incidence. Other instruments are available commercially to measure the shortwave (ultraviolet) or long-wave (infrared) components of solar and sky radiation.

3-6.1.1 Pyranometer. The most useful instrument for measuring solar radiation is the pyranometer. A typical instrument is manufactured by the Eppley Laboratory, Inc., Rhode Island (see Fig. 3.27). In this instrument a 3-cm disk is divided into six 60° sectors. Alternate sectors are painted with a stable black coating; the others are painted with a white nonhygroscopic borium sulfate paint. When exposed to solar radiation, the black surfaces become warmer than the white ones. A thermopile made of 48 plated-copper-on-constantan junctions produces a voltage of about 11 μV per W/m^2. The thermopile impedance is 350 ohms, and its output is linear to ±1% from 0 to 1400 W/m^2. The instrument time constant is about 3 to 4 s. Overall, the instrument exhibits a cosine response within ±2% for angles of incident from the vertical to 10° from the horizontal. The azimuth effect is negligible, i.e., the instrument can be rotated about a vertical axis without affecting the calibration. Temperature compensation of the output signal to ±1.5% over the range of −20 to +40°C is provided. The instrument's standard glass dome is precision Schott WG7 optical glass, which is transparent to wavelengths in the interval 280 to 2800 nm; a quartz dome is available that expands the effective spectral range from 200 to 4500 nm.

A pyranometer measures the total radiation incident on the hemisphere, i.e., a solid angle of 2π steradians. To separate the diffuse from the direct radiation, a shadow band (see Fig. 3.28) or an occulting disk may be used. In each case, the direct solar radiation is blacked out. However, with an occulting disk, an equatorial mount is necessary to keep the disk in the proper position. The shadow band is a simple device and is frequently used, but it does block out a portion of the diffuse radiation. Tables are available to correct for this shortcoming.

Pyranometers are available commercially to measure ultraviolet or infrared radiation. The ultraviolet radiometer houses a detector in a hermetically sealed selenium-barrier-layer cell that is covered by a quartz window. Band-pass interference filters are available to transmit in the four ultraviolet wavelength bands 295 to 385 nm, 295 to 325 nm, 325 to 355 nm, and 355 to 385 nm. The infrared radiometer isolates long-wave radiation from shortwave radiation by selective transmission of the long-wave component. This selectivity is achieved by means of a vacuum-deposited interference filter on the inner surface of a

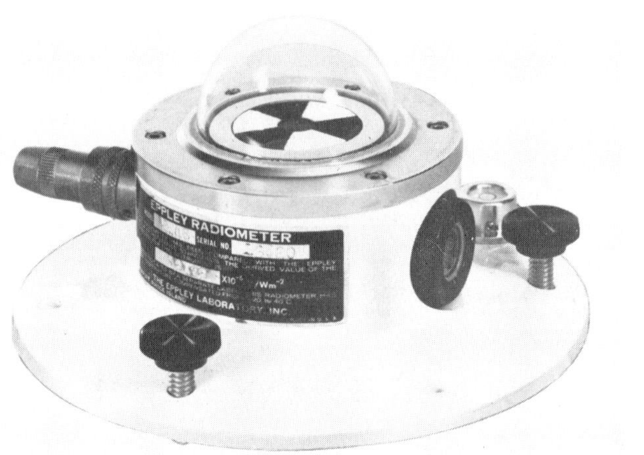

Fig. 3.27 Eppley black-and-white pyranometer for the measurement of global (sun and sky) radiation. (Eppley Laboratory, Inc.)

Fig. 3.28 Shadow band device. The band blocks out the direct solar radiation to the pyranometer (at center) to permit diffuse radiation measurement. (LI-COR, Inc.)

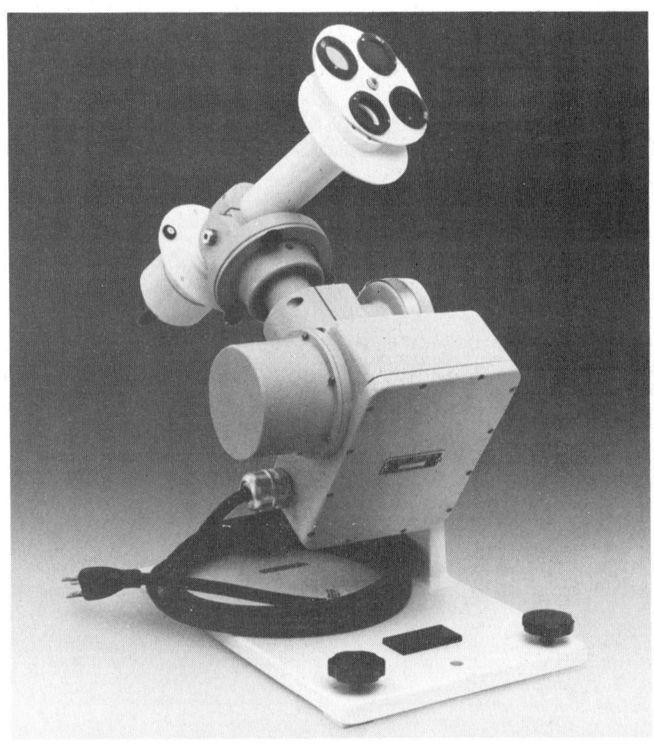

Fig. 3.29 Normal incidence pyrheliometer mounted on a solar tracker. (Eppley Laboratory, Inc.)

KRS-5 hemisphere. The filter transmits radiation only in the wavelength interval 4,000 to 50,000 nm.

3-6.1.2 Pyrheliometer. Direct solar radiation at normal incidence is measured by a pyrheliometer. One commercial instrument has a sensitive thermopile positioned at the end of a tube; an aperture at the other end limits the angle of view to 5.7°. The tube is blackened on the inside and chromium-plated on the outside. It is filled with dry air to atmospheric pressure and sealed at the viewing end by a quartz window. A manually rotatable disk with four holes can be mounted at the viewing end; filters can be inserted into the disk to measure solar radiation in selected wavelength bands. Since the pyrheliometer must track the sun to maintain normal incidence, it is usually supported by a motor-driven equatorial mount. Figure 3.29 shows the complete instrument as it appears in operation.

3-6.1.3 Net Radiometer. A net radiometer measures the difference between the total radiation impinging on a horizontal surface from above and the sum of the reflected (shortwave) radiation and the thermal (long-wave) radiation from the ground below. One of the earliest designs is based on a principle developed by Gier and Dunkle (1951). In their instrument, the net radiation flux is proportional to the output of a differential thermopile which measures the vertical temperature gradient across a horizontal Bakelite sheet. A schematic diagram of this net radiometer is shown in Fig. 3.30a. The entire sensing unit is mounted in an artificially ventilated housing containing a blower that produces a flow of air evenly distributed across the top and bottom of the flux plate (Fig. 3.30b). The airflow equalizes convective heat losses between the two surfaces and minimizes dust and dew accumulation. However, the net radiometer cannot be used during rainfall since wetting of the top surface causes an appreciable cooling that produces a spurious flux indication.

A radiometer that is sensitive to wavelengths of solar and terrestrial radiation from 300 to 60,000 nm was developed by Fritschen (1965). This instrument is a modification of an Australian radiometer developed by Funk (1959). The windows are polyethylene hemispheres and protect the sensor from wind and precipitation (Fig. 3.31). To prevent internal condensation and preserve the shape of the windows, a flow of dry air is maintained internally. The transducer is a 22-junction, temperature-compensated thermopile with a time constant of 60 s. The instrument sensitivity is 5 μV per W/m^2. Because the diameter of the sensor is only 6 cm, the instrument can

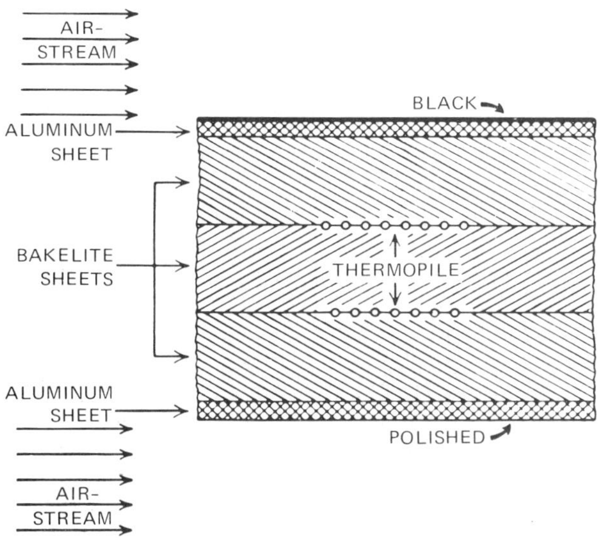

(a)

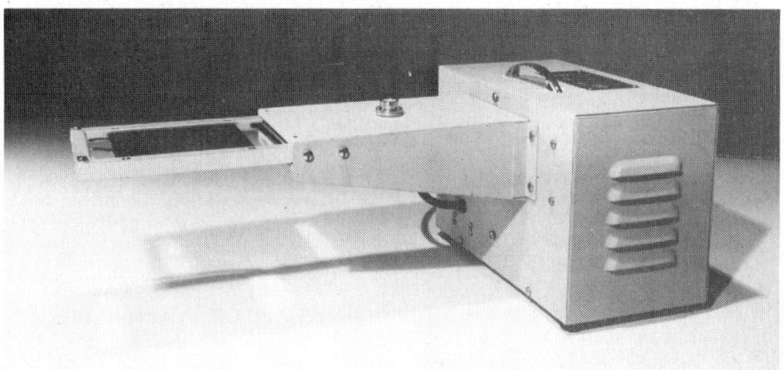

(b)

Fig. 3.30 (a) Schematic of a net radiometer. In this instance, the thermopiles are cemented between aluminum faceplates. (b) A net radiometer. (Teledyne Geotech)

measure net radiation near a surface without appreciable shading which alters the radiation balance. However, for most applications, the instrument should be 1 m above the surface.

3-6.1.4 Other Radiometers. There are numerous other kinds of solar-radiation instruments. For example, a bimetal actinometer contains a blackened bimetal strip that heats and bends as sunlight falls on it; the change in position can provide a measure of the solar radiation incident on a horizontal surface. The actinometer incorporates a system of compensating bimetals to correct for temperature changes and to keep the blackened sensor horizontal. Another well-known instrument is the Campbell–Stokes sunshine duration recorder. This instrument has a clear spherical glass ball mounted on an adjustable support and base. The ball acts as a lens to focus the sun's rays onto a specially prepared chart. As the solar image moves and if the incident radiation is sufficiently strong, a line is burned on the chart; the total length of this line is proportional to sunshine duration.

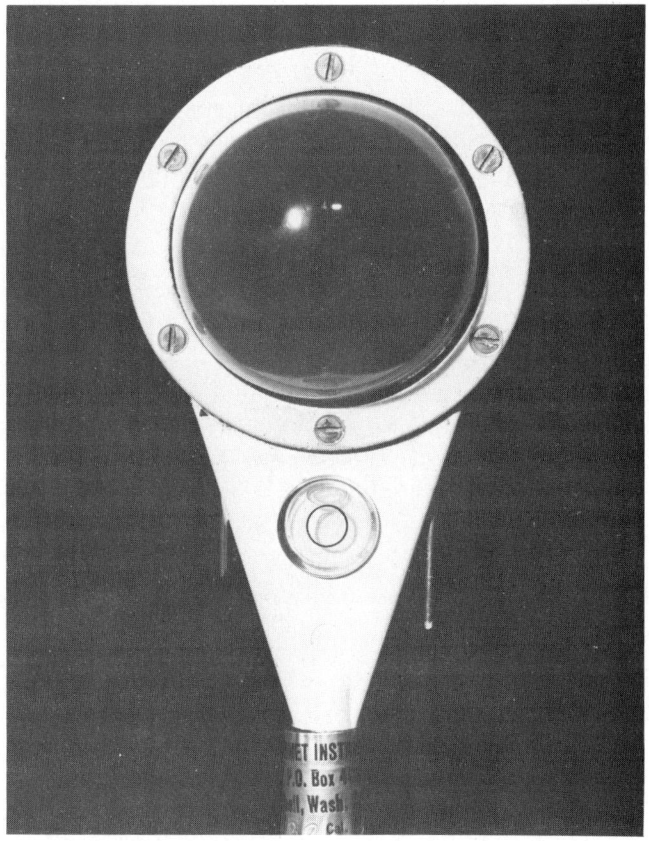

Fig. 3.31 Fritschen net radiometer. (Courtesy L. J. Fritschen)

3-6.2 Radiometer Response Characteristics

An ideal radiation sensor completely absorbs all incoming radiant energy and transforms it into a measurable quantity that is independent of wavelength, direction of incidence, and position on the sensor surface.

Suppose a radiation sensor is immersed in a radiation field whose intensity I is a function of wavelength (λ), azimuth (ϕ), and elevation (θ) angles as seen from the sensor surface, i.e., $I = I(\lambda,\phi,\theta)$. If the response of the sensor depends on these same variables, a sensor geometry function $G(\lambda,\phi,\theta)$ can be defined which relates its response to the position and spectral characteristics of any incoming radiation stream. With proper laboratory calibration techniques, G can be precisely determined for the sensor under consideration.

The output of the sensor is a measure of the radiation field which, to be useful, should be uniquely related to the character of that field. Sensor output is given by the following integral:

$$I_{out} = \int\int\int I(\lambda,\phi,\theta)G(\lambda,\phi,\theta)\, d\theta\, d\phi\, d\lambda \qquad (3.59)$$

Unambiguous results can be obtained for the determination of this field in only two cases. First, if the radiation field is that of a point source and its spectral characteristics are known, the integration over ϕ and θ involves only $G(\lambda,\phi,\theta)$. Since this latter function is known, the integration can be carried out. Consequently, for any given wavelength interval, I_{out} is directly proportional to the source intensity. The second case for which an unequivocal result can be obtained is for a sensor whose response is ideal. In this instance, the sensor geometry function G does not depend upon the wavelength or the azimuth and elevation angles. Hence the output integral becomes simply

$$I_{out} = G \int\int\int I(\lambda,\varphi,\theta)\, d\theta\, d\phi\, d\lambda \tag{3.60}$$

which is uniquely related to the radiation field. Any other circumstance involving an unknown but changing radiation field and a precisely known G may lead to a nonunique solution: two equal sensor outputs need not correspond to two identical radiation fields.

In the case of solar radiation measurements, we do not have a point source. Rather, the radiation field has both direct and diffuse components. Consequently, if we wish to measure (in the sense defined above) the characteristics of the radiation field, we must find sensors whose response is as nearly ideal as possible. Or, if this is not possible, we must standardize sensor geometry and attributes so that meaningful intercomparisons of data can be made. Consideration of these attributes and their measurement follows.

3-6.2.1 Calorimetric Radiometers. Although a variety of physical and chemical processes can be used to convert radiant energy into a measurable quantity, calorimetric techniques dominate. These techniques transform radiant energy into heat energy, and the measurement becomes one of heat quantities. Such a technique imposes no inherent restrictions on the spectral characteristics of the measured radiation, other than that of uniform absorption by the sensor in the spectral interval of concern.

Calorimetric radiometers, in essence, have only two elements: (1) a receiving surface to intercept the incoming radiant energy and transform it into heat and (2) some means of observing that heat by measuring the corresponding temperature effects. In its simplest form the radiometer can be envisioned as a sensor in contact with a heat conductor, which, in turn, is coupled to a heat sink. This conceptualized instrument is shown in Fig. 3.32. The sensor has a heat capacity C and is at a temperature T. We assume that the radiation impinging on the sensor is transformed without loss into heat, part of which serves to increase the temperature of the sensor while the remainder flows through a heat conductor of conductance K. This last is actually a lumped conductance and consists of all loss mechanisms combined; these include conduction, convection, and radiation processes. The conductor is coupled to the heat sink of the environment, whose heat capacity is infinite and whose temperature is that of the ambient, T_a.

To measure the radiant flux incident on the sensor, we measure the temperature of the receiver. The simple instrument shown above can be described by a linear first-order differential equation for the temperature of the sensor as a function of time. This equation can be written

$$I = C\frac{d(T - T_a)}{dt} + K(T - T_a) \tag{3.61}$$

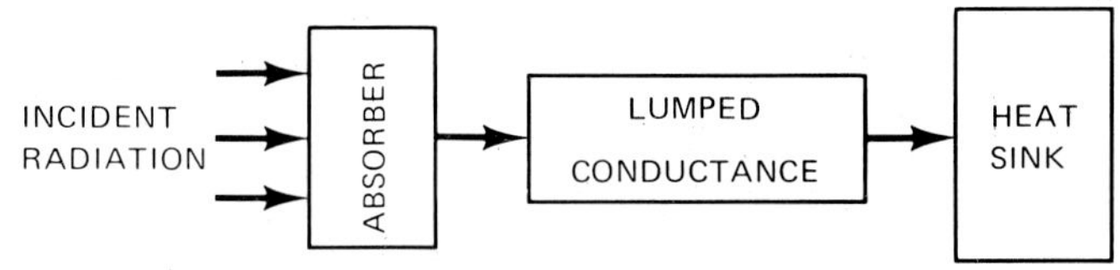

Fig. 3.32 Conceptual schematic for a calorimetric radiometer.

where I is the absorbed radiation, C is the heat capacity of the sensor, and K is the conductance of the conductor. The conductance term represents the rate at which heat is removed from the sensor. Thus, Eq. 3.61 above is a simple statement of the energy balance for the sensor, where I represents the incoming energy, a portion of which goes to heat the receiver (the first term on the right-hand side) and the remainder flows through the conductance to the heat sink (the second term on the right-hand side). Actually the conductance term is based on Newton's law of cooling, which states that the rate of heat loss is proportional to the difference in temperature between the body and its surroundings.

3-6.2.2 Sensor Response and Time Constant. Suppose at time t = 0, radiant power I is allowed to impinge on the sensor at a temperature T_a equal to that of the ambient. From Eq. 3.61, the sensor temperature after exposure to this constant field for a period of time t is

$$T = \frac{I}{K}\left[1 - \exp\left(-\frac{K}{C}t\right)\right] + T_a \tag{3.62}$$

where T_a is the ambient temperature and also the temperature of the heat sink. Note that, as t increases, the sensor approaches an equilibrium which can be found from Eq. 3.62 by letting $t \to \infty$. The equilibrium temperature is given by

$$T_e = \left(\frac{I}{K}\right) + T_a \tag{3.63}$$

a result independent of the heat capacity of the sensor. The quantity C/K is the time constant for the instrument which we will denote τ. Obviously C/K must have the dimensions of time, since the argument of the exponential function must be dimensionless. In terms of τ, we can write the above solution to the differential equation as

$$T = \frac{I}{K}\left[1 - \exp\left(-\frac{t}{\tau}\right)\right] + T_a \tag{3.64}$$

Thus τ is the time interval for the sensor temperature to reach 0.63 of the difference between its final and initial values for a stepwise change in radiant power, as can be readily shown if we set $t = \tau$ in the last equation.

Suppose we wish to find the change in sensor temperature if the instrument is shaded, as, for example, by the passage of a single cumulus cloud overhead. Suppose at time t = 0, the sensor temperature $T = T_0$; if the sensor is shaded suddenly, its temperature at a time t is given by

$$T = (T_0 - T_a) \exp\left(-\frac{K}{C}t\right) + T_a \qquad (3.65)$$

This expression is the solution of Eq. 3.61 for the initial conditions $T = T_0$ for $t = 0$, and $I = 0$ for all $t \geqslant 0$. In terms of the time constant $\tau = C/K$, which again appears as a natural parameter, we can rewrite Eq. 3.65 as

$$T - T_a = (T_0 - T_a) \exp\left(-\frac{t}{\tau}\right) \qquad (3.66)$$

And, as before, at the time $t = \tau$, $T - T_a = 0.37\ (T_0 - T_a)$, which is equivalent to an 0.63 response to a step-function decrease in radiant power.

The time constant is a measure of how quickly the sensor responds to changes in radiant-energy input. The smaller the value of the time constant, the faster the instrument approaches the new equilibrium value of sensor temperature after a change in the radiation field. For radiometers using thermojunctions to measure temperature changes, time constants range from as fast as 1 s to as long as 15 s; e.g., the Eppley black-and-white pyranometer has a time constant of 3 to 4 s, the normal incidence pyrheliometer 1 s, the Fritschen net radiometer 60 s, and the Funk net radiometer 30 s. A recording pyranometer using bimetallic strips may have a time constant of as much as 3 min. For this latter instrument, approximately 10 to 15 min must elapse before full output under equilibrium conditions is achieved.

We can measure response times by alternately shading and exposing the sensor, using a light source whose output passes through a shuttered aperture. A fast-response recorder—one capable of responding to changes more quickly than the sensor being studied—must be used to measure the instrument response. Usually this is an oscilloscope or a light-beam galvanometer.

3-6.2.3 Sensor Spectral Response. The radiation-absorbing surfaces of calorimetric instruments usually exhibit varying degrees of absorption as a function of incident wavelength. Figure 3.33, for example, compares the spectral reflectance of Parson's black with carbon black. Identical sensors using these two coatings can produce output signals that differ markedly when exposed to the same radiation field. Obviously the effective mean absorptance depends on the spectral characteristics of the incoming radiation in this case.

We can surmount the difficulty of finding a flat surface with uniform and precisely known absorptance characteristics over the spectrum of interest by shaping the sensor surface to form a blackbody cavity. Rays not fully absorbed at their first encounter with the receiving surface will be repeatedly reflected within the cavity until absorption does occur. In this way, the radiative loss back to the environment can be made exceedingly small. As a result, an

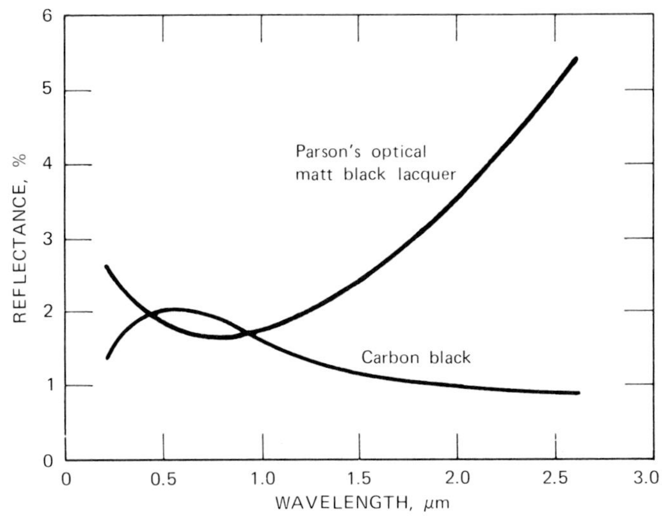

Fig. 3.33 Spectral response of carbon black and Parson's black. (From C. Fröhlich, Standards for Meteorological Radiation Measurements, in *Second Symposium on Meteorological Observations and Instrumentation*, pp. 267-271, American Meteorological Society, San Diego, Calif., 1972.)

effective absorption coefficient of near unity can be obtained for the spectrum of interest. However, owing to the means used to collimate the incoming radiation and to shield the blackbody cavity against losses to the environment, large thermal masses are introduced. These produce long time constants which limit the usefulness of these instruments to the measurement of radiation fields having long-time stability. Recent efforts have been directed to the development of rapidly responding blackbody instruments by making the sensor small. On the basis of this principle, instruments only 3.2 cm wide and 3.8 cm long have been constructed (Kendall, 1968).

One important factor that must be considered when applying a coating to the sensor surface is its stability against a change in its absorption characteristics. For example, recent instances have been noted in which the deterioration of Parson's black paint has resulted in a 20 to 30% loss in sensitivity for instruments using this coating. The degradation has been particularly pronounced for the combination of strong sunlight and warm temperatures. The rehabilitation of faulty data taken with degraded instruments is currently a joint project of the National Weather Service and the Department of Energy. Clearly, an annual calibration of all sensors should be made to ensure that such deterioration is not taking place.

3-6.2.4 Angle of Incidence. For a flat horizontal surface, the total flux intercepted by it is proportional to the projected area of the surface in the direction of the incident radiation stream. Since the projected area is equal to the product of the actual surface area and the cosine of the angle between the normal to the surface and the source, flat sensors exhibit a cosine response. For the solar beam, the response of a flat sensor is proportional to the cosine of the solar zenith angle. How well this relationship is attained in practice is shown in Fig. 3.34, in which the ratio of the

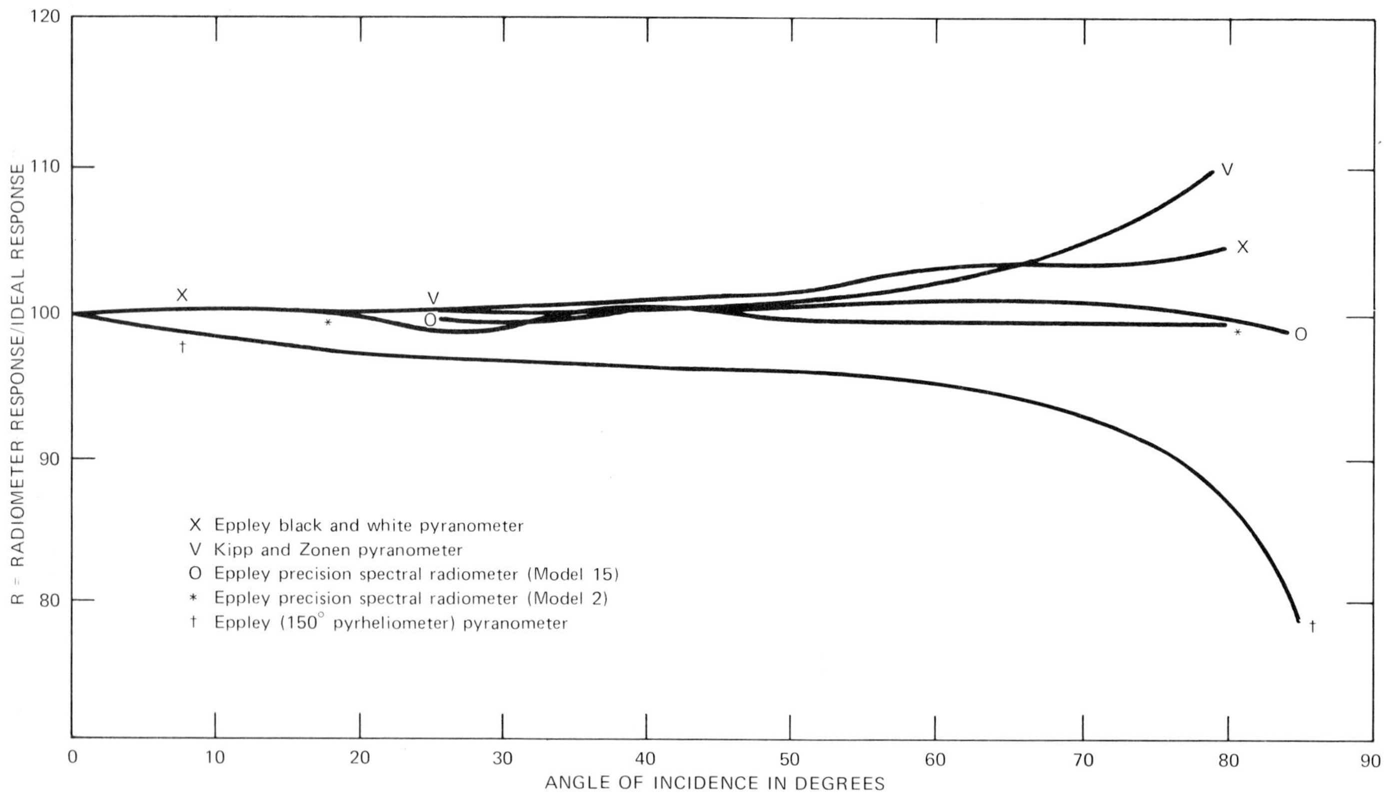

Fig. 3.34 Response characteristics of different radiometers compared to ideal cosine response as a function of angle of incidence. (From J. R. Latimer, Radiation Measurements, International Field Year for the Great Lakes Technical Manual, Series No. 2, 1978.)

radiometer response to the ideal response is plotted for a group of commercially available instruments (Latimer, 1978). This ratio is given by

$$R = 100 \frac{V_\theta}{V_0 \cos \theta} \tag{3.67}$$

where V_θ is the instrument response at an angle of incidence θ, and V_0 is the response at normal incidence ($\theta = 0$). Most of the sensors show ideal response (within 5%) out to the equivalent of a solar elevation angle of 20°.

Sensor output should be independent of azimuth. In other words, the output of an ideal sensor should show no change under constant radiation input conditions as the sensor is rotated about its vertical axis. For this uniformity of response to be achieved, most sensors have a circular geometry. Deviations from the ideal can be caused by optical distortions in the glass domes or bulbs that surround the sensor.

A stable (constant intensity) collimated light source and a turntable upon which the instrument can be mounted are required to measure the response of a sensor as a function of angle. The response of the instrument is measured for a series of zenith and azimuth angles.

3-6.2.5 Attitude Response. The response of an ideal sensor should be independent of its orientation with respect to the vertical. However, convection currents within the glass bulbs or domes arise owing to the difference in temperature between various elements of the sensor surfaces. As the sensor is tilted, these currents may become more effective in transferring heat from one surface to another. Complete inversion of the instruments can produce a decrease in sensitivity of up to 4% for some sensors. Under normal circumstances, the attitude-variation error is small in well-constructed instruments. However, the instrument still must be carefully leveled; otherwise, an apparent deviation from the cosine law will be experienced.

3-6.3 Some Factors Related to Measurement Accuracy

In view of the response characteristics of radiometers just considered, there are several factors related to the selection, exposure and siting of these instruments that affect the accuracy of solar radiation measurements. These factors are (a) the cosine response, (b) the attitude response, (c) deterioration of the sensor coating, (d) lack of adequate temperature compensation, (e) actually improper exposure and siting, and (f) dew and frost formation.

The first of these becomes important only for low sun-elevation angles at sunrise and sunset; at these times, errors of up to 5% may occur. Because this error is inherent it cannot be eliminated; however, a solar-attitude correction factor can be applied to the raw data to compensate for it, if necessary.

The attitude response is minimized by installing the sensor in a horizontal position. However, measurements of

Fig. 3.35 A pyranometer installed at an angle to simulate a solar collector. The black shield in the foreground limits the sensor's field-of-view to block radiation reflected from nearby objects as well as the upward-directed total radiation originating below the horizon.

the flux on a tilted surface may be necessary to evaluate solar collector performance. In this case, a pyranometer tilted to the appropriate angle and aligned in the proper direction can be used to make the measurements. Figure 3.35 shows a tilted pyranometer installed in just this way. For the utmost accuracy, it should be calibrated at this attitude.

Little can be done by the user to protect the sensor coating from degradation. Instead, he must rely on the manufacturer to provide a stable, long-lived coating. Nevertheless, frequent calibration checks should be carried out to detect any sensor deterioration as quickly as possible, if it should occur. Likewise, adequate temperature compensation to make the output signal independent of the ambient temperature is usually built into instruments supplied by reputable manufacturers. But, again, a failure of the compensation circuitry is detectable by frequent, simple tests with the instrument in a temperature-controlled chamber.

With regard to exposure and siting, a radiometer must be located so that its view of the sun is unobstructed at all times during the course of the study. For solar radiation monitors, this requirement means that a site must be selected which provides such an unobstructed view for the entire year. Otherwise, shadows produced by nearby objects such as poles or buildings can produce significant and unfillable gaps in the data. Consequently a pyranometer is best located on the roof of the tallest structure in the vicinity. However, because such an instrument sees all sources in its hemispherical field-of-view, the site horizon should be uncluttered and free of prominent radiators or reflectors. Figure 3.36 pictures such a roof-top facility housing a variety of radiometers to provide an accurate radiation data base for the Department of Energy. A net radiometer must be mounted over the surface at the particular site for which measurements of the net radiation are desired.

Fig. 3.36 Overall view of the University of Michigan solar radiation facility maintained in Ann Arbor, Mich., for the Department of Energy.

In northern latitudes, dew or frost may form and accumulate on the sensor's bulb. Heaters are available for some types of radiometers to prevent condensation. Without these heaters, the dew or frost ordinarily is removed daily during the morning routine cleaning of the bulb with a water-and-alcohol mixture.

3-6.4 Solar Radiation Standards

The International Pyrheliometric Scale 1956 (IPS) is the internationally accepted radiation scale for pyrheliometric measurements. This scale is maintained by the World Radiation Center in Davos, Switzerland, and is sponsored by the World Meteorological Organization. It is reproduced in the United States by the National Weather Service and the Eppley Laboratory. The preferred calibration techniques for radiometric instruments use the natural light sources of the sun and the sky. However, whatever procedure is used, the measurements ultimately must be traceable to a primary-standard radiometer. A direct comparison to a primary or working standard using the sun as a source is an accepted technique for the calibration of pyrheliometers. Note that the IPS is defined in terms of this type of sensor. An International Pyrheliometric Comparison was held in 1975 at the World Radiation Center for the purpose of comparing and recalibrating national primary-standard pyrheliometers; several groups from the United States participated to maintain or establish primary-standard instruments. A review article by Marchgraber (1970) treats instruments especially suitable as primary or working standards for radiation measurements important to meteorology.

A working-standard pyranometer is calibrated by the "shade" method. Here the pyranometer is used to measure first the total irradiance (direct plus diffuse) and then the diffuse component alone. The difference between these two measurements is a measure of the vertical component of the direct irradiance which is compared directly to a pyrheliometric determination of the same quantity. Subsequently other pyranometers can be calibrated by direct comparison in sunlight to this working standard.

Generally, all comparisons, for whatever purpose, are effected over a considerable period of time; thus, for example, 10-min-average irradiance values for two sensors might be compared for periods up to 30 d. In this way, a variety of sky conditions (light intensity, cloud cover) are sampled, leading to a more reliable calibration factor. Comparison calibrations should extend over a period of at least 5 d of precipitation-free weather, with the exposures of the sensors made as identical as possible.

Obviously calibration is an important process. Durrenberger and Brazel (1976) discuss the importance of and need for a better solar radiation data base for the United States, especially now because of the interest in solar energy. They point out that accurate radiation measurements are necessary to size solar-energy equipment properly and that an inadequate data base will waste economic resources. Unfortunately two significant sources of error are present in the National Weather Service (NWS) radiation data from 1957 to about 1972. The first arises from a systematic error of about 2.5% between the NWS and IPS 1956 scales; furthermore, in about one half of these data, an additional 7% error is present, owing to the use of Parson's optical black paint. The second error arises from the degradation of Parson's black in field instruments under conditions of strong sunlight and high temperatures; the degree of degradation is not uniform, and losses in sensitivity of up to 15% to 20% have been observed in some cases. Now there is an effort to rehabilitate the faulty data from the NWS pyranometer network so that errors will be kept to less than 5%.

3-7 PRESSURE MEASUREMENTS

Atmospheric pressure measurements are important in defining the state of the atmosphere. Some processes associated with power generation are pressure-dependent; therefore it is sometimes necessary to know the prevailing pressure to correct for differences from standard conditions (STP).

Atmospheric pressure for meteorological purposes is customarily measured in units of millibars, although other units related to the measurement of the height of a mercury column have also been used. However, the recent decision by the National Weather Service to adopt the international metric system (SI) means that a new unit for pressure must be introduced. In SI units, the standard pressure (force/area) unit is the pascal (Pa); for convenience, atmospheric pressures are better expressed in kilopascals (kPa). Because the unit is relatively new and also because many instruments will retain their obsolete scales (and, hence, be read in the old units), Table 3.4 is included to facilitate conversions from one system of units to the other.

3-7.1 Pressure Sensors

A variety of devices can be used to sense atmospheric pressure. However, of these, only two types find widespread application for meteorological purposes. These two are the mercury barometer and the aneroid barometer. Recently, solid-state semiconductor devices, including some with temperature compensation, have been developed with sufficient sensitivity to measure atmospheric pressure changes reliably. However, at the present time, not enough experience has been acquired using these units to assess their long-term stability and accuracy; consequently no "standard" sensor can be selected in this new category.

TABLE 3.4 Pressure Corresponding to One Standard Atmosphere†

101.325 kPa
1013.25 mb
760.000 mm Hg at 273.15°K
29.9213 in. Hg at 273.15°K
14.6959 lb/in.^{2}

†At standard gravity (9.80665 m/s^2).

Note: An error of 0.010 kPa is equivalent to 0.100 mb.

3-7.1.1 Mercury Barometer. The mercury barometer consists of a mercury-containing cistern that supports a vertical glass tube evacuated and sealed at its top end; the bottom end of the tube is open and is submerged in the mercury. Atmospheric pressure acts on the mercury surface in the cistern to force a column of mercury up into the tube. The length of the column is such that the pressure which it exerts on the mercury surface is equal to the pressure of the atmosphere on the same surface. Therefore a measurement of the length of the mercury column is a direct measurement of atmospheric pressure. It is important to note that the mercury barometer is a primary sensor because only the measurement of a length is necessary.

Some means must be provided to compensate for the changing level of mercury in the cistern as the column height varies in response to atmospheric pressure fluctuations. The Fortin barometer (Fig. 3.37) uses an adjustable cistern and an ivory peg as a zero-point fiducial. The mercury level can be raised or lowered by an adjusting screw until the mercury just contacts the tip of the peg. The height of the mercury column is read with the aid of a brass scale attached to the barometer-supporting assembly. Usually the column height is measured to 0.01 cm.

Various corrections must be applied to the height measurement to compensate for the effects of other environmental parameters which can affect the length of the mercury column. These parameters are the ambient temperature and the local acceleration of gravity at the station latitude and elevation. In addition, it is sometimes necessary to include an instrumental correction factor determined by calibration with reference to a standard barometer. Usually the temperature correction is the largest of these three factors.

The accuracy of the Fortin mercury barometer for a single corrected reading is ±0.020 kPa with a precision of ±0.010 kPa (World Meteorological Organization, 1971). Because of the fundamental nature of the meaurement, the mercury barometer is defined as the standard sensor against

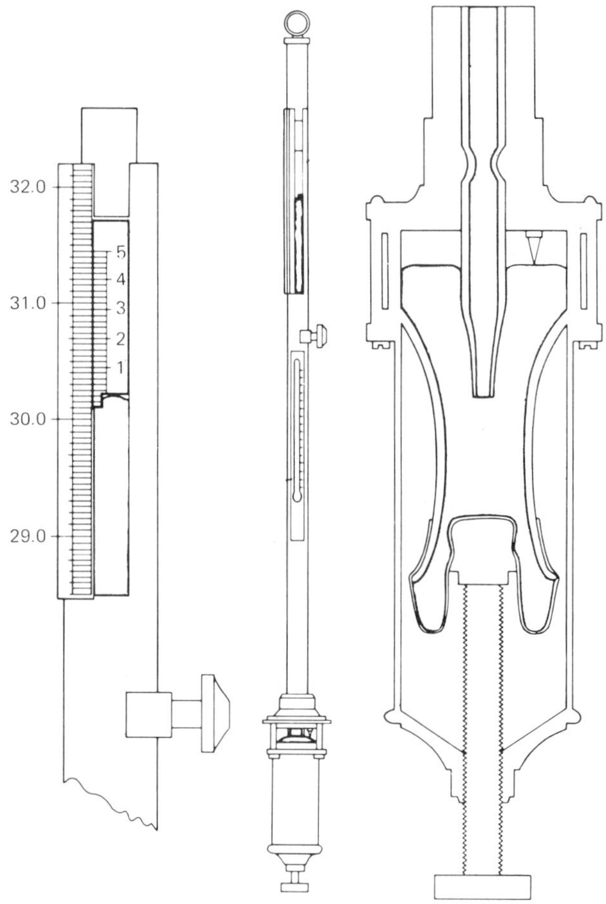

Fig. 3.37 Schematic drawing of a Fortin mercury barometer: (center) the entire barometer, (right) the barometer cistern, and (left) the vernier scale.

which all other types of pressure sensors are to be compared.

3-7.1.2 Aneroid Barometer. The pressure sensor of an aneroid barometer is an evacuated circular volume bounded by two thin metal disks. As the external atmospheric pressure (acting on the disks) changes, the disks flex and change their relative spacing. A mechanical or electrical linkage transmits this dimensional change to an appropriate transducer. An aneroid barometer may contain only one such sensing element, several identical elements stacked together, or a single bellows assembly which functions in a manner similar to that of the stacked array. These last two arrangements increase the sensitivity of the barometer as compared to the single-element instrument.

Aneroid barometers are convenient and easy to use: no correction need be applied for temperature or gravity. As a result, well-built units (precision aneroids) are found in widespread application. However, periodic calibration referenced to a mercury barometer is necessary. The accuracy of the aneroid barometer for a single reading is ±0.050 kPa with a precision of ±0.050 kPa (World Meteorological Organization, 1971).

3-7.1.2.1 Barograph. A recording barometer, called a barograph, is usually an aneroid barometer. In this instance the transducer is a mechanical linkage, one end of which is fixed to the bellows assembly and the other end to an ink pen. A rotating drum, driven by either a spring-powered or electric clock, moves under the pen, producing an inked trace that provides a permanent record of pressure vs. time (Fig. 3.38). Because the linkage is basically a lever mechanism, it is possible to magnify the change in the length of the bellows so that the pen actually moves through a greater distance than the corresponding change in height of a mercury column. For example, the pen deflection of the pictured instrument is 2.5 times that of a mercury barometer for the same pressure change. In this sense the aneroid barometer is more sensitive than a mercury barometer. Such an instrument can record pressure variations of very small magnitude and is called a microbarograph.

In addition to a zero-adjust mechanism, the microbarograph also contains one or two dashpots filled with oil. These units provide a viscous damping force to damp out rapid oscillations induced by either vibration or rapid pressure fluctuations caused by, for example, the opening and closing of a door.

3-7.1.2.2 NCAR Digital Barometer. There exists an ever-increasing need for a barometer particularly suited to long periods of unattended operation while it still maintains a meteorologically valid accuracy. To fill this need, Pike and Bargen (1976) at the National Center for Atmospheric Research (NCAR) developed an aneroid barometer whose sensor consists of a ceramic plate substrate sealed between two Ni-Span-C diaphragms (Fig. 3.39). Metalized areas on both surfaces of the substrate are one plate of a capacitor, the other plate of which is formed by the two metal diaphragms. The capacitance between the internal electrode and the diaphragms increases linearly with applied pressure. Low temperature-drift capacitance-measuring electronic circuitry is the transducer for this sensor.

The performance characteristics of their instrument are summarized in Table 3.5. Evaluation testing is now under way by a variety of users to determine more precisely the instrument's parameters, especially its long-term stability.

3-7.2 Exposure of Barometers

Normally, the measurement of any meteorological variable is made with the sensor responding to that variable completely immersed in the ambient atmosphere. This procedure poses a problem in the case of pressure measurements because wind blowing across the pressure sensor port induces variations in pressure at the port which are not related to the static atmospheric pressure. Consequently a barometer is usually installed in an enclosed room whose pressure is directly coupled to the ambient pressure. Care must be taken, however, to ensure that the barometer is not affected by the ventilating, heating, and/or air-conditioning equipment of the structure.

It is possible to construct a simple static pressure port vented to the ambient but connected to the barometer inside the structure. The pressure port overcomes the

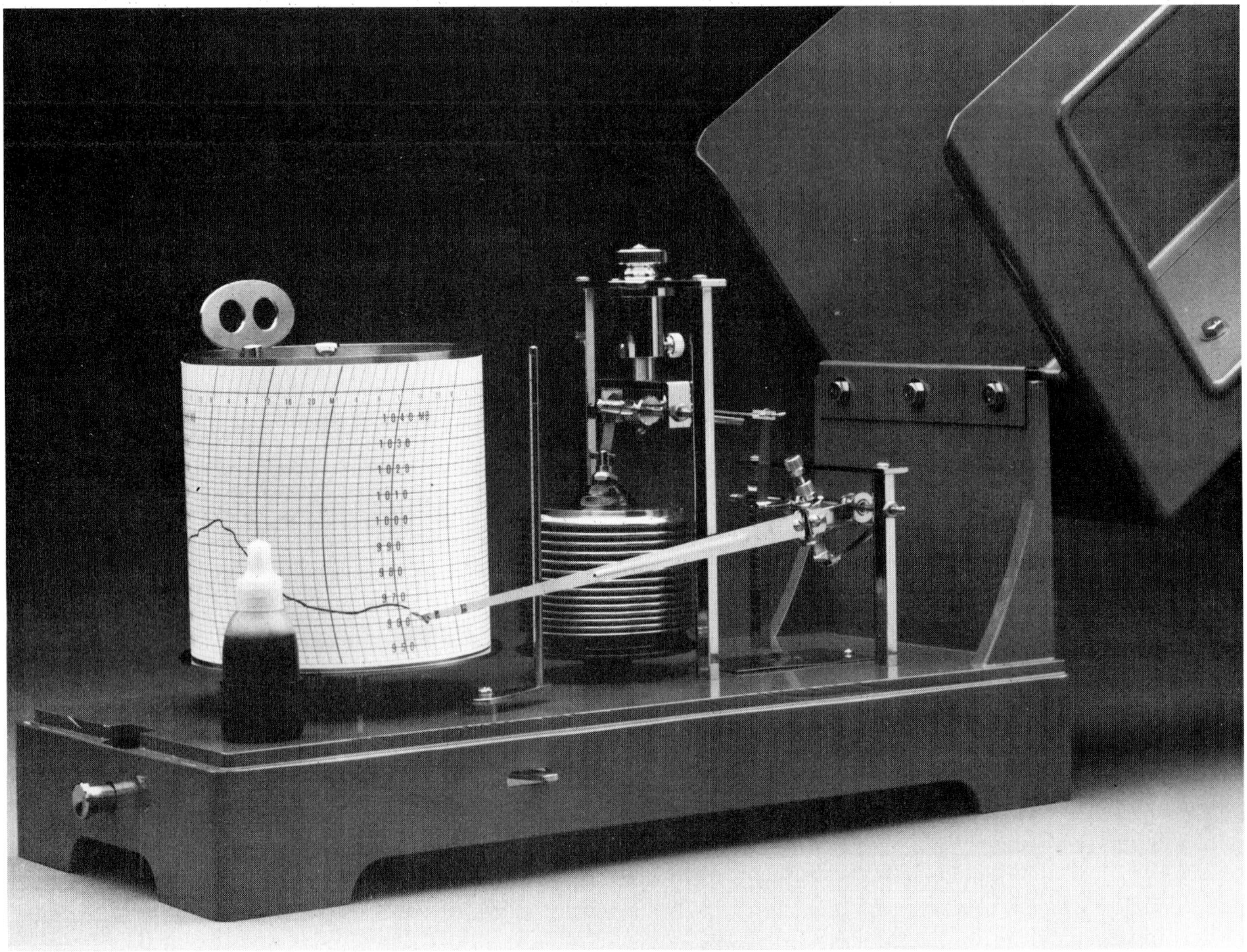

Fig. 3.38 An aneroid microbarograph. (WeatherMeasure Corp.)

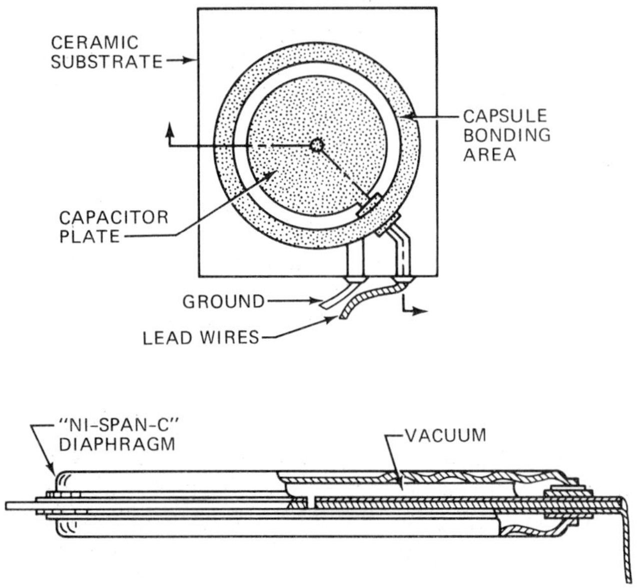

Fig. 3.39 The sensing element for the NCAR digital barometer. [From J. M. Pike and D. W. Bargen, The NCAR Digital Barometer, *Bull. Am. Meteorol. Soc.*, 57: 1106-1111 (1976).]

TABLE 3.5 Performance Characteristics of the NCAR Digital Barometer

Total pressure range: 60 to 110 kPa
Operating pressure range: ±5 kPa from set point
Averaging time: 64 s
Resolution: 0.001 kPa
Linearity: ±0.013 kPa
Hysteresis: ±0.005 kPa
Short-term repeatability (1 d): ±0.01 kPa
Long-term stability (1 yr): ±0.15 kPa (estimated)
Temperature coefficient: ±0.0070 kPa/°C

Size: 89 mm in diameter by 86 mm high
Weight: 454 g maximum
Power requirements: 12 ± 2 V dc, 10 mW typical
Data output format: 16-bit binary

Operating temperature range: −30°C to +50°C
Operating humidity range: 0 to 95°C noncondensing

deleterious effects that the building environment may have on the measurements and reduces the dynamic pressure head associated with the wind. The construction details for one such port are given in the *Manual of Barometry* (U. S. Department of Commerce, 1963). More recently, Miksad (1976) discusses a port that almost completely eliminates errors caused by dynamic wind loading.

3-8 ATMOSPHERIC MOISTURE MEASUREMENTS

The measurement of atmospheric moisture is of importance to the industrial and air-pollution meteorologist. For example, it is thought that the concentration of water vapor may affect the rate at which sulfur dioxide is converted to atmospheric sulfate and also that pollution photochemistry is influenced by the presence of water vapor. However, the sensors for atmospheric-moisture measurements are generally not as reliable as the sensors for other meteorological parameters; they require more maintenance and more frequent calibration. It is hoped that the continuing development of automatic dew-point sensors (Sec. 3-8.4) may alleviate these problems to some extent. Also, most moisture sensors have fairly long response times, on the order of several minutes. But the very recent development of the Lyman-alpha hygrometer (Sec. 3-8.5) with a response time measured in milliseconds has finally made available a sensor useful for moisture-flux measurements.

Relatively complete summaries of the underlying theory of and techniques for measuring atmospheric moisture are given by Middleton and Spilhaus (1953), Penman (1955), and Spencer-Gregory and Rourke (1957). A more recent paper by Wexler (1970) surveys the state of the art in hygrometry as applied to humidity measurements in gases near atmospheric pressure. Our emphasis here will be on those instruments most frequently used by meteorologists to measure humidity at the surface. These include the wet-bulb psychrometer, the hair hygrometer, the hygroscopic hygrometer, and the dew-point hygrometer.

3-8.1 Wet-Bulb Psychrometers

The reduction in temperature of a wetted surface owing to evaporational cooling can be used to determine the amount of moisture in the atmosphere. The ambient air temperature and pressure must also be known. This principle is used in the operation of the wet-bulb psychrometer, which consists of two temperature sensors, one of which is wrapped in clean muslin. The muslin is wetted with distilled water, and the instrument is placed in an airstream of from 3 to 5 m/s. The indicated temperature difference of the two sensors is easily converted to relative humidity or dew point through the use of psychrometric tables.

Commonly used temperature sensors for wet-bulb hygrometry include mercury-in-glass thermometers, resistance thermometers, thermocouples, and thermistors. Various means exist to produce an airstream for the wet-bulb thermometer. The sling psychrometer consists of two thermometers mounted side by side on a metal plate or frame that can be whirled by hand. The two thermometers mounted in this way also can be placed in an instrument shelter in front of a fan driven by a hand crank. Finally, spring-driven or electric motors can be used to create the airstream, as in the Assmann psychrometer.

3.8-2 Hair Hygrometers

For the last 150 yr, the hair hygrometer has been widely used for both industrial and meteorological purposes. It continues to be used because the sensor is simple in construction, lends itself easily to chart recording, and provides measurements that are of sufficient accuracy (about ±5%) for ordinary use in air temperatures above freezing. Human and animal hairs, after careful cleaning and degreasing in ethyl ether to ensure repeatability, elongate about 2.5% as the relative humidity changes from 0 to 100%. The relation between elongation and relative humidity is logarithmic in the relative humidity range 20 to 100%, i.e., $L = a + b \log RH$, where RH is the relative humidity, L is the length of the hairs, and a and b are constants. The loss in speed of response with temperatures below freezing may be a distinct shortcoming for some applications.

In the conventional hair hygrograph, the humidity sensor consists of a harp of human hairs held under slight tension. The harp drives a pen through a logarithmic-compensating linkage to produce an inked trace on a chart fixed to a clock-driven drum. Frequently a single drum is used to record the pen traces of both a humidity and a thermal sensor; this type of instrument is called a hygrothermograph (Fig. 3.35). Daily or weekly calibration checks are required. Human hairs can function well for several years in areas free from dust and salt but must be cleaned frequently in dirtier or polluted environments.

3-8.3 Hygroscopic Hygrometers

Salts, such as calcium chloride or lithium chloride, are hygroscopic and absorb water vapor when placed in a moist atmosphere. Various types of hygrometers based on this phenomenon have been devised. In these instruments the salt may be coated on a flat surface, such as a glass or plastic slide or may be impregnated on a fabric or a strand of fiberglass. The gravimetric hygrometer and the electrolytic hygrometer are examples of moisture-measuring instruments using hygroscopic salts.

3-8.3.1 Gravimetric Hygrometers. The actual amount of water vapor absorbed by the sensor is weighed by a sensitive balance built into this instrument. From this measurement the relative humidity can be determined if the temperature is known. This type of hygrometer is infrequently used for meteorological purposes.

3-8.3.2 Electrolytic Hygrometers. In this instrument the electrical resistance between two electrodes placed on a coated slide or between two points on an impregnated fiber is measured by a bridge, an ohmmeter, or some other electronic device. Because the conductivity of the unsaturated salt solution is a function of its moisture content, such resistance measurements can be related to the ambient relative humidity if the temperature is known. The radiosonde humidity element makes use of this principle.

A commonly used, modified form of the electrolytic hygrometer operates on the principle that for each value of water-vapor pressure of the surrounding air there is one temperature of the saturated solution at which equilibrium exists, i.e., the vapor pressure in the immediate vicinity of the lithium chloride salt surface equals that of the ambient vapor pressure. The sensor consists of a cylindrical tube covered by a wick of woven glass tape on which a bifilar wire filament is wound. The tape is impregnated with lithium chloride; the wires are connected to a low-voltage power source. The conductivity between the parallel wires depends on the amount of moisture absorbed by the hygroscopic lithium chloride. If the vapor pressure of the ambient atmosphere is greater than that in the immediate vicinity of the lithium chloride film, water vapor will be absorbed, the conductivity between the wires will increase, and more current will flow between them through the lithium chloride substrate. The augmented current flow raises the temperature of the sensor until the equilibrium vapor-pressure point is reached. If the vapor pressure of the ambient atmosphere is less than that surrounding the lithium chloride film, the reverse process takes place. Thus the sensor acts as an automatic self-regulating control mechanism in which the current flow between the heating wires is varied to maintain an equilibrium temperature established by the ambient vapor pressure. Now, if this equilibrium temperature is measured, it can be related to the ambient vapor pressure from the calibration curve for the hygrometer. Thus the measurement of moisture content becomes one of a temperature measurement. The thermal sensor can be a resistance thermometer or a thermistor, both of which allow a continuous record to be generated on a suitable electrical recording device.

3-8.4 Dew-Point Hygrometers

When a shiny metal surface exposed to the ambient air is cooled to a temperature below the dew point, moisture from the layer of air immediately adjacent to it condenses to form dew. The appearance of the surface changes markedly on the appearance of the condensate, and such a change is readily observed. If the temperature of the surface is measured at the precise instant when dew formation occurs, then this measurement is a direct determination of the dew point. The difference between the ambient air temperature and the dew point can be used in conjunction with the saturation vapor-pressure curve for air to determine the moisture content of the ambient sample.

On an operational basis, a manually operated dew-point apparatus is of little value. Thornthwaite and Owen (1940) tried to develop an automatic device that, though disappointing to them, embodies the basic ideas used in its modern counterparts. These instruments incorporate a mirrored surface illuminated by a collimated light source (Fig. 3.40). The surface can be heated or cooled by a thermoelectric device that forms the base for the mirror. As the mirror is cooled, the onset of dew formation is detected by the sudden appearance of a signal at the output of a photodetector that looks directly downward onto the mirror. The film of dew causes a nonspecular reflection component to appear in the illuminating beam of light that the photodetector can see. Immediately upon receiving the photodetector output signal by the control circuitry, mirror cooling ceases and heating starts, a process that ultimately evaporates the dew and causes the photodetector signal to disappear. At this point, mirror cooling is reinitiated. Thus the mirror temperature is automatically maintained at the dew point and is measured by a thermal sensor embedded in the mirror; the electrical signal from the mirror thermometer can be used as an input signal to a recording device. A second photodetector looks at the specularly reflected beam to generate a signal that compensates for the gradual accumulation of dust and other debris on the mirror's surface. A small air pump is used to draw ambient air continuously across the sensor's surface; it must be placed on the downstream side of the sensor.

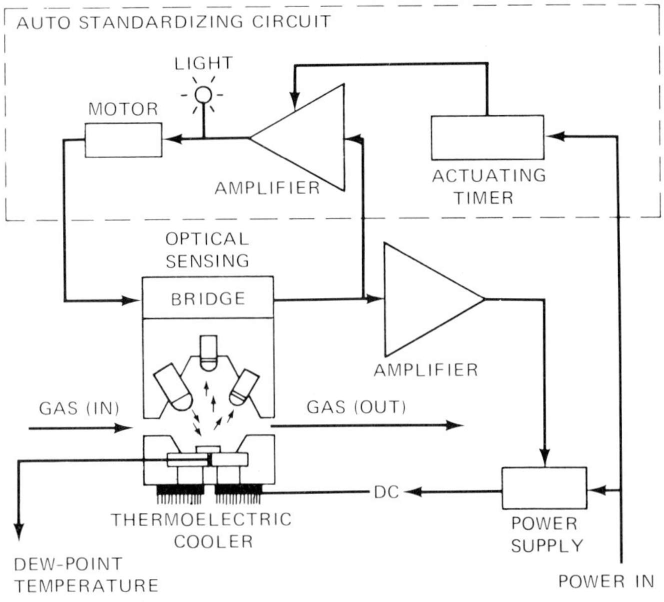

Fig. 3.40 Schematic of an automatic dew-point apparatus.

3-8.5 Lyman-Alpha Hygrometer

The Lyman-alpha hygrometer is based on the very strong selective absorption of Lyman-alpha radiation by water vapor. The instrument provides a direct measure of absolute humidity as a result. The theory, design, and construction of such a hygrometer are discussed by Buck (1976). Figure 3.41 shows the complete instrument, and its performance characteristics are listed in Table 3.6. Owing to the slow degradation of window materials used, a variable-path technique is employed to make the instrument self-calibrating. In its present stage of development,

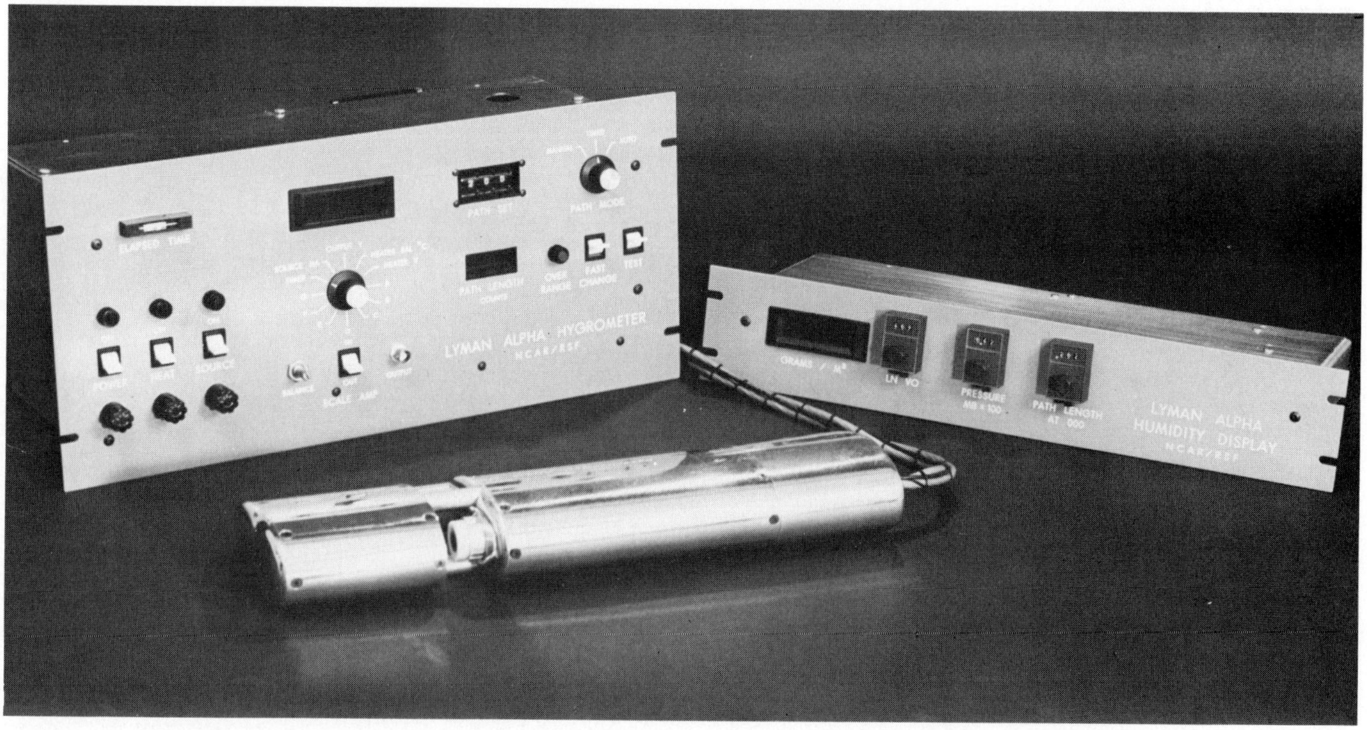

Fig. 3.41 The variable-path Lyman-alpha hygrometer. The sensor is in the foreground, control circuitry chassis on the left, and humidity-display panel on the right. [From A. L. Buck, The Variable-Path Lyman-Alpha Hygrometer and Its Operating Characteristics, *Bull. Am. Meteorological Soc.*, 57: 1113-1118 (1976).]

TABLE 3.6 Lyman-Alpha Hygrometer Performance Specifications

Time response, ms	12
Accuracy, °C dew point	1
Stability (up to 500 h), °C dew point	1
Resolution, °C dew point	<0.02
Operating range, °C dew point	−35 to 30
Noise, mV peak-to-peak	10
Gain change, percent/hour	0.8 typical
Environmental range	
Temperature, °C	−70 to +40
Pressure, mb	100 to 1100
Useful path length, cm	0.2 to 10
Path control accuracy, cm	±0.005
Effect of extraneous light, cosmic rays	Negligible
Source lifetime, h	>1000
Detector lifetime, h	>400

this instrument can be used successfully by qualified persons in studies that require its extensive capabilities; for routine observations, the use of other hygrometers should be considered first.

3-8.6 Hygrometer Calibration

Calibration techniques for hygrometers are of two types: primary and secondary. The primary techniques require only measurements of temperature and pressure while subjecting the calibration gas to changes in its moisture content, temperature, and/or pressure. For example, to obtain an air sample of known vapor content, we can mix known volumes of dry and saturated air at the same temperature. Alternately, we can saturate air at a given temperature and then raise its temperature to produce any required degree of saturation. These primary techniques require either fairly complex gas-handling systems or controlled-temperature environments for the duration of the hygrometer calibrations.

The secondary techniques all use the known vapor pressures of solutions to produce air samples containing various amounts of moisture. Table 3.7 lists the relative humidities attainable from a variety of saturated salt solutions at various temperatures; Table 3.8 lists the relative humidities obtainable from distilled water–glycerine mixtures at 25°C. The saturated salt solutions are particularly easy to use: a tray containing water and an excess of salt is placed in a sealed test chamber large enough to hold the hygrometer sensor. Some provision for circulating the air in the chamber and measuring its temperature is necessary. The accuracy of these secondary techniques is usually sufficient for most purposes.

3-9 PRECIPITATION MEASUREMENTS

3-9.1 Rain Gauges

Instruments for the measurement of precipitation (commonly called rain gauges) are of two general types: nonrecording and recording. In the first type the total amount of precipitation is determined by measuring the depth of a contained liquid column with a graduated

TABLE 3.7 Relative Humidity of Saturated Salt Solutions, %†

Temperature		Relative humidity of saturated salt solution, %‡				
°C	°F	KNO_3	NaCl	$Mg(NO_3)_2 \cdot 6H_2O$	$MgCl_2 \cdot 6H_2O$	LiCl
0	32	97	76	54	34	19
5	41	96	76	54	33	16
10	50	95	75	53	33	14
15	59	95	75	53	33	13
20	68	94	75	53	33	12
25	77	93	75	52	33	11
30	86	92	75	52	32	11
35	95	90	75	51	32	11
40	104	89	75	51	31	11

†(From W. E. K. Middleton and A. F. Spilhaus, 1953, *Meteorological Instruments*, p. 117, University of Toronto Press, Toronto, Canada.)

‡KNO_3 = potassium nitrate; NaCl = sodium chloride; $Mg(NO_3)_2 \cdot 6H_2O$ = magnesium nitrate; $MgCl_2 \cdot 6H_2O$ = magnesium chloride; LiCl = lithium chloride.

Table 3.8 Relative Humidity Obtained from Water–Glycerine Mixtures (25°C)†

Relative humidity, %	Glycerine (by weight), %	Specific gravity
10	95	1.245
20	92	1.237
30	89	1.229
40	84	1.216
50	79	1.203
60	72	1.184
70	64	1.162
80	51	1.127
90	33	1.079

†(From W. E. K. Middleton and A. F. Spilhaus, 1953, *Meteorological Instruments*, p. 117, University of Toronto Press, Toronto, Canada.)

dipstick. Because the measurement is usually made only once a day, only the accumulated total can be determined. The recording collector is equipped with electrical or mechanical devices to allow a continuous record of precipitation accumulation. In this instance, the record permits a determination of rate of accumulation as well as total accumulation. In both classes of instruments, rainfall is more accurately recorded than is the accumulation of snow, sleet, hail, or other frozen precipitation. A review article by Kurtyka (1953) describes almost 200 types of gauges that have been built and used.

3-9.1.1 Stick Rain Gauge. The most frequently used rain gauge is the stick gauge. It consists of an outer cylindrical container, a wide-mouthed funnel, and a smaller inner cylinder whose cross-sectional area is one-tenth that of the funnel's mouth (Fig. 3.42). Precipitation collected at the mouth is funneled into the small container to amplify the height of the liquid column; as a result, even small amounts (0.2 mm) can be measured with a graduated ruler. This gauge is suitable for those observations which require only an integrated measure over a long period of time, i.e., 12 to 24 h or longer.

3-9.1.2 Weighing Rain Gauge. A weighing rain gauge incorporates a scale mechanism to determine the weight of the water collected in its bucket system (Fig. 3.43). As the amount of catch increases, the bucket and its mounting platform are depressed. An attached pen shows the weight of the bucket (in units of amount of precipitation) on a drum-mounted chart that is driven by a clock mechanism. Depending on the rate of chart rotation, such a gauge permits a reasonably accurate determination not only of the total precipitation collected within any given time period but also of the rate of accumulation. A relatively simple calibration procedure makes use of weights that correspond to equivalent amounts of liquid water. The evaporation of the collected water can cause a cumulative error in the record, particularly if the water is allowed to stand over a long period of time. In some instances, inadequate temperature regulation of the weighing mechanism has produced spurious records indicating trace amounts of precipitation. However, neither of these two errors is serious if the gauge is tended on a daily basis.

3-9.1.3 Tipping-Bucket Rain Gauge. A tipping-bucket rain gauge incorporates a small bucket assembly that tips and empties when a given amount of water (usually 0.2 mm) fills it. In so doing, a second, identical bucket is presented to the orifice of the collector assembly, which, in turn, fills and tips, repositioning the first bucket. Each tip momentarily closes an electrical switch to generate an electrical impulse. This impulse can be used to activate an event pen, which, in turn, provides a continuous chart record of the occurrence in time of each unit of precipitation. Thus, in addition to total precipitation, it is possible to determine with high precision short-term accumulation rates, if the chart speed is fast enough. Alternately, the electrical impulses can drive an event counter to provide a measure of the total accumulation during a given time period. Evaporation and temperature changes produce no

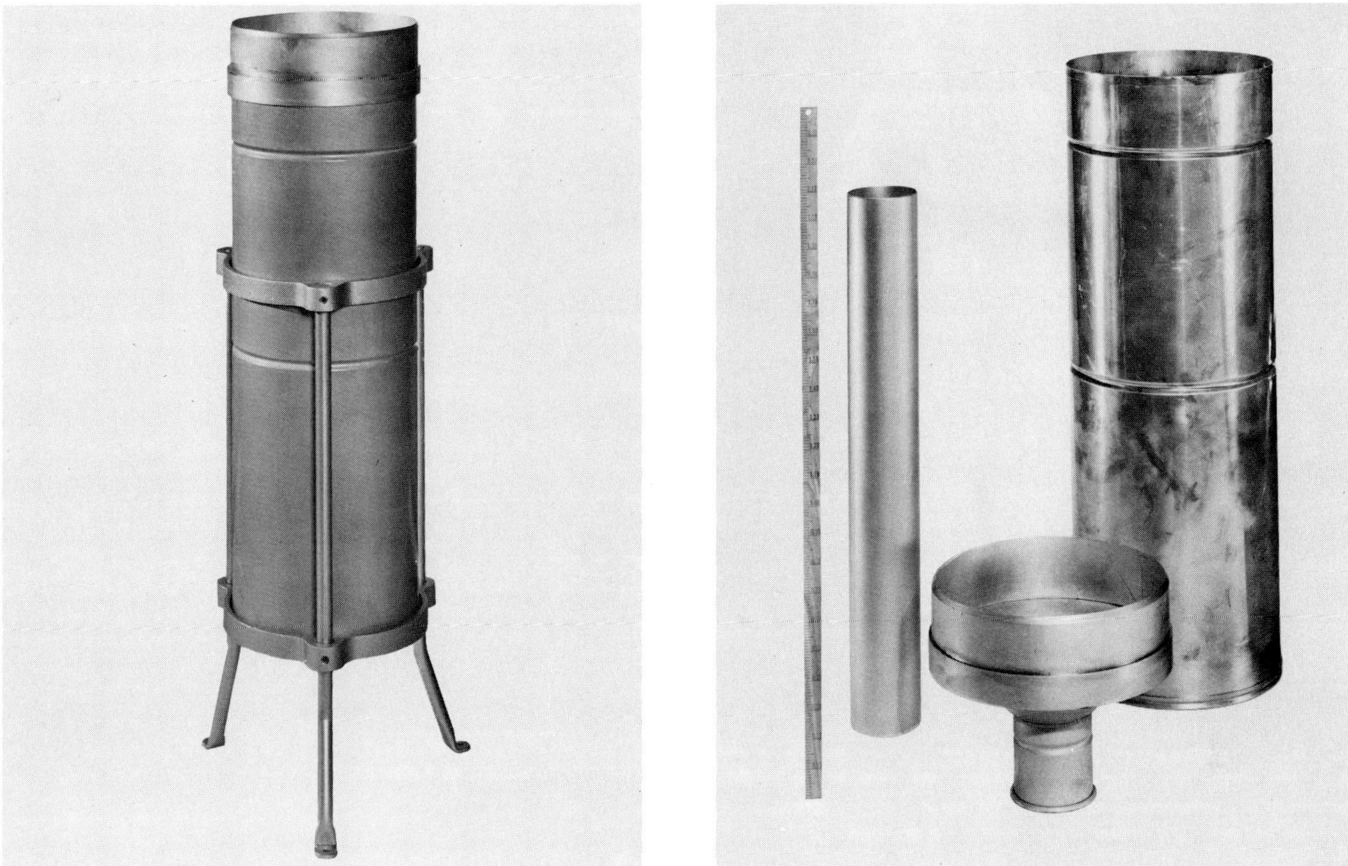

(a) (b)

Fig. 3.42 A standard rain gauge: (a) assembled; (b) components. The precipitation is collected in the wide-mouth funnel, which fits in the top of the large cylinder and passes into the narrow cylinder, which is mounted below the funnel. Precipitation accumulation is measured with the graduated rule. (Belfort Instrument Co.)

deleterious effects in this type of gauge. However, accumulations less than the amount required to tip the bucket go unrecorded.

3-9.2 Exposure of Rain Gauges

The major cause of inaccurate precipitation measurements is aerodynamic flow in the vicinity of the gauge mouth. Warnick (1956) established that any moderate to strong flow of air over a gauge standing vertically above the ground creates turbulent eddies that divert a significant portion of the precipitation away from its intake. In strong winds the gauge alone and upwind obstacles can produce deficiencies of from 10 to 30% in the measured precipitation. There are techniques which can be used to minimize this effect to some extent (Kurtyka, 1953), but it is impossible to eliminate this effect completely with any standard gauge.

The character of the precipitation itself may be of interest in certain studies, such as those relating to acid rain or to the determination of pollutant loadings. Usually these evaluations require considerably larger volumes of water than standard rain gauges provide. For this reason, several laboratories and Department of Energy contractors have designed and constructed collectors with very large inlet diameters and complex bucket-switching systems. These devices permit the separation of accumulated precipitation into short-term samples, each capable of containing several liters.

3-10 SURFACE-BASED REMOTE-SENSING TECHNIQUES

The capability to measure meteorological parameters at any selected point in the atmosphere from the earth's surface has been a long-sought-after goal. The concurrent development of solid-state electronics, devices such as the microprocessor which use this technology, and sensors that respond at a distance to the presence or fluctuations of naturally occurring atmospheric parameters have seemingly made this goal attainable. As these new sensors are used to study the characteristics of the particular atmospheric parameters to which they respond, the relationship between them and the conventional parameters (such as temperature, wind speed, and turbulent intensity) becomes better known; ultimately, these last can be derived from the first.

At the present time, much remote-sensing effort is still in the developmental phase. As a result, the equipment or the techniques are too sophisticated for routine deployment. However, two techniques have advanced to the stage of commercial availability: acoustic sounding and optical radar sounding of atmospheric structure. It is not by

Fig. 3.43 A weighing rain gauge. Precipitation enters the top of the gauge and passes into a bucket (not shown) that rests on the bucket platform. A cumulative record of the depth of precipitation is recorded on the clock-driven drum. (Belfort Instrument Co.)

accident that this development has taken place because these two techniques show promise of great utility when they are applied to the study of atmospheric dispersion in the boundary layer—a problem of importance in power production. Therefore we discuss these two techniques in some detail in the sections following.

Conventional radar and Doppler radar have been used for some time to monitor and elucidate the nature of convective storms. However, the utility of these radar techniques when applied to power-production-induced atmospheric perturbations is very limited, and the expense of acquiring and maintaining a meteorological radar usually cannot be justified. We do not discuss these techniques here for this reason. However, Lhermitte (1973) shows what Doppler radar can do. More recently, Serafin and Strauch (1978) describe real-time acquisition and processing of meteorological Doppler radar data.

We have limited our discussion of remote-sensing techniques to surface-based observations on purpose. Certainly, pollution phenomena associated with power-production processes have been observed from satellites, both manned and unmanned (see, for example, Randerson, 1977; Lyons and Pease, 1973; and Randerson, 1968). But, although regulatory agencies find these data helpful in their abatement activities, an individual user or producer is not likely to find these global-scale data as useful.

3-10.1 Acoustic Sounding

The acoustic radar† is a device used to measure both the temperature structure and the wind field of the boundary layer up to about 1 km in altitude. In operation, a sound pulse is transmitted upward through the atmosphere, and acoustic energy scattered by temperature fluctuations and wind turbulence is detected by an appropriately positioned receiver.

The fraction $\sigma(\theta)$ of the incident power scattered per unit incident flux into a unit solid angle from a unit volume can be written

$$\sigma(\theta) = (0.055)\lambda^{-1/3}\cos^2\theta \left[\frac{C_V^2}{c^2}\cos^2\left(\frac{\theta}{2}\right) + (0.13)\frac{C_T^2}{T^2}\right]\left(\sin\frac{\theta}{2}\right)^{-11/3} \qquad (3.68)$$

where θ is the scattering angle (measured from the initial direction of propagation of the transmitted wave), λ is the wavelength of sound for the average temperature T in the scattering volume, and c is the speed of sound in the undisturbed atmosphere (Little, 1969). The terms C_V and C_T are the velocity and thermal structure functions defined by

$$C_V^2 = \overline{\left(\frac{V(z) - V(z + \Delta z)}{(\Delta z)^{1/3}}\right)^2} \qquad (3.69)$$

and

$$C_T^2 = \overline{\left(\frac{T(z) - T(z + \Delta z)}{(\Delta z)^{1/3}}\right)^2} \qquad (3.70)$$

where V(z) and T(z) are the instantaneous wind speed and temperature at position z (in any direction). Note that C_V and C_T can be interpreted as the mean square of the velocity or temperature difference between two points in space a unit distance apart. In Eq. 3.68, two terms are shown to contribute to the scattering at any angle: one owing to fluctuations in wind velocity and the other to fluctuations in temperature. Note that the same equation also shows that the scattering is highly dependent on the scattering angle. No energy is scattered at 90° or 270°, because of the longitudinal nature of sound waves, and because that portion of the scattering owing to velocity fluctuations vanishes at 180°, i.e., at right angles and in the opposite direction to the initial direction of propagation.

Use is made of this last scattered-wave property to probe only temperature structure by operating the acoustic radar with the transmitter and receiver co-located. In this mode—called the monostatic mode—the acoustic radar transmits a short pulse of energy; the backscattered energy

---

†The term "acoustic radar" is a misnomer, since "radar" means "radio detection and ranging." More logically, a term like "acdar" (acoustic detection and ranging) or "sodar" (sound detection and ranging) should be used, and, in fact, both are. Because a clear preference for one or the other has not been established, we will continue to use the original phrase.

is detected by the receiver. Here, the backscattered power is given by the usual radar equation (McAllister et al., 1969)

$$P_r = P_t \frac{c\tau}{2} \frac{A}{R^2} L\sigma(\pi) \qquad (3.71)$$

where P_r = acoustic power returned
P_t = power transmitted
c = speed of sound
τ = duration of transmitted pulse
A = effective area of the receiving antenna
R = range to the target volume
L = round-trip attenuation factor

As mentioned above, only temperature fluctuations give rise to a backscatter signal. The size scale of these fluctuations is related to the wavelength of the transmitted signal and is equal to about one half the wavelength of the transmitted signal; for a 1600-Hz signal, the size scale is the order of only 10 cm.

A simplified functional diagram of an acoustic radar unit is shown in Fig. 3.44. Because only sound energy is involved in the acoustic radar, its circuits are much simpler than the analogs in conventional electromagnetic-pulsed radar. Operating and equipment parameters of a commercially available unit are given in Table 3.9.

In the monostatic mode, the amplitude of the echo signal vs. range and time is measured, amplified, and conveniently displayed on a facsimile recorder operating in synchronism with the receiver. Figure 3.45 shows the type of display produced in this mode of operation. There are two distinct processes that can give rise to a reflected signal: (1) turbulent mixing induced by wind shear where the temperature gradient is stable or slightly unstable and (2) mixing produced by thermally driven convective motions in an unstable atmosphere. The second of these two processes is characterized by a partial coherence in contrast to the first, which is completely without order. Both types of returns are identified in Fig. 3.45 as well as other interference signals and echoless regions. The convective thermal plumes indicate a nonuniform rise of thermally nonuniform parcels from the ground owing to surface heating.

Quantitative evaluation of monostatic data has proved difficult; neither the temperature nor temperature gradient is measurable. However, qualitative interpretation is possible. In Fig. 3.45, dark areas correspond to strong echoes coming from turbulent parcels with unstable lapse rates; white areas correspond to low-turbulence regions or neutral lapse rates. We can clearly see the rising top of the mixing layer as time passes; thus one use of this instrument is in the determination of mixing heights. Other applications include the study of site-related vertical mixing properties, the determination of inversion formation or break-up, and a heated-plume position and height monitor.

If the scattering volume that reflects the sound signal is in motion, then a Doppler frequency shift occurs in the scattered signal. This shift is related to the velocity of the medium as follows:

$$\Delta f = 2 \frac{v}{c} f \sin \frac{\theta}{2} \qquad (3.72)$$

where Δf = difference in frequency between the transmitted and received signals
f = original carrier frequency
v = velocity of the scattering element in the direction of propagation
c = phase velocity of the acoustic wave
θ = scattering angle

For a backscattered signal, $\sin(\theta/2) = 1$, and so, in the monostatic mode, the vertical wind-velocity profile can be measured directly if the frequency shift is determined. Such operation is possible as Beran, Little, and Willmarth (1971) show.

Another mode of operation, the bistatic mode, uses a transmitter and receiver separated by some distance. In this case, a measurement of the frequency shift in the received signal is used to determine the component of velocity of the scattering element in the direction of the bisector of the angle formed by the transmitter and receiver beam axes. If the transmitter is aimed vertically and operated in the monostatic mode, while a separate receiver operates in the bistatic mode, two components of the wind vector are measured simultaneously. From this measurement, a determination of the horizontal wind component in the plane formed by the two beam axes is possible. The addition of yet one other receiver, with its beam axis and that of the transmitter forming a plane perpendicular to the first, allows the total wind vector to be determined. In one such experiment, it was found that wind speed and direction can be measured with resolution equal to that of standard sensors (Beran and Clifford, 1972). In this case, the scattering volume at an altitude of 150 m was found to have a maximum horizontal dimension of 30 m and a maximum vertical depth of 34 m. The wind speed so determined is a spatial average over the volume of the scattering element; for a single pulse, the velocity resolution is ±0.5 m/s. The resolution can be improved (at the expense of spatial resolution) by averaging over a series of pulses; e.g., a 10-pulse average results in a resolution of ±0.15 m/s.

Several schemes exist to measure the entire wind velocity profile up to about 1 km. Two of these are (1) transmit a continuous wave from a vertically pointing transmitter and use a steerable narrow-beam receiver to look at different regions of the illuminated column and (2) operate a vertically pointing transmitter in a pulse mode and use a wide-beam receiver with spatial resolution defined by the pulse length and position by range-gating. With the aid of a dedicated minicomputer, sounder data can be analyzed in real time to provide a continually updated display of the wind profiles.

Strong surface winds, rain, and hail generate sufficient noise to saturate the sensitive audio receiver and, consequently, impose limitations on use in such inclement

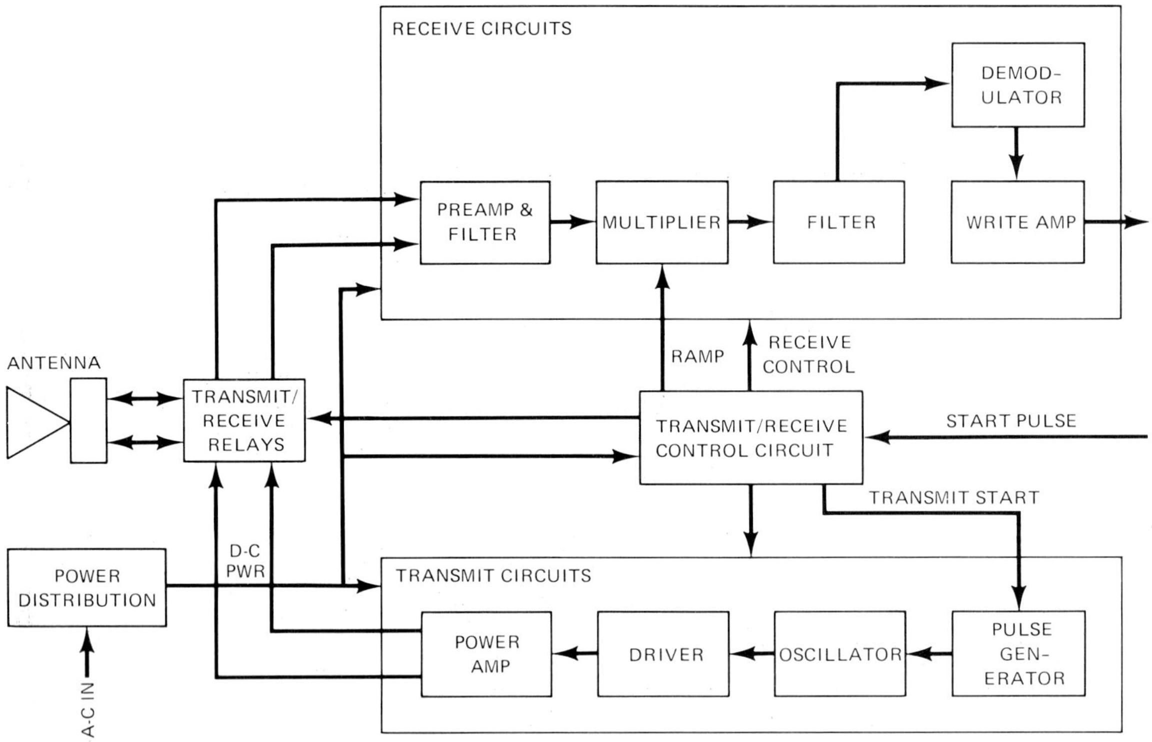

Fig. 3.44 Schematic block diagram for an acoustic sounder.

weather. Similarly operation in noisy environments may cause reception problems; however, the units have been operated successfully at airports where the passage of a plane overhead can blank the receiver for 2 to 3 min (Beran, 1971). Acoustic shields designed to attenuate undesirable side-lobe signals in the transmitted beam are available. These enclosures permit operation of the units in residential and urban neighborhoods with a minimum of audio disturbance. Additional information on problems relating to background noise is given by Hall, Westcott, and Simmons (1971).

In summary, acoustic radar can be used in a variety of ways to measure at a distance some properties of the atmosphere. Operation in the monostatic mode allows only a qualitative interpretation of the received signal; but gross features such as inversions and their change of height in time can be readily identified. In the bistatic mode, quantitative measurement of the wind-velocity profile up to altitudes of 800 m is possible.

A combined *r*adar-*a*coustic *s*ounding *s*ystem (RASS) is described by Marshall, Peterson, and Barnes (1972). In this system, acoustic pulses directed vertically upward are tracked by Doppler radar. The radar return signal is generated by localized changes in the atmospheric dielectric constant as the pulses of sound pass through the atmosphere. The atmospheric temperature profile is determined from an analysis of the radar signal which measures the velocity of the sound pulse. Only very powerful and highly sensitive radar systems can be used for the RASS because of the weak return generated.

3-10.2 Optical Sounding

Lidar units operate in a fashion analogous to that of microwave radar. Both units transmit pulses of electromagnetic energy, but the frequencies of these transmissions are in different regions of the electromagnetic spectrum. Radar transmits signals whose frequencies are in the microwave spectrum, whereas lidar frequencies are in the optical and the infrared. Lidar is sometimes called lasar radar or optical radar because of this close association. However, the more appropriate term "*lidar*"—an acronym for *li*ght *d*etection *a*nd *r*anging—is preferred because of its analogy with radar.

Principle uses of lidar are to measure (1) gaseous pollutant concentration, (2) cloud height and thickness, (3) wind velocity, and (4) dust or aerosol concentration. Now many of these applications involve complex equipment and extensive data-acquisition-and-processing systems. As a result, most of these are limited to research purposes only because of their complexity and cost. Eventually, some may become practical operational devices. But there are only two applications of real interest to the power-production community today: measurements of aerosol and gaseous pollutant concentrations. We will describe these as well as some promising current research.

A lidar unit consists of a transmitter and receiver whose axes are parallel. The transmitter is a pulsed laser; the receiver is a telescope–collimator–phototube combination. In operation the laser generates a short pulse of highly collimated coherent light that is transmitted upward through the atmosphere. This beam of light illuminates a column of the atmosphere. The transmitted pulse is

TABLE 3.9 Acoustic Radar Specifications

Transmit Pulse

Frequency, 1600 Hz
Duration, selectable 50, 100, 200 ms
Power input, 35 W
Repetition rate, 1 per 7 s (500-m scale); 1 per 14 s (1-km scale)

Receiver

Gain, 10^8
Gain compensation, proportional to time of echo return, with additional adjustment for short ranges
Bandwidth, less than 20, 40 and 80 Hz (between −3 dB points, 72 dB/octave cutoff) for 200-, 100-, and 50-ms pulses, respectively; automatically switched when pulse length is changed
Range, selectable 1 km or 500 m full scale, minimum 10 m due to 60-ms receiver dead time after pulse transmission
Resolution, 10 m

Antenna

Reflector, 1.2-m diameter paraboloid, 0.5-m focal length
Beam half-angle, 26°

Recorder

Writing technique, electric on conducting chart paper
Chart size, 15.2 cm wide by 22 m
Chart speed, 38.1 mm/h
Chart duration, 28 d

Sound Level (with Enclosure)

Transmitted horizontally, less than 60 dBA at 10 m
Maximum ambient for operation, 68 dBA for 1-km scale, 74 dBA for 500-m scale

Size and Weight

Electronics and recorder, 43 cm wide by 43 cm high by 15 cm diameter, 18 kg
Antenna and transducer, 133 cm diameter by 80 cm high, 23 kg

Power Input

115 V, 60 Hz, 50 W

scattered or absorbed and reradiated back to the receiver by a variety of processes, some of which are described subsequently. As a result, signals are received from the illuminated column to a height at which normal atmospheric attenuation reduces the return to the threshold level. The return signal passes through a narrow-band optical filter to reduce interference either from sky light or other sources and is detected by a photomultiplier. The output signal from this last device is a voltage whose amplitude can be measured vs. time. Ranging is accomplished by measuring the time interval since the transmission of the pulse. Because the pulse is electromagnetic in nature, its propagation velocity is that of the speed of light.

Two elastic scattering processes contribute to the reflection of the transmitted signal; these are Raleigh scattering and Mie scattering. Raleigh scattering takes place when light is reflected from bodies whose dimensions are much smaller than the wavelength of the incident light; in the atmosphere, the molecules of the constituent gases are responsible for Raleigh scattering. Mie scattering arises from reflections off bodies whose sizes are approximately equal to the wavelength of the transmitted signal; in the atmosphere, suspended aerosols and dust particles give rise to this process. For each of these two processes, the usual radar equation holds. A modification of this equation is necessary if a photomultiplier tube is used as a detector (Grams, 1966). The modified equation for the photoelectron emission rate per laser pulse from an optically thin layer of scatterers homogeneously distributed is

$$\frac{dn}{dt} = \eta W \frac{\lambda}{8\pi h} \frac{A}{R^2} K_t K_r K_a^2 N \sigma \tag{3.73}$$

where η = the quantum efficiency of the photocathode
W = transmitted energy (per pulse)
λ = wavelength
h = Planck's constant
A = area of receiving telescope
R = range of the scattering layer
K_t = efficiency of the transmitter
K_r = efficiency of the receiver
K_a^2 = round-trip attenuation factor
N = number density of scatterers
σ = radar cross section of an individual scatterer

As the preceding equation shows, the scattering cross section is proportional to the number density of scatterers, i.e., the number of molecules or aerosol particles per unit volume. In a dust-free atmosphere, a smoothly decreasing signal with altitude is to be expected from Raleigh scattering alone; the magnitude of such a signal provides a direct measure of atmospheric density as a function of altitude (range). Normally, the atmosphere is not dust-free. Instead, a sizable reflected signal is returned from the normal aerosol loading. As a result, these particles can be detected and their distribution and concentration measured as a function of altitude. For example, aerosol concentrations have been measured not only in the troposphere but also in the stratosphere and mesosphere by this technique (Fiocco and Smullin, 1963).

Scattering from the particulates in a power-plant plume allows lidar to monitor the plume position as well as the rate at which it is dispersing. An early study by Johnson (1969) used lidar for plume-position detection. More recent studies have shown the ability to differentiate between particulates from hot sources and dust by analyzing the depolarization of an initially polarized lidar beam (Derr, 1977).

An inversion aloft can trap aerosols below it and keep them in the mixed layer. Lidar can be used to detect and measure the location and altitude of these aerosols. Even more, a single fixed lidar can measure the vertical motions of these trapped aerosols and so monitor the diurnal variation of the mixing height (Uthe, 1972). A two- or three-dimensional profile of the aerosol cloud can be produced by a scanning lidar, either ground-based or airborne (Olsson et al., 1974; Grams, Patterson, and Wyman, 1975).

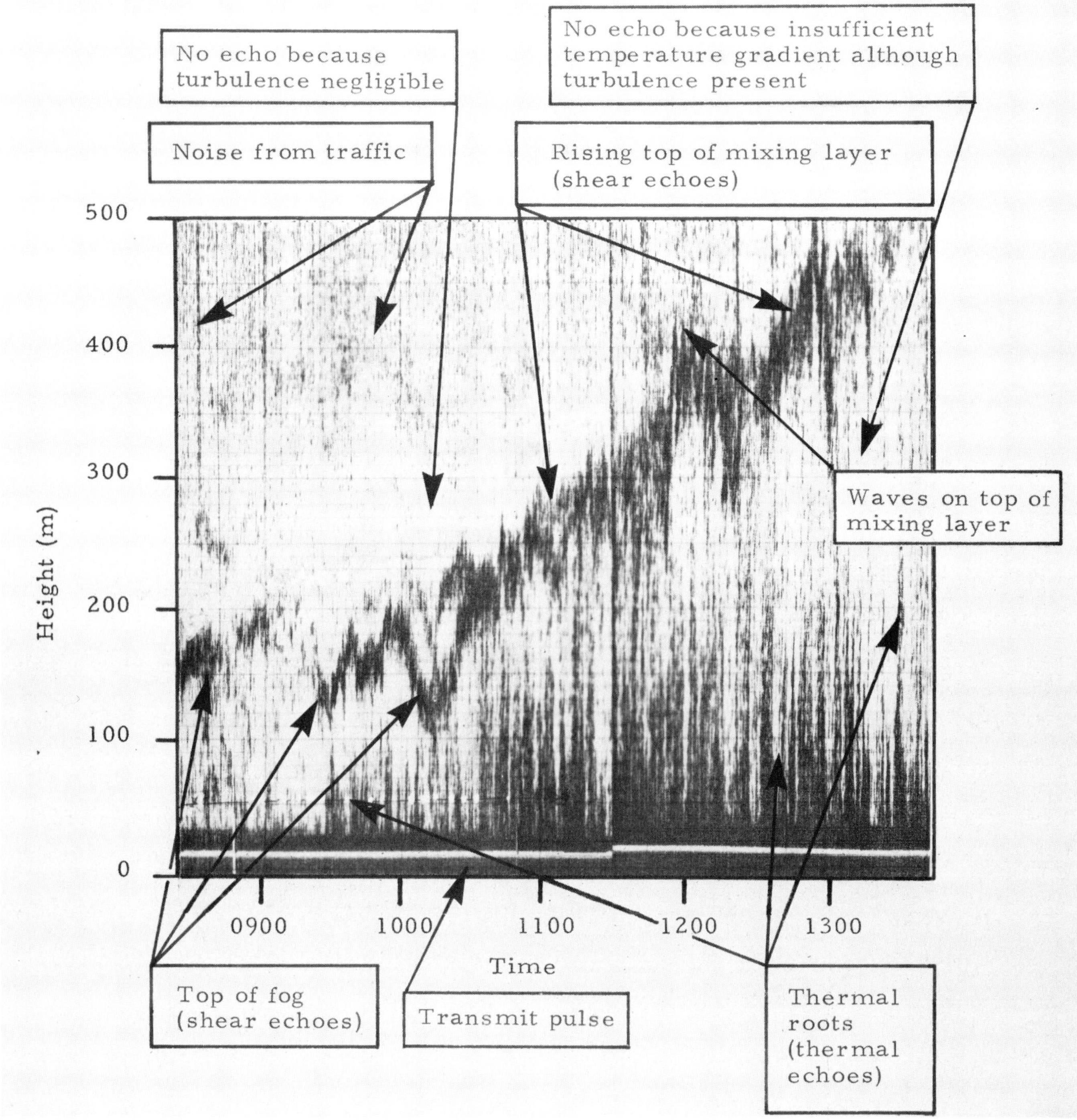

Fig. 3.45 Acoustic sounder record. The record shows the development of a deepening mixing layer evolving from early morning fog. (Courtesy of Aerovironment Inc.)

For both Raleigh and Mie scattering, the frequency of the reflected signal is the same as that of the transmitted signal. In contrast to these elastic scattering processes, Raman scattering is an inelastic process. In this case, the return is frequency-shifted owing to molecular transitions from one molecular vibration-rotation state to another. Since such structure transitions are unique to a given molecule, the frequency shift is specific to that molecule and can be used to identify the atmospheric constituent responsible for the scattering. For example, sulfur dioxide profiles in power-plant plumes have been detected and measured by this technique (Melfi, Brumfield, and Storey, 1973). Other examples may be found in the review article of Kildal and Byer (1971).

The analogy between optical and acoustic sounding techniques is set out by Russel et al. (1974), who used a monostatic lidar and a monostatic acoustic sounder to probe a common volume of the atmosphere; their paper contains several illustrations comparing the simultaneous outputs of these two sensors under a variety of meteorological conditions in an urban area.

The optical analogs to conventional electromagnetic radar are undergoing development. Such devices as FM–CW lidars and Doppler lidars have been built and tested. Their principles of operation are exactly analogous to that of the corresponding radar devices. The existing units are primarily research instruments used to further develop and refine the techniques. Schwiesow and Post (1977) discuss some of the uses which these devices may find when they become fully operational.

3-11 METEOROLOGICAL-DATA PROCESSING

A data-acquisition system gathers data from a variety of meteorological sensors whose outputs are usually in the form of analog electrical signals. These systems can have any degree of complexity, ranging from the very simple to the very complex. A simple system, for example, is a mechanical impulse counter that totals incoming pulses (as from a tipping-bucket rain gauge); a complex system might incorporate several remote, automated, multichannel, digital data-acquisition devices in communication with a central command- and data-processing unit. However, no matter what the degree of complexity, the primary function of the system is to process the incoming signals and convert them into suitable forms for subsequent analysis.

The ultimate inputs to the system are the environmental elements that impinge upon sensors designed to respond to them. As a result of this interaction, a given sensor undergoes a change of state that may or may not be converted to an electrical signal by a transducer. In the case of solar radiation, for example, the sensor can be a flat absorbing plate whose temperature changes as radiation illuminates it; a thermojunction transducer converts the temperature change to an electrical voltage that becomes the input signal to the data-acquisition system. On the other hand, a bimetal actinometer uses a sensor that undergoes a change in length in response to a change in insolation and, as a transducer, a simple mechanical linkage connected directly to a recording device. Consequently the operation of this instrument depends on only mechanical devices.

There is usually a wide range of available sensor/transducer combinations to measure a given meteorological variable, as shown by the last example. Therefore, during the initial design phase of a meteorological system, we must consider simultaneously the selection of all the sensors, the data-recording technique, and the data processing. To date, the steady increase in the capability of data-processing and computing equipment has been coupled with a steady decline in its cost. In view of this fact, it is legitimate to question whether the data should be hand-abstracted from chart recorders or whether a more sophisticated recording mechanism is justified, particularly when several instruments are involved. Since virtually all multiinstrument meteorological systems produce electrical voltage or current signals as outputs, only data-acquisition systems that require electrical signals as inputs will be considered in answering this question.

3-11.1 Chart Recorders

Analog voltages and currents are easily converted to shaft-position changes that lend themselves readily to chart recording. For transducers that have a binary voltage (i.e., "on" or "off") as an output, such as the tipping-bucket rain-gauge switch, time-of-event pens are often used to produce ink-trace records. As a result, one of the most common data records is the strip or circular chart on which one or more pens record the parameters of interest, usually as a function of time (Fig. 3.46). The recorders that produce these charts are highly developed devices, and well-constructed equipment of this type is very reliable: many recorders have run for periods as long as 10 or even 15 yr without major maintenance. Although these recorders have various internal mechanisms to convert electrical signals into pen movement, they can all provide good resolution of the recorded parameter with time, if run at sufficiently high chart speeds. Furthermore, the accessibility of the chart roll while still mounted on the recorder permits easy visual inspection of existing conditions as well as those over previous hours or days. This accessibility is advantageous if it is frequently necessary to evaluate recent meteorological changes and to act on this information quickly.

Fig. 3.46 A pair of strip-chart recorders mounted in a single cabinet. In this instance, a single motor drives both charts which permits the synchronized recording of two variables. (Esterline–Angus Instrument Co.)

Considerable effort is required to extract information from the chart records. Equipment is available to simplify the task, but such equipment should be considered as an aid to analysis rather than as a data-processing system. For example, data reduction can sometimes be facilitated by the addition of an integrating device operating in conjunction with the recorder. One such device can be used with a tipping-bucket rain gauge: a counter will automatically total events over a period of time while the recorder provides the detailed sequence-of-events record. In some recorders, an event pen can count directly on the record itself. More sophisticated integrators do exist, such as the mechanical ball-and-disk integrator (Moses and Kulhanek, 1962) and the electronic integrator that activates an event pen when a given total quantity of the measured parameter is attained.

Another means of facilitating data reduction involves alteration of either the sensor or the recording mechanism to give a slower pen response. In both cases, a reduction of detail is achieved to lessen the amount of information recorded. Sensors with long time constants (such as resistance thermometers of large mass) are often used to smooth out transients. Alternately, recorder response can be slowed either electrically or electromechanically. For example, Brown (1959) used servomotors operating through 1000 : 1 gear ratios to greatly slow the response of his recording system.

Hand reduction of chart data can be accelerated by the use of chart readers. These devices include mechanisms for the orderly advance and inspection of the chart by the operator and a simplified means of following the pen trace as the chart moves. In a typical system the operator follows the pen trace with a stylus or with wheel-driven vertical and horizontal cursors and, at selected intervals, the position of these last is converted into digital form and recorded automatically on punch cards or tape. Some chart readers can follow the trace photoelectrically and record the trace position automatically at set intervals of time or chart spacing.

3-11.2 Digital Data-Acquisition Systems

Currently there are a large number of automatic data-recording systems available commercially. The rapid advances in solid-state electronics of the past decade have produced circuitry that is extremely reliable, is able to take considerable abuse, requires very little power to operate, and, by past standards, is low in cost. With its use, considerable sophistication can be introduced into meteorological studies, and data-acquisition requirements that previously would have been considered impracticable are now routine.

Whether called a digital data-acquisition system or, more simply, a data logger, most of these systems contain a standard collection of circuits configured in much the same way. The elements of such a representative data-acquisition system are shown in Fig. 3.47. The incoming electrical signal passes first through a signal conditioner and filter whose functions are to tailor the signal to the input requirements of the multiplexer. This last unit is an electronic switch that sequentially connects each input channel to it for a specified time interval. When so connected, the signal from the selected channel passes through the multiplexer to the sample-and-hold circuitry,

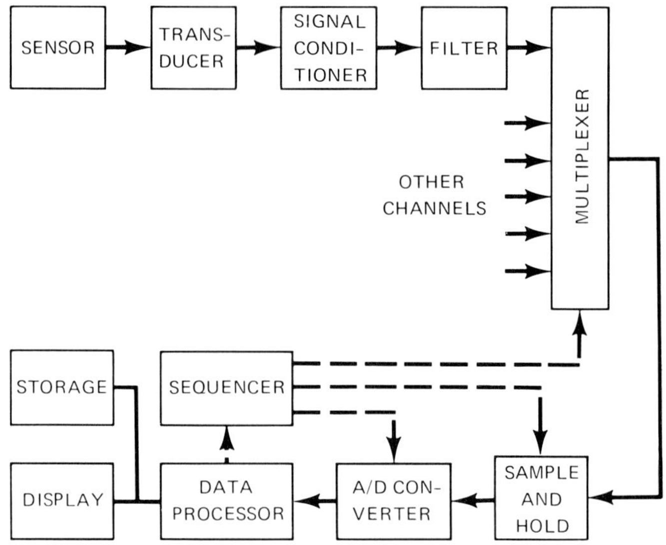

Fig. 3.47 Schematic block diagram for an analog-to-digital data acquisition system. The flow of data is shown by ——→; control signals follow the — — —→.

which quickly samples and maintains an unchanging output signal at the input of the analog-to-digital (A/D) converter. Here, the signal is finally converted to a digital form in which it can be fed to a computer, to a digital data transmitter, or to a digital data-recording device. To synchronize the operation of all these circuits and to control both the sequence in and frequency at which the data are acquired, a sequencer circuit controls the multiplexer, the sample-and-hold circuit, and the A/D converter. The sequencer, in turn, may be controlled by a minicomputer or microprocessor.

In general, once the signal has been converted to electrical form, all subsequent signal processing is done by electronic circuitry.

3-11.2.1 Signal Conditioner. The incoming signal passes first through a signal conditioner whose function is to alter and shape this signal into one suitable as input into the remaining circuitry. This circuit may perform the following functions, singly or in combination: (1) it can amplify the signal, (2) it can buffer the signal and effectively isolate the transducer from the data-acquisition system, and (3) it can convert a current signal into a voltage signal. Normally, each transducer is connected to an individual signal conditioner whose response can be specifically tailored to the particular form of the transducer's output signal. As a result, the output signals from the bank of signal conditioners can all be standardized to lie within the acceptable range of the multiplexer. Usually these signal conditioners are operational amplifiers.

3-11.2.2 Filter. A low-pass filter circuit usually follows that of the conditioner. The filter performs two functions: (1) it reduces the electrical noise in the signal and (2) it limits the bandwidth of the incoming signal. Why such a limitation is necessary is discussed in Sec. 3-11.3. Active filters, which use operational amplifiers as circuit elements, are preferred to passive resistance-inductance-capacitance filtering techniques. These active filters cause no insertion loss, minimize loading effects, and have response characteristics determined by temperature-stable resistors and capacitors.

3-11.2.3 Multiplexer. The multiplexer is an arrangement for time-sharing a single A/D converter; in this way, a separate converter is not needed for each channel. Since the A/D converter is usually the most costly part of the system (exclusive of a microprocessor or minicomputer that may control it), an analog multiplexer reduces the cost of the system. The multiplexer consists of a group of solid-state switches connected to the individual input channels and outputs connected to a common terminal. These switches can be activated by appropriate digital input codes generated by the sequencer. The switches are normally open; but under the control of the sequencer, they can be closed in any order for any specified interval of time. When a particular channel switch is closed, the input signal from that channel passes through the multiplexer and appears at the input to the sample-and-hold circuit. This switching can be made very rapidly, certainly in the order of microseconds or less. The sequences of switch selection can also be controlled, i.e., it is not necessary to go through all channels in order, but any switching order suited to the needs of the measurements can be generated. The rate at which any particular channel switch is activated determines the sampling frequency for that channel. Because of the speed at which the switch operates, a given channel can be sampled many times in a short period. Therefore a multiplexer allows considerable flexibility as to how frequently and in what sequence each channel is sampled.

3-11.2.4 Sample-and-Hold. The accurate conversion of a rapidly changing analog signal to digital form requires an A/D converter that responds in a time interval which is short compared to the temporal variation of the incoming signal. For a reduction in the severity of this requirement, especially for high-frequency signals and lower overall system costs, an inexpensive sample-and-hold circuit is inserted between the multiplexer and the A/D converter. The sample-and-hold circuit very quickly samples the output of the multiplexer under the control of the sequencer and maintains the signal level it senses at a constant value while the A/D converter is operating. In essence, a sample-and-hold circuit is a switch and a capacitor. With the switch closed the unit tracks the input signal, and the capacitor is always charged to the value of that signal. When the switch opens, the capacitor retains the instantaneous input voltage for some period of time. This unchanging voltage is the input signal to the A/D converter.

3-11.2.5 Analog-to-Digital Converter. The heart of the data-acquisition system is the A/D converter. Here, the analog electrical signal is converted into a digital form suitable for subsequent processing by digital equipment. An A/D converter relates the analog values to digital values by means of an appropriate code. There is a variety of techniques for generating this code, each of which gives rise to a particular form of A/D converter. Because of their complexity, the operation of these circuits will not be described in detail. It is sufficient to note that, as a result of the operation of an A/D converter, a binary-encoded signal which gives the magnitude of the corresponding analog signal appears at its output terminal. This output signal is finally routed to the output device which may process it further (a computer), transmit it to a remote location (a digital data transmitter), or store it (a digital data-recording device).

3-11.3 Signal Sampling

The principles of the signal-sampling process are shown in Fig. 3.48, which depicts a time-varying input signal (Fig. 3.48a) and a train of sampling pulses (Fig. 3.48b). These pulses represent the times at which the fast-acting switch in the multiplexer connects the analog input signal to the sample-and-hold circuit; their frequency, i.e., the sampling frequency, is ultimately set by the operator through the sequencer and is determined by the needs of the project. The output from this last circuit is shown in Fig. 3.48c. It is this step-wise varying signal that is converted to digital form by the A/D converter. A binary code is generated for the level of each step.

We can ask: why sample the signal at all? Presumably a given signal could be connected continuously to processing circuitry for full-time, real-time analysis. But such a procedure is extremely expensive because a single recorder or processor must be dedicated to just this one signal and, as a result, only a small fraction of its capability is utilized. On the other hand, if incoming signals are sampled periodically, a single data link can be used to transmit many channels of information to a single processor or recorder. Obviously this latter approach is more cost-effective, especially if a variety of signals from a large number of instruments is to be transmitted over any distance.

If the incoming continuous signal is sampled periodically and contains no frequency components higher than f, then the original signal can be recovered if it is sampled at a rate of at least 2f samples/s. A simple illustration of this requirement is shown in Fig. 3.49a, which shows a periodic signal sampled at a rate less than twice per cycle. If the sampled amplitudes are connected to construct a representation of the original signal (Fig. 3.49b), the reconstructed signal has a period that differs from the original. This process is called signal "aliasing." Obviously, as Fig. 3.49 shows, if the original waveform is sampled at least twice per cycle, the original frequency is retained in the reconstruction.

A second adverse effect of a too-infrequent sampling rate is "frequency folding." The spectrum of the sampled signal is centered at f_s with sidebands extending to $f_s \pm f$, where f_s is the sampling frequency and f the upper cutoff frequency for the original signal. If f_s is not greater than 2f, then the sampled-signal spectrum overlaps the high-frequency end of the original signal spectrum, i.e., the high-frequency components of the signal are folded into the

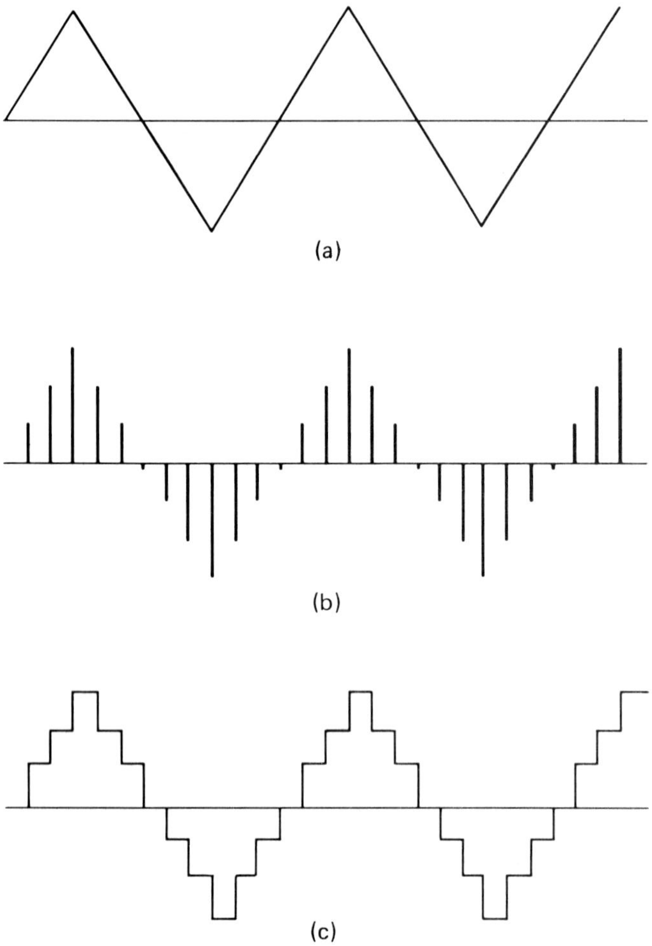

Fig. 3.48 Signal sampling. (a) The time-varying input signal; (b) the instantaneous voltage at the input to the sample-and-hold circuit; (c) the output voltage from this last circuit.

sampled-signal spectrum. These components cause distortion when an attempt is made to reconstruct the original signal. Frequency folding can be eliminated by either using the proper sampling frequency or filtering the original signal to eliminate the frequency components that lie above one half the sampling frequency. The low-pass filter discussed in Sec. 3-11.2.2 performs this function.

3-11.4 Selection and Use of a Data-Acquisition System

The advent of low-cost minicomputers and microprocessors has drastically lowered the cost of data-acquisition systems. Yet some of the systems, depending on capability, are still expensive items. However, in terms of the speed with which the data can be analyzed, the improved accuracy of the data-reduction process and the accompanying savings, these systems are worth the expenditure. The data-acquisition system incorporating a data processor trades a later-in-time analysis of the data for an on-the-spot, real-time analysis. One advantage of this is the system's ability to alert the user to changing conditions or malfunctioning sensors, either of which may require immediate attention or changes in the operation of the system.

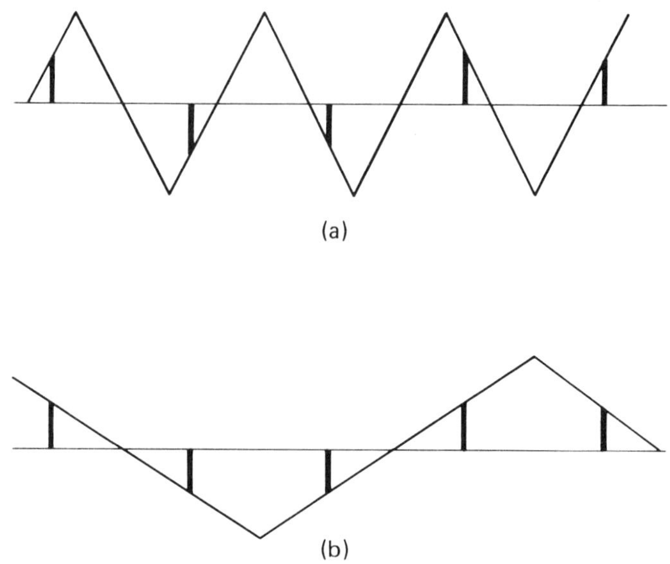

Fig. 3.49 Signal aliasing. (a) The original input waveform sampled at the indicated frequency; (b) the output waveform constructed from these samples.

Many prefer the use of strip-chart recorders to monitor the incoming data stream. The strip-chart recorder certainly has the advantage of an immediate visual presentation of the incoming data. (But it has a concurrent disadvantage: the data are in a form not suitable for computer processing, and a lengthy procedure is required to make them so.) On the other hand, in the case of the digital data-acquisition system (especially those which are rather small), the data are not immediately visible. Therefore it is essential that the system be designed so that the user can query the system at any time to find out if it is working properly and if the sensors are functioning as they should. This check can be accomplished either by a visual (though not necessarily hard copy) display of the analyzed data or by a series of data checks which exercise internally the components of the entire system. With a system of moderate size, it is always possible to generate strip-chart displays: incremental plotters can be used to plot the data with only a very short delay for analysis in a format similar to that of a strip-chart recorder; alternately, these data may be presented on a video monitor in a similar format.

Finally, the real-time capability to analyze the incoming data can be utilized to perform a series of data checks to ensure that not only is the acquisition system working but also that the sensors, which are the sources of the data, are functioning properly. Usually, some sort of quality-assurance test can be performed on the incoming data from a particular sensor to determine whether it is within acceptable limits and bears some appropriate relationship to other of the incoming data. In this way, self-consistency and cross-consistency can be evaluated in real time. Error flags can be inserted in the recorded output data to indicate where these consistency checks have failed.

The first time a data-acquisition system is put together, it is probably best to rely on an expert in the field of data processing for the short time actually required to specify the system, assemble it, and make it operational. Initially, there is a considerable amount of unfamiliar jargon and terminology that must be mastered by the neophyte before a proper appreciation of the potentialities of a given system can be attained. With continued use and exposure, the user becomes conversant with both the terminology and technology of digital data-acquisition systems. A high data-recovery rate is usually the result when this happens. But in the hands of an inexperienced user, a digital data-acquisition system may not perform its functions as designed, and a lower data-recovery rate is the result. In this last instance, the blame is generally placed on the system rather than on the user. When digital systems are used properly, data-recovery rates in excess of 95% are possible.

3-12 ACKNOWLEDGMENTS

Most of this chapter is based on an extensive rewrite of Chap. 6: Meteorological Instruments for Use in the Atomic Energy Industry, which appeared in *Meteorology and Atomic Energy—1968.* Because many people contributed to that original chapter and because after so long a period of time it is impossible to identify individual contributions, we wish to acknowledge the contributors to the 1968 publication: C. Ray Dickson, G. C. Gill, H. Moses, E. Robinson, M. E. Smith, and E. M. Wilkins. We also wish to thank C. M. Carnes for her diligent efforts at editing and typing the final version of the manuscript.

REFERENCES

Angell, J. K., and A. B. Bernstein, 1976, Evidence for a Reduction in Wind Speed on the Upwind Side of a Tower, *J. Appl. Meteorol.,* 15: 186-188.

Baynton, H. W., 1976, Errors in Wind Run Estimates from Rotational Anemometers, *Bull. Am. Meteorol. Soc.,* 57: 1127-1130.

Beran, D. W., 1971, Acoustics: A New Approach for Monitoring the Environment near Airports, *J. Aircr.,* 8: 934-936.

——, and S. F. Clifford, 1972, Acoustic Doppler Measurements of the Total Wind Vector, in *Second Symposium on Meteorological Observations and Instrumentation,* pp. 100-109, American Meteorological Society, San Diego, Calif.

——, C. G. Little, and B. C. Willmarth, 1971, Acoustic Doppler Measurements of Vertical Velocities in the Atmosphere, *Nature,* 230: 160-162.

Blanc, Milton L., 1958, *Climatological Investigation of Soil Temperature,* Technical Note 20, WMO 72, T.P. 28, World Meteorological Organization, Geneva, Switzerland.

Brazier, C. E., 1914, Recherches Expérimentales sur les Moulinets Anémométriques, *Ann. Bur. cent. Météorol. Fr.,* pp. 157-300.

Brevoort, M. J., and U. T. Joyner, 1935, *Experimental Investigation of the Robinson-Type Cup Anemometer,* Technical Report 513, National Advisory Committee for Aeronautics, Langley Memorial Aeronautical Laboratory.

Brown, R. M., 1959, An Automatic Meteorological Data Collecting System, *J. Geophys. Res.,* 64: 2369-2372.

Buck, A. L., 1976, The Variable-Path Lyman-Alpha Hygrometer and Its Operating Characteristics, *Bull. Am. Meteorol. Soc.,* 57: 1113-1118.

Corcoran, J. W., 1962, *Theoretical Analysis of Wind Vanes,* Report RD81, Beckman and Whitley, Inc., San Carlos, Calif.

——, and D. L. Esau, 1964, *Comparison of a Theoretical Model for Anemometer Cups with Experimental Data,* Meteorological Report 9, Beckman and Whitley, Inc., San Carlos, Calif.

Cramer, H. E., G. C. Gill, and F. A. Record, 1957, Heated-Thermocouple Anemometers and Light Bivanes—Massachusetts Institute of Technology, in *Exploring the Atmosphere's First Mile,* H. H. Lettau and B. Davidson (Eds.), pp. 233-242, Pergamon Press Ltd., London.

Deacon, E. L., 1951, The Over-Estimation Error of Cup Anemometers in Fluctuating Winds, *J. Sci. Instrum.,* 28: 231-234.

Derr, V. E., 1977, Lidar Techniques for Analyzing and Tracing Particulate Pollutants from Energy Production, in *Second National Conference on the Interagency R & D Program,* pp. 467-468, Washington, D. C.

Dines, W. H., 1887, A New Form of Velocity Anemometer, *Quart. J. Roy. Meteorol. Soc.,* 15: 95-99.

Durrenberger, R. W., and A. J. Brazel, 1976, Need for a Better Solar Radiation Data Base, *Science,* 193: 1154-1155.

Fiocco, G., and L. D. Smullin, 1963, Detection of Scattering Layers in the Upper Atmosphere (60–140 km) by Optical Radar, *Nature,* 199: 1275-1276.

Fritschen, L. J., 1965, Miniature Net-Radiometer Improvements, *J. Appl. Meteorol.,* 4: 528-532.

Fröhlich, C., 1972, Standards for Meteorological Radiation Measurements, in *Second Symposium on Meteorological Observations and Instrumentation,* pp. 267-271, American Meteorological Society, San Diego, Calif.

Fuchs, M., and C. B. Tanner, 1965, Radiation Shields for Air Temperature Thermometers, *J. Appl. Meteorol.,* 4: 544-547.

Funk, J. P., 1959, Improved Polyethylene-Shielded Net Radiometer, *J. Sci. Instrum.,* 36: 267-270.

Garbell, M. A., 1957, Fins for Meteorological Instruments, *J. Meteorol.,* 4: 82-90.

Gier, J. T., and R. V. Dunkle, 1951, Total Hemispherical Radiometers, *Trans. Am. Inst. Electr. Eng.,* 70: 339-343.

Gill, G. C., 1959, *Meteorological Instrumentation, Professional Notes, C. E., 199, Partial Bibliography on Anemometers and Calibration Devices: A Lecture by G. C. Gill Assisted by Floyd Elder,* University of Michigan, Ann Arbor, Mich.

——, 1973, The Helicoid Anemometer, *Atmosphere,* 11: 145-155.

——, 1975, Development and Use of the Gill UVW Anemometer, *Boundary-Layer Meteorol.,* 8: 475-495.

——, L. E. Olsson, and M. Suda, 1966, *Errors in Measurements of Wind Speed and Direction Made with Tower- or Stack-Mounted Instrument,* Publication 06973-1-P, Department of Meteorology and Oceanography, College of Engineering, University of Michigan, Ann Arbor, Mich.

Grams, G. W., 1966, Optical Radar Studies of Stratospheric Aerosols, Ph.D. Dissertation, Department of Meteorology, Massachusetts Institute of Technology.

——, E. M. Patterson, and C. M. Wyman, 1975, Airborne Laser Radar for Mapping Two-Dimensional Contours of Aerosol Concentration, *Opt. Quant. Electron.,* 7: 187-191.

Great Britain, Meteorological Office, 1956, Instruments for Surface Observations, in *Handbook of Meteorological Instruments, Part I,* Her Majesty's Stationery Office, London, England.

Hall, F. F., Jr., J. W. Wescott, and W. R. Simmons, 1971, Acoustic Echo Sounding of Atmospheric Thermal and Wind Structure, in *Remote Sensing of the Environment,* Seventh International Symposium, pp. 1715-1732, Willow Run Laboratories, University of Michigan, Ann Arbor, Mich.

Horst, T. W., 1973, Corrections for Response Errors in a Three-Component Propeller Anemometer, *J. Appl. Meteorol.,* 12: 716-725.

Humphrey, P. A., 1951, Atomic Energy Commission Meteorological Information Meeting, February 1–2, 1951, USAEC Report TID-399, pp. 14-16, NTIS.

Johnson, W. B., 1969, Lidar Observations of the Diffusion and Rise of Stack Plumes, *J. Appl. Meteorol.,* 8: 443-449.

Jones, J. I. P., and F. Pasquill, 1959, An Experimental System for Directly Recording the Statistics of the Intensity of Atmospheric Turbulence, *Quart. J. Roy. Meteorol. Soc.*, 85: 225-236.

Kendall, L. M., Sr., 1968, *The Standard Total-Radiation Absolute Radiometer*, National Aeronautics and Space Administration Technical Report 32-1263.

Kildal, H., and R. L. Byer, 1971, Comparison of Laser Methods for the Remote Detection of Atmospheric Pollutants, *Proc. IEEE*, 59: 1644-1663.

King, L. V., 1914, On the Convection of Heat from Small Cylinders in a Stream of Fluid: Determination of the Convection Constants of Small Platinum Wires with Applications to Hot-Wire Anemometry, *Philos. Trans. R. Soc., London, Ser. A*, 214: 373-432.

Kurtyka, J. C., 1953, *Methods of Measuring Precipitation for Use with Automatic Weather Station*, Report of Investigation No. 20, Illinois State Water Survey, Urbana, Ill.

Latimer, J. R., 1978, *Radiation Measurement*, International Field Year for the Great Lakes Technical Manual, Series No. 2.

Lhermitte, R., 1973, Meteorological Doppler Radar, *Science*, 182: 258-262.

Little, C. G., 1969, Acoustic Methods for the Remote Probing of the Lower Atmosphere, *Proc. IEEE*, 57: 571-578.

Lyons, W. A., and S. R. Pease, 1973, Detection of Particulate Air Pollution Plumes from Major Point Sources Using ERTS-1 Imagery, *Bull. Am. Meteorol. Soc.*, 54: 1163-1170.

MacCready, P. B., Jr., 1966, Mean Wind Speed Measurements in Turbulence, *J. Appl. Meteorol.*, 5: 219-225.

——, and H. R. Jex, 1964, Response Characteristics and Meteorological Utilization of Propeller and Vane Sensors, *J. Appl. Meteorol.*, 3: 182-193.

——, T. B. Smith, and M. A. Wolf, 1961, *Vertical Diffusion from a Low Altitude Line Source*, Vols. I and II, Meteorology Research, Inc., Altadena, Calif.

Marchgraber, R. M., 1970, The Development of Standard Instruments for Radiation Measurements, in *Meteorological Observations and Instrumentation*, S. Teweles and J. Giraytys (Eds.), pp. 302-314, American Meteorological Society, Boston, Mass.

Marshall, J. M., A. M. Peterson, and A. A. Barnes, Jr., 1972, Combined Radar-Acoustic Sounding System, *Appl. Opt.*, 11: 108-112.

McAllister, L. G., J. R. Pollard, A. R. Mahoney, and P. J. R. Shaw, 1969, Acoustic Sounding—A New Approach to the Study of Atmospheric Structure, *Proc. IEEE*, 57: 579-587.

McBean, G. A., 1972, Instrument Requirements for Eddy Correlation Measurements, *J. Appl. Meteorol.*, 11: 1078-1084.

Melfi, S. H., M. L. Brumfield, and R. W. Storey, Jr., 1973, Observations of Raman Scattering by SO_2 in a Generating Plant Stack Plume, *Appl. Phys. Lett.*, 22: 402-403.

Middleton, W. E. K., and A. F. Spilhaus, 1953, *Meteorological Instruments*, 3rd ed., University of Toronto Press, Toronto, Canada.

Miksad, R. W., 1976, An Omni-Directional Static Pressure Probe, *J. Appl. Meteorol.*, 15: 1215-1225.

Moses, H., and H. G. Daubek, 1961, Errors in Wind Measurements Associated with Tower-Mounted Anemometers, *Bull. Am. Meteorol. Soc.*, 43: 190-194.

——, and F. C. Kulhanek, 1962, Argonne Automatic Meteorological Data Processing System, *J. Appl. Meteorol.*, 1: 69-80.

Myers, R. F., 1952, A Low-Level Temperature Sounding for Routine Use, *Bull. Am. Meteorol. Soc.*, 33: 7-12.

Olsson, L. E., M. P. McCormick, W. P. Elliot, and S. H. Melfi, 1974, An Observational Study of the Mixing Layer in Western Oregon, *Atmos. Environ.*, 8: 241-252.

Ower, E., 1949, *The Measurement of Air Flow*, Chapman and Hall Ltd., London.

Patterson, J., 1926, The Cup Anemometer, *Trans. Roy. Soc., Can., Sec. III, Ser. III*, 20: 1-54.

Penman, H. L., 1955, *Humidity*, Reinhold Publishing Corp., New York.

Peterson, J. B., and S. H. J. Womack, 1937, *Electrical Thermometers for Aircraft*, National Bureau of Standards, Report 606, for National Advisory Committee for Aeronautics, National Aeronautics and Space Administration, Washington, D.C.

Pike, J. M., and D. W. Bargen, 1976, The NCAR Digital Barometer, *Bull. Am. Meteorol. Soc.*, 57: 1106-1111.

Randerson, D., 1968, A Study of Pollution Sources as Viewed by Earth Satellites, *J. Air Pollut. Control Assoc.*, 18: 249-253.

——, 1977, *Quantitative Analysis of Atmospheric Pollution Phenomenon, Skylab Explores the Earth*, NASA Technical Report SP-380, pp. 381-405.

Robinson, T. R., 1850, Description of an Improved Anemometer for Registering the Direction of the Wind and the Space Which It Traverses in Given Intervals of Time, *Trans. Roy. Irish Acad., Dublin*, 22: 155-178.

Russell, P. B., E. E. Uthe, F. L. Ludwig, and N. E. Shaw, 1974, A Comparison of Atmospheric Structure as Observed with Monostatic Acoustic Soundar and Lidar Techniques, *J. Geophys. Res.*, 79: 5555-5566.

Sanuki, M., 1950, Studies on Biplane Wind Vanes, Ventilator Tubes and Cup Anemometers, Part I, *Pap. Meteorol. Geophys. (Tokyo)*, 1: 81-132.

——, S. Kimura, and M. Baba, 1960, Wind Tunnel Experiment on Two-Directional Wind Vanes, *Pap. Meteorol. Geophys. (Tokyo)*, 11: 23-29.

Schrenk, O., 1929, Über die Trägheitsfehler des Schalenkreuz-Anemometers bei schwankender Windstärke, *Z. Tech. Phys.*, 10: 57-66.

Schubauer, G. B., and G. H. Adams, 1954, Lag of Anemometers, National Bureau of Standards, Washington, D.C., unpublished.

Schwiesow, R. L., and M. J. Post, 1977, Doppler Lidar for Measurement of Pollutant Transport, *Second National Conference on the Interagency R & D Program*, pp. 469-472, Washington, D. C.

Serafin, R. J., and R. Strauch, 1978, Meteorological Radar Signal Processing, in *Air Quality Meteorology and Atmospheric Ozone*, Report ASTM-STP-653, A. L. Morris and R. C. Barras (Eds.), pp. 159-182, American Society for Testing and Materials.

Sherlock, R. H., and M. B. Stout, 1931, *An Anemometer for a Study of Wind Gusts*, Engineering Research Bulletin No. 20, Office of Research Administration, University of Michigan, Ann Arbor, Mich.

Smith, F. B., 1961, An Analysis of Vertical Wind-Fluctuations at Heights Between 500 and 5,000 ft., *Quart. J. Roy. Meteorol. Soc.*, 87: 180-193.

Spencer-Gregory, H., and E. Rourke, 1957, *Hygrometry*, Crosby Lockwood and Sons, Ltd., London.

Spilhaus, A. F., 1934, *Analysis of the Cup Anemometer*, Meteorological Course, Professional Notes 7, Massachusetts Institute of Technology, Cambridge, Mass.

Swanson, R. N., and H. E. Cramer, 1965, A Study of Lateral and Longitudinal Intensities of Turbulence, *J. Appl. Meteorol.*, 4: 409-417.

Tanner, C. B., 1963, *Basic Instrumentation and Measurements for Plant Environment and Micrometeorology*, Soils Bulletin 6, University of Wisconsin, Madison, Wis.

Tapp, R. G., and R. C. Lile, 1978, A New Tethersonde System for Remote Measurement of Wind and Temperature, *Boundary-Layer Meteorol.*, 15: 465-479.

Television Digest Inc., 1979, *The Television Fact Book*, Television Digest, Inc., Washington, D. C.

Thompson, N., 1962, Intensities and Spectra of Vertical Wind Fluctuations at Heights Between 100 and 500 Ft. in Neutral and Unstable Conditions, *Quart. J. Roy. Meteorol. Soc.*, 88: 328-334.

Thornthwaite, C. W., and J. C. Owen, 1940, A Dew-Point Recorder for Measuring Atmospheric Moisture, *Mon. Weather Rev.*, 68: 315.

Trimmer, J. D., 1950, *Response of Physical Systems*, John Wiley & Sons, Inc., New York.

U. S. Department of Commerce and U. S. Department of Defense, 1972, *Federal Meteorological Handbook No. 5: Manual of Winds Aloft Observations,* GPO.*

U. S. Department of Commerce and U. S. Department of Defense, 1969, *Federal Meteorological Handbook No. 3: Manual of Radiosonde Observations,* GPO.*

U. S. Department of Commerce, Weather Bureau, 1963, *Manual of Barometry,* Vol. 1, 1st ed., GPO.*

Uthe, E. E., 1972, Lidar Observations of the Urban Aerosol Structure, *Bull. Amer. Meteorol. Soc.,* 53: 358-360.

Warnick, C. C., 1956, *Influence of Wind on Precipitation Measurements at High Altitudes,* Bulletin 10, University of Idaho Engineering Experiment Station, University of Idaho, Moscow, Idaho.

Wexler, A., 1970, Measurement of Humidity in the Free Atmosphere near the Surface of the Earth, in *Symposium on Meteorological Observations and Instrumentation,* pp. 262-282, American Meteorological Society.

World Meteorological Organization, 1971, *Guide to Meteorological Instruments and Observing Practice,* World Meteorological Organization, WMO No. 8, TP3, 4th ed., Secretariat of WMO, Geneva, Switzerland.

Chapter 3 covers research prior to August 1980.

*Available at the U. S. Government Printing Office, Washington, D. C.

4 MONITORING, SAMPLING, AND MANAGING METEOROLOGICAL DATA

Harold L. Crutcher, National Climatic Center

4-1 INTRODUCTION

Meteorological information is required during three reasonably distinct periods in the life history of any typical energy-generating facility: (1) during the selection of a site and the adaptation of the design and operating procedure to the site's special characteristics, (2) during routine operations, and (3) in the unlikely event of an accidental release of effluents. For these uses meteorological data must be gathered over long periods and retained indefinitely, and to do this requires a knowledge of the principles of monitoring, sampling, and managing data related to meteorology and to the pollutants released to the atmosphere during electrical power plant operation.

4-2 DATA MANAGEMENT

The detection and quantitative measurement of the various atmospheric processes is still an imperfect art; however, much useful information can be collected with proper planning. A field program to collect meteorological data should be designed to provide answers to the following 10 questions:

1. *What problem are you trying to solve?* What questions will you be expected to answer? What are the goals and objectives of your project?

2. *What are the time and space scales of the phenomena you are trying to detect?* Are you sampling micro-, meso-, or macroscale phenomena? Transient or stationary waves? For information on laying out an instrument network capable of resolving the desired atmospheric processes, see the paper by Culbertson and Willis (1964) and the recent article by Gaynor (1974).

3. *What is the required instrument response to detect the desired processes?* How much precision will be required of the instrumentation? Remember, the greater the precision needed, the greater the cost of the equipment. Will sudden temperature changes, wind, rain, dust, or pollution affect the performance of the instrument?

4. *How many sites must be instrumented to resolve the relevant information and what will be the cost?* Seinfeld (1972) discusses an objective technique to optimize instrument locations in a data network. Factors affecting the number of sites and their locations are that they should be easily accessible, they should be near power sources if power is required, they should be exposed properly to ensure accurate detection, and they may need to be protected against vandalism and theft.

5. *For what period of time will the project operate?* Will the data be collected for only part of one day or is a climatological record over several years required? What sampling period is required to identify the wavelengths of the relevant processes? Are instantaneous or average values required? Your processing capabilities are finite; so do not generate an indigestible amount of data.

6. *How many people will be required to operate and maintain the equipment?* How reliable is the equipment? If necessary, can it be left in the field, exposed to the weather elements, and survive? What standards are you establishing to ensure proper quality control?

7. *How will the data be collected and processed?* How will the data be reduced? Remember, other investigators may use your data. Also, more and more automatic sensing devices are being designed to provide a link between the data source and the computer. If a computer–instrument package is not used, who will convert the data to computer-compatible form? A discussion on the organization and collection of data is given by Sterling and Pollack (1968).

8. *How will the data be stored or filed?* Random access to a data file is desirable if possible. The most accessible files are usually those with simple and straightforward arrangements. When a data file is rather large, say 1000 to 2000 cards (over 80,000 characters of fully punched cards), the file should be transferred to a magnetic tape or disk file. Of course, data can be recorded directly on magnetic tape and thus eliminate the 80-character limitation that is characteristic of most punch cards.

9. *Who will analyze the data and at what cost?*

10. *Can you afford or justify the cost of the project?* If not, what are your alternatives? What can you sacrifice? Perhaps the space and temporal scales of the network are too large. Can you reduce the size of the project and still attain your goals?

If you can provide factual answers to these questions, you are on your way to a well-designed field program. Make

Crutcher: National Climatic Center, Environmental Data Services, National Oceanic and Atmospheric Administration, Asheville, North Carolina.

your decisions as objectively as possible, and try to avoid subjective decisions based on a majority vote of a committee. You may have to justify failures later.

4-3 QUALITY

The very act of making a product creates many by-products, some of which are desirable, recoverable (at a cost), and reusable and some of which are not. These by-products, such as pollution, vary in the degree to which they affect the quality of the surroundings. Though the quality of energy is the main thrust of a power-producing plant, the quality of all the environment must be considered in the total quality plan.

4-3.1 Management

Quality is not an intangible idea. It can be defined and quantified through either numbers or categories (attributes). The precision, the accuracy of measurements, the freedom from or the extent of "junk readings," and the number of interruptions (loss of record) are only a few of the attributes by which quality can be assessed. Quality also is a function of planning, reliability, sampling intervals, procedures, and representativeness. For example, order-of-magnitude estimates require only rather crude data, whereas detailed calculations may require very accurate, precise data. Different conditions require different planning strategies, different designs with different reliabilities, and different control modes. Improper control or elimination of this planning phase can totally remove or modify the quality sought. The measure of quality wanted or required is an a priori decision of management during the initial planning for the facility, during construction of the facility, and during the operation and maintenance of the facility. The quality pervades the entire facility (including all personnel), the power community, the producer, and the user.

Figure 1 describes a quality program showing the relationship among the following factors: quality management, quality design and reliability, quality assurance, and quality control.

4-3.2 Design and Reliability

Design and reliability must be an integral part of the planning process. The entire quality plan of the power plant must be specified. One small phase of this deals with the weather data. There are no manufacturing standards within the United States for meteorological instruments. Until such standards are formulated and approved, the requirements for each instrument must be specified by the user. In addition, there is no national requirement that the meteorological instrument output from one plant be comparable, compatible, or commensurate to that of another plant. Comparable, compatible, and commensurate data permit easier and less costly use of archived data.

There is no point in recording and archiving data if they will never be used to improve the product and to eliminate the unwanted by-products. Archiving generates operating costs and subsequent storage and servicing costs. The latter costs can be considerable over a period of time. Planning must also take into account the simple problem of space requirements.

It pays to archive good clean data. Sequences of bad data can so contaminate a study that the detection and removal of bad data, with subsequent reprocessing and analysis, ends in a cost three and four times that of a first run with good data. Also, the time lost can be considerable.

Consequently, management must dictate the quality of weather data required for its use in operation as backup, and for future planning. Requisite quality in the plant must be planned and specified. When only half-hour, one hour, or longer averages may be needed, there is no need to have a very fast response, and extremely precise and accurate instrument and recording system.

The reliability of the plans for providing the necessary information and the reliability of the instrumentation and its installation must be assessed. If not acceptable, the plans must be modified to provide the required reliability. Of course, specifications must be provided to a manufacturer and to those who install the equipment.

4-3.3 Assurance

The assurance phase of the quality program must ensure that instrument installation and exposure are correct, that instrument calibration records are what the manufacturer says they are, and that the output of the instruments and the storage and servicing of the records meet management's plans. Output accuracy and precision cannot be better than input. If the instrument exposure is bad, the data obtained are suspect. The assurance of quality is a monitoring process to ensure a specified input quality. Quality assurance, then, is mainly a check of procurement, installation, maintenance, and upkeep of the total sensing system.

4-3.4 Control

Quality control is the monitoring and control of a process to ensure the specified quality of the product. The product is the data used in real time, recorded, documented, and transmitted to archives. This process of obtaining relevant information requires, not separate and

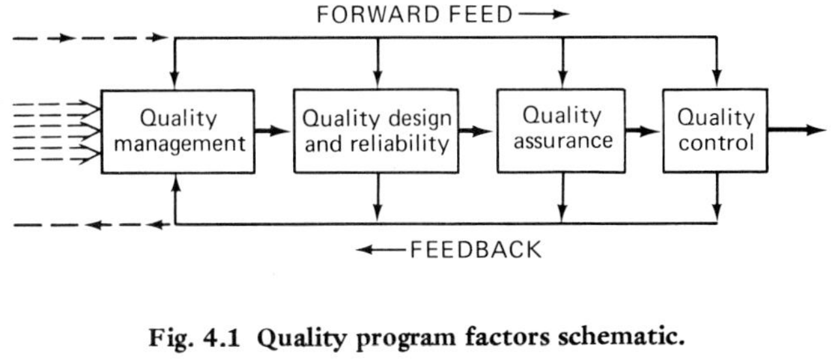

Fig. 4.1 Quality program factors schematic.

isolated operation, but open channels for forward feed and backfeed among the four phases shown in Fig. 1. Though a main channel is indicated, each phase should have continual, active contact with all other phases.

4-3.5 Management Policy

Because many effluents are released from power-production installations, ensuring atmospheric quality requires considerable monitoring, sampling, and communication. Jurgens and Rhodes (1976) discuss in some detail quality-assurance procedures appropriate to the meteorological system as well as to other monitoring systems. Their excellent paper forms the basis for the discussion that follows. Management policy for a total power-production quality program, especially the part dealing with atmospheric monitoring, should consider the following elements:

1. Objectives.
2. Organization and responsibility and the existence of or need to create manuals.
3. Plans and training.
4. Procurement, i.e., control, ordering, receiving, feedback, and corrective action.
5. Calibration and the existence of or need to create standards and procedures.
6. Internal checks.
7. Operation sampling, including sample handling and analysis.
8. Data recording, transmission, computation, and validation.
9. Preventive maintenance.
10. Reliability records and analysis.
11. Document control and documentation.
12. Configuration control.
13. On-site and performance audits.
14. Corrective actions whenever failure modes are detected in the monitoring process.
15. Statistical analysis.
16. Reporting.
17. Investigation.
18. Comparability between installations.
19. Costs.
20. Data archiving, storage, servicing, and recall.
21. Data publication (recorded data, summarized data, analyzed data, and statistical parameters).

Maintaining the history of a power installation is important. The record of weather data should include all information available from the incorporation of the weather requirements into the initial management policy through implementation (installation, maintenance, and servicing) to output. For example, when an instrument is replaced, both instrument and its replacement should be identified. Calibration records should be kept in the history (maintenance log). When the location of an instrument is changed, such as a change in height, the change should be noted. Photographs before and after installation or change provide an invaluable historical record. Also, it is helpful in later assessments to have in the history such data as the station's latitude, longitude, height above sea level, and height above ground.

4-4 ON-SITE METEOROLOGICAL PROGRAMS AT POWER-PRODUCTION INSTALLATIONS

4-4.1 Introduction

The preceding section discussed general meteorological guidelines. Some of the actual on-site data requirements are presented in this section.

Applicants for licenses to operate nuclear power plants are required by law (*Code of Federal Regulations,* Title 10, Part 50, Appendix E, "Emergency Plans for Production and Utilization Facilities") to include in their final safety analysis report their various plans for coping with emergencies. The plans must include criteria for determining when protective measures should be considered inside and outside the site boundary to protect health and safety and to prevent damage to property. In this regard the applicant must establish and maintain a meteorological program capable of rapidly assessing critical meteorological parameters.

Thus an on-site meteorological measurements program at a power plant site should be capable of providing the meteorological information required to make the following assessments:

1. A conservative assessment by the applicant and the regulatory staff of the potential dispersion of effluents to aid in evaluating the acceptability of a site and the adequacy of engineered safety features for a specific power plant.
2. An assessment by both the applicant and the regulatory staff of the maximum potential annual hazardous effluent output to the public which will result from the routine release of effluents.
3. A realistic assessment by the applicant and the regulatory staff of the potential dispersion of deleterious materials from, and the consequences of, a spectrum of accidental releases to aid in evaluating the environmental risk posed by a power plant.
4. A realistic assessment by the applicant and the regulatory staff of other environmental effects, such as fogging, icing (including snow), and salt drift from cooling towers, to aid in evaluating the environmental impact of a power plant.
5. A rapid, conservative assessment by the licensee and other appropriate persons of the consequences of an accidental extraordinary release of effluent material to the environment. The assessment should be used to provide early guidance to the licensee's emergency organization and to local, state, and federal agencies with responsibilities for coping with emergencies in determining the need for notification and participation of local, state, and federal

agencies and when appropriate measures should be taken to protect public health and safety and to prevent damage to property. On-site meteorological measurements should provide an adequate basis for short-distance atmospheric diffusion calculations. Meteorological data on the regional scale, mesoscale, and larger will be needed if it becomes necessary to make diffusion estimates for long distances. To ensure that the required data are readily available, the applicant should establish and maintain contact with the nearest National Weather Service Focal Point Air Pollution Meteorologist.†

Specific guidance for evaluating the potential consequences of design-basis accidental or extraordinary releases of effluents is given in appropriate safety guides. The basic transport diffusion model described in these guides can also be used to estimate the dispersion of particulates or other effluents in the atmosphere following the routine or accidental release of such materials. For an evaluation of short-term releases, the actual meteorological parameters measured during the release period should be used. For long-term releases the observed joint frequency distribution of wind speed and direction and atmospheric stability for the period should be used. If there are large variations in the rate of release, it may be necessary to subdivide the meteorological data into various categories of release rates.

Although there are differences in the specific types of meteorological information required for each of the preceding assessments, a single set of instruments can generally be used to obtain the basic data needed for all of them. For this reason, when a meteorological program for an initial site survey is established, careful consideration should be given to the operational needs of the plant for meteorological information. In particular, care should be taken to locate the stations at positions where the measurements will accurately represent the overall site meteorology and, if possible, where wind patterns will not be significantly influenced by plant structures.

The number of locations at which meteorological measurements are necessary will depend largely on the complexity of the terrain in the vicinity of the site. For example, the study of a hill–valley complex or a site near a large body of water would require a larger number of measuring points to determine airflow patterns and spatial variations of atmospheric stability. The minimum amount of meteorological data needed for a siting evaluation is considered to be the amount of data that is gathered on a continuous basis for a representative consecutive 12-month period. Only when regional climatic conditions for a year can be considered to be representative can a single station's data set for that year be considered to be representative. Even then, data from two full annual cycles or more are desirable. Data covering 5 to 10 years are even better.

The following sections describe verification of models and a suitable on-site meteorological program to provide the meteorological data needed to estimate potential radiation doses.

4-4.2 Verification of Models

Verification is, in itself, a rather controversial topic because of its many statistical facets and possible ramifications in cases where poor results are demonstrated and not intended. On completing the verification of a numerical model, for example, you should be able to answer the following questions: (1) Does the model demonstrate predictive skill? (2) If a change is made in the model to improve the forecast, can you demonstrate that the forecast improved significantly in a practical as well as a statistical sense? (3) Which components of the model explain most of the variance in the variables?

In numerical modeling, verification is usually understood to describe the process of comparing predictors with corresponding observations and using the results to evaluate the performance or skill of the model relative to some control or standardized prediction. Skill can be measured in various ways, e.g., with the skill score or Brier P-score or with such statistics as the percent correct, the mean absolute error, the root-mean-square error (RMSE), and by other indices. Control forecasts are usually selected as persistence, damped persistence (Lorenz, 1973), climatology, or other model forecasts of known skill, primarily because they all have the advantages of simplicity and stability.

Once the control has been specified, you can assemble the master data set or data file. Usually the data are collected within a grid network at positions that do not necessarily correspond with the grid points of the model. To transfer the observations to the model grid points, you should objectively analyze the observations in such a way as to specify values at the model grid points. Cressman (1959), Glahn, Hollenbaugh, and Lowry (1969), and Heimbach and Sasaki (1975) have developed procedures for objectively analyzing randomly spaced data. Having created a master data set, you can begin the model and the control simulations. Finally, you compare the control forecast and the model predictions against the master data set or observations. Note that you compare the two models against a standard set of observations. Your goal should be to attain an objective verification scheme that eliminates all elements of judgment in the verification. No matter how reasonable and realistic you think the simulations from your model are, the most conclusive and rewarding test of model performance is to compare the simulations with corresponding observations of the simulated phenomena.

More detailed discussions on verification can be found in a book by Panofsky and Brier (1958), a review of statistical procedures in meteorology by Julian and Murphy

†The name and address of the nearest Focal Point Air Pollution Meteorologist can be obtained by contacting the Air Pollution Meteorologist, Weather Analysis and Prediction Division, National Weather Service, National Oceanic and Atmospheric Administration, Silver Spring, Maryland 20910.

(1972), a paper by Bauer and Kutzbach (1974), and an interesting verification procedure used by Nappo (1974).

4-4.3 On-Site Program

In addition to providing meteorological data to estimate potential adverse effects to the public as a result of the routine or accidental release of potentially hazardous effluents to the environment, the on-site meteorological program data base can be used to assess other environmental effects.

4-4.3.1 Meteorological Parameters. Instrumentation at a minimum of two levels on a mast is required. Wind, temperature, and radiation data permit valid estimates of atmospheric transport and diffusion at the site. Wind speed and direction or wind components are required, and temperatures and temperature differences are needed. Measurements of the temperatures at one level and the temperature differences between two levels can be backed up by temperature measurements at the second level. Temperature-difference (ΔT) sensing is important because representative differences cannot be obtained by calculations from two different sensors. Each sensor has an output that is biased, censored, and, perhaps, even truncated in direct output or in output processing. Solar radiation measurements, direct and diffuse, should be supplemented by terrestrial radiation measurements. At sites where there is a potential for fogging or icing due to an increase in atmospheric moisture content caused by plant operation, instrumentation should be provided for measuring the dew point (or humidity) on the tower or mast. Measurements of cloudiness, visibility, and solar radiation may be necessary in conjunction with assessments of the cooling system.

Sites using brackish or saline water may require measurements of atmospheric air salt concentration and deposition. Also, sites with other unusual air-quality characteristics may require appropriate instrumentation and observation.

For realistic estimates of atmospheric transport and diffusion to a distance of 80 km from the plant site, additional information can be obtained, in part at least, from well-maintained meteorological systems, e.g., National Weather Service (NWS), military stations, and any other micrometeorological stations, if these existing stations are in representative locations.

4-4.3.2 Siting of Meteorological Instruments. The instrument tower or mast should be sited at approximately the same elevation as the finished plant grade and in an area where plant structures will have little or no influence on the meteorological measurements. Instrumentation should be installed on booms and located a minimum of two tower diameters from the tower on the booms to preclude substantial tower influence on the measurements. The lower set of instruments should sense wind speed and direction, temperature, dew point (where required) at an elevation of 10 m above the ground, and the ΔT between the lower level and the upper level. The upper set should sense wind speed and direction and temperature at the height at which material is released from the plant vent but should be positioned not less than 30 m, and preferably 60 m, above the lower sensor set. For stack releases a third set of sensors should be located at an elevation that will represent meteorological conditions at stack height so that diffusion conditions from stack releases can then be depicted. The solar and terrestrial radiation instruments should be placed in unobstructed locations. The distance between the tower and any obstruction should be at least ten times the height of the obstruction. The tower should also be placed directly upwind of the mean wind side of the stacks if possible or upwind off the major axis of the wind distribution, preferably not more than 45°.

4-4.3.3 Data Acquisition. Figure 2 is a flow diagram for data acquisition modified after George (1973). The recording media can be magnetic tapes (reels, cassettes, or cards), paper tapes or cards, or any medium from which data can be reproduced or extracted. Note the multiple junction point to the left of "direct manual entry" in the upper left. Data acquisition can be automatic, i.e., strip-charts or printed digital output. If automatic transcription readers are not available, however, the charts must be processed manually and visually to obtain data. Either analog (magnetic tape or strip-chart) or digital recording of data can be used as a basis for analysis. Use of both is recommended because strip-chart recording permits easy evaluation of sensing performance. For example, strip-chart recorders allow a fast check on the threshold and response of the wind vanes and anemometers, on the reasonableness of the ΔT between the dry- and wet-bulb temperatures, and on the general behavior of the radiation sensors. If you must wait for digital data to be plotted, considerable data can be lost before a faulty probe is detected and replaced. In lieu of redundant digital recorders, digital outputs can be supplemented by strip-chart recorders to minimize possible loss of data due to recorder malfunction. Recorders (analog or digital) for wind direction and speed and temperature difference (two temperatures or one temperature difference measurement on a tower or mast) should be located in the control room for use during plant operation. Accurate time checks and calibration marks should be made on the record once each 24 hr. The records should be documented (thoroughly identified) at the beginning and end of each tape, cassette, strip-chart, or any type of record.

4-4.3.4 Control Room Display. The data display from the primary meteorological system, which includes a time history (analog or digital printout) of wind direction and speed at each level and stability class or temperature difference with height, should be located in the control room for use during plant operation.

4-4.3.5 Instrument Accuracy.

Wind-direction accuracy desired for instantaneously recorded values is ±5° of azimuth. Boundary values are 0° and 360° for quality assurance (QA) screening.

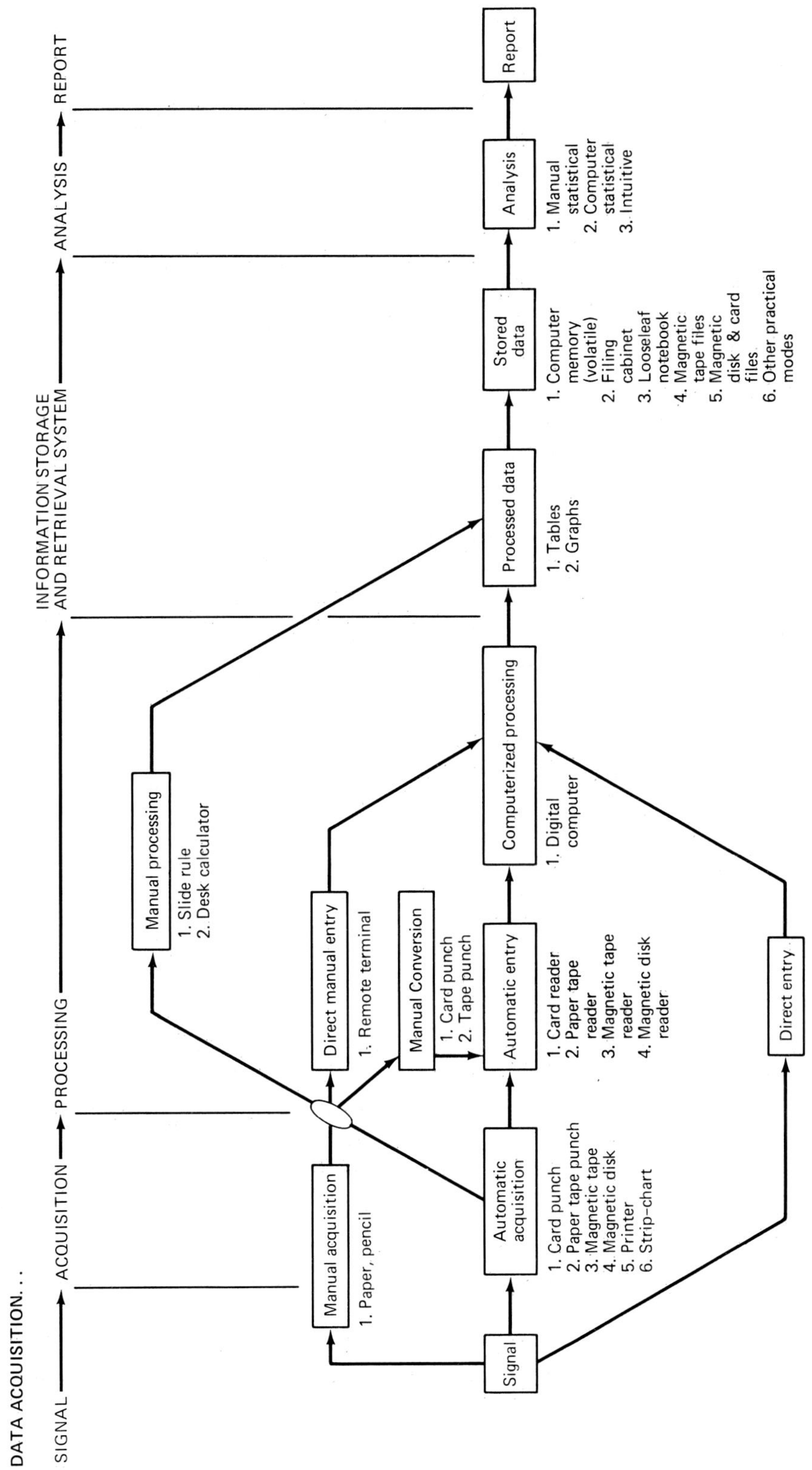

Fig. 4.2 Six basic steps in data systems. These same basic steps are used by all data systems, regardless of whether they are automated or manual or a combination of both. [Excerpted by special permission from *Chemical Engineering*, 80(28): 124 (Dec. 10, 1973) by McGraw-Hill, Inc., New York, N.Y. 10020.]

Wind-speed accuracy desired for time-averaged values is ±0.25 m/s for wind speeds less than 10 m/s, ±0.50 m/s for speeds over 10 m/s but less than 20 m/s, and ±1.00 m/s for speeds over 20 m/s. Starting-speed threshold of an anemometer is less than 0.5 m/s. Boundary values are 0 and 50 m/s, although an operational limit of 25 m/s can be accepted for initial QA screening

Temperature accuracy needed for time-averaged values is ±0.5°C. *Temperature-difference accuracy* from either the difference between averaged temperatures or the average temperature difference should be ±0.1°C. Boundary values are −100°C to 100°C, although values of −20°C to 45°C can be imposed for operational limits and QA screening. As pointed out previously, a separate probe system to measure temperature differences is needed; however, calculated temperature differences can be used when the probe values are not available. Boundary values depend on the thickness of the air layer. Nocturnal radiation over snow cover sometimes produces phenomenal vertical temperature gradients. For initial screening a suggested boundary value is ±50°C per 100 m. For individual months this difference would be less. Experience at a locale would permit modification of these limits.

Dew point accuracy required for time-averaged values is ±0.5°C. Boundary values are the same as for temperature.

Relative humidity is important. If it is measured, the instrumented output should be accurate and precise to ±2%, although the range is 0 to 100%.

The *pyramometer* accuracy desired is ±2,100 J/m^2 (≈±0.05 Langley). Boundary value checks are −21,000 to 105,000 J/m^2 (≈−0.50 to 2.50 Langleys) for QA screens. The Langley is solar radiation equal to 1 g · cal/min.

Archiving requirements for solar radiation data are being developed in laboratories of the National Oceanic and Atmospheric Administration (NOAA). Those who wish to record, transmit, and archive their solar radiation data in a format compatible with that being archived at the National Climatic Center should communicate with the staff at a NOAA Environmental Research Laboratory.

The *pyrgeometer* accuracy desired is ±2,100 J/m^2 (≈±0.05 Langley). Boundary value checks are +12,500 to +31,500 J/m^2 (≈+0.30 to 0.75 Langley) for QA screens.

The *pyrheliometer* accuract desired is ±2,100 J/m^2 (≈±0.05 Langley). Boundary value checks are −21,000 to 105,000 J/m^2 (−0.50 to 2.50 Langley) for QA screens.

Time lag of instruments is important because there are many state-of-the-art statistical calculations that demand low lag. The lag should be known and entered at the beginning and end of all records.

Measurement of precipitation by a recording rain gage, backed by a standard stick gage, should be within 3%, with a resolution of 0.25 mm.

Time should be accurate within ±5 min of actual time for all recording systems.

4-4.3.6 Instrument Maintenance and Servicing Schedules. Meteorological instruments should be inspected and serviced at a frequency which will ensure at least a 90% data recovery and which will minimize extended periods of outage. The use of redundant sensors and/or recorders, both digital and analog, can be another acceptable means of achieving the goal of 90% data recovery. An inventory of spare parts should be maintained. The instruments should be calibrated against a standard instrument at least semiannually. Quarterly calibration is required in some cases. Of course, maintenance must include all peripheral equipment, such as towers, supports, cables, and power supply, as well as shelters, fences, and buildings.

The system should be protected against lightning. Consideration should be given to protection of the system against other severe environmental hazards, e.g., icing, sand, salt, dust, air pollution, flooding.

A log of inspection, maintenance, and calibration should be maintained on site as a permanent record available for review.

4-4.3.7 Data Reduction and Compilation. All records should have prescribed time checks and random inspection time checks. Strip-charts must be documented at the beginning, at time checks during the recording, and at the end of the chart.

Wind, temperature, and humidity data should be averaged over a period of at least 15 min at least once each hour. This also should include the variance or standard deviation for the period. The selected periods can vary as required for a particular location.

The basic reduced wind data should be compiled into monthly (or seasonal) and annual joint frequency distributions of wind speed and wind direction by atmospheric stability class. Figure 3 gives an example of a suitable format for data compilation and reporting. Similar tables of joint frequency distribution should be prepared for all atmospheric stability classes. (See Table 1 for a possible classification of various atmospheric stability categories.) If the boxes on the form are made large enough, additional information can be given on the same sheet. For example, below the joint frequency distributions can be given such data as the percent of frequency within the stability class, the percent of frequency within the total data set, and the frequency divided by the total number of observations. If it is more convenient, the additional data can be recorded on separate forms, similar in format to Fig. 3. As a basic accounting check, the columns should be totaled both horizontally and vertically.

As an aid in assessing the environmental impact of plant operation, joint frequency distribution types of data summaries should be compiled to describe the frequency and extent of fogging and icing conditions caused by plant operation. Other environmental effects are important, too. For example, crop and forest damage can occur because of high SO_2 concentrations near fossil fuel plants. Fly ash

STATION ______ ______ ______
Town State Nation

LATITUDE
LONGITUDE
ALTITUDE (m) above sea level
HEIGHT (m) above ground ____ ____ ____

THICKNESS OF AIR LAYER
FOR STABILITY
LOWER ____________ UPPER ________
AVERAGING TIME ________________
SAMPLING FREQUENCY ____________

JOINT FREQUENCY DISTRIBUTION
OF WIND SPEED
AND WIND DIRECTION
(Per mille frequency)

PERIOD:

CRITERIA FOR THE TABLE SUBPERIOD OR SEASON
(SPRING, WINTER, ETC.)____________________

STABILITY CLASS____ TEMPERATURE (ΔT)____
PRECIPITATION TYPE____________________

NUMBER OF OBSERVATIONS
TOTAL FOR THE PERIOD____________________
TOTAL FOR THIS TABLE ____________________
EXTREME VALUE____________________

Degrees from north / Wind speed, m/sec	360-019	020-039	040-059	060-079	080-099	100-119	120-139	140-159	160-179	180-199	200-219	220-239	240-259	260-279	280-299	300-319	320-339	340-359	Undefined Directions	All Directions
>35																				
30																				
25																				
20																				
15																				
14																				
13																				
12																				
11																				
10																				
9																				
8																				
7																				
6																				
5																				
4																				
3																				
2																				
1																				
0																				
TOTALS																				

Fig. 4.3 Typical format for bivariate classification of meteorological data. The speed classifications, direction vectors, averaging time, and other details will vary from country to country and according to climate conditions.

damage to property can be significant close to the plant. Frequency distributions of these concentrations and other by-products associated with the wind should be possible to make. For example, fumigation episodes associated with the breakup of the nocturnal inversion can be studied. The distributions can be bivariate or trivariate on one page or a multivariate set including several bivariate forms.

In the evaluation of a site for a power plant, unique meteorological conditions at the site sometimes make it necessary or desirable to depart from the meteorological assumptions provided in safety guides. In these cases, when the data are reduced, it is necessary to analyze the joint frequency of persistent wind direction, wind speed, and atmospheric stability to determine appropriately conservative atmospheric diffusion factors for accidental and annual average releases of effluents. The data should be tabulated, and the assumptions and calculation procedures should be described. The probability distributions of χ/Q estimates for appropriate time periods should be presented.

4-4.3.8 Special Considerations. At some sites, because of complex flow patterns in nonuniform terrain, additional wind and temperature instrumentation and more comprehensive programs may be necessary. Measurements of precipitation and solar radiation are needed at all locations. Topographic maps should be used to support argument for or against alternative and/or additional sites.

Occasionally the unique diffusion characteristics of a particular site may warrant use of special meteorological instrumentation or studies, or both. Proposed studies of this nature should be described in the application for a construction permit.

4-4.3.9 Documentation. The program for on-site meteorological measurements should be fully documented

TABLE 4.1 Classification of Atmospheric Stability*

Stability classification	Pasquill categories	σ_θ,† deg	Temperature change with height, °C/100 m
Extremely unstable	A	25.0	<−1.9
Moderately unstable	B	20.0	−1.9 to −1.7
Slightly unstable	C	15.0	−1.7 to −1.5
Neutral	D	10.0	−1.5 to −0.5
Slightly stable	E	5.0	−0.5 to 1.5
Moderately stable	F	2.5	1.5 to 4.0
Extremely stable	G	1.7	>4.0

*These Pasquill (1961) categories were developed for climatic conditions in the British Isles. As experience has been gained over the past 20 years, changes to these are indicated (Skaggs and Robinson, 1976). There, the format used may vary from region to region, and Table 1 should be considered as a suggestion that some delineation in summary reports is needed.

†Standard deviation of horizontal wind-direction fluctuation over a period of 15 min to 1 hr. The values shown are averages for each stability classification.

in the safety analysis report. Documentation of the records themselves is discussed in more detail in the next section.

4-5 DOCUMENTATION

If a weather observation is made, it should be of some immediate use or some potential future use. If an observation is worth making, recording, and storing (archiving), it is worth identifying clearly so that those using it will know what it is and what it is all about. This is documentation. The record must be identified clearly, definitively, and completely so that it can be placed in the archives and serviced. To service a record means to store the record, to periodically or aperiodically verify its status, to retrieve it to answer a query, and to return it to storage after use.

Just as a misfiled book in a library is essentially lost, a badly identified record stands a good chance of being misfiled and therefore lost or even destroyed. Records assumed to be available may not be available. Space then would be occupied by useless material. Good documentation is a necessity to good housekeeping and vice versa.

4-6 RECORDING AND STORAGE MEDIA

Media for recording and storage are many and varied in form. As time goes on, new media forms and formats will become available, particularly in the miniaturization (microform) feature. These newer features permit more storage in less space. Today the main media are:

1. Paper for handwritten records in numbers, code, or abbreviated notes.
2. Chart paper for mechanically or electrically scribed information in analog or digital form. Various colored inks, codes, and symbols can be used.
3. Photographically recorded light traces.
4. Magnetically recorded analog and digital forms on magnetic tape or disks.
5. Punched cards, magnetic cards, or punched paper tape.
6. Microform film.

Good rag paper (linen) has a shelf life of many years. The paper generally used today is a sulphite paper that has a shelf life of about 10 years either for manuscript or for punched cards.

The shelf life of magnetic cards and tapes is not known. An acceptable shelf life beyond 10 years certainly has not been verified. Magnetic tape requires rewinding, and possibly rereading and rewriting each year (exercising), to maintain appropriate signal strength and discrimination. If this is not done, loss of information is inevitable. Cassettes of magnetic tape and disk assemblies are now coming into vogue. Their shelf life and usefulness (lack of trouble) also are unknown.

Laser-produced memories, bubble memories, and memories produced by other techniques have as yet not been demonstrated as viable, useful, and less costly than the forms mentioned previously. In addition, the necessary hardware and software to record, store, and read data have not been developed. Further, the costs are unknown, although costs of storage and servicing would be expected to be less than those required for the other forms. Reproduction of data might be another problem.

Microfilm, microfiche, and ultrafiche in color or in black and white are forms available to some extent for archiving. These forms in the past, especially where metals have been used, such as silver, have been subject to serious degradation due to ion movement within the film over a period of 20 years. Special care and "gold-washing" will prevent this degradation, sometimes called the "J" defect or Jupiter defect. The smaller the microform, the more serious the problem. Today's processing does not seem to be so affected, but continual watch and quality assurance during shelf life are required. The shelf life of microfilm in color is not known.

The shelf life of disk files, read-only memories, and core memories for data is unknown.

4-7 COSTS

Costs are controlled by the design and reliability of the system, as well as by the marketplace. Costs given here are approximate 1977 prices for presently available equipment sufficient to meet the minimum requirements of the Nuclear Regulatory Commission's Regulatory Guide 1.23 (formerly the USAEC Safety Guide 23). For the first two years these costs approximate one-third of a million dollars. These minimum costs do not include either office or storage space.

One tower, installation and equipment	$100,000
Annual maintenance cost	25,000
Annual cost of surveillance and quality assurance (including personnel and supplies, magnetic tape, paper, etc.)	50,000
Annual cost of data listings, etc. based on 15-min integrating intervals and automatic logging in digital form on magnetic tape, 13 parameters (channels) to a page, daily summaries	60,000

The cost of a mobile tower and equipment installation is essentially 50% of the cost of a permanent-type tower installation. Other costs remain essentially the same.

4-8 ARCHIVES

Many decisions have to be made regarding archiving. Each decision, of course, ramifies through other subsequent decisions. The first decision is whether to archive or not. If the decision is not to archive, then all records are destroyed once their operational aspect has been met. If the first decision is to archive, then the second decision is whether to archive in the owner's archives or in the National Archives. If the second decision is to archive in the owner's archives, then a decision must be made whether to archive in a transitory stage before going to final archives or to move directly to final archives. Whatever the decisions, there are costs involved to obtain, to prepare, and to move records to final archives for subsequent servicing. These include costs:

1. To procure, maintain, and replace equipment. These costs are extremely sensitive to the marketplace.
2. To procure, train, and maintain initial and replacement personnel necessary for the meteorological program. At 1977 prices, these costs are estimated to be a minimum of $50,000 per plant.
3. To obtain the data, provide quality control for the records (i.e., to monitor, record, and sample data), and document at every step.
4. To archive the data. Storage costs must be included in the plans of the primary facility to pay for the costs of receiving the records, entering them into storage, and storing them. There will be similar costs if records are moved to subsequent storage or archive depots.
5. To service the data. The data should be catalogued so that they are easily accessible.
6. To publish the data or specialized summaries.

These costs must be considered and included in every construction and operating budget. The servicing of the data can be on a cost basis to a customer (user) whether in house or outside. Each year of operation will add to the archive space required with the attendant costs of heat, lights, water, fire protection devices, and humidity controls and ventilation controls, as well as containers or devices to store the data media. If these costs are budgeted, then the individual power plant facility can function as its own archiving and servicing agent. If the decision is to use staging procedures on the way to a final archive, the budgeted costs can be transferred each year to the actual stage involved. Later they can be transferred to the owner's own central facility budget. This procedure is called front-end loading. That is, if a stage or a central facility is to be used, then for each block of data, a specified amount will be added to the operating budget of the central facility. Since space usually is limited at the initial facility, miniaturization (microform) should be used there, and the original documents should be moved to the central facility.

Shelving, containers, etc., require an initial cost of about $\$11/ft^3$ or $\$300/m^3$ (1977 prices). Annual storage costs are \$2.50 to $\$5.00/ft^2$, or \$30 to $\$60/m^2$. The lower costs are for space not protected by fire-suppression systems. Costs of storage can be substantially reduced by opting for microform storage, i.e., microfilm, microfiche, and ultrafiche. This lower storage cost helps compensate for the expense of preparing the microform and providing equipment to read it. The reduction in required space is 1000 to 1 for microfiche, which is limited to around 100 images on a card. The 1977 cost of such a card is about $0.10 per frame. The cost would be around $6 for a 60-frame (image) master card and $9 for one with 90 frames. Duplicate cards cost about $0.65. These costs presume that the documents are in order and are accepted. Ultrafiche, which takes even less storage space, costs $600 for a card containing 600 images, or about $1 per frame.

Once the documented data media are in final archives and front-end loading has been provided for the cost of archiving, costs for services do not require front-end loading. The user should pay for any further costs. If the user is within the company, he should pay the servicing costs. Of course, the company can also front-end load for this cost in its budgeting process. Manuscripts can be copied on currently available copying machines. Microfilm or microfiche is best prepared by commercial contract.

At some stage the decision may be made to offer the archived data of your facility or copies of the data to an operating central facility, a state, or a national organization for archiving and servicing. These data can be in original form or in compact summary and/or statistical form. For the last category, the General Services Administration maintains and services the archives of the national government. There are many subunits of the National Archives. The Library of Congress is one of these. For climate and weather data, the responsibility has been delegated to the National Climatic Center (NCC), Environmental Data Service (EDS), National Oceanic and Atmospheric Administration (NOAA), Department of Commerce (DOC). At the present time this facility is located at Asheville, N. C. Other NOAA storage facilities are the National Oceanographic Data Center (NODC), Washington, D. C., the National Geophysical and Solar Terrestrial Data Center (NGSDC), Boulder, Colo., and the National Marine Fisheries Service Data Centers managed from Washington, D. C. Data dealing with water densities, temperatures, salinities, waves, currents, and other oceanographic characteristics are stored at NODC. Earthquake or tremor records go to NGSDC.

The Environmental Protection Agency is developing its own separate data bank at which information including but not limited to meteorological data can be stored. This data bank is currently located at Research Triangle Park, N. C.

It is important to realize that the national facilities cannot accept ensembles of data unless these data are thoroughly documented and can be archived in catalogued form so that they will be easily accessible and serviceable. This cataloguing procedure should be followed for the first archive step even though it also may be the last. Good microform copy of summarized or listed data would be best. For entry into electronic computers, magnetic tape or disk media would be the best for either original data or processed data. Documentation should be given on the tapes and disks themselves as well as written on a label. Standard electronic computer formats, very important to standardization, would be required. Front-end loading, either by the originator or by the final archive facility, must be designed for, arranged for, and implemented.

REFERENCES

Bauer, K. G., and J. E. Kutzbach, 1974, Evaluation Standards for Dynamical Prediction Models, *J. Appl. Meteorol.*, 13: 505-506.

Cressman, G. P., 1959, An Operational Objective Analysis System, *Mon. Weather Rev.*, 87: 367-374.

Culbertson, B., and P. T. Willis, 1964, Observational Grid Spacing Effect on the Accuracy of a Physical Forecast Model, *J. Appl. Meteorol.*, 3: 541-546.

Gaynor, J. E., 1974, Optimum Averaging Times for Short-Term Wind Prediction in the Planetary Boundary Layer, *J. Appl. Meteorol.*, 13: 411-415.

George, T. J., 1973, Automatic Data-Acquisition, *Chem. Eng. (N.Y.)*, 80(28): 123-128 (Dec. 10, 1973).

Glahn, H. R., G. W. Hollenbaugh, and D. A. Lowry, 1969, *An Operationally Oriented Objective Analysis Program.* ESSA Technical Memorandum WBTM TDL-22, Environmental Science Services Administration, Silver Spring, Md.

Heimbach, James A., and Yoshikazu Sasaki, 1975, A Variational Technique for Mesoscale Objective Analysis of Air Pollution, *J. Appl. Meteorol.*, 14: 194-203.

Julian, P. R., and A. H. Murphy, 1972, Probability and Statistics in Meteorology: A Review of Some Recent Developments, *Bull. Am. Meteorol. Soc.*, 53: 957-965.

Jurgens, Robert B., and Raymond C. Rhodes, 1976, Quality Assurance and Data Validation for the Regional Air Monitoring System of the St. Louis Regional Air Pollution Study, in *Proceedings of the Conference on Environmental Modeling and Simulation*, Cincinnati, Ohio, April 20–22, 1976, Report EPA-600/9-76-016, pp. 730-735, Environmental Protection Agency, July 1976.

Lorenz, E. N., 1973, On the Existence of Extended Range Predictability, *J. Appl. Meteorol.*, 12: 543-546.

Nappo, C. J., 1974, A Method for Evaluating the Accuracy of Air Pollution Prediction Models, in *Proceedings of the Symposium on Atmospheric Diffusion and Pollution*, Santa Barbara, California, September 9–13, 1974, pp. 325-329, American Meteorological Society, Boston.

Panofsky, H. A., and G. W. Brier, 1958, *Some Applications of Statistics to Meteorology*, Pennsylvania State University, University Park, Penn.

Pasquill, F., 1961, The Estimation of the Dispersion of Windborne Material, *Meteorol. Mag.*, 90: 1-33.

Seinfeld, J. H., 1972, Optimal Location of Pollutant Monitoring Stations in an Airshed, *Atmos. Environ.*, 11: 847-858.

Skaggs, Donald L., and Elmer Robinson, 1976, A Comparison of Methods for Estimating Atmospheric Stability and Diffusion Coefficients, *J. Air Pollut. Control Assoc.*, 26: 888-891.

Sterling, T. D., and S. V. Pollack, 1968, *Introduction to Statistical Data Processing*, Prentice-Hall, Inc., Englewood Cliffs, N. J.

BIBLIOGRAPHY

Slade, D. H. (Ed.),1968, *Meteorology and Atomic Energy—1968*, USAEC Report TID-24190, Environmental Science Services Administration, Silver Spring, Md., NTIS.

Hilfiker, R. C., 1975, Exposure of Instruments, in *Air Pollution Meteorology*, pp. 6-29 to 6-38, USEPA Air Pollution Training Institute, Research Triangle Park, N. C.

Brooks, C. E. P., and N. Carruthers, 1953, *Handbook of Statistical Methods in Meteorology*, M. O. 538, Her Majesty's Stationery Office, London.

Chapter 4 covers research prior to August 1977.

ATMOSPHERIC BOUNDARY LAYER

Darryl Randerson, Weather Service Nuclear Support Office

5

5-1 FUNDAMENTALS

5-1.1 Introduction

Man and his power-production facilities exist in the atmospheric boundary layer, which occupies approximately the bottom kilometer of the atmosphere. Many of the effluents generated by man's activities are released directly into the atmospheric boundary layer and are transported, dispersed, and transformed there until they are finally removed. Anthropogenic and naturally generated effluents can be health hazards and cause millions of dollars in damage to property and vegetation each year. Thus it is little wonder that we should be concerned with achieving a better understanding of the role of boundary-layer processes in diluting and redistributing effluents. This concern has created an active interest in the search for mathematical and physical models that can account for the dynamic and thermodynamic processes of boundary-layer phenomena. Observational evidence and theoretical considerations have equipped us with a fundamental understanding of steady-state flow over relatively smooth surfaces, but little knowledge has been gained in the important areas of nonsteady flow over nonuniform surfaces. This aspect of boundary-layer diagnosis and prognosis offers unlimited opportunities for interrogation. Since today's theories help solve today's problems, this chapter deals primarily with what we have learned about the atmospheric boundary layer and how this knowledge can be put to practical use. The development of early theories of boundary-layer transport has been documented by Kampé de Fériet (1974).

5-1.2 Boundary-Layer Concept

The theory of boundary layers was first introduced into aerodynamics by Prandtl (1904), who used the concept to obtain approximate mathematical solutions to otherwise insoluble problems. A basic understanding of the concept of a boundary layer can be obtained from a development similar to that of Prandtl. Using Fig. 5.1, consider a rigid flat plate (or wall) of negligible thickness with a fluid flowing over it in a direction parallel to the plate. The velocity distribution in the fluid is indicated by the length of the arrows. To the left of the plate, let the speed profile of the approaching fluid be uniform. Apply a horizontal force to the top layer of the fluid to keep it moving at a constant, free-stream speed. As the fluid flows over the plate, surface friction and internal friction (vertical speed shear) retard the speed of the fluid near the rigid boundary. It is the accumulation of this low-energy part of the fluid that leads to the growth of a boundary layer. The region where the speed is reduced to less than the free-stream speed is known as the boundary layer. In Fig. 5.1, this region, δ, lies between the plate and the dashed line. For the purpose of mathematical analysis, the field of fluid flow can be divided into two regions: (1) a shallow boundary layer near the plate (the wall layer or surface layer) where the velocity distribution is determined by wall shear and surface roughness and (2) a region outside the boundary layer where frictional forces are negligibly small and the motion is driven by Coriolis and pressure forces (see Townsend, 1956). A region of overlap exists between these two layers, and dimensional arguments show that in this region the wind speed increases logarithmically with elevation (Sec. 5-3.3).

On a planetary scale and especially in the mid-latitudes, the flow of air is characterized by easily discernible, large, dynamic systems in which air motion is primarily determined by a balance between horizontal pressure gradients and the Coriolis effect. Contributions to the motion by surface friction, by viscosity, and by vertical momentum and heat exchange are neglected. From Chap. 2, recall that airflow occurring under these conditions and with the specified balance of forces is known as the geostrophic wind. Consequently, geostrophic flow occurs above an altitude where the effect of the surface of the earth is negligible, generally at an altitude of about 1 km. The atmospheric layer bounded by the ground and by the level at which geostrophic balance occurs can be regarded as the atmospheric boundary layer, i.e., the region of the atmosphere where the direct effect of the ground is detectable. In general, this definition does not cause any serious difficulties, and it works especially well when the momentum and heat exchange at the ground are significant. For example, with strong vertical wind shear, momentum is transferred toward the ground and a well-developed wind spiral (Sec. 5-3.6) forms, reflecting the transition of the geostrophic flow aloft to a different wind

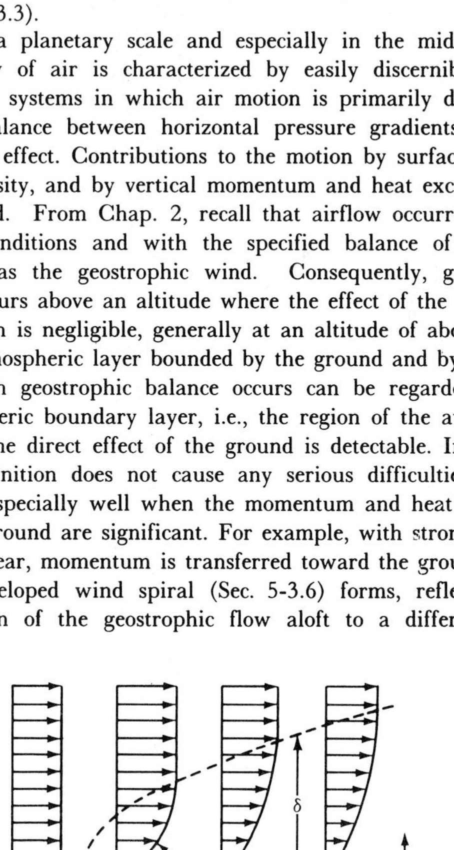

Fig. 5.1 Development of a boundary layer in a fluid as it flows over a smooth fixed plate. The arrows represent flow vectors; the dashed sloping line separates freely flowing fluid from that affected by the plate; δ is the depth of the boundary layer; and the vertically curved solid lines are speed (u) profiles.

Randerson: Weather Service Nuclear Support Office, National Oceanic and Atmospheric Administration, Las Vegas, Nevada.

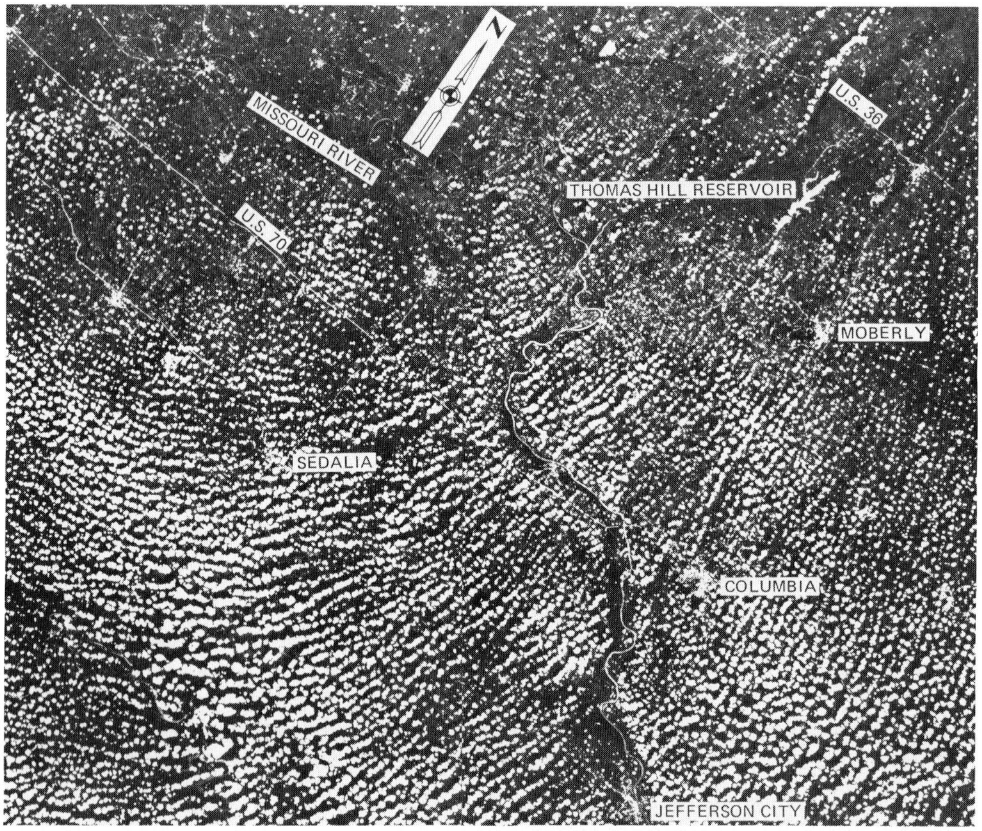

Fig. 5.2 Cloud streets over central Missouri, as viewed from Skylab-3 on Sept. 18, 1973, at approximately 1000 CST.

regime near the ground. From a different aspect, the boundary layer can be well defined when there is an upward heat flux at the ground, as occurs during the morning hours after sunrise. In this case a parcel of air, heated by contact with the ground, can be visualized to rise buoyantly to an altitude where the ambient temperature is the same or warmer than that of the parcel. Near this level a stable layer or even a temperature inversion can be formed, neatly defining the top of the atmospheric boundary layer. Another example of a meteorological phenomenon that portrays the top of the atmospheric boundary layer is the occurrence of cloud streets such as those pictured in Fig. 5.2. Analysis of such photographs has shown that when the cloud streets are about 2 km apart the boundary layer is approximately 1 km deep.

As we might expect, the atmospheric boundary layer is not always well defined. For example, when the atmosphere is stable and the heat flux is downward, turbulent exchanges are suppressed so that coupling between adjacent atmospheric layers is restricted. As a consequence, the flow field in this stable boundary layer is quite complex, with internally generated phenomena influencing the local wind. Conditions of this sort frequently occur over the interior of continents in winter.

The boundary layer is also poorly defined when the atmosphere is unstable over a deep layer, as in the desert regions in summer or in the vicinity of convective clouds and rain showers. When nonprecipitating clouds are present, the unstable boundary layer can be considered to be the layer between the ground and the base of the clouds.

A conceptual difficulty arises near the equator where the Coriolis effect is negligibly small so that a geostrophic approximation to the wind cannot be made. Consequently the height of the boundary layer is unclear; however, in many cases a temperature inversion exists near the 1-km level over large portions of the tropics. This inversion can be specified as the top of the boundary layer.

Observations have shown us that at any specific location the depth of the atmospheric boundary layer can vary from a few meters during windless nights (Businger and Arya, 1974) to several thousand meters in the middle of the afternoon. This diurnal cycle accounts for a large portion of the source of the energy which is generated and dissipated within the confines of the boundary layer and which results in the observed diurnal changes that occur in the wind and temperature profiles of the boundary layer. The time dependency of the atmospheric boundary layer is pursued further in Sec. 5-4.2.

5-1.3 Turbulence

The fluid flow represented in Fig. 5.1 is considered to be laminar, i.e., the fluid particles move along smooth, straight paths, parallel to the rigid plate. If we carefully introduced a liquid dye into this flow at an upstream point, the dye would follow a straight predictable line of motion and would tend to retain its shape for a considerable distance downstream, although it would gradually widen because of slow molecular diffusion. If we were to increase the fluid flow rate gradually, a transition speed would be reached where the behavior of the dye streak would become significantly altered from its laminar characteristics. This break up and rapid dispersion of the dye throughout the fluid would indicate to us that secondary motions were being generated in the fluid. These new motions are referred to as turbulence and are observed as irregular, chaotic motions that cause the fluid to become mixed.

The atmosphere can disperse gases and particulate matter rapidly because it is turbulent. In general, we can define a tur-

TABLE 5.1 Some Sources of Energy for the Generation of Turbulence

Microscale (< 10 km)	Mesoscale (10–1000 km)	Macroscale (> 1000 km)
Horizontal and vertical wind shear Thermal instability Surface roughness	Latent heat release in cloud formation Flow over topographical features, cities, islands, etc. Flow over surface with marked variations in surface character (e.g., land–sea breezes)	Horizontal instability of the zonal flow due to the latitudinal variation in solar heating

bulent fluid flow as one having the ability to disperse embedded gases and particles at a rapid rate. Experimental evidence tells us that the rate of turbulent dispersion is orders of magnitude greater than that accounted for by molecular diffusion. In fact, turbulence is the primary process by which momentum, heat, and moisture are fed into the atmosphere from the surface of the earth and then mixed in time and space. For a theoretical development of the concepts of atmospheric dispersion, see Chap. 6. Procedures for using models are described in Chap. 13.

Turbulence tends to manifest itself in eddies of all sizes, the largest eddies being limited by the dimensions of the fluid system and the very small-scale fluctuations being determined by the nonlinear terms in the equations of motion. Thus a continuous hierarchy exists from the largest down to the smallest eddies, with molecular diffusion occupying the bottom of the scale. There is a continuous transfer of kinetic energy from the largest eddies down to the smallest ones where the energy is dissipated.†

The mathematical details of the transition from laminar flow to turbulent flow regimes are not well understood. We do know from experiments that the transition usually begins with an initial or primary instability mechanism. This primary instability can generate secondary motions or eddies that become unstable. A sequence of these events can follow and generate localized turbulent eddies that arise at random locations at random times. These eddies can grow rapidly and merge with one another as they become larger and more numerous, a whole field of turbulent motion being created eventually. However, turbulence cannot maintain itself; it is dependent on its environment as an energy source. Table 5.1 illustrates the major sources of energy as a function of the scale of the motion. In general, Table 5.1 demonstrates that the creation of atmospheric turbulence depends on the magnitude of three factors: mechanical effects or the roughness of the ground, horizontal and vertical wind shear, and thermal instability.

Forced convection or mechanical turbulence and wind shear manifest themselves as horizontal and vertical turbulent fluctuations that are generated as air flows over rough ground. The faster the mean wind speed, the greater the contribution by mechanical effects at the ground, with the contribution decreasing as height increases. Thermal turbulence occurs in the form of fluctuations induced by the free convection of buoyant fluid parcels rising from the heated ground. For more details, see Chap. 2, Sec. 2-5.2.

The properties of atmospheric turbulence are of particular importance in understanding the phenomenon itself. Turbulence is a nonlinear, three-dimensional, continuum phenomenon. As a fluid flow, turbulence is diffusive, rotating, random motion. Turbulence is always dissipative and needs a continuous supply of energy to make up for that lost through the deformative action of viscous shear stresses. This dissipative property of turbulence represents a subtle distinction between such motions as random acoustic waves (and atmospheric gravity waves) and turbulent motions. Acoustic waves have insignificant viscous losses; i.e., they are essentially nondissipative (they can be dispersive), whereas turbulence is dissipative.

5-1.4 Mean Flow

Reynolds (1895) proposed an approach that reduced turbulent flow to a relatively manageable mathematical form and made direct mathematical treatment of turbulent motions possible. He suggested the total wind vector, **V**, could be thought of as consisting of two parts: a constant mean velocity, $\overline{\mathbf{V}}$, and a fluctuating (or turbulent) part, $\mathbf{V}'$, so that the real wind is given by

$$\mathbf{V} = \overline{\mathbf{V}} + \mathbf{V}' \tag{5.1}$$

or in component form as $u = \bar{u} + u'$, $v = \bar{v} + v'$, and $w = \bar{w} + w'$, where u, v, and w are the vector components along the x-, y-, and z-axes, respectively. Equation 5.1 tells us that to specify the turbulent part of the flow, $\mathbf{V}'$, we must first be able to identify a mean state of the flow, $\overline{\mathbf{V}}$. Knowledge of the mean flow is important to us for several reasons. First, the strength of the mean flow is directly related to the capacity of the atmosphere for diluting effluents. Second, the kinetic energy supply for turbulence lies in the large-scale mean motions. Third, in the atmospheric boundary layer, the mean flow is primarily determined by turbulence arising from frictional drag at the earth–atmosphere interface.

The mean wind flow can be determined from wind records such as those shown in Fig. 5.3. The procedure for temporally averaging wind-speed measurements taken at fixed locations is simple and makes use of the mean-value theorem of integral calculus. For the time average to be mathematically correct, the mean flow should be steady state over period Δt as illustrated in Fig. 5.3a. In this figure, $\overline{V}$ is selected so that the area bounded above $\overline{V}$ by trace V(t) is equal to the area below $\overline{V}$. In Figs. 5.3b and 5.3c, the fluctuations in the wind speed (turbulence) are of the same order of magnitude as the mean speed, a characteristic of atmospheric turbulence as distinguished from wind-tunnel turbulence, where $\overline{V}$ can be 100 to 1000 times V'. In Fig. 5.3b the mean wind speed can be defined with some confidence by the procedure portrayed in Fig. 5.3a. In Fig. 5.3c, how would you define a mean wind

†On the macroscale, energy is fed up-scale from synoptic storms to the general circulation.

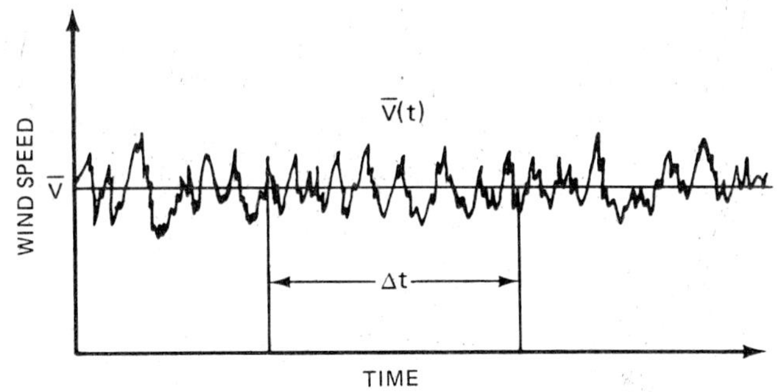

Fig. 5.3a Selecting $\overline{\mathbf{V}}$ by the mean-value theorem for integrals.

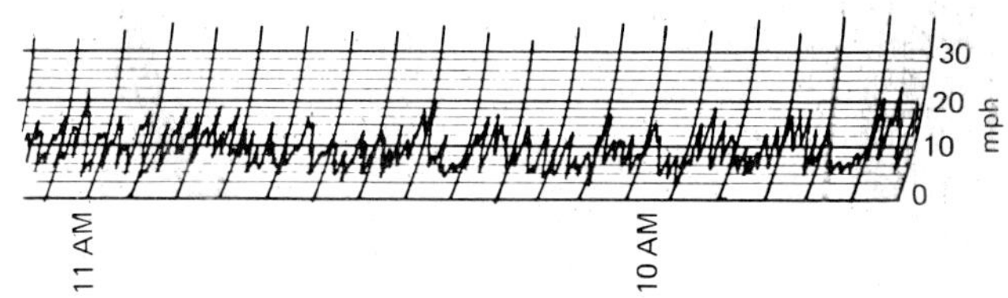

Fig. 5.3b Example of a rather steady mean wind speed with considerable fluctuation.

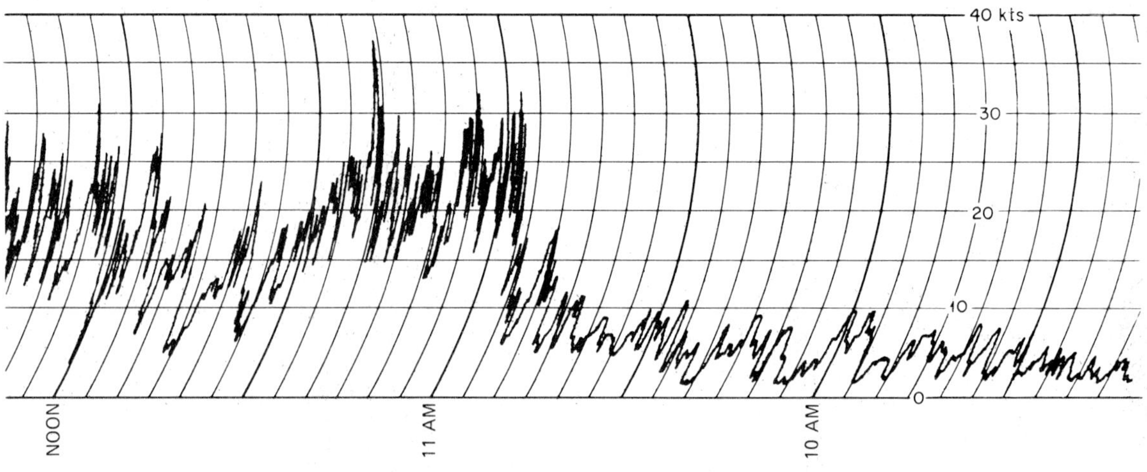

Fig. 5.3c Example of an abrupt change in wind speed.

speed for the period 1030 to 1130? This figure demonstrates that caution must be used in selecting the period of sampling because the use of a very long period could mask important changes taking place in the flow.

It is convenient to partition the instantaneous wind vector $\mathbf{V}$(x,y,z,t) at position (x,y), height, z, and time, t, into a mean flow vector $\overline{\mathbf{V}}$(x,y,z,t) and a superimposed turbulent velocity fluctuation $\mathbf{V}'$(x,y,z,t) so that

$$\mathbf{V}(x,y,z,t) = \langle \mathbf{V}(x,y,z,t) \rangle + \mathbf{V}'(x,y,z,t) \tag{5.2}$$

where the angular bracket represents an ensemble average operator. An ensemble average is formed, in principle, when the data record is long enough for us to presume a particular set of mean weather conditions has developed repeatedly so that all the possible minor variations of the variables have occurred. As a consequence, an ensemble average yields $\overline{\mathbf{V}'} = 0$. Equation 5.2 provides a procedure for defining the mean and turbulent parts of atmospheric flows; however, we generally cannot evaluate the ensemble mean because an extremely large number of realizations of the flow with the same meteorological conditions are not generally available. To estimate $\overline{\mathbf{V}}$(x,y,z,t), we assume that when it is possible to use one kind of average (ensemble, space, time) it does not matter which is used—the result will be the same. Such an assumption is called the ergodic hypothesis, and it permits us to replace the ensemble average with a time or space average over a realization of the flow. To apply this assumption, we must require that the statistics of the flow be invariant with time (stationary) and space (homogeneous) along at least one space coordinate.

The simplest case of a flow regime to which we can apply the ergodic theorem is one that is horizontally homogeneous and has a statistically stationary boundary layer. This type of boundary-layer concept is the one most commonly used in the simulation of atmospheric turbulence because the statistics of the flow vary only along the vertical axis. This condition is rarely realized in the atmospheric boundary layer because surface roughness, heat-transfer properties, and ensemble average temperatures of the earth must be invariant in time and in horizontal space. However, as an engineering approximation, the concept appears to be applicable in a local context to a wide variety of situations in which the terrain is relatively flat.

5-1.5 Deformation, Stress, and Viscosity

Fluids are deformed by forces that act on them. The relationship between these forces and the resulting strain or deformation in a fluid is of practical interest to us because the deformation of a fluid changes its flow characteristics. Owing to their molecular structure, all fluids offer at least some resistance to the deformation process. The property of resistance, viscosity, is most easily recognized in fluid flow near a boundary, where the flow ceases being uniform and becomes sheared as in Fig. 5.1. Recall that the flow in this figure is straight and parallel to the plate so that thin layers of faster-moving particles glide over thin layers of slower-moving particles. Because of the attraction of the molecules of a fluid for each other, the slower-moving molecules will tend to be dragged along by the faster-moving molecules so that the latter experience a force per unit area, a viscous stress or molecular stress, which retards their motion. Thus it is little wonder that the viscous stress is occasionally referred to as internal friction.

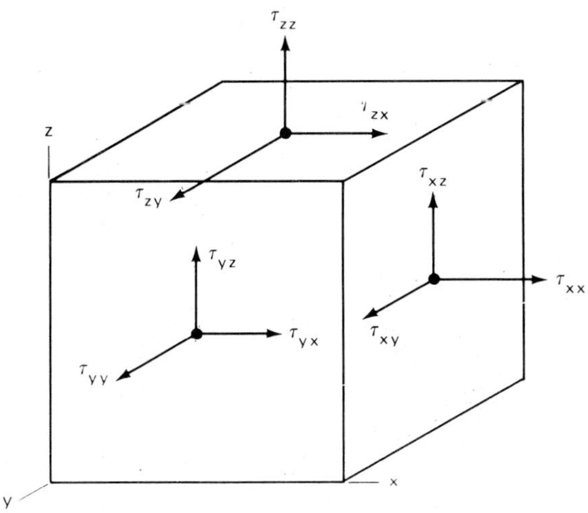

Fig. 5.4 Viscous stresses acting on the three surfaces of a fluid cube.

The viscous stresses acting on three faces of a cubic fluid element are shown in Fig. 5.4. Nine scalar quantities can occur on the surfaces of the element, but, to keep the discussion simple, we will consider only the x-directed stresses first. The three x-directed viscous stresses shown in Fig. 5.4 are the normal stress, τ_{xx}, and two tangential or shearing stresses, τ_{xy} and τ_{xz}. The first subscript, x, indicates that the stress acts on the plane surface that is perpendicular to the x-direction and the second subscript indicates the direction in which the shearing stress is acting. Thus τ_{xy} represents a force per unit area acting on a face normal to the x-axis with motion in the y-direction. The complete stress system requires nine scalar quantities for its description. All these quantities can be arranged into a stress matrix:

$$T_{ij} = \begin{bmatrix} \tau_{xx} & \tau_{xy} & \tau_{xz} \\ \tau_{yx} & \tau_{yy} & \tau_{yz} \\ \tau_{zx} & \tau_{zy} & \tau_{zz} \end{bmatrix}$$

Note that the stress tensor, T_{ij}, consists of nine components: three normal stresses, τ_{xx}, τ_{yy}, τ_{zz}, and six shearing stresses. The matrix is symmetrical about a diagonal from upper left to lower right so that the three stresses to the right of the diagonal are equal to the corresponding three stresses to the left of the diagonal. (Proof is given by Schlichting, 1955, p. 41.) Consequently there are actually only six independent viscous stresses we need to be concerned with—three normal stresses and three shearing stresses. How do these stresses deform our fluid cube in Fig. 5.4? Normal stresses act to stretch the cube by elongations in the x-, y-, and z-directions. Shearing stresses cause an angular displacement in the manner shown in Fig. 5.5. The deformation due to τ_{xz} is given by the angle, α, which is arctan $(\Delta x/\Delta z)$.

The viscous stresses acting on the surface of a fluid element are proportional to the rate at which the deformation is occurring. For atmospheric motions the simplest example of the deformation process is a vertical wind profile where straight, parallel, horizontal flow at every level is from the same direction but the speed increases steadily with height above the ground, as in Fig. 5.1. In this case only a shearing stress acts on a hypothetical fluid element. The shearing stress acts in the direction that tends to reduce the relative motion of adjacent layers of the fluid and is proportional to the vertical speed shear of the horizontal wind, $\partial \mathbf{V}/\partial z$. If we call μ the coefficient of viscosity, then the stress is given by

$$\tau = \mu \frac{\partial \mathbf{V}}{\partial z} \tag{5.3}$$

This equation, known as Newton's law of friction, forms the basis of our ideas on molecular viscosity. In Eq. 5.3, μ is a measure of the viscosity of the fluid and has the dimensions of $[M/(L \cdot T)]$. Equation 5.3 can be regarded as the mathematical definition of viscosity. The larger the value of μ (for constant $\partial \mathbf{V}/\partial z$), the more viscid the fluid. In an ideal fluid, $\mu = 0$; however, such a condition is an abstraction since all fluids possess some viscosity. For air at 20°C, $\mu = 1.8 \times 10^{-4}$ g/(cm·s). A simple method for calculating μ as a function of temperature is given in the *Handbook of Meteorology*, 1945, p. 447.

Knowledge of the stress is useful in finding mechanical energy that is produced by the wind at the rate $\tau \cdot \mathbf{V}$. Most of this energy production is near the ground where the wind shear is largest. For example, with $\tau = 2$ g/(cm·s^2) and a wind speed of 500 cm/s, the rate of mechanical energy production is 1000 g/s^3 or 1.0 W/m^2, after using the conversion factor, 1.0 W/m^2 = 10^3 g/s^3.

With the aid of Fig. 5.6, determine the stress acting on the surface of a cubic fluid element. For simplicity consider only the stress vector acting on the top and bottom faces, which are perpendicular to the z-axis. Assume that the x-direction coincides with the direction of the fluid flow. Specify a flow profile such as that shown on the left side of Fig. 5.6. Note that the flow is two-dimensional with no horizontal shear. There are two resultant stress vectors: τ_{zx} acting on the bottom face and $\tau_{zx} + (\partial \tau_{zx}/\partial z)\, dz$ acting on the top face. The resultant

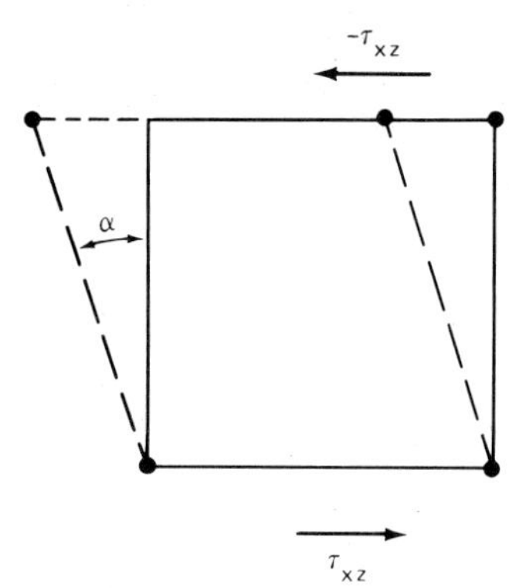

Fig. 5.5 Deformation by shearing stress τ_{xz}.

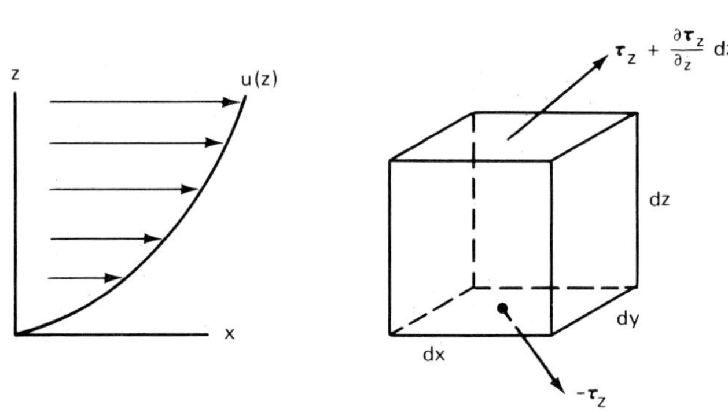

Fig. 5.6 Fluid flow profile and the resulting stresses acting on the dx · dy faces of a cubic fluid element.

stress acting on the dx · dy face is then the difference of the two stresses, or

$$\tau_{zx} + \frac{\partial \tau_{zx}}{\partial z} dx\, dy - \tau_{zx}\, dx\, dy = \frac{\partial \tau_{zx}}{\partial z} dx\, dy\, dz$$

Then the force per unit volume due to this one kind of stress becomes

$$\frac{\partial \tau_z}{\partial z} = \frac{\partial \tau_{zx}}{\partial z}\hat{\imath} + \frac{\partial \tau_{zy}}{\partial z}\hat{\jmath} + \frac{\partial \tau_{zz}}{\partial z}\hat{k} \tag{5.4}$$

or, for only the i component,

$$\frac{\partial \tau_{zx}}{\partial z} = \frac{\partial}{\partial z}\left(\mu \frac{\partial u}{\partial z}\right) = \mu \frac{\partial^2 u}{\partial z^2} \tag{5.5}$$

provided μ in Eq. 5.3 is independent of z. Similar terms can be obtained for the dx · dz and dy · dz faces to give the three components of the total surface force.

For the x component,

$$\hat{\imath}\left(\frac{\partial \tau_{xx}}{\partial x} + \frac{\partial \tau_{yx}}{\partial y} + \frac{\partial \tau_{zx}}{\partial z}\right) \tag{5.6}$$

For the y component,

$$\hat{\jmath}\left(\frac{\partial \tau_{xy}}{\partial x} + \frac{\partial \tau_{yy}}{\partial y} + \frac{\partial \tau_{zy}}{\partial z}\right) \tag{5.7}$$

For the z component,

$$\hat{k}\left(\frac{\partial \tau_{xz}}{\partial x} + \frac{\partial \tau_{yz}}{\partial y} + \frac{\partial \tau_{zz}}{\partial z}\right) \tag{5.8}$$

These three component equations represent the viscous force that should be added to the equations of motion for boundary-layer flow for an incompressible fluid.

The three normal stresses and three tangential stresses defined in Eqs. 5.6 to 5.8 can be expressed linearly; however, the derivation of these linear equations requires a long, complicated analysis [see Schlichting (1955) for details]. The final result is that the viscous stress in liquids and gases can be expressed in terms of the pressure, viscosity, and velocity gradients as follows:

For the tangential stresses,

$$\tau_{yx} = \mu\left(\frac{\partial u}{\partial y} + \frac{\partial v}{\partial x}\right) = \tau_{xy} \tag{5.9}$$

$$\tau_{yz} = \mu\left(\frac{\partial v}{\partial z} + \frac{\partial w}{\partial y}\right) = \tau_{zy} \tag{5.10}$$

$$\tau_{xz} = \mu\left(\frac{\partial w}{\partial x} + \frac{\partial u}{\partial z}\right) = \tau_{zx} \tag{5.11}$$

and for the normal stresses, where P represents static pressure,

$$\tau_{xx} = -P - \frac{2}{3}\mu\left(\frac{\partial u}{\partial x} + \frac{\partial v}{\partial y} + \frac{\partial w}{\partial z}\right) + 2\mu \frac{\partial u}{\partial x} \tag{5.12}$$

$$\tau_{yy} = -P - \frac{2}{3}\mu\left(\frac{\partial u}{\partial x} + \frac{\partial v}{\partial y} + \frac{\partial w}{\partial z}\right) + 2\mu \frac{\partial v}{\partial y} \tag{5.13}$$

$$\tau_{zz} = -P - \frac{2}{3}\mu\left(\frac{\partial u}{\partial x} + \frac{\partial v}{\partial y} + \frac{\partial w}{\partial z}\right) + 2\mu \frac{\partial w}{\partial z} \tag{5.14}$$

For incompressible flow the term $-(2\mu/3)(\partial u/\partial x + \partial v/\partial y + \partial w/\partial z)$ vanishes. Equations 5.9 to 5.14 are Stokes' law of friction for the flow of liquids and gases in which the stresses are proportional to the rate of strain of the fluid. Assuming that the pressure gradient outside the atmospheric boundary layer is essentially the same as that within, we can now assess the impact of the shearing stress on boundary-layer flow. As air flows over the ground, a shearing stress arises from frictional drag. The resulting stress is large enough to be important because the drag reduces the wind speed and thus reduces the Coriolis effect, which is directly proportional to the wind speed. Consequently the flow becomes ageostrophic, and the excessive pressure force causes the air to flow toward lower pressure until a balance is established between the shearing stress, the Coriolis effect, and the pressure gradient force. For simple flow regimes the shearing stress is at its maximum at the ground and decreases with height to approximately zero at the top of the boundary layer. Consequently the departure of the wind direction from that required by geostrophic conditions is also greatest at the surface and decreases with height.

5-1.6 Reynolds Stresses

The momentum of a body is defined as the product of its mass and its velocity and has the dimensions of (ML/T). In the flow of fluids, we usually express this product as the momentum per unit volume, or $\mathbf{M} = \rho\mathbf{V}$ with dimensions of $(M/L^2/T)$, where ρ is the air density. Since velocity is a vector quantity, momentum likewise must be a vector. Consequently momentum is described by three components. Momentum is transferred from one portion of a fluid to another by the motion of the fluid. The transfer of momentum per unit volume is referred to as the flux of momentum. For example, the upward flux of momentum associated with motion in the x-direction, M_{xz}, is expressed mathematically as $M_{xz} = \rho uw$.

In a fluid in turbulent motion, the fluctuating components give rise to additional stresses. To illustrate the consequence of this concept, consider a horizontal plane of area ΔA situated in a medium of turbulent flow with velocity components u, v, and w. The normal to area ΔA is imagined as parallel to the z-axis. Distances Δx and Δy are in the plane of ΔA. The mass of fluid passing through ΔA in time Δt is given by $\rho w \Delta A \Delta t$; thus the flux of momentum per unit volume in the z-direction is $M_{zz} = \rho w^2 \Delta A \Delta t$. The corresponding upward fluxes resulting from flow in the x- and y-directions are $M_{xz} = \rho uw \Delta A \Delta t$ and $M_{yz} = \rho vw \Delta A \Delta t$, respectively. If we use Eq. 5.1 and assume that density is constant and that density fluctuations can be neglected, the upward flux of x momentum per unit time resulting from the mean flow and from the fluctuating components is $M_{xz} = \rho uw \Delta A = \rho(\bar{u} + u')(\bar{w} + w')\Delta A$. Averaging gives

$$\overline{M}_{xz} = \rho(\bar{u}\bar{w} + \bar{u}w' + u'\bar{w} + u'w')\Delta A$$

and

$$\overline{M}_{xz} = \rho(\bar{u}\bar{w} + \overline{u'w'})\Delta A$$

since $\bar{\bar{u}} = \bar{u}$ and $\overline{u'} = \overline{v'} = \overline{w'} = 0$. Similar substitutions in the M_{yz} and M_{zz} equations give

$$M_{yz} = \rho(\bar{v}\bar{w} + \overline{v'w'})\Delta A$$

for the upward flux of y momentum per unit time and

$$M_{zz} = \rho(\bar{w}\bar{w} + \overline{w'w'})\Delta A$$

or

$$M_{zz} = \rho(\bar{w}^2 + \overline{w'^2})\Delta A$$

for the upward flux of z momentum per unit time.

These three fluxes have the dimensions of force acting on area ΔA so that dividing by ΔA yields force per unit area or stress. Now, since the flux of momentum per unit time through a plane surface is equivalent to an equal and opposite force exerted on the area by the surroundings, we conclude that the area under consideration is acted on by stresses $-\rho(\overline{w}^2 + \overline{w'^2})$, $-\rho(\bar{v}\bar{w} + \overline{v'w'})$, and $-\rho(\bar{w}\bar{u} + \overline{u'w'})$. The first of these three stresses is a normal stress and the other two are shearing stresses. Consequently, the superposition of turbulent fluctuations on the mean flow generates three additional stresses in the xy-plane,

$$\tau_{xz} = -\rho\overline{u'w'} \qquad \tau_{yz} = -\rho\overline{v'w'} \qquad \tau_{zz} = -\rho\overline{w'^2}$$

By analogy we have

$$\tau_{xx} = -\rho\overline{u'^2} \qquad \tau_{yx} = -\rho\overline{u'v'} \qquad \tau_{zx} = -\rho\overline{u'w'} \tag{5.15}$$

in the xz-plane and

$$\tau_{xy} = -\rho\overline{v'u'} \qquad \tau_{yy} = -\rho\overline{v'^2} \qquad \tau_{zy} = -\rho\overline{v'w'}$$

in the yz-plane. By symmetry, we actually have six new stresses due to turbulent fluctuations:

$$-\rho\overline{u'^2}, -\rho\overline{v'^2}, -\rho\overline{w'^2}, -\rho\overline{u'v'}, -\rho\overline{u'w'}, -\rho\overline{v'w'} \tag{5.16}$$

where, for example, $\rho u'w'$ represents the covariance of the u and w components of turbulence. These six terms are referred to as the Reynolds stresses of turbulent flow and must be added to the viscous stresses associated with the mean flow. The Reynolds stresses are advective terms and represent the advection of turbulence by turbulence. All the Reynolds stresses (or eddy stresses) are of the same order of magnitude except within the atmospheric boundary layer where the vertical variation is usually considerably greater than the horizontal variations. The significant difference between the Reynolds stress (e.g., $-\rho\overline{u'w'}$) and the viscous stress [e.g., $\mu(\partial u/\partial z)$] is that μ is a property of the fluid, whereas the fluctuations u' and w' are properties of the motion of the fluid and can be determined from the motion. At this point it should be obvious that the Reynolds stresses act in addition to the viscous stresses and are generally larger than the viscous stresses. The effect of the Reynolds stresses is to reduce the shear in a fluid in comparison to what it would be if only viscous stresses were present. The viscous stresses prevent the generation of extremely small scales of motion by dissipating kinetic energy into heat.

To appreciate the physical meaning of Eq. 5.15, consider a small horizontal area ΔA at elevation z where the x-momentum per unit volume is $\rho u = \rho(\bar{u} + u')$. Assume that the mean vertical motion is zero so that the volume of fluid flowing through ΔA in time Δt is due to the vertical fluctuation w' and is given by $w'\Delta A\Delta t$. Then the upward flux of x-momentum through ΔA is $\rho uw'\Delta A\Delta t$, the vertical flux per unit area per unit time is $\rho u(w') = \rho(\bar{u} + u')w' = \rho\bar{u}w' + \rho u'w'$, and the average vertical flux is then $\rho\overline{uw'} = -\rho\overline{u'w'}$ since $\overline{w'} = 0$. Consequently the $-\rho\overline{u'w'}$ component in Eq. 5.15 represents the advection of x-momentum downward by the vertical turbulent eddies provided $\partial u/\partial z > 0$. Through this transfer of momentum, the faster-moving air aloft tends to help maintain a steady-state flow against the effect of surface friction. Therefore, since we expect to see more guests (u′) with negative w′ (descending air) than with positive w′, we do not expect $\overline{u'w'}$ to vanish; in fact, it should be significantly different from zero. On the other hand, if the tangential stress $\overline{u'v'}$ does not vanish, the sign of the stresses depends on the gradient of the mean flow. Actually, we have no reason to associate gusts (or lulls) with a tendency for the wind to deviate in any specific direction so that the most likely result is $\overline{u'v'} = 0$. As for the normal stresses, $\rho u'^2$, etc., they represent the addition in dynamic pressure caused by the fluctuating components (see Eqs. 5.12 to 5.14). It is this increase in pressure that causes a pitot tube in a turbulent flow to read slightly higher than the actual mean velocity.

Experimental and observational evidence has shown that, when the wavelength of turbulent eddies is reduced to less than about a millimeter, the eddies dissipate by the conversion of their kinetic energy into heat energy by viscosity. Consequently there exists a layer of the lower atmosphere (the Kolmogorov microscale) whose thickness is of the order of the size of these turbulent eddies (see Sec. 5-1.8) in which the stress between the fluid and the boundary is due to viscosity alone and in which the Reynolds stresses are negligibly small in comparison to the viscous stresses. Outside this layer, if the flow is turbulent, there is another atmospheric layer in which the kinetic energy of the mean motion is converted into turbulent motion by the action of the Reynolds stresses and maintains the existence of the turbulence and the Reynolds stresses.

5-1.7 Eddy Viscosity and Mixing Length

Boussinesq (1877) was the first to propose that the effect of turbulent viscosity be accounted for through an analogy with Newton's law of friction (Eq. 5.3). Boussinesq introduced a mixing coefficient,† Λ, for the Reynolds stress in turbulent flow in the x-direction by setting

$$\tau_{xz} = \Lambda\frac{\partial\bar{u}}{\partial z} = \rho K_m\frac{\partial\bar{u}}{\partial z} \tag{5.17}$$

where K_m is the eddy viscosity and represents an eddy exchange process involving momentum. Note that $K_m = \Lambda/\rho$ and is analogous to $\nu = \mu/\rho$, the kinematic viscosity. Equation 5.17 has a disadvantage in that the mixing coefficient and the eddy viscosity are properties of the fluid, as is μ. For example, in the case of molecular diffusion, a quantity diffuses over

†Schmidt (1952) referred to the coefficient Λ as an "austausch," or exchange, coefficient.

a distance much larger than the mean molecular path length. As a result, it takes a very large number of molecular collisions to bring about molecular diffusion so that μ can be determined from the collision statistics (i.e., kinetic theory). By contrast, for turbulent diffusion only one or two excursions may convey a portion of the fluid far afield and set new boundaries for the diffusion of the fluid. Consequently K_m cannot be defined statistically. If an analogy to μ exists, it is in some other sense. Indeed, to explain laboratory-observed diffusion patterns, Deardorff and Willis (1975) demonstrated that K_m must be height and time dependent during convection mixing.

Using the analogy between molecular diffusion and turbulent mixing, Prandtl (1925) introduced the concept of a mixing length, ℓ, which corresponds to a molecular mean free path. Using a flow profile similar to the one in Fig. 5.6, where $u = u(z)$ and $\partial\bar{u}/\partial z > 0$, Prandtl visualized fluid parcels as coalescing together into lumps that move bodily and cling together for a given travel distance, both in the vertical and longitudinal directions. In addition, the lumps maintain their momentum parallel to the x-axis and conserve their momentum when rising from one level to another. He then assumed that, if a lump of fluid moved downward from a layer at $(z + \ell)$ with a mean speed $\bar{u}_2 = \bar{u}(z + \ell)$, it would be displaced by a distance $-\ell$ along the z-axis. If the lump of fluid retained its original momentum, its speed (u_1) in the new environment would be greater than the speed prevailing, since $\bar{u}_1 = \bar{u}(z)$ and $\partial\bar{u}/\partial z > 0$. The difference in the speeds is

$$\Delta u_1 = \bar{u}(z + \ell) - \bar{u}(z) = \ell \frac{\partial\bar{u}}{\partial z}$$

The last expression can be obtained by expanding the function $u(z + \ell)$ in a Taylor series and neglecting all high-order terms.† A similar expression can be obtained for lumps moving upward from a slow-moving medium to the faster-moving layer so that $\Delta u_2 = \bar{u}(z) - \bar{u}(z - \ell) = \ell\ \partial\bar{u}/\partial z$. The differences in the horizontal speed caused by the vertical motion of these lumps create u' so that the fluctuation can be written as

$$u' = \frac{1}{2}(\Delta u_1 + \Delta u_2) = \ell \frac{\partial\bar{u}}{\partial z} \tag{5.18}$$

Equation 5.18 simply states that ℓ represents the vertical distance traveled by a lump of fluid to make the difference between its original speed and the speed of its new environment equal to a fluctuation in the turbulent flow. We have assumed that the moving lump of fluid retains all of its original speed. From Eq. 5.15, $\tau_{xz} = -\rho\overline{u'w'}$; thus multiplying Eq. 5.18 by $\rho w'$, letting $K_m = \overline{w'\ell}$, and averaging, we obtain

$$\tau_{xz} = -\rho\overline{w'\ell}\frac{\partial\bar{u}}{\partial z} = -\rho K_m \frac{\partial\bar{u}}{\partial z} \tag{5.19}$$

but observational evidence tells us u', v', and w' are all of the same order of magnitude; thus Eq. 5.18 permits us to say $w' = -\ell(\partial\bar{u}/\partial z)$ and

$$\tau_{xz} = \rho\ell^2\left(\frac{\partial\bar{u}}{\partial z}\right)^2$$

†For another development, see Chap. 6, Sec. 1.3.

However, since the sign of τ must change with that of $\partial\bar{u}/\partial z$, the correct form of the equation is

$$\tau_{xz} = \rho\ell^2\left|\frac{\partial\bar{u}}{\partial z}\right|\frac{\partial\bar{u}}{\partial z} \tag{5.20}$$

The concept of a mixing length and its variation with height are controversial. One method is to use mixing lengths that approach constant values with increasing heights above the ground. A conceptual difficulty with this approach is that the turbulent fluxes far above the ground depend directly and instantaneously on the surface layer (see Sec. 5-2.5) characteristics; especially the heat flux. To overcome this difficulty, Busch, Chang, and Anthes (1976) suggest using a mixing length that is the scale of the energy-containing eddies. They assumed the time scale for the change of size of these eddies is inversely proportional to the strength of the mixing characterized by the local friction velocity, u_* (see Sec. 5-2.6) and the convective velocity, w_* (see Sec. 5-4.3). These investigators emphasize that this definition is arbitrary and may lead to overestimation of the wind shear, especially in weak wind and strong heating conditions.

Equation 5.20 is Prandtl's mixing-length hypothesis. Comparing this equation with that introduced by Boussinesq (Eq. 5.17) might lead us to conclude that little has been gained, but Eq. 5.20 is actually more suitable for calculating stress than is Eq. 5.17. Experimental evidence tells us that the resistance to flow introduced by turbulence is roughly proportional to the square of the speed of movement of the fluid. Observe that this result is obtained from Eq. 5.20 if the mixing length is independent of the magnitude of the speed. Another noteworthy point is that mixing length is still not a property of the fluid. The mixing length is a purely local function, but it is much easier to establish realistic assumptions about ℓ than about K_m. For these reasons Eq. 5.20 is of more general applicability than Eq. 5.17. However, according to Tennekes and Lumley (1972), mixing-length and eddy-viscosity models should be used only to generate analytical expressions for mean-speed profiles and for Reynolds stresses when these are needed for curve fitting in flows characterized by a single length and velocity scale.

Through Prandtl's mixing-length theory and his hypothesis of vertical conservation of momentum, analogous eddy-diffusion expressions can be derived for the vertical flux of sensible heat, H, and water vapor, E. Theory tells us that potential temperature, θ, and specific humidity, q, are conservative with respect to dry adiabatic motions. By analogy with Eq. 5.19 and the exchange coefficient concept, we obtain

$$H = \rho c_p\overline{w'\theta'} = -c_p A_{hz}\frac{\partial\bar{\theta}}{\partial z} = -\rho c_p K_h \frac{\partial\bar{\theta}}{\partial z} \tag{5.21}$$

where c_p is the specific gas constant (see Chap. 2 or Eq. 5.36) and $\overline{w'\theta'}$ is the vertical temperature flux. A shortcoming of Eq. 5.21 is encountered when the air layer in contact with the ground has a uniform potential temperature, thereby requiring H to vanish while it is actually still finite. A noteworthy point is that near sea level p ~ 1000 mb so that $\theta \sim T$. Consequently we can say $\partial\theta/\partial z \simeq \partial T/\partial z$ and $\overline{w'\theta'} \simeq \overline{w'T'}$ near sea level. Montgomery (1948) showed that $H = \rho c_p\overline{w'T'}$ can be used to represent sensible heat flux due to turbulent fluctuations with a time scale less than those resolved by macroscale measurements ($<10^3$ s). Similarly, for the vertical flux of water vapor,

$$E = \rho \overline{w'q'} = -A_{ez}\frac{\partial \bar{q}}{\partial z} = -\rho K_e \frac{\partial \bar{q}}{\partial z} \qquad (5.22)$$

Friehe and Schmitt (1976) used field data collected over water to develop expressions for $\overline{w'T'}$ and $\overline{w'q'}$ as functions of measured quantities. These expressions are

$$\overline{w'T'} \simeq 0.97 \times 10^{-3}\bar{V}\,\Delta T \qquad (m\cdot{}^\circ K/s)$$

for $0 < \bar{V}\,\Delta T < 25$, and

$$\overline{w'q'} \simeq 1.32 \times 10^{-3}\bar{V}\,\Delta q \qquad [(m/s)(g/m^3)]$$

where $\bar{V}$ is the mean wind speed (m/s) at a reference height of 10 m, ΔT(°C) is the difference between the surface temperature and the mean potential temperature at 10 m, and Δq is the difference between the water vapor density near the surface and at a height of 10 m. The surface value is obtained by assuming that the air is saturated at the surface temperature. These expressions are believed to be independent of wind speed, and the relation for $\overline{w'q'}$ is thought to be applicable for a wide range of stability. For stable conditions ($\bar{V}\ \Delta T < 0$), Friehe and Schmitt suggest

$$\overline{w'T'} = 0.0026 + 0.86 \times 10^{-3}\bar{V}\,\Delta T$$

For unstable conditions ($\bar{V}\ \Delta T > 25$) over water, Smith and Banke (1975) suggest

$$\overline{w'T'} = 1.46 \times 10^{-3}\bar{V}\,\Delta T$$

A procedure for estimating the temporal variation in $\overline{w'T'}$ is given in Busch, Chang, and Anthes (1976).

It is important to realize that the exchange coefficients K_m (or in component form K_x, K_y, K_z), K_h, and K_e are usually not equal. Businger et al. (1971) found the ratio K_h/K_m to be approximately 1.35 for neutral conditions in the surface layer. For stable conditions they found the ratio to range from 1.0 to 1.35, with an average value of 1.2. For additional discussion on K_h and K_m, see Lumley and Panofsky (1964, pp. 65-66). Although $K_h/K_m \simeq$ constant for near neutral conditions, it does not follow that K_h and K_m are each constants. In fact, all exchange coefficients are a property of the flow, not of the fluid, and they are not well behaved except in rather simple flows with linear profiles.

How are exchange coefficients specified? The numerous procedures for evaluating exchange coefficients range from estimates based on scalar analysis to lengthy computational procedures. The particular method usually depends upon the application of the derived values. Hanna (1968) described several different methods for calculating K_m and K_h, and O'Brien (1970) described a cubic polynomial fit for K_m for the surface layer up to the gradient-wind level. More recently, Shir (1973) and Estoque (1973) have described the numerical solutions obtained from models using different K_m and K_h formulations and Carl, Tarbell, and Panofsky (1973) suggested a K_m expression for very unstable conditions. In general, the K_m generalization proposed by Prandtl (1932) still appears to have applicability:

$$K_m = \ell^2\left[\left(\frac{\partial u}{\partial z}\right)^2 + \left(\frac{\partial v}{\partial z}\right)^2\right]^{1/2} \qquad (5.23)$$

where, for neutral stability only, ℓ is a mixing length given by Blackadar (1962) as $\ell = k(z + z_0)[1 + k(z + z_0)/\lambda]^{-1}$, where z_0 is the roughness length (see Sec. 5-3.3), k is the von Kármán constant (see Sec. 5-3.3), and $\lambda = 0.00027V_g/f$. Close to the ground, ℓ is linear with height so that $\ell = k(z + z_0)$ as a close approximation. In another approach, Gifford (1959) described a procedure for estimating K_y and K_z from aerial photographs of smoke plumes. Randerson, Garcia, and Whitehead (1971) used Gifford's formulas to obtain estimates of the vertical and horizontal exchange coefficients of a smoke plume photographed from a spacecraft. According to Gifford, the horizontal eddy diffusivity (K_y) can be estimated by two formulas,

$$K_y = \frac{\bar{u}Y_m^2 e^{1/3}}{2X_t} \qquad (5.24)$$

$$K_y = \frac{\bar{u}Y_m^2}{2X_m} \qquad (5.25)$$

where $\bar{u}$ is the mean wind speed along the major axis of the plume, X_t is the total length of a visible smoke plume, and X_m is the distance downwind from the source at which the maximum plume width, Y_m, occurs. Similar equations were formulated to provide estimates of the vertical eddy diffusivity:

$$K_z = \frac{\bar{u}Z_m^2 e^{1/3}}{2X_m} \qquad (5.26)$$

$$K_z = \frac{\bar{u}Z_m^2}{2X_t} \qquad (5.27)$$

where Z_m is the maximum half width of the plume at distance X_m. For additional information on procedures to estimate K_z, see the article by Draxler (1979).

Yu (1977) compared 14 parameterization schemes for calculating vertical exchange processes in the atmospheric boundary layer. Boundary-layer data from the O'Neill fifth period (Lettau and Davidson, 1953) and from Wangara day 32 (Clarke et al., 1971) were used. The most satisfactory simulation of wind profiles was produced by a model based on a turbulent energy equation proposed by Monin and Yaglom (1970). Temperature profiles were simulated best by the O'Brien (1970) K formula, especially during periods of convection.

5-1.8 Reynolds Number Similarity

In meteorological studies of the atmospheric boundary layer, much experimental work is done with models because the natural wind is difficult to control. Thus, for example, the full scale of the atmospheric flow field must be scaled down to maintain the proper characteristics of the natural flow in a wind-tunnel model, but otherwise it remains unchanged. If flows through two systems of different linear dimensions are to be similar, it is necessary to have a fixed ratio of the forces acting on a fluid parcel for all time at all geometrically similar points. This condition is referred to as the principle of similarity. In boundary-layer flow the two forces of primary concern are the viscous and inertial forces. The condition of similarity requires that the ratio of inertial force to viscous force be constant. In Fig. 5.6 the steady flow in the x-direction has an inertial force per unit volume equal to $\rho u(\partial u/\partial x)$. The viscous or internal frictional force per unit volume is, according to Eq.

5.5, $\mu(\partial^2 u/\partial x^2)$. For similarity the ratio of these forces must be constant:

$$\frac{\text{Inertial force}}{\text{Viscous force}} = \frac{\rho u(\partial u/\partial x)}{\mu(\partial^2 u/\partial x^2)} = \text{constant}$$

How do these two forces change as the magnitude of the terms varies? At a given point, the speed, u, is proportional to the free stream speed, V. The speed gradient is proportional to V/d, where d is a characteristic linear dimension. The denominator is proportional to V/d^2; thus the ratio, Re, is

$$\text{Re} = \frac{\rho V/d}{\mu V/d^2} = \frac{Vd}{\nu}$$

where $\mu/\rho = \nu$, the kinematic coefficient of viscosity. This equation yields a dimensionless number known as the Reynolds number (Re). Thus the natural and modeled flows are similar when the Reynolds number for both are equal.

Since Re is the ratio of the internal force to the viscous force, its value is as a measure of the relative importance of these forces. Thus we can deduce certain general characteristics about the fluid flow for large and small values of Re. Recall that in the boundary layer fast wind speeds tend to keep the air well mixed. Viscous forces tend to inhibit turbulence, however, since the effect of viscosity is to suppress irregular flow. Consequently, when Re is small (<1), the inertial forces do not dominate the viscous forces and the flow is laminar. When Re is large (>1), inertial forces prevail and the flow can be turbulent. This result was found by Reynolds (1883) in connection with his investigations of flow through pipes. Atmospheric flow characterized by Reynolds numbers greater than about 10^2 is usually turbulent.

How does one select a value for d in the atmosphere? For macroscale atmospheric flow, there is no obvious length associated with turbulence. In fact, it appears that turbulent motions can occur over a very wide range of sizes. However, for any value of d chosen for atmospheric flows, we are bound to conclude that Re will be very large because typical horizontal speeds are of the order of 10^2 to 10^3 cm/s and, for a temperature of 20°C, ν is approximately 0.15 cm^2/s. If d is small, say 10^4 cm, then Re = 10^7 to 10^8 in the atmosphere. Consequently we can conclude that atmospheric motions are generally turbulent. However, if we assume that turbulence due to eddy motion becomes small for some distance d = D above the ground, we can estimate the size of D through Reynolds number similarity by writing Re as DV/ν = constant. To specify the constant, assume that turbulent flows usually occur for Re > 10^2. Near the ground, V is of the order of 10^2 cm/s and ν is about 10^{-1} cm^2/s so that D < 10^{-1} cm. Thus the upper limit of D is of the order of 0.1 cm, which represents the smallest wavelength in turbulence or the Kolmogorov microscale.

5-1.9 Rossby Number

Another useful dimensionless parameter, the Rossby number, Ro, is the ratio of the inertial force to the Coriolis force. This ratio is expressed as Ro = V/fd, where f is the Coriolis parameter. For large values of Ro, it is generally assumed that the Coriolis effect can be neglected in our models. Large Ro occurs if f is small, as in the tropical latitudes, or if the dimensions of the model, d, are small, e.g., ≤ 100 km on a side. Consequently the Coriolis contribution is usually considered to be negligibly small for small-scale atmospheric processes. By contrast, if Ro is small, the earth's rotation becomes an important process. But what do large and small represent quantitatively? In the tropics, Ro has a magnitude of about 10^2, whereas for typical macroscale flow in the mid-latitudes it is approximately 10^{-1}. Haltiner (1971) recognized that a sufficient condition for the validity of the geostrophic approximation is that Ro ≲ 0.1. Thus, to neglect the Coriolis parameter, we could specify a large Ro as ≳ 0.1. Within 50 m of the ground (the surface layer), Ro is usually specified as

$$\text{Ro} = \frac{G}{f z_0}$$

where $G = (u_g^2 + v_g^2)^{1/2}$, the modulus of the geostrophic wind, and z_0 is the aerodynamic (surface) roughness length (see Sec. 5-3.3). The surface Ro is typically very large: 10^5 to 10^9.

5-1.10 Froude Number

The nondimensional ratio of the inertial force to the gravitational force is the Froude number:

$$\text{FR} = V^2/gd$$

where V is a representative wind speed, d is a characteristic length (such as the thickness of the atmospheric boundary layer), and g is the acceleration due to gravity. The Froude number is used in laboratory models whenever gravity is an important force (e.g., see Chap. 8, Sec. 8-5.7). The number can also be used as a measure of the importance of density stratification on flow over complex terrain.

5-2 GOVERNING EQUATIONS

5-2.1 Introduction

For a mathematical description of atmospheric motions, it is convenient to use a coordinate system that is fixed on the surface of the earth and rotates with the earth. The Cartesian coordinate system offers the desired features because it can be centered on the surface of the earth so that the vertical axis, z, is directed upward, opposite to gravity; the x-axis is in a plane perpendicular to z and is directed toward the east; and the y-axis is in the same plane as the x-axis but is directed toward the north. This coordinate system represents the framework in which the equations in this chapter will be presented.

5-2.2 Newton's Law of Motion

The fundamental principle governing atmospheric motions is Newton's second law, which states that the sum of all external forces acting on a body is equal to the product of the mass of the body times the acceleration produced. In fluid flow it is necessary to consider two classes of forces—body forces, **B**, and surface forces, **S**. Body forces are those which act throughout the fluid, e.g., gravitational force. Surface forces are those exerted by adjacent portions of the fluid on one another, e.g., the viscous stresses. Consequently, according to Newton's second law, the acceleration of a fluid parcel is given by the sum of the body and surface forces, per unit mass, acting on the parcel.

Newton's laws of motion apply to a nonaccelerating or inertial coordinate system. The Cartesian coordinate system is fixed with respect to the stars and is essentially an inertial reference frame for an earthbound observer; however, the system of coordinates rotates with the earth. Consequently, in an application of Newton's laws of motion, any point not at the center of rotation must experience a centripetal acceleration.

5-2.3 Equations of Motion

For atmospheric motions to occur, the forces per unit mass that can act on a fluid parcel are (1) pressure, **P,** (2) Newtonian gravitation, **G,** (3) friction, **F,** (4) viscous stress, **T,** and (5) centripetal acceleration resulting from the rotation of the earth, the Coriolis effect, **C.** Because these forces are balanced by the inertial force, **dV**/dt, the basic equation of motion is

$$\frac{d\mathbf{V}}{dt} = \mathbf{S} + \mathbf{B} = (\mathbf{P} + \mathbf{T} + \mathbf{F}) + (\mathbf{G} + \mathbf{C})$$

The component form of the equation of motion for incompressible, viscous flow is

$$\frac{Du}{Dt} = \frac{\partial u}{\partial t} + u\frac{\partial u}{\partial x} + v\frac{\partial u}{\partial y} + w\frac{\partial u}{\partial z} = -\frac{1}{\rho}\frac{\partial P}{\partial x} + fv + \nu\left(\frac{\partial^2 u}{\partial x^2} + \frac{\partial^2 u}{\partial y^2} + \frac{\partial^2 u}{\partial z^2}\right) \tag{5.28}$$

$$\frac{Dv}{Dt} = \frac{\partial v}{\partial t} + u\frac{\partial v}{\partial x} + v\frac{\partial v}{\partial y} + w\frac{\partial v}{\partial z} = -\frac{1}{\rho}\frac{\partial P}{\partial y} - fu + \nu\left(\frac{\partial^2 v}{\partial x^2} + \frac{\partial^2 v}{\partial y^2} + \frac{\partial^2 v}{\partial z^2}\right) \tag{5.29}$$

$$\frac{Dw}{Dt} = \frac{\partial w}{\partial t} + u\frac{\partial w}{\partial x} + v\frac{\partial w}{\partial y} + w\frac{\partial w}{\partial z} = -\frac{1}{\rho}\frac{\partial P}{\partial z} - g + \nu\left(\frac{\partial^2 w}{\partial x^2} + \frac{\partial^2 w}{\partial y^2} + \frac{\partial^2 w}{\partial z^2}\right) \tag{5.30}$$

$$\frac{\partial u}{\partial x} + \frac{\partial v}{\partial y} + \frac{\partial w}{\partial z} = 0 \tag{5.31}$$

Equations 5.28 to 5.30 are known as the Navier–Stokes (or primitive) equations† for frictionless, incompressible, viscous flow. The total derivative on the extreme left side of each equation represents the time rate of change of the speed of an individual particle of air. This term accounts for the change in velocity due to advection. The terms of the right side of each equation represent the pressure gradient force, Coriolis effect, gravitational force (z-direction only), and the force due to the viscous stresses. Finally, Eq. 5.31 is the continuity equation (see Chap. 2) for an incompressible atmosphere. This equation simply states that, if fluid is converging into (diverging out of) a region, the mass there must be increasing (decreasing) accordingly.

Robinson (1978) prepared a critical review of the use of the Navier–Stokes equations for predicting macroscale motions. He emphasized that the standard derivation of the Navier–Stokes equations is based on arguments that are strictly valid only in the trivial limit of macroscale uniformity and in the absence of thermal forcing. The issue raised by Robinson is important because its central theme is that we are unable to make deterministic statements about the future state of the atmosphere since the logical and mathematical tools available are not relevant to the problem. However, to simulate atmospheric flow, we must use the best mathematical and physical tools available and must use simplifying assumptions to make the mathematics manageable.

A special form of Eq. 5.30 can be attained for an inviscid fluid in which the vertical acceleration, Dw/Dt, is equal to zero:

$$0 = -\frac{1}{\rho}\frac{\partial P}{\partial z} - g \tag{5.32}$$

This is the hydrostatic equation (see Chap. 2), one of the best approximations of theoretical meteorology. It represents an atmospheric state of static equilibrium, with the net pressure force exactly balancing the force of gravity. From Eq. 5.32 it is apparent that at a given point in the atmosphere the pressure depends only on the air density and the thickness of the overlying atmosphere. The hydrostatic approximation can be applied safely to all meteorological phenomena except those in which an appreciable vertical acceleration occurs.

In a gas of constant pressure, temperature fluctuations cause density fluctuations. Consequently, when temperature becomes active enough to contribute to the creation of velocity fluctuations, a modification should be made in Eq. 5.30 (and in Eq. 5.35). This modification can be made by specifying $\rho = \bar{\rho} + \rho' = \bar{\rho}(1 + \rho'/\bar{\rho})$ and $P = \bar{P} + P'$, so that Eq. 5.32 becomes

$$-\frac{1}{\rho}\frac{\partial P}{\partial z} - g = \frac{1}{\bar{\rho}}\left(1 + \frac{\rho'}{\bar{\rho}}\right)^{-1}\frac{\partial}{\partial z}(\bar{P} + P')$$

After expanding $[1 + (\rho'/\bar{\rho})]^{-1}$ binomially, neglecting second-order terms, substituting $-g$ for $(1/\bar{\rho})(\partial\bar{P}/\partial z)$, and assuming that vertical pressure fluctuations can be neglected, we obtain

$$-\frac{1}{\rho}\frac{\partial P}{\partial z} - g = -g\frac{\rho'}{\bar{\rho}}$$

The fluctuating body force, $g\rho'/\bar{\rho}$, representing the buoyancy of an individual air parcel, should be used in Eqs. 5.30 and 5.35 when local vertical accelerations are important, as in thunderstorms and in volcanic eruption clouds (see Chap. 8, Sec. 8-2.1.1). Continuing with the assumption that the fluctuations in density result principally from thermal (as opposed to pressure) effects, we can show that $g\rho'/\bar{\rho} \simeq gT'/\bar{T}$ [see Spiegel and Veronis (1960) or Calder (1968)]. However, density fluctuations are frequently considered to be unimportant (the Boussinesq approximation), except when buoyant forces exist in the flow.

Generally we are interested in obtaining expressions for the mean flow and the accompanying fluctuations. To form the required time averages, substitute the component form of Eq. 5.1 in for each u, v, and w (and P) component of Eqs. 5.28 to 5.30:

$$\frac{D\bar{u}}{Dt} = -\frac{1}{\rho}\frac{\partial\bar{P}}{\partial x} + f\bar{v} + \nu\nabla^2\bar{u} - \frac{\partial}{\partial x}(\overline{u'u'}) - \frac{\partial}{\partial y}(\overline{u'v'}) - \frac{\partial}{\partial z}(\overline{u'w'}) \tag{5.33}$$

†These equations were derived by Navier (1822), Poisson (1830), Saint-Venant (1843), and Stokes (1849).

$$\frac{D\bar{v}}{Dt} = -\frac{1}{\rho}\frac{\partial \bar{P}}{\partial y} - f\bar{u} + \nu\nabla^2\bar{v}$$

$$- \frac{\partial}{\partial x}(\overline{v'u'}) - \frac{\partial}{\partial y}(\overline{v'v'}) - \frac{\partial}{\partial z}(\overline{v'w'}) \qquad (5.34)$$

$$\frac{D\bar{w}}{Dt} = -\frac{1}{\rho}\frac{\partial \bar{P}}{\partial z} - g + \nu\nabla^2\bar{w}$$

$$- \frac{\partial}{\partial x}(\overline{w'u'}) - \frac{\partial}{\partial y}(\overline{w'v'}) - \frac{\partial}{\partial z}(\overline{w'w'}) \qquad (5.35)$$

Observe that nine new "primed" terms have occurred in the time-averaging process. Thus the effect of turbulence is to introduce additional terms into the equations for mean flow. These new advective terms are the Reynolds stresses (Eq. 5.16). It is through Eqs. 5.33 to 5.35 that we strive to account for flow through the atmospheric boundary layer; however, these equations must satisfy the boundary conditions imposed on the atmospheric boundary layer. For example, above roughly 1 km, motion is due to macroscale pressure changes that set the outer boundary conditions on the atmospheric boundary layer. Normally the wind velocity at some height z = h is specified as the geostrophic velocity measured at h. However, flow near the ground depends on the topographic and surface influences of the area over which the flow develops. Consequently the bottom boundary conditions are a function of the condition of the ground.

We readily see that Eqs. 5.33 to 5.35 are nonlinear and that analytical solutions are difficult to attain without making simplifying assumptions. In fact, since Eqs. 5.33 to 5.35 vary temporally, five more equations are required to make the set complete. These five equations are expressions for the time changes in the Reynolds stress terms (see Donaldson, 1973a). Consequently, in attempting to simulate temporally varying flows, we have a choice of trying to proceed with only the averaged equations or with the full set of eight equations. With the latter approach, however, we are faced with calculating extremely complex relations at every grid point down to 10^{-3} s or less. With the averaged equations the flow fields are much simpler, but we have three more unknowns than equations or nine more averaged quantitives (e.g., $\overline{u'v'}$). As a result, crucial approximations are frequently made for the stress terms to make the equation set tractable (e.g., see Manton and Cotton, 1977). Efforts to make viable approximations for these parameters are often referred to as the closure problem of turbulence theory. From the use of high-order-closure models, we expect to obtain a better understanding of the physics of turbulence than is possible with eddy-diffusion approximations. This knowledge should lead to more reliable simulation of the mean wind field and, consequently, to better simulations of the atmospheric concentrations of pollutants released from manmade and natural sources. For an additional discussion on "closure," see Chap. 6 and Chap. 8, Sec. 8-2.3.

Simplified forms of the Navier–Stokes equations that describe a diurnally varying boundary layer are obtained by specifying a less complex flow regime. For example, the flow can be considered to occur over an infinite, flat surface of uniform roughness and temperature. Also, the flow is assumed to be dry and horizontally homogeneous with no radiative flux divergence. The mean quantities are only allowed to vary temporally along the vertical axis; thus, with the Boussinesq approximation,

$$\frac{\partial \bar{u}}{\partial t} = f(\bar{v} - \bar{v}_g) - \frac{\partial}{\partial z}(\overline{u'w'})$$

$$\frac{\partial \bar{v}}{\partial t} = -f(\bar{u} - u_g) - \frac{\partial}{\partial z}(\overline{v'w'})$$

and the temporal change in the mean potential temperature becomes (see Sec. 5-2.4)

$$\frac{\partial \theta}{\partial t} = -\frac{\partial}{\partial z}(\overline{w'\theta'})$$

Examples of how these expressions are used in numerical models of the atmospheric boundary layer are provided by Wyngaard and Coté (1974a), Pielke and Mahrer (1975), Yamada and Mellor (1975), Busch, Chang, and Anthes (1976), Brost and Wyngaard (1978), André et al. (1978), Lee and Kao (1979), Saab (1979), and Chang (1979). Geostrophic winds can be calculated (Endlich, 1968) from field data (Cornett and Randerson, 1977), estimated from observations, or specified empirically. The vertical profiles of geostrophic wind can be calculated each time step by means of a parabolic height interpolation as in the study by Yamada and Mellor (1975). On the bottom boundary, the variations can be required to follow Monin–Obukhov similarity (see Secs. 5-2.7 and 5-3.5). On the top boundary, all mean gradients can be held at zero. Parameterization of the turbulent energy budget at the top of the daytime atmospheric boundary layer has been described by Zeman and Tennekes (1977), and Carson and Richards (1978) reviewed procedures for estimating surface-level turbulent fluxes in stable conditions.

Different flow regimes can be derived from the Navier–Stokes equations, depending on the boundary conditions imposed and the assumptions posed about the field of flow. Excellent discussions of many different possible solutions to Eqs. 5.33 to 5.35 are given by Sutton (1953), Schlichting (1955), Plate (1971), and Estoque (1973). For discussions of experimental results from high-order-closure models, refer to the work of Bradshaw (1971), Deardorff (1972b, 1974a, 1974b), Wyngaard and Coté (1974a), Wyngaard (1975), Donaldson (1973a), Hoffert and Sud (1976), and others. As of 1979, closure models were changing so rapidly that it is beyond the scope of this chapter to attempt to assess the state of the art; however, some closure schemes are described in Chap. 6.

5-2.4 Heat Transfer

A solid or liquid surface acts as the bottom boundary to the atmosphere. This boundary can experience considerable temperature variation resulting from the absorption of short-wave solar radiation and the consequent emission of long-wave terrestrial radiation. Air in contact with the ground boundary interacts with the induced temperature field through the heat-transfer processes of conduction and convection (see Chap. 2). Heat can also be added to a parcel of air by such processes as the turbulent transfer of heat, condensation of water vapor, and radiation. However, Deardorff (1972a) showed that up to time periods of greater than 1 d these heating processes are of secondary importance.

To account for the interaction between the mean flow and the temperature field, we must define the heat transfer for a moving fluid parcel. For incompressible flow the thermal energy balance is determined by the internal energy, by the conduction and convection of heat, and by the generation of heat through friction. To simplify the mathematics, we neglect the contributions by solar and terrestrial radiation in the development of the heat-transfer equation.

From Chap. 2, recall that for a unit mass, the first law of thermodynamics can be written as

$$dH = c_p\,dT - \alpha dP \qquad (5.36)$$

where $c_p = 0.24$ cal/(g·°K) and α is the specific volume. This form of the first law of thermodynamics is useful because the increments dT and dP are easily measured.

Heat can be added to a fluid parcel by friction and from external sources by conduction. The amount of heat added by conduction (dH_c) into a unit volume in time dt can be obtained from

$$dH_c = b\left(\frac{\partial^2 T}{\partial x^2} + \frac{\partial^2 T}{\partial y^2} + \frac{\partial^2 T}{\partial z^2}\right) = b\nabla^2 T\,dt$$

where b is the thermal conductivity given by $b = \rho c_p \kappa$, where κ is the thermal diffusivity of the substance. This equation, which was developed by Fourier in 1822, represents the transfer of heat by conduction in an isotropic solid. The heat due to friction dH_f in time dt is given by $\mu\Phi$, where μ is the coefficient of viscosity and Φ expresses the energy dissipation by viscosity and is defined by

$$dH_f = \mu\Phi\,dt$$

where

$$\Phi = \left(\frac{\partial v}{\partial x} + \frac{\partial u}{\partial y}\right)^2 + \left(\frac{\partial w}{\partial y} + \frac{\partial v}{\partial z}\right)^2 + \left(\frac{\partial u}{\partial z} + \frac{\partial w}{\partial x}\right)^2 + 2\left(\frac{\partial u}{\partial x}\right)^2 + 2\left(\frac{\partial v}{\partial y}\right)^2 + 2\left(\frac{\partial w}{\partial z}\right)^2 - \frac{2}{3}\left(\frac{\partial u}{\partial x} + \frac{\partial v}{\partial y} + \frac{\partial w}{\partial z}\right)$$

Now the total heat added by conduction and friction is equal to dH in Eq. 5.36; thus the heat-transfer equation becomes

$$b\nabla^2 T\,dt + \mu\Phi\,dt = c_p\,dT - \alpha dp$$

where the properties c_p and b are assumed to be independent of temperature. A very simple form of this equation is obtained by neglecting dissipation by viscosity so that the heat transfer in an incompressible gas is

$$\rho c_p \frac{dT}{dt} = b\nabla^2 T \qquad (5.37)$$

In turbulent flow the definition of the temperature can be treated in a manner analogous to the wind so that the temperature can be regarded as the sum of the mean temperature, $\overline{T}$, and a fluctuating component, T′. Taking a time average of Eq. 5.37 and making use of the continuity equation, we obtain

$$\rho c_p \frac{d\overline{T}}{dt} = \frac{\partial}{\partial x}\left(K_x \frac{\partial \overline{T}}{\partial x} - \rho c_p \overline{u'T'}\right) + \frac{\partial}{\partial y}\left(K_y \frac{\partial \overline{T}}{\partial y} - \rho c_p \overline{v'T'}\right) + \frac{\partial}{\partial z}\left(K_z \frac{\partial \overline{T}}{\partial z} - \rho c_p \overline{w'T'}\right) \qquad (5.38)$$

Note the similarity between this equation and Eqs. 5.33 to 5.35. The $\overline{w'T'}$ type terms represent the eddy correlation (covariance) between temperature and vertical motion. In Eq. 5.38 the local rates of heat transfer are $[K_x(\partial\overline{T}/\partial x) - \rho c_p \overline{u'T'}]$, etc., so that the total rate of transfer is the sum of that arising from molecular conductivity ($K\partial\overline{T}/\partial x$ type of terms) and that coming from the eddy heat flux ($\rho c_p \overline{w'T'}$ type of terms). The eddy heat flux usually dominates the molecular-transfer terms. Consequently, if we neglect molecular conduction and assume horizontal uniformity in the atmosphere, Eq. 5.38 simplifies to

$$\frac{\partial \overline{T}}{\partial t} = \frac{\partial}{\partial z}(\overline{w'T'})$$

Substituting Eq. 5.21 for $\overline{w'T'}$ and assuming $\theta' \sim T'$ (P ~ 1000 mb) gives

$$\frac{\partial \overline{T}}{\partial t} = \frac{\partial}{\partial z}\left(K_h \frac{\partial \overline{T}}{\partial z}\right)$$

the familiar equation for heat conduction in a solid. Procedures for estimating the diurnal variation of the surface temperature from such a second-order equation were described by Brooks and Carruthers (1953), Panofsky and Brier (1965), and Lönnqvist (1962). DeWalle and Parmele (1974) described a method of calculating surface temperature variation by radiation budget techniques, and McCutchan (1979) proposed a procedure for estimating $\partial T/\partial t$ in mountainous terrain.

Now, returning to Eq. 5.38 and using the same assumptions, we can write

$$\frac{\partial(\rho c_p T)}{\partial t} = \frac{\partial(\rho c_p \overline{w'T'})}{\partial z}$$

where c_pT is enthalpy. Let $\rho c_p \overline{w'T'} = H$, the sensible heat flux, and specify the height where the heat flux is zero as h_0. Integration from z = 0 to z = h_0 yields

$$H = \int_0^{h_0} \rho c_p \frac{\partial T}{\partial t}\,dz \simeq \rho c_p \frac{\partial T}{\partial t} h_0 \qquad (5.39)$$

where h_0 is the height of the thermal boundary layer. This equation describes the heat flux throughout the boundary layer when $\partial T/\partial t$ and h_0 are known. The term ρc_p is approximately 300 cal/(m^3·°K) at sea level. Businger (1972) found Eq. 5.39 to yield values of H ranging from 8 W/m^2 to 16 W/m^2 for two consecutive 1-h periods shortly after sunrise. Hall, Edinger, and Neff (1975) found H = 4.5 W/m^2 for a 2-h period in late morning. For additional data, see Table 5.2.

TABLE 5.2 Some Measurements of the Vertical Flux of Sensible Heat from Various Field Programs†

Source	Range of values, W/m^2
Milford, Abdulla, and Mansfield (1979)‡	−19.0 to 163.0§
Spittlehouse and Black (1979)	−20 to 282
Ting and Hay (1975)	12 to 380
Lenschow (1973)¶	38 to 83
Donelan and Miyake (1973)	−12 to 8§
Hicks and Goodman (1971)	103 to 164
Pond et al. (1971)	9 to 32§
Sellers (1967)	−245 to 453
Swinbank (1964)	20 to 486
Monin and Obukhov (1954)	40 to 250
Cramer and Record (1953)¶	−29 to 121

†For a discussion of the horizontal advection of sensible heat, see the paper by Brakke, Verma, and Rosenberg (1978).

‡Observations from a powered glider.

§Range of values over water.

¶Report contains measurements of many other parameters.

5-2.5 Surface Layer

In a shallow layer of air near the ground, the frictional force and the horizontal shear stress are the dominant forces acting on the airflow. Furthermore, for unaccelerated mean motion over uniform terrain, observations tell us the pressure gradient and Coriolis forces are, individually, an order of magnitude smaller than the surface stress, τ_0; thus in the surface boundary layer the balance of forces can be approximated by the vector difference $\tau - \tau_0 = 0$, where τ is the viscous stress at the top of the layer. This vector expression states that the surface layer is characterized by (1) constant stress and (2) constant wind direction with height. Since we are discussing ensemble averages here, we are averaging the instantaneous flux distribution over a very large number of realizations. Moreover, if a variation in wind direction occurs, the vectorial aspect of the equation requires that there be a corresponding variation in the direction of the stress. Through what depth of atmosphere are these two characteristics of the boundary layer valid?

For unaccelerated motion over uniform terrain, $|(1/\rho)(\partial\tau/\partial z)|$ is of the order of 10^{-1} cm/s^2 and $\rho \simeq 10^{-3}$ g/cm^3 so that $\Delta\tau = 10^{-4}\,\Delta z$. Now define the depth of the surface boundary layer, h, as the height above the ground through which the magnitude of the viscous stress at the top of the boundary layer varies by less than 10%, or $\Delta\tau = 0.1\tau_0$. In other words, in the "constant stress" layer, the stress is within 10% of its surface value. Then, $0.1\tau_0 = 10^{-4}$ h, where Δz has been replaced by h, so that $h = 1000\,\tau_0$. Experimental evidence tells us τ_0 is of the order 1 to 10 $g/(cm \cdot s^2)$; thus h varies between 10 and 100 m. In general, the surface layer is often defined as the lowest 50 m of the atmospheric boundary layer where the vertical fluxes of heat and momentum are horizontally homogeneous and nearly constant with height. Haugen, Kaimal, and Bradley (1971) and Dyer and Hicks (1972) have found that, within the first 20 m of the surface layer, the vertical fluxes of heat and momentum are essentially constant (within ± 20%) provided the averaging period is long enough (>15 min). Using field data, Clarke (1970a) observed the depth of the surface layer to be approximately 90 m. In general, the depth of the surface layer is defined as $z < |L|$, where L is the Monin–Obukhov length scale defined in Sec. 5-2.7.

5-2.6 Scaling Parameters

From Eq. 5.20 it is implicit that the stress is proportional to the square of the mean wind speed. Then, if the viscous stress in the surface boundary layer is constant in magnitude and direction, we can define a friction velocity,† u_*, that is constant with height so that

$$u_*^2 = \frac{\tau}{\rho} \tag{5.40}$$

where τ and ρ are also assumed to be constant with height (but not time). Friction velocity is the relevant velocity scale for modeling diffusion in the surface layer. Arya (1972a) and Carson and Richards (1978) noted that knowledge of H, V(z), and $\Delta T(z)$ is required to determine u_*.

The surface stress, τ_0, can be approximated by the simple relationship, $\tau_0 = ½\rho_0 C_D \overline{V}^2$, where C_D is a drag coefficient, $½\,\rho_0\,\overline{V}^2$ is the dynamic head, and $\overline{V}$ is the mean wind speed at a height of a few meters. From the expression for τ_0 and Eq. 5.40, we find that $C_D = 2(u_*/\overline{V})^2$. Brook (1972) has proposed that $C_D = 0.08(I_u^2 + I_v^2 + I_w^2)$, where the I terms represent the intensity of the turbulence of each wind velocity component. From wind tower data, Smith and Banke (1975) proposed the overall regression formula for drag coefficient to be $C_D = (0.63 + 0.066\overline{V}_{10})10^{-3} \pm 0.23$, where $\overline{V}_{10}$ is the mean wind speed (m/s) at the 10-m level. For related material, see Sec. 5-2.8. In general, values of C_D range from approximately 10^{-3} over a smooth ocean to roughly 10^{-2} over a bushy canopy to 10^{-1} in a city environment. Pond, Fissel, and Paulson (1974) showed that an estimate of the 30-min averaged bulk surface stress could be calculated from $\tau = 1.53 \times 10^{-2}\overline{V}_{10}\,[g/(cm \cdot s^2)]$, where $\overline{V}_{10}$ (m/s) is the wind speed 10 m above the ocean surface. For additional information on bulk parameterization, see the paper by Liu, Katsaros, and Businger (1979).

Monin and Obukhov (1954) and Lettau (1949) took a more direct approach to specifying u_*. They started by considering horizontally homogeneous flow over an unlimited rough surface. The averaged characteristics of the flow were assumed to depend solely on vertical depth, z, so that the most important properties were the vertical fluxes of momentum, heat, and moisture. Thus the flux of momentum can be considered as a shearing stress. However, for convenience, instead of using shearing stress $\tau = -\rho\overline{u'w'}$, they defined a friction velocity according to Eq. 5.40 where $u_*^2 = -\overline{u'w'}$.‡

A similar approach can be used with the vertical flux of sensible heat to arrive at a scaling temperature, T_*, and a scale length, or stability length, L, assuming that we can ignore the radiant-energy flux. Table 5.2 shows that $|H|$ is of the order of 1 to 100 W/m^2. If the mean temperature, $\overline{T}$, varies with

†Also referred to as the mechanical-turbulence velocity scale, a measure of mechanical turbulence; a direct measure of the frictional force of the wind.

‡In a more complete form, $\mu_*^4 = (\overline{u'w'})^2 + (\overline{v'w'})^2$.

height, the general regime of turbulence is affected by the rate of production of buoyant or convective energy. Indeed, heat flux is the principal factor in the formation and maintenance of the unstable boundary layer. The heat released from the ground is distributed over much of the boundary layer, altering the temperature and generating turbulent kinetic energy through the buoyancy force.

As noted in Sec. 5-2.3, the vertical component of the equation of motion should contain the contribution by buoyancy in a medium of nonuniform temperature. Provided the temperature fluctuations remain small compared to $\overline{T}$, the buoyant force is represented by $g(T'/\overline{T})$ [or $g(\rho'/\overline{\rho})$ when pressure fluctuations are small] so that the buoyancy parameter $g/\overline{T}$ is introduced to the vertical component of the equation of motion (Eq. 5.35). A typical value of this parameter is 0.03 m/s. Whenever molecular viscosity and heat conduction can be neglected, $g/\overline{T}$ is the only dimensioned quantity of importance in all the equations for turbulent flow that are valid near the ground. In a broader sense, $g/\overline{T}$, H, u_*, and z can all be combined into unique parameters for defining the statistical characteristics of turbulence through similarity theory.

5-2.7 Similarity Theory

The Monin–Obukhov similarity theory for turbulence in a thermally stratified surface layer is now well known. To develop their similarity theory, Monin and Obukhov (1954) considered an idealized field of motion that is frequently approximated in the atmosphere near the ground. They assumed that all the statistical properties of the temperature and velocity fields do not vary with time and are homogeneous so that the operators $\partial/\partial t \equiv 0$ and $\partial/\partial x \equiv 0 \equiv \partial/\partial y$. In addition, the steady mean motion was considered to be unidirectional at all heights in the x direction so that $\mathbf{V} = V$, the $\nu\nabla^2$ terms were assumed to be negligibly small, the scale of the motion was considered to be small enough to omit the Coriolis effect, and the radiative heat inflow was neglected, $dT/dt = 0$. Consequently the equations of motion simplify to the form (see Calder, 1966)

$$\mu \frac{\partial V}{\partial z} - \rho\overline{u'w'} = \rho u_*^2 \tag{5.41}$$

$$-k\frac{\partial T}{\partial z} + k\overline{w'T'} = H \tag{5.42}$$

$$-\frac{\partial}{\partial z}\left(\frac{P}{\rho}\right) - \frac{\partial \overline{w'^2}}{\partial z} = -\frac{g}{\overline{T}}T' \tag{5.43}$$

Monin and Obukhov demonstrated that these three simultaneous coupled equations are necessary conditions for statistically steady and for horizontally homogeneous conditions in the surface layer. In addition, Obukhov (1946) and Monin and Obukhov showed that for microscale flow in the surface layer, the boundary conditions

$$\overline{w'T'} = \frac{H}{\rho c_p} = \text{constant} \tag{5.44}$$

and

$$-\rho\overline{u'w'} = \tau = \text{constant} \tag{5.45}$$

uniquely define the statistical characteristics of the turbulent flow predicted by Eqs. 5.41 to 5.43. In the surface layer where the turbulent fluxes are approximately constant at their surface values, $\rho_0 u_*$ and H_0, the similarity hypothesis states that the turbulence of the flow is determined solely† by the three parameters $g/\overline{T}$, u_*, and H over depth z. These governing surface-layer parameters yield a total of four independent velocity, temperature, and length scales. It is customary to use one velocity scale (u_*), one temperature scale (T_*), and two length scales (z and L). As we shall see, after nondimensionalizing with these scales, dimensional analysis predicts that the flow properties are functions of only z/L. Field data support this theory (e.g., see Businger et al., 1971).

Now look at L. Scale analysis allows us to combine the four surface-layer parameters to form the length scale, L:

$$L = -\frac{\rho c_p u_*^3 \overline{T}}{kgH} \tag{5.46}$$

where L is strongly dependent on wind speed since u_* depends on V. In fact, Eq. 5.46 tells us that if u_* doubles, with all the other forms held constant, L increases by a factor of 8.

Substituting from Eq. 5.44 into Eq. 5.46 yields

$$L = -\frac{u_*^3 \overline{T}}{gk\overline{w'T'}}$$

where L is the Monin–Obukhov length scale and contains only quantities that are approximately constant throughout the surface layer. This parameter is the dominant length scale governing diffusion and the flux-gradient relationships in the surface layer. The precise physical meaning of k is not fully understood, and its value is estimated by experiment (see Sec. 5-3.3). Note that the effects of water vapor have not been included in these expressions for L. If the effects of water vapor are included, Eqs. 5.44 and 5.46 become

$$\overline{w'T'} = \frac{H}{\rho c_p} + 0.61\,\overline{w'q'}\,\overline{T}$$

$$L = -\frac{u_*^3 \overline{T}_v}{gk(\overline{w'T'} + 0.61\,\overline{w'q'}\,\overline{T})}$$

where $\overline{T}_v$ is the mean virtual temperature. The meaning of Eq. 5.46 is discussed later in this section.

In a similar manner, a temperature scale, T_*, can be defined as

$$T_* = -\frac{H_0}{\rho c_p u_*} = -\frac{(\overline{w'T'})_0}{u_*} \tag{5.47}$$

where the minus signs have been included for convenience in Eqs. 5.46 and 5.47.

The preceding expressions for $\overline{w'T'}$, L, and T_* can be written in a more convenient form by defining the temperature flux (or the kinematic turbulent heat flux) Q as $H/\rho c_p$. Then we have

$$\overline{w'T'} = Q_0$$

$$L = -\frac{u_*^3 \overline{T}}{kgQ_0}$$

†Wyngaard (1973) discussed why other variables are not included among the governing group.

$$T_* = -\frac{Q_0}{u_*}$$

where Q_0 is the surface temperature flux and has the dimensions of m·°K/s. A typical summertime value for Q_0 is 0.25 m·°K/s.

All other quantities in Eqs. 5.41 to 5.43 are functions of the single independent variable z. However, this system of equations is not closed because the three Reynolds flux terms $\overline{u'w'}$, $\overline{w'T'}$, and $\overline{w'^2}$ are unknown.

On the basis of the similarity hypothesis, there is a unique functional relationship for the speed gradient:

$$\frac{\partial V}{\partial z} = f\left[u_*, T_*, \frac{g}{T}, z, \frac{\mu}{\rho}, \frac{k}{\rho c_p}\right]$$

All seven quantities in this equation involve only the dimensions of length, time, and temperature. Dimensional theory states that this relationship can be reduced to a universal one between a set of 4 = (7 − 3) independent nondimensional combinations of equations. For example, one combination is

$$\frac{z}{u_*}\frac{\partial V}{\partial z}, \frac{z}{L}, \frac{\rho u_* z}{\mu}, \frac{\rho c_p u_* z}{\mu}$$

and another is

$$\frac{z}{u_*}\frac{\partial V}{\partial z}, \frac{z}{L}, \frac{\rho u_* z}{\mu}, \frac{\mu c_p}{k}$$

Consequently we can define the relationship

$$\frac{kz}{u_*}\frac{\partial V}{\partial z} = \psi_m\left[\frac{z}{L}, \frac{\rho u_* z}{\mu}, \frac{\mu c_p}{k}\right]$$

where ψ denotes a universal function and k has been introduced on the left-hand side for consistency. Similarly, for temperature we obtain

$$\frac{kz}{T_*}\frac{\partial T}{\partial z} = \psi_h\left[\frac{z}{L}, \frac{\rho u_* z}{\mu}, \frac{\mu c_p}{k}\right]$$

where $\rho u_* z/\mu$ is a local Reynolds number. Now, if we restrict the flow to a given fluid, like air, then $\mu c_p/k$ becomes an absolute constant; and, if we specify the condition of fully developed turbulence (large Re), then the structure of the turbulence is independent of Re. For this special case, we have

$$\frac{kz}{u_*}\frac{\partial V}{\partial z} = \phi_m\left[\frac{z}{L}\right]$$

and

$$\frac{kz}{T_*}\frac{\partial T}{\partial z} = \phi_h\left[\frac{z}{L}\right]$$

where ϕ_m and ϕ_h are now universal functions of the single nondimensional variable z/L. Empirical expressions for ϕ_m and ϕ_h are described in Sec. 5-2.13.

5-2.8 Estimating L, H, and u_*

In Eq. 5.46, L is related to the height below which mechanically generated turbulence dominates the buoyant production of turbulence and above which the converse is true. During neutral conditions, |L| is approximately infinite (> 10^5 m), whereas L is negative during thermally unstable conditions (L > −5, extremely unstable) and positive during stable conditions (L < 10, very stable). Since |L| is seldom less than 10 m, thermal conditions in the lowest meter of the atmosphere are nearly neutral except for conditions with very slow wind speeds. If L is defined in terms of quantities that are approximately constant throughout the surface layer, it can be used as a quantity for characterizing the structure of the surface layer.

How is L determined? Basically the only measurements required to specify L are profiles of the local temperature and wind-speed gradients, say, between the 2-m and 10-m levels. Consequently it is convenient to recast Eq. 5.46 in terms of these two parameters. By using Eqs. 5.17, 5.21, and 5.40, we find that near the ground

$$L = \frac{u_* \overline{T}}{kg}\frac{\Delta V/\Delta z}{\Delta T/\Delta z} = \frac{u_* \overline{T}}{kg}\frac{\Delta V}{\Delta T}$$

when $K_m = K_h$ and where T is the absolute temperature and V is the wind speed. Tennekes and Lumley (1972) note that over a smooth surface $u_* \simeq 0.03\overline{V}$. Over wooded terrain, $u_* \simeq 0.05\overline{V}$. For near-neutral conditions over an open sea with wind speeds of about 8 m/s, Deacon (1973) found that $u_* \simeq 0.037V_{10}$, where V_{10} is the wind speed at the 10-m level. For conditions where direct measurement of the wind speed over the sea is not available, Hasse (1974) suggested that it is possible to estimate u_* from the geostrophic wind speed (V_g, based on surface pressure gradients) by $u_* \simeq 0.025V_g$. Errors involved in the evaluation of u_* from these simple equations are likely to be of the order of 20%. In a different study, Nieuwstadt (1978) made simultaneous use of both wind and temperature profiles to estimate u_* and T_* through least-squares methods.

A simple method for estimating L can be developed from the work of Liu and Durran (1977). They took Golder's (1972) nomograms that related estimates of L to the Pasquill–Gifford (P–G) stability categories (see Chaps. 1, 6, and 14), and, given z_0, they numerically derived the relation

$$L = (a_1 S + a_2 S^3)\, z_0^{-(b_1 - b_2|S| + b_3 S^2)^{-1}}$$

In this expression, S is a digital version of the P–G categories (see Table 5.3) and $a_1 = 0.00435$, $a_2 = 0.00372$, $b_1 = 0.503$, $b_2 = 0.231$, and $b_3 = 0.0325$. The final result is

$$\frac{1}{L} = a z_0^b$$

where the values of a and b are tabulated for each P–G category in Table 5.3. Methods of estimating z_0 are described in Secs. 5-3.3 and 5-3.4. Venkatram (1980) has proposed the simple empirical relation $L = Au_*^2$ for estimating L. In this expression, $A \propto -u_*/Q_0$. For stable conditions, $A \simeq 1100$ s^2/m. Irwin and Binkowski (1981) have developed an indirect procedure for finding L.

If direct measurements of H are possible, Eq. 5.44 can be solved easily. However, for accurate values of H, we should

TABLE 5.3 Values of 1/L as a Function of z_0, Calculated from the Liu–Durran Regression Equation†

P–G category	S	a	b	z_0: 0.01	1	0.1
					1/L	
A	−3	−0.1135	−0.1025	−0.18		−0.14
B	−2	−0.0385	−0.1710	−0.08		−0.06
C	−1	−0.0081	−0.3045	−0.03		−0.02
D	0	0	−0.5030	0		0
E	1	0.0081	−0.3045	0.03		0.02
F	2	0.0385	−0.1710	0.08		0.06

†Based on the data reported by Golder (1972). Values of z_0 and 1/L are in meters.

measure w′ and T′ directly with equipment that is capable of detecting turbulent energy fluctuations with periods less than 10 s and compute H from Eq. 5.44, where the quantity $\overline{w'T'}$ is of the order of ±0.1 to ±1.0 cm·°K/s. Bernstein and Young (1962) describe and discuss the measurement of H by both the direct eddy-correlation method (see Wesely, Thurtell, and Tanner, 1970) and by the energy balance analogy. An article detailing several techniques for determining surface layer heat fluxes has been prepared by Champagne et al. (1977). They compare the direct eddy-correlation method with two indirect methods, namely, the inertial-dissipation technique and the dissipation technique. Although the eddy-correlation method is direct, it generally involves the use of expensive equipment. Furthermore, if this equipment is on board a moving platform (e.g., aircraft or ship), the motions of the platform can severely contaminate the velocity signals. Moreover, Wyngaard (1973) noted that direct and indirect measurements require long averaging times and provide little relief from the problem of nonrepresentativeness. Wyngaard and Clifford (1978) found that the averaging time can be reduced by inferring surface-layer fluxes from structure parameters (see Sec. 5-2.14).

Dissipation rate methods are much less sensitive to motions of a platform. In addition, the dissipation rates of turbulent fluctuations can be measured at much higher frequencies than those of the platform motion; this minimizes contamination of the data. Champagne et al. (1977) found that over land both the direct and indirect measurements compared well in an unstable surface layer. For an overview of techniques for measuring surface fluxes over the ocean and for a discussion of the direct and dissipation methods of estimating $\overline{u'w'}$, see the article by Friehe and Gibson (1978).

5-2.9 Dimensionless Height z/L

In Eq. 5.46 the quantity $gH/c_p\rho\overline{T}$ represents the rate of production of turbulent energy, P_c, by convection. The rate of production of turbulent energy, P_m, by mechanical forces acting through depth z in neutrally stable air can be estimated near the ground as $u_*^3/kz \simeq u_*^2(\partial u/\partial z)$.† Using Eq. 5.46 and the ratio P_c/P_m, we can derive a dimensionless quantity, z/L, which expresses the ratio of convective to mechanical kinetic energy production in near-neutral air. From Eq. 5.46, we find that

$$-\frac{1}{L} = \frac{kgH}{u_*^3 c_p\rho\overline{T}} = \left(\frac{gH}{c_p\rho\overline{T}}\right)\left(\frac{k}{u_*^3}\right)$$

Multiply through by z and −1 to obtain the nondimensional expression

$$\frac{z}{L} = -\left(\frac{gH}{c_p\rho\overline{T}}\right)\left(\frac{kz}{u_*^3}\right) = -\frac{P_c}{P_m} \qquad (5.48)$$

For convenience, let

$$\zeta = \frac{z}{L} \qquad (5.49)$$

where ζ is a dimensionless height used to represent z/L. Values of ζ generally range from −5 to 5 where negative values represent unstable conditions and positive values stable conditions. Available evidence shows that when $\zeta > 0.2$, turbulence tends to cease.

Equation 5.48 is quite remarkable because of its implications. Remember that the mechanical production of turbulent energy comes from the shearing action of the mean flow, whereas the buoyant production of turbulent energy is the result of heating of the atmosphere adjacent to the ground. Also, the atmosphere is said to be stable if $\zeta > 0$, neutral if $\zeta = 0$, and unstable if $\zeta < 0$. Thus, if $\zeta < -1$, buoyant production exceeds mechanical production, and, if $\zeta > -1$, mechanical production exceeds buoyant production. For near neutral conditions $(-0.1 < \zeta < 0.1)$, the buoyant production vanishes and the turbulence is created mechanically by the mean wind shear. If $\zeta \gg 1$, then $H < 0$, corresponding to a downward flux of thermal energy into the ground from the air. In this case the buoyant production is negative, and the turbulence produced by the mechanical action of the mean-flow wind shear tends to be damped by the action of negative buoyant forces. In this stable state the structure of the turbulence behaves as if z is no longer important.

Returning to the definition of convective energy, the power per unit volume generated by convection is expressed as gH/c_pT. Near the ground, $g/c_pT \simeq 3.3 \times 10^{-5}\ m^{-1}$, after the proper manipulation of units. Consequently the atmosphere is a very inefficient heat engine when producing convective turbulence. For example, with a heat flux of 100 W/m^2 (see Table 5.2), convective effects generate turbulent energy at a rate of about 0.0033 W/m^3 so that, in a column of air 100 m deep, 0.33 W/m^2 is generated. In Sec. 5-1.5 the mechanical energy production was estimated at 1.0 W/m^2. Consequently, for the specified conditions, H would have to increase to more than 300 W/m^2 or the column depth must increase to more than 300 m (or some combination of both) before convective energy could dominate mechanical energy.

5-2.10 Local Free Convection

A free-convection-type state exists when ζ takes on large negative values. This state can be reached (1) if L takes on small negative values owing to an increase in the surface heat flux, H_0, or to a decrease in u_*, or (2) if z increases at fixed −L. In the first method we find from Eq. 5.46 that, as $u_* \rightarrow 0$ and $L \rightarrow 0$, a windless condition is approached

†The term u_*^3/kz is called the eddy dissipation rate, ϵ, for a neutrally stratified surface layer.

where all turbulent motion is caused by the buoyant force due to strong surface heating. Tennekes (1970) suggested that this state be termed local free convection. This condition never quite exists in the atmosphere since there is always some wind and thereby surface stress; however, for $\bar{u} < 7$ m/s and Q_0 as small as 0.1 m·°K/s, buoyant motions do tend to become the dominant energy source for turbulence. The second method of obtaining large negative values of ζ tells us that local free convection can be reached simply by going far enough above the ground but remaining in the atmospheric boundary layer. Consequently, of the four basic surface layer parameters ($g/\bar{T}$, u_*, H, and z) on which turbulence structure depends, u_* is no longer important for large $-\zeta$ values. For local free convection the governing surface variables become $g/\bar{T}$, Q_0, and z; thus from dimensional considerations it is possible to form only one each of the velocity, temperature, and length scales; namely, the convective velocity $u_f = (zgQ_0/\bar{T})^{1/3}$, temperature $T_f = Q_0/u_f$, and length z, where the subscript f has been used to indicate the local-free-convection state. Thus in the local-free-convection layer, turbulent velocities should scale with u_f, temperatures with T_f, and energy containing eddies with z. For examples of models of buoyancy-driven boundary layers, see the articles by Yamada and Mellor (1975) and by Zeman and Lumley (1976).

5-2.11 Richardson Number

Another nondimensional measure of whether or not turbulence will increase or decrease was proposed by Richardson (1920). The principle on which Richardson based his criteria is that a flow regime should remain turbulent provided the rate of supply of energy extracted from the mean flow by the Reynolds stresses is at least as great as the work that must be done to maintain turbulence against gravity. In the development of his criteria,† Richardson expressed the excess internal density of a rising eddy relative to the ambient air as $(g\rho \ell/T)(\partial T/\partial z) + \Gamma$ so that the rate of work against gravity by the upward eddy velocity w' per unit volume per unit time is $(g\rho w'\ell/T)(\partial T/\partial z) + \Gamma = (K_h \rho g/\theta)(\partial\theta/\partial z)$ where $K_h = w'\ell$ (as in Sec. 5-1.9) and where $(1/\theta)(\partial\theta/\partial z) = (1/T)(\partial T/\partial z) + \Gamma$, Γ being the dry adiabatic lapse rate (0.01°C/m). The work done by the Reynolds stresses per unit volume is $\tau(\partial u/\partial z) = \rho K_m(\partial u/\partial z)^2$. The condition for turbulence to increase was specified by Richardson as $K_m(\partial u/\partial z)^2 > (K_h g/\theta)(\partial\theta/\partial z)$ or $K_m/K_h > [(g/\theta)(\partial\theta/\partial z)]/(\partial u/\partial z)^2$, where $K_m(\partial u/\partial z)^2$ is the production of turbulence by vertical shear in the mean airflow and $(K_h g/\theta)(\partial\theta/\partial z)$ represents the rate of consumption of kinetic energy by buoyancy.

Now let

$$\mathrm{Ri} = \frac{(g/\theta)(\partial\theta/\partial z)}{(\partial u/\partial z)^2} \tag{5.50}$$

or in finite difference form

$$\mathrm{Ri} = \frac{g}{\theta}\left[\frac{\Delta\theta}{(\Delta u)^2}\,\Delta z\right]$$

where Ri is the (gradient) Richardson number, a useful stability parameter. This parameter can be thought of as a measure of the onset of turbulence. Originally, Richardson assumed that, if the atmosphere is neutrally stable, $K_m = K_h$ so that slightly turbulent motion will remain turbulent if $\mathrm{Ri} < 1$ and will be suppressed to laminar motion if $\mathrm{Ri} > 1$. Such an assertion implies the existence of a threshold or critical value of Ri which was originally taken as unity. However, perturbation theories usually yield critical Richardson numbers of about 0.25 (Taylor, 1931; Goldstein, 1931; Miles, 1961; Ludlam, 1967) while observations in wind tunnels and in the atmospheric boundary layer suggest that the critical value varies (Businger, 1973) with turbulence being certain for $\mathrm{Ri} < 0.15$ and absent for $\mathrm{Ri} > 0.5$.

For unstable thermal structures in the surface layer, Golder (1972) confirmed the hypothesis of Businger (1966), $\mathrm{Ri} = \zeta$, as a good overall approximation for unstable conditions, especially for $\zeta < -0.1$. Figure 5.7 shows the dependency of Ri on ζ for wind-tunnel data collected by Arya and Plate (1969). Near the neutral point the approximation weakens because for $\zeta \to 0$, $\mathrm{Ri} \simeq 0.74\zeta$. On the basis of the analysis of data collected in Kansas, a more general expression has been proposed by Businger et al. (1971) for unstable conditions

$$\mathrm{Ri} = \frac{(0.74\zeta)(1 - 15\zeta)^{1/4}}{(1 - 9\zeta)^{1/2}} \tag{5.51}$$

For stable stratifications, they recommend

$$\mathrm{Ri} = \frac{\zeta(0.74 + 4.7\zeta)}{(1 + 4.7\zeta)^2} \tag{5.52}$$

or, according to Golder (1972),

$$\phi_m = (1 - 15\mathrm{Ri})^{-1/4} \quad \text{for } \zeta < 0$$

$$\phi_m = (1 - 5\mathrm{Ri})^{-1} \quad \text{for } \zeta > 0$$

Simple calculations with Eq. 5.52 show that, as $\zeta \to \infty$, Ri approaches a limiting value of 0.21. Through a similar analysis, Webb (1970) found a limiting Ri of 0.20, if the thermal stability increases so that both the heat flux and the Reynolds stress approach zero at the same rate. Consequently, as $\zeta \to \infty$, the turbulence will diminish and the flow will become laminar so that the limiting value of Ri can be thought of as the critical Ri. However, Arya (1972b) warned that Ri may not have a unique limiting value and that it is likely to be a function of height. Clearly the final word about the critical Ri has not been said.

As an alternative to inverting Eq. 5.52, Webb (1970) suggested that for stable conditions the eddy diffusivities of heat and momentum are equal and the relationship between ζ and Ri is given by $\zeta = \mathrm{Ri}/(1 - 5\mathrm{Ri})$ or by $\mathrm{Ri} = \zeta/(1 + 5\zeta)$. More recently, Binkowski (1975) recommended using $\zeta = (1.35\ \mathrm{Ri})/(1 - 4.7\ \mathrm{Ri})$ during stable conditions. Applying this expression in the range $0 \leqslant \mathrm{Ri} \leqslant 0.19$, he reported excellent comparisons with corresponding values obtained from Eq. 5.52. For a comparison of Ri with the vertical dispersion parameter, σ_z, see Fig. 9.4 in Chap. 9.

An advantage of Ri is that it can be estimated directly from measurements of wind and temperature at two different levels. However, since there is no standard procedure for calculating Ri from equations similar to Eq. 5.50, calculated values of Ri should be accompanied with a brief description of the computation procedure, including documentation of the depth through which the vertical gradients are evaluated. Moreover, ordinary rawinsonde soundings smooth wind profiles so much that shal-

†See Sutton, 1953, pp. 152-153.

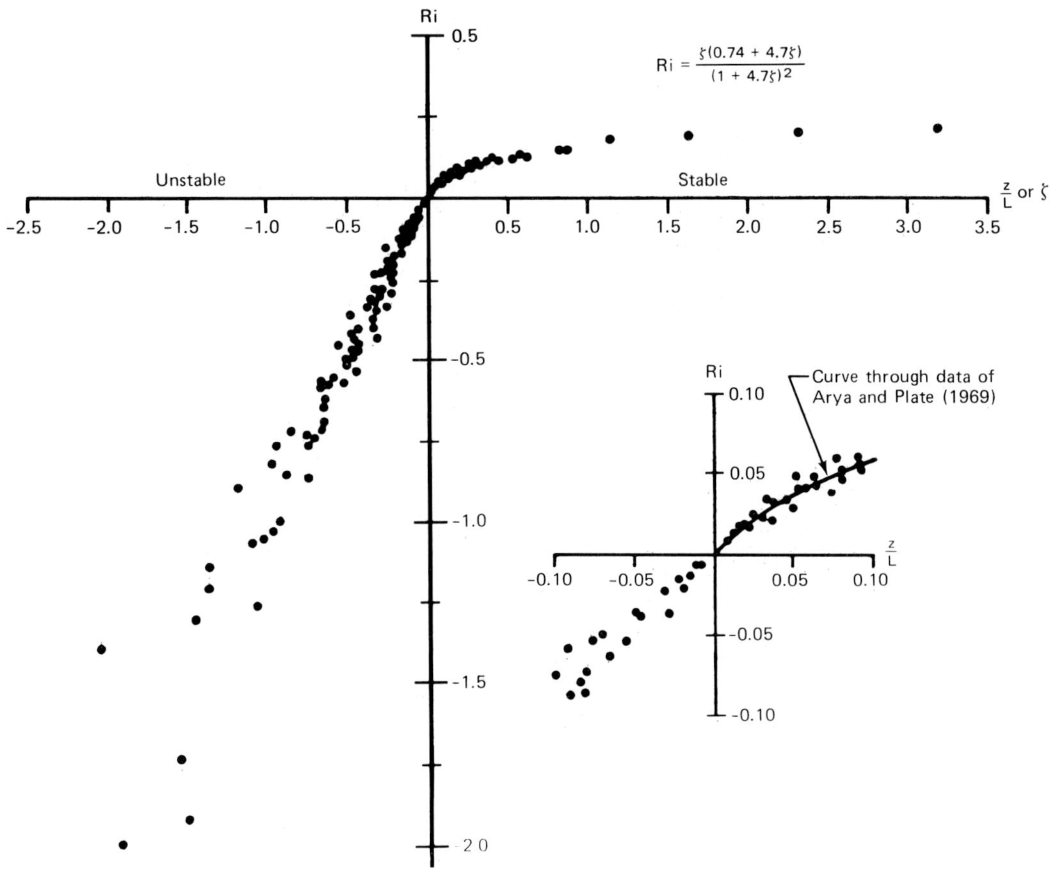

Fig. 5.7 The relationship between z/L and Ri for the whole range of atmospheric stability. Field data are from Businger et al. (1971); laboratory data in the stable area are from Arya and Plate (1969).

low layers of large vertical speed shear can be lost. Consequently, if Ri values are computed from rawinsonde data, they will be systematically too large because the vertical wind shears will be too small. Further, the vertical wind shear, $\partial u/\partial z$, should actually be written as the vector quantity $|\partial \mathbf{V}/\partial z|$ since both the u and v components of the wind are involved in the shear. If the wind direction varies by less than 5° over Δz, then the vector quantity can be treated as a scalar quantity to achieve quick estimates of $|\partial \mathbf{V}/\partial z|$. Small errors in wind velocity can lead to very large errors of Ri, especially for slow wind speeds. Peterson (1975) suggested using a bulk Ri for selecting wind profiles for near-neutral thermal stratifications. He views the gradient Ri, calculated from two low levels, as yielding anomalously large roughness lengths (z_0) and as not accounting for the stability of the upper parts of the surface layer. The bulk Ri is given by $gz\Delta T/(TV^2)$ where T is the temperature in °K at height z, V is the wind speed at height z, and ΔT is the temperature difference between that at height z and at a lower height. Schubert (1977) used the bulk Ri in a study of the momentum and heat transfer near the ground. In addition, Nickerson and Smiley (1975) and Benoit (1977) developed expressions for z/L as a function of the bulk Ri.

5-2.12 Richardson Flux Number

Since viscous dissipation plays an important role as a sink for turbulent energy, the condition $K_m = K_h$ is not a valid assumption for neutrally stable conditions (Businger et al., 1971; Pruitt, Morgan, and Lourence, 1973; Weber et al., 1975). Thus it is best to carry these two coefficients along and form another dimensionless number called the flux Richardson number, Rf, given by

$$Rf = Ri \frac{K_h}{K_m} \tag{5.53}$$

Assuming $T \simeq \theta$ for $p \simeq 1000$ mb (this replacement does not imply that the vertical gradients of T and θ are equal) and substituting $\overline{w'T'} = K_h \partial\theta/\partial z$ and $\overline{u'w'} = K_m \partial u/\partial z$ yields

$$Rf = \frac{g}{\overline{T}} \frac{\overline{w'T'}}{\overline{u'w'}\,(\partial u/\partial z)} = \frac{g}{\overline{T}} \left(\frac{H/\rho c_p}{\mu_*^2} \frac{\partial \bar{u}}{\partial z} \right) \tag{5.54}$$

From this form of the equation, assuming that $\partial u/\partial z = u_*/kz$ (see Eq. 5.62) and that L is given by Eq. 5.46, we see that

$$Rf = \frac{z}{L} = Ri \frac{K_h}{K_m} \tag{5.55}$$

This relationship holds only when the wind-speed profile is logarithmic or when the profile is not affected by buoyancy (except possibly for a contribution from z_0). Generally both Ri and Rf increase with height through the atmospheric boundary layer and thus are not really suitable for characterizing the overall stability of the boundary layer. Indeed, since the wind shear is large near the ground, Rf is always small near the surface, except when the air is motionless (Businger, 1973).

Furthermore, Eq. 5.55 illustrates that parameter z/L is fundamentally a Richardson number.

The dynamical significance of Rf is that it provides a measure of the relative importance of the buoyancy term relative to the shear-production term. Small values of Rf indicate that the flow is through nearly neutral air, with the wind velocity distribution in the vertical being described by a logarithmic profile (Eq. 5.60). This type of flow regime arises either because L is large (implying that H is small or u_* is large) or because z is small. Consequently in shear flows there always exists a layer near the ground in which the flow is characteristic of neutrally stratified flow, even in strongly stratified conditions.

When the value of Rf is negative, energy becomes available to the turbulence through the buoyancy forces and through the shearing stress. As the instability increases or as z becomes large, the effect of shear on the flow becomes progressively less important until an Rf value is reached where the presence of shear no longer significantly affects the dynamics of the flow. In other words, the flow is then driven entirely by buoyancy—the free-convection state. The transition from forced to free convection can be gradual.

Positive values of Rf indicate that the buoyancy forces are absorbing energy from the turbulence. Consequently the turbulence is suppressed.

In numerical models for turbulent flow, Yamada (1975) defined the critical Rf as a limiting value beyond which physically correct solutions were no longer possible. In this model all the turbulent moments were suppressed completely when the Rf exceeded the critical value. Results of the experiments yielded critical Rf numbers of 0.18 to 0.27. The values of the ratio of K_h to K_m ranged from 0.5 to 1.2 at the critical condition of stability.

5-2.13 Surface-Layer Profiles

It is generally recognized that the turbulent exchange processes in the atmospheric boundary layer contribute to the general circulation (e.g., see Charney and Eliassen, 1949; Petterssen, 1956). The lowest 50 m or so is of particular importance because it is this layer in which the interaction between the atmosphere and the surface is most strongly felt. It is through this layer that the fluxes of momentum, heat, and water vapor must be understood. In fact, a primary objective of micrometeorological research is to establish fluxes from a knowledge of mean profiles. As a consequence, much of the effort in micrometeorological research has been directed toward acquiring knowledge about the behavior of the vertical profiles of momentum, heat, and water vapor. Recent profile formulas are based on the similarity hypothesis proposed by Obukhov (1946) and first supported by field data in a paper by Monin and Obukhov (1954). Obukhov suggested that the average surface-layer wind and temperature profiles should depend only on the surface heat flux H_0, the buoyancy parameter $g/\overline{T}$, the surface shear stress τ_0, and height above the ground (see Eqs. 5.44 to 5.46).

In recent years several carefully controlled field experiments have been conducted over flat and relatively homogeneous terrain. Wind profile and heat flux data from these programs have been used to derive profile formulas for the surface layer. From a Kansas field experiment, Haugen, Kaimal, and Bradley (1971) and Businger et al. (1971) confirmed the following formulas for the wind and temperature profiles within the first 30 m of a stably stratified ($\zeta > 0$) surface layer

$$\phi_m = 1 + \beta\zeta \qquad (5.56)$$

$$\phi_h = 0.74 + \beta\zeta \qquad (5.57)$$

where $\phi_m = (kz/u_*)(\partial u/\partial z)$ is the dimensionless wind shear, $\phi_h = (kz/T_*)(\partial\overline{T}/\partial z)$ is the dimensionless temperature gradient, and $\beta = 5.0 \pm 0.2$ (Hicks, 1976). The conditions that lead to the differential forms of ϕ_m and ϕ_h actually are never met exactly in the atmospheric boundary layer; however, the use of these formulas can be regarded as a first step toward understanding the structure of turbulent boundary layers. Examples of the dependency of both ϕ_m and ϕ_h on ζ are illustrated by the generalized curves drawn in Figs. 5.8 and 5.9.

Carl, Tarbell, and Panofsky (1973) note that Eq. 5.56 loses its validity at levels where Ri > 0.15. In fact, Zdundkowski, Welch, and Paegle (1976) used $\phi_m = 5.1$ for

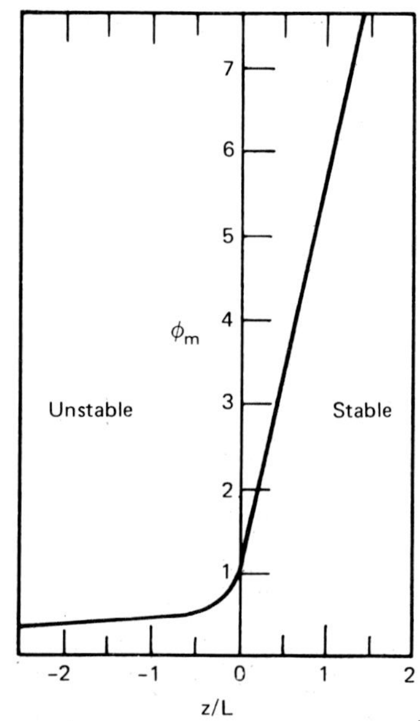

Fig. 5.8 Schematic of ϕ_m as a function of z/L.

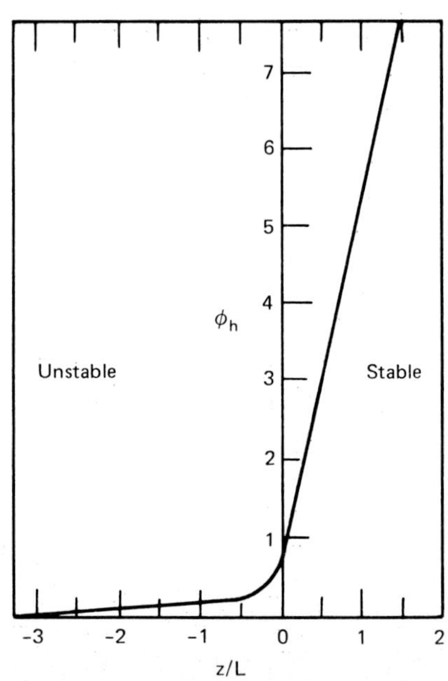

Fig. 5.9 Schematic of ϕ_h as a function of z/L.

$\zeta > 0.82$ and Eq. 5.56 for $\zeta < 0.82$. Furthermore, Hicks (1976) confirmed Eq. 5.56 for $\zeta < 0.5$, but he suggested

$$\phi_m = \left(8 - \frac{4.25}{\zeta} + \frac{1}{\zeta^2}\right)$$

for $0.5 < \zeta < 6$. For $\zeta > 6$, he proposed the linear relation $\phi_m = 0.76\zeta$. Data collected by Webb (1970) showed that $\phi_m \simeq \phi_h \simeq \phi_q \simeq 1 + \beta\zeta$ for $\zeta < 1$ and $\phi_m = 1 + \beta$ for $1 < \zeta < 6$. Clarke (1970b) also suggested that $\phi_m \simeq \phi_h \simeq \phi_q$ is a good approximation during stable conditions.

Although the Kansas data did not include numerous cases of unstable conditions containing large eddies, Businger et al. found that for an unstable surface layer, where ζ ranged from near 0 to -1,

$$\phi_m = (1 - 15\zeta)^{-1/4} \tag{5.58}$$

$$\phi_h = 0.74\,(1 - 9\zeta)^{-1/2} \tag{5.59}$$

Hicks (1976) reported that Eq. 5.58 provides an adequate description of the unstable surface layer wind profile up to $\zeta = -2$ and perhaps beyond this value. The validity of Eq. 5.58 and the KEYPS expression $\phi_m = (1 - \beta\zeta)^{-1/3}$ (Panofsky, 1961) for unstable conditions were tested by Carl, Tarbell, and Panofsky (1973) for $\beta = 16$. Both expressions fit the observations† well up to $\zeta = -10$. Therefore for unstable conditions ϕ_m is usually taken as $(1 - 16\zeta)^{-1/4}$, a relation strongly supported by Paulson (1970) and Dyer and Hicks (1970). In addition, Carl, Tarbell, and Panofsky noted that Eq. 5.59 slightly overestimated ϕ_h for small ζ.

For near-neutral stability, Businger et al. suggested that

$$\phi_m = 1.0 + 3.0\zeta + 10.2\zeta^2$$

$$\phi_h = 0.74 + 3.0\zeta + 9.2\zeta^2$$

Consequently, for neutral stability ($\zeta = 0$), $\phi_m = 1$, which yields a logarithmic wind profile (Sec. 5-3). Moreover, with $\zeta = 0$, $\phi_h = 0.74$, which indicates that $K_h/K_m = 1.35 = \phi_m/\phi_h$.

In an independent field experiment in South Carolina, Weber et al. (1975) found that the expressions based on the Kansas data fit the South Carolina data well. Noteworthy departures from the experiments of Businger et al. were found to occur above the 160-m level and to become increasingly worse with elevation. Weber et al. concluded that the simple formulas predicted by similarity theory for the surface layer could not be extended above the 160-m level, except as a first approximation.

Based on an analysis of data collected during a field experiment in Minnesota (Izumi and Caughey, 1976), Briggs (1978) found that for $\zeta < 0$

$$\phi_m = \frac{(1 - 1.6\zeta)[1 - 10(z/z_i)^2]}{(1 - 4\zeta)[1 + 5(z + z_i)^2]}$$

where z_i is the height of the mixed layer (see Sec. 5-4.2). Neglecting the contribution by z_i, Briggs found that

$$\phi_m = \frac{1 - 1.6\zeta}{1 - 4\zeta}$$

†All instruments below 30 m.

This expression is almost identical to the KEYPS profile (Sec. 5-3.5) for $\zeta < -0.3$. For large $|\zeta|$, ϕ_m approaches 0.4 for an asymptote instead of zero.

Using three different flux-gradient relationships, Lo and McBean (1978) examined the differences in the fluxes computed from the same set of data. Depending on the equations used, they noted that the computed flux may differ by 20 to 40% for unstable conditions and by much greater amounts for stable stratifications. Moreover, uncertainties in the von Kármán constant result in a difference of 17% in the computed value of u_*. Consequently, even though the Monin–Obukhov similarity theory can be used to compute surface fluxes from field data, the resulting fluxes are only as good as our knowledge of the flux-gradient relations. Yaglom (1977) reviewed the flux-profile relationships and noted that existing experimental data are insufficient for reliable determination of the universal functions entering these relationships.

5-2.14 Surface-Layer Structure Parameters

The concept of structure parameters can be used to interrelate the properties of turbulence with each other (see Lumley and Panofsky, 1964; Wyngaard, 1973). In addition, structure parameters are important in the study of the acoustic and optical properties of the atmosphere. A structure parameter, C_i, can be defined as the variance, D, of the difference between a transferable property, f_i:

$$D_i(r) = \overline{[f_i(r) - f_i(r + \Delta r)]^2} = C_i^2 r^{2/3}$$

where r is a position vector (Gaynor, 1977).† The subscript i is an index used to specify whether the structure parameter is for temperature, T; humidity, q; or speed, V. According to Bradley and Antonia (1979), these three structure parameters can be defined for turbulent shear flows as

$$C_T^2 = \frac{\overline{w'T'}}{\sigma_T(-\overline{u'w'})^{1/2}}$$

$$C_q^2 = \frac{\overline{w'q'}}{\sigma_q(-\overline{u'w'})^{1/2}}$$

$$C_v^2 = -\frac{\overline{u'w'}}{\overline{\Phi^2}}$$

where $\overline{\Phi^2}$ is the energy of the turbulence $(\overline{u'^2} + \overline{v'^2} + \overline{w'^2})$ and σ_T and σ_q are the standard deviations of temperature and humidity fluctuations, respectively. Tatarskii (1971) found $C_v^2 = 2\epsilon^{2/3}$, where ϵ is the rate of dissipation of kinetic energy. In the surface layer, ϵ is given by $u_*^2\partial u/\partial z + (g/\overline{T})\,\overline{w'T'}$. For unstable conditions, Wyngaard and Coté (1971) found that

$$\epsilon = \frac{u_*^3}{kz}\,(1 + 0.5|\zeta|^{2/3})^{3/2}$$

†An important difference between structure functions and autocorrelation functions is that structure functions involve only the differences between quantities. For this reason, structure functions have certain advantages over the correlation function as a practical measure of atmospheric variability. Therefore more precise analytical and empirical descriptions of boundary-layer fields can often be given by means of structure functions (e.g., see Panchev, 1971).

whereas for stable conditions the same equation is used but with $2.5|\zeta|^{1/3}$ replacing the $0.5|\zeta|^{2/3}$.

On the basis of the Monin–Obukhov similarity hypothesis, the structure parameters are thought to be uniquely determined by ζ. In fact, Bradley and Antonia (1979) pointed out that, when local-free-convection scaling is applicable, the expected behavior of these three parameters is

$$C_T^2 \text{ and } C_q^2 \propto (-\zeta)^{1/3}$$

$$C_v^2 \propto (-\zeta)^{-2/3}$$

On the basis of analysis of several different sets of data, Bradley and Antonia (1979) provided a rather complete analysis of structure parameters. They found that, for unstable conditions, C_T and C_v have a significant dependence on ζ. As $\zeta \rightarrow 0$, however, C_T and C_v approach values of approximately 0.4 and 0.09, respectively. By combining several sets of data, Bradley and Antonia found, for $-\zeta > 2$

$$C_T^2 = 0.0034 + 1.01\,(-\zeta)^{1/3}$$

as a regression equation with a goodness of fit of $r = 0.94$. However, for similar conditions, Wyngaard, Izumi, and Collins (1971) found that C_T^2 is related to Q_0 and is independent of L (and u_*). For convectively unstable conditions C_v^2 and ϵ are independent of height (e.g., Melling and List, 1980). Under such conditions, Kaimal et al. (1976) found a nondimensional value of C_v^2 of 1.5.

Departures from similarity theory can be attributed to the requirements of different sampling periods and averaging times. According to Bradley and Antonia, an approximation of the error involved in determining structure parameters (as a result of integration time) can be inferred from the recommendations of Sreenivasan, Chambers, and Antonia (1978). For example, with averaging times of 30 min, errors of nearly 30% are possible for C_T and nearly 25% for C_v and C_q. Kaimal (1973) also addressed the influence of averaging times on the estimation of structure parameters. He related C_v and C_T to Richardson numbers in a stable surface layer.

Wyngaard and Clifford (1978) made use of the principle that structure parameters can be obtained from path-averaged wave propagation measurements. For example, optical scintillation instruments can provide C_T^2. Through theoretical considerations, Wyngaard and Clifford suggested that, for dry, very unstable conditions ($-\zeta > 1$),

$$Q_0 \approx 0.5z(g/T)^{1/2}(C_T^2)^{3/4}$$

for the surface layer.

5-3 WIND PROFILES

5-3.1 Introduction

Normally the transport and dispersion of effluents emitted from power plants as well as from other man-made (and natural) pollutants are determined by the natural wind environment in the atmospheric boundary layer. Furthermore, airflow in the boundary layer can have a substantial impact on aerospace and aviation activities. In a very stable atmospheric boundary layer, extremely large vertical variations in the horizontal wind can occur. Fichtl (1972), Luers and Reeves (1973), and others have shown that aircraft passing through layers of strong wind shear can experience large changes in aerodynamic lift. Some investigators (Roberts, 1963; Melvin, 1971) have cited severe wind shear as the cause of long and short landings. The flow of air around tall structures should also be considered in building design (Owen, 1971). For example, the vortex shedding characteristics of buildings can be important in determining the location of plazas and open malls (Wise, 1971; Penwarden, 1972) as well as in influencing the dispersion of any effluent released from the building. Wind loads and resulting time-dependent pressures acting on the building are important to building design and to the expected lifetime of the building. Obviously knowledge of the behavior of airflow in the atmospheric boundary layer is required as input to many problems.

5-3.2 Historical Perspective

A historical outline on the development of wind profiles was provided by Sutton (1953) and more recently by Hanna (1969). Hanna also summarized theories for representing boundary-layer winds, and Tennekes (1973a) wrote an excellent paper on the contemporary development of the logarithmic wind profile and its implications with respect to classical theory. A comprehensive description of the calculation of wind profiles in the atmospheric boundary layer was given by Estoque (1973), who outlined the numerical approach for calculating boundary-layer profiles. Furthermore, Smith and Carson (1974) and Smith (1975) developed a scheme for estimating daytime wind profiles for the atmospheric boundary layer. The procedure permits the wind profile to vary with time of day and cloud amount so that the system has practical value at the site for which it was developed.

Hellmann (1915) was the first investigator to make a systematic measurement of wind profiles near the ground. He located anemometers at 5, 25, 50, 100, and 200 cm above the ground. Results of his experiments showed that the wind speed varied logarithmically with height according to $\log u_2 - \log u_1 = \eta \log z$ where η is a constant that was later shown to be dependent on the atmospheric thermal structure. Hellmann's discovery of the logarithmic change of wind speed with height was of fundamental importance. Indeed, the literature abounds with wind profile formulas. So much material is available that it is possible to discuss only those approaches and approximations that best illustrate past development and current trends of thought and effort.

The now classical laboratory experiments of Nikuradse (1933) of flow through sand-coated pipes established the existence of two regimes of flow in a turbulent boundary layer in which the effects of buoyancy were absent. Nikuradse proposed that aerodynamically smooth flow occurs when the vertical change in wind speed and τ are related to kinematic viscosity, ν. For heights above 30 ν/u_* the mean speed profile was found by experiment to take the form $\bar{u} = (u_*/k) \ln (u_*z/\nu) + B$ where parameter B varies according to the roughness of the sand grains and consequently is referred to as a roughness parameter, z_0. After Nikuradse's experiments, a number of meteorologists confirmed a linear relationship between u and ln z for a neutrally stable atmosphere and for different conditions of surface roughness.

5-3.3 A Surface-Layer Wind Profile

A classical procedure for developing the wind-profile equation for the surface boundary layer is to solve Eq. 5.40 for τ and substitute the resulting expression into Eq. 5.20 so that

$$\frac{\partial \bar{u}}{\partial z} = \frac{u_*}{\ell} \quad (5.60)$$

where mixing length, ℓ, is the only parameter on the right-hand side that can vary with height. Consequently Eq. 5.60 cannot be integrated with respect to height until we specify how ℓ varies with the height throughout the surface boundary layer. Near the ground mixing is bounded by the underlying surface. For points higher in the surface boundary layer, mixing is less inhibited by the ground, and eddies larger than those near the ground are permitted to form. A very simple assumption, incorporating this physical concept, can be expressed as $\ell = kz$ where k is the von Kármán constant. Incorporating this relation into Eq. 5.60 and integrating from z_0 to z yields

$$\bar{u} = \frac{u_*}{k} \ln \frac{z}{z_0} \quad (5.61)$$

the well-known logarithmic wind law that has become one of the cornerstones of micrometeorology. In this expression $\bar{u}$ is the mean wind speed at height z, and z_0 has been specified as the height above ground where $\bar{u} = 0$. Note that Eq. 5.61 is not applicable for $z < z_0$ and that it will yield an infinite speed shear if $z = 0$. Such a result is obviously undesirable. To keep the shear finite, a surface roughness length, z_0, is introduced so that $\ell = k(z + z_0)$. Integrating Eq. 5.60 with the specified substitution for ℓ yields

$$\bar{u} = \frac{u_*}{k} \ln \frac{z + z_0}{z_0}$$

This equation gives nonzero values for the eddy viscosity at the ground. For this reason it does not seem to be as valid as Eq. 5.61. For neutrally stable air flowing over homogeneous terrain, there is considerable experimental evidence (Panofsky, 1973) to verify Eq. 5.61. In fact, Thuillier and Lappe (1964), Carl, Tarbell, and Panofsky (1973), and Weber et al. (1975) have shown that the logarithmic profile is adequate up to heights of 150 to 160 m under neutral conditions. In addition, observations by Clarke (1970a) indicate Eq. 5.61 may be valid to 600 m during neutral conditions. More details about the logarithmic wind profile appear in interesting papers by Blackadar and Tennekes (1968) and by Tennekes (1973a). A tabulation of $\bar{u}/u_*$ values for various z's and z_0's is given in Chap. 8, Table 8.1.

Approximate values of z_0 and u_* are tabulated in Table 5.4. For a discussion on an experimental method for estimating z_0, see Lettau (1969). The dimensionless von Kármán constant was recently measured as 0.35 (Businger et al., 1971) for flow over smooth terrain. Over smooth surfaces, estimated values of k range from 0.33 (Tennekes, 1968) to 0.42 (Pruitt, Morgan, and Lourence, 1973), the uncertainty arising from errors (±10%) in the inferred friction velocity and from overestimates (20%) of the stress. By contrast, aerodynamic theory suggests that $2\pi k^2 = 1$ so that $k \simeq 0.4$ (Duncan, Thom, and Young, 1970, pp. 204 to 206 and 544 to 549).

TABLE 5.4 Representative Values of u_* and z_0 in the Constant Stress Layer†

Type of surface	z_0,‡ cm	u_*, cm/s k = 0.35	u_*, cm/s k = 0.40
Very smooth (ice, mud flats)	10^{-3}	14.3	16.4
Smooth sand	10^{-2}	17.7	20.2
Grass lawn up to 1 cm high	10^{-1}	23.0	26.3
Grass lawn up to 5 cm high	1	33.0	37.7
Thin grass up to 50 cm high	5	47.4	54.2
Thick grass up to 50 cm high	10	58.4	66.8
Woodland forest	20	76.0	86.9

†For $\bar{u}$ = 500 cm/s at z = 200 cm. For more information, see Table 8.1 in Chap. 8.

‡For urban areas, z_0 may be of the order of 100 to 300 cm.

5-3.4 Application

From a practical aspect an obvious question arises about Eq. 5.61. How do we arrive at values for u_* and z_0 for near-neutral conditions? A direct medium for answering this question is to determine u_* and z_0 from data such as those tabulated in Table 5.5 (also see Thuillier and Lappe, 1964). These wind-speed data were collected during periods of neutral thermal stability. By plotting the data on semilogarithmic paper, we can see in Fig. 5.10 that under such thermal conditions the wind speed tends to increase linearly with the logarithm of height. Thus we can conclude that the rate of change of wind speed with logarithmic height is constant or, mathematically, $\Delta u/(\ln z_2 - \ln z_1) = c = \Delta u/\ln (z_2/z_1)$. Assume that the wind speed vanishes at some nonzero height, $z_1 = z_0$, where z_0 is then simply the z-intercept for $u(z) = 0$ so that $u(z_2) = c \ln (z/z_0)$. The constant, c, is simply the slope of the u(z) line and depends partly on the vertical rate of transfer of momentum, τ_{xz}, and on air density, ρ, as in Eq. 5.15. For turbulent flow through pipes, laboratory experiments tell us that $c = (1/k)(\tau_{xz}/\rho)^{1/2} = u_*/k$ so that u_*/k is the slope of $u = u(z)$. Using linear regression ($y = mx + b$) with $\bar{u} = c \ln z + b$, where $b = -c \ln z_0$, yields $\bar{u}(z) = c \ln z/z_0$ and $z_0 = \exp(-b/c)$. The quantities u_* and z_0 can be obtained

TABLE 5.5 Wind Speed Data Vs. Height for Bushland, Texas†

Height, cm z	ln z	Wind speed (cm/s) at times (CST) 1700	1800	2000	1000
25	3.2	522	441	217	274
50	3.9	733	601	298	358
100	4.6	802	678	352	410
200	5.3	916	774	450	490
400	6.0	1069	920	541	569
800	6.7	1246	1082	665	622
1600	7.4	1409	1235	798	687

	Time (CST) 1700	1800	2000	1000
If k = 0.35				
c =	201.7	183.0	136.0	98.3
b =	−112.5	−151.0	−246.5	−33.7
u_* (cm/s) =	71.3	64.1	47.6	34.4
z_0 (cm) =	1.8	2.3	6.1	1.4

What values of u_* are obtained when k = 0.40?

†Data were collected by Franceschini (1959) at Bushland, Tex. (near Amarillo, Tex.) on December 10, 1959 (1700 and 1800 CST), December 11 (2000 CST), and December 15 (1000 CST).

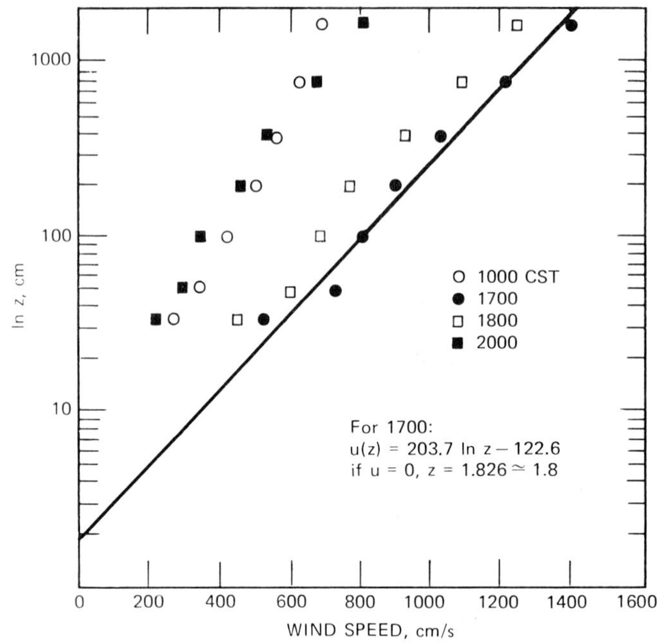

Fig. 5.10 Plot of the wind-speed data listed in Table 5.5. The straight line through the data for 1700 CST is a least-squares fit given by the equation on the bottom right-hand side of the figure. The fit yields a linear correlation coefficient of 0.99, indicating that the data are highly linear. Note that z_0 can be found by setting u(z) = 0 or approximated closely from the z-intercept.

from the wind data since z_0 is the intercept of the z axis when $\bar{u} = 0$. See Table 5.5 for calculated values of c, b, u_*, and z_0 for k = 0.35.

Other techniques for calculating u_* and z_0 are available, e.g., those of Robinson (1962) and Paulson (1970). By rearranging an equation developed by Paulson, we can determine z_0 from the expression $\ln z_0 = \ln z - \psi(\zeta) - k\bar{u}/u_*$, where $\psi(\zeta)$ is defined in Sec. 5-3.5. Paulson's expression allows us to make a linear extrapolation when $\ln z - \psi(\zeta)$ is plotted vs. $\bar{u}$ so that k/u_* is the slope and $\ln z_0$ is the intercept for $\bar{u} = 0$. Use of Paulson's relation for $\psi(\zeta)$ makes it possible to determine z_0 in unstable conditions. Experience tells us that simple estimates of z_0 can be obtained from $z_0 \simeq 0.1h$, where h is the height of the roughness elements, e.g., plant cover or buildings.

In retrospect, Eq. 5.61 describes a mean wind-speed profile in a neutrally stable surface boundary layer, provided the layer is characterized by small variations in stress, the flow is unaccelerated (steady state), the terrain is horizontally uniform, and the pressure and Coriolis forces are negligibly small. Together with u_*, z_0 incorporates the effect of small-scale surface irregularities on turbulent transfer processes in the surface layer.

Examples of the behavior of the mean speed and corresponding mean temperature profiles in the surface layer are presented in Fig. 5.11. Differences in the structure of the wind speed profiles are quite obvious. In general, results from numerous field expeditions indicate that for unstable conditions the increase in $\bar{u}$ with ln z is slower than that for a linear relationship, whereas for stable conditions it is faster (Fig. 5.12). Furthermore, experimental results show that for heights less than 10 m the speed profile for any stability category is given approximately by an equation of the form $\bar{u}(z) = A \ln z/z_0$ with the same value of z_0 for each case (see Monin, 1953).

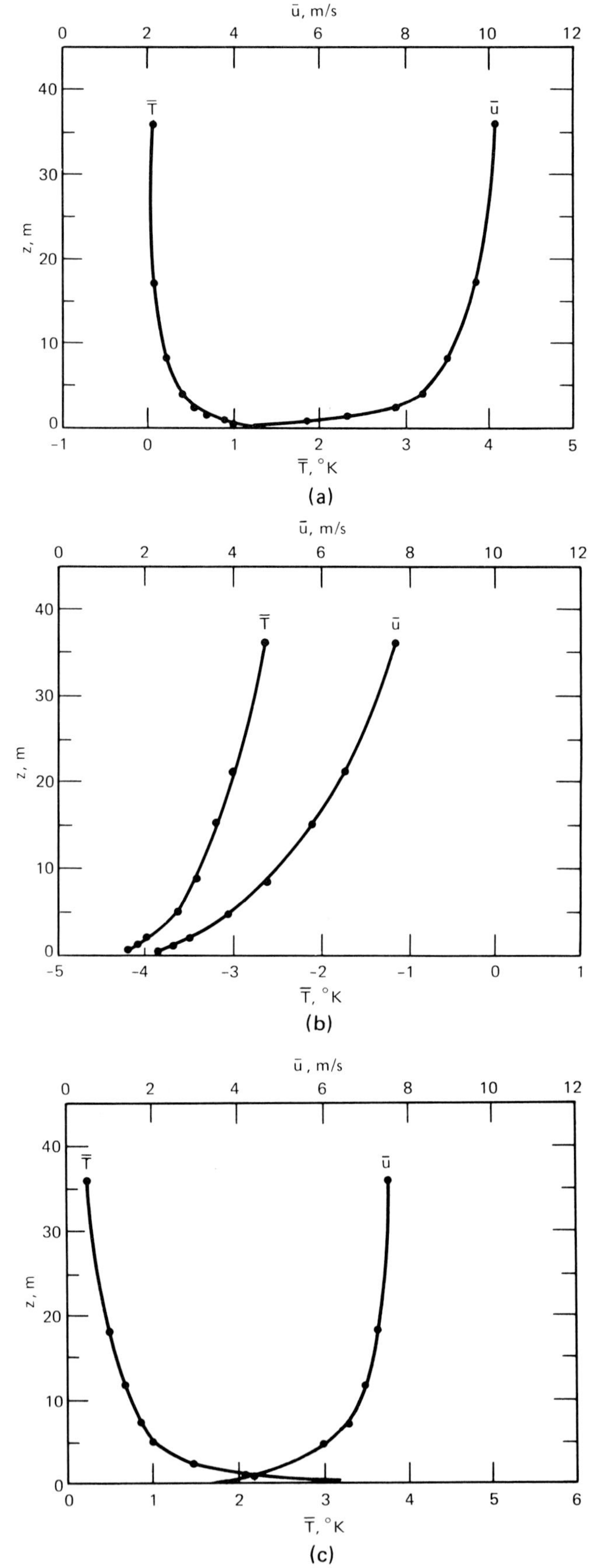

Fig. 5.11 Schematic representation of wind-speed and temperature profiles in the surface boundary layer (a) during neutral thermal stability, (b) during thermally stable conditions, and (c) during thermally unstable conditions.

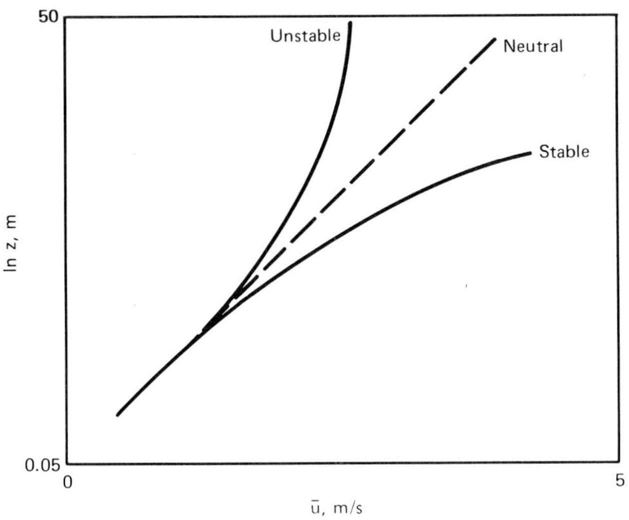

Fig. 5.12 Generalized wind-speed profiles as a function of thermal stability.

5-3.5 Log–Linear Profile

The next crucial step in developing the surface-layer wind profile was taken by Monin and Obukhov (1954), who accounted for the heating and cooling of the air from below, thereby attempting to unify the wind profile for different conditions of atmospheric stability. They proposed that

$$\frac{\partial \bar{u}}{\partial z} = \frac{u_*}{kz}\,\phi_m(\zeta) \qquad (5.62)$$

where the effects of stability are accounted for by the function $\phi_m(\zeta)$. Furthermore, they were able to remove the roughness length (essentially a constant of integration) from Eq. 5.61 by integrating between two assigned levels. As a result the unified wind profile can be expressed as a relation between differences of wind speed between two heights and the ratio of these heights. The central theme of their theory was to use the principles of similarity and thus specify one of the desired heights as the scale length L. In general, data were collected at heights of 2 and 15 m; however, L was obtained indirectly as a function of the vertical temperature gradient and an inverse function of $\bar{u}^2$. Omitting high-order terms and expanding the function $\phi_m(\zeta)$ in a power series, Monin and Obukhov obtained

$$\phi_m(\zeta) = 1 + \beta\zeta = \phi_m \qquad (5.63)$$

where β is a universal constant that was determined from their empirical data to be approximately 0.6. Substituting Eq. 5.63 into Eq. 5.62 yields the gradient form

$$\frac{\partial \bar{u}}{\partial z} = \frac{u_*}{kz}(1 + \beta\zeta) = \frac{u_*}{kz}\,\phi_m \qquad (5.64)$$

so that integrating with respect to z gives

$$\bar{u}(z) = \frac{u_*}{k}\left\{\ln\frac{z}{z_0} + \beta\frac{z}{L}\right\} \qquad (5.65)$$

where the term $(\beta/L)(z - z_0)$ has been approximated by $\beta z/L$, implying that Eq. 5.65 is valid for $z \gg z_0$. Equation 5.65 is known as the log–linear law because of the linearity of the $\beta z/L$ term. To evaluate β, plot z/L against $\bar{u}/u_*$.

For $\zeta < 0$, Eq. 5.65 occasionally appears in the literature in the more general form,

$$\bar{u}(z) = \frac{u_*}{k}\left[\ln\frac{z}{z_0} - \psi_m(\zeta)\right] \qquad (5.66)$$

and for temperature (when P $\simeq$ 1000 mb, otherwise use θ, θ_0, and θ_* for the T's),

$$\bar{T}(z) = T_0 + 0.74T_*\left[\ln\frac{z}{z_0} - \psi_h(\zeta)\right]$$

where height z has been scaled by z_0, L is given by Eq. 5.46, the orientation of $\bar{u}$ is taken to be the same as the surface stress, and diabatic effects are incorporated by introducing the function $\psi(\zeta)$. For neutral stratification, $\zeta = 0$ and $\psi(0) = 0$ so that the wind-speed profile specified by Eq. 5.66 reduces to the log law expressed in Eq. 5.61. In practice z_0 is small compared to L so that the universal function ψ_m is given by

$$\psi_m(\zeta) = \int_{z_0/L}^{z/L} \frac{1-\phi_m(\zeta)}{\zeta}\,d\zeta$$

When $z_0/|L| \ll 1$, then (see Paulson, 1970)

$$\psi_m(\zeta) = 2\ln\frac{1+x}{2} + \ln\frac{1+x^2}{2} - \tan^{-1}x + \frac{\pi}{2}$$

where

$$x = (1 - 15\zeta)^{1/4}$$

For ψ_h,

$$\psi_h(\zeta) = 2\ln\frac{1+y^2}{2}$$

where $y = (1 - \gamma\zeta)^{1/4}$. Businger et al. (1971) found γ to be 9, and Dyer and Hicks (1970) found γ to be 16. The differences in the γ's are due primarily to inaccuracies in measuring u_* and partly to the statistics of deriving the values for γ. For stable conditions $\zeta > 0$ or $L > 0$,

$$\bar{u}(z) = \frac{u_*}{k}\left\{\ln\frac{z}{z_0} + 4.7\zeta\right\}$$

and

$$\bar{T}(z) = T_0 + T_*\left\{0.74\ln\frac{z}{z_0} + 4.7\zeta\right\}$$

With functions ϕ_m and ϕ_h represented by integrals of the universal functions ψ_m and ψ_h, Busch, Chang, and Anthes (1976) and others noted that the fluxes of momentum and heat in the surface layer can be derived from

$$-\overline{u'w'} = \frac{u_*^2|\partial\bar{u}/\partial z|}{|\partial \mathbf{v}/\partial z|}$$

$$-\overline{v'w'} = \frac{u_*^2|\partial\bar{v}/\partial z|}{|\partial \mathbf{v}/\partial z|}$$

$$-\overline{w'T'} = -u_*T_* = Q_0$$

Under the assumption that the stress is in the direction of the shear, the vertical fluxes are related to the mixing length by first-order closure assumptions:

$$-\overline{u'w'} = K_{xz}\frac{\partial \bar{u}}{\partial z}$$

$$-\overline{v'w'} = K_{yz}\frac{\partial \bar{v}}{\partial z}$$

$$-\overline{w'T'} = K_h\frac{\partial \theta_c}{\partial z}$$

where $\partial\theta_c/\partial z = (\partial\theta/\partial z) - \gamma_c$ and $\gamma_c \approx 0.7°K/km$ (Deardorff, 1966). The inclusion of γ_c allows an upward heat flux in weakly stable conditions and it represents the countergradient heat flux (see Sec. 5-4.2).

Hicks (1976) reported that the height through which the log–linear law is valid appears to be a function of either wind speed or stability. He found that during conditions of slight stability a limit of $\zeta \simeq 0.5$ seems to be most appropriate; however, as wind speed decreases, this limit tends toward $z \simeq L$. Above $\zeta = 0.5$, the log–linear relationship is degraded slowly, tending toward but never reaching a purely logarithmic form before the initiation of the linear regime. In the intervening transition region, Hicks found the averaged wind profiles to be smooth with no evidence of a consistent step-function change. Above $\zeta \simeq 10$, Hicks suggested a linear dependence of wind speed on height, with the gradient approximately equal to $0.8\ u_*/kL$.

On the basis of analysis of data collected during a field experiment in Minnesota (Izumi and Caughey, 1976), Briggs (1978) found that for $\zeta < 0$

$$\bar{u}(z) = \frac{u_*}{k}\ln\frac{z}{z_0 D}$$

where $D = 1 - 4\zeta[1 + 5\,(z/z_i)^2]^{0.6}$. Neglecting the effect of the mixed layer (see Sec. 5-4.2) height, z_i, we have, for unstable conditions,

$$\bar{u}(z) = \frac{u_*}{k}\ln\left[\frac{z}{z_0}\,(1 - 4\zeta)^{0.6}\right]$$

On the basis of the idea of Zilitinkevich (1973), Ting and Hay (1975) noted that in a free-convection regime ($\zeta \lesssim -0.1$)

$$\bar{u}(z) = -5.31\,u_*^2\left(\frac{gzQ}{T_0}\right)^{-1/3} + C$$

where C is an integration constant. The significance of the term in parentheses is described in Sec. 5-4.3.

Over homogeneous terrain the wind direction in the surface boundary layer is usually taken as constant with altitude except in very stable conditions. In very stable air the wind direction can change dramatically between the ground and the 100-m level. Over complex terrain, Eqs. 5.61 and 5.65 and the assumption of constant wind direction with height may not be applicable. Furthermore, when cold and warm air advection processes are strong, caution is advised in applying Eqs. 5.61, 5.65, and the other wind-profile equations. Remember that all the wind-profile equations presented in this section are not time dependent; they are diagnostic equations, based on steady-state similarity theory applied to a whole surface layer that has been established for some time. However, even though the more general log–linear law is burdened with these restraints, it still works well up to 30 m above the ground, with no significant departures from the law up to the 150-m level except during stable conditions.

5-3.6 Ekman Layer Flow

If the pressure gradient and the Coriolis effect balance above the atmospheric boundary layer, a very simple flow field can develop within a boundary layer on a rotating surface. Within the boundary layer, the pressure gradient is approximately the same as above, but the Coriolis effect is reduced due to the frictional decay of the wind speed. Excess pressure gradient is then available to counteract boundary layer growth. However, when surface conditions are uniform over a large area, equilibrium can be established between the excess pressure gradient and the vertical wind shear near the boundary so that the boundary layer thickness remains constant. Observations such as those in Fig. 5.13 point toward this type flow regime. The top of the boundary layer is the gradient level where the stress is small relative to the pressure gradient and the Coriolis effect, and the flow is in geostrophic balance. Ekman (1902) was the first to demonstrate that the equations

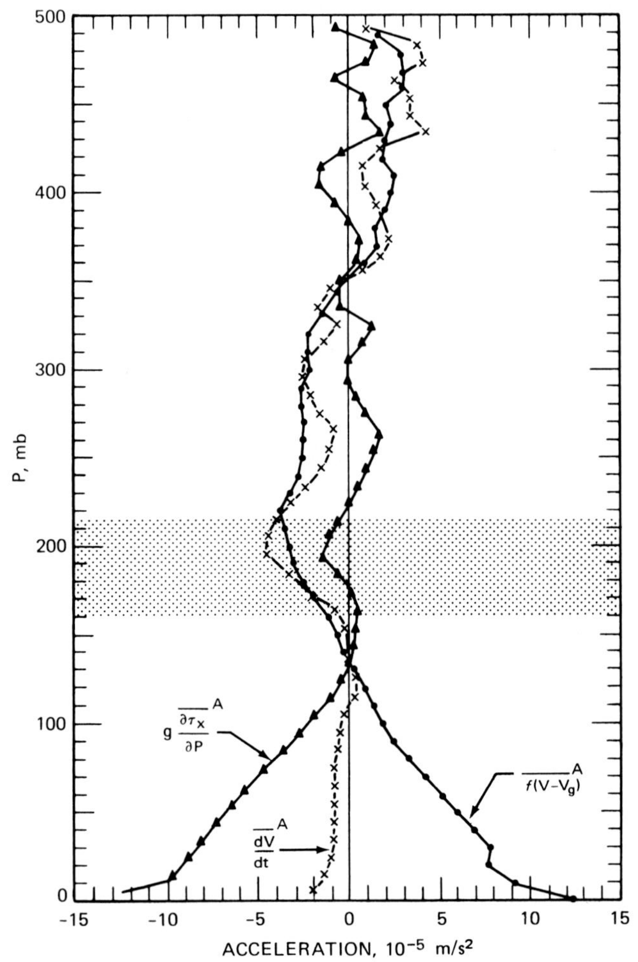

Fig. 5.13 Vertical profiles of total acceleration, $\overline{dV/dt}^A$; the excess of Coriolis over pressure gradient force per unit mass, $\overline{f(V-V_g)}^A$; and the frictional force in the direction of the mean surface wind, $\overline{g(\partial\tau_x/\partial P)}^A$, where A is a grid-scale area average on a P surface and P = 0 at sea level (Holland and Rasmusson, 1973).

of motion could be satisfied for these conditions if all acceleration terms were zero and if the flow was straight and parallel. He noted that these conditions imply a constant boundary-layer thickness. Consequently such a boundary layer on a rotating surface is referred to as the Ekman layer. Now let us consider the Ekman theory.

If we assume that (1) the horizontal mean wind shears are small compared to the vertical mean shears, (2) a balance of forces exists among the Coriolis, pressure-gradient, and eddy-viscosity terms at every level, and (3) there is a horizontal, unaccelerated flow that becomes geostrophic as $z \rightarrow \infty$, the steady-state equation of horizontal mean motion in component form is

$$\frac{\partial \bar{\tau}_{zx}}{\partial z} + \bar{\rho} f \bar{v} - \frac{\partial \bar{P}}{\partial x} = 0 \tag{5.67}$$

$$\frac{\partial \bar{\tau}_{zy}}{\partial z} - \bar{\rho} f \bar{u} - \frac{\partial \bar{P}}{\partial y} = 0 \tag{5.68}$$

Equations 5.67 and 5.68 can be combined into one complex expression by multiplying Eq. 5.68 by $i = (-1)^{-½}$ and adding. To express $\bar{\tau}$ in terms of the wind field, use Eq. 5.17 and specify an exchange coefficient, K, that is assumed to be independent of height; the complex equation of motion then becomes

$$\bar{\rho} K \frac{\partial^2}{\partial z^2} (\bar{u} + i\bar{v}) - \bar{\rho} f i (\bar{u} + i\bar{v}) - \frac{\partial \bar{P}}{\partial x} + i \frac{\partial \bar{P}}{\partial y} = 0$$

Now assume that the horizontal pressure gradient and density do not vary significantly with height above the surface so that they can be considered constants. Also, orient the x-axis in the direction of the wind at level z_g, the level where accelerations due to eddy-generated turbulence become small; thus $\partial \bar{P}/\partial x = 0$ and $\partial \bar{P}/\partial y = -\rho f u_g$, which yields

$$K \frac{\partial^2}{\partial z^2}(\bar{u} + i\bar{v} - \bar{u}_g) - i f(\bar{u} + i\bar{v} - \bar{u}_g) = 0 \tag{5.69}$$

This expression is a second-order, homogeneous, linear, partial differential equation with constant coefficients and with z as the only independent variable; thus the equation can be treated as an ordinary differential equation. Now if we assume that $\partial^2 \bar{u}_g/\partial z^2 = 0$, a solution to Eq. 5.69 is

$$\bar{u} + i\bar{v} - \bar{u}_g + C_1 e^{-(1+i)\chi z} + C_2 e^{(1+z)\chi z}$$

where $\chi = (f/2k)^{½}$.

Applying the boundary condition $\bar{u} + i\bar{v} = \bar{u}_g$ at $z = \infty$ shows that $C_2 = 0$, and applying the condition $\bar{u} + i\bar{v} = 0$ at $z = 0$ gives $C_1 = -\bar{u}_g$. Therefore the solution becomes

$$\bar{u} + i\bar{v} = \bar{u}_g (1 - e^{-\chi z} e^{-i\chi z})$$

or, using Euler's relation, $e^{-i\chi z} = \cos \chi z - i \sin \chi z$, and separating the real and imaginary parts, we obtain the steady-state expressions

$$\bar{u} = \bar{u}_g (1 - e^{-\chi z} \cos \chi z) \tag{5.70}$$

$$\bar{v} = \bar{u}_g e^{-\chi z} \sin \chi z \tag{5.71}$$

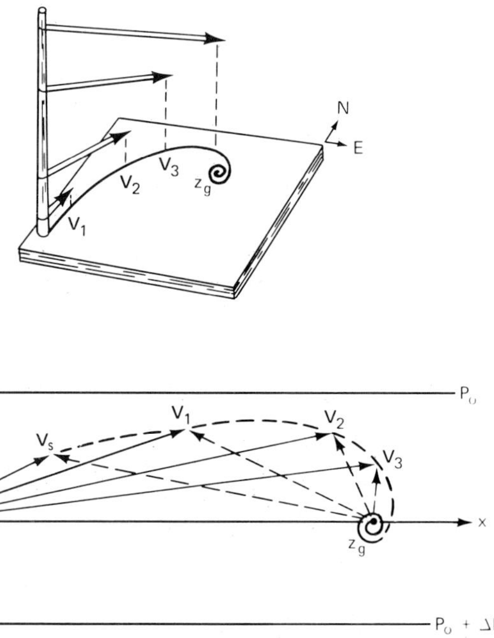

Fig. 5.14 The Ekman spiral.

The trigonometric functions in Eqs. 5.70 and 5.71 require that the wind direction vary continuously with elevation. A schematic representation of this solution is presented in Fig. 5.14. The vectors $\mathbf{V}_1$, $\mathbf{V}_2$, etc., represent the wind at level $z_s = 0$, z_1, etc., within the turning layer. Each dashed vector directed from point z_g to each $\mathbf{V}$ is the geostrophic departure at level z_s, z_1, etc. In Fig. 5.13, observe that the geostrophic departures decrease with height, ultimately vanishing at infinity since the solution spirals about z_g but never reaches it. The projection of the end points of the wind vectors onto a level surface forms a spiral having the property that the angle between the tangent to the curve and the geostrophic departure at any point on the curve is the same as that at height $z = 0$. This spiral curve is generally called the Ekman spiral because of its similarity to the hodograph of wind-driven ocean currents derived by Ekman (1902, 1905).

To determine how Eqs. 5.70 and 5.71 specify the turning of the wind with height, we need to consider the ratio of $\bar{v}$ to $\bar{u}$ which represents the tangent of the angle the wind makes with the isobars. At $z = 0$ the ratio is indeterminate; however, we can estimate this quantity by separately differentiating both components with respect to z and examining the ratio of the derivatives. We find that $\bar{v}/\bar{u} \rightarrow 1$ as $z \rightarrow 0$ so that, close to $z = 0$, the wind direction makes an angle of 45° with the isobars and points toward lower pressure (see Fig. 5.14). At a higher elevation where $z = \pi/\chi$, the v component in Eq. 5.71 vanishes, and the wind blows parallel to the isobars. Therefore the wind vector will turn clockwise, or veer, with increasing elevation in the Northern Hemisphere as long as the assumptions made in obtaining Eqs. 5.70 and 5.71 closely approximate real atmospheric conditions.

A calculated wind profile will depart from observations when the atmospheric conditions do not correspond to the basic assumptions used to arrive at Eqs. 5.70 and 5.71. For example, if strong cold- or warm-air advection is present in addition to friction, then the pressure gradient changes rapidly with height instead of remaining constant. In fact, cold-air advection causes the winds to back with elevation, whereas Ekman theory specifies veering. The flow of surface winds across isobars was studied by Hoxit (1974), who concluded that the cross-isobaric angle, α_0, increases considerably during warm-air advection and decreases during cold-air advection. More evidence of the effect of thermal winds on the actual wind veering in the atmospheric boundary layer has been reported by Lettau (1967) and Gray and Mendenhall (1973). Moreover, if the exchange coefficient varies with height, as observations (Wong and Brundidge, 1966) indicated, it will cause the wind to turn at a rate different from that required by Eqs. 5.70 and 5.71. Deardorff and Mahrt (1982) have shown how the use of a fixed exchange coefficient can lead to important inconsistencies when applying the Ekman approach in boundary layer models. For these reasons caution should be taken when using Eqs. 5.70 and 5.71.

For an interesting application of the Ekman theory to calculating surface winds in mountainous terrain, see the article by Ryan (1977). Ryan also proposed equations for correcting surface winds for sea breeze regimes, terrain slope, and diverting effects.

5-3.6.1 Transition to Ekman Layer Flow. Field data indicate that, with a large upward heat flux from the surface, small changes in mean wind and temperature occur with height above the surface layer. Consequently the mean profiles themselves may not be of much interest in the Ekman layer, but the bulk transfer coefficients of momentum and heat and the structure of the turbulence are important.

According to Zilitinkevich, Laikhtman, and Monin (1969), a straightforward extension of similarity theory to the Ekman layer leads to the relations

$$\frac{u_*}{G} = f_1\,(\text{Ro},\zeta_b)$$

$$\frac{T_*}{\Delta T} = f_2\,(\text{Ro},\zeta_b)$$

where Ro and G are defined in Sec. 5-1.9 and ζ_b is a bulk stability parameter of the form z/L. On the basis of the concepts of Rossby number similarity† suggested by Csanady (1967), Blackadar and Tennekes (1968) developed the principle of the transition from surface-layer flow to Ekman-layer flow. If we use Eqs. 5.67 and 5.68, substitute the geostrophic components for the pressure gradient, and consider a neutrally stable atmospheric boundary layer that is stationary, horizontal, and homogeneous, the steady-state equation of mean motion in component form becomes

$$\frac{\partial(\tau_{zx}/\rho)}{\partial z} + f(\bar{v} - \bar{v}_g) = 0 \qquad (5.72)$$

$$\frac{\partial(\tau_{zy}/\rho)}{\partial z} - f(\bar{u} - \bar{u}_g) = 0 \qquad (5.73)$$

Blackadar and Tennekes nondimensionalized these two equations by scaling the two stress components by ρu_*^2 and by defining $F_x = (\bar{u} - \bar{u}_g)/W$ and $F_y = (\bar{v} - \bar{v}_g)/W$ with $h = u_*^2/fW$, where h is a scale height of the neutral boundary layer, W is an unknown velocity scale, and F_x and F_y are the scaled velocity defects. In addition, they demonstrated that, if the turbulence scales of motion are determined in the same way as the scales of the velocity defects, the requirements for asymptotic similarity necessitate that W be the same order of magnitude as u_* so that outside the surface layer Eqs. 5.72 and 5.73 become

$$\frac{\bar{u} - u_g}{u_*} = F_x\left(\frac{zf}{u_*}\right)$$

and

$$\frac{\bar{v} - \bar{v}_g}{u_*} = F_y\left(\frac{zf}{u_*}\right) \qquad (5.74)$$

where u_*/f is the characteristic height of the Ekman boundary layer.

Equation 5.74 is known as the velocity-defect law. Tennekes (1973a) emphasized that Eq. 5.74 is valid only outside the surface layer, whereas equations of the form of

$$\frac{\bar{u}}{u_*} = F\,\frac{z}{z_0}$$

and

$$\frac{\bar{v}}{u_*} = 0 \qquad (5.75)$$

are valid only inside the surface layer. Thus Eqs. 5.74 and 5.75 should have a common region of validity which can be called the matched layer. In this layer, Eqs. 5.74 and 5.75 and their derivatives must agree with each other. The only way to achieve this matching is through the logarithmic wind profiles

$$\frac{\bar{u} - \bar{u}_g}{u_*} = \frac{1}{k}\ln\frac{zf}{u_*} + \frac{A}{k}$$

$$\frac{\bar{u}}{u_*} = \frac{1}{k}\ln\frac{z}{z_0} + \frac{B}{k}$$

which are valid simultaneously within the matched layer. Subtracting them from each other yields

$$\frac{\bar{u}_g}{u_*} = \frac{1}{k}\ln\frac{u_*}{fz_0} - \frac{A}{k} \qquad (5.76)$$

$$\frac{\bar{v}_g}{u_*} = -\frac{B}{k} \qquad (5.77)$$

where A and B are constants, such that, when $zf/u_* = 0.03$, $A \simeq 1$, and $B \simeq 5$; these values correspond to $z \simeq 100$ m. Equations 5.76 and 5.77 are referred to as the logarithmic drag law. In Eq. 5.76, u_*/fz_0 is a Rossby number (Ro). The practical utility of these expressions is limited; however, Agnew and Diehl (1978) have used similar expressions to estimate surface winds over the Beaufort Sea.

Wyngaard, Arya, and Coté (1974b) developed similar expressions through the use of a simple geostrophic drag law model for a convective boundary layer. They found that $\bar{v}_g = u_*^2/fz_i$ tends to be much smaller than $\bar{u}_g/\bar{u}_*$ so that the steady-state conditions

$$\frac{\bar{u}_g}{u_*} = \frac{1}{k}\ln\left(-\frac{L}{z_i}\right)$$

†Actually means Rossby number independence.

$$\frac{\Delta T}{T_*} = \frac{0.74}{k}\left[\ln\left(-\frac{L}{z_0}\right) - 1.0\right]$$

develop where z_i is the height of the lowest inversion base. In these expressions, $\bar{u}_g$ is assumed to equal $\bar{u}$ at the top of the surface layer. This assumption is based on the model prediction that, when $-z_i/L > 5$, an essentially shear-free convective layer exists with $\bar{u} = \bar{u}_g$ and $v_g = 0$. Above z_i, $\bar{v} \rightarrow \bar{v}_g$ and $\bar{u} = \bar{u}_g$. For additional information, see Sec. 5-4.3.

5-4 NUMERICAL SIMULATIONS AND OBSERVATIONS

5-4.1 Introduction

The concept of a model has been important in the modern development of many scientific disciplines. In a meteorological context, all models have one attribute in common: a set of partial differential equations that, when taken together, can be considered an explicit mathematical characterization of the model (see Chap. 15). The numerical aspects of the model generally refer to a set of equations approximately equivalent to the original differential equations of the model but obtained in a form specifically designed for numerical calculation. Typically, numerical models can be used either to diagnose or to prognose boundary-layer phenomena.

The implicitness of atmospheric processes makes it virtually impossible for a field program to study one process in isolation. Consequently one of the functions of modeling is to render the complexity of atmospheric processes more intelligible. Furthermore, many boundary-layer processes are not well understood, and thus the design of the model becomes difficult; however, a well-designed numerical model can provide empirical information that may permit better understanding of the physical processes included in the governing equations. The principal advantage of physically based numerical models is that they permit us to isolate the effects of individual processes or parameters. It is, of course, easy to run through a series of numerical experiments and vary one component of the model at a time to see how it affects the total results; however, if the experiment is to be more than a numerical exercise, the results must be compared with observed values of the real atmosphere. For this reason modeling should not proceed independent of field experimentation. Conversely, a field program that is not based on the understanding of theory derived from modeling experience could flounder and produce great masses of indigestible data. Finally, the ultimate use to which we put numerical models is determined to a large extent by the kind of models we design. If the model is to be a general research tool, formulated to reveal new information about atmospheric processes, it must be as comprehensive as possible (e.g., see Donaldson, 1973a). If the model is to be used operationally, however, it must be designed to accept field data automatically, analyze the data to obtain initial fields, and submit the initial conditions to the diagnostic and/or prognostic portions of the model within a time frame that satisfies the user's needs (e.g., see Cornett and Randerson, 1977). Thus operationally oriented models contain approximations to various physical factors to achieve a respectable solution in minimum time.

One of the fundamental needs in modeling the atmosphere on a global scale is to account for boundary-layer processes and to provide reasonable estimates of the depth and character of the mixed layer. Procedures for incorporating these processes into various general circulation models have been described by Clarke (1970b); Delsol, Miyakoda, and Clarke (1971); Deardorff (1972a), Smeda (1979); and others. Even air quality models need reliable procedures for assessing the effect of boundary-layer processes on the transport and dispersion of effluents released into the specified air shed (e.g., see Shir and Shieh, 1974; Lebedeff and Hameed, 1976).

5-4.2 Some Simulations and Observations

One important condition in solving problems pertaining to atmospheric dispersion is the depth of the mixed layer (z_i). Knowledge of z_i is useful because it appears to limit the scale of convective overturning motions and to act as a lid on the dispersion of material released within the mixed layer. As a length scale, z_i can be thought of as the height aboveground of the lowest temperature inversion.† The height scale of a neutral boundary layer is thought to be proportional to u_*/f (Blackadar and Tennekes, 1968; Deardorff, 1972b).

Using boundary-layer data from the field expedition near Hay, Australia (the "Wangara" data, described in detail by Clarke et al., 1971), Yamada (1976) concluded that the dynamic height scale u_*/f is not complete—a conclusion consistent with that of Deardorff (1972b, 1974a), Wyngaard, Coté, and Rao (1974), and Arya (1975). In fact, for thermally unstable conditions the depth of the mixed layer appears to be fixed only by the height of the lowest inversion base ($h = z_i$) above the ground. This hypothesis has been confirmed by data from a field experiment in northwestern Minnesota (Izumi and Caughey, 1976) where micrometeorological observations were taken to near the 1200-m level above the ground during August 1973. The analyzed data show that between sunrise and local noon the daytime mixed layer can grow rapidly in response to steadily increasing surface heat flux. This temporal evolution is consistent with that observed remotely by Noonkester (1976) near Wallops Island. In general, z_i increased after sunrise at rates of 0.5 to 9.7 m/min. This increase continued until nearly 1 hr before the maximum surface temperature was attained, and then z_i decreased slowly in the afternoon. Between 1300 and 1600 local time, the data collected by Izumi and Caughey demonstrate that z_i remains relatively constant as the surface heat flux (at 4 m) reaches its maximum value. The observed rate of growth of z_i between 1200 and 1600 agrees well with the numerical predictions of a model developed by Mahrt and Lenschow (1976). This model contains a constant surface temperature flux ($Q_0 = \overline{w'T'}$) between local noon and 1600 local time. After 1600, Kaimal et al. (1976) found the evening dissolution of z_i to be abrupt relative to the steady morning growth. In particular, they found that the heat flux throughout the entire layer turns negative rapidly (Fig. 5.15) so that the lowest part of the boundary layer stabilizes while the upper part remains relatively unstable. This rapid transition in z_i has been simulated nicely by Wyngaard (1975), who modeled the stable boundary layer. This model predicts the turbulent fluxes and profiles of mean quantities. Wyngaard allowed the model to develop a stable boundary layer from neutral initial conditions. At the end of the day when $Q_0 = 0$, the stable boundary layer was initiated by assuming a constant cooling rate of 0.2, 0.5, 1.0, and 2.0°K/h at a height equivalent to approximately 1 m. A

†The symbol z_i is also the height of the convecting region. Wyngaard and Coté (1974a) define z_i as the height where the virtual heat flux reaches its negative maximum. The mixed layer can also be specified as the layer through which the potential temperature is less than that of the heated layer at the surface that drives the convection.

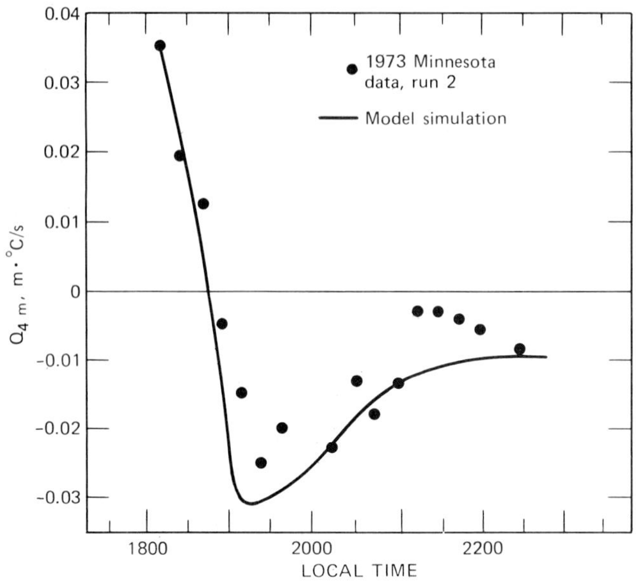

Fig. 5.15 Vertical heat flux, Q, at 4 m compared with a model simulation (Wyngaard, 1975).

striking result of the model is that, after an initial period of rapidly changing values, something approaching a steady state is obtained in a matter of hours. For example, the steady-state value for u_* (Fig. 5.16) is approached in 2 to 6 h and that for Q (Fig. 5.15) in 1 to 3 h. Interestingly the model actually predicts an overshoot of the asymptotic value of H which actually occurred in the verification data—the Minnesota observations. Incidentally the model simulation of the observed change in u_* (Fig. 5.16) from 0.4 to 0.1 m/s is encouraging.

In another numerical model Yamada and Mellor (1975) simulated z_i and other parameters over a 2-d period. The predictions made by the model have been verified with observations from the Wangara data. The variations in the structure of the atmospheric boundary layer predicted by this model are portrayed by Fig. 5.17. In this figure, observations indicate that the turbulence penetrated to near the 1600-m level in late afternoon and diminished rapidly after sunset. Note in particular that the authors distinguish between a mixed layer of low turbulence and one of high-intensity turbulence (shaded area). They define the mixed layer as the height at which the vertical

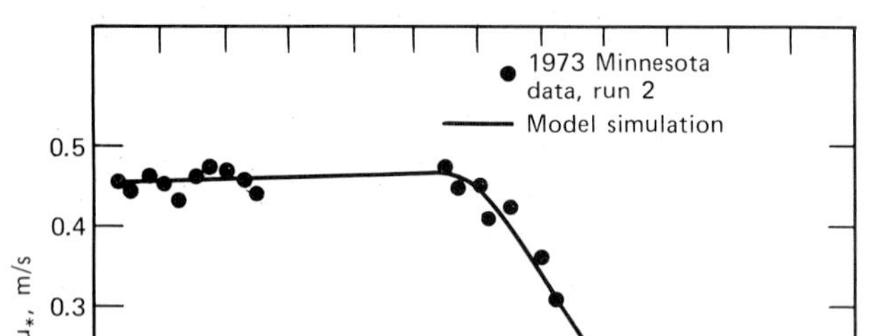

Fig. 5.16 Observed u_* values vs. calculated values (Wyngaard, 1975).

gradient of virtual potential temperature increases abruptly from nearly zero to a large value. In addition, they found the height of the neutrally stable atmospheric boundary layer to be approximately $0.3\ u_*/f$ but noted this to be a poor estimate for any given time.

Estimates of z_i as a function of t, H_0, and atmospheric lapse rate, γ, have been made in a number of investigations. For example, Betts (1973), Carson (1973), and Tennekes (1973b) gave results that support $z_i \propto \gamma^{-1/2}$. When $H_0 = 0$, however, Tennekes (1973b) showed that

$$z_i \simeq u_* \left(\frac{15T_0 t}{\gamma g}\right)^{1/3}$$

where g = 9.8 m/s, T_0 is the surface temperature in °K, t is in s, γ is in °K/m, and u_* is in m/s. This expression gives values of z_i close to those observed by Noonkester (1976) when t = 2 h, γ = 0.03 °K/m, and u_* = 0.33 m/s.

Laboratory experiments and field observations of the unstable surface layer indicate the presence of a countergradient in the sensible heat flux. Deardorff (1966) noted that, as $\partial\theta/\partial z$ goes from negative to positive while H_0 remains large and positive, K_h must go to infinity and then change signs so that the heat flux is countergradient. Such heat fluxes are a remarkable feature of convective turbulence and make application of eddy diffusivities hazardous.

Using acoustic sounder measurements, Schubert (1977) investigated the onset of what is called the momentum burst phenomena. Analysis of the sounder data indicate that the conditions necessary for the onset of this phenomena exist sometime during a clear afternoon when the heat flux changes sign and the ground cools so that the lowest part of the atmospheric boundary layer becomes stable. Before this situation develops, the entire boundary layer is in turbulent motion due to surface heating. A maximum *Rf* is reached just above the ground, turbulence is dampened, and a laminar layer forms. As the surface temperature drops, a strong wind shear develops, and Ri decreases below 0.25 so that the layer becomes conditionally unstable. Subsequently the laminar layer is eroded by turbulence from above, and, with a burst of momentum and heat, the mixed layer reaches the ground.

Field data show that the mixed layer flow is driven by buoyancy associated with a positive surface heat flux. The depth of the mixed layer usually increases by turbulent entrainment during daylight hours. Typically a buoyancy-driven mixed layer occurs during windless, sunny days. Such a layer usually consists of (1) a well-mixed convective region, (2) a thin entrainment layer above the mixed region, and (3) stably stratified air aloft. The entrainment of stable air into the well-mixed region occurs within a narrow region at the top of the mixed layer. Zeman and Tennekes (1977) have proposed methods for parameterizing this entrainment process.

Field data indicate that during convective conditions the surface layer extends to a height of about 0.1 z_i. Above this level the wind speed is approximately constant, and the wind direction turns slightly to the right (in the northern hemisphere) with increasing elevation.

5-4.3 Mixed Layer Parameters

When effluents are released into the atmosphere, mixing can result from mechanical friction or from thermal buoyancy. A mixed layer can be formed and maintained by mechanically and thermally generated turbulence. This layer is classified as

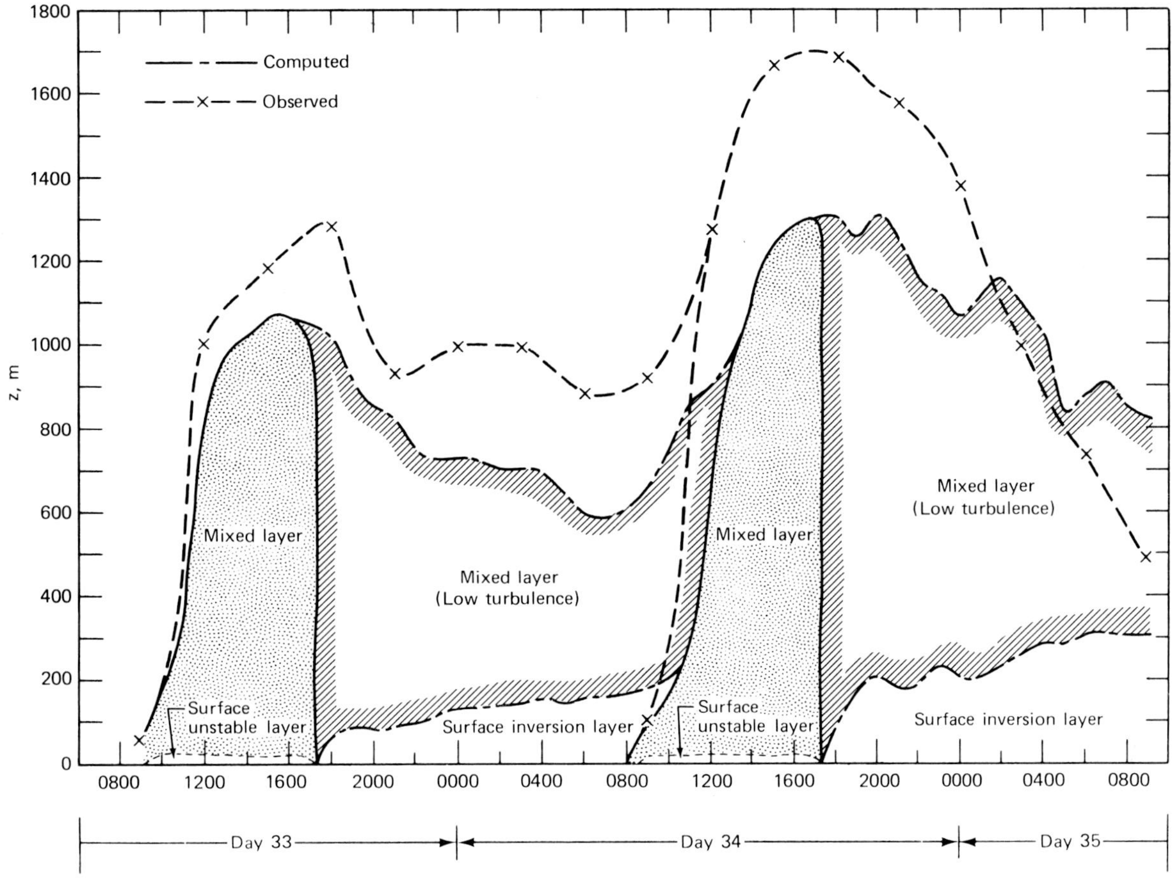

Fig. 5.17 Temporal evolution of the atmospheric boundary layer based on output from a model developed by Yamada and Mellor (1975).

either mechanical or convective, depending on which one dominates. When both mechanisms are active, the buoyancy-driven mixed layer usually dominates. The atmospheric boundary layer can be driven into a convective state when Q_0 is as small as 0.1 m°K/s and $\bar{u}$ = 5 m/s.

Willis and Deardorff (1974) demonstrated that time-dependent numerical models of the atmospheric boundary layer can be simplified if the time scale of the convective turbulence is too short to be significantly affected by the earth's rotation and if the time scale is little influenced by wind shear. Thus the mean wind ($\overline{V}$) could be considered as a mechanism for translating downstream the mixed-layer turbulence. For unstable conditions Deardorff (1970) showed that the governing parameters for mixed-layer turbulence are z_i and the convective velocity scale $w_* = (gz_iQ_0/\overline{T})^{1/3}$, where w_* is for the bulk of the convective layer. Over land during daytime, w_* is approximately 1 to 2 m/s. The scaling temperature for the mixed layer is given by $\theta_* = Q_0/w_*$. From his numerical experiments Deardorff (Lilly, 1968) deduced that the relevant Lagrangian time scale for an effluent to be carried through the whole mixed layer is given by z_i/w_*. The Eulerian aspects of this result have been confirmed with the Minnesota data by Kaimal et al. (1976). The practical implication is that z_i/W_* is the time scale needed, after release into the atmosphere, for an effluent to become well mixed throughout the depth of the mixed layer.

Willis–Deardorff scaling seems to be most appropriate when $z > 0.1z_i$ and when $|z_i/L| > 10$. Moreover, they showed that for z_i = 1.25 km and Q_0 = 0.2 m·°K/s, the time scale for convective turbulence is approximately 10 min, which is much smaller than the diurnal time scale of changes in Q_0, z_i, and in the pressure field driving the flow. Consequently the structure of the mixed layer should rapidly adjust itself in response to the changing boundary conditions.

A simple method for estimating w_* has been proposed by Venkatram (1978). He used a mixed-layer model to derive relationships between w_* and Q_0. According to Venkatram, the maximum (daily) convective velocity scale w_{*m} can be estimated by

$$w_{*m} = 4.74\, Q_m^{1/2}$$

where Q_m is the maximum surface temperature flux. This expression is useful because w_* can be estimated from Q_m with little knowledge of other boundary-layer variables. According to Briggs (1975),

$$Q_m = AR_i$$

where $R_i = \frac{2}{3} \sin \theta_m (1 - 0.8c)R$, where R_i is the incoming solar radiation, θ_m is the maximum solar elevation angle on the day of interest, c is the fractional cloudiness, and R is the solar constant. The constant A is a function of ground cover and varies from 0.25 for a crop canopy to 0.55 for a dry surface. Venkatram also developed the expression

$$w_{*m} = 1.12 \times 10^{-3}\, z_{im}$$

where z_{im} is the maximum mixed-layer depth in meters and the coefficient has the units of per second. These two estimates for w_{*m} are useful because w_{*m} determines the location of the maximum concentration of an effluent as well as its magnitude (Briggs, 1975).

From the definitions of w_* and L, we can write

$$-\frac{z_i}{L} = k\left(\frac{w_*}{u_*}\right)^3$$

For a given w_*, this expression requires $-z_i/L$ to get large rapidly as u_* decreases slowly. Also, this equation tells us that u_* need not be much less than w_* for a convective boundary layer to exist, namely, $z_i \gg |L|$ or $-z_i/L \gg 1$. For example, if $u_* = 0.5$ m/s and if $w_* = 1.5$ m/s, then $-z_i/L$ is roughly 10 for $k = 0.35$. Consequently the atmospheric boundary layer can be considered to be convective.

In Sec. 5-2.10 the length scales u_f and T_f were defined for local free convection. These scaling parameters can be combined with the mixed-layer parameters to give

$$\frac{u_f}{w_*} = \left(\frac{z}{z_i}\right)^{1/3}$$

$$\frac{T_f}{\theta_*} = \left(\frac{z}{z_i}\right)^{-1/3}$$

where z/z_i is a dimensionless length defining a position in the mixed layer. Consequently w_* and θ_* scale with z/z_i, a result predicted by the models of Deardorff (1972b) and of Wyngaard, Coté, and Rao (1974). These models predict that $\bar{u}/u_*$ and $\bar{v}/u_*$ are constant with height z/z_i for large $-z_i/L$ and for $z > 0.1z_i$. In fact, for $-z_i/L = 50$, the model of Wyngaard, Coté, and Rao predicts a very simple state where

$$\frac{\partial \overline{u'w'}}{\partial z} \simeq \frac{u_*^2}{z_i} \simeq -f\bar{v}_g$$

so that

$$\frac{\bar{v}_g}{u_*} \simeq -\frac{u_*}{fz_i}$$

and $\bar{u}_g \simeq \bar{u}$ over most of the convective layer. For the shear profiles the model yields sharply curved $\overline{u'w'}$ profiles for $z_i/L = 0$. For $-z_i/L > 10$, however, the model predicts $\overline{u'w'}$ to decrease monotonically with height, approaching zero at $z = z_i$ while $\overline{v'w'}$ vanishes over the entire layer. Recent analyses of field data have not confirmed these profiles (Kaimal et al., 1976). In fact, some of the data show $\overline{u'w'}$ and $\overline{v'w'}$ increasing with height, reaching a maximum at $z = 0.5z_i$ and decreasing above that height. Differences between model predictions and observations are attributed to inadequate averaging times for the field data.

Although L is not an important length scale in the local-free-convection region ($|L| < z < 0.1z_i$), nevertheless it is convenient to express local-free-convection predictions in terms of the usual surface layer scales and ζ (or z/L). According to Wyngaard, Coté, and Izumi (1971), some of these predictions are

$$\frac{\overline{w'^2}}{u_*^2} \approx 3.6\,(-\zeta)^{2/3}$$

$$\frac{\overline{\theta'^2}}{T_*^2} \approx 0.9\,(-\zeta)^{-2/3}$$

$$\frac{\overline{w'^2\theta'}}{u_*^2\,T_*} \approx -1.3\,\zeta^{1/3}$$

or in terms of mixed-layer parameters (Kaimal et al., 1976)

$$\frac{\overline{w'^2}}{w_*^2} \approx 1.8\left(\frac{z}{z_i}\right)^{2/3}$$

$$\frac{\overline{\theta'^2}}{\theta_*^2} \approx 1.8\left(\frac{z}{z_i}\right)^{-2/3}$$

Data indicate that, for convective conditions, these two predictions have validity up to $0.1z_i$. For the third-order moment

$$\frac{\overline{w'^2\theta'}}{w_*^2\,\theta_*} \approx 0.9\left(\frac{z}{z_i}\right)^{1/3}$$

Data show that this prediction is approached at $0.02z_i$. Above $0.02z_i$ the sign of $\overline{w'^2\theta'}$ is different. Similar results have been reported by Lenschow, Wyngaard, and Pennell (1980).

5-4.4 Dispersion in the Convective Boundary Layer

Willis and Deardorff (1978) used a laboratory model to study the dispersion of effluents from an elevated, continuous point source located in a convective boundary layer (see Fig. 5.18). For a source height $h_s \simeq 0.24\ z_i$ and at a downwind distance, x, they found that the plume first reaches the ground at $x \simeq 0.15\ z_i\ (\bar{V}/w_*)$. The term $\bar{V}/w_*$ can be thought of as a dimensionless stability parameter (Willis and Deardorff, 1976). The model also showed that the average plume concentration maximum descends to ground level at $x \simeq 0.4\ z_i\ (\bar{V}/w_*)$. Similar results were reported by Lamb (1978, 1979), who used a numerical model to study the effects of release height, h_s, on effluents released into a convective boundary layer. For $h_s > 0.25\ z_i$, Lamb showed that the plume centerline descended at a rate of about $0.5\ w_*$. For $h_s > 0.1\ z_i$, Lamb (1979) found that the maximum ground-level concentration is approximately $1.2\ S/h_s z_i\ \bar{V}$, where S is the rate at which material is emitted [particles/(volume · time)]. Note that

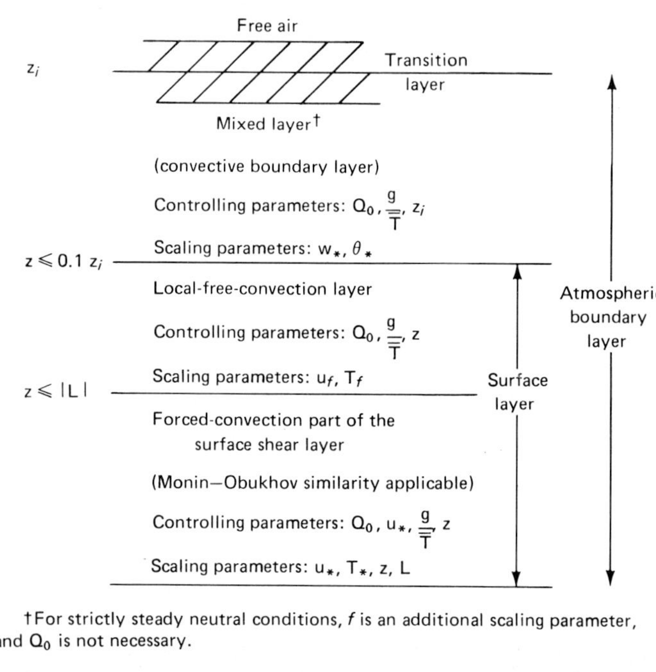

†For strictly steady neutral conditions, *f* is an additional scaling parameter, and Q_0 is not necessary.

Fig. 5.18 Conceptualized structure of the atmospheric boundary layer.

the maximum concentration is independent of L. Lamb also found that the maximum concentration at the ground occurs at $x \simeq 2\ h_s\ (\overline{V}/w_*)$. He emphasized that his results apply only to situations where dispersion is buoyancy-dominated and where diffusion in the direction of the mean flow is negligible compared to transport. These two conditions are met when $z_i/L < -10$, $h_s/L < -1$, and $1.2\ w_* \leqslant \overline{V} \leqslant 6w_*$. For additional discussion of dispersion in the convective layer, see the article by Venkatram (1979a).

It should be emphasized that, during convective conditions, vertical motion (both updrafts and downdrafts) can dominate the dispersion of an effluent. Updraft speeds appear to be faster than downdraft speeds. Consequently, if momentum is conserved, the downward flux of effluent passes through a larger horizontal area than in updrafts. Lenschow (1970) found that downdrafts are relatively nonturbulent, and thus plumes tend to remain intact until they approach the ground. These two regimes could be partitioned on the basis of the intensity of the turbulence. On the basis of this hypothesis, Venkatram and Kurtz (1979) argued that Gaussian plume models are not applicable during convective conditions and proposed a new model.

Field data indicate that during unstable conditions, the standard deviation of the u and v components of the wind (σ_u and σ_v) scale with w_*. In fact, when the atmosphere is very convective, σ_u and $\sigma_v = 0.6\ w_*$ throughout the mixed layer, all the way to the ground. Near the ground, the standard deviation of the vertical component of the wind, σ_w, is proportional to u_*. In the neutral surface layer $\sigma_w \simeq 1.25\ u_*$ (Panofsky, 1973) or $\sigma_w \simeq 1.3\ u_*$ (Merry and Panofsky, 1976) for near-neutral conditions. For stable conditions both σ_w and u_* become so small that their ratio is subject to much error. For neutral and stable conditions, however, Binkowski (1979) found that $\sigma_w = 1.28\ u_*$ and that $\sigma_u = 2.1\ u_*$ and $\sigma_v = 1.78\ u_*$. For unstable conditions, Panofsky et al. (1977) reported that $\sigma_u = \sigma_v = u_*\ (12 - 0.5z_i/L)^{1/3}$ and $\sigma_w = 1.3\ u_*\ (1 - 3z_i/L)^{1/3}$, whereas for stable conditions $\sigma_u = \sigma_v = 2.3\ u_*$. For fully convective conditions, Irwin (1979) attempted to combine field and laboratory data to find a relation between σ_w/w_* and z/z_i. If we use his information and let $\sigma_w/w_* = 0.2$ for $z/z_i = 10^{-3}$, an approximation for σ_w is

$$\sigma_w \sim 0.9\ w_* \left(\frac{z}{z_i}\right)^{0.2}$$

To relate turbulence to dispersion in simple air-quality models currently used in regulatory applications, perhaps use can be made of such parameters as σ_w/u_*, v/w_*, and z_i/w_*. For example, in Gaussian-plume models, values for σ_w and σ_y can be estimated from u_* and w_*.

During unstable conditions and near pollution sources, the horizontal deviations of the concentration, σ_x and σ_y, are proportional to $0.6\ w_*t$. At large distances, σ_x and σ_y may be proportional to $(w_*z_it)^{1/2}$ (which is analogous to molecular diffusion). Also, Willis and Deardorff (1976) reported that when convective turbulence is the dominant process, $\sigma_z = z_i\ f(t_*)$, where t_* is a nondimensional time (w_*t/z_i). However, Weber, McDonald, and Briggs (1977) noted that for stable conditions similarity theory predicts that, at a fixed distance, σ_z is a function of L. They suggest that for $\sigma_z < L$ the theory is approximated by $\sigma_z \propto L^{1/2}x^{1/2}$, where x is the downwind distance. For stable conditions, Venkatram (1980) has proposed a very simple method for calculating ground-level concentrations. More information on σ's is given in Chaps. 6, 9, and 14 and in the papers by Panofsky et al. (1977), Venkatram (1979b), and Draxler (1979).

5-4.5 Atmospheric Boundary-Layer Structure

To summarize the structure of the atmospheric boundary layer over land, refer to Fig. 5.18. Conceptually the atmospheric boundary layer is visualized as a three-layer structure determined by the parameters considered relevant to the turbulence in each layer. For example, in the forced convection part of the surface layer, wind shear plays a dominant role and Monin–Obukhov similarity applies. The similarity hypothesis states that the turbulence of the flow is determined solely by Q_0, u_*, $g/\overline{T}$, and z. Consequently the temperature and wind speed can be scaled with $T_* = Q_0/u_*$ and $u_*^2 = \tau/\rho$. Dimensionless parameters formed with T_* and u_* are universal functions of z/L. Furthermore, the wind shear layer is confined to a height range given by $z < |L|$.

In the local-free-convection layer, z continues to be a significant length scale, but τ_0 is no longer important, and u_* is important only for scaling momentum flux. Consequently the controlling parameters reduce to H_0, $g/\overline{T}$, and z so that the normalized scaling velocity and temperature are $u_f = (gzQ_0/\overline{T})^{1/3}$ and $T_f = Q_0/w_*$. According to the local-free-convection predictions by Wyngaard, Coté, and Izumi (1971), dimensionless groups formed with u_f and T_f should be constants. For very unstable conditions $(-z/L > 1)$ Kaimal et al. (1976) found that the Kansas data generally support these predictions. They also defined the upper limit of the free-convection layer as $0.1\ z_i$, a result supported by the modeling studies of Deardorff (1974a). If $-L > 0.1\ z_i$, the forced convection layer can be enveloped by free convection.

In the mixed layer, turbulence is considered to be insensitive to τ_0 so that z_i is the controlling length scale. Consequently the normalized scaling velocity and temperature for this layer become $w_* = (gz_iH_0/\overline{T})^{1/3}$ and $\theta_* = Q_0/w_*$. The momentum flux (u′w′) in this layer is controlled by u_* and the layer is well mixed down to $-L$. According to Deardorff (1972b) and to Wyngaard, Coté, and Rao (1974), dimensionless groups formed with w_* and θ_* should be functions of z/z_i. At night the air near the ground is normally stably stratified with little or no turbulence above a layer of continuous, mechanically produced turbulence. This mixed layer is usually 10 to 100 m deep but can be several 100 m deep on cloudy, windy nights. During daytime the air is normally stably stratified above the convective boundary layer, which is usually 500 to 2000 m deep in the afternoon. Above the mixed layer is the free air where friction and other surface effects are negligible and the geostrophic approximation can be used.

The observed characteristics of the atmospheric boundary layer depend strongly on thermal stability. On the basis of this property, Deardorff (1978) has tentatively proposed the following classification of stability regimes for the boundary layer:

(1) Unstable—buoyant for $z_i/L < -2$. During this regime the boundary layer is also called the mixed layer. In addition, $\partial\theta/\partial z \sim 0$ down to 0.1 to 0.2 z_i, $-\overline{u'w'}$ is 2 to 6 times greater than u_* during midday (Kaimal et al., 1976), and the mixed layer grows with time. The region $|L| < z_i < 0.1\ z_i$ is referred to as the free convection layer.

(2) Unstable—with shear for $-2 < z_i/L < 0$. This regime occurs frequently over tropical oceans where air–sea contrasts are small. In addition, this regime has a thicker surface-shear layer and a greater turning of the wind with

height than in regime 1. The mixed layer can also grow with time.

(3) Stable—continuous for $0 < z_i/L < 2$. This regime occurs over land at night or over the ocean when the air is slightly warmer than the sea surface.

(4) Stable—sporadic for $z_i/L > 2$. This regime is typical over land at night with clear to partly cloudy skies and weak to moderate geostrophic wind. Continuous turbulence occupies only a small fraction of z_i and a low-level nocturnal jet can occur.

The depth of the atmospheric boundary layer can be specified in a number of ways. For example, during neutral conditions it is a fraction of u_*/f; or it can be chosen as the height of a capping inversion during unstable conditions. It could also be specified as the first height above the ground where the local Ri > 0.25. However, remember, this height can vary temporally and spatially. The temporal variability of the height of the atmospheric boundary layer has been studied recently by Stull (1973), Yamada and Mellor (1975), Brost and Wyngaard (1978), and others. For a convective boundary layer entraining into neutrally stratified air, a time-dependent expression for this height has been proposed by Deardorff (1974c) for clear days ($\partial z_i/\partial t \simeq 0.2\ w_*$). The spatial variability over water has been reported by Agee and Lomax (1978). Many recent model studies strongly suggest that the depth of the convective boundary layer is determined by z_i.

5-4.6 Z-less Stratification

In a neutrally stable boundary layer, $L \rightarrow \infty$ so that the only length scale is z, which means that the turbulence scales with the distance z from the surface. Such a condition implies that the turbulent eddies are as large as they can be, i.e., as great as the distance to the surface. However, if the atmosphere begins to stabilize to a condition with a positive potential temperature gradient, the development of large eddy motions is suppressed. Consequently the energy-containing eddies will become much smaller than their counterparts during neutrally stable conditions. As the stability increases, the eddy sizes should be limited entirely by the stability and not by the distance from the surface. Then, what do the eddies scale with? The only other length scale available is L, which fixes the size of the eddies. Consequently, under strong stability, we expect only $g/\overline{T}$, Q_0, and u_* to characterize the turbulence structure. Thus the turbulent velocities scale with u_*, the temperatures with T_*, and the dominant eddy sizes with L so that we expect the fluctuating velocities, temperatures, and eddy sizes to be independent of height, z. In other words, we have a z-less stratification.

5-4.7 Nocturnal Boundary Layer

At night, radiative cooling of the ground and of the air in contact with it can generate a stably stratified boundary layer commonly referred to as the nocturnal boundary layer. Subsequent evolution of this strongly capped mechanically mixed layer appears to be governed by continued surface cooling, by turbulent transfer processes, and to some extent by the flux divergence of long wave radiation within the layer. Mechanical turbulence can occur in stably stratified flow near the ground because of the large vertical wind shear and relatively small Ri that can be developed there. This nocturnal mixed layer is shallower than the daytime mixed layer and varies more in depth. The nocturnal mixed layer is thought to range in depth from 3 to 300 m, the depth depending strongly on the upper-level wind speeds and on the downward heat flux at the surface. Furthermore, as with the daytime mixed layer (and the urban mixed layer), its top acts as a lid on diffusive processes; thus knowledge of the depth of the nocturnal boundary layer would be quite useful in computing the diffusion of material in stably stratified flow.

On the basis of the similarity theory, the height of the stable boundary layer, z_s, can be given by (Monin, 1970)

$$z_s \propto \frac{ku_*}{f}$$

More recently, Businger and Arya (1974) suggested that

$$z_s \propto \left(\frac{u_* L}{f}\right)^{1/2}$$

However, Yu (1978) concluded that dimensional scale height formulas such as these and others are only appropriate for stability conditions where u_*/fL ranges from 5 to 45, or for slight to moderate stability. Nieustadt and Driedonks (1979) analyzed the effects of advection on the nocturnal boundary layer. They found that advection can be neglected only if the horizontal velocity gradient is much less than $3 \times 10^{-5}\ s^{-1}$ and if the horizontal potential temperature gradient is much less than 3×10^{-5} °C/m. Under these conditions the thermal wind can also be neglected. They also emphasized that these restrictions are seldom met in reality.

Blackadar (1976) developed a model for predicting time-dependent wind speed and temperature profiles in the nocturnal boundary layer. The model also provides the vertical distribution of exchange coefficient that would be helpful in estimating diffusion of material released into the nocturnal boundary layer. In another study, Briggs (1977) proposed

$$z_s \simeq u_*^{3/2} C^{-1/2}$$

where u_* is the geometric average of the initial and steady-state values of friction velocity and C is $g/\overline{T}$ times the average cooling rate. Briggs reports that this model yields predictions consistent with the steady-state results from numerical models (Wyngaard, 1975).

When u_* and L are approximately constant, theory suggests that the height of the nocturnal boundary layer should be independent of time. However, observations indicate that z_s frequently increases with time during the night (Izumi and Barad, 1963). On the basis of these observations, Deardorff (1971) proposed that the rate of growth of the nocturnal boundary layer is given by

$$\frac{\partial z_s}{\partial t} = 0.025\, u_* \frac{1 - z_s}{0.35\, u_*/f}$$

Using similarity theory, Zilitinkevich and Monin (1974) suggested

$$\frac{\partial z_s}{\partial t} \propto \frac{ku_* L}{z_s}$$

and Khakimov (1976) proposed

$$\frac{\partial z_s}{\partial t} = 1.5\,(0.3u_* - f z_s)$$

By comparing some of these time-dependent equations with observations, Yu (1978) concluded that they were weakly significant in a statistical sense, explaining only 25% of the variance. More recently, Yamada (1979), Smeda (1979), and Zeman (1979) proposed simple prognostic expressions for simulating the development of the nocturnal boundary layer.

5-4.8 Change of Terrain and Wind Profiles

Up to this point, the profiles and concepts for the atmospheric boundary layer have been developed for steady-state flow over smooth surfaces. Such conditions are rare over land so that we can ask what happens to air as it flows from over a horizontally uniform surface to a nonuniform surface. Such a question creates a host of complexities, of which only a few have been investigated in depth; however, we do know that the effect of a change in terrain is, first, to modify the flow in the layer adjacent to the ground. In other words, an internal boundary layer develops within the atmospheric boundary layer. Above the internal boundary layer, the flow has the characteristics of the approaching flow regime, whereas, below, an inner layer seems to develop close to the ground. The inner layer adapts quickly (over distances of several kilometers) to the local surface roughness. Of course, since it is difficult to find topographical features that remain homogeneous for such distances, we would rarely find a fully adapted inner (turbulent) boundary layer.

Early evidence of the existence of an internal boundary layer was shown through the duct flow experiments conducted by Jacobs (1939). Investigation of the development of an internal boundary layer was pursued by Elliott (1958), who suggested using two different log-law profiles to describe the two different flow regimes. A log law of the form

$$u(z) = (u_{*s}/k)\ln(z/z_{0s})$$

was used to represent the wind profile over a smooth surface and $u(z) = (u_{*r}/k)\ln(z/z_{0r})$ was used for the flow over a rough surface. The abrupt change in surface roughness is represented by $m \equiv \ln(z_{0a}/z_{0b})$. When there is a marked change in roughness, say from a water surface to a rough coastline, $m \simeq -5$. For example, if $z_{0a} = 1$ cm and $z_{0b} = 100$ cm, then $m = -4.6$. Jensen (1978) showed that, for $m = -5$, a 70% change in surface stress can occur. Moreover, field measurements demonstrate that the height of the internal boundary layer grows slowly with increasing distance but the surface stress changes abruptly and then grows slowly to an equilibrium value. Calculations by Jensen (1978) indicate that, at downwind distances of 20 km, a 20% excess in surface stress still exists even though the internal boundary layer has grown to contain the whole atmospheric boundary layer.

Although Elliott's theory of the internal boundary layer is crude, his observations fit theoretical predictions quite well. Since Elliott's pioneering work, new theories have been proposed (Taylor, 1962; Panofsky and Townsend, 1964), but little improvement has been attained. Plate and Hidy (1967) did achieve slightly better results. Attempts to account for modification of the internal boundary-layer parameters by using a turbulent energy equation model (Bradshaw, Ferris, and Atwell, 1967) have not been very successful (Peterson, 1969); however, numerical experiments by Shir (1972) look promising and do offer improvements over previous theories because the physics of the problem can be studied better. For summaries of internal boundary-layer research, the reader is referred to Plate (1971); to the references in the paper by Shir (1972); to the review by Businger (1972), who also describes dust devils and other small-scale atmospheric phenomena; and to the developmental work by Rao, Wyngaard, and Coté (1974).

Field programs (e.g., Bradley, 1968) have been initiated to document changes in atmospheric diffusion characteristics that can occur as boundary-layer air flows from over a smooth to a rough surface (and from a cold surface to a warm one). Of particular interest has been flow from over water bodies to metropolitan areas. For example, along the southern shoreline of Long Island, N. Y., Raynor et al. (1975) studied the nature of the atmospheric dispersion of oil fogs released from a source located 2 to 5 km offshore. Only onshore flow conditions were studied. Results showed that wind profiles (to 16-m level) measured on the beach were representative of those over the ocean. Analysis of the available data showed that the height of the internal boundary layer, h_i, in meters, could be expressed as

$$h_i = \frac{u_*}{u}\left[\frac{F(|\Delta\theta|)}{|\Delta T/\Delta z|}\right]^{1/2}$$

where F is the fetch over the downwind surface (m) and $|\Delta T/\Delta z|$ is the absolute value of the lapse rate over the source region (°K/m^{-1}). This result is consistent with the wind tunnel observations made by Meroney, Cermak, and Yang (1975). They found that in the initial stages of development, h_i shows a half-power dependence on fetch. Knowledge of h_i is an important factor in estimating ground-level concentrations of pollutants. The value of h_i is used as the depth over which vigorous mixing is expected to occur. In terms of selecting a good micrometeorological site for studying the development of internal boundary layers, Bradley (1968) suggested one criterion to be a height-to-fetch ratio of 1 : 200.

At the Danish Atomic Energy Commission Research Establishment at Risø, Denmark, micrometeorological measurements (Panofsky and Petersen, 1972) have been made since early 1974. Peterson, Kristensen, and Su (1976) have used data from the Risø site to verify the theoretical wind predictions of the change-of-roughness models developed by Peterson (1969) and by Rao, Wyngaard, and Coté (1974). Results show that these second-order closure models predict a consistent shape for the downwind profiles, but the growth of the internal boundary layer is slower than that for observed data. In general, the model predictions compared poorly with observations. Consequently developments in this research area will continue to unfold (e.g., see Venkatram, 1977; Wood, 1978; and Mulhearn, 1978).

5-4.9 Effluents, Radiation, and the Atmospheric Boundary Layer

Recently a variety of numerical models have been formulated to incorporate solar and terrestrial radiative processes, as well as the radiative properties of particulates and gases. These types of models permit us to study the effects of urbanization and industrialization on the atmospheric boundary layer. Most of these models have been one-dimensional, but one of the earliest models to incorporate radiative effects into a physical boundary layer model was the three-dimensional model used by Atwater (1972, 1975). He concluded that the radiative flux divergence was not of primary importance in maintaining the urban heat island. He also noted that radiatively active pollutants such as aerosols can alter the temperature structure of the atmosphere less than other major urbanization factors. However, during daytime, radiatively active effluents contribute to increasing both the vertical stability and the near-surface concentrations when compared with nonreactive effluents. In a different model, Zdunkowski, Welch, and Paegle (1976) emphasized that earlier models disregarded the effects of water vapor on the attenuation parameters, and they used a more complete radiation treatment than Atwater (1975). From their one-dimensional model, they concluded that a dense layer of pollution can markedly affect temperature and wind profiles in the atmospheric boundary layer. Dense pollution can cause strong cooling of the air layer near the ground relative to clear air. Furthermore, air pollution can cause temperature inversions to intensify and grow and wind speeds to decrease due to increasing thermal stability. Similar results were reported by Ackerman (1977), who reported that the advection of pollution is important in modeling the boundary layer. He also found that the main effects of aerosols on urban climate is to increase the daytime temperatures by a few degrees. Complex radiative processes were found to contribute to feedback mechanisms for altering temperature profiles. However, Ackerman emphasized that, with regard to climate, pollutants in general and aerosols in particular have a fairly small effect on urban temperatures.

Viskanta, Bergstrom, and Johnson (1977) used a one-dimensional transport model to conclude that aerosols over an urban area can lead to a decrease of absorption by the ground and to an increase in absorption of solar radiation in the boundary layer. Hence a net warming of the boundary layer would be expected. However, the net effect of gaseous pollutants under a clear sky is to increase the absorption of thermal radiation by the ground and to cool the boundary layer. However, Venkatram and Viskanta (1977) advised caution in drawing premature conclusions from incomplete models, even though the results of their model are consistent with those of other studies. They emphasize that we need to learn more about the radiative properties of aerosols and that the use of eddy-diffusion models has limited relevance to the physics of the daytime convective boundary layer. Along similar lines, Bach (1976) noted that over urban areas the net effects of gaseous and particulate effluents on radiative transfer have not been established because of the uncertainty of the physical and chemical properties of aerosols. In particular, the absorbing and scattering properties of aerosols and particulates and the absorbing and emitting properties of gases (that interact primarily in the infrared spectrum) should be accounted for. Future modelers will continue to investigate these complicating processes.

LIST OF SYMBOLS

Symbols used frequently in this chapter are listed here. The equation or section number indicates the first appearance of the symbols. The noted units for each symbol are those which are typically used in the calculations discussed in the text, although other units may be used in specific instances.

Symbol	Definition
A	Area or a constant as in Eq. 5.71
a	Dimensionless constant
B	Dimensionless constant
B	Body force vector
b	Thermal conductivity, Eq. 5.37
C_D	Drag coefficient
C_i	Structure parameter, Sec. 5-2.14
C_1, C_2	Coefficients
C	Coriolis force vector
c_p	Specific heat for dry air at constant pressure, 0.24 cal/(g·°K), Eq. 5.21
D	Scale distance
d	Characteristic linear dimension, Eq. 5.16
E	Vertical flux of water vapor, Eq. 5.22
e	Subscript referring to water vapor, Eq. 5.22
F	Fetch
F	Vector force for friction
F_x, F_y	Components of the scaled velocity defects, Eq. 5.74
f	Coriolis parameter, or when used as a subscript to indicate the local-free-convection state
G	The modulus of the geostrophic wind
G	Newtonian gravitational vector
g	Gravitational acceleration, Eq. 5.26, or a geostrophic parameter when used as a subscript, Eq. 5.69
H	Vertical flux of sensible heat, Eq. 5.21
H_c	Heat transfer by conduction
H_f	Heat due to friction
H_m	Kinematic surface heat flux
H_0	Surface flux of sensible heat
h_i	Height of internal boundary layer
i	$(-1)^{1/2}$
K	Exchange coefficient, eddy-diffusivity coefficient
K_e	Exchange coefficient for water vapor, Eq. 5.22
K_h	Exchange coefficient for sensible heat, Eq. 5.21
K_m	Exchange coefficient for momentum, Eq. 5.23
K_x, K_y, K_z	Components for the exchange coefficients for momentum
k	Von Kármán's dimensionless constant, $\simeq 0.4$, Eq. 5.54
L	Monin–Obukhov length scale, Eq. 5.46
M	Momentum vector
P	Atmospheric pressure, Eq. 5.12
P	Pressure force vector
P_c	Production of turbulent energy by convection
P_m	Production of turbulent energy by mechanical forces
Q_0	Surface temperature flux, Sec. 5-2.7
q	Specific humidity, Eq. 5.22
R	Solar constant
Re	Reynolds number, Eq. 5.16
Rf	Richardson flux number, Eq. 5.53
Ri	Richardson number, Eq. 5.50
R_i	Incoming solar radiation
Ro	Rossby number
S	Stability parameter
S	Surface force vector
T	Atmospheric temperature
T_*	Scale temperature, Eq. 5.47
T_f	Scale temperature for free convection
T_{ij}	Stress tensor
t	Time
u, v, w	Components of the wind in the x-, y-, and z-directions, respectively, Eqs. 5.24 to 5.26
u′, v′, w′	Fluctuating components of the wind
$\overline{u}$, $\overline{v}$, $\overline{w}$	Mean wind components

u_* Friction velocity, Eq. 5.40
V Total wind vector, Eq. 5.1
V Characteristic wind speed
W A velocity scale
w_* Convective velocity scale, Sec. 5-4.3
w_{*m} Maximum convective velocity scale
X_m Distance downwind where the maximum plume width (Y_m), occurs, Eq. 5.25
x, y, z Positions in a Cartesian coordinate system oriented so that the x-axis is in the direction of the mean horizontal vector wind, the y-axis is crosswind, and the z-axis is vertical
Y_m Maximum plume width, Eq. 5.24
X_t Total length of a visible smoke plume, Eq. 5.27
Z_m Maximum half width of a plume at distance X_m, Eq. 5.27
z_i Depth of the mixed layer
z_{im} Maximum depth of the mixed layer
z_0 Roughness length, Eq. 5.61
z_s Height of the stable boundary layer
α Specific volume, Eq. 5.36
β A constant, Eq. 5.63
γ Lapse rate
ζ A dimensionless height used to represent z/L, Eq. 5.49
ℓ Mixing length, Eq. 5.18
η A constant
θ Potential temperature, Eq. 5.21
θ_m Maximum solar elevation angle
θ_* Scale temperature for the mixed layer
κ Thermal diffusivity
Λ Austausch or mixing coefficient, Eq. 5.17
λ A parameter given by $(27V_g \times 10^{-5})/f$
μ Coefficient of viscosity, 1.8 g/(cm·s), Eq. 5.3
ρ Atmospheric density
$\sigma_u,\sigma_v,\sigma_w$ Standard deviation of the u, v, and w components of the wind, respectively
$\sigma_x,\sigma_y,\sigma_z$ Standard deviation of the concentration in the horizontal (x and y) and vertical directions, respectively
τ Tangential stress on a unit area of fluid
τ_0 Surface layer tangential stress
ν Kinematic coefficient of viscosity, Eq. 5.16
Φ Energy dissipation by viscosity
ϕ_h Dimensionless temperature gradient, Eq. 5.57
ϕ_m Dimensionless wind shear, Eq. 5.56
χ A parameter given by $(f/2k)^{1/2}$, Eq. 5.70
ψ Universal function, Sec. 5-3.5
$'$ Primed terms represent turbulent fluctuations

REFERENCES

Ackerman, T. P., 1977, A Model of the Effect of Aerosols on Urban Climates with Particular Applications to the Los Angeles Basin, *J. Atmos. Sci.*, 34: 531-547.

Agee, E. M., and F. E. Lomax, 1978, Structure of the Mixed Layer and Inversion Layer Associated with Patterns of Mesoscale Cellular Convection During AMTEX 75, *J. Atmos. Sci.*, 35: 2281-2301.

Agnew, T., and L. Diehl, 1978, Prediction of Representative Surface Winds in the Beaufort Sea Computerized Support System, in *Conference on Weather Forecasting and Aviation Meteorology*, Oct. 16–19, 1978, Silver Spring, Md., pp. 111-116, American Meteorological Society, Boston.

André, J. C., G. DeMoor, P. Lacarrère, G. Therry, and R. DuVachat, 1978, Modeling the 24-Hour Evolution of the Mean and Turbulent Structures of the Planetary Boundary Layer, *J. Atmos. Sci.*, 35: 1861-1883.

Arya, S. P. S., 1972a, Comment on the Paper by P. A. Taylor: "A Note on the Log-Linear Velocity Profile in Stable Conditions," *Q. J. R. Meteorol. Soc.*, 98: 460-461.

——, 1972b, The Critical Condition for the Maintenance of Turbulence in Stratified Flow, *Q. J. R. Meteorol. Soc.*, 98: 264-273.

——, 1975, Geostrophic Drag and Heat Transfer Relations for the Atmospheric Boundary Layer, *Q. J. R. Meteorol. Soc.*, 101: 147-161.

——, and E. J. Plate, 1969, Modeling of the Stably Stratified Atmospheric Boundary Layer, *J. Atmos. Sci.*, 26: 656-665.

Atwater, M. A., 1972, Thermal Effects of Urbanization and Industrialization in the Boundary Layer: A Numerical Study, *Boundary-Layer Meteorol.*, 3: 229-245.

——, 1975, Thermal Changes Induced by Urbanization and Pollutants, *J. Appl. Meteorol.*, 14: 1061-1071.

Bach, W., 1976, Global Air Pollution and Climatic Change, *Rev. Geophys. Space Phys.*, 14: 429-474.

Benoit, R., 1977, On the Integral of the Surface Layer Profile-Gradient Functions, *J. Appl. Meteorol.*, 16: 859-860.

Bernstein, A. B., and J. A. Young, 1962, The Measurement of Vertical Gradients near the Ground, *J. Appl. Meteorol.*, 1: 458-470.

Betts, A. K., 1973, Nonprecipitation Cumulus Convection and Its Parameterization, *Q. J. R. Meteorol. Soc.*, 99: 178-196.

Binkowski, F. S., 1975, On the Empirical Relationship Between the Richardson Number and the Monin–Obukhov Stability Parameter, *Atmos. Environ.*, 9: 453-454.

——, 1979, A Simple Semiempirical Theory for Turbulence in the Atmospheric Surface Layer, *Atmos. Environ.*, 13: 247-254.

Blackadar, A. K. 1962, The Vertical Distribution of Wind and Turbulence Exchange in a Neutral Atmosphere, *J. Geophys. Res.*, 67: 3095-3102.

——, 1976, Modeling the Atmospheric Boundary Layer, in *Third Symposium on Atmospheric Turbulence, Diffusion, and Air Quality*, Oct. 19–22, 1976, Raleigh, N. C., pp. 46-49, American Meteorological Society, Boston.

——, and H. Tennekes, 1968, Asymptotic Similarity in Neutral Barotropic Planetary Boundary Layers, *J. Atmos. Sci.*, 25: 1015-1020.

Boussinesq, J., 1877, Essai sur la Théorie des Eaux courantes, *Mémoires Présentés par Divers Savants à l'Académie des Sciences de l'Institut de France*, Vol. 23, pp. 1-680.

Bradley, E. F., 1968, A Micrometeorological Study of Velocity Profiles and Surface Drag in the Region Modified by a Change in Surface Roughness, *Q. J. R. Meteorol. Soc.*, 94: 361-378.

——, and R. A. Antonia, 1979, Structure Parameters in the Atmospheric Surface Layer, *Q. J. R. Meteorol. Soc.*, 105: 695-705.

Bradshaw, P., 1971, Calculation of the Three-Dimensional Turbulent Boundary Layer, *J. Fluid Mech.*, 46: 417-445.

——, 1972, The Understanding and Prediction of Turbulent Flow, *Aeronaut. J.*, 76: 403-418.

——, D. H. Ferris, and N. P. Atwell, 1967, Calculation of Boundary-Layer Development Using the Turbulent Energy Equation, *J. Fluid Mech.*, 28: 593-616.

Brakke, T. W., S. B. Verma, and N. J. Rosenberg, 1978, Local and Regional Components of Sensible Heat Advection, *J. Appl. Meteorol.*, 17: 955-963.

Briggs, G. A., 1975, Plume Rise Predictions, *Lectures on Air Pollution and Environmental Impact Analyses*, pp. 59-105, American Meteorological Society, Boston.

——, 1977, Predictions of Nocturnal Mixing Layer Parameters, *Manuscript*, Atmospheric Turbulence and Diffusion Laboratory.

——, 1978, Atmospheric Turbulence and Diffusion Laboratory, personal communication.

Brook, R. R., 1972, The Measurement of Turbulence in a City Environment, *J. Appl. Meteorol.*, 11: 443-450.

Brooks, C. E. P., and N. Carruthers, 1953, *Handbook of Statistical Methods in Meteorology*, Her Majesty's Stationery Office, London.

Brost, R. A., and J. C. Wyngaard, 1978, A Model of the Stably Stratified Planetary Boundary Layer, *J. Atmos. Sci.*, 35: 1427-1440.

Busch, N. E., S. W. Chang, and R. A. Anthes, 1976, A Multilevel Model of the Planetary Boundary Layer Suitable for Use in Mesoscale Dynamic Models, *J. Appl. Meteorol.*, 15: 909-919.

Businger, J. A., 1966, Transfer of Heat and Momentum in the Atmospheric Boundary Layer, in *Proceedings of the Symposium on Arctic Heat Budget and Atmospheric Circulation*, Santa Monica, Calif., pp. 305-332, RAND Corp.

——, 1972, The Atmospheric Boundary Layer, in *Remote Sensing of the Atmosphere*, V. E. Derr (Ed.), Wave Propagation Laboratory, Environmental Research Laboratory, Boulder, Colo.

——, 1973, Turbulent Transfer in the Atmospheric Surface Layer, in *Workshop on Micrometeorology*, D. A. Haugen (Ed.), pp. 67-100, American Meteorological Society, Boston.

——, and S. P. S. Arya, 1974, Height of the Mixed Layer in the Stability Stratified Planetary Boundary Layer, *Adv. Geophys.*, 18A: 73-92.

——, J. C. Wyngaard, Y. Izumi, and E. F. Bradley, 1971, Flux Profile Relationships in the Atmospheric Surface Layer, *J. Atmos. Sci.*, 28: 181-189.

Calder, K. L., 1966, Concerning the Similarity Theory of A. S. Monin and A. M. Obukhov for the Turbulent Structure of the Thermally Stratified Surface Layer of the Atmosphere, *Q. J. R. Meteorol. Soc.*, 92: 141-146.

——, 1968, In Clarification of the Equations of Shallow Layer Thermal Convection for a Compressible Fluid Based on the Boussinesq Approximation, *Q. J. R. Meteorol. Soc.*, 94: 88-92.

Carl, D. M., T. C. Tarbell, and H. A. Panofsky, 1973, Profiles of Wind and Temperature from Towers over Homogeneous Terrain, *J. Atmos. Sci.*, 30: 788-794.

Carson, D. J., 1973, The Development of a Dry Inversion-Capped Convectively Unstable Boundary Layer, *Q. J. Roy. Meteorol. Soc.*, 99: 450-467.

——, and P. J. R. Richards, 1978, Modelling Surface Turbulent Fluxes in Stable Conditions, *Boundary-Layer Meteorol.*, 14: 67-81.

Champagne, F. H., C. A. Frieche, J. C. LaRue, and J. C. Wyngaard, 1977, Flux Measurements, Flux Estimation Techniques, and Fine-Scale Turbulence Measurements in the Unstable Surface Layer Over Land, *J. Atmos. Sci.*, 34: 515-530.

Chang, S. W-J., 1979, An Efficient Parameterization of Convective and Nonconvective Planetary Boundary Layers for Use in Numerical Models, *J. Appl. Meteorol.*, 18: 1205-1215.

Charney, J. G., and A. Eliassen, 1949, A Numerical Method of Predicting the Perturbations of the Middle Latitude Westerlies, *Tellus*, 1: 38-54.

Clarke, R. H., 1970a, Observational Studies of the Atmospheric Boundary Layer, *Q. J. R. Meteorol. Soc.*, 96: 91-114.

——, 1970b, Recommended Methods for the Treatment of the Boundary Layer in Numerical Models, *Austral. Meteorol. Mag.*, 18: 51-73.

——, A. J. Dyer, R. R. Brook, D. G. Reid, and A. J. Troup, 1971, *The Wangara Experiment: Boundary Layer Data*, Technical Paper No. 19, Commonwealth Scientific and Industrial Research Organization, Division of Meteorology and Physics, Australia.

Cornett, J. S., and D. Randerson, 1977, Verification of Some Numerical Models for Operationally Predicting Mesoscale Winds Aloft, *J. Appl. Meteorol.*, 16: 764-773.

Cramer, H. E., and F. A. Record, 1953, The Variation with Height of the Vertical Flux of Heat and Momentum, *J. Meteorol.*, 10: 219-226.

Csanady, G. T., 1967, On the Resistance Law of a Turbulent Boundary Layer, *J. Atmos. Sci.*, 24: 467-471.

Deacon, E. L., 1973, Geostrophic Drag Coefficients, *Boundary-Layer Meteorol.*, 5: 321-340.

Deardorff, J. W., 1966, The Counter-Gradient Heat Flux in the Lower Atmosphere and in the Laboratory, *J. Atmos. Sci.*, 23: 503-506.

——, 1970, A Three-Dimensional Numerical Investigation of the Idealized Planetary Boundary Layer, *Geophys. Fluid Dyn.*, 1: 377-410.

——, 1971, Rate of Growth of the Nocturnal Boundary Layer, in *Symposium on Air Pollution, Turbulence, and Diffusion*, H. W. Church and R. E. Luna (Eds.), pp. 183-190, Sandia Laboratories, Albuquerque, N. Mex.

——, 1972a, Parameterization of the Boundary Layer for Use in General Circulation Models, *Mon. Weather Rev.*, 100: 93-106.

——, 1972b, Numerical Investigation of Neutral and Unstable Planetary Boundary Layers, *J. Atmos. Sci.*, 29: 91-115.

——, 1974a, Three-Dimensional Numerical Study of the Height and Mean Structure of a Heated Planetary Boundary Layer, *Boundary-Layer Meteorol.*, 7: 81-106.

——, 1974b, Three-Dimensional Numerical Study of Turbulence in an Entraining Mixed Layer, *Boundary-Layer Meteorol.*, 7: 199-226.

——, 1974c, Three-Dimensional Study of the Height and Mean Structure of a Heated Planetary Boundary Layer, *Boundary-Layer Meteorol.*, 7: 81-106.

——, 1978, personal communication, Oregon State University.

——, and L. Mahrt, 1982, on the Dichotomy in Theoretical Treatments of the Atmospheric Boundary Layer, *J. Atmos. Sci.*, 39: 2096-2098.

——, and G. E. Willis, 1975, A Parameterization of Diffusion Into the Mixed Layer, *J. Appl. Meteorol.*, 14: 1451-1458.

Delsol, F., K. Miyakoda, and R. H. Clarke, 1971, Parameterized Processes in the Surface Boundary Layer of an Atmospheric Circulation Model, *Q. J. R. Meteorol. Soc.*, 97: 181-208.

DeWalle, D. R., and L. H. Parmele, 1974, Application of Error Analysis to Surface Temperature Determination by Radiation Budget Techniques, *J. Appl. Meteorol.*, 13: 430-434.

Donaldson, C. duP., 1973a, Construction of a Dynamic Model of the Production of Atmospheric Turbulence and the Dispersal of Atmospheric Pollutants, in *Workshop in Micrometeorology*, D. A. Haugen (Ed.), pp. 313-392, American Meteorological Society, Boston.

——, 1973b, *Atmospheric Turbulence and the Dispersion of Atmospheric Pollutants*, Report EPA-R4-73-016a, Environmental Protection Agency Report, available from NTIS as Report PB-225524.

Donelan, M., and M. Miyake, 1973, Spectra and Fluxes in the Boundary Layer of the Trade-Wind Zone, *J. Atmos. Sci.*, 10: 444-464.

Draxler, R. R., 1979, Estimating Vertical Diffusion from Routine Meteorological Tower Measurements, *Atmos. Environ.*, 13: 1559-1564.

Duncan, W. J., A. S. Thom, and A. D. Young, 1970, *Mechanics of Fluids*, 2nd ed., American Elsevier Publishing Co., New York.

Dyer, A. J., and B. B. Hicks, 1970, Flux-Gradient Relationships in the Constant Flux Layer, *Q. J. R. Meteorol. Soc.*, 96: 715-721.

——, and B. B. Hicks, 1972, The Spatial Variability of Eddy Fluxes in a Constant Flux Layer, *Q. J. R. Meteorol. Soc.*, 98: 206-212.

Ekman, V. W., 1902, Om jordrotationens inverkan på vindströmmar i hafvet, *Nyt. Mag. Naturvid*, 40: 37-63.

——, 1905, On the Influence of the Earth's Rotation on Ocean Currents, *Ark. Mat., Astron. Fys.*, 2: 1-52.

Elliott, W. P., 1958, The Growth of the Atmospheric Internal Boundary Layer, *Trans. Am. Geophys. Union*, 39: 1048-1054.

Endlich, R. M., 1968, Direct Computation of Geostrophic Winds from Observed Winds Using the Balance Equation, *J. Appl. Meteorol.*, 7: 994-1003.

Estoque, M. A., 1973, Numerical Modeling of the Planetary Boundary Layer, in *Workshop on Micrometeorology*, D. A. Haugen (Ed.), pp. 217-270, American Meteorological Society, Boston.

Fichtl, G. H., 1972, Wind Shear near the Ground and Aircraft Operations, *J. Aircr.*, 9: 765-770.

Franceschini, G. A., 1959, *Micrometeorological Observations over Non-Ideal Surfaces*, Scientific Report No. 1, Refs. 59-65, Texas A&M University, Department of Meteorology, College Station, Tex.

Friehe, C. A., and K. F. Schmitt, 1976, Parameterizations of Air–Sea Interface Fluxes of Sensible Heat and Moisture by the Bulk Aerodynamic Formulas, *J. Phys. Oceanogr.*, 6: 801-809.

——, and C. H. Gibson, 1978, Estimates of the Surface Fluxes over the Ocean, in *Turbulent Fluxes Through the Sea Surface, Wave Dynamics, and Prediction*, A. Favre and K. Hasselmann (Eds.), pp. 67-79, Plenum Publishing Corp.

Gaynor, J. E., 1977, Acoustic Doppler Measurement of Atmospheric Boundary Layer Structure Functions and Energy Dissipation Rates, *J. Appl. Meteorol.*, 16: 148-155.

Gifford, F. A., 1959, Smoke Plumes as Quantitative Air Pollution Indices, *Int. J. Air Pollut.*, 2: 42-50.

Golder, D., 1972, Relations Among Stability Parameters in the Surface Layer, *Boundary-Layer Meteorol.*, 3: 47-58.

Goldstein, S., 1931, On the Stability of Superimposed Streams of Fluids of Different Densities, *Proc. R. Soc. (London), Ser. A*, 132: 524-548.

Gray, W. M., and B. R. Mendenhall, 1973, A Statistical Analysis of Factors Influencing the Wind Veering in the Planetary Boundary Layer, in *Climatological Research*, the Hermann Flohn 60th Anniversary Volume, pp. 167-194, Meteorological Institute, University of Bonn, Bonn, Germany.

Hall, F. F., J. G. Edinger, and W. D. Neff, 1975, Convective Plumes in the Planetary Boundary Layer, Investigated with an Acoustic Echo Sounder, *J. Appl. Meteorol.*, 14: 513-523.

Haltiner, G. J., 1971, *Numerical Weather Prediction*, John Wiley & Sons, New York.

Hanna, S. R., 1968, A Method of Estimating Vertical Eddy Transport in the Planetary Boundary Layer Using Characteristics of the Vertical Velocity Spectrum, *J. Atmos. Sci.*, 25: 1026-1033.

——, 1969, *Characteristics of Winds and Turbulence in the Planetary Boundary Layer*, ESSA Technical Memorandum, Report ERLTM-ARL-8, Environmental Science Service Administration, Oak Ridge, Tenn.

Hasse, L., 1974, On the Surface to Geostrophic Wind Relationship at Sea and the Stability Dependence of the Resistance Law, *Beitr. Phys. Atmos.*, 47: 45-55.

Haugen, D. A., J. C. Kaimal, and E. F. Bradley, 1971, An Experimental Study of Reynolds Stress and Heat Flux in the Atmospheric Surface Layer, *Q. J. R. Meteorol. Soc.*, 97: 168-180.

——, J. C. Kaimal, and E. F. Bradley, 1978, Measuring Temperature Structure Parameter Profiles with an Acoustic Sounder, *J. Appl. Meteorol.*, 17: 895-899.

Hellman, G., 1915, Ü. d. Bewegung d. Luft in d. untersten Schichten d. Atm., *Met. Zeit.*, 32: 1-16.

Hicks, B. B., 1976, Wind Profile Relationships from the "Wangara" Experiment, *Q. J. R. Meteorol. Soc.*, 102: 535-551.

——, and H. S. Goodman, 1971, The Eddy-Correlation Technique of Evaporation Measurement Using a Sensitized Quartz-Crystal Hygrometer, *J. Appl. Meteorol.*, 10: 221-223.

Hoffert, M. I., and Y. C. Sud, 1976, Similarity Theory of the Buoyantly Interactive Planetary Boundary Layer with Entrainment, *J. Atmos. Sci.*, 33: 2136-2151.

Holland, J. Z., and E. M. Rasmusson, 1973, Measurements of the Atmospheric Mass, Energy, and Momentum Budgets over a 500-Kilometer Square of the Tropics, *Mon. Weather Rev.*, 101: 44-55.

Hoxit, L. R., 1974, Planetary Boundary Layer Winds in Baroclinic Conditions, *J. Atmos. Sci.*, 31: 1003-1020.

Irwin, J. S., 1979, Estimating Plume Dispersion: A Recommended Generalized Scheme, in *Fourth Symposium on Turbulence, Diffusion, and Air Pollution*, Jan. 15–18, Reno, Nev., pp. 62-69, American Meteorological Society, Boston.

——, and F. Binkowski, 1981, Estimation of the Monin–Obukhov Scaling Length Using On-Site Instrumentation, *Atmos. Environ.* (in press).

Izumi, Y., and M. Barad, 1963, Wind and Temperature Variations During Development of a Low-Level Jet, *J. Appl. Meteorol.*, 2: 668-673.

——, and S. J. Caughey, 1976, Minnesota 1973 Atmospheric Boundary Layer Experiment Data Report, Report AFCRL-TR-76-0038, Air Force Cambridge Research Laboratories, L. G. Hanscom Field, Mass.

Jacobs, W., 1939, Unformung eines turbulenten Geschwindigkeitsprofiles, *Z. Angew. Math. Mech.*, 19: 87-100 (translation in National Advisory Committee for Aeronautics, Technical Memorandum No. 951).

Jensen, N. O., 1978, Change of Surface Roughness and the Planetary Boundary Layer, *Q. J. R. Meteorol. Soc.*, 104: 351-356.

Kaimal, J. C., 1973, Turbulence Spectra, Length Scales, and Structure Parameters in the Stable Layer, *Boundary-Layer Meteorol.*, 4: 289-309.

——, J. C. Wyngaard, D. A. Haugen, O. R. Coté, Y. Izumi, S. J. Caughey, and C. J. Readings, 1976, Turbulence Structure in the Convective Boundary Layer, *J. Atmos. Sci.*, 33: 2152-2169.

Kampé de Fériet, J., 1974, Turbulent Atmospheric Diffusion: The First Twenty-Five Years, 1920–1945, *Adv. Geophys.*, 18A: 1-23.

Khakimov, I. R., 1976, The Wind Profile and the Thickness of the Neutrally Stratified Atmospheric Boundary Layer, *Izv. Acad. Sci., USSR, Atmos. Ocean. Phys.*, 12: (10).

Lamb, R. G., 1978, A Numerical Simulation of Dispersion from an Elevated Point Source in the Convective Planetary Boundary Layer, *Atmos. Environ.*, 12: 1297-1304.

——, 1979, The Effects of Release Height on Material Dispersion in the Convective Planetary Boundary Layer, in *Fourth Symposium on Atmospheric Turbulence, Diffusion, and Air Pollution*, Jan. 15–18, Reno, Nev., pp. 27-33, American Meteorological Society, Boston.

Lebedeff, S. A., and S. Hameed, 1976, Laws of Effluent Diffusion in the Steady-State Atmospheric Surface Layer in Stable and Unstable Conditions, *J. Appl. Meteorol.*, 15: 326-336.

Lee, H. N., and S. K. Kao, 1979, Finite-Element Numerical Modeling of the Atmospheric Boundary Layer, *J. Appl. Meteorol.*, 18: 1287-1295.

Lenschow, D. H., 1970, Airplane Measurements of Planetary Boundary Layer Structure, *J. Appl. Meteorol.*, 9: 874-884.

——, 1973, Two Examples of Planetary Boundary Layer Modification over the Great Lakes, *J. Atmos. Sci.*, 30: 568-581.

——, J. C. Wyngaard, and W. T. Pennell, 1980, Mean-Field and Second-Moment Budgets in a Baroclinic, Convective Boundary Layer, *J. Atmos. Sci.*, 37: 1313-1326.

Lettau, H., 1949, *Isotropic and Non-Isotropic Turbulence in the Atmospheric Surface Layer, Geophysical Research Papers, No. 1*, Report NP-1761, Air Force Cambridge Research Laboratories.

——, 1967, Small to Large-Scale Features of Boundary Layer Structure over Mountain Slopes, in *Proceedings of the Symposium on Mountain Meteorology*, E. R. Reiter and J. L. Rassmussen (Eds.), pp. 1-74.

——, 1969, Note on Aerodynamic Roughness—Parameter Estimation on the Basis of Roughness—Element Description, *J. Appl. Meteorol.*, 8: 828-832.

——, and B. Davidson, 1953, *Exploring the Earth's First Mile, Vol. II. Site Description and Data Tabulation*, Pergamon Press, New York.

Lilly, D. K., 1968, Models of Cloud-Top Mixed Layers Under a Strong Inversion, *Q. J. R. Meteorol. Soc.*, 94: 292-309.

Liu, M-K., and D. Durran, 1977, On the Prescription of the Vertical Dispersion Coefficient over Complex Terrain, in *Joint Conference on Applications of Air Pollution Meteorology*, Nov. 29–Dec. 2, 1977, Salt Lake City, Utah, pp. 172-175, American Meteorological Society, Boston.

Liu, W. T., K. B. Katsaros, and J. A. Businger, 1979, Bulk Parameterization of Air–Sea Exchanges of Heat and Water Vapor Including the Molecular Constraints at the Interface, *J. Atmos. Sci.*, 36: 1722-1735.

Lo, A. K., and G. A. McBean, 1978, On the Relative Errors in Methods of Flux Calculations, *J. Appl. Meteorol.*, 17: 1704-1711.

Lönnqvist, O., 1962, On the Diurnal Variation of Surface Temperature, *Tellus*, 14: 96-101.

Ludlam, F. H., 1967, Characteristics of Billow Clouds and Their Relation to Clear-Air Turbulence, *Q. J. R. Meteorol. Soc.*, 93: 419-435.

Luers, J. K., and J. B. Reeves, 1973, *Effect of Shear on Aircraft Landing*, Report NASA-CR-2287, George Marshall Space Flight Center, Huntsville, Ala., NTIS.

Lumley, J. L., and H. A. Panofsky, 1964, *The Structure of Atmospheric Turbulence*, Wiley–Interscience, New York.

Mahrt, L., and D. H. Lenschow, 1976, Growth Dynamics of the Convectively Mixed Layer, *J. Atmos. Sci.*, 33: 41-51.

Manton, M. J., and W. R. Cotton, 1977, Parameterization of the Atmospheric Surface Layer, *J. Atmos. Sci.*, 34: 331-334.

McCutchan, M. H., 1979, Determining the Diurnal Variation of Surface Temperature in Mountainous Terrain, *J. Appl. Meteorol.*, 18: 1224-1229.

Melling, H., and R. List, 1980, Characteristics of Vertical Velocity Fluctuations in a Convective Urban Boundary Layer, *J. Appl. Meteorol.*, 19: 1184-1195.

Melvin, W. W., 1971, Wind Shear on the Approach, *Shell Aviat. News*, 393: 16-21.

Meroney, R. W., J. E. Cermak, and B. T. Yang, 1975, Modeling of Atmospheric Transport and Fumigation at Shoreline Sites, *Boundary-Layer Meteorol.*, 9: 69-90.

Merry, M., and H. A. Panofsky, 1976, Statistics of Vertical Motion over Land and Water, *Q. J. R. Meteorol. Soc.*, 102: 255-260.

Miles, J. W., 1961, On the Stability of Heterogeneous Shear Flows, *J. Fluid Mech.*, 11: 284-290.

Milford, J. R., S. Abdulla, and D. A. Mansfield, 1979, Eddy Flux Measurements Using an Instrumented Powered Glider, *Q. J. R. Meteorol. Soc.*, 105: 673-693.

Monin, A. S., 1953, The Mechanism of the Heating of the Air in the Open Steppe, *Akad. Nauk SSSR, Inst. Geogr.*, pp. 100-123.

——, 1970, The Atmospheric Boundary Layer, *Annual Review of Fluid Mechanics*, pp. 225-250, Annual Reviews, Inc.

——, and A. M. Obukhov, 1954, Basic Regularity in Turbulent Mixing in the Surface Layer of the Atmosphere, *Tr. Geofiz. Inst., Akad. Nauk, SSSR, Sb. Statei*, 24: 163-167.

——, and A. M. Yaglom, 1970, *Statistical Fluid Mechanics: Mechanics of Turbulence*, Vol. I, MIT Press, Cambridge, Mass.

Montgomery, R. B., 1948, Vertical Eddy Flux of Heat in the Atmosphere, *J. Meteorol.*, 5: 265-274.

Mulhearn, P. J., 1978, A Wind-Tunnel Boundary-Layer Study of the Effects of a Surface Roughness Change: Rough to Smooth, *Boundary-Layer Meteorol.*, 15: 3-30.

Navier, C. L. M. H., 1822, Mémorie sur les Lois du Mouvement des Fluides, *Mém. Acad. Sci., Toulouse*, 6: 389-440.

Nickerson, E. C., and V. E. Smiley, 1975, Surface Layer and Energy Budget Parameterizations for Mesoscale Models, *J. Appl. Meteorol.*, 14: 297-300.

Nieuwstadt, F., 1978, The Computation of the Friction Velocity u_* and the Temperature Scale T_* from Temperature and Wind Velocity Profiles by Least-Square Methods, *Boundary-Layer Meteorol.*, 14: 235-246.

——, and A. G. M. Driedonks, 1979, The Nocturnal Boundary Layer: A Case Study Compared with Model Calculations, *J. Appl. Meteorol.*, 18: 1397-1405.

Nikuradse, J., 1933, Strömungsgesetze in rauhen Rohen, *Forschung.-Arb. Ing.-Wesen*, No. 361.

Noonkester, V. R., 1976, The Evolution of the Clear Air Convective Layer Revealed by Surface-Based Remote Sensors, *J. Appl. Meteorol.*, 15: 594-606.

O'Brien, J. J., 1970, A Note on the Vertical Structure of the Eddy Exchange Coefficient in the Planetary Boundary Layer, *J. Atmos. Sci.*, 27: 1213-1214.

Obukhov, A. M., 1946, Turbulence in an Atmosphere with a Nonuniform Temperature, English translation in *Boundary-Layer Meteorol.*, 2: 7-29 (1971).

Orlanski, I., B. B. Ross, and L. J. Polinsky, 1974, Diurnal Variation of the Planetary Boundary Layer, *J. Atmos. Sci.*, 31: 965-989.

Owen, P. R., 1971, Buildings in the Wind, *Q. J. R. Meteorol. Soc.*, 97: 396-413.

Panofsky, H. A., 1961, An Alternative Derivation of the Diabatic Wind Profile, *Q. J. R. Meteorol. Soc.*, 87: 109-110.

——, 1973, The Boundary Layer Above 30 m, *Boundary-Layer Meteorol.*, 4: 250-264.

——, and G. W. Brier, 1965, *Some Applications of Statistics to Meteorology*, Pennsylvania State University, University Park, Pa.

——, and E. L. Petersen, 1972, Wind Profiles and Change of Terrain Roughness at Risø, *Q. J. R. Meteorol. Soc.*, 98: 845-854.

——, H. Tennekes, D. H. Lenschow, and J. C. Wyngaard, 1977, The Characteristics of Turbulent Velocity Components in the Surface Layer Under Convective Conditions, *Boundary-Layer Meteorol.*, 11: 355-361.

——, and A. A. Townsend, 1964, Change of Terrain Roughness and the Wind Profile, *Q. J. R. Meteorol. Soc.*, 90: 147-155.

Panchev, St., 1971, *Random Functions and Turbulence*, Pergamon Press, Oxford.

Paulson, C. A., 1970, The Mathematical Representation of Wind Speed and Temperature Profiles in the Unstable Atmospheric Surface Layer, *J. Appl. Meteorol.*, 9: 857-861.

Penwarden, A. D., 1972, Wind Environments Around Tall Buildings, *Building Research Station Digest*, 141, Building Research Station Advisory Service, Garton, Watford, England, Her Majesty's Stationery Office, London.

Peterson, E. W., 1969, Modification of Mean Flow and Turbulent Energy by Change in Surface Roughness Under Conditions of Neutral Stability, *Q. J. R. Meteorol. Soc.*, 90: 561-576.

——, 1975, The Risø Profiles: A Study of Wind and Temperature Data from the 123-m Tower at Risø, Denmark, *Q. J. R. Meteorol. Soc.*, 101: 107-117.

——, L. Kristensen, and C-C. Su, 1976, Some Observations and Analysis of Wind over Nonuniform Terrain, *Q. J. R. Meteorol. Soc.*, 102: 857-869.

Petterssen, S., 1956, *Weather Analysis and Forecasting*, 2nd ed., Vol. I., McGraw-Hill Book Co., New York.

Pielke, R. A., and Y. Mahrer, 1975, Representation of the Heated Planetary Boundary Layer in Mesoscale Models with Coarse Vertical Resolution, *J. Atmos. Sci.*, 32: 2288-2308.

Plate, E. J., 1971, *Aerodynamic Characteristics of Atmospheric Boundary Layers*, AEC Critical Review Series, TID-25465, NTIS.

——, and G. M. Hidy, 1967, Laboratory Study of Air Flowing over a Smooth Surface onto Small Water Waves, *J. Geophys. Res.*, 72: 4627-4641.

Poisson, S. D., 1830, Sur les équations générales de l'Équilibre et du Mouvement des Corps solides, Élastiques, et Fluides, *J. l'École Polytechn*, 13: 1-174.

Pond, S., G. T. Phelps, J. E. Paquin, G. McBean, and R. W. Stewart, 1971, Measurements of the Turbulent Fluxes of Momentum, Moisture, and Sensible Heat over the Ocean, *J. Atmos. Sci.*, 28: 901-917.

——, D. B. Fissel, and C. A. Paulson, 1974, A Note on Bulk Aerodynamic Coefficients for Sensible Heat and Moisture Fluxes, *Boundary-Layer Meteorol.*, 6: 333-339.

Prandtl, L., 1904, Über flüssigkeitsbewegung bei sehr kleiner Reibung, in *Proceedings of the Third International Mathematics Congress*, Heidelberg.

——, 1925, Über die ausgebildete Turbulence, *Z. Angew. Math. Mech.*, 5: 136.

——, 1932, Meteorologische Anwendung der Stromungslehre, *Beitr. Phys. Atmos., Bjerkness Festschrift*, 19: 188-202.

Pruitt, W. O., D. L. Morgan, and F. J. Lourence, 1973, Momentum and Mass Transfers in the Surface Boundary Layer, *Q. J. R. Meteorol. Soc.*, 99: 370-386.

Randerson, D., J. G. Garcia, and V. S. Whitehead, 1971, Photogrammetric and Photometric Investigation of a Smoke Plume Viewed from Space, *J. Appl. Meteorol.*, 10: 1122-1131.

Rao, K. S., J. C. Wyngaard, and O. R. Coté, 1974, The Structure of the Two Dimensional Internal Boundary Layer over a Sudden Change of Surface Roughness, *J. Atmos. Sci.*, 31: 738-746.

Raynor, G. S., P. Michael, R. M. Brown, and S. SethuRaman, 1975, Studies of Atmospheric Diffusion from a Nearshore Oceanic Site, *J. Appl. Meteorol.*, 14: 1080-1094.

Reynolds, O., 1883, An Experimental Investigation of the Circumstances Which Determine Whether the Motion of Water Shall Be Direct or Sinuous, and of the Law of Resistance in Parallel Channels, *Philos. Trans., R. Soc. London, Ser. A*, 174: 935-982.

——, 1895, On the Dynamical Theory of Incompressible Viscous Fluids and the Determination of the Criterion, *Philos. Trans., London, Ser. A*, 186: 123-164.

Richardson, L. F., 1920, The Supply of Energy from and to Atmospheric Eddies, *Proc. R. Soc. (London), Ser. A*, 97: 354.

Roberts, C. F., 1963, A Preliminary Analysis of Some Observations of Wind Shear in the Lowest 100 Feet of the Atmosphere for Application to the Problem of the Control of Aircraft on Approach, *Vertical Wind Shear in the Lower Layers of the Atmosphere*, Technical Note 93, pp. 203-218, World Meteorological Organization, Geneva, Switzerland.

Robinson, G. D., 1978, Weather and Climate Forecasting as Problems in Hydrodynamics, *Mon. Weather Rev.*, 106: 448-457.

Robinson, S. M., 1962, Computing Wind Profile Parameters, *J. Atmos. Sci.*, 19: 189-190.

Ryan, B. C., 1977, A Mathematical Model for Diagnosis and Prediction of Surface Winds in Mountainous Terrain, *J. Appl. Meteorol.*, 16: 571-584.

Saab, A. E., 1979, Contribution a l'Étude du Transport Turbulent de Masse dans l'Átmosphèra, *Bull. Dir. Etud. Rech., Ser. A*, No. 4, Electricité de France, Paris.

Saint-Venant, B. de, 1843, Note à joindre au Mémoire sur la Dynamique des Fluides, *Compt. Rend.*, 17: 1240-1243.

Schlichting, H., 1955, *Boundary Layer Theory*, Pergamon Press, New York.

Schmidt, W., 1925, Der Massenaustausch in freier Luft and verwandte Erscheinungen, *Problems der Kosmischen Physik*, Vol. 7, Verlag von Henri Grand, Hamburg.

Schubert, J. F., 1977, Acoustic Detection of Momentum Transfer During the Abrupt Transition from a Laminar to a Turbulent Atmospheric Boundary Layer, *J. Appl. Meteorol.*, 16: 1292-1297.

Sellers, W. D., 1967, *Physical Climatology*, University of Chicago Press, Chicago.

Shir, C. C., 1972, A Numerical Computation of Air Flow over a Sudden Change of Surface Roughness, *J. Atmos. Sci.*, 29: 304-310.

——, 1973, A Preliminary Numerical Study of Atmospheric Turbulent Flows in the Idealized Planetary Boundary Layer, *J. Atmos. Sci.*, 30: 1327-1339.

——, and L. J. Shieh, 1974, A Generalized Urban Air Pollution Model and Its Application to the Study of SO_2 Distributions in the Saint Louis Metropolitan Area, *J. Appl. Meteorol.*, 13: 185-204.

Smeda, M. S., 1979, Incorporation of Planetary Boundary-Layer Processes into Numerical Forecasting Models, *Boundary-Layer Meteorol.*, 16: 115-129.

Smith, F. B., 1975, Turbulence in the Atmospheric Boundary Layer, *Sci. Prog. (Oxford)*, 62: 127-151.

——, and D. J. Carson, 1974, A Scheme for Deriving Day-Time Boundary-Layer Wind Profiles, *Meteorol. Mag.*, 103: 241-255.

Smith, S. D., and E. G. Banke, 1975, Variation of the Sea-Surface Drag Coefficient with Wind Speed, *Q. J. R. Meteorol. Soc.*, 101: 665-673.

Spiegel, E. A., and G. Veronis, 1960, On the Boussinesq Approximation for a Compressible Flow, *Astrophys. J.*, 131: 442-447.

Spittlehouse, D. L., and T. A. Black, 1979, Determining of Forest Evapotranspiration Using Bowen Ratio and Eddy Correlation Measurements, *J. Appl. Meteorol.*, 18: 647-653.

Sreenivasan, K. R., A. J. Chambers, and R. A. Antonia, 1978, Accuracy of Moments of Velocity and Scalar Fluctuations in the Atmospheric Surface Layer, *Boundary-Layer Meteorol.*, 14: 341-359.

Stokes, G. G., 1849, On the Theories of the Internal Friction of Fluids in Motion, and of the Equilibrium and Motion of Elastic Solids, *Trans. Cambridge Philos. Soc.*, 8: 287-319.

Stull, R. B., 1973, Inversion Rise Model Based on Penetrative Convection, *J. Atmos. Sci.*, 30: 1092-1099.

Sutton, O. G., 1953, *Micrometeorology*, McGraw-Hill Book Co., New York.

Swinbank, W. C., 1964, The Exponential Wind Profile, *Q. J. R. Meteorol. Soc.*, 90: 119-135.

Tatarskii, V. I., 1971, *The Effects of the Turbulent Atmosphere on Wave Propagation*, Israel Program for Scientific Translation, Report TT-68-50464, NTIS.

Taylor, G. I., 1931, Effect of Variation in Density on the Stability of Superimposed Streams of Fluids, *Proc. R. Soc. (London), Ser. A*, 132: 499-523.

Taylor, R. J., 1962, Small-Scale Advection and the Neutral Wind Profile, *J. Fluid Mech.*, 13: 529-539.

Tennekes, H., 1968, Outline of a Second-Order Theory of Turbulent Pipe Flow, *AIAA (Am. Inst. Aeronaut. Astronaut.) J.*, 6: 1735-1740.

——, 1970, Free Convection in the Turbulent Ekman Layer of the Atmosphere, *J. Atmos. Sci.*, 27: 1027-1034.

——, 1973a, The Logarithmic Wind Profile, *J. Atmos. Sci.*, 30: 234-238.

——, 1973b, A Model for the Dynamics of the Inversion Above a Convective Boundary Layer, *J. Atmos. Sci.*, 30: 558-567.

——, 1974, The Atmospheric Boundary Layer, *Phys. Today*, 27: 52-63.

——, and J. L. Lumley, 1972, *A First Course in Turbulence*, MIT Press, Cambridge, Mass.

Thuillier, R. H., and U. O. Lappe, 1964, Wind and Temperature Characteristics from Observations on a 1400-Foot Tower, *J. Appl. Meteorol.*, 3: 299-306.

Ting, C. L., and D. R. Hay, 1975, On the Heat Flux and Friction Velocity in Free Convection near the Ground, *J. Atmos. Sci.*, 32: 637-638.

Townsend, A. A., 1956, *The Structure of Turbulent Shear Flow*, Cambridge, University Press, New York.

Venkatram, A., 1977, Internal Boundary Layer Development and Fumigation, *Atmos. Environ.*, 11: 479-482.

——, 1978, Estimating the Convective Velocity Scale for Diffusion Applications, *Boundary-Layer Meteorol.*, 15: 447-452.

——, 1980a, The Relationship Between the Convective Boundary Layer and Dispersion from Tall Stacks, *Atmos. Environ.*, 14: 763-768.

——, 1980b, A Semi-Empirical Method to Compute Concentration Associated with Surface Releases in the Stable Boundary Layer, *Atmos. Environ.*, 16: 245-248.

——, and J. Kurtz, 1979, Models to Study the Short-Term Impact of Large Point Sources, in *Fourth Symposium on Atmospheric Turbulence, Diffusion, and Air Pollution*, Jan. 15–18, Reno, Nev., American Meteorological Society, Boston.

——, and R. Viskanta, 1977, Effects of Aerosol-Induced Heating on the Convective Boundary Layer, *J. Atmos. Sci.*, 34: 1918-1933.

Viskanta, R., R. W. Bergstrom, and R. O. Johnson, 1977, Radiative Transfer in a Polluted Urban Planetary Boundary Layer, *J. Atmos. Sci.*, 34: 1091-1103.

Webb, E. K., 1970, Profile Relationships: The Log-Linear Range and Extension to Strong Stability, *Q. J. R. Meteorol. Soc.*, 96: 67-90.

Weber, A. H., J. S. Irwin, J. P. Kahler, and W. B. Peterson, 1975, Atmospheric Turbulence Properties in the Lowest 300 Meters, Report EPA-600/4-75-004, Environmental Protection Agency.

——, K. R. McDonald, and G. A. Briggs, 1977, Turbulent Classification Schemes for Stable and Unstable Conditions, in *Joint Conference on Applications of Air Pollution Meteorology*, Nov. 29–Dec. 2, 1977, Salt Lake City, Utah, pp. 96-102, American Meteorological Society, Boston.

Wesely, M. L., G. W. Thurtell, and C. B. Tanner, 1970, Eddy Correlation Measurements of Sensible Heat Flux Near the Earth's Surface, *J. Appl. Meteorol.*, 9: 45-50.

Willis, G. E., and J. W. Deardorff, 1974, A Laboratory Model of the Unstable Planetary Boundary Layer, *J. Atmos. Sci.*, 31: 1297-1307.

——, and J. W. Deardorff, 1976, A Laboratory Model of Diffusion Into the Convective Planetary Boundary Layer, *Q. J. R. Meteorol. Soc.*, 102: 427-445.

——, and J. W. Deardorff, 1978, A Laboratory Study of Dispersion from an Elevated Source Within a Modeled Convective Planetary Boundary Layer, *Atmos. Environ.*, 12: 1305-1311.

Wise, A. F. E., 1971, Wind Effects Due to Groups of Buildings, *Philos. Trans. R. Soc. London, Ser. A*, 269: 469-485.

Wong, E. Y. J., and K. C. Brundidge, 1966, Vertical and Temporal Distributions of Heat Conductivity and Flux, *J. Atmos. Sci.*, 23: 167-178.

Wood, D. H., 1978, Calculation of the Neutral Wind Profile Following a Large Step Change in Surface Roughness, *Q. J. R. Meteorol. Soc.*, 104: 383-392.

Wyngaard, J. C., 1973, On Surface-Layer Turbulence, in *Workshop on Micrometeorology*, D. A. Haugen (Ed.), pp. 101-149, American Meteorological Society, Boston.

——, 1975, Modeling the Planetary Boundary Layer: Extension to the Stable Case, *Boundary-Layer Meteorol.*, 9: 441-460.

——, S. P. S. Arya, and O. R. Coté, 1974, Some Aspects of Convective Planetary Boundary Layers, *J. Atmos. Sci.*, 31: 747-754.

——, and S. F. Clifford, 1978, Estimating Momentum, Heat, and Moisture Fluxes from Structure Parameters, *J. Atmos. Sci.*, 35: 1204-1211.

——, and O. R. Coté, 1971, The Budgets of Turbulent Kinetic Energy and Temperature Variance in the Atmospheric Surface Layer, *J. Atmos. Sci.*, 28: 190-201.

——, and O. R. Coté, 1974, The Evolution of a Convective Planetary Boundary Layer: A Higher-Order-Closure Model Study, *Boundary-Layer Meteorol.*, 7: 289-308.

——, O. R. Coté, and Y. Izumi, 1971, Local Free Convection, Similarity, and the Budgets of Shear Stress and Heat Flux, *J. Atmos. Sci.*, 28: 1171-1182.

——, O. R. Coté, and K. S. Rao, 1974, Modeling the Atmospheric Boundary Layer, *Adv. in Geophys.*, 18: 193-211.

——, Y. Izumi, and S. A. Collins, 1971, Behavior of the Refractive Index Structure Parameter near the Ground, *J. Opt. Soc. Am.*, 61: 1646-1650.

Yaglom, A. M., 1977, Comments on the Wind and Temperature Flux-Profile Relationships, *Boundary-Layer Meteorol.*, 11: 89-102.

Yamada, T., 1975, The Critical Richardson Number and the Ratio of the Eddy Transport Coefficients Obtained from a Turbulence Closure Model, *J. Atmos. Sci.*, 32: 926-933.

——, 1976, On the Similarity Functions A, B, and C of the Planetary Boundary Layer, *J. Atmos. Sci.*, 33: 781-793.

——, and G. Mellor, 1975, A Simulation of the Wangara Atmospheric Boundary Layer Data, *J. Atmos. Sci.*, 32: 2309-2329.

——, 1979, Prediction of the Nocturnal Surface Inversion Height, *J. Appl. Meteorol.*, 18: 526-531.

Yu, T-W., 1977, A Comparative Study on Parameterization of Vertical Turbulent Exchange Processes, *Mon. Weather Rev.*, 105: 57-66.

——, 1978, Determining Height of the Nocturnal Boundary Layer, *J. Appl. Meteorol.*, 17: 28-33.

Zdundkowski, W. G., R. M. Welch, and J. Peagle, 1976, One-Dimensional Numerical Simulation of the Effects of Air Pollution on the Planetary Boundary Layer, *J. Atmos. Sci.*, 33: 2399-2414.

Zeman, O., 1979, Parameterization of the Dynamics of Stable Boundary Layers and Nocturnal Jets, *J. Atmos. Sci.*, 36: 792-804.

——, and J. L. Lumley, 1976, Modeling Buoyancy-Driven Mixed Layers, *J. Atmos. Sci.*, 33: 1974-1988.

——, and H. Tennekes, 1977, Parameterization of the Turbulent Energy Budget at the Top of the Atmospheric Boundary Layer, *J. Atmos. Sci.*, 34: 111-123.

Zilitinkevich, S. S., 1973, Shear Convection, *Boundary-Layer Meteorol.*, 3: 416-423.

——, D. L. Laikhtman, and A. S. Monin, 1969, Dynamics of the Atmospheric Boundary Layer, *Izv. Atmos. Oceanic Phys.*, 3: 297-333.

——, and A. S. Monin, 1974, Similarity Theory of the Atmospheric Boundary Layer, *Izv. Akad. Nauk. SSSR, Fiz. Atmos. Okeana*, 10: 587-599.

Chapter 5 covers research prior to February 1980.

THEORIES OF ATMOSPHERIC TRANSPORT AND DIFFUSION

6

S. K. Kao, University of Utah

6-1 INTRODUCTION

Diffusion of an inert trace constituent released in a fluid depends primarily on the turbulent motion, which mixes the constituent with the fluid, and on the mean velocity differences of the fluid, which increases the distance between particles and therefore the size of effective eddies of the turbulent motion. Since the larger eddies contain more kinetic energy, the rate of diffusion would generally increase with increasing diffusion time. Turbulent diffusion of a cluster of particles is thus scale dependent, and the turbulent motion involved is nonstationary. Turbulent diffusion in a shear flow can be very complex. The main purpose of this chapter is to analyze the effects of the turbulence characteristics and simple mean flow on the dispersion of a trace constituent in the atmospheric boundary layer.

There are essentially two approaches to the description of transport and turbulent diffusion: the Eulerian approach in which the transport and diffusion of pollutants are described relative to a fixed coordinate system and the Lagrangian approach in which the concentration is described by following fluid particles. Because more measurements are made at fixed locations, the Eulerian method is an effective way of predicting the pollution concentration. However, the Eulerian system of equations for turbulent diffusion in the Reynolds form is an open one; to solve the system of equations, we must use an appropriate closure scheme.

Because contaminants move with fluid particles, the Lagrangian approach is the natural way of describing pollutant distribution. Statistical analyses of particle movement (Lagrangian approach) have greatly contributed to the theories of diffusion. However, because of the complexity of the Lagrangian governing equations, the Lagrangian analyses have been confined mostly to diffusion in statistically stationary and homogeneous fields of turbulence, whereas predictions of the time-dependent pollution concentration have been accomplished mostly with the Eulerian governing equations. Since both Eulerian and Lagrangian methods are useful in describing the transport and diffusion of pollutants, both approaches are discussed here.

The chapter is organized to present first a general description of various approaches to transport and diffusion, then the governing equations in the Eulerian and the Lagrangian systems, and, finally, analyses of solutions to these equations under various assumptions ranging from simple to complex.

To give a detailed treatment and to cover an extensive literature on theories of transport and turbulent diffusion in the atmosphere's boundary layer would require many more pages than one chapter would permit. Because of limited space, only the basic theories of transport and diffusion are briefly presented, and a limited number of literature references are quoted.

6-2 THE EULERIAN APPROACH

To describe the transport and diffusion of pollutants in a fluid by measurements at fixed locations in a fixed coordinate system is generally referred to as the Eulerian approach. Since most measurements are made at fixed locations, they are readily available for diffusion analysis. Furthermore, because the governing equations for the transport and diffusion of pollutants in the Eulerian system are easier to handle than those in the Lagrangian system, the Eulerian method is an effective way of analyzing and predicting the concentration.

In the first part of this chapter the governing equations are analyzed, the closure schemes are discussed, and several diffusion models are presented.

6-2.1 The Governing Equations

For decades the Eulerian approach to the description of transport and turbulent diffusion has been centered on the "gradient transport model" in which turbulent transport is assumed to be in the direction of the mean concentration gradient in analogy to the molecular diffusion. This model has enjoyed a great deal of popularity because of its simplicity. However, analyzing the necessary conditions for this model to satisfy a simple mean-free-path type of random transport process, Corrsin (1974) has found that the length and time scales of the transport should be

Kao: University of Utah, Salt Lake City, Utah. Unfortunately, Dr. Kao died before he was able to review the chapter.

sufficiently uniform and small compared with the length and time scales of variations in the mean field gradients of the transferrable property. Moreover, from laboratory flow studies he has shown that for diffusion in the turbulent boundary layer some of the conditions are violated. Therefore it is necessary to deal with the problem of transport and turbulent diffusion with the use of the whole set of governing equations.

For the convenience of analysis, the governing equations—consisting of the diffusion equation, momentum equation, continuity equation, thermodynamic equation, and equation of state—are expressed in tensor notation for which the repeated indices denote summation over three components

$$\frac{\partial C}{\partial t} + u_j \frac{\partial C}{\partial x_j} = D \frac{\partial^2 C}{\partial x_j \partial x_j} + Q \tag{6.1}$$

$$\frac{\partial u_i}{\partial t} + u_j \frac{\partial u_i}{\partial x_j} = \frac{1}{\rho} \frac{\partial}{\partial x_k} \left[\mu \left(\frac{\partial u_i}{\partial x_k} + \frac{\partial u_k}{\partial x_i} \right) - \left(p + \frac{2}{3} \mu \frac{\partial u_j}{\partial x_j} \right) \delta_{ik} \right] - g\, \delta_{3i} - \epsilon_{ijk} f_j u_k \tag{6.2}$$

$$\frac{\partial \rho}{\partial t} + \frac{\partial}{\partial x_j} (\rho u_j) = 0 \tag{6.3}$$

$$c_v \rho \left(\frac{\partial T}{\partial t} + u_j \frac{\partial T}{\partial x_j} \right) = -\, p \frac{\partial u_j}{\partial x_j} + \kappa \frac{\partial^2 T}{\partial x_j \partial x_j} + S \tag{6.4}$$

$$p = R \rho T \tag{6.5}$$

where C denotes the concentration of an inert trace constituent, D is the molecular diffusion coefficient, μ is the coefficient of viscosity as defined in Chap. 5, δ_{ik} is the Kronecker delta tensor, ϵ_{ijk} is the alternating tensor, $f_1 = 2\Omega \sin \phi$ and $f_2 = 2\Omega \cos \phi$ are, respectively, the vertical and meridional components of the Coriolis parameter, R is the universal gas constant as defined in Chap. 2, and Q and S are the source or sink of the trace constituent and heat, respectively. Altogether there are seven equations and seven unknowns: u_1, u_2, u_3, C, ρ, p, T. In the preceding notation the subscripts 1, 2, and 3 represent the direction of the velocity component in the x, y, z directions, where x = 1, y = 2, and z = 3. Similar expressions are developed in Chap. 5.

To facilitate the analysis, let

$$\begin{aligned} u_i(x,t) &= \tilde{u}_i \\ p(x,t) &= p_b + \tilde{p} \\ \rho(x,t) &= \rho_b + \tilde{\rho} \\ T(x,t) &= T_b + \tilde{T} \\ C(x,t) &= C_b + \tilde{C} \end{aligned} \tag{6.6}$$

where p_b, ρ_b, T_b, and C_b are, respectively, the pressure, density, temperature, and concentration of the equilibrium state if the atmosphere were at rest, i.e., if $u_b = 0$, and where the terms under the tildas represent a physical state different from that at position **x** and time t (for additional information see Eq. 6.21). Consequently it follows from the momentum equation that

$$\frac{\partial p_b}{\partial x_i} = -\, g \rho_b \delta_{3i} \tag{6.7}$$

so that

$$p_b = p_b(x_3) \qquad \rho_b = \rho_b(x_3) \tag{6.8}$$

From the equation of state, we find that

$$p_b = R \rho_b T_b \tag{6.9}$$

and

$$T_b = T_b(x_3) \tag{6.10}$$

and, from the heat energy equation, that

$$\frac{\partial^2 T_b}{\partial x_3^2} = 0 \tag{6.11}$$

Thus for the equilibrium state

$$\begin{aligned} T_b(x_3) &= T_0 \left(1 - \frac{x_3}{H}\right) \\ p_b(x_3) &= p_0 \left(1 - \frac{x_3}{H}\right)^{gH/RT_0} \\ \rho_b(x_3) &= \rho_0 \left(1 - \frac{x_3}{H}\right)^{(gH/RT_0)-1} \end{aligned} \tag{6.12}$$

where H is the height at which $T_b = 0$ and T_0, ρ_0, and p_0 are, respectively, the temperature, density, and pressure at the surface.

Since in the atmosphere $\tilde{\rho}/\rho_b \ll 1$, it can be shown that the equation of continuity (Eq. 6.3) can be reduced to

$$\frac{\partial \tilde{u}_i}{\partial x_i} = 0 \tag{6.13}$$

The momentum (Eq. 6.2) can be reduced to

$$\frac{\partial \tilde{u}_i}{\partial t} + \tilde{u}_j \frac{\partial \tilde{u}_i}{\partial x_j} = -\frac{1}{\rho_b} \frac{\partial \tilde{p}}{\partial x_i} + g \frac{\tilde{\theta}}{T_b} \delta_{3i} - \epsilon_{ijk} f_j \tilde{u}_k + \frac{\mu}{\rho_b} \frac{\partial^2 \tilde{u}_i}{\partial x_j \partial x_j} \tag{6.14}$$

where θ is the potential temperature.

For an adiabatic stratification in T_b, the heat energy (Eq. 6.4) can be approximated by

$$\frac{\partial \tilde{T}}{\partial t} + \tilde{u}_j \frac{\partial \tilde{T}}{\partial x_j} = \frac{\kappa}{c_v \rho_b} \frac{\partial^2 \tilde{T}}{\partial x_j \partial x_j} + \frac{\tilde{S}}{c_v \rho} \tag{6.15}$$

where κ is the thermal diffusivity defined in Chap. 5, Sec. 5-2.4. Equation 6.4 can be further approximated by

$$\frac{\partial\tilde{\theta}}{\partial t} + \tilde{u}_j \frac{\partial\tilde{\theta}}{\partial x_j} = \frac{\kappa}{c_v\rho_b} \frac{\partial^2\tilde{\theta}}{\partial x_j \partial x_j} + \frac{\tilde{S}}{c_v\rho} \tag{6.16}$$

where θ is potential temperature.

For $u_i = 0$ and $C = C_b$, the diffusion equation becomes

$$\frac{\partial C_b}{\partial t} = D \frac{\partial^2 C_b}{\partial x_j \partial x_j} + Q \tag{6.17}$$

which is purely molecular diffusion and is much smaller than the turbulent diffusion. Therefore

$$C(x,t) = C_b(x,t) + \tilde{C}(x,t)$$

$$\approx \tilde{C}(x,t) \tag{6.18}$$

Thus the diffusion equation becomes

$$\frac{\partial\tilde{C}}{\partial t} + \tilde{u}_j \frac{\partial\tilde{C}}{\partial x_j} = D \frac{\partial^2\tilde{C}}{\partial x_j \partial x_j} + \tilde{Q} \tag{6.19}$$

The equation of state can be shown to be approximated by

$$\tilde{p} \cong R\rho_b\tilde{T} \cong R\rho_b\tilde{\theta} \tag{6.20}$$

The advantage of the new set of the governing Eqs. 6.13, 6.14, 6.16, 6.19, and 6.20 is that ρ_b and T_b in these equations are functions of height (x_3) only.

6-2.2 Equations for the Mean and Fluctuating Quantities

Atmospheric motion is generally in a turbulent state, and turbulent flows are irregular and random. Therefore, in dealing with turbulent flows, Eqs. 6.13, 6.14, 6.16, and 6.19 become partial differential equations whose dependent variables are random functions. Since random variables cannot be predicted precisely, none of these partial differential equations can be solved exactly. However, we can determine the statistical properties of the meteorological variables. Note that the turbulent transports of momentum, heat, and mass of contaminants result from the random fluctuations in the velocities and are many orders of magnitude greater than the molecular transports of the corresponding properties.

To analyze the turbulent motion and transports in a fluid, we find it convenient to express the instantaneous value of a physical quantity η as the sum of its time mean and fluctuating part (Reynolds, 1895)

$$\tilde{\eta}(x,t) = \bar{\eta}(x,t;\tau) + \eta'(x,t;\tau) \tag{6.21}$$

where the time mean $\bar{\eta}$ is defined as

$$\bar{\eta}(x,t;\tau) = \frac{1}{\tau} \int_{t-(\tau/2)}^{t+(\tau/2)} \tilde{\eta}(x,t')\, dt' \tag{6.22}$$

Thus

$$\bar{\eta}'(x,t;\tau) = 0 \tag{6.23}$$

Note that the time-averaged quantity previously defined is a function of time, position, and the averaging interval τ, and so are the measured meteorological quantities that are generally nonstationary functions. The choice of the averaging interval τ depends on the macroscale of the motion that you choose for the mean motion. In general, τ should be large enough so that an adequate range of turbulent motion can be included but small enough so that the macroscale features representing the mean flow can be excluded. Let

$$\begin{aligned} \tilde{u}_i &= \bar{u}_i + u_i' \\ \tilde{C} &= \bar{C} + C' \\ \tilde{p} &= \bar{p} + p' \\ \tilde{\theta} &= \bar{\theta} + \theta' \end{aligned} \tag{6.24}$$

Substituting the first of the preceding equations into the Eq. 6.13 gives

$$\frac{\partial\bar{u}_i}{\partial x_i} + \frac{\partial u_i'}{\partial x_i} = 0 \tag{6.25}$$

Taking time mean of Eq. 6.25 results in the continuity equation for the mean motion

$$\frac{\partial\bar{u}_i}{\partial x_i} = 0 \tag{6.26}$$

Subtraction of Eq. 6.26 from Eq. 6.25 yields the continuity equation for the turbulent motion

$$\frac{\partial u_i'}{\partial x_i} = 0 \tag{6.27}$$

Following the similar procedure, we obtain the rest of the governing equations for the mean and fluctuating concentration, momentum, potential temperature, and pressure:

$$\frac{\partial\bar{C}}{\partial t} + \bar{u}_j \frac{\partial\bar{C}}{\partial x_j} = D \frac{\partial^2\bar{C}}{\partial x_j \partial x_j} - \frac{\partial}{\partial x_j}(\overline{u_j'C'}) + \bar{Q} \tag{6.28}$$

$$\frac{\partial\bar{u}_i}{\partial t} + \bar{u}_j \frac{\partial\bar{u}_i}{\partial x_j} = -\frac{1}{\rho_b}\frac{\partial\bar{p}}{\partial x_i} + \frac{g}{T_b}\bar{\theta}\delta_{3i} - \epsilon_{ijk} f_j \bar{u}_k + \nu \frac{\partial^2\bar{u}_i}{\partial x_j \partial x_j} - \frac{\partial}{\partial x_j}\langle\overline{u_j'u_i'}\rangle \tag{6.29}$$

$$\frac{\partial \bar{\theta}}{\partial t} + \bar{u}_j \frac{\partial \bar{\theta}}{\partial x_j} = \hat{\kappa} \frac{\partial \bar{\theta}}{\partial x_j \partial x_j} - \frac{\partial}{\partial x_j}(\overline{u_j'\theta'}) + \frac{\bar{S}}{c_v \rho_b} \tag{6.30}$$

$$\bar{p} = R\rho_b \bar{\theta} \tag{6.31}$$

$$\frac{\partial C'}{\partial t} + \bar{u}_j \frac{\partial C'}{\partial x_j} = D \frac{\partial^2 C'}{\partial x_j \partial x_j} - u_j' \frac{\partial \bar{C}}{\partial x_j} - \frac{\partial}{\partial x_j}(u_j'C' - \overline{u_j'C'}) + Q' \tag{6.32}$$

$$\frac{\partial u_i'}{\partial t} + \bar{u}_j \frac{\partial u_i'}{\partial x_j} = - \frac{1}{\rho_b}\frac{\partial p'}{\partial x_i} + \frac{g}{T_b}\theta'\delta_{3i} - \epsilon_{ijk} f_j u_k' + \nu \frac{\partial^2 u_i'}{\partial x_j \partial x_j} - u_j' \frac{\partial \bar{u}_i}{\partial x_j} - \frac{\partial}{\partial x_j}(u_j'u_i' - \overline{u_j'u_i'}) \tag{6.33}$$

$$\frac{\partial \theta'}{\partial t} + \bar{u}_j \frac{\partial \theta'}{\partial x_j} = \hat{\kappa} \frac{\partial^2 \theta'}{\partial x_j \partial x_j} - u_j' \frac{\partial \bar{\theta}}{\partial x_j} - \frac{\partial}{\partial x_j}(u_j'\theta' - \overline{u_j'\theta'}) + \frac{S'}{c_v \rho} \tag{6.34}$$

$$p' = R\rho_b \theta' \tag{6.35}$$

where $\hat{\kappa} = \kappa/(c_v \rho_b)$ and $\nu = \mu/\rho_b$

Equations 6.28 to 6.31 and 6.26 are the governing equations in generalized Reynolds form for the mean quantities, similar to the corresponding equations (Eqs. 6.19, 6.14, 6.16, 6.20, and 6.13) for the instantaneous quantities with additional terms involving convergence of mean turbulent fluxes of fluctuating quantities, $\overline{u_j'C'}$, $\overline{u_j'u_i'}$, and $\overline{u_j'\theta'}$. Therefore a total of 19 unknowns are involved in the 7 governing equations for the mean quantities. To solve for the 7 mean quantities ($\bar{u}_1$, $\bar{u}_2$, $\bar{u}_3$, $\bar{C}$, $\bar{p}$, $\bar{\rho}$, $\bar{T}$), we need 12 new equations for the mean turbulent fluxes of the fluctuations of the concentration, velocities, and potential temperature. With the use of the equations for fluctuating quantities (Eqs. 6.32 to 6.35 and 6.27), it can be shown that the 12 equations for the rate of change of the mean turbulent fluxes of fluctuating concentration, velocity, and potential temperature take the forms of

$$\frac{\partial}{\partial t}\overline{u_j'C'} + \bar{u}_k \frac{\partial \overline{u_j'C'}}{\partial x_k} = - \frac{\partial}{\partial x_k}\overline{u_j'u_k'C'} - \overline{u_j'u_k'}\frac{\partial \bar{C}}{\partial x_k} - \overline{u_k'C'}\frac{\partial \bar{u}_j}{\partial x_k} - \frac{1}{\rho_b}\frac{\partial}{\partial x_j}\overline{p'C'} + \frac{1}{\rho_b}\overline{p'\frac{\partial C'}{\partial x_j}} + \frac{g}{T_b}\overline{\theta'C'}\delta_{3j} - \epsilon_{jki} f_k \overline{u'C'} + D\overline{u_j'\frac{\partial^2 C'}{\partial x_k \partial x_k}} + \nu\overline{C'\frac{\partial u_j'}{\partial x_k \partial x_k}} \tag{6.36}$$

$$\frac{\partial}{\partial t}\overline{u_j'u_i'} + \bar{u}_k \frac{\partial \overline{u_j'u_i'}}{\partial x_k} = - \frac{\partial}{\partial x_k}\overline{u_i'u_j'u_k'} - \overline{u_j'u_k'}\frac{\partial \bar{u}_i}{\partial x_k} - \overline{u_k'u_i'}\frac{\partial \bar{u}_j}{\partial x_k} + \frac{1}{\rho_b}\left(\overline{p'\frac{\partial u_j'}{\partial x_i} + p'\frac{\partial u_i'}{\partial x_j}}\right) - \frac{1}{\rho_b}\left(\frac{\partial}{\partial x_i}\overline{u_j'p'} + \frac{\partial}{\partial x_j}\overline{u_i'p'}\right) + \frac{g}{T_b}\left(\overline{u_j'\theta'}\delta_{3i} + \overline{u_i'\theta'}\delta_{3j}\right) - \epsilon_{j\ell m} f_\ell \overline{u_i'u_m'} - \epsilon_{i\ell m} f_\ell \overline{u_i'u_m'} + \nu\left(\overline{u_j'\frac{\partial^2 u_i'}{\partial x_k \partial x_k} + u_i'\frac{\partial^2 u_j'}{\partial x_k \partial x_k}}\right) \tag{6.37}$$

$$\frac{\partial}{\partial t}\overline{u_j'\theta'} + \bar{u}_k \frac{\partial \overline{u_j'\theta'}}{\partial x_k} = - \frac{\partial}{\partial x_k}\overline{u_k'u_j'\theta'} - \overline{u_j'u_k'}\frac{\partial \bar{\theta}}{\partial x_k} - \overline{u_k'\theta'}\frac{\partial \bar{u}_j}{\partial x_k} + \frac{g}{T_b}\overline{\theta'^2}\delta_{3j} - \frac{1}{\rho_b}\frac{\partial}{\partial x_j}\overline{p'\theta'} + \frac{1}{\rho_b}\overline{p'\frac{\partial \theta'}{\partial x_j}} - \epsilon_{jki} f_k \overline{u_i'\theta'} + \hat{\kappa}\overline{u_i'\frac{\partial^2 \theta'}{\partial x_k \partial x_k}} + \gamma\overline{\theta'\frac{\partial u_j'}{\partial x_k \partial x_k}} \tag{6.38}$$

$$\frac{\partial \overline{\theta'^2}}{\partial t} + \bar{u}_k \frac{\partial \overline{\theta'^2}}{\partial x_k} = - \frac{\partial}{\partial x_k}\overline{u_k'\theta'^2} - 2\overline{u_k'\theta'}\frac{\partial \bar{\theta}}{\partial x_k} - 2\hat{\kappa}\overline{\frac{\partial \theta'}{\partial x_k}\frac{\partial \theta'}{\partial x_k}} + \hat{\kappa}\frac{\partial^2 \overline{\theta'^2}}{\partial x_k \partial x_k} \tag{6.39}$$

$$\frac{\partial \overline{C'^2}}{\partial t} + \bar{u}_k \frac{\partial \overline{C'^2}}{\partial x_k} = - \frac{\partial}{\partial x_k}\overline{u_k'C'^2} - 2\overline{u_k'C'}\frac{\partial \bar{C}}{\partial x_k} - 2D\overline{\frac{\partial C'}{\partial x_k}\frac{\partial C'}{\partial x_k}} + D\frac{\partial^2 \overline{C'^2}}{\partial x_k \partial x_k} \tag{6.40}$$

$$\frac{\partial}{\partial t}\overline{\theta'C'} + \bar{u}_k \frac{\partial \overline{\theta'C'}}{\partial x_k} = - \frac{\partial}{\partial x_k}\overline{u_k'\theta'C'} - \overline{u_j'C'}\frac{\partial \bar{\theta}}{\partial x_j} - \overline{u_j'\theta'}\frac{\partial \bar{C}}{\partial x_j} + \hat{\kappa}\overline{C'\frac{\partial^2 \theta'}{\partial x_k \partial x_k}} + D\overline{\theta'\frac{\partial^2 C'}{\partial x_k \partial x_k}} \tag{6.41}$$

$$\overline{p'C'} = \rho_b R\, \overline{\theta'C'} \tag{6.42}$$

$$\overline{p'\theta'} = \rho_b R\, \overline{\theta'^2} \tag{6.43}$$

Now that we have a set of 24 equations for 24 mean quantities: $\overline{u_1}$, $\overline{C}$, $\overline{p}$, $\overline{\rho}$, $\overline{T}$, $\overline{u_j'C'}$, $\overline{u_i'u_j'}$, $\overline{C'\theta'}$, $\overline{u_j'\theta'}$, $\overline{\theta'^2}$, $\overline{C'^2}$, $\overline{p'C'}$ and $\overline{p'\theta'}$ with i, j = 1, 2, 3. However, in the 24 equations the following new variables appear: $\overline{u_i'u_j'u_k'}$, $\overline{u_j'u_k'C'}$, $\overline{u_j'u_k'\theta'}$, $\overline{C'^2u_k'}$, $\overline{u_k'\theta'C'}$, $\overline{u_k'\theta'^2}$. It is readily seen that when equations for the rate of change of these new variables are constructed more new variables will appear. Therefore, to solve the problem, we must close the system of equations with a certain scheme. One of the closure schemes is to express higher order correlation terms at a lower order correlation level; e.g., to express the third-order correlation $\overline{u_i'u_j'u_k'}$ in terms of the second-order correlation $\overline{u_k'u_i'}$. In recent years a number of turbulence closure theories have been developed. Applications that have been made to the atmosphere's boundary layers can be found in the works of Mellor (1973), Wyngaard, Coté, and Rao (1974), Launder (1975), Yamada and Mellor (1975), Lewellen and Teske (1976), André et al. (1976), Zeman and Lumley (1976), and Freeman (1977). Some of the closure schemes and applications are presented in the next several sections.

6-2.3 First-Order Closure Schemes and Gradient Transport Theory

One of the earliest first-order closure schemes was constructed on the concept of the mixing length model. This model is analogous to that of molecular transport of momentum; i.e., it assumes that a turbulent fluid consists of parcels or vortices of fluid that transport material, momentum, and heat from one location to another. In the process of transport, the fluid parcels or vortices retain their properties over a certain length before being mixed with the surrounding fluid (Boussinesq, 1877; Taylor, 1915; Prandtl, 1925).

To analyze the diffusion process, we first consider the turbulent transport of a physical quantity η in the one-dimensional case and assume that the time mean of the physical quantity is a function of the vertical coordinate z. Consider a fluid parcel consisting of a coherent conglomerate of fluid particles with a size of dissipation scale determined by the coefficient of kinematic viscosity ν and the rate of dissipation ϵ. Let this parcel be originated at level $z - \ell_{\eta z}$ carrying the physical quantity equal to the mean of the quantity at that level, namely, $\overline{\eta}(z - \ell_{\eta z})$. Assuming that this fluid parcel conserves its property during its travel to level z, we have the fluctuation in η at level z

$$\eta'(z) = \overline{\eta}(z - \ell_{\eta z}) - \overline{\eta}(z)$$

$$= - \ell_{\eta z} \frac{\partial \overline{\eta}}{\partial z} + \frac{1}{2} \ell_{\eta z}^2 \frac{\partial^2 \overline{\eta}}{\partial z^2} - + \ldots$$

$$= - \ell_{\eta x} \frac{\partial \overline{\eta}}{\partial z} \left(1 - \frac{1}{2} \frac{\ell_{\eta z}}{L_{\eta z}} + - \ldots\right) \qquad (6.44)$$

where

$$L_{\eta z} = \frac{\dfrac{\partial \overline{\eta}}{\partial z}}{\dfrac{\partial^2 \overline{\eta}}{\partial z^2}} \qquad (6.45)$$

is a characteristic length of the mean η-field. It is seen that only if $\ell_{\eta z} \ll L_{\eta z}$ can fluctuation η' be approximated by

$$\eta'(z) \approx - \ell_{\eta z} \frac{\partial \overline{\eta}}{\partial z}$$

where $\ell_{\eta z}$ is z component of the mixing length for η. Therefore the mean turbulent flux of η can be written

$$\overline{u_j'\eta'} = - \overline{w'\ell_{\eta z}} \frac{\partial \overline{\eta}}{\partial z}$$

Defining the vertical component of the eddy diffusivity of

$$K_{\eta z} = \overline{w'\ell_{\eta z}}$$

we can express the mean turbulent flux of η as

$$\overline{u_j'\eta'} = - K_{\eta z} \frac{\partial \overline{\eta}}{\partial z} \qquad (6.46)$$

which states that the mean turbulent flux of η is in the direction of the negative gradient of mean η.

Now consider the three-dimensional case. In this case, $\overline{\eta} = \overline{\eta}(x,t)$. Let $\ell_{\eta k}$ be the k component of the mixing length vector for η. If

$$\ell_{\eta k} \ll \frac{\dfrac{\partial \overline{\eta}}{\partial x_k}}{\dfrac{\partial^2 \overline{\eta}}{\partial x_k^2}} \qquad k = 1, 2, 3$$

then the fluctuation in η can be expressed

$$\eta' \cong - \ell_{\eta k} \frac{\partial \overline{\eta}}{\partial x_k}$$

and the mean turbulent flux of η can be written as

$$\overline{u_j'\eta'} = - \overline{u_j'\ell_{\eta k}} \frac{\partial \overline{\eta}}{\partial x_k}$$

$$= - K_{\eta jk} \frac{\partial \overline{\eta}}{\partial x_k} \qquad (6.47)$$

where

$$K_{\eta jk} \equiv \overline{u_j'\ell_{\eta k}} \qquad (6.48)$$

is the jk components of the eddy diffusivity of η.

With the use of this first-order closure scheme, the prediction equation for the mean concentration (Eq. 6.28), momentum (Eq. 6.29), and potential temperature (Eq. 6.30) can be written as

$$\frac{\partial \bar{C}}{\partial t} + \bar{u}_j \frac{\partial \bar{C}}{\partial x_j} = D \frac{\partial^2 \bar{C}}{\partial x_j \partial x_j} + \frac{\partial}{\partial x_j}\left(K_{Cjk} \frac{\partial \bar{C}}{\partial x_k}\right) + \bar{Q} \quad (6.49)$$

$$\frac{\partial \bar{u}_i}{\partial t} + \bar{u}_j \frac{\partial \bar{u}_i}{\partial x_j} = -\frac{1}{\rho_b}\frac{\partial \bar{p}}{\partial x_i} + \bar{\theta}\frac{g}{T_b}\delta_{3i} - \epsilon_{ijk} f_j \bar{u}_k + \nu \frac{\partial^2 \bar{u}_i}{\partial x_j \partial x_j} + \frac{\partial}{\partial x_j}\left(K_{mjk}\frac{\partial \bar{u}_i}{\partial x_k}\right) \quad (6.50)$$

$$\frac{\partial \bar{\theta}}{\partial t} + \bar{u}_j \frac{\partial \bar{\theta}}{\partial x_j} = \hat{\kappa} \frac{\partial^2 \bar{\theta}}{\partial x_j \partial x_j} + \frac{\partial}{\partial x_j}\left(K_{hjk}\frac{\partial \bar{\theta}}{\partial x_k}\right) + \frac{\bar{S}}{\rho c_v} \quad (6.51)$$

where K_C, K_m, and K_h are, respectively, the coefficients of the eddy diffusivity for concentration, momentum, and heat. Note that Eqs. 6.47 to 6.51, together with the equations of continuity and state, $\partial \bar{u}_i/\partial x_i = 0$ and $\bar{p} = R\rho_b\bar{\theta}$, form a closed system if the eddy diffusivities K_{Cjk}, K_{mij}, and K_{hjk} are known functions of the time mean variables, space, and time. However, the definition of eddy diffusivity (Eq. 6.48) does not offer such information. To seek a relation between the eddy diffusivity and the time mean variables, we must analyze the characteristics of the eddy diffusivity. Because turbulent diffusion is a consequence of turbulent motion, eddy diffusivity must be closely associated with the Reynolds stress tensor $\overline{u_j' u_k'}$. On the basis of this reasoning, it has been proposed that the mixing length, $\ell_{\eta k}$, be expressed as

$$\ell_{\eta k} = (t_\eta' u')_{Lk} \quad (6.52)$$

where u_{Lk}' is the k component of the Lagrangian turbulent velocity, and $t_{\eta Lk}'$ is the time scale for the mixing length $\ell_{\eta k}$. For small $t_{\eta Lk}'$, such that u_{LK}' remains unchanged, the mixing length $\ell_{\eta k}$ can be approximated by the product of the Eulerian turbulent velocity u_k' and time scale $t_{\eta k}'$. Thus

$$\ell_{\eta k} = (t_\eta' u')_k \quad (6.53)$$

and the eddy diffusivity can be written

$$K_{\eta jk} = \overline{u_j'(t_\eta' u')_k} \quad (6.54)$$

Therefore the eddy diffusivity can be expressed in terms of the Reynolds stress and a characteristic time.

Notice that, in Eq. 6.54, $K_{\eta jk}$ has the following nine components:

$$\begin{bmatrix} \overline{(t_\eta' u')_1 u_1'} & \overline{(t_\eta' u')_2 u_1'} & \overline{(t_\eta' u')_3 u_1'} \\ \overline{(t_\eta' u')_1 u_2'} & \overline{(t_\eta' u')_2 u_2'} & \overline{(t_\eta' u')_3 u_2'} \\ \overline{(t_\eta' u')_1 u_3'} & \overline{(t_\eta' u')_2 u_3'} & \overline{(t_\eta' u')_3 u_3'} \end{bmatrix}$$

Since, in general,

$$\overline{(t_\eta' u')_k u_j'} \neq \overline{(t_\eta' u')_j u_k'}$$

$$K_{\eta kj} \neq K_{\eta jk}$$

Therefore, unlike the Reynolds stress which has six components, the coefficient of eddy diffusivity $K_{\eta jk}$ generally has nine components. Diffusion equation (Eq. 6.49) can be written as

$$\begin{aligned} \frac{\partial \bar{C}}{\partial t} + \bar{u}_1 \frac{\partial \bar{C}}{\partial x_1} + \bar{u}_2 \frac{\partial \bar{C}}{\partial x_2} + \bar{u}_3 \frac{\partial \bar{C}}{\partial x_3} &= D\left(\frac{\partial^2}{\partial x_1^2} + \frac{\partial^2}{\partial x_2^2} + \frac{\partial^2}{\partial x_3^2}\right)\bar{C} \\ &+ \frac{\partial}{\partial x_1}\left(K_{C11}\frac{\partial \bar{C}}{\partial x_1}\right) + \frac{\partial}{\partial x_2}\left(K_{C22}\frac{\partial \bar{C}}{\partial x_2}\right) \\ &+ \frac{\partial}{\partial x_3}\left(K_{C33}\frac{\partial \bar{C}}{\partial x_3}\right) + \frac{\partial}{\partial x_1}\left(K_{C12}\frac{\partial \bar{C}}{\partial x_2}\right) \\ &+ \frac{\partial}{\partial x_2}\left(K_{C21}\frac{\partial \bar{C}}{\partial x_1}\right) + \frac{\partial}{\partial x_2}\left(K_{C23}\frac{\partial \bar{C}}{\partial x_3}\right) \\ &+ \frac{\partial}{\partial x_3}\left(K_{C32}\frac{\partial \bar{C}}{\partial x_2}\right) + \frac{\partial}{\partial x_1}\left(K_{C13}\frac{\partial \bar{C}}{\partial x_3}\right) \\ &+ \frac{\partial}{\partial x_3}\left(K_{C31}\frac{\partial \bar{C}}{\partial x_1}\right) + \bar{Q} \end{aligned} \quad (6.55)$$

Instead of deriving diffusion Eq. 6.55 through the analysis of the effect of Reynolds stress on the diffusion process as shown above, Hinze (1959) and Calder (1965) obtained a similar equation from the viewpoint of tensor structure.

To analyze the effect of shear flow on turbulent diffusion, let the turbulent velocity be decomposed into two components: a component of the intrinsic turbulent velocity denoted by u_b', which is defined as the turbulent velocity in a uniform, constant mean flow, and a component generated by the velocity shear of the mean flow, denoted by u_s' (Kao, 1976a). Thus the i component of the turbulent velocity can be expressed as

$$u_i' = u_{bi}' + u_{si}' \quad (6.56)$$

Invoking the generalized mixing length hypothesis (von Kármán, 1930), we can express the part of the turbulent velocity generated by the velocity shear of the mean flow as

$$u_{si}' = -\ell_{um}\frac{\partial \bar{u}_i}{\partial x_m} \quad (6.57)$$

In view of Eqs. 6.52 and 6.54, Eq. 6.57 can be written as

$$u_{si}' = -\left[t_u'(u_b' + u_s')\right]_m \frac{\partial \bar{u}_i}{\partial x_m} \quad (6.58)$$

Therefore Eqs. 6.56, 6.54, and 6.49 can be written, respectively,

$$u_i' = u_b' - [t_u'(u_b' + u_s')]_m \frac{\partial \bar{u}_i}{\partial x_m} \tag{6.59}$$

$$K_{\eta jk} = \overline{(u_b' + u_s')_j [t_\eta'(u_b' + u_s')]_k} \tag{6.60}$$

$$\frac{\partial \bar{C}}{\partial t} + \bar{u}_j \frac{\partial \bar{C}}{\partial x_j} = \frac{\partial}{\partial x_j}\left\{\overline{(u_b' + u_s')_j [t_C'(u_b' + u_s')]_k} \frac{\partial \bar{C}}{\partial x_k}\right\} + D \frac{\partial^2 \bar{C}}{\partial x_j \partial x_j} + \bar{Q} \tag{6.61}$$

Notice that u_{si}' can generally be expressed in terms of t', u_b', and $\bar{u}_i/\partial x_m$ by solving the set of equations (Eq. 6.58) for $i = 1, 2, 3$, and by specifying that $\overline{u_{bi}' u_{bj}'} = 0$ for $i \neq j$. Therefore K_{Cjk} can be expressed in terms of the variance of the intrinsic turbulent velocities, the velocity shear of the mean flow, and the characteristic time. Since the magnitude of the molecular diffusivity is much smaller than that of the eddy diffusivity, the terms involving the molecular diffusion in Eq. 6.61 can be neglected in the later analysis.

The equations for mean momentum (Eq. 6.50) and thermal energy (Eq. 6.51) can be written as

$$\frac{\partial \bar{u}_i}{\partial t} + \bar{u}_j \frac{\partial \bar{u}_i}{\partial x_j} = -\frac{1}{\rho_b}\frac{\partial \bar{p}}{\partial x_i} + \frac{g}{T_b}\bar{\theta}\delta_{3i} - \epsilon_{ijk} f_j \bar{u}_k + \nu \frac{\partial^2 \bar{u}_i}{\partial x_j \partial x_j} + \frac{\partial}{\partial x_j}\left\{\overline{(u_b' + u_s')_j [t_m'(u_b' + u_s')]_k} \frac{\partial \bar{u}_i}{\partial x_k}\right\} \tag{6.62}$$

and

$$\frac{\partial \bar{\theta}}{\partial t} + \bar{u}_j \frac{\partial \bar{\theta}}{\partial x_j} = \hat{\kappa} \frac{\partial^2 \bar{\theta}}{\partial x_j \partial x_j} + \frac{\partial}{\partial x_j}\left\{\overline{(u_b' + u_s')_j [t_\theta'(u_b' + u_s')]_k} \frac{\partial \bar{\theta}}{\partial x_k}\right\} + \frac{\bar{S}}{\rho c_v} \tag{6.63}$$

respectively.

6-2.3.1 Gradient Transfer Diffusion in a Uniform Mean Flow. Transport and turbulent diffusion in a uniform mean flow is the simplest diffusion model. This model yields a Gaussian solution and has been widely used to estimate the pollution distribution in the atmosphere.

For uniform mean flow, $u_{si}' = 0$ in Eq. 6.58. For convenience of analysis, let the x-axis be parallel to the mean flow of which the mean velocity $\bar{u} = U =$ constant. In so doing, the coordinate axes are parallel to the principal axes of the stress tensor of the turbulent motion. Therefore $\overline{u_i' u_j'} = 0$ for $i \neq j$, and Eq. 6.61 becomes

$$\frac{\partial \bar{C}}{\partial t} + U \frac{\partial \bar{C}}{\partial x} = \frac{\partial}{\partial x}\left(t_{Cx}' \overline{u_b'^2} \frac{\partial \bar{C}}{\partial x}\right) + \frac{\partial}{\partial y}\left(t_{Cy}' \overline{v_b'^2} \frac{\partial \bar{C}}{\partial y}\right) + \frac{\partial}{\partial z}\left(t_{Cz}' \overline{w_b'^2} \frac{\partial \bar{C}}{\partial z}\right) + \bar{Q} \tag{6.64}$$

where u, v, w are, respectively, the x, y, z components of the velocity. Since in a uniform mean flow the variances of the turbulent velocities are independent of position, they can be taken out of the differential signs so that Eq. 6.64 can be written as

$$\frac{\partial \bar{C}}{\partial t} + U \frac{\partial \bar{C}}{\partial x} = K_{xx} \frac{\partial^2 \bar{C}}{\partial x^2} + K_{yy} \frac{\partial^2 \bar{C}}{\partial y^2} + K_{zz} \frac{\partial^2 \bar{C}}{\partial z^2} + \bar{Q} \tag{6.65}$$

where $K_{xx} = t_{Cx}' \overline{u_b'^2}$, $K_{yy} = t_{Cy}' \overline{v_b'^2}$, and $K_{zz} = t_{Cz}' \overline{w_b'^2}$ are, respectively, the x, y, and z components of the eddy diffusivity. Equation 6.65, called the "gradient transfer diffusion equation" or "K-theory diffusion equation," was first investigated independently by Taylor (1915) and Schmidt (1925). It may be noticed that the diffusion equation (Eq. 6.65) is parabolic.

For diffusion from an instantaneous point source of source strength Q(gm) released initially at point (x_0, y_0, z_0) in an infinite medium, Eq. 6.65 has a solution of the form

$$\bar{C}(x,y,z,t) = \frac{\bar{Q}}{(2\pi t)^{3/2}(K_{xx}K_{yy}K_{zz})^{1/2}} \times \exp\left[-\frac{(x - x_0 - Ut)^2}{2K_{xx}t} - \frac{(y - y_0)^2}{2K_{yy}t} - \frac{(z - z_0)^2}{2K_{zz}t}\right] \tag{6.66}$$

Equation 6.66 indicates that the center of mass of the marked fluid particles would move at the speed of the mean flow and that the distribution of concentration of a trace constituent is Gaussian. One of the shortcomings of a parabolic type diffusion equation is its prediction that the trace constituent in the fluid will appear instantaneously everywhere (though not equally) if the trace constituent is introduced at some point in the fluid. Since such instantaneous propagation of particles is impossible, we must assume that the diffusion equation is correct only after a sufficiently long time has elapsed. This assumption is discussed further in Sec. 6-2.5.

In the presence of a plane boundary located at $z = 0$, because of the complete reflection of the marked fluid particles at the plane, the solution of Eq. 6.65 for diffusion from an instantaneous point source initially released at (x_0, y_0, z_0) is

$$\bar{C}(x,y,z,t) = \frac{\bar{Q}}{(2\pi t)^{3/2}(K_{xx}K_{yy}K_{zz})^{1/2}} \times \exp\left[-\frac{(x - x_0 - Ut)^2}{2K_{xx}t} - \frac{(y - y_0)^2}{2K_{yy}t}\right] \times \left\{\exp\left[-\frac{(z - z_0)^2}{2K_{zz}t}\right] + A\exp\left[-\frac{(z + z_0)^2}{2K_{zz}t}\right]\right\} \tag{6.67}$$

where A is a coefficient of absorption at the ground.

For diffusion in an infinite medium from a continuous fixed point source of strength $\overline{Q}$ (g/s) located at the origin of the coordinate system, Eq. 6.65 has a steady-state solution of the form

$$\overline{C}(x,y,z) = \frac{\overline{Q}}{4\pi(K_{yy}K_{zz}x^2 + K_{zz}K_{xx}y^2 + K_{xx}K_{yy}z^2)^{1/2}}$$

$$\times \exp\left\{-\frac{U}{2K_{xx}}\left[\frac{1}{K_{xx}^{1/2}}\left(\frac{x^2}{K_{xx}} + \frac{y^2}{K_{yy}} + \frac{z^2}{K_{zz}}\right)^{1/2} - x\right]\right\} \quad (6.68)$$

When $x \gg (y^2 + z^2)^{1/2}$, the preceding solution reduces to

$$\overline{C}(x,y,z) = \frac{\overline{Q}}{4\pi(K_{yy}K_{zz})^{1/2}x}\exp\left[-\frac{U}{4x}\left(\frac{y^2}{K_{yy}} + \frac{z^2}{K_{zz}}\right)\right] \quad (6.69)$$

In the presence of a plane boundary located at $z = 0$, the concentration distribution for diffusion from a continuous point source located at (0,0,H) is therefore

$$\overline{C}(x,y,z) = \frac{\overline{Q}}{4\pi(K_{yy}K_{zz})^{1/2}x}\exp\left(-\frac{Uy^2}{4K_{yy}x}\right)$$

$$\times \left\{\exp\left[-\frac{U(z-H)^2}{4K_{zz}x}\right] + \exp\left[-\frac{U(z+H)^2}{4K_{zz}x}\right]\right\} \quad (6.70)$$

For diffusion from an infinite line source on the y-axis in an infinite medium, the solution to Eq. 6.68 takes the form of

$$\overline{C}(x,z) = \frac{\overline{Q}}{2\pi(K_{xx}K_{zz})^{1/2}}K_0\left[\frac{U}{2}\left(\frac{x^2}{K_{xx}^2} + \frac{z^2}{K_{xx}^2}\right)^{1/2}\right]$$

$$\times \exp\left(\frac{Ux}{2K_{xx}}\right) \quad (6.71)$$

where K_0 is the modified Bessel function of the third kind.

When $x \gg z$ and $Ux \gg K_{xx}$, the preceding solution reduces to (e.g., Hinze, 1959)

$$\overline{C}(x,z) = \frac{\overline{Q}}{2(\pi K_{zz}Ux)^{1/2}}\exp\left(-\frac{Uz^2}{4K_{zz}x}\right) \quad (6.72)$$

for a uniform wind field.

6-2.3.2 Generalized Gradient Transfer Diffusion Equation in a Uniform Mean Flow. From the viewpoint of turbulence, the rate of dispersion of a cluster of particles generally depends on the cluster size and turbulence characteristics. As the cluster expands, larger turbulence eddies become effective in the dispersion. Since the larger eddies contain more energy, the rate of dispersion would increase with dispersion time; therefore turbulent dispersion is scale dependent. The generalization of the diffusivity concept is based on the hypothesis that the diffusivity is dependent on the wave vector k of the concentration field. Thus the spectral turbulent diffusivity coefficient K(k) can be defined as a function of the concentration wave vector (Berkowicz and Prahm, 1979).

Consider the case of one-dimensional diffusion of a material of concentration C(y,t) which may be expressed in Fourier series,

$$C(y,t) = \sum_{k=-\infty}^{\infty} C_k(y,t) \quad (6.73)$$

where

$$C_k(y,t) = A(k,t)e^{iky} \quad (6.74)$$

and C_k represents the Fourier components of the concentration mode $k = \pi n/L$, $n = \pm 0, 1, \ldots, \pm\infty$, and 2L is the length of the region to which the diffusion is limited, i.e.,

$$\frac{\partial}{\partial y}C(y,t) = 0 \quad (6.75)$$

at $y = \pm L$.

By analogy to the diffusion equation in gradient transfer theory, we assumed for each Fourier mode

$$\frac{\partial}{\partial t}C_k = K(k)\frac{\partial^2}{\partial y^2}C_k \quad (6.76)$$

Thus

$$C_k(y,t) = C_k(y,0)\,e^{-k^2K(k)t} \quad (6.77)$$

Note that the diffusion equation for the total concentration field cannot be written in a form similar to Eq. 6.76 where K(k) is not constant. However, if we introduce a turbulent diffusivity transfer function defined as

$$F(y,y') = \frac{1}{2L}\sum_{k=-\infty}^{\infty} K(k)\,e^{ik(y-y')} \quad (6.78)$$

it can be shown that the following equation describes the turbulent diffusion in terms of the spectral diffusivity coefficients

$$\frac{\partial}{\partial t}C(y,t) = \frac{\partial}{\partial y}\int_{-L}^{L} F(y,y')\frac{\partial}{\partial y'}C(y',t)\,dy' \quad (6.79)$$

Equation 6.79 is the generalized gradient transfer diffusion equation. The integral term on the right-hand side, taken with the minus sign, is the turbulent diffusivity flux which depends not only on local values of concentration gradients but also on values at remote points.

With the use of Eqs. 6.79, 6.73, and 6.74 and in view of the orthogonality relation

$$\int_{-L}^{L} e^{i(k-m)y}\,dy = \begin{cases} 2L & \kappa = m \\ 0 & k \neq m \end{cases} \quad (6.80)$$

we obtain

$$\sum_k \frac{\partial}{\partial t} C_k = \sum_k K(k) \frac{\partial^2}{\partial y^2} C_k \qquad (6.81)$$

which is equivalent to Eq. 6.76.

Note that, in the case where $K(k) = K_0$ = constant for all k, it follows that $F(y,y') = K_0 \delta(y - y')$ and Eq. 6.76 reduces to

$$\frac{\partial}{\partial t} C(y,t) = K_0 \frac{\partial^2}{\partial y^2} C(y,t) \qquad (6.82)$$

and, for an instantaneous point source of strength Q and located at y = 0, the solution to Eq. 6.82 is

$$C(y,t) = \frac{Q}{2L}\left[1 + 2\sum_{k=1}^{\infty} e^{-k^2 K_0 t} \cos ky\right] \qquad (6.83)$$

If $L \to \infty$, the summation in Eq. 6.83 can be replaced by integration with respect to k by

$$\frac{1}{2L}\sum_n \to \frac{1}{2\pi}\int dk$$

As a consequence, we obtain the well-known Gaussian solution

$$C(y,t) = \frac{Q}{\pi^{1/2}\sigma} \exp\left(-\frac{y^2}{2\sigma^2}\right) \qquad (6.84)$$

where $\sigma^2 = 2K_0 t$ is the solution for K theory.

To estimate the diffusivity coefficient K(k) for a cluster of particles of small dimension (10 cm to 600 m), we can express the standard deviation of its linear dimension as (Pasquill, 1974)

$$\sigma = \alpha \epsilon^{1/2} t^{3/2} \qquad (6.85)$$

where α is a dimensionless constant of the order of unity and ϵ is the rate of dissipation of turbulent kinetic energy. It follows from the definition of eddy-diffusivity coefficient that

$$K = \frac{1}{2}\frac{d\sigma^2}{dt} \qquad (6.86)$$

and from Eq. 6.85 that

$$K = \frac{3}{2}\alpha^{2/3}\epsilon^{1/3}\ell^{4/3} \qquad (6.87)$$

where $\ell = \sigma$ is the typical scale of the turbulence. To satisfy the condition that $K \to K_0$ as $k \to 0$, Eq. 6.87 can be expressed as (see Berkowicz and Prahm, 1979)

$$K(k) = \frac{K_0}{1 + Dk^{4/3}} \qquad (6.88)$$

where D is a constant specifying the width of the spectral diffusivity function and $D \sim \ell_m^{4/3}$, where ℓ_m is the size of the energy-containing eddy.

Using Eq. 6.88, $F(y,y')$ computed from Eq. 6.78 for D = 1.5 and 3.0 is shown in Fig. 6.1. The length of the computational region is 2L = 32 units. In the case of D = 0, $F(y,y')$ behaves like a δ distribution with a zero width, and the corresponding distribution becomes Gaussian. Using Eq. 6.88, we can express concentration field as

$$C(y,t) = \frac{Q_0}{2L}\left[1 + 2\sum_{k=1}^{\infty} \exp\left(-\frac{K_0}{1 + Dk^{4/3}} k^2 t\right) \cos ky\right] \qquad (6.89)$$

The concentration distribution computed with the use of Eq. 6.89 for D = 0, 1.5, and 3.0 is shown in Fig. 6.2. Note that, at the initial stage, the spread of the cloud with D = 1.5 and 3.0 is considerably smaller than that of the Gaussian (D = 0) profile. However, as diffusion time increases, the concentration distribution approaches the Gaussian distribution. Measurements of the optical density of a smoke plume viewed from space (Randerson, Garcia, and Whitehead, 1971) show concentration profiles similar to Fig. 6.2.

6-2.3.3 Transport and Turbulent Diffusion in a Linear Shear Flow. In the atmosphere and in the ocean, the mean velocity is rarely uniform. In fact, velocity shear generally exists in the boundary layer. There are two main physical effects of shear flows on turbulent diffusion: (1) the turbulent motion mixes the trace constituent with the fluid, and (2) the velocity differences of the mean flow increase the distance between particles and thus increase the size of effective turbulence eddies. Since larger eddies contain more kinetic energy, the rate of diffusion increases with increasing diffusion time.

As a first approximation to the diffusion in the atmospheric surface boundary layer, we consider the diffusion in a linear shear flow.

Let $\bar{u} = Az$, $\bar{v} = 0$, $\bar{w} = 0$, where $A = \partial\bar{u}/\partial z$ is the vertical velocity shear of the mean flow. In this case, the x, y, z components of the turbulent velocity can be expressed as Eq. 6.56

$$u' = u_b' - At_z' w_b'$$

$$v' = v_b'$$

$$w' = w_b'$$

and the diffusion equation, Eq. 6.55, can be written (Kao and Doran, 1976)

$$\frac{\partial \bar{C}}{\partial t} + Az\frac{\partial \bar{C}}{\partial x} = K_{xx}\frac{\partial^2 \bar{C}}{\partial x^2} + K_{yy}\frac{\partial^2 \bar{C}}{\partial y^2} + K_{zz}\frac{\partial^2 \bar{C}}{\partial z^2} - K_{xz}\frac{\partial^2 \bar{C}}{\partial x \partial z} + \bar{Q} \qquad (6.90)$$

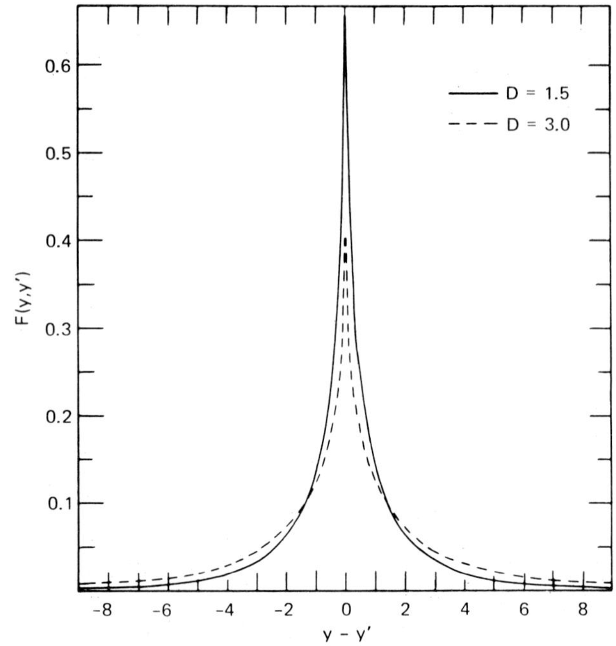

Fig. 6.1 The turbulent diffusivity transfer function. Equation 6.88 is used for the spectral diffusivity coefficients. F(y,y') is in units of $K_0/\Delta y$, where Δy is the length unit. [From R. Berkowicz and L. P. Prahm, Generalization of K Theory for Turbulent Diffusion. Part 1. Spectral Turbulent Diffusivity Concept, *Journal of Applied Meteorology*, 36: 266-272 (1979).]

where

$$
\begin{aligned}
K_{xx} &= t'_x \left(\overline{u_b'^2} + t_z'^2\, \overline{w_b'^2}\, A^2\right) \\
K_{yy} &= t'_y \overline{v_b'^2} \\
K_{zz} &= t'_z \overline{w_b'^2} \\
K_{xz} &= t'_z\, (t'_x + t'_z)\, \overline{w_b'^2}\, A = (t'_x + t'_z)\, K_{zz} A
\end{aligned}
\qquad (6.91)
$$

It will be shown that the term involving K_{xz}, which arises from $\overline{u'w'}$ in the linear shear flow, would affect the concentration distribution near the source.

Equation 6.90 can be cast in a dimensionless form by defining a characteristic time $T = A^{-1}$ and a characteristic length $L = K_{zz}^{1/2} A^{-1/2}$. Let the dimensionless time and coordinates be $\hat{t} = t/T$, $\hat{x} = x/L$, $\hat{y} = y/L$, and $\hat{z} = z/L$, and let

$$
\begin{aligned}
\beta &= \frac{K_{xx}}{K_{zz}} \\
\gamma &= \frac{K_{yy}}{K_{zz}} \\
\alpha &= \frac{\overline{u'^2}}{\overline{w'^2}} = \frac{\overline{u_b'^2} + t_z'^2\, \overline{w_b'^2}\, A^2}{\overline{w_b'^2}} \\
R_{u'w'} &= \frac{\overline{u'w'}}{\left(\overline{u'^2}\;\overline{w'^2}\right)^{1/2}} = -A\left(\frac{K_{zz}}{K_{xx}}\, t'_x t'_z\right)^{1/2}
\end{aligned}
\qquad (6.92)
$$

so that Eq. 6.90 can be written in dimensionless form as

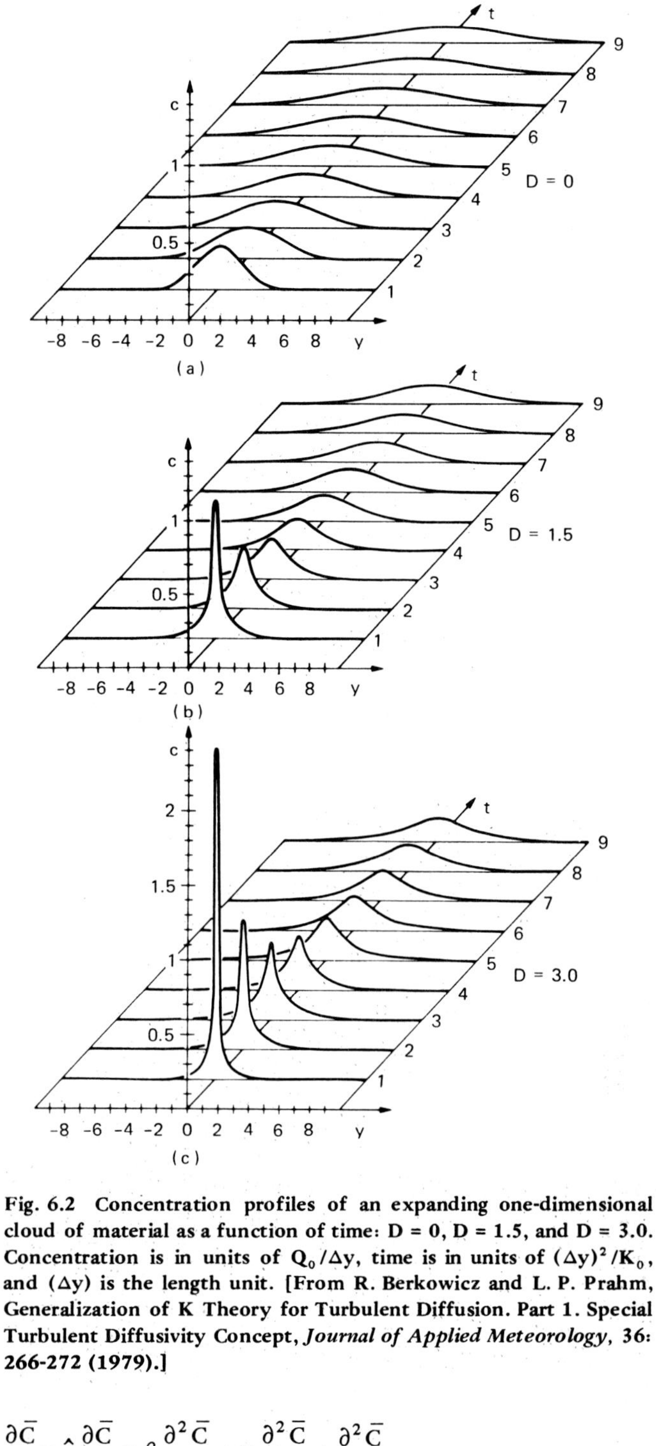

Fig. 6.2 Concentration profiles of an expanding one-dimensional cloud of material as a function of time: D = 0, D = 1.5, and D = 3.0. Concentration is in units of $Q_0/\Delta y$, time is in units of $(\Delta y)^2/K_0$, and (Δy) is the length unit. [From R. Berkowicz and L. P. Prahm, Generalization of K Theory for Turbulent Diffusion. Part 1. Special Turbulent Diffusivity Concept, *Journal of Applied Meteorology*, 36: 266-272 (1979).]

$$
\frac{\partial \bar{C}}{\partial \hat{t}} = \hat{z}\,\frac{\partial \bar{C}}{\partial \hat{x}} = \beta\,\frac{\partial^2 \bar{C}}{\partial \hat{x}^2} + \gamma\,\frac{\partial^2 \bar{C}}{\partial \hat{y}^2} + \frac{\partial^2 \bar{C}}{\partial \hat{z}^2}
- \alpha^{1/2}\left(1 + \frac{\beta}{\alpha}\right)\left|R_{u'w'}\right| \frac{\partial^2 \bar{C}}{\partial \hat{x}\,\partial \hat{z}} + \frac{\bar{Q}}{A} \qquad (6.93)
$$

For diffusion of an instantaneous point source of source strength $\hat{Q}$ released at the origin of the coordinate system at t = 0

$$
\bar{C}(\hat{x},\hat{y},\hat{z},0) = \hat{Q}\,\delta(\hat{x})\,\delta(\hat{y})\,\delta(\hat{z}) \qquad (6.94)
$$

Defining the Fourier transform of $\bar{C}(\hat{x},\hat{y},\hat{z},\hat{t})$ in the space of (k,p,q) as

$$F(k,p,q,\hat{t}) = \int\int\int_{-\infty}^{+\infty} \bar{C}(\hat{x},\hat{y},\hat{z},\hat{t})\, e^{i(k\hat{x}+p\hat{y}+q\hat{z})}\, d\hat{x}\, d\hat{y}\, d\hat{z} \qquad (6.95)$$

Fourier transform of Eq. 6.93 with the use of Eqs. 6.95 and 6.94 leads to

$$\frac{\partial F}{\partial \hat{t}} - k\frac{\partial F}{\partial q} = -(\beta k^2 + \gamma p^2 + q^2 - mkq) + \hat{Q} \qquad (6.96)$$

where

$$m = \alpha^{1/2}(1 + \beta/\alpha)|R_{u'w'}|$$

and

$$F(k,p,q,0) = \hat{Q} \qquad (6.97)$$

The method of bicharacteristics yields a solution to Eqs. 6.96 and 6.92:

$$F(k,p,q,\hat{t}) = \hat{Q}\exp\left\{-\left[k^2\left(\beta\hat{t} - \frac{1}{2}m\hat{t}^2 + \frac{1}{3}\hat{t}^3\right) + p^2\gamma\hat{t} + q^2\hat{t} + kq(\gamma^2 t - m\hat{t})\right]\right\} \qquad (6.98)$$

Taking the inverse Fourier transform of Eq. 6.93, we obtain

$$\bar{C}(\hat{x},\hat{y},\hat{z},\hat{t}) = \frac{\hat{Q}}{(2\pi\hat{t})^{3/2}\gamma^{1/2}\left(\beta - \frac{1}{4}m^2 + \frac{1}{12}t^2\right)^{1/2}} \exp\left\{-\frac{\hat{y}^2}{4\gamma\hat{t}} - \frac{\hat{z}^2}{4\hat{t}} - \frac{\left[x + \frac{1}{2}(m - \hat{t})\hat{z}\right]^2}{4\hat{t}\left(\beta + \frac{1}{12}\hat{t}^2 - \frac{m^2}{4}\right)}\right\} \qquad (6.99)$$

As can be seen from Eq. 6.99, the inclusion of the m term arises from the presence of K_{xz} in Eq. 6.90.

The preceding result can be compared with that of Carter and Okubo (1965), who consider a diffusion equation similar to Eq. 6.90 but without the term involving K_{xz}. For the model being discussed here, let us use values of $R_{u'w'} = -0.3$ and $\alpha = 3.5$, based on field experiments of Powell (1974), and $\beta = 20$ and $\gamma = 10$, based on experiments of Pasquill and Gifford (Islitzer and Slade, 1968). The solid curves in Fig. 6.3 show isopleths of mean concentration normalized to the maximum concentration at the center of the plume, with t/T = 15. The major axis of the mean concentration profile forms an angle of approximately 5.2° with the mean flow. The result for $\hat{y} = 0$ is shown, although the general characteristics are the same for all values of $\hat{y}$. The dashed curves in Fig. 6.3 correspond to the solution of Carter and Okubo (1965). Therefore the effect of the xz component of the eddy diffusivity, K_{xz}, is to rotate the principal axis of diffusion from the mean flow

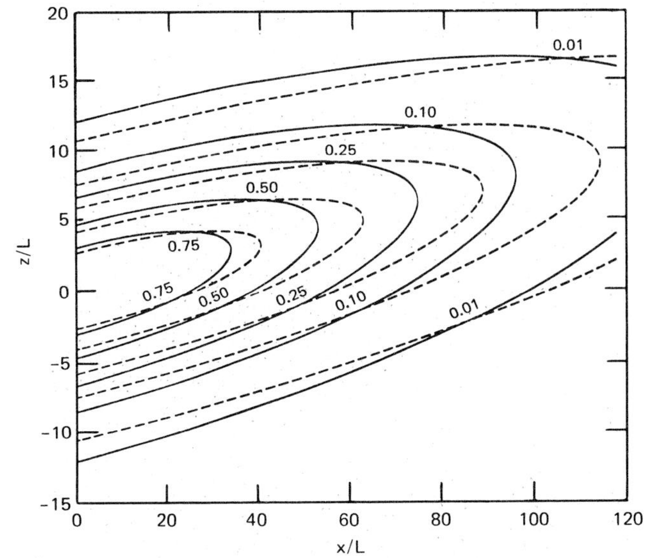

Fig. 6.3 Isopleths of normalized concentration at time t/T = 15. ——, present model; – – –, model of Carter and Okubo. [From S. K. Kao and J. C. Doran, Turbulent Diffusion in Linear Shear Flow, *Pure and Applied Geophysics*, 114: 357-363 (1976).]

upward, resulting in increasing concentration along the vertical but reducing concentration along the mean flow.

6-2.3.4 Transport and Turbulent Diffusion in the Surface Boundary Layer. Many pollution sources, such as motor vehicles and smoke stacks, are located near the earth's surface. Because of the strong velocity shear in the boundary layer, diffusion characteristics in the boundary layer are very different from those in a mean flow of uniform velocity or linear velocity shear as discussed in the previous sections.

Earlier studies of diffusion in the surface boundary layer proposed that, for diffusion in a shear flow with a mean velocity profile of the form of az^m, the eddy diffusivity would have the same form bz^n (W. G. L. Sutton, 1943; O. G. Sutton, 1953; Deacon, 1949; Davies, 1950, 1954; Rounds, 1955; F. B. Smith, 1957).

In a more general case (Yeh and Huang, 1975), for

$$\bar{u}(z) = az^m$$

$$K_{zz} = bz^n \qquad (6.100)$$

$$K_{yy} = H(x)z^p$$

the solution of diffusion equation

$$\bar{u}(z)\frac{\partial \bar{C}}{\partial z} = \frac{\partial}{\partial y}\left(K_{yy}\frac{\partial \bar{C}}{\partial y}\right) + \frac{\partial}{\partial z}\left(K_{zz}\frac{\partial \bar{C}}{\partial z}\right) + Q\delta(x)\delta(y)\delta(z) \qquad (6.101)$$

is

$$\bar{C}(x,y,z) = \frac{Q(a)^{1/2}}{(4\pi\zeta)^{1/2}} \exp\left[-\frac{a(y-y_s)^2}{4\zeta} - \frac{a(z^{2+m-n}+z_s^{2+m-n})}{b(2+m-n)^2 x}\right] \times \frac{(zz_s)^{(1-n)/2}}{b(2+m-n)x} \times I_{-\gamma}\left[\frac{2a(zz_s)^{(2+m-n)/2}}{b(2+m-n)^2 x}\right] \tag{6.102}$$

where Q is the source strength, $(0,y_s,z_s)$ is the coordinate of the point source, $I_{-\gamma}$ is the modified Bessel function of first kind of order $-\gamma = -(i-n)/(2+m-n)$, ζ is defined as

$$\zeta = \int_0^{\zeta} H(\eta)\, d\eta \tag{6.103}$$

and the parameters, a, b, m, n, p, and H(x) depend on the atmospheric stability and surface roughness.

For a source near the ground, the solution to Eq. 6.102 can be reduced to

$$\bar{C}(x,y,z) = \bar{C}_g(x,y;y_s,z_s=0) \times \exp\left[-\frac{a(z^{2+m-n}+z_s^{2+m-n})}{b(2+m-n)^2 x}\right] \tag{6.104}$$

where $\bar{C}_g$ is the concentration at ground for a point source at ground. The analytic concentration solution to Eq. 6.104 confirms Elliott's findings that the vertical concentration distribution for a point source at ground is of exponential form (Elliott, 1959).

Note that Eq. 6.102 is generally non-Gaussian. However, under special conditions, if we set m = n and write

$$\sigma_z = \left(2\frac{b}{a}x\right)^{1/2}$$

then, for a source at the ground ($z_s = 0$), Eq. 6.102 becomes

$$\bar{C} = \bar{C}_g(x,y) \exp\left(-\frac{z^2}{2\sigma_z^2}\right) \tag{6.105}$$

Equation 6.105 is now in the Gaussian form of Eq. 6.72.

The question is whether the power law for the velocity profile and that for the eddy diffusivity correspond with each other. To answer the question, let $\bar{u}(z) = az^m$ and substitute it into Eq. 6.61 to obtain

$$\frac{\partial\bar{C}}{\partial t} + \left(az^m - \frac{1-m}{am}z^{-m}\right)\frac{\partial\bar{C}}{\partial x} = t_x'(\overline{u_b'^2} + A\overline{w_b'^2}) \times \frac{\partial^2\bar{C}}{\partial x^2} + t_y'\overline{v_b'^2}\frac{\partial^2\bar{C}}{\partial y^2} - (1+A)A\frac{\overline{w_b'^2}}{am}z^{1-m} \times \frac{\partial^2\bar{C}}{\partial x\,\partial z} + \frac{A\overline{w_b'^2}}{am}z^{1-m}\frac{\partial^2\bar{C}}{\partial z^2} \tag{6.106}$$

In this equation, notice that the diffusion process in a shear flow with mean velocity az^m involves not only K_{zz} and K_{xz} but also a new term of the form $[(1-m)/am]z^{-m}(\partial\bar{C}/\partial x)$, affecting the advection of concentration in the downwind direction.

In the surface boundary layer, the mean velocity generally takes a log–linear form. In the following discussion, we will use a numerical technique to simulate the transport and diffusion† in a neutral surface boundary layer in which the mean velocity profile can be expressed as a log form (see Chap. 5)

$$\bar{u}(z) = \frac{u_*}{k}\ln\left(\frac{z+z_0}{z_0}\right) \tag{6.107}$$

where k = 0.4 is the von Kármán's constant, u_* is the friction velocity, and z_0 is the roughness length. It has been shown (Kao and Taylor, 1977) that in the neutral boundary layer Eq. 6.61 becomes

$$\frac{\partial\bar{C}}{\partial t} + \left[\frac{u_*}{k}\ln\left(\frac{z+z_0}{z_0}\right) + \frac{16\,k\,u_*^3}{\left(\overline{u_b'^2} + \frac{u_*^4}{\overline{w_b'^2}}\right)}\right]\frac{\partial\bar{C}}{\partial x} - ku_*\frac{\partial\bar{C}}{\partial z} = 16\,k\,u_*(z+z_0)\frac{\partial^2\bar{C}}{\partial x^2} - k\,u_*(z+z_0) \times \left[\frac{1}{\overline{w_b'^2}} + \frac{16}{\left(\overline{u_b'^2} + \frac{u_*^4}{\overline{w_b'^2}}\right)}\right]\frac{\partial^2\bar{C}}{\partial x\,\partial z} + ku_*(z+z_0)\frac{\partial^2\bar{C}}{\partial z^2} \tag{6.108}$$

Equation 6.108 can be nondimensionalized by defining a characteristic velocity, length, and time scale, respectively, as u_*(m/s), h (height of the surface boundary layer), and $T = hu_*^{-1}$. Equation 6.108 can be put in a dimensionless form as

$$\frac{\partial\bar{C}}{\partial\hat{t}} + \hat{u}(z)\frac{\partial\bar{C}}{\partial\hat{x}} + \hat{w}_e\frac{\partial\bar{C}}{\partial\hat{z}} = \hat{K}_{xx}(z)\frac{\partial^2\bar{C}}{\partial\hat{x}^2} + \hat{K}_{xz}(z)\frac{\partial^2\bar{C}}{\partial\hat{x}\partial\hat{z}} + \hat{K}_{zz}\frac{\partial^2\bar{C}}{\partial\hat{z}^2} \tag{6.109}$$

where the dimensionless variables and parameters are defined as

$$\hat{t} = t/T \quad \hat{x} = x/H \quad \hat{z} = z/h \quad \hat{u}(z) = \bar{u}(z)/u_* \quad \hat{w}_e = -ku_*$$

$$\hat{K}_{xx}(z) = 16k(z+z_0)/h \quad \hat{K}_{zz}(z) = k(z+z_0)/h \tag{6.110}$$

$$\hat{K}_{xz}(z) = -ku_*^2/h\left(\frac{1}{\overline{w_b'^2}} + \frac{16}{\overline{u_b'^2} + \frac{u_*^4}{\overline{w_b'^2}}}\right)(z+z_0)$$

†A similarity Lagrangian solution for diffusion in the surface boundary layer is given in Sec. 6-3.10.

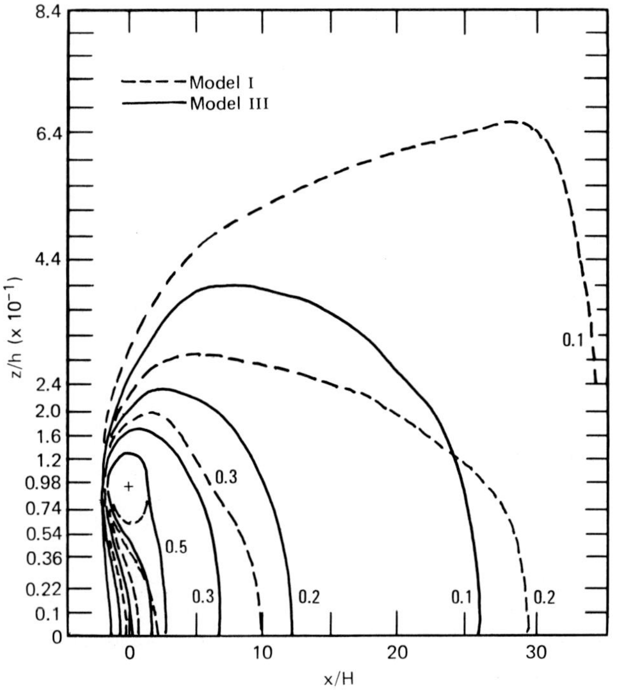

Fig. 6.4 Isopleths of normalized concentration for Models I (– – –) and III (——) with source located at 0.098. (From Kao and Taylor, 1977.)

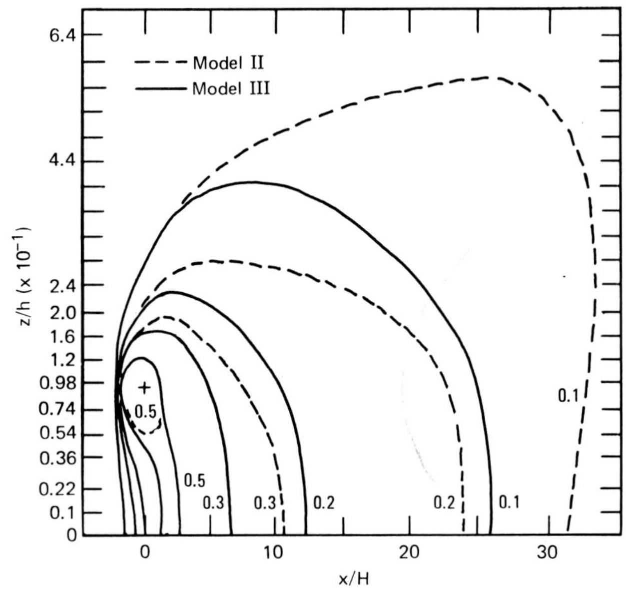

Fig. 6.5 Isopleths of normalized concentration for Models II (– – –) and III (——) with source located at 0.098. (From Kao and Taylor, 1977.)

Equation 6.109 has been numerically integrated for diffusion from a continuous point source located at $\hat{z} = 0.098$ (approximately $z = 5$ m), and the distribution of the normalized concentration is shown by the solid curves in Fig. 6.4. The dashed curves are the isopleths of concentration resulting from diffusion from a continuous point source at $\hat{z} = 0.098$ in a uniform mean flow with a mean velocity equal to that at $\hat{z} = 0.098$ in a logarithmic velocity profile (Eq. 6.107). It can be noted that, because of the strong velocity shear near the surface, the logarithmic velocity profile (Model III in Fig. 6.4) tends to diffuse more contaminants than the uniform mean velocity (Model I in Fig. 6.4). Also, the normalized concentration distribution in the surface boundary layer is compared with a logarithmic wind profile with and without the term involving $\hat{K}_{xz}$ as shown in Models III and II, respectively, in Fig. 6.5. It is apparent that the off-diagonal component, K_{xz}, diffuses more contaminants near the surface than do the diagonal components. In view of the important role played by the off-diagonal component of the eddy diffusivity in shear flow, it is necessary to include K_{xy} in the diffusion equation.

The effect of shear flow on diffusion was also investigated by Saffman (1962), Högström (1964), Gee and Davis (1963), F. B. Smith (1965), Tyldesley and Wallington (1965), Monin and Yaglom (1971), and others.

One of the methods of solving the diffusion equation for shear flow is the method of moments (Aris, 1956). In this procedure the mn-moment of concentration can be defined as

$$C_{mn}(z,t) = \int_{-\infty}^{\infty}\int_{-\infty}^{\infty} C(x,y,z,t)x^m y^n \, dx\, dy$$

Now, by applying the preceding integrations to the diffusion equation,

$$\frac{\partial C}{\partial t} + \bar{u}(z)\frac{\partial C}{\partial x} = K_{xx}(z)\frac{\partial^2 \bar{C}}{\partial x^2} + K_{yy}(z)\frac{\partial^2 \bar{C}}{\partial y^2} + \frac{\partial}{\partial z}\left[K_{zz}(z)\frac{\partial C}{\partial z}\right]$$

for $|m, n = 0, 1, 2, \ldots,$ we obtain

$$\frac{\partial C_{00}}{\partial t} - \frac{\partial}{\partial z}\left[K_{zz}(z)\frac{\partial C_{00}}{\partial z}\right] = 0$$

$$\frac{\partial C_{10}}{\partial t} - \frac{\partial}{\partial z}\left[K_{zz}(z)\frac{\partial C_{10}}{\partial z}\right] = \bar{u}(z)C_{00}$$

$$\frac{\partial C_{20}}{\partial t} - \frac{\partial}{\partial z}\left[K_{zz}(z)\frac{\partial C_{20}}{\partial z}\right] = 2\bar{u}(z)C_{10} + 2K_{xx}(z)C_{00}$$

and so forth. The resulting equations can be solved with appropriate initial and boundary conditions.

6-2.3.5 Transport and Turbulent Diffusion in the Planetary Boundary Layer. To study the transport and turbulent diffusion of pollutants in the planetary boundary layer, we consider the case where the mean velocity and temperature are horizontally homogeneous in the boundary layer, the governing equations (Eqs. 6.49 to 6.51) reduce to

$$\frac{\partial \bar{C}}{\partial t} + \bar{u}(z,t)\frac{\partial \bar{C}}{\partial x} + \bar{v}(z,t)\frac{\partial \bar{C}}{\partial y} = \frac{\partial}{\partial x_j}\left(K_{cjk}\frac{\partial \bar{C}}{\partial x_k}\right) + \bar{Q} \qquad (6.111)$$

$$\frac{\partial \bar{u}}{\partial t} = f(\bar{v} - v_g) + \frac{\partial}{\partial z}\left(K_{uzz}\frac{\partial \bar{u}}{\partial z}\right) \qquad (6.112)$$

$$\frac{\partial \bar{v}}{\partial t} = -f(\bar{u} - u_g) + \frac{\partial}{\partial z}\left(K_{vzz}\frac{\partial \bar{v}}{\partial z}\right) \qquad (6.113)$$

$$\frac{\partial \bar{\theta}}{\partial t} = \frac{\partial}{\partial z}\left(K_{\theta zz}\frac{\partial \bar{\theta}}{\partial z}\right) + \frac{\bar{S}}{\rho c_v} \qquad (6.114)$$

where $f = 2\Omega \sin \phi$ is the vertical component of the Coriolis parameter and u_g and v_g are the x and y components of the geostrophic wind velocity. For turbulent diffusion in the planetary layer, a series of models for the vertical component of the eddy diffusivity coefficient has been proposed (Blackadar, 1962; O'Brien, 1970; Pasquill, 1975; Smith and Matthews, 1975).

In the planetary boundary layer, it can be shown that on the right side of Eq. 6.111 the convergence of the mean turbulent flux of concentration is dominated by the vertical and lateral components of the flux. Thus Eq. 6.111 can be approximated by

$$\frac{\partial \bar{C}}{\partial t} + \bar{u}(z,t)\frac{\partial \bar{C}}{\partial x} + \bar{v}(z,t)\frac{\partial \bar{C}}{\partial y} = \frac{\partial}{\partial z}\left(K_{czz}\frac{\partial \bar{C}}{\partial z}\right) + K_{cyy}\frac{\partial^2 \bar{C}}{\partial y^2} + \bar{Q} \qquad (6.115)$$

In this section we will construct a planetary-boundary-layer diffusion model that takes into account the effects of the earth's rotation, thermal stratification, and stack height on concentration distribution.

If the lowest 10 m of the boundary layer is assumed to be within the surface layer where the wind and potential temperature take the form of (see Chap. 5, Sec. 5-3.5)

$$\bar{u} = \frac{u_{*0}}{k}\left[\ln\left(\frac{z + z_0}{z_0}\right) - \psi_m\left(\frac{z}{L}\right)\right] \qquad (6.116)$$

$$\bar{\theta} = \bar{\theta}_0 + 0.74\,\theta_{*0}\left[\ln\left(\frac{z + z_0}{z_0}\right) - \psi_h\left(\frac{z}{L}\right)\right] \qquad (6.117)$$

where L is the Monin–Obukhov length (see Chap. 5, Sec. 5-2.7)

$$L \equiv \frac{-c_p \rho T_0 u_{*0}^3}{kgH_0} \qquad (6.118)$$

with H_0 and ρu_{*0}^2 being the surface fluxes of heat and momentum, respectively. The functions ψ_m and ψ_h are integrals of universal functions ϕ_m and ϕ_h as given by Paulson (1970). The exchange coefficients can be computed from

$$K_{mzz} = u_{*0}^2\left[\left(\frac{\partial \bar{u}}{\partial z}\right)^2 + \left(\frac{\partial \bar{v}}{\partial z}\right)^2\right]^{-1/2} \qquad (6.119)$$

$$K_{hzz} = u_{*0}\,\theta_{*0} = -\frac{H_0}{c_p \rho} \qquad (6.120)$$

Above the surface layer in the planetary boundary layer, the vertical component of the eddy diffusivity for momentum K_m is assumed to be proportional to "the scale of energy-containing eddies, λ," and the square root of the kinetic energy of the turbulent motion per unit mass, e (Busch, Chang, and Anthes, 1976), i.e.,

$$K_{uzz} = K_{vzz} = K_{mzz} = \lambda^2\left[\left|\frac{\partial v}{\partial z}\right|^2 - \frac{g}{T_0}\frac{K_{\theta zz}}{K_{mzz}}\frac{\partial \bar{\theta}}{\partial z}\right]^{1/2} \qquad (6.121)$$

and the expression for the vertical component of the eddy diffusivity for sensible heat and containment can be expressed as

$$K_{\theta zz} = -K_{mzz}\frac{\phi_m}{\phi_h} \qquad (6.122)$$

$$K_{Czz} = \lambda^2 \frac{\phi_m}{\phi_h}\left|\frac{\partial \bar{v}}{\partial z}\right| \qquad (6.123)$$

where

$$\lambda = u_*\left|\frac{\partial \bar{v}}{\partial z}\right| \qquad (6.124)$$

is the scale of energy-containing eddies and u_* is the local frictional velocity. The expression for $K_{\theta zz}$ in Eq. 6.122 is derived from the Kansas experiment (Businger et al., 1971), where

$$\frac{K_{hzz}}{K_{mzz}} = \frac{\phi_m}{\phi_h} = \begin{cases} \dfrac{1.0 + 4.7\,zL^{-1}}{0.74 + 4.7\,zL^{-1}} & z/L > 0 \qquad (6.125) \\[2ex] \dfrac{1.35(T - 9\,zL^{-1})^{1/2}}{(1 - 15\,zL^{-1})^{1/4}} & z/L < 0 \qquad (6.126) \end{cases}$$

The vertical temperature profile observed at 0835 CST on Aug. 25, 1953, at O'Neil, Nebr., during the Great Plains Experiments (Stull, 1973) was used to specify the initial condition for potential temperature. The heat flux at the ground was specified by the function

$$\overline{w'\theta'} = \max \begin{cases} 0.25 \sin \dfrac{2\pi}{24h} & t > t_0 \\ -0.06 & t = t_0 \end{cases} \qquad (6.127)$$

Sunrise ($t = t_0$) was set at 0635 CST. The geostrophic wind components are assumed to be $u_g = 10$ m/s and $v_g = 0$. With the preceding specified initial conditions and values of K_{mzz} computed by Busch, Chang, and Anthes (1976),

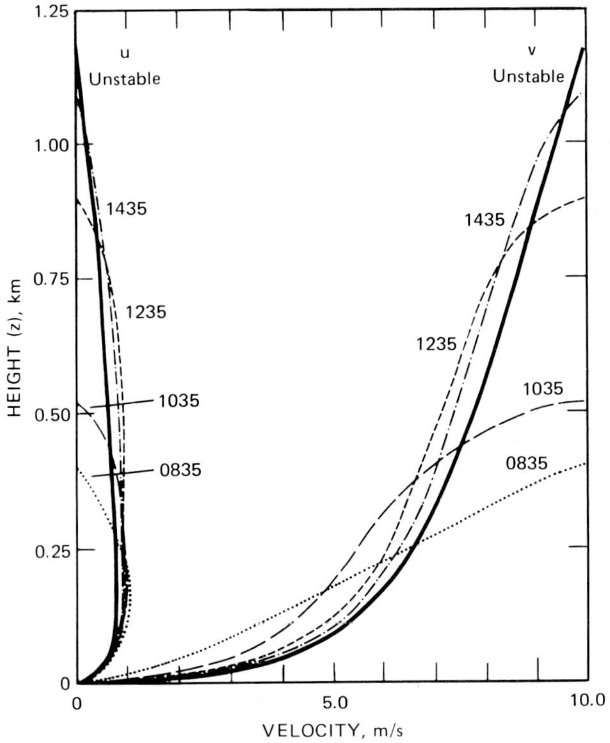

Fig. 6.6 The vertical profiles of the along- and cross-isobar components (u, v) of the velocity. [From S. K. Kao and Y. N. Yeh, A Note on the Effects of Stack and Inversion Heights on Diffusion of Pollutants in the Planetary Boundary Layer, *Atmospheric Environment*, 12: 397-407 (1979).]

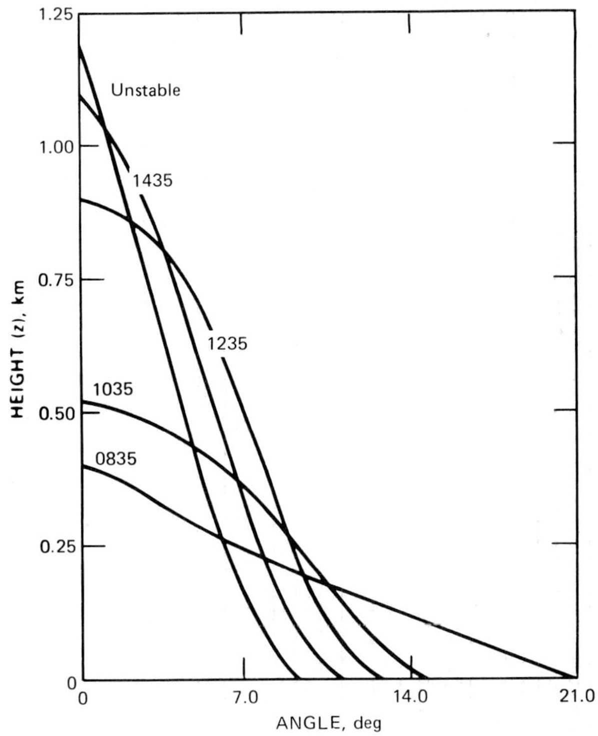

Fig. 6.7 The cross-isobar angle (α) at 0835, 1035, 1235, 1435 CST, and for the unstable case. [From S. K. Kao and Y. N. Yeh, A Note on the Effects of Stack and Inversion Heights on Diffusion of Pollutants in the Planetary Boundary Layer, *Atmospheric Environment*, 12: 397-407 (1979).]

Eqs. 6.112 and 6.113 were integrated to obtain the vertical profiles of $\bar{u}$ and $\bar{v}$ at 0835, 1035, 1235, and 1435 CST as shown in Fig. 6.6. In the same figure we have also plotted a speed profile for the unstable case. It can be noted that the speed shear of the mean speed decreases as the inversion height increases with increasing time. The vertical profiles of cross-isobar (or cross geostrophic wind) angle α as shown in Fig. 6.7 indicate that, at the surface, α decreases from 20.5° at 0835 to 11° at 1435 and to 9.5° for the unstable case. The vertical profiles of the vertical component of the coefficient of eddy diffusivity K_{Czz} for the inversion and unstable cases are shown in Fig. 6.8. Note that K_{Czz} generally increases with height near the surface, reaches its maximum value at half the inversion height, then decreases toward inversion height 400 m at 0835, 520 m at 1035, 900 m at 1235, and 1100 m at 1435; for the unstable case, K_{Czz} remains constant upward from its maximum.

To numerically integrate the diffusion equation with the use of the velocity and eddy-diffusivity profiles as shown in Figs. 6.6 and 6.8, we assume that there is no flux ($\partial\bar{C}/\partial z = 0$) at the surface ($z = 0$) and at the inversion height and that $K_{Cyy} = K_{Czz}$ (Kao and Yeh, 1979).

Parts (a) to (d) of Fig. 6.9 show the distribution of concentration relative to the source strength resulting from the transport and diffusion of contaminants from a continuous fixed point source located at $z = 100$ m above the ground with inversions located at $z = 400$ and 1100 m at 0835 and 1435, respectively. Figure 6.9a shows the distribution of the concentration relative to the source strength on a vertical plane passing through the source and parallel to the x-axis at 0835, whereas Fig. 6.9b shows the concentration distribution at the ground at 0835, which indicates that, because of the large velocity shear near the ground, contaminants released from the point source in the boundary layer tend to spread toward the ground with maximum ground concentration occurring near a downwind distance 400 m; the concentration decreases rapidly from the source toward the inversion height at 400 m. Parts (c) and (d) of Fig. 6.9 show the vertical yz profiles of the concentration relative to the source strength at downwind radial distances of 1 and 2 km, respectively. In these two figures it is shown that the maximum concentration shifts toward the right with increasing height (in the Northern Hemisphere), a consequence of the Coriolis effect on the wind which advects the contaminants.

Parts (a) and (b) of Fig. 6.10 show the concentration distribution similar to Figs. 6.9a and 6.9b except at 1435. It can be noticed that in these figures the vertical component of the eddy diffusivity increases as thermal stability decreases from 0835 to 1435 (inversion height increases). As a consequence, the cross-isobar angle decreases and, therefore, a reduction in the right shift of the

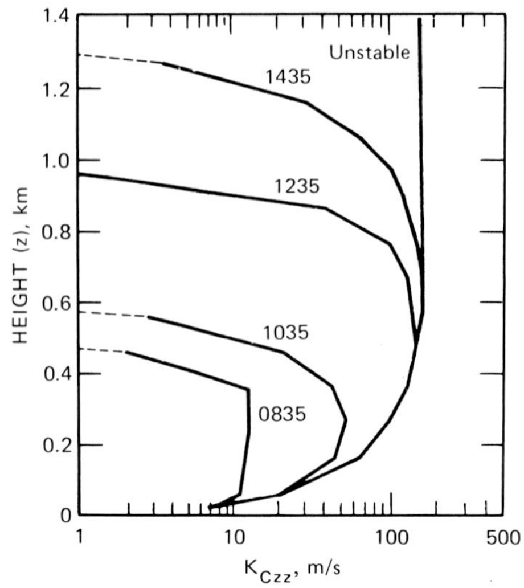

Fig. 6.8 The vertical profile of the vertical component of the coefficient of the eddy diffusivity at 0835, 1035, 1235, 1435 CST, and for the unstable case. [From S. K. Kao and Y. N. Yeh, A Note on the Effects of Stack and Inversion Heights on Diffusion of Pollutants in the Planetary Boundary Layer, *Atmospheric Environment*, 12: 397-407 (1979).]

maximum concentration with height occurs at 1435 as shown in Figs. 6.10c and 6.10d as compared with Figs. 6.9c and 6.9d at 0835.

Examining the diffusion characteristics in a uniform mean flow (Sec. 6-2.3.1), in a linear velocity shear flow (Sec. 6-2.3.3), in a surface boundary layer (Sec. 6-2.3.4), and in a planetary boundary layer (Sec. 6-2.3.5), we find that the concentration distribution resulting from diffusion from a continuous fixed source in the planetary boundary layer behaves very differently from concentrations in the uniform mean flow and in the surface boundary layer. The main differences are that in the planetary boundary layer the maximum concentration shifts clockwise (counterclockwise) in the Northern Hemisphere (Southern Hemisphere) with increasing height and that the concentration distribution is no longer Gaussian.

6-2.4 Higher Order Closure Scheme and Turbulent Diffusion

In the preceding sections several first-order closure schemes have been presented, and solutions to the diffusion equation have been analyzed. In spite of the shortcomings of the mixing length model, it provides a first approximation to the turbulent transfer of trace elements, momentum, and heat in boundary layers in which the velocity shear is strong. However, in more complex cases, e.g., in regions of maximum mean quantities where turbulence generally exists, the mixing length model fails to diffuse pollutants at the proper rate. In these cases it is necessary to use higher order closure schemes. In this section several higher order closure schemes will be presented. The procedure for obtaining a second-order closure is to express the double and triple correlation terms in the second-moment (Eqs. 6.36 to 6.39) in terms of the mean velocities and the Reynolds stress tensor as shown in the following discussion.

Extending Hanjalic and Launder's (1972) and Mellor's (1973) assumptions regarding the third- and second-order correlation terms in Eqs. 6.37 and 6.38 for the mean turbulent fluxes of momentum and temperature to the terms in Eq. 6.36 for the mean turbulent flux of trace constituents, we obtain

$$\overline{u_j'u_k'C'} = -\lambda_1 q\left(\frac{\partial \overline{u_j'C'}}{\partial x_k} + \frac{\partial \overline{u_k'C'}}{\partial x_j}\right) \tag{6.128}$$

$$\overline{u_j'C'} = -\lambda_2 q \frac{\partial \overline{C'^2}}{\partial x_j} \tag{6.129}$$

$$\overline{p'\frac{\partial C'}{\partial x_j}} = -\frac{q}{3\ell_1}\overline{u_j'C'} \tag{6.130}$$

$$\overline{p'C'} = \overline{p'\theta'} = \overline{p'u_i'} = 0 \tag{6.131}$$

$$\overline{u_i'u_j'u_k'} = -\lambda_3 q\left(\frac{\partial \overline{u_j'u_k'}}{\partial x_i} + \frac{\partial \overline{u_k'u_i'}}{\partial x_j} + \frac{\partial \overline{u_i'u_j'}}{\partial x_k}\right) \tag{6.132}$$

$$\overline{u_j'u_k'\theta'} = -\lambda_4 q\left(\frac{\partial \overline{u_j'\theta'}}{\partial x_k} + \frac{\partial \overline{u_k'\theta'}}{\partial x_j}\right) \tag{6.133}$$

$$\overline{u_k'\theta'} = -\lambda_5 q \frac{\partial \overline{\theta'^2}}{\partial x_k} \tag{6.134}$$

$$\overline{u_k'C'\theta'} = -\lambda_6 q \frac{\partial \overline{C'\theta'}}{\partial x_k} \tag{6.135}$$

$$\overline{p'\left(\frac{\partial u_i'}{\partial x_j} + \frac{\partial u_j'}{\partial x_i}\right)} = -\frac{q}{3\ell_2}\left(\overline{u_i'u_j'} - \frac{q^2}{3}\delta_{ij}\right) + Bq^2\left(\frac{\partial \bar{u}_i}{\partial x_j} + \frac{\partial \bar{u}_j}{\partial x_i}\right) \tag{6.136}$$

$$\overline{p'\frac{\partial \theta'}{\partial x_j}} = -\frac{q}{3\ell_3}\overline{u_j'\theta'} \tag{6.137}$$

$$2\hat{\nu}\,\overline{\frac{\partial u_i'}{\partial x_k}\frac{\partial u_j'}{\partial x_k}} = \frac{2}{3}\frac{q^3}{\Lambda_1}\delta_{ij} \tag{6.138}$$

$$2\hat{\kappa}\,\overline{\frac{\partial \theta'}{\partial x_k}\frac{\partial \theta'}{\partial x_k}} = 2\frac{q}{\Lambda_2}\overline{\theta'^2}, \quad 2D\,\overline{\frac{\partial C'}{\partial x_k}\frac{\partial C'}{\partial x_k}} = 2\frac{q}{\Lambda_2}\overline{C'^2} \tag{6.139}$$

where λ_n, ℓ_n, and Λ_n are constants and $q = (\overline{u_i'u_i'})^{1/2}$. If these equations are inserted into Eqs. 6.36 to 6.40, we obtain

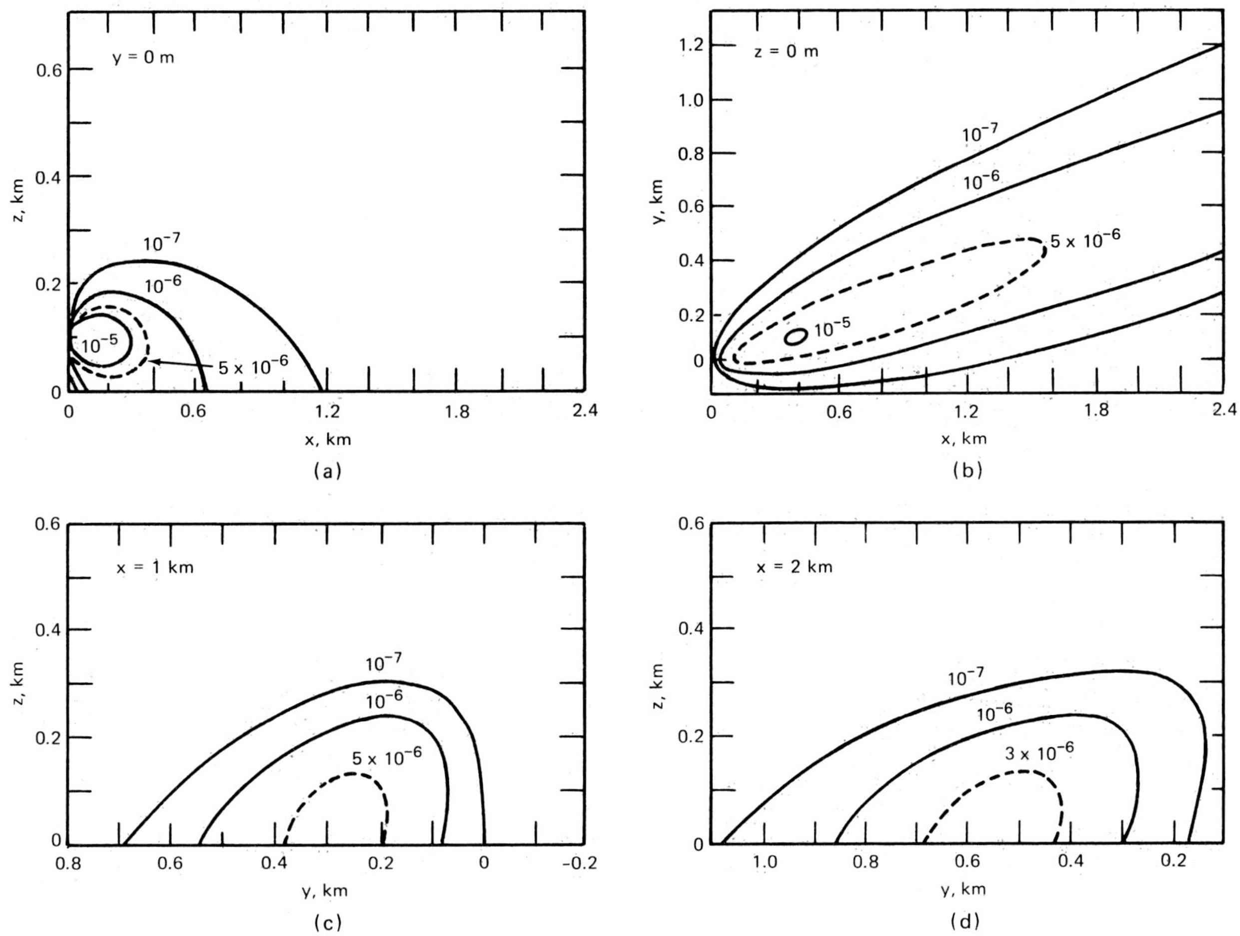

Fig. 6.9 Distribution of concentration relative to the source strength at 0835 CST resulting from the transport and dispersion from a fixed point source at height z = 100 m. Inversion height is 400 m.

$$\frac{D\overline{u_j'C'}}{Dt} = \frac{\partial}{\partial x_k}\left[\lambda_2 q\left(\frac{\partial \overline{u_j'C'}}{\partial x_k} + \frac{\partial \overline{u_k'C'}}{\partial x_j}\right) + D\,\overline{u_j'\frac{\partial C'}{\partial x_k}} + \hat{\gamma}\,\overline{\frac{C'\partial u_j'}{\partial x_k}}\right] + \epsilon_{ijk} f_k \overline{u_i'C'} - \overline{u_j'u_k'}\frac{\partial \bar{C}}{\partial x_k} - \overline{C'u_k'}\frac{\partial u_j}{\partial x_k} - \frac{q}{3\ell_3}\overline{u_j'C'} \tag{6.140}$$

$$\frac{D\overline{u_j'u_i'}}{Dt} = \frac{\partial}{\partial x_k}\left[\lambda_3 q\left(\frac{\partial\overline{u_j'u_k'}}{\partial x_j}+\frac{\partial\overline{u_k'u_i'}}{\partial x_j}+\frac{\partial\overline{u_i'u_i'}}{\partial x_k}\right) + \hat\gamma \frac{\partial\overline{u_i'u_j'}}{\partial x_k}\right] - f_k\left(\epsilon_{jkm}\overline{u_m'u_i'} + \epsilon_{ikm}\overline{u_m'u_j'}\right) - \overline{u_k'u_i'}\frac{\partial\bar u_j}{\partial x_k} - \overline{u_k'u_j'}\frac{\partial \bar u_i}{\partial x_k} - \frac{1}{T_b}\left(g_j\overline{u_i'\theta'} + g_i\overline{u_j'\theta'}\right) - \frac{q}{3\ell_2}\left(\overline{u_i'u_j'} - \frac{q^2}{3}\delta_{ij}\right) + Bq^2\left(\frac{\partial\bar u_i}{\partial x_j}+\frac{\partial \bar u_j}{\partial x_i}\right) - \frac{2}{3}\frac{q^3}{\Lambda_1}\delta_{ij} \tag{6.141}$$

$$\frac{D\overline{u_j'\theta'}}{Dt} = \frac{\partial}{\partial x_k}\left[\lambda_4 q\left(\frac{\partial\overline{u_j'\theta'}}{\partial x_k}+\frac{\partial\overline{u_k'\theta'}}{\partial x_j}\right)+\hat\kappa\,\overline{u_j'\frac{\partial\theta'}{\partial x_k}} + \hat\gamma\,\overline{\theta'\frac{\partial u_j'}{\partial x_k}}\right] - \epsilon_{jkm}f_k\overline{u_m'\theta'} - \overline{u_j'u_k'}\frac{\partial\bar\theta}{\partial x_k} - \overline{u_k'\theta'}\frac{\partial \bar u_j}{\partial x_k} - \frac{g_j}{T_b}\overline{\theta'^2} - \frac{q}{3\ell_3}\overline{u_j'\theta'} \tag{6.142}$$

$$\frac{D\overline{\theta'^2}}{Dt} = \frac{\partial}{\partial x_k}\left[(\gamma_5 q+\hat\kappa)\frac{\partial\overline{\theta'^2}}{\partial x_k}\right] - 2\overline{u_k'\theta'}\frac{\partial\bar\theta}{\partial x_k} - 2\frac{q}{\Lambda_2}\overline{\theta'^2} \tag{6.143}$$

$$\frac{D\overline{C'^2}}{Dt} = \frac{\partial}{\partial x_k}\left[(\lambda_2 q + D)\frac{\partial\overline{C'^2}}{\partial x_k}\right] - 2\overline{u_k'C'}\frac{\partial\bar C}{\partial x_k} - 2\frac{q}{\Lambda_3}\overline{C'^2} \tag{6.144}$$

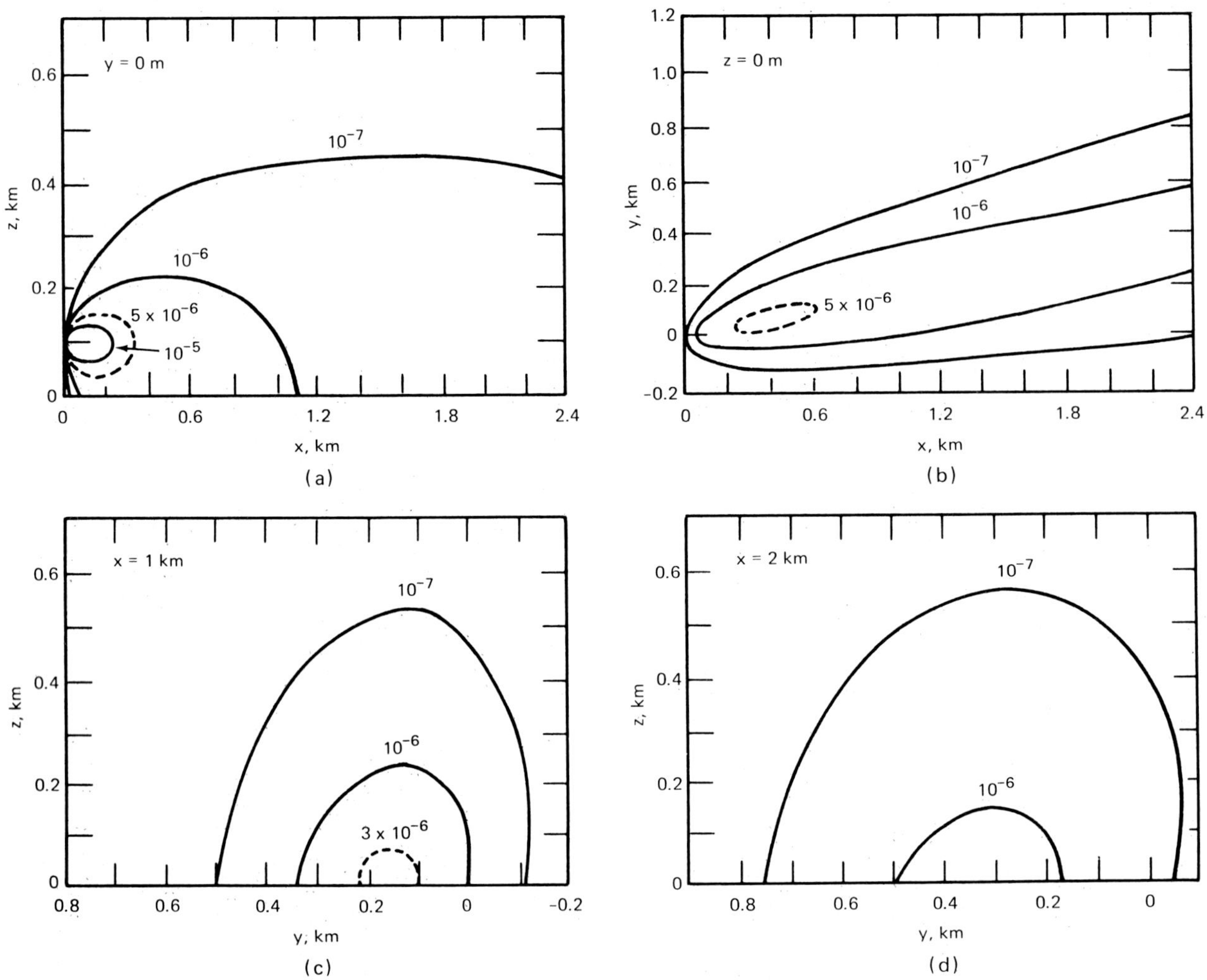

Fig. 6.10 Distribution of concentration relative to the source strength at 1235 CST resulting from the transport and dispersion from a fixed point source at height z = 100 m. Inversion height is 1100 m.

$$\frac{D\overline{\theta'C'}}{Dt} = \frac{\partial}{\partial x_k}\left(\lambda_6 q \frac{\partial\overline{\theta'C'}}{\partial x_k}\right) + \lambda_2 q \frac{\partial\bar{\theta}}{\partial x_k}\frac{\partial \overline{C'^2}}{\partial x_k} + \lambda_5 q \frac{\partial\overline{\theta'^2}}{\partial x_k}\frac{\partial \bar{C}}{\partial x_k} \qquad (6.145)$$

where operator

$$\frac{D}{Dt} \equiv \left(\frac{\partial}{\partial t} + \bar{u}_j \frac{\partial}{\partial x_j}\right) \qquad (6.146)$$

Therefore Eqs. 6.28 to 6.31 and 6.140 to 6.145, a total of 21 equations, form a closed system for 21 variables: $\bar{C}$, $\bar{u}_i$, $\bar{p}$, $\bar{\theta}$, $\overline{u_j C'}$, $\overline{u_i u_j}$, $\overline{u_j \theta'}$, $\overline{\theta'^2}$, $\overline{C'^2}$, and $\overline{\theta' C'}$. However, this modeling of third-order correlations requires many undetermined constants, only some of which can be determined from measurements in simple turbulent flows.

An alternative method is the so-called subgrid-scale scheme (e.g., Deardorff, 1972). This method uses a grid size that is sufficiently small to describe explicitly most of the turbulence structures. However, this scheme requires a three-dimensional simulation; the memory capacity of the present generation of computers limits the size of the domain of simulation.

Another closure scheme is based on the assumption that velocity is a Gaussian random variable such that the fourth-order correlations can be expressed in terms of the second-order correlations. This method is called the quasi-normal approximation (Millionschchikov, 1941; Proudman and Reid, 1954; Tatsumi, 1957). However, this method leads to an unphysical development of negative values of the energy density in the original energy-containing range (Ogura, 1962, 1963). This deficiency can be traced to an excessive growth of third-order correlations (Orszag, 1970) since one effect of fourth-order correlations, which is to limit the buildup of triple correlations (André, 1974), is lost in the quasi-normal approximation. This effect can be replaced by introducing an appropriate eddy-damping coefficient in the rate of change of third-order correlations (Orszag, 1970; Kraichnan, 1971; Leith, 1971).

One of the damping mechanisms is the clipping approximation which is based on the fact that triple correlations of turbulent quantities must verify realizability conditions derived from generalized Schwartz's inequalities (Orszag and Kruskal, 1968; Blanchet, 1970, 1975). For example, the triple correlation of the three turbulent quantities ξ', η', and ζ' can be expressed as

$$\overline{\xi'\eta'\zeta'} = \overline{\xi'(\eta'\zeta' + M)} \tag{6.147}$$

where M is a constant. Applying Schwartz's inequality to the right-hand side of the preceding equation, we obtain

$$\overline{\xi'\eta'\zeta'}^2 \leqslant \overline{\xi'^2}\ (\overline{\eta'^2\zeta'^2} + 2M\overline{\eta'\zeta'} + M^2) \tag{6.148}$$

If we minimize the right-hand side of the preceding inequality by letting

$$M = -\overline{\eta'\zeta'}$$

then

$$\overline{\xi'\eta'\zeta'}^2 \leqslant \overline{\xi'^2}\ (\overline{\eta'^2\zeta'^2} - \overline{\eta'\zeta'}^2) \tag{6.149}$$

It is seen from the preceding equality that the fourth-order correlation $\overline{\eta'^2\zeta'^2}$ imposes a limit on the development of triple correlation $\overline{\xi'\eta'\zeta'}$.

If we apply the quasi-normal approximation

$$\overline{\eta'^2\zeta'^2} = \overline{\eta'^2}\ \overline{\zeta'^2} + 2\overline{\eta'\zeta'}^2 \tag{6.150}$$

to the right-hand side of Eq. 6.149, we obtain by symmetry

$$|\overline{\xi'\eta'\zeta'}| \leqslant \min \begin{cases} [\overline{\xi'^2}\ (\overline{\eta'^2}\ \overline{\zeta'^2} + \overline{\eta'\zeta'}^2)]^{1/2} \\ [\overline{\eta'^2}\ (\overline{\zeta'^2}\ \overline{\xi'^2} + \overline{\zeta'\xi'}^2)]^{1/2} \\ [\overline{\zeta'^2}\ (\overline{\xi'^2}\ \overline{\eta'^2} + \overline{\xi'\eta'}^2)]^{1/2} \end{cases} \tag{6.151}$$

The clipping approximation consists of enforcing inequality (Eq. 6.151) on triple correlations.

We shall now apply the quasi-normal model modified by the clipping approximation to the planetary boundary layer. Here we consider again the system of equations given by Eqs. 6.28 to 6.30 and 6.36 to 6.41. We will neglect the molecular diffusion terms

$$\begin{array}{lll} D\dfrac{\partial^2\overline{C'^2}}{\partial x_k\partial x_k} & D\,\overline{u_i'\dfrac{\partial^2 C'}{\partial x_k\partial x_k}} & D\,\overline{\theta'\dfrac{\partial^2 C'}{\partial x_k\partial x_k}} \\ \nu\dfrac{\partial^2\overline{u_i'u_j'}}{\partial x_k\partial x_k} & \nu\overline{C'\dfrac{\partial^2 u_i'}{\partial x_k\partial x_k}} & \nu\overline{\theta'\dfrac{\partial^2 u_i'}{\partial x_k\partial x_k}} \\ \hat{\kappa}\dfrac{\partial^2\overline{\theta'^2}}{\partial x_k\partial x_k} & \hat{\kappa}\overline{C'\dfrac{\partial^2\theta'}{\partial x_k\partial x_k}} & \hat{\kappa}\overline{u_i'\dfrac{\partial\theta'}{\partial x_k\partial x_k}} \end{array} \tag{6.152}$$

since they are small compared to turbulent diffusion terms.

The molecular dissipation rate of Reynolds stresses will be estimated by

$$2\nu\overline{\frac{\partial u_i'}{\partial x_k}\frac{\partial u_j'}{\partial x_k}} = \frac{2}{3}\epsilon\delta_{ij} \tag{6.153}$$

where ϵ is the dissipation rate of eddy kinetic energy $\frac{1}{2}q^2 = \frac{1}{2}\overline{u_i'u_i'}$. By dimensional consideration, the dissipation rate ϵ can be estimated by

$$\epsilon = C_1\frac{q^3}{\lambda} \tag{6.154}$$

where λ is length scale of the energy-containing eddies. The molecular dissipation rate of concentration and temperature variances will be modeled by

$$\begin{aligned} 2D\overline{\frac{\partial C'}{\partial x_k}\frac{\partial C'}{\partial x_k}} &= C_2\frac{\epsilon}{q_2}\overline{C'^2} \\ 2\hat{\kappa}\overline{\frac{\partial\theta'}{\partial x_k}\frac{\partial\theta'}{\partial x_k}} &= C_3\frac{\epsilon}{q^2}\overline{\theta'^2} \end{aligned} \tag{6.155}$$

The turbulent transport terms will be taken as (Donaldson, 1973)

$$\begin{aligned} \frac{1}{\rho_b}\overline{p'u_i'} &= -C_4\,q\lambda\frac{\partial\overline{u_i'u_k'}}{\partial x_k} \\ \frac{1}{\rho_b}\overline{p'C'} &= -C_4\,q\lambda\frac{\partial\overline{u_k'C'}}{\partial x_k} \\ \frac{1}{\rho_b}\overline{p'\theta'} &= -C_4\,q\lambda\frac{\partial\overline{u_k'\theta'}}{\partial x_k} \end{aligned} \tag{6.156}$$

The correlation between pressure fluctuation and velocity, concentration, or temperature gradient will be modeled like the return-to-isotopy terms (Launder, 1975)

$$\begin{aligned} \frac{1}{\rho_b}\overline{p'\left(\frac{\partial u_i'}{\partial x_j} + \frac{\partial u_j'}{\partial x_i}\right)} &= -C_5\frac{\epsilon}{q^2}\left(\overline{u_i'u_j'} - \frac{1}{3}q^2\delta_{ij}\right) - C_6\left(G_{ij} - \frac{2}{3}G\,\delta_{ij}\right) \\ \frac{1}{\rho_b}\overline{p'\frac{\partial C'}{\partial x_i}} &= -C_7\frac{\epsilon}{q^2}\overline{u_i'C'} - C_8\,G_{iC} \\ \frac{1}{\rho_b}\overline{p'\frac{\partial\theta'}{\partial x_i}} &= -C_9\frac{\epsilon}{q^2}\overline{u_i\theta'} - C_{10}G_{i\theta} \end{aligned} \tag{6.157}$$

where G_{ij} is the total (dynamic and thermal) generation of Reynolds stress $\overline{u_i'u_j'}$

$$G_{ij} = -\left(\overline{u_i'u_k'}\frac{\partial \bar{u}_j}{\partial x_k} + \overline{u_j'u_k'}\frac{\partial \bar{u}_i}{\partial x_k}\right) + \frac{g}{T_b}(\overline{u_i'\theta'}\,\delta_{3j} + \overline{u_j'\theta'}\,\delta_{3i}) \qquad (6.158)$$

G is the total generation of eddy kinetic energy

$$G = -\overline{u_i'u_k'}\frac{\partial \bar{u}_i}{\partial x_k} + \frac{g}{T_b}\overline{u_i'\theta'}\,\delta_{3i} \qquad (6.159)$$

and G_{iC} and $G_{i\theta}$ are, respectively, the generation of concentration and heat flux $\overline{u_i'C'}$ and $\overline{u_i'\theta'}$.

$$G_{iC} = -\overline{u_k'C'}\frac{\partial \bar{u}_i}{\partial x_k} + \frac{g}{T_b}\overline{C'\theta'}\,\delta_{3i}$$

$$G_{i\theta} = -\overline{u_k'\theta'}\frac{\partial \bar{u}_i}{\partial x_k} + \frac{g}{T_b}\overline{\theta'^2}\,\delta_{3i} \qquad (6.160)$$

We shall now compute the third-order correlations between the turbulent velocity and/or concentration using the quasi-normal model modified by the clipping approximation. Again we shall neglect the molecular diffusion terms. As examples, the equations for $\overline{w'^3}$ and $\overline{w'C'^2}$ can be obtained after some algebra:

$$\frac{\partial \overline{w'^3}}{\partial t} + \bar{u}_k\frac{\partial \overline{w'^3}}{\partial x_k} = -\frac{\partial \overline{u_k'w'^3}}{\partial x_k} + 3\overline{w'^2}\frac{\partial \overline{u_k'w'}}{\partial x_k} - 3\,\overline{u_k'w'^2}\frac{\partial \bar{w}}{\partial x_k} + 3\frac{g}{T_b}\overline{w'^2\theta'} \qquad (6.161)$$

$$\frac{\partial \overline{w'C'^2}}{\partial t} + \bar{u}_k\frac{\partial \overline{w'C'^2}}{\partial x_k} = -\frac{\partial \overline{u_k'w'C'}}{\partial x_k} + \overline{C'^2}\frac{\partial \overline{u_k'w'}}{\partial x_k} + 2\,\overline{w'C'}\frac{\partial \overline{u_k'C'}}{\partial x_k} - \overline{u_k'C'^2}\frac{\partial \bar{w}}{\partial x_k} - 2\,\overline{u_k'w'C'}\frac{\partial \bar{C}}{\partial x_k} + \frac{g}{T_b}\overline{C'^2\theta'} \qquad (6.162)$$

$$\frac{\partial \overline{w'\theta'^2}}{\partial t} + \bar{u}_k\frac{\partial \overline{w'\theta'^2}}{\partial x_k} = -\frac{\partial \overline{u_k'w'\theta'^2}}{\partial x_k} + \overline{\theta'^2}\frac{\partial \overline{u_k'w'}}{\partial x_k} + 2\,\overline{w'\theta'}\frac{\partial \overline{u_k'\theta'}}{\partial x_k} - \overline{u_k'\theta'}\frac{\partial \bar{w}}{\partial x_k} - 2\,\overline{u_k'w'\theta'}\frac{\partial \bar{\theta}}{\partial x_k} + \frac{g}{T_b}\overline{\theta'^3} \qquad (6.163)$$

The quasi-normal approximation asserts that

$$\overline{u_k'w'^3} = 3\,\overline{u_k'w'}\;\overline{w'^2}$$

$$\overline{u_k'w'C'^2} = \overline{u_k'w'}\;\overline{C'^2} + 2\,\overline{u_k'C'}\;\overline{w'C'} \qquad (6.164)$$

$$\overline{u_k'w'\theta'^2} = \overline{u_k'w'}\;\overline{\theta'^2} + 2\,\overline{u_k'\theta'}\;\overline{w'\theta'}$$

Thus Eqs. 6.161 to 6.163 become

$$\frac{\partial \overline{w'^3}}{\partial t} + \bar{u}_k\frac{\partial \overline{w'^3}}{\partial x_k} = -3\,\overline{u_k'w'}\frac{\partial \overline{w'^2}}{\partial x_k} - 3\,\overline{u_k'w'^2}\frac{\partial \bar{w}}{\partial x_k} + 3\frac{g}{T_b}\overline{w'^2\theta'} \qquad (6.165)$$

$$\frac{\partial \overline{w'C'^2}}{\partial t} + \bar{u}_k\frac{\partial \overline{w'C'^2}}{\partial x_k} = -\overline{u_k'w'}\frac{\partial \overline{C'^2}}{\partial x_k} - 2\,\overline{u_k'C'}\frac{\partial \overline{w'C'}}{\partial x_k} - \overline{u_k'C'^2}\frac{\partial \bar{w}}{\partial x_k} - 2\,\overline{u_k'w'C'}\frac{\partial \bar{C}}{\partial x_k} + \frac{g}{T_b}\overline{\theta'C'^2} \qquad (6.166)$$

$$\frac{\partial \overline{w'\theta'^2}}{\partial t} + \bar{u}_k\frac{\partial \overline{w'\theta'^2}}{\partial x_k} = -\overline{u_k'w'}\frac{\partial \overline{\theta'^2}}{\partial x_k} - 2\,\overline{u_k'\theta'}\frac{\partial \overline{w'\theta'}}{\partial x_k} - \overline{u_k'\theta'}\frac{\partial \bar{w}}{\partial x_k} - 2\,\overline{u_k'w'\theta'}\frac{\partial \bar{\theta}}{\partial x_k} + \frac{g}{T_b}\overline{\theta'^3} \qquad (6.167)$$

In the clipping approximation the quasi-normal Eqs. 6.165 to 6.167 are forced to follow the realizability inequalities

$$\left|\overline{w'^3}\right| \leqslant (2\,\overline{w'^2}^{\,3})^{1/2}$$

$$\left|\overline{w'C'^2}\right| \leqslant [\overline{C'^2}(\overline{w'^2}\,\overline{C'^2} + \overline{C'w'}^2)]^{1/2} \qquad (6.168)$$

$$\left|\overline{w'\theta'^2}\right| \leqslant [\overline{\theta'^2}(\overline{w'^2}\,\overline{\theta'^2} + \overline{\theta'w'}^2)]^{1/2}$$

6-2.5 Diffusion Equation of Hyperbolic Type

As mentioned in Sec. 6-2.3.1, the parabolic, gradient transfer diffusion equation (Eq. 6.65) governs the dispersion of a trace constituent in a uniform flow and is an approximation to the rather complicated motion in turbulent flow. One of the shortcomings of the parabolic diffusion equation is its prediction that the trace constituent in the fluid will appear instantaneously everywhere (though not equally) if the trace constituent is introduced at some point in the fluid. Since such instantaneous propagation of particles is impossible, we must assume that the diffusion is correct only after a sufficiently long time has elapsed. This time depends naturally upon the velocity of propagation of the constituent, which in turn depends upon the mean mixing length of the constituent. Once the time required for the constituent to get to the point in question is exceeded, we can then presume that the diffusion equation applies. The partial differential equation that includes this effect can be expressed in a one-dimensional case as

$$\frac{\partial^2 C}{\partial t^2} + \eta \frac{\partial C}{\partial t} = v^{*2} \frac{\partial^2 C}{\partial y^2} \tag{6.169}$$

where v^* is the velocity of propagation of the plume boundary in the y-direction and η has a dimension of frequency. Equation 6.169 is of the hyperbolic type. Note that Eq. 6.169 reduces to the wave equation if $\eta \to 0$ and reduces to the parabolic diffusion equation if v^* and $f \to \infty$ and $v^{*2}/\eta \to$ constant.

For an instantaneous point source of strength Q released at y = 0 and t = 0, i.e.,

$$C(y,0) = Q\delta(y) \tag{6.170}$$

with the assumption

$$\left.\frac{\partial C(y,t)}{\partial t}\right|_{t=0} = 0$$

Eq. 6.169 has the following solution:

$$C(y,t) = Q\begin{cases} \dfrac{1}{4}\dfrac{\eta}{v^*} e^{-\frac{1}{2}\eta t}\left[I_0(Y) + \dfrac{1}{2} ft \dfrac{I_1(Y)}{Y}\right] & |y| < v^*t \\ \dfrac{1}{2} e^{-\frac{1}{2}\eta t} & y = \pm v^*t \\ 0 & |y| > v^*t \end{cases} \tag{6.171}$$

where

$$Y = \frac{1}{2}\frac{\eta}{v^*}(v^{*2}t^2 - y^2)^{\frac{1}{2}}$$

and I is the modified Bessel function.

For large diffusion time ($ft \gg 1$; $|y| \ll v^*t$), Eq. 6.171 becomes Gaussian

$$C(y,t) \approx \frac{Q}{\left(\dfrac{4\pi v^{*2} t}{\eta}\right)^{\frac{1}{2}}} \exp\left(-\frac{\eta y^2}{4v^{*2}t}\right)$$

Figure 6.11 showing a sketch of Eq. 6.171 indicates that, for $|y| \ll v^*t$, the solution is approximately Gaussian and that there are two wave fronts at $y = \pm v^*t$ and C = 0 beyond the wave fronts.

With applications of stochastic methods, the diffusion equation of hyperbolic type has also been derived and investigated by Goldstein (1951), Davies, Diamond, and Smith (1954), Monin (1959), and Scheidegger (1958).

6-3 THE LAGRANGIAN APPROACH

To describe the transport and diffusion of a trace constituent by following the marked fluid particles is generally referred to as the Lagrangian method. Since contaminants move with fluid particles, the Lagrangian method is the natural way of describing the pollution distribution. Most of the statistical theories of turbulent diffusion were developed on the basis of the Lagrangian approach.

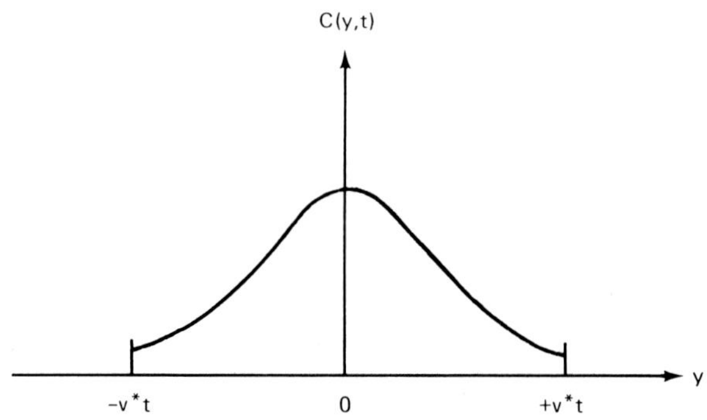

Fig. 6.11 One-dimensional solution of hyperbolic diffusion equation.

6-3.1 Lagrangian Description of Fluid Motion

Consider a particle moving with the fluid. We can identify it by a position vector **r** running from the origin of the coordinate system to the particle. As the particle moves, **r** changes as a function of time so that $d\mathbf{r}/dt$ represents the particle velocity and $d^2\mathbf{r}/dt^2$ represents its acceleration. In problems of diffusion we are interested in the concentration, C, of a certain contaminant. As long as we direct our attention to a single fluid element, no complication arises. However, there are an infinite number of elements in a fluid, and contaminant concentration is a field quantity; it varies from point to point, i.e., with coordinates x_1, x_2, x_3, and time t. Thus contaminant concentration can be expressed as

$$C = C(x_1,x_2,x_3,t) \tag{6.172}$$

if we consider physical quantities as functions of position (x_1,x_2,x_3) and time t as a continuous medium. These are called Eulerian variables.

Suppose we consider parcels of materials, such as contaminant concentration. The parcels are moving and at some instant t = 0, a given parcel is located at the point whose Cartesian coordinates are (a_1,a_2,a_3). This parcel has a certain concentration at time t, while some other parcel, located at some other point and having different initial coordinates, has a different concentration. In other words, at time t the concentration of a certain parcel can be considered a function of its initial coordinates (a_1,a_2,a_3) and time t. Thus

$$C = C(a_1,a_2,a_3,t) \tag{6.173}$$

The quantities (a_1,a_2,a_3) that identify a parcel of material are called Lagrangian (or material) coordinates. We denote the *ith* component of the Lagrangian particle position at time t, by

$$X_i = X_i(\mathbf{a},t;t_0) \qquad i = 1,2,3 \tag{6.174}$$

where t_0 is the initial time and $\mathbf{a}(a_1,a_2,a_3)$ is the initial position vector. Thus, at $t = t_0$, $X_i = a_i$. The i component of the Lagrangian velocity can be expressed

$$u_{Li}(\mathbf{a},t;t_0) = \left(\frac{\partial X_i}{\partial t}\right)_{\mathbf{a}} \tag{6.175}$$

and the *ith* component of the (Lagrangian) particle acceleration is

$$\frac{\partial u_{Li}}{\partial t} = \left(\frac{\partial^2 X_i}{\partial t^2}\right)_{\mathbf{a}} \tag{6.176}$$

The connection between the Lagrangian and the Eulerian velocities is given by

$$u_{Li}(\mathbf{a},t) = u_i[X(\mathbf{a},t;t_0),t] \tag{6.177}$$

so that

$$\frac{\partial u_{Li}}{\partial t} = \frac{\partial u_i}{\partial t} + (\mathbf{u}\cdot\nabla)u_i \tag{6.178}$$

where u_i is the *ith* component of the Eulerian velocity. The transformation from the Eulerian equations to the Lagrangian equations involves replacement of the independent variables (x_1,x_2,x_3,t) by $(X_1,X_2,X_3;t)$ and transformation from the unknown Eulerian functions $u_i(X_1,X_2,X_3,t)$ to the Lagrangian variables $X_1(a_1,a_2,a_3,t)$ in the preceding equations.

6-3.2 The Governing Equations in the Lagrangian System

In the Lagrangian system the diffusion equation, momentum equations, continuity equation, and heat-energy equation can be written, respectively, as (Kao, 1981)

$$\rho J\left(\frac{X_1,X_2,X_3}{a_1,a_2,a_3}\right) = \rho_0 \tag{6.181}$$

$$\begin{aligned}\left(\frac{\partial T}{\partial t}\right)_{\mathbf{a}} = -\frac{p}{C_p\rho_0}\Bigg\{&J\left[\frac{(\partial X_1/\partial t),X_2,X_3}{a_1,a_2,a_3}\right] + J\left[\frac{X_1,(\partial X_2/\partial t),X_3}{a_1,a_2,a_3}\right] + J\left[\frac{X_1,X_2,(\partial X_3/\partial t)}{a_1,a_2,a_3}\right]\Bigg\}\\ &+ \frac{\kappa}{C_p\rho_0}\Bigg\{J\left[\frac{\frac{\rho}{\rho_0}J\left(\frac{T,X_2,X_3}{a_1,a_2,a_3}\right),X_2,X_3}{a_1,a_2,a_3}\right] + J\left[\frac{X_1,\frac{\rho}{\rho_0}J\left(\frac{X_1,T,X_3}{a_1,a_2,a_3}\right),X_3}{a_1,a_2,a_3}\right]\\ &+ J\left[\frac{X_1,X_2,\frac{\rho}{\rho_0}J\left(\frac{X_1,X_2,T}{a_1,a_2,a_3}\right)}{a_1,a_2,a_3}\right]\Bigg\}\end{aligned} \tag{6.182}$$

where

$$J\left(\frac{M,Q,R}{a_1,a_2,a_3}\right) \equiv \begin{vmatrix} \frac{\partial M}{\partial a_1} & \frac{\partial Q}{\partial a_1} & \frac{\partial R}{\partial a_1} \\ \frac{\partial M}{\partial a_2} & \frac{\partial Q}{\partial a_2} & \frac{\partial R}{\partial a_2} \\ \frac{\partial M}{\partial a_3} & \frac{\partial Q}{\partial a_3} & \frac{\partial R}{\partial a_3} \end{vmatrix} \tag{6.183}$$

and $\rho_0 = \rho_0(a_1,a_2,a_3,0)$ is the density of the fluid parcel at the initial time.

Note that the nonlinear terms, involving advection, in the equations in the Eulerian system disappear in the Lagrangian system. However, terms involving the pressure force, the molecular diffusion, friction, and heat conduc-

(Text continues on the next page.)

$$\left(\frac{\partial C}{\partial t}\right)_{\mathbf{a}} = D\frac{\rho}{\rho_0}\left\{J\left[\frac{\frac{\rho}{\rho_0}J\left(\frac{C,X_2,X_3}{a_1,a_2,a_3}\right),X_2,X_3}{a_1,a_2,a_3}\right] + J\left[\frac{X_1,\frac{\rho}{\rho_0}J\left(\frac{X_1,C,X_3}{a_1,a_2,a_3}\right),X_3}{a_1,a_2,a_3}\right] + J\left[\frac{X_1,X_2,\frac{\rho}{\rho_0}J\left(\frac{X_1,X_2,C}{a_1,a_2,a_3}\right)}{a_1,a_2,a_3}\right]\right\} \tag{6.179}$$

$$\begin{aligned}\left(\frac{\partial^2 X_i}{\partial t^2}\right)_{\mathbf{a}} = &-\frac{1}{\rho_0}J\left(\frac{p,X_j,X_k}{a_1,a_2,a_3}\right) - \epsilon_{ijk}f_j\left(\frac{\partial X_k}{\partial t}\right)_{\mathbf{a}} - g\,\delta_{3i}\\ &+\frac{\mu}{\rho_0}\left(J\left\{\frac{\frac{\rho}{\rho_0}J\left[\frac{(\partial X_i/\partial t),X_2,X_3}{a_1,a_2,a_3}\right],X_2,X_3}{a_1,a_2,a_3}\right\} + J\left\{\frac{X_1,\frac{\rho}{\rho_0}J\left[\frac{X_1,(\partial X_i/\partial t),X_3}{a_1,a_2,a_3}\right],X_3}{a_1,a_2,a_3}\right\} + J\left\{\frac{X_1,X_2,\frac{\rho}{\rho_0}J\left[\frac{X_1,X_2,(\partial X_i/\partial t)}{a_1,a_2,a_3}\right]}{a_1,a_2,a_3}\right\}\right)\\ &+\frac{1}{3}\frac{\mu}{\rho_0}J\left(\frac{X_j,X_k,\frac{\rho}{\rho_0}\left\{J\left[\frac{(\partial X_i/\partial t),X_2,X_3}{a_1,a_2,a_3}\right] + J\left[\frac{X_1,(\partial X_i/\partial t),X_3}{a_1,a_2,a_3}\right] + J\left[\frac{X_1,X_2,(\partial X_i/\partial t)}{a_1,a_2,a_3}\right]\right\}}{a_1,a_2,a_3}\right)\end{aligned} \tag{6.180}$$

tion become nonlinear in the equations in the Lagrangian system.

In turbulent flow the effects of the molecular diffusion, friction, and heat conduction are generally small. If terms involving these effects are neglected, Eqs. 6.179, 6.180, and 6.182 may be written

$$\left(\frac{\partial C}{\partial t}\right)_a = 0 \tag{6.184}$$

$$\left(\frac{\partial^2 X_i}{\partial t^2}\right)_a = -\frac{1}{\rho_0} J\left(\frac{p, X_j, X_k}{a_1, a_2, a_3}\right) - \epsilon_{ijk} f_j \frac{\partial X_k}{\partial t} - g\,\delta_{3i} \tag{6.185}$$

$$\left(\frac{\partial T}{\partial t}\right)_a = -\frac{p}{C_p \rho_0}\left\{J\left[\frac{(\partial X_1/\partial t), X_2, X_3}{a_1, a_2, a_3}\right] + J\left[\frac{X_1, (\partial X_2/\partial t), X_3}{a_1, a_2, a_3}\right] + J\left[\frac{X_1, X_2, (\partial X_3/\partial t)}{a_1, a_2, a_3}\right]\right\} \tag{6.186}$$

Integration of Eq. 6.184 with respect to time yields

$$C(a_1, a_2, a_3, t) = C(a_1, a_2, a_3, 0) \tag{6.187}$$

indicating that the concentration of a fluid parcel remains the same for all time. It is because of this Lagrangian characteristic of fluid parcels that statistics of Lagrangian characteristics prove to be a powerful tool in the study of diffusion, as will be seen in later sections.

In view of the conservation of particle concentration, the prediction of particle movement is germain in the prediction of pollution. However, predicting particle movement with the Lagrangian equation of motion is not easy, because of the nonlinearity in the pressure-force term. For this reason, little study has been made of the Lagrangian dynamics, and the Lagrangian approach has mostly been used in the statistical analysis of diffusion. However, in some comparatively simple cases, particle dynamics has been investigated by Levin (1954), Chadam (1962), Kao and Wurtele (1959), Kao (1960, 1962), Wurtele (1960), and Krasnoff and Peskin (1971).

6-3.3 Transport and Turbulent Diffusion in a Uniform Mean Flow (Lagrangian Approach)

A Lagrangian approach to the dispersion of contaminants was first introduced by Taylor (1921) and was later generalized by Batchelor (1949) and others. The approach is based on the concept that the probability density of the random particle position is the distribution of the mean (Eulerian) concentration. To describe the transport and diffusion of pollutants in the Lagrangian system, we can proceed with the statistical analysis as follows:

Consider a marked fluid particle moving in a field of homogeneous and stationary turbulence. Let (x_1', x_2', x_3') be the location of the marked fluid particle at the initial instant t_0, and let (x_1, x_2, x_3) be the location at a later time t. The displacement $(x_1 - x_1', x_2 - x_2', x_3 - x_3')$ of the marked particle is then a random function of time interval $t - t_0$. If $b(x_1 - x_1', x_2 - x_2', x_3 - x_3', t; t_0)$ is the probability-density function of the particle displacement from point (x_1', x_2', x_3') at time t_0 to $(x - x_1', x_2 - x_2', x_3 - x_3')$ at time t, the probability $B(x_1, x_2, x_3, t)$ of the arrival of the marked fluid particle at point (x_1, x_2, x_3) at time t can be expressed as

$$B(x_1, x_2, x_3, t) = \iiint_{-\infty}^{+\infty} B(x_1', x_2', x_3', t_0) \times b(x_1 - x_1', x_2 - x_2', x_3 - x_3', t; t_0)\, dx_1'\, dx_2'\, dx_3' \tag{6.188}$$

which is equal to the mean concentration of marked fluid particles at time t. Equation 6.188 gives the relationship between the ensemble average field and the particle-displacement probability.

The condition for the conservation of the marked fluid particles requires that the integration of Eq. 6.188, over all space, equal the total mass of the marked fluid particles, i.e.,

$$\iiint_{-\infty}^{+\infty} B(x_1, x_2, x_3, t)\, dx_1\, dx_2\, dx_3 = Q \tag{6.189}$$

where Q is the total mass of the marked fluid particles and $B(x_1, x_2, x_3, t) = 1$ or 0 depending on whether (x_1, x_2, x_3) lies inside or outside the region of marked fluid particles at time t.

The ensemble average (see Chap. 5, Sec. 5-1.4) concentration $\langle C \rangle$ of the marked fluid particles can be expressed as

$$\langle C(x_1, x_2, x_3, t; x_1', x_2', x_3', t_0)\rangle = Q\, b(x_1 - x_1', x_2 - x_2', x_3 - x_3', t; t_0) \tag{6.190}$$

To describe the transport and diffusion of marked fluid particles, we are primarily interested in the movement of the center of mass of the particles and the rate of change in the mean-square particle distance from the center of mass. The former can be determined from the first moment of the particle displacement, and the latter can be calculated from the rate of change in the second moment of the particle displacement relative to the center of mass. If turbulence is stationary and homogeneous along the x_i-direction, let the probability density function of the displacement $X_i = x_i - x_i'$ of a particle after a time interval $t - t_0$ be given by $b(X_i, t; t_0)$, which is independent of the initial position x_i' of the particle. The first moment of the *ith* component of the particle displacement gives the *ith* component of the displacement of the center of mass of the particles,

$$\langle X_i(\mathbf{a}, t; t_0)\rangle = \int_{-\infty}^{+\infty} X_i(\mathbf{a}, t; t_0)\, b(X_i, t; t_0)\, dX_i \tag{6.191}$$

The *ith* component of particle displacement relative to the center of mass is

$$X_i'(\mathbf{a}, t; t_0) = X_i(\mathbf{a}, t; t_0) - \langle X_i(\mathbf{a}, t; t_0)\rangle \tag{6.192}$$

The *ith* component of the mean displacement relative to the center of mass can be expressed as the first moment of the *ith* component of the displacement relative to the center of mass

$$\langle X_i'(a,t;t_0)\rangle = \int_{-\infty}^{\infty} [X_i(a,t;t_0) - \langle X_i(a,t;t_0)\rangle]\, b(X_i,t;t_0)\, dX_i = 0 \qquad (6.193)$$

A most useful function that can be used as a measure of the dispersion of marked fluid particles is the mean-square displacement of particles relative to the center of mass. This function can be expressed as the second moment of the particle displacement relative to the center of mass, according to

$$\langle X_i'^2(a,t;t_0)\rangle = \int_{-\infty}^{\infty} [X_i(a,t;t_0) - \langle X_i(a,t;t_0)\rangle]^2\, b(X_i,t;t_0)\, dX_i \qquad (6.194)$$

Taylor (1921) considered turbulent diffusion of particles released continuously from a fixed point in a uniform mean flow in which the field of turbulence is statistically homogeneous and stationary. In this case, by ergodicity, an ensemble average of a set of realizations at a given time can be replaced by averaging a single realization for a sufficiently long time. The ensemble average of the position coordinates of the marked fluid particles released from the fixed point will be on the coordinate axis, oriented along the mean flow passing through the fixed point, and is equal to the product of the diffusion time and the velocity of the mean flow (Fig. 6.12). The *ith* component of the eddy-diffusivity coefficient, which can be defined as half the rate of change of the *ith* component of the mean-square particle displacement, can be expressed as

$$K_{Li}(t) \equiv \frac{1}{2}\frac{\partial}{\partial t}\langle X_i'^2(a,t;t_0)\rangle = \frac{1}{2}\frac{\partial}{\partial t}\overline{X_i'^2(a,t;t_0)} \qquad (6.195)$$

where the overbar denotes the time average over a single realization.

Without a loss of generality, the term on the right-hand side of the preceding expression can be written in terms of the x_i component of the turbulent velocity of the particle, so that

$$\frac{1}{2}\frac{\partial}{\partial t}\langle X_i'^2(a,t)\rangle = \langle X_i'(a,t)u_{Li}'(a,t)\rangle = \int_0^t \langle u_{Li}'(a,t')u_{Li}'(a,t)\rangle\, dt' \qquad (6.196)$$

Since the field of turbulence is homogeneous and stationary (see Appendix 6-A.2), the autocorrelation function depends only on the time difference between two times, i.e., $(t - t')$. Let $\tau = t - t'$, so that

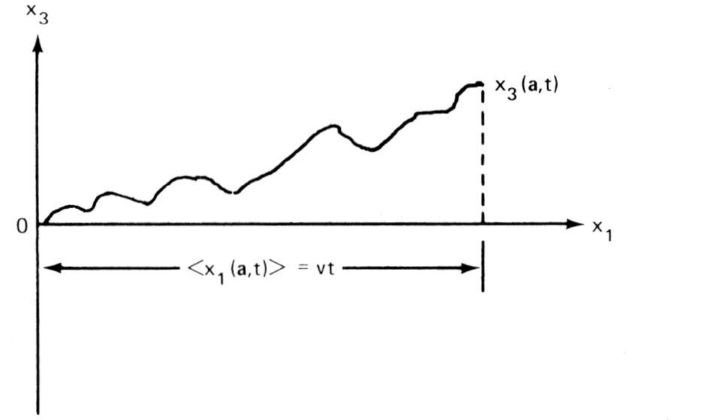

Fig. 6.12 Particle trajectory from a continuous fixed point source in a uniform mean flow.

$$\langle u_{Li}'(a,t)u_{Li}'(a,t')\rangle = \langle u_{Li}'(a,t)u_{Li}'(a,t+\tau)\rangle = \langle u_{Li}'^2\rangle R_{Li}(\tau) \qquad (6.197)$$

where

$$R_{Li}(\tau) = \frac{\langle u_{Li}'(a,t)u_{Li}'(a,t+\tau)\rangle}{\langle u_{Li}'^2(a,t)\rangle} \qquad (6.198)$$

is the Lagrangian autocorrelation coefficient of the *ith* component of the turbulent velocity. Consequently, using Eqs. 6.196 and 6.197, the rate of particle dispersion (Eq. 6.195) can be written as

$$K_{Li}(t) = \langle u_{Li}'^2\rangle \int_0^t R_{Li}(\tau)\, d\tau \qquad (6.199)$$

Integration of Eq. 6.199 with respect to time yields the mean-square particle displacement due to the *ith* component of the turbulent motion

$$\langle X_i'^2(t)\rangle = \langle X_i'^2(0)\rangle + 2\langle u_{Li}'^2\rangle \int_0^t\int_0^\eta R_{Li}(\tau)\, d\tau\, d\eta \qquad (6.200)$$

in which $\langle X_i'^2(0)\rangle = 0$ for particles released from a fixed source. Equation 6.200 is generally referred to as the Taylor diffusion theorem (Taylor, 1921).

Integrating Eq. 6.200 by parts, Kampé de Fériet (1939) obtained

$$\langle X_i'^2(t)\rangle = \langle X_i'^2(0)\rangle + 2\langle u_{Li}'^2\rangle \int_0^t (t-\tau)R_{Li}(\tau)\, d\tau \qquad (6.201)$$

For small diffusion time t, the rate of dispersion (Eq. 6.199) and mean-square particle displacement (Eq. 6.200) can be estimated by

$$K_{Li}(t) \approx \langle u_{Li}'^2\rangle t \qquad (6.202)$$

and

$$\langle X'^2(t)\rangle \approx \langle X_i'^2(0)\rangle + \langle u_{Li}'^2\rangle t^2 \qquad (6.203)$$

respectively. Equation 6.202 states that, for small diffusion time in a homogeneous, stationary field of turbulence in a uniform mean flow, the rate of dispersion can be estimated by the product of the variance of the velocity and the diffusion time. Furthermore, Eq. 6.203 says that the mean square of the displacement of particles from their mean position can be approximated by the product of the variance of the velocity and square of the diffusion time.

For large diffusion time, the rate of dispersion and the mean square of the displacement of particles from their mean position can be estimated by

$$K_{Li}(t) \approx \langle u_{Li}'^2 \rangle T_{Li} \tag{6.204}$$

$$\langle X_i'^2(t) \rangle \approx \langle X_i'^2(0) \rangle + 2\langle u_{Li}'^2 \rangle T_{Li} t \tag{6.205}$$

where

$$T_{Li} \equiv \int_0^\infty R_{Li}(\tau)\, d\tau \tag{6.206}$$

is the *ith* component of the Lagrangian integral time scale of the turbulent motion in the Lagrangian system. Equations 6.204 and 6.205 indicate that, for large diffusion time in a homogeneous, stationary field of turbulence in a uniform mean flow, the rate of dispersion can be estimated by the product of the variance of the velocity and the Lagrangian integral time scale. In addition, the mean square of the particle displacement from the mean position can be approximated by twice the product of the variance of the velocity, the Lagrangian integral time scale, and the diffusion time.

For turbulent diffusion in a coordinate system, where the coordinate axes are not parallel to the principal axes of diffusion, we need to know, in addition to $\langle X_i'^2(t) \rangle$, the mean of the product of different components of the particle displacement from their mean position

$$\langle X_i'(t) X_j'(t) \rangle = \langle [X_i(t) - \langle X_i(t) \rangle]\, [X_j(t) - \langle X_j(t) \rangle] \rangle$$

It can be shown that by following the procedure developed to arrive at Eq. 6.199 the rate of change of the ij component of the eddy diffusivity is

$$K_{Lij}(t) \equiv \frac{1}{2} \frac{\partial}{\partial t} \langle X_i'(t) X_j'(t) \rangle$$
$$= \frac{1}{2} (\langle u_{Li}'^2 \rangle \langle u_{Lj}'^2 \rangle)^{1/2} \int_0^t [R_{Lij}(\tau) + R_{Lji}(\tau)]\, d\tau \tag{6.207}$$

and

$$\langle X_i'(t) X_j'(t) \rangle = (\langle u_{Li}'^2 \rangle \langle u_{Lj}'^2 \rangle)^{1/2} \int_0^t \int_0^\eta [R_{Lij}(\tau) + R_{Lji}(\tau)]\, d\tau\, d\eta + \langle X_i'(0) X_j'(0) \rangle \tag{6.208}$$

where the coefficient of the cross correlation of the i and j components of the Lagrangian turbulent velocity.

$$R_{Lij}(\tau) \equiv \frac{\langle u_{Li}'(a,t) u_{Lj}'(a,t+\tau) \rangle}{(\langle u_{Li}'^2 \rangle \langle u_{Lj}'^2 \rangle)^{1/2}} \tag{6.209}$$

is the coefficient of the Lagrangian cross correlation of the i and j components of the Lagrangian velocity.

For small diffusion time, Eqs. 6.207 and 6.208 can be approximated by

$$K_{Lij}(t) \approx \langle u_{Lk}'(a,0) u_{Lj}'(a,0) \rangle t \tag{6.210}$$

and

$$\langle X_i'(t) X_j'(t) \rangle \approx \langle X_i'(0) X_j'(0) \rangle + \langle u_{Li}'(a,0) u_{Lj}'(a,0) \rangle t^2 \tag{6.211}$$

respectively.

For large diffusion time, Eqs. 6.207 and 6.208 can be estimated by

$$K_{Lij}(t) \approx \frac{1}{2} (\langle u_{Li}'^2 \rangle \langle u_{Lj}'^2 \rangle)^{1/2} (T_{Lij} + T_{Lji}) \tag{6.212}$$

and

$$\langle X_i'(t) X_j'(t) \rangle \approx \langle X_i'(0) X_j'(0) \rangle + (\langle u_{Li}'^2 \rangle \langle u_{Lj}'^2 \rangle)^{1/2} (T_{Lij} + T_{Lji}) t \tag{6.213}$$

where

$$T_{Lij} \equiv \int_0^\infty R_{Lij}(\tau)\, d\tau \tag{6.214}$$

is the Lagrangian integral time scale for the i and j components of the Lagrangian turbulent velocity.

6-3.3.1 Sutton's Lagrangian Autocorrelation Function and Diffusion Model. O. G. Sutton (1953) proposed a Lagrangian autocorrelation function for the turbulent motion in a stratified boundary layer. He reasoned that in the boundary layer the Lagrangian autocorrelation function must depend on the turbulence energy, $\langle v_i'^2 \rangle$, on the molecular viscosity, ν, and on the time lag, τ. From these parameters the simplest dimensionless quantity that can be formed is $\nu/\langle v_i'^2 \rangle \tau$. To satisfy the conditions that $R_L(0) = 1$ and $R_L(\infty) = 0$, Sutton obtained

$$R_{Li}(\tau) = \left(\frac{\nu}{\langle v_i'^2 \rangle \tau + \nu} \right)^n \qquad 0 \leqslant n \leqslant 1 \tag{6.215}$$

where the values of n for various degrees of thermal stratification are as follows:

Thermal Stratification, n

Large lapse	0.20
Neutral	0.25
Moderate inversion	0.33
Strong inversion	0.50

Substituting Eq. 6.215 into Eq. 6.200 and then carrying out the integrations, we find that

$$\langle X_i'^2(t)\rangle = \langle X_i'^2(0)\rangle + \frac{2\nu^n}{(1-n)(2-n)\langle v_i'^2\rangle} \times [(\nu + \langle v_i'^2\rangle t)^{2-n} - \nu^{2-n} - (2-n)\langle v_i'^2\rangle \nu^{1-n} t] \tag{6.216}$$

Since $0 < n < 1$, for sufficiently large diffusion time, the second term in Eq. 6.216 dominates. Thus

$$\langle X_i'^2\rangle \simeq \frac{2\nu^n}{(1-n)(2-n)\langle v_i'^2\rangle} (\langle v_i'^2\rangle t)^{2-n} \tag{6.217}$$

Sutton defined a virtual diffusion coefficient,

$$C_{xi}^2 = \frac{4\nu^n}{(1-n)(2-n)\bar{u}^2}\left(\frac{\langle v_i'^2\rangle}{\bar{u}^2}\right)^{1-n} \tag{6.218}$$

Equation 6.217 can be written as

$$\langle X_i'^2\rangle = \frac{1}{2} C_{xi}^2 (\bar{u}t)^{2-n} \tag{6.219}$$

Note that C_{xi}^2 is proportional to the relative intensity of turbulence to the $(1-n)$*th* power.

Comparing Eq. 6.129 with the result from the K theory,

$$\langle X_i'^2\rangle = 2K_{xi}t \tag{6.220}$$

we find that the two are equivalent only when $n = 1$.

6-3.4 The Probability Density of Lagrangian Velocity

In this section we show that in a homogeneous stationary field of turbulence the ensemble mean of the intensity of the Lagrangian turbulent velocity is equal to that of the Eulerian turbulent velocity. Let us use the concept that, in the field of homogeneous and stationary turbulence, a random function averaged over the entire space in the Lagrangian system is equal to that averaged over in the Eulerian system, i.e.,

$$\lim_{V\to\infty} \frac{1}{V} \iiint_V G[\mathbf{X}(a_1,a_2,a_3,t),t]\, da_1\, da_2\, da_3 = \lim_{V\to\infty} \frac{1}{V} \iiint_V G(x_1,x_2,x_3,t)\, dx_1\, dx_2\, dx_3 \tag{6.221}$$

Since we are interested in the Lagrangian and Eulerian velocities, let

$$G[\mathbf{X}(\mathbf{a},t),t] = \langle \exp[i\mathbf{k}\cdot\mathbf{u}_L'(\mathbf{a},t)]\rangle$$

$$G(\mathbf{x},t) = \langle \exp[i\mathbf{k}\cdot\mathbf{u}'(\mathbf{x},t)]\rangle$$

and take the ensemble average of both integrals in Eq. 6.221 to obtain

$$\lim_{V\to\infty} \frac{1}{V} \iiint_V \langle \exp[i\mathbf{k}\cdot\mathbf{u}_L'(a_1,a_2,a_3,t)]\rangle\, da_1\, da_2\, da_3 = \lim_{V\to\infty} \frac{1}{V} \iiint_V \langle \exp[i\mathbf{k}\cdot\mathbf{u}'(x_1,x_2,x_3,t)]\rangle\, dx_1\, dx_2\, dx_3 \tag{6.222}$$

Since the field of turbulence is homogeneous, the characteristic functions, i.e., the averaged exponential functions, are independent of position and can be taken out of the integrals. Therefore

$$\langle \exp[i\mathbf{k}\cdot\mathbf{u}_L'(\mathbf{a},t)]\rangle = \langle \exp[i\mathbf{k}\cdot\mathbf{u}'(\mathbf{x},t)]\rangle \tag{6.223}$$

implying that in homogeneous turbulence

$$\langle u_{Li}'^2\rangle = \langle u_i'^2\rangle \tag{6.224}$$

so that the variance of the Lagrangian velocities is equal to that of the Eulerian velocities. Therefore the probability densities of the Lagrangian and Eulerian turbulence fields are identical in homogeneous turbulence (Tennekes and Lumley, 1972).

6-3.5 Mean-Square Particle Displacement in Relation to Energy Spectrum

To gain an insight into the mechanism for turbulent diffusion, let us analyze the effect of the spectral energy of turbulent motion on the rate of dispersion and the mean-square particle displacement. First express the correlation function of the *ith* component of the Lagrangian velocity in terms of the Fourier transform of the normalized energy spectrum $F_{Li}(n)$ as

$$R_{Li}(\tau) = \int_0^\infty F_{Li}(n)\cos 2\pi n\tau\, dn \tag{6.225}$$

with

$$\int_0^\infty F_{Li}(n)\, dn = 1$$

Substitution of Eq. 6.225 into Eqs. 6.199 and 6.200 yields

$$K_{Li}(t) = \langle u_{Li}'^2\rangle \int_0^t \int_0^\infty F_{Li}(n)\cos 2\pi n\tau\, dn\, d\tau$$

$$= \langle u_{Li}'^2 \rangle t \int_0^\infty F_{Li}(n) \frac{\sin 2\pi nt}{2\pi nt} dn \qquad (6.226)$$

$$\langle X_i'^2(t) \rangle = \langle u_{Li}'^2 \rangle t^2 \int_0^\infty F_{Li}(n) \left(\frac{\sin \pi nt}{\pi nt}\right)^2 dn + \langle X_i'^2(0) \rangle \qquad (6.227)$$

For small diffusion time, Eqs. 6.226 and 6.227 can be approximated by

$$K_{Li}(t) \approx \langle u_{Li}'^2 \rangle t \int_0^\infty F_{Li}(n)\, dn = \langle u_{Li}'^2 \rangle t \qquad (6.228)$$

and

$$\langle X_i'^2(t) \rangle \approx \langle X_i'^2(0) \rangle + \langle u_{Li}'^2 \rangle t^2 \int_0^\infty F_{Li}(n)\, dn = \langle u_{Li}'^2 \rangle t^2 + \langle X_i'^2(0) \rangle \qquad (6.229)$$

respectively. Consequently, for small diffusion time, both the rate of dispersion and the mean-square particle displacement are affected by the turbulent motion of all frequencies.

For large diffusion time, it can be shown that Eqs. 6.226 and 6.227 can be approximated by

$$K_{Li}(t) \approx \frac{1}{4} S_{Li}(0) \qquad (6.230)$$

and

$$\langle X_i'^2(t) \rangle = \frac{t}{\pi} \langle u_{Li}'^2 \rangle \int_0^\infty F_{Li}\left(\frac{\zeta}{2\pi t}\right) \left(\frac{1-\cos \zeta}{\zeta^2}\right) d\zeta + \langle X_i'^2(0) \rangle$$

$$\approx \frac{t}{\pi} \langle u_{Li}'^2 \rangle F_{Li}(0) \int_0^\infty \left(\frac{1-\cos \zeta}{\zeta^2}\right) d\zeta + \langle X_i'^2(0) \rangle$$

$$\approx \frac{1}{2} S_{Li}(0) t + \langle X_i'^2(0) \rangle \qquad (6.231)$$

where $\zeta = 2\pi nt$ and $S_{Li}(0)$ is the energy spectrum at zero frequency. Therefore, for large diffusion time, both the rate of dispersion and the mean-square particle distance are affected by the spectral energy of the low-frequency eddies.

Hay and Pasquill (1959) noticed that the integrand of Eq. 6.231 has the same expression as the energy spectrum corrected for the effect of averaging over a time interval t (see Appendix, Eq. 6.A54).

$$F_{Lit}(n) = F_{Li}(n) \left(\frac{\sin \pi nt}{\pi nt}\right)^2 \qquad (6.232)$$

where $n = \omega/2\pi$. $F_{Lit}(n)$ is the observed normalized energy spectrum obtained from a wind-velocity recorder that has been averaged over a time interval t. If the averaging interval is selected to be equal to the diffusion time, Eq. 6.227 can be expressed as

$$\langle X_i'^2(t) \rangle = \langle u_{Li}'^2 \rangle t^2 \int_0^\infty F_{Lit}(n)\, dn + \langle X_i'^2(0) \rangle \qquad (6.233)$$

Since $\langle u_{Li}'^2 \rangle \int_0^\infty F_{Lit}(n)\, dn$ is the total turbulence energy in a velocity signal that has been subjected to a moving average over a time interval t, Eq. 6.233 can be written as

$$\langle X_i'^2(t) \rangle = \langle u_{Lit}'^2 \rangle t^2 + \langle X_i'^2(0) \rangle \qquad (6.234)$$

where $\langle u_{Lit}'^2 \rangle$ is the variance of the *ith* component of the Lagrangian velocity after being subjected to a moving average over a time interval t. As will be demonstrated in a later section, Eq. 6.233 is a very useful formula for predicting mean-square particle displacement and concentration of contaminants.

To seek a relationship between the rate of change of the ij component of the coefficient of eddy diffusivity, we consider Eq. 6.207. Since $[R_{Lij}(\tau) + R_{Lji}(\tau)]$ is an even function of τ

$$(\langle u'^2 \rangle \langle u'^2 \rangle)^{1/2} [R_{Lij}(\tau) + R_{Lji}(\tau)] = \int_0^\infty [C_{Lij}(n) + C_{Lji}(n)] \cos \pi 2\pi n\tau\, dn \qquad (6.235)$$

where $C_{Lij}(n)$ is the cospectrum of *ith* and *jth* components of the Lagrangian velocity. Substitution of Eq. 6.235 into Eqs. 6.207 and 6.208 yields

$$K_{Lij}(t) = \frac{1}{2} t \int_0^\infty [C_{Lij}(n) + C_{Lji}(n)] \left(\frac{\sin 2\pi nt}{2\pi nt}\right) dn \qquad (6.236)$$

and

$$\langle X_i'(t) X_j'(t) \rangle = \frac{1}{2} t^2 \int_0^\infty [C_{Lij}(n) + C_{Lji}(n)] \left(\frac{\sin \pi nt}{\pi nt}\right)^2 dn + \langle X_i'(0) X_j'(0) \rangle \qquad (6.237)$$

respectively. For small diffusion time, Eqs. 6.236 and 6.237 can be approximated by

$$K_{Lij}(t) \approx t \int_0^\infty [C_{Lij}(n) + C_{Lji}(n)]\, dn \approx 2 \langle u_{Li}' u_{Lj}' \rangle t \qquad (6.238)$$

and

$$\langle X_i'(t) X_j'(t) \rangle \approx t^2 \int_0^\infty [C_{Lij}(n) + C_{Lji}(n)]\, dn + \langle X_i'(0) X_j'(0) \rangle \approx 2 \langle u_{Li}' u_{Lj}' \rangle t^2 + \langle X_i'(0) X_j'(0) \rangle \qquad (6.239)$$

For large diffusion time, Eqs. 6.236 and 6.237 can be established by

$$K_{Lij}(t) \approx \frac{1}{8}\,[C_{Lij}(0) + C_{Lji}(0)] \qquad (6.240)$$

$$\langle X_i'(t)X'(t)\rangle \approx \frac{1}{4}\,[C_{Lij}(0) + C_{Lji}(0)]\,t \qquad (6.241)$$

6-3.6 Relation Between Eulerian and Lagrangian Velocities

From the analysis of the dispersion of the contaminants, we showed that the prediction of mean-square particle displacements in a field of statistically homogeneous and stationary turbulence depends on the velocity correlations of the Lagrangian turbulent velocities. In practice, however, the Eulerian velocities are readily measured, whereas the Lagrangian velocities, in particular the vertical component, are hard to determine. The question is therefore whether the Lagrangian velocities can be transformed to the Eulerian velocities. Mathematically, it involves coordinate transformation and has been formulated by Prandtl and Tietjens (1934), Gerber (1949), and Gifford (1955). The detailed transformation is very complex and proved to be impractical (Lumley, 1957). However, assuming that the energy spectrum can be completely determined in the inertial subrange, Corrsin (1953) obtained the energy spectra of the Eulerian and Lagrangian velocities,

$$S_{V_E}(n) = \begin{cases} B_E(\bar{u}\epsilon)^{2/3}\, n^{-5/3} & n \geqslant n_E \\ 0 & n < n_E \end{cases} \qquad (6.242)$$

$$S_{V_L}(n) = \begin{cases} B_L \epsilon n^{-2} & n \geqslant n_L \\ 0 & n < n_L \end{cases} \qquad (6.243)$$

where ϵ is the rate of dissipation, B_E and B_L are constants, and n_E and n_L are the lower limits of the Eulerian and Lagrangian energy spectra, respectively. Integration of the preceding energy spectra with respect to frequency over the whole frequency range gives the ratio β, between the Lagrangian and Eulerian characteristic time scales

$$\beta \equiv \frac{t_L}{t_E} = \frac{n_E}{n_L} = \left(\frac{3}{2}\right)^{3/2} \frac{B_E^{3/2}}{I\,B_L} \qquad (6.244)$$

where $I = \sigma_v/\bar{u}$ is the intensity of turbulence. Equation 6.244 can be written as

$$\beta I = \left(\frac{3}{2}\right)^{3/2} \frac{B_E^{3/2}}{B_L} \qquad (6.245)$$

which indicates that the product of the turbulence intensity and the ratio between the Lagrangian and Eulerian characteristic time scales in the inertial subrange is a constant. Various theoretical estimates of the ratio $\beta = t_L/t_E$ all lead to βI = constant (Saffman, 1962; Philip, 1967). The factor $B_E^{3/2}/B_L$ in Eq. 6.245 can be determined with use of the Eulerian and Lagrangian energy spectra given by Eqs. 6.242 and 6.243

$$\frac{B_E^{3/2}}{B_L} = \frac{[nS_{V_E}(n)]^{3/2}}{\bar{u}nS_{V_L}(n)} \qquad (6.246)$$

Using Eq. 6.246 and observed data, Pasquill (1968) found that $B_E^{3/2}/B_L \approx 0.23$ and that $\beta I \approx 0.35$ to 0.8.

Experimentally, it has been found that a similarity exists between the autocorrelation functions of the Eulerian and Lagrangian velocities in wind-tunnel experiments (Mickelsen, 1955), in the microscale (Gifford, 1955; Hay and Pasquill, 1957), and in the large-scale (Kao and Bullock, 1964; Kao, 1965) motion in the atmosphere. On the basis of Gifford's experience on simultaneous Eulerian and Lagrangian velocities (Gifford, 1962), Hay and Pasquill (1957) proposed that

$$R_L(\beta\tau) = R(\tau) \qquad (6.247)$$

where β is the ratio between the Lagrangian and Eulerian time scale.

It can be shown by integrating Eq. 6.247 with respect to τ from zero to infinity that

$$\beta = \frac{T_L}{T_E} \qquad (6.248)$$

where T_L and T_E are, respectively, the Lagrangian and Eulerian integral time scales. For atmospheric microscale turbulence, β has been experimentally found to be between 4 and 20 (Pasquill, 1974), whereas for the large-scale turbulence in the atmosphere it is found to be between 0.3 and 0.5 (Kao and Bullock, 1964; Kao, 1965, 1974; Murgatroyd, 1969).

With the use of Eq. 6.247 and the inverse of Eq. 6.225, it can be shown that

$$F_{Li}(n) = \beta F_i(\beta n) \qquad (6.249)$$

which establishes a relationship between the Lagrangian and Eulerian energy spectra.

The mean-square displacement (Eq. 6.227) can be shown to be expressed in terms of the Eulerian energy spectrum as

$$\langle X_i'^2(t)\rangle = \langle X_i'^2(0)\rangle + \langle u_i'^2\rangle t^2 \int_0^\infty \beta F_i(\beta n)\left(\frac{\sin \pi n t}{\pi n t}\right)^2 dn \qquad (6.250)$$

which can be further expressed as

$$\langle X_i'^2(t)\rangle = \langle X_i'^2(0)\rangle + \langle u_i'^2\rangle t^2 \int_0^\infty F_i(n) \left(\frac{\sin\frac{\pi n t}{\beta}}{\frac{\pi n t}{\beta}}\right)^2 dn \quad (6.251)$$

With the use of Eqs. 6.232 and 6.234, Eq. 6.251 can be written as

$$\langle X_i'^2(t)\rangle = \langle X_i'^2(0)\rangle + \left\langle u_i'^2, \frac{t}{\beta}\right\rangle t^2 \quad (6.252)$$

Therefore the mean-square distance at diffusion time t can be predicted by the sum of the initial mean-square distance and the product of the diffusion time squared and the variance of the Eulerian velocity after being subjected to a moving average over a time interval t/β.

Through analogous manipulation, Eq. 6.237 can be expressed in terms of the cospectra of the Eulerian velocities as

$$\langle X_i'(t)X_j'(t)\rangle = \frac{1}{2} t^2 \int_0^\infty [C_{ij}(n) + C_{ji}(n)] \left(\frac{\sin\frac{\pi n t}{\beta}}{\frac{\pi n t}{\beta}}\right)^2 dn + \langle X_i'(0)X_j'(0)\rangle \quad (6.253)$$

where $C_{ij}(n)$ is the Eulerian cospectrum of the *ith* and *jth* components of the velocities.

6-3.7 Gaussian Diffusion Model in a Uniform Mean Flow

As a measure of the diffusion of contaminants, we analyzed the mean-square displacement of marked fluid particles in the previous sections and showed that it is the second spatial moment of the particle displacement probability distribution. If the probability distribution is Gaussian, the second moment of the particle displacement would be sufficient to describe the concentration distribution. It has been shown that, if the Fourier components of a particle velocity are independent random variables, the probability distribution is Gaussian (Ogura, 1952; Bartlett, 1956). Although many successful theoretical treatises on turbulent flow assume that Fourier components of distant wave numbers are independent, Batchelor (1953) indicated that individual Fourier components are not independent. Although there has been no rigorous theoretical justification for the Gaussian distribution of the particle probability function, experimental data indicate that the Gaussian model provides a good first approximation to the concentration distribution.

To seek the concentration distribution resulting from diffusion from a continuous point source in a uniform mean flow, we assume that the probability distribution is Gaussian and consider diffusion from a continuous point source located at the origin of a Cartesian coordinate system. Let the coordinate axes be along the principal axes of the dispersion, so that $\langle X_i'(a,t)X_j'(a,t)\rangle = 0$ for $i \neq j$, and let the x-direction be along the mean velocity $\langle u_1\rangle = U$ of the fluid. Consider the continuous plume as consisting of an infinite number of puffs emitted from the source. The mean concentration from an ensemble average of instantaneous puffs of a trace constituent of strength $Q\,dt'$ emitted at $t = t'$ over a time interval dt' can be expressed as

$$d\langle C(x,y,z,t)\rangle = \frac{Q\,dt'}{(2\pi)^{3/2}\sigma_x(t-t')\sigma_y(t-t')\sigma_z(t-t')} \times \exp\left\{-\frac{[x-U(t-t')]^2}{2\sigma_x^2(t-t')} - \frac{y^2}{2\sigma_y^2(t-t')} - \frac{z^2}{2\sigma_z^2(t-t')}\right\} \quad (6.254)$$

where

$$\sigma_{xi}^2(t-t') = \langle X_i'^2(t-t')\rangle$$

For a continuous point source, which can be considered as consisting of an infinite number of puffs, the mean concentration from an ensemble average of instantaneous puffs can be obtained by integrating Eq. 6.254 from $-\infty$ to t. If the turbulent velocities are small as compared with the mean velocity, we can neglect turbulent diffusion in the direction of the mean velocity. Therefore the continuous plume can be considered as consisting of an infinite number of disks perpendicular to the direction of the mean flow. The mean concentration in a disk can be found by integrating Eq. 6.254 with respect to x from $-\infty$ to ∞, so that

$$\langle C(x,y,z,t)\rangle = \int_{-\infty}^{\infty} \frac{Q}{(2\pi)^{3/2}\sigma_x(t-t')\sigma_y(t-t')\sigma_z(t-t')} \times \exp\left\{-\frac{[x-U(t-t')]^2}{2\sigma_x^2(t-t')} - \frac{y^2}{2\sigma_y^2(t-t')} - \frac{z^2}{2\sigma_z^2(t-t')}\right\} \frac{dx}{U} \quad (6.255)$$

Since $t - t'$ is the interval between time of release of a disk and its arrival at x, the mean concentration can be expressed as

$$\langle C(x,y,z)\rangle = \frac{Q}{2\pi U\sigma_y(x)\sigma_z(x)} \times \exp\left[-\frac{y^2}{2\sigma_y^2(x)} - \frac{z^2}{2\sigma_z^2(x)}\right] \quad (6.256)$$

where

$$\sigma_{xi}^2(x) = \langle X_i'^2(t)\rangle$$

It can easily be shown that Eq. 6.256 satisfies the continuity equation so that

$$\iint_{-\infty}^{+\infty} U\langle C(x,y,z)\rangle\, dy\, dz = Q \quad (6.257)$$

To construct the solution for the mean concentration, we integrate Eq. 6.251 with respect to t′ from $-\infty$ to t and obtain

$$\langle C(x,y,z,t)\rangle = \int_{-\infty}^{t} \frac{Q}{(2\pi)^{3/2}\sigma_x(t-t')\sigma_y(t-t')\sigma_z(t-t')}$$

$$\times \exp\left\{-\frac{[x-U(t-t')]^2}{2\sigma_x^2(t-t')} - \frac{y^2}{2\sigma_y^2(t-t')} - \frac{z^2}{2\sigma_z^2(t-t')}\right\} dt' \tag{6.258}$$

By letting $t - t' = \tau$, the preceding equation becomes

$$\langle C(x,y,z)\rangle = \int_{0}^{\infty} \frac{Q}{(2\pi)^{3/2}\sigma_x(\tau)\sigma_y(\tau)\sigma_z(\tau)}$$

$$\times \exp\left[-\frac{(x-U\tau)^2}{2\sigma_x^2(\tau)} - \frac{y^2}{2\sigma_y^2(\tau)} - \frac{z^2}{2\sigma_z^2(\tau)}\right] d\tau \tag{6.259}$$

For small diffusion time, at a short distance from the source, $\sigma_x \approx \sigma_u t$, $\sigma_y \approx \sigma_v t$, $\sigma_z \approx \sigma_w t$, Eq. 6.258 can be integrated to yield (Frenkiel, 1953)

$$\langle C(x,y,z)\rangle = \frac{Q\sigma_u \exp\left(-\frac{U^2}{2\sigma_u^2}\right)}{(2\pi)^{3/2}\sigma_v\sigma_w r^2}$$

$$\times \left[1 + \left(\frac{\pi}{2}\right)^{1/2} \frac{Ux}{\sigma_u r} \exp\left(\frac{U^2x^2}{2\sigma_u^2 r^2}\right) \operatorname{erfc}\left(-\frac{1}{(2)^{1/2}}\frac{Ux}{ur}\right)\right] \tag{6.260}$$

where

$$r^2 = x^2 + \left(\frac{\sigma_u}{\sigma_v}\right)^2 y^2 + \left(\frac{\sigma_u}{\sigma_w}\right)^2 z^2$$

Figure 6.13 shows the nondimensional concentration distribution for $\sigma_u = \sigma_v = \sigma_w$ and $\sigma_u/U = 0.5$ in a section parallel to the mean flow and passing through the source.

For

$$\sigma_u/U \approx 0.1, \qquad r \approx x, \qquad \operatorname{erfc}\left[-\frac{1}{(2)^{1/2}}\frac{Ux}{\sigma_u r}\right] \approx 2$$

and Eq. 6.260 can be approximated by Eq. 6.256.

Under the assumption that there is total reflection of the contaminant at the boundary, the mean concentration due to a continuous point source over a flat surface can be shown to take the form of

$$\langle C(x,y,z)\rangle = \frac{Q}{2\pi U\sigma_y(x)\sigma_z(x)} \exp\left[-\frac{1}{2}\left(\frac{y}{\sigma_y}\right)^2\right]$$

$$\times \left\{\exp\left[-\frac{1}{2}\left(\frac{Z-H}{\sigma_z(x)}\right)^2\right] + \exp\left[-\frac{1}{2}\left(\frac{Z+H}{\sigma_z(x)}\right)^2\right]\right\} \qquad Z \geqslant 0 \tag{6.261}$$

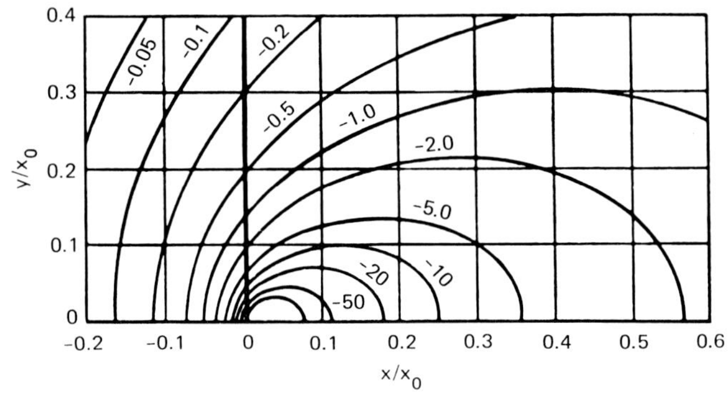

Fig. 6.13 Contours of constant mean concentration in meridional section across a continuous point source. Abscissa and ordinate are x/x_0 and y/x_0 (with $x_0 = Ut_L$, a suitable length scale). Figures or contours indicate values of nondimensional concentration, $\chi U x_0^2/q$. [From F. N. Frenkiel, Turbulent Diffusion: Mean Concentration Distribution in a Flow Field of Homogeneous Turbulence, *Advances in Applied Mechanics*, 3: 61-107 (1953).]

where H is the effective emission height which is equal to the sum of stack height and height of plume rise from the stack.

It can be shown that Eq. 6.261 satisfies the continuity equation so that

$$\int_0^{\infty}\int_{-\infty}^{\infty} U\langle C(x,y,z)\rangle \, dy\, dz = Q \tag{6.262}$$

Equation 6.261, which was first presented by O. G. Sutton (1947a, 1947b), has been widely used for describing concentration distribution resulting from dispersion of contaminants continuously emitted from an elevated point source. With Eq. 6.261, the ground concentration can be obtained by letting z = 0 so that

$$\langle C(x,y,0)\rangle = \frac{Q}{\pi U\sigma_y\sigma_z} \exp\left[-\frac{1}{2}\left(\frac{y}{\sigma_y}\right)^2 - \frac{1}{2}\left(\frac{H}{\sigma_z}\right)^2\right] \tag{6.263}$$

In addition, the maximum concentration at the ground occurs on the line (x,0,0) so that Eq. 6.263 becomes

$$\langle C(x,0,0)\rangle = \frac{Q}{\pi U\sigma_y\sigma_z} \exp\left(-\frac{1}{2}\frac{H}{\sigma_z}\right)^2 \tag{6.264}$$

Another application can be developed. For example, if a stable layer occurs at z = I, the concentration distribution can be shown to take the following form,

$$\langle C(x,y,z)\rangle = \frac{Q}{2\pi U\sigma_y\sigma_z} \exp\left[-\frac{1}{2}\left(\frac{y}{\sigma_y}\right)^2\right]\left\{\exp\left[-\frac{1}{2}\left(\frac{z-H}{\sigma_z}\right)^2\right]\right.$$

$$\left. + \exp\left[-\frac{1}{2}\left(\frac{z+H}{\sigma_z}\right)^2\right]\right\}$$

$$+\sum_{n=1}^{\infty}\left\{\exp\left[-\frac{1}{2}\left(\frac{z-H-2nI}{\sigma_z}\right)^2\right]\right.$$

$$+\exp\left[-\frac{1}{2}\left(\frac{z+H-2nI}{\sigma_z}\right)^2\right]$$

$$+\exp\left[-\frac{1}{2}\left(\frac{z-H+2nI}{\sigma_z}\right)^2\right]$$

$$\left.+\exp\left[-\frac{1}{2}\left(\frac{z+H+2nI}{\sigma_z}\right)^2\right]\right\}$$

$$\text{for } 0 \leqslant z \leqslant I \quad (6.265)$$

In this equation four terms usually are sufficient to approximate the concentration distribution not far from the source.

For dispersion from an infinite crosswind line source over a flat surface, the concentration distribution can be obtained by considering the line source as consisting of an infinite number of point sources. Thus we have

$$\langle C(x,z)\rangle = \int_{-\infty}^{\infty} \frac{Q}{2\pi U\sigma_y(x)\sigma_z(x)} \exp\left[-\frac{1}{2}\left(\frac{y-y'}{\sigma_y}\right)^2\right]$$

$$\left\{\exp\left[-\frac{1}{2}\left(\frac{z-H}{\sigma_z}\right)^2\right] + \exp\left[-\frac{1}{2}\left(\frac{z+H}{\sigma_z}\right)^2\right]\right\} dy'$$

$$= \frac{Q}{(2\pi)^{1/2}U\sigma_z(x)}\left\{\exp\left[-\frac{1}{2}\left(\frac{z+H}{\sigma_z}\right)^2\right]\right.$$

$$\left.+ \exp\left[-\frac{1}{2}\left(\frac{z-H}{\sigma_z}\right)^2\right]\right\} \quad (6.266)$$

The Gaussian distribution developed so far gives a first approximation to the lateral concentration distribution for dispersion of contaminants from an elevated continuous fixed source over a flat surface. However, for a ground source, the Gaussian model would not give a good description of the concentration distribution along the vertical. For dispersion of contaminants from a continuous fixed source over three-dimensional terrain, Kao (1976b) has shown that the concentration distribution takes the form of

$$\langle C(x,y,z)\rangle = \frac{Q}{2\pi U\sigma_y(x)\sigma_z(x)} \exp\left[-\frac{1}{2}\left(\frac{y}{\sigma_y}\right)^2\right]$$

$$\times\left(\exp\left[-\frac{1}{2}\left(\frac{z-H}{\sigma_z}\right)^2\right]\right.$$

$$\left.+ \exp\left\{-\frac{1}{2}\left[\frac{z+H-2\,h(x,y)}{\sigma_z}\right]^2\right\}\right) \quad (6.267)$$

where h(x,y) is the height of the terrain and is a function of the horizontal dimensions x and y. It can be shown that Eq. 6.267 satisfies the continuity condition for the conservation of marked fluid particles,

$$\int_{-\infty}^{\infty}\int_{h(x,y)}^{\infty} U\langle C(x,y,z)\rangle\, dz\, dy = Q \quad (6.268)$$

Strictly speaking, the concentration distributions developed apply to an environment in which the mean velocity is constant. Therefore they can at best be applied to regions far from the boundaries since, near boundaries, the velocity shear of the mean flow is generally strong.

Examining the Gaussian models for describing the concentration distribution, we find that the important parameters involved are the mean velocity of the fluid and the mean-square displacements $\sigma_y^2(x)$ and $\sigma_z^2(x)$. The latter are greatly influenced by the meteorological conditions in the atmosphere. Diffusion data over flat terrain under various stability conditions have been summarized in Chap. 9 and by Pasquill (1961, 1962, 1974) and Gifford (1961, 1976) and have been modified by Briggs (1974) with the inclusion of the buoyancy effect on the plume. Figure 6.14 shows Briggs's modified curves for σ_y and σ_z. The dashed portions of these curves indicate the tentative nature of the extrapolation beyond about 10 km, and the letters A through F denote atmospheric stability categories ranging from unstable to very stable (Table 6.1). More discussion and references on this topic can be found in Chap. 8. Additional information on Gaussian models is given in Chap. 13.

6-3.8 Transport and Diffusion from a Continuous Point Source in a Uniform Shear Flow

In Sec. 6-2.3, we discussed the transport and diffusion of pollutants from a fixed continuous point source in a uniform shear flow in the Eulerian system. In this section we consider the transport and diffusion of pollutants in a uniform shear flow in the Lagrangian system.

Let the mean velocity be along the x_1-direction and the shear of the mean velocity be $d\langle u_1\rangle/dx_2$ = constant. We assume that the Eulerian velocity field is homogeneous in all directions and the Lagrangian velocity field is stationary. Consequently, since $\langle u_2\rangle = \langle u_3\rangle = 0$, the velocity components of a particle are

$$u_{L1}(a,t) = \frac{d\langle u_1\rangle}{dx_2} X_2'(a,t) + u_{L1}'(a,t)$$

$$u_{L2}(a,t) = u_{L2}'(a,t) \quad (6.269)$$

$$u_{L3}(a,t) = u_{L3}'(a,t)$$

The position of the particle can be determined by integrating the preceding equations with respect to time to yield

$$X_1(a,t) = a_1 + \int_0^t \left[\frac{d\langle u_1\rangle}{dx_2} X_2'(a,t) + u_{L1}'(a,t)\right] dt'$$

$$X_2(a,t) = a_2 + \int_0^t u_{L2}'(a,t')\, dt' \quad (6.270)$$

$$X_3(a,t) = a_3 + \int_0^t u_{L3}'(a,t')\, dt'$$

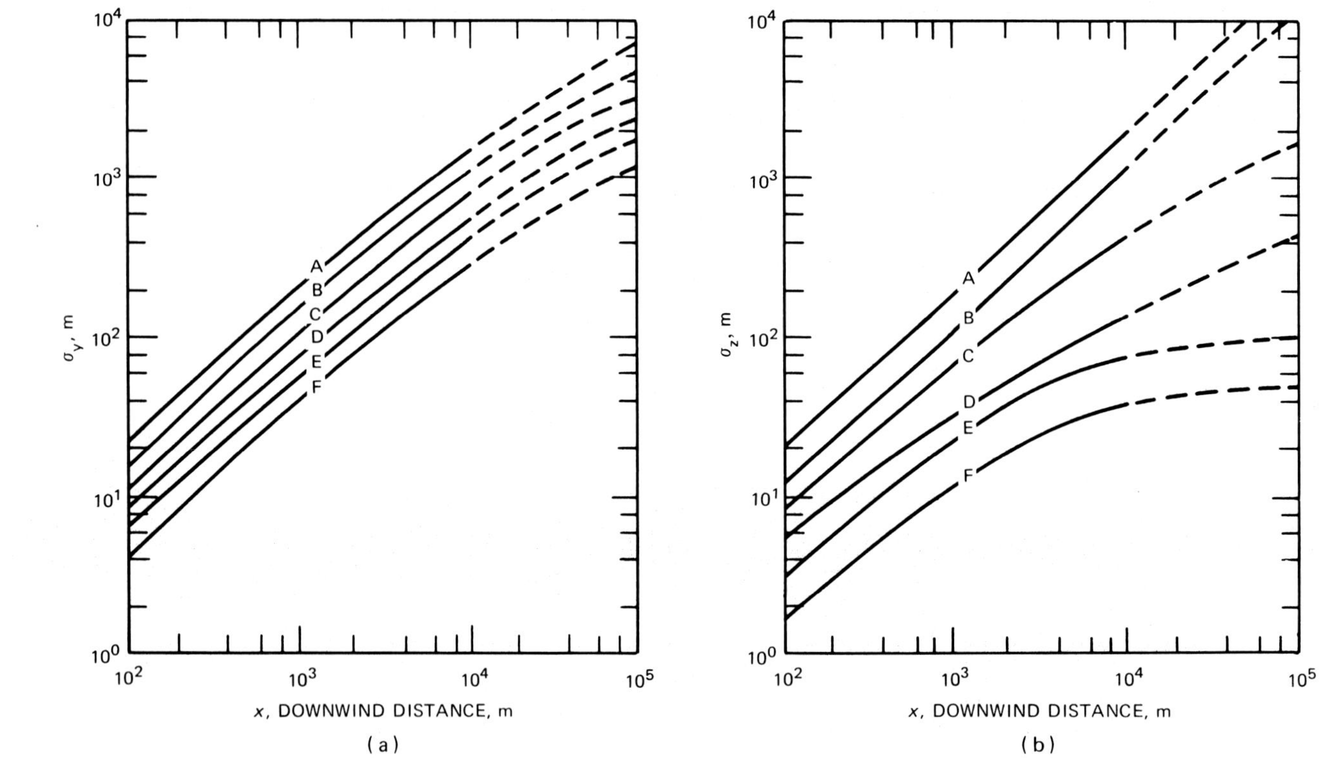

Fig. 6.14 Curves (a) of σ_y and (b) of σ_z based on Brigg's (1974) interpolation formulas for flow over open country. [See Table II of Hosker (1974).]

To describe the transport and diffusion of marked fluid particles, we are interested in the displacement of the particles and the rate of change in the mean-square particle displacement relative to the fixed source. The position of the marked fluid particles can be determined by taking the mean of the preceding equations, so that

$$\langle X_1(\mathbf{a},t)\rangle = a_1 + \frac{d\langle u_1\rangle}{dx_2} a_2 t$$
$$\langle X_2(\mathbf{a},t)\rangle = a_2 \tag{6.271}$$
$$\langle X_3(\mathbf{a},t)\rangle = a_3$$

Therefore the mean velocity at the initial position or the transport velocity of the center of mass is $a_2 \, d\langle u_1\rangle/dx_2$.

With the use of Eqs. 6.269 and 6.270, the rate of change of the members of the displacement tensor can be shown to take the forms of

$$\frac{\partial}{\partial t}\langle X_1^2(\mathbf{a},t)\rangle = 2\int_0^t \left\langle \left\{ \left(\frac{d\langle u_1\rangle}{dx_2}\right)^2 X_2(\mathbf{a},t)X_2(\mathbf{a},t') + \frac{d\langle u_1\rangle}{dx_2}[X_2(\mathbf{a},t)u'_{L1}(\mathbf{a},t') + X_2(\mathbf{a},t')u'_{L1}(\mathbf{a},t)] + u'_{L1}(\mathbf{a},t')u'_{L1}(\mathbf{a},t) \right\} \right\rangle dt' \tag{6.272}$$

$$\frac{\partial}{\partial t}\langle X_1(\mathbf{a},t)X_2(\mathbf{a},t)\rangle = \int_0^t \left\langle \left\{ \frac{d\langle u_1\rangle}{dx_2}[X_2(\mathbf{a},t)u'_{L2}(\mathbf{a},t) + X_2(\mathbf{a},t')u'_{L2}(\mathbf{a},t)] + u'_{L1}(\mathbf{a},t')u'_{L2}(\mathbf{a},t) + u'_{L1}(\mathbf{a},t)u'_{L2}(\mathbf{a},t') \right\} \right\rangle dt' \tag{6.273}$$

$$\frac{\partial}{\partial t}\langle X_2^2(\mathbf{a},t)\rangle = 2\int_0^t \langle u'_{L2}(\mathbf{a},t')u'_{L2}(\mathbf{a},t)\rangle \, dt' \tag{6.274}$$

$$\frac{\partial}{\partial t}\langle X_3^2(\mathbf{a},t)\rangle = 2\int_0^t \langle u'_{L3}(\mathbf{a},t')u'_{L3}(\mathbf{a},t)\rangle \, dt' \tag{6.275}$$

Integrating Eqs. 6.272 to 6.275 with respect to time, we obtain

$$\langle X_1^2(\mathbf{a},t)\rangle = a_1^2 + \left(\frac{d\langle u_1\rangle}{dx_2}\right)^2 \langle u'^2_{L2}(\mathbf{a},t)\rangle \left[\frac{2}{3}t^3\int_0^t R_{L22}(\tau)\,d\tau - t^2\int_0^t \tau R_{L22}(\tau)d\tau + \frac{1}{3}\int_0^t \tau^3 R_{L22}(\tau)\,d\tau\right] + \frac{d\langle u_1\rangle}{dx_2}\langle u'_{Li}(\mathbf{a},t)u'_{L2}(\mathbf{a},t)\rangle \left\{\int_0^t (t-\tau)^2 \times [R_{L12}(\tau) + R_{L21}(\tau)]\,d\tau + 2\int_0^t \tau(t-\tau)R_{L12}(\tau)\,d\tau\right\} + 2\langle u'^2_{L1}(\mathbf{a},t)\rangle \int_0^t (t-\tau)R_{L11}(\tau)\,d\tau \tag{6.276}$$

TABLE 6.1 Definition of Six Stability Classes for Use with the Pasquill–Gifford Curves†

Surface wind speed, m/s	Daytime insolation: Strong	Daytime insolation: Moderate	Daytime insolation: Slight	Nighttime conditions: Thin overcast or ≥4/8 cloudiness‡	Nighttime conditions: ≤3/8 cloudiness
<2	A	A–B	B		
2	A–B	B	C	E	F
4	B	B–C	C	D	E
6	C	C–D	D	D	D
>6	C	D	D	D	D

Stability classes:
A extremely unstable conditions
B moderately unstable conditions
C slightly unstable conditions
D neutral conditions§
E slightly stable conditions
F moderately stable conditions

†From Gifford, 1968.
‡The degree of cloudiness is defined as that fraction of the sky above the local apparent horizon which is covered by clouds.
§Applicable to heavy overcast, day or night.

$$\langle X_1(a,t)X_2(a,t)\rangle = a_1a_2 + \frac{d\langle u_1\rangle}{dx_2}\langle u'^2_{L2}(a,t)\rangle \times \int_0^t t(t-\tau)R_{L22}(\tau)\,d\tau + \langle u'_{L1}(a,t)u'_{L2}(a,t)\rangle \int_0^t (t-\tau) \times [R_{L12}(\tau)+R_{L21}(\tau)]\,d\tau \qquad (6.277)$$

$$\langle X_2^2(a,t)\rangle = a_1^2 + 2\langle u'^2_{L2}(a,t)\rangle \int_0^t (t-\tau)R_{L22}(\tau)\,d\tau \qquad (6.278)$$

$$\langle X_3^2(a,t)\rangle = a_3^2 + 2\langle u'^2_{L3}(a,t)\rangle \int_0^t (t-\tau)R_{L33}(\tau)\,d\tau \qquad (6.279)$$

Equations 6.276 to 6.279 were first obtained by Corrsin (1953).

The rate of change of the members of the displacement tensor can be expressed in terms of the diffusivity tensor $K_{Lij}(t)$ for diffusion in a uniform flow, or mathematically

$$\frac{\partial}{\partial t}\langle X_1^2(a,t)\rangle = 2\frac{d\langle u_1\rangle}{dx_2}\langle X_1(a,t)X_2(a,t)\rangle + 2K_{L11}(t) \qquad (6.280)$$

$$\frac{\partial}{\partial t}\langle X_1(a,t)X_2(a,t)\rangle = \frac{d\langle u_1\rangle}{dx_2}\langle X_2^2(a,t)\rangle + K_{L12}(t) + K_{L21}(t) \qquad (6.281)$$

$$\frac{\partial}{\partial t}\langle X_2^2(a,t)\rangle = 2K_{L22}(t) \qquad (6.282)$$

$$\frac{\partial}{\partial t}\langle X_3^2(a,t)\rangle = 2K_{L33}(t) \qquad (6.283)$$

Notice that as $t \to \infty$

$$\langle X_1^2(a,t)\rangle \to a_1^2 + \frac{2}{3}\left(\frac{d\langle u_1\rangle}{dx_2}\right)^2 \langle u_2'^2\rangle T_{L22}t^3 \qquad (6.284)$$

$$\langle X_1(a,t)X_2(a,t)\rangle \to a_1a_2 + \frac{d\langle u_1\rangle}{dx_2}\langle u_2'^2\rangle T_{L22}t^2 \qquad (6.285)$$

$$\langle X_2^2(a,t)\rangle \to a_2^2 + 2\langle u_2'^2\rangle T_{L22}t \qquad (6.286)$$

$$\langle X_3^2(a,t)\rangle \to a_3^2 + 2\langle u_2'^2\rangle T_{L33}t \qquad (6.287)$$

The Lagrangian integral time scale, T_{Lij}, is defined in Eq. 6.214.

In Eqs. 6.284 to 6.287, the Lagrangian velocity variance $\langle u'^2_{Li}\rangle$ has been replaced by the Eulerian velocity $\langle u_i'^2\rangle$. It is seen from the preceding expressions that at large diffusion time the diffusion rate along the direction of mean flow is proportional to product of the Lagrangian integral time scale, the velocity variance, the second power of the mean velocity shear, and the diffusion time. However, the diffusion rate along the direction perpendicular to the mean flow is proportional to the product of the Lagrangian integral time scale and the velocity variance.

To determine the concentration distribution in the linear shear flow, we see from Eqs. 6.270 and 6.271 that, for large diffusion time,

$$\frac{\langle [X_1(a,t)-\langle X_1(a,t)\rangle][X_2(a,t)-a_2]\rangle}{\{\langle [X_1(a,t)-\langle X_1(a,t)\rangle]^2\rangle\langle [X_2(a,t)-a_2]^2\rangle\}^{1/2}} \to \frac{(3)^{1/2}}{2} \qquad (6.288)$$

Therefore the isopleths of probability density of mean concentration can be expressed by ellipses of the form

$$\frac{(X_1-\langle X_1\rangle)^2}{\langle (X_1-\langle X_1\rangle^2)\rangle} - (3)^{1/2}\frac{\langle (X_1-\langle X_1\rangle)(X_2-a_2)\rangle}{[\langle (X_1-\langle X_1\rangle)^2\rangle\langle (X_2-a_2)^2\rangle]^{1/2}} + \frac{(X_2-a_2)^2}{\langle (X_2-a_2)^2\rangle} = \text{constant} \qquad (6.289)$$

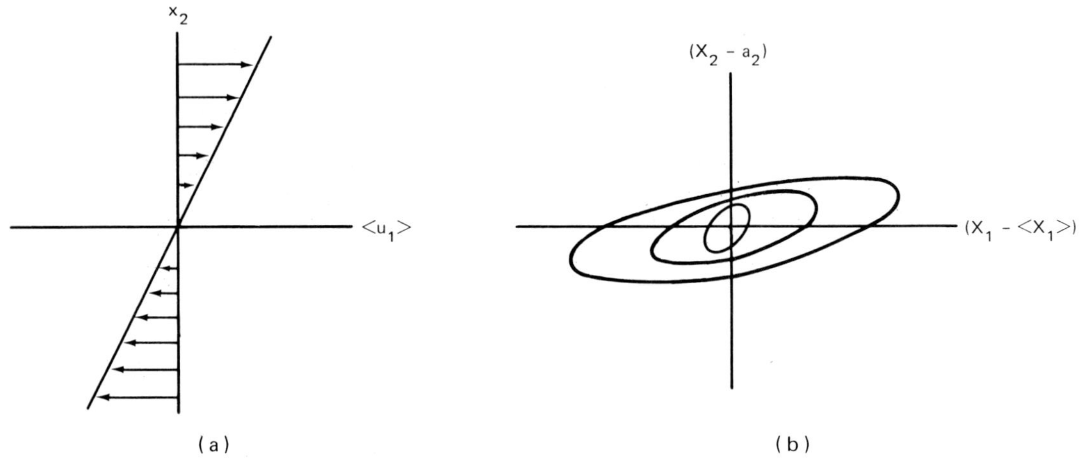

Fig. 6.15 (a) A mean velocity profile of uniform shear. (b) A dispersion pattern in the uniform shear flow. (From S. Corrsin, Remarks on Turbulent Heat Transfer, in *Proceedings of the Iowa Thermodynamics Symposium*, pp. 5-30, University of Iowa, Iowa City, 1953.)

The elliptical distribution of concentration is sketched in Fig. 6.15. The angle between the major axis of the ellipse and the x_1-direction is given by the expression

$$\arctan\left|\frac{\langle(X_2 - a_2)^2\rangle}{\langle(X_1 - \langle X_1\rangle)^2\rangle}\right|^{1/4} = \frac{(3)^{1/2}}{\dfrac{d\langle u_1\rangle}{dx_2}\,t} \qquad (6.290)$$

which decreases with increasing diffusion time.

6-3.9 Relative Diffusion

Analyses of relative diffusion are basic in the study of diffusion of particles from instantaneous sources. The problem of relative dispersion between particles is a complex one because the turbulent energy between particles generally increases with increasing distance between particles. Therefore turbulent motion between particles is nonstationary in nature, and the rate of dispersion between particles generally increases with diffusion time.

The statistical characteristics of dispersion between particles were first studied by Richardson (1926). However, formal mathematical foundation was established by Batchelor (1950, 1952) and by Brier (1950). Further investigations were carried out by Gifford (1959), Lin (1960a and b), Corrsin (1962), and Csanady (1966).

As a measure of the dispersion between particles, consider the characteristics of the mean-square particle displacement between particles. To do so, consider two marked fluid particles moving in a uniform mean flow in which the field of turbulence is statistically homogeneous and stationary. Let $X_{i\alpha}(a_\alpha,t)$ and $X_{j\beta}(a_\beta,t)$ be, respectively, the i and j position component of particles α and β at time t. Denote the i and j components of the relative position of particles α and β as

$$\begin{aligned} X_{ri}(t;a_\alpha,a_\beta) &= X_{i\alpha}(a_\alpha,t) - X_{i\beta}(a_\beta,t) \\ X_{rj}(t;a_\alpha,a_\beta) &= X_{j\alpha}(a_\alpha,t) - X_{j\beta}(a_\beta,t) \end{aligned} \qquad (6.291)$$

Furthermore, we can define the rate of change of the ensemble average of the relative particle position tensor as

$$\frac{\partial}{\partial t}\langle X_{ri}(t)X_{rj}(t)\rangle = \int_0^t [\langle u_{ri}(t)u_{rj}(t')\rangle + \langle u_{ri}(t')u_{rj}(t)\rangle]\,dt' \qquad (6.292)$$

where $u_{ri}(t)$ is the *ith* component of the relative particle velocity.

By letting $t' = t - \tau$, Eq. 6.292 can be written as

$$\frac{\partial}{\partial t}\langle X_{ri}(t)X_{rj}(t)\rangle = \int_0^t [\langle u_{ri}(t)u_{rj}(t-\tau)\rangle + \langle u_{ri}(t-\tau)u_{rj}(t)\rangle]\,d\tau \qquad (6.293)$$

Integration of this equation with respect to time gives the averaged relative particle distance tensor

$$\begin{aligned}\langle X_{ri}(t)X_{rj}(t)\rangle = \int_0^t \int_0^{t'} &[\langle u_{ri}(t')u_{rj}(t'-\tau)\rangle \\ &+ \langle u_{ri}(t'-\tau)u_{rj}(t')\rangle]\,d\tau\,dt' \\ &+ \langle X_{ri}(0)X_{rj}(0)\rangle \end{aligned} \qquad (6.294)$$

However, turbulence between particles is generally nonstationary so that the relative movement between particles is primarily affected by turbulent motions of the scale of the distance between particles. Furthermore, this distance changes with time. Consequently the correlation of relative particle velocities $\langle u_{ri}(t)u_{rj}(t-\tau)\rangle$ depends on both diffusion time and time lag τ—as does the correlation function of the relative particle velocities

$$R_{rij}(t,\tau) = \frac{\langle u_{ri}(t)u_{rj}(t-\tau)\rangle}{[\langle u_{ri}^2(t)\rangle\langle u_{rj}^2(t)\rangle]^{1/2}} \qquad (6.295)$$

Thus the averaged relative particle distance tensor can be written

$$\langle X_{ri}(t)X_{rj}(t)\rangle = \int_0^t \int_0^{t'} [\langle u_{ri}^2(t')\rangle\langle u_{rj}^2(t')\rangle]^{1/2} [R_{rij}(t',\tau) + R_{rji}(t',\tau)]\, d\tau\, dt' + \langle X_{ri}(0)X_{rj}(0)\rangle \quad (6.296)$$

Letting $i = j$, we obtain the *ith* component of the mean-square particle distance

$$\langle X_{ri}^2(t)\rangle = \langle X_{ri}^2(0)\rangle + 2 \int_0^t \int_0^{t'} \langle u_{ri}^2(t')\rangle R_{ri}(t',\tau)\, d\tau\, dt' \quad (6.297)$$

where the autocorrelation function for the relative particle velocities is defined as

$$R_{ri}(t,\tau) = \frac{\langle u_{ri}(t)u_{ri}(t-\tau)\rangle}{\langle u_{ri}^2(t)\rangle} \quad (6.298)$$

with no summation over i. Notice that in a field of homogeneous, stationary turbulence the variance of relative particle velocities is a function of diffusion time, whereas the variance of the velocities of a single particle is not. Therefore Taylor's diffusion theorem (Taylor, 1921), which applies to the movement of a single particle, does not apply to diffusion between particles.

To examine the time dependency of the variance of the relative particle velocities, let us express the relative particle velocities in forms of the velocities of individual particles

$$u_{ri}(t;a_\alpha,a_\beta) = u_{Li\alpha}(a_\alpha,t) - u_{Li\beta}(a_\beta,t) \quad (6.299)$$

Thereby the variance of the relative particle velocities can be written

$$\langle u_{ri}^2(t)\rangle = \langle u_{Li\alpha}'^2(t)\rangle + \langle u_{Li\beta}'^2(t)\rangle - 2\langle\langle u_{Li\alpha}'(t)u_{Li\beta}(t)\rangle$$
$$= 2[\langle u_{Li}'^2\rangle - \langle u_{Li\alpha}'(t)u_{Li\beta}'(t)\rangle] \quad (6.300)$$

For small diffusion time, if particles are released very close to each other, we find that Eq. 6.300 can be approximated by

$$\langle u_{ri}^2(t)\rangle \approx 0 \quad (6.301)$$

For large diffusion time, particles are so far apart that the velocity of one particle becomes independent from that of the other; thus Eq. 6.300 approaches

$$\langle u_{ri}^2(t)\rangle \rightarrow 2\langle u_i'^2\rangle \quad (6.302)$$

Therefore we expect that the variance of the relative particle velocities increases with increasing diffusion time until it reaches a limiting value of twice the variance of the Eulerian velocities.

For large diffusion time we find, from Eq. 6.297, that

$$\langle X_{ri}^2(t)\rangle \approx 2\langle u_i'^2\rangle \overline{T}_{ri} t + \langle X_{ri}^2(0)\rangle \quad (6.303)$$

where

$$\overline{T}_{ri} = \lim_{J\to\infty} \frac{1}{J} \int_0^J \int_0^\infty R_{ri}(t,\tau)\, d\tau\, dt \quad (6.304)$$

is the averaged integral time scale of the relative particle velocities.

For small and intermediate diffusion times, Batchelor (1956) applied the theory of turbulence and dimensional analysis to the determination of the mean-square particle distance $\langle X_{ri}(t)X_{ri}(t)\rangle$. He argued that in the inertial range the rate of relative diffusion, $d\langle X_{ri}X_{ri}\rangle/dt$ depends only on the initial particle distance $\langle X_{ri}(0)X_{ri}(0)\rangle^{1/2}$, the rate of turbulent-energy transfer, ϵ, and the diffusion time t, for small diffusion time $t < t^* \approx \epsilon^{-1/3}\langle X_{ri}(0)X_{ri}(0)\rangle^{1/3}$, but it depends on ϵ and t for intermediate time ($t > t^*$). Therefore, by dimensional consideration

$$\frac{d}{dt}\langle X_{ri}(t)X_{ri}(t)\rangle \propto \epsilon^{2/3}\langle X_{ri}(0)X_{ri}(0)\rangle^{1/3} t \qquad t < t^* \quad (6.305)$$

$$\frac{d}{dt}\langle X_{ri}(t)X_{ri}(t)\rangle \propto \epsilon t^2 \qquad t > t^* \quad (6.306)$$

Integrating Eqs. 6.305 and 6.306 yields

$$\langle X_{ri}(t)X_{ri}(t)\rangle - \langle X_{ri}(0)X_{ri}(0)\rangle \propto t^2 \qquad t < t^* \quad (6.307)$$

and

$$\langle X_{ri}(t)X_{ri}(t)\rangle \propto \epsilon t^3 \qquad t > t^* \quad (6.308)$$

respectively.

Elimination of t from Eqs. 6.308 and 6.306 gives

$$\frac{d}{dt}\langle X_{ri}(t)X_{ri}(t)\rangle \propto \epsilon^{1/3}\langle X_{ri}(t)X_{ri}(t)\rangle^{2/3} \qquad t > t^* \quad (6.309)$$

which is known as Richardson's 4/3 power law except that Eq. 6.309 applies to the inertial range. Obukhov (1941) applied Kolmogorov's second hypothesis for isotropic turbulence in the inertial range to relative diffusion and obtained a result similar to Eq. 6.309. Further investigations on relative diffusion have been made by Batchelor (1950, 1952), Batchelor and Townsend (1956), Hay and Pasquill (1959), and Smith and Hay (1961). Gifford (1959) confirmed the validity of the relative diffusion in the atmospheric boundary layer, and Frenkiel and Katz (1956) and Kellogg (1956) confirmed it for the stratosphere.

Assuming that the mean-square relative particle acceleration is statistically homogeneous and stationary, Lin (1960a, b) derived a 4/3 power diffusion formula that is free of the limitation to the inertial range. Therefore the 4/3 power diffusion formula is no longer limited to the inertial range. For this reason we present Lin's analysis. Denote the *ith* component of the relative acceleration between particles α and β as

$$a_{ri}(t;a_\alpha,a_\beta) = a_{Li\alpha}(a_\alpha,t) - a_{Li\beta}(a_\beta,t) \quad (6.310)$$

Integration of Eq. 6.310 with respect to time gives the relative velocity

$$u_{ri}(t+\tau) - u_{ri}(t) = \int_t^{t+\tau} a_{ri}(t')\, dt' \qquad (6.311)$$

The main objective of relative diffusion analysis is to seek the expression for the mean-square distance and the rate of diffusion between particles. To do so we consider

$$\frac{1}{2}\frac{d^2}{dt^2}\langle X_{ri}(t)X_{ri}(t)\rangle = \langle u_{ri}(t)u_{ri}(t)\rangle + \langle X_{ri}(t)a_{ri}(t)\rangle \qquad (6.312)$$

Here, repeated indices represent summation over three components.

To calculate $\langle u_{ri}(t)u_{ri}(t)\rangle$ in the preceding equation, we analyze

$$\frac{1}{2}\frac{d}{d\tau}\sum_{i=1}^{3}\langle [u_{ri}(t+\tau)u_{ri}(t)]^2\rangle = \int_t^{t+\tau}\langle a_{ri}(t+\tau)a_{ri}(t')\rangle\, dt' \qquad (6.313)$$

By changing of variables it can be shown that

$$\frac{1}{2}\frac{d}{d\tau}\sum_{i=1}^{3}\langle [u_{ri}(t+\tau)u_{ri}(t)]^2\rangle = \int_0^{\tau}\langle a_{ri}(t)a_{ri}(t+\tau')\rangle\, d\tau' = \langle a_{ri}a_{ri}\rangle\int_0^{\tau} R_{ra}(\tau')\, d\tau' \qquad (6.314)$$

where

$$R_{ra}(\tau) = \frac{\langle a_{ri}(t)a_{ri}(t+\tau)\rangle}{\langle a_{ri}(t)a_{ri}(t)\rangle} \qquad (6.315)$$

is the autocorrelation function of the relative particle accelerations.

For sufficiently large $\tau > \tau_1$, where $R_{ra}(\tau)$ converges to zero

$$\langle a_{ri}a_{ri}\rangle\int_0^{\tau_1} R_{ra}(\tau)\, d\tau \equiv T_{ra} \qquad (6.316)$$

Integrating Eq. 6.314, we obtain

$$\sum_{i=1}^{3}\langle [u_{ri}(t+\tau) - u_{ri}(t)]^2\rangle = 2T_{ra}\tau \qquad (6.317)$$

or

$$\langle u_{ri}(t+\tau)u_{ri}(t+\tau)\rangle + \langle u_{ri}(t)u_{ri}(t)\rangle - 2\langle u_{ri}(t)u_{ri}(t+\tau)\rangle = 2T_{ra}\tau \qquad (6.318)$$

Since for large τ, $\langle u_{ri}^2(t+\tau)\rangle \gg \langle u_{ri}^2(t)\rangle$ and $\langle u_{ri}(t)u_{ri}(t+\tau)\rangle \approx 0$,

$$\langle u_{ri}(t+\tau)u_{ri}(t+\tau)\rangle \approx 2T_{ra}\tau \qquad (6.319)$$

The term $\langle X_{ri}(t)a_{ri}(t)\rangle$ in Eq. 6.312 can be expressed as

$$\langle X_{ri}(t)a_{ri}(t)\rangle = \int_0^t\int_0^{\eta}\langle a_{ri}(t)a_{ri}(t')\rangle\, dt\, d\eta \qquad (6.320)$$

Letting $t' = t - \tau$, we can show that Eq. 6.320 can be reduced to

$$\langle X_{ri}(t)a_{ri}(t)\rangle = \langle a_{ri}a_{ri}\rangle\int_0^t \tau R_{ra}(\tau)\, d\tau \qquad (6.321)$$

Defining

$$\tau_2 = \frac{\int_0^{\tau_1}\tau R_{ra}(\tau)\, d\tau}{\int_0^{\tau_1} R_{ra}(\tau)\, d\tau} \qquad (6.322)$$

we obtain

$$\langle X_{ri}(t)a_{ri}(t)\rangle \approx T_{ra}\tau_2 \qquad (6.323)$$

With the use of Eqs. 6.319 and 6.323, Eq. 6.312 can be expressed as

$$\frac{1}{2}\frac{d^2}{dt^2}\langle X_{ri}(t)X_{ri}(t)\rangle \simeq \begin{cases} 2T_{ra}t + T_{ra}\tau_2 & t > \tau_1 \\ 2T_{ra}t & t \gg \tau_1 \end{cases} \qquad (6.324)$$

Integration of the preceding equation with respect to t gives the turbulent diffusivity

$$K_r(t) \equiv \frac{1}{2}\frac{d}{dt}\langle X_{ri}(t)X_{ri}(t)\rangle = T_{ra}t^2 \qquad t > \tau_1 \qquad (6.325)$$

One more integration of the preceding equation yields

$$\langle X_{ri}(t)X_{ri}(t)\rangle = \langle X_{ri}(0)X_{ri}(0)\rangle + \frac{2}{3}T_{ra}t^3 \qquad t > \tau_1 \qquad (6.326)$$

Elimination of t from Eqs. 6.324 and 6.325 gives the eddy diffusivity

$$K_r(t) \equiv \frac{1}{2}\frac{d}{dt}\langle X_{ri}(t)X_{ri}(t)\rangle = \left(\frac{2}{3}\right)^{2/3} T_{ra}^{1/3}\langle X_{ri}(t)X_{ri}(t)\rangle^{2/3} \qquad \tau \gg \tau_1 \qquad (6.327)$$

which gives Richardson's $\frac{4}{3}$ power law (Richardson, 1926).

Considering the growth of a three-dimensional Gaussian cluster in a field of homogeneous, isotropic turbulence, Smith and Hay (1961) took averages of Eq. 6.297 over all pairs of particles in the cluster and found that

$$\left(\frac{d}{dt}\langle X_{ri}X_{ri}\rangle^{1/2}\right)_{max} = \frac{2}{3}\beta\bar{u}i^2 \qquad (6.328)$$

which indicates that the maximum rate of the closure growth is proportional to the square of the turbulence intensity, $i = \langle u'^2\rangle^{1/2}/\bar{u}$.

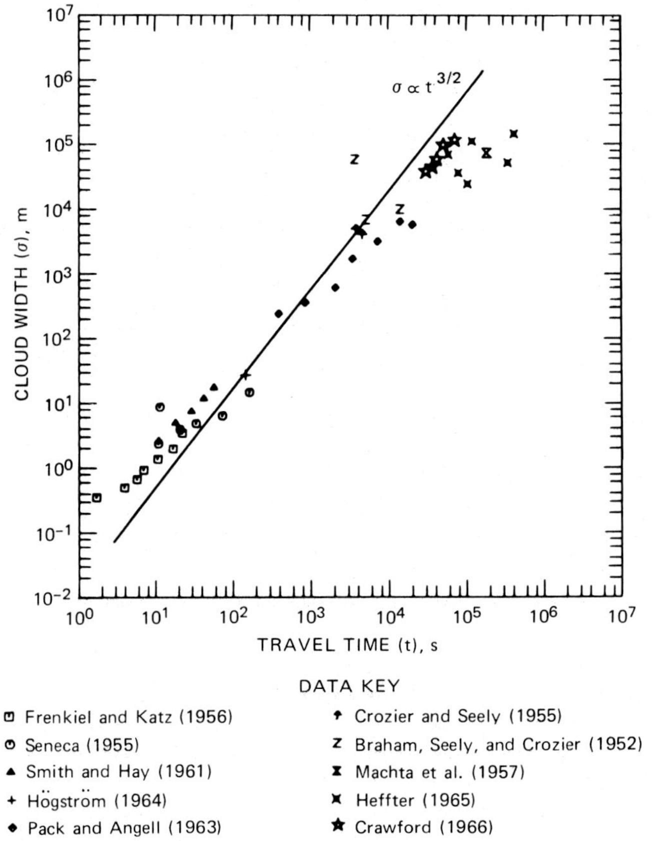

Fig. 6.16 Crawford's (1966) data on relative diffusion showing cloud width σ vs. travel time t, from which purely stratospheric points have been removed. [From F. A. Gifford, Jr., Tropospheric Relative Diffusion Observations, *Journal of Applied Meteorology*, 16: 311-313 (1977).]

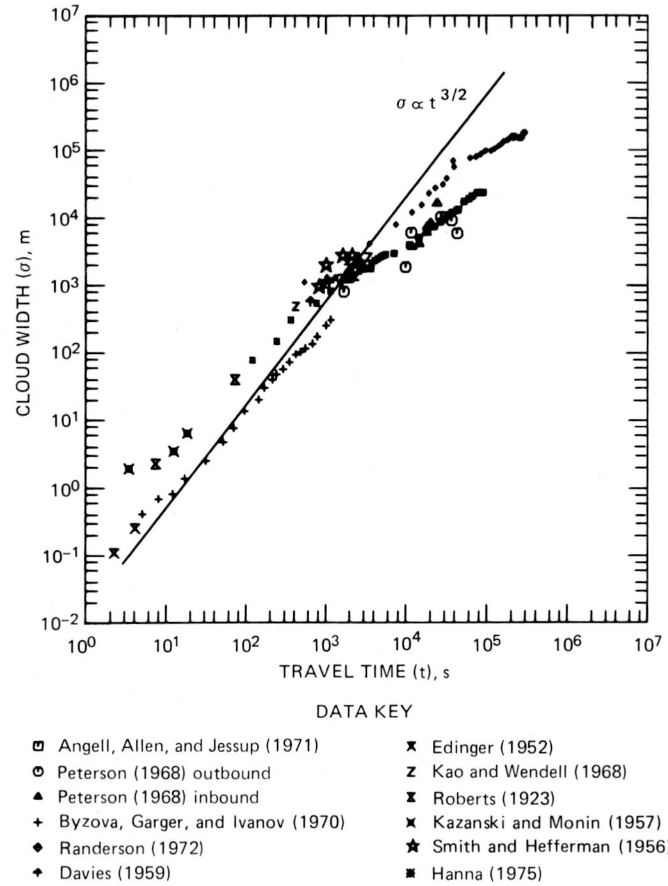

Fig. 6.17 Additional tropospheric relative diffusion data showing cloud width σ vs. travel time t. [From F. A. Gifford, Jr., Tropospheric Relative Diffusion Observations, *Journal of Applied Meteorology*, 16: 311-313 (1977).]

Gifford (1957a, 1957b) analyzed short-range, smoke-puff spreading data from several small-scale boundary-layer tracer experiments and found good evidence of $\langle X_{ri}(t)X_{ri}(t)\rangle \sim t^3$ for periods of a few tenths of a second corresponding to inertial subrange spreading. Hage (1964) and Crawford (1966) combined data from many atmospheric relative diffusion experiments and found that the mean-square particle distance is approximately proportional to $t^{2.4}$ (Fig. 6.16). Combining the relative diffusion data collected by a number of investigators (Smith and Hefferman, 1956; Kazanski and Monin, 1957; Davies, 1959; Edinger, 1952; Roberts, 1923; Kao and Wendell, 1968; Peterson, 1968; Byzova, Garger, and Ivanov, 1970; Angell, Allen, and Jessup, 1971; Randerson, 1972; and Hanna, 1975), Gifford (1977) found a t^3 spreading regime extending to somewhere between 1000 and 3000 s (Fig. 6.17). Then a slower regime appears. Analyzing the mean-square particle distance with the use of the General Circulation Model of the National Center for Atmospheric Research and using the wind data interpolated from the weather maps analyzed by the National Meteorological Center, Kao and his co-workers (Kao and Chi, 1976; Kao, 1974) found that for the large-scale relative diffusion in the troposphere and stratosphere the mean-square particle distance is approximately proportional to $t^{5/2}$.

6-3.10 Lagrangian Similarity Theory of Diffusion from a Continuous Ground Source

Most statistical theories of diffusion are based on the assumption that the field of turbulence is statistically homogeneous and stationary and that they apply to regions in which there is no velocity shear of the mean flow. Near a boundary surface, where the velocity shear of the mean flow is strong, turbulence is no longer homogeneous normal to the boundary so that the statistical theory of diffusion is no longer applicable. The Lagrangian similarity theory of diffusion was designed to describe the diffusion from a source in the surface boundary layer, which has a thickness of roughly a few tens of meters (see Chap. 5).

In an analysis of the rate of particle dispersion at the top of a smoke plume in the surface boundary layer, Monin (1959) proposed a similarity theory on the plume rise in relation to the vertical flux of momentum and sensible heat in the boundary layer. The similarity theory has been studied by many investigators, notably by Batchelor (1959, 1964) and Ellison (1959), who studied the neutral surface boundary layer, and by Gifford (1962), Panofsky and Prasad (1965), who studied the stratified boundary layer. Cermak (1963) and his collaborators, Yaglom (1965),

Pasquill (1966), and Klug (1968), have also made valuable contributions.

Two basic hypotheses in the Lagrangian similarity theory are (1) the rate of increase of the ensemble averaged particle displacement along the vertical depends primarily on the friction velocity, heat flux, and mean particle position, and (2) the rate of the ensemble averaged particle displacement in the horizontal is equal to the mean velocity at a certain level relative to the mean vertical position Z of the particles. Consequently, for diffusion from a ground source, we can express the rate of the mean vertical displacement as

$$\frac{\partial}{\partial t}\langle Z(\mathbf{a},t)\rangle = Au_*\,\psi\left(\frac{\langle Z\rangle}{L}\right) \tag{6.329}$$

and the rate of the mean horizontal displacement as

$$\frac{\partial}{\partial t}\langle X(\mathbf{a},t)\rangle = \bar{u}(B\langle Z\rangle) \tag{6.330}$$

where L is the Monin–Obukhov length, ψ is a universal function, and A and B are constants. In the surface boundary layer, the mean velocity takes the form of

$$\bar{u}(B\langle Z\rangle) = \frac{u_*}{k}\left[f\left(\frac{B\langle Z\rangle}{L}\right) - f\left(\frac{Bz_0}{L}\right)\right]$$

so that Eq. 6.330 becomes

$$\frac{\partial}{\partial t}\langle X(\mathbf{a},t)\rangle = \frac{u_*}{k}\left[f\left(\frac{B\langle Z\rangle}{L}\right) - f\left(\frac{Bz_0}{L}\right)\right] \tag{6.331}$$

Combining Eqs. 6.329 and 6.331 gives a relationship between $\langle X\rangle$ and $\langle Z\rangle$.

$$\langle X(\mathbf{a},t)\rangle = \frac{1}{kA}\int_0^{\langle Z\rangle} \frac{\left[f\left(\frac{B\langle Z\rangle}{L}\right) - f\left(\frac{Bz_0}{L}\right)\right]}{\psi\left(\frac{\langle Z\rangle}{L}\right)}\,d\langle Z\rangle \tag{6.332}$$

For the neutral case, $\psi = 1$, and $\bar{u}(z) = u_*k^{-1}\ln(Z/z_{0r})$ so that the preceding equation can be integrated to give

$$\langle X(\mathbf{a},t)\rangle = \frac{\langle Z\rangle}{kA}\left[\ln\frac{B\langle Z\rangle}{z_0} - 1 + \frac{z_0}{\langle Z\rangle}(1 - \ln B)\right] \tag{6.333}$$

To seek an expression for the ensemble averaged concentration distribution, we use dimensional analysis and obtain

$$\langle C(x,y,z,t)\rangle = \frac{Q}{\langle Z\rangle^3}\,\psi\left[\frac{x-\langle X\rangle}{\langle Z\rangle}, \frac{y}{\langle Z\rangle}, \frac{z-\langle Z\rangle}{\langle Z\rangle}\right] \tag{6.334}$$

For a continuous source, the mean concentration can be written

$$\langle C(x,y,z)\rangle = \int_0^\infty \frac{Q}{\langle Z\rangle^3}\,\psi\left(\frac{x-\langle X\rangle}{\langle Z\rangle}, \frac{y}{\langle Z\rangle}, \frac{z-\langle Z\rangle}{\langle Z\rangle}\right) dt \tag{6.335}$$

By a change of variable to $(x - \langle X\rangle)/\langle Z\rangle$, the prior integration can be expressed as

$$\langle C(x,y,z)\rangle = \int_0^\infty \frac{Q}{Au_*\langle Z\rangle^2}\,\psi\left(\frac{x-\langle X\rangle}{\langle Z\rangle}, \frac{y}{\langle Z\rangle}, \frac{z-\langle Z\rangle}{\langle Z\rangle}\right) \times \frac{d(x-\langle X\rangle)/\langle Z\rangle}{(x-\langle X\rangle)/\langle Z\rangle + d\langle X\rangle/d\langle Z\rangle} \tag{6.336}$$

Consider the centerline concentration at ground, $y = 0$, $z = 0$, $x \cong \langle X\rangle$. Thus the integral is essentially determined by values of $\langle Z\rangle$ near $\langle X\rangle = x$, and

$$\langle C(x,0,0)\rangle \propto \frac{Q}{Au_*\langle Z\rangle^2(d\langle X\rangle/d\langle Z\rangle)} \tag{6.337}$$

To obtain an axial, ground-level concentration as a function of x, we must carry out the integration in Eq. 6.332. For details see Gifford (1962). Table 6.2 shows a comparison of the ground concentration dependence on x from a continuous point source predicted by the similarity theory and observed at Project Prairie Grass (Gifford, 1962).

6-3.11 Fluctuations in Turbulent Diffusion

The basic problem of microscale fluctuations in turbulent diffusion has been investigated by Obukhov (1949), Corrsin (1951, 1952), and Batchelor (1959). In atmospheric diffusion, Gifford (1959, 1960, 1972) was the first to recognize the important effect of plume meandering on concentration fluctuation and to deal with the probability distribution of concentration and "peak-to-mean concentration ratio." In this connection investigations have also been made by Hilst (1957), Singer, Kazuhiko, and del Campo (1963), Hinds (1969), Barry (1971), and Knox and Pollack (1972).

When a quantity of tracer is released in a field of homogeneous turbulence, the instantaneously observable concentration $C(x_1,x_2,x_3,t)$ of the tracer, at time t after release and at a fixed point, (x_1,x_2,x_3) in the field, is a random variable. Let $B(C,x_1,x_2,x_3,t)$ be the probability distribution of the instantaneously observable concentration being less than C. It is seen that $B(0,x_1,x_2,x_3,t) = 0$, and $B \rightarrow 1$ as $C \rightarrow \infty$.

Quantitatively, meandering is defined as the random displacement, (s_1,s_2,s_3), of the center of mass of the tracer. Let $b(s_1,s_2,s_3,t)$ be the probability density, such that $b(s_1,s_2,s_3,t)\,ds_1\,ds_2\,ds_3$ is the probability of the center of mass being in the volume element $ds_1\,ds_2\,ds_3$ around the point (s_1,s_2,s_3).

The concentration observed at fixed distances from the center of mass, $C(x_1 - s_1,x_2 - s_2,x_3 - s_3,t)$ is also a random variable. Let us define $B(C,x_1 - s_1,x_2 - s_2,x_3 - s_3,t)$ as the probability of the observed concentration at $(x_1 - s_1,x_2 - s_2,x_3 - s_3,t)$ being less than C. Thus

$$B(C,x_1,x_2,x_3,t) = \iiint B(C,x_1 - s_1,x_2 - s_2,x_3 - s_3,t) \times b(s_1,s_2,s_3,t)\,ds_1\,ds_2\,ds_3 \tag{6.338}$$

TABLE 6.2 Comparison of the Dependence of the Ground-Level Concentration from a Continuous Point Source on Downwind Distance as Predicted by the Lagrangian Similarity Theory and Observed ($C \sim x^m$)†

Stability	⟨Z⟩/L	⟨X⟩, m	Observed m	Calculated m
Very stable	+0.1	100–200	−1.3	−1.3
Moderately stable	+0.01	400–800	−1.4	−1.6
Neutral	±0.001	400–800	−1.8	−2.0
Moderately unstable	−0.01	400–800	−2.6	−2.8
Very unstable	−0.1	400–800	−3.3	−3.0

†From F. A. Gifford, Jr., Diffusion in the Diabatic Surface Layer, *Journal of Geophysical Research*, 67: 3207-3212 (1962).

where the integration extends over all space.

Taking into account the effect of the large-scale meandering of plume elements in addition to the small scale randomness of concentration within elements, Gifford (1959, 1960) has developed a fluctuating plume model, which is equivalent to setting $C = C(x_1 - s_1, x_2 - s_2, x_3 - s_3, t)$ with probability one. In this model the lateral and vertical concentration distributions across a plume are assumed to be Gaussian, giving the expression for the instantaneous concentration,

$$\frac{C}{Q} = \frac{1}{2\pi\sigma_r^2 \bar{u}_1} \exp\left[-\frac{(x_2 - s_2)^2 + (x_3 - s_3)^2}{2\sigma_r^2}\right] \qquad (6.339)$$

where Q is the rate of emission from the source, s_2 and s_3 are the distances to the center of the instantaneous plume from the mean plume axis, and σ_r^2 is the second moment of the distributed tracer material about the centroid of the plume. Assuming that the distribution of s_2 and s_3 are Gaussian and that $\langle s_2^2\rangle = \langle s_3^2\rangle = \langle s^2\rangle$, the mean value of C/Q can be shown to be

$$\frac{\langle C\rangle}{Q} = \frac{1}{2\pi\bar{u}(\sigma_r^2 + \langle s^2\rangle)} \exp\left[-\frac{x_2^2 + x_3^2}{2(\sigma_r^2 + \langle s^2\rangle)}\right] \qquad (6.340)$$

The peak-to-mean concentration ratio can be shown to take the form of

$$\frac{C_{max}}{\langle C\rangle} = \left(1 + \frac{\langle s^2\rangle}{\sigma_r^2}\right) \exp\left[\frac{x_2^2 + x_3^2}{2(\sigma_r^2 + \langle s^2\rangle)}\right] \qquad (6.341)$$

If $x_2 = x_3 = 0$, i.e., on the mean plume axis, and since $\langle s^2\rangle \rightarrow$ constant, $C_{max}/\langle C\rangle \rightarrow 1$ for large diffusion time, then

$$\frac{C_{max}}{\langle C\rangle} = 1 + \frac{\langle s^2\rangle}{\sigma^2} \qquad (6.342)$$

On the basis of a long series of short-period observations of ground-level concentration values from continuous, isolated, elevated sources, Barry (1971) found that the peak-to-mean ratio represents two points on a probability distribution and suggested that the concentration frequency distribution takes an exponential form as follows:

$$F(C) = F(0) \exp\left[F(0) \frac{C}{\langle C\rangle}\right] \qquad (6.343)$$

where F(C) is the frequency of short-period concentrations (averaged over t_a) exceeding C, ⟨C⟩ is the concentration averaged over $t_b \gg t_a$, and F(0) is the frequency with which concentration exceeds the zero value at a given surface-level sampling point. With an application of Eq. 6.343, the peak-to-mean ratio can be shown to be

$$\frac{C_{max}}{\langle C\rangle} = -\frac{\left[\ln F(0) + \ln \frac{t_b}{t_a}\right]}{F(0)} \qquad (6.344)$$

Using techniques in turbulence analysis, Csanady (1967, 1973) derived a lognormal distribution function for concentrations in a plume. He also discussed the probability distribution and the variance of concentration fluctuations. To develop a lognormal distribution, consider a quantity of tracer with initial concentration C_0, released in a field of turbulence. Because the sampling volume is finite, any practical instrument will record the whole range of concentrations between zero and C. To describe this state of affairs, we can introduce an "intermittency factor," $\gamma(x_1, x_2, x_3, t)$, for the concentration, similar to that used by Townsend (1956) for turbulent energy, such that

$$1 - \gamma(x_1, x_2, x_3, t) = \lim_{C\rightarrow 0} B(C, x_1, x_2, x_3, t) \qquad (6.345)$$

In a moving coordinate system, we define an "intermittency factor" $\gamma(x_1 - s_1, x_2 - s_2, x_3 - s_3, t)$ such that

$$\lim_{C\rightarrow 0} B(C, x_1 - s_1, x_2 - s_2, x_3 - s_3, t) = 1 - \gamma(x_1 - s_1, x_2 - s_2, x_3 - s_3, t) \qquad (6.346)$$

Thus

$$\gamma(x_1, x_2, x_3, t) = \iiint B(x_1 - s_1, x_2 - s_2, x_3 - s_3, t) \times b(s_1, s_2, s_3, t)\, ds_1\, ds_2\, ds_3 \qquad (6.347)$$

To determine the functional form of the concentration probability distribution, $B(C, x_1 - s_1, x_2 - s_2, x_3 - s_3, t)$, let us consider the process of dilution of the concentration due to the mixing of the tracer with ambient fluid as the effect of a number of "diluting impulses"; each such impulse mixing with the tracer produces a decrease in concentration by a random factor $\eta_i < 1$, leading to a succession of concentration values between "impulses" as

$$C_0,\ C_0\eta_1,\ C_0\eta_1\eta_2,\ C_0\eta_1\eta_2\eta_3,\ \ldots,\ C_0\eta_1 \ldots \eta_n$$

Thus

$$\ln C = \ln C_0 + \ln \eta_1 + \ln \eta_2 + \ldots + \ln \eta_n \tag{6.348}$$

Equation 6.348 indicates that $\ln(C/C_0)$ is a sum of independent random variables and, by the Central Limit Theorem, tends to a normal distribution for a large number of diluting impulses. We can, therefore, take the concentration probability distribution for a continuous plume at ground level to be lognormal (Csanady, 1973; Frenkiel and Klebanoff, 1975) as

$$B(C, x_2, x_3, t) = \frac{1}{2}\left\{1 + \operatorname{erf}\left[\frac{\ln \dfrac{C}{C_s}}{(2)^{1/2}\sigma}\right]\right\} \tag{6.349}$$

where C_s is the median concentration and σ is the logarithmic standard deviation.

The probability density is therefore

$$b(C, x_2, x_3, t) = \frac{dB}{dc} = \frac{1}{(2\pi)^{1/2}\sigma\, C}\exp\left[-\frac{\left(\ln \dfrac{C}{C_s}\right)^2}{2\sigma^2}\right] \tag{6.350}$$

The mean and variance of the concentration at (x_2, x_3, t) are, respectively,

$$\langle C(x_2, x_3, t)\rangle = \int_{B=0}^{1} C\,\frac{dB}{dc}\,dc = C_s\, e^{\sigma^2/2} \tag{6.351}$$

$$\langle C'^2(x_2, x_3, t)\rangle = C_s^2\, e^{\sigma^2}(e^{\sigma^2} - 1) \tag{6.352}$$

The root mean square to the mean concentration ratio is therefore

$$\frac{\langle C'^2\rangle^{1/2}}{\langle C\rangle} = (e^{\sigma^2} - 1)^{1/2} \tag{6.353}$$

To derive an expression for the peak-to-mean concentration ratio, let the peak values be those which exceed 99% of the ensemble and denote them by C_{max}. Thus

$$B(C_{max}, x_2, x_3) = 0.99 \tag{6.354}$$

It can be shown with the use of Eqs. 6.351 and 6.352 that

$$\frac{C_{max}}{\langle C\rangle} = e^{\sigma(2.326 - 0.5\sigma)} \tag{6.355}$$

which indicates that the peak-to-mean ratio is a function of σ alone.

6-A APPENDIX: RANDOM FUNCTION, PROBABILITY DISTRIBUTION, AND SPECTRUM

6-A.1 Introduction

Motions in the atmosphere and oceans are generally turbulent. An important characteristic of turbulent motion is the diffusivity of turbulence, which causes rapid mixing and transport of momentum, mass, and heat in the fluid. It is this turbulence mechanism by which pollutants in the atmosphere and oceans can be effectively dispersed. In meteorology we are dealing with the entire spectrum of atmospheric motion, ranging from a scale as small as molecular motion to a scale as large as the diameter of the earth.

To define turbulent motion from the mean motion, we must select a space or time domain (scale) over which an averaging process is performed. Such an averaging process filters out the eddy (turbulent) motions of scales smaller than the selected space or time domain of the average. Motions of scales smaller than the domain of average contribute to the eddy motion, whereas those of larger scales contribute to the mean motion. It is the motion of the eddies which provides the mechanism for the turbulent dispersion of pollutants in the fluid, whereas the motion of the larger scales provides the transport mechanism for the pollutants. Therefore a knowledge of the characteristics of the energy spectrum of the turbulent motion in fluids is essential in the understanding and predicting turbulent diffusion of pollutants in fluids.

The distribution of the kinetic energy in the microscale motion varies considerably with boundary and meteorological conditions such as the thermal stratification and the distribution of velocity shear of the mean wind. Analyses of the energy spectrum of the large-scale motion in the free atmosphere generally have been limited to a space scale greater than 10^6 m and to a time scale greater than 24 h. The distribution of the kinetic energy in the mesoscale, 10^4 to 10^5 m, for a time scale of several hours is least known since it is smaller than the distance between current meteorological stations.

6-A.2 The Random Process

To describe the process of turbulent diffusion, we first examine the nature of turbulent motion. Figure 6.A1 shows typical traces of the horizontal wind speed and direction at 16 m and 120 m. Figure 6.A2 shows the vertical wind-direction trace taken from a typical bivane wind sensor at 10 m. These wind traces show that the fluctuations of wind speed and direction are of a random nature. For characteristics of the fluctuations of these turbulence quantities, it is

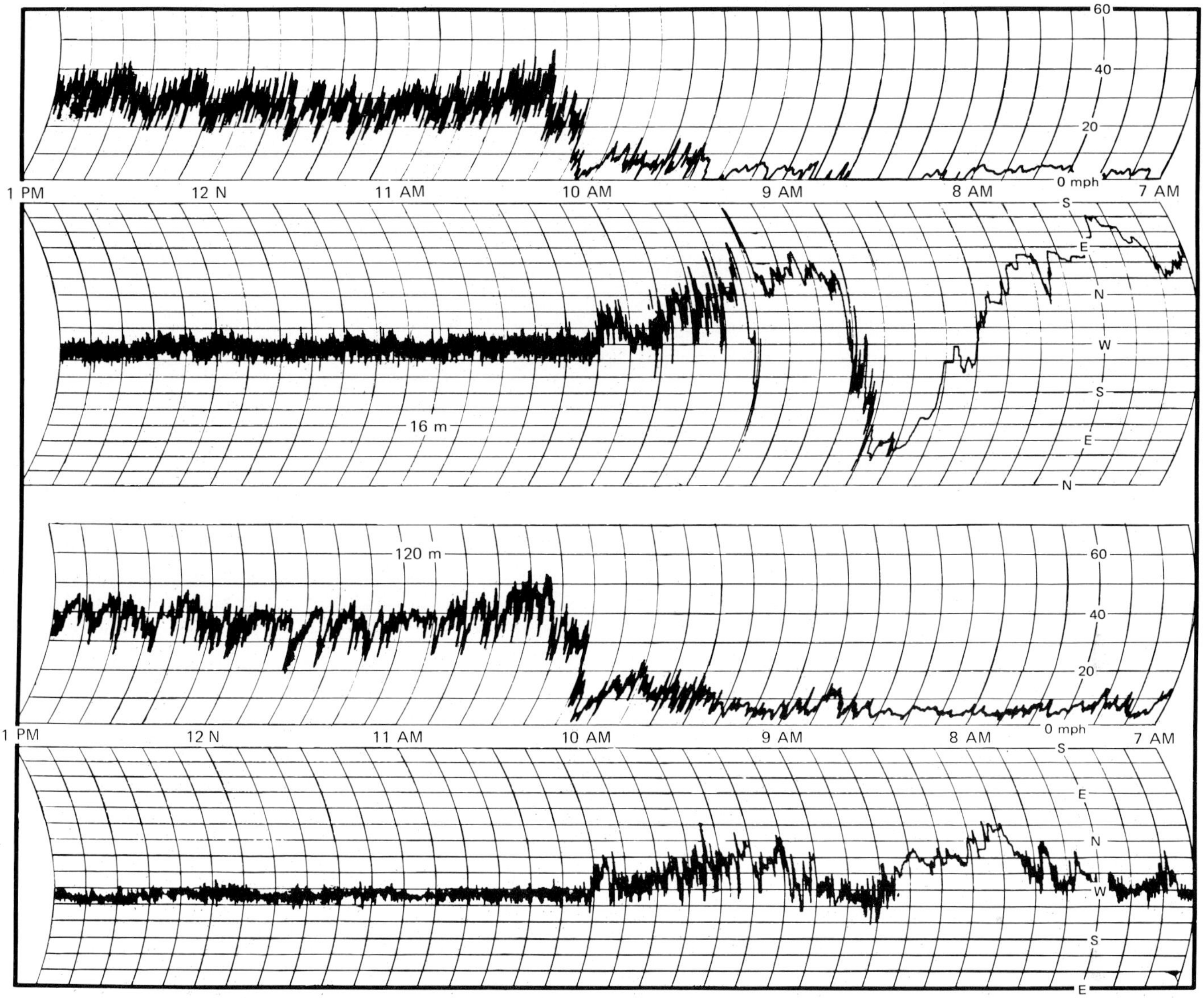

Fig. 6.A1 Typical horizontal wind speed and direction traces at 16 m and 120 m.

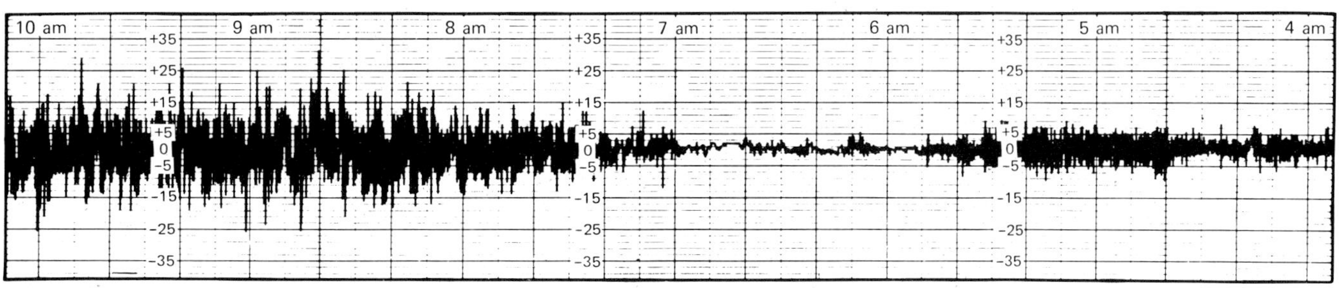

Fig. 6.A2 Example of a vertical wind-direction trace (chart speed, 3 in./h). The bivane sensor was mounted 10 m above the surface.

appropriate to calculate the statistical characteristics of these quantities.

The concept of a random process has been applied with considerable success to problems in turbulence and diffusion. A type of random process that has found frequent application in turbulence and diffusion studies can be described as being homogeneous, stationary, isotropic, and Gaussian. The property of homogeneity implies that the statistical properties derived from the random process do not vary in space, whereas that of stationarity means that the statistical properties do not vary with time. Isotropy can be defined by the condition that the statistical properties of the field are independent of rotation and reflection of the coordinate axes. A Gaussian random process is one in which the frequency distribution of values of the variable takes the Gaussian, or normal, curve.

It is convenient to study a random process with the application of the concept of a random variable introduced in the theory of probability. A variable is called random when, as a result of experimentation under some definite conditions, it can assume any one of a number of possible numerical values, which are distributed according to certain probability laws. In the statistical study of random processes, we are not interested in the investigation of the individual realizations of the random variable but of the statistical characteristics of a whole set of realizations.

6-A.3 Random Functions and Probability

Let $v(t)$ be a random function of time, and let m of n realizations of the random function at time t_1 have values less than v_1. Experiments show that, as n increases, the ratio m/n becomes stabilized and oscillates about a definite number. This number, which represents the probability of this event, can be designated by $P[v(t_1) < v_1]$, which is evidently a function of t_1 and v_1. Let the distribution function be

$$B(v_1,t_1) = P[v(t_1) < v_1] \tag{6.A1}$$

If $B(v_1,t_1)$ has a partial derivative with respect to v_1, then the probability density function can be defined by

$$b(v_1,t_1) = \frac{\partial}{\partial v_1} B(v_1,t_1) \tag{6.A2}$$

Therefore

$$b(v_1,t_1)\,dv_1 = P[v_1 < v(t_1) \leqslant v_1 + dv_1] \tag{6.A3}$$

expresses the probability of the occurrence of v in the interval $(v_1, v_1 + dv_1)$. To obtain the statistical characteristics of a random function, we first calculate the mean value of the function, which is defined by

$$\langle v(t)\rangle = \int_{-\infty}^{\infty} v(t)b(v,t)\,dv \tag{6.A4}$$

where the angle bracket denotes the ensemble average.

To characterize a random function more completely, we calculate the statistical moments of the random function around its mean. The *kth* order of such moment about the mean can be expressed as

$$M_k = \int_{-\infty}^{\infty} [v(t) - \langle v(t)\rangle]^k b(v,t)\,dv \tag{6.A5}$$

The first moment of a function gives the mean value of the function; the second moment provides the variance of the function, which measures the width-square of the function; the third moment provides a measure of asymmetry of the function.

For $k = 2$, we obtain the variance of $v(t)$,

$$\sigma_v^2(t) = M_2 = \langle [v(t) - \langle v(t)\rangle]^2\rangle \tag{6.A6}$$

The variance of velocity, which measures the intensity of turbulent motion, is a basic quantity in the analysis of turbulent motion and turbulent diffusion.

Since the third moment M_3 of a random function has a dimension of the third power of the random function, the coefficient of asymmetry, or skewness, can be defined as

$$S = \frac{M_3}{M_2^{3/2}} \tag{6.A7}$$

One of the statistical functions which proves to be very useful in an analysis of the energy spectrum of turbulence and its relationship to diffusion is the autocorrelation function of turbulent velocities, which is defined by

$$\langle v'(t_1)v'(t_2)\rangle = \int_{-\infty}^{+\infty}\int_{-\infty}^{+\infty} [v_1 - \langle v(t_1)\rangle]\,[v_2 - \langle v(t_2)\rangle] \times b(v_1,t_1;v_2,t_2)\,dv_1\,dv_2 \tag{6.A8}$$

where $v'(t)$ is the velocity deviation from its mean. In view of the autocorrelation function that depends on two values of time, it provides some insight into the dynamic process of the random function. Note that the autocorrelation function is symmetrical with respect to t_1 and t_2 and that for $t_1 = t_2 = t$ it gives the variance of the velocity.

For convenience of analyses, the normalized autocorrelation function, called the coefficient of correlation, defined by

$$R_v(t_1,t_2) = \frac{\langle v'(t_1)v'(t_2)\rangle}{\sigma_v(t_1)\sigma_v(t_2)} \tag{6.A9}$$

is commonly used.

To investigate the dependency between two random processes, $u(t)$ and $v(t)$, we consider the joint probability

$$B(u_1,t_1;v_2,t_2) = P[u(t_1) < u_1; v(t_2) < v_2] \tag{6.A10}$$

and the joint probability density

$$b(u_1,t_1;v_2,t_2) = \frac{\partial^2}{\partial u_1 \partial v_2} B(u_1,t_1;v_2,t_2) \quad (6.A11)$$

The cross-correlation function of the random processes u(t) and v(t) is defined by the joint moment,

$$\langle u'(t_1)v'(t_2)\rangle = \langle [u(t_1) - \langle u(t_1)\rangle] [v(t_2) - \langle v(t_2)\rangle] \rangle \quad (6.A12)$$

and the coefficient of the cross-correlation function is defined as

$$R_{uv}(t_1,t_2) = \frac{\langle u'(t_1)v'(t_2)\rangle}{\sigma_u(t_1)\sigma_v(t_2)} \quad (6.A13)$$

Note that the cross-correlation function is asymmetrical with respect to t_1 and t_2,

$$\langle u'(t_1)v'(t_2)\rangle = \langle v'(t_2)u'(t_1)\rangle \quad (6.A14)$$

Note also that, if the random processes are independent to each other, then

$$\langle u'(t_1)v'(t_2)\rangle = \langle u'(t_1)\rangle \langle v'(t_2)\rangle = 0 \quad (6.A15)$$

6-A.4 Stationarity and Ergodicity of Random Processes

A class of random functions that plays an important role in the theory of random functions as well as in turbulence and diffusion is the stationary random function. It is the random function for which the probability distribution is invariant with respect to the change of time reference, i.e.,

$$b(v_1,t_1) = b(v_1,t_1 + \tau) \quad (6.A16)$$

$$b(v_1,t_1;v_2,t_2) = b(v_1,t_1 + \tau;v_2,t_2 + \tau) \quad (6.A17)$$

Since τ is arbitrary, we can choose $\tau = -t_1$. Thus

$$b(v_1,t_1) = b(v_1) \quad (6.A18)$$

$$b(v_1,t_1;v_2,t_2) = b(v_1,0;v_2,t_2 - t_1) \quad (6.A19)$$

and the correlation coefficient becomes

$$R_v(\tau) = \frac{\langle v'(t)v'(t+\tau)\rangle}{\sigma_v^2(t)} \quad (6.A20)$$

where τ is the time lag. It is readily seen that the variance of a stationary random function is a constant.

Perhaps a most remarkable characteristic of a stationary random process is the ergodicity of the process, which states that the statistical characteristic of a stationary random process obtained by averaging a set of realizations of the process at a given time is equal to the characteristic obtained by averaging a single realization for a sufficiently long interval of time. Thus the correlation function can be expressed

$$R_v(\tau) = \lim_{T\to\infty} \frac{\frac{1}{T}\int_0^T v'(t)v'(t+\tau)\,dt}{\frac{1}{T}\int_0^T v'^2(t)\,dt} \equiv \frac{\overline{v'(t)v'(t+\tau)}}{\overline{v'^2(t)}} \quad (6.A21)$$

where the overbars denote the time averages of a single realization. This ergodicity of a stationary random process provides a great deal of simplification in the calculation of statistical characteristics of stationary random processes, as will be shown in the analyses of turbulence and diffusion.

6-A.5 Generalized Harmonic Analysis

The fact that random functions are neither periodic nor integrable prevents the applications of Fourier integral or series expansion in the ordinary manner. However, a random function can be expressed in terms of Fourier–Stieltjes integral as follows:

$$v(t) = \int_{-\infty}^{\infty} e^{i\omega t}\, dZ_v(\omega) \quad (6.A22)$$

where $Z(\omega)$ is a complex random function of angular frequency ω, which has the following characteristics:

$$\begin{aligned} &\langle Z_v(\omega)\rangle = 0 \qquad \text{for any value of } \omega \\ &\langle dZ_v(\omega_1)\, dZ_v^*(\omega_2)\rangle = 0 \qquad \omega_1 \neq \omega_2 \\ &\langle dZ_v(\omega)\, dZ_v^*(\omega)\rangle \equiv dG_v(\omega) \end{aligned} \quad (6.A23)$$

where the asterisk is designated to indicate the conjugate of the complex function.

Similar to the Fourier–Stieltjes integral for v(t) as previously shown, the correlation function random process can be expressed by

$$\sigma_v^2 R_v(\tau) = \int_{-\infty}^{\infty} e^{i\omega\tau}\, dG_v(\omega) \quad (6.A24)$$

If $R(\tau)$ tends to zero fast enough as $\tau \to \pm\infty$, such that $\int_{-\infty}^{\infty} |R(\tau)|\, d\tau$ is finite, then the correlation function can be expressed in terms of the Fourier integral

$$\sigma_v^2 R_v(\tau) = \int_{-\infty}^{\infty} e^{i\omega\tau}\, S_v(\omega)\, d\omega \quad (6.A25)$$

Therefore

$$G_v(\omega) = \int_{-\infty}^{\omega} S_v(\omega)\, d\omega \quad (6.A26)$$

or

$$S_v(\omega) = \frac{d}{d\omega} G_v(\omega) \quad (6.A27)$$

where $S_v(\omega)$ is called the power-spectral density or spectrum of the random function v(t) and $G_v(\omega)$ denotes the power-spectral contribution to the variance σ_v^2 from frequencies below ω.

The inverse Fourier transform of Eq. 6.A25 gives

$$S_v(\omega) = \frac{\sigma_v^2}{2\pi} \int_{-\infty}^{\infty} e^{-i\omega\tau} R_v(\tau)\, d\tau \qquad (6.A28)$$

For real random processes, $R_v(\tau) = R_v(-\tau)$; therefore $S_v(\omega) = S_v(-\omega)$ and

$$\sigma_v^2 R_v(\tau) = \int_{-\infty}^{\infty} S_v(\omega) \cos \omega\tau\, d\omega$$

$$= 2 \int_0^{\infty} S_v(\omega) \cos \omega\tau\, d\omega \qquad (6.A29)$$

and

$$S_v(\omega) = \frac{\sigma_v^2}{\pi} \int_0^{\infty} R_v(\tau) \cos \omega\tau\, d\tau \qquad (6.A30)$$

The normalized power spectrum is defined as $S_v(\omega)/\sigma_v^2$.

For $\tau = 0$, Eq. 6.A29 becomes

$$\frac{1}{2}\langle v'^2 \rangle = \int_0^{\infty} S_v(\omega)\, d\omega \qquad (6.A31)$$

which indicates that the kinetic energy of the motion is the resultant of the contribution of the power spectra of all frequencies.

It is convenient to express Eq. 6.A31 in the following form

$$\frac{1}{2}\langle v'^2 \rangle = \int_0^{\infty} \omega S_v(\omega)\, d \ln \omega \qquad (6.A32)$$

so that the area under the curve of the integrand, $\omega S_v(\omega)$, in each frequency range will give the contribution of the frequency range to the total variance. Since most published spectra in meteorological literature are expressed in terms of frequency $n = \omega/2\pi$, Eq. 6.A32 can be written as

$$\frac{1}{2}\langle v'^2 \rangle = \int_0^{\infty} n S_v(n)\, d(\ln n) \qquad (6.A33)$$

Consider the cross correlation of two random functions, u′(t) and v′(t). It is easily shown that

$$\langle u'(t_1)v'(t_2)\rangle = \int_{-\infty}^{\infty} \int_{-\infty}^{\infty} \exp\, [i(\omega_1 t_1 - \omega_2 t_2)]$$

$$\times \langle dZ_u(\omega_1)\, dZ_v^*(\omega_2)\rangle \qquad (6.A34)$$

For stationary random processes, $\langle u'(t_1)v'(t_2)\rangle = \langle u'(0)v'(t_2 - t_1)\rangle = \langle u'(t)v'(t + \tau)\rangle$ and

$$R_{uv}(\tau) = \frac{\langle u'(t)v'(t+\tau)\rangle}{\sigma_u(t)\sigma_v(t)} \qquad (6.A35)$$

where $\tau = t_2 - t_1$.

Since

$$\langle dZ_u(\omega_1)\, dZ_v^*(\omega_2)\rangle = \begin{cases} 0 & \omega_1 \neq \omega_2 \\ S_{uv}(\omega)\, d\omega & \omega_1 = \omega_2 = \omega \end{cases} \qquad (6.A36)$$

we have

$$\sigma_u\sigma_v R_{uv}(\tau) = \int_{-\infty}^{\infty} e^{i\omega\tau} S_{uv}(\omega)\, d\omega \qquad (6.A37)$$

and its inverse transform

$$S_{uv}(\omega) = \frac{\sigma_u\sigma_v}{2\pi} \int_{-\infty}^{\infty} e^{-i\omega\tau} R_{uv}(\tau)\, d\tau \qquad (6.A38)$$

Let

$$R_{uv}(\tau) = Re_{uv}(\tau) + Ro_{uv}(\tau) \qquad (6.A39)$$

$$S_{uv}(\omega) = C_{uv}(\omega) - iQ_{uv}(\omega) \qquad (6.A40)$$

where

$$Re_{uv}(\tau) = \frac{1}{2}[R_{uv}(\tau) + R_{uv}(-\tau)] \qquad (6.A41)$$

$$Ro_{uv}(\tau) = \frac{1}{2}[R_{uv}(\tau) - R_{uv}(-\tau)] \qquad (6.A42)$$

are, respectively, the even and odd part of the cross-correlation function, and

$$C_{uv}(\omega) = \frac{\sigma_u\sigma_v}{2\pi} \int_{-\infty}^{\infty} Re_{uv}(\tau) \cos \omega\tau\, d\tau \qquad (6.A43)$$

$$Q_{uv}(\omega) = \frac{\sigma_u\sigma_v}{2\pi} \int_{-\infty}^{\infty} Ro_{uv}(\tau) \sin \omega\tau\, d\tau \qquad (6.A44)$$

are, respectively, the cospectrum and quadrature spectrum. The normalized cospectrum and quadrature spectrum are defined as $C_{uv}(\omega)/\sigma_u\sigma_v$ and $Q_{uv}(\omega)/\sigma_u\sigma_v$, respectively.

The inverse transform of Eq. 6.A43 gives

$$\sigma_u\sigma_v Re_{uv}(\tau) = \int_{-\infty}^{\infty} C_{uv}(\omega) \cos \omega\tau\, d\omega \qquad (6.A45)$$

For $\tau = 0$, Eq. 6.A45 becomes

$$\langle u'v' \rangle = \int_{-\infty}^{\infty} C_{uv}(\omega)\, d\omega \qquad (6.A46)$$

Thus the covariance is the resultant of the contribution of the cospectra of all frequencies.

Figure 6.A3 shows the averaged spectrum of the horizontal wind speed at an average height of 100 m above the ground in the frequency range from 0.0007 to 900 cycles/h (Van der Hoven, 1957). Note that there are two

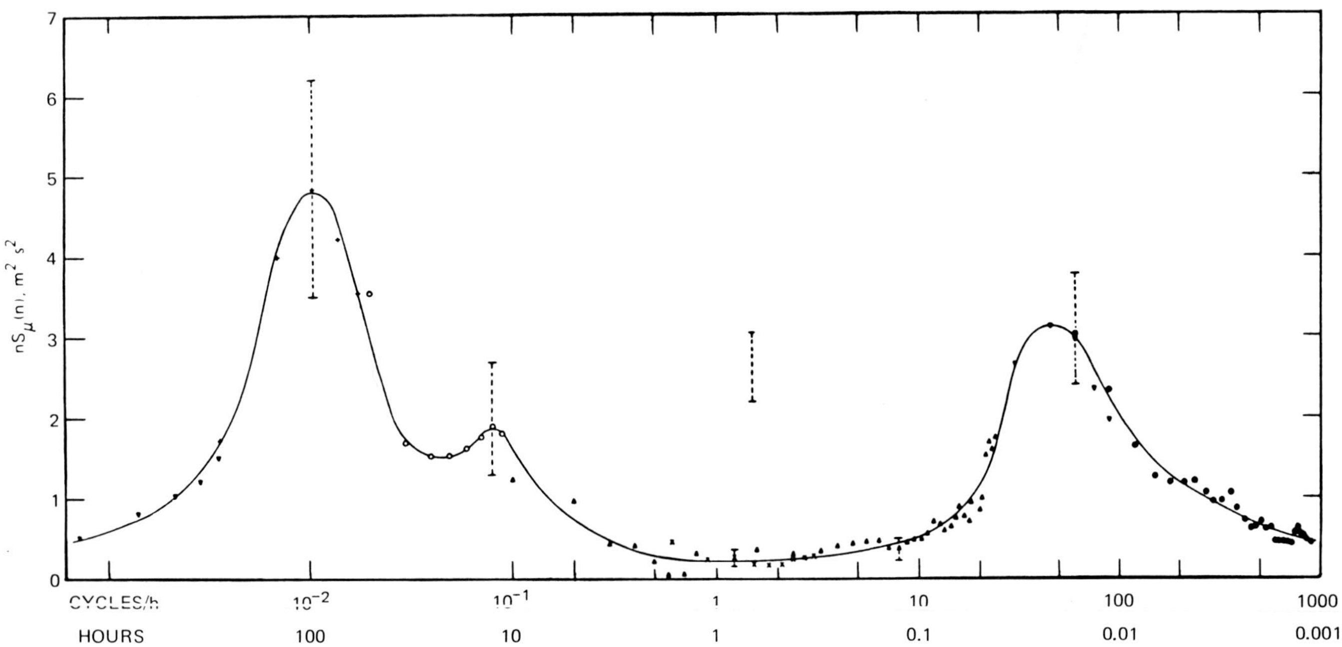

Fig. 6.A3 Spectrum of the horizontal wind speed at an average height of 100 m above the ground. [From I. Van der Hoven, Power Spectrum of Horizontal Wind Speed in the Frequency Range from 0.0007 to 900 Cycles Per Hour, *Journal of Meteorology*, 14: 160-164 (1957).]

major energy peaks, one near the frequency at 10^{-2} cycles/h (corresponding to a period of 4 d) and another at 50 cycles/h (corresponding to a period of 1 min), and that an energy gap occurs near the frequency at 1 cycle/h (corresponding to a period of 1 h).

Figure 6.A4 shows typical energy spectra of the vertical and lateral components of the wind for a series of experiments conducted under various thermal stratifications at 15 m (Walker, 1965). The energy spectra of the vertical component of the wind (Fig. 6.A4a) show that in stable conditions the spectrum shows small turbulence energy involved and a small energy peak occurring at rather high frequencies. During instability the warm ground introduces a thermal component into the turbulence, which results in the shifting of the peak to lower frequencies.

The energy spectra of the lateral component of the wind (Fig. 6.A4b) show that there are two energy peaks for all thermal stability conditions, one at a comparatively high frequency and one at lower frequencies. The high-frequency peaks are due to the effect of the surface roughness, whereas the low-frequency peaks are a contribution of the thermally induced eddies.

The normalized spectra for the power spectra and cospectra in the surface boundary layer for near neutral conditions (Fig. 6.A5, Kaimal et al., 1972) have been computed with the experimental data obtained from the 1968 Kansas experiment. In this figure, u, v, w denote, respectively, the along, lateral, and vertical components of the wind, and θ denotes the temperature. Note that the energy peaks of the longitudinal, lateral, and vertical components of the velocity shift from lower to higher frequency.

At both stations, above Salt Lake City, Utah (windward side of the mountains), and above Denver, Colo. (lee side of the mountains), the transport of westerly momentum below the mountain top (approximately 700 mb for Salt Lake City and 650 mb for Denver) is toward the south at practically all frequencies but toward the north in most frequencies at the highest levels. At both stations most of the contributions to momentum transports come from periods longer than 2 d. The relative contribution of the lowest frequencies increases markedly with increasing height, an indication that most of the momentum transport in the free atmosphere is accomplished by eddies of long periods.

6-A.6 Moving Average

In the treatment of geophysical data, an averaging or filtering concept is frequently used. This concept can be introduced directly in the measurement or in the interpretation. We are interested in deriving the expression for the power spectrum of the filtered data in terms of that of the original data. Let v(t) be the original random function, $\psi(t)$ be the averaging function, and $v_\psi(t)$ be the averaged function. By definition we have

$$v_\psi(t) = \int_{-\infty}^{\infty} v(t')\psi(t-t')\,dt' \qquad (6.A47)$$

Let the Fourier transform of v(t), $\psi(t)$, and $v_\psi(t)$ be $V(\omega)$, $\Psi(\omega)$, and $V_\psi(\omega)$, respectively.

The Fourier transform of Eq. 6.A47 yields

$$V_\psi(\omega) = V(\omega)\Psi(\omega) \qquad (6.A48)$$

The power spectrum of v(t), $\psi(t)$, and $v_\psi(t)$ can be expressed as

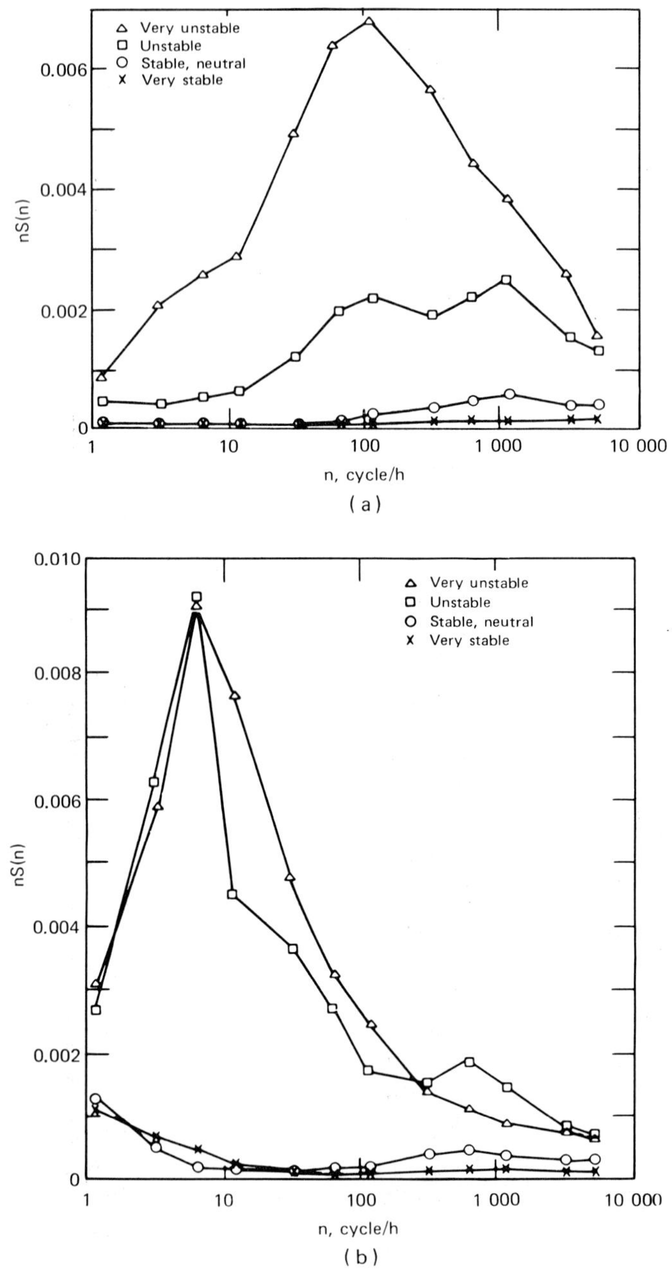

Fig. 6.A4 Spectra of the vertical (a) and horizontal (b) wind-direction fluctuations at 15 m grouped by atmospheric stability. [From E. R. Walker, A Particulate Diffusion Experiment, *Journal of Applied Meteorology*, 4: 614-621 (1965).]

$$S_v(\omega) = V(\omega)V^*(\omega)$$

$$S_\psi(\omega) = \Psi(\omega)\Psi^*(\omega)$$

and

$$S_{v\psi}(\omega) = V_\psi(\omega)V_\psi^*(\omega) \tag{6.A49}$$

respectively. With the use of Eqs. 6.A48 and 6.A49, the power spectrum of $v_\psi(t)$ can be written as

$$S_{v\psi}(\omega) = S_v(\omega)S_\psi(\omega) \tag{6.A50}$$

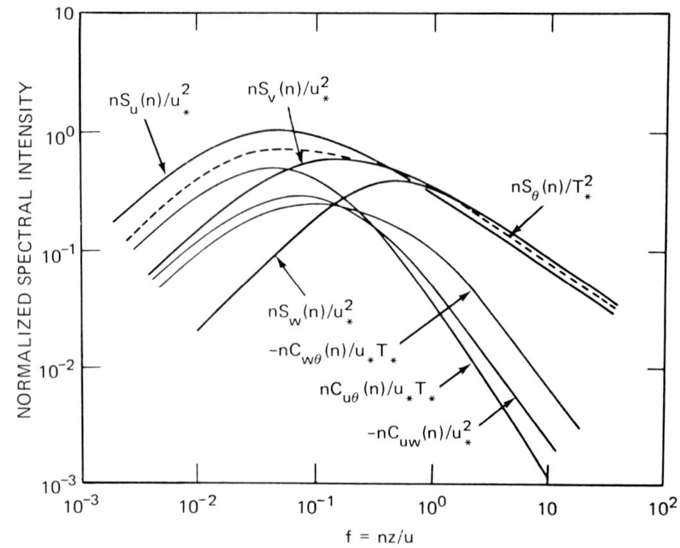

Fig. 6.A5 Spectra under near neutral atmospheric condition. [J. C. Kaimal et al., Spectral Characteristics of Surface-Layer Turbulence, *Quarterly Journal of the Royal Meteorological Society*, 98: 563-589 (1972).]

For a moving average over a time interval τ symmetrical about t, the averaging function can be expressed as

$$\psi(t;\tau) = \begin{cases} \dfrac{1}{\tau} & -\dfrac{\tau}{2} \leqslant t \leqslant \dfrac{\tau}{2} \\ 0 & |t| > \dfrac{\tau}{2} \end{cases} \tag{6.A51}$$

The Fourier transform of the averaging function is

$$\Psi(\omega;\tau) = \frac{1}{\tau}\int_{(-\tau/2)}^{(\tau/2)} e^{i\omega t}\,dt = \frac{\sin(\omega\tau/2)}{(\omega\tau/2)} \tag{6.A52}$$

and the power spectrum of the averaging function is

$$S_\psi(\omega;\tau) = \left[\frac{\sin(\omega\tau/2)}{(\omega\tau/2)}\right]^2 \tag{6.A53}$$

The power spectrum Eq. 6.A50 of the averaged function $v_\psi(t;\tau)$ can be expressed as

$$S_{v\psi}(\omega;\tau) = S_v(\omega)S_\psi(\omega;\tau) = S_v(\omega)\left[\frac{\sin(\omega\tau/2)}{(\omega\tau/2)}\right]^2 \tag{6.A54}$$

As shown in Sec. 6-3.6, Eq. 6.A54 is very useful in analyzing wind data for diffusion studies.

LIST OF SYMBOLS

Symbols used frequently in Chap. 6 are listed here.

a_i — i component of the initial Lagrangian coordinate of a fluid particle

a_{ri} i component of the acceleration between fluid particles
b Probability density function
B Probability distribution function
C Concentration of a trace constituent
C_{ij} Cospectrum of i and j components of the velocity
c_p Specific heat at constant pressure
c_p Specific heat at constant volume
D Width of the spectral diffusivity function
$f = 2\Omega \sin \phi$ Coriolis parameter
g Acceleration due to gravity
K_i i component of the eddy diffusivity in the Eulerian coordinate system
K_{ij} ij component of the eddy diffusivity in the Eulerian coordinate system
K_{Li} i component of the eddy diffusivity in the Lagrangian coordinate system
K_{Lij} ij component of the eddy diffusivity in the Lagrangian coordinate system
K_r Eddy diffusivity between fluid particles
l Mixing length
L Characteristic length
P Pressure
Q_{ij} Quadrature spectrum of the i and j components of the velocity
R Universal gas constant
R_i Autocorrelation coefficient (function) of the i component of the Eulerian velocity
R_{ij} Cross-correlation coefficient (function) of the i and j components of the Eulerian velocity
R_{Li} Autocorrelation coefficient (function) of the i component of the Lagrangian velocity
R_{Lij} Cross-correlation coefficient (function) of the i and j components of the Lagrangian velocity
R_{ri} Autocorrelation coefficient (function) of the i component of the Lagrangian velocity between fluid particles
R_{rij} Cross-correlation coefficient (function) of the i and j components of the Lagrangian velocity between fluid particles
Re_{ij} Even part of the cross-correlation function R_{ij}
Ro_{ij} Odd part of the cross-correlation function R_{ij}
S_{v_E} Power spectrum of Eulerian velocity v
S_{v_L} Power spectrum of Lagrangian velocity v_L
S_{ij} Cross spectrum of the i and j components of the velocity
t Time
T Temperature
T_E Integral time scale of the Eulerian velocity
T_L Integral time scale of the Lagrangian velocity of a single fluid particle
T_r Integral time scale of the Lagrangian velocity between particles
u_i, u_j, u_k Components of the Eulerian velocity
u_{Li} i component of the Lagrangian velocity
u_{ri} i component of the Lagrangian velocity between fluid particles
u_b' Intrinsic turbulent velocity
u_s' Turbulent velocity generated by shear flow
x_i, x_j, x_k Components of the Eulerian coordinates
X_i i component of the Lagrangian coordinates of a fluid particle
X_{ri} i component of the Lagrangian coordinates between fluid particles
ρ Density
ϕ Latitude
θ Potential temperature
Ω Angular velocity of the earth
σ Standard deviation
τ Time lag
ϵ Rate of dissipation
ν Coefficient of kinematic viscosity
ψ Universal function

REFERENCES

André, J. C., 1974, Irreversible Interaction Between Cumulants in Homogeneous, Isotropic, Two-Dimensional Turbulence Theory, *Phys. Fluids*, 17: 15-21.

——, G. De Moor, P. Lacarrère, and R. du Vachat, 1976, Turbulence Approximation for Inhomogeneous Flows, Part I. The Clipping Approximation, *J. Atmos. Sci.*, 33: 476-481.

Angell, J. K., P. W. Allen, and E. A. Jessup, 1971, Mesoscale Relative Diffusion Estimates from Tetroon Flights, *J. Appl. Meteorol.*, 10: 43-46.

Aris, R., 1956, On the Dispersion of a Solute in a Fluid Flowing Through a Tube, *Proc. R. Soc. (London), Ser. A.*, 235: 67-77.

Barry, P. J., 1971, A Note on Peak-to-Mean Concentration Ratios, *Boundary-Layer Meteorol.*, 2: 122-126.

Bartlett, M. S., 1956, *An Introduction to Stochastic Processes*, Cambridge University Press, London.

Batchelor, G. K., 1949, Diffusion in a Field of Homogeneous Turbulence, I. Eulerian Analysis, *Aust. J. Sci. Res.*, 2: 437-450.

——, 1950, The Application of the Similarity Theory of Turbulence to Atmospheric Diffusion, *Q. J. R. Meteorol. Soc.*, 76: 133-145.

——, 1952, Diffusion in a Field of Homogeneous Turbulence, II. The Relative Motion of Particles, *Proc. Cambridge Philos. Soc.*, 48: 345-362.

——, 1953, *The Theory of Homogeneous Turbulence*, Cambridge University Press, London.

——, 1956, *Surveys in Mechanics*, Cambridge University Press, London.

——, 1959, Some Reflections on the Theoretical Problems Raised at the Symposium, *Advances in Geophysics*, Vol. 6, pp. 449-452, F. N. Frenkiel and P. A. Sheppard (Eds.), Academic Press, New York.

——, 1964, Diffusion from Sources in a Turbulence Boundary Layer, *Arch. Mech. Stosow.*, 3: 661-670.

——, and A. A. Townsend, 1956, *Turbulent Diffusion, Surveys in Mechanics*, G. K. Batchelor and R. M. Davies (Eds.), Cambridge University Press, London.

Berkowicz, R., and L. P. Prahm, 1979, Generalization of K Theory for Turbulent Diffusion, Part I. Spectral Turbulent Diffusivity Concept, *J. Appl. Meteorol.*, 36: 266-272.

Blackadar, A. K., 1962, The Vertical Distribution of Wind and Turbulent Exchange in a Neutral Atmosphere, *J. Geophys. Res.*, 67: 3095-3102.

Blanchet, J., 1970, Étude Statistique d'un Système Fluide—Application à la Météorologie, *La Météorologie V*, Vol. 15, pp. 15-78.

——, 1975, Concepts Nouveaux sur la Mécanique de la Turbulence, *La Météorologie VI*, Vol. 1, pp. 63-73; *ibid.*, Vol. 2, pp. 47-56.

Boussinesq, J., 1877, Essai sur la Théorie des Eaux Courantes, *Mémoires Présentés par Divers Savants à l'Académie des Sciences de l'Institut de France*, Vol. 23, pp. 1-680, Paris.

Braham, R. B., B. K. Seely, and W. D. Crozier, 1952, A Technique for Tagging and Tracing Air Parcels, *Trans. Am. Geophys. Union,* 33: 825-833.

Brier, G. W., 1950, The Statistical Theory of Turbulence and the Problem of Diffusion in the Atmosphere, *J. Meteorol.,* 7: 283.

Briggs, G. A., 1974, *Diffusion Estimation for Small Emissions,* 1973 Annual Report, USAEC Report ATDL-106, Environmental Research Laboratories, Air Resources Atmospheric Turbulence and Diffusion Laboratory, Oak Ridge, Tenn., National Oceanic and Atmospheric Administration.

Busch, N. E., S. W. Chang, and R. A. Anthes, 1976, A Multilevel Model of the Planetary Boundary Layer Suitable for Use with Mesoscale Dynamic Models, *J. Appl. Meteorol.,* 15: 909-919.

——, J. C. Wyngaard, Y. Izumi, and E. F. Bradley, 1971, Flux-Profile Relationship in the Atmospheric Surface Layer, *J. Atmos. Sci.,* 28: 181-189.

Businger, J. A., 1973, Turbulent Transfer in the Atmospheric Surface Layer, in *Workshop on Micrometeorology, Boston, 1972,* D. A. Haugen (Ed.), pp. 67-98, American Meteorological Society, Boston, Mass.

Byzova, N. L., Ye. K. Garger, and V. N. Ivanov, 1970, Experimental Estimation of the Lagrangian Time Scale of Turbulence, *Izv. Atmos. Oceanic Phys.,* 6: 315-320.

Calder, K. L., 1965, On the Equation of Atmospheric Diffusion, *Q. J. R. Meteorol. Soc.,* 91: 514-517.

Carter, H. H., and A. Okubo, 1965, *A Study of the Physical Processes of Movement and Dispersion in the Cape Kennedy Area,* Chesapeake Bay Institute, Johns Hopkins University, Ref. 65-2.

Cermak, J. E., 1963, Lagrangian Similarity Hypothesis Applied to Diffusion in Turbulent Shear Flow, *J. Fluid Mech.,* 15(Part I): 49-64.

Chadam, J., 1962, *On a Theory of Turbulent Diffusion,* Report AFCRL-52-1107, Air Force Cambridge Research Laboratory.

Corrsin, S., 1951, On the Spectrum of Isotropic Temperature Fluctuations in an Isotropic Turbulence, *J. Appl. Phys.,* 22: 469-473.

——, 1952, Heat Transfer in Isotropic Turbulence, *J. Appl. Phys.,* 23: 113-118.

——, 1953, Remarks on Turbulent Heat Transfer, in *Proceedings of the Iowa Thermodynamics Symposium,* pp. 5-30, University of Iowa, Iowa City.

——, 1962, Turbulence: Experimental Methods, *Encyclopedia of Physics,* Vol. 8/2, S. Flugge and C. Trusdell (Eds.), pp. 524-590, Springer, Berlin.

——, 1974, Limitations of Gradient Transport Models in Random Walks and in Turbulence, *Advances in Geophysics,* Vol. 18A, pp. 25-60, F. N. Frenkiel and R. E. Munn (Eds.), Academic Press, New York.

Crawford, T. V., 1966, *A Computer Program for Calculating the Atmospheric Dispersion of Large Clouds,* USAEC Report UCRL-50179, University of California, Lawrence Radiation Laboratory.

Crozier, W. D., and B. K. Seely, 1955, Concentration Distributions in Aerosol Plumes Three to Twenty-Two Miles from a Point Source, *Trans. Am. Geophys. Union,* 36: 42-52.

Csanady, G. T., 1966, Accelerated Diffusion in the Skewed Shear Flow of Lake Currents, *J. Geophys. Res.,* 71: 411-420.

——, 1967, Concentration Fluctuations in Turbulent Diffusion, *J. Atmos. Sci.,* 24: 21-28.

——, 1973, *Turbulent Diffusion in the Environment,* 248p., D. Reidel Publishing Co.

Davies, D. R., 1950, Three-Dimensional Turbulence and Evaporation in the Lower Atmosphere, II. *Q. J. Mech. Appl. Math.,* 3: 64-73.

——, 1954, On Diffusion from a Continuous Point Source at Ground Level into a Turbulent Atmosphere, *Q. J. Mech. Appl. Math.,* 1: 168-178.

Davies, R. W., 1959, Large-Scale Diffusion from an Oil Fire, *Advances in Geophysics,* Vol. 6, pp. 413-414, F. N. Frenkiel and P. A. Sheppard (Eds.), Academic Press, New York.

——, R. J. Diamond, and T. B. Smith, 1954, *A Mathematical Treatment of Turbulent Diffusion,* Report OSR-TN-54-62, American Institute of Aerological Research.

Deacon, E. L., 1949, Vertical Diffusion in the Lowest Layers of the Atmosphere, *Q. J. R. Meteorol. Soc.,* 75: 89-103.

——, 1972, Numerical Investigations of Neutral and Unstable Planetary Boundary Layer, *J. Atmos. Sci.,* 29: 91-115.

Deardorff, J. W., 1970a, A Three Dimensional Numerical Investigation of the Idealized Planetary Boundary Layer, *Geophys. Astrophys. Fluid Dyn.,* 1: 377-410.

Donaldson, C. duP., 1973, Construction of a Dynamic Model of the Production of Atmospheric Turbulence and the Dispersal of Atmospheric Pollutants, *Workshop on Micrometeorology,* pp. 313-392, D. A. Haugen (Ed.), American Meteorological Society, Boston, Mass.

Edinger, J. G., 1952, A Technique for Measuring the Detailed Structure of Atmospheric Flow, Geophysics Research Paper No. 19, pp. 241-261, Air Force Geophysics Research Division, Cambridge, Mass.

Elliott, W. P., 1959, The Areas Within Concentration Isopleths Downwind of Continuous Point Source, *Int. J. Air Water Pollut.,* 2: 115-126.

Ellison, T. H., 1959, Meteorology, *Sci. Prog. (New Haven),* 47(187): 495-506.

Freeman, B. E., 1977, Tensor Diffusivity of a Trace Constituent in a Stratified Boundary Layer, *J. Atmos. Sci.,* 34: 124-136.

Frenkiel, F. N., 1953, Turbulent Diffusion: Mean Concentration Distribution in a Flow Field of Homogeneous Turbulence, *Adv. Appl. Mech.,* 3: 61-107.

——, and I. Katz, 1956, Studies of Small-Scale Turbulent Diffusion in the Atmosphere, *J. Meteorol.,* 13: 388.

——, and P. S. Klebanoff, 1975, On the Lognormality of the Small-Scale Structure of Turbulence, *Boundary-Layer Meteorol.,* 8: 173-200.

Gee, J. H., and D. R. Davies, 1963, A Note on Horizontal Dispersion from an Instantaneous Ground Source, *Q. J. R. Meteorol. Soc.,* 89: 542-546.

Gerber, R., 1949, Sur la Réduction á un Principe Variationnel des Équations du Mouvement d'un Fluide Visqueux Incompressible, *Proceedings of the Fourth Institute,* Université de Grenoble, p. 157.

Gifford, F. A., Jr., 1955, A Simultaneous Lagrangian–Eulerian Turbulence Experiment, *Mon. Weather Rev.,* 83: 293-301.

——, 1957a, Relative Atmospheric Diffusion of Smoke Puffs, *J. Meteorol.,* 14: 410-414.

——, 1957b, Further Data on Relative Atmospheric Diffusion, *J. Meteorol.,* 14: 475-476.

——, 1959, Statistical Properties of a Fluctuating Plume Dispersion Model, *Atmospheric Diffusion and Air Pollution,* Advances in Geophysics, Vol. 6, pp. 117-138, F. N. Frenkiel and P. A. Sheppard (Eds.), Academic Press, New York.

——, 1960, Peak to Average Concentration Ratios According to a Fluctuating Plume Dispersion Model, *Int. J. Air Water Pollut.,* 3: 253-260.

——, 1961, Use of Routine Meteorological Observations for Estimating Atmospheric Diffusion, *Nucl. Safety,* 2(4): 47-51.

——, 1962, Diffusion in the Diabatic Surface Layer, *J. Geophys. Res.,* 67: 3207-3212.

——, 1968, An Outline of Theories of Diffusion in the Lower Layers of the Atmosphere, in *Meteorology and Atomic Energy—1968,* David H. Slade (Ed.), USDOE Report TID-24190, p. 101, NTIS.

——, 1972, The Form of the Frequency Distribution of Air Pollution Concentrations, in *Proceedings of Symposium on Statistical Aspects of Air Quality Data,* pp. 3.1-3.17, Chapel Hill, N. C.

——, 1976, Turbulent Diffusion-Typing Schemes: A Review, *Nucl. Safety,* 17(1): 68-86.

——, 1977, Tropospheric Relative Diffusion Observations, *J. Appl. Meteorol.*, 16: 311-313.

Goldstein, S., 1951, On Diffusion by Discontinuous Movements and on the Telegraph Equation, *Q. J. Mech. Appl. Math.*, 4 (Part 2): 129-156.

Hage, K. D., 1964, *Particle Fallout and Dispersion Below 30 km in the Atmosphere*, USAEC Report SC-DC-64-1463, Sandia Corporation.

Hanjalic, K., and B. E. Launder, 1972, A Reynolds Stress Model of Turbulence and Its Application to Thin Shear Flows, *J. Fluid Mech.*, 52: 609-638.

Hanna, S. R., 1975, Relative Diffusion of Tetroon Pairs During Convective Conditions, *paper presented at First Conference on Regional and Mesoscale Modeling Analysis and Prediction*, Las Vegas, Nev., American Meteorological Society, Boston, Mass.

Hay, J. S., and F. Pasquill, 1957, Diffusion from a Fixed Source at a Height of a Few Hundred Feet in the Atmosphere, *J. Fluid Mech.*, 2: 299-310.

——, and F. Pasquill, 1959, Diffusion from a Continuous Source in Relation to the Spectrum and Scale of Turbulence, Atmospheric Diffusion and Air Pollution, *Advances in Geophysics*, Vol. 6, pp. 345-365, F. N. Frenkiel and P. A. Sheppard (Eds.), Academic Press, New York.

Heffter, J. J., 1965, The Variation of Horizontal Diffusion Parameters with Time for Travel Periods of One Hour or Longer, *J. Appl. Meteorol.*, 4: 153-156.

Hilst, G. R., 1957, The Dispersion of Stack Gases in Stable Atmospheres, *J. Air Pollut. Control Assoc.*, 7: 205-210.

Hinds, W. T., 1969, Peak-to-Mean Concentration Ratios from Ground Level Sources in Building Wakes, *Atmos. Environ.*, 3: 145-156.

Hinze, J. O., 1959, *Turbulence*, McGraw-Hill Book Co., New York.

Högström, U., 1964, An Experimental Study on Atmospheric Diffusion, *Tellus*, 16: 205-225.

Hosker, R. P., Jr., 1974, Estimates of Dry Deposition and Plume Depletion over Forests and Grassland, in *Physical Behavior of Radioactive Contaminants in the Atmosphere*, Symposium Proceedings, Vienna, 1973, pp. 291-309, STI/PUB/354, International Atomic Energy Agency, Vienna.

Islitzer, N., and D. Slade, 1968, Diffusion and Transport Experiment, in *Meteorology and Atomic Energy—1968*, pp. 177-188, D. Slade (Ed.), USDOE Report TID-24190, NTIS.

Kaimal, J. C., J. C. Wyngaard, Y. Izumi, and O. R. Cote, 1972, Special Characteristics of Surface-Layer Turbulence, *Q. J. R. Meteorol. Soc.*, 98: 563-589.

Kampé de Fériet, M. J., 1939, Les Fonctions Aléatoires Stationnaires et la Theorie Statistique de la Turbulence Homogène, *Ann. Soc. Sci. Brux. Ser. I*, 59: 145.

Kao, S. K., 1960, Nonlinear Oscillations of a Particle in a Long Wave in a Rotating Fluid, *J. Geophys. Res.*, 65: 279-286.

——, 1962, Nonlinear Oscillations of a Particle in a Long Wave in a Rotating Viscous Fluid, *Phys. Fluids*, 5: 639-643.

——, 1965, Some Aspects of the Large-Scale Diffusion and Turbulence in the Atmosphere, *Q. J. R. Meteorol. Soc.*, 91: 10-17.

——, 1974, Basic Characteristics of Global Scale Diffusion in the Troposphere, in *Advances in Geophysics*, Vol. 18B, pp. 15-32, F. N. Frenkiel and R. E. Munn (Eds.), Academic Press, New York.

——, 1976a, A Model for Turbulent Diffusion in Shear Flows, in *Third Symposium on Atmospheric Turbulence Diffusion and Air Quality*, Raleigh, N. C., Oct. 26–29, pp. 217-220, American Meteorological Society, Boston.

——, 1976b, A Model for Turbulent Diffusion over Terrain, *J. Atmos. Sci.*, 3: 157-158.

——, 1981, The Lagrangian Governing Equations of a Rotating Compressible Viscous Atmosphere, *Pure Appl. Geophys.*, 119: 9-15.

——, and W. S. Bullock, 1964, Lagrangian and Eulerian Correlations and Energy Spectra of Geostrophic Velocities, *Q. J. R. Meteorol. Soc.*, 90: 166-174.

——, and C. N. Chi, 1976, Statistical Characteristics of Three-Dimensional Particle Movement in NCAR General Circulation Model, *J. Atmos. Sci.*, 33: 1042-1049.

——, and J. C. Doran, 1976, Turbulent Diffusion in Linear Shear Flow, *Pure Appl. Geophys.*, 114: 357-363.

——, and D. M. Taylor, 1977, *Atmospheric Diffusion in the Surface Turbulent Boundary Layer*, ERDA Report COO-2455-16, NTIS.

——, and L. L. Wendell, 1968, Some Characteristics of Relative Particle Dispersion in the Atmospheric Boundary Layer, *Atmos. Environ.*, 12: 397-407.

——, and M. Wurtele, 1959, The Motion of a Parcel in a Constant Geostrophic Wind Field of Parabolic Profile, *J. Geophys. Res.*, 64: 765-777.

——, and Y. N. Yeh, 1979, A Note on the Effects of Stack and Inversion Heights on Diffusion of Pollutants in the Planetary Boundary Layer, *Atmos. Environ.*, 13: 873-878.

Kármán, Th. v., 1930, Mechanical Similarity and Turbulence, *Nachr. Ges. Wiss. Göttingen*, 58-76.

Kazanski, A. B., and A. S. Monin, 1957, The Form of Smoke Jets, *Izv. Atmos. Oceanic Phys.*, 8: 1020-1033.

Kellogg, W. W., 1956, Diffusion of Smoke in the Stratosphere, *J. Meteorol.*, 13: 241.

Klug, W., 1968, Diffusion in the Atmospheric Surface Layer: Comparison of Similarity Theory and Observation, *Q. J. R. Meteorol. Soc.*, 94: 555-562.

Knox, J. B., and R. I. Pollack, 1972, An Investigation of the Frequency Distributions of Surface-Air Pollutant Concentrations, USAEC Report UCRL-74063, Lawrence Livermore Laboratory.

Kraichnan, R. H., 1971, An Almost-Markovian Galilean-Invariant Turbulence Model, *J. Fluid Mech.*, 47: 513-524.

Krasnoff, E., and R. L. Peskin, 1971, The Lagrangian Model for Turbulent Diffusion, *Geophys. Astrophys. Fluid Dyn.*, 2: 123-146.

Launder, B. E., 1975, On the Effects of a Gravitational Field on the Turbulent Transport of Heat and Momentum, *J. Fluid Mech.*, 67: 569-581.

Leith, C. E., 1971, Atmospheric Predictability and Two-Dimensional Turbulence, *J. Atmos. Sci.*, 28: 145-161.

Levin, L. M., 1954, Size Distribution Function for Cloud-Droplets and Raindrops, *Dokl. Akad. Nauk SSSR*, 94(6): 1045-1048.

Lewellen, W. S., and M. E. Teske, 1976, Second-Order Closure Modeling of Diffusion in the Atmospheric Boundary Layer, *Boundary Layer Meteorol.*, 10: 69-90.

Lin, C. C., 1960a, On a Theory of Dispersion by Continuous Movements, *Proc. Natl. Acad. Sci.*, 46: 566-570.

——, 1960b, On a Theory of Dispersion by Continuous Movements, II. Stationary Anisotropic Processes, *Proc. Natl. Acad. Sci. U.S.A.*, 46: 1147-1150.

Lumley, J. L., 1957, *Some Problems Connected with the Motion of Small Particles in Turbulent Fluid*, Ph.D. Thesis, Johns Hopkins University, Baltimore, Md.

Machta, L., H. L. Hamilton, L. F. Hubert, R. J. List, and K. M. Nagler, 1957, Airborne Measurements of Atomic Debris, *J. Meteorol.*, 14: 165-175.

Mellor, G. L., 1973, Analytic Prediction of the Properties of Stratified Planetary Surface Layers, *J. Atmos. Sci.*, 30: 1061-1069.

Mickelsen, W. R., 1955, *An Experimental Comparison of the Lagrangian and Eulerian Correlation Coefficients in Homogeneous Isotropic Turbulence*, National Advisory Committee for Aeronautics, Washington, Technical Note No. 3570, National Aeronautics and Space Agency.

Millionschchikov, M., 1941, On the Theory of Homogeneous and Isotropic Turbulence, *C. R. (Dokl.) Acad. Sci. USSR*, 32: 615-618.

Monin, A. S., 1959, Smoke Propagation in the Surface Layer of the Atmosphere, *Atmospheric Diffusion and Air Pollution, Advances in Geophysics,* Vol. 6, pp. 29-40, 331-343, F. N. Frenkiel and P. A. Sheppard (Eds.), Academic Press, New York.

——, and A. M. Yaglom, 1971, *Statistical Fluid Mechanics of Turbulence,* pp. 364-373, The MIT Press, Cambridge, Mass.

Murgatroyd, R. J., 1969, Estimations from Geostrophic Trajectories of Horizontal Diffusivity in the Mid-Latitude Troposphere and Lower Stratosphere, *Q. J. R. Meteorol. Soc.,* 95: 40-62.

O'Brien, J. J., 1970, A Note on the Vertical Structure of the Eddy Exchange Coefficient in the Planetary Boundary Layer, *J. Atmos. Sci.,* 27: 1213-1215.

Obukhov, A. M., 1941, Distribution of Energy in the Spectrum of Turbulent Current, *Izv. Akad. Nauk SSSR, Ser. Geogr. Geofiz.,* 4–5: 453-466.

Ogura, Y., 1952, The Theory of Turbulent Diffusion in the Atmosphere, *J. Meteorol. Soc. Jpn.,* 30: 23.

——, 1962, Energy Transfer in Isotropic Turbulent Flow, *J. Geophys. Res.,* 67: 3143-3149.

——, 1963, A Consequence of the Zero-Fourth-Cumulant Approximation in the Decay of Isotropic Turbulence, *J. Fluid Mech.,* 16: 33-40.

Orszag, S. A., 1970, Analytical Theory of Turbulence, *J. Fluid Mech.,* 41: 363-386.

——, and M. D. Kruskal, 1968, Formulation of the Theory of Turbulence, *Phys. Fluids,* 11: 43-60.

Pack, D. H., and J. K. Angell, 1963, A Preliminary Study of Air Trajectories in the Los Angeles Basin Derived from Tetroon Flights, *Mon. Weather Rev.,* 91: 583-604.

Panofsky, H. A., and B. Prasad, 1965, Similarity Theories and Diffusion, *Int. J. Air Water Pollut.,* 9: 419-430.

Pasquill, F., 1961, The Estimation of the Dispersion of Windborne Material, *Meteorol. Mag.,* 90: 33-49.

——, 1962, Some Observed Properties of Medium-Scale Diffusion in the Atmosphere, *Q. J. R. Meteorol. Soc.,* 88: 70-79.

——, 1966, Lagrangian Similarity and Vertical Diffusion from a Source at Ground Level, *Q. J. R. Meteorol. Soc.,* 92: 185-195.

——, 1968, Some Outstanding Issues in the Theory and Practice of Estimating Diffusion from Sources, in *Symposium on the Theory and Measurement of Atmospheric Turbulence and Diffusion in the Planetary Boundary Layer,* USAEC Report SC-M-68-191, pp. 17-30, Sandia Laboratories.

——, 1974, *Atmospheric Diffusion,* Halsted Press and John Wiley & Sons, New York.

——, 1975, The Dispersion of Material in the Atmospheric Boundary Layer, *Lectures on Air Pollution and Environmental Impact Analysis,* D. A. Haugen (Ed.), pp. 1-34, American Meteorological Society, Boston, Mass.

Paulson, C. A., 1970, Mathematical Representation of Wind Speed and Temperature Profiles in the Unstable Atmospheric Boundary-Layer, *J. Appl. Meteorol.,* 9: 857-861.

Peterson, K. R., 1968, Continuous Point-Source Plume Behavior out to 160 Miles, *J. Appl. Meteorol.,* 7: 217-226.

Philip, J. R., 1967, Relation Between Eulerian and Lagrangian Statistics, Boundary Layers and Turbulence, *Phys. Fluids, Suppl.,* 69.

Powell, D. C., 1974, *Analyses of Parallel and Orthogonal Wind Components Along Lines of Towers over Homogeneous Desert,* Ph. D. Thesis, Department of Meteorology, University of Utah, Salt Lake City.

Prandtl, L., 1925, Über Die Ausgebildete Turbulence, *Z. Angew. Math. Mech.,* 5: 136-139.

——, and O. G. Tietjens, 1934, *Fundamentals of Hydromechanics and Aeromechanics,* United Engineering Trustees, Inc.

Proudman, I., and W. H. Reid, 1954, On the Decay of a Normally Distributed and Homogeneous Turbulent Velocity Field, *Philos. Trans. Roy. Soc. London, Ser. A,* 247: 163-189.

Randerson, D., 1972, Temporal Changes in Horizontal Diffusion Parameters of a Single Nuclear Debris Cloud, *J. Appl. Meteorol.,* 11: 670-673.

——, J. G. Garcia, and V. S. Whitehead, 1971, Photogrammetric and Photometric Investigation of a Smoke Plume Viewed from Space, *J. Appl. Meteorol.,* 10: 1122-1131.

Reynolds, O., 1895, On the Dynamical Theory of Incompressible Viscous Fluids and the Determination of the Criterion, *Philos. Trans. R. Soc., London, Ser. A,* 186: 123-164.

Richardson, L. F., 1926, Atmospheric Diffusion Shown on a Distance-Neighbour Graph, *Proc. R. Soc. (London), Ser. A.,* 110: 709.

Roberts, O. F. T., 1923, The Theoretical Scattering of Smoke in a Turbulent Atmosphere, *Proc. R. Soc. (London), Ser. A,* 104: 640-654.

Rounds, W., 1955, Solutions of the Two-Dimensional Diffusion Equations, *Trans. Am. Geophys. Union,* 36: 395.

Saffman, P. G., 1962, The Effect of Wind Shear on Horizontal Spread from an Instantaneous Ground Source, *Q. J. R. Met. Soc.,* 88: 382-393.

Scheidegger, A. E., 1958, The Random-Walk Model with Autocorrelation of Flow Through Porous Media, *Can. J. Phys.,* 36: 649-658.

Schmidt, W., 1925, Der Massenaustrausch in Freier Luft und Verwandte Erscheinungen, *Probleme der Kosmischen Physik,* Verlag von Henri Grand, Hamburg.

Seneca, J. 1955, Measures de diffusivite turbulenté sur des flocons de fumeé, *J. Sci. Meteorol.,* 7: 221-225.

Singer, I. A., I. Kazuhiko, and R. Gonzalez del Campo, 1963, Peak to Mean Pollution Concentration Ratios for Various Terrain and Vegetation Cover, *J. Air Pollut. Control Assoc.,* 13: 40-42.

Smith, E. J., and K. J. Hefferman, 1956, The Decay of the Ice Nucleating Properties of Silver Iodide Released from a Mountain Top, *Q. J. R. Meteorol. Soc.,* 82: 301-309.

Smith, F. B., 1957, The Diffusion of Smoke from a Continuous Elevated Point Source into a Turbulent Atmosphere, *J. Fluid Mech.,* 2: 49.

——, 1965, The Role of Wind Shear in Horizontal Diffusion of Ambient Particles, *Q. J. R. Meteorol. Soc.,* 91: 318-329.

——, and J. S. Hay, 1961, The Expansion of Clusters of Particles in the Atmosphere, *Q. J. R. Meteorol. Soc.,* 87: 82-101.

——, and S. A. Matthews, unpublished, 1975.

Stull, R. B., 1973, Inversion Rise Model Based on Penetrative Convection, *J. Atmos. Sci.,* 30: 1092-1099.

Sutton, O. G., 1947a, The Problem of Diffusion in the Lower Atmosphere, *Q. J. R. Meteorol. Soc.,* 73: 257-281.

——, 1947b, The Theoretical Distribution of Airborne Pollution from Factory Chimneys, *Q. J. R. Meteorol. Soc.,* 73: 426-436.

——, 1953, *Micrometeorology,* McGraw-Hill Book Co., New York.

Sutton, W. G. L., 1943, On the Equation of Diffusion in a Turbulent Medium, *Proc. R. Soc. (London), Ser. A,* 182: 48-75.

Tatsumi, T., 1957, The Theory of Decay Process of Incompressible Isotropic Turbulence, *Proc. R. Soc. (London), Ser. A,* 239: 16-45.

Taylor, G. I., 1915, Eddy Motion in the Atmosphere, *Philos. Trans. R. Soc. (London), Ser. A,* 215: 1-26.

——, 1921, Diffusion by Continuous Movements, *Proc. London Math. Soc., Ser. 2,* 20: 196-202.

Tennekes, H., and J. L. Lumley, 1972, *A First Course in Turbulence,* The MIT Press, Cambridge, Mass.

Townsend, A. A., 1956, *The Structure of Turbulent Shear Flow,* Cambridge University Press, London.

Van der Hoven, I., 1957, Power Spectrum of Horizontal Wind Speed in the Frequency Range from 0.0007 to 900 Cycles Per Hour, *J. Meteorol.,* 14: 160-164.

Walker, E. R., 1965, A Particulate Diffusion Experiment, *J. Appl. Meteorol.,* 4: 614-621.

Wyngaard, J. C., O. R. Cote, and K. S. Rao, 1974, Modeling the Atmospheric Boundary Layer, *Advances in Geophysics,* Vol. 18A, pp. 193-211, F. N. Frenkiel and P. A. Sheppard (Eds.), Academic Press, New York.

Wurtele, M. G., 1960, Lagrangian Dynamics and the Parcel Method in Atmospheric Models, *J. Meteorol.,* 17: 661-669.

Yaglom, A. M., 1965, Lagrangian Properties of Turbulence in a Diabatic Surface Layer and in Convective Jets, *Izv. Acad. Sci. USSR, Atmos. Oceanic Phys. (Engl. Transl.)*, 1: 94-99.

Yamada, J., and G. Mellor, 1975, A Simulation of Wangara Atmospheric Boundary Layer, *J. Atmos. Sci.*, 32. 2309-2329.

Yeh, G. T., and C. H. Huang, 1975, Three-Dimensional Air Pollutant Modeling in the Lower Atmosphere, *Boundary-Layer Meteorol.*, 9: 381-390.

Zeman, O., and J. Lumley, 1976, Modeling Buoyancy Driven Mixed Layers, *J. Atmos. Sci.*, 33: 1974-1988.

BIBLIOGRAPHY

Corrsin, S., 1959, Progress Report on Some Turbulent Diffusion Research, in *Advances in Geophysics*, Vol. 6, *Atmospheric Diffusion and Air Pollution*, pp. 161-164, F. N. Frenkiel and P. A. Sheppard (Eds.), Academic Press, New York.

——, 1963, Estimation of the Relation Between Eulerian and Lagrangian Scales in Large Reynolds Number Turbulence, *J. Atmos. Sci.*, 20: 115.

Deardorff, J. W., 1970a, A Three Dimensional Numerical Investigation of the Idealized Planetary Boundary Layer, *Geophys. Astrophys. Fluid Dyn.*, 1: 377-410.

——, 1970b, A Numerical Study of Three Dimensional Turbulent Channel Flow at Large Reynolds Numbers, *J. Fluid Mech.*, 41: 453-480.

Egan, B. A., and J. R. Mahoney, 1972, A Numerical Modelling of Advection and Diffusion of Urban Area Source Pollutants, *J. Appl. Meteorol.*, 11: 312-322.

Frenkiel, F. N., 1951, Frequency Distribution of Velocities in Turbulent Flow, *J. Meteorol.*, 8: 316-320.

——, 1952, Application of Statistical Theory of Turbulent Diffusion to Micrometeorology, *J. Meteorol.*, 9: 252-259.

——, 1955, Turbulent Diffusion, in *Proceedings of Eighth International Congress on Theoretical and Applied Mechanics*, 105-127.

Gee, J. H., 1964, A Further Note on Horizontal Dispersion from an Instantaneous Ground Source, *Q. J. R. Meteorol. Soc.*, 90: 478-480.

Gifford, Jr., F. A., and S. R. Hanna, 1971, Urban Air Pollution Modelling, in *Second International Clean Air Congress of the International Union of Air Pollution Prevention Associations*, pp. 1146-1151, Academic Press, New York.

Godson, W. L., 1958, The Diffusion of Particulate Matter from an Elevated Source, *Arch. Meteorol. Geophys. Bioklimatol., Ser. A*, 10: 305.

Haugen, D. A., 1966, Some Lagrangian Properties of Turbulence Deduced from Atmospheric Diffusion Experiments, *J. Appl. Meteorol.*, 5: 647-652.

Kaimal, J. C., 1974, Basic Characteristics of Global Scale Diffusion in the Troposphere, *Advances in Geophysics*, Vol. 18B, pp. 15-62, F. N. Frenkiel and R. E. Munn (Eds.), Academic Press, New York.

Kao, S. K., 1972, A Note on Eulerian–Lagrangian Time-Scale Transformation for the Large-Scale Atmospheric Turbulence, *J. Geophys. Res.*, 77: 7188-7189.

——, J. N. Paegle, and W. E. Normington, 1974, Mountain Effect on the Motion in the Atmosphere's Boundary Layer, *Boundary-Layer Meteorol.*, 7: 501-512.

Lumley, J. L., and H. A. Panofsky, 1964, *The Structure of Atmospheric Turbulence*, John Wiley and Sons, New York.

Mellor, G. L., and T. Yamada, 1974, A Hierarchy of Turbulence Closure Models for Planetary Boundary Layers, *J. Atmos. Sci.*, 31: 1791-1806.

Moses, H., and J. E. Carson, 1967, Stack Design Parameters Influencing Plume Rise, Paper 67-84, in *Fifth Annual Meeting of the Air Pollution Control Association*, Cleveland, Ohio.

Obukhov, A. M., 1949, Structure of Temperature Field in a Turbulent Flow, *Izv. Akad. Nauk SSSR, Ser. Geogr. Geofiz.*, 13: 58-69.

——, 1953, The Relation Between the Space- and Time-Correlation Function in a Turbulent Flow, *J. Meteorol. Soc. Jpn.*, 31: 355-369.

Ogura, Y., 1959, Diffusion from a Continuous Source in Relation to a Finite Observation Interval, Atmospheric Diffusion and Air Pollution, *Advances in Geophysics*, Vol. 6, pp. 149-159, F. N. Frenkiel and P. A. Sheppard (Eds.), Academic Press, New York.

——, 1962, Energy Transfer in a Normally Distributed and Isotropic Turbulent Velocity Field in Two Dimensions, *Phys. Fluids*, 5: 395-401.

Pasquill, F., 1971, Atmospheric Dispersion of Pollution, *Q. J. R. Meteorol. Soc.*, 97: 369-395.

Shir, C. C., 1973, A Preliminary Numerical Study of Atmospheric Turbulent Flow in the Idealized Planetary Boundary Layer, *J. Atmos. Sci.*, 30: 1327-1339.

Slade, D. H. (Ed.), 1968, *Meteorology and Atomic Energy—1968*, USDOE Report TID-24190, NTIS.

Smith, F. B., The Turbulent Spread of a Falling Cluster, Atmospheric Diffusion and Air Pollution, in *Advances in Geophysics*, Vol. 6, *Atmospheric Diffusion and Air Pollution*, pp. 193-210, F. N. Frenkiel and P. A. Sheppard (Eds.), Academic Press, New York.

Smith, M. E., and I. A. Singer, 1966, An Improved Method of Estimating Concentrations and Related Phenomena from a Point Source Emission, *J. Appl. Meteorol.*, 5: 631-639.

Sutton, O. G., 1932, A Theory of Eddy Diffusion in the Atmosphere, *Proc. R. Soc. (London), Ser. A*, 135: 143.

Tyldesley, J. B., and C. E. Wallington, 1965, The Effect of Wind Shear and Vertical Diffusion on Horizontal Dispersion, *Q. J. R. Meteorol. Soc.*, 91: 158-174.

Wyngaard, J. C., and O. R. Cote, 1971, The Budgets of Turbulent Kinetic Energy and Temperature Variance in the Atmospheric Surface Layer, *J. Atmos. Sci.*, 28: 190-201.

Yaglom, A. M., 1962, An Introduction to the Theory of Stationary Random Functions, Prentice-Hall, Inc., Englewood Cliffs, N. J.

Yamada, J., 1977, A Numerical Experiment on Pollutant Dispersion of Horizontally Homogeneous Atmospheric Boundary Layer, *Atmos. Environ.*, 11: 1015-1024.

Chapter 6 covers research prior to February 1980.

FLOW AND DIFFUSION NEAR OBSTACLES

7

R. P. Hosker, Jr., Atmospheric Turbulence and Diffusion Laboratory

7-1 INTRODUCTION

Obtaining reliable answers to questions regarding whether the behavior of effluent plumes near hills or buildings will create pollution problems—and, if so, how frequently—is difficult. Currently, relatively little guidance is available in an easy-to-apply handbook format; consequently, reference usually must be made to the technical literature.

This chapter attempts to consolidate existing information on flow and effluent diffusion near obstacles. Initially, flow patterns typically observed around bluff obstacles in general and buildings in particular are examined, and available data on the extent of the upwind influence of a body, the characteristics and size of the near-body flow, and the behavior of the far wake are summarized. A somewhat fundamental viewpoint that emphasizes the physics of the flow has been adopted, and the treatment often is somewhat theoretical. This mode of presentation reflects the method of study common in fluid dynamics literature.

Next, effluent dispersion in these complex flow fields is addressed; however, this subject is discussed in a much more applied fashion that reflects the traditional approach of the relevant literature. A number of rules of thumb and formulations are presented for estimating such things as appropriate stack heights and pollutant concentrations near buildings. However, main emphasis is placed on the flow fields because understanding their likely behavior in any particular case makes practical tasks, such as stack placement and design and pollution assessment, far easier.

Section 7-2 surveys flow characteristics common to many wake-related problems, even wakes behind bodies remote from ground surfaces. Sections 7-3 and 7-4 examine the response of near-body flows to velocity gradients in the incident wind and to the presence of a ground plane. The problem of central interest—a ground-mounted bluff obstacle immersed in a deep incident boundary layer—is considered in Sec. 7-5. Section 7-6 addresses the effects of the incidence angle of the wind and of the clustering of obstacles; and Sec. 7-7 examines the effects of changes in ambient turbulence and in atmospheric stability. Finally, Sec. 7-8 summarizes the available guidelines and methods for air-pollution-related calculations in regions aerodynamically affected by buildings or other obstacles.

This chapter presupposes some understanding of the basic principles and vocabulary of fluid mechanics; the texts of Batchelor (1967), Landau and Lifshitz (1959), and Schlichting (1960) are excellent background references.

Hosker: Atmospheric Turbulence and Diffusion Laboratory, National Oceanic and Atmospheric Administration, Oak Ridge, Tennessee.

Extensive use has been made of experimental data, many of which were generated in laboratory facilities. However, no discussion of the principles for successful modeling of flow near buildings is included because these principles are sufficiently explained in the literature. Interested readers should consult articles by Cermak and his colleagues (e.g., Cermak et al., 1966; Cermak, 1971, 1974), Isyumov, Jandali, and Davenport (1974), and Sundaram and his co-workers (e.g., Sundaram, Skinner, and Ludwig, 1970; Ludwig, Sundaram, and Skinner, 1971; McVehil, Ludwig, and Sundaram, 1967), among others. Snyder (1972) has reviewed these principles and has recently (1981) proposed fluid modeling guidelines that should prove useful in evaluating old test programs and in standardizing new ones.

Mention of applicable mathematical modeling techniques, both analytical and numerical, has deliberately been kept brief because this work seems to be a continually evolving art. Literally dozens of references are available; however, papers by Djurić and Thomas (1971), Hotchkiss (1971), Hirt and Cook (1972), Kiya and Arie (1972), Hunt and Mulhearn (1973), Frost, Harper, and Fichtl (1975), Ukeguchi, Okamoto, and Ohba (1975), Shieh, Frost, and Bitte (1977), Mason and Sykes (1979a), and Puttock and Hunt (1979) seem particularly interesting.

7-2 UNBOUNDED UNIFORM FLOW OVER BLUFF BODIES

7-2.1 Flow Near the Upwind Face

Some of the major features of flow near obstacles immersed in turbulent boundary layers are common to any high Reynolds number (Re) flow around a bluff body; see, for example, Birkhoff and Zarantonello (1957) and Wu (1972). This section reviews some of the information available on such phenomena. For simplicity, Fig. 7.1 illustrates the two-dimensional flow case; however, extension to three dimensions is conceptually easy. Upwind of a bluff object will be a "displacement zone" where incident fluid particles are first influenced by the presence of the obstacle. In this zone both fluid speed and direction will be affected as the flow attempts to travel around and over the object. As expected, pressure on the front surfaces of the obstacle will be higher than the ambient pressure as a result of the deceleration of the approach flow. Flow patterns in this upwind displacement region are fairly predictable using aerodynamic theory (e.g., Batchelor, 1967; Parkinson and Jandali, 1970; Bearman and Fackrell, 1975).

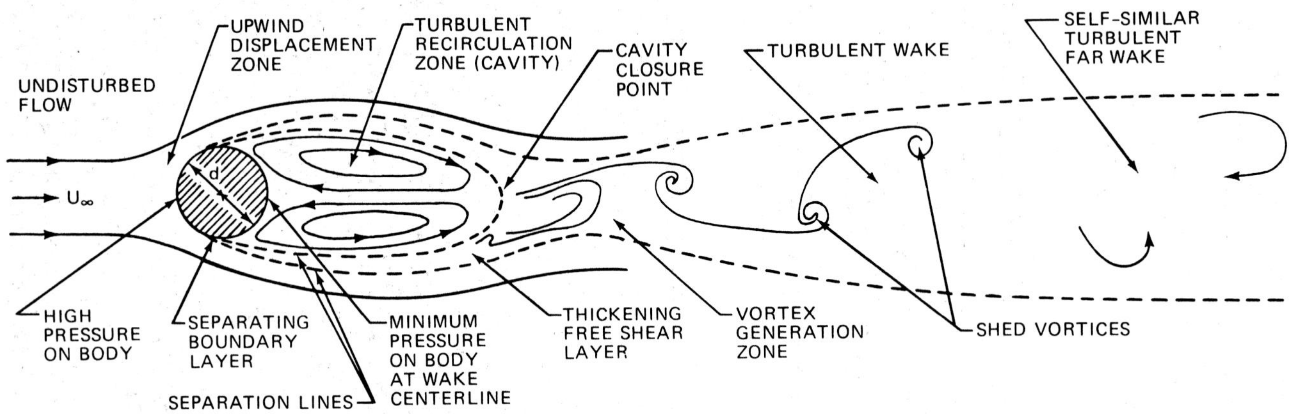

Fig. 7.1 Typical mean flow field associated with a two-dimensional bluff body immersed in a uniform high Reynolds number flow. (Based on T. Y. Wu, Cavity and Wake Flows, in *Annual Review of Fluid Mechanics*, Vol. 4, pp. 243–284, Annual Reviews, Inc., Palo Alto, Calif., 1972.)

7-2.2 Separation and the Near-Body Wake

Boundary layers that form along the exposed surfaces of a bluff obstacle generally will separate from these surfaces at locations determined either by geometry (i.e., at sharp edges) or by aerodynamic effects. When the separation lines are not fixed by geometry, their locations tend to fluctuate with variations in the overall flow pattern.

The separated boundary layers move out into the surrounding fluid as free shear layers. If the obstacle is sufficiently long, these layers may reattach to the surface at some downstream location and eventually be shed again at the end of the body. In either case, these separated layers curve inward toward the wake axis and enclose a cavity (or recirculation bubble) immediately downwind of the body. The cavity zone is characterized by low mean speed, high turbulence intensity, recirculation, relatively long residence times of fluid particles trapped within the bubble, and low, nearly constant pressure. As pointed out by Bearman (1972a), Bradshaw and Wong (1972), and Frost (1973), the steady-state cavity shape and pressure both result from the dynamic balance between the fluid withdrawn from the cavity by entrainment into the bounding shear layers and the fluid supplied to the cavity by the reverse flow that originates with the bifurcating shear layers at the point of cavity closure. Therefore, flow characteristics of the recirculation zone depend on the characteristics of the boundary layer shed by the obstacle and on the details of the turbulent mixing within the free shear layers. Such things are not easily predicted, and detailed experimental information is somewhat sparse because of the difficulty of measurements.

Uniform flow past a two-dimensional object has been relatively well researched; however, most of the research has concentrated on evaluating force-related terms, such as surface pressure distributions and aerodynamic drag coefficients (see Hoerner, 1965, for a thorough survey of such literature). For this reason, flow details, such as cavity size, often are not specified. In the case of a long thin plate of height H spanning an unbounded flow, the resulting standing eddy or bubble is between 2.0H and 2.8H in length (Arie and Rouse, 1956; Fail, Lawford, and Eyre, 1957; Castro, 1971; Bradbury and Moss, 1975; Sullerey, Gupta, and Moorthy, 1975). Within this cavity, reverse flow speeds approximately 40% of the incident wind speed and high turbulence intensities,† relative to free-stream wind speed, of 35% (streamwise and spanwise velocity components) and 60% (transverse velocity component) have been reported by Bradbury and Moss (1975).

If this long thin plate is thickened in the along-wind direction, it becomes a rectangular bar of thickness L. The recirculation zone length varies with the ratio L/H (discussed below). For L/H = 2.8 or more, the shear layers reattach to the side surfaces of the bar and finally separate again at the trailing edge (Nakaguchi, Hashimoto, and Mato, 1968; Bostock and Mair, 1972).

The recirculation region behind long circular or elliptical cylinders depends on the exact location of the boundary-layer separation lines along the cylinder surface; this location in turn depends on such parameters as Reynolds number, incident turbulence intensity and length scales, and cylinder surface roughness. No comprehensive survey of the variation of near-wake behavior associated with variations in these parameters has been reported. Similarly, flow behind two-dimensional objects other than plates, rectangular bars, and circular cylinders seems to remain largely unexplored.

Eddy characteristics behind axisymmetric and three-dimensional bodies immersed in uniform flow have been reported by several authors. For a thin square plate of side H mounted normal to the incident flow, a cavity length of 2.8H and a cavity width of 2.0H at the midlength point have been found (Fail, Lawford, and Eyre, 1957). The maximum reverse flow speed was about 60% of the free-stream wind. The wake boundary, which was roughly axisymmetric, was described by

†In this chapter, measurements within highly turbulent regions frequently will be cited. Many of the data were obtained in wind tunnels using heated-sensor anemometers. The reader should bear in mind that such measurements can be erroneous because of nonlinearities that arise in high turbulence intensity flows. Hinze (1959) suggests that turbulence with an intensity up to 25% may be explored if errors on the order of 10 to 20% can be tolerated. Methods to remove these errors exist (e.g., Comte-Bellot, 1976; Freymuth, 1978), but often they are not fully used. Newer measurement techniques bypass the problem entirely. Nevertheless, quantitative results within zones of strong turbulence intensity should be treated with caution; in particular, their indiscriminate use in critical applications is strongly discouraged.

$r/H = (x/H)^{1/4}$, where r is the radial distance of the boundary from the wake centerline and x is the distance downwind of the plate. Fail, Lawford, and Eyre (1957) remark that large changes in plate shape (e.g., squares, circles, triangles, and gear-like shapes) seem to have rather small effects on wake characteristics.

Calvert (1967a) studied near-wake flow behind cones with angles ranging from 0° (i.e., a circular cylinder) to 180° (i.e., a circular flat plate). Cavity length x_r, measured from the cone base, was found to increase linearly with increasing cone angle ϕ. For a 60° cone, the maximum reverse flow speed within the cavity was about 50% of the incident wind speed; the turbulence intensity, referred to as local mean speed, ranged as high as 65%. Near the center of the cavity, the local turbulence intensity was about 25%. As might be expected, the highest turbulence intensities were observed near the cavity closure point on the wake centerline and near the rear face of the cone. Roshko and Lau (1965), Calvert (1967a), and Sullerey, Gupta, and Moorthy (1975) have demonstrated similarity in the pressure and velocity distributions within the cavity region that develops behind simple obstacles such as plates and cones. They demonstrated these similarities by using suitable wake parameters to nondimensionalize the data.

The free shear layers (e.g., Fig. 7.1) that enclose the cavity are zones of high vorticity and usually are fully turbulent, even at moderate Reynolds numbers (Goldstein, 1965). Consequently, the vorticity diffuses into the neighboring field and thickens the shear layers, while the stresses within the shear layers become increasingly nonuniform with distance from the boundary-layer separation point (Wu, 1968, 1972). Therefore, the free shear layers become unstable and roll up into individual vortices that break loose and are advected downstream. This process occurs near the downwind closure point of the cavity region (Roshko and Fiszdon, 1967).

If the flow obstacle is rounded, so that the separation points on its surface are determined by aerodynamic force balances rather than by geometry alone, then the details of this wake vortex formation depend strongly on Reynolds number. For Reynolds numbers between 3×10^2 and 1×10^5, vortex shedding is periodic; however, as Reynolds number increases, the shedding eventually deteriorates into a fully turbulent flow where no periodic shedding is detected (Goldstein, 1965; Bearman, 1972a; Wu, 1972). The flow, even for two-dimensional obstacles, tends to be highly three-dimensional in this range. For Reynolds numbers larger than about 5×10^6, periodic vortex shedding and near-two-dimensionality reappear (Roshko and Fiszdon, 1967; Bearman, 1972a). For rounded obstacles bulk parameters such as drag coefficient C_D typically display large variations with Reynolds number (see Fig. 7.2). If, however, the flow separation occurs at the geometrically fixed sharp edges of a thin two-dimensional obstacle, such as a flat plate or a disk, the vortex shedding is more or less independent of Reynolds number, and bulk parameters, such as drag coefficient, are virtually constant for Reynolds numbers between approximately 10^3 and 10^7 (Hoerner, 1965; see Fig. 7.2).

Somewhat similar processes occur near thin three-dimensional bodies; for example, the flow behind an inclined disk has been described as a vortex "chain" consisting of interlocked vortex loops or rings (Calvert, 1967b) imbedded in a turbulent wake. If, on the other hand, the sharp-edged bluff obstacle has a finite downwind length, some interesting changes can take place. For example, with increasing downwind length to crosswind height ratio L/H, the drag coefficient of a simple bluff bar rises from about 2.0 at $L/H = 0$ (the thin two-dimensional plate value) to almost 3.0 for $L/H \simeq 0.6$; the coefficient then decreases with increasing L/H and reaches roughly 1.9 for $L/H > 1.2$ (Nakaguchi, Hashimoto, and Mato, 1968; Bostock and Mair, 1972; Bearman and Trueman, 1972; see Fig. 7.3a). Apparently, this abrupt rise in drag is caused by shear layer roll up and vortex formation immediately adjacent to the base of the obstacle in the critical L/H range; these vortices entrain fluid from the base region, generating a zone of very low pressure (Bearman and Trueman, 1972). Figure 7.4 indicates the roll-up process for several L/H ratios; at $L/H \simeq 0.6$, the vortices appear to be sweeping right across the rear face of the body. As might be expected, within this critical L/H range the flow pattern—and, hence, the pressure distribution and drag—is somewhat dependent on Reynolds number (Fig. 7.3b). These results suggest that the design and modeling of tall structures with thickness-to-width ratios near the critical value 0.6 should be approached with caution.

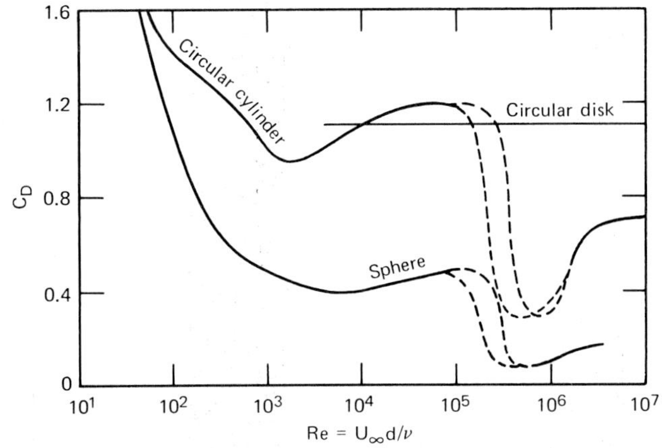

Fig. 7.2 Drag coefficient C_D as a function of Reynolds number (Re) for cylinder, sphere, and thin disk normal to uniform flow. All bodies have diameter d. Dashed curves show results from different wind tunnels. The abrupt decrease in C_D for $10^5 < Re < 10^6$ is due to boundary layers on the cylinder and sphere becoming turbulent and separating from obstacles farther aft than do laminar boundary layers. Therefore, the wake is narrower and drag is reduced. (From G. K. Batchelor, *An Introduction to Fluid Dynamics*, p. 341, Cambridge University Press, London and New York, 1967.)

In general, detailed predictions of the behavior of the free shear layer and the resultant wake characteristics near and just downwind of the recirculation zone are not available. The flow is unsteady and depends strongly on the obstacle geometry and incident flow characteristics.

7-2.3 The Far Wake

The turbulent wake that arises downwind of the cavity zone at high Reynolds numbers can be reliably predicted only at rather large distances (50 to 100 body diameters) downwind of the obstacle. At such distances the mean flow in the wake has become essentially parallel to the external flow, the flow speed has recovered to values approximately equal to the external wind speed, the pressure along the wake has returned to the ambient value, and the along-wind changes in wake structure are quite slow.

Once the wake has taken on these characteristics, then simple theoretical analyses (Hinze, 1959; Schlichting, 1960) indicate that the width (or height) of a two-dimensional far wake

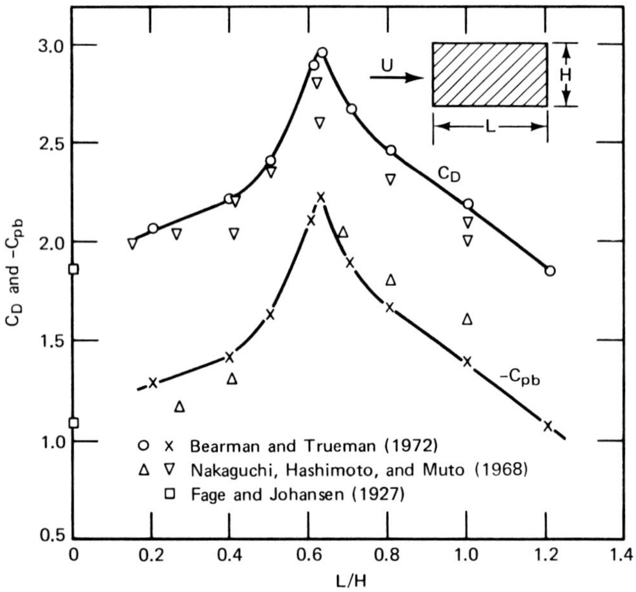

Fig. 7.3a Variation of drag (C_D) and rear face pressure (C_{pb}) coefficients with length-to-height ratio L/H for rectangular bars in uniform flow. [From P. W. Bearman and D. M. Trueman, An Investigation of the Flow Around Rectangular Cylinders, *Aeronaut. Q.*, XXIII(Pt. 3): 229-237 (1972).]

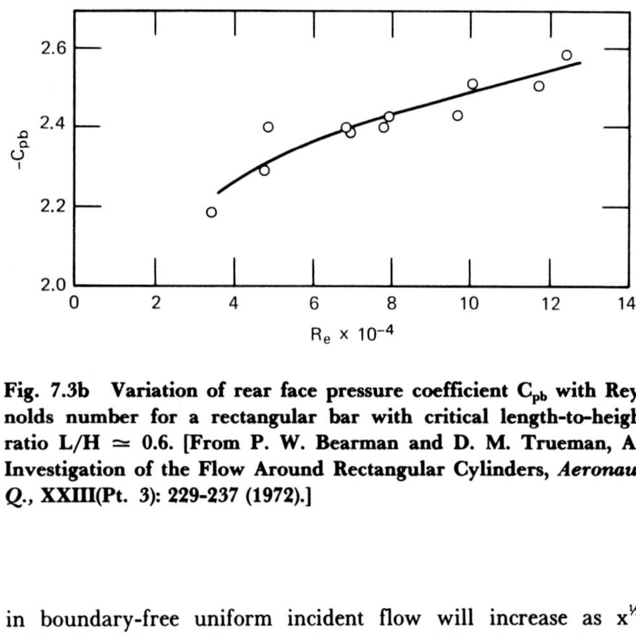

Fig. 7.3b Variation of rear face pressure coefficient C_{pb} with Reynolds number for a rectangular bar with critical length-to-height ratio L/H ≃ 0.6. [From P. W. Bearman and D. M. Trueman, An Investigation of the Flow Around Rectangular Cylinders, *Aeronaut. Q.*, XXIII(Pt. 3): 229-237 (1972).]

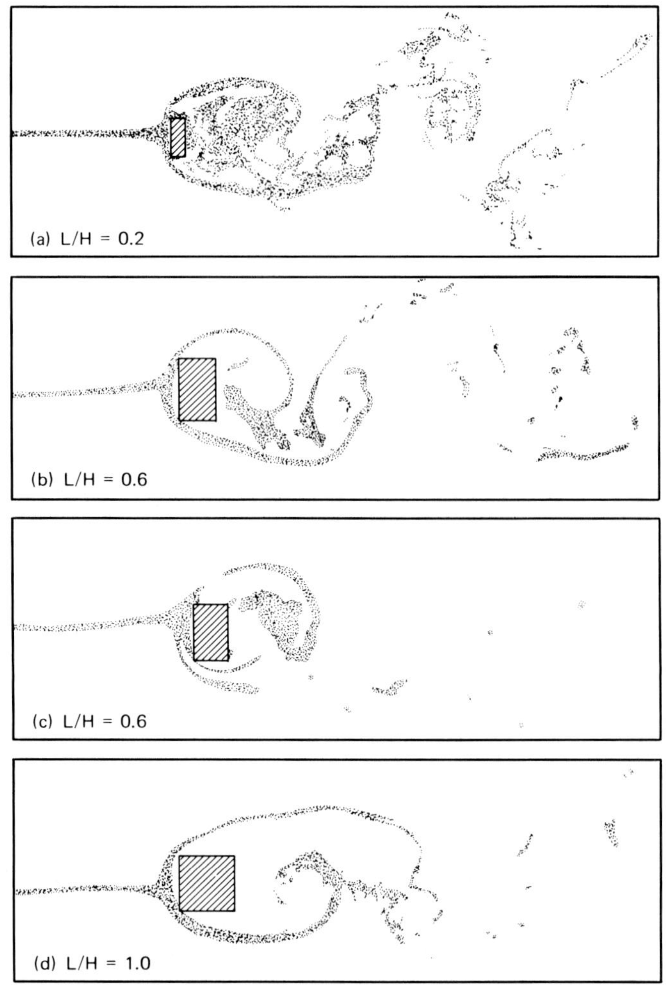

Fig. 7.4 Shear layer roll up and vortex formation behind two-dimensional rectangular bodies of differing along-width length to crosswind height ratios in uniform flow [traced from photographs in P. W. Bearman and D. M. Trueman, An Investigation of the Flow Around Rectangular Cylinders, *Aeronaut. Q.*, XXIII(Pt. 3): 229-237 (1972).]

in boundary-free uniform incident flow will increase as $x^{1/2}$, while the difference between the mean axial wake velocity and the free-stream velocity decreases as $x^{-1/2}$. Because the velocity fluctuations are proportional to this mean velocity difference (Tennekes and Lumley, 1972), they too decrease as $x^{-1/2}$. On the other hand, the width of a three-dimensional far wake should increase only as $x^{1/3}$ (Swain, 1929), and the axial velocity deficit and the velocity fluctuations in the wake should disappear as $x^{-2/3}$. Schlichting (1960) has demonstrated the excellent agreement between theory and experiment for the two-dimensional case of flow behind an infinitely long circular cylinder, and Halitsky (1968) has presented empirical equations in agreement with the theory for the three-dimensional case of a square flat plate normal to the incident flow.

7-3 EFFECT OF VELOCITY GRADIENT ON UNBOUNDED FLOW OVER BLUFF BODIES

The presence of a velocity gradient in the flow approaching a bluff body is known to have important effects. However, not much information is available describing the consequences of velocity gradient alone (as opposed to shear flows with turbulence and/or shear flows near ground planes) on the details of the resulting flow field.

7-3.1 Near-Body Flow and Wake Phenomena

Taylor (1917) and Tsien (1943) first examined the problem of laminar nonviscous shear flow past a two-dimensional cylinder with its axis across the flow (i.e., horizontal), perpendicular to the velocity gradient. They found that streamlines near the body are altered considerably from the uniform flow case, and they gave an expression for the shear-induced lift on the cylinder (see Batchelor, 1967, for a sketch of the resulting flow field). Tsien's (1943) work was quite general in that it included any two-dimensional obstacle that can be conformally mapped into a cylinder, such as a Joukowski airfoil.

Masch and Moore (1963) described the flow field near a vertical two-dimensional body with its axis parallel to the velocity gradient. If the incident wind speed increases with increasing height, then the stagnation pressure will be higher near the top of the front face than near the bottom. Consequently, this pressure gradient drives a downward-directed flow on the front face. The opposite situation exists on the rear of the body; the local wake pressure is lower at the top because of the higher winds; therefore, the induced flow is upward along the rear face. The downward transfer of fluid on the front face serves to lower the stagnation pressure near the top and to increase it near the bottom; the upward flow on the rear face causes exactly the opposite effect. Consequently, the drag near the top of the body is decreased by the induced flow, while the drag near the bottom is increased. Symes and Meroney (1970) found approximately a 7% net decrease in total drag for a finite-length cylinder in a shear flow. This description qualitatively agrees with Lighthill's (1956) work. Lighthill calculated zones of upward and downward velocity components; he found that the induced upward flow behind a cylinder is considerably weaker than the strong downwash just ahead of the body.

Three-dimensional work seems limited primarily to Lighthill's (1956) work on the sphere; he presents his results in the form of tabulated vorticity field components rather than flow schematics.

The presence of shear in the incident flow also can affect the boundary layers that form on an obstacle. Ahmad, Luxton, and Antonia (1976) examined the behavior of a turbulent boundary layer developing along a flat plate subjected to both uniform and sheared turbulent external flows. They concluded that, at least for moderate velocity gradients, the consequences of external shear are much less significant than the intensity and scale of the external turbulence, which had large effects on the boundary-layer structure. The shear was influential primarily in the outer regions of the boundary layer, where the local Reynolds stresses increased when the velocity gradient was positive and decreased when it was negative. In addition, near the outer edge of the boundary layer, they observed a thin zone where the local mean velocity was nearly constant with height and where the local Reynolds stresses were negligible. They attributed these effects near the interface with the outer flow to efficient turbulent mixing, which destroyed the shear locally. They suggested, however, that, if the shear was very large, it might dominate this mixing process so that the thin layer of constant velocity might not appear.

The implications of the preceding results for such phenomena as the mode and location of boundary-layer separation from bluff body surfaces are not clear and require further study. Gaster (1971) found that, at least at low Reynolds number, a shear flow past a long cylinder that has its axis parallel to the velocity gradient produced a vortex wake consisting of several discrete cells, each of which had its own characteristic shedding frequency (similar behavior was observed for a slightly tapered cylinder in a uniform flow field).

Maull and Young (1973) obtained the same results using a parallel-sided round-nosed bluff body in a shear flow at much higher Reynolds number than that used by Gaster (1971). They suggested that, on rounded bodies, vortex shedding tends to occur along the obstacle such that the local Strouhal number ($St = f\ell/U$, where f, ℓ, and U are the local values of vortex shedding frequency, obstacle dimension, and incident velocity, respectively) is a constant. The necessary spatial coherence of the shed vortices along the obstacle requires, on the other hand, that the shedding frequency be constant, at least over some finite spanwise distance. Eventually, however, the shedding process adjusts itself by an abrupt frequency jump, and a second vortex cell of a new characteristic frequency is formed. In this way the local Strouhal number can remain almost constant along the length of the obstacle. The local surface pressures naturally reflect this discrete structure. Because the cell boundaries are not geometrically fixed, they fluctuate with perturbations of the flow field (Stansby, 1976). In particular, the cell boundaries and shedding frequencies can be modified by artificially introducing longitudinal vorticity into the wake (Maull and Young, 1973).

However, in experiments with a two-dimensional thin flat plate (length parallel to the velocity gradient) where the separation points were fixed by geometry, Bradbury and Moss (1975) found no evidence of a cellular shedding structure; instead, the local Strouhal number was observed to vary weakly with position along the plate. In all cases the axes of the shed vortices were observed to be tilted with respect to the obstacle axis (shown in Fig. 7.5).

Kiya and Arie (1977a) have studied vortex shedding from an inclined thin flat plate that is placed horizontally across a shear flow and has its long axis perpendicular to the velocity gradient $\partial U/\partial z = \omega_0$. They found that, for a plate of height H in a flow with velocity components ($U_\infty + \omega_0 z$, 0, 0), such parameters as the Strouhal number, the velocity at the outer edge of the free shear layer, and the drag coefficient all varied approximately linearly with the shear parameter $\omega_0 H/U_\infty$. They suggested, however, that for symmetric bluff bodies, such as normal flat plates and cylinders that are oriented similarly, the flow properties, such as Strouhal number, should be even functions of the shear parameter because the bodies experience the same incident flow for $\pm\omega_0 H/U_\infty$. Further work on this topic is required.

Shear flow phenomena such as these can be important for understanding the response of tall structures, such as chimneys, to the periodic forces produced by vortex shedding (Scruton, 1967; Owen, 1971; and Cermak, 1974; among many others). Studies of these phenomena also help to elucidate the flow

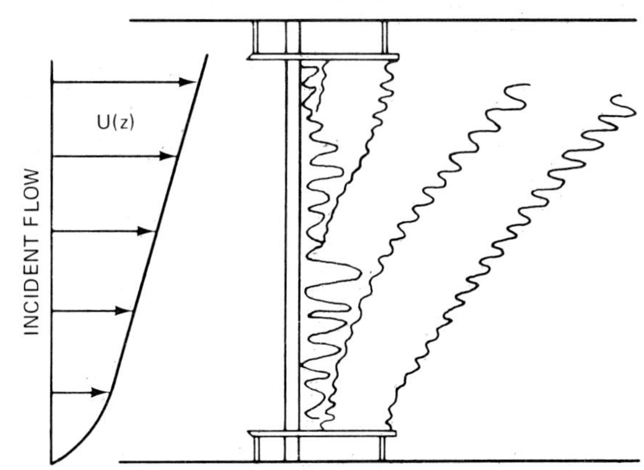

Fig. 7.5 Vortex shapes shed from stationary cylinder exposed to a shear flow, as shown by smoke visualization. [From P. K. Stansby, The Locking-On of Vortex Shedding due to the Cross-Stream Vibration of Circular Cylinders in Uniform and Shear Flows, *J. Fluid Mech.*, 74(Pt. 4): 641-665, Cambridge University Press, 1976.]

fields incident on other structures downwind of a vortex-shedding body.

The "locking-on" of the vortex shedding frequency to natural vibrational frequencies of bluff bodies is also a topic of considerable practical architectural interest (e.g., Owen, 1971; Mair and Maull, 1971; Bearman and Davies, 1975; and Griffin, 1978). However, the effect of shear on this phenomenon seems to have been studied only for circular cylinders (Stansby, 1976). Scruton and Rogers (1971) describe several construction techniques, such as helically wound strakes, by which the vortex shedding from certain bluff bodies may be modified or eliminated.

Little information exists on the effect of incident shear on the separation and reattachment of boundary layers on true three-dimensional objects. In fact, as pointed out by Mair and Maull (1971), relatively little work has been done on flow separation and vortex shedding near generalized three-dimensional bluff bodies in any kind of flow. More research into this fundamental area is needed if we are to understand and develop the potential for controlling such phenomena.

Information on the characteristics of the recirculation zone behind bluff bodies in a pure shear flow is largely unavailable. Bradbury and Moss (1975) found that the cavity behind a tall thin plate that has a width W and is aligned vertically (parallel to the velocity gradient) has a cavity-length-to-plate-width ratio, x_r/W, between 1.7 and 3.6. The ratio increases with the local incident velocity. They observed that at a given height the along-wind and spanwise vertical components of turbulence intensity, normalized by the local incident wind speed, were quite alike in both magnitude and dependence on downwind distance. However, both components were relatively strong functions of height. The transverse component of the turbulence intensity did not resemble the other components. It also showed the strongest dependence on height. Similar information for other types of bluff bodies would be highly useful and instructive, but it is not available.

7-3.2 The Far Wake

Kiya and Arie (1973) developed a theory for the laminar wake far behind a symmetric two-dimensional body that has its axis normal to the velocity gradient and is in a weakly sheared flow. They found that, just as in uniform flow, the wake width grew as $x^{1/2}$ and the velocity defect† decayed as $x^{-1/2}$. The wake was predicted to become increasingly asymmetric downwind; the wake width was larger on the lower velocity side of the flow, and the maximum velocity deficit migrated from the centerline behind the body toward the low-speed side. However, such a laminar wake is unstable at large Reynolds numbers. Kiya, Tamura, and Arie (1976) carried out a similar calculation for a turbulent wake behind a circular cylinder placed horizontally across a vertically varying shear flow with components $(U_\infty + \omega_0 z, 0, 0)$. Again it was found that the wake width increases as $x^{1/2}$ and the velocity defect decays as $x^{-1/2}$. Their predictions fit their experimental data for distances more than about 20 diameters downstream. Figure 7.6a shows the growing asymmetry of the theoretical and experimental values for the magnitude of the normalized wake velocity defect $|U_\infty + \omega_0 z - u|/U_\infty$. Figure 7.6b displays the profiles of measured longitudinal turbulence intensity at several downwind stations. The intensity is lower on the low-speed edge (bottom) of the flow; however, it dies out less quickly with changes in distance than on the higher speed edge (top) of the field.

Similar types of data are not available for the far-wake behavior behind three-dimensional bodies in unbounded shear flows.

†The difference between the velocity at a given location with and without the obstacle in place.

7-4 EFFECT OF A GROUND PLANE ON UNIFORM AND NEAR-UNIFORM FLOW OVER BLUFF BODIES

7-4.1 Flow Near the Upwind Face

If an inviscid incompressible fluid in uniform steady flow is incident on a bluff body, the resulting upwind velocity and pressure distributions can be calculated from potential flow theory. As long as the zero viscosity condition remains satisfied, the presence of a ground surface on which the obstacle rests serves only to introduce a plane of symmetry (between the model and its mirror image) into the problem. In practice, however, the viscosity of all real fluids produces boundary layers both on the exposed surfaces of the bluff body and on the ground plane. The retarding action of viscosity often is sufficient to prevent the fluid from completely traversing zones of adverse (opposing) pressure gradient; consequently, the flow separates.

Separation usually alters the flow patterns from those of potential flow. If, however, the exact separation points can be specified (e.g., for sharp-edged bluff bodies) and the paths of the separation streamlines downwind can be fairly accurately described, then an adjusted potential flow computation can be performed (e.g., Parkinson and Jandali, 1970; Kiya and Arie, 1977b).

If the ground surface extends very far upwind, the boundary layer induced by viscosity may be very thick relative to the obstacle height. In this case, the flow incident on the obstruction is fundamentally a shear flow; such examples are considered in Sec. 7-5. If the boundary layer is quite thin relative to the obstacle, then its effects will be confined to a region close to the ground.

Any boundary-layer velocity gradient near the ground surface will cause a downward-directed pressure gradient on the upwind faces of the obstacle. This pressure gradient in turn generates downward flow close to the body face and produces a recirculating upwind eddy or vortex near the ground and oriented across the flow.

Because of the pressure rise toward the body as the incident flow is decelerated, the main flow will separate from the ground plane some distance ahead of the body, passing above the frontal vortex and striking the obstacle somewhere just above the boundary-layer height. Because the boundary layer here is supposed to be thin, these effects should appear only as modest perturbations to the incident uniform mean flow (Fig. 7.7). Until recently, this type of flow field was common in many laboratory studies of flow near models of buildings and other obstacles; results of some of these studies are discussed in this section.

Leutheusser and Baines (1967) cite work by European authors on the methods of mounting bluff bodies on ground plates to achieve a good likeness of a potential flow in a wind tunnel. They suggest that the distance from the leading edge of the ground plate to the upwind face of the model should be approximately equal to the smaller of either the obstacle height or width in order to produce pressure distributions resembling a full model-and-mirror-image result.

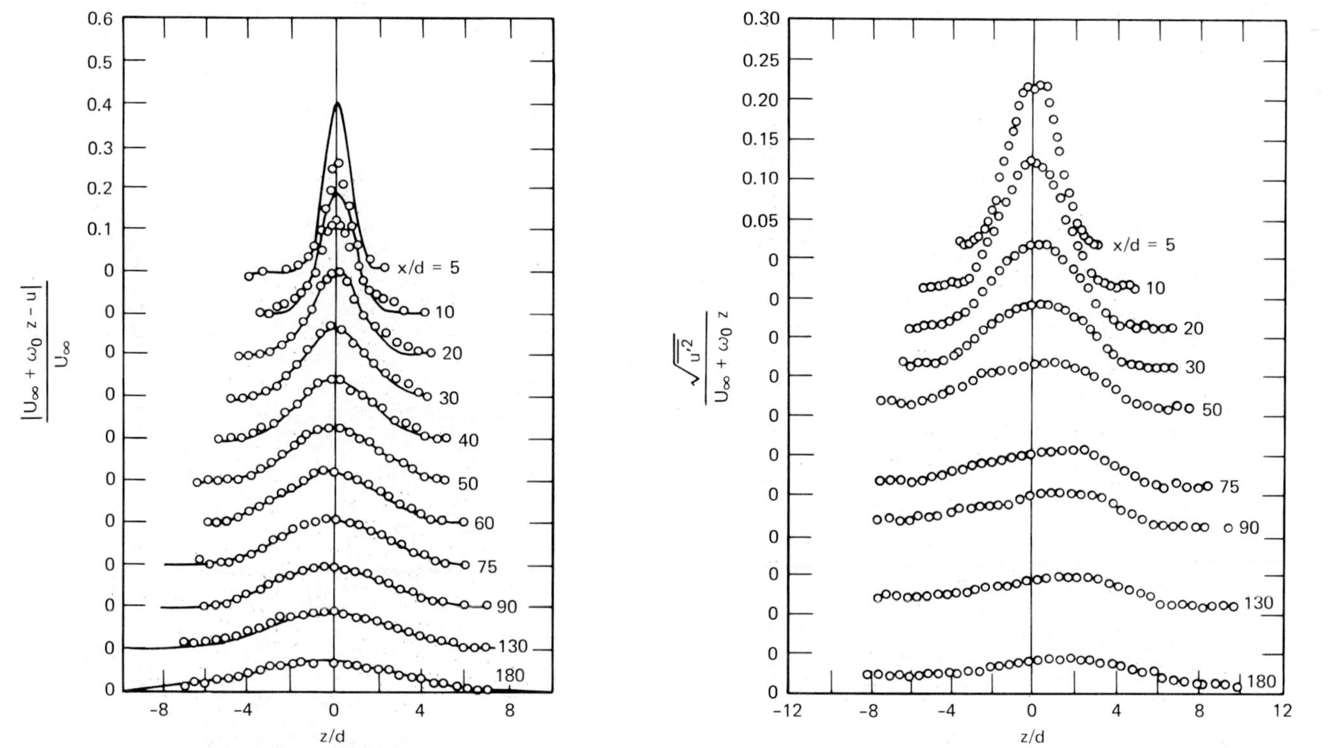

(a) Velocity defect profiles for several downwind locations x/d.

(b) Longitudinal turbulence intensity profiles relative to local incident wind speed for several x/d.

Fig. 7.6 Far-wake flow variables for a two-dimensional cylinder of diameter d mounted horizontally across a weak vertical shear flow $v = (U_\infty + \omega_0 z, 0, 0)$, such that $\omega_0 d/U_\infty \simeq 0.0235$. The solid lines in part (a) are theoretical predictions; the circles represent experimental values. [From M. Kiya, H. Tamura, and M. Arie, Turbulent Wakes Behind Two-Dimensional Bodies in a Uniform Shear Flow, *Bull. JSME (Jpn. Soc. Mech. Eng.)*, 19(129): 274–282 (1976).]

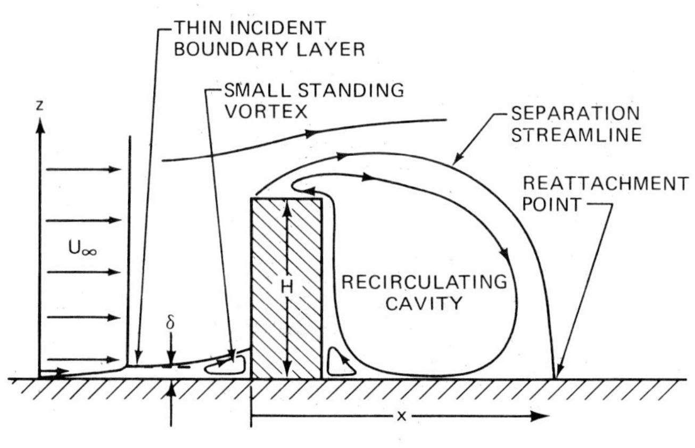

Fig. 7.7 Typical flow field near a strictly two-dimensional sharp-edged bluff body subjected to a nearly uniform incident flow where the depth of the incident boundary layer, δ, is much smaller than H. The concept that the cavity is bounded by the separation streamline should not be extrapolated to three-dimensional obstacles. (From J. C. R. Hunt, W. H. Snyder, and R. E. Lawson, Jr., Flow Structure and Turbulent Diffusion Around a Three-Dimensional Hill, in *Fluid Modeling Study on Effects of Stratification, Part I. Flow Structure,* Report EPA-600/4-78-041, U. S. Environmental Protection Agency, 1978.)

Arie and Rouse (1956) found that the presence of a ground plane downwind of the obstacle did not drastically affect the pressure distribution on the windward face of a normal flat plate. Evidently the upwind flow field depends primarily on the upwind conditions. Similar results were reported by Apelt and West (1975).

Castro and Robins (1977) examined near-uniform flow about a cube placed normal to the flow. They observed that the thin boundary layer along the upwind ground plane separated ahead of the block at a distance dependent on the boundary-layer thickness. The depth of even rather thin upwind boundary layers and the size of the upwind standing vortex significantly affected the vertical pressure distributions on a bluff body. However, at least for the somewhat thin layers in these tests, the horizontal variation of pressure across the upwind face of the cube was independent of the boundary-layer thickness.

7-4.2 Separation and the Near-Body Wake

In regions immediately adjacent to and downwind of an obstacle, the presence of a ground surface becomes considerably more significant to the flow. As pointed out by Arie and Rouse (1956), one of the major effects introduced by a ground plane is the severe restriction of vortex shedding and wake movement behind the body. The exact details of a given separation pattern, cavity flow, and near wake depend on the geometry and surface characteristics of the individual obstacles and on the nature of the approach flow; quantitative results are available for only a few thoroughly investigated examples.

*7-4.2.1 **Two-Dimensional Obstacles.*** Apelt, West, and Szewczyk (1973) studied a two-dimensional cylinder (diameter d) placed across a uniform flow; the cylinder was equipped along the wake centerline with wake splitter plates of various lengths. The wake width and drag were observed to decrease continuously as the splitter plate length, ℓ_s, increased from 0 to 1.0d. This was largely due to the flow separation point moving steadily aft on the cylinder. The roll up of the free shear layers shed by the body was hindered by the increasing length of the splitter plate. As this plate lengthened, the overall size of the vortex formation zone also increased in length and the transfer of fluid and vorticity from one side of the wake to the other was restricted. Gerrard (1966) reported similar phenomena.

By the time ℓ_s/d reached values in the range 5 to 7, the shed vortices were very small and apparently random, coupling of the flow across the wake had disappeared, and the wake envelope was narrow and steady. Reattachment of the separated streamline was observed ~5d from the cylinder center; this location did not vary with further increases in ℓ_s. Flow reversal in the wake cavity was observed between the cylinder and the reattachment point. Drag was virtually constant for $\ell_s/d \geqq 5$. Although the values determined in these experiments undoubtedly depend on the cylinder roughness and incident flow conditions, the general trends should be valid for other cases.

For thin sharp-edged obstacles where the separation points are determined by geometry, somewhat similar behavior can be expected. For example, Apelt and West (1975) found a coupled-vortex shedding pattern and oscillatory wake behind a long, thin, flat plate of height h normal to the flow. When a wake splitter plate was added, this behavior was sharply modified; for $\ell_s/h \geqq 3$, periodic vortex shedding was suppressed and the drag coefficient declined to a near-constant value.†

Arie and Rouse (1956) found that reattachment of the separation streamline behind a plate of height h occurred at $x_r/H = 16.8$, where $H \equiv h/2$ is the height of an equivalent fence above the ground surface. They observed that the separation streamline, as defined in Fig. 7.7, reached a maximum elevation $z/H \simeq 2.7$ near $x/H = 8$ (Fig. 7.8a). The recirculation zone enclosed by this streamline was nearly a semiellipse in cross section. The maximum turbulence intensity tended to remain at the height $z/H \simeq 2.5$, whereas the momentum flux maximum tended to follow the separation streamline (Fig. 7.8b).

Other experimenters found that x_r/H ranges from 11 (Baltaxe, 1967) to 14 (summary by Frost, 1973) for fence-like obstacles. The degree of test-section blockage in individual wind tunnels undoubtedly has an influence on the variability of the results (Castro and Fackrell, 1978).

This range of recirculation zone lengths—approximately 11 to 17 fence heights—may be compared with the results that show the cavity length for plates normal to unbounded flow typically is 2.0 to 2.8 plate heights (Sec. 7-2).

†Apelt and West (1975) point out that the drag coefficients they cite probably are inflated considerably by blockage effects. Their numerical values should not be used without applying corrections such as those suggested by Maskell (1963).

Evidently the presence of a ground plane dramatically increases the size of the cavity.

When a fence is thickened in the along-wind direction, it becomes a two-dimensional rectangular bar of height H and length L, and the flow situation becomes more complex. Locher (1969) noted that on such a bar the free shear layer that separated from the front edge would reattach to the upper surface of the bar if L/H was greater than 4.5; conversely, he observed no reattachment for L/H less than 2. When $2 < L/H < 4.5$, reattachment was intermittent and unstable.

Similar findings were reported by Arie et al. (1975). Although their work used bars that were either completely or partially immersed in boundary-layer incident flows, their results do not appear to be very sensitive to the boundary-layer depth relative to the bar. They found that for $L/H < 2.0$, the free shear layer curved over the top of the obstacle without contacting its upper surface; reattachment then occurred on the ground plane about 9H downwind of the front face of the bar. For L/H larger than 3.0, the shear layer reattached to the upper surface of the bar about 2.5H from its leading edge; the flow separated again at the trailing edge and traveled parallel to the ground surface. This secondary separation is somewhat like that at a rear-facing step. In this instance, reattachment of the final separation streamline to the ground plane occurred about 5H downstream of the rear face.

Comparable results behind rear-facing steps have been published by several authors; x_r/H was found to range between 5.8 (Bearman, 1965) and 7 (Tani, 1957). Intermediate values are quoted by Bradshaw and Wong (1972). Thus the rear-facing step has a recirculation zone somewhere between 5 and 7 step heights in length as compared to the two-dimensional fence cavities of 13 to 17H. This difference probably is caused by the decided upward deflection of the flow as it separates from a fence top compared to the step separation where the streamlines depart from the top parallel to the ground. The fence separation streamline must curve through a larger angle than the step separation streamline; consequently, a much longer recirculation zone can be expected.

For these two-dimensional cases, a number of authors (e.g., Mueller, Korst, and Chow, 1964; Plate, 1971) have successfully modeled the flow near the upper boundary of the recirculation zone. They take the isotach $U = U_\infty/2$ as the streamwise coordinate that divides a free jet from a lower region of near-zero velocity; this contour is identical to the separation streamline near the body, and it must be determined empirically. The predicted mean-velocity and shear-stress distributions within the resulting mixing zone along the upper cavity boundary are in good agreement with experimental results.

The behavior of the separated shear layer near its point of reattachment to the ground plane is important for an understanding of the dynamics of the flow within the cavity and of the near-wake region just downstream. Bradshaw and Wong (1972) have provided some insight into this behavior. At reattachment, the shear layer divides (Fig. 7.9). If the length of the recirculation zone is more than a few times the thickness of the initial boundary layer that separates from the two-dimensional body, then as much as one-half or more of the reattaching shear layer may be diverted upstream. This diverted fluid feeds the recirculation zone and balances the entrainment of fluid from the cavity into the bounding shear layer above. The large eddies in the layer that are responsible for most of the turbulent shear stress apparently are torn approximately in two near the reattachment point; consequently, the shear stress and the turbulent length scales decrease abruptly downwind of this

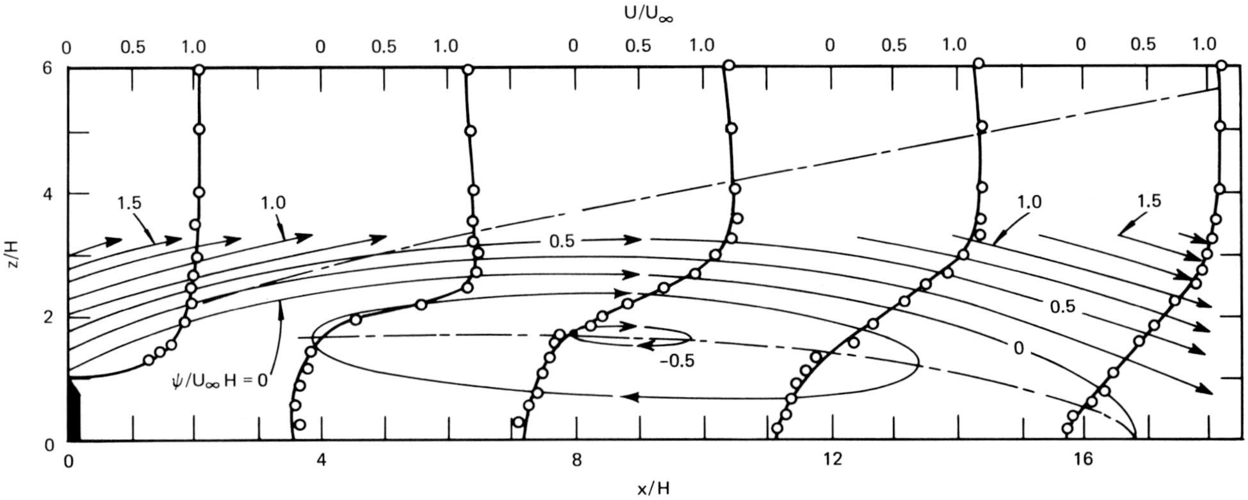

(a) Mean velocity profiles and flow streamlines ψ for several downwind stations.

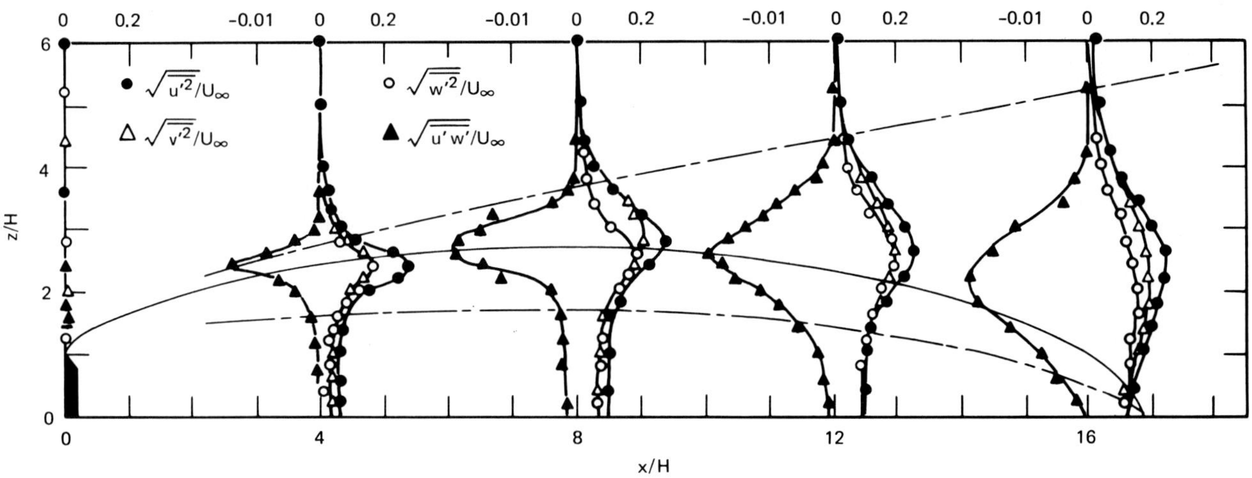

(b) Turbulence intensity relative to upstream wind speed for several downwind stations.

Fig. 7.8 Flow streamlines, mean velocity profiles, and turbulence intensity profiles behind a two-dimensional fence of height H on a ground plane with uniform incident flow. [From M. Arie and H. Rouse, Experiments on Two-Dimensional Flow over a Normal Wall, *J. Fluid Mech.*, 1(Pt. 2): 129-141, Cambridge University Press, 1956.]

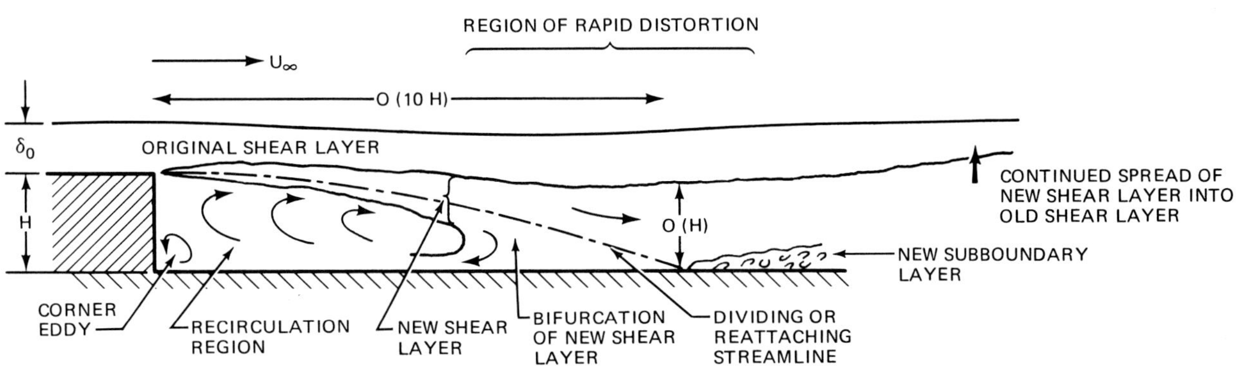

Fig. 7.9 Schematic of reattaching flow behind a two-dimensional step. [From P. Bradshaw and F. Y. F. Wong, The Reattachment and Relaxation of a Turbulent Shear Layer, *J. Fluid Mech.*, 52(Pt. 1): 113-135, Cambridge University Press, 1972.]

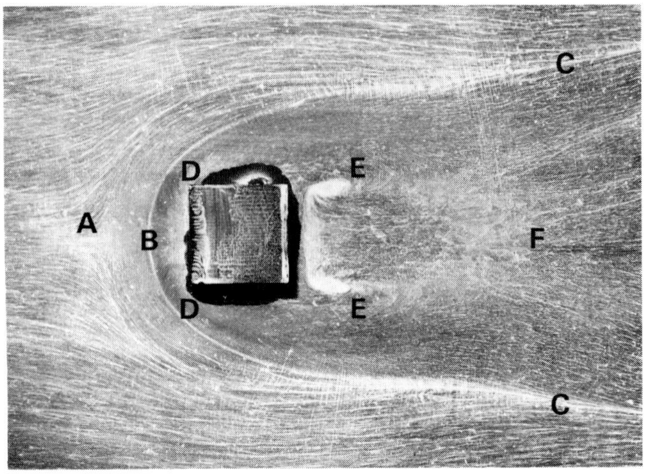

Fig. 7.10 View from above of near-ground flow pattern around a cube exposed to a nearly uniform ($\delta/H \simeq 0.1$) wind. Lettered features are described in text. [From I. P. Castro and A. G. Robins, The Flow Around a Surface-Mounted Cube in Uniform and Turbulent Streams, *J. Fluid Mech.*, 79(Pt. 2): 307-335, Cambridge University Press, 1977. Photograph courtesy of I. P. Castro and A. G. Robins.]

point. Therefore the turbulence just downstream of the cavity closure point is nearly independent of elevation above the surface (except for the new sub-boundary layer that immediately begins to develop). A considerable travel distance is required before a new boundary-layer-like flow is fully established.

Bradshaw and Wong (1972) point out two other factors that can affect the cavity flow regime. First, in-cavity flow, obstructions of significant size relative to the body can appreciably modify the pressure field near reattachment and, consequently, alter the behavior of the reattaching shear layer and the overall recirculating flow. Tani (1957) reported similar observations. This suggests that the presence of objects such as outbuildings or trees in the lee of a building can significantly affect its recirculation-zone characteristics. Extrapolation of data from laboratory to field or from site to site for such cases will require care. Second, the addition or removal of cavity fluid because of three-dimensionalities induced near the sides of an obstacle can significantly modify the recirculation-zone characteristics. This must be borne in mind when attempting to apply two-dimensional cavity data to real obstacles.

7-4.2.2 ***Three-Dimensional Obstacles.*** A photograph published by Castro and Robins (1977) is instructive with regard to the three-dimensional flow pattern around a cube-like object in a near-uniform incident flow (Fig. 7.10). Separation and lateral deflection of the main flow occur near point A in this figure. Point B indicates the maximum upwind extent of the small horizontally oriented standing vortex that is induced at the base of the obstacle by the thin boundary layer of the incident flow (see also Fig. 7.7, which shows a rough approximation of the centerline flow pattern). This vortex wraps itself around the obstacle and trails off downwind close to the ground as a counterrotating vortex pair, points C. If no incident boundary layer existed, this horseshoe vortex would not exist. Flow separation also occurs along the upwind vertical edges of the cube, points D; in this particular example, at least, reattachment to the sides does not take place. The shear layers shed at D roll up into vertically oriented counterrotating vortices near E; after interacting with the main flow somewhere near the roof height of the building, these vortices bend over and also trail off downstream as an elevated vortex pair (not visible in this ground-level visualization). The end of the cavity zone behind the cube appears near point F.

Probably the most complete study of cavity size behind sharp-edged three-dimensional obstacles was carried out by Evans (1957). To visualize the extent of the recirculation zone, Evans employed smoke in a nearly uniform incident flow. Using this technique, he systematically examined the flow behind block-like model buildings that had various lengths, widths, heights, and roofs (some roofs were pitched, some were not). His observations, a few of which are summarized here, were reported in tabular form and in a lengthy sequence of sketches.

Evans found that an increase in the along-wind building dimension L (relative to the height H) leads to a decrease in the length of the cavity. This decrease in cavity size continues until $L/H \gtrsim 2$; then the flow, which initially had separated at the upwind roof edge, reattaches to the flat roof. Final separation takes place at the trailing edge of the roof; the separation streamline leaves the body roughly parallel to the roof, and the cavity length becomes nearly independent of L/H. Evans also found that, as the crosswind dimension W increases (for a given L), both the width and length of the cavity increase. He observed that, at least in a near-uniform incident flow, the flow pattern near the top of the buildings was largely invariant with building height; the portion of air traveling over (rather than around) the top was nearly constant, while the amount of air displaced laterally around the building increased with H. The cavity length therefore increased with building height. Increases in pitch of a gable roof also increased the extent of the cavity. Evans' (1957) original report should be studied for information on the effects of roof overhang, building orientation, and more elaborate building configurations.

An empirical expression for the length of the recirculation zones behind flat-roof block buildings normal to the wind has been developed (Hosker, 1979) using Evans' (1957) data. Hosker reasoned that the cavity length x_r behind any building must depend on the three basic building dimensions, on the speed and vertical variations of the incident wind, on the intensity and scale of the incident turbulence, on the bulk stability of the incident flow, on the aerodynamic characteristics of the upwind surface, and on the fluid viscosity. For neutrally stable, uniform, low-turbulence flow over a short, smooth, flat ground plane upwind of a simple sharp-edged body, however, most of these dependencies should be irrelevant. For such a simple flow, x_r should be primarily a function of the building dimensions and the flow Reynolds number. If we assume that the Reynolds number is quite large, then we observe experimentally that the flow field no longer depends on the Reynolds number; consequently, by forming nondimensional groups from the remaining variables, we expect that $x_r/H = f(L/H, W/H)$.

The extent of the cavity should be measured from the separation point. Hence, two expressions, one for short ($L/H \lesssim 2$, for instance) and one for long ($L/H \gtrsim 2$) buildings, will be required because for short buildings the recirculation length is reckoned from the upwind face of the building, whereas for long buildings (where the flow has reattached to the roof) the cavity length is measured from the lee face.

Several limiting cases suggest themselves. For example, as the obstacle length and width both approach zero, the cavity should disappear. For very wide bodies (large W/H), the results should become independent of W/H (the two-

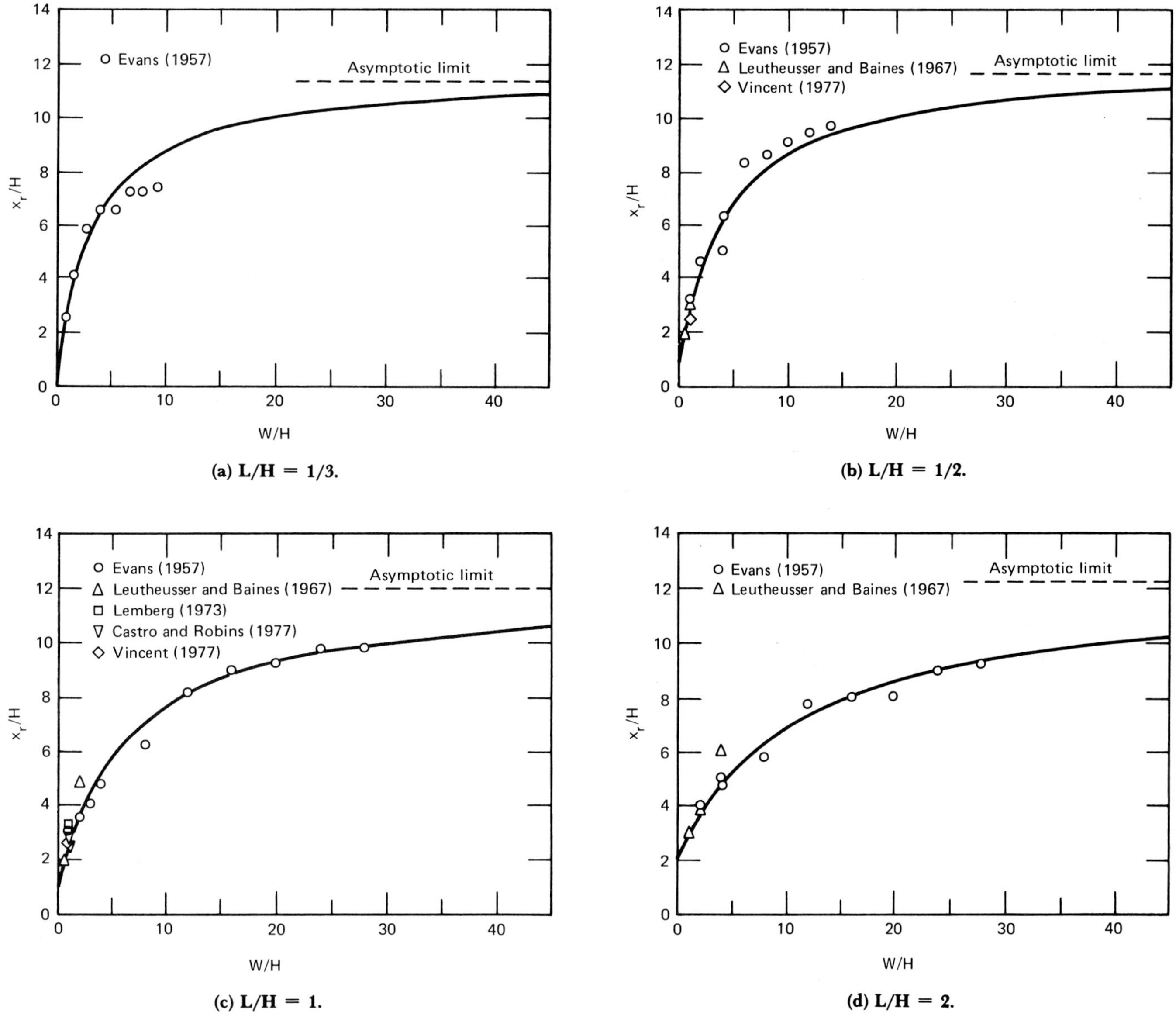

Fig. 7.11 Empirical expression (Eq. 7.1, solid line) for wake cavity length compared to data on rectangular prisms in nearly uniform flow. The expression is for bodies where $L/H \lesssim 2$, and the cavity length is measured from the upwind building face.

dimensional limit). In particular, for $W/H \rightarrow \infty$ and $L/H \rightarrow 0$, the short building expression should predict the two-dimensional fence results discussed earlier. For $W/H \rightarrow \infty$ and $L/H \rightarrow \infty$, the long building expression should approach the data on two-dimensional rear-facing steps.

The following equation, developed to fit Evans' (1957) data for short ($L/H \lesssim 2$) block buildings, meets most† of the pertinent requirements

$$\frac{x_r}{H} = \frac{L}{H} + \frac{A(W/H)}{1.0 + B(W/H)} \tag{7.1a}$$

where A and B are weak functions of L/H given by

$$A = -2.0 + 3.7\left(\frac{L}{H}\right)^{-1/3}$$

and (7.1b)

$$B = -0.15 + 0.305\left(\frac{L}{H}\right)^{-1/3}$$

Equation 7.1 is indicated in Figs. 7.11a to 7.11d, together with Evans' (1957) data; the agreement is generally within ±15% or better. Data from other authors are also shown and are predicted equally satisfactorily by Eq. 7.1. For $W/H \rightarrow 0$, $x_r/H \rightarrow L/H$; i.e., the cavity, which is measured from the

†This expression cannot be quite correct in form because, as $L/H \rightarrow 0$ for fixed W/H, $x_r/H \rightarrow$ constant independent of W/H, which is not true.

upwind building face, disappears. For $W/H \rightarrow \infty$ and $L/H \rightarrow 0$, $x_r/H \rightarrow 12.1$, which is toward the lower end of the range of two-dimensional fence data.

When $L/H \lesssim 2$, then Eq. 7.1 appears adequate for engineering estimates of x_r/H to within 20% or better. It should be borne in mind, however, that the expression is entirely empirical and is subject to revision as more data become available. In fact, Fig. 7.11 suggests some ranges of body sizes for which further information would be welcome.

For obstacles where $L/H \gtrsim 2$, Evans (1957) observed reattachment of the initial separation streamline onto the building roof; final separation then occurred at the trailing edge of the (flat) roof. The flow is independent of L/H and is rather reminiscent of flow behind a rear-facing step as the body width becomes large. If the cavity extent is measured from the final separation point (i.e., from the lee building face), then the expression (Hosker, 1979)

$$\frac{x_r}{H} = \frac{1.75(W/H)}{1.0 + 0.25(W/H)} \tag{7.2}$$

fits Evans' (1957) data relatively well—typically within ±15%—as is indicated in Fig. 7.12. As $W/H \rightarrow \infty$, $x_r/H \rightarrow 7.0$, which agrees reasonably well with the results behind two-dimensional steps (e.g., Tani, 1957; Bearman, 1965; Bradshaw and Wong, 1972; plotted in Fig. 7.12). Equation 7.2 should suffice for predicting x_r/H within ±25% for cases where $L/H \gtrsim 2$ or so.

Not as much information has been reported on the crosswind extent of the cavity region. Figure 7.13a contains data from previously cited authors on short buildings; the equation

$$\frac{2y_r}{W} = 1.1 + 1.7 \exp\left(-0.55 \frac{W}{H}\right) \tag{7.3}$$

where y_r is the maximum half-width of the cavity, appears to fit these points reasonably well. Note that $(2y_r/W) \rightarrow$ constant for $W/H \geqq 10$ and that $y_r \rightarrow 0$ as $W \rightarrow 0$. Not enough data are available to clarify any explicit dependence on L/H; indeed, all the results in Fig. 7.13a were observed behind relatively short bodies where $L/H \leqq 1$. Further investigation is needed to determine whether this relation is applicable for longer buildings.

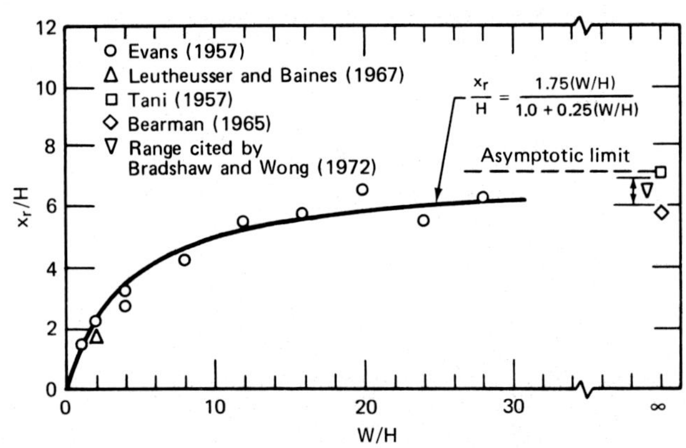

Fig. 7.12 Empirical expression (Eq. 7.2, solid line) for wake cavity length compared to data on rectangular prisms in nearly uniform flow. The expression is for bodies where L/H > 2, and the cavity length is measured from the lee building face.

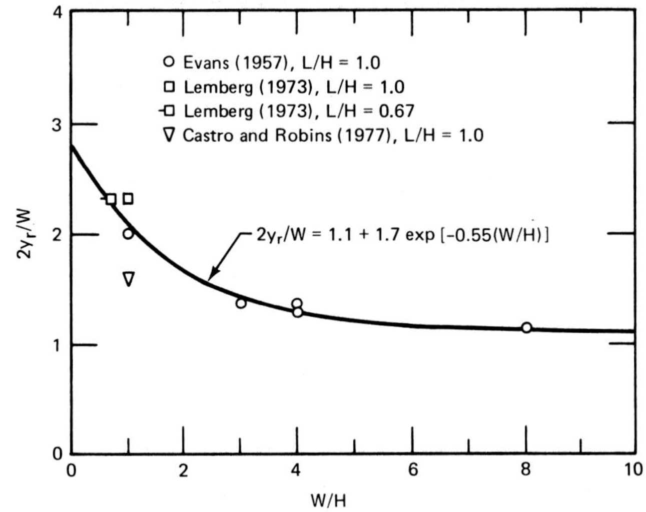

(a) Maximum crosswind extent of cavity.

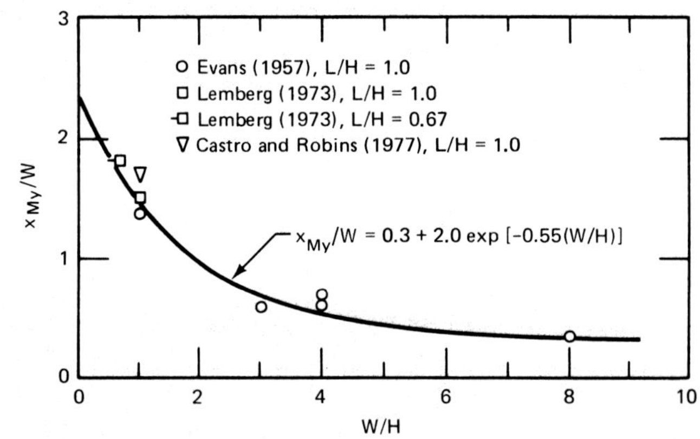

(b) Location of maximum crosswind extent of cavity, measured from upwind face.

Fig. 7.13 Maximum wake cavity width and its location for block-type bodies, L/H < 2, in near-uniform flow.

The location of maximum crosswind extent of the wake cavity, x_{My}, is presented in Fig. 7.13b; the expression

$$\frac{x_{My}}{W} = 0.3 + 2.0 \exp\left(-0.55 \frac{W}{H}\right) \tag{7.4}$$

gives a reasonable approximation to the short building data. Here $x_{My} \rightarrow 0.3W$ as $W/H \rightarrow \infty$ and $x_{My} \rightarrow 0$ as $W \rightarrow 0$, where x_{My} is measured from the separation point on the windward face of the obstacles. It is unknown whether a similar expression can be constructed for obstacles where $L/H > 2$ or so.

Finally, the maximum height, z_r, of the recirculation zone is sketched in Fig. 7.14. Significant scatter, some of which undoubtedly is caused by dependence on the width-to-height ratio W/H, appears in the data. Insufficient data are available to determine this effect. The equation

$$\frac{z_r}{H} = 1.0 + 1.6 \exp\left(-1.3 \frac{L}{H}\right) \tag{7.5}$$

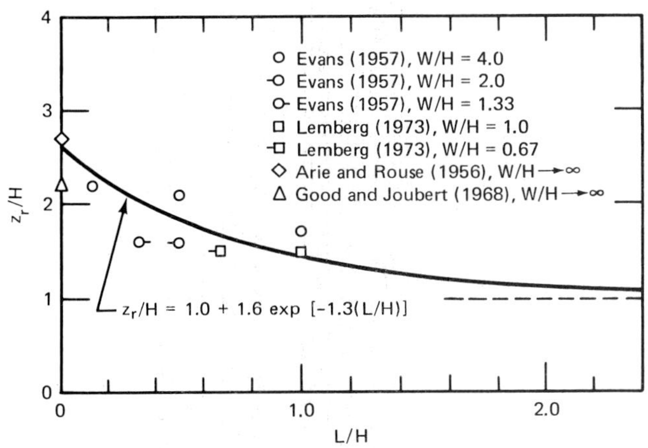

Fig. 7.14 Maximum wake cavity height behind block-type obstacles in near-uniform flow.

is suggested as a plausible rough estimate until more information is available. Note that, as $L/H \rightarrow 0$, $z_r/H \rightarrow 2.6$, which is close to the two-dimensional fence results; as $L/H \rightarrow 2$ or so, $z_r \rightarrow H$, which is consistent with the notion of flow reattachment to the roof, with final separation occurring at the body height H.

The location of the maximum cavity height is denoted by x_{Mz}. Arie and Rouse (1956) and Good and Joubert (1968) indicate that $x_{Mz}/H \sim 6$ to 8 for thin, two-dimensional fences. On the other hand, data from Evans (1957) and Lemberg (1973) suggest that, at least for the ranges $0.125 \leqq L/H \leqq 1$ and $0.67 \leqq W/H \leqq 4$, $x_{Mz}/H \sim 1.5$ to 2.5. No pattern could be discerned in these data. A systematic study of maximum cavity height and its dependence on building shape is needed.

Quantitative data on the recirculation zone behind finite-size rounded bodies mounted on a surface plane are rather sparse. Fiedler and Wille (1970) examined ground-mounted vertically oriented cylinders and cones of finite extent at relatively modest Reynolds numbers ($<10^5$). They found turbulence intensities, relative to local mean speed, of nearly 80% and turbulence length scales almost an order of magnitude smaller than are observed behind two-dimensional cylinders. Significant changes in the vortex-shedding characteristics also were observed. Fiedler and Wille postulated that the downwash into the near wake that occurs at the upper rear of the cylinder acts as an air curtain and forms a kind of wake splitter plate that restricts the transfer of vorticity across the wake centerline. Thus, the cavity region behind finite-length cylinders depends on the relative importance of the downwash, i.e., in some sense on the diameter-to-height ratio d/H, as well as on the usual quantities such as Reynolds number, surface finish, and incident flow properties.

Lemberg (1973) reported that, behind a vertically mounted circular cylinder of $d/H = 1.0$, the cavity length was $x_r/d = 2.7$; the maximum cavity width was $2y_r/d = 1.7$, which occurred at $x_{My}/d = 1.4$; and the maximum cavity height was $z_r/H = 1.1$, which occurred just above the rear edge of the cylinder at $x_{Mz}/d = 0.5$. For another cylinder where $d/H = 0.67$, Lemberg observed $x_r/d = 3.1$, $2y_r/d = 1.6$ at $x_{My}/d = 0.9$, and $z_r/H = 1.2$ at $x_{Mz}/d = 0.5$. In both cases x_r, x_{My}, and x_{Mz} were measured from the center of the body, and flow separation from the cylinder sides was observed about 100° from the upwind stagnation line. The flow over the cylinder roof separated in a sheet from the leading edge of the front rim. Additional measurements are needed, however, to establish empirically a predictive capability for the recirculation zone characteristics behind ground-plane-mounted rounded bodies.

7-4.3 The Far Wake

The far wake of a wide ground-mounted obstacle subjected to a nearly uniform incident flow has been experimentally and theoretically investigated by Mueller, Korst, and Chow (1964) and by Sforza and Mons (1970). They found that such wakes behave at large distances downwind ($x/H > 15$ or 20, for example) rather like developing boundary layers, at least for two-dimensional obstructions. For example, the mean velocity profiles for the two-dimensional case obey a similarity scaling, as do many boundary-layer flows. Mueller, Korst, and Chow (1964) showed that the so-called momentum and displacement thicknesses (e.g., Schlichting, 1960), which depend on integrals of the mean velocity profiles, agreed fairly well with the results appropriate to a developing turbulent boundary layer. The momentum thickness grew as $x^{0.8}$, whereas the ratio of displacement to momentum thicknesses approached a constant value of 1.3.

The data of Sforza and Mons (1970) suggest that the centerline decay of maximum wake velocity defect $(\Delta U/U_\infty)_{max}$ proceeds approximately as $x^{-0.95}$ for two-dimensional obstacles ($W/H \rightarrow \infty$, $L/H = 0.5$), as $x^{-1.0}$ for a fairly wide three-dimensional body ($W/H = 10$, $L/H = 0.5$), and as $x^{-2.6}$ for a block with a square upwind face ($W/H = 1.0$, $L/H = 0.16$). Their theory predicts a decay as $x^{-0.9}$ and agrees reasonably well with the two-dimensional data; however, the theory fails for the three-dimensional case. Sforza and Mons (1970) observed the vertical location of the maximum velocity deficit to rise approximately as $x^{0.5}$. They also found that the height of the line at which the velocity defect has diminished to one-half of its local maximum value (this is a measure of wake height) rises in the vertical roughly as $x^{0.4}$ for both two-dimensional and wide three-dimensional blocks. However, for the square-faced block, this line "necks down" initially and eventually grows more slowly as $x^{0.35}$. This initial narrowing of the vertical wake behind a block-like obstacle was ascribed to the effect of concentrated vortices being shed at roof level behind the obstacle. The crosswind wake behind a wide three-dimensional body increased as $x^{0.45}$.

Both the studies just cited must be viewed with some caution because the experiments, of necessity, involved a wake that was at least partially immersed in a boundary layer that developed along the ground surface. Possible interactions between the wakes and this boundary layer and the resultant effects on the wake velocity profiles and spreading characteristics are uncertain. Further work in this area is needed.

7-5 SHEAR FLOW NEAR BLUFF BODIES ON A GROUND PLANE

Certain important features of flow near an obstacle immersed in a turbulent shear layer result mainly from the presence of the shear. These features often are less familiar to aerodynamicists than those discussed in the preceding section. One example is the standing horizontally oriented vortex generated near the ground upwind of the obstacle. In the three-dimensional case, this vortex wraps around the body to trail off

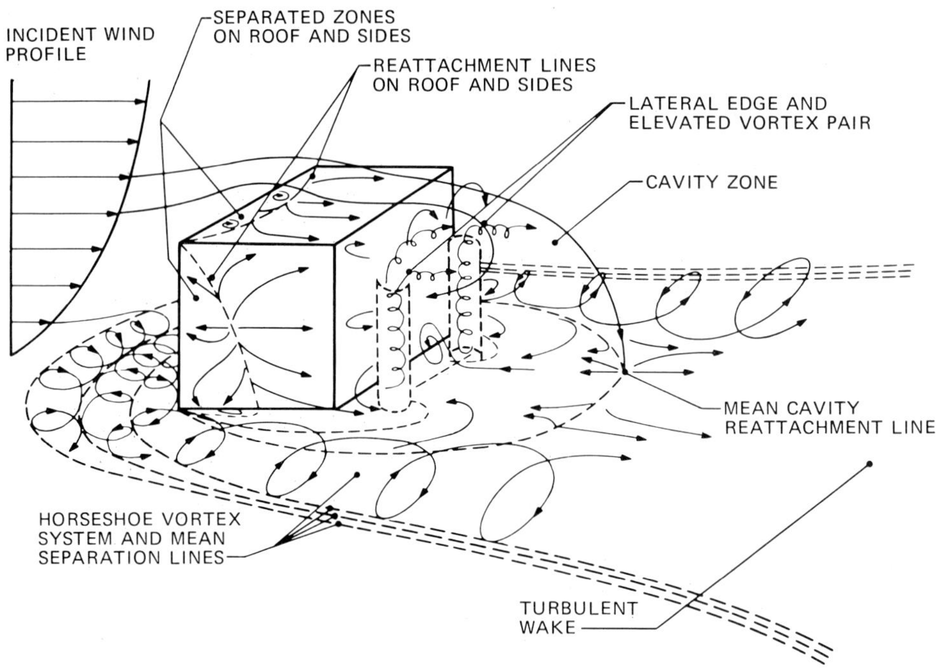

Fig. 7.15 Recent model of flow near a sharp-edged three-dimensional building in a deep boundary layer. [Based on H. G. C. Woo, J. A. Peterka, and J. E. Cermak, *Wind-Tunnel Measurements in the Wakes of Structures*, NASA Contractor Report NASA CR-2806, National Space and Aeronautics Administration, 1977; and on J. C. R. Hunt, W. H. Snyder, and R. E. Lawson, Jr., Flow Structure and Turbulent Diffusion Around a Three-Dimensional Hill, in *Fluid Modeling Study on Effects of Stratification, Part I. Flow Structure*, Report EPA-600/4-78-041, U. S. Environmental Protection Agency, 1978.]

downwind as a counterrotating vortex pair. This vortex was mentioned briefly in Sec. 7-4 as a relatively minor consequence of the thin incident boundary layers that accompany experimental modeling of any nominally uniform flow. In this section, however, the incident shear layer is deep relative to the obstacle, and the resulting vortex is a significant feature of the flow field. A fairly typical flow field for normal wind incidence on a block-like body is sketched in Fig. 7.15 (Woo, Peterka, and Cermak, 1977; Hunt et al., 1978).

7-5.1 Flow Near the Upwind Face

The important differences in the flow near the upwind face of an obstacle immersed in a uniform flow vs. a shear flow are illustrated in Figs. 7.16 and 7.17. Figure 7.16a indicates the expected flow near a two-dimensional wall subjected to a uniform wind profile; Figure 7.16b shows the upwind recirculation zone or eddy characteristic of an incident shear flow. Perhaps this latter sketch is an idealization because Jackson (1973) observed that, in some circumstances, the two-dimensional eddy could be induced to break up into a three-dimensional cellular system of vortices, the ends of which bend and rise over the obstacle to trail off downwind as counterrotating vortex pairs (Fig. 7.16c). The cell spacing seemed to be about six obstacle heights. Verification of this phenomenon would be useful.

Uniform flow approaching a cube-like object is sketched in Fig. 7.17a. Figure 7.17b shows a shear flow incident on the same object; streamlines enter the upwind eddy directly (Hunt et al., 1978; also see Fig. 7.15). Note that the upwind eddy appearing in this three-dimensional case is forced around the edges of the structure to trail off downwind on either side; viewed from above, this vortex rather resembles a horseshoe, hence, its name.†

*7-5.1.1 **Two-Dimensional Sharp-Edged Obstacles.*** For two-dimensional solid walls or fences immersed in shear flow (Fig. 7.16b), the incident flow typically separates from the ground surface one or two wall heights upwind (e.g., Baines, 1963; Good and Joubert, 1968; Arie et al., 1975). The exact separation location depends somewhat on the thickness of the shear flow, δ, relative to the obstacle height H. Also, the separation location probably depends on the shape of the incident velocity profile (Corke and Nagib, 1976). Upwind of the separation line the flow is directed toward the obstacle; downwind of the line the flow is away from the wall. The separation line is a line of near-zero speed; consequently material suspended in the flow, such as snow particles, will tend to deposit on the ground along this line (Baines, 1963).

On the face of the two-dimensional wall, a reattachment line (Fig. 7.16b) can be expected for z/H between 0.5 and 0.8, with 0.6 being the most commonly cited value (e.g., Baines, 1963; Good and Joubert, 1968; Arie et al., 1975; Ranga Raju, Loeser, and Plate, 1976). Above the reattachment line the main flow is incident on the wall, and the flow immediately adjacent to the wall is directed upward; below the line the flow of the recirculating eddy is directed downward near the face. In effect,

†Roper (1965), Norman (1972), and Werle (1975) found that under certain conditions the horseshoe vortex may consist of a system of up to seven alternating-rotation vortices whose diameter increases with closeness to the obstacle. Figure 7.22 shows three such vortices. In atmospheric flows, however, only one vortex seems to have been observed.

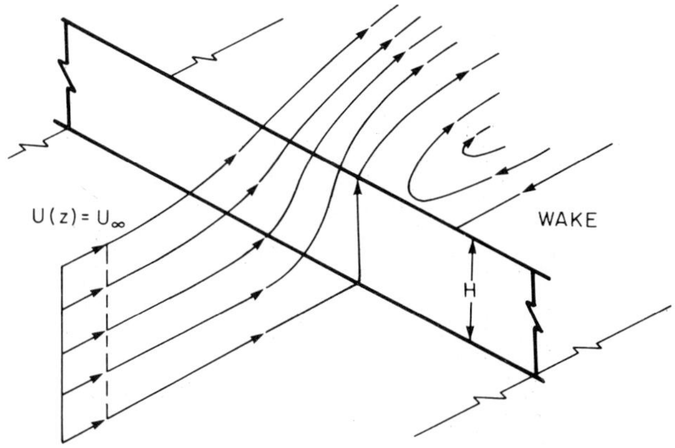

(a) Uniform incident flow ($\delta = 0$). (Based on W. D. Baines, Effects of Velocity Distribution on Wind Loads and Flow Patterns on Buildings, in *Proceedings of the Symposium on Wind Effects on Buildings and Structures,* National Physical Laboratory, Teddington, Middlesex, England, June 26–28, 1963, Vol. I, pp. 197-225, Her Majesty's Stationery Office, London, 1965.)

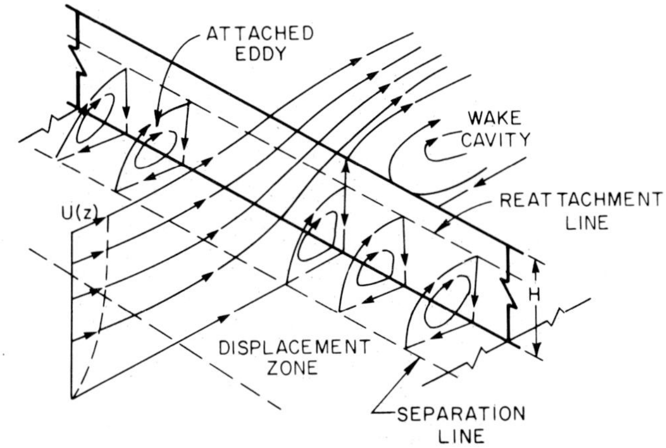

(b) Boundary-layer incident flow. [$\delta/H \sim 0(1)$]. (Based on W. D. Baines, Effects of Velocity Distribution on Wind Loads and Flow Patterns on Buildings, in *Proceedings of the Symposium on Wind Effects on Buildings and Structures,* National Physical Laboratory, Teddington, Middlesex, England, June 26–28, 1963, Vol. I, pp. 197-225, Her Majesty's Stationery Office, London, 1965.)

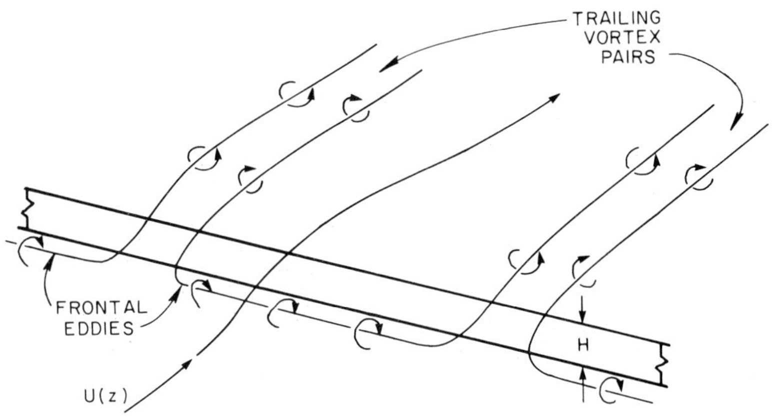

(c) Sketch of observed cellular eddy structure, boundary-layer incident flow. (Based on P. S. Jackson, *The Flow Around Obstacles in Boundary Layers,* Ph.D. Dissertation, University of Cambridge, Cambridge, England, 1973.)

Fig. 7.16 Flow patterns upwind of a two-dimensional wall for different incident flows.

the upwind eddy indicated in Fig. 7.16b is bounded by these separation and reattachment lines.

The extent of the overall region of influence ahead of a wall or fence (the upwind displacement zone of Fig. 7.1) varies—at least in the wind tunnel—according to $x_{p\infty}/H = 15(\delta/H)^{0.7}$ (Good and Joubert, 1968), where $x_{p\infty}$ is the distance upwind where the surface pressure field is first perturbed by the presence of the wall. However, this result gives displacement zones about an order of magnitude larger than those commonly observed in the atmosphere. It seems likely that the high turbulence intensities characteristic of the atmospheric boundary layer suppress the obstacle's influence until the flow is much closer to the vertical wall. Measurements near dense windbreaks suggest a displacement zone only 5H to 10H in front of the obstacle (van Eimern et al., 1964); Souster and Lee (1975) found upwind flow disturbances only to about 4H in deep power-law-type ($U \propto z^n$) wind-tunnel shear flows.

Frost (1973) presents a survey of velocity distribution, shear stresses, and pressure distributions upwind of two-dimensional walls. Some recent work can be found in Castro and Fackrell (1978). Kiya and Arie (1972), Kiya, Sakamoto, and Arie (1976), and Kiya and Arie (1976) have developed a theory for two-dimensional walls; their calculated pressure distributions and flow streamlines are in good agreement (Fig. 7.18) with observations for two-dimensional fences and surface-mounted half-cylinders of both circular and elliptical cross section. Empirical data are necessary, however, to evaluate certain constants of their theory for individual cases; i.e., their method is not entirely predictive. For details of the technique, the original papers should be consulted. Bitte and Frost (1976) have refined the method somewhat so that, when the empirical information necessary for the model is available, the technique yields excellent agreement with experimental data for flow over a fence.

Jackson (1975) has developed an analytical theory for turbulent boundary-layer flow approaching a two-dimensional hill of small slope. The problem is shown to be similar to that of a uniform inviscid flow incident on the hill. Agreement of the velocity and pressure fields with both field and wind-tunnel data is fairly good. Other examples of analytical and numerical solutions for flow near two-dimensional hills with various geometries abound in recent literature (e.g., Taylor and Gent, 1974; Jackson and Hunt, 1975; Deaves, 1976; Derickson and Meroney, 1977; Bouwmeester, Meroney, and Sandborn, 1978;

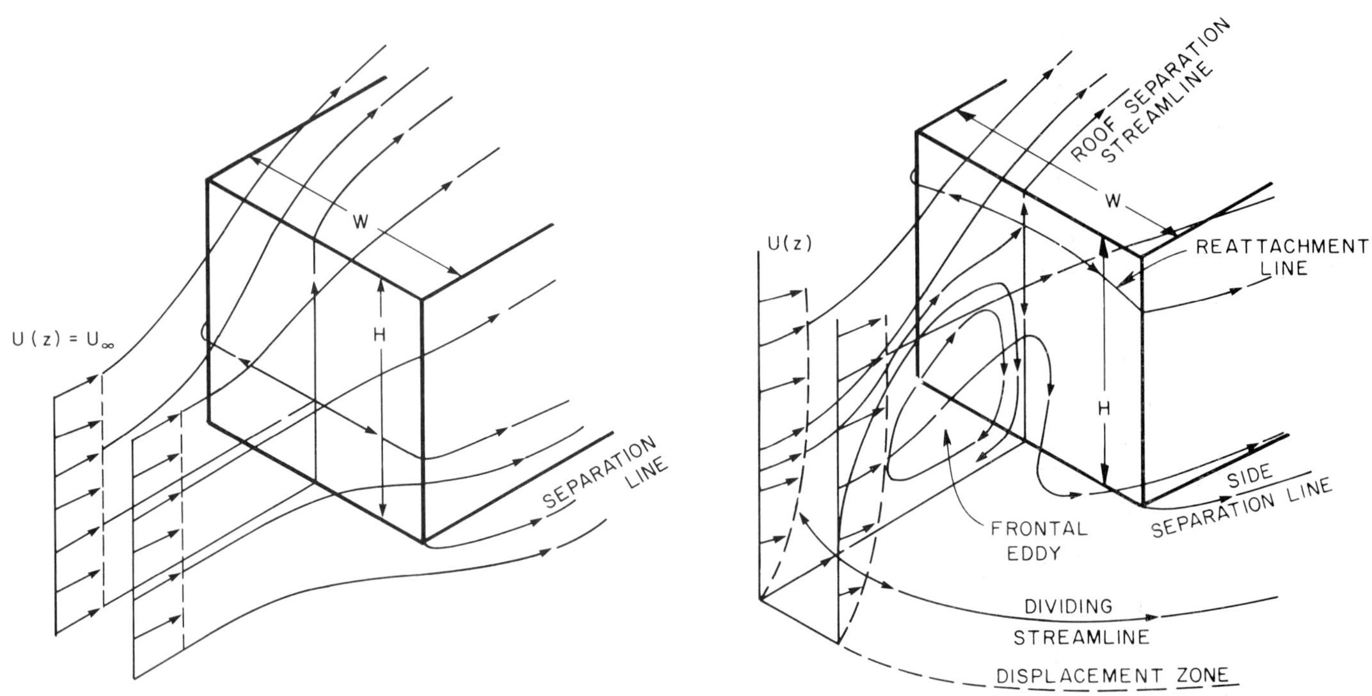

(a) Uniform incident flow. **(b) Boundary-layer incident flow.**

Fig. 7.17 Flow patterns upwind of a cube for different incident flows. (Based on W. D. Baines, Effects of Velocity Distribution on Wind Loads and Flow Patterns on Buildings, in *Proceedings of the Symposium on Wind Effects on Buildings and Structures,* National Physical Laboratory, Teddington, Middlesex, England, June 26–28, 1963, Vol. I, pp. 197-225, Her Majesty's Stationery Office, London, 1965.)

Mason and Sykes, 1978). Many of these studies stem from the current interest in wind energy conversion. Field data are more difficult to obtain, but Bradley (1980) has shown reasonably good agreement between some of these theories and observations taken over a wide hill. Close to the surface of the hill the wind speed increases by a factor related to the hill height-to-length ratio (Jackson and Hunt, 1975), and all three turbulence components increase as well. Consequently the local relative intensities of turbulence are approximately the same as their upwind values. Therefore pollutant dispersion near such a hill should be affected primarily by streamline convergence rather than by enhanced turbulent diffusion (Bradley, 1980). Other data have been reported by Sacré (1979). The interested reader should consult the specialized papers for more information.

7-5.1.2 ***Three-Dimensional Sharp-Edged Obstacles.*** In the more complex case of three-dimensional obstacles immersed in a shear flow (shown in Fig. 7.17b), current ability to predict the extent of the overall upwind displacement zone and the size and shape of the upwind near-body eddy is limited.

As the main flow approaches the obstacle, it decelerates longitudinally and accelerates laterally and/or vertically to pass around the body. This gross deflection of the incident flow strongly affects the shear-induced standing eddy. The size and strength of this vortex depend on the curvature of the incident velocity profile relative to the body height and on the width-to-height ratio of the obstacle.

Corke and Nagib (1976) conducted wind-tunnel studies on a block-like structure ($L/H = W/H = 0.52$) immersed in a family of power-law incident velocity profiles ($U \propto z^n$). The value of n covered the range typical of the atmosphere (Davenport, 1963). They found that, at least for heights below 0.75H, the streamwise mean velocity component near the upwind face was not very sensitive to the incident profile. However, the mean vertical component was quite responsive to changes in wind profile. In particular, the dividing point between upward and downward flow occurs at $z/H \simeq 0.53$ for $n \simeq 0.11$, at $z/H \simeq 0.72$ for $n \simeq 0.23$, and at $z/H \simeq 0.77$ for $n = 0.30$ and 0.35. The main flow lies above this point, and the frontal vortex is found below it; evidently, the diameter of the vortex increases with increasing wind profile exponent; the height of reattachment of the main flow onto the building face also increases with increasing wind profile exponent. This behavior is reflected in the vertical distribution of pressure on the upwind face. The horizontal pressure distributions show the lower pressures expected because of the lower velocities incident at a given height as the power-law exponent n increases; however, the shape is similar to that found in near-uniform flow.

For obstacles much taller than they are wide, the frontal vortex will extend to a height of about one building width on the upwind face (Baines, 1963). Above this height, the surface pressure will vary with height approximately as the square of the locally incident velocity; however, the pressure on the lower portion of the face, which contacts the standing vortex, will be nearly constant (Baines, 1963).

Typical mean pressure distributions on the upwind faces of generalized building types have been given by Baines (1963), Symes and Meroney (1970), and Castro and Robins (1977), among others. Pressure distributions on particular buildings, both model and full-scale, are common because these data are useful to architects; see, for example, the various proceedings of

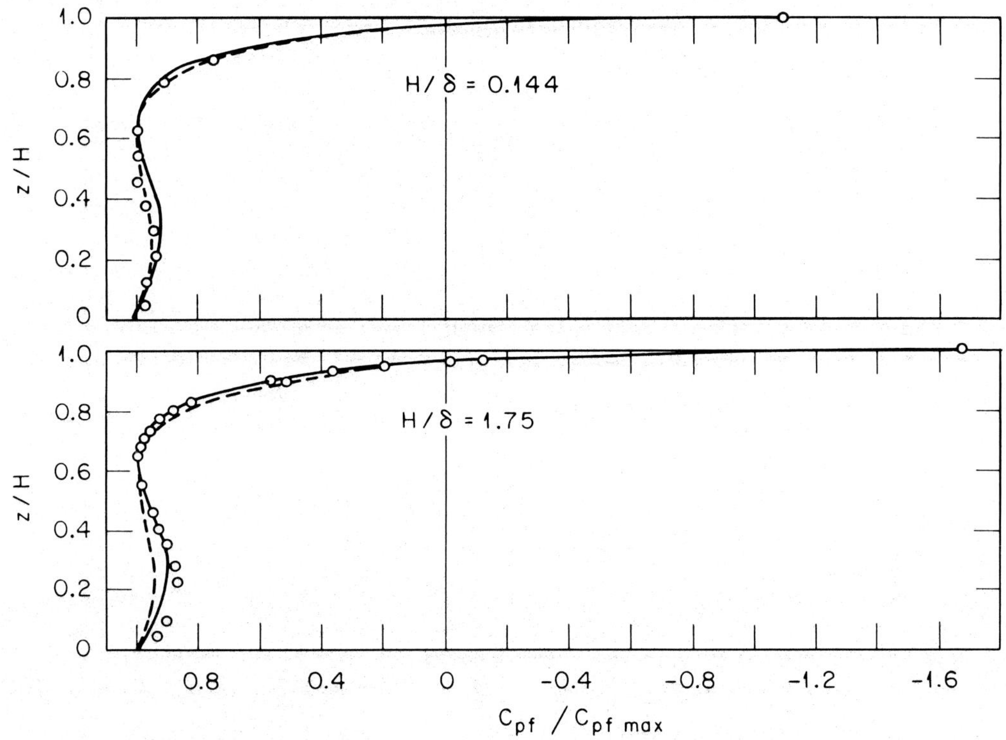

(a) Pressure coefficients on upwind face for deep and shallow boundary-layer flow. Solid line (theoretical) assumed maximum pressure occurred at z/H = 0.65; broken line assumed maximum pressure at z/H = 0.60.

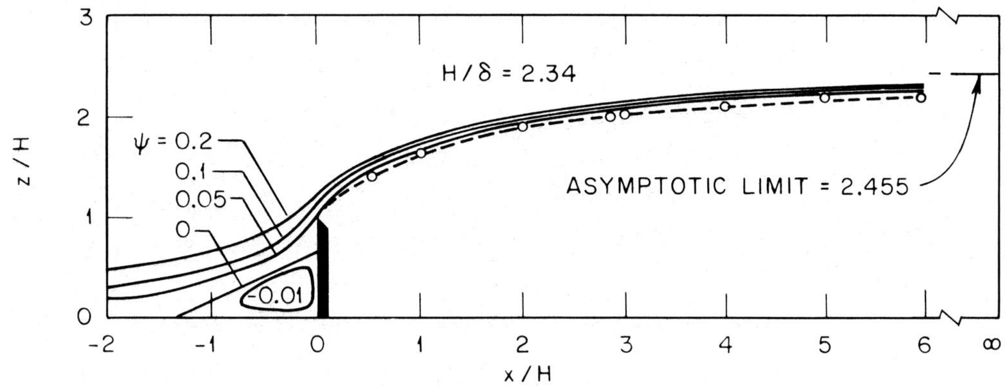

(b) Streamlines near fence [Calculations by Kiya and Arie (1972), using data of Good and Joubert (1968).]

Fig. 7.18 Comparison between theory and experiment for shear flows incident on a two-dimensional fence. Data are indicated by open dots.

the international symposia on *Wind Effects on Buildings and Structures* (1963, 1967, 1971, 1975) and their successors (e.g., *Fifth International Conference on Wind Engineering,* 1979).

The line of flow separation from the ground upwind of a three-dimensional body has not yet been universally specified. Frost and Shahabi (1977) found in field measurements that, at least for a fairly wide building, the separation line was approximately 0.9H in front of the obstacle. As in the two-dimensional case, particle deposition can be expected along this dividing streamline. Wilson and Netterville (1978) observed that perturbations in incident concentration profiles first appeared about 1.3H upwind of a model building.

Because the separation and reattachment points on the ground and the building face are fixed by aerodynamic force balances rather than by geometry, locations of the points will fluctuate with perturbations of the overall flow field (Hunt, 1975). Such perturbations can be introduced both by ambient turbulence or unsteadiness in the approach flow and by the feedback effect of turbulence generated in the wake of the obstacle (Norman, 1972). Hence the upwind vortex can be anticipated to be rather unsteady. As indicated in Fig. 7.15, this vortex interacts with the incident flow to wrap around the building and trail off downwind; therefore fluctuations in its strength and location will lead to perturbations of the wake flow and subsequently to feedback to the upwind vortex. Prediction of the location and characteristics of this vortex–wake interaction is likely to be possible only in a time-averaged sense.

Some theoretical work on neutrally stable flow approaching a three-dimensional hill has been published (e.g., Mason and Sykes, 1979a). Limited data from the field (Mason and Sykes, 1979b; Bradley, 1980) and laboratory (e.g., Hunt and Snyder, 1980) have recently become available for comparison. The reader should consult these four papers for details.

7-5.2 Separation and the Near-Body Wake

The separation process is likely to be dominated by locally generated pressure gradients and by surface features of the obstacle; consequently, at least for sharp-edged bodies, separation may not be very sensitive to the details of the incident velocity profile. The presence of shear and turbulence in the incident flow, however, may have strong effects on separation from rounded bodies, where the process depends on the locally determined force balance at the obstacle surface and is not dominated entirely by the geometry. The importance of turbulence in the incident flow is discussed in more detail later.

The influence of incident flow shear and turbulence on the free shear layers shed by an obstacle and on the reattachment process has been studied for a few special cases (e.g., Bradshaw and Wong, 1972; Chang, 1966; Corke and Nagib, 1976; Mueller, Korst, and Chow, 1964; Plate, 1971). Arie et al. (1975) were able to successfully model the flow within the shear layer shed by a two-dimensional block immersed in a deep boundary layer; however, their work is valid only at some distance upwind of the cavity closure point.

Cermak (1975) discussed the use of locally measured surface pressures to deduce the flow patterns near the surface. For example, if the flow is separated over an entire surface and the outer streamlines of the flow are not sharply curved, then the mean pressures on that surface will be less than the ambient, and they will be nearly constant. On the other hand, if separation is followed by reattachment further downwind, as in the case of a long building, the mean pressures on that surface will be nonuniform and will exhibit large fluctuations. Where surface pressure gradients are large, the near-surface flow will be along the gradient in the direction of decreasing pressure. When the surface pressure gradients are weak, however, the surface flow field is driven by the outer flow rather than by the local pressure gradient. When separation occurs, the near-surface flow will move toward the separation point or line, and the flow will travel away from stagnation locations.

7-5.2.1 Two-Dimensional Sharp-Edged Obstacles. Hunt (1974) pointed out that, for two-dimensional obstacles immersed in deep boundary layers, the extent of the cavity or recirculation region increases with increasing smoothness of the upwind terrain and increasing sharpness of the obstacle. He suggested the range $6 \leqq x_r/H \leqq 13$ as typical; however, values outside these limits have been reported.

Even for simple fences and fence-like objects, the cavity length seems to vary widely. For example, Woodruff and Zingg (1952) found cavity closure occurred about 10.7H downwind of a fence-like plate. In an extensive series of tests, Nagabhushanaiah (1961) found that behind model fences $x_r/H = 12$; depending on the incident flow conditions, the maximum cavity height ranged between 1.4H and 1.8H. Plate and Lin (1965) observed x_r/H values between 11 and 12 behind a wedge with its upwind face normal to the flow; Chang (1966) reported $x_r/H \simeq 12.8$ for a similar obstacle. Petryk and Brundrett (1967) observed $x_r/H \simeq 17$ behind a fence. de Bray (1971) reported reattachment lengths between 11H and 16H behind fences; maximum cavity heights were between 2H and 2.5H. In a rare field experiment, Caput et al. (1973) estimated $x_r/H \sim 4$ behind a long plastic fence; this value may be an underestimate, however, because a value of 5 or 6 appears to agree better with their particle deposition results. Raine and Stevenson (1977) observed $x_r/H = 9$ behind a fence; Sheih et al. (1978) reported that x_r/H ranged between 7.4 and 9.2.

There are several reasons for such variation in the x_r/H values. Raine and Stevenson (1977) pointed out that fence flow depends on properties such as incident turbulence intensity, the ratio of turbulent length scale to roughness length, and the ratio of fence height to roughness length. Castro and Fackrell (1978) demonstrated that laboratory data may be strongly influenced by both wind-tunnel blockage and the depth of the incident shear flow relative to the fence. Ogawa, Diosey, and Oikawa (1979) reported on the effect of stability changes on x_r/H. In view of the many variables, test-to-test scatter in the data is not surprising.

One possible explanation for the variability of x_r has been provided indirectly by Tani's (1952, 1958) observation that the characteristic length of the protected zone behind a windbreak depends logarithmically on the ratio of fence height to the aerodynamic roughness height of the upwind surface. If the recirculation zone behaves similarly,† then we can postulate $x_r/H \sim c \ln(H/z_0)$, where c is about 1.5. Field data of Caput et al. (1973) are overpredicted by perhaps 60% by this relation, and Raine and Stevenson's (1977) model results are underpredicted by about 25%. The data of Sheih et al. (1978) are at least consistent with the relation. However, the results of Ogawa, Diosey, and Oikawa (1979) are poorly reproduced by this relation. Instead, the expression‡ $x_r/H \sim 1.6\sqrt{Re_{turb}} \simeq 4[\ln(H/z_0)]^{1/2}$ appears to fit their field and laboratory data reasonably well. This latter relation underpredicts the field data of Caput et al. (1973) by about 100%; but the result of Raine and Stevenson (1977) is underpredicted by only 5%. Again, the data of Sheih et al. (1978) are consistent with the relation.

Neither of the preceding estimates for x_r/H takes account of other important variables, such as turbulence properties, atmospheric stability, or facility-induced changes. Currently too few data sets are available from which to establish a predictive relation. It would be useful if future experiments explicitly reported incident flow properties, such as u_*, z_0, shear layer

†The drag coefficient of an obstacle is a kind of bulk indicator of the properties of the wake cavity zone. It is interesting, therefore, that Ranga Raju, Loeser, and Plate (1976) state that the drag coefficient of a two-dimensional fence, referenced to the upwind friction velocity u_*, depends only on H/z_0, irrespective of H/δ. They suggest that only the portion of the incident boundary layer near the ground is significant in determining the separation and wake cavity characteristics. Their assertion is a confirmation of Jensen's (1958) criterion for wind-tunnel modeling: namely, that H/z_0 be equal in both model and prototype. However, Castro and Fackrell (1978) recently have challenged the conclusions of Ranga Raju, Loeser, and Plate; consequently, this issue is far from completely settled.

‡For a fence of height H, the so-called turbulent Reynolds number is defined as $Re_{turb}(H) \equiv U(H)H/K_m(H)$, where K_m is the turbulent diffusivity. If $K_m(z) = ku_*z$ is used, $Re_{turb} = k^{-2} \ln(H/z_0)$ for a neutral flow. Hunt (private communication) has pointed out that the adjective "turbulent" is something of a misnomer because the quantity is as much a measure of the mean wind speed at H and ez_0 as it is a measure of the incidence turbulence.

depth δ, stability parameters (e.g., Richardson number or Monin length), and turbulence intensity and length scales. Laboratory results should include an estimate of the blockage ratio and its likely impact on the data.

Estimation of x_r/H is not improved when two-dimensional blocks are considered, because the additional variable of along-wind length L is added to the problem. As in the case of nearly uniform incident flow, when L is sufficiently large, the separated flow can reattach to the flat roof of the block so that final separation at the roof trailing edge takes place with the flow nearly parallel to the ground (similar to the separation flow observed for the rear-facing step). The bounds for reattachment to the top of the block are not sharply defined. If the ratio $L/H < 1$, the flow probably is not reattached; for $L/H > 3$ or 4, the flow probably is reattached. For $1 < L/H < 3$, reattachment depends on the nature of the incident flow, especially on its turbulence (Castro, 1979), and on the characteristics of the block's surface.

For example, Arie et al. (1975) found that for $L/H = 2$, reattachment to the roof of a rectangular bar did not occur; the wake cavity length, measured from the upwind face, was about 9H. For $L/H = 4$, however, reattachment to the roof was observed, and the bubble length, measured from the rear face, was about 5H. In their field observations of flow near a cliff-like escarpment, Bowen and Lindley (1974) found that separation on the top persisted for several cliff heights from the edge; the separation zone reached a maximum height of about 0.2H relative to the top surface of the escarpment.

Castro (1979) observed roof reattachment on a bar with $L/H = 1$ when the incident boundary layer was allowed to develop over a rough ground plane. He found that, depending on the size of the bar relative to the boundary-layer depth, the cavity length, measured from the rear face, was between 8H and 10H. On the other hand, if the incident boundary layer developed over a smooth ground plane, reattachment to the roof was not observed, and the cavity length from the front face was between 9.2 and 10H (Castro and Fackrell, 1978). Counihan (1971) also found, for a two-dimensional block with $L/H = 1$, no reattachment on the roof; however, he observed a recirculation zone that was only about 5H long relative to the upwind face. Counihan, Hunt, and Jackson (1974) observed for a similar two-dimensional case that $x_r/H \simeq 7$ relative to the windward face.

Ogawa (1973) found $x_r/H = 10.5$ from the front face of a relatively thin block ($L/H = 1/6$). He also presented the data of several authors and suggested that x_r/H varies significantly with the turbulent Reynolds number. The relation $x_r/H \sim 0.21\ Re_{turb} \sim 1.3\ \ln(H/z_0)$ is a good approximation to Ogawa's numerical results, which in turn fit the data fairly well. Ogawa's (1973) data also are consistent with the idea that, besides the significant variation with H/z_0, x_r/H decreases as L/H increases, albeit weakly. This dependence on H/z_0 is in agreement with results obtained in near-uniform incident flows (discussed earlier) and with one of the hypotheses made above for the cavity region behind a thin two-dimensional fence. The values of x_r/H quoted by Ogawa (1973) are between 6 and 13 and agree with Hunt's (1974) summary.

For rear-facing steps with the flow attached to the "roof," Tani (1957) noted a cavity length of about 7H from the lee face of the step. Finally, Tillman (1945; cited by Bradshaw and Wong, 1972) reported the apparent extreme value of $x_r = 13.5H$ from the upwind face of a two-dimensional block where no roof reattachment was observed.

7-5.2.2 ***Three-Dimensional Sharp-Edged Obstacles.*** Addition of a finite crosswind dimension drastically complicates the near-wake flow, even for normal flow incidence. Along the upwind edges of the building, the flow cannot turn sharply enough to follow the roof and sides; consequently it separates from the obstacle. Air immediately adjacent to the roof and sides moves in the upwind direction in these separated zones, and a recirculating flow is formed. The boundary between it and the outer flow is a turbulent shear layer that curves inward toward the body; if the obstacle is sufficiently long, these shear layers contact the surfaces of the obstacle, and the flow reattaches to the roof and sides. When this occurs, the flow downwind of the reattachment line will be directed toward the rear of the structure. An examination of the literature suggests that reattachment usually will occur in atmospheric flows if the length-to-height ratio L/H is unity or larger and if the building is not too wide (Hosker, 1979). Reattachment may be delayed or may not occur if W/H is large.

Wilson (1979) has examined the roof recirculation zone in some detail. Let ζ_b be the smaller of H and W and let ξ_b be the larger. By defining the characteristic length scale as $\tilde{R} \equiv \zeta_b^{2/3}\xi_b^{1/3}$, Wilson found that the roof cavity extends a distance $L_c \simeq 0.9\tilde{R}$ from the upwind roof edge and reaches a maximum height of about $0.22\tilde{R}$ above the roof at a distance $\tilde{R}/2$ from the upwind edge (Fig. 7.19). This cavity is bounded above by a turbulent shear layer (zone II in Fig. 7.19) and above that by the roof wake (zone III). The upper edge of the shear layer, which begins near the maximum cavity height, is approximately determined by

$$\frac{Z_{II}}{\tilde{R}} \simeq 0.27 - 0.1\,\frac{x}{\tilde{R}} \tag{7.6}$$

and the roof wake boundary is approximated by

$$\frac{Z_{III}}{\tilde{R}} \simeq 0.28\left(\frac{x}{\tilde{R}}\right)^{1/3} \tag{7.7}$$

In both expressions, x is measured from the upwind roof edge and is greater than $\tilde{R}/2$. Additional field and laboratory work to validate these estimates would be welcome.

The horizontally oriented standing vortex in front of the building is forced by the incident wind to wrap around the sides of the object and to trail off near ground level as a counterrotating "horseshoe" vortex pair. In the lee of the obstacle, near its lateral edges, is a pair of vertically oriented vortices; these interact with the main flow field near roof level and also trail off downwind, in this case as an elevated counterrotating pair. The very complex flow field sketched in Fig. 7.15 has emerged in recent years from the work of Morkovin (1972) and his colleagues, Britter, Hunt, and Puttock (1976), Woo, Peterka, and Cermak (1977), and Hunt et al. (1978). Two significant features of this model should be noted: (1) the streamlines bounding the cavity zone are not the separation streamlines and (2) transfer of material into and out of the cavity is not limited to turbulent transport across this bounding interface. Instead, material can be advected into the recirculation zone along the entering streamlines and can be removed via the rising spiral vortices generated near the lateral lee edges of the obstacle. Earlier concepts of cavity flow, like the model of Fig. 7.7, are strictly applicable only to purely two-dimensional problems.

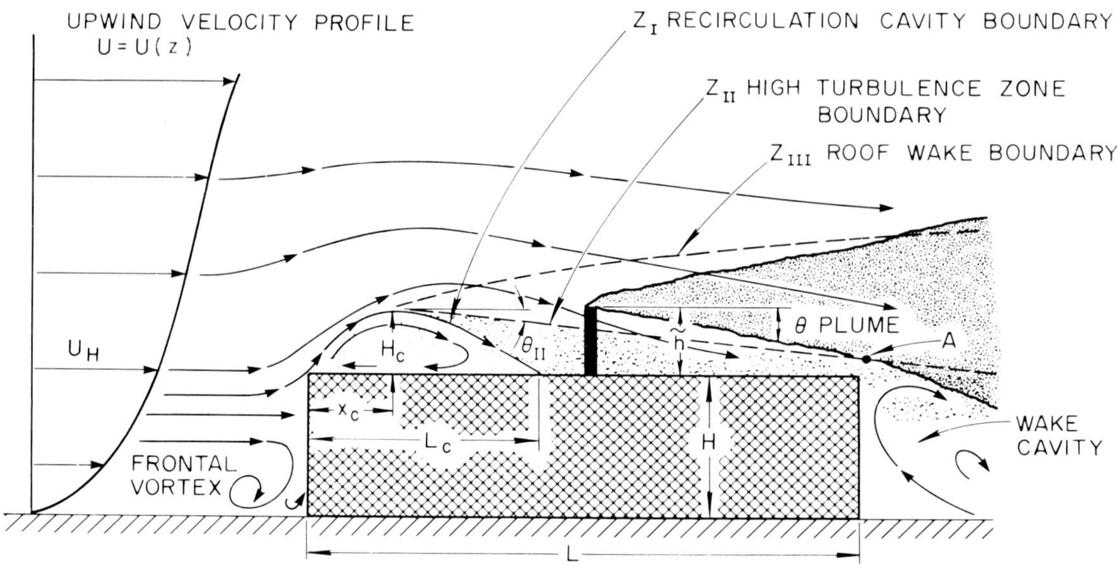

Fig. 7.19 Flow over center of long flat building roof for wind perpendicular to upwind face. [From D. J. Wilson, Flow Patterns over Flat-Roofed Buildings and Application to Exhaust Stack Design, *ASHRAE Trans.*, 85(Pt. 2): 284–295 (1979). Reprinted by permission of American Society of Heating, Refrigerating and Air-Conditioning Engineers, Inc.]

An improved understanding of the physics of the near-wake problem, such as the model indicated in Fig. 7.15, should eventually lead to better methods of predicting near-building winds and gustiness, air-pollutant buildup, and other practical problems. For some purposes, however, the concept of a more-or-less closed bubble may still be adequate. For example, one may merely need to estimate the approximate extent of a shelter zone near a windbreak, or one may wish to estimate the boundaries of the recirculation zone where buildup of pollutant concentrations from local emissions is likely. Furthermore, Vincent (1978) suggests that in a thick, turbulent boundary-layer flow (like the atmospheric surface layer, see Chap. 5), the simpler cavity concept may still be adequate because turbulent transfer across the shear layer probably dominates advective transport. In view of Fig. 7.15, however, this suggestion is somewhat controversial. Further study of the pertinent transfer processes clearly is required.

Information on the behavior of the horseshoe vortex is sparse. Ogawa (1973) and Frost and Shahabi (1977) performed a limited amount of flow visualization work during their field experiments; their results only verify the presence of organized vorticity near the edges of obstacles but do not provide data on the paths, velocities, or persistence of these vortices. Colmer's (1970) measurements also suggest such vortex structure in the wake but do not allow prediction of the vortex path.

In Colmer's (1970) work, there is some indication that the wake effectively rises off the ground as it moves downwind. Similar behavior has been reported in wind-tunnel tests (e.g., Lemberg, 1973; Hansen and Cermak, 1975). Penwarden and Wise (1975) studied the path of the line of maximum speed (y_{MS}) resulting from the horseshoe vortex wrapping around a thin (small L) model building. Relative to the wake centerline, their data agree fairly well with the relation $y_{MS}/W \simeq 0.5 + c\sqrt{x/W}$, where x is measured from the front face of the building, and $c \simeq 0.39$. The "constant" c must actually depend on H and L, especially when L is large and the flow reattaches to the sides, but not enough data are available to determine this dependence. The behavior and persistence of the horseshoe vortex require further study, especially with regard to the interaction of the vortex with the deep turbulent boundary-layer flows typical of the atmosphere.

As was true for the horseshoe vortex, little is known about the elevated vortex pair produced by the interaction of the vertically oriented lee edge vortices with the mean flow near roof level (Fig. 7.15). Again, Colmer's (1970) field measurements, the smoke releases of Frost and Shahabi (1977), and wind-tunnel experiments (e.g., Counihan, 1971) indicate the presence of these vortices. However, even qualitative descriptions of their behavior and interactions with the mean and cavity flows and with the trailing horseshoe vortex are not yet available.

Variations with changes in the incident flow of the turbulent shear layer bounding the cavity zone have been studied by Corke and Nagib (1976). As noted earlier, the diameter of the horseshoe vortex can be expected to increase with increasing power-law exponent n of the approach flow profile. The pressure coefficients on the lee face (based on U_∞) were found to decrease from about -1 to about -3 as n increased from 0.11 to 0.38; this decrease caused the separated shear layers near the sides of the obstacle to curve more sharply toward the wake centerline and thus to narrow the wake. Fluctuations in pressure on the downwind edges of the sides increase because of the increased proximity of these turbulent layers. On the other hand, the pressures on the roof are such that the greatest curvature of a shear layer shed from the roof leading edge occurs with the lowest value of n. The resultant interaction of the side and roof shear layers with each other, with the nearby vortices, and with the incident flow determines the size of the recirculation zone for any given case. Because all these flow features are turbulent and prone to meandering, the cavity will fluctuate in extent; hence, it probably is best considered in a time-averaged sense.

Different cavity behavior for similar obstacles has been reported; this is probably because of the importance of the incident flow in determining the paths of the separated shear layers bounding the cavity. For example, Yang and Meroney (1970) found a cavity length between 3H and 4H, measured from the upwind face, behind a cubical model. They also remarked that the ground-level transition zone between the reverse flow of the cavity and the downwind-directed near wake was about 2H in length. However, Counihan (1971) reported a cavity length of only 1 to 1.5H relative to the rear face of a cube because the flow had reattached to the roof and sides of the model. The differences between these two results probably are due to the different characteristics of the incident boundary-layer flows. Castro and Robins (1977) found that for sufficiently deep incident boundary layers, the flow always reattached to the top and sides of their cubical model; this reattachment occurred about L/4 from the upwind edges (see also Robins and Castro, 1977a, b). The end of the wake cavity was then about 1.5H from the lee face of the obstacle. Vincent (1978) recently attempted an explicit examination of the effect of δ/H on the cavity behavior. In instances of fairly deep boundary-layer flows ($\delta/H \simeq 8$), Vincent observed $x_r/H \simeq 1.8$ relative to the rear face of a cube; this length gradually decreased with increasing δ/H to a value of about 1.3 when $\delta/H \simeq 20$. Apparently the flow had reattached to the sides and roofs of his models. It is not clear, however, whether this change in cavity size resulted solely from δ/H effects or whether Vincent's experimental methods could have contributed to the change by introducing changes in the turbulence characteristics (such as relative eddy size and effective roughness length) of the approach flow. Further study is needed to resolve this question.

A variety of cavity sizes have been reported for noncubical objects. Dreher and Cermak (1973) state that the recirculation zone behind a typical peaked-roof house will extend 2H to 3H behind the lee face. Ogawa (1973) observed $x_r/H \simeq 7.5$ behind a fairly wide fence ($W/H = 16$) in the real atmosphere. Drivas and Shair (1974) found a cavity length of about 3H behind the rear face of a real building ($L/H = 1.4$, $W/H = 5.3$); apparently the flow reattached to the roof about L/2 behind the windward roof edge except close to a small rooftop penthouse. Huber and Snyder (1976) reported $x_r/H = 3.5$ from the upwind face of a model building ($L/H = 1$, $W/H = 2$); they observed a maximum cavity height and half-width of about 1.5H; both occurred at 1.5H from the windward face. Snyder and Lawson (1976) found cavity heights of 1.5H behind a tall thin building model ($L/H = 0.33 = W/H$) and 2.0H behind a somewhat wider structure ($L/H = 0.33$, $W/H = 2.0$). Field work by Frost and Shahabi (1977) indicated a cavity of similar size behind a moderately wide structure ($L/H = 0.75$, $W/H = 8.4$); on the average, x_r/H was about 3.8 relative to the rear face, while the cavity height reached about 2H at a distance 1.5H behind the building. Woo, Peterka, and Cermak (1977) conducted wind-tunnel tests to complement the work of Frost and Shahabi (1977). For a model building of modest width ($L/H = 0.75$, $W/H = 2.4$), the cavity extended about 3.5H from the rear face and attained a maximum height of about 2.5H near the model. The Woo, Peterka, and Cermak model of Frost and Shahabi's building ($L/H = 0.75$, $W/H = 8.4$) provided a cavity length of about 7H from the lee face, and a cavity height slightly greater than that obtained by Frost and Shahabi. Finally, Hatcher et al. (1978) recorded cavity lengths between 2.5H and 3.1H relative to the upwind surface of a model of a complex experimental reactor building ($L/H_{max} = 1.6 = W/H_{max}$).

Current data are insufficient to develop relations such as Eqs. 7.1 and 7.2 to estimate the extent of the cavity region behind even simple three-dimensional block structures immersed in deep boundary-layer flows. However, the work of Ranga Raju, Loeser, and Plate (1976) and of Ogawa (1973) and Tani (1952, 1958) suggests that the major dependence on the incident flow characteristics appears through the height-to-roughness ratio H/z_0; or, more specifically, through the logarithm of that ratio. That is, the gross features (e.g., cavity size) of a wake flow may not be very sensitive to the details of the approach flow. Hence, Eqs. 7.1 and 7.2, originally developed for near-uniform incident flows, were compared (Hosker, 1979) to the boundary-layer data cited above.

Figure 7.20 shows data where $L/H \leqq 1$ and/or where the investigators explicitly stated that the flow was not reattached to the roof and sides of the obstacle; the shaded area in this figure indicates values calculated using Eq. 7.1 ±25% and using $L/H = 0.75$ as a reasonable median value for the obstacle thickness ratio. The cavity extent x_r is given relative to the windward face, where separation occurs. Agreement of the data with Eq. 7.1 is fair. The scatter is substantially larger than in the uniform flow case for which Eq. 7.1 was developed. This probably results from the effects of the many different incident boundary layers. The worst disagreements between Eq. 7.1 and the data arise from Caput et al.'s (1973) field work, which is overpredicted by perhaps 6 to 8 fence heights (~150 to 200%), and from Frost and Shahabi's (1977) work, which is overpredicted by about 3 building heights (~67%).† The reason for the discrepancy between the field and laboratory results is unclear; however, it may be due to differences between the model and field incident wind characteristics. Nevertheless, Eq. 7.1 generally suffices to predict the cavity length behind simple block-like structures to within ±50%.

Figure 7.21 compares data where $L/H > 1$ and/or where the flow reattached to the obstacle roof and sides to values calculated from Eq. 7.2 ±25%. The cavity extent here is given relative to the lee face. The largest discrepancy occurs for the wind-tunnel data of Hatcher et al. (1978). These data were taken using a model of a peculiarly shaped reactor building. The field data of Drivas and Shair (1974) are overpredicted by about 1 building height (~33%). In general, however, Eq. 7.2 predicts cavity length fairly well for the case where the initially separated boundary layers reattach to the obstacle.

Both Fig. 7.20 and 7.21 indicate regions where additional data on cavity extent are needed. To facilitate future work toward understanding the scatter in the boundary-layer data, experimentalists should specify incident flow characteristics such as boundary-layer depth, roughness length z_0, and turbulence intensity and scale. Direct observations of separation and reattachment points would be useful. Field observations of cavity behavior seem rare; such observations should be encouraged so that further comparisons between the results of field observations and the results of laboratory work can be made.

7-5.2.3 ***Rounded Obstacles.*** As remarked earlier, the recirculation zone behind rounded structures depends on the exact location of boundary-layer separation from the obstacle's surface. Factors such as Reynolds number, incident flow turbulence scale and intensity, and details of the obstacle shape and surface roughness are important. Generalized studies are sparse.

Huber et al. (1976) examined the near wake of several two-dimensional ridges placed normal to a deep incident shear

†An interesting feature of the latter case is the fairly good agreement between Woo, Peterka, and Cermak's (1977) wind-tunnel simulation of Frost and Shahabi's (1977) field site and the predictions of Eq. 7.1.

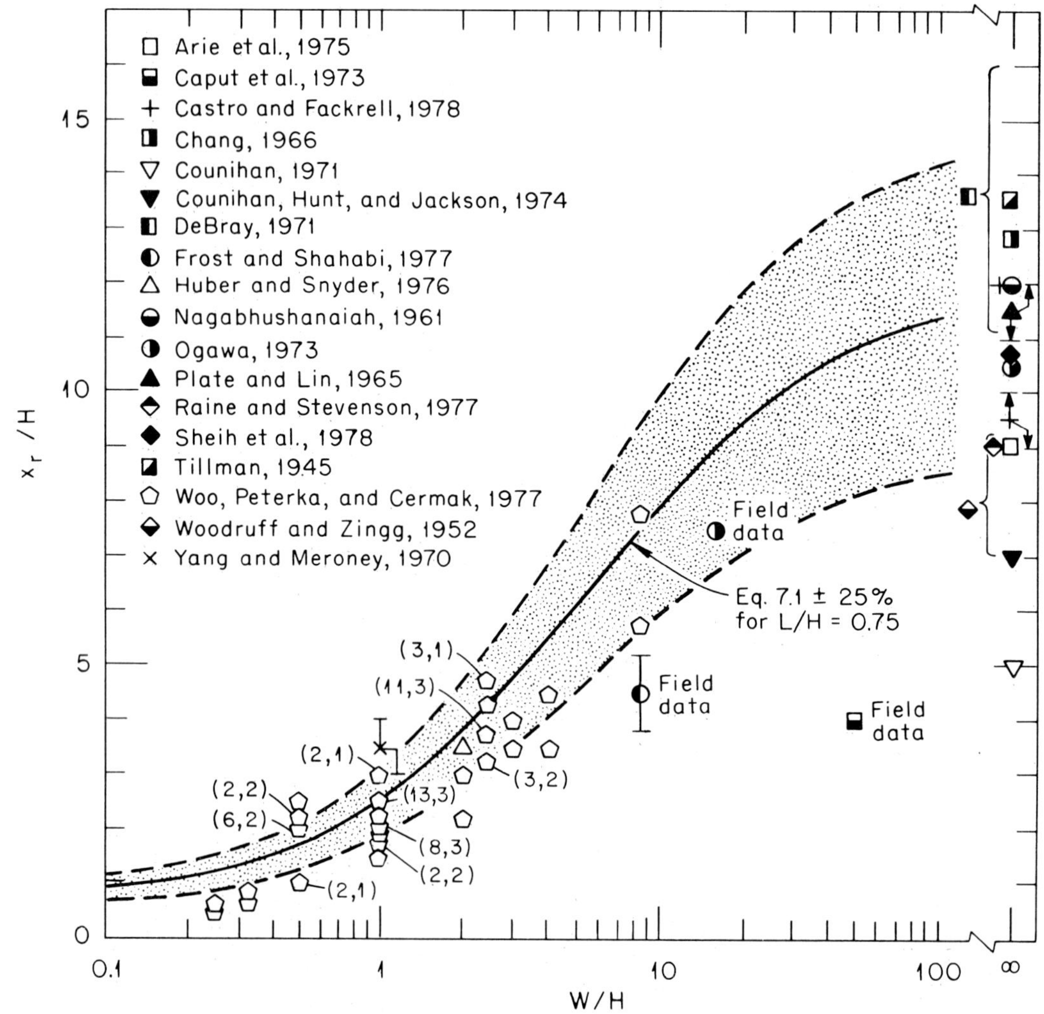

Fig. 7.20 Comparison of cavity length data taken in deep boundary layers to Eq. 7.1 ± 25%, for L/H = 0.75 (i.e., no reattachment). Number pairs near data points indicate number of points and number of test boundary layers, respectively.

flow. They observed, as expected, that the separation location on a bell-shaped ridge depended on the Reynolds number; however, in most cases, separation took place on the lee slope somewhat below the peak of the ridge. Consequently, the cavity was smaller than when separation was artificially triggered at the ridge's apex; such forced separation generated a cavity similar to that generated behind a triangular ridge for which separation occurred naturally at the ridge top (Fig. 7.22). In their forced separation tests, Huber et al. (1976) found a maximum cavity height of about 2H and a cavity length, measured from the ridge center (i.e., from the separation point), of approximately 10H. When the separation location was fixed, the flow patterns within the cavity were not very sensitive to the shape of the lee slope.

Bouwmeester, Meroney, and Sandborn (1978) also studied flow over two-dimensional ridges. Depending on the slope of the windward ridge face, separation upwind of the ridge sometimes was observed. They reported that lee-side separation depends on both upwind and downwind slopes. On rounded ridges, the actual separation point is a function of surface roughness, incident turbulence, and atmospheric stability. Cavity regions as large as 15H were observed; however, the cavity was smaller for ridges with gentle slopes. Their report should be consulted for details on wind profiles, turbulence levels, and cavity sizes. Field data are given by Sacré (1979) and Bradley (1980).

As in the case of sharp-edged bodies, boundary-layer flow around three-dimensional rounded obstacles is fundamentally different from the flow around their two-dimensional counterparts; the difference can be attributed to the presence of streamwise organized vorticity downwind of the obstacles. This vorticity together with the indeterminate location of separation that is typical of rounded objects, results in a complex flow pattern that varies according to the details of individual obstacles as well as with the incident flow characteristics. Nevertheless, qualitative descriptions of the major phenomena are possible.

For example, Gregory and Walker (1951), Hawthorne and Martin (1955), and Mochizuki (1961) all observed a horseshoe vortex wrapped around ground-mounted obstacles—a cylinder, a hemisphere, and a sphere, respectively. In addition, Gregory and Walker (1951) and Mochizuki (1961) reported an elevated vortex pair behind their respective obstructions. The elevated vortex pair resulted from the bending over at roof level of spiral vortices that were generated behind the sides of the bodies; except for the obvious geometric differences, the model of the flow is akin to that depicted in Fig. 7.15. Symes and Meroney

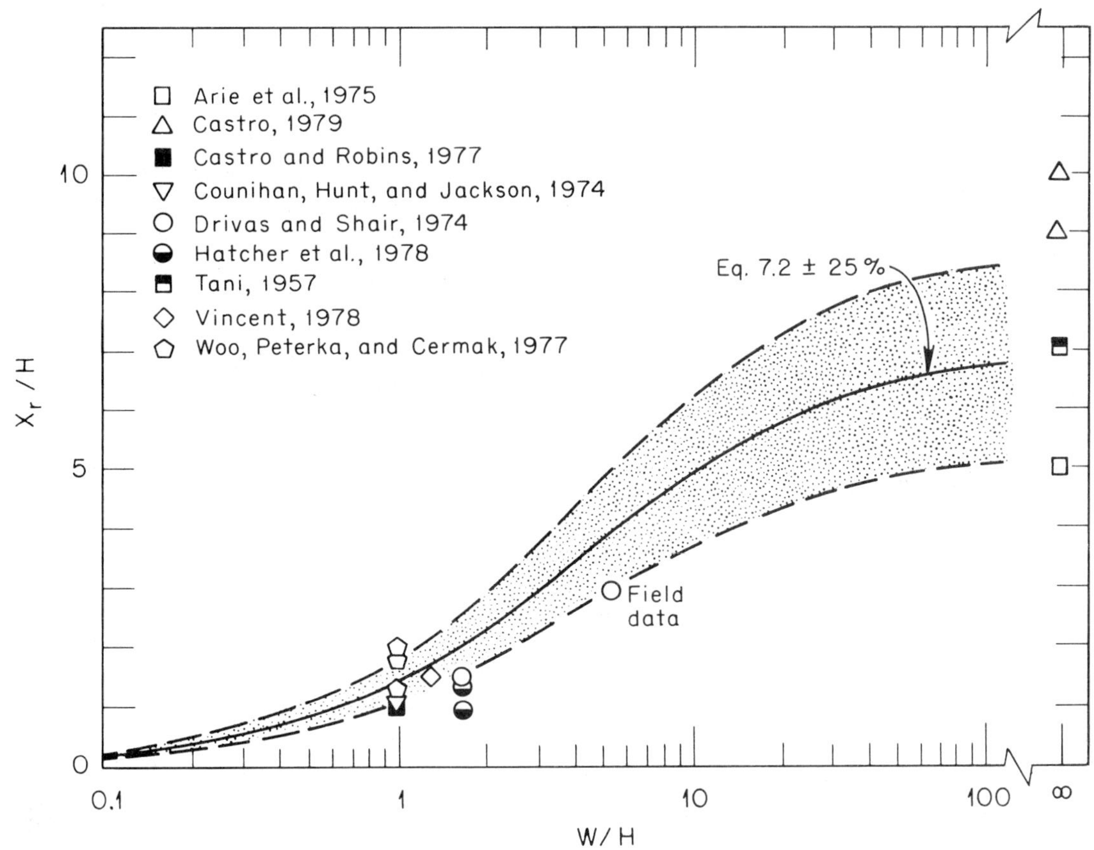

Fig. 7.21 Comparison of cavity length data taken in deep boundary layers to Eq. 7.2 ± 25%, for large L/H (i.e., reattachment observed).

(1970) observed similar behavior during their studies of cylinders and cone frustrums; they noticed, however, that for a sufficiently tapered cone (vertex angle greater than 19°), the horseshoe vortex was not generated and the character of the wake was changed significantly.

In the research cited in the previous paragraph, the details of the flow changed somewhat as Reynolds number changed. Nevertheless, as Sedney (1973) pointed out, the qualitative features seen in low-speed flows are similar to those observed in high-speed turbulent flow. For example, Iversen and Greeley's (1978) study of a volcanic cinder cone in both the field and the wind tunnel found trailing vortex effects evident in the scouring of loose material from the ground on both sides behind the cone. However, the cone frustrum equivalent to their volcano has a vertex angle of about 100°; this angle is much larger than in the cases that Symes and Meroney (1970) examined and found that the horseshoe vortex had vanished. The reason for this discrepancy is not known. Iversen and Greeley (1978) also noted a zone of deposition approximately one rim diameter in extent behind the crater base. This zone probably results from the slow mean winds characteristic of the cavity region. Figure 7.23a indicates the general situation for a hemisphere; Fig. 7.23b shows the flow near a more typical hill (Hunt and Snyder, 1980). The details can be expected to vary on a case-by-case basis because the separation line, the closeness of approach of the trailing vortices to the wake centerline, and the vortex roll up process itself are dependent on (at least) the Reynolds number. Field data on an isolated hill have been reported by Eliseev (1973) and Mason and Sykes (1979b).

Common cylindrical shapes have been studied by several authors. Halitsky (1968) reported on a reactor shell model consisting of a cylinder topped by a hemisphere. Separation occurred more than 90° along the side from the front stagnation line; the resulting cavity reached 2.3 diameters (about 1.8H) downwind of the center of the shell. Field observations by Probert, Hasoon, and Lee (1973) of large cylindrical oil tanks indicated that, even for the Reynolds number ranges typical in the atmosphere, the separation line moved rearward on the sides as the Reynolds number ranged from 4×10^6 to 27×10^6. They also found that the separation location varied with height so that the cavity was narrower at the top than at the bottom of the tanks. Whether this resulted from velocity shear in the incident flow, from secondary circulations, or from the finite height of the tanks is not clear.

Generally speaking, a case-by-case study is the safest approach to determining the near-wake structure of three-dimensional rounded obstacles immersed in a boundary layer.

7-5.3 The Far Wake

As remarked earlier, the far wake of two-dimensional ground-mounted obstacles behaves similar to a developing boundary layer. This seems to hold true even when the incident flow is itself a (deep) boundary layer. Although several authors (e.g., Townsend, 1965; Petryk and Brundrett, 1967) have worked on establishing a theoretical basis for understanding the two-dimensional far wake, the theory and data of Counihan, Hunt, and Jackson (1974) are particularly signifi-

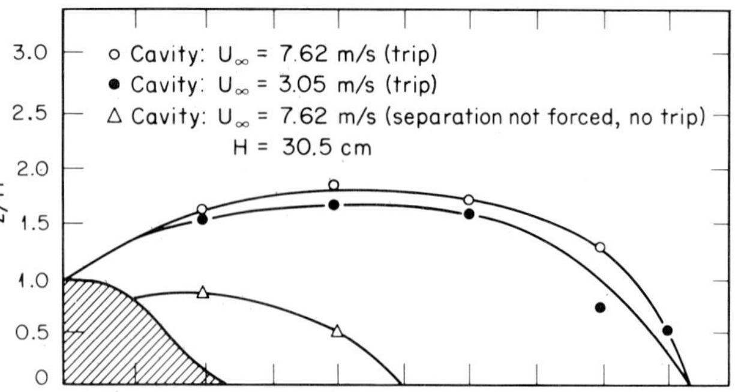

(a) Cavity extent in lee of Gaussian ridge.

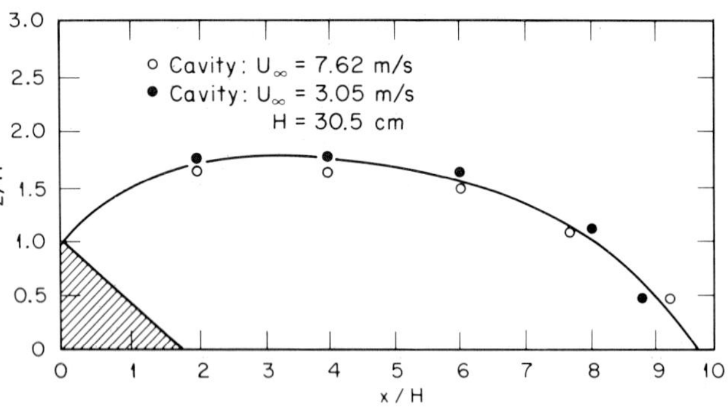

(b) Cavity extent in lee of triangular ridge.

Fig. 7.22 Cavity behind two-dimensional ridges immersed in deep boundary-layer flows. (From A. H. Huber, W. H. Snyder, R. S. Thompson, and R. E. Lawson, Jr., *Stack Placement in the Lee of a Mountain Ridge*, Environmental Sciences Research Laboratory Report EPA-600/4-76-047, U. S. Environmental Protection Agency, 1976.)

cant. They model the physical situation (Fig. 7.24) as follows: some distance downwind of the recirculation zone, close to the ground, is a layer where turbulent energy production is in local equilibrium with dissipation. Within this "wall" layer, the mean velocity defect has the usual logarithmic dependence on height (see Chap. 5); in addition, because the layer continues to develop and thicken, the mean velocity defect is, of necessity, a function of downwind distance. Above this zone is a "mixing layer" with a depth that is assumed small relative to the overall wake length; as in many wake and boundary layer cases, the flow in this layer is taken to be self-preserving. The mixing layer is capped by a third region that acts primarily as an inviscid perturbation to the incident flow field. The incident flow obeys a power-law profile with exponent n (typically, $0.1 \leqq n \leqq 0.4$).

Two important predictions of this theory are: (i) the mean velocity defects in both the wall and mixing layers diminish with downwind distance approximately as $(x/H)^{-1}$, and (ii) the overall wake depth grows slowly as $(x/H)^{1/(2+n)}$. The theory also assumes local equilibrium to derive rates of decay for the turbulent shear stress deficit in the wall layer, for the shear stress excess in the mixing layer, and for the mean-square turbulence intensity in the mixing layer.

Counihan, Hunt, and Jackson (1974) compared their own two-dimensional wake data as well as data of other authors to this theory; they found fair-to-good agreement in some respects for both laboratory and field experiments. For example, the mean velocity defect does follow the predicted self-preserving form (Fig. 7.25a), and its maximum does decay approximately as $(x/H)^{-1}$ (Fig. 7.25b). However, although the shear stress profiles appear to be self-preserving, their exact values are not predicted very well. The theory and the data appear to agree best for x/H in the approximate range of 7 to 21.

Recent work by Castro (1979) has verified the self-preserving nature of the mean velocity defect profiles at heights above the wall layer for moderate distances downwind ($10 \leqq x/H \leqq 30$). As predicted, the maximum velocity defect did decay in this region approximately as $(x/H)^{-1}$. However, the turbulent shear stress perturbation did not follow the predictions; instead, it died out no faster than $(x/H)^{-1}$; the exact decay rate seemed to depend on the turbulent Reynolds number of the incident flow. Near the ground, neither the mean velocity defect nor the perturbation shear stress obeyed the theory; instead they decayed more rapidly than in the mixing layer. Consequently, the recovery of the wake to a boundary layer-like profile proceeded most rapidly near the ground; far downwind ($x/H \simeq 50$), wake effects were only observed aloft. Castro suggested that the failures of the theory are probably associated with its assumptions about the turbulence. In particular, the theory uses an eddy viscosity formulation that is not borne out by the data. In fact, Castro remarked that no eddy viscosity assumption is likely to be satisfactory in view of the changes in turbulent structure as downwind distance increases. Additional theoretical and experimental work is required to improve our understanding of these flows and to provide a larger data base for model development and validation. Field data in particular are needed.

The far wake of a three-dimensional ground-mounted body immersed in a deep incident boundary layer is even more complex than that of a two-dimensional body because the presence of potentially persistent longitudinally oriented vortices generated by the body can strongly influence the flow far downwind. For example, a trailing counterrotating vortex pair can transport higher velocity air from above the wake down into the central portions of the wake and produce a sort of velocity overshoot; i.e., the measured speeds in the wake are higher than would be found at that same location if the obstacle were not present (see the far wake profile indicated in Fig. 7.23a). This effect seems to have been first reported by Jacobs (1938); however, he offered no real explanation for the phenomenon.

Some theoretical insight into the wake of three-dimensional obstacles was first provided by Hunt and his colleagues (Hunt and Smith, 1969; Hunt, 1970, 1971, 1974) via a similarity analysis akin to that used for two-dimensional obstacles (Counihan, Hunt, and Jackson, 1974). The theory ignores the presence of organized vorticity in the wake and assumes that the mean velocity defect drops off in the crosswind directions following a Gaussian curve. A power-law (exponent n) incident wind profile is assumed. The main predictions of this theory are: (i) the mean velocity defect decays as $(x/H)^{-(3+n)/(2+n)}$, and (ii) the mean-square turbulence intensity excess diminishes as $(x/H)^{-(4+n)/(2+n)}$. Lemberg (1973) carried out a related analysis that used a variable eddy viscosity in the wake rather than the constant value assumed by Hunt. Lemberg's results are similar to Hunt's; for example, the maximum mean velocity defect is predicted to decay as $(x/H)^{-(3+n)/2}$ and is expected to display a bell-shaped crosswind profile.

Hunt (1974) compared data from several sources with the theory. Figure 7.26a shows the mean velocity defect plotted in similarity variables to demonstrate the self-preserving nature of the wake profile; in this figure, the various broken lines indicate the best lines through the experimental results; the solid

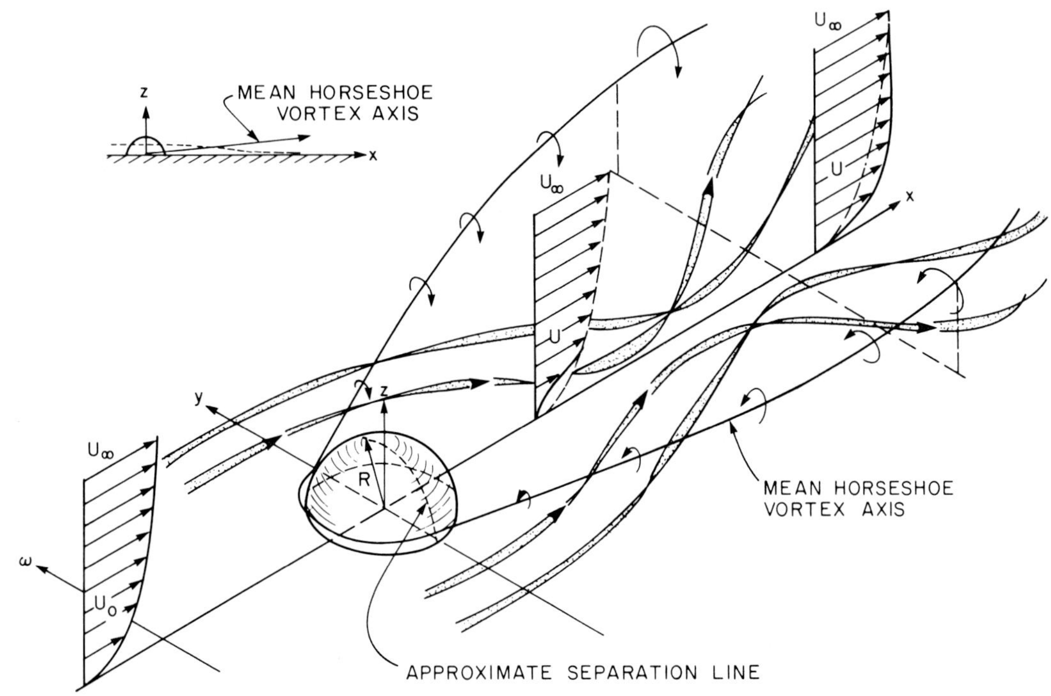

(a) Flow past hemisphere. (From A. C. Hansen and J. E. Cermak, Vortex-Containing Wakes of Surface Obstacles, Colorado State University Fluid Dynamics and Diffusion Laboratory Report CER75-76ACH-JEC16, Colorado State University, Fort Collins, Colo., 1975.)

(b) Flow past simple bell-shaped hill, with "cowhorn vortex" aloft. [From J. C. R. Hunt and W. H. Snyder, Experiments on Stably and Neutrally Stratified Flow over a Model Three-Dimensional Hill, *J. Fluid Mech.*, 96(Pt. 4): 671-704, Cambridge University Press, 1980.]

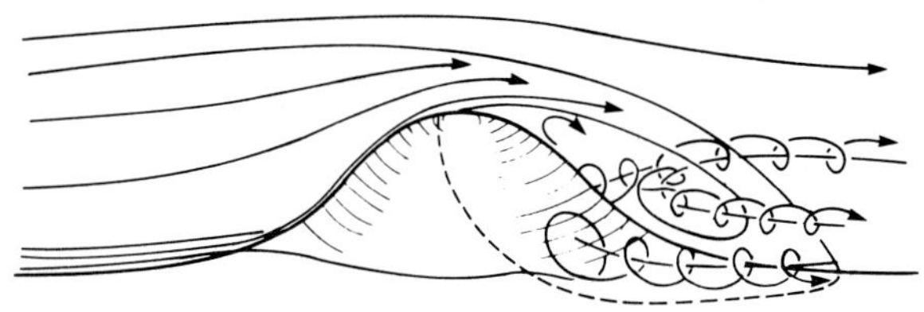

Fig. 7.23 Schematics of deep boundary-layer flow past rounded obstacles, showing lines of flow separation (dotted lines) and trailing vortices.

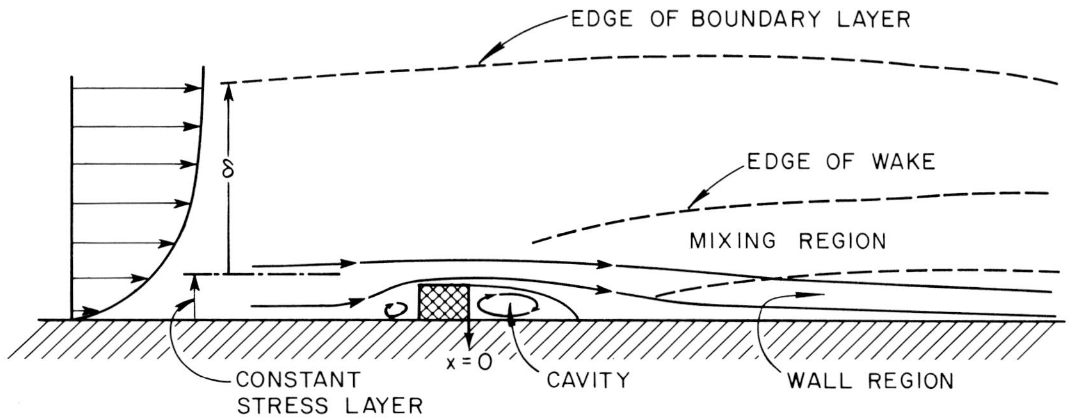

Fig. 7.24 Principal regions of flow and typical mean streamlines for a two-dimensional obstacle in deep turbulent boundary layer. [Based on J. Counihan, J. C. R. Hunt, and P. S. Jackson, Wakes Behind Two-Dimensional Surface Obstacles in Turbulent Boundary Layers, *J. Fluid Mech.*, 64(Pt. 3): 529-563, Cambridge University Press, 1974.]

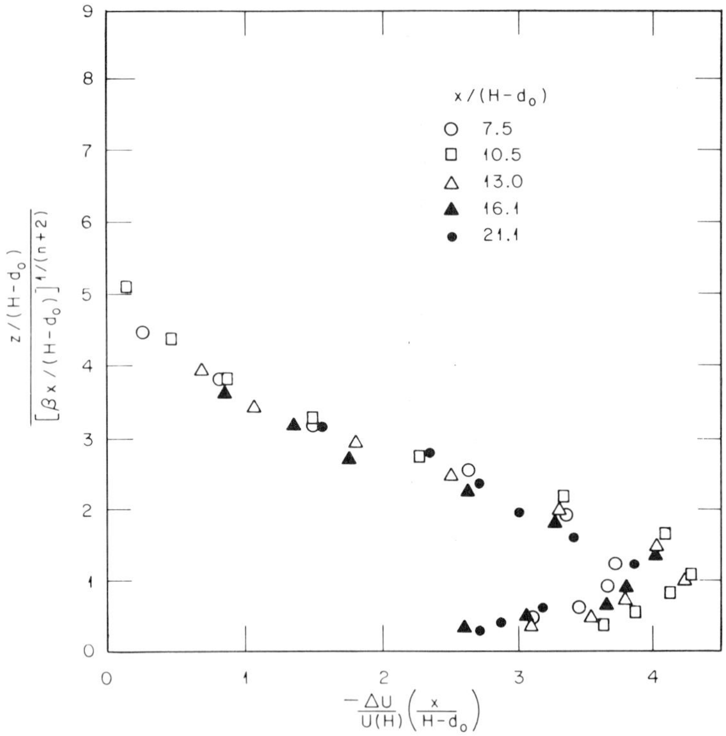

(a) **Mean velocity defect $\Delta U = U_{wake}(x,z) - U_{\infty}(z)$ as self-preserving profile behind two-dimensional block. Incident flow follows power law: $U_{\infty}(z) = U_0[(z - d_0)/\delta]^n$, with d_0 the zero plane displacement and δ the boundary-layer thickness. $\beta = 2ku_*/U_{\infty}(H) \simeq 0.5$, $n = 0.125$, $H/\delta \simeq 1/8$.**

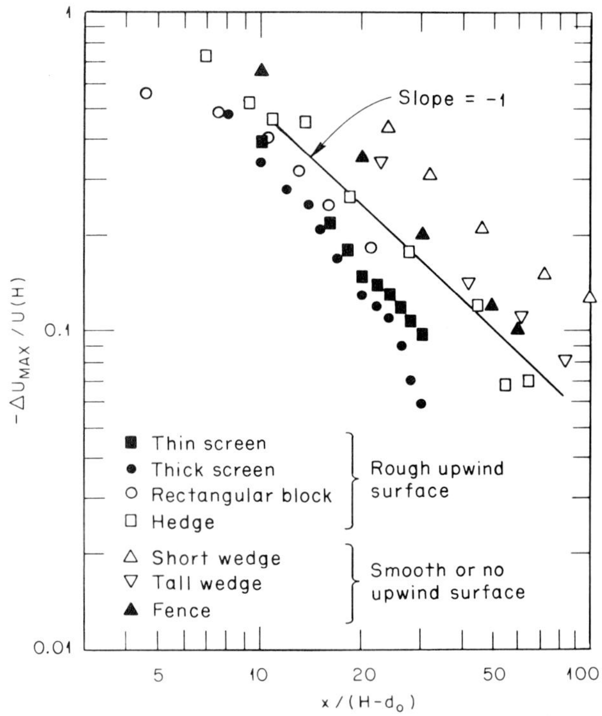

(b) **Maximum velocity defect as function of distance behind two-dimensional blocks.**

Fig. 7.25 Mean velocity defect behind two-dimensional obstacles in deep boundary-layer flows. [From J. Counihan, J. C. R. Hunt, and P. S. Jackson, Wakes Behind Two-Dimensional Surface Obstacles in Turbulent Boundary Layers, *J. Fluid Mech.*, 64(Pt. 3): 529-563, Cambridge University Press, 1974.]

curve is from the theory. The agreement is fair; Hunt (1974) suggested that the fit would be improved if he had used the variable eddy viscosity of Lemberg (1973). The least agreement between data and theory occurred for a cube placed at an angle to the incident wind; in this case, the likely presence of strong longitudinal vortices in the wake may account for the discrepancy.

Woo, Peterka, and Cermak (1977) also were able to demonstrate good agreement between theory and experiment. Figure 7.26b shows that the maximum mean velocity defect does decay approximately as $(x/H)^{-3/2}$; the actual exponents seem very close to the value (−1.58) found by Lemberg (1973). Figure 7.27 compares the predicted lateral spread of the wake to the data; the agreement with the theory's Gaussian assumption for crosswind spread is at best fair. In fact, Woo, Peterka, and Cermak (1977) were unable to fit their crosswind profile data to this hypothesis; they suggested that a strong horseshoe vortex in their wake may have been responsible. Peterka and Cermak (1975a) published lateral profiles of both mean velocity defect and turbulence intensity excess behind three-dimensional blocks. The horseshoe vortex present in their work induced "bumps" in the otherwise rather bell-shaped curves. The bumps appeared on both sides of the wake centerline at a distance of 1.0W to 1.5W. Similar data were reported by Kothari, Peterka, and Meroney (1979).

To account for the effects of organized vorticity in the wake of the three-dimensional obstacle, Hunt (to be published) has recently developed a small perturbation theory for such cases. In this theory, as described by Hansen and Cermak (1975), streamwise-oriented vortices serve to advect momentum from above downward into the wake on one side of the vortex; thus, for example, a counterrotating trailing vortex pair (e.g., Fig. 7.23a) can induce zones of somewhat higher than normal velocity on both sides of the wake centerline. The vortices involved can be those of the horseshoe system or those shed near roof level; multiple vortices are treated just by superposing their effects. The overall wake behavior is then taken as the superposition of the momentum wake theory discussed above, plus the contributions of the vortices.

Hunt's asymptotic results show that the mean velocity excess due to a vortex depends on the instantaneous vortex strength. Initially, the mean velocity excess increases slowly (as $x^{1/2}$) until it reaches a distance from the vortex origin that depends on the height of the vortex above the ground and the turbulent Reynolds number at that location. Further downstream this induced velocity excess begins to die out as $x^{-1/2}$. As Hansen and Cermak (1975) point out, the pure momentum wake always leads to velocity deficits in the wake; these velocity deficits decay about as $x^{-3/2}$. The vortex wake, however, induces velocity excesses on one side of each vortex, and these diminish very slowly as $x^{-1/2}$. In a wake where both phenomena appear, the vortex wake may dominate far downwind and there may be a region of velocity excess in the far wake. Whether this condition will actually occur depends, of

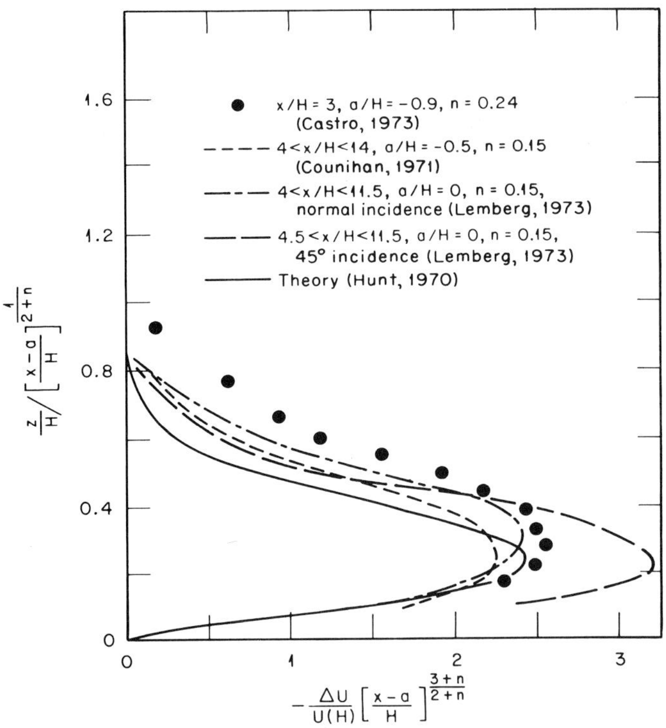

(a) Mean velocity defect $\Delta U = U_{wake}(x,z) - U_\infty(z)$ as self-preserving profile behind three-dimensional obstacle (cube). Incident flow follows power law of exponent n.

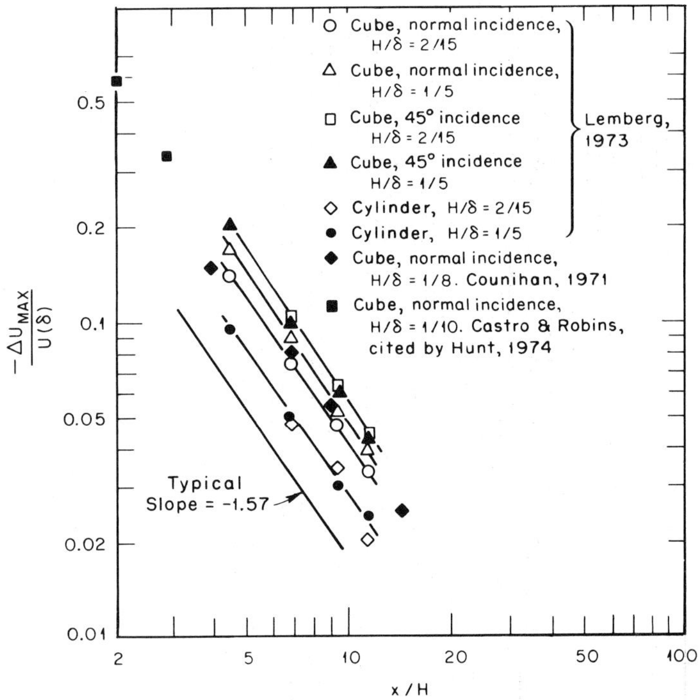

(b) Maximum velocity defect along wake centerline as a function of distance behind three-dimensional obstacles.

Fig. 7.26 Mean velocity defect behind three-dimensional obstacles in deep boundary layers. [From J. C. R. Hunt, *Wakes Behind Buildings*, presented at the Aeronautical Research Council's Atmospheric Environment Committee Meeting Oct. 23, 1974, ARC Report 35601, *Atmos.* 229, Aeronautical Research Council of Great Britain, Her Majesty's Stationery Office, London, 1974.]

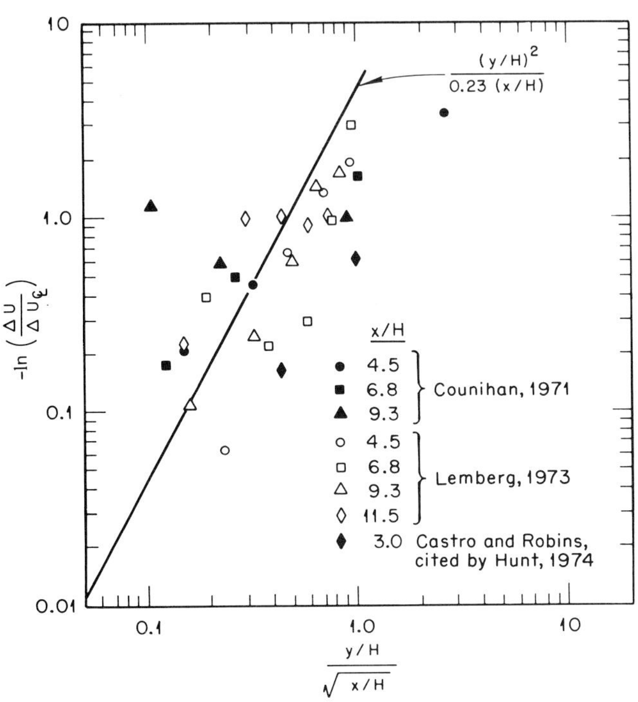

Fig. 7.27 Crosswind decay of mean velocity defect (relative to defect at z = 0 along wake centerline) behind cubes in deep boundary layers. [From J. C. R. Hunt, *Wakes Behind Buildings*, presented at the Aeronautical Research Council's Atmospheric Environment Committee Meeting Oct. 23, 1974, ARC Report 35601, Atmos. 229, Aeronautical Research Council of Great Britain, Her Majesty's Stationery Office, London, 1974.]

course, on the persistence and strength of the vortices present in any individual case.

Kothari, Peterka, and Meroney (1979) extended Hunt's combined momentum/vortex analysis to account for obstacle-induced perturbations to the profile of an incident scalar contaminant. Although the contaminant in Kothari, Peterka, and Meroney's (1979) research was actually the temperature of a moderately stable boundary layer flowing past isolated model buildings, the method that the researchers employed could easily be modified to examine the wake concentrations attributable to effluent being mixed downward from, for example, an elevated plume passing over a building from some upwind source. Here again the pure momentum wake leads to a perturbation that dies out approximately as $x^{-3/2}$, whereas the vortex-induced perturbation grows linearly with x and becomes most significant in the far wake.

Validation of this vortex wake theory requires evaluation of the instantaneous vortex strength. In any real turbulent flow, however, vortices meander according to the larger scales of the turbulence; the meander increases with turbulence intensity, and it increases in amplitude with increasing downwind distance. Consequently, measurements at any fixed location will provide only some type of average value for the vortex strength. This average value will be significantly less than the actual instantaneous strength, and it will indicate a decay rate greater than the actual value.

Hansen and Cermak (1975) proposed a theory to account for the meander effect. Their theory makes it possible to use experimental data to obtain the information required by Hunt's vortex wake theory. Among other things, Hansen and Cermak's (1975) work predicts that a trailing vortex pair will spread outward and upward with increasing downwind distance. This result agrees with the observations of Barker and Crow (1977) regarding vortices near the ground.

Data from many sources are available for comparison to the preceding theories on the wake of three-dimensional obstacles; Hunt's (1974) summary has already been cited. Wind-tunnel studies by Hansen, Peterka, and Cermak (1975), Peterka and Cermak (1975a), and Woo, Peterka, and Cermak (1977) all found that, as long as the block-like models examined were normal to the wind and immersed in a very deep, neutrally stable, incident boundary layer, the mean velocity defect decayed in approximate agreement with the momentum wake prediction $(x/H)^{-3/2}$. Colmer's (1970) field data behave similarly; however, a velocity overshoot at 23H downwind and 3H above ground suggests vortex influence. If Frost et al.'s (1977) method of removing upwind effects from their data is correct, their field data also display a momentum wake defect decay. Wakes in this category tend to be somewhat short (perhaps 10 to 20 building heights in downwind extent).

If a block-like structure is inclined to the wind, or if rounded objects such as a hemisphere are examined, or if the flow approaching a block-like obstacle is moderately stable, strong persistent vortices can be generated behind the obstacles and can dramatically alter the character of the far wake. Figure 7.28a illustrates this for a block building at an angle α to the incident wind. For such orientations a strong counterrotating vortex pair is generated near the upwind roof corner (Fig. 7.28b) and apparently it persists to great distances. In this case, velocity excesses were observed in the wake to at least 80H downwind.

Hansen and Cermak (1975) examined the flow behind a hemisphere (radius R) where a strong horseshoe vortex system was evident; they compared their data to the combined momentum-vortex wake theory. The vortices were observed to rise off the surface nearly linearly, $\bar{z}/R \simeq 0.031(x/R) - 0.08$, and to spread slowly relative to the centerline, $\bar{y}/R \simeq \pm 0.6(x/R)^{0.4}$. The vortex wake theory predicts an even slower rate of spread, $(x/R)^{0.15}$ to $(x/R)^{0.34}$. The maximum velocity defects are located outboard of the vortices. Figure 7.29 compares the Hansen and Cermak data to the combined momentum-vortex wake theory. As the figure shows, the theory seems able to predict the general shape of the profile, but it does not accurately predict the actual values. The gradual buildup and eventual decay of the wake velocity excess

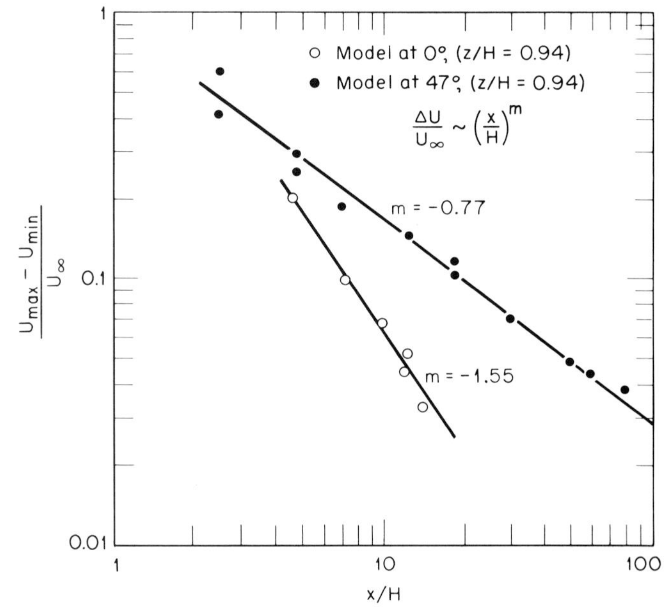

(a) Distance decay of maximum mean velocity difference behind three-dimensional block in a deep, neutrally stable boundary-layer flow for two wind directions.

(b) Visualization of roof corner vortex pair for wind incident at 47° to normal of long obstacle face.

Fig. 7.28 Significance of wind incidence angle to wake decay and near-building flow field. (From A. C. Hansen and J. E. Cermak, *Vortex-Containing Wakes of Surface Obstacles,* Colorado State University Fluid Dynamics and Diffusion Laboratory Report CER75-76ACH-JEC16, Colorado State University, Fort Collins, Colo., 1975. Photograph courtesy of J. E. Cermak.)

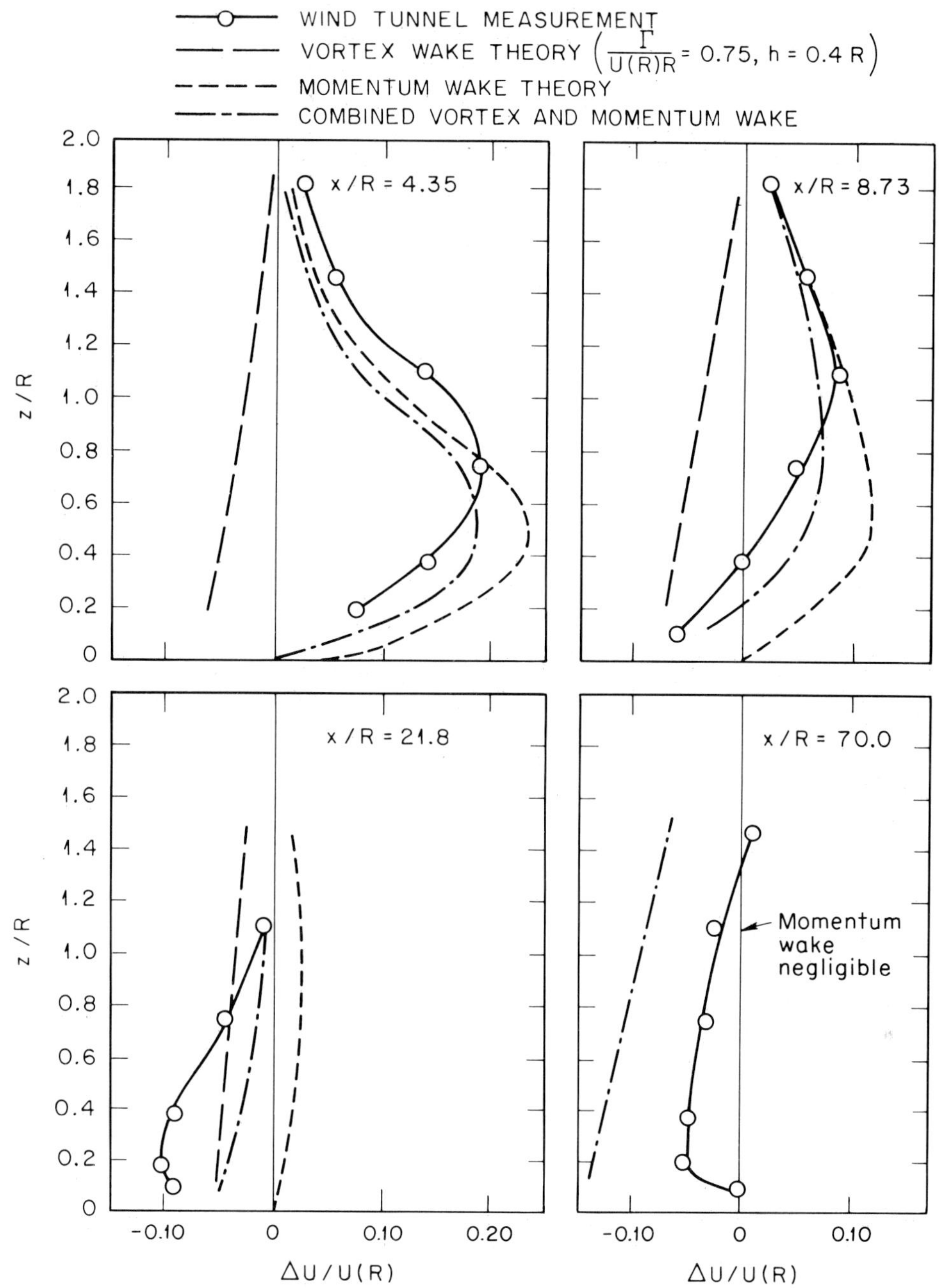

Fig. 7.29 Comparison of measured vertical profiles of velocity deficit $\Delta U \equiv U - U_\infty$ along wake centerline for a hemisphere (radius R) with various theoretical predictions. (From A. C. Hansen and J. E. Cermak, *Vortex-Containing Wakes of Surface Obstacles*, Colorado State University Fluid Dynamics and Diffusion Laboratory Report CER75-76ACH-JEC16, Colorado State University, Fort Collins, Colo., 1975.)

is shown in Fig. 7.30; the initial rise is slower than theory predicts, but the decay rate seems to agree with the prediction of $(x/R)^{-1/2}$. Notice, too, that this wake extends to over 80 building heights downwind.

If the incident flow is moderately stable, even a block-like obstacle normal to the wind can generate a wake influenced by vortices. Kothari, Peterka, and Meroney (1979) observed vortex-dominated wakes to at least 60H downwind of several building models. The mean velocity excess diminished at rates between $x^{-0.9}$ and $x^{-2.1}$. Apparently, the rates depended on building geometry. Perhaps flow reattachment (or its absence) to the roof and sides influences the result. The turbulent velocity fluctuations, on the other hand, dropped rapidly to background levels by 7.5 to 10H downwind; the decay rates ranged between $x^{-1.7}$ and $x^{-2.5}$; most of the rates were slightly faster than Hunt's theoretical estimate of x^{-2}. The mean temperature excess increased initially (out to a distance of 8 to 15H); then it began to die off slowly as $x^{-0.2}$ to $x^{-0.7}$; the rate at which it died off depended on building shape. This temperature wake was produced by the trailing horseshoe vortex that transferred the stable warm air aloft down toward the wake centerline; the temperature effect decayed more slowly and persisted farther downwind than the velocity wake. Figure 7.31 shows the excellent agreement observed between vertical pro-

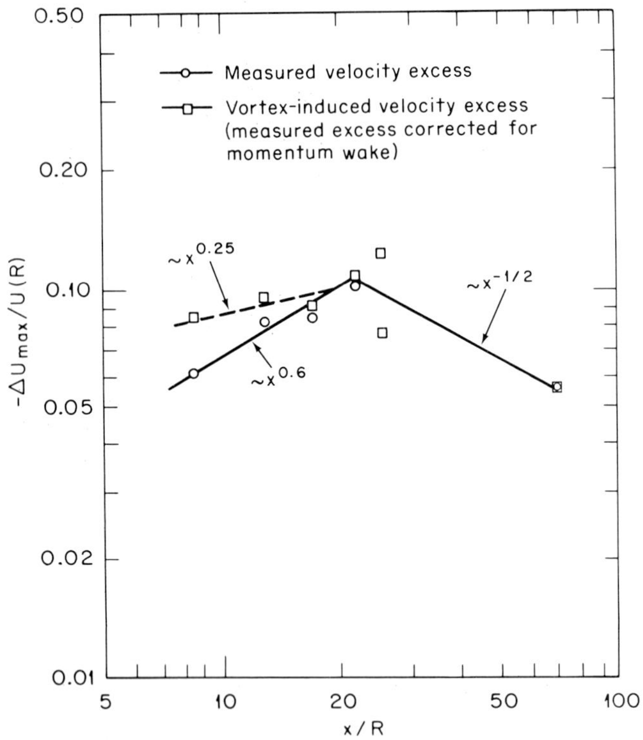

Fig. 7.30 Decay of maximum vortex-induced velocity perturbation in wake of hemisphere with radius R. (From A. C. Hansen and J. E. Cermak, *Vortex-Containing Wakes of Surface Obstacles*, Colorado State University Fluid Dynamics and Diffusion Laboratory Report CER75-76ACH-JEC16, Colorado State University, Fort Collins, Colo., 1975.)

files of mean temperature excess and the combined momentum/vortex wake theory. The variance of the turbulent temperature fluctuations actually became negative beyond 8 or 10H because the fluid being transported toward the surface was not only warmer but it was also less turbulent than the surrounding air. These temperature fluctuations did not return to background levels before 60H downwind.

Our understanding of the far wake behind a three-dimensional body can be summarized as follows: in wakes where organized vorticity is unimportant, the theoretical work of Hunt (e.g., 1974) and Lemberg (1973) explains the observations fairly well. When the data are plotted in the similarity variable format suggested, self-preserving velocity profiles are found. The mean velocity defect decays approximately as $(x/H)^{-3/2}$; the mean-square turbulence intensity excess disappears as $(x/H)^{-2}$. The crosswind velocity profile is approximately bell-shaped. The overall wake extent typically is over 10 to 20H for wakes where organized vorticity is unimportant; however, the overall wake extent may depend on the building aspect ratio W/H (as the separation zone does). When organized vorticity becomes important in the wake, as is the case for hemispheres, block buildings inclined to the incident wind, or block buildings in a stable flow, the wake can assume a vortex character. The vortices may induce a velocity excess in the wake; this excess is predicted to die out very slowly, as $(x/H)^{-1/2}$, and can persist to very large distances downwind. The vortices tend to rise slowly from the ground and spread gradually from the wake centerline. Depending on the ambient turbulence level, the detectable wake may extend to 50 or 100H downwind. A wake temperature excess also can be produced if the incident flow is stable. The theoretical work of Hunt (to be published), Hansen and Cermak (1975), and Kothari, Peterka, and Meroney (1979) provides a rough framework for understanding such combined momentum-vortex wakes, but further work is required to produce good quantitative predictions.

A few caveats are in order here. First, most of the data summarized above were obtained in laboratory experiments rather than in the field. Many of the investigators (e.g., Lemberg, 1973; Hansen, Peterka, and Cermak, 1975; Woo, Peterka, and Cermak, 1977) noted an apparent sensitivity of their results to the ratio of obstacle height to incident boundary-layer depth H/δ. A great deal more work is needed to elucidate or eliminate this factor. About all that can be said now is that H/δ in a model test should be made similar to that expected in the field. Secondly, it is well known (e.g., Saffman, 1973; Widnall, 1975) that the processes of vortex generation and decay are dependent on Reynolds number. Hence, wind-tunnel studies of vortex wake behavior, typically conducted at Reynolds numbers that are orders-of-magnitude smaller than those obtained in the atmosphere, must be viewed with caution until a better understanding of this effect is available. This may be particularly troublesome in modeling stable flows where the Reynolds number is usually smaller than normal to accommodate the thermal scaling of the flow (e.g., Snyder, 1972, 1981).

Both of these points suggest that more field data are vital, especially if the possible significance of vortex-type wakes is to be established. The work of Hansen and Cermak (1975) showed that vortex meander caused by ambient turbulence results in an apparently weak and rapidly decaying vortex as detected by a fixed sensor, even though the instantaneous vortex may be quite strong and persistent. This result means that vortices may be difficult to observe in the wakes of buildings in the field; their presence may have to be inferred from evidence such as velocity overshoot in the wake at large distances downwind (e.g., Colmer, 1970). Furthermore, the vortices can be expected to rise and spread from the wake centerline with increasing downwind distance; hence, measurements confined to the centerline may not be very useful. Perhaps future field work on building wake phenomena should use a fairly dense tower or tethered instrumented balloon array with both crosswind and along-wind sites, or, better, remote sensing instruments such as lidars or Doppler radars.

7-6 EFFECTS OF WIND INCIDENCE ANGLE AND CLUSTERING OF BUILDINGS

7-6.1 Wind Incidence

The preceding discussion has been almost exclusively devoted to problems where the wind is normally incident on one face of an obstacle. When the wind is not at right angles to the body, however, several interesting phenomena that are of interest to both structural engineers and air quality specialists appear. It is somewhat surprising, therefore, that, except as results of special investigations for particular structures, few quantitative data exist. Only a rather qualitative description of some of the important effects is attempted here.

Figure 7.32 is a top view of flow approaching a structure such as a tall building at an angle α. The flow separates at both upwind corners S and T, but it can reattach to the exposed side at some point R. This reattachment location is the result of an aerodynamic force balance and, consequently, depends on the instantaneous flow conditions as well as on the

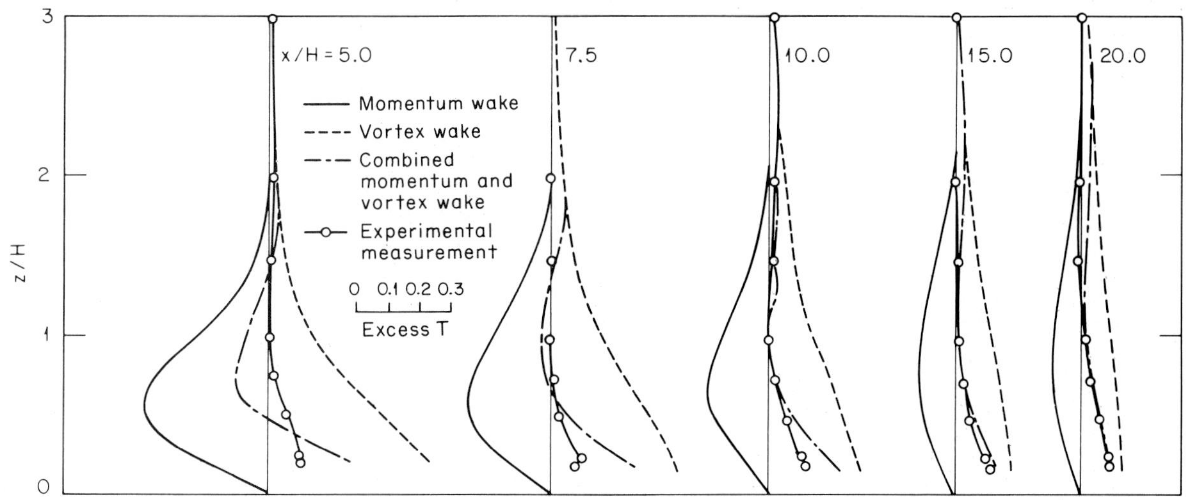

(a) L/H = 2.54, W/H = 0.78.

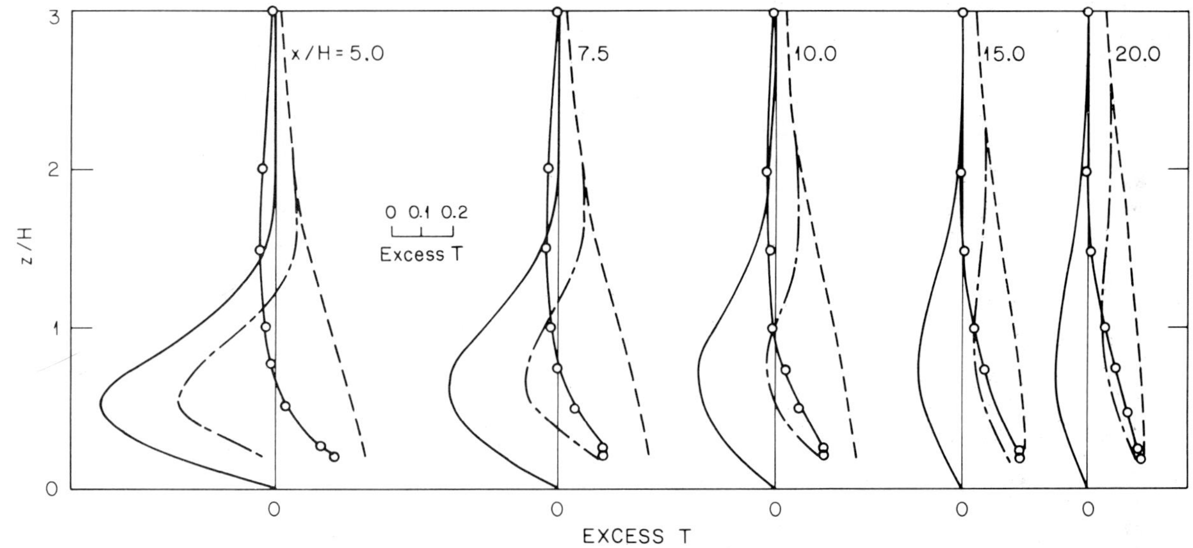

(b) L/H = 0.78, W/H = 2.54.

Fig. 7.31 Comparison of various theoretical profiles with wake centerline data on temperature excess behind a model building immersed in a stable, deep boundary-layer flow. (From K. M. Kothari, J. A. Peterka, and R. N. Meroney, *Stably Stratified Building Wakes*, NRC Report NUREG/CR-1247, Nuclear Regulatory Commission, 1979.)

angle of incidence α. The resulting unsteadiness generates constant fluctuations in the pressure loading on the exposed face. Figure 7.32 also indicates the anticipated velocity profiles at several locations along the building. Within the separated bubble, the flow (and transport of materials such as rain and air pollutants) can be upwind. Robertson (1975) found such a separated region followed by reattachment on two-dimensional prisms subjected to a low turbulence flow for $14° < \alpha < 45°$. Turbulent flows more characteristic of the atmosphere tend to promote reattachment; for example, Robertson et al. (1978) found wall separation and reattachment in a flow of 10% turbulence intensity for $9° < \alpha < 45°$.

Lemberg's (1973) studies suggest that the vertical extent of the flow perturbation introduced by a block-like building is less when the wind is not normal to one side; the somewhat streamlined shape of such a structure when it is approached from an angle makes it easier for the flow to pass around the sides of the building rather than be deflected over its top. Probably for the same reason, the upwind portion (at least) of the horseshoe vortex is not as large in diameter when the wind comes at an angle; the eddy region does not reach as high on the exposed faces as it would otherwise, and there is less downflow.

Many authors (e.g., Ostrowski, Marshall, and Cermak, 1967; Lemberg, 1973; Hansen and Cermak, 1975) have reported a strong counterrotating vortex pair generated near the upwind roof corner of a block-type structure for wind incidence angles near 45°. Cermak (1975, 1976) has pointed out that the vortex pair tends to destroy the roof recirculation region observed for normal flow incidence, while inducing very low pressures close to the upwind roof corner. Lack of a roof cavity is favorable for building ventilation because effluents released near the roof will be carried off; however, the low pressures can cause severe roof damage. Hansen and Cermak (1975) observed roof vortex formation for wind approach angles

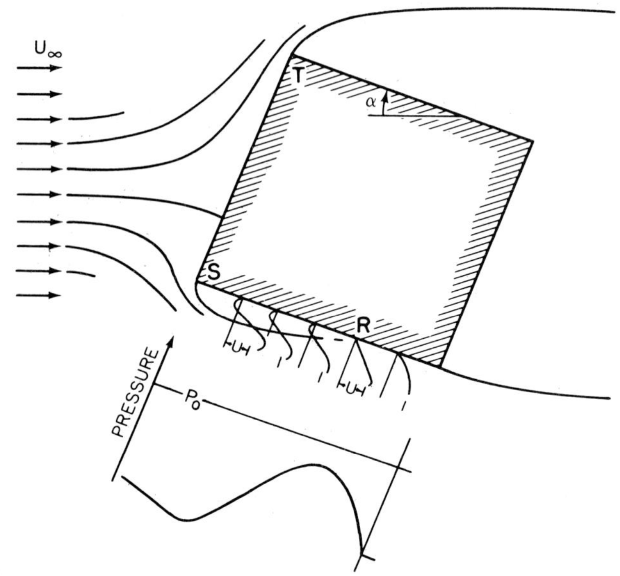

Fig. 7.32 Top view of flow incident at an angle α, showing separation and reattachment on exposed side as well as typical velocity profiles and surface pressure along that face. (Based on J. M. Robertson, Pressure Field at Reattachment of Separated Flows, in *Proceedings of Second U. S. National Conference on Wind Engineering Research,* Colorado State University, June 22–25, 1975, Colorado State University, Fort Collins, Colo., 1975.)

between 37° and 57°. They remarked that the instantaneous strength of these vortices seemed to be considerably greater than the ones they found in the horseshoe vortex near the ground. If the structure, in plan view, is not square, then the vortex generated along the longer side will be the stronger (Woo, Peterka, and Cermak, 1977).

Castro and Robins (1977) carried out a detailed study of the flow near a cube placed at 45° to the wind. Figure 7.33 shows their visualization of the patterns on the roof and ground surfaces. The horseshoe vortex wraps around the building from point A. The pair of roof corner vortices is generated at B. The vertically oriented lee-face spiral vortices, at C, are somewhat larger than observed for normally incident flow. The near wake is broadened and lengthened. The reattachment line behind the cavity is very complex and reaches further downwind under the elevated vortex pair than it does along the wake centerline. The trailing elevated vortices can be expected to transfer any effluent available aloft down into the wake, and, consequently, to produce higher ground-level concentrations (Barrett, Hall, and Simmonds, 1978; Yang and Meroney, 1970).

The influence of the trailing vortices on the far wake is not clear at this time. Castro and Robins (1977) found that the vortices apparently had dissipated within a few building heights downwind; they attributed this rapid disappearance to the very turbulent incident boundary-layer flow. This behavior agrees with Lemberg's (1973) data for buildings at an angle of flow. In fact, Lemberg's data show, if anything, less scatter around

Fig. 7.33 View from above of near-ground flow around a cube exposed to a nearly uniform wind incident at 45°. Lettered features are described in text. [From I. P. Castro and A. G. Robins, The Flow Around a Surface-Mounted Cube in Uniform and Turbulent Streams, *J. Fluid Mech.,* 79(Pt. 2): 307-335, Cambridge University Press, 1977. Photograph courtesy of I. P. Castro and A. G. Robins.]

his theoretical momentum wake profiles than was found for normal flow incidence. Yet Hansen and Cermak (1975) and Woo, Peterka, and Cermak (1977), among others, report vortex wake persistence out to 80H and beyond. Part of the discrepancy may result from the different turbulence characteristics of the various wind-tunnel flows. However, Hansen, Peterka, and Cermak's (1975) comment that, out to about 10H downwind, the wake behind an angled building is similar to the wake behind a building at normal incidence may be significant. Many of the studies cited above carried out measurements to only 10 or 12H or less; it is beyond this distance that the vortex wake seems to be important. Perhaps the vortices may have gone undetected near the body because the swirl velocities are only 3% to 5% of the free-stream wind speed (Hansen and Cermak, 1975; Woo, Peterka, and Cermak, 1977). The difficulty of observing meandering vortices of low apparent strength using fixed sensors has been discussed in Sec. 7-5.3. The question must be regarded as unsettled. However, further experimental work, particularly in the field, will be needed to resolve these uncertainties.

7-6.2 Clustering Effects

Except for very simple examples, such as two-dimensional bodies across a uniform wind, relatively little theoretical work seems to have been done on the general aerodynamic problem of flow near clusters of objects. Goldstein (1965) summarized the far wake similarity solution for a row of parallel rods. The mean velocity defect decays as $(x/\ell)^{-1}$, where ℓ is the distance between rod centers, and in the crosswind direction the defect varies sinusoidally with y/ℓ. The agreement between theory and experiment is felt to be good.

Close behind the rods, however, the flow is not well understood, and experiments must be relied on. Bearman and Wadcock (1973), among others, have studied the flow behind two-dimensional pairs of cylinders (diameter d) and flat strips (width d) placed side-by-side normal to the wind; the space S between the edges of these obstacles was varied systematically. They found that, for $0.1 \lesssim S/d \lesssim 1$, the near wake was both unsteady and asymmetric with one body having a larger cavity than the other at a given time. This phenomenon is bistable in that the wake does not favor either body but switches back and forth. For small spacings, $S/d \lesssim d$, there is also a mean repulsive force between the bodies. Roshko, Steinolfson, and Chattoorgoon (1975) found similar results for triangular-section rods which may be akin to square cylinders placed at 45° to the incident wind (see also Reinhold, Tieleman, and Maher, 1977). Roshko, Steinolfson, and Chattoorgoon (1975) suggest that the jet-like flow penetrating the gap may be the bistable cause of the wake asymmetry as the jet flips from one cavity region toward another. Bearman and Wadcock (1973) speculate that this effect, if it appears in the atmosphere, could lead to large fluctuating loads on adjacent structures solely because of the wake interaction between them. The implications for local effluent transport are unclear.

Zdravkovich and Pridden (1977) have systematically studied various permutations of cylinder pair location that range from side-by-side to tandem. Bistable flows were common; for certain orientations and spacings, discontinuous changes in flow pattern also were common. Further investigation, especially with three-dimensional bodies in boundary-layer winds, seems warranted.

Houghton and Carruthers (1976) discuss some of the structural problems (especially torsional loading) that may arise when one building is influenced by another. Peterka and Cermak (1975b) and Soliman (1976) also provide some useful discussions and experimental pressure coefficients.

A few general studies of building clusters have been reported; however, most of the studies have been conducted at a laboratory scale where the experimental techniques and instrumentation may have influenced the results. For example, in order to provide information on pedestrian-level winds for architectural purposes, Penwarden and Wise (1975) have treated the two-body problem of a high-rise building downwind of a lower structure. Figure 7.34a illustrates the eddies they observed in a typical test configuration; Figure 7.34b shows the ground-level flow patterns induced. If, as in Fig. 7.34c, the low building is moved upwind, the frontal portion of the horseshoe vortex seems to follow it. This not only elongates the ground area in contact with the recirculation zone but also weakens the eddy and destabilizes its position. However, in a similar test, Isyumov and Davenport (1975) observed highly turbulent, intermittent flow between model buildings rather than organized vortices. The incident boundary layer in Isyumov and Davenport's tests probably was a more realistic approximation of the atmospheric boundary layer than that of Penwarden and Wise. This may explain the differing results.

Britter and Hunt (1979) suggest that the quantitative results of both investigations be regarded with caution because the heated-sensor anemometers used in these studies cannot supply accurate data in highly turbulent reversing flows. Specifically, such anemometers inherently combine data on the mean and turbulent components of the flow, but they do not supply information on the direction of the mean flow. Consequently, results—such as those shown in Fig. 7.34d—that attempt to relate the wind speed at pedestrian level to speeds aloft or to speeds that occur when buildings are not present probably are not quantitatively precise; however, qualitatively, they may be quite useful. In any case, the results depend strongly on variables such as the relative building heights, spacing of the buildings, and the building aspect ratios. Reports by Wise (1971) and Penwarden and Wise (1975) can be consulted for additional data, although the caution recommended by Britter and Hunt (1979) should be observed.

Britter and Hunt (1979) conducted a useful study of flow patterns between a low upwind building and a tall downwind structure; theoretical estimates of the peak velocity generated near ground level between the buildings appear possible. The Britter and Hunt work should be expanded to additional cases.

Gandemer (1975, 1976) has presented some interesting sketches of flow patterns expected for other common building arrays. Figure 7.35 reproduces some of his work. His "comfort parameter" Ψ is the ratio of the sum of mean and root-mean-square (rms) turbulent wind speeds at the site of interest to that sum if the buildings are absent. Values of $\Psi > 1$ represent zones of locally higher mean wind speed and/or turbulence; values < 1 are sheltered areas. Gandemer's (1975, 1976) reports should be consulted for details; Britter and Hunt's (1979) caveat about regarding the quantitative values with caution applies here as well.

The ultimate in building clustering, the repetitive urban array, has been modeled by Hoydysh, Griffiths, and Ogawa (1974), among others. Figure 7.36a illustrates the recirculation zone observed in a two-dimensional street canyon perpendicular to the wind. The presence of along-wind gaps (i.e., streets) through the building rows interrupts this vortex and introduces corner vortices that interact with the main cavity vortex (Fig. 7.36b) to form a vertical spiral. The corner vortices

(a) Smoke visualization of eddy structure ahead of a tall building downwind of a low building.

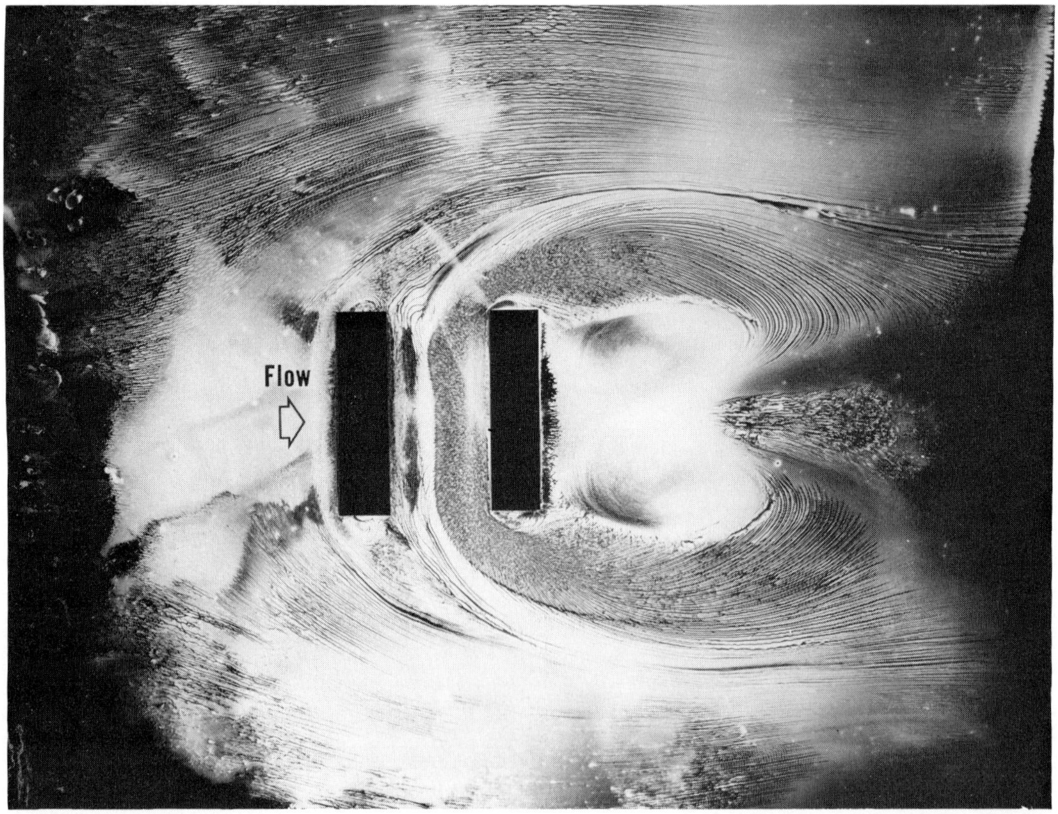

(b) Top view of surface flow patterns associated with (a). Solid line between buildings shows region of maximum wind speed in the frontal vortex; dashed line indicates path of maximum speed in the corner stream.

(Figure continues on next page.)

Figure 7.34 (continued)

(c) Same as (a) and (b), except that the low building is farther upwind. Line (solid) of peak frontal vortex speed also moves upwind, but peak corner speed line (broken) changes little.

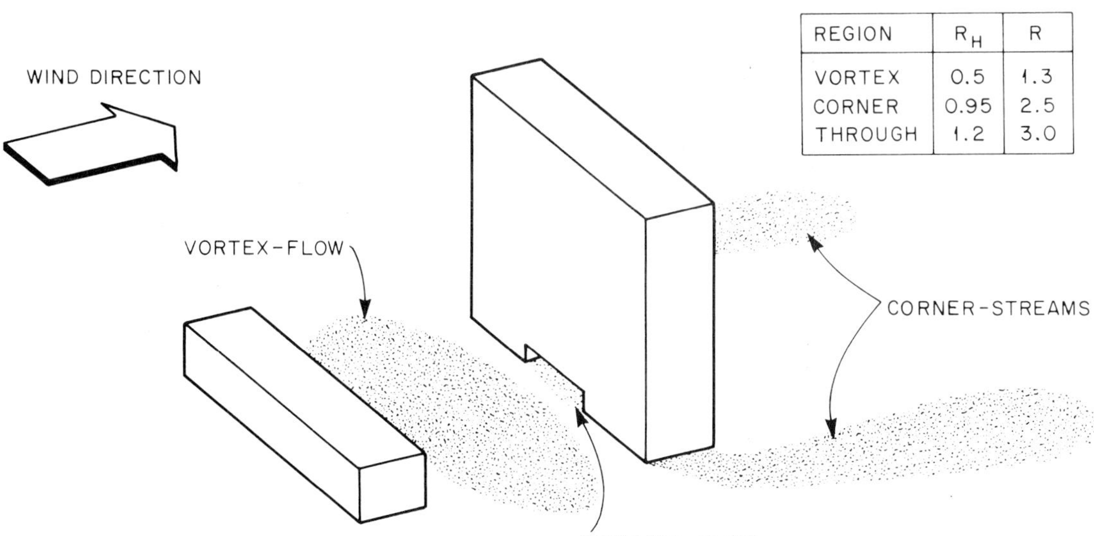

(d) Locations and likely maximum speed ratios R_H (speed at pedestrian height relative to unobstructed wind at roof height H) and R (speed at pedestrian height relative to wind at same height if buildings were absent) for zones of locally high wind near simple building cluster.

Fig. 7.34 Flow near simple building clusters. (From A. D. Penwarden and A. F. E. Wise, *Wind Environment Around Buildings,* Building Research Establishment Report, Her Majesty's Stationery Office, London, 1975. Photograph courtesy of A. D. Penwarden and A. F. E. Wise; Crown copyright.)

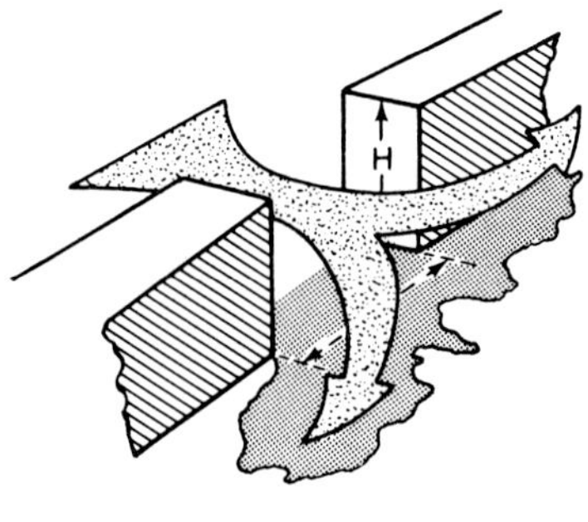

(a) Gap in a long obstruction. When gap width is of order one or two H, the speed ratio Ψ reaches a maximum value near 1.3 for normally incident winds.

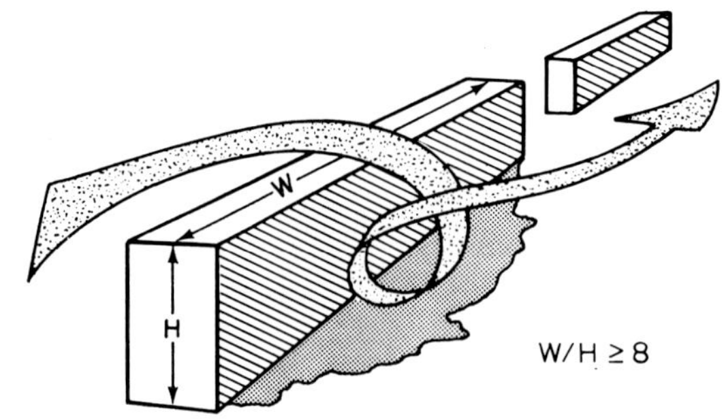

(b) Wind incident at an angle on a long obstruction creates a spiraling zone in the near wake; the speed ratio Ψ has a maximum value of about 1.4.

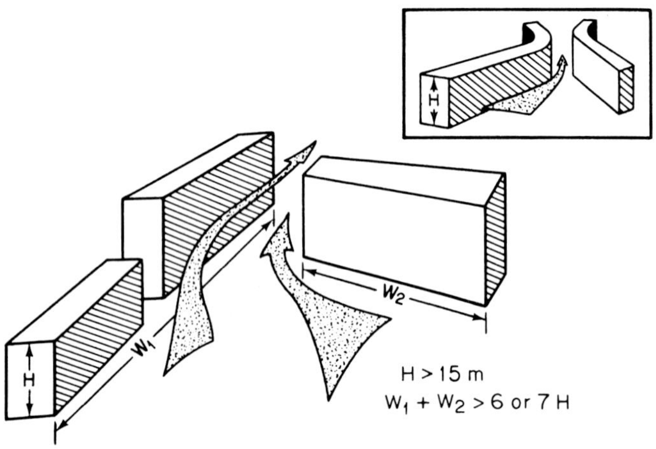

(c) Venturi effect. When gap opening is two or three times the average building height, the flow in the gap is a maximum $\Psi \simeq 1.3$ for $H \simeq 25$ m and $\Psi \simeq 1.6$ for $H \simeq 50$ m. A curved passageway (inset) is an excellent approximation to a true nozzle; for $H \simeq 15$ m, $\Psi \simeq 2.0$ in the "throat."

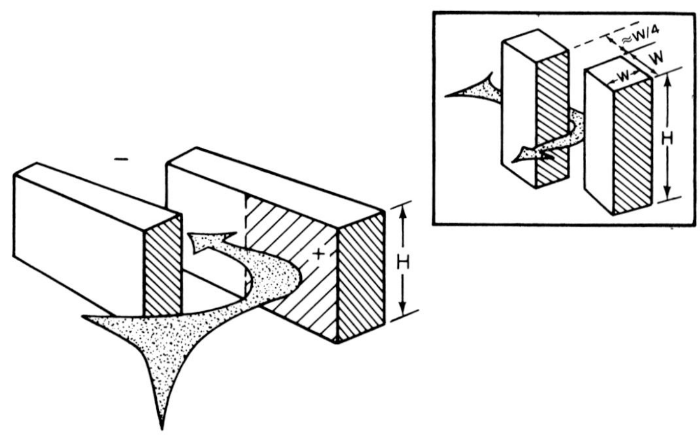

(d) Jetting between zones of high and low pressure. For $H \simeq 15$ m, $\Psi \simeq 1.2$; for $H \simeq 30$ m, $1.3 \leqslant \Psi \leqslant 1.6$. For tall towers (inset), with $H \simeq 90$ m, $\Psi \simeq 1.8$ when the spacing reaches the critical value of about $W/4$.

Fig. 7.35 Typical flow patterns and approximate wind speed ratios near building clusters. The speed ratio, or comfort parameter, Ψ, is given by the sum of the mean and rms turbulent wind speeds at pedestrian height relative to that sum if the buildings were absent. (From J. Gandemer, Inconfort du au vent aux abords des batiments: concepts ae'rodynamiques, Cahier du Centre Scientifique et Technique du Batiment, No. 170, Cahier 1384, Centre Scientifique et Technique du Batiment, Nantes, France, 1976.)

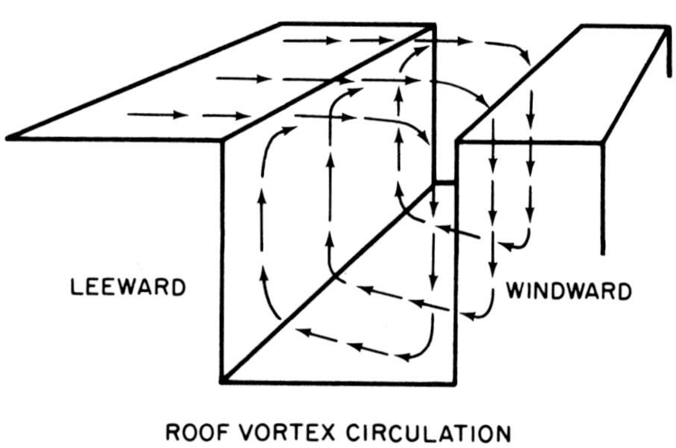

(a) Recirculation in a two-dimensional street canyon.

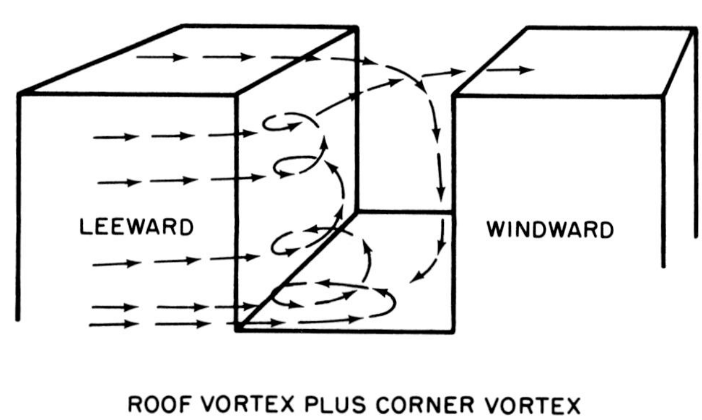

(b) Three-dimensional interaction of corner and cavity vortex near edge of a city block.

Fig. 7.36 Recirculation patterns in urban street canyons for normal wind incidence. (From W. G. Hoydysh, R. A. Griffiths, and Y. Ogawa, A Scale Model Study of the Dispersion of Pollution in Street Canyons, presented at the 67th Annual Meeting of the Air Pollution Control Association, Denver, June 9–13, 1974, Paper No. 74-157, 1974.)

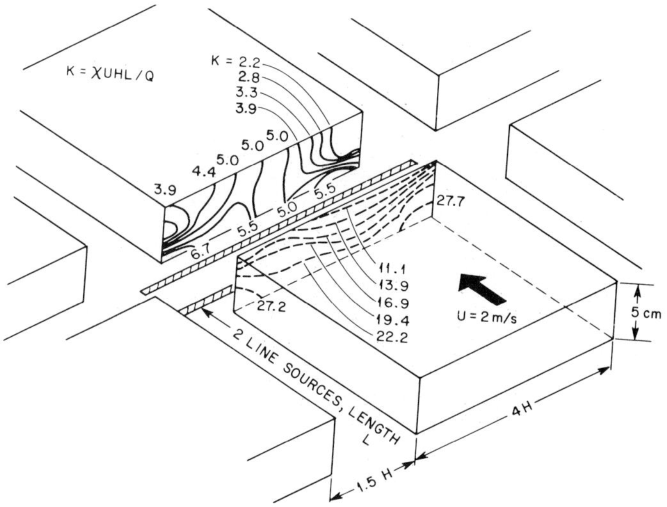

(a) **Wind normal to street canyon.**

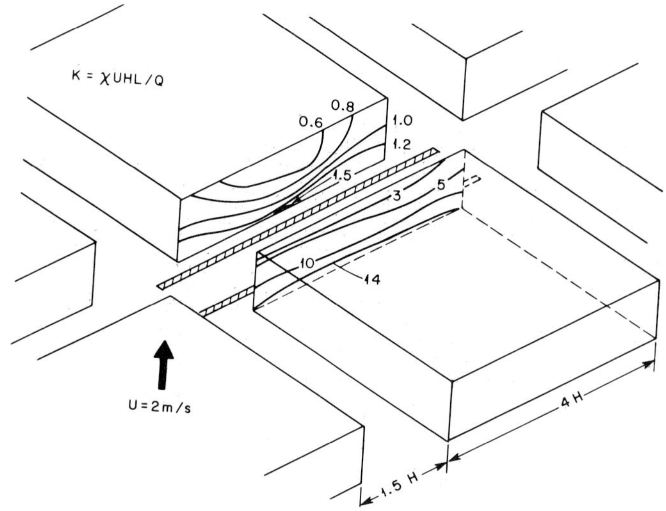

(b) **Wind at 45° angle to street canyon.**

apparently alternate in size so that behind a given block, only one vortex dominates the flow at any time.

These flow patterns are particularly important in pollutant dispersion. Cermak, Lombardi, and Thompson (1974) measured concentration fields for several angles of wind incidence in an urban-like array. Figure 7.37a shows the fairly symmetric isopleth patterns found when the wind is normal to the pollutant source in the street. The maximum concentrations occur near the rear corners of the upwind building because of the influence of the corner vortices; in fact, high concentrations are found over the entire rear face of the upwind structure because of the main cavity recirculation. Figure 7.37b shows the same situation when the wind approaches at a 45° angle; the concentrations are reduced by a factor of two or three, and they seem to vary primarily with height rather than lateral position. The greatest concentrations again occur on the rear face of the upwind building, but the strong corner effect is no longer evident. A single taller structure is introduced in Fig. 7.37c. In this case, concentrations are reduced from those in Fig. 7.37a by a factor of approximately two. Hoydysh, Griffiths, and Ogawa (1974) observed similar behavior and suggested that it might be useful to scatter high-rise buildings through urban regions to improve street-level ventilation. Pollutants apparently are moved upward in the spiral vortices in the wake of the high rises while the cleaner air aloft is diverted downward along the upwind faces of the buildings. The result is generally a reduction of effluent concentrations at heights below the mean roof level. One drawback, as pointed out by Wedding, Lombardi, and Cermak (1977), is that concentrations in the immediate lee of the tall buildings may be high. These authors also noted that placement of an urban array at 45° to the mean wind produced the best overall air quality. They suggested that city planners could take advantage of this by aligning new development at an angle relative to the likely wind direction for atmospheric conditions when urban ventilation and, hence, air quality would otherwise be poor.

Most cluster studies have dealt with very specific sites such as industrial or nuclear reactor complexes (e.g., Meroney, Cermak, and Chaudhry, 1968; Start et al., 1977; Allwine, Meroney, and Peterka, 1980); therefore, they are of somewhat limited use when discussing general principles. The main thrust of such investigations usually has been the mapping of downwind

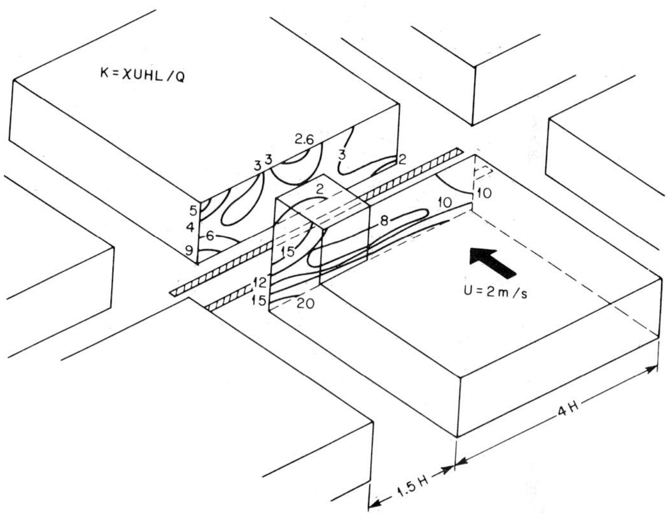

(c) **As in (a), but the upwind building is slightly modified.**

Fig. 7.37 Concentration coefficients K near walls of an urban street canyon. (From J. E. Cermak, D. J. Lombardi, and R. S. Thompson, *Applications of Physical Modeling to the Investigations of Air Pollution Problems in Urban Areas*, presented at the 67th Annual Meeting of the Air Pollution Control Association, Denver, June 9–13, 1974, Paper No. 74-160.)

effluent concentration patterns for different wind directions and source characteristics. Few studies have provided detailed flow visualizations and/or velocity measurements as well. An exception is the work of Hatcher et al. (1978), who modeled surface flow patterns as well as effluent concentrations around an experimental reactor complex. Figure 7.38 shows some of their results with and without the auxiliary buildings. Major effects on the flow patterns were found mainly when the outbuildings were upwind or close alongside the main reactor building. Concentrations were generally lower when the auxiliary buildings were present; however, the actual shape of the patterns was fairly sensitive to the details of the upwind terrain and building configuration.

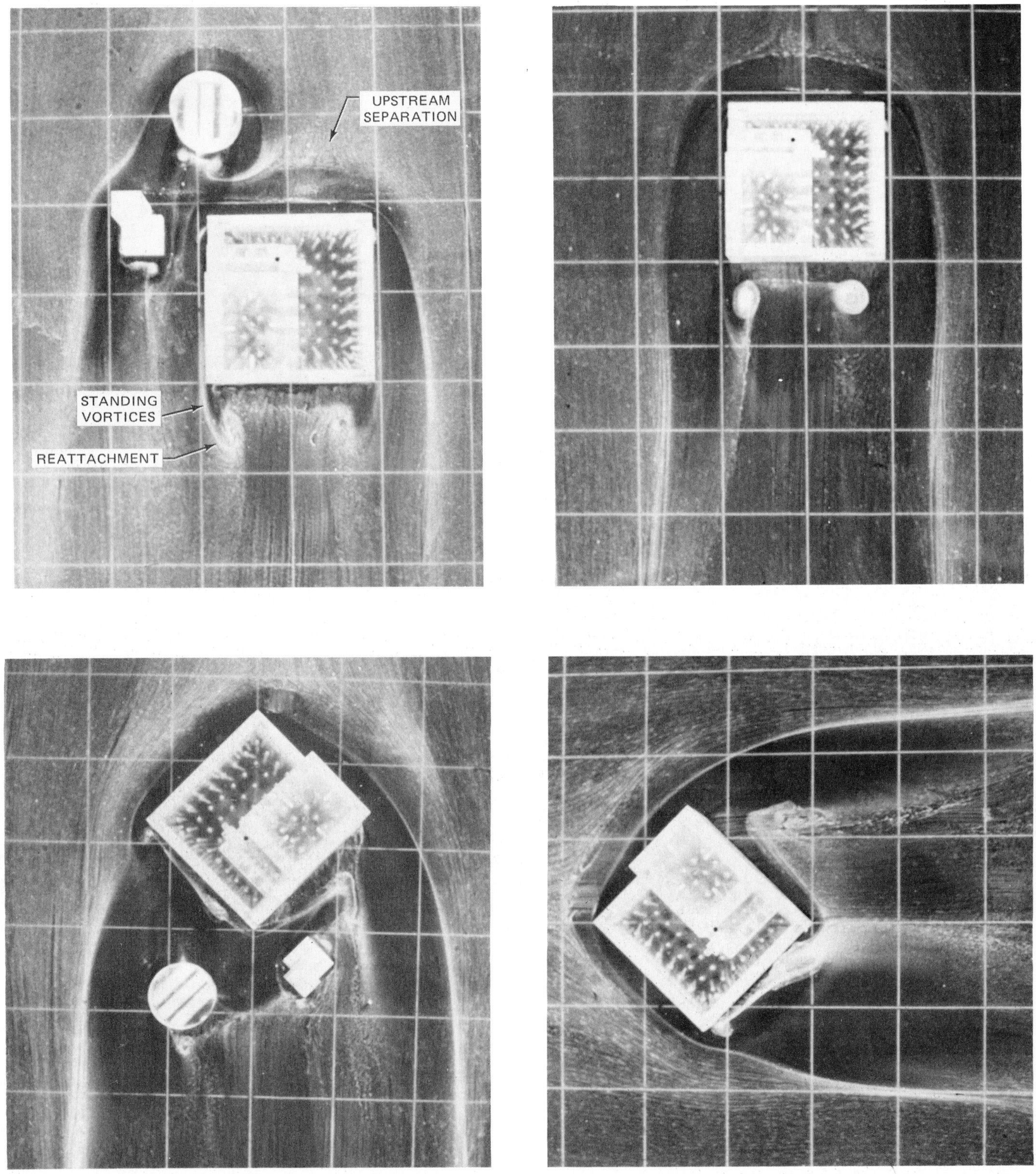

Fig. 7.38 **Surface flow patterns with and without auxiliary buildings, for two angles of wind incidence. (From R. V. Hatcher, R. N. Meroney, J. A. Peterka, and K. Kothari, *Dispersion in the Wake of a Model Industrial Complex,* NRC Report NUREG-0373, Nuclear Regulatory Commission, 1978.)**

Overall, the science of building cluster aerodynamics is still in its infancy. Little theoretical guidance is available. Although a large number of wind-tunnel and water-channel studies have been carried out, more are necessary to close the gaps in our present understanding of the phenomena involved. When the physical insights gained by such work are extrapolated to real building complexes, they must be accompanied by detailed measurements in the field; however, obtaining such measurements usually is difficult and costly. It is unlikely that anything but a qualitative predictive ability for flow near building clusters will be available in the foreseeable future; quantitative estimates will require study on a case-by-case basis.

7-7 EFFECTS OF AMBIENT TURBULENCE AND ATMOSPHERIC STABILITY ON FLOW NEAR OBSTACLES

This section discusses in some detail the effects of incident turbulence on the flow patterns around obstacles. The influence of the body on turbulence also is discussed. Current understanding of these phenomena seems sufficient to provide only rough guidance to anticipated effects in any specific case; detailed predictions must await further study.

7-7.1 Ambient Turbulence Effects

7-7.1.1 Effects of Turbulence on Flow Near an Obstacle. For many years it has been known that the presence of turbulence in a fluid flow can significantly alter the aerodynamic behavior of a body immersed in that flow. Vickery (1966) examined in detail the case of a two-dimensional square cylinder placed across the flow. He found significant differences between smooth and turbulent flows in the lift and drag behavior at various angles of wind incidence. He suggested that such changes resulted from the turbulent shear layer that was shed from the upwind body edges reattaching to the side of the cylinder that was partially exposed to the wind; he also suggested that the incident turbulence made possible or hastened such reattachment. The reattached flow then produced a smaller wake cavity, higher base pressure, and lower drag than would be seen otherwise. For large enough angles of attack, however, the flow must reattach to the exposed side regardless of the turbulence level; therefore, major shifts in the flow structure are no longer possible and turbulence can produce only rather small changes in the base pressure and drag.

Lee (1975) examined much the same problem and reported similar results. He suggested that the observed increase in base pressure (decrease in drag) with increasing turbulence actually results from a more rapid thickening of the separated shear layers. This thickening leads first to intermittent and then to complete reattachment to the body sides; to movement of the shear-layer roll up (vortex formation) region further downwind of the body (see Fig. 7.4 for a somewhat similar effect); and to weakening of the resultant vortices.

Bearman (1971a) carried out related experiments on thin square plates and circular disks, where the flow separation points are fixed by geometry. He reported that this case was unlike that of Vickery (1966) because the drag was higher in turbulent flow than in smooth flow. Bearman (1971a) also pointed out the dependence of the pressure field on the turbulent scale as well as the intensity of the turbulence. The results in a given case can be expected to depend in a complicated way on the shape of the body, its size relative to the turbulent length scale, and the intensity of the turbulence.

A number of studies on other obstacle types have been published (e.g., Barriga, Crowe, and Roberson, 1975; Katsura, 1974; Modi and El-Sherbiny, 1975; Robertson, 1975), although the possibilities for studies on obstacle types are far from being exhausted. Roberson et al. (1972) summarized their rather extensive experience as follows: (i) changes in incident turbulence level produce changes in the Reynolds stresses and flow patterns near any obstacle; (ii) for bodies for which flow reattachment is impossible (such as thin flat plates), drag will increase as the intensity of turbulence increases; (iii) bodies for which flow reattachment is possible can exhibit, depending on body shape, either increases or decreases in drag as the turbulence increases; (iv) bodies that are especially susceptible to turbulence effects tend to have crosswind-width-to-along-wind-length ratios on the order of unity; and (v) turbulence can significantly affect the unsteady as well as the mean aerodynamic forces on bodies. Scruton (1967) and Laneville, Gartshore, and Parkinson (1975), among others, have described the importance of such behavior in structural engineering and architecture.

Blessmann (1971) and Castro and Robins (1977) studied ground-mounted three-dimensional objects in both smooth and turbulent winds. Substantial changes were observed in building surface pressures as well as reattachment of the flow to the roof and sides of cubical building models, the latter being a feature commonly seen when deep simulated boundary layers are used as the incident wind.

The major effects of incident turbulence seem to be related to enhanced thickening of and mixing along the shear layers shed from windward surfaces of sharp-edged bodies, promoting flow reattachment whenever possible, with smaller wake cavities as the usual result. On bodies where reattachment cannot occur, such as thin plates, turbulence-enhanced mixing still takes place along the shear layers bounding the cavity and affects the pressure distribution in the near wake. Incident turbulence improves the mixing of high-momentum fluid from the outer flow through boundary layers down to the body surface. On rounded obstacles, this mixing helps the boundary layer remain attached to the surface against adverse pressure gradients; consequently, separation from the body is delayed. The net effect is akin to that produced by roughening the body surface or increasing the Reynolds number of the flow, procedures widely used in wind-tunnel testing (Goldstein, 1965).

7-7.1.2 Effect of an Obstacle on Incident Turbulence. Much of our current understanding of the influence of a bluff body on the turbulence approaching it results from experimental work of Bearman (1969, 1971a,b, 1972a, b), and from the theoretical development by Hunt (1971, 1972, 1973, 1975). The theoretical development was based on earlier research by Batchelor and Proudman (1954).

Consider a typical turbulent eddy of integral length scale Λ_x incident on an isolated tall building of width W (shown in Fig. 7.39a). The presence of "end effects" (streamline A) and vortices near ground level (streamline C) complicate the problem enormously. As a first approximation, only the central portion of the flow is considered (streamlines B); it is idealized as a uniform turbulent flow incident on a uniformly tall block. Several possibilities exist:

(i) Suppose $\Lambda_x \gg W$. Then, as the eddy approaches, the building sees, in effect, a change in the magnitude and direction of the entire incident wind. As Bearman (1972b) points out, the fluctuating velocity component in the x direction, u, is blocked and, therefore, will behave much like the mean velocity U; it will decrease along the streamline as the eddy nears the build-

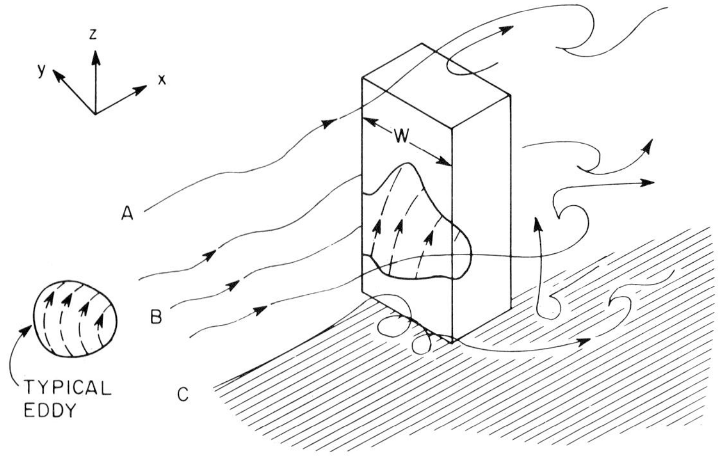

(a) **Streamlines around a tall block and distortion of a typical eddy as deduced from theory and observation. [From J. C. R. Hunt, Turbulent Velocities near and Fluctuating Surface Pressures on Structures in Turbulent Winds, in *Proceedings of the Fourth International Conference on Wind Effects on Buildings and Structures*, Heathrow (London), Sept. 8–12, 1975, pp. 309-320, Cambridge University Press, London, 1975.]**

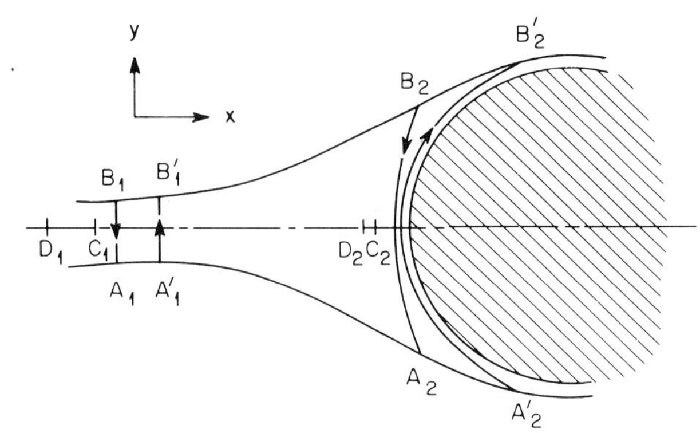

(b) **Flow near a two-dimensional cylinder showing longitudinal compression and transverse stretching of vortex lines, and piling up of vorticity near stagnation point. [From J. C. R. Hunt, A Theory of Turbulent Flow Round Two-Dimensional Bluff Bodies, *J. Fluid Mech.*, 61(Pt. 4): 625-706, Cambridge University Press, 1973.]**

Fig. 7.39 Modification of turbulent eddies incident on a bluff body.

ing. The crosswind component v, on the other hand, must increase in order to change the flow direction, while in this two-dimensional approximation, the vertical component w is unchanged. As $\Lambda_x/W \rightarrow \infty$, the root-mean-square values of the turbulence components near the stagnation streamline are described by (e.g., Bearman, 1972b)

$$\frac{u_{rms}}{U} \simeq \frac{u_{\infty rms}}{U_\infty} \tag{7.8a}$$

$$\frac{v_{rms}}{v_{\infty rms}} \simeq 2 - \frac{U}{U_\infty} \tag{7.8b}$$

$$w_{rms} \simeq w_{\infty rms} \tag{7.8c}$$

Equations 7.8 actually are based on flow near the stagnation streamline of a cylindrical obstacle. This provides an easily calculable method of quantifying the sideward deflection of flow produced by the finite body width without having to deal with the problem of flow separation near the edges of the original block body.

(ii) Now suppose $\Lambda_x \ll W$. An argument by Hunt (1973) demonstrates that the dominant feature may be the velocity perturbations resulting from the distortion of vorticity produced as the eddy stretches to reach around the obstacle. The argument goes as follows: the incident eddy travels toward the body with initial vector velocity $\mathbf{u}_\infty$ and vorticity $\boldsymbol{\omega}_\infty$. The obstacle's presence induces changes in the mean flow. These changes distort the vorticity and generate a perturbed velocity $\mathbf{u}_d$ in addition to $\mathbf{u}_\infty$. If $\Lambda_x \gtrsim W$, the change in $\boldsymbol{\omega}_\infty$ occurs over a distance of order W; therefore, we expect $|\mathbf{u}_d| \sim |W\boldsymbol{\omega}_\infty|$. On the other hand, if the eddy is small compared to the body, $\Lambda_x \ll W$, the vorticity is changed over a distance of order Λ_x; therefore, $|\mathbf{u}_d| \sim |\Lambda_x\boldsymbol{\omega}_\infty|$. Since $|\mathbf{u}_\infty| \sim |\Lambda_x\boldsymbol{\omega}_\infty|$, we have $|\mathbf{u}_d|/|\mathbf{u}_\infty| \sim W/\Lambda_x$ for $\Lambda_x/W \gtrsim 1$; for $\Lambda_x/W \ll 1$, we have $|\mathbf{u}_d|/|\mathbf{u}_\infty| \sim 1$. In other words, vorticity distortion affects only slightly the velocity of large eddies but can alter significantly the velocity of small eddies. The so-called "rapid distortion theory" (e.g., Batchelor and Proudman, 1954; Hunt, 1973) can be used to account for vorticity distortion. The theory assumes that if turbulence is disturbed rapidly enough by changes in the mean velocity field, then each eddy is distorted before it can exchange energy with any other eddy. This assumption places limitations (e.g., small turbulent intensity and relatively large scale) on the usefulness of the theory, but it permits linearization and solution of an otherwise nearly intractable problem. Physical implications of the theory can be understood by using Fig. 7.39b (Hunt, 1973). The vorticity $\boldsymbol{\omega} = \nabla \times \mathbf{u}$ associated with an eddy traveling along the stagnation streamline has components $(\omega_1,\omega_2,\omega_3)$; as the eddy approaches the body, vortex lines in the streamwise direction are attenuated, while those in the crosswind direction are stretched. Consequently, u and w will increase, but v will decrease. Vortex lines "pile up" immediately adjacent to the impenetrable body surface; u must go to zero, but v and w will increase. For $U/U_\infty \lesssim 0.5$, the theory predicts that, as $\Lambda_x/W \rightarrow 0$ along the stagnation streamline (Bearman, 1972b; Hunt, 1973),

$$\frac{u_{rms}}{u_{\infty rms}} \simeq \left[\frac{3}{4}\left(\frac{U_\infty}{U}\right) + \frac{3}{8}\left(\frac{U}{U_\infty}\right)\ln\left(\frac{4U_\infty}{U}\right) - \frac{1}{2}\right]^{1/2} \tag{7.9a}$$

$$\frac{v_{rms}}{v_{\infty rms}} \simeq \left[\frac{3}{4}\left(\frac{U_\infty}{U}\right)\ln\left(\frac{4U_\infty}{U}\right) - 1\right]^{1/2} \tag{7.9b}$$

$$\frac{w_{rms}}{w_{\infty rms}} \simeq \left[\frac{3}{4}\left(\frac{U_\infty}{U}\right) - \frac{3}{8}\left(\frac{U}{U_\infty}\right)\ln\left(\frac{4U_\infty}{U}\right) - \frac{3}{2}\right]^{1/2} \tag{7.9c}$$

Experimental data have been reported by Bearman (1972b) and Britter, Hunt, and Mumford (1979). Trends predicted by the theory seem to be fairly well verified, and the limiting cases of Eqs. 7.8 and 7.9 serve to bound the results.

The above work applies along the streamlines approaching the front stagnation area of the obstacle. Behavior along the sides will be different because the vortex lines will be stretched longitudinally but will be restricted in the crosswind direction; a reduction in u and an increase in v is anticipated (Hunt, 1975), at least away from the surface.

The theory provides some insight into the mechanisms of turbulence modification near buildings; in particular, it sheds light on the role of turbulence scale. In a practical sense, the theory can be applied only to rather tall structures at altitudes where the shear-induced frontal vortex is unimportant and where the end effects resulting from flow over the roof have not yet become significant. Additional research is needed to examine levels where both are important. Information provided by work such as the above should prove useful in understanding problems such as fluctuating loads on cladding and modifications to effluent diffusion rates induced by the presence of a building.

7-7.1.3 Effects of Ambient Turbulence on Diffusion. As described above, the presence of turbulence in the flow approaching a body can cause fundamental changes in the flow patterns near and behind the obstacles. Changes such as modifications to wake cavity size and shape, weakening of the vortex formation process, and enhancement of mixing and entrainment in the separated shear layers that bound the cavity have important consequences for effluent dispersion. The overall effects must depend on the details of the turbulence (e.g., its intensity and scale) as well as on the geometric characteristics of the bluff body itself. Relatively little work has been reported concerning the influence of changes in ambient turbulence on dispersion near bodies of architectural interest.

Humphries and Vincent (1976) studied the effects of variations of turbulence intensity and scale on the wake properties and characteristic retention time t_d of a pollutant released into the wake cavity of a thin disk suspended in a uniform flow. On the basis of dimensional analysis, they postulated that the size and shape of the cavity, the pressure coefficient on the rear of the disk, and the retention time all ought to depend on the dimensionless turbulence parameter $\Upsilon \equiv (u_{rms}/U_\infty)(\Lambda_x/D)$, where Λ_x is the integral length scale and D is the disk diameter. Their measurements of nondimensional retention time $\tau \equiv U_\infty t_d/D$, nondimensional cavity length x_r/D, cavity shape factor $f_s \equiv V_c/A_cD$ (V_c is the cavity volume; A_c is cavity surface area), and rear face pressure coefficient C_{Pb} are shown in Fig. 7.40; the measurements appear to confirm this hypothesis.

It is interesting to note that Bearman's (1971a) measurements seem to depend on a different form of turbulence parameter, $(u_{rms}/U_\infty)(\Lambda_x/D)^2$. The reason for this discrepancy apparently lies in the dimensional analyses of the different authors.

Humphries and Vincent (1976) also suggest that $\tau \propto (\Upsilon C_D)^{-1}$, where C_D is the overall drag coefficient for the disk. Because C_D is itself proportional to Υ, they find that $\tau \propto C_D^{-2}$, where C_D is measured under the turbulence conditions of interest. This estimate is verified by their data. Because C_D often can be measured with less difficulty than τ, their parameterization provides a convenient means for estimating pollutant retention behind a disk (see Sec. 7-8.2.1.3).

Further experiments are needed to establish whether such behavior is typical of other obstacle shapes and to determine the exponent of the dependence on Λ_x/D. However, such simple behavior cannot be expected for ground-mounted obstacles in turbulent shear flows because properties, such as τ and C_D, will depend on other variables such as the boundary-layer-thickness-to-building-height ratio.

Vincent (1977) considered ground-mounted cubes in near-uniform smooth and turbulent flows; he measured characteristic retention time t_d and cavity length x_r. In smooth flow he observed $\tau \equiv U_\infty t_d/H \simeq 6.8$. This compares to a value of between 7 and 8 for a square plate and to 7.4 for a circular disk; both the square plate and the circular disk were suspended in boundary-free uniform smooth flow.

In turbulent flow, however, Vincent observed a very complicated behavior (Fig. 7.41) of τ and x_r/H with the turbulence parameter Υ. He explained this behavior as follows: at low levels of free-stream turbulence, τ decreases sharply from its value in a perfectly smooth flow because of the enhanced turbulent mixing in the shear layers bounding the wake cavity. This increased entrainment reduces the pressure immediately behind the body and causes the cavity length to shrink. As Υ continues to increase, the flow, initially separated from the roof and sides, produces a smaller main cavity behind the body and areas of recirculation on the sides and roof near the upwind building edges. Larger Υ mixes effluent upwind from the wake cavity to the roof cavity and causes τ to rise, whereas the shear along the cavity boundary is less than when the flow does not reattach; consequently, x_r/H also increases. As Υ increases further, the steadily enhanced mixing in the shear layers causes τ to decline once again while continuing to reduce the wake pressure causing x_r/H to become nearly constant. Vincent (1977) remarks that although these explanations are speculative, it is important to note that the region of maximum variation in both τ and x_r/H corresponds to the level of turbulence where flow reattachment occurs and the flow patterns are greatly altered.

Vincent (1978) reported similar experiments with cubes using turbulent boundary layers as the incident flows. In this work the turbulence characteristics such as u_{rms}, Λ_x, and boundary-layer depth δ were fixed by the wind-tunnel technique for a particular test sequence and the model height H was varied. Unfortunately, in these experiments, it is impossible to separate the effects of boundary-layer-depth-to-height ratio from those of the turbulence alone. Thus, as δ/H increased in a given test flow, so did Λ_x/H, while u_{rms}/U_∞ remained constant. Vincent suggests that, at least for the range $\delta/H \geqq 8$, the resulting increase in Υ probably is responsible for most of the observed changes in nondimensional retention time τ and cavity length x_r/H. He argues that these changes are caused by the increasingly effective mixing along the shear layers bounding the cavity; the mixing removes material from the wake and produces smaller values of τ and lower wake pressures; this, in turn, generates shorter cavities.

Additional research repeating Vincent's (1978) experiments and using a contoured screen to generate a low-turbulence shear flow would be interesting because the effect of shear-induced phenomena (such as horseshoe vortex production) on the cavity characteristics could be separately evaluated. Experiments using constant-depth boundary-layer flows of variable turbulence characteristics and work on other simple building shapes also are needed.

7-7.2 Effects of Atmospheric Stability

The effects of atmospheric stability on flow patterns and effluent dispersion have been recognized for decades. Unstable conditions superpose convective turbulence on whatever

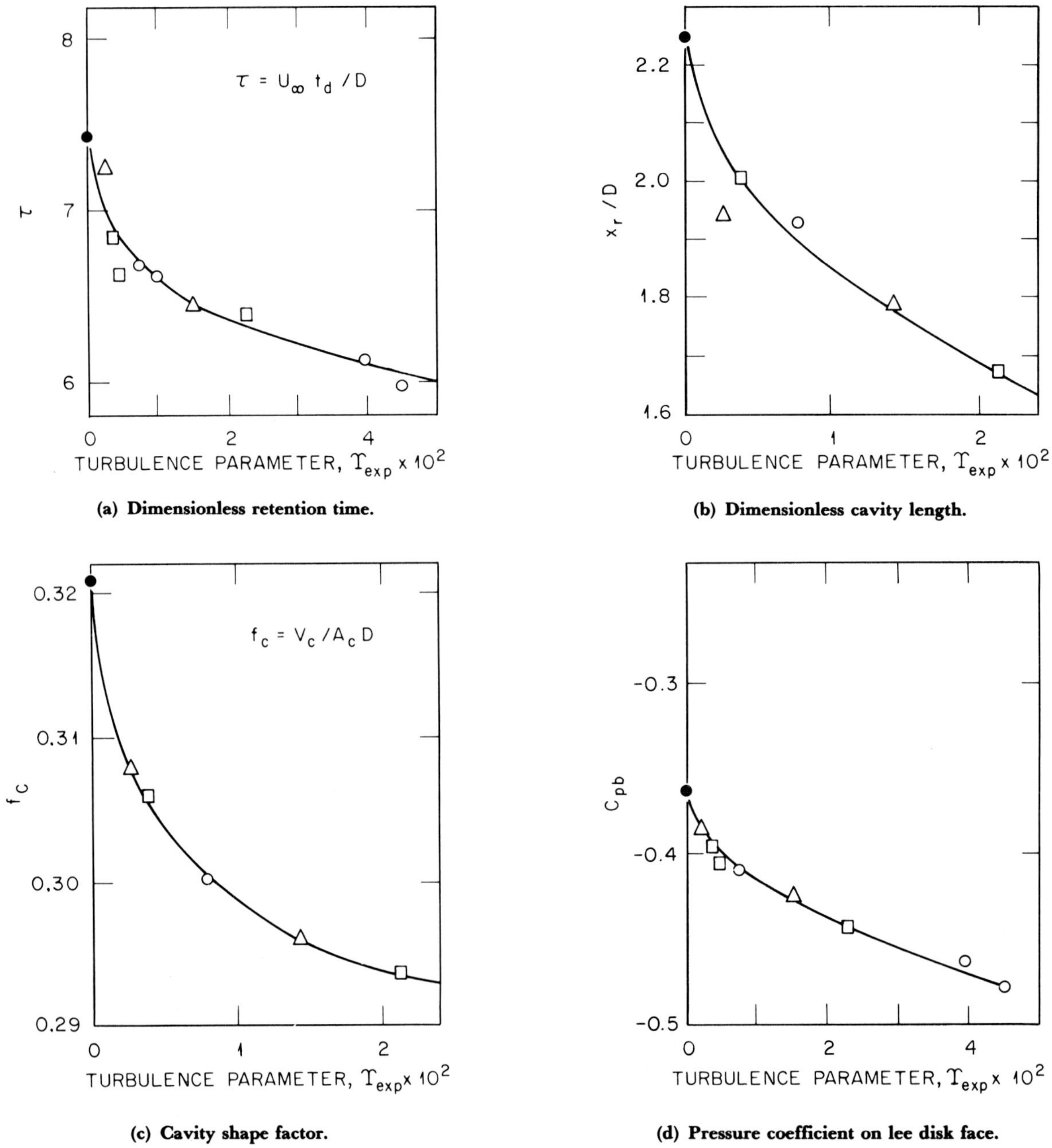

(a) Dimensionless retention time.

(b) Dimensionless cavity length.

(c) Cavity shape factor.

(d) Pressure coefficient on lee disk face.

Fig. 7.40 Dependence of wake-cavity properties on turbulence parameter $\Upsilon \equiv (u_{rms}/U_\infty)(\Lambda_x/D)$ for a circular disk. [From W. Humphries and J. H. Vincent, Experiments to Investigate Transport Processes in the Near Wakes of Disks in Turbulent Air Flow, *J. Fluid Mech.*, 75(Pt. 4): 737-749, Cambridge University Press, 1976.]

mechanically-induced turbulence already exists. Consequently, turbulent wakes can be expected to spread more rapidly, to decay to background flow conditions more rapidly, and to disperse effluent more quickly than in the case of neutral flow. On the other hand, stable stratification tends to dampen turbulent mixing, inhibiting wake growth and pollutant dispersion. Attention in this section will be confined to flow and dispersion near obstacles in stable atmospheres because undesirably high effluent concentrations are more likely in such a case.

7-7.2.1 Highly Stratified Flow Near Obstacles. Over the past thirty years a large body of literature dealing with stable flows over two- and three-dimensional obstacles has been published. The review by Long (1972) and the text by Turner (1973) provide useful introductions to this work.

A more recent article by Lin and Pao (1979) summarizes current information on wake phenomena in stratified flows. A large number of interesting effects are possible, depending on the degree of stratification in the flow (as measured by a Froude or Richardson number) and the size and shape of the obstacle. Specific effects that can develop include an upstream wake "blocking," lee waves, rotors, and abrupt collapse of the downwind vertical wake thickness some distance downwind of the body.

Much of this work concerns uniform stratified flows. Often the obstacle that is studied is suspended in the fluid away from

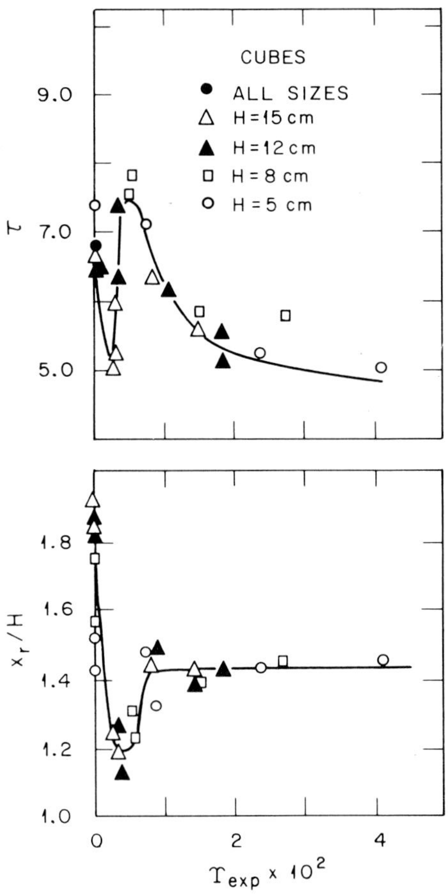

Fig. 7.41 Nondimensional effluent retention time $\tau \equiv U_\infty t_d/H$ and cavity length x_r/H as functions of the turbulence parameter $T \equiv (u_{rms}/U_\infty)(\Lambda_x/H)$ for cubes on a ground plane in near-uniform turbulent flow. [From J. H. Vincent, Model Experiments on the Nature of Air Pollution Transport near Buildings, *Atmos. Environ.*, 11(8): 765-774 (1977).]

boundaries. Stratified shear-flow studies frequently have dealt with obstacles such as two-dimensional mountains. Only a few studies have been published on three-dimensional stratified shear flow near simple obstructions such as those commonly encountered in meteorology and air-pollution problems.

The stability parameter most often used to classify theoretical and laboratory work on stratified wakes is the internal Froude number†

$$F_D \equiv \frac{U_\infty}{ND} \tag{7.10a}$$

where D is a characteristic body dimension (frequently the height H) and N is the Brunt-Väisälä frequency† (in rad/sec)

$$N \equiv \left(-\frac{g}{\rho}\frac{d\rho}{dz}\right)^{1/2} \tag{7.10b}$$

N is the angular frequency at which a fluid particle in a stable atmosphere would oscillate about its equilibrium location if initially displaced by an infinitesimal disturbance. When $F_D \gg 1$, buoyancy forces are negligible and the fluid behaves approximately as a neutral atmosphere. As F_D approaches unity, buoyancy effects become increasingly significant; for $F_D \rightarrow 0$, the stabilizing effect of the density gradient overwhelms the kinetic energy of the flow. This essentially limits the motion of the flow to horizontal planes and suppresses any turbulence or turbulent transport between these planes.

Lin and Pao (1979) summarized the characteristics of the flow around two-dimensional obstacles in strongly stratified fluids. A particularly intriguing feature is an upstream wake that appears when the relative motion of body and fluid is small. This effect, often referred to as "blocking" of the flow, is explained by the inability of a weakly energetic fluid particle to overcome the stabilizing force induced by a strong density gradient and rise over the obstacle. Instead, the particle is trapped ahead of the body and an upstream wake is formed. Details of the flow pattern formed seem to depend on the Reynolds number as well as on the Froude number. Downstream of the obstacle, lee waves are formed; these can lead to the recirculating eddies or "rotors" that frequently are observed behind mountain ranges (Scorer, 1958). At somewhat higher speeds, a turbulent wake may be generated behind the two-dimensional obstruction. This wake has been observed to initially spread in the vertical, but then it either approaches a constant wake depth, or actually collapses to a smaller asymptotic value. In unbounded fluids, at least, the entire wake has also been observed to oscillate vertically with a frequency $\simeq N/2\pi$ (Pao, 1973).

Laboratory studies of very low Froude-number flows past three-dimensional ground-mounted obstacles of some architectural and meteorological interest have been conducted by Liu and Lin (1976) and Riley, Liu, and Geller (1976). The latter authors compared some of their data to a numerical model of the flow patterns (shown in Fig. 7.42). As a fluid element approaches a hill, it slows down and begins to rise slightly; as it begins to pass around the obstacle it accelerates and begins to descend. Somewhere near the point of maximum obstacle diameter, the flow separates and the fluid elements tend to return to their equilibrium levels. The numerical work of Riley, Liu, and Geller (1976) did not allow for the effects of surface boundary layers or for the significant unsteadiness induced by vortex shedding in the separated wake; the authors attributed the discrepancies between their data and the numerical solutions to these shortcomings.

Brighton (1977, 1978) used both laminar-flow laboratory observations and a generalization of Drazin's (1961) theory in his study of very stable shear flows past obstructions similar to those examined by Riley, Liu, and Geller. In the limit, as $F_H \equiv U_\infty/NH \rightarrow 0$, where H is the obstacle height, the flow reduces to irrotational purely horizontal motion at each height above the ground; the solutions at each level can be calculated from ordinary potential flow theory. However, when $F_H \neq 0$, the flow is not entirely confined to planes of constant height; near the obstacle, small displacements of the constant density surfaces are generated by the combined effects of velocity shear, obstacle taper, and Coriolis forces.

Figure 7.43 illustrates the main phenomena observed in these highly stable flows. Upstream of the body (region A), the flow is nearly horizontal and, except near the top, is fairly well approximated by the potential flow solution appropriate to each level. Region B consists of a layer of thickness of order U_∞/N near the top of the obstacle; just upwind of the obstacle, the flow rises to pass over the body, whereas there is a strong

†The definitions of F_D and N vary slightly from author to author; caution is advised.

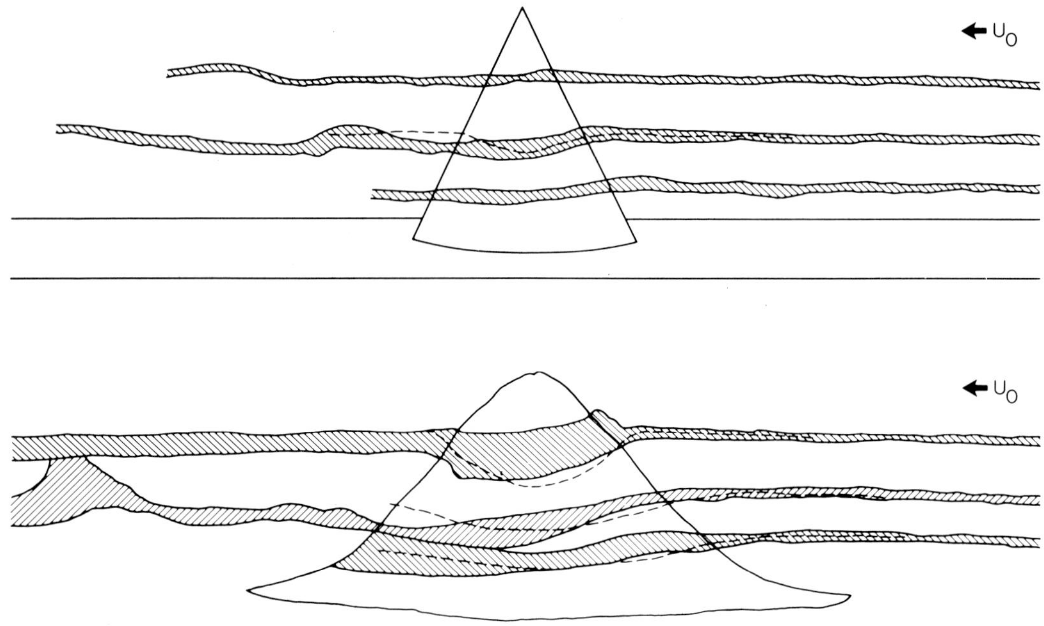

Fig. 7.42 Experimental (shaded) and calculated (dashed lines) flow patterns for stable flows near obstacles. (From J. J. Riley, H.-T. Liu, and E. W. Geller, *A Numerical and Experimental Study of Stably Stratified Flow Around Complex Terrain,* Report EPA-600/4-76-021, U. S. Environmental Protection Agency, 1976.)

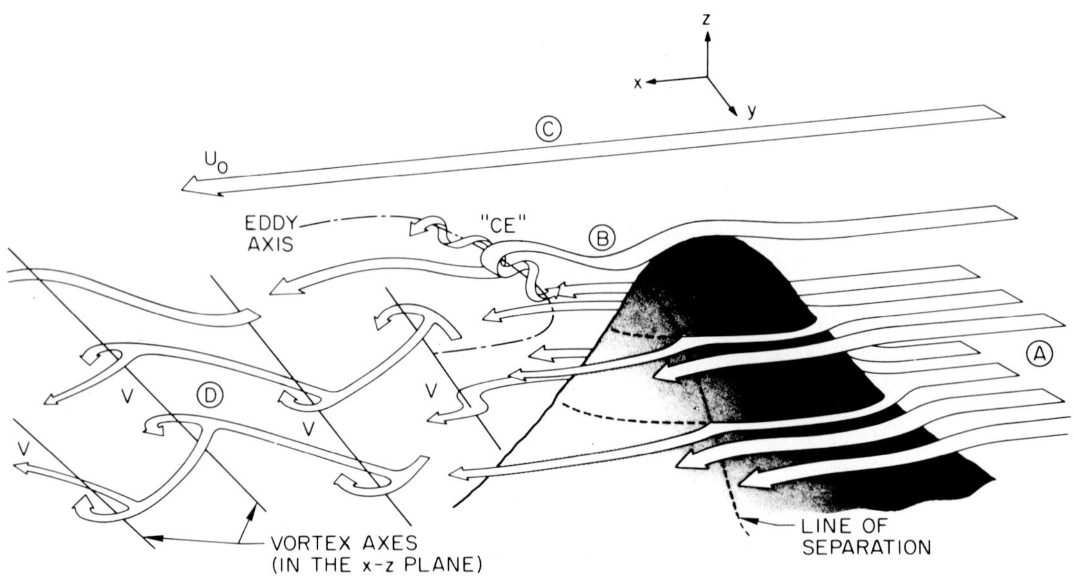

Fig. 7.43 Major flow patterns near a typical bluff obstacle under conditions of very strong stratification. [From P. W. M. Brighton, Strongly Stratified Flow Past Three-Dimensional Obstacles, *Q. J. R. Meteorol. Soc.,* 104(440): 289-307 (1978).]

downflow on the lee slope. Small-scale lee waves are generated a bit farther downwind. If the Froude number is high enough, an elevated vortex may appear under the crest of the first lee wave; the ends of this eddy trail off downwind on either side of the obstacle. Brighton (1977) called this phenomenon a "cowhorn eddy" because of its shape when viewed from above. In region C, well above the top of the body, there is essentially no disturbance to the incident flow. Region D is the separated wake zone; although vortex shedding does occur, the very strong stratification limits motion primarily to horizontal planes. Vortex shedding at one height apparently influences vortex shedding at other heights so that the vortex axes tend to be inclined from the vertical (see Fig. 7.5 for a somewhat similar result in a neutral shear flow). Brighton's (1977, 1978) publications should be consulted for details of the flow near specific obstacles (cylinder, cone, hemisphere).

Hunt, Snyder, and Lawson (1978) and Hunt and Snyder (1980) extended Brighton's (1977, 1978) work to somewhat larger Froude numbers. Figures 7.44 and 7.45 show the turbulent flow patterns observed near a bell-shaped three-

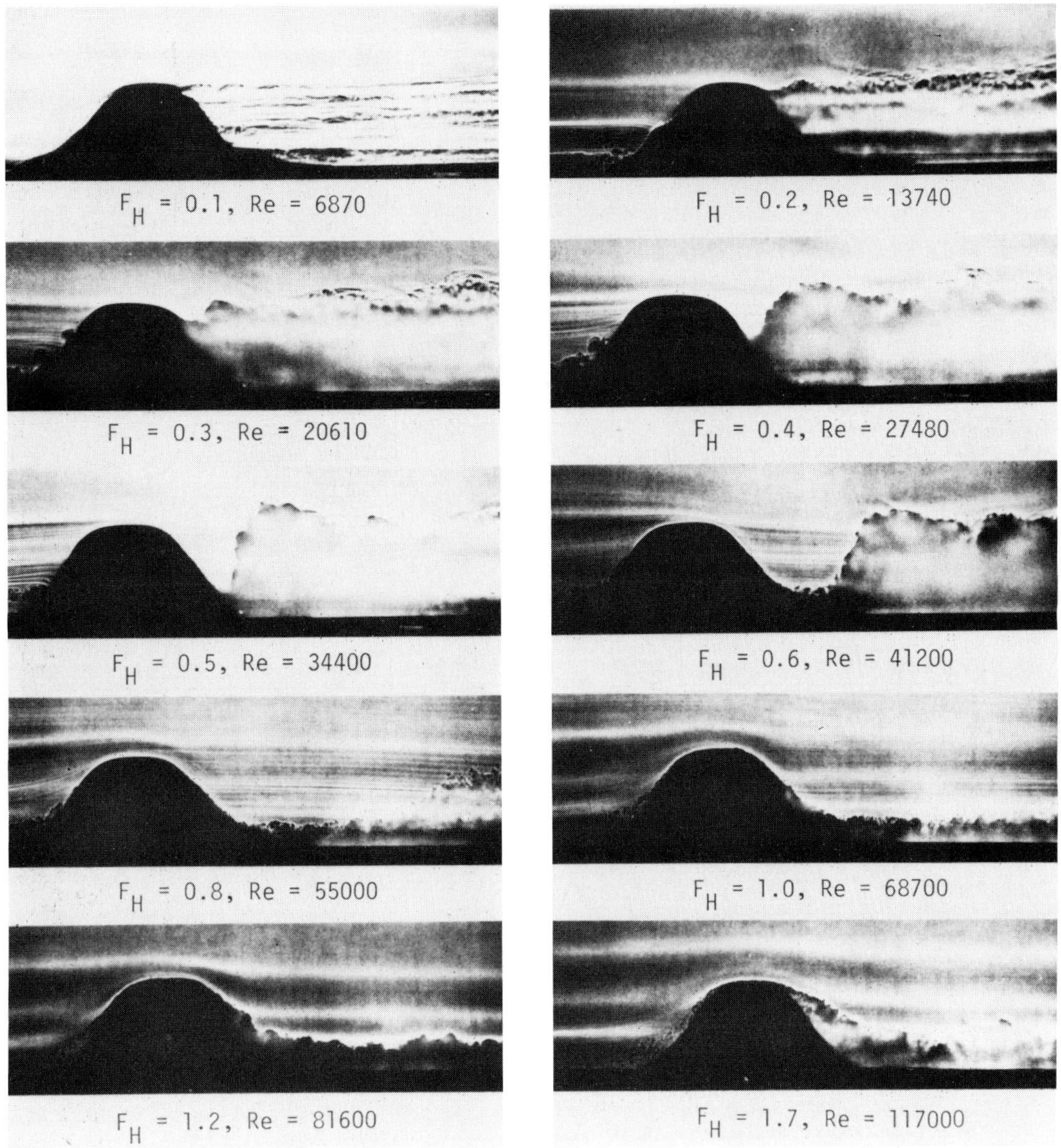

Fig. 7.44 Shadowgraphs of stratified flow near a three-dimensional hill of height H, where $F_D \equiv U_\infty/NH$. (From J. C. R. Hunt, W. H. Snyder, and R. E. Lawson, Jr., Flow Structure and Turbulent Diffusion Around a Three-Dimensional Hill, in *Fluid Modeling Study on Effects of Stratification, Part I. Flow Structure,* Report EPA-600/4-78-041, U. S. Environmental Protection Agency, 1978.)

dimensional hill. At the lowest Froude numbers, $F_H \leqq 0.2$, the flow behaves much as described by Riley, Liu, and Geller (1976); that is to say, it moves mostly in horizontal planes, rises slightly as it nears the hill, dips a bit as it passes around the hill, and, finally, rises once more as the streamlines separate from the hillside. Only a rather thin layer near the summit has sufficient kinetic energy to overcome the gravitational potential energy barrier and to pass over the hilltop. On the upwind part of the hill, a weak downslope flow on the centerline produces a weak and, perhaps, intermittent horseshoe vortex near the foot of the hill. Just downwind of the summit is a weak hydraulic jump, separation from the hillside perhaps triggered by lee wave effects, and a rather oscillatory wake. The latter feature induces oscillations (meander) in the upwind flow as well.

For a somewhat larger Froude number, $F_H \simeq 0.4$ for example, the layer of fluid that is near the hilltop and has enough energy to pass above the hill is thicker. As in the case with the lower Froude number, fluid parcels below this level must pass around the hill, again rising as they near the obstacle, but dipping more significantly than with lower Froude numbers as they move past the hill, and continuing to lose elevation as they move off downwind. The horseshoe vortex on the windward side remains intermittent, but it is much stronger than with smaller Froude numbers. On the lee side of the hill, separation is delayed to a point approximately halfway down the slope; a hydraulic jump and lee waves occur somewhat further downstream at hilltop height.

By the time $F_H \simeq 0.6$, lee-slope separation does not occur until near the base of the hill and the hydraulic jump is further yet downwind. When $F_H \simeq 0.9$, all of the approaching flow in the center plane is energetic enough to go over the top of the hill; the flow then passes down the entire lee slope without separation. As the Froude number increases to values larger than 1.0, occurrence of the boundary-layer separation commonly

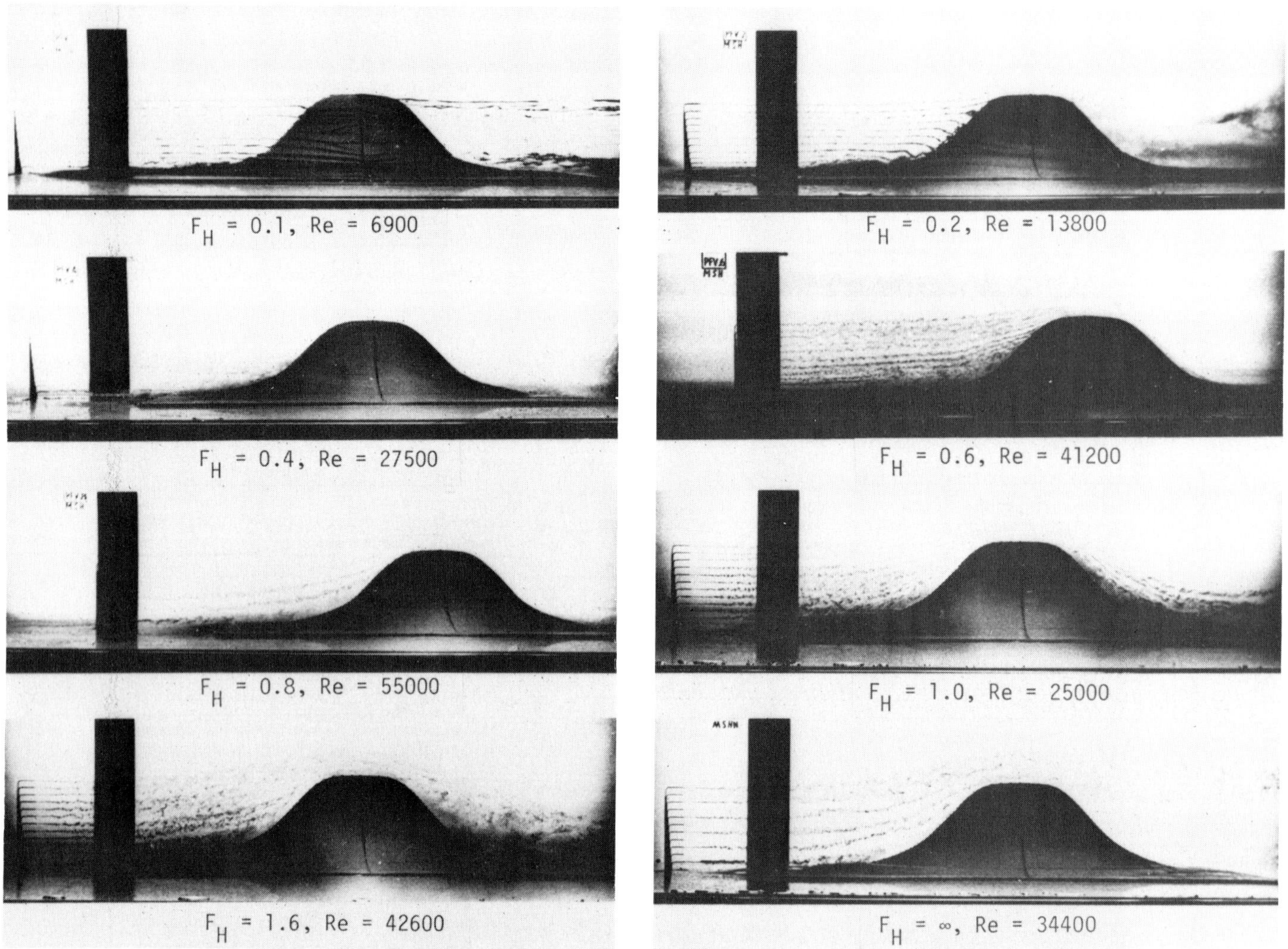

Fig. 7.45 Multi-tracer-stream visualizations of stratified flow near a three-dimensional hill of height H, $F_H \equiv U_\infty/NH$. (From J. C. R. Hunt, W. H. Snyder, and R. E. Lawson, Jr., Flow Structure and Turbulent Diffusion Around a Three-Dimensional Hill, in *Fluid Modeling Study on Effects of Stratification, Part I. Flow Structure,* Report EPA-600/4-78-041, U. S. Environmental Protection Agency, 1978.)

observed in near-neutral flows is possible near the hilltop; increases in F_H permit the resulting turbulent wake to grow both laterally and vertically. By the time $F_H \simeq 1.7$, this boundary-layer separation takes place just downwind of the summit; it produces a large recirculating region covering much of the lee slope; consequently, surface flow along the lee slope centerline is directed upslope. For this Froude number of about 1.7, the flow is similar to that of the neutrally stable case ($F_H \rightarrow \infty$) except that the streamlines are packed closer to the hilltop and they lose elevation more quickly downwind of the hill. In fact, stratification remains significant far downwind of the obstacle for Froude numbers as large as 10.

Observations of stratified flow past typical building shapes such as cubes and rectangular prisms have not been widely reported. The work by Kothari, Peterka, and Meroney (1979) is a notable exception. Highly detailed studies such as that of Hunt, Snyder, and Lawson (1978) would provide very useful information about the flow patterns near such commonly used shapes. In particular, the effects of stratified shear flow on the frontal horseshoe vortex, on reattachment to the roof and sides, and on the cavity and far wake regions need further investigation.

7-7.2.2 ***Effects of Atmospheric Stability on Near-Obstacle Dispersion.*** The effects of atmospheric stratification on effluent dispersion have been studied for many years. In situations where there are no barriers to the flow, the changes in turbulent mixing are accounted for by means of suitably adjusted dispersion parameters (e.g., σ_y and σ_z) in plume concentration formulations. Field work at several locations (e.g., Halitsky and Woodard, 1974; Sagendorf and Dickson, 1976; Wilson et al., 1976) indicates that the results in the stable case are highly site-dependent, primarily because of the wind-direction meander typical of the light winds that usually accompany strong stratification. The situation is further complicated by the presence of obstructions in the flow.

Here attention is restricted to laboratory studies in the hope of elucidating general trends. Further discussion of concentration fields near buildings can be found in Sec. 7-8.

Several features important for pollutant dispersion can be discerned from the previous description of near-obstacle flows under stable conditions. For example, wide obstructions (such as ridges or large buildings) may cause partial blocking of the incident flow and create a stagnant region with high air pollution potential just ahead of the body. The rise and subsequent dip of air parcel trajectories passing around hills may bring a plume near the ground. Lee wave-induced separation may mix effluents that manage to pass over the obstacle down to the surface, whereas rotors may recirculate such effluents. The tendency to constant wake height (or even wake collapse) and slow horizontal wake growth serves to limit downwind dispersion so that concentrations within the wake may be quite large. On the other hand, wake meander, whether produced by wake instabilities or by fluctuations of the incident flow direction, or both, will tend to reduce the downwind concentrations, at least in the time-averaged sense.

Yang and Meroney (1970) pointed out the consistently higher ground-level effluent concentrations behind individual buildings in stable rather than neutral conditions. They noted that within five building heights or so of their models, the effect of different stability conditions on dispersion was not significant because of the dominance of mechanically induced turbulence within this region. Farther downwind, however, suppression of this turbulence became important, and plume dispersion rates were reduced in both the vertical and crosswind directions. Hatcher et al. (1978) observed similar effects in their study of a reactor complex. In some cases effluent material would "puddle" in nearly stagnant regions behind the structures, producing locally high concentrations.

Hunt, Snyder, and Lawson (1978) studied the behavior of an effluent plume incident on a rather large-scale obstacle such as a hill. They suggested the following fairly quantitative guidelines:

1. As stratification becomes stronger (F_H smaller), plume thickness will become smaller and plume width will become larger. If the incident plume started from a level below that of the hilltop but is able to go over the crest, it will pass very close to the hill surface and will be contracted in the vertical and expanded in the horizontal.

2. Let h_* be the critical streamline height such that a plume released at a height less than h_* will impact the windward hill face and then pass around the hill. A plume released from a source higher than h_* will go over the rest. This critical height is approximated by the relation

$$h_* \simeq H(1 - F_H) = H - \frac{U_\infty}{N} \tag{7.11}$$

3. In moderately stratified flows ($0.4 < F_H < 1.5$), separation on the lee side of the hill is largely suppressed. If the plume release height is higher than h_* so that the plume clears the hill but is not much higher than $H(1 + F_H) = H + (U_\infty/N)$, the plume can pass down the lee slope very close to its surface. Notice that this possibility can be avoided by choosing an effluent release height greater than about $H(1 + 1.5)$, which is the familiar "2½ times" rule for stack heights.

4. In strongly stratified flows ($F_H < 0.3$, for example), the flow in the immediate lee of the hill consists of mainly horizontal recirculations. A plume emitted within this region at a height less than h_* would be carried to the lee surface. For less stable flows, the plume could be brought to the ground as a result of lee waves or rotors.

Figure 7.46 shows Hunt, Snyder, and Lawson's (1978) visualizations of a plume incident on a three-dimensional hill for a variety of stack heights and stabilities; these should be compared to Figs. 7.44 and 7.45.

Lin, Liu, and Pao (1974) studied a more complicated obstacle consisting of a pair of bell-shaped hills connected at their shoulders by a ridge. Figure 7.47 shows a typical result for the flow patterns for a Froude number $F_H \simeq 0.4$. The flow is partially blocked and brought to rest ahead of the ridge connecting the two hills. Fluid that is diverted around the hills dips significantly in the process, much as was observed for a single hill (Hunt, Snyder, and Lawson, 1978).

Further experimentation is needed to clarify the effects of stability on concentrations near simple shapes of engineering interest such as blocks and cylinders. Brighton's (1977, 1978) work is useful in this regard; however, the research should be repeated at higher Reynolds numbers to verify the results. Field studies on isolated structures are needed to confirm the modeling work. Such work eventually will allow formulation of rules of thumb to estimate stability effects on near-obstacle flow and concentration fields.

7-8 EFFLUENT DISPERSION CHARACTERISTICS AND CALCULATIONS

An understanding of the flow patterns near buildings and other flow obstacles should prove extremely useful in understanding and predicting effluent concentration fields near these objects. Numerous authors have attempted to use such insights to develop theoretical and numerical models of these fields. However, much of our current predictive capability continues to rest on field and fluid modeling experiments and on a number of rules of thumb arising from such experience. Many of these rules are cited in the following text. Further discussion and computational aids are given in a recent report (Hosker, 1982).

To permit comparison of the discussions of flow and concentration patterns, this section has been arbitrarily divided into examinations of situations where the effluent source is upwind (by a few building heights) of an obstacle, close to or on the obstacle, or downwind of the cavity behind the obstacle. Unfortunately, despite its obvious importance, surprisingly little information seems to be available on the last topic.

7-8.1 Isolated Sources Upwind of Obstacles

7-8.1.1 Individual Buildings. Previous discussions of the flow fields generally expected around obstacles provide some basis for understanding the concentration field near a building subjected to an isolated incident effluent plume (e.g., see Fig. 7.15). The rising flow near the upper one-third or so of the upwind face can divert pollutants upward, while the downwash generated by the frontal eddy over the lower two-thirds of the face can be expected to mix the effluent downward and increase concentrations near ground level. Some of the material will be carried off in the trailing horseshoe vortex. Pollutants that enter the cavity zone from the sides or from above will be fairly well mixed in that region; as a result, concentrations over the rear face of the building will be relatively uniform. If the building is sufficiently large in the along-wind direction (i.e., large L), the flow that has separated from the upwind roof edge may reattach and bring the effluent closer to roof level; in such an event, high effluent concentrations could result. However, if L is small, the roof-level flow will not reattach, and the incident plume will be deflected above roof level.

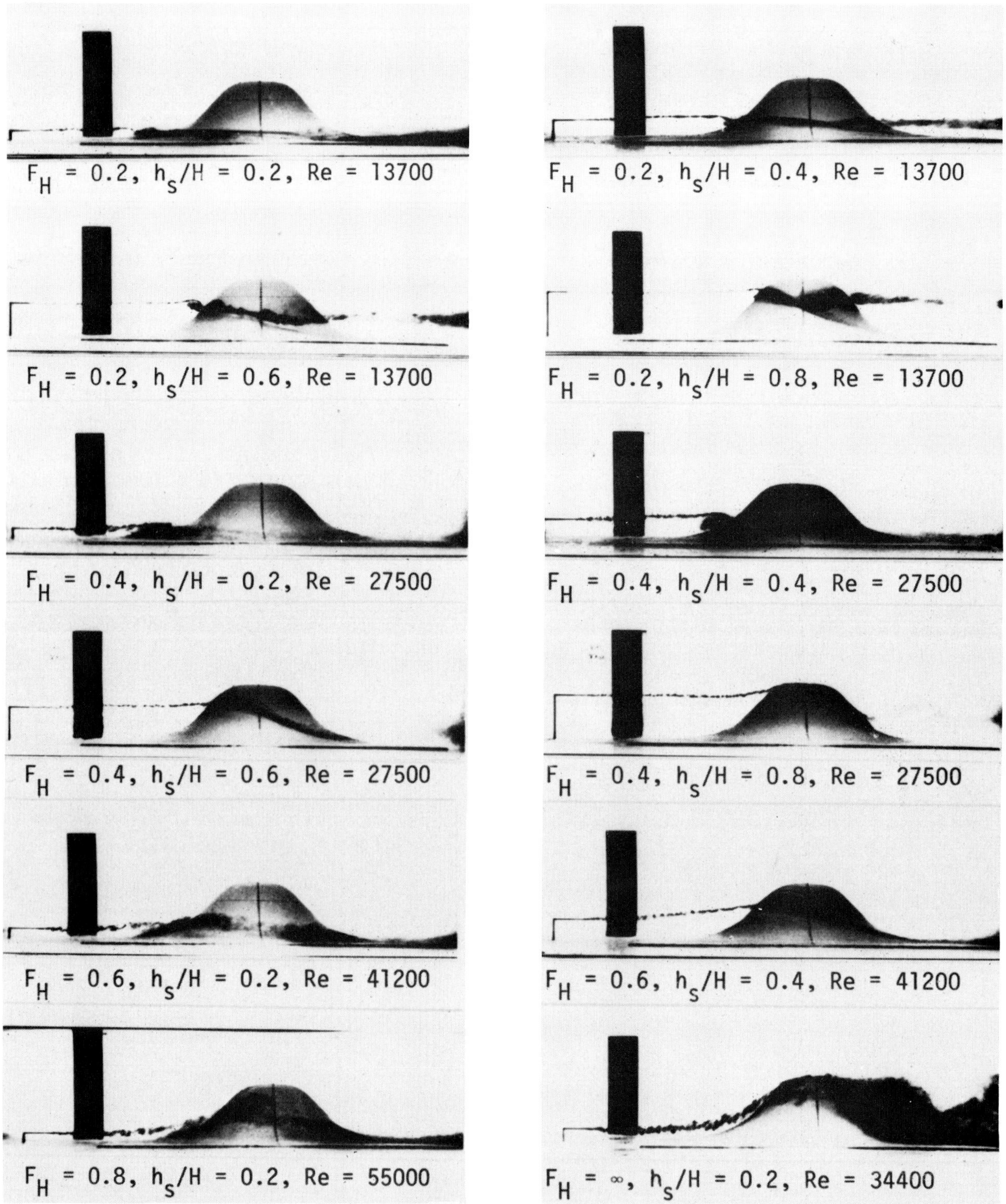

Fig. 7.46 Effluent plume incident on bell-shaped hill for several stack-to-hill height ratios h_s/H and several Froude numbers $F_H \equiv U_\infty/NH$. (From J. C. R. Hunt, W. H. Snyder, and R. E. Lawson, Jr., Flow Structure and Turbulent Diffusion Around a Three-Dimensional Hill, in *Fluid Modeling Study on Effects of Stratification, Part I. Flow Structure,* Report EPA-600/4-78-041, U. S. Environmental Protection Agency, 1978.)

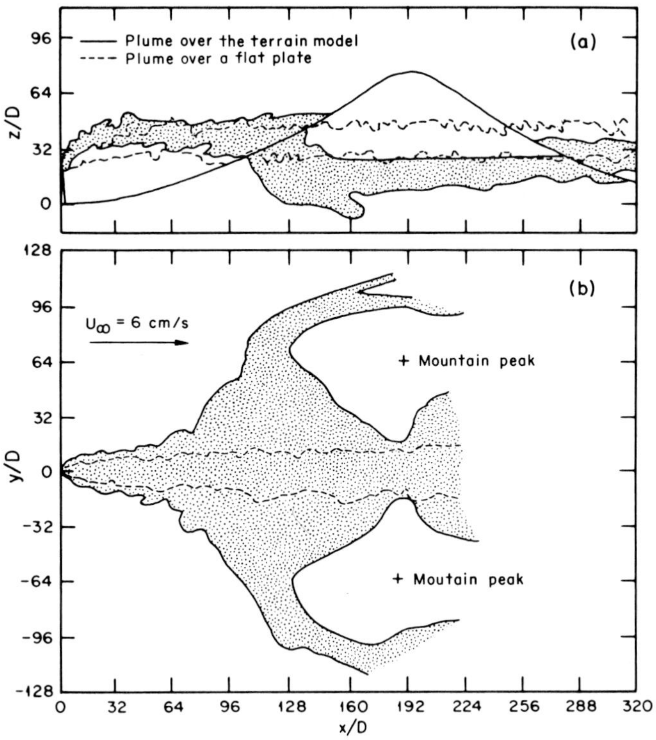

Fig. 7.47 Vertical (a) and lateral (b) spread of elevated plume in stratified flow near a pair of three-dimensional hills, height H. Observed plume (stippled area) can be compared to open-terrain plume (broken line). $F_H \simeq 0.42$, Re $\simeq$ 7920, $h_s/H \simeq 0.31$. (From J.-T. Lin, H.-T. Liu, and Y.-H. Pao, *Laboratory Simulation of Plume Dispersion in Stably Stratified Flows over a Complex Terrain,* Flow Research Report No. 29, prepared for U. S. Environmental Protection Agency, Flow Research, Inc., Kent, Wash., 1974.)

Consequently, concentrations on the roof will be similar to those found on the rear face.

The overall effect of the building, as viewed from a distance, is to induce a rapid enlargement and dilution of the incident plume. Ground-level concentrations near the building will be higher than would be observed if the building did not exist; however, further downwind the concentrations will be lower than would be observed if the building did not exist.

Work by a number of authors has provided quantitative confirmation of such expectations. For example, Wilson and Netterville (1978) found that the influence of a building on the concentration profiles extended only about 1.3H upwind of the building face. Just downwind of the structure they observed rather uniform concentrations and an elevated plume maximum about 1.1H above the ground (the initial plume height was 1H). Ground-level concentrations were increased locally by factors of two to five over the no-building values. As a useful rule of thumb for stack design, Wilson and Netterville (1978) suggested that the downwash effect generated by the building can be approximated by assuming that the concentration on all building surfaces is uniform and equal to the average of the concentrations expected at roof and at ground level with no building present. They also remarked that, at least for an incident plume released at the height of the "target" building, the concentration at ground level near the building may conservatively be taken equal to the maximum concentration intercepted by any point on the building surface.

Lucas (1972) has published a design technique for estimating the required height for an isolated stack so that hazardous or legislated effluent concentration limits are not exceeded on or near buildings located downwind of the source. The method requires a preliminary estimate of required source height using the best available open-terrain model. The height, width, and distance from the source of each downwind structure also are needed. Little explicit use is made of the known aerodynamic flow fields; however, consequences of the flow fields are reflected in the rules of thumb provided. For example, it is assumed that effluent will not be affected by surrounding structures if the stack is more than 2.5 times higher than all of the neighboring buildings. Additionally, it is assumed that the turbulent wake of a building rises to 1.5 times the building width or length, whichever is less, and extends downwind with gradually decreasing effectiveness to 10 building heights or half-widths, whichever is less. The Lucas (1972) paper should be consulted for full details.

Lucas' (1972) recommended procedure is as follows:

1. Let h_1 be a first approximation to the stack height. It can be selected by using standard open-terrain plume models (see Chap. 13) to determine the height required in the absence of neighboring buildings to ensure that the maximum predicted ground-level concentration is less than or equal to the safe and/or legal limit for the particular effluent. Standard conservative† practice should be followed.

2. For the ith downwind building ($i = 1, 2, 3, \ldots$), list its distance from the stack x_i, its height H_i, and its width normal to the wind W_i (if some buildings are smaller in both height and width than a building closer to the source, they may be ignored).

3. For each building, let ζ_i be the smaller of H_i and W_i, and let η_i be the smaller of H_i and $W_i/2$. Calculate an effective wake height for each building as $Z_{ei} \equiv H_i + 1.5\zeta_i$.

4. In the set of all buildings for which $x_i/\eta_i \leqq 5$, select the largest value of Z_{ei} as Z_{eM}.

5. Now consider the buildings for which $5 < x_i/\eta_i \leqq 10$; ignore those for which $Z_{ei} \leqq Z_{eM}$. For the remainder (if any), calculate the adjusted wake height $Z_{ai} \equiv Z_{eM} + (2.0 - 0.2x_i/\eta_i)(Z_{ei} - Z_{eM})$, and take the maximum value in this set as Z_{aM}.

6. Compare Z_{em} and Z_{aM}; call the larger Z_M.

7. Now consider the heights H_i for those buildings where $x_i \leqq 5h_1$ and the reduced heights $J_i \equiv H_i - 0.2x_i$ for those buildings where $5h_1 < x_i \leqq 20h_1$; the largest of any of these values is taken to be H_M.

8. Calculate the corrected stack height using

$$h_s = H_M + h_1\left(1 - \frac{H_M}{Z_M}\right) \tag{7.12}$$

Although Lucas (1972) suggested that this technique is appropriate for stacks of 30 m or less, he urged further comparison with experimental data to validate the method.

A somewhat similar scheme has been suggested by Wilson and Netterville (1976). They remarked that the most important effect of downwind buildings on an incident effluent plume is to mix the plume downward to produce a fairly uniform con-

†Conservative, as generally used in air-pollution meteorology, means that all calculating procedures and assumptions are selected to ensure that estimated concentrations will never be less than the actual values.

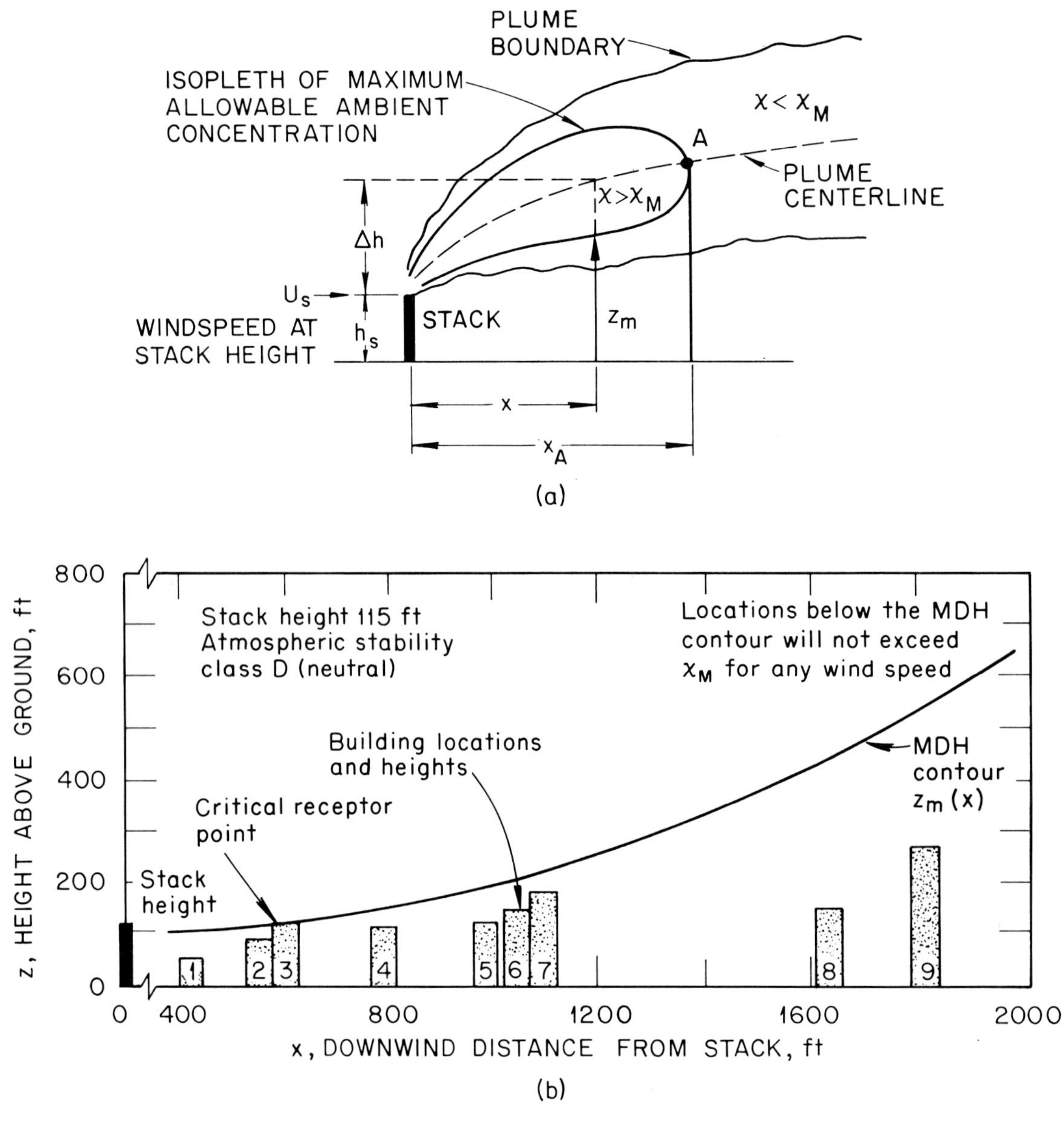

Fig. 7.48 Definition (a) and sample calculation result (b) for the minimum descent height (MDH), Z_m, of contour of maximum allowable concentration. [From D. J. Wilson and D. D. J. Netterville, Influence of Downwind High-Rise Buildings on Stack Design, *J. Air Pollut. Control Assoc.*, 26(10): 976-980 (1976).]

centration between roof level and the ground. Hence, for design purposes, the concentration at roof level should be maintained below the allowable limits regardless of the location of effluent receptors such as building air intakes. To ensure this, they introduced the concept of "minimum descent height," Z_m, of the maximum permissible concentration χ_M.

As indicated in Fig. 7.48a, Z_m is the height above the surface of the isopleth of maximum allowable concentration at a particular location x. This parameter is a function of the wind speed. In very light winds, the lower plume boundary will be high above the ground, but stronger winds will bend the plume over more sharply. However, strong winds also improve dilution and reduce concentrations in the plume. Consequently, the isopleth of maximum allowable concentration will first approach the ground as the wind speed at stack height U_s increases; the isopleth will reach a height of closest approach (Z_m) to the surface at some critical speed (U_A), and then it will recede from the ground as U_s continues to increase.

Proper stack height is calculated by iteration using an initial guess for the stack height h_s, a standard plume equation (see Chap. 13) and an appropriate estimate of plume rise (see Chap. 8). The minimum descent height (Z_m) is plotted as a function of downwind distance (Fig. 7.48b) on a graph that indicates the locations and heights of downwind structures. All stability categories should be considered. If Z_m intersects any existing buildings, or if Z_m is too low to permit future development of the area, then the initial stack height estimate should be increased and the procedure repeated. On the other hand, this method may be used by zoning authorities to estimate maximum permissible building heights for new construction downwind of existing stacks.

The actual calculation is tedious and best carried out on a computer.

1. Take, as a first approximation, h_s equal to the minimum stack height needed to produce acceptable air quality levels at all downwind surface locations in open flat terrain. All possible

atmospheric stabilities and wind speeds should be considered in this computation.

2. Select an atmospheric stability category j (j = 1, 2, ..., 6, where 1 corresponds to the usual "A" stability, 4 to "D," and so on). Choose the appropriate curves or functional forms for the dispersion parameters $\sigma_{yj}(x)$ and $\sigma_{zj}(x)$.

3. Choose a particular wind speed at stack height U_k, k = 1, 2, ...

4. Use a suitable plume rise estimation formula (from Chap. 8 for example) and the known effluent and ambient characteristics to calculate the plume rise Δh_k for that wind speed.

5. Solve the following equation for Z_{jk} at a number of downwind distances x_ℓ, ℓ = 1, 2, ...

$$\exp\left[-\frac{(Z_{jk} - h_s - \Delta h_j)^2}{2\sigma_{zj}^2}\right] + \exp\left[-\frac{(Z_{jk} + h_s + \Delta h_j)^2}{2\sigma_{zj}^2}\right] = \frac{2\pi \chi_M U_k \sigma_{yj}\sigma_{zj}}{Q} \tag{7.13}$$

Notice that for locations beyond the "end" of the plume (x_A in Fig. 7.48a), no real values will be possible for Z_{jk}, whereas for $x_\ell < x_A$, a pair of values will be obtained at each x_ℓ. The smaller value in each pair represents $Z_{mjk}(x_\ell)$ and is the minimum descent height for stability category j and wind speed U_k at the location x_ℓ.

6. Choose a new stack-height wind speed U_{k+1}, and repeat steps four through six until the entire wind-speed range expected in selected stability category j has been covered.

7. Examine the entire set of values $\{Z_{mjk}\}$ for each x_ℓ; the smallest value is $Z_{mj}(x_H)$ and represents the minimum descent height for any wind speed in stability category j at the location x_H.

8. Choose a new stability category, j + 1, for example, and repeat steps two through seven until all possible stability categories have been considered.

9. Examine the entire set of values $\{Z_{mj}(x_\ell)\}$ for each x_ℓ; the smallest of these is $Z_m(x_\ell)$ and represents the minimum descent height at location x_ℓ for any wind speed or stability category.

10. Plot Z_m as a function of downwind distance. Also plot the locations and heights of present and/or planned buildings (e.g., Fig. 7.48b). If the curve $Z_m(x)$ contacts any of these structures, the selected stack height h_s is too low. Increase the stack height and repeat steps two through ten until satisfactory results are obtained.

Wilson and Netterville (1976) also presented a simpler approximate procedure that tends to overestimate $Z_m(x)$ by 20% or more within five stack heights of the source, but that is within 2% of the correct value for $x/h_s \simeq 20$ or more. This method, although not conservative, may be useful in cases where the buildings of concern are some distance away from the isolated stack; also, this method can serve as a first approximation to Z_m in any case. The original paper should be consulted for details.

Wilson and Netterville (1976) made several useful comments about their procedure. First, and perhaps most important, they remarked on the uncertainties inherent in all air-pollution calculations. They recommended that the usual rural dispersion curves be used for σ_y and σ_z because the extra dilution provided by the larger dispersion parameters characteristic of built-up areas supplies a safety factor. Finally, they noted that, in practice, the critical atmospheric conditions for air pollution may be either an unstable atmosphere (e.g., looping plumes) or nocturnal plume trapping followed by morning fumigation, whereas the minimum descent height curve is usually closest to the ground under neutrally stable conditions. In other words, the stack designer must consider the full range of possibilities of effluent behavior, and not rely entirely on the technique just outlined.

Some intriguing theoretical work on this complex problem of estimating near-building pollutant concentrations resulting from upwind sources has been published by Hunt and his colleagues (Hunt and Mulhearn, 1973; Britter, Hunt and Puttock, 1976; Puttock and Hunt, 1979; Puttock, 1979). Their research has used both the statistical theory of turbulent diffusion and the classical diffusion equation approach. The results are mathematically somewhat complex, and, although both point and line effluent sources have been considered, the studies thus far have been limited mostly to two-dimensional obstacles. Nevertheless, some useful simplifications of the general theoretical problem have been suggested by this work.

For example, the statistical theory shows that in flows undergoing serious deformation, such as those around buildings, the cross-sectional area and resultant concentration of an incident plume are determined mainly by convergence or divergence of the flow streamlines; and they do not depend too strongly on the turbulence characteristics. Hence a semi-empirical eddy diffusivity solution may be at least as appropriate for calculation of dispersion around obstacles as it is for such calculations in pure shear flows, where it is commonly utilized (Britter, Hunt, and Puttock, 1976). The detailed statistical theory may be useful primarily for providing appropriate values for the eddy diffusivity itself (Hunt and Mulhearn, 1973). For flows constrained within horizontal planes (e.g., very strong atmospheric stratification), the concentration of effluent at the upwind stagnation point on a two-dimensional building should be nearly equal to that predicted for the same location in the absence of the building, as long as the effluent source is at least ten characteristic body dimensions upwind. A similar result is obtained for truly three-dimensional flows; however, in this case, the source need be only two building dimensions or more upstream (Britter, Hunt, and Puttock, 1976). These theoretical predictions (and their associated laboratory validations) support the previously outlined simplifications suggested by Lucas (1972) and Wilson and Netterville (1978). The theory suggests that behind the obstacle (at least for two-dimensional bodies) the concentration in the separated cavity zone is approximately constant and is equal to the average concentration along the cavity's separation streamline. This concentration is calculated by assuming that the separation streamline is an impenetrable barrier (Britter, Hunt, and Puttock, 1976; Puttock and Hunt, 1979). The outer flow, at least for the cases tested thus far, is reasonably well approximated by simple potential flow (Puttock and Hunt, 1979; Puttock, 1979). Some of these ideas have been used by Egan (1975) for estimating pollutant concentrations in complex terrain. However, further effort is needed, especially for the difficult extension of the theory to three-dimensional obstacles. The original papers should be reviewed for details of the computations.

7-8.1.2 ***Clusters of Buildings.*** Because of the enormous variations in flow patterns possible, even for relatively simple configurations, pollutant dispersion of an incident plume by a cluster of buildings can be described only in general terms. In particular, gross plume deflections near long walls, "jetting" effects near gaps, and a strong sensitivity to the wind direction can be expected. Wind-tunnel and/or field tests are virtually mandatory for any realistic assessment of a particular site.

Cagnetti (1975) has described an interesting field study of a comparatively simple reactor complex that illustrates some of the problems. To discuss his results, it is convenient to introduce the concentration coefficient (Halitsky, 1961, for example) defined by

$$K \equiv \frac{\chi U A}{Q} \tag{7.14}$$

where χ is the concentration, Q is the source strength, A is a characteristic area of the structure being examined, and U is the wind speed at the effective release height. Figures 7.49a, b, and c show isopleths of constant K for several wind speeds and directions. The patterns do not resemble the closed "footprint" characteristic of plumes in open terrain, but they are consistent in both shape and value. K ranges from about 0.02 along the front of building B to about 0.002 some 50 m behind its lee face. For the wind directions treated, the effluent appears to be simply deflected to the north and east by the upwind walls of the structure. Figures 7.49d and e show the consequences of a modest shift in wind direction and an increase in speed; the bulk of the flow is directed toward building C and the gap between structures B and C. The isopleth pattern here is much more reminiscent of a classical plume; whereas the K values, which are about 0.1 just behind building A and drop to 0.06 or 0.08 behind building C, are about an order of magnitude larger than the previous cases.

Cagnetti (1975) also included a limited amount of data (only two sampler arcs) on the far wake; at distances of 450 m or more downwind of the source, K appears (Fig. 7.50) to vary approximately as $x^{-1.7}$ without much dependence on wind direction. However, in experiment No. 9, K decreases much more rapidly (as $x^{-3.6}$); Cagnetti (1975) remarked on the "peculiar" nature of this particular test and postulated that during unstable conditions the extra dilution introduced by a building is somehow more efficient at low wind speeds. In view of the limited data set, this must be regarded as speculative. Cagnetti (1975) concluded his results by noting that the building cluster reduced concentrations from the anticipated open terrain values by a factor of four at 300 m from the source and by a factor of two at 550 m. By 800 to 1000 m downwind, the concentrations observed were about equal to those expected with no buildings downwind of the source.

7-8.2 Sources Close to Buildings

In this section emphasis will be directed toward the concentration fields generated by effluent sources close to or on buildings. As before, the flow patterns described earlier provide a qualitative understanding of the results for at least the relatively simple case of an isolated structure. Rules of thumb suitable for design use and descriptions of typical experimental results are included.

One important flow feature in this situation is the frontal eddy that induces downwash over approximately the lower two-thirds of the building and then wraps itself around the ends of the obstacle to trail off downwind. Effluent released from surface vents at mid-height or below on the windward face of a building will be carried downward to ground level and eventually will be swirled past the building sides. Some of this material will be advected and diffused into the cavity region, but much of it will be transported past the cavity to the far wake (Hinds, 1969; Meroney and Yang, 1970; Robins and Castro, 1977b).

On the upper third of the structure, flow will be directed upward and sideward; in the case of sharp-edged obstacles, the flow will separate from the windward edges of the roof and sides and will carry effluent from face vents on the upper portion of the obstacle out and away from other building surfaces. If, however, effluents are released within the resulting separated areas (e.g., on the roof), the subsequent ineffective ventilation and flow recirculation can lead to locally high concentrations. If the air intake system for the building is also within the separated zone, a potential health/safety hazard is created because the high concentrations may then be readily circulated throughout the structure. If devices such as heat exchangers are located in these zones of separation and recirculation, they will be prone to self-contamination (plume reingestion) that can seriously degrade their performance and life expectancy.

If the along-wind length of the obstacle is sufficiently large, the edge-separated flow will reattach to the roof and sides and will carry effluent from vents on the windward face closer to the building; however, it will provide better ventilation at outlets located in these reattached areas. Wilson (1976a; 1977a) noticed that side vents near the ground produced a somewhat higher effluent concentration at a given distance from the source than did more elevated vents; this probably is because of the lower wind speed closer to the ground. Flow that has reattached to the sides and top eventually separates again from the lee edges of those surfaces.

If the obstacle is at an angle to the wind, the flow will remain attached to the most exposed face; on the less exposed windward side, the flow may or may not separate and reattach. Effluent dispersion will vary accordingly. On the roof, a counterrotating vortex pair will be generated near the upwind corner; these vortices will transport (possibly clean) high-momentum air from above down to roof level. A roof cavity will not occur and effluents from roof vents will be swept away quickly. As a result rooftop concentrations will be kept relatively low.

Behind the structure, the notion of a more or less closed recirculating bubble bounded by a turbulent shear layer may remain a useful approximation for calculations (Vincent, 1977); however, the possibility of direct advection to and from this highly turbulent region must be recognized, i.e., a "leaky" bubble must be allowed. In atmospheric flows the cavity seems to be rather nonstationary (Frost and Shahabi, 1977), and it may even collapse intermittently (Meroney, 1982). Flow adjacent to the lee face of the obstacle will be more or less upward. Material released within the cavity will be mixed rapidly throughout the cavity's volume; therefore, contamination of any surfaces contacting the flow is likely (e.g., Meroney and Yang, 1971; Drivas and Shair, 1974; Huber and Snyder, 1976; Robins and Castro, 1977b). In particular, for the case of buildings that are thin in the along-wind direction (small L/H), the separated flow from the windward edges of the top and sides becomes part of the cavity; consequently, effluents can be expected to appear on the building's roof and sides as well as on its downwind face. Releases somewhat above ground level within the cavity will be mixed throughout the region more rapidly than emissions close to the ground (Drivas and Shair, 1974).

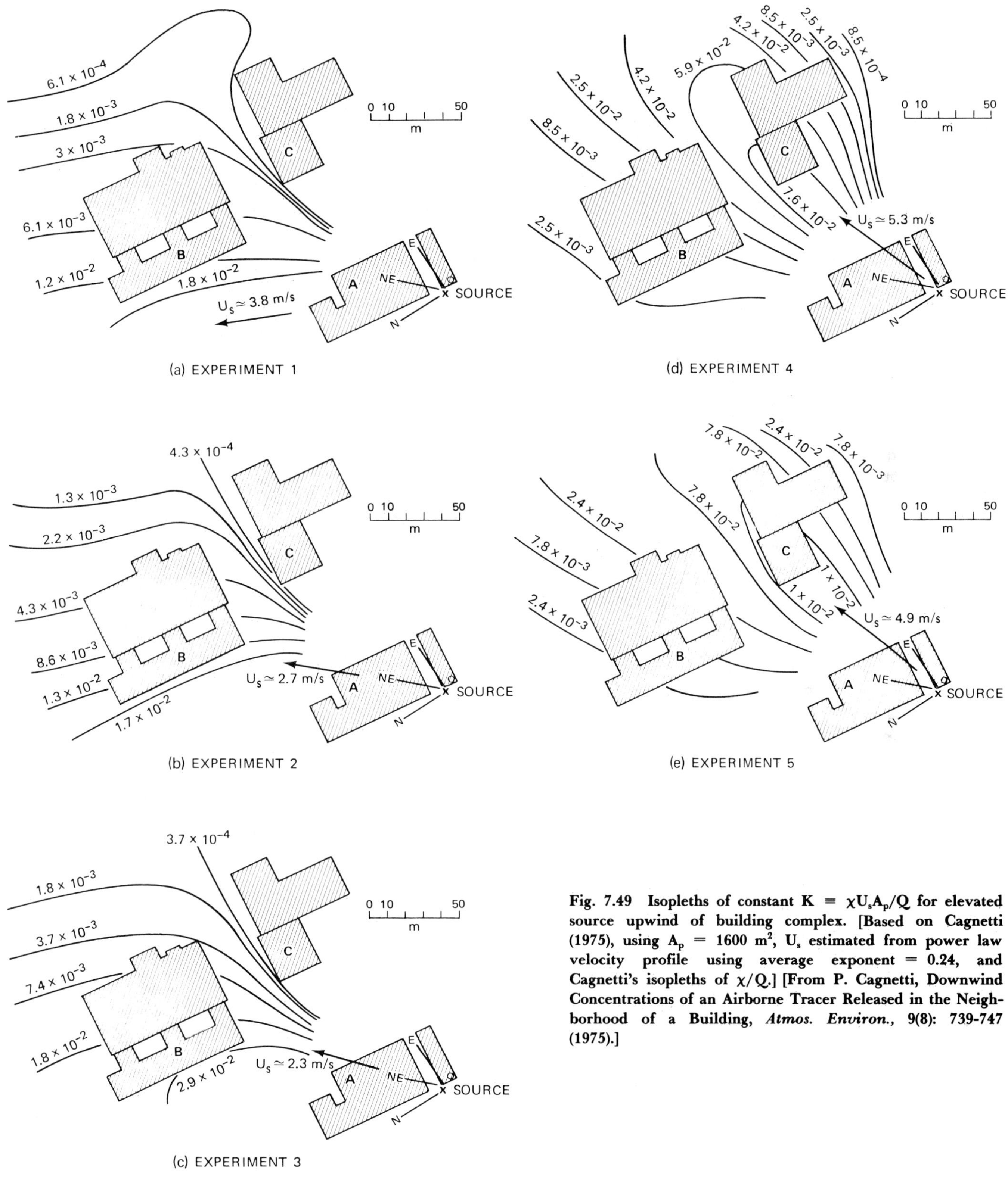

Fig. 7.49 Isopleths of constant $K \equiv \chi U_s A_p/Q$ for elevated source upwind of building complex. [Based on Cagnetti (1975), using A_p = 1600 m^2, U_s estimated from power law velocity profile using average exponent = 0.24, and Cagnetti's isopleths of χ/Q.] [From P. Cagnetti, Downwind Concentrations of an Airborne Tracer Released in the Neighborhood of a Building, *Atmos. Environ.*, 9(8): 739-747 (1975).]

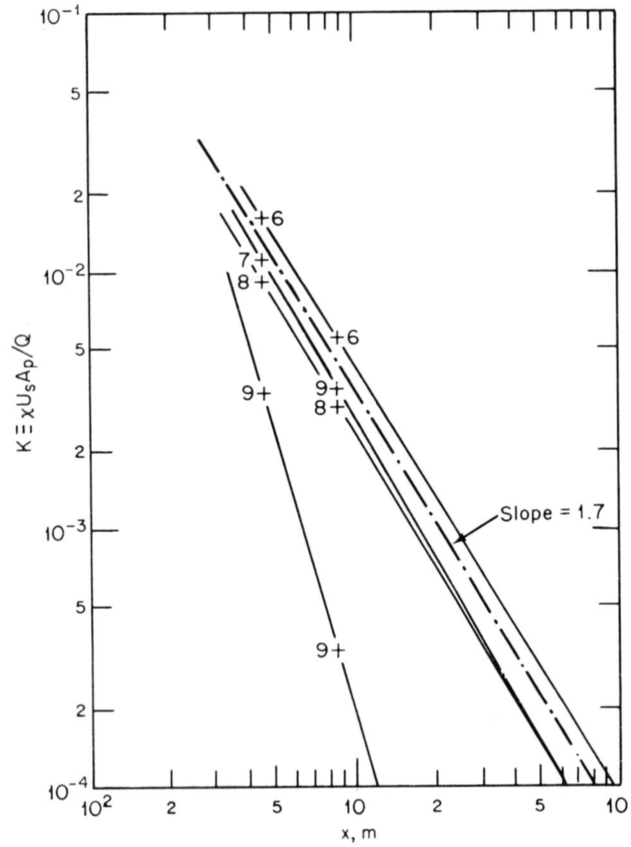

Fig. 7.50 Far wake behavior of concentration coefficient K ≡ $\chi U_s A_p/Q$ behind a building cluster for an upwind elevated source. [Based on P. Cagnetti, Downwind Concentrations of an Airborne Tracer Released in the Neighborhood of a Building, *Atmos. Environ.*, 9(8): 739-747 (1975).]

The curvature of the turbulent shear layer bounding the cavity from above induces downwash of the outer flow that is especially noticeable near the bubble closure. Material released above the cavity but subject to this flow can be advected close to the ground and can produce higher concentrations than would be expected if the building were absent. This process has important implications for stack height and placement decisions (discussed in the next section).

7-8.2.1 Concentration Fields Near Isolated Buildings.

7-8.2.1.1 Estimation of Effective Stack Height. Concentration patterns near an obstacle are intimately related to the characteristics of the effluent discharge and to the interaction of the discharge with local flows. Effluents released too close to the structure and/or at too low an exhaust speed w_e generally will produce high concentrations near the building. Therefore, successful stack or vent design must take account of building and chimney aerodynamics and effluent buoyancy. Some recent work incorporating such effects is outlined below. For additional information, the reader can refer to comprehensive survey reports by Wilson (1976a) and by Meroney (1982) and to a series of papers by Wilson (1976b, 1977a,b, 1979). Meroney's study provides useful discussion of the behavior of very dense and/or cold plumes (e.g., liquefied gases); behavior of such plumes is of growing concern in the energy and transportation industries.

In any stack design methodology, some attention must be directed toward the aerodynamic effect of the stack itself on the plume. Sherlock and Stalker (1941) described the production in strong wind of vertically oriented vortices shed by the stack and a longitudinally oriented counterrotating vortex pair generated at the stack outlet. These latter vortices transfer emitted material downward into the stack wake; this results in a lower effective release height for the material and it produces features such as the commonly observed blackening of chimney tops. Scorer (1968) mentioned installing flat, horizontal disks at the stack mouth. If the diameter of these disks is approximately equal to three stack diameters, they will prevent effluent from being drawn down into the stack wake cavity and thus will preserve the entire useful height of the stack. Similar devices are used on ship funnels; however, these rely on a fixed direction of wind incidence for their effectiveness. Such a convenience is seldom experienced in building design.

Downwash behind the stack probably is best avoided by keeping the efflux momentum significantly higher than that of the ambient flow. Sherlock and Stalker (1941) observed downwash distances of only one stack diameter or so for an efflux to stack-height speed ratio of $w_e/U_s = 1$. For $w_e/U_s = 1.5$, practically no downwash was observed. The latter value has become rather widely used, but it should be viewed as a lower bound for design work; that is to say, w_e/U_s should be maintained greater than 1.5 whenever possible to prevent downwash and to increase the momentum rise of the effluent plume (e.g., Turner, 1969; ASME, 1973).

The effluent density, another factor that can affect downwash, enters the problem as the square root of its ratio to the ambient density (e.g., Wilson, 1976a, b) and, in many cases, may have a rather modest effect on the value of w_e/U_s needed to avoid downwash. The effluent density of course may be important for very hot or cold and/or high or low molecular weight effluents.

Releasing material via a tall stack has been recognized for many years as the best way to avoid plume interference by the aerodynamic flow field around an obstruction. The "2 and 1/2 times" rule, which states that the stack outlet should be 2.5 times the height of neighboring flow obstacles above the ground, makes use of the approximation that the zone of influence of a hill or structure generally reaches less than 2.5H from the surface. Hence, material emitted above this height is ejected into a more or less unperturbed flow. As a result, reasonable travel distances and significant plume dilution can occur before the effluent diffuses to the ground.

Three things limit across-the-board use of this rule of thumb: economics, aesthetics, and the fact that, in many instances, the rule of thumb is needlessly conservative. As examples of this last factor, concentrations often are low enough at the stack exit so that the maximum amount of dilution possible is unnecessary to avoid objectionable levels of contamination, or the actual zone of disturbance above the obstacle is significantly lower than the rule would suggest. The latter situation is common for tall, thin structures (e.g., Snyder and Lawson, 1976).

Briggs (1973) has outlined a useful procedure for estimating the effective height of emission. The method incorporates corrections for stack downwash, building-induced disturbances to the flow, and buoyancy effects. Building model studies by Snyder and Lawson (1976) on neutrally buoyant stack emissions that do not experience downwash have verified that, for a short wide building and a tall thin building, Briggs's alternative to the "2 and 1/2 times" rule is adequate to avoid building-induced effects on effluent concentration. The procedure is as follows:

1. Compensate for stack-induced downwash by calculating a corrected release height h': if the stack outlet is not vertical, set $h' = h_s$; if it is vertical,

$$h' = h_s + 2\left(\frac{w_e}{U_s} - 1.5\right)d_i \tag{7.15}$$

where d_i is the inside diameter of the stack of height h_s. The factor of two that precedes the momentum-induced plume rise is a conservative value. Factors as large as three sometimes have been used (e.g., Briggs, 1975).

2. Calculate the release height h'' corrected for both building aerodynamic effects and stack-induced downwash. Let ζ_b be the smaller of the frontal building dimensions H or W. If the downwash-corrected release height $h' > H + 1.5\zeta_b$, the plume is above the region of building influence; set $h'' = h'$ and go to step (4). If $h' < H$, set

$$h'' = h' - 1.5\zeta_b \tag{7.16a}$$

If h' is between H and $H + 1.5\zeta_b$, set

$$h'' = 2h' - (H + 1.5\zeta_b) \tag{7.16b}$$

and go to step (3). The mathematical possibility that h'' is negative is not physically troublesome because h'' is an intermediate computational parameter and later steps ensure that the actual effective emission height is not negative.

3. The plume may remain aloft or may be entrained into the wake cavity and become essentially a ground-level source. If $h'' > 0.5\zeta_b$, the plume remains elevated; go to step (4). On the other hand, if $h'' < 0.5\zeta_b$, the plume is trapped in the cavity and should be treated as a ground source of initial cross-sectional area $\simeq \zeta_b^2$. Go to step (5).

4. Compensate for plume buoyancy by calculating the effective height of emission h_e. If the effluent is mostly ($\geqslant$ 98%) air and if its temperature is within 2 or 3°C of the ambient, buoyancy is negligible and $h_e = h''$; go to step (5). Otherwise, determine the relative density difference of the plume using

$$\Delta = \frac{\overline{m}_e}{\overline{m}_a}\frac{T_a}{T_e} - 1 \tag{7.17}$$

where subscripts e and a refer to effluent and ambient, respectively, T is temperature in degrees Kelvin, and $\overline{m}$ denotes the molecular weight (28.8 g/mole for air). The mean molecular weight of the effluent is determined from the mass fraction f_i of each of the component gases through the relation $\overline{m}_e^{-1} = \Sigma\, \overline{m}_i^{-1} f_i$. For many sources, the dominant component will be air; therefore $\overline{m}_e \simeq \overline{m}_a$. If the effluent contains water (liquid or vapor), evaporation or condensation may affect the buoyancy (Briggs, 1975, or Chap. 8).

If $\Delta > 0$, the plume is denser than air. Hoot, Meroney, and Peterka (1973) give a comprehensive treatment of the behavior of dense plumes and the concentrations that result. If $\Delta < 0$, the plume is buoyant. Evaluate the buoyancy flux $F_b = -g\dot{V}_e\,\Delta/\pi$, where $\dot{V}_e$ is the volume of effluent emitted per unit time. Then calculate the buoyant plume rise Δh using standard procedures (e.g., Briggs, 1969, 1975, or Chap. 8). Set $h_e = h'' + \Delta h$, and go to step (5).

5. Calculate downwind concentrations using standard plume dispersion formulations (Chap. 13); wherever a source height is required, use h_e as the effective height of emission.

Comparison of effective source heights computed in this way with the virtual source heights estimated by Robins and Castro (1977b) for their wind-tunnel tests suggests that Briggs's (1973) method may be quite conservative (i.e., gives low values for h_e), at least for a cubical building with a roof-centered stack. More data for a variety of building shapes, effluent release locations and conditions, and atmospheric conditions are needed to further test Briggs's method.

7-8.2.1.2 Concentrations on the Building Roof and Sides. If a stack must be placed on a building roof, and if reattachment of the flow to the roof can be expected, then Wilson's (1979) suggestions may prove useful for avoiding contamination of air intakes or other critical locations on the roof. The approximate size and position of the roof cavity and the boundary shear layer above it have been described in Sec. 7-5.2.2, and in Fig. 7.19. If the stack is high enough that the effluent plume's lower edge remains above the shear layer (zone II of Fig. 7.19), contamination of the roof is unlikely; if the plume clears the wake boundary as well, there is no danger of contamination. The required height depends on the rate of plume spread. Wilson suggested that it generally is adequate to draw a line of slope 0.2 in the upwind direction from a point on either the zone II or zone III boundary (depending on the safety margin needed) immediately above the air intake or other critical receptor nearest the lee edge of the roof. The effective stack exit should lie above this line. The 0.2 slope of the line corresponds to a standard deviation of wind vertical fluctuation angle σ_ϕ of perhaps 0.1 radian (5.7°), which seems reasonable; however, values greater than 0.3 radian (17.2°) have been observed over open country under some conditions (Lumley and Panofsky, 1964). Whether this higher value is appropriate to flow over a roof is uncertain.

The sensitivity of local concentrations to variations in parameters such as stack height h_s and efflux velocity w_e is of considerable interest. Smith (1975, 1978) conducted detailed field experiments on a small building fitted with a variable-height stack at the roof center; he determined concentrations on the roof, sides, and within 5H downwind for a variety of source and atmospheric conditions. The most prominent feature of his results is the strong dependence of roof concentrations on the ratio h_s/H. For example, a 10% increase in h_s/H caused a 67% decrease in roof concentration and a 25% decrease in concentration on the lee face. Smith also remarked on the varying sensitivity of the results to changes in other parameters. The most significant variables, in order of importance after h_s/H, are the efflux velocity ratio w_e/U_s, the angle of wind incidence α, the turbulence intensity σ_u/U_s, and the atmospheric stability as categorized by, for example, the Richardson number. Smith noted that roof concentrations were nearly an order of magnitude lower than the lee face concentration for all stack heights except flush vents. This suggests that roof-mounted air intakes generally will be better than side-mounted inlets whenever the stack height and/or efflux velocity are large enough to keep the emissions from direct contact with the roof.

Wilson (1976a, 1977b) has prepared a simple argument to demonstrate the sensitivity of roof concentrations to vent height. Effluent is emitted with negligible momentum from a short stack of height $\tilde{h}$ relative to the roof (i.e., $h_s = \tilde{h} + H$). Reattachment of the flow is assumed to occur upwind of the vent so that the wind affecting the effluent is directed toward the lee edge of the roof. A coordinate system is selected such that $\tilde{x} = 0 = \tilde{y} = \tilde{z}$ at the stack base. The effluent centerline concentration on the roof is then estimated from a standard plume equation; for example, $\chi \simeq Q(\pi\sigma_y\sigma_z U_s)^{-1} \exp(-\tilde{h}^2/2\sigma_z^2)$, so that the nondimensional concentration coefficient is given by $K \equiv \chi U_s A_p/Q \simeq A_p(\pi\sigma_y\sigma_z)^{-1} \exp(-\tilde{h}^2/2\sigma_z^2)$. For a flush vent, $\tilde{h} = 0$; therefore $K_0 \simeq A_p(\pi\sigma_{y0}\sigma_{z0})^{-1} U_s/U_H$, where the subscript 0 here indicates spreading appropriate to a roof-level release, and A_p is the projected frontal area of the building. After some algebra,

$$\frac{K}{K_0} \simeq \exp\left(\frac{-B\tilde{h}^2}{2\sigma_z^2}\right) \tag{7.18}$$

where B depends on the receptor location and the wind profile over the building. To examine the relative effectiveness of two stack heights $\tilde{h}_1$ and $\tilde{h}_2$, Eq. 7.18 gives

$$\frac{K_2}{K_1} \simeq \left(\frac{K_1}{K_0}\right)^{(\tilde{h}_2/\tilde{h}_1)^2 - 1} \tag{7.19a}$$

or, in terms of the dilution, $D \equiv K_e/K$,

$$\frac{D_2}{D_1} \simeq \left(\frac{D_1}{D_0}\right)^{(\tilde{h}_2/\tilde{h}_1)^2 - 1} \tag{7.19b}$$

K_e is the concentration coefficient at the vent exit $= (U_s/w_e)(A_p/A_e)$, where A_e is the exit area of the vent.

Suppose a short stack of height $\tilde{h}_1$ produces a dilution D_1 of, for instance, 800 compared to a dilution D_0 of 100 produced by a flush vent. If the stack height is increased by 50% to $\tilde{h}_2 = 1.5\tilde{h}_1$, then $D_2/D_1 \simeq 13.4$, or $D_2 \simeq 10{,}800$. However, if $\tilde{h}_1$ only produces a dilution D_1 of 200 compared to the flush vent value $D_0 = 100$, then the same percentage increase in stack height only gives $D_2/D_1 \simeq 2.4$, or $D_2 \simeq 476$. In other words, an increase in stack height produces significant decreases in roof-level concentration only if the stack being added to already is high enough to cause much more dilution than a flush vent. If values of K_1/K_0 or D_1/D_0 can be obtained from field or laboratory data for a particular building and vent height $\tilde{h}_1$, then the consequences of changes in vent height can be explored by using Eq. 7.19. A handbook of such ratios would be very helpful.

Wilson (1976a) has examined the case of a stack at the roof center of a model of a building where $W/H = L/H = 2.0$. A receptor is located at the center of the lee roof edge. The flow is reattached to the roof well ahead of the stack location. In this particular case, he found that the data were well fitted by the result $K/K_0 = \exp[-9.78(\tilde{h}/H)^2]$, where $\tilde{h} \leqq 0.5H$. The additional dilution caused by raising an effluent vent a height $\tilde{h}$ above the roof is then given by $D/D_{min} \simeq \exp[10(\tilde{h}/H)^2]$, where D_{min} is the minimum flush vent dilution (discussed below, see Eq. 7.22). If the building configuration and/or stack-receptor locations are changed, the numerical value of the factor B in Eq. 7.18 can be expected to change as well. Wind-tunnel and field data on a variety of building shapes, stack locations, and receptor sites are required before Eq. 7.18 can be widely used.

Halitsky (1963a,b) published an extensive study of concentrations on building side and roof surfaces for a variety of flush vent locations and exhaust conditions. Figures 7.51 a, b, c show the effects of changes in the effluent velocity ratio w_e/U_H. The highest roof concentrations naturally were obtained with the lowest value of w_e/U_H, 0.5; doubling this lowest value of w_e/U_H reduced the maximum effluent concentration at the roof by a factor of two to three, depending on receptor location. A further doubling of w_e/U_H, however, reduced the concentration only by a factor of 1.5 or less, with the main effect being felt close to the exhaust. Comparison of Figs. 7.51a and 7.51d shows the dramatic changes—especially above roof level—induced by installation of a relatively small stack. The modification reduced peak concentration almost by a factor of five, while the values over the whole roof became much more uniform. Figures 7.51e, f, g show the effects of different building geometries. Notice that in all but Fig. 7.51g, the rooftop concentration maxima are found upwind of the vent location. This result is due to the strong flow within the roof cavity of these models; the incident flow separated at the windward roof edge and apparently never reattached to the surface. Consequently, the flow near the roof was directed toward the upwind building edge and carried effluent from flush vents toward the front of the roof. On the other hand, the long, tall building in Fig. 7.51g experienced flow reattachment to the roof somewhere near the vent, and the quantity of material transported in the upwind direction was greatly reduced.

Halitsky's (1963a, b) experiments were conducted in a uniform, low-turbulence flow that was not very typical of the atmosphere. His results should therefore be compared with the results of Wilson's (1976a, b) work, which was accomplished in a simulated atmospheric surface layer. Figures 7.52 reproduce some of Wilson's (1976a, b) data. Zones of observed flow reattachment are indicated on the building surfaces. Upwind of the reattachment lines the near-surface flow is directed more or less toward the front of the structure, whereas the flow after reattachment is predominantly parallel to the incident wind direction. The K values shown in these figures are significantly higher than those cited by Halitsky (1963a, b). Tests by Wilson (1976a; 1977b) suggest that this results primarily from the much lower efflux velocity ratio used by Wilson ($w_e/U_H \simeq 0.3$ to 0.4, compared to 1.0 for Halitsky).

Figure 7.52a shows reattachment on the roof of the building. As discussed in earlier sections, this is a feature commonly observed in turbulent boundary layer winds. However, because reattachment took place considerably behind the vent, effluent was carried forward, and patterns similar to those in Figs. 7.51 a, b, c were generated. The somewhat squat building in Fig. 7.52b, however, experienced reattachment near the upwind roof edge; hence, the vent was subjected to only a rearward flow. The concentration pattern in Fig. 7.52b can be compared to that of Fig. 7.51e where the flow did not reattach. Similar behavior is apparent in Fig. 7.52c; the reattachment line was not given for this structure, but the isopleth pattern suggests that the line lies ahead of the exhaust location. Figure 7.51f is the no-reattachment counterpart of Fig. 7.52c. Finally, Fig. 7.52d shows reattachment just behind the vent; although most of the material was carried toward the lee edge, evidently some material was advected toward the front of the roof. Figure 7.52d can be compared to Fig. 7.51g which also had flow reattachment somewhere near the vent, but probably a bit further back, judging by the concentration patterns.

Wilson (1976a, b) also reported tests of a variety of flush roof vent locations. A few examples are shown in Figs. 7.53; Wilson (1976a) should be consulted for others. These data are potentially quite useful because exhausts often cannot be installed in the exact center of a building roof.

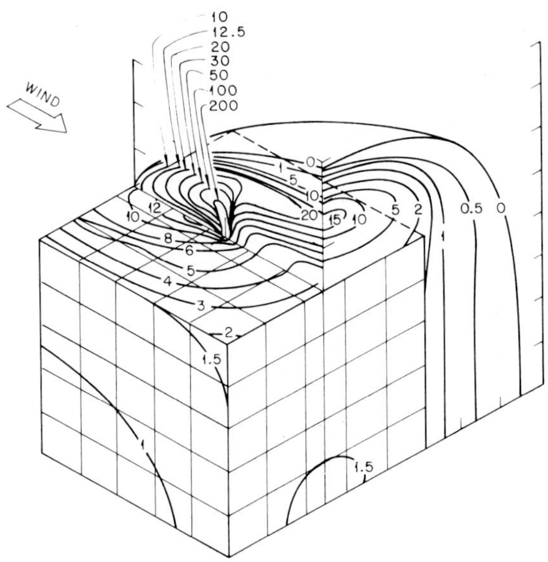

(a) Flush roof vent
L/H = W/H = 1.0
$w_e/U_H \simeq 1.0$

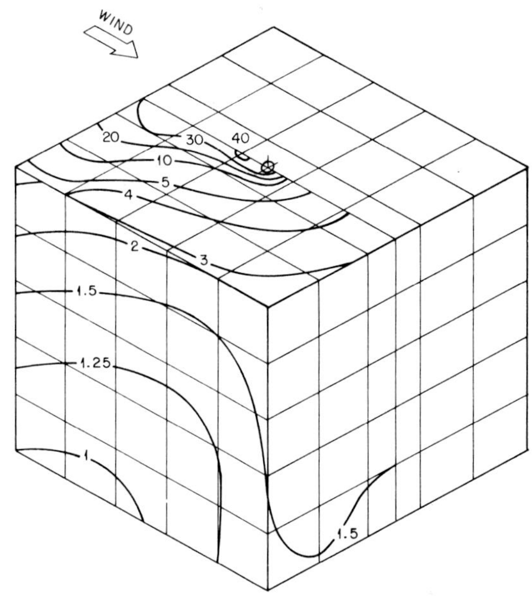

(b) Flush roof vent
L/H = W/H = 1.0
$w_e/U_H \simeq 0.5$

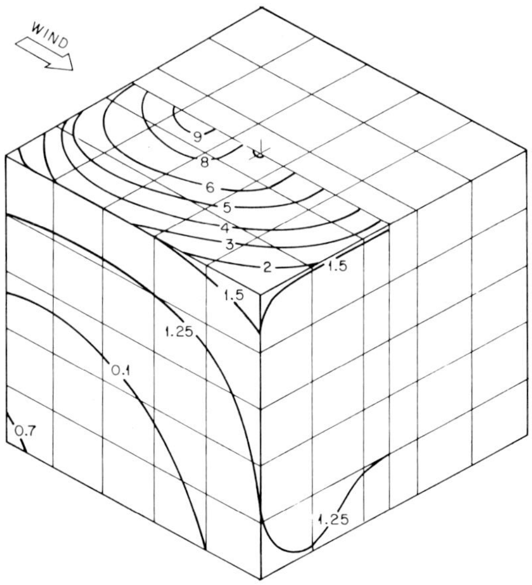

(c) Flush roof vent
L/H = W/H = 1.0
$w_e/U_H \simeq 2.0$

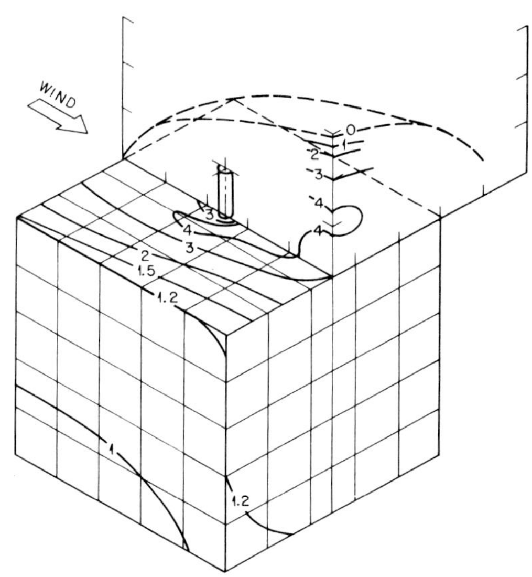

(d) Short roof stack
0.2H above roof
L/H = W/H = 1.0
$w_e/U_H \simeq 1.0$

(Figure continues on next page.)

Fig. 7.51 (continued)

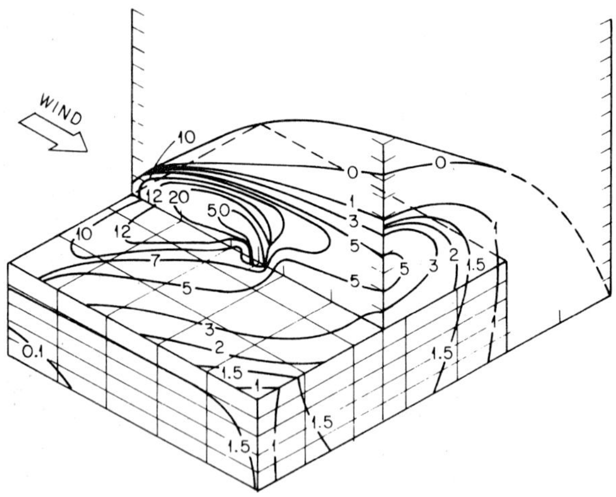

(e) Flush roof vent
L/H = W/H = 3.0
$w_e/U_H \simeq 1.0$

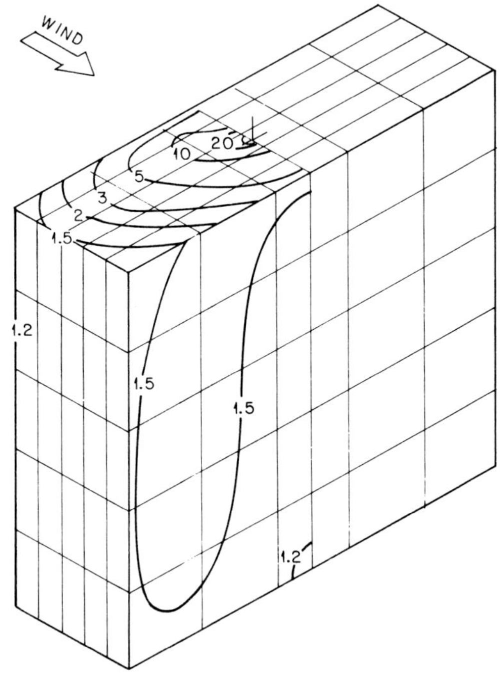

(f) Flush roof vent
L/H = 0.33; W/H = 1.0
$w_e/U_H \simeq 1.0$

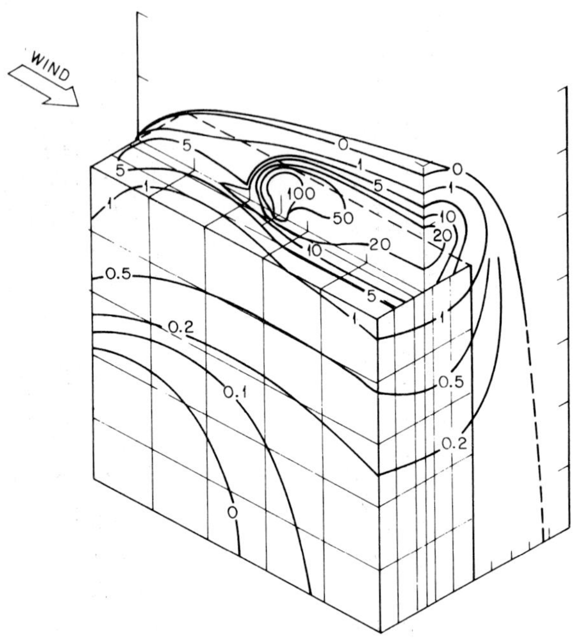

(g) Flush roof vent
L/H = 1.0; W/H = 0.33
$w_e/U_H \simeq 1.0$

Fig. 7.51 Concentration coefficient K contours for uniform incident wind and the effluent exit at roof center. [From J. Halitsky, *Gas Diffusion Near Buildings*, NYU Meteorology and Oceanography Geophysical Sciences Laboratory Report 63-3, New York University, College of Engineering, New York, 1963.]

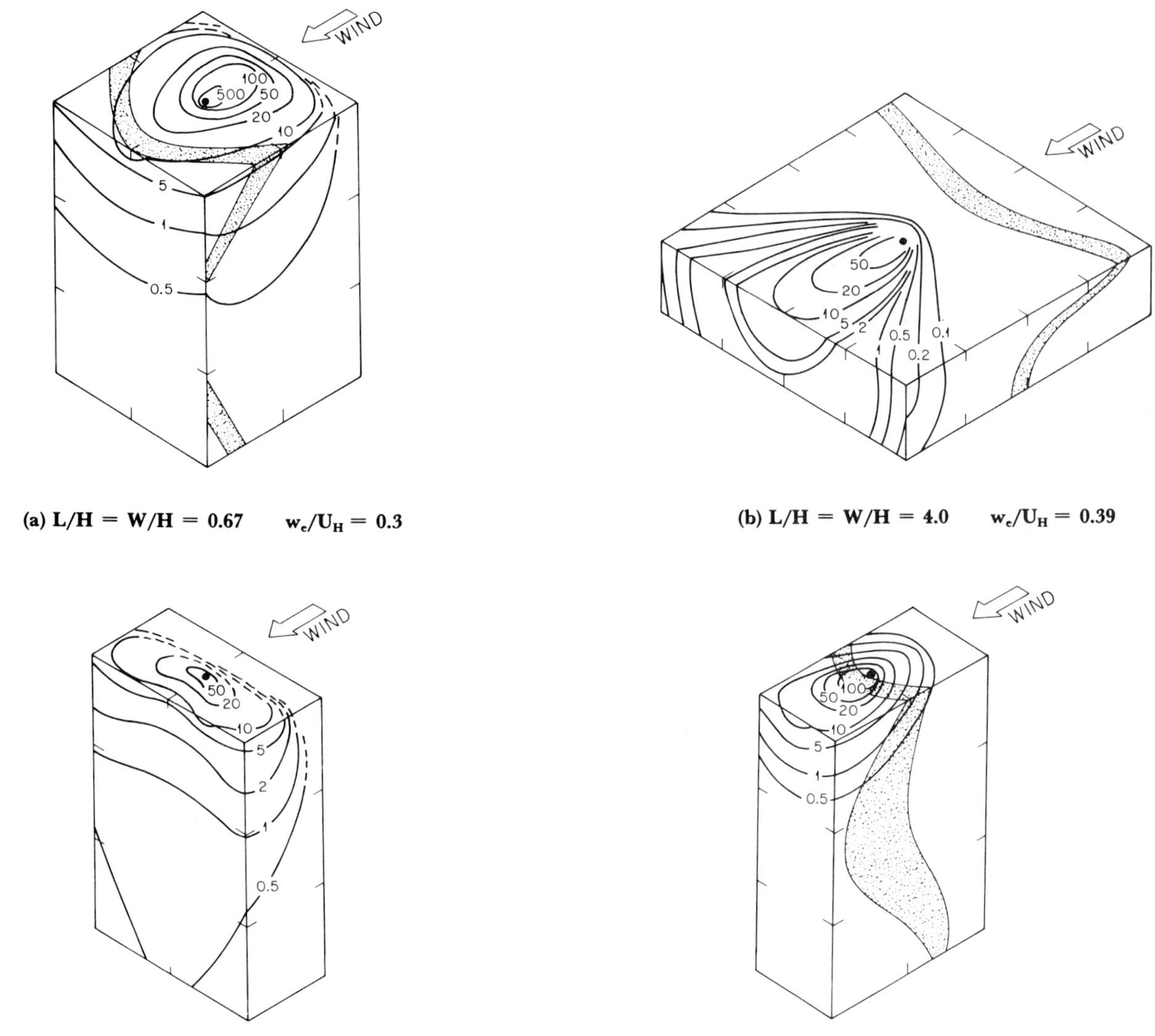

(a) L/H = W/H = 0.67 $w_e/U_H = 0.3$

(b) L/H = W/H = 4.0 $w_e/U_H = 0.39$

(c) L/H = 0.33; W/H = 0.67 $w_e/U_H = 0.29$

(d) L/H = 0.67; W/H = 0.33 $w_e/U_H = 0.29$

Fig. 7.52 Concentration coefficient $K = \chi U_H A_p/Q$ contours and zones of flow reattachment (shaded areas) for boundary-layer wind; effluents emitted at roof center from flush vents. [From D. J. Wilson, Contamination of Air Intakes from Roof Exhaust Vents, *ASHRAE Trans.*, 82(Pt.1): 1028-1038 (1976). Reprinted by permission of the American Society of Heating, Refrigerating, and Air-Conditioning Engineers, Inc. Figure also appears in D. J. Wilson, *Contamination of Building Air Intakes from Nearby Vents*, University of Alberta Department of Mechanical Engineering Report No. 1, University of Alberta, Edmonton, Alberta, Canada, 1976.]

Halitsky (1963a, b) may have been the first to suggest that, for a flush-vent-type source and a receptor somewhere on the surface of the same building, the dilution $D \equiv K_e/K$ is proportional to the square of the shortest distance between the vent and the measurement site. K_e is the concentration coefficient at the effluent vent, described above. Halitsky gave the relation

$$D \simeq \left[\tilde{\alpha} + 0.11\left(1 + \frac{\tilde{\alpha}}{5}\right)\left(\frac{s}{\sqrt{A_e}}\right)\right]^2 \qquad (7.20)$$

where s is the "stretched string" distance between the two points, A_e is the vent exit area, and $\tilde{\alpha}$ depends on the specific building/vent/atmospheric configuration. A lower bound on D is obtained by setting $\tilde{\alpha} = 1$. This expression is based, however, on tests conducted in a uniform, low-turbulence flow with an efflux-to-wind-speed ratio $w_e/U_H = 1$. More recently, similar tests on a more complex building resulted in an expression that is recommended by ASHRAE (1974) and explicitly uses this speed ratio:

$$D \simeq \left[4.66 + 0.147\left(\frac{s}{\sqrt{A_e}}\right)\right]^2\left(\frac{U_H}{w_e}\right) \qquad (7.21)$$

On the other hand, Wilson's (1976a, b; 1977a) tests in boundary layers used low values for w_e/U_H, 0.1 to 0.3 (depending on gas density); these values may be more typical of vents with rain deflectors or louvers. Wilson suggested that a lower bound to the dilution was

$$D_{min} \simeq 0.11\, K_e \frac{s^2}{A_p} \qquad (7.22a)$$

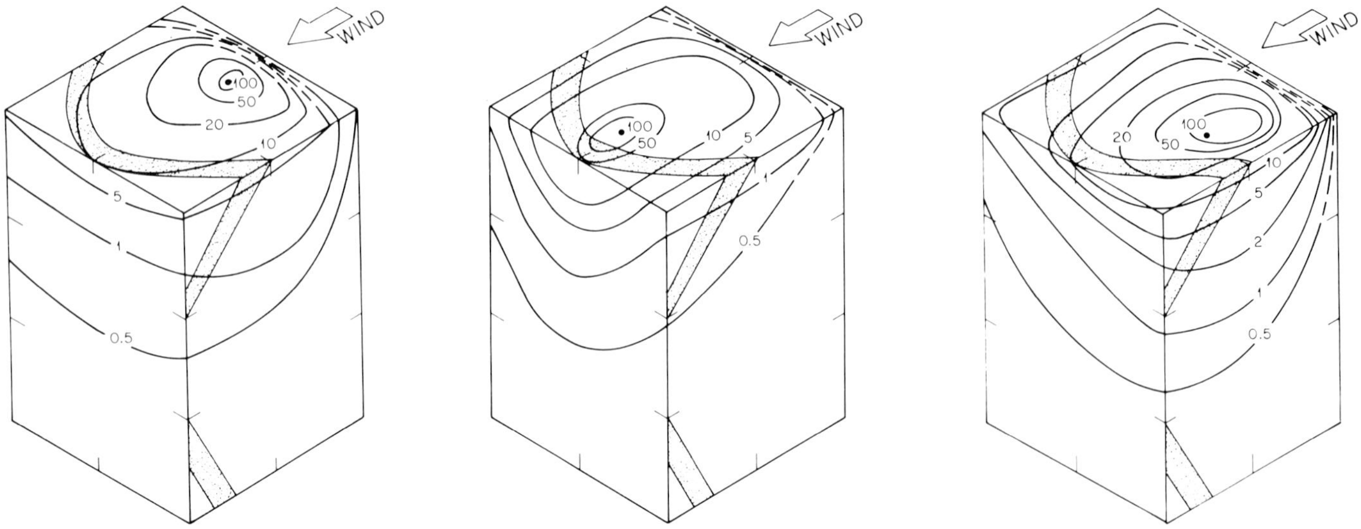

(a) Vent forward of center. **(b) Vent aft of center.** **(c) Vent to side of center.**

Fig. 7.53 Concentration coefficient $K = \chi U_H A_p/Q$ contours and zones of flow reattachment (shaded areas) for boundary-layer wind; effluents emitted from flush vents at different rooftop locations; $L/H = W/H = 0.67$ and $w_e/U_H \simeq 0.3$. (From D. J. Wilson, *Contamination of Building Intakes from Nearby Vents*, University of Alberta Department of Mechanical Engineering Report No. 1, University of Alberta, Edmonton, Alberta, Canada, 1976.)

An alternate convenient form is

$$D_{min} \simeq 0.11 \frac{U_H s^2}{\dot{V}_e} \tag{7.22b}$$

where $\dot{V}_e$ is the volume flow rate of effluent.

Plots of these equations together with data from several experiments indicate (Wilson, 1976a, b, 1977a; Halitsky, private communication) that Eq. 7.20, with $\tilde{\alpha} = 1$, provides a very conservative lower bound on dilution. On block-like structures, it may be an order of magnitude smaller than actual measurements. For a block-like structure, Eq. 7.22 offers less conservatism but, in most situations, still provides a lower bound to the data. However, it overpredicts (Halitsky, private communication) dilutions for block-like buildings at an angle to the wind and for a nonidealized structure (Clinical Center of the National Institute of Health in Bethesda). Equation 7.21 was specifically developed for this latter site.

In most cases, particularly for those involving mere nuisance contaminants (e.g., odors), Eq. 7.22 probably is an adequate lower bound to dilution. If the effluent is especially troublesome, however, an extra margin of safety can be included by using Eq. 7.20 with $\tilde{\alpha} = 1$. More data, particularly from field tests, would be helpful in validating this guideline.

In cases where the vent is somewhere on the sides of the building rather than on the roof, Wilson (1977a) found that Eq. 7.22 provides a good lower bound to the dilution, except when effluent emerges on the front of the structure and subsequently contaminates an air intake close to the ground on the building side. To compensate for this effect, Wilson suggested that the calculated minimum dilution be further reduced by a factor of five if any potential receptors were located at or below $H/5$ above the ground.

For more exact work, on-building concentration coefficients K often can be found in the literature (e.g., Halitsky, 1963a, b, 1968; Smith, 1975, 1978; Wilson, 1976a, b; and many others). The results are often, but not always, for very specific building configurations, such as a particular nuclear reactor. Nevertheless, substantial general guidance can be obtained from a careful study of these results and their variation with important parameters such as w_e/U_s and h_s/H. However, the experimental conditions—particularly for laboratory work—must be cautiously examined to assess the degree of validity of the data. For example, many early wind-tunnel studies were conducted in uniform speed, low turbulence flows that are not typical of the atmosphere; consequently, important phenomena such as flow reattachment to the model building roof or sides did not occur, and the resulting concentration patterns probably were considerably in error. A compendium of these data, together with a discussion of their general applicability and accuracy, would be most helpful.

Wilson (1976a) also made some useful estimates of the peak-to-mean concentration ratios that may be expected at roof or lee-face receptors as a result of effluent emitted at low speed from flush roof vents. The concentration probability distribution is assumed to be log-normal (see Csanady, 1973). Figure 7.54 indicates peak-to-mean ratios that may be exceeded a certain percentage of the time at a particular distance from the vent. Generally speaking, the fluctuating concentration will exceed the mean value by more than a factor of two only 10% of the time or less; it will exceed the mean by a factor of five only 5% of the time or less; and it will exceed the mean by a factor of ten only 1% of the time or less. Li, Meroney, and Peterka (1981) experimentally evaluated peak-to-mean concentration ratios behind a simple cube and found somewhat smaller values; the mean was exceeded by a factor of three only 1% of the time; the log-normal distribution, however, was followed fairly closely. Wilson (1976a) suggests that the minimum

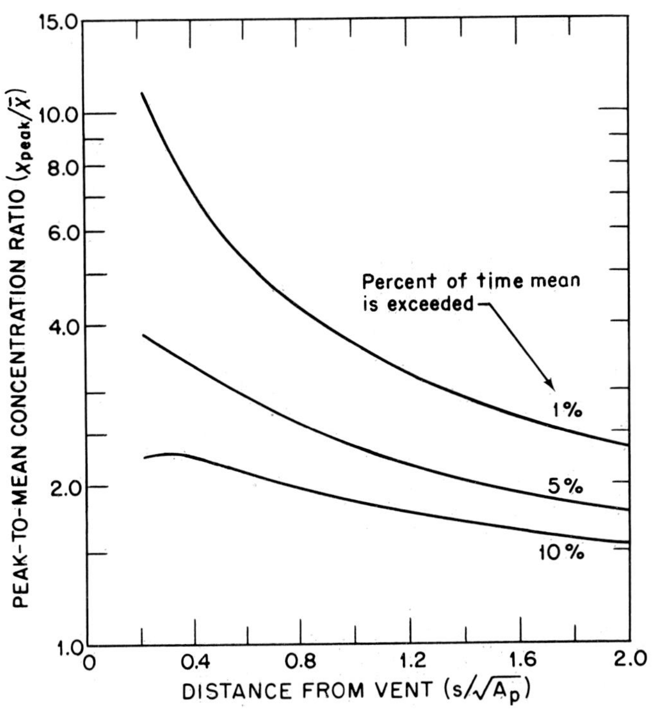

Fig. 7.54 Peak-to-mean concentration ratios for flush-mounted roof vents. (From D. J. Wilson, *Contamination of Building Air Intakes from Nearby Vents,* University of Alberta Department of Mechanical Engineering Report No. 1, University of Alberta, Edmonton, Alberta, Canada, 1976.)

dilution, as estimated from Eqs. 7.22, is a mean value suitable for averaging times of an hour or so. To estimate likely peak concentrations, he recommends

$$D_{peak} = f_S D_{min} \tag{7.23}$$

where the safety factor $f_S \simeq 0.2$ to 0.5 for nonhazardous but annoying effluents (e.g., kitchen odors); for toxic materials the safety factor is $f_S \simeq 0.1$. These results are all based on wind-tunnel studies; it is essential that field data be obtained to further validate these expressions.

7-8.2.1.3 Concentrations Within the Wake Cavity. As pointed out by Barry (1964) and Meroney (1982), most of the expressions proposed for concentration estimates within the cavity proper take the form

$$\chi \simeq \frac{Q}{cUA_p} \tag{7.24}$$

where the coefficient c is given values between 0.5 and 5.0. The effluent is presumed to be entirely captured in the wake. Evidently $K \equiv$ constant $(=1/c)$, with values between 0.2 and 2.0. A glance at Halitsky's (1963a, b) results (e.g., Figs. 7.51), where the concentrations on the lee face should be fairly representative of the cavity as a whole, indicates that these are indeed reasonable values. However, little understanding of the physical mechanisms that generate this result is conveyed, and, in particular, there is nothing to explain why the result might vary from one building to another.

Briggs's (1973) recommendation takes some account of the effective stack height and the building size. He defined the concentration coefficient in terms of the smaller building dimension ζ_b: $K \equiv \chi U \zeta_b^2/Q$. Consider the differential stack height $h' - H$, where h' is given by Eq. 7.15. If this is greater than $0.35\zeta_b$, then K generally is less than unity in the cavity. However, if $h' - H < 0.35\zeta_b$, K is typically 1.5 and perhaps as large as 3.0.

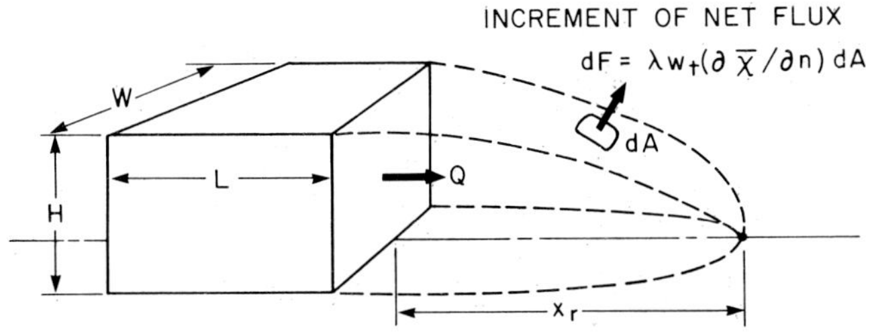

Fig. 7.55 Conceptual model of recirculating wake cavity with mass transfer confined to turbulent exchanges across the shear layer bounding the cavity. The cavity volume is V_c, and it is bounded by an active surface of area A_c. Effluent from a source of strength Q is fully entrained into the cavity.

A slightly more physical hypothesis is based on Vincent's (1977, 1978) ideas. As detailed in earlier sections, the cavity "bubble" is not, strictly speaking, a closed zone of recirculation with material transfer possible only across the turbulent shear layer bounding the region. Advection into and from the cavity can occur as well (see Fig. 7.15). Vincent's (1978) experiments suggest, however, that the bulk of the material transfer is conducted by the turbulent exchange processes across the boundary and only relatively minor contributions result from direct advection. Therefore, the cavity is modeled, at least to a first approximation, as a volume V_c bounded by an active surface of area A_c (Fig. 7.55). Material is emitted into this volume at a rate Q; it is mixed thoroughly to an average concentration $\bar{\chi}$ by the circulations within the bubble; and it is finally lost from the region because of turbulent transfer through the active surface. The net flux of matter through the boundary is approximated by $\lambda w_t \partial\bar{\chi}/\partial n$, where λ is the turbulent mixing length, w_t is a turbulent transport velocity, and n is the local normal to the surface A_c. The mass balance equation is

$$\frac{d}{dt}\left(\bar{\chi}V_c\right) \simeq Q - \lambda w_t \int_{A_c} \frac{\partial\bar{\chi}}{\partial n}\, dA$$

where $\bar{\chi}$ is the average concentration within the cavity. The integral is approximated by $\bar{\chi}A_c/\delta$, where δ is the thickness of the bounding shear layer across which the concentration changes from its internal value $(\bar{\chi})$ to its external value (0). The equation becomes

$$\frac{d\bar{\chi}}{dt} + \frac{\bar{\chi}}{t_d} \simeq \frac{Q}{V_c} \tag{7.25a}$$

where t_d is a characteristic diffusion time

$$t_d \equiv \frac{1}{w_t}\frac{\delta}{\lambda}\frac{V_c}{A_c} \tag{7.25b}$$

Case 1: constant source. At $t = 0$, $\bar{\chi} = 0$ and $Q =$ constant, then

$$\bar{\chi} \simeq \frac{Qt_d}{V_c}\left(1 - e^{-t/t_d}\right) \tag{7.26a}$$

Case 2: initial puff. At t = 0, $\bar{\chi} = \bar{\chi}_0$ and Q = 0, then

$$\bar{\chi} \simeq \bar{\chi}_0 \, e^{-t/t_d} \tag{7.26b}$$

Note that the latter result provides for a convenient means of evaluating t_d by injecting a known amount of material into the cavity and plotting the logarithm of the resulting concentration versus time. This procedure was carried out in field measurements by Drivas and Shair (1974). The steady-state limit of Eq. 7.26a is $\bar{\chi} \simeq Qt_d/V_c$, so that the average cavity concentration coefficient $\bar{K} \equiv \bar{\chi}U\,A_p/Q$ is

$$\bar{K} \simeq \frac{A_pUt_d}{V_c} \tag{7.27}$$

This expression can be interpreted as the ratio of the volume of air intercepted by the building in time t_d to the volume of air in the cavity. A more interesting interpretation is available if the cavity volume is approximated† by $V_c \sim A_px_r$, where x_r is the cavity length; then

$$\bar{K} \sim \frac{t_d}{x_r/U} \tag{7.28a}$$

That is, the cavity concentration coefficient is a measure of the characteristic diffusion time t_d relative to the characteristic "flushing" time of the cavity, x_r/U. In terms of the nondimensional time constant $\tau \equiv Ut_d/H$, the average coefficient is

$$\bar{K} \sim \frac{\tau}{x_r/H} \tag{7.28b}$$

This result may be more convenient to evaluate in some cases because τ and x_r/H may be available from experiments (e.g., Vincent, 1977, 1978) or from empirical formulations such as Eqs. 7.1 and 7.2. For a tall thin building or a short wide structure, the height H may not be the dimension most characteristic of the structure; in such cases $\sqrt{A_p}$ or Wilson's (1979) $\tilde{R}$ probably should be substituted for H in the above expressions. For example, in the field work of Drivas and Shair (1974), W/H = 5.3 and L/H = 1.4, not counting the penthouse on the roof. The velocity U was about 2.5 m/sec; the cavity reached about 3H (36 m) from the lee face of the building; and the characteristic diffusion time $t_d \simeq 60$ sec. Direct use of Eq. 7.28a gives $\bar{K} \simeq 4.2$, which seems somewhat large. If, however, we use $\tau = Ut_d/\sqrt{A_p}$ for this rather wide building, $\tau \simeq 5.2$ and therefore $\bar{K} \simeq 1.7$, which appears reasonable.

Before Eq. 7.27 and/or 7.28 can be used routinely, they must be checked experimentally. For example, direct evaluation of $\bar{K}$ using the limited data of Vincent (1978) and Eq. 7.28b gives a concentration coefficient that is approximately a factor of three larger than those cited by Barry (1964). A suitable adjustment factor may be required to compensate for the approximations implicit in the model and for the liberties taken with the turbulent flux parameterization. Direct measurements of $\bar{K}$, τ, and x_r/H are needed to verify this.

It must be emphasized that this simple model of mass transfer from a building's near wake is at odds with present understanding of the very complex advective and turbulent exchange processes that feed and drain recirculation zones (e.g., Sec. 7-5.2.2 and Fig. 7.15). Additional work is needed to elucidate the significance of the advective components in particular. In the meantime, the simple scheme just outlined must be regarded as highly tentative.

One last comment should be made about locating sources within the cavity: use of a moderately tall stack in this region can lead to surprising concentration levels. The work of Huber et al. (1976) on stacks behind a long model ridge illustrates this effect relatively well. In particular, they noted that a stack height ratio $h_s/H = 1.5$ produced the highest observed ground-level concentrations near the point of cavity closure. Evidently the effluent was caught in the upper portion of the recirculation pattern and carried to the surface. This is one situation where a shorter stack may provide lower concentrations than a tall one.

7-8.2.2 Concentrations Downwind of the Cavity ***Region.*** Briggs's (1973) method, outlined in the previous section, can be used to estimate an effective stack height for plumes emitted near buildings but not entrained completely in their wakes. This height is then used in the standard Gaussian plume expression for an elevated source. Briggs suggests that if the downwash-corrected stack height h′ is less than $H + 1.5\zeta_b$ (see Sec. 7-8.2.1.1), and if the release point is on the roof, or anywhere within about $\zeta_b/4$ of the building, then the plume is within the zone of building influence and should be treated as discussed earlier.

In cases of complete plume entrainment into the wake cavity, the methods generally used are based on simple modifications to the Gaussian plume equation for a ground source

$$\chi = \frac{Q}{\pi\sigma_y\sigma_zU} \tag{7.29}$$

along the plume centerline.

For example, one can account for the initial rapid mixing of the effluent into the wake by introducing a virtual source at some location upwind of the actual vent. A binormal distribution of material is assumed, with $\sigma_{y0} \simeq W/4.3$ and $\sigma_{z0} \simeq H/2.15$ (e.g., Turner, 1969) representing the dispersion parameters at the building. Next, evaluate the virtual locations x_{y0} and x_{z0} such that $\sigma_y(x_{y0}) = \sigma_{y0}$ and $\sigma_z(x_{z0}) = \sigma_{z0}$. Equation 7.29 is then employed using $\sigma_y = \sigma_y(x + x_{y0})$ and $\sigma_z = \sigma_z(x + x_{z0})$, where x is the physical distance between the building and the receptor site. Meroney (1982) has suggested virtual source locations for a number of likely stack-to-building height ratios.

A second method, introduced by Gifford (1960), uses the idea (attributed to Fuquay) of the building wake providing an initial plume dilution proportional to the product of wind speed and projected building area. Ground-level concentrations along the plume centerline are then given by

$$\chi = \frac{Q}{(\pi\sigma_y\sigma_z + cA_p)U} \tag{7.30}$$

The constant c was estimated on intuitive grounds (Gifford, 1975) to be between 1/2 and 2. The smaller value is the more conservative of the two and agrees relatively well with assorted test results (see Gifford, 1975, 1976).

†This may be most appropriate for flow that has reattached to the building and roof and sides.

A third method introduces total diffusion parameters Σ_y and Σ_z (Gifford, 1968, attributed to Davidson)

$$\chi = \frac{Q}{\pi \Sigma_y \Sigma_z U} \tag{7.31a}$$

where

$$\Sigma_y \equiv \left[\sigma_y^2 + \frac{cA_p}{\pi}\right]^{1/2} \tag{7.31b}$$

$$\Sigma_y \equiv \left[\sigma_z^2 + \frac{cA_p}{\pi}\right]^{1/2} \tag{7.31c}$$

The same range of c values applies as in the previous method.

In many cases the effluent plume will be neither completely elevated nor entrapped by the building wake; instead, a fluctuating partial entrainment in which portions of the plume are intermittently captured by the wake may occur. Johnson et al. (1975) introduced the so-called "split-h" model to account for the two different effective release heights generated by this intermittency. The model assumes that, in any given hour, a fraction ϵ of the plume will be entrained. The average concentration over an hour is then given by

$$\left(\frac{\chi}{Q}\right)_{ave} = \epsilon \left(\frac{\chi}{Q}\right)_{entr} + (1 - \epsilon) \left(\frac{\chi}{Q}\right)_{elev} \tag{7.32}$$

The concentration for the entrained portion is calculated from Eq. 7.31; a Gaussian crosswind distribution is assumed. The elevated concentration is computed from a standard Gaussian formulation with an effective source height estimated from Briggs's (1973) procedure (previously discussed). The entrainment fraction ϵ depends on parameters such as w_e/U_s, h_s/H, stack location, building geometry, atmospheric stability and turbulence intensity, and wind speed and direction. Johnson et al. (1975) found that, for the particular case of the Millstone nuclear reactor, w_e/U_s was the dominant influence, and they suggested an empirical form for ϵ that was based on w_e/U_s acting as such. However, their result probably is not appropriate for other sites. In fact, Thuillier and Mancuso (1980), using Johnson et al.'s form for ϵ, observed poor agreement with field data at a different site. Koga and Way (1979) have shown the strong dependence of partial entrainment on h_s/H, vent location, and w_e/U_s for a simple block structure. Experiments such as theirs may eventually produce schemes to estimate ϵ for a particular set of conditions, but this capability currently is not available.

As Gifford (1975, 1976) points out, all of these methods have two potential drawbacks. The first, which probably is not the fault of the model, is that the dispersion parameters generally used do not provide an adequate representation of the plume meander typically observed under light-wind stable conditions. In these circumstances, use of standard expressions or curves for σ_y and σ_z probably will provide significant overestimates of the concentration. Use of site-specific values for the dispersion parameters would alleviate this unwarranted conservatism.

The second shortcoming of the methods is inherent in their assumption that the dispersive characteristics of the atmosphere are unaltered by the presence of the building, which serves only to introduce some initial dilution. Far downwind the building effect is assumed negligible. This may or may not be true. As described in earlier sections, under some conditions wakes display a pronounced vortex character far downwind. This is particularly apt to occur for buildings at an angle to the wind. These highly persistent vortices typically exhibit rather low swirl velocities and considerable meander, and they tend to rise away from the ground with increasing downwind distance. Material entrained within these vortices will not spread in the normal fashion and may wind up as an elevated plume. Concentrations, especially aloft, may then drop off much more slowly with distance than the simple Gaussian formulation would predict. It is conceivable, then, that the Gaussian predictions will not be conservative under certain circumstances; indeed, they may predict concentrations lower than those actually available. Field programs are needed to attempt to detect such organized vorticity in the wake of isolated structures and to assess the vorticity's effect on concentration patterns. If significant vorticity effects are detected downwind of buildings, it will be necessary to use more complex models for wake diffusion that account for the effects of vortex flows on the concentration pattern (e.g., Kothari, Peterka, and Meroney, 1979).

Some attention has been devoted recently to development of improved expressions for σ_y and σ_z that better account for the effects of an obstacle on the concentration field. Huber and Snyder (1976) examined concentrations in the wake of a model building for various effluent release heights. They observed that emissions between ground level and roof height were dispersed rapidly in both the vertical and crosswind directions, whereas releases above roof level experienced only enhanced vertical dispersion. On the basis of this work, Huber (1977, 1979) recommended forms based on the concept of spread enhancement generated first by the building cavity and second by the decaying turbulence excess produced by the building (Huber and Snyder, 1976). For $3 \leqq x/H \leqq 10$, these forms are, for a squat building

$$\sigma_y' = 0.7\left(\frac{W}{2}\right) + 0.067(x - 3H)$$
$$\sigma_z' = 0.7H + 0.067(x - 3H) \tag{7.33a}$$

whereas for $10 \geqq x/H$, the forms are

$$\sigma_y' = \sigma_y(x + x_{y0})$$
$$\sigma_z' = \sigma_z(x + x_{z0}) \tag{7.33b}$$

Huber uses these in the Gaussian plume model, which, along the centerline, is simply

$$\chi = \frac{Q}{\pi \sigma_y' \sigma_z' U} \exp\left(\frac{-h_s^2}{2\sigma_z'^2}\right) \tag{7.34}$$

The virtual source locations x_{y0} and x_{z0} are calculated by requiring that $\sigma_y'(10W) = 0.7(W/2) + 0.5H$ and that $\sigma_z'(10H) = 1.2H$. The distance x is measured from the lee face of the obstacle. Figure 7.56 shows σ_y' and σ_z' for several building heights, where W is taken to be 2H. Note that atmospheric stability effects appear only for $x/H \geqq 10$.

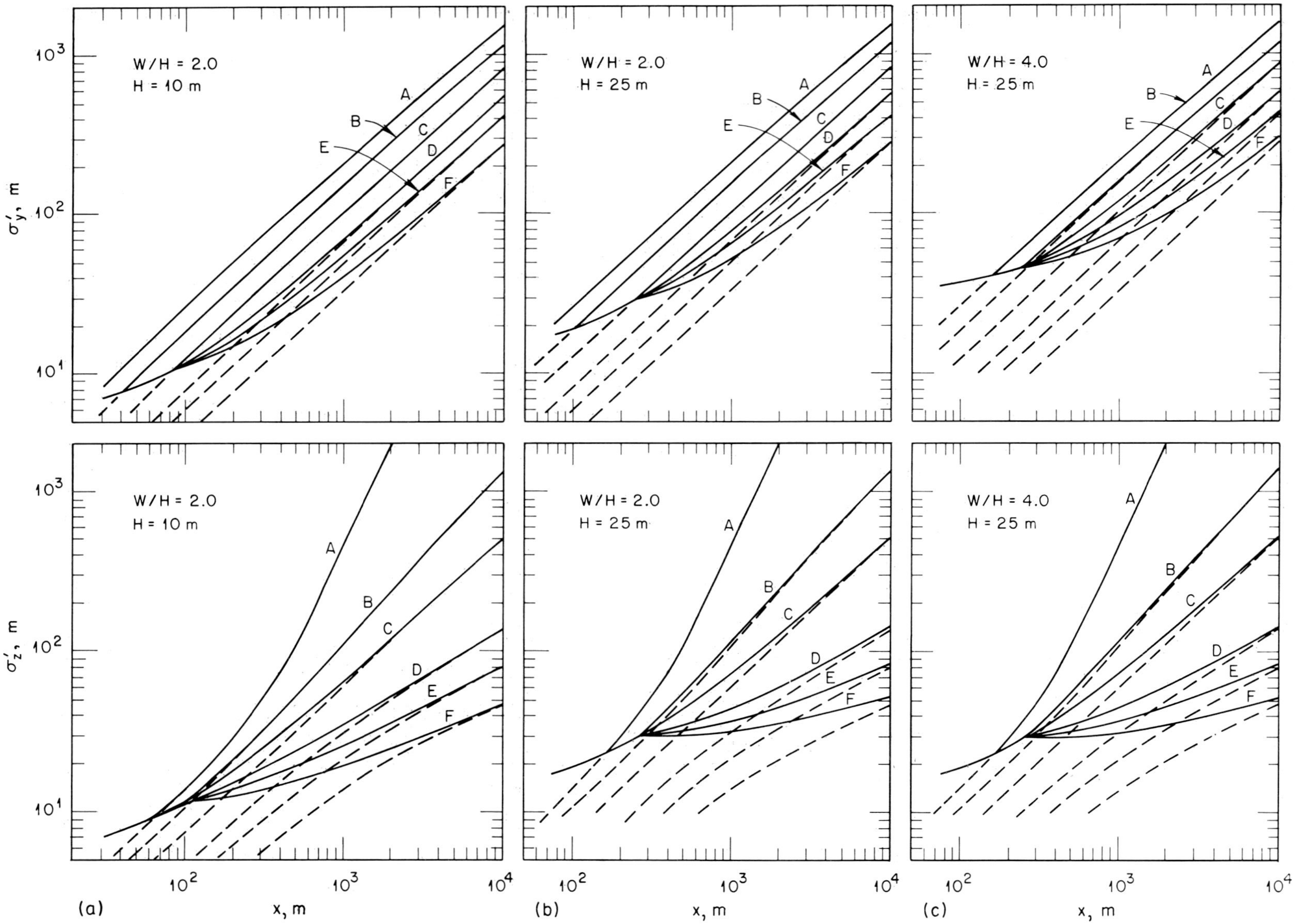

Fig. 7.56 Sample evaluation of building-wake-enhanced dispersion coefficients σ'_y and $\sigma'_z(x)$ (broken lines) compared to standard Pasquill–Gifford coefficients (solid lines) for buildings of different sizes and shapes (based on Huber's 1977 and 1979 recommendations). (Based on A. H. Huber, Incorporating Building/Terrain Wake Effects on Stack Effluents, in *Preprints of Joint Conference on Applications of Air Pollution Meteorology*, Salt Lake City, Nov. 29–Dec. 3, 1977, American Meteorological Society, Boston, Mass., 1977.)

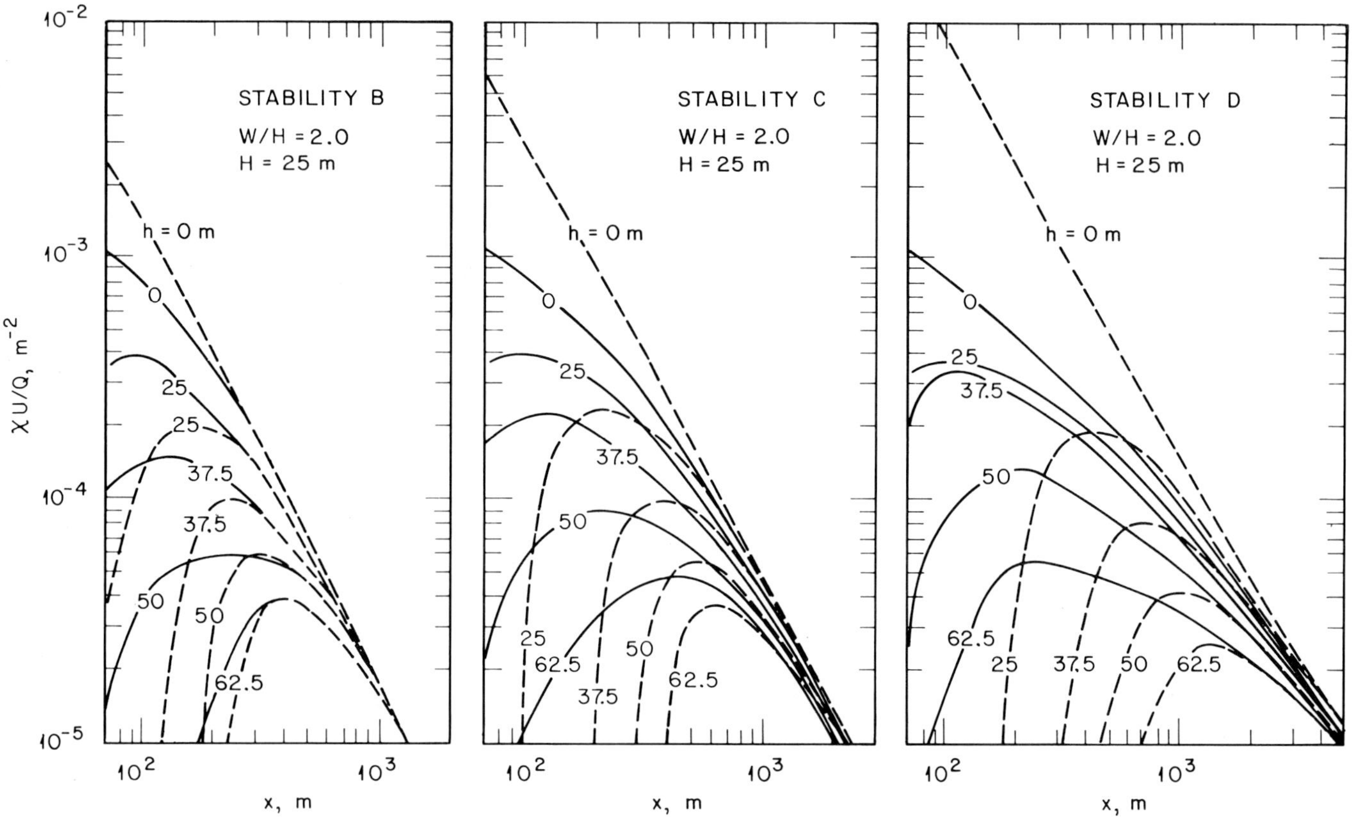

Fig. 7.57 Comparison of concentration estimates with (solid lines) and without (broken lines) building wake enhancement for various emission heights and atmospheric stability categories. (From A. H. Huber, *An Evaluation of Obstacle Wake Effects on Plume Dispersion*, presented at the Fourth Symposium on Turbulence, Diffusion, and Air Pollution, Reno, Nev., Jan. 15–18, 1979, American Meteorological Society, Boston, Mass.)

Figure 7.57 compares concentration estimates using the enhanced dispersion parameters σ_y' and σ_z' to those calculated using the standard forms for σ_y and σ_z. For effective source heights $h_e \leqq H$, both horizontal and vertical enhancement were assumed—i.e., σ_y' and σ_z' were used in Eq. 7.34. For effective stack heights $h_e > H$, only vertical enhancement was allowed—i.e., σ_y and σ_z' were used in the plume equation. Only for a ground-level source does enhancement produce lower concentrations everywhere. For all elevated sources considered, the enhanced dispersion model produced higher maximum ground-level concentrations at locations closer to the stack. The effects diminished with increasing stability and increasing stack height.

The agreement with Huber and Snyder's (1976) wind-tunnel data is fairly good (Fig. 7.58), especially for source heights $h_s \leqq 1.5H$. However, Hatcher et al. (1978) found that concentrations close behind a model reactor complex were underpredicted for elevated effluent releases; ground-level releases, however, were estimated fairly well. At least part of the discrepancy may result from surrounding structures that influence the wake of the main reactor building and modify the rates of wake spread and decay.

Agreement with field data seems to range from poor to good on a case-by-case basis (Fig. 7.59, from Huber, 1979). Part of the difficulty is that virtually all of the field work has involved clusters of buildings; therefore, it is difficult to determine appropriate single building heights and widths for use in the model, and the wake behavior may differ from that of an isolated obstacle. Also, the presence of other buildings between the source and receptor locations in some of these experiments may have diverted effluent material to other regions. Finally, none of the field studies determined the vertical plume height; only ground-level measurements of the horizontal spread were made. Consequently, it is impossible to say at this time whether or not the model assumptions (e.g., enhancement in both the horizontal and vertical parameters for low sources, but only the vertical parameter for higher sources) are verified in the field. As remarked before, field studies that determine both horizontal and vertical dispersion characteristics are urgently needed.

Concentration measurements behind simple isolated buildings in the field are rare and seem to have been restricted largely to the near-wake region (e.g., Smith, 1975, 1978). The work of Hinds (1969) appears to have been conducted to greater distances downwind than any other, and even it reached only to about $5\sqrt{A_p}$. Because the tracer material in his tests was released on the windward side of the building, the near-wake region was not very well mixed (as shown by his isopleth patterns), and the resulting rapid change of concentration with distance probably is not typical of that behind a building with a cavity-entrained plume.

Meroney and Yang (1970, 1971) examined the concentrations behind a cubical model to more than 30 building heights downwind. For ground-level releases, their concentration coefficient K decreased roughly as x to the −0.6 to −0.7 power. This decrease represents a much slower decay rate than would be estimated from the open-terrain Gaussian model, which,

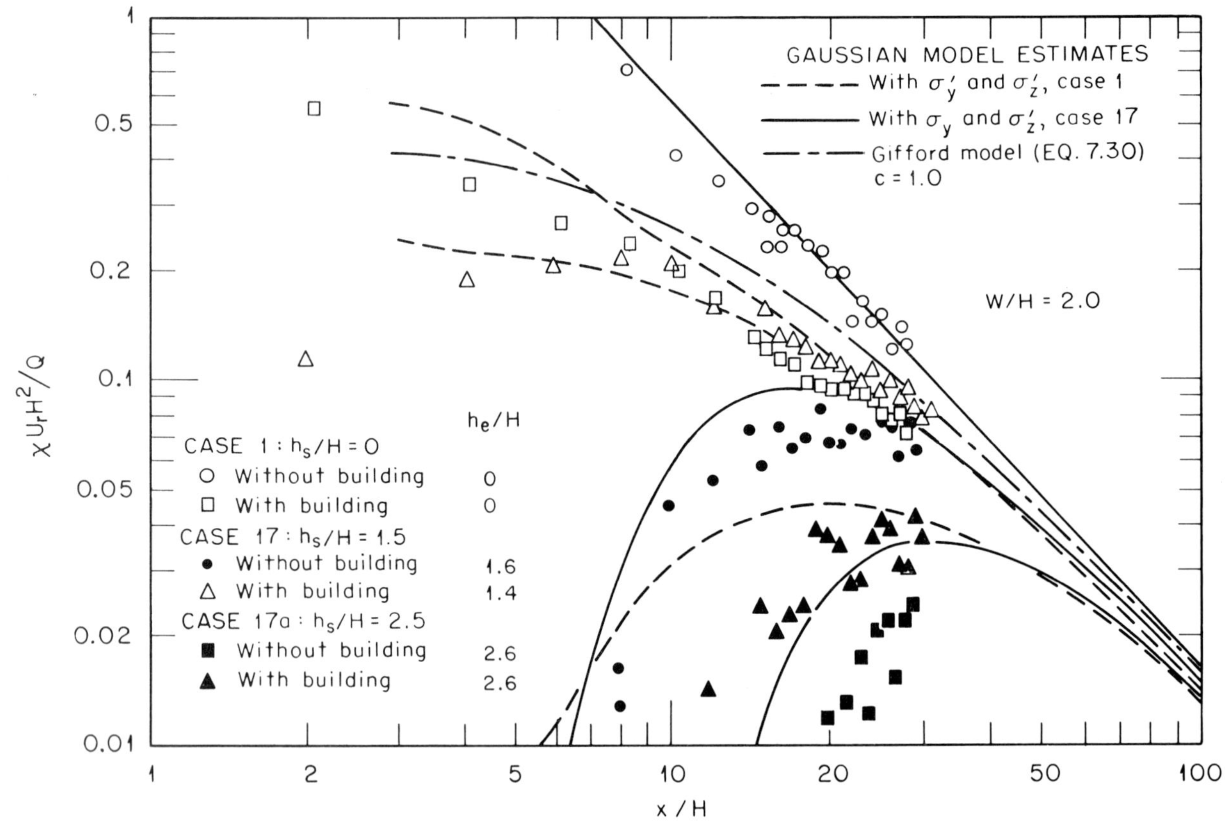

Fig. 7.58 Comparison of concentration coefficient K = $\chi U_r H^2/Q$ predicted from standard Gaussian model (solid lines), from Gifford's wake model (Eq. 7.30, dot–dash line), and from model using wake-enhanced dispersion parameters (broken lines) to wind-tunnel data for various stack-to-building height ratios, h_s/H. U_r is wind speed at 1.5 H. (From A. H. Huber, W. H. Snyder, R. S. Thompson, and R. E. Lawson, Jr., *The Effects of a Squat Building on Short Stack Effluents,* Environmental Sciences Research Laboratory Report EPA-600/4-80-055, U. S. Environmental Protection Agency, 1980.)

depending on stability, predicts powers of –1.3 to –1.6. Examination of Huber and Snyder's (1976) data in Fig. 7.58 indicates that, between 10H and 30H downstream of a ground-level source near a building, a line through the data points will behave roughly as $x^{-0.8}$. Beyond 30H, however, the wake-enhanced Gaussian model approaches a –1.3 power. On the other hand, Meroney and Yang (1971) carried out measurements to more than 50 building heights downwind with no sign of a similar increase in decay rate. Meroney and Yang (1971) hypothesize that this resulted from a very persistent horseshoe vortex that restricted lateral spread of the wake. The detailed flow measurements needed to confirm this explanation are not available.

Symes and Meroney (1970) reported concentration measurements behind a short cylinder and a sequence of cone frustrums of various tapers. The isopleth patterns varied somewhat with the body shape (Fig. 7.60), but the centerline ground-level concentrations beyond $x/H \simeq 10$ all were similar; they decayed as x to the –0.8 to –1.0 power. The measurements extend to about 35H downwind; a trend to more rapid change with distance is not evident in the data.

7-8.2.3 Clusters of Buildings. Nearly all field studies of diffusion in building wakes actually have involved clusters of structures typically at major industrial sites such as power stations. Consequently, it is often difficult to separate general trends from special effects generated by site-dependent flow interactions among and behind the obstacles. Such complicated flows may be at least partially responsible for the observed variations in far wake behavior in particular; such speculation can only be confirmed or repudiated by further measurements.

7-8.2.3.1 Near Wakes in Building Clusters. Because flow visualizations (smoke) and effluent releases usually are carried out on or immediately adjacent to some dominant structure such as a reactor containment vessel, much of the field work accomplished in the near wake region of building clusters probably is equally applicable to single building wakes. For example, Abbey (1976) and Start et al. (1977) described rather similar strong vertical mixing of smoke throughout the cavity of two dissimilar reactor buildings. During low-wind-speed stable conditions in particular, the building wake acts to convert a ground-level source into an effectively elevated one; plume concentration maxima appear aloft downwind of the obstacle.

Preliminary experiments with a pair of single cubes placed along the mean wind direction in a simulated atmospheric boundary layer have been described by Meroney, Peterka, and Kothari (1980). Effluent was emitted from the upwind cube at the roof center, using (alternately) a flush vent or a short stack, or from the center of the rear face. Higher concentrations were observed on the lee face of the downwind building than were found on its windward face. The effluent apparently was lifted up and out of the recirculation zone behind the upwind cube; consequently, concentrations on the upwind face of the downwind structure were reduced. This escaping effluent

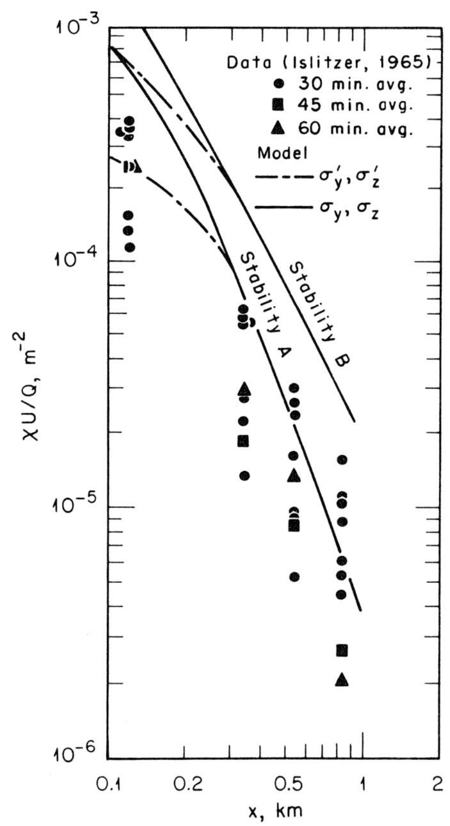

(a) Ground-level release. (From N. F. Islitzer, *Aerodynamic Effects of Large Reactor Complexes upon Atmospheric Turbulence and Diffusion*, USAEC National Reactor Testing Station Report IDO-12041, U. S. Weather Bureau, Idaho Falls, Idaho, 1965.)

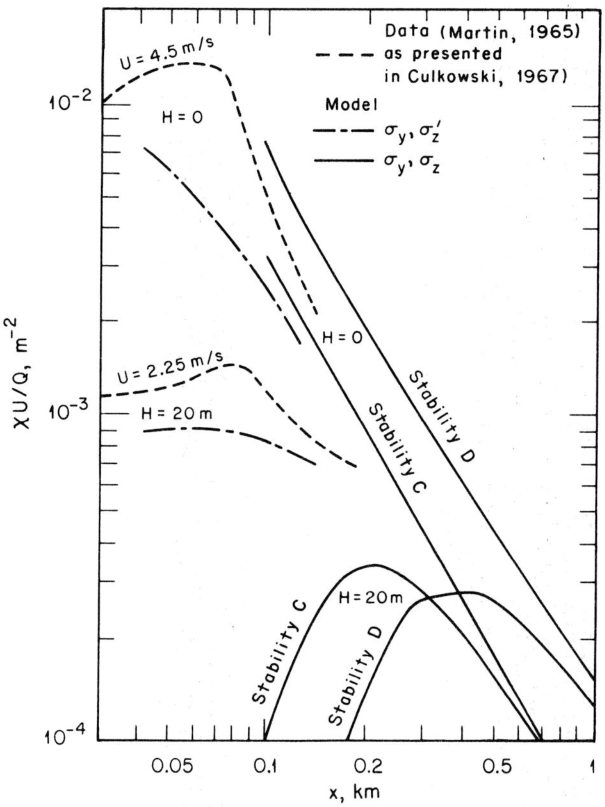

(b) Elevated release. (From J. E. Martin, The Correlation of Wind Tunnel and Field Measurements of Gas Diffusion Using Krypton-85 as a Tracer, Ph.D. Thesis and Report No. MMPP-272 of Michigan Memorial Phoenix Project, University of Michigan, Ann Arbor, Mich., 1965.)

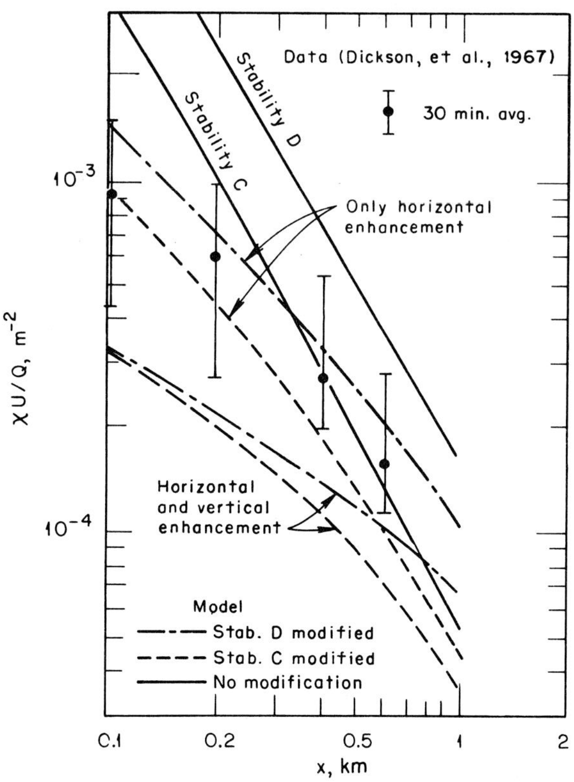

(c) Ground-level release, neutral and unstable condition. (From C. R. Dickson, G. E. Start, and E. H. Markee, Jr., Aerodynamic Effects of the EBR-II Containment Vessel Complex on Effluent Concentration, in the *Proceedings of the USAEC Meteorological Information Meeting,* Chalk River Nuclear Laboratories, Sept. 11–14, 1967, pp. 87-101, AECL Report 2787, Atomic Energy of Canada Ltd., Chalk River, Ontario, 1967.)

(Figure continues on next page.)

Fig. 7.59 (Continued)

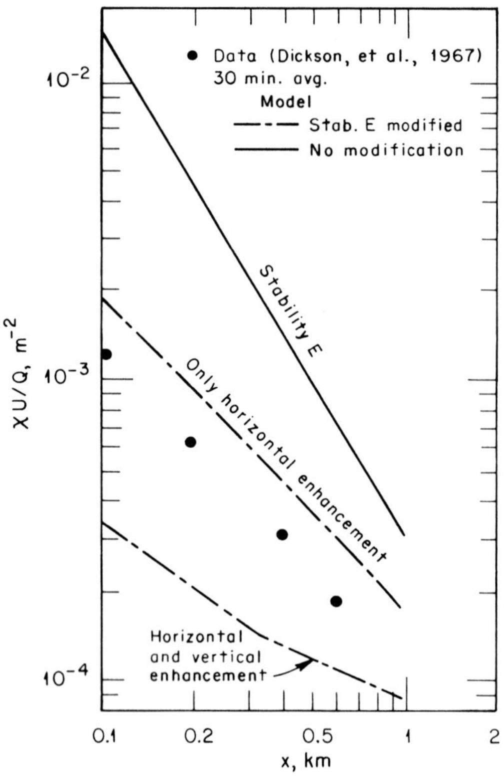

(d) Ground-level release, stable condition. (From C. R. Dickson, G. E. Start, and E. H. Markee, Jr., Aerodynamic Effects of the EBR-II Containment Vessel Complex on Effluent Concentration, in the *Proceedings of the USAEC Meteorological Information Meeting*, Chalk River Nuclear Laboratories, Sept. 11–14, 1967, pp. 87-101, AECL Report 2787, Atomic Energy of Canada Ltd., Chalk River, Ontario, 1967.)

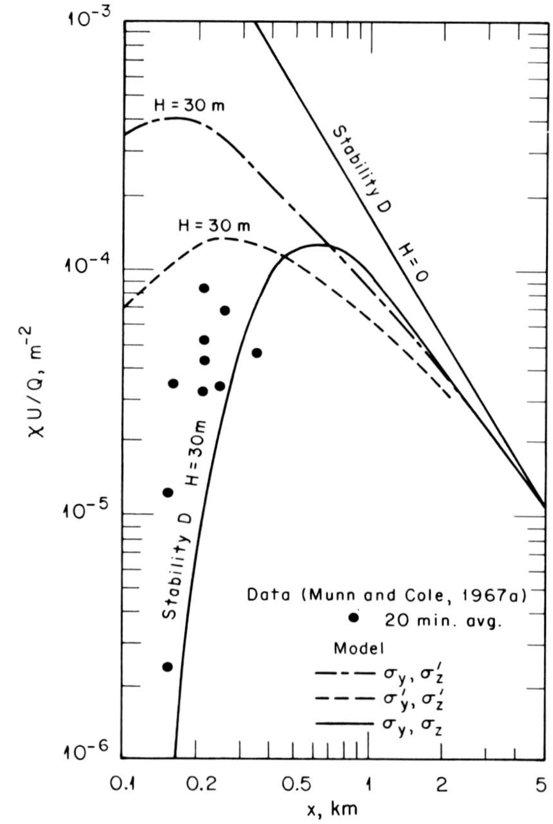

(e) Elevated release. [From R. E. Munn and A. F. W. Cole, Turbulence and Diffusion in the Wake of a Building, *Atmos. Environ.*, 1(1): 33-43 (1967).]

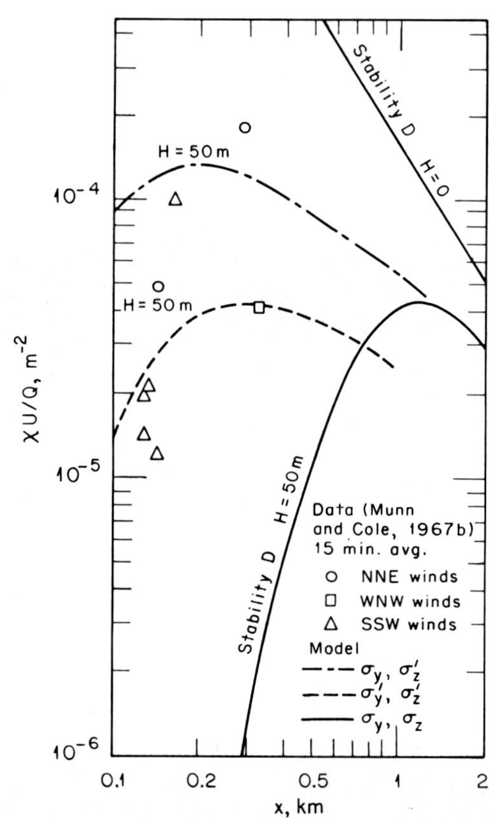

(f) Elevated release. [From R. E. Munn and A. F. W. Cole, Some Strong-Wind Downwash Diffusion Measurements at Douglas Point, Ontario, Canada, *Atmos. Environ.*, 1(5): 601-604 (1967).]

(Figure continues on next page.)

Fig. 7.59 (Continued)

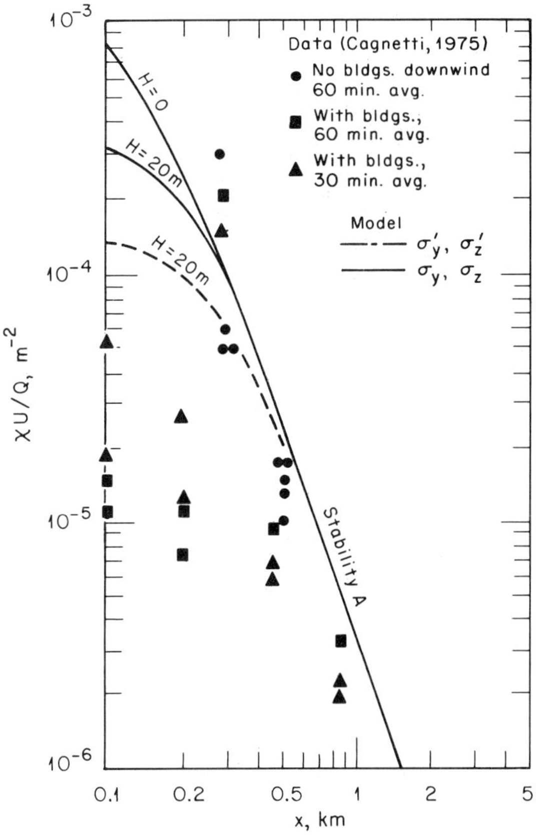

(g) Elevated release. [From P. Cagnetti, Downwind Concentrations of an Airborne Tracer Released in the Neighbourhood of a Building, *Atmos. Environ.*, 9(8): 739-747 (1975).]

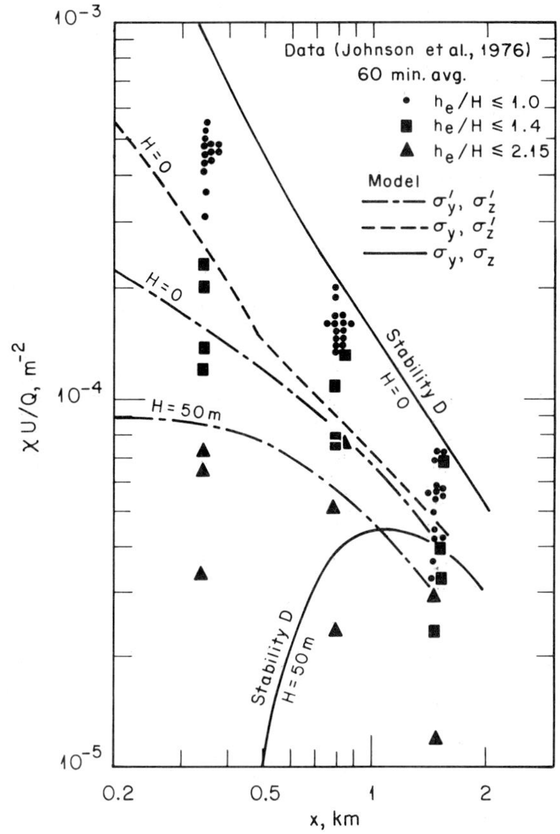

(h) Elevated release. (From W. B. Johnson, E. Shelar, R. E. Ruff, H. B. Singh, and L. Salas, *Gas Tracer Study of Roof-Vent Effluent Diffusion at Millstone Nuclear Power Station,* Report AIF/NESP-007b, Atomic Industrial Forum, Inc., Washington, D.C., 1975.)

Fig. 7.59 Comparison of model estimates with and without (solid lines) wake-induced dispersion enhancement to various field data (From A. H. Huber, *An Evaluation of Obstacle Wake Effects on Plume Dispersion,* presented at the Fourth Symposium on Turbulence, Diffusion, and Air Pollution, Reno, Nev., Jan. 15–18, 1979, American Meteorological Society, Boston, Mass.)

then was carried into the wake of the downwind cube via the shear layer generated by this structure. As a result, high concentrations formed on the lee face. Experiments of this type, combined with flow visualization procedures to help elucidate the flow patterns, can provide a better understanding of the general building cluster problem.

A number of investigators have noted critical windspeeds above which stack- or roof-emitted effluents were downwashed into the wake cavity (e.g., Davies and Moore, 1964; Martin, 1965; Munn and Cole, 1976a, b; Rodliffe and Fraser, 1971). A bimodal state, such that the plume remains aloft for low wind speeds but is fully entrained for other wind speeds, often is postulated for such cases. Lawson (1967) treated the data of Munn and Cole (1967a) in this manner by plotting their values of $\chi x^2/Q$ versus U to get two curves. The upper branch apparently corresponded to direct cavity entrainment and followed the empirical form $\chi x^2/Q \simeq 0.0136U$. In all cases corresponding to this high concentration line, wind speed U was greater than the effluent velocity w_e. The lower branch seemed to correspond to a plume that remained elevated, even though U was larger than w_e in a few cases. Lawson (1967), after a few simple assumptions, estimated that the maximum ground-level concentration for this branch occurred at $x_M = 0.32h_eU$ and was given by $\chi x_M^2/Q = 0.00576U$, which is about one-half the cavity entrainment value for a given wind speed. The wind-tunnel work of Meroney, Cermak, and Chaudhry (1968) demonstrated that, as U neared a critical speed, the downwash actually was intermittent first and then became permanent as U increased further. Attempts to use a simple model such as Lawson's (1967) for long-term dose estimates would require use of the wind-speed frequency distribution for the specific site to assess the percentage of time spent in either the high or low concentration state. The model of Johnson et al. (1975) requires similar information (see Eq. 7.32); however, it estimates concentrations in a slightly different way.

As expected, the turbulence in the near wake is significantly enhanced over the upwind value. Munn and Cole (1967a) measured σ_θ and σ_ϕ, the standard deviations of horizontal and vertical wind direction fluctuations, respectively. They observed $\sigma_{\theta w}/\sigma_{\theta\infty} \simeq 1.7$ and $\sigma_{\phi w}/\sigma_{\phi\infty} \simeq 2.3$ for five-second averages, and $\sigma_{\theta w}/\sigma_{\theta\infty} \simeq 1.6$ and

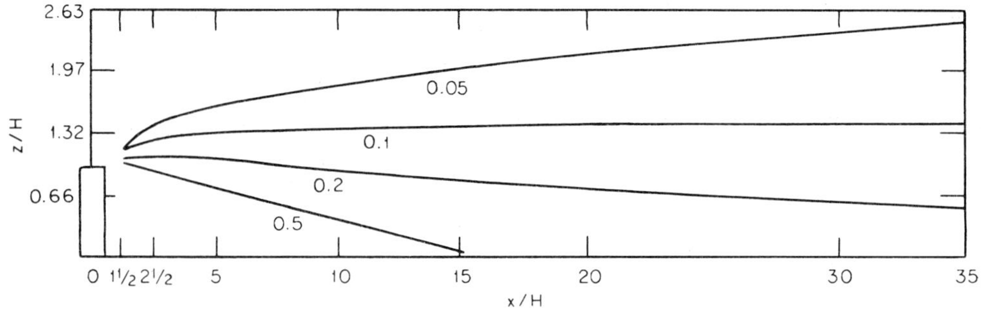

(a) Cylinder, diameter d = height H. Release from top rear of side.

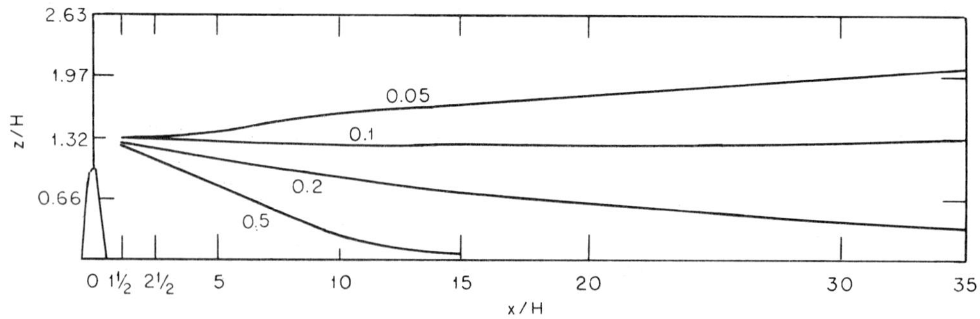

(b) Cone frustrum, cone tip angle ≃ 45°, base diameter = height. Release from top rear of side.

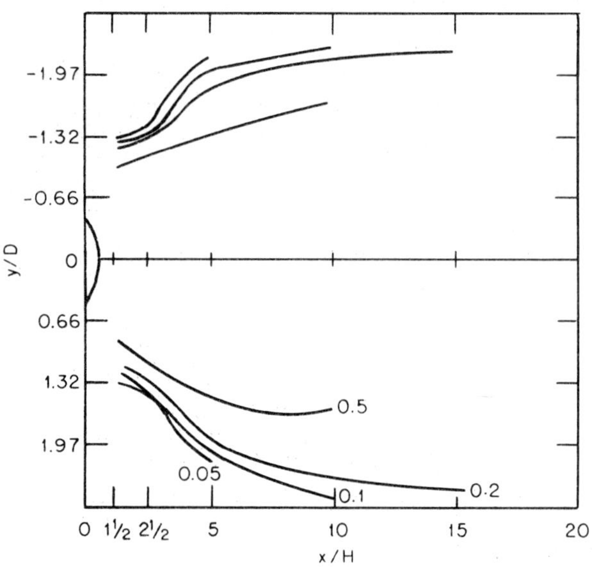

(c) Top view of cylinder release shown in (a).

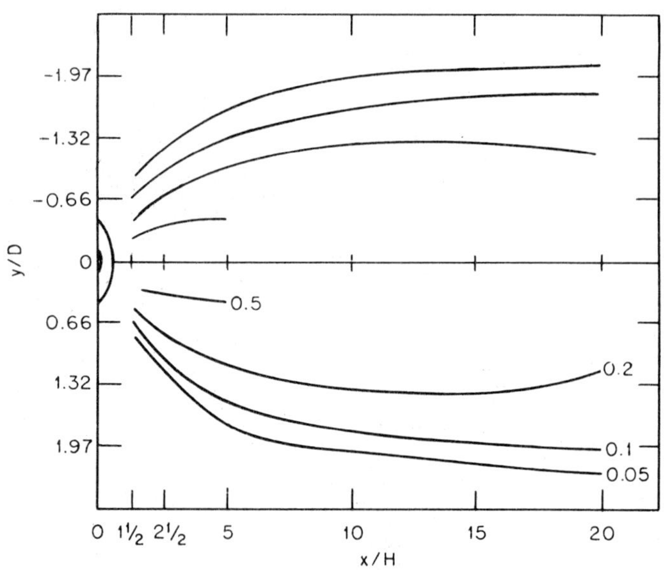

(d) Top view of cone release shown in (b).

Fig. 7.60 Typical nondimensional concentration coefficient ($K = \chi U A_p/Q$) isopleths behind cylinder and truncated cone. (From C. R. Symes and R. N. Meroney, *Cone Frustrums in a Shear Layer*, USAEC Report COO-2053-4, Colorado State University Fluid Dynamics and Diffusion Laboratory Report CER70-71CRS-RNM11, Colorado State University, Fort Collins, Colorado, 1970.)

$\sigma_{\phi w}/\sigma_{\phi\infty} \simeq 1.8$ for 100-second averages; the subscripts w and ∞ refer to wake and far upstream measurements. Evidently both σ_θ and σ_ϕ are significantly increased in the building wake, but, as shown by the five-second averages, more high-frequency turbulence energy is introduced in the vertical than in the horizontal. This extra energy accounts for the good vertical mixing commonly observed in the lee of buildings. The values obtained by Munn and Cole (1967a) probably are site-specific, and should not be extrapolated to other locations.

7-8.2.3.2 Far Wake Behavior Behind Building Clusters. Downwind of the cavity behind the structure that serves as an effluent release point, but within or close to the building cluster as a whole, the observed concentration isopleths display widely varied behavior. This is true especially when other large obstacles are present in the cluster. For example, Figs. 7.61 illustrate patterns produced during individual tests when effluent impinged on large cooling towers a short distance downwind of the release site (Start et al., 1977). Figures 7.61a and b were obtained for similar release and atmospheric conditions but with different wind directions, whereas Figs. 7.61c and d resulted from test runs with similar wind and stability conditions but using two different release points. Generalizations are risky; however, signs of significant wake behavior behind the cooling towers appear in all of these examples. In particular, small zones of concentration maxima are visible in the immediate lee of the towers. Perhaps this reflects strong mixing of initially elevated effluent down to the ground in the tower cavity. Some evidence of strong dilution between the cooling towers may be seen in Figs. 7.61b and d; a possible explanation of this dilution is flow "jetting" between the towers. It is clear that individual test results will vary widely from case to case and will change significantly with modest changes in wind characteristics and source location. Strong site-dependence is inevitable. Allwine, Meroney, and Peterka (1980), Hatcher and Meroney (1977), and Thuillier and Mancuso (1980) reported a similar sensitivity to test conditions in their wind-tunnel and field studies of various reactor complexes. Mathematical modeling of near-field behavior for such sites on an episode-by-episode basis will be extremely difficult, if not impossible.

On the other hand, Dickson, Start, and Markee (1969) presented *average* concentration isopleths for an ensemble of 15 tests (Fig. 7.62) for lapse and inversion conditions. Although significant initial broadening has occurred, these patterns are much more like those of a classical isolated plume. Simple modeling efforts may well be effective for predicting such ensemble average behavior [e.g., Huber's (1979) work discussed above].

Halitsky (1977) has developed a moderately complicated mathematical model for a building complex and has applied it to the EBR-II site studied by Dickson, Start, and Markee (1969). The model, based on measurements of unbounded uniform flow past a flat plate normal to the flow, predicts that the wake boundary grows in the horizontal and vertical as $x^{1/3}$ and that the mean velocity defect and turbulence intensity excess both decay as $x^{-2/3}$. These predictions do not agree with Counihan, Hunt, and Jackson's (1974) predictions for the wake behind an isolated building (see Sec. 7-5.3) even though the latter seem more appropriate to the problem. However, Halitsky's (1977) mean velocity defect prediction appears to agree well with Dickson, Start, and Markee's (1969) data. By adjusting a free parameter related to the size of the complex, the excellent agreement in concentration patterns shown in Fig. 7.63 was obtained. It may be that the field data do not have sufficient accuracy, resolution, and density to permit distinguishing between the theoretical formulations for wake behavior.

One interesting point is that the crosswind concentration distribution, matched to the data collected by Dickson, Start, and Markee (1969), was assumed by Halitsky to be parabolic rather than Gaussian, at least close to the reactor complex. Thuillier and Mancuso (1980), however, found that unimodal concentration distributions were generally well-fitted by a Gaussian curve. Further investigation of this would be welcome.

Abbey (1976) has summarized the observed decrease of concentration with downwind distance in the far wakes of building clusters. Figures 7.64a and b show the results of field tests and wind-tunnel studies, respectively, for neutral and unstable atmospheric conditions, whereas Fig. 7.64c combines the field and laboratory data for stable conditions. Beyond a few hundred meters downwind, a power-law dependence on distance is evident. For field tests under neutral and unstable conditions, the power-law exponent ranges between −1.1 and −1.8. The wind-tunnel tests, on the other hand, have exponents between −0.3 and −1.9. A recent laboratory study by Hatcher and Meroney (1977), not shown in these figures, found a decay rate of $x^{-1.8}$ with little dependence on stability. Halitsky's (1977) mathematical model for wake concentration seems to decrease as $x^{-1.2}$ and is in good agreement with Dickson, Start, and Markee's (1969) field value of $K \equiv \chi U A_p/Q \simeq 5(x/H)^{-1.3}$ at large values of x, where H is the containment vessel height. Under stable conditions, the field data in Fig. 7.64c behave as $x^{-1.3}$ or so, whereas both laboratory results decay much more slowly as $x^{-0.7}$. The Islitzer result in this figure decays very rapidly, as x^{-3}, but Abbey (1976) remarks that it is the result of only a single test and therefore may not be too reliable.

The reason for such discrepancies between wind-tunnel and field far-wake data is unknown; however, some speculation is possible. For example, under stable light wind conditions, wind meander is well known as an important contributor to plume dilution. Start et al. (1977) noted wind direction fluctuations so large that lateral plume spreading induced by the building complex being studied was nearly obscured. Other examples, such as Halitsky and Woodard's (1974) observations, are common in the literature. Such meander is virtually impossible to duplicate† in the laboratory; this may help to account for the relatively slow decrease of concentration with distance that is observed in the wind-tunnel studies. It may also be true that persistent vortex-containing wakes are more significant in the laboratory than in the field; if so, the resulting slow spread of the wake would inhibit plume dispersion and produce a lower concentration decay rate than would be observed in field experiments. The possible enhanced importance of such vortex phenomena in laboratory work could result from many factors, such as lower Reynolds number, incomplete simulation of the turbulence properties of the incident wind, and improper scaling of the model buildings relative to the incident boundary depth, turbulence scales, and other properties. Additional work to resolve this dilemma is needed; in the meantime, wind-tunnel data on far-wake concentration patterns behind building clusters must be viewed with caution.

(Text continues on page 317.)

†Bouwmeester, Kothari, and Meroney (1980), among others, have used time-weighting of wind-tunnel data to reproduce the effect of meander on average concentrations. The results have been promising.

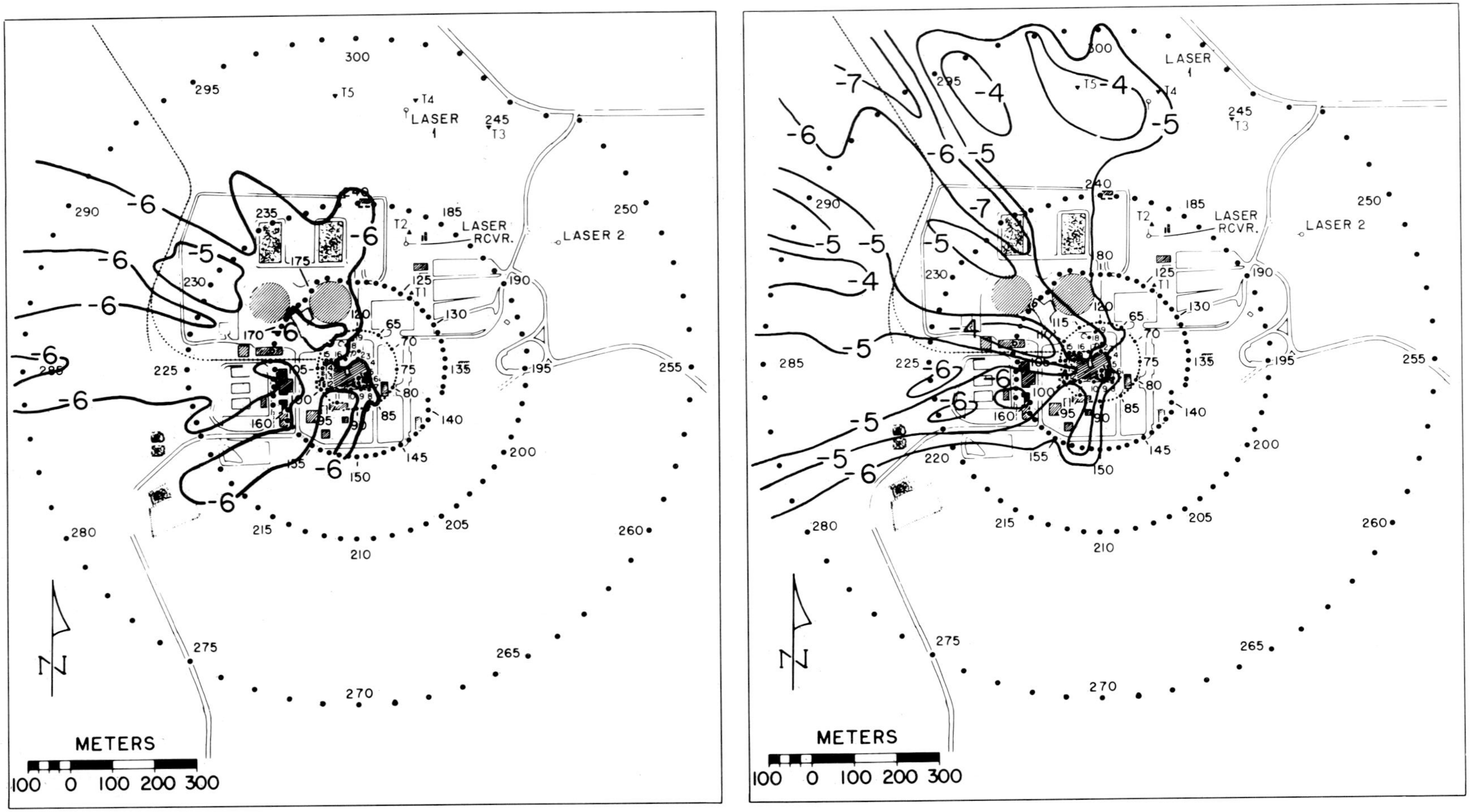

(a) Wind at release height from 161° at 1.8 m/s; $\sigma_\theta \simeq 3.2°$ at 10 m height.

(b) Wind at release height from 100° at 1.7 m/s; $\sigma_\theta \simeq 20°$ at 10 m height.

(Figure continues on next page.)

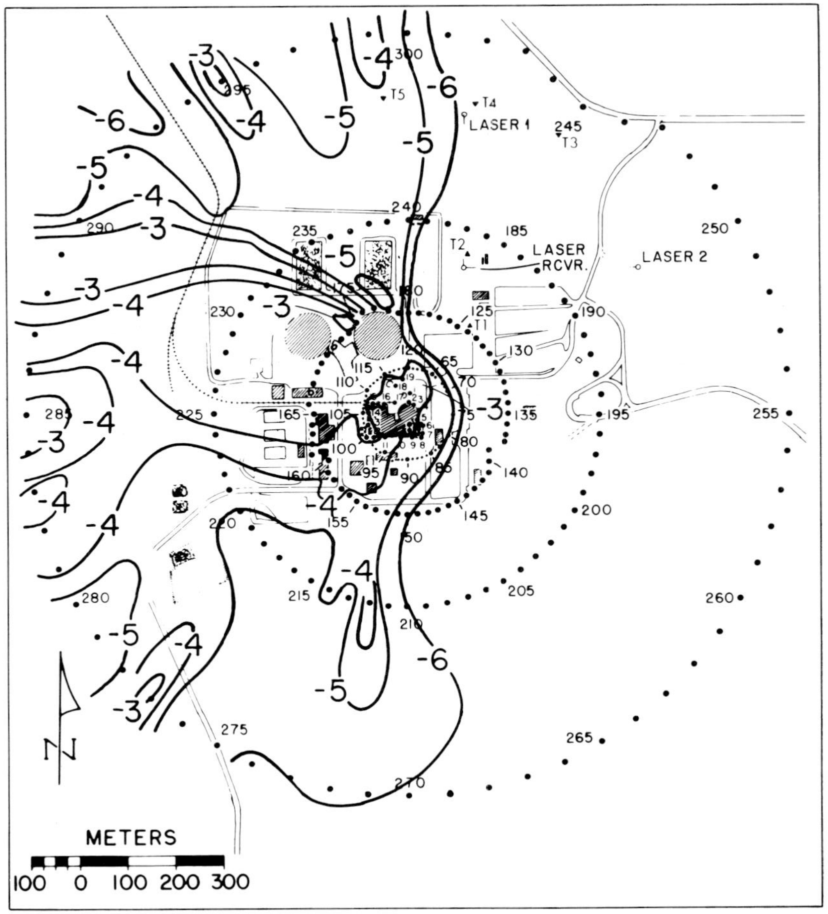

(c) Wind at release height from 110° at 0.9 m/s; $\sigma_\theta \simeq 19°$ at 10 m height.

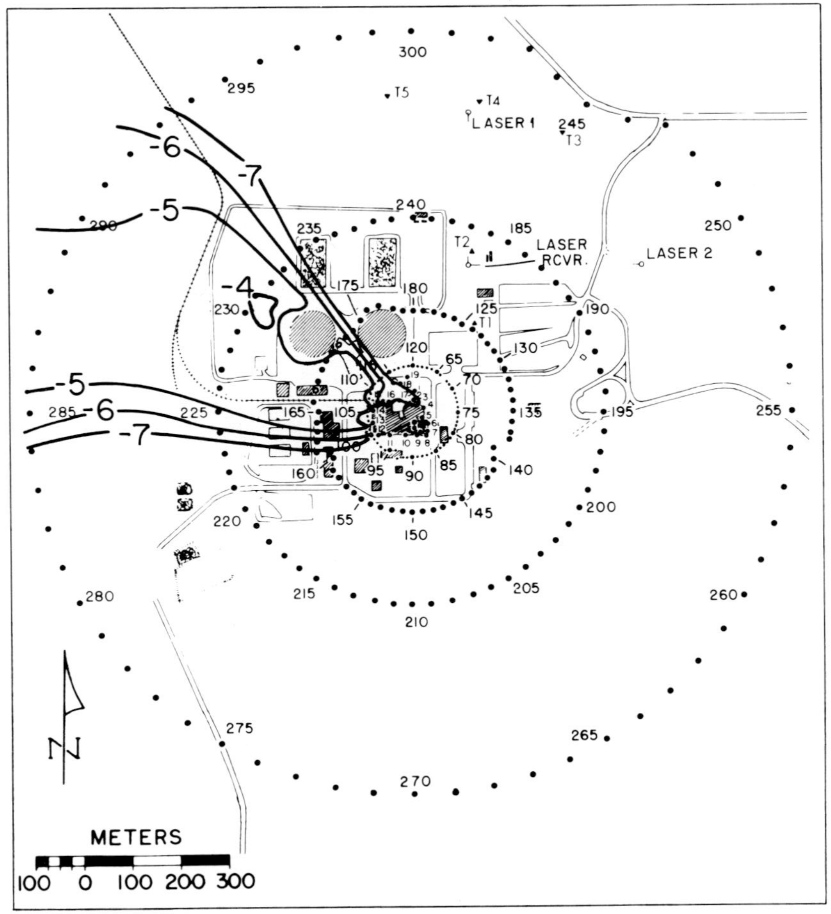

(d) Wind at release height from 110° at 2.5 m/s; $\sigma_\theta \simeq 19°$ at 10 m height [same conditions as (c) except for one different tracer and a different release point].

Fig. 7.61 Isopleths of $\chi U/Q$, in negative powers of ten, for tracers released under NRC stability category G. All releases were made from an auxiliary building roof adjacent to the containment building except for (c) which was made at the base of the containment building. (From G. E. Start, J. H. Cate, C. R. Dickson, N. R. Ricks, G. R. Ackermann, and J. F. Sagendorf, *Rancho Seco Building Wake Effects on Atmospheric Diffusion*, NOAA Technical Memorandum ERL-ARL-69, NOAA/ARL, Idaho Falls, Idaho, 1977.)

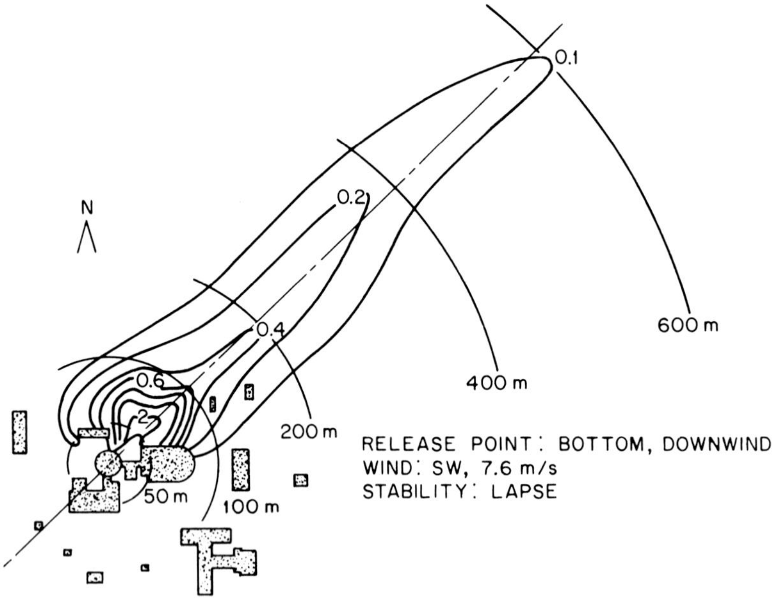

(a) K-mean isopleths of effluent released from containment building during lapse conditions.

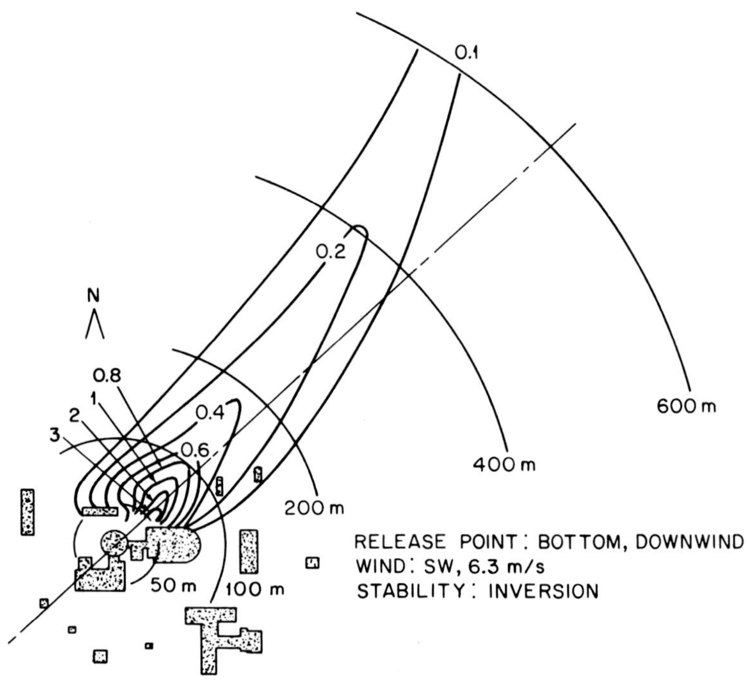

(b) K-mean isopleths of effluent released from containment building during inversion conditions.

Fig. 7.62 Ensemble-average isopleths of concentration coefficient K = $\chi UA_p/Q$, where A_p (= 665 m²) is taken to be the projected area of only the EBR-II containment building. (From C. R. Dickson, G. E. Start, and E. H. Markee, Jr., Aerodynamic Effects of the EBR-II Containment Vessel Complex on Effluent Concentration, in the *Proceedings of USAEC Meteorological Information Meeting,* Chalk River Nuclear Laboratories, Sept. 11–14, 1967, pp. 87-101, AECL Report 2787, Atomic Energy of Canada Ltd., Chalk River, Ontario, 1967.)

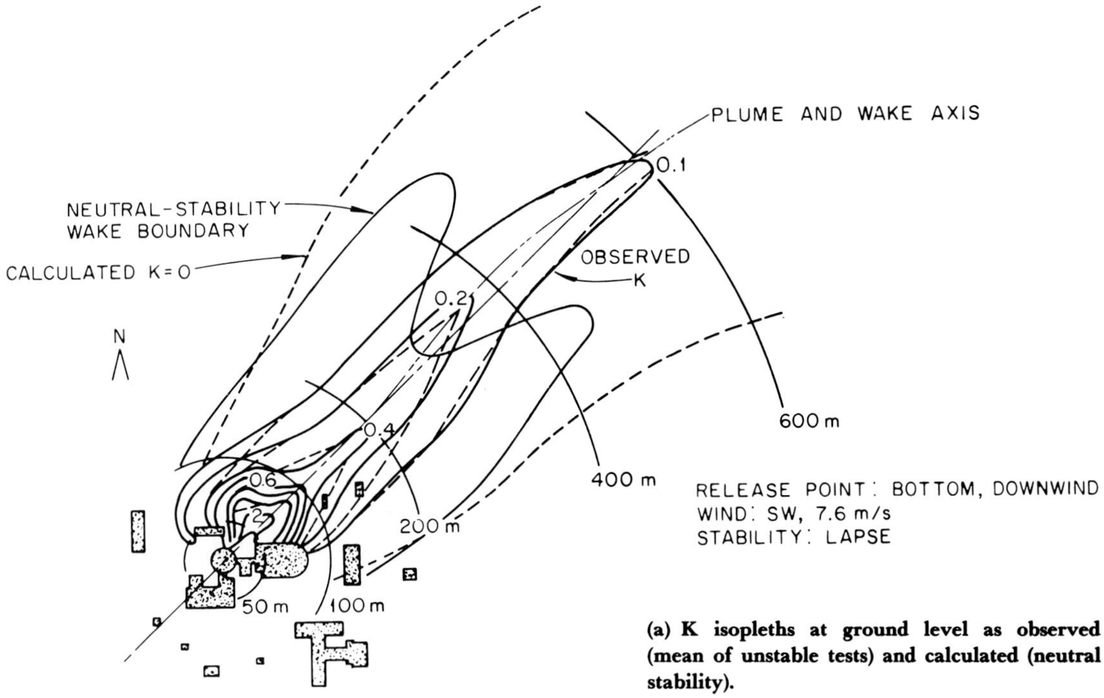

(a) K isopleths at ground level as observed (mean of unstable tests) and calculated (neutral stability).

(b) K isopleths at ground level as observed (mean of stable tests) and calculated (neutral stability).

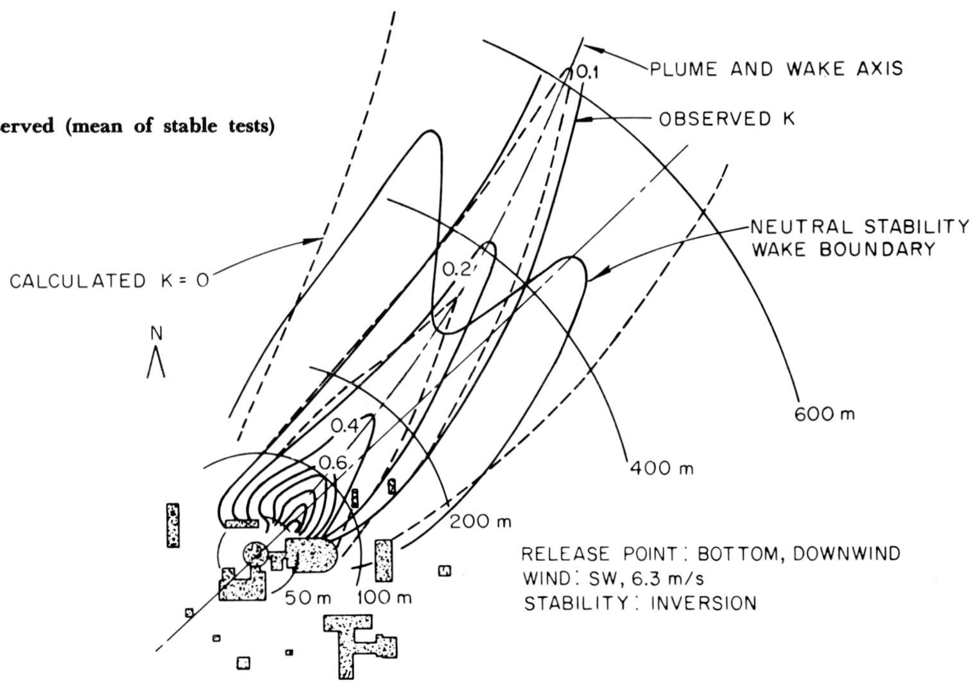

Fig. 7.63 Observed (Dickson, Start, and Markee, 1969; solid lines) and predicted (broken lines) concentration coefficient K contours behind a nuclear reactor complex. Zone where wake turbulence exceeds ambient atmospheric turbulence lies inside neutral-stability wake boundary line. [From J. Halitsky, Wake and Dispersion Models for the EBR-II Building Complex, *Atmos. Environ.*, 11(7): 577-596 (1977).]

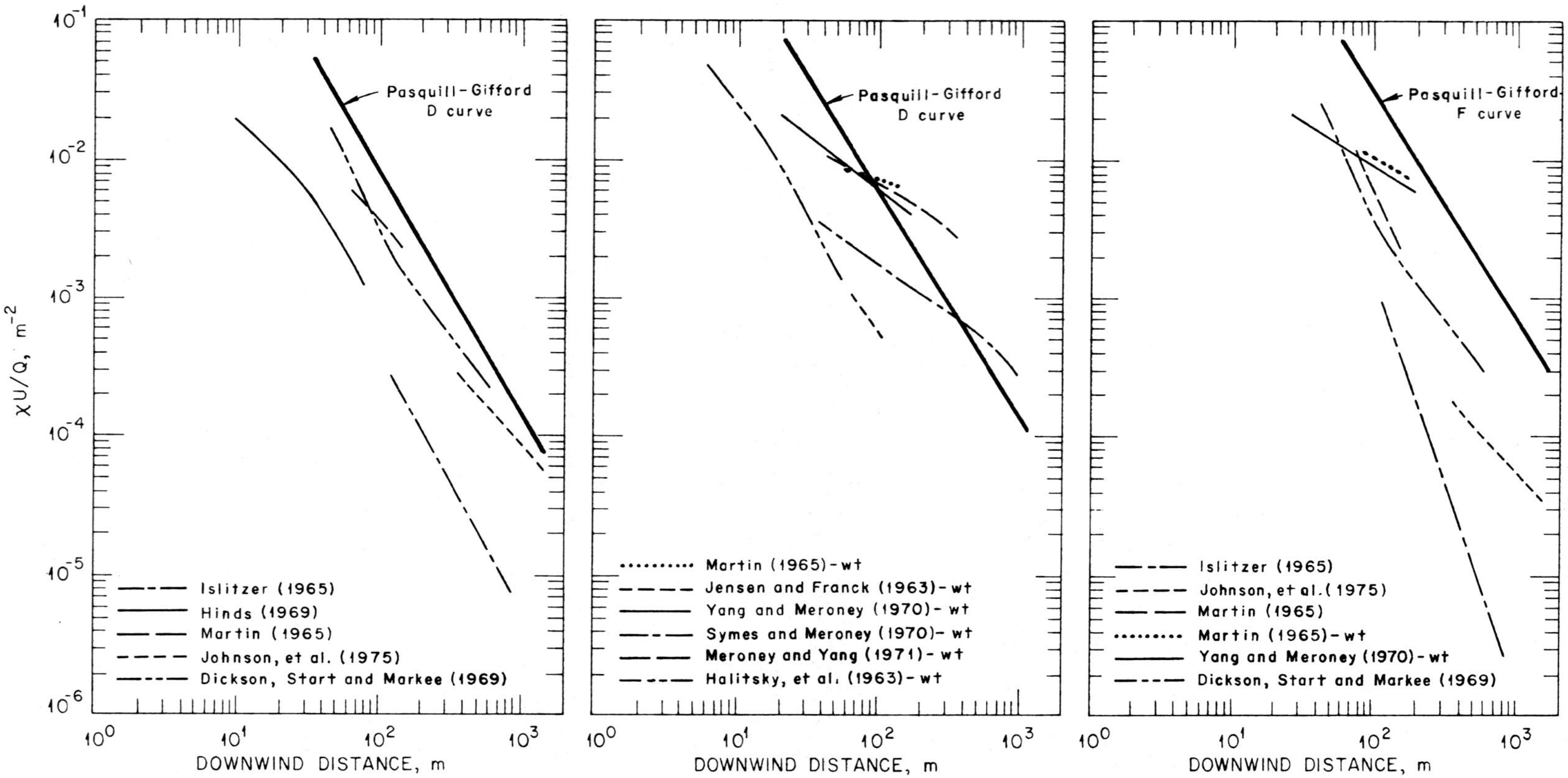

(a) Field tests in neutral and unstable conditions.

(b) Wind-tunnel (WT) tests in neutral and unstable conditions.

(c) Field and wind-tunnel tests in stable conditions.

Fig. 7.64 Summary of far-wake concentration decay with distance behind building clusters. Standard Pasquill-Gifford category D and F estimates are shown for comparison. (From R. F. Abbey, Jr., Concentration Measurements Downwind of Buildings: Previous and Current Experiments, in *Preprints of Third Symposium on Atmospheric Turbulence, Diffusion and Air Quality,* Raleigh, N.C., Oct. 19–22, 1976, pp. 247-254, American Meteorological Society, Boston, Mass., 1976.)

7-8.3 Sources Downwind of Buildings

Relatively little guidance is available for estimating the concentrations produced by a source located some distance downwind of an isolated building; therefore, this section is necessarily speculative. Suppose the source is located well back in the wake cavity. If it is a low or ground-level vent, then the effluent will travel toward the building's lee face and, in the process, it will be well stirred throughout the cavity. A higher stack at this location may have its effluent transported fairly directly to the surface by the downward curving shear layer bounding the cavity. If the cavity is fluctuating or collapsing, these building wake effects will be intermittently important. If the source is off the wake centerline, and if the horseshoe vortex system does not break down rapidly in the ambient turbulence, then effluent may be entrained in one of the vortices and carried off downwind; in this process, the effluent possibly could rise from the surface. If an elevated vortex pair is important, as may be true for buildings at an angle to the wind, then vortex entrainment and transport may be observed for an elevated source as well. In this case one might find effluent reaching the surface far downwind of the structure. Similar arguments may be appropriate for sources downwind of the cavity closure. One expects the enhancement to turbulent diffusion induced by the building wake to be the dominant effect, so that effluent from an elevated source is mixed to the ground more rapidly, whereas concentrations far downwind are lower than if the building was not present. If organized vorticity is present in the wake, however, it may be able to transfer rather concentrated material that is aloft down to the surface at large distances downwind.

Briggs (1973) suggested that his method of incorporating building wake effects into the effective stack height should be confined to cases where the stack is within $3\zeta_b$ downwind of the building—i.e., within or near the end of the cavity zone. If the effective stack height h_e is zero, Briggs's recommended model for computing downwind concentrations is essentially Eq. 7.30 with $c = 1$. If the plume remains aloft, he uses $\chi = 0$ for distances x such that $\sigma_z(x) < \sqrt{2/\pi}\ h_e$, whereas for $\sigma_z > \sqrt{2/\pi}\ h_e$,

$$\chi \simeq \frac{Q}{\pi\sigma_y\,(\sigma_z + \sqrt{2/\pi}\ h_e)U} \tag{7.35}$$

That is, dispersion behind the building is enhanced only in the vertical for an elevated release. No building contribution to dispersion is allowed by Briggs if the source is more than $3\zeta_b$ downwind.

Huber (1977, 1979) commented that the Huber and Snyder dispersion enhancement model (Eqs. 7.33 and 7.34) is best applied to sources within 2H of a building when $h_e < H + 1.5\zeta_b$. No recommendation was made for more distant sources.

In fact, despite its practical importance, little study seems to have been conducted on the concentration field produced by a source located downwind of the cavity zone of a building or building cluster. Barrett, Hall, and Simmonds (1978) studied the case of a single model building at an angle to the wind for a range of stack heights and downwind locations. They remarked that examination of the velocity and turbulence profiles behind the structure would suggest that the wake is virtually dissipated by $x/H \simeq 8$; however, the effects of the building on the concentration field can be observed out to at least 24H downwind. They attributed this result to downwash induced along the wake centerline by a persistent trailing horseshoe vortex. Figure 7.65 shows their maximum ground-level concentration coefficient $K_{max} \equiv \chi_{max}U_HH^2/Q$ as a function of nondimensional stack height h_s/H, for a range of stack locations downwind. Further study and field validation of these results are needed for a variety of building shapes and orientations. The work of Kothari, Peterka, and Meroney (1982) may be useful in this regard.

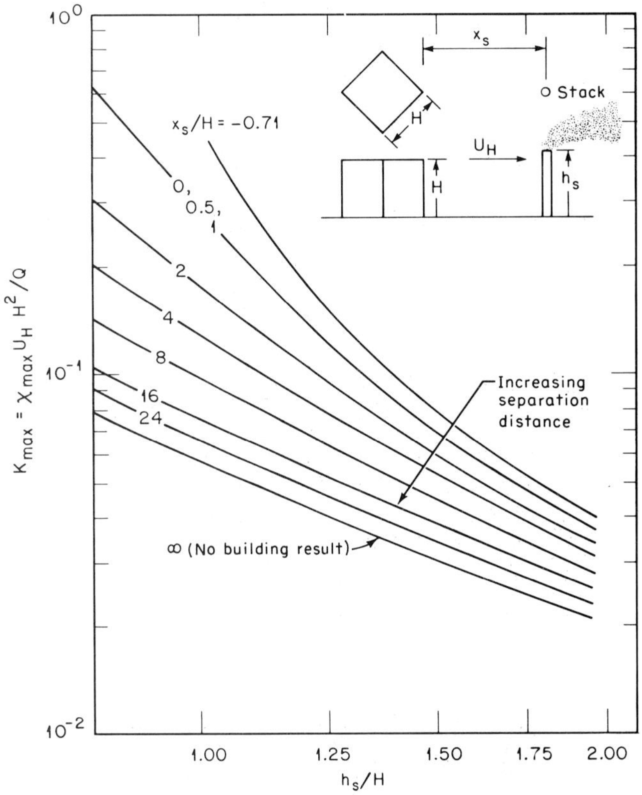

Fig. 7.65 Effect of building-stack separation on maximum ground-level concentration coefficient, for varying heights of a stack located downwind of a model building at an angle to the wind. (From C. F. Barrett, D. J. Hall, and A. C. Simmonds, Dispersion from Chimneys Downwind of Cubical Buildings: A Wind-Tunnel Study, in *Proceedings of the Ninth International Technical Meeting on Air Pollution Modeling and Its Application,* Toronto, Canada, Aug. 28–31, 1978, pp. 37–44, North Atlantic Treaty Organization, Committee on the Challenges of Modern Society, 1978.)

ACKNOWLEDGMENT

This work was carried out under agreements among the National Oceanic and Atmospheric Administration, the Department of Energy, and the Nuclear Regulatory Commission.

List of Symbols

Major symbols used in Chap. 7 are listed here; others are defined in the text as needed. An attempt has been made to make the nomenclature consistent throughout the chapter; consequently readers who refer to the original papers can expect changes in notation. Units are indicated in square brackets after each symbol.

Symbol	Definition
A_c	Surface area of wake cavity exposed to atmosphere $[L^2]$
A_e	Exit area of effluent vent $[L^2]$
A_P	Projected frontal area of an obstacle $[L^2]$
C_D	Drag coefficient, = drag force/$\left(\frac{1}{2}\rho U_\infty^2 A_p\right)$
C_P	Pressure coefficient, = $(p - p_\infty)/\left(\frac{1}{2}\rho U_\infty^2\right)$
c	Various constants
D	Diameter of a disk or sphere [L]
d	Diameter of cylinder [L]
F_b	Buoyancy flux of effluent, $= -g\dot{V}_e\Delta/\pi$ $[L^4t^{-3}]$
F_D, F_H	Froude number based on characteristic dimension D or H, $F_D = U_\infty/ND$
f	Vortex shedding frequency $[t^{-1}]$. Also, an arbitrary function
f_c	Cavity shape factor, $= V_c/A_cH$
g	Gravitational acceleration, 9.8 m/s^2 $[Lt^{-2}]$
H	Characteristic vertical dimension of an obstacle [L]
H_c	Maximum height of roof recirculation zone [L]
h	Height of a thin plate [L]
$\tilde{h}$	Height of stack relative to building roof [L]
h′	Effluent release height corrected for stack tip downwash [L]
h″	Effluent release height corrected for both stack tip downwash and building aerodynamic influence [L]
h_*	Critical source height for effluent to pass over a hill during stable atmospheric conditions [L]
h_e	Effective stack height, compensated for stack tip downwash, building aerodynamic effects, and plume buoyancy [L]
h_s	Stack height relative to ground [L]
Δh	Buoyant plume rise [L]
K	Concentration coefficient, $= \chi U A_p/Q$
K_e	Concentration coefficient at the effluent exit, $= (U_s/w_e)(A_p/A_e)$
K_m	Turbulent diffusivity for momentum $[L^2t^{-1}]$
k	von Kármán constant, $\simeq 0.4$
L	Characteristic dimension of an obstacle in the along-wind direction [L]
L_c	Length of roof recirculation zone [L]
ℓ	Any characteristic obstacle dimension [L]
$\overline{m}$	Molecular weight [M/mole]
N	Brunt-Väisälä frequency, $= [-(g/\rho)/(d\rho/dz)]^{1/2}$ $[t^{-1}]$
n	Exponent of power-law velocity profiles ($U \propto z^n$)
P	Local static pressure $[ML^{-1}t^{-2}]$
Q	Effluent source strength $[Mt^{-1}]$
R	Radius of hemisphere [L]
$\tilde{R}$	Wilson's (1979) characteristic dimension of an obstacle, $= \zeta_b^{2/3}\,\xi_b^{1/3}$ [L]
Re	Reynolds number, $= U_\infty\ell/v$
Re_{turb}	Turbulent Reynolds number, $= U_\infty\ell/K_m$
r	Radial distance of axisymmetric wake edge from wake centerline[L].
St	Strouhal number, $= f\ell/U$
s	Stretched string distance between source and receptor on the same building [L]
T	Absolute temperature [°K]
t_d	Characteristic retention time for effluent in a wake cavity [t]
U	Mean wind speed in the x direction $[Lt^{-1}]$
ΔU	Mean velocity defect, $= U - U_\infty$ $[Lt^{-1}]$
u,v,w	Fluctuating velocity components in the (x,y,z) directions, respectively $[Lt^{-1}]$
u_*	Friction velocity $[Lt^{-1}]$
V_c	Volume of wake cavity $[L^3]$
$\dot{V}_e$	Volume flow rate of effluent, $= w_eA_e$ $[L^3t^{-1}]$
v	Velocity vector, with components (u,v,w) $[Lt^{-1}]$
W	Characteristic dimension of an obstacle in the crosswind (horizontal) direction [L]
w_e	Effluent exit speed $[Lt^{-1}]$
w_t	Transport velocity across a turbulent shear layer $[Lt^{-1}]$
X_c	Distance from windward edge to location of maximum height of roof recirculation zone [L]
x,y,z	Along-wind, crosswind, and vertical coordinates, respectively [L]
$\tilde{x},\tilde{y},\tilde{z}$	Coordinates relative to base of a roof-mounted stack [L]
x_r	Length of recirculation zone behind an obstacle, measured from the flow separation point [L]
x_{yo}, x_{zo}	Virtual source location upwind of effluent vent, such that $\sigma_{yo} = \sigma_y(x_{yo})$, $\sigma_{zo} = \sigma_z(x_{zo})$ [L]
Z_e	Effective wake height, $= H + 1.5\,\zeta_b$ [L]
Z_m	Minimum descent height of maximum permissible concentration isopleth [L]
$Z_{I,\ II,\ or\ III}$	Boundaries of flow zones above a building roof (Fig. 7.19) [L]
z_o	Aerodynamic roughness length [L]
α	Angle of wind incidence [degrees or radians]
Γ	Circulation around trailing wake vortex, $= \oint \mathbf{v}\cdot d\boldsymbol{\ell}$ $[L^2t^{-1}]$
Δ	Relative density difference of effluent, $= (\rho_e - \rho_a)/\rho_a$
δ	Boundary layer or shear layer thickness [L]
ϵ	Fraction of time plume is entrained into a wake by aerodynamic downwash
ζ_b	Smaller of obstacle frontal dimensions W or H [L]
η	Smaller of the frontal dimensions H or W/2 [L]
Λ_x	Longitudinal integral length scale of turbulence [L]
λ	A turbulent mixing length [L]
v	Kinematic viscosity of air, $\simeq 0.15$ cm^2/sec $[L^2t^{-1}]$
ξ_b	Larger of obstacle frontal dimensions W or H [L]
π	3.14159 ...
ρ	Density $[ML^{-3}]$
σ_u	Standard deviation of longitudinal velocity fluctuation $[Lt^{-1}]$
$\sigma_y(x)$, $\sigma_z(x)$	Crosswind and vertical Gaussian plume dis-

	persion coefficients as functions of along-wind distances [L]
σ_{yo}, σ_{zo}	Plume dispersion coefficients corresponding to an upwind virtual source; typically, $\sigma_{yo} = W/4.3$, $\sigma_{zo} = H/2.15$ [L]
σ_θ	Standard deviation of horizontal fluctuations of wind direction [degrees or radians]
τ	Dimensionless retention time for effluent in wake cavity, $= U_\infty t_d/H$
Υ	Turbulence parameter, $= (u_{rms}/U_\infty)(\Lambda_x/H)$
χ	Effluent concentration $[ML^{-3}]$
Ψ	Gandemer's (1975) comfort parameter, = ratio of sum of mean and root-mean-square (rms) turbulent wind speeds at some location to that sum when no building is present
ψ	Stream function for two dimensions, such that $u = \partial\psi/\partial x$, $v = -\partial\psi/\partial y$ $[L^2t^{-1}]$
ω	Vorticity vector, $= \nabla \times \mathbf{v}$ $[t^{-1}]$

Subscripts

a	An ambient quantity
b	Quantity measured on back of an obstacle
e	An effluent quantity
f	Quantity measured on front of an obstacle
H	Value of quantity at roof height
i	Quantity for i*th* step or building
j	Indicates a particular stability category; in terms of P-G-T, j = 1 = "A", j = 4 = "D", etc.
M or max	A maximum value
m	A minimum value
r	A quantity characteristic of recirculation zone
rms	Root-mean-square value of quantity
s	Value of quantity at stack height
o	A constant or reference value
∞	An upstream or reference value

REFERENCES

Abbey, R. F., Jr., 1976, Concentration Measurements Downwind of Buildings: Previous and Current Experiments, in *Preprints of the Third Symposium on Atmospheric Turbulence, Diffusion and Air Quality*, Raleigh, N. C., Oct. 19–22, pp. 247-254, American Meteorological Society, Boston, Mass.

Ahmad, Q. A., R. E. Luxton, and R. A. Antonia, 1976, Characteristics of a Turbulent Boundary Layer with an External Turbulent Uniform Shear Flow, *J. Fluid Mech.*, 37 (Pt. 2): 369-396.

Allwine, K. J., R. N. Meroney, and J. A. Peterka, 1980, *Rancho Seco Building Wake Effects on Atmospheric Diffusion: Simulation in a Meteorological Wind Tunnel*, Nuclear Regulatory Commission, NRC Report NUREG/CR-1286, NTIS.

Apelt, C. J., G. S. West, and A. A. Szewczyk, 1973, The Effects of Wake Splitter Plates on the Flow Past a Circular Cylinder in the Range $10^4 < Re < 5 \times 10^4$, *J. Fluid Mech.*, 61(Pt. 1): 187-198.

——, and G. S. West, 1975, The Effects of Wake Splitter Plates on Bluff-Body Flow in the Range $10^4 < Re < 5 \times 10^4$, Pt. 2, *J. Fluid Mech.*, 71(Pt. 1): 145-160.

Arie, M., M. Kiya, H. Tamura, M. Kosugi, and K. Takaoka, 1975, Flow over Rectangular Cylinders Immersed in a Turbulent Boundary Layer, Part 2. Flow Patterns and Pressure Distributions, *Bull. JSME (Jpn. Soc. Mech. Eng.)*, 18(125): 1269-1276.

——, and H. Rouse, 1956, Experiments on Two-Dimensional Flow over a Normal Wall, *J. Fluid Mech.*, 1(Pt. 2): 129-141.

ASHRAE, 1974, Laboratories, in *Handbook and Product Directory, 1974, Applications*, Chap. 15, American Society of Heating, Refrigerating, and Air-Conditioning Engineers, Inc., New York.

ASME, 1973, *Recommended Guide For The Prediction of Airborne Effluents*, 2nd ed., American Society of Mechanical Engineers, New York.

Baines, W. D., 1963, Effects of Velocity Distribution on Wind Loads and Flow Patterns on Buildings, in *Proceedings of the Symposium on Wind Effects on Buildings and Structures*, National Physical Laboratory, Teddington, Middlesex, England, June 26–28, Vol. I, pp. 197-225, Her Majesty's Stationery Office, London, 1965.

Baltaxe, R., 1967, Air Flow Patterns in the Lee of Model Windbreaks, *Arch. Meteorol., Geophys., Bioklimatol., Ser. B*, 15(3): 287-312.

Barker, S. J., and S. C. Crow, 1977, The Motion of Two-Dimensional Vortex Pairs in a Ground Effect, *J. Fluid Mech.*, 82(Pt. 4): 659-671.

Barrett, C. F., D. J. Hall, and A. C. Simmonds, 1978, Dispersion from Chimneys Downwind of Cubical Buildings: A Wind-Tunnel Study, in *Proceedings of the Ninth International Technical Meeting on Air Pollution Modeling and Its Application*, Toronto, Canada, Aug. 28–31, pp. 37-44, North Atlantic Treaty Organization, Committee on the Challenges of Modern Society,

Barriga, A. R., C. T. Crowe, and J. A. Roberson, 1975, Pressure Distribution on a Square Cylinder at a Small Angle of Attack in a Turbulent Cross Flow, in *Proceedings of the Fourth International Conference on Wind Effects on Buildings and Structures*, Heathrow (London), Sept. 8–12, pp. 89-93, Cambridge University Press, London, 1977.

Barry, P. J., 1964, *Estimation of Downwind Concentration of Airborne Effluents Discharged in the Neighbourhood of Buildings*, AECL Report 2043, Atomic Energy of Canada, Ltd., Chalk River, Ontario, July 1964.

Batchelor, G. K., 1967, *An Introduction to Fluid Dynamics*, Cambridge University Press, London and New York.

——, and I. Proudman, 1954, The Effect of Rapid Distortion of a Fluid in Turbulent Motion, *Q. J. Mech. Appl. Math.*, 7(Part 1): 83-103.

Bearman, P. W., 1965, Investigation of the Flow Behind a Two-Dimensional Model with a Blunt Trailing Edge and Fitted with Splitter Plates, *J. Fluid Mech.*, 21(Part 2): 241-255.

——, 1969, Fluctuating Forces on Bluff Bodies in Turbulent Flow, in *Aerodynamics of Atmospheric Shear Flows*, Advisory Group for Aerospace Research and Development Conference Proceedings No. 48, pp. 24-1 to 24-13, North Atlantic Treaty Organization, 1970, National Aeronautics and Space Administration, Report Distribution and Storage Unit, Langley, Va.

——, 1971a, An Investigation of the Forces on Flat Plates Normal to a Turbulent Flow, *J. Fluid Mech.*, 46(Part 1): 177-198.

——, 1971b, Some Measurements of the Distortion of Turbulence Approaching a Two-Dimensional Body, in *Turbulent Shear Flows*, Advisory Group for Aerospace Research and Development Conference Proceedings No. 93, pp. 28-1 to 28-11, North Atlantic Treaty Organization, 1972, National Aeronautics and Space Administration, Report Distribution and Storage Unit, Langley, Va.

——, 1972a, Some Recent Measurements of the Flow Around Bluff Bodies in Smooth and Turbulent Streams, in *Proceedings of the Symposium on External Flows*, University of Bristol, July 4–6, pp. B.1-B.15, University of Bristol, England, 1972.

——, 1972b, Some Measurements of the Distortion of Turbulence Approaching a Two Dimensional Bluff Body, *J. Fluid Mech.*, 53(Pt. 3): 451-467.

——, and M. E. Davies, 1975, The Flow About Oscillating Bluff Structures, in *Proceedings of the Fourth International Conference on Wind Effects on Buildings and Structures*, Heathrow (London), Sept. 8–12, pp. 285-295, Cambridge University Press, London, 1977.

——, and J. E. Fackrell, 1975, Calculation of Two-Dimensional and Axisymmetric Bluff-Body Potential Flow, *J. Fluid Mech.*, 72(Pt. 2): 229-241.

——, and D. M. Trueman, 1972, An Investigation of the Flow Around Rectangular Cylinders, *Aeronaut. Q.* XXIII(Pt. 3): 229-237.

——, and A. J. Wadcock, 1973, The Interaction Between a Pair of Circular Cylinders Normal to a Stream, *J. Fluid Mech.*, 61(Pt. 3): 499-511.

Birkhoff, G., and E. H. Zarantonello, 1957, *Jets, Wakes, and Cavities*, in *Applied Mathematics and Mechanics Monographs*, F. N. Frenkiel and G. Temple (Eds.), Vol. 2, Academic Press, Inc., New York.

Bitte, J., and W. Frost, 1976, *Atmospheric Flow over Two-Dimensional Bluff Surface Obstructions*, Report NASA CR-2750, October, National Aeronautics and Space Administration, NTIS.

Blessmann, J., 1971, Pressures on Domes with Several Wind Profiles, in *Proceedings of the Third International Conference on Wind Effects on Buildings and Structures*, Tokyo, Sept. 6–9, pp. 317-326, Saikon Co., Ltd, Tokyo.

Bostock, B. R., and W. A. Mair, 1972, Pressure Distributions and Forces on Rectangular and D-Shaped Cylinders, *Aeronaut. Q.*, XXIII(Pt. 1): 1-6.

Bouwmeester, R. J. B., K. M. Kothari, and R. N. Meroney, 1980, *An Algorithm to Estimate Field Concentrations Under Nonsteady Meteorological Conditions from Wind Tunnel Experiments*, NRC Report NUREG/CR-1474, Nuclear Regulatory Commission, NTIS.

——, R. N. Meroney, and V. A. Sandborn, 1978, *Sites for Wind-Power Installations: Wind Characteristics over Ridges*, Colorado State University Fluid Dynamics and Diffusion Laboratory Report CER77-78RJBB-RNM-VAS51, Colorado State University, Fort Collins, Colo.

Bowen, A. J., and D. Lindley, 1974, Measurements of the Mean Wind Flow over Various Escarpment Shapes, in *Proceedings of the Fifth Australasian Conference on Hydraulics and Fluid Mechanics*, University of Canterbury, Christchurch, New Zealand, Dec. 9–13, Vol. I, pp. 211-219, University of Canterbury, New Zealand, 1975.

Bradbury, L. J. S., and W. D. Moss, 1975, Pulsed Wire Anemometer Measurements in the Flow Past a Normal Flat Plate in a Uniform Flow and in a Sheared Flow, in *Proceedings of the Fourth International Conference on Wind Effects on Buildings and Structures*, Heathrow (London), Sept. 8–12, pp. 485-496, Cambridge University Press, London, 1977.

Bradley, E. F., 1980, An Experimental Study of the Profiles of Wind Speed, Shearing Stress, and Turbulence at the Crest of a Large Hill, *Q. J. R. Meteorol. Soc.*, 106(447): 101-123.

Bradshaw, P., and F. Y. F. Wong, 1972, The Reattachment and Relaxation of a Turbulent Shear Layer, *J. Fluid Mech.*, 52(Pt. 1): 113-135.

Briggs, G. A., 1969, *Plume Rise*, AEC Critical Review Series, TID-25075, NTIS.

——, 1973, Diffusion Estimation for Small Emissions, in *Atmospheric Turbulence and Diffusion Laboratory 1973 Annual Report*, Publication No. ATDL-106, pp. 83-145, National Oceanic and Atmospheric Administration/Atmospheric Turbulence and Diffusion Laboratory, Oak Ridge, Tenn.

——, 1975, Plume Rise Predictions, in *Lectures on Air Pollution and Environmental Impact Analyses*, Chap. 3, pp. 59-111, Boston, Sept. 29–Oct. 3, American Meteorological Society, Boston, Mass.

Brighton, P. W. M., 1977, *Boundary Layer and Stratified Flows Past Obstacles*, Ph.D. Dissertation No. 10221, University of Cambridge, Cambridge, England.

——, 1978, Strongly Stratified Flow Past Three-Dimensional Obstacles, *Q. J. R. Meteorol. Soc.*, 104(440): 289-307.

Britter, R. E., and J. C. R. Hunt, 1979, Velocity Measurements and Order of Magnitude Estimates of the Flow Between Two Buildings in a Simulated Atmospheric Boundary Layer, *J. Ind. Aerodyn.*, 4(2): 165-182.

——, J. C. R. Hunt, and J. C. Mumford, 1979, The Distortion of Turbulence by a Circular Cylinder, *J. Fluid Mech.*, 92(Pt. 2): 269-301.

——, J. C. R. Hunt, and J. S. Puttock, 1976, Predicting Pollution Concentrations near Buildings and Hills, presented at the *Institution of Measurements and Control Conference on Systems and Models in Air and Water Pollution*, London, Sept. 22–24, pp. 7-1 to 7-15, Institution of Measurements and Control, London, U.K.

Cagnetti, P., 1975, Downwind Concentrations of an Airborne Tracer Released in the Neighbourhood of a Building, *Atmos. Environ.*, 9(8): 739-747.

Calvert, J. R., 1967a, Experiments on the Low-Speed Flow Past Cones, *J. Fluid Mech.*, 27(Pt. 2): 283-289.

——, 1967b, Experiments on the Flow Past an Inclined Disk, *J. Fluid Mech.*, 29(Pt. 4): 691-703.

Caput, C., Y. Belot, G. Guyot, C. Samie, and B. Seguin, 1973, Transport of a Diffusing Material over a Thin Wind-Break, *Atmos. Environ.*, 7(1): 75-86.

Castro, I. P., 1971, Wake Characteristics of Two-Dimensional Perforated Plates Normal to an Air Stream, *J. Fluid Mech.*, 46(Pt. 3): 599-609.

——, 1973, *An Experimental Investigation of the Flow Around a Surface Mounted Cube in a Uniform Free Stream*, CEGB Report R/M/N 687, Central Electricity Generating Board, England.

——, 1979, Relaxing Wakes Behind Surface-Mounted Obstacles in Rough Wall Boundary Layers, *J. Fluid Mech.*, 93(Pt. 4): 631-659.

——, and J. E. Fackrell, 1978, A Note on Two-Dimensional Fence Flows with Emphasis on Wall Constraint, *J. Ind. Aerodyn.*, 3(1): 1-20.

——, and A. G. Robins, 1977, The Flow Around a Surface-Mounted Cube in Uniform and Turbulent Streams, *J. Fluid Mech.*, 79(Pt. 2): 307-335.

Cermak, J. E., 1971, Laboratory Simulation of the Atmospheric Boundary Layer, *AIAA (Am. Inst. Aeronaut. Astronaut.) J.*, 9(9): 1746-1754.

——, 1974, Applications of Fluid Mechanics in Wind Engineering, presented at *Winter Annual Meeting of the ASME*, New York, Nov. 17–21, Freeman Scholar Lecture, American Society Mechanical Engineers, New York, 1974.

——, 1975, *Nature of Air Flow Around Buildings*, Colorado State University Fluid Dynamics and Diffusion Laboratory Report CER75-76JEC5, Colorado State University, Fort Collins, Colo.

——, 1976, Aerodynamics of Buildings, in *Annu. Rev. Fluid Mech.*, 8: 75-106, Annual Reviews, Inc., Palo Alto, Calif.

——, D. J. Lombardi, and R. S. Thompson, 1974, Applications of Physical Modeling to the Investigations of Air Pollution Problems in Urban Areas, presented at the *67th Annual Meeting of the Air Pollution Control Association*, June 9–13, Paper No. 74-160, Denver, Colo.

——, V. A. Sandborn, E. J. Plate, G. H. Binder, H. Chuang, R. N. Meroney, and S. Ito, 1966, *Simulation of Atmospheric Motion by Wind Tunnel Flows*, Colorado State University Fluid Dynamics and Diffusion Laboratory Report CER66JEC-VAS-EJP-RNM-GJB-HC-SI17, Colorado State University, Fort Collins, Colo.

Chang, S.-C., 1966, *Velocity Field in Separated Flow Behind a Model Hill*, M.S. Thesis, Colorado State University, Department of Civil Engineering, Fort Collins, Colo.

Colmer, M. J., 1970, *Some Full-Scale Measurements of the Flow in the Wake of a Hanger*, Royal Aircraft Establishment/Aeronautical Research Council C.P. No. 1166, replacing RAE Technical Report 70202-ARC32863, Her Majesty's Stationery Office, London, 1971.

Comte-Bellot, G., 1976, Hot-Wire Anemometry, in *Ann. Rev. Fluid Mech.*, 8: 209-231, Annual Reviews, Inc., Palo Alto, Calif.

Corke, T. C., and H. M. Nagib, 1976, *Sensitivity of Flow Around and Pressures on a Building Model to Changes in Simulated Atmospheric Surface Layer Characteristics*, IIT Fluids and Heat Transfer Report R76-1, May, Illinois Institute of Technology, Mechanics and Mechanical and Aerospace Engineering Department, Chicago, Ill.

Counihan, J., 1971, *An Experimental Investigation of the Wake Behind a Two-Dimensional Block and Behind a Cube in a Simulated Boundary Layer Flow*, Central Electricity Research Laboratories Note No. RD/L/N 115/71, Central Electricity Research Laboratories, Leatherhead, Surrey, England.

——, J. C. R. Hunt, and P. S. Jackson, 1974, Wakes Behind Two-Dimensional Surface Obstacles in Turbulent Boundary Layers, *J. Fluid Mech.*, 64(Pt. 3): 529-563.

Csanady, G. T., 1973, Turbulent Diffusion in the Environment, in *Geophysics and Astrophysics Monographs,* B. M. McCormac (Ed.), Vol. 3, D. Reidel Publishing Co., Boston, Mass.

Davenport, A. G., 1963, The Relationship of Wind Structure to Wind Loading, in *Wind Effects on Buildings and Structures,* Proceedings of Conference at National Physical Laboratory, Teddington, Middlesex, England, June 26–28, Vol. I, pp. 54-101, Her Majesty's Stationery Office, London, 1965.

Davies, P. O. A. L., and D. J. Moore, 1964, Experiments on the Behavior of Effluent Emitted from Stacks at or near the Roof Level of Tall Reactor Buildings, *Int. J. Air Water Pollut.*, 8(10): 515-533.

de Bray, B. G., 1971, *Protection by Fences*, presented at a Seminar on Wind Effects on Buildings and Structures, University of Auckland, New Zealand, May, Paper No. 9.

Deaves, D. M., 1976, Wind over Hills: A Numerical Approach, *J. Ind. Aerodyn.*, 1(4): 371-391.

Derickson, R. G., and R. N. Meroney, 1977, *A Simplified Physics Airflow Model for Evaluating Wind Power Sites in Complex Terrain*, presented at the Summer Computer Simulation Conference, Chicago, July 18–20, Colorado State University Fluid Dynamics and Diffusion Laboratory Report CER76-77RGD-RNM31, Colorado State University, Fort Collins, Colo.

Dickson, C. R., G. E. Start, and E. H. Markee, Jr., 1967, Aerodynamic Effects of the EBR-II Containment Vessel Complex on Effluent Concentration, in the *Proceedings of the USAEC Meteorological Information Meeting*, Chalk River Nuclear Laboratories, Sept. 11–14, pp. 87-101, AECL Report 2787, Atomic Energy of Canada, Ltd., Chalk River, Ontario.

——, G. E. Start, and E. H. Markee, Jr., 1969, Aerodynamic Effects of the EBR-II Reactor Complex on Effluent Concentration, *Nucl. Safety,* 10(3): 228-242.

Djurić, D., and J. C. Thomas, 1971, A Numerical Study of Convective Transport of a Gaseous Air Pollutant in the Vicinity of Tall Buildings, in *Proceedings of the Symposium on Air Pollution, Turbulence, and Diffusion*, New Mexico State University, Dec. 7–10, H. W. Church and R. E. Luna (Eds.), pp. 27-34, Sandia Laboratories, Albuquerque, N. Mex., March 1972.

Drazin, P. G., 1961, On the Steady Flow of a Fluid of Variable Density Past an Obstacle, *Tellus*, 13(2): 239-251.

Dreher, K. J., and J. E. Cermak, 1973, *Wind Loads on a House Roof,* Colorado State University, Fluid Dynamics and Diffusion Laboratory Report CER72-73KJD-JEC22, Colorado State University, Fort Collins, Colo.

Drivas, P. J., and F. H. Shair, 1974, Probing the Air Flow Within the Wake Downwind of a Building by Means of a Tracer Technique, *Atmos. Environ.*, 8: 1165-1175.

Egan, B. A., 1975, Turbulent Diffusion in Complex Terrain, *Lectures on Air Pollution and Environmental Impact Analyses*, Chap. 4, Boston, Sept. 29–Oct. 3, pp. 112-135, American Meteorological Society, Boston, Mass., 1975.

Eliseev, V. S., 1973, Stereophotogrammetric Investigation of the Air Flow in the Boundary Layer of the Atmosphere Above a Hill, in *Air Pollution and Atmospheric Diffusion*, M. E. Berlyand (Ed.), pp. 95-108, Halsted Press, New York, and Israel Program for Scientific Translations, Jerusalem, Israel, 1974.

Evans, B. H., 1957, *Natural Air Flow Around Buildings*, Texas Engineering Experiment Station Research Report No. 59, Texas A&M University, College Station, Tex.

Fage, A., and F. C. Johansen, 1927, *On the Flow of Air Behind an Inclined Flat Plate of Infinite Span*, Aeronautical Research Council Reports and Memoranda No. 1104, Her Majesty's Stationery Office, London.

Fail, R., J. A. Lawford, and R. C. Eyre, 1957, *Low-Speed Experiments on the Wake Characteristics of Flat Plates Normal to an Air Stream*, Aeronautical Research Council Reports and Memoranda No. 3120, Aeronautical Research Council of Great Britain, Her Majesty's Stationery Office, London, 1959.

Fiedler, H. E., and R. Wille, 1970, Some Observations in the Near Wake of Blunt Bodies, *AIAA (Am. Inst. Aeronaut. Astronaut.) J.*, 8(6): 1140-1141.

Fifth International Conference on Wind Engineering, 1979, Proceedings of Conference at Colorado State University, Fort Collins, July 8–14, Vols. I and II, Pergamon Press, Oxford and New York, 1980.

Freymuth, P., 1978, Extension of the Nonlinear Theory to Constant-Temperature Hot-Film Anemometers, *TSI Quart.*, 4(3): 3-7.

Frost, W., 1973, Review of Data and Prediction Techniques for Wind Profiles Around Manmade Surface Obstructions, in Advisory Group for Aerospace Research and Development Conference Proceedings No. 140, North Atlantic Treaty Organization, National Aeronautics and Space Administration's Report Distribution and Storage Unit, Langley, Va.

——, G. H. Fichtl, J. R. Connell, and M. L. Hutto, 1977, Mean Horizontal Wind Profiles Measured in the Atmospheric Boundary Layer About a Simulated Block Building, *Boundary-Layer Meteorol.*, 11(2): 135-145.

——, W. L. Harper, and G. H. Fichtl, 1975, Analysis of Atmospheric Flow over a Surface Protrusion Using the Turbulence Kinetic Energy Equation, *Boundary-Layer Meteorol.*, 8(3/4): 401-417.

——, and A. M. Shahabi, 1977, *A Field Study of Wind over a Simulated Block Building*, NASA Report CR-2804, National Aeronautics and Space Administration, NTIS.

Gandemer, J., 1975, Wind Environment Around Buildings: Aerodynamic Concepts, in *Proceedings of the Fourth International Conference on Wind Effects on Buildings and Structures*, Heathrow (London), Sept. 8–12, pp. 423-432, Cambridge University Press, London, 1977.

——, 1976, Inconfort du au Vent aux Abords des Batiments: Concepts Aérodynamiques, Cahiers du Centre Scientifique et Technique du Batiment, no. 170, Cahier 1384 (CSTB, Nantes, France). Available in translation as NBS Technical Note 710-9, March 1978.

Gaster, M., 1971, Vortex Shedding from Circular Cylinders at Low Reynolds Numbers, *J. Fluid Mech.*, 46(Pt. 4): 749-756.

Gerrard, J. H., 1966, The Mechanics of the Formation Region of Vortices Behind Bluff Bodies, *J. Fluid Mech.*, 25(Pt. 2): 401-413.

Gifford, F. A., 1960, Atmospheric Dispersion Calculations Using the Generalized Gaussian Plume Model, *Nucl. Safety,* 2(2): 56-59.

——, 1968, An Outline of Theories of Diffusion in the Lower Layers of the Atmosphere, in *Meteorology and Atomic Energy–1968*, pp. 65-116, D. Slade (Ed.), USAEC Report TID-24190, Chap. 3, Environmental Science Services Administration, NTIS.

——, 1975, Atmospheric Dispersion Models for Environmental Pollution Applications, *Lectures on Air Pollution and Environmental Impact Analyses*, Boston, Sept. 29–Oct. 3, Chap. 2, pp. 35-58, American Meteorological Society, Boston, Mass.

——, 1976, Turbulent Diffusion-Typing Schemes: A Review, *Nucl. Safety,* 17(1): 68-86.

Goldstein, S. (Ed.), 1965, Wakes, in *Modern Developments in Fluid Dynamics*, Vol. II, Chap. XIII, pp. 548-600, Dover Publications, New York.

Good, M. C., and P. N. Joubert, 1968, The Form Drag of Two-Dimensional Bluff-Plates Immersed in Turbulent Boundary Layers, *J. Fluid. Mech.*, 31(Pt. 3): 547-582.

Gregory, N., and W. S. Walker, 1951, *The Effect on Transition of Isolated Surface Excrescences in the Boundary Layer*, Aeronautical Research Council Reports and Memoranda No. 2779, pp. 1-10 and Figs. 1-20, Her Majesty's Stationery Office, London, 1956.

Griffin, O. M., 1978, A Universal Strouhal Number for the "Locking-On" of Vortex Shedding to the Vibrations of Bluff Cylinders, *J. Fluid Mech.*, 85(Pt. 3): 591-606.

Halitsky, J., 1961, Diffusion of Vented Gas Around Buildings, presented at the 54th Annual Meeting of Air Pollution Control Association, New York, Paper No. 61-35, June 11–15,

——, 1963a, *Gas Diffusion Near Buildings,* NYU Meteorology and Oceanography Geophysical Sciences Laboratory Report 63-3, New York University, College of Engineering, New York.

——, 1963b, Gas Diffusion Near Buildings, *ASHRAE (Am. Soc. Heating, Refrig. Air-Cond. Eng.) Trans.*, Paper No. 1855, 69: 464-485.

——, 1968, Gas Diffusion Near Buildings, in *Meteorology and Atomic Energy–1968,* D. H. Slade (Ed.), Chap. 5-5, pp. 221-255, USAEC Report TID-24190, Environmental Science Services Administration, NTIS.

——, 1977, Wake and Dispersion Models for the EBR-II Building Complex, *Atmos. Environ.*, 11(7): 577-596.

——, J. Golden, P. Halpern, and P. Wu, 1963, *Wind Tunnel Tests of Gas Diffusion from a Leak in the Shell of a Nuclear Power Reactor and from a Nearby Stack*, NYU Meteorology and Oceanography Geophysical Sciences Laboratory Report 63-2, New York University, College of Engineering, New York.

——, and K. Woodard, 1974, Atmospheric Diffusion Experiments at a Nuclear Power Plant Site Under Light Wind Inversion Conditions, in *Preprints of Symposium on Atmospheric Diffusion and Air Pollution*, Santa Barbara, Calif., Sept. 9–13, pp. 172-179, American Meteorological Society, Boston, Mass.

Hansen, A. C., and J. E. Cermak, 1975, *Vortex-Containing Wakes of Surface Obstacles*, Colorado State University Fluid Dynamics and Diffusion Laboratory Report CER75-76ACH-JEC16, Colorado State University, Fort Collins, Colo.

——, J. A. Peterka, and J. E. Cermak, 1975, *Wind-Tunnel Measurements in the Wake of a Simple Structure in a Simulated Atmospheric Flow*, Report NASA CR-2540, National Aeronautics and Space Administration, NTIS.

Hatcher, R. V., and R. N. Meroney, 1977, Dispersion in the Wake of a Model Industrial Complex, in *Preprints of Joint Conference on Applications of Air Pollution Meteorology*, Salt Lake City, Nov. 29–Dec. 2, pp. 343-346, American Meteorological Society, Boston, Mass.

——, R. N. Meroney, J. A. Peterka, and K. Kothari, 1978, *Dispersion in the Wake of a Model Industrial Complex*, NRC Report NUREG-0373, Nuclear Regulatory Commission, NTIS.

Hawthorne, W. R., and M. E. Martin, 1955, The Effect of Density Gradient and Shear on the Flow over a Hemisphere, *Proc. R. Soc. (London), Ser. A.*, 232: 184-195.

Hinds, W. T., 1969, Peak-to-Mean Concentration Ratios from Ground-Level Sources in Building Wakes, *Atmos. Environ.*, 3(2): 145-156.

Hinze, J. O., 1959, *Turbulence*, McGraw-Hill, New York, Toronto, and London.

Hirt, C. W., and J. L. Cook, 1972, Calculating Three-Dimensional Flows Around Structures and over Rough Terrain, *J. Comput. Phys.*, 10(2): 324-340.

Hoerner, S. F., 1965, *Fluid-Dynamic Drag*, Hoerner Fluid Dynamics, Brick Town, N. J.

Hoot, T. B., R. N. Meroney, and J. A. Peterka, 1973, *Wind Tunnel Tests of Negatively Buoyant Plumes*, Colorado State University Fluid Dynamics and Diffusion Laboratory Report CER73-74TGH-RNM-JAP-13, Colorado State University, Fort Collins, Colo.

Hosker, R. P., Jr., 1979, Empirical Estimation of Wake Cavity Size Behind Block-Type Structures, in *Preprints of Fourth Symposium on Turbulence, Diffusion, and Air Pollution*, Reno, Nev., Jan. 15–18, pp. 603-609, American Meteorological Society, Boston, Mass., 1979.

——, 1982, *Methods for Estimating Wake Flow and Effluent Dispersion Near Simple Block-Like Buildings*, NRC Report NUREG/CR-252, Nuclear Regulatory Commission, NTIS.

Hotchkiss, R. S., 1971, The Numerical Calculation of Three-Dimensional Flows of Air and Particulates About Structures, in *Proceedings of a Symposium on Air Pollution, Turbulence, and Diffusion*, New Mexico State University, Dec. 7–10, pp. 35-42, H. W. Church and R. E. Luna (Eds.), Sandia Laboratories, Albuquerque, N. Mex., March 1972.

Houghton, E. L., and N. B. Carruthers, 1976, *Wind Forces on Buildings and Structures: An Introduction*, Halsted Press, John Wiley & Sons, New York.

Hoydysh, W. G., R. A. Griffiths, and Y. Ogawa, 1974, *A Scale Model Study of the Dispersion of Pollution in Street Canyons*, presented at the 67th Annual Meeting of the Air Pollution Control Association, Denver, June 9–13, Paper No. 74-157,

Huber, A. H., 1977, Incorporating Building/Terrain Wake Effects on Stack Effluents, in *Preprints of Joint Conference on Applications of Air Pollution Meteorology*, Salt Lake City, Nov. 29–Dec. 2, pp. 353-356, American Meteorological Society, Boston, Mass.

——, 1979, An Evaluation of Obstacle Wake Effects on Plume Dispersion, presented at the *Fourth Symposium on Turbulence, Diffusion, and Air Pollution*, Reno, Nev., Jan. 15–18, American Meteorological Society, Boston, Mass.

——, and W. H. Snyder, 1976, Building Wake Effects on Short Stack Effluents, in *Third Symposium on Atmospheric Turbulence, Diffusion, and Air Quality*, Raleigh, N. C., Oct. 19–22, pp. 235-242, American Meteorological Society, Boston, Mass.

——, W. H. Snyder, R. S. Thompson, and R. E. Lawson, Jr., 1976, *Stack Placement in the Lee of a Mountain Ridge*, Environmental Sciences Research Laboratory Report EPA-600/4-76-047, U. S. Environmental Protection Agency, NTIS.

——, W. H. Snyder, R. S. Thompson, and R. E. Lawson, Jr., 1980, *The Effects of a Squat Building on Short Stack Effluents*, Environmental Sciences Research Laboratory Report EPA-600/4-80-055, U. S. Environmental Protection Agency, NTIS.

Humphries, W., and J. H. Vincent, 1976, Experiments to Investigate Transport Processes in the Near Wakes of Disks in Turbulent Air Flow, *J. Fluid Mech.*, 75(Pt. 4): 737-749.

Hunt, J. C. R., 1970, *Further Aspects and Theory of Wakes Behind Buildings and Comparison of Theory with Experiments*, CERL Report RD/L/R/1665, Central Electricity Research Laboratories, Leatherhead, Surrey, England.

——, 1971, The Effect of Single Buildings and Structures, *Philos. Trans. R. Soc. London, Ser. A*, 269: 457-467.

——, 1972, Some Theories for the Mean and Turbulent Velocity Distributions in Flows Around Bluff Bodies, in *Proceedings of a Symposium on External Flows*, University of Bristol, July 4–6, pp. C-1 to C-10, University of Bristol, England.

——, 1973, A Theory of Turbulent Flow Round Two-Dimensional Bluff Bodies, *J. Fluid Mech.*, 61(Pt. 4): 625-706.

——, 1974, *Wakes Behind Buildings*, presented at the Aeronautical Research Council's Atmospheric Environment Committee Meeting, Oct. 23, ARC Report 35601, Atmos. 229, Aeronautical Research Council of Great Britain, Her Majesty's Stationery Office, London.

——, 1975, Turbulent Velocities Near and Fluctuating Surface Pressures on Structures in Turbulent Winds, in *Proceedings of the Fourth International Conference on Wind Effects on Buildings and Structures*, Heathrow (London), Sept. 8-12, pp. 309-320, Cambridge University Press, London, 1977.

——, to be published. Cited in A. C. Hansen and J. E. Cermak, 1975.

——, C. J. Abell, J. A. Peterka, and H. Woo, 1978, Kinematical Studies of the Flows Around Free or Surface Mounted Obstacles: Applying Topology to Flow Visualization, *J. Fluid Mech.*, 86(Pt. 1): 179-200.

——, and P. J. Mulhearn, 1973, Turbulent Dispersion from Sources Near Two-Dimensional Obstacles, *J. Fluid Mech.*, 61(Pt. 2): 245-274.

——, and G. P. Smith, 1969, *A Theory of Wakes Behind Buildings and Some Provisional Experimental Results*, CERL Note RD/L/N31/69, Central Electricity Research Laboratory, Leatherhead, Surrey, England.

——, W. H. Snyder, and R. E. Lawson, Jr., 1978, Flow Structure and Turbulent Diffusion Around a Three-Dimensional Hill, in *Fluid Modeling Study on Effects of Stratification, Part I. Flow Structure*, Report EPA-600/4-78-041, U. S. Environmental Protection Agency, NTIS.

——, and W. H. Snyder, 1980, Experiments on Stably and Neutrally Stratified Flow over a Model Three-Dimensional Hill, *J. Fluid Mech.*, 96(Pt. 4): 671-704.

Islitzer, N. F., 1965, *Aerodynamic Effects of Large Reactor Complexes upon Atmospheric Turbulence and Diffusion*, USAEC National Reactor Testing Station Report IDO-12041, U. S. Weather Bureau, Idaho Falls, Idaho.

Isyumov, N., and A. G. Davenport, 1975, The Ground Level Wind Environment in Built-Up Areas, in *Proceedings of the Fourth International Conference on Wind Effects on Buildings and Structures*,

Heathrow (London), Sept. 8–12, pp. 403-422, Cambridge University Press, London, 1977.

——, T. Jandali, and A. G. Davenport, 1974, *Model Studies and the Prediction of Full Scale Levels of Stack Gas Concentration,* presented at 67th Annual Meeting of the Air Pollution Control Association, Denver, June 9–13, Paper No. 74-162,

Iversen, J. D., and R. Greeley, 1978, *Atmospheric and Wind Tunnel Experiments on the Amboy Crater Sand-Covered Lava Flow*, Iowa State University Engineering Research Institute Interim Report No. ISU-ERI-78235, Iowa State University, Ames, Iowa.

Jackson, P. S., 1973, *The Flow Around Obstacles in Boundary Layers*, Ph.D. Dissertation, University of Cambridge, Cambridge, England.

——, 1975, A Theory for Flow over Escarpments, in *Proceedings of the Fourth International Conference on Wind Effects on Buildings and Structures*, Heathrow, (London), Sept. 8–12, pp. 33-40, Cambridge University Press, London, 1977.

——, and J. C. R. Hunt, 1975, Turbulent Wind Flow over a Low Hill, *Q. J. R. Meteorol. Soc.*, 101(430): 929-955.

Jacobs, W., 1938, Strömung hinter einem einzelnen Rauhigkeitselement, *Ing.-Arch.*, IX: 343-355.

Jensen, M., 1958, The Model-Law for Phenomena in Natural Wind, *Ingenioren*, International Edition, 2(4): 121-128.

——, and N. Franck, 1963, *Model-Scale Tests in Turbulent Wind*, Part I. Danish Technical Press, Copenhagen.

Johnson, W. B., E. Shelar, R. E. Ruff, H. B. Singh, and L. Salas, 1975, *Gas Tracer Study of Roof-Vent Effluent Diffusion at Millstone Nuclear Power Station*, Report AIF/NESP-007b, Atomic Industrial Forum, Inc., Washington, D. C.

Katsura, J., 1974, Fluctuating Wind Pressure on the Side of the Surfaces on Models with Long Rectangular Sections, in *Proceedings of the Second USA–Japan Research Seminar on Wind Effects on Structures*, Kyoto, Japan, Sept. 9–13, Univ. of Hawaii Press, Honolulu, 1976.

Kiya, M., and M. Arie, 1972, A Free-Streamline Theory for Bluff Bodies Attached to a Plane Wall, *J. Fluid Mech.*, 56(Pt. 2): 201-219.

——, and M. Arie, 1973, A Perturbation Analysis of the Laminar Far Wake Behind a Symmetrical Two-Dimensional Body in a Uniform Shear Flow, *J. Fluid Mech.*, 61(Pt. 2): 305-321.

——, and M. Arie, 1976, Theory of Inviscid Shear Flow over Bluff Bodies Attached to a Plane Wall (Part 2, Application to Semielliptical Projection and Arbitrary-Shaped Body), *Bull. JSME (Jpn. Soc. Mech. Eng.)*, 19(131): 523-530.

——, and M. Arie, 1977a, An Inviscid Numerical Simulation of Vortex Shedding from an Inclined Flat Plate in Shear Flow, *J. Fluid Mech.*, 82(Pt. 2): 241-253.

——, and M. Arie, 1977b, An Inviscid Bluff-Body Wake Model Which Includes the Far-Wake Displacement Effect, *J. Fluid Mech.*, 81(Pt. 3): 593-607.

——, H. Sakamoto, and M. Arie, 1976, Theory of Inviscid Shear Flow over Bluff Bodies Attached to a Plane Wall, Part 1, Theoretical Formulation and Applications to Normal Plate and Semicircular Projection, *Bull. JSME (Jpn. Soc. Mech. Eng.)*, 19(131): 513-522.

——, H. Tamura, and M. Arie, 1976, Turbulent Wakes Behind Two-Dimensional Bodies in a Uniform Shear Flow, *Bull. JSME (Jpn. Soc. Mech. Eng.)*, 19(129): 274-282.

Koga, D. J., and J. L. Way, 1979, *Effects of Stack Height and Position on the Dispersion of Pollutants in Building Wakes*, IIT Fluids and Heat Transfer Report No. R79-2, Illinois Institute of Technology, Mechanics and Mechanical and Aerospace Engineering Department, Chicago, Ill.

Kothari, K. M., J. A. Peterka, and R. N. Meroney, 1979, *Stably Stratified Building Wakes*, NRC Report NUREG/CR-1247, Nuclear Regulatory Commission, NTIS.

Landau, L. D., and E. M. Lifshitz, 1959, *Fluid Mechanics*, Pergamon Press, London, and Addison-Wesley, Reading, Mass.

Laneville, A., I. S. Gartshore, and G. V. Parkinson, 1975, An Explanation of some Effects of Turbulence on Bluff Bodies, in *Proceedings of the Fourth International Conference on Wind Effects on Buildings and Structures*, Heathrow (London), Sept. 8–12, pp. 333-341, Cambridge University Press, London, 1977.

Lawson, T. V., 1967, Discussion of "Turbulence and Diffusion in the Wake of a Building," *Atmos. Environ.*, 1(2): 177-181.

Lee, B. E., 1975, The Effect of Turbulence on the Surface Pressure Field of a Square Prism, *J. Fluid Mech.*, 69(Pt. 2): 263-282.

Lemberg, R., 1973, *On the Wakes Behind Bluff Bodies in a Turbulent Boundary Layer*, Ph.D. Thesis and Boundary Layer Wind Tunnel Report No. BLWT-3-73, University of Western Ontario, London, Ontario, Canada.

Leutheusser, H. J., and W. D. Baines, 1967, Similitude Problems in Building Aerodynamics, *J. Hydraul. Div., Am. Soc. Civ. Eng.*, 93(HY3): 35-49.

Li, W. W., R. N. Meroney, and J. A. Peterka, 1981, *Wind-Tunnel Study of Gas Dispersion near a Cubical Model Building*, NRC Report NUREG/CR-2395, Nuclear Regulatory Commission, NTIS.

Lighthill, M. J., 1956, Drift, *J. Fluid Mech.*, 1(Pt. 0): 31-53.

Lin, J. -T., H. -T. Liu, and Y. -H. Pao, 1974, *Laboratory Simulation of Plume Dispersion in Stably Stratified Flows over a Complex Terrain*, Flow Research Report No. 29, prepared for the U. S. Environmental Protection Agency, Flow Research, Inc., Kent, Wash.

——, and Y. -H. Pao, 1979, Wakes in Stratified Fluids, in *Annu. Rev. Fluid Mech.*, 11: 317-338, Annual Reviews, Inc., Palo Alto, Calif.

Liu, H. -T., and J. -T. Lin, 1976, *Plume Dispersion in Stably Stratified Flows over Complex Terrain, Phase 2*, Report EPA-600/4-76-022, U. S. Environmental Protection Agency, NTIS.

Locher, F. A., 1969, *Some Aspects of Flow-Induced Vibrations of Hydraulic Control Gates*, Iowa Institute of Hydraulic Research, Report No. 116, University of Iowa, Iowa City.

Long, R. R., 1972, Finite Amplitude Disturbances in the Flow of Inviscid Rotating and Stratified Fluids over Obstacles, in *Ann. Rev. Fluid Mech.*, 4: 69-92, Annual Reviews, Inc., Palo Alto, Calif.

Lucas, D. H., 1972, Choosing Chimney Heights in the Presence of Buildings, in *Proceedings of the International Clean Air Conference*, Melbourne University, Australia, May, pp. 47-52, Clean Air Society of Australia and New Zealand, 1972.

Ludwig, G. R., T. R. Sundaram, and G. T. Skinner, 1971, *Laboratory Modeling of the Atmospheric Surface Layer with Emphasis on Diffusion*, Cornell Aeronautical Laboratory Report VC-2740-S-2, Cornell Aeronautical Laboratory (now CALSPAN), Buffalo, N.Y.

Lumley, J. L., and H. A. Panofsky, 1964, The Structure of Atmospheric Turbulence, Vol. 12 in *Monographs Texts in Physics and Astronomy,* R. E. Marshak (Ed.), Interscience Publishers, a Division of John Wiley & Sons, New York.

Mair, W. A., and D. J. Maull, 1971, Bluff Bodies and Vortex Shedding: A Report on Euromech 17, *J. Fluid Mech.*, 45(Pt. 2): 209-224.

Martin, J. E., 1965, *The Correlation of Wind Tunnel and Field Measurements of Gas Diffusion Using Krypton-85 as a Tracer*, Ph.D. Thesis and Report No. MMPP-272 of Michigan Memorial Phoenix Project, University of Michigan, Ann Arbor, Mich.

Masch, F. D., and W. L. Moore, 1963, Drag Forces in Velocity Gradient Flow, *Am. Soc. Civ. Eng. Trans.*, Paper 3394, 128(Pt. I): 48-64.

Maskell, E. C., 1963, *A Theory of the Blockage Effects on Bluff Bodies and Stalled Wings in a Closed Wind Tunnel*, Aeronautical Research Council Reports and Memoranda No. 3400, Her Majesty's Stationery Office, London, 1965.

Mason, P. J., and R. I. Sykes, 1978, A Simple Cartesian Model of Boundary Layer Flow over Topography, *J. Comput. Phys.*, 28(2): 198-210.

——, and R. I. Sykes, 1979a, Three-Dimensional Numerical Integrations of the Navier–Stokes Equations for Flow over Surface-Mounted Obstacles, *J. Fluid Mech.*, 91(Pt. 3): 433-450.

——, and R. I. Sykes, 1979b, Flow over an Isolated Hill of Moderate Slope, *Q. J. R. Meteorol. Soc.*, 105(445): 383-395.

Maull, D. J., and R. A. Young, 1973, Vortex Shedding from Bluff Bodies in a Shear Flow, *J. Fluid Mech.*, 60(Pt. 2): 401-409.

McVehil, G. E., F. R. Ludwig, and T. R. Sundaram, 1967, *On the Feasibility of Modeling Small Scale Atmospheric Motions*, Cornell Aeronautical Laboratory Report ZB-2328-P-1, Cornell Aeronautical Laboratory (now CALSPAN), Buffalo, N. Y.

Meroney, R. N., 1982, Turbulent Diffusion Near Buildings, in *Engineering Meteorology*, E. J. Plate (Ed.), Chap. 11, pp. 481-525, Elsevier Scientific Pubs. Co., Amsterdam.

——, J. E. Cermak, and F. H. Chaudhry, 1968, *Wind-Tunnel Model Study of Shoreham Nuclear Power Station Unit I for Long Island Lighting Company*, Colorado State University Fluid Dynamics and Diffusion Laboratory Report CER68-69RNM-JEC-FHC1, Colorado State University, Fort Collins, Colo.

——, J. A. Peterka, and K. M. Kothari, 1980, *Wind-Tunnel Measurements of Dispersion and Turbulence in the Wakes of Nuclear Reactor Plants*, NRC Report NUREG/CR-1475, Nuclear Regulatory Commission, NTIS.

——, and B. T. Yang, 1970, *Gaseous Plume Dispersion About Isolated Structures of Simple Geometry,* in Proceedings of the Second International Clean Air Congress of the International Union of the Air Pollution Prevention Association, Dec. 6–11, Washington, D. C., H. M. England and W. T. Beery (Eds.), Academic Press, New York, 1971.

——, and B. T. Yang, 1971, *Wind-Tunnel Study on Gaseous Mixing Due to Various Stack Heights and Injection Rates Above an Isolated Structure*, Colorado State University Fluid Dynamics and Diffusion Laboratory Report CER71-72RNM-BTY16, Colorado State University, Fort Collins, Colo.

Mochizuki, M., 1961, Smoke Observation on Boundary Layer Transition Caused by a Spherical Roughness Element, *J. Phys. Soc. Jpn.*, 16(5): 995-1008.

Modi, V. J., and S. El-Sherbiny, 1975, Wall Confinement Effects on Bluff Bodies in Turbulent Flows, in *Proceedings of the Fourth International Conference on Wind Effects on Buildings and Structures*, Heathrow (London), Sept. 8–12, pp. 121-132, Cambridge University Press, London, 1977.

Morkovin, M. V., 1972, An Approach to Flow Engineering via Functional Flow Modules, in *Beitr. Strömungsmech.*, A. Walz 65th Anniversary Volume, Deutsche Luft-und Raumfahrt Forschungsvericht No. 72-27.

Mueller, T. J., H. H. Korst, and W. L. Chow, 1964, On the Separation, Reattachment, and Redevelopment of Incompressible Turbulent Shear Flow, *Trans. ASME (Am. Soc. Mech. Eng.) J. Basic Eng., Ser. D*, 86(2): 221-226.

Munn, R. E., and A. F. W. Cole, 1967a, Turbulence and Diffusion in the Wake of a Building, *Atmos. Environ.*, 1(1): 33-43.

——, and A. F. W. Cole, 1967b, Some Strong-Wind Downwash Diffusion Measurements at Douglas Point, Ontario, Canada, *Atmos. Environ.*, 1(5): 601-604.

Nagabhushanaiah, H. S., 1961, *Separation Flow Downstream of a Plate Set Normal to a Plane Boundary*, Ph.D. Thesis, Colorado State University Department of Civil Engineering, Fort Collins, Colo.

Nakaguchi, H., K. Hashimoto, and S. Muto, 1968, An Experimental Study on Aerodynamic Drag of Rectangular Cylinders, *J. Jpn. Soc. Aeronaut. and Space Sci.*, 16(168): 1-5.

Norman, R. S., 1972, *On Obstacle Generated Secondary Flows in Laminar Boundary Layers and Transition to Turbulence*, Ph.D. Thesis, Illinois Institute of Technology Mechanical Engineering Department, Chicago, Ill.

Ogawa, Y., 1973, *Effects of Building and Thermal Boundary Layer On Diffusion*, Ph.D. Thesis, Hokkaido University Department of City Environmental Engineering, Sapporo, Japan.

——, P. G. Diosey, and S. Oikawa, 1979, The Effects of the Characteristics of the Oncoming Wind on the Flow Behind a Two-Dimensional Model Building, in *Proceedings of the Fifth International Conference on Wind Engineering*, Fort Collins, Colorado, July 8–14, Vol. I, Pergamon Press, Oxford, New York, 1980.

Ostrowski, J. S., R. D. Marshall, and J. E. Cermak, 1967, Vortex Formation and Pressure Fluctuations on Buildings, in *Proceedings of the International Research Seminar on Wind Effects on Buildings and Structures*, Ottawa, Canada, Sept. 11–15, Vol. I, pp. 459-484, University of Toronto Press, Toronto, 1968.

Owen, P. R., 1971, Buildings in the Wind, *Q. J. R. Meteorol. Soc.*, 97(414): 396-413.

Pao, Y. -H., 1973, Measurements of Internal Waves and Turbulence in Two-Dimensional Stratified Shear Flows, *Boundary-Layer Meteorol.*, 5(1/2): 177-193.

Parkinson, G. V., and T. Jandali, 1970, A Wake Source Model for Bluff Body Potential Flow, *J. Fluid Mech.*, 40(Pt. 3): 577-594.

Penwarden, A. D., and A. F. E. Wise, 1975, *Wind Environment Around Buildings*, Building Research Establishment Report, Her Majesty's Stationery Office, London.

Peterka, J. A., and J. E. Cermak, 1975a, Turbulence in Building Wakes, in *Proceedings of the Fourth International Conference on Wind Effects on Buildings and Structures*, Heathrow (London), Sept. 8–12, pp. 447-463, Cambridge University Press, London, 1977.

——, and J. E. Cermak, 1975b, *Adverse Wind Loading Induced by Adjacent Buildings*, Preprint No. 2456 of ASCE National Structural Engineering Convention, New Orleans, April 14–18, American Society of Civil Engineers.

Petryk, S., and E. Brundrett, 1967, *Recovery of a Turbulent Boundary Layer Disturbed by a Single Roughness Element*, University of Waterloo Department of Mechanical Engineering Research Report No. 4, July, University of Waterloo, Ontario, Canada.

Plate, E. J., 1971, The Aerodynamics of Shelter Belts, *Agric. Meteorol.*, 8(3): 203-222.

——, and C. W. Lin, 1965, *The Velocity Field Downwind from a Two-Dimensional Model Hill, Part 1*, Colorado State University Fluid Dynamics and Diffusion Laboratory Report CER65-EJP-14, Colorado State University, Fort Collins, Colo.

Probert, S. D., M. A. Hasoon, and T. S. Lee, 1973, Boundary Layer Separation for Winds Flowing Around Large Fixed-Roof Oil-Storage Tanks with Small Aspect Ratios, *Atmos. Environ.*, 7(6): 651-654.

Puttock, J. S., 1979, Turbulent Diffusion from Sources near Obstacles with Separated Wakes, Part II. Concentration Measurements near a Circular Cylinder in Uniform Flow, *Atmos. Environ.*, 13(1): 15-22.

——, and J. C. R. Hunt, 1979, Turbulent Diffusion from Sources near Obstacles with Separated Wakes, Part I. An Eddy Diffusivity Model, *Atmos. Environ.*, 13(1): 1-13.

Raine, J. K., and D. C. Stevenson, 1977, Wind Protection by Model Fences in a Simulated Atmospheric Boundary Layer, *J. Ind. Aerodyn.*, 2(2): 159-180.

Ranga Raju, K. G., J. Loeser, and E. J. Plate, 1976, Velocity Profiles and Fence Drag for a Turbulent Boundary Layer Along Smooth and Rough Flat Plates, *J. Fluid. Mech.*, 76(Pt. 2): 383-399.

Reinhold, T. A., H. W. Tieleman, and F. J. Maher, 1977, Interaction of Square Prisms in Two Flow Fields, *J. Ind. Aerodyn.*, 2(3): 223-241.

Riley, J. J., H. -T. Liu, and E. W. Geller, 1976, *A Numerical and Experimental Study of Stably Stratified Flow Around Complex Terrain*, Report EPA-600/4-76-021, U. S. Environmental Protection Agency, NTIS.

Roberson, J. A., C. -Y. Lin, G. S. Rutherford, and M. D. Stine, 1972, Turbulence Effects on Drag of Sharp-Edged Bodies, *J. Hydraul. Div., Am. Soc. Civ. Eng.*, Proceedings paper No. 9061, 98(HY7): 1187-1203.

Robertson, J. M., 1975, Pressure Field at Reattachment of Separated Flows, in *Proceedings of the Second U. S. National Conference on Wind Engineering Research*, Colorado State University, June 22–25, Paper IV-26, Colorado State University, Fort Collins, Colo., 1975.

——, J. B. Wedding, J. A. Peterka, and J. E. Cermak, 1978, Wall Pressures of Separation–Reattachment Flow on a Square Prism in Uniform Flow, *J. Ind. Aerodyn.*, 2(4): 345-359.

Robins, A. G., and I. P. Castro, 1977a, A Wind Tunnel Investigation of Plume Dispersion in the Vicinity of a Surface Mounted Cube, I. The Flow Field, *Atmos. Environ.*, 11(4): 291-297.

——, and I. P. Castro, 1977b, A Wind Tunnel Investigation of Plume Dispersion in the Vicinity of a Surface Mounted Cube—II. The Concentration Field, *Atmos. Environ.*, 11(4): 299-311.

Rodliffe, R. S., and A. J. Fraser, 1971, Measurements on the Release of Gaseous Activity from a Short Stack, *Atmos. Environ.*, 5(4): 193-208.

Roper, A. T., 1965, *The Wake Region of a Circular Cylinder in a Turbulent Boundary Layer*, M.S. Thesis, Colorado State University Department of Civil Engineering, Fort Collins, Colo.

Roshko, A., and W. Fiszdon, 1967, On the Persistence of Transition in the Near-Wake, in *Problems of Hydrodynamics and Continuum Mechanics,* pp. 606-616, Society for Industrial and Applied Mathematics, Philadelphia, Pa.

——, and J. C. Lau, 1965, Some Observations on Transition and Reattachment of a Free Shear Layer in Incompressible Flow, in *Proceedings of the 1965 Heat and Fluid Mechanics Institute*, pp. 157-167, Stanford University Press, Stanford, Calif.

——, A. Steinolfson, and V. Chattoorgoon, 1975, Flow Forces on a Cylinder near a Wall or near Another Cylinder, in *Proceedings of the Second U. S. National Conference on Wind Engineering Research*, Colorado State University, June 22–25, Paper IV-15, Colorado State University, Fort Collins, Colo., 1975.

Sacré, C., 1979, An Experimental Study of the Airflow over a Hill in the Atmospheric Boundary Layer, *Boundary-Layer Meteorol.*, 17(3): 381-401.

Saffman, P. G., 1973, Structure of Turbulent Line Vortices, *Phys. Fluids*, 16(8): 1181-1188.

Sagendorf, J. F., and C. R. Dickson, 1976, Diffusion Under Low Windspeed, Inversion Conditions, in *Preprints of Third Symposium on Atmospheric Turbulence, Diffusion, and Air Quality*, Raleigh, N. C., Oct. 19–22, pp. 277-282, American Meteorological Society, Boston, Mass.

Schlichting, H., 1960, *Boundary Layer Theory*, 4th ed., McGraw-Hill, New York, Toronto, and London.

Scorer, R. S., 1958, *Natural Aerodynamics*, Pergamon Press, New York, London, and Paris.

——, 1968, *Air Pollution*, Pergamon Press, Oxford, London, New York, Toronto, and Paris.

Scruton, C., 1967, Aerodynamics of Structures, in *Proceedings of the International Research Seminar on Wind Effects on Buildings and Structures*, Ottawa, Canada, Sept. 11–15, Vol. I, pp. 115-161, University of Toronto Press, Toronto, 1968.

——, and E. W. E. Rogers, 1971, Steady and Unsteady Wind Loading of Buildings and Structures, in *Philos. Trans. R. Soc. London, Ser. A*, 269(1199): 353-383.

Sedney, R., 1973, A Survey of the Effects of Small Protuberances on Boundary-Layer Flows, *AIAA (Am. Inst. Aeronaut. Astronaut. J.*, 11(6): 782-792.

Sforza, P. M., and R. F. Mons, 1970, Wall-Wake: Flow Behind a Leading Edge Obstacle, *AIAA (Am. Inst. Aeronaut. Astronaut.) J.*, 8(12): 2162-2167.

Sheih, C. M., P. J. Mulhearn, E. F. Bradley, and J. J. Finnigan, 1978, Pollutant Transfer Across the Cavity Region Behind a Two-Dimensional Fence, *Atmos. Environ.*, 12(12): 2301-2307.

Sherlock, R. H., and E. A. Stalker, 1941, *A Study of Flow Phenomena in the Wake of Smokestacks*, University of Michigan Department of Engineering Research Bulletin No. 29, University of Michigan Press, Ann Arbor.

Shieh, C. F., W. Frost, and J. Bitte, 1977, *Neutrally Stable Atmospheric Flow over a Two-Dimensional Rectangular Block*, NASA Contractor Report NASA-CR-2926, National Aeronautics and Space Administration, NTIS.

Smith, D. G., 1975, *Influence of Meteorological Factors upon Effluent Concentrations on and near Buildings with Short Stacks*, presented at 68th Annual Meeting of Air Pollution Control Association, Boston, June 15–20, Paper No. 75-26.2,

——, 1978, *Turbulent Dispersion Around a Building in the Natural Wind and Effluent Reentry*, Ph.D. Thesis, Harvard University School of Public Health, Boston, Mass.

Snyder, W. H., 1972, Similarity Criteria for the Application of Fluid Models to the Study of Air Pollution Meteorology, *Boundary-Layer Meteorol.*, 3(1): 113-134.

——, 1981, Guidelines for Fluid Modeling of Atmospheric Diffusion, Report EPA-600/8-81-009, U. S. Environmental Protection Agency, NTIS.

——, and R. E. Lawson, Jr., 1976, Determination of a Necessary Height for a Stack Close to a Building, *Atmos. Environ.*, 10(9): 683-691.

Soliman, B. F., 1976, *Effect of Building Group Geometry on Wind Pressure and Properties of Flow*, University of Sheffield Building Sciences Report BS29, University of Sheffield, England.

Souster, C. G., and B. E. Lee, 1975, *The Measurement of Airflow Patterns Upstream of Some Simple Building Shapes*, University of Sheffield Department of Building Science Report BS 26, July, University of Sheffield, England.

Stansby, P. K., 1976, The Locking-On of Vortex Shedding Due to the Cross-Stream Vibration of Circular Cylinders in Uniform and Shear Flows, *J. Fluid Mech.*, 74(Pt. 4): 641-665.

Start, G. E., J. H. Cate, C. R. Dickson, N. R. Ricks, G. R. Ackermann, and J. F. Sagendorf, 1977, *Rancho Seco Building Wake Effects on Atmospheric Diffusion*, NOAA Technical Memorandum ERL-ARL-69, National Oceanic and Atmospheric Administration/Aerospace Research Laboratory, Idaho Falls, Idaho.

Sullerey, R. K., A. K. Gupta, and C. S. Moorthy, 1975, Similarity in the Turbulent Near Wake of Bluff Bodies, *AIAA (Am. Inst. Aeronaut. Astronaut.) J.*, 13(11): 1425-1429.

Sundaram, T. R., G. T. Skinner, and G. R. Ludwig, 1970, *Simulation of Small-Scale Atmospheric Phenomena*, Cornell Aeronautical Laboratory Report VC-2697-A-1, Cornell Aeronautical Laboratory (now CALSPAN), Buffalo, N. Y.

Swain, L. M., 1929, On the Turbulent Wake Behind a Body of Revolution, *Proc. R. Soc. (London), Ser. A.*, 125: 647-659.

Symes, C. R., and R. N. Meroney, 1970, *Cone Frustrums in a Shear Layer*, USAEC Report COO-2053-4, Colorado State University Fluid Dynamics and Diffusion Laboratory Report CER70-71CRS-RNM11, Colorado State University, Fort Collins, Colo.

Tani, I., 1957, Experimental Investigation of Flow Separation over a Step, in *Proceedings of the Boundary Layer Research Symposium at Freiburg, Aug. 26–29, 1957*, H. Görtler (Ed.), pp. 377-386, Springer-Verlag, Berlin and Göttingen, 1958.

Tani, N., 1952, On the Mechanism and Effect of Shelter Hedge, Wind-Tunnel Test of the Model Hedge (Pt. 2), *J. Agric. Meteorol., Tokyo*, 7(3-4): 134-136.

——, 1958, On the Wind Tunnel Test of the Model Shelter Hedge, *Bull. Natl. Inst. Agric. Sci., Ser. A.*, 6: 1-80.

Taylor, G. I., 1917, Motion of Solids in Fluids when the Flow is not Irrotational, *Proc. R. Soc. (London), Ser. A.*, 93: 99-113.

Taylor, P. A., and P. R. Gent, 1974, A Model of Atmospheric Boundary Layer Flow Above an Isolated Two-Dimensional "Hill"; An Example of Flow Above "Gentle Topography," *Boundary-Layer Meteorol.*, 7(3): 349-362.

Tennekes, H., and J. L. Lumley, 1972, *A First Course in Turbulence*, MIT Press, Cambridge, Mass., and London, England.

Thuillier, R. H., and R. L. Mancuso, 1980, *Building Effects on Effluent Dispersion from Roof Vents at Nuclear Power Plants*, Report EPRI-NP-1380, Project No. 1073-1, Electric Power Research Institute, Palo Alto, Calif.

Tillman, W., 1945, British Ministry of Aircraft Production, Völkenrode Translation MAP-VG 34-45T (cited by Bradshaw and Wong, 1972).

Townsend, A. A., 1965, The Response of a Turbulent Boundary Layer to Abrupt Changes in Surface Conditions, *J. Fluid Mech.*, 22(Part 4): 799-822.

Tsien, H. S., 1943, Symmetrical Joukowsky Airfoils in Shear Flow, *Q. Appl. Math.*, 1: 130-148.

Turner, D. B., 1969, *Workbook of Atmospheric Dispersion Estimates*, Publication No. 999-AP-26, U. S. Department of Health, Education, and Welfare, Public Health Service.

Turner, J. S., 1973, *Buoyancy Effects in Fluids*, Cambridge University Press, London and Cambridge.

Ukeguchi, N., H. Okamoto, and R. Ohba, 1975, The Numerical Analysis of the Diffusion Around Structures, in *Proceedings of the International Clean Air Conference,* Rotorua, New Zealand, Feb. 17–21, Vol. 1, pp. 330-345, Clean Air Society of Australia and New Zealand, Ann Arbor Science Publishers, Ann Arbor, Michigan.

USEPA, 1978, *Guideline on Air Quality Models,* Environmental Protection Agency Office of Air Quality Planning and Standards Report EPA-450/2-78-027 (OAQRS 1.2-080), U. S. Environmental Protection Agency, Research Triangle Park, N. C.

van Eimern, J., R. Karschon, L. A. Razumova, and G. W. Robertson, 1964, *Windbreaks and Shelterbelts*, Technical Note No. 59, WMO-No.147.TP.70, Secretariat of the World Meteorological Organization, Geneva, Switzerland.

Vickery, B. J., 1966, Fluctuating Lift and Drag on a Long Cylinder of Square Cross-Section in a Smooth and in a Turbulent Stream, *J. Fluid Mech.*, 25(Pt. 3): 481-494.

Vincent, J. H., 1977, Model Experiments on the Nature of Air Pollution Transport near Buildings, *Atmos. Environ.*, 11(8): 765-774.

——, 1978, Scalar Transport in the near Aerodynamic Wakes of Surface Mounted Cubes, *Atmos. Environ.*, 12(6/7): 1319-1322.

Wedding, J. B., D. J. Lombardi, and J. E. Cermak, 1977, A Wind Tunnel Study of Gaseous Pollutants in City Street Canyons, *J. Air Pollut. Control Assoc.*, 27(6): 557-566.

Werle, H., 1975, Ecoulements décollés: étude phénoménologique a partir de visualizations hydrodynamiques, in Advisory Group for Aerospace Research and Development Conference Proceedings No. 168, Paper No. 39, North Atlantic Treaty Organization, NASA's Report Distribution and Storage Unit, Langley, Va., National Aeronautics and Space Administration.

Widnall, S. E., 1975, The Structure and Dynamics of Vortex Filaments, *Annu. Rev. Fluid Mech.*, 7: 141-165, Annual Reviews, Inc., Palo Alto, Calif.

Wilson, D. J., 1976a, *Contamination of Building Air Intakes from Nearby Vents*, University of Alberta Department of Mechanical Engineering Report No. 1, University of Alberta, Edmonton, Alberta, Canada.

——, 1976b, Contamination of Air Intakes from Roof Exhaust Vents, *ASHRAE Trans.*, 82(Pt. 1): 1028-1038.

——, 1977a, Dilution of Exhaust Gases from Building Surface Vents, *ASHRAE Trans.,* 83(Pt. 1): 168-176.

——, 1977b, Effect of Vent Stack Height and Exit Velocity on Exhaust Gas Dilution, *ASHRAE Trans.*, 83(Pt. 1): 157-167.

——, 1979, Flow Patterns over Flat-Roofed Buildings and Application to Exhaust Stack Design, *ASHRAE Trans.*, 85(Pt. 2): 284-295.

——, and D. D. J. Netterville, 1976, Influence of Downwind High-Rise Buildings on Stack Design, *J. Air Pollut. Control Assoc.*, 26(10): 976-980.

——, and D. D. J. Netterville, 1978, Interaction of a Roof-Level Plume with a Downwind Building, *Atmos. Environ.*, 12(5): 1051-1059.

Wilson, R. B., G. E. Start, C. R. Dickson, and N. R. Ricks, 1976, *Diffusion Under Low Windspeed Conditions near Oak Ridge, TN*, NOAA Technical Memorandum No. ERL-ARL-61, National Oceanic and Atmospheric Administration/Aerospace Research Laboratory, Idaho Falls, Idaho.

Wind Effects on Buildings and Structures, 1963, Proceedings of a Conference at National Physical Laboratory, Teddington, Middlesex, United Kingdom, June 26–28, Vols. I and II, Her Majesty's Stationery Office, London, 1965.

Wind Effects on Buildings and Structures, 1967, Proceedings of an International Research Seminar, Ottawa, Canada, Sept. 11–15, Vols. I and II, University of Toronto Press, Toronto, 1968.

Wind Effects on Buildings and Structures, 1971, Proceedings of Third International Conference, Tokyo, Sept. 6–9, Saikon Co. Ltd., Tokyo, 1971.

Wind Effects on Buildings and Structures, 1975, Proceedings of the Fourth International Conference, Heathrow (London), Sept. 8–12, Cambridge University Press, London, 1977.

Wise, A. F. E., 1971, Effects due to Groups of Buildings, *Philos. Trans. R. Soc. London, Ser. A*, 269: 469-485.

Woo, H. G. C., J. A. Peterka, and J. E. Cermak, 1977, *Wind-Tunnel Measurements in the Wakes of Structures,* NASA Contractor Report NASA CR-2806, National Aeronautics and Space Administration, NTIS.

Woodruff, N. P., and A. W. Zingg, 1952, *Wind-Tunnel Studies of Fundamental Problems Related to Windbreaks*, U. S. Department of Agriculture Soil Conservation Service Report SCS-TP-112, August

Wu, T. Y., 1968, Inviscid Cavity and Wake Flows, *Basic Developments in Fluid Dynamics*, Vol. 2, pp. 1-116, M. Holt (Ed.), Chap. 1, Academic Press, New York and London.

——, 1972, Cavity and Wake Flows, *Ann. Rev. Fluid Mech.*, 4: 243-284, Annual Reviews, Inc., Palo Alto, Calif.

Yang, B. T., and R. N. Meroney, 1970, *Gaseous Dispersion into Stratified Building Wakes*, Colorado State University Fluid Dynamics and Diffusion Laboratory Report CER70-71BTY-RNM-8, Colorado State University, Fort Collins, Colo.

Zdravkovich, M. M., and D. L. Pridden, 1977, Interference Between Two Circular Cylinders; Series of Unexpected Discontinuities, *J. Ind. Aerodyn.*, 2(3): 255-270.

Chapter 7 covers research prior to December 1980.

PLUME RISE AND BUOYANCY EFFECTS

Gary A. Briggs, Atmospheric Turbulence and Diffusion Laboratory

8

8-1 INTRODUCTION

Much has been learned in the past decade about the structure of the atmospheric boundary layer. This new knowledge eventually will lead to better and more complete diffusion models, including better models of plume rise and buoyancy effects. Some improvements already have been made; others await theoretical simplifications and sufficiently complete verifying data. Indeed, the turbulence structure of the atmospheric boundary layer is more complicated than we would like. However, because few parameters are required for a first-order description of this structure, development of reasonably simple and general models for application to diffusion may take place within the next decade. In the meantime, we must practice an incomplete state of the art and hope that some unknown phenomenon will not critically affect our diffusion calculations.

Plume rise remains an important part of diffusion modeling. Except in strong winds, plume rise increases the effective source height from 2 to 10 times the actual release height for typical elevated, buoyant sources. Thus plume rise can reduce the maximum ground concentration by a factor of 3 to 100 or even more if the effluent penetrates into an elevated stable layer.

Numerous plume rise formulas and models have been developed in the past 30 years. Indeed, these numerous formulas and models often are a great source of confusion because they give a very wide range of answers for a specific problem. Therefore, to avoid chance selection of a model, investigators must carefully choose among the models by considering their theoretical basis and the type of verifying data that support the models.

However, in some areas, plume rise is now rather well understood, and there is a consensus, or at least a majority opinion, on how to calculate it. This is true for the rise trajectory before final rise is obtained and for final rise in stable conditions. Calculating plume rise in both of these situations is relatively straightforward because the effect of ambient turbulence on plume rise is negligible in these cases. One problem area, however, that still needs further study is partial penetration of stable layers based above a buoyant source. This topic will be discussed in Sec. 8-4.3.

The great unresolved plume rise problem is final rise, or effective stack height (the rise may never actually terminate), when ambient turbulence is the most effective rise-limiting agent. Many theoretical solutions are offered (see Sec. 8-2.3.4), but adequate data for testing them are practically nonexistent. The available data base is inadequate because, except for observations made in stable conditions, the vast majority of plume rise observations show the plume trajectory still rising at the greatest distance of observation. This distance (the greatest distance of observation) generally falls far short of the distance where the investigator would most want to know the effective stack height (h_e)—the distance at which the maximum ground concentration (MGC) occurs. In Sec. 8-5.7, a plume rise formula for neutral conditions is verified indirectly by using the observed MGC to infer h_e (one definition of plume rise is $\Delta h = h_e - h_s$, where h_s is the source height). Verification using additional data is needed. However, an inadequate data base requires an extensive ground concentration monitoring network around an isolated source.

For convective conditions, which are typical of the daytime when the wind is less than about 7 m/s, the question of plume rise is still unresolved. Only a few theoretical models include the appropriate convective parameters suggested by recent boundary-layer research, and these models have not been adequately tested. Effects of the physical situation on the behavior of a buoyant plume in convective conditions are exceedingly complex (see Sec. 8-5.5). Because we do not know which effects are dominant, the plume rise formula for this case is tentative.

8-2 BASIC MODELS FOR PLUME RISE

Dozens of models for predicting plume rise have been proposed. Discussing the fine points of each one of these models is impossible in this section; therefore readers interested in such detail are referred to Briggs (1975a) for the most comprehensive review of available models.

In this section we will discuss the conservation equations that are used in common by almost all the models and the closure assumptions that are most frequently used to complete the models. More complete qualitative discussions of the plume rise process are given by Briggs (1969, Chap. 2) and by Scorer (1958, Chap. 8).

None of the equations in this chapter have any hidden dimensional constants. All constants are dimensionless, and all other quantities are physical parameters of the fluid or gravitational acceleration. Since all equations are dimensionally consistent, any consistent set of units can be used; therefore only generic units are indicated.

Briggs: Atmospheric Turbulence and Diffusion Laboratory, National Oceanic and Atmospheric Administration, Oak Ridge, Tennessee.
Present address: Meteorological Laboratory, Environmental Protection Agency, Research Triangle Park, N. C.

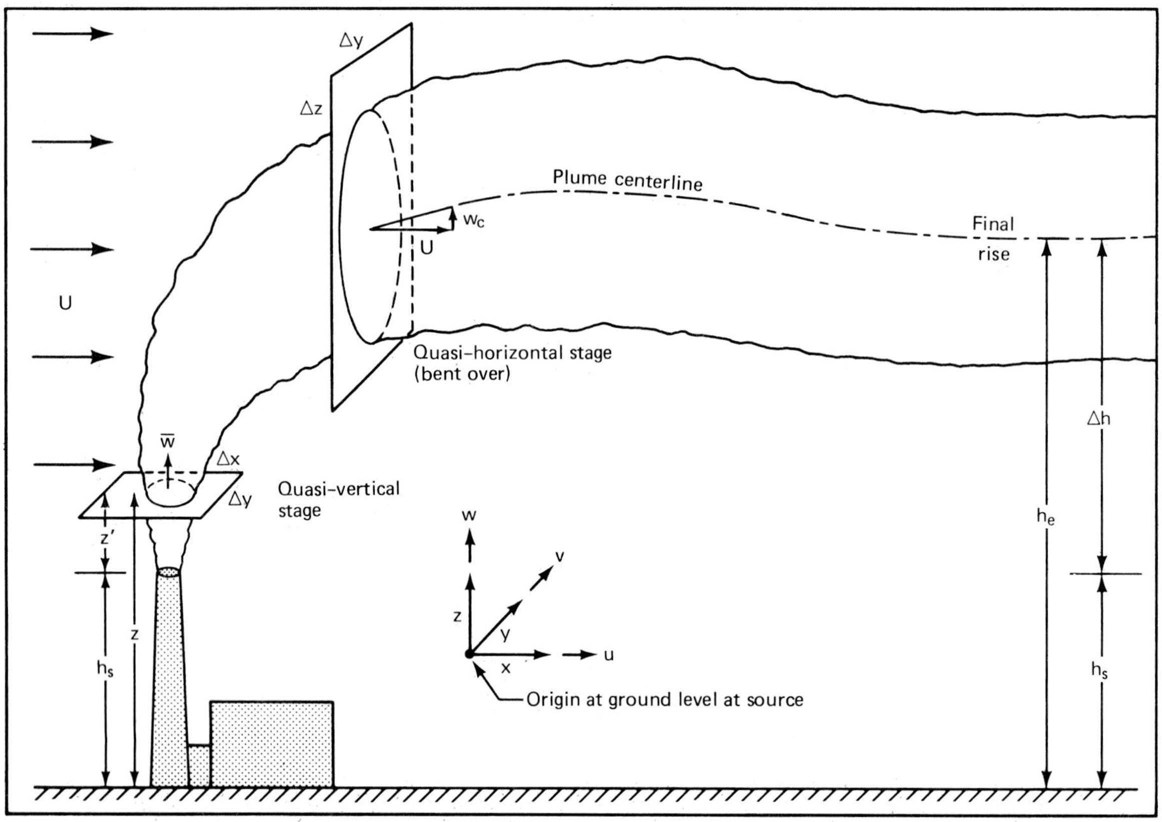

Fig. 8.1 Schematic drawing illustrating basic definitions for describing plume rise.

8-2.1 Conservation Equations

Because plumes are composed of fluid in motion, we will start our examination of the plume rise problem with some of the basic laws of fluid mechanics: conservation of mass, conservation of buoyancy, and conservation of momentum. To put these laws in a convenient form for describing plume trajectories, we integrate them over a cross section of the plume.

Most models consider a plume from a continuous source and assume that such a plume is continuous along its trajectory (i.e., its axis). The physical state of the plume can be described in bulk by integrating the conservation equations over the area of a plane intersecting the plume at some point along its axis. The resulting equations are for the flux of mass, potential density difference (i.e., buoyancy), and momentum through the plane of integration. For vertical plumes we use horizontal planes of integration, and they result in equations for the fluxes as a function of height above the source, z'. For bent-over plumes we use planes of integration perpendicular to the downwind direction, and they result in equations for the fluxes as a function of distance downwind of the source, x. For vertically emitted plumes that bend over in the wind, bending-over plumes, some models attempt to describe the complete trajectories by employing natural coordinates with the main coordinate tangent to the plume axis.

Some models consider the plume segmented into three-dimensional puffs (Moore, 1966). The conservation equations can then be integrated over the volume of each puff. This procedure avoids the complexity that results from changing the orientation of the plume axis in a bending-over plume. However, such models must specify the distance between the puffs, and they also must account for puff merger if the puffs expand into one another.

The following conservation equations, used in continuous-plume models, were derived in detail by Briggs (1975a). Only the results and the simplifying assumptions required to obtain the results are described here. We will use the convention in which z is the height above the ground and z' is the height above the source. Because dz' = dz, these are interchangeable. Figure 8.1 illustrates some of the basic geometric definitions used in this chapter.

*8-2.1.1 **Conservation of Mass.*** The statement that mass is neither created nor destroyed provides a useful auxiliary equation. For instance, it can be used to simplify the other conservation equations. For a point in a fluid, the equation is written

$$\frac{\partial \rho}{\partial t} = -\frac{\partial(\rho u)}{\partial x} - \frac{\partial(\rho v)}{\partial y} - \frac{\partial(\rho w)}{\partial z} \tag{8.1}$$

where ρ is the fluid density and u, v, and w are the fluid velocity components in the x, y, and z directions, respectively. We will use the convention that assumes z to be the vertical direction, x to be the downwind direction, and y to be the lateral direction. Equation 8.1, known as the continuity equation, says that the rate of change of mass within an element of space ($\Delta x \cdot \Delta y \cdot \Delta z$) is equal to the difference between the inflowing and outflowing flux of mass.

Equation 8.1 usually is simplified by using the Boussinesq approximation that density differences are unimportant except in the buoyant-force term (see Chap. 5, Sec. 5-2.3). This approximation is valid when, as is the case when plumes are modeled in water channels or tanks, the fluid is a liquid with small density differences. The approximation also is valid in the atmosphere as long as velocities are small when compared to the speed of sound and vertical scales of motion are small when compared to the scale height of the atmosphere (10 km for the earth's atmosphere). Although these conditions normally are satisfied, there are two notable exceptions: flow within a distance of a few tens of diameters of a rocket exhaust and plume rises from extremely large sources of buoyancy, such as volcanoes. However, observed plume rises from rocket exhausts do fit the $x^{2/3}$ law along with data from more conventional sources (Briggs, 1965), and observed plume rises from volcanoes do fit the conventional predictions for rise of buoyant plumes in stable air (Briggs, 1969). These predictions use the Boussinesq approximation. Apparently, at least for sources observed thus far, the distorting effect of nonconstant ρ has very little effect on plume rise. Assuming that $\rho \simeq$ constant, Eq. 8.1 becomes

$$\frac{\partial u}{\partial x} + \frac{\partial v}{\partial y} + \frac{\partial w}{\partial z} = 0 \tag{8.2}$$

A useful auxiliary equation for vertical plumes is formed by integrating Eq. 8.2 over the intersection of the plume with a horizontal plane. For an axisymmetric plume, the result is

$$\frac{dV_v}{dz} = 2\pi b_v v_e \tag{8.3}$$

where

$$V_v = {}_p\!\iint w\, dA_z \tag{8.4}$$

where V_v = the volume flux rising through the plume
b_v = a plume radius (see list of symbols)
v_e = the velocity of inflowing air at the plume perimeter
dA_z = an element of the plane of integration
p = integration just over the intersection of the plane with the plume

A similar equation could be written for a bent-over plume, but a different definition of V would better serve the purpose in that case.

The volume flux, V_v, will appear again when the potential density (buoyancy) equation is integrated; the volume flux will be needed to solve this equation and the momentum equation. However, Eq. 8.3 cannot be solved unless we know the entrainment velocity, v_e. Many models use Eq. 8.3 to predict V_v by making a closure assumption about how v_e relates to some other predicted quantity. Such assumptions can be checked in the laboratory by measuring $V_v(z)$ or by measuring the inflow velocity directly.

*8-2.1.2 **Effect of Compressibility**.* In an adiabatic process, no heat is added to the gas; thus any work the gas does in expanding is drawn from the internal energy. The first law for an adiabatic process is (see Chap. 2, Eq. 2.3)

$$0 = c_p\, dT - \frac{dp}{\rho} \tag{8.5}$$

Combining this expression with the hydrostatic equation (see Chap. 2, Sec. 2-2), we find the equation for air in adiabatic vertical motion is

$$\frac{\partial T}{\partial z} = (c_p\rho)^{-1}\frac{\partial p}{\partial z} = -\frac{g}{c_p} \tag{8.6}$$

where T = absolute temperature
p = the gas pressure
R = the gas constant in the ideal gas law
c_p = the heat capacity at constant pressure
g = gravitational acceleration

For air in earth's lower atmosphere, $g/c_p = 0.98°C/100$ m, the adiabatic lapse rate. Whenever air is mixed well, the temperature gradient closely approximates Eq. 8.6, except near a heated or cooled surface.

We can define a potential temperature, as was done in Chap. 2, that does not change when air is adiabatically lifted; i.e., the potential temperature is a conserved quantity when air is adiabatically lifted.

$$\theta = T\left(\frac{p_s}{p}\right)^{R/c_p} \tag{8.7}$$

where p_s is a standard pressure, usually defined as 1000 mb. For air, $R/c_p \simeq 0.286$. The relation between the ambient potential temperature gradient and the ambient temperature gradient $\partial T_a/\partial z$ is found by differentiating Eq. 8.7 and substituting the hydrostatic and ideal gas equations (subscript "a" denotes ambient values). The result is

$$\frac{\partial \theta_a}{\partial z} = \frac{\theta_a}{T_a}\left(\frac{\partial T_a}{\partial z} + \frac{g}{c_p}\right) \tag{8.8}$$

For well-mixed air, Eq. 8.6 applies and $\partial\theta_a/\partial z = 0$. This condition defines the neutral state of the atmosphere. If $\partial\theta_a/\partial z > 0$, any air lifted adiabatically will have its initial θ; therefore the air will become cooler and denser than its surroundings. Then hydrostatic force tends to push displaced air back to its original level, and the air is said to be stable. When $\partial\theta_a/\partial z < 0$, hydrostatic force tends to push displaced air away from its original level, and the air is referred to as unstable.

Normally it is assumed that no significant heat transfer takes place during plume rise other than that resulting from fine-scale mixing; therefore plume rise can be considered as an adiabatic process. This assumption is supported by the fact that plumes from power plants have been observed to rise and level off in stable air and to maintain a height constant within ±1% for up to 15 km downwind (Gartrell, Thomas, and Carpenter, 1961). If outgoing thermal radiation had cooled the plumes significantly, they would have lost altitude.

Loss of altitude can occur if a plume is sufficiently opaque. For example, the rate of radiation cooling of a flat blackbody plume at 288°K is 32 m/h multiplied by the average temperature difference between the plume and its radiation-emitting surroundings (both above and below) divided by the plume thickness. Consequently, if no mixing took place, a 100-m-thick blackbody plume in isothermal air ($\partial T_a/\partial z = 0$, $\partial\theta_a/\partial z = g/c_p$) could drop 32 m/h for every degree Celsius of temperature difference. However, the initial rise stage hardly would be affected because the time to maximum rise is approximately 4 min. Two other types of plumes that do not conserve potential temperature are highly radioactive plumes (Gifford, 1967) and plumes in which condensation or evaporation is taking place (Sec. 8-6.4).

For an adiabatically rising gaseous plume, we can now write a more accurate form of the continuity equation than Eq. 8.2 (recall that changes in fluid density were neglected in this equation). We will still assume a low Mach number so that dynamically induced changes in density can be neglected. We start with Eq. 8.1, assume stationarity ($\partial\rho/\partial t = 0$), and consider only the vertical derivative of density, $\partial\rho/\partial z$, to be important. By using the ideal gas equation of state (Chap. 2), Eq. 8.1 is easily put in terms of $\partial p/\partial z$ and $\partial T/\partial z$. The pressure within a plume closely matches the ambient pressure; therefore $\partial p/\partial z \simeq -g\rho_a$. For adiabatic motion, $\partial T/\partial z$ is given by Eq. 8.6. After making these substitutions, we make the approximations $T \simeq T_a$ and $p \simeq p_a$. The result is

$$\frac{\partial u}{\partial x} + \frac{\partial v}{\partial y} + \frac{\partial w}{\partial z} = \frac{w}{z_s} \tag{8.9a}$$

where

$$z_s = \frac{T_a}{(g/R) - (g/c_p)} \simeq 40\,\mathrm{m}\,\frac{T_a}{{}^\circ K} \simeq 10\,\mathrm{km} \tag{8.9b}$$

The length, z_s, is known as the scale height of the atmosphere.

The effect of this refinement on the plume volume flux can be seen by rederiving Eq. 8.3 from Eq. 8.9 instead of from Eq. 8.2. The result is

$$\frac{dV_v}{dz} = 2\pi b_v v_e + \frac{V_v}{z_s} \tag{8.10}$$

The last term is significant only if the plume rise is a significant fraction (1 km or greater) of z_s. If we neglect extrainment, integration of Eq. 8.10 gives $V_v = V_0 \exp(z/z_s)$. However, the expansion of the plume volume is not likely to have much effect on the centerline rise trajectory because the mass of the plume is unaffected by this expansion. The entrainment velocity is the factor that brings in new mass and dilutes plume quantities; therefore the v_e term of Eq. 8.10 is far more significant to trajectory calculations.

8-2.1.3 Conservation of Buoyancy. For most plumes, rise is primarily the result of buoyancy, which in turn results from density differences. Therefore it is important to account for how the total buoyancy of a plume changes as it rises.

Accounting for the buoyancy changes is relatively straightforward for plumes of incompressible fluids, such as in laboratory water channels and tanks. In this case Eq. 8.2 applies. We define ρ' as the density deficit of the plume so that $\rho = \rho_a - \rho'$. We then substitute this expression into Eq. 8.1, assume stationarity, and assume that ρ_a varies significantly only in the vertical. Integrating over a horizontal plane intersecting the plume and defining the edge of the plume by $\rho' = 0$, we find

$$\frac{dF}{dz'} = -sV_v \tag{8.11}$$

where, by definition,

$$F = {}_P\!\iint g\,\frac{\rho'}{\rho_a}\,w\,dA_z \tag{8.12}$$

and

$$s = \omega^2 = -\frac{g}{\rho_a}\frac{\partial\rho_a}{\partial z} \quad \text{(incompressible)} \tag{8.13}$$

In Eq. 8.11 we neglected a term involving $\partial(g/\rho_a)/\partial z$ because it is only of the order of ρ'/ρ_a multiplied by the other terms. Except for very close to the source, where V_v is relatively small, $\rho' \ll \rho_a$. The term F is the total volume flux of buoyant acceleration being carried by the plume and has the dimensions l^4/t^3. The buoyancy parameter, F, will appear in the equation for vertical momentum flux. The stability parameter, s, is equal to the restoring acceleration per unit vertical displacement felt by a parcel of incompressible fluid that is displaced. It has the dimensions t^{-2}.

For a compressible gaseous atmosphere, we use Eq. 8.9 instead of Eq. 8.2 and we proceed in a fashion similar as before. With the equation of state, the hydrostatic equation, and Eq. 8.8, we can express $\partial\rho_a/\partial z$ in terms of $\partial\theta_a/\partial z$. A number of terms cancel out, and the result is identical to Eqs. 8.11, 8.12, and 8.13 except that

$$s = \omega^2 = \frac{g}{\theta_a}\frac{\partial\theta_a}{\partial z} \quad \text{(compressible)} \tag{8.14}$$

In this case, s is the restoring acceleration per unit vertical displacement felt by a parcel of air that is adiabatically displaced. It is also the square of the Brunt-Väisälä frequency, ω, which is the natural frequency of vertical motions in gravity waves in the atmosphere.

We should note that, although mean molecular weight is a factor in computing the initial value of ρ', it does not enter into Eq. 8.14 because the ambient mean molecular weight, m_a, is assumed to be constant in the equation of state. This assumption is valid to a height of 90 km in the earth's atmosphere. For releases occurring above this altitude, the term $-(g/m_a)\partial m_a/\partial z$ should be added to the right side of Eq. 8.14.

Equations 8.11 to 8.14 are also true for bent-over plume as long as a horizontal plane completely intersects it and Eq. 8.4 is used to define V_v. However, it is more convenient to use a plane of integration perpendicular to the x-axis, particularly to simplify the momentum equation. We assume that s is approximately constant through the depth of the plume. Then we can derive an analog to Eqs. 8.11 and 8.12:

$$\frac{\partial F}{\partial x} = -s\,{}_P\!\iint w\,dA_x \tag{8.15}$$

where

$$F = {}_P\!\iint g\,\frac{\rho'}{\rho_a}\,u\,dA_x \tag{8.16}$$

and dA_x is an element on a plane perpendicular to the x-axis. Later, we will argue that, for a bent-over plume, u is approximated by the mean ambient wind, U. Assuming that U is also approximately constant through the depth of the plume, U can be brought outside the integral sign in Eq. 8.16 to provide one of the terms needed in the equation for vertical momentum.

We note that some authors refer to Eq. 8.11 or 8.15 as the "energy equation" simply because they base their derivation of the equation on the conservation of thermal energy. However, this label is difficult to support when the equation is applied to essentially incompressible fluids or to buoyancy resulting from molecular weight differences. It seems preferable to speak of it as the buoyancy conservation equation because this label is more general. Also, in a later section, we will discuss an energy conservation equation that is used in some plume rise models and, because it refers to kinetic rather than to thermal energy, is not at all equivalent to Eq. 8.11 or Eq. 8.15.

*8-2.1.4 **Conservation of Momentum.*** The momentum equation describes the effect of forces acting on a plume and therefore relates directly to the prediction of the plume trajectory. This equation often is presented without derivation, and, as a result, several different forms of it appear in the literature. For example, about half the models for bent-over plumes include a drag-force term, and half of them do not (Briggs, 1975a). Thus the form of this equation may be influenced by the intuition of the modeler. However, the equation can be derived from fairly basic principles. A detailed description of the derivation is given by Briggs (1975a). The mathematics will be treated briefly here, and the underlying assumptions will be discussed at some length.

Models for field-scale plumes universally neglect viscous forces. These forces are effective only at very small scales, e.g., of the order of 1 mm, and do not directly affect the large-scale motions of a plume. However, viscous force is directly related to the development and the dissipation of turbulence, which affects plume rise indirectly by diluting its buoyancy and upward momentum. Exact solution of the equations of motion requires the viscous force term. The problem is that with viscous force, turbulence develops and produces eddy motions on scales down to 1 mm and less; consequently the solution becomes quite intractable. Details of the turbulence motion are not needed to predict the mean trajectory of the plume because this motion generates no net force. We only need to know about the bulk effect of the turbulence on dilution of the plume. For example, it would be sufficient to know V as a function of z or of x. This problem will be discussed further in Sec. 8-2.3.

On the scale of the plume radius, the important forces acting on a plume are the accelerative pressure, which develops in response to the inertia of the ambient fluid in the way of the plume, and buoyancy force, which develops out of the hydrostatic pressure forces. The acceleration of a parcel of fluid resulting from pressure forces is just ρ^{-1} multiplied by the negative of the pressure gradient (the fluid accelerates toward lower pressure). Upward buoyant acceleration is just the sum of the acceleration resulting from the ambient hydrostatic pressure gradient, $-\rho^{-1}(\partial p_a/\partial z) = (\rho_a/\rho)g$, and gravitational acceleration, $-g$. This statement is merely a form of Archimede's principle. The resulting vector equation for the acceleration of a fluid parcel is

$$\frac{d\mathbf{v}}{dt} = -\rho^{-1}\nabla p' + g\frac{\rho'}{\rho}\mathbf{k} \qquad (8.17)$$

where $\mathbf{v}$ is the vector velocity, p' is the pressure perturbation ($p' = p - p_a$), $\rho' = \rho_a - \rho$, and $\mathbf{k}$ is the upward-directed unit vector.

The Lagrangian time derivative $d\mathbf{v}/dt$ can be expressed as partial derivatives:

$$\frac{d\mathbf{v}}{dt} = \frac{\partial \mathbf{v}}{\partial t} + u\frac{\partial \mathbf{v}}{\partial x} + v\frac{\partial \mathbf{v}}{\partial y} + w\frac{\partial \mathbf{v}}{\partial z} \qquad (8.18)$$

Taking a time average and assuming that the plume is stationary eliminates $\partial\mathbf{v}/\partial t$. The rest of the left side of Eq. 8.17 can be broken into three terms: the advection of $\mathbf{v}$ by the mean ambient wind, $U\,\partial\mathbf{v}/\partial x$; the gradient of the dynamic plume perturbation pressure; and a term involving the fluid vorticity, $\nabla\times\mathbf{v}$. For convenience, we assumed that the space derivatives of U are negligible, i.e., for practical purposes the ambient wind is constant in the environs of the plume.

The integrated form of Eq. 8.17 is simplified considerably if we assume, for the moment, that no turbulence exists outside the plume. If ambient turbulence is strong enough to affect the mean plume trajectory, it does so by diluting the plume rather than by exerting a net force on it. This dilution can be handled as a separate matter, e.g., through an entrainment assumption, and ambient turbulence can be ignored in the momentum equation. Then we can assume that the vorticity term $\nabla\times\mathbf{v}$ is zero outside the boundary of the plume; so it follows that the velocity there can be expressed by a potential

$$\mathbf{v} = \nabla\phi \text{ (outside the plume)} \qquad (8.19)$$

The theorem which states that any inviscid fluid that is initially in potential flow remains so until acted upon by a nonpotential force (Lamb, 1932, p. 18) is applicable. Because ∇p is a potential force, only the buoyancy term in Eq. 8.17 is a nonpotential force, and it (the buoyancy) is zero outside the plume. Therefore the flow outside the plume remains potential flow.

The mathematical consequence of all this is that, after integrating Eq. 8.17 over a plane intersecting the plume, terms interpretable as drag force cancel out (drag force is proportional to velocity squared multiplied by area). This result is consistent with the axiom that potential flow is frictionless. The only forces acting on the plume externally are pressure forces that are caused by the inertial resistance of the ambient fluid to acceleration caused by the advancing plume.

The presence of a turbulent wake outside the plume would challenge the above assumptions. However, such a wake would be finite in extent and could be included in the area of integration defining the plume. It could not be treated as a separate entity because a turbulent wake attached to the plume would entrain some plume material. This material would become part of the plume by definition (the buoyant plume is defined by $\rho' > 0$). Briggs (1975a) argues that, unlike a solid object, a rising plume cannot develop a wake by the separation of a high-vorticity viscous shear layer from a no-slip-condition solid boundary. Furthermore, laboratory measurements on rising thermals confirm the lack of vorticity outside the thermal (Tsang, 1971) and show that the streamlines there approximate potential flow (Richards, 1963). These arguments and measurements support using Eq. 8.19 and the consequent absence of drag terms from the momentum conservation equation.

Even with the previous assumptions, we obtain a somewhat complicated vector equation for momentum conservation by integrating Eq. 8.17 over the plume (Briggs, 1975a, Eq. 8.17). Fortunately, by making a few additional approximations, we can greatly simplify the vector equation for the idealized cases of nearly vertical or nearly horizontal plumes.

For a nearly vertical plume, we assume that the ambient wind has little effect and that $U = 0$; consequently, some terms are eliminated. For the vertical component of momentum, several terms are of the order of α^2 multiplied by other terms, where α is the entrainment coefficient defined by $\alpha = v_e/\overline{w}$ ($\overline{w}$ is a mean plume vertical velocity defined below). These terms can be neglected because experience has shown that $\alpha \simeq 0.1$. Therefore the momentum equation for vertical plumes reduces to

$$\frac{d(\overline{w}V_v)}{dz'} = \frac{F}{\overline{w}} \tag{8.20}$$

where V_v is defined by Eq. 8.4, F is defined by Eq. 8.12, and $\overline{w}$ is defined by

$$(\overline{w}V_v) = S_1 \, {}_{\infty}\!\iint w^2 \, dA_z \tag{8.21}$$

S_1 is a dimensionless profile shape factor defined by

$$S_1 = \left[\frac{{}_{P}\!\iint w(\rho'/\rho)dA_z \; {}_{P}\!\iint w \, dA_z}{{}_{P}\!\iint (\rho'/\rho)dA_z \; {}_{\infty}\!\iint w^2 \, dA_z} \right]^{1/2} \tag{8.22}$$

$S_1 = 1$ if $\rho' = 0$ everywhere. The integral over ∞ indicates integration over a plane extending far beyond the edge of the plume.

To derive Eq. 8.20 in such a simple form, we assume that the variation of ln S_1 with height is negligible compared to the variation of ln $(\overline{w}V_v)$ with height. This assumption is likely to be the case because S_1 is always of the order unity. Thus Eq. 8.20 is a valid approximation even if the flow is not similar (i.e., if it does not maintain constant profile shapes). The previous definition of $\overline{w}$ is convenient because it allows use of the top hat form of the plume equations (Morton, Taylor, and Turner, 1956) regardless of the real profile shapes of w and ρ'. For a buoyant vertical plume, the observations of Rouse, Yih, and Humphreys (1952) indicate that $S_1 = 0.92$, which is not far from unity.

For models that describe the bending-over stage of rise for a plume emitted vertically into a crosswind, Eq. 8.20 is approximately true when the plume axis is inclined at less than 45° from the vertical; however, S_1 changes more rapidly than in the case of no crosswind. The horizontal component of the momentum equation reduces to

$$\frac{d}{dz} \, {}_{\infty}\!\iint (u - U) \, w \, dA_z = 0 \tag{8.23}$$

This equation merely states that the total flux of horizontal momentum relative to U is constant. In analogy to Eq. 8.21, we can define a mean plume horizontal velocity $\overline{u}$ by

$$(\overline{u} - U)V_v = {}_{\infty}\!\iint (u - U) \, w \, dA_z \tag{8.24}$$

According to Eq. 8.23, the initial value of $(\overline{u} - U)V_v$ is conserved. Thus for a vertically oriented emission of volume flux V_0, we have

$$(\overline{u} - U)V_v = -UV_0$$

$$\frac{\overline{u}}{U} = 1 - \frac{V_0}{V_v} \tag{8.25}$$

For a bent-over plume, i.e., one whose axis is inclined at less than 45° above the horizontal, it makes sense to choose a plane of integration perpendicular to the mean wind direction. An element of this plane is denoted by dA_x. With this choice of plume orientation, several terms of the vertical momentum equation drop out. We neglect one of the other terms because it is of the order of V_0/V times the others; this ratio will be small for a bent-over plume. For convenience, we define a time $t = x/U$ so that $dx = U \, dt$. We can also use the potential flow relationship, $u = -\partial\phi/\partial x$, around the plume periphery (it is not valid inside the plume, where the flow is rotational). The result is

$$\frac{d}{dt} \left[{}_{P}\!\iint w \, dA_x - {}_{P}\!\oint \phi \, dy \right] = {}_{P}\!\iint g \frac{\rho'}{\rho} \, dA_x \tag{8.26}$$

where "${}_{P}\!\oint$" indicates integration around a path following the periphery of the plume on the A_x plane. An identical equation can be developed for the purely two-dimensional case, e.g., for a horizontal thermal released in a fluid at rest. This means that a mathematical basis can be provided for assuming the essential equivalence of rise of a bent-over continuous plume and the rise of a horizontal line thermal; we use the transformation $t = x/U$. The latter type of source has been used in several laboratory experiments (Richards, 1963, and Tsang, 1971).

The second term of Eq. 8.26 has a clear interpretation when we look at it in the context of two-dimensional flow. It represents the vertical component of the nonhydrostatic pressure force, p', acting on the plume at its interface with the ambient fluid. This pressure force develops because of the inertial resistance of the ambient fluid in the path of the rising plume, i.e., it results from the effective mass being accelerated outside the plume. The first term of Eq. 8.26 represents the vertical acceleration of fluid inside the plume. A term similar to this, but without the time derivative, appeared in Eq. 8.15.

For a bent-over plume, we define both an internal vertical momentum flux, M', and an effective vertical momentum flux, M, by

$$\frac{M'}{U} = {}_{P}\!\iint w \, dA_x \tag{8.27}$$

$$\frac{M}{U} = {}_{P}\!\iint w \, dA_x - {}_{P}\!\oint \phi \, dy \tag{8.28}$$

If similarity is maintained in the various motions of a plume as it rises and expands, the effective and internal vertical momentum fluxes will remain in constant ratio, i.e., $S_2 = M/M'$ will be constant. For a buoyant plume, this constant can be determined from the laboratory-measured streamlines of Richards (1963) by using the relation $\iint w \, dA_x = \oint \psi \, dz$, where ψ is the stream function describing two-dimensional flow. For these experiments, $S_2 \equiv 2.3$ (Briggs, 1975a).

The fact that S_2 is substantially greater than unity helps explain the discrepancy between observed radii of visible plumes and the effective plume radius consistent with observed plume trajectories and the momentum equation. The value of S_2 also will have a substantial effect on the predicted trajectories of plumes in stable air (Sec. 8-4.1.2).

To put Eq. 8.26 into convenient form, we multiply it by U. For a bent-over plume, $u \simeq U$, therefore

$${}_{P}\!\iint g \left(\frac{\rho'}{\rho} \right) U \, dA_x \simeq F$$

Then

$$\frac{dM}{dt} \simeq S_2 \frac{dM'}{dt} = F \tag{8.29}$$

This approximation is valid if the rate of change of ln S_2 is small when compared to that of ln M′ (this is likely). Equation 8.29 can readily be used with Eq. 8.15 for the variation of F if we keep in mind the transformation dt = dx/U.

Let us consider the adequacy of the approximation $u \simeq U$ for the bent-over plume. If we neglect terms of the order of (u − U)/U times other terms, the horizontal component of the momentum equation reduces to

$$\frac{d}{dx}\,{}_{P}\!\iint (u - U)U\, dA_x = 0 \qquad (8.30)$$

in analogy to Eq. 8.23 for quasi-vertical plumes. In analogy to Eq. 8.24, we can define a mean horizontal plume speed u_c by

$$(u_c - U)V = {}_{P}\!\iint (u - U)U\, dA_x \qquad (8.31)$$

However, we have not yet defined what is meant by the effective volume flux, V, for a bent-over plume. A convenient definition is

$$V = \frac{M}{w_c} \qquad (8.32)$$

where w_c is the rise velocity of the plume centerline moving down the plume axis at a horizontal speed u_c. In other words, w_c/u_c is the slope of the plume axis above the horizontal. For a vertically emitted plume, Eqs. 8.30 and 8.31 give an equation identical in form to Eq. 8.25, which is for a quasi-vertical plume:

$$(u_c - U)V = -UV_0 \qquad (8.33)$$

$$\frac{u_c}{U} = 1 - \frac{V_0}{V}$$

By visual observations, we can easily verify that, very early in the bent-over stage of rise, V becomes much larger than V_0 close to the source (V being approximately twice the cross-sectional area of the visible plume times U). Then u_c closely approaches U. Another visual estimate of u_c/U can be made from the slope of the plume centerline, w_c/u_c. We define the initial value of $M = w_cV$ as $M_0 = w_0V_0$, and substitute these definitions into Eq. 8.33

$$\left(\frac{u_c}{U}\right)^{-1} = 1 + \frac{(M_0/M)(w_c/u_c)}{w_0/U} \qquad (8.34)$$

For a nonbuoyant plume (F = 0), Eq. 8.29 tells us that $M = M_0$; thus Eq. 8.34 is simplified. As the trajectory slope becomes small compared to the initial velocity ratio, u_c approaches U. For a positively buoyant plume, M gets larger with travel time, and u_c/U approaches unity even closer.

*8-2.1.5 **Conservation of Kinetic Energy.*** The mass, buoyancy, and momentum equations constitute a complete set of conservation equations. Sometimes, however, a fourth equation is derived for kinetic energy. (In some literature on plume rise, the energy equation refers to thermal energy, not kinetic energy; see the last paragraph of Sec. 8-2.1.3). The kinetic energy equation is derived from the same starting point as the momentum equation, but both sides of the equation are multiplied by the plume-induced velocity before time averaging and integrating over a plane. If this velocity is time-averaged before it is used to multiply the momentum equation, an equation for the conservation of kinetic energy of mean plume-induced motion results. This equation introduces a term containing Reynolds stress, a variable that does not appear elsewhere in the conservation equations. If this velocity is not time-averaged first, an equation for the conservation of the total kinetic energy of plume-induced motion (mean plus turbulent) results. Turbulent kinetic energy appears as a new variable in this case. Both these forms of the kinetic energy conservation equation have been used in plume rise models (Briggs, 1975a, Sec. 2.4).

It is important to note that the energy conservation equation is not independent of the momentum conservation equation; it merely is an alternative. Which equation is used in a given model generally depends on which closure assumption is chosen. For example, if we assume that the entrainment velocity is related to $\overline{w}$, we generally use momentum conservation because it determines $(\overline{w}V_v)$; however, if we choose to relate entrainment velocity to the turbulent kinetic energy, we use the appropriate form of the kinetic energy conservation equation. Use of both momentum and kinetic energy equations in the same model does not obviate the need for a closure assumption, because inevitably an extra variable, such as turbulent kinetic energy, is introduced. Most plume rise models use only the momentum conservation equation (Briggs, 1975a, Table 1).

8-2.2 Initial Values

For most plumes, the most important source quantity affecting plume rise is the total buoyancy flux. To initialize Eq. 8.11, we must determine the value of F at the source, and we denote this value by πF_b. We can determine F_b by generalizing from either Eq. 8.12 or Eq. 8.16 and writing

$$(F)_0 \equiv \pi F_b = \iint g\frac{\rho_0'}{\rho_a}\, dV_0 \qquad (8.35)$$

If ρ_0' is constant,

$$\pi F_b = g\frac{\rho_0'}{\rho_a} V_0$$

where dV_0 is an element of the source volume flux and V_0 is the total source volume flux. For a round source of internal radius r_0 and an average exit velocity v_0, $F_b = g(\rho_0'/\rho_a)v_0r_0^2$.

We should note that, following convention, source parameters such as F_b are equal to the source flux divided by π. This convention was established because early plume rise models usually assumed round top hat profiles of plume quantities, and, therefore, π appeared on both sides of all flux conservation equations. Consequently, as a matter of convenience, all fluxes, including initial values, were divided by π. The buoyancy flux parameter, F_b, is defined in Eq. 8.35 so that it is consistent with this convention.

If the effluent is a heated gas composed mostly of air, we use the equation of state for an ideal gas and derive $\rho_0'/\rho_a = (T_0 - T_a)/T_0$. In some plume rise models, T_a, instead of T_0, appears in the denominator in the definition of F_b. As pointed out in Briggs (1972), this is equivalent to assuming that the buoyant force acts on a fluid of density ρ_0. Because a turbulent plume is composed of entrained fluid, the density of the fluid driven by the buoyant force should be approximated by ρ_a (this is assumed in Eq. 8.35). If the effluent temperature equals the ambient, but the effluent mean molecular weight is different from that of air, $\rho_0'/\rho_a = (m_a - m_0)/m_a$, where $m_a = 28.97$.

If the effluent is a heated gas with a heat capacity and mean molecular weight close to that of air, the buoyancy flux parameter can be related directly to the rate of sensible heat emission, Q_H, because $F_b = gQ_H/(\pi c_p \rho_a T_a)$. The term $g/(c_p \rho_a T_a)$ can be regarded as a constant at a given site elevation. However, note that from the ideal gas law $\rho_a T_a$ is proportional to ρ_a, which decreases with altitude. Thus, in the lower atmosphere, $\rho_a T_a$ decreases approximately 1% for 90 m of increased elevation. Near sea level, $F_b \simeq 3.7 \times 10^{-5} Q_H$ in m^4/s^3, if Q_H is expressed in cal/s. This is an especially useful relationship for open-combustion sources, such as flares and conflagrations. Most of the heat of combustion, except what is required to vaporize any water content in the fuel, goes into the sensible heat emission, but a fraction of it is lost to radiant heat (Schulze, 1978).

If the effluent is a heated gas with a different heat capacity or mean molecular weight than air, the total buoyant force present depends on the degree to which the heat of the effluent is conducted to entrained air. The formulas for F_b given by Briggs (1970 and 1975a) assume that all the heat is conducted to a large volume of entrained air. However, traverses of cooling tower plumes at several hundred meters downwind have revealed pockets of essentially uneroded effluent gas (Thompson, Norman, and Miller, 1976). This suggests that mixing down to fine scales, where appreciable molecular conduction can take place, takes considerable time. Therefore, at least in the early stage of rise, and perhaps for the whole plume trajectory, it is more appropriate to use Eq. 8.35 as it is, with $\rho_0'/\rho_a = 1 - (T_a m_0/T_0 m_a)$.

The next most important parameter to initialize is the total vertical momentum flux, $(\overline{w}V_v)$ or $M = w_c V$, which appears in Eqs. 8.20 and 8.29. (For a bent-over plume, $M = w_c V$, as V is defined in Eq. 8.32.) This quantity is particularly important if F_b is small. For a pure jet ($F_b = 0$), it is the primary source parameter. The initial value of $(\overline{w}V_v)$, denoted here by πF_m, is basically the vertical velocity of the effluent, w_0, times the effluent volume flux (flow rate). However, Eqs. 8.20 and 8.29 result after it is assumed that the plume density approximates ρ_a; this assumption may not be appropriate at the source. Because momentum is proportional to mass as well as to velocity, the source momentum flux is proportional to the source density. Therefore the initial value of $(\overline{w}V_v)$ or $(w_c V)$ is best given by

$$(\overline{w}V_v)_0 \equiv \pi F_m = \int\!\!\int \frac{\rho_0}{\rho_a} w_0 \, dV_0 \qquad (8.36)$$

If ρ_0 and w_0 are constant,

$$\pi F_m = \frac{\rho_0}{\rho_a} w_0 V_0$$

The volume flux parameter, V_v, may also need an initial value. The source value of V_v is less important than F_b or F_m for a turbulent plume because entrainment of ambient fluid typically increases the volume flux by 10^2 to 10^3. Therefore trajectory predictions are only slightly affected by setting $V_v = 0$ at $z' = x = Ut = 0$. If we initialize calculations with a more realistic value of V_v, we probably should use the source volume flux reduced to ambient density. In this way, the initial value of $\overline{w}$, which equals $(\overline{w}V_v)_0/V_0$, is equal to the weighted mean value of w_0. Thus, rather than simply V_0, the recommended initial value of V_v is

$$(V_v)_0 = \int\!\!\int \frac{\rho_0}{\rho_a} \, dV_0 \qquad (8.37)$$

If ρ_0 is constant,

$$(V_v)_0 = \frac{\rho_0}{\rho_a} V_0$$

Finally, we may need an initial value of the plume radius to use, for instance, in an entrainment relation like Eq. 8.3. Treating a plume as if it were emitted from a point source usually is adequate. If we want to take into account the finite source size, we can define a characteristic radius of a vertical plume by $b_v = (V_v/\pi\overline{w})^{1/2}$. Thus, for a round source with uniform density and efflux velocity, we have $(b_v)_0 = (\rho_0/\rho_a)^{1/2} r_0$.

8-2.3 Closure Assumptions

The equations for conservation of mass, buoyancy, and momentum have been written in convenient and simplified forms to describe plume rise, and the most appropriate initial values of each parameter have been determined. However, regardless of whether we use the momentum equation, the kinetic energy conservation equation, or both, we always have one fewer conservation equation than we have unknown dependent variables. This shortcoming is true of all models describing the rise of turbulent plumes. The problem occurs because we have neglected to compute the details of the turbulent motion (each and every turbulent eddy); instead, we have dealt with the problem in a bulk way, e.g., by integrating across the plume. This procedure at least leads to tractable equations. However, in integrating across the plume, we must introduce one new unknown. We therefore do not have a closed set of equations even though the original number of equations are retained. This difficulty is evident, for example, when integrating the continuity equation. In its simplest form, Eq. 8.2 is a relationship among u, v, and w, and these variables already appear in the vector momentum equation. For a vertical plume in a calm ambient, if we integrate Eq. 8.2 across a horizontal plane to the edge of an imaginary box that encloses the plume and the return flow so that $u = v = 0$ at the edge of this plane, all that results is the trivial assertion that $_\infty\int\!\int w \, dA_z$ is constant with height. The result tells us nothing about the plume. If instead we integrate over only the part of the plane intersected by the plume, we get Eq. 8.3. This equation tells us about the plume, but it also introduces a new unknown, v_e, the entrainment velocity.

The problem outlined above is known as the "closure problem." The equations cannot be solved for plume rise without one additional relationship. No relationship is available from known physical principles; we already have used the available conservation equations. The plume rise modeler must, at some point, invent an additional relationship, the "closure assumption," so that he can proceed with predicting plume rise.

Depending on which combination of conservation equations is chosen, the closure assumption can involve the entrainment velocity, the plume radius, the Reynolds stress, and/or the turbulent kinetic energy. Whatever unknown parameter is introduced, that parameter must be related to other variables that already are predicted by the chosen set of equations. The closure assumption has been made in dozens of different ways by different authors (the most compact summary of these assumptions is given in Table 1 of Briggs, 1975a). The closure assumption is the heart of the model, because the assumption used most distinguishes one model from another and allows a model to produce a prediction.

Some of these closure assumptions are well supported by comparisons with observations. Others lead to faulty predictions, and, for others, verifying observations are not available. In many cases, seemingly different plume rise models give very comparable predictions. However, in other cases the models disagree with one another very much, and we must avoid choosing models that yield very unrealistic predictions.

To avoid confusing alternatives, we will keep the following suggestions on closure assumptions as straightforward as possible. The most commonly used set of conservation equations, continuity–buoyancy–momentum, will be used. The simplest closure assumptions that give predictions consistent with observations of the most salient features of plume rise will be described. Where observations are lacking, incomplete, or ambiguous, as is the case when ambient turbulence affects plume rise, methods of predicting the rise will be suggested on the basis of simplicity, physical argument, and consistency of predicted ground concentrations with observational experience.

*8-2.3.1 **Vertical Plumes.*** When the continuity–buoyancy–momentum set of conservation equations is used to describe vertical plumes, the closure used most often is the "Taylor entrainment assumption,"

$$v_e = \alpha\overline{w} \tag{8.38}$$

where α is a dimensionless entrainment constant. This model was first proposed by G. I. Taylor (1945) and later expanded and published in the open literature by Morton, Taylor, and Turner (1956). Widely used, it is often taken for granted as the starting point for further work (see Briggs, 1975a, Table 1). The Taylor assumption is based on the idea that, because turbulence in the plume is generated by the velocity shear between the plume and the ambient fluid, all turbulent velocities, including the entrainment velocity, scale locally to the mean plume vertical velocity, $\overline{w}$.

As with any closure assumption, we would like to know how consistent the plume rise predictions are when compared with observations. Qualitatively, Eq. 8.38 performs quite well. At large values of z/D, where D is the source diameter, Eq. 8.38 leads to the predictions that $\overline{w} \propto F_m^{1/2}/z'$ for a jet, $\overline{w} \propto (F_b/z')^{1/3}$ for a buoyant plume, or $b_v \propto z'$ in either case. These relationships were long ago established in laboratory experiments. The predicted final rises for stable conditions, $\Delta h \propto (F_m/\omega^2)^{1/4}$ and $\Delta h \propto (F_b/\omega^3)^{1/4}$, also are verified by laboratory and field experiments.

One quantitative test of the model is the consistency of the best-fit values of α for different flow situations. Ricou and Spalding (1961) tested the prediction of Eqs. 8.3 and 8.38 that, for a vertical jet,

$$\frac{dV_v}{dz} = 2\pi b_v(\alpha\overline{w}) = 2\pi\alpha F_m^{1/2} \tag{8.39}$$

Here we define b_v dynamically with the equation

$$V_v = \pi\overline{w}b_v^2 \tag{8.40}$$

so that $\overline{w}b_v = (\overline{w}V_v/\pi)^{1/2}$, which, according to Eq. 8.20, is conserved for a jet because F = 0. The initial value of $\overline{w}V_v$ was defined as πF_m in Sec. 8-2.2. Ricou and Spalding measured the inflow rate through a cylinder surrounding the jet to determine dV_v/dz and found Eq. 8.39 to work with $\alpha = 0.080$. Hill (1972) confirmed this result in a similar experiment; however, he found that a reduced entrainment rate occurred in the flow establishment region defined by $z' < 10D$. For a buoyant plume, the measurement of Rouse, Yih, and Humphreys (1952) yielded Gaussian distributions of w and ρ', with $S_1 = 0.92$. The present model correctly predicts the observed $V_v(z')$ with $\alpha = 0.124$, and the observed $\overline{w}(z')$ with $\alpha = 0.125$.

The optimum values of α are more consistent for predicting the final rise in linearly stratified ambients. Here we compare predictions of the unaltered volume flux model with fallback (described in Sec. 8-4.1) with experiments made recently in the large water channel at the Environmental Protection Agency (EPA) Fluid Modeling Facility. In these experiments, the plume geometry was more completely defined than in any previous experiments. The centerlines of the stratified plumes are correctly predicted with $\alpha = 0.084$ for jets and with $\alpha = 0.074$ for buoyant plumes. Values resulting from these experiments at the EPA Fluid Modeling Facility are similar to the measured value by Ricou and Spalding; $\alpha = 0.08$ is a serviceable compromise.

An alternative model that gives similar qualitative predictions is the buoyancy and momentum conservation equation with the closure $b_v = \alpha' z'$ (Abraham, 1963). The same result is obtained from the more elaborate model of Priestley and Ball (1955); their model includes the kinetic energy conservation equation for mean motion and assumes that the Reynolds stress in that equation is proportional to $\overline{w}^2$. This relationship for b_v contrasts with the Morton, Taylor, and Turner (1956) model in that the latter predicts $b_v = 2\alpha z'$ for jets and $b_v = (6/5)\alpha z'$ for buoyant plumes in their rising stage. The models also differ somewhat qualitatively because the latter predicts a flaring of the plume as final rise is approached. (Abraham's model allows for negative entrainment of excess volume flux as $\overline{w}b_v^2$ diminishes.) For the rising stage, with the optimum values of α quoted above (0.08 and 0.125), $b_v = 0.16\,z'$ for jets and $b_v = 0.15\,z'$ for buoyant plumes. These results suggest that Abraham's model with $\alpha' = 0.155$ is more consistent with observations. However, for final rise in stable conditions, the Morton, Taylor, and Turner model with $\alpha = 0.08$ yields predictions more consistent with observations.

*8-2.3.2 **Bent-Over Plumes.*** The Taylor assumption, Eq. 8.38, is frequently used as a closure assumption for bent-over plumes. Because these plumes acquire most of the horizontal wind speed as they entrain ambient air, most of the velocity shear present results from the vertical component of plume motion, w_c. The vertical shear is the primary source of the plume's self-generated turbulence. Consequently, the assumption is made that $v_e = \beta w_c$, with

$$\frac{dV}{dx} = 2\pi b v_e = 2\pi b(\beta w_c) \qquad (8.41)$$

where β is a dimensionless entrainment constant. Beta is functionally similar to α, but it is much larger because the axis of a bent-over plume is perpendicular, rather than parallel, to its rise. This results in more efficient mixing with the ambient fluid. By defining the dynamic plume radius, b, with

$$V = \frac{M}{w_c} = \pi U b^2 \qquad (8.42)$$

and by using the transformations $dt = dx/U = dz_c'/w_c$, where z_c' is the centerline rise of the plume, we reduce Eq. 8.41 to

$$\frac{db}{dz_c'} = \beta \qquad (8.43)$$

if U is constant with height.

An auxiliary quantity, called here the internal radius, b′, is defined by

$$V' = \frac{M'}{w_c} = \pi U b'^2 \qquad (8.44)$$

$$\frac{db'}{dz_c'} = \beta' \qquad (8.45)$$

Note that b′ is the radius that occurs in the buoyancy conservation equation, Eq. 8.15.

Briggs (1975a) quotes various photographic data on plume depth (top elevation minus bottom elevation) versus plume rise that show $b''/z_c' = 0.45$ to 0.52, where b″ is the visible plume half-depth. Straight-line fits to plots of b″ vs. z_c' pass very near the origin, which suggests that $b \simeq 0$ and $b' \simeq 0$ at $z_c' = 0$. The best fit of predicted to observed plume trajectories was obtained with $\beta = 0.6$ (see Table 3 and Eq. 48 of Briggs, 1975a). This coefficient applies to the dynamic radius, which includes effective mass outside the plume. As for the internal radius, the streamlines measured by Richards (1963) suggest that $S_2 = (\beta/\beta')^2 = 2.3$ for buoyant plumes and, therefore, that $\beta/\beta' \simeq 1.5$. This ratio is consistent with the optimum fits of predicted rises for the rising stage, which depends on β, and for the leveled off stage in stable conditions, which depends on β', as reported by Fay, Escudier, and Hoult (1969). Briggs (1975a) concluded that $\beta' = 0.4$ for buoyant plumes.

For the trajectories of bent-over jets, Hoult and Weil (1972) found that $\beta = 0.6$ again gives the best overall fit of predictions to observations. The velocity ratio, $R = w_0/U$, ranged from 3 to 27 in the experimental data Hoult and Weil quoted. However, Briggs (1970) noted a correlation of the best-fit β with R while comparing results of six different laboratory experiments with the ⅓ law (Sec. 8-3.2). For R ranging from 1 to 40, observed centerlines for nonbuoyant plumes are fit well by

$$\beta = 0.4 + 1.2\,R^{-1} \qquad (8.46)$$

We do not understand why conditions at the stack, i.e., R, should affect the dimensionless entrainment rate far downwind of the stack; Eq. 8.46 is simply an empirical correlation. Entrainment may be affected by the relative intensity of vorticity engendered during the bending-over stage of rise. At axial distances of 15 to 70D in a jet, Kamotani and Greber (1972) found a strong self-preserving double vortex with R = 7.7, but they found only a very poorly organized double vortex with half that value of R.

When R < 1, the low pressure in the lee of a stack is strong enough to pull a jet downward and thereby cause downwash (see Sec. 8-6.2).

8-2.3.3 ***Bending-Over Plumes.*** About a dozen models that attempt to predict the whole trajectory in a continuous fashion (from vertical at the stack to quasi-horizontal far downwind) have been suggested for vertically emitted plumes in a crosswind (Briggs, 1975a, Table 1). About one-half these models consist of various interpolation schemes from $v_e = \alpha\overline{w}$, when there is no crosswind, to $v_e = \beta w_c$, when the plume becomes quasi-horizontal. To my knowledge, no side-by-side comparison of these various schemes has been made against the same set of observed trajectories; thus no basis exists for recommending any one of the models as superior.

In most cases, we have no need to calculate the plume trajectory at distances as small as the bending-over stage. For a jet, according to the ⅓ law, Eq. 8.56, the plume axis bends to 45° above the horizontal at $x \lesssim 0.4RD$, where D is the jet diameter. This normally is a very small distance. For a buoyant plume, according to the ⅔ law, Eq. 8.58, the plume axis attains 45° at $x \simeq 1.2\,F_b/U^3$. This distance also is rather small if U > 2 m/s. Only over a rather narrow range of wind speed would the rising trajectory of a plume be neither quasi-vertical nor quasi-horizontal; a wind speed in this low range might be hard to predict accurately.

8-2.3.4 ***Ambient Turbulence Effects.*** Experts are in general agreement that a turbulent plume's self-generated turbulence is dominant in causing entrainment near the source and that ambient turbulence, if present, becomes dominant only at some further stage downwind. The most common way of introducing the effect of ambient turbulence is by assuming that it makes some contribution to the entrainment velocity, v_e. Briggs (1975a, Table 2) summarizes three basically different closures that have been proposed in this regard. One is that $v_e \propto K/b$, where K is the ambient turbulent diffusivity; this approach is valid if b is much larger than the dominant scale of ambient turbulence, l. Another closure that often is used is $v_e \propto u'$, where u′ is the ambient turbulent root-mean-square velocity; this closure should be valid if b is of the same order as l. The third closure assumption is that $v_e \propto (\epsilon b)^{1/3}$, where ϵ is the ambient turbulent energy-dissipation rate. This approach is valid if b is much smaller than l because ϵ relates directly to the energy contained in the smaller eddies of the ambient turbulence, i.e., in the inertial subrange. (Note that v_e relates only to the growth of a plume relative to its axis, i.e., to relative diffusion rather than to total diffusion.)

These three closure assumptions lead to quite different asymptotic behavior predictions. Nevertheless, it is difficult to make a clear case for any of the predictions on the basis of comparison with observations because nearly 100% of the available observations either approximate the self-generated turbulence stage of rise or show leveling attributable to stable ambient stratification rather than to ambient turbulence. Then, too, measurements of ambient turbulence usually are totally lacking. Briggs (1975a) argues that the ϵ assumption is the most valid because, at least up to the distance at which the maximum ground concentration occurs, $b < l$ is the most likely of the three.

An alternative approach is to assume that plume rise is terminated rather suddenly when ambient turbulence breaks up the self-generated structure of a plume (this structure is described in some detail in Sec. 2.6 of Briggs, 1975a). Several models assume breakup occurs when the internal turbulent velocities, which are assumed to be proportional to w_c, become equal to u' (they are larger than u' in the first stage of rise). Briggs (1975a) suggested that only the smaller of the ambient eddies are likely to first break into and then disturb the plume structure. Because the small-eddy intensity relates directly to the turbulent dissipation rate for both plume and ambient turbulence, breakup might occur when the internal turbulent energy dissipation rate, which is assumed proportional to w_c^3/b, drops below the ambient value, ϵ. Specifically, it was proposed to terminate plume rise when

$$\epsilon = \frac{\eta w_c^3}{z_c'} \tag{8.47}$$

where η is a dimensionless constant estimated to be of the order of 1.5. Because w_c^3/z_c' diminishes rapidly with height in the first stage of rise, the predicted termination height is not very sensitive to the chosen value of η. Equation 8.47 is not really a closure assumption because the first stage closure assumption, Eq. 8.41, is used; however, it does account for the effect of ambient turbulence on terminating plume rise. An equivalent approach is to terminate rise when w_c is equal to ambient turbulent velocities of eddies of the same size as the plume diameter; these velocities are proportional to $(\epsilon z_c')^{1/3}$.

A different type of termination assumption may be successful in convective conditions when large-scale updrafts and downdrafts exist. This assumption states that downdrafts of velocity w_d bring the plume down to the ground at some touchdown distance, x_d, even though the plume is still rising with its first stage velocity relative to the downdraft. The maximum ground concentration is experienced at or near $x = x_d$. The mean rise at this distance is then the rise most relevant to defining the effective stack height. The mean rise is not affected by the velocity of the downdrafts, because the presence of the ground and the continuity equation result in the average $w = 0$ for the ambient motions. This model will be described in more detail in Sec. 8-5.2.

8-2.4 Summary of Principal Equations

The principal conservation equations and closure assumptions developed in Sec. 8-2.1 and 8-2.3 are summarized below. Refer to Sec. 8-2.2 for the determination of initial values.

8-2.4.1 Buoyancy Conservation. For a vertical plume the buoyancy conservation equation is

$$\frac{dF}{dz'} = -sV_v \tag{8.11}$$

For a bent-over plume, we have from Eqs. 8.15, 8.27, 8.32, and 8.44

$$\frac{dF}{dz_c'} = -sV' = -\frac{sV}{S_2} \tag{8.48}$$

Again we use the transformations $dt = dx/U = dz_c'/w_c$. For the vertical plume, z' is simply the height above the source, whereas, for the bent-over plume, z_c' refers to the height of the plume centerline above the source. For the rising stage of a buoyant plume, $S_2 \simeq 2.3$; however, its value has not been determined for other cases. The initial value of the buoyancy flux, F, is πF_b.

8-2.4.2 Momentum Conservation. For a vertical plume the vertical momentum conservation equation is

$$\frac{d(\overline{w}V_v)}{dz'} = \frac{F}{\overline{w}} \tag{8.20}$$

For a bent-over plume, we have, from Eqs. 8.29 and 8.32, the following equation that is identical in form to the equation for the vertical plume:

$$\frac{d(w_cV)}{dz_c'} = \frac{F}{w_c} \tag{8.49}$$

However, note the difference in the definitions of $\overline{w}$ and w_c. For a vertical plume, $\overline{w}$ is a carefully defined mean plume velocity that involves a shape factor S_1. For a bent-over plume, w_c is simply the centerline rise velocity $dz_c'/dt = Udz_c'/dx$. The initial value of the momentum flux, $\overline{w}V_v$ or w_cV, is πF_m.

8-2.4.3 Volume Flux. We must note also the differences in the definitions of the plume volume flux. In the case of a vertical plume, the plume volume flux is defined by Eq. 8.4, and, for convenience, we defined a dynamic radius b_v by

$$V_v = \pi \overline{w} b_v^2 \tag{8.40}$$

For a bent-over plume, the dynamic volume flux, V, and the internal volume flux, V′, are defined by Eqs. 8.28 and 8.32 and by $V = M/w_c$ and $V' = M'/w_c$, respectively. The dynamic and internal plume radii are defined by

$$V = \pi U b^2 \tag{8.42}$$

$$V' = \pi U b'^2 \tag{8.44}$$

Note that $V/V' = S_2$ and $b/b' = (S_2)^{1/2}$.

8-2.4.4 Closure Assumptions. The most widely used closure assumption is the Taylor entrainment hypothesis: the rate at which ambient air is drawn into the plume is proportional to the velocity shear between the plume and the ambient fluid, and this shear consists mainly of the plume's vertical velocity. For a vertical plume, from Eqs. 8.3, 8.38, and 8.40, we can write

$$\frac{dV_v}{dz'} = 2\pi\alpha b_v \overline{w} = 2\alpha(\pi \overline{w} V_v)^{1/2} \tag{8.50}$$

The best current value of the entrainment constant is $\alpha = 0.08$. For a bent-over plume, from Eqs. 8.41 and 8.42, we have

$$\frac{dV}{dz_c'} = 2\pi\beta b U = 2\beta(\pi U V)^{1/2} \tag{8.51}$$

Alternatively, we already have shown that, if U is constant with height, the above reduces to the simple linear relationship

$$\frac{db}{dz_c'} = \beta \tag{8.43}$$

$$b \simeq \beta z_c' \tag{8.52}$$

The best current value of the entrainment constant is $\beta = 0.6$ for a buoyant plume and $\beta = 0.4 + 1.2/R$ for a jet.

8-3 RISE NEAR THE SOURCE

Sometimes we need to predict the rise at distances where a plume is still in its first-stage rise, before ambient stability or turbulence have much effect. For instance, we might want to predict whether a plume will clear a nearby obstacle or to estimate what vertical velocities an aircraft would encounter in a plume. Finally, we might want to estimate how rapidly ambient air is entrained into a plume so that we can calculate its pumping potential, or perhaps its ability to concentrate ambient latent heat into a vertical column that might initiate a convective cloud.

In Sec. 8-4 we show that ambient stratification has very little effect on plume trajectories when $t < \omega^{-1}$, which ranges from 20 s in strong stratification to 100 s in weak stratification. Then we can neglect s in Eqs. 8.11 and 8.48 and state that $F \simeq \pi F_b$, which was its initial value. Furthermore, in Sec. 8-5 we show that buoyant plumes rise for a considerable distance before ambient turbulence retards the rise. This distance approximates 10 stack heights for conventional power plants.

8-3.1 Rise in a Calm

When $U \ll \overline{w}$, we can use the equations for vertical plumes to compute how vertical velocities and the plume volume flux change with height. In many cases we can use experimental information directly. For an extensive review of past work on plumes in calm surroundings, see Turner (1969).

8-3.1.1 Jets. A jet is, by definition, a plume with no initial buoyancy ($F_b = 0$). An extensive survey on jets was given by Pai (1954). For the incompressible axisymmetric jet in a calm fluid, Pai notes that it takes about 10 source diameters of travel for a similarity of flow profiles to become established. Hill (1972) noted that similar distance was required before the entrainment constant became constant. Once the jet has traveled beyond this required distance, it expands conically from an apex near the center of the source. We already have noted that experiments have shown $\alpha = 0.08$ for a vertical jet. For a point source, $b = 2\alpha z'$; so $b_v = 0.16z'$. (To derive $b_v = 2\alpha z$ is simple; from Eq. 8.20, $\overline{w}V_v$ is constant for a jet; then Eq. 8.39 can be readily integrated, $V_v = \pi\overline{w}b_v^2$ can be substituted, and terms can be canceled.) According to Pai (1954), the centerline mean velocity is 13.0 w_0r_0/z' for a round source with $\rho_0 = \rho_a$. More generally, it is given by $13F_m^{1/2}/z'$. From Eq. 8.20, we have $\overline{w}V_v = \pi F_m$, i.e., the initial value. By substituting $V_v = \pi\overline{w}b_v^2$, we find that $\overline{w} = F_m^{1/2}/b_v = 6.25F_m^{1/2}/z'$, or 0.48 times the centerline mean velocity. The total volume flux carried by the jet is $V = \pi b_v(\overline{w}b_v) = 0.50F_m^{1/2}z'$.

8-3.1.2 Transition to Buoyancy Domination. For a buoyant plume with $F = \pi F_b$, a constant, we can integrate Eq. 8.20 by first defining travel time for a vertical plume in terms of z′ and $\overline{w}$:

$$dt = \frac{dz'}{\overline{w}} \tag{8.53}$$

By substituting into Eq. 8.20 and integrating (at $t = 0$, $\overline{w}V_v = \pi F_m$), we find that

$$\frac{\overline{w}V_v}{\pi} = F_m + F_b t \tag{8.54}$$

Transition to a buoyancy-dominated plume then occurs at $t = F_m/F_b = w_0(\rho_0/\rho_0')/g$. To interpret this time in terms of height above the source, we use the approximation $b_v = 0.155z'$ in the term $\overline{w}V_v = \pi\overline{w}^2b_v^2$ in Eq. 8.54. Taking the square root of this equation, substituting $\overline{w} = dz'/dt$, integrating, and substituting $t = F_m/F_b$, we find that transition to buoyancy domination takes place at

$$z' \simeq 4\,F_m^{3/4}F_b^{-1/2} \tag{8.55}$$

8-3.1.3 Buoyant Plumes. The effect of the source momentum flux becomes increasingly negligible above the height given by Eq. 8.55. For a purely buoyant plume, Eq. 8.54 gives $\overline{w}V_v = \pi F_b t$. By using this relationship, we can easily solve the entrainment equation, Eq. 8.50, by substituting $dz' = \overline{w}\,dt$, multiplying both sides by $2\overline{w}V_v$, and integrating to get $V_v = \pi(8\alpha/5)^{1/2}F_b^{3/4}t^{5/4}$. This equation can be substituted back into $\overline{w}V_v$ to get $\overline{w}$ as a function of t, which can be integrated to find a transformation relation from t to z. By using $b_v = (V/\pi\overline{w})^{1/2}$, we find that $b_v = (6/5)\alpha z' = 0.15z'$ (recalling that $\alpha = 0.125$ best fits observations for this case). According to Rouse, Yih, and Humphreys (1952), the centerline vertical velocity is $6.9(F_b/z')^{1/3}$. For the same experimental data, we compute $S_1 = 0.92$ and $\overline{w} = 3.2(F_b/z')^{1/3}$, which is about 0.46 times the centerline vertical velocity. The total volume flux carried by the buoyant plume is $V = 0.225F_b^{1/3}z'^{5/3}$.

8-3.2 Rise in a Crosswind

Plumes emitted into the atmosphere usually bend over because of the wind. Except in very light winds, plumes bend over near the source and become only slightly inclined above the horizontal. In this case, plume rise models for bent-over plumes that assume horizontal transport with the mean wind speed are appropriate. In the models used in this section, we frequently use the transformation $x = Ut$ and the definitions, for a bent-over plume, $V = \pi Ub^2$ or $V' = \pi Ub'^2$.

8-3.2.1 Jets. The trajectory prediction for a bent-over jet can be derived rather directly. By the definition of a jet, $F_b = 0$; from the momentum conservation equation, w_cV is constant and equal to its initial value, πF_m. By using the simple closure of Eq. 8.52, we have $w_cV = (\pi\beta^2Uz_c^2)(dz_c/dt)$. If U is constant, this equation integrates to give

$$z_c' = \left(\frac{3F_mx}{\beta^2U^2}\right)^{1/3}$$

$$= \left(\frac{3}{4}\right)^{1/3}\left(\frac{RD}{0.4 + 1.2/R}\right)^{2/3}x^{1/3} \tag{8.56}$$

We have used $x = Ut$ and $R = w_0/U$. The second expression assumes a round source with $\rho_0 = \rho_a$ and $\beta = 0.4 + 1.2/R$.

Equation 8.56 is compared with observations in Fig. 8.2. Note that Eq. 8.56 fits observed jet centerlines rather well, whereas most of the observations of the velocity minimum or of the pressure, concentration, or temperature maximums are

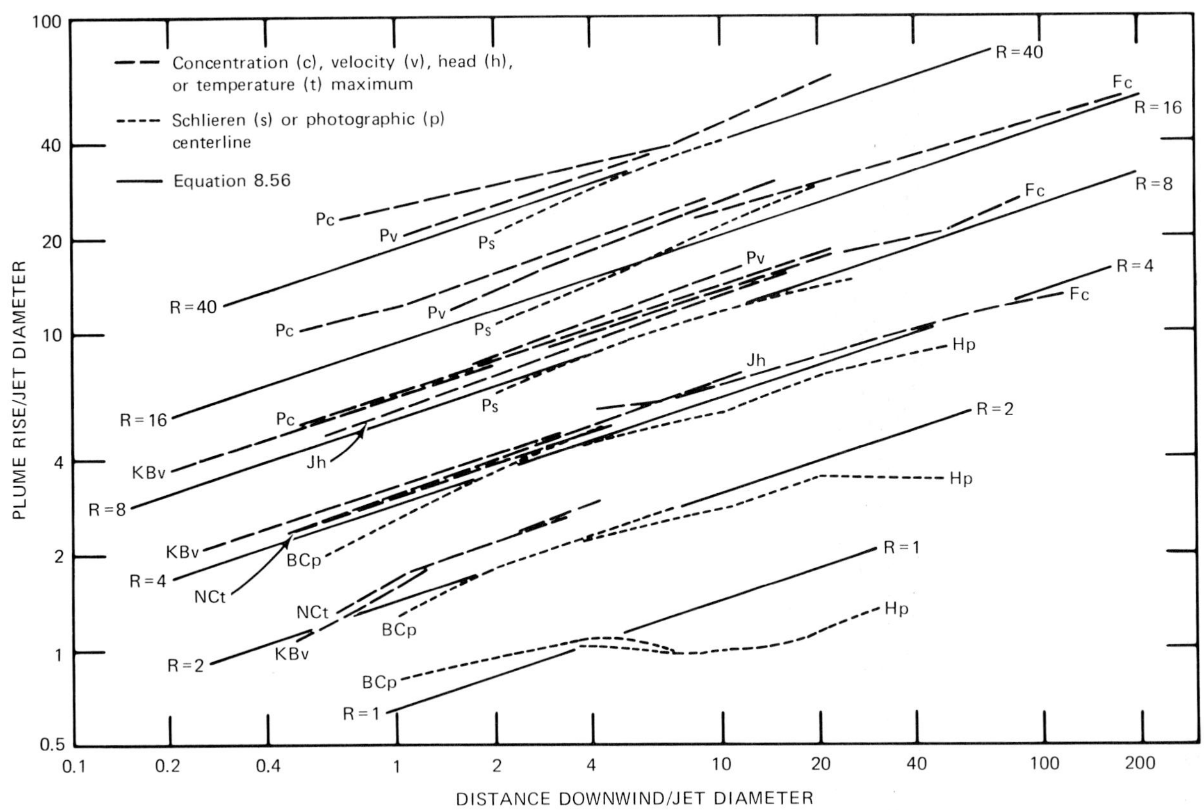

Fig. 8.2 Rise vs. distance for jets in a crossflow, R = w_0/U. Observations are as follows: BC, Bryant and Cowdrey (1955); F, Fan (1967); H, Halitsky, Randall, and Hackman (1950); J, Jordinson (Patrick, 1967); KB, Keffer and Baines (1963); NC, Norster and Chapman (Patrick, 1967); P, Patrick (1967).

slightly higher than the observed centerlines. At lower values of R, the observations of Halitsky, Randall, and Hackman (1950) show a leveling off that may reflect the effect of ambient turbulence. Photographs show more ragged, broken-up appearing plumes at the larger distances and smaller values of R. The centerline rises were estimated from the published photographs by using the median of the three higher Reynolds number runs and have been added to Fig. 3 of Briggs (1975a).

The ⅓ law, Eq. 8.56, seems to describe the observed trajectories well as close to the stack as 0.5D downwind (i.e., at the lee edge of a stack). This fit is surprising because flow similarity is not developed here, and the jets are more nearly vertical than horizontal at this point. Hoult and Weil (1972) suggested a formula in which $z_c' \propto x^{1/2}$ is very close to the stack in the quasi-vertical stage. Figure 8.2, however, indicates that, even when very close to the stack, $z_c' \propto x^{1/3}$ fits most of the observations better than $z_c' \propto x^{1/2}$; this fact may simply be regarded as fortuitous.

8-3.2.2 Transition to Buoyancy Domination. When F is constant and equal to its initial value, Eq. 8.49 integrates to $w_c V = \pi F_m + \pi F_b t$, which is the result obtained for vertical plumes in Eq. 8.54 (we made the substitution $w_c = dz_c/dt$). This relationship shows once again that a plume becomes buoyancy dominated in a time $t = F_m/F_b$, which typically is only about 5 s for plumes resulting from combustion. For a plume in a crosswind, this time represents very little downwind travel, usually <50 m.

Equation 8.54 can be integrated easily after substituting $w_c = dz_c'/dt$ and $V = \pi U b^2 = \pi \beta^2 U z_c'^2$. Substituting $t = x/U$, the result is

$$z_c' = \left(\frac{3}{\beta^2}\frac{F_m}{U^2}x + \frac{3}{2\beta^2}\frac{F_b}{U^3}x^2\right)^{1/3} \qquad (8.57)$$

This equation gives a transition from the $x^{1/3}$ law for bent-over jets to the $x^{2/3}$ law for bent-over buoyancy-dominated plumes. In Fig. 8.3 we see that Eq. 8.57 fits observations in this transition region fairly well. Note, however, that $\beta = 0.6$ was used with the F_b term, whereas $\beta = 0.4 + 1.2/R$ was used with the F_m term to be consistent with the asymptotic behaviors of jets and buoyant plumes. Furthermore, the z_c' given by Eq. 8.57 is multiplied by a factor 1.13 before plotting Fig. 8.3. This enhanced value of z_c' fits the model better because the observations in Fig. 8.3 represent the height of maximum concentration, which is shown in Fig. 8.2 to be somewhat above the plume centerline.

8-3.2.3 Buoyant Plumes. By dropping the F_m term from Eq. 8.57, we have the asymptotic prediction for the rise of buoyant bent-over plumes before ambient stratification or turbulence have much effect. Substituting $\beta = 0.6$, this is

$$z_c' = \frac{1.6 F_b^{1/3} x^{2/3}}{U} \qquad (8.58)$$

the well-known ⅔ law for buoyant plumes. This relationship first appeared implicitly in Sutton's (1950) formulation; it was explicitly derived by Csanady (1956), who also first demonstrated the consistency of the relationship with observations.

Since 1956 scientists have demonstrated many times that Eq. 8.58 agrees with laboratory plume data and with a substantial majority of field observations. Examples are discussed

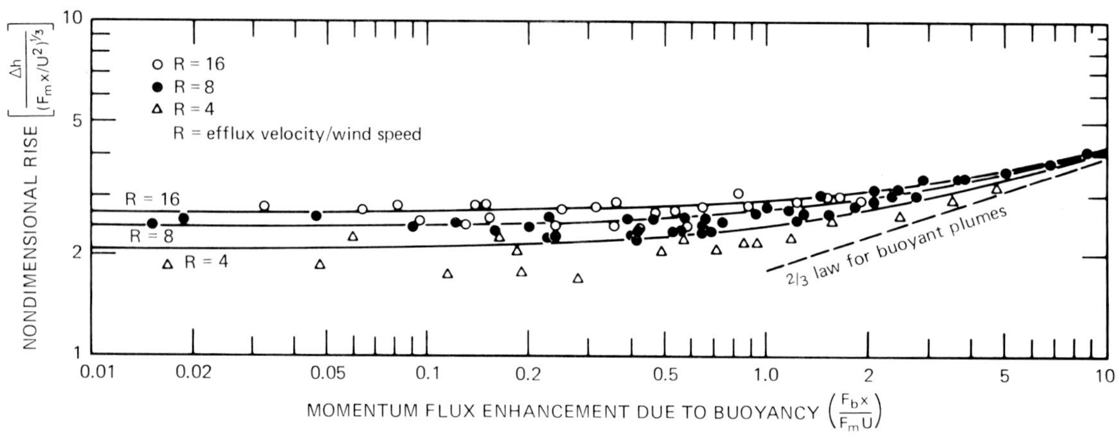

Fig. 8.3 Transition from the "⅓ law" for jets to the "⅔ law" for buoyant plumes vs. Eq. 8.57 times a factor of $(1.2)^{3/4}$. Points plotted are heights of concentration maximums (Fan, 1967).

in Briggs (1965, 1969, 1970); Fay, Escudier, and Hoult (1969); Bringfelt (1969a); and Hoult and Weil (1972). The optimum value of the dimensionless constant ranges from 1.2 to 2.6 in these various studies [these values are summarized in Table 3 of Briggs (1975a)]; however, 10 of the 16 cited optimum experimental values ranged only from 1.5 to 1.7; therefore 1.6 appears to be the best overall value. All the wind-tunnel studies are in agreement with this value. The more deviant values observed at certain field sites can be attributed to many possible causes such as stack-, building-, or terrain-induced downwash; terrain elevation variations; and lake- or river-induced convective cells.

Fay, Escudier, and Hoult's (1969) study gives some idea of the deviation of individual observations compared with Eq. 8.58, as opposed to the best average fits for each source. They found that ln $(z_c'/F_b^{1/3}U^{-1}x^{2/3})$ was distributed approximately normally. For field data involving dozens of diverse sources, their analysis shows a standard deviation corresponding to a factor of 1.55. This result implies that 16% of the dimensionless ratios were more than 1.55 times the median value and 16% were less than $1/1.55 = 0.65$ times the median. Laboratory data show much less scatter; the standard deviation of the logarithm of the ratio corresponds to a factor of 1.17.

The large scatter in field data can be attributed to many reasons. All the downwash, terrain, and body of water effects vary with wind speed and wind direction. Large eddies of the ambient turbulence affect individual plume observations. Finally, the best fit is sensitive to the measured value of U, and many of these measurements were affected by random turbulent fluctuations different from the fluctuations in U experienced by the plume itself.

Because wind speed generally increases with height, we would expect the best fit value of n to $z_c' \propto x^n$ to be somewhat less than 2/3 for a buoyant bent-over plume. According to the model of Murthy (1970), if $U \propto z_c'^p$, then $n = (2/3)/(1 + p)$. His wind-tunnel experiments with the profile $U \propto z_c'^{1/3}$ agreed fairly well with the model, with the observed n being about 0.45 instead of the predicted 0.50. Bringfelt (1969a) found that, somewhat contrary to expectations, $n = 0.73$ gave the best fit to his field observations in neutral conditions; $n = 0.62$ was the best fit for stable conditions. The lower value for stable conditions is consistent with the fact that the relative wind shear is larger in stable conditions. In a later analysis Bringfelt (1969b) noted that $n = 0.66$ was the best fit in the range $x = 250$ to 500 m, whereas $n = 0.50$ fit best in the range $x = 500$ to 1000 m. This result possibly reflects some tendency toward leveling. Most of these observations were of plumes from small, industrial sources.

Djurfors and Netterville (1978) also advanced a model accounting for wind shear for both close-in rise and final rise in stable conditions. Unfortunately, they could not verify this model with the field observations on hand because of a large amount of scatter.

8-3.2.4 ***Buoyancy-Generating Plumes.*** Although most plumes tend to conserve their buoyancy, it is conceivable to have a buoyancy-generating effluent (for example, a strongly radioactive plume). Gifford (1967) considered such plumes and assumed that the plume generated new buoyancy at a constant rate $\partial F_b/\partial t$. The buoyancy conservation equation, Eq. 8.48, is modified by this term:

$$\frac{dF}{dt} = -s(w_c V') + \frac{\partial F_b}{\partial t} \tag{8.59}$$

where we replaced dz_c' with $w_c\,dt$. In neutral conditions, or at small values of $s^{1/2}\,t$, we can disregard the s term and integrate Eq. 8.59 to $F = F_b + (\partial F_b/\partial t)t$. The momentum equation, Eq. 8.49, then integrates to

$$w_c V = F_m + F_b t + \frac{1}{2}\frac{\partial F_b}{\partial t}t^2 \tag{8.60}$$

By substituting $w_c = dz_c'/dt$ and $V = \beta^2 U z_c'^2$, Eq. 8.60 can be integrated to give an equation for z_c' vs. x that is similar to Eq. 8.57 except for the addition of the buoyancy-generating term. When the buoyancy-generating term dominates, the trajectory is simply linear:

$$z_c' = \left(\frac{\partial F_b/\partial t}{2\beta^2}\right)^{1/3} U^{-4/3}x \tag{8.61}$$

Contrary to the behavior of jets and buoyancy-conserving plumes, the buoyancy-generating plume is not expected to level off in stable conditions (unfortunately, no data are available to test model predictions). At large values of $s^{1/2}t$, the plume tends to be in buoyancy equilibrium with its surroundings; therefore $F \simeq 0$ and the two terms on the right of Eq. 8.59 balance each other. If we make the same substitutions as before and integrate, the result is

$$z_c' = \left[\frac{3(dF_b/\partial t)x}{\beta'^2 U^2 s}\right]^{1/3} \tag{8.62}$$

However, β' probably will be reduced from the rising stage value of 0.4 because stable stratification tends to dampen vertical mixing motions. In fact, we really do not know whether the Taylor entrainment assumption will continue to apply in this situation. (If entrainment ceased altogether at a time proportional to $s^{-1/2}$, the plume would continue to rise with a trajectory similar to that given by Eq. 8.61, but it would have a reduced constant.)

8-4 RISE LIMITED BY STABLE STRATIFICATION

During the night, air normally is stably stratified with little or no turbulence above a mechanically mixed layer near the ground. This layer of continuous turbulence usually is only 10 to 100 m deep (it may reach several hundred meters deep in cloudy high-wind conditions). Above this layer there are only infrequent outbursts of turbulence. These outbursts, called turbulent bursts or episodes, are similar in origin to clear-air turbulence, and they may occur, depending on the site and the synoptic situation, every few hours or not at all. Nevertheless, the normal state of the atmosphere above the ground layer is stable, and turbulent bursts are not known to cause high ground concentrations in comparison to other situations.

In the daytime air normally is stably stratified above the convective mixing layer. The convective layer usually is 500 to 2000 m deep; however, in the early morning when convection is just beginning, the layer can be shallower than 500 m. Occasionally, plumes from high stacks can penetrate into the stable air above a daytime convective mixing layer.

In stable air plume rise is limited for buoyancy-conserving plumes because of the negative buoyancy acquired by entraining ambient air and carrying it to levels of higher ambient potential temperature. The behavior of the momentum flux is revealed by substituting $dz' = \overline{w}\,dt$ or $dz_c' = w_c\,dt$ into the momentum and buoyancy conservation equations, differentiating the former (Eq. 8.20 or 8.49) with respect to t, and substituting (Eq. 8.11 or 8.48). The result is the buoyancy conservation equation

$$\frac{d^2(\overline{w}V_v)}{dt^2} = -\omega^2(\overline{w}V_v) \quad \text{(vertical plumes)}$$

$$\frac{d^2(w_cV)}{dt^2} = -\omega'^2(w_cV) \quad \text{(bent-over plumes)} \tag{8.63}$$

where $\omega = s^{1/2}$, the Brunt-Väisälä frequency, and $\omega' = (s/S_2)^{1/2}$. For the purpose of general discussion, define $\omega'' = \omega$ for vertical plumes and $\omega'' = \omega'$ for bent-over plumes. Similarly, define $M'' = \overline{w}V_v$ for vertical plumes and $M'' = w_cV$ for bent-over plumes. Equation 8.63 is the equation of a simple harmonic oscillator with a period $2\pi/\omega''$. Because V_v or V grows with time, while M'' oscillates, $\overline{w}$ or w_c must behave like a damped harmonic oscillator. The plume, as described by Eq. 8.63, should rise to an asymptotic height and approach it in the manner of a damped oscillator. (A remarkable thing about this result is that it is completely independent of any closure assumption; Eq. 8.63 is based on conservation principles alone.) One important conclusion that can be made from this equation is that stable stratification has very little effect on plume rise if $t \ll 1/\omega''$.

The validity of Eq. 8.63 at large values of $\omega''t$ is questionable. Observations show that plume height oscillations damp out very quickly and that often plumes approach an asymptotic height with no overshoot at all. The reason for this quick damping may be that wave drag carries away much of the vertical momentum as the plume approaches its final height (see Warren, 1960). The situation is analogous to that of an object floating on a water surface. If you lift the object slightly and let go, it will overshoot its original equilibrium depth somewhat, but it will quickly approach that depth again. Given an undisturbed water surface and Archimedean forces alone, the object would bob up and down forever, but the waves it makes quickly carry away its kinetic and potential energy. The object does experience Archimedean forces, but the forces are not a function of the object height alone, because the water surface adjacent to the object is responding to the vertical motion of the object itself, only with a phase lag.

Similar waves, called gravity waves, are produced in the atmosphere as a rising plume displaces the ambient air. The effect of these waves is not accounted for in the buoyancy equation, because ρ_a was assumed to be a function of height alone and not of time. Nonetheless, Eq. 8.63 apparently does hold well as far as the point of maximum rise because its trajectory predictions compare well with observations (see Figs. 8.4 and 8.5).

It is worthwhile to study visually plumes that rise and level off in the very early morning before convective-mixing-layer development. Often a slight overshoot of the final height is seen, and infrequently more than one oscillation is evident. Such plumes appear lumpy and chaotic—obviously very turbulent—in the rising stage; however, these features smooth out as the plume starts to level. The leveled part of the plume often is seen as smooth, horizontal, or slightly wavy filaments. Such an appearance indicates a complete lack of turbulence; these plumes can drift downwind tens of kilometers with no further increase in thickness. From such observations, we might conclude that published curves showing the vertical dispersion coefficient increasing with distance do not apply to elevated plumes leveled off in stable air. Lateral dispersion may occur, because stable stratification inhibits only vertical motions. Horizontal eddies that may be present can cause the plume to meander and spread laterally. Such eddies are not turbulence in the strict sense, because their motions are two-dimensional and there is no vortex stretching.

8-4.1 Rise Through a Constant Density Gradient

When s is approximately constant with height, as it often is in elevated layers and in approximately the lowest 100 m of the night atmosphere, Eq. 8.63 describes approximately a simple harmonic oscillation. Recognizing that at $t = 0$, we have $M'' = \pi F_m$ and $dM''/dt = \pi F_b$, the solution for constant s is

$$\frac{M''}{\pi} = F_m \cos(\omega''t) + \frac{F_b}{\omega''}\sin(\omega''t) \tag{8.64}$$

Thus the maximum rise, where $\overline{w}$ or w_c goes to zero, occurs at $\omega''t = \pi/2$ for a jet and equals π for a strongly buoyant

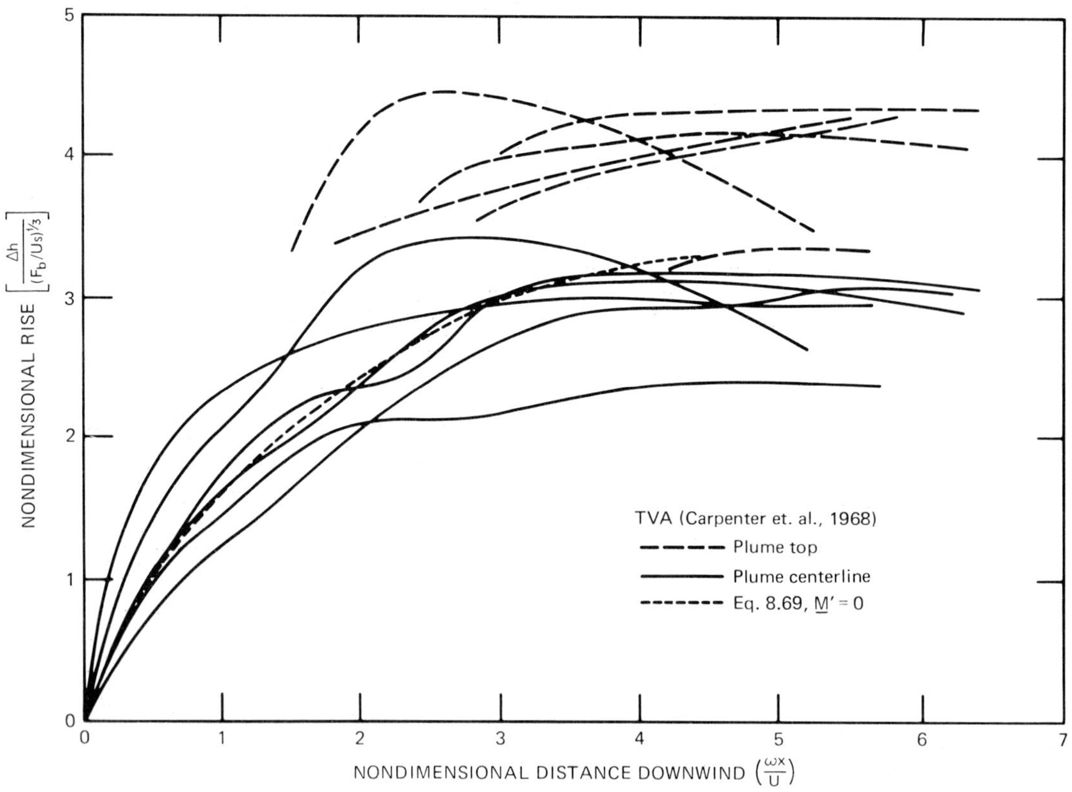

Fig. 8.4 Rise vs. distance in stable conditions for measurements at Tennessee Valley Authority power plants.

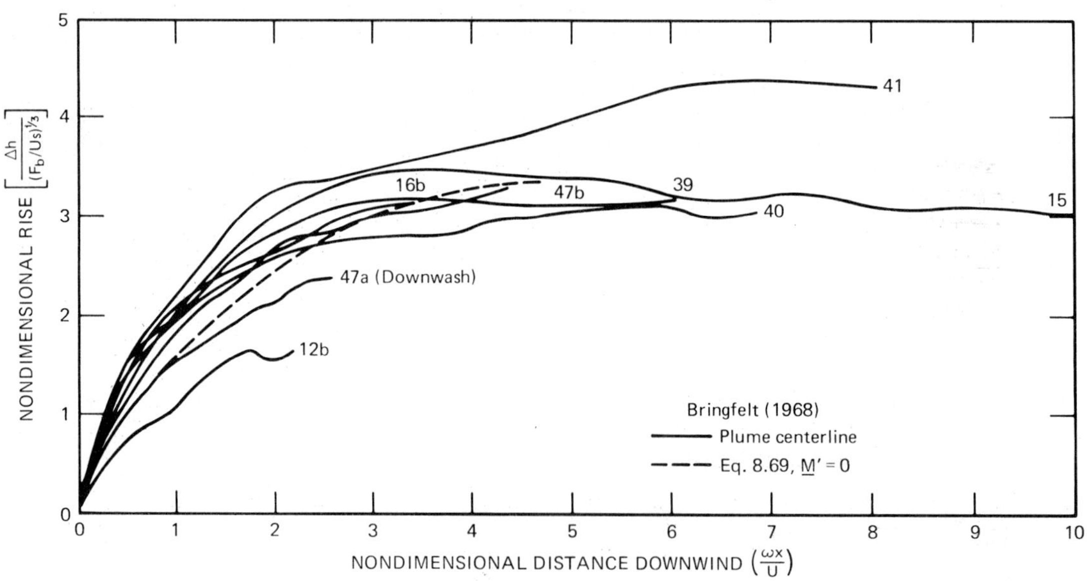

Fig. 8.5 Rise vs. distance in stable conditions for Bringfelt's (1968) measurements at small industrial sources.

plume. In the general case, maximum rise occurs at $\omega''t = \tan^{-1}(-\omega'' F_m/F_b)$.

When s is constant, we can reverse the procedure used to derive Eq. 8.63 to get

$$\frac{d^2F}{dt^2} = -\omega''^2 F$$

$$\frac{F}{\pi} = F_b \cos(\omega'' t) - \omega'' F_m \sin(\omega'' t) \qquad (8.65)$$

Thus the buoyancy flux also oscillates and is 90° out of phase with the momentum flux. A jet develops negative buoyancy immediately, whereas a buoyant plume starts to become negatively buoyant at $\omega' t = \pi/2$, which is exactly ½ of its time to maximum rise.

So far, all the conclusions made about plume rise into stable air have been made on the basis of conservation equations alone. To go further, we must invoke a closure assumption.

*8-4.1.1 **Vertical Plumes.*** Complete solutions for $\overline{w}$ and r as functions of z in vertical plumes can be obtained by using Eq. 8.64 with the Taylor entrainment hypothesis, Eq. 8.50, and by making the substitutions $dz' = \overline{w}\,dt$ and $V_v = \overline{w}b_v^2$. Unfortunately, this set of equations cannot be solved explicitly. Numerical solutions have been obtained by Morton, Taylor, and Turner (1956), Morton (1959), Morton and Middleton (1973), Hino (1962), and others.

For the pure cases of point-source buoyant plumes or jets, Briggs (1975a) suggested an explicit approximate solution that is especially useful when more complicated temperature stratifications are considered. This procedure is called the unaltered volume flux technique. The effect of stratification on the solution of V_v as a function of z′ is rather minor; for a buoyant plume, V_v is reduced by only 2% at half the maximum rise and by 16% at the maximum rise. Therefore, for the purpose of calculation, we assume that V_v is unaltered from its value as a function of z′ in a neutral atmosphere. When this simple form of V_v is substituted into Eq. 8.64, a solution for maximum rise is easily obtained and is only 2.5% less than the value given by the exact solution for a buoyant plume. The model cannot be expected to be more accurate than this because it does not take into account wave drag or the effect of stratification on entrainment.

As noted earlier, Eq. 8.4 predicts the maximum rise to occur at $\omega t = \tan^{-1}(-\omega F_m/F_b)$. Plumes are negatively buoyant at this point, therefore some settling back should occur. When we substitute this value of ωt into Eq. 8.65, we find that $F_{max} = -(F_b^2 + \omega^2 F_m^2)^{1/2}$ at the maximum rise (indicated by subscript max). The simplest way to estimate the final rise, or equilibrium height, is to assume that the plume settles back with no further mixing from the maximum rise to the height where F = 0. In other words, the plume settles back a distance equal to F_{max}/sV_{max}, in accordance with Eq. 8.11. Combined with the unaltered volume flux assumption, this method predicts an equilibrium height equal to ½ = 0.50 times the maximum rise for a jet and 13⁄16 = 0.81 times the maximum rise for a buoyant plume. Experimental values cited below give ratios of 0.55 and 0.74, respectively.

Experiments recently were conducted in a salt-water density stratified channel at the EPA Fluid Modeling Facility in Research Triangle Park, NC. These experiments, reported here for the first time, were done for both pure jets and buoyant plumes of various initial diameters and Reynolds numbers. The minimum Reynolds number required for flow similarity was established, and the rise of forced plumes in which both the buoyancy and the initial momentum are important was investigated by varying the dimensionless momentum ratio $\underline{M} = \omega F_m/F_b$.

For pure jets,

$$z'_{max} = 4.4\left(\frac{F_m}{\omega^2}\right)^{1/4}$$

$$z'_{eq} = 2.44\left(\frac{F_m}{\omega^2}\right)^{1/4} \qquad (8.66)$$

where z'_{max} is the top of the visible plume and z'_{eq} is the center height of the spreading out, stratified plume. The theoretically predicted coefficients in Eq. 8.66, using the above model with α = 0.08, are 5.0 and 2.5, respectively. The thickness of the spreading out plume is about $1.1\,(F_m/\omega^2)^{1/4}$; consequently, the bottom is at $1.9\,(F_m/\omega^2)^{1/4}$ and the top is at $3.0\,(F_m/\omega^2)^{1/4}$.

For very buoyant plumes,

$$z'_{max} = 7.2\left(\frac{F_b}{\omega^3}\right)^{1/4} - C$$

$$z'_{eq} = 5.3\left(\frac{F_b}{\omega^3}\right)^{1/4} - C \qquad (8.67)$$

Note that a correction term, C = 3D, was consistent with the data; this implies that the virtual point source of buoyancy is located three source diameters below the actual source height. With α = 0.08, the theoretically predicted coefficients for point sources in Eq. 8.67 are 6.3 and 5.1, respectively. Thus the unaltered volume flux model with fallback to equilibrium height makes good predictions of z'_{eq}, but it does not predict z'_{max} as well; however z'_{eq} is the level of concern to diffusion modeling. The thickness of the spreading out buoyant plume is about $1.2(F_b/\omega^3)^{1/4}$; therefore the bottom is at $4.7(F_b/\omega^3)^{1/4} - 3D$ and the top is at $5.9(F_b/\omega^3)^{1/4} - 3D$.

Dimensional analysis of the lower plume, where stability effects are negligible, predicts that the virtual point source of a vertical plume is located a distance D times some function of the source Froude number, Fr_0, below the top of the source. This function was not adequately defined in the preceding EPA experiment at Research Triangle Park; however, a function that did improve the data fit over the range of values explored was

$$C = \frac{3D}{1 + 0.01 Fr_0} \qquad (8.68)$$

where

$$Fr_0 = \frac{w_0^2}{gD(\rho_a - \rho_0)/\rho_0}$$

Typical values of Fr_0 for hot plumes range from 0.5 to 20.

Morton, Taylor, and Turner's (1956) theory predicts that, for a fixed value of F_b, plume rise will drop as momentum flux is increased until, at large values of dimensionless momentum flux ($\underline{M} = \omega F_m/F_b$), the rise begins to increase. Such behavior was observed in the EPA experiments. The drop in plume rise results from increased entrainment into the lower part of the plume as $\underline{M}$ increases. The theory predicts that the greatest loss of plume rise occurs in the vicinity of $\underline{M} \simeq 2$; the EPA experiment confirmed this prediction. However, the theory predicts a 20% decrease in (z'_{max} + C), whereas only a 10% decrease was observed. A 20% decrease was observed in (z'_{eq} + C). For $\underline{M} < 1$, the observed decrease in both z'_{max} and z'_{eq} was approximately 0.8 $\underline{M}^{3/4}(F_b/\omega^3)^{1/4}$. For most hot plumes, $\underline{M} \ll 1$; therefore this decrease is not very significant.

*8-4.1.2 **Bent-Over Plumes.*** The bent-over plume trajectory is rather easily obtained from Eq. 8.64 because $w_c = dz'_c/dt$, $t = x/U$, and V is a relatively simple function of z'_c. We assume the validity of Eq. 8.52 so that $V = \pi U b^2 = \pi U(\beta z'_c)^2$. Integrating Eq. 8.64 with respect to t, we get

$$z'_c = \left(\frac{3F_b}{\beta^2 Us'}\right)^{1/3} [\underline{M}' \sin(\omega' t) + 1 - \cos(\omega' t)]^{1/3} \tag{8.69}$$

where $\underline{M}' = \omega' F_m/F_b$. The $\underline{M}'$ term in Eq. 8.69 represents the effect of the initial efflux vertical momentum. Because the maximum rise occurs at $\omega' t = \tan^{-1}(-\underline{M}')$, we have

$$z'_{max} = \left(\frac{3F_b}{\beta^2 Us'}\right)^{1/3} [1 + (1 + \underline{M}'^2)^{1/2}]^{1/3} \tag{8.70}$$

With the assumption of no further mixing as the plume falls back to z'_{eq}, we have from Eq. 8.48 ($z'_{max} - z'_{eq}$) = $F_{max}/(sV'_{max})$, where $V'_{max} = U(\beta' z'_{max})^2$. This gives $z'_{eq}/z'_{max} = 2/3 = 0.67$ for a jet, whereas for a strongly buoyant plume $z'_{eq}/z'_{max} = 5/6 = 0.83$. The latter prediction is very close to the 0.84 asymptote predicted by using the entrainment assumption $db/dt = \beta|\overline{w}|$ favored by some modelers. (Fay, Escudier, and Hoult, 1969). With this assumption the plume grows as it gets into the damped harmonic oscillator stage of up and down motions; however, in the damped harmonic oscillator stage, wave drag effects are likely to be significant, and these effects are not accounted for in any model at present.

Equation 8.70 can be used to predict the rise of a bent-over jet into stable air by setting $\beta = 0.4 + 1.2/R$, $z'_{eq}/z'_{max} = 2/3$, and letting $\underline{M}'$ be very large. Unfortunately, no data currently exist against which to check this prediction. This case may be of little practical importance because the effluent density has to closely match that of air for $\underline{M}'$ to be large; roughly speaking, $\rho_0/\rho_a \ll 0.001\, w_0/(m/s)$.

For most hot plumes, $\underline{M}'$ is 0.1 or less, and the effect of initial momentum in Eq. 8.70 is negligible. When $\beta = 0.6$, $s' = s/2.3$, and $z'_{eq}/z'_{max} = 5/6$, the numerical coefficient in Eq. 8.70 becomes 2.8. This coefficient is in agreement with rises observed in a stably stratified wind tunnel (Hewett, Fay, and Hoult, 1971). However, on the basis of ten different analyses of observed rises, most of them of power plant plumes, Briggs (1975a) recommended

$$\Delta h = 2.6\left(\frac{F_b}{Us}\right)^{1/3} \tag{8.71}$$

The range of best-fit coefficients was 2.3 to 3.1, excluding the one set of observations in which stack downwash was a likely influence. However, 2.6 still may be slightly conservative because some of these observations were made at distances short of the maximum rise point. Much of the scatter in best-fit values may be due to difficulties in making accurate field observations. Briggs (1975a) cited a case in which two different research teams measured the rise of the same plume at the same time; however, the measurements were made at different distances in the leveled-off plume and by different techniques (lidar and vertical SO_2 profiles). The ratios of the rises observed by the two teams ranged from 0.7 to 1.4. Another difficulty is that the temperature gradient seldom is close to linear in the real atmosphere. In all the cited analyses, $\partial\theta_a/\partial z$ was taken as the average from the top of the stack to the top of the leveled-off plume.

Figures 8.4 and 8.5 show some observed trajectories of visible plumes from power plants and from small industrial sources. These trajectories are compared with the prediction of Eq. 8.69 with $\underline{M}'$ set at zero. The rises are made nondimensional, dividing by $(F_b/Us)^{1/3}$, and the distances downwind are made nondimensional, dividing by (U/ω). Equation 8.69 is plotted with $\underline{M}' = 0$ and $\omega/\omega' = (s/s')^{1/2} = S_2^{1/2} = 1.5$. This trajectory results in a maximum rise predicted at $x = \pi U/\omega' = 4.7(U/\omega)$. Five of the six centerline trajectories of the Tennessee Valley Authority (TVA) plumes reach a very flat maximum somewhere between 3 and $6(U/\omega)$. Four trajectories observed by Bringfelt (1968) do peak and settle back, and these maximize somewhat between 3 and $6.5(U/\omega)$. The wind-tunnel plume trajectories observed by Hewett, Fay, and Hoult (1971) are more uniform and have a rather flat peak in the rise at slightly beyond $x = 4U/\omega$.

8-4.2 Rise Through a Density Jump

Occasionally the atmosphere is near-neutral except for one or more thin layers through which the temperature rises (for example, see Simon and Proudfit, 1967). If a plume can penetrate such an inversion layer, the stability of the layer will prevent downward mixing; thus ground concentrations from the plume will be very low or nonexistent. On the other hand, if the plume cannot penetrate, upward diffusion will be prevented, and the material is trapped beneath the inversion. Consequently, whether or not such an inversion can be penetrated may be of critical concern.

We can approximate the temperature profile by assuming $\partial\theta_a/\partial z = 0$ except for a jump in potential temperature, $\Delta\theta_i$, at a height above the source z'_i. We define $\Delta\theta_i$ as the total change of potential temperature through the stable layer, $\int (\partial\theta_a/\partial z)\, dz$; and we define an inversion strength parameter $b_i = (g/\theta_a)\Delta\theta_i$.

If a plume penetrates the inversion, according to Eq. 8.11 or 8.48, its buoyancy flux is reduced by $b_i V_i$, where V_i is the volume flux of the plume at z_i, V_v for vertical plumes and V′ for bent-over plumes. (This result is obtained by integrating over z′ or z'_c through the inversion and holding the volume flux constant; because the volume flux approximates V_i, b_i results from the integral of s through the inversion layer.) Therefore, if $F > b_i V_i$ when the plume reaches the inversion, the plume remains positively buoyant after penetration and keeps on rising. However, if F is only slightly greater than $b_i V_i$, part of the plume may get trapped in the inversion because peripheral regions of the plume are less buoyant than the mean. If F is somewhat less than $b_i V_i$, very little penetration will occur, but the plume will still be buoyant relative to the air under the inversion. As a result, the plume may push against the inversion and resist downward mixing. In general, the degree of penetration depends on $F/b_i V_i$.

It is possible for a nonbuoyant plume, a jet, to punch through an inversion if it has sufficient momentum. However, the plume becomes negatively buoyant relative to the layer above the inversion, and it will settle back and sandwich itself somewhere in the inversion. Because of mixing with the upper layer, the plume will become buoyant relative to the layer below the inversion. Rise of the plume above the inversion will be proportional to w_i^2/b_i, where w_i is the vertical velocity of the plume at z_i', $\overline{w}$ for a vertical plume and w_c for a bent-over plume. The relative degree of mixing with the upper layer is then some function of $\overline{w}_i^2/b_i z_i'$.

8-4.2.1 ***Vertical Plumes.*** For a point-source buoyant plume rising in neutral air, $F = \pi F_b$, and the solution of Eqs. 8.20 and 8.50 is $V_v = (9\alpha/10)(32\alpha F_b/15)^{1/3} z'^{5/3} = 0.04 F_b^{1/3} z'^{5/3}$ if $\alpha = 0.08$. By substituting $z' = z_i'$ to get V_i, we predict penetration of the inversion if $V_i < F_b/b_i$, i.e.,

$$z_i' < \frac{6.9 F_b^{2/5}}{b_i^{3/5}} \qquad (8.72)$$

Figure 8.6 shows how laboratory-measured inversion heights required for penetration (Vadot, 1965) depend on a dimension-

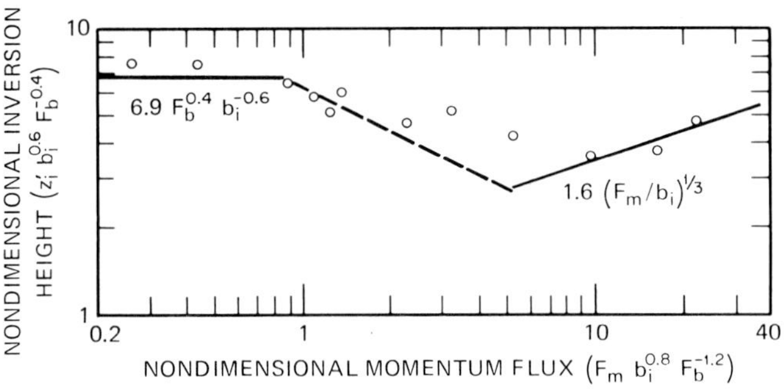

Fig. 8.6 Maximum inversion height for penetration vs. nondimensional momentum flux [data from Vadot (1965)].

less source momentum parameter. For the most buoyant cases, which appear on the left side of the figure, Eq. 8.72 works very well. If the Rouse, Yih, and Humphreys (1952) experimental value of V is used, the predicted coefficient in Eq. 8.72 is 4.9, which appears too low.

If a plume is momentum-dominated so that $V_v = F_m^{1/2}(2\alpha z')$, excess buoyancy above the inversion is predicted with $\alpha = 0.08$ if

$$z_i' < \frac{6.2(F_b/F_m^{1/2})}{b_i} \qquad (8.73)$$

If F_b results from heat and not from molecular weight, this criterion reduces simply to

$$T_0' r_0 > 0.16(T_0/T_a)^{1/2} \Delta\theta_i z_i'$$

Equation 8.73, represented by the dashed line in Fig. 8.6, appears to give a good prediction of inversion penetration at intermediate values of the momentum flux ratio.

For large values of momentum flux, $\overline{w} \propto F_m^{1/2}/z'$; thus the punch-through ability of the plume depends on $F_m/b_i z_i'^3$. Vadot's data indicate penetration if

$$z_i' < 1.6\left(\frac{F_m}{b_i}\right)^{1/3} \qquad (8.74)$$

However, if Eq. 8.73 is not satisfied, such a plume will be negatively buoyant above the inversion and will settle back into the inversion.

8-4.2.2 ***Bent-Over Plumes.*** Bent-over jets have very little inversion penetration ability. Vadot's experiments on vertical jets indicate that a predicted maximum rise of about $3z_i'$ is required for good penetration. Assuming $V = \pi U(\beta z_c')^2$ and $V' = \pi(\beta' z_c')^2$, and using $\beta = 0.6$ and $\beta' = 0.4$ as approximations, Eqs. 8.48 and 8.49 predict $z'_{max} \geqslant 3z_i'$ when

$$z_i' \leqslant 0.565\left(\frac{F_m}{U\beta\beta'}\right)^{2/5} b_i^{-1/5} \qquad (8.75)$$

$$\simeq 1.0\left(\frac{F_m}{U}\right)^{2/5} b_i^{-1/5}$$

(Equation 8.48 gives $F = 0$ below $z' = z_i'$; Equation 8.49 gives $w_c V = \pi F_m$ at $z' = z_i'$; above this height, Eq. 8.48 gives $F = -b_i V_i$, and Eq. 8.49 can be integrated and set so that $w_c V = 0$ at $z' = 3z_i'$.) This results in a small height that typically is less than 15 m. Even if Eq. 8.75 is satisfied, the jet will be denser than the air above the inversion and will settle back into the inversion layer.

Bent-over buoyant plumes have much more inversion penetration capacity than bent-over jets, but they do not have as much as vertical buoyant plumes (the latter entrain less ambient air). Using Eq. 8.48 with $V' = U(\beta' z_c')^2$ and $\beta' = 0.4$, we predict that a plume still is positively buoyant ($F > 0$) after penetrating an inversion if

$$z_i' < \beta'^{-1}\left(\frac{F_b}{U b_i}\right)^{1/2} = 2.5\left(\frac{F_b}{U b_i}\right)^{1/2} \qquad (8.76)$$

For a power plant this z_i' typically is of the order of 100 m.

However, a buoyant plume could partially penetrate a considerably higher inversion. Briggs (1975a) suggested a model for partial penetration based on approximating the plume as having a rectangular cross section extending from 0.5 to $1.5z_c'$ and a width equal to $\pi(\beta' z_c')^2/1.0\ z_c' = 0.5z_c'$. In this approximation, w effectively equals w_c throughout the plume cross section. Instead of Eq. 8.48, which treats the plume as if it all crosses the inversion at once, we use the following equation to derive the change in F that results from only the portion of the plume crossing the height z_i' at a given time:

$$\frac{dF}{dz_c'} = w_c^{-1}\frac{dF}{dt}$$

$$= -w_c^{-1} U b_i \int w\, dy$$

$$\simeq -U b_i (0.5 z_c') \qquad (8.77)$$

The resulting equilibrium height, where F goes to zero, is found by integrating Eq. 8.77 from $F = F_b$ at $1.5z_c' = z_i'$, where the plume is presumed to enter the inversion. The result is

$$\frac{z_{eq}'}{z_i'} = \frac{2}{3}\left(1 + \frac{28 F_b}{U b_i z_i'^2}\right)^{1/2} \qquad (8.78)$$

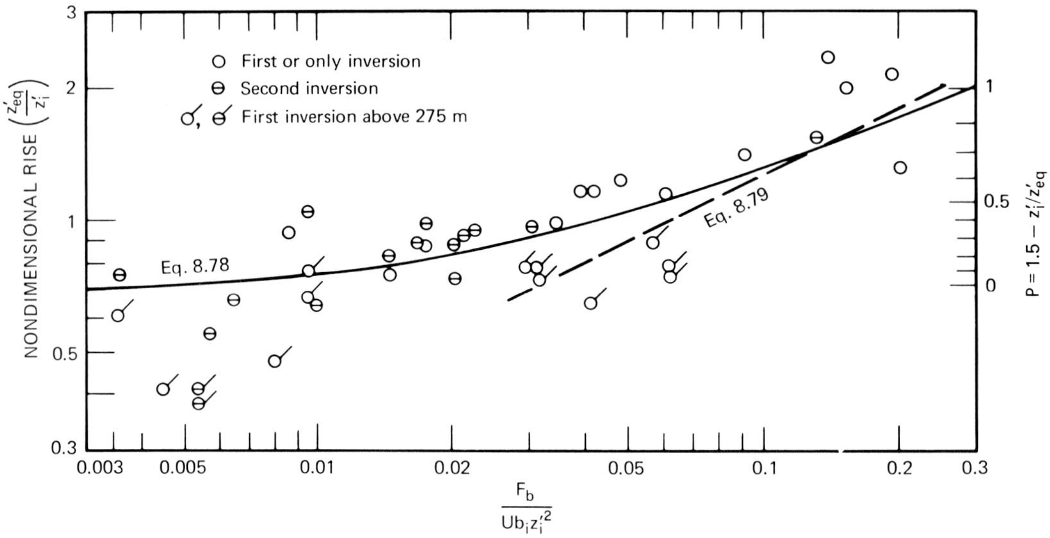

Fig. 8.7 Centerline rise divided by inversion height vs. nondimensionalized buoyancy flux [data from Simon and Proudfit (1967)]. Flags on the circles indicate $z'_i \geq 275$ m for the first inversion.

This model is compared in Fig. 8.7 with the data of Simon and Proudfit (1967) from the Ravenswood power plant in New York City. The heat emission used to compute F_b was reckoned from the load on only the largest unit. Two other stacks, each carrying 35% as much heat emission at full load, were operating. If any plume merger occurred, the buoyancy would be enhanced, but the plume would be widened also; the net effect in this case probably would be small. The stability of the air in between inversion layers is not reported, and this stability could significantly affect the final plume height. For most of the runs, two inversion layers were reported. When this layering occurred, $(b_i z_i'^2)$ was computed for both inversions and summed for the application of Eq. 8.78 to the upper inversion.

The figure shows fairly good agreement with Eq. 8.78, although the scatter is considerable, especially for runs in which $z'_i \geq 275$ m for the first inversion (flagged points). In cases where $z'_{eq} < 0.6z'_i$, the inversion probably has little influence on plume leveling; the air may be stable below the inversion. Complete, or nearly complete, penetration occurs when Eq. 8.76 is satisfied. A more conservative estimate that bypasses the need to know z'_i is given by the dashed line in Fig. 8.7; this estimate is expressed by

$$z'_{eq} = 4\left(\frac{F_b}{Ub_i}\right)^{1/2} \tag{8.79}$$

However, this equation applies only in the range $2/3 \leq z'_{eq}/z'_i \leq 2$.

Briggs (1975a) suggested a simple estimate of the fraction of the plume that penetrates the inversion, P. His estimate is based on the rectangular plume model:

$$P = 1.5 - \frac{z'_i}{z'_{eq}} \tag{8.80}$$

where, again $2/3 \leq z'_{eq}/z'_i \leq 2$. This assumption is untested. The remainder of the plume material $(1 - P)$ is assumed to diffuse downward and to reflect off the bottom of the inversion.

Recently, Manins (1979) completed an experiment that used a water channel to model partial penetration of an elevated inversion layer by buoyant plumes. The fraction of material trapped by the inversion, f, was defined by dividing the measured concentration integrated through the depth of the inversion layer with that integrated through the whole tank depth. Because the vertical concentration profile was measured only through the center of the stabilized plume, it is not known whether f correctly represents the distribution of the laterally integrated concentration. For f to correctly represent this distribution, the lateral dispersion would have to be the same above the inversion layer as it is within the layer, and that is unlikely. If the lateral dispersion is larger above the inversion, in the less stable fluid, f would be smaller than the value defined by this technique, and the penetration factor would be larger. The data show a fair amount of scatter but give a good fit to a simple theoretical formula developed by Manins, $f = 0.08(1 + p^{-1}) - p$, where $p = F_b/Ub_i z_i'^2$. If it is assumed that P is the same as $(1 - f)$, this result is less optimistic about penetration than Eqs. 8.78 and 8.80. How to resolve this discrepancy in results is not clear, especially since the quantity measured in the Ravenswood experiments, z'_{eq}, cannot be directly compared to the quantity measured in the modeling experiments.

8-4.3 Rise into an Elevated Stable Layer

Frequently, the lower atmosphere consists of a well-mixed, nearly adiabatic layer that is adjacent to the ground and is capped by a thick, stable layer. Examples are the mixed layer produced by daytime heating of the ground and marine inversions near coasts. Plume rise will be limited by the stable stratification of the upper layer; however, if the bottom of this layer is low enough, the plume can push completely into the stable layer. If the bottom of the layer is too high to allow complete penetration, the plume may still be able to partially penetrate it.

Let us approximate the stratification as neutral $(s = 0)$ below a height z'_b above the source and as a constant value of s above this height. A simple conservative approach is to calculate a predicted bottom elevation and a predicted top elevation of the stratified plume as if the upper stability, S, held throughout (because the air below z_b' is less stable, the actual rise will be greater). Then assume that the fraction of the plume above z_b penetrates the stable air, and the fraction below z_b is reflected downward.

Briggs (1975a) proposed a more complicated approach—based on a rectangular plume concept—to the problem. Although the method seems to be a more exact approach, it tends to overpredict the degree of penetration observed in the data cited in this section; on this basis, the simple approach outlined here seems preferable.

8-4.3.1 ***Vertical Plumes.*** With regard to vertical plumes, we can apply the approach of the preceding section to the results from the EPA Fluid Modeling Laboratory studies reported in Sec. 8-4.1.1. The bottom level for a stratified jet is $1.9\ (F_m/s)^{1/4}$ above the stack and the top level is $3.0\ (F_m/s)^{1/4}$ above the stack. If z_b' lies somewhere within this range, assume that the fraction of plume material penetrating the stable air is

$$P = \frac{(3.0 - z_b')/(F_m/S)^{1/4}}{1.1} \tag{8.81}$$

For a stratified buoyant plume, the bottom level is $4.7\ F_b^{1/4}s^{-3/8} - 3D$ and the top level is at $5.9\ F_b^{1/4}s^{-3/8} - 3D$. If z_b' lies somewhere within this range, assume that the fraction of plume material penetrating the stable air is

$$P = \frac{5.9 - (z_b' + 3D)/F_b^{1/4}S^{-3/8}}{1.2} \tag{8.82}$$

8-4.3.2 ***Bent-Over Plumes.*** No experimental information is available on rise of bent-over jets into stable air; however, bent-over jets are unlikely to have significant penetration abilities. Experimental measurements for bent-over buoyant plumes indicate that Eq. 8.71 gives good predictions of centerline rises. However, there has been little analysis of the bottom and top elevations of leveled-off plumes. Photographic observations of a rising plume indicate that the top and bottom are at about (1 ± 0.5) times the centerline height (Briggs, 1969; Bringfelt, 1969a). When these relative heights are used for a leveled plume, the method outlined in Sec. 8-4.3 predicts a penetration

$$P = \frac{3.9 - z_b'/(F_b/US)^{1/3}}{2.6} \tag{8.83}$$

$$P = 0 \text{ when } z_b' > 3.9(F_b/US)^{1/3}$$

$$P = 1.0 \text{ when } z_b' < 1.3(F_b/US)^{1/3}$$

Rough estimates of the percentage of plume material above z_b' are made here from inspection of the lidar cross sections of the Keystone plume reported by Johnson and Uthe (1969). On three days (Oct. 21, 22, and 24, 1968), periods occurred during which $z_b' > 0$ and adequate profile information existed. Determination of z_b' in this research is somewhat problematic. There was a terrain rise of between 69 m and 93 m between the plant site, where soundings were made, and the lidar site; this elevation difference is ignored, i.e., the altitude of the mixed layer is assumed constant. The profile determinations of

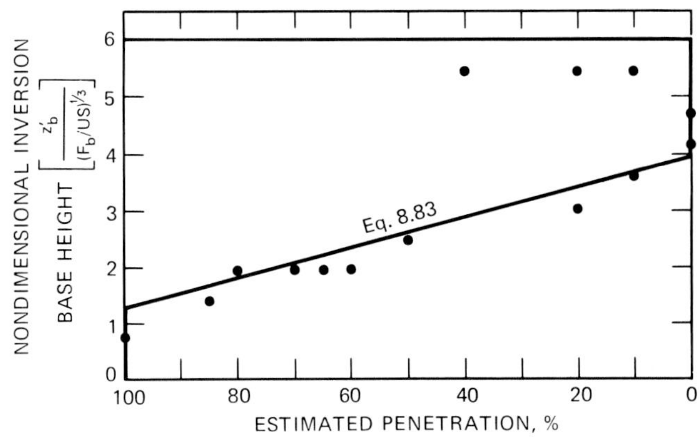

Fig. 8.8 Nondimensional inversion base height vs. estimated percentage penetration based on lidar cross sections [Keystone plume (Johnson and Uthe, 1969)].

mixed layer height for the three periods under consideration are fairly unambiguous; close agreement is exhibited among two or more of the following indices: minimum temperature, minimum wet-bulb temperature depression, top of wind-direction shear, and top of wind-speed shear. These data were determined from helicopter temperature and humidity profiles and from double theodolite pibals. The average of wind speed at the stack level and at the top of the plume is used for U. The stability parameter, S, was determined from the temperature difference between 245 m, at the stack top, and 650 m in an early morning sounding.

When Eq. 8.83 was used to calculate penetration and the results were compared with estimates shown in Fig. 8.8, Eq. 8.83 overestimated penetration by 10 to 15% in four cases; in three cases, Eq. 8.83 predicted no penetration, but 10 to 40% penetration was observed. For the remaining seven cases, the fit between the penetration calculated using Eq. 8.83 and the penetration observed is excellent. The more complicated model based on the rectangular plume given in Briggs (1975a) overpredicts penetration for these data.

8-4.4 Rise into Irregular Stability Profiles

Much more complex temperature profiles than those described so far sometimes occur. However complex, these profiles can be approximated with a series of end-to-end straight lines, i.e., by a series of layers with s constant in each layer. For the purpose of calculating plume rise, the fit to the actual profile can be approximate because the rise height generally depends only on the −⅓ or the −⅜ power of s. We can predict the approximate rise of the centerline of a buoyant plume through such stratification by integrating $dF/dz' = -sV_v$ or $dF/dz_c' = -sV'$ through each layer until $F = 0$; that is, we assume that the centerline rise equals approximately the equilibrium height. Let us designate $F = F_j$ when the plume reaches the bottom of the *jth* layer at $z' = z_j'$, and $s = s_j$ within this layer. For the zeroth layer, $z_0' = 0$ and $F_0 = \pi F_b$.

For a vertical plume we use the unaltered volume flux assumption given by $V_v = 0.04F_b^{1/3}z'^{5/3}$ (Sec. 8-4.2.1). Within the *jth* layer, we find that

$$F = F_j - 0.015s_jF_b^{1/3}\,(z'^{8/3} - z_j'^{8/3}) \tag{8.84}$$

We can use this equation to calculate F_{j+1} at $z' = z'_{j+1}$ until F goes negative. We find z'_{eq} in that layer by setting F = 0 and solving for $z' = z'_{eq}$ with Eq. 8.84.

For a bent-over plume, we use $V' = U(\beta' z'_c)^2$. When $\beta' = 0.4$, we get for the *jth* layer

$$F = F_j - 0.053 s_j U(z_c'^3 - z_j'^3) \tag{8.85}$$

The same procedure is used for finding $z'_c = z'_{eq}$.

If a question arises regarding whether to use Eq. 8.84 or Eq. 8.85 for a given layer, choose the equation that gives the greatest decrease in F through that layer. In general, transition from the vertical model to the bent-over model occurs at a height $z' \simeq 0.02F_b/U^3$, which is usually quite small.

8-5 RISE LIMITED BY AMBIENT TURBULENCE

8-5.1 Some Physical Considerations

8-5.1.1 Two Ways Turbulence Affects Effective Stack Height. When air is not stably stratified, it is easily stirred up by shearing motion or by heating from below. As a result, the air becomes filled with eddying motions that we call turbulence. These motions can limit the effective stack height of a rising plume in two different ways. First, if the motions are vigorous enough on a small scale, they can mix the plume with enough ambient air that both the buoyancy and the vertical momentum of the plume are greatly diluted. This causes the plume to approach an asymptotic height. Second, if large-scale turbulent eddies are vigorous enough, they might push some segments of the plume to the ground and cause a maximum ground concentration even before the mean plume centerline has started to level. In this case the maximum ground concentration averaged over an hour or so is a function of the plume rise at the distance that plume segments most frequently hit the ground, rather than of an asymptotic plume rise. Either of these two situations can dominate, depending on the source parameters and on the relative vigor of the large and small turbulent eddies.

8-5.1.2 Mechanical and Convective Turbulence. Turbulence has two different origins, and, as a result, displays different characteristics. Air blowing over and around roughness elements on the ground gets stirred up mechanically. The intensity of the turbulence depends on the wind speed and the size of the roughness elements (these include prominent terrain features, structures, and vegetation). Convective turbulence, on the other hand, is caused by ground-heated air expanding and becoming lighter than the air above it. This heated air rises while the heavier air above it descends. The intensity of convective turbulence depends predominantly on the rate at which sensible heat is transferred from the ground to the air and on the depth through which the overturning motions occur, called the mixing depth, z_i (the mixed layer often referred to here is defined by $0 \leqslant z \leqslant z_i$). The intensity of mechanical turbulence, especially for the small eddies, drops rapidly with height above the ground; hence the source height will affect plume rise when mechanical turbulence is the limiting factor. In contrast to this drop-off, the intensity of convective turbulence is relatively constant through most of the mixed layer.

8-5.2 The Breakup Model

The breakup model assumes that the mean plume rise effectively terminates before any significant fraction of the plume material reaches the ground by way of downdrafts. The termination of rise may occur slowly as ambient turbulence gradually breaks into the plume structure to dilute its buoyancy and upward momentum; or the plume may maintain its integrity for a fairly long time and then suddenly break up when the intensity of its internal turbulence decreases to below the level of the ambient turbulent intensity. No strong observational evidence is available to favor either of these views, but the latter scenario is more easily modeled and will serve adequately to predict ground concentrations.

A simple breakup assumption was given by Eq. 8.47, namely, that plume rise terminates when $\eta w_c^3/z'_c = \epsilon$, which is the ambient eddy dissipation rate; η was estimated to be 1.5, z'_c is the centerline rise of the bent-over plume in the rising stage, and $w_c = dz'_c/dt$. For a buoyant plume we substitute Eq. 8.58, with $x = Ut$, into Eq. 8.47 to get the termination time (differentiate Eq. 8.58 to get w_c); then we substitute the termination time back into Eq. 8.58 to get

$$\Delta h = \left(\frac{2}{3\beta^2}\right)^{3/5} \left(\frac{F_b}{U}\right)^{3/5} \left(\frac{\eta}{\epsilon}\right)^{2/5} \tag{8.86}$$

where $\Delta h = z'_c$ at the breakup point. For a jet we apply the same procedure to Eq. 8.56 to get

$$\Delta h = \beta^{-6/7} \left(\frac{F_m}{U}\right)^{3/7} \left(\frac{\eta}{\epsilon}\right)^{1/7} \tag{8.87}$$

Note the relatively weak dependence of Δh on η and ϵ, especially for the jet. This low sensitivity is fortunate because current estimates of these quantities are rather crude.

8-5.3 The Touchdown Model

In the touchdown model, we assume that large downward eddies bring a significant fraction of the plume material to the ground before the relative rise terminates. When we refer to relative rise, we mean the vertical displacement of the plume compared to that of the immediate ambient air, which itself may be rising or descending. We use Eq. 8.56 or Eq. 8.58 to describe the centerline relative rise, and we assume that this motion is superimposed on an average negative vertical velocity, w_d, in the ambient downdrafts. To compute the touchdown distance, $x = x_d$, we set

$$h_s + z'_b - w_d \frac{x}{U} = 0 \tag{8.88}$$

where z'_b is the relative rise of the bottom of the plume. This modeling concept is illustrated in Fig. 8.9. To get the mean plume rise at this distance, we substitute $x = x_d$ back into Eq. 8.56 or Eq. 8.58 for the centerline rise (the mean ambient vertical velocity is, of course, nearly zero). Because the plume is released into downdrafts only a fraction of the time, the mean rise, rather than the rise of plume segments caught in downdrafts, is appropriate for computing an effective source height for use in modeling the time-averaged ground concentration. The mean rise at $x = x_d$ is most appropriate for prediction of the maximum mean ground concentration.

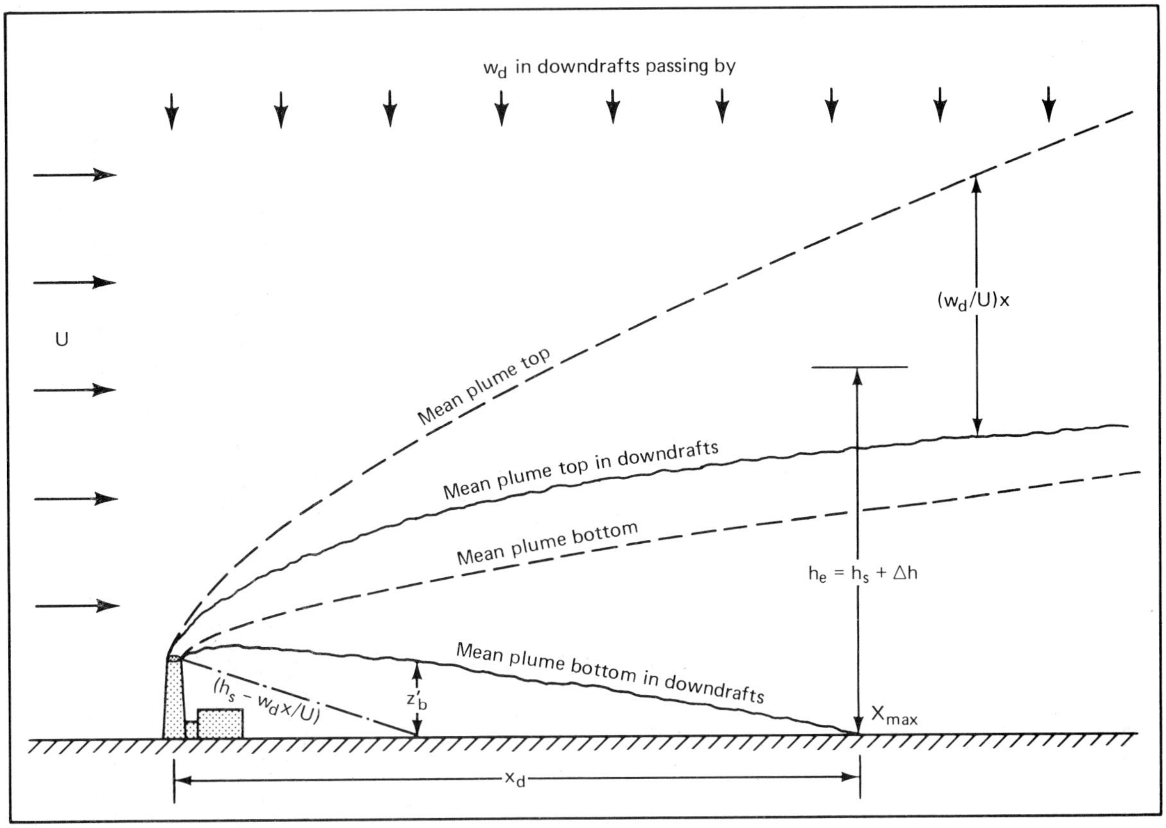

Fig. 8.9 Schematic drawing illustrating the "touchdown" model.

On the other hand, this model is easily adapted to predict the maximum instantaneous ground concentration. The internal radius of a plume segment is approximately $\beta' z_c' \simeq 0.4 z_c'$ for a buoyant plume. The volume dilution of an average plume segment is $\sim \pi U(0.4z_c')^2$, where z_c' is calculated from Eq. 8.58. Thus the instantaneous concentration in a plume segment is approximately $Q/\pi U(0.4z_c')^2$, where Q is the source release rate of the substance in question. The maximum such concentration to reach the ground occurs near $x = x_d$; therefore this is the distance to substitute into Eq. 8.58.

Photographic measurements for a buoyant plume indicate that $z_b' = 0.5 z_c'$ (Briggs, 1969; Bringfelt, 1969a). This substitution gives touch down at

$$h_s + 0.5 z_c' - w_d \frac{x}{U} = 0 \tag{8.89}$$

By substituting this value of x, which defines x_d, into Eq. 8.58 for z_c', we have Δh_d as the solution to

$$\frac{\Delta h_d}{h_s} = 1.6 \left(\frac{F_b}{U w_d^2 h_s} \right)^{1/3} \left(1 + 0.5 \frac{\Delta h_d}{h_s} \right)^{2/3} \tag{8.90}$$

where Δh_d is defined as z_c' at $x = x_d$.

With regard to a jet, little information is available on the height of the bottom or top of the plume as compared to the centerline. Photographs taken by Halitsky, Randell, and Hackman (1950), appear to indicate that the experiments with higher Reynolds number have a plume depth (top elevation minus bottom elevation) to z_c' ratio approximately equal to β in the early stage of rise, where $\beta = 0.4 + 1.2/R$ and $R = w_0/U$. If this geometry holds true for a jet in a downdraft so that $z_b' \simeq (1 - \beta/2) z_c'$, we have touch down at

$$h_s + \left(1 - \frac{\beta}{2} \right) z_c' - \frac{w_d x}{U} = 0 \tag{8.91}$$

By substituting this $x (= x_d)$ into Eq. 8.56 for z_c', we have $z_c' = \Delta h_d$ as the solution to

$$\frac{\Delta h_d}{h_s} = \left(\frac{3F_m}{\beta^2 U w_d h_s^2} \right)^{1/3} \left[1 + \left(1 - \frac{\beta}{2} \right) \frac{\Delta h_d}{h_s} \right]^{1/3} \tag{8.92}$$

Both Eq. 8.90 and Eq. 8.92 are cubic equations. However, some practical approximations can be made. For instance,

$$\Delta h_d \simeq 2.9 \left(\frac{F_b}{U w_d^2} \right)^{3/5} h_s^{2/5} \tag{8.93}$$

underestimates the value of $(h_s + \Delta h_d)$ in Eq. 8.90 by no more than 15% in the range $U = 0.1 F_b/(h_s w_d^2)$ to infinity; this range covers all practical windspeeds for sources for which the ground concentrations predicted by this model are significant, compared to those predicted for the neutral case. For jets,

$$\Delta h_d = 2.15 \beta^{-6/7} \left(1 - \frac{\beta}{2} \right)^{2/7} \left(\frac{F_m}{U w_d} \right)^{3/7} h_s^{1/7} \tag{8.94}$$

fits Eq. 8.92 very well; it underestimates the value of $(h_s + \Delta h_d)$ by no more than 10% for $U \geq 0.03(w_0^2/w_d)(r_0/h_s)^2$ [using a typical value of $(1 - \beta/2)^{2/3}\, \beta^{-2} \simeq 2.2$].

8-5.4 Rise Limited by Mechanical Turbulence

To predict plume rise in a turbulent atmosphere, we must estimate the turbulent eddy dissipation rate, ϵ, and typical downdraft velocities for the larger eddies, w_d. Turbulence quantities are most well-known in the neutral, mechanically driven, surface layer where they can be described by simple functions of the wind speed (see Chaps. 5 and 6). We will summarize these known relations, discuss their range of validity, and apply them to the estimation of plume rise.

8-5.4.1 ***Ambient Turbulent Eddy Dissipation Rate, ϵ, and Speed of Ambient Downdraft, w_d, in the Neutral Surface Layer.*** As described in Chap. 5, the primary quantity that scales mechanically-produced turbulent velocities in the surface layer is the friction velocity, u_*. Direct measurement of u_* requires fast-response, simultaneous measurement of u and w; however, u_* can easily be estimated from the mean wind speed, $\bar{u}$, at height z by using the long-established log profile law (this was derived in Chap. 5; see Eq. 5.61).

$$\bar{u} = \left(\frac{u_*}{k}\right) \ln \left(\frac{z}{z_0}\right) \tag{8.95}$$

where k is the von Kármán constant, z is the height above the ground, and z_0 is the roughness length characterizing the surface. Measurements indicate that the value of k is about 0.35 over very smooth surfaces and about 0.4 over more typical ground cover.

The roughness length, z_0, is of the order of 0.1 times the average height of protrusions from the surface; however, z_0 varies according to the spacing and shapes of the protrusions (z_0 is discussed in more detail in Chap. 5). Values of z_0 range from one centimeter or less for a short grass surface to one meter for a forest, and, for a highly developed urban area, values of z_0 can go up to four meters. For mixed countryside consisting of fields, woods, and scattered small buildings, z_0 is of the order of 0.3 m. Fortunately Eq. 8.95 is not very sensitive to the value of z_0; therefore an eyeball estimate usually is adequate for practical purposes. Table 8.1 gives an idea of the typical range of $\bar{u}/u_*$ based on Eq. 8.95.

TABLE 8.1 Ratios of $\bar{u}/u_*$ for Neutral Surface Layer

Site type	z_0	z = 10 m	z = 30 m	z = 100 m	z = 300 m
Urban	3 m	—	—	9	12
Surburban, forest	1 m	—	9	12	14
Mixed countryside	0.3 m	9	12	15	17
Tall grass, crops	0.1 m	12	15	18	21

The eddy dissipation rate in the neutral surface layer is given directly by

$$\epsilon = \frac{u_*^3}{kz} \tag{8.96}$$

Note the rapid decrease of ϵ with height.

The total variance (root-mean-square) of turbulent vertical velocities in the neutral surface layer is about $1.25u_*$ (Panofsky, 1973). Assessing w_d is more difficult because it represents only the downward velocities of the larger eddies; however, w_d will be some fraction of the total variance. If we assume that one-third to one-half of the turbulent energy is in an efficient eddy size range for displacing a plume downward, then $w_d = 0.7u_*$ to $0.9u_*$.

8-5.4.2 ***Vertical Extent of Surface Layer Relationships.*** The preceding relations apply to the surface layer, which is only the lowest part of the turbulent boundary layer (see Chap. 5, Sec. 5-2.5). The surface layer sometimes is defined as the layer in which the stress is within 20% of its surface value. Theory indicates that this height is proportional to u_* (Panofsky, 1973). For the Australian experiments reported by Clarke (1970), the depth of this layer in near-neutral cases was about 90 m, or u_* times 250 s.

The logarithmic velocity profile, Eq. 8.95, appears to hold to a somewhat greater height than that of 20% stress reduction. For the Dallas tower, with measurements to 430 m, the log law holds to 100 to 150 m in 5 m/s winds (Thuillier and Lappe, 1964); this height is 300 to 450 s times u_*. Clarke's (1970) observations show a good approximation to Eq. 8.95 to a height of 600 m, about 1700 s times u_*. Much of the discrepancy between these heights may be because Clarke used temperature gradient and wind speed to define his neutral class, whereas Thuillier and Lappe used only the temperature gradient.

Direct measurements of ϵ have not been made because turbulent energy dissipates into heat on very small scales; however, ϵ can be inferred from turbulence spectra or as the remainder in the energy balance equation. One instrument on the market indicates $\epsilon^{1/3}$ by using the high-frequency pressure fluctuations picked up by a pitot tube mounted on an airplane. Measurements by Herbert (1971) with this instrument showed that $\epsilon^{1/3}$ was approximately proportional to $z^{-1/3}$ in the layer from 300 m to 1200 m on cloudy days with $U \geq 7$ m/s (this should be neutral conditions). Clarke (1970) measured the mechanical production rate of turbulent kinetic energy, which equals ϵ if there is no advection or convective production. Clarke's measurements for near-neutral conditions show a good fit to Eq. 8.96 to over 1200 m.

The vertical velocity variances measured by Yokoyama (1971) averaged $1.16u_*$ through the maximum height of measurement, z = 313 m, where the local value of stress rather than the surface value is used to define u_*. Because stress induced by ground friction drops with height, the vertical velocity variance and w_d also drop with height. The Australian data indicate a 20% drop in u_* at about 160 m, or 450 s times u_*.

The foregoing information indicates that, in neutral conditions, surface-layer estimates of u_*/U, ϵ, and w_d are valid for low-level sources. For plume heights exceeding 100 m, the estimates are approximately valid only at high wind speed. We will show later that the neutral plume rise formulas apply to high sources only during high winds, except for brief transitional periods between stable and unstable conditions. Because little is known about the quantitative behavior of these quantities outside the surface layer, it seems reasonable to use the surface layer relationships at this time.

8-5.4.3 ***Plume Rise Estimates.*** To calculate the rise of a buoyant plume at breakup, we substitute Eq. 8.96 for ϵ into Eq. 8.86. Evaluating ϵ at the height of breakup, $z = h_s + \Delta h$, we find

$$\Delta h = 1.2\left(\frac{F_b}{U u_*^2}\right)^{3/5} (h_s + \Delta h)^{2/5} \tag{8.97}$$

where we have assumed $\beta=0.6$, $\eta=1.5$, and $k=0.4$. For simplicity we can approximate Eq. 8.97 by

$$\Delta h \simeq 1.54\left(\frac{F_b}{Uu_*^2}\right)^{2/3} h_s^{1/3} \tag{8.98}$$

As long as $\Delta h < 2h_s$, which usually is the case when rise in neutral conditions is of concern, this expression gives a conservative estimate of $h_e = h_s + \Delta h$ by no more than 10%; $\Delta h < 2h_s$ when U exceeds approximately $5(F_b/h_s)^{1/3}$. Equation 8.98 is not useful for gas turbines because, except in very high winds, $\Delta h > 2h_s$.

By comparing Eq. 8.97 with the touchdown plume rise model, we find that Eq. 8.90 predicts a smaller plume rise only if $\Delta h \gg h_s$ and $w_d > 0.88u_*$. Such a situation appears unlikely; consequently the breakup model represented by Eqs. 8.97 and 8.98 is the more conservative model, but only by a small margin. Both breakup and touchdown effects may be significant in this case.

To calculate the rise of a jet at breakup, we substitute Eq. 8.96 with $z = h_s + \Delta h$ into Eq. 8.87, so that

$$\Delta h = 0.93\left(\frac{F_m}{\beta^2 Uu_*}\right)^{3/7} (h_s + \Delta h)^{1/7} \tag{8.99}$$

where we have assumed $\eta=1.5$ and $k=0.4$; as for the rising stage of a jet, $\beta=0.4+1.2/R$ is suggested. By neglecting the effect of h_s, which is not very significant, we simplify Eq. 8.99 to

$$\Delta h \simeq \frac{0.9(U/u_*)^{1/2}F_m^{1/2}}{\beta U} \tag{8.100}$$

For a typical value $U/u_*=15$, Eq. 8.100 gives $\Delta h = 4.35DR/(1+3/R)$, which, in the range R = 1 to 32, is 0 to 50% higher than the early jet rise formula $\Delta h = DR^{1.4}$. The latter formula was proposed by Davidson in 1949 as an empirical fit to wind-tunnel data, but the turbulence conditions in the wind tunnel were not reported. When Eq. 8.99 is compared with the prediction of the touchdown model, we find that the breakup model gives a rise at least a factor of two lower; consequently the breakup model should be the more relevant model for a jet.

8-5.5 Rise Limited by Convective Turbulence

To estimate the effect of turbulence on plume rise, we need to estimate the eddy dissipation rate, ϵ, and an average downdraft velocity, w_d. Much has been learned in the past decade about turbulence in the convective boundary layer. Parameters most important in affecting convective turbulence are the mixing height above the ground, z_i, and the heat flux parameter, $H_* = (g/T)\overline{w'T'}$, which is equal to the sensible heat flux from the ground to the air times $(g/c_p\rho T)$. Hence the largest convective eddies scale to z_i and turbulence velocities scale to $w_* = (H_* z_i)^{1/3}$, the convective scale velocity (see Chap. 5, Sec. 5-4.3).

*8-5.5.1 **Ambient Turbulent Eddy Dissipation Rate, ϵ, and Speed of Ambient Downdraft, w_d, in the Convective Boundary Layer.*** The mean value of ϵ is relatively easy to estimate because theory shows that the rate of production of turbulent energy by heat convection is directly proportional to the sensible heat flux. At the surface, this rate equals H_*. Because heat flux drops off approximately linearly to zero at the top of the mixed layer, the mean production rate is approximately $0.5H_*$. Measurements show that, in most cases, ϵ is nearly constant with height in the convective boundary layer (this is not true close to the ground, where mechanical turbulence always dominates). For the Minnesota experiments, the measured $\bar{\epsilon}/H_* = 0.45$ to 0.85, with the higher values tending to occur in the late afternoon (Kaimal et al., 1976). Some unpublished measurements of $\bar{\epsilon}/H_*$ in East Tennessee in very convective, early afternoon conditions indicate a range of 0.23 to 0.72 with a mean value of 0.45 (Briggs, 1978). Overall, the experimental measurements indicate that $\bar{\epsilon}/H_* \simeq 0.5$ through the midday.

However, the portions of a plume that reach the ground in the highest concentrations are those brought down by downdrafts, and ϵ in downdrafts is less than $\bar{\epsilon}$ because of stable potential temperature gradients within downdrafts. Direct atmospheric measurements of ϵ in downdrafts alone are not available. However, Deardorff's numerical model of the convective boundary layer (Deardorff, 1974), which agrees well with observations of many of the turbulence characteristics, indicates that $\epsilon \simeq 0.5\bar{\epsilon}$ in the downdrafts in the middle of the mixed layer (Lamb, 1978). This value increases to $\epsilon \simeq 0.6\bar{\epsilon}$ in the lowest fourth of the mixed layer. For the purpose of determining Δh, we currently estimate that $\epsilon \simeq 0.5\bar{\epsilon} \simeq 0.25H_*$. This value is several times as large as the very tentative estimate suggested in Briggs (1975a).

With regard to the downdraft speeds, Deardorff's numerical experiment indicates that $w_d = 0.31w_*$ to $0.47w_*$ in the layer from $z/z_i = 0.1$ to 0.7 (Lamb, 1978); the earlier estimate that $w_d \simeq 0.4w_*$ appears to be a good compromise (Briggs, 1975a). Separate analysis of statistics pertaining to downdrafts has not been done in past field experiments, but the Minnesota experiments indicated total vertical velocity variances in the neighborhood of $0.6w_*$ in the middle of the mixed layer (Kaimal et al., 1976); two-thirds of this value seems a reasonable figure for w_d.

In the lowest tenth of the convective mixing layer, vertical velocities scale to $(H_* z)^{1/3}$. The numerical experiment gave a vertical velocity variance $\sigma_w \simeq 1.5(H_* z)^{1/3}$ (Lamb, 1978), whereas field measurements give $\sigma_w \simeq 1.3(H_* z)^{1/3}$ (Panofsky et al., 1977). Therefore, the downdraft velocities given by the numerical experiment might be slight overestimates; they were $w_d \simeq 0.9(H_* z)^{1/3}$.

*8-5.5.2 **Plume Rise Estimates.*** To calculate the rise of a buoyant plume at breakup, we substitute $\epsilon = 0.25H_*$ into Eq. 8.86. The result is

$$\Delta h = 3\left(\frac{F_b}{U}\right)^{3/5} H_*^{-2/5} \tag{8.101}$$

where we have assumed $\beta=0.6$ and $\eta=1.5$. A similar expression was suggested by Weil (1974a). If we compare Eq. 8.101 to the prediction of the approximate solution of the touchdown model, Eq. 8.93, we find that Eq. 8.101 gives the lowest rise as long as $w_d \leqslant 0.97(H_* h_s)^{1/3}$. This inequality is definitely the case when $h_s > 0.1z_i$ because $w_d \simeq 0.4w_*$ in the middle of the mixed layer. This inequality probably is the case in the lowest tenth of the mixing layer also because we estimate that $w_d \leqslant 0.9(H_* z)^{1/3}$. Once again the breakup and the touchdown models give similar estimates of Δh; this means that both mechanisms may significantly affect the rise. However, field data contain too much scatter to validate all the elements of a complex modeling approach that includes several mechanisms.

Equation 8.101 is recommended as simple and, perhaps, slightly conservative. Equation 8.101 also is more convenient to use than the touchdown model because it does not require knowing the mixing heights [nonetheless, Briggs (1980) concludes that in the daytime one must use some form of convective scaling, including z_i, to successfully model diffusion at distances greater than 10 min times U]. However, the touchdown model is, at this point, still an equally viable approach.

Venkatram (1980) has used an equation like Eq. 8.88, along with assumptions about the probability distribution of the impact point, in a model that predicts maximum average concentration vs. distance for elevated buoyant sources. The diffusion coefficients in this model are also based on application of research results using convective scaling, w_* and z_i. Venkatram compared the concentrations predicted by this model with those measured at two power plants in Maryland (Weil, 1977); on the basis of the usual diffusion modeling criterion that a prediction within a factor of two, for a 1-h average, is a successful prediction, the comparison was good.

More recently, Weil (1980a) tested Venkatram's model against a broader data base. The model worked quite well and compared favorably with the best of conventional Gaussian models for data from four of the six power plants. Venkatram's model, when applied to the plant that experienced very low wind speeds (less than 1 m/s), predicted concentrations nearly a factor of two higher than those shown by the data. Weil suggested that some penetration of the stable air above the mixed layer may have occurred; such penetration would account for the model's overprediction. At the largest plant, for which h_s = 305 m, the model predicted concentrations approximately a factor of ten lower than those shown by the data. By using trajectory calculations, Weil concluded that this plume was so buoyant that it bumped the top of the mixed layer before touch down occurred; he suggested a correction to Venkatram's model that successfully brought the predictions in line with the data.

Note that, in connection with Gaussian modeling, the touchdown model gives results very similar to those of Gaussian models if it is assumed that the plume trajectory follows the 2/3 law, Eq. 8.58, and that $\sigma_z \propto w^*(x/U)$. Weil (1979) has proposed a model that uses these assumptions along with more elaborate assumptions.

To calculate the rise of a jet at breakup, we substitute $\epsilon = 0.25H$ into Eq. 8.87:

$$\Delta h = 1.3\left(\frac{0.4 + 1.2}{R_s}\right)^{-6/7} \left(\frac{F_m}{U}\right)^{3/7} H_*^{-1/7} \tag{8.102}$$

By comparing this equation with the downdraft model for jets, Eq. 8.92, we find at least 35% less rise; therefore the breakup model definitely is more conservative for a jet.

The following suggestions on estimating H_*, which is $g/(c_p\rho_aT_a)$ times the sensible heat flux at the surface, H, are included to aid in making plume rise calculations for convective conditions. Briggs (1975a), citing data given by Weil (1974a), suggested that over a wide range of surface conditions $H = 0.4 \pm 0.15$ times R_s, the incoming solar radiation. The lower values of H/R_s tend to correspond to high evaporation surfaces such as well-watered crops. On the other hand, Weil (1977) used $H = 0.3R_s$, and Weil and Hoult (1973) used $H = 0.37R_s$. When the small exponents of H_* that appear in Eqs. 8.101 and 8.102 are considered, the difference in these estimates is not critical. Incoming solar radiation, R_s, is fairly easy to measure on-site with a pyrheliometer; the pyranometer (see Chap. 3, Sec. 3-6.1.1) is preferred because it measures global solar radiation. Briggs (1975a), lacking measurements such as those produced by the pyranometer, proposed using $R_s = (2/3) \sin \Theta_s (1 - 0.8C)S$, where Θ_s is the solar elevation angle, C is the fractional cloudiness, and S is the solar constant. The solar constant, S, times $g/(c_p\rho_aT_a)$ is 0.039 m^2/s^3 times 1013 mb/p_a (except at high altitudes, the latter ratio can be taken as unity.) A typical value of H_* on a clear summer day, when convective conditions are very likely to dominate, is 0.007 m^2/s^3. In dry conditions, $H_* \simeq 0.01$ m^2/s^3 can occur at most latitudes in summer. Values of H_* in deserts could be somewhat higher.

One caution needs to be made about applying convective plume rise formulas to low wind-speed conditions, especially for large buoyancy sources. In such cases, Eq. 8.101 or one of the downdraft models occasionally will predict a higher mean rise than the mixed layer height. However, because z_i usually exceeds h_s by some hundreds of meters, plumes rarely are able to penetrate above z_i significantly (to estimate penetration ability, refer to Sec. 8-4.3, and set $z_b' = z_i - h_s$). Thus z_i usually acts as a lid on the plume and suppresses plume rise. The limiting of plume rise in this manner apparently was observed in 4 out of 11 cases studied by Weil (1979), who was using lidar to map vertical profiles of the Morgantown power plant plume. He determined empirically that the centerline rise of these plumes was

$$\Delta h = 0.62(z_i - h_s) \tag{8.103}$$

for the single-stack cases. When two stacks were operating, he observed 20% greater rise (this was only one period of observation).

Information about climatological average values of z_i in the contiguous United States can be found in Holzworth (1972). In the summer these values range from about one to four kilometers at midday.

8-5.6 Some Practical Consequences

8-5.6.1 Mechanical-Stable Transition. The simplest way to estimate the transition point between one factor for limiting plume rise and another is to set the two appropriate plume rise formulas equal to each other. Setting Eq. 8.97 for high rise ($\Delta h \gg h_s$) equal to Eq. 8.71 for bent-over buoyant plumes in stable conditions yields transition at $U = 0.77 \times (U/u_*)_n^{3/4} F_b^{1/4} s^{1/8}$ [subscript n refers to the neutral profile, i.e., $(U/u_*)_n = k^{-1} \ln (z/z_0)$]. Interestingly, this speed is about 50 times the transition wind speed between the calm and bent-over stable cases, determined by the same method. In the event of $(U/u_*)_n = 15$ and an isothermal temperature profile, we find that transition from the stable to the neutral cases occurs at $U = 2.2F_b^{1/4}$, in mks units. For the low rise case, we use Eq. 8.98 instead of Eq. 8.97; this gives transition at $U = 0.73(U/u_*)_n^{4/5}(F_bh_s)^{1/5}$. For most sources this speed differs very little from the high-rise transition estimate. For large sources, these transition wind speeds are fairly high, and therefore neutral or mechanical turbulence limiting of plume rise at night is infrequent.

The transition point also can be expressed as a ratio of the Monin-Obukov length, $L = -u_*^3/kH_*$, to the effective stack height, $h_e = h_s + \Delta h$. By using the surface layer relation $s = (H_*/ku_*z)(0.74 + 4.7z/L)$ (Businger et al., 1971), we find that for large $\Delta h/h_s$, transition to neutral plume rise occurs when $L \geqslant 1.94(h_s + \Delta h)$. For $\Delta h/h_s = 1$, which is more typical, transition occurs when $L > 1.16(h_s + \Delta h)$. This criterion is not as practical as the one above, because L is hard to measure; however, it is satisfying in that L is a measure of the depth of mechanical turbulence undamped by the stabilizing effect of ground cooling.

8-5.6.2 ***Mechanical-Convective Transition.*** Because the breakup model may be the more conservative model for both mechanical and convective turbulence, transition between these two cases occurs simply when the respective values of ϵ are equal. According to the estimates used in Secs. 8-5.4.1 and 8-5.4.2, this transition occurs when $0.25H_* = u_*^3/kh_e$, or when $-L = 0.25h_e$, where $h_e = h_s + \Delta h$.

A practical estimate of the transition point for the high-rise case shows that transition occurs at $U = 0.7 \times (U/u_*)_n^{5/6} (F_bH_*)^{1/6}$. We then expect that, for the typical values $U/u_* = 15$ and $H_* = 0.005 m^2/s^3$, mechanical turbulence will dominate when $U \geqslant 2.8F_f^{1/6}$ (in mks units). Transition occurs at $u_* = 0.63(kH_*h_s)^{1/3}$ in the low-rise case; and, for the typical values cited in the high-rise case, mechanical turbulence will dominate when $U \geqslant 1.2h_s^{1/3}$. For a given source, the highest of these two estimates is the most relevant.

8-5.6.3 ***Maximum Ground Concentration (MGC).*** The simplest way to estimate the MGC is to assume that the plume expands symmetrically around its centerline at a height h_e above the ground; if the concentration is uniform within the plume cross section, the volume dilution at the moment of ground contact would be $U(\pi h_e^2)$; therefore the concentration would be $0.32Q/(Uh_e^2)$, where Q is the source strength. More often, however, a bi-Gaussian concentration distribution is assumed. With a constant ratio of vertical-to-lateral diffusion dimensions, σ_z/σ_y, the greatest ground concentration is

$$X_{max} = \frac{0.234(\sigma_z/\sigma_y)Q}{(Uh_e^2)} \tag{8.104}$$

There is some debate about whether σ_z/σ_y is constant or not, and the variability of σ_z/σ_y with distance affects the power of h_e derived in Eq. 8.104. However, in neutral and unstable conditions, the large turbulent eddies are not tall and skinny, nor are they flattened, as eddies might be when strong stable stratification is present. Because of this geometry, σ_z/σ_y can never be far from unity, and Eq. 8.104 should be approximately correct.

When $h_e = h_s + \Delta h$, and Δh is a function of wind speed, we find relatively low MGC at both fast wind speed—because of enhanced axial diffusion—and at slow wind speed—because the large Δh enhances the radial diffusion. The highest MGC, X_c, occurs at some intermediate wind speed, U_c, called the critical wind speed. In general, U_c can be determined for a given plume rise equation by setting the derivative, with respect to U, of $(Uh_e^2) = U(h_s + \Delta h)^2$ equal to zero. For instance, if $\Delta h \propto U^{-n}$, we find that the minimum dilution occurs at the wind speed that $\Delta h = h_s/(2n - 1)$. This value of Δh and the corresponding U are substituted back into Eq. 8.104 to predict X_c.

For the neutral case, this procedure can be applied to Eq. 8.97 by taking the total derivative of both sides while holding F_b, h_s, and U/u_* constant, substituting $d(Uh_e)^2 = h_e^2\, dU + 2Uh_e\, d\Delta h = 0$. All the differentials cancel, and the result is $\Delta h = (1/3)h_s$ at the critical wind speed (this is also true for the approximate rise formula, Eq. 8.98; it was chosen to give the same U_c and X_c). This gives

$$U_c = 2.15\left(\frac{U}{u_*}\right)^{2/3}\left(\frac{F_b}{h_s}\right)^{1/3} \tag{8.105}$$

$$X_c = \frac{0.043(u_*/U)^{2/3}Q}{F_b^{1/3}h_s^{5/3}} \tag{8.106}$$

where we have used $\sigma_z/\sigma_y \simeq 0.7$ for the neutral case (ASME, 1973). With $U/u_* \simeq 15$, Eq. 8.106 is consistent with observed MGC's in high-wind conditions at several Tennessee Valley Authority and Central Electricity Generating Board power plants (Briggs, 1975a, Table 7).

Because Eq. 8.101 for Δh is a simple power law, the procedure for the unstable case is straightforward. At the critical wind speed, $\Delta h/h_s = 5$, which is very much larger than it is in the neutral case. We then find that

$$U_c = 0.43F_bH_*^{-2/3}\, h_s^{-5/3} \tag{8.107}$$

$$X_c = \frac{0.015QH_*^{2/3}}{F_bh_s^{1/3}} \tag{8.108}$$

where we have used $\sigma_z/\sigma_y \simeq 1.0$ for the unstable case (ASME, 1973). Two things are remarkable about this solution. First, if $Q \propto F_b$, increased source strength has no effect on X_c. Second, increasing h_s reduces X_c, but it has much less effect than it does in the neutral case. However, increasing h_s may reduce U_c to a very small value so that the critical wind speed is never reached in practice. Also, note from Eq. 8.103 that if $5h_s > 0.62(z_i - h_s)$, or if $z_i < 9h_s$, the plume bumps the top of the mixed layer at a U value greater than that given by Eq. 8.107; thus Eq. 8.108 is invalidated. (To find the wind-speed value below which bumping occurs set Eq. 8.101 for Δh equal to Eq. 8.103 for Δh.)

In the bumping situation with no significant penetration of the overlying stable layer, there is no critical wind speed because X_{max} increases monotonically with decreasing U; according to Eq. 8.103, $(h_s + \Delta h)$ is limited to $(0.38h_s + 0.62z_i)$; therefore $X_{max} \propto U^{-1}$. However, because of the horizontal component of convective turbulent velocities, there is a minimum to the possible effective value of U in convective conditions. This minimum should be of the order of σ_u and σ_v, or about $0.6w_*$. By setting the bumping speed equal to $0.6w_*$, we can assume that, unless F_b exceeds $0.04H_*z_i^2(1 - h_s/z_i)^{5/3}$, bumping will rarely or never occur. Bumping and the consequent restricted dilution are more likely for smaller values of z_i; however, as $(z_i - h_s)$ becomes smaller, a larger fraction of plume material can penetrate into the overlying stable layer (see Sec. 8-4.3.2); consequently the amount of trapping possible is limited.

Another quantity of interest is the relative value of X_c and U_c for the neutral and unstable cases. We have, in the absence of bumping,

$$\frac{(X_c)_{unstable}}{(X_c)_{neutral}} = 0.35H^{2/3} \tag{8.109}$$

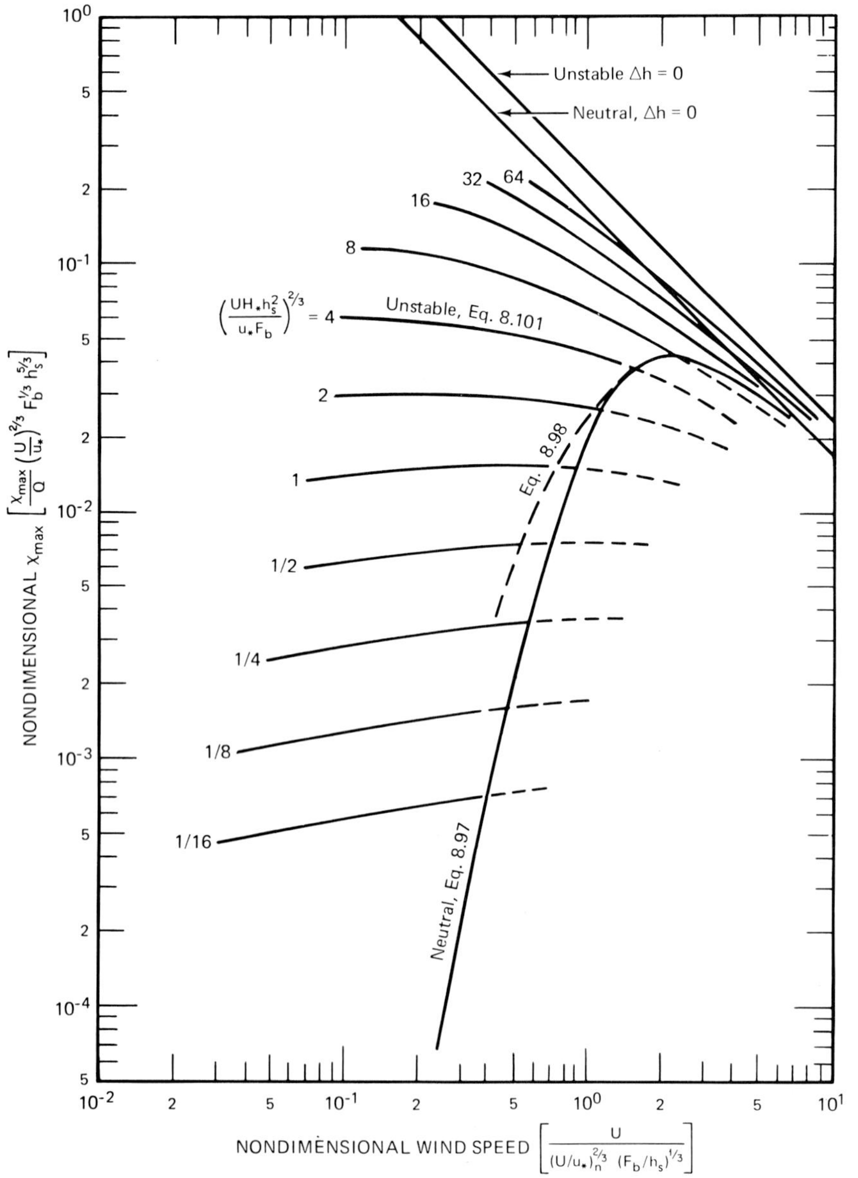

Fig. 8.10 Nondimensional maximum ground concentration vs. nondimensional wind speed predicted by Eq. 8.103 with various plume-rise equations for neutral and unstable conditions. The assumption is made that $\sigma_z/\sigma_y = 1.0$ for unstable conditions and 0.7 for neutral conditions.

$$\frac{(U_c)_{\text{unstable}}}{(U_c)_{\text{neutral}}} = 0.20H^{-2/3} \tag{8.110}$$

where $H = (U/u_*)_n H_* h_s^2/F_b$, a dimensionless heat flux parameter. Note that $H_* h_s^2/F_b$ is the ratio of buoyancy emitted by the ground within a radius equal to h_s to the buoyancy emitted by the stack; this ratio is a natural measure of relative effectiveness. When this ratio is larger than about 0.3, the unstable case is potentially worse than the neutral. However, U_c in this case is well over an order of magnitude smaller than U_c for the neutral case; for some sources it may be less than 1 m/s. Such slow speeds never are reached in convective conditions because the convective turbulent velocities themselves are of the order of 1 m/s.

Relative values of the MGC predicted for the neutral and unstable cases are shown as a function of nondimensionalized wind speed in Fig. 8.10. The MGC, as predicted by Eq. 8.104, also is nondimensionalized so that the neutral case is represented by a single curve for all sources. For the unstable case, a family of curves represents different values of H. To make the comparison realistic, we have limited the unstable curves in Fig. 8.10 to actual ranges of nondimensional U likely to be encountered by sources having the given value of H in summer conditions; this range was determined by first plotting $(F_b/h_s)^{1/3}$ vs. (F_b/h_s^2) for a variety of sources and then assuming that U itself ranges from 1 to 20 m/s. The unstable curves are dashed as they pass through the neutral curve because higher wind speeds are in the neutral domain. The degree to which a X_{max} curve drops below the no-plume-rise curve represents $(h_e/h_s)^2$; therefore plume rise is indirectly represented in the figure.

A typical value of $(U/u_*)_n H_*$ in summer conditions is 0.1 m^2/s^3. In such conditions, $H^{2/3}$ typically ranges from 2 to 20 for process emissions, from 2 to 4 for power plants, and from 0.07 to 0.5 for gas turbines. In the latter case, convective conditions are of no particular concern because the plume rise is many times h_s. For power plants, X_{max} in slow-wind-speed, convective conditions may be of the same order as X_{max} in fast-wind-speed, neutral conditions (see Sec. 8-5.7).

8-5.7 Comparison with Observations

Direct verification of any plume rise formula for neutral and unstable conditions is difficult. Photographs of plumes almost invariably show upward inclined plume trajectories at the distance where the plume becomes indistinguishable from the background, at least when a number of such trajectories are averaged. In unstable conditions, looping caused by up and down convective motions confuses the issue of whether or not the plume has leveled in the mean. Adequate measurement requires that a sufficient number (about a half dozen) of plume cross sections be taken and averaged at a distance near the maximum ground concentration while meteorological variables are simultaneously measured. Lidar is the most promising tool to map plume cross sections. Complete vertical profiles of temperature especially are needed because buoyant plumes from large sources often bump the top of the mixed layer at slow wind speeds. In such a case, turbulence is not the limiting factor.

Eleven periods of lidar observations were made at the Morgantown, Maryland, power plant, which has two stacks that are each 213 m high. Data from the observations were analyzed in some detail by Weil (1979), and, as was mentioned in Sec. 8-5.5.2, he concluded that during four of these periods the plume rise was limited by the capping effect of the stable air above the mixed layer, i.e., by bumping. In three of these periods, with only one stack operating, Eq. 8.103 was accurate within 5%. In another four of the periods, the plume appeared still to be rising at the greatest distance of the lidar cross sections, which ranged from 2.5 to 4.6 km. During only one of these 11 periods, all of which occurred in early autumn, did the plume rise appear to be limited by convective turbulence; in this period, the observed rise exceeded the prediction of Eq. 8.101 by 35%. However, the accuracy of the wind-speed measurement made during this run is questionable (Weil, 1980b). The rise during this period also exceeded the ⅔ law prediction, Eq. 8.58, by 34%; this suggests that U was overestimated. During the remaining two periods, the plume rise appeared to be limited by mechanical turbulence; U averaged near 12 and 14 m/s during these periods. For the period with a single stack operating, the observed rise was only 2% greater than that predicted by Eq. 8.97. For the period with two stacks operating, the rise was 16% greater than that predicted by Eq. 8.97 if the buoyancy from only one stack is used to calculate F_b. Briggs's (1974) method (see Sec. 8-6.1.1), when applied to this period, predicts a 13% rise enhancement as a result of two stacks operating.

Another experimental approach is to model plume rise in a turbulent ambient fluid in a wind tunnel or water channel. This technique has been successfully used for final rise of buoyant plumes in stable ambient fluids (see, for instance, Fan, 1967, or Hewett, Fay and Hoult, 1971). More recently, Wilson (1979) observed leveling of high-momentum, low-buoyancy plumes in a wind tunnel in neutral stratification, i.e., the plume rise evidently terminated because of ambient mechanical turbulence. This work was still in progress at the time of this writing. In principle, the technique also could be applied to plume rise in convective turbulence, but the scaling requirements are more difficult. To properly scale buoyancy, the researcher must maintain the same Richardson number or Froude number (see Chap. 5, Sec. 5-1.10) as occurs at full scale. In practice, because gravity cannot easily be adjusted for, this means that, when length scales are reduced by large factors, velocities must be reduced in proportion to the square-root of the length scale reduction factor. This reduction of velocity and length scales reduces the flow Reynolds numbers by a comparatively large factor; such a reduction means that the range of turbulent eddy scales is reduced proportionately. If the Reynolds numbers are too greatly reduced, the turbulence does not behave like full-scale turbulence; indeed, there may not even be turbulence flow, only laminar flow. To increase the Reynolds number, we must considerably increase density differences or gravity (such as by using a centrifuge), or we must increase the scale by using a larger facility.

An alternative to direct measurement of Δh is to test whether the average ground concentration is correctly predicted by using a given plume rise formula. Of course, this procedure involves diffusion modeling as well as plume rise modeling; consequently it depends on the correctness of both model components. The easiest case to test is for X_{max} because this is given in an approximate way by Eq. 8.104. Rather than designating σ_z and σ_y as functions of distance, only the ratio σ_z/σ_y needs to be specified. However, this method does require a rather extensive network of ground samplers, so that X_{max} is picked up.

This approach is employed in Fig. 8.11 to test our plume rise equations by using the X_{max} data of Moore (1974a) from the Northfleet and Tilbury power plants. The data are nondimensionalized for plotting so that data from both plants can be compared. Basically thirty-minute average values of X_{max} are reported for a number of wind-speed categories. Mean plant load and SO_2 emission also are reported for each category; these data are divided into unstable, slightly stable, and stable categories that are based on the temperature gradient on a nearby meteorological tower. Other quantities have to be estimated. By using $z_0 = 0.2$ m and the mean of the two stack heights (150 and 100 m), it is estimated that $(U/u_*)_n \simeq 2.5 \ln(h_s/z_0) \simeq 16$.

No measurements of H_* were obtained; however, by comparing the minimum-temperature lapse rate that defined the unstable category with lapse rates measured in the Minnesota experiment (Izumi and Caughey, 1976), it is estimated that $H_* \geqslant 25$ cm^2/s^3 for this category. By assuming a heat flux equal to 0.37 times the incoming radiation (Weil and Hoult, 1973), we estimate that the maximum H_* for a clear summer day in Great Britain is about 75 cm^2/s^3. Therefore to encompass the probable range of actual values, h_e for both $H_* = 25$ and 75 cm^2/s^3 is plotted in Fig. 8.11b.

Both plants had two stacks in operation; plume rise was calculated on the basis of heat emission from only one stack. This was accomplished by assuming Q_H equal to 1/6 of the load, whereas the total Q for both stacks was used. If any plume rise enhancement occurred as a result of merging of the two plumes, it would tend to lower the observed concentration and to increase the effective stack height. A 30-m bluff near the plant at Northfleet has a detrimental effect on the plume height. To account for the bluff, Hamilton (1967) suggested subtracting 30 m from the actual height. In this chapter we have taken a compromise position and made all calculations as if the stack height was 135 m, which is 15 m less than the actual stack height.

At low wind speeds, X_{max}/Q is much smaller for slightly stable cases than for unstable cases; only the unstable points are shown in Fig. 8.11 for $U < 6$ m/s. However, at 7 m/s the values for both stability categories come together strikingly in spite of the different lapse rate ranges; this suggests that the neutral regime is operative. On the basis of Sec. 8-5.6.1, we would estimate transition from stable to neutral at $U \simeq 7$ to 9 m/s for these two plants, whereas on the basis of Sec. 8-5.6.2,

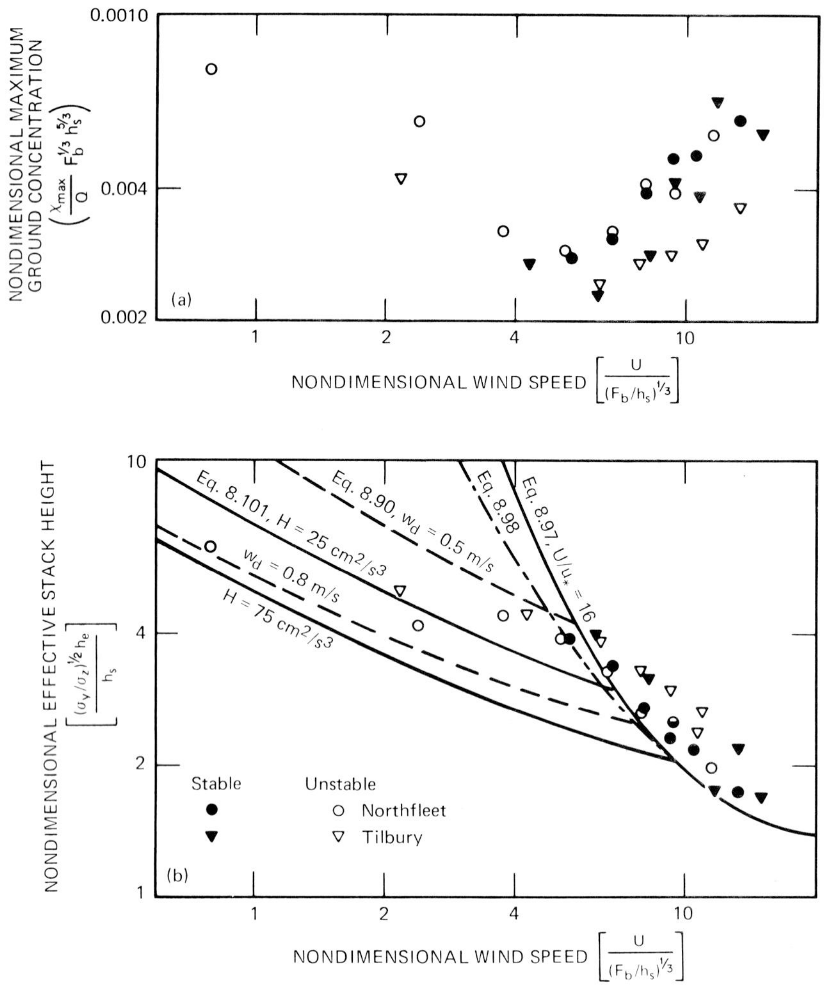

Fig. 8.11 Maximum ground concentration data of Moore (1974a) nondimensionalized by buoyancy flux and stack height (Northfleet stack height modified to 135 m). Effective stack height was determined from Eq. 8.104. The assumption is made that σ_z/σ_y = 1.0 for unstable conditions and 0.7 for neutral conditions.

we would estimate transition from unstable to neutral at $U = 6$ to 8 m/s; these estimates are consistent with the observations. At the higher wind speeds, points for both stability categories are clustered together, except for those for Tilbury unstable; for an unknown reason, the (X_{max}/Q) values are lower than the others. Points other than those for Tilbury unstable show a definite minimum in the vicinity of $U = 5(F_b/h_s)^{1/3} \simeq 6$ m/s with progressively higher values of X_{max}/Q in the higher wind-speed categories (U ranged from 1 to 18 m/s).

In Fig. 8.11b, the same X_{max}/Q data are transformed into a plot of h_e vs. U by using Eq. 8.104. The most striking feature of this plot is the abrupt kink in the points at $U = 6(F_b/h_s)^{1/3} \simeq 7$ m/s. This result suggests a transition between the unstable and neutral regimes at this point. As the plume rise formulas suggest, h_e has a much weaker dependence on U at slower wind speeds than at faster wind speeds. Equation 8.97 for neutral conditions fits the trend at $U \geqslant 7$ m/s very well and stays just on the low side of the data points. The simpler approximation, Eq. 8.98, fits just as well in this range. Excluding Tilbury unstable points, the equations appear to give a Δh about 20% lower than observed. Given the limitations of the data, especially the presence of two stacks, 1.2 is a reasonable factor of safety.

It is harder to evaluate the performance of the unstable rise formulas with these data because convective parameters were not measured. Equation 8.101 with $H_* = 25$ to 75 cm^2/s^3 produces a broad band, about a factor of 1.4 wide, which nearly encompasses the three points for $U < 5$ m/s. However, it does not encompass the substantial dip in X_{max}/Q that data from both plants show in the range $U = 5$ to 7 m/s. It is possible that these wind speeds are associated with more cloudiness and, therefore, smaller H_* than the slow wind speeds; even so, Eq. 8.101 may be overly conservative.

The range of predictions given by the touchdown model, Eq. 8.90, with $w_d = 0.5$ to 0.8 m/s, is shown by the dashed lines. (This range is based on $w_d = 0.4w_*$ and $w_*^3 = H_* z_i = 2$ to 8 m^3/s^3.) The band of predicted values is broader than that derived with Eq. 8.101; however, the curves have very much the same shape and they encompass all the points for which $U \leqslant 7$ m/s. Note that the point at the slowest wind speed, Northfleet at 1 m/s, represents a condition where the plume is likely to be bumping the top of the mixed layer. By taking $\Delta h = 2.4F_b^{1/3}U^{-1}x^{2/3} - (w_d/U)x$ for the top of the plume, we find a maximum Δh of $6F_b/(Uw_d^2)$; this exceeds 1000 m at Northfleet if $U < 2.5$ m/s and $w_d = 0.8$ m/s. Thus, at $U = 1$ m/s, the plume is likely to be restricted by the mixed layer lid, which increases X_{max} and decreases h_e. About all that can be concluded for the unstable case is that Eq. 8.101 is reasonably accurate but is probably conservative, perhaps by a factor of 1.2 to 1.5. Data with concurrent measurements of H_* or solar radiation and z_i are needed to properly verify the theoretical formulas available for this case.

8-6 SOME SPECIAL TYPES OF SOURCES

Several types of sources may produce plume rises substantially different from the predictions of the foregoing sections because of phenomena not accounted for so far. If two or more sources are in very close proximity, the plumes may merge and, thereby, increase the total vertical momentum and buoyancy present in one plume; this usually increases the plume rise over what the same sources would exhibit when isolated from one another. Plume rise from multiple sources is treated in Sec. 8-6.1.

If the effluent is emitted vertically from a stack too slowly, some or all of the effluent may be drawn into the low-pressure region on the lee side of the stack and, consequently, reduce plume rise. Called stack downwash, this situation is treated briefly in Sec. 8-6.2.

It has been suggested that the plume rise from gas turbines might require treatment different from that of conventional sources because of very high temperatures and, in some cases, unusually high exit velocities at the source. This possibility is discussed in Sec. 8-6.3.

Some rather elaborate models have been constructed for predicting the rise enhancement resulting from latent heat release in plumes containing large quantities of water vapor. The amount of rise enhancement caused by this effect is discussed in Sec. 8-6.4.

8-6.1 Plume Rise From Multiple Sources

In the past many power plants were constructed with lines of closely spaced stacks, each stack being fed by one or two generating units. Although many of these plants are still in operation, the trend in newer plants has been to build fewer and taller stacks or even to build just one very tall stack. By constructing power plants in this manner, ground concentrations are reduced by combining all the buoyancy to achieve a higher plume rise. Another example of a multiple source is two or more cooling towers in close proximity to one another; this construction is becoming more common. Process emissions sources that involve multiple stacks provide another example of the multiple source. The line source, such as a roof-peak vent or the active burn line of a forest fire, is another type of source that can be treated like a line of closely-spaced, individual sources. Extending the plume rise predictions for single sources to the case where plumes merge with one another and thereby combine their momentum and buoyancy is desirable for all these sources.

8-6.1.1 Briggs's Empirical Approach. The first attempt to calculate rise for multiple sources was by Briggs (1974), who used an empirical approach guided by theoretical limits. The data used were plume rises measured by TVA (Carpenter, Thomas, and Gartrell, 1968) at power plants with one, two, three, four, or nine equally spaced stacks operating in a line configuration. These sources were strongly buoyant. Nearly all the observations were made on bent-over plumes before leveling occurred or after leveling in stable conditions; consequently the observed rises were compared with either Eq. 8.58, the 2/3 law, or Eq. 8.71, for rise of buoyant plumes in stable stratification. In both of these equations, the plume rise for a single source, Δh_1, is proportional to $F_b^{1/3}$, where F_b is the buoyancy flux from one stack; therefore the maximum rise enhancement factor for N stacks would be $N^{1/3}$ if all the buoyancy completely combined. It was reasoned that the degree of plume merger, and, hence, of plume rise enhancement, should depend in some way on the ratio of Δh_1 to the spacing between the stacks, Δs. The observed plume rises were then compared to $\Delta h_1 E$, where E is a plume-rise enhancement factor with the assumed form $E = [(N+S)/(1+S)]^{1/3}$ and S is a dimensionless spacing factor that depends on $\Delta s/\Delta h_1$. Note that for a very large S, corresponding to large spacing, $E = 1$, whereas for small S, $E = N^{1/3}$. Three different expressions for S were tried, and the one that gave the most consistent agreement with data, with the least scatter, was

$$S = 6\left[\frac{(n-1)\Delta s}{N^{1/3}\Delta h_l}\right]^{3/2} \tag{8.111}$$

$$\Delta h_n = \Delta h_1 E = \Delta h_1 \left(\frac{N+S}{1+S}\right)^{1/3} \tag{8.112}$$

No dependence was found on the wind direction relative to the line of stacks. Therefore it was suggested that, for a cluster of sources, Eq. 8.111 and Eq. 8.112 could be used with $(N-1)s$ replaced by the greatest distance across the cluster.

A technique also was suggested for calculating the plume rise from a cluster of N buoyant sources in a light wind, when the rise is essentially vertical. Entrainment considerations were used to estimate that plume merger would occur near the height $z' = 4D/N^{1/2}$, where D is the diameter of the whole cluster. This height defines z'_m. The merged plume could then be treated as one plume with all the buoyancy combined, but coming from a virtual point source height $(1 - N^{1/2})z'_m$ below the actual source heights. This treatment would be appropriate only if z'_m is less than the rise for a single plume and if the predicted rise for the merged plume is larger than the single-plume rise. However, this approach for vertical plumes is completely untested, as appropriate data were not found.

8-6.1.2 Analytical Models. Three different mathematical models based on physical considerations of plume merger have been suggested. Briggs (1975a) assumed that bent-over plumes merge at the point $z'_c = \Delta s$ and that thereafter all buoyancy and momentum is combined into a single plume. The volume flux of the combined plume was assumed to be $V = U\beta^2 z'_c\,[z'_c + (N-1)\Delta s]$, instead of the $V = U\beta^2 z'^2_c$ that applies to a single plume. Otherwise, the same set of differential equations for z'_c was used. The approximate result for enhancement of rise in the 2/3 law stage was

$$E^3 = \frac{[2N + (N-1)(\Delta s/\Delta h_1)^3]}{[2 + 3(N-1)\Delta s]/\Delta h_1} \tag{8.113}$$

where $\Delta h_1 \geqslant \Delta s$.

The approximation $\Delta h_n \sim \Delta h_1$ was used in the denominator to avoid having to solve a cubic equation.

Another physical model for a line of chimneys was suggested by Murphy (1975). For the first stage of separate plumes, this model starts with the same set of differential equations as previously mentioned in the discussion of Briggs's model and sets $b = \beta z'_c$. Just after plume merger, Murphy applied a second-stage model in which the plume is treated as an infinite line source; b is dropped from both sides of the differential equations, making them two-dimensional, and the thickness of the plume is assumed to be $\pi\beta z'_c$ in this stage. In the third and final stage, Murphy assumed a round plume once again, but now all buoyancy and momentum were combined. The three stages are matched by assuming continuity in z'_c and dz'_c/dx and by choosing virtual origins for the second and third stages so that the continuity requirement is met at the transition points.

The result is

$$E = 1 \qquad (\text{for } p \leq 1)$$

$$E = \frac{1}{3p} + \frac{2}{3}p^{1/2} \qquad (\text{for } 1 \leq p \leq N^{2/3}) \tag{8.114}$$

$$E = N^{1/3} - \frac{N - 1}{3p} \qquad (\text{for } N^{2/3} \leq p)$$

where $p = [(3/2)\beta(\Delta h_1/\Delta s)N]/(N-1)$ and $\beta = 0.455$.

A quite different approach was developed by Anfossi et al. (1978). The authors claimed it to be applicable even to plumes of different buoyancy content originating from different source heights. If an unmerged plume from a shorter stack reaches the height of a taller stack, the two plumes are assumed to merge regardless of spacing between stacks. In situations where very large spacing occurs, Anfossi et al. use an interpolation in which the effective merger height for the merged plume is assumed to equal $(h_s + \Delta h_1)$ for the shorter stack; for very small spacing, Anfossi et al. assume the effective merger height to be $h_s + D$, where $D = \Delta s$ for two chimneys, or, when more than two chimneys are involved, D equals the distance between the outermost chimneys. The buoyancy flux for each source is then reduced so that the calculated rise for the single plume would be the same as for the separate plume, but the rise would start from this assumed merger height; these reduced buoyancies are then added and are treated as a single source emitted from the merger height. The predicted enhancement factor for a line of equally spaced, equally buoyant sources is

$$E = \frac{(1 + N^{1/3}\Delta h_1)/D}{(1 + \Delta h_1)/D} \tag{8.115}$$

However, when different sources in a complex have very different stack heights or buoyancies, the investigator must be cautious about assuming that all plumes will merge. Merger depends on many geometric and dynamic factors not accounted for in this method, including the wind speed and wind direction.

All the authors referenced in this section compared the predictions of their models, at least for $N = 2$, with the semi-empirical results of Briggs (1974) and concluded that, as different as these approaches seem, their predictions were very similar. Only Anfossi et al. refer to any other data base; however, that data base consisted of only six runs of observations, and few details are given. Field data usually evidence much scatter, which makes it difficult to verify predicted rise enhancements with an accuracy of more than 10 to 30%. Therefore the choice among the above four schemes is rather arbitrary; the choice could be based entirely on simplicity, e.g., for the appropriate source configuration, Eq. 8.115.

8-6.1.3 A Wind-Tunnel Study. Overcamp and Ku (1980) recently completed a wind-tunnel study of plume rise from one, two, three and four buoyant sources of equal strength and source height. These results conflict with Briggs's (1974) analysis of field data in that they found that rise enhancement was very strongly dependent on wind angle relative to the line of stacks, Θ. Essentially, they conclude that for $\Theta = 0°$ (wind parallel to the line of stacks) and for $\Theta = 30°$, complete plume merger occurs, with $E = N^{1/3}$, whereas for $\Theta = 60°$ and $\Theta = 90°$, no rise enhancement is observed ($E = 1$).

It easily can be argued that E must in some way depend on Δs and Δh_1 as well as on Θ. One approach, different from the previous ones, is to assume that E depends on the ratio of lateral stack spacing, $\Delta s \sin \Theta$, to the plume thickness at the time that the top of the downwind plume reaches the bottom of the upwind plume. Because plume thickness is proportional to rise, it follows that this condition is met when x is proportional to the longitudinal spacing between the stacks, $\Delta s \cos \Theta$. Therefore, for buoyant plumes following the 2/3 law, the previously mentioned ratio (see second sentence of this paragraph) is proportional to $\Delta s \sin \Theta/[L_b^{1/3} (\Delta s \cos \Theta)^{2/3}]$, where $L_b = F_b/U^3$. If this ratio is small, the lower plume is skewed only slightly off the centerline of the upper plume at the point that they first touch; therefore the lower plume will be pulled up into the upper (i.e., upwind source) plume and total merger will occur. If this ratio is large, the plumes miss each other at this point and collide at a much greater distance with more of a shoulder-to-shoulder orientation then. Consideration of the double vortex structure of each plume suggests that the plumes may simply push each other aside with no merger occurring in this case. Of course, in-between cases with gradual merger can occur.

An interesting feature of this approach is the relatively strong dependence of merger on Θ. Rise enhancement depends on $(\Delta s/L_b)^{1/3}$, which ranged only over a factor of two in Overcamp and Ku's (1980) two-stack experiments, while it depends on $\sin \Theta/(\cos \Theta)^{2/3}$. This prediction of a strong dependence on Θ seems consistent with the experimental results. An analysis of the results for two stacks indicates that no rise enhancement occurs if

$$S' = \frac{\Delta s \sin \Theta}{L_b^{1/3}(\Delta s \cos \Theta)^{2/3}} > 3.3 \tag{8.116}$$

For $S' < 2.3$, the wind-tunnel results indicate rise enhancement. However, for the present, users would be prudent to assume that this enhancement does not exceed the value of E given by one of the methods outlined in Sec. 8-6.1.2.

8-6.2 Stack Downwash

Wind blowing past any obstacle, such as a smokestack, induces a low pressure in the immediate wake on the lee side of the obstacle. If gas is emitted with insufficient momentum and buoyancy, it may be drawn into this low-pressure region and carried downward along the stack. When the gas buoyancy is small, theoretical considerations indicate that the occurrence and degree of downwash depends on the ratio of efflux velocity to cross-flow velocity at the top of the stack, $R = w_0/U$ (Briggs, 1969). The early wind-tunnel work of Sherlock and Stalker (1941) indicated that downwash is negligible if $R > 1.5$, and, although the shape of the stack exit has some influence, this guideline generally has been reliable. When $R = 1$, Sherlock and Stalker's observations show the bottom of the plume about one stack diameter below the rim.

Buoyancy was not a significant factor in Sherlock and Stalker's (1941) experiments because very high Froude numbers were used. The effect of low Froude number (high buoyancy) at the gas exit was investigated by Overcamp and Hoult (1971) in a wind-tunnel experiment with a model of a natural draft cooling tower of hyperbolic shape. Using a theoretical argument comparing the growth of the stack wake with the 2/3 law, Eq. 8.58, these authors predict a critical condition between downwash and no downwash when $L_b \propto D$ (however, to get this result, their Eqs. 15 and 16 have to be evaluated at either $x \propto L_b$ or $x \propto D$, and they neglect to mention this assumption). This prediction is substantiated by the wind-tunnel observations, with the critical value of the velocity ratio given by

$$R = 0.45\left[\frac{\rho_a w_0^2}{g(\rho_a - \rho_0)r_0}\right]^{1/3} \tag{8.117}$$

Note that the term inside the bracket on the right side of Eq. 8.117 is a source Froude number that is defined somewhat differently than in Eq. 8.68. For smaller values of R, strong interaction of the plume with the cooling tower wake is observed, and, for larger values, interaction is weak or absent.

For most sources other than natural draft cooling towers, the source Froude number is quite large compared to unity; in this case, $R > 1.5$ is a sufficient criterion to avoid stack downwash. Such downwash also can be prevented by shielding the plume from the stack wake; a disk one stack diameter in width placed around the rim of the stack appears to be adequate (Scorer, 1959).

Little guidance is available on how to calculate plume rise if downwash does occur. For a jet issuing from a vertical, straight-sided stack, Briggs (in Noll and Davis, 1976) suggests a rise

$$\Delta h = 2(R - 1.5)D \tag{8.118}$$

This suggestion was based on some very crude evaluations of photographs from early wind-tunnel experiments. The suggestion may be unduly conservative at $R = 1.5$, but the occurrence of negative plume rise at small values of R is observed. A simple correction based on R and D may be inadequate for buoyant plumes because, qualitatively, strongly interacting plumes are seen interacting with the stack wake for a long distance downwind. The lower part of the plume diffuses downward in the stack's turbulent wake. One simple correction could be based on the growth of the stack wake width, which grows in proportion to $(Dx)^{1/2}$ (Overcamp and Hoult, 1971). Some fraction of this could be subtracted from the 2/3 law up to the distance calculated for final rise. Unfortunately, observations currently available are inadequate for checking any such proposed downwash correction.

8-6.3 Plume Rise from Gas Turbines

Most plume rise observations in the literature were made on plumes from fossil fuel steam plants or from process emissions (primarily from the former). Empirical plume rise formulas are based entirely on these data, and theoretical models have been primarily tested against these kinds of data (some comparisons with laboratory modeling data also have been made). In comparison to fossil fuel steam plants, the stacks or vents for gas turbines are much shorter, the emissions are much hotter, and, in some cases, the exhaust velocities are much larger. Sometimes it is questioned whether the plume rise formulations recommended for conventional plants apply to gas turbines and whether gas turbine plumes are buoyancy or momentum dominated.

As for the latter question, simple theoretical considerations show that the vertical momentum in a rising plume becomes buoyancy dominated in a time F_m/F_b after the emissions leave the stack. Because the plume travels horizontally at a speed somewhat less than U, the travel distance to buoyancy domination for a hot plume is less than $(w_0 U \rho_0)/(g\rho_0') = (w_0 U T_a)/(g T_0')$. For gas turbines this distance ranges from 0.8 to 3 seconds times the wind speed and represents a very short distance. We can conclude that, even with very high efflux velocities, because of the high T_0', gas turbine plumes become buoyancy dominated very soon; consequently equations for buoyant plume rise are appropriate.

The question of how well standard formulas for buoyant plume rise predict the rise of gas turbine plumes can be answered by carefully considering the three papers summarized below. All these papers were presented in 1975 at the Sixty-Eighth Annual Meeting of the Air Pollution Control Association, and all include some observations of rise from gas turbines.

Moore and Robins (1975) compared some wind-tunnel modeling tests and some lidar observations of the Norwich gas turbine plume with three different models. These three models were (1) Moore's lumpy plume model (Moore, 1974b; Briggs, 1975b), (2) an older CERL formulation (Moore, 1974b), and (3) a simplified version of the pre-1975 Briggs model (Briggs, 1975b). Field data included some 6000 lidar scans made of the plume cross section at 250 to 1500 m downwind. Source parameters were: $h_s = 60$ m, $r_0 = 2.4$ m, $T_0 \simeq 718°K$, $w_0 = 40$ m/s, and $F_b \simeq 1350\ m^4/s^3$. For the field data the average variance about the means of the ratios of predicted to observed rise was about 24% for all three formulations tried. In the first 1000 m of travel, the mean plume trajectory followed Eq. 8.58 rather closely; it averaged 2% lower than the predicted rise. At 1500 m the mean plume rise dropped, on the average, to 16% below the 2/3 law prediction; this suggests some tendency toward leveling (the average rise velocity was about 0.2 m/s at this point). The wind-tunnel model of the same source at full load gave results similar to those just described, except that the rise tended to be about 10% higher than predictions from Eq. 8.58 and was even higher at $x \leq 150$ m full-scale. (The curves shown for Eq. 8.58 by the authors in their figures are about 60% too high for the F_b value calculated from the source parameter information; the stated Q values are also inconsistently high). At half load, with $R = w_0/U = 0.7$, downwash was observed, and the plume rise was about 20% less than that given by Eq. 8.58.

Another wind-tunnel study on a model of a gas turbine plant was done by Hoydysh, Diosey, and Yoshida (1975). The full-scale source parameters were: $h_s = 13.4$ m, stack area $= 50\ m^2$, $T_0 = 644°K$, $w_0 = 9.1$ m/s, and $F_b = 800 m^4/s^3$. In this experiment, the rise very close in—at $x = 40$ m full-scale—exceeded the prediction of Eq. 8.58 by about 20%; however, at larger distances—about $x = 200$ m full-scale—the rise tended to approach the prediction within ±10%. This result occurred only for wind speeds at which there was little or no downwash, in the range $R = 1.4$ to 2.7. In the tests with $R < 0.7$, downwash was substantial and large ground concentrations were measured at $x = 80$ m (full-scale).

Finally, Hoult et al. (1975) reported on another set of dual measurements in the field and in a wind-tunnel simulation. The full-scale source parameters were: $h_s = 14$ m, stack area $= 49\ m^2$, $T_0 = 722°K$, $w_0 = 10$ m/s, and $F_b \simeq 920\ m^4/s^3$. The field observations involved only one twenty-eight minute period of photographic observations. During this period the plume was observed to downwash about half the time in that the bottom of the plume descended below the rim of the stack and remained below it for a distance of several stack heights, even though $R = w_0/U$ was 1.56. For the purpose of calculating the plume centerline height, z_c', the bottom of the plume was considered to be at $z' = 0$ when it descended below the rim of the stack. Agreement between the wind-tunnel modeling of the plume and the field measurements was good. The median rise of the laboratory plumes was about 7% less than that of the full-scale plume. The scatter in both sets of observations was large and comparable. The median trajectory did parallel the 2/3 law, but the rise was about 30% less than that predicted by Eq. 8.58. This result probably was caused primarily by the downwash effect. The authors note that using (T_0'/T_a) in the definition of F_b, as had been proposed in some earlier plume rise models, did not result in the proper scaling of the wind-tunnel observations. Much better agreement with field observations was obtained with $-\rho_0'/\rho_a$ ($= T_0'/T_0$ for $m_0 = m_a$) in the definition of F_b, as is used here.

We can conclude that both field-observed and laboratory-modeled plume rises from gas turbines are consistent with the predictions of the standard plume rise equations for buoyant plumes (most notably, Eq. 8.58). However, plume rise is inhibited when downwash is present. At precisely what point downwash begins depends somewhat on the specific source geometry, the geometry of the surrounding buildings, and the wind direction. For the gas-turbine plumes observed by Moore and Robins (1975) and for those observed by Hoydysh, Diosey, and Yoshida (1975) considerable downwash occurred at $R = w_0/U = 0.7$, whereas little or none occurred for $R \gtrsim 1.4$. In contrast, Hoult et al. (1975) observed a 30% loss of plume rise at $R = 1.56$ for their building configuration. Their experience indicates that a potential downwash situation can be simulated quite well in a wind tunnel.

8-6.4 Plume Rise from Moist Plumes

This section deals primarily with the rise of plumes in which no significant amount of phase change takes place. Plumes that contain a large amount of water have the potential—depending on whether enough latent heat is acquired or released to change the phase of the water—to lose or to gain substantial buoyancy. The following discussion deals with the maximum effect that the phase change of water can have on plume rise (the question of plume visibility is not discussed).

8-6.4.1 Plumes with Liquid Water. If a plume contains liquid water in the form of small droplets that do not fall out quickly, and if this plume rises into and mixes with dry air, all of this liquid water could evaporate. This evaporation would absorb latent heat from the air and from the effluent gases; as a consequence the plume would cool and lose buoyancy. Several decades ago, some scrubbed plumes contained so many water droplets that evaporation often caused the plumes to become negatively buoyant and to descend to the ground very close to the stacks. Today's scrubbers generally are much more efficient at intercepting water droplets before the effluent leaves the unit.

If the mass flow rate of liquid water in the effluent, Q_w, can be determined (perhaps by mass balance methods), calculating the maximum possible loss of buoyancy flux is easy. The heat loss that can occur is LQ_w, where L is the latent heat of evaporation per unit mass. Part of the buoyancy that is lost in cooling the plume is balanced by a gain in buoyancy from the change in water density; water droplets are much denser than air, whereas water vapor is lighter than air at the same temperature. The negative buoyancy of water droplets divided by the buoyancy associated with LQ_w is

$$[(\rho_w - \rho_a)c_pT]/(\rho_wL) \simeq c_pT/L \simeq 1/8$$

where ρ_w is the density of liquid water. The positive buoyancy of water vapor divided by the buoyancy associated with LQ_w is negligible (about 0.0005 times c_pT/L). Thus, if it has not already been considered in calculating ρ_0,

$$(1/8gLQ_w)/(\pi c_p\rho_aT_a) = gQ_w/\pi\rho_a$$

should be subtracted from the calculated value of F_b to allow for the weight of the water droplets; eight times this amount should be subtracted from F_b after evaporation takes place.

Complete evaporation of visible mist often can be seen close to a wet source on days that are not damp or very cold. To calculate the nearest distance at which evaporation can take place, assume complete mixing with a volume $\pi U(\beta' z_c')^2$ of the ambient, where $\beta' = 0.4$. Then total evaporation can take place when $Q_w = \rho_{ws}(1 - RH)\pi U(\beta' z_c')^2$, where RH is the relative humidity of the air and ρ_{ws} is the absolute humidity of the air at saturation (ρ_{ws} is a function only of temperature). If this z_c' exceeds the predicted plume rise with no evaporation, we do not have to take evaporation into account. Otherwise, complete evaporation should take place at some point past this z_c' because it takes time for mixing to work down to small scales within the plume.

8-6.4.2 Plumes with Water Vapor. If a plume contains water vapor and rises into moist or very cold air, some of this water vapor may condense, release latent heat, warm the plume and thus increase the buoyancy of the plume. Most notably, this process might happen with cooling tower plumes because their effluent is nearly saturated with water vapor. The following discussion summarizes the conclusions drawn in an appendix on cooling tower plumes from Briggs (1975a).

Cooling tower plumes typically contain one to five times as much latent heat as sensible heat. If all of the water vapor in such a plume condensed, buoyancy would increase several times and plume rise would be augmented considerably. In reality, only a small fraction of the water vapor can condense because air has a holding capacity for water vapor and only the excess can condense. Furthermore, this holding capacity increases with temperature because, as excess water vapor condenses, the latent heat released warms the air and increases its holding capacity. This situation is illustrated in Fig. 8.12. The curved line represents the capacity of the air to hold water vapor, called the saturation absolute humidity, ρ_{ws}, vs. the air temper-

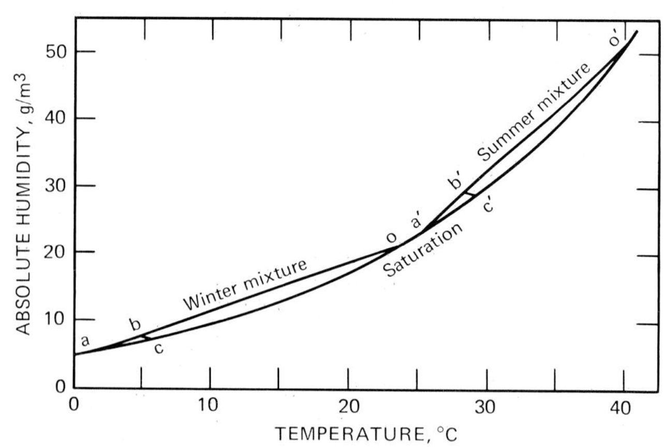

Fig. 8.12 Absolute humidity vs. air temperature. Curve represents saturation values. Straight lines represent state of various mixtures of saturated effluent with saturated ambient (100% relative humidity) for typical winter and summer cases.

ature. We assume a cooling tower plume that is mostly air and, initially, is saturated. The point o represents a typical source temperature and absolute humidity for the winter. The point o′ represents typical summer values. The points a and a′ represent winter and summer values, respectively, for the ambient air under saturation conditions (100% relative humidity). These are not typical values; instead, they represent conditions most favorable to latent heat release (in particular, a′ represents extremely muggy conditions). More typically, the relative humidity is 75% or less, and the plume quickly reevaporates as it rises and mixes with undersaturated air. For the winter case illustrated, the ratio of latent heat in the excess plume moisture, $\rho_{wo} - \rho_{wa}$, to sensible heat is 1.4. For the summer case, it is 3.8.

Points b and b′ represent the state of mixtures of plume and ambient air. The ratio of the length a − o (or a′ − o′) to the length a − b (or a′ − b′) is equal to the ratio of total volume of the mixed plume to the total volume of the original effluent. Because $V/V_0 \simeq (U/w_0)(0.4z_c'/r_0)^2$, this ratio quickly becomes large and the point b (or b′) moves close to point a (or a′). (This explains why the plume quickly desaturates if the relative humidity, ρ_{wa}/ρ_{ws} at T_a, is much less than 100%.) The points b and b′ lie above the ρ_{ws} curve; therefore the plume is supersaturated and some moisture will condense. Wigley (1975) showed that plume moisture will condense quickly until the amount of vapor remaining closely approaches the saturation point (100% relative humidity, or $\rho_w = \rho_{ws}$). However, the latent heat release warms the plume and the state of the mixture follows lines b − c and b′ − c′ [these lines have a slope of $-\rho_a c_p/L \simeq -0.5$ g/(m^3-°C)]. The resultant temperature increase is $T_c - T_b$ for the winter case; this increase is only about 0.35 times the initial temperature excess, $T_b - T_a$; therefore only about ¼ of the excess latent heat is releasable. For the summer case, the temperature boost is only about 0.25 times the initial temperature excess, and only about 1/15 of the excess latent heat is releasable.

We can conclude from Fig. 8.12 that, even in the most favorable circumstances (saturated environment), the boost in plume buoyancy because of the release of latent heat carried by the effluent is not substantial. Considering that, for rise into stable air, plume rise depends only on the ⅓ power, or less, of source buoyancy, the boost given to the plume rise by latent heat release probably will not exceed 10% over the rise resulting from sensible heat.

However, for a saturated plume rising through a saturated environment, such as through a cloud layer, one substantial modification to the plume equations (Weil, 1974b) is necessary. The stability parameter becomes

$$s \simeq \frac{g}{\theta_a}(1 + 11q_s)\left(\frac{\partial T_a}{\partial z} + \Gamma_m\right) \qquad (8.119)$$

where $q_s = \rho_{ws}/\rho_a$ and Γ_m is a modified wet adiabatic lapse rate. As a rule-of-thumb, $\rho_{ws} = 0.01$ kg/m^3 at 11°C and its value doubles with every additional 11°C; therefore the q_s term is not large. In a saturated ambient environment, such as a cloud, the temperature lapse usually is very close to the wet adiabatic lapse rate, which is approximated by

$$\Gamma_s = \frac{\Gamma_d(1 + 30q_s)}{1 + 157q_s} \qquad (8.120)$$

where $\Gamma_d = (g/c_p) = 0.0098$°C/m, which is the dry adiabatic lapse rate that is used in the definition of s in the unsaturated case. The modified wet lapse rate, Γ_m, is approximated by

$$\Gamma_m = \Gamma_s\left(1 + \frac{15q_s}{1 + 41q_s + 336q_s^2}\right) \qquad (8.121)$$

Thus, even if a cloud layer is well-mixed and $\partial T_a/\partial z = -\Gamma_s$, s is positive because of the increasing weight of water droplets carried by the plume (this presumes that the droplets don't have time to fall out of the fast-rising plume, while droplets in the ambient air do precipitate out). However, the difference between Γ_s and Γ_w is very slight. For practical purposes, a cloud layer in which $\partial T_a/\partial z \simeq -\Gamma_s$ is neutral, and a buoyant plume that reaches the bottom of this layer probably will rise to the top of the layer with little change in buoyancy. Briggs (1975a) noted that, in the Amos cooling tower study (Kramer et al., 1975), in the nine tests where the plumes reached an elevated cloud layer, the plumes rose into or among the clouds, but they did not penetrate substantially above the clouds; either the cloud layers were thin with very dry, stable air immediately above them, or the cloud layers were very deep so that the plumes were quite dilute and hardly distinguishable from the ambient air by the time they reached the top of the layer.

8-7 SUMMARY AND CONCLUSIONS

All analytical plume rise models are based on conservation relations and at least one closure assumption. The conservation relations are based on the assumptions that mass, potential density or buoyancy, and momentum are conserved. This chapter has shown how the equations expressing these properties for a point in a fluid are integrated over planes intersecting a plume in order to transform them into equations that describe the conservation of fluxes of quantities carried by the plume. Compressibility of air has little effect on the calculation of plume rise, particularly for rises of less than one or two kilometers. A fourth conservation equation, which is for plume kinetic energy, is used in some models; however, this equation really is an alternative form of the momentum conservation equation and is not independent of it.

Most models use the mass, buoyancy, and momentum conservation equations as their starting point. This approach introduces one new unknown, the entrainment velocity, v_e. (The introduction of at least one new unknown is an inevitable result of parameterizing the effect of turbulence in a bulk way; introducing this new unknown can be avoided only by computing every detail of the turbulent motion.) The equations can be solved by making some assumption about how v_e relates to quantities already in the equations; this assumption is called the closure assumption. The Taylor entrainment assumption, $v_e \propto \overline{w}$ or $v_e \propto w_c$, is the closure assumption most frequently used. This assumption is consistent with the observation that the plume radius expands linearly with rise height until final rise is approached. The Taylor entrainment assumption also leads to reasonably accurate predictions of rise speed and of final rise in stable air. For vertical plumes the constant of proportionality is 0.08. For bent-over plumes, it is 0.6 for buoyant plumes and $0.4 + (1.2/R)$ for nonbuoyant plumes, where $R = w_0/U$. These values are empirical best fits.

A different closure assumption is required when ambient turbulence, rather than stable stratification, becomes the most effective factor limiting plume rise. At some point ambient turbulence overwhelms the plume and causes more mixing than internally generated turbulence. Three different closures for v_e at this stage of rise are used in various models; these three closures relate v_e to an ambient eddy diffusivity, to the variance of turbulent velocity, or to the eddy dissipation rate, ϵ. We have expressed a preference for the latter approach because it is likely that the plume diameter is smaller than the most energetic ambient eddies. A gradually leveling trajectory can be calculated by assuming a second stage of rise in which $v_e \propto (\epsilon b)^{1/3}$, where b is an effective plume radius. However, for simplicity, we suggest that the first stage rise, given by $v_e \propto w_c$, be terminated when the internal dissipation rate of the plume matches ϵ. This concept has been called the breakup model. An alternative model, the touchdown model, in which the plume is pushed down to the ground by large downdrafts also has been explored.

These models have been applied to plume rise prediction for a number of idealized cases and have resulted, for the most part, in simple power law formulas. These formulas have been tested against the appropriate type of field observation or, in some cases, wind-tunnel or water-channel plume modeling data.

When there is almost no crosswind, plumes rise nearly vertically, and, as was established long ago, such plumes expand conically until final rise is approached. For negligible source diameter, jets expand with $b = 0.16z'$, whereas buoyant plumes expand with $b = 0.15z'$, where b is a characteristic plume radius defined by Eqs. 8.4, 8.21, 8.22, and 8.40. The total volume flux carried by a jet is $0.5F_m^{1/2}z'$, whereas for a buoyant plume it is $0.225F_b^{1/2}z'^{5/3}$, where the source terms F_m and F_b are defined in Sec. 8-2.2. These expressions are useful for computing the quantity of ambient air brought into a plume. In the absence of a crosswind, ambient turbulence probably will be too light to affect plume rise; therefore vertical plumes rise until they encounter stable stratification. Their rise through constant stratification is given by Eqs. 8.66 and 8.67; equations to determine the height and thickness of the plume as it spreads horizontally after rise has terminated, as determined by modeling experiments, also are given in Sec. 8-4.1.1. The ability of plumes to punch through thin inversions is described by Eqs. 8.72 to 8.74. A simple method for computing partial or full penetration into a thick, stable layer overlying a mixed layer is given in Sec. 8-4.3.1. Finally, a general method for calculating the equilibrium height of a buoyant plume rising vertically through irregular stable stratification is given with Eq. 8.84.

Most of the time sufficient wind exists to bend over plumes. This phenomenon occurs when U exceeds 1 m/s or so; the transition speed can be estimated by setting the appropriate formulas for rise in no crosswind and for bent-over plume rise equal to each other. The radius of bent-over plumes, similar to that of vertical plumes, is observed to increase linearly with rise above the source, z_c'. Ambient turbulence or stratification has no effect on the plume trajectory for considerable distance. Momentum conservation and the linear growth of plume radius for jets result in the ⅓ law,

$$z_c' = \left(\frac{3F_m}{\beta^2U^2}\right)^{1/3} x^{1/3} \tag{8.56}$$

Experimentally, $\beta = 0.4 + (1.2/R)$ results in a good fit to observed jet trajectories. For buoyant plumes the well-known 2/3 law results:

$$z_c' = \frac{1.6F_b^{1/3}x^{2/3}}{U} \tag{8.58}$$

The constant 1.6 is a median value among best fits to many separate experiments; it corresponds to $\beta = 0.6$.

If a bent-over plume rises through stable air, we can calculate the change in the buoyancy of the plume relative to its surroundings by using Eq. 8.48 with $V' = U(\beta' z_c')^2$, where β' is the entrainment constant for the internal mass of the plume (the constant β includes effective mass that is outside the plume). Final rise for a jet rising through constant stratification can be predicted by using Eq. 8.70, but we have no appropriate data to verify this prediction. However, estimates suggest that ρ_0' must be much less than 0.001 $\rho_a w_0/(m/s)$ in order for buoyancy to be negligible. The final rise of even lukewarm plumes will be buoyancy-dominated; in this case

$$\Delta h = 2.6\left(\frac{F_b}{Us}\right)^{1/3} \tag{8.71}$$

fits closely the final rise observed in many separate experiments. This result is only 7% lower than the theoretical prediction. Equation 8.75 gives an idea of the ability of a bent-over jet to penetrate a thin, elevated inversion; however, this ability turns out to be very slight. By using Eq. 8.76, we find that a typical buoyant plume can completely penetrate such an inversion if it is less than about 100 m above the stack. A buoyant plume can partially penetrate a considerably higher inversion. Some guidance is provided by Eqs. 8.78, 8.79, and 8.80, which are tested against one set of field observations. A simple method for computing partial or full penetration of a buoyant plume into a thick, stable layer overlying a mixed layer is given by Eq. 8.83; again, this method is tested against one set of field observations. Finally, a general method for calculating the equilibrium height of a buoyant, bent-over plume rising through irregular stable stratification is given with Eq. 8.85.

The effective stack height identified with a bent-over plume rising through turbulent air might be limited two ways. The smaller turbulent eddies might overcome the organized structure of the plume and break it up into a rapidly diffusing, essentially passive plume. Alternatively, the eddies larger than the plume might push it to the ground even though the plume is still rising relative to its immediate environment. This vertical transport could cause a maximum ground concentration to occur before the mean plume trajectory levels; therefore the mean height at this point would constitute an effective stack height. To make these predictions, both the magnitude of the small-eddy energy, which is determined by ϵ, and the average large-scale downdraft velocity, w_d, have to be estimated for both mechanical turbulence (neutral conditions) and for convective turbulence (unstable conditions). Experimental information for making these estimates is relatively new and is incomplete. According to current estimates, the breakup model appears to be slightly more critical, i.e., to give the lowest plume rise predictions. Plume rise for jets in turbulent conditions can only be theoretically estimated because no satisfactory data exist (see Eqs. 8.99, 8.100, and 8.102). Rise for buoyant plumes in mechanical turbulence is given by Eq. 8.97. Except for gas turbine plumes and ground sources, this equation is adequately approximated by the simpler equation

$$\Delta h \simeq 1.54\left(\frac{F_b}{Uu_*^2}\right)^{2/3} h_s^{1/3} \tag{8.98}$$

Very few plume rise observations have been carried out far enough to directly test predictions for final rise in neutral and unstable conditions; however, measured maximum ground concentrations are in satisfactory agreement with predictions made by using Eq. 8.98; usually, they are slightly lower. For buoyant plumes in convective turbulence, the equation

$$\Delta h = 3 \left(\frac{F_b}{U}\right)^{3/5} H_*^{-2/5} \tag{8.101}$$

appears to lead to somewhat too high predictions for maximum ground concentration, but present data are not adequate tests. In particular, a simultaneous measurement of sensible heat flux or solar radiation is needed to estimate H_*. The transition points between stable, neutral, and unstable regimes that limit plume rise can be estimated with formulas given in Secs. 8-5.6.1 and 8-5.6.2.

Some miscellaneous types of sources are covered in Sec. 8-6. Section 8-6.1 deals with multiple plumes that are close enough together that plume merger might occur and boost the rise ability. Equations 8.111 and 8.112 specify a semiempirical plume rise enhancement factor for such a situation. The factor is based on an analysis of data from lines of two, three, four, and nine stacks. Three alternative methods of calculating plume rise enhancement are discussed in Sec. 8-6.1.2. Some of the methods are simpler than others, but they all are compared by their authors to the same data analysis. A new analysis based on wind-tunnel data indicates a strong dependence of plume rise enhancement on wind direction relative to the line of stacks; a simple physical model explaining this sensitivity is proposed. Section 8-6.2 deals with stack downwash. Such downwash is avoided if $R > 1.5$, and Eq. 8.117 indicates that a smaller value of R is adequate for a source with a small Froude number. Loss of plume rise in the case of downwash is discussed briefly. Section 8-6.3 concerns the rise of gas turbine plumes and concludes that such plumes are highly buoyancy-dominated, even when the efflux velocity is large. Data from two field experiments and three modeling experiments are examined; and it is concluded that, unless the efflux velocity is low and some downwash of the plume occurs, the plume rise of gas turbine plumes is consistent with that of other sources of the same buoyancy flux. The occurrence and severity of the downwash depend on $R = w_0/U$, the building geometry, and the wind direction. Except in very humid conditions, evaporation occurs in plumes with water droplets; therefore a correction for the latent heat absorbed must be made to the buoyancy. The section also concludes that, in the case of plumes with large amounts of water vapor, only a small fraction of the latent heat can be released through condensation. A noticeable boost in the plume rise over that resulting from sensible heat can occur only in conditions of nearly 100% relative humidity, and then the boost amounts only to about 10%. Cloud environments, with a temperature gradient near the wet adiabatic lapse rate, are essentially neutral to a moist plume. Plumes that reach cloud layers tend to get caught in them because of drier, more stable air above them.

In reviewing the work already accomplished and considering future work needed in plume rise, we can conclude that, except for more subtle effects such as the influence of source diameter and wind shear, rise close to the source has been well examined. Final rise in stably stratified ambient fluid also has been well examined, except that few experiments have been done for nonbuoyant plumes (jets). More work is needed in investigating the partial penetration of plumes into stable air overlying neutral air (this stratification occurs frequently in the daytime). For all the cases discussed so far, physical modeling either in wind tunnels or in water channels is a proved tool, and modeling is often the cheapest and quickest method of investigation.

The area in greatest need of additional study is plume rise (or effective stack height) that is limited by ambient turbulence. Some work in progress indicates that wind-tunnel modeling of final rise in neutral conditions, i.e., mechanical turbulence, is feasible, but proper scaling with sufficiently high Reynolds numbers becomes increasingly difficult in the case of rise induced more by source buoyancy than by source momentum (see Sec. 8-5.7). Wind tunnels have not been adequate for modeling large-scale convective turbulence, but water-tank measurements have compared well with measurements in the atmospheric convective boundary layer when w_*-z_i scaling is applied. Such a tank might be capable of modeling the rise of a buoyant plume in convective turbulence if, as was done by Tsang (1971), a horizontal line thermal is used in place of a bent-over plume. The alternative to physical modeling is field measurements, which generally are more difficult and more expensive to make. Field observations using photographic techniques have not been adequate for defining plume rise at sufficiently large distances in neutral and unstable conditions. However, in recent years, lidar measurements of plumes have proved adequate and essential parameters like z_i and w_* have been measured in the same experiments. As more data using lidar and other remote sensing techniques are collected, less tentative modeling approaches to plume rise can be recommended.

LIST OF SYMBOLS

Symbol	Definition
b	Dynamic plume radius, Eq. 8.40, vertical plume; Eq. 8.42, bent-over plume
b′	Internal plume radius, bent-over plume, Eq. 8.44
b_i	Inversion strength parameter, $(g/\theta_a)\Delta\theta_i$
b_i'	Modified inversion strength parameter, b_i/S_2
c_p	Specific heat capacity at constant pressure
dA_x	Element of plane perpendicular to x-axis
dA_z	Element of plane perpendicular to z-axis
D	Source internal diameter, 2 r_0
F	Volume flux of buoyant acceleration ÷ π
F_b	Source value of F
F_m	Source value of $\overline{w}V$, momentum flux ÷ $\pi\rho_a$
Fr_0	Source Froude number, Eq. 8.68
g	Gravitational acceleration
h_e	Effective source height
h_s	Actual source height
Δh	Plume rise, final height of centerline of plume above source height, or $(h_e - h_s)$
H	Surface heat flux parameter, $(g/T)\overline{w'T'}$
$\mathcal{H}$	Dimensionless heat flux parameter, $(U/u_*)_n Hh_s^2/F_b$
k	The von Kármán constant, 0.35 to 0.4
L	The Monin–Obukhov length, $-u_*^3/kH$
L	Latent heat of evaporation per unit mass
m	Mean molecular weight
M	Internal vertical momentum flux of bent-over plume, Eq. 8.27
$\mathcal{M}$	Dimensionless momentum flux parameter, $\omega F_m/F_b$
$\mathcal{M}'$	Modified momentum flux parameter, $\omega' F_m/F_b$
N	Number of similar sources in proximity to each other

p_a	Ambient hydrostatic pressure
p	Fraction of plume material that penetrates elevated inversion or stable layer
Q_H	Sensible heat emission rate
r_0	Internal radius of round source
R	Velocity ratio, w_0/U
s	Ambient stability parameter, Eqs. 8.13 and 8.14
S_1	Dimensionless plume profile factor, Eq. 8.22
S_2	Ratio of effective to internal momentum flux of a bent-over plume
t	Travel time downwind of source, x/U
T	Absolute temperature
u	Fluid velocity component, x direction
U	Mean ambient wind speed
U_c	Critical wind speed at which X_{max} is maximum
v	Fluid velocity component, y direction
v	Vector velocity
v_e	Entrainment velocity, Eq. 8.3, vertical plume, $(\partial V/\partial x)/2b$, bent-over plume
V	Volume flux of vertical plume ÷ π. Dynamic volume flux of bent-over plume ÷ π; Eq. 8.32
V′	Internal volume flux of bent-over plume, $M/\overline{w}$
w	Fluid velocity component, z direction
$\overline{w}$	Mean vertical velocity in a vertical plume, Eq. 8.21
	Rise rate of centerline of bent-over plume, $\overline{u}\,d\overline{z}'/dx$
w*	Convective scale velocity, $(Hz_m)^{1/3}$
x	Distance downwind of source
X_c	X_{max} at the critical wind speed, U_c
X_{max}	Maximum average ground concentration
y	Lateral distance
z	Vertical distance above the ground
z′	Vertical distance above the source height
$\overline{z}'$	Centerline rise of bent-over plume above source
z_m	Mixing layer height above ground
α	Entrainment coefficient for vertical plume, $v_e/\overline{w}$
β	Entrainment coefficient for bent-over plume, dynamic radius, Eq. 8.43
β'	Entrainment coefficient for bent-over plume, internal radius, Eq. 8.45
ϵ	Ambient turbulent eddy dissipation rate
η	Dimensionless constant for internal turbulent dissipation rate
θ	Potential temperature, Eq. 8.7
$\Delta\theta_i$	Change in potential temperature through elevated inversion
ω	Brunt-Väisälä frequency of stratified ambient, $s^{1/2}$
ω'	Modified Brunt-Väisälä frequency, $(s/S_2)^{1/2}$
Subscript "a"	Ambient value
Subscript "n"	Neutral atmosphere
Subscript "o"	Source value
$_p\iint$	Integration over plume area only ($\rho' > 0$)
$_\infty\iint$	Integration over infinite plane
$_p\oint$	Integration around periphery of plume

REFERENCES

Abraham, G., 1963, *Jet Diffusion in Stagnant Ambient Fluid,* Delft Hydraulics Laboratory, Publication No. 29.

Anfossi, D., G. Bonino, F. Bossa, and R. Richiardone, 1978, Plume Rise from Multiple Sources: A New Model, *Atmos. Environ.,* 12: 1821-1826.

ASME, 1973, *Recommended Guide for the Dispersion of Airborne Effluents,* M. E. Smith (Ed.), American Society of Mechanical Engineers, New York.

Briggs, G. A., 1965, A Plume Rise Model Compared with Observations, *J. Air Pollut. Control Assoc.,* 15: 433-438.

——, 1969, *Plume Rise,* USAEC Critical Review Series, TID-25075, NTIS.

——, 1970, *Some Recent Analyses of Plume Rise Observations,* paper presented at the Second International Clean Air Congress of the International Union of Air Pollution Prevention Associations, Washington, D. C., Dec. 6–11, 1970.

——, 1972, Discussion of Chimney Plumes in Neutral and Stable Surroundings, *Atmos. Environ.,* 6: 507-510.

——, 1974, Plume Rise from Multiple Sources, in *Cooling Tower Environment—1974,* University of Maryland Adult Education Center, Mar. 4–6, 1974, Steven R. Hanna and Jerry Pell (Coordinators), pp. 161-179, CONF-740302, NTIS.

——, 1975a, Plume Rise Predictions, in *Lectures on Air Pollution and Environmental Impact Analyses,* Workshop Proceedings, Boston, Mass., Sept. 29–Oct. 3, 1975, pp. 59-111, American Meteorological Society, Boston.

——, 1975b, Discussion of a Comparison of the Trajectories of Rising Buoyant Plumes with Theoretical Empirical Models, *Atmos. Environ.,* 9: 455-462.

——, 1978, *Analysis of Data from the ETTEX Experiment,* unpublished. [See S. R. Hanna, C. J. Nappo, R. P. Hosker, and G. A. Briggs, 1974, *Description of the Eastern Tennessee Trajectory Experiment (ETTEX),* ATDL Contribution File No. 103, Atmospheric Turbulence and Diffusion Laboratory, National Oceanic and Atmospheric Administration, Oak Ridge.]

——, 1980, What Convective Scaling Does for Diffusion Data, in *Plume Diffusion Modeling,* S. R. Hanna, G. A. Briggs, F. A. Gifford, and C. J. Nappo, NOAA TM ERL-ARL-103, Atmospheric Turbulence and Diffusion Laboratory, Oak Ridge, Tenn. (ATDL File No. 80/24): 54-100.

Bringfelt, B., 1968, Plume Rise Measurements at Industrial Chimneys, *Atmos. Environ.,* 2: 575-598.

——, 1969a, A Study of Buoyant Chimney Plumes in Neutral and Stable Atmospheres, *Atmos. Environ.,* 3: 609-623.

——, 1969b, Author's Reply to Discussion of Plume Rise Measurements at Industrial Chimneys (see preceding entry), *Atmos. Environ.,* 3: 317-319.

Bryant, L. W., and C. F. Cowdrey, 1955, The Effects of Velocity and Temperature of Discharge on the Shape of Smoke Plumes from a Tunnel or Chimney, Experiments in a Wind Tunnel, *Proc. Inst. Mech. Eng. (London),* 169: 371-400.

Businger, J. A., J. C. Wyngaard, Y. Izumi, and E. F. Bradley, 1971, Flux–Profile Relationships in the Atmospheric Surface Layer, *J. Atmos. Sci.,* 28: 181-189.

Carpenter, S. B., F. W. Thomas, and F. E. Gartrell, 1968, *Full-Scale Study of Plume Rise at Large Electric Generating Stations,* Tennessee Valley Authority, Muscle Shoals, Ala.

Clarke, R. H., 1970, Observational Studies in the Atmospheric Boundary Layer, *Q. J. R. Meteorol. Soc.,* 96: 91-114.

Csanady, G. T., 1956, The Rise of a Hot Smoke Plume, *Aust. J. Appl. Sci.,* 7: 23-28.

Davidson, W. F., 1949, The Dispersion and Spreading of Gases and Dusts from Chimneys, in *Transactions of Conference on Industrial Wastes,* 14th Annual Meeting of the Industrial Hygiene Foundation of America, Pittsburgh, Pa., pp. 38-55.

Deardorff, J. W., 1974, Three-Dimensional Numerical Study of the Height and Mean Structure of a Heated Planetary Boundary Layer, *Boundary-Layer Meteorol.,* 7: 81-106.

Djurfors, S., and D. Netterville, 1978, Buoyant Plume Rise in Nonuniform Wind Conditions, *J. Air Pollut. Control Assoc.,* 28: 780-784.

Fan, L., 1967, *Turbulent Buoyant Jets into Stratified or Flowing Ambient Fluids,* Report KH-R-15, California Institute of Technology.

Fay, J. A., M. Escudier, and D. P. Hoult, 1969, *A Correlation of Field Observations of Plume Rise,* Publication No. 69-4, Fluid Mechanics Laboratory, Massachusetts Institute of Technology.

Gartrell, F. E., F. W. Thomas, and S. B. Carpenter, 1961, An Interim Report on Full Scale Study of Dispersion of Stack Gases, *J. Air Pollut. Control Assoc.,* 11: 60-65.

Gifford, F. A., 1967, The Rise of Strongly Radioactive Plumes, *J. Appl. Meteorol.,* 6: 644-649.

Halitsky, J., C. P. Randell, and M. Hackman, 1950, Communication on Dust Deposition from Chimney Stacks, *Proc. Inst. Mech. Eng. (London),* 162: 366-367.

Hamilton, P. M., 1967, Plume Height Measurements at Northfleet and Tilbury Power Stations, *Atmos. Environ.,* 1: 379-387.

Herbert, G. A., 1971, Energy Profiles of Small-Scale Wind Fluctuations, in *Atmospheric Transport and Diffusion in the Planetary Boundary Layer,* I. Van der Hoven (Ed.), Report NOAA-TM-ERL-ARL-32, pp. 10-17, National Oceanic and Atmospheric Administration.

Hewett, T. A., J. A. Fay, and D. P. Hoult, 1971, Laboratory Experiments of Smokestack Plumes in a Stable Atmosphere, *Atmos. Environ.,* 5: 767-789.

Hill, B. J., 1972, Measurement of Local Entrainment Rate in the Initial Region of Axisymmetric Turbulent Air Jets, *J. Fluid Mech.,* 51(4): 773-779.

Hino, M., 1962, *Ascent of Smoke in a Calm Inversion Layer of Atmosphere: Effects of Discharge Velocity and Temperature of Stack Gases,* Report TH-6201, Technical Laboratory, Central Research Institute of Electric Power Industry (Japan).

Holzworth, G. C., 1972, *Mixing Heights, Wind Speeds and Potential for Urban Air Pollution Throughout the Contiguous United States,* Report AP-101, Environmental Protection Agency.

Hoult, D. P., S. R. O'Dea, G. L. Touchton, and R. J. Ketterer, 1975, *Turbulent Plume in a Turbulent Cross Flow: Comparison of Wind Tunnel Tests with Field Observations,* paper presented at the 68th Annual Meeting of the Air Pollution Control Association, Boston, Mass., June 15–20, 1975 (preprint No. 75-49.1).

——, J. A. Fay, and L. J. Forney, 1969, A Theory of Plume Rise Compared with Field Observations, *J. Air Pollut. Control Assoc.,* 19: 585-590.

——, and J. C. Weil, 1972, Turbulent Plume in a Laminar Cross Flow, *Atmos. Environ.,* 6: 513-531.

Hoydysh, W. G., P. Diosey, and T. Yoshida, 1975, *Scaling Considerations for Simulating Plume Rise and Plume Dispersion from Gas Turbine Generators,* paper presented at the 68th Annual Meeting of the Air Pollution Control Association, Boston, Mass., June 15–20, 1975 (preprint No. 75-49.5).

Izumi, Y., and J. S. Caughey, 1976, Minnesota 1973 Atmospheric Boundary Layer Experiment Data Report, Report AFCRL-TR-76-0038, Environmental Research Paper No. 547, Air Force Cambridge Research Laboratories.

Johnson, W. B., and E. E. Uthe, 1969, *Lidar Study of Stack Plumes,* Final Report, Stanford Research Institute Project 7289, contract PH-22-68-33, for Department of Health, Education, and Welfare and National Air Pollution Control Administration, Division of Meteorology.

——, and E. E. Uthe, 1971, Lidar Study of the Keystone Stack Plume, *Atmos. Environ.,* 5: 703-724.

Kaimal, J. C., J. W. Wyngaard, D. A. Haugen, O. R. Coté, Y. Izumi, S. J. Caughey, and C. J. Readings, 1976, Turbulence Structure in the Convective Boundary Layer, *J. Atmos. Sci.,* 33: 2152-2169.

Kamotani, Y., and I. Greber, 1972, *Experiments on a Turbulent Jet in a Cross Flow,* paper presented at the 10th Aerospace Sciences Meeting, Jan. 17–19, 1972, San Diego, Calif., American Institute of Aeronautics and Astronautics.

Keffer, J. F., and W. D. Baines, 1963, The Round Turbulent Jet in a Cross-Wind, *J. Fluid Mech.,* 15: 481-497.

Kramer, M. L., D. E. Seymour, M. J. Butler, R. N. Kempton, P. J. Brennan, J. J. Conte, and R. G. Thompson, 1975, *John E. Amos Cooling Tower Flight Program Data: December 1974–March 1975,* prepared for the American Electric Power Service Corp. by Smith-Singer Meteorologists, Inc.

Lamb, Sir Horace, 1932, *Hydrodynamics,* 6th ed., Cambridge University Press, Cambridge, 1st American ed., 1945, p. 58, Dover Publications, New York.

Lamb, R. G., 1978, personal communication.

Manins, P. C., 1979, Partial Penetration of an Elevated Inversion Layer by Chimney Plumes, *Atmos. Environ.,* 13: 733-741.

Moore, D. J., 1966, Physical Aspects of Puff Models, *Int. J. Air Water Pollut.,* 10: 411-417.

——, 1974a, Observed and Calculated Magnitudes and Distances of Maximum Ground Level Concentration of Gaseous Effluent Material Downwind of a Tall Stack, in *Advances in Geophysics, Ser. B,* Vol. 18B, pp. 201-221, Academic Press, New York.

——, 1974b, A Comparison of the Trajectories of Rising Buoyant Plumes with Theoretical/Empirical Models, *Atmos. Environ.,* 8: 441-457.

——, and A. G. Robins, 1975, *A Comparison Between Calculated and Observed Full Scale and Model Gas Turbine Trajectories,* paper presented at the 68th Annual Meeting of the Air Pollution Control Association, Boston, Mass., June 15–20, 1975 (preprint 75-49.4).

Morton, B. R., 1959, The Ascent of Turbulent Forced Plumes in a Calm Atmosphere, *Int. J. Air Pollut.,* 1: 184-197.

——, and J. Middleton, 1973, Scale Diagrams for Forced Plumes, *J. Fluid Mech.,* 58(1): 165-176.

——, G. I. Taylor, and J. S. Turner, 1956, Turbulent Gravitational Convection from Maintained and Instantaneous Sources, *Proc. R. Soc. (London), Ser. A,* 234: 1-23.

Murphy, B. L., 1975, *Plume Rise from a Row of Chimneys,* paper presented at the 68th Annual Meeting of the Air Pollution Control Association, Boston, Mass., June 15–20, 1975 (preprint 75-04.7).

Murthy, C. R., 1970, On the Mean Path of a Buoyant Chimney Plume in Non-Uniform Wind, *J. Appl. Meteorol.,* 9: 603-611.

Noll, K. E., and W. T. Davis (Eds.), 1976, *Power Generation: Air Pollution Monitoring and Control,* p. 66, Ann Arbor Science Publishers, Ann Arbor, Mich.

Overcamp, T. J., and D. P. Hoult, 1971, Precipitation in the Wake of Cooling Towers, *Atmos. Environ.,* 5: 751-765.

——, and T. Ku, 1980, *A Wind Tunnel Study of Plume Rise from Two Stacks,* paper presented at the AMS/APCA Second Joint Conference on Applications of Air Pollution Meteorology, New Orleans, La., Mar. 24–27, 1980.

Pai, S. I., 1954, *Fluid Dynamics of Jets,* D. Van Nostrand Co., New York.

Panofsky, H. A., 1973, Tower Micrometeorology, in *Workshop on Micrometeorology,* pp. 151-176, American Meteorological Society, Boston, Mass.

——, H. Tennekes, D. H. Lenschow, and J. C. Wyngaard, 1977, The Characteristics of Turbulent Velocity Components in the Surface Layer Under Convective Conditions, *Boundary-Layer Meteorol.,* 11: 355-361.

Patrick, M. A., 1967, Experimental Investigation of the Mixing and Penetration of a Round Turbulent Jet Injected Perpendicularly into a Traverse Stream, *Trans. Inst. Chem. Eng. (London),* 45: 16-31.

Priestley, C. H. B., and F. K. Ball, 1955, Continuous Convection from an Isolated Source of Heat, *Q. J. R. Meteorol. Soc.,* 81: 144-157.

Richards, J. M., 1963, Experiments on the Motions of Isolated Cylindrical Thermals Through Unstratified Surroundings, *Int. J. Air Water Pollut.,* 7: 17-34.

Ricou, F. P., and D. B. Spalding, 1961, Measurements of Entrainment by Axisymmetrical Turbulent Jets, *J. Fluid Mech.,* 11: 21-32.

Rouse, H., C. S. Yih, and H. W. Humphreys, 1952, Gravitational Convection from a Boundary Source, *Tellus,* 4: 201-210.

Schulze, R. H. (Ed.), 1978, Plume Rise from Flares, in *Atmospheric Diffusion Notes,* Issue No. 4 (May), Trinity Consultants, P.O. Box 31481, Dallas, Texas 75231.

Scorer, R. S., 1958, *Natural Aerodynamics,* pp. 143-217, Pergamon Press, New York.

——, 1959, The Behavior of Chimney Plumes, *Int. J. Air Pollut.,* 1: 198-220.

Sherlock, R. H., and E. A. Stalker, 1941, *A Study of Flow Phenomena in the Wake of Smoke Stacks,* Engineering Research Bulletin No. 29, University of Michigan, Ann Arbor, Mich.

Simon, C., and W. Proudfit, 1967, *Some Observations of Plume Rise and Plume Concentration Distributed over New York City,* Paper 67-83, presented at the 60th Annual Meeting of the Air Pollution Control Association, June 11–16, 1967, Cleveland, Ohio.

Sutton, O. G., 1950, The Dispersion of Hot Gases in the Atmosphere, *J. Meteorol.,* 7: 307-312.

Taylor, G. I., 1945, *Dynamics of a Mass of Hot Gas Rising in the Air,* USAEC Report MDDC-919 (LADC-276), Los Alamos Scientific Laboratory, NTIS.

Thompson, D. W., J. M. Norman, and R. L. Miller, 1976, *Airborne Measurements of Turbulent Temperature and Velocity Fluctuations in Cooling Tower Plumes,* paper presented at the Third Symposium on Atmospheric Turbulence, Diffusion and Air Quality, Oct. 19–22, 1976, Raleigh, N. C., pp. 576-580, American Meteorological Society, Boston, Mass. (preprint).

Thuillier, R. H., and U. O. Lappe, 1964, Wind and Temperature Profile Characteristics from Observations on a 1400 ft Tower, *J. Appl. Meteorol.,* 3: 299-306.

Tsang, G., 1971, Laboratory Study of Line Thermals, *Atmos. Environ.,* 5: 445-471.

Turner, J. S., 1969, Buoyant Plumes and Thermals, *Annu. Rev. Fluid Mech.,* 1: 29-44.

Vadot, L., 1965, *Study of Diffusion of Smoke Plumes into the Atmosphere* (in French), Centre Interprofessionnel Technique d'Etudes de la Pollution Atmospherique, Paris.

Venkatram, A., 1980, Dispersion from an Elevated Source in a Convective Boundary Layer, *Atmos. Environ.,* 14: 1-10.

Warren, F. W. G., 1960, Wave Resistance to Vertical Motion in a Stratified Fluid, *J. Fluid Mech.,* 7: 209-229.

Weil, J. C., 1974a, *Comparison Between Measured and Model-Estimated Ground-Level SO_2 Concentrations Downwind from the Dickerson Power Plant,* Report PPSP-MP-11, Maryland Power Plant Siting Program.

—, 1974b, The Rise of Moist, Buoyant Plumes, *J. Appl. Meteorol.,* 13: 435-443.

—, 1977, *Evaluation of the Gaussian Plume Model at Maryland Power Plants,* Report PPSP-MP-16, Maryland Power Plant Siting Program.

—, 1979, *Assessment of Plume Rise and Dispersion Models Using Lidar Data,* Report PPSP-MP-24, Maryland Power Plant Siting Program.

—, 1980a, *Performance of Simple Models for Stack Plume Dispersion During Convective Conditions,* Report PPSP-MP-30, Maryland Power Plant Siting Program.

—, 1980b, personal communication, Environmental Center, Martin Marietta Corp., Baltimore, Md.

—, and D. P. Hoult, 1973, A Correlation of Ground-Level Concentrations of Sulfur Dioxide Downwind of the Keystone Stacks, *Atmos. Environ.,* 7: 707-721.

Wigley, T. M. L., 1975, Condensation in Jets, Industrial Plumes and Cooling Tower Plumes, *J. Appl. Meteorol.,* 14: 78-86.

Wilson, D., 1979, *Wind Tunnel Simulation of Plume Dispersion at Syncrude Mildred Lake Site,* Environmental Research Monograph 1979-1, Syncrude Canada Ltd., Edmonton, Alberta, Canada.

Yokoyama, O., 1971, An Experimental Study on the Structure of Turbulence in the Lowest 500 Meters of the Atmosphere and Diffusion in It, *Rep. Natl. Res. Inst. Pollut. Resour., Kawaguchi, Jpn.,* No. 2.

Chapter 8 covers research prior to November 1980.

DIFFUSION AND TRANSPORT EXPERIMENTS

9

Roland R. Draxler, Air Resources Laboratories

9-1 INTRODUCTION

9-1.1 Scope

The goals of diffusion experiments are diverse, and this diversity works against the simple comparison of experimental results. The purest diffusion experiments attempt to relate the dynamic and physical characteristics of the earth–atmosphere system to the mean and turbulent atmospheric structure and, in turn, to relate these characteristics to diffusion. In other experiments, instead of investigating the genesis of the mean and turbulent flow, effort is directed toward establishing the relation between a number of particular atmospheric states and the concomitant diffusion. Some experiments are mounted to establish the diffusion climate of a particular site. Finally, there are experiments designed to evaluate the effects of a particular pollutant-releasing process. These motivational differences are reflected in experimental techniques and the resulting data.

Most of this chapter is devoted to a review and comparison of many recent controlled-source diffusion experiments. A number of different subjective criteria were used in choosing the experiments to be discussed. Some experiments were included because of the broad range of meteorological conditions sampled, some for the high quality and large quantity of data, and others for interesting features of the technique relating meteorological data to diffusion data. Numerous small field studies that would be too repetitious were excluded. Data that appeared to be presented in interim form or that were difficult to obtain were also excluded.

The primary concern in this chapter is the turbulent diffusion of gases and aerosols. Such effects as plume rise, aerodynamics of flow around obstacles (Chap. 7), deposition (Chap. 12), washout (Chap. 11), and changes in the effluent along its atmospheric path (Chap. 10) are discussed only when necessary to the interpretation of observed pollutant concentrations and plume dimensions. With a few exceptions, experiments that included these complications are avoided in this chapter.

Draxler: Air Resources Laboratories, Environmental Research Laboratories, National Oceanic and Atmospheric Administration, Silver Spring, Maryland.

The experimental data are divided into two broad categories. The first category, the "classical" section (9-3.1), includes those experiments where passive tracers (no deposition) were used in a controlled surface or low-level elevated releases over homogeneous terrain. These experiments have been extensively documented in Chap. 4 of *Meteorology and Atomic Energy—1968,* by Islitzer and Slade. Very little additional data have been collected in this category since the 1968 publication. Some new analyses are presented which reduce the scatter between the observed diffusion and certain meteorological variables used as predictors. Recent developments have suggested that a universal function can be applied to a wide variety of meteorological situations to predict horizontal and vertical plume growth from measurements of wind-direction fluctuation.

The second major category also includes controlled releases of passive tracers but with emphasis on those "special cases" (9-3.2) excluded from the classical section. Experiments in this category were performed over areas of inhomogeneous terrain: coastal areas, mountains, urban areas, and forests. Also included are experiments with releases during very low wind speeds and those where sampling was conducted at long distances. Most of the recent experiments have been conducted in the special cases category. The complications of the special cases usually prohibited the massive data collection conducted for the classical experiments, and the results tend to be site specific in that other variables, in addition to the ones used in analyses of classical experiments, are required to explain the observed diffusion.

9-1.2 Definitions

The diffusion data presented in this chapter will usually be discussed in terms of instantaneous concentration, C, exposure, E, and average concentration, $\overline{C}$. These quantities are defined as

$$C = \frac{\text{mass of pollutant}}{\text{volume of air}} \tag{9.1}$$

$$E = \int_{t_0}^{t_0+t} C\,dt \tag{9.2}$$

$$\bar{C} = \frac{1}{t}\int_{t_0}^{t_0+t} C\,dt = \frac{E}{t} \tag{9.3}$$

where C in each case is the concentration at a given moment and t is the time required for plume passage. Dosage, another frequently used term, is defined as the product of exposure and wind speed. In computing $\bar{C}$, the volume of air used in the denominator of Eq. 9.1 is the volume of air sampled. If the emission rate is not continuous, the computed concentration will be different from the average concentration at the sampler for the duration of the plume passage. Emission rates, mass per unit time, will be denoted by Q. Line source emission, mass per unit length per unit time, will be denoted by q. Instantaneous emissions have no time units.

Frequently, sampling data along an arc are used to derive the horizontal standard deviation of the tracer

$$\sigma_y = D\left\{\frac{\Sigma[C_i\,(y_i - \bar{y})^2]}{\Sigma C_i}\right\}^{1/2} \tag{9.4}$$

where D = distance from the sampling arc to the source location

y_i = angular arc position of the sampler in radians

C_i = concentration at the sampler

i = a particular sampler location on the arc

The standard deviation can be computed from any measurement of tracer mass, not just concentration. A similar equation can be written for the vertical standard deviation of the tracer,

$$\sigma_z = \left(\frac{\Sigma C_i Z_i^2}{\Sigma C_i}\right)^{1/2} \tag{9.5}$$

where Z_i is the height above ground of the sampler. A fundamental assumption in this expression is that the plume centerline is on the ground.

For many experiments the distribution of tracer is assumed to be Gaussian, although analyses of some experimental data have shown this assumption to be valid only for ensemble averages and more frequently in the horizontal than the vertical dimension. If the distribution is Gaussian, then the standard deviation of the plume can be defined by

$$\sigma_y = \frac{Y_0}{2.15} \qquad \sigma_z = \frac{z_0}{2.15} \tag{9.6}$$

where Y_0 and z_0 are the horizontal and vertical distances from the peak concentration to the point where the concentration is $\frac{1}{10}$ of the peak value. With this assumption, σ_z is defined for a ground source with full reflection. By approximating the edge of the plume as the position where the concentration falls to $\frac{1}{10}$ the peak concentration, experimenters are able to determine the standard deviation, a more common variable, from a limited number of samplers by isoplething the concentration data.

In practice, extensive measurements of C are not at all common, because the relative complexity and expense of the real-time or sequential samplers required for such measurements usually prohibit their use in large numbers. The major diffusion experiments, those with numerous sampling arcs and many hundreds of samplers, usually involved measures of exposure (Eq. 9.2) rather than of average concentration. In these experiments the samplers were activated before the dispersing tracer arrived and were deactivated after the last of the tracer was assumed to have passed. This time interval is only estimated since real-time or sequential samplers are not in wide use. The reason for this sampling mode in experiments that are considered to describe continuous-source characteristics over finite time intervals is the inordinately large cost of the equipment necessary to activate and deactivate hundreds of samplers simultaneously during the emission of a continuous plume.

The total time required for the passage of all the material in a detached plume at a point downwind will generally be greater than the release interval because of the effects of along-wind turbulent diffusion and of the along-wind spreading introduced by vertical wind shear in a turbulent medium. Dividing the exposure by a time interval equal to the release time will indeed give an average concentration in cases of detached plumes, but the meaning of such average concentrations must be very precisely known if they are to be related, for instance, to meteorological indices. If the entire elongated detached plume is sampled, however, we would assume nearly the same level of exposure as would be obtained by sampling a continuous-source plume for a period of time equal to the release time of the detached plume.

9-2 BACKGROUND

9-2.1 Data Comparison

The comparison of data from diffusion experiments is complicated by differences between the experimental sites, by mode of tracer release, and by sampling methods. Some other important factors include:

1. *Limited extent of sampling grids.* The diffusion experiments can be divided into two types: (1) those in which the sampling was carried out on geometrical grids involving hundreds of samplers and (2) those in which sampling was carried out by a limited number of mobile sampling systems individually placed for each experiment. The former experiments usually have supplied the more comprehensive data. The data from dense surface sampling grids, except for a few percent of the observations, have usually been confined to distances of less than 10 km. This shortcoming is a consequence of funding and manpower limitations. However, the technical feasibility of surface sampling has been demonstrated to distances of over 1000

km (Cowan et al., 1976), and aircraft sampling using various tracers has been accomplished to distances in excess of 100 km (Peterson, 1968; Bench et al., 1978).

Vertical measurements of turbulence and diffusion are usually unavailable, even in the extensive horizontal grid experiments. Although the detailed description of the vertical plume distribution presents a difficult logistic and funding problem, a considerable amount of information can be obtained from the careful use of a limited number of towers and balloon- and aircraft-borne samplers.

In addition to limited vertical measurements, orientation of the grids in fixed directions frequently precluded sampling during conditions characterized by large angular changes in wind direction, conditions not described by theory but certainly within the realm of feasible experimental measurements.

2. *Diffusion during nonhomogeneous turbulence conditions.* Although homogeneous turbulence is called for by most diffusion theories, the atmosphere never approaches vertical homogeneity and infrequently approaches horizontal homogeneity. For example, radiational inversions are marked by strong vertical gradients of wind speed, wind direction, and turbulence. The limited angular extent of the grids and the paucity of elevated meteorological sensors and tracer samplers in most experiments limit the description of diffusion during just those times that diffusion is of greatest interest. Those experiments classified under special cases in this chapter, by definition, do not meet the requirement of homogeneous turbulence.

3. *Time interval of tracer release and sampling.* The release of pollutants from single sources in industry varies from virtually instantaneous releases to continuous releases over periods of years. The major experimental studies have, with few exceptions, been conducted for release and sampling times in the range of minutes to a few hours. Only rarely has an experimental series been designed to furnish data over a variety of release and sampling times.

4. *The effects of deposition.* The data from most diffusion experiments conducted using sources in the lowest hundred or so meters of the atmosphere can be affected by deposition of the tracer. Because of the difficulty of experimentally determining deposition, the study of deposition has not always been a part of a diffusion experiment.

Deposition will affect the absolute value of concentration, the rate of concentration decreases with distance, and the vertical concentration distribution. The importance of the effect of deposition varies with the type of tracer, atmospheric stability, and wind speed. Earlier experiments with SO_2 and aerosol tracers are particularly suspect. More recent experiments have used nondepositing gases, such as SF_6. Measurements of the horizontal dispersion parameter σ_y are not affected by deposition.

9-2.2 Effects of Averaging and Sampling

The value of any time dependent variable will be affected by the response characteristic of the measuring instrument, the duration of sampling, and any averaging during the sampling period. The importance of these factors in a particular experiment depends on the desired application of the results. Table 9.1 from Pasquill (1974) summarizes the effect of these three limiting factors upon the spectrum of the sampled quantity.

TABLE 9.1 Summary of Separate Filtering Effects of Sampling over Time s, Averaging over Time t, and of Instrument with Exponential Time-Constant α*

Reduction of spectral density, %	Frequency at which reduced spectral density applies as a result of		
	Limited sampling	Averaging	Instrument inertia
>95	<0.125/s	>0.81/t	>0.69/α
50	0.44/s	0.44/t	0.159/α
<5	>0.81/s	<0.125/t	<0.037/α

*From F. Pasquill, *Atmospheric Diffusion*, 2nd ed. (rev.), Ellis Horwood, Ltd., Chichester, England, 1974.

If the instrument time constant α (see Chap. 3), the sampling period s, and the averaging time t are known, then the frequencies (f) can be calculated for which the observed spectrum will be reduced <5%, 50%, and >95% of its original value. Table 9.1 shows which frequency fluctuations in the sampled quantity will not be fully recorded by the sampling system.

For instance, if an instrument with a time constant of 1.0 s is used to obtain continuous measurements of air concentration, concentration fluctuations with periods (1/f) less than 1.5 s will not be recorded (spectral density reduction >95%) and all those with periods greater than 27 s will be recorded (spectral density reduction <5%).

Several empirical studies (Stewart, Gale, and Crooks, 1958, and Cramer, 1959) established that the decrease in concentration with increased averaging time will be equal to

$$C_L = C_S \left(\frac{t_S}{t_L}\right)^P \tag{9.7}$$

where P = 0.2 and S indicates the shorter averaging time and L the longer averaging time.

Gifford (1960) indicated that the exponent P would vary with release height and distance from the source. Singer (1961) and Singer, Kazuhiko, and del Campo (1963) show dependence of P on stability and terrain. Hino (1968) confirmed P = 0.2 for sampling times less than 10 min and suggested P = 0.5 for sampling durations of 10 min to 5 h. Hino found P = 0.5 to be more applicable during unstable conditions. Figure 9.1, from Hino (1968), summarizes the data he used to obtain his results.

9-2.3 Diffusion Classifications

The primary objective of most diffusion experiments is to relate the observed diffusion to certain meteorological variables. The history of various diffusion typing schemes has been presented by Gifford (1976). Frequently, mea-

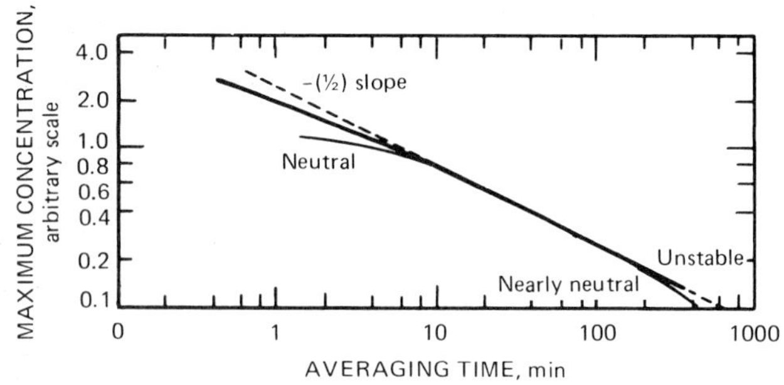

Fig. 9.1 Maximum concentration as a function of averaging time. [From M. Hino, Maximum Ground-Level Concentration and Sampling Time, *Atmos. Environ.*, 2: 149-165 (1968).]

sured meteorological data are limited at specific sites. Consequently, under these conditions, methods have been developed to estimate diffusion.

One of the earliest estimation methods, Sutton (1953), incorporates a variety of equations and a body of experimentally determined values of the diffusion coefficients.

In 1957 Cramer published a set of three graphs from which the plume dimensions σ_y and σ_z for distances up to 1.6 km could be determined directly from a knowledge of horizontal wind-direction standard deviation. The curves in these graphs were based on the Project Prairie Grass and the Round Hill data (Cramer, 1957). The completeness of this system (estimates could be made by simple reference to one or two graphs) and the simplicity with which it could be applied made this approach useful to an increasingly large audience.

Pasquill (1961) attempted to develop a simple system of estimating downwind concentrations. Pasquill suggested that both σ_y and σ_z could be estimated in accordance with the suggestions of Hay and Pasquill (1959). The Hay and Pasquill paper presented a convenient method of estimating both vertical and lateral cloud spread from measurements of wind-direction fluctuations made with a suitably responsive instrument. Recognizing that the data on vertical wind-direction fluctuation required by this method were not generally available, Pasquill suggested values for the appropriate degree of vertical spreading which could be estimated from stability considerations. In the absence of other meteorological data, he suggested that stability be estimated from wind speed and the amount of insolation. He also showed how the lateral spreading of the plume could be estimated from the range of the wind-direction trace for long (about 1 h) pollutant releases and suggested a series of wind-direction-range values to be used in lieu of actual wind measurements for short releases during steady wind-direction conditions. These direction-range values were related to the same estimates of stability used to infer the vertical spreading.

As pointed out by Pasquill and also by Gifford (1961), the values of plume height and width described by Pasquill can be expressed in terms of the diffusion coefficients σ_y and σ_z of the generalized Gaussian plume model. Gifford (1961) performed this conversion and presented the resulting curves. In Gifford's scheme, both σ_y and σ_z were extracted from the appropriate curves representing the various thermal-stability values.

Golder (1972) compared the various stability classification schemes using observed meteorological data and constructed a nomogram to extract the Monin–Obukhov length from an estimate of the Pasquill stability class (see Pasquill, 1961) and the roughness length.

When measurements of vertical fluctuations of the wind are unavailable, a new scheme for estimating vertical diffusion has been suggested by F. B. Smith (1973). The necessary nomograms are shown in Fig. 9.2. First, an estimate of the stability category is made (Fig. 9.2a); then σ_z for the appropriate distance at standard roughness z_0 = 10 cm is obtained (Fig. 9.2b); then σ_z is corrected for a departure from neutral conditions (Fig. 9.2c); and, finally, σ_z is corrected for a different roughness (Fig. 9.2d). These curves were derived from solutions of the two-dimensional diffusion equation for different profiles of $\bar{u}$ and K_z. Any stable layers aloft would limit the vertical growth, and hence the derived σ_z should not exceed the mixed-layer depth.

Another method used to derive σ_y and σ_z values from wind fluctuation measurements has been outlined by Draxler (1976). On the basis of some of the "classical" field experiments described in this chapter, a universal function was developed that satisfies the linear and square root growth limits required by Taylor's theory. The method differentiates between stable/unstable and ground/elevated sources and is based on the development of a universal function suggested by Pasquill (1971). This function can be derived from diffusion experiments by normalizing σ_y and σ_z by the appropriate wind-direction-fluctuation component and travel distance (x). With this universal function, f,

$$\sigma_y = \sigma_\theta \, x \, f\left(\frac{t}{t_i}\right) \tag{9.8}$$

$$\sigma_z = \sigma_\phi \, x \, f\left(\frac{t}{t_i}\right) \tag{9.9}$$

where t is the travel time and t_i is a constant related to the Lagrangian time scale. Draxler (1976) suggests that

$$f\left(\frac{t}{t_i}\right) = \left[1 + 0.9\left(\frac{t}{t_i}\right)^{\frac{1}{2}}\right]^{-1} \tag{9.10}$$

for horizontal diffusion, where t_i is equal to 300 s for ground sources and 1000 s for elevated sources. The wind-direction fluctuation is measured at the source height. Doran, Horst, and Nickola (1978b) found that the universal function was a better predictor of ground-level measured dispersion with elevated sources when the wind-direction fluctuation at ground level was used. Pasquill (1976) also gives similar values for the universal function but without considering the effects of wind speed. Draxler (1976) also found that Eqs. 9.9 and 9.10 can be used to predict vertical

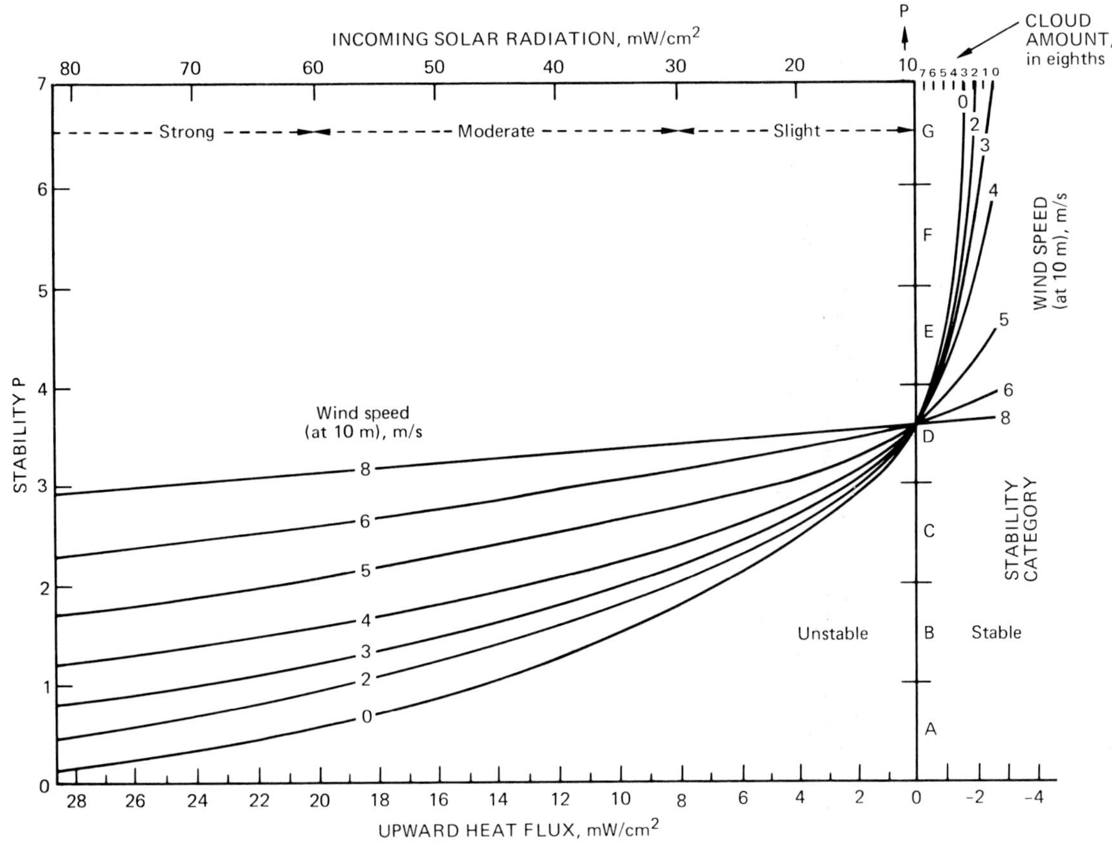

Fig. 9.2a Stability number P in terms of upward heat flux or cloud amount and wind speed for a roughness length of 10 cm. [From F. B. Smith, A Scheme for Estimating the Vertical Dispersion of a Plume from a Source Near Ground Level, in *Proceedings of the Third Meeting of the Expert Panel on Air Pollution Modeling*, Paris, Committee on the Challenge of Modern Society, NATO, Report 14, Brussels (1973).]

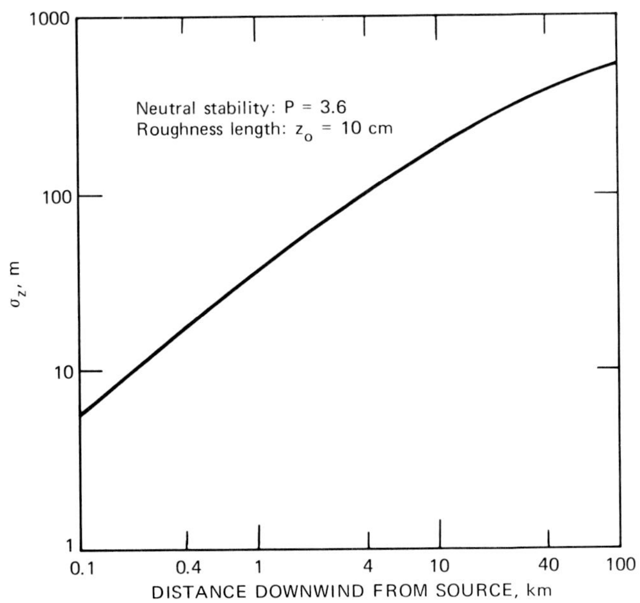

Fig. 9.2b Vertical dispersion during neutral conditions and a roughness length of 10 cm. [From F. B. Smith, A Scheme for Estimating the Vertical Dispersion of a Plume from a Source Near Ground Level, in *Proceedings of the Third Meeting of the Expert Panel on Air Pollution Modeling*, Paris, Committee on the Challenge of Modern Society, NATO, Report 14, Brussels (1973).]

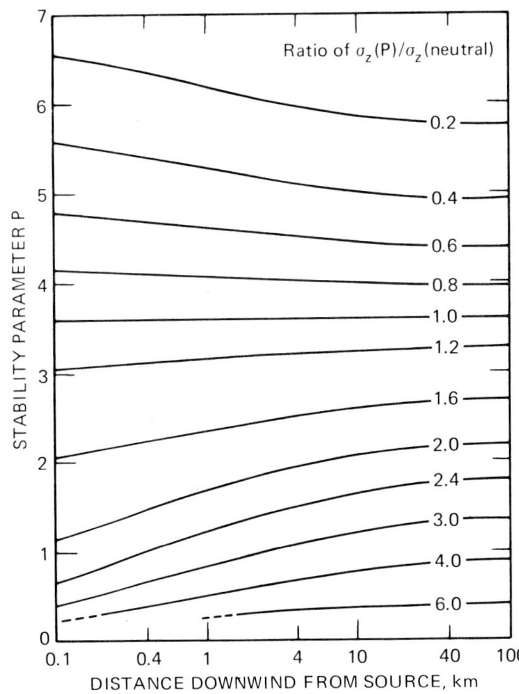

Fig. 9.2c Isopleths of $\sigma_z(P)/\sigma_z$ (neutral) independent of roughness. [From F. B. Smith, A Scheme for Estimating the Vertical Dispersion of a Plume from a Source Near Ground Level, in *Proceedings of the Third Meeting of the Expert Panel on Air Pollution Modeling*, Paris, Committee on the Challenge of Modern Society, NATO, Report 14, Brussels (1973).]

(Figure continues on the next page.)

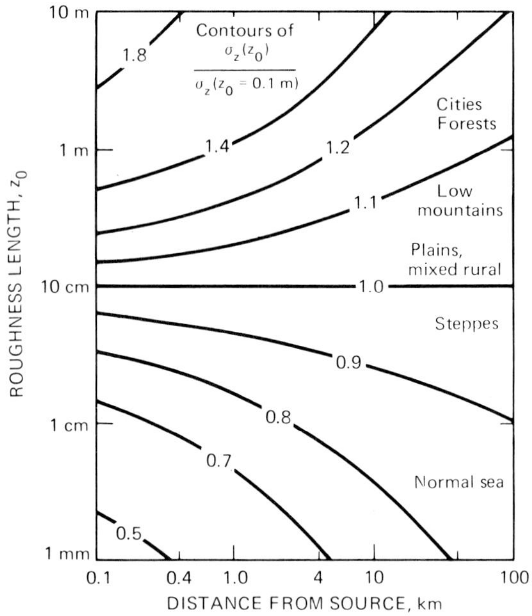

Fig. 9.2d Isopleths of $\sigma_z(z_0)/\sigma_z(z_0 = 0.1$ m) independent of heat flux. [From F. B. Smith, A Scheme for Estimating the Vertical Dispersion of a Plume from a Source Near Ground Level, in *Proceedings of the Third Meeting of the Expert Panel on Air Pollution Modeling,* Paris, Committee on the Challenge of Modern Society, NATO, Report 14, Brussels (1973).]

dispersion from an elevated source during unstable conditions when t_i was equal to 500 s.

9-2.4 Release and Sampling Techniques

The design of an atmospheric tracer-sampling system should include properties that make an experiment reproducible. Some essential properties include the use of a nontoxic material, one that is detectable at low concentrations and has little natural background; the release mechanism should provide for the release of a known amount, constant during the release period; the analysis method should be inexpensive and accurate; and the tracer should be conservative, such that any transit losses (absorption, chemical reactions, or settling) are negligible.

The most common tracer used for the experiments in this chapter are particulates, usually zinc cadmium sulfide, commonly known as a fluorescent particle (FP) because of its distinctive glow under ultraviolet light. A summary of the mechanics of using FP and some problems with the technique are discussed by Leighton et al. (1965). The FP sizes range from 1 to 5 μm with a mean diameter of about 2.5 μm. It is released by a blower, which mechanically separates the particles and ejects them aloft. Usually the spray is directed away from the ground to prevent excessive deposition.

Some sampling systems capture the particles on coated surfaces. One type known as a "Rotorod" consists of two thin vertical cylinders, coated with silicone grease and rotated about a common axis. This system samples about 40 L/min of air. The direction of rotation can be reversed, and another sample will be obtained on the rods. The sequential drum sampler collects particles on a coated drum that is rotated at a predetermined interval. Membrane filters are also used for particle collection.

The method of analysis is normally tedious, manual microscopic counting, although it is possible to measure fluorescence automatically with a photomultiplier tube if sufficient particles are present to overcome the noise from the background.

There is some doubt about the nature of the transit loss of FP. Several multiple tracer tests have been done (Niemeyer and McCormick, 1968; Ludwick, 1966; and Eggleton and Thompson, 1961) in which FP concentrations were compared to those of a gas. Generally the ratio of gas to particles increased with distance from the release point. Simpson (1961) computed deposition for FP for some diffusion trials in stable stratification and found losses of up to 50%. Absolute concentrations and vertical plume growth would be difficult to assess with a depositing tracer, although the horizontal distribution will be unaffected.

Other problems with FP tracers are a loss of fluorescence of the smaller particles in sunlight and high humidity. Dust, especially when sampling over long periods, may obscure the particles. Some experiments have released multiple colored FP, but sometimes it was difficult to identify the colors under a microscope.

Particles other than FP that have been used include lycopodium spores, water soluble dyes (uranine), and oil fog. Oil fog, produced by the incomplete combustion of a gas–oil mixture, is observed by photography or by transecting the plume with an optical densitometer. To use photography, we must make some assumptions in defining the edge of the plume since its definition depends on the nature of the background. Also Pasquill (1974) pointed out that absorption of solar radiation by the visible plume may produce buoyancy effects. Richardson (1920) first used smoke to measure diffusion, and Roberts (1923) developed the current tenets of opacity theory, a definition of the visible edge of the plume. Gifford (1957) generalized the methodology of determining dispersion coefficients from visible plume measurements. Gifford's method required no a priori assumptions regarding the plume growth and hence could be used to analyze relative diffusion observations.

A major group of recently developed tracers are gases. Field experiments have been performed with sulfur dioxide, Freon-12, and certain radioactive gases such as ^{85}Kr and ^{133}Xe. Some recent advances in electron capture gas chromatography (Simmonds et al., 1972; Lovelock, 1974; and Dietz, Coté, and Goodrich, 1976) make the use of halogenated compounds more attractive owing to the simplicity of analysis. One type, sulfur hexafluoride (SF_6), has been used in many experiments. Collins et al. (1965) and Turk et al. (1968) provide some background into the early use of SF_6 and the results of several field trials.

Gases are usually released from pressurized containers. Sampling techniques vary. Whole-air samplers are common. Bags of Saran or other suitable material are filled by a pump with a preset flow rate. Evacuated cylinders and cryogenic systems have also been used. Sampling systems for these gases have to be tested to ensure that the tracer does not leak through the bags or react with container walls.

The selection of a particular tracer for an experiment is based on several factors. The FPs were an early favorite because only simple technology was required to use this system. Many early tracer tests were sponsored by the Department of Defense, which let out the same equipment for many different tests. In all cases, cost of tracer and analysis was a factor. More recently, advances in gas chromatography permitted development of in-situ analyzers and measurements to much lower concentrations.

Experimenters working with low concentrations have to consider whether there is any natural background of the tracer. Each experiment has different requirements, and there are probably several tracers that can be used.

9-2.5 Summary of Data Sources

A summary of the tracer experiments used in this chapter is given in Tables 9.2 and 9.3.

9-3 EXPERIMENTAL DATA

9-3.1 Classical Diffusion Experiments

A reevaluation of many of the dispersion data included in this section has been completed by Horst, Doran, and Nickola (1979). They compared various diffusion theories to empirical data for nonbuoyant point sources up to 100 m in height over homogeneous terrain. These types of experiments are considered the "classical diffusion experiments."

9-3.1.1 Continuous Surface Releases. The term "surface release" has been used to classify those experiments in which the height of release was within a few meters of the surface. More detailed diffusion data are available from continuous-source surface experiments than from any other type. Measurements of vertical concentration distribution are sparse. However, if an estimate of the deposition can be made, effective vertical distribution data can be estimated from the measurements of the lateral spread and centerline or crosswind integrated concentration.

A number of problems arise during surface-release experiments. For example, during unstable (daytime) conditions, the emitted plume disperses rapidly with distance from the source so that measurement of tracer concentrations at the longer distances is difficult. Furthermore, the rapid vertical spreading rules out tower measurements of the vertical distribution of the tracer except at distances on the order of only a few hundred meters. Vertical distributions have been measured by aircraft- or tethered-balloon-borne sensors in such cases. However, the aircraft measurements, which are logistically the most flexible, provide only an instantaneous slice across the plume. Moreover, during extreme instability, such as might occur over a desert in summer, convective cells of sizes near the range of the total

TABLE 9.2 Classical Diffusion Experiments*

Location	Source type	No. of releases	Tracer type	Maximum sampling distance, km	Terrain
O'Neill, Nebr.	C-P-S	70	SO_2	0.8	Prairie
Round Hill, Mass.	C-P-S	20	SO_2	0.2	Flat with obstructions
United Kingdom	C-P-S	80	FP/spores	75	Downland
Hanford, Wash.	C-P-S	66	FP	25.6	Desert
Cape Canaveral, Fla.	C-P-S	76	FP	4.8	Dunes
Vandenburg Air Force Base, Calif.	C-P-S	109	FP	5.7	Foothills
Idaho Falls, Idaho	C-P-S	35	Dye	3.2	Desert
Porton, United Kingdom	C-P-E	19	Spores	0.5	Rolling
Idaho Falls, Idaho	C-P-E	16	Dye	1.8	Desert
Sweden	C-P-E	111	Oil fog	5	Low hills
Brookhaven, N. Y.	C-P-E		Oil fog	50	Low woods
Hanford, Wash.	C-P-E	16	FP	0.6	Desert
United Kingdom	I-P-S	10+	Spores	0.3	Downland
Dallas Tower, Tex.	I-L-E	37	FP	10	Rolling
Edwards Air Force Base, Calif.	I-P-S	43	Beryllium	24	Desert
New Mexico	I-P-E	36	Titanium	11 min	Altitude, 7 to 10 km
Dugway, Utah	I-P-S	33	Gunpowder	1.7	Flat desert
Pt. Arguello, Calif.	I-P-S	17	FP/dye	1	Rough terrain
Idaho Falls, Idaho	I-P-S	4		6	Desert
Hanford, Wash.	I-P-S	8	^{85}Kr	0.8	Desert

*The following abbreviations are used for the source configuration: C, continuous; I, instantaneous; P, point; L, line; E, elevated; and S, surface.

TABLE 9.3 Special Case Diffusion Experiments*

	Source type	No. of releases	Tracer type	Maximum sampling distance, km	Special features	
Shoreline					Release	Sampling
Long Island, N. Y.	C-P-S	11	Oil fog	7	Offshore	Offshore
Oceanside, Calif.	I-L-E I-P-S C-P-S	55	FP	15	Offshore	Onshore
Santa Barbara, Calif.	C-P-S	1	Hydrocarbons	3	Offshore	Offshore
Bolsa Island, Calif.	C-P-S	8	FP	1.3	Offshore	Offshore
Victoria, Tex.	I-L-E	17	FP	160	Offshore	Onshore
Corpus Christi, Tex.	I-L-E	9	FP	40	Offshore	Onshore
Rough					Topography	
Mt. Iron, Calif.	C-P-S	113	FP	13	Ridge–canyon foothills	
Jülich, F.R. Germany	C-P-E	44	CuS	10	Rhine River valley, Z_0 = 1.8 m	
Karlsruhe, F.R. Germany	C-P-E	25	HTO, hydrocarbons	5	Rhine River valley, Z_0 = 1.1 m	
Norway	C-P-S	53	FP, oil fog	13	Fjord and deep valley	
Loisach Valley, F.R. Germany	C-P-E	11	FP	16	Deep valley, Alp foothills	
Complex					Topography	
Huntington Canyon, Utah	C-P-S	11	SF_6	10	Steep-walled canyon	
Garfield, Utah	C-P-S	8	SF_6	5	Rough mountain	
Tonopah, Nev.	I-L-E	7	FP	40	Isolated ridge	
Twentynine Palms, Calif.	I-L-E	10	FP	10	Isolated ridge	
Forest					Height, m	Type
Brookhaven, N. Y.	C-P-S	72	Spores	0.1	13	Pines
Searcy, Ark.	C-L-S	131	FP	30	18	Deciduous
Chippewa, Minn.	I-L-E I-L-S I-P-S	24	FP	2	20	Deciduous
Mt. Rainier, Wash.	I-P-S C-P-S	31	FP	0.24	>30	Douglas fir
Charleston, S. C.	I-P-S	13	FP	0.6	30	Broadleaf trees
Turbo, Columbia	I-L-E	13	FP	1.2	42	Jungle
Urban						
St. Louis, Mo.	C-P-S	42	FP	16		
Ft. Wayne, Ind.	I-L-E	72	FP	16		
Low wind					Topography	
Oak Ridge, Tenn.	C-P-S	11	SF_6, CBr_2F_2	4	Tree height, 20 m; hill, 30 m	
Idaho Falls, Idaho	C-P-S	11	SF_6	0.4	Flat and open	
Three Mile Island, Pa.	C-P-S	11	SF_6	0.25	Flat and wooded	
Hanford, Wash.	C-P-S	5	^{85}Kr	1.2	Flat and open	

*The following abbreviations are used for the source configuration: C, continuous; I, instantaneous; P, point; L, line; E, elevated; S, surface.

plume length can cause significant departures from the assumed state of turbulent homogeneity and thus preclude simple relations between turbulence information and concurrent diffusion.

A variety of important effects must be considered when estimates of diffusion during stable situations are studied. The mean horizontal wind vector shows a decided shear in the lower levels of the atmosphere during inversions. Furthermore, the horizontal wind fluctuation characteristics vary considerably with sampling time and also with height. The problem of the various possible vertical configurations of horizontal wind fluctuations and their effect on diffusion near the surface has rarely been treated in a systematic way. During a strong surface-based nocturnal inversion, the horizontal wind fluctuations above 100 m, or even less, will be small, with the wind direction frequently appearing as an almost laminar trace on conventional wind-recording equipment. At the surface, however, wind fluctuations ranging from those which give a 10-min σ_θ value of a few degrees to those which touch momentarily at every point of the compass are possible (this same phenomena is also possible during the day with very light winds).

Rather large rates of deposition are possible with nonconservative tracers released from surface sources during stable conditions. Since much of the material from a surface source will remain within the region of the surface-roughness elements, the opportunity for plume depletion is large. This depletion lowers the downstream concentration values from those computed on the basis of turbulent diffusion and further modifies the shape of the vertical distribution of the effluent.

O'Neill, Nebraska (Project Prairie Grass), and Round Hill Field Station, Massachusetts. Project Prairie Grass, a comprehensive turbulence and diffusion research project, was carried out during the summer of 1956 over a large flat field in the Great Plains. A complete description of the experimental details and tabulations of diffusion and meteorological data has been presented by Barad (1958) and by Haugen (1959). Numerous papers based on these measurements have been written (notably, Cramer, 1957; Barad and Haugen, 1959; Haugen, 1966). Cramer related the diffusion data taken during the experiments to the concurrent wind fluctuation measurements and also summarized the results of a similar, although more limited, series of tests conducted at the Round Hill field station of the Massachusetts Institute of Technology.

The two studies present diffusion data for 90 field experiments: about 70 at the O'Neill site and 20 at the Round Hill station. Diffusion measurements at O'Neill were determined from sulfur dioxide gas released from 46 cm above the surface and sampled along semicircular arcs from 50 to 800 m from the source. The gas collectors were placed at a height of 1.5 m above the surface and were located along the arcs at 1 or 2° intervals. In addition, the vertical concentration distribution was available from six 20-m towers mounted along the 100-m arc. The tracer was released for 10 min in each experiment, sampling beginning before the arrival of the tracer at the first arc and ending after the last of the material has passed the last arc. The actual measurement was of exposure, but the authors assumed that the difference between the exposure and the average concentration computed for a 10-min sampling time was small and chose to use the average concentration in their discussions. Similar measurements were made at the Round Hill site at arcs located 50 m, 100 m, and 200 m from the source. Here a continuous plume was sampled for 10 min. Meteorological measurements at both sites included the mean wind speed, the wind direction determined at a height of 2 m near the gas source, and vertical profiles of the mean wind speed and temperature. The O'Neill site is unusually smooth ($z_0 < 0.01$ m) with an unobstructed upwind fetch of at least 1.0 km. The Round Hill site is considerably rougher ($z_0 > 0.1$ m) with roughness elements in the form of trees and small buildings. Differences in elevation of terrain on the order of 30 m were found within the first 0.5 to 1.0 km upwind from the test area.

A summary diagram of σ_y stratified by wind-direction fluctuation is presented in Fig. 9.3. Pasquill (1974, page 207) presents σ_z at the 100-m arc stratified by Richardson number (in Fig. 9.4). Haugen (1966) empirically determined the Lagrangian autocorrelation func-

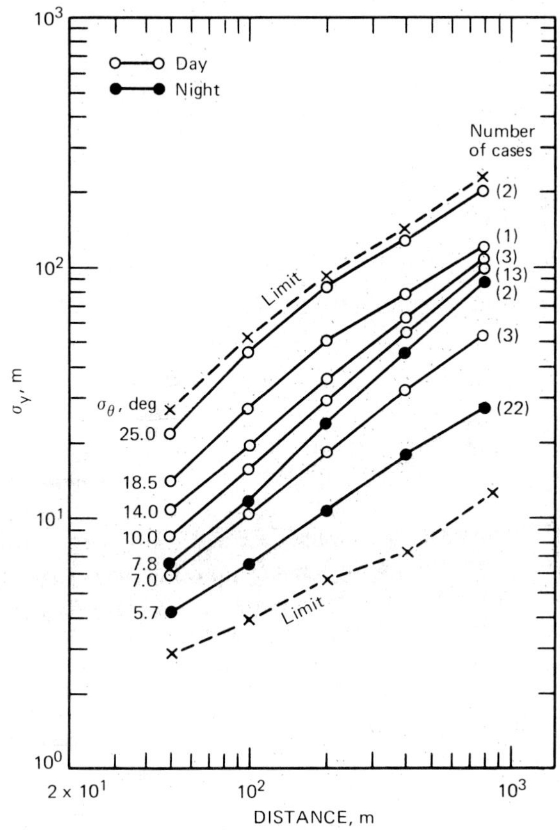

Fig. 9.3 Measured values of σ_y observed during the Prairie Grass experiments and stratified by average values. (From Islitzer and Slade, 1968).

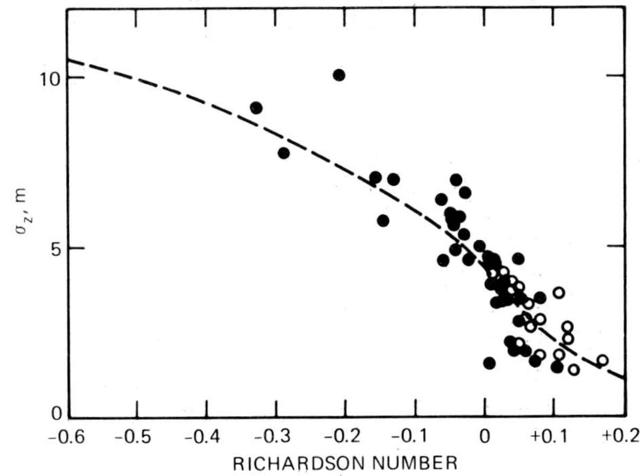

Fig. 9.4 Data on σ_z at 100 m from a continuous source at ground level. [From F. Pasquill, *Atmospheric Diffusion*, 2nd. ed. (rev.), Ellis Horwood, Ltd., Chichester, England, 1974.]

tion and estimated β, the ratio between the Lagrangian and Eulerian time scales, to be about 4. These diffusion trials firmly established the close dependence of plume width on the wind-direction fluctuations even over varying degrees of roughness as suggested by the Round Hill data.

Cramer, Record, and Vaughan (1958) determined the rates of plume growth under three broad stability categories of unstable, neutral, and stable. These growth rates are given in Table 9.4. The rate of growth of σ_y was found to be invariant with distance. The growth rate of σ_z increased with distance in stable stratification.

TABLE 9.4 Plume Growth Exponents from the Prairie Grass Experiment*

Stability	σ_y	σ_z
Unstable	0.8–0.9	1.0–2.0
Neutral	0.8	1.0
Stable	0.6	1.0–0.4

*Adapted from Cramer, Record, and Vaughan, 1958.

The dense sampling network provided the first quantitative empirical evidence that the horizontal distribution of material within the plume was Gaussian (Cramer, 1957); however, Elliott (1961) found that the vertical distribution usually deviated significantly from a Gaussian. These deviations could be due to the scale of turbulence over the rather short release time of 10 min being insufficient to provide a consistent average profile (see Horst, 1979).

Elliott's analysis consisted of fitting the concentration data to

$$\frac{C}{C_0} = \exp(-bZ^s) \tag{9.11}$$

where Z is the height of the concentration observation and b and s are constants determined from the plot. If the distribution is Gaussian, s = 2. The mean value of s for all experiments was found to be 1.49.

United Kingdom. This section includes several surface-release continuous-source experiments. The first, a short-range series (Hay and Pasquill, 1959), was designed to evaluate the factor β, the ratio of the Lagrangian and Eulerian autocorrelogram time scales, for use in the expression for the mean-square particle distribution

$$\overline{Y^2} = \overline{V'^2_{(T,t)}}\,T^2 \tag{9.12}$$

where T, the travel time, is given by $x/\bar{u}$ and t, the time over which $\overline{V'^2}$ is averaged, is given by T/β (V' is the lateral turbulent velocity). The arithmetic mean of β was found to be 4, similar to the Prairie Grass result.

Thompson (1965) described a set of 32 short-range trials at Porton to measure vertical diffusion within 200 m from a line source. He tested a two-dimensional diffusion model, where the vertical mixing coefficient was determined from the vertical wind fluctuation data. Also β was found to vary from 2 in near neutral conditions to 10 in stable conditions.

Another set of 40 longer range trials at Cardington from 4 to 60 km from a line source is described by Thompson (1966). He found β to vary from 1 to 15, from unstable to moderately stable stratification out to distances of 10 km.

Hanford, Washington (Project Green Glow-30 Series). During the period 1959–1962, a series of 66 diffusion experiments were carried out at Hanford, Wash. The results of 46 tests have been reported by Fuquay et al. (1963). Details of the early experiments in this series, as well as tabulations of diffusion and meteorological data, have been described by Barad and Fuquay (1962a). Other publications based on these experiments include discussions of tracer-deposition characteristics, computations of vertical diffusion (Simpson, 1961), the area contained within specified isopleths, tracer travel time (Elliott, Engelmann, and Nickola, 1961), diffusion during conditions of shear flow (Barad and Fuquay, 1962b), and other general descriptions of the experiment results (Fuquay, Simpson, and Hinds, 1964).

The fluorescent tracer, zinc sulfide, was sampled at 833 locations on arcs at 200, 800, 1 600, 3 200, 12 800, and 25 600 m from a source 1.5 m above the ground. The last two arcs were not operated during 26 of the 46 experiments analyzed. Drum samples at the last two arcs were used to obtain the time of arrival of the tracer at these distances.

Vertical diffusion, to heights of 62 m, was measured at the first four arcs by five towers on each arc. Spaced 8° apart, each tower supported 15 samplers. The test site was quite flat to gently rolling with sagebrush cover some 1 to 2 m high. The tracer, with water as a carrier, was released for a 30-min period through two fog generators. The field samplers were in operation well before and after the passage of the tracer cloud and thus provided measures of exposure rather than average concentration. The fluorescent particles were collected on filters, and their total mass was determined by a technique that involved irradiating the particles with an alpha-emitting isotope and counting the scintillations with an automatic electronic counter.

Figure 9.5 is a representation of the growth–travel time relation. The irregular curvature in the region $0.25 < \sigma_\theta\bar{u} < 0.50$ separates the curves into two sets: one set converging at the origin and the other converging at negative values. This irregularity is believed by Fuquay, Simpson, and Hinds (1964) to be caused by differences in the wind-direction shear associated with $\sigma_\theta\bar{u}$. The greatest shear was observed during stable atmospheric conditions that were associated with small values of $\sigma_\theta\bar{u}$. The shearing in stable atmospheres results in comparatively large values of the lateral spread, σ_y.

By material balance methods, Simpson (1961) computed rather large losses of tracer due to deposition during four releases into very stable atmospheres. These losses amounted to as much as 90% of the source during the most stable test. The effect of deposition on the air concentra-

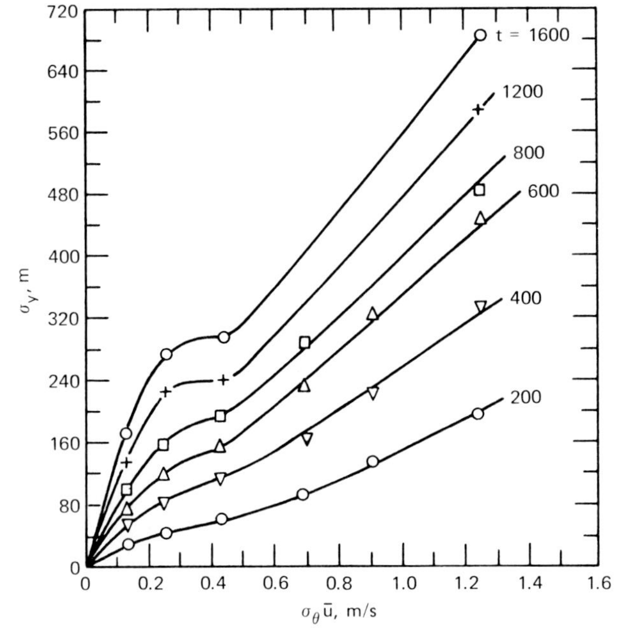

Fig. 9.5 Values of σ_y vs. $\sigma_\theta \bar{u}$ for various travel times. The irregularity in the curve is believed to be due to wind shear during very stable conditions. [From J. J. Fuquay, C. L. Simpson, and W. T. Hinds, Prediction of Environmental Exposure from Source Near the Ground Based on Hanford Experimental Data, *J. Appl. Meteorol.*, 3: 761-770 (1964).]

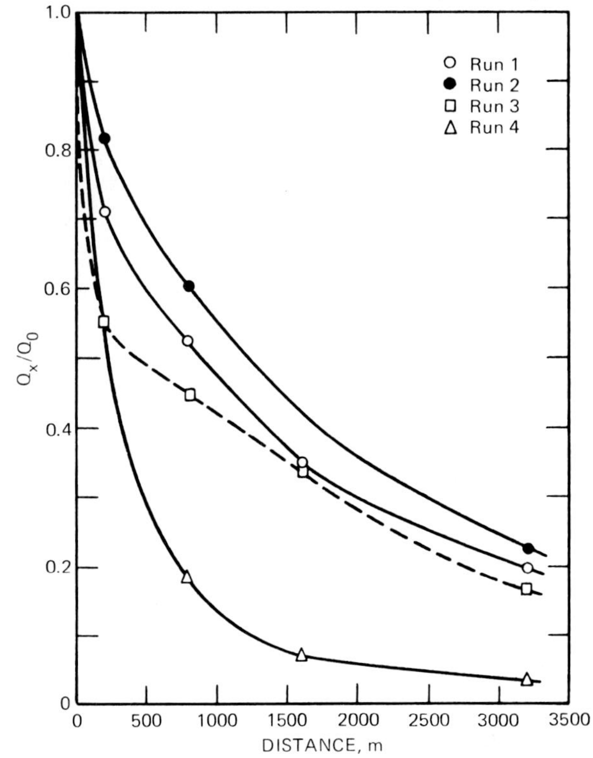

Fig. 9.6 Relative mass transport through vertical planes perpendicular to the average plume direction as observed during the Green Glow tests. The decrease in the ratio Q_x/Q_0 indicates the extent of deposition processes operating during these four tracer releases. (From Simpson, 1961).

tion is shown in Fig. 9.6, which relates the total mass transport, Q_x, past a certain distance to the total released tracer, Q_0.

Cape Canaveral, Florida (Project Ocean Breeze), and Vandenberg Air Force Base, California (Project Dry Gulch). Field diffusion programs were conducted at Cape Canaveral, Fla., and Vandenberg Air Force Base, Calif., during 1961 and 1962 (Haugen and Fuquay, 1963; Haugen and Taylor, 1963). These programs were undertaken to establish quantitative prediction equations for use as safety tools at these missile-test ranges. Tracer emissions were planned for periods of onshore flow which, at these seaside locations, were usually accompanied by daytime instability. In all, 76 experiments were carried out at Cape Kennedy and 109 at Vandenberg.

Fluorescent particles were released from ground-level sources for 30-min periods at both sites. The diffusion course at Cape Canaveral consisted of three concentric arcs 1.2, 2.4, and 4.8 km from the source. Samplers were positioned 1.5 and 4.5 m above the ground surface. The complexity of the terrain at Vandenberg led to sampling over two different courses. The first course, with samplers at arcs 2.3 and 5.7 km from the source, was oriented so that the flow from the source ascended in its path to the arcs. The other course consisted of three arcs at 0.85, 1.5, and 4.7 km with the radius connecting the midpoints of these arcs oriented approximately up a valley. Samplers were positioned at 1.5 m above the ground surface. The terrain at Cape Kennedy is fairly flat with thick vegetation extending to 5 m, whereas the Vandenberg site is quite rugged with elevations at the arcs varying by more than 60 m. Comprehensive supporting meteorological measurements were available at both sites. These measurements included σ_θ data for a variety of running averaging times and vertical temperature-difference data.

Surface Release Summary. The dispersion data from most of the previous experiments normalized as described by Eqs. 9.8 and 9.9 are shown in Figs. 9.7 to 9.9. The solid lines in all figures is Eq. 9.10. These graphs demonstrate that the surface source horizontal dispersion can be represented by a universal function to within a factor of 2 (also see Horst, Doran, and Nickola, 1979). One of the first indications that the dispersion data can be normalized to yield some universal function was apparent from Islitzer and Slade's (1968) presentation of various tracer test σ_y values normalized by σ_θ and shown as a function of distance. Vertical dispersion is more complex, and during unstable conditions the universal dispersion factor is too variable to exhibit any systematic decrease. In fact, statistical techniques to predict vertical growth are not recommended owing to the variation of vertical turbulence with height.

An analysis by Doran, Horst, and Nickola (1978a) indicated that the universal function should increase with sampling and averaging time. Most of the experiments, summarized by Figs. 9.7 and 9.8, were for times of less than 30 min. The authors indicated that values of dispersion function can be as much as 10 to 20% larger when a 1-h averaging time is used.

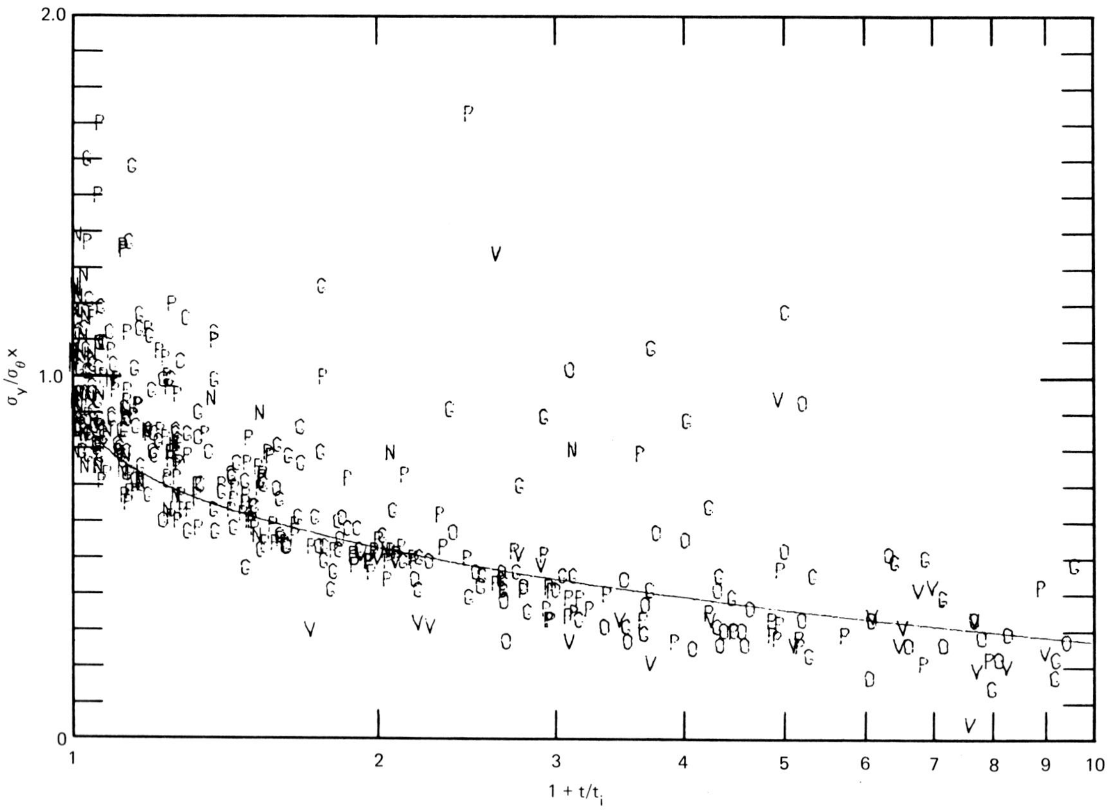

Fig. 9.7 Horizontal diffusion from a ground source during unstable stratification, with $\bar{t}_i$ = 300 s. —, Eq. 9.10. (From R. R. Draxler, *Diffusion from a Continuous Point Source,* March 1975, M.S. Thesis, The Pennsylvania State University, University Park, Pa., 1975.)

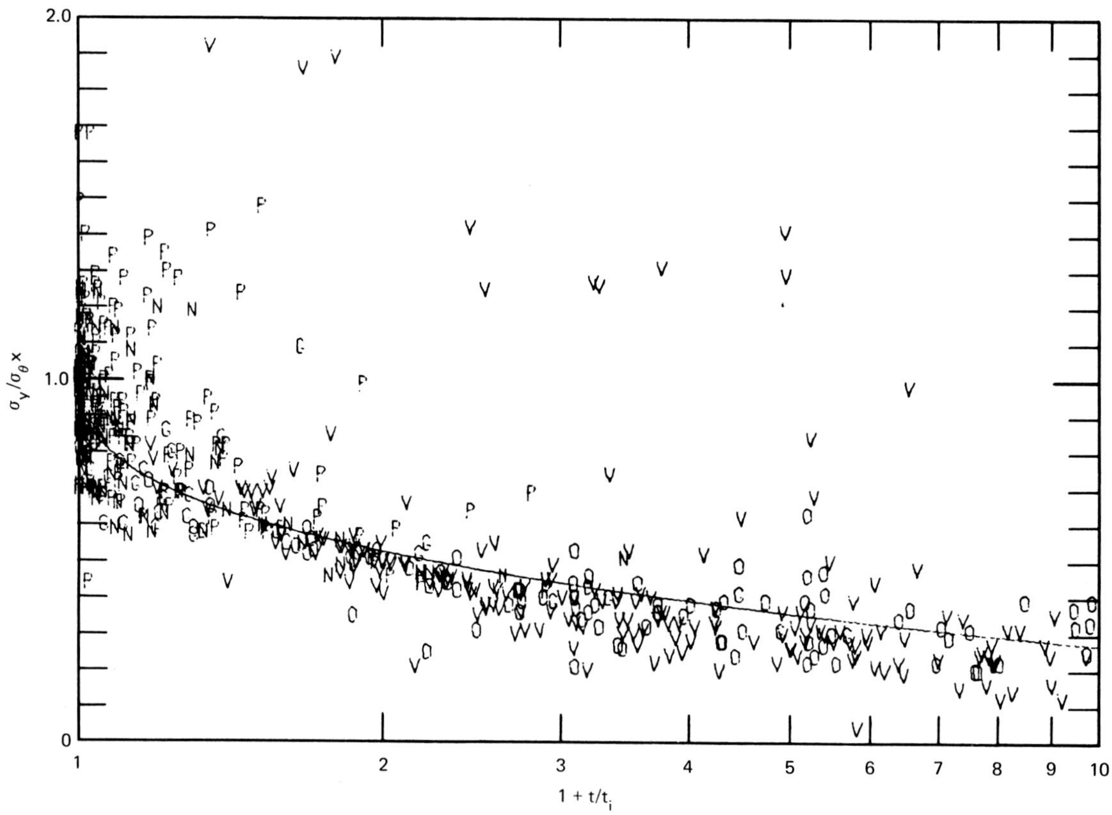

Fig. 9.8 Horizontal diffusion from a ground source during stable stratification, with $\bar{t}_i$ = 300 s. —, Eq. 9.10. (From R. R. Draxler, *Diffusion from a Continuous Point Source,* March 1975, M.S. Thesis, The Pennsylvania State University, University Park, Pa., 1975.)

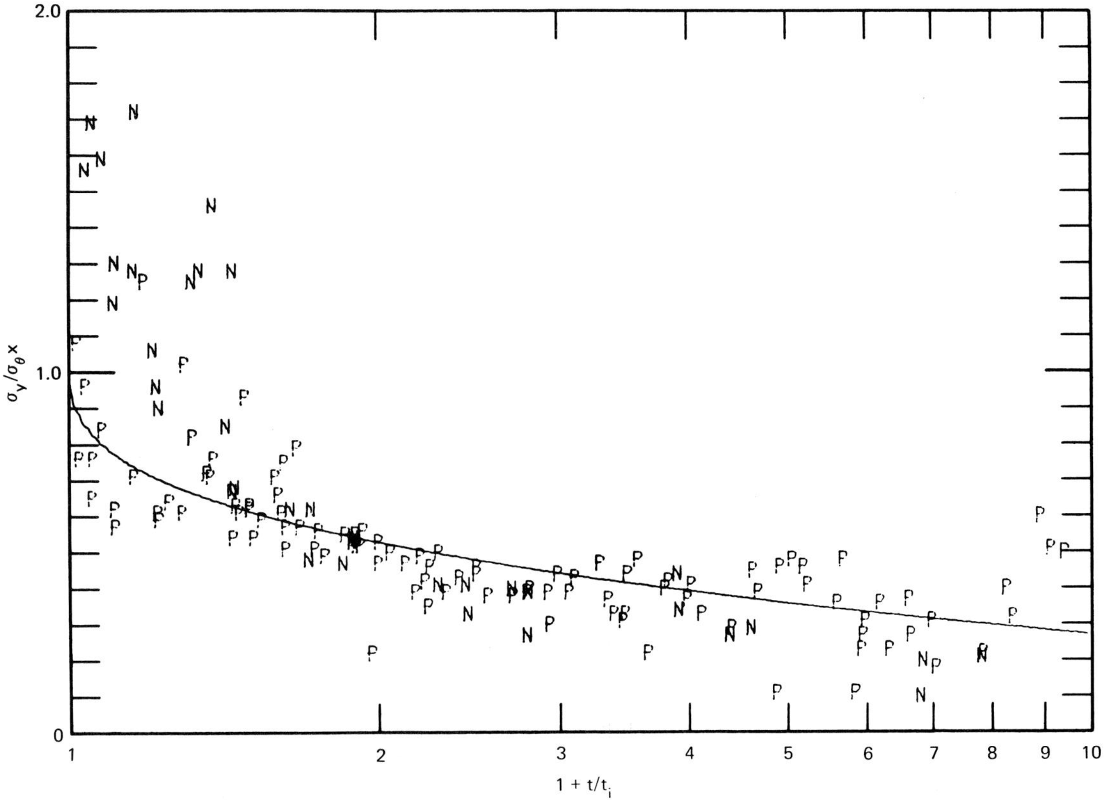

Fig. 9.9 Vertical diffusion from a ground source during stable stratification with $\bar{t}_i$ = 300 s. —, Eq. 9.10. (From R. R. Draxler, *Diffusion from a Continuous Point Source*, March 1975, M.S. Thesis, The Pennsylvania State University, University Park, Pa., 1975.)

*Idaho Falls, Idaho (Idaho National Engineering Laboratory).** A study of atmospheric diffusion and deposition from a surface source was reported by Islitzer and Dumbauld (1963). The grid consisted of six arcs of surface samplers extending 150 to 1800 m from the release point and five sampling towers 30 m high and 70 m apart installed at the 400-m arc to obtain mass-balance measurements for some releases. Direct measurements of deposition were also made at the 100- and 200-m arc on flat plates constructed to simulate a bare soil surface. The uranine dye tracer was released at a height of 1.0 m for 1-h periods during a wide range of stability conditions.

The measured lateral plume spreads during lapse conditions were related to the lateral wind-direction standard deviation for each measurement distance by an averaging time determined for β = 5.0, an experimentally determined average value.

Deposition velocities (V_d, defined as the airborne concentration divided by the rate of deposition, see Chap. 12) were determined for 11 of the diffusion tests. The method was based on computations involving the total amount of released material deposited between the source and 400 m (as determined from tracer material-balance measurements). The results along with the concurrent meteorological data are given in Table 9.5. Higher values of the deposition velocity during unstable conditions, as opposed to stable conditions, are evident in the results.

*Idaho National Engineering Laboratory (INEL) was formerly National Reactor Testing Station.

TABLE 9.5 Meteorological Data and Deposition Velocities Computed from Tracer Material-Balance Measurements at Idaho National Engineering Laboratory*

U-Unstable, S-Stable test	$\bar{u}_4$ m, m/s	Travel distance, m	Amount deposited, %	σ_z (measured), m	V_d, cm/s
U	6.3	400	54	14.6	9.2
S	4.9	400	35	9.5	2.3
U	6.0	200	32	13.5	8.0
U	4.7	200	24	13.2	4.4
U	6.2	200	22	13.7	5.4
U	6.0	400	48	13.4	6.8
U	3.9	400	32	12.2	2.4
S	4.7	400	38	10.0	2.6
S	2.6	400	10	5.0	0.2
S	2.5	400	14	5.5	0.2
U	8.4	400	37	17.8	8.9

*From N. F. Islitzer and R. K. Dumbauld, Atmospheric Diffusion-Deposition Studies over Flat Terrain, *Int. J. Air Water Pollut.*, 7: 999-1022 (1963).

9-3.1.2 Continuous Elevated Releases. Continuous elevated releases are of interest in practical applications since many real releases take place at some distance above the earth's surface. Measurements of axial concentration are difficult to obtain because of the problem of positioning samplers at a sufficient distance above the surface. Elevated-source, plume-center measurements have rarely been made from a grid of towers. Consequently axial concentration measurements have involved manually positioned samplers suspended from large balloons in the plume center or traverses by helicopters or light aircraft. Since the elevated plume will usually disperse toward the surface, the conventional surface sampling grids have been used successfully to infer the rates of crosswind spreading under all but very stable conditions. In some cases, particularly during strong inversions, estimates of diffusion are simplified by the comparative unimportance of surface deposition of the elevated plume material.

Porton, United Kingdom. Some of the first experiments on the problems of diffusion were initiated in 1921, the year that the Meteorological Department of the Chemical Defense Experimental Station at Porton, England, began operation. Some of the most important work done at the station centered about determinations of the shape and magnitude of the downwind lateral diffusion from continuously maintained sources. These early measurements, by hand-operated instrumentation, were adequate for describing the crosswind concentration distribution, the change of this distribution with distance, at least in the first few hundred meters, and the stability dependence of the magnitude of the spreading. A summary of these early measurements can be found on pp. 173–183 of *Atmospheric Diffusion* by Pasquill (1974).

Hay and Pasquill (1957) published some of the first results of their various experiments designed to relate atmospheric diffusion directly to concurrent wind fluctuations. A continous source of Lycopodium spores was released at 150 m above the ground and sampled at distances from 100 to 500 m downwind by a series of samplers suspended along the cable of a balloon. Hay and Pasquill gave the results of 10 such experiments, each lasting approximately 30 min. It was found that σ_ϕ, the vertical standard deviation of the wind fluctuation, measured in angular units at the height of the source, was remarkably similar to σ_z, the particle distribution, also measured in angular units, to distances of 500 m. This indicated to the authors that, to a fair approximation, the particles traveled in almost straight lines from the source to the samplers. From this fact they concluded that the Lagrangian correlation coefficient remained close to unity over the travel times of these experiments, the ratio σ_z/σ_ϕ varied from 0.94 to 1.25 in eight of the experiments, the other two experiments exhibited values outside this range owing to marked inhomogeneity in the turbulence. The arc average value of σ_z for essentially neutral conditions was about 3.9°.

Idaho Falls, Idaho, Idaho National Engineering Laboratory (INEL). The INEL test was a series of 16 releases of a fluorescent tracer, uranine dye in solution, made in unstable atmospheres from the top of a 46-m tower (Islitzer, 1961). The terrain at this site is fairly flat with desert characteristics. The tracer was released for 30-min periods over the same grid as the surface releases. The tower was instrumented at several levels with bivanes and anemometers.

One of the aspects of the INEL test was the direct evaluation of plume width from horizontal wind-direction-fluctuation data. The wind direction at the 46-m level was read as 5-s, end-to-end averages for the period of the test. The standard deviation of this series was computed, multiplied by the distance from the source, and plotted against the measured particle-distribution standard deviation values computed for each of the 16 releases at each of the 6 arcs. The result of this rather simple use of the wind-direction fluctuation is given by the expressions

$$\begin{aligned} \sigma_\theta x &= 1.23\ \sigma_y \\ \sigma_\phi x &= 1.23\ \sigma_z \end{aligned} \tag{9.13}$$

where the wind direction fluctuation data are 5-s averages determined at source height. The use of σ_y and σ_z values obtained from Eq. 9.13 in the Gaussian plume-concentration formula results in downstream computed surface-concentration values closely matching the observed at all distances.

Sweden. An extensive set of measurements of the spreading of very short plume segments released from elevated sources was reported by Högström (1964) and reanalyzed by Haeggblom, Gyllander, and Widemo (1969). The tests were conducted at a coastal and an inland site near Stockholm, Sweden. Oil-fog plumes were generated over a 30-s period and released from masts at heights of from 24 to 87 m above the surface. Photographs of the plume segments taken from a point immediately upwind of the release point yielded measures of the vertical and lateral dimensions of plume segments that were assumed to be equivalent to those of an instantaneously generated point-source puff situated at the midpoint of the plume segment. The use of a 30-s release time permitted photographic observation of the oil fog to much greater distances than would have been possible with true instantaneous releases.

The photographs were suitably enlarged, and the contours representing the visible edges were traced. These contours were assumed to represent some integrated value of concentration. The contours were usually elliptical. The lengths of the lateral and vertical axes of the ellipses were measured and related to standard deviations above the puff center by assumptions regarding the rate of change of the standard deviation with travel time. The lateral and vertical motions of the puff centers were also measured. In all cases the downwind distance of the center of the released material was estimated from the mean wind speed at the height of release and the elapsed time.

The usual experimental procedure consisted of the sequential release of four to six puffs over an interval of an hour. Each interval was considered as one experiment. In all, 430 puffs were released during 111 separate experiments. With few exceptions, all the puffs were released during conditions ranging from near neutral to very stable. Högström (1964) developed the stability parameter,

$$\lambda = \log\left(10^5\ \frac{\partial\theta/\partial z}{\overline{u}_f^2}\right) \tag{9.14}$$

where $\partial\theta/\partial z$ is the local vertical gradient of potential temperature between 30 and 122 m and $\bar{u}_f$ is the free wind speed. The wind speed was obtained from twice daily rawinsonde ascents at a point over 100 km from the point of observation. The correlation of this stability parameter with the measured data was greater than for a variety of combinations of other predictors including potential temperature gradient, wind-speed gradient, and actual wind speed, all measured within the height range of the releases.

The basic measurements were of σ_{y_r} and σ_{z_r} (Högström's notation), the lateral and vertical standard deviations of the mass distribution within a puff relative to its moving center, and σ_{y_c} and σ_{z_c}, the standard deviations, in the appropriate directions, of the puff centers. The dispersion of the puff centers was computed from groups of about 25 individual puffs for each of the stability–distance combinations within the range of the measurements. The values σ_y and σ_z, the standard deviations applicable to a continuous release of the length of one of the experiments (about 1 h), were computed from the relations

$$\sigma_y^2 = \sigma_{y_r}^2 + \sigma_{y_c}^2$$
$$\sigma_z^2 = \sigma_{z_r}^2 + \sigma_{z_c}^2 \tag{9.15}$$

Högström noted that all the terms in the preceding equations are stability dependent with the exception of σ_{y_c}. He suggests that the effect of stability is unimportant at the long wavelengths effective in the lateral dispersion of the puff centers. Further, the vertical diffusion results supported the theory of parabolic growth at large distances from the source. A summary of some of the dispersion data for a 1-h release at one site (Ågesta) by Haegglom, Gyllander, and Widemo (1969) is shown in Figs. 9.10 and 9.11. Note that the greater the release height the smaller the effect of stability on σ_y and the greater the effect on σ_z.

Brookhaven National Laboratory, New York. Figure 9.12 summarizes the results of atmospheric diffusion experiments conducted over a period of years at the Brookhaven National Laboratory. The Brookhaven site is situated in central Long Island, N. Y., a region characterized by terrain rising to no more than 100 m above sea level with most relief being less than 50 m. The ground cover consists of alternating open fields and areas of scrub oak and pine extending to 10 m. A 130-m meteorological tower is instrumented with wind vane and temperature instruments.

The diffusion trials at this site can be divided into three groups: (1) surface-release (2 m) diffusion-deposition tests to distances of about 100 m (Raynor and Smith, 1964), (2) a series of elevated-source (110 m) diffusion experiments using oil fog (M. E. Smith, 1956), and (3) aircraft sampling of the plume released from the Brookhaven reactor (e.g., Peterson, 1968).

The appropriately marked σ_y values in Fig. 9.12 were obtained during aircraft measurements of the ^{41}Ar in the reactor plume. The data were obtained during temperature

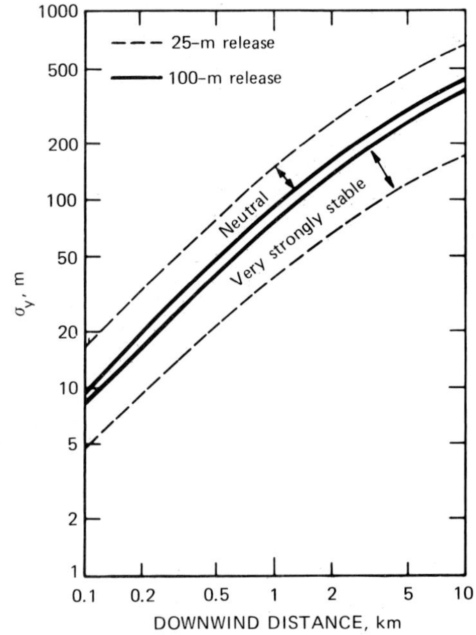

Fig. 9.10 Horizontal dispersion at Ågesta for 1 h at two different release heights. Neutral $\lambda = -\infty$; very strong stable $\lambda = 2.9$. (From L. E. Haegblom, C. H. Gyllander, and V. Widemo, *Atmospheric Diffusion, Investigations at Studsvik*, Nykoping, Sweden, 1969.)

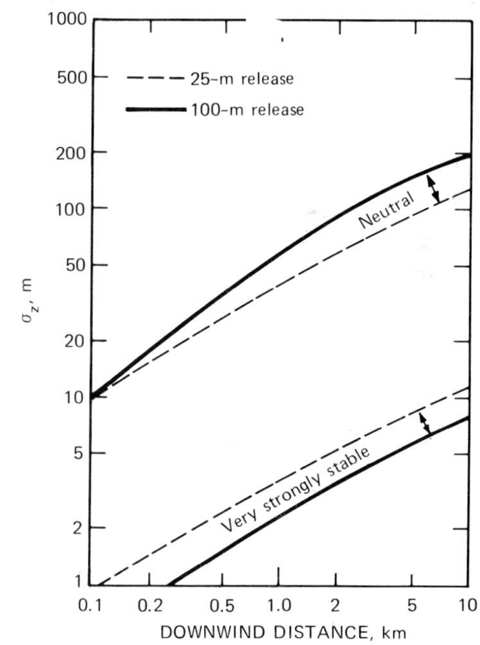

Fig. 9.11 Vertical dispersion at Ågesta for 1 h at two different release heights. Neutral $\lambda = -\infty$; very strong stable $\lambda = 2.9$. (From L. E. Haeggblom, C. H. Gyllander, and V. Widemo, *Atmospheric Diffusion, Investigations at Studsvik*, Nykoping, Sweden, 1969.)

inversions by equipping a light aircraft with dual Geiger-tube detectors mounted under the wings. The detectors were identical except that one permitted free passage of air through an aluminum-shield container and the other did not. Thus the former detector responded to both gamma and beta radiation from the ^{41}Ar, whereas the latter was effectively shielded against beta radiation. Individual disin-

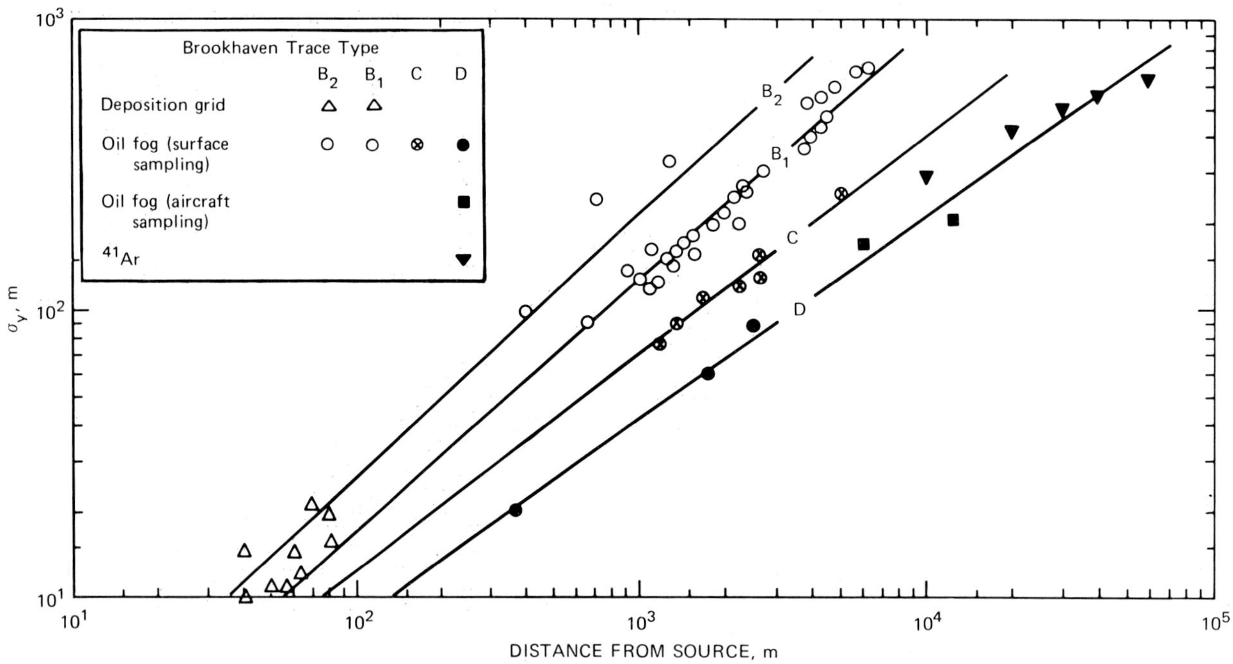

Fig. 9.12 Values of σ_y measured at Brookhaven National Laboratory in a variety of continuous-source experiments. (From Islitzer and Slade, 1968).

tegrations as detected by the Geiger tubes were recorded on a stereo tape recorder.

The σ_y data in Fig. 9.12 are stratified by the Brookhaven gustiness types. These types are subjective estimates of the lateral intensity of turbulence determined from analogue wind-direction recordings, which have been found to correspond in the mean to the values of σ_θ, the standard deviation of the horizontal wind-direction fluctuation, given in Table 9.6. No σ_θ value is given for type A since the wind direction as defined in this case is too variable for the computation of this statistic.

TABLE 9.6 Values of σ_θ and $\bar{u}$ Corresponding to the Brookhaven Trace Types*†

Trace type	σ_θ, deg	$\bar{u}$, m/s
A	(Too variable)	2
B_2	20	3
B_1	13	7
C	7	13
D	4	5

*Values are derived from measurements at 108 m with 1-h sampling time and a 6-s averaging interval.

†From Islitzer and Slade, 1968.

Hanford, Washington. Measurements of vertical diffusion in elevated plumes during various degrees of inversion intensity were made at Hanford (Hilst and Simpson, 1958). A fluorescent pigment was emitted from a 56-m elevation for 60-min periods simultaneously with a continuous plume of smoke used to observe the visual width of the dispersing particles. The vertical concentration distribution of the fluorescent pigment was obtained from an array of cylindrical impaction samplers held aloft by tethered blimps and centered in the visible smoke plume during the tests. Simultaneous observations at a variety of distances to 600 m from the source permitted the empirical determination of the distance dependency σ_z.

The horizontal spread of the smoke plume was photographed at 1-min intervals from an altitude of 1800 m. From these photographs both the instantaneous and the time-averaged horizontal widths of the plume were obtained. By making assumptions about the horizontal concentration distribution in a plume segment, Hilst (1957) deduced values for the short-period horizontal particle variance about the instantaneous-plume axis. This variance, when added to the meander, i.e., the variance of the plume center about the time-mean axis, gives the total horizontal particle-distribution variance (as in Eq. 9.15).

Elevated Release Summary. A summary of the horizontal dispersion from most of the previous experiments is shown in Fig. 9.13. The larger variation of σ_y may partly be due to variation in release height as well as stability if indications from the experiments in Sweden (Fig. 9.10) are representative. The applicability of the universal dispersion function (Eq. 9.10) to elevated releases has been confirmed by Doran, Horst, and Nickola (1978b) to be independent of a release height during neutral to stable conditions when the ground level σ_θ is used in Eq. 9.8.

Figure 9.14 shows a summary of σ_z as a function of distance. These data are plotted as the normalized dispersion function in Figs. 9.15 and 9.16. In these figures σ_z was normalized by the release-levels σ_ϕ measurement. The universal function (Eq. 9.10) should be applicable for all travel times where σ_z is less than the release height. However, during stable stratification, Draxler (1976) suggests a slightly different form for the dispersion function

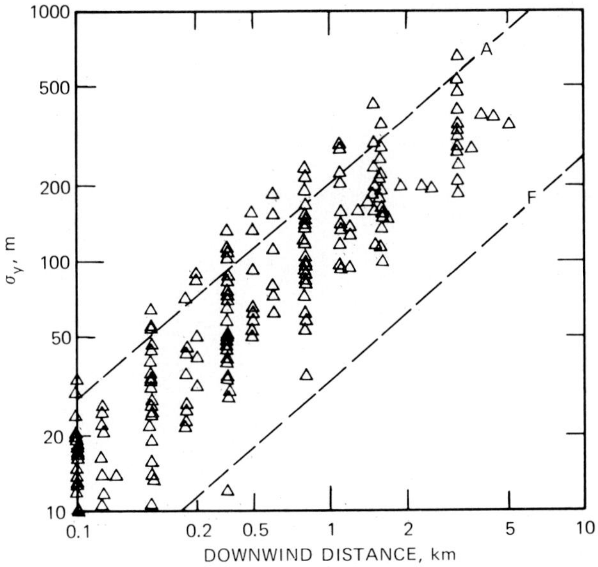

Fig. 9.13 Summary of horizontal dispersion from elevated sources; — — —, Pasquill–Gifford A and F stability dispersion curves.

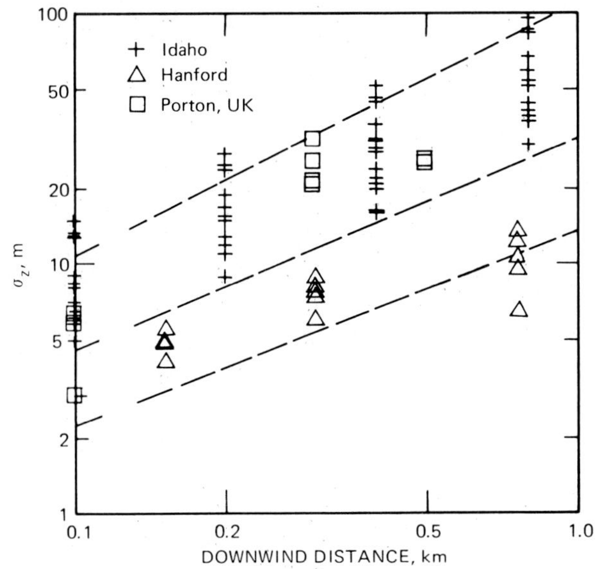

Fig. 9.14 Summary of vertical dispersion from elevated sources; — — —, Pasquill–Gifford B, D, and F stability dispersion curves.

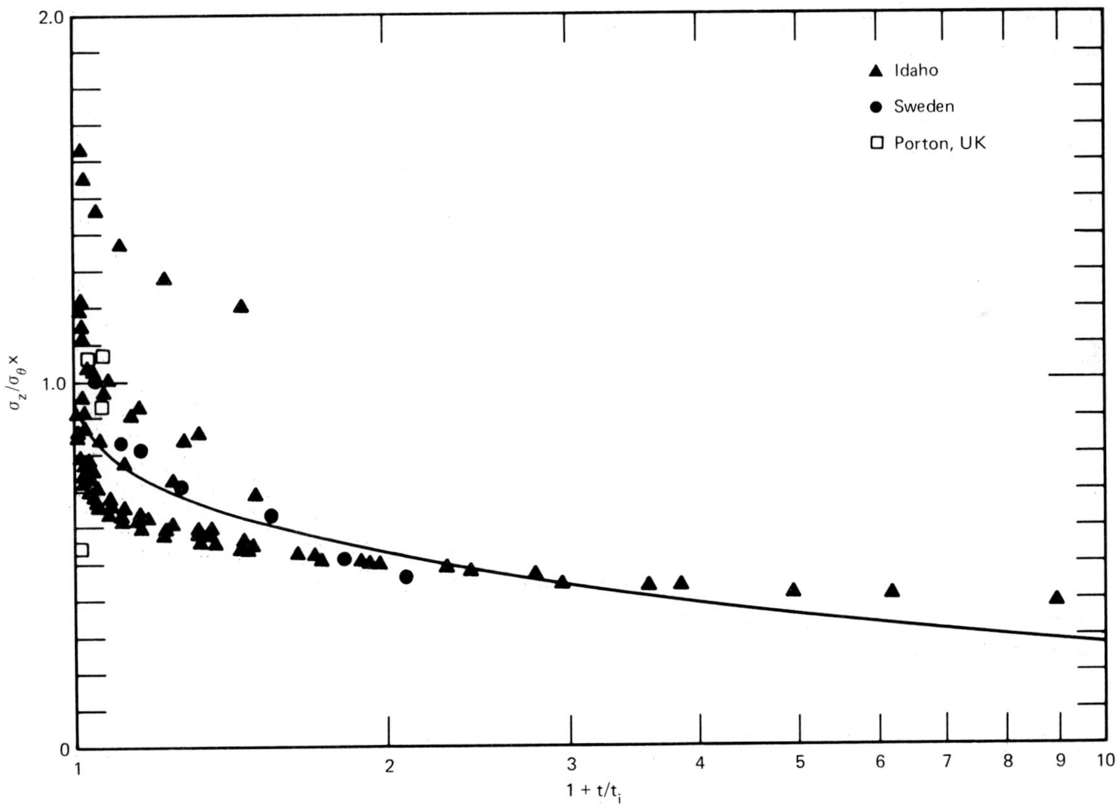

Fig. 9.15 Vertical diffusion from an elevated source during unstable stratification, with $\bar{t}_i$ = 500 s. —, Eq. 9.10. (From R. R. Draxler, *Diffusion from a Continuous Point Source, March 1975*, M.S. Thesis, The Pennsylvania State University, University Park, Pa., 1975.)

$$f\left(\frac{t}{t_i}\right) = \left[1 + 0.95\left(\frac{t}{t_i}\right)^{0.81}\right]^{-1} \tag{9.16}$$

which is shown as the solid line in Fig. 9.16.

9-3.1.3 Quasi-Instantaneous Releases. The practical evaluation of the diffusion parameters in the instantaneous point-source equation has received little attention when compared with the effort expended in determining the characteristics of continuous point-source plumes. Certainly part of the reason for the paucity of instantaneous-source experiments has been the traditional interest in the vast number of continuously emitting chimneys. By

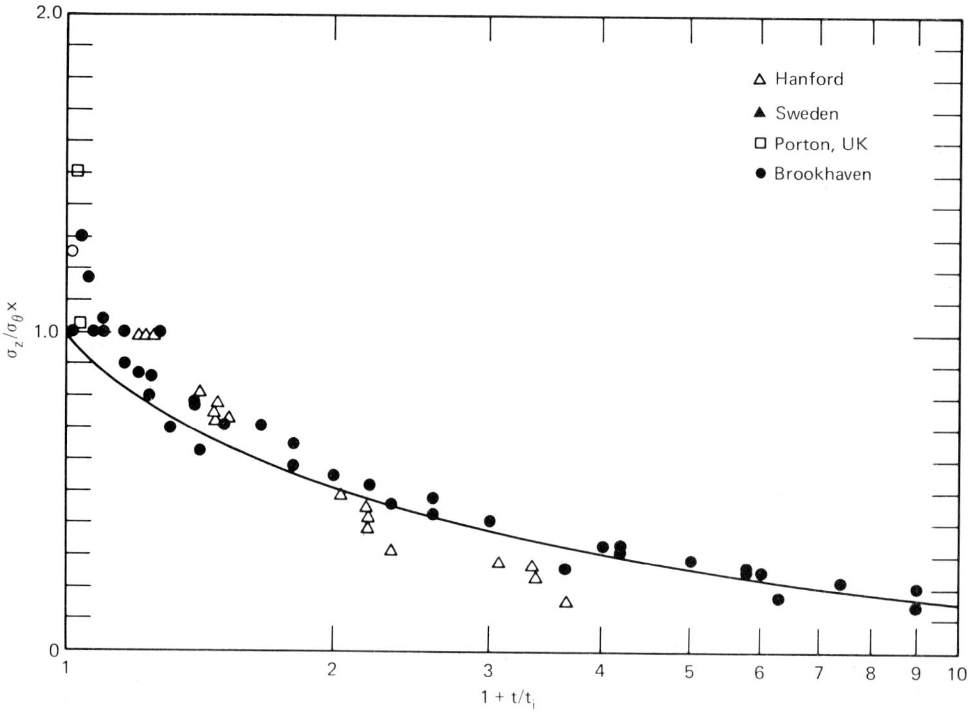

Fig. 9.16 Vertical diffusion from an elevated source during stable stratification, with $\bar{t}_i$ = 100 s. —, Eq. 9.16. (From R. R. Draxler, *Diffusion from a Continuous Point Source, March 1975*, M.S. Thesis, The Pennsylvania State University, University Park, Pa., 1975.)

comparison, instantaneous sources are usually associated with accidents and occasional military requirements.

The instantaneous mass distribution of a puff or plume in the instantaneous-source diffusion problems usually considered is a result of a finite number of discrete turbulent fluctuations. Turbulence theory deals statistically, not deterministically, with atmospheric motions. Therefore turbulence theory cannot be invoked to determine the details of an instantaneous observation of a diffusing pollutant, but the characteristics of an ensemble of puffs can be discussed in terms of turbulence theory.

United Kingdom. Experiments designed to relate to the diffusion of a puff have involved the use of the relation presented by Smith and Hay (1961),

$$\frac{\Delta\sigma_I}{\Delta X} = \frac{2}{3}\beta\, i^2 \qquad (9.17)$$

which relates the maximum spread of a cluster of dispersing particles to a meteorological variable. Here β is the Lagrangian–Eulerian time-scale factor, i is the intensity of turbulence along the appropriate coordinate, and σ_I is the standard deviation of instantaneous material distribution. Equation 9.17 applies strictly only in the case of homogeneous, isotropic turbulence conditions which are usually violated near the ground. Smith and Hay state, however, that Eq. 9.17 can be applied in an empirical sense during conditions of inhomogeneous or anisotropic turbulence.

Smith and Hay carried out two sets of experiments to evaluate Eq. 9.17. The first set consisted in catapulting charges of Lycopodium spores to heights of 2 m above the surface and making measurements of σ_y on crosswind arcs at distances to 300 m. Continuous measures of wind speed and direction were made during each experiment.

The particle spread, σ_y, had no obvious relation to the intensity of turbulence, but this is not surprising in view of the fact that the 3-min sampling time for the turbulence intensity included the effect of eddies much larger than those eddies which could possibly have contributed directly to the spread of the tracer.

A set of medium-range experiments are summarized in the Smith and Hay (1961) article. In these experiments, zinc cadmium sulfide particles were used as the tracer element. The particles were released from an aircraft flying on a crosswind track the center of which was several kilometers upwind of a sampling apparatus mounted on the cable of a captive balloon. Measurements of wind speed and vertical direction variation were made from instrumentation mounted on the cable.

The results of these two experiments are presented in Fig. 9.17. Equation 9.17 seems to furnish a fairly good estimate of the spread of a cluster of particles over a 10-fold range of the intensity of turbulence, a 1000-fold range of distance, and a 1000-fold range of cluster size.

Figure 9.17 showed that a value of 4.5 for β gives a reasonable fit to the observed data. Thus Eq. 9.17 can be rewritten

$$\frac{\Delta\sigma_I}{\Delta X} = 3\, i^2 \qquad (9.18)$$

Dallas Tower, Texas. A series of diffusion tests using elevated line sources released from low-flying aircraft was carried out in the vicinity of the 430-m Dallas television

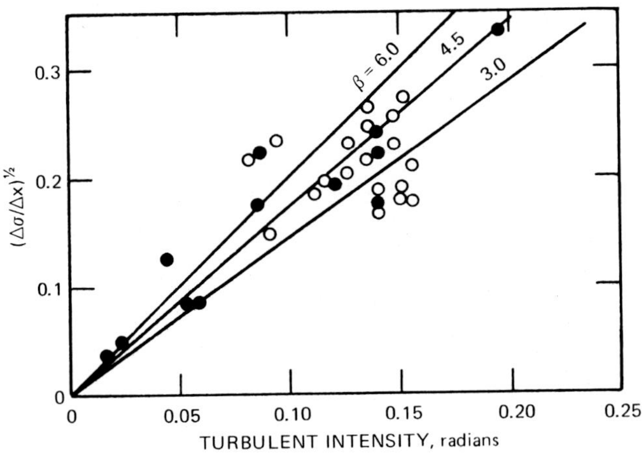

Fig. 9.17 Experimental values for the spread of clusters of particles. o, crosswind spread from instantaneous point sources at distances to 300 m; •, vertical spread from aircraft-released line sources at distances of 3 to 100 km; —, based on Eq. 9.17 for the indicated values of β, the Lagrangian–Eulerian time-scale factor. [From F. B. Smith and J. S. Hay, The Expansion of Clusters of Particles in the Atmosphere, *Q.J.R. Meteorol. Soc.*, 87: 82-101 (1961).]

tower at Cedar Hill, Tex. (MacCready, Smith, and Wolf, 1961). Thirty-seven tests were carried out in three 2-week intervals in April, June, and August 1961. Extensive use was made of wind and temperature measurements made at various levels of the tower. Mean-wind and turbulence information was obtained from wind vane and bivane instrumentation. Vertical sampling of the zinc cadmium sulfide tracer material was accomplished by sequential filter samplers at various levels on the tower and by rotorod samplers located on the ground 1.6 km apart to a distance of about 48 km downwind of the release line. During a portion of the tests, a crosswind rotorod line was added near the downwind end of the main ground sampler line. All releases were made at night during stable conditions. Source lines ranged between 14.5 and 42 km in length, between 1.6 and 11 km upwind of the tower, and between 110 and 320 m above the tower base. The primary objective of the program was to relate the measured diffusion characteristics of the cloud to the observed turbulence, wind velocity, and temperature in a quantitative manner.

The principal analysis effort was concerned with the explanation of the observed vertical cloud width on the tower, the distance from the release time to the first appearance of the cloud at the ground, and the location and magnitude of the maximum exposure at the ground. The basic theoretical relation investigated was that of Smith and Hay (1961), where i_w, the intensity of turbulence (vertical turbulence in this case), was estimated from tower measurements of σ_ϕ, the standard deviation of the vertical wind-direction distribution. The quantity σ_ϕ generally increases with the sample length up to some point where it becomes asymptotic to a maximum value. In the case of the Dallas tower data, it was found that almost all the increase occurred within 3 min.

The results of the tests showed that the vertical dimensions of the cloud were consistently larger than expected. One reason for this discrepancy was that the initial size of the cloud at release was appreciable because of the turbulence created by the releasing aircraft.

Dilution of the cloud from the ends of the release line was measured during nine of the tests in which adequate sampling was available along a crosswind line at the downwind end of the main sampler line. It was found that the cloud diluted laterally at a rate corresponding to

$$\frac{\Delta\sigma}{\Delta x} = i_v \tag{9.19}$$

where i_v was computed from the root-mean-square value of the horizontal direction variation over the period of interest. This rate is in marked contrast to the vertical spreading rate of $3\,(i_w)^2$ of Eq. 9.18.

Edwards Air Force Base, California. An atmospheric diffusion program (Project Sand Storm), aimed primarily at the development of quantitative statements of dilution rates of pollutants released as quasi-instantaneous volume sources, was carried out at the Air Force Rocket Propulsion Laboratory test facility at Edwards Air Force Base, Calif. (Taylor, 1965). All 43 experiments were carried out under unstable conditions. The sources consisted of solid-propellant rocket motors that contained known amounts of finely divided metallic beryllium in mixture with the propellant. Firing durations ranged from 2 to 8 s, and puffs were produced with initial visible diameters of 15 to 45 m. The tracer was collected on membrane filters located along circular arcs about the firing point. Originally the grid consisted of 10 such arcs extending from 100 to 2400 m. After 14 experiments all the samplers were consolidated on six arcs extending from 200 to 2400 m, thus increasing the sampler density.

Meteorological data were obtained from a 60-m tower located 60 m upwind of the release point. Temperature sensors, as well as high-response wind systems and bivanes, were mounted on the tower. Information on the visual appearance of the puffs and some measurements of puff dimensions were available from phototheodolites.

Since all the experiments were conducted under thermally unstable conditions, it was assumed that the vertical rate of growth was positively correlated with the lateral rate of growth. Therefore an attempt was made to use measured values of σ_θ^2, wind speed, and distance to estimate the normalized exposure. Regression analyses using the measured exposures and meteorological data yielded the result that the inclusion of the wind-speed and fluctuation data contributed little to the reduction of the variance obtained when distance was used as the sole criterion. The authors surmise that, since all the tests were conducted under thermally unstable conditions, the meteorological and diffusion data tended to occur within a very limited range. Had tests also been conducted under stable conditions, a greater range of data would have been observed, and it is quite likely that the correlations would have been higher.

New Mexico. Observations of smoke puffs were made by Kellogg (1956) and by Frenkiel and Katz (1956). In Kellogg's experiments, vials of titanium tetrachloride and water were attached to trains of balloons and exploded at predetermined altitudes by small charges of cordite set off by baroswitches. The release altitudes ranged from 7.1 to 18.9 km, and the resulting puffs of smoke were observed by phototheodolites. From the records of the phototheodolites, both the positions and visible sizes of the puffs were obtained. Eighteen usable sets of observations were reported in the form of graphs of visible diameter against time, up to a maximum time between 3 and 11 min. The smoke puffs observed by Frenkiel and Katz were released within the first 100 m above a water surface during unstable conditions. The puffs are generated by exploding small charges of gunpowder carried aloft by a tethered balloon. Positions and sizes of the smoke puffs were obtained from photographs at 1-s intervals, and the data on smoke-puff radius were tabulated by Frenkiel and Katz for 19 cases, with total durations of observation ranging from 7 to 20 s. The results from these New Mexico experiments are discussed in the summary.

Dugway, Utah. A series of bomblet-produced puffs, both gaseous and particulate released at the Dugway Proving Ground, Utah, were the subject of an intensive investigation by Cramer et al. (1964). Thirty-three different experiments were used in the analysis. These experiments consisted of small explosive releases carried out primarily during the night over densely instrumented sampling grids extending to about 1700 m from the source in the case of the particulates and to over 100 m for the gaseous-release series. Release times were either 3 or 26 s. Continuous records of wind speed and direction were available for each experiment and were used to furnish the values of σ_θ and $\bar{u}$ required by the models. No estimates of deposition or absorption were made. Since the efficiency of the various disseminators used in the tests varied widely, source strengths were estimated from measures of the flux of material through a dense array of towers surrounding and close to the source point.

In the interest of documenting the lateral and vertical plume spread observed during these experiments, average values of these data are presented in the summary diagrams (Figs. 9.18 and 9.19). The lines representing the average values of the two sets of longer range tests in each of the diagrams contain the corrections for initial finite puff size derived by the authors.

Point Arguello, California. Another experiment that can be used to furnish instantaneous-source diffusion information was conducted at the Naval Missile Facility at Point Arguello, Calif. (T. B. Smith et al., 1964). This experiment consisted of both quasi-instantaneous and longer releases of zinc cadmium sulfide and uranine dye over an irregular network of sampling stations at the coastal California site. The site itself is quite irregular with individual ridges and peaks rising to well over 300 m. Sampling was conducted along irregular lines surrounding the release point. It was not possible to obtain puff dimensions and exposures at a number of distances down-

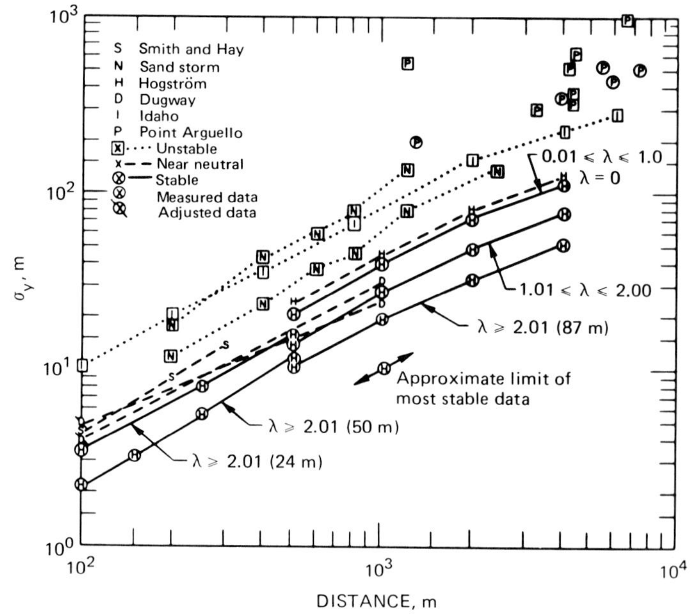

Fig. 9.18 Summary diagram for measured σ_{yI} data. Note that measured data have been differentiated from those to which some adjustment has been applied. Relative dispersion from the Swedish experiments by Högström has been included. (From Islitzer and Slade, 1968.)

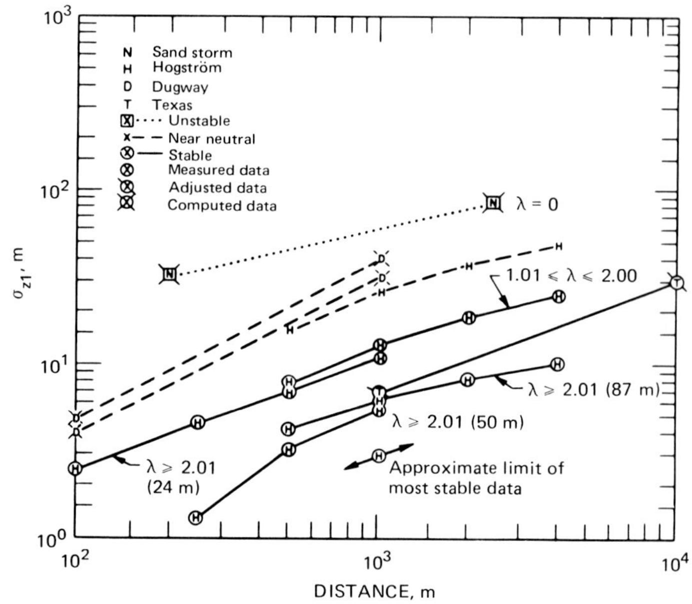

Fig. 9.19 Summary diagram for measured σ_{zI} data. Computed data have been added for the Sand Storm series because of the lack of information on instantaneous-source vertical spreading during unstable conditions. Relative dispersion from the Swedish experiments by Högström has been included. (From Islitzer and Slade, 1968.)

wind for each puff. However, the number of samplers was sufficient to obtain the width of a puff and an estimate of the maximum exposure at one distance for each release. The results of a number of 60-s releases, which were not complicated by release or sampling in the vicinity of a pronounced ridge, are presented in the summary Fig. 9.18.

Idaho Falls, Idaho. Islitzer and Markee (1964) reported on quasi-instantaneous releases that consisted of puffs of radioactive material emitted during tests at INEL in southeast Idaho. Release times for four tests were about 30 s, and the resulting material was sampled over instrumented grids to 750 m in three cases and to over 6000 m in the fourth case. The lateral puff spread was found to be more closely approximated by the square, rather than the first power (Eq. 9.19 of the Dallas Tower experiments), of the lateral turbulence intensity. It is interesting to note that T. B. Smith et al. (1964) also found such a relation. Although there is some doubt as to the sampling and averaging times to apply to the wind data to obtain the appropriate intensity of turbulence for Smith and Hay's (1961) Eq. 9.18, there appears to be merit in the use of this formulation for predicting puff dimensions directly from wind-fluctuation measurements.

Since the source strength was not accurately known in these experiments, only relative exposure data were obtained. The plume-width measurements for the one long-distance experiment are given in the summary Fig. 9.18.

Hanford, Washington. The data from eight experiments of puffs of ^{85}Kr released near the surface are reported by Nickola, Ramsdell, and Ludwick (1970). An analysis of these data by Nickola (1971) found that regardless of stability $\sigma_x > \sigma_y > \sigma_z$.

Instantaneous Source Summary. An interesting result from relative diffusion theory is the prediction that the particle variances from an instantaneous source, σ_{xI}^2, σ_{yI}^2, or σ_{zI}^2, initially increase as the third power of the travel time. Gifford (1957) reexamined individual runs from Kellogg's data and found that the stratospheric data tended to support a t^3 regime initially and to change to a t^2 regime for particle-variance growth after some characteristic time of travel. The upper tropospheric smoke studies, however, best followed a t^2 dispersion regime throughout all travel times. A similar t^3 regime was found for the smoke-width measurements given by Frenkiel and Katz (1956) for about 12 s with a definite transition to a t^2 law after that time. For Kellogg's stratospheric data, the transition time was usually about 3 min. Crawford (1966) concluded that a best fit of the relative dispersion data over all travel times showed $\sigma \propto t^{1.2}$ This was confirmed by Randerson (1972) for travel times between 1 and 10 h. A later analysis by Gifford (1977) which included some other tropospheric relative diffusion data seemed to indicate that a t^3 regime fit the data more closely at travel times of less than 3000 s and showed slower growth at later times. Careful examination of summary Figs. 9.18 and 9.19, where the data have been grouped by Högström's stability parameter λ, shows that the horizontal and vertical diffusion data as a function of distance have slopes of one or less.

9-3.2 Special Case Diffusion Experiments

The experiments in this section include all those where flow complications existed that are not adequately described by the classical diffusion theory. For instance, frequently the flow is influenced by the occurrence of multiple boundary layers after a change of terrain, such as near a coastal region, mountain or ridge-and-valley wind systems, forest and clearcut areas, and nearby cities. Some of the other experiments that have been included are in areas with simple topography or at least homogeneous rough terrain but are complicated owing to low wind speeds, excessive wind shear, and long-range plume travel. Comparison of results even in similar categories is difficult, usually because the objectives of the experiments were different and the nature of the data collected reflected those differences. Some attempts at obvious comparisons are made.

*9-3.2.1 **Shoreline Diffusion.*** Some of the experiments described in the previous sections also belong under the category of shoreline diffusion: Ocean Breeze, Dry Gulch, and possibly Mt. Iron. However, these experiments were not designed to determine the effect of the ocean on the diffusion environment but to answer some other operational problem of the sponsor. None of the tracer was released over water in any of these experiments.

The experiments described in this section can be divided into three types: (1) tracer release and tracer sampling conducted over water, (2) tracer release over water and sampling conducted inland, and (3) tracer release and sampling conducted inland.

Long Island, New York. In response to the possibility of locating electric power plants offshore, Brookhaven National Laboratory initiated a study that included tracer releases off the south shore of Long Island, N. Y. (Raynor et al., 1975).

The release point was a boat located 2 to 6 km off Tiana Beach. An oil-fog tracer was discharged 6 to 7 m above the surface during onshore winds. Another boat, with a photometric densitometer aboard, traversed the plume between the source point and the shore. An additional densitometer was aboard a van traveling along the beach. Plume-growth data were derived from photographs taken from the boat and from an aircraft used for meteorological measurements.

Two towers, one on the beach and the other on a parking lot overlooking the beach, were used to gather meteorological data representative of overwater values. The overwater roughness length was found to vary from 1×10^{-6} to 2×10^{-3} m.

Most of the tests were conducted with southwest winds which bring air from the mainland out over the cooler water and then across Long Island. The modification of the temperature structure of the marine air as it passed over land is shown in Fig. 9.20. The main control on the atmospheric stability was found to be the air–sea temperature difference.

During stable conditions (positive Richardson number), over half the σ_y values are less than those of the Pasquill F-curve, an indication that horizontal diffusion was substantially less over water than would be found over land in similar stability conditions. The unstable cases, although fewer, suggest that horizontal diffusion over water can be comparable to that over land in these conditions. The

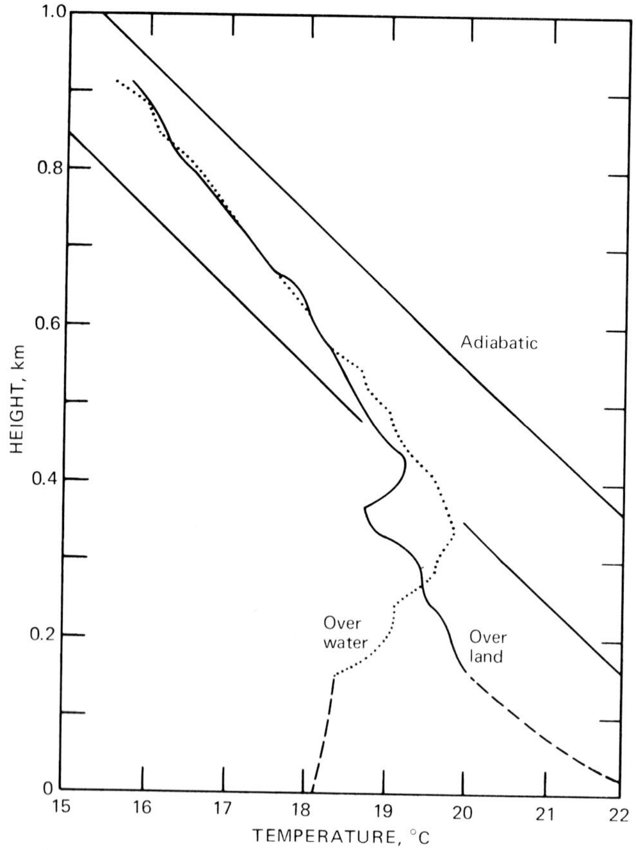

Fig. 9.20 Illustration of the modification of the overwater temperature profile as the marine air passes over land. [From G. S. Raynor et al., Studies of Atmospheric Diffusion from a Nearshore Oceanic Site, *J. Appl. Meteorol.*, 14: 1080-1094 (1975).]

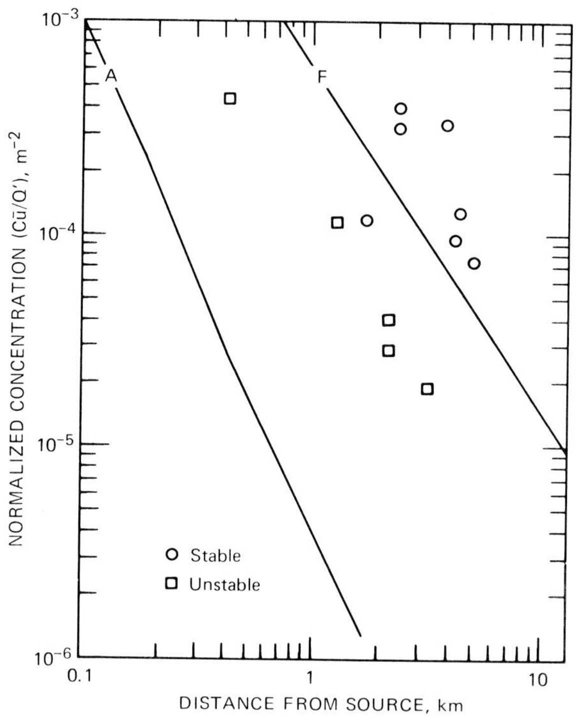

Fig. 9.21 Normalized concentration observed during the overwater Long Island diffusion trials. —, predictions based on Pasquill A and F stability curves. [From G. S. Raynor et al., Studies of Atmospheric Diffusion from a Nearshore Oceanic Site, *J. Appl. Meteorol.*, 14: 1080-1094 (1974).]

authors found σ_z tends to be less over water than over land in both stable and unstable conditions. These conclusions are supported by the concentration data of Fig. 9.21. The data points for σ_y and σ_z are shown in the summary graphs for this section.

Oceanside, California. A fluorescent-particle tracer diffusion study was conducted along the California coast between Oceanside and Del Mar. The project consisted of a number of different types of releases and sampling procedures designed to characterize diffusion in the shoreline region. A data summary and the results are given by T. B. Smith and Niemann (1969).

The topography of the test site consists of a coastal plain gradually increasing in elevation inland with a minimal vegetation cover. A 20- to 30-m bluff adjacent to the beach obstructs onshore flow.

The field program was divided into offshore and onshore releases. There were two sets of offshore releases, line-source and point-source. Line-source releases were conducted simultaneously from a boat and an aircraft; the point-source releases were done from the boat. In addition, 55 instantaneous point-source releases were made on land with the plume remaining over land. Most of the releases were made in the late afternoon. Eight additional offshore point-source releases (Smith and Beesmer, 1967) 90 km north of the Oceanside area are also reported by Smith and Niemann.

The onshore sampling network for the offshore releases extends 16 km crosswind and 15 km downwind. Several locations within this area were equipped for greater sampling resolution in both the horizontal and vertical. All sampling was conducted inland. The horizontal exposure data were fitted to a Gaussian curve to determine σ_y for a particular trial because the samplers were not located on uniform arcs.

One of the criteria for the selection of the test site was a consistent onshore flow. The area is characterized by a marine inversion usually based from 300 to 600 m with onshore flow below the inversion and a deep seaward flow aloft. The winds at the inversion base were less than at the inversion top.

There was considerable modification of the temperature profile as the air progressed inland. During the day, warming was found below the inversion with a slight cooling within the inversion due to mixing of cooler air from below. Generally the convection inland increased the height of the mixed layer during the day, but no change of mixed layer relative to sea level was exhibited during the evening. Areas adjacent to the coast experienced little diurnal variations of mixed-layer height.

Analysis of the turbulent velocity covariance showed large random eddies with periods of about 5 s generated either over the ocean or by the coastal discontinuity. These eddies degenerated into higher frequency turbulence inland and produced large variability in the total turbulent energy.

The effect of these eddies was also observed in the diffusion data. The σ_y values of the onshore point-source trials (Fig. 9.22) were evaluated by Smith and Niemann (1969) on the basis of a power law growth given by

$$\sigma_y(x) = (x/r)^\alpha \, \sigma_{y_r} \tag{9.20}$$

where σ_{y_r} is determined at some reference distance r. The average value of α was about 0.8 to the first arc and 1.4 to the second arc. The increasing turbulence inland accounted for part of the larger growth to the second arc. The largest increase was exhibited from 3 to 7 km from the source where the plume size was about the same as the horizontal scale of turbulence, 200 to 400 m.

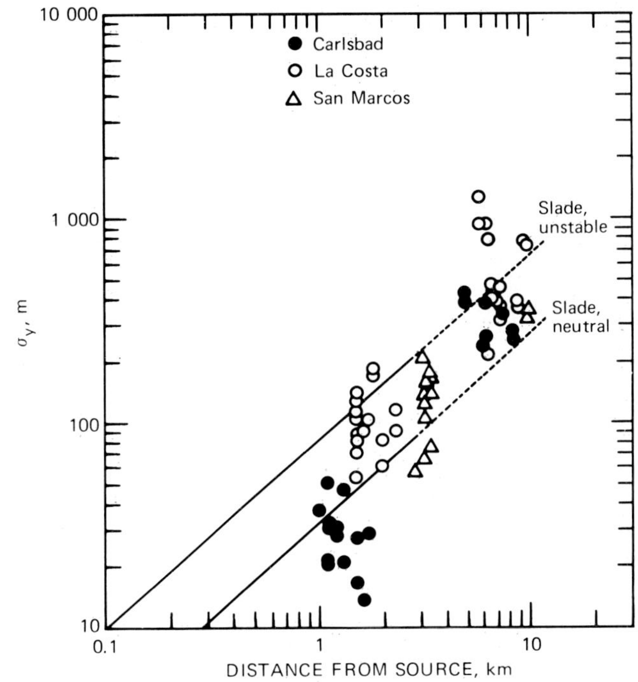

Fig. 9.22 Inland instantaneous point-source releases at varying distances from the coast. Solid and dashed lines are from Slade (1966) for instantaneous sources. (From Smith and Niemann, 1969.)

The offshore line-source releases showed rapid upward mixing before the plume reached shore. The vertical concentration profile was practically homogeneous. There was a slight upward displacement of the plume at the coastline. Further downwind the exposures were primarily determined by the depth of the mixed layer (Fig. 9.23).

In addition to the five offshore point-source releases at Oceanside, the eight releases from Bolsa Island, 1.3 km offshore, are shown in Fig. 9.24. The authors found very slow plume growth over water, roughly corresponding to the rate that would be observed under stable stratification over land. There was an abrupt transition at the shoreline to a more rapid growth representative of the inland meteorological environment. The authors recommend treating the overwater releases as two distinct regimes, with the overwater part of the trajectory being treated in the usual manner and a virtual point-source treatment when the plume moves inland.

Victoria, Texas. A summary of 17 elevated tracer releases offshore of the Texas Gulf coast near Corpus Christi is presented by Smith and Miller (1966). Two tracers were released simultaneously: fluorescent particles (FP) that were 2 to 4 μm in diameter and large particles (LP) of about 20 μm in diameter.

The terrain is flat, rising only 120 m in the first 120 km, then becoming gently rolling hills to 160 km inland. The initial vegetation of rice paddies changes to low crops and eventually light tree cover going inland. No significant flow obstacles exist in the area.

The tracers were released from two jet aircraft, one for each tracer. The flight path was a few kilometers offshore and parallel to the coast. Most of the releases were at a height of 90 m. The actual source configuration is difficult

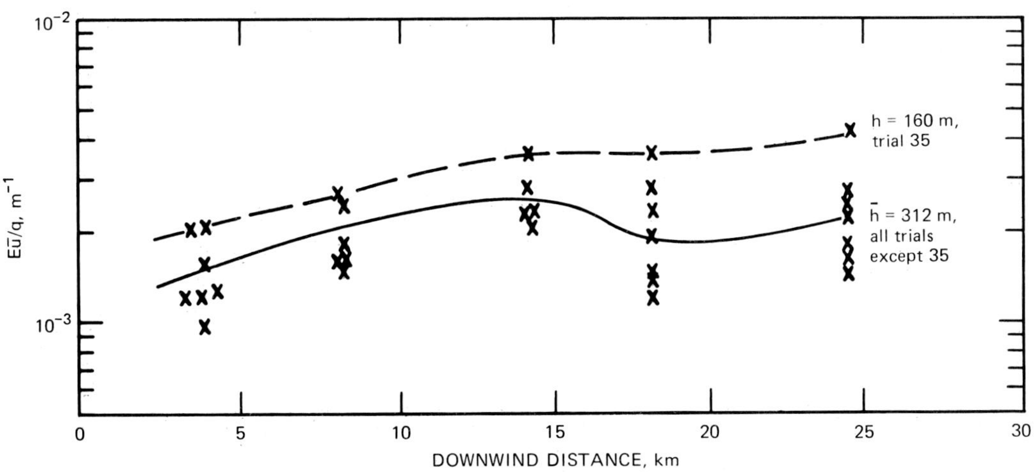

Fig. 9.23 Downwind exposure for the instantaneous line-source releases. The mean mixed depth was 312 m for all trials except trial 35 where much higher exposures were observed. (From Smith and Niemann, 1969.)

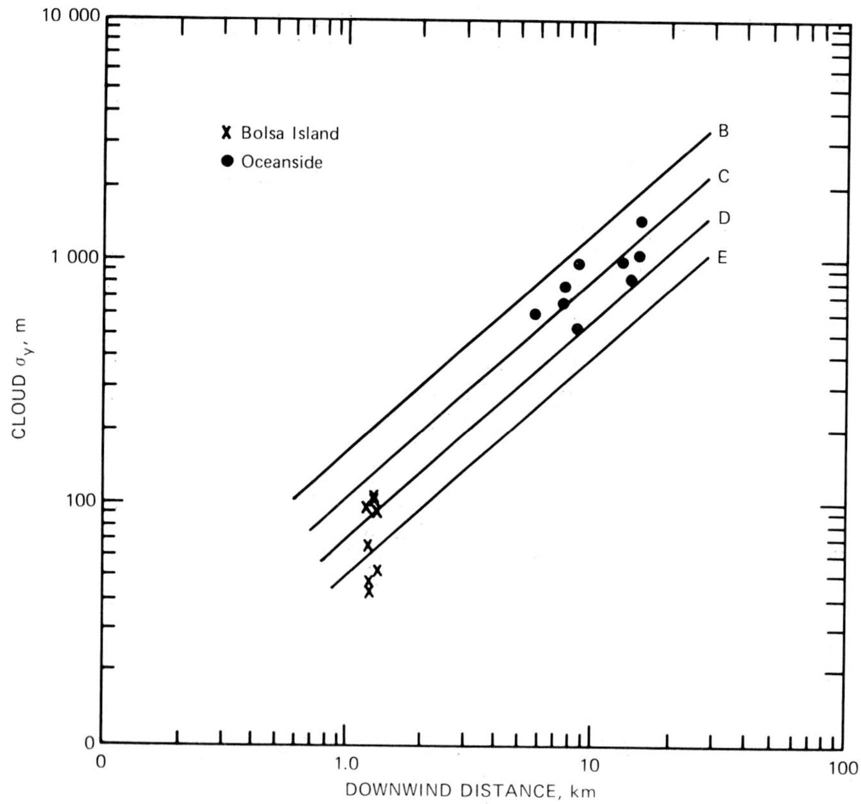

Fig. 9.24 Continuous overwater point-source releases at Bolsa Island and Oceanside. Plume growth at Bolsa Island is representative of overwater travel and at Oceanside the samples were taken several kilometers inland. —, Pasquill–Gifford curves. (From Smith and Niemann, 1969.)

to determine since the aircraft wake distributed the material nonuniformly in the vertical. The authors point out that the wake effect is in itself affected by the ambient turbulence.

Six main sampling lines were set up. Four of them were parallel to the coastline, about 160 km in length and spaced evenly from 40 to 180 km downwind of the release line. Another 16-km sampling line was set up along the beach. The last sampling line was normal to the others, through the midpoint of the release line going downwind. Two towers and a parafoil kite near the downwind end of the grid permitted limited vertical sampling.

The program was conducted during the summer when the synoptic pattern is such that the predominant flow is onshore (from the southeast) with the sea breeze manifesting itself only as variations in the wind speed. The water is sufficiently warm to produce a deep (200 to 600 m) adiabatic layer above the surface. Inland modification of the air resulted in a frictional shear layer 60 to 90 m deep near the coast below the neutral maritime air. This layer increased to a depth of 200 m inland. At night no real turbulent layer was formed.

Spectral analysis of the wind-fluctuation data showed that the peak turbulent energy of the maritime air was at about a wavelength of 210 m. Little energy is required to produce motions in the neutral marine air, and further analysis of the temperature data confirmed the presence of even lower frequency oscillations.

Tracer and visible smoke releases showed that there was rapid dispersal after release to the top of the neutral layer. However, the tracer concentrations were not uniform at this stage. It was found that the exposure of the inland samplers depended on the amount of material injected into the shear layer, and hence exposure correlated well with the depth of the neutral layer (Fig. 9.25). The LP arrived at the samplers as much as 2 h before the FP, probably because the large particles from the faster moving air aloft settled out before the smaller particles. Also, dispersion in the alongwind direction was found to be greater in light winds, possibly owing to retardation of some of the material by the terrain.

Corpus Christi, Texas. A set of four tracer experiments in various parts of the United States has been described in a summary volume by Smith and Wolf (1963). One of these experiments includes nine aircraft releases offshore of Corpus Christi, Tex. The coastal region of Corpus Christi is similar to that of the Victoria area.

The tracer, FP, was usually released at night from an aircraft flying parallel to the shoreline and several kilometers offshore. One sampling line extended downwind for 40 km near the release center. Three towers along the sampling line were used for meteorological measurements but not for vertical sampling.

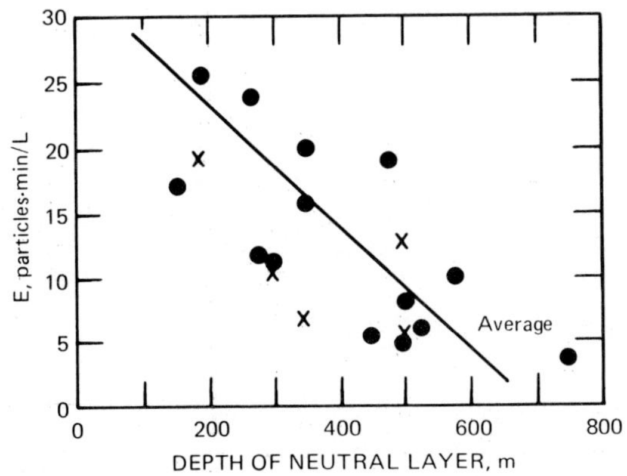

Fig. 9.25 Exposure over land from an instantaneous line-source release over water. (From Smith and Miller, 1966.)

Very warm water temperatures produced an adiabatic layer 300 to 500 m deep by late afternoon. During the evening the atmosphere above the water surface became slightly stable. The diurnal temperature change was small.

Although not observed by the meteorological sensors, a helical downward flow was presumed by the authors on the basis of the variability and rather fast appearance of the tracer at the surface samplers (15 to 30 m/min to the first arc, 2 km from release line). Owing to the fast downward transport, no model was found to compare favorably with the data.

Shoreline Summary. The problem of diagnosing the diffusion of atmospheric pollutants near coastlines and over water (i.e., very smooth terrain) has been studied by several authors. For example, Prophet (1961) presented a survey of diffusion and meteorological data collected from ocean and shoreline areas. He pointed out that the diffusion was primarily related to the air–water temperature difference and that the depth of the boundary layer (or the inland modification of the temperature profile) was proportional to the square root of the distance from the coast. Slade (1962) analyzed wind records for flow across the Chesapeake Bay. He found that during westerly flow, $\bar{u}$ increased and σ_θ decreased from the west to east side of the bay, which is the overwater trajectory. Table 9.7 summarizes some of Slade's results.

Van der Hoven (1967) reviewed atmospheric transport and diffusion experiments at coastal sites since 1961. Emphasis was placed on reactor siting studies. Van der Hoven also adapted a figure (Fig. 9.26) from Prophet to show the variation of vertical mixing coefficient with air–water temperature differences. He also presents a method, suggested by Prophet, of determining the distance downwind from the coast where the turbulent surface layer will intersect an elevated plume. Collins (1971) confirmed this method by some oil-fog tracer tests at a site in eastern Massachusetts. The growing inland boundary layer will contain a ground-level source until it intersects the marine inversion. Therefore only elevated sources at coastal sites during onshore flow will be initially dominated by the marine dispersion environment. Later these sources are subject to fumigation when the elevated plume intersects the growing inland boundary layer.

Hosker (1975) compares a number of models to the Brookhaven and Bolsa Island experiments. The best predictions of σ_y and σ_z were from those models which included the wind direction fluctuation (σ_θ).

Gifford (1976) suggested methods to account for the varying roughness of the sea surface and the effect of evaporation on stability estimates.

Fig. 9.27 represents the normalized exposure for all line-source releases over water with inland sampling. Results of these experiments demonstrate that the downwind exposure is inversely proportional to the depth of the mixing layer. The rate of vertical mixing was large enough that, in most cases, the material was distributed throughout the marine layer before reaching shore, suggesting a simple box model can be used to simulate inland exposures.

Some measurements of natural hydrocarbon emissions offshore of Santa Barbara, Calif., have been reported by Harrison and Maas (1976). Several passes were made across the plume up to 3.2 km downwind by a boat with a

TABLE 9.7 Average Wind-Direction Fluctuation and Wind Speed for Land and Water Trajectories*

Time	$(T_w - T_a)$†	Number of cases	Land σ_θ, deg	Land $\bar{u}$, mph	Water σ_θ, deg	Water $\bar{u}$, deg	Ratio of C_{water} to C_{land}
Day	>+2° F	52	10.4	12.9	5.8	18.7	1.5
	+2 to −7° F	25	10.3	11.6	5.3	16.5	1.6
	<−7° F	54	11.5	15.6	5.0	18.1	2.9
Night	>+2° F	118	9.5	7.9	5.2	15.9	0.9
	+2 to −7° F	30	10.1	11.4	5.1	16.6	1.6
	<−7° F	24	10.1	11.0	4.3	13.2	2.4
	Total or average	303	10.2	11.0	5.2	16.7	1.5

*From I. Van der Hoven, Atmospheric Transport and Diffusion at Coastal Sites, *Nucl. Safety*, 8(5): 490-499 (1967).

†T_w is the average 24-h water temperature, and T_a is the hourly average air temperature on the west shore.

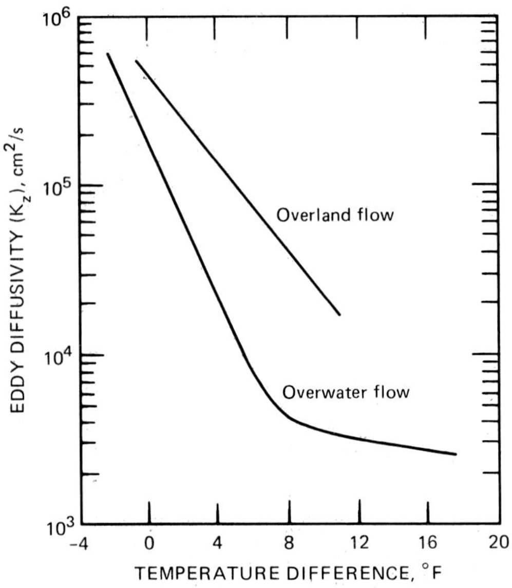

Fig. 9.26 Vertical eddy diffusivities computed from experimental observations of the height to which vertical mixing extends and the corresponding travel time. (From Prophet, 1961.)

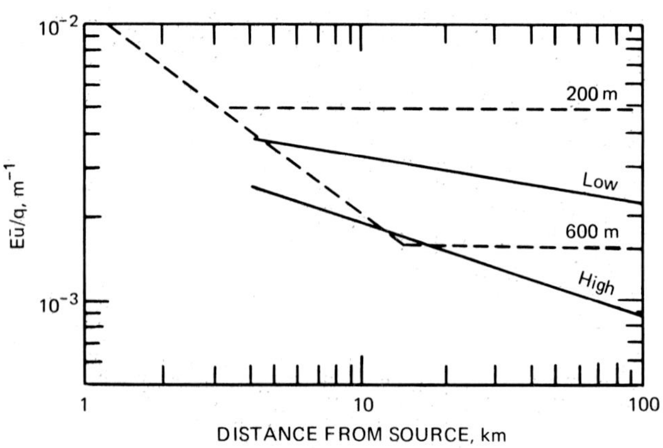

Fig. 9.27 Normalized exposure for composite of Victoria, Oceanside, and Corpus Christi aircraft line-source trials. "Low" and "High" classify experimental trials according to low and high mixing depths. The dashed line is calculated for a C stability leveling off at mixing depths of 200 and 600 m, the range between low and high.

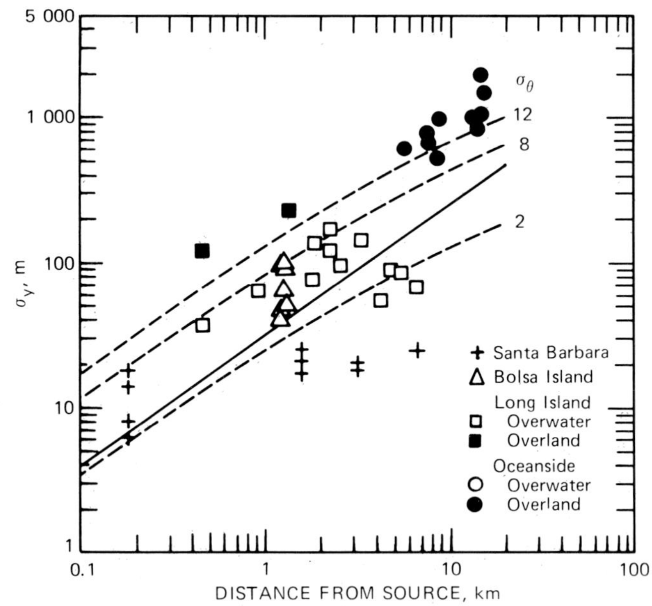

Fig. 9.28 Continuous point-source releases over water or in offshore flow situations. —, Pasquill–Gifford F stability; – – –, calculated by Eqs. 9.8 and 9.10 for the σ_θ values shown.

hydrocarbon analyzer to obtain atmospheric hydrocarbon cross sections. No meteorological observations were reported. The data points are included in summary Fig. 9.28, which presents the σ_y values observed over water and onshore for the overwater continuous point-source releases. The dashed lines represent predictions of σ_y based on Eqs. 9.8 and 9.10 which take into account wind-direction fluctuation. The σ_θ values used in the calculation of the curves in Fig. 9.28 represent the approximate range observed over water between stable (2°) and unstable conditions (8°) and during offshore or inland flow situations (12°). This figure shows that the overwater horizontal dispersion is related to the horizontal wind-direction fluctuations in the same way as over land, although σ_θ over water is much less than over land (see Gryming, Lyck, and Hedegaard, 1978).

The vertical dispersion coefficient, σ_z, was computed for the overwater continuous point-source releases assuming continuity (Eq. 9.6). This assumes that there was no deposition loss of tracer. The results are shown in Fig. 9.29 for Santa Barbara and Long Island, where only gases were used as tracers. The solid curve represents the Pasquill–Gifford F stability curve. The dashed line is the curve of σ_z obtained from

$$\sigma_z = (2K_z t)^{1/2} \tag{9.21}$$

where K_z, the overwater mixing coefficient, is obtained from Fig. 9.26 for an air–water temperature difference of 8°F, representative of the Long Island data. As many authors have pointed out, the air–water temperature difference is the primary factor affecting vertical diffusion over water.

Two experiments, Oceanside and Victoria, also collected sequential samples used to determine duration of plume passage. The duration of plume passage depends on the release duration and the alongwind dispersion, σ_x. Since these were both instantaneous line-source releases, σ_x can be obtained directly in units of minutes by taking the second moment of the concentrations from the sequential sampler. Figure 9.30 shows σ_x as a function of travel time from the source. The solid line is the least-square regression line

$$\sigma_{x_{min}} = 0.09\, t_{min}^{1.1} \tag{9.22}$$

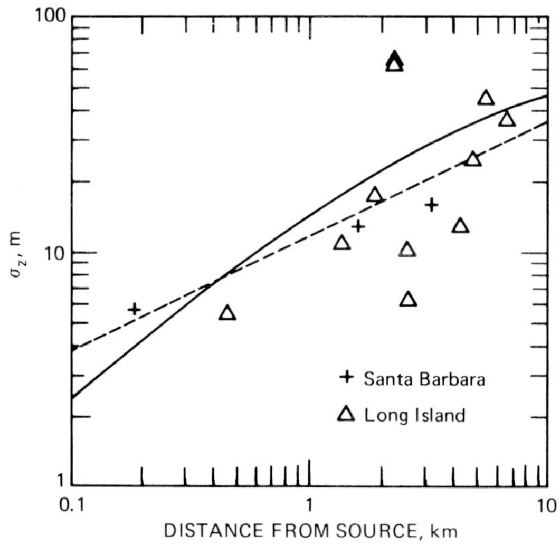

Fig. 9.29 Vertical diffusion at Santa Barbara and Long Island for overwater releases and sampling. —, Pasquill–Gifford F curve; – – –, σ_z by Eq. 9.21 with K_z from Fig. 9.26.

which has a slope that is larger than the usual slope of 0.8 observed for horizontal dispersion.

9-3.2.2 Rough Terrain Diffusion. The rough terrain experiments were conducted under a wide variety of conditions. In general, the experiments are limited to homogeneous terrain but with large roughness elements where σ_y and a roughness length still could be determined. Experimental sites with major flow obstacles which make estimates of σ_y difficult are included in the complex terrain section (9-3.2.3).

At some of the sites, major terrain elements exist off the experimental grid, but their effect on the observed dispersion has not been recorded. Most of the experiments have been performed in valleys and forests, over farmlands, and near urban centers. Some of these experiments could belong to other sections, but they were compiled here because the experiments were designed to measure the effect of increased roughness rather than the effects of large terrain features.

Mt. Iron, California. This area, near Pt. Arguello, Calif., was selected for FP tracer releases in rough terrain (Hinds, 1967). The site is a mountainous area south of Vandenburg Air Force Base, where the Dry Gulch program was conducted. The terrain configuration is of the ridge–canyon type with dense, low shrubs covering the ridges and grasses on the lowlands.

Horizontal winds were measured at the release point and in the outlying regions. Temperature profiles by wiresonde and hourly radiosonde releases were also made. There were 113 tracer releases of 15- to 30-min duration from two different source points. The sampling network was set up on existing roads and trails. Since the arcs were not concentric about either of the source points, σ_y values were estimated from isopleths of exposure.

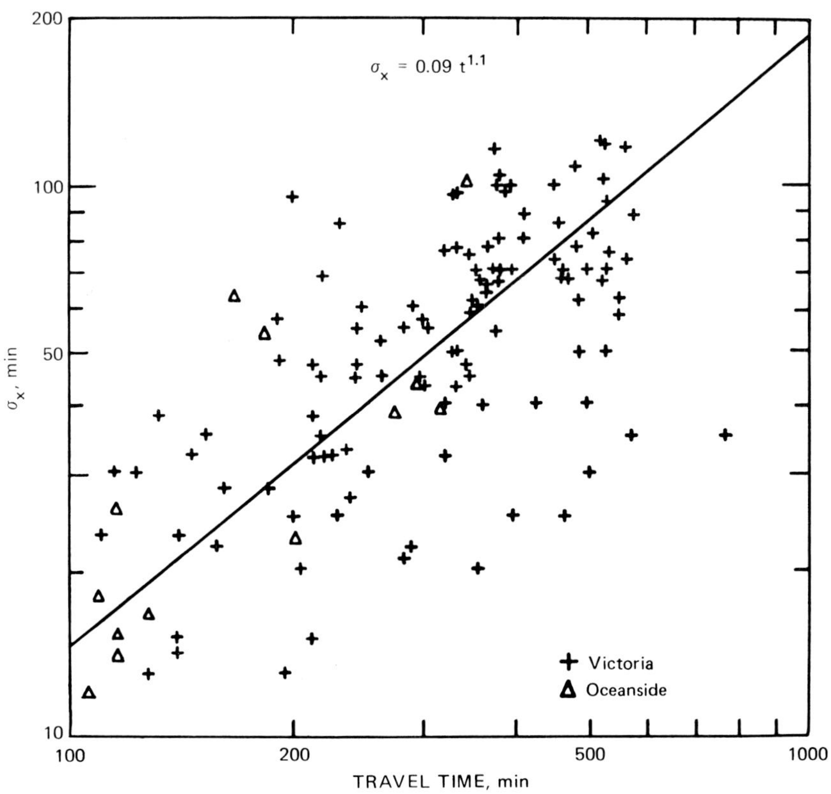

Fig. 9.30 Second moment of plume duration from sequential samples taken at Oceanside and Victoria. Regression equation is given (—) for σ_x and t in minutes.

One of the effects of rough terrain was to produce marked inhomogeneities in the wind field. Wind speeds were found to increase at ridge tops and decrease in the canyons. A characteristic of the marine climate of the region was a lack of any strong surface inversion at night; the nocturnal lapse rate was usually slightly stable or near neutral.

Comparing the different exposure patterns, Hinds categorized the plume configurations into three groups. The first contained 27% of the cases and was called "modified cigar" by the author. This type showed monotonically decreasing exposures. The second group, containing 67% of the cases, was called the "canyon low" and showed plumes interspersed with low exposures coincident with the canyon floors. The remaining 6% were similar to the canyon low but with more complex structure. The latter two types were found only at night.

The average ratio of peak exposures between the canyon and the downwind ridge was about 0.5 during the day and 0.4 at night, whereas the range of ratios was between 0.1 and 0.8. In all cases the ridge exposure was higher. No tracer was collected in the stagnant pockets of air in the canyon floors.

The comparison of normalized peak exposure as a function of distance for Dry Gulch, Hanford, and Mt. Iron is shown in Figs. 9.31 and 9.32. The major difference to be noted here is that the climate at Hanford is more continental and therefore very stable conditions may be present during the night hours.

The night tests (Fig. 9.31) show variations between summer and winter, but the seasonal variation is not so pronounced for the daytime tests (Fig. 9.32). During winter nights, when radiational cooling is more significant in rough terrain, exposures were higher over the smoother terrain (Dry Gulch and Hanford) where the drainage did not counterbalance the effects of stratification. Generally, the effects of terrain and climate differences are enhanced during nighttime dispersion. During the summer, when nighttime profiles were near neutral, the exposures were almost the same for the two sites.

From an analysis of all the diffusion data, the author concluded that σ_y fits a power law which is similar to that found by others, namely,

$$\sigma_y = 0.65x^{0.75} \tag{9.23}$$

and which is shown in Fig. 9.33.

Jülich, Federal Republic of Germany. Forty-four tracer experiments at the Jülich Nuclear Research Center have been reported by Vogt and Geis (1974). The test area is located in the Rhine River valley and is composed of forest, farming areas, and small urban centers. The roughness length was found to be 1.8 m at the release site. Copper sulphate aerosols tagged with ^{65}Cu (half-life 12.6 h) were released at heights of 50 m (28 trials) or 100 m (16 trials) and sampled up to 10 km downwind by battery-operated radio-controlled filters.

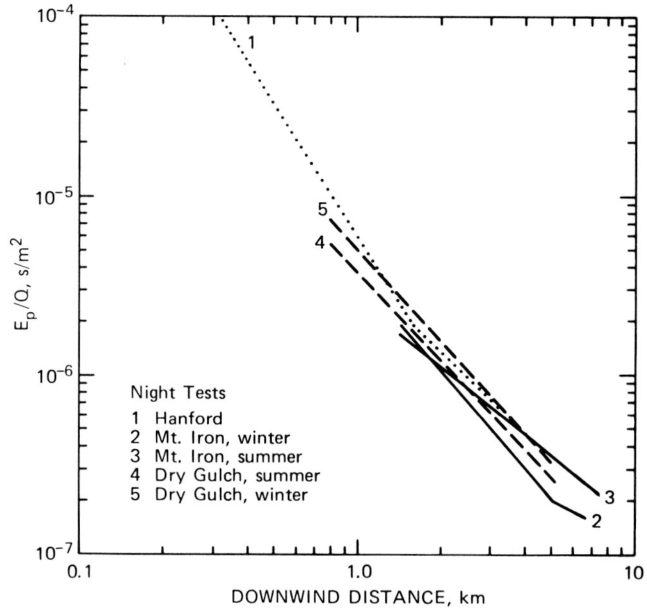

Fig. 9.31 Exposure averages from Mt. Iron, Dry Gulch, and Hanford nighttime tests. (From Hinds, 1967.)

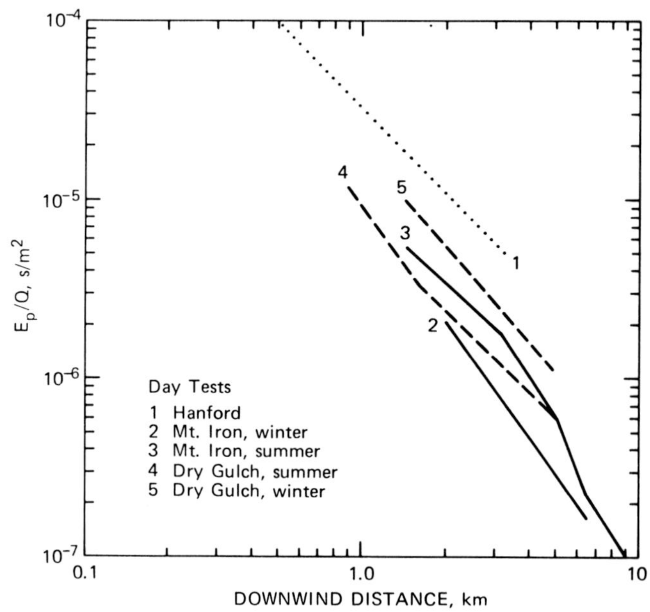

Fig. 9.32 Exposure averages from Mt. Iron, Dry Gulch, and Hanford daytime tests. (From Hinds, 1967.)

The authors used regression methods to determine the coefficients for

$$\sigma_y = p_y x^{q_y}$$
$$\sigma_z = p_z x^{q_z} \tag{9.24}$$

The coefficients are summarized in Table 9.8. Generally, the authors found that dilution is increased by rough terrain by a factor of 2 in unstable conditions, by a factor of 4 in neutral conditions, and by a factor of 2 to 3 during

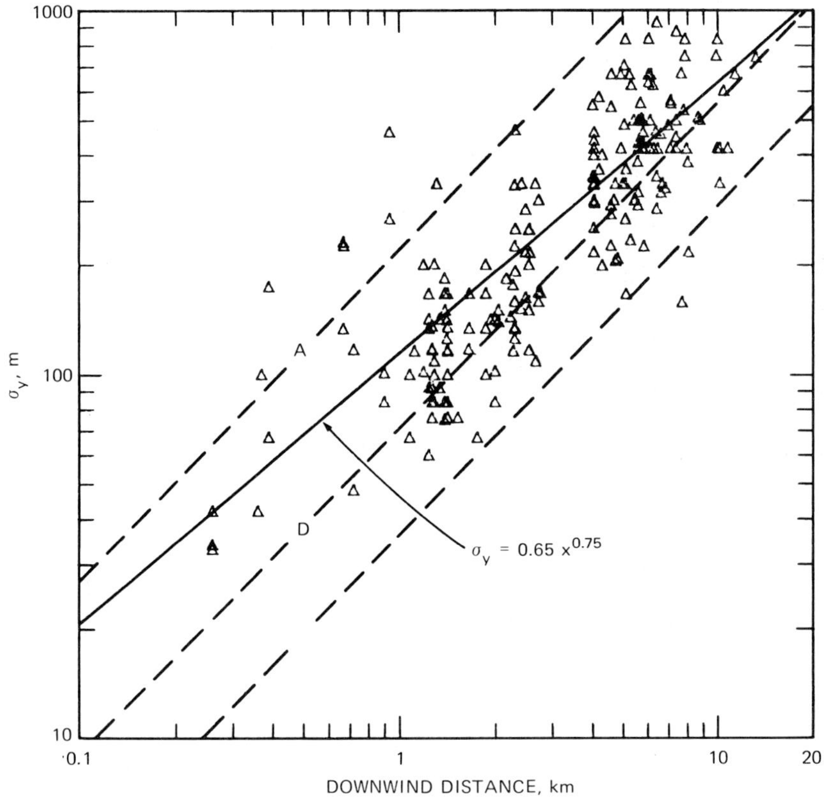

Fig. 9.33 Horizontal diffusion data from Mt. Iron. —, regression line; – – –, Pasquill–Gifford curves.

TABLE 9.8 Mean Values of the Diffusion Coefficients p_y, q_y, p_z, and q_z for the Different Diffusion Categories*

Diffusion categories	B	C	D	E	F
p_y	0.8685	0.7184	0.6248	1.691	5.382
q_y	0.8097	0.7837	0.7672	0.6211	0.5778
p_z	0.2222	0.2149	0.2048	0.1616	0.396
q_z	0.9680	0.9438	0.9358	0.8094	0.6183

*Data from K. J. Voght and H. Geis, *Tracer Experiments on the Dispersion of Plumes over Terrain of Major Surface Roughness,* Report Jul-1131-ST, Central Division for Radiation Protection, Kernsforschungsanlage Jülich GmbH, Jülich, Federal Republic of Germany, 1974.

stable stratification. The increased dilution was always greater during neutral stability.

Further analysis of the Jülich data presented by Vogt (1977) showed evidence for the release-height dependence of σ_z but not σ_y. Further, σ_z was found to be independent of release height (50 to 100 m) during stable stratification and independent of stability for the lower level release (50 m) during unstable and neutral conditions.

A summary of the 50-m release-height curves of σ_y and σ_z for Jülich is shown in Figs. 9.34 and 9.35. The large values of σ_y during E and F stabilities may be due to wind meander during the hour-long release.

Karlsruhe, Federal Republic of Germany. At the Karlsruhe Nuclear Research Center, about 25 experiments, mostly in neutral stability, are reported by Thomas et al. (1974) and Thomas and Nester (1974). The test area is located in the Rhine River valley, and the roughness length was measured to be 1.1 m. A fully instrumented 200-m meteorological tower collected data during the trials.

Tritiated water vapor and halogenated hydrocarbons were released from a height of 100 m for about 20 to 30 min. Sampling was conducted to 5 km from the source and analyzed by liquid scintillation spectrometry for tritiated water vapor and electron capture gas chromatography for the hydrocarbons.

The vertical dispersion results did not show the steep rise of σ_z during unstable conditions as exhibited by the Pasquill–Gifford curves. The authors attributed this to the release elevation. Most of the data were collected in unstable conditions, and the observed σ_z was larger than the corresponding one of Pasquill–Gifford for the same stability. The largest increases were observed during neutral stability when mechanical turbulence dominates.

In general, the σ_y values observed were larger than the corresponding Pasquill–Gifford values, but they increased with distance less rapidly, approaching those of Pasquill–Gifford at longer distances. The authors surmised that the larger initial dilution was due to the roughness elements, buildings and trees, at the release site.

Norway. Experiments to determine vertical diffusion in valleys and over water in northern Norway by Gotaas (1975) have been reported by Minott, Shearer, and Marker (1977). From a total of 53 tests, 7 were performed over

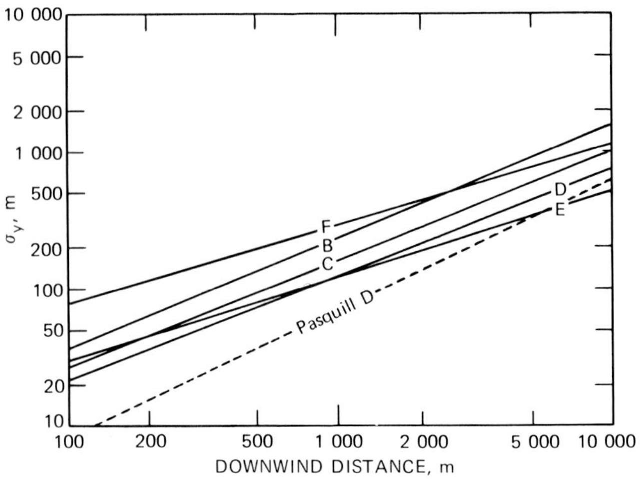

Fig. 9.34 Horizontal diffusion parameter as a function of the source distance for diffusion categories B to F according to the Jülich measurements at an emission height of 50 m. – – –, Pasquill D curve. (From K. J. Vogt, *A New System of Release Height Dependent Diffusion Parameters for the Gaussian Plume Model*, preprint of the International Clean Air Congress, Tokyo, Japan, 1977.)

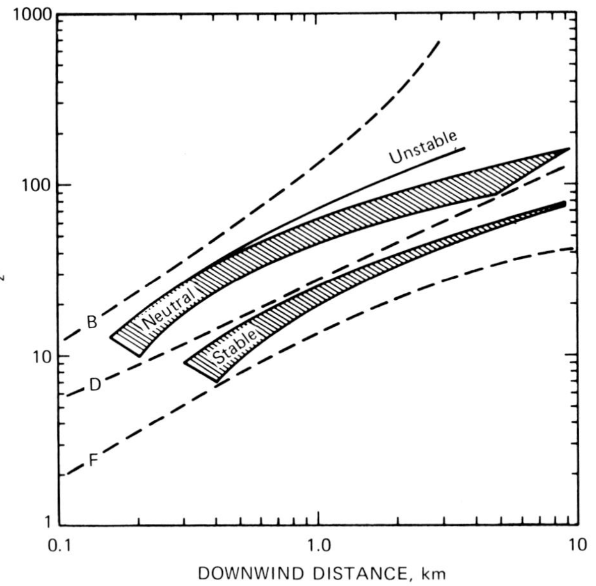

Fig. 9.36 Vertical diffusion parameter for the Norway experiments are shown in the shaded areas. – – –, Pasquill–Gifford σ_z curves. (From D. H. Minott, D. L. Shearer, and R. S. Marker, Report COO-4026-3, Traveler Research Corporation, Westerfield, Conn., 1977.)

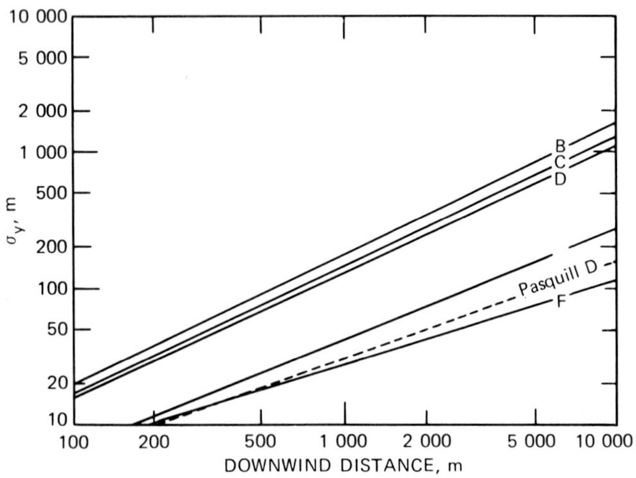

Fig. 9.35 Vertical diffusion parameter as a function of source distance for diffusion categories B to F according to the Jülich measurements at an emission height of 50 m. – – –, Pasquill–Gifford D curve. (From K. J. Vogt, *A New System of Release Height Dependent Diffusion Parameters for the Gaussian Plume Model*, preprint of the International Clean Air Congress, Tokyo, Japan, 1977.)

water in a fjord, the rest in valleys 2 to 5 km wide at the base and 6 to 8 km wide at the ridges. Ridge tops averaged about 500 m above the valley floors. Releases of FP or oil fog were from mid-valley and from one to several hours duration. Sampling was over the flat part of the valleys 3 to 13 km downwind. Overwater releases were conducted with FP which was sampled with rotorods that were suspended from 150-m high captive balloons. Some aircraft samples were also taken.

The results are summarized in Fig. 9.36. The vertical diffusion observed in Norway is similar to that of the Pasquill–Gifford curves for much smoother terrain during unstable and neutral conditions. There is some question whether very unstable conditions, such as categories A or B, would be observed in such a northern latitude. Vertical dispersion during stable conditions was greater than over smooth terrain. However, the adjacent complex terrain had no noticeable effect on the observed dispersion. The underlying surface seemed to be of most importance as is demonstrated by the overwater fjord dispersion trials, where the σ_z was similar to those observed in other overwater dispersion experiments.

Loisach Valley, Federal Republic of Germany. The results of 11 FP tracer dispersion tests in a mountain valley have been summarized by Reiter (1974, 1975) and Reiter and Sladkovic (1976). The test area is situated in southern Germany near the foothills of the Alps. The sampling grid, designed to measure horizontal dispersion, was placed in the flat basin of the valley. The valley is 9 by 2 km with the adjacent peaks rising up to 1000 m above the valley floor. The valley is mixed between urban, forest, and open fields with a roughness length of 0.44 m.

The FP was released from a hill 300 m above the valley for about 1 h, and samplers were placed on the valley floor 3 to 16 km downwind. All experiments were during unstable conditions.

Horizontal dispersion (Fig. 9.37) was similar to that over smooth terrain except when the plume started to impact against the valley walls; then the concentration of FP became nearly uniform across the valley (Fig. 9.38). The authors found that the best dispersion model was one that assumed full absorption rather than reflection of the FP at the valley walls.

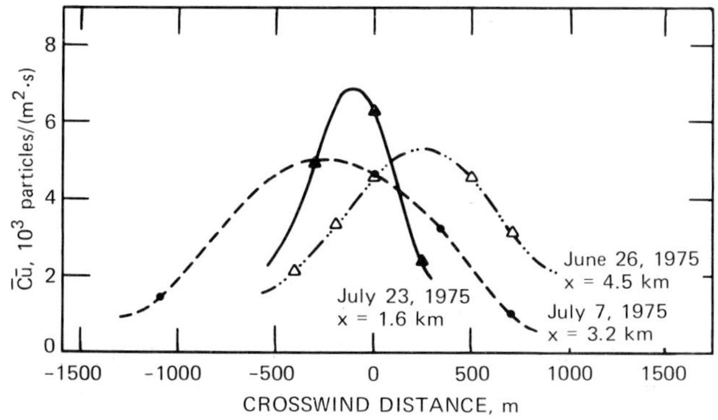

Fig. 9.37 Observed normalized concentration at Loisach Valley before the plume contacts the valley sides. (From R. Reiter and R. Sladkovic, *Boundary Layer Aerosol Transport Measurements in a Valley System*, Part III. *Final Technical Report*, European Research Office, U.S. Army, London, England, 1976.)

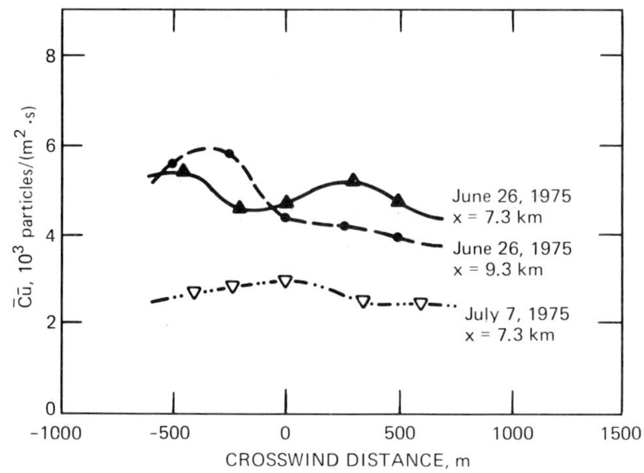

Fig. 9.38 Observed normalized concentration at Loisach Valley after the plume contacts the valley sides. (From R. Reiter and R. Sladkovic, *Boundary Layer Aerosol Transport Measurements in a Valley System*, Part III. *Final Technical Report*, European Research Office, U.S. Army, London, England, 1976.)

Rough Terrain Summary. Data analyses usually consisted of ordering the dispersion data by Pasquill stability classes. As several of the authors pointed out, this was not sufficient to provide accurate predictions of dispersion in rough terrain owing to the large scatter of σ_y and σ_z within a stability class. Part of the problem is that the Pasquill stability classification method does not provide any compensation to account for the increased roughness. Also, the method of computing the Pasquill stability class varied among the researchers, and large differences existed in source environment, ranging from relatively open terrain to complex building structures nearby.

The σ_y and σ_z values are summarized in Figs. 9.39 and 9.40. The experimental data are shown by average curves except Karlsruhe in which the individual points are plotted. The Karlsruhe data points are probably representative of the scatter of the other sites as well. Mt. Iron is not included in Fig. 9.39, but it falls between the limits of Pasquill A and E curves as seen in Fig. 9.33. For both σ_y and σ_z there is evidence that there is enhanced initial dispersion near the source due to local roughness elements. This enhanced dispersion has been noted previously by Bowne (1974) from diffusion data collected near urban and forested areas.

In general, there seems to be greater horizontal dispersion than shown by the Pasquill–Gifford curves during unstable conditions (Fig. 9.39). At Jülich and Karlsruhe the release times were from 20 to 60 min, whereas the Pasquill–Gifford curves are for 10-min releases. Dispersion would be greater for longer release times. This can be seen in the Jülich E and F curves on Fig. 9.34 where the authors attributed the large values of σ_y (greater than during unstable conditions) to the effect of wind-direction meander during the experiment. This effect is unlikely to be important in the E and F category Pasquill–Gifford curves because of the shorter release time. Additional support for this conclusion is seen in Fig. 9.33 where the Mt. Iron data (30-s release) exhibit the same range as the Pasquill–Gifford curves. Both Jülich and Karlsruhe were elevated releases, but the Jülich experiments (50- and 100-m release elevations) suggested little dependence of σ_y on release height.

The σ_z data (Fig. 9.40) are a little more straightforward in that, at least, during unstable conditions there seems to be more correspondence among the different experimental sites. Sampling duration should be less important for vertical dispersion; hence more direct comparison is possible. Jülich and Norway data seem to conflict during stable stratification, one suggesting enhanced dispersion over Pasquill–Gifford, the other not. The problem may be in definition of stability; the Norway classification was by vertical temperature gradient and the Jülich by surface weather data. The more extreme northern latitude of the Norway site complicates the comparison. Evidence from other experiments (see Urban and Forest sections) suggests that vertical dispersion is indeed enhanced over the Pasquill–Gifford for similar stability conditions because of the increased mechanical turbulence.

The predictive techniques used for σ_y in the classical section (9-3.1) can possibly be applied over rough terrain that is homogeneous. The increased turbulence should be incorporated into the wind-direction-fluctuation σ_θ, and hence in any predictive method that uses σ_θ. Predictions of σ_y, using Eqs. 9.8, 9.10, and the observed σ_θ, are compared with observations of σ_y at Mt. Iron and shown in Fig. 9.41. Although much scatter is present, most σ_y values are calculated to within a factor of 2 of the observed value.

9-3.2.3 Complex Terrain Diffusion. Atmospheric dispersion experiments in this section involve major variations of topography over the diffusion grid. Under these conditions the wind field is unlikely to be homogeneous, and local variations of wind speed and direction and terrain-induced vertical motions can influence the pollutant dispersion. Plume impaction on elevated terrain and separation and flow over terrain obstacles are some of the effects that

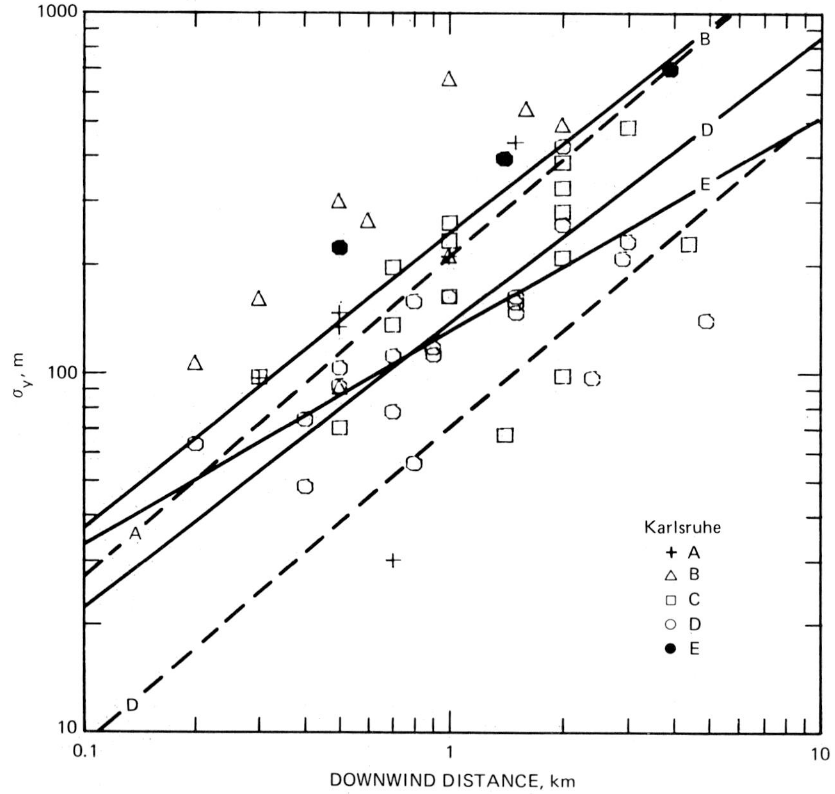

Fig. 9.39 Summary of the horizontal dispersion data. Plotted data points, Karlsruhe; —, Jülich; – – –, Pasquill–Gifford curves.

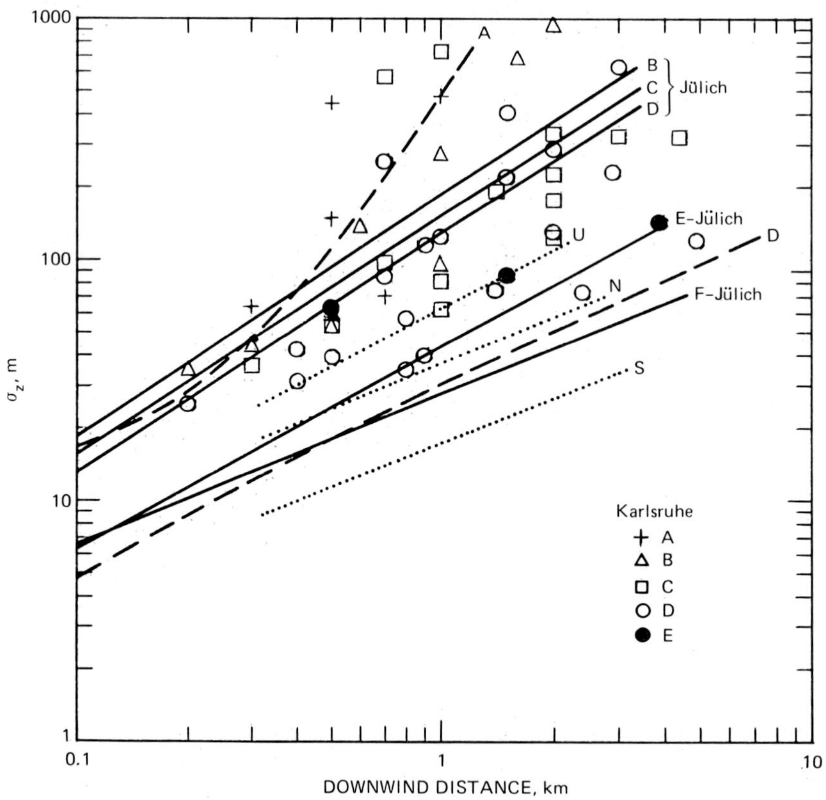

Fig. 9.40 Summary of the vertical dispersion data. Plotted data points, Karlsruhe; —, Jülich; . . ., U, N, and S summarize the Norway unstable, neutral, and stable curves, respectively; – – –, Pasquill–Gifford curves.

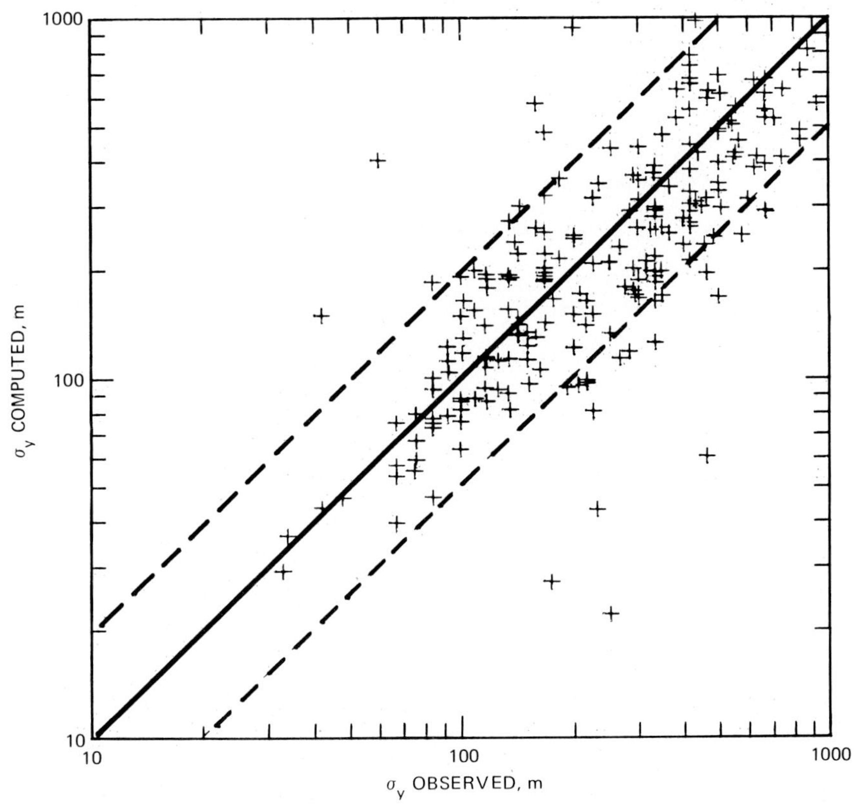

Fig. 9.41 Predictions of σ_y for Mt. Iron based on wind-direction fluctuation and Eqs. 9.8 and 9.10.

have to be studied. Also mountain–valley wind systems become more important for this group of experiments. Nocturnal downslope gradient flows, stagnation pockets in topographic depressions, and daytime upslope thermal flows have to be evaluated to identify their effect on dispersion.

Experimental limitations that restrict experiments in simple terrain are even more restrictive in complex terrain. These experiments are a first attempt to provide data to be used in solving problems of tremendous difficulty. Simple solutions that have been developed from classical experiments over homogeneous terrain, which assume spatial homogeneity of the winds, are not applicable over complex terrain. Although empirical corrections have been devised to allow for the increased dispersion in complex terrain, these tend to be site-specific and cannot be extrapolated with confidence to other sites where different terrain features may interact differently with the flow to produce different dispersion conditions. However, experiments conducted to date are of practical value in permitting better qualitative judgements on the likely effect of complex terrain features.

Huntington Canyon, Utah. A tracer diffusion experiment within the confines of a steep-walled canyon was conducted at Huntington Canyon, Utah, in March and April of 1973 (Start, Dickson, and Wendell, 1973; Start, Dickson, and Wendell, 1975). Eleven releases of SF_6 were made, six in neutral or lapse conditions, and five during stable stratification. The releases were continuous for 45 to 60 min from a 183-m stack about 10 km up the canyon during nonstable stratification and during stable stratification near the ground. The up-canyon release point was necessary because the winds reversed during the evening. Oil fog was also generated at the source for visual observation. Ground samples of about 1-h duration were taken with whole-air samplers. A helicopter was also used to sample the plume centerline located visually by the oil fog.

Meteorological data were collected to classify the trials by Pasquill stability. Generally, the meteorological environment was found to be too complex to allow simple predictive techniques. Three types of dispersive mechanisms were found to affect diffusion in the canyon: (1) turbulence generated at the ridge tops could be transported downward to increase diffusion in the canyon; (2) downslope density flows from feeder canyons could act with the drainage flow from the main canyon to produce helical circulations; and (3) wake turbulence in the lee of various terrain features also increased the mixing.

The authors compared their results to those obtained by using the flat-terrain methods. They found the observed dilution under all Pasquill classes to be similar to that calculated during unstable conditions for flat terrain. In neutral conditions the canyon dilution was greater than what would be predicted by flat-terrain methods by a factor of 5, and during stable stratification the canyon dilution was greater by a factor of 15. During the trials conducted in unstable conditions, the dilution was mainly produced by the downward transport of turbulence from the mountain tops. Occasionally the plume would impact the canyon sides for a few minutes, but hourly averages still showed enhanced dispersion by as much as a factor of 3 over that of flat terrain. During trials in stable conditions, the plume was diluted throughout the canyon.

The authors suggest that to predict tracer concentrations we must use a combination of joint wind direction, wind speed, and stability frequency statistics with the different dilution factors for the various stability classes. The results of the Huntington Canyon trials are shown in the summary diagrams.

Garfield, Utah. Start, Ricks, and Dickson (1974b) summarized eight releases of SF_6 from a smelter plant site near Garfield, Utah, on the slopes of the Oquirrh Mountains during June 1973. Terrain elevations range from 1440 to 2380 m mean sea level. Salt Lake City is the closest major metropolitan area. This experiment was similar to the Huntington Canyon trials except that the canyons are not so deep or so steep. A complication to the meteorological environment of the release site is the existence of a diurnal lake–valley breeze. All successful releases were conducted during moderately unstable stratification with onshore flow.

Seven 60-min releases of SF_6 were made through the 122-m stack while the smelter was in operation so that the plume was visible. An additional aerial release was made at 366 m. The 40 surface samplers were left on from 2 to 3 h. One-minute helicopter samples were also obtained.

Normalized concentrations as a function of distance are included in the summary. A comparison of these test results with the ones predicted by the flat-terrain equations show that, over the complex terrain, dilution is increased two to four times over that predicted for B and C stabilities.

In those experiments in which it could be computed (Fig. 9.42), σ_y was found to be about twice that given by the Pasquill–Gifford curves. Increased horizontal dispersion is due to lateral deflections of the plume around obstacles, looping plumes that spread out laterally at the surface when descending, and increased turbulence and wind shear in complex terrain.

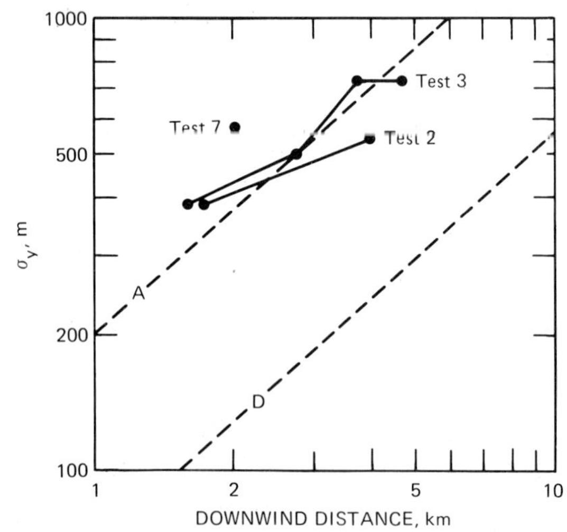

Fig. 9.42 Values of σ_y at Garfield for tests 2, 3, and 7. Stabilities were C for two of the tests and B for the other. – – –, Pasquill–Gifford curves. (From E. E. Start, N. R. Hicks, and C. R. Dickson, NOAA Technical Memo-ERL-51, Idaho Falls, Idaho, 1974b.)

Vertical growth of the plume is more complex than horizontal growth. Three types of growth were observed, depending on the vertical stability structure. First, a low, stable layer permitted the plume to be trapped in stagnant pockets between areas of higher elevation. This case was not observed during the SF_6 tracer releases, but it has been observed from the plant plume. Second, a higher stable layer enabled the plume to flow over terrain features, but rapid vertical mixing produced a uniform distribution beneath the lid. Third, when no lid was present, the plume mixed rapidly aloft, and low concentrations were observed at the ground.

Tonopah, Nevada. Tracer releases near Tonopah, Nev., were part of a program described by Smith and Wolf (1963). The mechanical details of the meteorological instrumentation, tracer release, and sampling network are similar to the Corpus Christi trials. The test site is dominated by a 305-m ridge that was perpendicular to the wind field during the trials. There is a gradual slope on both faces of the ridge. The tracer (FP) releases were made from 120 to 430 m above the ridge top 16 km upwind. The sampling line extended across the ridge.

There was considerable diurnal variation in the wind field due to the drainage effects of radiational cooling. During the evening hours the constriction in the depth of the layer as the air passed over the ridge increased the wind speeds. Downward flow was enhanced on the lee side. Tracer releases confirmed this analysis. Daytime releases found the highest exposure on the windward side; nighttime releases showed higher values to the lee side.

The authors surmised that the tracer, embedded in stable flow, experienced an upward transport with little downward mixing along the windward slope so that little tracer reached the surface. The indications are that a more complex turbulence field in the lee of the ridge resulted in a greater degree of vertical mixing and thus in higher surface exposure.

Twentynine Palms, California. Ten FP tracer releases conducted in complex mountainous terrain at the U. S. Marine Base at Twentynine Palms, Calif., have been summarized by T. B. Smith (1965). The ridge is 16 km long and about 300 m above the valley floor. The slope is rather gentle on the lee side. Some other small diffusion trials in complex terrain are also summarized in this volume.

Meteorological instrumentation consisted of four towers located on the ridge and on both slopes. The towers were equipped to measure component winds, temperature, and turbulence.

Line-source tracer releases, 17 km in length, were provided by two aircraft, one for yellow FP, the other for green FP. Releases were at ridge level or 155 m below the ridge top. Surface sampling was accomplished at 55 rotorod sites and vertically at the tower locations. Sequential samplers were located at the towers too.

As in the other programs, the tracer was found to descend on the lee slope. This downward transport was found to be due to combinations of wave flow and drainage. Wave flow over mountainous terrain was found to be significant in the downward transport of tracers when wind speeds above the ridge are 7 to 15 m/s. Wavelengths would be on the order of 5 to 25 km. Drainage occurred during the evening when wind speeds were 6 m/s or less. This condition increased the exposures on both slopes.

Complex Terrain Summary. These experiments all show the very significant effects of nonregular terrain on expected diffusion patterns. The classical methods, as well as those developed more recently for flat terrain grids, offer little help in estimating diffusion in the more complicated terrain.

A summary of results of the two point-source experiments is shown in Fig. 9.43. At Huntington and Garfield where the terrain slope is on the order of several hundred meters per kilometer, dispersion was greater than that over flat terrain for all stabilities. Generally, quantitative estimates of σ_y and σ_z could not be made. Results from other experiments (aircraft releases above ridge–valley systems) showed higher exposures on the windward side of the ridge during daytime releases and higher exposures on the leeward side during nighttime releases. However, when windspeeds are low (<6 m/s) at night, drainage flow can produce high exposures on both sides.

Other less comprehensive reports on transport and diffusion experiments are available. A few are listed here for reference. A summary of the Garfield and Huntington Canyon experiments is presented by Start, Dickson, and Ricks (1974a). Other FP and oil-fog tracer trials at three power plants located in areas of different complexity are summarized briefly by Hovind, Spangler, and Anderson (1974). The authors found the dispersion during stable conditions to be enhanced by a factor of 10 over that which would be observed over flat terrain, a result similar to that observed at Huntington Canyon. Over rough but open terrain, during neutral stability, dispersion was enhanced by a factor of 6. Meteorological measurements over mountainous terrain which were used to postulate a diffusion transport model have been presented by DeMarrais, Downing, and Meyer (1968). Clements and Barr (1976) discuss the mesa–canyon diffusion characteristics near Los Alamos, concentrating on the nocturnal drainage situation. They found that when the mesa wind was across the canyon, at 2 m/s or more, a roll eddy was observed within the canyon which caused considerable mixing and exchange with the mesa air aloft. Fosberg et al. (1976) present a modified Gaussian diffusion model for complex terrain based on wind-field divergence. The authors assume that the divergence δ, and the mean wind $\bar{u}$, are constant along the plume and this gives an exponential multiplier

$$\exp\left(-\frac{\delta x}{\bar{u}}\right) \tag{9.25}$$

to the traditional Gaussian solution. Estimates of divergence values ranged over several orders of magnitude (10^{-5}/s to 10^{-2}/s).

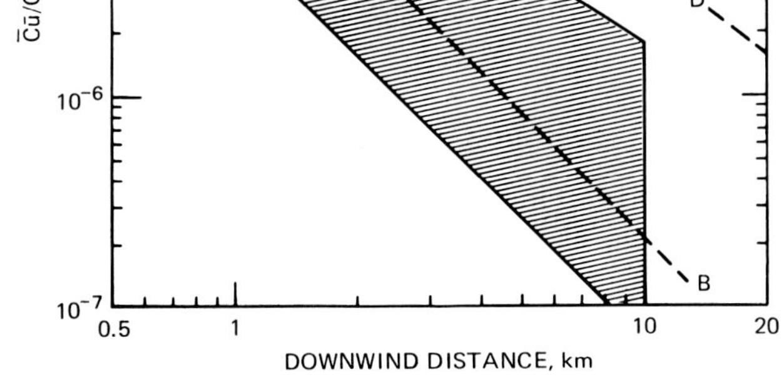

Fig. 9.43 A summary of downwind normalized concentration at Garfield and Huntington Canyon. The mean range between stable and unstable conditions of observed concentration is shown in the shaded areas. – – –, flat-terrain estimates based on Pasquill–Gifford curves.

In general, dispersion in complex terrain was found to be enhanced over that which would be predicted by flat-terrain dispersion equations. The effects of complex terrain were most pronounced at night and least during the day in unstable conditions. The enhanced dispersion during stable conditions is likely to be more site specific, and hence quantitative estimates should not be generalized. The nighttime drainage flows and wake obstacle turbulence depend on the topography of the site. The stable canyon trials were limited to the Huntington Canyon site, and it is probable that the dispersion at night is not so large as at Garfield. This may be partly confirmed by the diffusion data collected at Jülich, a rough site but with homogeneous topography. The authors found that, in unstable and neutral conditions, dispersion was enhanced by similar factors as at Huntington and Garfield, but at night the dispersion was enhanced only by about a factor of 2 compared to the factor of 15 observed at Huntington Canyon.

*9-3.2.4 **Forest Diffusion.*** The effect of a forest upon the diffusion of tracers depends on the degree to which the forest alters the meteorological environment and the location of release points with respect to the forest. Variations in forest density alter the contribution of mechanical turbulence to diffusion. Variations in crown density alter the thermal properties, and holes in the crown permit sunlight to penetrate to the forest floor. Therefore seasonal and spatial variations in diffusion can be expected. Such factors as different types of forests, dense undergrowth or lack of undergrowth, hilly terrain, swampy or dry areas, and differences in tracer release configurations must be considered when comparing results of tracer experiments.

Brookhaven, New York. Short-range pollen tracer experiments were conducted for several years at Brookhaven National Laboratory near Upton, N. Y., and the results have been summarized by Raynor, Hayes, and Ogden (1974). The data can be obtained from Raynor, Hayes, and Ogden (1972) and Raynor, Hayes, and Ogden (1973). The feasibility of using pollen spores as atmospheric tracers is discussed by Raynor et al. (1966).

The forest is comprised primarily of pines 10 to 13 m in height. The roughness length was determined to be 1 m. The undergrowth is sparse with few branches present below 3 m. All trials were conducted during the day when lapse rates outside the forest were unstable.

All tracer releases were outside the forest or just inside the interface. A total of 72 point-source pollen releases (15 μm in diameter) of 20- to 40-min duration were made anywhere from the edge to 60 m from the forest at heights of 1.75 to 14 m. In addition, there were 36 area-source releases and 3 distant-source releases.

A 70- by 100-m sampling grid of rotoslides was constructed, such that the first row of 7 samplers was 1 m outside the forest. The other rows were 10, 20, 40, 60, 80, and 100 m within the forest. Sixteen towers were constructed for vertical sampling up to 21 m. All sampling positions had deposition slides. All samplers were started and stopped simultaneously.

A summary of the meteorology of the Brookhaven forest stand has been given by Raynor (1971). Generally, when the airflow was toward the forest, there was little change in wind speed approaching the stand, but there was increased horizontal and vertical turbulence. The speed of the air that penetrated the trunk space was faster than in the crown area for about 60 to 80 m into the forest where equilibrium was reached and wind speeds changed little with height. This equilibrium distance was about 10 times the tree height. During the daytime, equilibrium wind speeds were about 20% of the winds above the forest. The daytime heating at the canopy top formed an inversion below the canopy, but because of the greater mechanical turbulence in the canopy there was considerable exchange of air through this level.

Some of the results of the point-source experiments are shown in Figs. 9.44 and 9.45. The forest results are compared with pollen tracer tests over open terrain conducted at a site nearby by Raynor, Ogden, and Hayes (1970).

The authors found that there was good consistency between experiments, especially since all were conducted during similar meteorological conditions. Upon approaching the forest the plume initially broadened both horizontally and vertically because of the increased turbulence. This effect was found to depend on foliage density at the forest edge. The flow was into the trunk space of the forest and over the top. Inside the forest the lateral spread was slower than that found during the open-terrain experiments. Although unstable lapse rates existed outside of the forest, heating of the canopy produced an inversion within the forest. Vertical dispersion was greater than during the open-terrain experiments. From visible tracer tests, subsidence was observed in open areas within the forest and upward motion in dense forest areas. Wind speeds within the forest were about 20% of those above the forest and changed little with height below the canopy after the initial

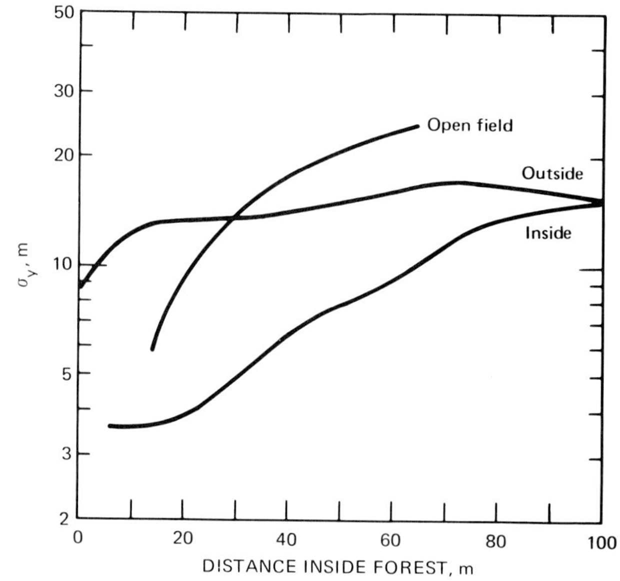

Fig. 9.44 Plume width as a function of distance inside the forest for point-source releases of ragweed pollen at a height of 1.75 m from 10 m outside the forest and 6 m inside the forest. For the open field test, plume width is a function of distance from the source. [From G. S. Raynor, J. V. Hayes, and E. C. Ogden, Particulate Dispersion into and Within a Forest, *Boundary Layer Meteorol.*, 7: 429-456 (1974).]

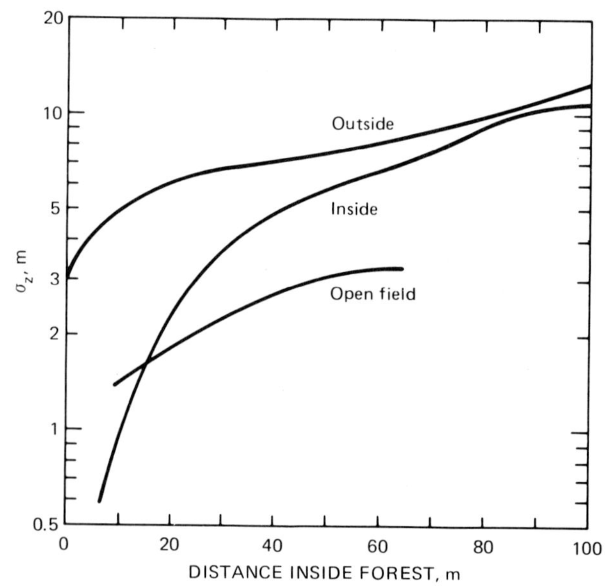

Fig. 9.45 Plume height as a function of distance inside the forest for point-source releases of ragweed pollen at a height of 1.75 m from 10 m outside the forest and 6 m inside the forest. For the open field test, the plume height is a function of distance from the source. [From G. S. Raynor, J. V. Hayes, and E. C. Ogden, Particulate Dispersion into and Within a Forest, *Boundary Layer Meteorol.*, 7: 429-456 (1974).]

60- to 80-m transition zone. Material was lost through impaction on the foliage at the forest edge.

Searcy, Arkansas. Bowne et al. (1969) conducted three different field diffusion programs near Searcy, Ark. A total of 131 tracer releases were designed to learn how complex vegetation patterns affect aerosol transport and diffusion within 30 km of the source. Also, the diffusion and deposition effects of an isolated woodlot and a line woodlot were to be evaluated.

The test grid was about 80 km northeast of Little Rock, Ark. The area is relatively flat, but hills to the northwest rise to about 60 m. Agriculture is predominant in the grid area with wooded areas dispersed throughout. The woodlot consisted of trees about 18 m in height with dense underbrush within the first meter. Meteorological measurements were made on six towers at three levels for the trajectory experiments. Three of these towers were located upwind of, within, and downwind of the woodlot.

A typical tracer experimental period usually consisted of four tracer releases, alternate yellow and green FP. Most of the 72 releases for the trajectory trials were made during the evening. An 800-m line source was generated by putting the dispenser on a truck and making six to eight passes in about 25 min. The 800-m release line for the isolated woodlot was located from 200 to 800 m upwind for 36 releases. The release line for the line woodlot was 1600 m in length, and 22 releases were made.

The sampling grid for the trajectory trials consisted of four arcs laid out on existing roads. The woodlot tests had a separate microgrid that included three 35-m towers and surface samplers out to 2 km. Both σ_y and σ_z were computed by the moment method and by the distance to the 10% concentration isopleth. The σ_y values were corrected by the authors to represent a point-source release.

The trajectory tests were designed to determine the effect of local terrain on long-distance (30 km) dispersion. In general, the terrain effects were larger with increasing stability. The plume tended to follow the channeling effects of terrain variations rather than the path indicated by wind directions from the tower sites. The trajectory trials were usually conducted early in the evening when the atmosphere became increasingly stable with time. Trajectories computed from the upper-level tower winds (90 m) best simulated the observed plume. The lateral dispersion data, adjusted to be comparable with point-source data, are shown in Fig. 9.46. Although most of the data were collected during neutral to stable conditions, most of the points lie within the neutral to unstable curves of Pasquill–Gifford for smooth terrain.

The woodlot trials indicated that horizontal dispersion (Fig. 9.47) was affected mainly by the characteristics of the woodlot. Larger σ_y correspond to the case with the leaves still on the trees. A slight reduction of the σ_y just inside the woodlot may be due to error in the collected concentration data.

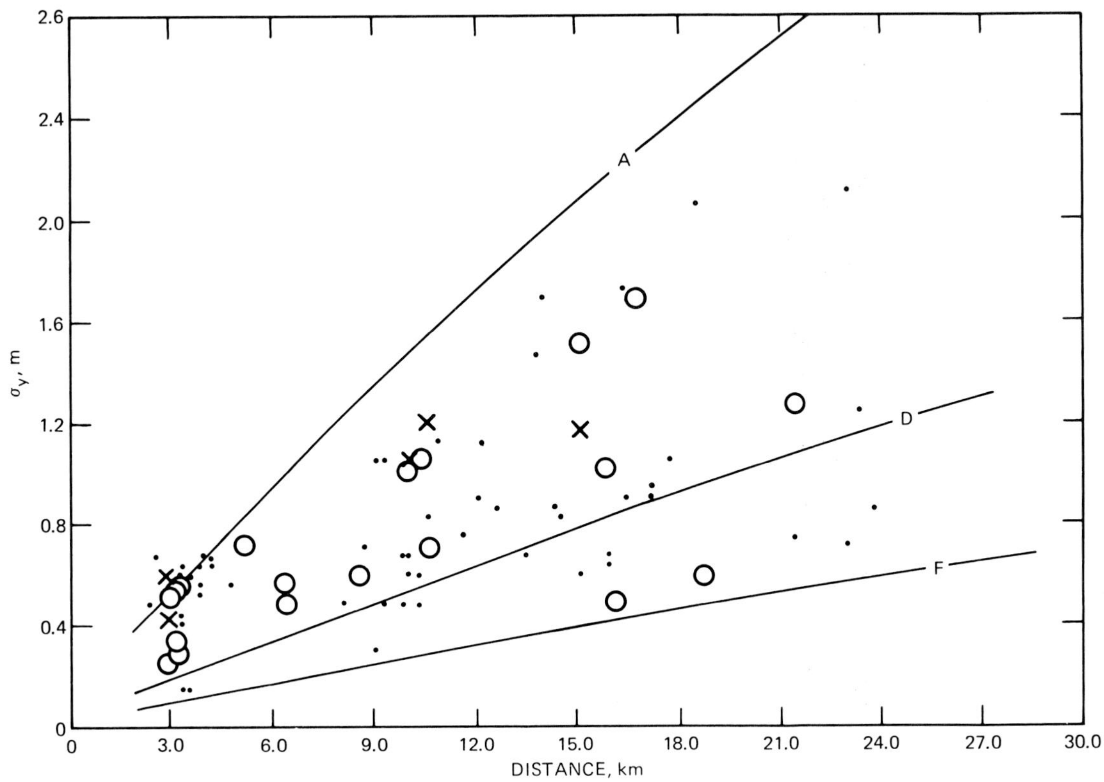

Fig. 9.46 Plume width as a function of distance for the Arkansas trajectory tests. —, the Pasquill–Gifford curves. Stability is determined by the temperature gradient: ○, unstable; X, neutral; ●, stable. (From N. E. Bowne et al., Report TRG-7236-343A, U.S. Army Dugway Proving Ground, Travelers Research Corporation, Hartford, Conn., 1969.)

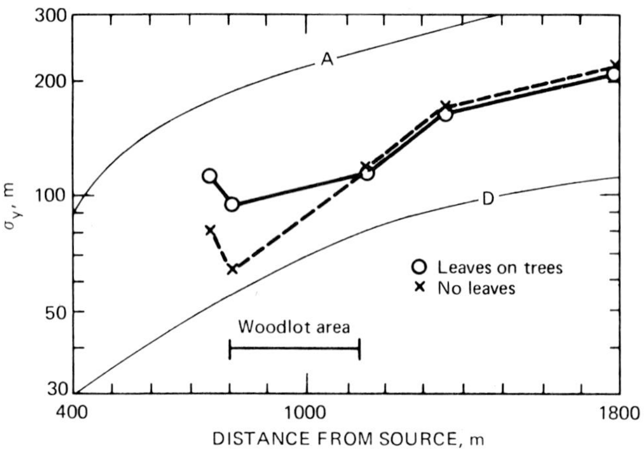

Fig. 9.47 Horizontal plume growth for the Arkansas woodlot tests. ——, Pasquill–Gifford curves for σ_y. (From N. E. Bowne et al., Report TRG-7236-343A, U. S. Army Dugway Proving Ground, Travelers Research Corporation, Hartford, Conn., 1969.)

Vertical diffusion (Fig. 9.48) was rapid, and the plume quickly grew to tree height once inside the woodlot and grew little thereafter. Vertical growth before entering the woodlot was substantial except in the most stable cases when the plume quickly expanded upon entering the woodlot. The line woodlot trials showed very high initial growth, believed to be due to the shelter effect from another upwind woodlot in the release line clearing. Other effects included a deficit of tracer in the lee of the woodlot. Streaming was observed during stable regimes; i.e., there were high dosage lobes on either side of the woodlot. Deposition computations give deposition velocities higher by a factor of approximately 2 during the day (10 vs. 4 cm/s).

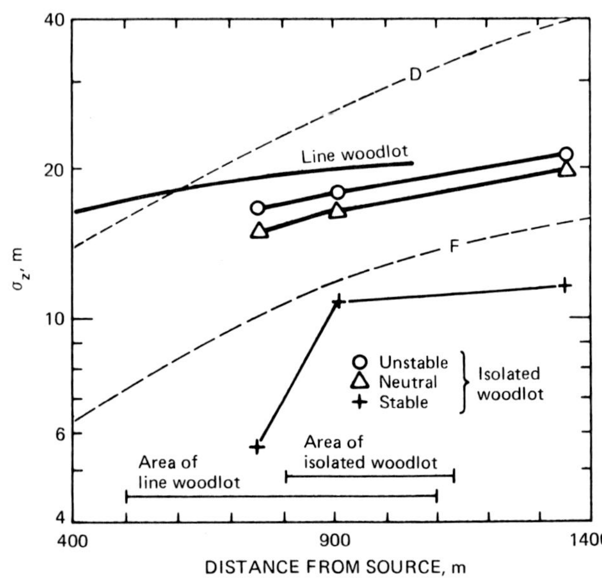

Fig. 9.48 Vertical plume growth for the Arkansas isolated woodlot tests by stability and for all stabilities for the line-woodlot trials. (From N. E. Bowne et al., Report TRG-7236-343A, U.S. Army Dugway Proving Ground, Travelers Research Corporation, Hartford, Conn., 1969.)

Chippewa National Forest, Minnesota. An FP tracer diffusion program at Chippewa National Forest has been summarized by Tourin and Shen (1969). The test site, located in North Central Minnesota about 32 km from Bemidji, is flat and forested mostly by deciduous species with crowns at about 20 m. The diffusion grid was 500 by 2000 m. Trials were conducted after sundown in summer and winter. Ten 44-m meteorological towers were set up along the long axis of the grid at 200-m intervals; these were also equipped with FP sampling equipment.

Three types of releases were planned: aircraft releases of FP along the main axis of the grid and simultaneous instantaneous, ground line-source releases (winter trials) and instantaneous point-source releases (summer trials). The sampling points for the aircraft releases were in the main grid at the meteorological tower locations. Sampling for the ground releases was centered in a microgrid, located near the center of the main grid. The sampling was done at 15-min intervals.

The effect of the forest on dispersion was found to be dependent on the season. In the summer night, the increased foliage in the crown became the primary radiating surface. Therefore stable lapse rates above the crown and unstable lapse rates below the crown were frequently observed. In the winter, radiational cooling occurred well below the crown, and neutral stratification, possibly due to greater momentum exchange between the above- and below-crown air, was predominant.

Results from the aerial releases showed somewhat faster downward transport in winter. Peak concentrations at ground level were usually observed within 30 min. A typical sequence of 15-min concentration profiles is shown in Fig. 9.49. The FP gradient was usually downward in the first sampling period and reversed to become upward during the second and subsequent sampling periods. The rate of concentration decrease below canopy, greater in winter than in summer, was attributed to upward transport rather than to advection.

The ground-burst FP clouds were observed to travel in a direction to the left of (see Table 9.9) or along the prevailing wind direction above the canopy. The greatest variation occurred during stable stratification in summer.

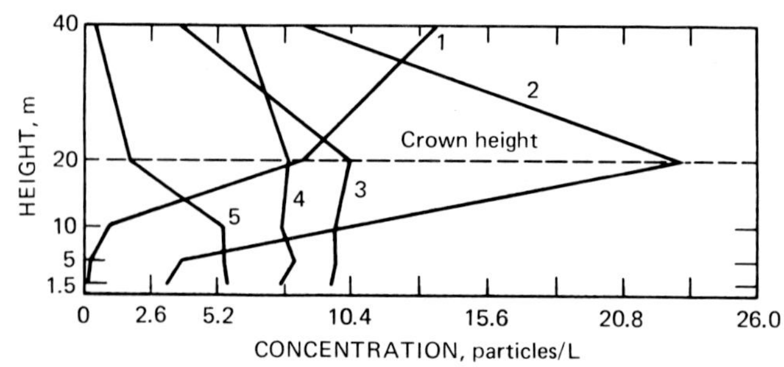

Fig. 9.49 Vertical profiles of average particle concentration for successive 15-min intervals during Test 203. (From M. H. Tourin and W. C. Shen, Diffusion Studies in Deciduous Forest, R.N. 3004, *Deciduous Forest Diffusion Study*, Vol. 1, Applied Science Division, Litton Systems, Inc., Minneapolis, Minn., 1969.)

TABLE 9.9 Counterclockwise Plume-Direction Shift from Above-Canopy Winds

	Stable	Neutral
Summer	58°	1°
Winter	22°	12°

Results from the point-source trials are shown as σ_y and σ_z values in Figs. 9.50 and 9.51. The FP samples were collected at only very short distances from the source. Predictions of plume height based on the wind-direction fluctuations [Eq. 9.9 with $f(t/t_i) = 1$] are shown as solid lines. The σ_y predictions are grossly overestimated (and not shown), possibly because of the large wind-direction meander and rather random flows observed at night with frequent intervals of wind speeds below the anemometer stall speed. Vertical plume-growth estimates from observed vertical wind-direction fluctuations provide reasonable values of σ_z.

Mt. Rainier, Washington. Short-range forest tracer experiments summarized by Fritschen et al. (1970) and Fritschen and Edmonds (1976) were conducted in the Charles Lathrop Pack Demonstration forest, 45 km southeast of Tacoma, Wash. The forest consists mostly of Douglas Fir, 20 to 30 m in height, and is located on the slopes of Mt. Rainier. The site is on an alluvial terrace above the confluence of two rivers. Variations in elevation at the site are ±2.5 m. Also, a mountain–valley wind system exists in the area. Daytime upslope winds were from a clear-cut area into the forest.

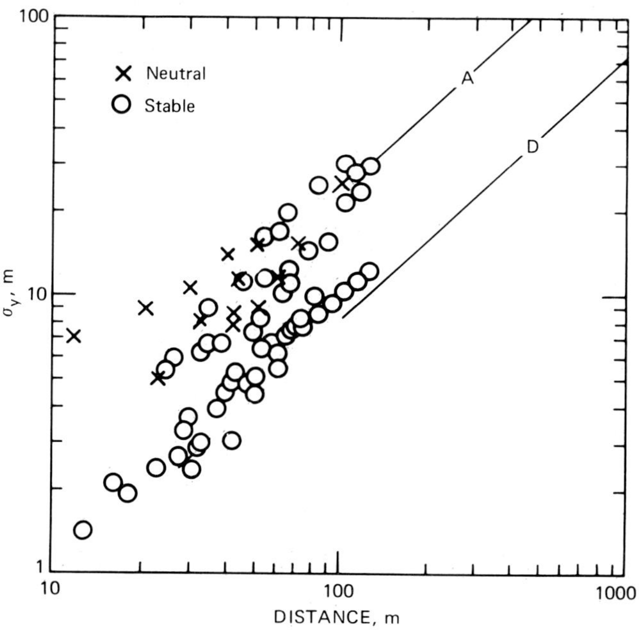

Fig. 9.50 Horizontal plume growth at Chippewa during the summer; —, Pasquill–Gifford dispersion curves. (From M. H. Tourin and W. C. Shen, Diffusion Studies in Deciduous Forest, R.N. 3004, *Deciduous Forest Diffusion Study*, Vol. 1, Applied Science Division, Litton Systems, Inc., Minneapolis, Minn., 1969.)

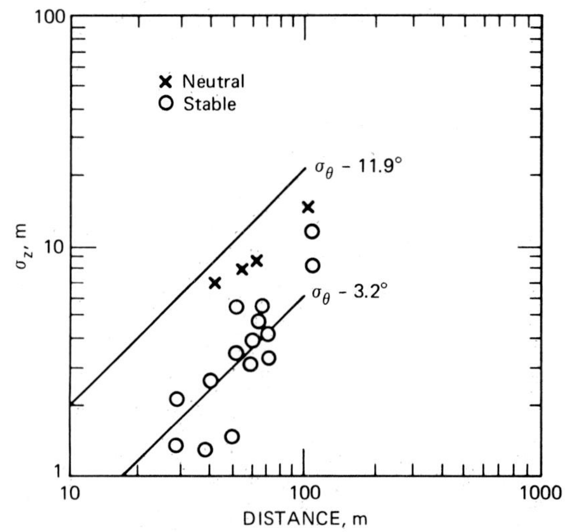

Fig. 9.51 Vertical plume growth at Chippewa during the summer; —, $\sigma_z = \sigma_\phi x$ as indicated.

The primary tracer was FP, but some trials were conducted with Fomés spores. Several simultaneous releases showed a high correlation between the two tracers. Releases were designed for several locations including: within the forest, at the clear-cut forest interface, and within the clear-cut area. Release heights varied from 1 to 20 m.

A parallelogram sampling grid consisted of samplers in nine rows at 30-m intervals, the first at the interface of the forest and the clear-cut area. The samplers in each row were also spaced at 30-m intervals. The samplers were located at a height of 1 m, and several 30-m towers had vertical arrays of rotorods. Two microgrids were established at the interface and within the forest (15-m spacing). Several different release configurations were tried, including instantaneous and continuous point sources.

Analysis of the meteorological data provided insight into the diffusion within the forest environment. The vertical temperature profile was found to be the most important factor in affecting forest dispersion. Solar absorption at the canopy tops produced instability in the upper crown and resulted in a stable layer below, restricting tracer cloud growth. Canopy openings and large trees acted as chimneys, venting the tracer from the trunk space. Radiational cooling from the canopy top produced an inversion within the canopy and isothermal conditions below the canopy. During the early evening a slight unstable layer would form above the warmer ground.

Generally, the vertical wind shear was directly proportional to the forest density. Ground-level wind-speed jets developed under canopies without undergrowth. The equilibrium wind speed under the canopy was found to be 20 to 30% of the speed at the top of the canopy. During the day, upslope winds reached equilibrium within 60 m (2 to 3 tree heights) of the interface. Drainage winds were in equilibrium to the canopy top, 10 tree heights from the interface, with slight acceleration near the top. An example of a within-forest dispersion trial at night during drainage flow is shown in Fig. 9.52. This case is typical of other trials in

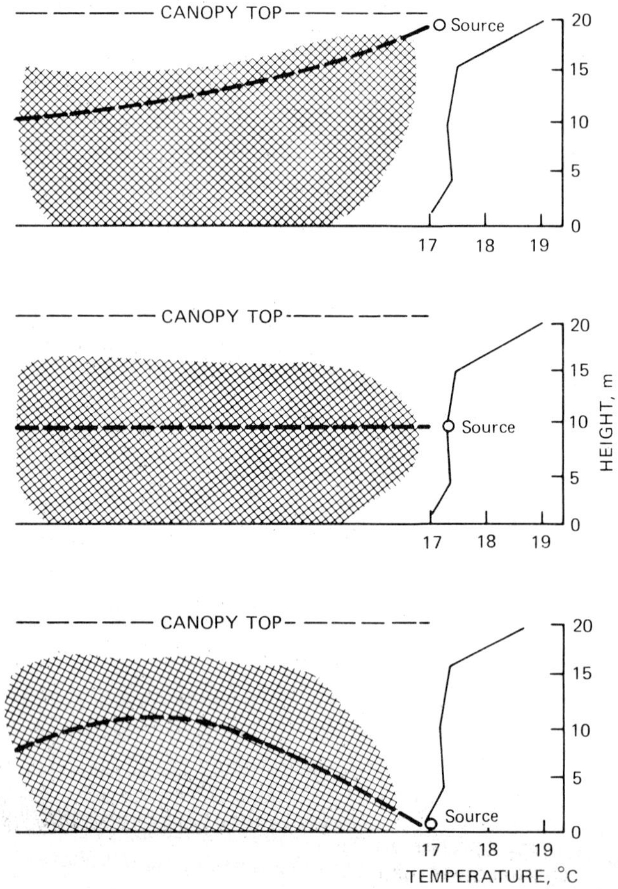

Fig. 9.52 Vertical extent of FP plumes shown in hatched area for three release heights. — — —, plume center line. (From L. J. Fritschen et al., Report TR-ECOM-68-G8-3, U.S. Army Electronics Command, Atmospheric Sciences Laboratory, Ft. Huachuca, Ariz., 1970.)

that the tracer is distributed rapidly beneath the canopy regardless of release height. In some cases the tracer would mix up rapidly, be trapped below the canopy, and drift for miles with little deposition. Lateral dispersion at night was larger than during the daytime owing to the extreme wind-direction fluctuations. In the daytime the vertical mixing was as rapid as the nighttime situation.

Microgrid releases also indicated day/night differences. Releases in the clear-cut area showed less vertical and little lateral growth when compared to those in the forest at night. Daytime releases in the clear-cut area showed enhanced vertical growth over those in the forest but less lateral growth. Releases at the interface and in the clear-cut area showed that the tracer that passed over the canopy penetrated into the canopy about 200 m into the forest. For the large grid the maximum concentration decreased with the cube of distance.

Measurements of deposition velocity ranged from 0.07 to 0.46 cm/s and indicated that the greatest losses occurred as follows:

1. During the afternoon in the clear-cut area.
2. At night at the downwind forest interface, where an acceleration, and hence greater turbulence, occurs.
3. Generally at night when there is an inversion at the crown and an unstable layer above the forest floor.

Charleston, South Carolina. Thirteen tracer releases were performed during the summer of 1964 at Francis Marion National Forest near Charleston, S. C., to determine the diffusion under a tropical jungle canopy. The results are reported by the Meteorology Research Laboratory (1968) in three volumes.

The area was selected because it was the closest representation to a tropical rain forest in the continental United States. The terrain is flat, a vertical range of less than 4.5 m within a 4-km radius of the site. In the summer it is hot and humid and the diurnal changes of temperature are small. Furthermore, there is a dominance of broadleaf trees of at least 30 m in height.

The releases consisted of instantaneous ground-level FP bomblets (usually 30 g) of several colors. In addition, smoke bombs for visual tracing were used.

A (6 by 6) sampling grid was established with rotorod samplers at heights of 1.5 and 9.0 m at 100-m intervals. Three 28-m towers at the west, east, and south portions of the grid had samplers at the 28-m level. A (5 by 5) microgrid at the center had samplers at the 1.5-m level spaced every 25 m in an orthogonal network. Some long-range samples were obtained from mobile units along nearby roads so that the sampling was extended to distances of 3 to 6 km. Some leaf samples and deposition plates were also taken. A 59-m tower at the center of the grid was used for bivane and temperature measurement.

Analysis of the meteorological data showed the roughness length to be 1.25 m. The gradient Richardson number was used to describe the stability. Although most of the below-canopy Richardson numbers indicated stable stratification, the total dosage was much lower than expected. Part of this reduction can be attributed to fallout.

Seventy percent of the tracer was deposited on the ground within the first 25 m, but only another 10% was deposited during the next half hour of travel. The authors concluded that the depletion after the large initial fallout would be insignificant in comparison to the other diffusion processes. Otherwise, cloud widths (Fig. 9.53) were found to be similar to the open-terrain values during unstable stratification. The speed of the FP cloud was about 11% of the above-canopy wind speed and the direction of travel 23° to the left of the above-canopy winds.

When plotted as a function of distance rather than time, vertical mixing was found to be more rapid under the canopy than it would be for open terrain, presumably owing to the very low wind speeds usually observed under the canopy. As in other forest diffusion trials, the transport out of the forest depended upon the vertical temperature profile.

Turbo, Colombia. Thirteen FP tracer releases above a jungle in northwestern Colombia, 32 km south of the Gulf of Uraba, have been summarized by the Bendix Corporation (1963). The tests were conducted during January, February, and March 1962. In these months, there is little rain and persistent northerly trade winds are dominant. The

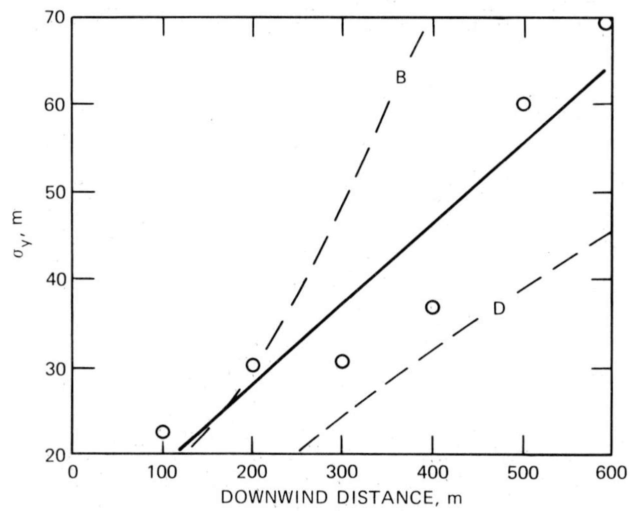

Fig. 9.53 Horizontal plume growth at Francis Marion National Forest. —, best fit through the data; – – –, Pasquill–Gifford curves. [From Meteorology Research Laboratory, *Diffusion Under a Jungle Canopy*, Vols. I(176), II(124), and III(154), Mexpar, Inc., Falls Church, Va., 1965, copyright American Standard, Inc.]

jungle canopy is about 42 m high, uniform, flat, and with no holes. There is considerable undergrowth below the main canopy.

The program was designed to determine the diffusion processes around the canopy and to measure the tracer penetration into the canopy and the tracer distribution below the canopy. Of the thirteen trials, nine were at night, seven of which were cloudy nights.

An aircraft was used to produce a 16-km line source 72 to 92 m above ground. Daytime releases were at 0900 LST; nighttime releases, at 1900 LST. Sampling was conducted along a north–south line of 12 towers spaced at 100-m intervals perpendicular to the release line. On the ends and at the midpoint of the line, sampling was done above the canopy as well as below. Rotorods were reversed and changed to give half-hour dosages. In addition, the two endpoint towers were equipped with meteorological instruments.

Tracer results showed that the FP maximum at ground level occurred about 30 min after the maximum above the canopy. The diffusion through the canopy depended upon the turbulence, lapse rate, and forest density. The decrease of the tracer within the forest after this time was a result of upward transport through the canopy top and not advection out of the grid. The rate of this decrease was related to the above-canopy wind speed and lapse rate. Above-canopy wind speeds were found to decrease to the canopy top (42 m). At 15 to 20 m, they were 5% of the 62-m wind speed and at 2 m were 1 to 2% of the 62-m value. The authors concluded that FP concentrations at ground level were one-eighth what they would have been had there been no trees.

Forest Diffusion Summary. Solutions to turbulent diffusion problems near forested areas require careful analysis of the local meteorological environment. Location of the source and receptors with respect to the forest is important. Tracer plumes approaching a forest from a cleared area will broaden both horizontally and vertically before entering the forest. This broadening depends upon the density of the forest stand. The plume may divide at the interface, and some of the material may pass aloft. Subsequent material exchanges through the canopy depend on the temperature and wind-speed profiles through the canopy and whether any canopy openings exist (i.e., irregular spacing or gaps).

Wind speed decreases on entering the forest stand. Speeds in the denser crown area are much less than those in the trunk space. The wind-speed profile becomes uniform with height after about 10 tree-heights distance within the forest. In general, wind speeds near ground level were found to be from 5 to 50% of the above-canopy value, the upper limit (50%) being reached during leafless conditions. Brown et al. (1969) developed an index to classify different forests on the basis of the ratio of the above-to-below canopy equilibrium wind speed during neutral conditions (Fig. 9.54). These ratios ranged from 4 for a deciduous forest in winter to 8 to 12 in summer and up to 24 for a tropical rain forest. Also, below-canopy wind speeds are usually so small that accurate turbulence measurements were difficult to obtain. Below-canopy plume-travel direction is to the left (33 to 78°) of the above-canopy wind direction.

Tracer releases below the canopy remain for the most part below the canopy. The factors controlling below-canopy diffusion and mass transfer out of the canopy are the temperature gradient and the wind speed. The below-canopy values of these parameters depend on the leaf density. A sample by Raynor (1971) of a 24-h sequence of temperature profiles for a forest and an adjacent open area is shown in Fig. 9.55. This figure shows how the canopy becomes the prime radiating surface so that during the daytime (1600 LST) heating of the canopy produces unstable lapse rates above and an inversion below the canopy. At night (0100 LST) radiation from the canopy produces an inversion above and usually isothermal conditions below the canopy.

A summary of the horizontal diffusion parameter, σ_y, for some of the experiments described in this section is shown in Fig. 9.56. Results indicate that lateral plume growth within the forest was less than the growth over open terrain during the day because of more stable conditions below the canopy. Nighttime trials seemed to give exceptionally large growth due to very large fluctuations of wind direction; hence little diurnal variation of σ_y in the forest was observed.

A summary of the vertical growth parameter, σ_z, is shown in Fig. 9.57. In all cases, vertical diffusion was found to be more rapid in the forest than over open terrain during similar stabilities, with tracer rapidly mixing up to the canopy which tended to act as a lid to prohibit further growth. Tracer concentrations tended to become uniform with height after about 1 km of travel. During the daytime the below-canopy inversion did not always prevent tracer

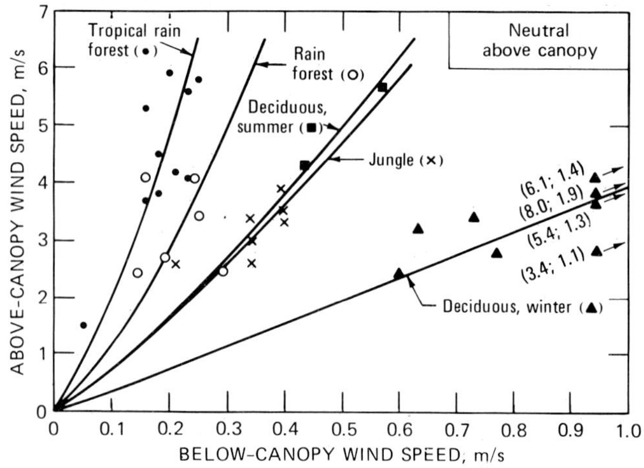

Fig. 9.54 Comparison of the above-to-below canopy wind speeds for various forests during neutral conditions. Coordinates are given in parentheses for data points outside the range of the abscissa scale. (From R. M. Brown et al., Report UT-2408-P-1, Cornell Aeronautical Laboratory, Buffalo, N.Y., 1969.)

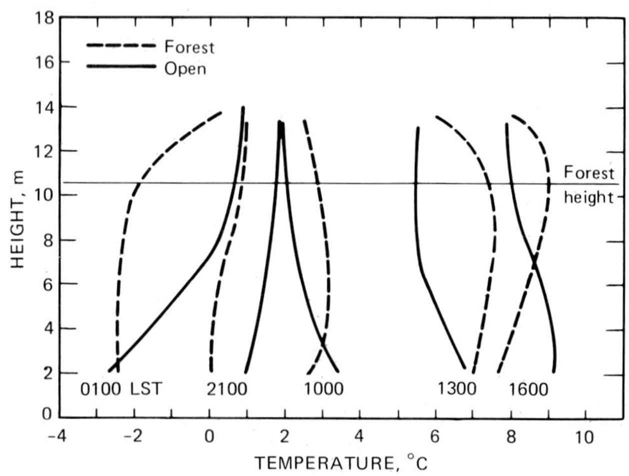

Fig. 9.55 Typical temperature profiles in the open and at 60 m into the forest. [From G. S. Raynor, Wind and Temperature Structure in a Coniferous Forest and a Contiguous Field, *For. Sci.*, 17: 351-363 (1971).]

penetration due to greater mechanical turbulence in the canopy. Nighttime vertical mixing was larger than during the daytime because of less stable conditions under the canopy. The aircraft-release dispersion tests indicated tracer penetration of the canopy within 30 min and a reversal of the concentration gradient from downward to upward.

Waldron (1971) reanalyzed some of these experiments and determined that the tracer growth rates could be summarized in the form

$$\sigma = kx^a \tag{9.26}$$

where a is equal to 0.82 for lateral growth and 2.03 for vertical growth.

9-3.2.5 Urban Diffusion. Although there are not so many experiments reported in this section as in some of the others, the modification of the dispersion environment by a city is understood well enough to permit predictions of dispersion with reasonable confidence. The thermal and roughness characteristics of a city and their effect on dispersion have been discussed by Pasquill (1970) and by Gifford (1972).

St. Louis, Missouri. A series of 42 point-source FP tracer releases between 1963 and 1965 at St. Louis, Mo., has been summarized by McElroy and Pooler (1968) and analyzed by Pooler (1966) and McElroy (1969). The city is relatively flat with variations in terrain between 140 and 170 m above mean sea level. Climatological data were

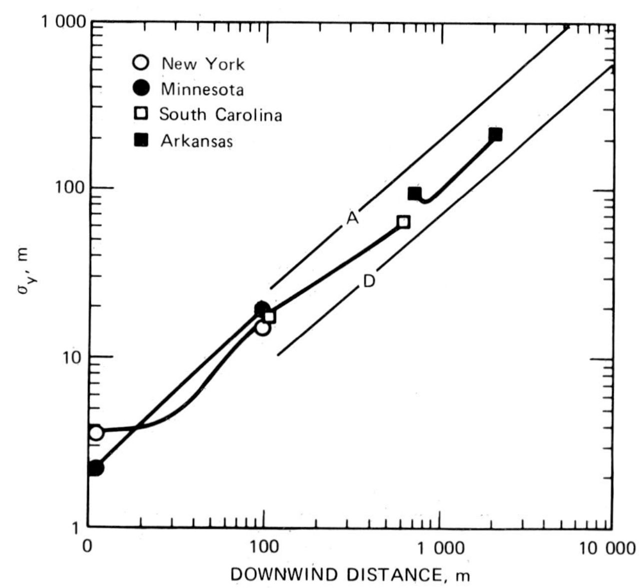

Fig. 9.56 Summary of horizontal dispersion for within-forest releases. —, Pasquill–Gifford curves.

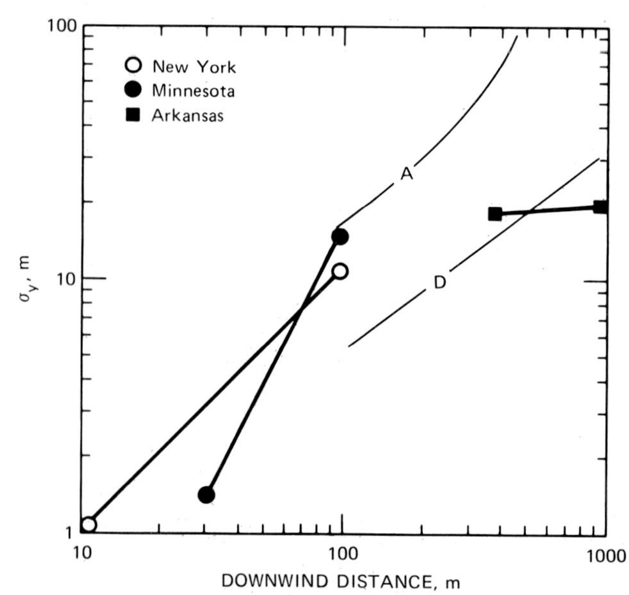

Fig. 9.57 Summary of vertical dispersion for within-forest release. —, Pasquill–Gifford curves.

collected from three stations set up with wind and temperature sensors at the periphery of the city. During the experiment, additional meteorological data were collected from three levels on a TV tower. Pibal and tetroon releases were made from the tracer dissemination point. Radiosonde ascents were made in the downtown area. A fast-response anemometer was installed at a height of 2 m near the release site.

Releases were made from two sites: one at Forest Park (ground level) and the other on top of a three-story building. The sampling arcs were oriented to encompass most directions except the northeast quadrant. The releases were about 1-h duration. Only 28 releases provided data for unstable conditions, 8 of which were in the evening. The arcs for Forest Park were at 0.8, 3.2, and 7.3 km downwind, and the building source had arcs at 2.1, 4.0, and 8.1 km downwind. Later, an additional arc at 16.1 km was added for both release points. For nine trials the vertical dosage profile was measured from a tethered balloon. Because of the large wind shear at night, the data proved to be inconclusive; therefore σ_z was computed only from surface sampling data. Vaughan and McMullen (1968) concluded that the deposition of tracer at St. Louis was insignificant by examining the size distribution of the tracer at the sampling arcs.

McElroy and Pooler's evaluation of the sampling data indicated that the major effect of the city is to increase the initial dispersion. The initial vertical (20 to 30 m) and horizontal (50 to 60 m) tracer cloud dimensions were closely approximated by the nearby buildings. Eventually, when the cloud grows larger than the eddies created by the obstructions, the plume growth approaches that of open country. Horizontal spread (Fig. 9.58) is somewhat larger than in open country for a particular stability. Vertical growth (Fig. 9.59) is even more enhanced (by about one Pasquill category). McElroy and Pooler suggest this increase is due to the relatively more important effect of convective motions over the city.

Predictions of σ_y made by Eqs. 9.8 and 9.10, although not strictly applicable to nonhomogeneous turbulence, result in an overprediction of only about 17% (solid line) with all values within a factor of 2 of the observed data (Fig. 9.60). This overprediction may be caused by the unrepresentative nature of the initial turbulence measurements in predicting downwind values where the plume size has become larger than the eddies created by the local obstructions.

Fort Wayne, Indiana. To isolate the effects of an urban area upon aerosol diffusion, Hilst and Bowne (1966) conducted a field program at Ft. Wayne, Ind., during 1964 and 1965. Bowne (1974) discussed some of the results of this program. Ft. Wayne is a moderately industrialized city about 10 km wide surrounded by flat farmland. The greatest variation in terrain is 60 m within 48 km of Ft. Wayne.

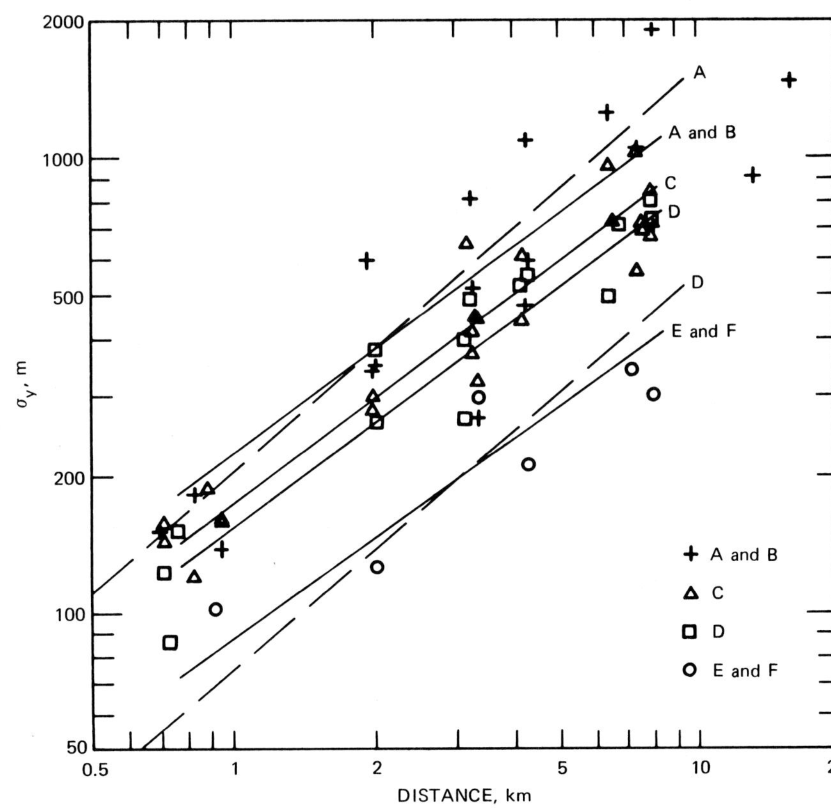

Fig. 9.58 Horizontal diffusion observed at St. Louis. —, best fit through the data; — — —, corresponding Pasquill–Gifford open-terrain curves.

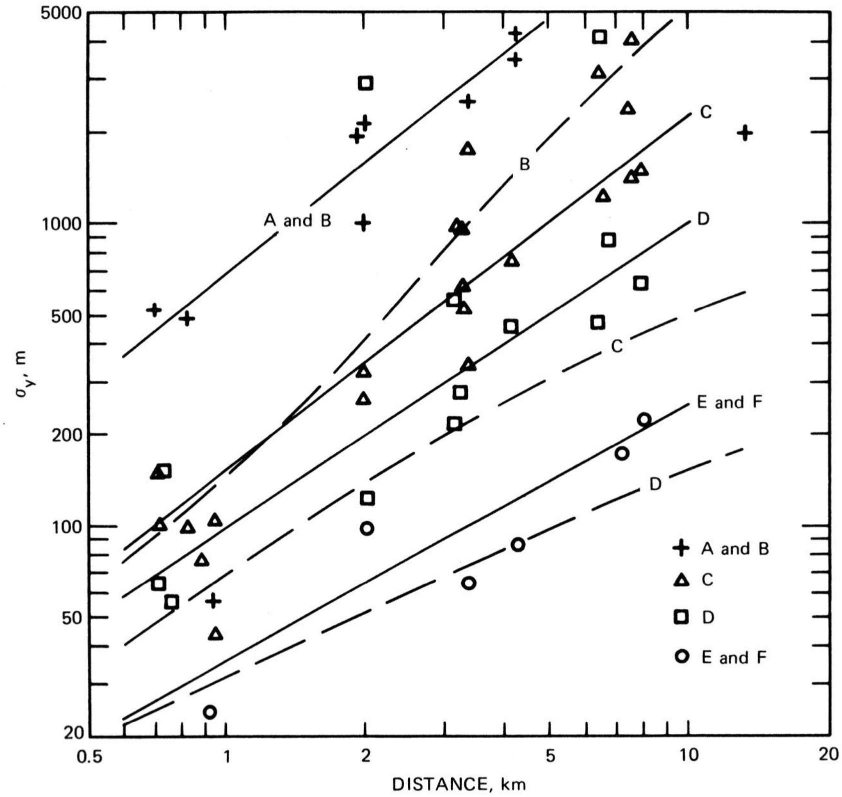

Fig. 9.59 Vertical dispersion from surface sampling at St. Louis. ——, best fit through the data; – – –, corresponding Pasquill–Gifford open-terrain curves.

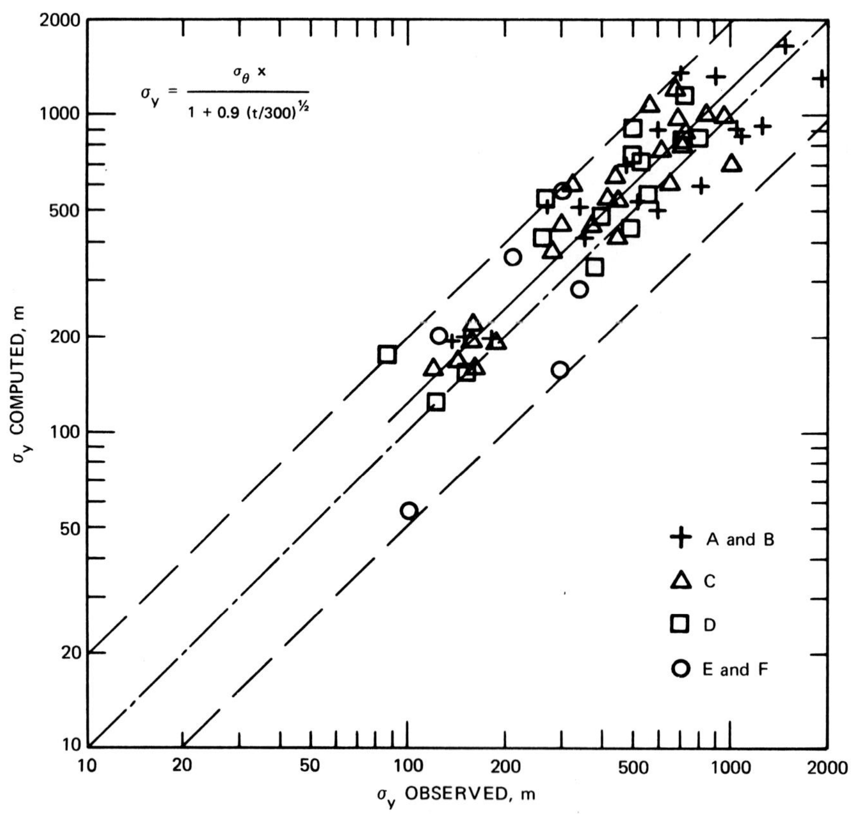

Fig. 9.60 Predictions of σ_y using Eqs. 9.8 and 9.10. ——, best fit through the data points, with about 17% above the one-to-one correspondence line (– · –); – – –, a factor of 2 above and below the one-to-one correspondence line.

Two releases of FP (yellow and green) were made in each trial by separate aircraft northwest of the city at altitudes of 91 to 214 m. A total of 36 trials were conducted, all after 1800 LST. The sampling system consisted of 250 rotorods located along 5 arcs. One arc was upwind of Ft. Wayne; three were within the city; and the last one was downwind in the rural area. The inner-city arcs were extended into the rural area. In addition, 25 sequential samplers were scattered throughout the grid. Three vertical sampling grids were constructed; one upwind of the city on a 244-m TV tower, another on a 63-m telephone tower within the city, and one downwind on a 488-m-high captive balloon.

Tracer results indicate that, because of the roughness and thermal properties of the city, there was a 30 to 50% increase in vertical mixing over the city as compared to the rural area. Exposures measured in the rural area were correspondingly higher than in the city. The authors suggest that this may in part be due to a toroidal circulation around the city with downward transport over the rural areas.

Results of vertical dispersion measurements at the TV tower and balloon are shown in Fig. 9.61. All the data points within 5 km were collected on the TV tower, upwind of the city, by varying the distance of the release line from the tower. The placement of the solid lines for D stability and E and F stabilities in Fig. 9.61 is admittedly very subjective. The lack of any D stability data above 200 m is due to the limitation of vertical measurement capability rather than lack of vertical growth. Although there is a large scatter, Fig. 9.61 confirms the increased vertical growth above an urban area in comparison to that observed over rural areas.

The authors identified four basic diffusion patterns, labeled categories I to IV. These categories represent a plume that stayed above the ground (I), slow downward motion of the cloud axis (II), rapid downward motion of the axis (III), and a thoroughly mixed layer (IV). These classes were found to be dependent on upwind stability, stability increasing with the lower category numbers.

Predictions of σ_z from an elevated source before the plume intersects the ground can be made by using the standard deviation of the vertical wind-direction fluctuation, σ_ϕ, and Eqs. 9.9 and 9.16. The results of these predictions are compared with Ft. Wayne data in Fig. 9.62. Although much scatter is still evident, two-thirds of the points lie within a factor of 2 of the one-to-one correspondence line.

Summary of Urban Dispersion. The alteration of the environment by an urban area leads to anomalies in turbulence and other meteorological parameters related to dispersion. The two primary factors are the increased roughness and different thermal properties of the city. Pasquill (1970) suggested that a roughness length from 1 to 3 m is appropriate for most cities. During neutral stratification he found that the vertical dispersion over a city could be accounted for by similarity theory when the increased roughness was taken into consideration. In general, a 100-fold increase in roughness length (0.03 m to 3 m) produced a fourfold increase in σ_z at short distances and a twofold increase at longer distances (5 km).

The heat island of an urban area is produced by the increased thermal capacity, anthropogenic heat sources, and the loss of evaporation from vegetation. These factors tend to keep the urban area warmer than the surrounding rural area.

The effect on dispersion is most pronounced at night when stable air entering the city becomes modified and an isothermal layer forms below a growing boundary layer.

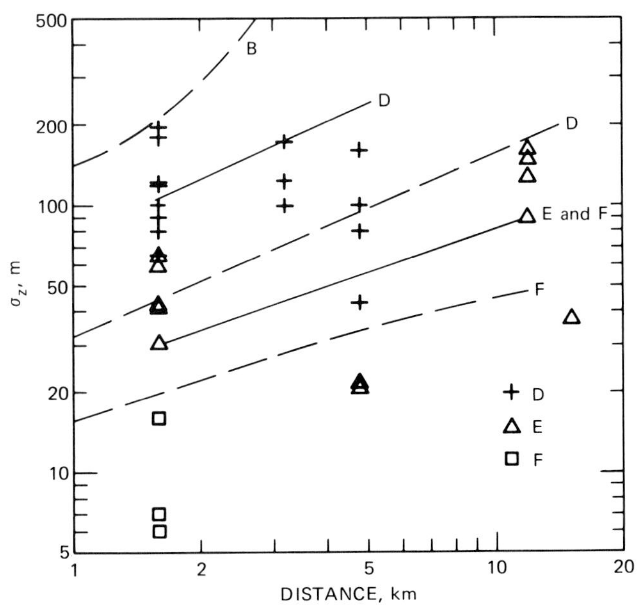

Fig. 9.61 Vertical diffusion observed at Ft. Wayne. —, best fit for D and E and F stabilities; – – –, Pasquill–Gifford open-terrain curves.

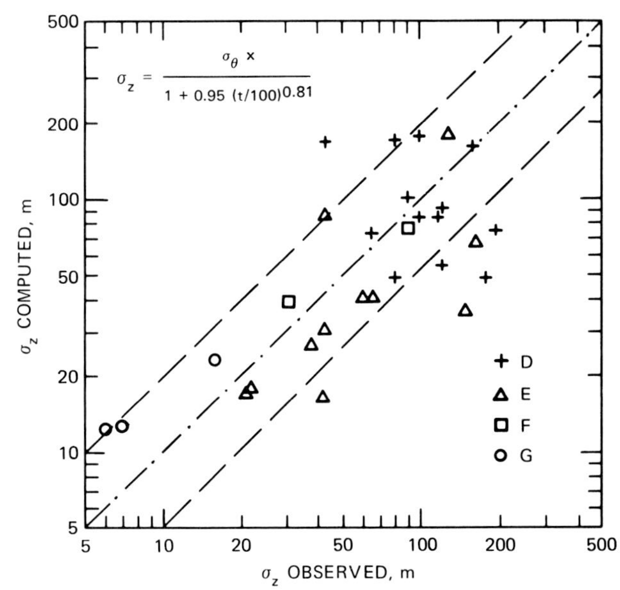

Fig. 9.62 Predictions of σ_z based upon the vertical wind-direction fluctuation at Ft. Wayne. – · –, one-to-one correspondence line; – – –, a factor of 2 above and below the one-to-one correspondence line.

This boundary layer may be several hundred meters deep; a value of 400 m is typical for New York City (Gifford, 1972). Gifford (1972) identified three regimes of city vs. rural dispersion. First, during windy and sunny conditions dispersion over rural and city areas would be similar. Second, during windy nights the rural area would have stable stratification in the lower layer and the city would have a mixed layer. Third, during calm nights convection cells may form over the city.

Angell et al. (1971) discussed the results of tetroon flights at Columbus, Ohio. They found that for tetroons flown at 200-m heights there was an anticyclonic turning as the tetroon passed over the downtown area. This turning was greater during more stable conditions. Stable stratification tended to be associated with winds from the southern sector. The anticyclonic turning was attributed to increased frictional forces and the formation of a meso-high with upward motion over the city. As the air flowed into the city, there was a 20% decrease in wind speed during stable conditions. This decrease in wind speed was also observed at Ft. Wayne. Although wind speed decreases, the turbulent velocities remain the same in the urban and rural areas, which leads to a corresponding increase in the turbulent intensities in the city (Gifford, 1972). During the daytime, wind speeds in the rural and city areas are about the same.

Summaries of the horizontal and vertical dispersion measurements conducted over cities are shown in Figs. 9.63 and 9.64. Also included in the summary figures are values of σ_y and σ_z for six releases of FP reported by D. Smith (1967) at Johnstown, Pa., conducted in the autumn during clear calm nights. The horizontal dispersion parameter, σ_y (Fig. 9.63), during stable stratification exhibits a size and growth similar to that of open country during neutral stratification (Pasquill–Gifford D), a reasonable result considering the adiabatic mixed layer that forms

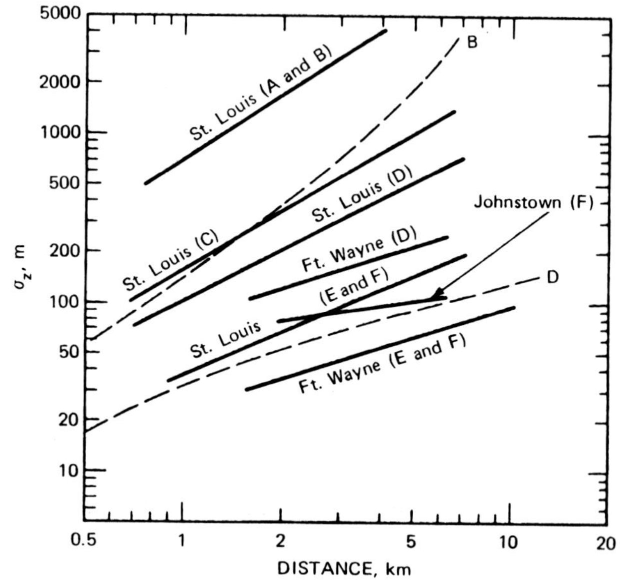

Fig. 9.64 Summary of vertical dispersion over urban areas. – – –, Pasquill–Gifford curves.

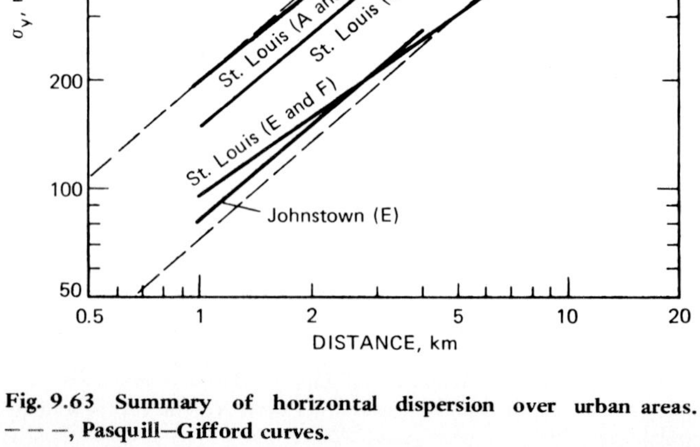

Fig. 9.63 Summary of horizontal dispersion over urban areas. – – –, Pasquill–Gifford curves.

over cities at night. Horizontal dispersion during the day in the city seems similar to that of open country for similar stabilities. Otherwise there is some evidence to indicate somewhat larger initial values of σ_y due to the effect of nearby buildings.

Vertical dispersion values (Fig. 9.64) should be viewed with caution. Ft. Wayne and Johnstown σ_z values are from observations, but the St. Louis values have been computed from measurements of surface concentration. Also, Johnstown and St. Louis were continuous surface tracer releases, but Ft. Wayne was an elevated instantaneous line source. The aircraft wake effect may be responsible for some of the initial dispersion, and, as a compensatory effect, the elevation of the release decreased the effect of turbulence generated by the buildings, a dominant feature in the surface-based releases. The release elevation may explain the lower values of σ_z at Ft. Wayne during stable stratification. Note that the D stability curve for Ft. Wayne was during nighttime releases, whereas many of the St. Louis D stabilities were daytime releases. We can conclude there is considerable enhancement of vertical dispersion in the city at night corresponding to what would be observed in adjacent rural areas. The flatter slope of the σ_z curve at Johnstown was due to a low inversion that limited the height of the mixed layer. The urban boundary layer grew downwind of the source to a maximum of about 150 m. D. Smith (1967) reported that the tracer concentrations were uniform below the inversion. Unstable vertical growth at St. Louis is large compared with open-country Pasquill–Gifford curves for similar stabilities. Pasquill (1970) suggested that these values may be too large and attributed the overestimation to a problem with the analytic procedure in computing σ_z.

Other urban experiments include some vertical dispersion trials at St. Louis, Winnipeg, and Minneapolis reported by Stanford University (1953) and analyzed by Waldron

(1964). Some horizontal diffusion trials at New York City have been reported by Megonnell (1958). The results of Hilst and Bowne (1966) were the subject of a wind-tunnel simulation by Chaudry and Cermak (1971). They were able to simulate the exposures measured at Ft. Wayne.

In summary, diffusion experiments over urban areas have found an increase in mixing over the city as compared with rural areas. Releases within the city have shown the initial dispersion to be related to the size of nearby buildings. At longer distances, plume growths approached those for the open country except for larger vertical plume spread.

9-3.2.6 Low-Wind-Speed Diffusion. Diffusion estimates during low-wind-speed stable conditions frequently are the most important because the standard methods for computing concentrations give the highest concentrations during these conditions. Large horizontal fluctuations in wind direction are usually observed at night, but because these fluctuations occur near the threshold value of most anemometers, earlier investigators were uncertain about the actual contribution of these fluctuations to the dispersion. Recently many investigators have confirmed these nocturnal fluctuations, and a climatology of bivane data during low wind speeds has been presented by Leahey and Halitsky (1973).

Oak Ridge, Tennessee. Wilson et al. (1976) reported on eleven tracer tests made during July and August of 1974. All were during low wind speeds, usually <0.5 m/s, and neutral to stable stratification. The test site is on a peninsula of the Clinch River about 25 km south of Oak Ridge, Tenn. The area is heavily wooded, and nearby ridge tops rise 30 m above the release point. It was cloudy for most of the trials, and half of them had either fog or rain.

Two tracers were used, SF_6 and CBr_2F_2 (dibromodifluoromethane). They were released at a height of 1 m for 1 h for nine trials and at 30.5 m for two trials. Oil fog was also used for visual tracking. Whole air bag samplers at 0.76 m height were placed on concentric circles 100 and 200 m from the release point. A partial arc was located at a distance of 400 m from the source. Additional samplers were placed along roads and along the river bank to obtain some samples to 4 km. The samplers on the inner arcs could be activated by remote control. Some aerial bag samples were obtained from a helicopter. Meteorological measurements were made from five towers and an orthogonal set of lasers for low wind-speed measurements. Because of the low wind speeds, no turbulence measurements were possible.

Tracer results indicated that dilution was increased with increasing stability. Figure 9.65 shows the mean range of observed σ_y values. The horizontal spread is about six times larger than the Pasquill–Gifford values (D to F stability) owing to the wind-direction meander that occurred during the hour-long release. The Pasquill–Gifford curves apply for release times of about 10 min, and therefore this type of meander is not taken into account.

The vertical dispersion parameter, σ_z, was computed from the observed surface and helicopter-collected air concentrations. For most trials the tracer filled the valley below the ridge tops (making σ_z computation difficult and non-Gaussian), and any tracer that mixed higher aloft would be affected by large wind shears and would be transported in different directions from the tracer below the ridgetops. Figure 9.66 shows the mean range of σ_z within the shaded area. The large initial dispersion is consistent with findings of other experiments in forested areas. At this site the canopy tops were at about 20 m. The growth rate with distance is slower than that which would be observed over open terrain, presumably owing to the limiting effect of the above-canopy inversion. From oil-fog releases and tracer concentration data, the authors surmised that for most

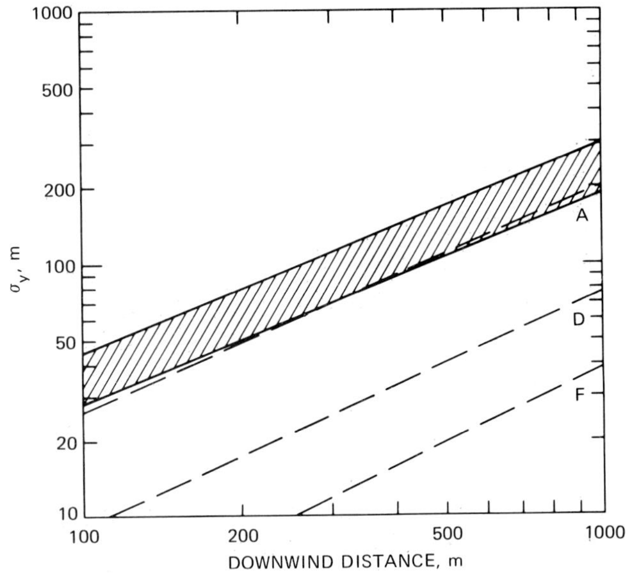

Fig. 9.65 Horizontal dispersion parameter measured near Oak Ridge during stable low-wind-speed conditions. — — —, Pasquill–Gifford curves.

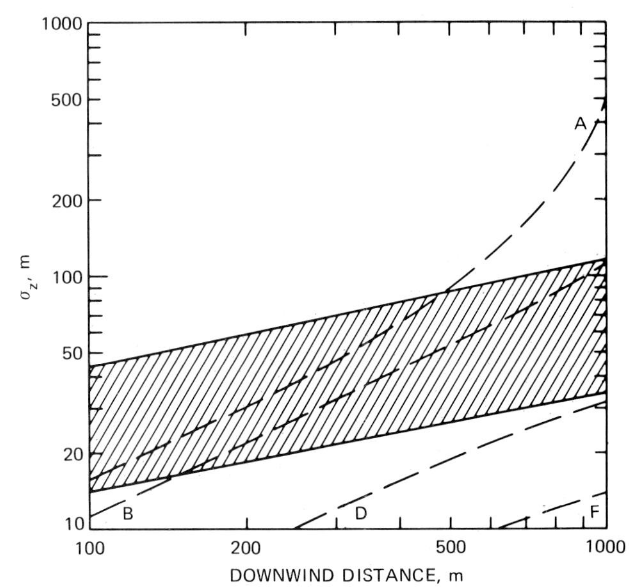

Fig. 9.66 Vertical dispersion parameter computed for the experiment near Oak Ridge during stable and low-wind-speed conditions. — — —, Pasquill–Gifford curves.

trials the tracer was well-mixed below the canopy. The enhanced vertical mixing (about 5.7 times greater than that which would be observed over open terrain during similar stabilities) was attributed to the increased roughness. However, under low wind speeds over smooth terrain, Sagendorf and Dickson (1974) found σ_z to be comparable to the Pasquill–Gifford curves.

Idaho Falls, Idaho. Eleven tracer releases of SF_6 at INEL have been described by Sagendorf and Dickson (1974). The site is a flat desertlike area located in southeastern Idaho. All releases were made during stable stratification, usually early in the morning, when wind speeds were less than 2 m/s. Large fluctuations of wind direction were observed during the release periods.

The SF_6 was released from a 1.5-m elevation for about 1 h. An oil-fog cloud was also generated along with the SF_6 for visual observations. The sampling grid extended 360° around the release point at three distances: 100, 200, and 400 m. Whole air bag samples were later analyzed by an electron capture gas chromatograph. During some tests, samples to heights of 9 m were collected from eight towers at the 200-m arc. Meteorological data were collected on a 61-m tower at the 200-m arc.

The wind-direction meander was reflected in the SF_6 concentration data. During some tests the plume was observed on the entire 360° arc. A distinctive vertical tilt of the plume was also observed. When the wind direction was relatively steady, the right-hand portion of the plume was elevated by 5 to 6 m. During these cases the wind direction veered with height. A small (5 m) knoll 800 m northeast of the release point may have affected tracer behavior. Small stationary "puddles" of tracer were sometimes observed. Vertical dispersion was similar to what would be observed over open terrain during very stable conditions.

Comparison of the observed concentration data at 200 m during Test 14 to those calculated by the standard Gaussian equation, using various methods to determine σ_y and σ_z, is shown in Fig. 9.67. In the "standard" method, σ_y and σ_z were taken from the Pasquill–Gifford stable curves. These curves do not account for increased wind-direction meander with light winds and hence overestimate the concentrations for the hour-long release. On the average, for all trials the standard Gaussian technique resulted in overprediction of about a factor of 8. The other method, which provided the best representation of the observed data for most trials, used 2-min values of $\bar{u}$ and σ_θ to compute σ_y by a regression equation developed earlier at INEL. Similar values of σ_y could be obtained from the Pasquill–Gifford curves. These 2-min values of σ_y, $\bar{u}$, and appropriate σ_z were used to calculate 2-min concentrations that were averaged over the entire release period.

Three Mile Island, Pennsylvania. A tracer study was conducted at Three Mile Island Nuclear Station, south of Harrisburg, Pa., to test various diffusion models during periods of stable stratification and slow wind speeds (General Public Utilities, 1972; Halitsky and Woodward, 1975). Three Mile Island is a relatively flat area in the Susquehanna River.

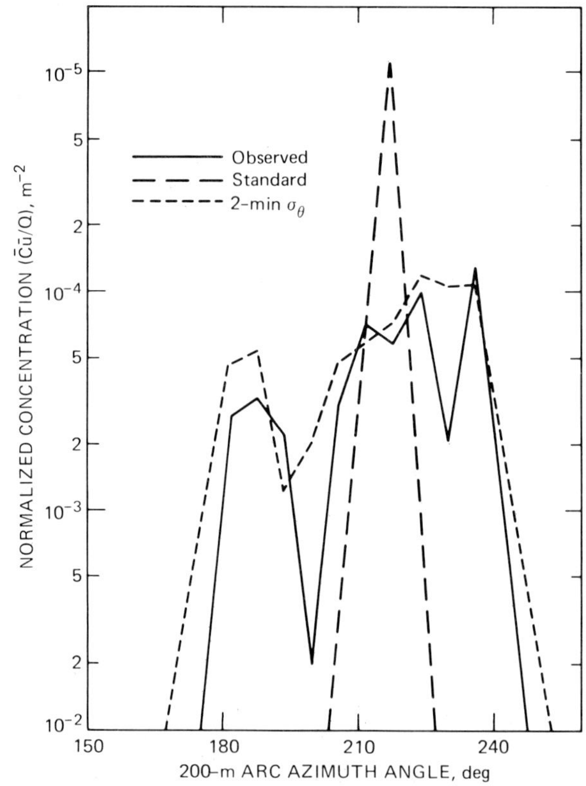

Fig. 9.67 Observed concentrations and model predictions along the 200-m arc at Idaho National Engineering Laboratory. (From J. F. Sagendorf and C. R. Dickson, *Diffusion Under Low Wind Speed Inversion Conditions*, NOAA Technical Memo-ERL-ARL-52, Air Resources Laboratories, Silver Spring, Md., 1974.)

Two types of SF_6 releases of about 45-min duration were planned. One group consisted of five releases from a height of 1.3 m in an open field and another of five releases adjacent to the containment building. The samplers were arranged in a circle of about 92-m radius for the open-field trials. A 244-m radius circle was used for the building trials. A later test attempted more complete vertical sampling.

In most trials, wind directions varied over 150°. The maximum concentration in the open field corresponded to the azimuth with the longest wind-direction persistence. The releases in the building wake sometimes produced peak concentrations askew from the persistent wind direction. Concentrations in the building wake were considerably smaller than those in the open-field trials. Releases in the building wake caused the tracer to transport upward to the approximate height of the building before diffusing downward.

The concentrations from the open-field trials were best predicted by a model that accounted for the large wind-direction fluctuations. When σ_y and σ_z were based on the Pasquill–Gifford curves (usually F stability), peak concentrations were overpredicted by a factor of 22. But when σ_y was replaced by a sector average, i.e., the concentrations were assumed to be uniform in a sector equal to the extreme range of observed wind directions during the test, the model overpredicted the peak by only 27%.

Hanford, Washington. A data compilation of many different kinds of diffusion experiments at Hanford has been summarized by Nickola (1977). An analysis of several of these experiments during slow wind and stagnation conditions has been reported by Nickola (1972), Nickola and Ludwick (1972), and Nickola, Clark, and Ludwick (1975).

The releases of ^{85}Kr were for 30 min from the 0.3-m level. Sampling was made with a Geiger–Müller counter at a height of 1.5 m. Data sampled were 10-s mean concentrations for 22 min, centered on the release period at distances out to 1.2 km.

The normalized exposures for some of the trials are shown in Fig. 9.68. Four trials during stable stratification showed similar exposures although two were during steady wind conditions and two were during stagnation. The authors suggest that one explanation may be that the cases with faster wind speeds also had less wind-direction fluctuation. A solid line representing several trials during neutral and unstable stratification is included in Fig. 9.68 for reference.

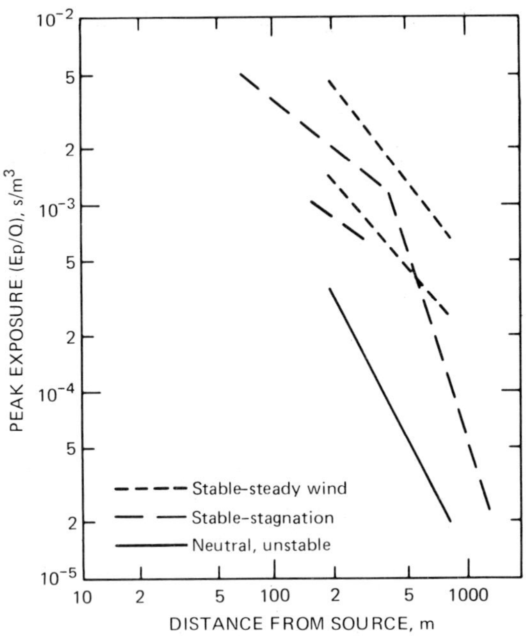

Fig. 9.68 Observed peak concentrations at Hanford. (From P. W. Nickola, Report BNWL-1651, Pt. 3, pp. 123-125, Battelle Northwest Laboratories, Richland, Wash., 1972.)

Low-Wind-Speed Diffusion Summary. In all the reported diffusion trials under stable low-wind-speed conditions, the observed concentrations were found to be lower than concentrations calculated with the Gaussian model and with σ_y and σ_z taken from the stable Pasquill–Gifford curves. When the calculations accounted for the short-term fluctuations in wind direction, concentration predictions were much closer to the observed values.

Van der Hoven (1976) analyzed low-wind-speed diffusion data and found that the lowest concentrations correspond to sites with the roughest terrain. In addition to the experiments in this section, he included some diffusion trials conducted in Louisiana. He partitioned the experimental sites into three types: I, flat and unobstructed terrain (Idaho and Washington); II, relatively flat but forested areas (Louisiana and Pennsylvania); and III, rough, hilly, and forested (Tennessee). These results are summarized in Fig. 9.69 which shows the range of the mean normalized peak concentrations for the three roughness types. Figure 9.69 shows that the lowest peak concentrations were observed at those sites with the roughest terrain, which also had the largest fluctuation in wind direction, although no relation between wind-direction fluctuation and concentration could be determined because of the large scatter. The solid lines in Fig. 9.69 represent the concentration calculated from the Pasquill–Gifford curves of σ_y and σ_z and could result in overprediction of concentration by as much as a factor of 500 at the rough sites. The misuse of the Pasquill–Gifford curves, intended for short-term dispersion estimates, can lead to serious errors in calculated concentrations.

9-3.2.7 Long-Range Diffusion. Most experiments discussed in the previous sections were tracer experiments in which sampling was conducted in the range of a few

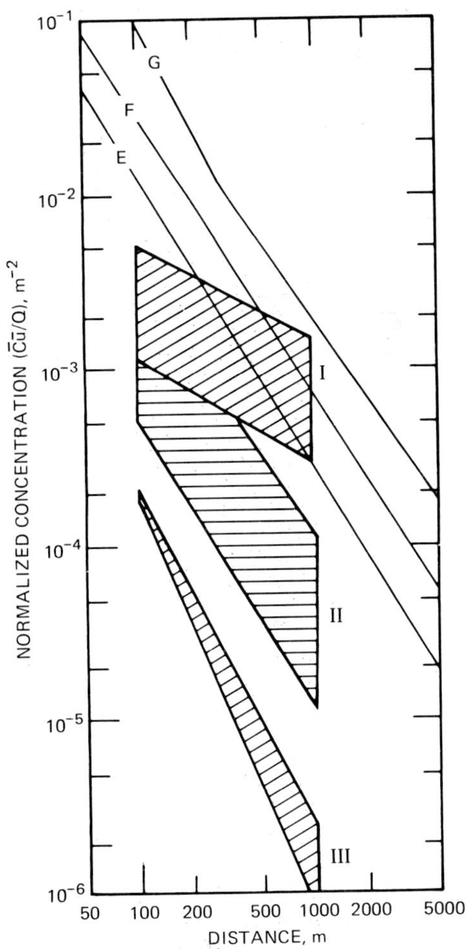

Fig. 9.69 Summary of low wind dispersion data grouped by roughness of the test site. —, concentrations calculated from Pasquill–Gifford E, F, and G stabilities. [From I. Van der Hoven, A Survey of Field Measurements of Atmospheric Diffusion Under Low-Wind Speed Inversion Conditions, *Nucl. Safety*, 17(2): 223-230 (1976).]

kilometers. Sampling beyond a few tens of kilometers becomes prohibitive because of the complicated logistics, increased cost, and the large amount of tracer that must be released to overcome the background concentrations. Also, there may be some bias in previous long-range data that have been collected. For most experimenters, the meteorological situations chosen were the ones in which the plume is well-defined so as to permit complete sampling. Situations with complicated meteorology and large wind shears would have been excluded. Although Heffter (1965) summarized many long-range "experiments," quite a few of them were samples taken from nuclear weapons tests, and most of these data are either classified or difficult to obtain. Some additional experiments have been conducted recently, but many are limited to only a single release.

In general, for most of the longer range experiments, wind shear becomes more important than the turbulence-generated dispersion. Therefore extrapolation of short-range dispersion data to longer distances can be misleading. Pasquill (1962) qualitatively assessed the role of alongwind and crosswind vertical wind shear in enhancing dispersion observed at a particular level. He summarized the process into two steps. First, shear produces spread when vertical mixing is slow, and, second, the shear-induced spread is later observed at a given level when mixing becomes vigorous. Several tracer experiments were conducted with FP released from the ground and aircraft. Pasquill found that there was a small variation with height of the direction and speed of travel of the particles when compared to the actual wind profile, except when small vertical mixing let the wind shear affect the tracer. Upon examination of shorter range diffusion data, Pasquill (1969) determined that the shear-induced dispersion becomes important at about 12 km from ground sources and at about 5 km from elevated sources.

Saffman (1962) estimated the contribution of wind shear to dispersion by solving the diffusion equation with an effective diffusivity that depended on the interaction of wind shear and vertical transport. He assumed a linear wind profile and a vertical diffusivity constant with height. The diffusion was divided into two regimes. Close to the source when the vertical growth is still important,

$$\sigma^2 = \frac{1}{25}\,\psi^2 K_z t^3 \tag{9.27}$$

where ψ (s^{-1}) is the slope of the wind-speed profile and K_z is the vertical mixing coefficient. At distances ($t > h^2/K_z$) from the source when further vertical growth is inhibited by an upper boundary, the asymptotic solution will apply

$$\sigma^2 = \frac{1}{60}\,\frac{\psi^2 h^4 t}{K_z} \tag{9.28}$$

where h is the height of the boundary layer. Equations 9.27 and 9.28 show that initially the dispersion grows by $t^{3/2}$ and later by $t^{1/2}$. The initial dispersion is directly proportional to the vertical mixing due to the growing plume's entering layers of additional wind shear. The latter phase is similar to the two-phase shear process described by Pasquill (1962), and hence the growth is indirectly proportional to the vertical mixing.

Tyldesley and Wallington (1965) used a method similar to Saffman except their solutions were derived through numerical techniques. They essentially confirmed Saffman's analysis. They divided diffusion into four regimes shown in Fig. 9.70: simple turbulent diffusion, a plume growing into a shear layer ($\sigma \propto X^{3/2}$), transition zone ($\sigma \propto X$), and a final zone where an upper boundary limits the shear effect ($\sigma \propto X^{1/2}$). By assuming that the depth of the layer h = 2.15 σ_z and that the plume width w = 4.3 σ_y, they derived from Eqs. 9.27 and 9.28 the plume widths shown in Table 9.10 for the three wind-shear diffusion regimes shown in Fig. 9.70.

F. B. Smith (1965) also investigated the problem of shear-induced spread. He derived an equation describing the dispersion through a statistical approach in which he assumed that the lateral and vertical turbulence was correlated. His results also confirmed those of previous investigators.

Peterson (1968) sampled an ^{41}Ar plume from the 107-m stack of the BNL reactor. Sampling was conducted by aircraft to 300 km downwind. Most of the plume travel was over water, and the atmosphere was neutral between 100 and 900 m. He found the plume growth to be almost linear with distance or travel time. These regression lines are shown in Figs. 9.71 and 9.72.

Csanady (1969) solved the diffusion equation with an effective diffusivity based on wind shear. He assumed an Ekman wind profile in the boundary layer, and the results were divided into two regimes. In the first regime, the plume is still growing; in the second regime, at longer travel times, the cloud grows much larger than the Ekman layer (global diffusion). In the second case,

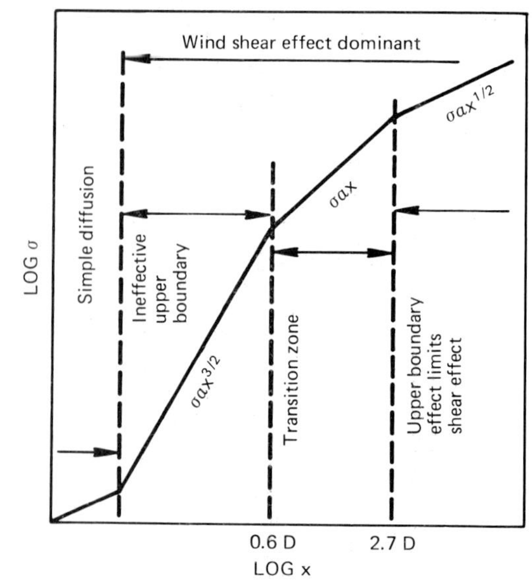

Fig. 9.70 Schematic diagram to illustrate the four phases of shear-influenced dispersion. [From J. B. Tyldesley and C. F. Wallington, The Effect of Wind Shear and Vertical Diffusion on Horizontal Dispersion, *Q.J.R. Meteorol. Soc.*, 91: 158-174 (1965).]

TABLE 9.10 Plume Width for Wind-Shear-Determined Dispersion*

Region	Applicable distance†	Plume width‡
Upper boundary ineffective	<0.63 D	$0.48 \frac{\Delta V'}{u} X$
Transition zone	>0.63 D <2.66 D	$0.48 \frac{\Delta V}{u} X$
Upper boundary limits shear effect	>2.66 D	$4.3 \left(\frac{2 K_z x}{u}\right)^{1/2}$
		$K = \frac{(\Delta V)^2 L^2}{120 K_z}$

*From J. B. Tyldesley and C. F. Wallington, The Effect of Wind Shear and Vertical Diffusion on Horizontal Dispersion, *Q. J. R. Meteorol. Soc.*, 91: 158-174 (1965).

$†D = \frac{\bar{u}h^2}{2 K_z}$

‡ΔV′, wind shear over depth 0.58 h. ΔV, wind shear over depth h.

$$\sigma_y^2 = 2 K_c t \qquad K_c = 0.2146 \frac{u_g^2}{f} \tag{9.29}$$

where f is the Coriolis parameter and u_g is the geostrophic wind speed. Csanady's solution of the turbulent and shear values of σ_y for typical values of $u_g = 15$ m/s and $f = 10^{-4}$/s is shown in the summary (Fig. 9.72).

Randerson (1972) extracted σ_y from values from airborne radiological samples taken from a Plowshare nuclear cratering experiment in Nevada in 1968. He identified three growth regimes in the data. The first, to about 0.5 h of travel, the growth was proportional to $t^{1/2}$. The second, to about 15 h of travel, the growth was proportional to $t^{1.2}$, and the third, after 15 h of travel, was again proportional to $t^{1/2}$. Randerson also tested Csanady's (1969) wind-shear model for σ_y and found that it predicted slightly larger values than those observed. The overestimate is probably due to the fact that the Csanady model is based on the classical Ekman profile with a 45° angle between the surface and geostrophic winds, thereby giving cross-isobar velocities roughly three times greater than usually observed.

Brown, Cohen, and Smith (1972) used an airborne densitometer to determine plume growth at a power plant on the north shore of Long Island, N. Y. Sampling flights

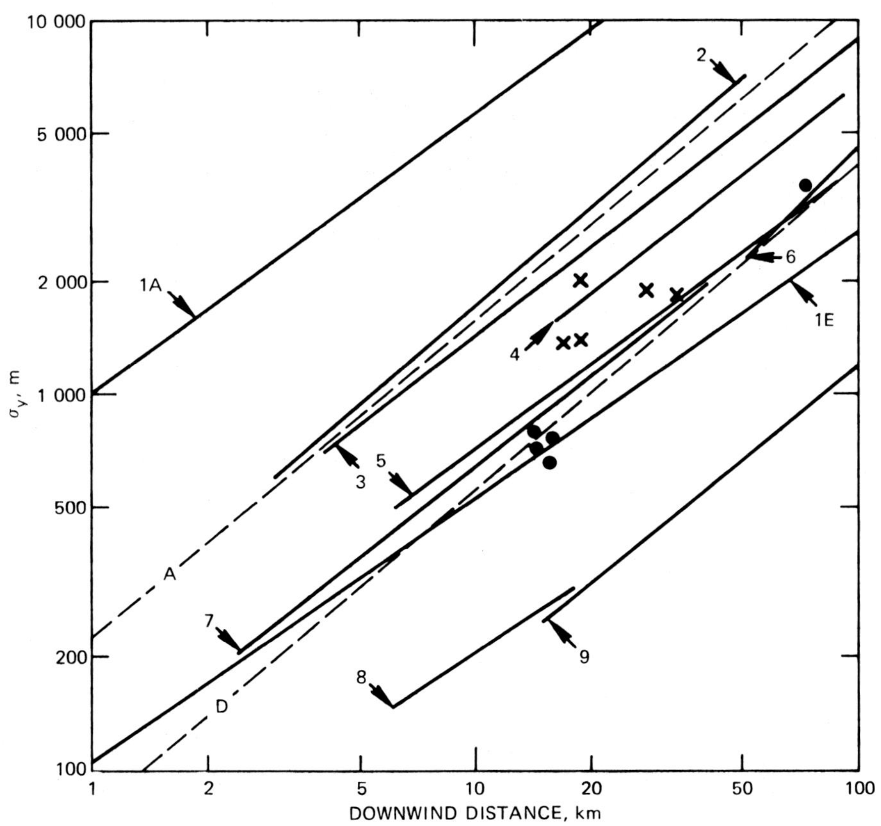

Fig. 9.71 Long-range observed horizontal spread for: 1A, Hemibach, Super, and McPartland (1975); stability A curve; 1E, stability E curve; 2, Clements (1979); 3, Pasquill (1974); 4, Bench et al. (1978); 5, Ferber and List (1973), ground-level samples; 6, Peterson (1968); 7, Brown and Michael (1975); 8, Ferber and List (1973), instantaneous aircraft samples; 9, Brown, Cohen, and Smith (1972), aircraft samples; ●, Pasquill (1962), neutral; x, Pasquill (1962), unstable; – – –, Pasquill–Gifford short-range A and D stability curves.

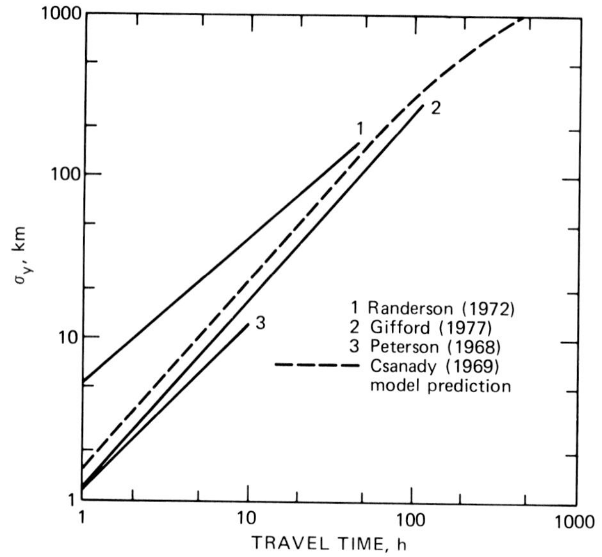

Fig. 9.72 Very long-range horizontal dispersion, observed and model predictions.

were made in the winter when conditions were stable near the ground but were neutral aloft. Horizontal standard deviations were relatively short term but integrated vertically through the depth of the plume. To eliminate effects of the buoyant source, only data >20 km downwind are summarized in Fig. 9.71.

Some additional flights at the power plant were summarized by Brown and Michael (1975). These flights were made when conditions were either stable or neutral. Several aircraft passes were made at each level to obtain a time average and at several levels to obtain horizontal growth through the entire depth of the plume at each distance. These data are also summarized in Fig. 9.71. The authors found that greater horizontal spread occurred during stable conditions because of large wind shears (about 0.1°/m).

A 3-h release of SF_6, CF_2Br_2, $C_2F_2Br_2$, and ^{85}Kr from a 76-m stack at INEL has been summarized by Ferber and List (1973). Ground-level sampling was accomplished by bag samplers 6, 18, 47, and 87 km downwind. Aircraft samples were obtained by an on-board electron capture gas chromatograph at several altitudes over some of the arcs. The results of the SF_6 sampling data are shown in Fig. 9.71. Very strong winds during the experiment indicated a Pasquill D stability.

Pasquill (1974) presented a reanalysis of some long-distance dispersion data by Richardson and Proctor (1925), Braham, Seely, and Crosier (1952), and Pasquill (1955, 1956). He found that σ_y grew in proportion to $X^{0.8}$ and did not show the square-root limit of growth at large distances. His summary is also shown in Fig. 9.71.

Heimbach, Super, and McPartland (1975) presented some dispersion data of silver iodide tracer released from a 91-m tower on a 61-m hill in rugged Montana countryside. Aircraft samples were taken as far out as 200 km. Plume widths were adjusted to represent 1-h averages from collected wind-fluctuation data. The extreme limits of his data, A and E stability, are shown in Fig. 9.71.

Gifford (1977) replotted available tropospheric relative diffusion data. A line that passes through most of the data is proportional to $t^{1.2}$. This summary curve, which includes some data previously mentioned in this section, is included in Fig. 9.72.

Bench et al. (1978) summarized data from a 4-h release of SF_6 and deuterated methane from the 62-m stack of the Savannah River Plant near Augusta, Ga. Samples were collected at ground level and by aircraft out to 94 km. Meteorological conditions during the experiment were unstable. Both aircraft and ground SF_6 samples are combined and shown in Fig. 9.71.

A 3-h ground-level release at INEL of SF_6, methane 20 and 21, and C_6F_{12} and C_8F_{16} has been reported by Clements (1979). Ground-level samples were taken at 3, 50, and 90 km. These data are shown in Fig. 9.71 for the SF_6 tracer. Other long-range experimental data not discussed in this section are available from Telegadas et al. (1980), Pendergast et al. (1979), and Ferber et al. (1981).

The summary Figs. 9.71 and 9.72 show that none of the data seem to approach the theoretical square-root growth with distance. The data in Fig. 9.71 stratified by stability are not shown after 100 km downwind because the significance of a single stability is hard to evaluate when the travel time approaches one-half the diurnal cycle. The data in Fig. 9.71 still show considerable stratification according to stability; however, differences in sampling duration between instantaneous aircraft and long-period ground-level samples provide as much variation in the plume dimensions as does the stability.

Data taken at distances greater than 100 km are included in Fig. 9.72 and plotted by travel time. It is presumed that all these data are affected by significant wind shear. Csanady's (1969) σ_y values for typical values of geostrophic wind speed (15 m/s) and Coriolis parameter (10^{-4}/s) are also included for comparison. Estimates of long-range dispersion are difficult to make, and any approach that considers wind shear at these distances is likely to provide more realistic results than those from extrapolation of short-range data beyond their range of applicability.

9-4 CONCLUDING REMARKS

Although considerable progress has been made in recent years in organizing dispersion data from sites with relatively simple terrain and meteorology, current interest centers about those sites with special features. In these cases, simple techniques are not always adequate to predict dispersion within accepted standards, and any extrapolation of results from different experimental sites should be viewed with caution. However, the dispersion data from these sites, with some exceptions, generally indicated increased dispersion over that which would be found above flat terrain during similar meteorological conditons.

REFERENCES

Angell, J. K., D. H. Pack, C. R. Dickson, and W. H. Hoecker, 1971, Urban Influence on Nighttime Air Flow Estimated from Tetroon Flights, *J. Appl. Meteorol.*, 10: 194-204.

Barad, M. L. (Ed.), 1958, *Project Prairie Grass, A Field Program in Diffusion, Geophysical Research Papers*, No. 59, Vols. I and II, Report AFCRC-TR-58-235, July 1958, (ASTIA Documents AD-152572 and AD-152573), Air Force Cambridge Research Center.

——, and J. J. Fuquay (Eds.), 1962a, *The Green Glow Diffusion Program, Geophysical Research Papers*, No. 73, Vols. I and II, USAEC Report HW-71400 (Report AFCRL-62-251(I), [January 1962 and Report AFCRL-62-251(II), April 1962], Air Force Cambridge Research Laboratories.

——, and J. J. Fuquay, 1962b, Diffusion in Shear Flow, *J. Appl. Meteorol.*, 1: 257-264.

——, and D. A. Haugen, 1959, A Preliminary Evaluation of Sutton's Hypothesis for Diffusion from a Continuous Point Source, *J. Meteorol.*, 16: 12-20.

Bench, H. L., et al., 1978, *Heavy Methane—SF_6 Tracer Test Conducted at the Savannah River Plant, December 10, 1975*, Report DP-1469, Savannah River Laboratory, Aiken, S. C.

Bendix Systems Division of the Bendix Corporation, 1963, *Jungle Canopy Penetration*, Vols. I, II, and III, Ann Arbor, Mich.

Bowne, N. E., 1974, Diffusion Rates, *J. Air Pollut. Control Assoc.*, 24: 832-835.

——, H. D. Entrekin, K. W. Smith, and G. R. Hilst, 1969, *Aerosol Diffusion over Woodlot Complexes*, Report TRC-7236-343A, U. S. Army Dugway Proving Ground, Travelers Research Corporation, Hartford, Conn.

Braham, R. R., B. K. Seely, and W. D. Crosier, 1952, A Technique for Tagging and Tracing Air Parcels, *Trans. Am. Geophys. Union*, 33: 825.

Brown, R. M., L. A. Cohen, and M. E. Smith, 1972, Diffusion Measurements in the 10—100 km Range, *J. Appl. Meteorol.*, 11: 323-334.

——, and P. Michael, 1975, Measured Effect of Shear on Plume Dispersion, in *Proceedings of the Second American Meteorological Society Conference on Air Pollution*, pp. 246-250, Santa Barbara, Calif.

Brown, R. A., G. E. McVehil, R. L. Peace, and R. W. Coakley, 1969, *Characterization of Forest Vegetation Analogs*, Report UT-2408-P-1, Cornell Aeronautical Laboratory, Buffalo, N. Y.

Chaudhry, F. H., and J. E. Cermak, 1971, *Wind Tunnel Modeling of Flow and Diffusion over an Urban Complex*, Desert Test Center, Fort Douglas, Utah.

Clements, W. E. (Ed.), 1979, *Experimental Design and Data of the April 1977 Multitracer Atmospheric Experiment at the Idaho National Engineering Laboratory*, Los Alamos Scientific Laboratory Report, Los Alamos, N. Mex., in preparation.

——, and S. Barr, 1976, Atmospheric Transport and Dispersion at a Site Dominated by Complex Terrain, in *Third Symposium on Atmospheric Turbulence, Diffusion and Air Quality*, pp. 430-435, Oct. 19—22, 1976, Raleigh, N. C., American Meteorological Society.

Collins, G. F., 1971, Predicting "Sea-Breeze Fumigation" from Tall Stacks at Coastal Locations, *Nucl. Safety*, 12(2): 110-114.

——, F. E. Bartlett, A. Turk, S. M. Edmonds, and H. L. Mark, 1965, A Preliminary Evaluation of Gas Air Tracers, *J. Air Pollut. Control Assoc.*, 15: 109-112.

Cowan, G. A., D. G. Ott, A. Turkevich, L. Machta, G. J. Ferber, and N. R. Daly, 1976, Heavy Methanes as Atmospheric Tracers, *Science*, 191: 1048-1050.

Cramer, H. E., 1957, A Practical Method for Estimating the Dispersal of Atmospheric Contaminants, in *Proceedings of the First National Conference on Applied Meteorology*, pp. C-33 to C-55, American Meteorological Society, Hartford, Conn.

——, 1959, Engineering Estimates of Atmospheric Dispersal Capacity, *Am. Ind. Hyg., Assoc. J.*, 20: 183-189.

——, G. M. DeSanto, R. K. Dumbauld, P. Morgenstern, and R. N. Swanson, 1964, *Meteorological Prediction Techniques and Data System*, Report GCA-64-3-G, Geophysical Corporation of America, Bedford, Mass.

——, F. A. Record, and H. C. Vaughan, 1958, *The Study of the Diffusion of Gases in the Lower Atmosphere*, Report AFCRC-TR-58-239, Department of Meteorology, Massachusetts Institute of Technology.

Crawford, T. V., 1966, *A Computer Program for Calculating the Atmospheric Dispersion of Large Clouds*, Lawrence Radiation Laboratory, USAEC Report UCRL-50179, University of California, Livermore, Calif.

Csanady, G. T., 1969, Diffusion in an Ekman Layer, *J. Atmos. Sci.*, 26: 414-426.

DeMarrais, G. A., G. L. Downing, and H. E. Meyer, 1968, Transport and Diffusion of an Aerosolized Insecticide in Mountainous Terrain, Report ERLTM-ARL-6, Silver Spring, Md.

Dietz, R. N., E. A. Cote, and R. W. Goodrich, 1976, Air Mass Movements Determined by Real-Time Frontal Chromatography of Sulfur Hexafluoride in *Measurement, Detection and Control of Environmental Pollutants*, Symposium Proceedings, Vienna, 1976, pp. 277-299, STI/PUB/432, International Atomic Energy Agency, Vienna.

Doran, J. C., T. W. Horst, and P. W. Nickola, 1978a, Variations in Measured Values of Lateral Diffusion Parameters, *J. Appl. Meteorol.*, 17: 825-831.

——, T. W. Horst, and P. W. Nickola, 1978b, Experimental Observations of Lateral and Vertical Dispersion Characteristics on Source Height, *Atmos. Environ.*, 12(12): 2259-2263.

Draxler, R. R., 1975, *Diffusion from a Continuous Point Source, March 1975*, M.S. Thesis, The Pennsylvania State University, University Park, Pa.

——, 1976, Determination of Atmospheric Diffusion Parameters, *Atmos. Environ.*, 10: 99-105.

Eggleton, A. E. J., and N. Thompson, 1961, Loss of Fluorescent Particles in Atmospheric Diffusion Experiments by Comparison with Radioxenon Tracer, *Nature*, 192: 935-936.

Elliott, W. P., 1961, The Vertical Diffusion of Gas from a Continuous Source, *Int. J. Air Water Pollut.*, 4: 33-46.

——, R. J. Engelmann, and P. W. Nickola, 1961, *Area-Dosage Relationships and Time of Tracer Arrival in the Green Glow Program*, Air Force Surveys in Geophysics, No. 134, Report AFCRL-468, Air Force Cambridge Research Laboratories.

Ferber, G. J., and R. J. List, 1973, Long-Range Transport and Diffusion Experiments, Air Resources Laboratories, Silver Spring, Md., unpublished.

——, K. Telegadas, J. L. Heffter, C. Ray Dickson, R. N. Dietz, and P. W. Krey, 1981, *Demonstration of a Long-Range Atmospheric Trace System Using Perfluorocarbons*, NOAA Technical Memorandum ERL-ARL-101, Air Resources Laboratories, National Oceanic and Atmospheric Administration, Silver Spring, Md.

Fosberg, M. A., D. G. Fox, E. A. Howard, and J. D. Cohen, 1976, Nonturbulent Diffusion Processes in Complex Terrain, *Atmos. Environ.*, 10: 1053-1055.

Frenkiel, F. N., and J. Katz, 1956, Studies of Small-Scale Turbulent Diffusion in the Atmosphere, *J. Meteorol.*, 13: 388-394.

Fritschen, L. J., C. H. Driver, C. Avery, J. Buffo, R. Edmonds, R. Kinerson, and P. Schiess, 1970, *Dispersion of Tracers into and Within a Forested Area*, Report TR-ECOM-68-G8-3, U. S. Army Electronics Command, Atmospheric Sciences Laboratory, Ft. Huachuca, Ariz.

——, and R. Edmonds, 1976, Dispersion of Fluorescent Particles into and Within a Douglas Fir Forest, in *Proceedings of Atmospheric Surface Exchange of Particulate and Gaseous Pollutants Symposium*, pp. 280-301, Richland, Wash.

Fuquay, J. J., C. L. Simpson, M. L. Barad, and J. H. Taylor, 1963, Results of Recent Field Programs in Atmospheric Diffusion, *J. Appl. Meteorol.*, 2: 122-128.

——, C. L. Simpson, and W. T. Hinds, 1964, Prediction of Environmental Exposures from Sources near the Ground, Based on Hanford Experimental Data, *J. Appl. Meteorol.*, 3: 761-770.

General Public Utilities (GPU), 1972, *Atmospheric Diffusion Experiments with SF_6 Tracer Gas at Three Mile Island Nuclear Under Low Wind Speed Inversion Conditions,* Report filed as Amendment 24 to Three Mile Island Unit 1 FSAR, AEC Docket 50-289.

Gifford, F. A., 1957, Relative Atmospheric Diffusion of Smoke Puffs, *J. Meteorol.*, 14: 410-414.

——, 1960, Peak to Average Concentration Ratios According to a Fluctuating Plume Dispersion Model, *Int. J. Air Pollut.*, 3: 253-260.

——, 1961, Use of Routine Meteorological Observations for Estimating Atmospheric Dispersion, *Nucl. Safety,* 2(4): 47-51.

——, 1972, Atmospheric Transport and Dispersion over Cities, *Nucl. Safety,* 13(5): 391-402.

——, 1976, Turbulent Diffusion-Typing Schemes: A Review, *Nucl. Safety,* 17(1): 68-86.

——, 1977, Tropospheric Relative Diffusion Observations, *J. Appl. Meteorol.*, 16: 311-313.

Golder, D. G., 1972, A Comparison of Stability Parameters, *Boundary-Layer Meteorol.*, 3: 47-58.

Gotaas, Y., 1975, *Atmospheric Dispersion in Valleys* (Part 1), Interim Report UM-34, Norwegian Defense Research Establishment, Kjeller, Norway.

Gryming, S. E., E. Lyck, and K. Hedegaard, 1978, Short-Range Diffusion Experiments on Unstable Conditions over Inhomogeneous Terrain, *Tellus,* 30: 392-403.

Haeggblom, L. E., C. H. Gyllander, and U. Widemo, 1969, *Atmospheric Diffusion, Investigations at Studsvik,* AE-379, Aktiebolaget Atomener 61, Nykoping, Sweden.

Halitsky, J., and K. Woodward, 1975, Atmospheric Diffusion Experiments at a Nuclear Power Plant Site Under Light Wind Inversion Conditions, in *Proceedings of the Second Conference on Atmospheric Turbulence and Diffusion,* pp. 172-175, Santa Barbara, Calif., American Meteorological Society, Boston, Mass.

Harrison, P. R., and S. J. Maas, 1976, *Monitoring of Natural Seeps in the Santa Barbara Channel Off Coal Point,* Report MRI-76-R-1408, Meteorological Research Institute, Inc., Altadena, Calif.

Haugen, D. A. (Ed.), 1959, *Project Prairie Grass, A Field Program in Diffusion,* Geophysical Research Papers, No. 59, Vol. III. Report AFCRC-TR-58-235 (ASTIA Document AD-217076), Air Force Cambridge Research Center, Mass.

——, 1966, Some Lagrangian Properties of Turbulence Deduced from Atmospheric Diffusion Experiments, *J. Appl. Meteorol.*, 5: 646-652.

——, and J. J. Fuquay (Eds.), 1963, *The Ocean Breeze and Dry Gulch Diffusion Programs,* Vol. I. USAEC Report HW-78435 [Report AFCRL 63-791(I)], Air Force Cambridge Research Laboratories, Mass., and Hanford Atomic Products Operation, Washington.

——, and J. H. Taylor (Eds.), 1963, *The Ocean Breeze and Dry Gulch Diffusion Programs,* Vol. II. Report AFCRL-63-791, Air Force Cambridge Research Laboratories, Mass.

Hay, J. S., and F. Pasquill, 1957, Diffusion from a Fixed Source a Few Hundred Feet in the Atmosphere, *J. Fluid Mech.*, 2 (Part 3): 299-310.

——, and F. Pasquill, 1959, Diffusion from a Continuous Source in Relation to the Spectrum and Scale of Turbulence, *Advances in Geophysics,* Vol. 6, pp. 345-365, F. N. Frenkiel and P. A. Sheppard (Eds.), Academic Press, Inc., New York.

Heffter, J. L., 1965, The Variation of Horizontal Diffusion Parameters with Time for Travel Periods of One Hour or Longer, *J. Appl. Meteorol.*, 4: 153-156.

Heimbach, Jr., J. A., A. B. Super, and J. T. McPartland, 1975, Dispersion from an Elevated Source over Coldstrip, Montana, presented at the *68th Annual Meeting of the Air Pollution Control Association, June 14–20, 1975,* Boston, Mass.

Hilst, G. R., 1957, The Dispersion of Stack Gases in Stable Atmospheres, *J. Air Pollut. Control Assoc.*, 7: 205-210.

——, and N. E. Bowne, 1966, *A Study of the Diffusion of Aerosols Released from Aerial Line Sources Upwind of an Urban Complex,* Vols. I and II. Final Report to Dugway Proving Ground, The Travelers Research Center, Inc., Hartford, Conn.

——, and C. L. Simpson, 1958, Observations of Vertical Diffusion Rates in Stable Atmospheres, *J. Meteorol.*, 15: 125-126.

Hinds, W. T., 1967, Diffusion over Coastal Mountains of Southern California, PNL Annual Report for 1967 to the USAEC Division of Biology and Medicine, Vol. II. *Physical Sciences,* Part 3, pp. 19-53, *Atmospheric Sciences,* D. W. Pearce (Ed.), Richland, Wash.

Hino, M., 1968, Maximum Ground-Level Concentration and Sampling Time, *Atmos. Environ.*, 2: 149-165.

Högström, U., 1964, An Experimental Study on Atmospheric Diffusion, *Tellus,* 16: 205-251.

Horst, T. W., 1979, Lagrangian Similarity Modeling of Vertical Diffusion from a Ground-Level Source, *J. Appl. Meteorol.*, 18: 733-740.

——, J. C. Doran, and P. W. Nickola, 1979, *Evaluation of Empirical Atmospheric Diffusion Data,* Report NUREG/CR-0798 (PNL-2599) Pacific Northwest Laboratories.

Hosker, R. P., 1975, A Comparison of Estimation Procedures for Overwater Plume Dispersion, in *Symposium on Atmospheric Diffusion and Air Pollution,* pp. 281-288, Santa Barbara, Calif., American Meteorological Society, Boston, Mass.

Hovind, E. L., R. C. Spangler, and A. J. Anderson, 1974, The Influence of Rough Mountainous Terrain upon Plume Dispersion from an Elevated Source, in *Symposium on Atmospheric Diffusion and Air Pollution,* Santa Barbara, Calif., pp. 214-216, American Meteorological Society, Boston, Mass.

Islitzer, N. F., 1961, Short-Range Atmospheric-Dispersion Measurements from an Elevated Source, *J. Meteorol.*, 18: 443-450.

——, and R. K. Dumbauld, 1963, Atmospheric Diffusion-Deposition Studies over Flat Terrain, *Int. J. Air Water Pollut.*, 7: 999-1022.

——, and E. H. Markee, 1964, Puff Diffusion Measurements from Reactor Destruction Tests, presented at the *American Meteorological Society National Conference on Micrometeorology,* Oct. 13–16, 1974, Salt Lake City, Utah.

——, and D. H. Slade, 1968, Diffusion and Transport Experiments, in *Meteorology and Atomic Energy—1968,* D. H. Slade (Ed.), USAEC Report TID-24190, NTIS, Washington, D. C.

Kellog, W. W., 1956, Diffusion of Smoke in the Stratosphere, *J. Meteorol.*, 13: 241-250.

Leahey, D. M., and J. Halitsky, 1973, Low Wind Turbulence Statistics and Related Diffusion Estimates from a Site Located in the Hudson River Valley, *Atmos. Environ.*, 7: 49-61.

Leighton, P. A., W. A. Perkins, S. W. Grinnel, and F. X. Webster, 1965, The Fluorescent Particle Atmospheric Tracer, *J. Appl. Meteorol.*, 4: 334-348.

Lovelock, J. E., 1974, The Electron Capture Detector Theory and Practice, *J. Chromatogr.*, 99: 3-12.

Ludwick, J. D., 1966, Xenon-133 as an Atmospheric Tracer, *J. Geophys. Res.*, 71: 4743-4748.

MacCready, P. B., T. B. Smith, and M. A. Wolf, 1961, *Vertical Diffusion from a Low Altitude Line Source,* Vols. I and II. Meteorology Research, Inc., Altadena, Calif.

McElroy, J. L., 1969, A Comparative Study of Urban and Rural Dispersion, *J. Appl. Meteorol.*, 8: 19-31.

——, and F. Pooler, 1968, *The St. Louis Dispersion Study,* Vols. I and II. Report APTDL-68-12, U. S. Department of Health, Education, and Welfare, Durham, N. C.

Megonnell, W. H., 1958, *Interstate Atmospheric Transport of Tracer Particles on the N. Y.–N.J. Metropolitan Area,* U. S. Department of Health, Education, and Welfare, Public Health Service, Cincinnati, Ohio.

Meteorology Research Laboratory, 1968, *Diffusion Under a Jungle Canopy,* Vols. I. (176), II. (124), and III. (154), Melpar, Inc., Falls Church, Va.

Minott, D. H., D. L. Shearer, and R. S. Marker, 1977, *Development of Vertical Dispersion Coefficients for Deep-Valley Terrain,* Report COO-4026-3, Travelers Research Corporation, Wethersfield, Conn.

Nickola, P. W., 1971, Measurements of the Movement, Concentrations, and Dimensions of Clouds Resulting from Instantaneous Point Sources, *J. Appl. Meteorol.,* 10: 962-973.

——, 1972, *A Comparison of Exposures Resulting from Inert Gas Releases with Exposures Predicted with the Aid of Pasquill's Diffusion Categories,* Battelle Northwest Laboratories Annual Report, Report BNWL-1651, Pt. 3, pp. 123-125.

——, J. V. Ramsdell, Jr., and J. D. Ludwick, 1970, *Detailed Time-Histories of Concentrations Resulting from Puff and Short-Period Releases of an Inert Radioactive Gas,* Report BNWL-1272, Battelle Pacific Northwest Laboratories.

——, and J. D. Ludwick, 1972, *Preliminary Concentration Measurements of Puffs Released During Near Calm Winds,* Battelle Northwest Laboratories Annual Report, Report BNWL-1651, Part 3, Battelle Northwest Laboratories.

Niemeyer, L. E., and R. A. McCormick, 1968, Some Results of Multiple-Tracer Diffusion Experiments at Cincinnati, *J. Air Pollut. Control Assoc.,* 18: 403-405.

Pasquill, F., 1955, Preliminary Studies of the Distribution of Particles at Medium Range from a Ground Level Point Source, *Q. J. R. Meteorol. Soc.,* 81: 636.

——, 1956, Meteorological Research at Porton, *Nature,* 166: 1148.

——, 1961, The Estimation of the Dispersion of Windborne Material, *Meteorol. Mag.,* 90: 33-49.

——, 1962, Some Observed Properties of Medium Scale Diffusion in the Atmosphere, *Q. J. R. Meteorol. Soc.,* 88: 70-79.

——, 1969, The Influence of the Turning of Wind with Height on Crosswind Diffusion, *Philos. Trans. R. Soc., London, Ser. A,* 265: 173-181.

——, 1970, Prediction of Diffusion over an Urban Area: Current Practice and Future Prospects, in *Proceedings of Symposium on Multiple-Source Urban Diffusion Models,* A. C. Stern (Ed.), pp. 3-1 to 3-25, U. S. Environmental Protection Agency, Research Triangle Park, N. C.

——, 1971, Atmospheric Dispersion of Pollution, *Q. J. R. Meteorol. Soc.,* 97: 369-395.

——, 1974, *Atmospheric Diffusion,* 2nd ed., Halsted Press, New York.

——, 1976, *Atmospheric Dispersion Parameters in Gaussian Plume Modeling,* U. S. Environmental Protection Agency, Report EPA-600/4-76-0306, Research Triangle Park, N. C.

Pendergast, M. M., A. L. Boni, G. J. Ferber, and K. Telegadas, 1979, *Measured Weekly ^{85}Kr Concentrations Within 150 km of the Savannah River Plant (March 1975 Through August 1976),* Report DP-1486, Savannah River Laboratory.

Peterson, K. R., 1968, Continuous Point Source Plume Behavior Out to 160 Miles, *J. Appl. Meteorol.,* 7: 217-226.

Pooler, F., Jr., 1966, A Tracer Study of Dispersion over a City, *J. Air Pollut. Control Assoc.,* 16: 677-681.

Prophet, D. T., 1961, *Survey of the Available Information Pertaining to the Transport and Diffusion of Airborne Material over Ocean and Shoreline Complexes,* Technical Report 89 (AD-258958), Aerosol Laboratory, Stanford University, Stanford, Calif.

Randerson, D., 1972, Temporal Changes in Horizontal Diffusion Parameters of a Single Nuclear Debris Cloud, *J. Appl. Meteorol.,* 11: 670-673.

Raynor, G. S., 1971, Wind and Temperature Structure in a Coniferous Forest and a Contiguous Field, *Forest Sci.,* 17: 351-363.

——, L. A. Cohen, J. V. Hayes, and E. C. Ogden, 1966, Dyed Pollen Grains and Spores as Tracers in Dispersion and Deposition Studies, *J. Appl. Meteorol.,* 5: 728-729.

——, J. V. Hayes, and E. D. Ogden, 1972, *Experimental Data on Particulate Dispersion into and Within a Forest,* Part I. *Dispersion from Upwind Point Sources,* Informal Report BNL-17750, Brookhaven National Laboratory.

——, J. V. Hayes, and E. C. Ogden, 1973, *Experimental Data on Particulate Dispersion into and Within a Forest,* Part II. *Dispersion from Upwind Area Sources,* Informal Report BNL-18064, Brookhaven National Laboratory, Upton, N. Y.

——, J. V. Hayes, and E. C. Ogden, 1974, Particulate Dispersion into and Within a Forest, *Boundary-Layer Meteorol.,* 7: 429-456.

——, P. Michael, R. M. Brown, S. SethuRaman, 1975, Studies of Atmospheric Diffusion from a Nearshore Oceanic Site, *J. Appl. Meteorol.,* 14: 1080-1094.

——, E. C. Ogden, and J. V. Hayes, 1970, Dispersion and Deposition of Ragweed Pollen from Experimental Sources, *J. Appl. Meteorol.,* 9: 885-895.

——, and M. E. Smith, 1964, *A Diffusion—Deposition Tracer System,* Report BNL-850 (T-343), Brookhaven National Laboratory, Upton, N. Y.

Reiter, R., 1974, *Boundary Layer Aerosol Transport Measurements in a Valley System,* Part I. Final Technical Report, European Research Office, U. S. Army, London, England.

——, 1975, *Boundary Layer Aerosol Transport Measurements in a Valley System,* Part II. Final Technical Report, European Research Office, U. S. Army, London, England.

——, and R. Sladkovic, 1976, *Boundary Layer Aerosol Transport Measurements in a Valley System,* Part III. Final Technical Report, European Research Office, U. S. Army, London, England.

Richardson, L. F., 1920, Some Measurements of Atmospheric Turbulence, *Philos. Trans. R. Soc. London, Ser. A,* 221: 1-28. 1-28.

——, and D. Proctor, 1925, Diffusion over Distances of Ranging from 3 to 86 Kilometers, *Mem. R. Meteorol. Soc.,* 1: 1-16.

Roberts, O. F. T., 1923, The Theoretical Scattering of Smoke on an Independent Atmosphere, *Proc. R. Soc. (London), Ser. A,* 104: 640-654.

Saffman, P. G., 1962, The Effect of Wind Shear on Horizontal Spread from an Instantaneous Ground Source, *Q. J. R. Met. Soc.,* 88: 382-393.

Sagendorf, J. F., and C. R. Dickson, 1974, *Diffusion Under Low Windspeed Inversion Conditions,* NOAA Technical Memo-ERL-ARL-52, Air Resources Laboratories, Silver Spring, Md.

Simmonds, P. G., G. R. Shoemaker, J. E. Lovelock, and H. C. Lord, 1972, Improvements in the Determination of Sulfur Hexafluoride for Use as a Meteorological Tracer, *Anal. Chem.,* 44: 860-863.

Simpson, C. L., 1961, *Some Measurements of the Deposition of Matter and Its Relation to Diffusion from a Continuous Point Source in a Stable Atmosphere,* Report NW-69292, Rev., Hanford Atomic Products Operation, Hanford, Wash.

Singer, I. A., 1961, The Relation Between Peak and Mean Concentrations, *J. Air Pollut. Control Assoc.,* 11: 336-341.

——, I. Kazuhiko, and R. Gonzalez del Campo, 1963, Peak to Mean Pollutant Concentration Ratios for Various Terrain and Vegetation Cover, *J. Air Pollut. Control Assoc.,* 13: 40-42.

Slade, D. H., 1962, Atmospheric Dispersion over Chesapeake Bay, *Mon. Weather Rev.,* 90: 217-224.

——, 1966, Summary of Measurements of Dispersion from Quasi-Instantaneous Sources, *Nucl. Safety,* 7(2): 225-229.

Smith, D., 1967, *Tracer Study in an Urban Valley (Johnstown, Pa.),* M.S. Thesis in Meteorology, The Pennsylvania State University.

Smith, F. B., 1965, The Role of Wind Shear in Horizontal Diffusion of Ambient Particles, *Q. J. R. Meteorol. Soc.,* 91: 318-329.

——, 1973, A Scheme for Estimating the Vertical Dispersion of a Plume from a Source Near Ground Level, in *Proceedings of the Third Meeting of the Expert Panel on Air Pollution Modeling,* Paris, North Atlantic Treaty Organization, Committee on the Challenge of Modern Society, Report 14, Brussels.

——, and J. S. Hay, 1961, The Expansion of Clusters of Particles in the Atmosphere, *Q. J. R. Meteorol. Soc.,* 87: 82-101.

Smith, M. E., 1956, The Variation of Effluent Concentrations from an Elevated Point Source, *A. M. A. Arch. Indust. Health*, 14: 56-68.

Smith, T. B., 1965, *Diffusion Study in Complex Mountainous Terrain*, Vol. I. Report MRI-65-FR-236, Meteorology Research Inc., Altadena, Calif.

——, and K. M. Beesmer, 1967, *Bolsa Island Meteorological Investigation*, Report MRI-67-FR-650, Meteorology Research, Inc., Altadena, Calif.

——, E. K. Kauper, S. Berman, and F. Vukovich, 1964, *Micrometeorological Investigation of Naval Missile Facility, Point Arguello, Calif.*, Vol. I. *Analysis*, Vol. II. *Data Supplement*, Report MRI-64-FR-167, Meteorology Research, Inc., Altadena, Calif.

——, and R. L. Miller, 1966, *Victoria Diffusion Trials*, Vol. I. Report MRI-66-FR-374, Meteorology Research, Inc., Altadena, Calif.

——, and B. L. Niemann, 1969, *Shoreline Diffusion Program, Oceanside, California*, 3 Vols., Technical Report, Report MRI-64-FR-860, Meteorology Research, Inc., Altadena, Calif.

——, and M. A. Wolf, 1963, *Vertical Diffusion from an Elevated Line Source over a Variety of Terrains*, Part A, Final Report, Report MRI-63-FR-71, Meteorology Research, Inc., Altadena, Calif.

Stanford University Aerosol Laboratory and the Ralph M. Parsons Co., 1953, *Behavior of Aerosol Clouds Within Cities*, Quarterly Reports 1 to 6, Report AD-31509, Stanford, Calif.

Start, G. E., C. R. Dickson, and N. R. Ricks, 1974a, Effluent Dilutions over and Within Mountain Canyons, in *Symposium on Atmospheric Diffusion and Air Pollution*, pp. 226-232, Santa Barbara, Calif., American Meteorological Society, Boston, Mass.

——, N. R. Ricks, and C. R. Dickson, 1974b, *Effluent Dilutions over Mountainous Terrain*, NOAA Technical Memo-ERL-ARL-51, Idaho Falls, Idaho.

——, C. R. Dickson, and L. L. Wendell, 1973, *Diffusion in a Canyon Within Rough Mountainous Terrain*, NOAA Technical Memo-ERL-ARL-38, Idaho Falls, Idaho.

——, C. R. Dickson, and L. L. Wendell, 1975, Diffusion in a Canyon Within Rough Mountainous Terrain, *J. Appl. Meteorol.*, 14: 333-346.

Stewart, N. G., H. J. Gale, and R. N. Crooks, 1958, The Atmospheric Diffusion of Gases Discharged from the Chimney of the Harwell Reactor BEPO, *Int. J. Air Pollut.*, 1: 87-102.

Sutton, O. G., 1953, *Micrometeorology*, McGraw-Hill Book Company, N. Y.

Taylor, J. H., 1965, *Project Sand Storm, An Experimental Program in Atmospheric Diffusion*, Environmental Research Papers, No. 134, Report AFCRL-65-649, Air Force Cambridge Research Laboratories, Cambridge, Mass.

Telegadas, K., G. J. Ferber, R. R. Draxler, M. M. Pendergast, A. L. Boni, J. P. Hughes, and J. Gray, 1980, *Measured Weekly and Twice Daily Krypton-85 Surface Air Concentrations Within 150 km of the Savannah River Plant (March 1975 Through September 1977)*, Final Report, NOAA Technical Memorandum ERL-ARL-80, National Oceanic and Atmospheric Administration, Silver Spring, Md.

Thomas, P., W. Hubschmann, L. A. Konig, H. Schuttelkopf, S. Vogt, and M. Winter, 1974, *Experimental Determination of the Atmospheric Dispersion Parameters over Rough Terrain*, Report KFK-2285, Part 1, Gesellschaft für Kernforschung mbH, Karlsruhe, Federal Republic of Germany.

——, and K. Nester, 1974, *Experimental Determination of the Atmospheric Dispersion Parameters over Rough Terrain*, Part 2, Report KFK 2286, Gesellschaft für Kernforschung mbH, Karlsruhe, Federal Republic of Germany.

Thompson, N., 1965, Short-Range Vertical Diffusion in Stable Conditions, *Q. J. R. Meteorol. Soc.*, 91: 175-183.

——, 1966, The Estimation of Vertical Diffusion over Medium Distances of Travel, *Q. J. R. Meteorol. Soc.*, 92: 270-276.

Tourin, M. H., and W. C. Shen, 1969, *Diffusion Studies in a Deciduous Forest, R.N. 3004 Deciduous Forest Diffusion Study*, Vol. I. Applied Science Division, Litton Systems, Inc., Minneapolis, Minn.

Turk, A., S. M. Edmonds, H. L. Mark, and G. F. Collins, 1968, Sulfur Hexafluoride as a Gas–Air Tracer, *Environ. Sci. Tech.*, 2: 44-48.

Tyldesley, J. B., and C. F. Wallington, 1965, The Effect of Wind Shear and Vertical Diffusion on Horizontal Dispersion, *Q. J. R. Meteorol. Soc.*, 91: 158-174.

Van der Hoven, I., 1967, Atmospheric Transport and Diffusion at Coastal Sites, *Nucl. Safety*, 8(5): 490-499.

——, 1976, A Survey of Field Measurements of Atmospheric Diffusion Under Low-Wind-Speed Inversion Conditions, *Nucl. Safety*, 17(2): 223-230.

Vaughan, L. M., and R. W. McMullen, 1968, *The Physical Analysis of Particle Size Distributions from Field Samples Obtained During St. Louis Fluorescent Particle Tracer Experiments*, Technical Report 145, Metronics Association, Inc., Palo Alto, Calif.

Vogt, K. J., 1977, *A New System of Release Height Dependent Diffusion Parameters for the Gaussian Plume Model*, preprint of the International Clean Air Congress, Tokyo, Japan.

——, and H. Geis, 1974, *Tracer Experiments on the Dispersion of Plumes over Terrain of Major Surface Roughness*, Report Jul-1131-ST, Central Division for Radiation Protection, Kernforschungsanlage Jülich GmbH, Jülich, Federal Republic of Germany.

Waldron, A. W., 1964, *An Analysis of the 1953 City Diffusion Trials*, presented at the National Conference on Micrometeorology, Salt Lake City, Utah.

——, 1971, *Expected Instantaneous Surface Line Source and Point Source Behavior in a Forest Environment*, Desert Test Center, Ft. Douglas, Utah.

Wilson, R. G., G. E. Start, C. R. Dickson, and N. R. Ricks, 1976, *Diffusion Under Low Windspeed Conditions near Oak Ridge, Tennessee*, NOAA Technical Memorandum ERL-ARL-61, Air Resources Laboratories, Silver Spring, Md.

Chapter 9 covers research prior to February 1979.

ATMOSPHERIC CHEMISTRY OF INORGANIC EMISSIONS

10

K. H. Abel, J. A. Young, and L. A. Rancitelli, Pacific Northwest Laboratory

10-1 INTRODUCTION

Combustion has been important to man since the beginning of civilization. As civilizations have grown, man's utilization of combustion has accelerated and changed technologically. The present massive scale of combustion used to supply the energy needs of our complex society has raised concern regarding the length of time fossil-fuel reserves, especially petroleum, will last. Moreover, the rate at which modern society consumes fossil fuels has produced concern about alteration of environmental parameters and about the long-term consequences of these changes, either beneficial or harmful.

Introduction of various acidic species, including SO_2 and NO_x, from combustion and other industrial sources has resulted in regional acidification of precipitation in both Europe and the United States (Proceedings of the First International Symposium on Acid Precipitation in the Forest Ecosystem, 1976; Galloway, Likens, and Edgerton, 1976; Cogbill and Likens, 1974; Likens and Bormann, 1974). There is also a question as to whether the well-documented increase of carbon dioxide in the atmosphere (Keeling et al., 1976a and 1976b; Stuiver, 1978; Woodwell, 1978; Plass, 1959; Callendar, 1958) since the beginnings of the industrial revolution may alter or already have begun to alter worldwide climate (Plass, 1959; McCormick and Ludwig, 1967; Barrett, 1975; Matthews, Kellogg, and Robinson, 1971; Inadvertent Climate Modification, 1971). The release of large quantities of normally trace inorganic species has accompanied the emission of carbon dioxide from anthropogenic activities. The elevated concentration of trace elements in aerosols over urban and industrial areas is now well documented (Winchester, 1973; Bogen, 1973; Gladney et al., 1974; Rahn and Winchester, 1971; Gordon, Zoller, and Gladney, 1973; Peirson et al., 1973; Harrison and Winchester, 1971; John et al., 1973; Brar et al., 1970; Tabor and Warren, 1958; Breeding et al., 1976; Young and Alkezweeny, 1977). Elemental enrichments are also noted in aerosols collected at locations remote from industrial activities (Duce, Hoffman, and Zoller, 1975; Zoller, Gladney, and Duce, 1974; Rahn and Winchester, 1971; Rancitelli and Perkins, 1970; Murozumi, Chow, and Patterson, 1969; Maerhaut and Zoller, 1977), implying either a measurable global anthropogenic effect or that some natural process gives rise to aerosols of chemical characteristics similar to those arising from anthropogenic activities. The measurement of discrete industrially derived particles long distances from industrial sources by Folger (1970) and Parkin, Phillips, and Sullivan (1970) indicates anthropogenic activities may be important. However, the trace-element enrichments observed in volcanic emissions by Lepel, Stefansson, and Zoller (1978) and Buat-Menard and Arnold (1978) indicate that natural processes may also be responsible for the elemental enrichments on aerosols observed in remote regions. The massive introduction of combustion by-products (Bertine and Goldberg, 1971) may be disrupting natural biogeochemical cycles, especially if the chemical or physical forms being introduced anthropogenically are different from those normally cycling through the environment. There is also considerable disagreement whether aerosol concentrations at remote locations are increasing.

Direct cause and effect environmental relationships attributable to inorganic emissions from combustion or industrial activities are difficult to establish beyond any reasonable doubt due to the complexity of natural systems and our relative ignorance concerning the interaction pathways therein. Aside from certain notable exceptions such as the mercury poisoning in Minamata, Japan, evidence of *direct* ecological consequences of significant human scale is rare. More often a regional long-term exposure such as the acidification of precipitation in the northeastern United States or Scandinavia occurs. It is then difficult to differentiate natural variability within the ecosystem from subtle ecological effects due to the exposure. Selected environmental parameters may be altered, such as pH during the acid rainfall, but overall effects are not easily detected. Only in certain ecosystems, i.e., weakly buffered Scandinavian lakes, have detectable deleterious consequences been noted. More often the natural system has the capacity to adjust or reequilibrate to the impact. Acid precipitation, for example, will be relatively rapidly equilibrated via carbonate or other acid–base systems in most biogeochemical systems (Eaton, Likens, and Bormann, 1973). However, this reequilibration process implies accelerated weathering in soils which means fertility losses may occur over long time periods of decades to centuries. Thus it is difficult to discern natural variability

Abel, Young, and Rancitelli: Pacific Northwest Laboratory, Richland, Washington.

and changes in ecosystems from those being introduced or accelerated via introduction of anthropogenic agents or activity.

To understand the influence of anthropogenic activities, one must identify the various anthropogenic sources. One very significant source is the inorganic emissions from coal combustion, and this will be the focus of this chapter. Our examination will begin by describing the origin and levels of the inorganic species in coal, theories regarding chemical behavior during combustion and ash formation, the resulting chemical and physical properties of fly ash and other fugitive emissions, and the ecological significance of these anthropogenic additions to natural biogeochemical cycles. Finally, we hope to point out during our examination areas where information is lacking and where there are needs for further research.

The task is immense since it brings together various aspects of chemistry, biology, biochemistry, geochemistry, and physics. Owing to limitations of space and/or human frailty, justice may not be done to all topics. The reader is referred to other chapters herein and to the recent open literature for further information.

10-2 HISTORICAL BACKGROUND

Presently, coal combustion supplies more than 40% of U. S. electrical needs. In 1970, 60% of the coal mined was used by electrical utilities, and only two-thirds of that coal underwent cleaning prior to combustion. Various estimates presently available predict approximate doubling of coal utilization by 1985 to more than 1 billion tons/yr.

Natural gas and petroleum had been replacing coal because of their cleaner, more stable combustion characteristics. Oil combustion had been attractive in eastern U. S. locations because of cheaper transportation costs via water and the much lower particulate emission levels. Coals typically average 10% ash, whereas residual oils are much lower in ash content (0.05 to 0.1%).

Mroz (1976) examined emissions from an oil-fired power plant and found that oil combustion emitted 50% less sulfur oxide and 90% less ash on a per megawatt basis, when compared to one selected coal combustion facility. However, oil combustion emissions have been shown to be higher in nickel, vanadium, and organic matter than those from coal combustion. Also, emissions from oil combustion are often more acidic than those from coal combustion, since coal fly ash is alkaline in nature and neutralizes some of the acidic species produced by combustion. In oil combustion the ash content is much lower; therefore associated neutralization is also significantly less.

However, the oil embargo of the early 1970s and the subsequent upward surge in foreign crude oil prices coupled with other factors have stopped or slowed the replacement of coal by oil. These other factors include the depletion of domestic reserves of crude oil and natural gas, the hesitancy to develop offshore domestic petroleum reserves until environmental problems can be assessed or overcome, and the reluctance of certain segments of the public and scientific community to accept nuclear generation facilities. These same factors have stimulated a concerted effort to develop domestic coal reserves and alternative energy sources such as geothermal, solar, shale oil, and fusion. However, these alternative sources require significant technological development before wide-scale utilization can occur and/or may not be applicable in all locations or on a large enough scale to supply a significant portion of our future energy needs.

Coal combustion is the largest single source of electrical energy and is likely to continue as such, so we have chosen to focus on it. Other sources such as geothermal, shale oil, solar, and fusion should be examined since they also have environmental risks associated such as H_2S, volatile metals, and brine disposal in geothermal, and fugitive particulates and organics in shale oil. However, coverage of these other sources is not within the scope of this chapter. Suffice it to say that each energy source has its unique faults environmentally, and a detailed examination needs to be made of each so that well-founded decisions can be made regarding the scope of application for each potential energy source.

Figures 10.1 and 10.2 (U. S. Atomic Energy Commission Office of Planning, 1973) indicate typical projected growths of the various energy sources through the year 2000. A recent update in Project Independence (1977) contains more detailed projections under 12 alternative demand and supply options up to the year 1990. Recent surveys (*Electrical World,* 1977) project a slower growth in energy use into and through the 1980s, averaging approximately 4% per year, down significantly from the earlier rates of 7% in the late 1960s. These projections are subject to even further downward projection if conservation and

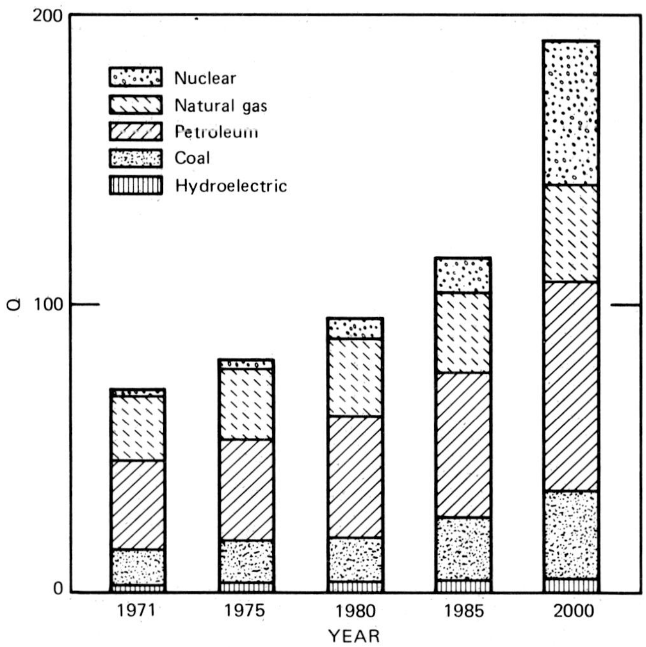

Fig. 10.1 Energy sources to meet total energy needs. Q = 10^{15} Btu.

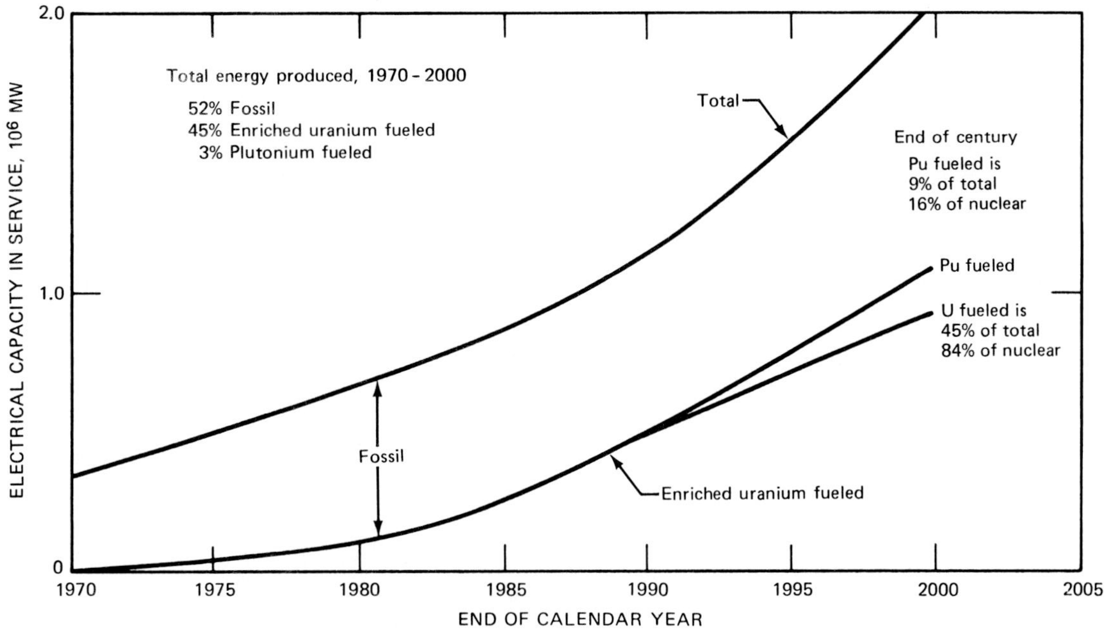

Fig. 10.2 Projections of U. S. electrical capacity.

increased efficiency of utilization continue to be implemented.

Since coal is not a clean fuel, the increased coal utilization projected has given cause for environmental concern. In fact, legislation enacted to prevent environmental degradation has mandated development of alternative conversion techniques, improved combustion conditions, and/or improved postcombustion emission control. Approaches include solvent refining, liquefaction, and gasification of coal to convert it to a cleaner and more attractive fuel; fluidized-bed combustion; and other altered firing conditions which decrease NO_x and/or SO_x emissions; and finally, improved emission control of both particulates via higher efficiency electrostatic precipitators and/or wet scrubbers, and gaseous emission control, specifically for removal of SO_x and NO_x.

10-3 INORGANIC EMISSIONS: ORIGIN IN COAL

Inorganic elemental emissions from coal combustion may originate in the coal matrix in several different forms. The presence of different chemical forms is significant since the forms may have dissimilar behavior during combustion, thus possibly affecting postcombustion chemical forms and concentration distributions. Second, the various forms may have differing susceptibilities for removal during any coal beneficiation applied prior to combustion. As an example, inorganic sulfur that occurs predominantly as pyrites in most coals can be readily separated by simple cyclone or other density separations applied prior to combustion, whereas the organic sulfur forms are much more difficult to remove.

Several classifications have been applied to describe the origin and forms of inorganically and organically associated elements in coal. To briefly summarize: An element in coal can be (1) a trace constituent incorporated into the living biological matter from which the coal body originated; (2) adsorbed to or incorporated into the organic matrix after deposition via percolating waters before, during, or after coal formation; (3) associated with inorganic detrital minerals deposited simultaneously or interbedded with the parent biogenic material; (4) associated with authigenic minerals formed during diagenesis of plant matter and/or from percolating groundwaters; or (5) incorporated into or sorbed onto inorganic mineral phases present in the coal deposit. Finally, inorganic matter can be mixed into the coal during the mining process from the overburden or seat earth associated with the coal deposit.

10-3.1 Organic Associations

Generally less than 10% of the trace elements are associated with the organic fraction in coals. However, specific elements such as Ge, B, Be, Br, and Ga can be associated primarily with the organic phase, as is shown in Table 10.1. This tabulation is a compilation from several sources and shows affinity rankings defined as association with the organic matter based on a density separation.

Zubovic (1966a and 1966b) analyzed coals from the Eastern Interior Province of the United States. The coals were from several different coal beds and differing depths within the same bed. Gluskoter (1975) compiled data on four Illinois coals (Eastern Interior Province). Filby, Shah, and Sawtter (1977) also analyzed an Eastern Interior U. S. coal, specifically Kentucky-14 bituminous coal. Horton and Aubrey (1950) used three British coals of different rank

TABLE 10.1 Average "Organic Affinity" of Selected Elements in Coal

Zubovic (1966a)	Horton and Aubrey (1950)	Filby, Shah, and Sawtter (1977)	Gluskoter (1975)
Ge (87)†	Ge (100)†	Br (70)†	Ge‡
Be (82)	V (100)	Hg (53)	B
Ga (79)	Be (75–100)	As (48)	Be
Ti (78)	Ga (75–100)	Sr (29)	Ti
B (77)	Ti (75–100)	Ba (24)	Ga
V (76)	B (75–100)	Tb (23)	Sb
Ni (59)	Mo (50–75)	Sb (21)	P
Cr (55)	Zn (50)	Na (20)	Co
Co (53)	Cr (0–100)	Co (17)	Se
Y (53)	Ni (0–75)	Sc (12)	Ni
Mo (40)	Co (25–50)	Lu (9)	Cu
Cu (34)	Cu (25–50)	Fe (6)	Cr
Sn (27)	Sn (0)	Cs (4)	Mn
La (3)		Th (3)	Mo
Zr (0)		Cr, K, Rb, Eu, Sm, Cs, Ni (0)	Zr
			Pb
			Cd
			Zn
			Ag

†Indicates percent associated with organic fraction.
‡Relative ranking.

from the same coal seam. Table 10.1 indicates considerable variability. Comparison between studies is difficult because of differences in elements analyzed; however, the elements Ge, Be, V, Ga, Ti, and B appear highly correlated with the organic fraction. Elements to a lesser degree associated with the organic fraction are Ni, Cr, Mo, Co, Cu, Zn, Sb, Hg, and As. Finally, Cr, K, Rb, Eu, Sm, Ce, Zr, and Th show minimal association with the organic density fraction separated in these studies.

The fraction of a trace element that is organically associated may have a higher volatility during combustion than the fraction associated with the inorganic mineral. The presence of both organic and inorganic forms of an element thus creates the possibility of a bimodal distribution for an element in the ash phase and gases after combustion. This will be discussed in more detail in later sections.

10-3.2 Inorganic Associations

The elements associated with inorganic phases in coal contribute the largest fraction of the coal ash. Inorganic mineral phases can constitute up to 50% by weight in low-grade coals but generally average approximately 10%. The specific minerals as well as the total inorganic content vary significantly. Numerous environmental parameters associated with the coal body during deposition of the original organic matter and throughout its subsequent alteration influence the ultimate inorganic mineral content and distribution. The major inorganic elements present in coal are the same as those in silicate rocks: Si, Al, Ca, Mg, Fe, S, Na, K, Cl, and Ti. The mineralogy of the inorganic matter has been examined rather extensively (Nelson, 1953; O'Gorman and Walker, 1971; Rao and Gluskoter, 1973; Kemezys and Taylor, 1964; Ball, 1935; Gauger, 1936; Gluskoter, 1967 and 1975). The predominant mineral types (O'Gorman and Walker, 1971) found are:

Clays: illite, kaolinite, montmorillonite
Carbonates: calcite, siderite, dolomite, ankerite
Sulfides: pyrite, marcasite, sphalerite

along with small amounts of other minerals including quartz, gypsum, hematite, rutile, apatite, barite, and felspar to name only a few. The presence of numerous minerals in a given coal and the relative abundances of the various minerals may play an important role during combustion. For example, pyrites are oxidized above 600°K, whereas carbonates decompose above 500°K. Alkali salts begin to vaporize appreciably above 1350°K and silica and alumina above 1900 and 2500°K, respectively. These differing volatilities and high-temperature stabilities can result in multiple behaviors for an element present in more than one mineral. The variation in mineralogy and mineral content in coal may well influence the elemental distribution in the fly ash particulates emitted from combustion.

The average elemental composition of coal is difficult to define. For example, coals have been found to vary greatly in their composition from one geographical area to the next, from coal bed to coal bed, and even within a given coal bed. Numerous authors have examined the elemental composition of coal and subsequent coal ash or fly ash released through combustion. Goldschmidt (1937) and, more recently, Abernathy, Peterson, and Gibson (1969) reported on trace elements in coal ashes. Horton and Aubrey (1950) reported on the minor elements and their distribution in some British coals. Zubovic (1966a and 1966b) has examined minor element distributions and controlling geochemical factors in Interior U. S. coals. In addition, Nicholls (1968) has examined the geochemistry of coal. O'Gorman and Walker (1972) reported on mineral matter and trace elements in coals. A great deal has been published by the Illinois Geological Survey concerning trace elements in coals (Gluskoter and Ruch, 1971; Ruch, Gluskoter, and Kennedy, 1971; Ruch, Gluskoter, and Shimp, 1973; Kuhn, 1973; Gluskoter and Lindahl, 1973). These authors examined Na, Cl, Hg, Cd, and other volatiles as well as other trace elements. Extensive coverage of the origin and distribution of mineral matter is contained in a book edited by Babu (1975). Furthermore, this book contains information on qualitative and quantitative trace-element analytical techniques, on the fate of trace elements during coal preparation and conversion, and on environmental toxicological effects of trace metals. To estimate the significance of elemental mobilization via coal and other fossil-fuel combustion, Bertine and Goldberg (1971), Vaughan et al. (1975), and Magee, Hall, and Varga (1973) have compiled the estimated average compositions shown in Table 10.2.

TABLE 10.2 Elemental Composition of Coal (parts per million)†

	A	B	C		A	B	C
Li	65	6–50		Ta			0.25
Be	3	0.6–3.1	1.2	Te			0.5
B	75	20–100		Th			1.5
Na	2,000		150	Tl			0.125
Mg	2,000			W			12.5
Al	10,000			As	5	0.5–70	5
P	500			Se	3	0.5–3.9	1
S	20,000			Rb	100		1
Cl	1,000			S	500	5–180	
K	1,000			Y	10	7–22	
Ca	10,000			Mo	5	1–6	10
Si	5	1–20	1	Ag	0.5		0.12
Ti	500			Cd		0.1	0.5
V	25	20–40	10	Sn	2	0.1–2	20
Cr	10	10–20	0.3	Ba	500	20–600	
Mn	50	4–70	50	La	10	3–11	
Fe	10,000		1,500	Ce	11.5		
Co	5	4–7	0.5	Pr	2.2		
Ni	15	10–25	20	Nd	4.7		
Cu	15	10–20	50	Sm	1.6		
Zn	50	5–20	15	Eu	0.7		
Ga	7	1–7	12.5	Gd	1.6		
Ge	5	1–20	30	Tb	0.3		
Ra	0.05			Ho	0.3		
Hg	0.012	0.07–0.21	0.5	Er	0.6		
Pb	25	1.2–35	10	Tm	0.1		
Bi	5.5	1	5	Yb	0.5	1–10	
U	1.0			Lu	0.7		
Sb			1				

†A, an average compilation combining coal and lignite (Bertine and Goldberg, 1971).
B, range of regional average concentrations in coal (Magee, Hall, and Varga, 1973).
C, compilation of an estimated Western coal (Vaughan et al., 1975).

10-4 INORGANIC EMISSIONS: POSTCOMBUSTION ELEMENTAL DISTRIBUTION IN PARTICULATES

Contemporary models describing inorganic elemental behavior during the combustion process are based on observations of elemental concentrations and distributions in postcombustion gases and particulates. These observations will be discussed prior to the presentation of combustion models. In addition, the findings from investigation of emissions from both dry and wet control devices are detailed below.

10-4.1 Electrostatic Precipitation

To describe the behavior of inorganic matter during coal combustion, Klein et al. (1975), Kaakinen et al. (1975), Block and Dams (1976), Gladney (1974), and Lyon (1977) investigated the balance of masses for various constituents. For most utility systems these investigations have shown that, relative to bottom ash, the precipitator fly ash is generally enriched with potentially volatile elements. Also, in many instances, the smaller fly ash particulates contain greater concentrations of these "volatiles" than the larger fly ash particulates (Davison et al., 1974; Kaakinen et al., 1975; Campbell et al., 1978a; Smith, Campbell, and Nielson, 1979; Campbell et al., 1978b). Facilities where this fractionation does not occur have also been noted by Gladney et al. (1976) and Vaughan et al. (1975). Figures 10.3, 10.4, and 10.10 illustrate the range of variations in the elemental composition of fly ash vs. particulate size encountered at three different combustion facilities. The fly ash was obtained from the respective pollution control devices at the generating stations and physically separated by particle size prior to analysis. The dramatic variation of fly ash chemical composition at the four facilities implies that multiple parameters including coal, coal form, firing conditions, boiler type, and emission control device and the conditions of its application may play significant roles in determining the forms and distributions of inorganic species released from combustion.

Table 10.3 further illustrates the partitioning that may occur as a result of combustion. First, as mentioned above,

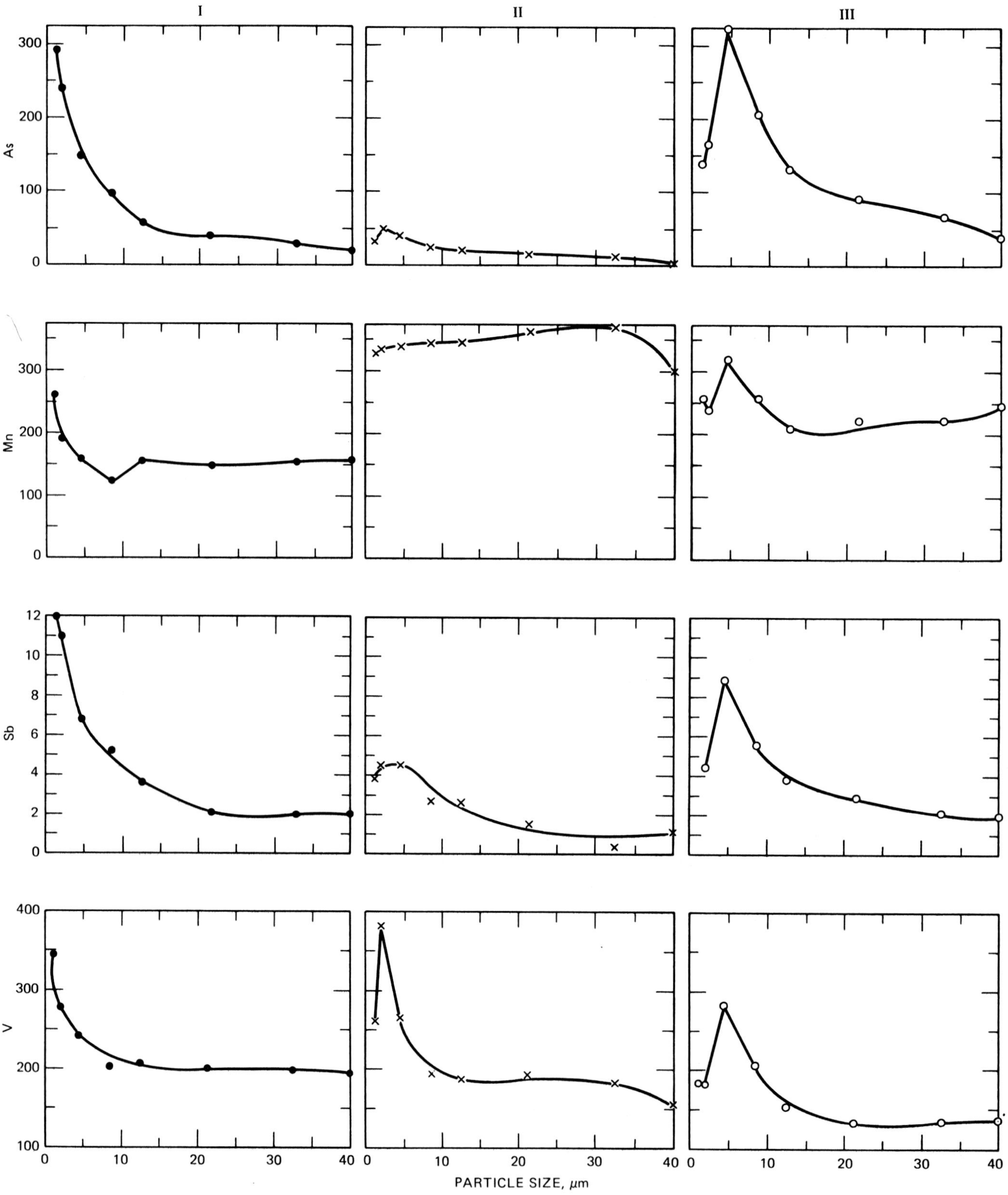

Fig. 10.3 Elemental composition of fly ash from three selected power plants as a function of particle size. Elemental composition is in units of parts per million except as noted. (From B. E. Vaughan et al., *Review of Potential Impact on Health and Environmental Quality from Metals Entering the Environment as a Result of Coal Utilization—A Battelle Energy Report*, Battelle Memorial Institute, Richland, Wash., 1975.)

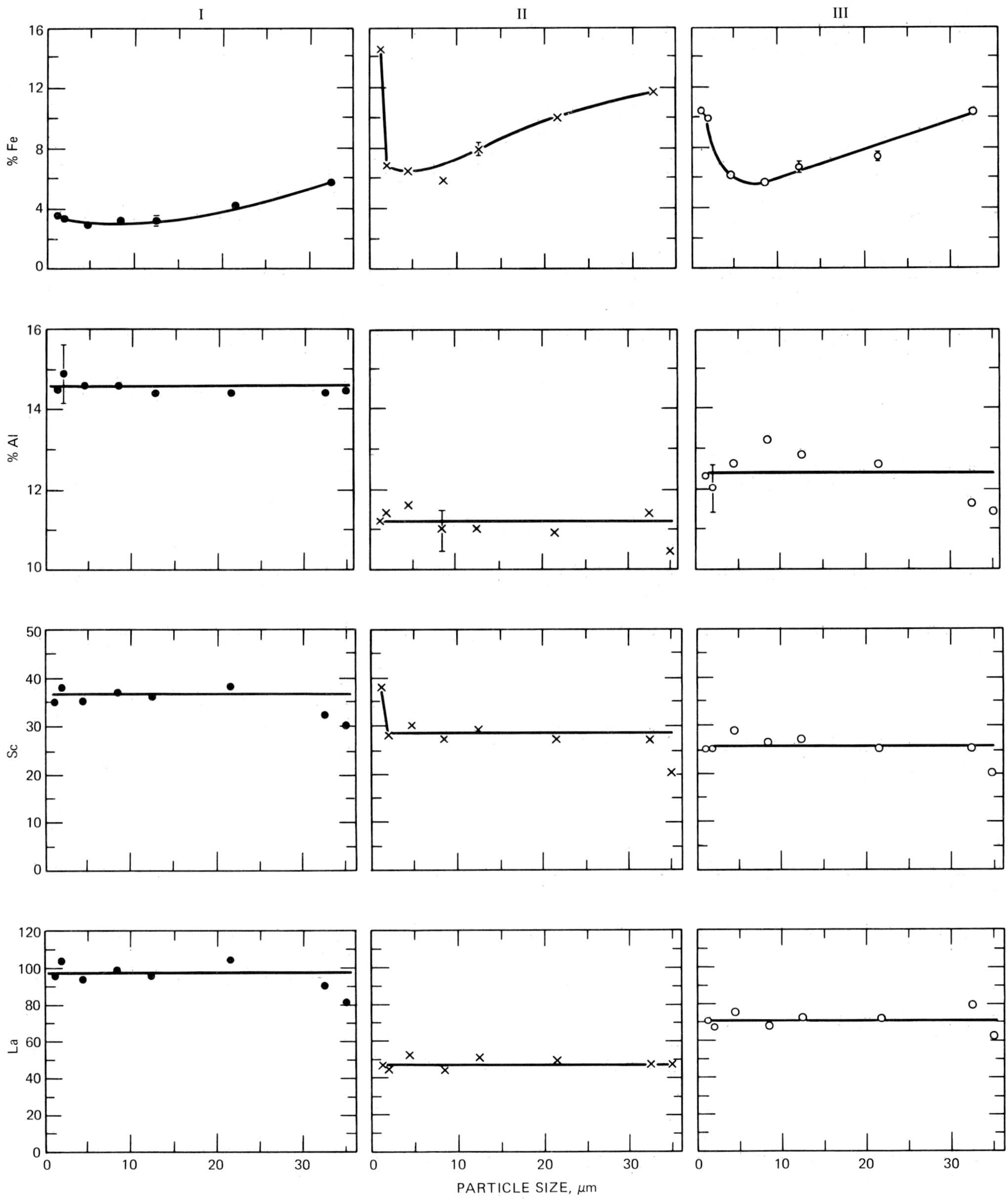

Fig. 10.4 Elemental composition of fly ash from three selected power plants as a function of particle size. Elemental composition is in units of parts per million except as noted. (From B. E. Vaughn et al., *Review of Potential Impact on Health and Environmental Quality from Metals Entering the Environment as a Result of Coal Utilization—A Battelle Energy Report,* Battelle Memorial Institute, Richland, Wash., 1975.)

TABLE 10.3 Elemental Comparison of Centralia Coal, Bottom Ash, and Fly Ash Concentrations (parts per million except as noted)

	Coal	Bottom ash	Fly ash†		
			A	B	C
Al (%)	2.12	10.7	11.3	12.4	12.7
As	1.3	0.65	5.7	11	11
Ba	290	1320	1790	1550	1580
Br	3.1	5.6	0.66	7.1	7.9
Cl (%)	0.02	0.1	0.1	0.1	0.1
Co	4.6	16	26	31	32
Cr	33	51	69	76	82
Cs	0.19	1.5	2.6	2.2	2.2
Cu (%)	0.01	0.04	0.04	0.04	0.04
Eu	0.64	3.1	3.4	3.6	3.4
Fe (%)	0.65	3.03	3.53	3.73	3.97
Hf	4.5	15	25	20	26
La	19	68	86	84	84
Mg (%)	0.4	1.7	1.4	2.9	2.3
Mn	92	620	570	560	630
Na (%)	0.24 (0.21)	1.14 (1.3)	1.0 (0.90)	1.18 (1.2)	0.91 (0.98)
Sb	0.46	0.21	1.8	2.0	2.5
Sc	0.19	0.80	1.1	1.1	1.1
Se	2.3	6.2	7.2	7.2	12.4
Sm	3.1	11	14	13	14
Sr	570	2670	3280	3060	2870
Ta	1.8	9.1	8.6	12	9.1
Tb	0.51	2.2	3.0	3.2	2.9
Th	2.6	11	15	14	15
Ti (%)	1.5	2.0	1.7	1.7	1.9
V	56	227	316	400	384
Yb	2.6	8.3	12	10	12

†The fly ash composition in the three columns represents samples from three fields of an electrostatic precipitator proceeding downstream from inlet to outlet.

notice the depletion of certain potentially volatile elements (As, Sb, V, . . .) in the bottom ash. Second, the data indicate that the chemical composition of the fly ash varies within the electrostatic precipitator. Remember that the fly ash composition shown in the three columns represents fly ash retained in the three different fields of an electrostatic precipitator. The samples from left to right proceed downstream from inlet to outlet. An increase in "volatile" element concentrations is noted (As, Br, Sb, and Se) as one progresses downstream, indicating that the particulates most difficult to remove (i.e., smallest) also have the highest concentration of these volatile inorganics. This increase in volatile element concentrations may also indicate condensation is occurring with progression through the electrostatic precipitator. The boiler system represented by these samples generates bottom ash infrequently. Therefore the bottom ash composition is integrated over a long time interval, thereby minimizing the possibility of a reliable mass balance between input coal, bottom ash, precipitated fly ash, and escaping material. The variation of the fly ash composition within the electrostatic precipitator further complicates the construction of a mass balance unless one has available a weighting factor for the relative amounts of fly ash retained in each field of the precipitator. This piece of information is site specific and generally available. However, it has not been taken into account in past mass balance studies to our knowledge.

The above data coincide with those of several investigators (Lyon, 1977; Davison et al., 1974; Kaakinen et al., 1975; Block and Dams, 1976; Campbell et al., 1978a; Smith, Campbell, and Nielson, 1979) and reinforce their postulate that the inorganic elements can be classified into approximately three groups based on combustion behavior and postcombustion chemical distributions. Group A elements (Al, Ba, Ca, Ce, Co, Cr, Cs, Dy, Eu, Fe, Hf, K, La, Mg, Mn, Na, Nb, Rb, Sc, Si, Sm, Sr, Ta, Tb, Th, Ti, V, Y, Yb, and Zr) show minimal enrichment in fly ash over bottom ash and are postulated as being nonvolatile during combustion. They form a relatively uniform melt or slag which is partitioned into the fly ash or bottom ash depending on furnace characteristics. The Group B elements (As, Ba, Br, Cd, Cl, Co, Cr, Cu, Ga, Hg, I, In, Mo, Mn, Nd, Ni, Pb, S, Sb, Se, Sn, Zn, ^{210}Po, and ^{210}Pb) are those which often appear significantly enriched in fly ash compared to bottom ash. These elements are thought to be volatilized partially or totally in the combustion zone and carried out of the combustion zone as gases. Consequently, they have no opportunity to condense on the bottom ash and are depleted in this phase. On cooling of the combustion gases downstream from the combustion zone, these elements are thought to condense or adsorb on the matrix formed by the nonvolatile Group A elements. The third group of elements are those which remain almost totally in the gas phase after combustion including Hg, Cl, Br, and F. These elements may be classified as the "extremely" volatile elements. Finally, there is a group of elements that display behavior intermediate between the nonvolatile and semivolatile groups A and B (Al, Ca, Ce, Cr, Cs, Ga, K, Mg, Mn, Mo, M, Na, Nb, Ni, Rb, Ti, V, and W). These elements are classified as intermediate since they are difficult to classify into either group. These groupings are a composite of several investigators in which there are differences in classification for certain elements. Thus an element listed in more than one group indicates differences in findings between authors. These differences can be ascribed to variations in combustion parameters, to possible different elemental forms in the original coals, to differing pollution control devices, and to analytical errors.

10-4.2 Wet Scrubbing

Venturi or wet scrubbing is a relatively recent innovation in emission control at large stationary combustion sources. The wet contact process accomplishes particulate control and also is capable of emission control for certain gaseous species including sulfur oxides. Sulfur oxide removal is facilitated by use of an alkaline buffer, either supplied by added carbonates or the natural alkalinity of fly ash supplemented with carbonate. Because of the alkaline nature of the system and the high dissolved and suspended solids present in the aqueous media, scaling, sludge buildup, and wear have been problems encountered with wet control. There is presently debate whether wet systems can have the reliability factor necessary for application at large central generation facilities.

Because wet scrubbing has only recently been applied at large generation facilities, little is known regarding the chemistry of the treated gas stream. Ragaini and Ondov (1976) and O'Toole et al. (1978) have noted that Venturi scrubbing of combustion gases gives rise to a different effluent particle size distribution than produced from dry emission control systems. The particulates emitted from Venturi systems also have a different chemical composition and differing physical properties. Both Ragaini and Ondov (1976) and O'Toole et al. (1978) note a bimodal particle size distribution after Venturi scrubbing, as shown in Fig. 10.5. Chemical composition and electron microscopy indicate that the particles in the region of 0.5 μm are predominately fly ash particles with an amorphous coating of evaporated carbonate-based scrubber liquor. The particles in the region <0.1 μm appear to be generated during passage through the Venturi system. These particles probably result from entrainment of scrubber liquor as a fine mist, with incomplete removal in the demisting stage. The

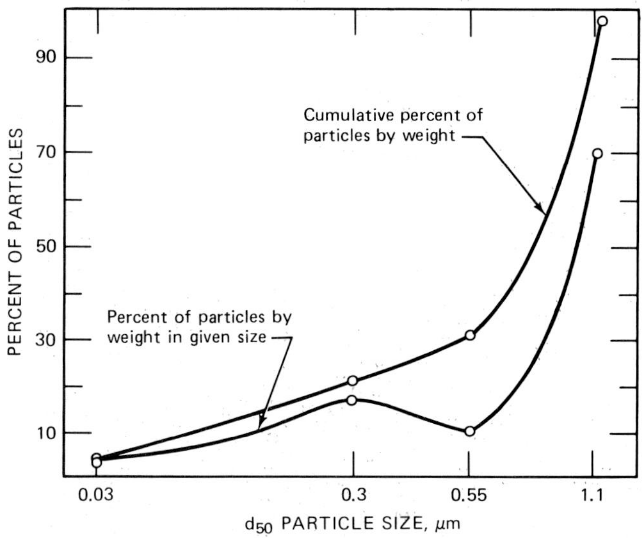

Fig. 10.5 Particulate distribution observed in the stack after Venturi scrubbing. (From J. J. O'Toole et al., *Potential for Gaseous and Heavy Metal Contamination from Energy Extraction Processes in the Northern Great Plains and the Consequent Uptake and Turnover in Range Ecosystems*, Report IS-4733, Ames Laboratory, 1978.)

evaporation of the mist during gas stream reheat prior to emission from the stack would be expected to give rise to the extremely fine particles, 0.1 μm. In general, the aerosols emitted from wet scrubbing are extremely small (1 μm) and, therefore, have potentially greater environmental impact.

The chemical composition vs. particle size encountered in the stack after passage through a Venturi scrubber system is shown in Table 10.4. The variation of several elements across the particle size spectrum is shown graphically in Fig. 10.6. Figure 10.7 illustrates elemental correlations found at 90, 95, and 99% confidence levels. Arsenic and selenium, both volatile elements, correlate at the 99% level of confidence, whereas Sr and Ca correlate at the 95% level, undoubtedly related to entrainment of scrubber solution. Other correlations at 99% level are Al and Mn, Ce and Na, Ce and Co, Hf and Sm, and Th and Ti reflecting the fly ash matrix elements. The correlation of Ca and V is not completely understood. They may both have the scrubber liquor as their predominant source.

These correlations and size-composition data are somewhat similar to those observed in dry systems. The dramatic increase of Se and As concentrations with decreasing particle size is noteworthy. The smallest particles are of concern since they may be respirable and also have the potential for remaining suspended in the atmosphere for long periods of time. Consequently, a higher concentration of these two potentially biologically toxic elements increases their potential for environmental damage. The oxidation state of the Se and As associated with the particles is not presently known. The oxidation state, along with the mobility of Se and As from the particles, needs to

TABLE 10.4 Elemental Composition Vs. Particle Size of Colstrip Unit #2 Stack Particulates as Determined by In-Stack Impactor (parts per million except where noted) after O'Toole et al. (1978)

	Stage 4	Stage 5	Stage 6	Nuclepore backup filter
d_{50} (micrometers)	1.1	0.55	0.3	0.3
Al (%)	16.1 ± 0.1	8.93 ± 0.02	6.6 ± 0.02	4.04 ± 0.01
As	77 ± 14	82 ± 4	145 ± 7	166 ± 3
Ca (%)	29.3 ± 1.4	19.7 ± 0.7	24.7 ± 0.7	23.1 ± 0.5
Ce	126 ± 5	71 ± 2	66 ± 2	62 ± 2
Co	75 ± 2	12.5 ± 0.3	5.4 ± 0.8	9.24 ± 0.12
Cr				88 ± 2
Cs	2.1 ± 1.0	0.96 ± 0.33	27 ± 0.6	4.1 ± 0.6
Eu	2.2 ± 0.3	1.7 ± 0.1	1.1 ± 0.1	0.48 ± 0.03
Fe (%)	1.53 ± 0.10	1.22 ± 0.21	1136 ± 0.22	1.38 ± 0.02
Ga	230 ± 30	130 ± 10	150 ± 10	140 ± 10
Gd		17 ± 3		19 ± 1
Hf	7.7 ± 1.1	6.0 ± 0.5	4.6 ± 0.5	0.77 ± 0.11
K (%)	0.53	0.13	0.18	0.19 ± 0.09
La	100 ± 20	47 ± 7	49 ± 7	16 ± 2
Mn	1750 ± 60	1050 ± 20	895 ± 21	714 ± 10
Na	1.11 ± 0.02	0.29 ± 0.04	0.34 ± 0.05	0.19 ± 0.02
Ni				42 ± 12
Rb				17 ± 5
Sb	21 ± 3	7 ± 2	20 ± 3	19 ± 2
Sc	19.5 ± 0.2	16.5 ± 0.4	12.3 ± 0.7	2.69 ± 0.02
Se	112 ± 10	96 ± 6	330 ± 20	450 ± 20
Sm	12 ± 1	8.8 ± 0.4	7.8 ± 0.5	2.5 ± 0.1
Sr (%)	1.09 ± 0.05	0.71 ± 0.02	0.84 ± 0.03	0.79 ± 0.02
Tb		0.82 ± 0.22	0.62 ± 0.23	0.15 ± 0.06
Th	37 ± 2	20 ± 2	22 ± 1	3.4 ± 0.2
Ti (%)	0.65 ± 0.1	0.43 ± 0.06	0.46 ± 0.07	0.23 ± 0.04
V	600 ± 30	240 ± 10	430 ± 10	390 ± 10
W	2.8 ± 2.1	1.2 ± 0.5	2.1 ± 0.7	3.1 ± 0.3
Zn	580 ± 30	240 ± 10		118 ± 3

be characterized before the environmental significance can be determined since toxicity depends on oxidation state and availability from the particle.

10-5 INORGANIC EMISSIONS: PHYSICAL CHARACTERISTICS OF PARTICULATES

Physical examination has indicated that fly ash particles from dry precipitator systems can be divided into several morphological groups. Fisher et al. (1978) identified 11 different morphological classes based on opacity and shape considerations:

A. Amorphous, nonopaque
B. Amorphous, opaque
C. Amorphous, mixed opaque and nonopaque
D. Rounded, vesicular, nonopaque
E. Rounded, vesicular, mixed opaque and nonopaque
F. Angular, lacy, opaque
G. Nonopaque, cenosphere
H. Nonopaque, plerosphere
I. Nonopaque, solid sphere
J. Opaque sphere
K. Nonopaque sphere with crystals
L. Combined particle classes B and F

A simple physical magnetic separation of fly ash from electrostatic precipitators has shown that the small magnetic fraction of the fly ash contains very high transition metal concentrations (Rancitelli and Abel, 1974). For two typical fly ashes, Tables 10.5 and 10.6 compare the original or nonseparated fly ash with the two fractions obtained via magnetic separation. These two fractions also have different physical characteristics when observed under a light microscope. The magnetic fraction is predominantly composed of opaque spheres and particles, whereas the nonmagnetic fraction is composed primarily of nonopaque, pale blue, red, or golden brown colored spheres. The presence of these particles of differing physical and chemical characteristics further indicates the nonequilibrium nature of the combustion process and increases the difficulty in modeling the overall chemical behavior of the inorganic species during combustion.

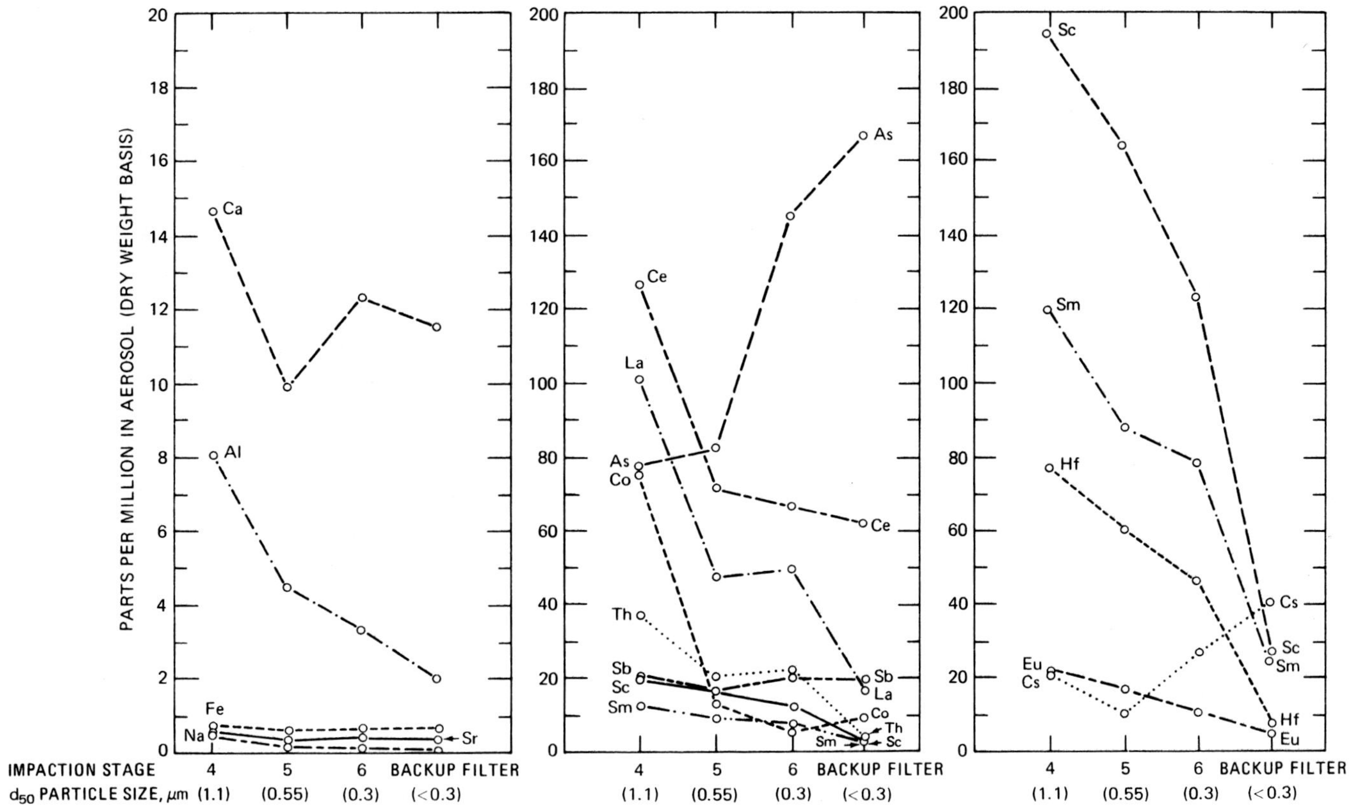

Fig. 10.6 Elemental concentrations vs. particle size observed after Venturi scrubbing. (From J. J. O'Toole et al., *Potential for Gaseous and Heavy Metal Contamination from Energy Extraction Processes in the Northern Great Plains and the Consequent Uptake and Turnover in Range Ecosystems,* Report IS-4733, Ames Laboratory, 1978.)

CORRELATIONS AT 90% LEVEL OF CONFIDENCE

Al: Ce, Co, Eu, Na, Sm, Th, Ti
As: -Eu, -Hf, -Sc
Ca: Fe, Ga, Sb
Ce: Ga, La, Sr
Co: Ga, Mn, Sr
Cs: Se
Eu: La, Mn, -Se, Sm, Ti
Fe: Sb, Sr
Ga: Mn, V
Hf: La, -Se, Th, Ti
La: Mn, Na, Sm
Mn: Na, Th, Ti
Na: Sr
Sb: V
Sc: -Se, Th, Ti
Sr: V

CORRELATIONS AT 95% LEVEL OF CONFIDENCE

Al: La
Ca: Sr
Ce: Hf, Mn
Co: Na
Eu: Hf, Sc
Ga: Na, Sr
La: Th, Ti
Sc: Sm
Sm: Th, Ti

CORRELATIONS AT 99% LEVEL OF CONFIDENCE

Al: Mn
As: Se
Ca: V
Ce: Na
Hf: Sm
Th: Ti

Fig. 10.7 Elemental correlations in sized particulates from a Venturi scrubber. (From J. J. O'Toole et al., *Potential for Gaseous and Heavy Metal Contamination from Energy Extraction Processes in the Northern Great Plains and the Consequent Uptake and Turnover in Range Ecosystems,* Report IS-4733, Ames Laboratory, 1978.)

TABLE 10.5 Chemical Composition Vs. Magnetic Characteristics of Coal Ash (parts per million except as noted)

	Original coal ash	Magnetic fraction	Nonmagnetic fraction
Al (%)	8.08	10.9	9.12
Ba	670	1300	870
Br		1.7	1.6
Cl (%)	0.06	0.06	0.06
Co	19	57	14
Cu (%)	0.03	0.3	0.3
Cr	41	78	47
Cs	0.26	0.81	1.4
Eu	2.3	2.9	1.9
Fe (%)	2.54	15.16	1.70
Hf	12	16	12
Hg		4.1	
K (%)	0.39	0.31	0.44
La	50	60	56
Mn	340	1060	290
Na (%)	0.52	0.54	0.51
Rb	7	6	8
Sb	0.40	0.37	0.30
Sc	22	30	22
Se	3.2	2.1	1.6
Sm	9.6	11.2	9.6
Sr	1370	2160	1550
Ti (%)	1.1	1.9	1.2
Ta	6.3	7.3	6.7
Tb	1.2	1.7	1.4
Th	9.0	7.0	8.3
V	190	350	220

TABLE 10.6 Chemical Composition Vs. Magnetic Properties of Centralia Precipitator Fly Ash (parts per million except as noted)

	Original fly ash	Magnetic fraction	Nonmagnetic fraction
Al (%)	11.3		
As	5.7	8.6	7.5
Ba	1790	930	1380
Br	0.66	7.9	7.3
Cl (%)	0.1		
Co	26	105	24
Cr	69	158	68
Cs	2.6		2.2
Cu (%)	0.04		
Eu	3.4	7.0	3.4
Fe (%)	3.53	45.0	3.12
Hf	25	9.4	24
La	86	54	84
Mg (%)	1.4		
Mn	570		
Na (%)	1.0 (0.90)	0.38	1.02
Sb	1.8	5.3	1.7
Sc	1.1	0.71	1.0
Se	7.2	5.4	7.1
Sm	14	8.9	14
Sr	3280	1670	3170
Ta	8.6	5.5	8.4
Tb	3.0	2.0	2.9
Th	15	7.5	14
Ti (%)	1.7		
V	316		
Yb	12	8.7	13

10-6 INORGANIC SPECIES: MODELING THE COMBUSTION PROCESS

Vaporization–condensation models have been developed to describe the behavior of elemental species during combustion and to explain the resultant inorganic chemical compositions observed. Flagan (1977) conceptualizes the fly ash particle formation process as shown in Fig. 10.8.

Two initial routes, breakup and vaporization, are proposed to occur, leading to the formation of the large and fine particle sizes, respectively. The breakup process is illustrated in more detail in Fig. 10.9. Initially, volatiles are vaporized, and a residual char is left behind. Further oxidation occurs on the surface and then internally. The internal burning may cause even further breakup of the char particle. During the combustion process temperatures are elevated enough to cause melting of the ash or inorganic mineral phases. Ash droplets may form on the char surface as the char burns more completely. Further char combustion reduces surface area and brings the ash droplets into contact allowing coalescence and growth of droplets. These droplets can be larger than the original inorganic mineral inclusions. As an overall result of char breakup and droplet coalescence, more than one ash particle can be produced from each coal particle, and resultant ash particles can be larger than the original mineral bodies.

A good deal of research has centered on the formation of submicron particles during combustion for several reasons. They are the most difficult to remove from combustion gas streams, are often enriched in many of the potentially toxic trace elements, have potentially longer atmospheric residence times, and pose a respiratory hazard.

Smith, Campbell, and Felix (1979) suggest that formation of the submicron particles may be a result of several complex processes. Four potential mechanisms are cited:

1. Formation by homogeneous nucleation. This is the mechanism of Flagan (1977) in which particles are formed from completely vaporized material.

2. Formation by heterogeneous condensation (Flagan, 1977) onto smaller particles. These particles are primarily unoxidized carbonaceous matter present under the reducing combustion conditions.

3. Formation from submicron mineral inclusions in the coal with further condensation of volatile matter after combustion.

4. Formation by the bursting of bubbles produced during rapid gas release under combustion conditions. Coagulation of smaller fragments can occur as well as condensation of volatilized material.

Working backward from the composition of single fly ash particles, Davison et al. (1974) constructed a simple

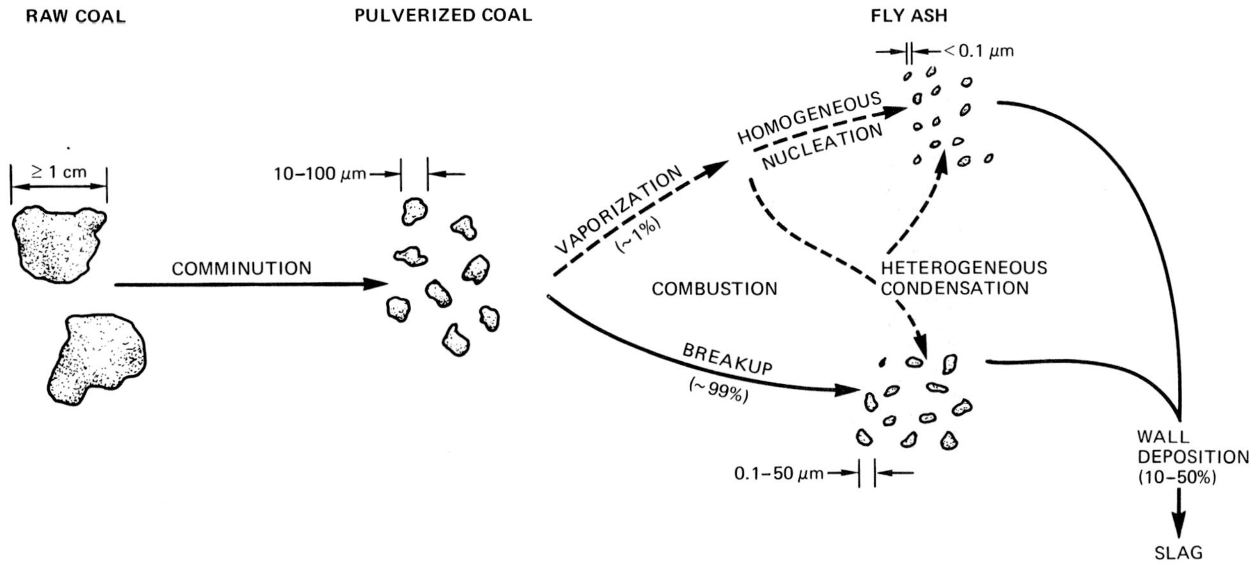

Fig. 10.8 Processes that contribute to fly ash particle formation. (From R. C. Flagan, *Ash Particle Formation in Pulverized Coal Combustion,* presented at Combustion Institute Western State Section, Apr. 18–19, 1977, Pittsburgh, Pa.)

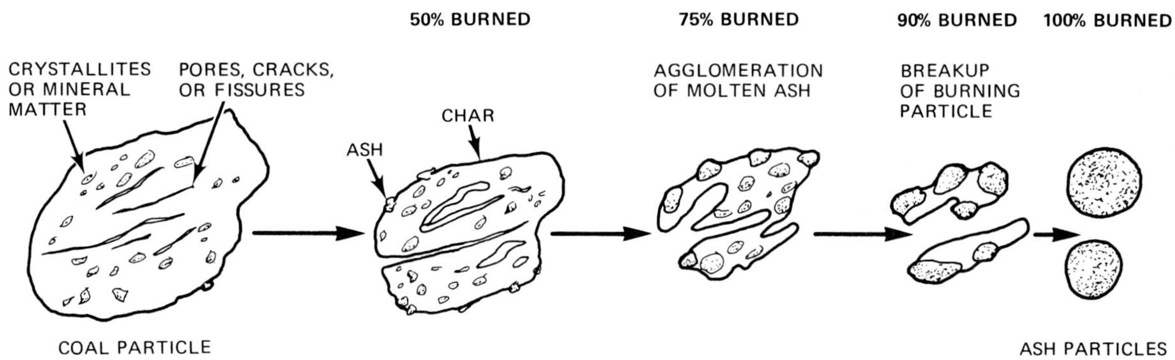

Fig. 10.9 Breakup mechanism for the formation of large fly ash particles. (From R. C. Flagan and S. K. Friedlander, Particle Formation in Pulverized Coal Combustion—A Review, in *Proceedings of Symposium on Aerosol Science and Technology 82nd National Meeting, American Institute of Chemical Engineering, Aug. 29–Sept. 1, 1976,* Atlantic City, N. J.)

mathematical model describing vaporization–condensation chemistry during combustion. They assumed an element in fly ash was distributed in two ways: (1) one fraction uniformly dispersed throughout the spherical aluminosilicate glass matrix matter, and (2) the second fraction uniformly deposited in a condensation layer on the surface of the matrix matter. The concentration of an element in the fly ash varies with particle diameter according to

$$C = C_m + \frac{6C_s}{\rho D}$$

where C = mean bulk concentration observed
C_m = concentration of an element in the matrix
C_s = concentration of an element in the surficial condensed layer
ρ = particle density
D = particle diameter

Flagan and Friedlander (1976) suggested another model in which the concentration should be proportional to D^{-2}. The Flagan–Friedlander model obviously produces a much more pronounced elemental enrichment for smaller particle sizes.

Smith, Campbell, and Felix (1979) found both models were inadequate in describing the elemental distributions in submicron particles for the sized fly ash they analyzed. They observed elemental concentrations that increased with decreasing diameter until approximately 1.0 μm. However, particles smaller than 1.0 μm were generally seen to show no further increase in concentrations for the volatile elements, as shown in Fig. 10.10.

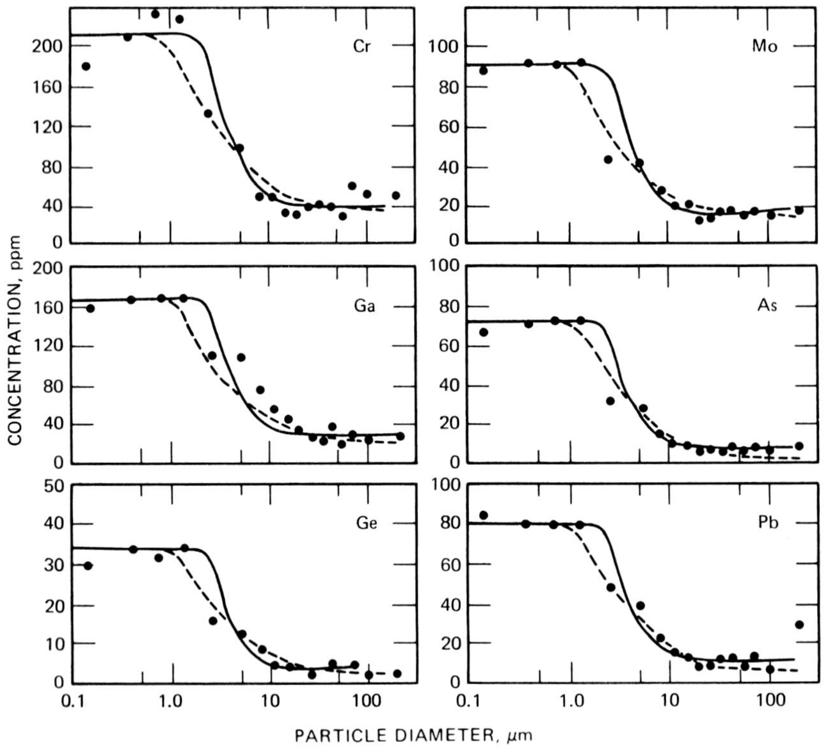

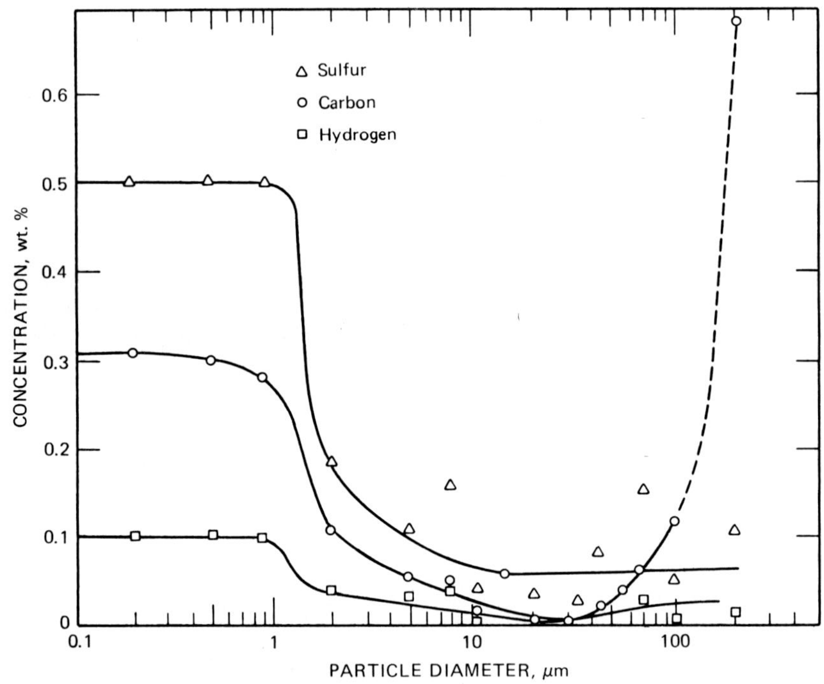

Fig. 10.10 **Concentration vs. particle size data for selected trace elements and hydrogen, carbon, and sulfur. These results show that the major elements in coal, hydrogen, carbon, and sulfur behave similarly to volatile trace elements. Additionally, the results show that the contribution of soot to the formation of submicron particles is negligible. The increase in carbon concentration for the largest particle sizes is ascribed to incomplete combustion and the formation of char. (From R. D. Smith, J. A. Campbell, and W. D. Felix, Atmospheric Trace Element Pollutants from Coal Combustion, in *Proceedings of American Institute Mining Engineering Meeting, Feb. 18–22, 1979,* New Orleans, La.)**

Smith, Campbell, and Felix (1979) and Smith, Campbell, and Nielson (1979) further refine the vaporization–condensation model and show that

$$C = \frac{C_sD^3 + (C_m - C_s)(D - 2L)^3}{D^3}$$

where L is the condensed layer thickness, D − 2L is the diameter of the matrix prior to condensation, and L is independent of particle diameter (Davison et al., 1974). If L is proportional to D^{-1} (Flagan and Friedlander, 1976),

$$C = \frac{C_sD^3 + (C_m - C_s)(D - f/D)^3}{D^3}$$

where $D \leqslant f^{1/2}$ (in micrometers), f = a constant, and L = f/2D.

The preceding vaporization–condensation models are unrealistic since they assume that all submicron particles are formed completely by condensation. Smith, Campbell, and Felix (1979) feel that the bursting process followed by coagulation may be the major source of particles in the submicron range. As a first approximation, these particles would have elemental concentrations that are independent of particle size. The larger particles are formed from molten mineral inclusions and have a condensed layer on the surface. The bulk concentration for a given particle size is then

$$C = X_1C_b + (1 - X_1)\frac{C_sD^3 + (C_m - C_s)(D - f/D)^3}{D^3}$$

where C_b is the concentration in the particles formed by bursting and coagulation and X_1 is the fraction of the particles of diameter D due to bursting and coagulation.

As shown earlier an element can be present in several chemical forms in coal. These forms may have widely varying volatilities during combustion. The volatility of elements bound in inorganic minerals in coal may also be diffusion limited during combustion. The diffusion-limiting process would lead to increased concentration in larger particle sizes. The vaporization–condensation model can be modified by adding factors R, the distance to the surface through which diffusion can occur, and V, the fraction of the element which is in the diffusion-controlled mineral inclusions.

The concentration in the fly ash then becomes

$$\begin{aligned} C = {} & [\rho_sC_sD^3 + V_mC_m(D - f/D - 2R)^3 \\ & + \rho_mC_m(1 - V) - \rho_sC_s(D - f/D)^3] \\ & \times [(\rho_m - \rho_s)(D - f/D)^3 + \rho_sD^3]^{-1} \end{aligned}$$

where ρ_s and ρ_m are the densities of the surficially condensed and matrix matter, respectively.

The presence of the surficial enriched layer postulated from observed particle size chemistry and mathematical modeling has been confirmed analytically. Linton et al. (1976, 1977), Pueschel (1976), Campbell et al. (1978, 1979) used a combination of several analytical techniques, including ion microprobe mass spectrometry and auger electron spectrometry, and determined surficial elevations in concentrations for S, Fe, K, Na Li, Pb, Tl, Mn, Cr, and V. Table 10.7 shows results of this surface characterization and includes information on the mobility or leachability of this surface layer.

TABLE 10.7 Elemental Surface Characteristics of Fly Ash (Linton et al., 1977)

Element	Surficial enrichment	Surface depletion upon leaching	Leachability
Al, Ti, Si	No	Insignificant	Low
S, Fe, K, Na, Li	Yes	Insignificant	Moderate to high
Pb, Tl, Mn, Cr, V	Yes	Observable	Moderate to high
Ca, Mg	No	Not significant	Moderate

In summary, the sophistication of the modeling process continues to improve. The presence of dissimilar elemental concentration vs. particle size distributions in fly ash from different combustion units indicates the chemistry of combustion is a function of multiple factors. These factors can range from the chemical composition and chemical forms of elements in the coal to numerous conditions during combustion, including firing parameters and physical configuration of the combustion unit. Also of importance may be the emission control technique applied—wet or dry—and numerous parameters associated with the type of emission control, such as temperature, contact time, and contact geometry.

10-7 GASEOUS INORGANIC EMISSIONS

The gaseous inorganic pollutants emitted during fossil-fuel combustion include CO_2, SO_x, NO_x, and lesser amounts of species such as CO, fluorides, mercury, and selenium. The oxides of carbon, sulfur, and nitrogen have been the focus of considerable research and controversy. An overview of the formation, atmospheric and environmental chemistries, and future environmental implications of these oxides will be presented in this chapter.

10-7.1 Carbon Dioxide

Traditionally CO_2 has not been considered a pollutant. However, since CO_2 absorbs outgoing infrared radiation from the earth, there has been considerable concern in recent years that the warming of the atmosphere caused by the increased atmospheric CO_2 concentrations resulting from man's activities will have undesirable consequences in future years. The atmospheric CO_2 concentration has increased from about 290 ppm in the late 19th century

(Callendar, 1958) to 335 ppm in 1978. The concentration increased by 1.5 ppm between 1977 and 1978 (*Nature*, 1979). If fossil-fuel consumption continues to increase at the present rate, atmospheric CO_2 concentration may double by the year 2020 (Woodwell et al., 1978).

This increase has often been attributed primarily to the burning of fossil fuels, with the kilning of limestone to produce calcium oxide in the manufacture of cement adding another 1 to 2% (Keeling, 1973). However, several investigators have recently concluded that the terrestrial biosphere is a significant source of CO_2 (in the following discussions the term biosphere will refer to the terrestrial biosphere only). Traditionally, the biosphere has often been considered to be a sink for CO_2, mainly because such a sink is needed to balance the CO_2 budget (Revelle and Shapero, 1978; Siegenthaler and Oeschger, 1978). It has been suggested that the increased rate of photosynthesis caused by the increased atmospheric CO_2 concentrations could cause the biosphere to act as a CO_2 sink. However, Broecker, Li, and Peng (1971), Woodwell et al. (1978), and others have concluded that the effect of increased rates of photosynthesis on atmospheric concentrations is small.

Several investigators have made estimates of the amount of CO_2 released by the biosphere, but these estimates have differed considerably. The increase is believed to be due primarily to the clearing of forests, which contain about 90% of the CO_2 fixed in the biosphere (Whittaker and Likens, 1973). The clearing of tropical forests is believed to be causing most of the release at the present time. The oxidation of dead organic material in the soil following the clearing of forests and the cultivation of virgin soil is also believed to result in the release of significant amounts of CO_2.

Bolin (1977) estimated from land use surveys that the annual rate of release of CO_2 is 0.7×10^{15} g from forests and 0.3×10^{15} g from soil organic matter. Bohn (1978), however, estimated that 1 to 2×10^{15} g of CO_2 is being released annually by the oxidation of soil organic matter alone. These rates are considerably less than the 5.0×10^{15} g of CO_2 currently being released annually by fossil fuels (Baes et al., 1977). However, other investigators using similar methods have arrived at higher estimates of the rate of CO_2 released by the biosphere (Hutchinson, 1954; Adams, Mantovani, and Lundell, 1977; Woodwell, 1978; Woodwell et al., 1978). Woodwell et al. (1978) estimated that the annual release could be anywhere between 2 and 18×10^{15} g and considered the best estimate to be 4 to 8×10^{15} g.

Stuiver (1978) calculated the rate of release of CO_2 by the biosphere from changes in the $^{14}C/C$ and $^{13}C/^{12}C$ ratios in tree rings since 1820. Because of isotopic fractionation during CO_2 fixation, the $^{13}C/^{12}C$ ratio is lower in the biosphere and in fossil fuels than in the atmosphere. Carbon-14 has a 5700-yr half-life; so the $^{14}C/C$ ratio in fossil fuels is negligible, but it is similar in the atmosphere and the biosphere. Therefore fossil-fuel burning will decrease the $^{14}C/C$ and $^{13}C/^{12}C$ ratios in the atmosphere, but the release of CO_2 by the biosphere will decrease only the $^{13}C/^{12}C$ ratio significantly. Stuiver (1978) estimated that 110×10^{15} g of CO_2 was released by the biosphere between 1850 and 1950, with most of the release occurring between 1860 and 1930. Fossil-fuel burning released 55×10^{15} g of CO_2 from 1850 to 1950 (Keeling, 1973). However, Stuiver's calculations indicated that the release by the biosphere was slower from 1950 to 1970. Since fossil-fuel burning released 55×10^{15} g of CO_2 from 1950 to 1970, this would indicate that fossil fuels were responsible for most of the CO_2 released during this period. However, Stuiver's data show considerable scatter, and the $^{13}C/^{12}C$ ratio in tree rings are influenced by environmental factors; therefore his estimates have fairly large error limits. Farmer (1979) has pointed out that temperature at the time of CO_2 fixation affects the $^{13}C/^{12}C$ ratio in the biosphere. If changes in the average temperature since 1820 have had a significant effect on the $^{13}C/^{12}C$ ratios that Stuiver measured in tree rings, then his estimates of CO_2 release will have to be corrected for this effect. The disagreement between various investigators concerning the magnitude, or even the existence, of the CO_2 release by the biosphere indicates that considerably more work is needed to determine more accurately the CO_2 flux to or from the biosphere.

Although more, and possibly considerably more, than 200×10^{15} g of CO_2 has been released into the atmosphere since the middle of the 18th century, the amount of CO_2 in the atmosphere has only increased by about 100×10^{15} g. The ocean is the only major CO_2 sink that has been identified. Since the ocean contains about 55 times as much CO_2 as the atmosphere, it would appear to be an adequate sink for excess CO_2. However, only about 1% of CO_2 dissolved in surface waters is present as CO_2 gas; the rest is in the form of HCO_3^- or CO_3^{2-} ions. As a result, a 10% increase in the atmospheric CO_2 concentration will produce only about a 1% increase in the total dissolved CO_2 in the surface ocean (Broecker, Li, and Peng, 1971). Therefore CO_2 must be transported to the deep ocean if the ocean is to serve as a significant sink for CO_2. Even when the atmosphere has reached equilibrium with the entire ocean, only about 85% of excess CO_2 will have entered the ocean (assuming carbonate sediments do not dissolve) (Broecker, Li, and Peng, 1971).

The surface waters require only about 1 yr to equilibrate with excess CO_2 added to the atmosphere (Broecker, Li, and Peng, 1971). However, the transport of the CO_2 to the deep ocean is very slow. The warm surface waters are separated from the cold deep ocean by a 1000-m-thick stable layer, known as the main thermocline, which inhibits vertical mixing. At high latitudes, cold temperatures and ice formation produce cold water of high salinity which sinks to form the intermediate, deep, and bottom waters of the ocean. Excess CO_2 in the atmosphere and the surface waters may reach the deep ocean either through the main thermocline or in these sinking waters. However, the time required for the surface ocean to mix

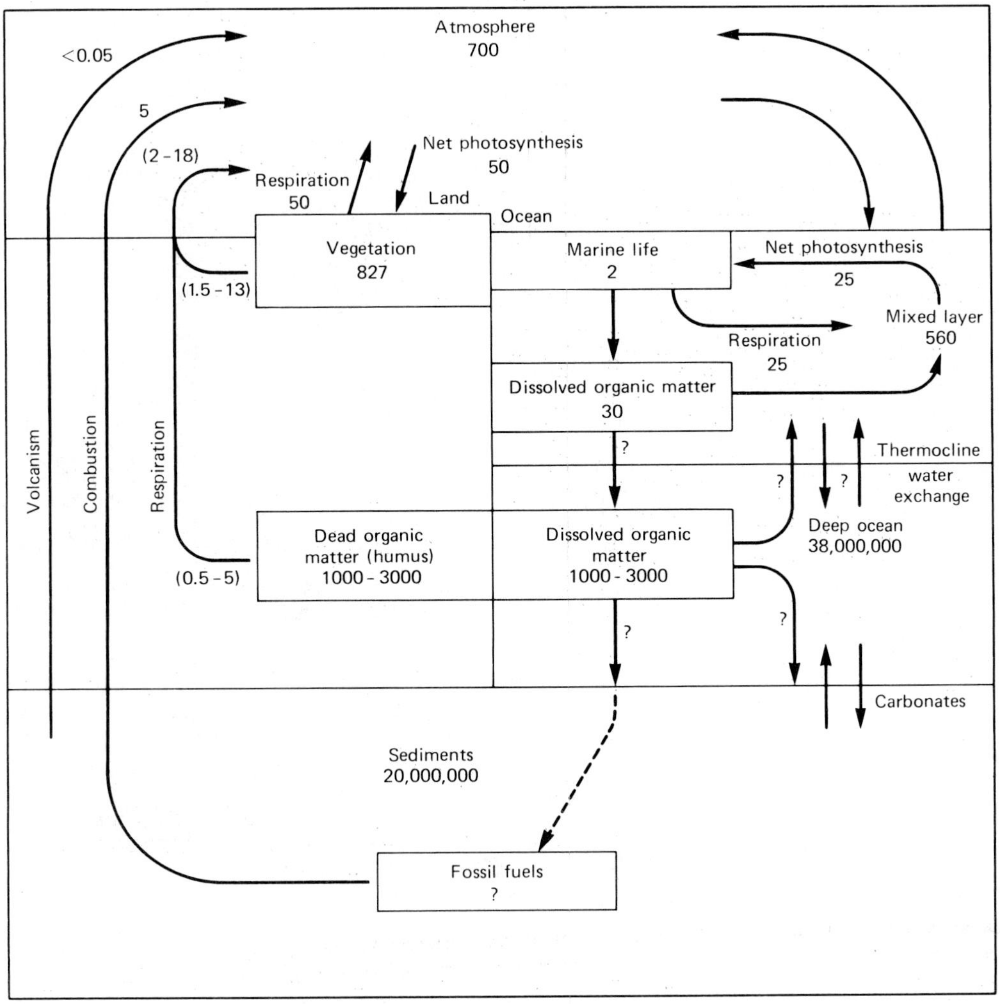

Fig. 10.11 Global carbon cycle showing movement in units of 10^{15} g. [From G. M. Woodwell, The Carbon Dioxide Question, *Sci. Am.*, 238: 34-43 (1978).]

with the deep ocean is still very long, about 1000 yr (Broecker, 1963).

Several investigators have estimated the rate at which the oceans have removed excess CO_2 from the atmosphere. They have generally agreed that the oceans could have removed an amount of CO_2 nearly equal to the difference between the amount of CO_2 that has been released into the atmosphere by fossil-fuel burning and the amount of excess CO_2 that remains in the atmosphere. However, if the biosphere has been a significant source of CO_2, then the calculated rates of removal become inadequate to explain present atmospheric concentrations. Siegenthaler and Oeschger (1978), for example, calculated rates of uptake of CO_2 by the ocean using a box diffusion model in which the biosphere, atmosphere, and surface ocean were considered to be well-mixed boxes and transfer to the deep ocean was assumed to result from eddy diffusion alone. Their calculations agreed best with measured atmospheric CO_2 and ^{14}C concentrations and oceanic ^{14}C concentrations if the terrestrial biosphere was assumed to be a sink for CO_2. They concluded that the biosphere could not be a significant CO_2 source if the ocean is the only major sink.

It seems clear that, if the biosphere is releasing CO_2 at a rate comparable to the rate of release by fossil fuels, either the ocean is taking up CO_2 at a rate faster than currently accepted oceanic mixing rates would allow or there are other significant sinks for CO_2. Measurements have recently been made of the concentrations of excess CO_2 in the ocean (Chen and Millero, 1979; Brewer, 1978). When more such measurements become available, it may be possible to determine more accurately the rate of uptake of CO_2 by the ocean.

Woodwell (1978) has constructed a global CO_2 balance sheet showing the size of the various CO_2 reservoirs and estimates of rates of flow between them (Fig. 10.11). It can be seen that there are several reservoirs of sufficient size to act as significant CO_2 sinks. Bohn (1978) has suggested that uncultivated soils could be accumulating dead biomass at a significant rate. The existence of fossil-fuel deposits is evidence that such accumulation has occurred in the past. Bohn estimates that organic (peat) soils alone are removing 0.4×10^{15} g of CO_2 per year. It has also been suggested that the sinking of dead organic material in the ocean may remove enough carbon from the surface ocean to increase the rate of removal of CO_2 by the ocean significantly. It seems clear that much more accurate knowledge of the fluxes between the various CO_2 reservoirs is required before accurate predictions of future atmospheric CO_2 concentrations can be made.

Several investigators have used atmospheric models to calculate the warming of the atmosphere that will result from a given increase in the atmosphere CO_2 concentration. These models can only give estimates of the warming because they contain many approximations and because the warming produced by CO_2 will be modified in a poorly understood manner by feedback from changes in parameters such as humidity, cloudiness, atmospheric stability, and ice cover. Schneider (1975) has critically examined several models and concluded that the best estimate of the global warming of the atmosphere that would be produced by doubling the CO_2 concentration is 1.5 to 3°K. Perhaps the most widely used estimate is that of Manabe and Wetherald (1975) who used a radiative-convective model to calculate that doubling the CO_2 concentration would raise the average surface temperature by 2.9°K. Their model also indicates that the surface temperature at high latitudes would be increased by two or three times this much because of the greater stability of the atmosphere at high latitudes and because the heating will decrease the ice cover and therefore increase the surface albedo.

At the present time, there is considerable uncertainty as to the overall effect warmer temperatures would have on mankind in general. One major worry is that food production will suffer. However, although crops may suffer in some regions and different crops may have to be planted in others, there will probably be regions where crops will benefit; so no one knows what the net effect will be. The higher precipitation rates resulting from the increased evaporation caused by the elevated temperatures might be expected to benefit food production.

Another major worry is that the warmer temperatures will melt enough ice at high latitudes to raise the sea level and cause major flooding. Mercer (1978) contends that the warmer temperatures caused by a doubling of the CO_2 concentration could cause the West Antarctic ice sheet to deglaciate rapidly because it is grounded below sea level. This would raise the average sea level by 5 m. However, he also believes that the much larger East Antarctic ice sheet would not undergo rapid melting because it is grounded above sea level.

At the present time the general consensus seems to be that the increasing CO_2 concentrations do not necessarily present a serious problem and could even be beneficial. Therefore there seems to be no urgent need to curtail CO_2 release at the present time. So far the increased CO_2 concentrations have not even been able to counteract the presumably natural cooling trend that has been occurring since about 1940. However, Broecker (1975) has studied past temperature cycles using $^{18}O/^{16}O$ ratios in Antarctic ice cores and concludes that a natural warming trend should soon start, which would be amplified by increased CO_2 concentrations. It is clear that considerably more work needs to be done to better understand the CO_2 cycle and to better understand the effect of warmer temperatures upon the overall climate so that possible serious consequences can be identified before it is too late to prevent them by curtailing net CO_2 release.

10-7.2 Carbon Monoxide

Carbon monoxide (CO) is an important atmospheric gas not only because of its effect on human health at the relatively high concentrations often found in urban atmospheres but also because its reaction with the hydroxy radical, (OH),

$$CO + OH \rightarrow CO_2 + H \qquad (10.1)$$

$$H + O_2 \rightarrow HO_2 \qquad (10.2)$$

plays an important role in limiting atmospheric OH concentrations (Wofsy, 1976; Crutzen and Fishman, 1977). The OH radical acts as an atmospheric cleansing agent, reacting with trace gases such as CH_4, H_2S, SO_2, CH_3Cl, $CH_xCl_yF_z$, and CH_yBr (Levy, 1971; McConnell, McElroy, and Wofsy, 1971; Wofsy, McConnell, and McElroy, 1972; Sze, 1977). There has been recent concern that an increase in CO concentrations caused by man's activities could lower OH concentrations and thereby increase the concentrations of these trace gases.

10-7.2.1 *Atmospheric CO Concentrations.* Carbon monoxide was first detected in the atmosphere relatively recently (Migeotte, 1949) and was initially assumed to be due almost entirely to the incomplete oxidation of carbon during fossil-fuel combustion. However, natural sources of CO have since been discovered which are comparable or larger than anthropogenic sources. The CO distribution in the marine troposphere is shown in Fig. 10.12 (Seiler and Zankl, 1976). The CO concentrations in the Northern Hemisphere average considerably higher than those in the Southern Hemisphere, indicating that there are significant anthropogenic and/or continental sources. North of the intertropical convergence zone typical CO concentrations over the oceans are 0.15 to 0.22 ppmv, whereas south of it concentrations drop rapidly to 0.08 ppmv (Peters and Jouvanis, 1979). In the stratosphere the CO concentrations decrease fairly rapidly to about 0.05 ppm in the first 1 or 2 km above the troposphere and then remain fairly constant at higher altitudes (Ehhalt et al., 1974; Seiler, 1974).

10-7.2.2 *CO Sources.* Estimates by Seiler and Zankl (1976) and Zimmerman et al. (1978) of the magnitudes of the various CO sources and sinks are summarized in Table 10.8. Seiler and Zankl (1976) estimate the total anthropogenic CO emission to be about 7.0×10^{14} g/yr, with 70 to 80% being contributed by automobile exhaust

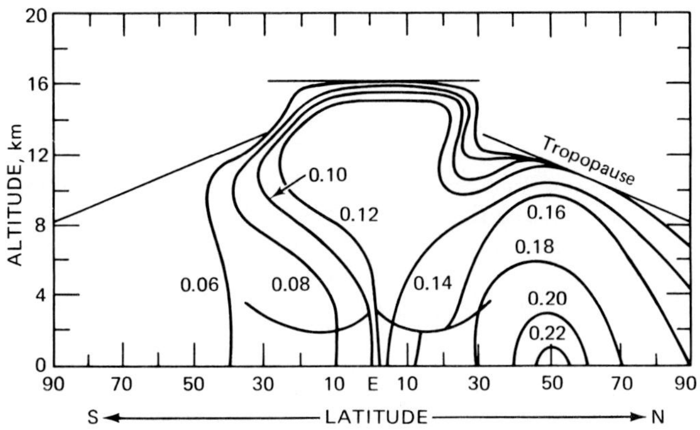

Fig. 10.12 General carbon monoxide distribution in marine air in the troposphere and lower stratosphere. Mixing ratios are in parts per million. (From W. Seiler and H. Zankl, Man's Impact on the Atmospheric Carbon Monoxide Cycle, in *Environmental Biogeochemistry*, Vol. 1, J. O. Nriagu (Ed.), Ann Arbor Press, Ann Arbor, Mich., 1976.)

TABLE 10.8 Sources and Sinks of Tropospheric Carbon Monoxide†

	10^{14} g of CO per year	Percent
Source		
Fossil-fuel burning	6.4	21–42
Nonfossil-fuel burning	0.6	2–4
Oceans	0.4	1–3
CO emissions from plants	0.2–2.0	1–6
Terpene and isoprene emissions from plants	4.2–13.3	28–43
CH_4	3.3–8.3	22–27
Total	15.1–31.0	
Sinks		
Soil uptake	4.5	8–23
Flux to stratosphere	1.1	2–6
Oxidation by OH	13.9–48.7	71–90
Total	19.5–54.3	

†Data from Seiler and Zankl (1976) as reported by Zimmerman et al. (1978).

and 20 to 30% being contributed by combustion in stationary sources and by industrial processes (Seiler, 1974).

The major natural CO sources are believed to be the oceans, plants, volcanoes, forest fires, and the gas-phase oxidation of CH_4, terpenes, and isoprene by OH. Several investigators measured CO supersaturations in the oceans and concluded that the oceans act as a CO source because of microbiological CO production (Swinnerton et al., 1969, 1970; Seiler and Junge, 1970; Lamontagne, Swinnerton, and Linnenbom, 1971; Seiler and Schmidt, 1974; Swinnerton and Lamontagne, 1974). Swinnerton and Lamontagne (1974) indicate that the CO concentration in the ocean corresponds to an equilibrium air-phase concentration more than an order of magnitude higher than actual atmospheric concentrations. Seiler and Zankl (1976) estimate that the oceans release 0.4×10^{14} g of CO per year into the atmosphere.

Several investigators have measured CO emanation from algae and higher plants, but their estimates of global production have varied widely (Krall and Tolbert, 1957; Wilks, 1959; Siegel, Renwick, and Rosen, 1962; Chapelle, 1962; Bidwell and Frazer, 1972; Seiler and Giehl, 1977). Bidwell and Frazer (1972) measured rates of uptake of ^{14}C-labeled CO by plants corresponding to 7 to 70×10^{14} g/yr globally. However, Seiler and Giehl (1977) point out that their method measured CO consumption but not CO production; therefore it did not provide a measure of the net CO production or consumption rate. From their measurements of CO emanation from plants, Seiler and Giehl (1977) concluded that the net global CO production by plants is 0.5×10^{14} g/yr.

The oxidation of methane (CH_4) by the OH radical is believed to constitute a major source of atmospheric CO. Methane oxidation begins with the reaction

$$CH_4 + OH \rightarrow CH_3 + H_2O \qquad (10.3)$$

The methyl radical then reacts to form formaldehyde, which decomposes to form CO. Zimmerman et al. (1978) estimate that 3.3 to 8.3×10^{14} g of CO per year is produced by this process.

The oxidation of CH_4 by OH constitutes the only known tropospheric sink for CH_4. Since CO consumes OH by the reaction in Eq. 10.1, increased CO concentrations might be expected to lower OH concentrations and thereby increase CH_4 concentrations. Methane has an infrared absorption band at 7.7 μm; thus increased CH_4 concentrations would lead to a heating of the atmosphere (Wang et al., 1976). Methane is also involved in photochemical reactions in the stratosphere which tend to lower ozone concentrations.

However, Hameed, Pinto, and Stewart (1979) concluded that the effect of CO on the OH concentrations depends on the $NO_x(NO + NO_2)$ concentrations. If NO concentrations are high, the HO_2 produced by Eq. 10.2 is converted back to OH by the reaction

$$HO_2 + NO \rightarrow NO_2 + OH \qquad (10.4)$$

The NO_2 also produces OH by the reactions

$$NO_2 + h\nu \rightarrow NO + O \qquad (10.5)$$

$$O + O_2 \rightarrow O_3 \qquad (10.6)$$

$$O_3 + h\nu \rightarrow O_2 + O(^1D) \qquad (10.7)$$

$$O(^1D) + H_2O \rightarrow 2OH \qquad (10.8)$$

Thus, for high NO_x concentrations, the injection of CO into the atmosphere results in a net production OH. Hameed, Pinto, and Stewart (1979) calculate that an increase in CO concentration decreases the OH concentration at NO_x concentrations below 0.23 ppbv but increases the OH concentrations at NO_x concentrations above 0.23 ppbv. If this is true, then CO emissions at remote locations where the NO_2 concentrations are about 0.1 ppbv (Noxon, 1975, 1978) would be expected to lower OH concentrations and increase CH_4 concentrations, but CO emissions in urban areas where NO_x concentrations are higher might raise OH concentrations. Carbon monoxide and CH_4 should therefore be monitored at remote locations to see whether there are trends in their concentration.

Zimmerman et al. (1978) have measured the rates of terpene and isoprene emission from plants. They estimate that the oxidation of these compounds in the atmosphere by OH produces 4.3 to 13.4×10^{14} g of CO per year, an amount comparable to that produced by either fossil-fuel burning or CH_4 oxidation.

10-7.2.3 ***CO Sinks.*** The uptake of CO by soil microorganisms, losses into the stratosphere, and oxidation by OH are believed to be the primary sinks for tropospheric CO. Ingersoll, Inman, and Fisher (1974) measured the uptake of CO by soils exposed to 100 ppm of CO. They estimated that global uptake of CO by soils is 13×10^{14} g/yr. However, when Leibl and Seiler (1976) measured the uptake by soils exposed to CO concentrations more typical of tropospheric concentrations (0.25 ppbv), they found smaller uptake rates and estimated a global rate of uptake by soils of 4.5×10^{14} g of CO per year. They found that the air above soil reached an equilibrium CO concentration that increased from 0.002 ppbv at 20°C to 0.18 ppbv at 52°C.

The major mechanism for the removal of CO from the atmosphere is now believed to be the reaction in Eq. 10.1 with OH. From the atmospheric concentrations of CO and CH_4 and from the rates of reaction of OH with CO and CH_4, it can be shown that four molecules of CO are oxidized to CO_2 by OH for every one molecule of CH_4 that is oxidized to CO by OH (Zimmerman et al., 1978). Zimmerman et al. (1978) estimate that the global consumption rate of CO by this reaction is 13.9 to 48.7×10^{14} g/yr. The OH radical may therefore act as a net sink for CO.

10-7.2.4 ***CO Residence Time.*** Robinson and Robbins (1967) calculated an atmospheric residence time of 2.7 yr for CO, assuming that anthropogenic sources are the only major CO sources. However, the realization that there are major natural sources has led to shorter calculated residence times, and the residence time is now believed to be around 0.1 to 0.2 yr (Weinstock and Niki, 1972; Seiler, 1974; Peters and Jouvanis, 1979). Weinstock (1969) derived a residence time of 0.1 yr for ^{14}CO from measurements of its atmospheric concentration and production rate.

10-7.3 Nitrogen Oxides

There are several gaseous oxides of nitrogen, but only nitrous oxide (N_2O), nitric oxide (NO), and nitrogen dioxide (NO_2) are important in the atmosphere. Nitric acid vapor is also found in the atmosphere, and the nitrate ion is found in association with particulate material and in rainfall.

10-7.3.1 Nitrous Oxide. N_2O is primarily produced biologically during the nitrogen cycle. It is an important gas because of the role it plays in chemistry of ozone in the atmosphere. Nitrous oxide is the main source of NO in the stratosphere through the reaction

$$N_2O + O(^1D) \rightarrow 2NO \tag{10.9}$$

The NO then proceeds to destroy ozone catalytically by means of the cyclic reactions

$$\begin{aligned} NO + O_3 &\rightarrow NO_2 + O_2 \\ NO_2 + O &\rightarrow NO + O_2 \end{aligned} \tag{10.10}$$

These reactions represent the major natural sink for stratospheric ozone. Recently, concern has developed that increased fertilizer use and fossil-fuel combustion will increase atmospheric N_2O concentrations, with a consequent decrease in the stratospheric ozone concentrations. However, uncertainties in the atmospheric residence time of N_2O, the magnitude of the natural sources and sinks of N_2O, and the amount of N_2O released into the atmosphere because of increased fertilizer use and combustion have made it difficult to predict the magnitude of the ozone depletion.

10-7.3.1.1 Atmospheric N_2O Concentrations. Nitrous oxide is the most abundant oxide of nitrogen in the atmosphere. Reported concentrations have ranged from around 250 to 600 ppbv. Recent measurements indicate that there is little systematic variation in N_2O concentration with latitude or altitude in the troposphere (Tyson, Arvesen, and O'Hara, 1978; Pierotti and Rasmussen, 1977), indicating that N_2O sources and sinks are rather uniformly distributed. The concentrations decrease rapidly in the stratosphere, particularly in the first 15 km above the tropopause, because of Eq. 10.9. Above 40 km, Ehhalt et al. (1974) found values lower than 10 ppbv.

Disagreement exists concerning the average tropospheric N_2O background concentration. Hahn and Junge (1977) consider that 260 ppbv represents the best estimate, whereas Rasmussen and Pierotti (1978) favor a value of 325 ppbv. Rasmussen and Pierotti (1978) suggest that the disagreement is the result of a bias between the different experimental techniques for measuring N_2O. They note that measurements using thermal conductivity detectors and concentrations calculated from measurements of the N_2O/CO_2 ratio, both of which require sample preconcentration, have almost always given values less than 300 ppbv. However, measurements using infrared absorption spectra, electron capture, or helium ionization detectors have given concentrations greater than 300 ppbv.

Interlaboratory comparisons between laboratories using the different techniques should be carried out to resolve this discrepancy so that trends in N_2O concentrations can be determined from measurements at different laboratories. Craig, Weiss, and Dowd (1976) reported a 1.5% increase in N_2O concentrations over the Pacific Ocean during the decade ending in 1976.

10-7.3.1.2 N_2O Sources. Biological Nitrogen Cycle. Most atmospheric N_2O is produced by the biological nitrogen cycle by nitrification and denitrification and, to a lesser extent, during the uptake of nitrites and nitrates by phytoplankton and higher plants. The major processes occurring during the biological cycle are:

Nitrogen fixation:

$$N_2 \rightarrow X \rightarrow NH_3 \rightarrow RN \qquad (10.11)$$

Nitrogen uptake:

$$NO_3^- \rightarrow NO_2 \rightarrow X \rightarrow N_2O \rightarrow NH_2OH \rightarrow NH_3 \rightarrow RN \qquad (10.12)$$

Ammonification:

$$RN \rightarrow X \rightarrow NH_3 \qquad (10.13)$$

Nitrification:

$$NH_3 \rightarrow NH_2OH \rightarrow X \rightarrow N_2O \rightarrow NO_2^- \rightarrow NO_3^- \qquad (10.14)$$

Denitrification:

$$NO_3^- \rightarrow NO_2^- \rightarrow X \rightarrow N_2O \rightarrow N_2 \qquad (10.15)$$

where X represents mostly unknown intermediates and RN represents biological nitrogen compounds.

Biological fixation occurs when various bacteria, blue-green algae, and symbiotic associations of microorganisms and higher plants, such as legumes, convert N_2 into NH_3 and then into organic nitrogen compounds. Phytoplankton and higher plants may also convert nitrates into organic nitrogen (nitrogen uptake), probably releasing some N_2O during the process. The metabolism of various microbial strains may in turn convert the nitrogen in these organic compounds back into ammonia (ammonification). Bacteria then oxidize the NH_3 through an unknown intermediate into N_2O, which is in turn converted into nitrites and then into nitrates (nitrification). In anoxic environments a variety of microorganisms uses nitrates instead of O_2 in their metabolism by converting the nitrates into nitrites and then, through an unknown intermediate, into N_2O and N_2 (denitrification).

Oceans. There has been disagreement on whether the ocean acts as a source or a sink for atmospheric N_2O. McElroy et al. (1976) have claimed that the consumption of N_2O by bacteria in the oceans results in the oceans' acting as a sink for 60 Tg of atmospheric N_2O per year (Tg = 10^{12} g). However, measurements of N_2O supersaturation in ocean surface waters (Hahn, 1974a, 1974b, 1975; Yoshinara, 1973; Rasmussen et al., 1976) have led most investigators to regard the oceans as a source of N_2O (Hahn and Junge, 1977; Cohen and Gordon, 1978).

There has also been disagreement whether N_2O production in the oceans results primarily from denitrification, as it does on land, or from nitrification. According to Hahn (1974a, 1975) denitrification is the main source of N_2O in the oceans, but in a later paper Hahn and Junge (1977) concluded that nitrification is the process most likely to produce N_2O in oxygenated seawater, even though denitrification could not be ruled out. Cohen and Gordon (1979) pointed out that the measurements that showed N_2O supersaturation in the Atlantic were carried out in areas where oxygen concentrations were too high for denitrification to be likely to occur. They concluded from their measurements of minimum N_2O concentrations in seawater with oxygen concentrations low enough for denitrification to occur that the reduction of N_2O to N_2 during denitrification results in the net consumption of N_2O during denitrification. They concluded that N_2O production in the oceans occurs during nitrification.

Industrial Fertilizer. The use of industrial fertilizers has been increasing at a rate of about 6% annually. Hardy and Havelka (1975) have pointed out that if this continues fertilizer use will increase from 40 Tg of N per year in 1974 to 200 Tg of N per year in the year 2000. This is comparable to the 250 Tg of N per year that has been estimated to be fixed annually by natural processes (Liu et al., 1977). Some of the fertilizer nitrogen will be converted fairly rapidly into N_2O and N_2 by denitrification, but the rest will be added to the rather large organic nitrogen pool and will be converted to N_2 and N_2O on a much slower time scale. However, there exists no consensus on the fraction of the fertilizer nitrogen that will be released as N_2O in a given period of time. McElroy, Wofsy, and Yung (1977) estimated that approximately one-third of the nitrogen will be released into the atmosphere within a decade, but other investigators believe that a significantly smaller fraction will be released. Crutzen and Ehhalt (1977) consider that 5 to 20% is the best estimate for the fraction that is released within a relatively short time, whereas Bolin and Arrhenius (1977) estimated that only 6% will be released within a decade. Hutchinson and Mosier (1979) measured the N_2O emission from a cornfield and found that the rate of N_2O release was only about 2% of the rate of nitrogen addition in fertilizer. If nitrogen fertilizer is used at a rate of 200 Tg per year in the year 2000 and if 30% of the nitrogen is released as N_2O, then the annual release of N_2O from fertilizer will equal about 4% of the present atmospheric inventory. But if only 2% of the nitrogen is released as N_2O, then the annual release will equal only about 0.2% of the present inventory. Clearly the rate of conversion of fertilizer nitrogen to N_2O must be known more accurately before the effects of fertilizer use on atmospheric N_2O and O_3 concentrations can be estimated reasonably accurately.

Combustion. Weiss and Craig (1976) and Pierotti and Rasmussen (1976) have measured N_2O concentrations in stack gases from coal, gas, and oil-fired electric power plants that were 10 to 100 times background concentrations, with the stack gases from coal and oil-fired power plants showing considerably higher N_2O concentrations

TABLE 10.9 N_2O Balance of the Troposphere (Hahn and Junge, 1977)

	10^{12} g of N_2O (N)	
	Likely value	Range
Source		
Marine	45	16–160
Land (soil)	16	6–65
Industrial N fertilizers	13	6–20
Direct man-made (mainly combustion)	2.5	1–4
Fresh water (lakes and rivers)	3	0–25
Lightning	?	10–55
Sum of all known sources	80	40–330
Sinks		
Stratospheric processes	13	9–18
Tropospheric processes [photolysis and reactions with $O(^1b)$, negative ions, OH, and HO_2]	2	1–3
Sum of all known sinks	15	10–21
Total source (and sink) estimated from inventory and residence time	135	80–300
Additional sinks required for balance	120	70–280

than those of gas-fired power plants. Pierotti and Rasmussen (1976) estimated that 1.9 Tg of N_2O(N) is produced annually by coal burning, whereas 0.14 Tg of N_2O(N) is produced annually by gas burning. Weiss and Craig (1976) estimated that 1.2 Tg of N_2O(N) is produced annually by coal burning and 0.3 Tg of N_2O(N) by oil burning. They point out that, although the estimated annual release by coal and oil burning is only 0.11% of the atmospheric inventory, such a release is large enough to account for up to half of the increase in the atmospheric concentrations reported by Craig, Weiss, and Dowd (1976).

Seinfeld (1975) suggested that the oxidation of organic amines or ammonia in automobile catalytic converters may produce N_2O by the reaction

$$2NH_3 + O_2 \rightarrow N_2O + 3H_2O \qquad (10.16)$$

Pierotti and Rasmussen (1976) measured elevated N_2O in the exhausts of cars equipped with catalytic converters, but not in the exhaust of cars without them. Weiss and Craig (1976) estimated that the global production of N_2O(N) by catalytic converters was 1.9 Tg in 1976.

10-7.3.1.3 N_2O Sinks. The stratosphere is generally believed to be the only major sink for atmospheric N_2O, although McElroy et al. (1976) have postulated that the oceans act as a major sink. Measurements by Johnston and Selwyn (1975) of the cross section for the photodissociation of N_2O indicate that the photolytic destruction of N_2O in the troposphere is negligible. However, the estimated rate of N_2O destruction in the stratosphere is far smaller than the rate of production at the earth's surface. The estimates of Hahn and Junge (1977) of the most likely values for the sources and sinks of N_2O are shown in Table 10.9. Also shown are their estimates of the range of uncertainty of the values and the total rate of addition and removal of N_2O from the atmosphere calculated from the atmospheric inventory and residence time. If these estimates are reasonably accurate, there clearly must be an unknown major sink for N_2O. This is a main reason McElroy et al. (1976) believe the oceans must act as a N_2O sink.

There has been recent evidence that desert areas (specifically the Sahara Desert) or dry dust particles suspended in the atmosphere may act as a sink for atmospheric N_2O. The proposed mechanism involves the adsorption of N_2O on silicate particles followed by the thermal or photolytic decomposition of the N_2O. The presence of water inhibits the adsorption, and therefore the decomposition, of N_2O. Schutz et al. (1970), Junge et al. (1971), and Pierotti, Rasmussen, and Chatfield (1978) have measured N_2O concentrations in air of Sahara origin which were 15, 10, and 5% below normal at Tenerife (300 km west of Africa), the eastern Atlantic, and the eastern tropical Pacific, respectively. Rebbert and Ausloos (1978) measured the thermal and photolytic decomposition of N_2O in 1-L flasks containing five dry sands of different origin and N_2O at low pressure [13.3 Pa (0.1 torr)]. They found that the half-life of N_2O in contact with Tunisian sand at 296°K was 350 ± 35 days. The half-life decreased when more sand was added or when the flask was exposed to light with wavelengths greater than 280 nm, but it increased significantly when water vapor was present. Alyea, Cunnold, and Prinn (1978), using a simple mass flow model, estimated that, if 10% of the N_2O that arrives near (1 m) the desert surface is destroyed, this would account for the generally accepted 15- to 20-yr atmospheric residence time of N_2O.

Pierotti, Rasmussen, and Chatfield (1978) have suggested that the destruction of N_2O in dust clouds over and downwind of the Sahara might be more important than the destruction at the desert surface. Because of intense heating at the desert surface, vertical mixing is very intense and dust clouds are carried to altitudes of 5 to 7 km. These dust clouds are transported clear across the Atlantic, where they are easily identifiable (Prospero and Ness, 1977). Pierotti, Rasmussen, and Chatfield (1978) measured depressed atmospheric N_2O concentrations in Saharan dust clouds but considered that the concentration decrease was too small to support firm conclusions, considering the limited number of measurements. It is evident that further research is required to determine whether the desert areas act as major sinks for N_2O.

10-7.3.1.4 N_2O Residence Time. There has also been disagreement over the atmospheric residence time, T, of N_2O. If a steady state exists, T = M/Q = M/S, where M is the mass in the atmosphere, Q is the source strength, and S is the removal rate. Unfortunately, values for T, Q, and S are very difficult to obtain directly. However, there are several indirect methods for estimating the residence time. Hahn and Junge (1977) have pointed out that, for very short time periods, ΔT, M varies according to

$$\frac{\Delta M}{\Delta T} = Q - S$$

Since M/T fluctuates about zero and Q and S are greater than zero and approximately equal to M/T,

$$\frac{\Delta M}{\Delta T} < \frac{\bar{M}}{T} \quad \text{or} \quad T < \frac{\Delta T \bar{M}}{\Delta M}$$

From the measured trends in the N_2O concentration at four locations, Hahn and Junge (1977) estimated that the N_2O residence time must be less than 6 yr.

Hahn and Junge (1977) also estimated the N_2O residence time from the fluctuations in the measured concentrations at three locations. The total variance, σ_t^2, of a given set of data should be related to the variance of the actual concentrations, σ_d^2, and to the variance due to methodological errors, σ_m^2, by the equation

$$\sigma_t^2 = \sigma_d^2 + \sigma_m^2$$

Junge (1974) has shown that the residence times of several tropospheric gases are related to their mean standard deviations, $\tilde{\sigma}_d$, by the empirical relationship

$$T = \frac{0.14}{\tilde{\sigma}_d}$$

with $\tilde{\sigma}_d$ given in percent and T in years. Junge (1974) states that this relationship must be stochastic and that the constant 0.14 may be off by as much as a factor of 3 for an individual gas. Using this relationship and the variances in the N_2O concentrations at three stations, Hahn and Junge (1977) estimated that the upper limit for the N_2O residence time is 11 yr. However, recent measurements of atmospheric N_2O concentrations by Yoshinara (1973), Rasmussen et al. (1976), and others have shown smaller variations, leading to longer calculated residence times. The general consensus now seems to be that the N_2O residence time is between 10 and 20 yr.

10-7.3.1.5 Ozone Reduction by N_2O. Widely varying predictions have been made of the decrease in stratospheric ozone concentrations that would result if fertilizer use increased 6% per year until the year 2000 and then remained constant after that. McElroy et al. (1976) have estimated that ozone concentrations could be reduced by about 20% or more by the year 2025, assuming that the oceans act as a major sink for N_2O. However, other investigators have arrived at much smaller estimates because they assumed that the oceans act as an N_2O source, making fertilizer produced N_2O a much smaller fraction of the total N_2O production. Estimated ozone reductions also become smaller if it is assumed that much of the fertilizer nitrogen enters the large organic nitrogen pool and is denitrified over a long period of time rather than being denitrified rather rapidly. It has been estimated that ozone reduction will be 1.5 to 9% by 2025 if 50% of the fertilizer is denitrified immediately (Crutzen, 1976a), 10% by 2025 if 60% of the fertilizer is denitrified immediately (Sze and Rice, 1976), less than 2% by 2000 and 20% by 2100 if 30% is denitrified immediately (Crutzen and Ehhalt, 1977), and 1% by 2025 if only 10% is denitrified immediately (Liu et al., 1976; Sze and Rice, 1976).

The general consensus (except for McElroy) seems to be that the ozone reduction produced by increased fertilizer use will only be a few percent by the turn of the century, although larger reductions could occur on a longer time scale. More work needs to be done to determine the magnitude of N_2O sources and sinks and the rate of conversion of fertilizer nitrogen to N_2O before more accurate predictions of ozone reduction can be made.

10-7.3.2 Nitric Oxide and Nitrogen Dioxide. Chemical and photochemical reactions cause rapid interconversion between nitric oxide (NO) and nitrogen dioxide (NO_2) in the atmosphere. Noxon (1975, 1978) has measured tropospheric background NO_2 concentrations of about 0.1 ppbv. The denitrification of soils introduces NO into the atmosphere, and the oxidation of atmospheric N_2 by lightning introduces NO_2. However, it has been estimated that in 1975, 47% of the total global NO_x (NO + NO_2) emission originated from fuel consumption and that 28% of the total was released during electric power generation. Anthropogenic NO_x emissions are believed to be about 20×10^{12} g/yr as N. During combustion, nitrogen oxides are produced both from the fuel nitrogenous matter and by the oxidation of N_2 in the air.

Nitrogen oxide emissions from stationary combustion sources are currently regulated by the Environmental Protection Agency (EPA). The emission standards and the primary and secondary ambient air-quality standards for nitrogen oxides, SO_2, and particulates are shown in Tables 10.10 and 10.11.

10-7.3.3 NO_X Photochemistry. The importance of NO and NO_2 in the atmosphere is due to the role they play in the production of photochemical smog. The ozone (O_3) concentration in the atmosphere is controlled by the following three reactions:

$$NO_2 + h\nu \xrightarrow{k_1} NO + O \tag{10.16}$$

$$O + O_2 \xrightarrow{k_2} O_3 \tag{10.17}$$

$$NO + O_3 \xrightarrow{k_3} NO + O_2 \tag{10.18}$$

This series of reactions is cyclical, there being no net chemical change. Since they are much faster than other reactions involving these species in the troposphere, a photostationary state should be formed such that

$$[O_3] \simeq \frac{[NO_2\ k_1]}{[NO]\ k_3}$$

Therefore the O_3 concentration should be proportional to the NO_2/NO ratio and the light intensity (through k_1).

However, Eschenroeder, Martinez, and Nordick (1972) found that the monitoring data of the Los Angeles Air Pollution Control District did not obey this relationship very well. Calvert (1976) and Bilger (1978) concluded that the discrepancy resulted from inhomogeneities in the air sampled because (1) the measuring instruments sampled over a finite period of time, averaging the concentrations of different air parcels having different composition, and (2) air parcels having different composition mixed too soon before sampling for a photostationary state to be achieved in the mixture.

TABLE 10.10 New Source Standards of Performance for Fossil-Fuel-Fired Steam Generators†

Pollutant	Standard
Particulate matter	0.10 lb per million Btu heat input, maximum 2 h average
	20% opacity (except that 40% opacity is permissible for not more than 2 min in any hour)
Sulfur dioxide	0.80 lb per million Btu heat input, maximum 2 h average when liquid fossil fuel is burned
	1.2 lb per million Btu heat input, maximum 2 h average when solid fuel is burned
Nitrogen oxides	0.20 lb per million Btu heat input, maximum 2 h average, expressed as NO_2 when gaseous fossil fuel is burned
	0.30 lb per million Btu heat input, maximum 2 h average, expressed as NO_2 when liquid fossil fuel is burned
	0.70 lb per million Btu heat input, maximum 2 h average, expressed as NO_2 when solid fossil fuel (except lignite) is burned

†Source· EPA Regulations Title 40, *Code of Federal Regulations,* Part 60.42 (40 CFR 60.42) to 40 CFR 60.44 and Air Quality Criteria for Sulfur Oxides (1967).

TABLE 10.11 National Primary and Secondary Ambient Air-Quality Standards†

Pollutant	Type of standard	Averaging time	Frequency parameter	Concentration g/m^3	Concentration ppm
Carbon monoxide	Primary and secondary	1 h	Annual maximum‡	40,000	35
		8 h	Annual maximum	10,000	9
Hydrocarbons (nonmethane)	Primary and secondary	3 h (0600–0900)	Annual maximum	160§	0.24§
Nitrogen dioxide	Primary and secondary	1 yr	Arithmetic mean	100	0.05
Photochemical oxidants	Primary and secondary	1 h	Annual maximum	160	0.08
Particulate matter	Primary	24 h	Annual maximum	260	
		24 h	Annual geometric mean	75	
	Secondary	24 h	Annual maximum	150	
		24 h	Annual geometric mean	60¶	
Sulfur dioxide	Primary	24 h	Annual maximum	365	0.14
		1 yr	Arithmetic mean	80	0.03
	Secondary	3 h	Annual maximum	1300	0.5

†Source: EPA Regulations 40 CFR 50 and Air Quality and Stationary Source Emission Control (1975).

‡Not to be exceeded more than once per year.

§As a guide in devising implementation plans for achieving oxidant standards.

¶As a guide to be used in assessing implementation plans for achieving the annual maximum 24-h standard.

Although the reactions in Eqs. 10.16 to 10.18 by themselves result in no net production of ozone, many other reactions take place which result in the oxidation of NO to NO_2, causing the O_3 concentration to rise. The hydroxyl radical (OH) is produced by

$$O_3 \xrightarrow{h\nu} O_2 + O(^1D) \quad (10.19)$$

$$O(^1D) + H_2O \rightarrow 2OH \quad (10.20)$$

and other reactions in the atmosphere. This radical, and to a lesser extent O and O_3, reacts with hydrocarbons, CO, aldehydes, and other hydrocarbon oxidation products to initiate free radical chain reactions which produce hyperoxy (HO_2), alkylperoxy (RO_2), and acylperoxy ($RCOO_2$) radicals. These oxidize NO to NO_2 by the reactions

$$HO_2 + NO \rightarrow HO + NO_2 \quad (10.21)$$

$$RO_2 + NO \rightarrow RO + NO_2 \quad (10.22)$$

$$RCOO_2 + NO \rightarrow RCO_2 + NO_2 \quad (10.23)$$

In an urban atmosphere the concentrations of O_3 are negligible during the night because the reaction in Eq. 10.16 does not take place in the absence of sunlight. During the morning the large increase in the amount of vehicular traffic causes the concentrations of primary contaminants such as CO, NO, and hydrocarbons to increase. However, the production of secondary contaminants such as O_3 and NO_2 does not occur to an appreciable extent until about 2 h after dawn. In the next 1 or 2 h, however, photochemical reactions accelerate and much of the NO becomes oxidized to NO_2, NO_2 concentrations reach a peak, and O_3 concentrations rise. Reactions such as

$$HO_2 + NO \xrightarrow{M} HNO_3 \quad (10.24)$$

$$HO + NO_2 \xrightarrow{M} HNO_3 \quad (10.25)$$

$$HNO_3 + \text{aerosol} \rightarrow \text{nitrates} \quad (10.26)$$

which convert NO_2 to nitrates then cause the NO_2 concentrations to decrease, although the NO_2/NO ratio continues to increase for a couple of hours, causing the O_3 concentrations to continue to rise (Calvert, 1976). When NO concentrations become sufficiently low, NO_2 reacts with $RCOO_2$ radicals to form peroxyacyl nitrates ($RCOO_2NO_2$) (Hecht, Seinfeld, and Dodge, 1974).

$$RCOO_2 + NO_2 \rightarrow RCOO_2NO_2 \text{ (PAN)} \quad (10.27)$$

Nitrates have been shown to have a deleterious effect on respiratory functions and on human health in general (Shy, 1975; French, 1975). The peroxyacyl nitrates are particularly potent eye irritants and phytotoxicants. Most of the NO_x is eventually converted to HNO_3 or PAN, with the PAN/HNO_3 ratio varying from 1 to 3 in urban atmospheres (Spicer, 1977).

The concentration of ozone reaches a maximum sometime after noon and then gradually declines for several hours. The afternoon traffic maximum injects additional NO into the atmosphere, lowering the NO_2/NO ratio and therefore further lowering the O_3 concentration (Haagen-Smit and Wayne, 1976).

Many of the photochemical smog reactions that occur in urban atmosphere also occur at lower concentrations in rural atmospheres, even in the relative absence of anthropogenic pollutants. The oxidation of naturally produced CH_4, CO, H_2CO, and H_2 by OH produces HO_2 and CH_3O_2 which oxidize NO to NO_2 by reactions in Eqs. 10.21 and 10.22 (Crutzen, 1973; Chameides and Walker, 1973, 1976; Chameides and Stedman, 1977). Chameides and Stedman (1977) concluded from their photochemical and transport model that one-third of the O_3 found in the troposphere is due to local photochemical production and two-thirds originates from the stratosphere. They also estimated that two-thirds of the O_3 destruction in the troposphere is due to photochemical processes, whereas only one-third is due to loss at the ground.

Several investigators have found in smog chamber experiments and in the results of photochemical models of urban or rural atmospheres that the maximum O_3 concentrations for a given hydrocarbon or NO_x concentration occur at a specific hydrocarbon/NO_x concentration ratio (Korth, Rose, and Stahman, 1964; Altshuller et al., 1967; Glasson and Tuesday, 1970; Hecht, Seinfeld, and Dodge, 1974). In the model of Hecht, Seinfeld, and Dodge (1974), high O_3 concentrations did not develop at very high hydrocarbon/NO_x ratios, because the NO_x was removed as stable products. At low ratios the hydrocarbon oxidation process did not produce enough free radicals to oxidize NO rapidly. The investigators concluded that the O_3 levels in the 1969 Los Angeles atmosphere could be reduced more effectively by lowering hydrocarbon levels than by lowering NO_x levels but that lowering both hydrocarbon and NO_x levels would be relatively ineffective in lowering O_3 concentrations.

The photochemical model of Stewart, Hameed, and Pinto (1977) gave maximum O_3 concentrations at an NO_x concentration of 0.5 ppbv for tropospheric background air. Since Noxon (1975, 1978) measured background NO_2 concentrations of about 0.1 ppbv, this indicates that increased NO_x emissions, such as by power plants, would raise O_3 concentrations in tropospheric background air.

NO_x Photochemistry in Power Plant Plumes. Fossil-fuel-fired power plants release NO_x mostly in the form of NO. Hydrocarbon emissions are also relatively low; so the oxidation of NO to NO_2 is slow. Because of the resulting low NO_2/NO ratio, the O_3 concentrations in power plant plumes are generally significantly below ambient concentrations for several miles downwind of the power plant (Davis, Smith, and Klauber, 1974; Hegg et al., 1977; Miller et al., 1978). However, several investigators have also observed the

development of elevated O_3 concentrations at greater distances downwind (Davis, Smith, and Klauber, 1974; Gillani et al., 1978; Keifer, Klauber, and Davis, 1977; Miller et al., 1978). Davis, Smith, and Klauber (1974) concluded that air-quality standards for O_3 are probably exceeded regularly in the plumes of large power plants during daylight hours in the summer. However, other investigators have failed to find elevated O_3 concentrations in power plant plumes (Ogren et al., 1976; Hegg et al., 1977). Miller et al. (1978) concluded that the development of elevated O_3 concentrations in a plume depends on whether or not mixing of the plume with environmental air produces nonmethane hydrocarbon (NMHC) to NO_x ratios in the plume that are favorable for O_3 production. Most studies have shown that O_3 concentrations go through a maximum when the NMHC/NO_x ratio is about 8 (Dimitriades, 1977). If the ratio in the plume is closer to this value than the ratio in the surrounding air, elevated O_3 concentrations would be expected to develop in the plume. Miller et al. (1978) found elevated O_3 concentrations in a power plant plume over Lake Michigan when the NMHC/NO_x ratio in the surrounding air was about 30. Mixing of this air with a power plant plume containing high NO_x and low NMHC concentrations would be expected to produce a NMHC/NO_x ratio in the plume that was more favorable for O_3 production than in the ambient air.

10-7.3.4 Atmospheric Nitrates and Nitric Acid. The oxidation of NO_x to nitric acid or nitrates, followed by wet or dry deposition, results in the removal of NO_x from the atmosphere. Until recently there have been only a few measurements of atmospheric nitric acid concentrations. Spicer (1977) measured concentrations ranging from 0.5 to 10 ppbv in Los Angeles, St. Louis, and Dayton, whereas Okita et al. (1976) measured concentrations from 0.2 to 8 ppbv in Tokyo. Okita et al. (1976) and Huebert and Lazrus (1978) measured nitric acid concentrations at remote locations ranging from 0.0 to 0.7 ppbv and <0.02 to 0.3 ppbv, respectively. Huebert and Lazrus (1978) found no variation in concentration with latitude from 70°N to 24°S, and little variation with altitude from 0.2 to 8 km, although concentrations were somewhat higher above the atmospheric boundary layer than in it.

Particulate nitrate concentrations generally are less than 1 ppbm in the atmospheric boundary layer, as shown in Table 10.12, although they may be higher in polluted locations (Okita et al., 1976). Concentrations at remote locations in the boundary layer are generally higher at marine than at continental locations because of the conversion of nitric acid to nitrates by the reaction

$$HNO_3\ (gas) + Cl^-\ (solid) \xrightarrow{H_2O} HCl\ (gas) + NO_3^-\ (solid) \qquad (10.28)$$

Above the boundary layer, concentrations are generally less than 0.2 ppbm at remote continental and marine locations (Huebert and Lazrus, 1978).

TABLE 10.12 Particulate Nitrate in the Atmospheric Boundary Layer

Location	Concentration range, ppbm	Best estimate of typical value, ppbm
Marine	0.07–1.1	0.8
Continental	0.06–0.6	0.1
Urban	0.15–3.5	1.4

10-7.4 Sulfur Oxides

Anthropogenic emissions of sulfur oxides are of such magnitude that they may be comparable to natural production levels. However, in heavily industrialized regions such as Europe and the northeastern United States, anthropogenic emissions are much higher than natural emissions. The primary anthropogenic sources of sulfur oxides are stationary combustion facilities, such as electric power plants. The relative magnitudes of the various anthropogenic sources of SO_2 are listed in Table 10.13.

TABLE 10.13 Sulfur Oxide Emissions (NEDS† Data for 1972)

Source	Emissions, 10^{12} g/yr	Percentage of total emissions
Stationary fuel combustion sources		
Electric power plants	15.8	54.3
Industrial	4.5	15.3
Commercial and residential	2.1	7.1
Industrial processes	6.1	21.1
Other stationary sources	0.06	0.2
Transportation	0.6	2.0

†National Environmental Data Service.

Anthropogenic sulfur is emitted primarily as SO_2. However, in the atmosphere the SO_2 becomes oxidized to form sulfuric acid and acid sulfates, which cause major environmental problems by raising the acidity of precipitation. In the past few years, evidence has been accumulating that the sulfates are also more harmful to human health than is SO_2 (Shy and Finklea, 1973; CHESS, 1974). Sulfate particles become trapped in the lungs, where they may cause respiratory disease. The dry deposition velocity of SO_2 on the earth's surface is about an order of magnitude greater than that of sulfate particles (Garland, 1978); therefore the rapid deposition of SO_2 and its conversion to sulfates cause SO_2 concentrations to decrease relatively rapidly downwind of a source. However, sulfate particles may be transported long distances and cause environmental problems hundreds or even thousands of kilometers downwind. The construction of taller stacks to reduce local pollutant concentrations has led to the dispersal of environmental problems due to sulfates over greater distances.

Sulfate concentrations and precipitation acidities downwind of SO_2 sources depend on the rate at which SO_2 is oxidized to sulfuric acid. Therefore considerable effort has been expended in recent years to determine the factors that control the SO_2 oxidation rate. The oxidation rate varies widely, depending on atmospheric conditions, because oxidation may result from many different processes, including homogeneous reactions with gaseous atmospheric pollutants, reactions in aqueous solutions, and reactions on particle surfaces. In the following sections the mechanisms for SO_2 oxidation which are considered to be the most important are discussed.

10-7.4.1 SO_2 Oxidation Pathways.

10-7.4.1.1 Homogeneous Gas-Phase Oxidation. The following discussion of gas-phase SO_2 oxidation is derived largely from the excellent review of the subject by Calvert et al. (1978). There are innumerable gas-phase reactions by which SO_2 may be oxidized in an impure atmosphere, but only a very few of these involving free radicals appear to produce most of the oxidation. Neither the direct oxidation of SO_2 by O_2, O_3, or NO_x nor the oxidation of excited states of SO_2 appears to play a major role in SO_2 oxidation.

The initial gas-phase oxidation of SO_2 in a plume from an electric power plant or other industrial operation is believed to result primarily from two reactions. The first of these involves the photolysis of NO_2 early in the plume dilution. At this time the O_2 and NO_2 concentrations are relatively low and the SO_2 concentration is high so that SO_2 can compete successfully with NO, NO_2, and O_2 for the oxygen atoms produced by the photolysis of NO_2:

$$NO_2 + h\nu \rightarrow O + NO \qquad (10.29)$$

$$O + SO_2 + M \rightarrow SO_3 + M \qquad (10.30)$$

$$SO_3 + H_2O \rightarrow H_2SO_4 \qquad (10.31)$$

Calvert et al. (1978) have estimated that for NO_2, NO, SO_2, and O_2 concentrations of 50, 500, 500, and 1×10^5 ppm, respectively, this reaction should produce an SO_2 oxidation rate of 1.4%/h in midday sunlight.

A similar early peak in the oxidation rate should result from the reaction of SO_2 with the OH produced by the photolysis of HONO. The HONO may be produced either by the homogeneous reaction of NO, NO_2, and H_2O or by heterogeneous reactions involving water condensation or particle surfaces:

$$NO + NO_2 + H_2O \rightarrow 2HONO \qquad (10.32)$$

$$HONO + h\nu \rightarrow OH + NO \qquad (10.33)$$

$$OH + SO_2 + M \rightarrow HOSO_2 + M \qquad (10.34)$$

$HOSO_2$ is an unstable free radical that reacts further to form sulfuric acid, peroxysulfuric acid, and other compounds that eventually lead to sulfate. Calvert et al. (1978) have estimated that for HONO, NO_2, SO_2, and CO concentrations of 10, 20, 500, 500, and 1 ppm, respectively, this reaction should produce an SO_2 oxidation rate of 1.2%/h in midday sunlight.

The rates of both of these reactions decrease rapidly with plume dilution; so we might expect the rate of gas-phase oxidation to decrease rapidly with travel distance. In fact, several investigators have measured very low rates of SO_2 oxidation in power plant plumes in relatively unpolluted environments (e.g., Newman, Forrest, and Manowitz, 1975; Lusis and Wiebe, 1976; Forrest and Newman, 1977). However, if the plume is diluted with urban air containing hydrocarbons, ozone, NO_x, and other pollutants, the rapid chain reactions characteristic of photochemical smog will take over, and the reactions of SO_2 with the radicals OH, HO_2, and CH_3O_2 will begin to dominate the gas-phase oxidation of SO_2. The rates of reaction of these species with SO_2 are fairly well known, but, except for a few measurements of OH, their concentrations have not been measured in the atmosphere; therefore it is possible only to estimate the rates at which they oxidize SO_2. It has been estimated that in a highly polluted, sunlight-irradiated lower troposphere typical OH concentrations of 7×10^6 mol/cm^3 (1.6×10^{-7} ppm) would be expected (Calvert and McQuigg, 1975) and that in a relatively clean lower troposphere somewhat lower concentrations of 1×10^6 mol/cm^3 (2.3×10^{-8} ppm) would be expected (Crutzen and Fishman, 1977). Using these concentrations, Calvert et al. (1978) estimated that the rate of oxidation of SO_2 by OH might be as high as 2.7%/h in a dirty atmosphere and typically 0.4%/h in a clean troposphere. However, the few direct measurements of OH in the atmosphere have yielded higher values than these, suggesting that the rate of oxidation of SO_2 by OH could be even higher. Wang et al. (1975) reported peak OH concentrations in Detroit of about 5×10^7 mol/cm^3. Davis et al. (1979) measured average midday OH concentrations of 9.5×10^5 mol/cm^3 in the Four Corners area near the Arizona–New Mexico border.

Payne, Steif, and Davis (1973) estimated that the reaction

$$HO_2 + SO_2 \rightarrow OH + SO_3 \qquad (10.35)$$

has a rate constant, $k = (8.7 \pm 1.3)\ 10^{-16}$ cm^3 mol^{-1} s^{-1}. According to Demerjian, Kerr, and Calvert (1974), a typical HO_2 concentration that might be expected in a highly polluted, sunlight-irradiated region of the lower troposphere is about 6×10^9 mol/cm^3. This leads to an SO_2 oxidation rate of about 1.9%/h by this reaction. However, in a clean atmosphere the HO_2 concentration and therefore the SO_2 oxidation rate by HO_2 may be an order of magnitude lower.

Sulfur dioxide is oxidized by the CH_3O_2 radical by the reactions

$$CH_3O_2 + SO_2 \rightarrow CH_3O + SO_3 \qquad (10.36)$$

$$CH_3O_2 + SO_2 \rightarrow CH_3O_2SO_2 \qquad (10.37)$$

Thermochemical arguments indicate that $CH_3O_2SO_2$ may decompose readily into CH_3O and SO_3 after a short delay (Calvert et al., 1978). Whitbeck et al. (1976) estimated that the sum of the rate constants for these two reactions is $(5.3 \pm 2.5) \times 10^{-15}$ cm^3 mol^{-1} s^{-1}. It is expected that in a highly polluted atmosphere the concentration of CH_3O_2 should be about 10^9 mol/cm^3 (Demerjian, Kerr, and Calvert, 1974). This would give an SO_2 oxidation rate of 2%/h. In the fairly clean troposphere, rates of 0.02%/h would be expected (Crutzen and Fishman, 1977).

10-7.4.1.2 SO_2 Oxidation in Aqueous Solutions.

Uncatalyzed Oxidation by O_2. When SO_2 dissolves in water, it hydrolyzes to form HSO_3^- and SO_3^{2-}. Bisulfite (HSO_3^-) is the most common species in the pH range of atmospheric droplets (3 to 6), but SO_3^{2-} is much more readily oxidized by O_2 than is HSO_3^-; so it is primarily SO_3^{2-} that is oxidized. Lowering the pH of a solution decreases the hydrolysis of SO_2 and thereby greatly decreases its solubility. Lowering the pH also decreases the fraction of the dissolved SO_2 that is present as SO_3^{2-}; therefore the oxidation of SO_2 is minimal in strongly acid solutions. It is generally believed that the diffusion of SO_2 from the gas phase into the water droplets is fast compared to the subsequent oxidation so that oxidation is the rate-controlling step.

The exact mechanism for uncatalyzed oxidation of SO_2 in aqueous solution is still a matter of discussion. The most widely accepted mechanism is the free radical chain reaction, which was proposed by Backstrom (1934):

Chain initiation: $$SO_3^{2-} + M^+ \rightarrow SO_3^{-\cdot} + M\cdot \qquad (10.38)$$

$$SO_3^{-\cdot} + O_2 \rightarrow SO_5^{-\cdot} \qquad (10.39)$$

Chain propagation:

$$SO_5^{-\cdot} + HSO_3^- \rightarrow HSO_5^{-\cdot} + SO_3^- \quad (pH \leqslant 7) \qquad (10.40)$$

$$SO_5^{-\cdot} + SO_3^{2-} \rightarrow SO_5^{2-} + SO_3^{-\cdot} \quad (pH \geqslant 7) \qquad (10.41)$$

$$SO_3^{2-} + HSO_5^- \rightarrow HSO_4^- + SO_4^{2-} \quad (pH \leqslant 7) \qquad (10.42)$$

Oxidation: $$SO_3^{2-} + SO_5^{2-} \rightarrow 2SO_4^{2-} \quad (pH\ 7) \qquad (10.43)$$

$$SO_3^{-\cdot} + SO_5^{-\cdot} \rightarrow S_2O_6^{2-} + O_2 \qquad (10.44)$$

Termination:

$$SO_5^{-\cdot} + \text{inhibitor} \rightarrow \text{nonreactive products} \qquad (10.45)$$

The inhibitor can be any number of inorganic or organic compounds (Schroeter, 1966). The preceding equations, however, predict an oxygen dependence in acid solution that has not been observed experimentally. Also, the initiation of the chain by a metal ion is much the same as the catalysis of the chain by the metal ion. Larson, Horike, and Harrison (1978) proposed a more complex, modified Backstrom mechanism that does not require initiation by a metal ion and is not dependent on the oxygen concentration.

It is generally believed that the uncatalyzed oxidation of SO_2 by O_2 in aqueous solution is unimportant in the atmosphere. Bielke, Lamb, and Muller (1975) estimated that in a cloud with a liquid water content of 0.1 g/m^3 the SO_2 oxidation rate would decrease from 1.5%/h at a pH of 6 to 10^{-6} %/h at a pH of 3. Bielke and Gravenhorst (1978) conclude that uncatalyzed oxidation is unimportant below a pH of about 6. However, Hegg and Hobbs (1978) state that uncatalyzed oxidation by O_2 or O_3 is competitive with catalyzed oxidation for $3 \leqslant pH \leqslant 4.5$ in the rural case and for $3 \leqslant pH \leqslant 5$ in the urban case.

Uncatalyzed Oxidation by Strong Oxidizing Agents.

Ozone. Penkett, Jones, and Eggleton (1975) and Larson, Horike, and Harrison (1978) found that the uncatalyzed oxidation of sulfite was first order with respect to O_3 if the sulfite was kept in excess. The rate of oxidation of SO_2 by O_3 increases with pH but not so rapidly as does the rate of oxidation by O_2 (Penkett, Jones, and Eggleton, 1975; Scott and Hobbs, 1967; Larson, Horike, and Harrison, 1978). Hegg and Hobbs (1978) concluded that at an O_3 concentration of 50 ppm, oxidation by O_3 should exceed that for O_2 below a pH of 4.2 and therefore should be of importance in water droplets in the atmosphere. The temperature dependence of the oxidation of SO_2 by O_2 and O_3 is weak; so the relative importance of O_2 and O_3 oxidation should not vary with temperature. It should be mentioned, however, that Barrie and Georgii (1975) measured rates of oxidation by O_3 in distilled water at a pH of about 4 which were around 10^2 lower than those measured by Penkett (1972) or Penkett, Jones, and Eggleton (1975). This discrepancy must be cleared up before the importance of uncatalyzed oxidation by O_3 can be evaluated.

H_2O_2. Penkett, Jones, and Eggleton (1975) showed that the oxidation of SO_2 by H_2O_2 is very fast even in the pH ranges of 3 to 5. Model calculations by Levy (1973) indicate that the H_2O_2 concentration in the lower troposphere should be about 1 ppb. The oxidation of SO_2 by dissolved H_2O_2 could be an important reaction in the atmosphere if H_2O_2 concentrations are at this level. However, actual measurements of H_2O_2 concentrations in clean and polluted atmospheres need to be made before the importance of the reaction can be evaluated.

10-7.4.1.3 SO_2 Oxidation by O_2 in the Presence of Catalysts. The oxidation of SO_2 by O_2 in water solutions is catalyzed by transition metals of the fourth period, especially manganese and iron. Catalyzed oxidation may occur either in solutions of deliquesced aerosols with catalyst concentrations of 10^{-1} to $10^{-3}M$ or in fog or cloud droplets having catalyst concentrations of 10^{-4} to $10^{-8}M$. The discussion in the following section is limited to the latter. Brimblecombe and Spedding (1974) found that the rate of oxidation of S(IV) in a solution with a pH of 4 was 10^3 faster in the presence of $10^{-6} M$ Fe^{3+} than in the absence of catalysts. Most investigators have found that the rate of oxidation was proportional to the concentrations of both the metal ion and the S(IV) species used (e.g., Fuller and Crist, 1941; Karraker, 1963; Brimblecombe and Spedding, 1974) and was independent of the O_2 concentration.

Considerable disagreement exists concerning the relative efficiencies of the various catalysts. Junge and Ryan (1958) found that the order was $Mn^{2+} > Cu^{2+} > Fe^{3+}$, but Bracewell and Gall (1967) reported that the order was $Fe^{3+} > Mn^{2+} > Cu^{2+}$, and Barrie and Georgii (1976) reported that the order was $Mn^{2+} > Fe^{2+} \gg Fe^{3+}$. Bassett and Parker (1951) reported that the anion had a definite effect on the catalytic efficiency, which may explain to some extent the disagreement.

Barrie and Georgii (1976) found that the rate of oxidation in a $MnCl_2$ solution with a pH of 2 to 4.5 increased by a factor of 5 to 10 when the temperature was increased from 8 to 25°C but that the temperature dependence was largely eliminated when $FeCl_3$ or $FeCl_2$ was added to the solution. However, the oxidation rate in this mixed catalyst solution was about an order of magnitude greater than for the same molar concentration of either catalyst alone. This synergistic effect has also been observed by Hoather and Goodeve (1934), Penkett, Jones, and Eggleton (1975), and Lenelle (1975) but not by Johnstone and Coughanowr (1958).

Since a large number of inorganic and organic compounds have been found which inhibit the oxidation of SO_2 in aqueous solution, presumably by radical chain termination, it is not certain whether catalysis or inhibition will predominate in the atmosphere. Betz (1977) found that the rate of SO_2 oxidation in rainwater collected in the Frankfurt area was one to two orders of magnitude faster than the rate measured in catalyst-free water. However, Larson (1976) found that the oxidation rate in urban rainwater was about the same as that in clean water of the same pH (5.5) whereas the oxidation rate in rural rainwater was 30% slower than that in clean water. It therefore seems likely that the importance of catalyzed oxidation may depend sensitively on the composition of the cloud drops.

The mechanism of metal catalyzed SO_2 oxidation is still a matter of discussion. Most recent studies have proposed a free radical chain mechanism similar to that for the uncatalyzed reaction (Brimblecombe and Spedding, 1974; Barrie and Georgii, 1976) although other mechanisms have been proposed, such as the autocatalysis of metal sulfite or bisulfite complexes (Bassett and Parker, 1951; Prichett, 1960; Freiberg, 1975).

The mechanism initiating the chain reaction is uncertain, but the simplest mechanism would be the metal-ion oxidation of the sulfite in the Backstrom mechanism, with the metal ion being reoxidized by a chain radical, such as SO_5^-, or by another metal ion. Such a reaction could explain the synergism between Mn^{2+} and Fe^{3+} or Fe^{2+}, since both Fe^{3+} and Fe^{2+} will oxidize the manganous ion. However, several investigators have preferred to invoke the formation of a metal sulfite or bisulfite complex that acts as the initial radical producing agent (e.g., Barrie and Georgii, 1976).

10-7.4.1.4 SO_2 Oxidation on Particle Surfaces.

Carbon. Several investigators showed that SO_2 becomes oxidized on the surfaces of various types of particles. Novakov, Chang, and Harker (1974) and Barbaray, Contour, and Mouvier (1977) reported that SO_2 is adsorbed and oxidized by soot particles, which are basically carbon with a graphite-like structure. Tartarelli et al. (1978) and Novakov, Chang, and Harker (1974) reported that the oxidation increased with humidity, and Tartarelli et al. (1978) reported that it increased with the acidity of the particles. According to Barbaray, Contour, and Mouvier (1977), the SO_2 is adsorbed and oxidized at active sites of oxygen complexes on the carbon. Novakov, Chang, and Harker (1974) believe that the adsorption and oxidation of SO_2 by soot particles (which constitute about 50% of urban particulate emissions) play a major role in urban atmospheres. However, data on the rate of oxidation of SO_2 by soot particles in the atmosphere are needed before the importance of this process can be evaluated.

Metal Oxides. Metal oxide particles have also been shown to oxidize SO_2. Hulett et al. (1972) showed that SO_2 is oxidized by CaO, MgO, MnO_2, and Fe_2O_3 particles. Urone et al. (1968) reported that SO_2 is oxidized very rapidly by finely ground particles of Fe_2O_3 and less rapidly by Fe_3O_4, $CaCO_3$, Cr_2O_3, V_2O_5, PbO, PbO_2, CaO, and Al_2O_3. Urone et al. (1968) reported that at least for V_2O_5 the oxidation was not very sensitive to humidity. According to Chun and Quon (1973), the oxidation of SO_2 on the surfaces of nonhygroscopic particles is not a true catalytic reaction. After the active sites on the particles have become occupied by the products of SO_2 oxidation, they are no longer available for subsequent oxidation; so the metal oxides have a definite, limited capacity. Chun and Quon (1973) reported that the capacity of Fe_2O_3 was essentially independent of humidity and was too low to account for more than a small fraction of the sulfate found in urban air.

Metal Salts. The rate of oxidation of SO_2 by hygroscopic or deliquescent particles of metal compounds has been shown to increase with humidity. However, the oxidation by wettable particles at higher humidities is presumably due to catalysis by metal ions dissolved in the surface aqueous film rather than to the particles themselves. The rate of oxidation by deliquescent particles increases rapidly with humidity above the deliquescence point as the particle completely dissolves. Haury, Jordan, and Hoffman (1978), for example, reported that the oxidation rate and capacity of $MnSO_4$ increased slowly with relative humidity below 70% but increased rapidly at higher humidities. The oxidation capacity of the droplets of $MnSO_4$ solution was found to be four times that of the $MnSO_4$ crystals.

10-7.4.1.5 Stabilization of S(IV) in Metal Complexes. Hansen et al. (1974, 1976) have recently proposed that the formation of metal–sulfite complexes in atmospheric aerosols may stabilize sulfur in the +4 oxidation state and that these complexes may be responsible for respiratory irritation. More than 20 cations form metal–sulfite complexes (Sillen and Martell, 1964), but probably only those of copper and iron are of environmental significance (Dasgupta, Mitchell, and West, 1979). Smith et al. (1976) and Eatough et al. (1977, 1978) reported S(IV) in aerosol from lead and copper smelters and from coal- and oil-fired power plants which they believe is in the form of Fe^{3+}–sulfite complexes. Eatough et al. (1978) also believe that sulfite complexes of Zn^{2+}, Pb^{2+}, and organic complexes may also be present in the aerosol. Hilton et al. (1979) reported that they have generated Fe^{3+}–sulfite aerosols in the laboratory. However, Dasgupta, Mitchell, and West (1979) disputed the stability of the Fe^{3+}–sulfite complexes. They and others reported that Fe^{3+} and sulfite react to form Fe^{2+}, $S_2O_6^{2-}$, and/or sulfate (Bassett and Parker, 1951; Higginson and Marshall, 1957; Pollard, Hanson, and Nickless, 1961; Karraker, 1963). Dasgupta, Mitchell, and West concluded that the formation of stable metal complexes in the atmosphere is unlikely. The reason for this disagreement must be discovered before the environmental significance of these complexes can be assessed.

10-7.4.1.6 Summary of SO_2 Oxidation. Calvert et al. (1978) estimated that the rate of homogeneous gas-phase oxidation of SO_2 by O and OH in a stack plume should initially be about 3%/h but that the rate of oxidation by these species should decrease rapidly as the plume is diluted. However, if the plume is diluted with urban air, oxidation by the HO, HO_2, and CH_3O_2 produced by photochemical smog should cause the reaction rate to increase again to as high as about 5%/h. Eggleton and Cox (1978) estimated similar rates of oxidation of SO_2 by HO, HO_2, and RO_2 in urban air (2.2 ± 6.5%/h). These estimated rates of homogeneous gas-phase oxidation are comparable to the actual rates of SO_2 oxidation that have been measured in the urban environment. Wilson et al. (1976), for example, observed an increase in the rate of SO_2 oxidation in a plume from about 1.5%/h from 10 to 21 km downwind to 5%/h from 32 to 45 km downwind. Because of this correspondence between estimated gas-phase oxidation rates and measured oxidation rates, gas-phase oxidation is now generally believed to be an important, if not the most important, oxidation route for SO_2. However, measurements of the atmospheric concentrations of O, HO, HO_2, and RO_2 are needed before the importance of homogeneous gas-phase oxidation can be firmly established.

The catalyzed oxidation of SO_2 by transition metals such as iron and manganese in aqueous solution may be important in situations where the catalyst concentrations are high, such as in urban and stack plumes and urban fogs. However, more laboratory and field studies are needed to determine the rates of these reactions. In the atmosphere the rate of catalyzed oxidation may depend sensitively on the relative concentrations of catalysts and inhibitors; so laboratory measurements should replicate as closely as possible actual atmospheric conditions.

The relative importance of SO_2 oxidation on particle surfaces has not yet been determined. Experiments designed to measure the rates of oxidation by actual atmospheric aerosols at different humidities are required.

10-7.4.2 Global Sulfur Cycle. Several investigators have constructed global sulfur budgets to examine the importance of anthropogenic sulfur emissions. The budget of Kellogg et al. (1975) is shown in Fig. 10.13. The quantity of sulfur released during the decay of biological materials is the most uncertain component of these budgets. In the budgets proposed by Friend (1973), Kellogg et al. (1975), and Robinson and Robbins (1972), the biogenic inputs were postulated at levels that balanced the sulfur budget and had little scientific basis.

Biological decay is believed to emit sulfur primarily in the form of hydrogen sulfide (H_2S), although Rasmussen (1974) and Lovelock, Maggs, and Rasmussen (1972) recently indicated that dimethyl sulfide, methyl mercaptans, and dimethyl disulfide may be the primary forms emitted. Friend (1973) estimated that 58 Tg/yr of H_2S is released by the terrestrial biosphere and that 48 Tg/yr is released by the oceans. However, Ostlund and Alexander (1963) indicated that the escape of significant amounts of H_2S from the ocean is unlikely because H_2S should be rapidly oxidized by seawater. Slatt et al. (1978) concluded that H_2S production by biological decay in the surface microlayer could result in the release of amounts of H_2S compatible with the estimate of Friend (1973).

There has been considerable debate concerning the rate and mechanism of the oxidation of H_2S in the atmosphere. According to Slatt et al. (1978), the only reaction that has been quantitatively investigated and shown to be sufficiently rapid to account for the substantial removal of H_2S is the surface-catalyzed reaction with ozone (Cadle and Ledford, 1966):

$$H_2S + O_3 \rightarrow SO_2 + H_2O \qquad (10.46)$$

However, the reaction

$$H_2S + OH \rightarrow HS + H_2O \qquad (10.47)$$

may also be important.

Calculated atmospheric residence times for H_2S have ranged from 2 h (Cadle and Ledford, 1966) to 3.5 d (Friend, 1973). Slatt et al. (1978) concluded from measurements of atmospheric H_2S concentrations at several locations that typical atmospheric background H_2S concentrations are 0.045–0.140 ppb at terrestrial locations and 0.005–0.050 ppb at marine locations and estimated that the average global background concentration is 0.015–0.075 ppb.

Carbon disulfide and carbonyl sulfide are two other minor species present in the sulfur cycle. Sources and sinks for these two gases are not well known. The CS_2 may be of

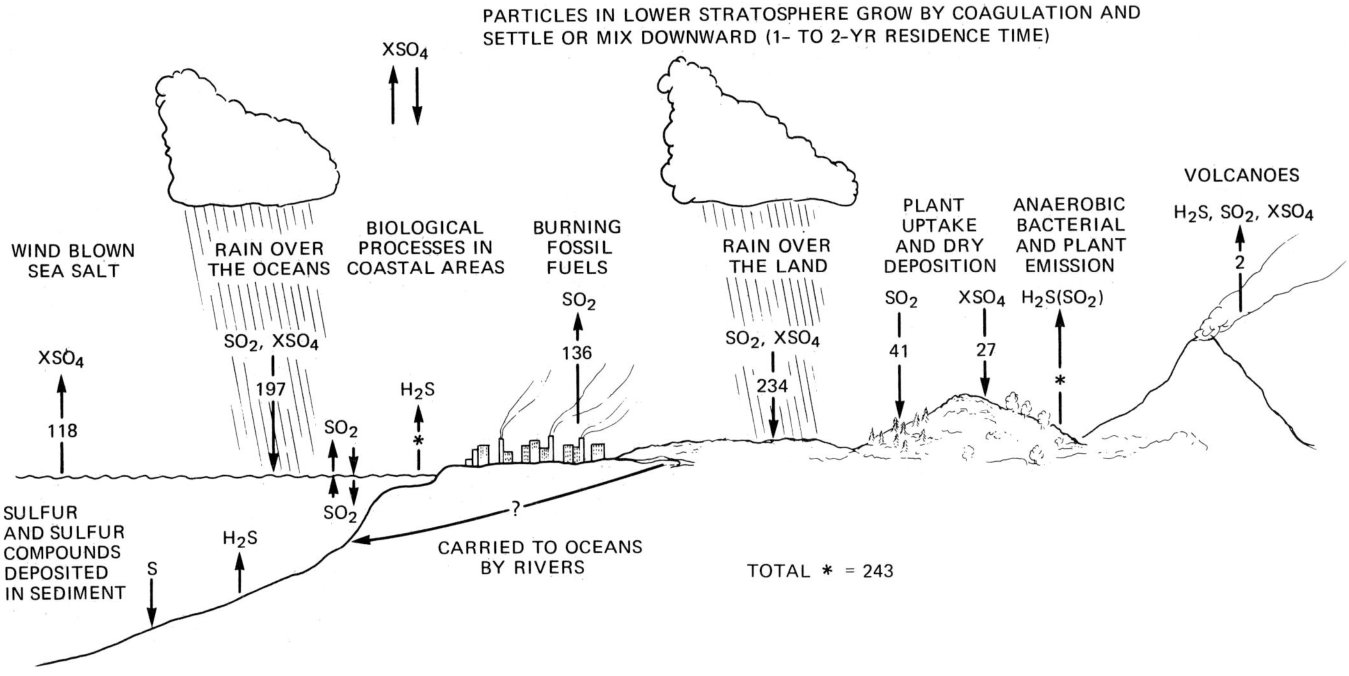

Fig. 10.13 Sources and sinks of atmospheric sulfur compounds. Units are 10^{12} g, calculated as sulfate per year. [From W. W. Kellogg, R. D. Cadle, E. R. Allen, A. L. Lazrus, and E. A. Martell, The Sulfur Cycle, *Science*, 175: 587-595 (1975).]

natural origin since Lovelock (1974) found uniform concentrations in Atlantic seawater from 50°N to 65°S and Sandalls and Penkett (1977) found that the troposphere and the ocean appeared to be near equilibrium.

Carbonyl sulfide may be produced by photolysis of CS_2 and/or from natural sources such as volcanoes, fumaroles, and biological processes (Stoiber et al., 1971; Crutzen, 1976b) and/or from industrial sources. Recently Crutzen (1976b) indicated that COS may be a significant source of sulfur for the stratospheric sulfate layer. Carbonyl sulfide is relatively inert in the troposphere but subject to oxidation in the stratosphere. One of the oxidation products is sulfate. Since the sulfate layer affects the earth's radiation balance and since there are indications that COS may be a by-product during fuel conversion (Gibson, Hammons, and Cameron, 1974) or during synthetic fuels production, Crutzen (1976b) suggests further examination be made to inventory COS sources and sinks.

10-7.4.3 Acid Rain. Acid rain has been recognized as possibly the major environmental problem caused by power plant emissions. The pH of precipitation in large parts of the eastern United States, southwestern Canada, and western Europe now has an annual average of 4 to 4.5, and the acidity is increasing. In the eastern United States and Europe, about two-thirds of the acidity is due to SO_2 emissions and one-third to NO_x, although the proportion of the acidity due to NO_x has been increasing in recent years (Ottar, 1978; Likens et al., 1979). In the western United States, however, two-thirds of the acidity is due to NO_x because the automobile contributes more to pollution than do power plants or heavy industry (Spectrum, 1979). The majority of the acidity due to sulfur results from the precipitation scavenging of sulfate, although a small but significant fraction results from absorbed SO_2 (Hales and Dana, 1979). The acidity of precipitation also depends on the availability of ammonia, which neutralizes the acidity due to nitrates and sulfates.

Greenland ice from snow that originated 180 yr ago has a pH of 6 to 7.6, indicating that precipitation was nearly neutral before the industrial revolution. The acidification of precipitation in Europe was first noticed in the 1950s in samples from the European Air Chemistry Network, which showed a central area of highly acid precipitation (pH 3 to 4) which was expanding year by year (Oden, 1968). Because of the concern aroused by these measurements, the Organization for Economic Cooperation and Development (OECD) approved a program in 1972 to investigate the Long Range Transport of Air Pollutants (LRTAP), with special emphasis on the acidification of precipitation. This study showed that acid precipitation has continued to spread until today it encompasses nearly all of northeastern Europe (Fig. 10.14). In northern Europe the sulfate concentration in precipitation rose markedly from 1955 to 1963 and has remained nearly constant since then. The concentration of nitrate in precipitation, however, has increased quite steadily at a rate of about 4%/yr since 1955 (Likens et al., 1979).

The changes in the acidity of precipitation that have occurred in the United States are much less certain because of the lack of continuous, high-density monitoring data. From 1959 to 1966, the U. S. Public Health Service and then the National Center for Atmospheric Research

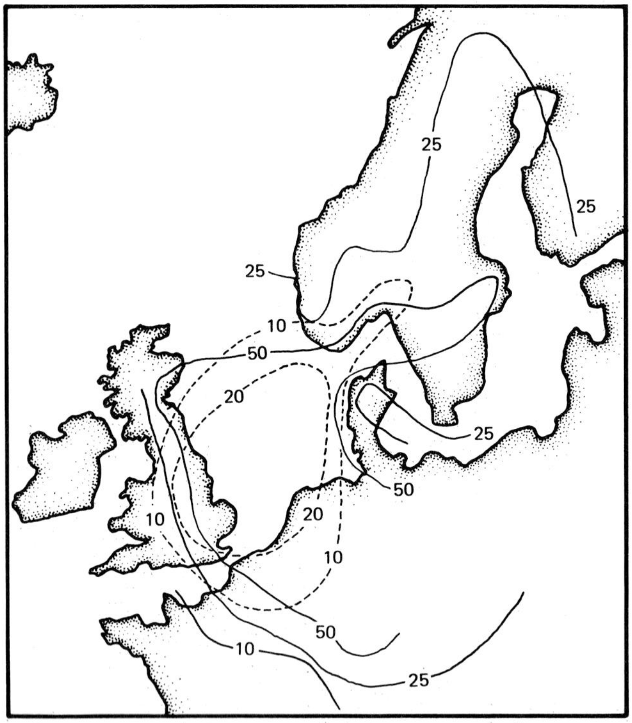

Fig. 10.14 Acidity of precipitation in northwestern Europe in microequivalents of hydrogen ions per liter in 1956 (dotted lines) and 1974 (solid lines.) [From G. E. Likens, R. W. Wright, J. N. Galloway, and T. J. Butler, Acid Rain, *Sci. Am.*, 241: 43-51 (1979).]

operated a precipitation chemistry network consisting of around 30 stations. This network showed that areas of acid precipitation were widespread in the eastern United States, with the most acid precipitation falling on New England, New York, and Pennsylvania. Precipitation chemistry measurements by several investigators show that the acidity of precipitation has increased since that time and that the area receiving acid precipitation has spread westward and southward (Fig. 10.15). The most rapid increase in precipitation acidity has been in the Southeast (Likens et al., 1979). Nitrate concentrations in precipitation have increased severalfold during the past decade or so. Sulfate concentrations increased by 60 to 65% between 1955 and 1965. The rate of increase slowed after 1966 but has recently increased again (Nisbet, 1975; Hales, 1978).

The long-range transport of sulfates and nitrates has resulted in acid rain far downwind of SO_2 emission sources. For example, it has been estimated that 77% of the acidity in precipitation in Sweden is due to sources outside of Sweden, mostly in Great Britain and western Europe (Likens et al., 1979).

Dry deposition of sulfur and nitrogen oxides also contributes to the acidification of soils and surface waters. The relative importance of wet and dry deposition is uncertain, however, because it is difficult to obtain precise measurements of dry deposition rates. Hales (1978) concluded from budget studies of several investigators that 10 to 50% of atmospheric sulfur is removed by dry deposition. However, the trajectory model of Eliassen (1978) indicates that about 65% of the sulfur deposition in Europe results from dry deposition. Fisher et al. (1978) also concluded from model calculations that dry deposition of sulfur is greater than wet deposition in Europe. Model calculations by McNaughton (1980) indicate that in August and October 1977 about 85% of the total sulfur deposition in the northeastern United States resulted from dry deposition. The rate of dry deposition of sulfur decreases more rapidly downwind of a source than does wet deposition because vertical mixing rapidly decreases ground-level sulfur concentrations, which determine dry deposition rates. Wet deposition depends on concentrations at higher elevations in the mixed layers. Sulfur dioxide also becomes converted to sulfate, which has about an order of magnitude lower deposition velocity than SO_2 but is more readily scavenged by precipitation than is SO_2.

The wet and dry depositions of sulfur and nitrogen oxides have lowered the pH of soils, lakes, and rivers over large areas, causing injury to plants and animals (Barrett and Brodin, 1955; Oden, 1968; Lazrus et al., 1974; Likens et al., 1979). Fresh waters with pH values below 5 are now found in southern Sweden, southern Norway, Great Britain,

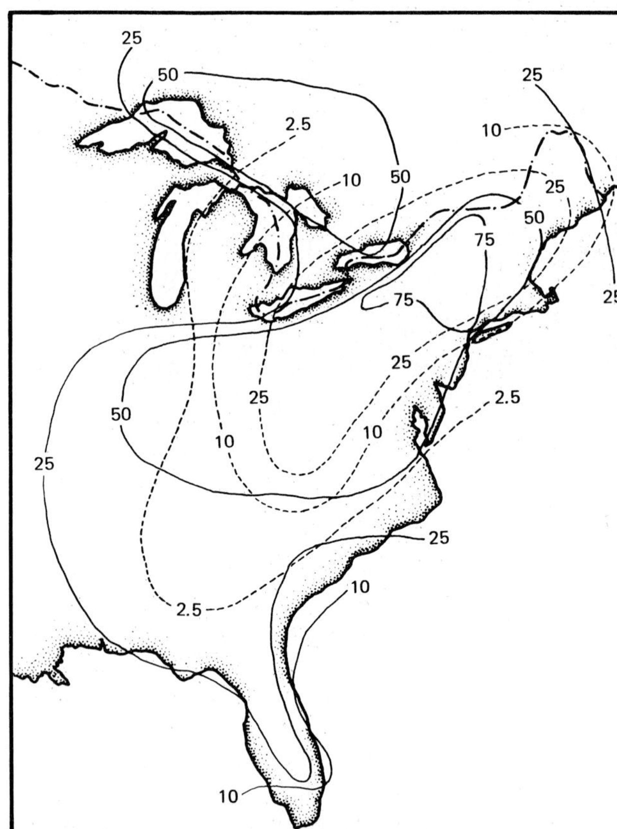

Fig. 10.15 Concentration of hydrogen ions in microequivalents per liter in precipitation in eastern North America in 1955–1956 (dotted lines) and 1975–1976 (solid lines.) [From G. E. Likens, R. W. Wright, J. N. Galloway, and T. J. Butler, Acid Rain, *Sci. Am.*, 241: 43-51 (1979).]

the Netherlands, East Germany, Nova Scotia, Ontario, northern New England, the Adirondacks, and Florida. Fish populations have been virtually wiped out or severely reduced in some of these regions. Freshwater ecosystems that are underlain by siliceous types of bedrock, such as granite, are especially sensitive to acid inputs because these rock types are resistant to weathering. Therefore the surface waters have low ion concentrations and poor ability to neutralize or buffer acid (Likens et al., 1979).

10-7.4.4 Large-Scale Measurement Programs. Because of the growing concern over the environmental effects of acid rain over considerable distances from emission sources, several large-scale programs have been or are being carried out to measure the concentration, transport, transformation, and deposition of sulfur and other pollutants. A major goal of these programs has been to develop and verify computer models that can be used to predict concentrations and wet and dry deposition rates from emission rates. Three of these programs are briefly discussed here.

LRTAP. During the Long Range Transport of Air Pollution (LRTAP), the atmospheric concentrations of SO_2 and sulfate and the acidity and sulfate concentrations in precipitations were measured on a daily basis at about 70 stations in Great Britain and western Europe from 1972 to 1975. Concentrations were also measured by aircraft during certain periods. Ammonium and nitrate ions were measured in some precipitation samples.

A number of investigators have compared the measured concentrations and deposition rates with those calculated with numerical models. Eliassen (1978) used a Lagrangian model to calculate the atmospheric concentrations and dry deposition rates of SO_2 and sulfate and the wet deposition rate of sulfate. In this model Eliassen assumed that sulfur was well mixed vertically through a 1000-m mixed layer. Horizontal dispersion was simulated by the calculation of trajectories every hour (every 2 h in one version of the model) for the SO_2 emitted from each grid point of a 127-km grid using the 850-mb winds. Sulfate and SO_2 concentrations along the trajectories were calculated with the following assumptions: (1) 0.05 of the emitted SO_2 is immediately oxidized to sulfate, and the remainder is oxidized at a rate of 3.5×10^{-6}/s, (2) SO_2 decays at a rate of 10^{-5}/s during dry periods and 4×10^{-5}/s during wet periods, and (3) sulfate decays at a rate of 4×10^{-6}/s during both wet and dry periods. It is assumed that the rate of conversion of SO_2 to sulfate increases during precipitation, causing the rate of SO_2 decay to increase and compensating for the loss of sulfate in precipitation. Dry deposition rates for SO_2 and sulfate were then calculated from the concentration fields by using a deposition velocity of 0.8 cm/s for SO_2 and 0.2 cm/s for sulfate. Wet deposition of sulfate was calculated from precipitation rates and an empirical relation relating atmospheric sulfate concentrations and sulfate concentrations in precipitations. Wet and dry deposition ratios calculated by Eliassen (1978) for sulfur for 1974 are shown in Figs. 10.16 and 10.17.

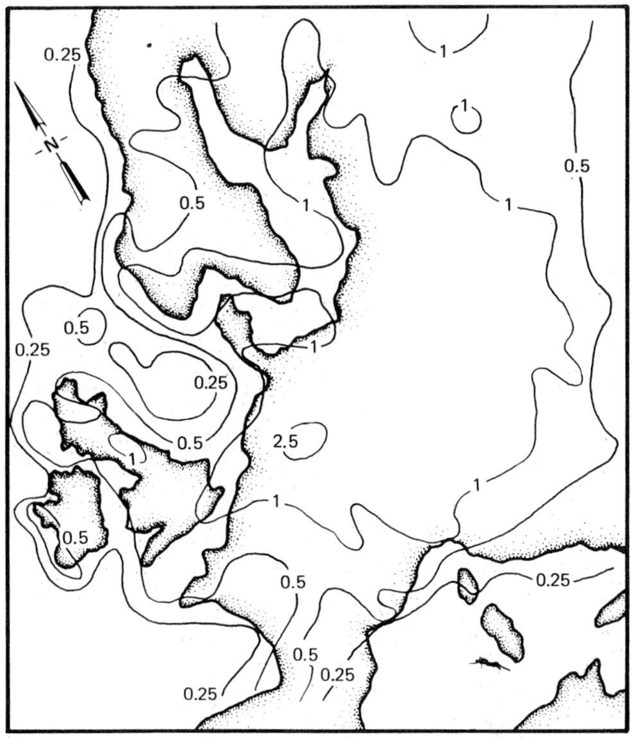

Fig. 10.16 Estimated wet deposition of sulfur (grams of sulfur per cubic meter) in 1974 in northwestern Europe. [From A. Eliassen, The OECD Study of Long Range Transport of Air Pollutants: Long Range Transport Modeling, *Atmos. Environ.*, 12: 479-487 (1978).]

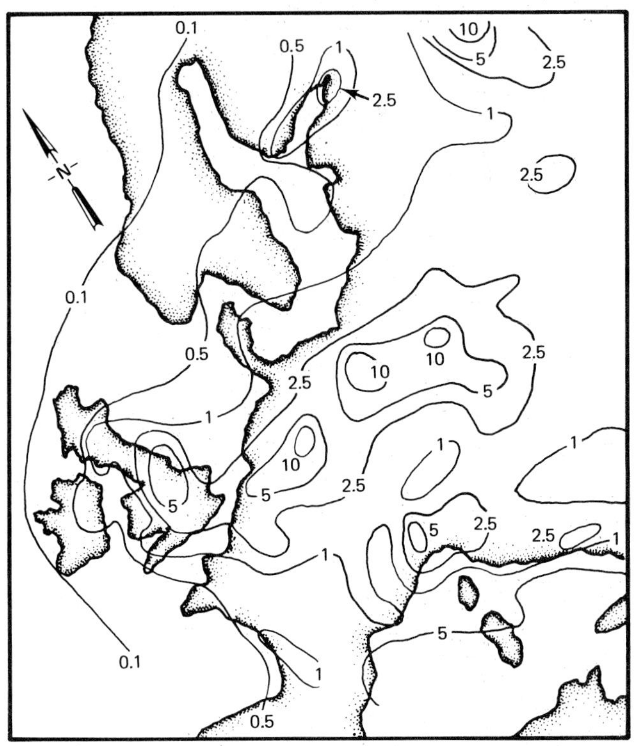

Fig. 10.17 Estimated total dry deposition of sulfur (grams of sulfur per cubic meter) in 1974 in northwestern Europe. [From A. Eliassen, The OECD Study of Long Range Transport of Air Pollutants: Long Range Transport Modeling, *Atmos. Environ.*, 12: 479-487 (1978).]

According to Eliassen this simple one-layer advection model cannot predict in detail the SO_2 and sulfate concentrations on specific days, but the calculated annual mean concentrations agree fairly well with observed values. Eliassen believes that the estimates of dry deposition should be accurate within ±40% and that the estimates of total deposition should be accurate within ±50%.

SURE. The Sulfate Regional Experiment (SURE) consists of an extensive program of measurements of atmospheric pollutants in the northeastern United States conducted from 1976 to 1980. Funded by the Electric Power Research Institute (EPRI), the program concentrated particularly on secondary pollutants, such as sulfate. The basic elements of the program are (1) ground-level measurements of SO_2, NO_x, O_3, sulfate, sulfite, nitrate, chloride, ammonium, trace elements, water-soluble organics, and particle mass at 54 stations in the northeastern United States (all these pollutants are measured at only nine of the stations); (2) measurements from airplanes; (3) construction of emission inventories for oxides of sulfur, oxides of nitrogen, total emitted particulate matter, and nonmethane hydrocarbons; and (4) development of a model to predict concentrations from emissions (Perhac, 1978).

MAP3S. The Multi-State Atmospheric Power Production Pollution Study (MAP3S) is a study of the concentrations, transport, transformations, and deposition of atmospheric, energy-related (AER) pollutants and is being conducted in the northeast quarter of the United States. Since coal is the most abundant fossil-fuel resource, this study focuses on the effects of emissions from coal-fired electric power plants, with particular emphasis on sulfur oxides. This study includes (1) the derivation of emission inventories for sulfur and nitrogen oxides, hydrocarbons, and particles; (2) the characterization of the physical and chemical properties of AER pollutants; (3) the determination of the spacial and temporal distribution of AER pollutants; (4) the determination of the rates of processes which determine the vertical and horizontal transport of AER pollutants; (5) the identification and determination of the rates of chemical and physical transformation processes of AER pollutants; (6) the determination of the rates of processes governing the wet and dry deposition of AER pollutants on the earth's surface and the determination of the effects of AER pollutants on precipitation chemistry, especially acid–base relationships; (7) the determination of the effects of AER pollutants on weather and climate; and (8) the development, verification, and demonstration of numerical models that will make possible the accurate assessment of atmospheric transport, transformation, and deposition of AER pollutants; and (9) the evaluation of various strategies for generating power while minimizing atmospheric pollution (MacCracken, 1978).

McNaughton (1980) used the Pacific Northwest Laboratory (PNL) regional model (Powell et al., 1978) to calculate atmospheric SO_2 and sulfate concentrations and sulfur depositions in the northeastern United States during August and October 1977. These results were then compared to the atmospheric concentrations measured by SURE and the sulfur depositions measured on the MAP3S precipitation chemistry network (Dana, 1979). In the PNL model, trajectories were calculated hourly from each of the 65 largest sulfur sources in the northeastern United States, using the average wind for the layer from 100 to 1000 m. It was estimated that these sources comprise about 65% of the total sulfur emissions. Diffusion, the oxidation of SO_2 to sulfate, and wet and dry deposition were calculated by using coupled mass conservation equations. Dry deposition velocities of 1 and 0.1 cm/s were used for SO_2 and sulfate, respectively. Wet removal of SO_2 and sulfate was given by $0.005\ P(t)$ and $0.232\ P(t)^{0.625}$ (Scott, 1978), respectively, where P(t) is the precipitation rate in millimeters per hour. A diurnal stability cycle was specified giving unstable conditions in the afternoon, neutral conditions in the morning and late afternoon, and stable conditions at night. Mixing height was given by a sinusoidally varying cycle representing the building from the surface of a stable nocturnal layer and a daytime mixed layer. The atmospheric SO_2 and sulfate concentrations and the sulfur depositions in precipitation calculated from the PNL model agreed fairly well with the measured values. Atmospheric SO_2 and sulfate concentrations from the model averaged about 70% of measured concentrations, and sulfur depositions in precipitation derived from the model averaged about 50% of measured depositions (at four stations). This is reasonable since the model simulations used only about 65% of the actual emissions. The correlation between individual atmospheric concentration measurements and concentrations derived from the model was about 0.6.

According to the model, about 98% of the total SO_2 deposition was by dry deposition, but only about 15% of the sulfate deposition was by dry deposition. The August precipitation generally removed more sulfate per event than did the October precipitation. The large August deposition values were associated with convective storms.

10-7.5 Other Trace Inorganic Volatile Species

Other minor inorganic volatile emissions have been noted from combustion sources. Mercury emissions have been measured by several authors, and mass balances indicate 80 to 100% of the mercury input in coal escapes emission control and is released primarily as elemental mercury vapor (Diehl et al., 1972; Billings et al., 1973; Kalb, 1975; Klein et al., 1975; Anderson and Smith, 1977).

Selenium has also been identified in flue gases. In a mass balance, Andren and Klein (1975) indicated that 10 to 20% of the selenium introduced in the coal escaped the emission control system (electrostatic precipitation). Of the selenium emitted, more than 90% was found in the vapor state as elemental selenium. The hazard posed by selenium to the environment is dependent on the environmental chemistry of the receiving soils (Lakin, 1973). In acidic soil the selenium is subject to oxidation to selenite and would be of low solubility and bioavailability. In alkaline soil areas the selenium burden from a combustion facility should be considered and evaluated since oxidation to the more toxic selenate may occur. The toxological effects of selenium and other inorganic materials are addressed in Chap. 17.

Halide species also are likely to escape dry emission control technology. Of primary concern are fluorides since they are toxic to biological systems. Atmospheric emissions of fluoride from combustion were estimated (*Fluorides*, 1971) at 1.5×10^{10} g in the United States in 1968, approximately the same as from the manufacture of aluminum or from open hearth furnaces. Recently Gladney et al. (1978) indicated boron release from coal-fired plants may also be substantial.

10-8 ABATEMENT TECHNOLOGY

Abatement technology has evolved to meet the required standards. Since the early part of the century, the evolution of improved electrostatic precipitators has increased particulate removal efficiencies until 99% removal can at the present time be achieved. Technology has also been developed for improved precipitation during combustion of lower grade, low sulfur, western coals.

The promulgation of the sulfur oxide standard has led to developments along serveral different pathways. One such development is the pretreatment of coal to remove sulfur physically or chemically, as exemplified by density separation of pyritic sulfur and solvent refining, respectively. Blending of coals to achieve a lower sulfur content and fluidized-bed combustion are other alternatives. Finally, treatment of flue gases after combustion is another option being undertaken. A number of wet and dry sulfur oxide removal techniques are in various stages of development. Venturi scrubbing, coupling fly ash removal and SO_x removal, is one option currently being applied at several power generation facilities. However, it has not proven reliable at all sites.

Two other sulfur oxide removal processes undergoing considerable use and development are the lime/limestone and double alkali processes. The lime/limestone process involves wet scrubbing the flue gas stream with a lime or limestone slurry. This alkaline slurry is effective in sulfur oxide removal; however, large quantities of an insoluble sludge are created, and significant abrasion, erosion, and scaling problems are often encountered. The dual or double alkali process uses a soluble alkali to effectuate SO_x removal from the flue gas. Generally, sodium or ammonium salts are used ($NaOH$, $NaCO_3$, $NaHCO_3$) which produce soluble products and thus decrease the erosion, abrasion, and scaling encountered during the lime/limestone process. The alkali sorbent can then be regenerated by treatment with a calcium salt which precipitates the calcium sulfite and calcium sulfate for disposal.

Disposal of the wastes generated by a wet scrubbing system and reliability pose the largest problems at present. The long-term environmental significance of scrubber wastes is uncertain. Also, there are indications that the Venturi system may be less efficient than dry electrostatic precipitation for removing small particles. In fact, small particulates appear to be generated by aerosol evaporation as the combustion gases pass through the Venturi system. Since these generated particulates are high in volatile and hazardous trace elements and are in the respirable range (<3 μm), they may remain suspended in the atmosphere for longer than suspension times for particles that were removed.

The potential problems associated with fly ash or sludge disposal are just beginning to be recognized. Fly ash has been considered an inert nontoxic medium, and disposal is accomplished by burial at the mine site in the many recent mine-mouth generating stations in the West. However, the discovery that surface enrichment of many volatile and potentially toxic trace elements occurs, coupled with the apparent leachability of the elements on aqueous contact, raises serious questions concerning groundwater contamination after disposal. The long-term leaching characteristics of sludge generated by wet chemical cleanup, the actual leaching into the aqueous phase from the fly ash during contact in Venturi systems, and the solubility of gas-phase species in scrubbing liquors are uncertain. Detailed examination needs to be made of trace-element behavior during wet cleanup to ensure that selective solubilization, if it occurs, does not pose any environmental problems. Extrapolating the ash formation rate of 13 metric tons/h for a 310-MW combustion facility over the lifetime of the facility, approximately 30 yr, indicates that 3.4×10^6 metric tons of ash will be generated and disposed of. The behavior of this reservoir of inorganic species *must* be characterized under the various potential environmental conditions that might be encountered.

If a constituent is present at a concentration of 100 ppm in the fly ash, all of which is readily leachable, and the maximum permissible concentration in water were 1 mg/L, then 3.4×10^8 m^3 of H_2O could be contaminated. Further, if an aquifer depth of 10 m is assumed, this aquifer could be contaminated over 34 km^2. Although this set of circumstances is not likely, it indicates the magnitude of the potential problem.

Preliminary research has indicated that most fly ash is alkaline in nature with an equilibrium pH on aqueous contact of greater than 11. At high pH Theis and Wirth (1977) observed a general decrease in leaching on aqueous contact as pH increased for As, Cd, Pb, Cr, Ni, Cu, and Zn. At pH above 12, arsenic solubility increased, attributable to unavailability of metal ions for precipitation of arsenate. Dreesen et al. (1977) reported similar behavior for metal species but observed that species generally anionic in character such as Mo, Fe, Se, B, and As were more readily leachable at higher pH values.

10-9 ENVIRONMENTAL IMPACT OF TRACE INORGANIC EMISSIONS

Fortunately, there appears to be no significant or drastic environmental degradation associated with the inorganic emissions from coal combustion over the short term. One must therefore assess the significance of the long-term or chronic exposure (see Chap. 17) to low levels of trace species. The overall mobilization of the trace elements via combustion has been shown to be similar to that accomplished by natural weathering for many of the trace elements (Klein et al., 1975).

Bertine and Goldberg (1971) compared mobilization via combustion of fossil fuels to that occurring naturally via geochemical weathering. They extrapolated average coal composition to worldwide combustion figures. Their findings indicated that mobilization via combustion was comparable to natural weathering for many elements (i.e., within two orders of magnitude). Natural weathering was estimated from riverine discharge and composition and also via oceanic sedimentation rate and sediment composition. These estimates make no allowances for relative volatilities and thus enrichment of volatile species in effluents from combustion. The authors speculate that under these additional considerations the anthropogenic mobilization of volatile elements may in fact be greater than mobilization produced naturally via weathering.

Establishing system-level effects in the environment attributable to these chronic exposures is difficult. In one watershed study Andren, Lindberg, and Bate (1975) showed that the ecosystem retained Pb, Cd, Cu, Zn, and Hg on a mass balance comparing aerosol input to stream outflow. These elements were also enriched in atmospheric aerosols presumably related to the several coal-fired facilities in the vicinity of the site. However, the fly ash contribution to the ecosystem trace-element levels via rain and aerosol deposition did not prove to be statistically significant.

Observation of environmental levels of trace inorganics in soils and/or biota in the environs of a coal-fired facility has likewise proven somewhat inconclusive. Horton and Dorsett (1976) observed elevation of only mercury in soils and scandium in plants near a facility in South Carolina, whereas Lyon (1977) could observe no enrichments in soils near the Allen Steam Plant in Tennessee. Klein and Russell (1973) observed elevation in very sandy soils of Ag, Cd, Cs, Cr, Cu, Fe, Hg, Ni, Ti, and Zn. They also observed elevation of Cd, Fe, Ni, and Zn in vegetation. Cannon and Swanson (1975) observed that soil levels of Hg, As, Sc, Te, Cd, and Pb decreased rapidly with distance from generation facilities at the Four Corners area (New Mexico–Arizona). Vegetation near the plant had not, however, accumulated significant quantities of the same elements.

Overall systems effects are difficult to predict. Vaughan et al. (1975) attempted to assess the long-term effects of a hypothetical facility and found under their parameter choices that Cd, Mo, Cu, and W might accumulate in vegetation. However, their assessment did not permit prediction for several trace elements of potential concern, including Ga, Ra, Te, Th, Tl, and V.

Also of interest regarding trace metal interactions at the system level is the work of Watson et al. (1976). They assessed the effects of smelter effluents, specifically Pb, Zn, Cd, and Cu, and found biological accumulation greatest in the forest floor litter. It also appeared that microbial decomposition of this litter was inhibited. This in turn decreased normal nutrient cycling processes. These types of effects are the most difficult to assess owing to the natural variability within environmental systems. However, the long-term significance of such environmental disturbances can be extremely important.

This points up the need for a great deal of research in terms of environmental significance. Van Hook and Shults (1977) point out specific areas where such research is needed.

10-10 FUTURE CONSIDERATIONS

The increased use of coal for combustion projected for the period until the turn of the next century, coupled with our present lack of adequate understanding regarding the long-term overall environmental significance of trace-element emissions or major inorganic emissions (NO_x, SO_x, CO_2), necessitates that such a scientific assessment and/or reassessment be undertaken.

This entails long-term investigation, and several programs such as SURE (Perhac, 1978), MISST (Wilson, 1978; Gillani, 1978; Gillani et al., 1978), MAP3S (MacCracken, 1978), and LRTAP (Ottar, 1976) have been designed in this direction. It is also imperative that the major constituents such as SO_x, NO_x, and CO_2 be assessed since these are producing documented visible environmental consequences at the present time as evidenced by acidification of precipitation and increasing atmospheric CO_2 levels. It is also apparent that current models and mass balances do not close for carbon. This indicates error in estimates in one or several of the cycling compartments and/or failure to take into account all the relevant interactions. Investigation of the long-term behavior, transport, and exchange of carbon in the ocean reservoir in both organic and inorganic phases is needed since it now appears that the oceanic medium is the long-term controlling agent of carbon dioxide content in the atmosphere. Also, these facets of the oceanic system are not well understood at present. The Department of Energy has instituted a program toward this end (Workshop, 1977).

With regard to minor element emission, there is a good deal of data available or becoming available regarding bulk concentration in fly ash. However, important environmental considerations not yet understood are the detailed chemical speciations or at least biological availability of these trace elements from the fly ash matrix. On bulk chemical consideration the reservoir of most trace species in soils is much greater than that being added via atmospheric influx of fly ash materials. However, bulk considerations are very deceptive if the trace elements in the injected fly ash are in a different chemical or physical form and consequently are more bioavailable and/or more biotoxic as may be the case.

Inorganic speciation may well be a variable factor dependent on such parameters associated with combustion as input coal composition and elemental distribution, combustion unit type and firing conditions, and, finally, emission control technique. If such is the case, there may be a methodology that will minimize environmental significance of inorganic emissions.

The prediction of environmental interactions and ultimate consequences, if any, is extremely difficult under conditions where trace inorganic species are being added at low levels over long periods of time. Ecologists and biologists have been further hampered by the lack of knowledge concerning the forms of added species. Thus a very high priority must be identification of discrete chemical species and long-term laboratory and field studies of the ecological behavior and fate of these specific combustion-related species.

The environmental significance of nonfugitive emissions, i.e., precipitated ash, bottom ash, process waters, and sludges from wet precipitation or flue gas cleanup, is also a major concern. Such emissions are of greater volume than the fly ash emitted to the atmosphere and also will not be dispersed over so wide an area upon disposal. Of definite concern are the species that are mobile at relatively alkaline pHs associated with aqueous contact for most fly ash.

REFERENCES

Abernathy, R. F., M. J. Peterson, and F. H. Gibson, 1969, *Spectrochemical Analysis of Coal Ash for Trace Elements*, Report of Investigation 7281, U. S. Department of the Interior, Bureau of Mines.

Adams, J. A. S., M. S. Mantovani, and L. L. Lundell, 1977, Wood Versus Fossil Fuel as a Source of Excess Carbon Dioxide in the Atmosphere: A Preliminary Report, *Science*, 196: 54-56.

Air Quality Criteria for Sulfur Oxides, 1967, U. S. Public Health Service, Department of HEW.

Air Quality and Stationary Source Emission Control, 1975, *National Research Council Report to U. S. Committee on Public Works, Series No. 944*, GPO, Washington.

Altshuller, A. P., S. L. Kopczynski, W. A. Lonneman, T. L. Becker, and R. Stater, 1967, Chemical Aspects of the Photooxidation of the Propylene–Nitrogen Oxide System, *Environ. Sci. Technol.*, 1: 899-914.

Alyea, F. N., D. M. Cunnold, and R. G. Prinn, 1978, Meteorological Constraints on Tropospheric Halocarbon and Nitrous Oxide Destructions by Siliceous Surfaces, *Atmos. Environ.*, 12: 1009-1011.

Anderson, W. L., and K. E. Smith, 1977, Dynamics of Mercury at Coal-Fired Power Plant and Adjacent Cooling Lake, *Environ. Sci. Technol.*, 11: 75-80.

Andren, A. W., and D. H. Klein, 1975, Selenium in Coal-Fired Steam Plant Emissions, *Environ. Sci. Technol.*, 9: 856-858.

——, S. E. Lindberg, and L. C. Bate, 1975, *Atmospheric Input and Geochemical Cycling of Selected Trace Elements in Walker Branch Watershed*, ERDA Report ORNL/NSF/EATC-13, Oak Ridge National Laboratory, NTIS.

Babu, S. P., 1975, Trace Elements in Fuel, *Adv. Chem. Ser.*, Volume 141, 216 pages.

Backstrom, H., 1934, The Chain Mechanism in the Autooxidation of Sodium Sulfite, *Z. Phys. Chem.*, B25: 99-121.

Baes, C. F., H. E. Goeller, J. S. Olsen, and R. M. Rotty, 1977, Carbon Dioxide and Climate: The Uncontrolled Experiment, *Am. Sci.*, 65: 310-327.

Ball, C. G., 1935, *Contributions to Study of Coal: Mineral Matter of No. 6 Bed Coal at West Frankfort, Franklin County, Illinois*, Investigation No. 33, Illinois State Geological Survey.

Barbaray, B., J. P. Contour, and G. Mouvier, 1977, Sulfur Dioxide Oxidation over Atmospheric Aerosol: X-Ray Photoelectron Spectra of Sulfur Dioxide Absorbed on V_2O_5 and Carbon, *Atmos. Environ.*, 11: 351-356.

Barrett, E. W., 1975, Inadvertent Weather and Climate Modification, *CRC Crit. Rev. Environ. Control*, 6: 15-90.

——, and G. Brodin, 1955, The Acidity of Scandanavian Precipitation, *Tellus*, 7: 251-255.

Barrie, L. A., and H. W. Georgii, 1975, An Experimental Investigation of the Absorption of Sulfur Dioxide by Cloud and Rain Drops Containing Heavy Metals, *Berichte des Institute fur Meteorologie der Universitat Frankfurt a.M.*, Nr. 28.

——, and H. W. Georgii, 1976, An Experimental Investigation of the Absorption of Sulfur Dioxide by Water Drops Containing Heavy Metal Ions, *Atmos. Environ.*, 10: 743-749.

Bassett, H., and W. Parker, 1951, The Oxidation of Sulfurous Acid, *J. Chem. Soc.*, 4: 1540-1560.

Bertine, K. K., and E. D. Goldberg, 1971, Fossil Fuel Combustion and the Major Sedimentary Cycle, *Science*, 173: 233-235.

Betz, M., 1977, *Untersucken uber die Absorption und Oxidation von Schwefel Dioxid in Naturlichem Regenwasser*, Diplomarbeit Universitat, Frankfurt.

Bidwell, R. D., and D. E. Frazer, 1972, Carbon Monoxide Uptake and Metabolism by Leaves, *Can. J. Bot.*, 50: 1435-1439.

Bielke, S., and G. Gravenhorst, 1978, Heterogeneous SO_2-Oxidation in the Droplet Phase, *Atmos. Environ.*, 12: 231-239.

——, D. Lamb, and J. Muller, 1975, On the Uncatalyzed Oxidation of Atmospheric SO_2 by Oxygen in Aqueous Systems, *Atmos. Environ.*, 9: 1083-1090.

Bilger, R. W., 1978, The Effect of Admixing Fresh Emissions on the Photostationary State Relationship in Photochemical Smog, *Atmos. Environ.*, 12: 1109-1118.

Billings, C. E., A. M. Succo, R. W. Conoglio, and R. A. Harley, 1973, Mercury Balance on a Large Pulverized Coal-Fired Furnace, *J. Air Pollut. Control Assoc.*, 23: 773-777.

Block, C., and R. Dams, 1976, Study of Fly Ash Emission During Combustion of Coal, *Environ. Sci. Technol.*, 10: 1011-1017.

Bogen, J., 1973, Trace Elements in Atmospheric Aerosol in the Heidelberg Area: Measured by Instrumental Neutron Activation Analysis, *Atmos. Environ.*, 7: 1117-1125.

Bohn, H. L., 1978, On Organic Soil Carbon and CO_2, *Tellus*, 30: 472-475.

Bolin, B., 1977, Changes of Land Biota and Their Importance for the Carbon Cycle, *Science*, 196: 613-615.

——, and E. Arrhenius, 1977, Nitrogen—An Essential Life Factor and a Growing Environmental Hazard, *Ambio*, 6: 95-105.

Bracewell, J. M., and D. Gall, 1967, The Catalytic Oxidation of Sulfur Dioxide in Solution at Concentrations Occurring in Fog Droplets, in *Proceedings of the Symposium on the Physio-Chemical Transformation of Sulfur Compounds in the Atmosphere and the Formation of Acid Smogs*, Mainz, Germany.

Brar, S. S., D. M. Nelson, J. R. Kline, P. F. Gustafson, E. L. Kanabrocki, and C. E. Moore, 1970, Instrumental Analysis for Trace Elements Present in Chicago Area Surface Air, *J. Geophys. Res.*, 75: 2939-2945.

Breeding, B. J., et al, 1976, Measurements of Atmospheric Pollutants in the St. Louis Area, *Atmos. Environ.*, 10: 181-194.

Brewer, P. G., 1978, Direct Observation of the Oceanic CO_2 Increase, *Geophys. Res. Lett.*, 5: 997-1000.

Brimblecombe, P., and D. J. Spedding, 1974, The Catalytic Oxidation of Micromolar Aqueous Sulfur Dioxide—I (Oxidation of Dilute Solutions of Iron(III)), *Atmos. Environ.*, 8: 937-945.

Broecker, W. S., 1963, Radioisotopes and Large Scale Oceanic Mixing, *The Sea*, 2: 88-108.

——, 1975, Climatic Change: Are We on the Brink of a Pronounced Global Warming, *Science*, 198: 460-463.

——, Yuan-Hui Li, and Tsung-Hung Peng, 1971, Carbon Dioxide—Man's Unseen Artifact, in *Impingement of Man on the Oceans*, pp. 287-324, Wiley-Interscience, Inc., New York.

Buat-Menard, P., and M. Arnold, 1978, The Heavy Metal Chemistry of Atmospheric Particular Matter Emitted by Mount Etna Volcano, *Geophys. Res. Lett.*, 5: 245-248.

Cadle, R. D., and M. Ledford, 1966, Reaction of Ozone with Hydrogen Sulfide, *Int. J. Air Water Pollut.*, 10: 25-30.

Callendar, G. S., 1958, On the Amount of Carbon Dioxide in the Atmosphere, *Tellus*, 10: 243-248.

Calvert, J. G., 1976, Test of the Theory of Ozone Generation in Los Angeles Atmosphere, *Environ. Sci. Technol.*, 10: 248-256.

——, and R. D. McQuigg, 1975, The Computer Simulation of the Rates and Mechanisms of Photochemical Smog Formation, *Int. J. Chem. Kinet.*, 1: 113-154.

——, Fu Su, J. W. Bottenheim, and O. P. Strausz, 1978, Mechanism of Homogeneous Oxidation of Sulfur Dioxide in the Troposphere, *Atmos. Environ.*, 12: 197-226.

Campbell, J. A., J. C. Laul, K. K. Nielson, and R. D. Smith, 1978a, A Separation and Chemical Characterization of Finely Sized Flyash Particles, *Anal. Chem.*, 50: 1032-1040.

——, R. D. Smith, K. K. Nielson, and K. H. Abel, 1978b, *Analysis of Submicron Flyash Particles*, paper presented at the American Chemical Society National Meeting, Miami Beach, Fla.

——, R. D. Smith, and L. E. Davis, 1978, Application of X-Ray Photoelectron Spectroscopy to the Study of Flyash, *Appl. Spectrosc.*, 32: 316-319.

——, R. D. Smith, L. E. Davis, and K. L. Smith, 1979, Characterization of Micron-Sized Flyash Particles by X-Ray Photoelectron Spectroscopy (ESCA), *Sci. Tot. Environ.*, 12: 75-85.

Cannon, H. L., and V. E. Swanson, 1975, *Contributions of Major and Minor Elements to Soils and Vegetation by the Coal-Fired Four Corners Power Plant, San Juan, New Mexico*, U. S. Geological Survey Open File Report No. 75-170, U. S. Geological Survey, Denver, Colo.

Chameides, W. L., and D. H. Stedman, 1977, Tropospheric Ozone: Coupling Transport and Photochemistry, *J. Geophys. Res.*, 82: 1787-1793.

——, and J. C. G. Walker, 1973, A Photochemical Theory of Tropospheric Ozone, *J. Geophys. Res.*, 78: 8751-8760.

——, and J. C. G. Walker, 1976, A Time-Dependent Photochemical Model for Ozone Near the Ground, *J. Geophys. Res.*, 81: 413-420.

Chapelle, E. W., 1962, Carbon Monoxide Oxidation by Algae, *Biochem. Biophys. Acta*, 62: 45-62.

Chen, G., and F. J. Millero, 1979, Gradual Increase in Oceanic CO_2, *Nature*, 277: 205-206.

CHESS (Community Health and Environmental Surveillance System), 1974, *Health Consequences of Sulfur Dioxide: A Report from CHESS, 1970–1971*, Report EPA-650/1-74-004, National Environmental Research Center, Research Triangle Park, N. C.

Chun, K. C., and J. E. Quon, 1973, Capacity of Ferric Oxide Particles to Oxidize Sulfur Dioxide in Air, *Environ. Sci. Technol.*, 7: 532-538.

Cogbill, C. V., and G. E. Likens, 1974, Acid Precipitation in the Northeastern United States, *Water Resour. Res.*, 10: 1133-1137.

Cohen, Y., and L. I. Gordon, 1978, Nitrous Oxide in the Oxygen Minimum of the Eastern Tropical North Pacific: Evidence for Its Consumption During Denitrification and Possible Mechanisms for Its Production, *Deep-Sea Res.*, 25: 509-524.

——, and L. I. Gordon, 1979, Nitrous Oxide Production in the Ocean, *J. Geophys. Res.*, 84: 347-353.

Craig, H., R. F. Weiss, and W. L. Dowd, 1976, paper presented at the American Geophysical Union Symposium on the Terrestrial Nitrogen Cycle and Possible Atmospheric Effects.

Crutzen, P. J., 1973, Gas-Phase Nitrogen and Methane Chemistry in the Atmosphere, in *Physics and Chemistry of the Upper Atmosphere*, B. M. McCormac and D. Reidel (Eds.), Dordrect, Netherlands.

——, 1976a, Upper Limits on Atmospheric Ozone Reductions Following Increased Application of Fixed Nitrogen to the Soil, *Geophys. Res. Lett.*, 3: 169-172.

——, 1976b, The Possible Importance of CSO for the Sulfate Layer in the Stratosphere, *Geophys. Res. Lett.*, 3: 73-76.

——, and D. H. Ehhalt, 1977, Effects of Nitrogen Fertilizers and Combustion on the Stratospheric Ozone Layer, *Ambio*, 6: 112-117.

——, and J. Fishman, 1977, Average Concentrations of OH in the Troposphere, and the Budgets of CH_4, CO, H_2, and CH_3CCl_3, *Geophys. Res. Lett.*, 4: 321-324.

Dana, M. T. (Ed.), 1979, *The MAP3S Precipitation Chemistry Network: Second Periodic Summary Report (July 1977–June 1978)*, DOE Report PNL-2829, Battelle Pacific Northwest Laboratory, NTIS.

Dasgupta, D. K., P. A. Mitchell, and P. W. West, 1979, Study of Transition Metal Ion–S(IV) Systems, *Atmos. Environ.*, 13: 775-782.

Davis, D. D., W. Heaps, D. Philen, and T. McGee, 1979, Boundary Layer Measurements of the OH Radical in the Vicinity of an Isolated Power Plant Plume: SO_2 and NO_2 Chemical Conversion Times, *Atmos. Environ.*, 13: 1197-1203.

——, G. Smith, and G. Klauber, 1974, Trace Gas Analysis of Power Plant Plumes via Aircraft Measurements: O_3, NO_X and SO_2 Chemistry, *Science*, 186: 773.

Davison, R. L., D. F. S. Natusch, J. R. Wallace, and C. A. Evans, 1974, Trace Elements in Flyash: Dependence of Concentration on Particle Size, *Environ. Sci. Technol.*, 8: 1107-1113.

Demerjian, K. L., J. A. Kerr, and J. G. Calvert, 1974, The Mechanism of Photochemical Smog Formation, *Adv. Environ. Sci. Technol.*, 5: 443-444.

Diehl, R. C., E. A. Haltman, H. Schultz, and R. J. Haren, 1972, *Fate of Mercury in the Combustion of Coal*, U. S. Department of the Interior, Bureau of Mines Managing Coal Wastes & Pollution Program, Technical Progress Report 54.

Dimitriades, B., 1977, Oxidant Control Strategies. Part 1. Urban Oxidant Control Strategy Derived from Existing Smog Chamber Data, *Environ. Sci. Technol.*, 11: 81-88.

Dreesen, D. R., E. S. Gladney, J. W. Owens, B. L. Perkins, C. L. Wienke, and L. E. Wangen, 1977, Comparison of Levels of Trace Elements Extracted from Flyash and Levels Found in Effluent Waters from Coal-Fired Power Plant, *Environ. Sci. Technol.*, 11: 1017-1019.

Duce, R. A., G. L. Hoffman, and W. H. Zoller, 1975, Atmospheric Trace Metals at Remote Northern and Southern Hemisphere Sites: Pollution or Natural? *Science*, 187: 59-61.

Eaton, J. S., G. E. Likens, and F. H. Bormann, 1973, Throughfall and Stemflow Chemistry in a Northern Hardwood Forest, *J. Ecol.*, 61: 495-508.

Eatough, D. J., L. D. Hansen, R. M. Izatt, and N. F. Mangelson, 1977, Analysis of Sulfite Species in Aerosols, *ACS Abstracts, Division of Environmental Chemistry*, 173rd Annual Meeting, New Orleans, pp. 57-60.

——, et al., 1978, The Formation and Stability of Sulfite Species in Aerosols, *Atmos. Environ.*, 12: 263-272.

Eggleton, A. E. J., and R. A. Cox, 1978, Homogeneous Oxidation of Sulfur Compounds in the Atmosphere, *Atmos. Environ.*, 12: 225-230.

Ehhalt, D. H., L. E. Heidt, R. H. Lueb, and N. Roper, 1974, Vertical Profiles of CH_4, H_2, CO, N_2O, and CO_2 in the Stratosphere, in

Proceedings of the Third Conference on the Climatic Impact Assessment Program, Report DOT-TSC-OST-74-15, pp. 153-160, U. S. Department of Transportation.

Electrical World, 1977, *28th Annual Electrical Industry Forecast*, pp. 48-58.

Eliassen, A., 1978, The OECD Study of Long Range Transport of Air Pollutants: Long Range Transport Modelling, *Atmos. Environ.*, 12: 479-487.

Eschenroeder, A. Q., J. R. Martinez, and R. A. Nordick, 1972, *Evaluation of a Diffusion Model for Photochemical Smog Simulation*, General Research Corporation, Santa Barbara, Calif., Final Report, EPA Contract No. 68-02-0336.

Farmer, J. G., 1979, Problems in Interpreting $\delta^{13}C$ Tree-Ring Records, *Nature*, 279: 229-231.

Filby, R. H., K. R. Shah, and C. A. Sawtter, 1977, A Study of Trace Element Distribution in a Coal Conversion Process Using Neutron Activation Analysis, in *Proceedings of the 1976 International Conference on Modern Trends in Activation Analysis*, pp. 664-674, International Atomic Energy Agency, Vienna.

Fisher, G. L., et al., 1978, Physical and Morphological Studies of Size-Classified Coal Flyash, *Environ. Sci. Technol.*, 12: 447-451.

Flagan, R. C., 1977, *Ash Particle Formation in Pulverized Coal Combustion*, presented at Combustion Institute Western State Section, Apr. 18-19, Pittsburgh, Pa.

——, and S. K. Friedlander, 1976, Particle Formation in Pulverized Coal Combustion—A Review, in *Proceedings of Symposium on Aerosol Science and Technology 82nd National Meeting, American Institute of Chemical Engineering*, Aug. 29–Sept. 1, Atlantic City, N. J.

Fluorides, 1971, National Academy of Sciences Committee on Biologic Effects of Atmospheric Pollutants, Washington.

Folger, D. W., 1970, Wind Transport of Land-Derived Mineral, Biogenic and Industrial Matter over the North Atlantic, *Deep-Sea Res.*, 17: 337-352.

Forrest, J., and L. Newman, 1977, Further Studies on the Oxidation of Sulfur Dioxide in Coal-Fired Power Plant Plumes, *Atmos. Environ.*, 11: 465-474.

Freiberg, J., 1975, The Mechanism of Iron Catalyzed Oxidation of SO_2 in Oxygenated Solutions, *Atmos. Environ.*, 9: 661-672.

French, J., 1975, *Recent Epidemiological Studies with Respect to Nitrogen Oxide,* presented at the Scientific Seminar on Automotive Pollutants, Washington.

Friend, J. P., 1973, The Global Sulfur Cycle, in *Chemistry of the Lower Atmosphere*, Chap. 4, S. I. Rasool (Ed.), Plenum Publishing Corp., New York.

Fuller, E. C., and R. H. Crist, 1941, The Rate of Oxidation of Sulfite Ions by Oxygen, *J. Am. Chem. Soc.*, 63: 1644-1650.

Galloway, J. H., G. E. Likens, and E. S. Edgerton, 1976, Acid Precipitation in the Northeastern United States: pH and Acidity, *Science*, 194: 722-724.

Garland, J. A., 1978, Dry and Wet Removal of Sulphur from the Atmosphere, *Atmos. Environ.*, 12: 349-362.

Gauger, A. W., 1936, Coal and Its Mineral Matter with Reference to Carbonization, *Blast. Furn. Steel Plant*, 24(4,5,6): pp.310-312 (April), pp. 406-409 (May), pp. 508-510 (June).

Gibson, C. R., G. A. Hammons, and D. S. Cameron, 1974, Environmental Aspects of El Paso's Burnham I Coal Gasification Complex, in *Symposium on Environmental Aspects of Fuel Conversion Technology*, St. Louis, Mo., May 13, 1974, Report EPA-650/2-74-118, pp. 91-100.

Gillani, N. V., 1978, Project MISST: Mesoscale Plume Modelling of the Dispersion, Transportation and Ground Removal of SO_2, *Atmos. Environ.*, 12: 569-588.

——, R. B. Husar, J. D. Husar, D. E. Patterson, and W. E. Wilson, Jr., 1978, Kinetics of Particulate Sulfur Formation in a Power Plant Plume out to 300 km, *Atmos. Environ.*, 12: 589-599.

Gladney, E. S., 1974, *Trace Element Enrichment from Coal-Fired Plants: A Study of the Chalk Point Electric Generating Station*, Ph.D. Thesis, University of Maryland, College Park, Md.

——, J. A. Small, G. E. Gordon, and W. H. Zoller, 1976, Composition and Size Distribution of In-Stack Particulate Material at a Coal-Fired Power Plant, *Atmos. Environ.*, 10: 1071-1077.

——, L. E. Wangen, D. B. Curtis, and E. T. Jurney, 1978, Observation on Boron Release from Coal-Fired Power Plants, *Environ. Sci. Technol.*, 12: 1085-1086.

——, W. H. Zoller, A. G. Jones, and G. E. Gordon, 1974, Composition and Size Distribution of Atmospheric Particulate Matter in Boston Area, *Environ. Sci. Technol.*, 8: 551-557.

Glasson, W. A., and C. S. Tuesday, 1970, Inhibition of Atmospheric Photo-Oxidation of Hydrocarbons by Nitric Oxide, *Environ. Sci. Technol.*, 4: 37.

Gluskoter, H. J., 1967, Clay Minerals in Illinois Coals, *J. Sediment. Petrol.*, 37: 205-214.

——, 1975, Mineral Matter and Trace Elements in Coal, *Adv. Chem. Ser.*, 141: 1-22.

——, and P. C. Lindahl, 1973, Cadmium Mode of Occurrence in Illinois Coals, *Science*, 101: 264-266.

——, and R. R. Ruch, 1971, Chlorine and Sodium in Illinois Coals as Determined by Neutron Activation Analysis, *Fuel*, 50: 65-76.

Goldschmidt, V. M., 1937, Rare Elements in Coal Ashes, *Ind. Eng. Chem.*, 27: 1100-1102.

Gordon, G. E., W. H. Zoller, and E. S. Gladney, 1973, Abnormally Enriched Trace Elements in the Atmosphere, in *Proceedings of 7th Conference on Trace Substances in Environmental Health*, pp. 167-174, University of Missouri, Columbia, Mo.

Haagen-Smit, A. J., and L. G. Wayne, 1976, Atmospheric Reactions and Scavenging Processes, *Air Pollut.*, 1: 235-288.

Hahn, J., 1974a, The North Atlantic Ocean as a Source of Atmospheric N_2O, *Tellus*, 26: 160-168.

——, 1974b, Nitrous Oxide in Air and Seawater over the Atlantic Ocean, in *The Changing Chemistry of the Oceans*, D. Dryssen and D. Jagner (Eds.), pp. 53-69, John Wiley & Sons, Inc., New York.

——, 1975, N_2O Measurements in the North Atlantic, *"Meteor" Forschungsergebnisse, Reihe A*, 16: 1-14.

——, and C. Junge, 1977, Atmospheric Nitrous Oxide: A Critical Review, *Z. Naturforsch.*, 32a: 190-214.

Hales, J. M., 1978, Wet Removal of Sulfur Compounds from the Atmosphere, *Atmos. Environ.*, 12: 389-399.

——, and M. T. Dana, 1979, Regional-Scale Deposition of Sulfur Dioxide by Precipitation Scavenging, *Atmos. Environ.*, 13: 1121-1132.

Hameed, S., J. P. Pinto, and R. W. Stewart, 1979, Sensitivity of the Predicted $CO-OH-CH_4$ Perturbation to Tropospheric NO_x Concentrations, *J. Geophys. Res.*, 84: 763-768.

Hansen, L. D., et al., 1974, Transition Metal—SO_3 Complexes: A Postulated Mechanism for Synergistic Effects of Aerosols and SO_2 on the Respiratory Tract, *Trace Substances in Environmental Health*, 8: 293-297, University of Missouri Press.

——, L. Whiting, D. J. Eatough, T. E. Jensen, and R. M. Izatt, 1976, Determination of Sulfur(IV) and Sulfate in Aerosols by Thermometric Methods, *Anal. Chem.*, 48: 634-638.

Hardy, R. F. W., and J. D. Havelka, 1975, Nitrogen Fixation Research: A Key to World Food, *Science*, 188: 633-643.

Harrison, P. R., and J. W. Winchester, 1971, Area-Wide Distribution of Lead, Copper, and Cadmium in Air Particulate from Chicago and Northwest Indiana, *Atmos. Environ.*, 5: 863-880.

Haury, G., S. Jordan, and C. Hoffman, 1978, Experimental Investigation of the Aerosol Catalyzed Oxidation of SO_2 Under Atmospheric Conditions, *Atmos. Environ.*, 12: 281-287.

Hecht, T. A., J. H. Seinfeld, and M. C. Dodge, 1974, Further Development of Generalized Kinetic Mechanism for Photochemical Smog, *Environ. Sci. Technol.*, 8: 327-339.

Hegg, D. A., and P. V. Hobbs, 1978, Oxidation of Sulfur Dioxide in Aqueous Systems with Particular Reference to the Atmosphere, *Atmos. Environ.*, 12: 241-253.

——, P.V. Hobbs, L. F. Rodke, and H. Harrison, 1977, Reactions of Ozone and Nitrogen Oxides in Power Plant Plumes, *Atmos. Environ.*, 11: 521-526.

Higginson, W. C. E., and J. W. Marshall, 1957, Equivalence Changes in Oxidation-Reduction Reactions in Solution: Some Aspects of the Oxidation of Sulphurous Acid, *J. Chem. Soc.*, 447-458.

Hilton, C. M., J. J. Christiansen, D. J. Eatough, and L. D. Hansen, 1979, Fe(III)–S(IV) Aerosol Generation and Characterization, *Atmos. Environ.*, 13: 601-605.

Hoather, R. C., and C. F. Goodeve, 1934, The Oxidation of Sulphurous Acid—III. Catalysis by Manganous Sulphate, *Trans. Faraday Soc.*, 30: 1149-1156.

Horton, J. H., and R. S. Dorsett, 1976, Effect of Stack Releases from a Coal-Fired Powerhouse on Minor and Trace Element Contents of Neighboring Soil and Vegetation, *Environmental Chemical and Cycling Process Symposium*, Augusta, Ga.

Horton, L., and K. V. Aubrey, 1950, The Distribution of Minor Elements in Vitrain: Three Vitrains from the Barnsley Seam, *J. Soc. Chem. Ind., London (Suppl. 1)*, pp. S-41-48.

Huebert, B. J., and A. L. Lazrus, 1978, Global Measurements of Nitric Acid Vapor and Particulate Nitrate, *Geophys. Res. Lett.*, 5: 577-580.

Hulett, L. D., T. A. Carlson, B. A. Fish, and J. L. Durham, 1972, Studies of Sulfur Compounds Adsorbed on Smoke Particles and Other Solids by XPS, in *Determination of Air Quality*, pp. 179-185, Plenum Press, New York.

Hutchinson, G. E., 1954, in *The Earth as a Planet*, p. 390, University of Chicago Press, Chicago.

Hutchinson, G. L., and A. R. Mosier, 1979, Nitrous Oxide Emissions from an Irrigated Cornfield, *Science*, 205: 115-117.

Inadvertent Climate Modification: Report of the Study of Man's Impact on Climate (SMIC), 1971, The MIT Press, Cambridge, Mass.

Ingersoll, R. R., R. E. Inman, and W. R. Fisher, 1974, Soil's Potential as a Sink for Atmospheric Carbon Monoxide, *Tellus*, 26: 151-158.

John, W., R. Kaifer, K. Rahn, and J. J. Wesolowski, 1973, Trace Element Concentrations in Aerosols from the San Francisco Bay Area, *Atmos. Environ.*, 7: 107-118.

Johnston, H. S., and G. S. Selwyn, 1975, New Cross Sections for the Absorption of Near Ultraviolet Radiation by Nitrous Oxide, *Geophys. Res. Lett.*, 2: 549-551.

Johnstone, H. F., and D. R. Coughanowr, 1958, Absorption of Sulfur Dioxide from Air, *Ind. Eng. Chem.*, 50: 1169-1172.

Junge, C. E., 1974, Residence Time and Variability of Tropospheric Trace Gases, *Tellus*, 26: 477-488.

——, B. Buckholt, K. Schutz, and R. Beck, 1971, N_2O Measurements in Air and Seawater over the Atlantic, *"Meteor" Forschungsergebnisse, Reihe B*, 6: 1-11.

——, and T. G. Ryan, 1958, Study of SO_2 Oxidation in Solution and Its Role in Atmospheric Chemistry, *Q. J. R. Meteorol. Soc.*, 84: 46-55.

Kaakinen, J. W., R. M. Jorden, M. H. Lawasani, and R. E. West, 1975, Trace Element Behavior in Coal-Fired Power Plant, *Environ. Sci. Technol.*, 9: 862-869.

Kalb, G. W., 1975, Total Mercury Mass Balance at a Coal-Fired Power Plant, *Adv. Chem. Ser.*, 141: 154-174.

Karraker, D. G., 1963, The Kinetics of the Reaction Between Sulfurous Acid and Ferric Ion, *J. Phys. Chem.*, 67: 871-874.

Keeling, C. D., 1973, Industrial Production of Carbon Dioxide from Fossil Fuels and Limestone, *Tellus*, 25: 174-190.

——, et al., 1976a, Atmospheric Carbon Dioxide Variations at Mauna Loa Observatory, Hawaii, *Tellus*, 28: 538-551.

——, J. A. Adams, Jr., C. A. Ekdahl, Jr., and P. R. Guenther, 1976b, Atmospheric Carbon Dioxide Variations at the South Pole, *Tellus*, 28: 552-564.

Keifer, W. S., G. M. Klauber, and D. D. Davis, 1977, paper presented at the 174th National Meeting of the American Chemical Society, Division of Environmental Chemistry, Chicago.

Kellogg, W. W., R. D. Cadle, E. R. Allen, A. L. Lazrus, and E. A. Martell, 1975, The Sulfur Cycle, *Science*, 175: 587-595.

Kemezys, M., and G. H. Taylor, 1964, Occurrence and Distribution of Minerals in Some Australian Coals, *J. Inst. Fuel*, 37: 389-397.

Klein, D. H., et al., 1975, Pathways of Thirty-Seven Trace Elements Through a Coal-Fired Power Plant, *Environ. Sci. Technol.*, 9: 973-979.

——, and P. Russell, 1973, Heavy Metals: Fallout Around a Power Plant, *Environ. Sci. Technol.*, 7: 357-358.

Korth, M. W., A. H. Rose, Jr., and R. C. Stahman, 1964, Effects of Hydrocarbon to Oxides of Nitrogen Ratios on Irradiated Auto Exhaust, Part 1, *J. Air Pollut. Control Assoc.*, 14: 168-175.

Krall, A. R., and N. E. Tolbert, 1957, A Comparison of the Light-Dependent Metabolism of Carbon Monoxide by Barley Leaves with That of Formaldehyde, Formate and Carbon Dioxide, *Plant Physiol.*, 32: 321-326.

Kuhn, J. K., 1973, Trace Elements in Whole Coal Determined by X-Ray Fluorescence, *Norelco Rep.*, 20: 7-10.

Lakin, H. W., 1973, Selenium in Our Environment, *Adv. Chem. Ser.*, 123: 96-112.

Lamontagne, R. A., J. W. Swinnerton, and V. J. Linnenbom, 1971, Nonequilibrium of Carbon Monoxide and Methane at the Air–Sea Surface, *J. Geophys. Res.*, 76: 5117-5121.

Larson, T. V., 1976, *The Kinetics of Sulfur Dioxide Oxidation by Oxygen and Ozone in Atmospheric Hydrometeors*, Ph.D. Thesis, University of Washington, Seattle.

——, N. R. Horike, and H. Harrison, 1978, Oxidation of Sulfur Dioxide by Oxygen and Ozone in Aqueous Solution: A Kinetic Study with Significance to Atmospheric Rate Processes, *Atmos. Environ.*, 12: 1597-1611.

Lazrus, A. L., J. P. Lodge, Jr., and B. W. Gandrud, 1974, Acidity of U. S. Precipitation, paper presented at the 1974 Spring American Geological Union Meeting.

Leibl, K. H., and W. Seiler, 1976, CO and H_2 Destruction at the Soil Surface, in *Microbial Production and Utilization of Gases*, H. G. Schlegel (Ed.), pp. 215-230, E. Goltze K. G., Kottingen, West Germany.

Lenelle, Y., 1975, Contribution a L'Etude de L'Oxidation des Sulfites dans la Pluie, paper presented at the European Sulfur Symposium, Ispra, Italy.

Lepel, E. A., K. M. Stefansson, and W. H. Zoller, 1978, The Enrichment of Volatile Elements in the Atmosphere by Volcanic Activity: Augustine Volcano 1976, *J. Geophys. Res.*, 83: 6213-6220.

Levy, H., 1971, Normal Atmosphere: Large Radical and Formaldehyde Concentrations Predicted, *Science*, 173: 141-143.

——, 1973, Photochemistry of Minor Constituents in the Troposphere, *Planet. Space Sci.*, 21: 575-591.

Likens, G. E., and F. H. Bormann, 1974, Acid Rain: A Serious Regional Problem, *Science*, 184: 1176-1179.

——, R. W. Wright, J. N. Galloway, and T. J. Butler, 1979, Acid Rain, *Sci. Am.*, 241: 43-51.

Linton, R. W., A. Loh, D. F. S. Natusch, C. A. Evans, Jr., and P. Williams, 1976, Surface Predominance of Trace Elements, *Airborne Part. Sci.*, 191: 852-854.

——, P. Williams, C. A. Evans, Jr., and D. F. S. Natusch, 1977, Determination of the Surface Predominance of Toxic Elements in Airborne Particles by Ion Microprobe, Mass Spectrometry, and Auger Electron Spectrometry, *Anal. Chem.*, 49: 1514-1520.

Liu, S. C., R. J. Cicerone, T. M. Donahue, and W. L. Chameides, 1976, Limitation of Fertilizer Induced Ozone Reduction by the

Long Lifetime of the Reservoir of Fixed Nitrogen, *Geophys. Res. Lett.*, 3: 157-160.

——, R. J. Cicerone, T. M. Donahue, and W. L. Chameides, 1977, Sources and Sinks of Atmospheric N_2O and the Possible Ozone Reduction Due to Industrial Fixed Nitrogen Fertilizers, *Tellus*, 29: 251-263.

Lovelock, J. E., 1974, CS_2 and the Natural Sulfur Cycle, *Nature*, 248: 625-626.

——, R. J. Maggs, and R. A. Rasmussen, 1972, Atmospheric Dimethyl Sulfide and the Natural Sulfur Cycle, *Nature*, 237: 452-453.

Lusis, M. A., and H. A. Wiebe, 1976, The Rate of Oxidation of Sulfur Dioxide in the Plume of a Nickel Smelter Stack, *Atmos. Environ.*, 10: 793-798.

Lyon, W. S., 1977, *Trace Element Measurements at the Coal-Fired Steam Plant*, CRC Press Inc., Cleveland, Ohio.

McConnell, J. C., M. B. McElroy, and S. C. Wofsy, 1971, Natural Sources of CO_2, *Nature*, 233: 187-188.

McCormick, R. A., and J. H. Ludwig, 1967, Climate Modification by Atmospheric Aerosols, *Science*, 156: 1358-1359.

MacCracken, R. C., 1978, MAP3S: An Investigation of Atmospheric Energy Related Pollutants in the Northeastern United States, *Atmos. Environ.*, 12: 649-660.

McElroy, M. B., J. W. Elkins, S. C. Wofsy, and Y. L. Yung, 1976, Sources and Sinks for Atmospheric N_2O, *Rev. Geophys. Space Phys.*, 14: 143-149.

——, S. C. Wofsy, and Y. L. Yung, 1977, The Nitrogen Cycle: Perturbations Due to Man and Their Impact on Atmospheric N_2O and O_3, *Philos. Trans. R. Soc. London*, 277B: 159-181.

McNaughton, D. J., 1980, Initial Comparison of SURE/MAP3S Sulfur Oxide Observations with Long-Term Regional Model Predictions, *Atmos. Environ.*, 14: 55-63.

Maerhaut, W., and W. H. Zoller, 1977, Determination of the Chemical Composition of the South Pole Aerosol by Instrumental Neutron Activation Analysis, *J. Radioanal. Chem.*, 37: 637-650.

Magee, E. M., H. F. Hall, and G. M. Varga, Jr., 1973, *Potential Pollutants in Fossil Fuels*, EPA Project Report, Contract No. 68-02-0629.

Manabe, S., and R. T. Wetherald, 1975, The Effects of Doubling the CO Concentration on the Climate of a General Circulation Model, *J. Atmos. Sci.*, 32: 3-15.

Matthews, W. H., W. W. Kellogg, and G. D. Robinson (Eds.), 1971, *Man's Impact on Climate*, The MIT Press, Cambridge, Mass.

Mercer, J. H., 1978, West Antarctic Ice Sheet and CO_2 Greenhouse Effect: A Threat to Disaster, *Nature*, 271: 321-325.

Migeotte, M. V., 1949, On the Presence of Atmospheric Gases from Infrared Telluric Bands, in *The Atmospheres of the Earth and Planets*, pp. 284-287, University of Chicago Press, Chicago.

Miller, D. F., A. J. Alkezweeny, J. M. Hales, and R. N. Lee, 1978, Ozone Formation Related to Power Plant Emissions, *Science*, 202: 1186-1188.

Mroz, E. J., 1976, *The Study of the Elemental Composition of Particulate Emissions from an Oil-Fired Power Plant*, Ph.D. Thesis, University of Maryland, College Park, Md.

Murozumi, M. T., J. Chow, and C. Patterson, 1969, Chemical Concentrations of Pollutant and Aerosols, Terrestrial Dusts and Sea Salts in Greenland and Antarctic Snow Strata, *Geochim. Cosmochim. Acta*, 33: 1247-1294.

Nature, 1979, 279: 467.

Nelson, J. B., 1953, *BCURA Monthly Bulletin*, 17: 41.

Newman, L., J. Forrest, and B. Manowitz, 1975, The Application of an Isotope Ratio Technique to a Study of the Atmospheric Oxidation of Sulfur Dioxide in the Plume from a Coal-Fired Power Plant, *Atmos. Environ.*, 9: 969-977.

Nicholls, G. D., 1968, The Geochemistry of Coal-Bearing Strata, in *Coal and Coal-Bearing Strata*, D. C. Murchison and T. S. Westoll (Eds.), pp. 269-307, American Elsevier Publishing Co., New York.

Nisbet, I., 1975, *Sulfates and Acidity in Precipitation Air Quality and Stationary Source Control*, NAS/NAE Report to U. S. Senate Committee on Public Works, Series No. 94-4, GPO, Washington.

Novakov, T., S. G. Chang, and A. B. Harker, 1974, Sulfates as Pollution Particulates: Catalytic Formation on Carbon (soot) Particles, *Science*, 186: 259-261.

Noxon, J. F., 1975, Nitrogen Dioxide in the Stratosphere and Troposphere Measured by Ground Based Absorption Spectroscopy, *Science*, 189: 547.

——, 1978, Tropospheric NO_2, *J. Geophys. Res.*, 83: 3051-3057.

Oden, S., 1968, *The Acidification of Air and Precipitation and Its Consequences on the Natural Environment*, Swedish Natural Science Research Council, Stockholm, Ecological Research Committee Bulletin No. 1, unpublished study.

O'Gorman, J. V., and P. L. Walker, Jr., 1971, Mineral Matter Characteristics of Some American Coals, *Fuel*, 50: 135-151.

——, and P. L. Walker, Jr., 1972, *Mineral Matter and Trace Elements in U. S. Coals*, Report to Office of Coal Research, U. S. Department of the Interior, Washington.

Ogren, J. A., D. L. Blumenthal, W. H. White, T. W. Tesche, M. A. Yocke, and M. K. Liu, 1976, *Determination of the Feasibility of Ozone Formation in Power Plant Plumes*, Report EA-307, Electric Power Research Institute, Palo Alto, Calif.

Okita, T., S. Morimoto, M. Izawa, and S. Konno, 1976, Measurement of Gaseous and Particulate Nitrates in the Atmosphere, *Atmos. Environ.*, 10: 1085-1089.

Ostlund, H. G., and J. Alexander, 1963, Oxidation Rate of Sulfide in Seawater, A Preliminary Study, *J. Geophys. Res.*, 68: 3995-3997.

O'Toole, J. J., et al., 1978, *Potential for Gaseous and Heavy Metal Contamination from Energy Extraction Processes in the Northern Great Plains and the Consequent Uptake and Turnover in Range Ecosystems*, Report IS-4733, Ames Laboratory, NTIS.

Ottar, B., 1976, Monitoring the Long-Range Transport of Pollutants: The OECD Study, *Ambio*, 6: 203-206.

——, 1978, Study on Long Range Transport, *Atmos. Environ.*, 12: 445-460.

Parkin, D. W., D. R. Phillips, and R. A. L. Sullivan, 1970, Airborne Dust Collections over the North Atlantic, *J. Geophys. Res.*, 75: 1782-1793.

Payne, W. A., L. J. Steif, and D. D. Davis, 1973, A Kinetics Study of the Reaction of HO_2 with SO_2 and NO, *J. Am. Chem. Soc.*, 95: 7614-7619.

Peirson, D. H., P. A. Cawse, L. Salmon, and R. S. Cambray, 1973, Trace Elements in the Atmospheric Environment, *Nature*, 241: 252-256.

Penkett, S. A., 1972, Oxidation of SO_2 and Other Atmospheric Gases by Ozone in the Troposphere, *Nat. Phys. Sci.*, 240: 105-106.

——, B. M. R. Jones, and A. E. J. Eggleton, 1975, Rate of Oxidation of Sodium Sulfite Solutions by Oxygen and Ozone, paper presented at the European Sulfur Symposium, Ispra.

Perhac, R. M., 1978, Sulfate Regional Experiment in Northeastern United States: The "SURE" Program, *Atmos. Environ.*, 12: 641-648.

Peters, K. P., and A. A. Jouvanis, 1979, Numerical Simulation of the Transport and Chemistry of CH_4 and CO in the Troposphere, *Atmos. Environ.*, 13: 1443-1462.

Pierotti, D., and R. A. Rasmussen, 1976, Combustion as a Source of Nitrous Oxide in the Atmosphere, *Geophys. Res. Lett.*, 3: 265-267.

——, and R. A. Rasmussen, 1977, The Atmospheric Distribution of Nitrous Oxide, *J. Geophys. Res.*, 82: 5823-5832.

——, R. A. Rasmussen, and R. Chatfield, 1978, Continuous Measurements of Nitrous Oxide in the Troposphere, *Nature*, 274: 574-576.

Plass, G. N., 1959, Carbon Dioxide and Climate, *Sci. Am.*, 201: 41-47.

Pollard, F. H., P. Hanson, and G. Nickless, 1961, Chromatographic Studies on the Oxidation of Sulfurous Acid by Ferric Ion in Aqueous Acid Solution, *J. Chromatogr.*, 5: 68-73.

Powell, D. C., D. J. McNaughton, L. L. Wendell, and R. L. Drake, 1978, *A Variable Trajectory Model for Regional Assessments of Air Pollution from Sulfur Compounds*, Report PNL-2734, Battelle Pacific Northwest Laboratory.

Pritchett, P. W., 1960, The Catalyzed Reaction of Oxygen with Sulfurous Acid and Its Effect, Ph.D. Thesis, University of Delaware, Newark.

Proceedings of the First International Symposium on Acid Precipitation in the Forest Ecosystem, 1976, Northeastern Forest Experimental Station, Upper Derby, Pa.

Project Independence: U. S. and World Energy Outlook Through 1990, 1977, GPO.

Prospero, J. M., and R. T. Nees, 1977, Dust Concentrations in the Atmosphere of the Equatorial North Atlantic: Possible Relationship to the Sahelian Drought, *Science*, 196: 1196-1198.

Pueschel, R. F., 1976, Aerosol Formation During Coal Combustion: Condensation of Sulfates and Chlorides in Flyash, *Geophys. Res. Lett.*, 3: 651-653.

Ragaini, R. C., and J. M. Ondov, 1976, Trace Element Emissions from Western U. S. Coal-Fired Power Plants, in *Proceedings of International Conference on Modern Trends in Activation Analysis*, Munich, Germany.

Rahn, K. A., and J. W. Winchester, 1971, *Sources of Trace Elements in Aerosols—An Approach to Clean Air*, Tech. Report, USAEC Contract AT(11-1)-1705.

Rancitelli, L. A., and K. H. Abel, 1974, Magnetic Studies of Centralia Flyash, in *Pacific Northwest Laboratory Annual Report for 1973 to USAEC Division of Biomedical and Environmental Research*, USAEC Report BNWL-1850, Pt. 2, pp. 126-128, Battelle Pacific Northwest Laboratory, NTIS.

——, and R. W. Perkins, 1970, Trace Element Concentrations in the Troposphere and Lower Stratosphere, *J. Geophys. Res.*, 75: 3055-3064.

Rao, C. P., and H. J. Gluskoter, 1973, *Occurrence and Distribution of Minerals in Illinois Coals*, Illinois Geological Survey Circular 476.

Rasmussen, R. A., 1974, Emission of Biogenic Hydrogen Sulfide, *Tellus*, 26: 254-260.

——, and D. Pierotti, 1978, Interlaboratory Calibration of Atmospheric Nitrous Oxide Measurements, *Geophys. Res. Lett.*, 5: 353-355.

——, D. Pierotti, J. Krasnec, and B. Halter, 1976, *Report on the Cruise of the Alpha Helix Research Vessel, March 5–20, 1976*, NSF Grant No. OCE-75-04688 A03.

Rebbert, R. E., and P. Ausloos, 1978, Decomposition of N_2O over Particulate Matter, *Geophys. Res. Lett.*, 5: 761-764.

Revelle, R. R., and D. C. Shapero, 1978, Energy and Climate, *Environ. Conserv.*, 5: 81-91.

Robinson, E., and R. C. Robbins, 1967, *Sources, Abundance and Fate of Gaseous Atmospheric Pollutants*, Stanford Research Institute Project PR-6755, prepared for the American Petroleum Institute, New York.

——, and R. C. Robbins, 1972, Emissions, Concentrations, and Fate of Gaseous Atmospheric Pollutants, in *Air Pollution Control*, Part II, pp. 1-93, W. Strauss (Ed.), Wiley-Interscience, New York.

Ruch, R. R., M. J. Gluskoter, and E. J. Kennedy, 1971, *Mercury Content of Illinois Coals*, Illinois Geological Survey, Environmental Geological Notes, No. 18.

——, H. J. Gluskoter, and M. F. Shimp, 1973, *Occurrence and Distribution of Potentially Volatile Trace Elements in Coal*, Illinois Geological Survey, Environmental Geological Notes, No. 73.

Sandalls, F. J., and S. A. Penkett, 1977, Measurements of Carbonyl Sulfide and Carbon Disulphide in the Atmosphere, *Atmos. Environ.*, 11: 197-199.

Schneider, S. H., 1975, On the Carbon Dioxide—Climate Confusion, *J. Atmos. Sci.*, 32: 2060-2066.

Schroeter, L. C., 1966, *Sulfur Dioxide: Application in Foods, Beverages, and Pharmaceuticals*, p. 56, Pergamon Press, Oxford.

Scott, B. C., 1978, Parameterization of Sulfate Removal by Precipitation, *J. Appl. Meteorol.*, 17: 1375-1389.

Scott, W. D., and P. V. Hobbs, 1967, The Formation of Sulfate in Water Droplets, *J. Atmos. Sci.*, 24: 54-57.

Schutz, K., C. Junge, R. Beck, and B. Albrecht, 1970, Studies of Atmospheric N_2O, *J. Geophys. Res.*, 75: 2230-2246.

Seiler, W., 1974, The Cycle of Atmospheric CO, *Tellus*, 26: 116-135.

——, and H. Giehl, 1977, Influence of Plants on the Atmospheric Carbon Monoxide, *Geophys. Res. Lett.*, 4: 329-332.

——, and C. Junge, 1970, Carbon Monoxide in the Atmosphere, *J. Geophys. Res.*, 75: 2217-2226.

——, and V. Schmidt, 1974, New Aspects of CO and H_2 Cycles in the Atmosphere, in *International Conference on Structure, Composition and General Circulation of the Upper and Lower Atmospheres and Possible Anthropogenic Perturbations*, Vol. 1, pp. 192-222.

——, and H. Zankl, 1976, Man's Impact on the Atmospheric Carbon Monoxide Cycle, in *Environmental Biogeochemistry*, Vol. 1, J. O. Nriagu (Ed.), Ann Arbor Press, Ann Arbor, Mich.

Seinfeld, J. H., 1975, *Air Pollution: Physical and Chemical Fundamentals*, McGraw-Hill Book Co., New York.

Shy, C. M., 1975, *Health Rationale for the Existing Nitrogen Oxide Emission Standard*, presented at the Scientific Seminar on Automotive Pollutants, Washington, D. C.

——, and J. F. Finklea, 1973, Air Pollution Effects Community Health, *Environ. Sci. Technol.*, 7: 204-208.

Siegel, S. M., G. Renwick, and L. A. Rosen, 1962, Formation of Carbon Monoxide During Seed Germination and Seedling Growth, *Science*, 137: 683-684.

Siegenthaler, U., and H. Oeschger, 1978, Predicting Atmospheric Carbon Dioxide Levels, *Science*, 199: 388-395.

Sillen, L. G., and A. E. Martell, 1964, Stability Constants of Metal-Ion Complexes, *J. Chem. Soc. London*, Special Publication No. 17.

Slatt, B. J., D. F. S. Natusch, J. M. Prospero, and D. L. Savoie, 1978, Hydrogen Sulfide in the Atmosphere of the Northern Equatorial Atlantic Ocean and Its Relation to the Global Sulfur Cycle, *Atmos. Environ.*, 12: 981-991.

Smith, R. D., J. A. Campbell, and W. D. Felix, 1979, Atmospheric Trace Element Pollutants from Coal Combustion, in *Proceedings of American Institute Mining Engineering Meeting, February 18–22*, New Orleans, La.

——, J. A. Campbell, and K. K. Nielson, 1979, Concentration Dependence upon Particle Size of Volatilized Elements in Flyash, *Environ. Sci. Technol.*, 13: 553-558.

Smith, T. J., D. J. Eatough, L. D. Hansen, and N. F. Mangelson, 1976, The Chemistry of Sulfur and Arsenic in Airborne Copper Smelter Particulates, *Bull. Environ. Contam. Toxicol.*, 6: 651-659.

Spectrum, *Environment*, 1979, 21(10): 21.

Spicer, C. W., 1977, Photochemical Atmospheric Pollutants Derived from Nitrogen Oxides, *Atmos. Environ.*, 11: 1080-1095.

Stedman, D. H., and J. O. Jackson, 1975, The Photostationary State in Photochemical Smog, *Int. J. Chem. Kinet.*, 1: 493-501.

Stewart, R. W., S. Hameed, and J. P. Pinto, 1977, Photochemistry of Tropospheric Ozone, *J. Geophys. Res.*, 82: 3134-3140.

Stoiber, R. E., D. C. Leggett, T. F. Jenkins, R. P. Murrmann, and W. I. Rose, Jr., 1971, Organic Compounds in Volcanic Gas from Santiaguito Volcano, Guatemala, *Bull. Geol. Soc. Am.*, 82: 2299-2302.

Stuiver, M., 1978, Atmospheric Carbon Dioxide and Carbon Reservoir Changes, *Science*, 199: 253-258.

Swinnerton, J. W., and R. A. Lamontagne, 1974, Carbon Monoxide in the South Pacific Ocean, *Tellus*, 26: 136-142.

——, V. J. Linnenbom, and C. H. Cheek, 1969, Distribution of Methane and Carbon Monoxide Between the Atmosphere and Natural Waters, *Environ. Sci. Technol.*, 3: 836-838.

——, V. J. Linnenbom, and R. A. Lamontagne, 1970, Ocean: A Natural Source of Carbon Monoxide, *Science*, 167: 984-986.

Sze, N. D., 1977, Anthropogenic CO Emissions: Implications for the Atmospheric $CO-OH-CH_4$ Cycle, *Science*, 195: 673-675.

——, and H. Rice, 1976, Nitrogen Cycle Factors Contributing to N_2O Production from Fertilizers, *Geophys. Res. Lett.*, 3: 343-346.

Tabor, E. C., and W. V. Warren, 1958, Distribution of Certain Metals in the Atmosphere of Some American Cities, *Arch. Ind. Health*, 17: 145-151.

Tartarelli, R., R. P. Davini, F. Morelli, and P. Corsi, 1978, Interactions Between SO_2 and Carbonaceous Particulates, *Atmos. Environ.*, 12: 289-293.

Theis, T. L., and J. L. Wirth, 1977, Sorptive Behavior of Trace Metals on Fly Ash in Aqueous Systems, *Environ. Sci. Technol.*, 11: 106-110.

Tyson, B. J., J. C. Arvesen, and D. O'Hara, 1978, Interhemispheric Gradients of CF_2Cl_2, $CFCl_3$, CCl_4 and N_2O, *Geophys. Res. Lett.*, 5: 535-538.

Urone, P., H. Lutsep, C. M. Noyes, and J. F. Parcher, 1968, Static Studies of Sulfur Dioxide Reactions in Air, *Environ. Sci. Technol.*, 2: 611-618.

U. S. Atomic Energy Commission Office of Planning, 1973, data presented at Division of Biomedical and Environmental Research Directors' Meeting, Germantown, Md., November 5–7.

Van Hook, R. I., and W. D. Shults (Eds.), 1977, Effects of Trace Contaminants from Coal Combustion, in *Proceedings of a USERDA-Sponsored Workshop*, Knoxville, Tenn., August 2–6.

Vaughan, B. E., et al., 1975, *Review of Potential Impact on Health and Environmental Quality from Metals Entering the Environment as a Result of Coal Utilization—A Battelle Energy Program Report*, Battelle Memorial Institute, Richland, Wash.

Wang, C. C., L. I. Davis, C. H. Wu, S. Japar, H. Niki, and B. Weinstock, 1975, Hydroxyl Radical Concentrations Measured in Ambient Air, *Science*, 189: 797-800.

Wang, W. C., Y. L. Yung, A. A. Lacis, T. Mo, and J. E. Hansen, 1976, Greenhouse Effects Due to Man-Made Perturbations of Trace Gases, *Science*, 194: 685-690.

Watson, A. P., R. I. Van Hook, D. R. Jackson, and D. E. Reichle, 1976, *Impact of a Lead Mining-Smelting Complex on the Forest Floor Litter Anthropod Fauna in the New Lead Belt Region of Southeast Missouri*, ERDA Report ORNL/NSP/EATC-30, Oak Ridge National Laboratory, NTIS.

Weinstock, B., 1969, Carbon Monoxide: Residence Time in the Atmosphere, *Science*, 166: 224-225.

——, and H. Niki, 1972, Carbon Monoxide Balance in Nature, *Science*, 176: 290-292.

Weiss, R. F., and H. Craig, 1976, Production of Atmospheric Nitrous Oxide by Combustion, *Geophys. Res. Lett.*, 3: 751-753.

Whitbeck, M. R., J. W. Bottenheim, S. Z. Levine, and J. G. Calvert, 1976, A Kinetic Study of the CH_3O_2 and $(CH_3)_3CO_2$ Radical Reactions by Kinetic Flash Spectroscopy, Abstracts of 12th Informal Conference on Photochemistry, U. S. National Bureau of Standards, Gaithersburg, Md., pp. K1-1, K1-5.

Whittaker, R., and G. Likens, 1973, in *Primary Productivity of the Biosphere*, Springer-Verlag, Berlin, 305 pages.

Wilks, S. S., 1959, Carbon Monoxide in Green Plants, *Science*, 129: 964-966.

Wilson, W. E., 1978, Sulfates in the Atmosphere: A Progress Report on Project MISST, *Atmos. Environ.*, 12: 537-548.

——, R. J. Charlson, R. B. Husar, K. T. Whidbey, and D. Blumenthal, 1976, Sulfates in the Atmosphere, paper presented at the 69th Meeting of the Air Pollution Control Association, Portland, Oreg.

Winchester, J. W., 1973, Trace Metal Associations in Urban Airborne Particulates, *Bull. Am. Meteorol. Soc.*, 54: 94-97.

Wofsy, S. C., 1976, Interactions of CH_4 and CO in the Earth's Atmosphere, *Ann. Rev. Earth Planet. Sci.*, 4: 401-409.

——, J. C. McConnell, and M. B. McElroy, 1972, Atmospheric CH_4, CO, and CO_2, *J. Geophys. Res.*, 77: 4477-4493.

Woodwell, G. M., 1978, The Carbon Dioxide Question, *Sci. Am.*, 238: 34-43.

——, R. H. Whittaker, W. A. Reiners, G. E. Likens, C. C. Delwiche, and D. B. Botkin, 1978, The Biota and the World Carbon Budget, *Science*, 199: 141-146.

Workshop on the Global Effects of Carbon Dioxide from Fossil Fuels, 1977, U. S. Department of Energy Study Group on the Global Environmental Effects of Carbon Dioxide, Miami Beach, Florida, Report CONF-770385.

Yoshinara, T., 1973, *Nitrous Oxide in the Sea*, Ph.D. Thesis, Dalhousie University, Halifax, N.S.

Young, J. A., and A. J. Alkezweeny, 1977, Trace Element Concentrations Downwind of Milwaukee and Chicago, in *Proceedings of the 4th Joint Conference on Sensing of Environmental Pollutants*, Paper No. 227, New Orleans, La.

Zimmerman, P. R., R. B. Chatfield, J. Fishman, P. S. Crutzen, and P. L. Hanst, 1978, Estimate on the Production of CO and H_2 from the Oxidation of Hydrocarbon Emissions from Vegetation, *Geophys. Res. Lett.*, 5: 679-682.

Zoller, W. H., E. S. Gladney, and R. A. Duce, 1974, Atmospheric Concentrations and Sources of Trace Metals at the South Pole, *Science*, 183: 198-200.

Zubovic, P., 1966a, *Physicochemical Properties of Certain Minor Elements as Controlling Factors in Their Distribution in Coal*, *Adv. Chem. Ser.*, 55: 221-231.

——, 1966b, Minor Element Distribution in Coal Samples of the Interior Coal Province, *Adv. Chem. Ser.*, 55: 232-241.

Chapter 10 covers research prior to April 1980.

11 PRECIPITATION SCAVENGING

W. George N. Slinn, Pacific Northwest Laboratory

11-1 INTRODUCTION

As a result of natural atmospheric cleansing processes, individual aerosol particles and gas molecules have finite residence times in the atmosphere. Some estimates for these residence times are given in Table 11.1. The three dominant atmospheric cleansing processes are precipitation scavenging, dry deposition, and physical/chemical transformations. General criteria about the relative importance of these three cleansing processes cannot be established because, as will be seen, the cleansing efficiencies depend sensitively on specific pollutant and meteorological variables. In this chapter, the emphasis is on precipitation scavenging although, when appropriate, some comments will be made about the other two major cleansing or removal processes.

By precipitation scavenging (or wet removal, wet deposition, or washout) is meant the removal of pollutants or other trace constituents from the atmosphere by various types of precipitation (rain, snow, hail, graupel, etc.). The suffix "out" will be used synonymously for the word "scavenging"; e.g., rainout will mean rain scavenging. If a distinction is desired between in-cloud and below-cloud scavenging (although normally there is no major distinction), then no technical jargon will be used. Fig. 11.1 summarizes some of the terminology to be used in this chapter and illustrates some of the variables upon which washout depends.

One of the goals of this chapter is to demonstrate the influence of precipitation scavenging on the residence times shown in Table 11.1. A more pressing goal, however, is to provide sufficient theory and data about precipitation scavenging to permit estimates for health and environmental impact assessments of pollution releases to the atmosphere, especially during the production of electrical power. In general, the user is assumed to have data about the precipitation and pollutant. Because these data may be meager, some formulas presented require only a minimum amount of additional information. Although variations of pollutant properties within the atmosphere are mentioned, the reader should go to Chap. 10 and its references for more complete information. Similarly, the reader is referred to Chap. 12 for a more complete discussion of dry deposition. Indeed, the reader will frequently be referred elsewhere even for more complete descriptions of precipitation scavenging since space restrictions on this chapter force the omission of many details. In an attempt to compensate for space restrictions and omissions, reliance is made on many figures and the old Chinese proverb: one picture is worth more than ten thousand words. And if there are complaints concerning too many formulas, a response could be: one formula can be worth more than ten thousand pictures!

Slinn: Pacific Northwest Laboratory
Present address: P. O. Box 999, Richland, WA 99352

It may be useful to give some general references on precipitation scavenging. Four conference proceedings that contain many details are those edited by Styra, Garbalyauskas, and Luyanas (1970), Engelmann and Slinn (1970), Semonin and Beadle (1977), and Pruppacher, Semonin, and Slinn (1983). Some of the earliest precipitation scavenging studies were reported by Chamberlain (1953), McCulley et al. (1956), Greenfield (1957), and May (1958). Reviews and surveys include those by Vaughan and Perkins (1961), Facy (1962), Junge (1963), Zimin (1964), Makhon'ko (1966), and Engelmann (1968). Junge's text and Engelmann's contribution to the previous edition of this book (Engelmann, 1968) remain as valuable reference works.

As a cautionary note, the user should realize that, typically, precipitation scavenging estimates are not very reliable. Long-term (e.g., seasonal) average estimates using washout ratios for total mass scavenged may be accurate to within a factor of two to four if conditions are comparable to the cases for which data are available, but for short-term estimates (e.g., an accidental release of radioactivity) to expect more than order-of-magnitude accuracy is not realistic. Therefore corresponding assessments of health and environmental impacts cannot be made more accurately. It is hoped that sufficient detail is presented in this chapter to demonstrate specific topics that currently limit the accuracy of predictions, to entice further study of these topics, and to provide a framework for extrapolations to the removal of pollutants that may be of interest in the future. As a corollary to present uncertainties, the user is urged to check current literature in this continuously developing field (see, especially, current issues of *Atmospheric Environment, Izvestia—Atmospheric and Oceanic Physics, Journal of Atmospheric Sciences,* and *Journal of the Japanese Meteorological Society* and *Tellus B*).

11-2 GENERAL FORMULATION AND AVERAGES

11-2.1 The Convection-Diffusion Equation

The fundamental equation in precipitation scavenging studies is the general conservation equation. If $\chi(\mathbf{r},t)$ is an appropriate concentration of the pollutant in air (e.g., χ da may be the number of particles per unit volume and whose radii are from a to a + da), this conservation equation is

$$\frac{\partial \chi}{\partial t} = -\nabla \cdot (\mathbf{flux}) + G(\chi) - L(\chi) \quad (11.1)$$

where G and L represent gain and loss rates per unit volume, respectively. A form of this conservation equation that may be more familiar can be seen by dividing the flux into contributions from (1) Brownian (or molecular) diffusion, $-D\nabla\chi$;

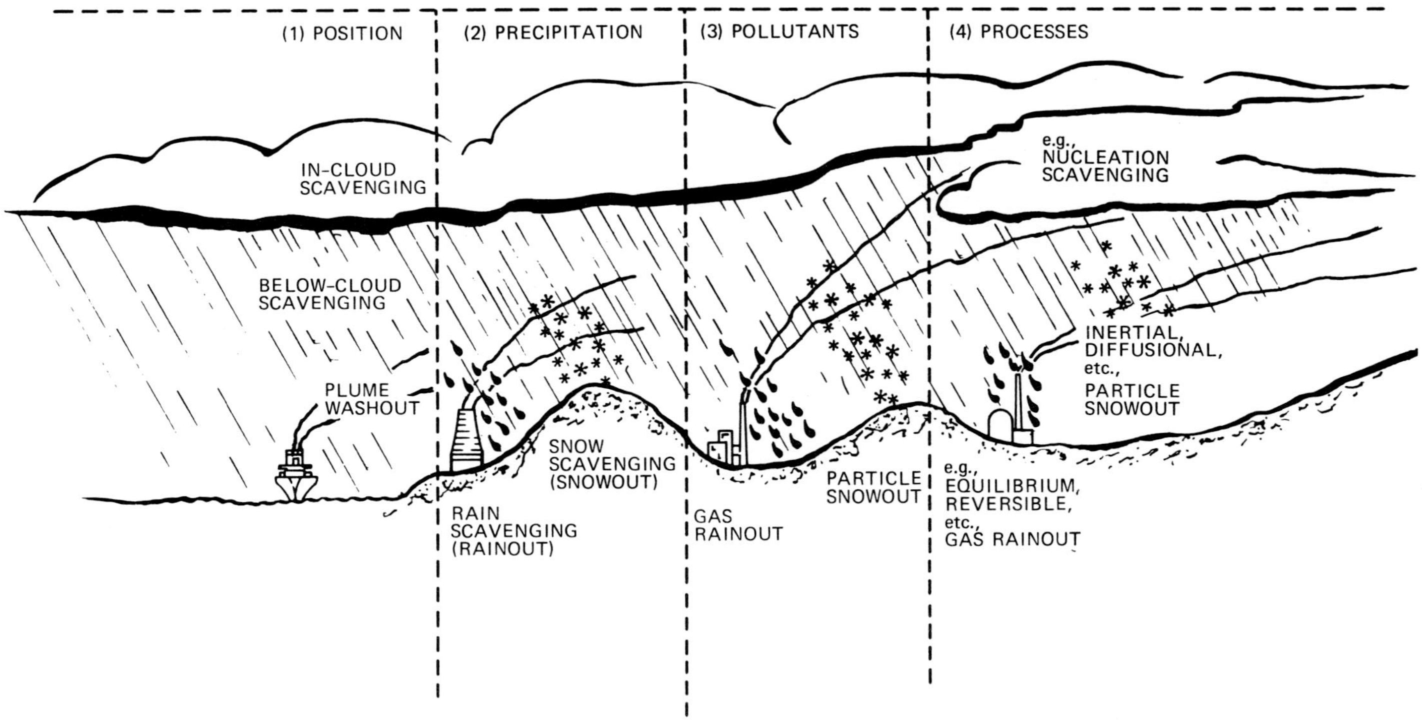

Fig. 11.1 Examples of terminology used in this chapter. [From Slinn (1977b).]

TABLE 11.1 Estimates for Average Residence Times in the Troposphere[a,b]

Material	Vinyl chloride[c] (CH_3I)	SO_2	0.1- to 1.0-μm particles (e.g., SO_4^{2-} and H_2O)	DDT	HT PCB's[d]	CH_4 CCl_3F[e]	CO_2 N_2O	^{85}Kr	O_2
Residence time	~hours	~days	~1 wk	~3 mo	~1 yr	1–10 yr	~10 yr	~10 yr	~10^4 yr

[a]Estimates are based on Junge (1972, 1974), Singh, Salas, and Cavanagh (1977), and Rasmussen et al. (1978).

[b]If "residence times" are considered more carefully (e.g., Bolin and Rodhe, 1973), then three different times can be identified: age, transit time, and turnover time. Some appreciation for the differences can be gained from noting that the turnover time and the average transit time are equal for steady-state reservoirs and that this common time is equal to the average age for well-mixed reservoirs.

[c]Similarly for ethylene dichloride and trichlorethylene.

[d]Polychlorinated biphenyls.

[e]Jesson, Meakin, and Glasgow (1977) estimate the tropospheric residence times for fluorocarbon-11 (CCl_3F) and fluorocarbon-12 (CCl_2F_2) to be 10 to 20 yr.

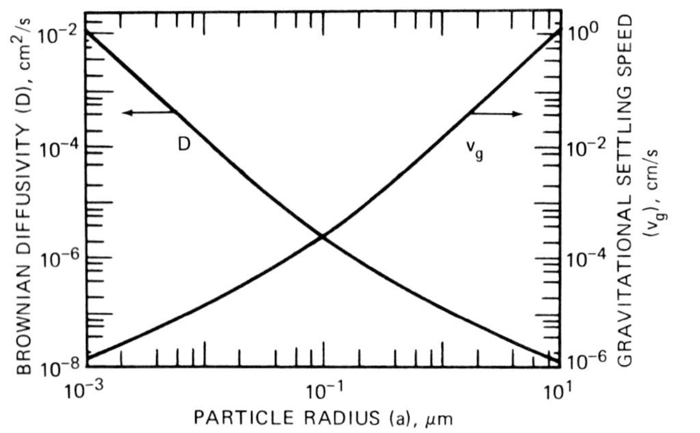

Fig. 11.2 The diffusivity, D, and gravitational settling speed, v_g, as functions of particle radius, a. The results are for spheres of unit specific density in air at 1 atm and 20°C. The gravitational settling speed and particle stopping time, τ, are related by $v_g \simeq \tau g$, where g is the acceleration caused by gravity. (Data from N. A. Fuchs, *The Mechanics of Aerosols,* Pergamon Press, New York, 1964.)

TABLE 11.2 Brownian Diffusivity and Gravitational Settling Speed for Particles of Unit Density in Air at 1 atm and 20°C†

Particle radius (a), μm	Diffusivity (D), cm²/s	Settling speed (v_g), cm/s
1×10^{-3}	1.28×10^{-2}	1.31×10^{-6}
2×10^{-3}	3.23×10^{-3}	2.62×10^{-6}
5×10^{-3}	5.24×10^{-4}	6.63×10^{-6}
1×10^{-2}	1.35×10^{-4}	1.37×10^{-5}
2×10^{-2}	3.59×10^{-5}	2.91×10^{-5}
5×10^{-2}	6.82×10^{-6}	8.64×10^{-5}
1×10^{-1}	2.21×10^{-6}	2.24×10^{-4}
2×10^{-1}	8.32×10^{-7}	6.73×10^{-4}
5×10^{-1}	2.74×10^{-7}	3.47×10^{-3}
1×10^{0}	1.27×10^{-7}	1.28×10^{-2}
2×10^{0}	6.10×10^{-8}	4.93×10^{-2}
5×10^{0}	2.38×10^{-8}	3.02×10^{-1}
1×10^{1}	1.38×10^{-8}	1.21×10^{0}

†Another important aerosol parameter is the particle stopping time, $\tau = v_g/g$, where $g = 980$ cm/s². (Based on N. A. Fuchs, *The Mechanics of Aerosols,* Pergamon Press, New York, 1964.)

(2) any drift velocity, **d**, with respect to the air (e.g., caused by gravity or diffusiophoresis); and (3) motion with the air with (three-dimensional) velocity **v**. Further, if the air is treated as incompressible (i.e., $\nabla \cdot \mathbf{v} = 0$), Eq. 11.1 becomes the convection-diffusion equation

$$\frac{\partial \chi}{\partial t} + \mathbf{v} \cdot \nabla \chi = \nabla \cdot (D \nabla \chi - \mathbf{d}\chi) + G(\chi) - L(\chi) \tag{11.2}$$

Figure 11.2 and Table 11.2 give data for the Brownian diffusivity, D, and gravitational settling speed, v_g, as functions of particle size. The restrictions on Eq. 11.2, to meet requirements of local thermodynamic equilibrium and to ignore influences of the pollutant on the fluid motion, are not significant for practical applications to be considered here.

Equation 11.2 is the starting point for most precipitation scavenging studies. Succinctly, the goal of this chapter is to specify the contribution of the scavenging to the loss rate per unit volume, L. However, in general, Eq. 11.2 is too complicated as it stands, and various simplifications, averages, and approximations must be introduced. The remainder of Sec. 11-2 indicates some of these simplifications.

11-2.2 Turbulence and Time Averages

In Eq. 11.2, **v** is the average velocity of the air molecules, the average taken over time periods long compared to the average time between molecular collisions ($\sim 10^{-9}$ s) and over distances large compared to the mean free path for molecules (~ 0.1 μm at STP). As a consequence of performing these averages, terms enter Eq. 11.2 that describe drift ($\mathbf{d}\chi$) and diffusion ($D\nabla\chi$) of the trace constituent. Usually, however, it is also convenient to average **v** over larger time and space scales, especially if the flow is turbulent; consequently, additional drift and diffusion terms enter. Unfortunately, though, universally applicable time and space scales for averaging turbulent fluctuations are not available; for example, a global modeler's turbulence can be a micrometeorologist's mean flow! That is, the distinction between turbulence and mean flow depends on the time and space scales of interest for the particular problem under study.

For applications in precipitation scavenging, suppose there is a minimum time period of interest, T. For example, T may represent the response time of an instrument that measures the concentration of the pollutant. Then for some problems, all fluid motions with characteristic time scale shorter than T may be considered as turbulent. If all terms in Eq. 11.2 are separated into mean values during this chosen time period (these means will be identified with a tilde, ˜) and into fluctuations about this mean (identified by ′), then an average of Eq. 11.2 leads to

$$\frac{\partial \tilde{\chi}}{\partial t} + \tilde{\mathbf{v}} \cdot \nabla \tilde{\chi} = \nabla \cdot (-\tilde{\boldsymbol{F}} - \tilde{\mathbf{d}}\tilde{\chi}) + \tilde{G} - \tilde{L} \tag{11.3}$$

in which $\tilde{G}$ and $\tilde{L}$ may be highly complex, e.g., for the case of chemical reactions nonlinear in concentrations. In Eq. 11.3, the (negative of the) average flux is

$$-\tilde{\boldsymbol{F}} = -\widetilde{\mathbf{v}' \cdot \nabla \chi'} + D\nabla\tilde{\chi} - \widetilde{\mathbf{d}'\chi'} \tag{11.4}$$

As discussed in earlier chapters, a frequently used parameterization for the "turbulent diffusion" contribution to the flux is the gradient-flux ansatz

$$\widetilde{\mathbf{v}' \cdot \nabla \chi} = -\mathbb{K} \cdot \nabla \tilde{\chi} \tag{11.5}$$

where $\mathbb{K}(T)$ is known as the (second-order tensor) turbulent diffusivity. Limitations on this assumption are discussed elsewhere (e.g., Corrsin, 1974). In response to the criticism that this parameterization sometimes fails because it is unable to describe observed fluxes in the direction of the gradient of the mean concentration, it can be argued that, when this occurs, an inappropriate choice has been made for the averaging time, T.

11-2.3 Dry Deposition and Resuspension

The main emphasis in this chapter will be to develop parameterizations for the precipitation scavenging contribution to the mean loss rate per unit volume, $\tilde{L}$, appearing in Eq. 11.3. However, a few comments about dry deposition and resuspension may be useful. For one reason, wet and dry removal contributions to a pollutant's atmospheric residence time will be compared. Also, in some cases it is convenient to formulate washout not as a loss rate per unit volume but as a boundary condition (with a wet deposition velocity) similar to parameterizations for dry deposition and resuspension.

To see these parameterizations for dry deposition and resuspension, imagine a convection-diffusion equation, similar to Eq. 11.3, written for the pollutant's mean concentration, $\tilde{C}$, in the bounding medium (soil, water, vegetation, etc.). One inviolable boundary condition is equality of normal components of the flux (subscript, n) at the interface (subscript, i). From Eqs. 11.3 and 11.4 and for the case of negligible turbulence at the interface (note that v_n must be zero at an impermeable interface), this first boundary condition becomes

$$\left[D\frac{\partial\tilde{\chi}}{\partial n} - (\tilde{d}_n\tilde{\chi} + \widetilde{d_n'\chi'})\right]_{air,i} = \left[D\frac{\partial\tilde{C}}{\partial n} - (\tilde{d}_n\tilde{C} + \widetilde{d_n'C'})\right]_{medium,i} \quad (11.6)$$

Incidentally, to the author's knowledge, no one has yet determined what corrections, if any, are needed for molecular diffusivities near boundaries where the assumed randomness of molecular impacts is clearly violated by the proximity of the boundary (Fuchs, 1964; Slinn and Shen, 1971). The second terms on both sides of Eq. 11.6 describe any mean drift of the pollutant caused, for example, by gravity or by thermophoresis or diffusiophoresis. The last terms on both sides of the equation can be related to particle drift caused by nonuniformities in the turbulence (e.g., Caporaloni et al., 1975).

The required second boundary condition for the two coupled (second-order) convection-diffusion equations in the two media is parameterized in a number of ways. The simplest way is for the case when the surface is a perfect sink for the pollutant; then $\tilde{\chi}_i = 0$ and the equations are decoupled. In the more general case, two general procedures are to specify the concentration discontinuity at the interface (in terms of a Henry's law constant or a resuspension factor or other parameter) or to specify the magnitude of the flux (in terms of deposition and resuspension velocities). Table 11.3 lists some of the parameterizations that have been used. Clearly, as discussed in more detail in Chapter 12, much research must yet be performed to interpret these parameterizations in terms of underlying physical and chemical processes controlling interfacial exchange. For purposes in this chapter, the net (positive, downward) flux will be parameterized in terms of a dry deposition velocity, v_d, and a resuspension velocity, v_r:

$$F_i = v_d\tilde{\chi}_i - v_r\tilde{C}_i \quad (11.7)$$

Incidentally, sometimes it is convenient to parameterize dry deposition as a loss rate per unit volume, e.g., within vegetative canopies (Slinn, 1982b); in contrast, as mentioned earlier, sometimes it is convenient to parameterize precipitation scavenging in terms of a wet deposition velocity. The latter will be briefly illustrated in the next section and described in more detail later in the chapter.

11-2.4 Vertical Averages and the Wet Deposition Velocity

In the time-averaged convection-diffusion equation, Eq. 11.3, precipitation scavenging enters as a time-averaged loss rate per unit volume, $\tilde{L}$. With this formulation it is possible, at least in principle, to describe, for example, variations in the pollutant's vertical distribution that are caused by variations in precipitation intensity with height. However, for practical applications, such vertical variations normally can be ignored (or are ignored for want of adequate information), and what is sought is the vertically averaged reduction in concentration (usually as a function of travel time or downwind distance) and/or the wet flux of the contaminant to the earth's surface. This flux itself is a vertical average of the removal rate; in a real sense, the precipitation performs the integration. Thus, in summary, there is practical value in considering various vertical averages of Eq. 11.3.

The simplest case to consider is that of aerosol particles captured irreversibly by precipitation elements. Then the rate of removal depends linearly on the pollutant's air concentration and is independent from the number of particles previously captured: $L(\mathbf{r},t;a) = \psi(\mathbf{r},t;a)\chi(\mathbf{r},t;a)$. Further, as the simplest illustration, suppose the scavenging rate, ψ, is independent of position, $\mathbf{r}$ (especially, independent of height, z). For this case the removal term in the convection-diffusion equation, Eq. 11.3, can be eliminated if a new dependent variable $\tilde{\zeta} = \tilde{\chi}\exp(\int\tilde{\psi}\,dt)$ is introduced. The resulting equation describes diffusion of $\tilde{\zeta}$ without wet removal. If this equation can be solved, then resubstitution for $\tilde{\chi}$ leads to

$$\tilde{\chi}(\mathbf{r},t) = G(\mathbf{r},t;\sigma_y,\sigma_z)\exp\left(-\int\tilde{\psi}\,dt\right) \quad (11.8)$$

where G represents some plume formula (e.g., for a Gaussian plume). Furthermore, if $\tilde{\psi}$ is independent of time, the removal term for a steady-state plume then becomes $\exp(-\tilde{\psi}x/\bar{u})$, where $x/\bar{u}$ is the downwind travel time. If the removal rate is not independent of height, then, as a first approximation, both a time- and space-averaged removal rate can be used in Eq. 11.8. The alternative, i.e., retaining the spatial variation of $\tilde{\psi}$, obviously leaves a difficult equation to solve.

But independent of these considerations, a vertical average of the removal rates appears naturally because the wet flux to the surface is the sum of the removal from all volume elements aloft. In particular, if horizontal variations in $\tilde{\chi}$ and $\tilde{\psi}$ are ignorable over the horizontal distance traversed by hydrometeors during their fall, then the wet flux of pollutant particles to the earth's surface can be approximated by

$$(+\downarrow) \quad W = \int_0^\infty dz\,\tilde{L} = \int_0^\infty dz\,\tilde{\psi}\,\tilde{\chi} \quad (11.9)$$

where $(+\downarrow)$ signifies the sign convention for positive flux (i.e., down). Similar to the case of dry deposition, Eq. 11.9 can be used to define a wet deposition velocity (Junge, 1963)

$$v_w = \frac{W}{\tilde{\chi}_0} \quad (11.10)$$

i.e., the wet deposition velocity is the wet flux divided by the surface-level air concentration. For example, if aerosol particles are uniformly distributed in the mixed layer and if vertical

TABLE 11.3 Parameterizations for the Second Interfacial Boundary Condition for Cases of Nonzero Flux†

Options	Interrelations and comments
Specification of an equilibrium jump discontinuity:	
For gases,	
Liquid absorber, simple solutions	
$\tilde{\chi}_i = H\tilde{C}_i$	The Henry's law constant, $H = \alpha^{-1}$, where α is the solubility coefficient. Applicable only for portion of gas that is in simple solution.
Other absorbers and reactive gases	
$\tilde{\chi}_i = (\alpha_*)^{-1}\tilde{C}_i$	α_* = air/surface-medium partition coefficient.
For particles,	
$\tilde{\chi}_i = J\tilde{C}_i$	J = jump discontinuity parameter.
$\tilde{\chi}_i = K_*\tilde{G}$	$\tilde{G} \simeq \delta\tilde{C}_i$ is the approximate "surface concentration," where δ is the effective depth to which the pollutant has penetrated into the surface medium; therefore $K_* \simeq J/\delta$.
$\tilde{\chi}_b = K\tilde{G}$	K is known as the resuspension factor; note that use of the bulk concentration in the atmosphere (at a suitable reference height), $\tilde{\chi}_b$, rather than the concentration at the interface, $\tilde{\chi}_i$, can severely restrict the practical usefulness of K since upwind sources can contribute to $\tilde{\chi}_b$ (but not to $\tilde{\chi}_i$ at equilibrium).
Specification of the magnitude of the common dry flux, *F*:	F is the magnitude of the net flux, D-R, where D is the dry deposition flux and R is the resuspension flux.
For gases,	
$F = k_a(\tilde{\chi}_b - \tilde{\chi}_i)$	k_a is the transfer velocity in the atmosphere from where the bulk concentration is measured to the interface; for the measurement at height h and mean eddy diffusivity K_z in this interval, $k_a \simeq K_z/h$.
$F = k_s(\tilde{C}_i - \tilde{C}_b)$	Similarly, k_s is the net transfer velocity in the surface medium to where the bulk concentration is measured.
$F = k_0(\tilde{\chi}_b - \tilde{\chi}_*)$	$\tilde{\chi}_*$ is the concentration in the atmosphere that would be in equilibrium with the bulk concentration in the surface medium; e.g., $\tilde{\chi}_* = H\tilde{C}_b$. k_0 is the overall transfer velocity. From the first two expressions for *F*, one obtains $k_0^{-1} = k_a^{-1} + Hk_s^{-1}$; this shows the overall transfer resistance, k_0^{-1}, to be like the sum of electrical resistances in series (Liss and Slater, 1974; Liss and Slinn, 1983).
For particles,	
$D = v_d^c\tilde{\chi}_i$	v_d^c is Calder's deposition velocity; v_d^c can have any value less than infinity (Calder, 1961).
$D = v_d^{ch}(\tilde{\chi}_h - \tilde{\chi}_i)$	v_d^{ch} is Chamberlain's deposition velocity; h is a convenient reference height; a resistance analogy can be developed (Chamberlain, 1960).
$F = v_d\tilde{\chi}_i - v_r\tilde{C}_i$	At equilibrium, $F = 0$, and then $\tilde{\chi}_i = (v_r/v_d)\tilde{C}_i = J\tilde{C}_i$ or $J = v_r/v_d$ (Slinn, 1976b).
$F = v_d\tilde{\chi}_i - \Lambda\tilde{G}$	Λ is a resuspension rate; $\Lambda \simeq v_r/\delta$ (Amato, 1971).
$F = \Omega\tilde{A} - \Lambda\tilde{G}$	Ω is a similar deposition rate, where $\tilde{A}$ is the average atmospheric concentration of the pollutant. For example, if h_d is the average height to which the pollution has mixed, then $\tilde{A} = h_d\tilde{\chi}_i$ and $\Omega = v_d/h_d$ (Slinn, 1976b).

†Based on W. G. N. Slinn, Formulation and a Solution of the Diffusion, Deposition, Resuspension Problem, *Atmos. Environ.*, 10: 763-768, 1976.

variations in air density and removal rate are ignored, then Eqs. 11.9 and 11.10 yield

$$v_w = H\tilde{\psi} \tag{11.11}$$

where H is the height of the mixed layer. As an illustration of this result, if H = 1 km and $\psi = 10^{-4}\,s^{-1}$, then $v_w = 10$ cm/s. For another illustration of Eq. 11.10, consider the familiar Gaussian plume (including a term for wet removal)

$$\tilde{\chi} = \frac{\dot{Q}}{2\pi\sigma_y\sigma_z\bar{u}} \exp\left[-\left(\frac{y^2}{2\sigma_y^2} + \frac{\psi x}{\bar{u}}\right)\right] \times \left\{\exp\left[-\frac{(z-h)^2}{2\sigma_z^2}\right] + \exp\left[-\frac{(z+h)^2}{2\sigma_z^2}\right]\right\} \tag{11.12}$$

where $\dot{Q}$ is the source strength (e.g., in units of g/s). By evaluating Eq. 11.9 and substituting the result into Eq. 11.10, the wet deposition velocity becomes

$$v_w = \sqrt{\frac{\pi}{2}} \exp\left(\frac{-h^2}{2\sigma_z^2}\right) \sigma_z\tilde{\psi} \tag{11.13}$$

which becomes similar to the result given by Eq. 11.11 when $\sigma_z \simeq H$ and $H >> h$, where h is the effective stack height.

In contrast to the case of particles, however, it is usually not an adequate approximation to assume that gases are irreversibly captured by hydrometeors; reversible gas scavenging will be considered in Sec. 11-4. However, for both particles and gases, the formulation in terms of a wet deposition velocity has practical value. Indeed, in the case of gases that form simple solutions in rainwater, the gas concentration, $\tilde{\kappa}_0$, in surface-level raindrops can usually be approximated by

$$\tilde{\kappa}_0 = \alpha\tilde{\chi}_0 \tag{11.14}$$

where values for α, the solubility coefficient, are given in a later section (see Table 11.8). The wet flux is then $W = \tilde{\kappa}_0 p_0$, where p_0 is the surface-level precipitation flux [$cm^3/(cm^2 \cdot s)$ or cm/s], and therefore the wet deposition velocity is

$$v_w = \frac{W}{\tilde{\chi}_0} = \alpha p_0 \quad \text{(simple gases)} \tag{11.15}$$

i.e., the solubility coefficient multiplied by the precipitation rate.

Similarly, the wet deposition velocity for particles can be written in terms of measured values for the washout or scavenging ratio

$$w_r = \left(\frac{\tilde{\kappa}}{\tilde{\chi}}\right)_0 \quad \text{(dimensionless)} \tag{11.16}$$

through the use of

$$v_w = \frac{W}{\tilde{\chi}_0} = \frac{\tilde{\kappa}_0 p_0}{\tilde{\chi}_0} = w_r p_0 \tag{11.17}$$

Measured values for the washout ratios for both particles and gases will be discussed in Sec. 11-3.6; in the meantime, the parameterization of precipitation scavenging in terms of a wet deposition velocity (viz., Eq. 11.17) will be used to estimate residence times.

11-2.5 Reservoir Residence Times

In Sec. 11-2, so far, averages of the convection-diffusion equation have been taken over scales used to define turbulence and over scales characteristic of a single storm. The goal of this last subsection of Sec. 11-2 is to sketch some concepts about "reservoir residence times" for atmospheric trace constituents. Thereby, still larger-scale averages will be needed, with the needed size dictated by the size of the reservoir and/or the magnitude of the "pollutant's" residence time.

To define a pollutant's residence (or turnover) time in some reservoir (e.g., the mixed layer of the atmosphere, the northern troposphere, etc.), we first integrate the convection-diffusion equation over the volume of the reservoir. With the use of Gauss's divergence theorem, we obtain the obvious statement of conservation of mass

$$\frac{\partial Q}{\partial t} = -\oint \mathbf{F}\cdot\hat{n}\,dS + \int G\,dV - \int L\,dV \tag{11.18a}$$

where $Q = \int \chi\,dV$ is the total quantity of the pollutant in the reservoir.

If the net inflow through the reservoir's boundaries $\langle -\oint \mathbf{F}\cdot\hat{n}\,dS\rangle$ is written as the difference between the inflow (I) and outflow (O), if we use the symbols P (production) for the volume integrated gain term and R (removal) for the total loss rate, and if for later convenience we now divide Eq. 11.18a by Q, then Eq. 11.18a takes the obvious form

$$\frac{1}{Q}\frac{\partial Q}{\partial t} = \frac{(P+I)}{Q} - \frac{(R+O)}{Q} \tag{11.18b}$$

Equation 11.18b relates seven different natural time constants. With term-by-term identification, Eq. 11.18b can be written in the form

$$\frac{1}{\tau_N} = \left(\frac{1}{\tau_P} + \frac{1}{\tau_I}\right) - \left(\frac{1}{\tau_R} + \frac{1}{\tau_O}\right) \equiv \frac{1}{\tau_G} - \frac{1}{\tau_D} \tag{11.18c}$$

in which the following abbreviations have been used for the subscripts: N = net, P = production, I = inflow, R = removal, O = outflow, G = growth, and D = decay. Equation 11.18c simply states that the net growth rate of Q is the growth rate (in turn equal to the rates of production and inflow) less the decay rate (in turn equal to the rates of removal and outflow).

For steady-state conditions, the net growth rate is zero. This net growth rate can be zero because both the growth and decay rates vanish, but the more interesting case of zero net growth is when there is a balance between growth and decay. When this balance occurs, the common time constant is called the residence or turnover time, τ. Thus, for steady-state conditions,

$$\frac{Q}{(P+I)} \equiv \tau_G = \tau_D \equiv \frac{Q}{(R+O)} \tag{11.18d}$$

TABLE 11.4 Application of Eq. 11.20 to Estimate Tropospheric Residence Times for Certain Atmospheric Trace Constituents†

Material	Vinyl chloride	SO_2	H_2O	0.1- to 1-μm particles	DDT	Fluoro-carbons	^{85}Kr
τ_d	≫1 d	~10 d	~10 d	≧1 mo	~3 mo	≫3 yr	≫10 yr
τ_w	≫1 d	~2 mo	~10 d	~10 d	1–10 yr	≫3 yr	≫10 yr
τ_c	~hours	1-3 d			≫1 yr	≫3 yr	≫10 yr
τ_p	~3 yr‡					~3 yr‡	~10 yr§
τ¶	~hours	1-3 d	~1 wk	~1 wk	~3 mo	~3 yr	~10 yr

†See Table 11.1.

‡The physical transformation considered here is physical transport from the troposphere to the stratosphere and subsequent chemical reaction in the stratosphere.

To illustrate the use of Eq. 11.18d, consider the case when the reservoir is the entire troposphere. Suppose the pollution's outflow to the earth's surface, by wet and dry deposition, is described via

$$O = \int (v_w + v_d)\chi_0 \, dA \tag{11.19}$$

where v_w and v_d are average wet and dry deposition velocities. Suppose, also, that there are volumetric chemical and physical loss terms that are first-order in the pollution's concentration: $L = (\lambda_c + \lambda_p)\chi$. Then Eq. 11.18d gives for the residence time

$$\frac{1}{\tau} = \frac{1}{\tau_w} + \frac{1}{\tau_d} + \frac{1}{\tau_c} + \frac{1}{\tau_p} \tag{11.20}$$

where $\tau_p = \lambda_p^{-1}$, $\tau_c = \lambda_c^{-1}$, $\tau_d = h_d/v_d$, and $\tau_w = h_w/v_w$, in which the height scales are given by $h = \int [\chi(z)/\chi(O)]\, dz$. Table 11.4 demonstrates how estimates for the individual removal rates in Eq. 11.20 can be used to estimate the residence time (or overall removal rate, τ^{-1}) for a number of species, some of which were previously listed in Table 11.1.

The estimates shown in Table 11.4 are quite crude. To refine these estimates, we need many improvements: to account for unsteady and inhomogeneous conditions in reservoirs (Schwartz, 1979; Slinn, 1980), to account for the stochastic nature of the removal and transformation processes (Rodhe and Grandell, 1972, 1981; Slinn, 1982a, and see Sec. 11-5), and to define the dependence of the removal rates on properties of the particulate and gaseous species. As an illustration of this last point, Figure 11.3 shows an estimate of the dependence of the tropospheric residence time for particles on their size. Surveying information leading to estimates for the τ_w shown in this figure is one of the objectives of the next section.

In Sec. 11-3, describing precipitation of particles, and Sec. 11-4, discussing washout of gases, the tilde (˜), indicating time average values, is suppressed to simplify notation.

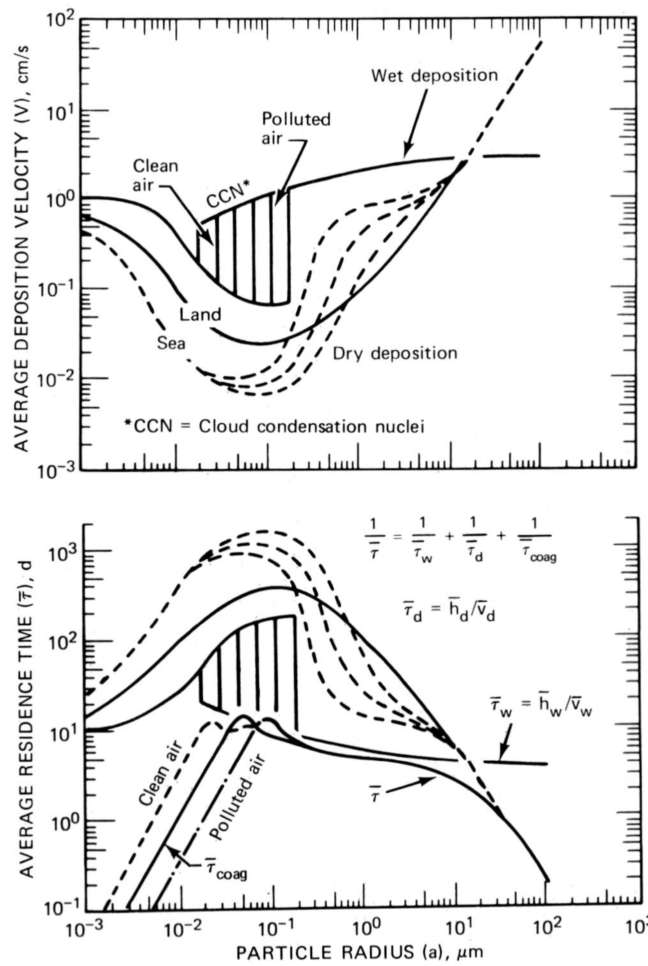

Fig. 11.3 A qualitative indication of how the tropospheric residence time of particles depends on their size. Estimates for the particle dry deposition velocities were obtained from the theory described by Slinn and Slinn (1980) and Slinn (1982a,b); estimates for the wet deposition velocities were based on the data obtained by Cawse (1974) and Gatz (1975a) (see Sec. 11-3); and estimates for the contribution to the residence times from interparticle coagulation (whereby particles of a given size "disappear," but can remain in the atmosphere attached to other particles) were based on Junge (1963) (see Sec. 11-3). [From W. G. N. Slinn, 1983a, Air-to-Sea Transfer of Particles, Chapter 6 of *Air–Sea Exchange of Gases and Particles*, P. S. Liss and W. G. N. Slinn (Eds.), NATO Advanced Study Institute Series, D. Reidel Publishing Co., Dordrecht, Holland (Kluwer Academic Publishers, Higham, Mass.).]

11-3 PRECIPITATION SCAVENGING OF PARTICLES

11-3.1 Collision, Retention, and Collection Efficiencies

If we assume that the chance for capture of an individual aerosol particle by hydrometeors is independent from the number of particles already captured, the loss rate per unit volume, L, can be written as

$$L(\mathbf{r},t;a) = \psi(\mathbf{r},t;a)\,\chi(\mathbf{r},t;a) \tag{11.21}$$

where ψ is the scavenging rate. Particle size is characterized by the parameter a (e.g., a = particle radius). By consideration of the number of particles swept up by each hydrometeor during a time interval Δt, through integration over all hydrometeor sizes, and by division by Δt, the scavenging rate for both in- and below-cloud scavenging is seen to be

$$\psi = \int_0^\infty \epsilon E(a,\lambda)A_x[v_t(\ell) - v_g(a)]N(\ell;\mathbf{r},t)\,d\ell \tag{11.22}$$

where v_g (usually negligible) is the gravitational settling speed of the particles (see Fig. 11.2); ℓ characterizes the size of the hydrometeor (e.g., drop radius or circumscribed radius for a snowflake); λ is a characteristic capture length scale for the hydrometeor (e.g., drop radius); N, v_t, and A_x are hydrometeor number density, terminal speed, and effective cross-sectional area, respectively; and E and ϵ are the particle/hydrometeor collision and retention efficiencies, respectively. The product ϵE is known as the collection efficiency.

Figure 11.4 illustrates the collision efficiency. One way to interpret the collision efficiency for an obstacle of effective cross-sectional area A_x is to consider EA_x to be the effective collisional cross-sectional area. This area is similar to the familiar "cross section" in nuclear physics. For a spherical drop of radius R, $A_x = \pi(R + a)^2$, where a is the particle radius. For another view of the collision efficiency, consider an area, A_c, which is far upstream from the obstacle and from which emanating particles will collide with the obstacle (see Fig. 11.4). Then the collision efficiency can be defined via $E = A_c/A_x$. Some authors prefer to use the square root of this area ratio, referred to as the linear collision efficiency. A collision efficiency of unity means that all particles in the geometric volume swept out by a hydrometeor as it falls (relative to the particles) will collide with the hydrometeor. The collision efficiency can be larger than unity, e.g., if electrical interactions are large, but usually $E < 1$.

One way to calculate the collision efficiency is to determine the total rate of collisions between particles and the hydrometeor (i.e., integrate the flux of particles to the hydrometeor over the surface area of the hydrometeor) and to normalize the result by the total flow of particles past an area in the upstream flow equal to the effective cross-sectional area of the hydrometeor. For a drop and for the flux independent of position on the drop's surface, this procedure would give

$$E = \frac{(\text{flux})4\pi R^2}{\chi_0(v_t - v_g)\pi(R+a)^2} \simeq \frac{4(\text{flux})}{v_t\chi_0} \tag{11.23}$$

where χ_0 is the upstream concentration of particles, (flux) is the rate of collection divided by the surface area, and $v_t - v_g$ is the relative speed between the drop and the particle. One way to evaluate the collision efficiency experimentally, under the assumption that the retention efficiency is unity, is to determine the number of particles, N, collected by an individual hydrometeor as it falls a length L through a chamber in which the particle number concentration is χ_0. For a raindrop this gives for the collection efficiency

$$\epsilon E = \frac{N}{\chi_0\pi(R + a)^2 L} \tag{11.24}$$

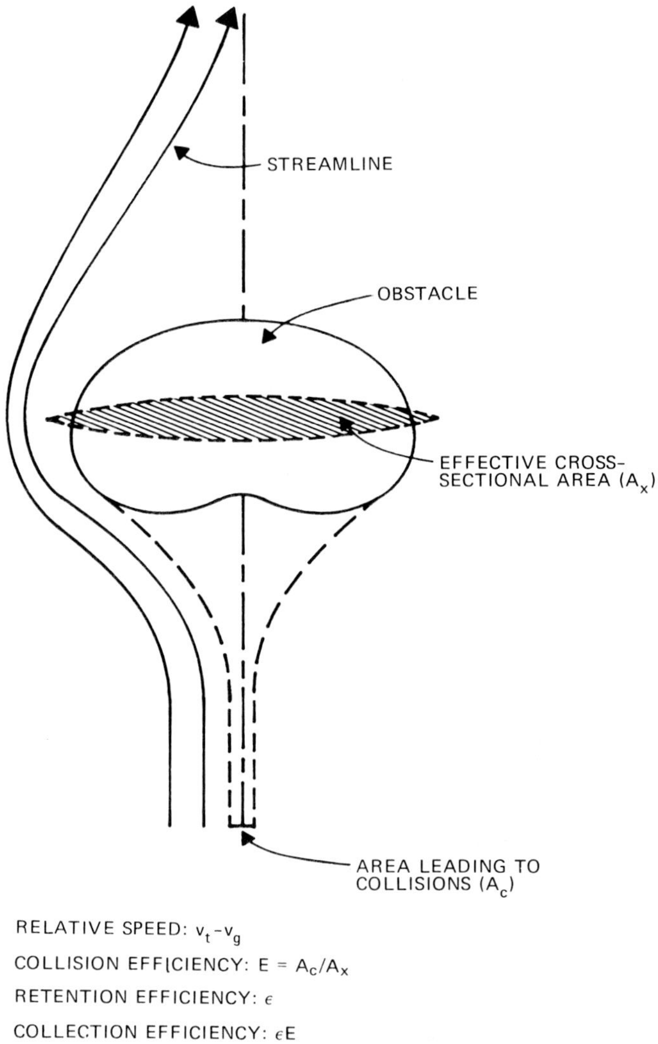

Fig. 11.4 A schematic defining the collision efficiency. The retention efficiency is almost invariably taken as unity (see discussion in text).

The equivalence of Eqs. 11.23 and 11.24 for the case $\epsilon = 1$ can be seen by setting $L = (v_t - v_g)T$, where T is the fall time, and then by suitably defining the average flux to the hydrometeor. However, whether or not the two methods outlined by Eqs. 11.23 and 11.24 give comparable results depends on the realism of the assumption that the retention efficiency, ϵ, is unity.

The realism of the assumption that the retention efficiency is unity remains a major uncertainty in precipitation scavenging predictions. As stated well by Hidy (1970):

> One of the ghosts in the scavenging-theory closet is the question of particle or droplet bounce during collisions. This effect, measured by the retention efficiency, is difficult to evaluate because of the lack of a theory for near-particle interactions and few experiments from which to evaluate the efficiency. There are indications that collisions between solid particles and droplets or between particles and ice crystals have a retention efficiency essentially unity. Therefore, in the absence of any better information, this efficiency has to be taken as unity in any theoretical model.

Section 11-3.5.3 discusses further the retention efficiency problem; some dramatic uncertainties in the collection efficiency may indicate that the retention efficiency "ghost" has returned to haunt us. In the meantime, we proceed with the assumption that $\epsilon = 1$ and therefore equate the collection efficiency to the collision efficiency. The next two sections address specification of the collision efficiency, first for particle/drop collisions and then for particle/ice crystal collisions.

11-3.1.1 Particle/Drop Collision Efficiency. Many physical processes can contribute to particle/drop collisions. One way to identify the different processes is to identify the nondimensional groups (e.g., Reynolds number, Stokes number, etc.) that appear in the governing equations when these equations are nondimensionalized. The governing equations can be for the motion of a single particle or for a continuous stream of particles in the neighborhood of a single drop; through a classical set of equations known as the Langevin, Fokker–Planck, and Smoluchowski equations, it is possible to demonstrate equivalence between the single-particle and the continuum formulation (e.g., see Fuchs, 1964, or Slinn and Shen, 1971). In the nondimensionalization of the governing equations, the choice is made here to nondimensionalize velocities by the terminal speed of the drop, v_t, and lengths by the drop's radius, R. This choice of velocity and length scales causes no difficulties in the present discussion, but the choice can lead to deceptive results if perturbation solutions to the equations are sought (Slinn, 1974a, 1977b).

For steady-state conditions (i.e., a coordinate system fixed to a nonaccelerating drop) and with the same symbols for dimensionless quantities as when they possessed dimensions, the governing nondimensional equations include:

1. The momentum equation for the (incompressible) fluid:

$$\mathbf{v} \cdot \nabla \mathbf{v} = -\frac{1}{\mathrm{Eu}} \nabla p + \frac{1}{\mathrm{Re}} \nabla^2 \mathbf{v} \tag{11.25}$$

where pressure is nondimensionalized with its upstream value, p_0; the Euler number, $\mathrm{Eu} = \rho_0 v_t^2/p_0$, is the ratio of dynamic to static pressure; and the Reynolds number, $\mathrm{Re} = Rv_t/\nu$, is the ratio of rates of convection to diffusion of vorticity.

2. The convective diffusion equation for the particles:

$$\mathbf{v} \cdot \nabla \chi = \nabla \cdot \left[\frac{1}{\mathrm{Pe}} \nabla \chi - \mathbf{d}\chi\right] \tag{11.26}$$

where the Péclét number, $\mathrm{Pe} = Rv_t/D = \mathrm{Re\,Sc}$, is the ratio of rates of convection to diffusion of the particles and the Schmidt number, $\mathrm{Sc} = \nu/D$, is the ratio of the diffusivities for vorticity and for particles.

3. The momentum equation for the particles, which can be approximated by

$$\mathbf{u} \cdot \nabla \mathbf{u} = -\frac{\mathbf{d}}{\mathrm{St}} + \frac{R}{v_t^2} \mathbf{a} \tag{11.27}$$

where $\mathbf{u} = \mathbf{v} + \mathbf{d}$ is the total velocity of the particles; $\mathbf{d}$ is the drift velocity relative to the fluid; $\mathbf{a}$ represents other (dimensional) forces per unit mass on the particle (e.g., caused by electrical charges); and the Stokes number, s or $\mathrm{St} = \tau v_t/R$, is the ratio of the particle stopping time, τ (see Fig. 11.2), to the characteristic time scale of the flow field, R/v_t.

The preceding nondimensional equations show that the particle/drop collision efficiency can depend on many parameters. Two other nondimensional groups not explicit in Eqs. 11.25 to 11.27 are the ratio of particle radius to drop radius, $\kappa = a/R$, and the ratio of dynamic viscosities for water and air, $V = \mu_w/\mu_a$, which enters when circulation within the drop is considered. Other dimensionless groups enter if there are heat and moisture fluxes in the air: (1) a first-order estimate for particle drift caused by a heat flux $\mathbf{q}$ (the "thermophoretic" drift arises from more violent molecular impacts on one side of the particle than on the other) is $\mathbf{q}/(5p)$, where p is the pressure and (2) a first-order estimate for particle drift caused by a vapor flux $\dot{m}_v$ (the "Stefan flow" drift arises from drag of the particle by the vapor flux) is $\dot{m}_v/\rho_a$, where ρ_a is the density of air. Corrections to these estimates are available in the literature (e.g., Slinn and Hales, 1971). However, even with the absence of phoretic processes (including electrophoresis), the collision efficiency obviously can be extremely complicated—it can be expected to be a function of, at least, six dimensionless groups:

$$E = E(\mathrm{Re}, \mathrm{Eu}, \mathrm{Sc}, \kappa, V, \mathrm{St}) \tag{11.28}$$

Unfortunately, there is not sufficient space in this chapter to elaborate on methods that have been used to deduce functional forms for these dependencies; however, a 16-page summary is given in Slinn (1977b). The resulting tentative suggestion for the particle/drop collision efficiency is shown in Fig. 11.5. The theoretical and experimental bases are sufficient for the diffusional and inertial portions for the results in Fig. 11.5 to be considered reliable to within a factor of about 2 or 3. However, there are very significant (order-of-magnitude) uncertainties about the collision efficiency for particles in the size range $0.1 \lesssim a \lesssim 1\ \mu m$. These and other uncertainties are discussed in Sec. 11-3.5.3.

11-3.1.2 Particle/Ice Crystal Collision Efficiency. In contrast to substantial effort devoted to the particle/drop collision efficiency problem, relatively few studies have been devoted to particle/ice crystal collisions. One reason is the huge variety of crystal types; another is the lack of even approximate descriptions of the flow fields about (and through) ice crystals. In spite of this dearth of knowledge, the particle size dependence of the collision efficiency can be expected to display the three features shown in Fig. 11.5 for the particle/drop collision efficiency: the dominance of (1) molecular diffusion for particles of radii $a \lesssim 0.1\ \mu m$, (2) inertial impaction for particles with $a \gtrsim 1\ \mu m$, and, in the absence of significant phoretic effects, (3) interception for particles of intermediate size range. Figure 11.6 demonstrates a semiempirical expression for the collision efficiency, *E*, between particles and ice crystals. Through appropriate choice of the capture length scale λ (see the inset in Fig. 11.6), it is possible to force the curves through most of the available collection efficiency data. However, recent results showing an order-of-magnitude larger ξ for rimed vs. unrimed crystals (Murakami, Hiramatsu, and Magono, 1981) have not yet been interpreted theoretically (Slinn, 1983b).

11-3.2 Approximations for the Scavenging Rates

Presumably it is clear already that there are significant uncertainties about precipitation scavenging rates; more uncertainties are described later. Because of these uncertainties, it appears to be consistent to introduce a number of approximations into the formulation. In this section, approximations will be introduced first for rain and then for snow scavenging; in later sections, the results of these approximations are compared with some data. To initiate the presentation, we simplify Eq. 11.22 for the scavenging rate by taking the retention effi-

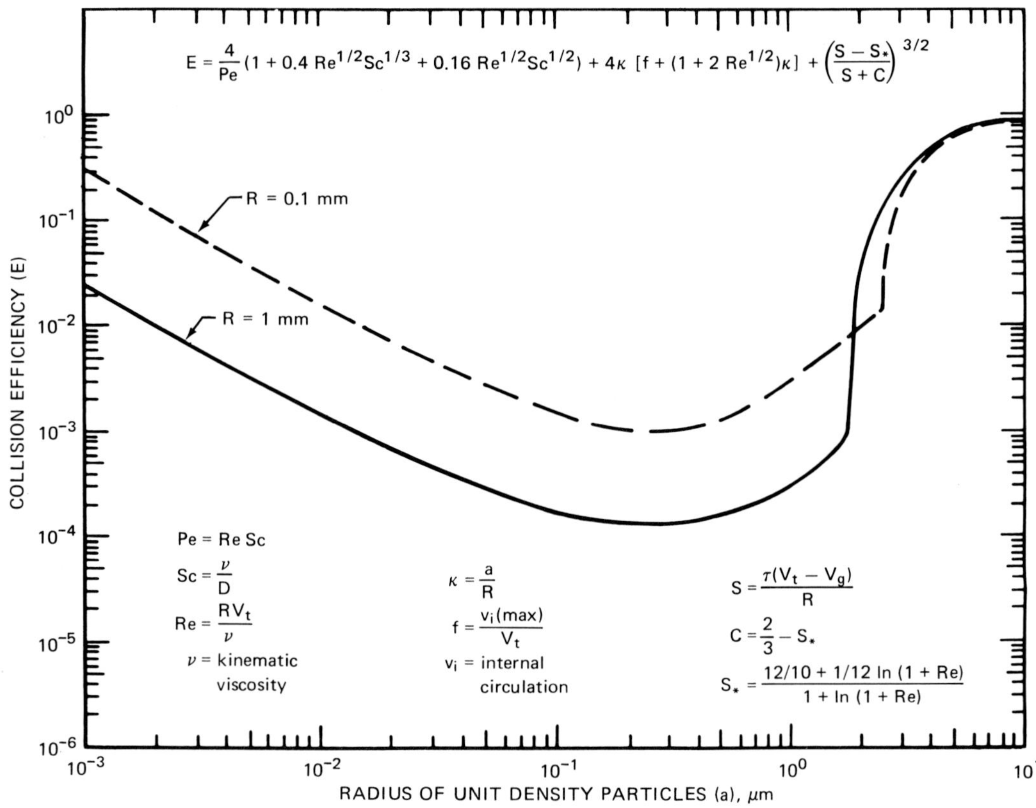

Fig. 11.5 A semiempirical expression for the particle/drop collision efficiency (Slinn, 1977b, 1977c); the expression shown at the top of the figure is evaluated for two drop radii, R. The contribution to *E* from Brownian diffusion ($a \lesssim 0.1\ \mu m$) includes the evaluation by Zimin (1964) of the effect of internal circulation. The crude estimate of the contribution to *E* from interception ($0.1 \lesssim a \lesssim 2\ \mu m$) uses the internal circulation results (see Fig. 11.60) obtained by LeClair et al. (1972). The contribution to *E* from impaction ($a \gtrsim 2\ \mu m$) uses a semiempirical fit to the numerical results obtained by Beard and Grover (1974). The minimum in *E* for $a \simeq 0.5\ \mu m$ reflects both the decreasing diffusivity for larger aerosol particles and the relative inefficiency of collection by interception. For particles of mass density ρ_p different from 1, the inertial contribution to the collision efficiency should be scaled by $(\rho_w/\rho_p)^{1/2}$ where $\rho_w = 1\ g/cm^3$; e.g., for $\rho_p = 4\ g/cm^3$, the critical Stokes number, s_*, for a 1-mm drop occurs at $a \simeq 1.7(1/4)^{1/2} \simeq 0.85\ \mu m$.

ciency to be unity and by ignoring the gravitational settling speed of the pollutant particles. The result for the scavenging rate is

$$\psi(\mathbf{r},t) = \int_0^\infty E(a,\lambda)\, A_x v_t(\ell)\, N(\ell;\mathbf{r},t)\, d\ell \qquad (11.29)$$

where, again, λ is a characteristic capture length for the hydrometeor and ℓ characterizes the hydrometeor size, e.g., $\ell = R$, the drop radius.

11-3.2.1 ***Rain Scavenging of Particles.*** To evaluate Eq. 11.29 for particle scavenging, we need to know the settling speeds and number distribution for the drops. Figure 11.7 shows data for the settling speeds along with a simple empirical expression that fits a portion of the data; many other empirical expressions are available (e.g., Best, 1950; Beard and Pruppacher, 1969). Empirical expressions to fit raindrop number distributions are much more difficult to obtain. As illustrated in Fig. 11.8, for some stable frontal storms, the Marshall–Palmer distribution (e.g., see Mason, 1971)

$$N(D) = N_0 \exp(-\epsilon D) \qquad (11.30)$$

where D is the raindrop diameter, provides a reasonable fit to average number distributions using $N_0 = 0.08/cm^4$ and $\epsilon = 41p^{0.21}/cm$, where p is the rainfall rate in mm/h. However, Fig. 11.9 shows that the parameters N_0 and ϵ can have great variability even during the passage of a single storm. More dramatically, Figs. 11.10 and 11.11 indicate that drop distributions for convective storms are sometimes not even qualitatively similar to the decaying exponential given by Eq. 11.30. As Fig. 11.12 shows, many other functions have been proposed to fit the number distributions for drops, but almost no guidance is available to recommend a specific distribution and its parameters for use in scavenging predictions.

As a result of these uncertainties about the drop size distribution, the uncertainties mentioned earlier (e.g., the retention and collision efficiencies) and the uncertainties yet to be described (e.g., electrical charges and particle characteristics and their modifications in the atmosphere), the following approximation to Eq. 11.29 has been advanced (Slinn, 1977b). First, the similarity is noted between Eq. 11.29 and the defining expression for the rainfall rate

$$p(\mathbf{r},t) = \int_0^\infty \frac{4}{3}\pi R^3 v_t N(R;\mathbf{r},t)\, dR \qquad (11.31)$$

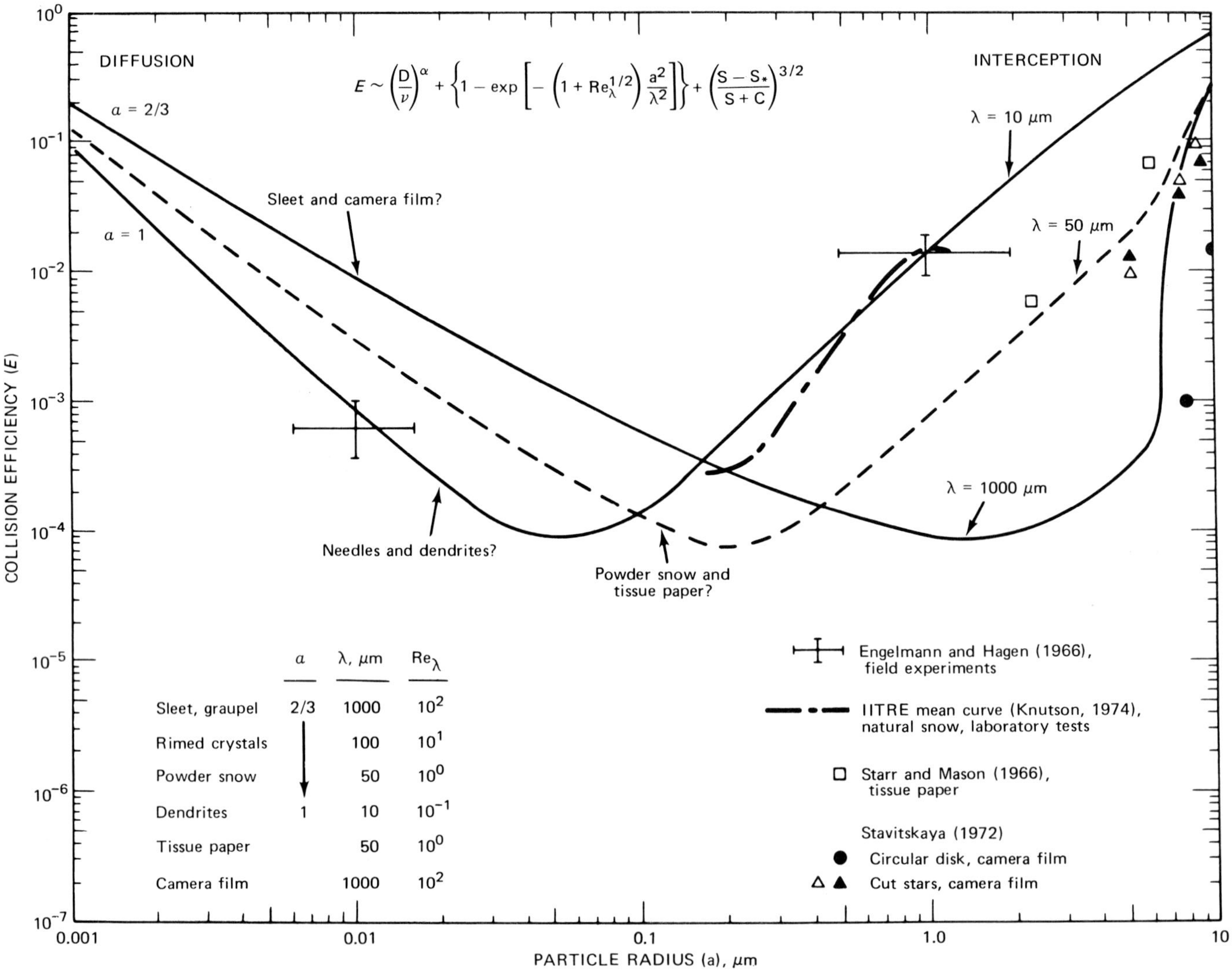

Fig. 11.6 A tentative suggestion for the particle/ice crystal collision efficiency, *E*, as a function of particle size and characteristic capture length, λ, of the ice crystal. Substantially more research is needed before the particle/ice crystal collision efficiency can be put on a firm basis. [Based on W. G. N. Slinn, Some Approximations for the Wet and Dry Removal of Particles and Gases from the Atmosphere, *J. Water, Air, Soil Poll.*, 7: 513-543 (1977).]

Approximations to Eq. 11.29 are now made by evaluating E using the mass mean raindrop radius, R_m, and by multiplying the integrand in Eq. 11.29 by R and by dividing the integral by R_m. These approximations lead to the approximate expression for the rain scavenging rate

$$\psi_r(a;\mathbf{r},t) = \frac{c\, p(\mathbf{r},t)\, \overline{E}(a,R_m)}{R_m} \tag{11.32}$$

where c is a numerical factor of order unity and the collision efficiency is evaluated by using a mean drop size. In comparisons with data collected during frontal storms, it appears adequate to take $c \simeq \frac{1}{2}$ and to use the result given by Mason (1971) for fairly steady rains

$$R_m = 0.35\,\text{mm}\left(\frac{p}{1\text{mm/h}}\right)^{1/4} \tag{11.33}$$

An alternative to the approximations used here to obtain Eq. 11.32 from Eq. 11.29 is to use an empirical raindrop number distribution in Eq. 11.29, to assume $\overline{E}$ to be independent from R, and to evaluate the resulting integral (e.g., see Slinn, 1971a, or Scott, 1977). However, as was suggested earlier, the many uncertainties (e.g., in the adequacy of the drop number distribution) suggest that the method used here is preferable. This suggestion is fortified by comparisons with data in Sec. 11-3.5.

11-3.2.2 Snow Scavenging of Particles. The development of a similar approximation for particle scavenging by snow is not so simple. One approximation is described elsewhere (Slinn, 1977b); this chapter advances a different and simpler approximation. Figure 11.13 shows data for the mass, m, of various crystal types as a function of their maximum linear dimension, D_c (e.g., diameter of circumscribed sphere). Functional forms for the empirical curves shown fit to the data of Fig. 11.13 are given in Table 11.5. To develop the approx-

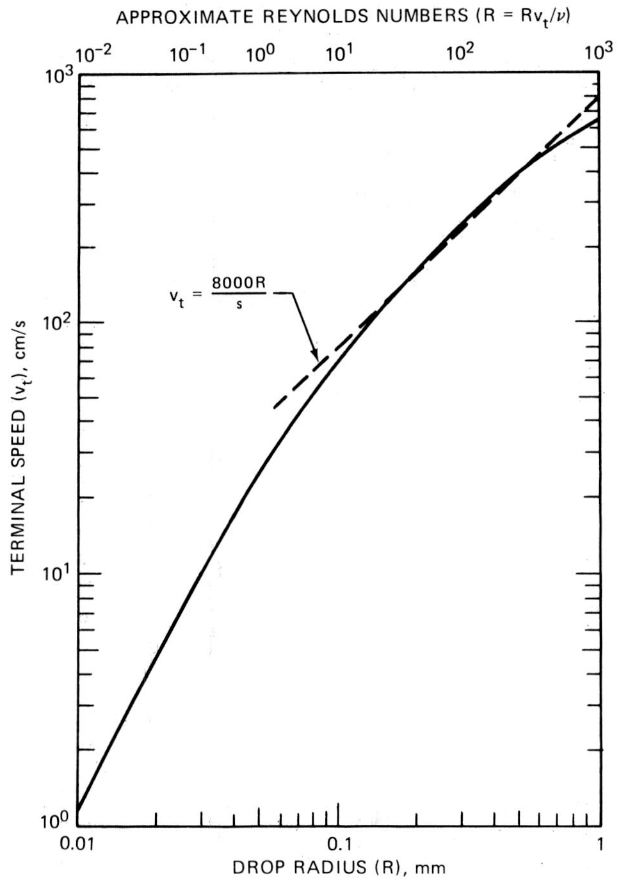

Fig. 11.7 Terminal velocities of water drops in still air (pressure 1013 mb; temperature 20°C). (Data from B. J. Mason, *The Physics of Clouds,* 2nd ed., Clarendon Press, Oxford, 1971.)

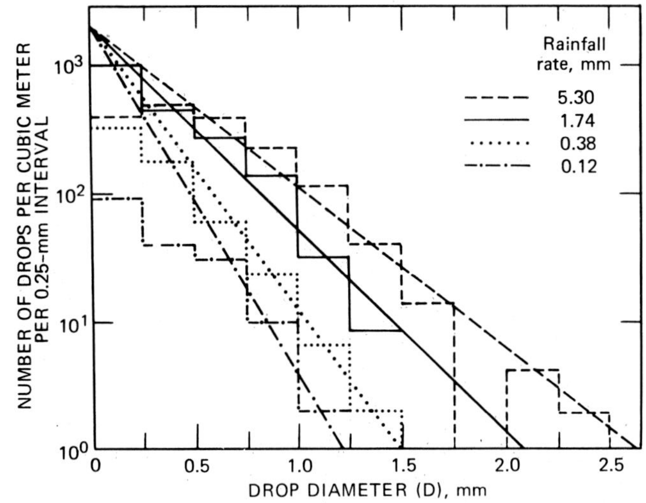

Fig. 11.8 Number distributions for raindrops from stable frontal cloud systems exhibiting marked radar melting bands but otherwise little structure. Straight lines are fits using the Marshall–Palmer distribution, Eq. 11.30. (Based on B. J. Mason, *The Physics of Clouds,* 2nd ed., Clarendon Press, Oxford, 1971.)

imate expression for the snow scavenging rate, first notice the similarity between the expression for the snowfall rate (rainwater equivalent)

$$p(\mathbf{r},t) = \frac{1}{\rho_w} \int_0^\infty m(D_c)v_t(D_c)N(D_c;\mathbf{r},t)dD_c \tag{11.34}$$

where ρ_w is the density of water, and the approximate expression for the snow scavenging rate

$$\psi_s(a;\mathbf{r},t) \simeq \frac{\pi}{4} E(a,\lambda) \int_0^\infty D_c^2 v_t(D_c)N(D_c;\mathbf{r},t)\, dD \tag{11.35}$$

in which λ is the characteristic capture length shown in Fig. 11.6. By comparing these two expressions and using the data of Table 11.5, we get the approximation

$$\psi_s(a;\mathbf{r},t) = \frac{\gamma p(\mathbf{r},t)E(a,\lambda)}{D_m} \tag{11.36}$$

where γ is a dimensionless constant of order unity and D_m is the characteristic length given in the right-hand column of Table 11.5.

Unfortunately, few data are available to test the accuracy of the snow scavenging rate given by Eq. 11.36. Figure 11.14 shows the predicted variation of ψ_s with precipitation rate for different crystal types for the case $E = 1$ and $\gamma = \pi/4$ and gives values for γ chosen to fit data obtained by Engelmann et al. (1966). Further studies of snow scavenging of particles are definitely needed. Comparisons between Eq. 11.36 and predictions made by others are described in Sec. 11-3.5.

11-3.3 Particle-Average Scavenging Rates

The collection efficiency describes the capture of a single particle by a single hydrometeor. The scavenging rate, Eq. 11.29, is a weighted average of the collection efficiency over all hydrometeor sizes. The result was approximated by Eq. 11.32 for rain scavenging and by Eq. 11.36 for snow scavenging. Similarly, for practical applications, it is almost always desirable to develop approximations for the particle-average scavenging rate. This average is taken over the distribution of particle sizes. This task is discussed in the remainder of the section.

Except in fundamental studies, rarely is it of interest to know the scavenging of the *number* of aerosol particles of a specific size. Instead, practical interest is usually on the scavenging of some moment of the number distribution corresponding, for example, to the distribution of radioactivity borne by the particles or to the total mass of the particles. To develop the formulation for particle-average scavenging rates, let f(a) describe the probability density function for the number distribution; e.g., if $\chi(a)\, da$ is the number of aerosol particles per unit volume, whose radii are between a to a + da, then

$$f(a) = \frac{\chi(a)}{N_T} \tag{11.37}$$

where $N_T = \int_0^\infty \chi(a)\, da$ is the total number density of particles. Then the normalized, particle-average removal rates are

$$\langle {}^n\psi \rangle = \frac{\int_0^\infty \psi(a)a^n f(a)\, da}{\int_0^\infty a^n f(a)\, da} \tag{11.38}$$

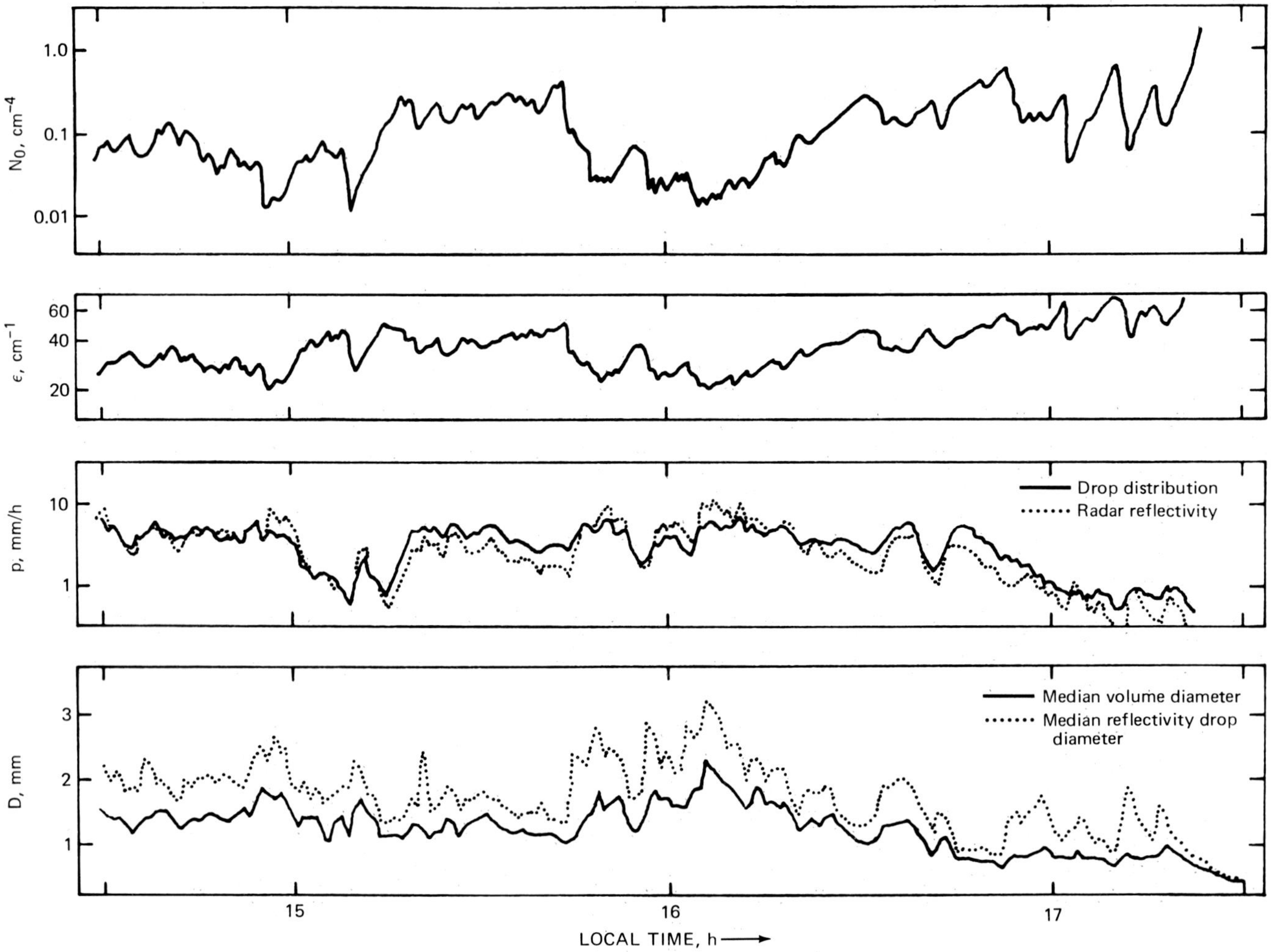

Fig. 11.9 Variations in the parameters N_0 and ϵ of the Marshall–Palmer distribution, Eq. 11.30, during a large-scale rain event in western Europe. [Based on A. Waldvogel, The N_0 Jump of Raindrop Spectra, *J. Atmos. Sci.*, 31: 1067-1078 (1974).]

In Eq. 11.38, the case n = 2 would be appropriate for the removal of radioactivity distributed on particles proportional to the particle surface area; the case n = 3 would be appropriate for the removal rate for total pollutant mass; for some cases it may be necessary to replace a^n in Eq. 11.38 by more general functions of a.

One of the significant consequences of this rather mundane analysis is that the particle mass rainout (or snowout) rate for a polydisperse aerosol can be much greater than might at first be expected for particles with specified mean size (Slinn, 1971a; Dana and Hales, 1976). This difference is illustrated in Fig. 11.15, which shows the results of numerical integrations of Eq. 11.38 for f(a) given by a lognormal distribution (geometric mean particle size, a_g, and geometric standard deviation, σ_g) and for a lognormal raindrop size distribution with indicated parameters. The collision efficiency used in $\psi(a)$ in Eq. 11.38 was essentially as given in Fig. 11.5 but with the interception term taken to be 3a/R.

In particular, these results show that the mass rainout rate for a polydisperse aerosol with $0.1 < a_g < 1$ μm can be much larger than for a monodisperse aerosol ($\sigma_g = 1$) with a_g in the same range. This observation has been advanced to explain why the mass rainout rates shown in Fig. 11.16 do not display a strong dependence on mean particle size. Thus, although the rainfall rate dependence of $\langle {}^3\psi_r \rangle$ is roughly as expected (e.g., using Eq. 11.32 in Eq. 11.38), the particle size dependence is much weaker than would have been expected from the dependence of the collision efficiency on particle size shown in Fig. 11.5. The results are more in accord with the mass-average rainout rates shown in Fig. 11.15 although unavailability of information on particle size at the time of the scavenging precludes a firm comparison. That is, although particle size information was obtained during equipment testing, it is necessary to determine particle size during the scavenging measurements if a complete comparison is to be made. One of the complications is particle growth because of water vapor condensation; this is discussed in the next section. Another complicating feature of these specific studies is the occurrence of simultaneous dry deposition (see Wolf and Dana, 1969; Slinn, 1974b, 1977b).

11-3.4 Changes in the Scavenging Rates via Particle Modifications

The last paragraph in the preceding section mentioned one topic in the general area of particle modifications in the atmosphere, namely, particle growth by water vapor condensation. Other modifications are Brownian diffusional coagulation and

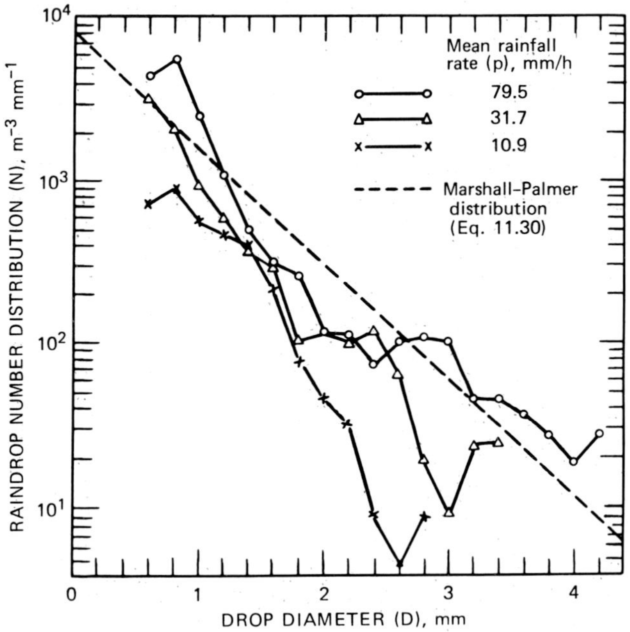

Fig. 11.10 Number distributions for raindrops from convective storms. Dashed curve is the Marshall–Palmer distribution, Eq. 11.30, for p = 79.5 mm/hr. [Based on A. N. Dingle and K. R. Hardy, The Description of Rain by Means of Sequential Raindrop-Size Distributions, *Q. J. R. Meteorol. Soc.*, 88: 301-314 (1962).]

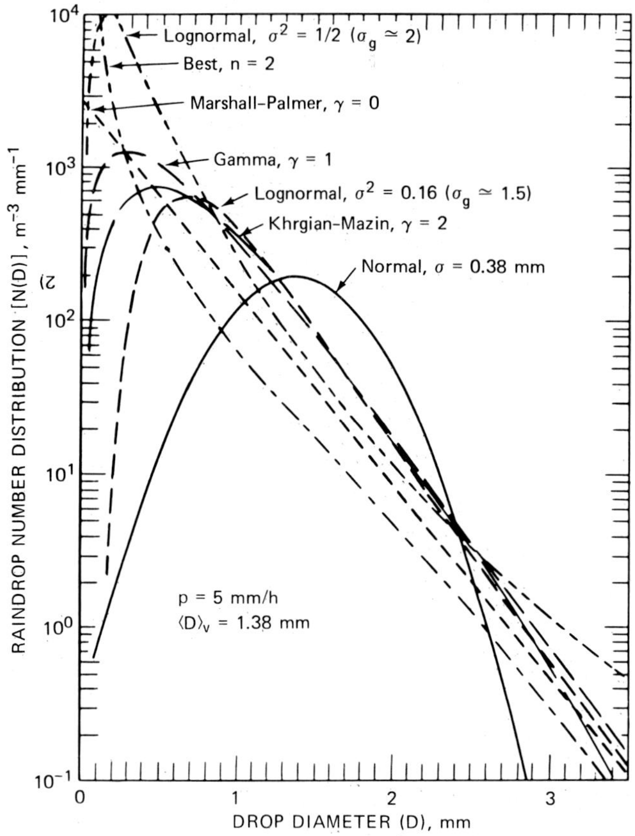

Fig. 11.12 A comparison of some empirical formulas proposed by various authors for raindrop number distribution function, all evaluated at the same mean-volume drop diameter and rainfall intensity. [From W. G. N. Slinn, Some Approximations for the Wet and Dry Removal of Particles and Gases from the Atmosphere, *J. Water, Air, Soil Poll.*, 7: 513-543 (1977).]

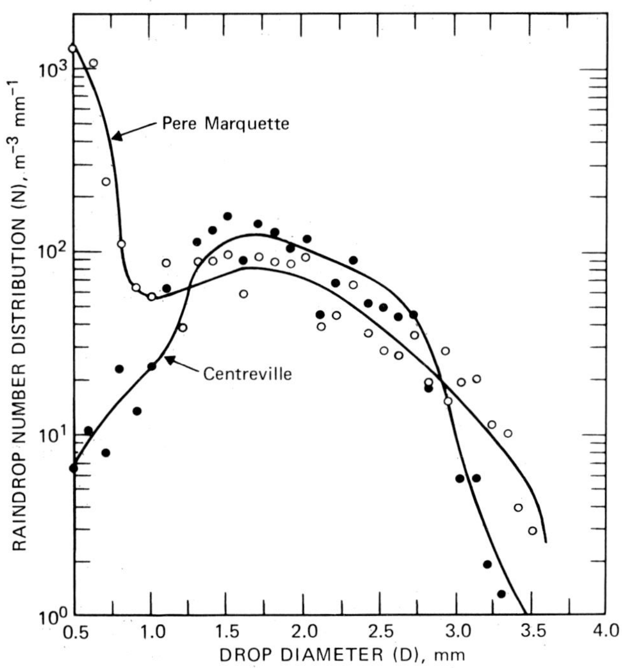

Fig. 11.11 Average number distributions for raindrops from convective storms near St. Louis, Mo. Data for the Pere Marquette site, relatively distant from urban and industrial pollution, are a composite of 10 samples with p = 14.6 mm/h; data for the Centreville site, closer to St. Louis, are a composite of 16 samples with p = 14.4 mm/h. Note that a (straight-line) Marshall–Palmer distribution would not be even qualitatively similar to either curve. [Based on R. Cataneo, Surface Measurements of Condensation Nuclei and Raindrop Distributions, *Summary Report of METROMEX Studies*, F. A. Huff (Ed.), Report ISWS-73-R174, pp. 84-88, Illinois State Water Survey, 1973.]

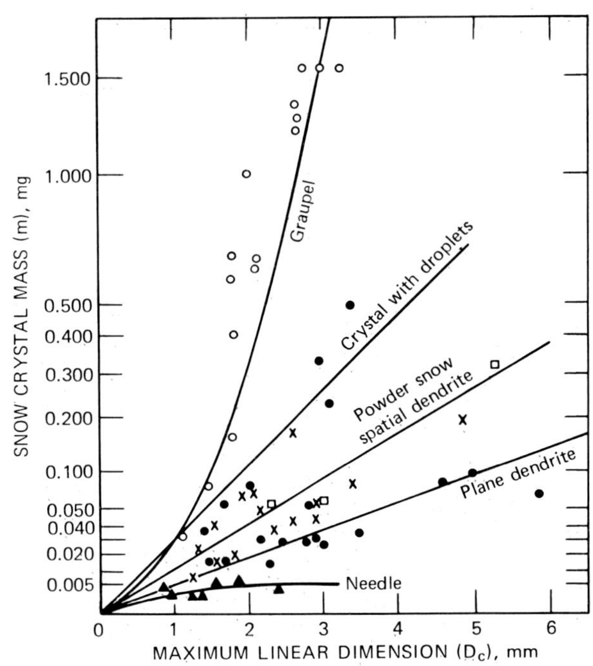

Fig. 11.13 Mass as a function of maximum linear dimension for various ice crystals and aggregates. Table 11.5 gives functional forms of the empirical curves shown. (Based on B. J. Mason, *The Physics of Clouds*, 2nd ed., Clarendon Press, Oxford, 1971.)

TABLE 11.5 Empirical Relations Between Masses and Maximum Linear Dimensions of Ice Crystals† and the Characteristic Length Scale D_m to Be Used in the Snow Scavenging Rate, Eq. 11.36‡

Crystal type	Mass,§ g	D_m, cm
Graupel particles	$6.5 \times 10^{-2} D_c^3$	1.4×10^{-2}
Rimed plates and stellar dendrites	$2.7 \times 10^{-3} D_c^2$	2.7×10^{-3}
Powder snow and spatial dendrites	$1.0 \times 10^{-3} D_c^2$	1.0×10^{-3}
Plane dendrites	$3.8 \times 10^{-4} D_c^2$	3.8×10^{-4}
Needles	$2.9 \times 10^{-5} D_c$	3.8×10^{-3}

†See Fig. 11.13.
‡Data from Mason (1971).
§D_c, in cm, is the diameter of the sphere that just contains the crystal; for needles, D_c = length of needle.

changes in chemical composition (and therefore wettability) caused, for example, by the adsorption and chemical transformation of trace gases. Because of modifications such as these, particle scavenging rates can be significantly modified. In Secs. 11-3.4.1 to 11-3.4.3, some of these modifications will be mentioned; however, the discussion cannot be complete here and should be viewed more as an attempt to alert the user to potential complications.

11-3.4.1 ***Particle Growth by Water Vapor Condensation.*** Because of the apparent strong dependence of the particle/hydrometeor collision efficiency on particle size (see Figs. 11.5 and 11.6), it is important to account for particle growth by water vapor condensation. As is discussed in many cloud physics texts, water vapor condensation is promoted by soluble constituents in the particle (Raoult's law) and is inhibited by the increase in equilibrium vapor pressure over a curved surface (Kelvin effect). Figure 11.17 shows, for the indicated salts, the resulting (Kohler) curves of the equilibrium particle (or droplet) radius for specified saturation ratio. Equilibrium sizes for $(NH_4)_2SO_4$ particles exposed to saturated and subsaturated atmospheres are shown in Fig. 11.18. Finally, Fig. 11.19 shows the equilibrium size of other nuclei; the curves (identified in the figure legend) for samples of aerosol particles collected from the atmosphere are especially significant. Typically, particles found in the atmosphere have an internally mixed chemical composition; the resulting curves show that for a relative humidity (RH) less than about 90%, particle growth is relatively modest ($\lesssim$ factor of 2). Consequently, as far as precipitation scavenging calculations are concerned, particle growth probably can be ignored for RH $\lesssim$ 90%, given the existence of so many other uncertainties. However, in the typical case during precipitation, with RH > 90%, particle growth can be substantial.

An essential feature not revealed in the equilibrium curves of Figs. 11.17 to 11.20 is the time required for particle growth. For example, for a relative humidity of 99.5%, Fig. 11.20 shows the relative mean growth rate, $\delta(t)$, and standard deviation about this mean for twelve different salts. As another example of time dependent particle growth, Fig. 11.21 shows the condensational growth for a few different sizes of NaCl particles subjected to indicated supersaturations. Of course, analytical expressions for the growth rates of individual, isolated particles are available (e.g., Mason, 1971). However, these expressions are complicated, and they require the supersaturation as input data; normally, the supersaturation would not be known well for precipitation scavenging predictions. As

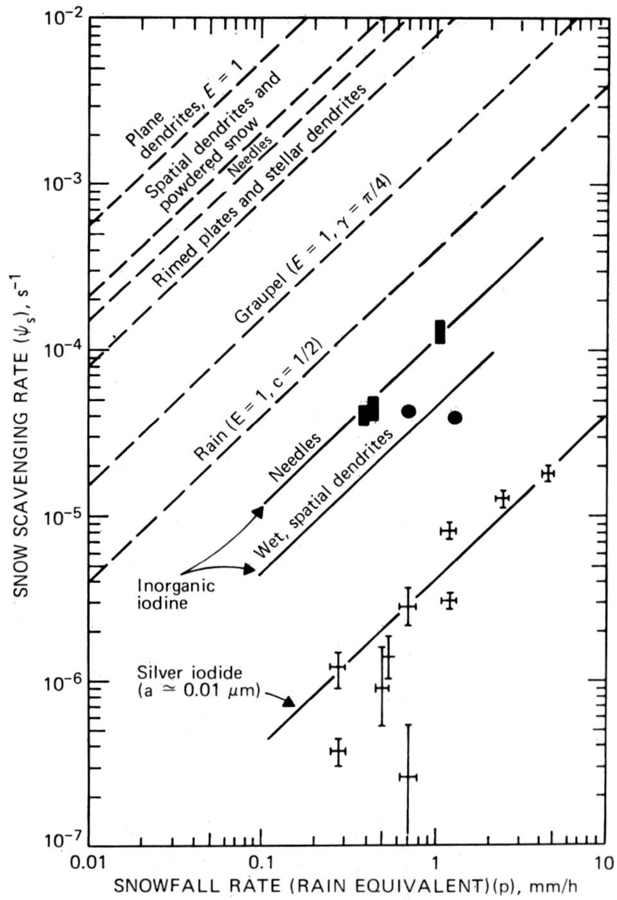

Fig. 11.14 Estimates for the snow scavenging rates as functions of precipitation intensity. The dashed curves, with $E = 1$ and $\gamma = \pi/4$, suggest the dependence on crystal type. Different values for γE were chosen to fit the experimental data of Engelmann et al. (1966): 7.3×10^{-3} for process-plant iodine (possibly dissolved in plume droplets) scavenged by needles; 1.8×10^{-4} for silver iodide particles scavenged by various crystal types, usually powdered snow or spatial dendrites. To obtain E from these data for plotting in Fig. 11.6, the value $\gamma = 1/3$ was chosen for snow scavenging of silver iodide particles.

another complicating feature, the presence of other particles will influence the saturation as available water vapor is shared among a distribution of droplet sizes. For example, Fig. 11.22 shows how the supersaturation, initially specified by a moderate updraft of 60 cm/s, reaches a maximum and then subsides as droplets grow within a cloud. Updrafts an order of magnitude smaller and larger than this value are not uncommon in frontal and convective clouds, respectively. Thus, in summary, accounting for particle growth for precipitation scavenging predictions is not a simple undertaking.

At the time of this writing, a number of research groups are attempting to develop convenient parameterizations of particle growth for precipitation scavenging calculations. An essential feature is that, since water vapor condensation on particles is a surface phenomenon, the relative growth rate decreases with decreasing surface to volume ratio: $4\pi a^2 da/dt \sim 4\pi a^2(1/a)$. Consequently, if the supersaturation is unknown, the particle size should at least have the form,

$$a(t) = a_0 \left[1 + \frac{t}{\tau_g}\right]^{1/2} \quad (a \geq a_c) \tag{11.39}$$

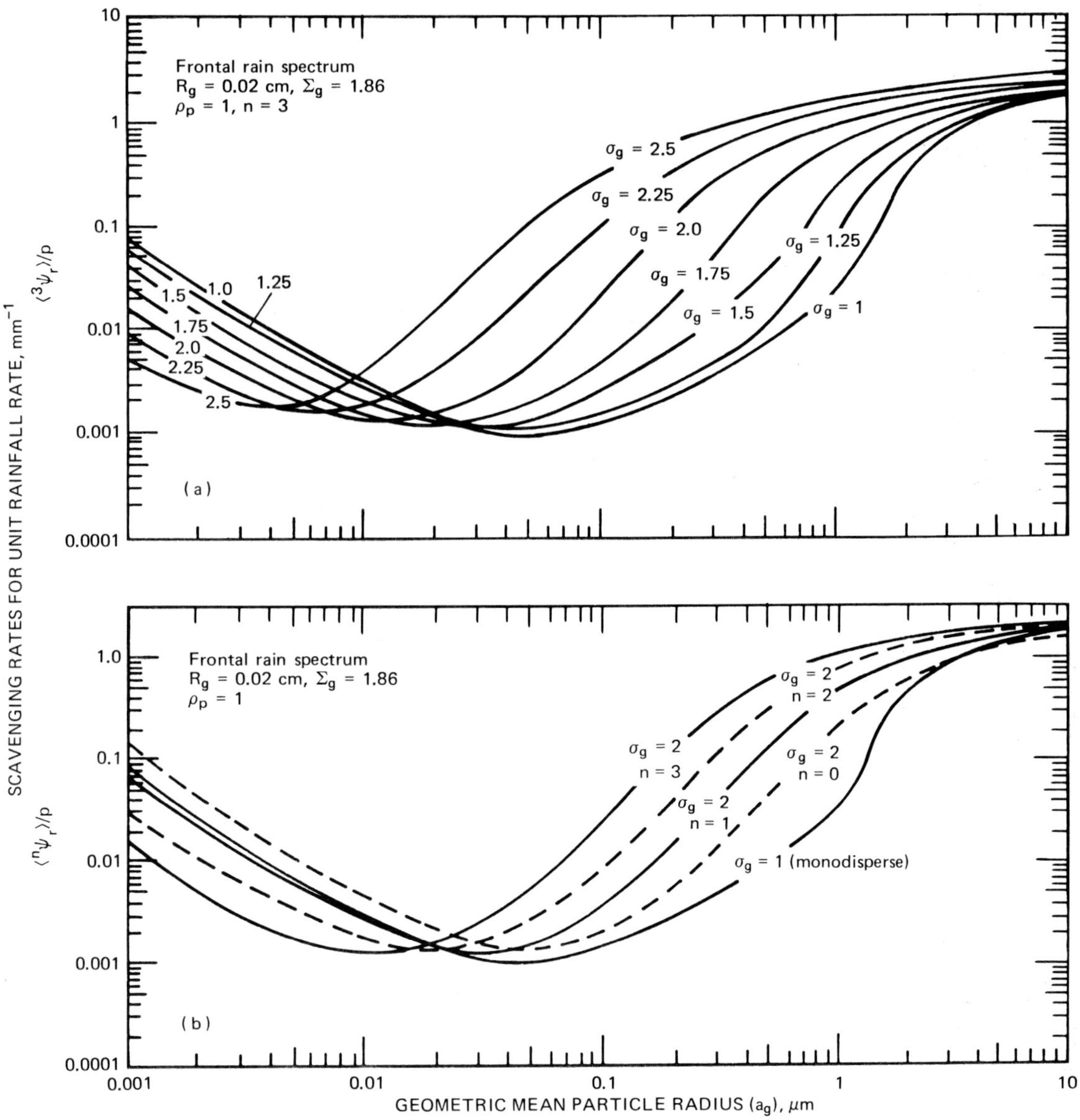

Fig. 11.15 Illustrations of particle average scavenging rates. (a) n of Eq. 11.38 is fixed at the value 3 (appropriate for the mass scavenging rate for spherical particles), and consequences of a broader particle size distribution (larger σ_g) are shown. (b) σ_g is fixed, and n is varied. [Based on M. T. Dana and J. M. Hales, Statistical Aspects of the Washout of Polydisperse Aerosols, *Atmos. Environ.*, 10: 45-50 (1976).]

where a_c is a critical size (see Fig. 11.17) below which particle growth is negligible for the existing (typically unknown) saturation ratio and where τ_g, the characteristic time constant for condensational growth, should at least display a particle size dependence, e.g.,

$$\tau_g = (10^2\ \mathrm{s})\ (a_0/1\ \mu\mathrm{m})^2 \qquad (11.40)$$

Although much more elaborate (and accurate) formulas for the growth rate are available (e.g., Mason, 1971), the necessary input information would usually be unknown. Also, in addition to this case of growth of particles that can act as cloud condensation nuclei, it may be necessary for a specific pollution release to include the possibility that the particles can act as ice nuclei.

A number of authors have considered the influence on scavenging of particle growth by water vapor condensation. Junge (1963) gives the following estimates for ϵ_n, the fraction of the total aerosol mass that would be active as cloud condensation nuclei: 0.5 for continental aerosols of high concentrations; 0.8 for continental aerosols of low concentrations; 0.9 to 1.0 for maritime aerosols and tropospheric aerosols with a total particle concentration less than 200 to 300/cm^3. Hicks (1966) suggested that most bomb debris fallout particles in the troposphere could be active as cloud condensation nuclei. Storebo and Dingle (1974) numerically integrated the equations describing condensational growth of particles and predicted the resulting particle removal caused by rain falling through the particle cloud as the cloud was advected up and past a small ridge. In Sec. 11-3.4.3, a method to account for both particle growth and attachment processes is outlined.

*11-3.4.2. **Attachment Processes Within Plumes and Clouds.*** Many potentially significant processes can influence the scavenging of particles other than particle inertia, intercep-

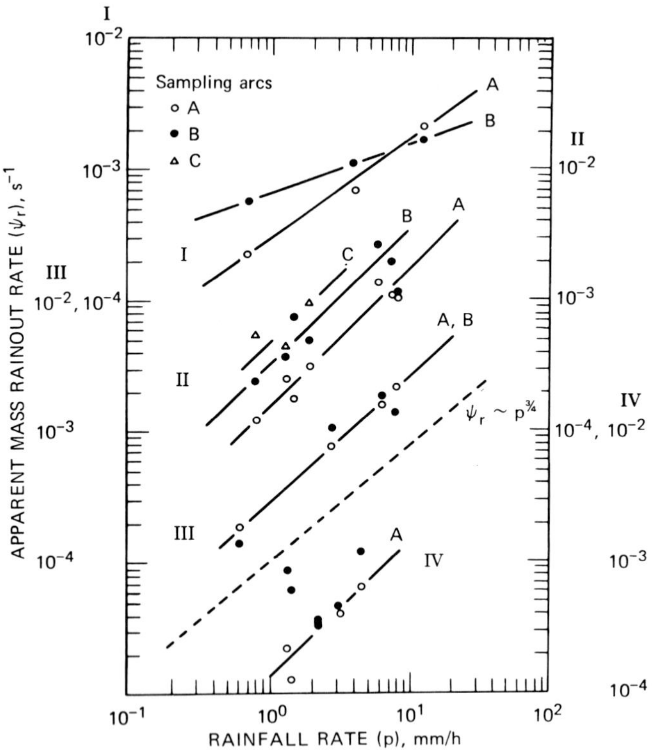

Fig. 11.16 The apparent rain scavenging rate for aerosol particles measured by Dana and Wolf (1968), Wolf and Dana (1969), Dana, Wolf, and du Plessis (1970), and Dana (1971, 1972). All tests were during fairly steady (prefrontal) rain. Sampling arcs A, B, and C were located 50, 100, and 150 ft downwind from a 26-ft tower. Curves I: rhodamine particles, mass median diameter (mmd) = 11.2 μm. Curves II: fluorescein particles, mean $a^2\rho_p$ = 35 $\mu m^2 \cdot g/cm^3$. Curves III: rhodamine particles, mmd = 1.2 μm. Curve IV: rhodamine particles, mmd = 4.8 μm. Different depositions on different arcs suggest the presence of simultaneous dry deposition; the polydispersity of the aerosol size distribution may explain why the removal rates are substantially greater than the rates that would be predicted on the basis of the particle/drop collision efficiency given in Fig. 11.5. [From W. G. N. Slinn, Some Approximations for the Wet and Dry Removal of Particles and Gases from the Atmosphere, *J. Water, Air, Soil Poll.*, 7: 513-543 (1977).]

tion, diffusion, and growth by water vapor condensation. In this section the emphasis will be on the attachment of the contaminant to other aerosol particles (including droplets and ice crystals). When attachment occurs, the effective collection efficiency between the contaminant and hydrometeor can be drastically altered; for example, if a 0.1-μm particle attaches to a 10-μm cloud droplet, then the effective collision efficiency for the contaminant could increase by about three orders of magnitude from the value for 0.1-μm particles given in Fig. 11.5. After some of the attachment processes are described, methods for incorporating their effects into scavenging predictions are outlined.

Brownian diffusional coagulation is an attachment process of predominant interest for contaminant particles with large diffusivities [i.e., particle radii a ≪ 1 μm (see Fig. 11.2)] and for three types of host particles: (1) cloud or plume droplets and ice crystals, (2) the background aerosol, and (3) other contaminant particles, i.e., self-coagulation. For large host particles such as cloud droplets, the attachment rate (Junge, 1963) is

$$\alpha(a) = 4\pi D(a) r N \tag{11.41}$$

where r is the host particle radius (or other characteristic dimension) and N is their number density. For example, for cloud droplets with r ≃ 10 μm and N ≃ $10^3/cm^3$ (liquid water content of about 4 g/m^3) then α (in s^{-1}) is numerically equal (approximately) to 10D with D in cm^2/s. Therefore, according to Fig. 11.2, the characteristic attachment time, α^{-1}, is ⩽ 15 min for particles with radii a ⩽ 10^{-2} μm. Consequently, if particles smaller than about 0.01 μm attach this

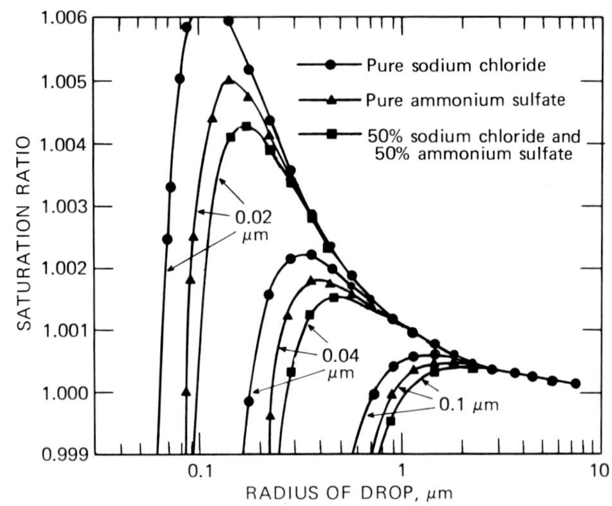

Fig. 11.17 The equilibrium solution droplet radius as a function of the saturation ratio for different initial (dry) particles of the indicated salts with the indicated dry radii, r_d. [Based on C. S. Chen, Evaluation of the Vapor Pressure over an Aerosol Particle, *J. Atmos. Sci.*, 31: 847-849 (1974).]

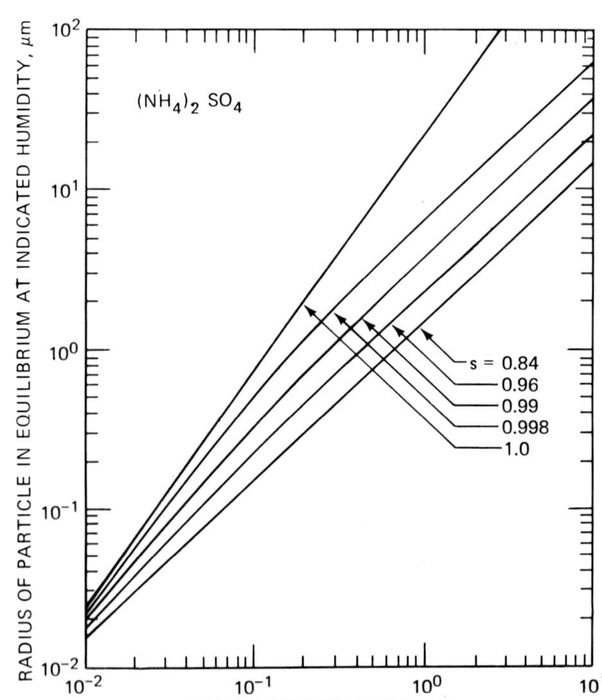

Fig. 11.18 Equilibrium radii of ammonium sulfate particles as a function of their dry sizes and the ambient saturation ratio, s. [Based on J. W. Fitzgerald, Approximate Formulas for the Equilibrium Size of an Aerosol Particle as a Function of Its Dry Size and Composition and the Ambient Humidity, *J. Appl. Meteorol.*, 14: 1044-1049 (1975).]

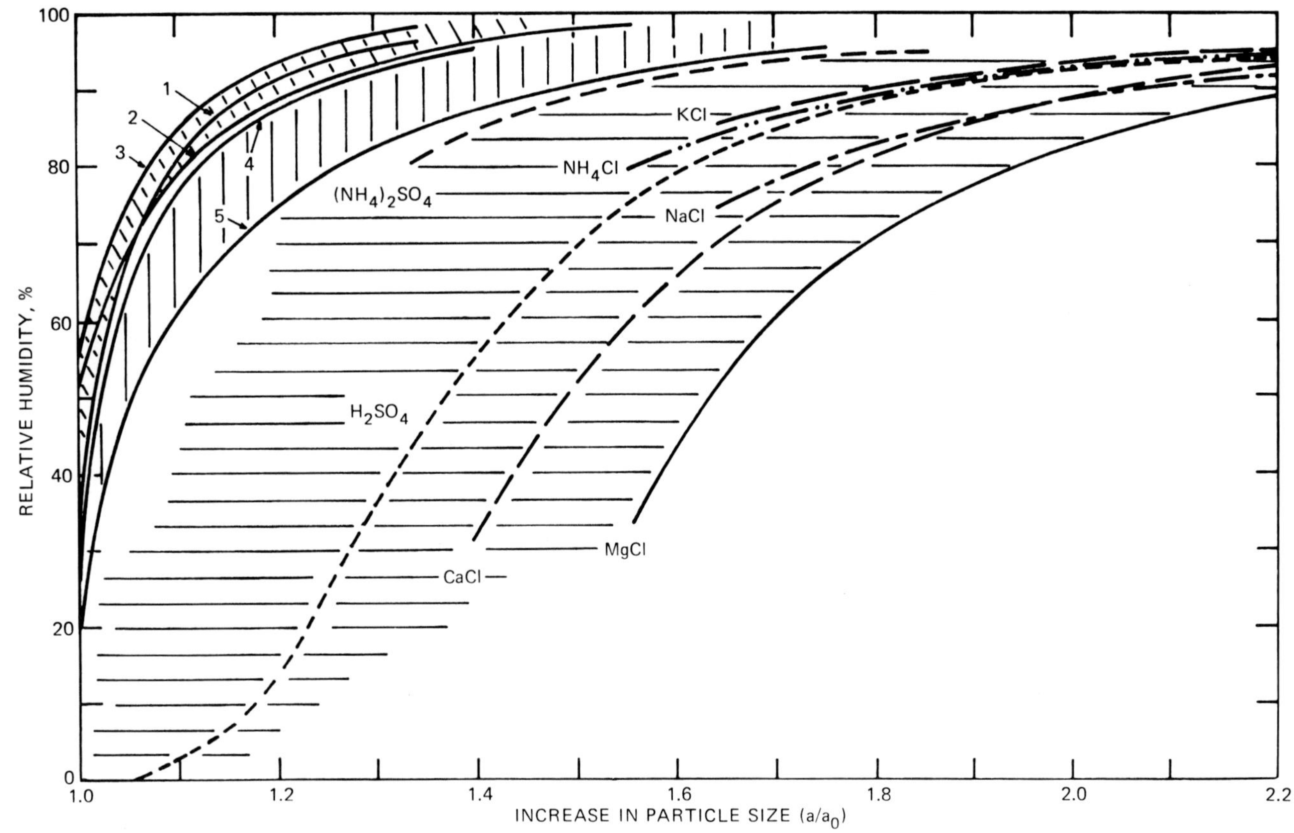

Fig. 11.19 Relative increase in particle size caused by water condensation on particles of initial radius $a_0 \gtrsim 0.1\ \mu m$ as a function of relative humidity, for different chemical compositions of the particles. (1), diagonal hatching, average growth curves for individual particles: curve 1, after Junge (1952); curves 2 and 3 after Charlson, Pueschel, and Ahlquist (1969), for p = 2 and 3, respectively, where the scattering coefficient $b \sim a^p$ in which a is particle radius. (2), vertical hatching, average growth curves of samples of atmospheric aerosol particles: curves 4 and 5 for particle density ρ_p = 1.0 g/cm³ and 2.5 g/cm³, respectively. (3), horizontal hatching, curves representing external mixtures of 60% pure salt and 40% insoluble matter (calculated curves). [Based on P. Winkler, The Growth of Atmospheric Aerosol Particles as a Function of the Relative Humidity—II. An Improved Concept of Mixed Nuclei, *Aerosol Sci.*, 4: 373-387 (1973).]

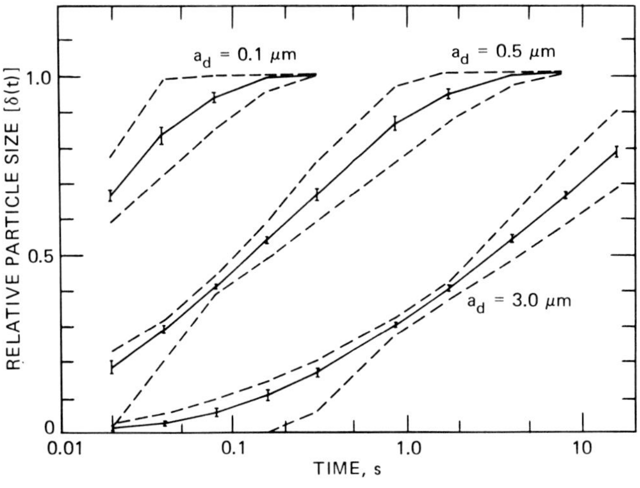

Fig. 11.20 The relative increase of particle radii, δ(t), for 12 different salts as a function of time since exposure to air at 37°C and 99.5% relative humidity (in the human bronchial tree). The relative increase of each particle's radius is $\delta(t) = [a(t) - a_d]/[a(\infty) - a_d]$, where a_d is the initial dry radius and $a(\infty)$ is the equilibrium radius, $t \to \infty$. The salts considered and their equilibrium radii for the case a_d = 0.5 μm are as follows: $AlCl_3$, 1.9 μm; $CoCl_2 \cdot 2H_2O$, 1.6 μm; $CoCl_2 \cdot 6H_2O$, 1.5 μm; KNO_3, 1.7 μm; $MgCl_2$, 1.95 μm; NaCl, 2.05 μm; $NaNO_3$, 1.8 μm; NaOH, 2.35 μm; $Ni(NO_3)_2 \cdot 6H_2O$, 1.4 μm; $Pb(NO_3)_2$, 0.95 μm; $UO_2(NO_3)_2 \cdot 6H_2O$, 1.05 μm; $ZnSO_4 \cdot 7H_2O$, 0.95 μm. ——, mean increase for all salts; I, standard deviation about the mean; - - - -, maximum and minimum values of the growth for all salts considered for the three initial particle radii shown. It is concluded that for practical applications in precipitation scavenging, differences in relative growth rates for these different salts could be ignored. [Based on G. A. Ferron, The Size of Soluble Aerosol Particles as a Function of the Humidity of the Air. Application to the Human Respiratory Tract, *J. Aerosol Sci.*, 8: 251-267 (1977).]

rapidly to cloud droplets and if particles larger than about 0.1 μm rapidly grow to cloud droplet size, it appears that, after a short time in clouds and in most plumes, only the particles in the size range $0.01 < a < 0.1$ μm would have an effective collision efficiency smaller than unity.

When a calculation is made for the attachment of a contaminant to host particles of size comparable to the mean free path of air molecules, λ_m ($\simeq 0.1$ μm at STP), it is necessary to correct Eq. 11.41 to account for noncontinuum aspects of diffusion (Junge, 1963; Brock, 1970). If the number of chemically similar host aerosol particles per unit volume whose radii are from r to r + dr is n(r) dr, the resulting distribution of attached contaminant particles or gases among this host aerosol size distribution is given approximately by the "attachment spectrum"

$$A(r) = \frac{r^2}{1 + (r/\lambda_m)} n(r) \tag{11.42}$$

Figure 11.23 shows an example of Eq. 11.42 for three model aerosol distributions; Fig. 11.24 shows both attachment spectra and attachment rates, according to different theories, for attachment of radioactive ions to the indicated particle distribution; Fig. 11.25 compares a number of theories against most available data; and Fig. 11.26 indicates the rate of attachment of I_2 and SO_2 to particles. One of the major consequences of radioactive and/or chemically reactive gas attachment to aerosol particles (e.g., radioactive I_2 attachment to smoke from an

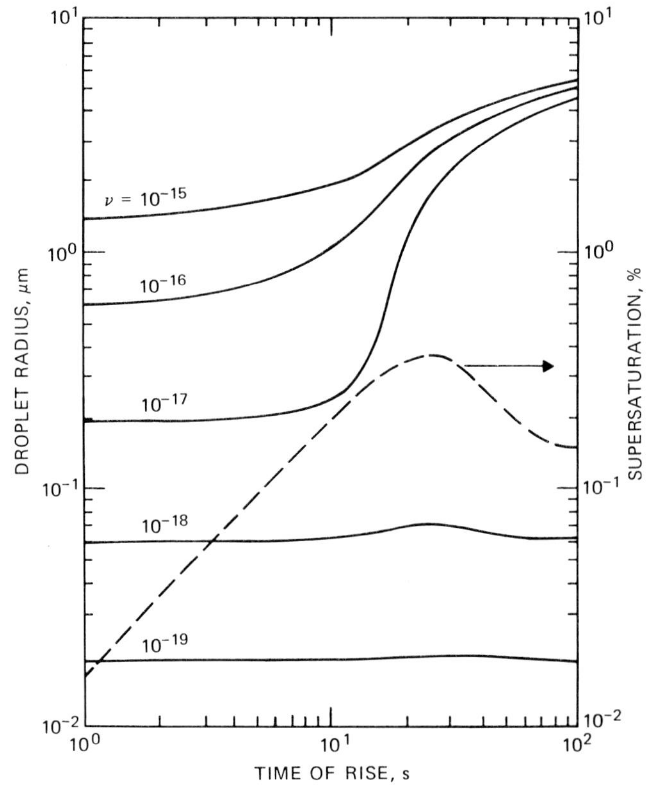

Fig. 11.22 The growth of cloud droplets on NaCl particles of different sizes (ν = particle mass in moles) when carried above cloud base in a steady updraft of 60 cm/s. Original analysis by Howell (1949); see also Mordy (1959). (Based on B. J. Mason, *The Physics of Clouds*, 2nd ed., Clarendon Press, Oxford, 1971.)

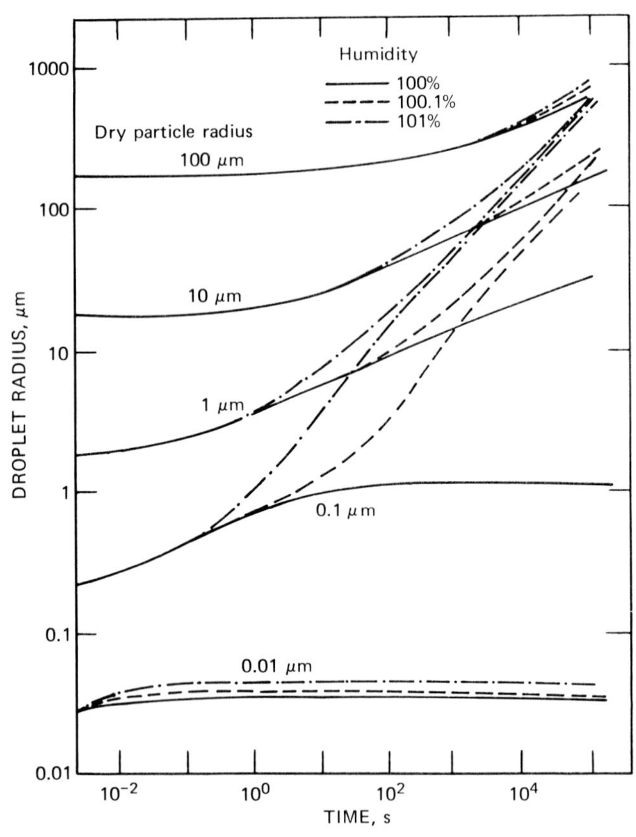

Fig. 11.21 Condensational growth of NaCl solution drops at a pressure of 760 mm Hg and T = 20°C for indicated initial particle sizes and humidities. [Based on E. L. Aleksandrov, L. M. Levin, and Y. S. Sedunov, Condensation Growth of a Solution Drop, *Izv., Atmos. Ocean. Phys.*, 3: 872-880 (1967).]

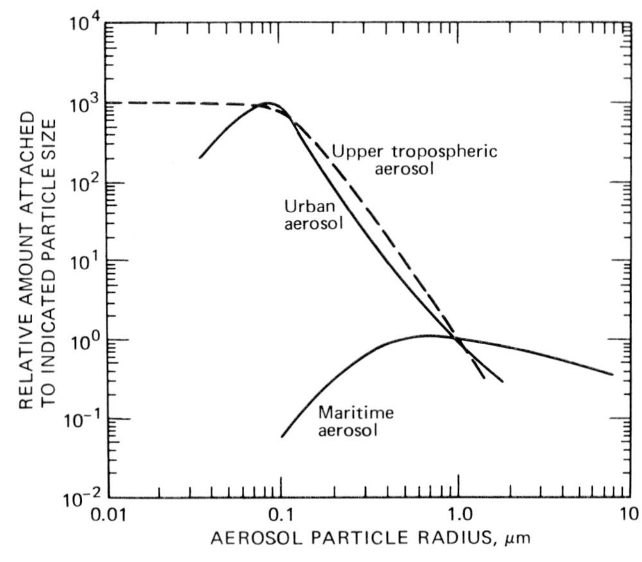

Fig. 11.23 Relative attachment spectra (e.g., for the attachment to aerosol particles of a reactive gas such as I_2) for different model aerosol particle number distributions. The curves are normalized to the value unity for a 1-μm particle. [Results from J. R. Brock, Attachment of Trace Substances on Atmospheric Aerosols, in *Precipitation Scavenging (1970)*, AEC Symposium Series, Richland, Wash., June 2–4, 1970, Rudolf J. Engelmann and W. George N. Slinn (Coordinators), pp. 373-380, CONF-700601, NTIS, 1970.]

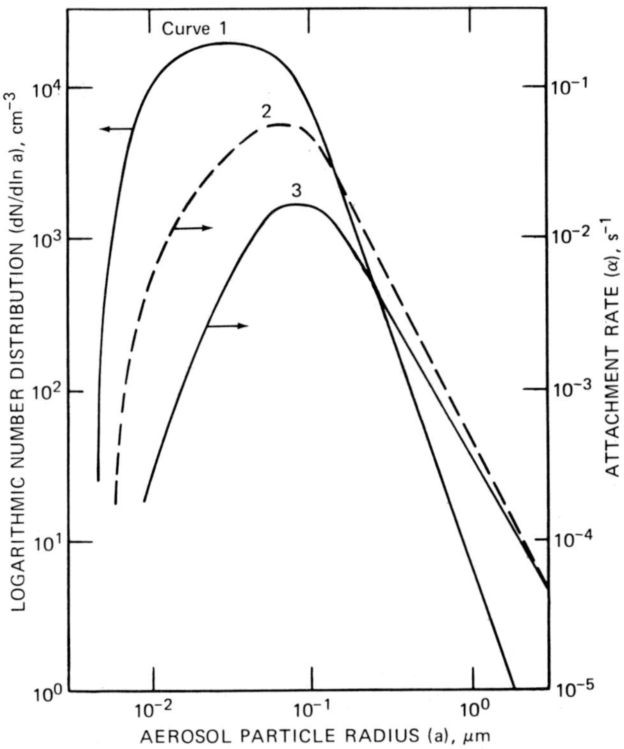

Fig. 11.24 Model attachment spectra and rates for attachment of radon and thoron decay products to aerosol particles. Curve 1: typical continental aerosol particle number distribution characterized by 17,850 particles/cm^3, 16,800 Aitken particles/cm^3, 1050 large particles/cm^3, and 1 giant particle/cm^3. Curve 2: relative attachment rate according to Eq. 11.41 (Smoluchowski's solution). Curve 3: the attachment rate according to Lassen and Rau (1960) accounting for non-continuum aspects of the attachment process, as in Eq. 11.42. The relative distribution of the radioactivity among the particles is given as in the figure; the rate of attachment is given by values on the right-hand ordinate. (Based on C. E. Junge, *Atmospheric Chemistry and Radioactivity*, Academic Press, New York, 1963.)

accidental fire) is that the gas can be much more slowly scavenged, especially if it attaches predominantly to, for example, particles with radii in the range $0.01 \lesssim a \lesssim 0.1$ μm. One of the major complications in predicting attachment rates and spectra is the great variability in the "natural" aerosol particle size distribution (see Fig. 11.27).

Self-coagulation of a contaminant aerosol is described by the Smoluchowski equation, which is difficult to solve analytically (e.g., see Hidy and Brock, 1972). In applications, it may also be necessary to include condensation of gases on the particles (Levin and Sedunov, 1967; Pich, Friedlander, and Lai, 1970; Heisler, Friedlander, and Husar, 1973; Junge, 1975). Figure 11.28 illustrates theoretical studies of both water vapor condensation and self-coagulation of a model aerosol; Fig. 11.29 shows an attempt to model measured volume distributions of a Los Angeles aerosol. From these and previous figures, it is seen that (1) self-coagulation has most effect on particles of radius $a \lesssim 0.01$ μm, (2) gases condense primarily on particles in the radius range $0.1 \lesssim a \lesssim 1$ μm (since, usually, most particle surface area is in this size range), and (3) the results of water vapor condensation can be a dramatic increase in the concentration of particles of size $a \gtrsim 1$ μm.

In total, a bewildering number of microphysical processes can influence the attachment of contaminant particles to cloud and plume particles. Figure 11.30 displays the results of many

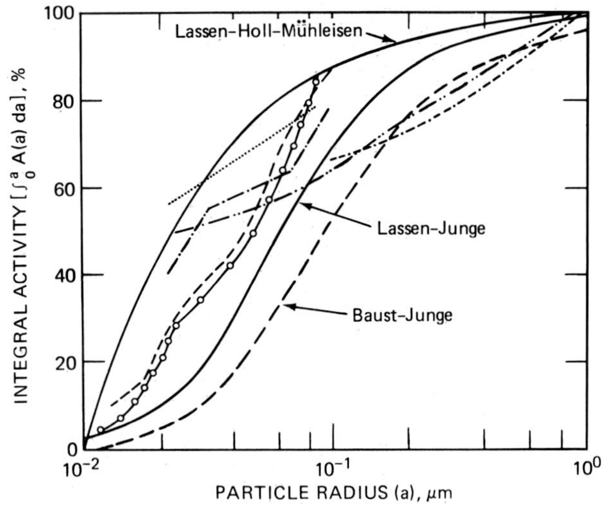

Fig. 11.25 The integral attachment spectra for radioactive primary ions attaching to model and measured aerosol particle size distributions. —— (upper curve), Lassen's (1961) attachment formula applied to the Holl-Mühleisen (1955) aerosol particle distribution model; —— (lower curve), Lassen's attachment formula applied to the Junge (1962) aerosol distribution model; — — —, Baust (1967) attachment coefficient applied to Junge's aerosol distribution model; —·—, Schumann's (1963a) measurements of natural radioactivity on aerosol particles; —··—··—, Schumann's (1963b) measurements of particles tagged with ThB; ··········, measurements by Bricard (1965); -·-·-·-, measurements by Wieser (1966); ---, measurements by Mohen and Stierstadt (1963); -o-o-, measurements by Havlovic and Snihs (1971). [Based on V. Havlovic and J. O. Snihs, Distribution of Natural Radioactivity on the Environmental Submicronic Aerosols, *Tellus*, 23: 441-454 (1971).]

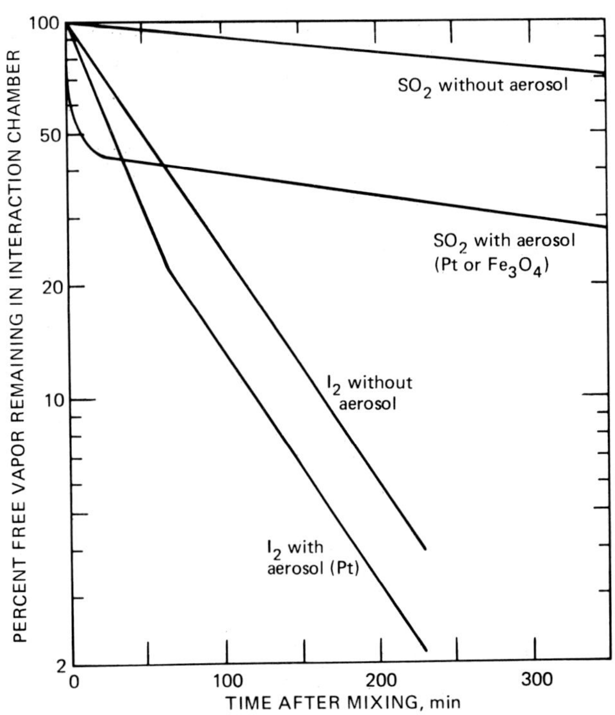

Fig. 11.26 Attachment of SO$_2$ and I$_2$ on aerosol particles. Initial aerosol particle number density, about 10^4/cm^3; maximum particle surface area is for particles of radii in the range 0.01 to 0.1 μm (particles were generated by the exploding wire technique); initial concentration of SO$_2$, 19 ppm; initial I$_2$ concentration, 1 ppm; the "without aerosol" curves indicate attachment to the walls of the chamber. [Based on Smith, Walker, and Stettler (1967).]

of these processes. Rather than describe details of these processes here, we refer the reader to the original publications (Greenfield, 1957; Facy, 1962; Junge, 1963, 1975, 1977; Slinn and Hales, 1971; Slinn, 1971b, 1974c; Dingle and Lee, 1973; Williams, 1977; Kerrigan and Rosinski, 1977) and add the fol-

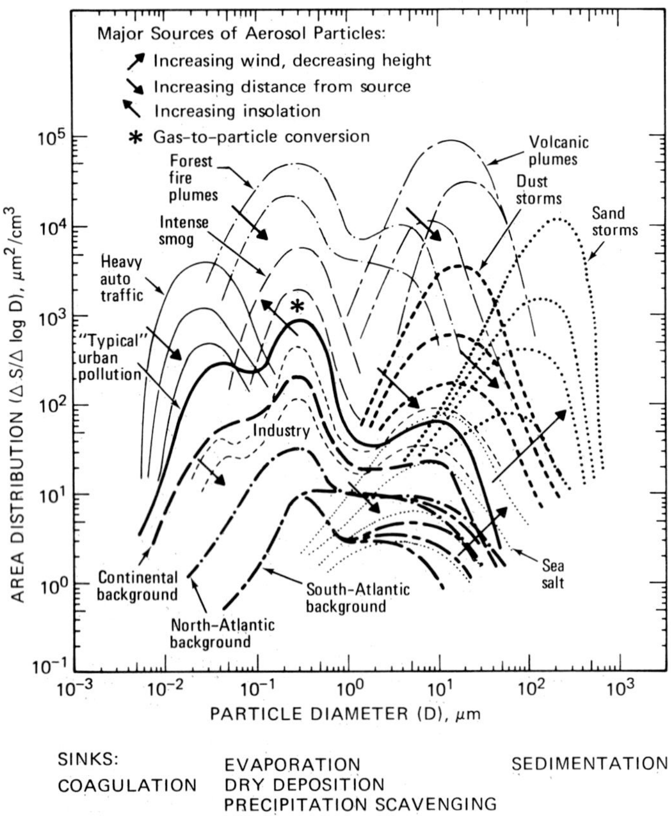

Fig. 11.27 An illustration of some of the variability that can be found in aerosol particle "size distributions" (here the surface area of all aerosol particles in a unit volume) for particles near the surface of the earth. [From W. G. N. Slinn, Sources and Surface-Area Distributions of Atmospheric Particles, *Atmos. Environ.*, in press (1983).]

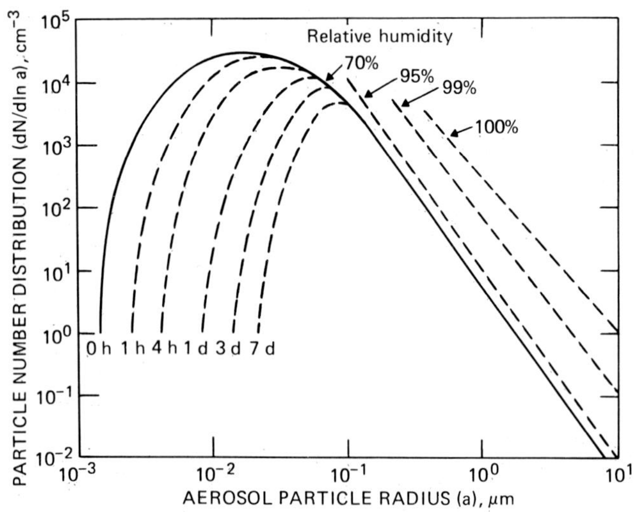

Fig. 11.28 An illustration of modifications in the aerosol particle number distribution caused by both self-coagulation and by water vapor condensation at the indicated relative humidities. [Based on Junge (1964).]

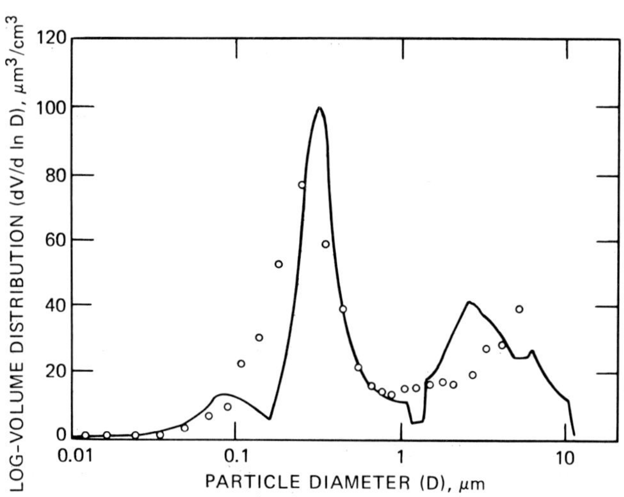

Fig. 11.29 Calculated and measured (points) volume distributions of the Los Angeles aerosol, averaged from 0900 to 2000 on Sept. 3, 1969. The peak in the calculated distribution from 0.1 to 1.0 μm results from condensed organics. The soil dust and marine components cause the mode at 2.6 μm. [Based on S. L. Heisler, S. K. Friedlander, and R. B. Husar, The Relationship of Smog Aerosol Size and Chemical Element Distributions to Source Characteristics, *Atmos. Environ.*, 7: 633-649 (1973).]

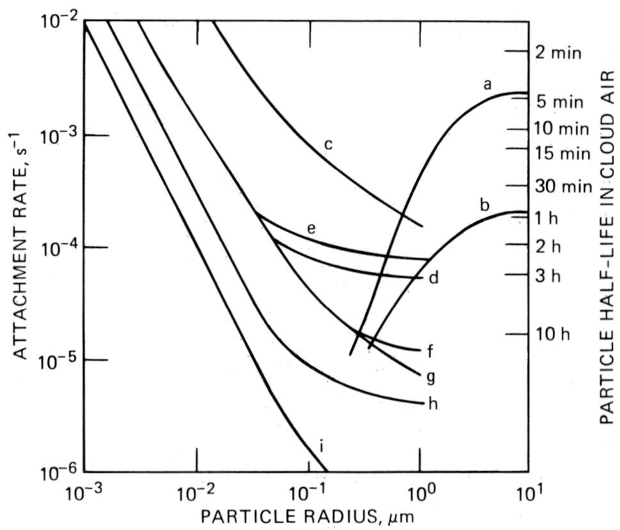

Fig. 11.30 Calculated rates of attachment of aerosol particles within plumes and clouds as a function of particle size. Curves a and b are for inertial capture by raindrops for rain rates of 10 and 1 mm/hr, respectively. Curve c represents attachment in an external electric field with $E_0 = 3 \times 10^5$ volts/m and particles in bipolar charge equilibrium. Curve d represents estimates for attachment by turbulent diffusion and convection, and curves e and f represent the corrections to d caused by phoretic effects. Curve g represents the attachment rate by Brownian diffusion and convection. Curve h represents attachment to cloud droplets, considering turbulent diffusion without convection. Curve i represents attachment of particles to raindrops, considering Brownian diffusion and convection for a rain rate of 10 mm/hr. For curves c to h, the cloud (or plume) droplet concentration is taken to be 300/cm³. [From A. L. Williams, Analysis of In-Cloud Scavenging Efficiencies, in *Precipitation Scavenging (1974)*. AEC Symposium Series, Champaign, Ill., Oct. 14–18, 1974, Richard G. Semonin and Robert W. Beadle (Coordinators), pp. 258-275, CONF-741003, NTIS, 1977.]

TABLE 11.6 Some Numerical Codes Containing Precipitation Scavenging Calculations

References	Brief description
Peterson and Crawford (1970)	Large-cloud diffusion code with precipitation scavenging subroutine; simple exponential removal term with removal rate $\sim p^{\alpha}$; comparisons against measured ^{131}I deposition after nuclear test "Cabriolet."
Dana et al. (1973); Hales, Dana, and Wolf (1977)	Primary emphasis on describing reversible scavenging of gases from industrial plumes; adequately predicts wet deposition as measured in a number of field studies; user orientated and available for use by others.
Dingle and Lee (1973)	Numerical study (1) of attachment and removal of particles from a stratiform cloud by rain from an overriding cloud and (2) of particle growth and removal from a convective cloud.
Storebo and Dingle (1974)	Numerical study of condensational growth of particles within lifted clouds and scavenging by rain from an overriding cloud.
Heffter and Ferber (1975)	Large-scale plume model incorporating scavenging via washout ratios and long-term average precipitation.
Baum, Wong, and Dolan (1974)	User-orientated program to calculate in-cloud scavenging of particles, especially for scavenging and resulting doses from nuclear bomb tests in the atmosphere.
Ritchie, Brown, and Wayland (1976)	Research-oriented numerical study for assessing consequences of nuclear reactor accidents; simple exponential removal term but incorporates important considerations of probability and spatial structure of rain.
Wendell and Powell (1976)	Simple exponential removal terms within a large-scale "puff model" for transport and diffusion.
Norment (1977)	Subroutine in a large transport and diffusion code. In- and below-cloud scavenging of particles, as functions of particle size, rain rate (as a function of height), and accounting for condensational growth and Brownian diffusional coagulation of particles.
Lange and Knox (1977)	Incorporates a simple exponential removal term in a large transport and diffusion code.
Watson, Barr, and Allenson (1977)	Simple description of scavenging within a thorough analysis of meteorological fields (velocities and precipitation) and their probability distributions.

lowing comments: (1) there is not unanimity (e.g., see Beal, 1972; Tennekes and Woods, 1973) about the influence of turbulence on coagulation (note that both droplets and particles are smaller than the turbulence microscale $\nu/u_* \simeq 100\ \mu m$); (2) the value for the electric field strength, $E_0 = (3 \times 10^5$ V/m), chosen to display electric effects in Fig. 11.30 may be appropriate for cumulonimbus clouds, but the same value is large compared with typical electric fields in other clouds; (3) thermophoretic contributions are probably most significant during cloud glaciation and evaporation; and (4) nucleation scavenging is not displayed explicitly in Fig. 11.30. In summary, we emphasize that, even if the host of details about attachment processes is not appreciated, it is essential to appreciate, at least, that aerosol particles in the atmosphere are not static but continuously evolve and that their evolution can dramatically influence their removal.

11-3.4.3 Removal Rates for Modified Particles. At the outset it should be acknowledged that it is conceptually feasible to develop expressions and computer programs that completely describe particle coagulation, attachment, condensational growth, and removal of particles within plumes and clouds. Table 11.6 identifies some numerical codes that contain varying complexities of precipitation scavenging calculations; one code uses more than 1500 FORTRAN statements to describe scavenging and needs more than 10^5 words of large core memory. However, in a thorough analysis, values must be prescribed for various parameters, such as cloud updraft, plume entrainment, electric field strengths, turbulence, supersaturations, temperatures, collection efficiencies, particle size distributions, chemical compositions, etc. Unfortunately, uncertainties in these parameterizations are normally large, and therefore it is the author's opinion that elaborate formulations of the processes will usually provide no more accurate predictions than can be obtained with simple formulations. Consequently, and with appreciation for space limitations, only simple models will be described in this chapter.

One of the simplest models for in-plume and in-cloud scavenging describes the two-stage process: attachment of the contaminant to the condensed water at a rate α, and then removal of the condensed water by larger hydrometeors at a rate β (Facy, 1962; Davis, 1970, 1972; Slinn 1971b, 1974c; Dingle and Lee, 1973). For this simple model, it is assumed that microphysical and large-scale meteorological processes are stationary and that space-average values of α and β can be defined. Thus, α and β are treated as constant, and this restricts the generality of the results (Hales, 1978b). However, the model is capable of providing a useful framework for interpreting experimental results.

To formulate the model, let χ_a be the concentration of the contaminant (e.g., per unit size interval and per unit volume) in the air, i.e., unattached to condensed water. If this contaminant attaches to the cloud water at rate α and if, in addition, allowance is made for the direct scavenging of the contaminant by large hydrometeors at a rate γ, then

$$\frac{D\chi_a}{Dt} = -\alpha\chi_a - \gamma h(t - t_i)\chi_a \tag{11.43}$$

where $h(t - t_i)$ is the Heaviside or unit step function, starting at the time of the onset of precipitation, $t = t_i$, and where D/Dt is the total (or Lagrangian) derivative. Similarly, the concentration, χ_c, of the contaminant attached to cloud water is governed by

$$\frac{D\chi_c}{Dt} = \alpha\chi_a - \beta h(t - t_i)\chi_c \tag{11.44}$$

where, again, β is the rate at which the condensed cloud water is scavenged by larger hydrometeors. The total contaminant concentration is $\chi_t = \chi_a + \chi_c$, and the rate at which the total contaminant is removed from unit volume is $L = \gamma\chi_a + \beta\chi_c$.

To obtain approximate solutions to Eqs. 11.43 and 11.44, we assume that meteorological conditions experienced by the contaminant change slowly compared to attachment and removal rates α, β, and γ. Therefore, α, β, and γ being treated as constants, the total time derivative is replaced by $\partial/\partial t$. The solution to the resulting equations for χ_t subject to $\chi_t(t = 0) = \chi_0$ is $\chi_t(t \leqslant t_i) = \chi_0$ and for $t \geqslant t_i$:

$$\frac{\chi_t}{\chi_0} = \left[\frac{\beta - \gamma}{\beta - (\alpha + \gamma)}\right] e^{-\alpha t}\, e^{-\gamma(t - t_i)} + \left\{1 - \left[\frac{\beta - \gamma}{\beta - (\alpha + \gamma)}\right] e^{-\alpha t_i}\right\} e^{-\beta(t - t_i)} \tag{11.45}$$

A number of special cases of Eq. 11.45 are of interest. For the usual case that the attachment rate, α, and the cloud water removal rate, β, are both much larger than γ, then Eq. 11.45 simplifies to

$$\frac{\chi_t}{\chi_0} = \frac{\beta}{\beta - \alpha} e^{-\alpha t} + \left\{1 - \frac{\beta}{\beta - \alpha} e^{-\alpha t_i}\right\} e^{-\beta(t - t_i)} \qquad (\alpha,\beta \gg \gamma) \tag{11.45a}$$

Also of interest is the case where there is a long time between cloud formation and rain initiation. Then, with $\alpha t > \alpha t_i \gg 1$, Eq. 11.45a simplifies to

$$\chi_t = \chi_0 \exp[-\beta(t - t_i)] \qquad \begin{bmatrix}\alpha,\beta \gg \gamma \\ \alpha t > \alpha t_i \gg 1\end{bmatrix} \tag{11.45b}$$

For the case described by Eq. 11.45b, the overall scavenging rate is just the rate at which the condensed water is removed from the cloud. Then from Eq. 11.32, $\beta \equiv \psi = [cp(z)\,E(\bar{r},R_m)]/R_m(z)$, where $\bar{r}$ is the mean size of the condensed water particles and $p(z)$ is the (height dependent) rain rate within the cloud. Similarly, Eq. 11.36 can be used for β for regions of the cloud where precipitation is in the form of ice crystals. Finally, for the case where the contaminant is placed in a precipitation cloud (viz, $t_i = 0$), Eq. 11.45 yields

$$\frac{\chi_t}{\chi_0} = \frac{\beta}{\beta - \alpha} e^{-\alpha t} - \frac{\alpha}{\beta - \alpha} e^{-\beta t} \qquad \begin{bmatrix}\alpha,\beta \gg \gamma \\ t_i = 0\end{bmatrix} \tag{11.45c}$$

For the cloud water removal rate, β, rapid compared with the attachment rate, α, Eq. 11.45c yields

$$\chi_t = \chi_0 \exp(-\alpha t) \qquad \begin{bmatrix}\beta \gg \alpha \gg \gamma \\ t_i = 0\end{bmatrix} \tag{11.45d}$$

Conversely, for $\alpha \gg \beta$ (and $\gamma,\beta \gg \gamma$ and $t_i = 0$),

$$\chi_t = \chi_0 \exp(-\beta t) \qquad \begin{bmatrix}\alpha \gg \beta \gg \gamma \\ t_i = 0\end{bmatrix} \tag{11.45e}$$

Thus the slower of the two processes limits the rate of removal and can be interpreted as the overall scavenging rate. This result is qualitatively illustrated in Fig. 11.31 for the case where attachment is by Brownian diffusional coagulation with cloud droplets. It is the author's experience that usually the cloud water removal rate is the rate-limiting stage of the removal process and therefore Eq. 11.45e (or Eq. 11.45b) is more generally applicable.

One of the disadvantages of the preceding analysis, however, is that condensational growth of particles is not displayed explicitly. Indeed, for condensational growth, it is not that particles attach to condensed water but that water attaches to the particles. This problem can be partially solved, as was done for Fig. 11.31, by indicating the time for particle growth to a specific cloud droplet size (e.g., 10 μm). This results in a pseudo attachment rate to describe condensation. However, this technique, alone, does not account for the removal of particles that have not completed their condensational growth. One way to remedy this is to return to Eq. 11.43 and allow γ to be time dependent. Alternatively, consider the contaminant to be distributed among three components (free, f; growing, g; hosted,

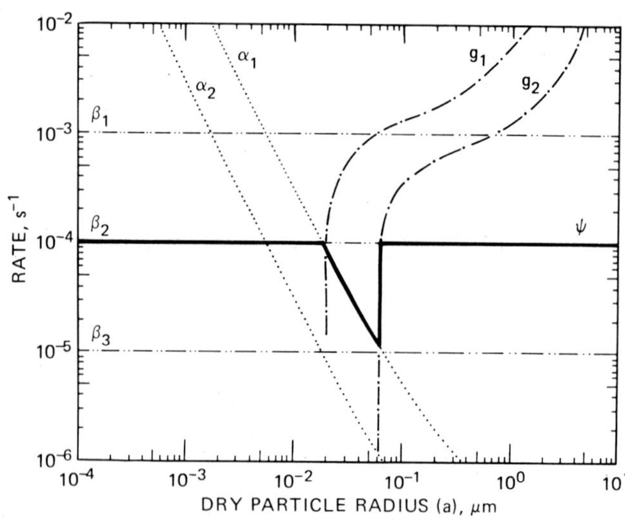

Fig. 11.31 A qualitative plot of the dependence of the overall in-cloud scavenging rate, ψ (solid curve), on particle size. For this multi-rate model, ψ is the slowest of the various rates: α = attachment rate (shown here only for Brownian coagulation and for α_1 the values N = 200 cloud droplets/cm^3 and R = 10 μm were used); β = cloud water removal rate; and g = reciprocal of the time for condensational growth to 10 μm. From considerations such as these, it is deduced that, usually, the cloud water removal rate is the rate-limiting step of the removal process and therefore $\psi \simeq \beta$. This estimate, however, neglects the likelihood that the more active the particle is as a cloud condensation nucleus (e.g., larger size) then the more rapidly will the particle be removed from a cloud. Therefore, ψ is expected to increase with particle size larger than ~0.1 μm, e.g., as indicated in Fig. 11.3. [Based on Slinn (1971b) and W. G. N. Slinn, Rate-Limiting Aspects of In-Cloud Scavenging, *J. Atmos. Sci.*, 31: 1172-1173 (1974).]

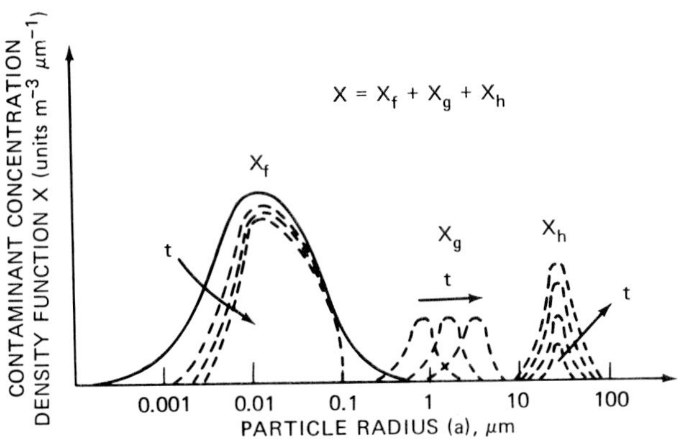

Fig. 11.32 A schematic illustration of the variation with increasing time, t, of three portions (free, f; growing, g; and hosted, h) of the total aerosol particle population. [From W. G. N. Slinn, Some Approximations for the Wet and Dry Removal of Particles and Gases from the Atmosphere, *J. Water, Air, Soil Poll.*, 7: 513-543 (1977).]

h) as sketched in Fig. 11.32. Then the rate at which the total contaminant is scavenged can be obtained simply by evaluating the rate at which each component is scavenged. For example, the mass average rain scavenging rate (and similarly for other moments) can be approximated by (Slinn, 1977c)

$$\langle {}^3\psi_r\rangle = \frac{cp(z)}{R_m}\Bigg(F_{f0}\exp(-\bar{\alpha}t)\bar{E}\left[\bar{a}_f\exp\left(\frac{7}{2}\sigma_f^2\right),\ R_m\right] + F_{g0}\bar{E}\left\{\left[\bar{a}_{g0}^2\exp(7\sigma_{g0}^2) + \frac{t}{100s}(\mu m)^2\right]^{1/2},\ R_m\right\} + F_{f0}[1-\exp(-\bar{\alpha}t)]\bar{E}(\bar{r}, R_m)\Bigg) \quad (11.46)$$

where c = a numerical factor $\simeq$ 0.5
$p(z)$ = height-dependent rainfall rate
R_m = mass-mean raindrop size
F_{f0} = initial fraction of "free" particles, unattached to condensed water
$\bar{\alpha}$ = average rate of attachment of the contaminant to the condensed water
$\bar{E}(a,R_m)$ = mean collection efficiency of the free particles assumed to have a lognormal size distribution with parameters $\bar{a}_f$ and σ_f
$F_{g0} = 1 - F_{f0}$ = fraction of the contaminant involved in condensational growth to size at time t assumed to be given by (compare Eq. 11.39) $a(t) = [\bar{a}_{g0}^2\exp(7\sigma_{g0}^2) + (t/100\,s)\mu m^2]^{1/2}$
$\bar{r}$ = mean radius of the host (cloud or plume) particles

Although Eq. 11.46 may look rather opaque, it is actually quite transparent. It shows that the contributions to the effective collection efficiency (all terms following cp/R_m) change with time, t, since entering the cloud or plume, as the free particles (initial fraction = F_{f0}) attach to host particles (of mean radius $\bar{r}$) at an average rate $\bar{\alpha}$ and as the mean size of the growing particles (initial fraction = F_{g0} = 1 − F_{f0}) increases in size. To obtain Eq. 11.46, a specific growth rate, a time independent attachment rate, and lognormal particle size distributions (parameters $\bar{a}_f, \sigma_f; \bar{a}_{g0}, \sigma_{g0}$) were assumed. In spite of this apparent complexity, it is hoped that Eq. 11.46 adequately demonstrates the general nature of this type of formulation and will suggest to the reader how to develop similar formulas for practical problems of interest. The case of snow scavenging leads to an expression essentially the same as Eq. 11.46 except that γ/D_m replaces c/R_m and E replaces E.

Fig. 11.33 shows a specific evaluation of Eq. 11.46 using $E(a,R_m)$ as given in Fig. 11.5 and $E(\bar{r},R_m) = 1$. One result demonstrated in this figure is that the rainout rate usually increases for a broader particle size distribution (increasing σ). A second result demonstrated is that, for time large compared to the attachment and growth times, the overall removal rate becomes close to the value for the cloud water removal rate. That this case of $\psi \simeq \beta$ frequently occurs in practice is suggested from an analysis of experimental data.

11-3.5 Interpretations of Some Scavenging Field Data

In previous sections, many approximations have been introduced during the development of approximate formulas to describe precipitation scavenging of aerosol particles. The adequacy of these approximations can best be judged through comparisons of predicted and measured scavenging rates. In this section, some of these comparisons will be made. Unfortunately, however, the comparisons cannot be complete, because the number of illuminating field experiments, especially for snow scavenging, is insufficient.

An approximate expression for the particle rain scavenging (rainout) rate was given by Eq. 11.32, namely,

$$\psi_r(a;\mathbf{r},t) = \frac{cp(\mathbf{r},t)\bar{E}(a,R_m)}{R_m} \quad (11.32)$$

where, again, c is a numerical factor near unity, p is the rain rate, $\bar{E}$ is the particle/drop collection efficiency (compare Fig.

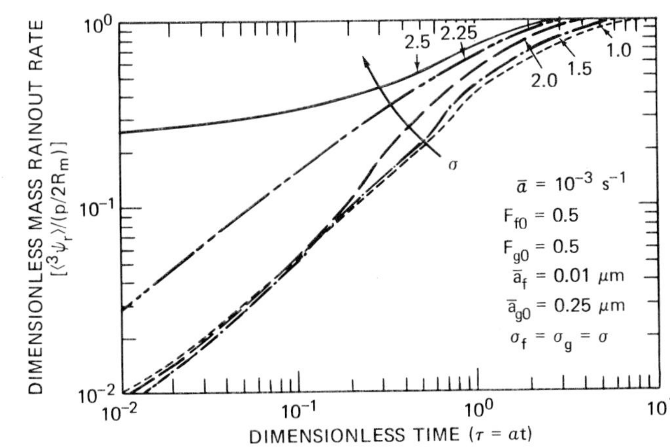

Fig. 11.33 An evaluation of Eq. 11.46 demonstrating that the effective rate at which particle mass is scavenged can increase dramatically with increasing polydispersity of the aerosol particle size spectrum (i.e., increasing σ) and increasing time, τ, for particle growth by water vapor condensation and for particle attachment to plume or cloud droplets (or ice crystals). [Based on W. G. N. Slinn, Some Approximations for the Wet and Dry Removal of Particles and Gases from the Atmosphere, *J. Water, Air, Soil Poll.*, 7: 513-543 (1977).]

11.5) based on the mean drop radius, and, at least for fairly steady rains, it was suggested in Eq. 11.33 that

$$R_m = 0.35 \text{ mm}\left(\frac{p}{1 \text{ mm/h}}\right)^{1/4} \tag{11.33}$$

Figure 11.16 demonstrated that the rainfall rate dependence of ψ_r, with or without the use of Eq. 11.33, is fairly adequately described by Eq. 11.32. For frontal storms, the average of the results obtained by Dana (1970) gives $\psi_r/\bar{E} = (1.6 \text{ mm}^{-1})p$ which is fairly well represented by Eq. 11.32 with c = ½ and a reasonable value for R_m. For convective storms, the average of the results given by Vali (1977) is $\psi_r/\bar{E} = 2.5 \times 10^{-4} \text{ s}^{-1} [p/(1 \text{ mm/h})]^{0.8}$ which is fairly well represented by Eq. 11.32 using Eq. 11.33 for R_m and c = ⅓. Alternatively, we could use c = ½ and a correspondingly larger numerical factor in the expression for R_m; however, the present accuracy of neither the data nor the theory can justify the pursuit of such refinements.

Similarly, an approximate expression for the particle snowout rate was given by Eq. 11.36,

$$\psi_s(a;\mathbf{r},t) = \frac{\gamma p(\mathbf{r},t)\, E(a,\lambda)}{D_m} \tag{11.36}$$

where γ = numerical factor near unity
p = precipitation rate (rainwater equivalent)
E = particle/ice crystal collection efficiency based on the capture length scale λ (see Fig. 11.6)
D_m = characteristic (volume/area) length scale, listed for different crystal types and ice aggregates in Table 11.5.

In contrast to the case of rain scavenging, however, where there is a factor of about two uncertainty between data and approximate theory (when the uncertainty in the collection efficiency is suppressed), there are very great uncertainties, even about the general form of the removal rate, for snow scavenging. This statement is illustrated by Table 11.7, which shows empirical expressions proposed by Knutson and Stockham

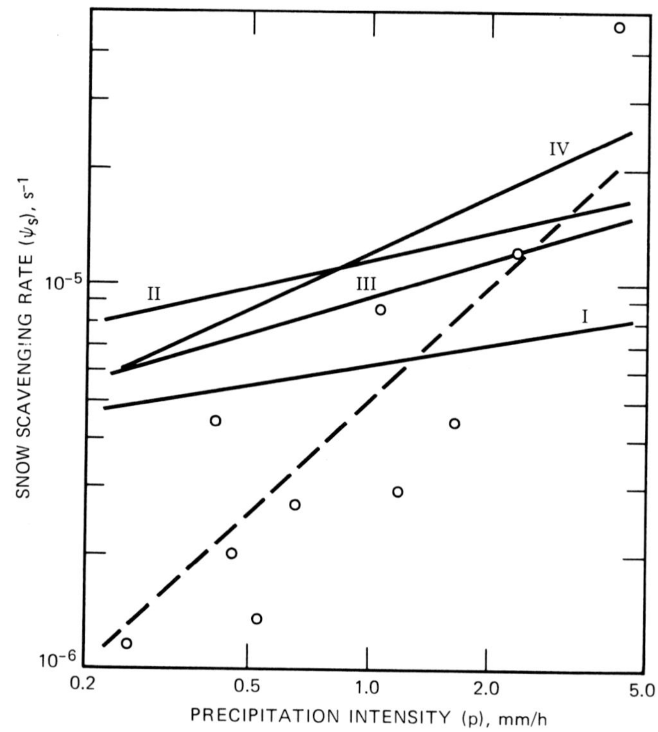

Fig. 11.34 Comparisons of empirical expressions for the rate of scavenging of particles by snow against data from Wolf and Dana (1969). Curves I–IV are the Knutson and Stockham formulae given in Table 11.7 using D_p = 0.5 μm. The dry mass mean diameter of the AgI particles was actually about 0.012 μm, ranging in size from about 0.002 to 0.05 μm (Hagen, 1967). The dashed curve "through" the data points (circles) is given by Eq. 11.36 as in Fig. 11.14. (Based on Knutson and Stockham, 1977.)

(1977) for the snow scavenging rate for different ice aggregates. In particular, the substantially different dependence on precipitation rate between the expressions in Table 11.7 and Eq. 11.36 should be noticed. The source of this difference appears to be the strong dependence of the mean crystal size on precipitation rate assumed by Knutson and Stockham. Although, as was mentioned earlier, there are insufficient data to establish firmly the more appropriate expressions, Fig. 11.34 suggests to the author that Eq. 11.36 contains the more appropriate dependence of ψ_s on precipitation rate.

The preceding discussion, however, does not address the important question of the adequacy with which the collection efficiencies are known. Here, again, insufficient field data are available; in the present author's opinion, the best available data are shown in Fig. 11.35. The measurements were taken by Radke, Eltgroth, and Hobbs (1978) as the opportunities occurred during a 3-yr period. The method used was to obtain airborne measurements of particle size distributions before and after the aerosol was scavenged by rain showers. A total of eight such opportunities were realized. For four cases, the aerosol was in plumes from coal-fired power plants; the scavenging times were 960, 600, 400, and 265 s. For three other cases, "natural" aerosol was scavenged for 1800, 1100, and 480 s. Not included to form the average shown in Fig. 11.35 is an eighth case—scavenging of aerosol from a volcanic maar by graupel—in this case, the collection efficiency was found to be greater than unity for all particles except those with 0.3 ≲ a ≲ 0.7 μm. To obtain the collection efficiencies, the (small) differences in the (typically large) numbers of parti-

TABLE 11.7 Empirical Equations for the Snow Scavenging Rate for Four Types of Ice Aggregates†

Type of ice aggregate	Empirical equation‡
I Lump graupel	$\psi_s = 0.21\, p^{0.188} f(T,D_p)$§
II Densely rimed dendrites	$\psi_s = 0.40\, p^{0.266} f(T,D_p)$
III Densely rimed radiating assemblages of dendrites	$\psi_s = 0.32\, p^{0.305} f(T,D_p)$
IV Aggregates of unrimed radiating assemblages of dendrites or dendrites	$\psi_s = 0.41\, p^{0.57} f(T,D_p)$

†Formulas from Knutson and Stockham (1977).

‡Units are ψ_s (s), p (mm/h), T (°F), and D_p (μm); the empirical correlations were derived from data for monodisperse particle diameter in the range $0.25 \leqslant D_p \leqslant 3.5$ μm and snowflake circumscribed diameter in the range $1 \leqslant D_c \leqslant 10$ mm.

§$f(T,D_p) = \text{antilog}\,(-3.17 - 0.0178T + 3.41 \log D_p - 0.973 \log^2 D_p - 7.20 \log^3 D_p)$.

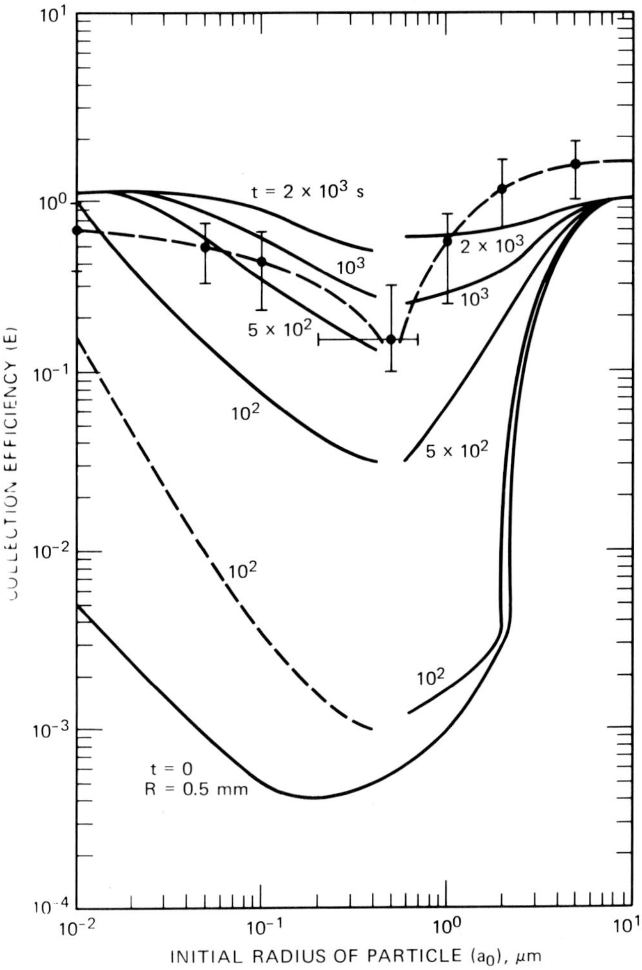

Fig. 11.35 An interpretation of the measured collection efficiency of Radke, Eltgroth, and Hobbs (1978) as a function of particle size. The data curve shows the mean (•) and standard deviation about the mean (⫯) for eight sets of data collected over a 3-yr period. The solid curve at the bottom of the figure (labeled t = 0) is the collection efficiency from Fig. 11.5 for mean raindrop radius R = 0.5 mm. The dashed curve labeled t = 10^2 (s) is obtained assuming E = E(0) exp ($-\alpha t$) + 1.0 [1 − exp ($-\alpha t$)] with the attachment rate, α, for Brownian coagulation with 10^3 droplets/cm^3 of radius r = 10 μm. The solid curves at a < 0.5 μm labeled 10^2, 5 × 10^2, 10^3, and 2 × 10^3 s are obtained by using the attachment rate labeled as curve c in Fig. 11.30. The solid curves at a > 0.5 μm are obtained by assuming that the particles grew by water vapor condensation to the size (and correspondingly increased collection efficiency) given by a(t) = [a_0^2 + (t μ^2/100 s)]$^{1/2}$.

cles before and after scavenging were found and then were used with measured hydrometeor site distributions to deduce E. Important details about the individual cases should be studied in the original publication; to the present author's knowledge, these details have not yet been explained. In Fig. 11.35 are also shown theoretical curves (see legend) that suggest to the author that condensational growth of particles larger than ≃0.5 μm and attachment of smaller particles to cloud or plume droplets may explain the substantially greater values for the collection efficiency than would have been expected for inert particles on the basis of Fig. 11.5. Further comments about remaining uncertainties in the interpretation of these data will be made in Sec. 11-3.7.

Less definitive as to particle size dependence but extremely valuable data on particle scavenging by convective storms are shown in Fig. 11.36. Burtsev, Burtseva, and Malakhov (1970) released radioactively tagged aerosol particles into different zones of a number of cumulonimbus storms. In analogy with particle sizes from exploding wires, the authors assumed that the particle mass mean diameter from exploding shells was about 0.2 μm. However, Dingle (1978) does not have great confidence in the validity of this analogy. Details about the tracer injections into the storms are given by the authors (Burtsev et al., 1970) as follows:

> The aerosol was injected into two regions of the clouds: (a) into the region of cloud-borne droplets, at an altitude of 3.5–4.5 km from the cloud base (subzero-temperature region); (b) into the precipitation zone, at an altitude of 0.5–1.5 km from the cloud base. As a rule, the lower boundary of the raining cloud dissolves, and this zone is either a zone of pure precipitation, or a zone containing both precipitation particles and cloud elements. The radioactive aerosol was injected by means of antiaircraft shells with AgI as crystallizing agent.

From this description it may be correct to infer that AgI ice nuclei were released simultaneously with the radioactively tagged aerosol, and this could have influenced the scavenging of those particles released in the "cloud droplet zone." However, this uncertainty has comparatively minor influence on the value of the results. The dashed curve in Fig. 11.36 shows the theoretical predictions of the scavenging rate based on Eqs. 11.32 and 11.33 (or Eq. 11.46); different values for the mass-average collection efficiency, $\bar{E}$, could be chosen to obtain better fits to the data for scavenging from different parts of the storm. The relatively good fit to the data suggests to the author that most of the tracer particles acted as cloud condensation (or freezing) nuclei, grew to cloud droplet size (presumably more rapidly and larger in the "cloud droplet zone"), and were scavenged with an efficiency comparable to the efficiency with

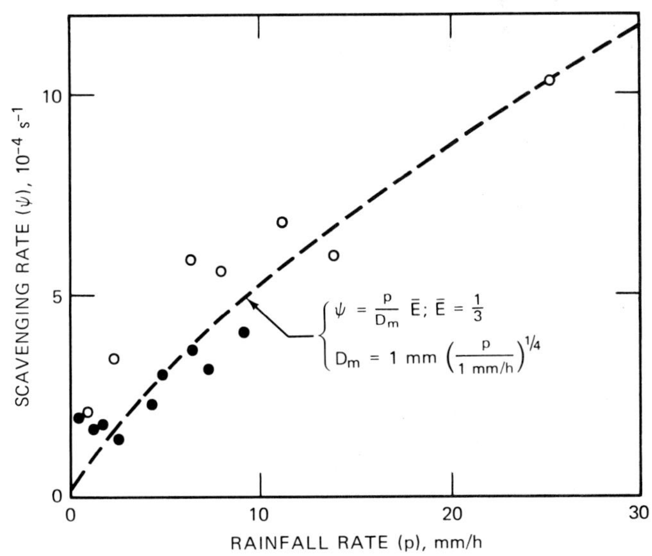

Fig. 11.36 A comparison of the theoretical prediction of Eq. 11.32 (dashed curve) against experimental data obtained by Burtsev, Burtseva, and Malakov (1970) for the in-cloud scavenging of tracer mass released into the top of the rain shaft (•) and into the region of "cloud drops" (O) of a cumulonimbus cloud. [Based on W. G. N. Slinn, Some Approximations for the Wet and Dry Removal of Particles and Gases from the Atmosphere, *J. Water, Air, Soil Poll.*, 7: 513-543 (1977).]

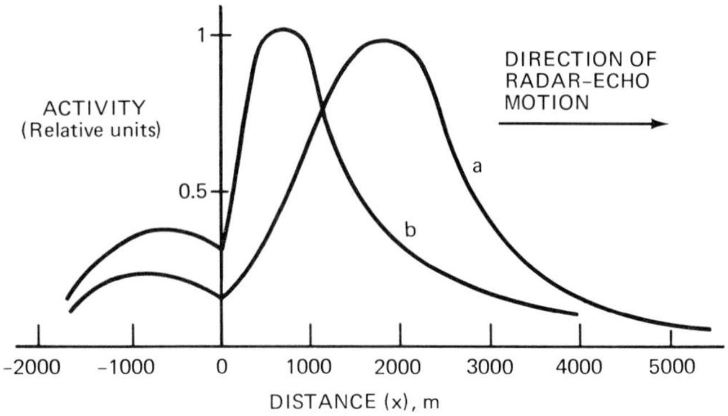

Fig. 11.37 The average, for 16 radioactive tracer releases, of the maximum specific activity of precipitation collected as a function of distance from the shell-explosion epicenter in the direction of the radar-echo motion. Curve a is for 7 tracer injections into the "cloud-droplet" zone, and curve b is for 9 injections into the top of the rain shafts of cumulonimbus storms. Notice that in the latter case the average downwind distance to the maximum concentration was only 0.6 to 0.7 km (the average removal rate was 2×10^{-4}/s and an average of about 65% of the injected tracer was deposited on the sampling grid), whereas for injection into the cloud-droplet zone the maximum concentration occurred at about 2 km (the average removal rate was 6×10^{-4}/s and about 85% of the tracer was deposited). (Based on Burtsev, Burtseva, and Malakhov, 1970.)

which a cloud removes its condensed water. Dingle (1978) suggests that the relatively rapid removal may be caused by the presence of a substantially greater proportion of supermicron particles in the initial release than was assumed by the authors.

Gatz (1977a) reviews approximately 100 field studies of the consequences of injecting tracer aerosol particles into storms. Figure 11.37 shows average downwind (and upwind) distributions of the tracer for the 16 convective storms injected by Burtsev, Burtseva, and Malakhov (1970); on the average, 65% of the tracer injected into the rain shaft and 85% injected into the cloud droplet zone were deposited within about ½ h after injection. Parts a, b, and c of Fig. 11.38 show, respectively: (a) the total precipitation, (b) the ^{210}Po deposition, and (c) the ^{32}P deposition for a cloud propagating to the northeast across the sampling grid. The ^{210}Po was in the form of a polonium nitrate; the ^{32}P was a chromium phosphate powder. The different chemical natures of these compounds may explain the differences in their removal, but probably the major cause was that the polonium was injected (using "antihail" rockets) into the forward, growing part of the storm whereas the phosphorus was injected into the raining, central and rear portions. For the deposition patterns and estimated removal rates for convective storms in which tracers were released into the updraft and entrained into the cloud at midlevels, the reader is referred to the original literature (Dingle, Gatz, and Winchester, 1969; Semonin, 1972; Dingle, 1977). Figure 11.39 shows the deposition pattern for a tracer release within a frontal storm (Atkinson et al., 1972).

For a summary of the substantial effort devoted to these tracer studies, the reader is referred to Gatz (1977a). In this chapter, an attempt is made to interpret the results for practical applications. For an estimate of the scavenging of tracer particles released into or entrained by storms, it appears that the best available procedure is to assume that:

1. Overall removal efficiency is about the same (~10 to ~90%) as the storm's efficiency for removing cloud water, ϵ_{cw} (see Fig. 11.40).

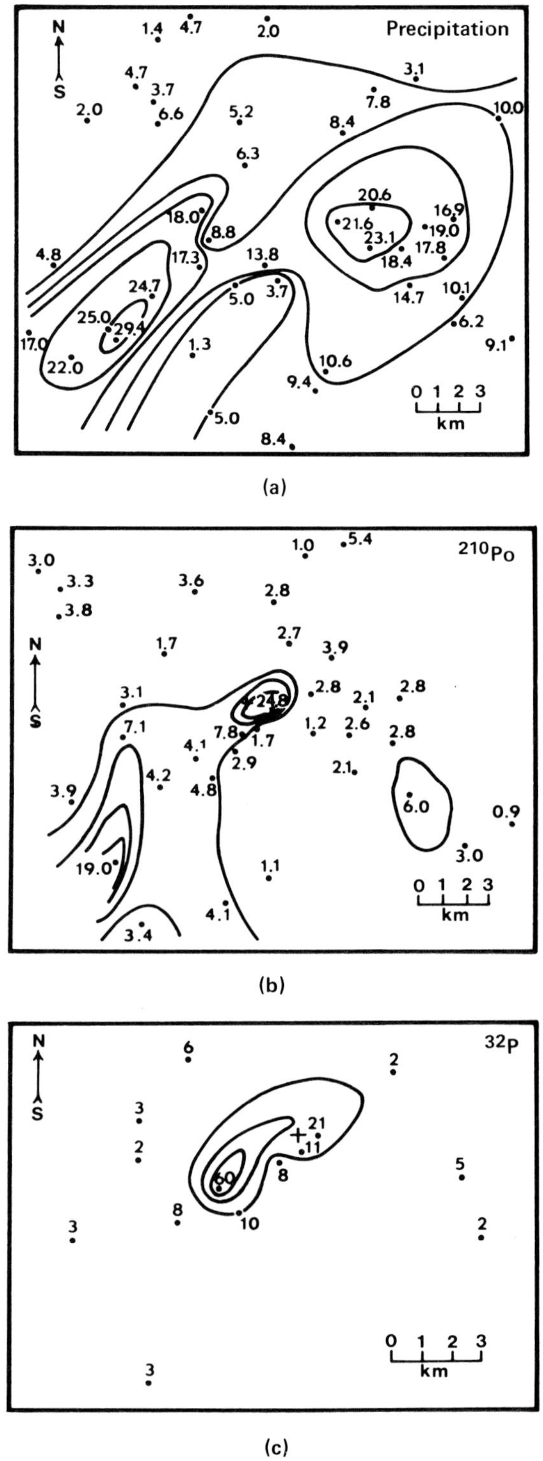

Fig. 11.38 Distributions of precipitation, ^{210}Po, and ^{32}P obtained by Shopauskas et al. (1970) on 5 July 1966. Part (a) shows the total precipitation in mm that fell on the test area during the 15 to 25 min of the storm's presence. The cumulonimbus cloud moved northeastward across the grid; the wind was about 20 m/s at 220° at the tracer injection altitude. Part (b) shows the relative distribution of deposited ^{210}Po. This tracer was injected by four rockets that exploded 3.8 km above the point marked + into the forward part of the cloud, 10 minutes before rain commenced, 5 minutes before, at rain commencement, and 2 min after. The authors estimate that it required only about 2 min for the tracer to attach to the cloud droplets. Part (c) shows the relative distribution of ^{32}P injected by seven rockets into the central and rear portions of the cloud. [Based on Shopauskas et al., 1970.]

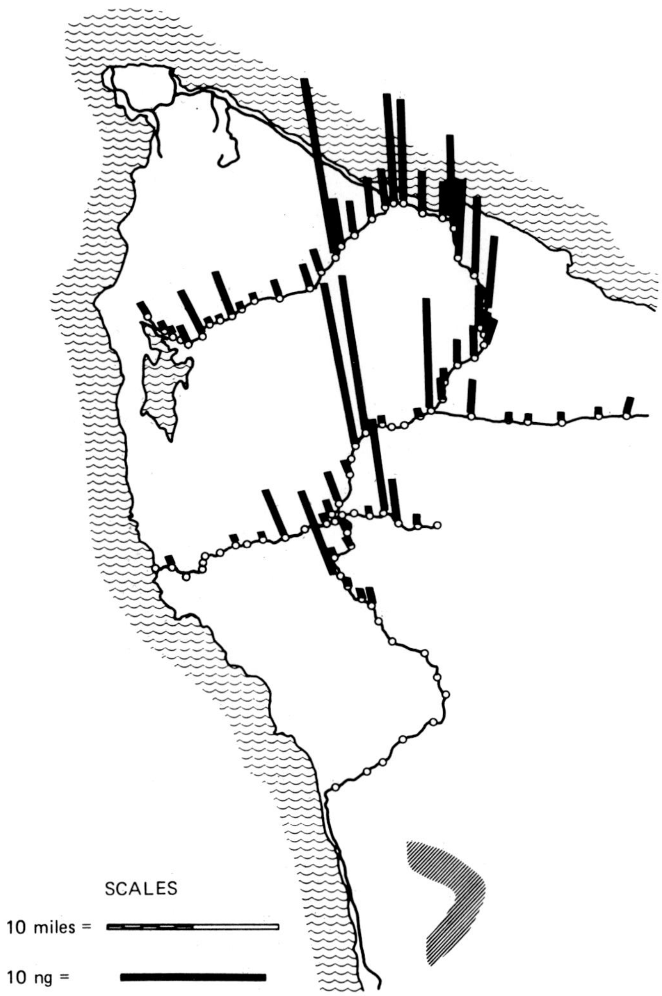

Fig. 11.39 Deposition pattern of an indium aerosol released on Mar. 11, 1971, from within a frontal storm by burning pyrotechnic flares on an aircraft on the flight path shown in the bottom right of the figure. In total, 450 g of indium (mass median diameter ≃ 0.5 μm) was released during the 7-min burn of the flares. Release altitude (5000 ft) was at the freezing level where the winds were 30 m/s at 180°; at 1000 mb the winds were 10 m/s at 130°. The data can be interpreted by using a rapid (~1/min) attachment rate and a rate-limiting, cloud-water-removal rate of ~1/h. It appears that the maximum of the deposition pattern occurred beyond the sampling grid, i.e., beyond about 80 km from the tracer release area. For other results in this test series with frontal storms, see Atkinson et al. (1972), Slinn (1973), Davis, Young, and Thorp (1974), and Davis and Young (1977). (From Slinn, 1977a.)

2. Attachment and/or nucleation processes are efficient, and therefore the rate-limiting stage of the removal process is the removal of the cloud water.

3. Overall removal rate as a function of altitude, z, depends on the rainfall rate, apparently (see Fig. 11.36) as given approximately by

$$\psi(z) = \frac{\epsilon_{cw}\, p(z)}{2R_m} \tag{11.47}$$

where the z-dependence of the rainfall rate should be noticed (see Kessler, 1969, and Machta, 1977; see Fig. 11.41).

4. The deposition patterns depend on storm dynamics and wind fields; the results can be surprisingly complex (Summers and Renick, 1971; Dingle, 1977).

Admittedly these guidelines leave many aspects of the scavenging unspecified, but that, in itself, fairly adequately describes the present state of knowledge in this subject area. More recent analyses of scavenging in terms of precipitation efficiencies (more appropriate than Eq. 11.47) are given elsewhere (Slinn, 1983a,b).

11-3.6 Particle Washout Ratios

In the previous sections, the emphasis was on particle scavenging *rates;* in this section emphasis is on scavenging or washout *ratios* (Junge, 1963; Gatz, 1966; Engelmann, 1970). As was mentioned in Sec. 11-2.4, the washout ratio (or precipitation scavenging concentration factor) is the ratio of the pollutant concentration, κ_0, in surface-level precipitation to the concentration, χ_0, in surface-level air:

$$w_r = \left(\frac{\kappa}{\chi}\right)_0 \tag{11.48}$$

Also, since the wet flux, W, of the pollutant to the surface is $\kappa_0 p_0$, then the relation between the wet deposition velocity, $v_w = W/\chi_0$, and the washout ratio, w_r, is simply

$$v_w = \frac{W}{\chi_0} = \frac{\kappa_0 p_0}{\chi_0} = w_r p_0 \tag{11.49}$$

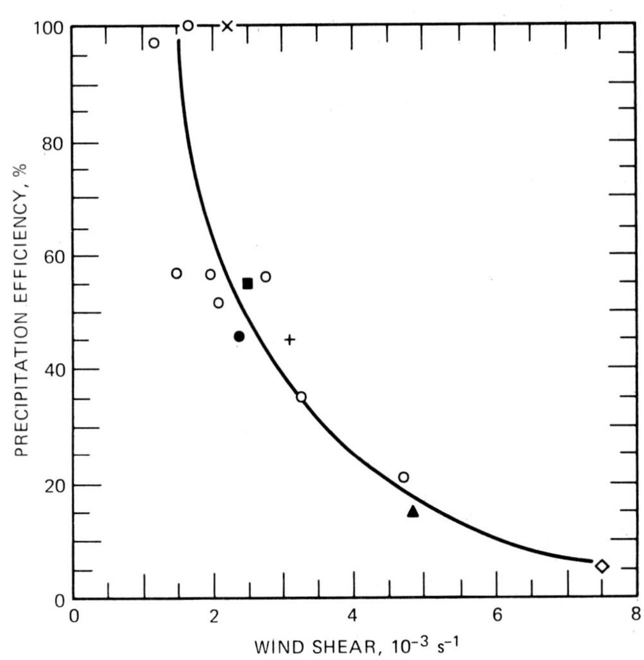

Fig. 11.40 Precipitation efficiency (ratio of precipitation out to water vapor inflow) as a function of the vertical wind shear from cloud base to cloud top for several cumulonimbus clouds. [Based on J. D. Marwitz, Precipitation Efficiency of Thunderstorms on the High Plains, in preprints of the Third Conference on Weather Modification, Rapid City, S. D., pp. 245-247, American Meteorological Society, Boston, Mass., 1972, and G. B. Foote and J. C. Frankhauser, Airflow and Moisture Budget Beneath a Northeast Colorado Hailstorm, *J. Appl. Meteorol.*, 12: 1330-1353 (1973).]

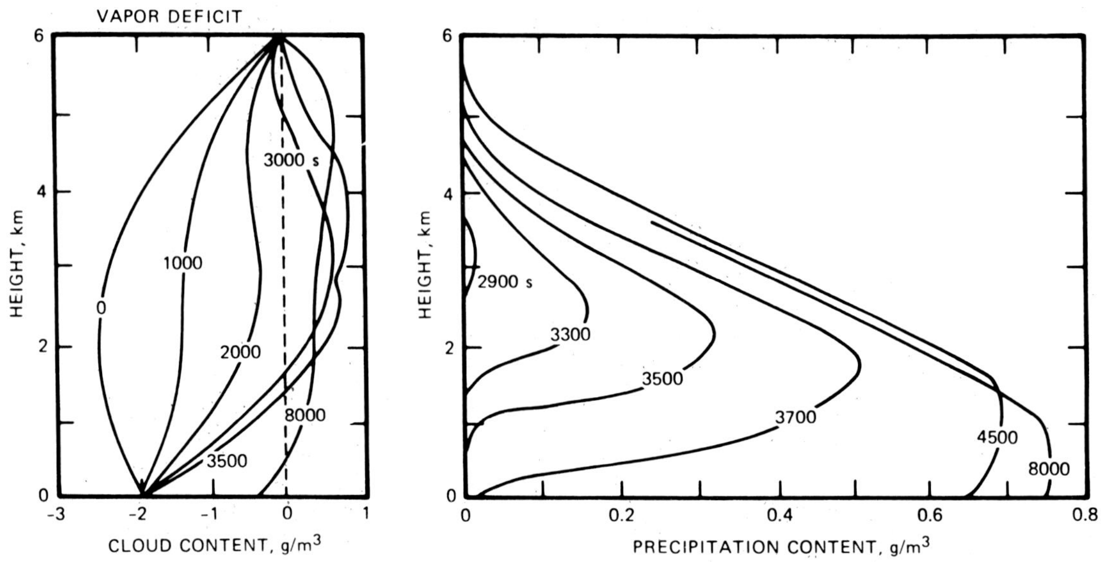

Fig. 11.41 Calculated values for the vertical profiles of water vapor and cloud water (left) and precipitation (right) in an initially unsaturated circulation with weak updrafts. For other examples see Kessler (1969) and Norment (1977). (Based on E. Kessler, On the Distribution and Continuity of Water Substance in Atmospheric Circulations, *Meteorol. Monogr.*, Vol. 10, No. 32, American Meteorological Society, Boston, Mass., 1969.)

where p_0 is the surface-level precipitation rate. For example, if $w_r = 3.6 \times 10^5$ and $p_0 = 1$ mm/h, then $v_w = 10$ cm/s.

To relate particle scavenging rates to washout ratios, the wet flux is calculated by integrating the loss rate per unit volume:

$$W = \int_0^\infty dz\, L = \int_0^\infty dz\, \psi\chi \qquad (11.50)$$

For example, for rain scavenging the use of Eq. 11.32 in Eqs. 11.49 and 11.50 leads to

$$v_w = w_r p_0 = c \int_0^\infty dz\, \frac{p(r,t)}{R_m}\, \frac{\chi(r,t)}{\chi_0}\, \overline{E}\,(a,R_m) \qquad (11.51)$$

Alternatively, if only the z-dependence of p and χ are significant, if we take c = ½, and if the effective collection efficiency (accounting for particle modifications in the atmosphere) is used, then Eq. 11.51 becomes

$$w_r = \frac{hE_{eff}}{2R_m} \qquad (11.52)$$

where the characteristic height from which the pollution is removed is defined via

$$h = \int_0^\infty dz \left[\frac{p(z)}{p_0}\right]\left[\frac{\chi(z)}{\chi_0}\right] \qquad (11.53)$$

For example, for h = 1 km, $2R_m = 1$ mm, and $E_{eff} = 0.36$, then $w_r = 3.6 \times 10^5$.

Typically, experimental values for particle washout ratios have magnitudes of about 10^6 when the nondimensional values are ratios of mass densities (i.e., mass/unit volume ÷ mass/unit volume). Unfortunately, though, washout ratios reported in the literature possess (sometimes implicitly) an assortment of units (such as cubic centimeters per cubic meter, grams of air per grams of water, etc.) and therefore have magnitudes near 10^0, 10^3, or 10^6. Here, dimensionless ratios of mass densities (not mass mixing ratios) will be used, and their magnitudes will be near 10^6.

Actually, the enormous concentrating ability of precipitation (which is one of the prime reasons for interest in precipitation scavenging) can be seen from a simple physical argument. A single 1-mm raindrop can be the result of the coalescence of about 10^6 10-μm cloud droplets. In turn, if each cloud droplet formed as the result of water condensation on a pollutant particle, then the particles that were originally distributed in about 10^4 cm^3 of air (a typical number density of cloud droplets is 10^2 droplets/cm^3) is physically concentrated by a factor of about 10^6 within the volume of a single drop. A similar concentration factor is also appropriate for snow scavenging; this suggests that the ice phase within clouds is as efficient at scavenging as is the liquid phase (see also Koenig, 1977). However, if the pollution is constrained beneath the clouds, then the washout ratio could be substantially smaller than 10^6. This result can be seen by splitting the integral of Eq. 11.51 into heights beneath and within clouds. The main difference between the contributions from these two regions is that, beneath the cloud base, it is expected that there would be less particle growth and, therefore, a smaller collection efficiency (however, see Fig. 11.35).

Figure 11.42 shows the type of results that initially generated interest in washout ratios. The data are for total bomb-debris fission products (Gedeonov et al., 1970). The authors found

$$w_r = \left(\frac{\kappa}{\chi}\right)_0 = (1.0 \pm 0.3) \times 10^6 \qquad (11.54)$$

to be the average (± one standard deviation) for measurements of both natural and bomb-debris radioactivity, which were made throughout the world. Figure 11.43 displays some of the variations in the bomb-debris data.

Some of the variations of the data in Fig. 11.43 can be understood from Junge's scavenging model (Junge, 1963, 1975), see Fig. 11.44. If the average pollutant concentration entering the storm is $\overline{\chi}$ and if a fraction ϵ_i of the pollutant enters the cloud water, then the pollutant concentration in the cloud water is $\epsilon_i\overline{\chi}$ units per volume or $\epsilon_i\overline{\chi}/L$ units per unit mass of cloud water, where L is the condensed water content of the cloud ($0.1 \lesssim L \lesssim 10$ g/cm^3). If this same concentration is in the precipitation reaching the ground (e.g., assuming no significant evaporation or condensation of water and that drops grow by coagulation of equally polluted drops), then the concentration, κ_0, in surface-level precipitation would also be $\epsilon_i\overline{\chi}/L$ units per unit mass of water or, on a volume basis,

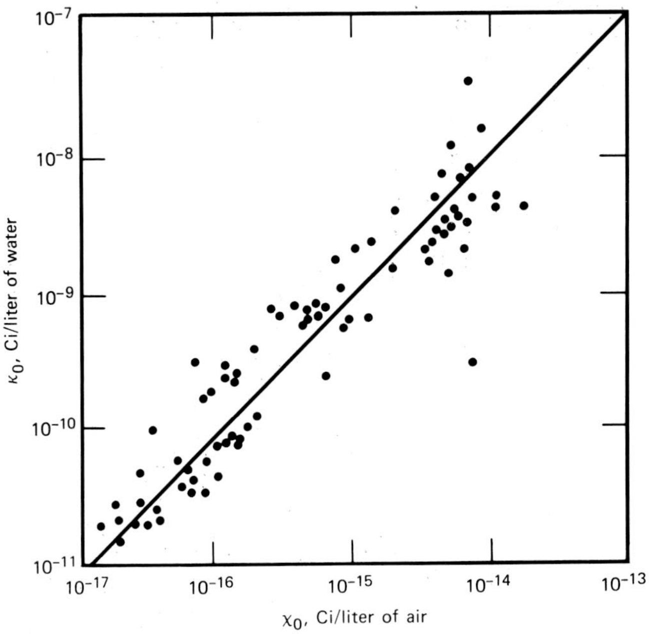

Fig. 11.42 An example of the correlation between measured concentrations of total fission products in precipitation, κ_0, and their concentrations in surface-level air, χ_0. Notice that the best fit for the washout ratio, viz., $w_r = (\kappa/\chi)_0 = (1.0 \pm 0.3) \times 10^6$, appears equally adequate over 4 orders of magnitude of variation in κ and χ. (Based on Gedeonov et al., 1970.)

$$w_r = \left(\frac{\kappa}{\chi}\right)_0 = \frac{\rho_w \epsilon_i}{L} \frac{\overline{\chi}}{\chi_0} \tag{11.55}$$

where ρ_w is the mass density of water (1 g/cm^3). Notice that with $\epsilon_i\overline{\chi}/\chi_0 \simeq 1$ and $L \simeq 1$ g/m^3, Eq. 11.55 yields $w_r \simeq 10^6$. Some of the variations in the washout ratios shown in Fig. 11.43 might be explained in terms of Eq. 11.55 if (1) ϵ_i is larger for fresh bomb debris than for aged debris; (2) the condensed water content of clouds is larger for larger precipitation rates; and (3) account is taken of the vertical distribution of the pollutant entering the storm. This last item appears to be especially significant when estimates for washout ratios are desired for particles associated with urban pollution. It is to be kept in mind that, for bomb debris initially injected into the stratosphere, typically $\overline{\chi}/\chi_0 > 1$; in contrast, for pollutants with surface-level sources, typically $\overline{\chi}/\chi_0 < 1$. Consequently, it is expected that the washout ratios for urban particles would be less than 10^6.

Some of the major dependencies of washout ratios for trace metal aerosol particles on such variables as particle size, rain-

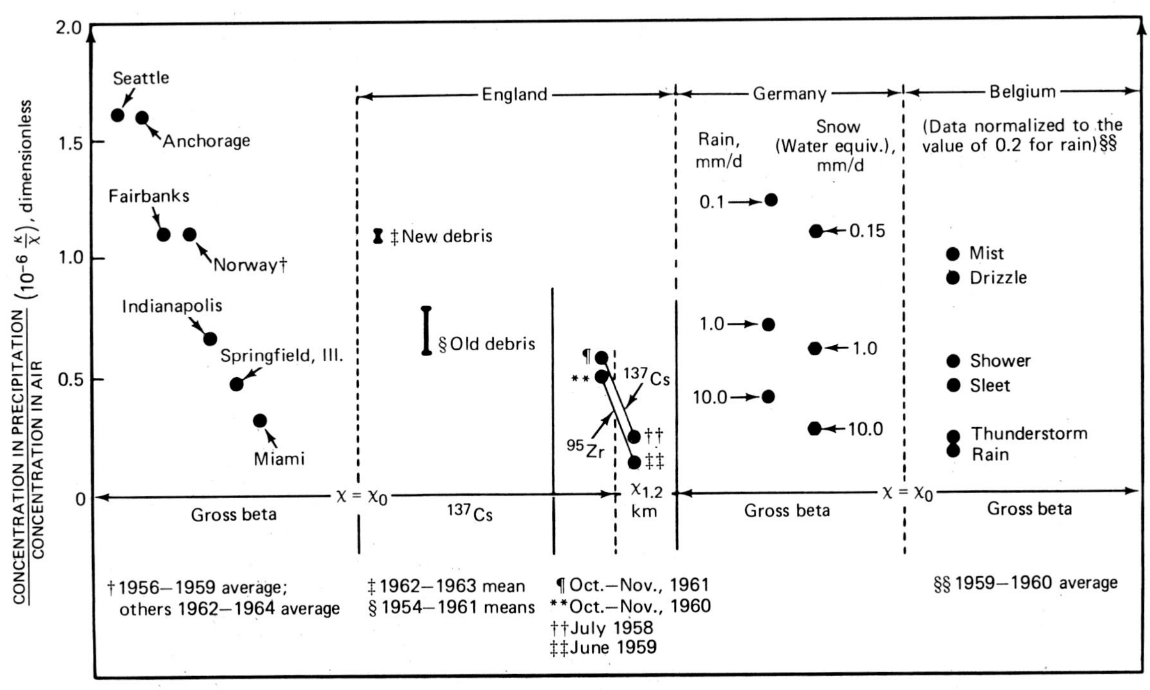

Fig. 11.43 An indication of some of the sources of variation in the bomb-debris washout ratios, namely: geographic location, debris age, height of measurement of χ, and precipitation intensity. Data from Junge (1963) and Engelmann (1970). [From W. G. N. Slinn, Some Approximations for the Wet and Dry Removal of Particles and Gases from the Atmosphere, *J. Water, Air, Soil Poll.*, 7: 513-543 (1977).]

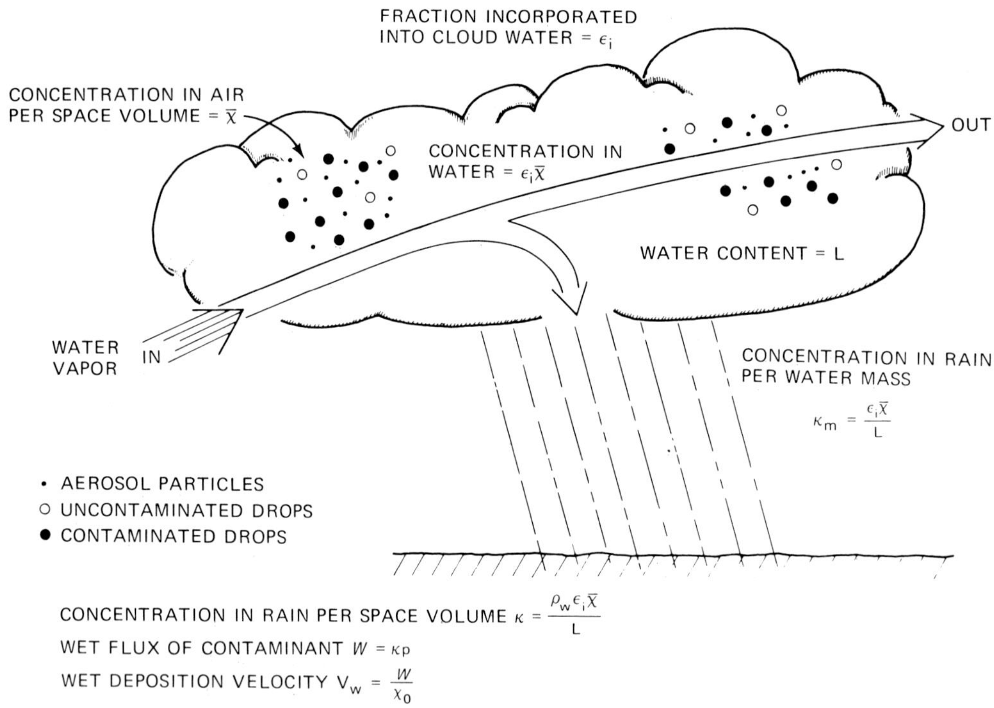

Fig. 11.44 A schematic illustration for one derivation of the washout ratio, $w_r = (\kappa/\chi)_0$. [Based on W. G. N. Slinn, Some Approximations for the Wet and Dry Removal of Particles and Gases from the Atmosphere, *J. Water, Air, Soil Poll.*, 7: 513-543 (1977).]

fall rate, and storm type can be interpreted on the basis of Eq. 11.52, viz.,

$$w_r = \frac{h}{2R_m}\bar{E}_{eff}[a(\tau),R_m] \tag{11.56}$$

where the dependence of particle size on nondimensional time, τ, has been made explicit. For snow scavenging, Eq. 11.36 (rather than 11.32) should be used in Eq. 11.50; for mass-average washout ratios, the effective collection efficiency is as in Eq. 11.46. For bomb-debris radionuclides, Fig. 11.45 shows that there is a relatively weak dependence of w_r on rainfall amount; this weak dependence for similar contaminants scavenged under similar meteorological conditions is suggested in Eq. 11.56 by the weak dependence of R_m on p (e.g., $R_m \sim p^{1/4}$). Figure 11.46 displays washout ratios for various trace elements obtained by Cawse (1974), mainly for frontal storms. These data are precipitation-weighted values, and therefore results from a few storms with large precipitation can have a dominant influence on the reported values. Also, the sampling method collected both wet and dry deposition. In contrast to the results shown in Fig. 11.46, mostly for frontal storms, Fig. 11.47 shows precipitation-weighted washout ratios obtained by Gatz (1975a, 1975b) for convective storms. Figure 11.48 suggests, on the basis of Eq. 11.56, that the larger values of w_r found for frontal storms is caused by the longer times available for particle growth (and therefore larger collection efficiencies) in frontal storms compared to those in convective storms. However, the difference may be caused by differences in the (poorly known) effective height, h (see Eq. 11.53) from which the pollution is scavenged and by differences in contributions from below- and in-cloud scavenging. Also, the term "frontal storms" does not adequately describe their variety; for example, even stable warm fronts frequently contain rainbands that could be caused by convective activity. Thus, in summary, for practical applications using available data for washout ratios is preferable to using Eq. 11.56 (or similar expressions) although the theory can help guide extrapolations.

It is important to mention some of the limitations in the use of particle washout ratios and/or wet deposition velocities. First, for urban pollutants and χ measured in urban areas, the dependence of w_r on total rainfall is neither so weak nor so predictable as is the case for bomb-debris radionuclides shown in Fig. 11.45. This statement is illustrated in Fig. 11.49. Also, as seen in Fig. 11.47, values of w_r differ substantially (factors of 2 to 10) for frontal vs. convective storms. Furthermore, as displayed in Fig. 11.50, there are order-of-magnitude differences in the washout ratios for different storms of the same type. Actually, however, these variations are to be expected; precipitation scavenging is a complicated process, and it would be naive to expect that all meteorological variations could be parameterized in a single variable. The fact that the washout ratio does show some regularity and therefore does have some practical value no doubt follows because the ratio represents the average over many variables, e.g., particle and precipitation size distributions, vertical distributions, and sometimes even storm types. If enough averages are taken of almost any variable, its variability will be suppressed! In contrast, if some of the averages are not taken, then one sees (as in Fig. 11.51) that there are large variations in w_r as a function of time during an individual storm. In turn, as shown in Fig. 11.52, these variations in w_r foretell of even larger variations in the wet deposition velocity $v_w = pw_r$. Thus, although long-time (e.g., seasonal) average values for the washout ratio for a particular location may be accurate to within about a factor of 2 to 4 provided the data base for a particular pollutant is extensive, it is to be expected, on the other hand, that short-term (e.g., accident case) predictions based on washout ratios are no more accurate, and indeed are likely to be less accurate, than the

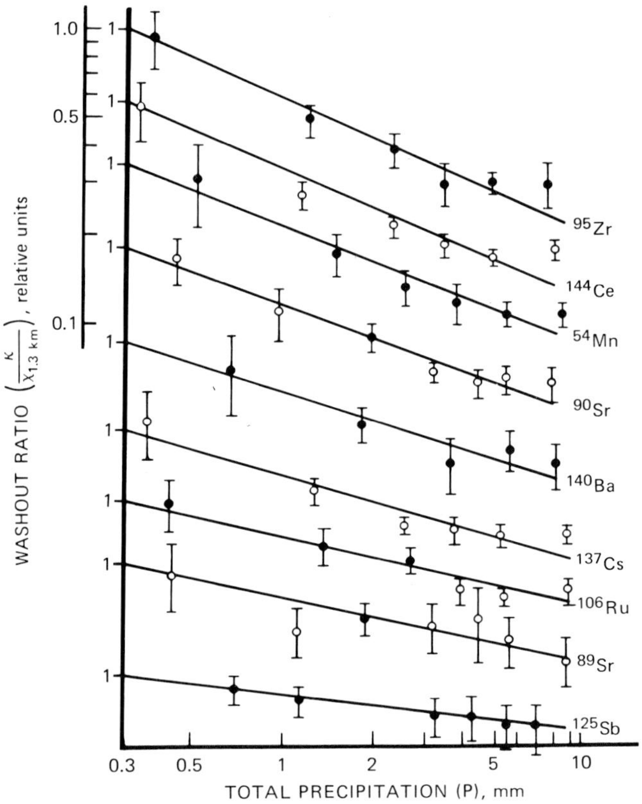

Fig. 11.45 Monthly average κ/χ values for a number of radionuclides as a function of total precipitation. The air concentration, χ, was measured at the cloud-layer altitude (typically 1.3 km); the 24-h precipitation totals, P, were calculated by dividing the monthly precipitation by the number of rain days in the month. The curves are displaced vertically to improve clarity; for all curves the 0.1 to 1.0 ordinate is appropriate. Original data were taken from a number of Harwell reports; graph redrawn from Makhon'ko, Avramenko, and Makhon'ko (1970) in which the original references can be found.

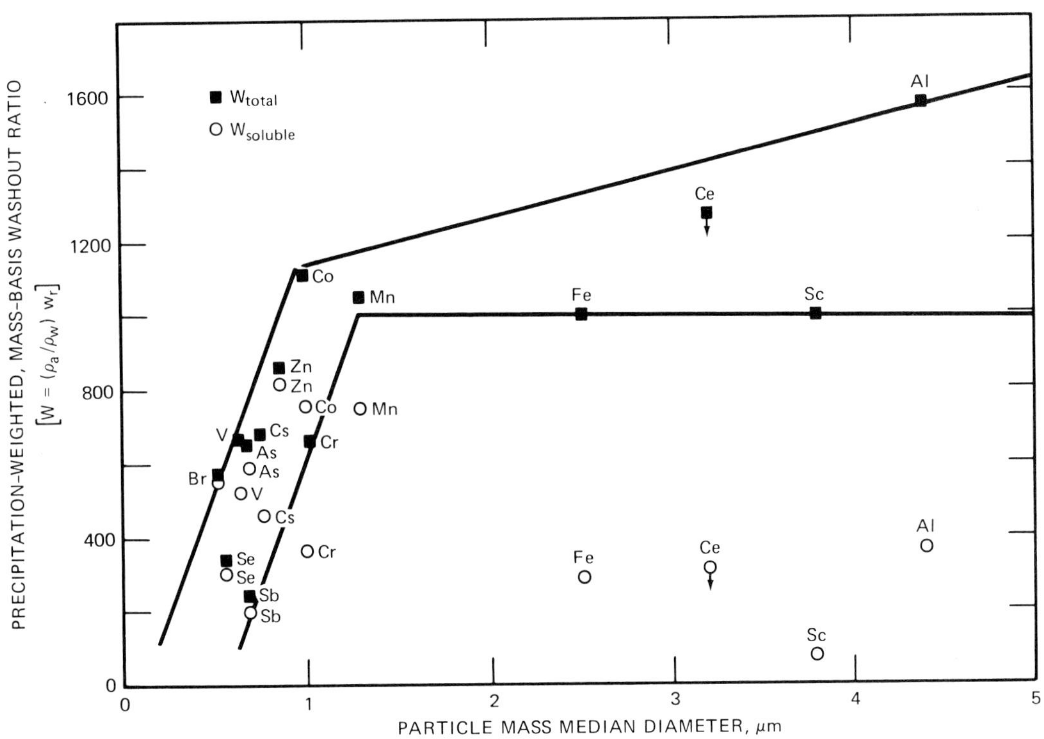

Fig. 11.46 Variations of precipitation-weighted washout or scavenging ratios with particle mass median diameter. Data obtained by Cawse (1974) at Chilton, U.K., during July–December, 1973; much of the data is probably for frontal storms. The solid squares represent washout ratios computed from total (soluble plus insoluble) concentrations in precipitation. Open circles represent w_r values calculated using only the soluble concentrations. The precipitation samples included contributions from both dry and wet deposition. Downward pointing arrows indicate upper limit value. Washout ratios on a mass basis are about 3 orders of magnitude smaller than those on a volume basis: $\rho = 1.23 \times 10^{-3}$ g/cm^3; $\rho_w = 1.0$ g/cm^3. (Based on D. F. Gatz, Wet Deposition Estimates Using Scavenging Ratios, in *First Specialty Symposium on Atmospheric Contributions to the Chemistry of Lake Water,* Sept. 28–Oct. 1, 1975, pp. 2-29, International Association for Great Lakes Research, 1975.)

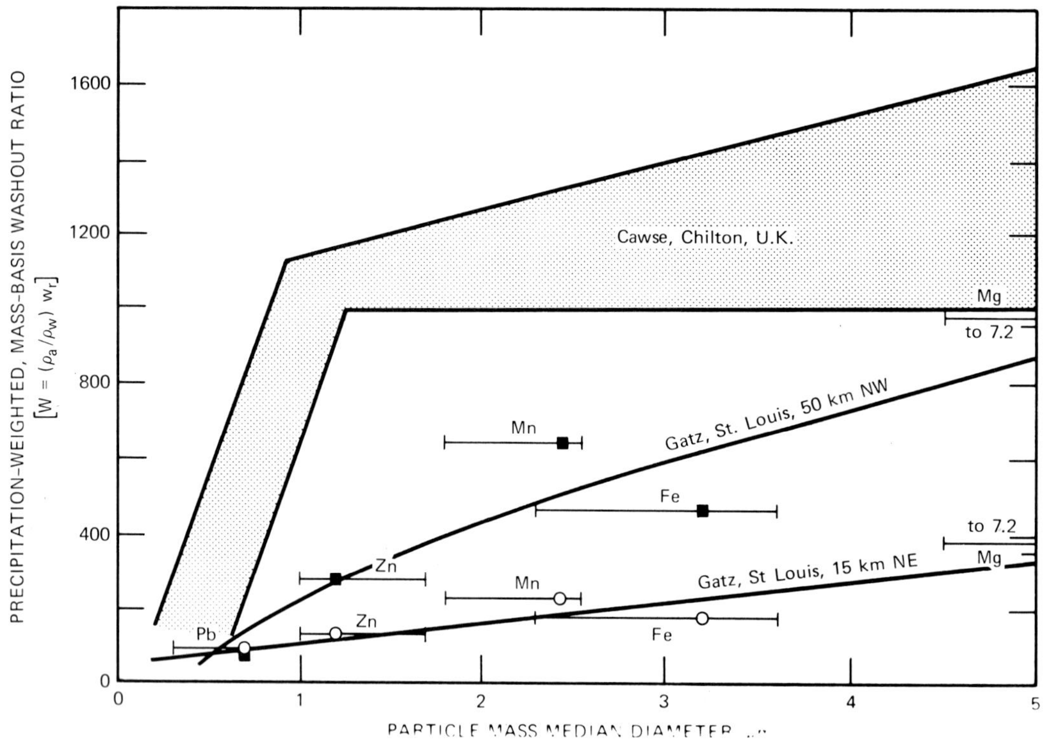

Fig. 11.47 Comparison of rainfall-weighted washout ratios as a function of particle size for Chilton (monthly samples, including dry deposition, obtained by Cawse, 1974) and for St. Louis (daily samples, excluding dry deposition, for convective storms, obtained by Gatz, 1975a). (Based on D. F. Gatz, Wet Deposition Estimates Using Scavenging Ratios, in *First Specialty Symposium on Atmospheric Contributions to the Chemistry of Lake Water*, Sept. 28–Oct. 1, 1975, pp. 2-29, International Association for Great Lakes Research, 1975.)

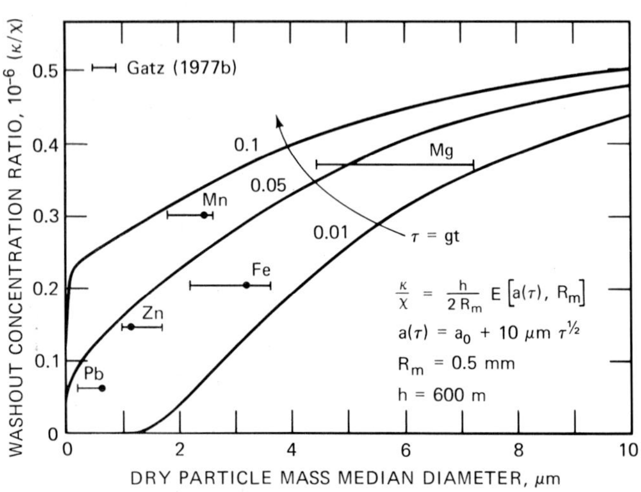

Fig. 11.48 Comparison of the predictions of Eq. 11.56 against rainfall-weighted washout ratios obtained by Gatz (1977b) for various trace metals scavenged by convective storms. In the evaluation of Eq. 11.56, different growth times (or growth rates) were considered, E was as given in Fig. 11.5, and the characteristic height h (see Eq. 11.53) was chosen solely on the basis of improving the fit to the data. [Based on W. G. N. Slinn, Some Approximations for the Wet and Dry Removal of Particles and Gases from the Atmosphere, *J. Water, Air, Soil Poll.*, 7: 513-543 (1977).]

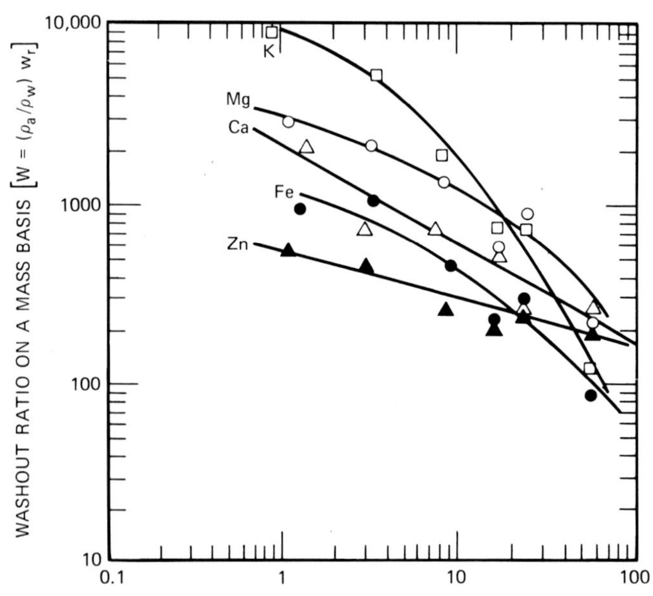

Fig. 11.49 Variations of washout ratios with amount of daily precipitation. The washout ratio on a mass basis is about 3 orders of magnitude smaller than that on a volume basis (ρ_a = 1.23×10^{-3} g/cm^3; ρ_w = 1.0 g/cm^3). (Based on D. F. Gatz, Wet Deposition Estimates Using Scavenging Ratios, in *First Specialty Symposium on Atmospheric Contributions to the Chemistry of Lake Water*, Sept. 28–Oct. 1, 1975, pp. 2-29, International Association for Great Lakes Research, 1975.)

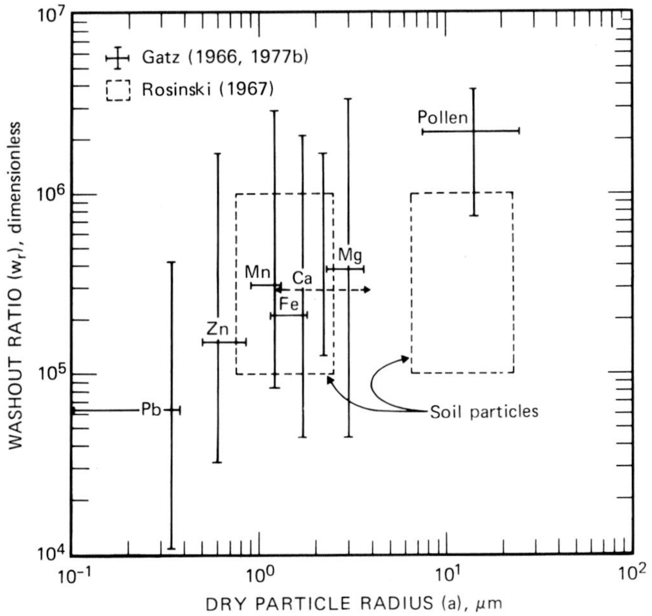

Fig. 11.50 Variations of washout ratios for different aerosol particles for different convective storms. The horizontal variation of the data obtained by Gatz (1977b) reflects different mass median diameters measured for the material at different sites; the vertical bar occurs at the mass median radius appropriate for St. Louis. The vertical variation spans approximately from 5 to 95% on a cumulative frequency distribution of measured washout ratios. An indication of the source of the variability in Rosinski's data is shown in Figure 11.51. [From W. G. N. Slinn, Parameterizations for Resuspension and for Wet and Dry Deposition of Particles and Gases for Use in Radiation Dose Calculations, *Nucl. Saf.*, 19(2): 205-219 (1978).]

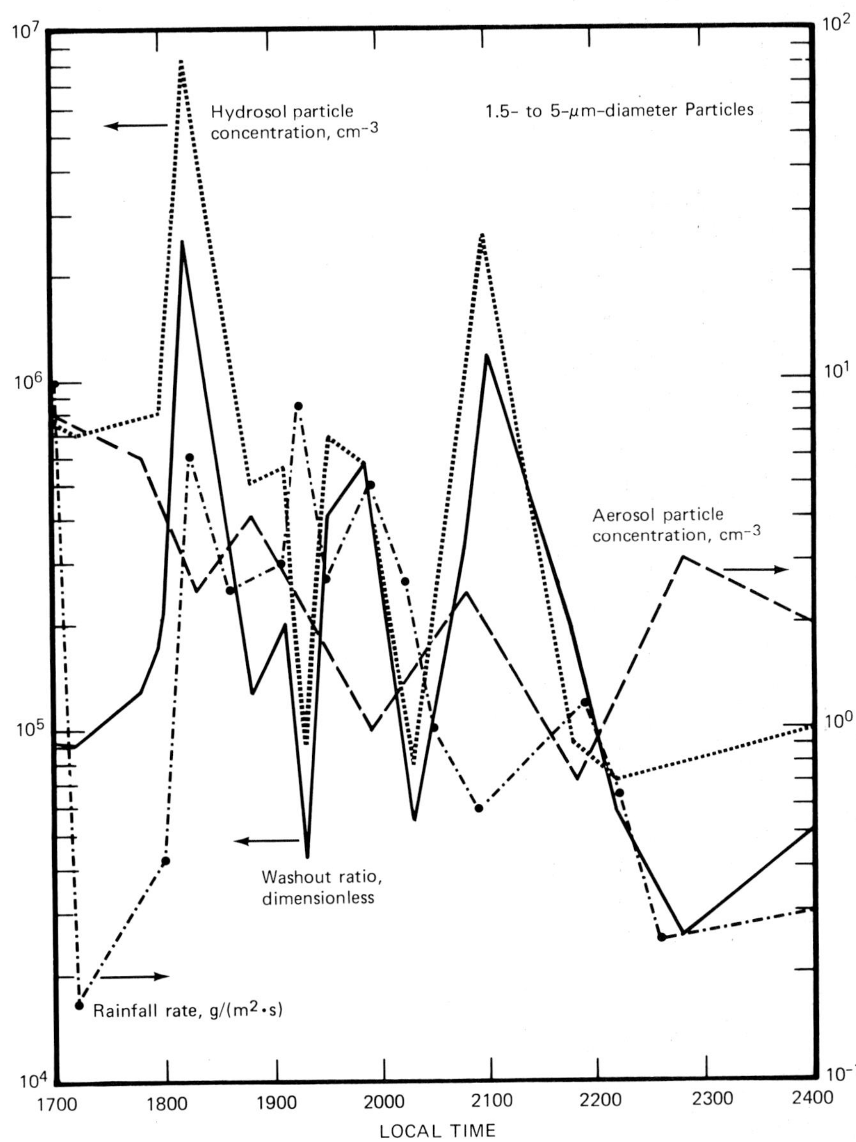

Fig. 11.51 Variation in κ, χ, w_r, and p during a single convective storm on June 30, 1966. Data were taken from Rosinski (1967). [From W. G. N. Slinn, Parameterizations for Resuspension and for Wet and Dry Deposition of Particles and Gases for Use in Radiation Dose Calculations, *Nucl. Saf.*, 19(2): 205-219 (1978).]

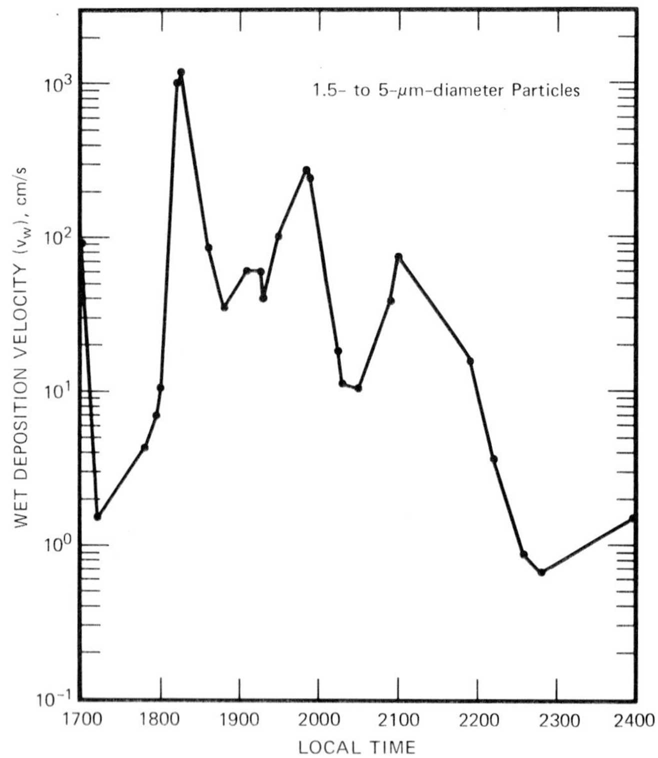

Fig. 11.52 Variations in the wet deposition velocity derived from the data shown in Fig. 11.51. Data were taken from Rosinski (1967). [From W. G. N. Slinn, Parameterizations for Resuspension and for Wet and Dry Deposition of Particles and Gases for Use in Radiation Dose Calculations, *Nucl. Saf.*, 19(2): 205-219 (1978).]

(sometimes) order-of-magnitude calculations based on scavenging rates (Slinn, 1977b and 1978). In summary, if one knows or has predictions for the vertical distribution of a pollutant, particle size distributions, characteristics of the precipitation, general features of the storms, etc., this information (or any part of it) should be used by calculating scavenging rates and not ignored by employing washout ratios.

11-3.7 Some Major Uncertainties About Particle Scavenging

At many places in this section on particle scavenging, the existence of major uncertainties has been mentioned. An extensive amount, comparable to this entire section, could be written about the collision efficiency for submicron particles or about the retention efficiency problem or about snow scavenging. However, detailed descriptions have been avoided because of space limitations and because of a desire to present a coherent picture of the entire field. Nevertheless, space should be devoted, at least, to calling attention to some of the uncertainties and to mentioning topics requiring further investigation. The order of the presentation will follow the order in which the topics have already appeared.

Early work on the retention efficiency problem was performed by McCulley et al. (1956) and by Pemberton (1960). McDonald (1963) extended the generality of Pemberton's calculations and showed that the collection efficiency could be reduced drastically for nonwettable particles (contact angle, $\theta = 180°$) and reduced substantially for all partially wettable particles ($0 < \theta < 180°$); McDonald's results, using Fonda and Herne's calculations (Herne, 1960) for the collision efficiency, are shown in Fig. 11.53. Other studies of the retention efficiency problem include those by Engelmann (1963) who presented a surface energy model of coalescence and Esmen (1973) who interpreted the scavenging of soluble vs. insoluble particles on the basis of differences in retention efficiency. Berg, Gaukler, and Squier (1965) presented important results showing that the rate at which particles are wetted depends on the wettability of the particle and on the voltage difference between the drop and particle. A good source for more current work (e.g., El-Shimi and Goddard, 1974; Miller and Ruckenstein, 1974) is the *Journal of Colloid and Interface Science*. Stulov, Murashkevich, and Fuchs (1978) suggest that all collisions result in capture unless incoming particles collide with previously captured particles residing on the drop's surface. However, if the previously captured particles are wettable, presumably they will not remain on the drop's surface. In summary, to the author's knowledge, there has not yet been a definitive study that necessitates relinquishing the convenient assumption that the retention efficiency is unity for particles smaller than about 10 μm. The reason, however, may be that other uncertainties are so large as to preclude a definitive statement about the retention efficiency.

Some of these other uncertainties are about the collision efficiency. For example, Fig. 11.54 shows a dramatic decrease in the collection efficiency with increasing drop size. This

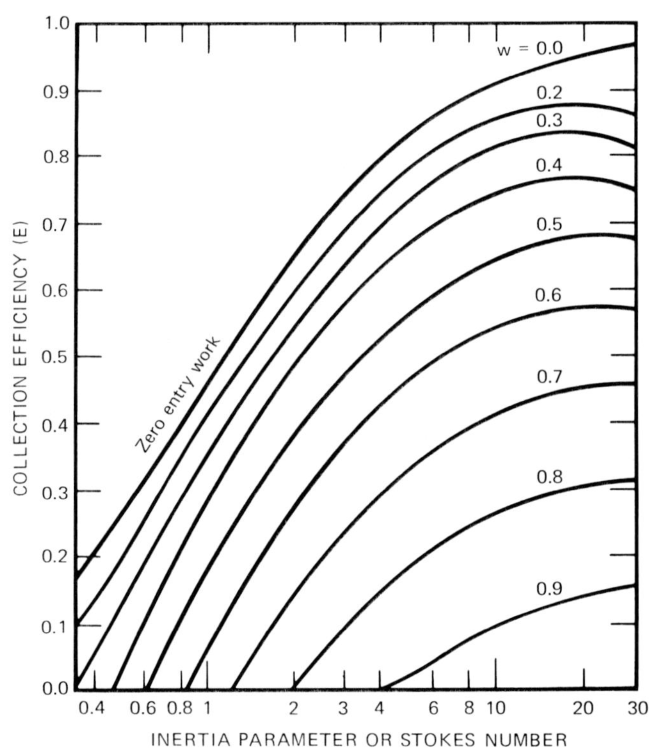

Fig. 11.53 Collection efficiency as a function of the Stokes' number (or impaction parameter) for different particle wettabilities, defined by the parameter $w = v_{pn}/v_t$, where $v_{pn} = (4\sigma \cos \beta / a\rho_p)^{1/2}$ in which σ = surface tension, $\beta = \pi - \theta$, θ = contact angle ($\theta = \pi$, completely nonwettable particle), a = particle radius, and ρ_p = particle density. The zero entry-work curve is for $\theta = \pi/2$, and McDonald (1963) suggests that the same curve is also appropriate for $\theta < \pi/2$. $\theta = 0$ is for completely wettable particle; work is done on particles as they enter a drop if $\theta < \pi/2$. [From J. E. McDonald, Rain Washout of Partially Wettable Insoluble Particles, *J. Geophys. Res.*, 68: 4993-5003 (1963).]

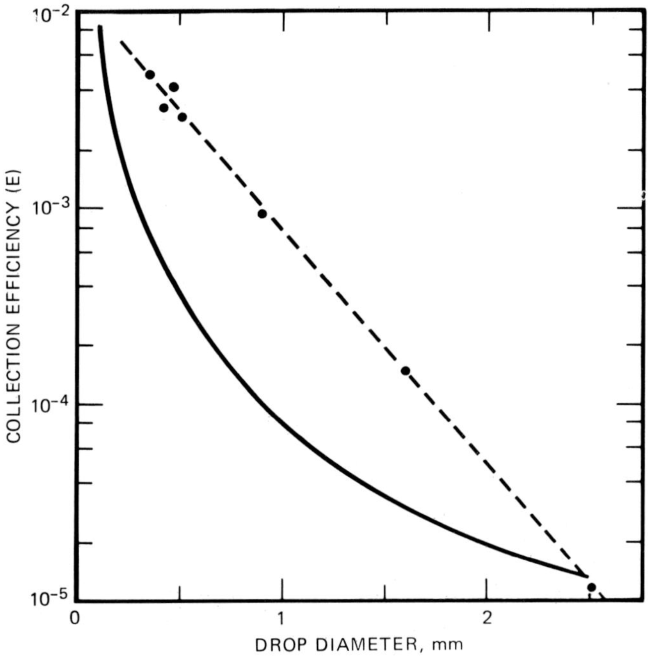

Fig. 11.54 Measured dependence of collection efficiency on drop size. Collected particles were *Bacillus subtilis* spores, which are rod shaped with a diameter of 0.7 μm and a length of 1.2 μm. •, experimental results; ———, experimental resolution. Because the drops were not at thermal equilibrium as they fell through the scavenging chamber, thermophoresis may have influenced the collection efficiency. The dramatic decrease of E with R is not consistent with results found by Starr and Mason (1966) for slightly larger particles. (From Adam and Semonin, 1970.)

decrease is not consistent with the collision efficiency shown in Fig. 11.5 or with current models for the retention efficiency since, for large drops, the energy available for particle penetration into the drop would be larger. In this same study, Adam and Semonin (1970) investigated the influence on the collection efficiency of electrical charges (see Fig. 11.55), and therefore the results shown in Fig. 11.54 do not appear to be explainable on the basis of electrical interactions. A number of authors (e.g., Englemann, 1965; Slinn, 1967; Berg, 1970) have suggested that the cause of the decrease in E with increasing R is wake shedding at large Reynolds numbers, but the numerical studies by Grover, Pruppacher, and Hamielec (1977)—see Fig. 11.57—do not predict such a dramatic dependence on drop size as is shown in Fig. 11.54.

Figure 11.56 illustrates dramatically the uncertainties about submicron particle/drop collection efficiencies: from this illustration, the reader may conclude that essentially nothing is known about the collection efficiency for submicron particles! Studies by Pruppacher and his colleagues (see Fig. 11.57 and references in the legend) suggest that thermophoresis could have had a significant influence on particle capture during some of the laboratory studies leading to the results shown in Fig. 11.56. Thus it may be that, if humidity in scavenging chambers is kept low to suppress particle growth, thermophoretic capture is enhanced over values predicted for normal atmospheric conditions when there is precipitation; on the other hand, if the humidity is maintained near 100%, particle growth can be of great importance. There are not corresponding uncertainties about particle/ice crystal collection efficiencies, almost certainly because there are substantially fewer data.

Theoretical studies of the collision efficiency for particles in the radius range $0.1 < a < 1$ μm provide some guidance but are not complete. Fuchs (1964) and Slinn (1977a) admit to the crudeness of their calculations of the interception term. The conclusions made in the important numerical studies by Beard (1974)—see Fig. 11.56—can be questioned because of his assumption that once particles enter a drop's wake they would be captured by the drop. Calculations that include electrical effects (e.g., see Fig. 11.58 and Zebel, 1968) can show dramatic changes in the collision efficiencies, but the value of such calculations is severely restricted by a lack of data for realistic charges and electrical fields. Rosenkilde (1976) has surveyed available data and the influence of electrical effects on attachment of particles to precipitation embryos within clouds. In summary, the present author's opinions are that, beneath clouds, electrical effects are usually insignificant (since natural aerosol particles likely possess a Boltzmann distribution of charges and electric fields are relatively weak) and that, within clouds, the details about electrical effects can be ignored since the data are consistent with the assumption that, by one mechanism or another, the particles are attached to cloud particles. The numerical calculation of collisions between particles and simple ice crystals (Pitter, 1977) suggests that inertial contribution can be described by the relation shown in Fig. 11.6

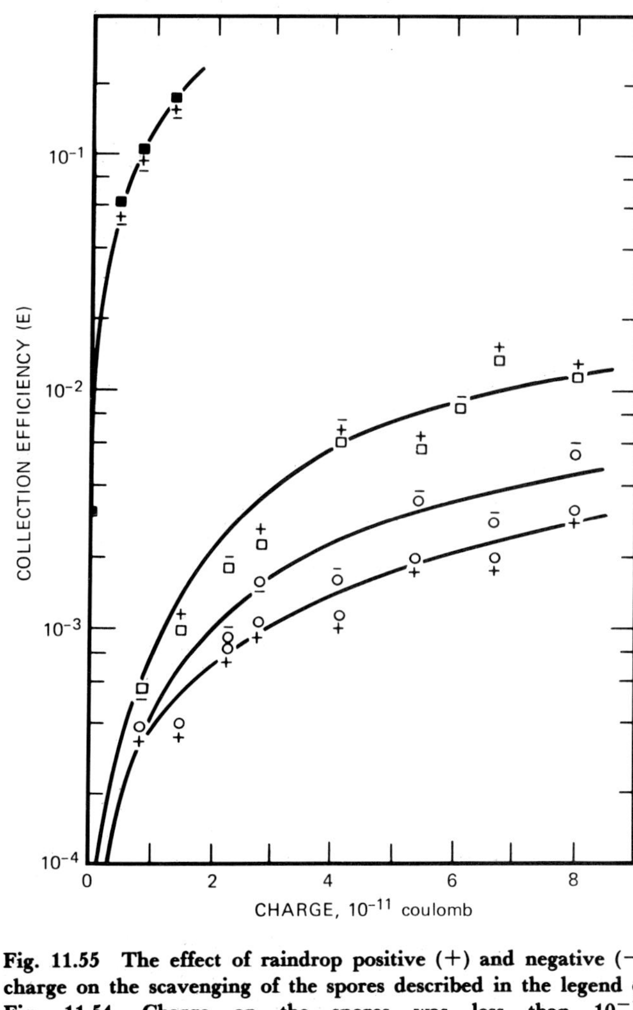

Fig. 11.55 The effect of raindrop positive (+) and negative (−) charge on the scavenging of the spores described in the legend of Fig. 11.54. Charge on the spores was less than 10^{-16} coulomb. ■, 0.4-mm diameter; □, 2.8-mm diameter; ○, 3.7-mm diameter. Typical charges on raindrops from frontal storms are in the range from 10^{-12} to 10^{-14} coulomb. (From Adam and Semonin, 1970.)

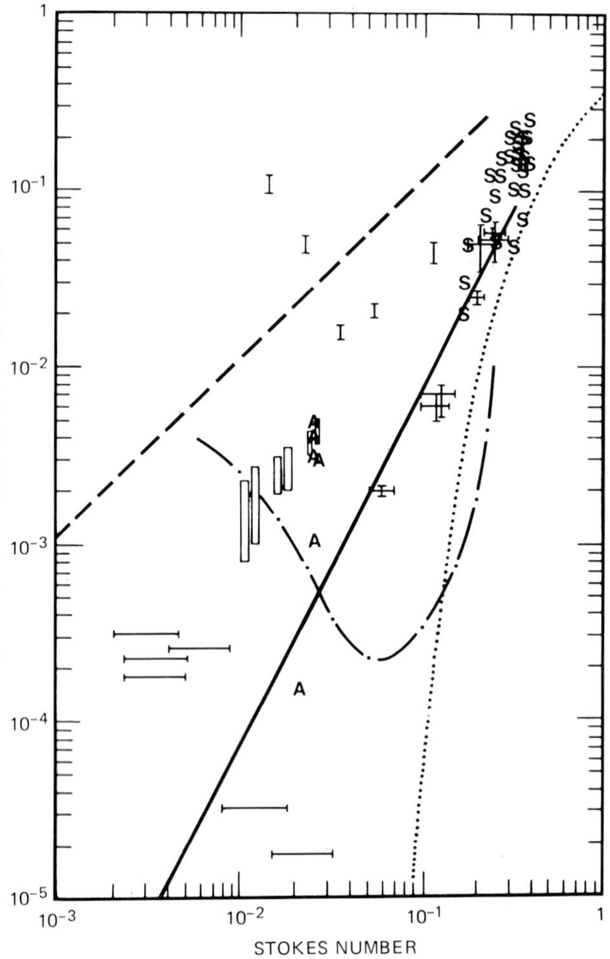

Fig. 11.56 An illustration of the uncertainties about the collision and collection efficiencies for submicron particles; if other data were included, it could be seen that there are more than four orders of magnitude in the uncertainty. Some details about the data shown are as follows. S (Starr and Mason, 1966): collection of spores of radii 2.25, 2.6, and 6.4 μm by drops ranging in radii from 0.1 to 1 mm; only drops with R ≲ 0.3 μm reached terminal velocity in the apparatus. ╪ (Walton and Woolcock, 1960): stationary drop collecting methylene blue particles of mean radii 1.25 and 2.5 μm. A (Adam and Semonin, 1970): *Bacillus subtilis* spores, rod-shaped with a diameter of 0.7 μm and a length of 1.2 μm (see also Fig. 11.54). ▯ (Sood and Jackson, 1969): *Bacillus subtilis* spores scavenged by 0.5- to 2.6-mm-radius drops, falling at more than 95% of their terminal velocity. ├──┤ (Kerker and Hampl, 1974): 0.15- to 0.6-μm-radius AgI particles scavenged by 0.94- to 2.54-mm-radius drops; results extrapolated to drop terminal velocity. I (Lai, Dayan, and Kerker, 1978): drop radius = 0.62 mm; AgCl particle mean radii ranging from 0.15 to 0.45 μm. The following theoretical curves are some of many that could be displayed. — · — (Beard, 1974: numerical results (see text). ...., exp (− 1/s), an empirical curve suggested by Berg (1970) and used by Slinn (1975b) as an extrapolation from an exact solution for potential flow at large Stokes' number. ———, (¾)s^2, an analytical result using Oseen's uniformization technique for low Stokes' number and for Stokes' flow about a sphere. — — —, $\sqrt{4/3}$ s, similarly, but for potential flow about a sphere (Slinn, 1975b). From results such as these, it is to be concluded that the collection efficiency for 0.1- to 1-μm particles is essentially unknown; see, however, Fig. 11.57. [Based on M. Kerker and V. Hampl, Scavenging of Aerosol Particles by a Falling Drop and Calculation of Washout Coefficients, *J. Atmos. Sci.*, 31: 1368-1376 (1974).]

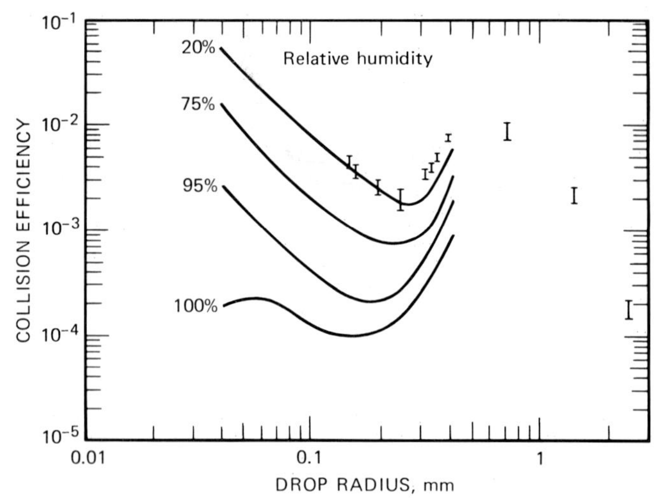

Fig. 11.57 Numerical results for the collision efficiency by Grover, Pruppacher, and Hamielec (1977) and experimental results for the collection efficiency by Wang and Pruppacher (1977) as functions of the drop size for 0.25-μm-radius particles. Numerical results are for uncharged particles and drops. The experimental results (I) were obtained at a relative humidity of (23 ± 2)% and a temperature of (22 ± 2)°C; the numerical results are for relative humidities of 20%, 75%, 95%, and 100% and predict that thermophoresis can play a significant role in laboratory studies of collection efficiency. Also see Slinn and Hales (1971). [Based on P. K. Wang and H. R. Pruppacher, An Experimental Determination of the Efficiency with Which Aerosol Particles Are Collected by Water Drops in Subsaturated Air, *J. Atmos. Sci.*, 34: 1664-1669 (1977).]

and that electrical contributions are as given by Kraemer and Johnston (1955). However, to the author's knowledge, no one has yet described the flow field about (and through) realistic snow flakes.

In earlier sections, comments have already been made with respect to uncertainties about hydrometeor size distributions, particle modifications in the atmosphere, and various attachment processes within clouds and plumes. As a single but very important example, consider again the results obtained by Radke, Eltgroth, and Hobbs (1978), shown in Fig. 11.35. Although at first glance the theory may appear to be capable of explaining the main features of the data, very significant problems remain. One problem is that the attachment rate, chosen to obtain the theoretical curves for particles less than 0.5 μm radius, is very much larger (greater than an order of magnitude larger) than would be expected to apply. It may be, however, that it is not Brownian coagulation but electrical effects that dominate the attachment within plumes. Or, as another possibility, the larger-than-expected scavenging of these particles may be caused by the scavenging of gases (whose unavail-

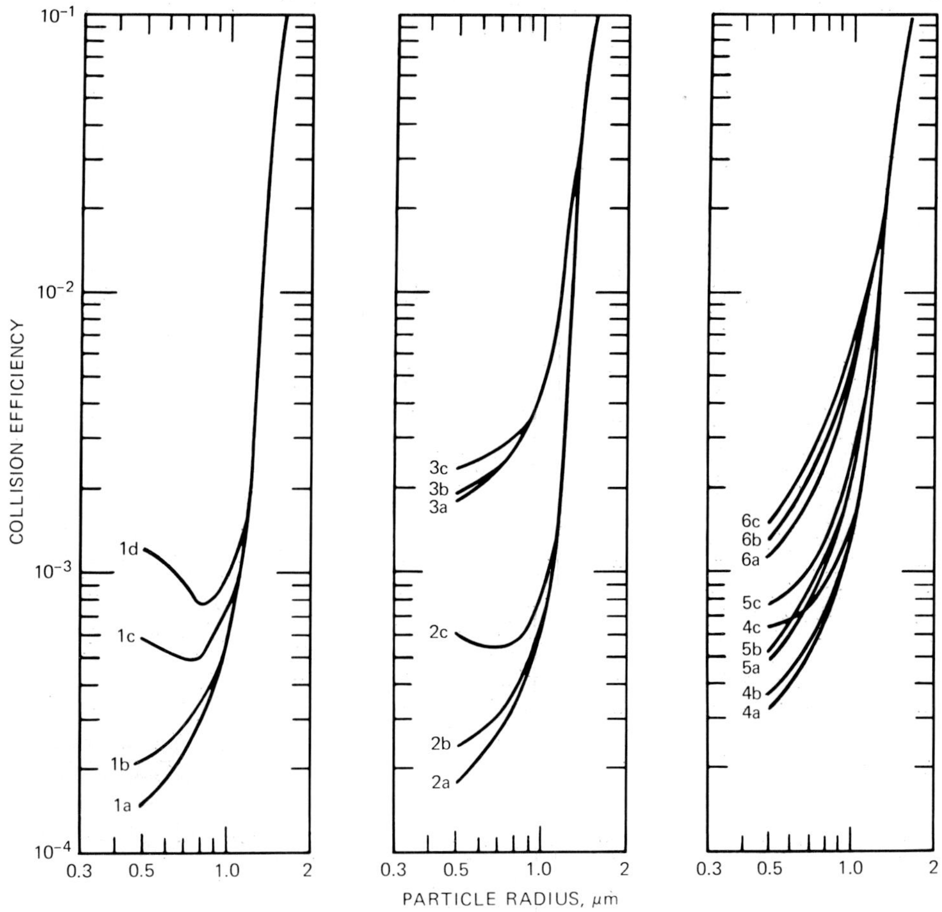

Fig. 11.58 Numerical results obtained by Grover, Pruppacher, and Hamielec (1977) for the dependence of the collision efficiency (for a drop of radius R = 0.173 mm falling at its terminal velocity in air at 10°C and 900 mb) on particle size, background electric field, E_0, and relative humidity. Particle density = 2 g/cm^3; electrical charges $Q = qr^2$ where q is the parameterized surface charge density, r is drop or particle radius in cm, and Q is charge in esu. In each set of curves (1, 2, 3, etc.) a, b, c, and d correspond to relative humidities of 100, 95, 75, and 50%, respectively. For curves 1, $q_a = 0$, $q_R = 0$, and $E_0 = 0$; for curves 2, $q_a = \mp 0.2$ esu/cm^2, $q_R = \pm 0.2$ esu/cm^2, and $E_0 = 0$; for curves 3, $q_a = \mp 0.2$ esu/cm^2, $q_R = \pm 0.2$ esu/cm^2, and $E_0 = 0$; for curves 4, $q_a = 0$, $q_R = 0$, and $E_0 = 5 \times 10^4$ V/m; for curves 5, $q_a = 0$, $q_R = 0$, and $E_0 = 1 \times 10^5$ V/m; and for curves 6, $q_a = 0$, $q_R = 0$, and $E_0 = 3 \times 10^5$ V/m. [Based on S. N. Grover, H. R. Pruppacher, and A. E. Hamielec, A Numerical Determination of the Efficiency with Which Spherical Aerosol Particles Collide with Spherical Water Drops Due to Inertial Impaction and Phoretic and Electrical Forces, *J. Atmos. Sci.*, 34: 1655-1663 (1977).]

ability for gas-to-particle conversion may cause an artificial scavenging of particles). It may be that thermophoretic effects are significant; it may be that some particles smaller than 0.5 μm (chosen arbitrarily for the theoretical curves in Fig. 11.35) grew by condensation; it may be that electrical effects are enhanced after particles pass through electrostatic precipitators; etc. It may even be that the data are misleading! These possibilities illustrate some of the uncertainties that exist in the field of particle scavenging.

In summary, there are many uncertainties about precipitation scavenging of particles. In the earlier sections, an attempt was made to circumvent these uncertainties and present semiempirical results for scavenging rates and ratios for interpretation of the data. These semiempirical results are recommended for use in practical applications, with error estimates derived from the present difference between theory and data. However, because of the present uncertainties, users should search the current scientific literature for improvements.

11-4 RAIN SCAVENGING OF GASES

In Sec. 11-3, the focus was on the scavenging of particles by rain and snow. We now turn to rain scavenging (rainout) of gases. Snow scavenging of gases can usually be ignored unless there is a peculiar reason for affinity of the gas by the ice crystal lattice. Of course, if the snow is melting or if snowflakes capture cloud or plume droplets that contain some pollutant gas, the gas can be brought to the earth's surface by the snow. However, these cases are rather rare and, indeed, can be treated in a manner similar to rain scavenging. Therefore the emphasis will be on gas rainout.

In some respects, gas rainout is simpler to understand and to describe than is particle rainout; in other respects, gas rainout is more complicated. The simplifications follow because the rate at which a gas is transferred to a drop (the analog of the collision efficiency) is known reasonably well; in many cases the rate is rapid enough to allow the assumption that the gas concentration in the drop, κ, has reached the value which is in equilibrium with the existing air concentration, χ, provided χ does not change too rapidly. However, complications follow because the air concentration can vary and because it is usually an extremely complicated problem to know what concentration is in equilibrium, not only with the gas of interest but also with a host of other chemical species in the drop. This problem is an analog to the particle retention efficiency problem. As a consequence of these complications, rainout of gases usually must be treated differently for different plume geometries, different gas species, and different "background" concentrations of trace chemicals in the rain. Clearly, a complete presentation cannot be given here; however, an attempt will be made to present enough details so that salient features are apparent. For more complete discussions the reader can consult Vaughan and Perkins (1961), Junge (1963), Postma (1970), and Hales (1972, 1978a). In the following subsections we address the following topics: gas transfer to drops; gas solubilities in water; transfer within drops; scavenging rates and washout ratios; reactions within drops; and some general aspects about precipitation chemistry.

11-4.1 Gas Transfer to Drops

The transfer of a gas through the atmosphere to a drop (or to an aerosol particle) can be described in a manner similar to that used in Sec. 11-3.4.2 to describe Brownian diffusion coagulation. If the drop is stationary and retains all gas molecules impinging on it, the gas concentration, χ, in the neighborhood of an isolated drop is governed by the diffusion equation

$$D\nabla^2\chi = 0 \qquad (11.57)$$

subject to the boundary conditions: $\chi(r \rightarrow \infty) = \chi_b$, where χ_b is the "bulk" concentration, far from the drop, and $\chi(r = R) = 0$ (perfect sink), where R is the drop radius and D is the molecular or Brownian diffusivity of the gas in air. The solution to this boundary value problem is $\chi = \chi_b (1 - R/r)$. From this solution, the magnitude of the flux to the drop

$$|\Gamma| = |D\nabla\chi|_{r=R} = \frac{D\chi_b}{R} \qquad (11.58)$$

In other words, the nondimensional flux† (known as the Sherwood number), for mass transfer to a stationary sphere, is

$$Sh = \frac{|\Gamma|}{(D\chi_b/R)} = 1 \qquad (11.59)$$

For a moving sphere, the steady-state convective diffusion equation $\mathbf{v}\cdot\nabla\chi = D\nabla^2\chi$ is substantially more difficult to solve. Fuchs (1964) has reviewed attempts to find analytic solutions to this equation. For present purposes, it will be adequate to use the Frössling correlation (e.g., Bird, Stewart, and Lightfoot, 1964; Slinn, 1967) to account for the "ventilation" effect of the motion. This procedure yields

$$Sh = 1 + 0.4\, Re^{1/2} Sc^{1/3} \qquad (11.60)$$

†Some authors nondimensionalize the flux with $D\chi_b/2R$ which yields, for this case, Sh = 2.

where $Re = v_t R/\upsilon$ is the Reynolds number for a drop falling at its terminal speed, v_t, through air of kinematic viscosity υ and where $Sc = \upsilon/D$ is the Schmidt number. It might be noticed that Eq. 11.60 is the same expression as was used to describe the flux of particles of diffusivity D to a drop; the relationship between the flux and the collection efficiency is simply

$$E = \frac{4\pi R^2 Sh(D\chi_b/R)}{\pi R^2 \chi_b v_t} = \frac{4Sh}{Pe} \qquad (11.61)$$

where $Pe = Re\,Sc = v_t R/D$ is the Péclét number.

Another convenient way to describe the transfer of the gas through the atmosphere to the drop is in terms of a gas-phase (as opposed to a liquid-phase) transfer velocity. Similar to the definition of the dry deposition velocity, if the flux is divided by the air concentration far from the drop (viz., χ_b), then the transfer velocity through the atmosphere becomes

$$k_a = \frac{|\Gamma|}{\chi_b} = D\frac{Sh}{R} = \left\{Sc\,\frac{Sh}{Re}\right\} v_t \qquad (11.62)$$

For Sh as given by Eq. 11.60, for low molecular weight gases with $Sc \simeq 1$, and for drops with $Re \gtrsim 10$ and $v_t \simeq (8000/s)\,R$, Eq. 11.62 yields $k_a \simeq 10$ cm/s, independent of drop size. This is a relatively rapid transfer velocity. However, whether or not the drop will accept the delivery of the gas at this rate depends on the solubility of the gas in water, on chemical reactions that may convert the gas to a less volatile form, and on the transfer of the gas within the drop. These topics will now be addressed.

11-4.2 Gas Solubilities in Water

In Sec. 11-4.1 it was assumed that the drop was a perfect sink for the gas. This assumption allowed a simple estimation to be made for the gas phase transfer velocity, k_a. However, for most gases, the boundary condition at the drop's surface should not be $\chi(r = R) = 0$ but $\chi(r = R) = \chi_i$, where the concentration at the interface, in air, χ_i, depends on the solubility of the gas in the drop and therefore on the amount of gas already captured by the drop. For example, if the drop is fully saturated with the gas (that is, at a concentration in equilibrium with the concentration in the atmosphere), then the net flux will vanish; indeed, should the concentration of the gas in the air then fall to a lower value (e.g., as the drop emerges from beneath a plume) the gas can desorb from the drop (Hales, 1972). In a later subsection (see Eq. 11.70), desorption will be described analytically; first, it is useful to describe the solubilities of gases in water.

Theoretical predictions of gas solubilities in water suffer from lack of understanding of the liquid state; however, an application of some basic physical principles can assist in the understanding of solubilities. The rate at which molecules in the vapor phase collide with unit area of the solution is the result familiar from kinetic theory: $n_g\bar{v}/4$, where n_g is the number density of the solute molecules in the atmosphere and $\bar{v} = (8kT/m)^{1/2}$ is the mean thermal speed of these molecules (kT is their thermal energy). At equilibrium, this flux is equal in magnitude to the flux to the atmosphere from the liquid state, and (following Hughes, 1971) this common magnitude is given by $\nu n_\sigma f(\epsilon/kT)$, where ν is the frequency with which the solute molecules vibrate with respect to their solvent neighbors,

n_σ is the surface concentration of the "clusters" that contain solute molecules, and $f(\epsilon/kT)$ is the probability that a solute molecule possesses a total energy, ϵ, sufficient to ensure its escape. Equating the magnitudes of these two fluxes and introducing suitable estimates for ν, n_σ, and $f(\epsilon/kT)$ leads to the result (Hughes, 1971)

$$\frac{n_g}{n_s} = \frac{\chi_b}{\kappa_b} = \frac{3}{(s-1/2)!}\left[\frac{\epsilon}{kT}\right]^{s-\frac{1}{2}} \exp\left[-\frac{\epsilon}{kT}\right] \quad (11.63)$$

in which n_s is the bulk number density of solute molecules in solution (and therefore the ratio of number densities, n_g/n_s, is the same as the ratio of the bulk mass concentrations χ_b/κ_b). This ratio, n_g/n_s or χ_b/κ_b, is known as the Ostwald absorption coefficient; in this chapter, its inverse $\kappa_b/\chi_b \equiv \alpha$ will be called the solubility coefficient.

In Eq. 11.63, s is a parameter indicating the number of solvent molecules coupled to each solute molecule. It can be expected that the value of s is much larger for large polar molecules (e.g., DDT) than for small nonpolar molecules (e.g., CH_4). Therefore the inverse of Eq. 11.63 (i.e., the solubility coefficient, α) can be expected to be (and is) much larger for DDT than CH_4. Also, the result given by Eq. 11.63 is in general agreement with the empirical result for the temperature dependence of the solubility coefficient

$$\ln \alpha^{-1} = \ln \frac{\chi_b}{\kappa_b} = a - b \ln T - \frac{c}{T} \quad (11.64)$$

This result can also be obtained from the thermodynamic argument that, at equilibrium, the chemical potentials are equal. For a number of gases, Hughes (1971; p. 46) gives values for the constants a,b,c appearing in Eq. 11.64 and relates these constants to specific heats and changes in enthalpy.

For practical calculations of the concentrations of gases in water, tabulated values for the solubility coefficients are used. Unfortunately, though, there is a confusing array of different methods by which solubilities are tabulated. In Table 11.8 are listed some solubility coefficients defined via (one form of) Henry's law

$$\kappa_b = \alpha \chi_b \quad (11.65)$$

which gives the equilibrium concentration of the gas in water (κ_b, in units, for example, of mol/liter of g/cm^3 or number of molecules/cm^3) in terms of the concentration of the gas in the atmosphere (χ_b, measured in the same units as κ_b). Notice that α is dimensionless and the concentrations are to be given in dimensions of "per volume" rather than "per mass." That is, one possible source of confusion is that some tabulated values are given on a mass basis (e.g., moles per unit mass of water divided by moles per unit mass of air); such values are about three orders of magnitude smaller than values to be used in this chapter. An advantage of α, the water-to-air partition coefficient as defined by Eq. 11.65, is that the larger this "solubility coefficient" is then the greater the mass of gas that will be absorbed at equilibrium.

The comments in the previous paragraph, however, only begin to reveal the confusion this author has experienced while trying to read tabulated values of the solubility coefficients. Most tables give one or another form of the Henry's law constant. Henry's law states that the mass of a gas that dissolves in a definite volume of liquid is directly proportional to the vapor pressure of the gas above the solution. The proportionality "constant" (known as the Henry's law constant) depends on the gas, the liquid, and the temperature. As an illustration of the dependence on the type of gas: at 0°C and 760 mm pressure, 1 liter of distilled water dissolves 0.0489 liter of O_2, 1.713 liters of CO_2, 78.789 liters of SO_2, or 1176 liters of NH_3. One of the major sources of confusion with tabulated values of the Henry's law constant, *H*, is the variety of units used. For example, if Henry's law is written as $p_A = HX_{A\ell}$, where $X_{A\ell}$ is the mole fraction of the gas A in the liquid [i.e., $X_{A\ell}$ = moles of A in solution divided by (moles of A in solution plus moles of the solvent)], then *H* has the units of pressure and is reported in atmospheres, mm Hg, dynes/cm^2, Pa, etc.! Another source of confusion (although the resulting numerical difference is usually of little practical significance if $X_{A\ell}$ is small, as it is for most gases) is that some authors take Henry's law in the form $p_A = H^* x_{A\ell}$, where $x_{A\ell}$ is the mole *ratio* of the gas A in solution (i.e., moles of A in solution divided by moles of the solution). Other authors define the Henry's law constant via $p_A = X_{A\ell}/H'$, and still other authors use other than mole fractions or mole ratios for the concentration of the gas in the liquid. The footnote to Table 11.8 provides details on how to convert *H*, reported in pressure units, to the dimensionless (volume basis) solubility coefficient, α.

The values for the solubility coefficients given in Table 11.8 can be used to predict only the *equilibrium* amount of the *dissolved* gas in solution. For example, if, in addition, the dissolved gas reacts chemically (e.g., $CO_2 + H_2O \rightarrow HCO_3^+ + OH^-$), further considerations must be made to evaluate the concentrations of the reaction products. In a later section, this topic will be briefly reviewed; first, it is useful to determine under what conditions equilibrium will be attained.

11-4.3 Gas Transfer Within Drops and the Overall Gas Transfer Velocity

Henry's law, e.g., in the form of Eq. 11.65, describes the equilibrium gas concentration in a well-mixed drop. It can be demonstrated (Slinn, 1983a) that it is an acceptable approximation to use Henry's law also to describe the jump in concentration at the interface (subscript i) between air and drop, even for otherwise nonequilibrium conditions; that is,

$$\kappa_i = \alpha \chi_i \quad (11.66)$$

However, whether or not the equilibrium concentration is established within the drop (during the time, for example, when a drop traverses a plume) depends on the rate at which the gas is transferred to the interior of the drop. For a stagnant drop of radius R, the only internal transfer mechanism is by molecular diffusion. From an appropriate diffusion equation, it is seen that the characteristic diffusion time constant $\tau_D \simeq R^2/D$. Since the diffusivity of gases in stagnant water is very small ($\sim$1 cm^2/d), this time constant is, relatively, very large for large drops: if R = 1 mm, then $\tau_D \simeq (10^{-2}\,cm^2)/(10^{-5}\,cm^2/s) = 10^3$ s. Even for an 0.1-mm drop, the time constant is about 10 s, during which time the drop would fall about 10 m. For drops smaller than about 0.1 mm = 100 μm (e.g., plume and cloud droplets), then diffusion within even stagnant drops is rapid enough to allow the assumption that the equilibrium concentration of the gas within each drop is attained in time intervals short compared with those of interest.

TABLE 11.8 Gas Solubilities in Distilled Water†

Gas	Solubility
Noble Gases (20 to 25°C)	
He	9.4×10^{-3}
Ne	1.1×10^{-2}
Ar	3.6×10^{-2}
Kr	6.3×10^{-2}
Xe	1.4×10^{-1}
Rn	2.3×10^{-1}
Inorganic Gases	
CF_4 (25°C)	4.9×10^{-3}
SF_6 (25°C)	5.5×10^{-3}
N_2 (25°C)	1.6×10^{-2}
H_2 (25°C)	1.8×10^{-2}
CO (25°C)	2.6×10^{-2}
O_2 (25°C)	3.1×10^{-2}
NO (25°C)	4.6×10^{-2}
NO_2 (25°C)	2.4×10^{-1}
O_3 (25°C)	3.2×10^{-1}
N_2O (20°C)	6.7×10^{-1}
COS (13.5°C)	6.7×10^{-1}
CO_2 (25°C)	8.3×10^{-1}
CO_2 (20°C)	9.5×10^{-1}
CO_2 (10°C)	1.2×10^{0}
CO_2 (0°C)	1.6×10^{0}
H_2S (10°C)	3.4×10^{0}
Cl_2 (15°C)	3.3×10^{0}
SO_2 (25°C)	3.0×10^{1}
Inorganic Gases (Cont.)	
SO_2 (18°C)	4.2×10^{1}
SO_2 (0°C)	7.4×10^{1}
HI (0°C)	4.9×10^{2}
HCl (0°C)	5.1×10^{2}
HBr (0°C)	6.1×10^{2}
HNO_3 (25°C)	1.2×10^{3}
NH_3 (25°C)	1.5×10^{3}
Some Organic Gases (at 25°C)‡	
Alkanes	
Methane	3.7×10^{-2}
Ethane	4.9×10^{-2}
Propane	3.5×10^{-2}
Isobutane	2.1×10^{-2}
Isopentane	1.8×10^{-2}
4-Methyloctane	2.5×10^{-3}
Alkenes	
Ethene	1.1×10^{-1}
1-Butene	3.3×10^{-2}
1-Octene	2.6×10^{-2}
Dienes	
Butadiene	3.3×10^{-1}
1,4-Pentadiene	2.1×10^{-1}
Alkynes	
Propyne	2.2×10^{0}
1-Butyne	1.3×10^{0}
Organic Gases (Cont.)	
Monoaromatics	
Benzene	4.5×10^{0}
Toluene	3.7×10^{0}
n-Pentylbenzene	4.0×10^{0}
Polynuclear aromatics	
Naphthalene	5.8×10^{1}
Biphenyl	8.8×10^{1}
Anthracene	4.1×10^{2}
Pyrene	2.1×10^{3}
Halogenated alkanes and alkenes	
Chloromethane	2.6×10^{0}
Carbon tetrachloride	1.2×10^{0}
Tetrachloroethene	1.1×10^{0}
1,2-Dibromomethane	7.7×10^{1}
Halogenated aromatics	
Chlorobenzene	7.1×10^{0}
Fluorobenzene	3.9×10^{0}
Pesticides	
Lindane	7.7×10^{3}
Aldrin	8.8×10^{2}
Dieldrin	2.3×10^{3}
DDT	4.7×10^{2}
Methylparathion	1.2×10^{5}
Ronnel	1.2×10^{3}
Leptophos	9.2×10^{3}
Fenitrothion	$\sim 10^{6}$

†Solubility coefficients typically decrease with increasing temperature (e.g., see the data in this table for CO_2) and with the presence of other constituents (e.g., for O_2 the solubility in seawater is about 80% of its value in fresh water at the same temperature). Solubility data listed by Heines and Peters (1974), Liss and Slater (1974), and Junge (1977) are relisted in Slinn et al. (1978). The data listed in this table are from Slinn et al. (1978), from Schwartz et al. (1983), from the excellent review by Mackay and Shiu (1981), and from the text of W. Gerrard, *Solubility of Gases and Liquids*, Plenum Press, N.Y., 1976. In the conversion of data given in terms of Henry's law constant, H (defined via $p_A = HX_{A\ell}$, where p_A is the pressure of gas A above the solution and $X_{A\ell}$ is the mole fraction of gas A in the liquid), to find the (volume basis) solubility coefficient, α, the following relationships are used:

1. $$p_A = \rho_A \frac{RT}{M_A} = HX_{A\ell} = H \frac{M_A/M_A}{(M_A/M_A) + (M_\ell/M_\ell)}$$

where $\rho_A \equiv \chi_A$ is the mass density of the gas above the solution, R is the universal gas constant, T is the absolute temperature, M_A is the molecular weight of the gas, M_A is the mass of gas A in solution, M_ℓ is the mass of the liquid, and M_ℓ is the molecular weight of the liquid ($\simeq 18$ g for water).

2. $M_\ell = \rho_\ell V_\ell$, where ρ_ℓ is the density of the liquid ($\simeq 1$ g/cm^3 for water) and V_ℓ is the volume of the liquid.

3. $M_A = \kappa_A V_\ell$, where κ_A is the mass concentration of the gas in the liquid.

Putting these relationships together and assuming that the mole fraction is small (i.e., $M_A/M_A \ll M_\ell/M_\ell$), we find the result

$$\kappa_A = \alpha\chi_A = \left[\frac{\rho_\ell RT}{M_\ell H}\right] \chi_A$$

which defines the relationship between α and the Henry's law constant defined in terms of mole fractions. If, instead, the Henry's law constant is defined in terms of mole ratios [i.e., $p_A = H^* x_{A\ell}$, where $x_{A\ell} = (M_A/M_A)/(M_\ell/M_\ell)$], no approximations are needed to reach the result $\alpha = (\rho_\ell RT)/(M_\ell H^*)$. Alternatively, if as in Gerrard (already cited), the solubility is given as mole ratio at 1 atm pressure, $x_{A\ell}$, then

$$\alpha = \frac{\rho_\ell RT}{M_\ell} \frac{x_{A\ell}}{1 \text{ atm}}$$

Finally, if Henry's law constant is given via $p = H_c \kappa$, with H_c in units of Pa/(mol/m^3), then use of the ideal gas law leads to $\alpha = \kappa/x = RT/H_c$.

For convenience to the reader, the following numerical values are presented: for water, $\rho_\ell \simeq 1$ g/cm^3, $M_\ell \simeq 18$ g/mol, and, depending on the units used for pressure in Henry's law, $R = 82.05$ atm cm^3/°K $= 6.24 \times 10^4$ mm Hg cm^3/°K $=$ 8.314 Pa m^3/°K $= 2.45 \times 10^3$ in Hg cm^3/°K, all per (gram) mole.

‡For estimates of uncertainties in these values (generally $\sim \pm 10\%$), see Mackay and Shiu (1981).

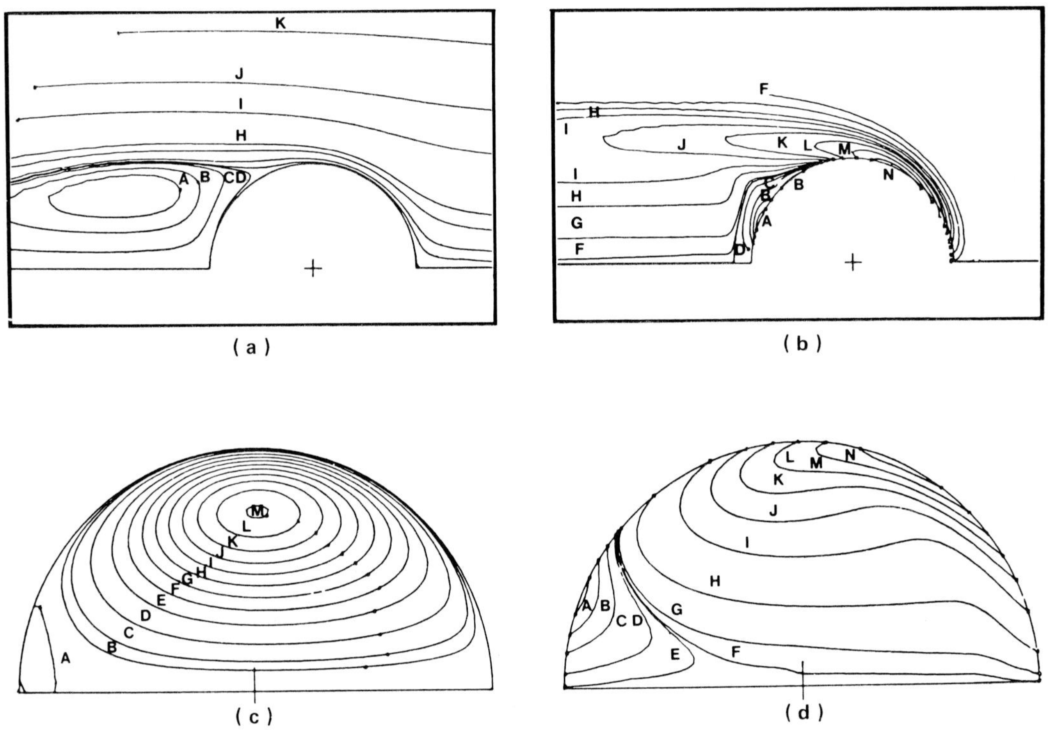

Fig. 11.59 Calculated flow fields (a,c) and vorticity fields (b,d) outside and inside a circulating water sphere for Re = Rv_t/ν = 150. [Based on B. P. LeClair, A. E. Hamielec, H. R. Pruppacher, and W. D. Hall, A Theoretical and Experimental Study of the Internal Circulation in Water Drops Falling at Terminal Velocity in Air, *J. Atmos. Sci.*, 29: 728-740 (1972).]

For drops larger than about 0.1 mm, a fortunate simplification is available. In these drops, internal circulations develop, driven by the viscous drag on the drop's surface as it falls at its terminal velocity. Figures 11.59 and 11.60 illustrate this phenomenon (LeClair et al., 1972). Typically the induced velocity (which can also influence particle collection; see Fig. 11.5) is a few percent of the drop's terminal velocity. Therefore convective mixing proceeds with a characteristic time constant $\tau_c \simeq R/(1\%\ v_t)$ and, with $v_t = (8000/s)R$, this results in mixing during $\tau_c \simeq 10^{-2}$ s. Consequently, both for small and large drops, mixing within drop is relatively rapid.

To see the consequence of this discussion, let the flux of gas to a drop be written in terms of an overall transfer velocity, k_0, via

$$|\Gamma| = k_0[\chi_b - \chi_*] \tag{11.67}$$

where χ_* is the gas concentration in air that would be in equilibrium with the existing gas concentration in the bulk of the liquid phase, κ_b. For gases that obey Henry's law, $\chi_* = \kappa_b/\alpha = H\kappa_b$. In addition, define gas and liquid phase transfer velocities via $|\Gamma| = k_a[\chi_b - \chi_i]$ and $|\Gamma| = k_\ell[\kappa_i - \kappa_b]$, respectively, where the subscript i identifies concentrations at the interface. Then, using $\kappa_i = \alpha\chi_i$, rewriting Eq. 11.67 as

$$|\Gamma| = k_0[(\chi_b - \chi_i) + \alpha^{-1}(\kappa_i - \kappa_b)] \tag{11.68}$$

(since $\chi_i - \alpha^{-1}\kappa_i \equiv 0$) and using the defining equations for k_a and k_ℓ, we obtain for the overall transfer velocity

$$k_0^{-1} = k_a^{-1} + (\alpha k_\ell)^{-1} \tag{11.69}$$

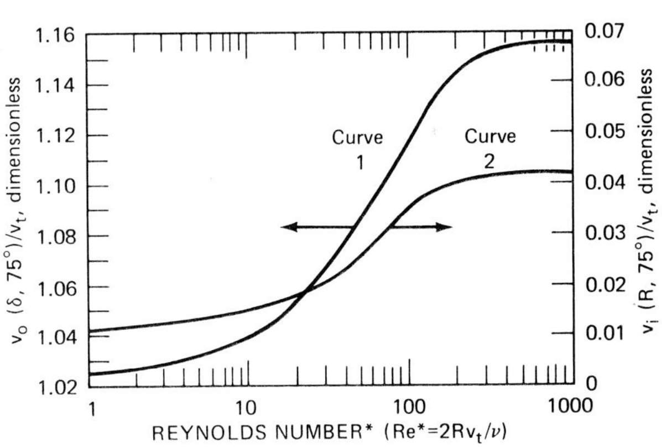

Fig. 11.60 Variation of the tangential velocity with Reynolds number as obtained by LeClair et al. (1972). Curve 1 (left-hand ordinate), velocity at the edge of the boundary layer at 75° from the upstream axis. Curve 2 (right-hand ordinate), velocity at the surface of a circulating water sphere falling through air and at 75° from the upstream axis. [Based on B. P. LeClair, A. E. Hamielec, H. R. Pruppacher, and W. D. Hall, A Theoretical and Experimental Study of the Internal Circulation in Water Drops Falling at Terminal Velocity in Air, *J. Atmos. Sci.*, 29: 728-740 (1972).]

This result displays the total resistance to transfer, k_0^{-1}, to be the sum of the transfer resistance in the atmosphere, k_a^{-1}, plus the transfer resistance in the drop, $(\alpha k_\ell)^{-1}$. Earlier, see Eq. 11.62, it was seen that for low molecular weight gases and for drops with Re $\gtrsim$ 10, then $k_a \simeq 10$ cm/s, independent of drop size. Also, from the previous paragraph, it was seen that, for drops with internal circulation, k_ℓ is a few percent of v_t; for example, $k_\ell \simeq 1\% \, v_t \simeq (80/s)R$, which varies from about 1 to 10 cm/s as R varies from 0.1 to 1 mm. Consequently, for low molecular weight gases with small solubility coefficients (e.g., CO, CO_2, CH_4, etc.; see Table 11.8), the resistance to gas transfer is similar in the two phases (both transfer velocities are a few percent of v_t). For low molecular weight gases with large solubility coefficients, however (e.g., SO_2, NH_3, HTO), $\alpha \gg 1$, and the resistance in the liquid phase $(\alpha k_\ell)^{-1}$ is negligible compared to the resistance in the atmosphere. In this case, the overall transfer velocity, k_0, can be approximated by the transfer velocity in the atmosphere, k_a. For large molecular weight gases such as DDT (with $Sc = \nu/D \gg 1$) with substantial solubility ($\alpha \gg 1$), again it is possible for the transfer resistances in the two media to be of comparable magnitude (Slinn et al., 1978).

11-4.4 Gas Rainout, Plume Washdown, and Washout Ratios

We will now examine how the significance of the results in the previous section depends on the vertical distribution of the gaseous plume, and, to simplify notation, we will drop the subscript "b" (for bulk). For any vertical profile of the pollutant's concentration in air, $\chi(z)$, the instantaneous flux to a drop of radius R leads to a change in the (bulk) gas concentration in the drop according to

$$\frac{d}{dt}\left(\frac{4}{3}\pi R^3 \kappa\right) = k_0 4\pi R^2 \left[\chi(z) - \frac{\kappa}{\alpha}\right] \tag{11.70}$$

From this equation it is seen that the characteristic time for the drop to reach its equilibrium concentration, $\alpha\chi$, is

$$\tau_e = \frac{\alpha R}{3k_0} \tag{11.71}$$

Consider now the case of low molecular weight gases with low solubility coefficients (e.g., CO, CO_2, CH_4, etc.). In this case, if k_0 is approximated by αk_ℓ and k_ℓ by a few percent of $v_t \simeq (8000 \, s^{-1})R$, then $\tau_e \simeq 10^{-2}$ s; that is, in time intervals less than essentially all time intervals of interest, these gases will have reached their equilibrium concentration $\alpha\chi(z)$. On the other hand, for low molecular weight gases with large solubility coefficients (e.g., SO_2, NH_3, HTO), that is, for gases whose transfer is controlled by transfer through the atmosphere, then $k_0 \simeq k_a \simeq v_t/Re^{1/2}$ and $\tau_e \simeq \alpha R Re^{1/2}/(3 v_t)$. During this time interval a drop would fall a distance $\lambda_e = \tau_e v_t \simeq \alpha Re^{1/2} R/3$. In Table 11.9 are displayed these equilibration length scales, λ_e, for different gases and two drop sizes. From these results it is seen that the amount of these gases that will be dissolved in the rain depends on the vertical distribution of the gaseous plume. For example, Fig. 11.61 shows the concentration of SO_2 dissolved in different size drops as they fall through an elevated plume.

For cases where $\chi(z)$ varies too rapidly for drops to attain their equilibrium concentration of gas, then the description of gas scavenging can be quite formidable. Numerical solutions to the resulting equations for SO_2 scavenging are available (Hales, 1978a). In addition, an analytic solution to Eq. 11.70 for a Gaussian plume whose vertical distribution is uninfluenced by the rain is also available (see Dana et al., 1973 or Hales, 1978a). The preliminary analysis by Slinn (1974d) shows that a plume is redistributed in the atmosphere by the

TABLE 11.9 Approximate Fall Distances, 10 λ_e, in Meters, for Drops to Attain Their Equilibrium Concentration, $\alpha\chi_0$, of the Indicated Gases for Distilled Water

Gas	Drop radius, mm	
	0.1	1.0
CO_2, CH_4	10^{-1}	10^0
DDT (gas)	10^{-1}	10^1
NH_3	10^0	10^2
HTO	10^2	10^4

absorption/desorption phenomenon; however, the effective plume "washdown" velocity, αp (where p is the rain rate), is normally quite small. Figure 11.62 illustrates this plume washdown effect.

On the other hand, if drops attain their equilibrium concentration, gas rainout becomes almost trivial to describe. Thus, if the gas concentration in a drop immediately equilibrates with the ambient air concentration, then no gas will be scavenged from aloft, and the gas can even desorb from the rain. If χ varies slowly in distances comparable to the equilibration length (e.g., CH_4 in a plume greater than about 1 m in width; DDT well mixed in a layer deeper than about 100 m; HTO well mixed throughout the troposphere), then at the surface of the earth, the gas concentration in the drop will be $\kappa_0 = \alpha\chi_0$, or the washout ratio is simply the solubility coefficient:

$$w_r = \left(\frac{\kappa}{\chi}\right)_0 = \alpha \quad \text{(at equilibrium)} \tag{11.72}$$

Correspondingly, the wet flux is given by

$$W = \kappa_0 p_0 = \alpha p_0 \chi_0 = v_w \chi_0 \quad \text{(at equilibrium)} \tag{11.73}$$

where p is the precipitation rate and the wet deposition velocity is

$$v_w = \frac{W}{\chi_0} = \alpha p_0 \quad \text{(at equilibrium)} \tag{11.74}$$

For example, for dissolved NH_3 and $p_0 = 1$ cm/h, then $v_w \simeq 10^3 \times (1 \text{ cm/h}) \simeq 1$ cm/s. Notice that this wet deposition velocity is, relatively, very small compared with v_w for particles (see Fig. 11.52) and yet v_w for NH_3 is much larger than for many gases (see the solubility coefficients in Table 11.8).

There are, however, a number of appropriate caveats. One is that the "solubilities" of gases in water may be drastically influenced by the presence of other species. For example, Richards et al. (1983) found approximately two orders of magnitude more S(IV) (e.g., as in SO_2) in Los Angeles cloud water than would be expected on the basis of Henry's law. They speculate that the SO_2 formed adducts with carbonyl compounds such as formaldehyde (HCHO) in the cloud water. As another example, Pankow, Isabelle, and Asher (1983) found one to three orders of magnitude more of a number of hydrocarbons in Portland rainwater than would be expected on the basis of Henry's law, which nevertheless was found to give reasonable results for remote areas (Atlas and Giam, 1981). The second major caveat is that gases in rain can be chemically con-

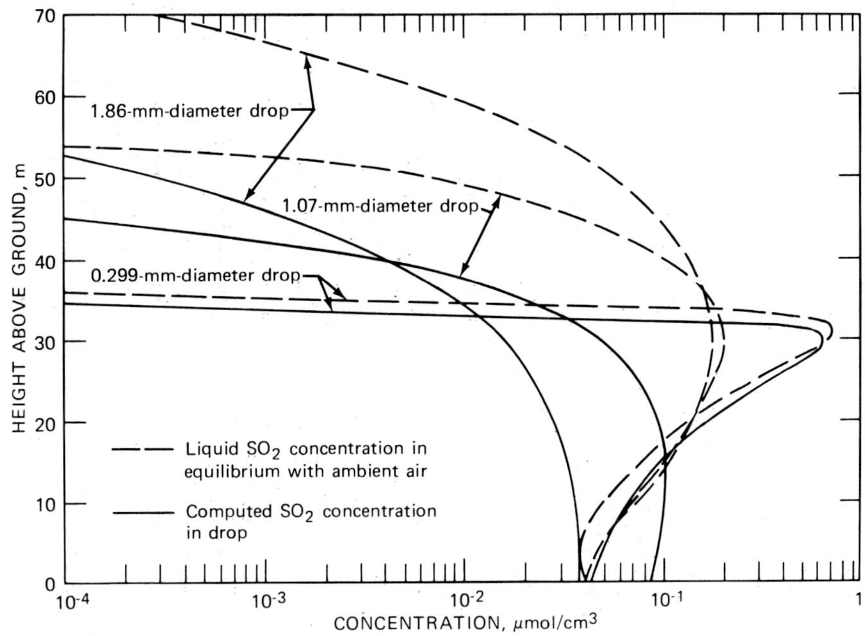

Fig. 11.61 Concentration of SO_2 in different size drops as they fall through an elevated plume as calculated by Hales. The dashed curves give indications of the SO_2 air concentrations experienced by the drops as they fall (at different angles under the influence of a mean wind, $\bar{u}$) through the plume. Transfer of SO_2 to each drop is assumed to be gas phase limited (i.e., $k_0 \simeq k_a$). Notice that, although the concentration in the largest drop considered substantially lags the equilibrium concentration, by the time this drop reaches the ground it possesses an SO_2 concentration near the value for local equilibrium. (Based on M. T. Dana, J. M. Hales, W. G. N. Slinn, and M. A. Wolf, *Natural Precipitation Washout of Sulfur Compounds from Plumes*, EPA Report EPA-R3-73-047, Ecological Research Series, Environmental Protection Agency, 1973.)

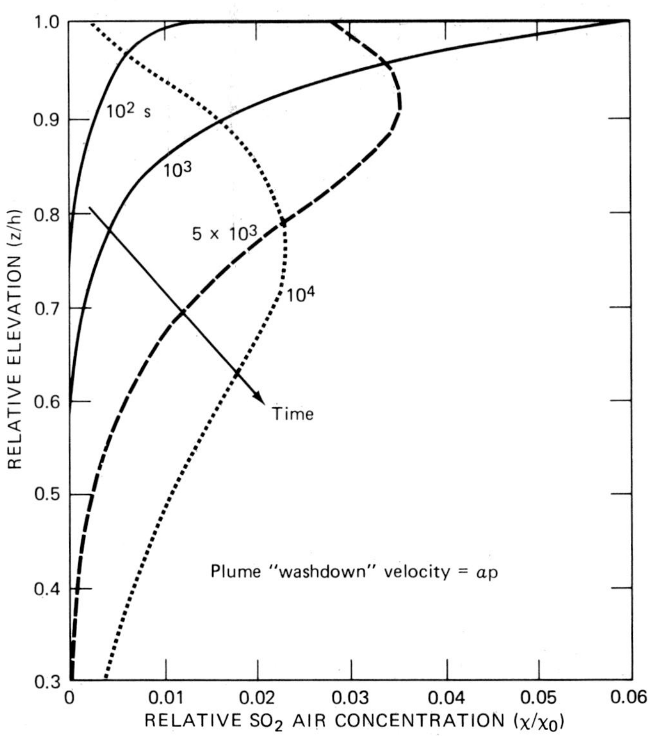

Fig. 11.62 The influence on an elevated SO_2 plume of the absorption/desorption of the gas by raindrops. As time advances, the plume is washed downward; the "washdown" velocity of the plume's centroid is approximately αp, where α is the solubility coefficient of the gas in rain and p is the rain rate. [From W. G. N. Slinn, The Redistribution of a Gas Plume Caused by Reversible Washout, *Atmos. Environ.*, 8: 233-239 (1974).]

verted to less volatile species [e.g., S(IV) as SO_2 to SC(VI) as SO_4^{2-}], leading, for example, to acidic rain. We now turn to some aspects of this complicated problem of chemical reactions in cloud water and rainwater.

11-4.5 Chemical Reactions in Drops

So far in this review of gas scavenging, we have considered only the portion of the gas that is in simple solution in the rainwater. In reality, when molecules of a specific gas enter a drop, some molecules remain in solution whereas others, to a greater or lesser extent, depending on the gas and the chemical constituents of the water, react reversibly or irreversibly with the water molecules and possibly with the other trace constituents in the drop. Because the rates of these reactions and the degree to which they proceed depend sensitively on the particular gas and the presence of other species, it is clear that the rainout of gases and their reaction products can become enormously complicated: each gas must be considered separately, but to evaluate its removal we must also consider the influence of all other species present. Perhaps understandably, this total problem has never been solved. Here the goal will merely be to indicate some of the considerations that are needed and to refer the reader to appropriate scientific literature.

For practical applications of precipitation scavenging, it is extremely convenient to separate the chemical reactions within drops into two categories: those that proceed much more rapidly than the shortest time scale of interest, and those that proceed more slowly. The shortest time scale of interest would typically be the time required for a drop to fall through a plume, of the order of one second. For those reactions that proceed much more rapidly than this time scale (e.g., essentially all ionization reactions), we can usually assume that the reactions have proceeded to reach equilibrium, although (depending on solubility) the drop may not have time to become saturated with the gas (e.g., for SO_2, see Garland, 1978). Some of these ionization reactions are considered in the next subsection. Following that, some slower, oxidation reactions are discussed.

11-4.5.1 Ionization Reactions and Equilibrium. As an example of a fast ionization reaction, consider the dissociation of water

$$H_2O \underset{k_r}{\overset{k_d}{\rightleftarrows}} H^+ + OH^- \tag{11.75}$$

At equilibrium, the dissociation and recombination fluxes are equal and therefore

$$k_d[H_2O] = k_r[H^+][OH^-] \tag{11.76}$$

where [ ] symbolizes concentrations measured in moles per liter (mol/liter), and Eq. 11.76 essentially defines the dissociation and recombination rate constants k_d and k_r. (Note that their dimensions are not the same.) At equilibrium, the concentrations are also related by the equilibrium constant, K, and therefore

$$K = \frac{[H^+][OH^-]}{[H_2O]} = \frac{k_d}{k_r} \tag{11.77}$$

The equilibrium constant for this reaction has the value 1.82×10^{-16} mol/liter at 25°C (or $K_w \equiv [H^+][OH^-] \simeq 1.0 \times 10^{-14}$ mol²/liter²) and is about an order of magnitude smaller at 0°C. Eigen and DeMaeyer (1955) measured k_r to be $[(1.4 \pm 0.2) \times 10^{11}]$/(mol·s/liter) and therefore from Eq. 11.77, $k_d = 2.5 \times 10^{-5}$/s. In other words, if initially the water was un-ionized, H^+ (or H_3O^+) ions would be produced at an initial rate governed by

$$\frac{d[H^+]}{dt} = k_d[H_2O] \tag{11.78}$$

and, therefore, ionization equilibrium would be attained in much less than a millisecond. Similar dissociation and recombination rates are applicable for all species (e.g., NH_4^+, HS^-, and SO_3^-) whose ion diffusion rates are similar to the diffusion of H_3O^+. Larger ions diffuse more slowly in water, and therefore equilibrium is attained more slowly; but still, typically, ionization reactions proceed at a rate fast compared with the time required for a drop to fall through a plume.

As a consequence of the rapidity of ionization reactions, it is reasonable to assume that ionization equilibrium is attained in time intervals short compared with time intervals of interest. Therefore it is theoretically possible to calculate the concentration of ions in raindrops. For example, for pure water at equilibrium, $[H^+] = [OH^-]$, and from Eq. 11.77 we obtain $[H^+] = [1.0 \times 10^{-14}\ \text{mol}^2/\text{liter}^2]^{1/2} = 10^{-7}$ mol/liter or from

$$pH = -\log[H^+] \tag{11.79}$$

we obtain the familiar result for pure water: pH = 7.0 at 25°C (and pH = 7.48 at 0°C). However, it is quickly seen that even the calculation of the equilibrium concentration of ions is an extremely difficult task since so many ionization reactions can occur.

As an example, following Junge (1963), consider the dissolution and ionization of CO_2. From Henry's law we have

$$[CO_2]_\ell = \alpha_{CO_2}[CO_2]_g \tag{11.80}$$

where Junge used the solubility coefficient $\alpha_{CO_2} = 0.8$ at 10°C. In the absence of other species, the following reactions can occur:

$$(CO_2)_\ell + H_2O \rightleftarrows H_2CO_3 \rightleftarrows H^+ + HCO_3^-$$

$$HCO_3^- \rightleftarrows H^+ + CO_3^{2-}$$

$$H_2O \rightleftarrows H^+ + OH^-$$

At equilibrium, charge balance requires

$$[H^+] = [OH^-] + [HCO_3^-] + 2[CO_3^{2-}] \tag{11.81}$$

Also, at equilibrium the concentrations of the species are dictated by the equilibrium constants. At 10°C these constants have the following values (Junge, 1963):

$$K_1 = \frac{[H^+][HCO_3^-]}{[H_2CO_3]} = 3.4 \times 10^{-7}\ \text{mol/liter} \tag{11.82}$$

$$K_2 = \frac{[H^+][CO_3^{2-}]}{[HCO_3^-]} = 3.2 \times 10^{-11}\ \text{mol/liter} \tag{11.83}$$

$$K_3 = [H^+][OH^-] = 3.6 \times 10^{-15}\ \text{mol}^2/\text{liter}^2 \tag{11.84}$$

Given the gas phase concentration of CO_2, these five equations, Eqs. 11.80 to 11.84, can be solved for the five unknowns $[CO_2]_\ell$, $[H^+]$, $[OH^-]$, $[HCO_3^-]$, and $[CO_3^{2-}]$. For $[CO_2]_g = 1.34 \times 10^{-5}$ mol/liter (i.e., ≃ 300 ppm), the results are $(CO_2)_l$ = 0.7 mg/liter, (HCO_3^-) = 0.14 mg/liter, $(CO_3^{2-}) = 0.19 \times 10^{-5}$ mg/liter, and $[H^+] = 2.5 \times 10^{-6}$ mol/liter or pH = 5.6.

As pointed out by Junge (1963), although pH values in the range 5 to 6 are quite common in unpolluted atmospheres, CO_2 is probably not the controlling factor because small amounts of NH_3 will have an effect on the pH comparable to that of CO_2. Thus consider the dissolution and dissociation of NH_3:

$$[NH_3]_\ell = \alpha_{NH_3}[NH_3]_g \tag{11.85}$$

$$(NH_3)_\ell + H_2O \rightleftarrows NH_4OH \rightleftarrows NH_4^+ + OH^- \tag{11.86}$$

with $\alpha_{NH_3} = 2.8 \times 10^3$ at 0°C and

$$K_4 = \frac{[NH_4^+][OH^-]}{[NH_4OH]} = 2.0 \times 10^{-5}\ \text{mol/liter} \tag{11.87}$$

Then the $CO_2 + NH_3$ system (with an appropriate change in the electrical neutrality equation, Eq. 11.81) gives for $[NH_3]_g = 1.75 \times 10^{-10}$ mol/liter (= 3 μg/m³): $(NH_4OH) = 1.2 \times 10^{-3}$ mg/liter, $(NH_4) = 2.6$ mg/liter,

$(NH_3)_\ell = 0.4\ \mu g/m^3$, $(H_2CO_3) = 0.71$ mg/liter, $(HCO_3^-) =$ 3.3 mg/liter, $(CO_3^{2-}) = 10^{-3}$ mg/liter, and pH = 7.0.

From these considerations, Junge (1963) concludes that the mineral acids H_2SO_4, HCl, HNO_3, etc., are largely responsible for lowering the pH below the neutral value 7.0 at 25°C (or 7.5 at 0°C). However, it is extremely difficult to predict the resulting pH: it would be necessary to know the concentrations of all anions and cations in the rain. Alternatively, these concentrations could be measured or, most simply, the pH of rains could be measured. These alternatives have been pursued and are briefly described in a later section dealing with precipitation chemistry. Now, however, we desire to conclude this section on ionization equilibrium by responding to the inverse question: if the acidity of the rain were known, what would be the equilibrium concentrations of a dissolved gas and its ionization products?

As an example, consider SO_2 and its ionization products. Equilibrium between SO_2 in the gaseous and liquid phases is described by Henry's law:

$$[SO_2]_\ell = \alpha_{SO_2}[SO_2]_g \tag{11.88}$$

and equilibrium between dissolved SO_2 and bisulfite ion, namely,

$$(SO_2)_\ell + H_2O \rightleftarrows H_2SO_3 \rightleftarrows H^+ + HSO_3^- \tag{11.89}$$

is specified by the equilibrium constant

$$K_5 = \frac{[H^+][HSO_3^-]}{[H_2SO_3]} = 1.3 \times 10^{-2}\ \text{mol/liter} \tag{11.90}$$

at 25°C. ($K_5 = 2.3 \times 10^{-2}$ mol/liter at 0°C; see Hales, 1978a). Bisulfite is also ionized via $HSO_3^- \rightleftarrows H^+ + SO_3^{2-}$. However, at a pH less than 5, less than a percent of the sulfur is present as sulfite (SO_3^{2-}), and therefore this ionization is usually ignored in scavenging calculations. (For details, see Barrie, 1978.) Until the next subsection, we will ignore the relatively slow oxidation reactions that form SO_4^{2-}.

Now, in response to the question posed, assume that the rain contains a background or initial concentration [Ex] of univalent anions from a strong acid or univalent cations from a strong base. [Ex] is positive for a strong acid and negative for a strong base.† This concentration can be related to the initial hydrogen ion concentration by the condition of electrical neutrality

$$[H^+]_0 = [Ex] + [OH^-]_0 \tag{11.91}$$

and the equilibrium ionization of water

$$K_w = [H^+]_0[OH^-]_0 \tag{11.92}$$

That is, Eqs. 11.91 and 11.92 yield

$$[Ex] = [H^+]_0 - \frac{K_w}{[H^+]_0} \tag{11.93}$$

Hales (1978a) suggests that under most conditions the second term in Eq. 11.93 can be ignored, leaving

$$[Ex] \simeq 10^{-pH_0} \tag{11.94}$$

With the dissolution and ionization of SO_2 according to Eqs. 11.88 and 11.89 and the new condition of electroneutrality,

$$[H^+] = [HSO_3^-] + [Ex] + \frac{K_w}{[H^+]} \tag{11.95}$$

these equations can be solved for $[HSO_3^-]$. The result (Hales and Sutter, 1973; Hill and Adamowicz, 1977) for the total concentration of sulfur compounds, $[S] = [SO_2]_\ell + [HSO_3^-]$, is

$$[S] = \alpha_{SO_2}[SO_2]_g + \frac{1}{2}\{-[Ex] + ([Ex]^2 + 4\alpha_{SO_2}K_5[SO_2]_g)^{1/2}\} \tag{11.96}$$

Figure 11.63 shows a plot of this result; similar results are available for NH_3, HNO_2, and HNO_3 (e.g., Schwartz et al., 1983). Such results can be used to define appropriate enhanced-solubility coefficients, α_*, e.g., via

$$[S] = \alpha_*(SO_2)_g \tag{11.97}$$

which lead to the washout ratio for total sulfur, $w_r = \alpha_*$, which depends on $[SO_2]_g$. Some illustrative values are shown in Table 11.10; for the temperature dependence of this α_*, see Barrie (1981). However, we have here assumed that equilibrium conditions prevail (note that for $\alpha_* \sim 10^4$ and dependent on $[SO_2]_g$, then the "e-fold" equilibration length is quite

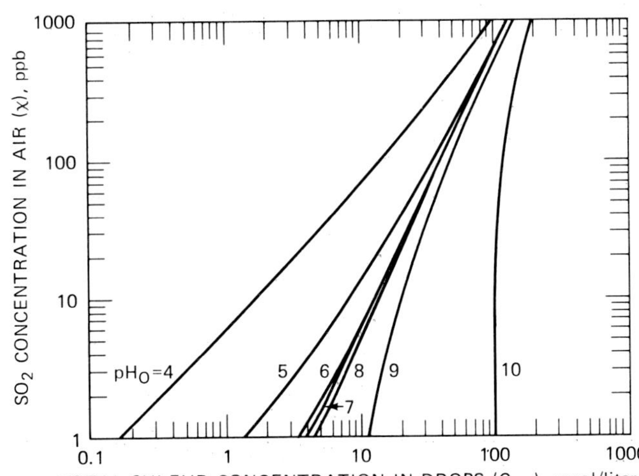

Fig. 11.63 Total sulfur compound concentration in drops of indicated pH as a function of the SO_2 air concentration (see Eq. 11.96) ignoring the relatively slow oxidation of HSO_3^- to SO_4^{2-}. [Based on F. B. Hill and J. F. Adamowicz, A Model for Rain Composition and the Washout of Sulfur Dioxide, *Atmos. Environ.*, 11: 917-928 (1977).]

†A strong acid, for example, donates all of its protons to solution; thus the equilibrium dissociation constant for a strong acid (and similarly for a strong base) satisfies K ≫ 1. Examples of strong acids are HF, HCl, H_2SO_4, HNO_3, etc. In contrast, at a pH of less than 5, many weak acids such as H_2CO_3 do not completely dissociate.

TABLE 11.10 Enhanced Solubility Coefficient, α_*, or Washout Ratio, w_r, for SO_2†

pH	χ, ppbv	α_*, dimensionless‡
5.0	1	3.7×10^4
	3	3.0×10^4
	10	2.2×10^4
	30	1.5×10^4
6.0	1	9.0×10^4
	3	5.5×10^4
	10	3.2×10^4
	30	1.9×10^4

†Adapted from Slinn et al. (1978).

‡Ratio of moles of total dissolved sulfur (other than SO_4^{2-}) per unit volume to moles (per unit volume) of SO_2 in air, for water at indicated pH at 20°C and for the indicated air concentrations of SO_2 (from Hales and Sutter, 1973; Dana, Battelle–Nothwest, personal communication).

large: ~1 km for a 1-mm-radius drop if $[SO_2]_g \simeq 10$ ppb, see Garland, 1978), and we have considered only fast ionization reactions. To describe the total amount of, for example, sulfur removed by rain, we must also consider the relatively slow oxidation of sulfite to sulfate.

11-4.5.2 Oxidation Reactions and Reaction Rates. Fast ionization reactions were illustrated in the previous section: sulfur reactions were emphasized; Postma (1970) illustrates the rapid ionization reactions that occur when elemental iodine is dissolved in water. In this section, slower reactions will be considered. As was discussed earlier, the distinction between slow and rapid depends here on the time scale dictated by the particular scavenging event of interest: seconds to minutes for drops falling through plumes (the latter for a plume spread through the depth of a mixed layer) and minutes to hours for scavenging within clouds (from unstable convective to stable frontal storms). In this section, the emphasis will be again on sulfur scavenging.

For a description of SO_2 scavenging by raindrops, the following equations are used (Hill and Adamowicz, 1977):

1. Henry's law at the air/water interface (Eq. 11.66).
2. Equation 11.70 to describe the flux of SO_2 to each drop [with the overall transfer velocity, $k_0 \simeq k_a$, the transfer through the atmosphere (see remarks in Sec. 11-4.4), and k_a related to the Sherwood number via Eq. 11.62].
3. Electroneutrality, which becomes, with the oxidation of bisulfite to sulfate (see Eq. 11.95),

$$[H^+] = [HSO_3^-] + 2[SO_4^{2-}] + [Ex] + \frac{K_w}{[H^+]} \tag{11.98}$$

4. A rate equation for the production of SO_4^{2-}

$$\frac{d[SO_4^{2-}]}{dt} = k_{ox}[HSO_3^-] \tag{11.99}$$

where k_{ox} is the (assumed first-order reaction) rate constant for the oxidation.

Solutions to this set of equations are shown in Figs. 11.64(a) and (b).

Although calculations of the type illustrated in Fig. 11.64 are interesting, they are of limited practical value. One limitation is from their focus on the time during which raindrops fall (rather than on the typically longer times available for chemical reactions within clouds); a more serious limitation, for both raindrops and cloud drops, results from significant deficiencies in knowledge about the reaction rates. As just one example of these deficiencies, during the past decade there have been at least 100 reports on the rate of aqueous-phase oxidation of S(IV) to S(VI), and quite likely another decade of progress will be needed before this oxidation rate is reliably known. Recent results (e.g., Martin and Damschen, 1981; Hegg and Hobbs, 1982; Maahs, 1983) appear to confirm the proposal by Penkett et al. (1979) that, for cloud water with pH below about 5.5, hydrogen peroxide (H_2O_2) is a more important oxidant for S(IV) than is ozone (O_3) for expected concentrations of these oxidants.

To illustrate the last statement, it is convenient (following Schwartz et al., 1983) to introduce the following preliminaries. The oxidation of all cloud-water or rainwater sulfur in the fourth oxidation state (e.g., HSO_3^-) to the sixth oxidation state (e.g., SO_4^{2-}) is indicated schematically, e.g., for the case of oxidation by O_3, by

$$O_3 + S(IV) \rightarrow S(VI) + O_2 \tag{11.100}$$

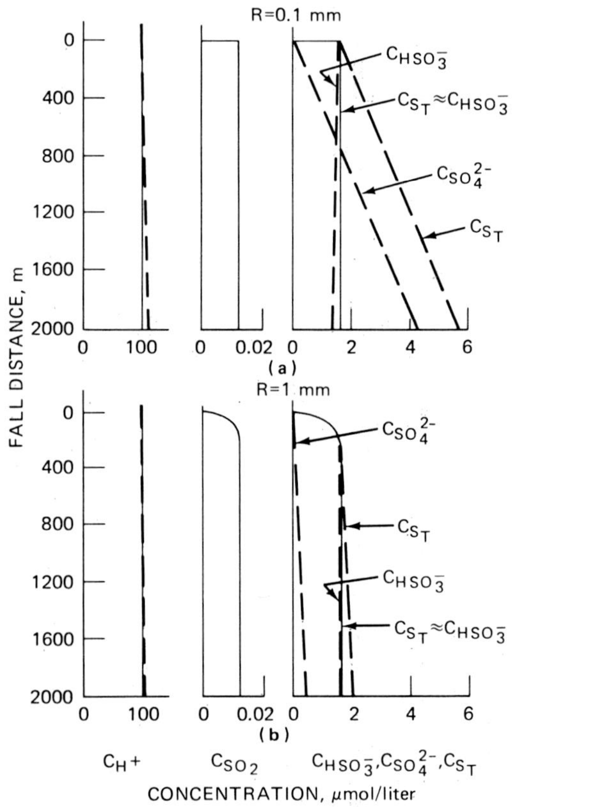

Fig. 11.64 Concentration transients in (a) a small drop (R = 0.1 mm) and (b) in a large drop (R = 1 mm) as they fall through a uniformly mixed layer with the air concentration of SO_2 = 10 ppb and background pH of drops = 4.0. Solid curves are for the HSO_3^- oxidation rate $k_{ox} = 1 \times 10^{-3}$/s. [Based on F. B. Hill and J. F. Adamowicz, A Model for Rain Composition and the Washout of Sulfur Dioxide, *Atmos. Environ.*, 11: 917-928 (1977).]

The rate of this oxidation is taken to be given by

$$R = -\frac{d}{dt} S(IV)$$

$$= k_{ox}[S(IV)][O_3(aq)] \quad mol/(liter \cdot s) \qquad (11.101a)$$

where k_{ox} represents the aqueous-phase rate coefficient. If we assume that the aqueous-phase concentrations are in equilibrium with the ambient gas-phase concentrations, then through use of the enhanced-solubility coefficient for SO_2 as given in Eq. 11.97, we can write Eq. 11.101a in terms of the gas-phase concentrations via

$$R = k_{ox}\, \alpha_{O_3}\, \alpha^{*}_{S(IV)}\, \chi_{O_3}\, \chi_{SO_2} \qquad (11.101b)$$

But if this oxidation rate of S(IV) (e.g., within clouds of liquid water content L in grams per square meter) is to be compared with clear-air oxidation of SO_2 (e.g., by the hydroxyl radical, OH), it is appropriate to recognize that the aqueous-phase reaction occurs only within the fraction L/ρ_w ($\sim 10^{-6}$) of the available volume, where ρ_w is the density of water. It is therefore useful to define, also, the effective oxidation rate

$$r = \frac{LR}{\langle \rho_w \chi_{SO_2} \rangle} \qquad (11.102)$$

These rates, evaluated with assumed values for gas-phase O_3 and H_2O_2, are shown in Fig. 11.65 and indicate that, if air concentrations of H_2O_2 are as large as 1 ppb ($\sim 1.5\ \mu g/m^3$), then H_2O_2 dominates both aqueous-phase and gas-phase oxidation of S(IV), assuming reasonable values for the average liquid water contents for clouds and their frequency of occurrence in temperate latitudes (Möller, 1980; Oppenheimer, 1983).

However, a number of associated caveats remain. First, there are, as yet, no reliable data for the concentrations of H_2O_2 in air or in cloud water—a situation that likely will soon be rectified. Second, the field studies by Hegg and Hobbs (1982), apparently confirming the rapid oxidation of S(IV) in clouds, give results for relatively short-time-scale wave clouds and in relatively clean air in the western United States; these authors therefore caution against the extrapolation of their results to other situations. Third, for longer-lived (stratiform) clouds in more polluted areas (in the eastern United States), Schwartz et al. (1983) found a wide range of values for the (interstitial) SO_2 unincorporated into the cloud water. These authors concluded: "The observation of measurable SO_2 concentration in cloud (air) suggests that aqueous-phase oxidation reactions converting SO_2 or SO_3^- to sulfate are not always sufficiently fast to cause the immediate disappearance of SO_2 from the interstitial air. In addition, there is the caveat that perhaps can be summarized best by quoting the report by Richards et al. (1983) dealing with Los Angeles cloud water: "The rate-data...predict that when the H_2O_2 concentration is greater than 40 μmol/liter (as measured), the lifetime of S(IV) is less than one second at the pH observed for this sample; so the presence of 10 μmol/liter of S(IV), after the cloud water stood for one hour, confirms that the reaction between H_2O_2 and S(IV) has been inhibited."

11-4.6 Precipitation Chemistry and Applications

In the previous section, some of the difficulties of calculating the chemical composition of rain were displayed. An alternative approach that can be used for scavenging predictions is to rely on available data of precipitation chemistry. Thus, from available data for the concentration, κ_0, of reaction products in rain, average effective washout ratios, $w_r = (\kappa/\chi)_0$, can be formed, where χ_0 is the concentration of the precursor gas in surface-level air. Some examples are shown in Table 11.11.

There is, however, a major difficulty with using results of the type shown in Table 11.1: it is not yet clear how much of the precipitation-borne, for example, SO_4^{2-} is derived from SO_2 oxidation within the storm vs. the amount derived from the direct scavenging of SO_4^{2-} that entered the storm [e.g., as H_2SO_4 or $(NH_4)_2SO_4$]. This difficulty is well illustrated by results reported by Hidy et al. (1983). For all data available from the SURE project, when they took the ratio of SO_4^{2-} in precipitation to SO_4^{2-} in surface-level air, they found the average scavenging ratio $w_r = (0.6 \pm 0.1) \times 10^6$; but, if they divided the SO_4^{2-} concentration in precipitation, instead, by the airborne particulate SO_4^{2-} plus S as SO_2, then the result was $w_r = (0.09 \pm 0.02) \times 10^6$. Similarly for NO_3^-: if the scavenging ratio was based only on the airborne particulate nitrate, then $w_r = (7.7 \pm 2.9) \times 10^6$; but, by adding to the denominator also the N from airborne NO_x and HNO_3, then $w_r = (0.09 \pm 0.04) \times 10^6$. It therefore seems clear that the

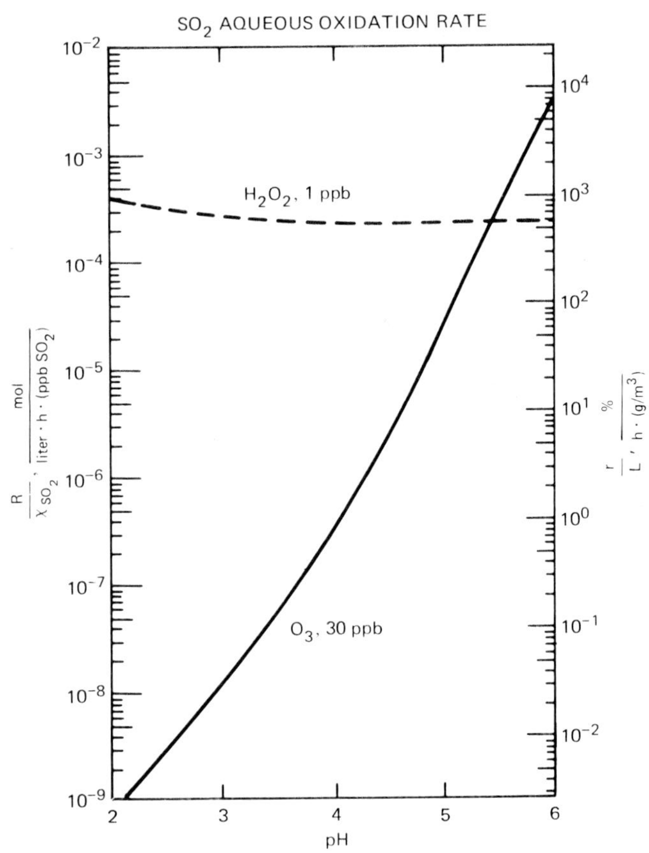

Fig. 11.65 Rate of aqueous-phase oxidation of S(IV), by O_3 (30 ppb) and H_2O_2 (1 ppb), as a function of solution pH. R/χ_{SO_2} represents aqueous reaction rate per parts per billion of gas-phase SO_2; r/L represents rate of reaction referred to gas-phase SO_2 partial pressure, per g/m³ liquid-water volume fraction. [Adapted from S. E. Schwartz, P. H. Daum, M. R. Hjelmfelt, and L. Newman, Cloudwater Acidity Measurements and Formation Mechanisms—Experimental Design, in *Precipitation Scavenging, Dry Deposition, and Resuspension,* Vol. 1, H. R. Pruppacher, R. G. Semonin, and W. G. N. Slinn (Coords.), Elsevier, New York, 1983.]

TABLE 11.11 Effective Scavenging (or Washout) Ratios for Some Reactive Species

Precursor gas	Mass ratio	w_r, dimensionless†
NH_3	$(NH_4^+)/(NH_3)$	$(1.2 - 2.2) \times 10^5$‡
NO_2§	$(NO_3^-)/(NO_2)$	$(1.8 - 4.0) \times 10^5$‡
SO_2	$(SO_4^{2-})/(SO_2)$	$(0.5 - 1.5) \times 10^5$‡
SO_2	$(SO_4^{2-})/(SO_2)$	$(0.1 - 1.0) \times 10^5$¶
SO_2	$(SO_4^{2-})/(SO_2)$	$(1.5, 3.0) \times 10^5$**

†Ratio of reaction product mass density in rain to precursor-gas mass density in air.

‡From Junge (1963) for various locations in Europe.

§See the text for comments about special problems in identifying the precursor of scavening NO_3^-.

¶Estimated from available data for the St. Louis METROMEX project; convective storm scavenging.

**Estimated from July 1972 to June 1973, National Air Sampling Network data for urban and nonurban sites, respectively, in northeastern United States (Dana, Battelle–Northwest, personal communication).

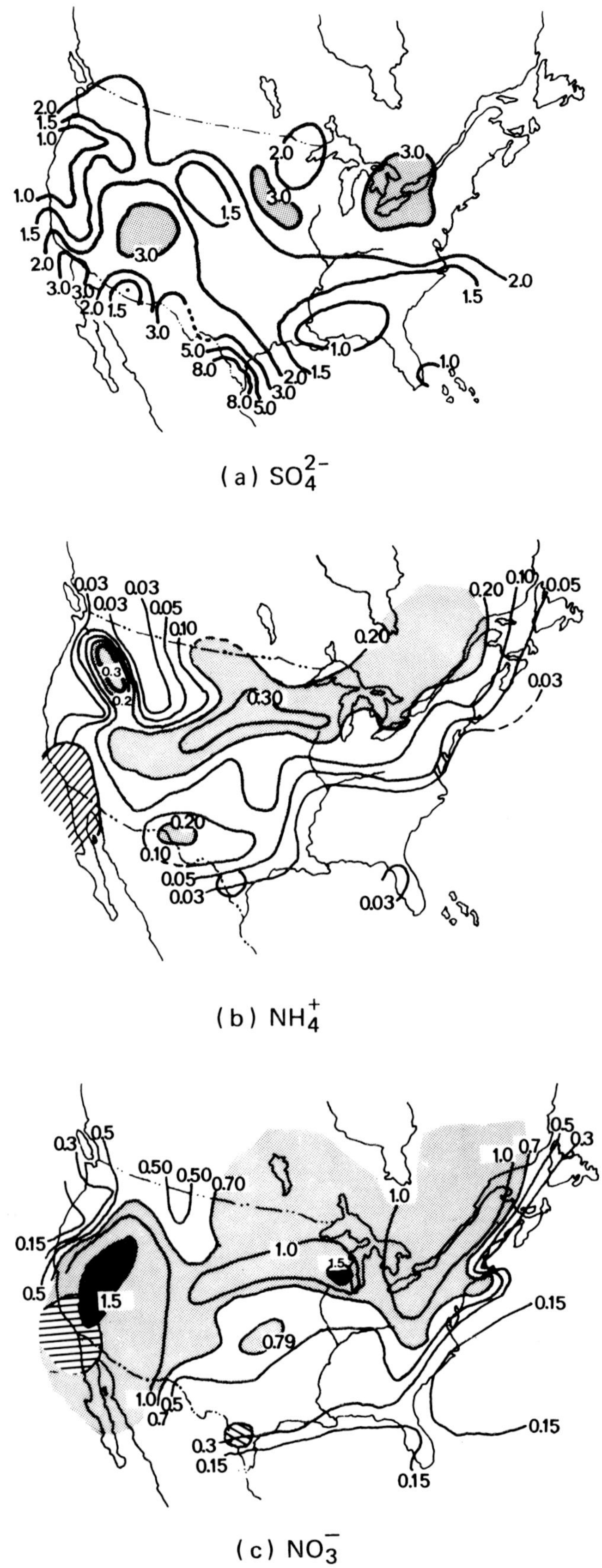

Fig. 11.66 Average concentrations in rain of (a) SO_4^{2-}, (b) NH_4^+, and (c) NO_3^- (mg/liter) measured for (a) during July 1955 to June 1956 and for (b) and (c) during July to September 1955. (Based on C. E. Junge, *Atmospheric Chemistry and Radioactivity*, Academic Press, N. Y., 1963.)

use of precipitation chemistry data for predictions of the scavenging of reactive species is not without pitfalls; note, for example, that for HNO_3 at 25°C, $w_r = \alpha = 5 \times 10^6$ (see Table 11.8).

Besides this use of precipitation chemistry data for scavenging predictions, however, the data are extremely valuable for other purposes. Although the focus of this chapter has been on how precipitation scavenging cleanses the atmosphere, there is great concern about the simultaneous pollution of the precipitation. The resulting phenomenon known as *acid rain* (better: acidic deposition) can perturb ecological balances, damage industrial products, and hasten the weathering of art treasures (e.g., see Dochinger and Seliga, 1976). In this closing section of the chapter, a few comments will be devoted to this alternative view of precipitation scavenging.

The collection of precipitation chemistry data in the United States has suffered from a lack of continuity. The first extensive network was a 60-station network operated from July 1955 to June 1956 (Junge and Gustafson, 1956). Figure 11.66 shows data from this network. Using these data and the data from a network established during 1965 and 1966 by Lodge et al. (1968), Cogbill and Likens (1974) estimated the average acidity of precipitation during these periods; their results and measured pH values for 1972 and 1973 are shown in Fig. 11.67.

The estimates shown in Fig. 11.67 have had considerable influence stimulating popular interest in "acid rain," but there are significant questions about these estimates. Two major questions are: (1) how accurate is the calculated pH (Liljestrand and Morgan, 1979) and (2) how representative are the data (Stensland and Semonin, 1982)? Figure 11.68 illustrates a response to the second question. Figure 11.68a illustrates Stensland and Semonin's estimate of the 1955–1956 pH distribution derived from the Junge data, using estimates for the concentrations of magnesium (which were not measured in the Junge network). Notice that there are differences between Fig. 11.68a and 11.67a but the differences are not too dramatic. What is dramatic, however, are the concentrations of calcium (and one would expect similar concentrations for magnesium) measured in the 1955–1956 Junge network (see Fig. 11.68b): these calcium values are by a factor of about 5 higher than normal

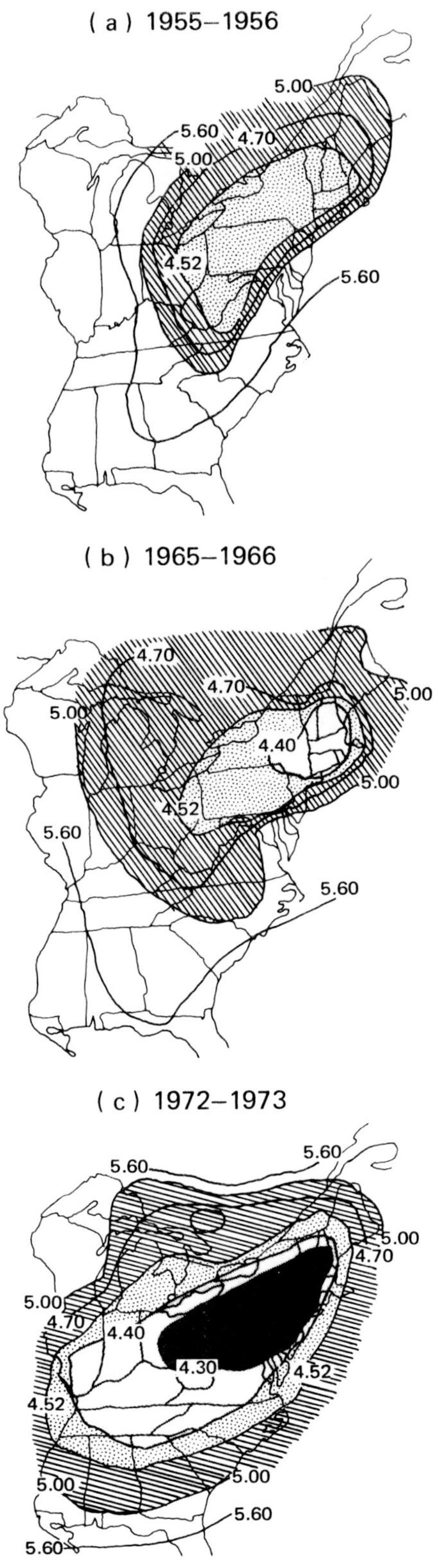

Fig. 11.67 Distribution of predicted, (a) and (b), and observed, (c), pH during the indicated time periods. The predicted values are based on measured concentrations of ionic species in rain (Junge, 1958; Junge and Werby, 1958; Lodge et al., 1968; Gambell and Fisher, 1966; Pearson and Fisher, 1971). [Based on C. V. Cogbill, The History and Character of Acid Precipitation, in Eastern North America, in L. S. Dochinger and T. A. Seliga (Eds.), *Proceedings of the First International Symposium on Acid Precipitation and the Forest Ecosystem,* May 12–15, 1975, General Technical Report NE-23, pp. 363-370, USDA Forest Service, 1976.]

values during the 1970s (Stensland and Semonin, 1982). These authors expect that the high concentrations of air- and precipitation-borne calcium (and presumably magnesium), commonly attributed to be derived from wind erosion of soil, were caused by the major drought that occurred over the central and lower Great Plains during the early and mid-1950s; in particular, in December 1955 and in February, March, and April 1956, severe dust storms were reported throughout the Great Plains states. These authors then estimated what the pH might have been in 1955–1956 (Fig. 11.68c) if the calcium and magnesium had not been anomalously high, and this result is compared with the current distribution of the pH of precipitation, shown in Fig. 11.68d. From this comparison, as opposed to the comparison available from Fig. 11.67, one would be hard pressed to conclude that precipitation in the northeastern United States has become more acidic during the past quarter century, although it is clear that the precipitation is more acidic in the Northeast than in the Great Plains. It also seems clear that more data, past and future, could help constrain speculations.

A more extensive data base is available for northern Europe; Fig. 11.69 identifies the precipitation chemistry network stations. It can be seen from Fig. 11.70 that there have been significant increases in the NO_3^{2-}–N and NH_4^-–N depositions at some stations during the past 20 yr; it is expected that the source of NO_3^{2-} is NO_x gases from various industries, high-temperature engines, and oil-based power plants (Odén, 1976). The sources of the NO_x (and SO_2) pollution are essentially all "upwind" releases and range from those nearby to those which occurred about a week previously. In turn, Scandinavian sources pollute countries downwind. However, pH decreases are of special concern in the Scandinavian countries because the buffering capability of the soil is relatively weak. As a result, fish and forests in these countries are suffering from this inadvertent modification of the natural chemical balance (Overrein, 1976). Similar consequences are of concern for the northeastern United States and southeastern Canada, but this concern should be tempered by the fact that the only trend in this acidic deposition during the past quarter century, which is statistically significant, is this increase in concern (Hansen and Hidy, 1982; Bilonick and Nichols, 1983).

11-5 SUMMARY COMMENTS

In this survey of precipitation scavenging, unfortunately a great deal of material has been omitted. It is estimated that at least 10^5 pages have been written about precipitation scavenging during the past 30 yr. Therefore, if this chapter were a balanced survey, the compaction would be more than 1000-fold. Thus even a balanced survey would possess many omissions. Now, in this final section, rather than to attempt a further compaction, the plan is to review briefly how the surveyed material can be used for scavenging predictions and what topics require further investigation.

11-5.1 Review for Applications

There are many different types of precipitation scavenging problems. Twelve types that are relatively easy to distinguish are for particles and for gases, for short- and long-term (e.g., accidental and routine) releases, and for local ($\lesssim$100 km), regional ($\sim 10^2$ to 10^3 km), and global spatial scales. The number of types are doubled, to 24, if a distinction is drawn

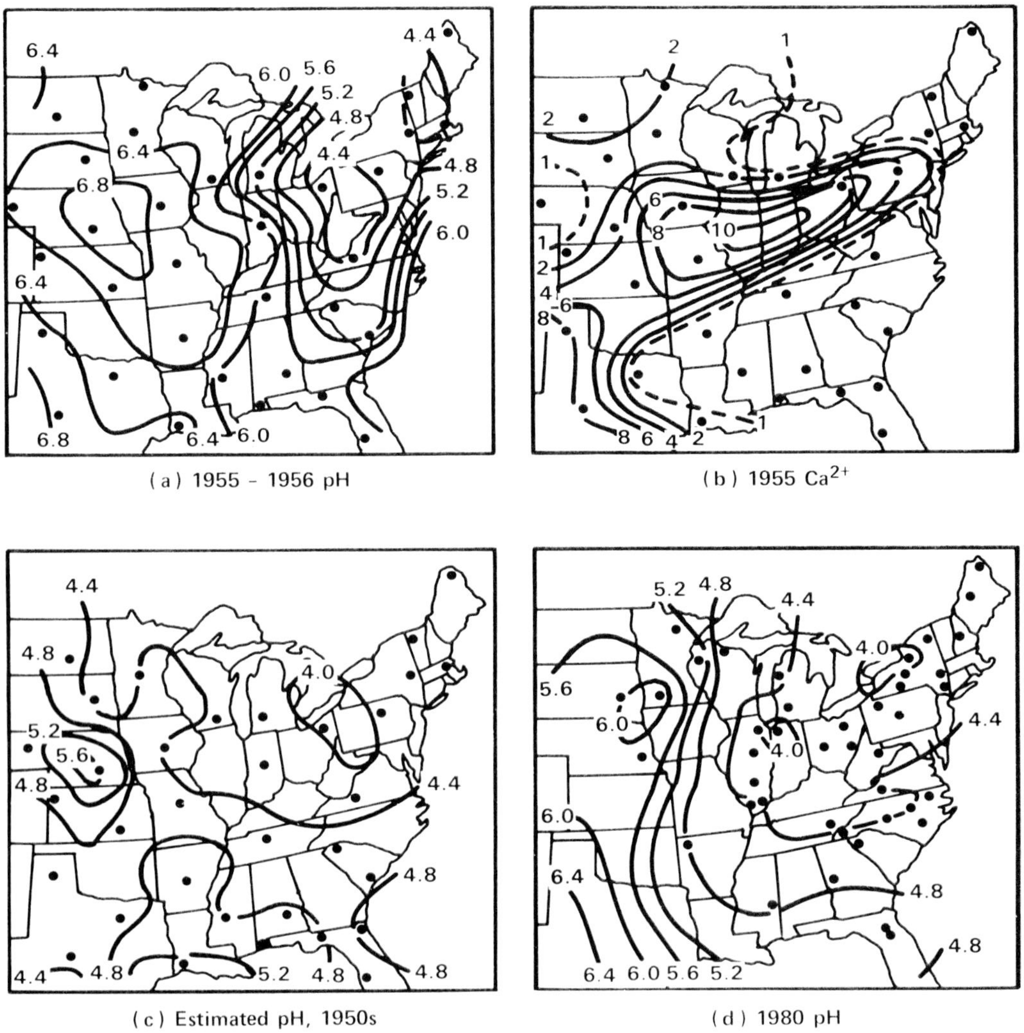

Fig. 11.68 (a) Estimated 1955–1956 pH distribution of precipitation derived from the Junge data, using estimated magnesium concentrations and an empirical correction factor. (b) The calcium concentration (mg/liter) for December 1955 from the Junge network. (c) The pH distribution for 1955–1956 after correcting for assumed anomalously high concentrations of calcium and magnesium. (d) Distribution of median pH values of precipitation from the NADP network as of September 1980. [Adapted from G. J. Stensland and R. G. Semonin, Another Interpretation of the pH Trend in the United States, *Bull. Am. Meteorol. Soc.*, 63: 1277–1284 (1982).]

between scavenging by stable vs. unstable storms. The present author resists another doubling, to 48, to differentiate between below- and in-cloud scavenging; the formulas presented in this chapter generally do not need to rely on this differentiation. This section presents methods for predicting scavenging for all 24 cases.

Predicting the possible consequences of a postulated nuclear accident can illustrate the case of local deposition of particles from a short-term release. Diagnostic studies of a particular accident would be relatively straightforward if appropriate meteorological data were available; these studies would be similar to the interpretations of tracer deposition, described in Sec. 11-3.5. Note, however, that transport of particles to neighboring cells of a convective storm is but poorly understood (Dingle, 1977). In contrast to these relatively straightforward diagnostic studies, forecasts of precipitation scavenging would be very difficult, even for forecasts only a few hours into the future. The reason for this difficulty is, of course, the difficulty of forecasting details about precipitation, such as probability of occurrence, aerial extent, and precipitation rate and type. The range of possible wet deposition locally is from zero (if there is no precipitation) to essentially complete removal (e.g., if all the released particles are entrained into the updraft of a convective cloud). In this latter example, essentially all particles could be deposited within about 10^3 s after ingestion by the cloud and onto an area dictated by the ensuing motion of the cloud and by the trajectories of the hydrometeors. Within this range of possibilities, the probability of a specific deposition pattern (e.g., on a nearby urban area) depends on the probability that the wind is in the direction of concern, on how much diffusion will dilute the pollutant's concentration, and on the probability that precipitation will occur during the time when the material is over the area of concern. To the author's knowledge, Watson, Barr, and Allenson (1977) have made the most complete analysis of this type of problem, and the interested reader should refer to their examples. Two of the important factors in their study are estimates for the probability of rain (e.g., Fig. 11.71) and for the probability of the aerial extent of the precip-

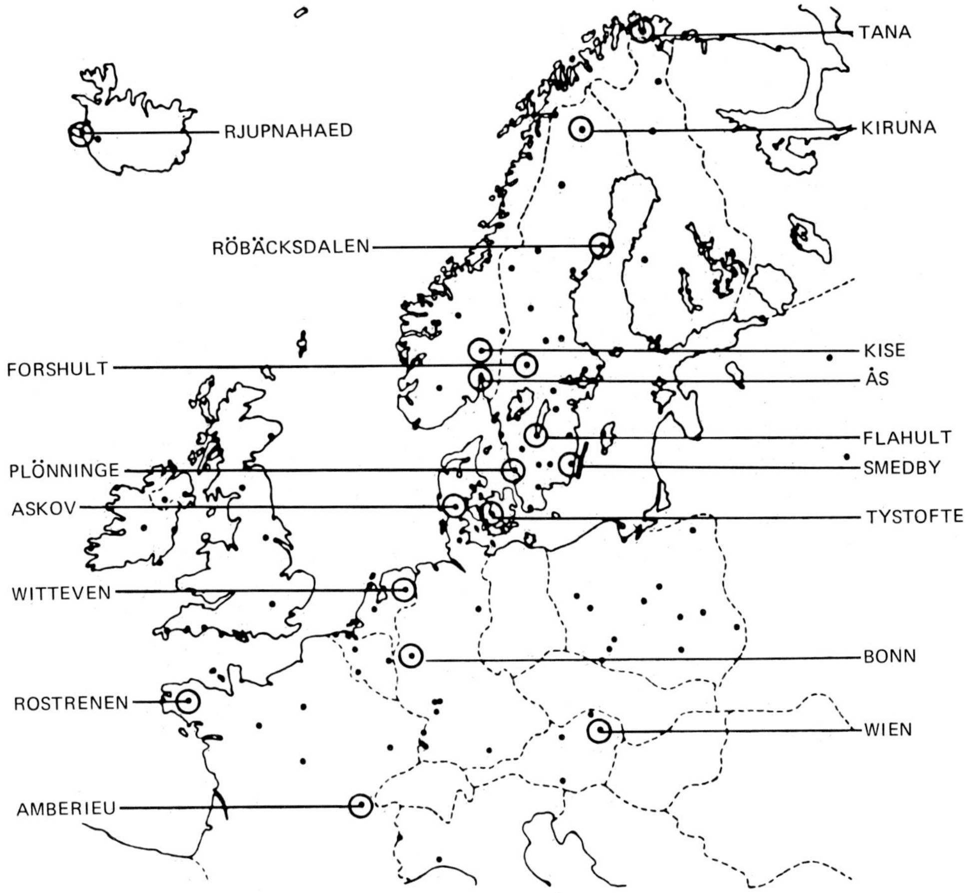

Fig. 11.69 The European precipitation chemistry network. Each dot represents a sampling station for precipitation and air; long-term data are available for the circled stations. [Based on S. Odén, The Acidity Problem—An Outline of Concepts, in L. S. Dochinger and T. A. Seliga (Eds.), *Proceedings of the First International Symposium on Acid Precipitation and the Forest Ecosystem,* May 12–15, 1975, General Technical Report NE-23, pp. 1-36, USDA Forest Service, 1976.]

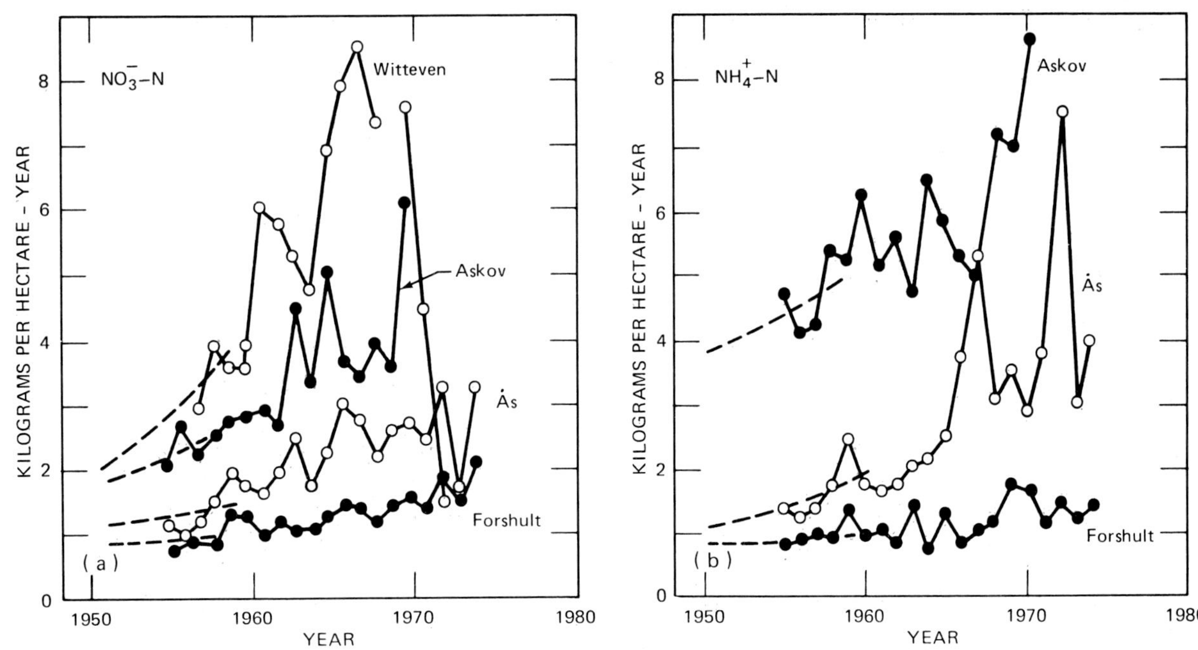

Fig. 11.70 Time records of the wet deposition of nitrogen as (a) nitrate and (b) ammonium ion at some stations of the European network (compare Fig. 11.68). [Based on S. Odén, The Acidity Problem—An Outline of Concepts, in L. S. Dochinger and T. A. Seliga (Eds.), *Proceedings of the First International Symposium on Acid Precipitation and the Forest Ecosystem,* May 12–15, 1975, General Technical Report NE-23, pp. 1-36, USDA Forest Service, 1976.]

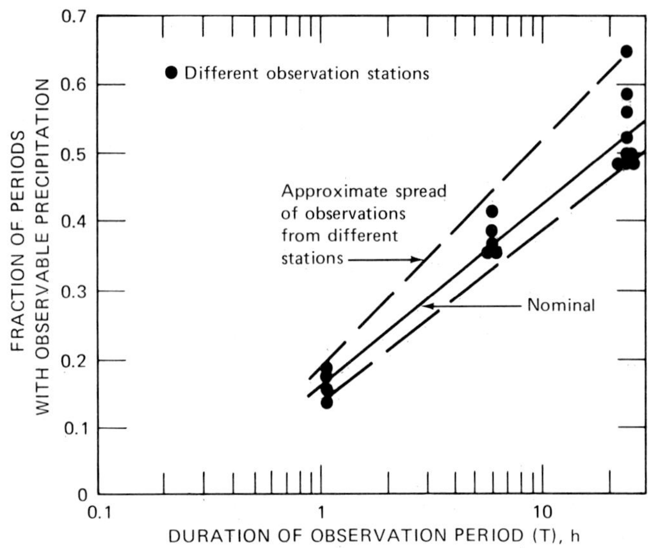

Fig. 11.71 Estimated probability of precipitation in Germany during the indicated observation period. [Based on Watson, Barr, and Allenson (1977).]

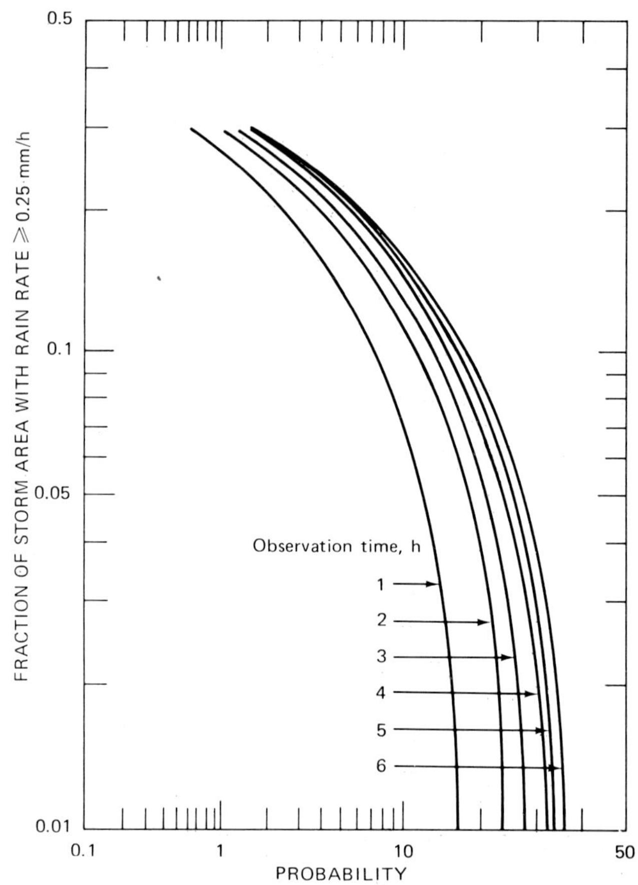

Fig. 11.72 Estimated probability of rain (with rate ⩾ 0.25 mm/h) in an area greater than the indicated fraction of the total storm area, during the indicated observation time, T. [Based on Watson, Barr, and Allenson (1977).]

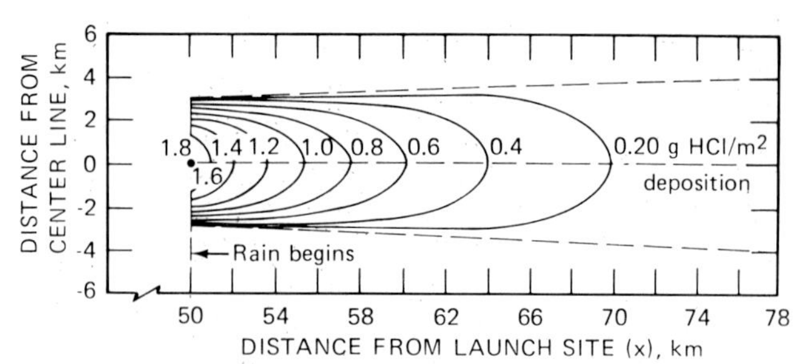

Fig. 11.73 Predicted "acid footprint" resulting from irreversible, below-cloud rain scavenging (p = 20 mm/h; ū = 7.5 m/s) of HCl released from the burn of a proposed rocket booster for the space shuttle vehicle. [Based on Pellett (1977).]

itation (Fig. 11.72). Typically it is not easy to obtain such data; when they are obtained, the scavenging rates that can be used are those given by Eq. 11.32 (for rain scavenging as a function of particle size), by Eq. 11.36 (similarly, but for snow scavenging), and by Eq. 11.38 or 11.46 (particle-average scavenging rates for polydisperse and active aerosol particles).

Local wet deposition from a short-term release of a gaseous pollutant is strongly dependent also on specific properties of the gas and on its vertical distribution in the atmosphere. How to treat these complications for gases that form simple solutions in rain (reversible scavenging) was described in Sec. 11-4.4; sometimes a numerical integration of Eq. 11.70 would be necessary (see Dana et al., 1973). For very reactive gases, such as HCl or I_2, however, irreversible scavenging can be assumed; in such cases the analysis proceeds in the same manner as for particles but with the collection efficiency (e.g., in Eq. 11.32) replaced by (4/Pe)Sh, where Pe is the Péclét number and Sh is the Sherwood number. As an example, Fig. 11.73 shows the predicted "acid footprint" of HCl, deposited in rain downwind of a burn of a proposed rocket booster for the space-shuttle vehicle.

Long-term average deposition, within about 100 km of a source of either particles or gases, is simpler to describe or to predict than short-term (e.g., accidental) releases because average meteorological data can be used. However, care must be taken to account properly for dry vs. wet periods. Thus, in the (unlikely) event that precipitation is continuous at the annual-average rate, then a simple Gaussian (or similar) plume model could be used, multiplied by a simple "exponential decay" term to describe scavenging (see Eq. 11.8). But for discontinuous precipitation, more pollution would be available for removal farther downwind (if removal by dry deposition is slower than by wet deposition). As a result, the "decay term," caused by scavenging, need not be a simple exponential. Rodhe (1980) illustrates this and suggests approximate methods to account for different removal rates during wet and dry periods. Also note, as illustrated in Fig. 11.74, that the wind rose for precipitation periods is normally not the same as the annual-average wind rose; further, what is obviously needed is a precipitation-weighted wind rose. As a result, as shown in Fig. 11.75, mean pollutant deposition patterns can be substantially different in shape from the shape of mean air concentration isopleths. For gases, local deposition near continuous releases is again complicated by the possibility of desorption of gases from rain. If account is taken for this phenomenon, e.g., see Fig. 11.76, then reasonably good predictions of reality can be obtained, although such complications as unknown gas reaction rates can cause uncertainties (see Hales, Dana, and Wolf, 1977). Beyond about 10 stack heights downwind, however, the vertical distribution of most gases (see Sec. 11-4.4 and Table 11.9) will be sufficiently diffuse to allow the assumption, especially for long-term average predictions, that the gas concentration in the rain is in equilibrium with the local surface-level air concentration. Then the most convenient description of the wet removal flux, W, is in terms of the wet deposition velocity, v_w (see Eq. 11.73): $W = v_w \chi_0$. This parameterization can then be used in the same manner as the parameterization for the dry removal flux, $D = v_d \chi_0$.

It may be useful to repeat that the wet deposition velocity parameterization for precipitation scavenging can be used for both particles and gases. For particles, this parameterization is always appropriate, regardless of the time and space scales of interest (see Sec. 11-3.6), although descriptions in terms of removal rates may be more convenient and accurate if appropriate information is available (see the closing remarks of Sec. 11-3.6). For particles, $v_w = w_r p_0$, where p_0 is the surface-level

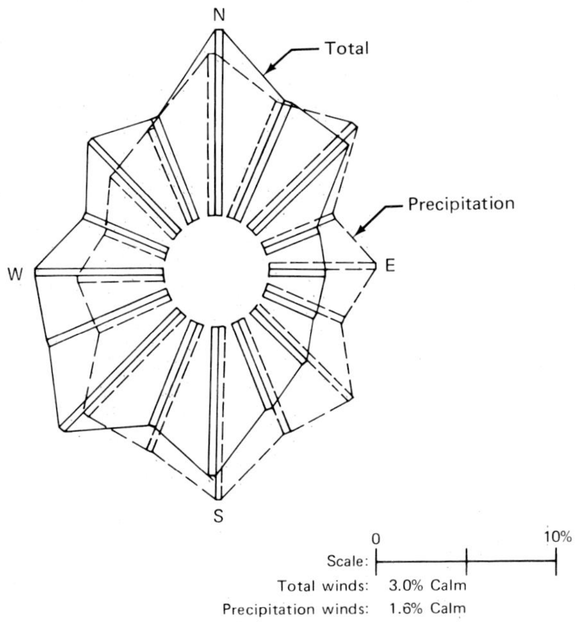

Fig. 11.74 Directional distribution (wind rose) of annual average, surface-level wind (all speeds) for Sudbury, Ontario, 1955–1972 (solid lines) and similarly during precipitation events (dashed lines) except only for April–September periods. Analysis performed by the Atmospheric Environment Service, Canada; results supplied through the courtesy of S. Stevens, Air Resources Branch, Ministry of the Environment, Ontario.

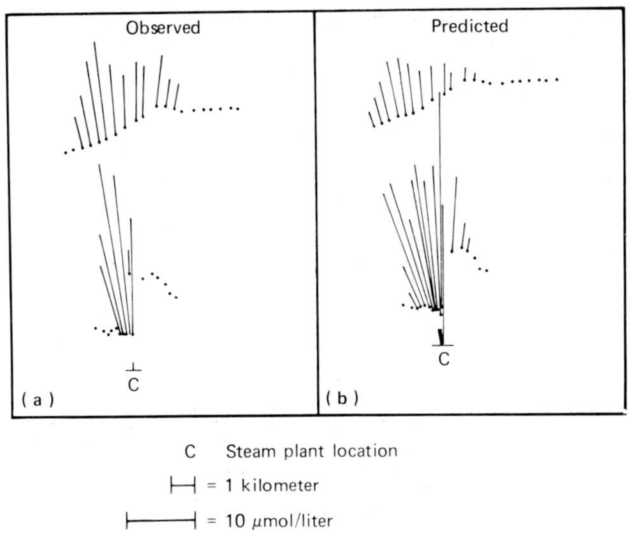

Fig. 11.76 Observed and predicted SO_2 concentrations in rain near a coal-fired power plant in Centralia, Wash. The EPAEC model calculations were based on assumptions that transport of the SO_2 through the atmosphere was rate limiting (i.e., well-mixed drops) and that the oxidation rate corresponded to an SO_2 half life of 15 min. [Based on Hales, Dana, and Wolf (1977).]

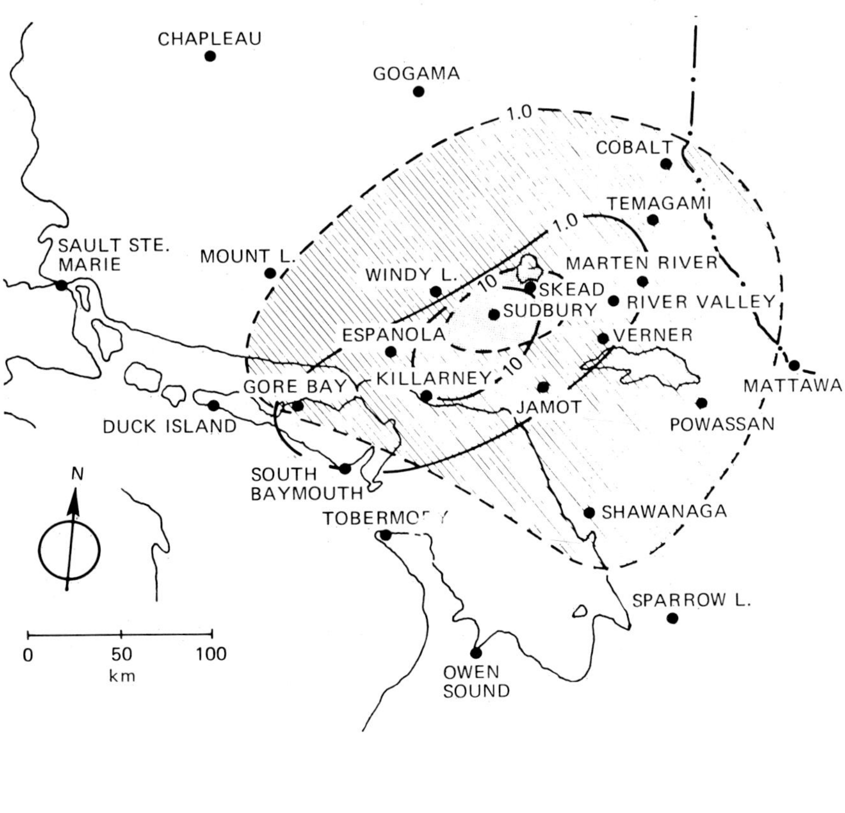

Fig. 11.75 Measured total nickel deposition rate in northern Ontario, near the Sudbury smelter. Solid lines reflect geometric-mean deposition prior to the installation of a 381-m "super stack"; dashed lines are geometric means after installation. Units are 10^{-9} g/(cm^2·day). [Based on Muller and Kramer (1977).]

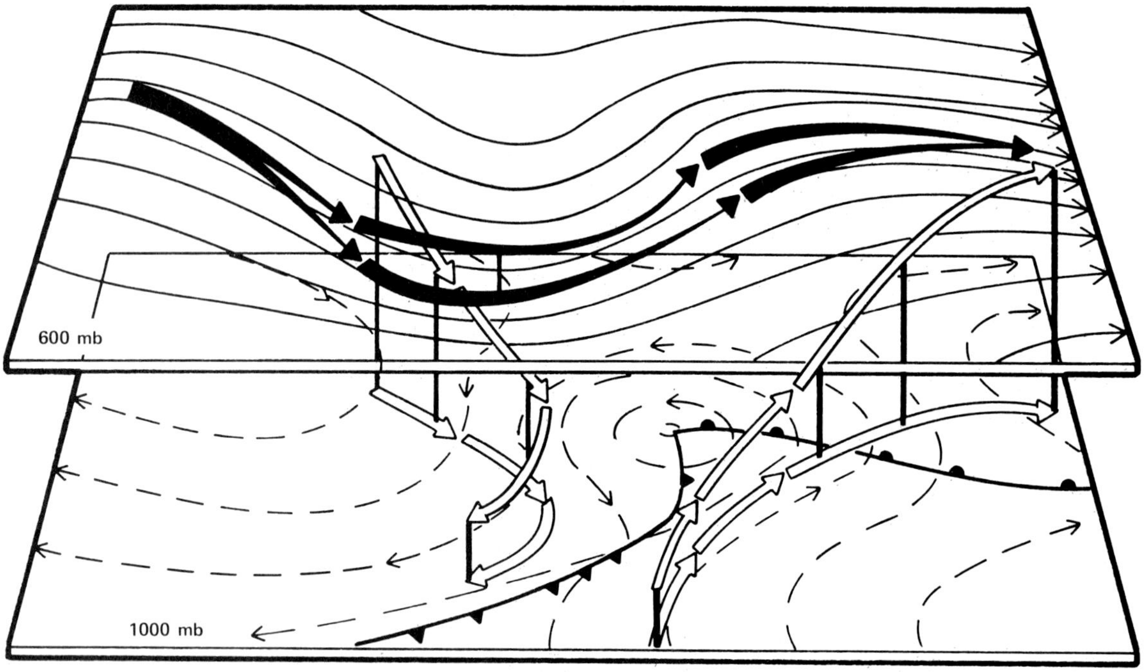

Fig. 11.77 A schematic indication of airflows associated with an extratropical cyclone in the Northern Hemisphere. It is typical for precipitation to occur both where the warm, typically moist air ascends at the warm front (●●●) and where the cold front (▲▲▲) overtakes and lifts the moist air. It is therefore clear that the deposition patterns for pollution scavenged by such a precipitation system will depend strongly on exactly how the pollution becomes involved in the precipitation. The solid arrows aloft (arrow lengths are proportional to wind speeds) indicate a typical pattern of the jet stream. (Figure derived from a similar figure in E. Palmen and C. W. Newton, *Atmospheric Circulation Systems*, Academic Press, 1959.)

precipitation rate and w_r represents the washout ratios as given in Figs. 11.42 to 11.49. The wet deposition velocity parameterization for gases, however, should be used cautiously unless the plume is well mixed in the vertical (see Table 11.9). When it is well mixed, then for gases that form simple solutions in water, $v_w = \alpha p_0$, where alpha is for the appropriate solubility coefficients (see Table 11.8). For reactive gases, $v_w = \alpha_* p_0$, where α_* is for an appropriate enhanced-solubility coefficient.

Regional-scale (i.e., $\sim 10^2$ to 10^3 km) wet deposition following short-term releases of either particles or gases is again complicated by uncertainties in winds, diffusion, and precipitation. For example, even for the relatively simple case of frontal precipitation occurring during the time of release, the resulting deposition patterns can be significantly different depending on the location of the fronts and on the atmosphere's stability (e.g., see Fig. 11.77). For the more complicated and more likely cases that the pollutant is transported for some time before encountering a precipitating system, then there are the difficult problems associated with analyzing or predicting pollutant trajectories and diffusion about these trajectories. These problems will not be addressed here (see Danielsen, 1974) except to mention that, even after 24-hr travel time, it is not uncommon for end points of isentropic trajectories through precipitating systems to be more than 500 km different from predictions using isobaric trajectories (e.g., see Davis and Wendell, 1976). In contrast to these difficulties of predicting the pollutant's air concentration, the tasks of predicting the wet removal should the pollutant enter a storm are relatively straightforward. The removal can be parameterized in terms of removal rates or wet deposition velocities. Indeed, regardless of initial particle size, wettability, etc., it appears to be an accurate estimate to specify that the efficiency with which particles are removed by a storm, ϵ, is the same as the efficiency with which the storm removes its condensed cloud water, ϵ_{cw} (Slinn, 1982c). For gases that form simple solutions in water and for problems at these large spatial scales, it is almost certainly acceptable to set the gas concentration in surface-level rain, κ_0, equal to the value in equilibrium with the surface-level air concentration, χ_0; namely, $\kappa_0 = \alpha\chi_0$. For reactive gases, it is necessary to solve the difficult problem of how much of the gas has been converted to its reaction products before it enters the first storm, while it is within the storm, before it enters the second storm, etc. Thus, in summary, the most difficult aspects of analyzing or predicting wet deposition at large spatial scales following a short-term release of a pollutant are associated with atmospheric dynamics and atmospheric chemistry.

The case of regional-scale deposition from routine releases of pollutants is being actively investigated by a number of research groups. Many of these investigations are concerned with the long-range transport and deposition of sulfur compounds released from fossil-fuel-fired stations generating electricity. A valuable general reference to the appropriate scientific literature is the composite edition of *Atmospheric Environment*, Vol. 12, Nos. 1–3, which contains papers presented at the 1977 Dubrovnik conference on "Sulfur in the Atmosphere." Theoretically the long-term average problem is simpler to solve than the short-term release case because average trajectories and average chemical reaction rates can be used. However, this presupposes that these averages are known. So far, what has been done to obtain average trajectories is to evaluate many (typically, isobaric) trajectories and, from these calculations, to evaluate the average trajectory and the "diffusion"

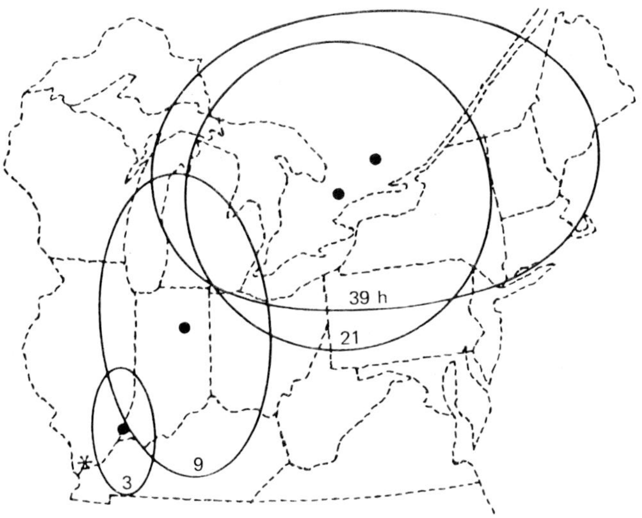

Fig. 11.78 Computed mean location (•) and one standard deviation (solid curves) for 40 puffs of pollutants, after the indicated time period since release from the position indicated by *. Advection is computed using a mean wind in a 2-km boundary layer and evaluating pollutant position every 3 h. (Based on C. M. Sheih, Application of a Lagrangian Statistical Trajectory Model to the Simulation of Sulfur Pollution over Northeastern United States, in *Proceedings of the Third Symposium on Atmospheric Turbulence, Diffusion, and Air Quality,* Raleigh, N. C., October 19–22, 1976, pp. 311-317, American Meteorological Society, Boston, Mass., 1976.)

about this average. Note that this diffusion reflects the spread of individual (for example, 72 hr) trajectories from the mean trajectory; typically, smaller scale diffusion of the pollutant about an individual trajectory is ignored. Figure 11.78 illustrates the type of results obtained for 40 trajectories from Apr. 1 to 20, 1972. Similarly, average chemical reaction rates, e.g., for the oxidation of SO_2, are obtained from measurements of oxidation rates for specific test periods, e.g., see Fig. 11.79. Generalizing such results to specify "average" reaction rates is at present a tenuous undertaking. Nevertheless, predictions for

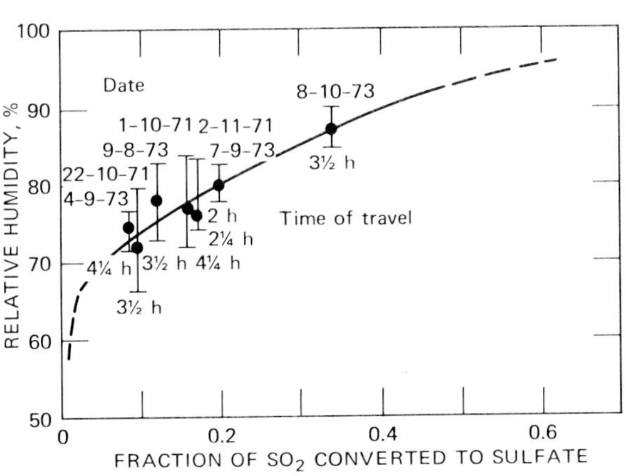

Fig. 11.79 Inferred fraction of SO_2 converted to sulfate as a function of relative humidity. Measurements were made along flight paths off the east coast of the United Kingdom. [Based on F. B. Smith and R. D. Hunt, Meteorological Aspects of the Transport of Pollution over Long Distances, *Atmos. Environ.,* 12: 461-478 (1978).]

regional-scale transport and deposition of routinely released pollutants have been made using these methods, e.g., see Fig. 11.80. More confidence in the realism of these modeling exercises should result from studies now in progress.

Finally, for global-scale deposition resulting from short- or long-term releases of particles or gases, we return to the problem mentioned at the outset of this chapter: estimating the residence times of atmospheric trace constituents. At the end of Sec. 11-2, an outline was given of the steady-state case, with balance between a pollutant's production and removal. Here, to end this chapter, some aspects of the unsteady case will be demonstrated, and this will reveal some needed thrusts for future research.

Earlier, in Sec. 11-2.5, the residence (or turnover time), τ, was estimated by assuming steady conditions and by using τ = (Total Burden or Quantity, Q) divided by (Production plus Inflow Rates, P + I) or divided by (Removal plus Outflow Rates, R + O). For example, for atmospheric sulfur, the mixing ratio is about 1 ppb, and, since the mass of the troposphere is about 4×21^{21} g, then Q = 4 Tg. Also, natural and anthropogenic sources of sulfur sum to about 200 Tg/yr. Consequently, for the tropospheric residence time for sulfur compounds, we have $\tau \simeq 4$ Tg/(200 Tg/yr) $\simeq$ 1 week. Similar estimates are appropriate for the turnover time for water vapor (Junge, 1963) and for submicron particles (e.g., see Lambert, Sanak, and Polian, 1983). But, especially for atmospheric trace constituents whose tropospheric residence times are as short as a week and for species whose source distributions are nonuniform, rarely do steady conditions prevail (Schwartz, 1979). For applications, it is therefore appropriate to examine the unsteady case.

For the unsteady case, consider a puff release of some pollution and suppose that there is no production of the pollution in the atmosphere. Assume that the pollution can be removed by wet and dry processes, as well as by first-order chemical and physical transformations. If we average the convection-diffusion equation (Eq. 11.3) over a volume large enough to contain the pollution (so that the advection term is zero; otherwise, obvious additional terms will enter the resulting equation) and if we divide the resulting equation by the total amount of pollution, q(t), still airborne at time t [i.e., $q(t) = \int \chi \, dV$], then we can define the overall removal rate, τ^{-1}, via

$$\frac{1}{\tau} = -\frac{1}{q}\frac{\partial q}{\partial t} = \frac{1}{\tau_w} + \frac{1}{\tau_d} + \frac{1}{\tau_c} + \frac{1}{\tau_p} \qquad (11.103)$$

where the separate contributions to the removal processes are similar to those in Eq. 11.20. In one interpretation Eq. 11.103 describes the time evolution of the pollution as if it were all located at the centroid of its spatial distribution.

The root problem in the application of Eq. 11.103 is that most removal processes are random in time, i.e., stochastic. Thus, if data were available for q, say for a species subjected only to wet and dry removal (illustrated in Fig. 11.81a), then Eq. 11.103 could be used to derive τ (see Fig. 11.81b). Note that this overall removal rate, τ^{-1}, varies with time. If data were available for an ensemble of realizations of the type illustrated in Fig. 11.81a, then as suggested in Figs. 11.81c and 11.81d, a probability density function, $f(\tau)$, for τ could be obtained. From this, a mean residence time could be defined as the ensemble mean:

$$[\tau] = \int_0^\infty \tau \, f(\tau) \, d\tau \qquad (11.104)$$

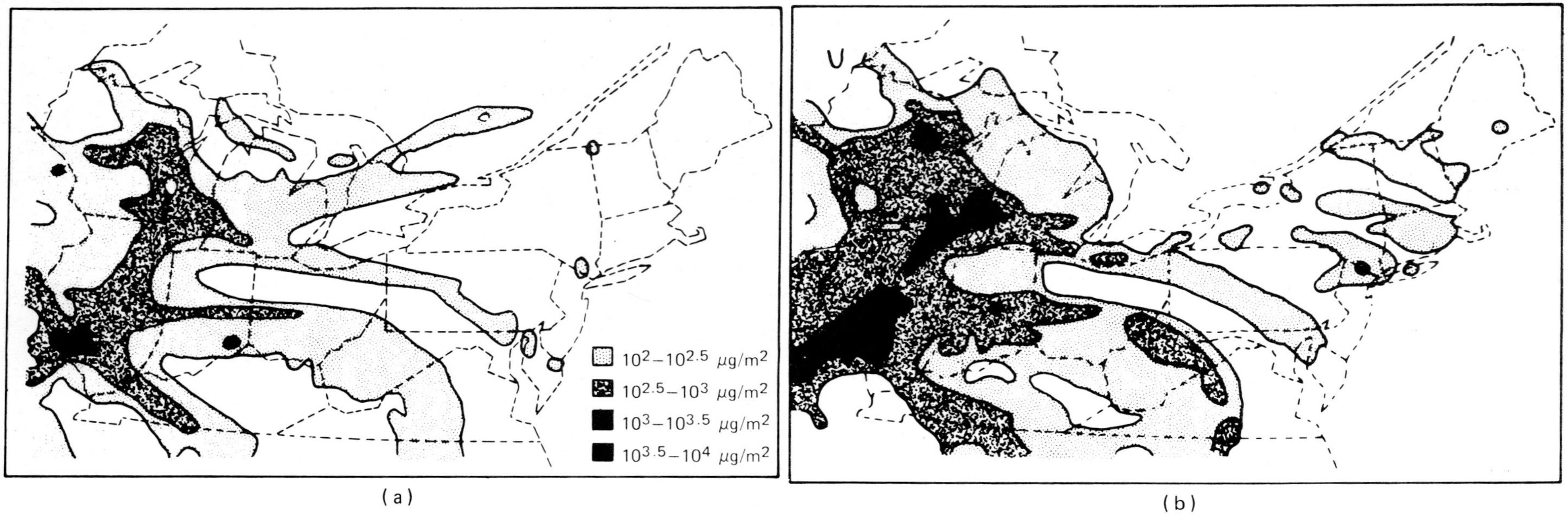

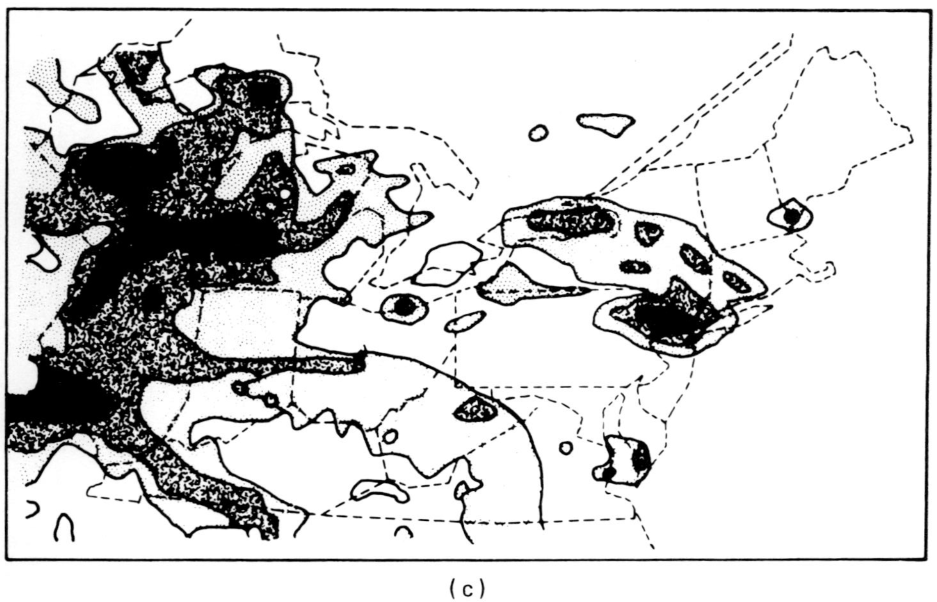

Fig. 11.80 Predicted sulfate deposition in micrograms per square meter for (a) dry removal, (b) wet and dry removal assuming average precipitation, and (c) wet and dry removal assuming real time precipitation, for an 18-d test period (Apr. 2–20, 1974), from a single 10^5 ton/yr SO_2 source near St. Louis, Mo. The motion of an expanding puff from a 400-m effective stack height was calculated using interpolations from synoptic radiosonde wind data. Values for parameters used in the model were as follows: dry deposition velocities $v_d(SO_2)$ = 1 cm/s and $v_d(SO_4^{2-})$ = 0.1 cm/s; collection efficiencies (in Eq. 11.32) $\bar{E}(SO_2)$ = 1.0 and $\bar{E}(SO_4^{2-})$ = 0.1; mean drop diameter = 2 mm; and an SO_2 oxidation rate = 5%/h. More realistic analysis should include changes in oxidation rate with humidity (especially within clouds), the finite dissolution and conversion of SO_2 in drops as a function of rain pH, and changes in SO_2 and SO_4^{2-} dry deposition velocities with surface and atmospheric conditions. [From L. L. Wendell, D. C. Powell, and R. L. Drake, A Regional Scale Model for Computing Deposition and Ground Level Air Concentration of SO_2 and Sulfates from Elevated and Ground Sources, in *Proceedings of the Third Symposium on Atmospheric Turbulence, Diffusion, and Air Quality*, Raleigh, N. C., October 19–22, 1976, pp. 318-324, American Meteorological Society, Boston, Mass., 1976.)

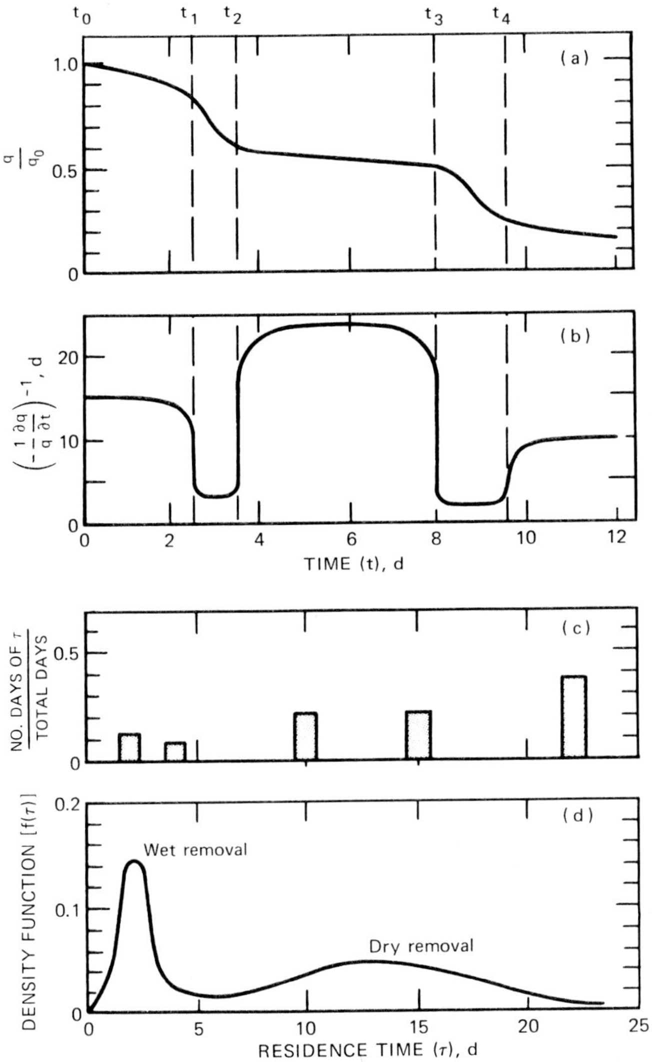

Fig. 11.81 A schematic indication of how data for q(t)/q(0) could be used to deduce the probability density function f(τ), yielding the mean residence time (τ). [From Slinn et al., Some Aspects of the Transfer of Atmospheric Trace Constituents Past the Air–Sea Interface, *Atmos. Environ.*, 12: 2055-2087 (1978).]

In reality, however, sufficient data are not available for q, and various other methods are used to estimate [τ]. For example, Junge (1963) summarizes estimates for τ, for submicron particles, based on seven studies of the natural and anthropogenic radioactivity attached to these particles. For these seven studies, the mean value for τ was 18.9 d; the standard deviation about this mean was 13.4 d! More recent reviews (Martell and Moore, 1974; Lambert et al., 1983)—but still of insufficient data—give, for submicron particles in the troposphere, [τ] ≃ 1 week, which can also be deduced from a simple analysis of Eq. 11.103.

To illustrate current studies of Eq. 11.103, consider the case (which is probably appropriate for 0.1- to 1-μm particles) when the dominant removal process is wet deposition, with the first-order removal rate $\tau_w^{-1} = \psi$. Then Eq. 11.103 obviously gives

$$\frac{\partial q}{\partial t} = \psi q \rightarrow q(t) = q(0)\exp\left(-\int_0^t \psi(t')dt'\right) \qquad (11.105)$$

In this equation the time variation is as clocked by the particles moving with the air (i.e., a Lagrangian rather than an Eulerian coordinate system).

As time advances, the removal rate can vary from near zero (dry deposition has been ignored) to large values (e.g., 100%/h) as the particles pass in and out of storms. However, the time the particles experience precipitation (e.g., in an updraft of a cumulonimbus storm or in a rainband of a frontal storm) may be quite short. If both the time and the scavenging rate are small, the integral in Eq. 11.105 over "precipitation period i" can be approximated by

$$\int_{t_i}^{t_{i+1}} \psi\, dt' \simeq \epsilon_i \ll 1 \qquad (11.106)$$

where ϵ_i is obviously the fraction, of the total number of particles of a given size class, which is scavenged by storm i.

With the approximation given by Eq. 11.106, Eq. 11.105 can be written in the form

$$\frac{q}{q_0} = \exp\left(\sum_{i=1}^{n} \epsilon_i\right) \rightarrow \ln\frac{q}{q_0} = -\left(\sum_{i=1}^{n} \epsilon_i\right) \qquad (11.107)$$

where n is the total number of storms experienced by the particles during time t. By this technique we have managed to change the integral over the stochastic variable ψ (i.e., an untractable infinite sum of the random variables ψΔt) into a finite sum of random variables (which, nevertheless, still is not trivial to evaluate). If ϵ_i is not small (but the storms are still short compared to the total time scale of interest), then as indicated in Fig. 11.82 the obvious generalization to Eq. 11.107 is

$$\frac{q(t)}{q(0)} = \prod_{i=1}^{n}(1-\epsilon_i) \rightarrow \ln\frac{q}{q_0} = \sum_{i=1}^{n}\ln(1-\epsilon_i) \qquad (11.108)$$

Equations 11.107 and 11.108 give relations between the stochastic variables q(t)/q(0) and ε. An additional stochastic variable is n, the number of storms that the particles encounter during time t. If the statistics of ε and n are known, there are a number of ways to use these equations to deduce statistics about q/q_0. However space restrictions here force the omission of details (for these, see Slinn, 1982a, 1983a). In outline, one method to proceed is to use the "law of large numbers" (if the number of storms is "sufficiently" large), which states that, subject only to weak restrictions on the probability density function for ε, the sum (say given by Eq. 11.108, for ln q/q_0) will have a normal or Gaussian distribution. In other words, q(t)/q(0) will have a lognormal distribution. From this distribution (or directly from the sums in Eq. 11.108—using the result that the mean of the sum is the sum of the means of independent variables), we can obtain the mean value of q/q_0, about which there is the lognormal distribution:

$$\frac{q}{q_0} = \exp\left[\bar{\nu}t\ln(1-\bar{\epsilon})\right] \simeq \exp\left[-\bar{\epsilon}\bar{\nu}t\right] \qquad (11.109)$$

in which $\bar{\epsilon}$ is the average efficiency with which particles are scavenged and $\bar{\nu}$ is the average (Lagrangian) frequency at which the particles encounter storms. Thereby, we can define the contribution of wet removal, $\bar{\tau}_w$, to the tropospheric residence time via

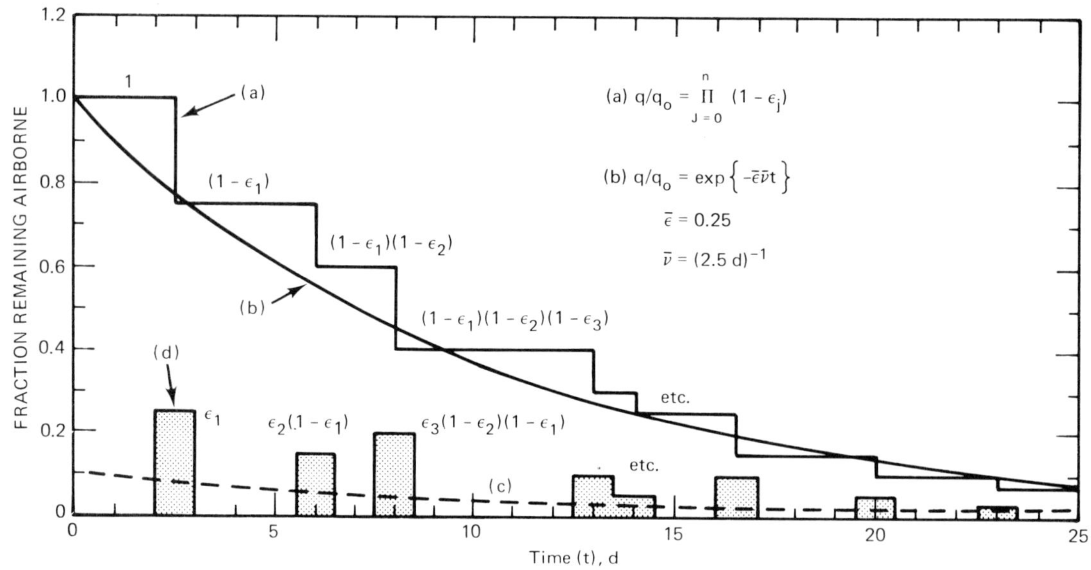

Fig. 11.82 A schematic indication of the stochastic nature of precipitation scavenging. [From Slinn et al., Some Aspects of the Transfer of Atmospheric Trace Constituents Past the Air–Base Interface, *Atmos. Environ.*, 12: 2055-2087 (1978).]

$$\left[\frac{q}{q_0}\right] = \exp\left[\frac{t}{\bar{\tau}_w}\right] \rightarrow \bar{\tau}_w = (\bar{\epsilon}\,\bar{\nu})^{-1} \quad (11.110)$$

As an example of Eq. 11.110, if $\bar{\epsilon} = ½$ and if $\bar{\nu} = 2.5$ d in the winter and $\bar{\nu} = 5$ d in the summer, then

$$\bar{\tau}_w = \begin{cases} 5\ \text{d, winter} \\ 10\ \text{d, summer} \end{cases} \quad (11.111)$$

The values given in Eq. 11.111 are, however, not very reliable. Much additional study is needed before reliable values are available for the average scavenging efficiency of storms (for reviews see Slinn, 1982c, 1983b) and for the average (Lagrangian) frequency of storms (for recent studies see Vickers and Slinn, 1979; Hamrud, Rodhe, and Grandell, 1981; Small and Sampson, 1982). Results more general than those outlined here have been developed by Rodhe and Grandell (1981).

But it would be inappropriate to end this chapter with Eq. 11.111, with its focus on averages. Precipitation scavenging is a stochastic process, and, just as life would be very dull if average conditions always prevailed, so too the essence of scavenging is in the variations about the mean, which likely have a "long-tailed" lognormal distribution. For example, although the mean tropospheric residence time of sulfate is probably about a week, Fig. 11.83 suggests how a deep storm at lower latitudes can inject sulfate (and other pollutants) into the stratosphere, resulting in a residence time of many months and leading to the observed pollution in the Arctic (e.g., Barrie, Hoff, and Duggupaty, 1981). Such events can lead to "global-scale finger pointing," with the western Europeans pointing at pollution sources in North America, the Soviets pointing at sources in western Europe, and the North Americans completing the circle by pointing at European and Asian pollution in Alaska and the Northwest Territories. But there is the hope that awareness of unity might yet result.

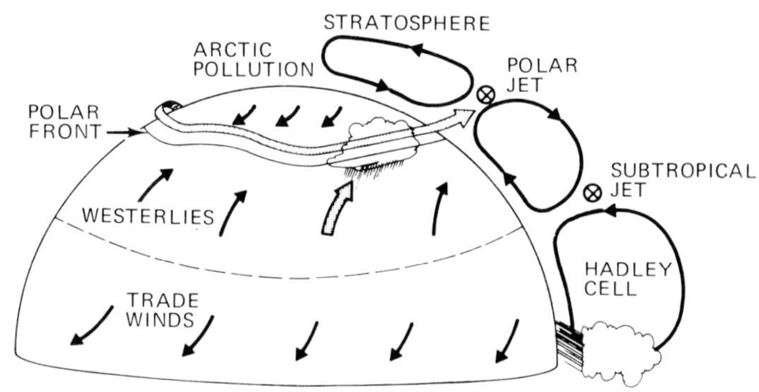

Fig. 11.83 A suggestion of how pollution, not scavenged by a storm, can subsequently be transported in the stratosphere and eventually can contribute to pollution in the Arctic (e.g., during subsidence of stratospheric air during the Arctic spring). [From W. G. N. Slinn, Estimates for the Long-Range Transport of Air Pollution, *J. Water, Air, Soil Pollut.*, 18: 45–64 (1982).]

11-6 ACKNOWLEDGMENTS

The following individuals are given my sincere thanks for the guidance, ideas, arguments, and pleasure they have given me during the past 15 yr: R. J. Engelmann, M. A. Wolf, M. T. Dana, J. Rosinski, and O. Vittori. Also, much of what I have learned is from C. E. Junge, A. N. Dingle, and D. F. Gatz. None of this would have been possible, however, without the continuous and appreciated financial support of R. W. Beadle, project monitor first with the U. S. Atomic Energy Commission, then the Energy Research and Development Administration, and now the Department of Energy. Special acknowledgments are due A. N. Dingle, J. M. Hales, L. A. Barrie, H. Rodhe, and D. Randerson for suggesting significant improvements to earlier drafts of this chapter. This work was performed under DOE Contract EY-76-S-06-2227 TA No. 27.

LIST OF SYMBOLS

Although all symbols are defined in the text, it may be useful to list those used frequently, along with their dimensions (L = length, T = time, M = mass, Q = charge; "units" refers to any convenient measure of pollutant amount, e.g., μg, moles, dpm, etc.). In cases of multiple uses of a single symbol, the number of the first equation in which the symbol has its specified use is also listed.

Symbol	Definition
a	Aerosol particle radius, L
a_g	Geometric mean particle radius, L
A	Area, L^2
A(r)	Attachment spectrum, units/L^4
c	Empirical constant $\simeq$ 0.5, dimensionless
C	Pollutant concentration in bounding medium, units/L^3
d	Drift velocity of pollutant with respect to the fluid velocity, L/T
D	Brownian or molecular diffusivity of particle or gas molecule in air, L^2/T, Eq. 11.2
D_c	Maximum linear dimension (e.g., diameter of circumscribed sphere) for ice crystals or aggregates, L
D_m	Characteristic volume/area length scale for ice particles, L, Eq. 11.36
D	Magnitude of dry flux, i.e., unassociated with precipitation, units/L^2T
E	Particle/drop collision efficiency, dimensionless. E is also the collection efficiency when the retention efficiency is unity.
E	Particle/ice crystal collision efficiency, dimensionless
Eu	$\rho_0 v_t^2/p_0$ = Euler number, ratio of dynamic to static pressures, dimensionless
Flux, F	Magnitude of convective and diffusive flux of contaminant, units/L^2T
g	Gravitational acceleration, L/T^2
G	Source (or gain) term, units/L^3T
h	Effective stack height, L, Eq. 11.12
h	Characteristic height from which pollution is scavenged, L, Eq. 11.52
h_d	Characteristic dry removal height scale, L
h_w	Characteristic wet removal height scale, L
H	Height of mixed layer, L
H	Henry's law constant, dimensionless
k_a	Transfer velocity (or conductance) through the atmosphere, L/T
k_l	Transfer velocity (or conductance) through the liquid, L/T
k_0	Overall transfer velocity, L/T
K_z	Coefficient of turbulent diffusivity for diffusion in the vertical direction, L^2/T
$\mathbb{K}$	Turbulent or eddy diffusivity tensor, L^2/T
l	Characteristic dimension of hydrometeor, L
L	Sink (or loss) term, units/L^3T
N	Number density function, number/L^4
p	Precipitation rate or flux (if for snow, then rainwater equivalent), L/T or L^3/L^2T
P	Total precipitation during a specified time period, L
Pe	Rv_t/D = Péclét number; the ratio of convective to diffusive transport, dimensionless
q	Total quantity of contaminant remaining in the atmosphere, units
$\dot{Q}$	Source strength of pollutant, units/T
r	Host particle radius, L
$\vec{r}$	Position vector, L
R	Drop radius, L
R_m	Volume or mass mean drop radius, L
Re	Rv_t/ν = Reynolds number; the ratio of convection to diffusion of vorticity, dimensionless
s,St	$\tau v_t/R$ = Stokes' number; the ratio of characteristic length (or time) scales for particle motion in a fluid, dimensionless
S_*	Critical Stokes' number below which there is no contribution to collisions from inertial impaction, dimensionless
Sc	ν/D = Schmidt number; the ratio of diffusivities for vorticity and for the pollutant, dimensionless
Sh	Sherwood number, nondimensional mass flux
$\bar{u}$	Mean wind speed, L/T
u	Particle velocity, L/T
v	(u,v,w) = fluid velocity, L/T
v_d	Dry deposition velocity, L/T
v_g	Gravitation settling speed for particles, L/T
v_i	Internal circulation for drop, L/T
v_r	Resuspension velocity, L/T
v_t	Terminal speed for hydrometeors, L/T
v_w	Wet deposition velocity, L/T
V	μ_w/μ_a = ratio of dynamic viscosities for water and air, dimensionless
w_r	κ/χ, washout ratio, dimensionless
W	wet flux, i.e., flux associated with precipitation, units/L^2T
α	H^{-1}, solubility coefficient, dimensionless, Eq. 11.14
α	Attachment rate, T^{-1}, Eq. 11.41
α_*	Enhanced solubility coefficient, dimensionless
β	Cloud water removal rate, T^{-1}
γ	Rate for direct scavenging of particles, T^{-1}, Eq. 11.43
γ	Empirical constant, dimensionless, Eq. 11.36
Γ	Flux of gas to drop, units/L^2T
ϵ	Retention efficiency, dimensionless
ϵ_{cw}	Cloud water removal efficiency, dimensionless
ϵ_n	Nucleation efficiency, dimensionless
κ	Pollution concentration in rainwater, units/L^3, Eq. 11.14
κ	a/R = interception parameter, dimensionless, Eq. 11.28
λ	Characteristic capture length scale of hydrometeor, L
λ_e	Equilibration length scale, L
ν	Coefficient of kinematic viscosity for air, L^2T
ρ	Mass density, M/L^3
σ	Standard deviation
τ	Particle stopping time, T, see Table 11.2 and definition of Stokes' number
$\tau_{d,w,c,p}$	Characteristic times for pollutant removal by dry, wet, chemical, and physical processes, respectively, T
τ_e	Equilibration time for drop to saturate with gas, T
τ_g	Characteristic time for particle growth by water vapor condensation, T
ψ	Precipitation scavenging rate coefficient, T^{-1}
ψ_r	Rain scavenging rate, T^{-1}
ψ_s	Snow scavenging rate, T^{-1}

χ	Pollutant concentration in air, units/L^3
χ_*	The pollutant concentration in air that would be in equilibrium with bulk concentration in other medium, units/L^3

Subscripts

a	Air, atmosphere
b	Bulk
c	Chemical
cw	Cloud water
d	Dry
g	Gravitational, growth
i	Interface
m	Mass-mean
n	Normal
o	Surface level, overall, upstream
p	Physical, particle
r	Rain, resuspension
s	Snow
t	Terminal
w	Wet, water
z	Vertical component

Other Symbols and Abbreviations

$\tilde{\xi}$	Average of ξ during chosen time interval
ξ'	The (turbulent) fluctuations of ξ about the mean value $\tilde{\xi}$
$\bar{\xi}$	Average of ξ over some variable
$\langle \xi \rangle$	Average of ξ over particle size
$\langle n_\xi \rangle$	*nth* moment average over particle size
$\lvert \xi \rvert$	Magnitude of ξ
mmd	Mass mean diameter
[ ]	Concentration in moles per liter
$(\downarrow +)$	Defining direction for positive component

REFERENCES

Adam, J. R., and R. G. Semonin, 1970, Collection Efficiencies of Raindrops for Submicron Particulates, in *Precipitation Scavenging (1970)*, AEC Symposium Series, Richland, Wash., June 2–4, 1970, Rudolf J. Engelmann and W. George N. Slinn (Coordinators), pp. 151-160, CONF-700601, NTIS.

Aleksandrov, E. L., L. M. Levin, and Y. S. Sedunov, 1967, Condensation Growth of a Solution Drop, *Izv., Atmos. Oceanic Phys.*, 3: 872-880.

Amato, A. J., 1971, *A Mathematical Analysis of the Effects of Wind on Redistribution of Surface Contamination*, USAEC Report WASH-1187, NTIS.

Atkinson, D. G., et al., 1972, Precipitation Scavenging of Tracers Released into Frontal Storms, in *Pacific Northwest Laboratory Annual Report for 1971 to the USAEC Division of Biology and Medicine. Vol. II. Physical Sciences. Part I. Atmospheric Sciences*, C. L. Simpson et al., USAEC Report BNWL-1651 (Pt. 1), pp. 1-15, Battelle Pacific Northwest Laboratories.

Atlas, E., and C. S. Giam, 1981, *Science*, 211: 163-165.

Barrie, L. A., 1978, An Improved Model of Reversible SO_2—Washout by Rain, *Atmos. Environ.*, 12: 407-412.

——, 1981, The Prediction of Rain Acidity and SO_2 Scavenging in Eastern North America, *Atmos. Environ.*, 15: 31-42.

——, S. Beilke, and H. W. Georgii, 1977, SO_2 Removal by Cloud and Fog Drops as Affected by Ammonia and Heavy Metals, in *Precipitation Scavenging (1974)*, ERDA Symposium Series, Champaign, Ill., Oct. 14–18, 1974, Richard G. Semonin and Robert W. Beadle (Coordinators), pp. 151-166, CONF-741003, NTIS.

——, R. M. Hoff, and S. M. Daggupaty, 1981, The Influence of Mid-Latitudinal Pollution Sources on Haze in the Canadian Arctic, *Atmos. Environ.*, 15: 1407-1420.

Baum, S., P. W. Wong, and P. J. Dolan, 1974, *NUCROM: A Model of Rainout from Nuclear Clouds* (Final Report on Contract DNA001-73-C-0148), Stanford Research Institute, Menlo Park, Calif.

Baust, E., 1967, Die Anlagerung von radioaktiven Atomen und Ionen an Aerosolteilchen, *Z. Phys.*, 199: 187-206.

Beal, S. K., 1972, Turbulent Agglomeration of Suspensions, *Aerosol Sci.*, 3: 113-125.

Beard, K. V., 1974, Experimental and Numerical Collision Efficiencies for Submicron Particles Scavenged by Small Raindrops, *J. Atmos. Sci.*, 31: 1595-1603.

——, 1977, Rain Scavenging of Particles by Electrostatic–Inertial Impaction and Brownian Diffusion, in *Precipitation Scavenging (1974)*, ERDA Symposium Series, Champaign, Ill., Oct. 14–18, 1974, Richard G. Semonin and Robert W. Beadle (Coordinators), pp. 183-194, CONF-741003, NTIS.

——, and S. N. Grover, 1974, Numerical Collision Efficiencies for Small Raindrops Colliding with Micron Size Particles, *J. Atmos. Sci.*, 31: 543-550.

——, and H. R. Pruppacher, 1969, A Determination of the Terminal Velocity and Drag of Small Water Drops by Means of a Wind Tunnel, *J. Atmos. Sci.*, 26: 1066-1072.

Berg, T. G. O., 1970, Collection Efficiency in Washout by Rain, in *Precipitation Scavenging (1970)*, AEC Symposium Series, Richland, Wash., June 2–4, 1970, Rudolf J. Engelmann and W. George N. Slinn (Coordinators), pp. 169-186, CONF-700601, NTIS.

——, T. A. Gaukler, and L. A. Squier, 1965, Kinetics of Wetting in Washout of Dust, in *Radioactive Fallout from Nuclear Weapons Tests*, AEC Symposium Series, Germantown, Md., Nov. 3–6, 1964, Alfred W. Klement, Jr. (Ed.). pp. 281-300, CONF-765, NTIS.

Best, A. C., 1950, Empirical Formulae for the Terminal Velocity of Water Drops Falling Through the Atmosphere, *Q. J. R. Meteorol. Soc.*, 76: 302-310.

Bilonick, R. A., and D. G. Nichols, 1983, Temporal Variations in Acid Precipitation over New York State—What the 1965–1979 USGS Data Reveal, *Atmos. Environ.*, 17: 1063-1072.

Bird, R. B., W. E. Stewart, and E. N. Lightfoot, 1964, *Transport Phenomena*, John Wiley & Sons, New York.

Bolin, B., and H. Rodhe, 1973, A Note on the Concepts of Age Distribution and Transit Time in Natural Reservoirs, *Tellus*, 25: 305-308.

Bricard, J., 1965, Action of Radioactivity and of Pollution upon Parameters of Atmospheric Electricity, in *Problems of Atmospheric and Space Electricity*, S. C. Coroniti (Ed.), Elsevier, Amsterdam.

Brock, J. R., 1970, Attachment of Trace Substances on Atmospheric Aerosols, in *Precipitation Scavenging (1970)*, AEC Symposium Series, Richland, Wash., June 2–4, 1970, Rudolf J. Engelmann and W. George N. Slinn (Coordinators), pp. 373-380, CONF-700601, NTIS.

Burtsev, I. I., L. V. Burtseva, and S. G. Malakhov, 1970, Washout Characteristics of a ^{32}P Aerosol Injected into a Cloud, in *Atmospheric Scavenging of Radioisotopes*, proceedings of a conference in Palanga, USSR, June 7–9, 1966, B. Styra, Ch. A. Garbalyauskas, and V. Yu. Luyanas (Eds.), pp. 242-250, translated from *Issledovanie Protsessov Samoochischeniya Atmosfery ot Radioaktivnykh Izotopov* by D. Lederman; available as TT-69-55099 from the National Technical Information Service.

Calder, K. L., 1961, Atmospheric Diffusion of Particulate Material, Considered as a Boundary Value Problem, *J. Meteorol.*, 18: 413-416.

Caporaloni, M., F. Tampieri, F. Trombetti, and O. Vittori, 1975, Transfer of Particles in Nonisotropic Air Turbulence, *J. Atmos. Sci.*, 32: 565-569.

Cataneo, R., 1973, Surface Measurements of Condensation Nuclei and Raindrop Distributions, in *Summary Report of METROMEX*

Studies, F. A. Huff (Ed.), Report ISWS-73-R174, pp. 84-89, Illinois State Water Survey, Urbana, Ill.

Cawse, P. A., 1974, *A Survey of Atmospheric Trace Elements in the United Kingdom,* Report AERE-R-7669, Atomic Energy Research Establishment, Harwell, England. See also Gatz (1975a).

Chamberlain, A. C., 1953, *Aspects of Travel and Deposition of Aerosols and Vapour Clouds,* Report AERE-R-1261, Atomic Energy Research Establishment, Harwell, England.

——, 1960, Aspects of the Deposition of Radioactive and Other Pollutants, *Int. J. Air Pollut.,* 3: 63-88.

Charlson, R. J., F. F. Pueschel, and N. C. Ahlquist, 1969, The Integrating Nephelometer as an Instrument for Studying Atmospheric Aerosols, in *Proceedings of the Seventh International Conference on Condensation and Ice Nuclei,* Prague and Vienna, pp. 292-298.

Chen, C. S., 1974, Evaluation of the Vapor Pressure over an Aerosol Particle, *J. Atmos. Sci.,* 31: 847-849.

Cogbill, C. V., 1976, The History and Character of Acid Precipitation in Eastern North America, in *Proceedings of the First International Symposium on Acid Precipitation and the Forest Ecosystem,* May 12–15, 1975, Columbus, Ohio, L. S. Dochinger and T. A. Seliga (Eds.), USDA Forest Service General Technical Report NE-23, pp. 363-370, U. S. Department of Agriculture, NTIS.

——, and G. E. Likens, 1974, Acid Precipitation in the Northeastern United States, *Water Resources Res.,* 10: 1133-1137.

Corrsin, S., 1974, Limitations of Gradient Transport Models in Random Walks and in Turbulence, *Adv. Geophys.,* 18A: 25-60.

Dana, M. T., 1970, Scavenging of Soluble Dye Particles by Rain, in *Precipitation Scavenging (1970),* AEC Symposium Series, Richland, Wash., June 2–4, 1970, Rudolf J. Engelmann and W. George N. Slinn (Coordinators), pp. 137-147, CONF-700601, NTIS.

——, 1971, Washout of Soluble Dye Particles, in *Pacific Northwest Laboratory Annual Report for 1970 to the USAEC Division of Biology and Medicine. Vol. II. Physical Sciences. Part I. Atmospheric Sciences,* C. L. Simpson et al., USAEC Report BNWL-1551 (Vol. II)(Pt. 1), pp. 63-67, Battelle Pacific Northwest Laboratories.

——, 1972, Washout of Rhodamine Dye Particles, in *Meteorology and Atomic Energy—1968,* David H. Slade (Ed.), USAEC Report TID-24190, pp. 36-44, Environmental Science Services Administration, NTIS.

——, 1978, Battelle Pacific Northwest Laboratory, personal communication.

——, and J. M. Hales, 1976, Statistical Aspects of the Washout of Polydisperse Aerosols, *Atmos. Environ.,* 10: 45-50.

——, J. M. Hales, W. G. N. Slinn, and M. A. Wolf, 1973, *Natural Precipitation Washout of Sulfur Compounds from Plumes,* Report EPA-R3-73-047, Ecological Research Series, Environmental Protection Agency.

——, and M. A. Wolf, 1968, Experimental Studies in Precipitation Scavenging, in *Pacific Northwest Laboratory Annual Report for 1967 to the USAEC Division of Biology and Medicine. Vol. II. Physical Sciences. Part 3. Atmospheric Sciences,* C. L. Simpson et al., USAEC Report BNWL-715-3, pp. 128-140, Battelle Pacific Northwest Laboratories.

——, M. A. Wolf, and L. A. DuPlessis, 1970, Field Experiments in Precipitation Scavenging, in *Pacific Northwest Laboratory Annual Report for 1969 to the USAEC Division of Biology and Medicine. Vol. II. Physical Sciences. Part I. Atmospheric Sciences,* C. L. Simpson et al., USAEC Report BNWL-1307 (Pt. 1), pp. 77-81, Battelle Pacific Northwest Laboratories.

Danielsen, E. F., 1974, Review of Trajectory Methods, *Adv. Geophys.,* 18B: 73-94.

Daum, P. H., S. E. Schwartz, and L. Newman, 1983, Studies of the Gas- and Aqueous-Phase Composition of Stratiform Clouds, in *Precipitation Scavenging, Dry Deposition, and Resuspension,* Proceedings of the Chamberlain Conference, Vol. 1, Santa Monica, Calif., No. 29–Dec. 3, 1982, H. R. Pruppacher, R. G. Semonin, and W. G. N. Slinn (Coordinators), Elsevier, New York.

Davis, W. E., 1970, An Interpretation of the In-Cloud Scavenging of Particulates, in *Precipitation Scavenging (1970),* AEC Symposium Series, Richland, Wash., June 2–4, 1970, Rudolf J. Engelmann and W. George N. Slinn (Coordinators), pp. 107-110, CONF-700601, NTIS.

——, 1972, A Model for In-Cloud Scavenging of Cosmogenic Radionuclides, *J. Geophys. Res.,* 77: 2159-2165.

——, and L. L. Wendell, 1976, Some Effects of Isentropic Vertical Motion Simulation in a Regional Scale Quasi Lagrangian Air Quality Model, in *Proceedings of the Third Symposium on Atmospheric Turbulence, Diffusion, and Air Quality,* Oct. 19–22, 1976, Raleigh, N. C., pp. 403-406, American Meteorological Society, Boston, Mass.

——, and J. A. Young, 1977, Results of In-Cloud Tracer Releases in Frontal Storms, in *Precipitation Scavenging (1974),* ERDA Symposium Series, Champaign, Ill., Oct. 14–18, 1974, Richard G. Semonin and Robert W. Beadle (Coordinators), pp. 337-361, CONF-741003, NTIS.

——, J. A. Young, and J. M. Thorp, 1974, Progress Report on In-Cloud Scavenging in Frontal Storms, in *Pacific Northwest Laboratory Annual Report for 1973 to the USAEC Division of Biomedical and Environmental Research. Part 3. Atmospheric Sciences,* C. L. Simpson et al., USAEC Report BNWL-1850(Pt.3), pp. 109-111, Battelle Pacific Northwest Laboratories.

Dingle, A. N., 1975, Response to "Wet Deposition Estimation Using Scavenging Ratios" by D. F. Gatz, in *Proceedings of the First Speciality Symposium on Atmospheric Contribution to the Chemistry of Lake Waters,* Sept. 28–Oct. 1, 1975, International Association for Great Lakes Research.

——, 1977, Scavenging and Dispersal of Tracer by a Self-Propagating Convective Shower System, in *Precipitation Scavenging (1974),* ERDA Symposium Series, Champaign, Ill., Oct. 14–18, 1974, Richard G. Semonin and Robert W. Beadle (Coordinators), pp. 395-424, CONF-741003, NTIS.

——, 1978, The University of Michigan, personal communication.

——, and K. R. Hardy, 1962, The Description of Rain by Means of Sequential Raindrop-Size Distributions, *Q. J. R. Meteorol. Soc.,* 88: 301-314.

——, D. F. Gatz, and J. W. Winchester, 1969, A Pilot Experiment Using Indium as Tracer in a Convective Storm, *J. Appl. Meteorol.,* 8: 236-240.

——, and Y. Lee, 1973, An Analysis of In-Cloud Scavenging, *J. Appl. Meteorol.,* 12: 1295-1302.

Dochinger, L. S., and T. A. Seliga, 1976 (Eds.), *Proceedings of the First International Symposium on Acid Precipitation and the Forest Ecosystem,* May 12–15, 1975, Columbus, Ohio, General Technical Report NE-23, USDA Forest Service, NTIS; many of the papers were also published in 1977 issues of *J. Water, Air Soil Pollut.*

Eigen, M., and L. DeMaeyer, 1955, Untersuchungen uber die Kinetik der Neutralisation, *Z. Elektrochem.,* 59: 986-993.

El-Shimi, A., and E. D. Goddard, 1974, Wettability of Some Low Energy Surfaces 1. Air/Liquid/Solid Interface, *J. Colloid Interface Sci.,* 48: 242-248.

Engelmann, R. J., 1963, A Surface Energy Model in the Raindrop Wetting of Scavenged Particles, USAEC Report HW-SA-3545, Hanford Laboratories, Richland, Wash. See also, *Rain Scavenging of Particles,* USAEC Report HW-79382, Hanford Atomic Products Operation, NTIS.

——, 1965, Rain Scavenging of Zinc Sulphide Particles, *J. Atmos. Sci.,* 22: 719-724.

——, 1968, The Calculation of Precipitation Scavenging, in *Meteorology and Atomic Energy—1968,* D. H. Slade (Ed.), USAEC Report TID-24190, pp. 208-221, Environmental Science Services Administration, NTIS.

——, 1970, Scavenging Predictions Using Ratios of Concentrations in Air and Precipitation, in *Precipitation Scavenging (1970),* AEC Symposium Series, Richland, Wash., June 2–4, 1970, Rudolf J. Engelmann and W. George N. Slinn (Coordinators), pp. 475-486, CONF-700601, NTIS.

——, et al., 1966, Washout Coefficients for Selected Gases and Particulates, USAEC Report BNWL-SA-657, Battelle Pacific Northwest Laboratories; see also related USAEC Reports BNWL-SA-155 and BNWL-SA-798.

——, and D. I. Hagen, 1967, Precipitation Scavenging Studies, in *Pacific Northwest Laboratory Annual Report for 1966 to the USAEC Division of Biology and Medicine. Vol. II. Physical Sciences. Part I. Atmospheric Sciences,* D. W. Pearce and M. R. Compton, USAEC Report BNWL-481 (Pt. 1), pp. 43-52, Battelle Pacific Northwest Laboratories, NTIS.

——, and W. G. N. Slinn, 1970 (Coords.), in *Precipitation Scavenging (1970),* AEC Symposium Series, Richland, Wash., June 2–4, 1970, Rudolf J. Engelmann and W. George N. Slinn (Coordinators), CONF-700601, NTIS.

Esmen, N. A., 1973, Atmospheric Scavenging of Soluble and Insoluble Matter, *Sci. Total Environ.,* 2: 181-189.

Facy, L., 1962, Radioactive Precipitation and Fallout, *Nuclear Radiation in Geophysics,* H. Israel and A. Krebs (Eds.), pp. 202-204, Springer-Verlag, Berlin.

Ferron, G. A., 1977, The Size of Soluble Aerosol Particles as a Function of the Humidity of the Air. Application to the Human Respiratory Tract, *J. Aerosol Sci.,* 8: 251-267.

Fitzgerald, J. W., 1975, Approximate Formulas for the Equilibrium Size of an Aerosol Particle as a Function of Its Dry Size and Composition and the Ambient Relative Humidity, *J. Appl. Meteorol.,* 14: 1044-1049.

Foote, G. B., and J. C. Fankhauser, 1973, Airflow and Moisture Budget Beneath a Northeast Colorado Hailstorm, *J. Appl. Meteorol.,* 12: 1330-1353.

Fuchs, N. A., 1964, *The Mechanics of Aerosols,* Pergamon Press, N. Y.

Gambell, A. W., and D. W. Fisher, 1966, *Chemical Composition of Rainfall in Eastern North Carolina and Southern Virginia,* Geological Survey Water Supply Paper 1535-K, NTIS.

Garland, J. A., 1978, Dry and Wet Removal of Sulphur from the Atmosphere, *Atmos. Environ.,* 12: 349-362.

Gatz, D. F., 1966, *Deposition of Atmospheric Particulate Matter by Convective Storms,* Ph.D. Thesis and USAEC Report COO-1407-6, University of Michigan.

——, 1975a, Wet Deposition Estimates Using Scavenging Ratios, in *Proceedings of the First Specialty Symposium on Atmospheric Contributions to the Chemistry of Lake Water,* Sept. 28–Oct. 1, 1975, pp. 21-29, International Association for Great Lakes Research.

——, 1975b, Pollutant Aerosol Deposition into Southern Lake Michigan, *J. Water, Air Soil Pollut.,* 5: 239-251.

——, 1977a, A Review of Chemical Tracer Experiments on Precipitation Systems, *Atmos. Environment,* 11: 945-954.

——, 1977b, Scavenging Ratio Measurements in METROMEX, in *Precipitation Scavenging (1974),* ERDA Symposium Series, Champaign, Ill., Oct. 14–18, 1974, Richard G. Semonin and Robert W. Beadle (Coordinators), pp. 71-87, CONF-741003, NTIS.

Gedeonov, L. I., Z. G. Gritchenko, F. M. Flegontov, and M. I. Zhilkina, 1970, Coefficient of Radioactive-Aerosol Concentration in Atmospheric Precipitation, in *Atmospheric Scavenging of Radioisotopes,* proceedings of a conference in Palanga, USSR, June 7–9, 1966, B. Styra, Ch. A. Garbalyauskas, and V. Yu. Luyanas (Eds.), pp. 150-156, translated from *Issledovanie Protsessov Samoochischeniya Atmosfery ot Radioaktivnykh Izotopov* by D. Lederman; available as TT-69-55099 from the National Technical Information Service.

Greenfield, S. M., 1957, Rain Scavenging of Radioactive Particulate Matter from the Atmosphere, *J. Meteorol.,* 14: 115-123.

Grover, S. N., H. R. Pruppacher, and A. E. Hamielec, 1977, A Numerical Determination of the Efficiency with Which Spherical Aerosol Particles Collide with Spherical Water Drops Due to Inertial Impaction and Phoretic and Electrical Forces, *J. Atmos. Sci.,* 34: 1655-1663.

Hagen, D. I., 1967, Precipitation Scavenging of Sub-Micron Particles. Part B—Comparison of Theory with Field Experiments, in *Proceedings of the USAEC Meteorological Information Meeting,* Chalk River, Ontario, Sept. 11–14, 1967, C. A. Mawson (Ed.), Report AECL-2787, pp. 541-552, Chalk River Nuclear Laboratories.

Hales, J. M., 1972, Fundamentals of the Theory of Gas Scavenging by Rain, *Atmos. Environ.,* 6: 635-659.

——, 1978a, Wet Removal of Sulfur Compounds from the Atmosphere, *Atmos. Environ.,* 12(1-3): 389-399.

——, 1978b, Battelle Pacific Northwest Laboratory, personal communication.

——, M. T. Dana, and M. A. Wolf, 1977, Application of the EPAEC Scavenging Model to Calculations for Industrial Plumes, in *Precipitation Scavenging (1974),* ERDA Symposium Series, Champaign, Ill., Oct. 14–18, 1974, Richard G. Semonin and Robert W. Beadle (Coordinators), pp. 625-638, CONF-741003, NTIS.

——, and S. L. Sutter, 1973, Solubility of Sulfur Dioxide in Water at Low Concentrations, *Atmos. Environ.,* 7: 997-1001.

Hamrud, M., H. Rodhe, and J. Grandell, 1981, A Numerical Comparison Between Lagrangian and Eulerian Rainfall, *Tellus,* 33: 235-241.

Hansen, D. A., and G. M. Hidy, 1982, Review of Questions Regarding Rain Acidity Data, *Atmos. Environ.,* 16: 2107-2126.

Havlovic, V., and J. O. Snihs, 1971, Distribution of Natural Radioactivity on the Environmental Submicronic Aerosols, *Tellus,* 23: 441-454.

Heffter, J. L., and G. A. Ferber, 1975, *A Regional to Continental Scale Transport, Diffusion and Deposition Model, Part II: Diffusion-Deposition Models,* Technical Memorandum ERL-ARL-50, Air Resources Laboratories, National Oceanic and Atmospheric Administration, NTIS.

Hegg, D. A., and P. V. Hobbs, 1982, Measurements of Sulfate Production in Natural Clouds, *Atmos. Environ.,* 16: 2663-2668.

Heines, T. S., and L. K. Peters, 1974, The Effect of Ground Level Absorption on the Dispersion of Pollutants in the Atmosphere, *Atmos. Environ.,* 8: 1143-1153.

Heisler, S. L., S. K. Friedlander, and R. B. Husar, 1973, The Relationship of Smog Aerosol Size and Chemical Element Distributions to Source Characteristics, *Atmos. Environ.,* 7: 633-649.

Herne, H., 1960, The Classical Computation of the Aerodynamic Capture of Particles by Spheres, in *Aerodynamic Capture of Particles,* pp. 26-34, E. G. Richardson (Ed.), Pergamon Press, New York.

Hicks, B. B., 1966, Nucleation and the Wet Removal of Fallout, *J. Appl. Meteorol.,* 5: 169-174.

Hidy, G. M., 1970, Theory of Diffusive and Impactive Scavenging, in *Precipitation Scavenging (1970),* AEC Symposium Series, Richland, Wash., June 2–4, 1970, Rudolf J. Engelmann and W. George N. Slinn (Coordinators), pp. 355-372, CONF-700601, NTIS.

——, and J. R. Brock, 1972, *Dynamics of Aerocolloidal Systems,* Pergamon Press, New York.

——, R. Countess, H. Collins, and S. Machiraju, 1983, *Precipitation-Scavenging Chemistry for Sulfate and Nitrate from the SURE and Related Data,* Final Report, DOE Report EPRA-EA-1914, Vol. 2, Electric Power Research Institute, NTIS.

Hill, F. B., and J. F. Adamowicz, 1977, A Model for Rain Composition and the Washout of Sulfur Dioxide, *Atmos. Environ.,* 11: 917-928.

Hindman, E. E., P. V. Hobbs, and L. F. Radke, 1975, Airborne Investigations of Aerosol Particles from a Paper Mill, Paper 75-32.9 in *Proceedings of the 68th Annual Meeting of the Air Pollution Control Association,* Boston, Mass., June 15–20, 1975; see also P. V. Hobbs, L. F. Radke, and E. E. Hindman, 1976, An Integrated Airborne Particle-Measuring Facility and Its Preliminary Use in Atmospheric Aerosol Studies, *J. Aerosol Sci.,* 7: 195-211.

Holl, W., and R. Mühleisen, 1955, On the Equilibrium of Ionization in Air Containing Nuclei, *Geophys. Pure Appl.,* 31: 115-118.

Howell, W. E., 1949, The Growth of Cloud Droplets in Uniformly Cooled Air, *J. Meteorol.,* 6: 134-149.

Hughes, E. A. M., 1971, *The Chemical Statics and Kinetics of Solutions,* Academic Press, New York.

Jesson, J. P., P. Meakin, and L. C. Glasgow, 1977, The Fluorocarbon–Ozone Theory-II. Tropospheric Lifetimes—An Estimate of the Tropospheric Lifetime of CCl_3F, *Atmos. Environment,* 11: 499-508.

Junge, C. E., 1952, Die Konstitution des Atmosphärischen Aerosols, *Ann. Meteorol.,* 5 (Beiheft): 1-55.

——, 1958, The Distribution of Ammonia and Nitrate in Rainwater over the United States, *Am. Geophys. Union, Trans.,* 39: 241-248.

——, 1962, Radioactive Aerosols, in *Nuclear Radiation in Geophysics,* pp. 169-201, H. Israël and A. Krebs (Eds.), Springer-Verlag, Berlin.

——, 1963, *Atmospheric Chemistry and Radioactivity,* Academic Press, New York.

——, 1964, *The Modification of Aerosol Size Distributions in the Atmosphere,* Final Technical Report on Contract No. DA 91-591-EVC 2979, Report AD445 873, U. S. Department of the Army, European Research Office, NTIS.

——, 1972, The Cycle of Atmospheric Trace Gases—Natural and Man Made, *Quart. J. Roy. Meteorol. Soc.,* 98: 711-729.

——, 1974, Residence Time and Variability of Tropospheric Trace Gases, *Tellus,* 26: 477-488.

——, 1975, Processes Responsible for the Trace Content in Precipitation, in *Proceedings of the Snow and Ice Symposium,* Grenoble, UGGI Assembly.

——, 1977, Basic Considerations About Trace Constituents in the Atmosphere as Related to the Fate of Global Pollutants, presented at the 169th ACS National Meeting, Philadelphia, Apr. 7–11, 1975, Symposium on Fate of Pollutants in the Air and Water Environment; published in *Advances in Environmental Science and Technology,* I. H. Suffet (Ed.), Vol. B, Pt. 1, pp. 7-25, Wiley-Interscience, New York, 1977.

——, and P. E. Gustafson, 1956, Precipitation Sampling for Chemical Analysis, *Bull. Am. Meteorol. Soc.,* 37: 244-245.

——, and R. Werby, 1958, The Concentration of Chloride, Sodium, Potassium, Calcium, and Sulphate in Rain Water over the United States, *J. Meteorol.,* 15: 417-425.

Kallend, A. S., A. R. W. Marsh, J. H. Pickles, and M. V. Proctor, 1983, Acidity of Rain in Europe, *Atmos. Environ.,* 17: 127-137.

Kerker, M., and V. Hampl, 1974, Scavenging of Aerosol Particles by a Falling Drop and Calculation of Washout Coefficients, *J. Atmos. Sci.,* 31: 1368-1376.

Kerrigan, T. C., and J. Rosinski, 1977, Scavenging of Aerosol Particles in Severe Convective Storms, in *Precipitation Scavenging (1974),* ERDA Symposium Series, Champaign, Ill., Oct. 14–18, 1974, Richard G. Semonin and Robert W. Beadle (Coordinators), pp. 466-493, CONF-741003, NTIS.

Kessler, E., 1969, On the Distribution and Continuity of Water Substance in Atmospheric Circulations, *Meteorological Monographs,* Vol. 10, No. 32, American Meteorological Society, Boston, Mass.

Knutson, E. O., and J. D. Stockham, 1977, Aerosol Scavenging by Snow: Comparisons of Single-Flake and Entire-Snowfall Results, in *Precipitation Scavenging (1974),* ERDA Symposium Series, Champaign, Ill., Oct. 14–18, 1974, Richard G. Semonin and Robert W. Beadle (Coordinators), pp. 195-207, CONF-741003, NTIS.

Koenig, L. R., 1977, Cloud Glaciation Characteristics Important to Parameterizing Scavenging, in *Precipitation Scavenging (1974),* ERDA Symposium Series, Champaign, Ill., Oct. 14–18, 1974, Richard G. Semonin and Robert W. Beadle (Coordinators), pp. 672-694, CONF-741003, NTIS.

Kraemer, H. F., and H. F. Johnston, 1955, Collection of Aerosol Particles in the Presence of Electric Fields, *Ind. Eng. Chem.,* 47: 2426-2434.

Lai, K. Y., N. Dayan, and M. Kerker, 1978, Scavenging of Aerosol Particles by a Falling Water Drop, *J. Atmos. Sci.,* 35(4): 674-682.

Lambert, G., J. Sanak, and G. Polian, 1983, Mean Residence Time of the Submicrometer Aerosols in the Global Troposphere, in *Precipitaion Scavenging, Dry Deposition, and Resuspension,* Proceedings of the Chamberlain Conference, Santa Monica, Calif., Nov. 29–Dec. 3, 1982, H. R. Pruppacher, R. G. Semonin, and W. G. N. Slinn (Coordinators), Elsevier, New York.

Lange, R., and J. B. Knox, 1977, Adaptation of a Three-Dimensional Atmospheric Transport-Diffusion Model to Rainout Assessments, in *Precipitation Scavenging (1974),* ERDA Symposium Series, Champaign, Ill., Oct. 14–18, 1974, Richard G. Semonin and Robert W. Beadle (Coordinators), pp. 732-758, CONF-741003, NTIS.

Lassen, L., 1961, Die Anlagerung von Zerfallsprodukten der natürlichen Emanationen an elektrisch geladene Aerosole, *Z. Phys.,* 163: 363-376; also see Junge, 1963, *Atmospheric Chemistry and Radioactivity,* pp. 222-226, Academic Press, New York.

——, and G. Rau, 1960, Die Anlagerung radioaktiver Atome an Aerosole (Schwebestoffe) im Grössenbereich 0.7–5 μm (Radius), *Z. Physik,* 161: 339-345; also see Junge, 1963, *Atmospheric Chemistry and Radioactivity,* pp. 222-226, Academic Press, New York.

LeClair, B. P., A. E. Hamielec, H. R. Pruppacher, and W. D. Hall, 1972, A Theoretical and Experimental Study of the Internal Circulation in Water Drops Falling at Terminal Velocity in Air, *J. Atmos. Sci.,* 29: 728-740.

Lee, Y., and A. N. Dingle, 1974, *Numerical Models for Precipitation Scavenging,* COO-1407-52, Department of Atmospheric and Oceanic Science, University of Michigan, NTIS.

Levin, L. M., and Yu. S. Sedunov, 1967, Kinetics of the Formation of a Cloud Spectrum, *Izv., Atmos. Oceanic Physics,* 3: 419-433.

Liljestrand, H. M., and J. J. Morgan, 1979, Error Analysis Applied to Indirect Methods for Precipitation Acidity, *Tellus,* 31: 421-431.

Liss, P. S., and P. G. Slater, 1974, Flux of Gases Across the Air–Sea Interface, *Nature,* 247: 181-184.

——, and W. G. N. Slinn (Eds.), 1983, *Air–Sea Exchange of Gases and Particles,* NATO Advanced Study Institute Series, D. Reidel Publishing Co. (Kluwer Acad. Pub., Higham, Mass.) Dordrect, Holland.

Lodge, J. P. Jr., K. C. Hill, J. B. Pate, E. Lorange, W. Basbergill, A. L. Lazrus, and G. S. Swanson, 1968, *Chemistry of United States Precipitation,* National Center for Atmospheric Research, Boulder, Colo.

Maahs, H. G., 1983, Measurements of the Oxidation Rate of Sulfur (IV) by Ozone in Aqueous Solution and Their Relevance to SO_2 Conversion in Nonurban Tropospheric Clouds, *Atmos. Environ.,* 17: 341-345.

McCulley, C. R., et al., 1956, Scavenging Action of Rain on Airborne Particulate Matter, *Ind. Eng. Chem.,* 48: 1512-1516.

McDonald, J. E., 1963, Rain Washout of Partially Wettable Insoluble Particles, *J. Geophys. Res.,* 68: 4993-5003.

Machta, L., 1977, Vertical Distribution of Water for Precipitation in a Zonally Averaged Model, in *Precipitation Scavenging (1974),* ERDA Symposium Series, Champaign, Ill., Oct. 14–18, 1974, Richard G. Semonin and Robert W. Beadle (Coordinators), pp. 662-671, CONF-741003, NTIS.

Mackay, D., and W. Y. Shiu, 1981, A Critical Review of Henry's Law Constants for Chemicals of Environmental Interest, *J. Phys. Chem. Ref. Data,* 10: 1175-1199.

McKay, H. A. C., 1971, The Atmospheric Oxidation of Sulfur Dioxide in Water Droplets in the Presence of Ammonia, *Atmos. Environ.,* 5: 7-14.

Makhon'ko, K. P., 1966, Spontaneous Removal of Radioactive Dust from the Lower Troposphere (Review), *Izv., Atmos. and Oceanic Phys.,* 2: 304-314.

——, A. S. Avramenko, and E. P. Makhon'ko, 1970, Washout of Radioactive Isotopes and Chemical Compounds from the Atmosphere, in *Atmospheric Scavenging of Radioisotopes,* proceedings of a conference in Palanga, USSR, June 7–9, 1966, B. Styra, Ch. A. Garbalyauskas, and V. Yu. Luyanas (Eds.), pp. 174-184, translated from *Issledovanie Protsessov Samoochischeniya Atmosfery ot Radioaktivnykh Izotopov* by D. Lederman; available as TT-69-55099 from the National Technical Information Service.

Martell, E. A., and H. E. Moore, 1974, Tropospheric Aerosol Residence Times—A Critical Review, *J. Rech. Atmos.,* 8: 903-910.

Martin, L. R., and D. E. Damschen, 1981, Aqueous Oxidation of Sulfur Dioxide by Hydrogen Peroxide at Low pH, *Atmos. Environ.,* 15: 1651-1621.

Marwitz, J. D., 1972, Precipitation Efficiency of Thunderstorms on the High Plains, *Preprint of the Third Conference on Weather Modification,* pp. 245-247, Rapid City, S. D., American Meteorological Society, Boston, Mass.

Mason, B. J., 1971, *The Physics of Clouds, 2nd ed.,* Clarendon Press, Oxford.

May, F. G., 1958, The Washout of Lycopodium Spores by Rain, *Quart. J. Roy. Meteorol. Soc.,* 84: 451-458.

Miller, C. A., and E. Ruckenstein, 1974, The Origin of Flow During Wetting of Solids, *J. Colloid Interface Sci.,* 48: 368-373.

Miller, J. M., 1977, A Statistical Evaluation of the U. S. Precipitation Chemistry Network, in *Precipitation Scavenging (1974),* ERDA Symposium Series, Champaign, Ill., Oct. 14–18, 1974, Richard G. Semonin and Robert W. Beadle (Coordinators), pp. 639-661, CONF-741003, NTIS.

——, and R. G. de Pena, 1972, Contribution of Scavenged Sulfur Dioxide to the Sulfate Content of Rain Water, *J. Geophys. Res.,* 77: 5905-5916.

Mohnen, V., and K. Stierstadt, 1963, Die Verteilung der natürlichen Radioaktivität auf das Grössenspektrum des natürlichen Aerosols, *Z. Phys.,* 173: 276-293.

Möller, D., 1980, Kinetic Model of Atmospheric SO_2 Oxidation Based on Published Data, *Atmos. Environ.,* 14: 1067-1076.

Mordy, W., 1959, Computations of the Growth by Condensation on a Population of Cloud Droplets, *Tellus,* 11: 16-44.

Muller, E. F., and J. R. Kramer, 1977, Precipitation Scavenging in Central and Northern Ontario, in *Precipitation Scavenging (1974),* ERDA Symposium Series, Champaign, Ill., Oct. 14–18, 1974, Richard G. Semonin and Robert W. Beadle (Coordinators), pp. 590-601, CONF-741003, NTIS.

Murakami, M., C. Hiramatsu, and C. Magono, 1981, Observation of Aerosol Scavenging by Falling Snow Crystals at Two Sites of Different Heights, *J. Meteorol. Soc. Jpn.,* 59; 763-771.

Norment, H. G., 1977, A Precipitation Scavenging Model for Nuclear Fallout Research, in *Precipitation Scavenging (1974),* ERDA Symposium Series, Champaign, Ill., Oct. 14–18, 1974, Richard G. Semonin and Robert W. Beadle (Coordinators), pp. 695-731, CONF-741003, NTIS.

Odén, S., 1976, The Acidity Problem—An Outline of Concepts, in *Proceedings of the First International Symposium on Acid Precipitation and the Forest Ecosystem,* pp. 1-36, May 12–15, 1975, Columbus, Ohio, General Technical Report NE-23, USDA Forest Service, NTIS; many of the papers were also published in 1977 issues of *J. Water, Air Soil Pollut.*

Oppenheimer, M., 1983, The Relationship of Sulfur Emissions to Sulfate in Precipitation, *Atmos. Environ.,* 17: 451-460.

Overrein, L. N., 1976, A Presentation of the Norwegian Project "Acid Precipitation—Effects on Forest and Fish," in *Proceedings of the First International Symposium on Acid Precipitation and the Forest Ecosystem,* pp. 37-42, May 12–15, 1975, Columbus, Ohio, General Technical Report NE-23, USDA Forest Service, NTIS; many of the papers were also published in 1977 issues of *J. Water, Air Soil Pollut.*

Pankow, J. F., L. M. Isabelle, and W. E. Asher, 1983, *Trace Organic Compounds in Rain,* submitted for publication; reprint available from the Oregon Graduate Center, 19600 N.W. Walker Rd., Beaverton, Oreg. 97006.

Pearson, F. J., Jr., and D. W. Fisher, 1971, *Chemical Composition of Atmospheric Precipitation in the Northeastern United States,* Geological Survey Water Supply Paper 1535-P, NTIS.

Pellett, G. L., 1977, Washout of HCl and Application to Solid Rocket Exhaust Clouds, in *Precipitation Scavenging (1974),* ERDA Symposium Series, Champaign, Ill., Oct. 14–18, 1974, Richard G. Semonin and Robert W. Beadle (Coordinators), pp. 437-465, CONF-741003, NTIS.

Pemberton, C. S., 1960, Scavenging Action of Rain on Non-Wettable Particulate Matter Suspended in the Atmosphere, in *Aerodynamic Capture of Particles,* pp. 168-178, E. G. Richardson (Ed.), Pergamon Press, New York.

Penkett, S. A., B. M. R. Jones, K. A. Brice, and A. E. J. Eggleton, 1979, The Importance of Atmospheric Ozone and Hydrogen Peroxide in Oxidising Sulfur Dioxide in Cloud and Rainwater, *Atmos. Environ.,* 13: 123-137.

Peterson, K. R., and T. V. Crawford, 1970, Precipitation Scavenging in a Large-Cloud Diffusion Code, in *Precipitation Scavenging (1970),* AEC Symposium Series, Richland, Wash., June 2–4, 1970, Rudolf J. Engelmann and W. George N. Slinn (Coordinators), pp. 425-431, CONF-700601, NTIS.

Pich, J., S. K. Friedlander, and F. S. Lai, 1970, The Self-Preserving Particle Size Distribution for Coagulation by Brownian Motion III. Smoluchowski Coagulation and Simultaneous Maxwellian Condensation, *J. Aerosol Sci.,* 1: 115-126.

Pitter, R. L., 1977, A Numerical Study of Scavenging by Ice Crystals, Final Report on NSF Grant ATM75-22640, Oregon Graduate Center, Beaverton, Oreg., NTIS.

Postma, A. K., 1970, Effect of Solubilities of Gases on Their Scavenging by Raindrops, in *Precipitation Scavenging (1970),* AEC Symposium Series, Richland, Wash., June 2–4, 1970, Rudolf J. Engelmann and W. George N. Slinn (Coordinators), pp. 247-260, CONF-700601, NTIS.

Pruppacher, H. R., R. G. Semonin, and W. G. N. Slinn (Coordinators), 1983; *Precipitation Scavenging, Dry Deposition, and Resuspension,* Proceedings of the Chamberlain Conference, Santa Monica, Calif., Nov. 29–Dec. 3, 1982, Elsevier, New York.

Radke, L. F., M. W. Eltgroth, and P. V. Hobbs, 1978, Precipitation Scavenging of Aerosol Particles, in *Preprints of the Conference on Cloud Physics and Atmospheric Electricity,* Issaquah, Wash., July 31–August 4, 1978, pp. 44-48, American Meteorology Society, Boston, Mass.

Rasmussen, R. A., D. Pierotti, T. F. Biddleman, and P. Wilkniss, 1978, Halogenated Hydrocarbons, in *Tropospheric Transport of Pollutants and Other Substances to the Ocean,* J. M. Prospero (Ed.), National Academy of Sciences, Washington, D. C.

Richards, L. W., J. A. Anderson, D. L. Blumenthal, J. A. McDonald, G. L. Kok, and A. L. Lazrus, 1983, Hydrogen Peroxide and Sulfur (IV) in Los Angeles Cloud Water, *Atmos. Environ.,* 17; 911-914.

Ritchie, L. T., W. D. Brown, and J. R. Wayland, 1976, *Effects of Rainstorms and Runoff on Consequences of Nuclear Reactor Accidents,* ERDA Report SAND 76-0429, Sandia Laboratories, NTIS.

Rodhe, H., 1980, Estimate of Wet Deposition of Pollutants Around a Point Source, *Atmos. Environ.,* 14: 1197–1199.

——, and J. Grandell, 1972, On the Removal Time of Aerosol Particles from the Atmosphere by Precipitation Scavenging, *Tellus,* 24: 442–454.

——, and J. Grandell, 1981, Estimates of Characteristic Times for Precipitation Scavenging, *J. Atmos. Sci.,* 38: 370-386.

Rosenkilde, C. E., 1976, *Electrical Aspects of Rainout,* Final Report on Subcontract No. 7465605, January 1976, Lawrence Livermore Laboratory, NTIS.

Rosinski, J., 1967, Insoluble Particles in Hail and Rain, *J. Appl. Meteorol.,* 6: 1066-1074.

Schumann, G., 1963a, Grössenspektra radioaktiver Aerosole, *Staub,* 23: 412-416.

——, 1963b, Investigation of Radon Daughters, *J. Geophys. Res.,* 68: 3867-3869.

Schwartz, S. E., 1979, Residence Times in Reservoirs Under Non-Steady-State Conditions: Application to Atmospheric SO_2 and Aerosol Sulfate, *Tellus,* 31: 530-547.

——, P. H. Daum, M. R. Hjelmfelt, and L. Newman, 1983a, Cloudwater Acidity Measurements and Formation Mechanisms—Experimental Design, in *Precipitation Scavenging, Dry Deposition, and Resuspension,* Proceedings of the Chamberlain Conference, Vol. 1, Santa Monica, Calif., Nov. 29–Dec. 3, 1982, H. R. Pruppacher, R. G. Semonin, and W. G. N. Slinn (Coordinators), Elsevier, New York.

Scott, B. C., 1977, Parameterization of Sulfate Removal by Precipitation, PNL-SA-6657, Battelle Pacific Northwest Laboratory, NTIS.

Scott, W. D., and P. V. Hobbs, 1967, The Formation of Sulfate in Water Droplets, *J. Atmos. Sci.*, 24: 54-57.

Semonin, R. G., 1972, Tracer Chemical Experiments in Midwest Convective Storms, in *Preprints of the Third Conference on Weather Modification,* Rapid City, S. D., pp. 83-87, American Meteorology Society, Boston, Mass.

——, and R. W. Beadle (Coords.), 1977, *Precipitation Scavenging (1974),* ERDA Symposium Series, Champaign, Ill., Oct. 14–18, 1974, CONF-741003, NTIS.

Sheih, C. M., 1976, Application of a Lagrangian Statistical Trajectory Model to the Simulation of Sulfur Pollution over Northeastern United States, in *Proceedings of the Third Symposium on Atmospheric Turbulence, Diffusion, and Air Quality,* Oct. 19–22, 1976, Raleigh, N. C., pp. 311-317, American Meteorological Society, Boston, Mass.

Shopauskas, K. K., et al., 1970, Washout of Radioisotopes Injected into a Cloud from Data of Ground Observations, in *Atmospheric Scavenging of Radioisotopes,* proceedings of a conference in Palanga, USSR, June 7–9, 1966, B. Styra, Ch. A. Garbalyauskas, and V. Yu. Luyanas (Eds.), pp. 233-241, translated from *Issledovanie Protsessov Samoochischeniya Atmosfery ot Radioaktivnykh Izotopov* by D. Lederman; available as TT-69-55099 from the National Technical Information Service.

Singh, H. B., L. J. Salas, and L. A. Cavanagh, 1977, Distribution, Sources, and Sinks of Atmospheric Halogenated Compounds, *J. Air Pollut. Control Assoc.*, 27: 332-336.

Slinn, S. A., and W. G. N. Slinn, 1980, Predictions for Particle Deposition on Natural Waters, *Atmos. Environ.*, 14: 1013-1016.

Slinn, W. G. N., 1967, Precipitation Scavenging of Submicron Particles Part A—Theory, in *Proceedings of the USAEC Meteorological Information Meeting,* Chalk River, Ontario, Sept. 11–14, 1967, C. A. Mawson (Ed.), Report AECL-2787, pp. 527-541, Chalk River Nuclear Laboratories.

——, 1971a, Numerical Explorations of Washout of Aerosol Particles, in *Pacific Northwest Laboratory Annual Report for 1970 to the USAEC Division of Biology and Medicine. Vol. II. Physical Sciences. Part I. Atmospheric Sciences,* C. L. Simpson et al., USAEC Report BNWL-1551 (Vol. II)(Pt. 1), pp. 75-81, Battelle Pacific Northwest Laboratories.

——, 1971b, Aerosol Particle Size Dependence of the "Rainout Rate," in *Pacific Northwest Laboratory Annual Report for 1970 to the USAEC Division of Biology and Medicine. Vol. II. Physical Sciences. Part I. Atmospheric Sciences,* C. L. Simpson et al., USAEC Report BNWL-1551 (Vol. II)(Pt. 1), pp. 3-8, Battelle Pacific Northwest Laboratories.

——, 1973, In-Cloud Scavenging Studies, in *Pacific Northwest Laboratory Annual Report for 1972 to the USAEC Division of Biomedical and Environmental Research. Vol. II. Physical Sciences. Part 1. Atmospheric Sciences,* C. L. Simpson et al., USAEC Report BNWL-1751 (Pt. 1), pp. 76-79, Battelle Pacific Northwest Laboratories.

——, 1974a, Analytical Investigations of Inertial Deposition of Small Aerosol Particles from Laminar Flows onto Large Obstacles, Parts A & B, in *Pacific Northwest Laboratory Annual Report for 1973 to the USAEC Division of Biomedical and Environmental Research. Part 3. Atmospheric Sciences,* C. L. Simpson et al., USAEC Report BNWL-1850(Pt.3), pp. 179-196, Battelle Pacific Northwest Laboratories.

——, 1974b, A Postulated Estimate of Wet Versus Dry Deposition Downwind of a Point Source of Pollution, in *Pacific Northwest Laboratory Annual Report for 1973 to the USAEC Division of Biomedical and Environmental Research. Part 3. Atmospheric Sciences,* C. L. Simpson et al., USAEC Report BNWL-1850(Pt.3), pp. 121-132, Battelle Pacific Northwest Laboratories.

——, 1974c, Rate-Limiting Aspects of In-Cloud Scavenging, *J. Atmos. Sci.*, 31: 1172-1173.

——, 1974d, The Redistribution of a Gas Plume Caused by Reversible Washout, *Atmos. Environ.*, 8: 233-239.

——, 1975a, Atmospheric Aerosol Particles in Surface Level Air, *Atmos. Environ.*, 9: 763-764.

——, 1975b, Analytical Investigations of Inertial Deposition of Small Aerosol Particles from Laminar Flows onto Large Obstacles, Parts C and D, in *Pacific Northwest Laboratory Annual Report for 1974 to the USAEC Division of Biomedical and Environmental Research. Part 3. Atmospheric Sciences,* C. L. Simpson et al., USERDA Report BNWL-1950 (Pt. 3), pp. 150-167, Battelle Pacific Northwest Laboratories.

——, 1976a, Dry Deposition and Resuspension of Aerosol Particles—A New Look at Some Old Problems, in *Atmosphere–Surface Exchange of Particulate and Gaseous Pollutants (1974),* ERDA Symposium Series, Richland, Wash., Sept. 4–6, 1974, R. J. Engelmann and G. A. Sehmel (Coords.), pp. 1-40, CONF-740921, NTIS.

——, 1976b, Formulation and a Solution of the Diffusion Deposition, Resuspension Problem, *Atmos. Environ.*, 10: 763-768.

——, 1976c, Natural Atmospheric Cleansing Processes, pp. 193-199 in *Proc., Fourth Nat. Conf. on Fire and Forest Meteorology,* Nov. 16–18, 1976, St. Louis, Missouri, D. H. Baker, and M. A. Fosberg (Coords.), General Technical Department Report RM-32, USDA Forest Service, Rocky Mountain Forest and Range Experiment Station, Fort Collins, Colo.

——, 1977a, Precipitation Scavenging: Some Problems, Approximate Solutions, and Suggestions for Future Reseach, in *Precipitation Scavenging (1974),* ERDA Symposium Series, Champaign, Ill., Oct. 14–18, 1974, Richard G. Semonin and Robert W. Beadle (Coordinators), pp. 1-60, CONF-741003, NTIS.

——, 1977b, Proposed Definitions for Precipitation Scavenging, in *Precipitation Scavenging (1974),* ERDA Symposium Series, Champaign, Ill., Oct. 14–18, 1974, Richard G. Semonin and Robert W. Beadle (Coordinators), pp. 813-818, CONF-741003, NTIS.

——, 1977c, Some Approximations for the Wet and Dry Removal of Particles and Gases from the Atmosphere, *J. Water, Air Soil Pollut.*, 7: 513-543.

——, 1978, Parameterizations for Resuspension and for Wet and Dry Deposition of Particles and Gases for Use in Radiation Dose Calculations, *Nucl. Saf.*, 19(2): 205-219 (1978).

——, 1980, Relationships Between Removal Processes and Residence Times for Atmospheric Pollutants, *AIChE Symp. Ser.*, 196(76): 185-203.

——, 1982a, Estimates for the Long-Range Transport of Air Pollution, *J. Water, Air, Soil Pollut.*, 18; 45-64.

——, 1982b, Predictions for Particle Deposition to Vegetative Canopies, *Atmos. Environ.*, 16; 1785-1794.

——, 1982c, Some Influences of the Atmospheric Water Cycle on the Removal of Atmospheric Trace Constituents, in *Atmospheric Chemistry,* E. D. Goldberg (Ed.), pp. 57-90, Dahlem Konferenzen 1982, Springer-Verlag, New York.

——, 1983a, Air-to-Sea Transfer of Particles, Chapter 6 of *Air–Sea Exchange of Gases and Particles,* P. S. Liss and W. G. N. Slinn (Eds.), NATO Advanced Study Institute Series, D. Reidel Pub. Co., Dordrecht, Holland (Kluwer Acad. Pub., Higham, Mass.).

——, 1983b, A Potpourri of Deposition and Resuspension Questions, in *Precipitation Scavenging, Dry Deposition, and Resuspension,* Proceedings of the Chamberlain Conference, Vol. 2, Santa Monica, Calif., Nov. 29–Dec. 3, 1982, H. R. Pruppacher, R. G. Semonin, and W. G. N. Slinn (Coordinators), Elsevier, New York.

——, 1983c, Sources and Surface-Area Distributions of Atmospheric Particles, *Atmos. Environ.*, 17 (in press).

——, and J. M. Hales, 1971, A Reevaluation of the Role of Thermophoresis as a Mechanism of In- and Below-Cloud Scavenging, *J. Atmos. Sci.*, 28: 1465-1471.

——, L. Hasse, B. B. Hicks, A. W. Hogan, D. Lal, P. S. Liss, K. O. Munnich, G. A. Sehmel, and O. Vittori, 1978, Some Aspects of the Transfer of Atomospheric Trace Constituents Past the Air–Sea Interface, *Atmos. Environ.*, 12: 2055-2087.

——, and S. F. Shen, 1971, Brownian Diffusion of a Nonuniform Gas, *J. Stat. Phys.*, 3: 291-306.

——, et al., 1978, Some Aspects of the Transfer of Atmospheric Trace Constituents past the Air–Sea Interface, *Atmos. Environ.*, 12(11): 2055-2087.

Small, M. J., and P. J. Sampson, 1982, Mathematical Simulation of Lagrangian Precipitation and Associated Sulfur Wet Deposition, in *A Speciality Conference on Atmospheric Deposition*, E. P. Frederick (Ed.), pp. 273-285, Air Pollution Control Association, APCA Speciality Conference Proceedings SP-49, Publishers Choice Book Mfg. Co., Mars, Penn.

Smith, B. M., R. L. Walker, and L. E. Stettler, 1967, Interactions of Airborne Vapors with Particulates, in *Health Physics Division Annual Progress Report for Period Ending July 31, 1967,* USAEC Report ORNL-4168, Oak Ridge National Laboratory, NTIS.

Smith, F. B., and R. D. Hunt, 1978, Meteorological Aspects of the Transport of Pollution over Long Distances, *Atmos. Environ.*, 12(1-3): 461-477.

Sood, S. K., and M. R. Jackson, 1969, Scavenging Study of Snow and Ice Crystals, Report IITRI-C-6105-9, Illinois Institute of Technology Research Institute, NTIS.

Starr, J. R., and B. J. Mason, 1966, The Capture of Airborne Particles by Water Drops and Simulated Snow Crystals, *Q. J. R. Meteorol. Soc.*, 92: 490-499.

Stavitskaya, A. V., 1972, Capture of Water-Aerosol Drops by Flat Obstacles in the Form of Star Shaped Crystals, *Izv., Akad. Nauk. Atmos. Oceanic Phys.*, 8: 768-774.

Stensland, G. J., and R. G. Semonin, 1982, Another Interpretation of the pH Trend in the United States, *Bull. Am. Meteorol. Soc.*, 63: 1277-1284.

Storebo, P. B., and A. N. Dingle, 1974, Removal of Pollution by Rain in a Shallow Air Flow, *J. Atmos. Sci.*, 31: 533-542.

Stulov, L. D., F. I. Murashkevich, and N. Fuchs, 1978, The Efficiency of Collision of Solid Aerosol Particles with Water Surfaces, *J. Aerosol Sci.*, 9: 1-6.

Styra, B., Ch. A. Garbalyauskas, and V. Yu Luyanas (Eds.), 1970, *Atmospheric Scavenging of Radioisotopes,* proceedings of a conference in Palanga, USSR, June 7-9, 1966, translated from *Issledovanie Protsessov Samoochischeniya Atmosfery ot Radioaktivnykh Izotopov* by D. Lederman; available as TT-69-55099 from NTIS.

Summers, P. W., and J. H. Renick, 1971, Case Studies of the Physical Effects of Seeding Hailstorms in Alberta, in *Proceedings of the International Conference on Weather Modification,* Canberra, Australia, Sept. 6-16, 1971, American Meteorological Society, Boston, Mass.

Tennekes, H., and J. D. Woods, 1973, Coalescence in a Weakly Turbulent Cloud, *Q. J. R. Meteorol. Soc.*, 99: 758-763.

Vali, G., 1977, Washout in High Plains Thundershowers, in *Precipitation Scavenging (1974)*, ERDA Symposium Series, Champaign, Ill., Oct. 14-18, 1974, Richard G. Semonin and Robert W. Beadle (Coordinators), pp. 494-502, CONF-741003, NTIS.

Vaughan, L. M., and W. A. Perkins, 1961, The Washout of Aerosol Particles and Gases by Rain, Technical Report No. 88, Report NP-10217, Aerosol Laboratory, Stanford University, NTIS.

Vickers, D., and W. G. N. Slinn (Eds.), 1979, *Estimates for Wet and Dry Removals' Contributions to the Residence Time for Atmospheric Pollutants in the Eastern United States*, USDOE Report SR-0980-6, Oregon State University, NTIS.

Waldvogel, A., 1974, The N_0 Jump of Raindrop Spectra, *J. Atmos. Sci.*, 31: 1067-1078.

Walton, W., and A. Woolcock, 1960, The Suppression of Airborne Dust by Water Spray, in *Aerodynamic Capture of Particles,* pp. 129-153, E. G. Richardson (Ed.), Pergamon Press, New York.

Wang, P. K., and H. R. Pruppacher, 1977, An Experimental Determination of the Efficiency with Which Aerosol Particles Are Collected by Water Drops in Subsaturated Air, *J. Atmos. Sci.*, 34: 1664-1669.

Watson, C. W., S. Barr, and R. E. Allenson, 1977, *Rainout Assessment: The ACRA System and Summaries of Simulation Results,* ERDA Report LA-6763, Los Alamos Scientific Laboratory, NTIS.

Wendell, L. L., and D. C. Powell, 1976, An Examination of the Effects of Real Time Versus Time-Averaged Precipitation Data on SO_2 and SO_4 Removal in the Regional Assessment Model, in *Pacific Northwest Laboratory Annual Report for 1975 to the USERDA Division of Biomedical and Environmental Research,* ERDA Report BNWL-2000(Pt. 3), pp. 224-229, Battelle Pacific Northwest Laboratory, NTIS.

——, D. C. Powell, and R. L. Drake, 1976, A Regional Scale Model for Computing Deposition and Ground Level Air Concentration of SO_2 and Sulfates from Elevated and Ground Sources, in *Proceedings of the Third Symposium on Atmospheric Turbulence, Diffusion, and Air Quality,* Oct. 19-22, 1976, Raleigh, N. C., pp. 318-324, American Meteorological Society, Boston, Mass.

Wieser, P., 1966, Über die Grössenverteilung des natürlichen radioaktiven Aerosols, *Atompraxis,* 12: 294-296.

Williams, A. L., 1977, Analysis of In-Cloud Scavenging Efficiencies, in *Precipitation Scavenging (1974)*, ERDA Symposium Series, Champaign, Ill., Oct. 14-18, 1974, Richard G. Semonin and Robert W. Beadle (Coordinators), pp. 258-275, CONF-741003, NTIS.

Winkler, P., 1973, The Growth of Atmospheric Aerosol Particles as a Function of the Relative Humidity-II. An Improved Concept of Mixed Nuclei, *Aerosol Sci.*, 4: 373-387.

Wolf, M. A., and M. T. Dana, 1969, Experimental Studies in Precipitation Scavenging, in *Pacific Northwest Laboratory Annual Report for 1968 to the USAEC Division of Biology and Medicine. Vol. II. Physical Sciences. Part I. Atmospheric Sciences,* C. L. Simpson et al., USAEC Report BNWL-1051(Pt. 1), pp. 18-25, Battelle Pacific Northwest Laboratories.

Young, J. A., T. M. Tanner, C. W. Thomas, and N. A. Wogman, 1977, Entrainment of Tracers into Convective Clouds at 10,000 to 13,500 Feet near St. Louis, in *Precipitation Scavenging (1974)*, ERDA Symposium Series, Champaign, Ill., Oct. 14-18, 1974, Richard G. Semonin and Robert W. Beadle (Coordinators), pp. 362-381, CONF-741003, NTIS.

Zebel, G., 1968, Capture of Small Particles by Drops Falling in Electric Fields, *J. Colloid Interface Sci.*, 27: 294-304.

Zimin, A. G., 1964, Mechanisms of Capture and Precipitation of Atmospheric Contaminants by Clouds and Precipitation, in *Problems of Nuclear Meteorology,* I. L. Karol' and S. G. Malakhov (Eds.), USAEC Rept. AEC-tr-6128, pp. 139-182, translated from *Voprosy Yadernoi Meteorologii* (Russian), NTIS.

Chapter 11 covers research prior to June 1978.

DEPOSITION AND RESUSPENSION

George A. Sehmel, Pacific Northwest Laboratory

12

12-1 INTRODUCTION

The transport of particles and gases from the atmosphere to environmental surfaces and from those environmental surfaces back into the atmosphere is the subject of this chapter. When mass is transported from the atmosphere to an environmental surface, the removal process is described as *deposition*. The transport from the environmental surface back into the atmosphere is known as *resuspension*. These two mass-transfer processes are important in minimizing airborne concentrations at breathing height (deposition removes airborne pollutants) or in predicting airborne concentrations arising from surface contamination (resuspension).

These mass-transfer processes are of most interest for noxious airborne pollutants such as plutonium, sulfur compounds, lead, arsenic, and other materials. Because these particles can be inhaled, the potential hazards of deposition and resuspension to man must be known and minimized if necessary. Determination of these hazards requires that airborne concentrations be predicted from validated models that describe the transport of toxic material downwind. Unfortunately man is most often exposed to the hazard of inhaling these particles in the bottom 2 m of the atmosphere. This region of the atmosphere is generally neglected by models used to simulate the downwind transport and diffusion of pollutants. This chapter demonstrates how this neglect has limited our predictive capability and summarizes reported values of air–surface exchange mass-transfer rates described by dry deposition velocities and resuspension rates.

Currently, deposition and resuspension models have equal applicability to both natural and man-generated particles and gases. We cannot distinguish between the mass transfer behavior of two different particles of different chemical composition or solubility. Most of our knowledge of particle behavior cannot even distinguish between the behavior of different particle shapes. To simplify the physics and mathematics so that the aerodynamic behavior of a particle can be described, we usually assume particles to be spherical. For various gases, deposition rates are a function of chemical reactivity and solubility. Deposition rates are influenced by water solubility over water surfaces or within plant surfaces.

Although many experiments have been conducted to determine the physics of particle and gas deposition, the results from these field experiments have not been generalized even within an order of magnitude, because of experimental uncertainties and limited data. Most experiments either have not been conducted under completely controlled conditions or have possibly been interpreted by too simple theoretical models. In field experiments, to interpret or predict downwind concentrations at the ground, modelers have related the deposition removal rate to assumed atmospheric diffusion coefficients.

To interpret deposition measurements, the modeler either had to assume a deposition removal rate or had to specify a vertical diffusion parameter. Usually neither of these parameters was measured independently. Independent predictions of the deposition removal rate are presented in this chapter.

The reinsertion processes discussed here consider both resuspension and suspension. Resuspension describes material deposited from the atmosphere and then subsequently reentrained or resuspended into the atmosphere. Suspension describes the deposition of particles on a surface by some process, such as an industrial spill, and the subsequent insertion of these particles into the atmosphere. Since the combined processes are usually referred to as resuspension, this term will be used in this chapter to describe both resuspension and suspension. Another reason for referring to these two processes by one term is that at the present time we cannot distinguish between the subsequent behavior of particles deposited from the atmosphere or from an industrial spill. A complicating factor is that pollutant particles on the ground most probably lose their individual identity by becoming attached to host particles. When the pollutant particle is transported downwind, it is usually attached to this host particle (Sehmel, 1976b, 1978b) rather than being transported as a discrete pollution particle.

Resuspension occurs both as a result of wind and the mechanical disturbance of the soil (Hereim and Ritchie, 1976) but more readily by the latter. Mechanical stresses, such as walking or digging, can raise particles above ground level into the main airstream so that they can be more rapidly transported downwind than they could by general wind resuspension. Mechanical stress also adds more energy to the soil surface than simple wind resuspension so that the soil is more readily resuspended. On the basis of limited

Sehmel: Pacific Northwest Laboratory, Richland, Washington.

data, some estimates have been made of wind-induced resuspension rates as well as mechanical resuspension factors. These values are discussed in Secs. 12-4 and 12-5.

Agricultural erosion studies are briefly reviewed even though they have limited applicability to respirable toxic particle resuspension since they investigate bulk transport by soil erosion. The results of agricultural research are reported (Skidmore and Woodruff, 1968) in terms of a soil erodibility index. For further information the reader is referred to Chepil whose many papers are listed by Orgill (1977). From these studies, predictions of average soil loss in a field can be estimated in units of surface depth loss per unit time.

Airborne soil can change or alter expected airflow characteristics at the air–surface interface because the energy exchange between air and airborne particles alters the wind speed as a function of height (Bagnold, 1941). This change in the wind-speed profile from a logarithmic distribution is not discussed here, however. Although a change in wind speed can occur for large airborne dust loadings, the effects on resuspension and deposition have not been quantified.

A summary of deposition and resuspension research has been published by Engelmann and Sehmel (1976). That volume, as well as this chapter and its references, will probably lead the interested reader to most published results on these topics.

12-1.1 Transport Models

Although both deposition and resuspension measurements have been made and some correlating bases exist for predicting mass transfer at the air–surface interface, prediction models need much improvement. Many models have been developed to describe the combined deposition, resuspension, diffusion, and transport of an airborne plume. However, insufficient and inadequate data are available to validate these models. Furthermore, most modeling work compares predictions from different models rather than attempts to validate individual models. These model comparisons do not significantly improve our predictive capabilities in describing downwind transport.

Multibox resistance models are often used as a mathematical convenience to simplify the continuum mass-transfer processes (Calder, 1961) by including an initial box in the lower 1 mm assumed in chemical engineering models and additional boxes for incremental distances above 1 mm. These models describe mass transfer within the air–surface interface in terms of mass-transfer resistances. These models have been developed by Bennett, Hill, and Gates (1973), Bennett and Hill (1975), Caporaloni et al. (1975), Garland (1976a), Liss and Slater (1976), Markee, Andrews, and Jubach (1977), Murphy (1976), Sehmel and Hodgson (1978), Slinn (1976a, 1976b, 1977), and Slinn et al. (1978).

Multibox models are used by ecologists interested in the effect of pollutants on plants rather than in its inhalation hazard to man. Boxes are used to describe different leaf areas, such as the stomatal resistance to gas mass transfer. In this case, increased gas mass transfer occurs during daylight while stomata are open; decreased gas mass transfer occurs during darkness while the stomata are closed.

An integration method has not been validated for summing deposition on individual plant leaves from the canopy top to bottom. The multibox models require that calculations of flux and concentration gradient equality be made at the interfaces between boxes, but data to validate these calculations are not available to establish reference points between the boxes.

Data gathered from the multibox models can be used to predict the effects of pollutants on vegetation and structural materials as well as on man. This chapter, however, is mainly concerned with predicting airborne concentrations that are potentially hazardous to man.

When multibox resistance models are applied to diffusion and transport models, usually only two boxes can be used: an upper box described by standard meteorological diffusion and a lower box described by either a deposition velocity or a resuspension rate. Unfortunately these models have not been validated. In the absence of validated models, predictions of deposition velocities and resuspension rates are those reported in current literature and summarized in this chapter.

12-1.2 Deposition Velocity

Deposition velocity is a mass-transfer boundary condition at the air–surface interface in atmospheric diffusion and transport models. It includes the multibox resistances describing the total mass-transfer effects within the lowest few meters of the atmosphere. Chamberlain and Chadwick (1953) defined the deposition velocity as the ratio of the deposition flux divided by the airborne pollutant concentration per unit volume at some height above that surface layer. Thus the deposition velocity is usually reported in units of centimeters per second.

Historically the deposition velocity, v_d, has been defined as a deposition flux, F, divided by an airborne concentration, χ. For a depositing polydispersed aerosol and sometimes for a monodisperse aerosol,

$$v_d = -\frac{F}{\chi} \tag{12.1}$$

(The minus sign is required since the flux downward is a negative quantity whereas the deposition velocity is defined as positive.) For monodispersed particles the deposition velocity, $K_{1\ m}$, is defined as

$$K_{1m} = -\frac{F}{C} \tag{12.2}$$

where C is the airborne concentration of monodispersed particles. The deposition velocity nomenclature K, rather than v_d, reflects that deposition velocities are a function of particle diameter. The subscript 1 m indicates a reference pollutant concentration at a 1-m reference height.

For measurement of the airborne concentration of a pollutant required in Eqs. 12.1 and 12.2, a reference height of about 1 to 1.5 m is used for land surfaces, whereas 10 to 15 m is used for ocean surfaces. For all reference heights and upwind sources, the detailed physics required to describe the integrated mass transfer within these lowest layers is not known (Doran, 1977). However, approximations and predictions of deposition mass-transfer rates are given in Sec. 12-3.4.

12-1.3 Resuspension Coefficients

Resuspension was described in the early literature (Langham, 1971) by a resuspension factor, RF, which is the ratio of airborne contaminant concentration per unit air volume, χ, divided by the contaminant surface concentration per unit area on the ground, G:

RF (units of m^{-1})

$$= \frac{\chi\ (\text{airborne concentration, } m^{-3})}{G\ (\text{surface concentration, } m^{-2})} \qquad (12.3)$$

The airborne concentration is measured at some height above the ground surface concentration sample. Neither the height of air sampling nor the depth of surface soil sampling is specified, although breathing height and a 1-cm depth sample in the ground might be used.

The magnitude of the resuspension factor, which has ranged in experimental studies from about 10^{-10} to 10^{-2} m^{-1} (9×10^{-11} to 3×10^{-4} m^{-1} for wind resuspension and 1×10^{-10} to 4×10^{-2} m^{-1} for mechanical stresses from man's activities), depends in part on the sampling techniques used to measure particle concentrations. The definition given to "surface concentration" must be precisely stated since the term can mean anything from the pollutant in the first millimeter depth in the soil per unit area to that in the first several centimeter depth per unit area. Similarly the definition of "airborne concentration" is not often precise because the measurement height above the surface is not specified. Different measurement heights and particle sizes will certainly change the resuspension factor.

As Langham (1971) noted, although the resuspension factor is useful as an index to relate air concentrations to surface concentrations, it is "aesthetically nauseating and simpleminded." One deficiency is that the airborne concentration is assumed to be related to the surface concentration on the soil at a person's feet rather than upwind. Another deficiency is that resuspension factors are applicable only for the conditions for which they were determined. At the present time no general method has been devised to extrapolate or to predict resuspension factors for any situation. Even the taking of average resuspension factors is a risky concept because of the many orders of magnitude uncertainty in the resuspension factor. A third deficiency is that most resuspension factors have not been determined under conditions in which the resuspension source characteristics were carefully controlled. Sources are usually not uniformly distributed across an area, and the particle size on the ground has not been known.

Unfortunately resuspension factors do not describe particulate resuspension rates nor can they realistically be used to predict downwind air exposure and ground deposition. For downwind calculations, resuspension rates and airborne particle concentration profiles are needed so that the resuspension term can be dimensionally consistent with particle diffusion deposition models. For use in these models, a group of suspension (resuspension) rates, S, must be evaluated with units of the fraction resuspended per unit time. Suspension rates, S, rather than resuspension rates, are determined experimentally. Thus the total resuspension rate S is

$$S = f(S_{air}, S_{salt}, S_{sc}, S_W, S_M) \qquad (12.4)$$

The three modes of soil transport as a function of particle size are S_{air}, S_{salt}, and S_{sc}, where S_{air} is the suspension rate for particles moving in true airborne suspension, S_{salt} is the suspension rate for particles moving in saltation, and S_{sc} is the suspension rate for particles moving in surface creep. These three variables must be considered, although one may dominate, since soil particle movement occurs by different mechanisms (Bagnold, 1941, 1960) for different sized particles. Wind resuspension rates, S_W, from all three modes of transport and mechanically caused resuspension rates, S_M, must also be considered. Each S_i must be evaluated, and general correlations for the S_i's must be established before a reasonable model can be developed for predicting particle resuspension and translocation.

When the contaminant surface concentration is known, the upward resuspension flux can be determined by multiplying the surface concentration by the resuspension rate. Conceptually this calculation is simple. In reality the physics are complex. One complication arises in determining the surface contamination level. For example, assume that (1) the resuspension rate is constant and (2) the total loss to air is small. Even in this case resuspension fluxes could change with time since the surface contamination source could also change owing to washing, weathering, and penetrating more deeply into the soil. Consequently, depending on the soil depth used to define the source strength, widely different resuspension fluxes could be calculated. If the source strength is measured to include all contaminants below the surface, the calculated resuspension flux would remain constant. If only the surface contaminant is measured, the calculated resuspension fluxes would increase with time.

Conceptually, both resuspension rates and factors are related only to the contaminant that the wind stress can act on, which might be within the top millimeter or so of a soil surface. However, to determine the contaminant surface concentration experimentally, the contaminant must usually be calculated as a function of varying sampling depths. Often only the top 1 cm of soil is sampled. This presents no problem if the contaminant is only on the top surface of that 1-cm section because only this portion can normally be resuspended. However, if the contaminant is uniformly distributed throughout the top 1 cm, the estimate of the pollutant available for resuspension is too large. Other experiments have evaluated contaminant surface concentrations at even greater depths.

This inconsistency in determining the resuspension factor is also caused by sampling all particle sizes. These sampling techniques distort resuspension physics because they ignore how airborne concentrations of a particular particle size are related to the contaminant surface concentrations in that same particle-size range on the ground surface.

12-1.4 Concentration Prediction Accuracy

Many models are used to calculate the downwind transport, diffusion, and deposition of particles and gases. In general, the transport–diffusion models can be modified to include the effects of surface mass transfer. Some of the modified models include Sutton's equations (Chamberlain, 1955), Pasquill's model (1962, 1974), Pasquill's equations (Brook, 1968), a Gaussian plume model using surface depletion (Horst, 1977a, 1977b), Gaussian plume models by others (Csanday, 1955, 1957; Van der Hoven, 1963; Trevino, 1972; Mills, Dahlman, and Olson, 1974; Scriven and Fisher, 1975; Shreffler, 1975; Amato, 1976; Overcamp, 1976; Travis, 1976; Healy, 1974, 1977a; Vögt, 1977; Davis, 1978), Slinn's theoretical models (1976a, 1977, 1978), and finite difference models (Draxler and Elliott, 1977). In addition, surface soil transport by saltation and surface creep has been considered by Amato (1976). These models will not be specifically described here, because there is much uncertainty as to which one, if any, of them has been validated.

Gaussian plume models have often been used to predict the transport of particles over great distances. These predictions are a function of the atmosphere diffusion coefficients. However, Crawford et al. (1977, 1978) and others point out that the experimental conditions on which atmospheric dispersion curves (Pasquill, 1974) were developed are often overlooked or neglected in such predictions. The experimental data used in the models to define these curves were gathered from near-surface release experiments conducted on a smooth (roughness coefficient of approximately 1 cm), flat area for distances of less than 1 km and sampling times of 3 to 10 min. Modelers, however, have forgotten or overlooked the tentative nature of these dispersion curves and have extrapolated the dispersion curves to elevated release heights and to distances of over 200 km from the source.

If the uncertainty in predicting from different diffusion and transport models were resolved, a recommendation for a validated model including air–surface mass transfer might be possible. Mathematical models inherently have levels of uncertainty attached to them (American Meteorological Society 1977 Committee on Atmospheric Turbulence and Diffusion, 1978). For highly idealized conditions (i.e., uniform terrain, steady meteorology, source and ambient parameters measured carefully by research-grade instruments), the observed maximum downwind ground-level concentration is expected to be within 10 to 20% of that calculated for a surface-level source and within 20 to 40% for an elevated source. A factor of 2 accuracy is the best that can be expected in most urban-scale dispersion model predictions. Additional but unspecified uncertainties are suggested when dry deposition and resuspension are considered in model predictions. In using Gaussian plume models, Crawford et al. (1977, 1978) suggested tentative accuracy statements in estimating airborne concentrations which were based largely on scientific judgment. However, there are not enough data on which to base a reliable statistical estimate of model predictions. Nevertheless, for a highly instrumented flat-field site from which previous data and meteorological and airborne concentrations are available, it might be possible to estimate the ground-level airborne concentration from a continuous point source at downwind distances of less than 10 km to within a factor of 0.2 for the plume centerline concentration.

The accuracy in predicting airborne concentrations is expected to increase as averaging times increase. For a specific hour and downwind receptor point, the accuracy depends on calculating the exact plume trajectory. For flat terrain, relatively steady meteorological conditions, and distances of less than 10 km, airborne concentrations for an individual case might be estimated to within a factor of about ±10. The ensemble of a large number of air concentration calculations for a specific receptor point compares favorably with the ensemble of air concentration measurements at that point. Accuracies are about a factor of ±4 for monthly and seasonal averages. For an annual average concentration estimate, the accuracy is about a factor of ±2 for distances of 10 to 100 km in relatively flat terrain.

For complex terrain, a few experiments indicate departures of more than a factor of 10 from the estimates of the Pasquill–Gifford dispersion curves. For the complex terrain situation, the data are insufficient to generalize the results and to document accuracy. When the effects of surface mass transfer are included in addition to diffusion calculations, there appears to be no valid model that could be recommended for complex terrain. Even models for flat terrain show uncertain predictions. Nevertheless, as models become validated in the future, data will be needed on which to base surface mass-transfer calculations. The published data for deposition and resuspension are the focus of attention in this chapter.

Although there is uncertainty at the present time as to which transport model to use, the user is required to base predictions on current models. To make matters more difficult, the user is often limited in the complexity of the model he chooses. Since most model users are not only limited by computer capacity but also by costs, model predictions using modifications of the Pasquill (1962) diffusion curves are often used.

12-1.5 Concentration Profiles

Rather than discussing each transport model separately (see Chap. 13), this chapter discusses only the expected air concentration profiles that models should predict. These profiles are the airborne pollutant concentrations as a function of height aboveground. Figure 12.1 shows idealized, two-dimensional, airborne pollutant concentrations as a function of height for four different situations. Part (a) of Fig. 12.1 shows airborne concentrations for resuspension from a local source. In this case airborne concentrations are maximum at ground level and decrease with height. Downwind from that local source, airborne concentrations change, as suggested in part (b) of Fig. 12.1, and show a depletion near ground level due to particle deposition. Obviously, large deposition rates produce more depletion than low deposition rates. For both high and low deposition rates, a maximum concentration occurs at some height above ground level.

This maximum airborne concentration at an elevated height is also typical of elevated pollutant releases. For example, part (c) of Fig. 12.1 shows airborne concentration profiles for an upwind elevated source for both unstable and stable atmospheres. In a stable atmosphere, there is minimal vertical mixing, and the airborne pollutant plume does not diffuse as rapidly.

Airborne pollutant concentration profiles resulting from an upwind surface release are shown in part (d) of Fig. 12.1 for equal total deposition. Both stable and unstable conditions are considered. Airborne concentrations at ground level are less for the unstable case since pollutants mix upward more rapidly than in a stable atmosphere.

These profiles can be modified by precipitation, as well as by dry deposition. Wet removal is discussed in Chap. 11. However, in an analysis of the hazards of respirable airborne pollutants, only dry deposition can always be relied on to remove airborne pollutants near an industrial

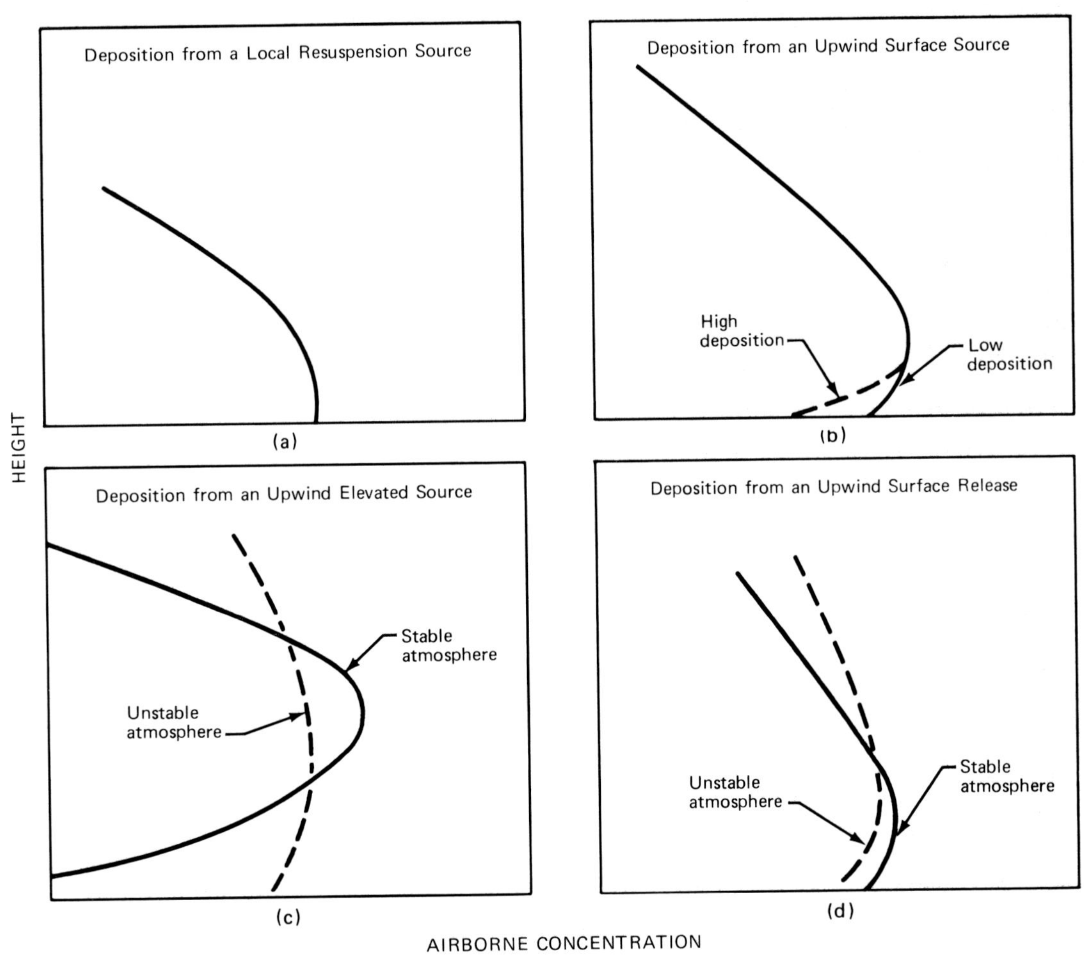

Fig. 12.1 Idealized airborne pollutant concentrations as a function of height for both resuspension and deposition sources.

released source. Dry deposition is always cleansing the atmosphere near ground level, whereas removal by precipitation is intermittent. Predictions (Marenco and Fontan, 1976) indicate dry deposition is more important than wet deposition for removing airborne pollutants from anthropogenic sources. This is not to contradict previous literature that suggested wet deposition was the primary removal mechanism for cleansing the atmosphere. Early literature indicating that wet deposition was more efficient referred only to the removal of radioactive particles from the stratosphere.

12-1.6 Respirable Particles

In air–surface mass-transfer exchanges, the most important particles to consider for both deposition and resuspension are those which can be inhaled and which potentially cause damage in the lung (see Chap. 16). In general, the respirability of a particle depends on its diameter. Particles less than 3.5 μm in diameter are most often considered to be respirable, but some data indicate that particles as large as 15 μm can be inhaled. The size of an inhaled particle determines where it will be deposited in the tracheobronchial tree and the relative hazard. Smaller particles are deposited in the alveoli; larger particles in the neighborhood of 15 μm or greater are denuded from the air by the nasal passageway.

Respirable particles (Volchok, Knuth, and Klemman, 1972) are the prime concern of this chapter; however, because pollutants are also transported downwind on nonrespirable particles, we will be looking at the air–surface mass transfer of both respirable and nonrespirable particles. The resuspension of pollutant particles transported on nonrespirable particles can occur if particles of respirable size become detached from the host particles and are resuspended (Sehmel, 1978a).

12-2 OTHER VARIABLES INFLUENCING AIR–SURFACE MASS-TRANSFER PROCESSES

The previous discussion has indicated many of the uncertainties and qualifications in accurately predicting airborne concentrations and mass transfer at the air–surface interface. Despite these uncertainties, much is known about mass-transfer processes. These processes are discussed in this chapter, but first we need to consider the other variables that influence mass transfer.

12-2.1 Soil Erosion

Because pollutant particles deposited on soil will probably become resuspended while attached to larger host soil particles, resuspension is also concerned with the physics of soil erosion. The main differences between research on noxious particle resuspension and that on soil erosion are in the mass of material transported and the particle diameters studied. Research into resuspension phenomena involves the study of transporting either toxic particles in the respirable diameter range or respirable particles on nonrespirable host soil particles. By contrast, research on soil erosion is directed toward determining bulk soil loss [densities ranging from about 0.9 to 2.8 g/cm^3 (Chepil, 1950b)].

Although resuspension rates and resuspension factors are probably influenced by different soil types, no data base exists for determining the effects of soil types or soil size distribution on resuspension. Nevertheless, some soil erosion concepts that might have the greatest influence on resuspension physics are discussed here.

Soil erosion becomes important to resuspension when vegetation surfaces are disturbed. Wind erosion alone was no problem in the Great Plains so long as native grasses protected the soil. It was after poor farming techniques that wind and drought transformed the Great Plains into a gigantic dust bowl in the 1930s.

The influence of wind alone must not be underestimated. It has been estimated that wind erosion of the total earth surface contaminates the atmosphere (Travis, 1976) with 10^4 to 10^6 tons of soil per day. This windblown soil accounts for approximately 9% of the total mass of atmospheric aerosol (Hidy and Brock, 1970).

Another factor to be considered in resuspension is total soil erosion over thousands of years, which can be estimated in many areas by the depth to which roots or rocks are exposed. In Greenland the average erosion in exposed areas was 9 to 37 m per thousand years (Judson, 1968). In vegetated areas, surface losses were less, only 3 m per thousand years and even less in Utah and California where root exposure depths indicate erosion between 0.02 to 0.1 m per thousand years. An estimate for continent-wide erosion is 0.024 m per thousand years. These average surface-depth erosion rates, ranging from 6×10^{-11} to 1×10^{-7} cm/s, correspond to an average resuspension rate of from 6×10^{-11} to 1×10^{-7} fraction resuspended per second for the top 1 cm of soil.

One problem in relating such soil erosion data to resuspension is that total erosion has usually been measured over wide source areas rather than as a function of small, local source areas. Although data on soil erosion and visibility (Hagen and Woodruff, 1973, 1975) indicate that erosion occurs in these local source areas, the data do not directly evaluate the resuspension rates of local contaminants.

Results from agricultural experience, however, can be used to suggest what might occur in resuspension processes. For "dry" soil with no vegetation, wind erosion occurs if either the wind speed is approximately 4 to 7 m/s, or greater, or the more than 65% of the soil particles are 2 mm or less in diameter (Van Doren, 1944). To reduce soil erosion, farmers alternate strips of crop and fallow land. This practice decreases soil erosion by controlling the abrasion of local soil by upwind soil. Alternating strips of cultivated land is more effective when crops are standing. For instance, standing wheat stubble is more effective than flattened wheat stubble. Stubble in rows perpendicular to the wind direction also control wind erosion more effectively than rows parallel to the wind direction.

Soil erosion and, consequently, resuspension is not uniform across a field. Soil erosion rates are least near the windward side of the field and increase with downwind distance. Erosion rates increase with downwind distance because the abrasion of soil increases as soil is moved from upwind. This avalanching effect increases with downwind distance. Because of these soil erosion characteristics, the resuspension rates of respirable particles must be determined as a function of location in the field, height and density of land cover, and land usage.

Some data suggest that avalanching is a critical consideration in determining resuspension rates only for certain soil particle sizes. By a dry-sieving analysis of a 5-cm-deep surface soil sample, Chepil (1945c) found that erosive material less than 0.8 mm in diameter tends to accumulate at the lee sides of fields. A change in wind direction would, of course, cause a corresponding drift accumulation in the opposite direction. These phenomena suggest resuspension rates will change when the wind direction changes; however, theories are needed for predicting the attachment and movement of pollutant particles across fields.

Another limitation in using the results of agricultural erosion in resuspension studies is that they make no attempt to describe the respirable material that becomes airborne. Results on erosion are reported in terms of a wind erosion equation (Woodruff and Siddoway, 1965; Skidmore and Woodruff, 1968; Skidmore, 1976), which is a product of a soil erodibility index, a climatic index, a soil ridge roughness factor (Chepil, 1950a), the unsheltered field length, and an equivalent quantity of vegetation. The product of these terms is proportional to the annual soil loss in tons per acre or depth loss per unit time. This equation can predict soil losses up to about 340 tons/(acre · yr), which corresponds to about a predicted maximum average surface loss of 4 cm/yr or 1×10^{-7} cm/s.

Abnormal wind stresses, such as during a dust devil, also cause significant soil erosion, although their effects on the resuspension of pollutants have not been thoroughly investigated. During dust devils, soil concentrations of up to about 5 g/m^3 have been measured at heights 1 to 2 m aboveground (Sinclair, 1976). Even particles up to 350 μm in diameter have been transported. Although dust devils can transport large concentrations of soil, dust devils occur infrequently and therefore transport less soil than general dust storms. The physics of dust-devil transport are discussed by Sinclair (1976), and further discussion here will be omitted since the transport of toxic materials by dust devils had not been quantified experimentally.

12-2.1.1 *Soil Types.* Surface soils are described in terms of their particle diameter ranges—either sand, silt, or clay—and their mineral and organic matter content. Sand ranges from 50 to 2000 μm in diameter; silt ranges from 2 to 50 μm; and clay is less than 2 μm in diameter (Gillette, 1976). The percentage of sand, silt, and clay particles in the soil is referred to as the texture; the degree of aggregation of these materials is referred to as structure. In a non-aggregated soil a random distribution of individual particles exists, with smaller grains of silt and clay filling openings between larger grains of sand.

Soils consisting of a mixture of both coarse and fine particles are more easily eroded (Newman et al., 1976), as is the case for most sand dunes where the particle size ranges from 0.1 to 0.55 mm in diameter (Chepil, 1945b). Studies of soil mixtures have shown that particles less than 0.42 mm in diameter erode easily; those from 0.42 to 0.84 mm are more resistant. Material from 0.84 to 6.4 mm in diameter is highly resistant to erosion. A last classification includes all particles with diameters greater than 6.4 mm. When these largest particles erode, often 90% of the material moves within about 0.3 m of ground level.

12-2.1.2 *Surface Crust.* The resistance of the surface crust to soil erosion depends on the time since the last windstorm. If there has been a long interval between windstorms, a much higher wind speed is required to erode the soil since a surface crust can form which serves as a protective barrier against wind abrasion (Hinds and Sauer, 1976). This crust can be either biotic or clay cementation. If soil particles are eroded, the particles can cause abrasion of the surface crust and expose the more highly erosive soil beneath. This abrasive action is more rapid on the leeward side of a field, where the concentration of eroding particles is greatest.

A surface crust forms (Chepil, 1951b) when silt and clay particles smaller than 20 μm in diameter become dispersed during rain and subsequently dry. The crust depth seldom exceeds a thickness of about 0.2 cm. After a series of rains, the small silt and clay particles are carried downward, coarser particles being left at the top surface. Since some of these coarser particles can be subject to more rapid erosion even while the surface is drying, surface resuspension characteristics change continually, even for soils undisturbed except by natural phenomena.

12-2.1.3 *Soil Transport Classification.* Soil movement can occur on several scales. On a small scale, many sand grains tend to move simultaneously in a series of intermittent bursts confined to small surface areas. After movement ceases, the area remains quiescent until another short burst occurs (Sutherland, 1967). As wind speed increases, particle bursts become more active until particles are suspended and transported into the air. During these bursts, there may be either a gain or a loss of electrostatic charge on individual particles as they roll along the surface (Masironi and Fish, 1967).

Large-scale erosion has been described by three terms, which express continual soil movement as a function of surface stresses: saltation, surface creep, and suspension. Differences in motion between these three are gradual. How the soil is transported depends on the diameter of the particle and the wind speed or wind turbulence. Saltation describes particles, ranging in diameter from about 100 to 500 μm (Newman et al., 1976), which jump or bounce within a layer close to the air–surface interface. Particles transported by surface creep range in diameter from about 500 to 1000 μm. The particles move by a sliding or rolling motion and are pushed along the ground by wind stresses and momentum exchange from the impact of smaller particles transported in saltation. Particles smaller than about 100 μm in diameter move by suspension. The amount of soil eroded by saltation, surface creep, and suspension varies greatly for different soils (Chepil, 1965; Oksza-Chocimowoski, 1976): suspension moves from 3 to 40% by weight, saltation 50 to 75%, and surface creep 5 to 25%.

12-2.1.4 *Saltation.* Of the three types of large-scale erosion just described, saltation (Chepil, 1945a) moves the greatest amount of soil. Saltating particles rotate at speeds from about 200 to 1000 revolutions/s and are ejected almost vertically into the airstream. The velocity of this vertical ejection is comparable to the friction velocity, u_* (Owen, 1964). While gaining considerably horizontal momentum from the wind, saltating particles descend with rather flat trajectories—at an angle of between 6 and 12° from horizontal—to strike the soil surface and thereby cause avalanching. After striking the surface, saltating particles either saltate again or cause surface creep. In either case, avalanching can cause some particles to move in true suspension.

Saltation motions indicate that sand is transported not by a diffusion process but rather through the impact of saltating sand grains, which act as projectiles. Since this motion cannot be described by eddy diffusion, existing meteorological transport and diffusion models have limited ability to describe the transport caused by saltating particles. Model development and validation are needed to describe saltation transport.

One way to minimize the eroding effects of saltation flows is to increase the widths of short vegetation stubble in a field (Chepil, 1945a). A 0.6-m-wide strip of short stubble can trap about 50% of saltating particles; a 3-m-wide strip can trap between 90 and 98% of saltating particles; and a 10-m-wide strip can trap over 99% of saltating particles. The height of the stubble also minimizes erosion. An 8-cm stubble traps about 60% of saltating particles, compared to a 15-cm stubble which traps about 85% of saltating particles.

Soil ridges are less efficient at controlling saltation than stubble of equal height. Because the ridges are widely spaced, saltating particles bounce off one or more times before finally being trapped within furrows. The collection efficiency of ridges depends on the ridge size (Chepil, 1945a). Ridges 6 cm high and 23 cm wide were found to be about 50% as effective as 15-cm-high wheat stubble in collecting saltating particles; ridges 13 cm high and 46 cm wide were about 77 to 83% as effective, and ridges 30 cm high were 85 to 92% as effective.

12-2.1.5 *Suspension.* Suspension describes the movement of soil particles below about 100 μm in diameter. This fine dust is lifted by the wind when saltating particles strike the surface. If it were not for saltating movement, this fine dust would not be eroded readily. Dust less than 50 μm and especially less than 10 μm in diameter is extremely resistant (Chepil, 1943) to wind erosion. However, when this fine soil is mixed with coarser particles capable of saltation, it is easily transported by suspension.

12-2.1.6 *Threshold Speed.* The wind speed necessary to begin soil erosion, termed the threshold speed, has been defined and depends on a complicated set of factors. One is whether a surface crust has formed on an exposed field (Chepil, 1945b). The wind speed required to initiate soil erosion in any given season is determined in part by this crust. The erosion of soil with a surface crust can be initiated by wind speeds of 6 to 13 m/s at an elevation of 0.3 m. Once that crust is broken, a bare field exposed to the continued action of wind erodes when the wind speed reaches 6 to 7 m/s at an elevation of 0.3 m.

According to Chepil (1945b), the lowest wind speed required to produce soil movement has been designated as the minimal fluid threshold speed, and the wind speed required to move the largest sand grains, as the maximal fluid threshold speed. Bagnold (1941) used the term "initial fluid threshold" speed as the wind speed required to initiate movement of the predominant sand dune material.

Determining threshold speeds is complicated by the movement of upwind saltating particles. During saltation the minimal impact threshold speed is the wind speed required to initiate soil movement by the force of descending soil particles carried in saltation rather than by direct surface wind stresses. After erosion has started, it will spread downwind by the bombardment of particles moving in saltation. The saltating particles cause movement in other parts of the field which under an equal air speed without bombardment would have remained at rest. A minimal threshold speed for a field depends on the minimal threshold speed for the most erosive upwind spots in the field. Thus the wind speed required to initiate soil erosion in a field of saltating particles is equal to the minimal impact speed rather than to the minimal fluid threshold speed.

Small threshold speeds occur for particles ranging in diameter from 0.1 to 0.15 mm. For smaller particle diameters the threshold speed increases with a decrease in particle size. For particle diameters larger than 0.15 mm, the threshold speed increases. For mixtures of soil particle sizes, the threshold speed is smaller than that required to erode only the largest particle. In other words the threshold speed increases directly with average particle size and decreases with a wider range of particle diameters in the soil (Chepil, 1943). Bagnold (1941) indicates that particles above approximately 100 μm will saltate and be carried aloft only if the wind speed becomes significantly greater than the threshold speed.

12-2.1.7 *Horizontal Soil Fluxes.* The results of soil erosion rates are important to resuspension studies since pollutants are often transported on eroding soil particles. The results of this research are often reported in terms of a horizontal soil particle flux, with particular emphasis on the transport of larger particles. O'Brien and Rindlaub (1936) developed the original relationship between the amount of sand transported and wind speed. These measurements were made at Clatsop Beach at the mouth of the Columbia River. At this location the sand has a mean diameter of about 0.02 cm. Their results have been adjusted to the amount of sand drift, Q, in kg/(m width • h). Results show

$$Q = 0.79u^3 \tag{12.5}$$

for wind speeds (u in m/s) greater than 6 m/s at an elevation of 1.5 m. Bagnold (1941) found similar results for the movement of sand during dust storms in the Egyptian desert. For a mean sand particle diameter of 0.025 cm, the amount of sand transported is expressed by the relationship

$$Q = 5.2 \times 10^5 \ (u - u_T)^3 \tag{12.6}$$

where u is the wind speed in m/s at a height of 1 m and u_T is the threshold speed equal to 4 m/s. It is obvious from comparing coefficients that horizontal fluxes are site specific.

Both expressions show that the horizontal flux of sand is proportional to wind speed, u^3. Similar results are reported in the wind erosion equation developed by Chepil (1951a). In wind-tunnel experiments the soil movement rate was a function of the friction velocity, u_*^3. However, wind-tunnel results indicated the total horizontal flux of sand was a function of the friction velocity, u_*^5.

Although O'Brien and Rindlaud (1936) originally determined that sand movement was a function of wind, the concept was independently studied in more detail by Bagnold (1941) and Chepil (1951b). Bagnold, who did the more extensive early work, was concerned with deposits of individual nonadhering grains larger than about 80 μm in diameter. He conducted his research on Sahara sand. Chepil also studied soil erosion and focused on particles less than 0.4 mm in diameter. Thus both Bagnold and Chepil were concerned with the movement of nonrespirable particles rather than respirable particles. Their research is important, however, because respirable particles are also transported on nonrespirable particles. Sehmel (1978a, 1978c) has shown there can be a greater flux of airborne plutonium on nonrespirable particles than on respirable particles.

Building on the findings of these investigators, Hsu (1971) developed a correlation for predicting sand transport as a function of mean sand particle diameter for data published between 1936 and 1963. In this correlation the rate of sand transport is proportional to the Froude number, $Fr^{1.5}$. The Froude number, Fr, is given by $Fr = u^2/gD$ and represents the ratio of inertial force, u^2/D, to gravitational force. With this correlation between wind speed (u), gravitational acceleration (g), and the mean sand grain diameter (D), the horizontal soil flux in kg/(m • h) can be scaled.

12-2.1.8 *Sand Grain Models.* Wind stresses on individual sand grains have been estimated by several investigators through sand grain models. In the modeling studies, force balances, including frictional drag, form drag, gravitation, and the Bernoulli effect, have been considered (Akiba, 1975; Andres, 1972; Chepil, 1950a, 1959, 1965; Ford, 1957; Kawamura, 1975; Malina, 1941; and Slinn, 1976b). The theoretical models have limited applicability to resuspension physics over a wide surface area since model results require integration over the entire surface area, a procedure involving formidable uncertainties. In addition, models of surface forces on individual particles need some means by which to describe the effects of avalanching probabilistically. These models will not be described further in this chapter.

12-2.2 Airborne Soil Concentrations

Airborne dust loadings tend to modify the air–surface mass transfer and exchange processes. Airborne soil particles can interact with airborne pollutant material as well as modify air velocities and exchange processes. They can agglomerate with pollutant particles or can react chemically in gas–solid reactions. In either case, airborne pollutants can be modified in their physical and chemical structure while airborne. These modifications could lead to changes in the transport properties of the pollutant.

Experimental field measurements of airborne soil concentrations have been organized to show the volume distributions for particles of different diameters. The volume distribution is given by

$$\frac{\Delta V}{\Delta \ln D} = \frac{\pi D^4}{6} \frac{\Delta N}{\Delta D} \tag{12.7}$$

in cubic micrometers per cubic meter. The total airborne particle volume is the integral of Eq. 12.7. The total dust loading can be calculated from the total volume and average soil density; a density of 2 g/cm^3 was assumed.

The volume distributions to be discussed are shown in Fig. 12.2 for data obtained at the Hanford site near Richland, Wash. (Sehmel, 1975b, 1977a). The volume distributions are for wind erosion without mechanical disturbance in a semiarid area. Figure 12.2 shows the airborne volume distributions determined by two sampling techniques. Airborne particle volume distributions were determined with an optical particle counter for particle diameters smaller than 5 μm. The shaded area is the experimental range. The upper limit for this diameter range corresponds to the maximum dust storm condition mea-

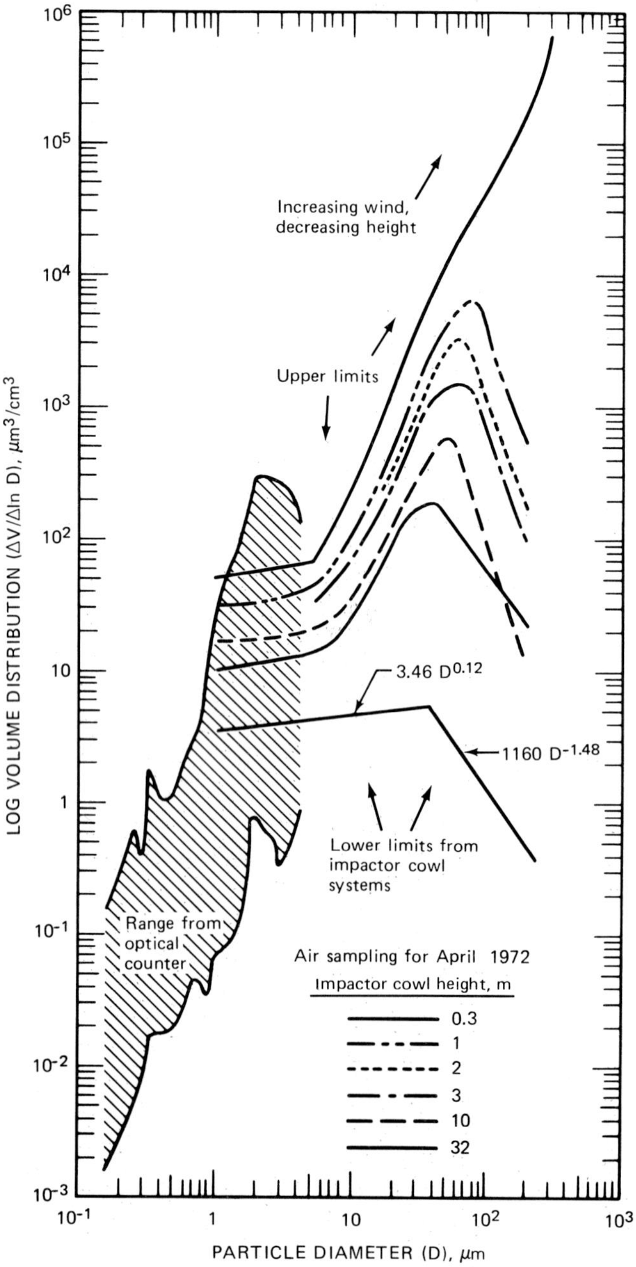

Fig. 12.2 Airborne particle volume distributions at Hanford, Wash.

sured, which corresponded to a maximum visibility of 1.4 km. The lower limit was for a clear day. Larger nonrespirable particles have been sampled with an opened-cavity collector (Sehmel, 1973a), and in this case the airborne volume distribution was a function of height. The maximum concentration measured was at 0.3 m above ground level. The volume distributions and concentrations decreased as the sampling height was increased.

Limits of airborne-soil mass loadings can be estimated from Fig. 12.2. For particle diameters less than 10 μm, the upper limit is 0.7 mg/m^3, and the lower limit is 7.7 μg/m^3. For particle diameters larger than 10 μm, the upper limit is 232 mg/m^3, and the lower limit is 20 μg/m^3. Thus a clean airstream might be one in which the airborne mass loading is in the neighborhood of 20 μg/m^3 (Sehmel, 1977b). In comparison, a dust loading of 110 mg/m^3 is barely tolerable for breathing (Stewart, 1967), whereas the maximum dust concentration measured in a dust devil is approximately 5 g/m^3 (Sinclair, 1976).

12-2.3 Deposition Variables

As Table 12.1 shows, many factors influence dry deposition rates: meteorological variables, properties of the depositing pollutant, and surface variables.

12-2.3.1 *Meteorological Variables.* One meteorological variable that influences dry deposition is aerodynamic surface roughness, z_0, which is used by micrometeorologists to assist in a description of the wind-speed profile above different surfaces. For near-neutral atmospheric stability, the logarithm of wind speed is described by (see Chap. 5)

$$u = \frac{u_*}{k} \ln \frac{z + z_0}{z_0} \qquad (12.8)$$

where u = the measured wind speed
z = the measured height aboveground
u_* = the friction velocity
k = von Kármán's constant, which is approximately 0.4 (see Chap. 5)

For a surface of greater geometric roughness, the height is adjusted to a zero-displacement plane, d, within the canopy. In this case the relationship is

$$u = \frac{u_*}{k} \ln \frac{z - d + z_0}{z_0} \qquad (12.9)$$

The researcher using these equations can adjust the quantities d and z_0 until straight lines are obtained on semilogarithmic paper. Thus these d and z_0 values have no physical meaning other than an empirical data fit. Often d is about three-fourths of the canopy height, whereas z_0 might range from 10^{-4} to 10^2 cm (flat plate with d = 0 to a forest with d = 7 m, respectively). As a rule of thumb, the friction velocity might be a few percent of the average wind speed (see Chap. 5).

The aerodynamic roughness height, z_0, is an extrapolated ordinate at the z intercept. Although z_0 is useful in micrometeorology for momentum transfer by the air, it is not known whether there is a difference between the aerodynamic roughness height for momentum transfer by air and roughness heights for particle and gas mass transfer (Brutsaert, 1975). Calculations of zero-plane displacement have the same unknowns as those for aerodynamic roughness. Zero-plane displacements for mass must be determined and compared to those for heat and momentum transfer.

In micrometeorology, diffusion parameters are used to describe the transport of pollutants even though the effects of vegetative or structural canopies on diffusion coefficients

TABLE 12.1 Some Factors Influencing Dry Deposition Removal Rates

Micrometeorology variables	Depositing material		Surface variables
	Particles	Gases	
Aerodynamic roughness	Agglomeration	Chemical activity	Accommodation
Mass transfer	Diameter	Diffusion	Exudates
Particles	Density	Brownian	Trichomes
Gases	Diffusion	Eddy	Pubescence
Heat	Brownian	Partial pressure	Wax
Momentum	Eddy equal to	in equilibrium	Biotic surfaces
Atmospheric stability	Particle	with surface	Canopy growth
Diffusion, effect of	Momentum	Solubility	Dormant
Canopy	Heat		Expanding
Diurnal variation	Effect of canopy on		Senescent
Fetch	Diffusiophoresis		Canopy structure
Flow separation	Electrostatic effects		Areal density
Above canopy	Attraction		Bark
Below canopy	Repulsion		Bole
Friction velocity	Gravitational settling		Leaves
Inversion layer	Hygroscopicity		Porosity
Pollutant concentration	Impaction		Reproductive structure
Relative humidity	Interception		Soils
Seasonal variation	Momentum		Stem
Solar radiation	Physical properties		Type
Surface heating	Resuspension		Electrostatic properties
Temperature	Shape		Leaf, vegetation
Terrain	Size		Boundary layer
Uniform	Solubility		Change at high winds
Nonuniform	Thermophoresis		Flutter
Turbulence			Stomatal resistance
Wind velocity			Nonbiotic surfaces
Zero-plane displacements			pH effects on
Mass transfer			Reaction
Particles			Solubility
Gases			Pollutant penetration and distribution in canopy
Heat			
Momentum			Prior deposition loading
			Water

are poorly known. This lack of knowledge about the air–surface interface is one limiting factor in applying mass-transfer boundary conditions to transport and diffusion equations. Because the effect of canopies on diffusion cannot be predicted, the assumption is usually that diffusion parameters, such as those of Pasquill, are applicable to all situations.

Another important meteorological factor is the effect of terrain on dry deposition removal. In the past, attempts have been directed only toward measuring mass transfer across a uniform terrain. However, it is rather difficult to find what might be called a uniform terrain in the world since it is the exception rather than the rule. Limited effort has been made to determine the effects of nonuniform terrain on air–surface mass transfer; they can be predicted only from models that have not been validated.

12-2.3.2 ***Pollutant Properties.*** The properties of the depositing material which influence deposition are shown in columns 2 and 3 of Table 12.1. The removal of particles from the atmosphere is accelerated by gravitational settling. The parameters that determine gravitational settling are particle diameter (squared) and particle density. Usually small changes in temperature and relative humidity hardly affect calculations of particle settling velocities; however, hygroscopic particles can absorb sufficient water to change the particle mass, density, and shape.

One important consideration seldom addressed in transport calculations is the effect of particle shape and size on airborne transport and surface mass transfer. The usual assumption is that particles are spherical. However, many particles released from industrial sources are neither spherical nor solid. For instance, fly ash can contain hollow spheres and aggregates of many smaller individual particles. Although attempts have been made to describe the settling velocities of aggregates, no general method exists to predict their settling velocities. Studies have shown that the settling velocity of aggregates can have a settling velocity as small as 0.01 of that for a solid particle of equal mass (Sehmel, 1956).

Another pollutant property of interest is the effects of diffusion on both particles and gases. For very small particles (diameters $<0.1\ \mu m$), the Brownian diffusion coefficient increases as the particle diameter decreases. In the limit as the particle diameter decreases, the Brownian diffusion coefficient for a particle approaches that for a gas. Brownian diffusion is extremely important in the final stage of small particle and gas mass transfer. In fact, within 1 mm of the deposition surface, Brownian diffusion is more important than particle inertia, particle eddy diffusion, and gravitational settling in causing final deposition.

For larger particles (diameters >1 μm), eddy diffusivity becomes important in causing deposition, as does particle inertia and gravitational settling. For rough surfaces, particle eddy diffusivity cannot be predicted. However, for a smooth surface, Sehmel (1971) showed that an effective particle eddy diffusivity can be much greater than the eddy diffusivity of air momentum.

For very small distances close to a surface, the effective eddy diffusivity for a particle can be an order of magnitude greater than that for air momentum transfer. This effective diffusivity is caused in part by the inertia of a particle being approximately 10^3 greater than an equal volume of air. The magnitude of this increased diffusivity makes the uncertainty in deciding to use the diffusivity of momentum or heat at larger heights a minor uncertainty of only about 1.4 (Businger et al., 1971).

The atmospheric eddy diffusivity can also be altered by a forest canopy. The diffusivity must change as a function of height as canopy elements interact with particle and gas pollutants. Nevertheless, it is the combined effects of diffusion and gravitational settling which bring particles and gases down through the canopy; however, neither of these changes can be predicted at the present time.

Possibly the most important factors influencing gas deposition or mass transfer are gas solubility in water (either open water or water on vegetation surfaces) and gas concentration in the water built up from prior deposition (Whitman, 1923; Bird, Stewart, and Lightfoot, 1960). If the gas reacts in the water phase, there could be no back pressure or resistance to additional surface mass transfer. However, if the gas dissolves without irreversible chemical reaction in the water, mass transfer can decrease as the amount of material deposited increases. Thus the mass-transfer rates should be a function of gas solubility in water.

12-2.3.3 *Surface Properties.* Many surface variables affecting deposition are listed in column 4 of Table 12.1. These include the accommodation or adhesion of particles as they strike surfaces (Romney et al., 1963). For instance, if a particle strikes a sticky or wet surface, it would not be so quickly reentrained as if it had hit another less accommodating surface. Deposition can also be affected by changed canopy structure at strong wind speeds, as when grass blades bend from their vertical orientation in strong winds. The effects of wind speed on canopy geometry and subsequent changes in deposition cannot be predicted from experimental field data.

Deposition rates for gases could be strongly influenced by the physiological state of vegetation. Gas mass transfer in vegetation is a function of whether the stomata on the leaf surfaces are open or closed, i.e., open during the day and closed at night (Bennett, Hill, and Gates, 1973; Bennett and Hill, 1975).

Few data have been collected to determine the penetration of depositing pollutants through a canopy. In some field experiments the deposited pollutant has been collected only from upper surfaces of the vegetation canopy. Other experiments have shown a deposition flux to the underlying surface. In wind-tunnel experiments with simple surfaces, Sehmel and Hodgson (1978) have demonstrated that penetration is a function of wind speed, particle diameter, and canopy structure. However, there are too few data to generalize the relative canopy penetration during deposition.

12-2.4 Resuspension Variables

Probably even more variables influence resuspension than deposition. Many of these are listed in Table 12.2 under these classifications: particle properties, soil properties, particle–soil interaction, surface properties, topography, and meteorological variables. Some variables in each column are discussed briefly.

12-2.4.1 *Particle Properties.* One of the most important variables influencing resuspension is particle diameter, shown in column 1 of Table 12.2. When this variable is being considered, the question really is, "What does particle diameter mean in terms of resuspension source characteristics?" At the present time the effects of the diameter of deposited pollutant particles on resuspension cannot be predicted.

The complicating factor in determining source characteristics is that, since pollutants become attached to host surface soil particles, source characteristics must be determined by the pollutant's distribution on the host particle. The methods used to determine particle size distributions in collected soil samples can change the pollutant distribution and even the distribution of host soil particles. Changes in particle attachment, for instance, can be expected in either liquid or air elutriation. Rosinski et al. (1976) have shown that respirable particles are released when a particle is dropped into liquid. Sehmel (1978a) has shown that respirable particles are released when a soil particle is dropped on a solid surface. In both methods, smaller respirable particles are released from host soil particles.

12-2.4.2 *Soil Properties.* The properties of soil that affect particle resuspension are listed in column 2 of Table 12.2. An overriding factor is the soil moisture content. As long as soil is saturated, resuspension will not occur; as the surface layer dries, however, resuspension can occur. The increased resuspension rate with decreasing surface soil moisture content is unknown.

The size and distribution of soil particles also influence resuspension. Soils are usually classified by the diameter of the particles comprising them; silt, clay, and sand identify

TABLE 12.2 Variables That Influence Resuspension

Particle properties	Soil properties	Particle–soil interaction	Surface properties	Topography	Meteorological variables
Chemical reactivity	Moisture content	Attraction forces	Areas of erodible surface	Agricultural	Air density, affected by
Density	Ratio of erodible to nonerodible fractions	Molecular forces Van der Waals forces	Cohesiveness of particles	Broken	Humidity
Hydrophobic		Chemical binding forces (Valence forces)	Binding action of materials	Desert	Pressure
Hydrophylic					
Particle diameter	Soil-size frequency distribution	Electrostatic forces	Surface moisture	Diurnal	Temperature
Respirable	Specific gravity	Magnetic forces	Cover	Flat	Soil burden carried
Nonrespirable	Structure	Capillary forces at the surface of aggregates filled with liquid	Obstructions	Forested	Influence on air velocity
Frequency Distribution	Organic matter	Chemical reaction	Roughness, large scale	Hills	Avalanching
Shape	Lime content	Contact area effect on adhesion	Mechanical turbulence	Humidity	Saltation
Surface roughness	Texture	Particle shape	Overall sheltering	Seasonal	Surface moisture content
	Temperature	Particle size	Roughness, small scale	Temperature	Dew
		Nature of particle and surface	Orientation of obstacles	Undulating	Frost upheaval
		Surface roughness	Percent of area covered by nonerodible aggregates or obstacles	Uniform	Relative humidity
		Adhesion of agglomerates vs. individual particles		Urban	Temperature
		Crystallization of dissolved substances	Sheltering of individual particles	Valley	Surface dry
		Deposition of suspended colloidal particle	Stability of aggregates against abrasion and disintegration by moisture and weathering	Water	Surface drying
					Surface frozen
		Hardening binders			
					Wind speed
		Liquid bridges	Temperature		Wind removal forces
		Mineral bridges	Topographic features		Mean wind speed
		Relative humidity	Broken		Frequency, period, and intensity of gusts penetrating to the surface
		Temperature	Uniform		
		Time of contact	Vegetative cover		Transient and steady drag
		Viscous surface coating	Live vegetation		Vertical wind profile
			Plant residue		Stability
					Vertical turbulent exchange
					Surface roughness
					Temperature stratification
					Wind direction

three types. Soil erosion and resuspension rates can be altered by the proportion of each type of soil and the abrading action of soil carried from upwind.

This alteration can also be a function of time. After precipitation and drying occur, a surface crust of salts can form. This crust tends to minimize soil erosion until it is broken by mechanical action or by the upheaval of soil caused by thawing and freezing (Caine and Morin, 1976).

12-2.4.3 ***Particle–Soil Interactions.*** Many interactions between particle and soil can influence the attachment of pollutant particles to host soil particles. These interactions are listed in column 3 of Table 12.2. Although soil–surface interactions have been measured (Corn and Stein, 1965; Corn, 1966; Walker and Fish, 1965; Punjrath and Heldman, 1972a, 1972b) in laboratory experiments for very simple surfaces, the results cannot be generalized to the more complex interactions in the environment.

Researchers investigating resuspension can, however, use some information about soil erosion. For instance, even though the effects of surface properties on resuspension are unknown, the greater the resistance of the canopy to wind erosion, the smaller the resuspension from underlying soil. The experience of wheat farmers substantiates the fact that vegetation tends to decrease soil erosion. We also know that soil erodes least at the upwind edge of a field and that erosion increases downwind across the field, caused in part because of the avalanching effect by soil particles as they strike and abrade downwind surfaces. This avalanching effect, which fans out across a field, has not been addressed in the literature on resuspension.

Studies also determined that resuspension can occur from vegetation as well as from soil. For instance, Aylor (1976) determined that large numbers of particles were resuspended from corn. Similar work has been done by Chamberlain (1970, 1974, 1975), Heinemann, Vögt, and Angeletti (1976a), Heinemann et al. (1976b), and Peters and Witherspoon (1972).

12-2.4.4 ***Topographical Variables.*** Although little is known about the effect of regional topography on resuspension other than visibility differences caused by airborne dust (Orgill and Sehmel, 1976), the effect of topography may influence average resuspension rates. Most resuspension literature discusses only desert or semiarid areas, rather than humid areas. Soil erosion data apply primarily to agricultural areas.

Bagnold's (1941) studies of erosion in the Sahara Desert are typical of the bulk of literature on erosion. However, the studies have limited applicability to resuspension within most of the United States because sand from the Sahara Desert is made up of larger particles than sand found in most areas of the United States where resuspension might occur. Bagnold also reported that he saw sand moving so close to ground level during a dust storm that only a man's head and shoulders were visible above the dust storm. Particles moving in this fashion are clearly not respirable particles and not the principal concern in studies of resuspension.

Although some data on soil erosion in agricultural areas might be applied in resuspension studies, few data have been collected to assess the resuspension that can occur in urban areas. Some data do show (Newman et al., 1976) that material is "refloated" by traffic on city streets. These suspended materials are primarily abraded tire and pavement particles and recycling pollutant particles. Only limited data quantify the reflotation rate (Sehmel, 1973c); see Figs. 12.11 and 12.12 discussed in Sec. 12-5.3. However, sometimes this urban resuspension is discussed in the literature as fugitive dust. A review of the literature on fugitive dust emissions is not included here.

12-2.4.5 ***Meteorological Variables.*** Although many meteorological variables are assumed to influence resuspension, their effects have not, for the most part, been quantified. During the development of the meteorological field, emphasis was mainly directed toward describing transport and diffusion with a mathematical boundary condition, or black-box region, which described mass transfer at the air–surface interface. In this region, especially the last few millimeters from the surface, wind stresses can cause resuspension. A detailed description of the black box is normally, of necessity, neglected in model calculations of transport and diffusion above the air–surface mass transfer. As indicated in Sec. 12-2.3.1, the air–surface mass transfer is even neglected in the mathematical concept of friction velocity and aerodynamic surface roughnesses, which describe wind stress.

The physics of turbulent wind stresses next to the surface are still being studied. In the literature on fluid flow, an undisturbed laminar boundary layer is no longer considered to occur next to a smooth surface. Air turbulence always penetrates to the surface intermittently and unpredictably, especially over rough surfaces. Turbulence can penetrate in a burst directed toward the ground or in a small dust devil. The author has seen small dust devils, on the order of 10 cm in height and spaced about 1 m apart, resuspend zinc sulfide tracer from an asphalt surface. These dust devils may have been caused by solar heating rather than by turbulence from the average air speed.

12-3 DEPOSITION PROCESSES

Field measurements of plume depletion and particle and gas dry deposition velocities are summarized in this section. Deposition velocities and fractional plume depletion are considered because airborne pollutant plumes are depleted at the air–surface interface by the dry deposition of particles and gases.

One variable affecting deposition and plume depletion is the available surface area, which is always greater in a vegetative canopy. The increased surface area is described by a leaf-area index, i.e., the canopy projected leaf area divided by the ground area covered by a plant stem. Leaf-area indexes are about 0.5 for natural grasslands, up to 3 for irrigated and fertilized pastures (Knight, 1973), 3 to 5 for mature row crops, 4 to 6 for deciduous forests, and even larger for evergreen forests (Whittaker and Woodwell, 1967). The total area index for a forest, including trunk and branch area, can be as large as 10.

Although these surface areas are important in determining the deposition rates, the overall effects of surface area on deposition velocities and plume depletion have not been quantified. Since field results cannot be generalized, model-derived dry deposition velocities are recommended for meteorological transport and diffusion models.

12-3.1 Experimental Plume Depletion

Plume depletion of particles and gases by dry deposition processes can be either determined experimentally or predicted from theoretical diffusion and transport meteorological models; however, the data base is not satisfactory for either method. Existing experimental plume-depletion data, as a function of distance, are summarized in Table 12.3. Although meteorological models attempting to predict deposition should agree with the experimental data, the models often lack validation. Until validation occurs, there is little reason to select one predictive model over another. One stumbling block in validating the models is that the particle diameters used in the initial plume-depletion experiments were unknown or were not necessarily recorded in the literature. Since deposition velocities are a function of particle diameter, determining the diameter is crucial for validating transport models.

Experimental determinations have shown that a particle plume depletes with up to 97% of the particles being removed across a distance of 3200 m (Simpson, 1961), 65% between 45 and 91 m (Sehmel and Hodgson, 1976), and 80% at a distance of 200 m (Start, 1970a, 1970b). Data for iodine, a gas, indicate plume depletion is only up to 3% at a range of 380 m (Zimbrick and Voillequé, 1969). A comparison of these depletion percentages and the distances over which depletion occurred shows that these data cannot be generalized. This experimental data base is too limited to generalize plume-depletion characteristics for particles and gases. To predict plume depletion for other situations, transport and diffusion meteorological models must be used even though the meteorological models need extensive experimental validation.

12-3.2 Dry Deposition Velocities for Particles

The best existing method to predict plume depletion is by the particle dry deposition velocity. Dry deposition velocities for many materials and various deposition surfaces including soils are summarized in Table 12.4. In the table, the velocities for different materials are listed as a function of particle diameter. The results of studies in which particle diameters were not reported are arranged at the end of the table in alphabetical order by each investigator. If possible, the range of experimental deposition velocities for each experiment is presented rather than the "average" deposition velocities. This range is critical in emphasizing the uncertainties in many deposition experiments and in our ability to predict deposition mass-transfer rates. The deposition velocities reported range over 5 orders of magnitude from a minimum of 10^{-3} cm/s to 180 cm/s. It is expected, and is shown if known, that the deposition velocity is greater than the gravitational settling velocity, v_t.

Most data gained through field experiments, however, cannot be extrapolated to other situations because the variables have not been adequately controlled. At the present time, for instance, field deposition data do not differentiate between experiments as a function of any particle property other than particle diameter. Furthermore, deposition results are often reported when a particle diameter distribution was studied rather than a single or a narrow particle diameter range. Although we know that

TABLE 12.3 Experimental Plume Depletion as a Function of Distance

Depositing material	Particle diameter if specified, μm	Air speed, m/sec u	u_*	Atmospheric stability	Deposition surface	Deposition distance range, m	Plume depletion, %	Deposition velocity, cm/s	Reference
Uranine	1 median	2.5–8.4			Sagebrush	200	22–32		Islitzer and Dumbauld, 1963
						400	10–54		
ZnS					Sagebrush	200	0–1		Nickola and Clark, 1976
						842	2.8–5.7		
Uranine	0.7	1.5			Sagebrush, 1.2–1.5 m high	45–91	65		Sehmel and Hodgson, 1976
ZnS					Sagebrush	200	19–45		Simpson, 1961
						800	40–82		
						1600	58–93		
						3200	78–97		
ZnS					Cheat grass	3200	70–75		Simpson, 1964
Uranine		3–8				To 200	80		Start, 1970a, 1970b; Van der Hoven, 1970
Iodine			0.4	Unstable	Pasture	100–380	3.0	1.0	Zimbrick and Voillequé, 1969
			0.6	Unstable	Pasture	100–380	1.1	0.7	
			0.6	Unstable	Pasture	100–300	1.1	0.6	
			0.6	Unstable	Pasture	100–380	1.0	1.2	
			0.4	Stable	Pasture	50–300	1.3	0.1	
			0.1	Stable	Pasture	50–100	0.7	0.2	

TABLE 12.4 Dry Deposition Velocities for Particles

Depositing material	Particle size if specified		Air speed, cm/s		Atmospheric stability	Deposition surface	Deposition velocity, cm/s	Reference
	Diameter, μm	Other	u	u*				
Particles	0.03–30			10–200		Z_0 from 0.001–10 cm	10^{-3}–40	Sehmel and Hodgson, 1978
Pollen	20	v_t = 1.5 cm/s				Microscope slides in grassland	4.5 (3 x v_t)	Raynor, 1976
	32–35	3.3					9.9	
	90–100	20					60	
Pollen	20	v_t = 1.5 cm/s				Microscope slides in forest	1.5 (1 x v_t)	Raynor, 1976
	32–35	3.3					3.3	
	90–100	20					20	
Lycopodium spores	32		110–920			Glass plate	0.7–3.5	Gregory, 1950
Lycopodium spores	32	v_t = 1.4–2.1 cm/s	104–740	12–94		Grassland	0.7–3.5	Chamberlain, 1966
Lycopodium spores	32					Moss bags	9 x (grass v_d)	Clough, 1975
Oleic acid	0.5					Moss bags	15 x (grass v_d)	Clough, 1975
Uranine	2		10^2–10^3			Pine and oak shoots	0.003–10	Belot and Gauthier, 1975
	5						0.02–30	
	10						0.1–60	
Plutonium		0.3 MMD*				Tray	7.3	Stewart, 1963
		16–29 MMD					3.2	
Rhodamine B	0.8					Snow	<0.6–19	Nickola and Lee, 1975
ZnS	5 MMD					Snow	0–31	
ZnS	5					Sagebrush	1.5–5.4	Nickola and Clark, 1976
$CuSO_4$	3–7			8–34		Paper	0.005–0.06	Horbert, Vögt, and Angeletti, 1976
						Smooth metal	0.007–0.08	
						Rough metal	0.006–0.007	
						Grass	0.015–0.15	
						Clover	0.018–0.15	
						Soil	0.01–0.05	
ZnCdS	3	v_t = 0.13 cm/s				Artificial surface in forest	0.1–0.5	Fritschen and Edmonds, 1976
ZnS	2.5				Stable	Sagebrush	0.5	Simpson, 1961
Natural aerosol	1–10		600			Grass sward	0.8	Clough, 1973
Blowing field		<6 m, most ≈1 μm					9.8	Porch and Lovill, 1976
Plutonium		1–2 AMAD†	0.4			Leaf area	4×10^{-3}–8×10^{-3}	Cataldo, Klepper, and Craig, 1976
$Pu(NO_3)_4$		0.5–1.7	1			Single bean leaf	2.6×10^{-3}–2.7×10^{-2}	Craig, Klepper, and Buschbom, 1976
Pu compounds		0.8–1.8	1			Plant canopy	3.5×10^{-3}–1.2×10^{-2}	Craig, Klepper, and Buschbom, 1976
Uranine	<1					Sand	0.03–0.3	Islitzer and Dumbauld, 1963
^7Be	<1					Ocean with rain	<0.6–5.3	Silker, 1976
Au colloid	0.8					Bean leaf	4×10^{-3}	Vaughn, 1976
Particles	0.05–0.1			5.5–20		Field	Increases with u*, 0.1–1.1	Wesely et al., 1977
Beta–gamma Kiwi effluent	Median	σ_g				Resin-coated tray at 0.8 m height at distance, m		Henderson and Fultyn, 1965
	0.003–0.6					4900	0.4–0.8	
	0.006	16				9800	2–85	

TABLE 12.4 (Continued)

Depositing material	Particle size if specified: Diameter, μm	Particle size if specified: Other	Air speed, cm/s: u	Air speed, cm/s: u*	Atmospheric stability	Deposition surface	Deposition velocity, cm/s	Reference
	0.1	8				39000	0.1–3	
	0.1–0.4					1220	0.2–0.4	
	0.6	4				1220	0.07–0.50	
	0.2–1.4					2440	0.8–110	
	0.02–0.2					9800	0.3	
	0.02–4					1220	0.4–2.5	
	0.4–6					610	7–110	
	0.1–8					1220	0.9–100	
	2.8	4				4900	0.3–3	
	2.7–6					2440	0.4–5	
	3.9	3				2440	0.03–3	
	2.3–4					1220	0.5–6	
	5	4				4900	48–180	
Pb, auto exhaust	Aggregated					Whole leaves	0.008–0.12	Little and Wiffen, 1977
						Leaf disks	0.001–0.17	
	Aggregated exhaust, 0.04–0.01					Moss bags	0.1–2	
						Bare soil	0.005–0.05	
						Filter paper	0.003–0.1	
						Moss bags near highway	0.7–39	
	Nonaggregated (aggregated)					Grass	0.13 (0.019)	
						Soil under grass	0.015 (0.007)	
						Bare soil	0.035 (0.008)	
						Grass and soil	0.14 (0.026)	
Trace elements						Filter paper at 1.5 m	<0.4–1	Cawse and Peirson, 1972
							0.05–4.8	Cawse, 1974
							<0.01–>3	Cawse, 1975
Trace elements						Filter paper at 1.5 m	0.06–13	Cawse, 1976
Na							0.2–4.3	
Mg							0.6–>3	
Al							0.9–2.7	
Cl							0.2–6.3	
K							0.6–13	
Ca							0.4–1.4	
Sc							0.6–2.5	
Ti							0.7–2.2	
V							0.2–<0.7	
Cr							0.6–6.8	
Mn							0.4–0.9	
Fe							1.0–2.5	
Co							0.3–1.9	
Ni							0.7–<2	
Cu							<0.6–1.1	
Zn							0.4–4.5	
As							<0.1–<0.6	
Se							0.1–0.6	
Br							0.1–2.6	
Rb							<0.9–3.0	
Ag							<0.3–1.1	
Cd							<0.4–>8	
In							0.3–>6	
Sb							0.06–<0.4	
I							<0.3–<2	
Cs							0.2–0.6	
La							>0.6–3.5	

(Table continues on the next page.)

TABLE 12.4 (Continued)

Depositing material	Particle size if specified: Diameter, μm	Particle size if specified: Other	Air speed, cm/s: u	Air speed, cm/s: u_*	Atmospheric stability	Deposition surface	Deposition velocity, cm/s	Reference
Ce							0.5–1.9	
Sm							0.6–4.1	
Eu							0.3–>2	
Th							0.7–2.2	
Trace elements						Open bucket	0.2–6.4	Crecelius et al., 1978
Cl							6.4	
K							2.6	
Ca							2.6	
Br							2.3	
Fe							0.37	
Mn							0.59	
Ti							0.67	
Cr							0.60	
Pb							0.38	
Se							0.24	
As							0.70	
V							0.24	
Cu							3.0	
Ni							4.0	
Zn							1.8	
^{7}Be							0.96	
RaB and RaC						Filter paper	0.05	Bullas published by Chamberlain, 1960
Field tests			Deposition increases with u	Independent of lapse rate				Convair, 1960
^{137}Cs						Water	0.09 ± 0.06	
						Soil	0.04 ± 0.05	
						Grass	0.2–0.5	
						Sticky paper	0.2–0.4	
^{103}Ru						Water	2.3 ± 1.0	
						Soil	0.1 ± 0.2–<0.6 ± 0.4	
						Grass	0.02–0.8	
						Sticky paper	0.2–2.0 ± 1.0	
^{95}Zr–^{95}Nb						Water	5.7 ± 3.4	
						Soil	2.9 ± 2.7	
						Sticky paper	1.4 ± 0.7	
^{141}Ce						Sticky paper	0.7 ± 0.6	
^{127}Te, ^{129}Te						Sticky paper	0.7 ± 1.3	
Pb, auto exhaust						Teflon	0.29	Davidson, 1977
Pb				Light	Stable	Snow	0.16–2.0	Dovland and Eliassen, 1976
Sulfate				Light	Stable	Snow	0.2–2.9	
Fission products						Desert vegetation	3	Fuquay, 1957
Fluorides						Forage	0.3 ± 0.02	Israel, 1977
						Vertical limed filter paper	0.49 ± 0.19	
Beta activity						Sticky plate	1.0	Makhonko, 1970
Mixed fission products						Grass	0.04–0.3	Megaw and Chadwick, 1956
						Grass substrate	0.02–0.17	
						Filter paper	0.03–0.16	
Ba						Grass	0.001–0.003	
						Substrate	0.001–0.006	
Sr						Grass	0.002–0.004	
						Substrate	0.003–0.01	
Mixed fission products, Ba, Sr						Oats		
						Ears	0.001–1.2	
						Stalks	0.004–1.8	
						Water	Function of particle diameter	Möller and Schumann, 1970; Schumann, 1975

TABLE 12.4 (Continued)

Depositing material	Particle size if specified		Air speed, cm/s		Atmospheric stability	Deposition surface	Deposition velocity, cm/s	Reference
	Diameter, μm	Other	u	u*				
						Water	Function of particle diameter and u	Sehmel and Sutter, 1974
Pb						Pluviometer	0.13	Servant, 1976
Atmospheric fallout						Fallout collectors		Small, 1960
						Annual	0.5–0.9	
						Monthly	0.2–3.4	
Atmospheric fallout						Sticky paper	0.07	Stewart et al., 1957
RaD						Tank	0.05–0.5	Styro and Shalaveyus, 1966
Natural aerosol						Great Lakes	0.8–7.6	Whelpdale, 1976
Lead, bromine, zinc						Urban St. Louis,		Young, 1978
						1973	2.2 ± 1.4	
						1975	1.8 ± 0.7	

*MMD = mass median diameter.
†AMAD = activity median aerodynamic diameter.

particle diameter greatly influences deposition velocities, other variables such as wind speed do also. Convair (1960) has shown that field deposition velocities increase with wind speed, even though in these field tests the deposition velocities were independent of atmospheric stability. Sehmel and Hodgson's (1978) model supports these tests by predicting that deposition velocity is independent of atmospheric stability.

Determination of deposition velocities from artificial surfaces in the field presents other problems. Artificial collection surfaces, such as fallout collectors, are often used to determine deposition velocities because of the formidable task of determining deposition velocities in the field. For fallout collectors to be useful in estimating plume depletion, they should have many of the characteristics of the surfaces being simulated. However, when deposition is directly measured on environmental surfaces, the question always arises as to whether or not enough surface has been sampled to obtain a representative deposition sample. When vegetation is being sampled, the total deposition rates must be determined as a function of height throughout the canopy as well as on the soil surface below the canopy.

Although much deposition velocity data are presented in Table 12.4 for both artificial and natural surfaces, little is known about the relationship between deposition on fallout collectors and deposition on natural surfaces. Currently the results from fallout collectors indicate only changes with downwind distance and time rather than the total deposition on natural surfaces.

Diffusion and transport models have been used to infer deposition velocities from limited field data. Predictions from these models are suspect at the present time because so little is known about mass-transfer processes at the air–surface interface and about dispersion coefficients as a function of height. Until data are sufficiently defined for these parameters, the uncertainties inherent in making deposition velocity predictions from diffusion and transport models must always be considered. The uncertainties are present since even diffusion models without deposition have limited validation data, even to 10 km (Crawford et al., 1977, 1978).

12-3.3 Dry Deposition Velocities for Gases

Values for the deposition velocities of gases are summarized in Table 12.5. Gases are listed alphabetically rather than in the order of their measured deposition velocities because the range of deposition velocities in any experiment is so wide. In fact, uncertainties in the deposition velocities for a single field experiment can range at least 1 order of magnitude. Dry deposition velocities for all gases range 4 orders of magnitude, from 2×10^{-3} up to 26 cm/s, only 1 order of magnitude less than the range for particle dry deposition, which is from 10^{-3} to 180 cm/s. Thus the range of deposition velocities for gases is comparable to that for particles.

Reported deposition velocities measured in field experiments will be considered to show the great range even for individual gases. The deposition velocities for iodine and SO_2 have been studied extensively because these gases could be effluents from nuclear (iodine) and nonnuclear (SO_2) industries. Even though studied extensively, these deposition velocities are so widely scattered that they could not be confidently predicted. The deposition velocity of iodine ranges 3 orders of magnitude, from 0.03 to 26 cm/s; the deposition velocity of SO_2 ranges 2 orders of magnitude, from 0.04 to 7.5 cm/s.

Although the deposition velocity of particles cannot be predicted with certainty, there is some indication that the deposition velocities for gases may depend on atmospheric stability (Bunch, 1968). Whelpdale and Shaw (1974) have shown a dependency on atmospheric stability, but the uncertainties are not stated in the deposition velocities and an inconsistency in the data is unexplained. They report that the deposition velocities to grass and snow for neutral conditions are greater than those for stable conditions. However, the deposition velocity to a water surface for a neutral atmosphere is greater than that for a stable atmosphere.

12-3.3.1 ***Water Solubility.*** Some uncertainties about the deposition velocity for gases are a function of surface conditions. For example, some gases are more water soluble than others; other gases can react within the water phase. These differences in solubility can cause a gas back pressure or equilibrium gas pressure to develop at the receptor surface which can resist dry deposition from the atmosphere. Some of these gas transfer problems are considered for water deposition by Slinn et al. (1978). Hill (1971) has completed one of the more extensive experimental studies of gas mass transfer onto potential environmental surfaces. His data have been recalculated to determine the deposition velocities shown in Table 12.5.

12-3.3.2 ***Gas Reactions.*** An important uncertainty in predicting deposition velocities is the reaction of gases with condensation nuclei and other small particles in the atmosphere during transport. After reaction, the gas loses its identity and is transported as the host particle with which the gas initially reacted. Consequently mass-transfer rates will be altered rapidly as gases are modified while airborne. Once the gas has become attached to a particle, Brownian transport and the deposition velocity decrease because the Brownian diffusion coefficient of a particle is much less than that of a gas.

12-3.4 Predicted Dry Deposition Velocities

Sehmel and Hodgson (1978) predicted deposition velocities, $K_{1\ m}$, as a function of particle diameter from 10^{-2} to 10^{2} μm, of friction velocities from 10 to 200 cm/s, of aerodynamic roughness heights from 10^{-3} to 10 cm, of particle densities from 1 to 11.5 g/cm^3, and of atmospheric stabilities for Obukhov's lengths from −10 to +10 m (unstable and stable atmospheres, respectively). These predictions were based on wind-tunnel-derived correlations for surface mass-transfer resistances for depositing particles. Predictions indicate that deposition velocities can range over several orders of magnitude from about 10^{-3} up to 10 cm/s. Moreover, they increase as roughness height increases, usually as friction velocity increases, and they are nearly independent of atmospheric stability.

Since field deposition experiments abound in uncertainties, it is encouraging that deposition velocity predictions are in the same range as those determined in field experiments. The validity of the deposition velocity model was also supported in a field experiment (Sehmel, Sutter, and Dana, 1973). For a polydispersed aerosol the predicted deposition velocity of 0.17 cm/s compared favorably with the experimental measurement of 0.21 cm/s. Model predictions also indicate the functional dependency of deposition velocity on the several controlling parameters. For a concentration reference height of 1 m and a constant friction velocity of 30 cm/s, predicted deposition velocities, $K_{1\ m}$, are shown in Fig. 12.3 as a function of aerodynamic surface roughness and particle density. In all cases the predicted deposition velocities are greater than the particle's gravitational settling velocity, i.e.,

$$K_{1\ m} \geqslant v_t \tag{12.10}$$

The gravitational settling velocity increases proportionally with particle density and the square of particle diameter. Only in the particle diameter range from about 0.1 to 1 μm is the deposition velocity nearly constant for a selected surface roughness, particle density, and friction velocity. For particle diameters larger than about 1 μm, deposition velocities increase because of an increase in eddy diffusion and gravitational settling. For large particles, deposition velocities approach their respective gravitational settling velocity. The deposition velocity of particles less than about 0.1 μm increases with decreasing particle diameter because of Brownian diffusion. Figure 12.3 shows the lower limits for deposition velocities calculated only from Brownian diffusion. On the left-hand side of the figure, a lower limit is shown for distances of 1 cm and 0.01 cm. For the calculation of these lower limits, only Brownian diffusion was assumed to cause mass transfer below the respective distance, whereas both Brownian diffusion and atmospheric diffusion were assumed to cause mass transfer for greater distances. These lower limits are a function of each distance. Thus Brownian diffusion is a controlling factor. When the controlling diffusion distance decreases from 1 cm to 0.01 cm adjacent to the deposition surface, the lower limit for deposition velocities increases nearly 2 orders of magnitude. Obviously predicted deposition velocities for small particles are dependent on Brownian diffusion adjacent to the deposition surface.

Upper limits for dry deposition velocities are also shown in Fig. 12.3. For these calculations the surface resistance to mass transfer within 1 cm of the surface was assumed to be zero. Between 1 cm and 1 m, deposition velocities were calculated by including only atmospheric diffusion (Sehmel, 1970, 1973b) and gravitational settling. For particle diameters less than 1 μm, the upper limit is nearly constant and decreases from 1.1 cm/s at 1 μm to 1.08 cm/s at 10^{-3} μm. Calculations were simplified by using the ground surface area rather than the total canopy surface area as the deposition surface. For the upper limit, deposition velocities increase for particle diameters greater than 2 μm and approach their respective terminal settling velocity. However, most surfaces are not smooth, and there

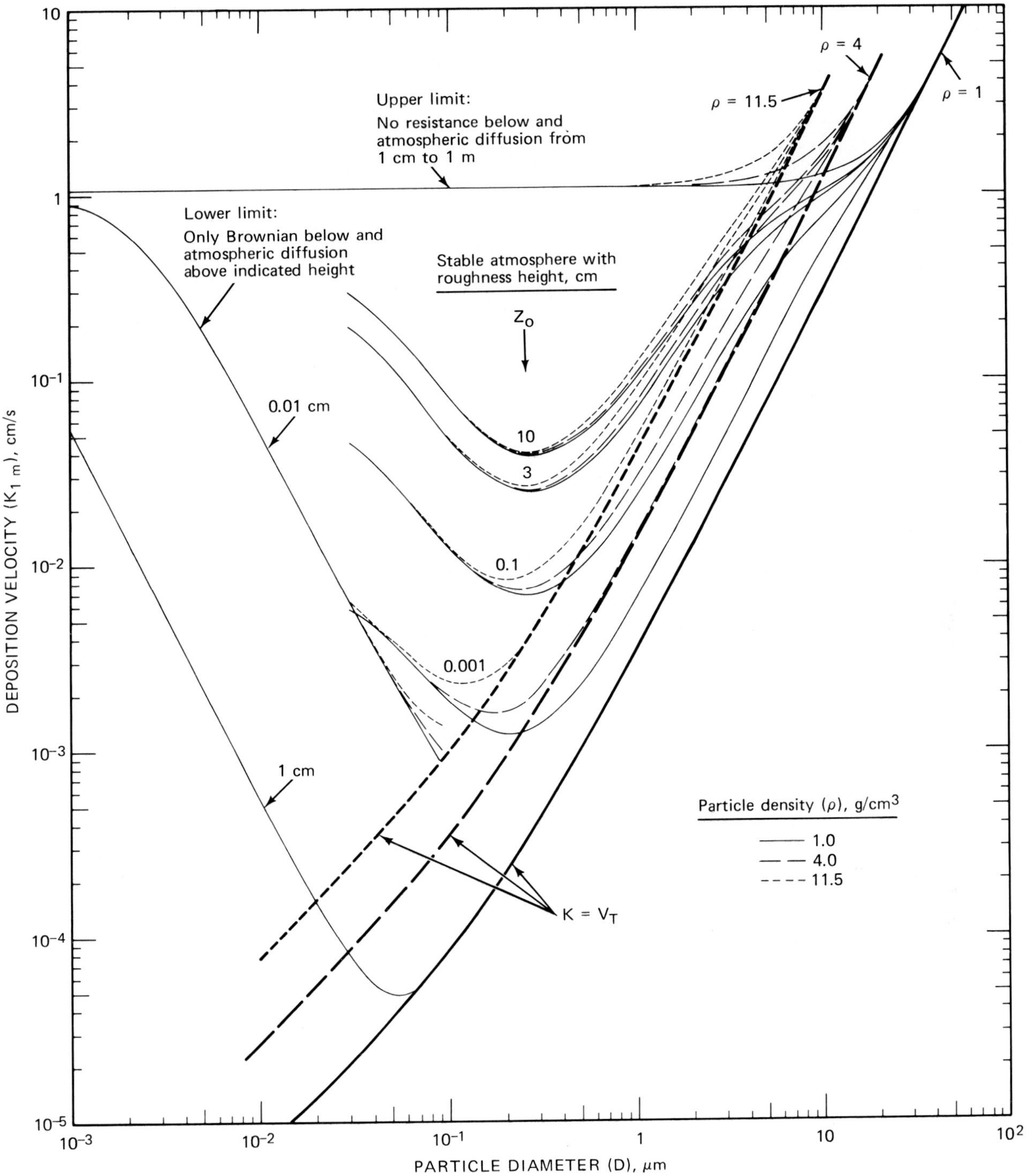

Fig. 12.3 Predicted deposition velocities at 1 m for $u_* = 30$ cm/s and particle densities of 1, 4, and 11.5 g/cm^3.

is greater surface area available in the canopy than on the underlying ground surface.

Figures 12.4 to 12.7 show predicted deposition velocities for a range of friction velocities and as functions of particle diameter, aerodynamic surface roughness, and particle density. Although predictions are shown only for a stable atmosphere, further predictions by Sehmel and Hodgson (1978) indicate that atmospheric instability increases the deposition velocities. The increase is small compared to the effects of particle diameter, friction velocity, and aerodynamic surface roughness. The deposition velocity of particles with diameters outside the range from 0.1 to 1 μm always increases with an increase in friction velocity. There is a minimum deposition velocity of 0.025 cm/s, however, which occurs for a friction velocity in the range of 20 to 30 cm/s. It is not known whether this minimum is physically real or only reflects uncertainties in the prediction techniques.

TABLE 12.5 Dry Deposition Velocities for Gases

Depositing material	Deposition surface	Deposition velocity, cm/s	Reference
Cl_2	Alfalfa	1.8–2.1	Calculated from Hill, 1971
CO_2	Alfalfa	0.3	Calculated from Hill, 1971
Dimethyl sulfide		Maximum rates	Judeikis and Wren, 1977
	Adobe clay soil	0.28	
	Sandy loam soil	0.064	
Fluorides	Forage	1.9 ± 0.5	Israel, 1977
	Vertical limed filter paper	0.36 ± 0.06	
HF	Field crops	3.1 ± 0.6	Israel, 1974
	Alfalfa	1.6	
HF	Alfalfa	3.5	Calculated from Hill, 1971
H_2S		Maximum rates	Judeikis and Wren, 1977
	Adobe clay soil	0.016	
	Sandy loam soil	0.015	
	Fe_2O_3	0.38	
^{85}Kr	Grass	Upper limit, 2.3×10^{-11}	Voillequé, Adams, and Echo, 1970
Iodine	Grass	Upper limit, 7.2 ± 0.9	
Iodine	Water	Increases with u, 0.2–1.2	Allen and Neff, 1975
Iodine	Activated-charcoal fallout plate		Bunch, 1968; Zimbrick and Voillequé, 1969; Markee, 1971
	Lapse	0.38–0.87	
	Inversion	0.03–0.30	
Iodine	Snow, neutral	0.34 median (0.085–0.41)	Bunch, 1968
Iodine	Open-range crested wheat grass	Median ± 1 σ in one test (range)	Bunch, 1968
	Growing	0.85 (0.033–4.2)	
	Damp	0.15 (0.033–0.66)	
	Damp	0.10 (0.019–0.225)	
	Dusty, dry	0.14 (0.13–0.15)	
	Dusty, dry	0.28, 0.54	
Iodine	Mixed pasture grass	Median ± 1 σ in one test (range)	Bunch, 1968
	Unstable		
	Dry	1.2 ± 1.6 (0.66–3.4)	
	Dry	1.0 ± 1.2 (0.42–3.2)	
	Growing	0.61 ± 1.6	
	Growing	0.79 ± 2.0 (0.15–6.2)	
	Neutral		
	Growing	0.79 ± 1.5 (0.14–2.0)	
	Stable		
	Green	3.3 ± 3.3 (1.4–26)	
	Freshly mown	1.7 ± 2.6 (0.21–6.2)	
	Freshly mown	0.41 (0.26–4.8)	

TABLE 12.5 (Continued)

Depositing material	Deposition surface	Deposition velocity, cm/s	Reference
	Freshly mown	0.23 (0.025–0.5)	
	Green	0.087 (0.021–0.41)	
Iodine	Grass	1.2–3.8	Chamberlain and Chadwick, 1953
Iodine	Grass	0.9–1.6	Chamberlain and Chadwick, 1966
Iodine	Field $u_* = 15$ to 48 cm/s $z_0 = 1$ to 5 cm		Chamberlain, 1960
	Soil + grass	1.1–2.6	
	Leaves	0.3–1.3	
	Paper leaves	0.6–2.0	
	Paper in Petri dish	0.3–0.9	
Iodine	Field grasses	0.3–0.6	Cline, Wilson, and Hungate, 1965
Iodine	Field		Convair, 1960
	Grass	0.5–4.0	
	Sticky paper	0.02–0.9	
Iodine	SL-1 accident		Gifford and Pack, 1962
	1 km	0.25	
	8.5 km	0.21	
	67 km	0.23	
Iodine	Pasture grasses	0.4–1.1	Hawley and Markee, 1964
Iodine	Routine releases, Hanford	2.8	Parker, 1956
Iodine	Grass ($u_* = 10$ to 69 cm/s)	0.3–2.7	Heinemann, Vögt, and Angeletti, 1976a
Iodine	Grass	0.12–8.0	Heinemann et al., 1976b
	Average dry	0.3–2.8	
	Average damp	0.9–6.3	
	Clover	1.0–4.2	
Iodine	Pasture grasses	0.2–1.2, function grass density	Markee, 1967
Iodine	Field	2.2	Megaw and Chadwick, 1956
Iodine with HIO_3, HIO_4, and organic iodides	Soil	2.7×10^{-2}	Miyamaga, Kasai, and Imai, 1973
	Sand	2.5×10^{-3}	
	Water	1.0×10^{-3}	
	"Leaf area"		
	Pine needle	2.0×10^{-3}	
	Leaf beet	2.1×10^{-3}	
	Spinach	1.1×10^{-3}	
	Chinese cabbage	7×10^{-4}	
Iodine	Windscale accident		Stewart and Crooks, 1958
	North England	0.30	
	South England	0.11	
Iodine	Grass	0.12–6.9	Vögt et al., 1973
Iodine	Mixed pasture grass	2.1, 2.4	Zimbrick and Voillequé, 1969
Methyl iodide	Pasture grass	1.4×10^{-4} – 2.4×10^{-3}	Atkins, Chadwick, and Chamberlain, 1967
Methyl iodide	Activated-charcoal fallout plate	0.12	Bunch, 1968
Methyl iodide	Mixed pasture grass	10^{-4}–10^{-2}	Bunch, 1968
Methyl iodide	Grass	(0.9% of that for molecular iodine)	Heinemann et al., 1976b
Methyl iodide	Mixed pasture grass	(<0.05% for molecular iodine)	Zimbrick and Voillequé, 1969
NO	Grass	Minus to 0.9	Droppo et al., 1976
NO	Alfalfa	0.1	Calculated from Hill, 1971
NO_2	Alfalfa	1.9	Calculated from Hill, 1971

(Table continues on the next page.)

TABLE 12.5 (Continued)

Depositing material	Deposition surface	Deposition velocity, cm/s	Reference
NO_X	Alfalfa and oats	0.5	Rasmussen, Taheri, and Kabel, 1974, calculated from Tingey, 1968
NO_X	Grass	Minus to 0.02	Droppo et al., 1976
O_3	Juniper bush	2.0	Calculated from Aldaz, 1969
	Alaska tundra	0.7	
	Sand or dry grass	0.5	
	Grass, Australia	0.4	
	Grass, Nebraska	0.2	
	Snow	0.1	
	Water, fresh	0.07	
	Water, sea	0.04	
	Water, distilled	0.02	
	Clean Mylar	0.002	
O_3	Grass	Minus to 0.03	Droppo et al., 1976
O_3	Grass	0.2–0.6	Droppo and Doran, 1978
O_3	Short dry grass	0.09–0.9	Calculated from Galbally, 1968
O_3	Near neutral		Galbally, 1971
	Dry soil mass		
	z_0 ~0.1 cm	1.3–1.7	
	z_0 ~0.4 cm	0.7	
	z_0 ~0.9 cm	0.9–1.2	
	Tundra		
	z_0 ~1 cm	1.4	
	Vegetation		
	z_0 ~1.6	0.7–1.5	
O_3	Soil 4% H_2O	1.76	Garland, 1976a
	Soil 27% H_2O	0.84	
	Peat 43%	1.39	
	Peat 74%	0.46	
	Grass	0.55	
	Grass + SO_2	0.74	
	Fuller's earth	0.73	
	Kaolin	0.49	
	Talc	0.33	
	Talc + exposure ozone	0.16	
	Sand	0.14	
	$CaCO_3$	0.22	
	$CaCO_3$ + SO_2	0.38	
O_3	Alfalfa	1.7	Calculated from Hill, 1971
PAN	Alfalfa	0.8	Calculated from Hill, 1971
SO_2	Grass	Minus to 2.4	Droppo et al., 1976
SO_2	Snow	0.07–0.38	Dovland and Eliassen, 1976
SO_2	Grass, $u_* = 0.07$ to 0.33 m/s	0.2–2.06	Garland et al., 1974
SO_2	Forest	<2	Garland, 1976a
	Fresh water	0.46	
	Short grass	0.55	
	Medium grass	0.77–1.9	
	Calcareous soil	1.1	
SO_2	Ice	0.10–0.17	Garland, 1976b
SO_2	Alfalfa	2.7	Calculated from Hill, 1971
SO_2		Maximum rates	Judeikis and Stewart, 1976
	Cement	2.5	
	Ready-mixed cement	2.0	
	Exterior stucco	1.8	
	Cement	1.6	
	Exterior stucco	0.86	
	Adobe clay soil	0.66	
	Sandy loam soil	0.65	
	Asphalt	0.04	
SO_2		Maximum rates	Judeikis and Wren, 1977

TABLE 12.5 (Continued)

Depositing material	Deposition surface	Deposition velocity, cm/s	Reference
	Adobe clay soil	0.92	
	Sandy loam soil	0.60	
	Fe_2O_3	3.9	
SO_2	Great Britain	1.3	Meetham, 1950
SO_2	B stability		Owers and Powell, 1974
	u = 2.6 m/s		
	Grass	0.5–0.7	
	Water	0.7–1.1	
	D stability		
	u = 5.2 m/s		
	Upwind grass	2.1–2.8	
	Hedge	4.9	
	To 30-m downwind hedge	1.4–1.8	
	D stability		
	u = 1.7 to 1.9 m/s		
	Grass	0.7 ± 0.1	
	Water	0.5 ± 0.1	
SO_2	Grass, $u_* = 0.2$ to 0.4 m/s	0.27–1.5	Shepherd, 1974
SO_2	Wheat	0.1–0.8	Unsworth and Fowler, 1976
	17-m forest	<0.6	
SO_2	Grass		Whelpdale and Shaw, 1974
	Lapse	2.4	
	Neutral	2.6	
	Stable	0.5	
	Snow		
	Lapse	1.6	
	Neutral	0.52	
	Stable	0.05	
	Water		
	Lapse	4.0	
	Neutral	2.2	
	Stable	0.16	
SO_2	Urban St. Louis		Young, 1978
	1973	3.8 ± 1.7	
	1975	6.2 ± 1.3	
ThB	Grass		Chamberlain, 1962
	Gas on 0.08-μm	2.6	
	nuclei	0.077	

Estimates of both the aerodynamic surface roughness height, z_0, and the air friction velocity, u_*, are needed to predict deposition velocities. Aerodynamic surface roughness is about 0.15 of the vegetation and physical roughness height (Plate, 1971). This simple relationship does not attempt to describe any changes in surface roughness which occur as the wind speed changes, e.g., a field of long grass becoming smooth during high wind speeds. The aerodynamic surface roughness of different surfaces at different wind speeds is given in Table 12.6, along with calculated friction velocities based on Eq. 12.8 for the average air speed at a height of 2 m. The magnitude of the friction velocity ranges from 3 to 15% of the air velocity at that height.

12-4 RESUSPENSION FACTORS

The resuspension factors reported in this section can be used to make conservative estimates of the potential hazards through inhalation. However, these factors were determined in experiments without differentiation between respirable and nonrespirable airborne particles.

12-4.1 Resuspension Factors Caused by Wind

Resuspension factors attributed to wind rather than to mechanical stresses are given in Table 12.7. These values have been either reported in or calculated from the

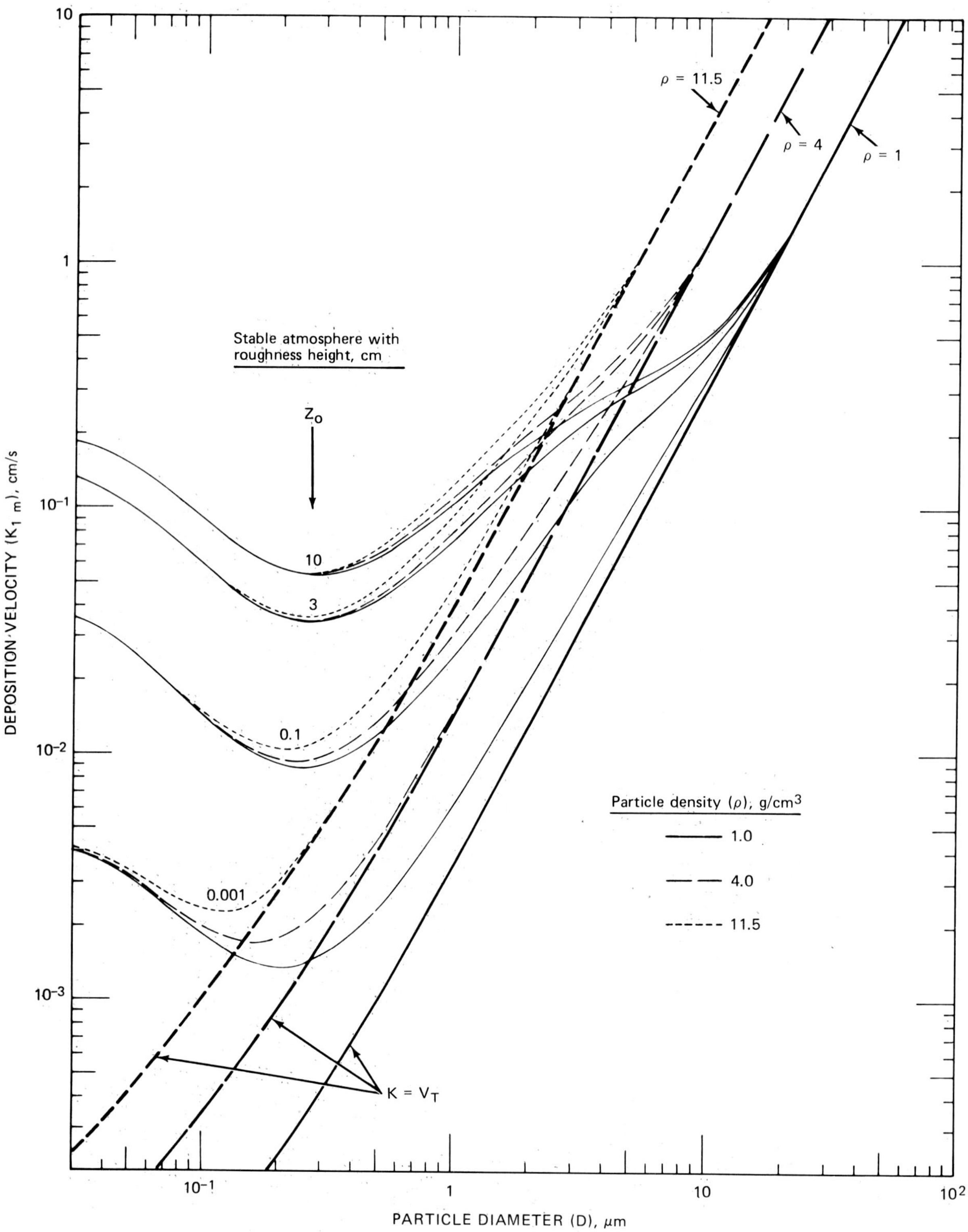

Fig. 12.4 Predicted deposition velocities at 1 m for $u_* = 10$ cm/s and particle densities of 1, 4, and 11.5 g/cm^3.

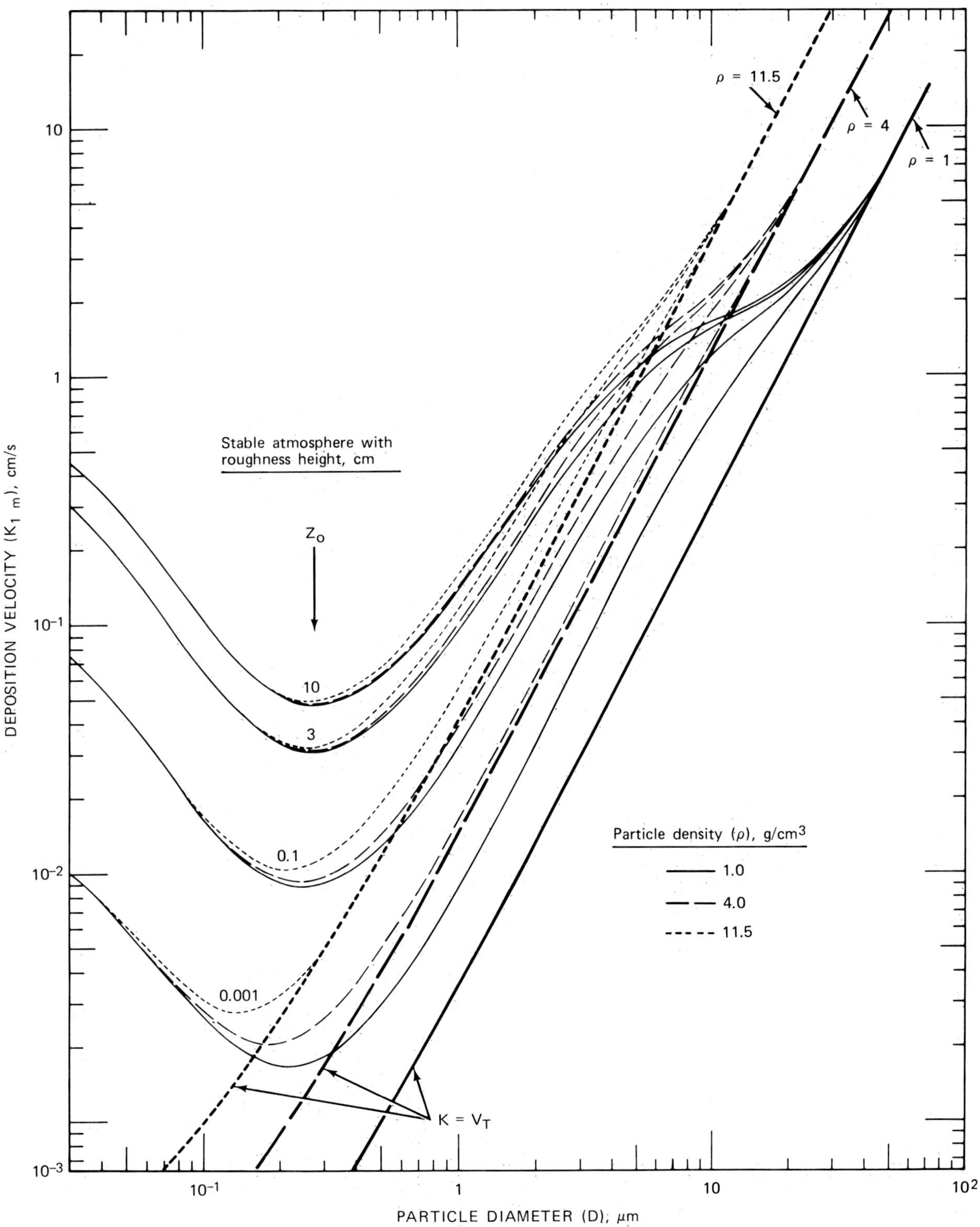

Fig. 12.5 Predicted deposition velocities at 1 m for $u_* = 50$ cm/s and particle densities of 1, 4, and 11.5 g/cm^3.

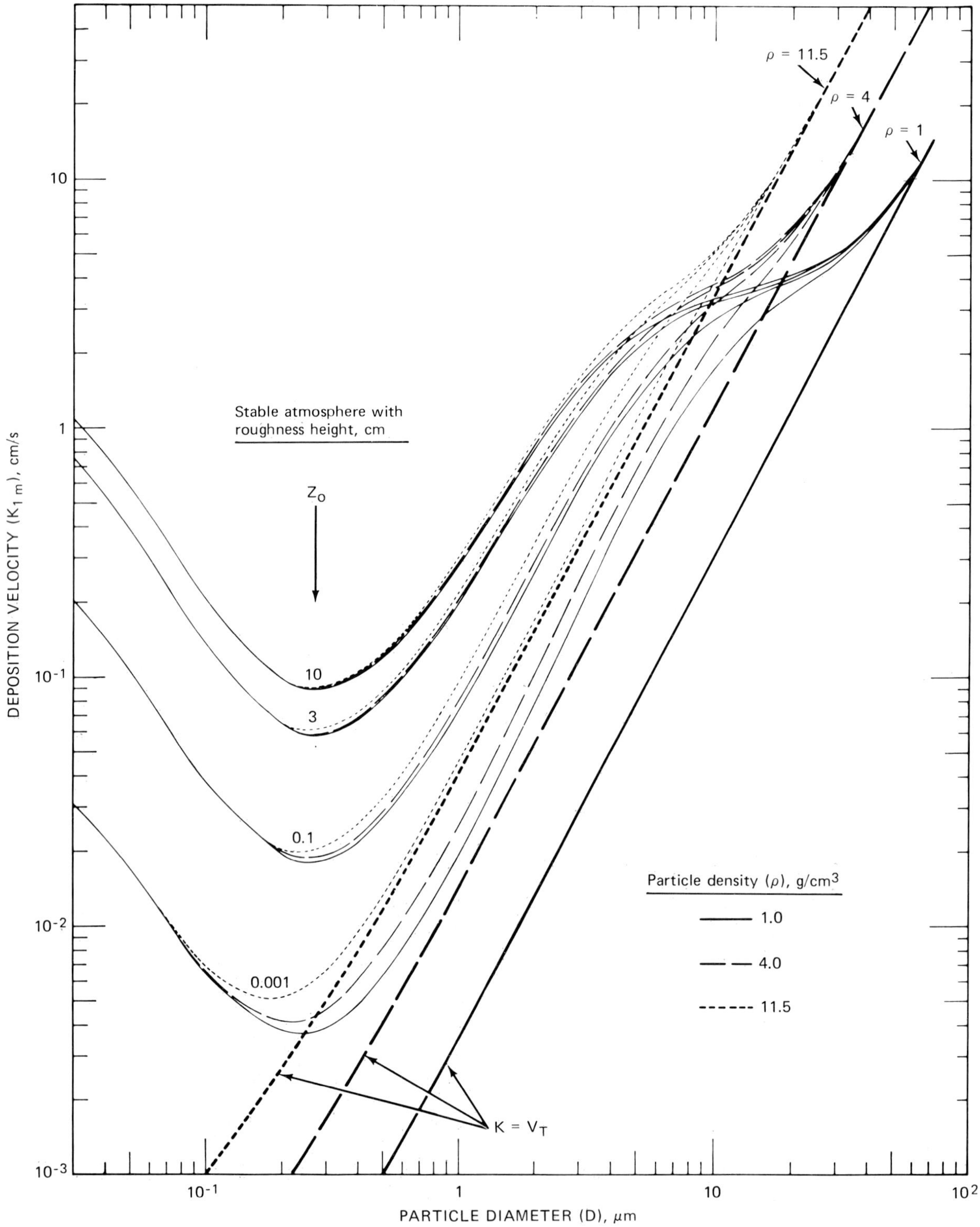

Fig. 12.6 Predicted deposition velocities at 1 m for $u_* = 100$ cm/s and particle densities of 1, 4, and 11.5 g/cm^3.

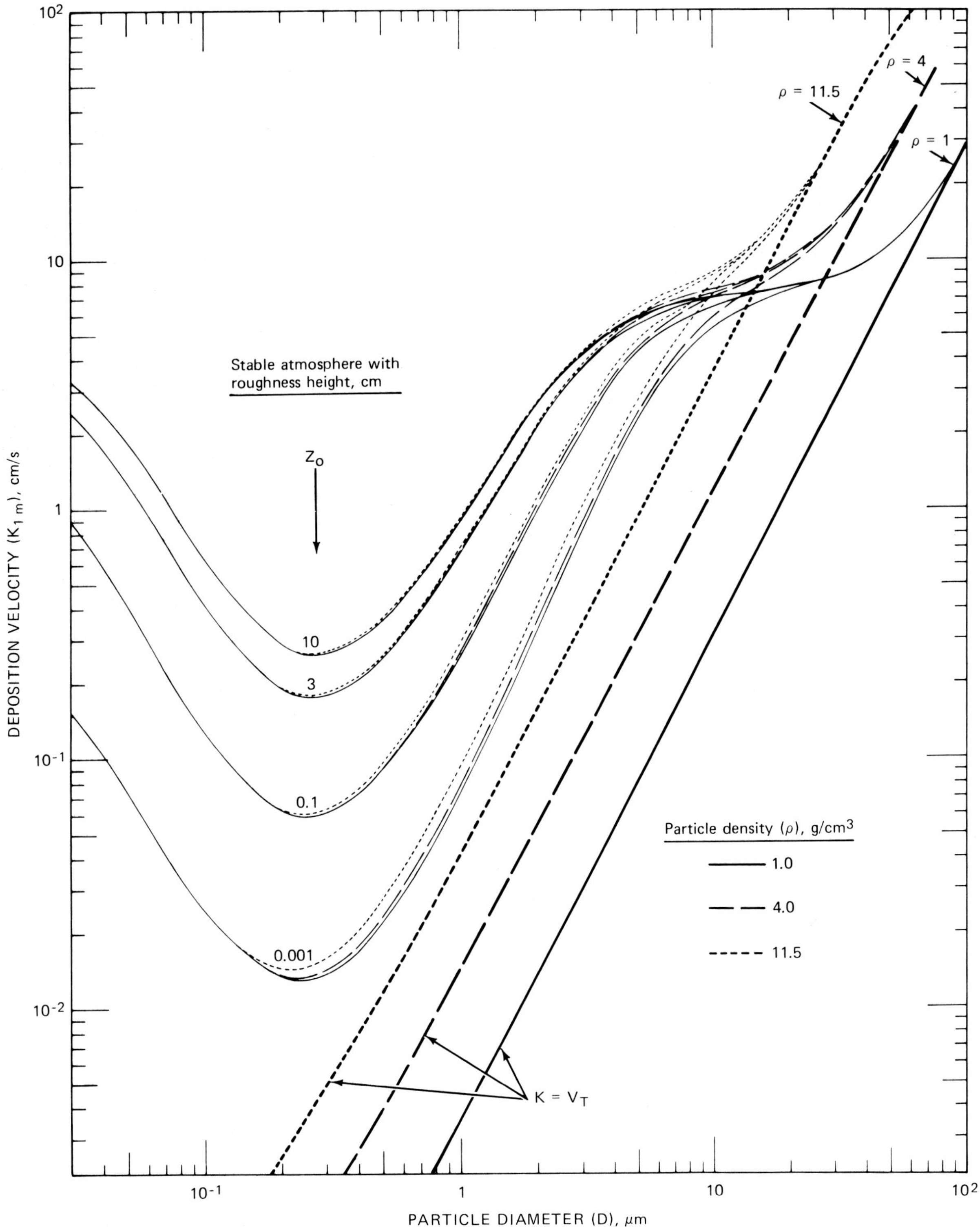

Fig. 12.7 Predicted deposition velocities at 1 m for u_* = 200 cm/s and particle densities of 1, 4, and 11.5 g/cm^3.

TABLE 12.6 Calculated Friction Velocities for Typical Aerodynamic Surface Roughness Lengths

Surface	z_0, cm†	For u = 1 m/s at 2 m: u_*, m/s	u_*/u, %	For u = 5 m/s at 2 m: u_*, m/s	u_*/u, %	For u = 20 m/s at 2 m: u_*, m/s	u_*/u, %	For u = 40 m/s at 2 m: u_*, m/s	u_*/u, %
Smooth mud flats, ice	0.001	0.033	3.3	0.16	3.3	0.66	3.3	1.31	3.3
Smooth snow on short grass	0.005	0.038	3.8	0.19	3.8	0.76	3.8	1.51	3.8
Smooth sea	0.02	0.043	4.3	0.22	4.3	0.87	4.3	1.74	4.3
Level desert	0.03	0.045	4.5	0.23	4.5	0.91	4.5	1.81	4.5
Snow surface, lawn to 1 cm	0.1	0.053	5.3	0.26	5.3	0.97	4.9	2.10	5.3
Mown grass									
1.5 cm	0.2	0.058	5.8	0.29	5.8	0.98	4.9	2.32	5.8
3.0 cm	0.7	0.071	7.1	0.35	7.1	1.01	5.0	2.82	7.1
4.5 cm									
u_2 m = 2m/s	2.4	0.090	9.0	0.45	9.0	1.04	5.2	3.61	9.0
u_2 m = 6–8 m/s	1.7	0.084	8.4	0.42	8.4	1.03	5.1	3.35	8.4
To 5 cm	1–2	0.075	7.5	0.38	7.5	1.51	7.5	3.02	7.5
		0.087	8.7	0.43	8.7	1.73	8.7	3.46	8.7
To 60 cm	4–9	0.102	10.2	0.51	10.2	2.04	10.2	4.07	10.2
		0.127	12.7	0.64	12.7	2.54	12.7	5.08	12.7
Long grass									
60 to 70 cm									
u_2 m = 1.5 m/s	9.0	0.13	12.7	0.64	12.7	2.54	12.7	5.09	12.7
u_2 m = 3.5 m/s	6.1	0.11	11.4	0.57	11.4	2.27	11.4	4.54	11.4
u_2 m = 6.2 m/s	3.7	0.10	10.0	0.50	10.0	2.00	10.0	3.99	10.0
Fully grown root crops	14	0.15	14.7	0.73	14.7	2.93	14.7	5.87	14.7

*Using 0.4 for von Kármán's constant and a zero-displacement length.
†Based on Priestley, 1959, and Slade, 1968.

literature. Data are listed in alphabetical order by author rather than in order of increasing magnitude because resuspension factors can range up to 3 orders of magnitude in one experiment. The range of resuspension factors rather than an average is listed to emphasize that most field experiments have been conducted with a resuspension source that was not uniform. In the field experiments, both large and small particles were resuspended, and particles were not uniformly distributed across the source. Because of these conditions, a single average resuspension factor for an individual experiment has an uncertainty of at least 1 to 2 orders of magnitude. In addition to this range in an individual experiment, resuspension factors range widely between experiments. In field experiments conducted to determine the influence of wind on resuspension, resuspension factors ranged from 10^{-11} to 10^{-4} m^{-1}, a range undoubtedly influenced by many of the factors presented in Table 12.2.

The extent to which wind speed influences resuspension factors is also indicated in Table 12.7. Iranzo and Salvador (1970) reported that airborne plutonium concentrations increased when wind speeds in Palmares, Spain, were above about 10 m/s. When the average wind speed is below 5 m/s, Stewart (1967) reported that resuspension factors tended to increase with wind speed to the third power. However, caution must be exercised in applying Stewart's data to other than slow wind-speed conditions because Sehmel (1977a) determined in tracer experiments that resuspension becomes more rapid for wind speeds above, not below, about 5 m/s.

Conceptually, resuspension factors should be a function of both the resuspension source area present and the sampling height. Experimentally there appears to be no significant difference between resuspension factors for a 1-m^2 source or a 1300-m^2 source. This anomalous behavior is unexpected, especially since little resuspension is expected for wind speeds less than 5 m/s.

That resuspension factors are also a function of the sampling height for airborne concentrations is borne out by the Maralinga trials data (Stewart, 1967) shown in Table 12.7. Airborne concentrations were found to decrease about 1 order of magnitude as the sampling height increased from 0.3 to 0.6 m.

Although a scientifically defensible basis for selecting a resuspension factor value is desirable, there is uncertain justification for the method of selecting reasonably conservative values of wind-induced resuspension factors. This uncertainty results from the fact that, in the original

TABLE 12.7 Resuspension Factors from Wind Resuspension Stresses

Location	Source material	Wind-speed dependency	Resuspension factor, m^{-1}	Reference
Nevada Test Site, Schooner	^{181}W		10^{-7}–6×10^{-5}	Anspaugh et al., 1970
Nevada Test Site, GMX	Plutonium			Anspaugh et al., 1975
Near center			3×10^{-10}	
Near edge			2×10^{-9}	
New York, average air concentrations and surface deposition	Fallout		5×10^{-9}–2×10^{-6}	Calculated from Bennett, 1976
New York	^{238}U		5×10^{-8}	Calculated by Bennett, 1976
United Kingdom	^{238}U		5×10^{-9}	Hamilton calculated by Bennett, 1976
Palmares, Spain	Plutonium, 10-d samples	$u < 9.7$ m/s	1.4×10^{-9}–7.8×10^{-6}	Calculated from Iranzo and Salvador, 1970
Nevada Test Site, dusty rural air	Plutonium		7×10^{-6}	Langham, 1971
Hanford	Inert tracer, <1 μm	Increases	1.7×10^{-10}–2.7×10^{-7}	Sehmel and Lloyd, 1975
Rocky Flats	Plutonium		10^{-9}–10^{-5}	Sehmel and Orgill, 1973
Maralinga trials	Uranium	$u < 5$ m/s		Stewart, 1967
	At 0.3 m height		3×10^{-4}	
	At 0.6 m height		1×10^{-5}	
Paving stones	U_3O_8, 2-m² source	$u < 5$ m/s	8×10^{-9}–6×10^{-6}	Stewart, 1967
Sandy, desert grass	U_3O_8, 9-m² source		1×10^{-8}–5×10^{-5}	
Sandy soil with charred debris	^{210}Po	RF u^3		Stewart, 1967
	1-m² source	$u < 5$ m/s	2×10^{-9}–9×10^{-6}	
	1300-m² source	$u < 5$ m/s	7×10^{-8}–1×10^{-5}	
Monte Bello Islands, Hurricane trials	Fallout	$u < 5$ m/s		Stewart, 1967
	16-m source, lightly vegetated sand and rock		1×10^{-6}–1×10^{-5}	
	Near tower shot		1×10^{-8}–2×10^{-7}	
	Near road, no disturbance		1×10^{-8}–1.5×10^{-6}	
Cleared sandy soil	Yttrium chloride, aqueous solution	$u < 5$ m/s		Stewart, 1967
	1-m² source		9×10^{-11}–5×10^{-8}	
	9-m² source		5×10^{-10}–6.2×10^{-7}	

experiments, airborne concentrations were measured only under slow wind-speed conditions when resuspension factors are expected to be less, not in fast wind-speed conditions.

Anyone using Table 12.7, therefore, is advised to rely on site-specific data only from locations similar to the ones he is studying; they may be the best available estimates.

This recommendation carries a caveat, however. New York fallout data in Table 12.7 do not suggest that fallout resuspension factors should be used for determining local resuspension. The fallout resuspension factor was calculated by assuming that airborne concentrations were caused by resuspension rather than by stratospheric fallout. A "true" average resuspension factor for surface fallout might be less than the fallout range of 10^{-9} to 10^{-6} m^{-1} for New York. Therefore a better index of an average resuspension factor might be for ^{238}U resuspended from natural uranium in surface soil, where the resuspension factors range from about 10^{-9} to 10^{-8} m^{-1}.

12-4.2 Airborne Concentration Half-Life

The length of time since a contaminant was deposited is another variable that can influence airborne concentrations at a resuspension site. The change in airborne concentration with time near a resuspension site is reported in terms of an airborne concentration half-life, the apparent time required for the average airborne concentration to reduce to one-half of the original value. This length of time depends at least on average wind speeds and varies according to the availability of surface contamination to be resuspended. Although originally assumed to be constant, the half-life is now considered to change with time.

Half-lives summarized in Table 12.8 range from 35 d to years. This variability in the half-lives of even the same source material is probably caused by differences in the variable conditions, such as the wind speed, vegetative cover, time since deposition, and sampling technique at each test site. Some of the half-life data have been combined into overall models (Anspaugh et al., 1974;

TABLE 12.8 Half-Life Times for Decrease of Airborne Concentrations near Resuspension Sites

Location	Source material	Duration of experiment	Half-life for airborne concentration	Half-life for retention on foliage, d	Reference
Nevada Test Site, Plumbbob			Decreases with time		Shreve, 1958
Nevada Test Site	Plutonium	15 months 5 to 15 months	No change at 230 m Concentration decreased by only a factor of 1 to 10 at 760 m		Healy, 1974; comparison of Wilson, Thomas, and Stannard, 1960, and Olafson and Larson, 1961
Nevada Test Site, Yucca Flat			Concentrations increased from 1956 to 1958		Mork, 1970
Russia	Alpha and beta	2 to 4 yr >4 yr	Decreased 2 to 3 orders of magnitude Practically constant		Izrael, 1971
Nevada Test Site, Plumbbob	Plutonium	1 to 5 months	35 d		Wilson, Thomas, and Stannard, 1960
Nevada Test Site	Plutonium		35 d		Langham, 1971
Nevada Test Site, Schooner	^{181}W	1.8 months	38 d		Anspaugh et al., 1973
Nevada Test Site, Baneberry	Gross gamma	9 months	76 d		Anspaugh et al., 1973
Hanford Area aged source	^{137}Cs, ^{238}Pu, ^{239}Pu	33 months	5 months to years		Sehmel, 1977a, 1977b
Hanford	Inert tracer	8 months	>8 months		Sehmel, 1977a
Rocky Flats after soil disturbance	Plutonium		9 months		Sehmel and Orgill, 1974
Field test	Solutions ^{85}Sr, ^{51}Cr, ^{214}Pb, and 1-μm-diameter polystyrene			49 in winter, 19 in summer (after 7 weeks)	Chadwick and Chamberlain, 1970

Kathren, 1968; Oksza-Chocimowoski, 1976, 1977), which are consistent with Shreve's (1958) original observation that the half-life decreases with time. These models describe an apparent half-life decrease with time by equations that describe different exponential changes (in half-life) with time.

The half-life may not always decrease with time. Half-lives for airborne concentrations at the Plumbbob nuclear test site were originally reported by Wilson, Thomas, and Stannard (1960) to range from 35 to 40 d. More recent results indicate a great uncertainty about these half-lives. In the original study starting in May 1957, the 35-d half-life was determined for an air sampling time period of less than 5 months. In the summer of 1958, Olafson and Larson (1961) measured airborne concentrations intermittently for times up to 15 months. In comparing the results from these two studies, Healy (1974) found that, during the 15-month time period, no appreciable change was reported in airborne concentrations within the data scatter at a site 230 m from the original test location. This result could indicate that resuspension was occurring at an average rate continuously during that 15-month period if the source were uniform with distance. At a site 760 m from the test location, airborne concentrations decreased by a factor of only 1 to 10. Thus changes in the air concentration may depend on proximity to the source and possibly local surface contamination levels.

Although the half-life concept appears to be questionable from experimental measurements of airborne concentrations, there is theoretical justification to expect some half-life to occur for resuspension source availability. Surface contamination will become attached to soil particles and migrate into the soil structure with time. This migration and coverage by vegetation and upwind soil erosion would make the source less available for resuspension. Data are insufficient, however, to characterize the half-life adequately. The possibility also exists that the surface contamination that does not migrate could become more readily available for resuspension by degradation processes, possibly biodegradation (Wildung and Garland, 1977) and solubility with time.

The half-life for airborne concentrations is also related to the changes in contamination on vegetation surfaces as a function of time, as experiments measuring the changes in surface contamination level on plant foliage have shown.

Chadwick and Chamberlain (1970), for example, reported a difference in the half-lives of pollutants retained on foliage between winter and summer months. As shown in Table 12.8, the half-life was greater during the winter.

12-4.3 Resuspension Factors Caused by Mechanical Disturbances

Resuspension occurs as a result of mechanical stresses caused by man's activities as well as by simple wind-induced resuspension. Resuspension factors measured during walking, working, driving, etc., are summarized in Table 12.9. Because of the range between the resuspension factors for each experiment, the results reported in each reference are listed together. Some judgment was used in arranging the data. In general, larger resuspension factors are listed first even if there are smaller values within the data grouping for the reference.

Resuspension factors for mechanical resuspension stresses range from 10^{-10} to 10^{-2} m^{-1}. This range overlaps the 10^{-11} to 10^{-4} m^{-1} values for simple wind resuspension. The upper range of resuspension factors for laboratory experiments is 2 orders of magnitude greater than the 10^{-4} m^{-1} for outdoor experiments.

Resuspension factors are included for both laboratory and outdoor experiments. Results from laboratory experiments are included since these data can be an index to an upper limit for resuspension factors in outdoor situations. In laboratory experiments the upper limit results from containing all resuspended particles in an unventilated room. Since, under these conditions, there is no wind to carry resuspended particles upward other than the air movement produced by the mechanical disturbance itself, data from laboratory experiments must be used cautiously in determining resuspension factors.

Much work remains in determining resuspension factors for both wind and mechanical resuspension (Sehmel, 1977a; Cowherd and Mann, 1976). Additional data are still needed for extrapolation or interpolation to determine the effect of the chemical or physical properties of resuspended material on resuspension. Few data in either Table 12.7 or 12.9 can be used to determine the effects of pollutant particle diameter on resuspension.

12-5 RESUSPENSION RATES

Sehmel (1977a) measured resuspension rates experimentally using tracer particles and mass balances of airborne tracer. In these experiments, resuspension rates were measured as a function of wind speed as well as respirable and nonrespirable particle diameters. Mechanical resuspension was measured for vehicular traffic on asphalt and an area vegetated with cheatgrass and for pedestrian traffic on asphalt. Sehmel's use of mass balance techniques to determine resuspension rates corrects a deficiency in early tracer resuspension data reported by Healy (1955, 1977b) and Healy and Fuquay (1958, 1959). In that early research, airborne concentrations of suspended tracers were estimated with air samplers located only at ground level. Later a diffusion model was used to calculate apparent resuspension rates. Since both methods can produce inaccurate data, only resuspension rates determined from mass balance techniques are discussed in this section.

12-5.1 Wind-Induced Resuspension

Airborne concentrations produced by wind stresses have been reported as a function of both wind speed and friction velocity. Resuspension data summarized in Table 12.10 show that air concentrations increase from the 1.1 to the 6.4 power of wind speed or friction velocity. However, agricultural studies by Bagnold (1941, 1960) and Chepil (1945a) indicate that fields erode as a function of air speed to the third power. The differences in rate might be attributed to the size of soil being considered in each study. The agricultural study focused on the erosion of nonrespirable particles, whereas the resuspension studies summarized in Table 12.10 focused on respirable particles. The wind speed on which airborne concentrations of respirable particles depend shows a great uncertainty.

Although the wind-speed dependency reflects resuspension physics, resuspension rates are needed for a boundary condition in meteorological transport models.

The particle resuspension rates summarized in Table 12.11 have either been calculated by using transport and diffusion models in interpreting measured concentrations at only one height or have been measured by integrating airborne concentrations as a function of height. Although the data determined from concentrations measured as a function of height with controlled resuspension sources are least questionable, resuspension rates for both experimental techniques range from about 10^{-12} to 10^{-4} fraction resuspended per second. This 8-order magnitude uncertainty is slightly less than the 11 orders of magnitude uncertainty for resuspension factors.

Average wind-induced resuspension rates for tracer particles have been measured (Sehmel, 1977a; Sehmel and Lloyd, 1975) for resuspension from three surfaces: an asphalt surface, a cheatgrass area, and an area lightly vegetated with cheatgrass and sagebrush. For average wind speeds of 1 to 4 m/s, wind resuspension rates for the asphalt surface ranged from 5×10^{-9} to 6×10^{-8} fraction resuspended per second. For average wind speeds of 1 to 5 m/s at the cheatgrass area, wind resuspension rates ranged from 5×10^{-9} to 6×10^{-8} fraction resuspended per second.

Resuspension rates (RR) of tracer particles from a lightly vegetated portion of the Hanford site are shown in Fig. 12.8. The rates are for submicrometer particles as deposited, and they are listed as a function of aerodynamic particle diameter when resuspended. The rates ranged from about 10^{-11} to 10^{-7} fraction resuspended per second.

TABLE 12.9 Resuspension Factors from Mechanical Resuspension Stresses

Location	Source material	Resuspension stress	Resuspension factor range, m^{-1}	Reference
Unventilated room	Beryllium	Vigorous sweeping	1×10^{-2}–4×10^{-2}	Mitchell and Eutsler, 1967
Small unventilated room	Alpha floor contamination	People walking	3×10^{-4}–2×10^{-2}	Calculated from Brunskill, 1967
Changeroom with no ventilation	Alpha on coveralls	Two to four people changing coveralls	$<2 \times 10^{-4}$–1.7×10^{-2}	Calculated from Brunskill, 1967
Changeroom, concrete floor, 9 air changes/h	10% "loose" alpha floor contamination	Four to six people moving in room	2×10^{-4}–3×10^{-3}	Brunskill, 1967
Simulated work	Asbestos contaminated coat	Machining operations	1.2×10^{-3}–5.3×10^{-3}	Carter, 1970
		Stacking sheets	2.0×10^{-3}–4.2×10^{-3}	
Room, concrete floor	Plutonium facility	No circulation	1×10^{-5}–2×10^{-4}	Glauberman, Bootmann, and Breslin, 1967
		Fan air stress	3×10^{-4}–3×10^{-3}	
		Fan and dolly movement	4×10^{-3}–1.5×10^{-2}	
		No circulation after tests	5×10^{-4}–1×10^{-3}	
Room, concrete floor	Uranium facility	No circulation	7×10^{-5}–4×10^{-4}	
		Fan air stress	3×10^{-5}–2×10^{-4}	
		Dolly movement	1×10^{-4}–2×10^{-4}	
		Fan and dolly movement	2×10^{-4}–1×10^{-3}	
Maralinga trials	Fallout	Road survey, at 1–2 d	$<1 \times 10^{-8}$–$>2 \times 10^{-6}$	Stewart, 1967
		Cab landrover, 5th h	6.4×10^{-5}	
		Cab landrover, 8th h	2.5×10^{-5}	
	Uranium	Dust stirred at 0.3 m height	1×10^{-3}	Stewart, 1967
	Plutonium	Vehicle dust at 0.3 m height	3×10^{-7}–7×10^{-4}	
		Pedestrian dust at 0.3 m height	1.5×10^{-6}–3×10^{-4}	
Particles settled in room	ZnS	Vigorous work, sweeping	1.9×10^{-4}	Fish et al., 1967
	MMD = 3.1 μm	Walking	3.9×10^{-5}	
	$\sigma_g = 1.9$	Light work	9.4×10^{-6}	
	CuO	Light sweeping	7.1×10^{-4}	
	MMD = 2.0 μm			
	$\sigma_g = 2.3$			
Civil defense trial	^{131}I	Bomb-recovery operation with brick/plaster dust		Stewart, 1967
		Enclosed space	4×10^{-5}–2×10^{-4}	
		Open area	2×10^{-6}	
Nevada Test Site	Plutonium	Extensive vehicular traffic	7×10^{-5}	Langham, 1971
Laboratory room	Plutonium oxide	No movement	2×10^{-8}	Jones and Pond, 1967
		14 steps/min	10^{-5}	
		36 steps/min	5×10^{-5}	
	Plutonium nitrate	No movement	2×10^{-8}	
		14 steps/min	10^{-6}	
		36 steps/min	5×10^{-6}	
Monte Bello Islands, Hurricane trails	Fallout	Road survey from back landrover		Stewart, 1967
		4th d	8×10^{-7}–3×10^{-5}	
		7th d	7×10^{-7}–4×10^{-6}	
		At tailboard, 7th d	1.6×10^{-5}–3.1×10^{-5}	
Australian desert, "Totem," 1953	Fallout gamma	Walking survey	3×10^{-7}	Stewart, 1967
		Vehicle survey		
		At tailboard	2×10^{-6}	

TABLE 12.9 (Continued)

Location	Source material	Resuspension stress	Resuspension factor range, m^{-1}	Reference
Contaminated field	Plutonium	Inside vehicle	2×10^{-5}	Calculated from Milham et al., 1976
		Downwind tractor	$5 \times 10^{-8}-1.0 \times 10^{-6}$	
		In tractor cab	2.4×10^{-7}	
		Fertilizing	7.9×10^{-9}	
		Subsoiling	5×10^{-6}	
		Planting, light disking	$1.8 \times 10^{-8}-7.9 \times 10^{-6}$	
		Mowing	1.3×10^{-5}	
Tractor cab in contaminated field	Plutonium	Tractor operations	$4 \times 10^{-9}-2 \times 10^{-7}$	Milham et al., 1976
New York	Fallout plutonium	Average air concentration/surface concentration	$5 \times 10^{-9}-2 \times 10^{-6}$	Calculated from Bennett, 1976
Palmares, Spain	Plutonium	Yearly average farming	$1.2 \times 10^{-10}-3.8 \times 10^{-7}$	Calculated from Iranzo and Salvador, 1970
Contaminated sludge	Plutonium	Rototilling	5.6×10^{-8}	Calculated from Myers et al., 1976

TABLE 12.10 Air Concentration Dependency on Wind Speed

Location	Source material	Concentration increases with wind speed to power n: u^n	Concentration increases with wind speed to power n: u_*^n	Reference
Nevada Test Site, Baneberry	Gross gamma			Anspaugh et al., 1973
Nevada Test Site, GMX	Plutonium		2.2	Anspaugh et al., 1976
Nevada Test Site, Yucca Flat	Plutonium	Increases a factor of 10 from u <2 m/s to u = 5 to 7 m/s		Mork, 1970
Rocky Flats	Plutonium	2.1		Sehmel and Orgill, 1973
Rocky Flats	Plutonium	5.9		Sehmel and Lloyd, 1976
Hanford				
On-site	Plutonium	1.0–1.5		Sehmel, 1977b
Off-site	Plutonium			Sehmel, 1977a
	<5 m/s	1.1		
	>5 m/s	3.0–9.3		
Hanford	Tracer	1.0–4.8		Sehmel, 1977a
Eroding field	Soil particle diameter, μm			Sehmel and Richmond, 1978
	0.4–0.6		2.6–3.4	
	0.6–0.7		5.0	
	2.7–2.9		6.4	
Eroding agricultural field	Soil		6.38	Shinn et al., 1976; Anspaugh et al., 1975
Nevada Test Site, desert pavement			2.1	

TABLE 12.11 Resuspension Rates

Location	Source material	Resuspension rate, s^{-1}, determined from: Calculation using transport model and concentration at only one height	Integration of measured concentrations	Reference
Corn leaves	Pollen		10^{-6} to 10^{-4}	Calculated from Aylor, 1976
Nevada Test Site	Plutonium	2.7×10^{-12} to 4.8×10^{-10}		Anspaugh et al., 1975
Nevada Test Site, GMX	Plutonium	10^{-12} to 10^{-10}		Healy, 1974
Nevada Test Site	Plutonium	1×10^{-10} to 6×10^{-8}		Healy, 1974, calculated from Wilson, Thomas, and Stannard, 1960
Hanford Site	ZnS	1×10^{-8} to 3.5×10^{-6}		Healy, 1974, recalculation of Healy, 1955, and Healy and Fuquay, 1958
Eroding field from top 1 cm depth	Soil			Calculated from Skidmore and Woodruff, 1968
0.5 ton/(acre • month)		2×10^{-9}		
40 tons/(acre • month)		2×10^{-7}		
310 tons/(acre • month)		1×10^{-6}		

They were measured during selected wind speeds and increased with the 1.0 to 4.8 power of wind speed.

Average wind-induced tracer resuspension rates of both respirable and nonrespirable particles are shown in Fig. 12.9. Average resuspension rates for nonrespirable particles were nearly independent of time and were on the order of 10^{-11} fraction resuspended per second. Average resuspension rates for respirable particles ranged from about 10^{-11} to 10^{-7} fraction resuspended per second and did not decrease with time.

Resuspension rates cannot be predicted as a function of the diameters of the source particles or the physical and chemical properties of the particles. There are too few data even to predict differences between resuspension rates for the same pollutant for surfaces with different vegetation coverage.

12-5.2 Correlation for Wind-Induced Resuspension Rates

Resuspension rates measured in several other experiments (Sehmel, 1975a) are plotted in Fig. 12.10 as a function of aerodynamic surface roughness height. In this figure, resuspension rates range over 7 orders of magnitude, from 10^{-10} to 10^{-3} fraction resuspended per second. For an estimate of fractional removal per year, these rates need to be compared with the 3.2×10^7 seconds in a year. Figure 12.10 is a guideline for estimating resuspension rates for other aerodynamic roughness heights. However, this correlation with aerodynamic surface roughness should be used with extreme caution. For accuracy, resuspension rates should be determined as a function of physical parameters other than aerodynamic surface roughness alone. Until better correlations are developed, however, this correlation does give some justification for estimating resuspension rates.

12-5.3 Resuspension by Vehicular Traffic

Resuspension rates for vehicular traffic on an asphalt road are shown in Fig. 12.11 as a function of vehicle speed (Sehmel, 1973c). The resuspension rate was calculated by determining the fraction of the tracer resuspended each time the vehicle was driven along the road. The fraction of tracer resuspended each time the vehicle passed ranged from 10^{-5} to 10^{-2}. Resuspension rates did increase with vehicle speed. Also, when vehicles were driven on the lane containing the tracer, the resuspension rates were greater than when the vehicle was driven on the lane adjacent to the tracer lane. For vehicle speeds above 32 km/h (20 mph) on the tracer lane, resuspension rates for both car and truck passage were found to be comparable.

These rates were also found to depend on the time the tracer particles had been on the asphalt road. After the tracer had been on the road for 4 d, the particle resuspension rates decreased, as Fig. 12.12 shows. Even considering the influence of time, resuspension rates were still greater when the vehicle was driven through the tracer lane than when it was driven on the adjacent lane.

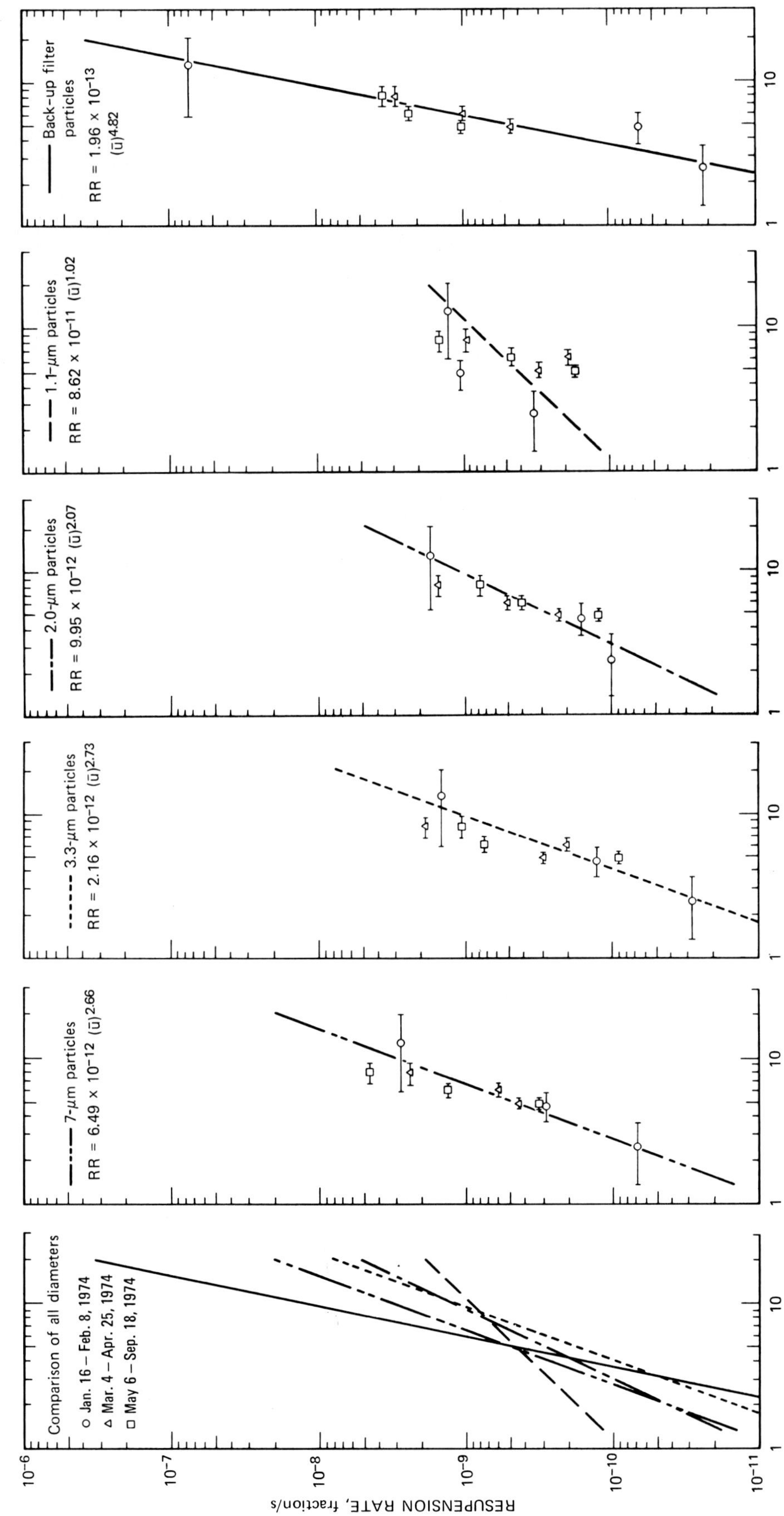

Fig. 12.8 Tracer resuspension rates as a function of particle diameter and wind speed increments.

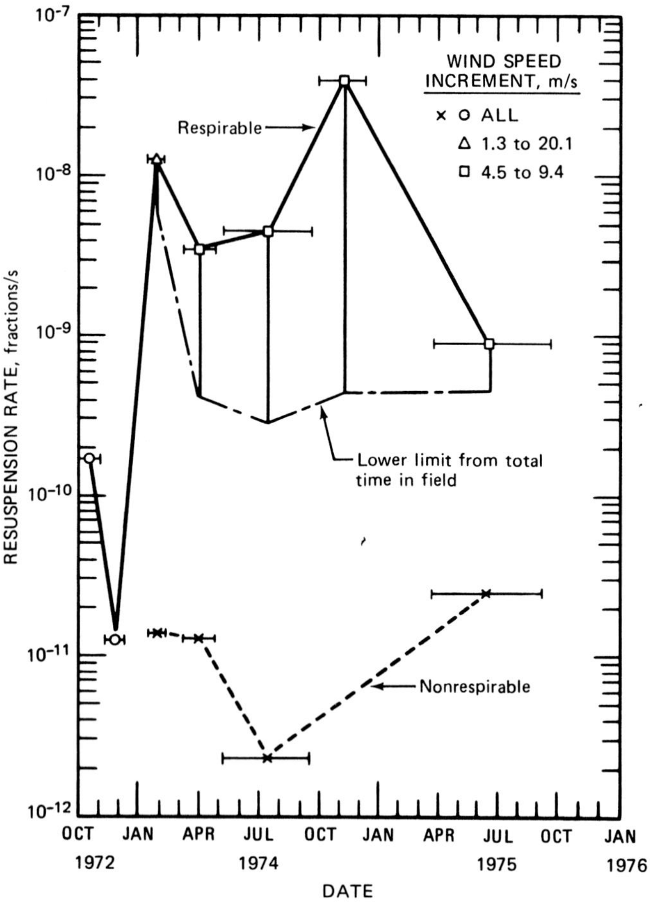

Fig. 12.9 "Average" wind caused tracer resuspension rates (lightly vegetated desert on the Hanford site).

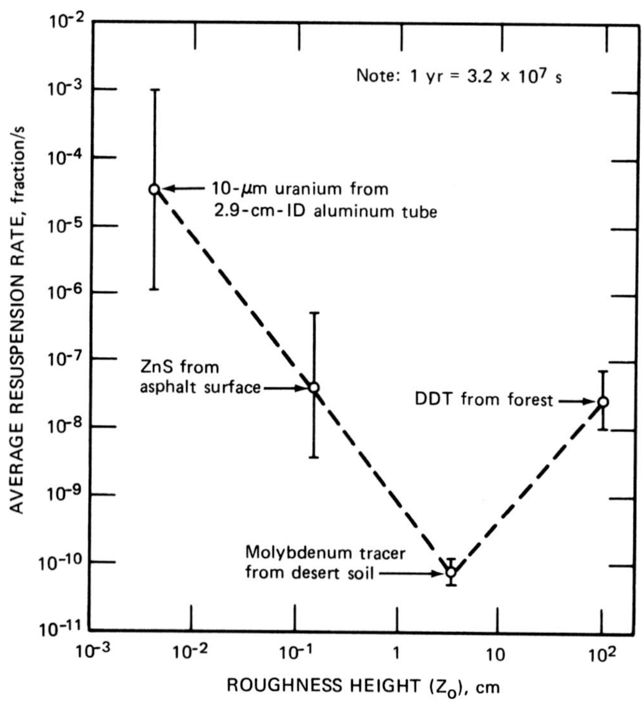

Fig. 12.10 Initial correlation of wind caused tracer resuspension rates.

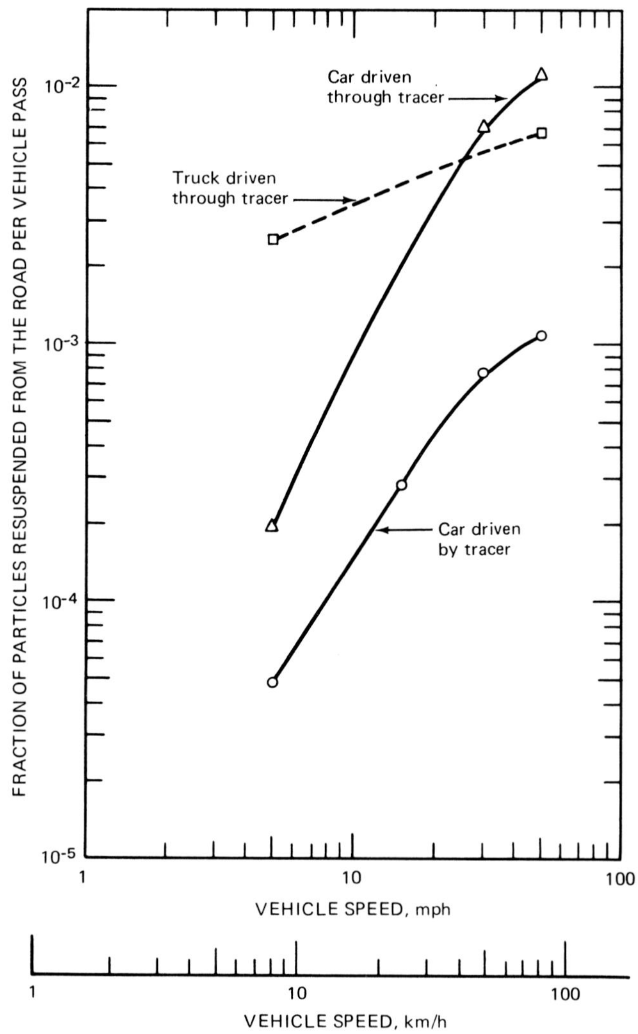

Fig. 12.11 Rates of tracer particle resuspension caused by vehicle passage over an asphalt road 4 d after particle deposition.

Sehmel (1976a) also determined that truck traffic in a cheatgrass area caused less resuspension than truck traffic on an asphalt road because the vegetation prevented part of the turbulence generated by the truck from reaching the ground and resuspending the tracer. Figure 12.13 compares the resuspension rates for both surfaces.

Although increased resuspension rates are expected at higher speeds because of turbulence, Sehmel found that the resuspension rates decreased for the cheatgrass road as speed increased from 8 to 48 km/h (5 to 30 mph). This decrease can be attributed to the sequence of truck speeds used in the experiment. The initial resuspension experiment was conducted at 8 km/h (5 mph). At this low speed the most readily resuspended tracer particles were removed from the cheatgrass. Only when the truck speed was increased from 48 to 64 km/h (30 to 40 mph), with the consequent increase in air turbulence, did resuspension rates increase.

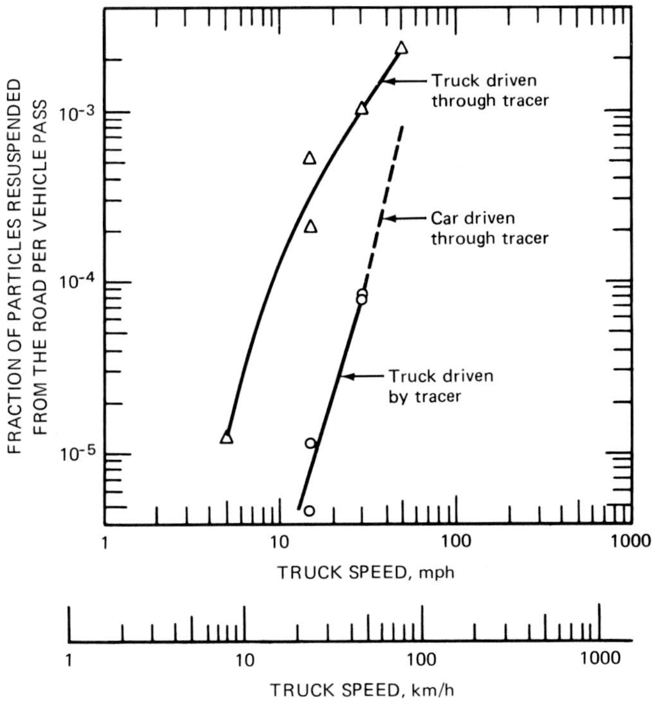

Fig. 12.12 Rates of tracer particle resuspension caused by vehicle passage over an asphalt road 4 d after particle deposition.

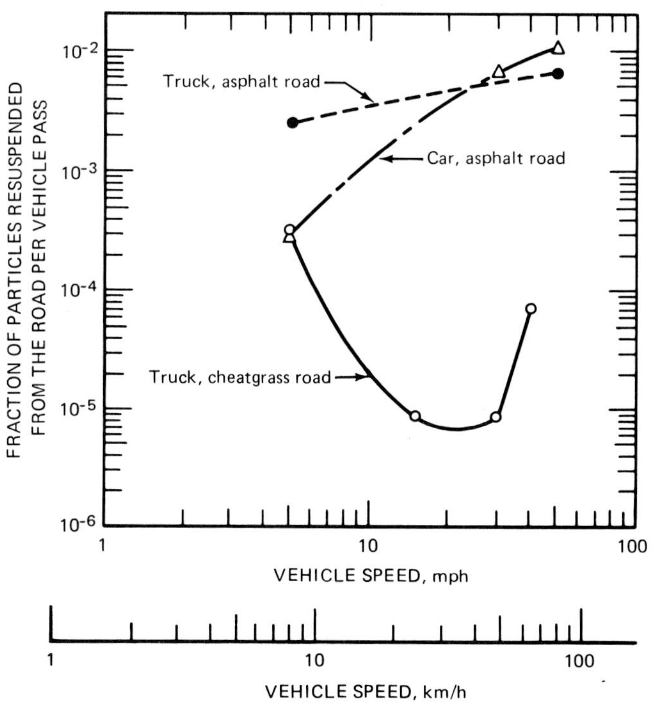

Fig. 12.13 Rates of tracer particle resuspension caused by vehicle passage over asphalt and cheatgrass roads.

12-5.4 Resuspension by Pedestrian Traffic

Resuspension caused by pedestrian traffic was determined as the fraction of tracer particles resuspended each time a person walked the length of a tracer lane. A man walking along a tracer lane on an asphalt road caused a fraction from 1×10^{-5} to 7×10^{-4} of the tracer to be resuspended with each pass.

12-6 CONCLUSIONS

Mass transfer across the air–surface interface is important for all toxic materials, regardless of their size. If the materials are respirable, predicting the concentrations that might potentially be inhaled is crucial. These concentrations can be calculated from atmospheric diffusion and transport models that include a boundary condition for mass transfer across the air–surface interface. The problem is that many of these models either have not been validated or have been validated only under limited conditions. Models have not been validated, for instance, for the case of significant air–surface mass transfer. Once these models are validated, deposition velocities and resuspension rates will be available. Until then, they may not be the best available method of predicting downwind concentrations.

Although many experiments have been conducted and theoretical models have been developed to predict the effects of particle and gas deposition since the literature was reviewed in *Meteorology and Atomic Energy—1968* (Slade, 1968), our ability to predict surface mass-transfer deposition rates from those field experiments has not improved significantly, with one exception. A model developed by Sehmel and Hodgson (1978) to predict deposition velocities is considered to so improve our capability to predict deposition velocities that it should be used until either more refined and definitive deposition experiments are completed or improved predictive models are developed from the existing data.

Accurate predictions of surface depletion of airborne pollutant plumes from anthropogenic sources require that the best available estimates for dry deposition velocities be used. However, the relative significance of dry deposition has only recently been appreciated in the literature. Earlier deposition experiments were focused on the wet removal of nuclear fallout debris from the stratosphere rather than on dry deposition near ground level. The fact that fallout is removed primarily by wet deposition has overshadowed a more important consideration: the key role of pollutant release height in determining the relative effectiveness of dry vs. wet removal processes. Since most energy sources release pollutant near ground level, studies of dry deposition may be more significant than studies of wet deposition in predicting the removal of pollutants from the atmosphere.

A limitation of early field experiments was that experimental variables were not adequately controlled, i.e., the particle size distribution was either not known or not reported. However, deposition velocities are a function of particle diameter. Consequently most field-determined mass-transfer data should be interpreted with qualification. The deposition velocities determined in these field experiments ranged from 10^{-3} up to 20 cm/s, a range also predicted by Sehmel and Hodgson's empirical model.

The model predicts deposition velocities as a function of particle diameter, friction velocity, aerodynamic surface roughness, and particle density. These predictions emphasize that the distribution of pollutant particle size must also be known, but the model does not include any geometrical parameter. The model does include a minimum deposition velocity of 0.025 cm/s. For smaller particles and gases, deposition velocities increase because of Brownian diffusion. For larger particles, deposition velocities increase because of gravitational settling and eddy diffusion.

Resuspension, the other mass-transfer process discussed in this chapter, has been receiving increased attention in the last several years, as is evidenced in the 1974 conference proceedings (Engelmann and Sehmel, 1976) which considers both resuspension and dry deposition. Resuspension can be a continuing problem after toxic materials are deposited on environmental surfaces. After the material is on the ground, it will be recycled into the atmosphere to be inhaled by man. This chapter has demonstrated that resuspension physics are so poorly defined that much research is needed.

One problem is that resuspension coefficients have not been adequately defined. One resuspension coefficient relates airborne concentrations to local surface-contamination levels but does not describe either the vertical flux from resuspension or the total downwind flux. Any calculation using an "average" resuspension factor must be qualified as being uncertain within 2 to 3 orders of magnitude because resuspension factors, which cannot be predicted, have ranged 2 to 3 orders of magnitude even in a single field experiment. The second resuspension coefficient describes the resuspension rate, i.e., the fraction of material resuspended per unit time. If the surface contamination level is known, the vertical flux from resuspension is the product of resuspension rates and source concentrations, using this coefficient. Resuspension rates measured for both wind and mechanical stresses indicate that mechanical stresses can cause more material to be resuspended at one time than does wind, but the effect is short-lived because mechanical stresses usually last over less time and space.

Deposition and resuspension research results are equally applicable to pollutants from both nuclear and nonnuclear energy sources. Differences in these air–surface mass transfer processes cannot be distinguished as a function of chemical properties. Dry deposition can be predicted as a function of particle size, but resuspension cannot.

Experimental deposition and resuspension research are needed to define the transport and mass-transfer processes more adequately. In future studies of these processes, the experimental conditions must be controlled and defined, and the experimental techniques used and the theoretical interpretation of the results must be validated. Model and experimental validation are the crux and requirement for future research of air–surface mass-transfer processes.

12-7 APPLICATIONS

This chapter has summarized the literature on dry deposition of particles and gases and resuspension of particles. In addition, data describing dry deposition and resuspension have been tabulated. However, use of the summarized data must always be qualified since the data cannot be generalized. It is hoped that generalizations will be forthcoming, but until such generalizations are formed, the data must be used with caution. When applying these data, the user must be aware of their limitations, even when, as is often the case, available data are too limited to indicate guidance.

Possibly one of the most important limitations is that particle size has been inadequately reported in much of the literature. Nevertheless, there are different particle diameters that might be of greatest interest, depending on the user's viewpoint regarding application. For instance, pollution depositing on environmental surfaces, including biological and ecological surfaces, might indicate that one particle size was a larger problem area to the biota than another particle size. These particles are usually larger than respirable. In this summary, however, the respirable particle size range is emphasized.

12-7.1 Particle Size

Normally the diameter of respirable particles is less than about 3.5 μm, whereas the diameter of inhalable particles is less than 15 μm. If we specify that our prime interest is particles that are respirable and inhalable to 15 μm, then the next analysis step is to estimate or measure the particle size distribution. Although the particle size distribution is usually estimated from literature sources, here only direct experimental particle size measurements will be considered.

Ideally, particle collection is by isokinetic sampling, which describes the collection of airborne particulates while not disturbing the motion of the approaching air. If the air sampling velocity (vector) does not match the velocity (vector) of the approaching air, particle inertia will cause relative particle motion with respect to air motion. Hence particle collection will deviate from the motion of the free air. In determining the particle size, we need to be aware of possible particle-sampling biases introduced by non-isokinetic air sampling; otherwise, resulting predictions would also show a bias that is not properly qualified for particle size uncertainty.

As a hypothetical example of bias to illustrate the point, consider the measurement of airborne plutonium arising from particle resuspension. The airborne plutonium consists of two different monodispersed particle diameters of either 1 μm (a respirable particle) or 100 μm in diameter (a nonrespirable particle). For this hypothetical size distribution, predictions are to determine if airborne concentrations are within (that is, less than) a maximum permissible airborne concentration as given by health standards (ICRP, 1959; ERDA, 1977). The true air concentration of resuspended respirable plutonium particles (1 μm) is assumed to be at the limit set by the maximum permissible airborne concentration guide; i.e., this maximum permissible concentration, $MPC_{168\ h}$, is 6×10^{-14} μCi/cm^3 (ERDA, 1977).

In many applications for plutonium, the entire size distribution of airborne particles is characterized by a single sample containing only respirable particles. For such a case the total airborne concentration could be increased by air sampling for both 1-μm-diameter particles and 100-μm-diameter particles. Thus the predicted airborne hazard based on the total sample would increase because of the increased plutonium collection. However, if the absolute concentration of 1-μm particles remained at the maximum permissible airborne concentration, the real airborne hazard would not be increased by simply including both respirable and nonrespirable particles in the sample. Similarly, in many applications, the entire size distribution of airborne particles is characterized by a single particle diameter, for instance, the activity median aerodynamic diameter (AMAD). For such a case, the AMAD could be increased by air sampling for both 1-μm diameter particles and 100-μm-diameter particles. Thus the predicted airborne hazard based on the AMAD would decrease because of the increased average particle diameter. However, if the absolute concentration of 1-μm particles remained at the maximum permissible airborne concentration, the real airborne hazard would not be decreased by simply including both respirable and nonrespirable particles in the sample. Therefore anyone who uses the data must be aware of, at least, some limitations and problems in particle sampling and data interpretation. Some of these problems will be considered below.

12-7.2 Resuspension Source Characterization

In calculating resuspension, we are frequently faced with the problem of how to characterize the source material on the ground. In other words should the toxic material be characterized as a function of particle size, or should it be characterized simply as the gross amount of pollutant on the ground? Both approaches have serious problems when we consider existing knowledge. For instance, consider some problems in determining the toxic particle size distribution on the ground (assuming that the toxic material is on the ground rather than on vegetation). Even if pollutant particle size distributions were known when the pollutant was deposited, we must consider whether the resuspension-controlling particle size is the toxic material size or the agglomerate size of toxic material particles attached to host soil particles.

Although particle size must have an effect on resuspension, neither the theoretical nor experimental data base for resuspension has yet been developed with enough precision to allow us to predict the particle size dependency with confidence. For instance, we can hypothesize whether the toxic material is in the respirable size range or whether the total material should be used to characterize the resuspension source strength for the surface soil. From a theoretical standpoint, we would like to predict resuspension as a function of the particle size distribution on the ground surface. Even on this theoretical basis, practical problems exist in quantifying the pollutant size distribution on the ground. For example, there are at least two approaches: (1) analyze a surface soil sample in the laboratory and (2) use a portable wind tunnel to suspend surface material for subsequent sampling at the wind-tunnel outlet. Neither approach is strictly comparable to resuspension by either wind or local mechanical stresses since (1) a single surface soil sample is representative only if the entire available surface is uniform and (2) there may not be a quantitative one-to-one correspondence for particles on the ground and in the air.

Since considerable effort is currently being expended in collecting surface soil samples for subsequent laboratory analysis of the particle size distribution, let us consider some theoretical problems with this approach. The problems considered are associated with modifying the parent size distribution in the laboratory. Dry sieving through screens is one technique for determining particle size distributions. Release of smaller particles from host particles by impact on the screens might be similar to release of these particles from host particles during impact from dropping. For instance, particles of 0.2- to 0.8-μm diameters are released (Sehmel, 1978a) when a 1-mm-diameter particle is dropped 51 cm to impact on a solid surface. Similar particle releases by wet sieving are expected to alter the apparent particle size of distribution. Particles have been shown (Rosinski et al., 1976) to become detached from host particles when the host particle is dropped into liquid. Thus, when determining particle size distributions, we should be aware of the technique used for particle size analysis; it could significantly alter the parent size distribution. However, considering existing technology, we may have little choice of the techniques used. Although optical and electron microscopy are also available, these techniques are very expensive to use on a routine basis, and the data are always qualified since only a very small sample is usually viewed.

Another problem is that source-sampling strategies require the assumption of a uniform resuspension source or the experimental determination of the areal variation in source strength. For instance, should a surface sample be selected under vegetation or between vegetation, between rocks or on top of rocks, or are there more representative sampling sites that can be selected based on either known or expected deposition patterns?

An economic problem as well as a scientific problem arises when we consider nonhomogeneous resuspension sources. The inference is that costs usually indicate that only relatively small surface samples can be analyzed. However, I feel that costs for particle size determination of a small 1-cm^2 or 100-cm^2 surface sample would be more profitably expended by additional gross surface samples to define more accurately the areal distribution of the source strength.

Another practical problem in defining source strength is soil sampling depth. For instance, should the sampling depth be 1 mm, 1 cm, 5 cm, or some other depth? Various depths as well as samplings have been used as a function of depth. If the total source strength is desired, the sampling depth is not too important (except for the greater cost involved in analyzing for toxic material in a greater soil volume) as long as the greatest portion of toxic material is within the sampling depth. However, if we are interested in determining the source strength available for immediate resuspension, only the surface material should be sampled.

Surface source strength changes with time because toxic material migrates into the soil. After some time, the logarithm of concentration might be expected to decrease linearly with depth. However, if the toxic material is leached by precipitation or if the source has been covered by clean soils, a maximum soil concentration could occur at some distance below the surface. In this case, concentrations would not exhibit a simple exponential decrease with depth, and the surface concentration could not be determined by extrapolating concentrations determined at greater depths. Although various problems exist in characterizing resuspension surface source strengths, the practical solution appears to be a sampling depth of about 1 mm to 1 cm. The 1-mm sampling depth might be obtained by a vacuuming technique, and the 1-cm depth might be obtained by a scooping technique.

12-7.3 Estimation of Deposition Velocity

In estimating deposition velocities, we must be aware of the basic assumptions required for applying air–surface mass-transfer rates described by dry deposition velocities. Often the estimator has assumed that the deposition velocity is constant at 1 cm/s for all situations considered. This simplification is a mathematical convenience that often does not occur in the real world.

Dry deposition velocities are a function of the many variables discussed in this chapter, but here dry deposition velocities are emphasized to be a function of particle diameter, even for gases. The assumption is that gases become attached to airborne condensation nuclei particles, and, as a result, the gas diffusion coefficient is altered. The pure gas has a higher Brownian diffusion coefficient than gas sorbed on particles since sorbed gas has a Brownian diffusion coefficient corresponding to the host airborne particle. The Brownian diffusion coefficient of a particle is less than that for a gas and decreases as the particle size increases. Therefore, for particle diameters less than 0.1 μm, the final deposition process within the 1 mm adjacent to a deposition surface is often controlled by Brownian diffusion.

Next, consider the single deposition velocity describing an airborne particle size distribution. An effective deposition velocity, v_d, is needed to account for the size distribution. To calculate an effective deposition velocity, we must first know the deposition velocity, $k_{1\ m}$, for each particle size within the airborne particle size distribution. Thus the effective deposition velocity at any one location is the summation over particle diameters, i, of

$$v_d = \frac{\Sigma (K_{1\ m})_i\, \chi_i}{\chi} \tag{12.11}$$

In addition, the effective deposition velocity is a function of downwind distance since each particle size is depleted from the airborne plume with different mass-transfer rates. That is, the particle plume will show relatively more depletion at the air–surface interface for all particle diameters outside the range of 0.1 to 1 μm than for particles within that range. To determine either the average airborne concentration or the effective deposition velocity at some downwind location, we should calculate airborne concentrations for each particle size. Subsequently, airborne concentrations are summed for each particle size, and the total average concentration is determined.

In general, however, these summations have been neglected by most estimators. This neglect may not be justifiable in all applications unless the resulting concentration predictions are conservative for the situation being considered. Even if the estimator considers that the conservative prediction of airborne concentrations is calculated by neglecting plume depletion at breathing height, he makes the assumption that atmospheric transport models have been validated with data for most situations. However, there has been limited transport model validation even to 3 to 10 km (Crawford et al., 1977, 1978, and American Meteorological Society 1977 Committee on Atmospheric Turbulence and Diffusion, 1978). In addition, a basic requirement in the assessment of potentially dangerous conditions is that realistic estimates of airborne concentrations can be calculated, but such an assessment is realistic only when plume depletion by dry deposition is considered.

In retrospect, then, the point is that we must be aware of the preceding assumptions and limitations that go into the computation and use of deposition velocity. Furthermore, other variables that can contribute to changing the deposition velocity include atmospheric stability, wind speed and direction, aerodynamic surface roughness, friction velocity, modification of these parameters by local terrain, and diurnal and seasonal variations, in addition to the other variables listed in Table 12.1, including gravitational settling, particle size and shape, and gas solubility in water. These areas will not be discussed further, but they must be considered for each individual situation, even if their effects on resuspension and deposition cannot always be predicted.

12-7.4 Illustrations

Several applications of the data summarized in this chapter are presented here as potential guidelines. Examples are given for calculating deposition velocities, resuspension factors, and resuspension rates. Theoretical atmospheric transport model predictions of downwind airborne concentrations, based on deposition velocities and resuspension rates, could also be used to estimate airborne concentrations. For details on atmospheric transport, see Chaps. 6, 13, and 14.

12-7.4.1 ***Deposition.*** First, let us consider dry deposition. Assume that we are interested in predicting the deposition of a 1-μm-diameter monodispersed particle, having a density of 1 g/cm^3, over sagebrush when the particle release height is within 1 m of the ground. Instead of estimating the deposition velocities to calculate the plume depletion, in this case we can go directly to the first and third entries of Table 12.3 to estimate the plume depletion. Use of the 0.7-μm data for a 1-μm particle diameter is based on deposition velocities (see Figs. 12.3 to 12.7), which are nearly independent of particle diameters between 0.1 and 1 μm. Thus Table 12.3 predicts 22 to 32% plume depletion within 200 m, 10 to 54% plume depletion within 400 m, and 65% plume depletion within 91 m.

As an illustration of estimating a dry deposition velocity, assume the same conditions except that the particle diameter is now 5 μm. If we are fortunate enough to find similar experimental conditions in Table 12.4, the deposition velocity could be predicted. However, a completely similar experiment is not listed in Table 12.4, except possibly for 5-μm zinc sulfide at deposition velocities from 1.5 to 5.4 cm/s (Nickola and Clark, 1976).

For illustration purposes the deposition velocity will be selected from Figs. 12.3 to 12.7. To use these figures, we need both the aerodynamic surface roughness, z_0, and the friction velocity, u_*. Obviously, if experimental surface-roughness data were available for the situation being considered, that experimental surface roughness should be used. The friction velocity is estimated from Table 12.6. Although the listed surfaces do not include sagebrush, a surface roughness between 5 and 10 cm might be estimated from the long-grass data. A surface roughness of 5 cm and a wind speed of 5 m/s at 2 m are assumed. In this case, Table 12.6 suggests that the friction velocity is of the order of 0.6 m/s. The 0.6 m/s or 60 cm/s is used to determine the deposition velocities from Figs. 12.3 to 12.7. However, a friction velocity of 60 cm/s is not shown. A deposition velocity for 60 cm/s could be determined by interpolation (friction velocities between 50 and 100 cm/s), or we might assume that friction velocities of 50 and 60 m/s are comparable. In the latter case the deposition velocity for a friction velocity of 50 cm/s is used.

Looking at Fig. 12.5 for a friction velocity of 50 cm/s and a 5-μm particle diameter, we select a deposition velocity for a unit particle density and an aerodynamic surface roughness of 5 cm. For these conditions, a deposition velocity of 1 cm/s is predicted. For a smaller diameter of 0.3 μm, the deposition velocity is estimated to be 0.04 cm/s. Thus the deposition velocity for a particle diameter of 5 and 0.3 μm differs by a factor of 25.

From the preceding examples, we can see some difficulties that arise in calculating and predicting deposition velocities and, therefore, the importance of always providing the user with qualifying information on the assumptions used in the prediction calculations.

12-7.4.2 ***Resuspension: Vehicular.*** Now let us consider particle resuspension by vehicular traffic. Assume that an equal number of particles from 1 to 100 μm in diameter are on an asphalt road. Figures 12.11 to 12.13 indicate that no data are available to permit us to estimate the effects of different particle sizes on particle resuspension. Thus, in calculations, resuspension rates are assumed to be independent of particle diameter.

Then assume that resuspension occurs on the same day particles were deposited and that sedans are traveling at 25 m/s (55 mph). From the limited data in Fig. 12.11, which shows resuspension rates as a function of vehicular speed, the fraction resuspended from the road per vehicle pass is approximately 10^{-2}. Thus the fraction of material remaining on the road after one vehicle pass is 0.989, and the fraction remaining after n vehicle passes is $(0.989)^n$. Only 50% of the source would remain on the road after 53 sedans had driven through the tracer, and only 1% would remain after 352 sedans had driven through the tracer. However, at urban speeds of 56 km/h (35 mph), the fraction of particles resuspended from the road per sedan pass is 8.2×10^{-3}. Thus 84 sedan passes are required for 50% source removal. After 559 sedans had passed, only 1% of the source would remain.

12-7.4.3 ***Resuspension: Wind.*** Airborne concentrations from resuspension can be calculated from either resuspension factors (Tables 12.7 and 12.9) or from resuspension rates (Figs. 12.8 to 12.10 and Table 12.11). Possible applications of both resuspension factors and resuspension rates will be discussed.

In practice you may be interested in several types of resuspension predictions, including predictions of airborne concentrations at breathing height for people within contaminated areas, predictions of downwind airborne concentrations from resuspension, and predictions of changes in surface source contamination levels. Calculations using resuspension factors will be considered first. Resuspension factors, m^{-1}, are defined as the airborne concentration per cubic volume of air, m^{-3}, divided by the surface contamination per unit area, m^{-2}, below the airborne concentration measurement site. Resuspension factors are most useful for predicting airborne concentrations above contaminated surfaces; however, resuspension factors are not rates and are not useful in predicting changes with time in surface contamination levels.

The prediction of an average resuspension factor must always be qualified by uncertainty, since no correlation for predicting resuspension factors exists. As shown in

Table 12.7, the range of the wind-caused resuspension factor is from 9×10^{-11} to 3×10^{-4} m^{-1}—a range of 6 to 7 orders of magnitude. Even within one experimental location, the range can be over 3 orders of magnitude. If you are faced with having to estimate a resuspension factor, try to find a similar situation in Table 12.7, but use the number with caution.

12-7.4.4 ***Resuspension: Mechanical.*** Resuspension factors can be estimated for mechanical stresses from Table 12.9. However, there is no general correlation for predicting resuspension factors. These resuspension factors range over 8 orders of magnitude, from 1.2×10^{-10} up to 4×10^{-2} m^{-1}. Even within a single experiment, resuspension factors can range over 2 orders of magnitude. Thus uncertainties in predicting resuspension factors are very large. Resuspension factors cannot be predicted as a function of the source characteristics, particle size, particle density, or amount of mechanical stress producing resuspension.

12-7.4.5 ***Wind-Induced Resuspension Rates.*** The fraction, G/G_0, of a resuspension source remaining after time t is

$$G/G_0 = e^{-St} \tag{12.12}$$

To obtain a feel for the source strength as a function of time and resuspension rates, defined as the fraction of surface contaminant removed per second, compare the rate to the 3.2×10^7 s equal to 1 yr. If a resuspension rate were constant at 1×10^{-8} s^{-1}, 27% of the resuspension source would be depleted in a year. At a constant resuspension rate of 1×10^{-11} s^{-1}, only 0.03% of the resuspension source would be depleted in a year. For resuspension rates in the order of 10^{-11} to the 10^{-8} s^{-1}, material will continue to resuspend for many years. Similarly, if it is assumed that weathering does not occur, the resuspension source strength will also remain essentially constant for many years.

Resuspension rates caused by wind stresses can be estimated from Figs. 12.8 to 12.10 and from Table 12.11. Resuspension-rate variability will be considered first. For instance, in Table 12.11 resuspension rates for general surface sources range from 10^{-12} to 10^{-6} s^{-1}. The resuspension rates are of interest when they are compared to the resuspension rates calculated from agricultural erosion considerations. For instance, soil erosion rates of 0.5, 40, and 300 tons/(acre · month) were selected. These erosion rates were converted to average resuspension rates by assuming a 1-cm source depth for eroding soil. On the basis of this assumption, resuspension rates increased from 2×10^{-9}, 2×10^{-7}, and 1×10^{-6} s^{-1} for the three soil erosion rates, respectively. These resuspension rates are within the same orders of magnitude as those measured for pollutant resuspension. It is not known whether this similarity indicates a valid approach for calculating average resuspension rates for pollutants. The validity is questioned since erosion rates are determined for total soil movement rather than for the respirable particle size range for resuspension. Nevertheless, if you need to calculate resuspension rates for field conditions, you might justifiably calculate average resuspension rates from agricultural erosion equations until resuspension rates are experimentally determined or generalizations are made.

Although resuspension rates calculated from agricultural erosion data are in the range of resuspension-rate measurements shown in Figs. 12.8 to 12.10, a recommended procedure is required. If any of the experimental conditions are similar to your conditions, the best approach is to use resuspension rates measured for the similar conditions. This similarity will usually be the exception. For other conditions the recommendation is that resuspension rates be selected from Fig. 12.9, which gives the temporal variability of wind-caused tracer resuspension rates. A range of resuspension rates could be selected from the data shown in Fig. 12.10 for use in sensitivity analysis. Admittedly, this correlation is based only on the aerodynamic surface roughness height. There are many other variables discussed in this chapter that also should influence resuspension rates, but their effects are unknown.

Our ability to predict resuspension rates is limited by great uncertainty in the selected values, and, unfortunately, too few data exist for developing a more refined generalized resuspension-rate correlation. Much research is needed to develop a better generalized resuspension-rate correlation.

In any theoretical modeling approach using either resuspension rates or deposition velocities, you must be concerned with informing the reader about the model validity and the uncertainties in model predictions. Too many decisions are being made on the basis of numerical values predicted from models without considering the uncertainties in the numerical values. Even though we are using theoretical models at the forefront of science, we must be scientifically honest in our predictions. Uncertainties in those predictions must be communicated to the scientific community, to the legislative bodies, to those satisfying legal regulations, and especially to the general public. In the environmental pollution area, we must have the courage to continue to use and apply the most valid models, and we must also have the courage to conclude that existing modeling approaches may not be the best for predicting airborne concentrations. We must also have the courage to develop and to use, although not as theoretically satisfying, pragmatic approaches that will yield required answers.

ACKNOWLEDGMENT

This paper is based on work performed under U. S. Department of Energy contract EY-76-C-06-1830.

LIST OF SYMBOLS

Symbols used in Chap. 12 are listed here. The dimensions mass, length, and time are abbreviated as M, L, and T, respectively. Equation numbers indicate where the symbol first appears or where additional clarification can be found.

C	Airborne concentration of monodisperse particles (L^{-3}), Eq. 12.2
d	Zero-displacement plane (L), Eq. 12.9
D	Particle diameter (L), Eq. 12.7
F	Deposition flux (units L^{-2} T^{-1}), Eq. 12.1
G	Pollutant concentration on ground per unit area but to an unspecified depth (units L^{-2}), Eq. 12.3
k	von Kármán's constant (dimensionless), Eq. 12.8
$K_{1\,m}$	Deposition velocity for a single particle diameter with the airborne concentration measured at an elevation of 1 m (LT^{-1}), Eq. 12.2
N	Number of particles (L^{-3}), Eq. 12.7
Q	Soil drift perpendicular to the wind direction (ML^{-1} T^{-1}), Eq. 12.5
RF	Resuspension factor (L^{-1}), Eq. 12.3
S	Suspension or resuspension rate (fraction resuspended/T), Eq. 12.4
S_{air}	Suspension rate for particles moving in true airborne suspension air (fraction resuspended/T), Eq. 12.4
S_{salt}	Suspension rate for particles moving in saltation (fraction salt resuspended/T), Eq. 12.4
S_M	Suspension rate caused by mechanical stresses (fraction resuspended/T), Eq. 12.4
S_{sc}	Suspension rate for particles moving in surface creep (fraction resuspended/T), Eq. 12.4
S_W	Suspension rate caused by wind stresses (fraction resuspended/T), Eq. 12.4
t	Time (T), Eq. 12.12
u	Air velocity (LT^{-1}), Eq. 12.5
u_T	Threshold velocity for soil transport (LT^{-1}), Eq. 12.6
u_*	Friction velocity (LT^{-1}), Eq. 12.8
V	Particle volume (L^3), Eq. 12.7
v_d	Deposition velocity for a polydisperse aerosol determined in the field (LT^{-1}), Eq. 12.1
v_t	Monodisperse particle gravity settling velocity (LT^{-1}), Eq. 12.10
z	Height above ground (L), Eq. 12.8
z_0	Aerodynamic surface roughness (L), Eq. 12.8
χ	Airborne concentration of polydisperse particles (units L^{-3}), Eq. 12.1
ρ	Particle density (ML^{-3}), Fig. 12.3

REFERENCES

Akiba, M., 1975, *Size of Sand Grain and Critical Wind Velocity at Which Sand Grains Begin to Move,* Report NASA-TT-F-16190, available from the National Aeronautics and Space Administration, Washington, D. C.

Aldaz, L., 1969, Flux Measurements of Atmospheric Ozone over Land and Water, *J. Geophys. Res.,* 74: 6943-6946.

Allen, M. D., and R. D. Neff, 1975, Measurements of Deposition Velocity of Gaseous Elemental Iodine on Water, *Health Phys.,* 28: 707-715.

Amato, A. J., 1976, Theoretical Resuspension Ratios, in *Atmosphere–Surface Exchange of Particulate and Gaseous Pollutants (1974),* ERDA Symposium Series, Richland, Wash., Sept. 4–6, 1974, pp. 891-905, CONF-740921, NTIS.

American Meteorological Society 1977 Committee on Atmospheric Turbulence and Diffusion, 1978, Accuracy of Dispersion Models, *Bull. Am. Meteorol. Soc.,* 59(8): 1025-1026.

Andres, R. M., 1972, A Simulation of Saltation Dynamics with Application to Terrestrial and Martian Model Surfaces, No. 10, in *Proceedings of the 1972 Heat Transfer and Fluid Mechanics Institute,* R. B. Landis and G. J. Hardeman (Eds.), pp. 163-181, Stanford University Press.

Anspaugh, L. R., P. L. Phelps, G. Holladay, and K. O. Hamby, 1970, Distribution and Redistribution of Airborne Particulates from the Schooner Cratering Event, in *Proceedings of the Fifth Annual Health Physics Society Midyear Topical Symposium,* Idaho Falls, Nov. 3–6, 1970, Vol. 2, pp. 428-446.

——, P. L. Phelps, N. C. Kennedy, and H. G. Booth, 1973, Wind-Driven Redistribution of Surface-Deposited Radioactivity, in *Environmental Behavior of Radionuclides Released in the Nuclear Industry,* Symposium Proceedings, Aix-en-Provence, 1973, pp. 167-184, STI/PUB/345, International Atomic Energy Agency, Vienna.

——, P. L. Phelps, N. C. Kennedy, H. G. Booth, R. W. Goluba, J. M. Reichman, and J. S. Koval, 1974, Resuspension Element Status Report, in *The Dynamics of Plutonium in Desert Environments,* P. B. Dunaway and M. G. White (Eds.), USAEC Report NVO-142, pp. 221-310, Nevada Operations Office, NTIS.

——, P. L. Phelps, N. C. Kennedy, J. H. Shinn, and J. M. Reichman, 1976, Experimental Studies on the Resuspension of Plutonium from Aged Sources at the Nevada Test Site, in *Atmosphere–Surface Exchange of Particulate and Gaseous Pollutants (1974),* ERDA Symposium Series, Richland, Wash., Sept. 4–6, 1974, pp. 727-743, CONF-740921, NTIS.

——, J. H. Shinn, P. L. Phelps, and N. C. Kennedy, 1975, Resuspension and Redistribution of Plutonium in Soils, *Health Phys.,* 29: 571-582.

Atkins, D. H. F., R. C. Chadwick, and A. C. Chamberlain, 1967, Deposition of Radioactive Methyl Iodide to Vegetation, *Health Phys.,* 13: 91-92.

Aylor, D. E., 1976, Resuspension of Particles from Plant Surfaces by Wind, in *Atmosphere–Surface Exchange of Particulate and Gaseous Pollutants (1974),* ERDA Symposium Series, Richland, Wash., Sept. 4–6, 1974, pp. 791-812, CONF-740921, NTIS.

Bagnold, R. A., 1941, *The Physics of Blown Sand and Desert Dunes,* Methuen & Company, London.

——, 1960, The Re-Entrainment of Settled Dusts, *Int. J. Air Pollut.,* 2: 357-363.

Belot, Y., and D. Gauthier, 1975, Transport of Micronic Particles from Atmosphere to Foliar Surfaces, Chap. 40 in *Heat and Mass Transfer in the Biosphere, Part 1: Transfer Processes in the Plant Environment,* pp. 583-591, Scripta Book Co., Halsted Press, New York.

Bennett, B. G., 1976, Transuranic Element Pathways to Man, in *Transuranium Nuclides in the Environment,* Symposium Proceedings, San Francisco, pp. 367-383, STI/PUB/410, International Atomic Energy Agency, Vienna.

Bennett, J. H., and A. C. Hill, 1975, Interactions of Air Pollutants with Canopies of Vegetation, *Responses of Plants to Air Pollution,* Academic Press, New York.

——, A. C., Hill, and D. M. Gates, 1973, A Model for Gaseous Pollutant Sorption by Leaves, *J. Air Pollut. Control Assoc.,* 23(11): 957-962.

Bird, R. B., W. E. Stewart, and E. N. Lightfoot, 1960, *Transport Phenomena,* 2nd ed., pp. 652-680, John Wiley & Sons, New York.

Brook, A. J., 1968, *The Effect of Deposition on the Concentration of Windborne Material,* Report AHSB(S)-R-157, United Kingdom Atomic Energy Authority, Health and Safety Branch, Risley, Lancs., England.

Brunskill, R. T., 1967, The Relationship Between Surface and Airborne Contamination, in *Surface Contamination,* B. R. Fish (Ed.), pp. 93-105, Pergamon Press, New York.

Brutsaert, W., 1975, The Roughness Length for Water Vapor, Sensible Heat, and Other Scalars, *J. Atmos. Sci.,* 32: 2028-2031.

Bunch, D. F. (Ed.), 1968, *Controlled Environmental Radioiodine Tests,* Progress Report 3, USAEC Report IDO-12063, Idaho Operations Office, NTIS.

Businger, J. A., J. C. Wyngaard, Y. Izumi, and E. F. Bradley, 1971, Flux-Profile Relationships in the Atmospheric Surface Layer, *J. Atmos. Sci.,* 28: 181-189.

Caine, N., and P. Morin, 1976, *The Significance of Frost Action and Surface Soil Characteristics to Wind Erosion at Rocky Flats, Colorado,* Oct. 1, 1975–May 30, 1976, Second Progress Report, Institute of Arctic and Alpine Research, University of Colorado, Boulder, Colo.

Calder, K. L., 1961, Atmospheric Diffusion of Particulate Material, Considered as a Boundary Value Problem, *J. Meteorol.,* 18: 413-416.

Caporaloni, M., F. Tampieri, F. Trombetti, and O. Vittori, 1975, Transfer of Particles in Monisotropic Air Turbulence, *J. Atmos. Sci.,* 32: 565-568.

Carter, R. F., 1970, *The Measurement of Dust Levels in a Workshop Environment,* Report AWRE-028/70, United Kingdom Atomic Energy Authority, Atomic Weapons Research Establishment, Aldermaston, Berks., England.

Cataldo, D. A., B. L. Klepper, and D. K. Craig, 1976, Fate of Plutonium Intercepted by Leaf Surface: Leachability and Translocation to Seed and Root Tissues, *Transuranium Nuclides in the Environment,* Symposium Proceedings, pp. 291-301, STI/PUB/410, International Atomic Energy Agency, Vienna.

Cawse, P. A., 1974, *A Survey of Atmospheric Trace Elements in the United Kingdom, 1972–1973,* Report AERE-R-7669, United Kingdom Atomic Energy Authority, Atomic Energy Research Establishment, Harwell, Oxfords., England.

——, 1975, *A Survey of Atmospheric Trace Elements in the United Kingdom: Results for 1974,* Report AERE-R-8038, United Kingdom Atomic Energy Authority, Atomic Energy Research Establishment, Harwell, Oxfords., England.

——, 1976, *A Survey of Trace Elements in the United Kingdom: Results for 1975,* Report AERE-R-8398, United Kingdom Atomic Energy Authority, Atomic Energy Research Establishment, Harwell, Oxfords., England.

——, and D. H. Peirson, 1972, *An Analytical Study of Trace Elements in the Atmospheric Environment,* Report AERE-R-7134, United Kingdom Atomic Energy Authority, Atomic Energy Research Establishment, NTIS.

Chadwick, R. C., and A. C. Chamberlain, 1970, Field Loss of Radionuclides from Grass, *Atmos. Environ.,* 4: 51-56.

Chamberlain, A. C., 1955, *Aspects of Travel and Deposition of Aerosol and Vapour Clouds,* Report AERE-HP/R-1261, United Kingdom Atomic Energy Authority, Atomic Energy Research Establishment, Harwell, Berks., England.

——, 1960, Aspects of the Deposition of Radioactive and Other Gases and Particles, *Int. J. Air Pollut.,* 3: 63-88.

——, 1962, *Transport of Particles Across Boundary Layers,* Report AERE-M-1122, United Kingdom Atomic Energy Authority, Atomic Energy Research Establishment, Harwell, Oxfords., England.

——, 1966, Transport of Lycopodium Spores and Other Small Particles to Rough Surfaces, *Proc. R. Soc. (London), Ser. A,* 296: 45-70.

——, 1970, Interception and Retention of Radioactive Aerosols by Vegetation, *Atmos. Environ.,* 4: 57-78.

——, 1974, Mass Transfer to Bean Leaves, *Boundary-Layer Meteorol.,* 6: 447-486.

——, 1975, Pollutant in Plant Canopies, in Heat and Mass Transfer in the Biosphere, Pt. 1, *Transfer Processes in the Plant Environment,* P. A. de Vries and N. H. Afgan (Eds.), pp. 561-582, Scripta Publishing Co., Washington, D. C.

——, and R. C. Chadwick, 1953, Deposition of Airborne Radioiodine Vapor, *Nucleonics,* 8: 22-25.

——, and R. C. Chadwick, 1966, Transport of Iodine from Atmosphere to Ground, *Tellus,* 18(2): 226-237.

Chepil, W. S., 1943, *Soil Erosion by Wind,* Soil Research Laboratory, Dominion Department of Agriculture, Report of Investigations, Swift Current, Saskatchewan, Canada.

——, 1945a, Dynamics of Wind Erosion: I. Nature of Movement of Soil by Wind, *Soil Sci.,* 60: 305-320.

——, 1945b, Dynamics of Wind Erosion: II. Initiation of Soil Movement, *Soil Sci.,* 60: 397-411.

——, 1945c, Dynamics of Wind Erosion: V. Cumulative Intensity of Soil Drifting Across Eroding Fields, *Soil Sci.,* 61: 257-263.

——, 1950a, Properties of Soil Which Influence Wind Erosion: I. The Governing Principle of Surface Roughness, *Soil Sci.,* 69: 149-162.

——, 1950b, Methods of Estimating Apparent Density of Discrete Soil Grains and Aggregates, *Soil Sci.,* 70: 351-362.

——, 1951a, Properties of Soil Which Influence Wind Erosion: IV. State of Dry Aggregate Structure, *Soil Sci.,* 72: 387-401.

——, 1951b, Properties of Soil Which Influence Wind Erosion: V. Mechanical Stability of Structure, *Soil Sci.,* 72: 465-478.

——, 1959, Equilibrium of Soil Grains at the Threshold of Movement by Wind, *Soil Sci. Soc. Am., Proc.,* 23: 422-428.

——, 1965, Transport by Wind: Transport of Soil and Snow by Wind, *Agric. Meteorol.,* 6: 123-132.

Cline, J. F., D. O. Wilson, and F. P. Hungate, 1965, Effect of Physical and Biological Conditions on Deposition and Retention of ^{131}I on Plants, *Health Phys.,* 11: 713-717.

Clough, W. S., 1973, Transport of Particles to Surfaces, *Aerosol Sci.,* 4: 227-234.

——, 1975, The Deposition of Particles on Moss and Grass Surfaces, *Atmos. Environ.,* 9: 1113-1119.

Convair, 1960, *Fission Products Field Release Test II,* Report NARF-60-10T (FZK-9-149; AFSWC-TR-60-26), U. S. Air Force Nuclear Aircraft Research Facility, Division of General Dynamics Corp., Ft. Worth, Tex., NTIS.

Corn, M., 1966, Adhesion of Particles, in *Aerosol Science,* pp. 359-392, C. N. Davies (Ed.), Academic Press, New York.

——, and F. Stein, 1965, Re-Entrainment of Particles from a Plane Surface, *Am. Ind. Hyg. Assoc. J.,* 26: 325-336.

Cowherd, C., Jr., and C. O. Mann, 1976, *Quantification of Dust Entrainment from Paved Roads,* paper presented at the 69th Annual Meeting of the Air Pollution Control Association, June 27–July 1, 1976, Portland, Oreg.

Craig, D. K., B. L. Klepper, and R. L. Buschbom, 1976, Deposition of Various Plutonium-Compound Aerosols onto Plant Foliage at Very Low Wind Velocities, in *Atmosphere–Surface Exchange of Particulate and Gaseous Pollutants (1974),* ERDA Symposium Series, Richland, Wash., Sept. 4–6, 1974, pp. 244-263, CONF-740921, NTIS.

Crawford, T. V., et al., 1977, *Atmospheric Transport of Radionuclides,* USAEC Report DP-MS-77-116, Savannah River Laboratory, NTIS.

——, et al., 1978, Atmospheric Transport of Radionuclides, F. O. Hoffman (General Chairman), *Proceedings of a Workshop on the*

Evaluation of Models Used for the Environmental Assessment of Radionuclide Releases, Gatlinburg, Tenn., Sept. 6–9, 1977, USAEC Report CONF-770901, Oak Ridge National Laboratory, NTIS.

Crecelius, E. A., D. E. Robertson, K. H. Abel, D. A. Cochran, and W. C. Weimer, 1978, Atmospheric Deposition of ^{7}Be and Other Elements on the Washington Coast, in *Pacific Northwest Laboratory Annual Report for 1977 to the DOE Assistant Secretary for Environment,* DOE Report PNL-2500(Pt.2), pp. 7.25-7.26, Pacific Northwest Laboratory.

Csanady, G. T., 1955, Dispersal of Dust Particles from Elevated Sources, *Aust. J. Phys.,* 8: 545-550.

——, 1957, Dispersal of Dust Particles from Elevated Sources, II. Limitations of the Approximate Theory, *Aust. J. Phys.,* 10: 558-564.

Davidson, C. I., 1977, The Deposition of Trace Metal-Containing Particles in the Los Angeles Area, *Powder Technol.,* 18: 117-126.

Davis, W. E., 1978, Comparison of Dry Deposition in Single- and Multi-Layer U. S. Scale Assessment Models, in *Pacific Northwest Laboratory Annual Report for 1977 to the DOE Assistant Secretary for Environment,* DOE Report PNL-2500(Pt.3), pp. 2.25-2.26, Pacific Northwest Laboratory, NTIS.

Doran, J. C., 1977, Limitations on the Determination of Deposition Velocities, in *Pacific Northwest Laboratory Annual Report for 1976,* DOE Report BNWL-2100-3, pp. 55-58, Battelle Pacific Northwest Laboratory, NTIS.

Dovland, H., and A. Eliassen, 1976, Dry Deposition on a Snow Surface, *Atmos. Environ.,* 10: 783-785.

Draxler, R. R., and W. P. Elliott, 1977, Long-Range Travel of Airborne Material Subjected to Dry Deposition, *Atmos. Environ.,* 11: 35-40.

Droppo, J. G., and J. C. Doran, 1978, Dry Deposition of Gaseous Pollutants from Coal-Fired Power Plants, in *Pacific Northwest Laboratory Annual Report for 1977 to the DOE Assistant Secretary for Environment,* DOE Report PNL-2500(Pt.3), pp. 1.27-1.29, Pacific Northwest Laboratory, NTIS.

——, D. W. Glover, O. B. Abbey, C. W. Spicer, and J. Cooper, 1976, *Measurement of Dry Deposition of Fossil Fuel Plant Pollutants,* Report EPA-600/4-76-056, Environmental Protection Agency, NTIS.

Energy Research and Development Administration, 1977, *ERDA, Manual of Operations,* Chap. 0524, "Standards for Radiation Protection," Appendix A, Table II.

Engelmann, R. J., and G. A. Sehmel (Coordinators), 1976, *Atmosphere–Surface Exchange of Particulate and Gaseous Pollutants (1974),* ERDA Symposium Series, Richland, Wash., Sept. 4–6, 1974, CONF-740921, NTIS.

Fish, B. R., R. L. Walker, G. W. Royster, and J. L. Thompson, 1967, Redispersion of Settled Particles, in *Surface Contamination,* B. R. Fish (Ed.), pp. 75-81, Pergamon Press, New York.

Ford, E. F., 1957, The Transport of Sand by Wind, *Trans. Am. Geophys. Union,* 38(2): 171-174.

Fritschen, L. J., and R. Edmonds, 1976, Dispersion of Fluorescent Particles into and Within a Douglas Fir Forest, in *Atmosphere–Surface Exchange of Particulate and Gaseous Pollutants (1974),* ERDA Symposium Series, Richland, Wash., Sept. 4–6, 1974, pp. 280-301, CONF-740921, NTIS.

Fuquay, J. J., 1957, *Meteorology as Related to Waste Disposal and Weapons Tests,* USAEC Report HW-47721-A, Hanford Atomic Products Operation, NTIS.

Galbally, I., 1968, Some Measurements of Ozone Variation and Destruction in the Atmospheric Surface Layer, *Nature,* 218: 456-457.

——, 1971, Ozone Profiles and Ozone Fluxes in the Atmospheric Surface Layer, *J. R. Meteorol. Soc.,* 97: 18-29.

Garland, J. A., 1976a, Dry Deposition of SO_2 and Other Gases, in *Atmosphere–Surface Exchange of Particulate and Gaseous Pollutants (1974),* ERDA Symposium Series, Richland, Wash., Sept. 4–6, 1974, pp. 212-227, CONF-740921, NTIS.

——, 1976b, Discussions: Dry Deposition on a Snow Surface, *Atmos. Environ.,* 10: 1033.

——, D. H. F. Atkins, C. J. Readings, and S. J. Caughey, 1974, Deposition of Gaseous Sulphur Dioxide to the Ground, *Atmos. Environ.,* 8: 74-79.

Gifford, F. A., Jr., and D. H. Pack, 1962, Surface Deposition of Airborne Material, *Nucl. Safety,* 3(4): 76-89 (June).

Gillette, D. A., 1976, Production of Fine Dust by Wind Erosion of Soil: Effect of Wind and Soil Texture, in *Atmosphere–Surface Exchange of Particulate and Gaseous Pollutants (1974),* ERDA Symposium Series, Richland, Wash., Sept. 4–6, 1974, pp. 591-609, CONF-740921, NTIS.

Glauberman, H., W. R. Bootmann, and A. J. Breslin, 1967, Studies of the Significance of Surface Contamination, in *Surface Contamination,* B. R. Fish (Ed.), pp. 169-178, Pergamon Press, New York.

Gregory, P. H., 1950, Deposition of Airborne Particles on Trap Surfaces, *Nature,* 66: 487.

Hagen, L. J., and N. P. Woodruff, 1973, Air Pollution from Duststorms in the Great Plains, *Atmos. Environ.,* 7: 323-332.

——, and N. P. Woodruff, 1975, Particulate Loads Caused by Wind Erosion in the Great Plains, *J. Air Pollut. Control Assoc.,* 25: 860-861.

Hawley, C. A., Jr., and E. H. Markee, Jr., 1965, Controlled Environmental Radioiodine Tests, in *Radioactive Fallout from Nuclear Weapons Tests,* AEC Symposium Series, Germantown, Md., Nov. 3–6, 1964, Alfred W. Klement, Jr. (Ed.), pp. 821-835, CONF-765, NTIS.

Healy, J. W., 1955, *A Preliminary Estimate of Wind Pickup and Impaction of Particles,* USAEC Report HW-35542, Hanford Atomic Products Operation, NTIS.

——, 1974, *A Proposed Interim Standard for Plutonium in Soils,* USAEC Report LA-5483-MS, Los Alamos Scientific Laboratory, NTIS.

——, 1977a, A Review of Resuspension Models, in *Proceedings of the Nevada Applied Ecology Group Symposium on the Dynamics of Transuranics in Terrestrial and Aquatic Environments,* DOE Report NVO-178, pp. 211-222, NTIS.

——, 1977b, *An Examination of the Pathways from Soil to Man for Plutonium,* DOE Report LA-6741-MS, Los Alamos Scientific Laboratory, NTIS.

——, and J. J. Fuquay, 1958, Wind Pickup of Radioactive Particles from the Ground, in *Proceedings of the Second United Nations International Conference on the Peaceful Uses of Atomic Energy, Geneva, 1958,* Vol. 18, pp. 291-295, United Nations, New York, 1958.

——, and J. J. Fuquay, 1959, Wind Pickup to Radioactive Particles from the Ground, in *Progress in Nuclear Energy, Health Physics,* Series XII, Vol. 1, pp. 427-436, Pergamon Press, New York.

Heinemann, K., K. J. Vogt, and L. Angeletti, 1976a, Deposition and Biological Half-Life of Elemental Iodine on Grass and Clover, in *Atmosphere–Surface Exchange of Particulate and Gaseous Pollutants (1974),* ERDA Symposium Series, Richland, Wash., Sept. 4–6, 1974, pp. 136-152, CONF-740921, NTIS.

——, K. M. Stoeppler, K. J. Vogt, and L. Angeletti, 1976b, *Studies on the Deposition and Release of Iodine on Vegetation,* ERDA Report ORNL-tr-4313, NTIS.

Henderson, R. W., and R. V. Fultyn, 1965, *Radiation Measurements of the Effluent from the Kiwi B4D-202 and B4E-301 Reactors,* USAEC Report LA-3397, Los Alamos Scientific Laboratory, NTIS.

Hereim, A. T., and B. Ritchie, 1976, Resuspended Bacteria from Desert Soil, in *Atmosphere–Surface Exchange of Particulate and Gaseous Pollutants (1974),* ERDA Symposium Series, Richland, Wash., Sept. 4–6, 1974, pp. 835-845, CONF-740921, NTIS.

Hidy, G. M., and J. R. Brock, 1971, An Assessment of the Global Sources of Tropospheric Aerosols, in the *Second International Clean Air Congress,* H. M. Englund and W. T. Beery (Eds.), pp. 1088-1097, Academic Press, New York.

Hill, A. C., 1971, Vegetation: A Sink for Atmospheric Pollutants, *J. Air Pollut. Control Assoc.,* 21(6): 341-346.

Hinds, W. T., and R. H. Sauer, 1976, Soil Erodibility, Soil Erosion, and Revegetation Following Wildfire in a Shrub–Steppe Community, in *Atmosphere–Surface Exchange of Particulate and Gaseous Pollutants (1974),* ERDA Symposium Series, Richland, Wash., Sept. 4–6, 1974, pp. 571-590, CONF-740921, NTIS.

Horbert, M., K. J. Vögt, and L. Angeletti, 1976, *Studies on the Deposition of Aerosols on Vegetation and Other Surfaces,* ERDA Report ORNL-tr-4314, available from Oak Ridge National Laboratory, Oak Ridge, Tenn.

Horst, T. W., 1977a, *The Estimation of Air Concentrations Due To the Diffusion, Deposition, and Resuspension of Contaminants by the Wind,* DOE Report PNL-2426, Pacific Northwest Laboratory, NTIS.

——, 1977b, A Surface Depletion Model for Deposition from a Gaussian Plume, *Atmos. Environ.,* 11: 41-46.

Hsu, S. A., 1971, Wind Stress Criteria in Eolian Sand Transport, *J. Geophys. Res.,* 76: 8684-8686.

International Commission on Radiological Protection (ICRP), 1959, Publication 2, *Report of Committee II on Permissible Dose for Internal Radiation,* p. 82, Recommendations of the International Commission on Radiological Protection, Pergamon Press, New York.

Iranzo, E., and S. Salvador, 1970, *Inhalation Risks to People Living near a Contaminated Area,* presented at the Second International Congress of the International Radiation Protection Association, May 3–8, 1970, Brighton, England.

Islitzer, N. F., and R. K. Dumbauld, 1963, Atmospheric Diffusion-Deposition Studies over Flat Terrain, *Int. J. Air Water Pollut.,* 7: 999-1022.

Israel, G. W., 1974, Deposition Velocity of Gaseous Fluorides on Alfalfa, *Atmos. Environ.,* 8: 1329-1330.

——, 1977, Accumulation of Gaseous and Particulate Fluorine Compounds, *Atmos. Environ.,* 11: 183-188.

Izrael, Yu. A., 1971, *Behavior and Migration of Radioisotopes Contaminating the Ground,* USSR State Committee for Uses of Atomic Energy, Moscow, USSR, USAEC Report UCRL-tr-10549, NTIS.

Jones, I. S., and S. F. Pond, 1967, Some Experiments To Determine the Resuspension Factor of Plutonium from Various Surfaces, in *Surface Contamination,* B. R. Fish (Ed.), pp. 83-92, Pergamon Press, New York.

Judeikis, H. S., and T. B. Stewart, 1976, Laboratory Measurement of SO_2 Deposition Velocities on Selected Building Materials and Soils, *Atmos. Environ.,* 10: 769-776.

——, and A. G. Wren, 1977, Deposition of H_2S and Dimethyl Sulfide on Selected Soil Materials, *Atmos. Environ.,* 11: 1221-1224.

Judson, S., 1968, Erosion of the Land, or What's Happening to Our Continents? *Am. Sci.,* 56: 356-374.

Kathren, R. L., 1968, *Towards Interim Acceptable Surface Contamination Levels for Environmental PuO_2,* USAEC Report BNWL-SA-1510, Battelle Pacific Northwest Laboratory, NTIS.

Kawamura, R. R., 1975, *Motion of Sand Grains by Wind,* Report NASA-TT-F-16191, National Aeronautics and Space Administration, Washington, D. C.

Knight, D. H., 1973, Leaf Area Dynamics of a Shortgrass Prairie in Colorado, *Ecology,* 54: 891-896.

Langham, W. H., 1971, Plutonium Distribution as a Problem in Environmental Science, in *Proceedings of Environmental Plutonium Symposium,* E. B. Fowler, R. W. Henderson, and M. F. Milligan (Eds.), USAEC Report LA-4756, p. 9, Los Alamos Scientific Laboratory, NTIS.

Liss, P. S., and P. G. Slater, 1976, Mechanism and Rate of Gas Transfer Across the Air–Sea Interface, in *Atmosphere–Surface Exchange of Particulate and Gaseous Pollutants (1974),* ERDA Symposium Series, Richland, Wash., Sept. 4–6, 1974, pp. 354-368, CONF-740921, NTIS.

Little, P., and R. D. Wiffen, 1977, Emission and Deposition of Petrol Engine Exhaust Lead. I. Deposition of Exhaust Lead to Plant and Soil Surfaces, *Atmos. Environ.,* 11: 437-447.

Makhonko, K. P., 1970, *The Deposition of Radioactive Dust and Its Elimination from the Atmosphere by Precipitation,* USSR Reports on Natural and Fallout Radioactivity, USAEC Report AEC-tr-7128, pp. 79-111, NTIS.

Malina, F. J., 1941, Recent Developments in the Dynamics of Wind-Erosion, *Trans. Am. Geophys. Union,* 22: 262-287.

Marenco, A. and J. Fontan, 1976, Influence of Dry Deposition on the Residence Time of Particulate Pollutants in the Troposphere, in *Atmosphere–Surface Exchange of Particulate and Gaseous Pollutants (1974),* ERDA Symposium Series, Richland, Wash., Sept. 4–6, 1974, pp. 54-61, CONF-740921, NTIS.

Markee, E. H., Jr., 1967, A Parametric Study of Gaseous Plume Depletion by Ground Surface Absorption, in *USAEC Meteorological Information Meeting,* Chalk River Nuclear Laboratories, Atomic Energy of Canada Ltd., Chalk River, Ontario, Canada, Sept. 11–14, 1967, C. A. Mauson (Ed.), Report AECL-2787, pp. 602-614, Atomic Energy of Canada Ltd.

——, 1971, *Studies of the Transfer of Radioiodine Gas to and from Natural Surfaces,* Report ERL-ARL-29, Environmental Research Laboratories, Air Resources Laboratories, NTIS.

——, L. Andrews, and R. Jubach, 1977, *On the Behavior of Dry Gaseous Effluent Deposition and Attendant Plume Depletion,* presented at the American Meteorological Society and Air Pollution Control Association Joint Conference on Applications of Air Pollution Meteorology, Nov. 29–Dec. 2, 1977, Salt Lake City, Utah, available from the American Meteorological Society, Boston, Mass.

Masironi, L. A., and B. R. Fish, 1967, Direct Observation of Particle Re-Entrainment from Surfaces, in *Surface Contamination,* B. R. Fish (Ed.), pp. 55-59, Pergamon Press, New York.

Meetham, A. R., 1950, Natural Removal of Pollution from the Atmosphere, *J. R. Meteorol. Soc.,* 76: 359-371.

Megaw, W. J., and R. C. Chadwick, 1956, *Some Field Experiments on the Release and Deposition of Fission Products and Thoria,* Report AERE-HP-M-114, United Kingdom Atomic Energy Authority, Atomic Energy Research Establishment, Harwell, Berks., England, NTIS.

Milham, R. C., J. F. Schubert, J. R. Watts, A. L. Boni, and J. C. Corey, 1976, Measured Plutonium Resuspension and Resulting Dose from Agricultural Operations on an Old Field at the Savannah River Plant in the Southeastern United States of America, in *Transuranium Nuclides in the Environment,* Symposium Proceedings, San Francisco, pp. 409-421, STI/PUB/410, International Atomic Energy Agency, Vienna.

Mills, M. T., R. C. Dahlman, and J. S. Olson, 1974, *Ground Level Air Concentrations of Dust Particles Downwind from a Tailings Area During a Typical Windstorm,* USAEC Report ORNL-TM-4375, Oak Ridge National Laboratory, NTIS.

Mitchell, R. N., and B. C. Eutsler, 1967, A Study of Beryllium Surface Contamination and Resuspension, in *Surface Contamination,* B. R. Fish (Ed.), pp. 349-352, Pergamon Press, New York.

Miyamaga, I., A. Kasai, and K. Imai, 1973, A Preliminary Experiment on the Deposition of Gaseous Radioiodine onto Environmental Materials, in *Environmental Behavior of Radionuclides Released in the Nuclear Industry,* Symposium Proceedings, Aix-en-Provence, 1973, pp. 157-165, STI/PUB/345, International Atomic Energy Agency, Vienna.

Möller, U., and G. Schumann, 1970, Mechanisms of Transport from the Atmosphere to the Earth's Surface, *J. Geophys. Res.*, 75: 3013-3019.

Mork, H. M., 1970, *Redistribution of Plutonium in the Environs of the Nevada Test Site,* USAEC Report UCLA-12-590, University of California, NTIS.

Murphy, C. E., 1976, *The Calculation of Deposition Velocities for ^3HHO, $^{14}CO_2$, $^{131}I_2$, $CH_3{}^{131}I$ on Vegetated Surfaces,* DOE Report DPST-LWR-76-163, Savannah River Laboratory, NTIS.

Myers, D. S., W. J. Silver, D. G. Coles, K. C. Lamson, D. R. McIntyre, and B. Mendoza, 1976, Evaluation of the Use of Sludge Containing Plutonium as a Soil Conditioner for Food Crops, in *Transuranium Nuclides in the Environment,* Symposium Proceedings, San Francisco, pp. 311-323, STI/PUB/410, International Atomic Energy Agency, Vienna.

Newman, J. E., M. D. Abel, P. R. Harrison, and K. J. Yost, 1976, Wind as Related to Critical Flushing Speed Vs. Reflotation Speed by High-Volume Sampler Particulate Loading, in *Atmosphere–Surface Exchange of Particulate and Gaseous Pollutants (1974),* ERDA Symposium Series, Richland, Wash., Sept. 4–6, 1974, pp. 466-496, CONF-740921, NTIS.

Nickola, P. W., and G. H. Clark, 1976, Field Measurement of Particulate Plume Depletion by Comparison with an Inert Gas Plume, in *Atmosphere–Surface Exchange of Particulate and Gaseous Pollutants (1974),* ERDA Symposium Series, Richland, Wash., Sept. 4–6, 1974, pp. 74-86, CONF-740921, NTIS.

——, and R. N. Lee, 1975, Direct Measurements of Particulate Deposition on a Snow Surface, in *Pacific Northwest Laboratory Annual Report for 1974,* DOE Report BNWL-1950-3, pp. 185-188, Battelle Pacific Northwest Laboratory, NTIS.

O'Brien, M. P., and B. D. Rindlaub, 1936, The Transportation of Sand by Wind, *Civ. Eng.*, 6(5): 325.

Oksza-Chocimowoski, G. V., 1976, *Resuspension Models Review,* Report ORP/LV-76-11, Environmental Protection Agency, NTIS.

——, 1977, *Generalized Model of the Time-Dependent Weathering Half-Life of the Resuspension Factor,* Report ORP/LV-77-4, Environmental Protection Agency, NTIS.

Olafson, J. H., and K. H. Larson, 1961, *Plutonium, It's Biology and Environmental Persistence,* USAEC Report UCLA-501, University of California, NTIS.

Orgill, M. M., 1977, Application of Soil Erosion Data to Particle Wind Resuspension, in *Pacific Northwest Laboratory Annual Report for 1976,* DOE Report BNWL-2100-3, pp. 79-84, Battelle Pacific Northwest Laboratory, NTIS.

——, and G. A. Sehmel, 1976, Frequency and Diurnal Variation of Dust Storms in the Contiguous United States, *Atmos. Environ.*, 10: 813-825.

Overcamp, T. J., 1976, A General Gaussian Diffusion-Deposition Model for Elevated Point Sources, *J. Appl. Meteorol.*, 15: 1167-1171.

Owen, P. R., 1964, Saltation of Uniform Grains in Air, *J. Fluid Mech.*, 420: 225-242.

Owers, M. J., and A. W. Powell, 1974, Deposition Velocity of Sulphur Dioxide on Land and Water Surfaces Using a ^{35}S Tracer Method, *Atmos. Environ.*, 8: 63-67.

Parker, H. M., 1956, Radiation Exposure from Environmental Hazards, in *Proceedings of the International Conference on the Peaceful Uses of Atomic Energy, Geneva, 1955,* Vol. 13, pp. 305-310, United Nations, New York.

Pasquill, F., 1962, Deposition of Airborne Material, in *Atmospheric Diffusion,* pp. 226-239, Van Nostrand Reinhold Co., Cincinnati.

——, 1974, *Atmospheric Diffusion,* 2nd ed., Halsted Press, New York.

Peters, L. N., and J. P. Witherspoon, 1972, Retention of 44–88 μ Simulated Fallout Particles by Grasses, *Health Phys.*, 22: 261-266.

Plate, E. J., 1971, *Aerodynamic Characteristics of Atmospheric Boundary Layers,* TID-25465, U. S. Atomic Energy Commission, NTIS.

Porch, W. M., and J. E. Lovill, 1976, *Aerosol Deposition and Suspension During a Texas Dust Storm,* ERDA Report UCRL-78011, Lawrence Livermore Laboratory, NTIS.

Priestley, C. H. B., 1959, *Turbulent Transfer in the Lower Atmosphere,* p. 21, University of Chicago Press, Chicago.

Punjrath, J. S., and D. R. Heldman, 1972a, Influence of Air Turbulence on Re-Entrainment of Small Particles from Flat Surfaces, *Trans. Am. Soc. Agric. Eng.*, 15: 559-562.

——, and D. R. Heldman, 1972b, Mechanisms of Small Particle Re-Entrainment from Flat Surfaces, *Aerosol Sci.*, 3: 429-440.

Rasmussen, K. H., M. Taheri, and R. L. Kabel, 1974, *Sources and Natural Removal Processes for Some Atmospheric Pollutants,* Report EPA-650/4-74-032, Environmental Protection Agency, NTIS.

Raynor, G. S., 1976, Experimental Studies of Pollen Deposition to Vegetated Surfaces, in *Atmosphere–Surface Exchange of Particulate and Gaseous Pollutants (1974),* ERDA Symposium Series, Richland, Wash., pp. 264-279, CONF-740921, NTIS.

Romney, E. M., R. G. Lindberg, H. A. Hawthorne, B. G. Bystrom, and K. H. Larson, 1963, Contamination of Plant Foliage with Radioactive Fallout, *Ecology*, 44: 343-349.

Rosinski, J., G. Langer, C. T. Nagamoto, and J. S. Bogard, 1976, Generation of Secondary Particles from Single Large Soil Particles upon Impact, in *Atmosphere–Surface Exchange of Particulate and Gaseous Pollutants (1974),* ERDA Symposium Series, Richland, Wash., Sept. 4–6, 1974, pp. 638-647, CONF-740921, NTIS.

Schumann, G., 1975, *The Process of Direct Deposition of Aerosols at the Sea Surface,* presented at the Interdisciplinary Symposium on Marine Meteorological Aspects at the XVI General Assembly of the International Union of Geodesy and Geophysics, Grenoble, France.

Scriven, R. A., and B. E. A. Fisher, 1975, The Long Range Transport of Airborne Material and Its Removal by Deposition and Washout. I. General Considerations. II. The Effect of Turbulent Diffusion, *Atmos. Environ.*, 9: 49-68.

Sehmel, G. A., 1956, *The Agglomeration of Solid Aerosol Particles,* M. S. Thesis, University of Illinois, Urbana, Ill.

——, 1970, Particle Deposition from Turbulent Air Flow, *J. Geophys. Res.*, 75: 1766-1781.

——, 1971, Particle Diffusivities and Deposition over a Horizontal Smooth Surface, *J. Colloid Interface Sci.*, 37: 891-906.

——, 1973a, An Evaluation of a High-Volume Cascade Particle Impactor System, in *Second Joint Conference on Sensing of Environmental Pollutants,* pp. 109-115, Instrument Society of America, Pittsburgh, Pa.

——, 1973b, Particle Eddy Diffusivities and Deposition Velocities for Isothermal Flow and Smooth Surfaces, *Aerosol Sci.*, 4: 125-138.

——, 1973c, Particle Resuspension from an Asphalt Road Caused by Car and Truck Traffic, *Atmos. Environ.*, 7: 291-301.

——, 1975a, Initial Correlation of Particle Resuspension Rates as a Function of Surface Roughness Height, in *Pacific Northwest Laboratory Annual Report for 1974,* ERDA Report BNWL-1950-3, pp. 209-212, Battelle Pacific Northwest Laboratory, NTIS.

——, 1975b, Airborne Dust Size Distributions as a Function of Wind Speed, in *Pacific Northwest Laboratory Annual Report for 1974,* ERDA Report BNWL-1950-3, pp. 213-215, Battelle Pacific Northwest Laboratory, NTIS.

——, 1976a, Particle Resuspension from Truck Traffic in a Cheatgrass Area, in *Pacific Northwest Laboratory Annual Report for 1975,* ERDA Report BNWL-2000-3, pp. 96-98, Battelle Pacific Northwest Laboratory, NTIS.

——, 1976b, *Airborne ^{238}Pu and ^{239}Pu Associated with the Larger Than "Respirable" Resuspended Particles at Rocky Flats During July 1973,* ERDA Report BNWL-2119, Battelle Pacific Northwest Laboratory, NTIS.

——, 1977a, *Transuranic and Tracer Simulant Resuspension,* DOE Report BNWL-SA-6236, Battelle Pacific Northwest Laboratory, NTIS.

——, 1977b, *Radioactive Particle Resuspension Research Experiments on the Hanford Reservation,* DOE Report BNWL-2081, Battelle Pacific Northwest Laboratory, NTIS.

——, 1978a, Resuspension and Transport of Respirable Particles on Nonrespirable Particles, in *Pacific Northwest Laboratory Annual Report for 1977 to the DOE Assistant Secretary for Environment,* DOE Report PNL-2500(Pt.3), pp. 2.6-2.8, Pacific Northwest Laboratory, NTIS.

——, 1978b, *Plutonium Concentrations in Airborne Soil at Rocky Flats and Hanford Determined During Resuspension Experiments,* DOE Report PNL-SA-6720, Pacific Northwest Laboratory, NTIS.

——, 1978c, *Airborne Plutonium-239 and Americium-241 Concentrations Measured from the 125-Meter Hanford Meteorological Tower,* DOE Report PNL-SA-6771, Pacific Northwest Laboratory, NTIS.

——, and W. H. Hodgson, 1976, Field Deposition Velocity Measurements to a Sagebrush Canopy, in *Pacific Northwest Laboratory Annual Report for 1975,* ERDA Report BNWL-2000-3, pp. 89-91, Battelle Pacific Northwest Laboratory, NTIS.

——, and W. H. Hodgson, 1978, *A Model for Predicting Dry Deposition of Particles and Gases to Environmental Surfaces,* DOE Report PNL-SA-6721, Pacific Northwest Laboratory, NTIS.

——, and F. D. Lloyd, 1975, Initial Particle Resuspension Rates—A Field Experiment Using Tracer Particles, in *Pacific Northwest Laboratory Annual Report for 1974,* ERDA Report BNWL-1950-3, pp. 203-207, Battelle Pacific Northwest Laboratory, NTIS.

——, and F. D. Lloyd, 1976, Resuspension of Plutonium at Rocky Flats, in *Atmosphere–Surface Exchange of Particulate and Gaseous Pollutants (1974),* ERDA Symposium Series, Richland, Wash., Sept. 4–6, 1974, pp. 757-779, CONF-740921, NTIS.

——, and M. M. Orgill, 1973, Resuspension by Wind at Rocky Flats, in *Pacific Northwest Laboratory Annual Report for 1972,* USAEC Report BNWL-1751(Pt.1), pp. 15-22, Battelle Pacific Northwest Laboratory, NTIS.

——, and M. M. Orgill, 1974, Resuspension Source Change at Rocky Flats, in *Pacific Northwest Laboratory Annual Report for 1973,* USAEC Report BNWL-1850-3, pp. 212-214, Battelle Pacific Northwest Laboratory, NTIS.

——, and K. L. Richmond, 1978, Airborne Soil Concentrations, in *Pacific Northwest Laboratory Annual Report for 1977 to the DOE Assistant Secretary for the Environment,* DOE Report PNL-2500(Pt.3), pp. 2.8-2.9, Pacific Northwest Laboratory, NTIS.

——, and S. L. Sutter, 1974, Particle Deposition Rates on a Water Surface as a Function of Particle Diameter and Air Velocity, *J. Rech. Atmos.,* 8: 911-920.

——, S. L. Sutter, and M. T. Dana, 1973, Dry Deposition Processes, in *Pacific Northwest Laboratory Annual Report for 1972,* USAEC Report BNWL-1751(Pt.1), pp. 43-49, Battelle Pacific Northwest Laboratory, NTIS.

Servant, J., 1976, Deposition of Atmospheric Lead Particles to Natural Surfaces in Field Experiments, in *Atmosphere–Surface Exchange of Particulate and Gaseous Pollutants (1974),* ERDA Symposium Series, Richland, Wash., Sept. 4–6, 1974, pp. 87-95, CONF-740921, NTIS.

Shepherd, J. G., 1974, Measurements of the Direct Deposition of Sulphur Dioxide onto Grass and Water by the Profile Method, *Atmos. Environ.,* 8: 69-74.

Shinn, J. H., N. C. Kennedy, J. S. Koval, B. R. Clegg, and W. M. Porch, 1976, Observations of Dust Flux in the Surface Layer for Steady and Nonsteady Cases, in *Atmosphere–Surface Exchange of Particulate and Gaseous Pollutants (1974),* ERDA Symposium Series, Richland, Wash., Sept. 4–6, 1974, pp. 625-637, CONF-740921, NTIS.

Shreffler, J. H., 1975, Numerical Experimentation with Particles Having Non-Zero Terminal Velocity in the Atmospheric Surface Layer, *Boundary-Layer Meteorol.,* 9: 191-204.

Shreve, J. D., Jr., 1958, *Summary Report, Test Group 57,* USAEC Report ITR-1515, Energy Research and Development Administration, NTIS.

Silker, W. B., 1976, Air to Sea Transfer of Marine Aerosol, in *Atmosphere–Surface Exchange of Particulate and Gaseous Pollutants (1974),* ERDA Symposium Series, Richland, Wash., Sept. 4–6, 1974, pp. 391-398, CONF-740921, NTIS.

Simpson, C. L., 1961, *Estimates of Deposition of Matter from a Continuous Point Source in a Stable Atmosphere,* USAEC Report HW-69292, Hanford Atomic Products Operation, NTIS.

——, 1964, Deposition Measurements at Hanford, in *Conference on AEC Meteorological Activities,* I. A. Singer, M. E. Smith, and E. W. Bierly (Eds.), USAEC Report BNL-914, pp. 141-143, Brookhaven National Laboratory, NTIS.

Sinclair, P. C., 1976, Vertical Transport of Desert Particulates by Dust Devils and Clear Thermals, in *Atmosphere–Surface Exchange of Particulate and Gaseous Pollutants (1974),* ERDA Symposium Series, Richland, Wash., Sept. 4–6, 1974, pp. 497-527, CONF-740921, NTIS.

Skidmore, E. L., 1976, A Wind Erosion Equation: Development, Application, and Limitations, in *Atmosphere–Surface Exchange of Particulate and Gaseous Pollutants (1974),* ERDA Symposium Series, Richland, Wash., Sept. 4–6, 1974, pp. 452-465, CONF-740921, NTIS.

——, and N. P. Woodruff, 1968, *Wind Erosion Forces in the United States and Their Use in Predicting Soil Loss,* Agricultural Handbook No. 346, U. S. Department of Agriculture, GPO.

Slade, D. H. (Ed.), 1968, *Meteorology and Atomic Energy—1968,* TID-24190, U. S. Atomic Energy Commission, NTIS.

Slinn, W. G. N., 1976a, Formulation and a Solution of the Diffusion–Deposition–Resuspension Problem, *Atmos. Environ.,* 10: 763-768.

——, 1976b, Dry Deposition and Resuspension of Aerosol Particles—A New Look at Some Old Problems, in *Atmosphere–Surface Exchange of Particulate and Gaseous Pollutants (1974),* ERDA Symposium Series, Richland, Wash., Sept. 4–6, 1974, pp. 1-40, CONF-740921, NTIS.

——, 1977, Some Approximations for the Wet and Dry Removal of Particles and Gases from the Atmosphere, *Water, Air, Soil Pollut.,* 7: 513-543.

——, 1978, Parameterizations for Resuspension and for Wet and Dry Deposition of Particles and Gases for Use in Radiation Dose Calculations, *Nucl. Safety,* 19(2): 205-219 (March–April).

——, L. Hasse, B. B. Hicks, A. W. Hogan, D. Lal, P. S. Liss, K. O. Munnich, G. A. Sehmel, and O. Vittori, 1978, Some Aspects of the Transfer of Atmospheric Trace Constituents Past the Air–Sea Interface, *Atmos. Environ.,* 12: 2055-2087.

Small, S. H., 1960, Wet and Dry Deposition of Fallout Materials at Kjeller, *Tellus,* 12(3): 308-314.

Start, G. E., 1970a, Comparative Diffusion and Deposition of Uranine Dye, Molecular Iodine Gas, and Methyl Iodide Gas, in *Proceedings of the Fifth Annual Health Physics Society Midyear Topical Symposium,* Idaho Falls, Nov. 3–6, 1970, Vol. 2, pp. 463-473.

——, 1970b, *Diffusion and Deposition Studies,* Report ERLTM-ARL 20, Environmental Science Services Administration, Air Resources Laboratories, NTIS.

Stewart, K., 1963, The Particulate Material Formed by the Oxidation of Plutonium, C. M. Nichols (Ed.), in *Progress in*

Nuclear Energy, Technology and Safety, Series IV, Vol. 5, pp. 535-579, Pergamon Press, New York.

——, 1967, The Resuspension of Particulate Material from Surfaces, in *Surface Contamination,* B. R. Fish (Ed.), pp. 63-74, Pergamon Press, New York.

Stewart, N. G., and R. N. Crooks, 1958, Long-Range Travel of the Radioactive Cloud from the Accident at Windscale, *Nature,* 182: 627-628.

——, R. G. D. Osmond, R. N. Crooks, and E. M. Fisher, 1957, *The World-Wide Deposition of Long-Lived Fission Products from Nuclear Test Explosions,* Report AERE-HP/R-2354, United Kingdom Atomic Energy Authority, Atomic Energy Research Establishment, Harwell, Berks., England.

Styro, B. I., and S. S. Shalaveyus, 1970, Deposition Rate of RaD in the Atmosphere, in *Proceedings of the Atmospheric Scavenging of Radioisotopes,* Palanga, USSR, June 7–9, 1966, Report TT-69-55099 (CONF-66073), p. 79, NTIS.

Sutherland, A. J., 1967, Proposed Mechanism for Sediment Entrainment by Turbulent Flows, *J. Geophys. Res.,* 72: 6183-6194.

Tingey, D. T., 1968, *Foliar Absorption of Nitrogen Dioxide,* M. A. Thesis, University of Utah, Salt Lake City, Utah.

Travis, J. R., 1976, A Model for Predicting the Redistribution of Particulate Contaminants from Soil Surfaces, in *Atmosphere–Surface Exchange of Particulate and Gaseous Pollutants (1974),* ERDA Symposium Series, Richland, Wash., Sept. 4–6, 1974, pp. 906-943, CONF-740921, NTIS.

Trevino, G., 1972, *On the Natural Redistribution of Surface Contamination,* USAEC Report WASH-1225, U. S. Atomic Energy Commission, NTIS.

Unsworth, M. H., and D. Fowler, 1976, Field Measurements of Sulphur Dioxide Fluxes to Wheat, in *Atmosphere–Surface Exchange of Particulate and Gaseous Pollutants (1974),* ERDA Symposium Series, Richland, Wash., Sept. 4–6, 1974, pp. 342-353, CONF-740921, NTIS.

Van der Hoven, I., 1963, A Diffusion-Deposition Model for Particulate Effluents from Ground-Tested Nuclear Engines, *Int. J. Air Water Pollut.,* 7: 1023-1032.

——, (Ed.), 1970, *Atmospheric Transport and Diffusion in the Planetary Boundary Layer,* Report ERLTM-ARL-20, Environmental Science Services Administration, Air Resources Laboratories, NTIS.

Van Doren, C. E., 1944, The Effect of Cloddiness of Soils on Their Susceptibility to Wind Erosion, *Agron. J.,* 36: 859-864.

Vaughan, B. E., 1976, Suspended Particle Interactions and Uptake in Terrestrial Plants, in *Atmosphere–Surface Exchange of Particulate and Gaseous Pollutants (1974),* ERDA Symposium Series, Richland, Wash., Sept. 4–6, 1974, pp. 228-243, CONF-740921, NTIS.

Vögt, K. J., 1977, Empirical Investigations of the Diffusion of Waste Air Plumes in the Atmosphere, *Nucl. Technol.,* 34: 43-47.

——, K. Heinemann, W. Matthes, G. Polster, M. Stoeppler, and L. Angeletti, 1973, *Propagation of Pollutants in the Atmosphere and Stress on the Environment,* USAEC Report BNWL-TR-204, Battelle Pacific Northwest Laboratory, NTIS.

Voillequé, P. G., D. R. Adams, and J. B. Echo, 1970, Transfer of Krypton-95 from Air to Grass, *Health Phys.,* 19: 835.

Volchok, H. L., R. H. Knuth, and M. T. Klemman, 1972, The Respirable Fraction of Plutonium at Rocky Flats, *Health Phys.,* 23: 395-396.

Walker, R. L., and B. R. Fish, 1965, *Adhesion of Particles to Surfaces in Liquid and Gaseous Environments,* USAEC Report ORNL-TM-1228, Oak Ridge National Laboratory, NTIS.

Wesely, M. L., B. B. Hicks, W. P. Dannevik, S. Frisella, and R. B. Husar, 1977, An Eddy-Correlation Measurement of Particulate Deposition from the Atmosphere, *Atmos. Environ.,* 14: 561-563.

Whelpdale, D. M., 1976, Dry Deposition over the Great Lakes, in *Atmosphere–Surface Exchange of Particulate and Gaseous Pollutants (1974),* ERDA Symposium Series, Richland, Wash., Sept. 4–6, 1974, pp. 382-390, CONF-740921, NTIS.

——, and R. W. Shaw, 1974, Sulphur Dioxide Removal by Turbulent Transfer over Grass, Snow, and Water Surfaces, *Tellus,* 26(1-2): 196-204.

Whitman, W. G., 1923, The Two-Film Theory of Gas Absorption, *Chem. Metall. Eng.,* 29: 146-148.

Whittaker, R. H., and G. M. Woodwell, 1967, Surface Area Relations of Woody Plants and Forest Communities, *Am. J. Bot.,* 54: 931-939.

Wildung, R. E., and T. R. Garland, 1977, *The Relationship of Microbial Processes to the Fate and Behavior of Transuranic Elements in Soils, Plants, and Animals,* DOE Report PNL-2416, Pacific Northwest Laboratory, NTIS.

Wilson, R. H., R. G. Thomas, and J. N. Stannard, 1960, *Operation Plumbbob: Biomedical and Aerosol Studies Associated with a Field Release of Plutonium,* Nevada Test Site, May–October 1957, USAEC Report WT-1511, Sandia Corporation, NTIS.

Woodruff, N. P., and F. H. Siddoway, 1965, A Wind Erosion Equation, *Soil Sci. Soc. Am., Proc.,* 29: 602-608.

Young, J. A., 1978, The Rates of Change of Pollutant Concentrations Downwind of St. Louis, in *Pacific Northwest Laboratory Annual Report for 1977 to the DOE Assistant Secretary for Environment,* DOE Report PNL-2500(Pt.3), pp. 1.35-1.39, Pacific Northwest Laboratory, NTIS.

Zimbrick, J. D., and P. G. Voilleqúe, 1969, *Controlled Environmental Radioiodine Tests at the National Reactor Testing Station, Progress Report No. 4,* USAEC Report IDO-12065, Idaho Operations Office, NTIS.

Chapter 12 covers research prior to May 1979.

13

DIFFUSION MODELING: PRINCIPLES OF APPLICATION

Sumner Barr and William E. Clements, Los Alamos National Laboratory

13-1 INTRODUCTION

In preparing this chapter the authors had considerable concern over the prospect of writing another summary of conventional diffusion methodology in the wake of the large number of excellent reviews that have been made in the last decade. Instead, we attempted to call attention to the wide range of modeling concepts and techniques and to point out how these methods have been recently used in the variety of settings where practical assessments must be done. We chose not to list or discuss the well-known or most often used models by their "trade names" but rather to address classes of modeling approach by their mathematical characteristics and their physics.

The last few years have been marked by a growing realization that the traditional modeling assumptions have a limited applicability in the assessments of real sites with their complicating factors of topography, structures, forests, water bodies, and time-dependent boundary-layer structure. This chapter addresses these peculiar regimes that make up a majority of assessment cases and attempts to summarize the adaptations of traditional diffusion methods and the entirely new techniques that have evolved to address them. When possible, we try to offer a measure of the confidence one might have in the results of a given technique applied to a problem.

The most frequently used modeling concept for practical applications is the Gaussian plume, and it has served as a vehicle for many useful and innovative studies. The Gaussian model has been formulated in many ways, from methods simple enough to use arithmetic done in the implementer's head to complex computer algorithms. The non-Gaussian diffusion concepts to be mentioned in this chapter will include:

—Statistical theory of diffusion.
—Method of moments.
—Monte Carlo and other stochastic methods.
—Particle-in-Cell (PIC).
—K-Theory.
—Similarity theory.
—Higher order closure methods.

Barr and Clements: University of California, Los Alamos National Laboratory, Los Alamos, New Mexico.

13-1.1 Gaussian and Non-Gaussian Diffusion Models

Conceptually, a diffusion model is a statement of the budget of contaminant or tracer material released to the atmosphere. Given some particular source of material, the questions of primary interest are:

1. Where does it go?
2. How rapidly does it dilute in getting there?
3. How rapidly and by what mechanisms is it removed from the atmosphere?

The answers to these questions are extremely complex and depend on the interacting factors of source effects, atmospheric structure, and elements of the underlying terrain. Although a considerable amount of research is necessary to completely address the fundamental questions, enough is known about each to implement models for practical estimates. The adequacy of the existing models for certain problems depends largely on the application. This section will mainly address itself to the second question, the rate and nature of dilution.

To avoid confusing fundamental concepts with complicated mathematics, we shall try to state the diffusion problem as simply as possible. Material released from a source occupies a certain volume and when mixed with air has a concentration value. As the material resides in the air for more time, it mixes with more air, the volume grows, and by this process the concentration decreases. Because the mixing tends to take place near the edges of the volume, the concentration is not generally uniform over the volume but decreases toward the edges. In the following sections we shall explore a number of techniques aimed at describing the concentration profile across a cloud of contaminant. The form of the profile, however, can be divorced from the problem of growth or dilution rate with time or travel distance, and for many applications the dilution rate may well be the most sensitive consideration.

The most widely used class of diffusion models assumes that the profile of concentration across a puff or plume satisfies the equation of a Gaussian distribution. The choice is by no means arbitrary but has a basis in classical diffusion theory and statistics. In particular, the classical Fickian diffusion represents solutions to the parabolic diffusion equation with constant diffusivity. These solutions are characterized by the Gaussian curve. The number of limitations that this rationale places on practical wind and turbulence fields indicates the need for another fundamental basis for the Gaussian assumption. The central limit theorem of statistics provides this argument for time-averaged concentrations. Finally, experimental diffusion studies support the use of the Gaussian distribution to describe concentration profiles in the lower atmosphere under certain conditions.

Besides the Gaussian models, there is a wide variety of other techniques available to estimate dispersion. Many of these are complex enough to require computers. However, computers are now sufficiently available that, if a method can provide a better solution to a practical problem that needs the best possible answer, it should not be passed up for want of a computing machine. Some concepts are simpler than the Gaussian, and for certain problems it is difficult to justify more than the simplest calculations. For example, under conditions of limited vertical mixing and a complicated area source configuration, as might be found in an urban complex, lateral boundary effects may become small and a uniform concentration within a characteristic volume determined by the size of the city and the depth of the mixed layer may be the most appropriate choice. Another example of a computationally simple model with useful results is the "top-hat" buoyant plume model (Briggs, 1970). The fundamental concept of the uniform concentration profile is to estimate the dimensions of the domain and the flux of source material into the volume and we have the concentration field specified.

Another major approach is through the solution of differential equations which are taken to describe diffusion in a particular mean flow configuration. The classical diffusion theory has its roots in the parabolic diffusion equation

$$\frac{\partial C}{\partial t} + \mathbf{V} \cdot \nabla C = \frac{\partial}{\partial x} K_x \frac{\partial C}{\partial x} + \frac{\partial}{\partial y} K_y \frac{\partial C}{\partial y} + \frac{\partial}{\partial z} K_z \frac{\partial C}{\partial z} \qquad (13.1)$$

and consists of attempts to solve Eq. 13.1 under a variety of transport wind and eddy-diffusivity fields. The solutions may be approached analytically by various approximation techniques or numerically. A variety of elegant analytical solutions to Eq. 13.1 are summarized in Sutton (1953), Pasquill (1974), and Monin and Yaglom (1971). Approximate solutions have been obtained by expansion techniques. Saffman (1962), Gee and Davies (1963), and Tyldesley and Wallington (1965) have used a method of expansion of the concentration field in its statistical moments first introduced by Aris (1956). These results allowed the solution for features of the concentration field in shear flows.

The application of random walk techniques to atmospheric diffusion has been attempted a number of times with promising results, although these methods have not been extensively pursued in models for practical application. Thompson (1971) and Hall (1975) describe numerical experiments with random walk methods that yield good agreement with traditional methods in the cases where the traditional techniques are applicable. The value of the random walk procedure is that it is not restricted to a particular mean wind or temperature specification. The techniques apply with equal facility to arbitrarily specified time and space variations of transport wind and temperature.

Analytical calculations on selected Markov random walk sequences have yielded a number of useful and interesting results, including those of Monin (1955, 1959), Scheidegger (1958), and Smith (1973) who derived hyperbolic diffusion equations of the form

$$\frac{\partial^2 C}{\partial t^2} + \frac{1}{\tau} \frac{\partial C}{\partial t} = -w^{*2} \frac{\partial^2 C}{\partial z^2} \qquad (13.2)$$

where t is a time scale and w^* is the velocity of propagation of the plume boundary in the z-direction. This expression taken as the governing equation of dispersal offers several conceptual advantages over the parabolic Eq. 13.1, including primarily a finite propagation rate of airborne material.

A similarity method as applied by Monin (1959), Gifford (1962), Cermak (1963), and Pasquill (1966) has proven useful for modeling diffusion in parallel shear flow near the ground. Similarity hypotheses assert that flow properties such as mean wind, turbulence, and concentration distributions are universal functions of only the governing variables of the particular flow. The Lagrangian similarity hypothesis seeks to express the spreading of material from a given source in a manner that follows the material from source to receptor in such a way that the statistical properties of a fluid particle and its motion depend only on a velocity scale and a length (or time) scale.

Consider an ensemble of particles originating from a point at ground level. Each particle after some time will be located at a different point in space from its neighbors. However, according to the similarity ideas, the mean height and second moment can be described in terms of only a scale factor and the time of travel.

The authors mentioned above noted the success of the Eulerian similarity theories introduced by Monin and Obukhov (1954) for describing momentum flux in the surface layer under a variety of stabilities and adopted the friction velocity, u_*, and the Monin–Obukhov length $L = -u_*^3/[k(g/T)(H/c_p\rho)]$ as the fundamental velocity and length scales (see Chap. 5 for details). Then the mean vertical and longitudinal positions of an ensemble of "particles" released at ground level are described by

$$\frac{d\bar{z}}{dt} = k u_* \, \phi\left(\frac{\bar{z}}{L}\right) \qquad (13.3)$$

$$\frac{d\bar{x}}{dt} = \bar{u} = \frac{u_*}{k}\left[f\left(\frac{\bar{z}}{L}\right) - f\left(\frac{z_0}{L}\right)\right] \qquad (13.4)$$

where ϕ and f are universal functions. Kao (1959) and Panofsky and Prasad (1965) offer the interpretation of $(d\bar{z}/dt)$ as $(\overline{w'^2})^{1/2}$. With this interpretation, the function ϕ can be obtained from empirical or theoretical studies as reported by Lumley and Panofsky (1964) and Cramer (1967). The function f has been studied extensively over the full range of stabilities that occur in the surface layer with useful mathematical approximations for limited ranges of stability. The function f is pictured in Fig. 13.1, whereas Fig. 13.2 shows an example of the function $\phi(z/L)$. The

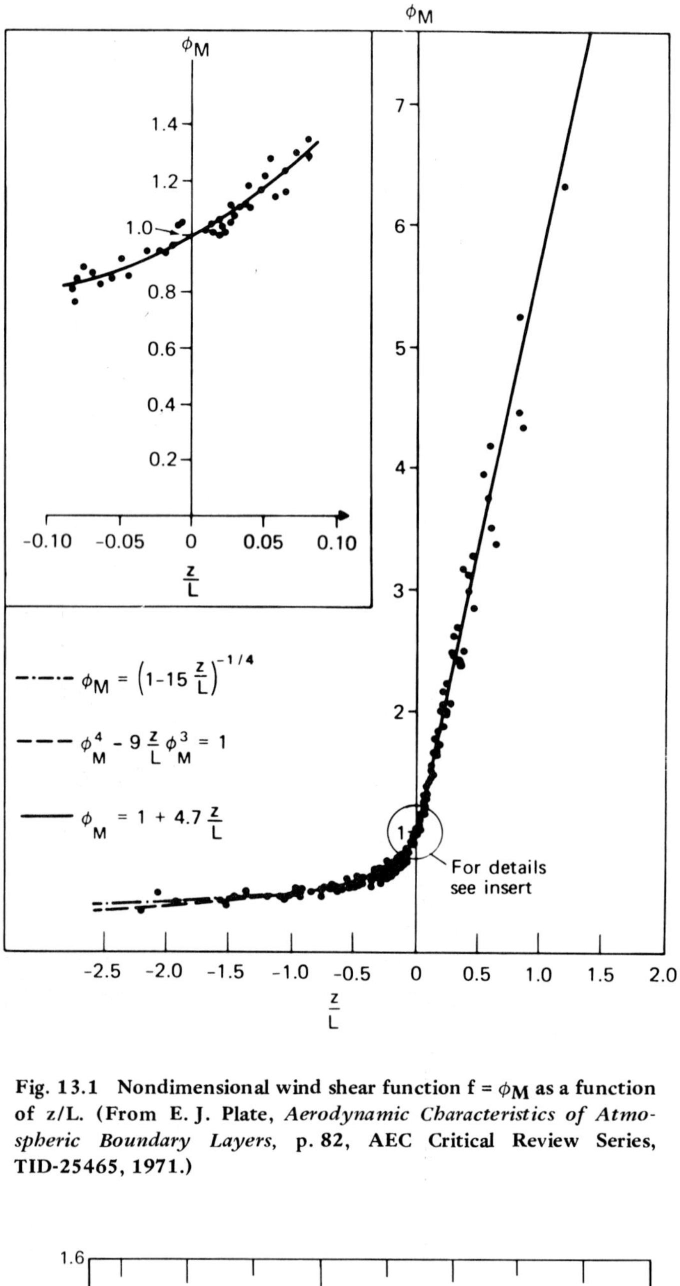

Fig. 13.1 Nondimensional wind shear function f = ϕ_M as a function of z/L. (From E. J. Plate, *Aerodynamic Characteristics of Atmospheric Boundary Layers*, p. 82, AEC Critical Review Series, TID-25465, 1971.)

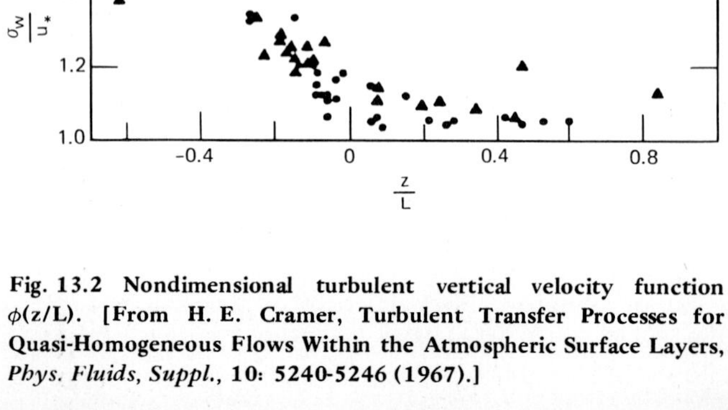

Fig. 13.2 Nondimensional turbulent vertical velocity function ϕ(z/L). [From H. E. Cramer, Turbulent Transfer Processes for Quasi-Homogeneous Flows Within the Atmospheric Surface Layers, *Phys. Fluids, Suppl.*, 10: 5240-5246 (1967).]

theory produces estimates of vertical diffusion rates for associated downstream travel distance. With the additional hypothesis that the probability of a particle's position relative to the ensemble mean position is also a universal function (e.g., Gaussian), we can obtain estimates for the concentration as a function of downwind distance.

Similarity theory for vertical diffusion has relied heavily on surface-layer formulations and therefore is valid only for relatively short travel distances where the diffusion of low-level sources does not exceed the surface-layer boundaries.

It is valuable to place the various diffusion calculation schemes in a context that can guide us in the selection of a model class for a given problem. We can work under the following rationale. Given the complete velocity field, including scale fluctuations, and the source structure, the diffusion becomes merely a computational tallying or bookkeeping problem. Seldom do we know enough about the velocity field or source function; then the computational concepts and tools get quite complicated.

The theoretical modeling problem is to choose those assumptions about the turbulent flow which most closely correspond to the particular atmospheric situation under investigation (e.g., homogeneous, axisymmetric, steady, linear shear, and isotropic). The practical modeling problem becomes one of attempting to simulate the most important aspects of velocity, turbulence, and source factors and of establishing tally methods.

The selection of a modeling concept depends in great part on the use to which the product will be put. There are two main classes of usage: direct evaluation of concentrations for design, regulation, safety, or economics and models used as components in systems studies. This includes multisource applications, source through effects studies, deposition, transformation, and urban planning. Among the strengths and weaknesses of each technique that should be considered are ease of computations (manual, computer, special computer needs, or time) and relevance to the actual situation being simulated (e.g., for shear flows, time-dependent flows, long-distance transport, and light wind wake flows).

In cases where pollutant emissions are regulated by state or federal agencies, the methods of estimating air-quality impact should be consistent among the responsible organizations. This consistency provides a basis of comparison among regulatory agencies, industry, and the public for assessing pollution concentrations and emission control needs. The Clean Air Act of 1967 and its amendments of 1970 and 1977 provide for application of models in order to achieve air-quality standards. To have a consistent basis of evaluation, the U. S. Environmental Protection Agency (1977) has established a set of guidelines on air-quality models. The guidelines focus on Gaussian plume algorithms and address source factors and data requirements. They include consideration of the complications added by long travel time and nonuniform surface conditions. Recommendations of specific codes and authors are available in the guideline document.

13-2 SELECTED RESULTS FROM STATISTICAL DIFFUSION THEORY

In the interest of consistency and in an effort to keep this chapter somewhat self-contained, we will summarize selected theoretical developments. We will cover mainly those analyses which form the basis of the modeling studies surveyed throughout this chapter. The reader who desires to study greater breadth and depth of turbulent diffusion theory is referred to Chap. 6 or any of several excellent texts including Pasquill (1974), Tennekes and Lumley (1972), Sutton (1953), Csanady (1973), and Monin and Yaglom (1971).

13-2.1 Single-Particle Analysis

The basis of the statistical theories of diffusion is the study of particles of fluid as they travel in the turbulent flow. This reference point that follows a fluid particle is referred to as the Lagrangian system, and most of the results derived from statistical theories, therefore, require information on the Lagrangian properties of the fluid turbulence.

The pioneering study in this area was done by Taylor (1921) in an analysis that provides the foundation for much of our current understanding of turbulent diffusion including theoretical development and guidance for practical models. Taylor restricted his analysis to homogeneous, stationary turbulence and analyzed the continuous motion of a particle introduced into this flow. In one dimension a very fundamental statement about the particle is that its distance from the origin is due to its accumulated motion since release. Without loss of generality this makes it possible to subtract the mean velocity and consider only the turbulent departures from the mean. This leads to the expression for the distance from the origin

$$y(t) = \int_0^t v'(t')\,dt' \qquad (13.5)$$

where v' is the fluctuating turbulent velocity. Using Eq. 13.5, we can derive useful information by repeating the release and tracking process a large number of times and by preserving the second moment of the particle distribution by squaring both sides of Eq. 13.5 then averaging over the ensemble of repeated experiments. This leads to

$$\overline{y^2}(t) = 2\overline{v'^2} \int_0^t \int_0^{t'} R_L(\xi)\,d\xi\,dt' \qquad (13.6)$$

where the overbar denotes the ensemble average and the function $R_L(\xi)$ is the Lagrangian autocorrelation

$$R_L(\xi) = \frac{\overline{v'(t)v'(t+\xi)}}{\overline{v'^2}} \qquad (13.7)$$

Theories and measurements of turbulence have provided useful information on the form of $R_L(\xi)$. In particular, it must be unity at $\xi = 0$ and must approach zero at very large ξ. These two properties yield useful information from Eq. 13.6 for the limits of the second moment for small and large travel times. At small times, $R_L(\xi)$ is approximately constant at unity, and Eq. 13.6 can be simply integrated to give

$$\overline{y^2}(t) \approx \overline{v'^2}\,t^2 \qquad (13.8)$$

At long times, since $R_L(\xi)$ converges to zero, the inner integral in Eq. 13.6 reaches a fixed value, K, and

$$\overline{y^2}(t) \approx 2Kt \qquad (13.9)$$

It can be shown that the constant, K, is the same as the eddy-diffusivity coefficient defined in Eq. 13.1. Adopting the notation $\sigma_y = (\overline{y^2})^{1/2}$ for the standard deviation for the particle distribution, we see in Eqs. 13.8 and 13.9 the often quoted limits of linear growth ($\sigma_y \propto t$) at early travel time and $\sigma_y \propto t^{1/2}$ at long travel times. Equation 13.6 may be converted to a form that yields insight into the role of eddy sizes by using the Fourier transform relation between $R_L(\xi)$ and the Lagrangian energy spectrum

$$\overline{y^2} = \overline{v'^2}\,t^2 \int_0^\infty F(n)\,\frac{\sin^2(\pi n t)}{(\pi n t)^2}\,dn \qquad (13.10)$$

Equation 13.10 expresses the particle position variance in terms of a filtered spectrum. All scales of motion contribute to the spread at all times; however, the second factor in the integral determines a weighting function that emphasizes certain ranges of eddy size. The reader should sketch the function $[\sin^2(\pi n t)]/(\pi n t)^2$. At small t, a wide range of frequencies, n, contribute to the integral, whereas at large t, only small n (large eddies) have significant weighting in the calculation. We can interpret the last result in the following way. Close to a continuous source, all scales of turbulence contribute to plume spread. Since the plume is narrow, small eddies serve to deform and tear it apart while large eddies account for a meander of the entire plume, thus producing an effectively wider time-averaged plume. At long downwind distance the small eddies merely move material around within a wide plume while the large eddies still provide the wandering plus significant local deformation. We shall see in the next section that these roles are different when diffusion is considered relative to a local centroid of a plume or puff.

13-2.2 Relative Diffusion

Although the single-particle diffusion theory proposed by Taylor is probably the single most influential theoretical development in turbulent diffusion, it was recognized quite early that many realistic problems require information on diffusion relative to the centroid of a wandering puff or plume. Brier (1950) and Batchelor (1952) investigated the relative diffusion problem in terms of the simplest ensemble—a pair of particles. Taylor's method applied to the velocity difference between the two particles leads to the expression

$$\overline{Y_i(t)Y_j(t)} = \overline{Y_i(t_0)Y_j(t_0)} + \int_{t_0}^t \int_{t_0}^t \overline{w_i(t')w_j(t'')}\,dt'\,dt'' \qquad (13.11)$$

where now the capital letter notation refers to particle separations in the i and j coordinate directions and the w terms are velocity differences between the two particles $[w_i = (v_{1i} - v_{2i})]$. Notice that there are actually more terms required to express the covariance terms in Eq. 13.11, although mathematical symmetry reduces the terms to six. For analysis purposes, we can select only one component (e.g., $i = j = 1$) and expand the velocity difference terms in the integral and Eq. 13.11 becomes

$$\overline{Y^2(t)} = \overline{Y^2(t_0)} + 2\overline{v'^2} \int_{t_0}^{t} \int_{t_0}^{t} R_L(t'' - t')\, dt'\, dt'' - 2 \int_{t_0}^{t} \int_{t_0}^{t} \overline{v_1'(t')v_2'(t'')}\, dt'\, dt'' \qquad (13.12)$$

The subscripts on the velocities here refer to particles 1 and 2. The major differences between Eqs. 13.12 and 13.6 are the dependence on the initial particle separation and the final integral term that depends on a correlation between one particle at one time and the other at another time. The latter term has been a major block to the progress of the Taylor approach to relative diffusion because it is difficult to evaluate for several reasons. It is nonstationary since, in general, the relative velocities change as the separation changes. Hence the relative diffusion rate depends on an additional time scale.

We can draw some conclusions regarding the effective scale of eddies for growth of an ensemble of particle pairs relative to a local centroid. Corssin (1962) introduces the idea of efficiency or diffusivity per unit kinetic energy to account for the fact that eddies much larger than $\overline{Y^2}$ will displace pairs of fluid points together whereas eddies smaller than $\overline{Y^2}$ will serve to separate particles. Hence the small eddies have a greater efficiency. On the other hand, the role of the small eddies is suppressed in a manner similar to fixed point diffusion, and we are left with the somewhat intuitive result that eddies on the size scale of the particle ensemble (or puff) are the most effective at spreading the puff. In view of the difficulties associated with solving Eq. 13.11 or 13.12 directly, several authors have turned to Kolmogorov's similarity theory for quantitative insight. Batchelor (1950) began by postulating the relative rate of particle separation by eddies in the locally isotropic range of turbulence that is known to exist at high Reynolds numbers. In such a flow it is possible to express the growth rate as an exclusive function of travel time, initial separation, viscosity, and turbulent energy dissipation. Dimensional analysis can then be used to determine basic power law dependencies on the relevant variables. By varying his restrictions on Y_0 and t, Batchelor derived a number of useful predictions including the early time linear growth $(\overline{y^2})^{1/2} \propto t$ and a late time result $(\overline{y^2})^{1/2} \propto {}^1\!/_2$. Most significantly, for an intermediate scale where growth is independent of initial size, this analysis produces an accelerated diffusivity where $(\overline{y^2})^{1/2} \propto t^{3/2}$, a result initially deduced on empirical grounds by Richardson (1926). The conditions for the prediction are that travel time is great enough that the initial particle separation (cluster size) is negligible but not so large that turbulent eddies outside the inertial subrange would have a significant role in cluster growth. The broad extent of the inertial subrange in the atmosphere suggests that the $t^{3/2}$ regime may be a viable one in practical diffusion cases. Gifford (1977) has recently collected observational evidence of this diffusion regime in the atmosphere for travel times in the range of 300 to 1000 s.

Smith and Hay (1961) address the problem of relative diffusion of a three-dimensional Gaussian cluster of particles. They relate the relative velocities of particle pairs to the velocities of particles relative to the cluster mean velocity. An extension of the hypothesis of a simple scale relationship between Lagrangian and Eulerian covariance functions to refer to the particle cluster leads to a growth function in terms of a filtered Eulerian spectrum. This amounts to a neglect of the nonstationarity of the relative velocity autocovariance on the basis that changes in the Lagrangian time scale due to cluster growth are small compared with the original scale governed by the cluster. A functional form of the spectrum is introduced (i.e., the transform of an exponential autocovariance) and integrated. The entire resulting function has a relatively complicated mathematical form but Smith and Hay note that

$$\left(\frac{d\sigma}{dx}\right)_{max} = \frac{2}{3}\beta i^2 \qquad (13.13)$$

where β is the ratio of Lagrangian to Eulerian time scales and i is the intensity of turbulence, $(\overline{u'^2})^{1/2}/\bar{u}$. The maximum growth rate is an excellent representation of the more detailed growth function to travel distances up to 100 times the initial cluster size.

13-2.3 Lagrangian–Eulerian Relationships

The statistical theories of diffusion, discussed in Sec. 13-2.2, are based on the evolving time history of marker particles in turbulent flows and therefore reflect a fundamentally Lagrangian variable. Most of the observational work in atmospheric boundary-layer turbulence has been based on time series data from a tower or other fixed sensor. Following Gifford (1967), we shall refer to the latter system of measurement as Eulerian fixed point. In order to make use of the readily available turbulence data in conjunction with the theoretical guidance on statistical diffusion, some relationship must be established between the turbulence structure as determined by the two systems. Gifford (1955) performed an experiment in which vertical turbulent energy spectra were simultaneously calculated from the fluctuations of a tower-mounted bivane at 300 ft and neutral balloons floating at the same level. He observed that the spectra were similar in shape but that the tower's (Eulerian fixed point) spectral peaks were displaced toward higher frequency. Hay and Pasquill (1959) made use of this scale difference in a practical method for predicting cloud growth from tower-observed turbulence. The scale relationship was expressed in terms of the autocorrelation functions as

$$R_L(\beta t) = R_E(t) \qquad (13.14)$$

where R_L is the Lagrangian and R_E is the Eulerian autocorrelation function.

Equation 13.14 is consistent with the shifted peak of the turbulence spectra since the Fourier transform relationships between autocorrelation and spectrum yields

$$nS_L(n) = \beta nS_E(\beta n) \qquad (13.15)$$

where S is the turbulent spectral density function and n is frequency. The scale ratio, β, may be directly observed as done by Gifford (1955), Angell (1964), and Angell et al. (1971), where balloon tracks represented the Lagrangian velocity fluctuations and fixed sensors were used for the Eulerian measurements. Also, several authors have used Taylor's (1921) diffusion theory to infer β values from tracer cloud growth and tower wind fluctuations. Using Taylor's theorem, in spectral form, Eq. 13.10, Hay and Pasquill (1959) combine Eqs. 13.15 and 13.10 to obtain

$$\overline{y^2} = \overline{v'^2}t^2 \int_0^\infty S_E(n) \left\{\frac{\sin [\pi n(t/\beta)]}{(\pi nt/\beta)}\right\}^2 dn \qquad (13.16)$$

and make use of the fact that the integral in Eq. 13.16 is identical in form to the spectrum of a fixed-point wind record that has been averaged over the time t/β. Hence

$$\overline{y^2} = \langle\overline{v'^2}\rangle_{t/\beta}\, t^2 \qquad (13.17)$$

represents a shorthand expression of Eq. 13.16 that shows more clearly the basic variables in estimating β from a fixed-point wind record and a diffusing plume. The parameter t/β is determined as the required averaging interval on the Eulerian velocity variance to achieve a balance of Eq. 13.17. Values of β obtained by this method and the direct comparison method are in the range of 2 to 12 and depend on several elements of atmospheric structure including turbulence intensity and stability. The dependence is consistent with the theoretical estimates discussed in the next few paragraphs.

13-2.4 Theoretical Approaches to the Lagrangian–Eulerian Relationship

Under the hypothesis that the Lagrangian and Eulerian turbulence spectra are similar in structure and differ only by a scale factor β, the theoretical approaches to defining β fall into three general classes as outlined by Pasquill (1974). Two of the concepts are discussed here.

The concept introduced by Ogura (1953) and used by Gifford (1955, 1967) suggests that the Eulerian frequency of velocity fluctuations at a fixed point is due to two linearly additive components. One is simply the advection of the assumed invariant eddy structure past the observation point, and the other component reflects the production and dissipation of eddies. By hypothesis, the latter term is identified with the Lagrangian frequency, and this allows the calculation of β by use of known turbulence functions. Following this approach, Gifford (1967) derived an inverse relation between β and the intensity of turbulence, i. Additional regression relations for β showed a very weak height dependence and a significant stability dependence where β ranged from 13 for very stable to near unity for unstable conditions.

The other concept summarized by Pasquill is ascribed to Corssin (1963) and involves the representation of the Lagrangian and Eulerian spectral forms, either by their respective inertial subrange forms or some other satisfactory relation (see Chap. 6). The results of this approach lead to the relation βi = constant with values in the range of 0.35 to 0.8. Figure 13.3 from Gifford (1967) shows the behavior of β with temperature gradient.

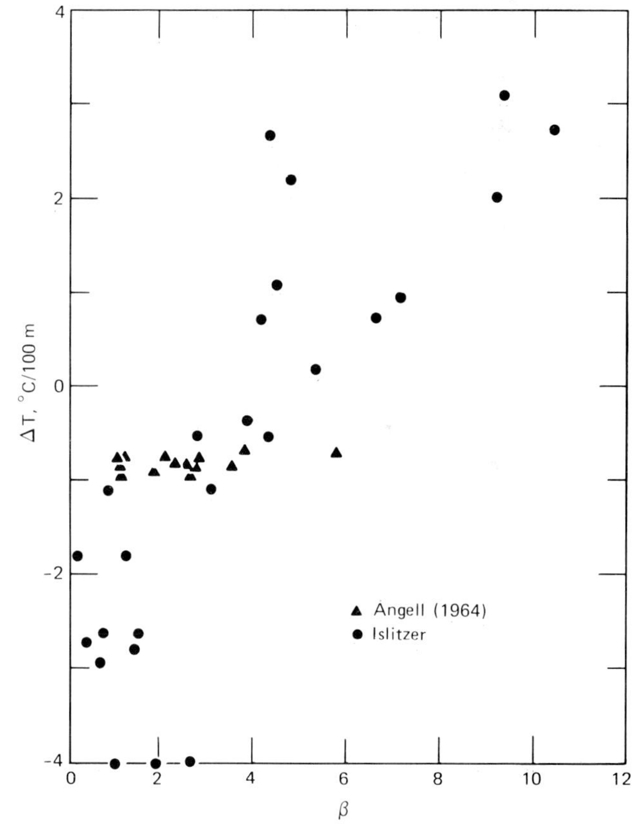

Fig. 13.3 **Observations of the Lagrangian/Eulerian time scale, β, plotted as a function of vertical temperature gradient. (From F. A. Gifford, Variation of the Lagrangian–Eulerian Time Number Scale Relationship with Stability, in *Proceedings of Joint U. S. Canadian Atomic Energy Commission's Conference on Micrometeorology*, pp. 485-499, Chalk River, Canada, 1967.**

13-3 GAUSSIAN PLUME MODELS

As described in Sec. 13-1, the basis of the Gaussian plume model is the hypothesis of the form of distribution of material within a plume. The two major areas of consideration are the derivation of formulas for different source configurations from the fundamental hypotheses and the determination of the appropriate input parameters. The results of model calculations can be no better than the basic hypotheses; hence a systematic development of practical formulas with special attention paid to the assumptions will provide us a basis for estimating the validity and accuracy of an application.

13-3.1 Derivation of Diffusion Formulas

We shall adopt as a starting point the Gaussian plume expression for a continuous point source of strength Q [g s^{-1}] located at $(0,y_0,h)$ in a uniform transport wind, u, with no material boundaries:

$$C(x,y,z) = \frac{Q}{2\pi\sigma_y\sigma_z\bar{u}} \exp\left[-\frac{(y-y_0)^2}{2\sigma_y^2} - \frac{(z-h)^2}{2\sigma_z^2}\right] \tag{13.18}$$

The diffusion parameters σ_y and σ_z are the lateral and vertical standard deviations of the assumed Gaussian distribution of concentration. They are functions of travel distance. Equation 13.18 is very familiar to readers who have had any acquaintance with atmospheric diffusion. The equation follows quite directly by a time integration of an instantaneous point source, but we have avoided that expression since the instantaneous mathematical "puff" has virtually no real physical counterpart. Although puffs exist in nature, individually they defy mathematical expression and should be modeled only in an ensemble mean sense. Equation 13.18 represents a mean picture of a plume for which instantaneous realizations may be patchy, meandering, and otherwise decidedly non-Gaussian.

As a fundamental working hypothesis, Eq. 13.18 contains certain implicit assumptions and limitations. First, it conserves mass. The reader may verify this fact by integrating the equation over all y and z using the standard integral

$$\int_0^\infty e^{-r^2x^2}\,dx = \frac{(\pi)^{1/2}}{2r} \tag{13.19}$$

(To achieve a finite mass effluent, assume a pulse of duration t = 0 to t = n and balance that known release amount against the integral of Eq. 13.18 over the corresponding values of x.)

The diffusion in the orthogonal directions is considered independent, but the distribution is jointly as well as separately Gaussian. This implies that these formulas may not characterize diffusion well when the off-diagonal terms of the diffusivity tensor (see Sec. 13-4) are important, such as in shear flow. In practice, however, if shear in the x-z plane only is involved, the neglect of the off-diagonal terms has little effect over moderate to large travel distances because x-directed diffusion is negligible.

The formulas for other source configurations can be produced by integrating Eq. 13.18 in space as indicated in Table 13.1, using for convenience the standard integral, Eq. 13.19, when appropriate. Reflecting boundary effects are handled by means of the method of images described by Sutton (1953) and others. Mathematical derivations and discussions of diffusion formulas, including the forms summarized in Table 13.1, have recently been reviewed by Drake (1979), and the reader interested in details is referred to that work. The reader may readily derive simplified formulas for ground-level (z = 0) concentrations from elevated sources by direct substitution into the appropriate Eq. 13.18 and the equations in Table 13.1. Detailed discussion of Gaussian plume formulas for a variety of source and receptor geometries are available in several references including Turner (1969), Pasquill (1974), and Sutton (1953). Although the processes of deposition at the

TABLE 13.1 Formulas for Selected Source Geometrics

Source configuration	Variable of integration	Limits	Formula
Continuous point			Eq. 13.21
Continuous point with reflecting surface			$C = \frac{Q}{2\pi\sigma_y\sigma_z\bar{u}} \exp\left[-\frac{(y-y_0)^2}{2\sigma_y^2} - \frac{(z-h)^2}{2\sigma_z^2} - \frac{(z+h)^2}{2\sigma_z^2}\right]$
Continuous line	y	$-\infty$ to $+\infty$	$C = \frac{Q}{(2\pi)^{1/2}\sigma_z\bar{u}} \exp\left[-\frac{(z-h)^2}{2\sigma_z^2} - \frac{(z+h)^2}{2\sigma_z^2}\right]$
Finite line	y	$-a$ to $+a$	$C = \frac{Q}{(2\pi)^{1/2}\bar{u}\sigma_z} \exp\left[-\frac{(h-z)^2}{2\sigma_z^2} - \frac{(h+z)^2}{2\sigma_z^2}\right]\left\{\text{erf}\left[\frac{a+y}{(2)^{1/2}\sigma_y}\right] + \text{erf}\left[\frac{a-y}{(2)^{1/2}\sigma_y}\right]\right\}$
Continuous area	x,y	$0 \leqslant x < 2a,\ -a \leqslant y \leqslant a$ evaluated at $x \geqslant 2a$	$C = \int_0^{2a} \frac{Q\,dx_0}{2(2\pi)^{1/2}\bar{u}\sigma_z(x_0)} \exp\left\{-\frac{1}{2}\left[\frac{z^2}{\sigma_z^2(x_0)}\right]\right\} \left\{\text{erf}\left[\frac{y+a}{(2)^{1/2}\sigma_y(x_0)}\right] + \text{erf}\left[\frac{a-y}{2\sigma_y(x_0)}\right]\right\} dx_0$

ground are very complex (see Chap. 12), it is often expedient to assume simple, limiting cases such as either total deposition of any airborne material that reaches the ground or, conversely, zero deposition, the so-called "reflection" boundary condition. The method of images is a useful and convenient method of accounting for the boundary conditions. An imaginary source located symmetrically with respect to the real source and bounding plane is assumed. The concentration is then taken to be the sum of the contributions from the real and the imaginary sources. The presence of a finite mixing depth, H, can be accounted for in the manner introduced by Holland (1953) and further described by Turner (1969). The Gaussian distribution in the vertical dimension is assumed until the distance x_L, where σ_z grows to 0.47 H. Beyond $2x_L$, a uniform vertical concentration profile between the ground and H is adopted, whereas between x_L and $2x_L$, concentrations are determined by interpolation.

To use the Gaussian plume model to make practical estimates, we must select the appropriate diffusion parameters σ_y and σ_z for travel distance, surface conditions, and meteorological conditions under consideration. This has typically involved the determination of a functional form for $\sigma_y(x)$ and $\sigma_z(x)$, where x is selected as the travel distance or travel time. The functions clearly depend on atmospheric stability; so a second major parameterization effort has been to relate the cloud growth $[\sigma_y(x), \sigma_z(x)]$ curves to various available measures of stability. An excellent survey of the development of cloud growth parameters is given by Gifford (1976). He summarizes several of the more popular turbulence typing schemes including those introduced by the Brookhaven National Laboratory (BNL), Pasquill, Tennessee Valley Authority (TVA), and Cramer. Most of these methods use meteorological indicators to establish one of five to seven different classes of atmospheric turbulence and relate it to the cloud growth by an empirical function for each category. The BNL method has five categories based on the total range of wind fluctuations in a 1-h period, and the related cloud growth refers to equivalent duration concentration samples from a nonbuoyant plume released at 108 m above flat rural fields and forests. The method suggested by Pasquill, and later modified slightly by Gifford, is the most frequently cited technique and uses six categories that may be discriminated by wind speed, cloud cover, and solar radiation as shown in Table 13.2. Table 13.3 relates the cloud growth categories for the Pasquill–Gifford (PG) scheme and the standard deviation of wind direction, σ_θ. The plume growth functions shown in Figs. 13.4 and 13.5 were empirically derived from 3-min mean concentration patterns within 800 m of a ground-level point source of nonbuoyant tracer. Turner (1964) amended the meteorological selection scheme to include sun angle, cloud cover, and cloud height. This has been shown by Golder (1972) to depart somewhat from the PG categories. The TVA performed many observations on buoyant tall stack emissions from power plants and derived a stage of growth curves for six categories selected on the basis of temperature lapse rate at plume height. The TVA data extended to 10 km from the source but represent quite a different measurement base from the other typing schemes. The observation times are shorter (2 to 5 min), the source is elevated and buoyant, and the meteorology is at a different altitude (lapse rates in the TVA method are all neutral to stable, although they are undoubtedly related to "unstable boundary layers" through superadiabatic lapse rates below and ground-initiated convention at plume height). Briggs (1974) has developed a series of curves that interpolate between the PG and the TVA methods, thereby providing a consistent estimation scheme for σ_y and σ_z over flat terrain from distances <1 km to 10 km. Briggs' curves are shown in Figs. 13.6 and 13.7, and the equivalent formulas are listed in Table 13.4. Cramer (1957) has favored azimuth wind direction fluctuations as a more direct measure of the turbulence that accounts for diffusion. The equivalent growth curves, derived for 30- to 60-min concentrations for short distances (<1 km), make use of power laws, $\sigma_y = ax^p$, $\sigma_z = bx^q$, where parameters a, b, p, and q depend on stability.

TABLE 13.2 Relation of Turbulence Types to Weather Conditions

A: Extremely unstable conditions — D: Neutral conditions†
B: Moderately unstable conditions — E: Slightly stable conditions
C: Slightly unstable conditions — F: Moderately stable conditions

Surface wind speed, m/sec	Daytime insolation: Strong	Daytime insolation: Moderate	Daytime insolation: Slight	Nighttime conditions: Thin overcast or ⩾ 1/8 cloudiness‡	Nighttime conditions: ⩽ 3/8 cloudiness
<2	A	A	B		
2	A–B	B	C	E	F
4	B	B–C	C	D	E
6	C	C–D	D	D	D
>6	C	D	D	D	D

†Applicable to heavy overcast, day or night.

‡The degree of cloudiness is defined as that fraction of the sky above the local apparent horizon which is covered by clouds.

TABLE 13.3 Relationships Between Pasquill–Gifford Stability Category and the Standard Deviation of the Horizontal Wind Direction σ_θ

Pasquill stability categories	σ_θ
A, extremely unstable	25.0°
B, moderately unstable	20.0°
C, slightly unstable	15.0°
D, neutral	10.0°
E, slightly stable	5.0°
F, moderately stable	2.5°

Inconsistencies among the various stability classification schemes for estimating plume growth have prompted efforts to reanalyze the classification methods. A workshop organized by the American Meteorological Society (Hanna et al., 1977) offered a consistent set of recommendations for the derivation and application of formulas for lateral

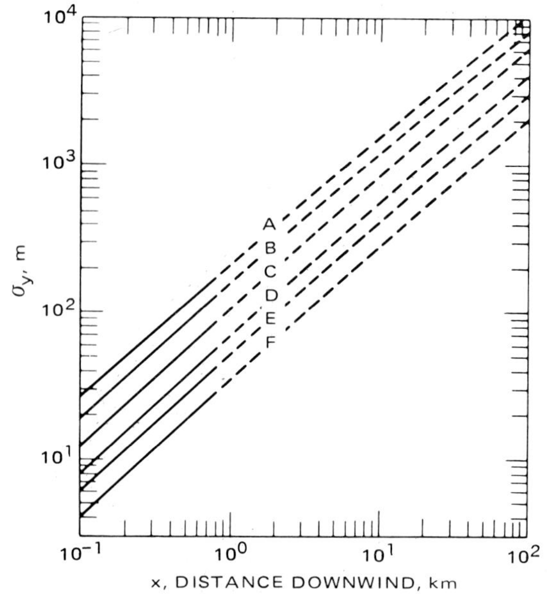

Fig. 13.4 Lateral diffusion coefficient, σ_y, vs. downwind distance for Pasquill–Gifford turbulence types. [From F. A. Gifford, Turbulent Diffusion-Typing Schemes: A Review, *Nucl. Saf.*, 17(1): 71 (January–February 1976).]

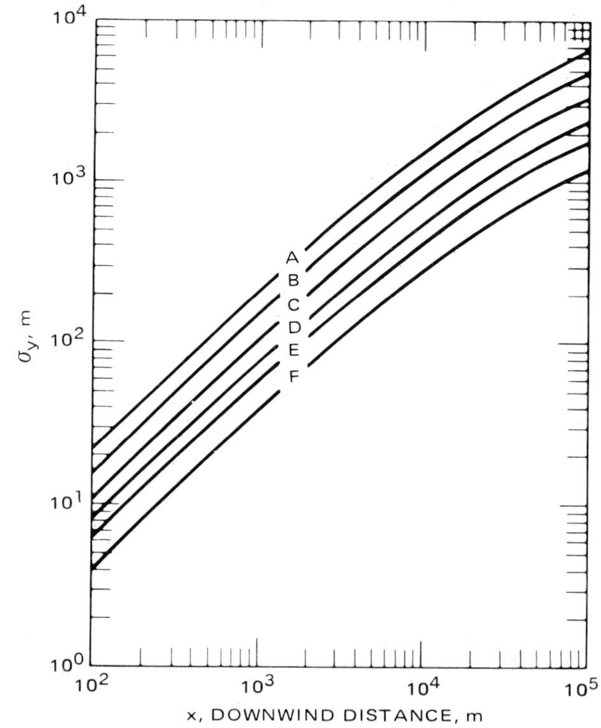

Fig. 13.6 Curves of σ_y based on recommendations by Briggs for flow over open country. [From F. A. Gifford, Turbulent Diffusion-Typing Schemes: A Review, *Nucl. Saf.*, 17(1): 74 (January–February 1976).]

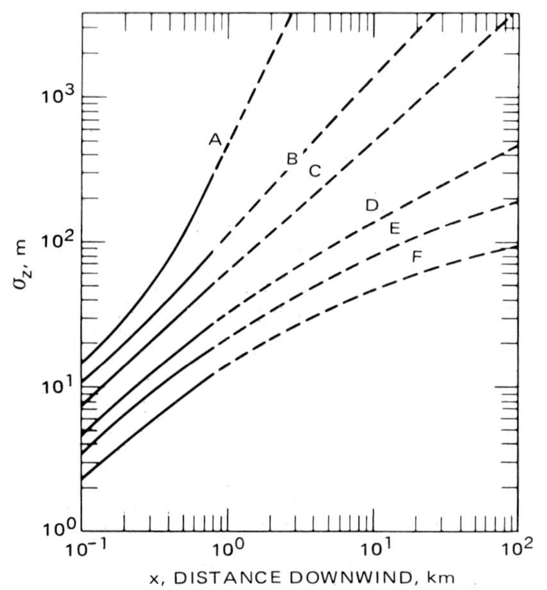

Fig. 13.5 Vertical diffusion coefficient, σ_z, vs. downwind distance for Pasquill–Gifford turbulence types. [From F. A. Gifford, Turbulent Diffusion-Typing Schemes: A Review, *Nucl. Saf.*, 17(1): 71 (January–February 1976).]

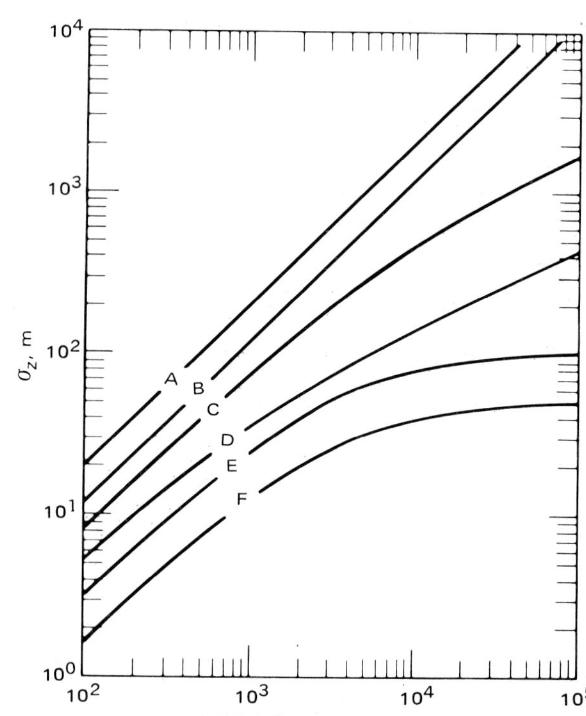

Fig. 13.7 Curves of σ_z based on recommendations by Briggs for flow over open country. [From F. A. Gifford, Turbulent Diffusion-Typing Schemes: A Review, *Nucl. Saf.*, 17(1): 74 (January–February 1976).]

and vertical plume growth (σ_y and σ_z). An appeal to fundamental principles and analysis of a broad data base guided the workshop participants to the recommendations that cover σ_y and σ_z formulas for different meteorological and roughness conditions and measurement requirements. The recommended formulations rely on turbulence statistics as in the practical expression

$$\frac{\sigma_y}{\sigma_\theta x} = f(x) \tag{13.20}$$

TABLE 13.4 Formulas Recommended by Briggs for $\sigma_y(x)$ and $\sigma_z(x)$ ($10^2 < x < 10^4$ m; Open-Country Conditions)

Pasquill type	σ_y, m	σ_z, m
A	$0.22x(1 + 0.0001x)^{-1/2}$	$0.20x$
B	$0.16x(1 + 0.0001x)^{-1/2}$	$0.12x$
C	$0.11x(1 + 0.0001x)^{-1/2}$	$0.08x(1 + 0.0002x)^{-1/2}$
D	$0.08x(1 + 0.0001x)^{-1/2}$	$0.06x(1 + 0.0015x)^{-1/2}$
E	$0.06x(1 + 0.0001x)^{-1/2}$	$0.03x(1 + 0.0003x)^{-1}$
F	$0.04x(1 + 0.0001x)^{-1/2}$	$0.016x(1 + 0.0003x)^{-1}$

or on basic boundary-layer scale parameters for daytime unstable conditions

$$\frac{\sigma_y}{h} = F_m\left(\frac{w^*}{V}\frac{x}{h}\right) \tag{13.21}$$

where σ_θ = standard deviation of horizontal wind angle
w^* = convective velocity scale
V = mean velocity for the mixed layer of height h
x = downwind travel distance

Suggestions are made for adjusting the conventional PG σ_y curves for sampling time, surface roughness, source height, and terrain effects. The application of σ_z curves requires careful consideration of the source height and buoyancy. Available methods for σ_z parameterization are less broadly applicable than σ_y. Techniques have a limited range of stability, release height, and travel distance for which they are valid. The workshop recommends more rigorous data inputs than are commonly used, including wind and gross turbulence statistics at source height elements of surface heat and momentum flux and surface roughness. Additional research is needed to derive general methods for diffusion parameterization and climatologies of the important parameters. Recent work in these directions is discussed by Draxler (1976; also Chap. 9), Pasquill (1976), Weber (1976), and Randerson (1979).

There is, of course, theoretical guidance for pollutant cloud growth rates for idealized turbulence field (i.e., homogeneous) in the work of Taylor (1921), Batchelor (1950), and others as summarized in Sec. 13-2. The accuracy of operational diffusion models is discussed in an American Meteorological Society (1978) committee position paper.

13-3.2 Quasi-Instantaneous Sources

It is mathematically straightforward to write a Gaussian description of the concentration field in puff emanating from an instantaneous source. Although useful in a computational sense, instantaneous point source calculations are not often used in practice owing to the difficulty in obtaining input parameters and properly interpreting the results. Most measurements of tracer from quasi-instantaneous sources are integrated over the entire cloud passage to yield a quantity with units of concentration × time. Formally, the integration produces an equation identical to that for the average concentration from a continuous point source, although the interpretation of the terms is different. The continuous source is expressed as a rate $[MT^{-1}]$, whereas the short-term release is a total amount $[M]$. Two other major points must be considered. First, the diffusion coefficients $\sigma_y(x)$ and $\sigma_z(x)$ are different for the two sources. In the case of a puff, the growth is due to the relative diffusion of the puff material about its centroid. The large turbulent eddies will transport the puff intact and contribute to a meandering path.

Furthermore, as the puff grows, it will incorporate more of the larger eddies into its growth mode rather than the meander. The mean concentration pattern from the continuous source represents the contributions of all eddies at all stages of its growth. These characteristics are evident from the theoretical results described in Sec. 13-2. In particular, the results of analysis of the separation rate of particle pairs carried out by Batchelor (1950) suggest an accelerated growth rate ($\sigma \propto t^{3/2}$) for intermediate travel times.

The second major point follows from the meandering behavior of quasi-instantaneous puffs. The trajectories of individual puffs will not generally follow the mean wind. Hence, although model puff concentrations may describe a statistical mean puff, it is really an infrequent occurrence. The meander of puffs about the mean position can be estimated from the lateral wind fluctuations. Cramer et al. (1964) suggest that the envelope of puff trajectories can be estimated by the lateral boundaries of a continuous source plume under the same meteorological conditions.

The growth of puffs is most frequently analyzed with respect to their own center of gravity and therefore is appropriately considered from the point of view of relative diffusion discussed in Sec. 13-2.2. Cramer et al. (1964) have developed Eq. 13.13 into a practical model for quasi-instantaneous sources of arbitrary source emission time, τ (provided $\tau \ll$ travel time to the receptor). Cramer noticed that for travel distances of about 100 m vertical growth for instantaneous and continuous sources is effectively the same. Hence the limited source duration effects are to be found mainly in the lateral growth. Assuming, as Smith and Hay do, that the initial source duration dominates over the growth effects, Cramer suggests a dependence,

$$\sigma_y(\tau) \propto \tau^{1/5} \tag{13.22}$$

This is based on estimates of variations of σ_y with sampling time and is consistent with spectral characteristics of the lateral turbulence component. The lateral diffusion coefficients in Cramer's model are merely normalized by a factor $\tau^{1/5}$ and used in the Gaussian plume formulation.

Högström (1964) performed an extensive series of experiments on puffs and separately determined the spread relative to the puff centers (σ_r) and the meandering of the puffs (σ_c). These were combined into a total spread through

$$\sigma^2 = \sigma_c^2 + \sigma_r^2 \tag{13.23}$$

applied to lateral and vertical dimensions. Högström presents data which indicate that beyond about 100 m travel distance the puff vertical growth rate is approximately the same as the total "quasicontinuous" growth rate. Near the source the fluctuation of puffs dominates the total vertical diffusion, whereas downstream the relative diffusion plays an increasingly greater role. The horizontal spreading is somewhat different. Within 0.5 km the puff growth rate exceeds the total rate of spread; beyond that distance, σ_{yr} grows much more slowly with distance. This is consistent with observations by Högström of a gap in lateral spectral energy at periods of about 1 h.

These results support the suggestion of Pasquill (1971) that power law expressions such as Eq. 13.22 cannot fully describe the temporal behavior of puff growth since they do not account for the spectrum of turbulence. That is, there are several independent time scales in the quasi-instantaneous source problem including source emission time, travel time, and characteristic spectral frequencies. In spite of the various theoretical complications, the formulation suggested by Cramer has a reasonably solid empirical basis and offers a good way to account for finite and variable source emission times.

13-4 NUMERICAL DIFFUSION MODELS

Numerical techniques have recently become very popular in formulating transport and diffusion problems, partly due to their flexibility in applications where the mean wind field or diffusivity function vary spatially. It is an oversimplification to refer to numerical models as one class of diffusion modeling since there are a large number of numerical methods applied in a variety of ways to diffusion problems (for more details see Chap. 14). However, here we shall refer to the use of finite differences or finite element methods addressed toward solving the fundamental equation for conservation of mass of a particular pollutant, shown in one dimension

$$\frac{\partial \bar{C}}{\partial t} + \frac{\partial}{\partial x}\left(\bar{u}\bar{C}\right) = \frac{\partial}{\partial x}(\overline{u'C'}) + S \tag{13.24}$$

where $\bar{u}$ and $\bar{C}$ indicate a local time average (e.g., 1 h) of the velocity and concentration, u' and C' are turbulent fluctuations from the local average, and S represents a source/sink function that embodies a great deal of physics, chemistry, and computational complexity in models where it is applied. Discussions and development of Eq. 13.24 are available, for example, in Johnson, Sklarew, and Turner (1976). In practice, the velocity field is supplied independently, either through specification, observation, or by a numerical module that calculates the velocity field from Navier–Stokes or mass conservancy formulations.

13-4.1 Formulation of Important Terms

The formulation of the turbulent flux term is of central interest to the diffusion estimation community. Traditionally, numerical models have adopted an eddy-diffusivity formulation of the form

$$\overline{u'C'} = K_x \frac{\partial \bar{C}}{\partial x} \tag{13.25}$$

It has been pointed out by Freeman (1976), Seltzer (1972), Calder (1965), and others that, in three dimensions, mathematical and physical consistency requires that the eddy diffusivity, K, be a second-order tensor, K_{ij}, containing nine terms. Commonly, only the diagonal terms are used, opening the possibility of miscalculation in pollutant plume configurations where the off-diagonal terms may be important (e.g., shear flows).

Expressions of the form of Eq. 13.25 may be viewed as "first-order closure," i.e., a quadratic turbulent flux term $\overline{u'C'}$ is directly related to a gradient of the mean field for purposes of computation. This is an approximation that may be quite satisfactory in some applications and quite limited in others. As a rule of thumb, if the "fundamental parameter," K, must become a complicated function of time, space, and the concentration field in order to describe the field, then the "K-theory" formulation is too simplistic for that problem. The flux term may readily be expressed as a function of higher order fluctuations eliminating the necessity of the diffusivity approximation. This is the basis for higher order closure methods as described by Donaldson (1972), Mellor and Yamada (1974), and others where a hopefully less-critical approximation is made to eliminate the higher order fluctuation terms. In practice, consistency requirements demand solution of a fairly large number of simultaneous partial differential equations. The method has been successfully applied in a number of turbulent flows including the atmospheric boundary layer (Yamada and Mellor, 1975; Wyngaard and Cote, 1974).

The advection term, $\mathbf{V} \cdot \nabla \bar{C}$, requires a velocity field for its evaluation. In some applications a simple field is satisfactory or even desired. There are many examples of very useful analyses based on constant winds in time and space, winds that depend only on height above ground, or a simple variation in the horizontal with no vertical dependence. However, there are many applications where complex advection fields are fundamental to the problem and must be reasonably approximated in time and space. In those cases a separate calculation using a method appropriate to the problem must be applied to estimate the velocity field. For example, in other sections of this chapter, we describe methods applicable to terrain obstacles, shoreline circulations, vegetative canopies, and urban complexes.

13-4.2 Method of Solution

Equation 13.24 with its terms appropriately formulated is amenable to solution by a variety of numerical techniques, the most common of which is the finite-difference method. Differentials in the governing equations are expressed as differences over preselected grid distances. An excellent example to use to become familiar with the concept of finite differences is the simplified transport equation

$$u \frac{\partial C}{\partial x} = K \frac{\partial^2 C}{\partial y^2} \tag{13.26}$$

where u and K are constant values. Equation 13.26 can be analytically solved to provide a basis of comparison for simple numerical exercises.

In one simple finite-difference form, Eq. 13.26 becomes

$$C_{i+1,j} - C_{i,j} = \frac{K}{u} \frac{\Delta x}{(\Delta y)^2} (C_{i,j+1} - 2C_{i,j} + C_{i,j-1}) \tag{13.27}$$

Equation 13.27 can be evaluated manually with a modest computational effort or coded on a small computer or even a programmable hand calculator. The interested reader can gain useful insights into finite-difference methods by exercising Eq. 13.27. For example, try $u = K = 1$, $\Delta y = 1/4$, and $dx = 1/32$; then $\Delta x = 1/16$. (Compare $(K/u)[\Delta x/(\Delta y)^2] - 1/2$ in each case). The calculations can be done quickly by hand for a point source. This exercise introduces the idea of numerical stability. For values of $(K/u)[\Delta x/(\Delta y)^2]$ larger than $1/2$, the step-by-step computed values of C increase without bound and without physical significance. In practice, this usually dictates a constraint on the step size of Δx. One problem that has plagued some practical numerical diffusion estimates is computational diffusion. The advection term in Eq. 13.27 should merely translate the pulse of concentration with no change in shape. However, due to truncation errors, the numerical application of the advection term tends to spread material with time. Long and Pepper (1976) have performed a comparison test on several popular differencing methods for the effect of computational diffusion. In some cases this incidental effect can exceed the specifically formulated diffusion. The conditions for numerical stability are derived and discussed in many texts on numerical methods, and it should serve here to stress that this is a fundamental concern in selecting a finite-difference formulation for diffusion problems. In view of the long-time concern, several methods have been developed to avoid or control numerical instability by using information at several steps in the X dimension. These must also be carefully analyzed for the particular problem at hand since side effects may be introduced such as:

1. Compatibility. The finite-difference equation may be equivalent to a fundamentally different kind of equation than the differential equation it is approximating (e.g., one may obtain a damped wave difference equation when trying to simulate a diffusion equation).

2. Computational diffusion. One method in currently extensive use is the PIC method modified to handle diffusion calculations (Sklarew, Fabrick, and Prager, 1972). In this method the diffusion effects are combined with the advection term through the definition of a diffusion velocity

$$v_d = -\frac{K}{C} \nabla C \tag{13.28}$$

yielding an equation of the form of a simple advection equation. The mass of pollutant is characterized by an ensemble of discrete particles. Concentration gradients required to calculate v_d are derived from the tally of particles in adjacent Eulerian cells. Particles are then transported directly for the next time step. Johnson, Sklarew, and Turner (1976) report the elimination of artificial diffusion by PIC methods. For further discussion the reader is referred to Sklarew, Fabrick, and Prager (1972) and Lange (1978).

13-5 STOCHASTIC DIFFUSION MODELS

Problems of dispersion of particles (real or theoretical) in the atmosphere are well suited to analysis by stochastic techniques. A puff or plume is modeled by an ensemble of fluid "particles" that have trajectories described in a stepwise fashion according to a particular stochastic formulation. The simplest sequence of steps is the random walk where each new step of a particle trajectory is randomly chosen from a distribution of possible steps independently of the previous history of particle motion. A common model used to aid the visualization of this process is to use the flip of a coin to determine positive or negative steps in a one-dimensional, single velocity system. Consider, for example, 10 such steps for each particle trajectory and then repeat for a large number of particles. The distribution so obtained is called the "Bernoulli distribution." Markov processes retain a memory of at least one step prior to the one under consideration. This adds mathematical complexity but also has the potential of adding realism.

Stochastic methods may be used in analytical derivations of partial differential diffusion equations or used directly in calculating diffusion by sampling the appropriate distributions, the latter being referred to as Monte Carlo techniques. Csanady (1973) derives the conventional parabolic diffusion equation based on the assumption of a random walk. Monin and Yaglom (1971) describe a more complex derivation, originally proposed by Monin (1955), that accounts for a finite propagation rate of material away from a source. By making use of a Markovian (two-dimensional) random function of position and velocity as well as an explicit requirement of a finite step-to-step velocity, Monin develops a hyperbolic diffusion equation. In particular, the equation is a familiar differential equation from mathematical physics called the "telegrapher's equation." Similar results are derived independently by Scheidegger (1958) for application to flow through a porous medium.

The Monte Carlo approach to atmospheric diffusion generally requires a computer, but numerical routines can be formulated to solve reasonable problems using only a fraction of the machine time required by finite-difference techniques. Thompson (1971) describes a series of Monte Carlo modeling exercises that evolve from the simplest random walk with no mean wind up through a flow over a hill. In the more complex model, Thompson introduces a turbulence field that exhibits Eulerian temporal velocity correlation and through which the particles travel, although he does not attempt to describe the Lagrangian autocorrelation of the moving particles. He stresses the adaptability of the method to varying topography, buoyancy, and anisotropy and produces three-dimensional plume growth patterns that compare favorably with known results.

Hall (1975) describes and compares two Monte Carlo models in a one-dimensional (vertical) application above a reflecting ground surface. He introduced a height dependence of the Eulerian scale and the intensity of turbulence and estimates vertical concentration profiles at selected downwind locations out to 3.2 km. The simplest of Hall's models is, again, the random walk where the particle velocity at each step is selected randomly from a Gaussian distribution with a zero mean and a unit standard deviation. Some Lagrangian autocovariance is introduced in the second model which is a Markov chain in particle velocities,

$$v_{i+1} = \alpha v_i + r_i \tag{13.29}$$

The parameter r_i is a random variable with zero mean and given standard deviation and α is a constant value estimated by

$$\alpha = \exp\left(-\frac{\Delta t}{\tau_L}\right) \tag{13.30}$$

where Δt is the selected time step and τ_L is the Lagrangian scale of turbulence. This choice of parameters yields physically reasonable autocorrelation functions. Hall makes use of the scale ratio, β, between Lagrangian and Eulerian time scales to utilize tower-based micrometeorological results in the particle track diffusion formulation. He points out the need for additional empiricism on turbulence structure in order to apply the technique. This is not necessarily a drawback. It is valuable in many cases to have a computational method that does not require data to be reduced to a single parameter for application. Watson and Barr (1976) attempt to incorporate additional structure of the velocity autocorrelation through a multiterm recursion relation

$$v_j = \sum_{i=1}^{i-1} \beta_{ji}\, v_i + r_j \tag{13.31}$$

In Eq. 13.31 r_j is the same as r_i defined in Eq. 13.29 and v is the fluctuating part of the appropriate velocity component. A matrix of β_{ij} terms is developed from the autocorrelation function. In preliminary applications the autocorrelation is derived by Fourier transform of a filtered Lagrangian spectrum. Several composite spectra have been worked up for free atmosphere applications without boundary effects, and the model has been applied to puff-type volume sources over travel times of 5 to 10 h. Alsmiller et al. (1978) modified the model to apply to a continuous near-ground point-source release and compared the results with the plume measurements of Peterson (1968) to travel distances of 290 km.

Monte Carlo methods are computationally very flexible and rather easily accommodate to time and space varying transport wind and turbulence fields as well as attendant physical processes such as deposition, transformation, and precipitation scavenging. However, as Hall points out, they require more data or empirical relations than some other diffusion models. In the proper application the reader should be aware of this option of a technique that can utilize all the relevant data which he can provide. In return, the method provides statistics of the diffusion process not available in other methods and, it is to be hoped, can use the increased data to produce more confidence in the model output.

13-6 CONCENTRATION FLUCTUATIONS

Mean concentration patterns from many source configurations have received considerable attention as described in the previous sections. However, an increasing number of practical environmental applications require information on the fluctuations of concentration about the mean. Certain species of vegetation are very tolerant to chronic exposures of low levels of atmospheric pollutants but are quite sensitive to brief periods of high concentration.

The stochastic nature of turbulence causes the concentration at some fixed location downwind of a point source to fluctuate with time. A continuous recording infinite-response instrument placed at this location might produce a trace like that shown in Fig. 13.8. If a specific time interval T is selected, then one can compute $\overline{C}(T)$ and talk about fluctuations about $\overline{C}$. Also, the probability distribution of fluctuations about $\overline{C}$ can be determined.

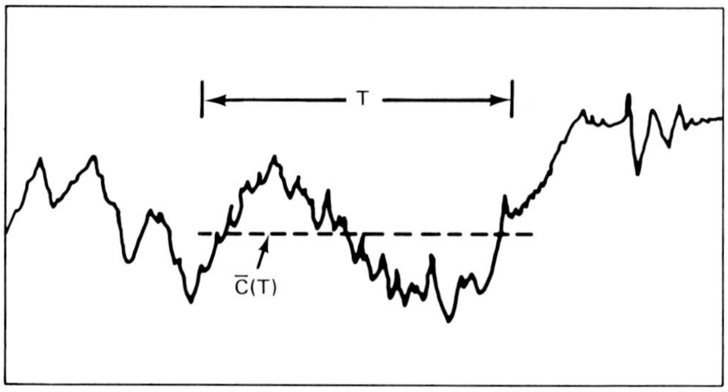

Fig. 13.8 Schematic trace of hypothetical pollutant concentration at a fixed sampler.

Since the trace in Fig. 13.8 represents a continuous record of an infinite-response instrument, all scales of fluctuations are observed and can be included in calculations. In many experiments or monitoring programs, average values over some time period τ are obtained. This might result in the data of Fig. 13.8 to be observed as shown in Fig. 13.9. Now in the same examination period, T, one still has fluctuations about $\overline{C}$, but some of the details of the fluctuations have been lost in the averaging process.

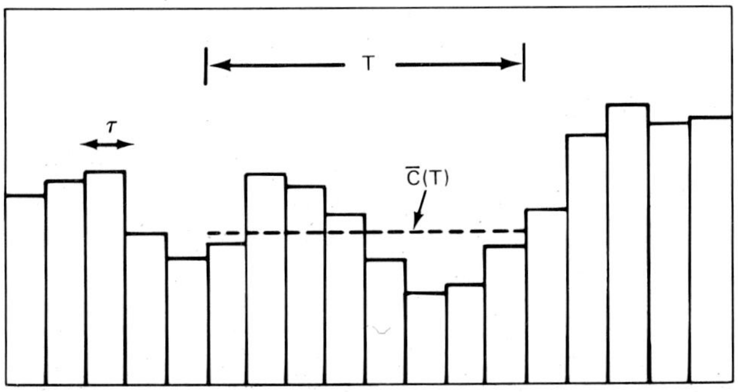

Fig. 13.9 Trace of Fig. 13.8 averaged over steps of duration (τ).

The closer τ gets to T, the less is known about the fluctuations. It should be clear that $\bar{C}$ is a function of T and that the probability distribution function for fluctuations and $\bar{C}$ for a given turbulence field depends on τ and T.

In response to the need for additional fluctuation information, several studies have been performed (Hinds, 1969; Gifford, 1959; Hilst, 1957; Singer, 1961; Singer, Imai, and Del Campo, 1963) to determine ratios of peak-to-mean concentration. Properties of this parameter can be related to source-receptor geometry, sampling time, and atmospheric stability as summarized in Gifford (1968). Gifford (1972) discusses the form of the frequency distribution of fluctuations for several source configurations.

Csanady (1967) surveys the concentration fluctuation problem in some depth from several viewpoints and is able to set a context for comparison of several other theoretical models. Using techniques familiar to turbulence analysis, Csanady derives a conservation equation for the variance of concentration fluctuations, s, and solves the partial differential equation for the special case of radial symmetry. The solution is based on a self-similar radial profile of the concentration variance with plume travel distance. One result of the analysis is a ratio of the concentration variance to the square of the mean concentration, $s/\bar{C}^2$, close to unity near the plume axis. The ratio increases outward to large values at the plume edge.

The value of $\overline{C^2}/\bar{C}^2 = 1$ supports the hypothesis of a Poisson-like distribution for concentration fluctuations near the plume axis since this distribution yields $s/\bar{C}^2 = 1$. On the fringes of a plume, Csanady introduces an intermittency term and suggests

$$F(C) = qe^{-q(C/M)} \tag{13.32}$$

as a distribution function.

Barry (1971) points out that the peak-to-mean ratio represents only two points on a probability distribution and suggests the use of the entire distribution function to convey more information. Barry adopts an exponential probability distribution function based on extensive empirical evidence compiled by him and earlier by Gartrell (1966)

$$F(C) = F(0) \exp\left[-F(0)\frac{C}{M}\right] \tag{13.33}$$

where F(C) is the frequency of short-term concentrations (averaged over t_p) exceeding $\bar{C}$, M is the concentration averaged over a time $t_m \gg t_p$, and F(0) is the frequency of nonzero concentrations, i.e., a measure of the time a sampling station is affected by a plume. Notice that Eq. 13.33 is identical in form to Eq. 13.32. The term F(0) is equivalent to the intermittency factor adopted by Csanady. The peak-to-mean ratio can be calculated as a special case of Eq. 13.33

$$\frac{P}{M} = \frac{[\ln F(0) + \ln(t_m/t_p)]}{F(0)} \tag{13.34}$$

Another derived functional form of concentration distributions is lognormal. If we simply make the hypothesis that the concentration at some time step is related to the concentration at the previous time step by a random multiplicative factor and integrate the process over many steps invoking the Central Limit Theorem, we can derive the lognormal probability density function,

$$F(C) = \frac{1}{C(2\pi\sigma_\ell^2)^{1/2}} \exp\left[-\frac{(\ln C - M)^2}{2\sigma_\ell^2}\right] \tag{13.35}$$

where M is the mean and σ_ℓ^2 is the variance of lnC.

Although the derivation is convenient, it does not imply any fundamental atmospheric properties. Rationales have been adopted for application of the function to air pollution. Csanady (1973) suggests the concept of "dilution impulses" to represent the stepwise random fluctuations. Gifford (1972) advises that the function is more relevant to urban pollution with its myriad of fluctuations due to random sources and meteorology than to the isolated plume. The derivation mentioned above breaks down at near-zero concentration and is appropriate only for small departures from an average concentration. The most likely concentration at a given location near an isolated point source is zero, a fact that challenges the generality of the lognormal distribution for such sources.

The lognormal and exponential distributions are fundamentally different in shape. The exponential distribution has its maximum ordinate at zero concentration, whereas the lognormal distribution approaches zero at the origin. Csanady (1973) suggests that a data set should be screened for zero concentrations and reflect only the presence of a plume in the vicinity of a sampler. By processing air concentration measurements in this way, he finds a quite general applicability of the lognormal probability density function for short-term air concentrations. The problem of presence or absence of material at a sampler is addressed by an intermittency factor which is the fraction of the total sampling interval that some contaminant was present. Indications are that, even near a plume axis, intermittency is in the range of 50% (±20%) depending on distance from source, stability, and roughness. Away from the axis the presence of positive concentrations is much lower.

Once a functional form for the probability density function of concentration is adopted, much of the earlier work on peak-to-mean ratios can be assessed and unified to some extent. There are two parameters in Eq. 13.35 that govern the distribution. These are the mean, M, and variance, σ, of the logarithm of the concentration. The properties of Eq. 13.35 are well documented and allow calculation of the mean $[e^{M} + (\sigma^2/2)]$, the variance $[e^{\sigma_\ell^2 + 2M}(e^{\sigma_\ell^2} - 1)]$, or a selected percentile. Csanady (1973) adopts a definition for the peak concentration as that value which exceeds 99% of the ensemble of observations. The peak-to-mean ratio then depends only on σ_ℓ (neglecting intermittency) as shown in Fig. 13.10. Values of

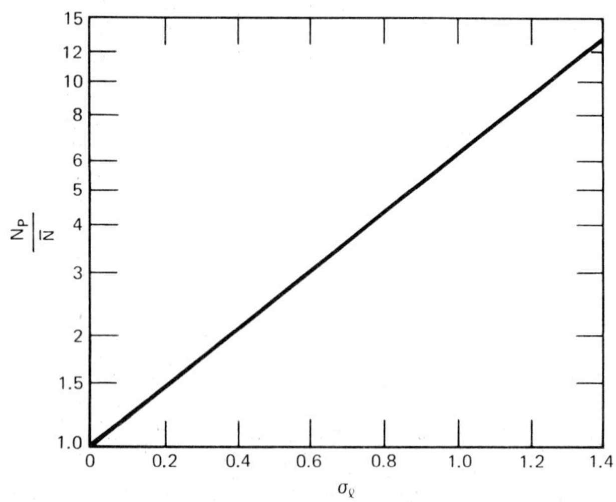

Fig. 13.10 Peak-to-mean ratios vs. log standard deviation. (From G. T. Csanady, *Turbulent Diffusion in the Environment*, D. Reidel Publishing Co., Dordrecht-Holland, Boston, Mass., 1973.)

M will vary greatly, depending on source strength and location of the sampling site. However, there is hope that σ_ℓ might be a more general function of meteorology and setting. Csanady suggests a range of σ_ℓ values from 0.2 for a stable atmosphere and smooth terrain up to 0.7 for rough terrain. These values correspond to the range of peak-to-mean ratios typically reported near plume axes.

The physical causes for the fluctuation of concentration at a sampler are the large-scale meandering of plume elements and the small-scale randomness of concentration within elements. Gifford (1959, 1960) has developed a fluctuating plume model that allows explicit estimates of the meandering effect. In this model disk-shaped plume elements are considered to meander about the mean plume axis. The lateral and vertical concentration profiles across a plume element are assumed to be Gaussian, giving the expression for the instantaneous concentration

$$\frac{C}{Q'} = \frac{1}{2\pi\overline{Y^2}u} \exp\left[-\frac{(y - D_y)^2 + (z - D_z)^2}{2\overline{Y^2}}\right] \qquad (13.36)$$

where $\overline{Y^2}$ is the second moment of distributed material about the centroid of a local plume element, and D_y and D_z are distances from the mean plume axis to the instantaneous plume centroid. In addition, D_y and D_z are assumed to be Gaussian about the mean axis with variance D^2. The model has the property that the mean concentration is identical to that of the conventional mean plume model.

The partitioning of the diffusion process into two scales generates two time scales for the assessment of effects. This important implication comes out of the formulation of Eq. 13.36, although the scales are not explicitly applied in Gifford's (1959, 1960) calculations other than to estimate asymptotic properties of peak-to-mean ratios for short and long travel times. In preparing data for comparison, Gifford uses a ratio of time scales, t_A/t_p, where t_A and t_p are the sampling periods for the average and peak, respectively. The observed peak-to-average concentration ratios depend quite strongly on the two sampling times, particularly in the event of ground-level concentrations near an elevated source. In this case the sampler is on the fringes of the plume, and convective eddies account for large fluctuations in the plume axis. Another practical problem producing large concentration fluctuations is the building wake dispersion problem. Hinds (1969) reports peak-to-mean ratios up to 30 in wake flow when t_A/t_p is 200.

The discussion up to this point deals largely with concentration fluctuations in a single plume from a well-defined source. The variation of concentrations in an urban area adds the uncertainty of time and space variable sources. This problem is less amenable than the isolated source problem to the kinematic analysis techniques described above, although Gifford (1972) offers a physical rationale for a lognormal distribution of urban air pollution. Larsen (1969) presents a mathematical summary of extensive urban air monitoring data for several pollutants in six cities. He concludes that "concentrations are approximately lognormally distributed for all pollutants in all cities for all averaging time." This relationship along with the second conclusion, which expresses the median concentration as a power law of averaging time, allows for generalization of the concentration variations. Larsen has adopted a characteristic graphic presentation, illustrated in Fig. 13.11,

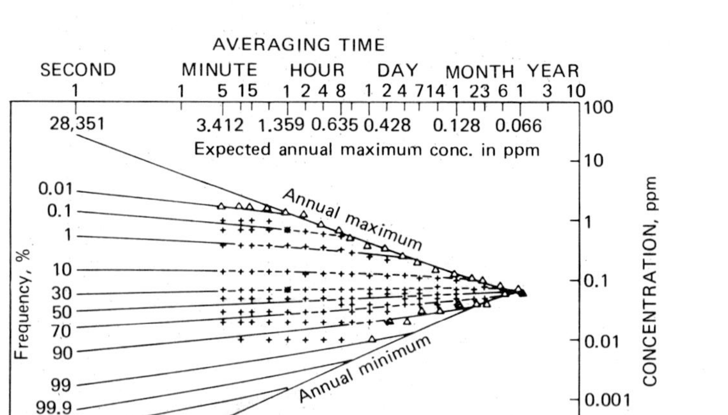

Fig. 13.11 Concentration vs. averaging time and frequency for NO_x in Washington, D. C. [From R. I. Larsen, A New Mathematical Model of Air Pollutant Concentration Averaging Time and Frequency, *J. Air Pollut. Control Assoc.*, 19: 24-30 (1969).]

that demonstrates the full range of properties of a particular pollutant and city combination.

The empirical support for the lognormal distribution provided by Larsen's studies along with the work described above suggests that this function may withstand additional scrutiny and prove to be useful in the estimation of urban pollution fluctuations. The exponential distribution has been the most useful in analyzing isolated point-source concentrations which have a signature quite different from the urban pollution due primarily to the high incidence of zero values. The research in this area is still incomplete, and the practical demands for information on fluctuations have only begun to be recognized. In his 1967 analysis, Gifford makes the point that the functional form selected for the distribution is likely to be somewhat less important than the proper meteorological description of the functional parameters.

13-7 DIFFUSION IN THE UNSTABLE PLANETARY BOUNDARY LAYER

The atmospheric boundary layer under strong insolation and light-to-moderate wind speeds exhibits unstable thermal stratification and associated convection currents. An easily observed consequence of the convection is the increase in the magnitude of wind fluctuations near the ground. Also, dilution of effluent plumes proceeds much more rapidly in unstable conditions than in other conditions as is characterized in Figs. 13.4 and 13.5. The danger in extrapolating curves like the A-Stability in Fig. 13.4 is obvious when we observe the impossibly large vertical dimensions of plumes for travel distances of more than a few kilometers. In considering diffusion in unstable conditions, it is necessary to view the entire unstable boundary layer and its associated overlying stable layer as a system, frequently referred to as the mixed layer (see Chap. 5). Recent studies of the dynamics and structure of the unstable boundary layer by Deardorff (1972, 1974), Wyngaard and Cote (1974), Clarke et al. (1971), Yamada and Mellor (1975), and others have provided the basis of a comprehensive picture of the time and space evolution of the mixed layer.

Figure 13.12 is offered as a brief schematic view of the diurnal character of the mixed layer with a hypothetical surface source of pollution and the presence of clouds as a reminder of the complexity of the system in practical applications.

The important features of the mixed layer for the purposes of diffusion estimation are:

1. The depth of the layer.
2. The magnitude of vertical and lateral turbulent fluctuations in the layer.
3. The profile of turbulence with height.

Scaling studies treat the depth of the convective mixed layer, z_i, as a fundamental quantity. It depends on time of day, intensity of surface heat flux, and thermal stability of the atmosphere outside the layer. Holzworth (1967) derived mixing depths for the entire United States from radiosonde records by some straightforward and simple assumptions about boundary-layer mixing. These first-order estimates have proven very useful for practical pollution problems. More recent efforts have concentrated on the dynamics of the mixed layer including the structure of the turbulence and the temporal development of the layer. This work, summarized by Tennekes (1973, 1974), Wyngaard (1973), and Deardorff (1973), is providing an additional depth of insight required to describe the details of plume behavior within the mixed layer.

Briefly, the mixed layer is driven by heating at the ground surface which produces unstable temperature stratification in the lowest layers and allows vigorous convection to occur. The convection promotes very efficient flux of virtually every transportable quantity in the vicinity of this surface layer. Convective plumes then traverse the quasi-adiabatically stratified interior of the mixed layer quite efficiently. At the top of the layer the plumes encounter an interface with more stably stratified air.

The processes that take place at the interface include a penetration of the stable air by an inertial overshoot and an exchange of energy from relatively large vertical convective

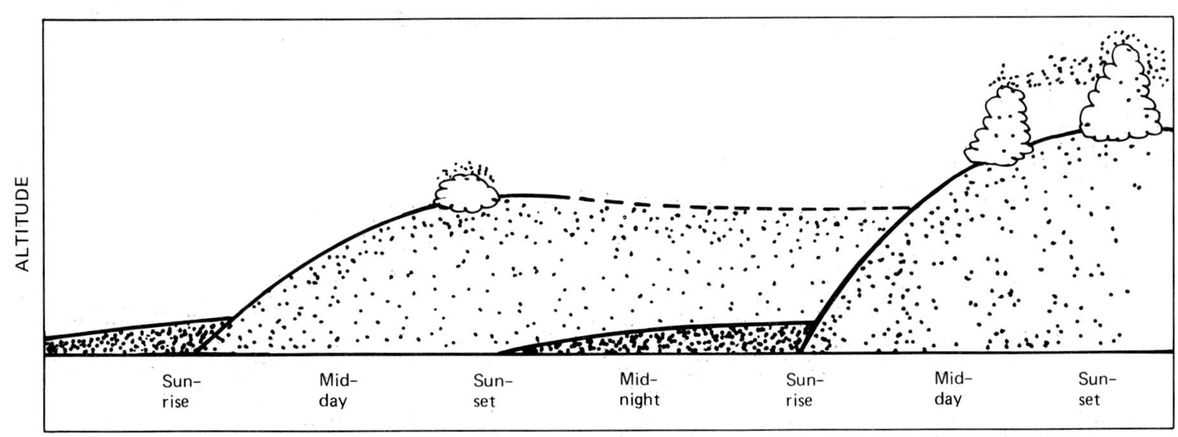

Fig. 13.12 Schematic diagram of diurnal pattern of mixed layer and associated pollution.

scales to smaller scales of turbulence in all directional components. In addition, moisture is also transported to the interface by boundary-layer mixing and tends to condense into the base of cumulus cloud layers at this elevation. Although Tennekes (1974) makes the point that cloud processes above the interface are not boundary-layer phenomena, they are nonetheless effective in transporting pollutants.

Some of the properties of pollution transport in the mixed layer appear to be a tendency to concentrate material into a secondary maximum near the top of the layer at relatively short travel distances (e.g., Deardorff and Willis, 1974). Further downstream the material spreads from the upper and lower boundaries of the mixed layer toward the center, producing a uniform concentration profile in the vertical direction. This characteristic of the mixing process is not modeled by the Gaussian plume approach in the intermediate travel range, although the methods described in Sec. 13-3 would produce uniform profiles at travel distances within a factor of 2 of the position reported by Deardorff and Willis.

13-8 STAGNATION ENVIRONMENT

The condition of very low mean wind speed and stable temperature gradient represents one of the poorest conditions for dilution of atmospheric emissions. The problem is compounded by the inapplicability of the traditional modeling techniques under these limiting conditions. Although an extension of predictors of the type shown in Figs. 13.4 and 13.5 indicate very diminished diffusivity with ever-increasing thermal stability, observations show that beyond a threshold stability the average spread of material actually increases. The particular wind speed–stability levels at which the minimum dilution rates occur and the character of the diffusivity function in the vicinity of the minimum are site dependent and probably depend on wind direction at a given site. The cause of the enhanced dilution under low wind speed and stable conditions is the presence of a low frequency meander of wind direction. The vertical component of turbulence, σ_w, appears to continue to diminish under the influence of thermal stability, whereas the low frequency fluctuations keep σ_θ above a site-dependent threshold level.

The practical need for study of the low wind speed and stable diffusion characteristics is by no means trivial. At most locations a significant fraction of the time winds are less than 2 m/s. Furthermore, they generally represent the worst cases of diffusion for a particular facility; hence the estimates contribute heavily to design concepts. If generalized estimation methods routinely yield overconservative results, expensive overdesign may result. However, lack of conservatism may result in damage to the environment. The answer is an adequate understanding of this condition to produce reasonable estimates of the dilution of atmospheric emissions.

There have been several modeling approaches to the low wind speed diffusion problem, all based on Gaussian plume concepts with some method of azimuth averaging used to account for the meandering effects. From the point of view of the turbulent energy spectrum, the low wind speed turbulence structure is expected to be quite different from other cases. With light winds and associated weak shears, the small-scale mechanical turbulence should be weak. The peak energy might be expected for periods equivalent to the low frequency meandering. Predictors based implicitly or explicitly on some generalized form of the turbulence spectrum might be expected to fail. Leahey and Halitsky (1973) attempted to account for the spectral effects by using the method of Hay and Pasquill discussed in Sec. 13-2 in which, as in Eq. 13.17,

$$\sigma_y(x) = x(\sigma_A)_{t,(x/\beta\bar{u})} \quad (13.37)$$

$$\sigma_z(x) = x(\sigma_E)_{t,(x/\beta\bar{u})} \quad (13.38)$$

In Eqs. 13.37 and 13.38, σ_A and σ_E are azimuth and elevation angle standard deviations sampled over time t, the time for which plume dimensions are desired. The smoothing interval, $x/\beta u$, is an advective time scale modulated by the Lagrangian to Eulerian time scale ratio, β, a parameter explicitly required in the Hay–Pasquill analysis. Using a series of bivane records, mean wind, and temperature profiles covering 153 h of data over three seasons in the Hudson Valley, Leahey and Halitsky (1973) derived a climatology of σ_y and σ_z for a 1.1-km travel distance. In a series of tables, they show the equivalent Pasquill–Gifford dilution category (Table 13.2) for the same travel distance as a function of wind speed and direction and vertical temperature gradient. The weakest diffusion estimated by the Hay–Pasquill method corresponds to Pasquill–Gifford category D, and there is a high frequency of vigorous dilution cases resulting from meandering under stable lapse rates.

Wilson et al. (1976) used a Gaussian plume model to interpret a set of experimental data. The diffusion coefficients, σ_y and σ_z, were deduced directly from ground-level and helicopter concentration measurements under stable conditions. They were compared with estimates derived using the methods of Sec. 13-3 with stability classes selected by temperature lapse rate for travel distances of 100, 200, and 400 m from a continuous point source. In general, observed concentrations were about 20 times lower than would be estimated from the lapse rate selection procedure. The difference could be assigned to both σ_y and σ_z being greater than predicted. The authors explain the enhanced vertical diffusion in terms of vegetative and topographic roughness of the site near Oak Ridge, Tenn. Discrepancies between observed and estimated concentrations tended to become greater as the stability increased, an indication that the lapse method of diffusion prediction is of questionable value under the conditions of the test. In fact, any method that fails to account for low frequency meandering wind direction is likely to give poor results under very stable light-wind conditions.

Sagendorf and Dickson (1976) describe two modeling techniques to account for the role of low frequency horizontal meander. One method, termed split-sigma, recognizes that vertical diffusion agrees with the Pasquill–Gifford estimates and becomes very weak with strong stable thermal stratification, whereas time-averaged σ_y is better estimated from the azimuth wind fluctuations. The second correction, the segmented plume concept, addresses the bimodal spectral energy through perceiving the finite duration release as a series of short-term (e.g., 2 min) plumes. The short-term lateral spread is given by the empirical formula

$$\sigma_y = 0.017\sigma_\theta x^{0.87} \tag{13.39}$$

where σ_y and x are in meters and σ_θ is in degrees. The meandering is accounted for by the integrating effects of a large number of separate plume positions. Again, the Gaussian plume model provides the vehicle for the adaptive alterations in diffusion estimators, apparently with success for the scales on which it was applied. Sagendorf and Dickson report excellent agreement on a 200-m arc with plumes covering more than 200° azimuth. At longer travel distances the time and space coherence of the meandering components of wind may break down and reduce the applicability of the method. The Sagendorf–Dickson paper is a good example of adaptation of a simple model to a fairly complex diffusion problem. The governing mechanisms were identified, found to be compatible with the Gaussian plume, and incorporated as a direct and simple method.

The low wind speed stable diffusion problem can be expected to be very site dependent and does not necessarily imply significant meandering. The existence of a slight but quasi-uniform large-scale topographic slope may produce drainage winds that are less than 2 m yet very steady in response to the local driving force. Clements and Barr (1976) report winds of 0.5 to 1.0 m/s which are steady for several hours of the night and occur on a majority of nights of the year.

13-9 TURBULENT DIFFUSION IN URBAN AREAS

The estimation of transport and diffusion in urban areas depends on the basic methodology described in Secs. 13-2 to 13-6. Indeed, the urban pollution problems are among those of most concern to public health officials and are the basis for a great deal of the research leading to practical estimation methods. There has been no attempt to make this section an exhaustive survey of urban air-pollution modeling since the subject involves a very extensive literature. Also, there have been a number of excellent surveys prepared recently to which the reader is referred for in-depth discussions of model application and testing [e.g., Hanna (1975), Moses (1969), and Johnson, Sklarew, and Turner (1976)].

13-9.1 Factors Affecting Urban Air Pollution

There are a number of aerodynamic, thermodynamic, and chemical factors that characterize the urban dispersion problem. One of the most significant factors affecting the utility of modeling approaches is the complexity of the source configuration. Cities are affected by isolated point sources at ground level and at elevated levels. They are superimposed on quasi-uniform, low-level residential sources and localized intense area sources associated with industrial activity. Added to these are an array of line sources produced by time-dependent traffic flow and the background of material transported from nearby cities. Of course, for particular pollutants, one or more of the source types may be dominant, but the list of potential sources serves as an example of the need for high-quality source inventories.

In addition to the increased complexity of source configuration, the urban complex alters the basic meteorological structure governing the transport and diffusion, including mixing depth, mean wind direction and speed, temperature structure, and related turbulence. Also, altered mechanisms of chemical and physical transformation and wet and dry deposition are fundamental characteristics of the urban problem.

It has been well established for many years that cities have surface temperatures several degrees warmer than the surrounding countryside. This urban heat island was documented empirically by Bornstein (1968), Findlay and Hirt (1969), Runnels, Randerson, and Griffiths (1972), and Oke and East (1971) and simulated in heat budget models by Myrup (1969) and Wagner and Yu (1972). Although the mechanics of the heat island is a challenging problem in micrometeorology, the consequences of the phenomenon are of primary interest in the pollutant transport problem. One such consequence is an airflow circulation created by low-level confluence of cool, clean, nonurban air to replace the rising heated air over the city. Several numerical models have been developed to estimate wind circulation and mixed layer depth associated with the urban heat island including those of Bornstein (1972), Vukovich (1971), Delage and Taylor (1970), and Leahey (1969). Superimposed on a moderately weak external wind, the circulation tends to tilt downwind with altitude and give patterns of subsidence and ascent as suggested in Fig. 13.13 presented by Vukovich (1971). The motion and temperature fields are closely interrelated and produce regions of stability and instability. Figure 13.14 from Clarke (1969) shows an alongwind cross section of stable layers over Cincinnati as well as the unstable air over the city.

The METROMEX program is a cooperative multiagency field program aimed primarily at investigating inadvertent weather modification produced by the St. Louis urban complex. Some of the results are directly applicable to transport and dispersion including the wind fields (Ackerman, 1974) and the modification of the capping inversion, which delineates the mixed layer (Spangler and Dirks, 1974). Figure 13.15 shows that the mixed layer tends to deepen over the city. Also, of course, there is evidence that the combined effects of particulates as cloud condensation nuclei, altered stability, and vertical velocities may produce a suggested increase in the precipitation. This, in turn,

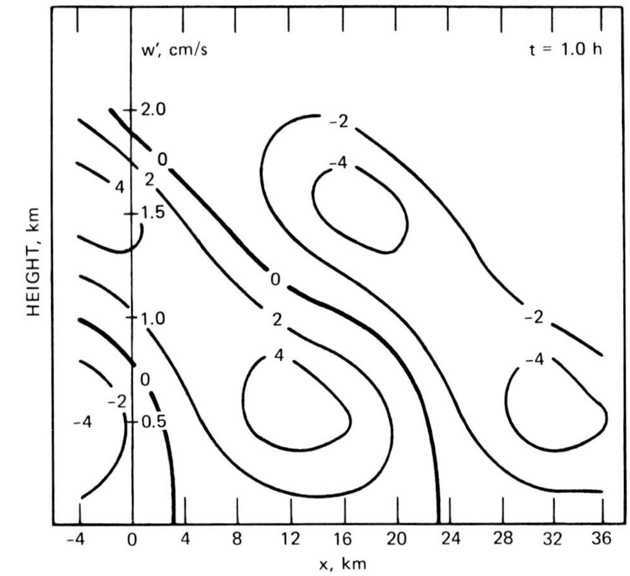

Fig. 13.13 An x–z cross section of vertical wind component in urban heat island. [From F. M. Vukovich, Theoretical Analysis of the Effect of Mean Wind and Stability on a Heat Island Circulation Characteristic of an Urban Complex, *Mon. Weather Rev.*, 99: 919-926 (1971).]

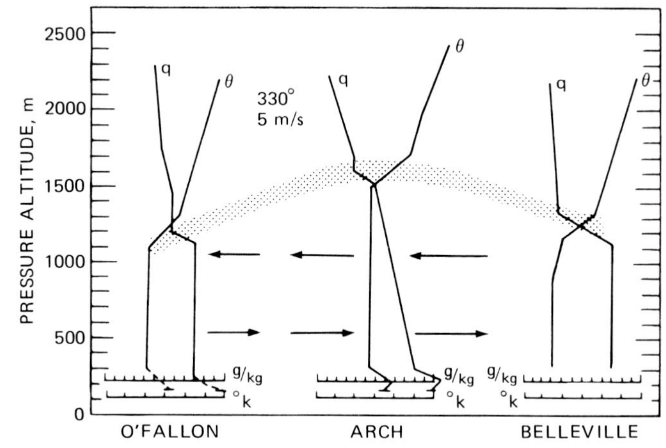

Fig. 13.15 Inversion height over St. Louis, Mo., July 2, 1971. [From T. C. Spangler and R. A. Dirks, Meso-Scale Variations of the Urban Mixing Height, *Boundary-Layer Meteorol.*, 6: 423-441 (1971).]

serves to alter the scavenging of airborne pollutants. The interactions are very complicated, and the role of these mechanisms as a sink for urban pollutants has not been quantified.

One other fundamental feature of cities that should be singled out is the roughness on the scale of building sizes. This will increase friction and retard mean wind flow and will also increase the turbulence levels and affect the plume-growth parameters. Gifford (1976) suggests a set of plume-growth parameters for urban transport as shown in Figs. 13.16 and 13.17. Because many practical problems require estimates on scales equivalent to or smaller than the characteristic spacing of the building roughness elements, a separate set of assessment models has been devised for this domain.

13-9.2 Models for Use in the Urban Setting

We have reviewed the phenomena that characterize the urban diffusion problem. They represent more physics and chemistry than we are able to reproduce with confidence in detailed mathematical models. Therefore most of the assessments to date have been made from simplifying assumptions. In fact, it is not clear that the addition of mathematical or physical complexity improves the quality of estimates in a general sense. Moses (1969) defines four vital components of an urban pollution model:

1. Source inventory.
2. Network of meteorological observations.
3. Network of pollutant monitoring stations.
4. Mathematical algorithm describing processes that transform source flux to observations at receptor locations.

Another important parameter in urban pollution models is city size (Oke, 1973; Gifford and Hanna, 1973). Moses further classified models based on the use of the

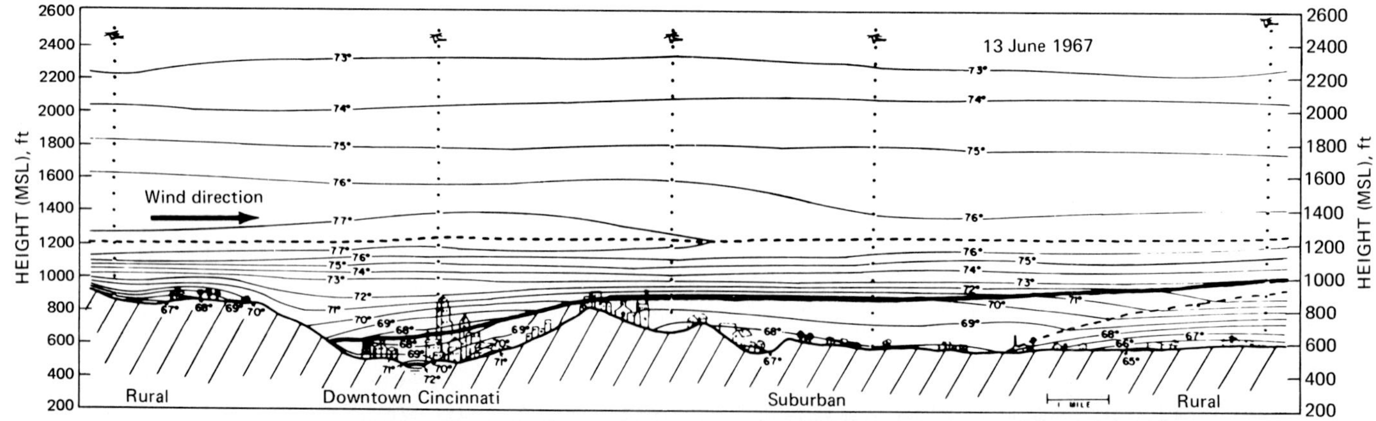

Fig. 13.14 Cross section of temperature over Cincinnati, Ohio, June 13, 1967. [From J. F. Clarke, Nocturnal Urban Boundary Layer over Cincinnati, Ohio, *Mon. Weather Rev.*, 97: 582-589 (1969).]

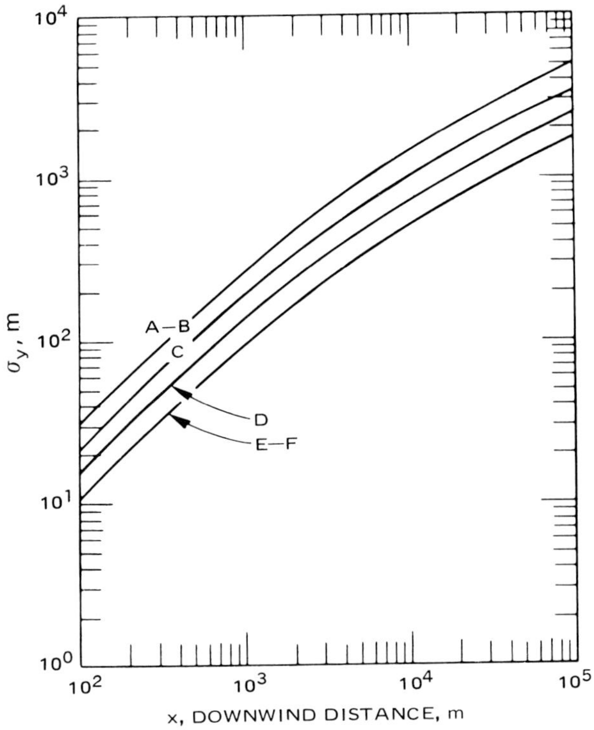

Fig. 13.16 Curves of σ_y based on recommendations by Briggs for flow over urban areas. [From F. A. Gifford, Turbulent Diffusion-Typing Schemes: A Review, *Nucl. Saf.*, 17(1): 79 (January–February 1976).]

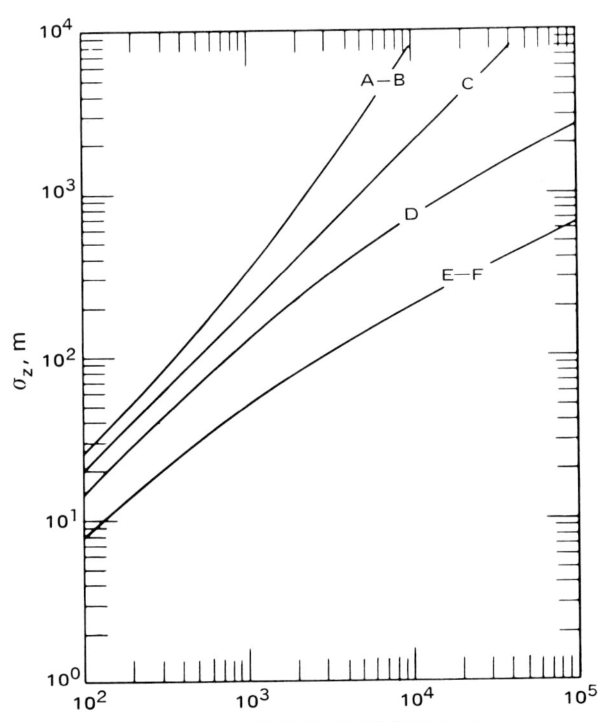

Fig. 13.17 Curves of σ_z based on recommendations by Briggs for flow over urban areas. [From F. A. Gifford, Turbulent Diffusion-Typing Schemes: A Review, *Nucl. Saf.*, 17(1): 79 (January–February 1976).]

observed pollution distributions as tests of fundamentally meteorological models (primary items 1, 2, and 4) or the empirical basis of the model (items 3 and 4). Hanna (1975) summarizes a series of models based on a sequence of increasing complexity of the mathematical algorithm (4) and points out strengths and weaknesses in the applicability of each class. His sequence starts with the rollback concept in which (4) is a simple proportion between (3) and the gross effluent (1) and includes regression algorithms and analytical and numerical computational methods.

In many urban assessment problems, it is satisfactory to treat the urban complex in an integral sense and to neglect the time and space morphology of source, sink, transport, and transformation effects. The advantage, as Hanna points out, is that the model can be formulated in an analytical framework suitable for rapid hand calculations. This simplified geometry is the basis for the box model that has been adopted by a number of authors. Because of inhomogeneous, nonlinear phenomena such as photochemical smog reactions in the urban pollution problem, there exists some controversy on the use of these simplified models (Lamb and Seinfeld, 1974; Hameed, 1974).

Lettau (1970) formulates a somewhat generalized box model that expresses the budget of pollutant in a simplified geometry including the boundary layer over a city. In this budget the time rate of change of pollutant concentration is balanced by the horizontal flux divergence associated with the mean wind, the turbulent flux through the top of the box, the surface based source of material, and interior sources and sinks. Such a mass balance problem reduces to a first-order differential equation that he then expresses as

$$\frac{dC}{dt'} = \frac{Q}{hf^*} - C \qquad (13.40)$$

where C = the average concentration in the box
Q = the average surface emission rate
h = an arbitrarily selected height of the box region
t' = a dimensionless time scale = f^*t
f^* = a "flushing frequency" (s^{-1}) which is related to ventilation rate

A solution of Eq. 13.40 is

$$C = e^{-t'}\left(C_0 + \int_0^{t'} e^{t'} \frac{Q}{hf^*}\, dt'\right) \qquad (13.41)$$

where C_0 is an initial concentration.

Two more definitions that allow evaluation of Eq. 13.40 are

$$f^* = \frac{u_v}{D} + \frac{1}{hC}\,((\overline{w'C'})_h) \qquad (13.42)$$

where

$$u_v = \frac{D}{C}(\nabla \cdot \overline{w}C) \qquad (13.43)$$

and D is the city diameter. The parameter of f^* has values in the range of 1 h^{-1} but tends to follow the wind speed.

With the rate of turnover, or flushing of urban air on the order of a few hours, the resolution requirements on source rate are governed by the basic frequency, f^*. Lettau performs a simple calculation with a typical diurnal source forcing function $Q(t)$ and ventilation, $u_v(t)$. He estimates peak concentrations in a 40-mile-diameter city to be two to four times larger than a 10-mile-wide city. Lettau's model reduces to the familiar form of simpler steady-state box models without interior sources as shown by Hanna (1975). Also, the ventilation parameter that Holzworth (1967) has estimated climatologically for many cities across the United States can be shown to be equivalent to Lettau's $(hf^*)^{-1}$.

The next stage of complexity in meteorologically based models is the multiple source Gaussian plume model. Although in concept the patterns of air concentration resulting from an array of sources could be obtained by manual superposition of many hand-calculated isopleths for individual sources, the logistics of the required calculations strongly favor a computer algorithm. Models that appear to handle the combination of major point sources and background area sources are described by Gifford and Hanna (1973). Elevated, buoyant point sources are handled by a point-source Gaussian plume formula (e.g., Eq. 13.16). Integrated area source contributions are summed for a series of finite area blocks in the upstream direction from a receptor point:

$$C = \left[\frac{\left(\frac{2}{\pi}\right)^{1/2}\left(\frac{\Delta x}{2}\right)^{1-b}}{ua(1-b)}\right]\{Q_0 + \sum_{i=1}^{n} Q_i\,[(2i+1)^{1-b} - (2i-1)^{1-b}]\} \quad (13.44)$$

where x is the alongwind block dimensions and a and b are meteorological diffusion parameters describing vertical plume dimensions through $Z(x) = ax^b$.

Table 13.5, adapted from Moses (1969) with several modifications, describes a number of modeling activities for many cities. The table is not a complete listing but is only indicative of the types of model applications and the methods chosen for testing models. Models have been tested in case studies with special air-quality measurements and in long-term comparisons with data routinely collected by local and federal air-quality monitoring systems.

The majority of entries in Table 13.5 depend on Gaussian plume or puff elements. Recently, the numerical techniques described in Sec. 13-4 have been applied to the urban problem (Randerson, 1970; Shir and Shieh, 1974), particularly when the wind field is distorted because of the combined effects of topography and urbanization or in small-scale problems where building structures force highly distorted wind fields. There is an entire subclass of intraurban diffusion problems in which the local effects of building wakes, street canyons, highway cuts, overpasses, airports, and shopping centers must be assessed. The modeling approaches for these problems are summarized by Johnson, Sklarew, and Turner (1976).

The remaining major class of the urban modeling problem includes the chemical transformation of pollutants within the urban complex. The classic problem in chemical transformation is the production of photochemical smog, although many other chemical reactions have important implications in urban air quality. Johnson et al. summarize five photochemical models based on techniques varying in complexity from dimensional analysis to detailed trajec-

TABLE 13.5 Summary of Selected Urban Modeling Studies

Investigator	Location	Modeling concept	For model comparison
Frenkiel	Los Angeles, Calif.	Gaussian	
Turner	Nashville, Tenn.	Gaussian	1036 24-h readings, 2710 2-h readings } SO_2
Clarke	Cincinnati, Ohio	Integrated Gaussian	33 d SO_2, NO_x
Fortak	Bremen, Germany	Gaussian plume	1 yr SO_2
Hilst and Bowne	Ft. Wayne, Ind.	Gaussian plume	21 experiment periods (days), 70 releases fluorescent particles
Pooler	St. Louis, Mo.	Gaussian plume	43 experiments fluorescent particles
Miller and Holzworth	Washington, D. C.	Integrated Gaussian	36 test days NO_x
	Los Angeles, Calif.		31 test days NO_x
	Nashville, Tenn.		41 test days SO_2
Davidson	New York, N. Y.		9 experiment periods, 5 d per period SO_2
Koogler, Sholtes, Davis, and Harding	Jacksonville, Fla.		12 d, 111 observations SO_2
Panofsky and Prasad	Johnstown, Pa.	Box model	Particulates
Hilst, Badgley, Yocum, and Bowne	Connecticut	Gaussian plume	None predicted; HC, CO, NO_x, SO_2 particulates
Roberts et al.	Chicago, Ill.	Integrated Gaussian puff	Hourly readings, 5 sites SO_2
Duewer et al.	San Francisco Bay area	Particle-in-Cell including chemical reactions	CO: 24 hourly readings at 12 stations HC: 24 hourly readings at 10 stations NO: 24 hourly readings at 6 stations NO_2: 24 hourly readings at 9 stations O_3: 24 hourly readings at 15 stations

tories and 16 chemical reactions. MacCracken et al. (1978) include a set of 48 chemical reactions with a PIC transport and diffusion code.

13-10 COMPLEX TERRAIN MODELING

13-10.1 Complications Added by Terrain

It is overly simplistic from a meteorological point of view to define a single category of complex terrain to contrast with homogeneous terrain. Complications to the classical analysis described in Sec. 13-3 may be introduced by everything from rolling hills to steep mountains as well as spatial variations in surface heating rates due to albedo, moisture, or vegetative differences. At present, there is no generalized method of calculating diffusion and transport in complex terrain, but one purpose of this section is to identify the types of complicating factors that should be accounted for in preparing and interpreting diffusion estimates. Recommendations for formulating a complex terrain modeling program as an integrated part of a broad study program are summarized in a DOE/EPA workshop proceedings (Barr et al., 1977).

Perhaps the major uncertainty introduced by the terrain is the alteration of the mean transport of a plume or puff by strongly space- and time-dependent local circulations. These motion systems depend on the scale of the terrain features, their height, and height gradients as well as the large-scale atmospheric structure of temperature and wind speed and direction. The particular application may introduce some scale requirements, e.g., through some relationship between source dimensions and terrain structure.

Some assessment should be made of the importance of the terrain effect before investing the effort necessary to formulate and run complicated terrain models. For example, if the dimensions of an isolated topographic obstacle are small compared with plume dimensions when the plume encounters the obstacle, the feature is likely to have only a minor influence on plume deflection or growth. Wind fields in a large valley or basin may be distinctly governed by topography but well documented empirically by a series of observation sites. Unless the dynamics of the winds are important to the application, the observed winds may suffice for air-quality estimation. However, careful judgment must be exercised before extrapolating one or two surface-based wind observation sets to an entire region. Topographic variations are likely to have more effect under light wind, stable conditions than in the daytime and under vigorous gradient winds.

The transport wind may be affected by the alteration of a free stream flow around or over a terrain obstacle or the local circulations induced by differential heating on a system of sloping surfaces. The diurnal cycle of some circulations may account for anomalous trajectories. For example, a weak nocturnal downslope drainage wind may transport a cloud of contaminant several kilometers during the course of the night and the cloud's path may reverse in 1 or 2 h after the transition to upslope with increasing insolation. Another complication induced by terrain is the production of a nonhomogeneous, nonstationary distribution of turbulent energy. Preferred regions of eddy formation and subsequent decay lead to regions of differing dilution rates.

Finally, in obstacle flow there are stagnation zones where the mean flow may be nearly absent but weak turbulent fluctuations are present. These special light wind cases transcend the hypotheses of the traditional diffusion models and must be treated separately as described in Sec. 13-8.

13-10.2 Deflections of a Free Stream Flow by Terrain Obstacles

The deflection of a fluid flow around an obstacle has long been an interesting but perplexing problem in hydrodynamics studies (see Chap. 7). Atmospheric flows around topographic features include most of the possible complications being three dimensional, nonsteady, nonhomogeneous, density stratified, and multiscaled. A timely discussion of some fundamental aerodynamics involved along with environmental applications is presented by Scorer (1978).

Egan (1975) presents a rather extensive review on the subject of turbulent diffusion in complex terrain. His discussions include airflow phenomena, mathematical and physical models, and diffusion experiments, all pertaining to complex terrain situations. Egan stresses the difference between flow deformation which will deform a plume while not altering the "centerline concentration" and terrain enhanced turbulent diffusivity which does alter centerline concentrations.

We should not expect a magic solution to this difficult problem in view of these complexities, but several lines of effort have proved very useful and will be introduced here. The methods will be categorized as empirical–analytic and dynamic.

The empirical–analytical techniques, including the methods of Wendell (1972), Dickerson (1978), and Sherman (1978), use objective analysis of mesoscale networks of wind observations on regions of complex terrain. Because most of the available meso-networks are ground-based towers, the results from them would necessarily reflect boundary-layer flow fields. Wendell uses a $1/r^2$ weighting factor to interpolate a continuous horizontal wind field from such a network of discrete point sensors. Performing the analysis over a climatological record length of a year, he is able to characterize the main features of the flow in a broad flat valley. Dickerson and Sherman perform an objective analysis on similar network observations under the constraint of mass conservation with a variety of hypotheses on the depth of flow. Both methods have been used in conjunction with computer diffusion codes to estimate transport with dilution. Similar approaches have been reported by Ackerman (1974), Randerson and Cornett (1974), and Cornett and Randerson (1977) who, in addition, considered extended detailed records of upper winds.

Ryan (1977) has developed a model for surface winds in mountainous terrain in which the effects of individual local flows (mountain–valley winds, slope winds, land–sea breezes, etc.) add vectorially to the larger scale wind system to produce the resultant wind. Formulations of the local effects are mostly empirically derived. Topography is used to generate sheltering and diverting modifications to the local wind components through sheltering and diversion terms. Although Ryan's model is specific for a site and, to some extent, meteorological conditions, it is a novel attempt to tie some of the complicating terrain effects together.

The majority of practical sites are not equipped with extensive observation networks. Therefore it is most desirable to estimate the role of terrain on the transport flow domain from the more fundamental inputs of topographic structure, free stream flow, and gross stability. Techniques adapted from classical hydrodynamics are proving useful in this endeavor. Anderson (1971), Hino (1968), and Scholtz and Brouckaert (1978) have used linearized flow theory to obtain the potential flow over an arbitrary terrain. The method is most useful for shallow obstacles with gradual slopes and an essentially two-dimensional flow (e.g., a single layer flow capped by an inversion).

We will briefly summarize the conditions for application of potential flow theory. Irrotational flows can be expressed in terms of a velocity potential, and the velocity is given as the gradient of ϕ, or, $\mathbf{V} = \nabla\phi$. If, in addition, the flow is incompressible with constant density, the equation of continuity is

$$\nabla \cdot \mathbf{V} = 0 \tag{13.45}$$

Substitution for the velocity into Eq. 13.45 gives the Laplace equation

$$\nabla^2 \phi = 0 \tag{13.46}$$

If the real fluid flow satisfied the requirements leading to Eq. 13.45, an extensive mathematical formalism is available for solving Eq. 13.45 under a variety of boundary conditions. Perhaps the most significant requirement for irrotationality is that viscosity be negligible. This condition need not preclude application to viscous fluids so long as the theory is applied to those regions and aspects of the flow which are not dominated by frictional effects. Close to terrain surfaces where frictional retardation is dominant or where boundary-layer separation and other wake effects determine the structure of the wind field, we should not rely heavily on potential flow models. However, many important features of wind field deformation and deviation by terrain obstacles can be estimated from solutions of Eq. 13.46. A solution to the linearized potential flow problem is given by

$$\phi(x,y,z) = \frac{U}{2\pi}\int_{-\infty}^{\infty}\int_{-\infty}^{\infty} h(\xi,y)\frac{y-\xi}{r^3}\,d\xi\,dy \tag{13.47}$$

where $r = [(x-\xi)^2 + (y-y)^2 + z]^{1/2}$, h is the height above the ground, and

$$u(x,y,z) = U + \frac{\partial\phi}{\partial x} \tag{13.48}$$

$$v(x,y,z) = \frac{\partial\phi}{\partial y} \tag{13.49}$$

Many practical sites do not satisfy the requirements of Eq. 13.47 because of large amplitudes and gradients of the underlying surface topography, essential three dimensionality, strong local forcing functions of differential heating, and frictional influences.

Scholtz and Brouckaert (1978) obtain a linear potential flow model by neglecting the Coriolis effect and assuming that the coupling between the surface flow and the geostrophic wind above is through the pressure gradient rather than momentum transfer. First-order influences of topography, land–sea breezes, slope and valley winds, and the large-scale pressure gradient are predicted by this model.

Several applications of nonlinear hydrodynamics to airflow over complex terrain have provided promising results. However, the inclusion of nonlinear terms of numerical models tends to increase the computational burden and force compromises between spatial resolution and computing costs. In order to maintain a balance between detailed representation of the underlying boundary and time required to perform the calculations, innovative efforts have been devoted to deriving simplified, yet valid, models of the governing flow phenomena. The simplifications cover a variety of techniques including:

1. Layer integrals.
2. Scale analysis to eliminate negligible terms.
3. Separation of variables.
4. Formulating a two-dimensional model in either the x-y or x-z plane depending on the particular physical phenomena under investigation.

Houghton and Kasahara (1968) introduced the use of the shallow fluid equations of oceanography to meteorological application, and Tingle and Bjorklund (1973) have adapted the theory to a model for complex terrain transport. The applicability of the shallow fluid equations requires only that the depth of the fluid layer be much less than the wavelength of the generated waves in the flow. This permits finite amplitude solutions that are somewhat more general than those from linearized wave theories. Basically, the shallow fluid approximation allows a variable depth to the effective fluid layer but neglects vertical velocities. Tingle and Bjorklund use a two-layer system where the lower relatively diverse layer may be taken to represent a region beneath an elevated inversion. Ohmstede (1979) presents a system of material layer equations that are more generally applicable to the atmosphere but retain valuable properties of the shallow fluid equations.

An important class of phenomena in stratified fluid flow in general and in the atmosphere in particular is the internal gravity wave. Basically, in stratified fluids, vertically displaced parcels of fluid are subject to a buoyant restoring force thus giving rise to the gravity wave phenomenon. Mountain waves are one classical example of internal gravity waves, but the fundamental physics are operative on all scales of motion including those relevant to boundary-layer transport and diffusion. Raymond (1972) has developed a two-dimensional (x-z plane) model for flow over a ridge that allows heating or cooling on the ridge. Figure 13.18a shows Raymond's calculations for flow over a small ridge under the specified temperature and wind profiles shown in Fig. 13.18b.

Among the models that utilize simplifications to a Navier–Stokes representation of the flow for cost and ease of computation is that of Fosberg, Marlatt, and Krupnak (1976). In addressing the horizontal wind field, they neglect the nonlinear advective terms and express the time rate of change of horizontal divergence and the vertical component of vorticity as functions of thermal and frictional driving forces and topographic boundary conditions.

Hirt, Nichols, and Romero (1975) describe a two-dimensional time-dependent model (SOLA) based on the full Navier–Stokes equations including inertial terms, friction, and external body forces. SOLA can be applied in either the x-z or x-y plane. The numerical formulation is based on the marker-and-cell (MAC) methods. Figure 13.19 is the asymptotic flow field produced by SOLA for the same ridge as described in Fig. 13.18 with a hypothetical "plume" of tracer particles overlaid to show the distortion produced by flow in the lee of the ridge. Notice the effect of friction in producing the separated flow and subsequent lee side eddy not present in Fig. 13.18.

Although two-dimensional models applied in either the x-y plane or x-z plane undoubtedly aid greatly in identifying important phenomena in topographic airflows, there are many fundamentally three-dimensional effects that must be studied. Anthes and Warner (1974) describe a three-

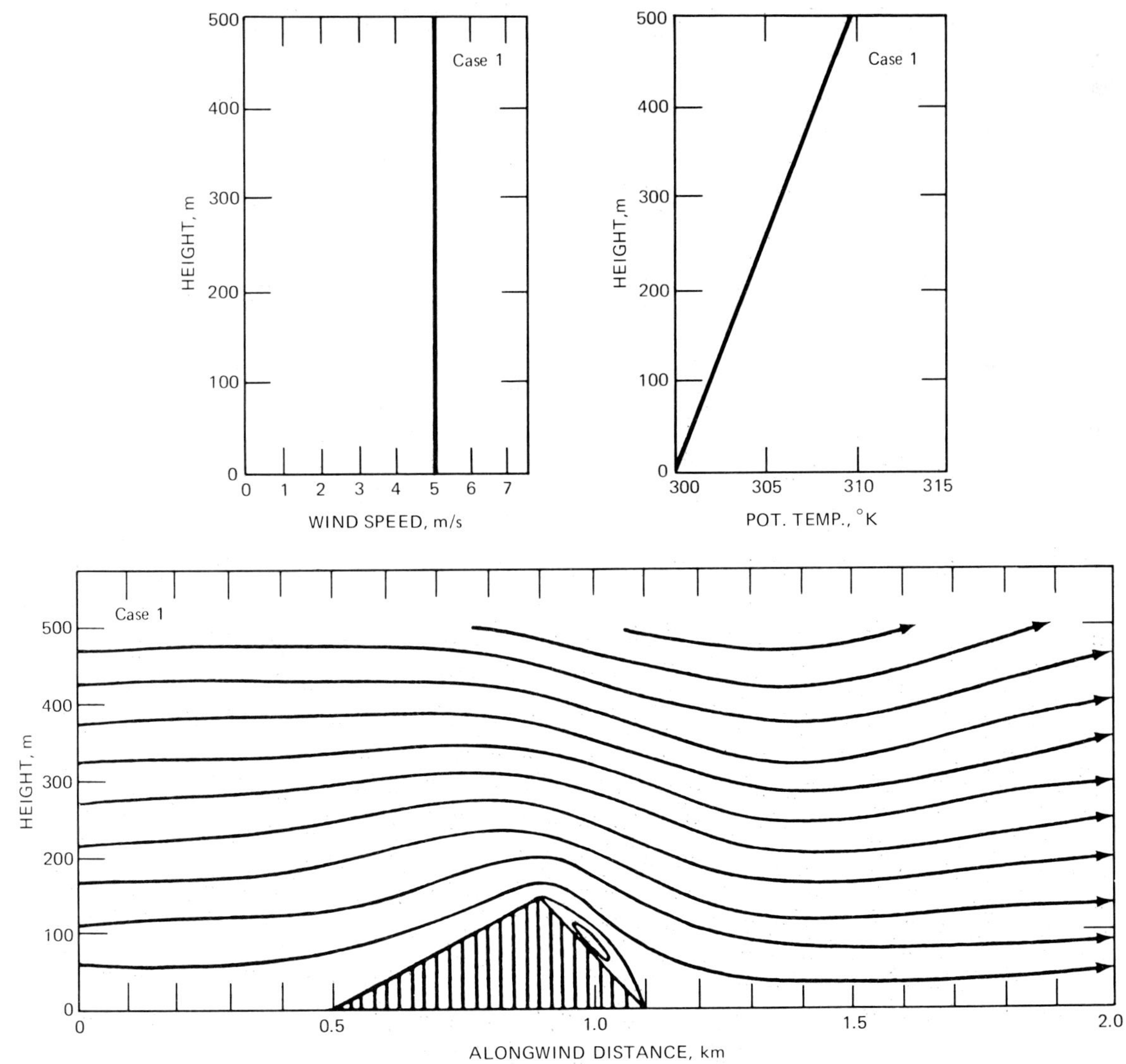

Fig. 13.18 Streamlines over hogback produced by Raymond's model for wind and temperature conditions indicated. [From D. J. Raymond, Calculation of Airflow over an Arbitrary Ridge Including Diabatic Heating and Cooling, *J. Atmos. Sci.*, 29: 837-843 (1972).]

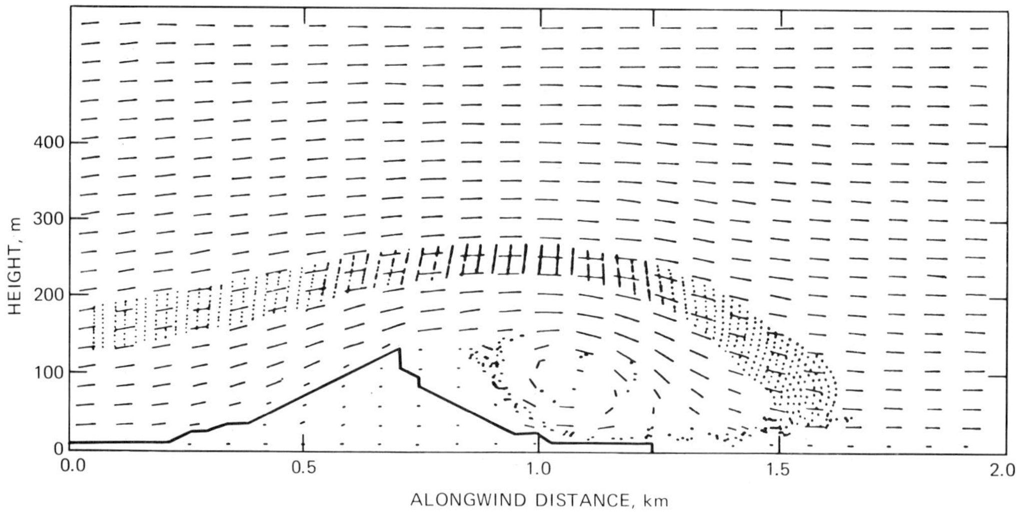

Fig. 13.19 Modified airflow over the same ridge as in Fig. 13.18 with frictional code, SOLA.

dimensional mesoscale grid model in a terrain-following (sigma) coordinate system that was applied to the Tularosa Basin of southern New Mexico at a grid spacing of 5 km. The governing equations include horizontal momentum equations with Coriolis and friction terms, mass continuity, thermodynamics, and the hydrostatic approximation. The number of layers in the vertical direction is variable, but in the application cited six layers were used. Yamada (1978) describes a similar model under development that makes use of second-moment closure equations to describe vertical turbulent fluxes of momentum, heat, and water vapor. Hirt and Cook (1972) and Hotchkiss and Hirt (1972) introduce a fully three-dimensional Navier–Stokes representation for flow of incompressible fluid around arbitrarily complex obstacles. The method faces severe limitations on spatial resolution in the context of present capabilities of computer speed and memory.

13-10.3 Thermally Driven Circulations

A major role of terrain irregularities in local wind and turbulence fields is in the initiation of thermal gradients which result in local circulation systems. We can separate these thermally driven circulations into two major classes, shallow and deep-convective. The shallow flows are characterized by slope winds, canyon flows, and valley-plain winds, whereas deep convection typically has vertical and horizontal dimensions of the same magnitude. Examples of the latter would include orographically induced cumulus convection.

A classical introduction to shallow local wind influences is given by Defant (1951). Land–sea breezes will be discussed in Sec. 13-12. Basic models of slope winds include those of Petkovsek and Hocevar (1971), Fleagle (1950), and Prandtl (1942) which are each slightly different simplifications of the Navier–Stokes equation.

Orville (1964) developed a model of upslope winds based on equations developed by Ogura (1962) for vorticity, rate of change of potential temperature, and continuity for an incompressible fluid. Orville points out that his two-dimensional approach may limit the generality of the model. A numerical model of the local circulation in a valley is presented by Thyer (1966), and Gleeson (1951) discussed the cross-valley winds forced by the differential heating of the slopes.

Although the role of local wind in atmospheric transport and dispersion for complex terrain situations is being recognized as being increasingly important, the basic physics of these problems is still not well understood.

In the time since the authors completed the draft of this chapter, research activity has led to significant development in several of the areas covered in the chapter. Techniques in numerical and stochastic modeling have been extended and refined. Also, the study of the mixed layer, complex terrain, and coastal environments have benefitted from experimental, theoretical, and laboratory simulation research supported by DOE, EPA, EPRI, DOD, and other agencies. In general, concepts in those areas which are explored qualitatively in this chapter are rapidly acquiring a more quantitative basis in the recent literature. The interested reader is urged to pursue the open literature since 1980 in boundary layer and diffusion meteorology and fluid mechanics to keep abreast of progress in the areas described.

13-10.4 Complex Terrain Diffusion Models

The main bulk of diffusion models for complex topography are empirical models derived from field tracer experiments of the type discussed in Chap. 9. There are several examples of computer generated models and a few examples of models based on analyses of wind and turbulence without tracers.

The first and most frequently occurring class of models are the empirical models characterized by Start, Dickson, and Wendell (1975) and Smith (1965). Here, Gaussian plume models are adopted for comparison with observed tracer concentrations. Start et al. have used the Pasquill–

Gifford parameterization for the diffusion coefficients, and Smith et al. adopted the Smith–Hay formula (see Sec. 13-3.2) for quasi-instantaneous sources. Comparison with the Pasquill–Gifford parameters shows, not surprisingly, a more rapid dilution when a plume interacts with terrain obstacles than that estimated by using the equivalent categories for flat terrain. The turbulence energy that affects the dilution has a number of sources additional to those considered in the original formulation for flat terrain. The departures are, however, systematic. This suggests that the simple methods, suitably extended, can continue to be useful in some areas of terrain complexities. In Huntington Canyon, Utah, Start, Dickson, and Wendell (1975) found the ratio of estimated (C_e) to observed (C_o) concentrations to depend on stability category as follows: Class C. . . $C_e/C_o = 1.4$, Class D. . . $C_e/C_o = 5.3$, Class F. . . $C_e/C_o = 18.7$. In unstable flows, thermally produced convection would appear to dominate the diffusive process, and this would not depend heavily on underlying topography. On the other hand, under very stable thermal stratification, gravity flows occur in complex terrain. Start, Dickson, and Wendell (1975) suggest that enhanced turbulence under stable conditions can occur owing to drainage flows from small feeder canyons and also to vortices shed from projections into the main drainage flow. Hovind, Spangler, and Anderson (1974) also used the Pasquill–Gifford parameters at three sites in the southwestern United States and suggest accounting for enhanced turbulence by means of a two-step change in Pasquill–Gifford stability category from F to C–D. The ratio of calculated to observed concentrations becomes a function of travel distance, decreasing from about 8 at 3 km to unity at about 12 km.

Although simple estimation schemes may work well for survey purposes, functional detail probably requires a more detailed parameterization, i.e., some direct characterization of the turbulence field. Smith et al. (1964), in dealing with releases over only a few minutes, have chosen to describe the growth rate of the diffusing cloud in terms of a directly measured turbulence intensity through Eq. 13.30. For longer duration releases, they apply the method of Hay and Pasquill (1959) described in Sec. 13-2.

Hinds (1970) reports a purely empirical model based on a multivariate regression of tracer data collected at Vandenberg Air Force Base. It is, of course, strictly valid only for the location and conditions of the experiments. Hinds stresses that, although his derived equation produced the best fit, the scatter in the data was such that it was difficult to justify more than a simple power law dependence of exposure on travel distance varying from $x^{-2.2}$ for unstable to $x^{-1.5}$ for very stable conditions.

A frequent occurrence noted by Hinds was the reduced exposure in canyons when a plume crossed the canyon. The average ratio of canyon to downwind ridge exposure was about 0.5. Finally, the effect of topography was emphasized in stable conditions and reduced under unstable conditions. This suggests that under stable thermal stratification the influence of drainage wind components and mechanical disturbances produces turbulence energetics quite different from the smooth horizontal plane. On the other hand, in a mixing regime dominated by thermal convection, the underlying terrain may make little difference.

A very interesting approach to a practical complex terrain diffusion model is offered by MacCready, Baboolal, and Lissaman (1974). They derive an approximate expression,

$$\frac{C}{Q} = \frac{1}{2.76\, L(\sigma_v \sigma_w x)^{1/2}} \tag{13.50}$$

for the centerline concentration from statistical diffusion concepts, L being the longitudinal Eulerian scale length of turbulence, derived from the LO-LOCAT (Jones et al., 1970) data set as 180 m. The lateral and vertical turbulent velocity parameters, σ_v and σ_w, are estimated in meters per second from the variance of terrain height and meteorological inputs. Figure 13.20, adapted from MacCready, Baboolal, and Lissaman (1974), demonstrates a relationship between the standard deviations of turbulence velocity fluctuations and terrain height. They also suggest an objective method for estimating s_{tr}, the standard deviation of terrain height variations.

In view of the logical role of terrain roughness-induced turbulence in atmospheric diffusion, extensions of the work described by MacCready et al. are a useful component of complex terrain studies.

Kao, Lee, and Smidy (1975) derive K-Theory parameters from a network of ground-based wind stations in

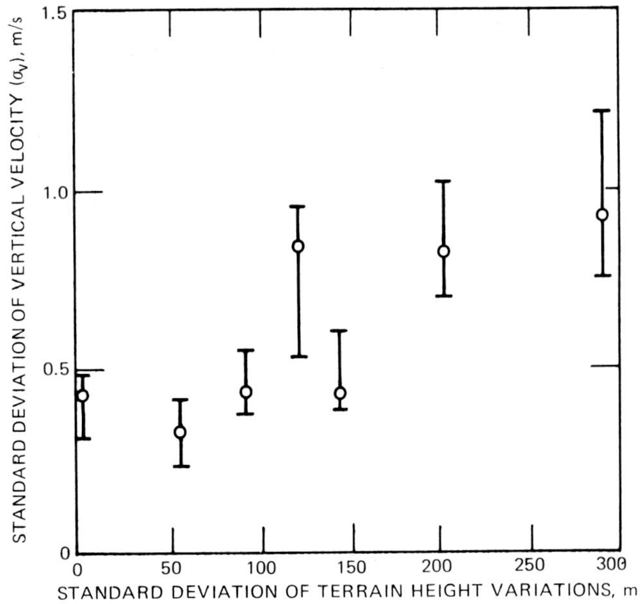

Fig. 13.20 Velocity fluctuations vs. standard deviation of topographic height variations, σ_{TR}. (From P. B. MacCready, Jr., L. B. Baboolal, and P. B. S. Lissaman, Diffusion and Turbulence Aloft over Complex Terrain, in *Symposium on Atmospheric Diffusion and Air Pollution*, Sept. 9–13, 1974, Santa Barbara, Calif., pp. 218-225, American Meteorological Society, Boston, Mass., 1974.)

the Salt Lake Valley. They derive a series of trajectories and calculate the statistics of the spread of these trajectories. They then estimate eddy diffusivities from the asymptotic relationship

$$K_{ii} = \frac{d}{dt}(\overline{x_i^2}) \tag{13.51}$$

where the subscript i represents east-west components and north-south components rather than along and across the mean wind. The diffusivities representative of the surface layer (10 to 100 m) varied diurnally and seasonally with early morning minima of 5 to 10 m^2/s and mid-afternoon to evening maxima of 60 to 100 m^2/s. Because of the data processing techniques, these parameters represent mesoscale motions and do not include contributions from the fluctuations with periods less than 10 min. This study represents a technique not described before in this section, the empirical trajectory analysis. Below, we will discuss the trajectory method of particle tracing in numerical complex terrain models. Whether empirical or computational, the analysis based on the vantage point of a moving particle is a Lagrangian approach, fundamentally different from the classical Eulerian K-Theory. Final results, however, should not depend on the particular method of particle tracking.

The third class of model applied to complex terrain is the numerical computational model. There are two major components to these models: a module that calculates the transport wind field as a function of time and space and a module that accounts for the microscale diffusion of particle clusters in the variable mean wind field. The mean field calculations typically include an undulating lower boundary characterizing the terrain, some driving wind either at a great height or an upstream boundary, and an equation that characterizes the physical interaction. Equations that have been useful in particular applications are discussed in Sec. 13-10.2 and vary from simple statements of mass conservation through complete primitive forms of the Navier–Stokes equations. The dispersion modules include the classical diffusion equation; Gaussian plumes, puffs, or disks; and PIC and Monte Carlo methods. Conceptually, we should be able to mix and match the modules, although the logistics of existing codes often tie the user to the author's choice of techniques. A few examples of modular transport and diffusion methods are indicated below.

Lantz et al. (1972) use an equation derived from mass conservation,

$$\nabla \cdot K\nabla\phi = q \tag{13.52}$$

where ϕ is the velocity potential, K is a "potential coefficient" introduced to account for height variations in a viscous flow, and q is a source term included to account for the topographic driving force. Their diffusion module is the Eulerian heat equation in three dimensions.

Hino (1968) reports a two-layer numerical model formulated in the sense of traditional boundary-layer theory described by Schlichting (1968). The outer, free layer is treated as an inviscid potential flow, whereas the viscous effects are confined to an inner boundary layer. The thinness of the inner layer allows simplifications in the equations. The diffusion in Hino's model is, again, the classical heat equation with a variable three-component diffusivity vector. The diffusion coefficients are either derived from similarity theory (K_z) or specified independently (K_y). Hino applied his model to a particular site with a north-south ridge of elevation from 300 to 680 m and compares concentrations from a 400-m-tall source with a wind-tunnel simulation.

Dickerson (1978), Sherman (1978), and Lange (1978) use a pair of models for the same purpose. The wind field is interpolated from a network of observations through a mass-consistent wind field model. This then forms the transport wind field for a PIC model that fulfills the diffusion role. For more details see Chap. 14.

13-11 DIFFUSION IN VEGETATIVE CANOPIES

The role of large roughness elements such as trees has only been tentatively addressed, but it is clear that a number of important complications to transport and diffusion of trace materials are introduced by vegetation. A large fraction of land area surrounding power plants and other industrial operations is tree covered. Even in semiarid locations, there is a high chance of coherence between near-ground plume travel and forested areas. For instance, river valleys which are preferred paths for nocturnal drainage flows and mountain slopes which may tend to intercept elevated plumes are generally tree covered.

The exchange of airborne material through vegetative canopies is of interest in a number of practical problems. An extensive literature (e.g., Lemon, 1963, 1967; Inoue, 1963; Uchijima and Wright, 1964) deals with quasi-horizontal homogeneity and treats one-dimensional vertical diffusion within and above crop canopies. A primary motivation of this work is the need to know exchange rates of CO_2 as they relate to crop growth. The modeling basis of the one-dimensional work is a mathematical statement of flux divergence

$$\frac{d}{dz}(F_z) = S(z) \tag{13.53}$$

where S(z) is a source function and F_z represents the vertical flux of the quantity of interest often expressed as

$$F_z = K_z \frac{\partial C}{\partial z} \tag{13.54}$$

The vertical diffusivity K_z has frequently been determined by analogy with momentum flux which has prompted a great deal of collateral work with wind speed profiles in canopies (Cionco, 1965, 1972a; Raynor, 1971; Thom, 1972; Plate and Quraishi, 1965; Baynton et al.,

1965; and others). Bergen (1976) presents a rather thorough review of work on momentum balance in forest canopies that includes an extensive bibliography.

The complications introduced by specific source configurations (e.g., point and line sources within or outside the canopy) have forced a slower development of models to handle their diffusion.

A number of field experiments have been performed such as those summarized by Waldron (1971) and those reported by Raynor, Hayes, and Ogden (1974a) and Tourin and Shen (1969). Diffusion parameters $\sigma_y(x)$ and $\sigma_z(x)$ can be derived from these experiments, but they are not easily generalized. The number of fundamental parameters in canopy diffusion is greatly expanded over the classical boundary-layer diffusion problem. For example, several scale parameters are essential to the canopy problem including canopy height, density of vegetative elements, and distance from the forest edge. Because of a significant reduction of wind speed within canopies, there are clear differences in velocity scale depending on whether material is transported within or above the roughness elements. There are hydrodynamic perturbations producing nonparallel flow, particularly near the boundaries of woodlots. Finally, with extensive surface areas available, deposition and biological uptake of particles and gases are dominant and very poorly understood processes.

We will consider modeling approaches for three major regions associated with canopies, each characterized by different physical processes. The tracing of a plume from a finite source will usually involve one or more of these canopy domains. The three zones discussed here are the region within the canopy, the region above the roughness elements, and the leading edge of a canopy.

13-11.1 Diffusion Processes Within the Canopy

The main features of this region are a major reduction in mean wind, extensive turbulence generated by the roughness elements, and enhanced deposition due to a large available surface area. There is also turbulence dissipation in canopies with the net effect of a drastic change in scale of turbulence between the canopy and the air layer above. Several empirical studies have established the basic structure of turbulence within crop and forest canopies. There are good indications that these turbulence features can be generalized and incorporated in the methods of Hay and Pasquill (1959) or Smith and Hay (1961).

Cionco (1972b) summarizes the dimensionless turbulence intensity, i, for a variety of canopies and looks for relationships of i with height, canopy type, and density. The values of i range from 0.4 to 0.6 with a weak upward trend for increasing density of roughness elements. Cionco also finds the turbulence intensity to increase with leaf area, with a maximum in the crown area of the forest. For simple canopies such as crops or trees with quasi-uniform leaf distribution, a uniform function of i with height within the canopy is a good approximation. Above the roughness elements, i decreases similarly to a simple boundary layer. One feature of turbulence within moderate to dense canopies is a relatively large scale meander with light mean winds which shows up in large values of i.

Spectral features of turbulence in forests have been documented by Allen (1968), McBean (1968), and Petit, Trinite, and Valentine (1976). Turbulence spectra for a model canopy in a wind tunnel are reported by Meroney (1968). From these studies the following picture of canopy turbulence emerges. The effects of eddies present in the atmosphere above the canopy are impressed into the forest, producing a maximum spectral density at periods of about 20 to 30 s close to the period of mechanical turbulence outside the canopy. In addition, with relatively strong winds just above the trees, a secondary higher frequency peak of turbulent energy (with wavelength equivalent to tree spacing) is produced by mechanical interaction with the trees. This energy appears to be locally dissipated because the secondary peak is missing at lower levels and only the larger scale eddies are observed near the ground. This produces the meandering wind drift character in the lower trunk area of forests. When the spectra of Petit, Trinite, and Valentine (1976), McBean (1968), and Allen (1968) are similarly normalized and plotted on a composite diagram (Fig. 13.21), they display some useful similarities. The peaks occur at 0.05 Hz (0.05 cycle/s), and the curves fall off at about the same rate on both the high and low frequency sides. Allen's spectra are slightly sharper than the others, but they also tend to broaden with height within the canopy. The method of Hay and Pasquill discussed in Sec. 13.2 can be applied to the composite spectrum of Fig. 13.21 to yield the plume-growth curve for a continuous point source shown in Fig. 13.22. In the range of travel times from 10 s to 20 min, the plume spreading rate ($\sigma_y \propto t^{0.7}$) is slower than in an open boundary layer. This is consistent with the observations of Raynor, Hayes, and Ogden (1974a) who found relatively slow lateral dispersion of tracer aerosol within canopies.

Whereas in the open boundary layer, lateral dispersion is more effective than vertical spread, especially in stable conditions, the roughness elements in the forest tend to enhance the vertical component of turbulence. The intensity and scale of vertical turbulent fluctuations tend to be comparable with those of the horizontal eddies, and plume growth follows the same pattern.

13-11.2 Tracer Plumes Above the Canopy

Several authors (Baynton et al., 1965; Petit, Trinite, and Valentine, 1976) have addressed the problem of pollutant sources of various geometries above a quasi-uniform vegetative canopy. Basically, the problems have been handled in three parts: diffusion in the open atmosphere above the canopy, interchange between the below-canopy and above-canopy zones, and dispersion within the roughness elements.

The open air component is handled by the appropriate methods for the source and meteorology discussed in Sec. 13-3 and the subcanopy dispersion as described above. The interchange has been addressed by Baynton et al. (1965) through the concept of penetration ratio, a ratio of

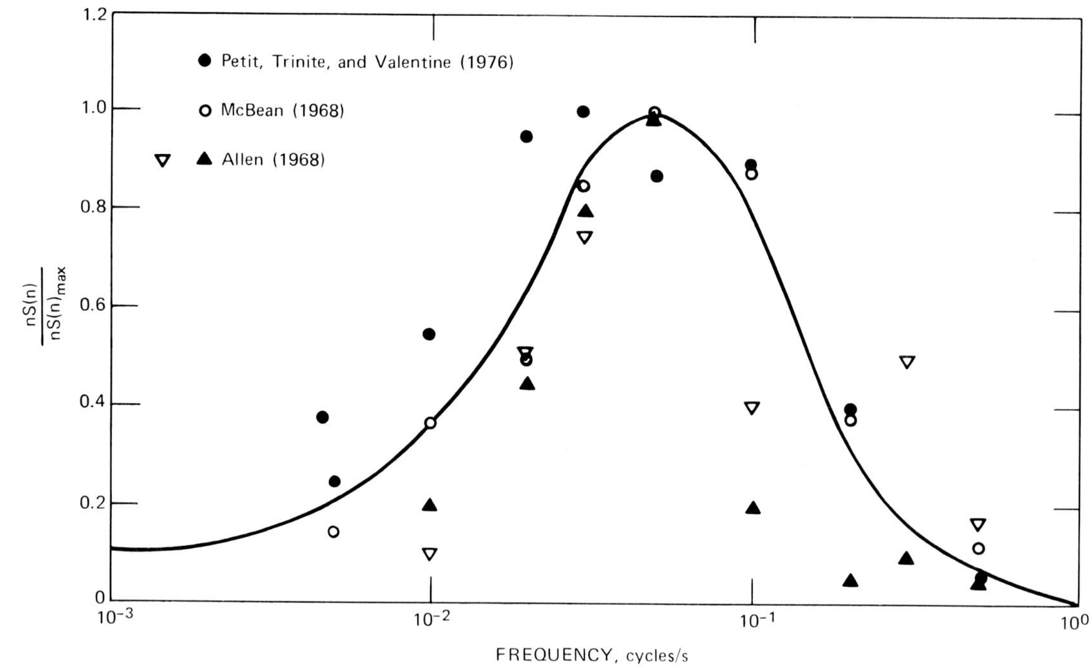

Fig. 13.21 Composite spectrum of horizontal velocity components within a forest canopy.

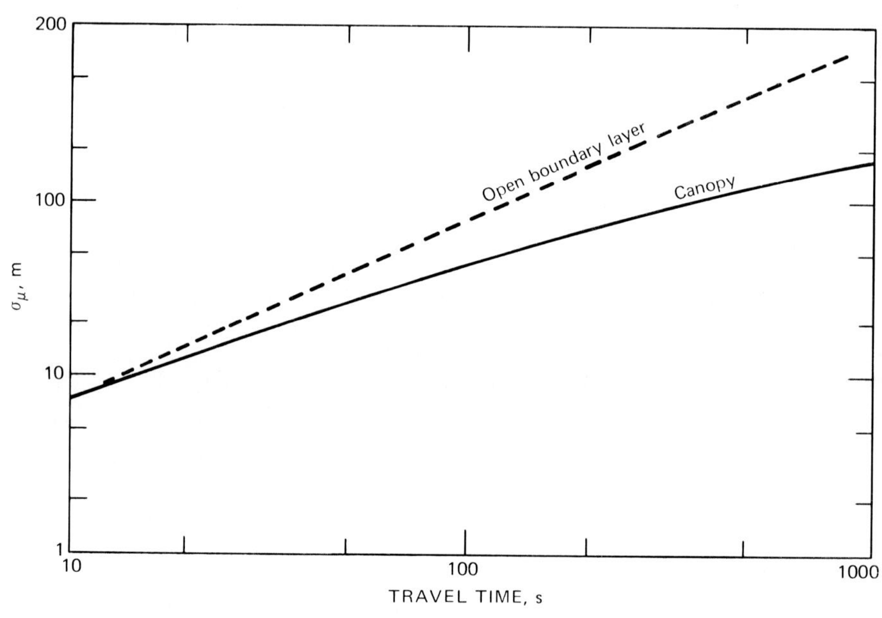

Fig. 13.22 Plume-growth curve within forest canopy derived from the spectrum in Fig. 13.21.

time-integrated concentrations within the canopy and overhead. The values of the ratio were related to lateral turbulence intensity through a regression analysis giving a predictive formula for exchange into one particular jungle canopy. Empirical determinations of a similar parameter for a wide range of canopy types are needed to broaden the basis for estimating exchange rates.

In the case of more slowly varying concentration fields, the flux of pollutant at the top of the vegetation may be evaluated through flux-gradient relationships with observed concentration gradients and various methods for estimating diffusivity.

13-11.3 Tracer Plumes at the Leading Edge of a Canopy

Few applications deal with uniform canopies of infinite extent. In practical cases, we must account for phenomena near the edges of forests. Rapid deceleration of the mean flow by trees tends to produce extensive turbulence and to force airflow up over the forest at a leading edge. These processes both contribute to rapid dispersal of pollutant that enters a wooded area from the windward side. After a transition region that Barr (1971) found to be about 20 to 30 tree heights, the diffusion becomes characteristic of the uniform canopy discussed above.

13-12 TURBULENT DIFFUSION MODELS FOR COASTAL ENVIRONMENTS

Many industrial operations are located for practical reasons on or near the coastlines of lakes or oceans. These coastal areas present unique atmospheric phenomena that must be considered in transport and diffusion calculations for such environments. Many of these phenomena and their relationship to atmospheric dispersion are poorly understood at the present, and only a few appropriate data sets are available with which to validate models. In this section we will briefly discuss some of the more important atmospheric processes associated with coastal environments and point out some of the experimental and modeling efforts in this field. For a more detailed coverage of this subject, the interested reader is referred to the excellent review article by Lyons (1975) on which much of the material in this section is based.

It should be pointed out at the onset that there are many different types of coastal environments, each with its individual problems. Any attempt to generalize these would be foolish. To have any credibility, assessments of dispersion problems in these environments must incorporate as much site-specific information as possible.

13-12.1 Unique Considerations in Coastal Areas

Because the heat capacity of water is greater than that of crustal soils and rocks and because of the surface mixing in oceans and lakes, the temperature of oceans and lakes remains relatively constant compared with land areas. This simple fact results in a number of phenomena that can affect atmospheric transport and dispersion.

Whenever the temperature of the land and that of adjacent water are sufficiently different, a local flow pattern is set up by the thermally induced pressure gradients. At night the land will cool faster than the water and there will tend to be an offshore flow in the air layer near the surface, whereas the opposite pattern may set up during the day. These flows may be theoretically explained by means of the circulation theorem (Holton, 1972), which predicts a less often experienced return flow in the upper layers. Hence, besides the obvious local diurnally changing transport field, we must also be aware of the possibility of pollutants' getting caught up in the return flow.

On a larger scale, continental air masses that pass over large bodies of water undergo significant modification. Phillips (1972) has shown that about half of the near-surface temperature changes in an air mass occurs within the first few minutes of travel over a body of water. When warm air is advected over colder water, the lower layers tend to become thermally stabilized and turbulence intensities drop. This effect is enhanced by the fact that the surface roughness of oceans or lakes is fairly uniform and usually less than that of the surrounding topography. Conversely, when there is cold air advection over warmer water, turbulence intensities increase owing to enhanced thermal convection. If the advected cold air is also dry, then this convection can be increased because of evaporation and the subsequent greater buoyancy of the moist surface air. In these situations pollutant plumes may find a rapid change in the dispersion properties of the transporting air mass as it crosses coastal boundaries.

Convection due to cold air advection over warm water can at times produce clouds and precipitation offshore. Besides the introduction of precipitation scavenging, these regions may themselves alter the stability and thereby present still another stability regime.

Although oceans and lakes are temporally more stable than land areas, there can exist large spatial variations in the surface temperature of the water, especially near shore. The determination of surface water temperature in most studies has not been done in detail; however, this temperature variation should be kept in mind in the interpretation of results so that one does not always use too simplified a model for the water body.

The surface roughness of oceans and lakes in general is more uniform and of less magnitude than that over land. Therefore, in addition to the conditional thermal modification of turbulence, the mechanical component of turbulence is generally less than that found at an inland site. Of course, stormy conditions can negate this simple rule, but under those conditions the role of surface roughness is reduced with respect to other generators of turbulence.

The preceding discussion points out that, in crossing the boundary between land and water, discontinuities exist in both the mechanical and the thermal components of turbulence. Most present transport and diffusion models are not well suited to cope with this situation.

Although a variety of coastal processes can affect pollutant plumes, of particular interest are plume trapping and fumigation because they can result in higher concentration at ground level. Consider the situation wherein cold stable air is blowing from the water onto a warmer land area. Under these conditions, a thermal internal boundary layer (TIBL) (Lyons, 1975) will develop capped by a stable layer. Figure 13.23 is a schematic of this situation. Plumes with effective release heights within the TIBL will be trapped as in any other elevated inversion situation. Plumes that are effectively injected above the mixing layer will develop according to the stability of the parent air mass. If at some distance downwind the elevated plume intersects the top of the TIBL, then rapid fumigation will occur and concentrations will increase at the surface. The converse situation can be imagined for warm air advection over colder waters wherein a stable boundary layer would be capped by an unstable air mass. It should be obvious that dispersion models in coastal environments should take into consideration the TIBL.

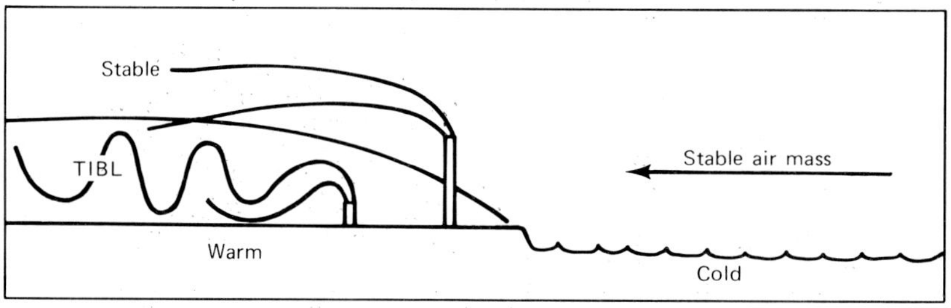

Fig. 13.23 Schematic diagram of plume interaction with the thermal internal boundary layer (TIBL).

13-12.2 Models Used in the Coastal Environment

The complexity of the coastal meteorological environment is no different from that of complex terrain or urban areas and, hence, the models are in a rather primitive state.

Some efforts have been directed at the problem of fumigation from a point source. A method for estimating maximum surface concentration in a fumigation zone has been suggested by Van der Hoven (1967). In this model the intersection of the plume centerline and the top of the TIBL is estimated from an empirically derived nomogram. Input requirements are the mean wind speed in the boundary layer and the potential temperature difference in the overlying stable air mass. Lyons and Cole (1973) have used the steady-state Gaussian plume equation to address the fumigation problem. They divided the region of dispersion near the land–water boundary into three zones: (1) stable zone above TIBL where standard Gaussian calculations are made; (2) the region between the estimated intersection of the top and bottom of the plume and the top of the TIBL (in this zone the concentration below the TIBL is considered to be constant in the vertical, Gaussian in the horizontal, and a modified σ_y in the fumigation zone); (3) the region within the TIBL in which the plume is trapped in a mixed layer of variable height and the distance x at which σ_y is evaluated is modified to account for the fact that the turbulence within the TIBL has not affected the plume for its entire travel distance. This point-source model has been the building block for a regional multisource model (GLUMP) (Lyons, 1975). GLUMP incorporates plume rise and multiple vertical atmosphere layers with independent meteorological and dispersion parameters. Other modeling efforts the reader should be aware of are those of Bierly and Hewson (1963) and of Mahoney and Egan (1971).

Most of these models work well under certain conditions and, if they can be loaded with the proper inputs, they may be useful in some predictions. However, the complexity of coastal environments strongly suggests that without detailed knowledge of site-specific phenomena all models are very limited.

13-12.3 Experimental Data

One of the first over-the-water diffusion experiments was one using oil fog by Woodcock and Wyman (1947). Raynor et al. (1974b) have conducted a series of field experiments off the shore of Long Island. Their results show significant differences between dispersion properties over land and water with the most common fact being that turbulence intensities over water are usually lower than those over land. This data set is probably the most comprehensive one available that is directed at coastal problems. Lyons (1975) discusses tests of the GLUMP model on field data.

13-12.4 Summary

Coastal regions present unique meteorological problems that result in plume behavior different from that at inland sites. Models are limited, and only a few good data sets are available by which to evaluate the models. This state of affairs is not surprising considering the meteorological complexity of the coastal environment. Much more must be learned about the associated phenomena, and appropriate data sets are badly needed. Extreme caution should be exercised in applying any existing modeling technique without an understanding of the meteorological problems of a given site.

LIST OF SYMBOLS

Symbols in Chap. 13 which appear infrequently and the meaning of which are clear from the context have been omitted from this list.

C	Concentration
c_p	Specific heat at constant pressure
D	Distance from mean plume axis to instantaneous centroid
E	Unnormalized turbulent spectral density function
f*	Flushing frequency
g	Acceleration due to gravity
H	Vertical heat flux
h	Height of source above ground, height above ground, or height of box model
i	Intensity of turbulence
K	Eddy diffusivity
k	Von Karman's constant
L	Monin–Obukhov length or longitudinal Eulerian scale length of turbulence

ℓ_0	Initial separation of particle pairs
M = C	Mean concentration
n	Frequency
Q	Source strength
R	Correlation coefficient
S	Source or sink term
S_L, S_E	Lagrangian and Eulerian turbulent spectral density functions, respectively
s	Variance of concentration fluctuations
T	Absolute temperature or averaging period
t	Time
u,v,w	Velocity components along the x-, y-, and z-axes
u', v', w'	Fluctuating components of velocity along the x-, y-, and z-axes
u_*	Friction velocity
V	Wind velocity vector
x,y,z	Rectangular coordinates, x usually being along the mean wind and z being vertical
T	Relative displacement of particles
β	Ratio of Lagrangian to Eulerian time scales
ξ	Time lag
ρ	Density
σ	Standard deviation
τ	Time scale

REFERENCES

Ackerman, B., 1974, METROMEX: Wind Field over St. Louis in Undisturbed Weather, *Bull. Am. Meteorol. Soc.*, 55: 93-95.

Allen, L. H., Jr., 1968, Turbulence and Wind Speed Spectra Within a Japanese Larch Plantation, *J. Appl. Meteorol.*, 7: 73-78.

Alsmiller, F. S., R. G. Alsmiller, H. W. Bertini, and C. L. Begovitch, 1978, *Monte Carlo Simulation of Turbulent Atmospheric Transport and Comparisons with Experimental Data*, USAEC Report ORNL/TM-6291, Oak Ridge National Laboratory, NTIS.

American Meteorological Society, 1978, Accuracy of Dispersion Models, *Bull. Am. Meteorol. Soc.*, 59: 1025-1026.

Anderson, G. E., 1971, Mesoscale Influences on Wind Fields, *J. Appl. Meteorol.*, 10: 377-386.

Angell, J. K., 1964, Measurements of Lagrangian and Eulerian Properties of Turbulence at a Height of 2500 Feet, *Q. J. R. Meteorol. Soc.*, 90: 57-71.

——, D. H. Pack, W. H. Hoecker, and N. Delver, 1971, Lagrangian–Eulerian Time-Scale Ratios Estimated from Constant Volume Balloon Flights Past a Tall Tower, *Q. J. R. Meteorol. Soc.*, 97: 87-92.

Anthes, R. A., and T. T. Warner, 1974, *Prediction of Mesoscale Flows over Complex Terrain*, Report ECOM-5532, U. S. Army Atmospheric Science Laboratory, White Sands Missile Range, N. Mex.

Aris, R., 1956, On the Dispersion of a Solute in a Fluid Flowing Through a Tube, *Proc. R. Soc. (London), Ser. A.*, 235: 67-77.

Barr, S., 1971, A Modeling Study of Several Aspects of Canopy Flow, *Mon. Weather Rev.*, 2: 122-126.

——, R. E. Luna, W. E. Clements, and H. W. Church (Eds.), 1977, *Workshop on Research Needs for Atmospheric Transport and Diffusion in Complex Terrain*, Sept. 28–30, 1976, Albuquerque, N. Mex., DOE Report CONF-7609160, NTIS.

Barry, P. J., 1971, A Note on Peak-to-Mean Concentration Ratios, *Boundary-Layer Meteorol.*, 2: 122-126.

Batchelor, G. K., 1950, The Application of the Similarity Theory of Turbulence to Atmospheric Diffusion, *Q. J. R. Meteorol. Soc.*, 76: 133-145.

——, 1952, Diffusion in the Field of Homogeneous Turbulence, II, The Relative Motion of Particles, *Proc. Cambridge Philos. Soc.*, 48: 345-362.

Baynton, H. W., W. G. Biggs, H. L. Hamilton, P. E. Sherr, and J. B. Worth, 1965, Wind Structure in and Above a Tropical Forest, *J. Appl. Meteorol.*, 4: 670-675.

Bergen, J. D., 1976, Air Flow in Forest Canopies, A Review of Recent Research in Modeling the Momentum Balance, in *Fourth National Conference on Fire and Forest Meteorology*, Nov. 16–18, 1976, St. Louis, Mo.

Bierley, E. W., and E. W. Hewson, 1963, Atmospheric Diffusion Studies near a Lake Shore, *J. Appl. Meteorol.*, 2: 390-396.

Bornstein, R. D., 1968, Observations of the Urban Heat Island Effect in New York City, *J. Appl. Meteorol.*, 7: 575-582.

——, 1972, Two Dimensional Simulations of Night Time Flow of a Stable Planetary Boundary Layer over a Rough Warm City, in *Proceedings of Conference on Urban Environment and Second Conference on Biometeorology*, Oct. 31–Nov. 2, 1972, Philadelphia, Pa., pp. 89-94, American Meteorological Society, Boston, Mass.

Brier, G. W., 1950, The Statistical Theory of Turbulence and the Problem of Diffusion in the Atmosphere, *J. Meteorol.*, 7: 283-290.

Briggs, G. A., 1970, *Diffusion Calculation for Small Emissions*, Contribution Number 36, Atmospheric Turbulence and Diffusion Laboratory, Oak Ridge, Tenn.

——, 1974, *Diffusion Estimation for Small Emissions*, Report ATDL-106, NOAA, Atmospheric Turbulence and Diffusion Laboratory, Oak Ridge, Tenn.

Calder, K., 1965, On the Equation of Atmospheric Diffusion, *Q. J. R. Meteorol. Soc.*, 91: 514-517.

Cermak, J. E., 1963, Lagrangian Similarity Hypothesis Applied to Diffusion in Turbulent Shear Flow, *J. Fluid Mech.*, 15(Part I): 49-64.

Cionco, R. M., 1965, A Mathematical Model for Air Flow in a Vegetative Canopy, *J. Appl. Meteorol.*, 4: 517-522.

——, 1972a, A Wind Profile Index for Canopy Flow, *Boundary-Layer Meteorol.*, 2: 255-263.

——, 1972b, Intensity of Turbulence Within Canopies with Sample and Complex Roughness Elements, *Boundary-Layer Meteorol.*, 2: 453-465.

Clarke, J. F., 1969, Nocturnal Urban Boundary Layer over Cincinnati, Ohio, *Mon. Weather Rev.*, 97: 582-589.

Clarke, R. H., A. J. Dyer, R. R. Brook, D. G. Reid, and A. J. Troup, 1971, *The Wangara Experiment: Boundary Layer Data*, Tech. Paper 19, Division of Meteorological Physics, CSIRO, Aspendale, Victoria, Australia.

Clements, W. E., and S. Barr, 1976, Atmospheric Transport and Dispersal at a Site Dominated by Complex Terrain, in *Third Symposium on Atmospheric Turbulence, Diffusion, and Air Quality*, Raleigh, N. C., October 19–22, 1976, CONF-761003, pp. 430-435, American Meteorological Society, Boston, Mass.

Cornett, J. S., and D. Randerson, 1977, Verification of Some Numerical Models for Operationally Predicting Mesoscale Winds Aloft, *J. Appl. Meteorol.*, 16: 764-773.

Corssin, S., 1962, Theories of Turbulent Diffusion, in *Mechanique de la Turbulence*, Centre National de la Recherche Scientifique, pp. 27-52, Paris, France.

——, 1963, Estimation of the Relation Between Eulerian and Lagrangian Scales in Large Reynold's Number Turbulence, *J. Atmos. Sci.*, 20: 115-119.

Cramer, H. E., 1957, A Practical Method for Estimating the Dispersal of Atmospheric Contaminants, in *Proceedings of the First National Conference on Applied Meteorology*, October 1957, Hartford, Conn., pp. C33-C55, American Meteorological Society, Boston, Mass.

——, 1967, Turbulent Transfer Processes for Quasi-Homogeneous Flows Within the Atmospheric Surface Layer, *Phys. Fluids, Suppl.*, 10: 5240-5246.

——, G. M. De Santo, K. R. Dumbald, P. Morgenstern, and R. N. Swanson, 1964, *Meteorological Prediction Techniques and Data System,* Report GCA-64-3-G, Geophysical Corporation of America, Bedford, Mass.

Csanady, G. T., 1967, Concentration Fluctuations in Turbulent Diffusion, *J. Atmos. Sci.,* 24: 21-28.

——, 1973, *Turbulent Diffusion in the Environment,* D. Reidel Publishing Co., Dordrecht-Holland, Boston, Mass.

Deardorff, J. W., 1972, Numerical Investigation of Neutral and Unstable Planetary Boundary Layers, *J. Atmos. Sci.,* 29: 91-115.

——, 1973, Three-Dimensional Numerical Modeling of the Planetary Boundary Layer, in *Workshop on Micrometeorology,* D. A. Haugen (Ed.), pp. 271-311, American Meteorological Society, Boston, Mass.

——, 1974, Three-Dimensional Numerical Study of the Height and Mean Structure of a Heated Planetary Boundary Layer, *Boundary-Layer Meteorol.,* 7: 81-106.

——, and G. E. Willis, 1974, Physical Modeling of Diffusion in the Mixed Layer, in *Symposium on Atmospheric Diffusion and Air Pollution,* Sept. 9–13, 1974, Santa Barbara, Calif., pp. 387-391, American Meteorological Society, Boston, Mass.

Defant, F., 1951, Local Winds, in *Compendium of Meteorology,* pp. 655-672, American Meteorological Society, Boston, Mass.

Delage, Y., and P. A. Taylor, 1970, Numerical Studies of Heat Island Circulations, *Boundary-Layer Meteorol.,* 1: 201-226.

Dickerson, M. H., 1978, MASCON—A Mass Consistent Atmospheric Flux Model with Complex Terrain, *J. Appl. Meteorol.,* 17: 241-253.

Donaldson, C. duP., 1972, Construction of a Dynamic Model of the Production of Atmospheric Turbulence and the Dispersion of Pollutants, in *Workshop in Micrometeorology,* D. A. Haugen (Ed.), pp. 313-391, American Meteorological Society, Boston, Mass.

Drake, R. L., 1979, Mathematical Models for Atmospheric Pollutants, in *Governing Equations for Atmospheric Trace Constituents and Solution Techniques,* DOE Report EPRI-EA-1131 (App. C), Battelle Pacific Northwest Laboratories, NTIS.

Draxler, R. R., 1976, Determination of Atmospheric Diffusion Parameters, *Atmos. Environ.,* 10: 99-105.

Egan, B. A., 1975, Turbulent Diffusion in Complex Terrain, in *Lectures on Air Pollution and Environmental Input Analyses,* Sept. 29–Oct. 3, 1975, pp. 112-135, American Meteorological Society, Boston, Mass.

Findlay, B. F., and M. S. Hirt, 1969, An Urban-Induced Meso-Circulation, *Atmos. Environ.,* 3: 537-542.

Fleagle, R. G., 1950, A Theory of Air Drainage, *J. Meteorol.,* 7: 227-232.

Fosberg, M. A., W. E. Marlatt, and L. Krupnak, 1976, *Estimating Air-Flow Patterns over Complex Terrain,* Research Paper RM-162, U. S. Department of Agriculture, Forest Service, Ft. Collins, Colo.

Freeman, B. E., 1976, Tensor Diffusivity of a Trace Constituent in a Stratified Boundary Layer, in *Third Symposium on Atmospheric Turbulence, Diffusion, and Air Quality,* Raleigh, N. C., October 19–22, 1976, CONF-761003, pp. 213-216, American Meteorological Society, Boston, Mass.

Gartrell, F. E., 1966, *Control of Air Pollution from Large Thermal Power Stations,* Revue Mensuelle 1966 de la Soc. Belge des Ingenieurs et das Industriels Bruxelles, pp. 1-12.

Gee, J. H., and D. R. Davies, 1963, A Note on Horizontal Dispersion from an Instantaneous Ground Source, *Q. J. R. Meteorol. Soc.,* 89: 542.

Gifford, F. A., 1955, A Simultaneous Lagrangian–Eulerian Turbulence Experiment, *Mon. Weather Rev.,* 83: 293-301.

——, 1959, Statistical Properties of a Fluctuating Plume Dispersion Model, in *Atmospheric Diffusion and Air Pollution,* F. N. Frenkiel and P. A. Sheppard (Eds.), pp. 117-137, Academic Press, New York.

——, 1960, Peak-to-Average Concentration Ratios According to a Fluctuating Plume Dispersion Model, *Int. J. Air Water Pollut.,* 3: 253-260.

——, 1962, Diffusion in the Diabatic Surface Layer, *J. Geophys. Res.,* 67: 3207-3212.

——, 1967, Variation of the Lagrangian–Eulerian Time Number Scale Relationship with Stability, in *Proceedings of Joint U. S. Canadian Atomic Energy Commission's Conference on Micrometeorology,* pp. 485-499, Chalk River, Canada.

——, 1968, An Outline of Theories of Diffusion in the Lower Layers of the Atmosphere, in *Meteorology and Atomic Energy, 1978,* D. H. Slade (Ed.), TID-24190, pp. 65-116, U. S. Atomic Energy Commission, NTIS.

——, 1972, The Form of the Frequency Distribution of Air Pollution Concentrations, in *Proceedings of the Symposium on Statistical Aspects of Air Quality Data,* Nov. 9–10, 1972, Chapel Hill, N. C., pp. 3.1-3.7.

——, 1976, Turbulent Diffusion-Typing Schemes: A Review, *Nucl. Safety,* 17(1): 68-86 (January–February 1976).

——, 1977, Tropospheric Relative Diffusion Observations, *J. Appl. Meteorol.,* 16: 311-313.

——, and S. R. Hanna, 1973, Modelling Urban Air Pollution, *Atmos. Environ.,* 7: 131-136.

Gleeson, T. A., 1951, On the Theory of Cross-Valley Winds Arising from Differential Heating of the Slopes, *J. Meteorol.,* 8: 398-405.

Golder, D., 1972, Relations Among Stability Parameters in the Surface Layer, *Boundary-Layer Meteorol.,* 3: 57-58.

Hall, C. D., 1975, The Simulation of Particle Motion in the Atmosphere by a Numerical Random-Walk Model, *Q. J. R. Meteorol. Soc.,* 101: 235-244.

Hameed, S., 1974, Modelling Urban Air Pollution, *Atmos. Environ.,* 8: 555-561.

Hanna, S. R., 1975, Urban Diffusion Problems, in *Lectures on Air Pollution and Environmental Impact Analyses,* D. A. Haugen (Ed.), American Meteorological Society, Boston, Mass.

——, 1976, Relative Diffusion of Tetroon Pairs During Convective Conditions, *J. Appl. Meteorol.,* 15: 588-593.

——, G. A. Briggs, J. Deardorff, B. A. Egan, F. A. Gifford, and F. Pasquill, 1977, AMS Workshop on Stability Classification Schemes and Sigma Curves—Summary of Recommendations, *Bull. Am. Meteorol. Soc.,* 58(12): 1305-1309.

——, C. J. Nappo, R. P. Hosker, and G. A. Briggs, 1974, *Description of the Eastern Tennessee Trajectory Experiment,* ATDL File No. 74-103, NOAA, Atmospheric Turbulence and Diffusion Laboratory, Oak Ridge, Tenn.

Hay, J. S., and F. Pasquill, 1959, Diffusion from a Continuous Source in Relation to the Spectrum and Scale of Turbulence, in *Atmospheric Diffusion and Air Pollution,* F. N. Frenkiel and P. A. Sheppard (Eds.), pp. 331-365, Academic Press, New York.

Hilst, G. R., 1957, The Dispersion of Stack Gases in Stable Atmospheres, *J. Air Pollut. Control Assoc.,* 7: 205-210.

Hinds, W. T., 1969, Peak-to-Mean Concentration Ratios from Ground Level Sources in Building Wakes, *Atmos. Environ.,* 3: 145-156.

——, 1970, Diffusion over Coastal Mountains of Southern California, *Atmos. Environ.,* 4: 107-124.

Hino, M., 1968, Computer Experiment on Smoke Diffusion over a Complicated Topography, *Atmos. Environ.,* 2: 541-558.

Hirt, C. W., and J. L. Cook, 1972, Calculating Three-Dimensional Flows Around Structures and over Rough Terrain, *J. Comput. Phys.,* 10: 324-340.

——, B. D. Nichols, and N. C. Romero, 1975, *SOLA—A Numerical Solution Algorithm for Transient Fluid Flows,* ERDA Report LA-5852, Los Alamos Scientific Laboratory, NTIS.

Högström, U., 1964, An Experimental Study on Atmospheric Diffusion, *Tellus,* 16: 205-251.

Holland, J. Z., 1953, *A Meteorological Survey of the Oak Ridge Area,* USAEC Report ORO-99, U. S. Atomic Energy Commission.

Holton, J. R., 1972, *An Introduction to Dynamic Meteorology,* Academic Press, New York.

Holzworth, G. C., 1967, Mixing Depths, Wind Speeds and Air Pollution Potential for Selected Locations in the United States, *J. Appl. Meteorol.,* 6: 1039-1044.

Hotchkiss, R. S., and C. W. Hirt, 1972, Particulate Transport in Highly Distorted Three-Dimensional Flow Fields, in *Proceedings of Summer Simulation Conference,* June 14–16, 1972, San Diego, Calif., American Institute of Aeronautics and Astronautics, New York.

Houghton, D. P., and A. Kasahara, 1968, Nonlinear Shallow Fluid Flow over an Isolated Ridge, *Commun. Pure Appl. Math.,* 21: 1-23.

Hovind, E. L., T. C. Spangler, and A. J. Anderson, 1974, The Influence of Rough Mountainous Terrain upon Plume Dispersion from an Elevated Source, in *Symposium on Atmospheric Diffusion and Air Pollution,* Sept. 9–13, 1974, Santa Barbara, Calif., pp. 214-217, American Meteorological Society, Boston, Mass.

Inoue, E., 1963, On the Turbulent Structure of Airflow Within Crop Canopies, *J. Meteorol. Soc. Jpn.,* 41: 317-325.

Johnson, W. B., R. C. Sklarew, and D. B. Turner, 1976, Urban Air Quality Modeling, in *Air Pollution, Vol. I, Air Pollutants, Their Transformation and Transport,* 3rd ed., A. C. Stern (Ed.), pp. 504-562, Academic Press, New York.

Jones, J. W., R. H. Mielke, and G. W. Jones, 1970, *Low Altitude Atmospheric Turbulence, LO-LOCAT Phase Three,* Report AFFDL-TR-70-10, Boeing Company report to Air Force Flight Dynamics Laboratory, Air Force Systems Command, Wright-Patterson Air Force Base, Ohio.

Kao, S. K., 1959, Turbulent Transfer in the Boundary Layer of a Stratified Fluid, *J. Meteorol.,* 16: 497-503.

——, H. N. Lee, and K. I. Smidy, 1975, An Analysis of the Topographical Effect on Turbulence and Diffusion in the Atmosphere's Boundary Layer, *Boundary-Layer Meteorol.,* 8: 323-334.

Lamb, R. G., and J. H. Seinfeld, 1974, Comments on "A Simple Dispersion Model for the Analysis of Chemically Reactive Pollutants," *Atmos. Environ.,* 8: 527-529.

Lange, R., 1978, ADPIC—A Three-Dimensional Particle-in-Cell Model for the Dispersal of Atmospheric Pollutants and Its Comparison to Regional Tracer Studies, *J. Appl. Meteorol.,* 17: 320-329.

Lantz, R. B., K. H. Coats, and C. V. Kloepfer, 1972, A Three Dimensional Numerical Model for Calculating the Spread and Dilution of Air Pollutants, in *Proceedings of Symposium on Air Pollution, Turbulence, and Diffusion,* H. W. Church and R. E. Luna (Eds.), pp. 43-55, Las Cruces, New Mex.

Larsen, R. I., 1969, A New Mathematical Model of Air Pollutant Concentration Averaging Time and Frequency, *J. Air Pollut. Control Assoc.,* 19: 24-30.

Leahey, D. M., 1969, *A Model for Predicting the Variation of Pollution Within the Urban Heat Island,* Contribution No. 78 of the Geophysical Sciences Laboratory, Department of Meteorology and Oceanography, New York University, New York.

——, and J. Halitsky, 1973, Low Wind Turbulence Statistics and Related Diffusion Estimates from a Site Located in the Hudson River Valley, *Atmos. Environ.,* 7: 49-61.

Lemon, E. R., 1963, *The Energy Budget at the Earth's Surface, Part II,* Report No. 72, U. S. Army Electronic Proving Ground, Fort Huschuca, Ariz.

——, 1967, Aerodynamic Studies of CO_2 Exchange Between the Atmosphere and the Plant, in *Harvesting the Sun: Photosynthesis in Plant Life,* Academic Press, New York.

Lettau, H. H., 1970, Physical and Meteorological Basis for Mathematical Models of Urban Diffusion Processes, in *Proceedings of Symposium on Multiple Source Urban Diffusion Models,* A. C. Stern (Ed.), U. S. Environmental Protection Agency, Research Triangle Park, N. C., pp. 2.1-2.26.

Long, P. E., and D. W. Pepper, 1976, A Comparison of Six Numerical Schemes for Calculating the Advection of Atmospheric Pollution, in *Third Symposium on Atmospheric Turbulence, Diffusion, and Air Quality,* Raleigh, N. C., October 19–22, 1976, CONF-761003, pp. 181-186, American Meteorological Society, Boston, Mass.

Lumley, J. L., and H. A. Panofsky, 1964, *The Structure of Atmospheric Turbulence,* pp. 126-140, Interscience Publishers, New York.

Lyons, W. A., 1975, Turbulent Diffusion and Pollutant Transport in Shoreline Environments, in *Lectures on Air Pollution and Environmental Input Analyses,* Sept. 29–Oct. 3, 1975, pp. 136-208, American Meteorological Society, Boston, Mass.

——, and H. S. Cole, 1973, Fumigation and Plume Trapping on the Shore of Lake Michigan During Stable Onshore Flow, *J. Appl. Meteorol.,* 12: 484-510.

McBean, G. A., 1968, An Investigation of Turbulence Within the Forest, *J. Appl. Meteorol.,* 7: 410-416.

MacCraken, M. C., D. J. Wuebbles, J. J. Walton, W. H. Duewer, and K. E. Grant, 1978, The Livermore Regional Air Quality Model: I Concept and Development, *J. Appl. Meteorol.,* 17: 254-272.

MacCready, P. B., Jr., L. B. Baboolal, and P. B. S. Lissaman, 1974, Diffusion and Turbulence Aloft over Complex Terrain, in *Symposium on Atmospheric Diffusion and Air Pollution,* Sept. 9–13, 1974, Santa Barbara, Calif., pp. 218-225, American Meteorological Society, Boston, Mass.

Mahoney, J. R., and B. A. Egan, 1971, A Mesoscale Numerical Model of Atmospheric Transport Phenomena on Urban Areas, in *Proceedings of the Second International Clean Air Congress,* Dec. 6–11, 1970, Washington, D. C., International Union of Air Pollution Prevention Association, Washington.

Mellor, G., and T. Yamada, 1974, A Hierarchy of Turbulence Closure Models for Planetary Boundary Layers, *J. Atmos. Sci.,* 31: 1791-1806.

Meroney, R. N., 1968, Characteristics of Wind and Turbulence in and Above Model Forests, *J. Appl. Meteorol.,* 7: 780-788.

Monin, A. S., 1955, Diffusion with a Finite Speed, *Izv. Akad. Nauk SSSR, Ser. Geofiz.,* 3: 234-248.

——, 1959, Smoke Propagation in the Surface Layer of the Atmosphere, in *Atmospheric Diffusion and Air Pollution,* F. N. Frenkiel and P. A. Sheppard (Eds.), *Adv. Geophys.,* 6: 331-343, Academic Press, New York.

——, and A. M. Obukhov, 1954, Basic Regularity in Turbulent Mixing in the Surface Layer of the Atmosphere, *Tr. Inst. Geofiz., Akad. Nauk SSSR,* 24: 163-187.

——, and A. M. Yaglom, 1971, *Statistical Fluid Mechanics,* J. L. Lumley (Ed.), The MIT Press, Cambridge, Mass.

Moses, H., 1969, Mathematical Urban Air Pollution Models, Report ANL/ES-RPY-001, Argonne National Laboratory.

Myrup, L. O., 1969, A Numerical Model of the Urban Heat Island, *J. Appl. Meteorol.,* 8: 908-918.

Ogura, Y., 1953, The Relation Between the Space and Time Correlation Functions in a Turbulent Flow, *J. Meteorol. Soc. Jpn.,* 31: 355-369.

——, 1962, Convection of Isolated Masses of Buoyant Fluid: A Numerical Calculation, *J. Atmos. Sci.,* 19: 492-502.

Ohmstede, W. D., 1979, *The Dynamics of Material Layers,* Report ASL-TR-0036, Atmospheric Sciences Laboratory, White Sands Missile Range, N. Mex., 38 pp.

Oke, T. R., 1973, City Size and the Urban Heat Island, *Atmos. Environ.,* 7: 769-779.

——, and C. East, 1971, The Urban Boundary Layer in Montreal, *Boundary-Layer Meteorol.,* 2: 411-437.

Orville, H. D., 1964, On Mountain Upslope Winds, *J. Atmos. Sci.,* 21: 622-633.

Panofsky, H. A., and B. Prasad, 1965, Similarity Theories and Diffusion, *Int. J. Air Water Pollut.,* 9: 419-430.

Pasquill, F., 1966, Lagrangian Similarity and Vertical Diffusion from a Source at Ground Level, *Q. J. R. Meteorol. Soc.,* 92: 185.

——, 1971, Atmospheric Dispersion of Pollution, *Q. J. R. Meteorol. Soc.*, 97: 369-395.

——, 1974, *Atmospheric Diffusion,* 2nd ed., John Wiley & Sons, New York.

——, 1976, *Atmospheric Dispersion Parameters in Gaussian Plume Modeling,* Part II, EPA-600/4-76-0306, U. S. Environmental Protection Agency, 44 pp.

Peterson, K. R., 1968, Continuous Point Source Plume Behavior Out to 160 Miles, *J. Appl. Meteorol.*, 7: 217.

Petit, C., M. Trinite, and P. Valentine, 1976, Study of Turbulent Diffusion Above and Within a Forest-Application in the Case of SO_2, *Atmos. Environ.*, 10: 1057-1063.

Petkovsek, Z., and A. Hocevar, 1971, Night Drainage Winds, *Arch. Meteorol., Geophys. Bioklimatol., Ser. A,* 20: 353-360.

Phillips, D. W., 1972, Modification of Surface Air over Lake Ontario in Winter, *Mon. Weather Rev.*, 100: 662-670.

Plate, E. J., 1971, *Aerodynamic Characteristics of Atmospheric Boundary Layers,* AEC Critical Review Series, TID-25465, NTIS.

——, and A. A. Quaraishi, 1965, Modeling the Velocity Distributions Inside and Above Tall Crops, *J. Appl. Meteorol.*, 4: 400-408.

Prandtl, L., 1942, *Fuhrer durch die stromungslehre,* Braunschweig, Vieweg, and Sohn.

Randerson, D., 1970, A Numerical Experiment in Simulating the Transport of Sulfur Dioxide Through the Atmosphere, *Atmos. Environ.*, 4: 615-632.

——, 1979, Review Panel on Sigma Computations, *Bull. Am. Meteorol. Soc.*, 60: 682-683.

——, and J. S. Cornett, 1974, Verification of a Mesoscale Wind-Prediction Model Derived from a Horizontal Vorticity Theorem, in *Proceedings of the Fifth Conference on Weather Forecasting and Analysis,* Mar. 4–7, 1974, St. Louis, Mo., pp. 126-131, American Meteorological Society, Boston, Mass.

Raymond, D. J., 1972, Calculation of Airflow over an Arbitrary Ridge Including Diabatic Heating and Cooling, *J. Atmos. Sci.*, 29: 837-843.

Raynor, G. S., 1971, Wind and Temperature Structure in a Coniferous Forest and a Contiguous Field, *For. Sci.*, 17: 351-363.

——, J. V. Hayes, and E. C. Ogden, 1974a, Particulate Dispersion into and Within a Forest, *Boundary-Layer Meteorol.*, 7: 429-456.

——, P. Michael, R. M. Brown, and S. Sethuraman, 1974b, A Research Program on Atmospheric Diffusion from an Oceanic Site, in *Symposium on Atmospheric Diffusion and Air Pollution,* Sept. 9–13, 1974, Santa Barbara, Calif., pp. 289-295, American Meteorological Society, Boston, Mass.

Richardson, L. F., 1926, Atmospheric Diffusion Shown on a Distance-Neighbor Graph, *Proc. R. Soc. (London), Ser. A,* 110: 709-737.

Runnels, R. C., D. Randerson, and J. F. Griffiths, 1972, An Observational Study of Winter Temperatures in the Urban Area of Houston, Tex., *Int. J. Biometeorol.*, 16: 119-129.

Ryan, B. C., 1977, A Mathematical Model for Diagnosis and Prediction of Surface Winds in Mountainous Terrain, *J. Appl. Meteorol.*, 16: 571-584.

Saffman, P. G., 1962, The Effect of Wind Shear on Horizontal Spread from an Instantaneous Ground Source, *Q. J. R. Meteorol. Soc.*, 88: 382-393.

Sagendorf, J. F., and C. R. Dickson, 1976, Diffusion Under Low Windspeed, Inversion Conditions, in *Third Symposium on Atmospheric Turbulence, Diffusion, and Air Quality,* Raleigh, N. C., October 19–22, 1976, CONF-761003, pp. 272-282, American Meteorological Society, Boston, Mass.

Scheidegger, A. E., 1958, The Random-Walk Model with Autocorrelation of Flow Through Porous Media, *Can. J. Phys.*, 36: 649-658.

Schlichting, H., 1968, *Boundary-Layer Theory,* 6th ed., McGraw-Hill Book Co., New York.

Scholtz, M. T., and C. J. Brouckaert, 1978, Modeling of Stable Air Flow over a Complex Region, *J. Appl. Meteorol.*, 17: 1249-1257.

Scorer, R. S., 1978, *Environmental Aerodynamics,* John Wiley & Sons, New York.

Seltzer, J. P., 1972, *A Study of the Tensor Diffusion Equation,* Ph.D. Thesis, Drexel University, Philadelphia, Pa.

Sherman, C. A., 1978, A Mass-Consistent Model for Wind Fields over Complex Terrain, *J. Appl. Meteorol.*, 17: 312-319.

Shir, C. C., and L. J. Shieh, 1974, A Generalized Urban Air Pollution Model and Its Application to the Study of SO_2 Distributions in the St. Louis Metropolitan Area, *J. Appl. Meteorol.*, 13: 185-204.

Singer, I. A., 1961, The Relationship Between Peak and Mean Concentrations, *J. Air Pollut. Control Assoc.*, 11: 336-341.

——, K. Imai, and R. G. Del Campo, 1963, Peak-to-Mean Pollutant Concentration Ratios for Various Terrain and Vegetation Cover, *J. Air Pollut. Control Assoc.*, 13: 40-42.

Sklarew, R. C., A. J. Fabrick, and J. E. Prager, 1972, Mathematical Modeling of Photochemical Smog Using the PICK Method, *J. Air Pollut. Control Assoc.*, 22: 865-869.

Smith, F. B., 1973, The Basic Equations for Finite-Velocity One-Dimensional Random-Walk Diffusion, Report Meteorological Office 14, Turbulence and Diffusion Note No. 33.

——, and J. S. Hay, 1961, The Expansion of Clusters of Particles in the Atmosphere, *Q. J. R. Meteorol. Soc.*, 87(374): 549-561.

Smith, T. B., 1965, *Diffusion Study in Complex Mountainous Terrain,* Report MRI-65-FR-236, U. S. Army Chemical Corps, Dugway Proving Ground, Utah, Meteorology Research, Inc., Altadena, Calif.

——, E. K. Kauper, S. Berman, and F. Vukovich, 1964, *Micrometeorological Investigation of the Naval Missile Facility,* Point Arguello, Calif., Report MRI-64-FR-167, Meteorology Research, Inc., Altadena, Calif.

Spangler, T. C., and R. A. Dirks, 1974, Meso-Scale Variations of the Urban Mixing Height, *Boundary-Layer Meteorol.*, 6: 423-441.

Start, G. E., C. R. Dickson, and L. L. Wendell, 1975, Diffusion in a Canyon Within Rough Mountainous Terrain, *J. Appl. Meteorol.*, 14: 333-346.

Sutton, O. G., 1953, *Micrometeorology,* McGraw-Hill Book Co., New York.

Taylor, G. I., 1921, Diffusion by Continuous Movements, *Proc. R. Soc. (London), Ser. A,* 164: 196-212.

Tennekes, H., 1973, Similarity Laws and Scale Relations in Planetary Boundary Layers, in *Workshop on Micrometeorology,* D. A. Haugen (Ed.), pp. 177-216, American Meteorological Society, Boston, Mass.

——, 1974, The Atmospheric Boundary Layer, *Physics Today,* 27: 52-63.

——, and J. L. Lumley, 1972, *A First Course in Turbulence,* The MIT Press, Cambridge, Mass.

Thom, A. S., 1972, Momentum, Mass and Heat Exchange of Vegetation, *Q. J. R. Meteorol. Soc.*, 98: 124-134.

Thompson, R., 1971, Numeric Calculation of Turbulent Diffusion, *Q. J. R. Meteorol. Soc.*, 97: 93-98.

Thyer, N. H., 1966, A Theoretical Explanation of Mountain Valley Wind by a Numerical Method, *Arch. Meteorol., Geophys. Bioklimatol., Ser. A.*, 15: 318-348.

Tingle, A. G., and J. R. Bjorklund, 1973, *Study and Investigation of Computer Algorithms for the Solution of the Shallow-Fluid Equations as a Means of Computing Terrain Influences on Wind Fields,* Report TR-73-302-01, H. E. Cramer Company, Salt Lake City, Utah.

Tourin, M. H., and W. C. Shen, 1969, *Deciduous Forest Diffusion Study, Financial Report, Volume I, Diffusion in a Deciduous Forest,* Applied Science Division, Litton Systems, Inc., Minneapolis, Minn.

Turner, D. B., 1969, *Workbook of Atmospheric Dispersion Estimates,* U. S. Department of Health, Education, and Welfare, Public Health Service, Cincinnati, Ohio.

Turner, F. B., 1964, A Diffusion Model for an Urban Area, *J. Appl. Meteorol.*, 3: 83-91.

Tyldesley, J. B., and C. E. Wallington, 1965, The Effect of Wind Shear and Vertical Diffusion on Horizontal Dispersion, *Q. J. R. Meteorol. Soc.*, 91: 158.

Uchijima, Z., and J. L. Wright, 1964, An Experimental Study of Air Flow in a Corn Plant–Air Layer, *Bull. Natl. Inst. Agric. Sci., (Japan), Ser. A.*, 11: 19-65.

U. S. Environmental Protection Agency, 1977, *Interim Guideline on Air Quality Models*, OAQPs No. 1.2-080, Oct. 1977, 85 pp.

Van der Hoven, F., 1967, Atmospheric Transport and Diffusion at Coastal Sites, *Nucl. Safety*, 8(5): 490-499.

Vukovich, F. M., 1971, Theoretical Analysis of the Effect of Mean Wind and Stability on a Heat Island Circulation Characteristic of an Urban Complex, *Mon. Weather Rev.*, 99: 919-926.

Wagner, N. K., and T. W. Yu, 1972, Heat Island Formation: A Numerical Experiment, in *Proceedings of Conference on Urban Environment and Second Conference on Biometeorology*, Oct. 31–Nov. 2, 1972, Philadelphia, Pa., pp. 83-88, American Meteorological Society, Boston, Mass.

Waldron, A. W., 1971, Expected Instantaneous Surface Line Source and Point Source Behavior in a Forest Environment, Technical Note, Desert Test Center, Fort Douglas, Utah.

Watson, C. W., and S. Barr, 1976, *Monte Carlo Simulation of the Turbulent Transport of Airborne Contaminants*, USAEC Report LA-6103, Los Alamos Scientific Laboratory.

Weber, A. H., 1976, Atmospheric Dispersion Parameters in Gaussian Plume Modeling, Part I, Report EPA-600/4-76-030a, 59 pp.

Wendell, L. L., 1972, Mesoscale Wind Fields and Transport Estimates Determined from a Network of Wind Towers, *Mon. Weather Rev.*, 100: 565-578.

Wilson, R. B., G. E. Start, C. R. Dickson, and N. R. Ricks, 1976, Diffusion Under Low Wind Speed Conditions near Oak Ridge, Tennessee, in *Third Symposium on Atmospheric Turbulence, Diffusion, and Air Quality*, Raleigh, N. C., October 19–22, 1976, CONF-761003, pp. 269-276, American Meteorological Society, Boston, Mass.

Woodcock, A. H., and J. Wyman, 1947, Convective Motion in Air over the Sea, *Ann. N. Y. Acad. Sci.*, 38: 749-777.

Wyngaard, J. C., 1973, On Surface Layer Turbulence, in *Workshop on Micrometeorology*, D. A. Haugen (Ed.), pp. 101-149, American Meteorological Society, Boston, Mass.

——, and O. R. Cote, 1974, The Evolution of a Convective Planetary Boundary Layer—a Higher-Order Closure Model Study, *Boundary-Layer Meteorol.*, 7: 298-302.

Yamada, T., 1978, A Three-Dimensional Numerical Study of Complex Atmospheric Circulations Produced by Terrain, in *Proceedings of Conference on Sierra Nevada Meteorology*, June 19–21, 1978, Lake Tahoe, Calif., American Meteorological Society, Boston, Mass.

——, and G. Mellor, 1975, A Simulation of the Wangara Atmospheric Boundary Layer Data, *J. Atmos. Sci.*, 32: 2309-2329.

Chapter 13 covers research prior to July 1980.

MODELING REGIONAL TO GLOBAL AIR POLLUTION

14

Joseph B. Knox and John J. Walton, Lawrence Livermore National Laboratory

14-1 INTRODUCTION

During the past decade there has been an increasing awareness of and concern for the effect, or potential effect, of man's activities on the environment. Some scientists suggest that man is on the verge of being able, if not already so, to influence his global environmental quality. Since the publication of *Meteorology and Atomic Energy—1968*, the need for both regional assessments and the resolution of global environmental issues has accelerated. New research programs have begun to examine the chronic emissions from energy generation and their potential effects on regional to global scales. When man injects chemically active species into the atmosphere, the treatment of the chemical evolution entails consideration of a large set of chemical and photochemical reactions. For example, there is evidence (T. A. Cahill, University of California at Davis, personal communication, 1975) that much of the anthropogenic respirable aerosol results from the transformation of sulfur dioxide (a product of the combustion of most fossil fuels) to sulfate in the presence of other reactive or photochemical pollutants. An SO_2 conversion rate as high as 7.5% per hour has been reported in some heavy smog conditions encountered in Los Angeles. This one example of regional air pollution illustrates the complexity of the general problem, namely, that emissions from one type of source can be affected by reactive species originating from other sectors of human activity, e.g., transportation. Even before the passage of the Clean Air Act of 1970, planners used fuel substitution and tall stacks to alleviate local air-quality problems. The effects of such control procedures are now gradually emerging: emission of SO_2 from tall stacks leads to (1) longer residence times in the atmosphere, (2) the opportunity for more complete conversion to sulfate, and (3) increased contribution of such sources to background at extended scales (Altshuller, 1976). It is becoming clear that to control air pollution we need to develop, validate, and implement regional assessment techniques based on the best possible physical and chemical understanding.

Our consideration of these more complex problems has been facilitated by the evolution of large, high-speed computers. The computer industry has responded to the needs of the scientific community with increasingly sophisticated machines, and the scientific community has reacted by seeking to address more challenging problems. This process has given birth to a wide range of computer simulation models. Among these are:

1. Parameterized transport models in one, two, and three dimensions which describe transport of nonreactive pollutants over scales ranging from a few kilometers to global; models are Eulerian, Lagrangian, and hybrids of the two.
2. Transport models that describe the evolution of chemically active species with reaction sets of varying degrees of complexity.
3. Models in which transport processes evolve subject to some of the feedback mechanisms that occur in the real atmosphere. (Models in this classification are beyond the scope of this chapter.)

The preceding considerations tend to emphasize the meteorological processes as they relate to the possible adverse effects of various means of power production. If we look at the atmosphere's "weather machine" as a complex system of energy balance, we see another use of our knowledge of atmospheric energy flows, i.e., the tapping of certain alternate energy resources. Specifically we may consider the requirements for the siting of solar energy systems, which are affected by cloud cover, and of wind energy systems, which capture energy from the planetary boundary layer. The development of siting techniques in support of these alternative energy resources is now part of regional meteorological research.

This chapter outlines some of the numerical models that can be used, and, where possible, presents some of the results obtained. In this discussion, we will progress in order of model dimensionality rather than domain covered, with examples from regional, continental, or global scales.

14-2 POLLUTANTS OF CONCERN

When power is generated from fossil fuels, the following primary pollutants are produced: oxides of nitrogen, SO_2, particulates, and, of course, CO_2 and water vapor (DeHaven and Woodfill, 1973; Davis, Smith, and Klauber, 1974). Carbon monoxide and hydrocarbons, although still believed

Knox and Walton: Lawrence Livermore National Laboratory, University of California, Livermore, California.

to be important, are not present in the large quantities that they are in internal combustion emissions. NO_x is a significant primary pollutant in its own right; in addition, mixtures of NO_x and reactive hydrocarbons in the presence of sunlight form oxidant, primarily ozone (O_3) (Leighten, 1961). Depending on the type of fuel used, SO_2 has come under increased scrutiny because of its position at the beginning of the chain of reactions leading to sulfate aerosol and acid rain (Beilke, Lamb, and Müller, 1975; Altshuller, 1976). For a discussion of the health effects of these species, the reader is directed to *Proceedings of the Conference on Health Effects of Air Pollutants*, 1973; *Research and Development Related to Sulphates in the Atmosphere*, 1975; and Chap. 16, this volume. These species are all of regional significance; while the SO_2/sulfate chain, however, may extend to continental scales (Altshuller, 1976). For a discussion of the chemistry associated with these species, the reader is referred to Chap. 10 and to Altshuller (1976), Beilke, Lamb, and Müller (1975), and Demerjian, Kern, and Calvert (1974).

One combustion product, CO_2, is not usually considered a pollutant on regional scales but may well have global consequences (*Energy and Climate*, 1977). Hydrogen sulfide (H_2S), a species associated with geothermal power, constitutes a problem at short range because of its odor, its toxicity, and its secondary pollutant, SO_2, which converts to sulfates a few hundred kilometers downwind (Junge, 1963; Cadle and Ledford, 1966). This regional air-quality problem is discussed in greater depth in Chap. 18.

Another type of fuel used in increasing quantities is nuclear fuel. From the nuclear fuel cycle including consumption and reprocessing, there are continuous small releases of ^{85}Kr, tritium, and ^{14}C. Releases from reprocessing plants can be about 50 times those from individual reactors. In addition, ^{85}Kr and tritium are of concern on the global scale. Details on the materials associated with nuclear power production and their effects on man are discussed in Chap. 16.

14-3 DEVELOPMENT OF THE BASIC EQUATION FOR POLLUTANT TRANSPORT

We wish to develop an equation with which we can follow the history of trace species from their release into the atmosphere until their exit from our domain of interest. A general equation for N chemically interacting species will be presented which may then be simplified for inert species or those undergoing simple decay.

Consider the physical processes in an arbitrary cartesian coordinate system. The single symbol x represents a point in this system, and the use of a Greek subscript, e.g., x_α, represents one of the coordinates, depending on the value of $\alpha(= 1, 2, \text{ or } 3)$. Similarly the components of the wind velocity are denoted as u_α. Further, when a Greek subscript is repeated in our equation, a summation is implied. The concentration of species i at time t is $c_i(x,t)$, where i ranges over the total number of species, N. The equations are written for the mixing ratio, $\chi_i = c_i/\rho$, where ρ is the air density at point x. This is a more natural form when dealing with scales where ρ may change significantly, e.g., over large vertical distances. Concentrations may be mass or molecular density. For brevity the time and space arguments will usually be omitted. The processes affecting species i that will be treated are advection and photochemical processes, which are denoted by $R_i(\chi_1, \ldots, \chi_n, T)$; $T(x,t)$ is the temperature. There may also be sources and sinks, represented by $s_i(x,t)$. The chemical term R_i, in general, contains photodissociation terms of the form $J_{ij}\chi_j$ and reaction terms of the form $Q_{ijk}\chi_j\chi_k$ (no summation), where J_{ij} is the photodissociation rate coefficient and Q_{ijk} is the chemical rate coefficient. Termolecular (three species) reactions may also occur. Molecular diffusion has not been included because its effect is less than the effects of other processes in the lower stratosphere and troposphere. Specifically, eddy diffusion, to be described shortly, is orders of magnitude greater. On the basis of these processes, N equations describing the conservation of the N species are

$$\frac{\partial}{\partial t}(\rho\chi_i) + \partial_\alpha(\rho u_\alpha \chi_i) = R_i(\chi_1, \ldots, \chi_N, T) + s_i(x,t) \qquad i = 1, N \qquad (14.1)$$

where the symbol ∂_α represents the derivative $\partial/(\partial x_\alpha)$. In using only this equation with specified winds and temperature, we assume that the species are at sufficiently low concentrations that they do not feedback on terms in the driving equation. Specifically the Navier–Stokes equation of motion and the energy equation are not affected by the presence of the pollutants. One feedback mechanism may be considered. If changes in trace species are great enough to affect radiation transport, photochemical rates may change. The radiative energy balance may then be sufficiently disturbed to cause temperature changes and hence changes in chemical reaction-rate coefficients. However, in most local and regional models, temperature and solar intensity are assumed known.

We now have an equation for χ_i that can be solved, in principle, given the sources, sinks, chemical and photochemical processes, and the atmospheric advection. With the atmospheric velocity field, we encounter a major obstacle. The atmosphere exhibits complex turbulent motions that are not easy to describe in detail. However, we do know that the general characteristics of the motion can be described as the sum of an average field $\bar{u}_\alpha$ plus a random variation u'_α,

$$u_\alpha = \bar{u}_\alpha + u'_\alpha \qquad (14.2)$$

In like manner the species mixing ratios can be described as

$$\chi_i = \bar{\chi}_i + \chi'_i \qquad (14.3)$$

where air density is assumed not to be perturbed. In Eqs. 14.2 and 14.3 by definition, $\overline{u'} = \overline{\chi'} = 0$. We have now divided the velocity and concentration variables into a mean within a sampling period plus a random variation. Substituting for u_α and χ_i in Eq. 14.1 and averaging,

$$\frac{\partial}{\partial t}(\rho\overline{\chi}_i) + \partial_\alpha(\rho\overline{u}_\alpha\overline{\chi}_i) + \partial_\alpha(\rho\overline{u'_\alpha\chi'_i}) = R_i(\overline{\chi}_1 + \chi'_1, \ldots, \overline{\chi}_N + \chi'_N) + s_i \qquad i = 1,N \qquad (14.4)$$

where terms of the form $Q_{ijk}\overline{\chi'_j\chi'_k}$ appear in R_i. These terms present considerable theoretical difficulties and are not included in most current models; for an analysis of the significance of these terms, the reader is referred to Hilst et al. (1973). The average of the product of the perturbations is called the turbulent flux,

$$f_\alpha = \rho\overline{u'_\alpha\chi'_i} \qquad (14.5)$$

These terms have long been troublesome when dealing with turbulent phenomena. If we generate a set of equations for f_α, still higher order terms result, e.g., $\overline{u'u'\chi'_i}$. Although these higher order closure problems are being considered by some investigators (e.g., Hilst et al., 1973), most modeling efforts use the simplest closure approximation given by K-theory (Monin and Yaglom, 1971). The basis of this concept is the idea that the flux is proportional to the gradient of the mean concentration field,

$$f_\alpha = -K_{\alpha\beta}\rho\partial_\beta\overline{\chi}_i \qquad (14.6)$$

The quantities $K_{\alpha\beta}$ are the elements of a symmetric second-rank cartesian tensor, and, because of their position in the transport equations, they are referred to as eddy-diffusion coefficients. In many analyses the off-diagonal terms are considered to be zero, although this is not always the case (Reed and German, 1965). With these assumptions and approximations, Eq. 14.4 reduces to

$$\frac{\partial(\rho\overline{\chi}_i)}{\partial t} + \partial_\alpha(\rho\overline{u}_\alpha\overline{\chi}_i) = \partial_\alpha(K_{\alpha\beta}\rho\partial_\beta\overline{\chi}_i) + \overline{R}_i(\overline{\chi}_1, \ldots, \overline{\chi}_N) + s_i(x,t) \qquad i = 1, N \qquad (14.7)$$

In rectangular cartesian coordinates, when the eddy-diffusion tensor is diagonal and $u_\alpha = (u,v,w)$,

$$\frac{\partial(\rho\overline{\chi}_i)}{\partial t} + \frac{\partial}{\partial x}(\rho\overline{u}\overline{\chi}_i) + \frac{\partial}{\partial y}(\rho\overline{v}\overline{\chi}_i) + \frac{\partial}{\partial z}(\rho\overline{w}\overline{\chi}_i) = \frac{\partial}{\partial x}\left(\rho K_x \frac{\partial\overline{\chi}_i}{\partial x}\right) + \frac{\partial}{\partial y}\left(\rho K_y \frac{\partial\overline{\chi}_i}{\partial y}\right) + \frac{\partial}{\partial z}\left(\rho K_z \frac{\partial\overline{\chi}_i}{\partial z}\right) + \overline{R}_i(\overline{\chi}_1, \ldots, \overline{\chi}_N) + s_i(x,t) \qquad i = 1,N \qquad (14.8)$$

In the applications that follow, the bars are dropped in Eqs. 14.7 and 14.8, and it is understood that the subscript i implies N equations, one for each species (i).

The equation obtained by integrating Eq. 14.7 over some volume also provides a basis for model development and gives a different perspective of the physical processes.

Consider a volume (V) in space bounded by a surface (S) within which Eq. 14.7 describes the concentration of N species. This volume may be anything from a large single cell encompassing the entire region of interest to one cell in a three-dimensional array of cells subdividing the region. In the latter case, transport will take place between the cells themselves and between boundary cells and the domain outside. We may even represent the structure analytically within each cell. For the present, we consider the Eulerian problem of the time rate of change of species within this stationary volume governed by the equation

$$\int_V \frac{\partial}{\partial t}(\rho\chi_i)\,dV = -\int_V \partial_\alpha(\rho u_\alpha\chi_i)\,dV + \int_V \partial_\alpha(\rho K_{\alpha\beta}\partial_\beta\chi_i)\,dV + \int_V R_i\,dV + \int_V s_i\,dV \qquad (14.9)$$

In Eq. 14.9 the advective term has been placed on the right-hand side since we are considering a material balance within a fixed volume. With the divergence theorem, Eq. 14.9 becomes

$$\int_V \frac{\partial(\rho\chi_i)}{\partial t}\,dV = -\int_S (\rho u_\alpha\chi_i)\,n_\alpha\,dS + \int_S (\rho K_{\alpha\beta}\partial_\beta\chi_i)\,n_\alpha\,dS + \int_V R_i\,dV + \int_V s_i\,dV \qquad (14.10)$$

Here n_α represents the components of the *outward* unit normal vector of the area increment dS. Equation 14.10 states that the rate of change of material within a volume (V) is due to the combination of advective and diffusive fluxes $(\rho u_\alpha\chi_i)$ and $(-\rho K_{\alpha\beta}\partial_\beta\chi_i)$ across the boundaries of the volume plus the net effects of reactions R_i and sources and sinks s_i within the volume (Fig. 14.1).

All processes shown in Fig. 14.1 are important, and under certain conditions any one process or combination of processes may dominate. Since computational power is not unlimited, the modeler must make decisions based on his particular problem as to which processes are to be emphasized and treated in most detail.

Source terms must be available in detail that is commensurate with the refinement of the model grid. A simple box model will require only total emission rates; a finely zoned two- or three-dimensional model must have sources defined with considerable spatial detail. Similarly a

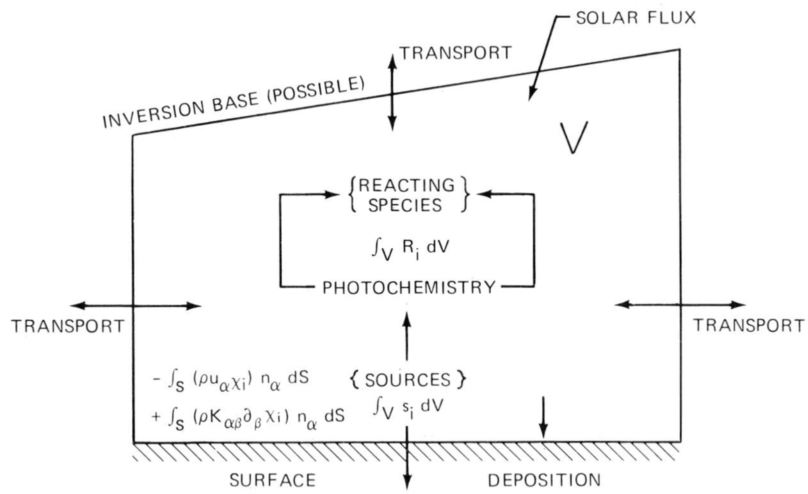

Fig. 14.1 The processes most commonly treated in transport models.

model that is used to calculate a regional concentration does not require the temporal detail of one in which the effects of rush-hour automobile traffic are to be determined. The magnitude of surface deposition is prescribed on the basis of known surface conditions and may become large but will generally not vary greatly over model domain.

Turbulent diffusion is generally small compared with advective transport and is numerically tractable. Advective transport, on the other hand, presents a problem when it comes to treatment with finite difference schemes. For example, the error term in a finite-difference approximation to the advection term can have the form of a diffusion that is orders of magnitude larger than that due to turbulence. This problem can be emphasized when strong local sources are present; here the artificial diffusion blurs out concentration distributions. More elaborate schemes may have the property of not maintaining positivity, which would be a problem if chemical reactions were being calculated. Still more sophisticated schemes may present a problem when it is desirable to couple them with complex chemical kinetics calculations. The problem is obviated if a Lagrangian calculation (following an air parcel) is done, but then the parcel trajectory must be treated with special care since this controls which sources inject into the parcel.

When chemical reactions are important, the governing equations cannot be solved independently since the chemical terms provide a coupling. Chemical and photochemical terms may have rates that differ by many orders of magnitude. This difference, called stiffness, means that a simple time integration scheme will either calculate fast reactions incorrectly or will require an unacceptably large number of time steps. Sophisticated time integration schemes have been developed to deal with stiff equations. In most cases it is difficult to use these time integration schemes and the more complex advection prescriptions together.

Solar radiation, which affects the photochemical terms, is itself the result of a complex calculation requiring information on the constituents of and radiative transport properties in the atmosphere. Fractional cloud cover, which can play a significant role, may be particularly hard to measure and model. Most current models use a prescribed solar intensity.

There are some rules of thumb that should be kept in mind.

—Increasing dimensionality and refinement of zoning may improve the treatment of transport.

—Increased zoning, using more computer storage, generally reduces the chemistry that can be treated.

—Lagrangian (parcel) calculations do not suffer the artificial diffusion problems experienced in Eulerian (fixed) grids.

We have not attempted to discuss the numerical techniques used to solve the equations in the succeeding sections of this chapter. This would merit a volume in itself. The interested reader is directed to Long and Pepper (1976) for a discussion of numerical advection schemes. Discussions of methods used in integrating stiff systems of equations may be found in Gear (1971) and Willoughby (1974).

14-4 GENERAL APPLICATIONS: SINGLE BOX, 1-D, 2-D, AND 3-D MODELS

Equations 14.1 through 14.8 presented in the foregoing section provide a convenient basis for pollutant-prediction models. Meteorological data required by the models may be provided either by observation or by a concurrently running meteorological analysis or prediction code. Analytic integration of the equations is possible only under severe simplifying assumptions. For nonreactive species, these procedures have provided a number of useful solutions for such cases as continuous point, line, and area sources (plume solutions) and instantaneous sources (puff solutions), discussed in Chaps. 6 and 13.

We are interested in solutions that, using more data, provide pollutant predictions for less specialized situations. Because of the complexity of this problem, we must resort to computer simulation. By indicating various approximations, we summarize a few of the current models.

This chapter is *not* intended as an exhaustive critical survey of all models currently available. There are a number of such surveys (Stern, 1968; Lamb, Badgley, and Rossano, 1973; state of California, 1974; Mathis and Grose, 1973). Models will be described only to illustrate some application of the equations, our purpose being to provide a general description of how such models are constructed.

14-4.1 The Single-Box Model (Zero-Dimensional Model)

The simplest model that can be devised using Eq. 14.10 is a single box of uniformly mixed species. The horizontal boundaries of the box are usually located so that all significant sources are included within the domain. In the vertical the box extends from the surface of the ground up through the well-mixed layer, which on the regional scale may be limited by a temperature inversion at height H. The vertical motion of the inversion base causes a volume change in the model. Because species within the volume are considered to be uniformly mixed, a detailed source distribution pattern is not required. Uniform boundary transport or ventilation is assumed. The coarseness of the representation may not warrant inclusion of diffusive transport across the boundaries. As in most models, sink terms are a flux of material $V_{di}\chi_i$ due to a deposition velocity V_{di} or simple decay $\alpha_i\chi_i$. A distinction is made between deposition and decay because the former represents material loss through interaction with the surface (i.e., settling and/or loss through contact with or filtering by vegetation), whereas the latter, decay, may describe radioactive decay or highly parameterized chemical transformation.

Under the following conditions of a uniformly mixed region with base dimensions $\delta x, \delta y$, a constant mixing depth H, a uniform wind u in the x-direction, and constant air density ρ, Eq. 14.10 becomes

$$\delta x \delta y\, H \frac{\partial c_i}{\partial t} = u \delta y\, H(c_{ia} - c_i) + \delta x \delta y\, H(R_i + s_i) \qquad (14.11)$$

(c_{ia} ambient upwind concentration) or

$$\frac{\partial c_i}{\partial t} = \frac{u}{\delta x}(c_{ia} - c_i) + R_i + s_i \qquad (14.12)$$

For the steady-state case in which the ambient concentration is assumed to be zero and there are no reactions,

$$c_i = \frac{s_i \delta x}{u} \qquad (14.13a)$$

When the volume source s_i is replaced with the area source flux q_i ($s_i = q_i/H$)

$$c_i = \frac{q_i}{u}\left(\frac{\delta x}{H}\right) \qquad (14.13b)$$

Simple deposition or decay terms do not significantly complicate the steady-state solution; with

$$s_i = \frac{q_i}{H} - \frac{V_{di}}{H} c_i + \alpha_i c_i$$

$$c_i = \left(\frac{q_i}{H} + \frac{u}{\partial x} c_{ia}\right)\left(\frac{u}{\partial x} + \frac{V_{di}}{H} + \alpha_i\right)^{-1} \qquad (14.13c)$$

These equations hardly warrant the name "model," because they are just a starting point. For example, if the reactive expression, R, contains terms no more complicated than $c_i c_j$, the coupled set of algebraic equations may have an analytic solution or may be solved iteratively. Time-dependent equations are also readily written. Hanna (1973) developed a box model based on the integration of the Gaussian equation for area sources. In this model a simple set of chemical reactions is incorporated to treat pollutants in the Los Angeles basin. The equation, in our notation, is

$$H \delta x \frac{\partial c_i}{\partial t} = q_i\, \delta x - uHc_i + R_i H\, \delta x \qquad (14.14)$$

This model can be made more realistic by making assumptions regarding the character of the wind and concentration profiles [e.g., $u(z) = u_0 (z/z_0)^n$, $c_i = a_i + b_i \log z/z_0$ as was done by MacCracken et al. (1978) in a more complex model]. Or, we may know something about transport through the inversion or convergence of the wind field around the box.

These simple models provide us with a convenient avenue for exploring the significance of a time-varying inversion height. For example, suppose that our source data are in terms of the total mass of species (i) emitted per second in the volume. Differentiation of the total mass $M_i = \delta x \delta y\, H(t)\, c_i$ gives

$$\frac{\partial M_i}{\partial t} = \partial x \partial y H \frac{\partial c_i}{\partial t} + \partial x \partial y c_i \frac{\partial H}{\partial t}$$

When Eq. 14.11 is substituted in the above,

$$\frac{\partial M_i}{\partial t} = \frac{u(M_{ia} - M_i)}{\delta x} + R_i + s_i + \frac{M_i}{H}\frac{\partial H}{\partial t} \qquad (14.15)$$

where R_i and s_i refer to mass rates and M_{ia} is the mass that would occupy the box at ambient concentration. The last term in Eq. 14.15 accounts for mass change due to motion of the inversion base. For a complete discussion of the effects of variable inversion lids on simple box models, the reader is directed to Tennekes (1976).

The models discussed thus far have dealt with a volume fixed in space. We now consider a Lagrangian single-box model in which the volume moves with the air motion. In this case eddy diffusion may play a significant role, depending on the source configuration. The reaction term represents chemical or photolytic processes operative within the moving volume; the source term is a function of time depending on the pollutant sources along the trajectory of the volume.

A variant of this idea, developed by Gelinas and Walton (1973), considered the concentration at the center of a plume of reactive materials being dispersed by eddy diffusion. The equation for species (i) in a volume V(t) is

$$\frac{dc_i}{dt} = R_i - \left[\frac{\partial \ln V(t)}{\partial t}\right] [c_i - c_{ia}] + s_i \tag{14.16}$$

The logarithmic derivative of V(t) accounts for exchange between the volume and its surroundings. It was shown that, if the volume considered was an infinitesimal element at the center of a Gaussian puff with standard deviations $(\sigma_x, \sigma_y, \sigma_z)$, then

$$\frac{\partial \ln V(t)}{\partial t} = \frac{\dot{\sigma}_x}{\sigma_x} + \frac{\dot{\sigma}_y}{\sigma_y} + \frac{\dot{\sigma}_z}{\sigma_z} \tag{14.17a}$$

$$= \frac{K_x}{\sigma_x^2} + \frac{K_y}{\sigma_y^2} + \frac{K_z}{\sigma_z^2} \tag{14.17b}$$

where the K's are obtained from a scale-dependent diffusion formula. The same result is obtained by recognizing that the second term on the right-hand side of Eq. 14.10 is the time rate of change of concentration due to diffusion only and by treating the equation at puff center ($x = y = z = 0$). In this example the box becomes a point at which the characteristics of the puff are calculated.

The principal advantages of single-box models are that few data are required and calculations are very rapid because variables are needed in only one zone. A less apparent advantage comes from the fact that little storage is required for transport terms; thus the single-box model can be thought of as a smog chamber with simple source/sink and transport terms added, which makes it possible to treat very elaborate sets of reactions. Gelinas and Walton (1973) in treating plume dispersion in the stratosphere used 40 reactions among 12 active chemical species. Such models permit rapid testing of meteorological parameters and sets of reactions and reaction-rate coefficients. If the meteorology and chemistry are sufficiently simple, no machine calculations may be required.

By treating the various processes in more detail, the intent is to obtain results that warrant the added effort and cost in return for higher resolution solutions or solutions based on more detailed physics and chemistry.

14-4.2 One-Dimensional Models

If, in Eqs. 14.7, 14.8, or 14.10, we assume that the dependent variables vary only with height, Eq. 14.8 becomes,

$$\frac{\partial(\rho\chi_i)}{\partial t} + \frac{\partial}{\partial z}(\rho u_z \chi_i) = \frac{\partial}{\partial z}\left(\rho K_z \frac{\partial \chi_i}{\partial z}\right) + R_i + s_i \tag{14.18}$$

If the domain is subdivided into J zones bounded by z_j, z_{j+1}, Eq. 14.10 becomes a set of J equations

$$\int_{z_j}^{z_{j+1}} \frac{\partial(\rho\chi_i)}{\partial t}\, dz = -F_{j+1} + F_j - f_{j+1} + f_j + \int_{z_j}^{z_{j+1}} R_i\, dz + \int_{z_j}^{z_{j+1}} s_i\, dz \qquad j = 0, J-1 \tag{14.19}$$

In Eq. 14.19 the advective and diffusive fluxes are, respectively,

$$F_j = \rho u_z(z_j)\, \chi_i(z_j) \tag{14.20a}$$

and

$$f_j = -\left(\rho K_z \frac{\partial \chi_i}{\partial z}\right)_{z=z_j} \tag{14.20b}$$

In practice, the advective terms do not usually appear. Their appearance would imply some flow through the upper and lower boundaries, flow convergence relative to the sides of the column, or a combination of the two. As a practical matter, these terms have been excluded from most models.

Consider now the question of boundary conditions. In the single-box models, these boundary conditions were not dignified with that name; the fact remains, however, that, by specifying a flow into the box, uc_{ia}, and a surface deposition, we imposed flux conditions at the boundaries of the simple domain. At each boundary either the concentration or its gradient (alternatively its flux) must be prescribed. This task is not always easy; it requires knowledge that may not be available. For example, for the single-box model, if the upwind concentration, c_{ia}, is unknown, even the simple steady-state Eq. 14.13c cannot be solved. For models in one dimension, poor boundary conditions cause errors that propagate throughout the computational domain and can defeat the original purpose of obtaining more complete results by using a more complex model. The most suitable conditions are dictated by the specific problem and by the data available.

Consider now two models: (1) a Eulerian for treating the vertical distribution of a globally averaged trace species and (2) a Lagrangian for treating pollution on a local scale.

The one-dimensional (vertical) model of Chang (1974) has the governing equation, after substituting c_i/ρ for χ_i in Eq. 14.18,

$$\frac{\partial c_i}{\partial t} = \frac{\partial}{\partial z}\left[K_z \rho \frac{\partial c_i/\rho}{\partial z}\right] + R_i + s_i \qquad (14.21)$$

The correction for changing air density is explicit in the diffusive transport term because this model covers a large vertical range (0 to 55 km). Chang emphasizes that, rather than thinking of this as a globally averaged model, we should consider it "as an approximate tool for studying the relative importance of transport and chemical kinetics processes . . . ." This model is designed to study the response characteristics of the stratosphere. The source term s_i is included so that perturbing injection of trace species can be made.

The choice of boundaries is dictated by the need either to have reliable estimates of boundary fluxes or to minimize model sensitivity to these conditions. A lower boundary at the earth's surface requires treatment of tropospheric processes. Specifying boundary fluxes at the tropopause is even more troublesome because of the difficulty of observing conditions there. Furthermore, if fluxes are fixed there, downward transport of species injected by stratospheric sources would be limited, or, worse, fixing concentrations at the tropopause would turn the troposphere into an infinite sink. Chang resolved this problem by placing the boundary at the surface and making the troposphere a buffer by using a large K_z in this region. This condition further separates the uncertain lower boundary conditions from the main region of interest, the stratosphere.

One-dimensional global models do not simulate the actual processes governing large-scale vertical transport included in two- and three-dimensional models; so all vertical transport must be lumped into the one parameter, K_z. There are at least as many K_z profiles suggested as there are one-dimensional global modelers, the effective transport simulated by each having been validated against some set of observations.

Radiation transport can be calculated on the bases of species concentration, which feeds back onto photochemical rates and, through the time- and height-varying temperature, to changes in chemical reaction-rate coefficients.

Models of this type have been used to study the coupling of vertical transport and chemical and photochemical processes that determine the structure of the atmosphere. They have been used also to determine the stratospheric role of such pollutants as nitric oxides from supersonic transports and chlorofluoromethanes used in aerosol cans and refrigeration.

An example of a Lagrangian model is the one developed by Eschenroeder and Martinez (1971, 1973). This model, called DIFKIN, simulates a column of air extending from the surface to an inversion lid through which there is assumed to be no flow (i.e., a zero-flux upper boundary condition). The depth of the column can vary as the inversion and topography change relative to each other in time and space. The column is zoned in z to permit calculation of species profiles resulting from the combined effects of chemistry and diffusive transport. Assuming constant air density ρ, the governing equation is

$$\frac{dc_i}{dt} = \frac{\partial}{\partial z}\left(K_z \frac{\partial c_i}{\partial z}\right) + R_i \qquad (14.22)$$

where it is understood that the total time derivative appears because the column is transported by the mean wind. The air density does not appear because its variation is negligible in the boundary layer.

The volume source term s_i is replaced by a time-dependent lower boundary flux condition that results from knowing where the column is at any time and what the sources are at each point along its trajectory. Clearly this model presents fewer problems in defining boundary conditions. At the same time, the user must exercise care in determining the column trajectories since an error here might carry the column over different source areas.

The K_z profile consists of three segments: (1) a linear increase from ground level to approximately 75 m, where $K_z = (50/60)\ (U + 5)\ m^2/s$ and U is the layer wind speed; (2) a layer of constant K_z; and (3) a linear decrease from approximately 140 m up to the inversion base.

The meteorological inputs required are wind velocity and inversion-base height. Good wind data are important since trajectories are constructed in hourly segments; these trajectories determine the sources.

The modelers intentionally kept the reaction set simple, but it could be expanded in the future as knowledge of photochemical processes advances. The variation of solar radiation as a function of time has been included.

In developing models such as these, within the constraint of one dimension, the developer/user largely determines the degree of detail. As long as sufficient data exist to add some other process, it too can be incorporated. The models are comparatively inexpensive in terms of computer time and data requirements; they provide useful information as long as their limitations are understood.

One-dimensional models can be made applicable to a number of problems, from urban to global scale. They usually provide vertical structure and make limited demands for data and computation time, one exception being diurnal calculations with large reaction sets (Wuebbles and Chang, 1975).

However, limitations exist. For example, in a global model the diffusion coefficient must in some way represent all scales of transport in the atmosphere. Also, if we are really trying to look at a column, there will always be some transport relative to the sides which may not be represented. A possible answer to these and other problems is to add another space dimension.

14-4.3 Two-Dimensional Models

In addition to adding another spatial dimension, two-dimensional models represent the advective transport in a more complete manner. In the one-dimensional Eulerian models, in most cases the advective term vanishes for reasons of mass continuity. The two-dimensional model permits explicit representation of circulation patterns; such variables as topography and inversion base height can also be treated in some detail.

Unlike the one-dimensional models, which generally describe vertical structure, the two-dimensional models simulate processes in either horizontal or vertical planes. Local and regional models (scales of a few hundred kilometers) are usually zoned in the horizontal plane, and global models, by latitude and altitude.

We now discuss briefly three models: one Lagrangian model on a regional scale and two Eulerian models, one regional and one global.

The first model, 2BPUFF (Crawford, 1966), was developed to model the growth of the deposition from a cloud of nuclear debris. In addition to diffusive transport, radioactive decay and wet and dry deposition are present. The wind transports the cloud as a whole along the cloud-center trajectory. Changes in wind speed and direction with altitude can be treated by subdividing the initial cloud into layers and performing a separate calculation for each.

The cloud is considered to be a right-circular cylinder. The governing equation is

$$\frac{dc}{dt} = K_r \left(\frac{\partial^2 c}{\partial r^2} + \frac{1}{r}\frac{\partial c}{\partial r}\right) + \frac{\partial}{\partial z}\left(K_z \frac{\partial c}{\partial z}\right) + s \qquad (14.23)$$

where K_r is a function of the characteristic horizontal dimension of the puff (scale-dependent diffusion). The term s may contain radioactive decay, $-\alpha c$, throughout the volume of the cloud. If the cloud is in contact with the surface, a deposition term is present.

It may seem that 2BPUFF and DIFKIN are basically the same, but they describe different phenomena. The DIFKIN column was selected so that it would cross specified receptor points. However, it was assumed to contain species with the same characteristics as in its surroundings, meaning that horizontal gradients are assumed to be small and hence there is no horizontal transport by diffusion processes. The 2BPUFF column, on the other hand, is a puff of material quite distinct from its surroundings; hence horizontal gradients and transport are included. Also, with DIFKIN it is known where the observation will be made; so parcel trajectories are back-tracked to determine the source areas crossed in its passage. In contrast, with 2BPUFF we know where the parcel starts and want to calculate the concentration along or under the prescribed cloud trajectory.

The vertical boundaries of the model are the surface and the inversion at the top of the convective layer (usually a few kilometers) or, in some applications, the tropopause. Possible horizontal growth is unlimited, but eventually concentrations will become so small that the problem need be run no longer. This duration depends on the meteorological conditions that define the horizontal and vertical diffusion coefficients K_r and K_z. The grid in which Eq. 14.23 is solved expands to include the whole cloud as it grows, vertical growth being limited by the surface and the inversion. Because we are dealing with a distinct cloud, the boundary condition in the r-direction is $c = 0$. When the pollutant reaches one of the vertical boundaries, a zero-flux boundary condition replaces the zero concentration, or at the ground surface a deposition flux boundary condition is used.

Initial cloud conditions are Gaussian distributions about the cloud center in both the r- and z-directions. The vertical diffusion coefficient, K_z, is similar in shape to that used by Eschenroeder and Martinez (1973) but extends over the deeper range of the convective layer on the troposphere. Also K_z is a prescribed function of z and the time t. The horizontal diffusion coefficient K_r depends on the horizontal cloud dimension (Richardson, 1926; Batchelor, 1950). The form is

$$K_r(z,t) = \alpha[\epsilon(z)]^{1/3}\,[\sigma_r(z,t)]^{4/3} \qquad (14.24)$$

where α is a dimensionless constant or order 1, ϵ is the atmospheric turbulent energy dissipation rate, and σ_r is a measure of the radial dimension of the cloud. For more discussion of this expression and its significance, see Walton (1973).

The second model (LIRAQ) is Eulerian and is used to calculate the horizontal spread and the chemical changes experienced by pollutants in a region with a characteristic dimension of 100 km (MacCracken et al., 1978). There are two versions of the model: LIRAQ-1 and LIRAQ-2. LIRAQ-1 computes the transport of nonreactive species with emphasis on fine resolution (1 km) and with an advanced advective transport algorithm. LIRAQ-2 computes the transport and chemical evolution of photochemically active species on a coarser grid (5 km) with a simpler transport algorithm. LIRAQ was designed to make maximum use of available regional data. The concentrations were assumed to have logarithmic vertical profiles (ρ = constant).

$$c_i(x,y,z,t) = a_i(x,y,t) + b_i(x,y,t)\ln\frac{z}{z_0} \qquad (14.25)$$

between the surface and the inversion base at H(x,y,t), where a_i and b_i are given in terms of $\bar{c}_i$ (the vertical average concentration), sources, sinks, and the surface vertical diffusion coefficient $K_z(z = z_0)$. The wind velocity has a power-law profile

$$u_x = u = u_0\left(\frac{z}{z_0}\right)^n \qquad u_y = v = v_0\left(\frac{z}{z_0}\right)^n \qquad (14.26)$$

where the subscription 0 refers to a reference level. A third assumption, that the horizontal diffusion coefficient is a weak function of z, is made to obtain the vertical average equation.

The LIRAQ governing equation describes the behavior of the vertically averaged concentration in a zone with horizontal dimensions $\delta x, \delta y$ and height $H(x,y,t)$. This equation is obtained by integrating Eq. 14.8 in z and defining $\bar{f}$,

$$\bar{f} = \frac{1}{H}\int_0^H f(z)\,dz \tag{14.27}$$

Assuming $K_x = K_y = K$, then

$$\frac{\partial}{\partial t}(H\bar{c}_i) = -\frac{\partial}{\partial x}[H\bar{c}_i\bar{u}(1+\beta_i)] - \frac{\partial}{\partial y}[H\bar{c}_i\bar{v}(1+\beta_i)] + \frac{\partial}{\partial x}\left(K\frac{\partial}{\partial x}H\bar{c}_i\right) + \frac{\partial}{\partial y}\left(K\frac{\partial}{\partial y}H\bar{c}_i\right) - T_i + H\bar{s}_i + H\bar{R}_i(\bar{c}_1, \ldots, \bar{c}_N) \tag{14.28}$$

The β_i terms in Eq. 14.28 appear in the relationship between the average of a product and the product of the averages, e.g., $\overline{uc_i} = \bar{u}\bar{c}_i\,(1+\beta_i)$. They are obtained analytically by comparing $(1/H)\int_0^H uc\,dz$ with $\bar{u}\bar{c}$, and their value is

$$\beta_i = \frac{1}{n+1}\frac{b_i}{\bar{c}_i} \tag{14.29}$$

The T_i represents all the terms, advective and diffusive, which account for transport through the not totally impervious inversion base. In T_i a volume correction term of the type described in Sec. 14-4.1 is important since during the day convection can eat away at the inversion base and in doing so can cause entrainment into the volume. The inversion can reform at lower levels at night and isolate some of the pollutant from the volume of interest. The last two terms $H\bar{s}_i$ and $H\bar{R}_i$ are the net effect of sources and sinks and chemical reactions in the volume. For some species the sinks are in the form of a deposition flux.

The horizontal eddy-diffusion coefficient (K) accounts for the transport due to flow scales, which are smaller than the zone size $(\delta x, \delta y)$, and is related to the zone size and the average energy dissipation rate, $K_x = K_y = k = \alpha_1 \epsilon^{1/3}(\sigma_x)^{4/3}$. The surface K_z used in a_i and b_i is proportional to the reference-level wind speed and a standard surface roughness. Finally, for photochemistry, the ambient temperature and the solar intensity are required. The types of reactions used in the chemistry portion of LIRAQ-2 are:

Photolysis	Aldehyde
$O - O_3 - NO_x$	NO_x with organic free radicals
$HO_x - NO_x - HNO_x$	Other organic radical
HO_x	Sulfur–sulfate
Alkene	
Alkane	

Mass consistent (i.e., nondivergent) regional wind fields, an inventory of surface and elevated sources, topography, and time-varying inversion-base heights are needed as inputs to the model.

As usual the lateral boundary conditions pose a problem. It is desired that the area just outside the region behave in somewhat the same manner as that inside. At the same time, if all sources are turned off, concentrations should go to some background value. The problem was handled by assuming that the concentration, c_0, just outside a boundary was

$$c_0 = [c_a^2 + (pc_b)^2]^{1/2} \tag{14.30}$$

The terms in Eq. 14.30 are c_a, an ambient concentration characteristic of the region; c_b, the concentration in the boundary zone; and $0 \leqslant p \leqslant 1$. In this way the region outside reflects (weakly, depending on p) what is going on inside while maintaining some ambient level greater than c_a.

The third two-dimensional model was developed by Widhopf (Widhopf and Taylor, 1974) to study stratospheric meridional ozone distributions over the globe. This model differs from the others we have described in several ways. First, since the calculation is done in the meridional plane, the equations are most conveniently written in spherical coordinates. Second, following Reed and German (1965), off-diagonal elements of the eddy-diffusion tensor are used to treat the large-scale turbulent transport.

A modified form of Eq. 14.7 is used

$$\rho\frac{\partial\chi_i}{\partial t} + \rho\frac{v}{r}\frac{\partial\chi_i}{\partial\phi} + \rho w\frac{\partial\chi_i}{\partial z} = \frac{1}{r}\frac{\partial}{\partial\phi}\left(\rho K_{\phi z}\frac{\partial\chi_i}{\partial z} + \frac{\rho}{r}K_{\phi\phi}\frac{\partial\chi_i}{\partial\phi}\right) + \frac{\partial}{\partial z}\left(\rho K_{zz}\frac{\partial\chi_i}{\partial z} + \frac{\rho}{r}K_{z\phi}\frac{\partial\chi_i}{\partial\phi}\right) + 2\frac{\rho}{r}\left(K_{zz}\frac{\partial\chi_i}{\partial z} + K_{z\phi}\frac{1}{r}\frac{\partial\chi_i}{\partial\phi}\right) - \frac{\rho}{r}\tan\phi\left(K_{\phi z}\frac{\partial\chi_i}{\partial z} + K_{\phi\phi}\frac{1}{r}\frac{\partial\chi_i}{\partial\phi}\right) + R_i + s_i \tag{14.31}$$

In Eq. 14.31, v and w are the meridional and vertical components of the mean winds, ϕ is the latitude, and z is the vertical distance from the surface of the earth at $R_e(r = z + R_e)$. The mean winds, temperature, air density, and solar intensity as well as the eddy-diffusion tensor are needed as data to drive the model.

Chemistry (R_i) was added to the model in steps: (1) the first set of reactions related O, O_2, and O_3 (the Chapman cycle), (2) reactions involving the oxides of nitrogen (NO_x), and (3) hydrogen species reactions. In this way the relative effects of added trace species could be evaluated. This process helped the modelers in assessing the species transport terms, which, because of the lack of data, were not available in some regions.

The Widhopf model has a lower boundary at 10 km and extends to 52.5 km; the troposphere is excluded from the calculation. At the lower boundary the concentrations of O_3 and N_2O and the mass mixing ratios of HNO_3, NO, and NO_2 are specified. The remaining species are assumed to be in photochemical equilibrium at the lower boundary. At the upper boundary, O_3 is also assumed to be in photochemical equilibrium. The mass mixing ratios of NO_2, N_2O, and HNO_3 are continued with a second-order extrapolation at this boundary.

In these two-dimensional models, significant physical processes have been represented in some detail; for example, fairly extensive sets of photochemical reactions have been included in the regional and global models. Some of the questions we might expect to address are: In the global model, are we looking at an "orange section" or some kind of average? In 2BPUFF, how adequately can we treat shear flow? In LIRAQ we have treated vertical structure analytically; how well does this represent the real physical situation?

Clearly, if we are concerned with the adequacy of the assumptions required to make a two-dimensional model treat a three-dimensional world, we can consider three-dimensional models. Then, all components of Eqs. 14.7, 14.8, or 14.10 are treated fully, and our model is as complete as are the assumptions on which the equations are based.

Before this step is taken, serious consideration must be given to the demands that are going to be put on the computer system and data network. For example, if a two-dimensional model treats the chemistry and transport of 10 species on a 20 by 20 horizontal mesh, we are led to the simultaneous solution of 4000 ordinary differential equations. If the region is then zoned in the vertical with 10 levels, we now must solve 40 000 ordinary differential equations, thereby increasing running times and data-storage demand. Furthermore, if we are trying to overcome the use of the analytic vertical wind profile in LIRAQ, we must have the observed winds over the region. These data are frequently sparse, and the problem increases greatly when we ask for vertical structure, which may be available from only one or two meteorological towers in the region. Finally, the same problem applies for species concentration measurements that are required for initialization. If the effort is warranted and trade-offs can be made (e.g., fewer reacting species), full application of the equations can be made.

14-4.4 Three-Dimensional Models

To the best of our present knowledge, there are no purely Lagrangian models in three dimensions. The first Eulerian model presented is that of Reynolds, Roth, and Seinfeld (1973). Equation 5 in the cited work, derived by Reynolds, is equivalent to our Eq. 14.8, with $K_x = K_y = K_H$, $K_z = K_V$ and the air density, ρ, held constant. Of particular interest in this reference is the discussion of the conditions for validity of the governing equation. The inquisitive reader is directed to other related works (Lamb, 1973; Lamb and Seinfeld, 1973).

The Reynolds model has been applied to Los Angeles. Because Eulerian models are extremely complex, computation time and data storage have been reduced by adjusting the boundary to exclude as many source-free grid squares as possible from the calculation. In treating the spatially varying surface height h(x,y) and space- and time-varying inversion-base height H(x,y,t), the authors have made the change of vertical variable

$$z \rightarrow \zeta = \frac{z - h(x,y)}{\Delta H(x,y,t)} \tag{14.32a}$$

where

$$\Delta H(x,y,t) = H(x,y,t) - h(x,y) \tag{14.32b}$$

The boundaries are thus, in a sense, fixed not only in the horizontal but also in the vertical ($\zeta = 0 \rightarrow 1$). The equation resulting from the substitution of Eqs. 14.32a and 14.32b into Eq. 14.8 is quite complicated, including derivatives of h and ΔH with respect to x,y and also for ΔH with respect to t. The equation is simplified by the assumption that the spatial derivatives of h and ΔH are small, giving

$$\begin{aligned}\frac{\partial}{\partial t}(\Delta H c_i) = &- \frac{\partial}{\partial x}(u \Delta H c_i) - \frac{\partial}{\partial y}(v \Delta H c_i) - \frac{\partial}{\partial \zeta}(W \Delta H c_i) \\ &+ \frac{\partial}{\partial x}\left(K_H \Delta H \frac{\partial c_i}{\partial x}\right) + \frac{\partial}{\partial y}\left(K_H \Delta H \frac{\partial c_i}{\partial y}\right) \\ &+ \frac{\partial}{\partial \zeta}\left(\frac{K_v}{\Delta H}\frac{\partial c_i}{\partial \zeta}\right) + R_i \Delta H + s_i \Delta H\end{aligned} \tag{14.33a}$$

where $W = w - \zeta/\partial \Delta H/\partial t$. Equation 14.33a looks much like Eq. 14.28 in Sec. 14-4.3 in which a vertical average had been performed, giving the $(1 + \beta_i)$ terms and T_i representing a lumped vertical transport term. Equation 14.33a is differenced on a 2-mile by 2-mile grid in the horizontal and in five vertical levels.

The horizontal eddy-diffusion coefficient, K_H, is a constant in this model. The vertical diffusion coefficient, K_V, is similar to that of Eschenroeder and Martinez (1973), i.e., it increases linearly through the first two levels of the model, is constant over the next two levels, and decreases linearly through the top level as the inversion base is approached. As in DIFKIN, the constant value of K_V through levels 3 and 4 (and hence the slopes above and below) is proportional to the horizontal wind speed in the zones. Source emissions for this model are discussed in detail by Roth et al. (1974). Time-varying winds and inversion-base height are required in a nondivergent form and are discussed by Reynolds, Roth, and Seinfeld (1973). Topography is of course required in the ΔH terms of the equation.

The chemistry in this model consists of 15 reactions among 10 species. Differential equations are integrated in time for the concentration of NO, NO_2, O_3, and hydrocarbons. The species O, OH, HO_2, RO_2, and NO_3 are assumed to be in pseudo-steady state, whereas HNO_2 is considered to be in steady state.

A flux boundary condition is used at the surface, equating the diffusive flux normal to the topography to a specified surface flux. Conditions at the inversion base and at the sides of the region are statements of the continuity of the normal component of the mass flux at these boundaries. For outflow the normal component of the diffusive flux is zero. Inflow is at a concentration equal to that just outside the boundary.

Recently another three-dimensional Eulerian model has been developed by Shir and Shieh (1974). The model was applied to diagnose the SO_2 distributions during 25 consecutive days in the St. Louis, Mo., metropolitan area. In this model the equivalent of Eq. 14.33a did not contain ΔH and $W = w$, the vertical component of the wind; so the concentration equation for SO_2 took the more classical form,

$$\frac{\partial c}{\partial t} = -\left(\frac{\partial uc}{\partial x} + \frac{\partial vc}{\partial y} + \frac{\partial wc}{\partial z}\right) + K_H \left[\frac{\partial}{\partial x}\left(\frac{\partial c}{\partial x}\right) + \frac{\partial}{\partial y}\left(\frac{\partial c}{\partial y}\right)\right] + \frac{\partial}{\partial z}\left(K_z \frac{\partial c}{\partial z}\right) + R + s \qquad (14.33b)$$

Using meteorological and source inventory data, this three-dimensional model diagnosed the temporal and spatial concentration distributions resulting from multiple-point and area sources. The strength of the article by Shir and Shieh lies not only in its explanation of the model but also in its verification of the model simulations for the 25-d period.

The third model, Atmospheric Diffusion Particle in Cell (ADPIC), is a Particle-in-Cell (PIC) model. PIC models are hybrid models that are partially Eulerian and partially Lagrangian in nature. They solve the three-dimensional advection-diffusion equation in its flux conservative form for a given mass-conservative advection field, with the pollutant concentration represented statistically by embedded Lagrangian marker particles in a Eulerian grid (Welch et al., 1965; Amsden, 1966; Sklarew, Fabrick, and Prager, 1972). The ADPIC model developed by Lange (1973, 1978) can calculate the time-dependent distribution of pollutants under such conditions as strongly distorted advection wind fields, calm conditions, space-variable surface roughness, wet and dry deposition, radioactive decay, and space- and time-variable diffusion parameters.

The ADPIC model represents the concentration field by a distribution of discrete particles at points r_α^p (p is the index of a specific particle). The computational cycle is composed of a Eulerian step (without change in time), in which a pseudovelocity is generated using the concentration field, followed by a Lagrangian step (advancing in time), in which the particles are moved by the pseudovelocity field. The change in particle locations by the pseudovelocity field gives a new concentration distribution, thus completing the computational cycle.

The pseudovelocity, V_α, is a mathematical artifice that can be derived using Eq. 14.7, written without the species index (i). We consider the region to be one of constant air density, ρ, in order that the equation can be written in terms of the concentration c. The species subscript is dropped because the particle approach does not lend itself readily to treatment of a number of chemically active species. For this reason the reaction term R is usually simple decay, and s contains sources plus surface deposition and washout. Equation 14.7 becomes

$$\frac{\partial c}{\partial t} = -\partial_\alpha (u_\alpha c) + \partial_\alpha (K_{\alpha\beta} \partial_\beta c) + R + s \qquad (14.34)$$

Collecting the partials with respect to x_α and replacing $K_{\alpha\beta}\partial_\beta c$ with $[c(K_{\alpha\beta}/c)\, \partial_\beta c]$,

$$\frac{\partial c}{\partial t} = -\partial_\alpha \left[c\left(u_\alpha - \frac{K_{\alpha\beta}}{c}\, \partial_\beta c\right)\right] + R + s \qquad (14.35)$$

The term $-(K_{\alpha\beta}/c)\, \partial_\beta c$ is the diffusivity velocity, and

$$V_\alpha = u_\alpha - \frac{K_{\alpha\beta}}{c}\, \partial_\beta c \qquad (14.36)$$

is the pseudovelocity. The equation governing c now becomes

$$\frac{\partial c}{\partial t} = -\partial_\alpha (V_\alpha c) + R + s \qquad (14.37)$$

which is still exact.

The pollutant is represented by a distribution of particles at the points $r_\alpha^p(t)$, where p runs over the total number of particles still in the problem domain. The advective winds, u_α, are specified in the grid; for example, u_α is obtained from an objective analysis model on a regional scale (see Sec. 14-5). In the Eulerian part of the cycle, the gradient field $\partial_\beta c$ is generated, giving the diffusive velocity $-(K_{\alpha\beta}/c)\, \partial_\beta c$. The pseudovelocity, V_α, is calculated using Eq. 14.36 and can be interpolated to each particle location; so now a velocity V_α^p is associated with each particle. In the Lagrangian step each particle moves, in time δt, to a new location,

$$r^p_\alpha(t + \delta t) = r^p_\alpha(t) + V^p_\alpha \, \delta t \tag{14.38}$$

Diffusivity coefficients are used that are valid for single puff or continuous plume releases that can grow in size from sources on the order of meters to debris clouds in the kilometer scale. For time periods on the order of 12 h and geographical radii on the order of 100 km, turbulent diffusivities are scale dependent (Batchelor, 1950; Walton, 1973), and Eq. 14.24 is used. For the vertical diffusivity coefficient in the surface layer, the turbulent diffusion due to both the mechanical shear induced by the drag of the earth's surface on the atmosphere and the vertical thermal convection within the atmosphere is taken into account through the parameter $\delta = z/L$ (Monin and Obukhov, 1954; Obukhov, 1971), where

$$L = -\frac{u_*^3 C_p \rho T}{kgQ} \tag{14.39}$$

and u_* = friction velocity
C_p = specific heat of air at constant pressure
ρ = density of air
T = temperature
Q = heat flux
g = gravitational acceleration
(see Chap. 5, Sec. 2.7)

L can be found from its functional relationship to the Richardson number, R_i, as in Businger et al. (1971). The formula for K_z is

$$K_z = u_* \, kz \frac{1}{\phi_m} = \left(\frac{kz}{\phi_m}\right)^2 \frac{\partial \bar{u}}{\partial z} \tag{14.40}$$

where k is the von Karman constant, $\phi_m = (kz/u_*)(\partial\bar{u}/\partial z)$ is a dimensionless wind shear, and $\bar{u}$ is again the average horizontal wind speed. Above the surface layer, it is assumed that K_z remains constant up to the mixing or inversion-base height, where it sharply drops to a minimum.

Owing to the part-Eulerian, part-Lagrangian makeup of ADPIC, the boundary conditions are broken into two separate conditions, one imposed on the Eulerian velocity field and one on the Lagrangian particles. Both sets must be consistent with each other. The two basic boundary conditions imposed on the pseudovelocity field, V_α, are constant mass flux, (cV_α) = const., corresponding to inflow and outflow of particles, and zero mass flux, $(cV_\alpha) = 0$, corresponding to reflection of particles from an impermeable boundary. There are intermediate cases as, for example, deposition of particles at the topography, in which case a deposition velocity is specified. The concentration field is sufficiently smooth by the time it reaches the outflow boundary that the boundary velocities can be specified by postulating a constant flux of particles through the boundary.

14-5 WIND FIELD ANALYSIS

In developing and using a complex model, the modeler is faced with information that is not completely adequate, from crudely defined source inventories to incomplete reaction sets with uncertain rate coefficients. One of the meteorological variables that can cause errors is the wind field, which in most two- and three-dimensional regional models is the major transport mechanism. In addition to being sparse, when wind field data are interpolated onto a model grid, the resulting winds may have nonphysical divergences. For primitive equation forecast models, this means that improper initialization will occur (Paegle and Paegle, 1976). For air-quality models, erroneous concentration fields will reflect this characteristic (Dickerson, 1978). Although the modeler may not be able to obtain better data, there are techniques he can use to make the wind field input physically consistent. We will discuss two approaches to the problem of providing wind field input to air-quality models. The first approach is based on the conservation of mass, and the second, on the conservation of mass and total energy. The idea is to improve wind fields interpolated from observed data by using physical constraints that are known to govern the system.

The first approach (Sherman, 1978) is based on the work of Sasaki (1958, 1970a, 1970b). It requires that the deviation of the adjusted winds from the observed winds be minimized in the least-squares sense while satisfying the condition of nondivergence.

In three dimensions the function is defined as

$$E(u,v,w,\lambda) = \int_V \Big[\alpha_1^2 (u - u^o)^2 + \alpha_1^2 (v - v^o)^2 + \alpha_2^2 (w - w^o)^2 + \lambda \left(\frac{\partial u}{\partial x} + \frac{\partial v}{\partial y} + \frac{\partial w}{\partial z}\right)\Big] dx\, dy\, dz \tag{14.41}$$

where x,y = horizontal directions
z = vertical direction
u,v,w = adjusted velocity components in the x,y,z directions
u^o, v^o, w^o = corresponding observed variables (w^o is taken to be 0 since it is rarely available)
$\lambda(x,y,z)$ = Lagrange multiplier
α_i = Gauss precision moduli reflecting observation errors and/or deviations of the observed field from the desired adjusted field

The associated Euler–Lagrange equations, which with the nondivergence constraint minimize Eq. 14.41, are

$$2\alpha_1^2 (u - u^o) + \frac{\partial \lambda}{\partial x} = 0 \tag{14.42a}$$

$$2\alpha_1^2 (v - v^o) + \frac{\partial \lambda}{\partial y} = 0 \tag{14.42b}$$

$$2\alpha_2^2 (w - w^o) + \frac{\partial \lambda}{\partial z} = 0 \tag{14.42c}$$

with the nondivergence constraint

$$\frac{\partial u}{\partial x} + \frac{\partial v}{\partial y} + \frac{\partial w}{\partial z} = 0 \qquad (14.42d)$$

Differentiating and using Eq. 14.42d gives

$$\frac{\partial^2 \lambda}{\partial x^2} + \frac{\partial^2 \lambda}{\partial y^2} + \left(\frac{\alpha_1}{\alpha_2}\right)^2 \frac{\partial^2 \lambda}{\partial z^2} = -2\alpha_1^2 \left(\frac{\partial y^o}{\partial x} + \frac{\partial v^o}{\partial y} + \frac{\partial w^o}{\partial z}\right) \qquad (14.43)$$

Boundary conditions become conditions on either λ or its normal derivative $(\partial\lambda/\partial n) = (\partial\lambda/\partial x)$, $(\partial\lambda/\partial y)$, $(\partial\lambda/\partial z)$. In the first case, if $\lambda = 0$, an open or flow-through condition pertains. If $(\partial\lambda/\partial n) = 0$, then no adjustment of the field takes place on the boundary.

Given the solution to Eq. 14.43, Eqs. 14.42a to 14.42c provide u, v, and w in terms of known quantities.

In a second approach (Ohmstede, 1976; Meyers et al., 1976), an energy-conservation equation is added to that for mass conservation. This approach treats equations integrated in the vertical through layers separated by material interfaces, the integral of a variable α being denoted by $\langle\alpha\rangle$. Use is also made of moments of the variables, the *nth* moment of α being denoted by $\langle\alpha/n\rangle$.

The form of the mass-conservation condition includes possible density variation

$$\frac{\partial\langle p\rangle}{\partial t} + \frac{\partial\langle m\rangle}{\partial x} + \frac{\partial\langle n\rangle}{\partial y} = 0 \qquad (14.44a)$$

in which ρ is air density and m and n are the x and y components of momentum. The other equation is for energy, E,

$$\frac{\partial\langle E\rangle}{\partial t} + \frac{\partial\langle U(E+p)\rangle}{\partial x} + \frac{\partial\langle V(E+p)\rangle}{\partial y} + p_1 \frac{\partial H}{\partial t} = C_p \zeta_0^S + k\nabla^2 K \qquad (14.44b)$$

In Eq. 14.44b, E includes kinetic energy (K) and potential and internal energy. The x and y velocity components are U and V, p is the pressure, and p_1 is the pressure at a material interface located at height H. The air—earth heat exchange is ζ_0^S. With the inclusion of the internal energy, a state equation must also be supplied.

The functional to be minimized now contains the zero and first-order moments of the momentum equation and the energy equation. The resulting set of equations to be solved contain up to thirteen fields: three residual fields, five Lagrange multipliers, the field for the variation of the depth of the layer, observed fields of the layer depth, first-order moments of the momentum, and terrain-height variation.

14-6 MODEL VERIFICATION

An essential adjunct to numerical modeling is verification. By verification we mean the verification of complex models through the use of independently determined parameters in the calculations and a comparison of the results to actual data. Clearly, the limitations and solution sensitivity of regional- to global-scale models should be examined and understood as fully as possible before they are used in assessments contributing to important policy decisions. In our view, model verification must be a continuing activity in order to test a model over a wide spectrum of occurrences and to develop statistics on model performance under full operational or field conditions. Intermodel comparisons and verifications require parallel calculations for common data bases for many occurrences as in the verification of five mesoscale wind-prediction models carried out by Cornett and Randerson (1977) and as in the air-pollution model verified by Shir and Shieh (1974). The current status of research is such that each investigator tends to verify his model on separate geographical or regional data bases, and then reports his findings. Because regional, and to some extent global, modeling is relatively recent, in-depth comparisons using common data are lacking.

Some areas important in model verification include, but are not limited to, the following:

—Inherent accuracy of the mathematical framework of making the calculations.
—Accuracy of available input data and model sensitivity to input uncertainties.
—Realism of representation of important processes.
—Appropriateness of assumptions in the model when applied to real situations.
—Simple experimental checks (e.g., the performance of the chemistry submodel against smog chamber results).
—Performance of the model under full field tests.

Many of the concerns that we raise in this section on model verification have been identified and discussed by Pasquill (1974). Our purpose is to describe areas in which some significant progress has been or is being achieved. Most of the above questions imply a comparison of model solutions with data. In general, emission inventory information, model resolution, and the spatial resolution of the field data need to be consistent for such comparisons to be meaningful. Prescribing source-term information by county, obtaining the model solution on a 1- to 5-km grid in the horizontal plane, and then comparing the results to street-corner sampling data is an example of perfect inconsistency.

In the sections that follow, we will use various illustrative results as examples of the preceding aspects of verifications.

14-6.1 Inherent Accuracy of the Mathematical Framework

One of the traditional tests in theoretical meteorology for model validation is the comparison of the model results to closed-form analytical solutions. Such a test is a measure

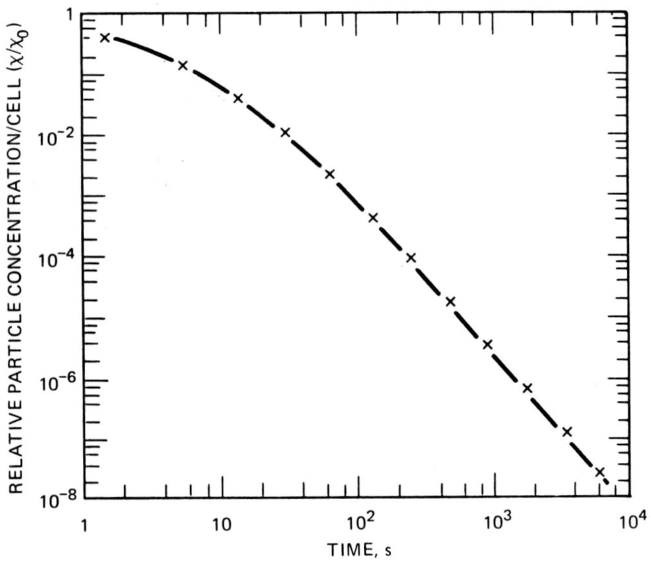

Fig. 14.2 Relative cloud center concentrations for puff diffusion in simple shear flow. ——, analytical solution. x, ADPIC.

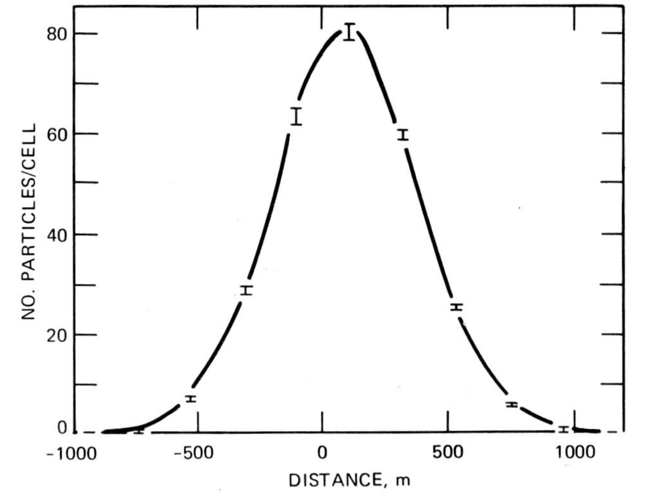

Fig. 14.3a Cross section of cell concentration vs. distance, T = 6254 s. ——, analytical solution. I, ADPIC.

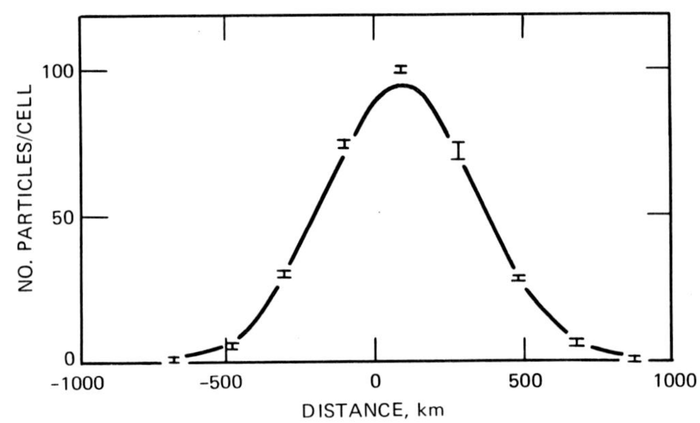

Fig. 14.3b Cross section of cell concentration vs. distance, T = 6328 s. ——, analytical solution. I, ADPIC.

of the inherent accuracy of the solution method. For example, Molenkamp (1968), Crowley (1968), and Long and Pepper (1976) have examined a number of prescriptions for the advection term in the transport equation. Ideally such comparisons are a part of the modelers' selection of a method of solution and hence normally occur at early stages of model development. Nonetheless, it is appropriate to include this topic in a discussion of model verification.

Extensive verification of the three-dimensional transport–diffusion model ADPIC has been achieved using closed-form analytical solutions for both puff and continuous-point sources in constant wind fields as well as those with wind shear (Lange, 1973). These tests indicate that the PIC models with the pseudo-velocity approach can be quite accurate. ADPIC was found typically to be within 5% of the analytical solution for extended periods of integration. This type of comparison can be seen in Fig. 14.2 which shows ADPIC vs. the analytical solution for the relative cloud center concentration for puff-type initial conditions and simple shear flow. The comparison of the puff shape at two times is shown in Figs. 14.3a and 14.3b. Computational experience indicates that even these reasonably small errors can be further reduced by finer Eulerian zoning and/or by increasing the number of Lagrangian marker particles. However, solution accuracies of the order of 5% or less appear to be quite adequate. Given such accuracies in the mathematical framework, benefits from improved knowledge of regional transport and diffusion processes can be realized.

In an attempt to overcome some of the problems inherent in finite difference techniques, some investigators are now adopting the method of finite elements for use in atmospheric flow problems (Strang and Fix, 1973; Gresho, Lee, and Sani, 1976). Using a finite set of local basis functions, they convert the governing partial differential equation to a finite system of ordinary differential equations. Some of the benefits that may accrue are high accuracy on coarsely meshed domains, easy and accurate representation of both complex terrain and boundary conditions, and inherent computational stability resulting from satisfaction of important global conservation laws (e.g., kinetic energy and mean square vorticity).

14-6.2 Accuracy of Input Data

Modelers and users of air-quality simulation models should be aware of the accuracy of input data when seeking solutions to assessment problems (Ellsaesser, 1960; Culbertson and Willis, 1964; Pasquill, 1974). It is unreasonable to expect model results to be better than the quality of the input information except in fortuitous cases that should be recognized as such and not portrayed as normal model performance. Clearly this is a problem for the modeler almost from the beginning because uncertainties in input

TABLE 14.1 Estimated Uncertainties of Various Data Available in the San Francisco Bay Area*

Input measurements	Horizontal resolution, km	Temporal resolution, h	Estimated uncertainty
Surface wind	~10	3	Speed 20%; direction 20°
Inversion height	~50	3–12	~50%
Surface concentrations	~20	1	20% or 1 unit, whichever is larger
Mobile sources	1	1	25%
Stationary point sources	0.5	1	20%
Stationary area sources	1	1	x2†
Basic photolysis	~35‡	1	30%
Primary reaction coefficients			<50%
Minor reaction coefficients			< x 10
Reactivities of hydrocarbon input			< x 2
Topographic height	1		25 m

*Based on MacCracken and Sauter (1975).
†Measurement is estimated to be accurate to within a factor of 2.
‡Separation distance of pyranometer measurement stations.

data will color the appraisal of the model as soon as verification is started. In regard to regional air-quality models, estimates have been made of these uncertainties in input information to the LIRAQ-2 model for reactive pollutant simulation. These uncertainty estimates are shown in Table 14.1. The Bay Area Air Pollution Control District (BAAPCD) is a primary source of hourly average values of oxidant, NO, NO_2, CO_2, and hydrocarbon. It should be noted that these station measurements are point measurements, whereas most models predict average concentrations over some volume of space, in the case of LIRAQ-2, 5- by 5-km grid elements. This difference is particularly important when the monitoring station is located near a major point source. Since air-quality measurements for a 1-h averaging time are estimated to be accurate to 20%, it would be unreasonable to expect model solutions to be more accurate than that figure in evaluating a large suite of simulation results. Such uncertainties in the input to complex models should not, in principle, detract from their use in assessment activities. One technique is to compare control runs to experimental model runs in which single changes in source inventory or control strategy are assumed. Solution differences should have significance, provided the calculational framework has sufficient accuracy and the framework is reasonably free of noise. In more-complex nonlinear models, and probably in three-dimensional models, this latter aspect of the solution may be important since noise has the potential for masking the effect being evaluated, either partially or totally. Noise might even be interpreted as the effect in some cases. Techniques for treating noise are starting to emerge in the modeling of global climate where global perturbations of climate are being evaluated in controlled runs (Warshaw, 1973; Chervin, Gates, and Schneider, 1974).

Another noteworthy point is that some of the error in observed winds need not be passed along to the air-quality simulation model. Errors in wind measurement and in exposure of wind sensors can lead to fictitious fields of convergence and divergence in interpolated wind fields which can seriously flaw solutions to the concentration prediction equation. For example, 100% errors in the concentration field can occur if unadjusted wind data are used (Dickerson, 1976). If regional mass-consistent wind field models of the type discussed in Sec. 14.5 are used, the divergence in the interpolated wind field can be reduced from 10^{-3} s^{-1} to 10^{-10} s^{-1} or less. Experience indicates that this procedure of error control is a necessity in Eulerian regional air-quality simulations.

As an illustration of what can be done, we present a sample set of results using the MATHEW code from Sherman (1978). The data were from the Idaho National Engineering Laboratory (INEL). The processing was performed on a 25 by 33 by 28 grid with horizontal spacing of 4.3 km and vertical spacing of 50 m.

Figures 14.4 through 14.7 illustrate the process by which each data set is derived and adjusted. The horizontal wind measurements shown in Fig. 14.4 are interpolated (with $1/r^2$ weighting) across the grid at a constant height of 50 m above the topography. The resulting interpolated field is plotted in Fig. 14.5. The lowest elevation in the grid is used as the reference level, and horizontal wind components are specified at each grid point, the vertical variation being determined by a power law within the surface layer; the interpolated field at 50 m is shown in

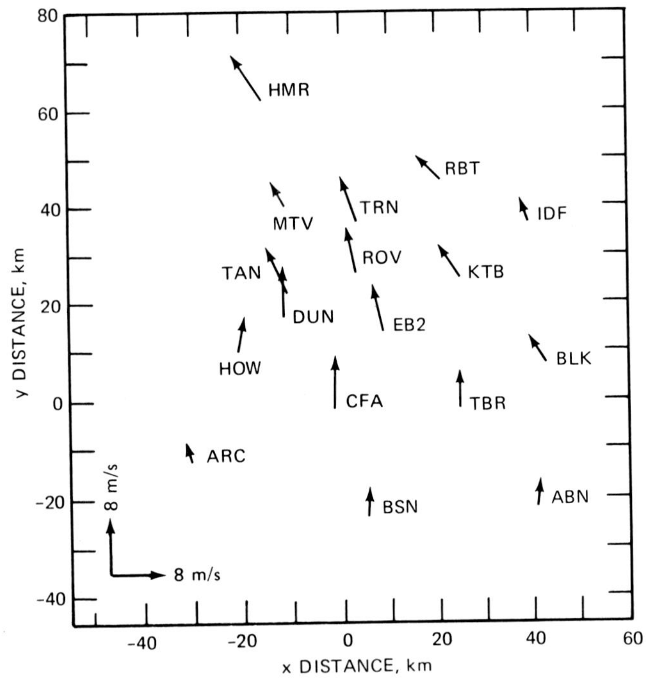

Fig. 14.4 INEL tower wind measurements adjusted to 50 m above the surface for 1330 MDT.

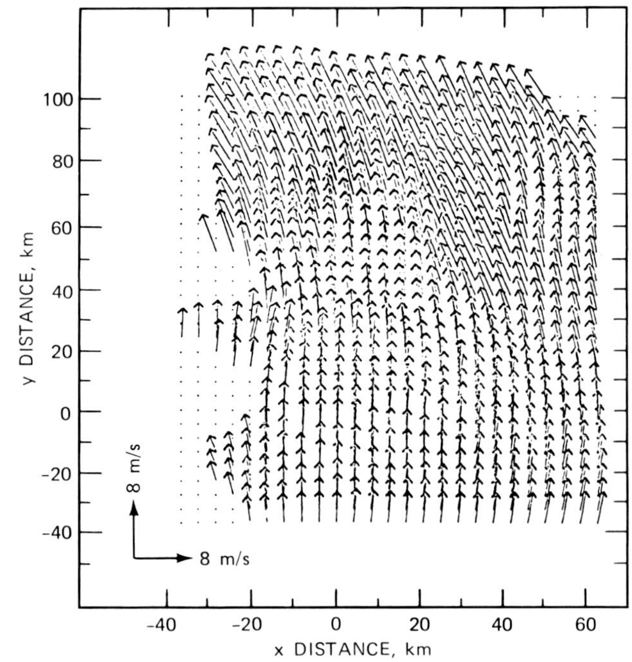

Fig. 14.6 INEL interpolated wind field 450 m above the reference level for 1330 MDT.

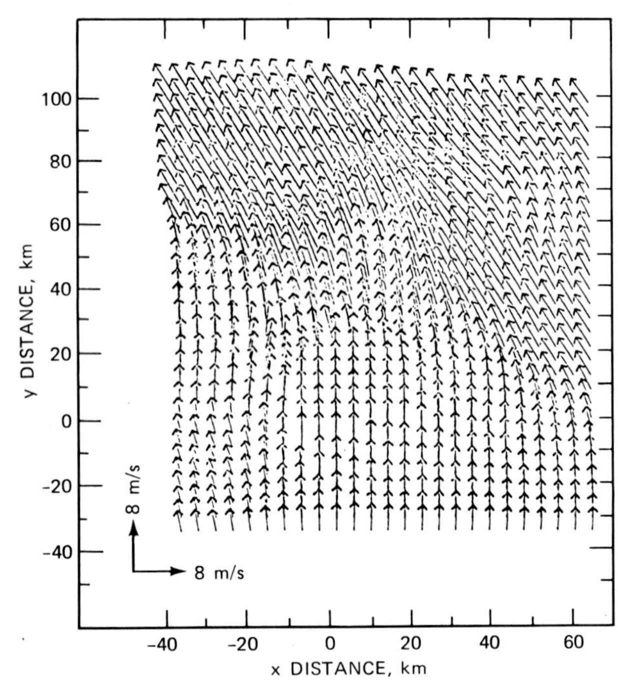

Fig. 14.5 INEL interpolated wind field 50 m above the topography for 1330 MDT.

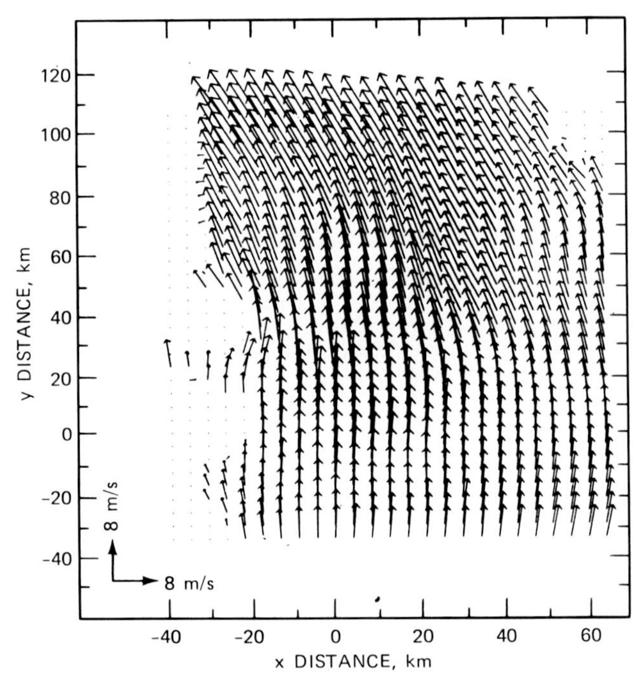

Fig. 14.7 INEL adjusted horizontal wind field 450 m above the reference level for 1330 MDT.

Fig. 14.5. The resulting interpolated horizontal velocity field at 450 m above the reference level is shown in Fig. 14.6. The entire three-dimensional velocity field is adjusted using the MATHEW model. Figure 14.7 shows the adjusted horizontal velocities at the 450-m level. Areas where no wind vectors are plotted represent grid points that are below the prescribed topography. The original and resulting divergences at the 450-m level, shown in Figs. 14.8 and 14.9, illustrate the effectiveness of this approach in reducing the residual divergence. The divergence values are reduced some 12 orders of magnitude in the mass-adjusted field.

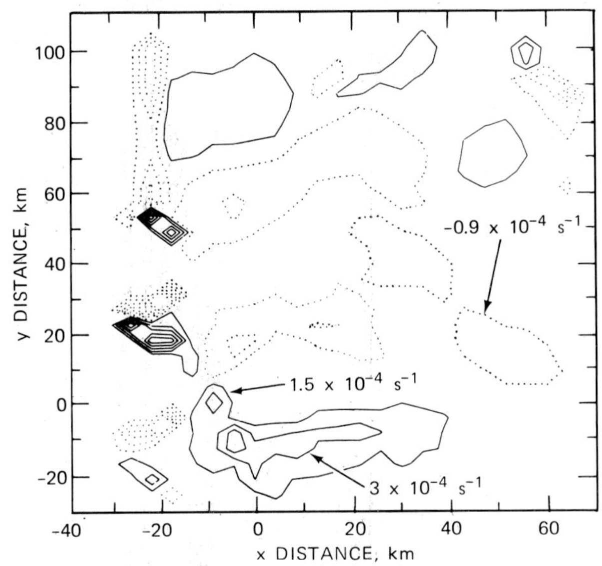

Fig. 14.8 INEL unadjusted divergences at 450 m above the reference level for 1330 MDT.

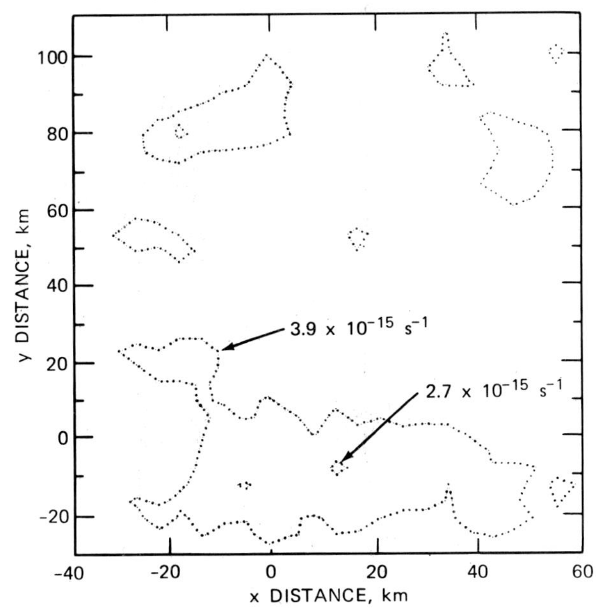

Fig. 14.9 INEL adjusted divergences at 450 m above the reference level for 1330 MDT.

14-6.3 Accuracy of Representation of Important Processes

Assuming the modeler has achieved consistency in resolution of the source inventory, resolution of the model, and the spatial resolution required for an assessment, there remain questions about how well the important processes are represented. For example, there are questions about

—Realism of the vertical eddy profile used in the simulation.

—Stability of the calculated regional mass-consistent wind field to decreasing $\Delta x, \Delta y$.

—Solution sensitivity to the inclusion of new chemical species or reactions.

An example of the latter is given below.

Currently, reasonably complete kinetic calculations for regional air pollution involve 40 to 50 species and nearly 100 reactions. The model developer is forced by the limitations of even the largest computers to simplify such reaction sets in the construction of multidimensional models. One technique is to ignore those reactions with rates less than about 1% of net competing processes as determined from smog chamber tests. These techniques may not yield a small enough set, in which case the modeler has resorted to lumped organic mechanisms. Use of a detailed mechanism for a single compound or other simplifications must be defended by comparison with the net observed process in smog chamber experiments. The effects of omitted reactions are often explored through the use of sensitivity calculations. There is, however, a lingering concern in depending on smog chamber data because data are obtained at relatively high initial concentrations, the radiation environment is often not well controlled, and wall reactions may play a significant role. Furthermore, the data from different chambers do not yield reproducible results. Under such circumstances, full field-scale verification studies are required to achieve a regional air-pollution model that can be used with reasonable confidence.

14-6.4 Appropriateness of Model Assumptions

Normally the burden of justifying key assumptions is left to the individual modeler to argue in the course of the development and presentation of his model. However, it is rare for the validity of key assumptions to be tested in full field studies. Usually the evaluation of errors so induced is included in the comparison of the overall model results to measurements from case studies. In complex models it is difficult to argue that this plan is unreasonable. On the other hand, it is reasonable to suspect that the case for confidence in models would be strengthened by suitable tests of key assumptions in actual field tests. We illustrate this aspect of model validation by (A) a discussion of the three key assumptions that make the concentration prediction equation integrable in LIRAQ and (B) field tracer tests of the assumption in LIRAQ that the inversion base is a lid on upward diffusion in subregions of the San Francisco Bay where the zones are remote from elevated heat sources.

A. From Sec. 14-6.3, it will be recalled that the authors (MacCracken et al., 1978) made three approximations to integrate the basic equation to obtain one for the vertically averaged concentration.

1. It was assumed that horizontal winds follow a simple power-law profile

$$u = u_0 \left(\frac{z}{z_0}\right)^n$$

and

$$v = v_0 \left(\frac{z}{z_0}\right)^n$$

The choice of n = 1/7 for neutral stability conditions was reported by Hess (1959). Experience indicates that the layer under the marine inversion is very nearly neutral during the period of evolution of photochemical pollutants. However, the subsidiary assumption, that there is no turning of the wind with height, is not supported by data; a point that was recognized and acknowledged by the authors.

2. The concentrations were considered to have logarithmic profiles

$$c_i(x,y,z,t) = a_i(x,y,t) + b_i(x,y,t) \ln \left(\frac{z}{z_0}\right)$$

This approximation was based on the assumption that, although vertical diffusion in the subinversion layer is not the predominant factor (horizontal advection and some chemical reaction terms are greater), it is a persistent process that tends to bring the profile above z_0 toward vertical equilibrium, satisfying

$$\frac{\partial}{\partial z}\left(K_z \frac{\partial c_i}{\partial z}\right) = 0$$

(sources above z_0 being considered negligible). The subsidiary assumption, related to assumption 1, that turning in the horizontal wind is a weak function of z gave a $K_z \sim z$ (Pasquill, 1974), hence the logarithmic profile. A more serious assumption is that chemical interactions do not affect the profile shape.

3. The final approximation, that horizontal eddy diffusion is only weakly dependent on z, also goes back to assumption 1 for its support. Using scale-dependent diffusion arguments, the authors obtained a horizontal eddy-diffusion coefficient

$$K_H(z) \sim z^{n-1/3} \simeq z^{-1/5}$$

when n = 1/7 in the simple power-law profile.

B. To support the assumption that a well-developed inversion forms a lid on upward transport of material, MacCracken and Sauter (1975, Appendix 7-3) performed a series of experiments. In these studies, fluorescent particles were released above the inversion, and traverses were made at successively lower altitudes after each one in which the cloud was detected.

The following qualitative conclusions could be drawn from the successful runs:

1. Downward spreading was quite restricted, no more than about 100 m in travel time of approximately 1 h.
2. Horizontal spreading was in conformity with the expected stabilities.

Hence no unexpected mechanisms operated to accelerate transport through the inversion, and it was concluded that vertical transport across the inversion could be modeled by a statistical transport theory with parameters appropriate to the strength of the inversion.

These general conclusions are supported by lidar observations of aerosol concentrations through marine temperature inversions. Lidar data show that the inversion base and the top of the particulate-loaded boundary layer are nearly coincident (Uthe and Russell, 1974).

14-6.5 Experimental Checks

The potential list of experimental checks on parameters that determine the relative importance of processes represented in numerical models is very large, and we will indicate only a few of these. For reactive pollutant models, one can use

—Smog chamber results to confirm the reaction rates and the specific reactions that are included in simulations.
—Small-scale experiments on the absorption rate of sulfur dioxide or ozone by plants under varying conditions of plant development.
—Wind tunnel and other small experiments designed to measure the deposition velocity of particulates of various sizes and simulated surface roughness.
—Small-scale field experiments to measure the absorption rates and desorption rates of surface layer tritium with plant or vegetation water reservoirs.
—Water tow tanks to simulate the dispersion of pollutants in the boundary layer in the presence of complex terrain.

It is beyond the scope of this chapter to extend this list or to explain in detail the above experiments and their results. However, the preceding list is representative of the interaction between the modeling community and the experimenters in the quest to define processes and their controlling parameters, which must be included in reasonably complete regional assessments. Many of these parameters and processes are important for assessment because they represent processes of transfer of pollutants from one reservoir (the air) to the soil or to food chains.

Because of the importance of these processes in determining exposure of man to many substances in his environment, the need for appropriately designed small-scale experiments will be long lasting.

14-6.6 Overall Performance of Models

Information on the overall performance of regional air-pollutant models has been long in coming, owing in part to the relatively large resources required to gather the necessary regional air-quality and meteorological data. During the past several years, the MATHEW/ADPIC models have been verified against several regional tracer studies. These studies have included methyl iodide tracer studies at the Idaho National Engineering Laboratory (INEL), Idaho Falls, Idaho, and ^{41}Ar plumes at the Savannah River Plant (SRP), South Carolina. The agreement between measurements (made in two and three dimensions) and calculations has been most encouraging; 60% of the time, calculated concentrations are within a factor of 2 of the observed values. This level of comparison is impressive because of the long distances and hence large range of concentrations involved. In these studies, measurements were taken at distances of 4 to 80 km from the source and included high-volume surface samplers for the methyl iodide and surface and airborne measurements of gamma energy from ^{41}Ar. Figure 14.10 shows the correspondence of the measured and calculated concentrations, normalized to a unit rate release, from ranges of a few kilometers to as far as 80 km and in the presence of complex terrain. In general, the factor-of-2 agreement 60% of time reported pertains mainly to comparison under or in the central portions of the tracer plume. The larger errors occur on the sides of the tracer plume where slight displacements in the plume can result in relatively large fractional errors but reasonably small absolute errors. A relatively small error, say a few degrees in the transport wind vector, can result in errors in the concentration field of at least a factor of 2 in regional problems. Hence the model performance documented to date appears to be reasonable in comparison to the expected input errors to regional models.

Another example is LIRAQ, a Eulerian regional air-quality model that incorporates mass-consistent advection, eddy diffusion, and photochemical kinetics. This model presently exists in two versions. LIRAQ-1 treats the transport of pollutants without representing detailed photochemical kinetics. LIRAQ-2 treats rather complex chemical reaction sets (15 active species and 50 reactions) in detail. It uses a modified Gear package to solve the large sets of coupled ordinary differential equations with a high-order implicit method (Gear, 1971; Hindmarsh, 1974). Typical domains for LIRAQ-1 and LIRAQ-2 are shown in Fig. 14.11.

The raw data processed by the LIRAQ submodels for subsequent input to the main program (MacCracken, 1975) are:

Geographic and topographic features of the region of interest
Hourly observations of species concentrations
Meteorological observations, e.g.,
- —Wind fields for transport
- —Inversion-base heights
- —Atmospheric transmission coefficients that affect photodissociation rates

Source inventories, which consist of
- —Mobile, hourly
- —Population, by census, daily
- —Point, by grid square
- —Airport, time independent

The pre-processing submodels and the main-code structure, including the executive processing routines, are shown in Figs. 14.12 and 14.13.

A 24-h simulation problem (LIRAQ-2) using a 20 by 20 mesh presently takes approximately 1 h of 7600 computer time (Fig. 14.14). For example, it requires about 11 min to run a 12-h problem from 0600 PST (approximately 30 min after sunrise) to 1800 PST (approximately 30 min before sunset). Sunrise (i.e., sunrise until 10 min after sunrise) requires approximately 4 min, and similar time is taken at sunset owing to the effects of rapidly varying photodissociation coefficients and species concentrations. Meteorological updates, which result in discontinuity in transport coefficients, also cause a slowing of the code at the time of their occurrence. Some sample results obtained from this model are shown in Figs. 14.15 to 14.22 (Duewer, MacCracken, and Walton, 1978). Figures 14.15 to 14.18 are contour plots of ozone surface concentration calculated by the LIRAQ-2 photochemical model for various times of the day for a high-oxidant day in the San Francisco Bay Area.

At 0700 PST (Fig. 14.15), the concentrations of ozone are quite low, of the order of parts per billion. At 1000

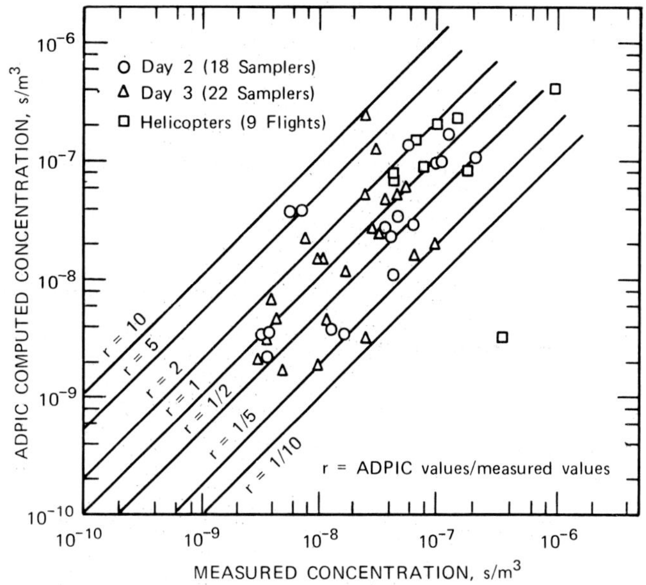

Fig. 14.10 Measured and computed relative concentrations, normalized to a continuous unit rate release, for three ^{41}Ar plumes at the Savannah River Plant (Lange, 1978).

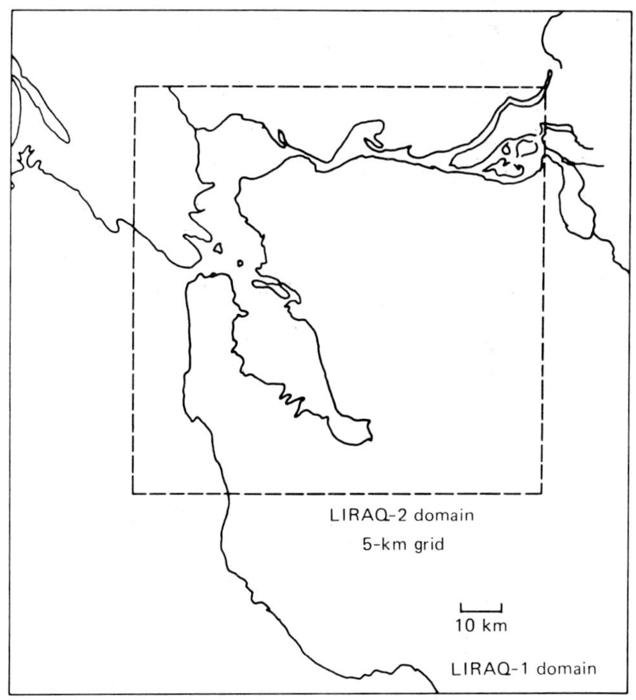

Fig. 14.11 Illustrative domains of calculation for LIRAQ-1 and LIRAQ-2.

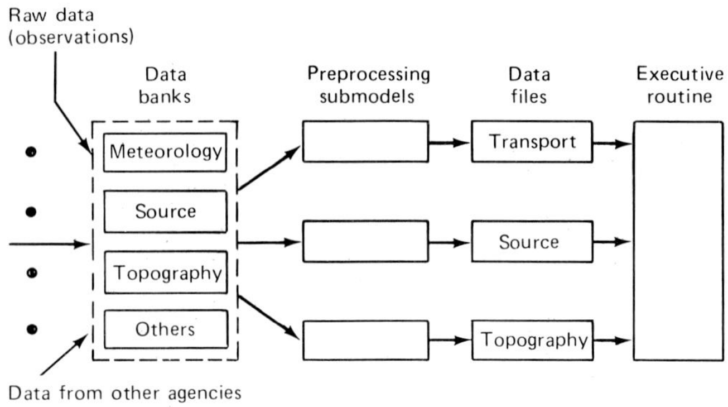

Fig. 14.12 The preprocessing submodels present tabulated or raw observational data in a form usable by the operational model.

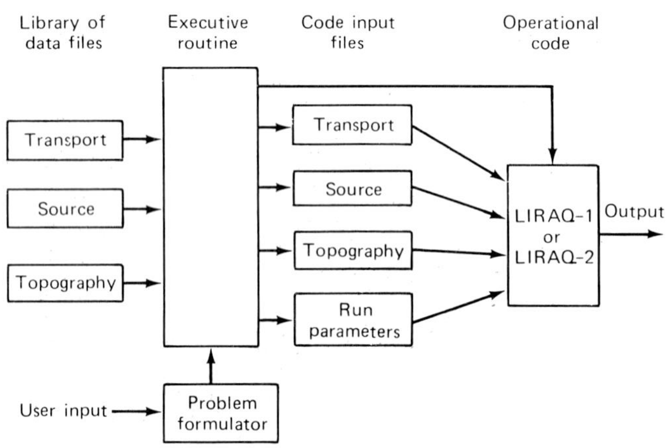

Fig. 14.13 The executive routine selects and modifies the data files as required by the user, on the basis of input from the problem formulator, and then executes the desired operational code.

(Fig. 14.16), the various stationary and mobile sources of hydrocarbons (HC) and oxides of nitrogen (NO_x) have been active for several hours, injecting these species into a still comparatively thin mixing layer. Here the action of the sun has begun to generate increased levels of ozone, and weak westerly winds have moved the material somewhat to the east. By 1300 (Fig. 14.17), the photochemistry, operating in an increasingly thick mixing layer, has almost reached its maximum effect (ozone concentrations of 0.1 to 0.13 ppm), and advection has carried significant amounts of ozone to the east. At 1600 (Fig. 14.18), photochemical production of ozone is slowing. Winds from the west begin cleaning out the region along the coast, although high concentrations still persist in the interior valleys.

Time histories of surface and average ozone concentrations and actual observations are given for specific locations in Figs. 14.19 to 14.22. Information describing the correlation of observation and model results appears in the upper right-hand corner of these plots. Comparisons with observation play a vital role in the process of tuning a limited number of model parameters. These plots portray the process of ozone buildup through the combined action of HC and NO_x with sunlight and advective transport and the later cleaning out due to advection.

The LIRAQ model performance as discussed and represented above is not atypical of other case studies of regional oxidant distributions conducted for verification purposes. The performance of the model in these case

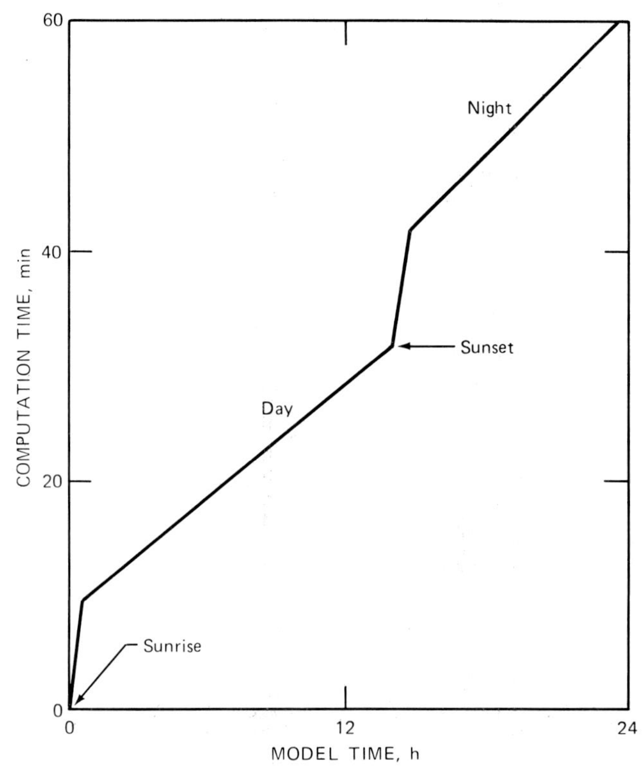

Fig. 14.14 A 24-h simulation requires approximately 1 h of computer time, unequally distributed through the period of simulation.

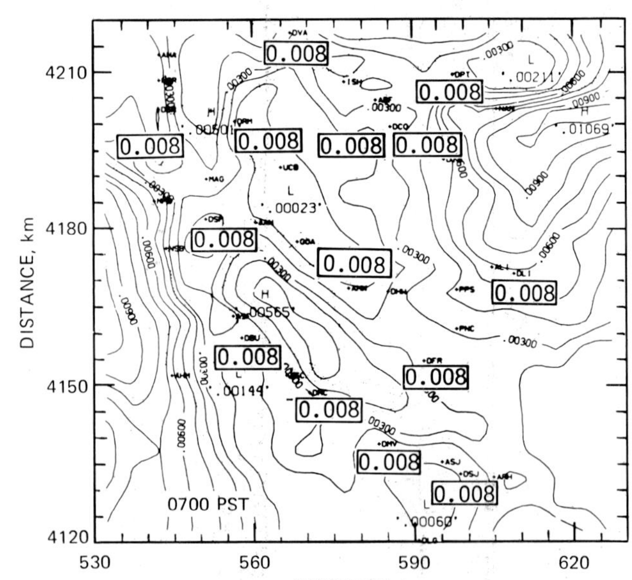

Fig. 14.15 Ozone surface concentration (ppm) at 0700.

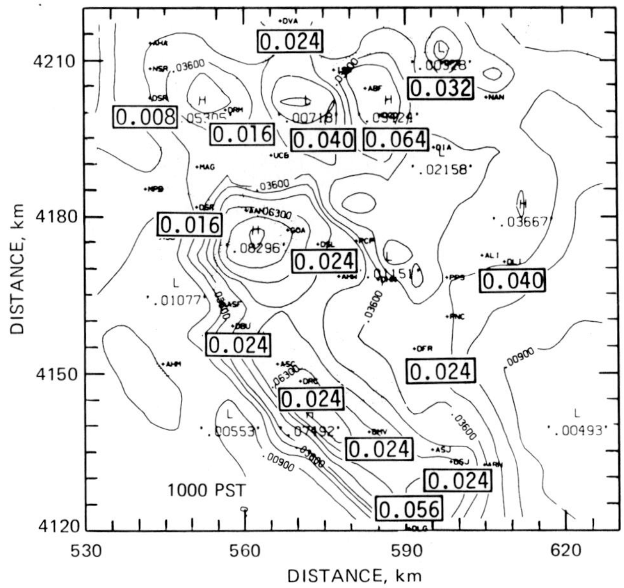

Fig. 14.16 Ozone surface concentration (ppm) at 1000.

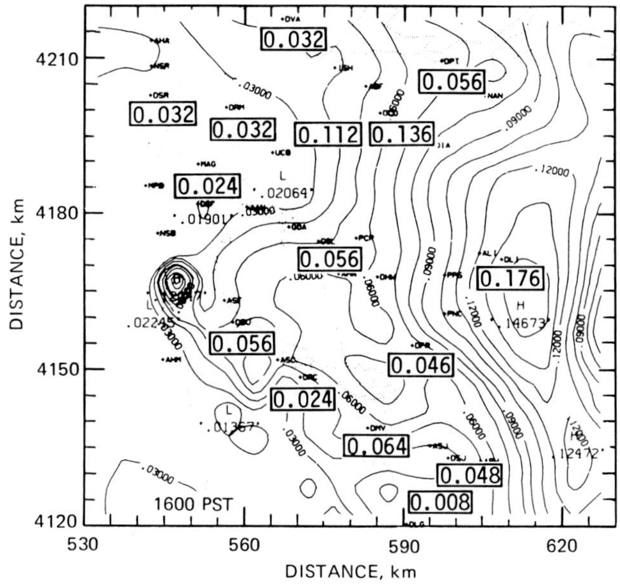

Fig. 14.18 Ozone surface concentration (ppm) at 1600.

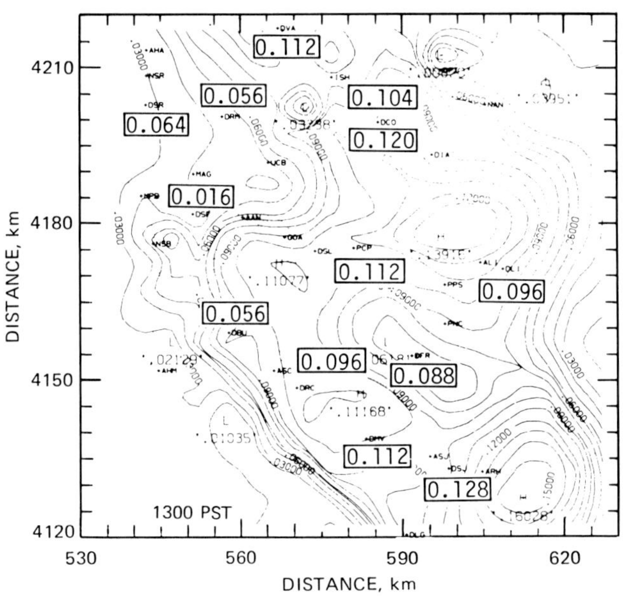

Fig. 14.17 Ozone surface concentration (ppm) at 1300.

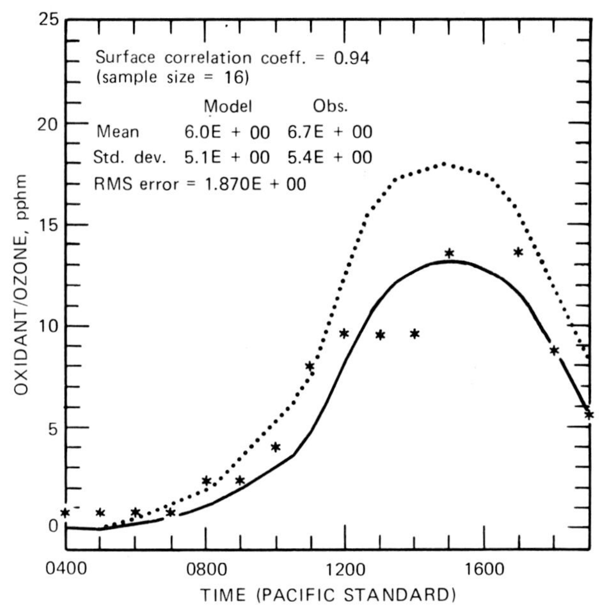

Fig. 14.19 Comparison of computed average (——) and surface (. . . .) concentrations (pphm) with observations (*) at Livermore.

studies is not inconsistent with the estimated accuracy of the air-quality input data used for comparison purposes. It will be noted that the calculated time history of oxidant at two of the stations presented was ahead, in time, of that observed. At the present time it is not known if this is primarily due to simplifications in the source inventory or to residual artificial diffusion in the computational procedures of the model. Much more extensive testing of the model in this region and others may well be required before this issue is resolved.

The concepts, validation results, and discussion given in this section are in no way to be interpreted as being the last word on verification. Regional modeling and its use in assessments on these scales is in its beginnings; thus the verification of such models is expected to be a current research topic for many more years until a confident basis of understanding and effects estimation emerges which can contribute to the important policy decisions. At any time it can only be expected that the best technical tools and methods be used in contributing to decisions. But in parallel, as new problems and applications surface, the methods must be updated and improved as results emerge from the continuing areas of regional environmental research. It is perhaps pertinent to conclude this chapter with

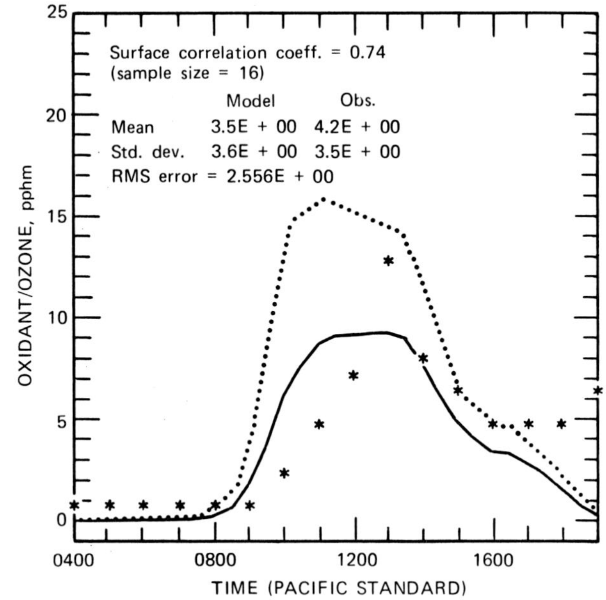

Fig. 14.20 Comparison of computed average (—) and surface (. . . .) concentrations (pphm) with observations (*) at San Jose.

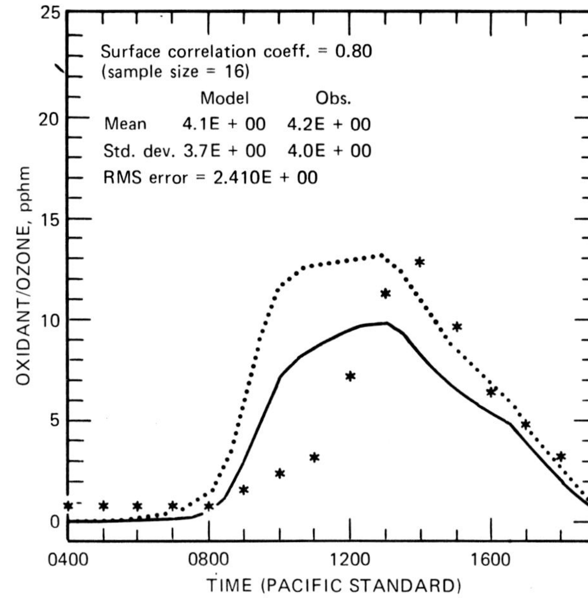

Fig. 14.22 Comparison of computed average (—) and surface (. . . .) concentrations (pphm) with observations (*) at Mountain View.

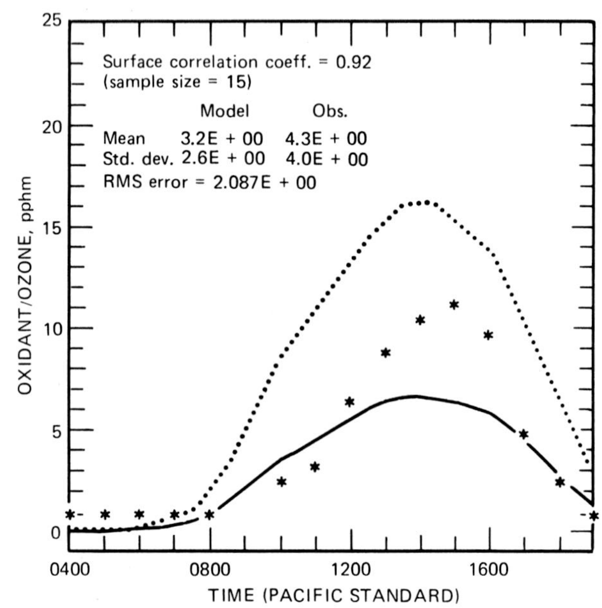

Fig. 14.21 Comparison of computed average (—) and surface (. . . .) concentrations (pphm) with observations (*) at Fremont.

brief discussions of two current problems: the fate of pollutants in regions of complex terrain and the regional simulation of sulfur dioxide and sulfate.

14-7 FATE OF POLLUTANTS IN REGIONS OF COMPLEX TERRAIN

Experience in attempting to regulate air quality in regions of complex terrain has proved to be somewhat unsatisfactory. Under regulatory pressure for decisions, simple models, like the Gaussian plume model, have been pressed into service under conditions for which they were never designed, including nonuniform and rough terrain, nonhomogeneous turbulent fields, and non-steady-state wind flows. Regulatory conservatism has been sought through the use of a plume impaction model despite the violation of the fundamental assumptions of the model. Recent tracer studies performed by the National Oceanic and Atmospheric Administration (NOAA) (Sagendorf and Dickson, 1974) suggest that such oversimplifications have led to overestimates of the surface air concentrations in and near complex terrain. Supportive evidence also comes from water tow-tank experiments (Lin, Liu, and Pao, 1974), which show, through flow visualization, the hydrodynamic flow of plumes around obstacles.

Still more recently, at the Seventh International Technical Meeting on Air Pollution Modeling and Its Applications (1976), several papers were presented on the fate of pollutants in the presence of complex terrain. The approaches included (1) an estimate of the regional flow field from a potential flow model, (2) a stream function model fitting actual wind observations, and (3) a transport–diffusion model driven by the input from a diagnostic flow model. More experience and several field tracer studies are now needed for intercomparison of model performance and model improvement. It is, however, pertinent to show one such regional simulation to illustrate the potential of this kind of modeling for assessing the fate of pollutants in the presence of complex terrain.

The use of a three-dimensional time-dependent model, such as MATHEW/ADPIC, for calculating regional temporal–spatial pollutant distributions in the presence of complex terrain is illustrated for the Indian Point operating releases during a typical fall day with diurnally reversing boundary layer flows. The topography of this region is

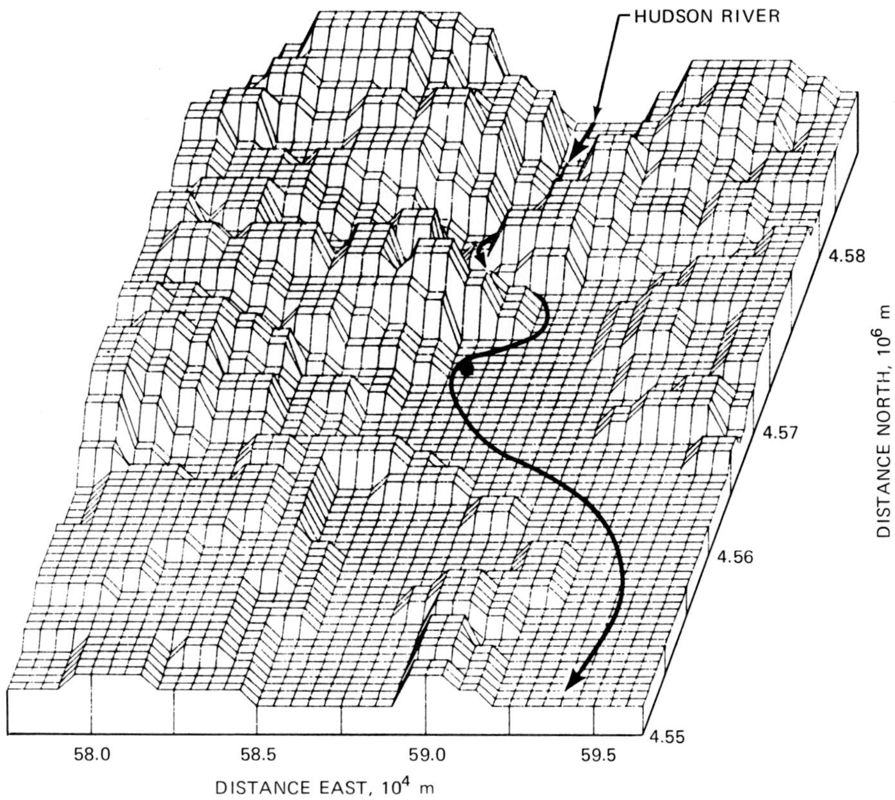

Fig. 14.23 Computer generated topography for the Hudson River Valley.

shown in Fig. 14.23. Figures 14.24 to 14.31 show the projection of the marker particles for selected times onto the horizontal plane, with the Hudson River in background display, during the diurnal cycles (Lange et al., 1976).

In this example, ADPIC modeled the transport and diffusion of the total pollutant plume by generating about 90,000 tracer particles for the typical 24-h period.

Examples of integrated surface air concentrations and isotopic deposition patterns from the period of simulation are shown in Figs. 14.32 to 14.34; these deposition patterns are largely determined by the regional topography and its relationships to the source location.

14-8 REGIONAL- AND CONTINENTAL-SCALE SIMULATION OF SULFUR DIOXIDE AND SULFATE

Recent examination of the nation's energy options has reemphasized the important role that coal may play in the future. Energy planning on a scale of even 50 yr involves tremendous investments. To plan for an adequate, reliable energy supply with acceptable environmental consequences requires the wise use of scientific information and the focused acquisition of knowledge where gaps exist. At the present time, utilities and governmental agencies are planning and designing research programs aimed at providing the technical data and understanding necessary for the judicious implementation of air-quality standards for sulfur dioxide and sulfates. Models of several different types and spatial resolutions are required to interpret the observed data, to provide a verified basis of simulation of regional- and global-scale studies, and to assess exposure of humans to increased sulfur concentrations. To meet these needs requires that many processes be simulated in space and time with resolution consistent with the intended assessment. For example, the sulfate imported into urban areas from rural power plant sites might be estimated by trajectory or sequential puff models of subcontinental scale; regional and urban models, however, may be required to obtain high-resolution exposure and deposition patterns to match regional data networks and effects studies. In principle, for consistency of assessment and model capability, a hierarchy of models may be required.

Before briefly discussing the spectrum of required models, we will reflect on our air-quality experience since 1970.

In the United States, with the passage of the Clean Air Act and the adoption of Ambient Air Quality Standards, we committed our populace to a national objective of improving air quality. This objective has been pursued in part by requiring tall stacks on major new strong SO_2 stationary sources. The Environmental Protection Agency (EPA) has attributed the documented decreasing annual SO_2 and suspended particulate loading in our urban surface air to implementation or movement toward compliance with these standards (EPA, 1974). At the same time the tall stacks provide longer airborne residence times for SO_2, permitting increased sulfate formation and transport (EPA,

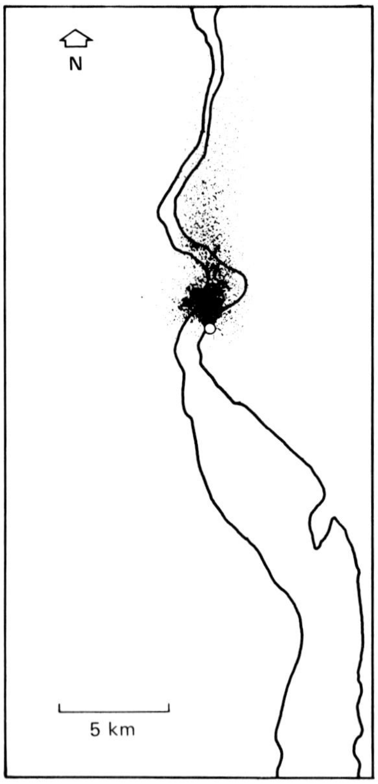

Fig. 14.24 ADPIC particle distribution, Hudson River Valley, 2000 EST.

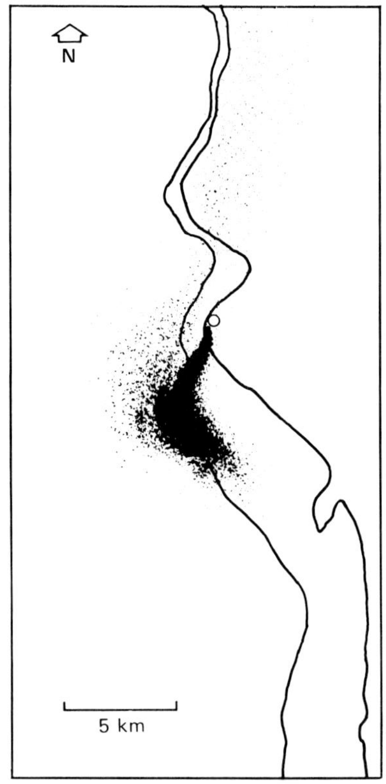

Fig. 14.26 ADPIC particle distribution, Hudson River Valley, 2300 EST.

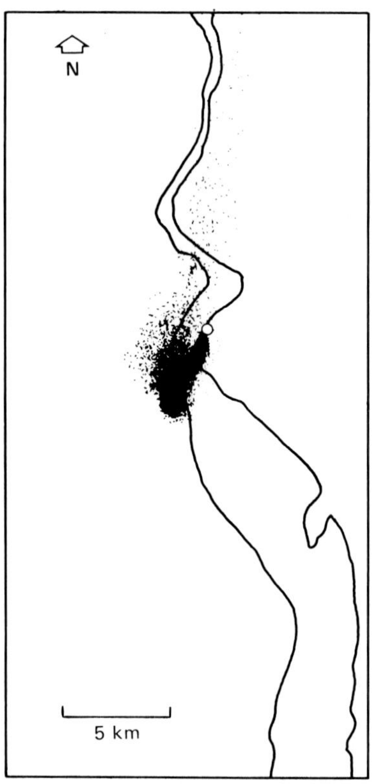

Fig. 14.25 ADPIC particle distribution, Hudson River Valley, 2200 EST.

Fig. 14.27 ADPIC particle distribution, Hudson River Valley, 0100 EST.

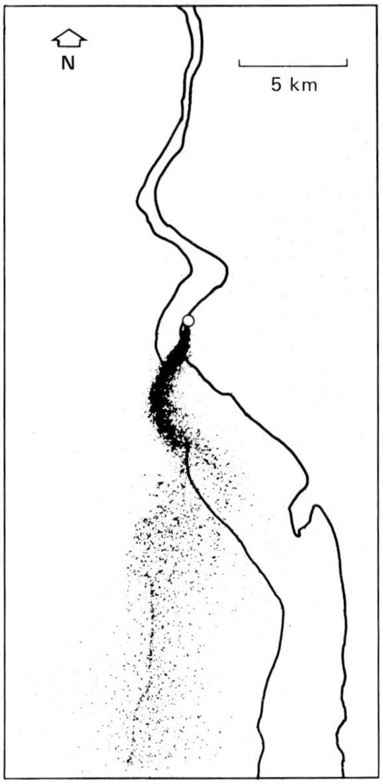

Fig. 14.28 ADPIC particle distribution, Hudson River Valley, 0700 EST.

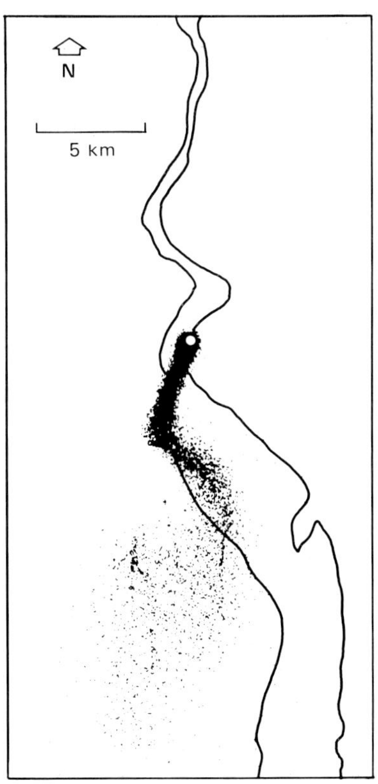

Fig. 14.29 ADPIC particle distribution, Hudson River Valley, 1100 EST.

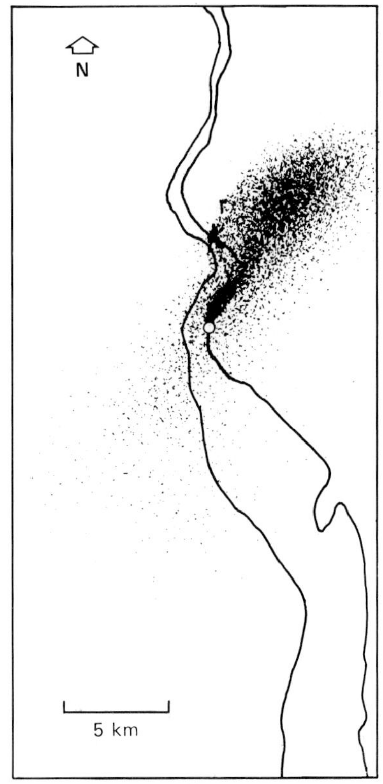

Fig. 14.30 ADPIC particle distribution, Hudson River Valley, 1200 EST.

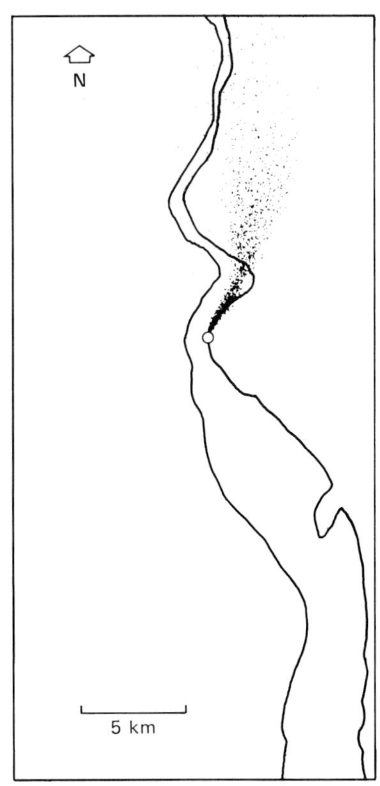

Fig. 14.31 ADPIC particle distribution, Hudson River Valley, 1700 EST.

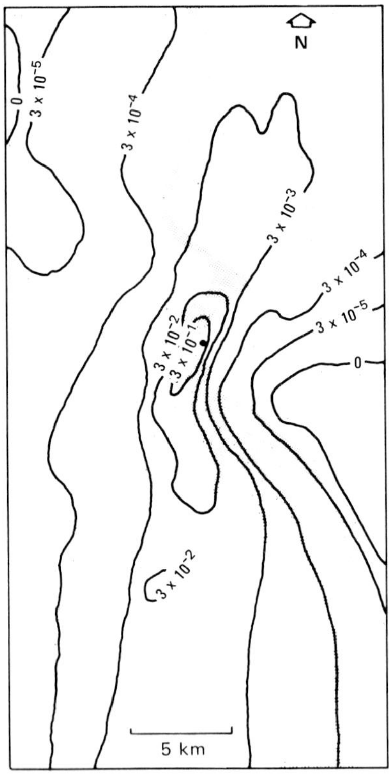

Fig. 14.32 The 48-h time-integrated surface air concentrations (s^2/m^3) for inert gas from a continuous unit surface release in the Hudson River Valley.

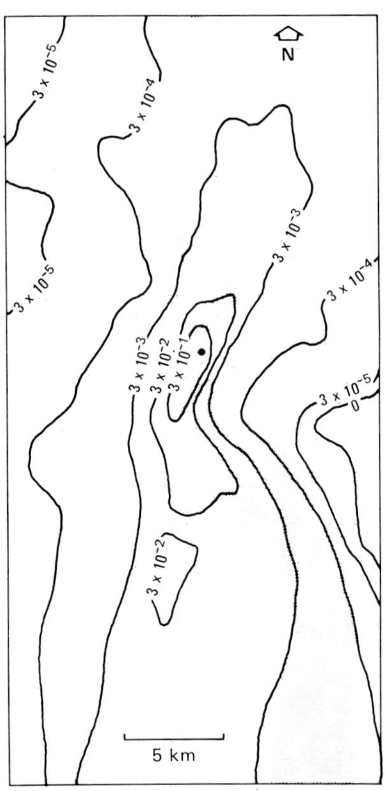

Fig. 14.33 The 48-h time-integrated surface air concentrations (s^2/m^3) for ^{131}I from a continuous unit surface release in the Hudson River Valley. The ^{131}I half-life = 6.9 x 10^5 s and the deposition velocity = 0.005 cm/s.

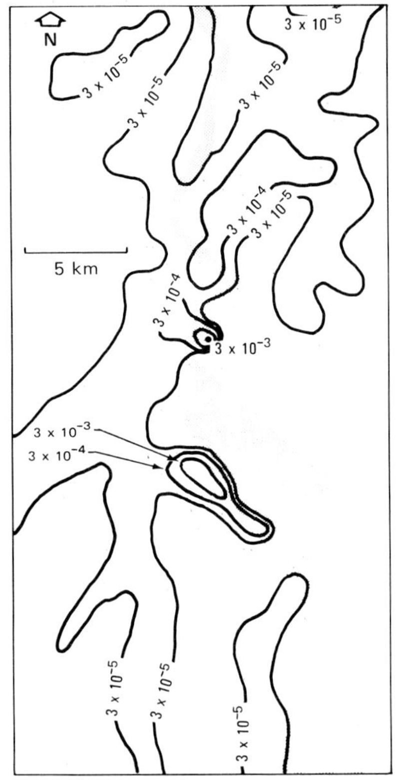

Fig. 14.34 A 48-h surface deposition (s/m^2) for ^{131}I from a continuous unit surface release in the Hudson River Valley. The ^{131}I half-life = 6.9 x 10^5 s and the deposition velocity = 0.005 m/s.

1975). The presence of sulfate contributes to the problem of acid rain on a regional scale (Likens and Bormann, 1974; Likens, 1976). In addition, increased particle loading has reportedly reduced solar radiation to the earth's surface by about 10% in rural areas (Bolin and Charlson, 1976). It is suggested that the growing season may be shortened by about 2 weeks owing to lower ground temperatures. This is an unexpected and unanticipated consequence of the implementation of the air-quality standards. These examples illustrate the complexities of technically striving for conservation of our global air resources.

In our opinion new models treating transport and transformation on subcontinental scales will be required to evaluate problems of this complexity. The lack of such models is being addressed by developing such programs as MAP3S by the Department of Energy, SURE by the Electric Power Research Institute, and STATE by the Environmental Protection Agency.

The models either developed in the past or now in development may be categorized as follows:

- —Sequential puff models or trajectory models, including transport, diffusion growth of the puff, deposition, and first-order transformation.
- —Simple regional box or mixed-layer models.
- —Modern coupled transport kinetic models suitable for relatively high-resolution regional distributions.
- —Reactive plume models for sulfur and sulfates applicable to ranges of 50 to 100 km.

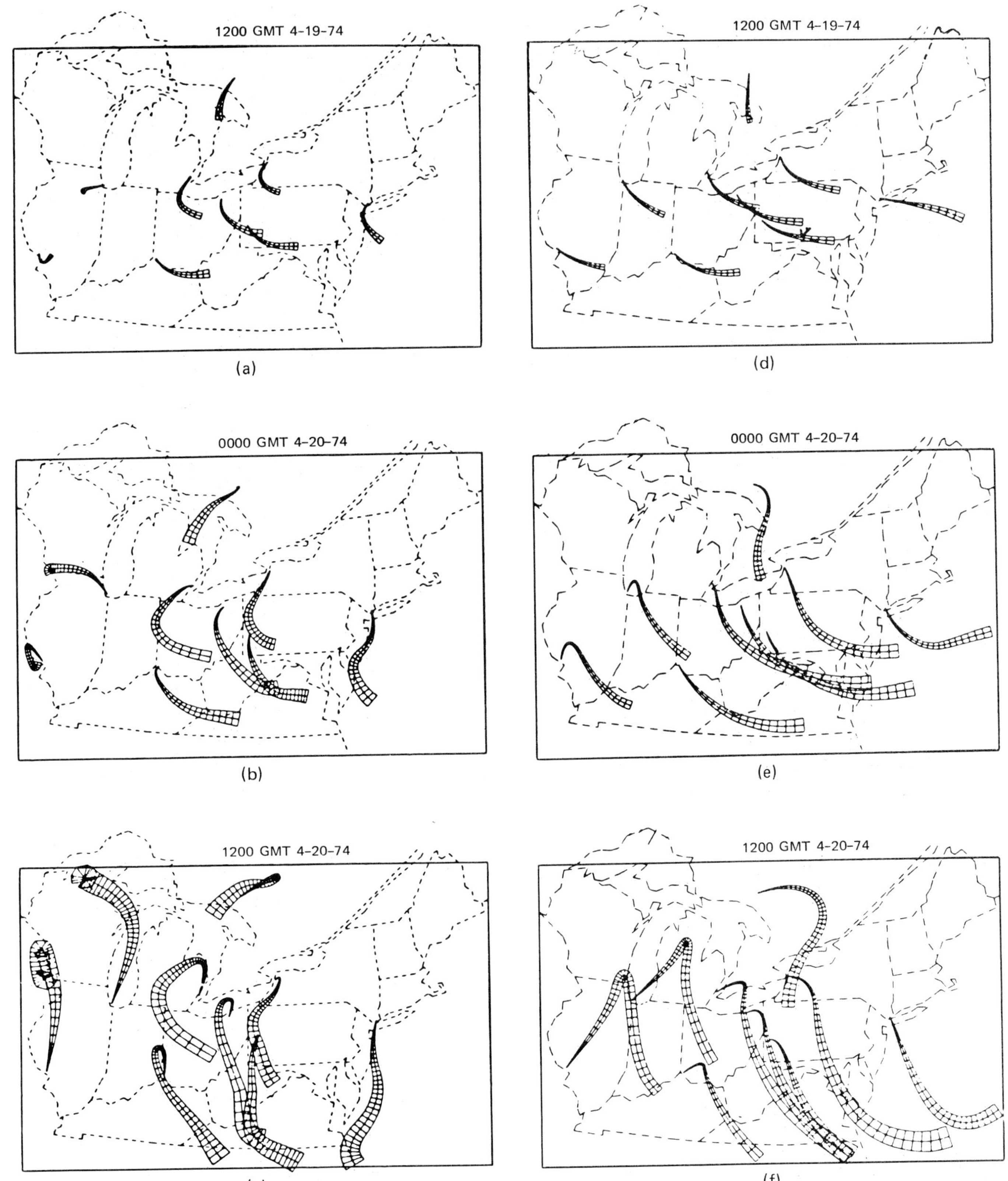

Fig. 14.35 Trajectory model results showing a streaker trace through particles ejected hourly from nine point sources. Plume width is calculated using stability-dependent diffusion coefficients. Parts (a) to (c) and (d) to (f) each show a 24-h sequence at 12-h intervals. Parts (a) to (c) show results using winds averaged over the 100- and 1000-m layers, and parts (d) to (f) show results using the winds at 850 mb (Wendell, Powell, and Drake, 1976).

TABLE 14.2 Sequential Puff or Trajectory and Simple Box or Mixed-Layer Models

Type	Reference	Purpose
Sequential puff or trajectory	Start and Markee (1967) Bolin and Persson (1975) Heffter and Taylor (1975) Heffter and Ferber (1975) Rao, Lague, and Egan (1976) Sheih (1976) Wendell, Powell, and Drake (1976) SURE (1976)	Extended range puff concentrations
Simple box or mixed layer	Hilst (1967) Reiquam (1970) Meyers and Cederwall (1975)	Regional scale; low-resolution SO_2 concentrations

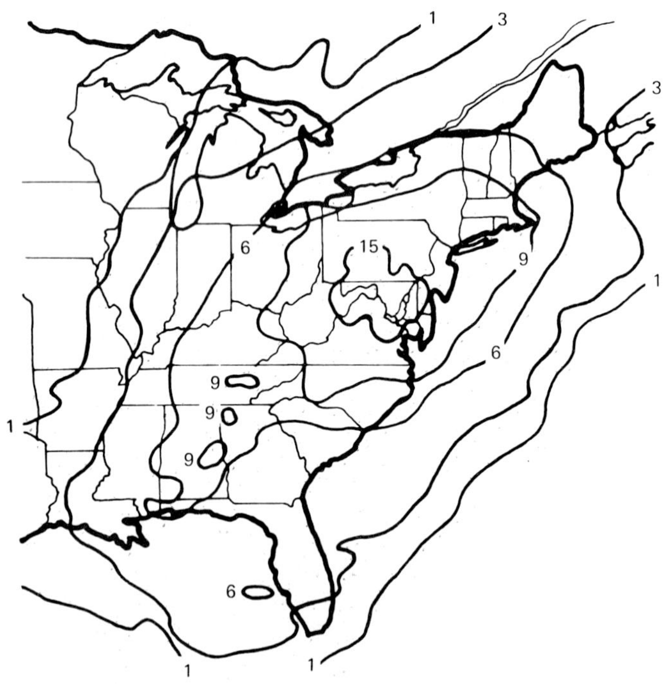

Fig. 14.36 Mean surface concentrations ($\mu g/m^3$) of sulfate (SO_4) from trajectory model calculation involving 74 point sources for October 1974 using winds at 12-h intervals, uniform vertical mixing over 1000-m depth, in-stack conversion of 2%, conversion rate of SO_2 to SO_4 of 5% per hour, and deposition velocities for SO_2 and SO_4 of 3 and 0.3 cm/s, respectively (Meyers and Cederwall, 1975).

The reader interested in the state of the art of the first two classes is referred to the sources in Table 14.2. An examination of the spatial resolution associated with the models by Wendell, Powell, and Drake (Fig. 14.35) and Meyers and Cederwall (Fig. 14.36) shows that they have different potential roles: the model by Wendell, Powell, and Drake can be used to estimate the contribution of a large point source to a downstream intermediate-range regional background or to the Lagrangian time history of pollutants in the moving puff (provided the source term is known along the trajectory).

The model by Meyers and Cederwall is useful for low spatial resolution estimates of surface air sulfate concentration for about 74 major finite sources. For instance, the model would not be suited for assessment studies of community health effects in an urban area in the form illustrated in Fig. 14.36.

This area is developing so rapidly that it would be premature to report it in more depth. It is to be expected that a rich literature will develop in response to this national need.

ACKNOWLEDGMENT

This work was performed under a contract with the U. S. Department of Energy.

REFERENCES

Altshuller, A. P., 1976, Regional Transport and Transformation of Sulfur Dioxide to Sulfates in the U. S., *J. Air Pollut. Control Assoc.*, 26(4): 318.

Amsden, A. A., 1966, *The Particle-in-Cell Method for the Calculation of the Dynamics of Compressible Fluids*, USAEC Report LA-3466, Los Alamos Scientific Laboratory, NTIS.

Batchelor, G. K., 1950, The Application of the Similarity Theory of Turbulence to Atmospheric Diffusion, *Q. J. R. Meteorol. Soc.*, 76: 133-146.

Beilke, S., D. Lamb, and J. Müller, 1975, On the Uncatalyzed Oxidation of Atmospheric SO_2 by Oxygen in Aqueous Systems, *Atmos. Environ.*, 9: 1083-1090.

Bolin, B., and R. J. Charlson, 1976, On the Role of the Tropospheric Sulfur Cycle in the Shortwave Radiative Climate of the Earth, *Ambio*, 5: 47-54.

——, and C. Persson, 1975, Regional Dispersion and Deposition of Atmospheric Pollutants with Particular Application to Sulfur Pollution Over Western Europe, *Tellus*, 24: 281-310.

Businger, J. A., J. C. Wyngaard, Y. Izumi, and E. F. Bradley, 1971, Flux Profile Relationships in the Atmospheric Surface Layer, *J. Atmos. Sci.*, 28: 181.

Cadle, R. D., and M. Ledford, 1966, The Reaction of Ozone with Hydrogen Sulfide, *Int. J. Air Water Pollut.*, 10: 25-30.

Chang, J. S., 1974, Simulations, Perturbations, and Interpretations, in *Proceedings of the Third Conference on the Climatic Impact*

Assessment Program, Report DOT-TSC-OST-74-15, pp. 330-341, U. S. Department of Transportation.

Chervin, R. M., W. L. Gates, and S. H. Schneider, 1974, The Effect of Time Averaging on the Noise Level of Climatological Statistics Generated by Atmospheric General Circulation Models, *J. Appl. Sci.*, 31: 2216-2219.

Cornett, J. S., and D. Randerson, 1977, Verification of Some Numerical Models for Operationally Predicting Mesoscale Winds Aloft, *J. Appl. Meteorol.*, 16: 764-773.

Crawford, T. V., 1966, *A Computer Program for Calculating the Atmospheric Dispersion of Large Clouds*, USAEC Report UCRL-50179, Lawrence Livermore Laboratory, NTIS.

Crowley, W. P., 1968, Numerical Advection Experiments, *Mon. Weather Rev.*, 96: 1-11.

Culbertson, B., and P. T. Willis, 1964, Observational Grid Spacing Effect on the Accuracy of a Physical Forecast Model, *J. Appl. Meteorol.*, 3: 541-546.

Davis, D. D., G. Smith, and G. Klauber, 1974, Trace Gas Analysis of Power Plant Plumes Via Aircraft Measurement: O_3, NO_x, and SO_2 Chemistry, *Science*, 186: 733-736.

DeHaven, J. C., and B. M. Woodfill, 1973, *Cost and Effectiveness of Strategies for Reducing Emissions from Fixed Sources: The San Diego Air Basin 1975*, Report R-1363-SD, The RAND Corporation.

Demerjian, K. L., J. A. Kern, and J. G. Calvert, 1974, The Mechanism of Photochemical Smog Formation, in *Advances in Environmental Science and Technology*, Vol. 4, J. N. Pitts, Jr., and R. L. Metcalf (Eds.), Wiley–Interscience, New York.

Dickerson, M. H., 1978, A Mass-Consistent Flux Model for Regions with Complex Topography, *J. Appl. Meteorol.*, 17: 241-253.

Duewer, W. H., M. C. MacCracken, and J. J. Walton, 1978, The Livermore Regional Air Quality Model. II. Verification and Sample Application in the San Francisco Bay Area, *J. Appl. Meteorol.*, 17: 273-311.

Ellsaesser, H. W., 1960, *Wind Variability*, Report AWS-TR-105-2, U. S. Air Force.

Energy and Climate, 1977, Studies in Geophysics, National Academy of Sciences, Washington, D. C.

Environmental Protection Agency, 1974, *Monitoring and Air Quality Trends Report, 1973*, Report EPA-450/1-74-007, available from NTIS as Report PB-240484.

——, 1974, *Position Paper on Regulation of Atmospheric Sulfates*, Report EPA-450/2-75-007, available from NTIS as Report 245760.

Eschenroeder, A. Q., and J. R. Martinez, 1971, *Further Development of the Photochemical Smog Model for the Los Angeles Basin*, General Research Corp., Report GR-1-191.

——, and J. R. Martinez, 1973, Concepts and Applications of Photochemical Smog Models, *Adv. Chem. Ser.*, 113: 101-168.

Gear, C. W., 1971, *Numerical Initial Value Problems in Ordinary Differential Equations*, Prentice-Hall, Englewood Cliffs, N. J.

Gelinas, R. J., and J. J. Walton, 1973, Dynamic-Kinetic Evolution of a Single Plume of Interacting Species, *J. Atmos. Sci.*, 31(7): 1807-1813.

Gresho, P. M., R. L. Lee, and R. L. Sani, 1976, Modeling and Planetary Boundary Layer Using the Galerkin Finite-Element Method, in *Third Symposium on Atmospheric Turbulence, Diffusion, and Air Quality*, Raleigh, N. C., October 19–22, pp. 82-89, available from the American Meteorological Society, Boston, Mass.

Hanna, S. R., 1973, A Simple Dispersion Model for the Analysis of Chemically Reactive Pollutants, *Atmos. Environ.*, 7: 1-15.

Heffter, J. L., and G. J. Ferber, 1975, *A Regional-Continental Scale Transport, Diffusion and Deposition Model, Part II: Diffusion-Deposition Models*, NOAA-TM-ERL-ARL-50, National Oceanic and Atmospheric Administration.

——, and A. D. Taylor, 1975, *A Regional-Continental Scale Transport, Diffusion and Deposition Model, Part I: Trajectory Model*, NOAA-TM-ERL-ARL-50, National Oceanic and Atmospheric Administration.

Hess, S. L., 1959, *Introduction to Theoretical Meteorology*, Holt, Rinehart, & Winston, New York.

Hilst, G. R., 1967, An Air Pollution Model of Connecticut, in *Proceedings of the IBM Scientific Computing Symposium on Water and Air Resource Management*, Yorktown Heights, N. Y., October 23–25, pp. 251-274, available from IBM Data Processing Division, White Plains, N. Y.

——, D. duP. Donaldson, M. Teske, R. Contiliano, and J. Feiberg, 1973, *The Development and Preliminary Application of an Invariant Coupled Diffusion and Chemistry Model*, Report NASA-CR-2295, Aeronautical Research Associates of Princeton, Inc.

Hindmarsh, A. C., 1974, *GEAR: Ordinary Differential Equation System Solver*, USAEC Report UCID-30001, Rev. 3, Lawrence Livermore Laboratory, Computer Documentation Report, NTIS.

Junge, C. E., 1963, *Air Chemistry and Radioactivity*, Academic Press, New York.

Lamb, R. G., 1973, Note on the Application of K-Theory to Diffusion Problems Involving Nonlinear Chemical Reaction, *Atmos. Environ.*, 7: 257.

——, F. I. Badgley, and A. T. Rossano, Jr., 1973, *A Critical Review of Mathematical Modeling Techniques for Air Quality with Relation to Motor Vehicle Transportation*, Report PB-224656, University of Washington, NTIS.

——, and J. H. Seinfeld, 1973, Mathematical Modeling of Urban Air Pollution: General Theory, *Environ. Sci. Technol.*, 7(3): 253-261.

Lange, R., 1973, *ADPIC: A Three-Dimensional Computer Code for the Study of Pollutant Dispersal and Deposition Under Complex Conditions*, USAEC Report UCRL-51462, Lawrence Livermore Laboratory, NTIS.

——, 1978, ADPIC: A Three-Dimensional Transport-Diffusion Model for the Dispersal of Atmospheric Pollutants and Its Validation Against Regional Tracer Studies, *J. Appl. Meteorol.*, 17: 320-329.

——, M. H. Dickerson, K. R. Peterson, C. A. Sherman, and T. J. Sullivan, 1976, *Particle-in-Cell vs. Straight-Line Airflow Gaussian Calculations of Concentration and Deposition of Airborne Emissions Out to 70 km for Two Sites of Differing Meteorological and Topographical Character*, USAEC Report UCRL-52133, Lawrence Livermore Laboratory, NTIS.

Leighton, P. A., 1961, *Photochemistry of Air Pollution*, Academic Press, New York.

Likens, G. E., 1976, Acid Precipitation, *Chem. Eng. News*, 54(48): 29-44 (November 22).

——, and F. H. Bormann, 1974, Acid Rain: A Serious Regional Environmental Problem, *Science*, 184: 1176-1179.

Lin, J., H. Liu, and Y. Pao, 1974, *Laboratory Simulation of Plume Dispersion in Stably Stratified Flows over Complex Terrain*, Report 29, Flow Research, Inc., Seattle, Wash., for the Environmental Protection Agency.

Long, P. E., and D. W. Pepper, 1976, A Comparison of Six Numerical Schemes for Calculating the Advection of Atmospheric Pollution, in *Third Symposium on Atmospheric Turbulence, Diffusion, and Air Quality*, Raleigh, N. C., October 19–22, pp. 181-187, available from the American Meteorological Society, Boston, Mass.

MacCracken, M. C. (principal investigator), 1975, *User's Guide to the LIRAQ Model: An Air Pollution Model for the San Francisco Bay Area*, USAEC Report UCRL-51983, Lawrence Livermore Laboratory, NTIS.

——, and G. D. Sauter (Eds.), 1975, *Development of an Air Pollution Model for the San Francisco Bay Area, Final Report to the National Science Foundation*, USAEC Report UCRL-51920, Vols. 1 and 2, Lawrence Livermore Laboratory, NTIS.

——, D. J. Wuebbles, J. J. Walton, W. H. Duewer, and K. E. Grant, 1978, The Livermore Regional Air Quality Model: I. Concept and Developments, *J. Appl. Meteorol.*, 17: 254-272.

Mathis, J. J. and W. L. Grose, 1973, *A Review of Methods of Predicting Air Pollution Dispersion,* Report NASA AP-322, National Aeronautics and Space Administration.

Meyers, R. E., and R. T. Cederwall, 1975, Fossil Pollutant Transport Model Development, in *Brookhaven National Laboratory Regional Energy Studies Program, Annual Report for Fiscal Year 1975,* P. F. Palmedo (Ed.), USAEC Report BNL-50478, pp. 46-61, Brookhaven National Laboratory, NTIS.

——, R. T. Cederwall, W. D. Ohmstede, and W. aufm Kampe, 1976, Transport and Diffusion Using a Diagnostic Mesoscale Model Employing Mass and Total Energy Conservation Constraints, in *Third Symposium on Atmospheric Turbulence, Diffusion, and Air Quality,* Raleigh, N. C., October 19–22, pp. 90-97, available from the American Meteorological Society, Boston, Mass.

Molenkamp, C. R., 1968, Accuracy of Finite-Difference Methods Applied to the Advection Equation, *J. Appl. Meteorol.*, 7: 160-167.

Monin, A. S., and A. M. Obukhov, 1954, The Basic Laws of Turbulence Mixing in the Surface Layer of the Atmosphere, *Tr. Inst. Geofiz., Akad. Nauk SSSR* (Works of the Geophysics Institute of the Academy of Sciences USSR), No. 24(151): 163-187.

——, and A. M. Yaglom, 1971, *Statistical Fluid Mechanics: Mechanics of Turbulence,* Vol. 1, Chap. 5, The MIT Press, Cambridge, Mass.

Obukhov, A. M., 1971, Turbulence in an Atmosphere with a Non-Uniform Temperature, *Boundary-Layer Meteorol.*, 2: 7-29.

Ohmstede, W. D., 1976, Diagnostic Analysis of Terrain Effects on Atmospheric Phenomena Using a Variational Approach, in *Proceedings of the Seventh Technical Exchange Conference, El Paso, Tex., November 30–December 3,* pp. 81-94, GPO.

Paegle, J., and J. N. Paegle, 1976, On Geopotential Data and Ellipticity of the Balance Equation: A Data Study, *Mon. Weather Rev.*, 104: 1279-1288.

Pasquill, F., 1974, *Atmospheric Diffusion,* 2nd ed., Halsted Press, New York.

Proceedings of the Conference on Health Effects of Air Pollutants, 1973, Assembly of Life Sciences, National Academy of Sciences–National Research Council, prepared for the Committee on Public Works, U. S. Senate, October 3–5, Serial No. 93-15.

Proceedings of the Seventh International Technical Meeting on Air Pollution Modeling and Its Applications, Airlie, Va., Sept. 7–10, 1976, Report PB-270799, pp. 81-94, NTIS.

Rao, K. S., J. S. Lague, and B. A. Egan, 1976, An Air Trajectory Model for Regional Transport of Atmospheric Sulfates, in *Third Symposium on Atmospheric Turbulence, Diffusion, and Air Quality,* Raleigh, N. C., October 19–22, pp. 325-331, available from the American Meteorological Society, Boston, Mass.

Reed, R. J., and K. E. German, 1965, A Contribution to the Problem of Stratospheric Diffusion by Large-Scale Mixing, *Mon. Weather Rev.*, 93(5): 313-321.

Reiquam, H., 1970, An Atmospheric Transport and Accumulation Model for Airsheds, *Atmos. Environ.*, 4: 233-247.

Research and Development Related to Sulphates in the Atmosphere, 1975, Hearings Before the Subcommittee on the Atmosphere of the Committee on Science and Technology, U. S. House of Representatives, July 9–14, No. 39.

Reynolds, S. D., P. M. Roth, and J. H. Seinfeld, 1973, Mathematical Modeling of Photochemical Air Pollution. I. Formulation of the Model, *Atmos. Environ.*, 7: 1033-1061.

Richardson, L. F., 1926, Atmospheric Diffusion Shown on a Distance-Neighbor Graph, *Proc. Roy. Soc. (London), Ser. A,* 110: 709-737.

Roth, P. M., P. J. W. Roberts, M. Liu, S. D. Reynolds, and J. H. Seinfeld, 1974, Mathematical Modeling of Photochemical Air Pollution. II. A Model and Inventory of Pollutant Emissions, *Atmos. Environ.*, 8: 97-130.

Sagendorf, J. F., and C. R. Dickson, 1974, *Diffusion Under Low Windspeed, Inversion Conditions,* NOAA-TM-ERL-ARL-52, National Oceanic and Atmospheric Administration, NTIS.

Sasaki, Y., 1958, An Objective Analysis Based on the Variational Method, *J. Meteorol. Soc. Jpn.*, 36: 77.

——, 1970a, Some Basic Formalisms in Numerical Variational Analysis, *Mon. Weather Rev.*, 98: 875.

——, 1970b, Numerical Variational Analysis Formulated Under the Constraints Determined by Longwave Equations and Low-Pass Filter, *Mon. Weather Rev.*, 98: 884.

Sheih, C. M., 1976, Application of a Lagrangian Statistical Trajectory Model to the Simulation of Sulfur Pollution over Northeastern United States, in *Third Symposium on Atmospheric Turbulence, Diffusion, and Air Quality,* Raleigh, N. C., October 19–22, pp. 311-317, available from the American Meteorological Society, Boston, Mass.

Sherman, C. A., 1978, A Mass-Consistent Model for Wind Fields Over Complex Terrain, *J. Appl. Meteorol.*, 17: 312-319.

Shir, C. C., and L. J. Shieh, 1974, A Generalized Urban Air Pollution Model and Its Application to the Study of SO_2 Distributions in the St. Louis Metropolitan Area, *J. Appl. Meteorol.*, 13: 185-204.

Sklarew, R. C., A. J. Fabrick, and J. E. Prager, 1972, Mathematical Modeling of Photochemical Smog Using the PICK Method, *J. Air Pollut. Control Assoc.*, 22: 865-869.

Start, G. E., and E. J. Markee, Jr., 1967, Relative Dose Factors from Long Period Point Source Emissions of Atmospheric Pollutants, in *USAEC Meteorological Information Meeting,* Chalk River, Ontario, Canada, September 11–14, C. A. Mawson (Ed.), Report AECL-2787, pp. 59-76, Atomic Energy of Canada Ltd.

State of California Air Resources Board Evaluation and Planning, 1974, *Air Quality, Land Use, and Transportation Models: Evaluation and Utilization in the Planning Process.*

Stern, A. C., 1968, *Air Pollution,* 2nd ed., Part II, pp. 187-347, Academic Press, New York.

Strang, G., and G. Fix, 1973, *An Analysis of the Finite Element Method,* Prentice-Hall, Englewood Cliffs, N. J.

SURE, 1976, *Design of the Sulfate Regional Experiment (SURE),* Final Report, Vol. I, Report EPRI EC-129, Environmental Research and Technology, Inc., Concord, Mass.

Tennekes, H., 1976, Observations on the Dynamics and Statistics of Single Box Models with a Variable Inversion Lid, in *Third Symposium on Atmospheric Turbulence, Diffusion, and Air Quality,* Raleigh, N. C., October 19–22, pp. 397-402, available from the American Meteorological Society, Boston, Mass.

Uthe, E. E., and P. B. Russell, 1974, Experimental Study of the Urban Aerosol Structure and Its Relation to Urban Climate Modification, *Bull. Am. Meteorol. Soc.*, 55: 115-121.

Walton, J. J., 1973, Scale Dependent Diffusion, *J. Appl. Meteorol.*, 12(3): 547-549.

Warshaw, M., 1973, *The Statistical Analysis of Simulated Climate Change,* Report R-1298-DOT, U. S. Department of Transportation.

Welch, J. E., F. H. Harlow, J. P. Shannon, and B. J. Daly, 1965, *The MAC Method,* USAEC Report LA-3425, Los Alamos Scientific Laboratory, NTIS.

Wendell, L. L., D. C. Powell, and R. L. Drake, 1976, A Regional Scale Model for Computing Deposition and Ground Level Air Concentration of SO_2 and Sulfates from Elevated and Ground Sources, in *Third Symposium on Atmospheric Turbulence, Diffusion, and Air Quality,* Raleigh, N. C., October 19–22, pp. 318-329, available from the American Meteorological Society, Boston, Mass.

Widhopf, :al Experiments on the ributions Using a Parameterized ... *Proceedings of the Third Conference on the Climatic ... ssessment Program,* Report DOT-TSC-OST-74-15, p. 30, 376-389, U. S. Department of Transportation.

Willoughby, R. A. (Ed.), 1974, *Stiff Differential Systems,* Plenum Publishing, New York.

Wuebbles, D. J., and J. S. Chang, 1975, Sensitivity of Time-Varying Parameters in Stratospheric Modeling, *J. Geophys. Res.,* 80(18): 2637-2642.

Chapter 14 covers research prior to February 1978.

15 ATMOSPHERIC EFFECTS OF ENERGY GENERATION

Steven R. Hanna, Atmospheric Turbulence and Diffusion Laboratory

15-1 INTRODUCTION

When energy is generated, gases and particulate matter are usually emitted into the atmosphere. Some of their effects are treated in other chapters. The effects of excess heat and moisture released to the atmosphere as the result of energy generation are discussed in this chapter. Specific methods for estimating such effects as cloud formation or drift deposition are outlined.

All energy produced by man must ultimately pass through the atmosphere to space. Thus even the waste heat released to the deep ocean by the diffuser pipes from a power plant will slowly be lost to the atmosphere by radiation, conduction, and evaporation; although the time scale for this loss may be weeks or months. In contrast, the waste heat released to the atmosphere through cooling towers can rise within minutes to elevations of several kilometers. The time scale for the heat loss from cooling ponds is intermediate between those for cooling towers and ocean releases.

15-1.1 Source Characteristics of Cooling Towers

The cooling tower is the source that is most thoroughly treated in this chapter. We will be concerned with the energy only after it leaves the mouth of the tower and refer the reader to other reports (e.g., Woodson, 1971; or Chan and Golay, 1977) for discussions of the detailed engineering aspects of the tower. There are two major types of cooling towers in use today: mechanical draft and natural draft. Photographs of these two types are given in Figs. 15.1 and 15.2. In the mechanical-draft tower, large fans force the vertical airflow, whereas in the natural-draft tower, the great size of the tower causes a vertical airflow to develop with no fan. These are so-called "wet" towers, in which heat exchange is accomplished by evaporation from countless water droplets, generated by splashing warm water over barriers. There are also a few "dry" towers, which exchange heat on the same principle as an automobile radiator (Boyack and Kearney, 1972) but which are usually impractical for an energy generating facility because of the large size required. At some locations, hybrid "dry—wet" towers may be practical (Patel, Croley, and Cheng, 1974), where the system is operated in the dry mode as much as possible to conserve water.

Generally natural-draft cooling towers are built in groups of two to four, with spacings of 200 to 500 m. Mechanical-draft towers are traditionally built in lines or "banks," in which 10 to 20 fan units are located in one 200-m-long structure. Recently some mechanical draft units have been arranged into doughnut-shaped structures, presumably to increase the plume rise from the group.

Typical characteristics of the initial plumes from these towers are given in Table 15.1. Vertical speeds, temperatures, and liquid-water content have been measured at cooling-tower mouths at several locations by Shofner et al. (1974). Models generally assume that the vertical speed is constant across the plume, although the vertical speed cross section in Fig. 15.3 shows quite a bit of systematic variation. The initial liquid-water content in Table 15.1 is only an estimate, based on a few measurements by Environmental Systems Corporation (1977) at the mouth of the Chalk Point cooling tower. It is generally agreed, though, that the plume is saturated as it leaves the tower.

Most of the liquid water in the plume is pure condensed water. However, a small fraction of the liquid water is "drift," which consists of drops with the chemical composition of the circulating water in the tower. To conserve water and chemicals and to minimize environmental damage, manufacturers try to reduce the drift loss. In some instances the salts or water-treatment chemicals in the water can harm vegetation in the vicinity of the cooling towers (see Sec. 15-2.3). Holmberg (1974) points out that by installing drift eliminators the manufacturers can keep the drift loss down to as low as 0.001% of the circulating water rate.

For the typical natural-draft source in Table 15.1, about 200 g/s of drift water is emitted from the tower. Since the size of the drift drop determines its settling speed, there has been great interest in programs in which the drift-drop size distribution is measured at operating cooling towers. Some of these distributions are given in Fig. 15.4.

Hanna: Atmospheric Turbulence and Diffusion Laboratory, National Oceanic and Atmospheric Administration, Oak Ridge, Tennessee.

Present address: Environmental Research and Technology, Inc., Concord, MA 01742.

Fig. 15.1 Photograph of mechanical-draft cooling tower. (Courtesy of J. D. Holmberg, The Marley Cooling Tower Company.)

Fig. 15.2 Photograph of natural-draft cooling towers of the General James M. Gavin Power Plant of the Ohio Electric Company. (Courtesy of M. L. Kramer, Meteorological Evaluation Services, Inc.)

TABLE 15.1 Characteristics of the Initial Plumes from Wet Cooling Towers

Type	Radius, m	Vertical speed, m/s	Temperature excess, °C	Liquid-water content, g liquid water/g air
Mechanical	5–10	10	10–20	0.0001–0.001
Natural	20–40	5	10–20	0.0001–0.001

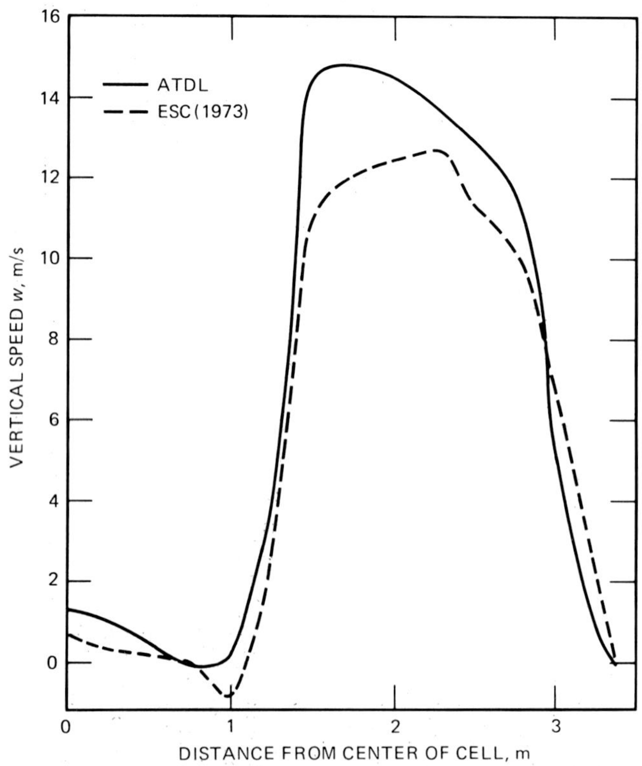

Fig. 15.3 Average updraft speed measured at the mouth of the No. 6 cell of the K-31 mechanical-draft cooling tower at Oak Ridge, Tenn., during the period June 25–June 29, 1973. The solid and dashed curves represent data taken by the Atmospheric Turbulence and Diffusion Laboratory (—) and the Environmental Systems Corporation (– – –), respectively.

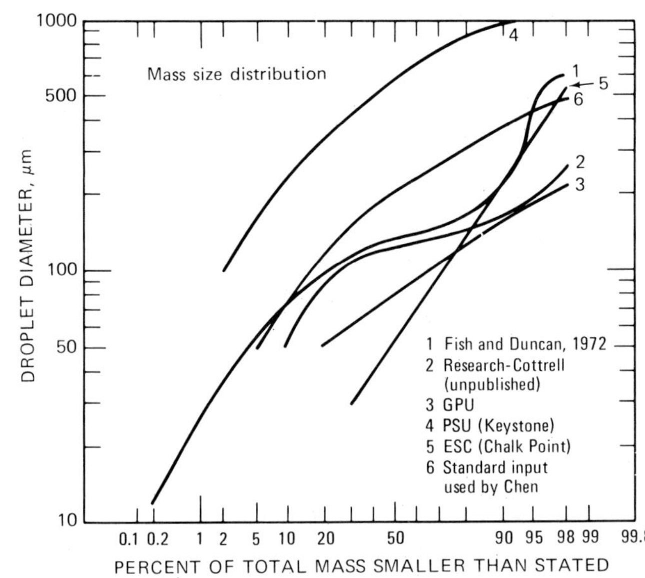

Fig. 15.4 Examples of cumulative mass distributions of drift droplets for natural-draft cooling towers. [From N. C. J. Chen and S. R. Hanna, Drift-Modeling and Monitoring Comparisons, *Atmos. Environ.*, 12: 1725-1734 (1978).]

The median drift-drop radius is on the order of 100 μm, but some large drops with radius of 1000 μm can occur. Where there are holes in the drift eliminators of mechanical draft towers, drops with radius 2 to 5 mm can escape.

15-1.2 Source Characteristics of Cooling Ponds

In geographical areas where land is not at a premium, it is often less expensive for an industry or utility to construct a cooling pond rather than a cooling tower. A rough rule is that 1 or 2 acres of pond surface are required for each megawatt of electricity produced. There are a number of large cooling ponds in operation, such as the ones at the Four Corners plant in New Mexico and at the Baldwin and Dresden plants in Illinois. Figure 15.5 is a photograph of the Dresden pond.

The pond loses heat to the atmosphere at the air–water interface by conduction, radiation, and evaporation, which all tend to increase as the surface temperature increases. Unfortunately there are not many measurements (however, see Hadlock, 1979) of the heat budget at a cooling pond. It is clear, however, that a cooling pond has a much larger time scale than a cooling tower, implying that it is not so sensitive to brief periods of high air temperatures. Designers generally prefer to force water in a large pond to flow around a series of baffles or dikes, ensuring that it has as much time as possible to lose its heat before being ingested into the utility's cooling system again. Inlet pipes are often located below skimmer walls, so that colder bottom water must be ingested. The surface temperature of a heated pond can be up to 20°C warmer than the surface temperature of a natural pond of the same size.

15-1.3 Source Characteristics of Spray Ponds

The rate of heat loss from a cooling pond can be increased by spraying the water into the air and thus increasing the surface area of water exposed to the air. Spray units are about 10 to 20 m apart, and spray water is about 10 m high (Ryan, 1975; Kelley, 1971). A few spray ponds are in operation, but maintenance of the sprayers is a problem. Wilber (1978) reports on measurements in which the dependence of cooling rates on wind speed and direction are determined but unfortunately provides no measurements of source characteristics (e.g., drop size distribution and evaporation rate).

Fig. 15.5 Photograph of Dresden cooling pond. (Courtesy of B. B. Hicks, Argonne National Laboratory.)

15-1.4 River, Lake, and Ocean Cooling

If there is a large source of cool water nearby, it makes sense to put the hot cooling water from the industrial process directly into the water. To increase the time and space scales of the heat-exchange process, the hot water is dispersed from diffuser pipes below the surface. Surface temperature increases of as much as 12°C near the source are reported by Csanady (1970) in the warm plume from a power plant on the Great Lakes. High temperatures are maintained in rivers for a relatively long time because of the smaller volume of water available for mixing. As a result, there are now stringent Environmental Protection Agency regulations regarding the discharge of waste heat into rivers. The heat is eventually lost to the atmosphere at the air—water interface over a period of several days through conduction, radiation, and evaporation.

15-1.5 Heat Releases over Large Areas

Power parks or energy centers have been considered in which 20 000 to 50 000 MW(e) (the symbol e indicates electrical production) of energy will be produced on one 10- by 10-km site. Waste-heat release would be through 50 to 100 natural-draft cooling towers. Presumably there would be great savings in costs of construction, maintenance, safeguards, and transmission lines for such a centralized facility.

Any large urban surface currently emits 10 to 100 W/m^2 of waste power, or about 5 to 50% of the incoming solar energy flux density. Weinberg and Hammond (1970) predict that, by the end of the 21st century, the waste energy flux, averaged over the globe, will be about 1 W/m^2. The potential climatic effects of this waste heat must be estimated in time to alter the progression of energy usage in case serious atmospheric effects are predicted.

15-2 SUMMARY OF CURRENT OBSERVED EFFECTS

Often there is confusion about whether an effect is only conjectured or has actually been observed. In this section observed effects are reviewed.

15-2.1 Sun Shading by Condensed Plumes

When the sun is shaded by condensation downwind of a cooling pond or tower, the energy balance at the underlying surface is altered. A reduction in the solar energy received by plants, animals, and all other substances can cause significant changes, such as the mildewing of a painted surface, an increase in fungus on a crop, or a reduction in skin cancer in humans. Over several years or decades, changes in radiation or cloudiness can cause changes in the natural vegetation types occupying an area. On the basis of the number of reports in the literature, this effect is presently of greatest concern to Europeans as shown by the work on shading reported by Junod et al. (1974) and Bøgh (1974). In Switzerland the government regulations state that "a mean maximum duration of shadow of 15 min/d is tolerable in built-up areas around cooling towers" (Junod et al., 1974, p.240).

The condensed water causing the shading is due to emissions of heat and water from the cooling system. At current power plants (energy outputs up to 2500 MW), the median visible plume length is about 250 to 500 m during the summer and 500 to 1000 m during the winter (e.g., Junod et al., 1974; Thomson et al., 1977; Coleman and Crawford, 1978). During very humid environmental conditions, which occur 10 to 30% of the time depending on location, cumulus or stratus clouds are observed up to 50 km downwind of the power plant. Results from Bøgh's (1974) analysis of the Biblis power station, where light and heat absorption by the plume in three wavelength ranges are accounted for, show that the expected shadowing at nearby villages is 2 to 5 min/d.

15-2.2 Ground Fog

When the visible plume reaches the surface, ground fog results, posing an obvious hazard to nearby traffic. Furthermore, vegetation and structures can also be affected by an increase in the incidence of ground fog. As Drake (1975) explains, fog is often observed within a few hundred meters downwind of spray ponds and cooling ponds, but it mainly occurs during atmospheric conditions that favor the formation of natural fog (e.g., cold, calm mornings). Ground fog within a few hundred meters of mechanical-draft cooling towers is also common, owing to the tendency of the plume to be brought to the ground by downwash behind the towers when the wind speed exceeds 3 to 5 m/s. Hanna (1974a) reports that downwash occurs about 40% of the time at the cooling towers at the Oak Ridge Gaseous Diffusion Plant. But because of the strong buoyant forces in the downwashed plumes, they are observed to lift off the ground at a distance of about 200 m from the towers, after they escape from the aerodynamic cavity region downwind of the towers.

Ground fog due to downwash from natural-draft cooling towers occurs very infrequently. Bøgh (1974) reported that the visible plume sometimes touched the ground at German electric production facilities with natural-draft cooling towers that are over 100 m tall. Also, D. W. Thomson of Pennsylvania State University has noted the Keystone plume touching the ground on a few isolated occasions (private communication, 1976).

Measurement programs should be initiated around existing electric production facilities with cooling ponds or towers to quantify observations of ground fog. There are several instruments available for sensing fog, most of which operate on the principle of attenuation of a light beam. It is necessary to install several dozens of these instruments and to operate them for several years to ensure that a statistically significant climatology is determined. A. Baille of the French National Institute of Agronomic Research reports that this type of experiment is now under way in France at a location where a cooling tower is under construction. The cooling tower is expected to produce a plume that will impinge on a high plateau on which grapes are grown. A similar climatological study has begun in the Rhine River Valley in West Germany, where it is possible that waste heat from power plants may alter the fragile agricultural balance.

15-2.3 Drift Deposition

Ground fog and drift deposition have received the greatest attention in environmental impact reports on cooling towers. Impurities such as fungicides or salts in the circulating cooling water can be carried out the top of the tower by water drops formed as the water splashes over the heat-exchange surfaces. There have been a few preliminary observations taken of the drift deposition around operating cooling towers. For example, Laulenian and Ulansky (1979) used sensitive paper and deposition pans to measure drift deposition near the power plant at Pittsburg, Calif. Taylor, Hanna, and Parr (1978) measured concentrations of drift chemicals in soil surrounding the Paducah, Ky., Gaseous Diffusion Plant. Overcamp, Israel, and Pringle (1976) reported measurements (taken during the summer of 1975) of NaCl deposition downwind of the Chalk Point tower, where 22 samplers were located from 0.18 to 3.23 km downwind of the tower. The Chalk Point program also measured drift deposition in December 1975, but the results are uncertain because of the splashing of the drift drops as they hit the sensitive paper (Environmental Systems Corporation, 1976). Other tests in June 1977 produced more satisfactory results (Environmental Systems Corporation, 1977; Meyer and Stanbro, 1977).

The main purpose of the 1977 test at Chalk Point was to determine the relative contributions of the drift from the cooling tower and from the stack plume. The stack plume contains saltwater drops because a scrubber is used to remove SO_2 from the plume. Meyer and Stanbro (1977) report that the maximum sodium deposition rate observed 0.5 km and 1.0 km from the cooling tower is 1000 kg/(km^2·month) and 400 kg/(km^2·month), respectively, with median drop diameters of 340 μm and 260 μm, respectively. By dyeing the cooling tower water pink, it was determined that about 90% of the total drift deposition is from the cooling tower at a distance of 0.5 km, with a decrease to about 50% at a distance of 1.0 km. Davis (1979) summarizes these Chalk Point data in a final project report. It is concluded that significant salt deposition occurs at Chalk Point only on the plant property (distances less than 0.4 km). The maximum depositon, once both units are operable, is expected to be 5800 kg/(km^2·month) with off-site drift deposition indistinguishable from the natural background.

Alkezweeny et al. (1974) measured chromate deposition downwind of the mechanical-draft cooling towers at the Oak Ridge Gaseous Diffusion Plant. Maximum chromate deposition rates are about 250 kg/(km^2·month) 30 m from the tower. Taylor et al. (1973) measured chromium in grass, tobacco, and soil downwind of these towers and found that concentrations significantly above background occurred within about 500 m of the towers. The measurements are difficult to interpret because of the complexity of the source, which consists of three separate banks of towers.

Taylor and Parr (1978) have continued their studies of drift interception by plants to determine the time scales of drift retention and the differences among various plant species. A portable aerosol generator was used to simulate drift drops. It was found that broad-leafed trees intercepted about 72% of the drift, whereas grass intercepted only about 24%. Three days after a one-shot drift application, the amount of drift mass (chromate is used as a tracer) remaining on the plant decreased by about 50 to 90%. However, some contamination remained after 8 to 12 weeks.

Simulated saline water drift was used for similar studies made by Francis (1977) and McCune et al. (1977). Among the various trees and shrubs that were tested, hemlock and dogwood seemed the most sensitive to drift water.

Lauver et al. (1978) and Armbruster et al. (1978) have been monitoring sodium and chloride levels in trees, shrubs, and field crops around the Chalk Point natural-draft cooling tower for several seasons. They report no accumulations above background levels in the plant species studied. However, Rochow (1978) reports damage to trees within 100 m of the Palisades mechanical-draft cooling towers where the plume often downwashes and impinges directly on the adjacent trees. Talbot (1979) reviews the potential biological impacts of cooling-tower salt drift by listing the threshold concentrations at which various floral species are harmed. As he points out, however, it may not be valid to extrapolate the results of laboratory experiments to the field.

15-2.4 Interference with Aircraft

Because cooling towers and ponds are sometimes located near airports, the question has come up whether or not moist plumes can affect aircraft. The aircraft of several research groups, including Pennsylvania State University (Thomson, Norman, and Miller, 1976 and Thomson et al., 1977, for the Keystone Steam Plant), Meteorology Research, Inc. (Woffinden, Harrison, and Anderson, 1976, Chalk Point Steam Plant), and Battelle Pacific Northwest Laboratory (Wolf, 1976, Rancho Seco Nuclear Plant) have been used to measure the characteristics of moist plumes. No one reported icing in the plumes, although this is probably due to the short duration of plume penetration. Turbulence in the plume is reported as being moderate to severe but is also of short duration. An abrupt updraft of about 5 m/s is encountered at either edge of the plume at a height of about one tower diameter above the tower mouth. A record of plume vertical speed sensed by an aircraft flying near the tower mouth at Keystone is given in Fig. 15.6 (Thomson, Norman, and Miller, 1976).

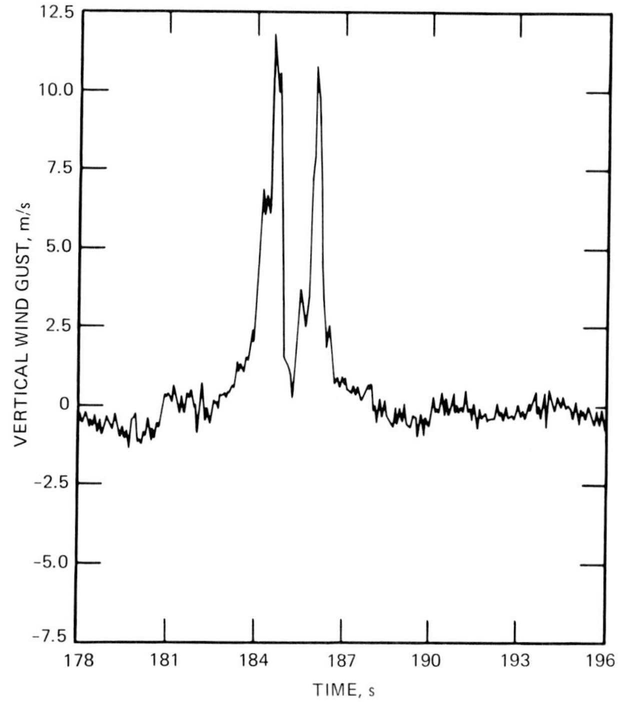

Fig. 15.6 Record of plume vertical speed sensed by an aircraft flying through two plumes at an elevation 90 m above the cooling towers at Keystone, Pa. (From D. W. Thomson, J. M. Norman, and R. L. Miller, Airborne Measurements of Turbulent Temperature and Velocity Fluctuations in Cooling Tower Plumes, in *Proceedings of the Third Symposium on Atmospheric Turbulence, Diffusion, and Air Quality*, Raleigh, N. C., American Meteorological Society, Boston, Mass., pp. 576-580, 1976.)

Because of limited visibility, Tennessee Valley Authority pilots do not fly helicopters in the condensed plume at Paradise Steam Plant (Slawson, Coleman, and Frey, 1974). To determine the possible extent of interference with aircraft, existing records of in-plume accelerations and visibilities should be compared with the flight specifications of aircraft which might by chance have to fly through the plume. Also, on a few occasions, experimental flights through the plume should be made by various commercial and military aircraft to estimate the actual accelerations and visibility losses.

15-2.5 Interactions with Chemical Plumes

Often cooling towers are built at fossil-fired power plants, where SO_2 and particles are emitted. If the plumes merge, chemical reactions may lead to the production of sulfates. It is known that the SO_2—sulfate oxidation process occurs more rapidly in a humid environment. In a study at Keystone several years ago, conducted by the Illinois Institute of Technology Research Institute (Stockham, 1971), it was determined that acidity in the stack plume increased when a plume containing SO_2 merged with a cooling-tower plume. Similar results, also based on the Keystone plant, are reported by Thomson, Norman, and Miller (1976).

Stacks are usually higher than adjacent cooling towers, but measurements indicate the total heat output from stacks to be less than that from cooling towers. Consequently the final plume elevation from both sources is nearly the same, about 500 m above the ground, and it is not surprising that the stack and cooling-tower plumes are sometimes observed to merge. Stockham (1971) used cascade impactors mounted on a helicopter to sample droplets in the merged Keystone plumes. He found that the number ratio of acid drops at pH 4 to 5 to neutral drops at pH 6 to 7 increased from 0.05 to 3.0 as relative humidity increased from 50% to 95%. The smallest drops had the highest acidity. Thomson, Norman, and Miller (1976) also reported that the smallest drops, those found in the last stage of the impactor, were associated with the highest acid concentrations.

15-2.6 High Winds

Very large buoyant plumes can concentrate vorticity existing in the region around the plume and cause waterspouts and other tornado-like vortices to form. At currently operating electricity producing facilities, vortices are seldom observed. However, Thomson (private communication, 1976) reports observing a small whirl dangling from the cooling-tower plume at Keystone. Hanna and Gifford (1975) show that vortices will be much more likely at large energy centers (20 000 to 50 000 MW), if they are built. Vortices were observed in the "Meteotron" oil burner experiment by Dessens (1964) and by Taylor et al. (1973) in large Australian bushfires. Orgill (1977) studied reports of these and several other fire whirlwinds in his comparison between fire plumes and cooling-tower plumes. He concluded that fire whirlwinds were common during large fires but that the power developed by a large fire was much greater than the power developed at current cooling towers. Vortices (e.g., dust devils) are not often observed during natural convection, where buoyancy is usually too weak, or at single cooling towers, where buoyancy overwhelms the tendency to concentrate vorticity.

The energy characteristics discussed in this section and the next section are summarized in Table 15.2.

15-2.7 Changes in Cloudiness, Precipitation, and Temperature

Every wet cooling tower produces a visible plume or cloud to some degree. Even cooling ponds and dry towers produce clouds when atmospheric conditions are right. For example, Everett and Zerbe (1977) report stratus clouds occasionally extending a few kilometers downwind of the Dresden cooling pond. Visible cooling-tower plumes average a few hundred meters in length, although the distribution is very broad, with short 50-m plumes on sunny summer afternoons, and long 50-km plumes on rainy days (Hanna, 1974a; Thomson et al., 1977; and Environmental Systems Corporation, 1977). The last two reports give visible plume outlines during a wide variety of weather conditions. Large cumulus clouds have been observed over cooling towers by Auer (1976) and Stinson and Brown (1976), but there are no instances in which a thunderstorm has been definitely tied to cooling-tower origin. The most common large cloud produced by a cooling tower is a flat, narrow stratus cloud extending for many kilometers, usually located just beneath the main stratus deck of a large rain system. Some good photographs of clouds produced by cooling towers and other industrial processes are given in the review article by Hosler and Landsberg (1977).

Current power plants, with maximum waste energy outputs of 5000 MW, are occasionally observed to generate light precipitation. Light to moderate snowfalls downwind of cooling towers were reported by Culkowski (1961) in Oak Ridge and Kramer et al. (1976) and Otts (1976) in Charleston, W. Va. The snow in Charleston fell in a plume-shaped sector at distances from about 5 km to about 50 km downwind of the John E. Amos Plant. Moore (1974) reports no significant difference in rainfall rates measured around cooling towers in England. Furthermore, there are no reports of precipitation caused by spray or cooling ponds. The statistical problems in determining a significant variation in rainfall are so great that it is doubtful that a prediction of a small increase (e.g., 5%) in rainfall could be verified. Many years of records at a large network of

TABLE 15.2 Energy Characteristics of Natural and Man-Made Sources

Source	Area	Time duration	Total power, MW	Observations
Surtsey Volcano (Bourne, 1964)	1 km^2	Several months	100 000	Continuous cloud; waterspouts
Australian Bushfire (Taylor et al., 1973)	50 km^2	Several hours	100 000	Cumulus cloud; convergence
Booster Rocket Test (Morris, 1968)	300 m^2	150 s	148 000	Cumulus cloud
Oil burners (Dessens, 1964)	3.2 km^2	Several hours	700	Cumulus cloud; dust devils
Oil fires (Davies, 1959)		Day	10 000	Large plume
Large city (Peterson, 1969)	10^3 km^2	Continuous	100 000	Effects on local climate
Thunderstorm (Hanna and Swisher, 1971)	10 km^2	Hour	180 000	2.5 cm/h rain
Power park	5 to 100 km^2	Continuous	100 000	

stations are required. A few mesoscale precipitation networks of this type have begun operation around cooling towers, e.g., in Michigan (see Ryznar, Baker, and Moses, 1976) and in Georgia (see Patrinos, Chen, and Miller, 1979), but it will be years before the data can be properly evaluated. Patrinos, Chen, and Miller (1979) state that there is "preliminary positive evidence of rainfall modification in the target area," which is an area within about 25 km of the cooling towers at Plant Bowen. A raingauge network around the cooling pond at Baldwin, Ill., has been set up by the Illinois State Water Survey, but results are not yet published.

Landsberg (1977) suggests in a short note that significant changes in precipitation due to waste-heat emissions from a single power plant will be detectable only during specific case studies at distances less than about 1 km from the plant. Dana and Wolf (1977) calculated that the changes in precipitation rate due to scavenging of cooling-tower plume drops by natural rain would be detectable only to distances on the same order as the figure given by Landsberg. Apparently studies of precipitation effects at current power plants should involve a dense network of raingauges in the area within about 5 km of the plant, and single rain events should be chosen for investigation.

In urban areas large amounts of heat are released from broad surface area sources. The heat release rate is about 100 W/m^2 in highly built-up areas, or 100 000 MW for a 1000-km^2 area. Increases in precipitation due to heat releases (on the order of 10^5 MW) from urban areas have been reported. Changnon, Semonin, and Huff (1976) have summarized their 1971–1974 METROMEX data from St. Louis and conclude that "the city is a focal point for both rain initiation and rain enhancement under conditions when rain is likely." A 25% increase in summer rain and greater percentage increases in thunderstorms and hail are found in an area from the edge of the city to about 30 km downwind of the city. The maxima are located to the east-northeast of the city, in agreement with the dominant synoptic wind direction. Observed average summer rainfall in the St. Louis area is given in Fig. 15.7 showing the maximum near Edwardsville and Collinsville, Ill. The orographic lifting caused by the Ozark Hills and the moisture source in the Mississippi–Missouri river bottomlands complicate the pattern, but the statistical analysis by Changnon, Semonin, and Huff (1976) shows that the high precipitation north-northeast of the city is statistically significant on the 5% level. They conclude that the precipitation increase is due partly to the added heat from the city and partly to the effects of added giant nuclei which accelerate the drop coalescence process. Huff and Vogel (1978) further analyzed the METROMEX data and concluded that the effect of the urban area on summer rain downwind of St. Louis was about two times greater than the effect of the bluffs and about three times greater than the effect of the hills.

The most obvious meteorological effect of urban areas is the heat-island effect, or increase of temperature near the urban surface. This effect is due to the greater ability of the urban concrete and asphalt to absorb and hold the sun's radiation. Landsberg and Maisel (1972) observed the buildup of the urban heat island in the new city of Columbia, Md., as it grew from farmland to a population of 25 000. Since surface temperatures are easy to measure, there are many reports of the urban heat island, and Peterson (1969) has summarized the ones published before that date. He finds that the magnitude of the urban heat island varies from about 5°C on calm nights to nearly zero on fair afternoons.

Aircraft observations of several meteorological parameters in the mixed layer over and downwind of St. Louis are summarized by Shea (1976) in Fig. 15.8. It is seen that St. Louis has a significant influence on Aitken nucleii, visibility, mixing height, turbulence, temperature, and humidity. Auer (1977) shows that visibility can deteriorate to less than 5 km at distances of several tens of kilometers downwind of the city. Again, the differences between a city and a power park should be underlined, to prevent Fig. 15.8 from being used to estimate the effects of a power park. The power park does not release as many pollutant particles and gases as does the city, and the waste heat from a power park is likely to rise higher because of the cooling-tower plumes. However, if there are industrial sources in the power park or if the waste heat is released from a broad area source such as a cooling pond, the analogy may be more valid.

Satellite photographs are an important tool for analyzing regional weather modification due to power plants and other man-made sources but have not yet been applied to the problem in a quantitative way. Other remote sensors, such as radars, lasers, and acoustic sounders, would be useful in the study of weather modification around steam plants.

15-3 ESTIMATING THE EFFECTS OF ENERGY GENERATION AT A SINGLE LARGE POWER PLANT

Before discussing large power parks or other complex sources of energy, it is reasonable to begin with the simplest source—a single large power plant. In a fossil-fired plant, about 60% of the total heat generation is lost to the atmosphere. In a nuclear plant, the loss is about 70%. Since a typical size for a large power plant is 1000 MW(e), about 1500 to 2000 MW is thus released to the atmosphere by the cooling system. This rate of energy release is approximately equivalent to the release of energy by a small thunderstorm or to the latent energy input by evaporation from a wet field with an area of 20 km^2 (Hanna and Swisher, 1971). Thus the energy release from a single, large power plant is on the same scale as mesoscale natural processes.

15-3.1 Direct Cooling by Lakes, Oceans, and Rivers

If sufficient cold water is available, it is reasonable to use it directly for cooling a power plant. The various engineering techniques for dispersing the heated effluent in

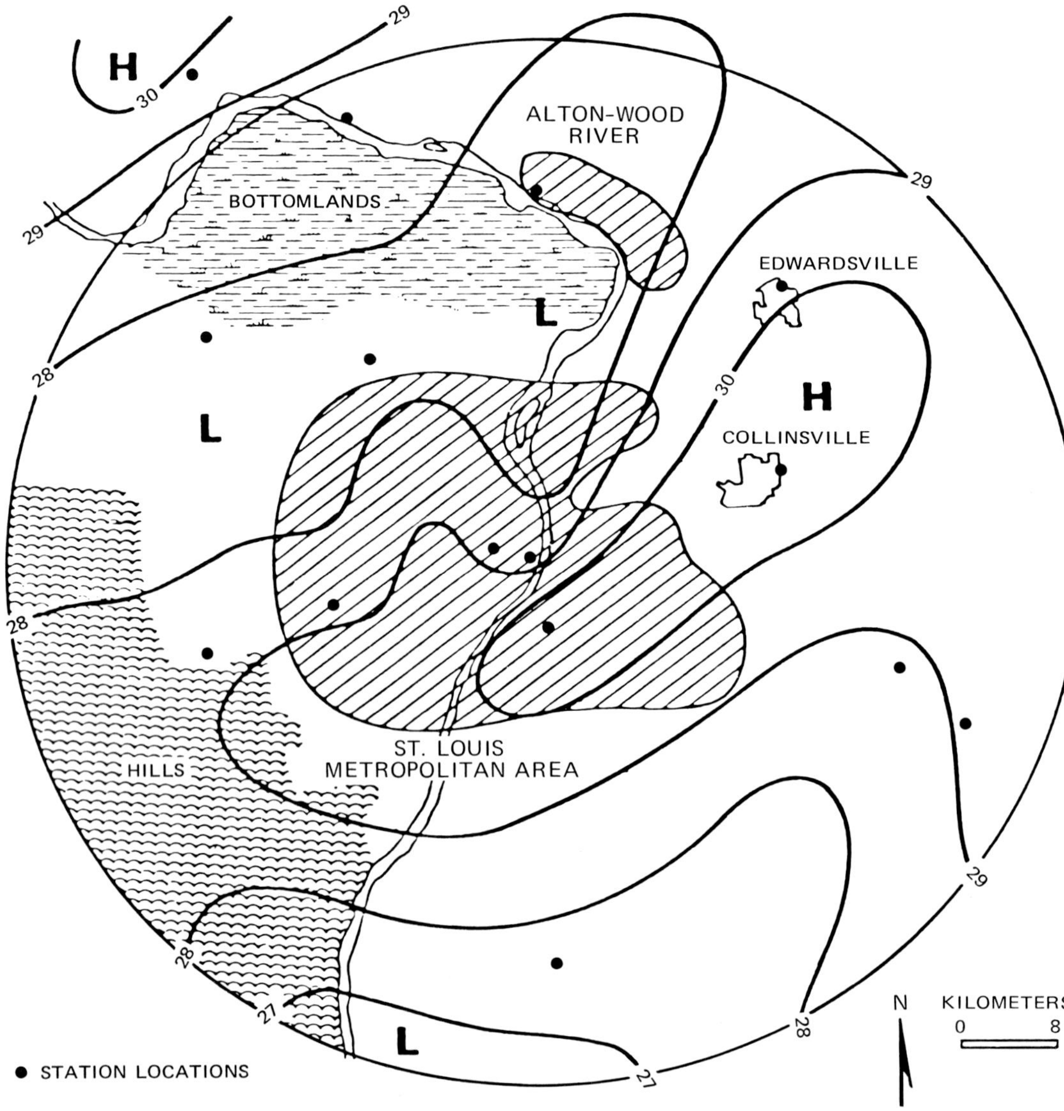

Fig. 15.7 Average summer rainfall (cm) in the region around St. Louis, 1941–1968 period. [From S. A. Changnon, Jr., R. G. Semonin, and F. A. Huff, A Hypothesis for Urban Rainfall Anomalies, *J. Appl. Meteorol.*, 15: 544-560 (1976).]

the water body and the various models for heated effluent dispersion are summarized by Policastro (1972). In this chapter, however, we will be concerned only with the heat as it is being emitted into the atmosphere.

Within a few hundred meters of the power plant, Csanady's (1970) observations show that the heated water plume can have a maximum excess temperature of a few degrees Celsius. More typical values of excess temperature are a few tenths of a degree. If ΔT is the excess temperature, then the excess radiation loss, ΔR, due to the heated plume is given by the equation:

$$\Delta R = \epsilon\sigma[(T_s + \Delta T)^4 - T_s^4] \tag{15.1}$$

where T_s is the surface temperature of the water far from the thermal plume, and the Stefan–Boltzman constant, σ, equals 5.67×10^{-8} W/(m$^2 \cdot$°K^4). The term ϵ is the emissivity, which is usually 0.95. The latent and sensible heat fluxes are altered more indirectly, because of the dependence of the diffusivities, K, on the excess temperature. Using the gradient transfer assumption, we can write the following approximations to the excess sensible heat, ΔH, and latent heat, ΔLE, fluxes due to the heated plume:

$$\Delta H = c_p \rho K \frac{\Delta T}{\Delta z} \tag{15.2}$$

$$\Delta LE = \rho L K \frac{\Delta q}{\Delta z} \tag{15.3}$$

where Δz = layer over which the temperature excess occurs
ΔT, Δq = excess surface temperature and water vapor content
c_p, L, ρ = specific heat of air at constant pressure, latent heat of vaporization, and air density, respectively

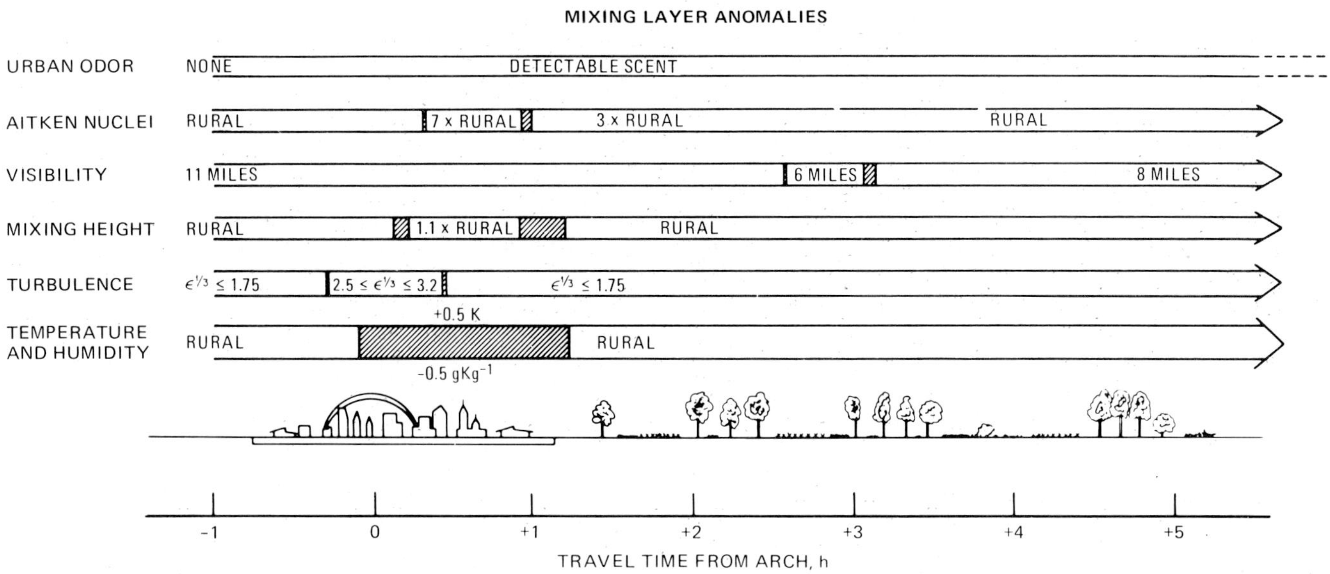

Fig. 15.8 Summertime mixing layer anomalies over and downwind of St. Louis expressed in terms of travel time from the city center. (From D. W. Shea, Observations of the Regional Influence on Thermodynamic Properties and Aerosol Patterns in the Plume Downwind of St. Louis, prepared for the Environmental Protection Agency and the National Science Foundation, Department of Atmospheric Science, College of Engineering, University of Wyoming, Laramie, 1976.)

To illustrate the use of these formulas, assume the following typical values for the lowest 1 m of the boundary layer over a hot-water plume.

$\Delta T = 0.2°K$ $\Delta q = 0.0002$ $\rho = 1.15 \text{ kg/m}^3$

$\Delta z = 1 \text{ m}$ $K = 0.05 \text{ m}^2/\text{s}$ $L = 2.25 \times 10^6 \text{ J/kg}$

$\epsilon = 0.95$ $c_p = 1000 \text{ J/kg°C}$

Then $\Delta R = 1.2 \text{ W/m}^2$, $\Delta H = 11.6 \text{ W/m}^2$, and $\Delta LE = 26 \text{ W/m}^2$. It would take a total water surface area of about 50 km^2 to dissipate the waste heat from a 1000 MW(e) power plant under these conditions [area = 2000 MW/ (38.8 W/m^2)].

The gross effects of thermal discharges on Lake Michigan are treated by Carson (1976). He assumes that the effluents are well-mixed into a 4.65×10^{18} g mass of water. The maximum temperature increase of the lake owing to a waste heat input of 1500 MW for 1 yr then is estimated by Carson to equal 0.003°C (total heat input/ mass$\cdot c_{pw}$), where c_{pw} is the specific heat for water and it is assumed that all the waste heat remains in the lake for the 1-yr period. The estimated warming of Lake Michigan is too small to be measurable. More exact estimates of warming in specific locations and other lakes or oceans depend on a good knowledge of the volume of water available for mixing and the amount of mixing that actually takes place.

Rivers, on the other hand, are usually well-mixed because of their constant motion. If the flow rate of a river is M_R(g/s), the waste-heat input is W (watts), and the specific heat of water is c_{pw}, then the temperature increase for complete cross-stream mixing is

$$\Delta T = \frac{W}{c_{pw} M_R} \qquad (15.4)$$

The radiation and the sensible and latent heat fluxes to the atmosphere from the river can then be calculated by Eqs. 15.1 to 15.3 or some other set of suitable equations developed for estimating the heat exchange over power surfaces.

15-3.2 Cooling Ponds

A cooling pond is usually a closed system, with hot water discharged into one end of the pond, cooled as it flows across the pond, and then withdrawn at the other end for another pass through the power plant. Most of the engineering of cooling ponds is guided by field experience (trial and error) rather than scientific concepts. Drake (1975), in his review and evaluation of the literature on cooling ponds and spray ponds, concludes: "The sets of data on spray ponds and small, hot ponds required to formulate and evaluate models are nonexistent." However, B. B. Hicks (personal communication, 1977) has stated that progress can be made by applying classical oceanic meteorology concepts. A few good theoretical models of cooling ponds do exist. The primary application of the models is to estimate the equilibrium surface temperature of the pond. Ryan, Harleman, and Stolzenbach (1974) formulated an expression for the heat loss from a pond during free-convection conditions. Free convection often exists near the surface of the pond owing to its excess surface temperature. Observations and modeling of the Dresden pond by Yamada (1979) show temperatures of 18°C and 0°C at the surface and a height of 20 cm, respectively, on a cold February morning. The heat budget or balance equation for a well-mixed steady-state cooling pond can be expressed as

$$Q = \epsilon \sigma T_s^4 + LE + H \qquad (15.5)$$

where Q is the imposed heat loading, which includes the waste heat, the incoming long-wave radiation, and the solar radiation. The outgoing long-wave radiation and the latent and sensible heat fluxes are on the right-hand side of Eq. 15.5. Because it is assumed that all fluxes are constant, an averaging time of one day to a few days is implied. In a measurement program the waste-heat loading is known from the plant operating record; the radiation fluxes can be measured by radiometers; and the latent and sensible fluxes can be determined from mean gradients or by eddy correlation techniques, as mentioned in Chap. 5

$$H = -\rho c_p K_z \frac{\partial \theta}{\partial z} \qquad LE = -\rho L K_z \frac{\partial q}{\partial z} \tag{15.6}$$

or

$$H = \rho c_p \overline{w'T'} \qquad LE = \rho L \overline{w'q'} \tag{15.7}$$

The technique in Eq. 15.7 is considered the best since the flux is measured directly. However, if fast-response instruments are not available, Hicks (1977a) showed that Eq. 15.6 can be used by substituting a reasonable form for the eddy diffusivity K (see Chap. 5), which is a function of the surface stress, height, and stability.

Hicks (1975) analyzed the heat budget over averaging times of a day or longer using Eq. 15.5 and suggested that every cooling pond will come to a limiting surface temperature determined by the maximization of the evaporation rate. His method works best if the total heat load of the pond is more than about 500 W/m^2. The basis of the method is that, over a saturated surface, the Bowen ratio $\beta = H/LE$ is a function only of the surface temperature (Priestly and Taylor, 1972).

$$\beta = 0.8 \frac{c_p}{L} \left(\frac{\partial q_s}{\partial T_s}\right) - 0.2 \tag{15.8}$$

where $\partial q_s/\partial T_s$ is the rate of change of the saturated mixing ratio with surface temperature. Hicks solves Eq. 15.5 for LE, sets its derivative with respect to T_s equal to zero, and obtains

$$Q - \epsilon\sigma T_s^4 = \frac{-4\epsilon\sigma T_s^3 (1 + \beta)}{\partial\beta/\partial T_s} \tag{15.9}$$

This equation can be solved to give the equilibrium surface temperature T_s. Notice that Eq. 15.9 implies that H and LE are given by the relations

$$H = \frac{-4\epsilon\sigma T_s^3 \beta}{\partial\beta/\partial T_s} \tag{15.10}$$

$$LE = \frac{-4\epsilon\sigma T_s^3}{\partial\beta/\partial T_s} \tag{15.11}$$

Hicks assumes ϵ equals 0.95 and proposes the following interpolation formulas for c_p/L and $\partial q_s/\partial T_s$:

$$\frac{c_p}{L} = p(1 + 0.00115\, T_s') \times 0.00066 \tag{15.12}$$

$$\frac{\partial q_s}{\partial T_s} = \frac{25\ 029}{(237.3 + T_s')^2} \exp\,[17.269\, T_s' (237.3 + T_s')] \tag{15.13}$$

where p is in millibars and T_s' is in degrees Celsius. The prime is used to distinguish this temperature from the temperature previously defined in the units °K. The derivatives $\partial q_s/\partial T_s$ and $\partial q_s/\partial T_s'$ are equivalent.

The method for estimating T_s described previously is tested with measurements at the 1-km^2 Dresden cooling pond and at a small 10-m^2 pond simulator by Hicks (1975). A comparison of predicted temperatures with observed temperatures in Fig. 15.9 shows good agreement. It is important to note this method suggests the equilibrium temperature is independent of wind speed.

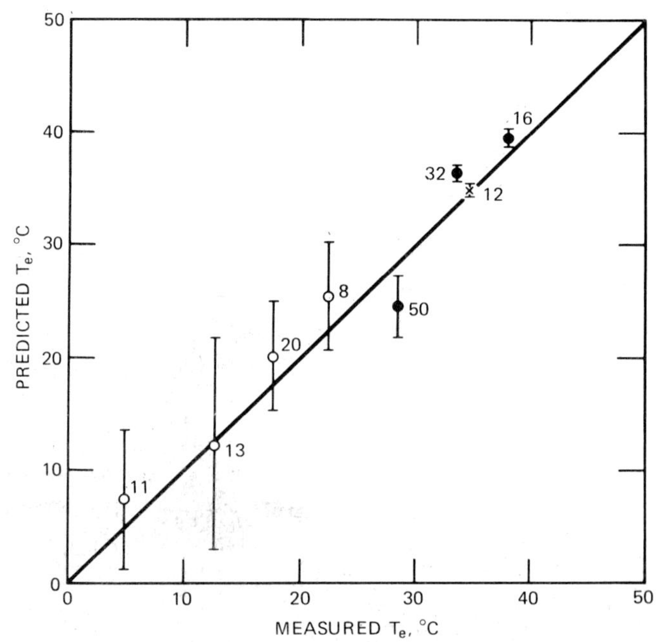

Fig. 15.9 A comparison between measured values of cooling pond water surface temperature and predictions based on Hicks's (1975) technique. The ○ represents evaluations made over a large cooling pond and the + represents the results of an eddy correlation experiment in a large cooling pond. Standard error bars and the number of contributing values are shown.

The major atmospheric effect of cooling ponds is the creation of fog. Currier, Knox, and Crawford (1974) define a fog index number

$$FIN = \frac{\Delta T}{e_s - e_a} \quad \frac{°F}{mb} \tag{15.14}$$

where ΔT is the surface-to-air temperature difference (in °F) and e_s and e_a are the saturated and ambient vapor pressures (in millibars) of the air. Values of FIN are plotted as a function of fog occurrence in Fig. 15.10 and show that little fog occurs when FIN < 10°F/mb (5.6°K/mb) and fog nearly always occurs when FIN > 100°F/mb (56°K/mb). However, a lot of scatter is shown in Fig. 15.10.

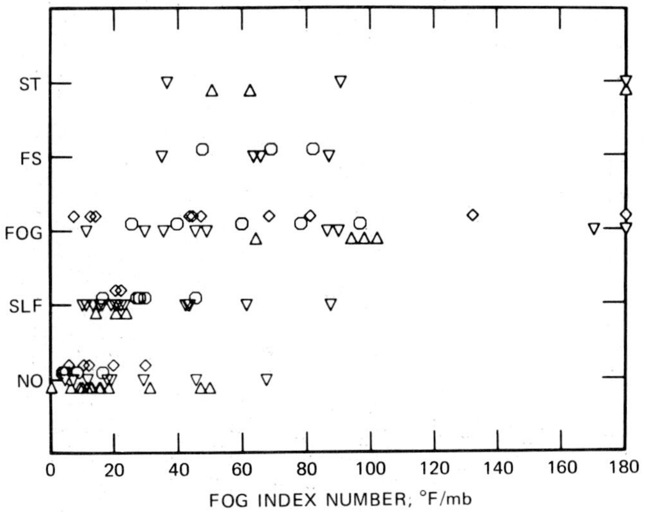

Fig. 15.10 Fog occurrence using the Fog Index Number (FIN) defined by Eq. 16.13. [Adapted from E. L. Currier, J. B. Knox, and T. V. Crawford, Cooling Pond Steam Fog, *J. Air Pollut. Control Assoc.*, 24: 860-864 (1974).] Wind speed classes are (△) less than 0.5 m/s, (▽) 0.5 to 2 m/s, (○) 2.5 to 5 m/s, and ◇ greater than 5 m/s. The NO means no fog; SLF, slight fog; FOG, fog; FS, fog and stratus; and ST, stratus.

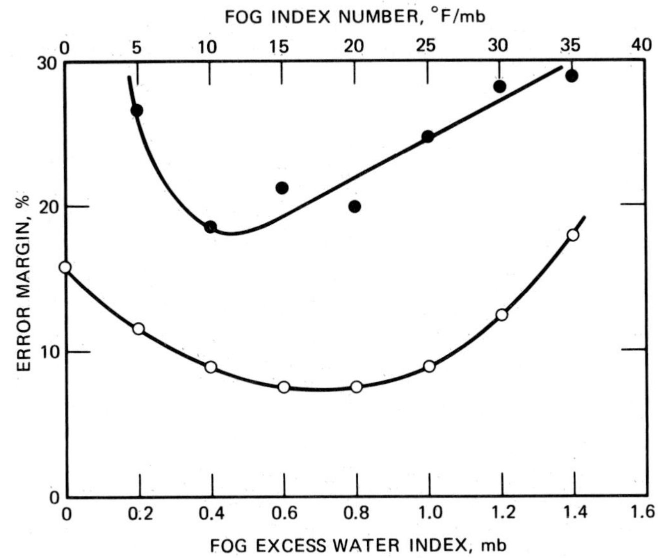

Fig. 15.12 Fog occurrence using fog indices defined by Eqs. 15.13 and 15.14. (Adapted from B. B. Hicks and J. D. Shannon, *A New Cooling Pond Index*, Radiological and Environmental Research Division, Annual Report ANL-76-88, pp. 104-107, Argonne National Laboratory. FEWI, ○; FIN, ●.

The scatter is reduced if the number FEWI defined by Hicks (1977b) is used

$$\mathrm{FEWI} = \frac{e_s(T_s) + e_a}{2} - e_s\,\frac{T_s + T_a}{2} \quad \mathrm{mb} \qquad (15.15)$$

In Fig. 15.11, the same fog data as in Fig. 15.10 are plotted as a function of FEWI. Some fog is expected when FEWI > 0.5 mb, and fog is virtually assured if FEWI > 1.0 mb. Figure 15.12 shows the error margin resulting if various values of FIN and FEWI are picked as a fog–no-fog criterion (Hicks and Shannon, 1977). This again demonstrates that FEWI is more valuable as a predictor than FIN

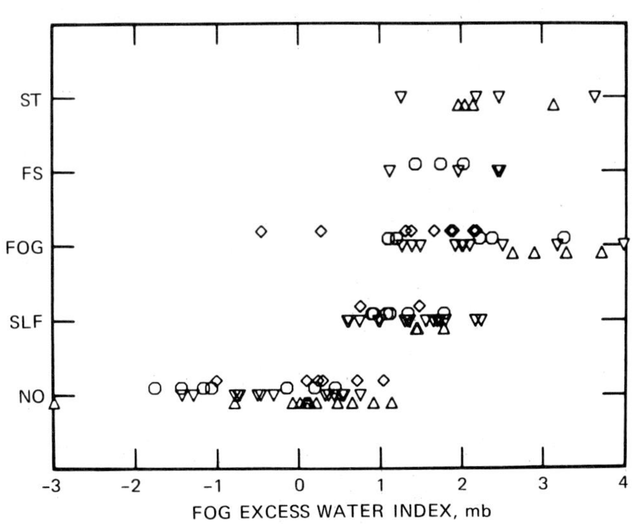

Fig. 15.11 Fog occurrence using the Fog Excess Water Index (FEWI) defined by Eq. 15.14. [Adapted from B. B. Hicks, The Prediction of Fog over Cooling Ponds, *J. Air Pollut. Control Assoc.*, 27: 140-142 (1977b)]. Same wind speed classes and legend as in Fig. 15.10.

and that best results are obtained if a value of FEWI equal to 0.7 mb is chosen as a fog criterion.

Drift deposition and icing are observed to be problems at cooling ponds only at downwind distances less than 100 or 200 m. Possibly because these distances are still within the plant boundaries, there have been no good measurements or modeling efforts directed toward this problem. The same can be said for spray ponds. The engineering aspects of the spray modules are discussed by Drake (1975) and Wilber (1978), but apparently no one is thoroughly studying their atmospheric effects.

15-3.3 Cooling Towers

The research area of atmospheric effects of cooling towers consumes much more time and money than the other areas discussed so far. For this reason, the theories are more complete and the data are more extensive in this section. To grasp the research problems better, the reader can refer to several reviews. The Energy Research and Development Administration report, CONF-740302, *Cooling Tower Environment—1974* contains 30 papers that cover topics ranging from visible plume length to social problems to biological effects. McVehil and Heikes (1975) conducted a critical review of cooling-tower plume modeling and drift measurement for the American Society of Mechanical Engineers. This document is especially useful because it compares the predictions of many different visible plume models against some observations. Carson (1976) documented and briefly discussed 191 references concerning evaporative cooling systems. *Atmospheric Environment* (Vol. 8, No. 4, 1974) is devoted to a conference on cooling-tower effects in England. Model "Olympics," in which all available cooling-tower plume models are compared with a common set of input data, have been conducted by Chen and Hanna (1978) and Policastro, Dunn, and Breig (1978).

15-3.3.1 Plume Rise. The material in Chap. 8 permits plume rise to be calculated for a variety of sources. In this section the plume rise theory is extended to handle initially saturated sources. In estimating the atmospheric effects of cooling towers, the plume rise must be known since the plumes typically do not level off until they have traveled several hundred meters vertically. The total buoyancy of the cooling tower at a fossil-fired power plant is at least three or four times the buoyancy of the stack plume. An excellent summary of the physical basis of theories of cooling-tower plume rise is given in Appendix B of Briggs (1975).

Downwash is often a factor in lowering the trajectory of the plume near the source. Overcamp and Hoult (1971), on the basis of a towing tank experiment with models of natural-draft towers, found that strong downwash (ground contact 2 to 4 tower heights downwind) occurs if

$$\frac{w_0}{U} \leqslant 0.45\ Fr_0^{2/3} \tag{15.16}$$

where $Fr_0 = w_0/(R_0 g\ \Delta T/T_0)^{1/2}$. In this expression U is wind speed (m/s), g is the acceleration of gravity (m/s^2), and Fr_0, w_0(m/s), R_0(m), and T_0(°K) are the initial plume Froude number, vertical speed, radius, and temperature, respectively. The Froude number for cooling towers is on the order of unity, and the initial vertical speed w_0 is about 5 m/s. Consequently it is estimated that downwash will occur for wind speeds greater than about 10 m/s. However, there are only a few observations of plumes touching the ground at natural-draft towers. Possibly their hyperbolic shape keeps natural-draft towers from experiencing downwash problems. Still, there is a slight bending down of the plume during high winds, and empirical factors to account for this dip in the trajectory at specific sites are suggested by Slawson, Coleman, and Frey (1974) and Bøgh (1974).

Because of their bulky shape, mechanical-draft towers are plagued with downwash problems. Hanna (1974a) estimates that downwash occurs about 40% of the time at the mechanical draft towers near Oak Ridge. But he also points out that the buoyant plumes "lift off" once they escape from the aerodynamic cavity region at a distance of about 200 m from the towers. The final effect of downwash on plume rise has not yet been studied, although intuitively we would predict a decrease in plume rise.

Most models assume that the plume is initially saturated, with a liquid-water content of about 0.001 g of liquid water per gram of air (g/g). Wigley's (1975) theoretical predictions suggest that initial supersaturations of 150 to 170% are possible in small jets or scrubbed plumes. But he predicts a natural-draft plume to be initially supersaturated by only a few percent and then undersaturates in a few seconds. Thus it appears that supersaturation need not be considered in cooling tower plume calculations.

Evaporative cooling-tower plumes by definition carry large amounts of latent heat. During hot weather the plume may carry 80% latent heat and 20% sensible heat. However, as Briggs (1975) points out, only a small fraction of this latent heat is released. A convenient way of seeing the relative amounts of latent and sensible heat is through examination of the psychrometric chart (Fig. 15.13). On this figure the energy change represented by a certain distance on the horizontal axis is roughly equal to the energy change represented by an equal distance on the vertical axis. For example, $\Delta T = 10°C$ and $\Delta q = 0.004$ g/g; each represent an energy difference of 10 J/g.

Hypothetical initial plume conditions (subscript po) and ambient conditions (subscript e) are shown on the figure. The initial ratio of latent to sensible heat equals about 2.0. In the absence of thermodynamic effects, the temperature and specific humidity of any mixture of ambient and plume air would fall along the straight line connecting the initial plume and ambient points. The amount of condensation at any T would then be the vertical distance between the straight and curved lines. Yet as latent heat is released, T increases, and consequently the total latent heat release is less than that given by the distance between the two lines. As an example, consider a mixture indicated on the figure by subscript pl which is 50% plume air and 50% ambient air. If thermodynamic changes are not accounted for, the 50–50 mixture would be at the midpoint of the straight line with a supersaturation of 0.0042 g/g and a potential fractional latent heat release of $[(0.0042\ g/g)\ (V/V_0)]/0.027\ g/g = 0.31$. But

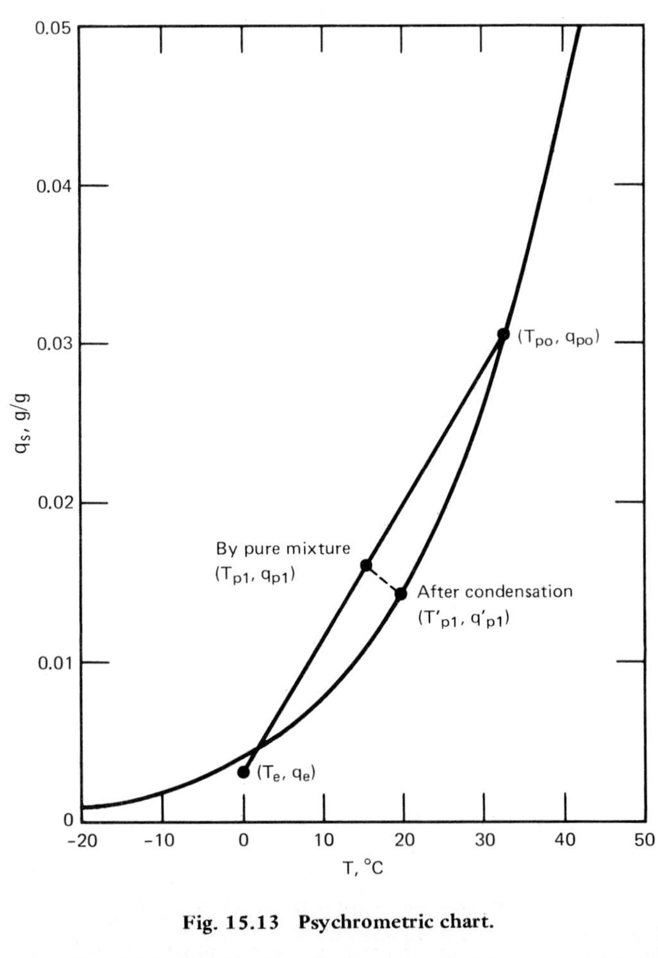

Fig. 15.13 Psychrometric chart.

warming by latent heat release brings the mixture to the primed point on the curved line, where the mixture is just saturated and about 0.0017 g/g of water has been condensed, amounting to a fractional latent heat release of $[(0.0017\ g/g)\ (V/V_0)]/0.027\ g/g = 0.13$. In this example less than 20% of the initial latent heat flux is released, and plume rise (proportional to the one-third power of the initial plume heat flux) increases by no more than 6%. Briggs (1975) deduces that the plume rise enhancement due to latent heat release is no more than 20% in any case, and that current measurements of cooling-tower plume rise are not sufficiently accurate to detect this difference. Thus plume rise calculations need not account for the latent heat flux. It is therefore recommended that the following formulas be used for estimating plume rise Δh from cooling towers:

$$\Delta h = 2.6 \left(\frac{F}{Us}\right)^{1/3} \qquad U > 1\ \text{m/s} \tag{15.17}$$

$$\Delta h = 5\,\frac{F^{1/4}}{s^{3/8}} \qquad U \leqslant 1\ \text{m/s} \tag{15.18}$$

A dividing line of U equal to 1 m/s is chosen because for typical buoyancy fluxes and stabilities, the two formulas (Eqs. 15.17 and 15.18) produce nearly equal values of Δh at U equal to 1 m/s. The trajectory near the source is given by

$$z = 1.6\,F^{1/3} u^{-1} x^{2/3} \left(1 + \frac{2F_m U}{Fx}\right) \tag{15.19}$$

where

$$\text{momentum flux } F_m = w_0^2 R_0^2 \quad \frac{m^4}{s^2}$$

$$\text{buoyancy flux } F = \frac{g}{T_{vpo}} (T_{vpo} - T_{veo}) \times w_0 R_0^2 \quad \frac{m^4}{s^3} \tag{15.20}$$

$$\text{and stability } s = \frac{g}{T_e}\left(\frac{\partial T_e}{\partial z} + \Gamma_d\right)$$

The dry adiabatic lapse rate, Γ_d, equals g/c_p, or 0.0095 °K/m. These formulas are fully derived by Briggs (1975) and are developed further in Chap. 8.

For saturated plumes rising in a saturated environment, Weil (1974) shows that the definition of the stability factor in Eqs. 15.17 and 15.18 should include the modified wet adiabatic lapse rate Γ_m rather than the dry adiabatic lapse rate Γ_d. Thus, for a saturated environment, the buoyant plume rise is unimpeded for an ambient lapse rate equal to the wet adiabatic lapse rate.

Most power plants have two to four natural-draft towers, spaced about 200 m apart. Mechanical-draft towers are in banks containing 10 to 20 cells. It is therefore necessary to account for the possibility of the merging of multiple plumes. Briggs (1974) analyzed some Tennessee Valley Authority data from multiple stacks and derived the semiempirical enhancement factor E_N

$$E_N = \left(\frac{N + S}{1 + S}\right)^{1/3} \tag{15.21}$$

where the dimensionless spacing factor S is defined by

$$S = 6.0 \left[\frac{(N-1)\,\Delta X}{N^{1/3}\,\Delta h}\right]^{3/2}$$

In these expressions N is the number of stacks, Δh is the plume rise from one stack, ΔX is the distance between stacks, and E_N is the ratio of the plume rise from N stacks to that from one stack. For a typical power plant with four natural-draft towers,

$N = 4 \qquad \Delta X = 200\ \text{m} \qquad \Delta h = 500\ \text{m} \qquad E_N = 1.17$

For a typical power plant with a single bank of 50 mechanical draft cells,

$N = 50 \qquad \Delta X = 10\ \text{m} \qquad \Delta h = 100\ \text{m} \qquad E_N = 1.80$

The plume rise enhancement is barely noticeable for the natural-draft towers, but it amounts to an 80% increase for the mechanical-draft towers. It can be concluded that the enhancement factor E_N should be applied to estimating plume rise from mechanical-draft towers, but it is not important for multiple natural-draft towers.

Unfortunately there are not very many observations of final plume rise from cooling towers that can be used for testing the formulas in this section. There are numerous studies of the initial stages of plume rise near the tower where the plume is still visible. But, to detect the final plume rise, either the plume must be condensed up through its point of maximum rise, or sophisticated temperature and humidity instruments on an aircraft or a remote sensor are needed. The most complete set of data available were reported by Kramer et al. (1975), who recorded the plume rise at the John E. Amos, West Virginia plant. The plume was visible from a small aircraft, and vertical profiles of ambient temperature, dew point, and wind speed were obtained. The observed plume rise and the results of several analytical plume rise formulas are listed in Table 15.3, which is taken from the report by Hanna (1976). The initial plume radius and vertical speed are 40 m and 4.4 m/s, respectively. The initial plume temperature is based on an empirical formula given by Kramer et al. (1975)

$$T_{po} = [297.4 + 0.635\,(T_d - 273)] \times [1 + 0.01\,(1 - RH)] \tag{15.22}$$

where T_d is the dew point in degrees kelvin and RH is the fractional relative humidity (i.e., RH = 0.5 when the relative humidity is 50%). There are three towers 200 m apart at Amos, but the heat output from only one of them is used in the calculations in Table 15.3. It is seen in the table that the average observed plume rise, 750 m, is about halfway between the average plume rise calculated using the sensible heat flux from one unit, 540 m, and the average plume rise calculated using the sum of the sensible and

TABLE 15.3 Observed and Calculated Plume Rise at John E. Amos Plant

Run	$\overline{U}$, m/s	$\Delta T/\Delta z$ over depth of plume, 10^3 °C/m	Observed plume rise, m	Analytical plume rise, m		
				Sensible (one unit)	Sensible and latent (one unit)	E_N applied to sensible (one unit)
1	14	13.5	560	250	450	270
6	11	3.0	850	430	770	490
8	9	9.8	370	250	450	270
10	5	2.1	820	590	1000	700
11	0	6.1	880	980	1500	
12	6	1.2	1060	720	1280	870
15	3	10.7	1120	780	1200	
16	6	7.2	1160	380	680	430
17	8	6.7	360	410	730	460
18	6	5.6	910	410	730	460
19	0	5.9	800	910	1400	
24	6	4.0	760	500	900	580
28	5	21.8	360	310	570	340
31	0	11.6	500	780	1200	
35	3	9.8	360	480	880	550
44	8	1.9	910	510	910	590
45	7	5.6	910	390	700	440
47	0	3.8	800	910	1400	
48	7	7.6	670	340	610	380
		Average	750	540	920	

latent heat fluxes from one unit, 920 m. The relative position of the observed plume rise is in agreement with the conclusion above that only a small fraction of the latent heat is released. The multiple plume enhancement factor, E_N, is about 1.2, not enough to increase the calculated plume rise to the observed level. The observed plume rise in the Amos data set might be best simulated by some combination of latent heat and multiple plume enhancement. For example, these data indicate that the effective heat flux is about 1.5 times the sensible heat flux. Refinements in prediction techniques require more accurate observations.

15-3.3.2 Ground-Level Fog Formation. In most environmental impact statements (Collins, Case, and Mule, 1971) and in the review by Hanna (1974b), it is recommended that the standard Gaussian plume formula, as described in Chaps. 6 and 13, be used for calculating ground-level fog formation due to cooling towers. But since the Gaussian formula assumes that the background concentration of water vapor is constant, this formulation is valid only for a well-mixed environment; e.g., a windy or convective day with a deep mixing layer. In this method, the source strength of excess water, Q, is

$$Q = \pi w_0 R_0^2 (q_{po} - q_{eo} + Q_{HO} + Q_{CO}) \tag{15.23}$$

where Q_{HO} and Q_{CO} are the initial hydrometeor and cloud-water contents of the plume. The ground-level concentration of excess water, Δq, calculated using the Gaussian plume formula is then compared with the saturation deficit $q_s - q_e$ of the ambient air and the difference yields the fog concentration. Visibilities in meters can be derived from the fog concentrations through use of Trabert's (1901) formula:

$$\text{Visibility} = \frac{2\overline{D}}{\omega} \tag{15.24}$$

where $\overline{D}$ is the mean drop diameter (μm) and ω is liquid-water content in grams per cubic meter in $\{\omega = 1150 [\Delta q - (q_s - q_e)]\}$.

Ground-fog problems do not usually occur with the well-mixed atmosphere assumed in the preceding derivation. Fog is most likely during stable conditions, when the atmosphere, by definition, is not well-mixed. When model estimates based on the above assumptions are applied to stable conditions, they are often questioned, for there are no observations of ground fog caused by cooling towers anywhere except in the area within a few hundred meters of the towers. A reason why the moisture does not reach the ground could be that there is usually a vertical gradient of water vapor in the atmosphere, with the highest value at the ground as shown in Fig. 15.14. The flux of moisture is naturally from high to low values of water vapor. To reach the ground, the water in the cooling tower plume would have to move against this gradient.

To simulate plume-water diffusion at distances downwind, where the maximum excess water in the plume is less than the difference between the natural water content at the surface and the plume elevation, "K" diffusion theory or second-order closure is required. The appropriate K coefficients and techniques are described in Chaps. 5 and 6.

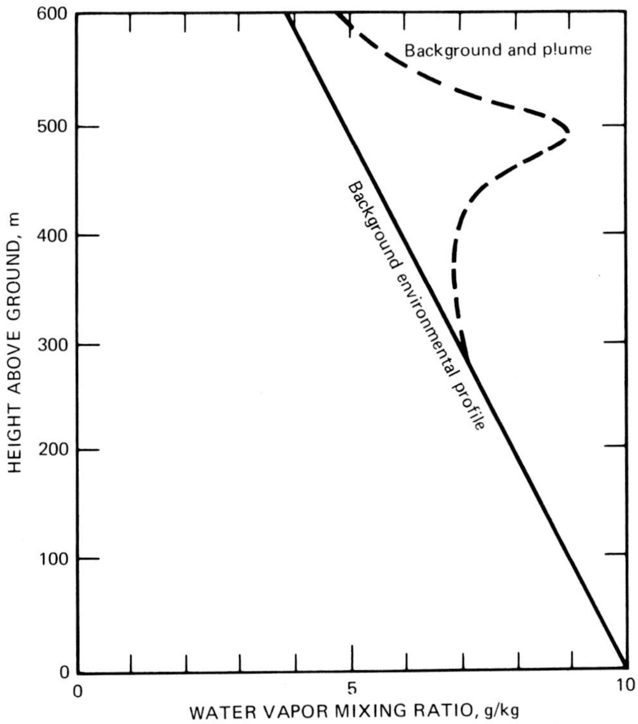

Fig. 15.14 Typical vertical profile of environmental water vapor mixing ratio, with the contribution of the cooling-tower plume superimposed.

The problem of fog near the cooling towers during high-wind downwash conditions is more easily treated. It is assumed that the moisture from the cooling tower is uniformly distributed in a vertical cross section across the cavity region. The concentration of excess water ($q - q_e$) in the cavity is calculated from the formula

$$q - q_e = \frac{Q}{AU} \tag{15.25}$$

where A is the cross-sectional area of the cavity. In a study of fogging at mechanical-draft cooling towers, Hanna and Perry (1973) found best results when A equals the height of the tower bank times its length.

15-3.3.3 Drift Deposition. The subject of drift deposition usually occupies a large number of pages in environmental impact statements. Near the power plant at Chalk Point, Md., there is great concern among tobacco growers that the saltwater drift from the towers will harm their crop (Pell, 1974). As in all areas of cooling-tower effects, good comprehensive field observation programs are difficult to find. But in spite of the lack of observations, there has been a proliferation of drift deposition models. In *Cooling Tower Environment—1974*, descriptions are given of the models developed by Schrecker, Wilber, and Shofner; Slinn; Pena and Hosler; Hanna; Roffman and Grimble; Laskowski; and Israel and Overcamp. McVehil and Heikes (1975) critically reviewed these and other drift deposition models. Currently, Chen and Hanna (1978) and Policastro, Dunn, and Breig (1978) are also reviewing and comparing drift models. The models must account for plume rise, the settling speed of the drift drops relative to the plume, and the possible evaporation of the drift drop. As shown by the drift drop size distribution in Fig. 15.4, most drift drops have diameters greater than about 100 μm, meaning that their fall speed is greater than 0.5 m/s, and that they will fall through the plume fairly rapidly.

Chen and Hanna (1978) provided the arbitrary set of initial conditions listed in Table 15.4 to 10 drift modelers. The models range from the K-model of Roffman and Grimble (1974) and the ballistic trajectory model of Hosler, Pena, and Pena (1972) to the Gaussian model of Israel and Overcamp (1974). A few of the predicted drift deposition curves are given in Fig. 15.15, which illustrates the wide variability in predicted drift deposition. The maximum predicted deposition rates for all the models studied range over two orders of magnitude and the locations of the maximum vary from 0.3 to 3.4 km from the source. However, it is reassuring that about two-thirds of the models agree within a factor of 3. The main reason for the variability is the differing assumptions for initial release height of the drift drops. One model assumes the release height to be the top of the tower, thus yielding high deposition rates close to the tower. Another model releases the drops at the height of maximum plume rise, thus yielding relatively low deposition rates and a maximum relatively far from the tower.

Overcamp and Israel (1979) performed a sensitivity test with their model and found that the initial drift drop size distribution had the greatest effect on downwind drift deposition. Thus it is very important to make good measurements of the drops at the tower mouth. The drift

TABLE 15.4 Parameters Used in the Drift Calculations in the Model Comparison by Chen and Hanna (1978)

Saturated plume initially
Plume temperature at exit from tower, T_{p_0} = 305°K
Ambient temperature near exit from tower, T_{e_0} = 275°K
Tower height = 100 m
Tower exit diameter = 60 m
Efflux velocity = 4.3 m/s
Amount of circulating water = 31.5 m^3/s

Drift rate = 2 x 10^{-5}
Water salinity = 3.45%
Wind speed = 4.3 m/s
Calculate salt deposition rate for a sector of 22.5°
Frequency of wind direction which blows toward the sector = 1
Ambient relative humidity = 70% (constant with height)
Isothermal ambient atmosphere (slightly stable atmosphere)

Drop size distribution:

Diameter interval, μm	Mass mean diameter for interval, μm	Mass fraction
0–100	50	0.05
100–200	150	0.3
200–300	250	0.4
300–400	350	0.15
400–500	450	0.075
500–600	550	0.025

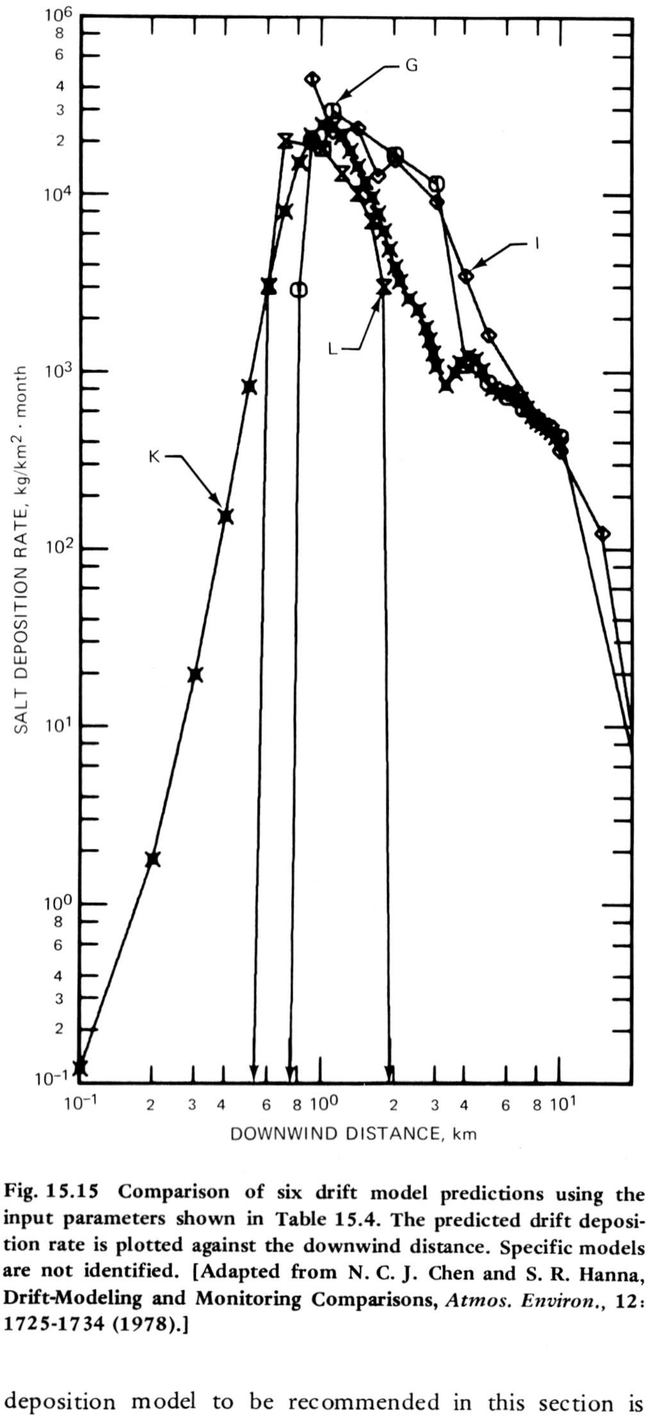

Fig. 15.15 Comparison of six drift model predictions using the input parameters shown in Table 15.4. The predicted drift deposition rate is plotted against the downwind distance. Specific models are not identified. [Adapted from N. C. J. Chen and S. R. Hanna, Drift-Modeling and Monitoring Comparisons, *Atmos. Environ.*, 12: 1725-1734 (1978).]

deposition model to be recommended in this section is based on the suggestion by Hanna (1974b) that drift drops with diameters greater than 200 μm should be treated using the tilted Gaussian plume model (Csanady, 1955). The division at this diameter is made because the settling speed of drops with diameters of 200 μm is about 0.72 m/s, which is roughly equivalent to turbulent vertical speed fluctuations. Other settling speeds for drops of 50, 100, 400, and 600 μm, are 0.06, 0.27, 1.62, and 2.47 m/s, respectively (Neuberger, 1957). As Van der Hoven (1968) points out, the large drops fall through the turbulence without being dispersed. The initial trajectory of the large drops and settling of the plume of small drops are relative to the axis of the plume, which is assumed to rise at a rate given by the equation for bent-over plume rise (Eq. 15.18.)

At a downwind distance where the height given by Eq. 15.18 equals the final plume rise estimated by Eqs. 15.16 or 15.17, the plume height is set equal to this value for all calculations further downwind. The concepts of the tilted plume and the ballistic trajectory are shown in Figs. 15.16 and 15.17.

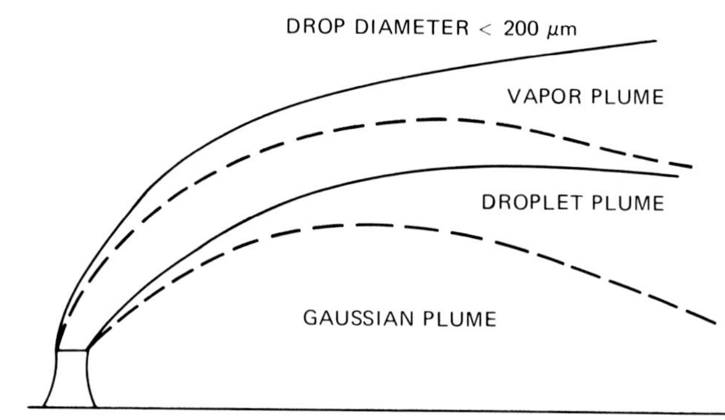

Fig. 15.16 Schematic drawing of technique for calculating drift deposition for small drops (D < 200 μm).

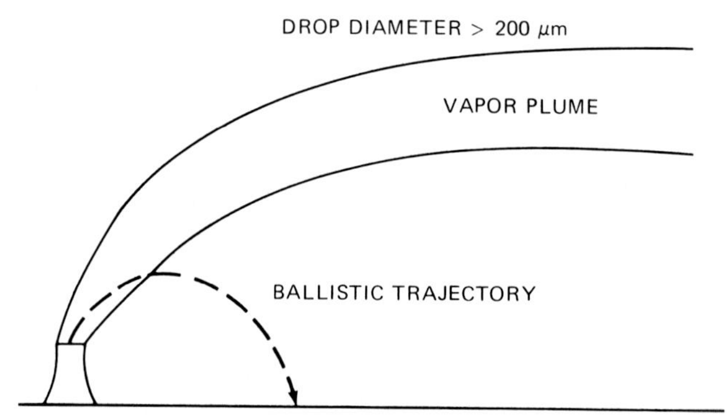

Fig. 15.17 Schematic drawing of technique for calculating drift deposition for large drops (D > 200 μm).

The large drop spectrum is usually assumed to be divided into about 10 classes, and the trajectory is calculated for drop sizes at the border of each class. Since time averages of a day or more are desirable, the deposition is assumed to be uniform across 22½° sectors, to conform with the conventional 16-point wind rose.

In all such trajectory models, the drop remains in a buoyant environment until it breaks away from the lower edge of the plume. The fraction of drift in any drop size class that breaks away from the plume in a downwind distance increment, Δx, can be assumed to equal $V_g(\Delta x/u)/2R$, which is the ratio of the distance fallen in that increment to the diameter of the plume. This assumption implies that the initial distribution of drift drops across the tower diameter, $2R_0$, is uniform. As distance from the tower increases, the fraction of drift that breaks away in each Δx is calculated until the sum of the fractions for any drop size class equals unity, at which point it is assumed that there are no drift drops in that class left in the plume. Settling speed, V_g (in centimeters per second), is given by the following analytical approximations to the data given by Engelmann (1968, p. 212)

$$
\begin{aligned}
V_g &= 3.2 \times 10^5 D^2 && D < 0.0093 \text{ cm} \quad \text{(Stokes' law)} \\
V_g &= 6816 D^{1.177} && 0.0093 \text{ cm} < D < 0.068 \text{ cm} \\
V_g &= 2155 D^{0.746} && 0.068 \text{ cm} < D < 0.26 \text{ cm} \\
V_g &= 1077 D^{0.224} && 0.26 \text{ cm} < D
\end{aligned}
\qquad (15.26)
$$

These analytical speeds are always within 10% of the actual speed for drift drops in the range $0.001 < D < 0.2$ cm. Formulas for the plume radius are given in Sec. 15-3.3.4. At the point the drop breaks away from the plume, it enters the ambient environment, which is assumed to have zero vertical speed. The height of this breakaway point equals the plume height minus the plume radius. If the environment is saturated, the drop does not evaporate, but in an unsaturated environment the drop will lose mass at the rate (Hosler, Pena, and Pena, 1972).

$$\frac{dm}{dt} = -\pi\delta(q_d - q_e)D\rho \mathrm{Sh} \qquad (15.27)$$

where m = mass, g
Sh = Sherwood number = $2 + 0.552\,\mathrm{Re}^{1/2}/(\delta/\nu)^{1/3}$
δ = diffusion coefficient of water vapor (~0.24 cm^2/s)
q_d = specific humidity at drop surface
ρ = air density, g/cm^3
ν = kinematic viscosity of air (~0.15 cm^2/s)
Re = DV_g/ν, droplet Reynolds number

The difference in specific humidities $(q_d - q_e)$ can be estimated from a relationship given by Mason (1971)

$$\frac{q_d - q_e}{q_e} = \frac{\{1 + i(M_0/M_s)[m_s/(m - m_s)]\}^{-1} - \mathrm{RH}}{\mathrm{RH} + (L^2\delta M \mathrm{Sh}\rho q_e)/(R_g T^2 k \mathrm{Nu})} \qquad (15.28)$$

where Nu = Nusselt number = $2 + 0.552\,\mathrm{Re}^{1/2}/(\kappa/\nu)^{1/3}$
RH = relative humidity (on scale of 0 to 1.0)
k = thermal conductivity of air, which equals 0.24 J/(m · °K · s) at 0°K and 0.32 J/(m · °K · s) at 100°C
$\kappa = k/c_p\rho$
m_s = mass of solute, g
M_0 = molecular weight of water (18 g/mol)
M_s = molecular weight of solute
R_g = universal gas constant [8.31 J/(mol · °K)]
i = van't Hoff factor (usually equal to about 2.0 for sea salt, according to Roffman et al., 1973)

Of course, as the mass of the drop changes, its settling speed also changes. In computer applications of these formulas, it is necessary to hold the settling speed constant over a certain time step Δt, calculate a new mass at the end of the time step based on evaluation of Eqs. 15.27 and 15.28, and then repeat the process using the new mass and its corresponding settling speed.

Drops with diameters less than 200 μm are treated using the tilted plume model, which is based on work by Csanady (1955). The droplet spectrum is divided into about five classes in the small drop range. All the drift mass in a class is assumed to behave like the drop with median diameter in that class. The distance the drop plume for each class has settled relative to the main plume equals $V_g t$, where the settling speed V_g is obtained from Eq. 15.26. The drift deposition equals the settling speed V_g times the surface concentration, which is estimated using the Gaussian plume formula (Chaps. 6 and 13).

Validation of drift deposition models has been hampered by the lack of drift deposition observations. Some observations by Alkezweeny et al. (1976) at Oak Ridge, Laulenian and Ulansky (1979) at Pittsburg, Calif., and Overcamp, Israel, and Pringle (1976) and Israel, Overcamp, and Pringle (1977) at Chalk Point have been reported but are not sufficiently accurate or complete to form a basis for model comparison. One problem is that the drift rate from modern towers is so small (0.001% of the circulating water rate) that surface deposition 1 km from the tower is just barely measurable. Another problem is that it is necessary to deploy large numbers of samplers.

During June 1977, the Dye Tracer Experiment was conducted at Chalk Point, Md. (Meyer and Stanbro, 1977; and Environmental Systems Corporation, 1977). The purpose of the experiment was to determine the relative contributions to total drift deposition of water from the cooling tower and from the scrubber-equipped stack. Pink uranine dye was put into the cooling-tower water, and no makeup water was added or blowdown water removed from the cooling-tower basin for the duration of the experiment (about 6 h). Drift samplers were placed on arcs at downwind distances of 0.5 and 1.0 km from the tower. It was necessary to conduct the experiment at night because sunlight affected the uranine dye. The sodium deposition pattern from the cooling tower observed by Meyer and Stanbro (1977) at the downwind distance of 0.5 km is given in Fig. 15.18, and the drop size distributions at the mouth of the Chalk Point tower are given in Table 15.5. About 90% of the drift deposition on that arc is due to the cooling tower. The average sodium deposition over a 30° arc was about 1000 kg/(km^2 · month). On the arc at a distance of 1 km, the average Na deposition was about 400 kg/(km^2 · month). Mass median drop diameters on the two arcs were 340 μm and 260 μm, respectively.

The following source and ambient terms were measured at Chalk Point at the time of the experiment by Environmental Systems Corporation (1977) and Meyer and Stanbro (1977).

Na emission : 1.86 g/s
Tower height = 124 m (isothermal with height)
R_0 = 27 m w_0 = 4.5 m/s
T_{p0} = 35.6°C T_a = 20°C
U = 8 m/s (at plume height)
Wind-direction variability ≅ 30°
Relative humidity = 93% (at plume height)

There was some variation with height of temperature, wind speed, and relative humidity near the ground, but at plume height these variables were fairly uniform.

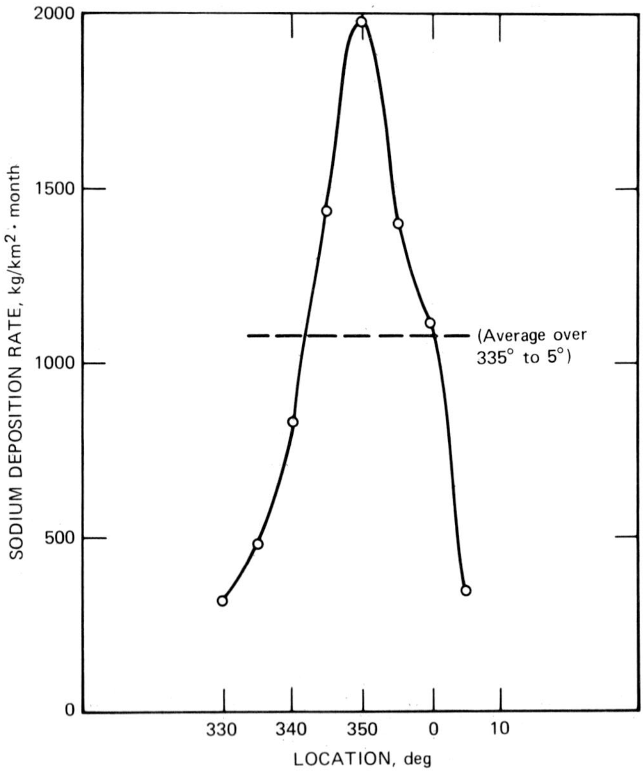

Fig. 15.18 Distribution of observed sodium deposition rate 0.5 km from the Chalk Point cooling tower during the Dye Tracer Experiment. (From J. H. Meyer and W. D. Stanbro, *Fluorescent Dye, a Novel Technique to Trace Cooling Tower Drift*, presented at the Fourth Joint Conference on Sensing Environmental Pollutants, November 1977.) The average on the arc (– – –) is 1080 kg/km^2 · month.

TABLE 15.5 Drop Size Distribution at Mouth of Chalk Point Tower*

Drop diameter range, μm	Percent of total mass	Drop diameter range, μm	Percent of total mass
10–30	13.8	350–400	0.9
30–50	28.8	400–450	0.7
50–70	13.4	450–500	0.6
70–90	8.9	500–600	1.0
90–110	6.2	600–700	1.0
110–130	4.2	700–800	0.9
130–150	3.4	800–900	0.8
150–180	3.8	900–1000	0.7
180–210	3.0	1000–1100	0.6
210–240	2.3	1100–1200	0.4
240–270	1.7	1200–1300	0.1
270–300	1.3	1300–1400	0.1
300–350	1.4		

*Environmental Systems Corporation, 1977.

The preceding information is all that is needed to operate most drift deposition models. Hanna, Pike, and Seitter (1978) describe the application of a simple drift deposition model to this data set, where it is assumed that the nearly saturated environmental conditions preclude evaporation of the drift drops. The application of the above-mentioned model, assuming no evaporation, yields predictions of Na drift deposition of 3600 and 800 kg/(km^2 · month) averaged over 30° arcs at downwind distances of 0.5 and 1.0 km, respectively. These figures are about a factor of 3 higher than the observed Na deposition. Predicted mass median drop diameters on the two arcs are 470 μm and 290 μm, which are within 40% of the observed values.

15-3.3.4 Visible Plume and Cloud Growth. The geometry of the condensed portion of the cooling-tower plume is important because of questions regarding sun shading, aesthetic effects, and possible weather modification. Condensation will occur whenever the concentration of excess water in the plume exceeds the ambient saturation deficit ($q_s - q_e$). The simplest model for estimating the end of the visible plume (Hanna, 1974a) assumes that the saturation mixing ratio in the plume is the same as that in the ambient air; i.e., the plume temperature can be assumed to be equal to the ambient temperature. Then the end of the visible plume is the point where the initial flux of water in the plume, $V_0 Q_0$, is just equal to the flux of water necessary to saturate the plume, $V(q_s - q_e)$. The ratio of volume flux of the effective moisture plume at height z to initial volume flux is given by Hanna (1972)

Windy:

$$\frac{V}{V_0} = \left[1 + 0.28\,\frac{z}{R_0}\left(\frac{U}{w_0}\right)^{1/2}\right]^2 \qquad (15.29)$$

Calm:

$$\frac{V}{V_0} = \left(1 + 0.11\,\frac{z}{R_0}\right)^2 \qquad (15.30)$$

The constants were changed slightly from the original paper to account for the "peak factor" explained in the next section. The ratio (U/w_0) is used in Eq. 15.29 so that the initial flux of momentum equals the flux of momentum in the bent-over plume as z approaches zero. Thus we obtain the following equations for the height and length of the visible plume (no downwash assumed):

Windy

$$\text{Height } h = 3.6\,R_0\left(\frac{w_0}{u}\right)^{1/2}\left[\left(\frac{q_0}{q_s - q_e}\right)^{1/2} - 1\right] \qquad (15.31)$$

$$\text{Length } l = 3.4\,\frac{R_0^{3/2}U^{3/4}w_0^{3/4}}{F^{1/2}}\left[\frac{q_0^{1/2}}{(q_s - q_e)^{1/2}} - 1\right]^{3/2} \qquad (15.32)$$

Calm:

$$\text{Height } h = 9.R_0\left[\left(\frac{q_0}{q_s - q_e}\right)^{1/2} - 1\right] \qquad (15.33)$$

Since the saturation deficit $q_s - q_e$ is assumed constant, these equations should be applied only in atmospheric layers where $q_s - q_e$ does not change much with height. These formulas were originally developed using observations from Oak Ridge, but they have also been tested in

TABLE 15.6 Calculated and Observed Visible Plume Length and Height at Oak Ridge (from Hanna, 1974a)

Date	Observed plume length (l'_0), m	Calculated plume length (l'_c), m	Observed plume height (h'_0), m	Calculated plume height (h'_c), m
Dec. 29, 1972	Calm		150	160
Jan. 10, 1973	500	490	200	520
Jan. 11, 1973	Calm		200	380
Jan. 31, 1973	Calm		175	240
Feb. 1, 1973	250	240	200	280
Feb. 5, 1973	125	170	100	190
Feb. 6, 1973	100	100	75	140
Feb. 7, 1973	200	140	200	260
Feb. 26, 1973	Calm		75	140
Mar. 9, 1973	Calm		75	120
Mar. 12, 1973	Calm		50	60
Mar. 20, 1973	500	1000	500	720
Apr. 8, 1973	150	170	100	190
Apr. 19, 1973	100	160	100	180
Apr. 26, 1973	1500	3600	500	600
May 10, 1973	Calm		400	430
May 17, 1973	700	1040	500	1220
	r(observed, calculated) = 0.98		r(observed, calculated) = 0.87	
	$l'_c/l'_0 = 1.2$		$h'_c/h'_0 = 1.55$	

model comparisons by McVehil and Heikes (1975) and Briggs (1975). In the Oak Ridge calculations, the moisture and buoyancy fluxes from all the cells in a large bank of cooling towers were combined. The observed and estimated visible plume lengths and heights are listed in Table 15.6. Correlations are quite good, but the calculated heights are about 50% too great.

When environmental conditions change markedly with height or when cloud physics processes become important, the preceding simple model for visible plume geometry must be abandoned. In particular, when cloud development or weather modification by cooling tower plumes is of interest, a thermodynamic cloud growth model is necessary. Such a model was developed by Hanna (1976) for analysis of clouds observed by Kramer et al. (1975) at the John E. Amos power plant in West Virginia. Except for the cloud microphysics parameterizations, it is similar to cooling tower plume models developed by Weil (1974), Wigley and Slawson (1971), and Hanna (1972). The model is constructed so that it agrees with known relationships developed by Briggs (1969, 1975) for the rise of buoyant stack plumes at small heights, and models developed by Weinstein (1970) and Kessler (1969) for the growth of natural continental cumulus clouds at large heights. The model is steady-state and one-dimensional, meaning that all parameters are a function only of height.

The entrainment functions used in the model are based on Briggs' (1975) latest work. He finds that best agreement with observations of stack plumes is obtained when it is assumed that the effective radii R_m, R_t, and R_w of the momentum, temperature, and moisture plumes, respectively, are not the same. It is argued that the effective momentum plume is broader than the temperature plume because the ambient air above the rising plume must be accelerated. Also, the effective moisture plume is assumed to have a smaller cross section than the temperature plume to account for the presence of lumps or eddies of saturated air. The ratio of the cross sectional area of the temperature plume to that of the moisture plume, E_w, is defined as the "peak factor," and is found by Slawson, Coleman, and Frey (1974), Meyer et al. (1974), and Briggs (1975) to equal about 2.0. The ratio of the cross-sectional area of the momentum plume to that of the temperature plume, E_m, is found to be about 2.25. The radii and entrainment rates $O = (1/V)(dV/dz)$ are given by the following expressions:

Vertical plume:

$$O_m = O_t = 0.15 \qquad O_w = 0.107$$

$$\frac{\partial R_m}{\partial z} = \frac{\partial R_t}{\partial z} = 0.15 - \frac{R_t(g/T_p)(T_p - T_e)}{2w^2}$$

$$\frac{\partial R_w}{\partial z} = 0.71\,\frac{\partial R_t}{\partial z} \tag{15.34}$$

Bent-over plume:

$$O_m = 1.2 \qquad O_t = 0.8 \qquad O_w = 0.57$$

$$\frac{\partial R_t}{\partial z} = 0.4 - \frac{R_t}{2U}\frac{\partial U}{\partial z}$$

$$\frac{\partial R_w}{\partial z} = 0.71\,\frac{\partial R_t}{\partial z}$$

$$\frac{\partial R_m}{\partial z} = 1.5\,\frac{\partial R_t}{\partial z} \tag{15.35}$$

The dividing line for the use of the bent-over or vertical plume entrainment rates is a wind speed, U, equal to 1 m/s. As a plume rises and the ambient wind speed changes, it is possible for the entrainment rate to switch back and forth between those appropriate for bent-over and vertical plumes.

In Briggs' plume rise theories, it is assumed that there is a "top-hat" (uniform) distribution of temperature, moisture, etc. in the plume. This assumption works well in practice, but recent observations indicate that the true distribution is more Gaussian in shape and can be bimodal. Airplane observations of the crosswind distribution of temperature in the Chalk Point plume were reported by Woffinden, Harrison, and Anderson (1976). The published raw data were assembled by Hanna, Rao, and Hosker (1977) into the graphical cross section given in Fig. 15.19.

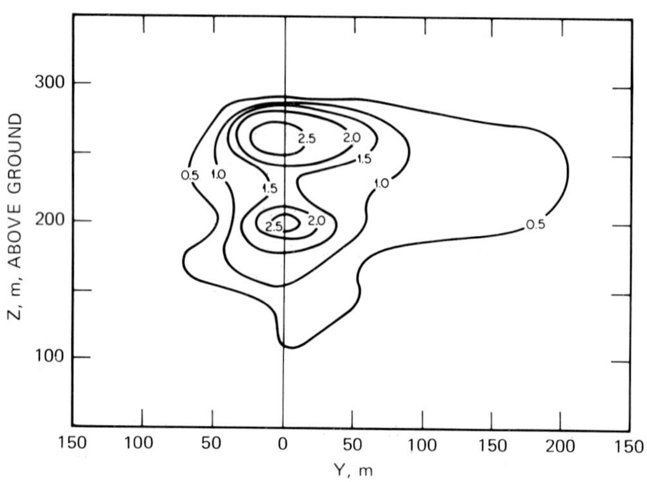

Fig. 15.19 Cross section of temperature excess, ΔT (°K), in the Chalk Point plume, June 22, 1976, 150 m downwind, traverses 19 to 27. The plume is bent over at this stage.

Also, as Briggs points out, there is a double-vortex circulation in the plume with upward motions relative to the plume axis near the center, and downward motions at the plume edges. The tangential speed of this secondary circulation averages about 2 m/s (Hanna, Pike, and Seitter, 1978) and may affect the breakaway of drift drops (Chen and Jung, 1978).

Assuming top hat distributions across the plume, it is possible to derive the following basic equations for the growth of cooling tower plumes:

Equation of Motion

$$\frac{\partial}{\partial z}\left(\frac{w^2}{2}\right) = \frac{g}{E_m}\left[\frac{T_p\,(1 + 0.61 q_p/E_w) - T_e\,(1 + 0.61 q_e)}{T_p\,(1 + 0.61 q_p/E_w)} - \frac{Q_c + Q_h}{E_w}\right] - \frac{O_m w^2}{R_m} \qquad (15.36)$$

The liquid water is broken down into cloud water, Q_c(g/g), and hydrometeor water, Q_h(g/g). The term on the left is the vertical acceleration of the plume, where steady-state conditions are assumed ($\partial w/\partial t = 0$). The terms on the right are the buoyant acceleration, the drag due to liquid-water drops, and the drag due to entrainment of environmental air.

Equation for Temperature Change

$$\frac{\partial T_p}{\partial z} = -\left[\frac{L}{c_p E_w}\frac{\partial q_{ps}}{\partial z}\right] - \frac{g}{c_p} - \frac{O(T_p - T_e)}{R} + \left[\frac{L_i}{c_p E_w}\frac{Q_h + Q_c}{\Delta z}\right] - \left[\frac{L}{c_p E_w}\frac{O_w(q_p - q_e)}{R_w}\right] \qquad (15.37)$$

The terms in brackets apply only when the plume is saturated. The expression Δz is the vertical height increment in the integration process. The terms on the right side of Eq. 15.39 are the heat gain due to latent heat release, the heat loss due to dry adiabatic expansion, the entrainment rate, the heat gained by a phase change of liquid water to ice, and the heat lost as the plume evaporates water in order to saturate the entrained air.

Equation for Water Vapor Change

$$\text{Unsaturated: } \frac{\partial q_p}{\partial z} = \frac{-O_w(q_p - q_e)}{R_w} \qquad (15.38)$$

$$\text{Saturated: } \frac{\partial q_p}{\partial z} = \frac{\partial q_{ps}}{\partial z} \qquad (15.39)$$

Equation for Cloud-Water Change

$$\frac{\partial Q_c}{\partial z} = \frac{-\partial q_{ps}}{\partial z} - \frac{10^{-3}\,(Q_c - 0.0005)}{w} - \frac{0.00522 Q_c\,(1000 Q_h)^{0.875}}{w} - \frac{O_w\,(q_p + Q_c - q_e)}{R_w} \qquad (15.40)$$

The terms on the right-hand side of Eq. 15.40 are the gain due to condensation, the losses to hydrometeor water due to conversion and coalescence (after Kessler, 1969), and the losses due to entrainment. Cloud water is used before hydrometeor water to saturate the entrained air.

Equation for Hydrometeor Water Change

$$\frac{\partial Q_h}{\partial z} = \frac{10^{-3}\,(Q_c - 0.0005)}{w} + \frac{0.00522 Q_c\,(1000 Q_h)^{0.875}}{w} - \frac{4.5 Q_h\,(1000 Q_h)^{0.125}\cos\,[\arctan\,(w/U)]}{w R_w} - \frac{O_w Q_h}{R_w} + \frac{K_2}{\Delta z} \qquad (15.41)$$

The physical meaning of the terms on the right-hand side are the gain from cloud water due to conversion and

TABLE 15.7 Environmental Data from Rancho Seco, Calif. (Wolf, 1976)

Height above tower, m	2/17/75			2/18/75			2/20/75		
	T_e*	T_d†	U‡	T_e	T_d	U	T_e	T_d	U
0	280.0	270.9	5.0	283.0	274.4	3.4	283.2	275.4	9.1
44	279.5	270.6	5.0				282.8	274.8	9.0
120	278.8	270.9	5.0	281.5	273.7	2.8	282.0	274.7	8.8
197	278.2	270.0	5.0						
273	278.9	268.9	5.6	280.4	272.3	2.2	280.6	273.4	8.2
425	279.2	265.3	7.2	279.5	269.9	1.9	279.2	271.2	7.1
578	278.4	266.8	9.4	278.5	267.1	1.5	277.9	271.3	6.0
654				277.8	267.2	1.8			
730	278.0	264.7	11.0	278.0	268.6	2.4	276.6	270.1	5.0
807				278.1	267.7	2.9			
883				278.0	266.5	3.0	276.1	269.0	4.5

*Dry bulb temperature, K.
†Dew point, K.
‡Wind speed, from pilot balloons, m/s.

coalescence, the loss due to rainout, the loss due to entrainment, and a final correction term in the numerical process which is used to saturate the entrained air if all the cloud water has been evaporated.

Any consistent equation for the saturated specific humidity can be used. Best agreement with observations of q_{ps} and T are given by the Goff–Gratch formulas (WMO International Meteorological Tables, 1966).

Saturation specific humidity

(a) $T_1 = 273.16°K < T < 373°K$

$$\ln q_{ps} = 2.303 \Bigg[10.79574\left(1 - \frac{T_1}{T}\right) + 1.50474 \times 10^{-4}\left(1 - 10^{-8.2969[(T/T_1)-1]}\right) + 0.42873 \times 10^{-3}\left(10^{4.76955[1-(T_1/T)]}\right) - 1\Bigg] - 5.028 \ln\frac{T}{T_1} - \ln p + 1.335 \qquad (15.42)$$

(b) $T < T_1 = 273.16°K$

$$\ln q_{ps} = 2.303 \Bigg[-9.09685\left(\frac{T_1}{T} - 1\right) + 0.87682\left(1 - \frac{T}{T_1}\right)\Bigg] - 3.56654 \ln\frac{T_1}{T} - \ln p + 1.335 \qquad (15.43)$$

In these formulas p is atmospheric pressure in millibars. Equation 15.42 is nearly the same as Eq. 15.13 but agrees with observations a little better.

Equations 15.34 through 15.43 are a complete set and may be solved on a digital computer to yield vertical profiles of R_t, T_p, w, Q_p, Q_c, and Q_h. To assure numerical stability, a computer program should use smaller height steps near the source; e.g., the current program uses steps of 0.01, 0.1, and 1.0 m height for heights less than 10 m, between 10 and 100 m, and above 100 m, respectively. It is important to recognize that the cloud physics parameterizations were derived from observations of natural clouds, and that they may not be true for cooling tower plumes. However, observations of condensation in cooling tower plumes are not yet sufficient for the derivation of independent cloud physics parameterizations.

To execute the plume and cloud growth program described above, the following input parameters must be provided: initial plume radius, vertical speed, temperature, and liquid-water content; and vertical profiles to at least 1000 m of ambient temperature, dew point, and wind speed. If there are multiple towers, it is assumed that neighboring plumes merge when their radii equal half of the distance between them.

To illustrate the application of the model, data from the Rancho Seco, Calif., towers, reported by Wolf (1976), are used. There are two natural-draft cooling towers, with radii 30 m, which are separated by 140 m. The ambient profiles of temperature, dewpoint, and wind speed are listed in Table 15.7, for three runs. Wolf (1976) measured these variables using a light aircraft. Initial vertical speed, w_0, of the plume on three separate days is 4.4, 3.6, and 3.9 m/s, respectively. Initial temperatures are calculated from Eq. 15.22. On the basis of Wolf's aircraft measurements in the plume (about 100 m above the tower mouth), the initial cloud and hydrometeor water contents, Q_c and Q_h, are set equal to 0.0005 and 0.0001 g/g, respectively. The predicted and observed excess temperature and water profiles are listed in Table 15.8, showing the capabilities of the model. Predicted vertical profiles of excess temperature, vertical speed, and liquid-water content are shown in Fig. 15.20 for February 17.

This model has also been tested using data from the John E. Amos cooling towers in West Virginia (Hanna, 1976) where cloud height was the primary variable. In another application, detailed cloud physics measurements by Auer (1976) in a cloud over a refinery were compared with the model predictions as reported by Hanna (1977a).

TABLE 15.8 Observed (Wolf, 1976) and Estimated Temperature and Water Content Excesses at Rancho Seco

Date	Height above tower, m	$(T_p - T_e)$, K		$(q_p + Q_c + Q_h - q_e)$, g/g	
		Observed	Model	Observed	Model
Feb. 17	120	7.2	2.3	0.0051	0.0033
	136	5.6	2.0	0.0036	0.0031
	151	5.0	1.6	0.0031	0.0028
	181	1.9	1.2	0.0018	0.0025
	273	2.3	−0.3	0.0021	0.0018
	349	0.7	Top at 340 m	0.0011	
Feb. 18	197	2.2	1.3	0.0022	0.0032
	273	2.1	0.7	0.0024	0.0024
	425	0.8	−0.1	0.0018	0.0019
	578	−0.3	−0.5	0.0008	0.0017
	730	−0.5	Top at 610 m	0.0004	
	883	−2.1		0.0004	
	974	−3.8		0	
Feb. 20	120	3.0	1.3	0.0019	0.0026
	151	3.0	0.9	0.0020	0.0020
	181	1.4	0.7	0.0008	0.0017
	273	0.9	0.4	0.0006	0.0012
	425	0	0.1	0.0004	0.0011
	486	1.0	0	0.0007	0.0009
	730	−0.4	Top at 720 m	−0.0001	

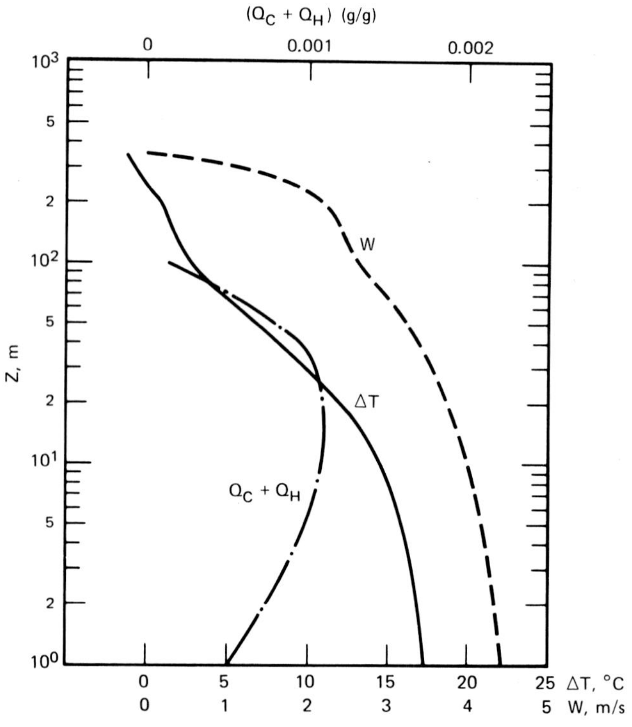

Fig. 15.20 Predicted vertical profiles of excess temperature ΔT, vertical speed w, and liquid water content $Q_c + Q_h$ for the Rancho Seco February 17 data in Table 16.7.

The observed and predicted values of cloud base and cloud height are in agreement with 10%. The above-mentioned model can be used to estimate the visible plume length and height, but for visible plumes with expected dimensions less than about 200 m, the simpler model described at the beginning of this section is preferable.

15-3.3.5 Interactions of Moist Plume with SO_2 Plume. In Sec. 15-2.5, it was reported that acid drops have been found in the Keystone, Pa., cooling tower plumes when the stack plume and the cooling tower plumes merge. The conversion of SO_2 to sulfates is accelerated when moisture is present. However, no models of this phenomena have yet been suggested. There is basic research under way on the subject of chemical reactions in turbulent mixtures, but no one has developed a useful applied model. Donaldson and Hilst (1972) point out that chemical reactions in inhomogeneously mixed fluids, such as a stack plume merging with a cooling tower plume, may proceed more slowly than reactions in a completely mixed fluid. Measurements which have been obtained by Thomson (1976) in the Keystone plume may permit the development of applied models of this phenomenon.

15-3.4 Dry Cooling Towers

The same models can be used to analyze dry towers as wet towers. The initial conditions must be changed so that the plume dew point equals the ambient dew point, and there is no liquid water present initially. Experiments with the model described in Sec. 15-3.3.4 suggest that, for the same total (sensible plus latent) initial heat flux, both the dry and the wet tower are likely to generate a cloud at the top of the plume (Hanna and Gifford, 1975). By the time the plume reaches a height of about 500 m, where the cloud base often occurs, it consists of over 90% ambient air. Thus the plume acts like a lifting mechanism for the ambient air, and condensation occurs near the lifting condensation level, no matter what the plume source is. On days when little condensation occurs, the total plume rise from the dry plume is greater than that for the wet plume since the latent heat in the wet plume has not yet been released.

Koenig, Murray, and Tag (1978) applied their cloud growth model to determine the differences between clouds generated by dry and wet cooling towers with the same initial heat flux. They consider three cases: (1) all latent heat, (2) equal sensible and latent heat, and (3) all sensible heat. The predicted profiles of clouds after 60 min of simulated time show that a much larger and taller cloud is formed as the fraction of sensible heat increases. For case 1 (all latent), there is much moisture available, but it is not used if the weakly buoyant plume does not rise much above the heat source. For case 3 (all sensible), there is no moisture in the plume, but it is strongly buoyant and eventually reaches the lifting condensation level where the environmental air which has been entrained by the rising plume condenses. Lee (1979) reached the same conclusions using his numerical model and by considering releases from single cooling towers and groups of cooling towers.

15-4 EFFECTS OF ENERGY GENERATION IN MESOSCALE POWER PARKS

From the observations and theoretical calculations in the previous sections, it appears that the largest operating energy centers (about 5000 MW dissipated) do not affect plants, animals, or structures in their vicinity to any great extent. It is true that long dense plumes are observed, and that drift deposition and fog are problems within about 1 km of a cooling tower or pond. But long-lasting, significant changes in weather are not observed beyond this distance.

Planners in the United States and in the USSR state that future power plants will have many units within a relatively small area. For example, a 50,000-MW(e) power park is being proposed by Gulf States Utilities for a 10-km^2 site on the Mississippi River at Baton Rouge. The concentration of power plants into a power park results in easier security, construction, maintenance, storage, and transmission. However, the rejection of 100 000 MW of waste heat over an area of 10 to 100 km^2 may cause serious atmospheric effects. Since there are no power parks of this size in existence, it is necessary to use models or natural analogs to estimate their effects.

Briggs (1974) suggested a criterion for determining whether a given buoyant source with characteristic vertical speed V_B would concentrate vorticity (cause whirlwinds) in an ambient fluid with characteristic tangential velocity V_∞

$$\frac{V_\infty}{V_B} > 0.9 \qquad \text{No concentration of vorticity. The whole flow swirls.}$$

$$0.15 < \frac{V_\infty}{V_B} < 0.9 \qquad \text{Vorticity can be concentrated.} \qquad (15.44)$$

$$\frac{V_\infty}{V_B} < 0.15 \qquad \text{No concentration of vorticity. Buoyancy is dominant.}$$

Briggs (1974) uses a numerical experiment by Deardorff (1972) to estimate V_∞, stating that "the maximum possible value of V_∞ is about $0.8(Hz_i)^{1/3}$, where z_i is the height of the mixing layer (usually of the order of 1 km) and the natural surface heat flux H equals $g H_0/(c_p \rho T)$, where H_0 is the average sensible heat flux at the ground." The value of V_∞ is about 1.7 m/s for a convective day ($H = 10^{-2} m^2/s^3$ and $z_i = 10^3$ m), for a scale R equal to z_i, where R is the radius of the convective element. For smaller scales. $V_\infty \sim 6(Hz_i)^{1/3}(R/z_i)$. Table 15.9 summarizes the application of this criterion to several buoyant sources. Observations agree with the predictions for the oil burners (whirlwinds observed), the brush fires (occasional whirlwinds), and the Saturn V firing (no whirlwinds). Strong natural convection occasionally gives rise to whirlwinds in the form of dust devils or tornados. On only one occasion has a whirlwind been reported at an operating cooling tower (D. Thomson, personal communication, 1976). It is important to note that vorticity concentration is predicted by this criterion for power parks. Of course we will have to wait until someone builds a power park before this prediction can be tested.

Many of the atmospheric effects of power parks can be roughly estimated using simple dispersion models. For example, let us assume that the excess water from a power park is mixed uniformly into a pie-shaped sector of angular width $22\frac{1}{2}°$ and depth z_i. The formula for the concentration of excess water at a distance x downwind is then

$$X = \frac{Q}{Uz_i(2\pi x/16)} \qquad (15.45)$$

Assume that the mixing depth z_i is 300 m, the wind speed U is 3.7 m/s, and the source strength Q is 4×10^7 g/s (typical of a 40 000 MW power park). Then the concentrations X at distances 20, 50, and 100 km downwind are estimated to be 0.004, 0.0010, and 0.00079 g/g, respectively. Hanna and Perry (1973) found that the probability of the saturation deficit at Oak Ridge, Tenn., being less than 0.001 g/g is 32%. Hence it is predicted that the visible plume will be greater than 50 km long 32% of the time.

More detail is introduced by considering the individual plumes and somehow parameterizing their merging. Hanna (1977b) constructed a climatology of cooling tower plumes from energy centers by using the one dimensional plume and cloud growth model described in Sec. 15-3.3.3 as a basis, and assuming that neighboring plumes merge at a height where their radius equals half the distance between the cooling towers. Four types of sources were assumed, with the following characteristics:

A. 10^3 MW single tower: This tower, with 30-m radius, is typical of natural draft cooling towers currently in operation. With the use of Eq. 15.22, initial vertical speed is 4.4 m/s and initial plume temperature is calculated from the environmental temperature and relative humidity. The plume is initially saturated and initial cloud-water and hydrometeor water contents are both equal to 0.001 g/g.

B. 10^4 MW single tower: The energy output is slightly larger than that from the largest existing power plants. The extra energy is obtained by keeping all parameters except

TABLE 15.9 Calculation of the Vorticity Concentration Parameter for Several Sources*

Source	Radius (R), m	Buoyancy flux, F (m^4/s^3)	Heat flux, H (m^2/s^3)	Mixing height (z_i), m	V_B, m/s	V_∞, m/s	V_∞/V_B
	Source environment						
Single cooling tower, sensible heat	25	3500	10^{-2}	10^3	5.2	0.32	0.06
Cluster of 20 cooling towers	500	70 000	10^{-2}	10^3	5.2	1.7	0.33
Sensible heat	5000	70 000	10^{-2}	10^3	2.4	1.7	0.71
Strong natural convection	500	2500	10^{-2}	10^3	1.7	1.7	1.0
Oil burners (Dessens, 1964)	125	6000	10^{-2}	10^3	3.6	1.6	0.44
Saturn V (Morris, 1968)	10	1 300 000	10^{-2}	10^3	51.0	0.13	0.0025
Australian bushfires (Taylor et al., 1973)	10^3	900 000	10^{-2}	10^3	13.8	1.7	0.12

*[From S. R. Hanna and F. A. Gifford, Meteorological Effects of Energy Dissipation at Large Power Parks, *Bull. Am. Meteorol. Soc.*, 56: 1069-1076 (1975).]

initial plume radius the same as in the first case. Initial plume radius is increased to 91 m.

C. 10^5 MW single tower: The energy output is equal to that proposed for a large energy center. The additional energy is obtained by keeping all parameters except initial plume radius the same as in the first case but increasing the initial plume radius to 300 m. Since the energy flux is so concentrated, the results of this analysis represent the "worst case."

D. 10^5 MW energy park: There are 100 cooling towers, each with the characteristics of case A, divided into 25 groups of 4. The groups are assumed to be spaced on a square grid, 1 km apart. The four towers in each group are in a square array, 200 m apart. Merging of plumes is assumed to occur when the radius of a plume equals one half of the distance between the sources. At that point, the energies of the plumes are assumed to add linearly, which is accomplished by increasing the plume radius to equal $(N)^{1/2}R$, where N is the number of plumes merging and R is the radius of a plume before merger.

The model was applied for all four sources to 241 rawinsonde observations from Nashville, Tenn., covering the months of January, April, July, and October 1974. Estimates of average annual plume rise from the four sources are 580, 1180, 2460, and 780m, respectively. A cloud forms at the top of the plume from the 10^3 MW tower 39% of the time. This fraction increases to 95% for the 10^5 MW source, as shown in Table 15.10. For the 10^5-MW power park, the average plume rise and cloud frequency is in between those for the 10^3- and 10^4-MW single towers, because plume merger in the power park occurs only a fraction of the time. It is concluded that merger is usually avoided if the towers are separated by a distance greater than the typical plume rise from a single tower (500 to 1000 m).

TABLE 15.10 Frequency of Cooling Tower Cloud Occurrence Estimated Using Nashville Radiosonde Observations*

Month (all 1974)	10^3 MW	10^4 MW	10^5 MW	10^5 MW power park
January	0.41	0.70	0.97	0.59
April	0.34	0.59	0.96	0.43
July	0.51	0.76	0.97	0.59
October	0.31	0.54	0.89	0.39
All runs	0.39	0.64	0.95	0.50

*[From S. R. Hanna, Predicted Climatology of Cooling Tower Plumes from Energy Centers, *J. Appl. Meteorol.*, 15(*11*): 1232-1233 (1977b).]

The application of a similar model to multiple cooling towers in a 50 000-MW(e) power park was reported by Lee (1976). His model has slightly different assumptions regarding entrainment and cloud microphysics than the model just described and in Sec. 15-3.3.3, but the results are essentially the same. His assumption regarding plume merging, which very crucially affects the final plume or cloud, is the same as the preceding one. To estimate the entrainment rate for a bent-over plume, he uses the methods described by Ooms (1972), who considers both the vertical and horizontal components of entrainment. If s is the coordinate axis tangent to the plume axis at any location, ϕ is the angle between the axis s and the horizontal, and $v = (w^2 + u^2)^{1/2}$ is the plume velocity along s, then the conservation of momentum equations (analogous to Eq. 15.36) for the vertical and horizontal directions are

$$\left(\frac{\partial}{\partial s}\right)(v^2 R^2 \sin \phi) = gR^2\left[\frac{T_{vp} - T_{ve}}{T_{vp}} - (Q_h + Q_c)\right]$$

$$- \frac{C_D}{\pi} R(U \sin \phi)^2 \cos \phi \qquad (15.46)$$

$$\left(\frac{\partial}{\partial s}\right)(v^2 R^2 \cos \phi) = \left(\frac{C_D}{\pi}\right) b(U \sin \phi)^2 \; |\sin \phi| \qquad (15.47)$$

The value of the drag coefficient C_D for a circular cylinder is assumed to be 0.3. The entrainment assumption is

$$\left(\frac{\partial}{\partial s}\right) vR^2 = 0.30R \; [w^2 + (u - U)^2]^{\frac{1}{2}} \qquad (15.48)$$

This reduces to the assumption in the earlier model if u equals U, the s axis is vertical, and the coefficient 0.3 is changed to 0.4.

Lee's (1976) cloud microphysics parameterizations can account for measured drift drop size distributions, as given by, e.g., Shofner et al. (1974), rather than imposing the Marshall–Palmer drop size distribution proposed by Kessler (1969). The details are given in Lee's paper. He also assumes that all hydrometeor drops fall from the plume when their settling speed is at least as great as the plume updraft speed. Consequently the rainfall rates that are predicted by this model are based on a very weak assumption. This problem led Hanna (1976) to conclude that rainfall rates predicted by models should not be reported until good observations are available that can be used to develop a plume rainout mechanism. As an example of the output of Lee's (1976) model, predicted vertical profiles of water content above a hypothetical 50 000-MW(e) power park near Shreveport, La., are given in Fig. 15.21. The total liquid-water content is seen to reach about 0.003 g/g.

Another popular method of modeling involves the use of diffusivity or K coefficients. Due to waste heat emissions, Pandolfo and Jacobs (1978) apply their three-dimensional model to predict mesoscale climatic changes near Basel, Switzerland. This model includes radiative transfers and a detailed surface heat budget. Hane (1978) looks at changes in a squall line as it passes over a power park dissipating 65 000 MW(e) of waste heat. He finds that maximum vertical velocities and liquid-water contents in the storm are changed by less than 10% by the power park. Bhumralkar (1976) has applied his model of flow over a heated island (Bhumralkar, 1973) to the problem of atmospheric effects of power parks. In this method, turbulent fluxes of quantities in a given direction are proportional to K times the gradient of the mean quantity in that direction. The governing equations are solved at several grid points located in a two-dimensional downwind horizontal and vertical coordinate system. In Bhumralkar's model there are 16 vertical levels on an expanding grid and 18 horizontal points at 1-km intervals.

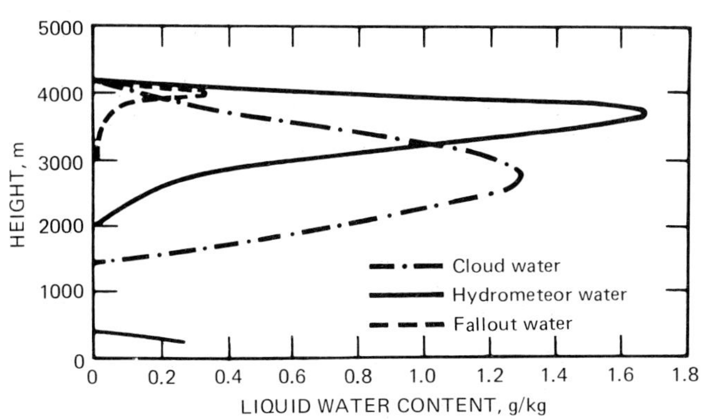

Fig. 15.21 Predicted variation of microphysical parameters with height, average sounding of 0000 GMT, July, Shreveport, La. (From J. Lee, A Numerical Simulation of Atmospheric Convection Caused by Heat Dissipation at Large Power Centers, in *Proceedings of the Third Symposium on Atmospheric Turbulence, Diffusion, and Air Quality*, Raleigh, N. C., available from the American Meteorological Society, Boston, Mass., 1976.)

A major problem is always the specification of K's, which are functions of the vertical wind and temperature gradients. There is also the question of how to introduce the waste heat from the power park into the model. Bhumralkar (1973) introduces the perturbations due to the power park as differential potential temperature and specific humidity values at the grid points at elevations of 225 m to 1200 m and horizontal distances of −2, −1, 0, +1, and +2 km. The grid system, perturbations, and some model predictions are shown in Fig. 15.22. The initial magnitudes of ΔT and Δq are 0.7°C and 0.0011 g/g. At subsequent times the perturbations are only at heights 225 and 425 m (shown in Fig. 15.22) and their magnitudes are reduced to 0.032°C and 0.0005 g/g. The use of the perturbation method is clearly a highly arbitrary scheme in this application to cooling tower plumes, although it is more reasonable for heated islands, urban areas, or natural cloud formation. At a power park the heat input is actually in the form of a buoyant vertical flux, but this type of initial condition is impossible to implement in most models.

Models of atmospheric mixed layers over urban areas can be altered for application to power parks. For example, McElroy (1971) constructed a two-dimensional model using K coefficients and tested it using observations over Columbus, Oh. It correctly simulates the magnitude of the urban heat island and also indicates the slight convergence of winds around an urban area.

More complex models are under development for application to the problems of plume and cloud formation from multiple sources. These include both first-order (K-models) and second-order closure schemes. The K-model by Hill (1974) accounts for the development and interaction of multiple clouds. Lewellen et al. (1976) have described a second-order-closure model for convective systems with phase changes. Rao and Hosker (1978) apply a two-dimensional second-order-closure model to the power park problem. It is assumed that clouds are shallow and nonprecipitating, and that the total heat release of 20 000 MW covers an area of 20 km^2. The average latent and sensible heat fluxes from the surface owing to the power

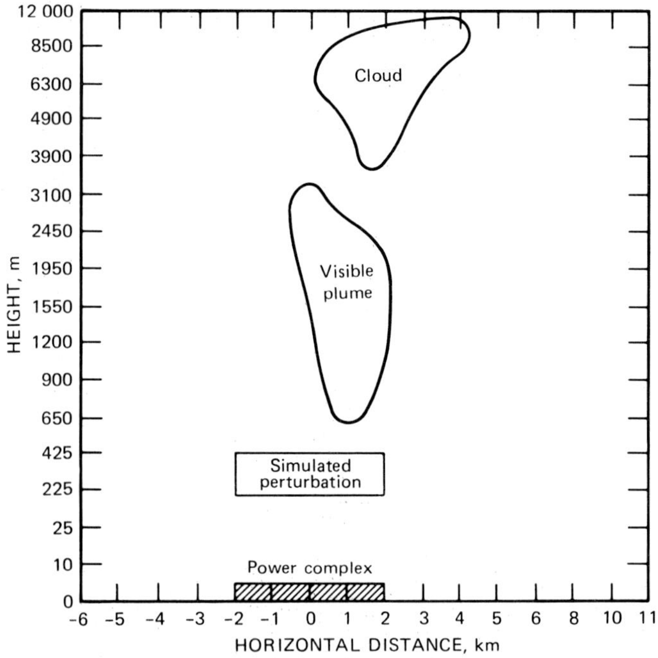

Fig. 15.22 Predicted plume and cloud formation after 15 min in the control experiment. (From C. M. Bhumralker, Weather Modification Caused by Waste Heat Rejected into the Atmosphere from Cooling Towers at Large Power Parks, in *Proceedings of the Third Symposium on Atmospheric Turbulence, Diffusion, and Air Quality*, Raleigh, N. C., pp. 581-585, available from the American Meteorological Society, Boston, Mass., 1976.)

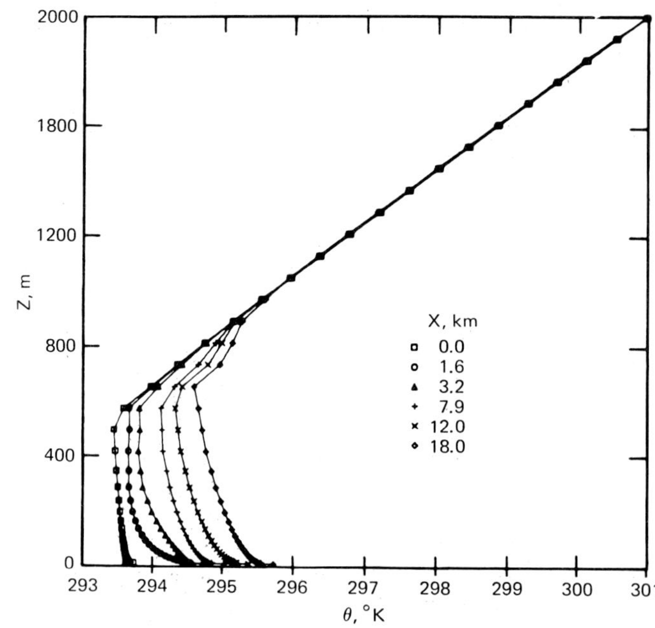

Fig. 15.23 Calculated vertical profiles of mean potential temperature as a function of downwind distance from a power park. (From K. S. Rao and R. P. Hosker, A Numerical Study of Meteorological Effects of Waste Heat and Moisture Releases from Hypothetical Power Parks, in *Proceedings of the Second AIAA/ASME Thermophysics and Heat Transfer Conference*, Palo Alto, Calif., available from the Atmospheric Turbulence and Diffusion Laboratory, Oak Ridge, Tenn., 1978.)

park are 800 W/m^2 and 200 W/m^2, respectively. Predicted temperature profiles as a function of distance downwind from the edge of the heat source are given in Fig. 15.23. The initial profile is assumed to be typical of a convective afternoon, and ambient relative humidity is 75%. It is seen that the mixed layer increases in depth by about 200 m and the mean mixed layer temperature increases by about 2°C as the air passes over the power park. Figure 15.24 shows the calculated isopleths of mean specific liquid water content in the convective cumulus clouds initiated by the power park for an ambient relative humidity of 90%. In its present form, this model is most applicable to cooling ponds, which are essentially area sources. To treat cooling towers, it is necessary to account for plume rise and multiple plume merging.

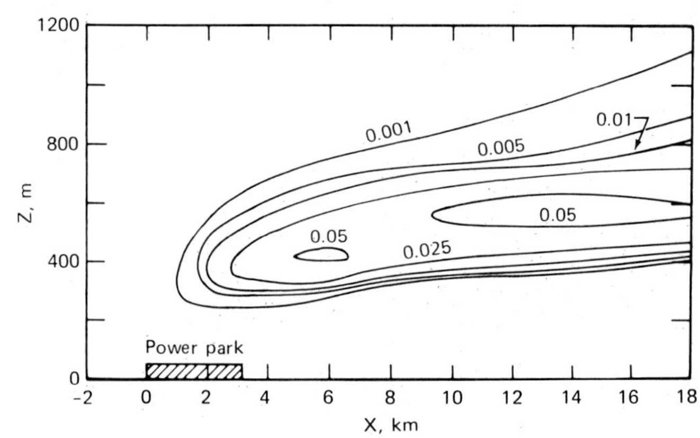

Fig. 15.24 Calculated isopleths of mean specific liquid water content in the convective cumulus clouds spawned by the power park (RH = 90%). (From K. S. Rao and R. P. Hosker, A Numerical Study of Meteorological Effects of Waste Heat and Moisture Releases from Hypothetical Power Parks, in *Proceedings of the Second AIAA/ASME Thermophysics and Heat Transfer Conference*, Palo Alto, Calif., available from the Atmospheric Turbulence and Diffusion Laboratory, Oak Ridge, Tenn., 1978.)

15-5 EFFECTS OF ENERGY GENERATION ON A SYNOPTIC OR GLOBAL SCALE

Energy generation processes produce obvious atmospheric effects at scales up to about 100 m. Long visible plumes, temperature excesses, and cloud formation are effects which have been discussed in previous sections of this chapter. As yet, no effects have been discerned on synoptic or global scales. There is a level of energy generation, however, which will produce effects on these larger scales, and it is the job of the researcher to identify this critical point.

Weinberg and Hammond (1970) calculate that the energy released artificially by man is now 5×10^{12} W, most of which reaches the atmosphere. They expect that man's ultimate energy release, based on 20×10^9 people, each using 2×10^4 W, will be about 4×10^{14}W. Spread over a global area of 5×10^{14} m^2, this amounts to an energy flux of 0.8 W/m^2. Table 15.11 compares man's energy fluxes with some natural energy fluxes. The incoming solar flux is about 440 times man's ultimate energy flux. A quick calculation using the formula Flux $\propto T^4$ shows that the global temperature will increase

TABLE 15.11 Energy Production per Unit Area of Natural and Artificial Processes at Various Scales

	Natural production, W/m^2		Artificial production, W/m^2	
Area, m^2	Event	Rate	Type of use	Rate
5×10^{14}	Dissipation of kinetic energy	5	Man's ultimate energy production	0.8
	Solar-energy absorption by atmosphere	25		
	Solar flux at top of atmosphere	350		
10^{12}	Cyclone latent heat release (1 cm of rain per day)	200	Northeastern United States ultimate production (10^8 people, 20 kW each)	2.0
	Cyclone kinetic energy production	30		
10^8	Thunderstorm		Suburban area (400 persons per km^2, 10 kW per capital) capital)	4
	Kinetic energy production	100		
	Latent heat release (1 cm of rain per 30 min)	1500	Super energy center or city	1000
			Agroindustrial complex	100
	Evaporation from lake	100		
10^4	Tornado kinetic energy production	10^4	Cooling pond (1000 kW per 2 acres)	130
			Cooling tower (10^6 kW per 10^4 m^2)	10^5

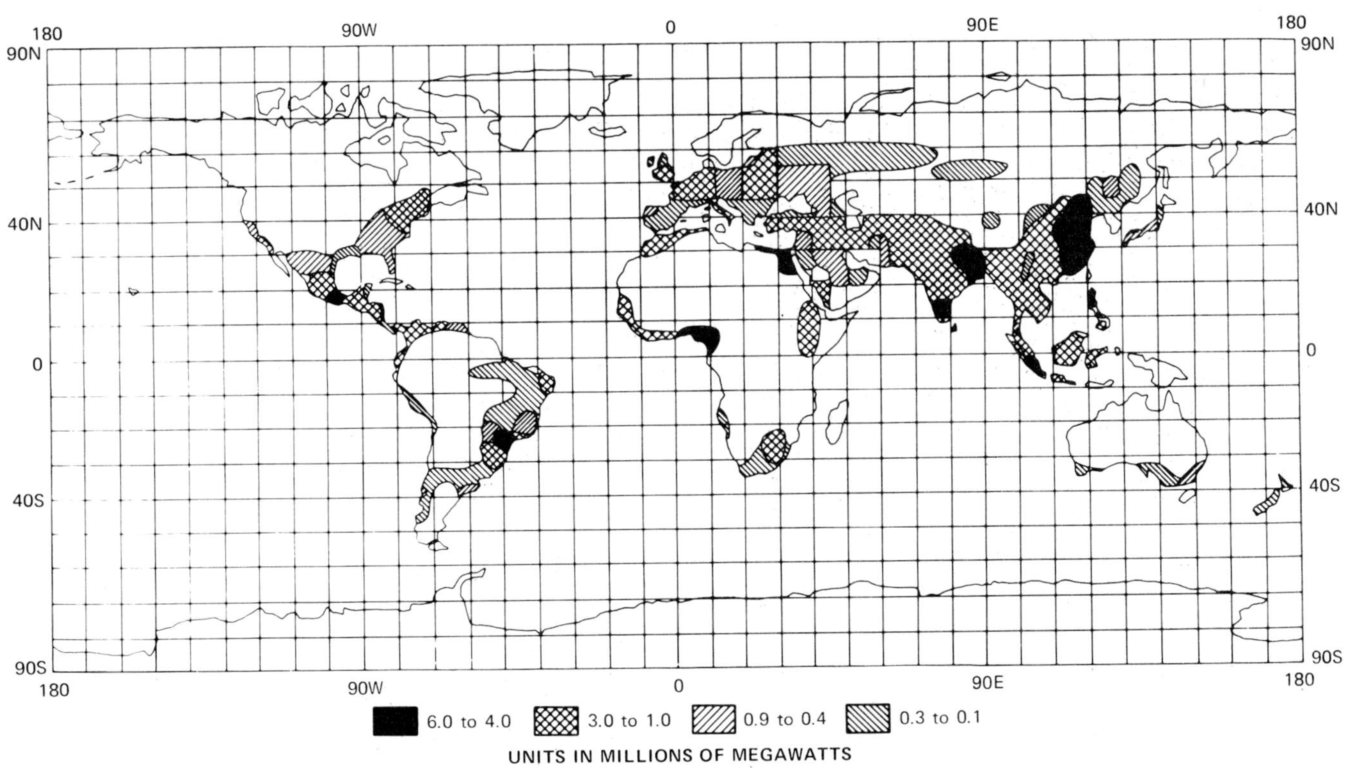

Fig. 15.25 Geographical distribution of expected levels of thermal pollution. [From W. M. Washington, Numerical Climate Change Experiments: The Effect of Man's Production of Thermal Energy, *J. Appl. Meteorol.*, 18: 3-16 (1972).]

by one-fourth of a degree centigrade due to the additional heat input by man. However, man's ultimate input is close to the kinetic energy dissipation rate, 5 W/m^2. Meteorologists are not agreed on the best method of determining the effects of this excess heat.

Washington (1972) collaborated with Weinberg and Hammond to produce Fig. 15.25, in which the total ultimate global production of 3×10^{14} W (assuming 15 kW per capita usage) is distributed according to population. There are several areas on the globe where man's energy flux is as much as 30 to 40 W/m^2, or about 20% of the average solar flux reaching the surface. Washington then applied the NCAR general circulation model using these heat fluxes as part of the surface energy balance. Other input and boundary conditions were typical of January. It was found that anomalies in the January average temperature of as much as 10°C occurred at different parts of the globe, but the locations of these anomalies were not correlated with the locations of the heat input. It was concluded that "the thermal pollution effect is probably small compared to the natural fluctuations of the model."

Washington and Chervin (1979) have continued this work, assuming that the energy flux from the entire eastern megalopolis of the United States in the year 2000 will equal the current flux from Manhattan (90 W/m^2). January and July temperature increases near the heat source were predicted to be 12°C and 3°C, respectively, with significant changes in soil moisture and precipitation. It must be recognized that the assumed heat input is probably an overestimate, and the resulting temperature anomalies are also overestimates.

Work is also under way at the International Institute for Applied Systems Analysis (Murphy et al., 1976) using the general circulation model of the United Kingdom Meteorological Office. In their experiment the 3×10^{14} W was assumed to be all released from two 3° latitude by 3 to 4° longitude rectangles. One rectangle is located off the coast of Japan, and the other is located either off the coast of the British Isles or off the coast of western Africa. A heat flux of 375 W/m was added to the sensible heat in the model over these areas. The results, given in terms of 40-d averages, show that most of the variability is not significant, but is due to the natural vagaries of climate. A few significant deviations occur, such as a surface pressure trough just east of the parks, and temperature increases of 5°C over the parks. No maps of global changes are reproduced here because of the question of whether the results are significant. Future models will be more realistic, accounting for such features as air–sea interaction, ocean circulations, and the interactions between cloudiness changes and other meteorological fields.

REFERENCES

Alkezweeny, A. J., D. W. Glover, R. N. Lee, J. W. Sloot, and M. A. Wolf, 1974, Measured Chromium Distributions Resulting from Cooling Tower Drift, in *Cooling Tower Environment—1974,* ERDA Symposium Series, University of Maryland Adult Education Center, Mar. 4–6, 1974, Steven R. Hanna and Jerry Pell (Coordinators), pp. 558-572, CONF-740302, NTIS, 1975.

Armbruster, J. A., C. L. Mulchi, L. W. Douglass, and P. C. Wolf, 1978, Response of Field Crops to Salt Drifts from a Natural Draft Cooling Tower, in *Proceedings of the Symposium on Cooling Tower Environment,* May 2–4, 1978, pp. 179-196, Water Resources Research Center Special Report No. 9, sponsored by the Power Plant Siting Program, Maryland Department of Natural Resources; publisher, Water Resources Research Center, University of Maryland, College Park, Md.

Atmos. Environ., 1974, 8(4).

Auer, A. H., 1976, Observations of an Industrial Cumulus, *J. Appl. Meteorol.,* 15: 406-413.

——, 1977, The Evaluation of Visibility Deterioration and Aerosol Spectra in the St. Louis Urban Plume, *J. Air Pollut. Control Assoc.,* 27: 1188-1191.

Bhumralkar, C. M., 1973, An Observational and Theoretical Study of Atmospheric Flow over a Heated Island, *Mon. Weather Rev.,* 101: 719-745.

——, 1976, Weather Modification Caused by Waste Heat Rejected into the Atmosphere from Cooling Towers at Large Power Parks, in *Proceedings of the Third Symposium on Atmospheric Turbulence, Diffusion, and Air Quality,* Raleigh, N. C., pp. 581-585, American Meteorological Society, Boston, Mass.

Bøgh, P., 1974, Experience with Combined Wind Tunnel–Plume Model Analysis of Cooling Tower Environmental Impact, in *Cooling Tower Environment—1974,* ERDA Symposium Series, University of Maryland Adult Education Center, Mar. 4–6, 1974, Steven R. Hanna and Jerry Pell (Coordinators), pp. 265-289, CONF-740302, NTIS, 1975.

Bourne, A. G., 1964, Birth of an Island, *Discovery,* 25: 16-19.

Boyack, B. E., and D. W. Kerney, 1973, *Plume Behavior and Potential Environmental Effects of Large Dry Cooling Towers,* Final Report, USAEC Report Gulf-GA-A12346, Gulf General Atomic Co.

Briggs, G. A., 1969, *Plume Rise,* AEC Critical Review Series, TID-24635, NTIS.

——, 1974, Plume Rise from Multiple Sources, in *Cooling Tower Environment–1974,* ERDA Symposium Series, University of Maryland Adult Education Center, Mar. 4–6, 1974, Steven R. Hanna and Jerry Pell (Coordinators), pp. 161-179, CONF-740302, NTIS, 1975.

——, 1975, Plume Rise Predictions, *Lectures on Air Pollution and Environmental Impact Analyses,* pp. 59-111, American Meteorological Society, Boston, MA.

Carson, J. E., 1976, Atmospheric Effects of Waste Heat Discharges into a Large Lake, in *Proceedings of the Second Federal Conference on the Great Lakes,* pp. 306-318.

——, 1976, *Atmospheric Impacts of Evaporative Cooling Systems,* Report ANL/ES-53, Argonne National Laboratory, NTIS.

Chan, J., and M. W. Golay, 1977, Comparative Performance Evaluations of Current Design Evaporative Cooling Tower Drift Eliminators, *Atmos. Environ.,* 11: 775-781.

Changnon, S. A., Jr., R. G. Semonin, and F. A. Huff, 1976, A Hypothesis for Urban Rainfall Anomalies, *J. Appl. Meteorol.,* 15: 544-560.

Chen, N. C. J., and S. R. Hanna, 1978, Drift-Modeling and Monitoring Comparisons, *Atmos. Environ.,* 12: 1725-1734.

——, and L. Jung, 1978, A Mathematical Model of Drift Deposition from a Bifurcated Cooling Tower Plume, *Atmos. Environ.,* 12: 1969-1980.

Coleman, J. H., and T. L. Crawford, 1978, Characterization of Cooling Tower Plumes from Paradise Steam Plant, in *Proceedings of the Symposium on Cooling Tower Environment,* May 2–4, 1978, pp. 1131-1150, Water Resources Research Center Special Report No. 9, sponsored by the Power Plant Siting Program, Maryland Department of Natural Resources; publisher, Water Resources Research Center, University of Maryland, College Park, Md.

Collins, G. F., C. R. Case, and C. J. Mule, 1971, *Cooling Tower Effects (Fogging); Vermont Yankee Generating Station,* Research Corporation of New England, Hartford, Conn., Project 5189.

Csanady, G. T., 1955, Dispersal of Dust Particles from Elevated Sources, *Aust. J. Phys.*, 8(4): 545-550.

——, 1970, Waste Heat Disposal in the Great Lakes, in *Proceedings of the Thirteenth Conference on Great Lakes Research*, Buffalo, N.Y., Apr. 1–3, pp. 388-396. Available at the International Association for Great Lakes Research, Ann Arbor, Mich.

Culkowski, W. M., 1961, An Anomalous Snow at Oak Ridge, Tennessee, *Mon. Weather Rev.*, 90(5): 194-196.

Currier, E. L., J. B. Knox, and T. V. Crawford, 1974, Cooling Pond Steam Fog, *J. Air Pollut. Control Assoc.*, 24: 860-864.

Dana, M. T., and M. A. Wolf, 1977, *Rainfall Enhancement Due to Scavenging of Cooling Tower Condensate*, USAEC Report BNWL-2295, Battelle Pacific Northwest Laboratory.

Davies, R. W., 1959, Large Scale Diffusion from an Oil Fire, in *Advances in Geophysics*, Vol. 6, F. N. Frenkiel and P. A. Sheppard (Eds.), pp. 413-415, Academic Press, New York.

Davis, E. A., 1979, *Environmental Assessment of Chalk Point Cooling Tower Drift and Vapor Emissions*, Report JHU-PPSP-CPCTP-28, The Johns Hopkins University, Applied Physics Laboratory, Laurel, Md.

Deardorff, J. W., 1972, Numerical Investigation of Neutral and Unstable Planetary Boundary Layers, *J. Atmos. Sci.*, 29: 91-115.

Dessens, J., 1964, Man-Made Thunderstorms, *Discovery*, 25: 40-44.

Devine, J. C., Jr., 1974, The Forked River Program: A Case Study in Saltwater Cooling, in *Cooling Tower Environment–1974*, ERDA Symposium Series, University of Maryland Adult Education Center, Mar. 4–6, 1974, Steven R. Hanna and Jerry Pell (Coordinators), pp. 509-557, CONF-740302, NTIS, 1975.

Donaldson, C. Du. P., and G. R. Hilst, 1972, Effect of Inhomogeneous Mixing on Atmospheric Photochemical Reactions, *Environ. Sci. Technol.*, 9: 812-816.

Drake, R. L., 1975, *A Review and Evaluation of Information on the Thermal Performance of Ultimate Heat Sinks: Spray Ponds and Cooling Ponds*, NRC Report BNWL-B-446, Battelle, Pacific Northwest Laboratories.

Engelmann, R. J., 1968, The Calculation of Precipitation Scavenging, in *Meteorology and Atomic Energy—1968*, D. H. Slade (Ed.), pp. 208-221, USAEC Report TID-24190.

Environmental Systems Corporation, 1976, *Chalk Point Cooling Tower Project Seasonal Test Data for the Period December 15–19, 1975*, Report PPSP-CPCTP-8, ESC, Knoxville, Tenn., Maryland Power Plant Siting Program, Chalk Point Cooling Tower Project.

——, 1977, *Cooling Tower Drift Dye Tracer Experiment*, Report PPSP-CPCTP-12, prepared by Environmental Systems Corporation.

Everett, R. G., and C. A. Zerbe, 1977, *Winter Field Program at the Dresden Cooling Ponds*, ERDA Report ANL-76-88, Part IV. in *Radiology and Environmental Research Division Annual Report, Atmospheric Physics*, January–December 1976, pp. 108-113.

Fish, B. R., and J. R. Duncan, 1972, Measurements of Liquid Droplet Emission from Cooling Towers and Process Stacks, in *Sixty-Fifth Annual Meeting of the Air Pollution Control Association*, Miami Beach, Paper 72-35.

Francis, B. A., 1977, *Effects of Simulated Saline Cooling Tower Drift on Woody Species*, Report PPSP-CPCTP-17, Water Resources Research Center, University of Maryland.

Gifford, F. A., Jr., 1968, An Outline of Theories of Diffusion in the Lower Layers of the Atmosphere, in *Meteorology and Atomic Energy—1968*, D. H. Slade (Ed.), USAEC Report TID-24190, pp. 65-116.

Hadlock, R. K., 1979, Meteorological Influence on Hot Cooling Pond Thermal Performance, in *Fourth Symposium on Turbulence, Diffusion, and Air Pollution*, Jan. 15–18, Reno, Nev., pp. 661-664, American Meteorological Society, Boston, Mass.

Hanna, S. R., 1972, Rise and Condensation of Large Cooling Tower Plumes, *J. Appl. Meteorol.*, 11: 793-799.

——, 1974a, Meteorological Effects of the Mechanical Draft Cooling Towers of the Oak Ridge Gaseous Diffusion Plant, in *Cooling Tower Environment—1974*, ERDA Symposium Series, University of Maryland Adult Education Center, Mar. 4–6, 1974, Steven R. Hanna and Jerry Pell (Coordinators), pp. 291-306, CONF-740302, NTIS, 1975.

——, 1974b, Fog and Drift Deposition from Evaporative Cooling Towers, *Nucl. Safety*, 15(2): 190-196.

——, 1976, Predicted and Observed Cooling Tower Plume Rise and Visible Plume Length at the John E. Amos Power Plant, *Atmos. Environ.*, 10: 1043-1052.

——, 1977a, Comments on "Observations of an Industrial Cumulus," *J. Appl. Meteorol.*, 15(11): 1232-1233.

——, 1977b, Predicted Climatology of Cooling Tower Plumes from Energy Centers, *J. Appl. Meteorol.*, 16: 880-887.

——, 1978, A Simple Drift Deposition Model Applied to the Chalk Dye Tracer Experiment, in *Proceedings of the Symposium on Cooling Tower Environment*, May 2–4, 1978, pp. 291-306, Water Resources Research Center Special Report No. 9, sponsored by the Power Plant Siting Program, Maryland Department of Natural Resources; publisher, Water Resources Research Center, University of Maryland, College Park, Md.

——, and F. A. Gifford, 1975, Meteorological Effects of Energy Dissipation at Large Power Parks, *Bull. Am. Meteorol. Soc.*, 56: 1069-1076.

——, and J. Pell (Coordinators), *Cooling Tower Environment—1974*, ERDA Symposium Series, University of Maryland Adult Education Center, Mar. 4–6, 1974, CONF-740302, NTIS.

——, and S. G. Perry, 1973, *Meteorological Effects of the Cooling Towers at the Oak Ridge Gaseous Diffusion Plant*, Part I. ATDL Report 86, Atmospheric Turbulence and Diffusion Laboratory, Oak Ridge, Tenn.

——, M. Pike, and K. Seitter, 1978, Secondary Motions in Cooling Tower Plumes, *J. Appl. Meteorol.*, 17: 1068-1071.

——, K. S. Rao, and R. P. Hosker, 1977, *Report on ATDL Research on Meteorological Effects on Thermal Energy Releases, Oct. 1, 1976–Sept. 30, 1977*, Atmospheric Turbulence and Diffusion Laboratory, Oak Ridge, Tenn.

——, and S. D. Swisher, 1971, Meteorological Effects of the Heat and Moisture Produced by Man, *Nucl. Safety*, 12: 114-122.

Hane, C. E., 1978, The Application of a 2D Convective Cloud Model to Waste Heat Release from Proposed Nuclear Energy Centers, *Atmos. Environ.*, 12: 1839-1848.

Hicks, B. B., 1975, On the Limiting Surface Temperature of Exposed Water Bodies, *J. Geophys. Res.*, 80(36): 5077-5081.

——, 1977a, A Study of Heat Transfer Processes Above a Cooling Pond, *Water Resour. Res.*, 13: 901-908.

——, 1977b, The Prediction of Fog over Cooling Ponds, *J. Air Pollut. Control Assoc.*, 27: 140-142.

——, and J. D. Shannon, 1977, *A New Cooling Pond Index*, Radiological and Environmental Research Division Annual Report ANL-76-88, pp. 104-107, Argonne National Laboratory.

Hill, G. E., 1974, Factors Controlling the Size and Spacing of Cumulus Clouds as Revealed by Numerical Experiments, *J. Atmos. Sci.*, 31: 646-673.

Holmberg, J. D., 1974, Drift Management in the Chalk Point Cooling Tower, in *Cooling Tower Environment—1974*, ERDA Symposium Series, University of Maryland Adult Education Center, Mar. 4–6, 1974, Steven R. Hanna and Jerry Pell (Coordinators), pp. 128-146, CONF-740302, NTIS, 1975.

Hosler, C. L., and H. E. Landsberg, 1977, The Effect of Localized Man-Made Heat and Moisture Sources on Mesoscale Weather Modification, in *Energy and Climate*, pp. 96-105, National Academy of Sciences, Washington, D. C.

——, J. Pena, and R. Pena, 1972, *Determination of Salt Deposition Rates from Drift from Evaporative Cooling Towers*, Department of Meteorology, The Pennsylvania State University, University Park, Pa.

Huff, F. A., and S. A. Changnon, Jr., 1972, Climatological Assessment of Urban Effects on Precipitation at St. Louis, *J. Appl. Meteorol.*, 11: 825-943.

——, and J. L. Vogel, 1978, Urban, Topographic, and Diurnal Effects on Rainfall in the St. Louis Region, *J. Appl. Meteorol.*, 17: 565-577.

Israel, G. W., and T. J. Overcamp, 1974, Drift Deposition Model for Natural Draft Cooling Towers, in *Cooling Tower Environment—1974*, ERDA Symposium Series, University of Maryland Adult Education Center, Mar. 4–6, 1974, Steven R. Hanna and Jerry Pell (Coordinators), pp. 614-628, CONF-740302, NTIS, 1975.

——, T. J. Overcamp, and W. J. Pringle, 1977, A Method to Measure Drift Deposition from Saline Natural Draft Cooling Towers, *Atmos. Environ.*, 11, 123-130.

Junod, A., R. J. Hopkirk, D. Schneiter, and D. Haschke, 1974, Meteorological Influences of Atmospheric Cooling Systems as Projected in Switzerland, in *Cooling Tower Environment—1974*, ERDA Symposium Series, University of Maryland Adult Education Center, Mar. 4–6, 1974, Steven R. Hanna and Jerry Pell (Coordinators), pp. 239-264, CONF-740302, NTIS, 1975.

Kelley, R. B., 1971, Large Scale Spray Cooling, *Ind. Water Eng.*, 8: 18-20 (Aug./Sept. 1971).

Kessler, E., 1969, On the Distribution and Continuity of Water Substance in Atmospheric Circulations, *Meteorol. Monogr.*, 10, American Meteorological Society, Boston, Mass.

Koenig, L. R., 1976, Analogs for Waste Heat Dissipation Effects, in *Atmospheric Effects of Nuclear Energy Centers Program*, Annual Technical Progress Report for July 1975–September 1976, A. A. Patrinos and H. W. Hoffman (Eds.), ERDA Report ORNL/TM-5778, pp. 84-115, Oak Ridge National Laboratory.

——, and C. M. Bhumralkar, 1974, *On Possible Undesirable Atmospheric Effects of Heat Rejection from Large Electric Power Centers*, Report R-1628-RC, Rand Corp., Santa Monica, Calif.

——, F. W. Murray, and P. M. Tag, 1978, Differences in Atmospheric Convection Caused by Waste Energy Rejected in the Forms of Sensible and Latent Heats, *Atmos. Environ.*, 12: 1013-1020.

Kramer, M. L, D. E. Seymour, M. J. Butler, R. N. Kempton, P. J. Brennan, J. J. Conte, and R. G. Thomson, 1975, *John E. Amos Cooling Tower Flight Program Data, December 1974–March 1975*, Smith–Singer Meteorologists, Inc., Amityville, N. Y., American Electric Power Service Corp., Canton, Oh.

——, D. E. Seymour, M. E. Smith, R. W. Reeves, and T. T. Frakenberg, 1976, Snowfall Observations from Natural Draft Cooling Tower Plumes, *Science*, 193: 1239-1241.

Landsberg, H. E., and T. N. Maisel, 1972, Micrometeorological Observations in an Area of Urban Growth, *Boundary-Layer Meteorol.*, 2: 365-370.

——, 1977, Rainfall Variations Around a Thermal Power Station, *Atmos. Environ.*, 11: 565.

Laskowski, S., 1974, Mathematical Transport Model for Salt Distribution from a Saltwater Natural-Draft Cooling Tower, in *Cooling Tower Environment—1974*, ERDA Symposium Series, University of Maryland Adult Education Center, Mar. 4–6, 1974, Steven R. Hanna and Jerry Pell (Coordinators), pp. 598-613, CONF-740302, NTIS, 1975.

Laulenian, N. S., and S. L. Ulansky, 1979, Cooling Tower Drift: Comprehensive Case Study, in *Pacific Northwest Laboratories Annual Report for 1978 to the Department of Energy, Part 3. Atmospheric Sciences*, pp. 226-230, Pacific Northwest Laboratories.

Lauver, T. L., C. R. Curtis, G. Patterson, and L. W. Douglass, 1978, The Effects of Saline Cooling Tower Drift on Seasonal Variations of Sodium and Chloride Concentrations in Native Perennial Vegetation, in *Proceedings of the Symposium on Cooling Tower Environment*, May 2–4, 1978, pp. 149-162, Water Resources Research Center Special Report No. 9, sponsored by the Power Plant Siting Program, Maryland Department of Natural Resources; publisher, Water Resources Research Center, University of Maryland, College Park, Md.

Lee, J., 1976, A Numerical Simulation of Atmospheric Convection Caused by Heat Dissipation at Large Power Centers, in *Proceedings of the Third Symposium on Atmospheric Turbulence, Diffusion, and Air Quality*, Raleigh, N. C., pp. 563-570, American Meteorological Society, Boston, Mass.

——, 1979, Potential Weather Modification Caused by Waste Heat Release from Large Dry Cooling Towers, *J. Heat Transfer*, 101: 164-168.

Lewellen, N. S., D. A. Oliver, M. E. Teske, and G. G. Williamson, 1976, *Status Report on a Low-Level Atmospheric Turbulence Model for Marine Environment*, Report 289, prepared for U. S. Department of the Navy by Aeronautical Research Associates of Princeton, Inc., Princeton, N. J.

McCune, D. L., G. H. Silberman, R. H. Mandl, L. H. Weinstein, P. C. Freudenthal, and P. A. Giardina, 1977, Studies on the Effects of Saline Aerosols of Cooling Tower Origin on Plants, *J. Air Pollut. Control Assoc.*, 27: 319-324.

McElroy, J. L., 1971, An Experimental and Numerical Investigation of the Nocturnal Heat Island over Columbus, Ohio, Ph.D. Thesis, Department of Meteorology, Pennsylvania State University, University Park, Pa.

McVehil, G. E., and K. Heikes, 1975, *Cooling Tower Plume Modeling and Drift Measurement*, Ball Brothers Research Corp., Boulder, Colo., American Society of Mechanical Engineers.

Mason, B. J., 1971, *The Physics of Clouds*, Clarendon Press, Oxford, England.

Meyer, J. H., T. W. Eagles, L. C. Kohlenstein, J. A. Kagan, and W. D. Stanbro, 1974, Mechanical Draft Cooling Tower Visible Plume Behavior: Measurements, Models, Predictions, in *Cooling Tower Environment—1974*, ERDA Symposium Series, University of Maryland Adult Education Center, Mar. 4–6, 1974, Steven R. Hanna and Jerry Pell (Coordinators), pp. 307-352, CONF-740302, NTIS, 1975.

——, and W. D. Stanbro, 1977, *Fluorescent Dye, a Novel Technique to Trace Cooling Tower Drift*, presented at Fourth Joint Conference on Sensing Environmental Pollutants, New Orleans, Johns Hopkins University, Applied Physics Laboratory, Laurel, Md.

Moore, D. J., 1974, Recent Central Electricity Generating Board Research on Environmental Effects of Wet Cooling Towers, in *Cooling Tower Environment—1974*, ERDA Symposium Series, University of Maryland Adult Education Center, Mar. 4–6, 1974, Steven R. Hanna and Jerry Pell (Coordinators), pp. 205-220, CONF-740302, NTIS, 1975.

Morris, D. G., 1968, Initiation of Convective Clouds Due to Static Firing of the Saturn V First Stage, *Bull. Am. Meteorol. Soc.*, 49: 1054-1058.

Murphy, A. H., A. Gilchrist, W. Hafele, G. Kromer, and J. Williams, 1976, *The Impact of Waste Heat Release on Simulated Global Climate*, Report RM 76-79, International Institute for Applied Systems Analysis, Laxenburg, Austria.

Murray, F. W., L. R. Koenig, and P. M. Tag, 1978: Numerical Simulation of an Industrial Cumulus and Comparisons with Observations, *J. Appl. Meteorol.*, 17: 655-688.

Neuberger, H. H., 1957, *Introduction to Physical Meteorology*, College of Mineral Industries, The Pennsylvania State University, University Park, Pa.

Ooms, G., 1972, A New Method for the Calculation of the Plume Path of Gases Emitted by a Stack, *Atmos. Environ.*, 6: 899-909.

Orgill, M. M., 1977, *The Fire Analog: A Comparison Between Fire Plumes and Energy Center Cooling Tower Plume*, USDOE Report PNL-2433, Battelle Pacific Northwest Laboratories.

Otts, R. E., 1976, *Locally Heavy Snow Downwind from Cooling Towers*, NOAA Technical Memorandum NWS-ER-62, National Weather Service, Eastern Region.

Overcamp, T. J., G. W. Israel, and W. J. B. Pringle, 1976, Drift Droplet Deposition Measurements from a Brackish-Water Natural Draft Cooling Tower, in *Proceedings of the Third Symposium on Atmospheric Turbulence, Diffusion, and Air Quality,* Raleigh, N. C., pp. 586-592, American Meteorological Society, Boston, Mass.

——, and D. P. Hoult, 1971, Precipitation in the Wake of Cooling Towers, *Atmos. Environ.,* 5: 751-765.

——, and G. W. Israel, 1979, Sensitivity Analysis of a Salt Deposition Model for Natural Draft Cooling Towers, *Atmos. Environ.,* 13: 61-70.

Pandolfo, J. P., and C. A. Jacobs, 1978, Computer Simulation of Meso-Scale Meteorological Effects of Alternative Waste Heat Disposal Methods, in *Proceedings of the Conference on Waste Heat Management and Utilization,* Miami Beach, Center for Environment and Man, Inc., Hartford, Conn.

Patel, V. C., T. E. Croley II, and M. S. Cheng, 1974, Optimum Design of Wet–Dry Combination Cooling Towers for Power Plants, in *Cooling Tower Environment—1974,* ERDA Symposium Series, University of Maryland Adult Education Center, Mar. 4–6, 1974, Steven R. Hanna and Jerry Pell (Coordinators), pp. 24-57, CONF-740302, NTIS, 1975.

Patrinos, A. A. N., N. C. J. Chen, and R. L. Miller, 1979, Spatial Correlations of Monthly Rainfall: Applications in Climatology and Weather Modification Experiments, *J. Appl. Meteorol.,* 18: 719-732.

Pell, J., 1974, The Chalk Point Cooling Tower Project, in *Cooling Tower Environment—1974,* ERDA Symposium Series, University of Maryland Adult Education Center, Mar. 4–6, 1974, Steven R. Hanna and Jerry Pell (Coordinators), pp. 88-127, CONF-740302, NTIS, 1975.

Pena, J. A., and C. L. Hosler, 1974, Influence of the Choice of the Plume Diffusion Formula on the Salt-Deposition-Rate Calculation, in *Cooling Tower Environment—1974,* ERDA Symposium Series, University of Maryland Adult Education Center, Mar. 4–6, 1974, Steven R. Hanna and Jerry Pell (Coordinators), pp. 573-584, CONF-740302, NTIS, 1975.

Peterson, J., 1969, *The Climate of Cities: A Survey of Recent Literature,* U. S. Department of Health, Education and Welfare, Public Health Service, Raleigh, N. C.

Policastro, A. J., 1972, State of the Art of Analytical Modeling of Heated Effluent Dispersion in Large Lakes, in *Proceedings of the Fifteenth Conference on Great Lakes Research,* Madison, Wis., Apr. 5–7, 1972, pp. 642-654. Available at the International Association for Great Lakes Research, Ann Arbor, Mich.

——, W. E. Dunn, and M. Breig, 1978, Evaluation of Theory and Performances of Salt Drift Deposition Models for Natural Draft Cooling Tower, in *Proceedings of the Second AIAA/ASME Thermophysics and Heat Transfer Conference,* Palo Alto, Calif., Environmental Statements Project, Argonne National Laboratory, Argonne, Ill.

Priestly, C. H. B., and R. J. Taylor, 1972, On the Assessment of Surface Heat Flux and Evaporation Using Large-Scale Parameters, *Mon. Weather Rev.,* 100: 81-92.

Rao, K. S., and R. P. Hosker, 1978, A Numerical Study of Meteorological Effects of Waste Heat and Moisture Releases from Hypothetical Power Parks, in *Proceedings of the Second AIAA/ASME Thermophysics and Heat Transfer Conference,* Palo Alto, Calif., Atmospheric Turbulence and Diffusion Laboratory, P. O. Box E, Oak Ridge, Tenn.

Rochow, J. J., 1978, Compositional Structure and Chemical Changes to Forest Vegetation from Fresh Water with Cooling Tower Drift, in *Proceedings of the Symposium on Cooling Tower Environment,* May 2–4, 1978, pp. 119-138, Water Resources Research Center Special Report No. 9, sponsored by the Power Plant Siting Program, Maryland Department of Natural Resources; publisher, Water Resources Research Center, University of Maryland, College Park, Md.

Roffman, A., et al., 1973, *The State of the Art of Saltwater Cooling Towers for Steam Electric Generating Plants,* USAEC Report WASH-1244, Westinghouse Electric Corporation.

——, and R. E. Grimble, 1974, Drift Deposition Rates from Wet Cooling Systems, in *Cooling Tower Environment—1974,* ERDA Symposium Series, University of Maryland Adult Education Center, Mar. 4–6, 1974, Steven R. Hanna and Jerry Pell (Coordinators), pp. 585-597, CONF-740302, NTIS, 1975.

Rotty, R. M., 1975, *Waste Heat Disposal from Nuclear Power Plants,* National Oceanic and Atmospheric Administration Technical Memorandum ERL-ARL-47.

Ryan, P., 1975, *Spray Cooling,* Bechtel Power Co., San Francisco.

——, D. R. F. Harleman, and K. D. Stolzenbach, 1974, Surface Heat Loss from Cooling Ponds, *Water Resour. Res.,* 10(5).

Ryznar, E., D. G. Baker, and H. Moses, 1976, Coastal Meteorological Networks to Determine Effects of Nuclear Plant Cooling Systems, *Bull. Am. Meteorol. Soc.,* 57(12): 1441-1447.

Sanderson, M., and R. Gorski, 1978, The Effect of Metropolitan Detroit–Windsor on Precipitation, *J. Appl. Meteorol.,* 17: 423-427.

Schrecker, G. O., R. Wilber, and F. Shofner, 1974, Prediction and Measurement of Airborne Particulate Concentrations from Cooling-Device Sources and in the Ambient Atmosphere, in *Cooling Tower Environment—1974,* ERDA Symposium Series, University of Maryland Adult Education Center, Mar. 4–6, 1974, Steven R. Hanna and Jerry Pell (Coordinators), pp. 455-482, CONF-740302, NTIS, 1975.

Shea, D. M., 1976, *Observations of the Regional Influence on Thermodynamic Properties and Aerosol Patterns in the Plume Downwind of St. Louis,* Environmental Protection Agency, and the National Science Foundation, Department of Atmospheric Science, College of Engineering, University of Wyoming, Laramie.

Shofner, F., G. O. Schrecker, T. Carlson, and R. Webb, 1974, Measurement and Interpretation of Drift Particle Data, in *Cooling Tower Environment—1974,* ERDA Symposium Series, University of Maryland Adult Education Center, Mar. 4–6, 1974, Steven R. Hanna and Jerry Pell (Coordinators), pp. 427-454, CONF-740302, NTIS, 1975.

Slawson, P. R., J. H. Coleman, and J. W. Frey, 1974, Some Observations on Cooling-Tower Plume Behavior at the Paradise Steam Plant, in *Cooling Tower Environment—1974,* ERDA Symposium Series, University of Maryland Adult Education Center, Mar. 4–6, 1974, Steven R. Hanna and Jerry Pell (Coordinators), pp. 147-160, CONF-740302, NTIS, 1975.

Slinn, W. G. N., 1974, An Analytical Search for the Stochastic-Dominating Process in the Drift-Deposition Problem, in *Cooling Tower Environment—1974,* ERDA Symposium Series, University of Maryland Adult Education Center, Mar. 4–6, 1974, Steven R. Hanna and Jerry Pell (Coordinators), pp. 483-500, CONF-740302, NTIS, 1975.

Stinson, J. R., and C. D. Brown, 1976, *Inadvertent Cloud Formation and Modification by Refineries and Power Plants,* Report SRA 76-01, Stinson Research Association, No. 1 Third Place, Suite 603, Long Beach, Calif.

Stockham, J., 1971, *Cooling Tower Study,* Contract No. CPA-22-69-112, Illinois Institute of Technology Research Institute, Technology Center, Chicago, Ill., Environmental Protection Agency Pollution Control Office, Durham, N. C.

Talbot, J. J., 1979, A Review of Potential Biological Impacts of Cooling Tower Salt Drift, *Atmos. Environ.,* 13: 395-406.

Taylor, D. F., and D. T. Williams, 1968, Severe Storm Features of a Wildfire, *Agric. Meteorol.,* 5: 311-318.

Taylor, F. G., Jr., S. R. Hanna, and P. D. Parr, 1978, *Cooling Tower Drift Studies at the Paducah, Kentucky Gaseous Diffusion Plant,* USDOE Report ORNL/TM-6131, Oak Ridge National Laboratory, Oak Ridge, Tenn.

——, L. K. Mann, R. C. Dahlman, and F. L. Miller, 1974, Environmental Effects of Chromium and Zinc in Cooling-Water Drift, in *Cooling Tower Environment—1974,* ERDA Symposium Series, University of Maryland Adult Education Center, Mar. 4–6, 1974, Steven R. Hanna and Jerry Pell (Coordinators), pp. 408-426, CONF-740302, NTIS, 1975.

——, and P. D. Parr, 1978, *Interception and Retention of Simulated Cooling Tower Drift by Vegetation,* Oak Ridge National Laboratory.

Taylor, R. J., S. T. Evans, N. K. King, E. T. Stephens, D. R. Packham, and R. G. Vines, 1973, Convective Activity Above a Large Scale Brushfire, *J. Appl. Meteorol.,* 12: 1144-1150.

Thomson, D. W., 1976, *Environmental Impact of Evaporative Cooling Tower Plumes,* ERDA Report, Pennsylvania State University, Department of Meteorology, University Park, Pa.

——, J. M. Norman, and R. L. Miller, 1976, Airborne Measurements of Turbulent Temperature and Velocity Fluctuations in Cooling Tower Plumes, in *Proceedings of the Third Symposium on Atmospheric Turbulence, Diffusion, and Air Quality,* Raleigh, N. C., American Meteorological Society, Boston, Mass., pp. 576-580.

——, J. M. Norman, T. N. Chin, and R. L. Miller, 1977, *Airborne Studies of Natural Draft Cooling Tower Plumes: Meteorological Profiles and a Summary of In-Plume Turbulent Temperature and Velocity Fluctuations Progress Report,* Department of Meteorology, Pennsylvania State University, University Park, Pa.

Thorarinsson, S., and B. Vonnegut, 1964, Whirlwinds Produced by the Eruption of Surtsey Volcano, *Bull. Am. Meteorol. Soc.,* 45: 440-444.

Trabert, W., 1901, Die Extinction des Lichtes in Einem Truben Medium (Schweite in Wofken), *Meteorol. Z.,* 18: 518-520.

Van der Hoven, I., 1968, Deposition of Particles and Gases, in *Meteorology and Atomic Energy—1968,* D. H. Slade (Ed.), USAEC Report TID-24190, pp. 202-208, NTIS.

Washington, W. M., 1972, Numerical Climate Change Experiments: The Effect of Man's Production of Thermal Energy, *J. Appl. Meteorol.,* 11: 768-772.

——, and R. M. Chervin, 1979, Regional Climatic Effects of Large-Scale Thermal Pollution: Simulation Studies with the NCAR General Circulation Model, *J. Appl. Meteorol.,* 18: 3-16.

Weil, J., 1974, The Rise of Moist Buoyant Plumes, *J. Appl. Meteorol.,* 13: 435-443.

Weinberg, A. M., and R. P. Hammond, 1970, Limits to the Use of Energy, *Am. Sci.,* 58: 412-418.

Weinstein, A. I., 1970, A Numerical Model of Cumulus Dynamics and Microphysics, *J. Atmos. Sci.,* 27: 246-255.

Wigley, T. M. L., 1975, Condensation in Jets, Industrial Plumes, and Cooling Tower Plumes, *J. Appl. Meteorol.,* 14: 78-86.

——, and P. R. Slawson, 1971, On the Condensation of Buoyant, Moist, Bent-Over Plumes, *J. Appl. Meteorol.,* 10: 259-263.

Wilber, K. R., 1978, *Examination of the Sensitivity of Spray Cooling to Wind Speed and Direction,* presented at the American Meeting of the Cooling Tower Institute, Environmental Systems Corporation, Knoxville, Tenn.

World Meteorological Organization, 1966, Introduction to Tables 4.6 and 4.7, *International Meteorological Tables,* Report WMO-188-TP-94.

Woffinden, G. J., J. A. Anderson, and P. R. Harrison, 1976, *Aircraft Survey, Chalk Point Cooling Tower Plume,* Report 76R-1910, the Meteorological Research Institute, Altadena, Calif., for the Maryland Power Plant Siting Program.

——, P. R. Harrison, and J. A. Anderson, 1976, *Cooling Tower Plume Survey,* 4 Vols., Report PPSP-PCTP-15, Meteorology Research, Inc., Altadena, Calif.

Wolf, M. A., 1976, Natural Draft Cooling Tower Plume Characteristics Determined with Airborne Instrumentation, in *Pacific Northwest Laboratory Annual Report for 1975 to the USERDA Division of Biomedical and Environmental Research,* Part 3, *Atmospheric Sciences,* ERDA Report BNWL-2000(Pt.3), pp. 281-288.

Woodson, R. D., 1971, Cooling Towers, *Sci. Am.,* 224(5): 70-78.

Yamada, T., 1979, An Application of a Three Dimensional, Simplified Second Moment Closure Numerical Model to Study Atmospheric Effects of a Large Cooling Pond, *Atmos. Environ.,* 13: 693-704.

Chapter 15 covers research prior to December 1979.

RADIOACTIVE CLOUD DOSE CALCULATIONS

16

J. W. Healy, Los Alamos National Laboratory

16-1 INTRODUCTION

Previous chapters have described the methods of estimating atmospheric dispersion as well as deposition and resuspension. These chapters illustrate methods of estimating the air concentration from a source or a series of sources. For radioactive materials the next step is to estimate the radiation dose (energy absorbed) to the body or to specific organs that can result from exposure to the materials emitted in the estimated concentration. To define whether the release is reasonable, we can use the estimated dose in comparison with standards to estimate the effects attributable to the release. In recent years studies of the practicality of reducing the dose and the use of cost–benefit studies to assess the effects of an operation have become important in deciding appropriate releases. These studies require integration of a large number of disciplines, including the meteorologist.

The radiation dose is the common link among the potential damages from radiations or radioactive materials because it expresses the amount of energy deposited in a tissue, and this energy is the primary cause of damage. The existence of such a link for radiation is in marked contrast to chemically toxic materials or chemical carcinogens where each can act by a different mechanism. Thus, if we can estimate the radiation dose to the body or to a specific organ, we can estimate the potential future harm; however, for the chemical materials, independent tests must be made and judgments drawn for each. This is an overgeneralization of the usefulness of the radiation-dose concept because there may be some factors, such as the radiation-dose distribution through a tissue or organ, which are not well defined at the present time. In general, however, the radiation-dose concept in expressing possible damage is the basis of all radiation protection.

In this chapter we will provide methods for estimating the dose following dispersion and deposition of radioactive materials. Some introductory material on dose units and characteristics of radioactive material will be included for those unfamiliar with the field. A brief description of current standards and methods of estimating risk are also included for orientation but not in sufficient detail for use. The reader should refer to more detailed discussions on these points as given in the references.

Healy: Los Alamos National Laboratory, Los Alamos, New Mexico.

There are many different ways of calculating some of the doses. For example, complex computer codes can integrate the radiation flux and scatter over large volumes. These codes can provide accurate estimates once the source configuration is known. However, for predictive studies it is difficult, if not impossible, to define the release of the contaminant well enough in advance to allow full advantage of the accuracy of such codes. These codes are, however, very useful for parametric or probability studies, given the wide range of meteorological variables that are possible. The use of the computer in dose calculation will not be discussed further in this chapter, since the procedures should be derived and applied by one knowledgeable in the field. Instead, the objective is to provide some insight into methods of calculation and analytical approximations that provide reasonable answers given the necessary approximations.

16-2 SOURCES OF AIRBORNE MATERIALS

An important concept in the relationship between releases and the potential doses to a receptor is the so-called "source term." This term is the basis for all calculations and includes the quantities and types of materials emitted to the atmosphere, the position of emission relative to the receptor, the time period over which emission occurs, and the physical and chemical nature of the materials emitted. All are important in estimating the ultimate fate of the contaminant in the environs; however, the last item, the chemical and physical nature, is frequently the least known.

Releases can be characterized as routine or accidental. For routine releases the source term can be predicted from knowledge of past operations and measurements of the emissions. For accidents, however, it is necessary to predict, often from small-scale laboratory experiments, the quantities and characteristics of the materials involved. Since accidents are by definition unpredictable, except on a statistical basis, and the information on the quantities emitted and the environmental behavior of the radioactive materials are frequently poorly known, the source term for accident conditions may be very uncertain.

In the nuclear industry it is the general practice to reduce emissions to "as low as practicable" (ALAP). This term indicates that operations should not proceed as based on the limits in radiation-protection recommendations and regulations, but that every effort should be taken to reduce the emissions to a level that is economically and practically possible. The International Commission on Radiological Protection (ICRP) (1977) has used "as low as reasonably achievable" (ALARA) for this concept, and this term is also used by the U. S. Nuclear Regulatory Commission. In many cases, such efforts have resulted in emissions that are too low to be measured in the environs by conventional methods, and the doses received are estimated by calculations of dispersion and the resulting dose rather than by measurement.

The quantities and types of radioactive materials reaching the atmosphere are a function of the type of installation considered, the specific design, and the operating mode. The actual source term can only be determined from the characteristics of the particular installation. The following description of sources is general and intended to provide orientation rather than specific numbers for use.

16-2.1 Classification of Airborne Materials

Atmospheric contaminants can be in the form of gases or particles. Gases are monomolecular distributions of the contaminant formed by evaporation or sublimation, and particles are agglomerates of molecules in the solid or liquid state.

In general, gases will disperse in the same manner as the atmosphere in which they are carried, although at very high concentrations heavier gases may have a net settling force owing to their higher density. The deposition characteristics of a gas will depend on its chemical reactivity and on the difference between the boiling point of the contaminant and the ambient temperature of surfaces. Noble gases, such as xenon or krypton, will deposit to only a small extent on solid surfaces, although they can be dissolved in liquids. They are eliminated from the atmosphere primarily by radioactive decay. Other gases will deposit with varying degrees of efficiency. For example, when in the form of iodine vapor, elemental ^{131}I will deposit very strongly on many types of surfaces, including vegetation. Such gases will be eliminated from the atmosphere by deposition, by reaction with or adsorption on natural aerosol particles in the atmosphere, by photochemical reactions resulting in a more reactive compound, by washout with precipitation, or by biological processes. Since the removal rates of these processes depend on the chemical form of the gas, accurate predictions of interactions in the environs require data on the characteristics of the contaminant that influence deposition and their changes with time in the particular environmental conditions existing. Only a few materials have been studied to any appreciable degree; therefore such estimates require empirical verification.

Particles can be liquids or solids. When in liquid form, they are referred to as mists or, in high concentrations of submicron particles, as fogs. Solid particulates are dusts formed mechanically by grinding, crushing, and drilling; fumes are formed by combustion, sublimation, and condensation; smokes are usually formed by combustion of organic material. More information on the characteristics of these particulate classes is given in Chap. 17.

The general term "aerosol" is used to denote these disperse systems in air in analogy to the term "hydrosol" applied to disperse systems in water. The fate of aerosols after they are released to the atmosphere depends to a large degree on the aerodynamic properties of the aerosol which, in turn, depend on the particle size, shape, and density. Among the more important processes influencing such particles are gravitational settling, agglomeration, and impaction.

Gravity will act on particles to provide a net settling force. Depending on the surface area per unit mass and shape of these small particles, the resulting air resistance after a short time will produce a constant settling velocity when the force of gravity is exactly balanced by the drag forces. This velocity is referred to as the terminal velocity of the particle. In the atmosphere the forces exerted on small particles by turbulent motions are large compared with the force of gravity, and consequently such particles can remain suspended for long periods of time.

An important concept in describing the behavior of particles in the atmosphere is that of aerodynamic diameter, that is, the diameter of a spherical particle of unit density that has the same terminal velocity as the particle in question. The aerodynamic diameter provides an indication of the inertial properties of the particle that define not only the terminal velocity but also the impaction probabilities. When a stream of air flows around an obstacle in its path, the inertia of the particles contained in the stream causes a resistance to any change in motion. If the particle inertia is sufficiently high as compared to the flow direction change, the particle will not be carried by the air stream around the object but will impact on the object. The efficiency of impaction is defined as the ratio of the number of particles which impact on the object to the number of particles which would have passed through the same area if the object had not been present. The overall collection efficiency of the object will depend on both the impaction efficiency and the fraction of the impacting particles that are retained. Inertial forces are believed to play an important part in the deposition of larger particles from a turbulent medium such as the atmosphere. Small particles or particles of lower density will have lower impaction efficiencies than large or dense particles. Theory predicts that the inertial forces for small particles are so low that collection by direct impact becomes negligible but that collection is enhanced for very small particles by diffusion. Deposition of very small particles may also occur owing to electrostatic, thermal, and other forces.

Particles in the atmosphere will increase in size by agglomeration when the number of particles per unit volume is very high, as in the case of an aerosol generated by volatilization in a limited space. In a freshly formed fume, the particle sizes are very small so that agitation by Brownian movement causes frequent collisions and the formation of larger particles through agglomeration. As the particle size increases, the rate of collision and agglomeration decreases, and the larger particles are removed by gravitational settling or, in an agitated atmosphere, by impaction on surfaces. Such flocs of particles frequently have bulk densities considerably lower than the theoretical density of the parent material. This process places an upper limit of about 10^5 to 10^6 particles/cm^3 on the number of particles which can exist in a "stable" aerosol (Magill, Holden, and Ackley, 1956). Agglomeration rate, shape, and bulk density may be affected by the composition of the fluid in which the process occurs (i.e., the presence of steam can have a marked effect). The agglomeration of radioactive material may be different from the agglomeration of stable material because the radiation field associated with the radioactive material will produce ions that can result in charged particles and thus affect the forces of attraction between particles.

16-2.2 Characteristics of Radioactive Materials

The behavior of radioactive contaminants in the atmosphere is governed by their chemical and physical nature. That is, the dynamics of transport, deposition, and later resuspension will be nearly the same for these materials as for their nonradioactive counterparts. Some reservation is required here since the ionization produced by the radiation from the radioactive particle may change the electrical charge on the particle. Further, in large concentrations, such as could conceivably occur in a major accident, Gifford (1967) has shown that the heat produced by the absorption of the radiation in the air will cause increased rise of the plume.

16-2.2.1 Radioactive Decay. A radioactive atom is characterized by an unstable nucleus that tends to a more stable condition through the emission of a nuclear particle and/or a quantum of energy. If a charged particle is emitted, the resulting atom has a different atomic number and is therefore a different element. Such disintegrations occur according to the laws of chance. The decrease in the number of radioactive atoms in a time period, dt, is therefore

$$dN = -\lambda_R N \, dt \tag{16.1}$$

where N is the number of radioactive atoms at any time t and λ_R is a proportionality constant, called the disintegration constant, with the units of reciprocal time. λ_R is a constant for a given radioactive species and represents the fraction of the atoms present which disintegrate per unit time. If Eq. 16.1 is integrated using the condition that $N = N_0$ at zero time, the familiar exponential radioactive-decay equation results.

$$N = N_0 \exp(-\lambda_R t) \tag{16.2}$$

From Eq. 16.2 it can be shown that one-half of the radionuclei originally present will decay in a time given by

$$T_R = \frac{\ln 2}{\lambda_R} = \frac{0.693}{\lambda_R} \tag{16.3}$$

where T_R is the half-life of the radionuclide and is characteristic of the particular isotope.

In many cases the original radionuclide (parent) decays to a nucleus that is also unstable (daughter). The number of atoms of the daughter at some time after starting with an initial quantity of the parent (N_1^0 atoms) is given by

$$N_2 = N_1^0 \frac{\lambda_{R1}}{\lambda_{R2} - \lambda_{R1}} [\exp(-\lambda_{R1} t) - \exp(-\lambda_{R2} t)] \tag{16.4}$$

where the subscripts 1 and 2 refer to the parent and daughter nuclei, respectively. Equation 16.4 indicates that, if the half-life of the parent is very long with respect to that of the daughter (i.e., if λ_{R1} is small with respect to λ_{R2}), an equilibrium condition is reached after some time, depending on the half-life of the daughter, where the number of atoms of N_2 are such that the disintegration rate of the daughter equals the disintegration rate of the parent. Since the source of supply of the daughter atoms is the disintegration of the parent, the daughter atoms are, on the average, decaying at the same rate at which they are being supplied. Such a condition is known as secular equilibrium between the parent and the daughter,

$$\lambda_{R1} N_1 = \lambda_{R2} N_2 \tag{16.4a}$$

where $\lambda_{R2} \gg \lambda_{R1}$.

If the parent has a longer half-life than the daughter, $\exp(-\lambda_{R2} t)$ becomes negligible in comparison to $\exp(-\lambda_{R1} t)$ after some period of time. Equation 16.4 then reduces to

$$N_2 = N_1^0 \frac{\lambda_{R1}}{\lambda_{R2} - \lambda_{R1}} \exp(-\lambda_{R1} t) = N_1 \frac{\lambda_{R1}}{\lambda_{R2} - \lambda_{R1}} \tag{16.5}$$

If expressed as disintegration rate rather than number of atoms, Eq. 16.5 becomes

$$\lambda_{R2} N_2 = (\lambda_{R1} N_1) \frac{\lambda_{R2}}{\lambda_{R2} - \lambda_{R1}} \tag{16.6}$$

or in terms of activity

$$A_2' = \frac{\lambda_{R2}}{\lambda_{R2} - \lambda_{R1}} A_1' \tag{16.6a}$$

where the activity of the daughter is greater than that of the parent by $\lambda_{R2}/(\lambda_{R2} - \lambda_{R1})$. The condition expressed by Eq. 16.6 is known as transient equilibrium because the ratio of the amount of daughter to the parent is a constant, although the absolute quantities of the two radioisotopes present are continuously decreasing with time at a rate determined by the decay constant of the parent. Note that the disintegration rate of the daughter exceeds that of the parent once this equilibrium is established.

Chains in which radioactive daughters produce radioactive second daughters, third daughters, etc., can extend to many members. In the chain originating with ^{238}U, for example, there are 10 steps before stable ^{206}Pb is reached. Bateman (1910) gave equations for solving such a radioactive chain of any length with the condition that only the parent material is present at time zero. These are summarized as

$$N_n = N_1^0 \, [C_1 \exp(-\lambda_{R1} t) + C_2 \exp(-\lambda_{R2} t) + \ldots + C_n \exp(-\lambda_{Rn} t)] \quad (16.7)$$

$$\text{where } C_1 = \frac{\lambda_{R1}\lambda_{R2} \cdots \lambda_{R(n-1)}}{(\lambda_{R2} - \lambda_{R1})(\lambda_{R3} - \lambda_{R1}) \cdots (\lambda_{Rn} - \lambda_{R1})}$$

$$C_2 = \frac{\lambda_{R1}\lambda_{R2} \cdots \lambda_{R(n-1)}}{(\lambda_{R1} - \lambda_{R2})(\lambda_{R3} - \lambda_{R2}) \cdots (\lambda_{Rn} - \lambda_{R2})}$$

$$C_3 = \frac{\lambda_{R1}\lambda_{R2} \cdots \lambda_{R(n-1)}}{(\lambda_{R1} - \lambda_{R3})(\lambda_{R2} - \lambda_{R3}) \cdots (\lambda_{Rn} - \lambda_{R3})}$$

16-2.2.2 Activation. Radioactive materials can be produced by nuclear reactions caused by bombarding the nucleus with subatomic particles, such as energetic alpha particles, deuterons, or electrons, or with neutrons of all energies, including very low ones. Since a nuclear reactor, which produces power through nuclear reactions, provides a very large number of neutrons during the chain reaction, radioactive materials will be produced in structural members as well as in coolants, gases, or any other material in the vicinity of the reactor core.

Activation reactions in reactors can be caused by thermal neutrons (i.e., neutrons with kinetic energy equivalent to that caused by thermal energies) or by fast neutrons (i.e., neutrons with energy above thermal). Most of the thermal-neutron reactions are of the type in which the nucleus captures a neutron and immediately emits a gamma photon; thus the end result is a nucleus of the same element as the parent with a mass of one mass unit greater (n,γ reaction). With fast neutrons the reactions are more likely to result in a prompt emission of a particle, such as an alpha particle or one or more neutrons.

Such reactions can be written symbolically as, for example, the reaction of a thermal neutron with ^{40}Ar:

$$^{40}\text{Ar} + \text{n} \rightarrow h\nu + {}^{41}\text{Ar} \xrightarrow[1.8\ \text{h}]{\beta^-} {}^{41}\text{K}$$

This expression indicates that an ^{40}Ar nucleus has absorbed a neutron, produced an atom of ^{41}Ar which, then, emitted a quantum ($h\nu$) of gamma radiation. Argon-41 is unstable and decays by emission of a beta particle with a half-life of 1.8 h to a stable ^{41}K. In many cases the nucleus resulting from such a reaction is stable so that no radioactive material results.

The measure of the probability of a nucleus capturing a particle and undergoing a reaction is the activation cross section designated as σ_A. The cross section will vary with the energy of the bombarding particle and is different for each nucleus being bombarded. Cross sections are expressed in units of barns, which have the dimensions of 10^{-24} cm^2.

For a material exposed to a given neutron flux, nv [neutrons/(cm^2·s)], the number of interactions, and thus the net number of atoms of radioactive material produced (dN_R), in time dt is

$$dN_R = (nv)\, \sigma_A N_T \, dt - \lambda_R N_R \, dt \quad (16.8)$$

where N_T is the number of target atoms exposed and λ_R is the radioactive-decay constant. The second term accounts for the number of radioactive atoms lost by radioactive decay. Solution of this equation with the condition that the number of radioactive atoms present at the start of the irradiation is zero gives

$$N_R = \frac{(nv)\, \sigma_A N_T}{\lambda_R} \, [1 - \exp(-\lambda_R t)] \quad (16.9)$$

Thus the number of atoms of radioactive material will increase at a decreasing rate, dependent on the half-life of the radioactive material, until an equilibrium is attained between the rate of formation and the rate of decay. At equilibrium

$$N_R \lambda_R = (nv)\, \sigma_A N_T \quad (16.10)$$

In other words, at equilibrium the rate of decay of the material produced is equal to the rate at which it is formed.

16-2.2.3 Fission. Fission is a special type of nuclear reaction in which the nucleus of the bombarded atom splits into two pieces with the liberation of energy in the form of heat and radiations. Fission can be induced in most elements with an atomic number above about 30 by bombarding with sufficiently energetic particles. Only elements with atomic number above 90 can be fissioned by neutrons of low or moderate energy (<10 MeV) with appreciable cross sections. Some isotopes are fissionable with thermal neutrons (^{235}U, ^{233}U, and ^{239}Pu are most important), whereas others, such as ^{238}U, require neutrons of 1 MeV or greater energy. Of particular interest is the fact that the fission process produces neutrons that in themselves can produce further fissions. Thus it is possible to arrange a mass of fissionable material so that these additional neutrons produce a chain reaction. Because the fission cross section is high for slow neutrons, a thermal reactor is designed to produce further fissions by absorbing much of the energy of the neutrons before they enter the fuel. In a fast reactor the neutrons are utilized with only slight degradation of energy so that the cross sections for fission are lower but the parasitic captures in structural or other materials are also lower. This permits more efficient use of the neutrons and greater freedom in the choice of construction materials.

In the fission process the splitting of the nucleus is asymmetrical so that two fragments of different mass are formed. In ^{235}U, for example, only about 0.01% of the fissions result in a symmetrical division of the nucleus producing two fragments of identical size.

The average frequency of occurrence of each mass chain is referred to as the fission-yield curve and varies to some extent with the energy of the bombarding neutrons and the mass and type of the fissionable nucleus. The total fission yields add to 200% since the basis is the yield per 100 atoms fissioned, i.e., each fission will yield one atom on the lower part of the mass scale and one atom on the higher part.

When ^{239}Pu fissions, the low mass portion of the fission-yield curve is displaced about four mass units from ^{235}U, but the higher mass region is only slightly affected. This has considerable significance in reactor operation because a portion of the power may be produced by ^{239}Pu generated earlier in the cycle from ^{238}U. As a result, the quantities of strontium and other fission products of similar mass number may be lower than would be estimated from ^{235}U fission, and the quantities of ruthenium and fission products on the upper end of the lower mass curve may be higher.

The total amount of energy released per fission is about 200 MeV. Thus the production of 1 W of thermal power in a reactor corresponds to a fission rate of 3.1×10^{10} fissions/s, and a burst of 10^6 joules (1 MW·s) corresponds to a total of 3.1×10^{16} fissions. Since the energy of the explosion of 1 ton of TNT is taken as 4.2 joules (10^9 cal) (Glasstone, 1962), the fission of 1.3×10^{20} atoms will release an amount of energy equivalent to the explosion of 1 ton of TNT.

The formulas for computing the quantities of individual fission products present are similar to those for neutron activation and decay. For the first member of the chain,

$$\lambda_{R1} N_1 = \gamma F \, [1 - \exp(-\lambda_{R1} t)] \exp(-\lambda_{R1} T) \qquad (16.11)$$

where $\lambda_{R1} N_1$ = the disintegration rate of the first member
γ = the fission yield of the chain
F = the fission rate in fissions per second
t = the time over which the fission rate occurs
T = the time since the fission rate stopped

Similarly, for the second member

$$\lambda_{R2} N_2 = \frac{\gamma F \lambda_{R1} \lambda_{R2}}{\lambda_{R2} - \lambda_{R1}} \left\{ \frac{[1 - \exp(-\lambda_{R1} t)]}{\lambda_{R1}} \exp(-\lambda_{R1} T) - \frac{[1 - \exp(-\lambda_{R2} t)]}{\lambda_{R2}} \exp(-\lambda_{R2} T) \right\} \qquad (16.12)$$

where λ_{R1} and λ_{R2} are the decay constants of the first and second members. These equations assume that the nuclide is removed only by radioactive decay. In some cases, particularly in reactors with high neutron fluxes, some fission products with high neutron-capture cross sections may be eliminated by capture and transmutation to another element.

Inspection of Eqs. 16.11 and 16.12 indicates that the short-lived nuclides will soon build to an equilibrium level where the rate of decay is equal to the rate of formation by fission. As in the radioactive-decay equations, the time required to reach such an equilibrium depends on the half-life of the nuclide produced. The longer-lived nuclides, such as ^{90}Sr or ^{137}Cs, will continue to increase with the life of the core as the reactor operates. As a result, the mixture of fission products in a reactor is weighted much more heavily toward the longer-lived materials than are the fission products from a nuclear excursion or a nuclear detonation.

Way and Wigner (1951), by considering the statistics of fission and the resulting fission products, derived a useful rule of thumb for the quantities of radioactive materials produced. In their formulation the beta disintegration rate for times in excess of 10 s following a single fission is given by

$$\beta(t) = 5.2 \times 10^{-6} t_d^{-1.2} \qquad (16.13)$$

where $\beta(t)$ is the number of beta particles emitted per second and t_d is the time since fission in days.

16-2.3 Nuclear Power

The development of nuclear reactors for the production of electricity has provided the United States and other countries with an alternative to fossil fuels for the generation of electricity. The application of methods for calculating dose to populations surrounding such plants has played an important part in deriving emission limits for the plants as well as in assessing their real risk. For orientation a brief discussion on the essentials of nuclear power follows.

16-2.3.1 Fuel Cycle. The term "fuel cycle" has been adopted for nuclear energy work because the operation of a reactor produces nuclides that can fission and, after separation and fabrication into fuel elements, can be cycled back into the reactor to produce more power. The fuel cycle starts with the mining of uranium or, in an alternate cycle not presently used to any extent, thorium. The uranium ores are refined to produce uranium oxide that is converted to uranium hexafluoride, a gaseous compound at slightly elevated temperatures. With this compound an isotopic separation step is used to enrich the uranium in ^{235}U, normally 0.72% of the uranium, to levels of 2 to 6%. The ^{235}U is the isotope that will fission with slow neutrons and produce the chain reaction. The enrichment is necessary to overcome the parasitic loss of neutrons in the present light-water reactors.

After the uranium is enriched, it is again converted to oxide, and fuel pellets are produced and placed in zirconium cylinders to serve as fuel rods in the reactor. Great care is used in assembling these rods and placing them in the reactor because a rupture of the cladding will release fission products to the cooling water. Although a reactor can operate safely with some cladding defects, if the ruptures become too numerous, the reactor must be shut down and the damaged elements removed to avoid spreading radioactive materials through the remainder of the power-generation equipment or to the environment.

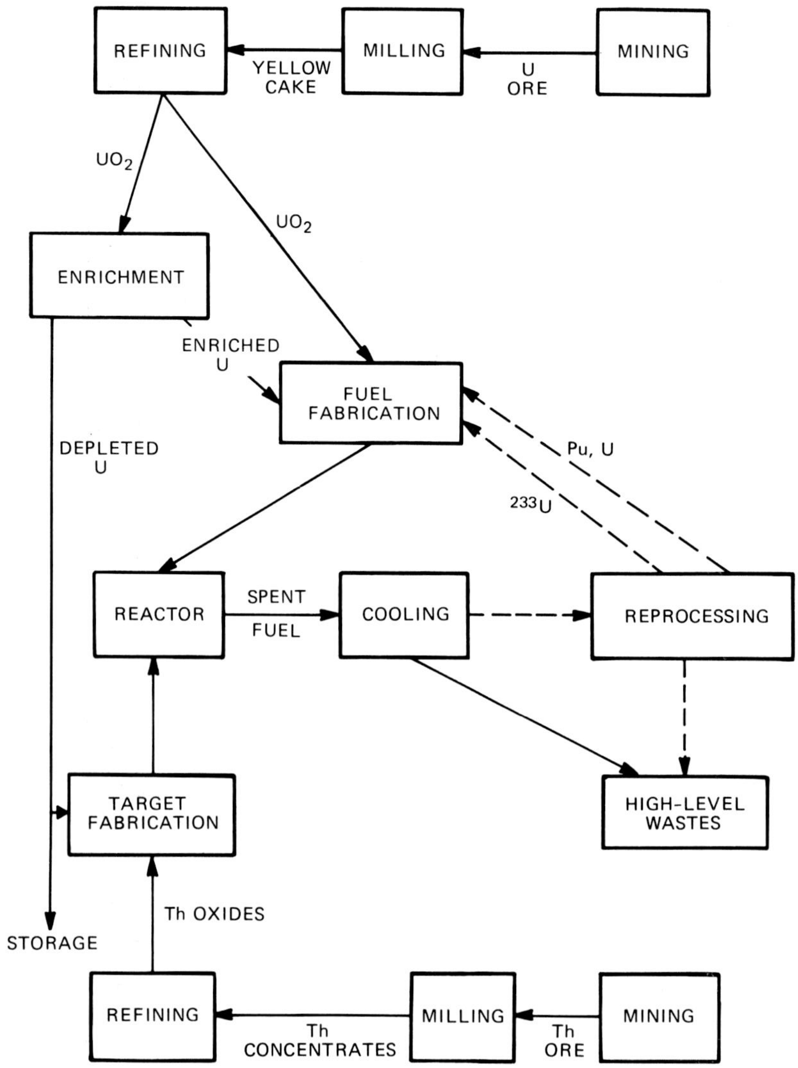

Fig. 16.1 Nuclear fuel cycles.

After the fuel has operated for its useful life, it is removed from the reactor, the zirconium cladding is removed, and the fuel is dissolved for chemical separation. Such separation provides three main streams: the waste products, uranium at an enrichment level lower than in the entering fuel, and plutonium produced by the capture of a neutron in the ^{238}U. Since plutonium is also fissionable, the latter two streams provide a source of fuel that can be recycled to the reactor. At the time of writing, the recovery of unused uranium and plutonium has been banned in the United States, and the spent fuel elements are considered as wastes.

An alternate fuel cycle is to use the plutonium in a fast breeder reactor. By capturing the excess neutrons in ^{238}U to form plutonium, the breeder reactor produces more fuel than it uses. It is operated with fast neutrons that have been little degraded in energy to avoid loss of neutrons by capture in the reactor structure. In a breeder reactor, the energy potential of the uranium can be more completely used, thus increasing by large factors the usefulness of the uranium available. In addition, the tails, or uranium depleted in ^{235}U, from the isotope separation plant can be used to breed more plutonium.

A different fuel cycle starting with thorium is also possible. The irradiation of natural thorium, which consists almost entirely of ^{232}Th, produces ^{233}U, which is also fissionable and can be used to fuel reactors. This cycle has not been applied to any extent in power reactors but is still under investigation. These fuel cycles are shown in Fig. 16.1.

16-2.3.2 Airborne Wastes from Fuel-Element Fabrication. Uranium and thorium are mined in underground operations or in open pits. Since the uranium and thorium and their daughter products are radioactive, some radioactive wastes will be produced at all steps in the chain of producing fuel. Both uranium and thorium have very long half-lives, and thus the mass associated with a given quantity of radioactivity is relatively high. Although this tends to minimize radiation problems with respect to the unirradiated fuel, the large quantities handled and the presence of the relatively short-lived daughter products, particularly in the early stages of refining, provide a potential for airborne release of some radioactive material.

Airborne dusts containing the parent materials and daughters are generated in the operations, but little is released to the atmosphere. The chief problem in the operation is with the daughter-product radon and its decay products, because radon is a gas and is easily expelled to the atmosphere. Extensive ventilation is required to limit the concentration in the mines to tolerable levels.

Uranium mill operations include crushing, grinding, and, in some mills, roasting. All these operations are potential sources of airborne dusts or gaseous daughter products. In most cases equipment to clean the air is provided along with forced ventilation of the areas.

A problem in recent years has been the tailings piles left after the separation of the uranium from the ore. The piles contain all the daughter products of uranium, including radium. Although the tailings piles contain no radioactivity that has not been present naturally, the removal and treatment of the ore have caused an increased mobility in pickup and spread through the atmosphere of the radionuclides and increased availability of the radon gas generated by the decay of radium. In some cases the tailings have been moved to building sites to serve as fill, and the radon gas emitted has caused high accumulations of radon in these buildings.

After the milling operation the concentrates are subjected to further chemical and metallurgical purification to produce uranium compounds. Since the majority of the long-lived daughter products have been removed earlier, airborne materials from these operations consist primarily of dusts or fumes of uranium or thorium generated by mechanical or chemical means. These are controlled by conventional industrial ventilation and air-cleaning equipment. Similarly the fuel-fabrication steps are primarily mechanical, and dusts are essentially the uranium or thorium compounds involved in the process.

The enrichment step is accomplished by converting the uranium to UF_6 (which is gaseous at a slightly elevated temperature) and processing this compound through molecular barriers to concentrate the ^{235}U. The UF_6 is very reactive chemically and, if released, combines with water in the air to produce a cloud of UO_2F_2 and HF. Since impurities can affect the process, care is taken to minimize the possibilities of leaks. Accidental releases are possible, however, in which case the UF_6 would be converted to small particles of uranium compounds and fluoride gases that could disperse downwind. If a release occurs at the product end of the process, the higher concentrations of ^{234}U and ^{235}U will result in more radioactivity per unit mass (relative to natural uranium) and a somewhat greater radiation hazard. In the event of a large release, the chemical toxicity of the fluorine compounds may well be the major immediate hazard.

16-2.3.3 Nuclear Reactors. The nuclear reactor is the energy-producing step in the cycle. Radioactive products that could be released to the atmosphere during reactor operation include the fissionable material and the fission and activation products. The actual releases during normal operation depend on the integrity of the fuel cladding and the cooling system, and wide variations are possible depending on the reactor type and use. For the purpose of this chapter, it is useful to classify reactors by the relation between the cooling system and the potential paths to the atmosphere as direct cycle or indirect cycle. In the direct cycle the coolant circulates to drive the electrical-generating system directly, with any entrained gases discharged through the turbine condenser, frequently to holding systems to allow decay of the short-lived gases. In the indirect cycle the coolant passes through a heat exchanger and produces steam or other heated gas in the secondary loop which drives the energy-recovery system.

Although corrosion and activation products must be considered a potential source for release of radioactive materials during routine operations, the major problem is in maintaining the integrity of the fuel and cladding so that fission products do not escape to the coolant and then to the wastes or the environs. During irradiation structural changes take place in the uranium oxides and cause the fuel to swell, thus producing cracks and breaks in the fuel materials. All these changes tend to place stresses on the fuel cladding so that eventually the cladding may break or lose its integrity. The fuel is designed to operate for a given period with only minimal failures, but even so the escaping fission-product gases may make a contribution to the radioactivity released in the effluents during normal operations.

The direct-cycle plants are best illustrated by the present boiling-water reactors (BWRs). In these plants the steam flows directly from the reactor to the turbine, after which it is condensed and returned to the reactor. A continuous demineralization system is used to remove most of the fission or corrosion products in the coolant. In these direct-cycle plants, some of the gaseous and volatile radioactive materials associated with the steam are carried with the gases from the turbine condenser and exhausted to the atmosphere with holdup to allow decay. Experience has shown that the phase separation at the surface of the boiling water in a BWR is an efficient method of decontaminating the steam for all except the noble gases. The majority of the more troublesome fission products remain in the water system to be removed by the ion exchangers. Even in the event of serious fuel-cladding failures in the present plants, the release to the steam is low and has been controlled by changing the faulty fuel rods, treating the exit air, or lowering the operating temperature of the faulty fuel rods.

In the reactors using the indirect cycle, the primary system is completely contained and any radioactive gases generated are withdrawn for storage and decay and subsequently released. In pressurized-water reactors (PWRs), the volume of noncondensable gases is reduced because the hydrogen and oxygen produced by radiolytic decomposition are removed in a catalytic recombiner. In the reactors that use liquid sodium as a coolant, the capacity of the sodium to dissolve and retain fission products even at the high temperatures involved is quite good. In particular the sodium appears to retain the radioiodines with high efficiency.

In any of these systems, leaks of the primary coolant with their associated dissolved fission products or activation products produce a potential source of airborne particulates and gases that must be removed by the air-cleaning system before being released to the atmosphere. It is difficult to ensure that a water system will remain leakproof at high pressure. Any leakage will result in flashing of the steam with dissolved materials suspended in the air of the buildings.

Accidents in nuclear reactors, as with other devices, are postulated to range over a wide spectrum of consequences, with the more serious accidents of lower probability than the more trivial ones (U. S. Nuclear Regulatory Commission, 1975). In a serious reactor accident, it is the potential release of radioactivity, in particular the fission products, that is of concern. The nuclear reactor is unique among power-generating devices in that the release of energy does not stop immediately on damping of the chain reaction because of the presence of the fission products that provide energy through radioactive decay. The rate of energy release decreases with time after the reactor is shut down, but there is a significant period of time when continued cooling is required, especially for high-power-density reactors, to prevent melting of the fuel and the release of a portion of the fission products to the containment vessel.

Safeguards against fission-product release, or other accident mechanisms, are designed to avoid the accident through redundancy in design and special systems to maintain cooling and to limit the consequences by containment of the radioactive materials. The task of a large number of engineers and specialists is the design and review of such systems.

The most serious accident at a power reactor was that at Three Mile Island, Unit 2, near Harrisburg, Pa., in 1979. Releases to the atmosphere consisted almost entirely of noble gases resulting from fission. The consequences to the public are predicted to be about one possible cancer developing in future years. However, the reactor containment building was heavily contaminated and resulted in large recovery costs.

For general studies of reactor accidents, the estimates of consequences arise from analysis of predicted events following an accident. Meteorological dispersion, deposition, and resuspension estimates play a major role in such analyses.

A major study of reactor accidents is the Rasmussen report (U. S. Nuclear Regulatory Commission, 1975) prepared by a group of engineers and scientists under the direction of Dr. Norman Rasmussen of MIT. This study estimates not only the consequences but also the probability of an accident having given consequences through fault-tree analysis. In general, the conclusions are that major accidents have very low probability.

16-2.3.4 Separation Plants. After discharge from the reactor, the fuel contains the fission products from the fission reaction as well as a large fraction of unused uranium and a significant quantity of plutonium. In the thorium cycle the thorium contains ^{233}U produced by a neutron reaction in ^{232}Th. If the uranium and/or plutonium can be separated from the fission products (which are highly radioactive and provide neutron-absorbing materials that are detrimental to the chain reaction), the uranium or plutonium can be recycled to provide additional fuel for the reactor.

In present-day separation plants, the fuel is dissolved in acid after the cladding has been removed. This solution is then passed through solvent extraction columns to remove the fission products and to separate the uranium and plutonium.

The major source of emission to the atmosphere is during the dissolution step. Here the ^{85}Kr, a noble gas with a 10.76-yr half-life, is released from the fuel. This nuclide emits only a low-energy beta particle with a minimum of accompanying gamma radiation. If cooling times (time between discharge from the reactor and dissolving) are less than 100 to 200 d, then ^{133}Xe will accompany the ^{85}Kr. This nuclide does have significant gamma emission, but such short cooling times have not been in common use.

Other nuclides of interest include ^{129}I, an isotope of iodine with a half-life of 1.6×10^7 yr, of interest because of its possible long-term buildup in the environs; ^{14}C with a half-life of 5730 yr; and ^{3}H with a half-life of 12.4 yr. The ^{14}C and ^{3}H are of interest because of their association with biological processes. If short cooling times are used, ^{131}I with a half-life of 8 d is of interest because of its potential incorporation into milk with transfer to children. Other nuclides, including a general spectrum of the fission products and actinides, can be expected to become airborne in small quantities by resuspension from contaminated surfaces in the plant and to enter the plant ventilation system. Various removal devices are used to minimize the discharge of these nuclides to the atmosphere. The dissolver gases are scrubbed or treated with silver zeolite beds to remove the iodines, and high-efficiency filters are used to remove the particulates. Treatment schemes for the removal of ^{85}Kr have been developed using cryogenic or solution–scrubber techniques.

However, at the present time, reprocessing of fuel is not permitted in the United States because of potential problems with diversion of plutonium and proliferation of nuclear weapons. If reprocessing is resumed in the future, it is possible that different processing schemes will be used to minimize the diversion of fissile material. This, in turn, may mean that any separation plant may be different from the plants used in the past.

16-2.4 Other Sources

The widespread use of radionuclides in research, industry, and medicine has resulted in a variety of potential sources ranging from a large number of users of small quantities of a few radionuclides to a few large laboratories where many different radionuclides are produced and used in large quantities. Each of these installations, at least potentially, can discharge radioactive materials into the atmosphere resulting in a multiple-source problem.

The use of radionuclides and reactors for auxiliary power sources and in propulsion systems has added a new dimension in accident potential because these systems may exist in unusual environments or may operate under unusual stresses. The safety problems are further complicated by the uncertainties in predicting the dispersion of the material (particularly for space missions) and in evaluating the biological effects of the various fuel forms.

Radioactive materials are also dispersed to the atmosphere from processes other than nuclear as a result of natural radioactive materials in the process material. One such process that has gained considerable attention is the burning of coal to generate electricity. Eisenbud and Petrow (1964) pointed out that, in terms of accumulation in the body and radiation dose to organs, the airborne effluent from coal-fired plants could equal or exceed the corresponding doses from nuclear power plants, depending upon the uranium and thorium content of the coal. McBride et al. (1978) analyzed the total radiation doses received over a 50-yr period from a 1% fly-ash release from coal containing 1 ppm of uranium and 2 ppm of thorium along with typical releases from nuclear power plants. They found that the individual bone doses to people in the vicinity of such a release were higher than those from the BWR. The whole-body and lung doses from the coal plant were about the same as those for a PWR but about one-half of those for a BWR. Since the radium content of coal and the amount of fly ash vary, the preceding estimates will not hold for all plants. Miller and Beck (1979) found average concentrations in U. S. coals of 1.7 μg/g of uranium, 4.5 μg/g of thorium, and 1700 μg/g of potassium. Their analyses indicate that the total contribution to dose from the coal fuel cycle is lower than that from the nuclear fuel cycle if both use the best available technology.

16-3 RADIOLOGICAL CONCEPTS

16-3.1 Radiations

The ionizing radiations encountered in radioactive effluents released to the atmosphere are alpha radiation, beta radiation, and gamma radiation. Neutron radiation problems are largely confined to the reactor or to a few sources used for calibration of instruments or various tests.

16-3.1.1 Alpha Radiation. Alpha radiations are particles that are physically identical to the nuclei of helium atoms and are ejected from the nucleus of certain radioactive atoms with kinetic energies of 4 to 10 MeV. Alpha particles have a mass of approximately 4 atomic mass units (6.642×10^{-24} g) and a positive charge of 2, resulting in strong interactions with atoms in the material they encounter. As a result of these interactions, the alpha-particle energy is imparted to the electrons of the atoms of the material traversed with consequent ion-pair production (usually a negative electron and a positive-charged remainder). The alpha particle produces a large number of such ion pairs per unit of track length (30 000 to 100 000 ion pairs per centimeter in air), resulting in a rapid loss of kinetic energy and consequently a short range. Alpha particles emitted from most radioactive materials will travel only 2 to 8 cm in air, depending on their energies. Because alpha particles of these energies will generally be stopped by the inert surface layer of skin, alpha emitters without accompanying beta or gamma radiation present little or no problem of external radiation but can produce serious damage if taken inside the body where a protective layer does not exist.

16-3.1.2 Beta Radiation. Beta radiations are electrons emitted from the nucleus of a radioactive atom with speeds approaching that of light and with energies of 0.02 to 3.2 MeV. Because of their lower mass and charge, interactions between beta particles and the electrons of matter to produce ionization are more infrequent than for alpha particles ($\leqslant$200 ion pairs per centimeter in air). The lower rate of energy loss (compared to alpha particles) permits a longer range, up to several meters in air and several centimeters in tissue. External beta radiation is important because it will penetrate the protective layer of the skin and produce damage to the underlying sensitive tissues if the kinetic energy is above about 100 keV. It will not penetrate to the deeper-seated organs of an animal as large as man, and damage is largely confined to the surface layers of the body, including exposed organs such as the eyes. Beta radiation gives rise to electromagnetic radiation, due to the slowing of the high-speed electrons in material (bremsstrahlung), but generally such radiation is significant only with very intense and energetic beta-ray sources with the electrons absorbed in a high atomic number material. Owing to the limited range of the beta radiation, the dose rate is sensitive to the local concentration, and wide variations can be encountered because of local shielding, pockets developed by increased deposition from eddies, or by later translocation and concentration by runoff from a rain storm.

A similar particle with a positive charge is emitted from some radionuclides. This particle is called a positron and behaves similar to the beta ray in passing through matter. However, when the positron energy is expended, it will combine with an electron to produce an annihilation reaction. In this reaction, both the positron and the electron are converted to 0.51-MeV gamma photons. These gamma rays are a mark of the presence of positrons, and the reaction itself is a classic demonstration of the equivalence of energy and matter.

16-3.1.3 Gamma and X Radiations. Gamma radiations are electromagnetic rays emitted from the nucleus of radioactive atoms. They are indistinguishable from X rays produced by rapid slowing down of electrons or from photons produced from orbital-electron shifts, and both X rays and gamma rays are included in dose calculations. Interactions of gamma photons passing through matter are less frequent than for charged particles, and thus their penetrating ability is much greater. Such interactions occur at random with the drop-off in intensity of the primary beam following an exponential relationship complicated by reentry of some scattered components back into the beam. The relaxation length (distance to decrease the intensity by a factor of 1/e) for typical gamma rays (~0.7 MeV) is on the order of 100 m for air, 50 cm for water, and 5 cm for lead. As a result the gamma-ray dose at any location is a function of the distribution of the radioactive material over a relatively large area or volume.

Gamma radiation is absorbed in matter by three different reactions, with the relative importance of each reaction dependent on the energy of the gamma radiation and the atomic number of the absorbing material. For low energy photons and low atomic-number materials, the absorption is primarily by photoelectric effect. In this reaction the total energy of the gamma photon is absorbed by one of the electrons in the atoms of an absorber, and the electron is ejected with an energy equal to the energy of the gamma ray minus the binding energy of the electron. The energy of the electron is dissipated by ionization in the surrounding medium. At moderate energies the gamma photon undergoes a reaction called the Compton effect. Here the photon collides with a bound electron and undergoes an elastic collision. A portion of the energy passes to the electron where it is again dissipated. However, the remaining fraction remains with the photon, which leaves the scene of the collision at some angle compared to its initial direction. The fraction of the energy taken by the electron and the photon depends on the angle of collision and the angle of scatter. The third reaction has a threshold energy of 1.02 MeV and consists of the photon converting to a positron and an electron in the force field of a nucleus. The threshold arises from the mass-energy equivalence of an electron or positron of 0.51 MeV. Any photon energy above the threshold is transferred as kinetic energy of the particles.

16-3.2 Radiation Units

A number of specialized units for expressing ionizing radiation have been derived. Since the damage from such radiation is primarily due to the release of energy in tissue, many of the radiation units are closely related to energy deposition.

16-3.2.1 Rad. The rad is the unit used to express the absorbed dose of ionizing radiation and is defined as the deposition of 100 ergs/g of any absorbing material (International Commission on Radiological Units and Measurements, 1962). The amount of energy delivered to a body will depend on the nature and degree of the interaction between the radiation and the body irradiated. Thus it is possible to have different absorbed doses in different materials subjected to the same radiation flux as, for example, when the same masses of lead and tissue are exposed to a gamma field.

The International Standards Organization (ISO) has defined a new unit of absorbed dose, the "gray" (International Commission on Radiological Protection, 1972). This unit represents the absorption of 1 J/kg, which is equivalent to 100 rads.

16-3.2.2 Roentgen. The roentgen (R) is the unit of exposure. The terms "exposure" and "exposure rate" have a special meaning in radiation use or protection. The exposure applies only to X or gamma radiation and is defined as the quotient of dQ/dm, where dQ is the absolute value of the total charge of the ions of one sign produced in air when all of the electrons liberated by photons in a volume element of air having mass of dm are completely stopped in air (International Commission on Radiological Units and Measurements, 1971). The unit of exposure is given by

$$1\,\mathrm{R} = 2.58 \times 10^{-4}\ \mathrm{coulomb/kg}$$

Since the roentgen is a measure of the interaction of gamma radiation and air, the absorbed dose (in rads) will vary in different materials from the same exposure (in roentgens). With moderate-energy gamma rays (0.2 to 3 MeV), an exposure of 1 R will result in an absorbed dose in muscle of about 0.87 rad.

The use of the word exposure as a special unit has the possibility of confusion since it has a narrow, defined meaning, as given previously, in addition to a more general meaning in the sense of being exposed to a condition. Exposure is used in both senses here, with the meaning intended to come from the context.

16-3.2.3 Dose Equivalent. All radiations do not produce identical biological effects for a given amount of energy delivered to tissue (i.e., for equal absorbed doses). In general, radiations with high specific ionization along their tracks will produce a greater effect, but the quantitative degree of difference will depend on the biological change chosen for study. The relative biological effectiveness (RBE) is used to compare the effectiveness of absorbed doses of radiation delivered in different ways. The RBE should be used only in radiobiological work, and the quality factor (QF) should be used for the linear-energy-transfer (LET) dependent factor for radiation-protection purposes (International Commission on Radiological Units and Measurements, 1962). Where appropriate, provision is also made for other factors, such as the distribution factor. The dose equivalent (DE) is now defined as the product of the absorbed dose, the quality factor, the dose distribution factor, and other necessary modifying factors. The concept of quality factor has a limited usefulness since the biological effectiveness depends on many factors, including the type and degree of biological damage taken as an end point, the absorbed dose rate, the fractionation of the dose, and the conditions of delivery. It is not possible to determine an accurate value of the quality factor for one given radiation as compared to another which will describe all effects, such as lethality, carcinogenic action, and damage to specific tissues.

The dose equivalent is numerically equal to the product of the dose in rads and an agreed-on conventional value of the quality factor which describes the effects of concern. The standard of comparison is X or gamma radiation having a linear energy transfer in water of 3.5 keV/μm and delivered at a rate of about 10 rads/min.

The unit of dose equivalent is the rem. In addition to the uncertainties introduced by the concept of the quality factor, the rem assumes values of the quality factor arrived at by convention rather than by direct measurement. Such values are normally chosen on the basis of long-term effects that may occur over the lifetime of the individual and are therefore applicable to long-term effects rather than the more pronounced immediate effects from a large absorbed dose. Values of QF as a function of LET are available. For general use in radiation protection, the values in Table 16.1 are recommended (National Council on Radiation Protection and Measurements, 1971). However, these values should not be used for immediate effects from high doses.

TABLE 16.1 Practical Quality Factors

Internal radiation	QF
X rays, gamma rays, electrons or positrons, energy >0.03 MeV	1
Electrons or positrons, energy <0.03 MeV	1
Neutrons, energy <10 keV	3
Neutrons, energy >10 keV	10
Protons	1–10
Alpha particles	1–20
Fission fragments and recoil nuclei	20

The ISO has also defined a new unit for the dose equivalent, the sievert, which is equal to 100 rems (International Commission on Radiological Protection, 1977).

In the case of exposure to mixed radiations, the dose equivalent is assumed to be equal to the sum of the dose equivalents (in rems) of each of the radiations involved.

16-3.2.4 ***Dose Commitment.*** The dose commitment is a concept frequently used in estimating exposure to radionuclides. As its name indicates, it is the total dose that an individual or population will receive over a defined period of time as a result of an exposure to, or the release of, a given radionuclide. For example, an annual lifetime dose-commitment value would refer to the total dose that an individual would receive when the doses from the quantity of radionuclide taken into the body in 1 yr are integrated over his lifetime. For short-lived radionuclides, with half-lives much less than 1 yr, the annual dose commitment approximates the annual dose. However, for long-lived radionuclides that are well retained in the body, the annual dose is only a fraction of the annual lifetime dose commitment. For example, if we assume that a radionuclide with a very long half-life is retained completely in the body for a lifetime period of 50 yr (frequently assumed for occupational exposure), then the annual dose that can be accumulated in any 1 yr is one-fiftieth of the annual dose commitment.

The population dose commitment is similar to the dose commitment except that the total dose to all members of the public following a release of radionuclides is summed for a defined time. The environmental dose commitment is similar to the population dose commitment except that the doses are summed over the half-life of the radionuclide. For very long-lived radionuclides, this can involve integrations over thousands to millions of years, a difficult and very uncertain task.

16-3.2.5 ***Curie.*** The curie (Ci) is the unit of activity. The activity is given by $\Delta N/\Delta T$, where ΔN is the number of nuclear transformations that occur in the quantity of material of concern in time Δt. One curie is that quantity of a radioactive nuclide in which the number of disintegrations (transmutations) per second is exactly 3.7×10^{10}.

The curie is related to the mass of a radioactive material through the half-life and the number of atoms per unit mass. The number of disintegrations per unit time can be obtained from Eq. 16.1. The mass of material associated with 1 Ci varies widely. For example, 1 Ci of ^{24}Na (half-life of 14.8 h) weighs 10^{-15} g, 1 Ci of ^{226}Ra (half-life of 1622 yr) weighs 1 g, and 1 Ci of ^{238}U (half-life of 4.5×10^9 yr) weighs 3×10^6 g. The relation between the curies and the mass of a specific radionuclide is given by

$$q = \frac{\lambda_R M_R N_0}{A(3.7 \times 10^{10})} \qquad (16.14)$$

where q = number of curies

λ_R = disintegration constant (fraction of the atoms that decay per second) $= 0.693/T_{1/2}$

M_R = mass of radionuclide, g

N_0 = Avogadro's number: 6.03×10^{23} atoms/(g·mol)

A = atomic weight of the radionuclide

Frequently there is some confusion about the use of the term "curie" when it is applied to mixtures of radionuclides. In this chapter the term will be used to describe the *total* disintegration rate of the mixture, that is, the *sum* of the activities contributed by the specific radionuclides of the mixture when more than one radionuclide is involved. Many radionuclides emit more than one gamma ray per disintegration or a beta particle and one or more gamma rays per disintegration, and care must be exercised when determining the radiation flux from a source (in curies) or when determining the source (in curies) that produces a given flux.

The relationship between the activity (Ci) and the radiation dose rate (rads/s) is dependent on the energy of the radiation emitted, the type of radiation emitted, the geometrical pattern between the radioactive material and the receptor, and the amount of absorbing material between the radioactive material and the receptor.

The ISO has recently defined a new unit of activity, the becquerel (Bq), which is 1 dis/s (International Commission on Radiological Protection, 1977). This is in accord with the ISO principle of expressing all units on a common basis using fundamental terms.

16-3.3 Radiation Measurements

Radiation measurements fall into two main categories: measurement of radiation fields and measurement of activity. In the first category the objective is to obtain indications of the dose rate or total dose that is potentially or actually received by people. In the second category the objective is to measure the amount of radioactive contaminant present so that potential doses from radionuclides taken into the body (internal emitters) can be estimated. The instruments used in each case have similar operating principles, although the actual configuration may differ, depending on the purpose.

In measurements of radioactivity the concern is primarily with the rate of emission of radiation with the actual weight of the radioactive material of little or no interest. Thus the proper unit for expressing the results is in curies since these values can be compared with the standards. The sensitivity of such measurements, as related to the appropriate standards for radioactive materials, is generally much greater than for nonradioactive materials, and the measurements can usually be made with greater ease. The ease of measurement provides a basis for controlling radioactive pollutants, which is excellent compared to most nonradioactive pollutants.

Although a number of different instruments are available for measuring radiation, we will consider the broad classes of ionization chambers, particle counters, spectroscopy equipment, photographic film, and thermoluminescent dosimeters (TLD's).

An ionization chamber in its simplest form consists of a pair of charged electrodes that collect ions formed within their electrical field. Most of the ionization occurring in the gas is the result of secondary interactions in the chamber walls; consequently the chamber response is governed by the wall material. Properly designed and calibrated chambers are usually used for dose or dose-rate measurements because they provide an indirect measure of the energy deposited in the chamber.

Counters are designed to record the passage of a single particle, whether primary or secondary, through the sensitive volume of the counter. Gas-filled counters may simply collect the ionization produced by the radiation and amplify it to produce a pulse; internal amplification may occur by the acceleration of the electrons and their subsequent collisions with molecules in the gas to produce a pulse proportional to, but greater than, the initial pulse (proportional counter); or a large amplification may occur by providing conditions such that the pulse spreads through the entire sensitive volume and causes a complete discharge between the electrodes (Geiger–Mueller counter). A scintillation counter registers electrical pulses from a phototube. These pulses are initiated by the small flash of light produced by the interaction of the radiation with the phosphor or crystals in the vicinity of the tube.

Counters are usually used for determining the activity present by measuring the number of particles or photons emitted by the radioactive material. Scintillation or proportional counters can be used to determine the dose or dose rates if the equipment is calibrated for the energy and type of radiation. Geiger counters are frequently used as radiation detectors because of their sensitivity and can be used to measure the dose rate if they are calibrated for the energy and type of radiation to be measured.

Spectroscopy measurements provide a measure of the energy of the radiation as well as the quantity. For gamma radiation, which is emitted in discrete energies during the decay of a nuclide, such information provides a means of identification of the radionuclide and the activity. Detectors are thallium-activated sodium iodide crystals or intrinsic detectors, such as a drifted germanium diode. In general, the intrinsic detectors give better resolution of the individual photons than the sodium iodide, but the sodium iodide crystal can be made larger and more sensitive. For alpha particles the separation of energy can be accomplished with liquid scintillation counting, but the most precise work is done with separated samples of the alpha emitter and an intrinsic detector. This system is widely used in radiochemistry and permits a direct measure of the chemical yield by using a tracer isotope that decays at a different alpha energy than the one of interest. In all these systems, electronic equipment to sort and register the pulses by sizes is needed. In fact, the measurement of gamma spectra with a number of different energies requires the use of a minicomputer for efficient operation.

Photographic film will darken on exposure to radiation, and such films have been widely used to indicate the individual dose received by workers. At the present time a large number of measurements of worker exposure are made using TLD's. These devices store dislocations caused by the interaction of the radiation in the crystal structure. Heating relieves these dislocations, resulting in a light that can be measured.

There are a number of publications on measurement techniques. A recent publication that is devoted particularly to the measurement of environmental radiation has been issued by the National Committee on Radiation Protection and Measurements (1976).

16-3.4 Effects of Radiation

The effects of a given dose of radiation depend on a large number of factors, including:

—The type of radiation
—The penetrating power of the radiation
—The magnitude of the absorbed dose
—The rate at which the dose is delivered
—The radiosensitivity of the organism for the effect involved
—The portion of the organism irradiated and the specific organs involved

The qualitative aspects of radiation effects are well known from past experience with human exposure and from extensive studies with animals, but the quantitative relationship between a dose delivered under given conditions and the resulting biological change is not as well defined for man. Only a few of the effects will be discussed here to illustrate the magnitudes of doses involved.

A massive exposure delivered in a relatively short time can produce immediate effects, which in their most severe form can cause the death of the organism. It is estimated that an X- or gamma-radiation dose of 400 to 500 rads delivered to a major portion of the human body in a short time interval will result in the death of about 50% of those persons so exposed within a few months (Braestrup and Wyckoff, 1958). Most mammals have about the same degree of sensitivity to the dose required to give 50% lethality in 30 d (LD 50), ranging from perhaps 275 rads for swine to 800 rads for the rabbit (Patt and Brues, 1954). Sublethal exposure of humans to above 100 to 200 rads will produce nausea, vomiting, diarrhea, and malaise with the symptoms developing earlier for the higher exposures (National Council on Radiation Protection and Measurements, 1974). A loss of hair might occur about 2 weeks after an exposure of greater than 300 rads (Nickson and Bane, 1959).

Skin reactions will develop if the dose is sufficiently high. Such effects are of primary importance in exposure to beta radiation or low-energy gamma radiation which can deliver very high doses to the skin with comparatively small doses to the remainder of the body. Doses of 400 to 1500 rads to the skin will result in an erythema (reddening) in several days and possible loss of hair in about 2 weeks (Nickson and Bane, 1959). The hair will return, but there is a possibility that it will be gray. At the higher end of this dose range, there may be dry scaling of the surface layers in about 3 weeks. Above 1500 rads an erythema will develop, and raw moist areas will form. Above 5000 to 7000 rads these areas may develop into areas that heal very slowly or not at all.

Late effects of radiation exposure are those which occur some time after the exposure or as a result of continued low-level exposure over a period of time. These effects are detected by a statistical review of the incidence of the effect in a population exposed compared to a similar population not exposed. That is, these effects are not an invariable result of exposure, as with the acute effects, but rather express themselves as an increase in probability of the individual's contracting a specific ailment. The degree of increase in probability depends on the radiation dose, the time of exposure, the organs exposed, and other parameters.

The increased incidence of certain malignancies with radiation exposure has been well documented both in experimental animals and in man. The ingestion of radium by dial painters in New Jersey between 1916 and 1925 claimed 41 victims over a period of about 20 yr (Furth and Lorenz, 1954). Bone sarcomas developed in individuals, a few of whom had over 20 μg of radium in their bodies. The increased incidence of leukemia among the Japanese surviving the nuclear detonations over Hiroshima and Nagasaki (Wald, 1958) and among certain patients treated with radiation (Court-Brown and Doll, 1957) has indicated that an increased incidence of leukemia can be produced by single exposures on the order of 100 rads. Similar results have been noted in a study of American radiologists who received radiation over a period of time during the days before the institution of present protection methods (March, 1950). Similarly, numerous skin cancers have developed among early workers with X rays where the majority of the exposure was to the hands. Direct evidence on the production of malignancies by lower doses is not available.

Both the United Nations Scientific Committee on the Effects of Atomic Radiation (UNSCEAR) (United Nations, 1972, 1977) and the Advisory Committee on the Biological Effects of Ionizing Radiations of the National Academy of Sciences (BEIR) (National Academy of Sciences–National Research Council, 1972) have prepared reviews of the long-term effects of radiations. The BEIR report, in particular, produced numerical estimates of the numbers of cancers produced by a given dose of radiation assuming a linear relationship between dose and cancer production, a direct relation between dose to a particular critical tissue and cancer induction, and no effect of the rate at which radiation was delivered. These numbers were qualified to be maximum values but are frequently used as actual cancers. In the Rasmussen study (U. S. Nuclear Regulatory Commission, 1975) of reactor accidents, factors were included on these estimates to provide for a lower effect at lower dose rates.

Radiation is also a mutagenic agent. Such agents can produce changes in the germ cells which can be passed to children and may produce abnormalities in future generations (World Health Organization, 1957). These effects will not appear in the person irradiated but may result in harmful changes in the future populations. Since such changes result from intermixing of the genes, an important criterion (in addition to the exposure to the individual) is the number of individuals exposed. The BEIR Committee (National Academy of Sciences–National Research Council, 1972) also provided estimates of the numbers of mutations that would be produced by irradiation of the gonads using the same assumptions as for the carcinogenic effect.

A third form of possible damage from radiation is that which can occur to the fetus. Particularly in the early stages of development, the fetal cells are in the process of dividing to produce the various structures of the body, and damage to these cells can produce abnormalities in the infant.

Another effect that occurs later is cataract induction. Cataracts have been noted from single clinical exposures of over 175 rads (Merriam and Focht, 1958). With the radiation delivered over a period of 3 months, the lowest exposure associated with a cataract was 550 rads. Life shortening from radiation exposure has been noted in animals where no specific injury was responsible for the death. Evidence for such an effect in man has been presented for the early radiologists (Warren, 1956) but has been questioned owing to the difficulty of obtaining an adequate control group for comparison (Lewis, 1957).

16-3.5 Background Radiation

Ionizing radiation of all types has been a part of man's environment from his beginning. Neutrons, as well as gamma and electron radiation, result from cosmic radiation; external gamma radiation along with alpha-, beta-, and gamma-emitting internal emitters result from primordial radionuclides and their daughters; and a small contribution results from the production of radioactivity in the lighter elements (beryllium through chlorine) from the nuclear reactions of the cosmic rays primarily with the atmosphere. In addition to providing a continual radiation dose to man, these background radiations will register on radiation-detection instruments and will require subtraction of this component if only the quantities of man-made radionuclides or their radiations are desired. The materials presented below are largely derived from a report by the National Council on Radiation Protection and Measurements (1975).

Cosmic rays result both from the sun and from the intergalactic radiation incident on the solar system. Since their intensity is affected by the magnetic fields of the sun and the earth, there is a variation with time and a major variation with latitude. When these high-energy particles strike the atmosphere, they are absorbed in nuclei, producing secondary radiations emitted from the struck nucleus. This results in an increase in dose at higher elevations because of the unattenuated secondary radiations. The dose rate to people at sea level is about 29 mrems/yr outdoors; whereas at Denver, Colo. (1600-m elevation), the dose is about 50 mrems/yr, and at Leadville, Colo. (3200-m elevation), the dose rate is 125 mrems/yr. Aircraft travel will add a small additional dose. (This is small because of the time at altitude.) It is estimated that a jet flight of 5-h duration at an altitude of 12 km will result in about 2.5 mrems to the whole body. However, a 10-h polar flight will result in about 10 mrems due to the higher cosmic-ray intensities in the higher latitudes and the longer flight time.

The primordial radioactive nuclides result from long-lived radioactive elements that were formed early in the history of the earth and have not had time to decay. The nuclides that make the most important contribution to the dose to man are ^{40}K, ^{238}U and its daughters, ^{235}U and its daughters, and ^{232}Th and its daughters. The heavy-element chains are given in Tables 16.2 to 16.4, which show the daughter products and their half-lives and radiations. Potassium-40 is an essential element for the body, and its concentration in various tissues is biologically controlled to the quantity needed. Although the dose to various organs is different, in general the dose rate contributed by ^{40}K ranges from 6 mrems/yr in cortical bone to 19 mrems/yr in soft tissue. In addition, the gamma radiation from ^{40}K in rocks and soils can result in an external radiation averaging about 8 mrems/yr, with variation possible across the country depending on the local concentrations in the vicinity.

External dose rates from these nuclides vary widely depending on the concentrations of these primordial nuclides in rocks and soil. Thus, in the Atlantic and Gulf Coastal Plains, the average level is given as 15 to 35 mrads/yr (taken for this purpose as milliroentgens per year), with an average of about 23 mrads/yr. For middle America the range is from 35 to 75 mrads/yr, with an average of 46 mrads/yr. However, in the Front Range area at Denver, Colo., the range is 75 to 140 mrads/yr, with an average of 90 mrads/yr. Higher levels exist elsewhere in the world. For example, in the monazite sand regions of Kerala, India, about 16 000 to 17 000 people are likely to be receiving an external dose exceeding 500 mrems/yr and several hundred people could be receiving doses of 2000 mrems/yr or greater (Gopal-Ayengar et al., 1972). Deposits of monazite sands also occur in Brazil with elevated radiation levels (Eisenbud, 1972).

This summary of background levels is intended to show the variations that can occur. Internal doses were largely omitted, although they can be significant for some organs. The small contributions from nuclides produced by cosmic rays or the contributions from fallout from nuclear tests were not included. However, the information does indicate

TABLE 16.2 Thorium Chain

Isotope	$T_{1/2}$	Daughter	Principal radiations,* MeV		
			Alpha†	Beta‡	Gamma‡
^{232}Th	1.405×10^{10} yr	^{228}Ra	4.00		
^{228}Ra	5.75 yr	^{228}Ac		0.010	
^{228}Ac	6.13 h	^{228}Th		0.376	0.915
^{228}Th	1.9131 yr	^{224}Ra	5.40		0.002
^{224}Ra	3.66 d	^{220}Rn	5.67		0.009
^{220}Rn	55.6 s	^{216}Po	6.29		
^{216}Po	0.15 s	^{212}Pb	6.78		
^{212}Pb	10.64 hr	^{212}Bi		0.100	0.117
^{212}Bi	60.55 min	^{208}Tl	6.04 (35.93%)		0.003
		^{212}Po	(64.07%)	0.717	0.281
^{208}Tl	3.07 min	^{208}Pb		0.561	2.37
^{212}Po	3.05×10^{-7} s	^{208}Pb (Stable)	8.78		

*Branches <1% not included.
†Average energy.
‡Total energy emitted per disintegration.

TABLE 16.3 Uranium Chain

Isotope	$T_{1/2}$	Daughter	Principal radiations,* MeV		
			Alpha†	Beta‡	Gamma‡
^{238}U	4.468 x 10^9 yr	^{234}Th	4.20		
^{234}Th	24.10 d	^{234}Pa		0.044	0.008
^{234}Pa	1.17 min	^{234}U		0.82	0.011
^{234}U	2.445 x 10^5 yr	^{230}Th	4.76		
^{230}Th	7.7 x 10^4 yr	^{226}Ra	4.66		
^{226}Ra	1600 yr	^{222}Rn	4.77		0.006
^{222}Rn	3.8235 d	^{218}Po	5.49		
^{218}Po	3.05 min	^{214}Pb	6.00		
^{214}Pb	26.8 min	^{214}Bi		0.22	0.23
^{214}Bi	19.9 min	^{214}Po		0.63	2.03
^{214}Po	1.64 x 10^{-4} s	^{210}Pb	7.69		
^{210}Pb	22.3 yr	^{210}Bi		0.007	0.002
^{210}Bi	5.012 d	^{210}Po		0.39	
^{210}Po	138.378 d	^{206}Pb (Stable)	5.31		

*Branches <1% not included.
†Average energy.
‡Total energy emitted per disintegration.

TABLE 16.4 Actinium Chain

Isotope	$T_{1/2}$	Daughter	Principal radiations,* MeV		
			Alpha†	Beta‡	Gamma‡
^{235}U	7.1 x 10^8 yr	^{231}Th	4.28		0.140
^{231}Th	25.52 h	^{231}Pa		0.078	0.017
^{231}Pa	3.248 x 10^4 yr	^{227}Ac	4.92		0.030
^{227}Ac	21.773 yr	^{223}Fr	4.19 (1.38%)		
		^{227}Th	(98.62%)	0.010	
^{227}Th	18.718 d	^{223}Ra	5.67		0.010
^{223}Fr	21.8 min	^{223}Ra		0.345	0.051
^{223}Ra	11.434 d	^{219}Rn	5.70		0.087
^{219}Rn	3.96 s	^{215}Po	6.81		0.054
^{215}Po	1.780 x 10^{-3} s	^{211}Pb	7.39		
^{211}Pb	36.1 m	^{211}Bi		0.453	0.050
^{211}Bi	2.13 m	^{207}Tl	5.55		0.043
^{207}Tl	4.77 m	^{207}Pb (Stable)		0.493	0.002

*Branches <1% not included.
†Average energy.
‡Total energy emitted per disintegration.

that some populations receive much higher doses than others and that background radiation must be considered when the significance of small additional doses is being assessed. Details on these and other doses in the United States are given in a report by the National Council on Radiation Protection and Measurements (1975a).

16-3.6 Radiation Limits

Shortly after the discovery of X rays in 1895, evidence of the harmful effects of indiscriminate exposure to radiation began to accumulate, and recommendations for limitations of exposure were made. In 1928 the ICRP was organized under the auspices of the Second International Congress on Radiology. In 1929 the Advisory Committee on X-Ray and Radium Protection was organized to develop recommendations concerning radiation protection in the United States. After World War II, the Advisory Committee was reorganized as the National Committee on Radiation Protection (NCRP) with representatives from professional societies and governmental agencies, as well as individual experts. The NCRP is now known as the National Council on Radiation Protection and Measurements. The recommendations developed by the group have served as the basis for most radiation-protection programs and, later, for rules and codes adopted by the various regulatory agencies in the United States. In 1959 the Federal Radiation Council (FRC) was formed to advise the President on radiation matters and to provide guidance for all federal agencies in the formulation of standards of protection against radiation damage. These responsibilities were assigned to the Environmental Protection Agency when the government was reorganized in 1970.

Each group has written lengthy reports that include a discussion of the factors taken into consideration in arriving at the recommendations and limits. These reports must be read fully and carefully to understand the recommendations as they were intended by the authors. Only deliberate study of the complete reports can provide the basis for judgment on the adequacy of a program on radiation protection. When limits are enumerated, they are intended as values for application in the light of existing conditions and not as design criteria for an apparatus or activity. All groups have stressed the fact that there are multiple sources and modes of exposure and have assumed a linear relationship between dose and effect. All groups have recommended keeping exposure to the minimum practical and have called for a balance of risk and benefit to justify those exposures which cannot be avoided. A balance between benefit and risk is difficult because they cannot be adequately reduced to a common denominator and considerable uncertainty is involved on both sides of the "scale." In the final analysis, professional judgment is required, and the basis for the judgment should be sound.

16-3.6.1 Occupational Exposure. The original recommendations of the NCRP and the ICRP were aimed at controlling the exposures of individuals working with radiation and were primarily based on effects that had been observed in early radiation workers. After World War II, a large body of information became available as a result of the animal experiments performed in connection with the Manhattan Project. Review of these data, plus continued review of the extensive work since that time, has led to the present recommendations for the control of occupational exposure.

The NCRP (National Council on Radiation Protection and Measurements, 1971) currently provides a prospective annual limit for whole-body occupational exposure of 5 rems in any 1 yr. The critical organs for purposes of this limit are the gonads, the lens of the eye, and red bone marrow. This is the limit to be used in planning work. However, there will be occasions when the actual dose exceeds this value, and a retrospective limit is given that indicates that no deviation from sound protection is implied if the retrospective dose equivalent does not exceed 10 to 12 rems for increments well distributed over time or even 15 rems for exceptionally well-distributed increments. Long-term accumulation should not exceed 5 rems multiplied by the number of years beyond age 18. Maximum permissible doses (MPD's) of 15 rems in any 1 yr to unlimited areas of the skin and 75 rems to the hands, of which not more than 25 rems shall be received in any one quarter, are included. The forearms are limited to 30 rems in any 1 yr, with no more than 10 rems in any one quarter. The MPD for all other organs is 15 rems in any 1 yr.

Recently the ICRP (International Commission on Radiological Protection, 1977) proposed an occupational limit of 5 rems/yr for whole-body radiation, with the dose to individual organs limited according to the relative risk of irradiating these organs to a level where the total risk does not exceed the risk presented by 5 rems/yr whole-body dose. For the individual organ the limit is obtained by the use of weighting factors applied to the dose to each organ or tissue, with the sum of the product of the weighting factor and the dose not to exceed 5 rems/yr. The values of the weighting factors, W_t, for individual organs are given in Table 16.5. A weighting factor of 0.06 is used for each of

TABLE 16.5 Weighting Factors for Individual Organs or Tissues in ICRP Dose Limit System

Tissue	Weighting factor
Gonads	0.25
Breast	0.15
Red bone marrow	0.12
Lung	0.12
Thyroid	0.03
Bone surfaces	0.03
Remainder	0.30

the remaining five organs or tissues receiving the highest dose, and the exposure of all remaining tissues can be neglected. Note that the stomach, small intestine, upper large intestine, and lower large intestine are treated as four different organs.

Regulations for the occupational exposure of nuclear energy workers have been derived from past recommendations of the NCRP (National Committee on Radiation Protection and Measurements, 1958), the ICRP (International Commission on Radiological Protection, 1965), and the FRC (Federal Radiation Council, 1960), with adaptations to make the recommendations suitable for regulations. The regulations (U. S. Atomic Energy Commission, 1975) limit occupational doses to 1.25 rems per calendar quarter to the whole body, head and trunk, active blood-forming organs, lens of the eye, or gonads; to 18¾ rems per calendar quarter to hands and forearms or feet and ankles; and 7.5 rems per calendar quarter to the skin of the whole body. In addition, the total dose to a given age, N, cannot exceed 5(N−18) rems. However, such regulations can change, and therefore the latest rules should be consulted.

Limitations for internal emitters have been derived by the NCRP (National Committee on Radiation Protection and Measurements, 1959) and the ICRP (International Commission on Radiological Protection, 1960) based on the radiation dose delivered to specific organs or, in the case of bone, on a biological comparison with the effects of radium (Langham and Healy, 1973; Healy, 1975). In the derivation of limits for internal emitters, the maximum permissible quantity in an organ is calculated from the organ size; the energy release in the organ; and, where appropriate, the distribution of the nuclide in the organ. This quantity of radioactive material provides a secondary limitation because it is derived from the radiation actually received by the organ. As guides to the conditions that could result in exposure of the organ to the MPD, tertiary numbers called maximum permissible concentrations (MPC's) of the radionuclide in air and water are derived by considering the uptake and translocation of the nuclide in the various organs. Such numbers are not applicable to environmental concentrations because of the many pathways through food chains that can result in much greater doses than the direct use of the air or water.

The current situation on these internal emitters is uncertain because of increased knowledge on the behavior of the nuclides in the body and improved models for estimating this distribution, particularly following inhalation (International Commission on Radiological Protection, 1966, 1972). The ICRP is currently reestimating the maximum permissible intake for the major nuclides using the dose to the bone surface for bone-seekers rather than the previous comparison with radium (International Commission on Radiological Protection, 1977). The NCRP is also reevaluating its position on internal emitters, and changes can be expected from both of these efforts.

16-3.6.2 Exposure of Populations. Individuals not working with radiation can be exposed in a variety of ways. These include the radiation from cosmic rays or natural radioactive materials in the earth, medical or dental uses of X rays, television tubes, radium dials, fallout from nuclear detonations, waste disposal from nuclear power plants, accidents involving radiation sources, etc. An estimate of the magnitude of some of these sources to the population as a whole is given in Table 16.6 (Environmental Protection Agency, 1976). The person-rem is a cumulative measure of the total radiation received by the population and is obtained by summing the products of the dose to groups of individuals by the numbers of individuals in each group. From the linear no-threshold theory, this will give a measure of the total potential damage.

All groups that have made recommendations for allowable doses to the public have recommended values lower than those for workers, and in recent years there has been a tendency in the regulations to limit doses based on the practicality of achieving doses lower than the recommended limits.

The ICRP (International Commission on Radiological Protection, 1977) recommends a limit of 0.5 rem/yr for critical groups with the evaluation for internal emitters using the weighting factors in Table 16.5. The critical group is defined as a group small enough to be relatively homogeneous with respect to age, diet, and those aspects of behavior which affect the doses received, and it is also the group that is expected to receive the highest doses. The ICRP also stresses the need to obtain an optimum net benefit from the practice causing the radiation by careful analysis of the costs, including potential harm, and the benefits obtained from the practice.

The NCRP (National Council on Radiation Protection and Measurements, 1971) states that the whole-body dose limits for the critical organs of an individual not exposed occupationally shall not exceed 0.5 rem/yr, in addition to natural background and medical exposures. An additional limitation for controlling possible genetic effects provides that the dose equivalent to the gonads for the population of the United States as a whole from all sources of radiation, with the exception of natural background and medical uses, should not exceed an average of 0.17 rem per person.

16-3.6.3 Emergency Radiation Limits. The radiation limits given earlier are applicable to routine situations that are amenable to control of the exposure. Exposure to higher radiation levels does not mean that damage will necessarily result. In emergency situations, where the radiation levels are high, it may be impossible to control the

TABLE 16.6 Summary of Current Dose Data†

	External		Internal	
Source	Individual dose, mrems/yr	Population dose, person-rems/yr	Individual dose, mrems/yr	Population dose, person-rems/yr
Ambient ionizing radiation		9.7×10^6		
Cosmic radiation	40.9–45	9.7×10^6		
Ionizing component	28–35.3	9.2×10^6		
Neutron component	0.33–6.8	4.9×10^5		
Worldwide radioactivity				
Tritium			0.04	9.2×10^3
Carbon-14			1.0	
Krypton-85	4×10^{-4}	80		
Terrestrial radiation	30–95		18–25	
Potassium-40	17		16–19	
Tritium			4×10^{-3}	
Carbon-14			1.0	
Rubidium-87			0.6	
Polonium-210	13[a]		2–3	
Radon-222	25[b]		3.0	

(Table continues on the next page.)

TABLE 16.6 (Continued)

Source	External: Individual dose, mrems/yr	External: Population dose, person-rems/yr	Internal: Individual dose, mrems/yr	Internal: Population dose, person-rems/yr
Technologically enhanced natural radiation				2.73×10^6
Uranium mill tailings			140–14 000[c]	2.5–70 000[d]
Radon in natural gas			54[e]	2.73×10^6
Radon in liquified petroleum gas			0.9–4.0	30 000
Fallout	~2[f]			
Uranium fuel cycle		2014		
Mining and milling			4.5×10^{-2}[g]	2.5
Fuel enrichment	0.17[h]	14	4.8–8.0[i]	14[j]
Fuel fabrication			2×10^{-3}[k]	3[k]
Power reactors				
BWR	54 max.[m]	1552[n]		
PWR	<1 max.[m]	~155[n]		
Research reactors				
Transportation				
Nuclear power industry		100[o]		
Radioisotopes		<170[o]		
Reprocessing and spent-fuel storage	5.8[p]	23[p]		
Radioactive waste disposal				
Federal facilities		~1.96		
ERDA	13–320[m]	8×10^{-7}–1.96[n]		
Department of Defense	<0.01			
Accelerators	0.04–4[m]	0.42–65		
Radiopharmaceuticals, production and disposal	0.2	0.083[q]		
Medical radiation				
X radiation	20[r]			
Radiopharmaceuticals				3.3×10^6[s]
Occupational and industrial radiation				
BWR	1230[t]			
PWR	1080[t]			
All occupations	0.80[u]			
Consumer products		~6100		
Television	0.025–0.043[v]			
Timepieces		~6100		

†From Environmental Protection Agency, *Radiological Quality of the Environment*, Report EPA-520/76-010, Office of Radiation Programs, 1976.

[a]Uranium-238 series.
[b]Thorium-232 series.
[c]Lung dose.
[d]Lung, rem/yr.
[e]Tracheobronchial dose.
[f]50-yr dose commitment divided by 50.
[g]Average individual lung dose within 80 km.
[h]Maximum potential exposure.
[i]Maximum potential exposure to lungs.
[j]Cumulative exposure within 40-mile radius.
[k]Average individual lung dose within 80 km.
[m]Fence-line boundary dose.
[n]Within a radius of 80 km.
[o]Estimated for 1973.
[p]For Nuclear Fuel Service Plant.
[q]Based on data from five institutions.
[r]Millirads per year (genetically significant dose).
[s]Estimated 1980 dose.
[t]Average occupational exposure per year.
[u]Average exposure for all occupations and 3.7 radiation workers/1000 persons in United States.
[v]5 cm from TV set; units of milliroentgens per hour.

exposure to any preset level, either because the radiation is received before action is instituted or because action involving radiation exposure is required to save lives or otherwise control the situation. Firm limits to be followed in all such situations cannot be derived because any exposure under these conditions involves a judgment as to the value of a given action as opposed to the possible damage that would result from the exposure. In a study of this situation, the NCRP (National Committee on Radiation Protection and Measurements, 1962) states that it is not possible to spell out every circumstance and that it is certainly not the business of any committee to assign values to tasks during an emergency and to the radiation exposure involved in doing so. The NCRP defines a radiation emergency as an accident that threatens to expose people to more than 25 R in 1 week. It does not include reconstruction activities following a peacetime accident after the hazard to the general population has passed.

Limits per se are not pertinent to emergency conditions during which the risk of acute exposure and immediate harm is present. Instead, such situations must be guided by individuals acquainted with the risks involved and the importance of various tasks.

16-4 EXTERNAL BETA RADIATION

Because beta radiation has a limited range (1 to 10 m in air), the beta dose depends mainly on the concentration of radioactive materials in the immediate vicinity of the individual. Any variation in the local concentration will result in corresponding variation in the dose. Thus the prior estimates of the beta doses can be regarded only as order-of-magnitude estimates of the average dose, not the maximum dose, particularly in those cases where the radioactive materials settle in close proximity to, or on, the person.

16-4.1 Beta Dose from Passing Clouds

A cloud of beta emitting radioactive materials surrounding a receptor will constitute a source of beta radiation. The importance of this source will depend on the energies of the radiations emitted, the size of the cloud, the concentration in the cloud, the time of cloud passage, and its proximity to a receptor.

16-4.1.1 Beta Dose from Infinite Clouds. A spherical cloud with a radius equal to the range of the highest energy beta particles and containing a uniform concentration of a beta-emitting isotope would be considered an infinite cloud for a receptor at the center because any additional beta emitters beyond the cloud dimensions would not alter the flux of beta particles to the receptor. For such a cloud the beta dose rate can be calculated from equilibrium considerations, provided the receptor volume is small and does not perturb the radiation field.

For an infinite uniform cloud containing χ curies of beta radiation per cubic meter, the beta dose rate in air at the center is given by

$$ {}_{\beta}D'_{\infty} = \frac{\chi \bar{E}_{\beta}(1.6 \times 10^{-6})(3.7 \times 10^{10})}{(1293)(100)} \tag{16.15} $$

$$ = 0.458\ \bar{E}_{\beta}\chi \tag{16.16} $$

where ${}_{\beta}D'_{\infty}$ = beta dose rate from an infinite cloud, rads/s
$\bar{E}_{\beta}$ = average beta energy per disintegration, MeV/dis
1.6×10^{-6} = number of ergs per million electron volts
χ = concentration of beta-emitting nuclide in the cloud, Ci/m^3
3.7×10^{10} = disintegration rate per curie, dis/(s·Ci)
1293 = density of air at standard temperature and pressure (STP), g/m^3
100 = energy absorbed per gram of absorbing material per rad, erg/(g·rad)

If the air is at other than standard density, Eq. 16.16 can be corrected by the ratio of the air densities, thus

$$ {}_{\beta}D'_{\infty} = 0.458\ \bar{E}_{\beta}\chi \frac{\rho_0}{\rho} \tag{16.17} $$

where ρ_0 is the density of air at STP (g/m^3) and ρ is the density of air at reference conditions (g/m^3). Although applicable throughout, this correction factor for air density will not be repeated in subsequent equations.

By the time a cloud or plume has reached the receptor, the dimensions of the cloud are generally on the order of tens of meters or greater. Under these conditions the cloud can be considered to be infinite in extent for any point within the cloud, and the error introduced by this approximation will be well within the uncertainties caused by other factors.

For these conditions of infinite cloud size, if $\chi(x,y,z,t)$ is the instantaneous concentration of the cloud at the receptor location at time t (in seconds) after the release, the external beta dose rate ${}_{\beta}D'_{\infty}(x,y,z,t)$ to the receptor at any time t after the release is given by

$$ {}_{\beta}D'_{\infty}(x,y,z,t) = 0.458\ \bar{E}_{\beta}\chi(x,y,z,t) \tag{16.18} $$

where $\chi(x,y,z,t)$ has been corrected for depletion by ground deposition and radioactive decay en route. The total direct beta dose ${}_{\beta}D_{\infty}(x,y,z)$ due to cloud passage is given by

$$ {}_{\beta}D_{\infty}(x,y,z) = 0.458\ \bar{E}_{\beta}\psi(x,y,z) \tag{16.19} $$

where $\psi(x,y,z)$ is the concentration time integral (Ci·s/m^3) at the receptor location and is corrected for depletion en route,

$$ \psi(x,y,z) = \int_{t_a}^{t} \chi(x,y,z,t)\, dt $$

t_a = time of cloud arrival, seconds since release
t = time of cloud departure, seconds since release

The meteorological formulas generally used to determine the cloud concentration or concentration time integral are developed for nondepositing source material and do not usually include a correction for ground deposition. When the source material is radioactive, correction factors must be included to account for the depletion by radioactive decay en route to the receptor. When a single radionuclide is involved, the correction is simply

$$\left(\frac{Q_x}{Q_0}\right)_r = \exp\left(-\lambda_R \frac{x}{\bar{u}}\right) \tag{16.19a}$$

where $(Q_x/Q_0)_r$ = reduction in quantity between emission and receptor by radioactive decay
x = downwind distance between the point of release and the receptor location, m
$\bar{u}$ = average wind speed during transit, m/s
λ_R = radioactive-decay constant, s^{-1}

When the source term includes several radionuclides, a single exponential equation is inadequate. The decay function can be calculated exactly by the sum of the individual decay factors, or an empirical value, such as a power function, can be used. In this case the correction factor can be indicated symbolically as

$$\left(\frac{Q_x}{Q_0}\right)_r = f_r(t)$$

The correction factor for depletion due to ground deposition is given by Briggs et al. (1968) as

$$\left(\frac{Q_x}{Q_0}\right)_d = \exp\left\{-\left(\frac{2}{\pi}\right)^{1/2}\left(\frac{v_d}{\bar{u}}\right)\int_0^x \exp\left[\frac{(-h^2/2\sigma_z^2)}{\sigma_z}\right]\right\} dx \tag{16.20}$$

where $(Q_x/Q_0)_d$ = reduction in quantity between emission and receptor by deposition
v_d = deposition velocity of the contaminant, m/s
$\bar{u}$ = mean wind speed, m/s
h = height of release, m
σ_z = standard deviation of concentration in the vertical, m

The total depletion correction term is the product of the two factors:

$$\left(\frac{Q_x}{Q_0}\right) = \left(\frac{Q_x}{Q_0}\right)_r \left(\frac{Q_x}{Q_0}\right)_d$$

In this chapter, in which meteorological dispersion formulas are used to calculate radiation dose values, the symbols for cloud concentration (χ) and cloud concentration time integral (ψ) at the receptor location are assumed to include the appropriate depletion correction factors.

The value of $\bar{E}_\beta$ in Eqs. 16.15 and 16.19 is the average kinetic energy of the beta particles released per disintegration. Beta particles are emitted from radioactive material with a continuous spectrum of energies ranging from zero to a maximum value characteristic of the specific material. The average energy of the beta-ray spectrum varies from about 25 to 45% of the maximum energy for negative electrons, depending on the atomic number of the radionuclide and the maximum energy (Hine and Brownell, 1956). In a recent report of data on radionuclides in routine releases from nuclear fuel-cycle facilities, tabulations are given of the maximum and average energy of each beta particle and the overall average for each nuclide (Kocher, 1977). In view of the accuracy with which the other parameters are known, it is usually adequate to consider the average beta energy for a specific radionuclide to be one-third of the maximum beta-ray energy. In a cloud composed of a number of radionuclides, it is necessary to obtain an overall average beta energy by considering the fraction of each species present and its average energy of emission.

Equations 16.15 and 16.19 describe the beta dose rate and the beta dose resulting from an essentially infinite cloud where media of different absorbing characteristics are not present. In the actual case the presence of the body in the cloud causes a perturbation to the radiation field. Owing to the very limited range of beta particles in tissue, the beta dose to a body surrounded by a beta-emitting cloud will generally be limited by geometrical considerations to irradiation from half the cloud. Protruding organs, such as the nose, ears, and fingers, may, however, receive radiation from greater than a 2π solid angle. For this reason the beta dose or dose rate at the surface of a body immersed in an infinite cloud will be about one-half of the equilibrium beta dose (or dose rate) of the full cloud. In addition, the surface of the earth provides a further perturbation in the radiation field. The dose rate at the surface is only one-half of that in an infinite cloud but increases to the full infinite-cloud dose at an elevation equal to the maximum range of the beta particles involved. Thus dose rates will vary from the head to the foot of an individual, with the highest dose rate at the head. This factor varies from 0.5 at the ground to 1.0 at heights greater than the range of the beta radiation. This correction factor will not be included in subsequent formulas, although specific corrections can be made for particular elements of tissue if the radiation energy and height of concern are defined.

The surface body dose from beta emitters in the infinite cloud can be approximated as

$$_\beta D_\infty(x,y,z) = (0.5)(0.458)\,\bar{E}_\beta \psi(x,y,z) \tag{16.21}$$

$$= 0.23\,\bar{E}_\beta \psi(x,y,z) \tag{16.22}$$

where $\psi(x,y,z)$ is the concentration time integral for the cloud ($Ci\cdot s/m^3$) corrected for depletion by ground deposition and radioactive decay en route. If the source (release of radioactive material) is known in terms of power units, e.g., $P_{\beta 1}$ (MeV/s) or $P_{\beta 2}$ (ergs/s), the beta dose (rads) to the surface of the body from the passing infinite cloud can be given by

$$_{\beta}D_{\infty(x,y,z)} = \frac{(0.5)(1.6 \times 10^{-6})}{1293 \times 100}\left[\frac{\psi_{(x,y,z)}}{Q_x}\right] P_{\beta 1} \qquad (16.23)$$

$$= 6.2 \times 10^{-12}\left[\frac{\psi_{(x,y,z)}}{Q_x}\right] P_{\beta 1} \qquad (16.24)$$

where $\psi_{(x,y,x)}/Q_x$ is the time-integrated dilution factor (s/m^3) and $P_{\beta 1}$ is the source term (MeV/s) emitted by the entire cloud at the time of release.

When the source term $P_{\beta 2}$ is in units of ergs per second, the equation for the beta dose becomes

$$_{\beta}D_{\infty(x,y,z)} = 3.86 \times 10^{-6}\left[\frac{\psi_{(x,y,z)}}{Q}\right] P_{\beta 2} \qquad (16.25)$$

Similar equations can be written for other beta power units; for example, to use watts, multiply Eq. 16.25 by 10^{-7}.

In Eqs. 16.21 to 16.25, the concentration and energy release rates are dependent on the source characteristics and the amount of radioactive decay and ground deposition that has occurred between the time of release and the time of exposure. Whether the release occurs over a short period of time or is continuous, the change in quantity due to radioactive decay during the time of actual passage of the cloud is frequently small compared to the change during transit between the release point and the receptor. In this case a single correction term has been applied to ψ to correct for the radioactive decay.

If the release consists of a mixture of radionuclides and the time of exposure to the radiation from a passing cloud is long with respect to the half-life of the radioactive material, it is desirable to correct the equation for decay during passage. In this case the source term in the equation for ψ is not constant, and the integration must include a term describing the time variation of the source, i.e., $\bar{E}_{\beta}\psi$ or $(\psi/Q)P_{\beta}$, must be replaced by $\int (E_{\beta}\chi)\,dt$ or by $\int (P_{\beta}\chi/Q_x)\,dt$. This expression can be evaluated from considerations of the individual radionuclides present: (1) by fitting an arbitrary mathematical function to the rate of decay and integrating or (2) by calculating the dose rate at various times during cloud passage and averaging over the exposure time.

16-4.1.2 Depth Dose. The beta dose calculated in the preceding section is the dose received at the surface of the body or clothing. To produce damage, the radiation must arrive at the sensitive tissues underneath by penetrating the clothing and the normal inert layer of skin. A complete estimate of the effects of a beta-ray dose requires evaluation of the dose to the underlying tissues.

From measurements of the distribution of dose in air around a point source of beta radiation, Loevinger, Japha, and Brownell (1956) derived an empirical equation for a point source that permits the dose to be calculated as a function of both the depth of penetration and the energy of the beta radiation. Their integration of this expression to give the dose at depths from an infinite slab source can be used to estimate the depth dose in the body. This estimate is an approximation because Loevinger's equations were derived for a homogeneous medium containing both source and receptor. According to the authors,

$$_{\beta}D_d = {}_{\beta}D_{\infty}\alpha\left\{c^2\left[3 - \exp\left(1 - \frac{\nu d}{c}\right) - \frac{\nu d}{c}\left(2 + \ln\frac{c}{\nu d}\right)\right] + \exp(1 - \nu d)\right\} \qquad (16.26)$$

where the quantity in square brackets equals zero when $d > c/\nu$.

In this equation $_{\beta}D_d$ is the dose at a mass depth of d(g/cm^2) in the body from a surface dose of $_{\beta}D_{\infty}$, and c is an energy-dependent parameter with the following values:

$$\begin{aligned} c &= 2 && 0.17\ \text{MeV} < E_0 < 0.5\ \text{MeV} \\ &= 1.5 && 0.5\ \text{MeV} \leqslant E_0 < 1.5\ \text{MeV} \\ &= 1 && 1.5\ \text{MeV} \leqslant E_0 < 3\ \text{MeV} \end{aligned} \qquad (16.26a)$$

where E_0 is the maximum beta energy emitted and α is a normalizing constant given by

$$\alpha = [3c^2 - (c^2 - 1)\,e]^{-1} \qquad (16.26b)$$

The values of α for various values of c are

c	α
1	0.333
1.5	0.297
2	0.260
3	0.190

In Eq. 16.26, ν is an apparent absorption coefficient derived from measurements in polystyrene with a correction for the electron density of tissue. It is expressed in units of square centimeters per gram so that the product νd is dimensionless. This absorption coefficient is given by

$$\nu = \frac{18.6}{(E_0 - 0.036)^{1.37}} \qquad (16.26c)$$

An additional term to correct for anomalous spectra with certain beta emitters has been omitted in Eq. 16.26c because the correction is not of significance for the accuracy of these calculations.

The dose at a given depth in the tissue will be strongly dependent on the energy of the beta particles emitted. For a mixture of different emitters, it is necessary to evaluate the dose from each emitter separately or to calculate for various energy groupings and weight the result by the fractional occurrence of each group. Figure 16.2 illustrates the dose rate to be expected at various depths in tissue from exposure to a cloud with a concentration time integral (ψ) of 1.0 $Ci{\cdot}s/m^3$ and with the beta energies shown. For a sustained concentration of 1.0 Ci/m^3, Fig. 16.2 will give the dose rate in rads per second from a cloud of infinite dimension outside the body. The depth is expressed in

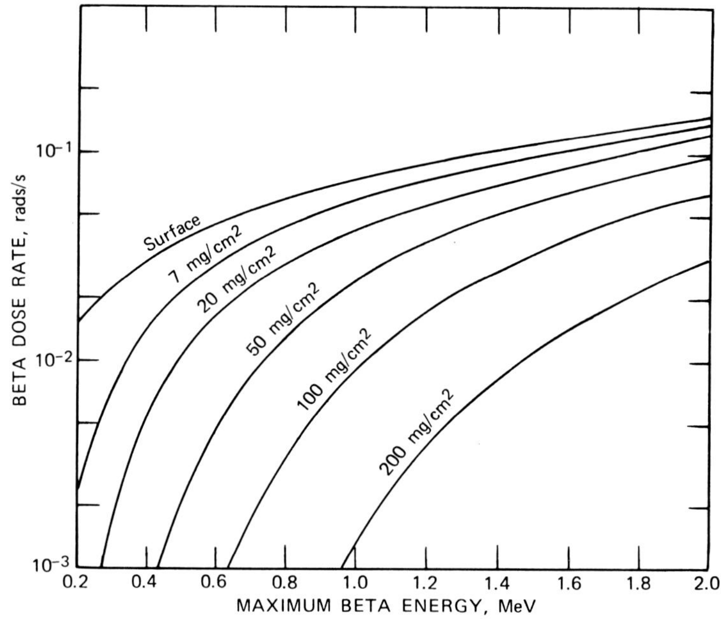

Fig. 16.2 Beta dose rates at various depths in tissue produced by an infinite cloud containing 1 Ci/m^3 of a beta emitter with the maximum energy as given.

milligrams per square centimeter, a unit that expresses the mass available for absorbing radiation to the thickness given. For tissue with a density of 1, a thickness of 100 mg/cm^2 corresponds to a linear depth of 1 mm. The value of c given previously produces discontinuities in the curve because of the step change in value at 0.5 and 1.5 MeV. For the curves given in Fig. 16.2, the values of c were interpolated linearly between the individual values and their energy.

The thickness of the inert layer of the epidermis varies over the body but is usually considered to be 7 mg/cm^2. For most calculations the surface dose given by Eqs. 16.21 and 16.26 will estimate the dose for the unclothed part of the body to within the accuracy of the knowledge of the local concentration.

16-4.2 Beta Dose from Material Deposited on the Ground

During the passage of the cloud, a portion of the radioactive material will be deposited on the ground by the mechanisms of dry deposition, rainout, or washout. This material can consist of any of the radioactive materials in the cloud except the noble gases. Ideally the pattern of deposition calculated from the dispersion equations will consist of a central line of high deposition, with the amount per unit area decreasing on either side in a Gaussian distribution similar to the pattern of concentration in the cloud. In practice, there will be localized variations in the dose rate due to (1) runoff and puddling of rain water, (2) increased deposition caused by eddies near obstructions, (3) distortion of the deposition pattern caused by variable winds, (4) early deposition of large particles that cannot remain airborne, (5) increased accumulation of radioactive materials in oils or greases, and (6) other effects. The present knowledge of the detailed diffusion in a particular terrain is not adequate to calculate these localized variations in any detail. The calculations of the radiation field above the ground provide the general area dose rates with the localized variations due to these factors superimposed on this general field. In dry weather, differences in the rate of dry deposition due to particle sizes, physical form of the material, and other factors will cause fractionation; thus the composition of the deposited material may vary with distance. In general, the beta dose from the separate fractions with different deposition constants must be evaluated separately and then summed.

The material deposited on the surface can be considered as a thin source that, because of the short range of the beta particles, is effectively infinite in the horizontal plane. Under these conditions the variation in the dose rate above the ground is given by Loevinger, Japha, and Brownell (1956) as

$$_{\beta}D'_{\infty}(b,t) = 0.0296\,\nu\bar{E}_{\beta}\alpha\omega\left\{\left[c\left(1+\ln\frac{c}{b\nu}\right) - \exp\left(1-\frac{b\nu}{c}\right)\right] + \exp\,(1-b\nu)\right\} \quad (16.27)$$

where the expression in the brackets equals zero when $b > c/\nu$. The term $_{\beta}D'_{\infty}(b,t)$ is the beta dose rate (rads/s) at height b (g/cm^2) above an infinite plane on which beta-emitting material has been deposited (1 cm^3 of air at STP weighs 1.293×10^{-3} g), ω is the surface activity (Ci/m^2), and values of the other parameters are as given previously.

Where the cloud is a Gaussian plume of radioactive material from the release of a single radionuclide at a

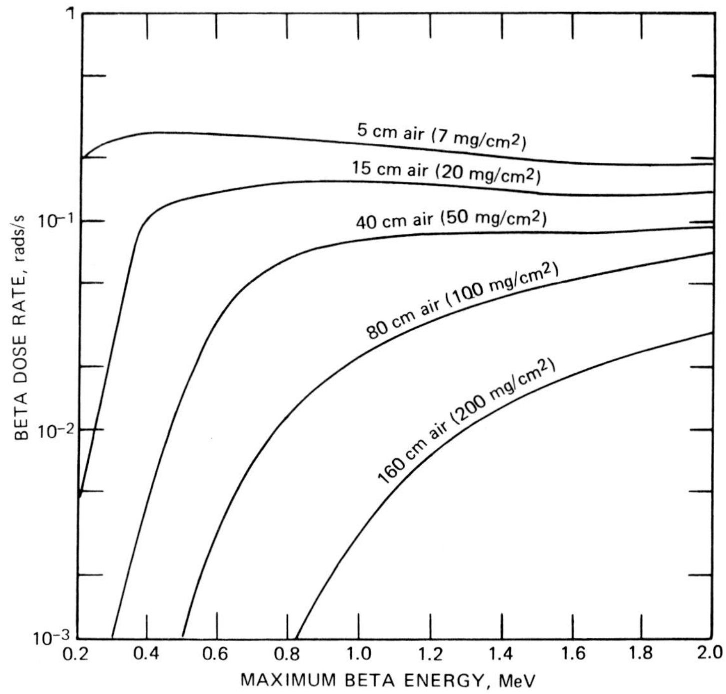

Fig. 16.3 Beta dose rates above a plane surface with a deposition of 1 Ci/m² of a beta emitter with the maximum energy as given. Curves are labeled with the depth of air in terms of length and total mass per unit area.

constant rate of Q_0' (Ci/s) from an elevated point h (m) above the ground plane, the concentration of the established plume at a location x,y,0 (with correction for depletion en route), is given in curies per cubic meter by

$$\bar{\chi}_{(x,y,0)} = \frac{Q_0'}{\pi \bar{u} \sigma_y \sigma_z} \left(\frac{Q_x}{Q_0}\right)_d \exp\left(-\lambda_R \frac{x}{\bar{u}}\right) \times \exp -\left(\frac{y^2}{2\sigma_y^2} + \frac{h^2}{2\sigma_z^2}\right) \qquad (16.28)$$

The activity deposited on the surface, $\omega_{(x,y,0)}$ (Ci/m²) is given by

$$\omega_{(x,y,0)} = \bar{\chi}_{(x,y,0)} v_d \int_0^{t_d} \exp(-\lambda_R t)\, dt \qquad (16.29)$$

where v_d is the deposition velocity (m/s) (see Chap. 12), and t_d is the time interval of discharge (s). The net activity on the ground after the first part of the plume reaches the point (x,y,0) will be given by

$$\omega_{(x,y,0)} = \bar{\chi}_{(x,y,0)} v_d \frac{1 - \exp(-\lambda_R t_d)}{\lambda_R} \qquad (16.30)$$

At times after the last of the plume has passed the point, the deposited activity (of a single radionuclide) will again decay according to the simple exponential, and the activity per unit surface area will be given by

$$\omega_{(x,y,0,t)} = \bar{\chi}_{(x,y,0)} v_d \frac{[1 - \exp(-\lambda_R t_d)] \exp(-\lambda_R t)}{\lambda_R} \qquad (16.31)$$

where t is the time interval (s) after the last of the plume has passed the point (x,y,0).

For the special case of an instantaneous release of Q_0 curies, i.e., $t_d \to 0$ and the quantity $[1 - \exp(-\lambda_R t_d)] \to \lambda_R t_d$, Eq. 16.31 becomes

$$\omega_{(x,y,0,t)} = \frac{Q_0 v_d (Q_x/Q_0)_d}{\pi \bar{u} \sigma_y \sigma_z} \exp\left[-\lambda_R \left(t + \frac{x}{\bar{u}}\right)\right] \times \exp -\left(\frac{y^2}{2\sigma_y^2} + \frac{h^2}{2\sigma_z^2}\right) \qquad (16.32)$$

where t is the time interval (s) after the center of the cloud has passed the point (x,y,0), Q_0 is the total quantity released, and Q_x is the quantity remaining at point d after deposition.

Variations in the beta dose rate will again be apparent when the material is deposited on surfaces other than the ground. For example, the dose rate from fission products on tall bushes, on buildings, or in regions of very rough terrain will be different from that calculated by Eq. 16.27, which is based on a plane surface.

Estimates of the dose rate above an infinite plane for different beta maximum energies are given in Fig. 16.3. The assumptions for the value of c given previously are also used in this figure.

16-4.3 Beta Dose Rates from Materials Deposited on the Body

Radioactive materials deposited directly on the skin can deliver relatively high beta doses because of their proximity to sensitive tissues. Although some gamma dose is usually associated with the beta dose, the longer range of the gamma radiation results in spreading the absorbed energy over a larger volume, whereas the short range of the beta particles results in intense energy deposition in the first few millimeters of the skin.

The physical form of the deposited material can differ from a uniform coating over the entire surface to the random distribution of a few small, very intensely radioactive particles, depending on the characteristics of the material in the passing cloud. These single particles are usually found close to the point of discharge or, in the event of a contaminated surface, in dust resuspended by the action of the winds. Such particles act as discrete sources of radiation and provide very-high dose rates to the tissue in the immediate vicinity of the particle. One can estimate dose rates of 10^6 rads/h at distances of 10 μm from a 1-μCi particle. The beta dose rate, however, drops rapidly with distance from the particle, and thus the total amount of tissue involved is very small. The significance of such calculated doses in terms of biological effects does not have a theoretical basis for interpretation. However, for alpha emitters, it has been shown that the effects of such focal radiation are less than the effects that would be produced if the same amount of radioactive material were uniformly distributed through the organ (Bair, Richmond, and Wachholz, 1974; National Council on Radiation Protection and Measurements, 1975b; National Academy of Sciences—National Research Council, 1976; Healy, Richmond, and Anderson, 1974), presumably because of the fewer numbers of cells irradiated. Although data are not as complete for beta emitters, they indicate the same pattern (Cember, 1964). Since the permissible limits are calculated on the basis of a uniform distribution through the tissue, the information indicates that such particles are of no special concern unless the activity is so great that an actual burn occurs.

The amount of radioactive material deposited on the body will vary over the body surfaces. It is to be expected that the greatest deposition will occur on the hair and on any oily spots on the body. The feet and legs may also become contaminated if the individual moves around in the contaminated area. The estimation of the quantities of material deposited on the body can be made only with poor accuracy; however, for an average estimate, it can be assumed that the deposition velocity for the transfer of radioactive material from the passing cloud to the body is the same as for the transfer of material to the ground. The beta dose rate to the body can be estimated from Eq. 16.27, where b is the depth within the body rather than the height in air above the ground.

In estimates of beta doses, the shielding effects and protection provided by clothing can be taken into account. The maximum beta dose will be delivered to the exposed surfaces of the skin. Clothing will provide attenuation of the beta-radiation dose because of its shielding effect and because the clothing will hold the deposited material away from the skin.

16-5 EXTERNAL GAMMA RADIATION

Although the principles and methods of calculating the dose from extended sources of gamma radiation are well known and straightforward, the solution for actual cases is the most complicated of the cloud-dosage calculations. As a result, the gamma dose from a passing cloud can be evaluated with good accuracy only for specific conditions.

With extended sources, such as with a cloud or with ground deposits, the gamma dose in air at a given point is dependent on the location of the radioactive materials for distances up to several hundred meters. The radiation field is therefore more uniform than with beta radiation because local variations in the material distribution are averaged over a much larger area or volume.

The use of the equations in this chapter should be approached with considerable caution because of the many assumptions and approximations that are necessary to obtain numerical solutions. There are many factors in a specific application which can cause significant differences between measured doses or dose rates and those calculated by the formulas presented.

16-5.1 The Point Source

Calculations of the gamma dose from an extended source start with consideration of the radiation received at the receptor from a differential area or volume that can be regarded as a point source. This radiation is then integrated over the entire source, with account being taken of the geometry of the source, variation in concentration, attenuation by interactions of the photons with matter in the path between source and receptor, and scattering of radiation from material outside the direct path to the receptor.

For a point source the geometrical considerations provide an inverse-square decrease in the dose with distance. Because the total radiation emitted (minus that absorbed) will pass through a sphere of radius equal to the distance between source and receptor, the flux passing through any area at that distance can be determined if the initial emission is known.

For a narrow, collimated beam falling on an absorber, the gamma radiation is absorbed according to

$${}_{\gamma}D = {}_{\gamma}D_0 \exp(-\mu x) \tag{16.33}$$

where ${}_{\gamma}D$ is the gamma-ray dose (rads), ${}_{\gamma}D_0$ is the gamma-ray dose (rads) that would be received without a shield of thickness x (m or g/cm^2), and μ is the total gamma-ray absorption coefficient (m^{-1} or cm^2/g). The absorption coefficient, μ, is a function of the specific shield material and energy of the radiation. A shield of thickness $x = 1/\mu$ would attenuate the beam by a factor of 1/e; this thickness is designated as the relaxation length. The product of μx is frequently used as a dimensionless thickness of absorber. The units of μ may be reciprocal length (linear absorption coefficient) or reciprocal mass per unit area (mass absorption coefficient).

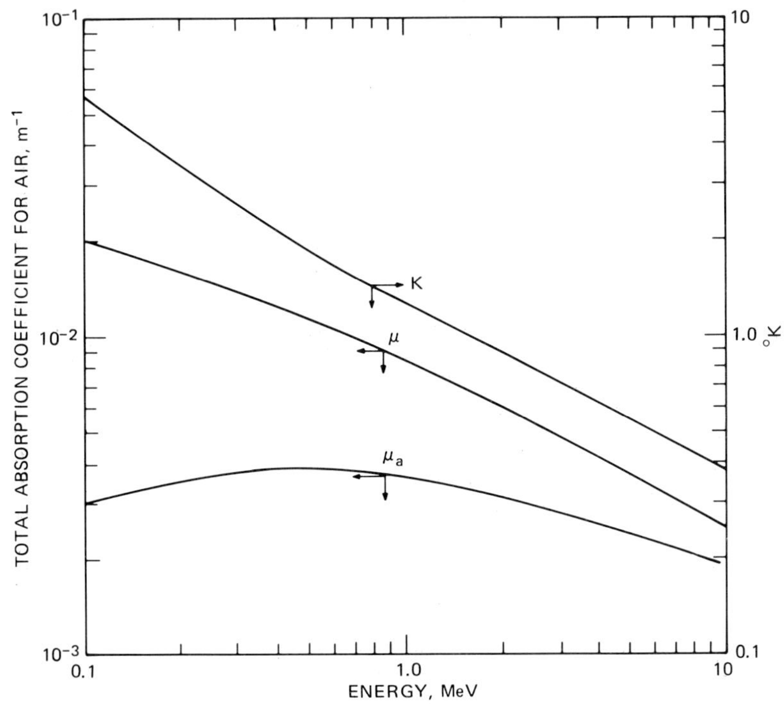

Fig. 16.4 Absorption coefficients and values of the estimated buildup constant for air at STP.

In the Compton scattering process, which predominates in the energy ranges and absorber atomic numbers of interest in cloud gamma-dose calculations, only a portion of the energy is absorbed and transferred to an electron that is absorbed close to its point of origin. The remainder is carried by scattered photons that, although removed from the beam under the conditions for which Eq. 16.33 is written, can contribute to the gamma dose received at some other point. The total cross section for this process is divided into two parts: (1) μ_a represents the fraction of the energy in the recoil electrons and thus represents the fraction of the energy absorbed near the site of the interaction, and (2) μ_s represents the fraction of the energy contained in the scattered photons. Thus $\mu = \mu_a + \mu_s$.

Contributions from scattered radiation in air or other materials not directly between the source and the receptor will cause the gamma dose calculated from Eq. 16.33 to be low for broad beams. Buildup factors to account for the increased radiation have been computed for several configurations and shielding materials (Goldstein and Wilkins, 1954). For air, Gamertsfelder (1960) has used a buildup factor of the form $(1 + k\mu x)$, where k is equal to $(\mu - \mu_a)/\mu_a$. This factor gives the correct value of energy absorption in an infinite radioactive medium but overestimates the scatter component from sources close to the receptor and underestimates the scatter from distant sources.

The gamma dose rate to air at a distance r meters from a point source of q (Ci) can then be written as

$${}_\gamma D'_r = \frac{\mu_a q\,(3.7 \times 10^{10})\,\bar{E}_\gamma\,(1.6 \times 10^{-6})(1 + k\mu r)\exp(-\mu r)}{4\pi r^2\,(1293)\,(100)} \qquad (16.34)$$

$$= \frac{0.0364\,\mu_a q \bar{E}_\gamma\,(1 + k\mu r)\exp(-\mu r)}{r^2} \qquad (16.34a)$$

where ${}_\gamma D'_r$ = dose rate (rads/s) at a distance r(m)
μ_a = energy absorption coefficient for air, m^{-1}
μ = total absorption coefficient for air, m^{-1}
$k = (\mu - \mu_a)/\mu_a$
$\bar{E}_\gamma$ = average gamma energy emitted at each disintegration, MeV/dis
3.7×10^{10} = disintegrations per second per curie
1.6×10^{-6} = ergs per million electron volts
1293 = density of air at STP, g/m^3
100 = ergs per gram per rad

The gamma dose rate to tissue will be higher than that to air by the ratio of the electron densities in the two mediums. Over the range of energies from 0.1 to 5.0 MeV, this ratio can be approximated as 1.11 for soft tissue. The gamma dose rate to tissue is therefore given by

$${}_\gamma D'_r = \frac{0.0404\,\mu_a q \bar{E}_\gamma\,(1 + k\mu r)\exp(-\mu r)}{r^2} \qquad (16.35)$$

Figure 16.4 gives values of the gamma absorption coefficients for air and the associated values of k as a function of gamma energy. If the air density differs from the standard conditions of temperature and pressure used in developing the relationship, corrections must be made when these coefficients are applied; namely, the ratio (ρ_0/ρ) should be used as a correction factor for values of μ and μ_a.

16-5.2 Cloud Gamma Dose

For clouds with dimensions that are small compared to the range of the gamma radiation, a calculation of the gamma dose at a given point must take into account the radiation received from various parts of the cloud. The complexities of geometry, absorption, and buildup make a complete solution difficult and require the use of special integrals. Approximate solutions have been obtained by neglecting absorption or buildup. Holland (1958) has calculated the dose, assuming that all particles of the cloud pass the receptor with the mean wind velocity and isotropic turbulence so that $\sigma_y = \sigma_z$, where σ_y represents the standard deviation of width and σ_z the standard deviation of height for the Gaussian-shaped cloud. The standard deviations for an instantaneous release were used. His calculations have been presented in the form of nomograms. A more detailed discussion on the limitations of the standard deviations is given in Chaps. 6 and 13. Couchman (1960) has described a computer technique based on the Sutton form of the diffusion equations which provides for attenuation, buildup, and nonisotropic diffusion. Gamertsfelder (1960) has provided a general method for an isotropic cloud which will be used in the following discussion.

16-5.2.1 Infinite Cloud. If the dimensions of a homogeneous cloud of gamma-emitting radioisotopes are large compared to the distance that the gamma rays travel, an equilibrium condition occurs similar to that noted for beta radiation. Under these conditions the rate of energy absorption per unit volume is equal to the rate of energy release per unit volume, and the gamma dose to air will be given by Eq. 16.18 or 16.19 if $\bar{E}_\beta$ is replaced by $\bar{E}_\gamma$, the average gamma energy released per disintegration. A correction factor of 1.11 for the ratio of electron densities of tissue to those for air converts this to dose to tissue.

$$_\gamma D'_\infty = 0.507\ \bar{E}_\gamma \chi \frac{\rho_0}{\rho} \tag{16.36}$$

where the dose rate is in rads per second and ρ_0/ρ is the correction factor that can be applied when the air is at other than standard conditions. (This factor will generally not be repeated in subsequent equations except in instances where it is required for clarity.)

If the receptor is at ground level, the gamma dose calculated with Eq. 16.36 must be corrected by a factor of 0.5 because the radiation is received from only one-half of the cloud owing to the presence of the ground. For an actual cloud the assumption of an infinite cloud of gamma-emitting material can be made only when the dimensions of the cloud are large compared to $1/\mu$. For an infinite cloud the value of the cloud concentration $\bar{\chi}_{(x,y,0,t)}$ (Ci/m^3) can be substituted for χ in Eq. 16.36 so that

$$_\gamma D'_{\infty(x,y,0,t)} = 0.25\ \bar{E}_\gamma \bar{\chi}_{(x,y,0,t)} \tag{16.37}$$

where $_\gamma D'_{\infty(x,y,0,t)}$ is the gamma dose rate (rads/s) to a person located at (x,y,0) at time t and $\sigma \gg 1/\mu$.

Equation 16.37 will yield the gamma dose rate for an established plume from the emission of gamma-emitting material at a constant rate of Q'_0 (Ci/s) if the plume concentration $\bar{\chi}_{(x,y,0)}$ (Ci/m^3) is used. If the concentration time integral $\psi_{(x,y,0)}$ (Ci·s/m^3) is used, the dose received during the cloud passage is given by

$$_\gamma D_{\infty(x,y,0)} = 0.25\ \bar{E}_\gamma \psi_{(x,y,0)} \tag{16.38}$$

Consistent with previous definitions, the value $\bar{\chi}$ and ψ must include the appropriate correction factors for depletion en route, i.e., $(Q_x/Q_0)_d$ and/or $(Q_x/Q_0)_r$.

16-5.2.2 Finite Cloud. Near the source of the plume, the gamma doses and dose rates will be less than those estimated by the infinite-cloud assumption. However, at positions close to the source of an elevated release before the cloud has reached the ground, the concentration will be small or even zero. Under these conditions the plume will be a gamma source, and much higher doses than would be estimated from the ground-level concentration will result. The following treatment of the smaller cloud is from work by Gamertsfelder (1960).

In terms of the reference system given in Fig. 16.5, the general equation for the instantaneous isotropic cloud concentration, $\chi_{(x,y,z,t)}$ (Ci/m^3), can be written as

$$\chi_{(x,y,z,t)} = \frac{Q_0 (Q_x/Q_0)}{(2\pi)^{3/2}\sigma_I^3} \times \exp - \left[\frac{(m^2 + r^2 - 2\,mr\cos\phi)}{2\sigma_I^2}\right] \tag{16.39}$$

where Q_0 is the total number of curies of the radionuclide in the cloud at the time of release, Q_x/Q_0 is a fraction of the original source material which has not been deposited on the ground or removed by radioactive decay en route, and σ_I is the standard deviation of concentration for an instantaneous release.

In Fig. 16.5 consider the gamma dose rate to a receptor on the ground at point (x_1, y_1, 0) and at time t where the gamma dose rate is due to the presence of q curies in a volume element of a cloud containing a total of Q curies and when the center of the cloud is moving with a speed u (average speed $\bar{u}$) in the x–z plane and is located at the point ($\bar{u}t$, 0, h) at time t. For an isotropic cloud

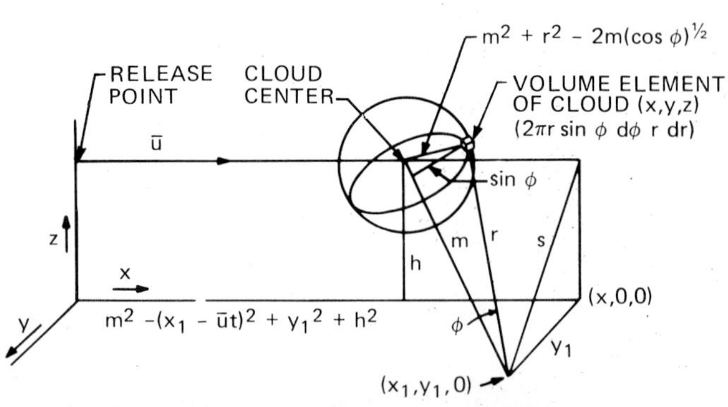

Fig. 16.5 Coordinate system for cloud gamma-dose calculations.

consisting of a single puff release, values of $\sigma_{xI} = \sigma_{yI} = \sigma_{zI} = \sigma_I$ are equal. If perfect reflection at the ground plane is assumed, the gamma dose rate to a receptor located at $(x_1, y_1, 0)$ from that part of the cloud which is reflected from the ground is the same as the dose received if the cloud passed through the ground plane. Since the cloud concentration is symmetrical with respect to the line from $(x_1, y_1, 0)$ to the cloud center at $(\bar{u}t, 0, h)$, the dose rate from the entire cloud at time t is obtained by replacing the incremental source term q in Eq. 16.39 with the instantaneous cloud concentration times the elemental volume $\chi_{(x,y,z)}$ dV (which is equivalent) and integrating the volume element 2π r sin ϕ dϕ r dr over the entire cloud volume. The gamma dose rate to the receptor from the cloud at time t is given by

$$ {}_\gamma D'(x_1,y_1,0,t) = 0.0404\mu_a \chi_{(x,y,z,t)} \bar{E}_\gamma \times \int_0^\infty \int_0^\pi \frac{(1 + k\mu r)\exp(-\mu r)}{r^2} 2\pi r^2 \sin\phi \, d\phi \, dr \quad (16.40) $$

Replacing $\chi_{(x,y,z,t)}$ in Eq. 16.40 by the equivalent in Eq. 16.39 gives

$$ {}_\gamma D'(x_1,y_1,0,t) = \frac{0.0404\,\mu_a \bar{E}_\gamma Q_0 (Q_x/Q_0)}{(2\pi)^{3/2}\sigma_I^3} \times \int_0^\infty \int_0^\pi (1 + k\mu r)\exp(-\mu r)\exp\left[-\frac{(m^2 + r^2 - 2\cos\phi)}{2\sigma_I^2}\right] \times 2\pi \sin\phi \, d\phi \, dr \quad (16.41) $$

so that

$$ {}_\gamma D'(x_1,y_1,0,t) = \frac{0.0404\,\mu_a \bar{E}_\gamma Q_0 (Q_x/Q_0)}{(2\pi)^{1/2}\sigma_I} \times \int_0^\infty (1 + k\mu r)\exp(-\mu r) \left\{\frac{\exp[-(m-r)^2/2\sigma_I^2] - \exp[-(m+r)^2/2\sigma_I^2]}{mr}\right\} dr \quad (16.42) $$

If σ_I and the cloud source depletion correction factors for decay and ground deposition are considered to be constants when the center of the cloud is at a distance $\bar{u}t$ from the point of release for the effective time of cloud passage, the equation for the gamma dose can be written

$$ {}_\gamma D_{(x,y,0)} = \frac{0.0404\,\mu_a \bar{E}_\gamma Q_0 (Q_x/Q_0)}{2\pi\sigma_I} \times \int_0^\infty \int_0^\infty (1 + k\mu r)\exp(-\mu r) \times \frac{\{\exp[-(m-r)^2/2\sigma_I^2] - \exp[-(m+r)^2/2\sigma_I^2]\}}{mr} dr \, dt \quad (16.43) $$

If I_1 and I_2 are defined as

$$ I_1 = \frac{\bar{u}}{4(2\pi)^{1/2}\mu\sigma_I} \int_0^\infty \int_0^\infty \frac{\exp(-\mu r)}{mr} \left\{\exp\left[-\frac{(m-r)^2}{2\sigma_I^2}\right] - \exp\left[-\frac{(m+r)^2}{2\sigma_I^2}\right]\right\} dr \, dt \quad (16.43a) $$

and

$$ I_2 = \frac{\bar{u}}{4(2\pi)^{1/2}\mu\sigma_I} \int_0^\infty \int_0^\infty \frac{\mu \exp(-\mu r)}{m} \left\{\exp\left[-\frac{(m-r)^2}{2\sigma_I^2}\right] - \exp\left[-\frac{(m+r)^2}{2\sigma_I^2}\right]\right\} dr \, dt \quad (16.43b) $$

Eq. 16.43 can then be written as

$$ {}_\gamma D_{(x,y,0)} = \frac{0.1616\,\mu\mu_a \bar{E}_\gamma Q_0 (Q_x/Q_0)(I_1 + kI_2)}{\bar{u}} \quad (16.44) $$

Values of I_1 and I_2 were determined by graphical and numerical integrations. Values of I_1 and I_2 are given in Figs. 16.6 and 16.7 for various values of $\mu\sigma_I$. In these figures, values of I_1 and I_2 are presented as a function of $\mu\sigma_I$ and are given for various values of μs, where $\mu s = \mu(y^2 + h^2)^{1/2}$ and s is the distance between the receptor and the center line of the cloud path. Dashed lines show values of I_1 and I_2 as functions of fixed ratios between s and σ_I and thus permit interpolation of I_1 and I_2 for values of μs other than those given.

The values of I_1 and I_2 can be applied to the continuous source if the instantaneous values of cloud width (σ_I) are replaced by values for the continuous plume (σ).

Equation 16.44 was developed for an isotropic single puff release of Q_0 curies of radioactive material. If the release were to occur at a constant rate (Q_0') over a period of time (t_d) so that $Q_0' t_d = Q_0$, then the governing meteorological variables to determine the concentration time integral, ψ, would be $\bar{u}$, σ_y, and σ_z. These sigma values would be larger than the instantaneous puff values, σ_I, because they would include the effects of meander of the plume at the receptor over the time interval of the release. Since the equation for the concentration time integral for the plume is identical in form to that for the puff release, the only difference in the computed values of ψ would be due to the differences in the wind speed and the plume spread during the long and short time intervals. Thus in both cases the dose is proportional to ψ.

The gamma dose rate (${}_\gamma D'$), which is the time derivative of the dose (${}_\gamma D$), will be proportional to the time derivative (Q_0') of the source (Q_0) if the plume has attained an equilibrium growth condition. Therefore Eq. 16.44 can be used to calculate the gamma dose rate from an established plume if the source release rate (Q_0') is used in place of the total source (Q_0) and if sigma values appropriate to a continuous source are used. This gives

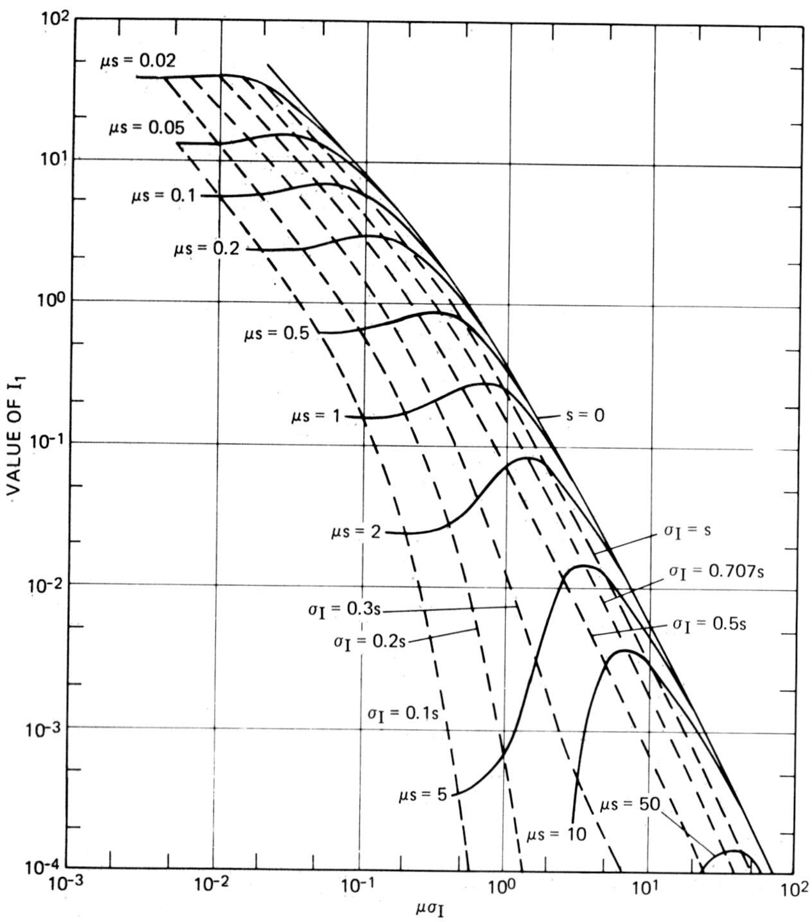

Fig. 16.6 Values of the I_1 integral (see Eq. 16.43a).

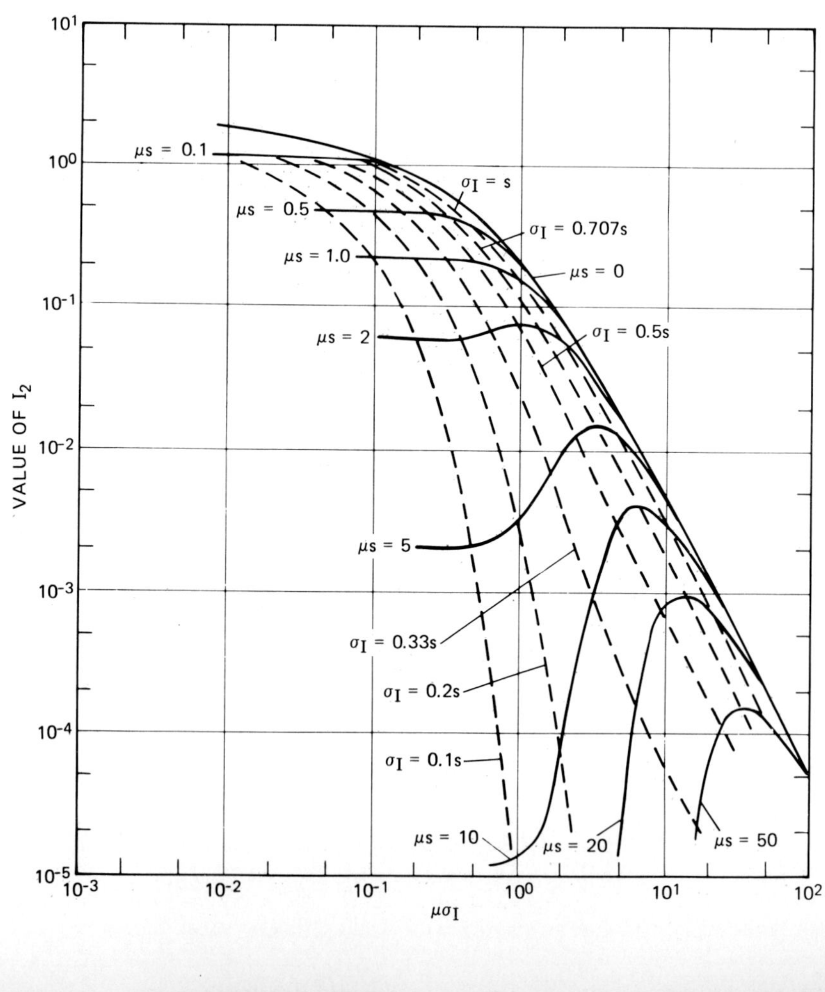

Fig. 16.7 Values of the I_2 integral (see Eq. 16.43b).

$$_{\gamma}D'_{(x,y,0)} = \frac{0.1616\ \mu\mu_a \bar{E}_{\gamma} Q'_0\ (Q_x/Q_0)(I_1 + kI_2)}{\bar{u}} \quad (16.45)$$

where $_{\gamma}D'_{(x,y,0)}$ is the gamma dose rate (rads/s) from the established plume to a receptor on the ground plane at a point x (m) downwind, y (m) perpendicular to the plume center line in the horizontal plane and where Q'_0 is the source release rate at the time of release (Ci/s).

It is frequently more convenient to consider the radioactive decay in the cloud to be represented by a power function. In this case the product $\bar{E}_{\gamma}Q_0$ can be replaced by the gamma power, P_{γ}, if suitable absorption coefficients can be assumed. Conversion of the constants in Eq. 16.44 gives

$$_{\gamma}D_{(x,y,0)} = \frac{4.36 \times 10^{-12}\ \mu\mu_a P_{\gamma 1}}{\bar{u}} (I_1 + kI_2) \quad (16.46)$$

where $P_{\gamma 1}$ is the total energy release rate in the cloud by gamma emission at the time of cloud passage (MeV/s), i.e., it is corrected for depletion en route, and $_{\gamma}D$ is the gamma dose in rads. If $P_{\gamma 2}$ is in watts, Eq. 16.46 becomes

$$_{\gamma}D_{(x,y,0)} = 27.3\ P_{\gamma 2} \frac{\mu\mu_a}{\bar{u}} (I_1 + kI_2) \quad (16.47)$$

and the same depletion correction factors used previously are applicable.

Values of $(I_1 + kI_2)$ can be computed as a single term, I_T, if the gamma energy and thus the absorption coefficients can be fixed. If it is assumed that the effective gamma energy in a cloud consisting of fission products is 0.7 MeV, then $\mu = 9.7 \times 10^{-3}$ and $\mu_a = 3.8 \times 10^{-3}$. The values of $I_T = (I_1 + kI_2)$ for this condition are given in Fig. 16.8.

The preceding equations apply to an isotropic cloud ($\sigma_{yI} = \sigma_{zI}$ or $\sigma_y = \sigma_z$). Since many of the present studies of diffusion result in a nonisotropic cloud formulation, particularly with inversions, a method of calculation is required. Gamertsfelder (1960) noted that, if $\sigma_y > \sigma_z$, the value of I_T for σ_z small with respect to the height of the cloud center line is the same as if $\sigma_y = \sigma_z$. However, for a cloud that is effectively infinite in height (σ_z = 700 m), the value of I_T should be reduced by the fraction σ_z/σ_y to account for the reduced concentration caused by the horizontal spread. Gamertsfelder suggests that a correction factor at intermediate distances can be obtained by drawing a curve with a correction factor on the abscissa and σ_z on the ordinate. The value of the correction factor should be unity for 0.7-MeV gamma rays at $\sigma_z \leqslant 0.035h$, and the curve should smoothly approach the value of σ_z = 700 m. This curve is shown in Fig. 16.9.

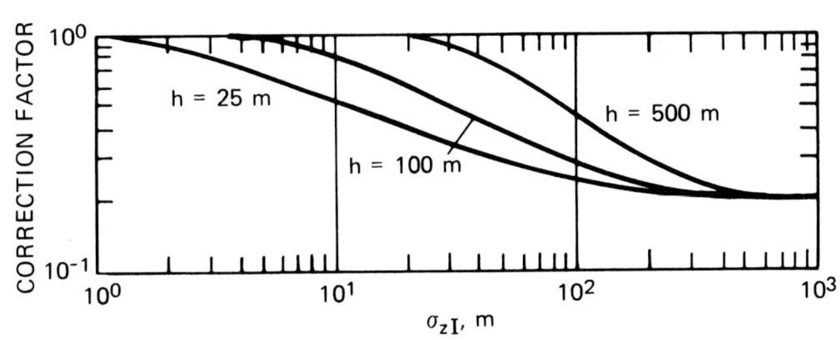

Fig. 16.9 Correction curves for gamma dose from nonisotropic clouds.

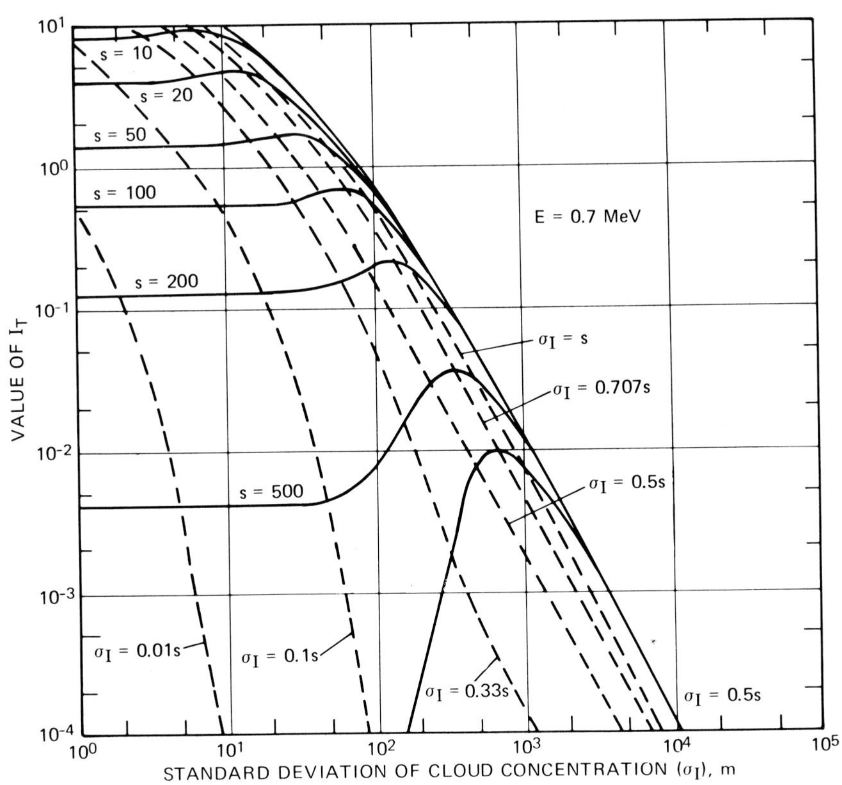

Fig. 16.8 Values of the I_T integral for $\bar{E}_{\gamma}$ = 0.7 MeV.

An alternative approach to the problem of correcting for a nonisotropic cloud which gives comparable answers is to use an average value of σ given by

$$\sigma_{av} = (\sigma_y \sigma_z)^{1/2} \qquad (16.48)$$

16-5.2.3 Ratio of Finite to Infinite Cloud Dose. Since it is relatively simple to estimate the gamma dose or dose rate from a cloud large enough to be considered infinite, it is frequently more convenient to apply correction factors to the dose calculations for infinite clouds to determine the dose from a cloud of finite size. The concentration time integral at a point on the path of the center of a cloud of finite size is given by

$$\psi_{(x,0,h)} = \frac{Q_x}{\pi \bar{u} \sigma_y \sigma_z} \qquad (16.49)$$

If the concentration of an infinite cloud is numerically equal to the concentration time integral of the finite cloud, i.e., $\chi_\infty = \psi$, and the exposure time to the infinite cloud is 1 s, the ratio of the gamma doses from the two clouds can be calculated by using Eqs. 16.38, 16.44, and 16.48, so that

$$\frac{_\gamma D_{(x,y,0)}}{_\gamma D_{\infty(x,y,0)}} = 2.03\ \mu\mu_a \sigma_{av}^2\ (I_1 + kI_2) \qquad (16.50)$$

The values of this ratio are given in Fig. 16.10 for 0.7-MeV gamma photons and values of the cloud standard deviation. Also included are values for the center-line dose and the dose at various distances from the center line in units of the cloud width, as measured by the standard deviation. Note that for smaller plumes the gamma dose at a distance from the plume is controlled more by the penetrating nature of gamma radiation than by the concentration at the location. Figure 16.11 illustrates this effect by comparing the ratio of the gamma dose at various distances from the center line to the center-line dose and comparing these values to the Gaussian function that describes the decrease in concentration with distance from the plume. It is not until the plume width reaches a standard deviation of about 1000 m that the dose is reasonably described by the concentration function.

The ratio of finite to infinite cloud gamma dose at the center line of the plume is given for three different gamma energies in Fig. 16.12. The spread in ratios is only a factor of 2 for energies that differ by a factor of 20.

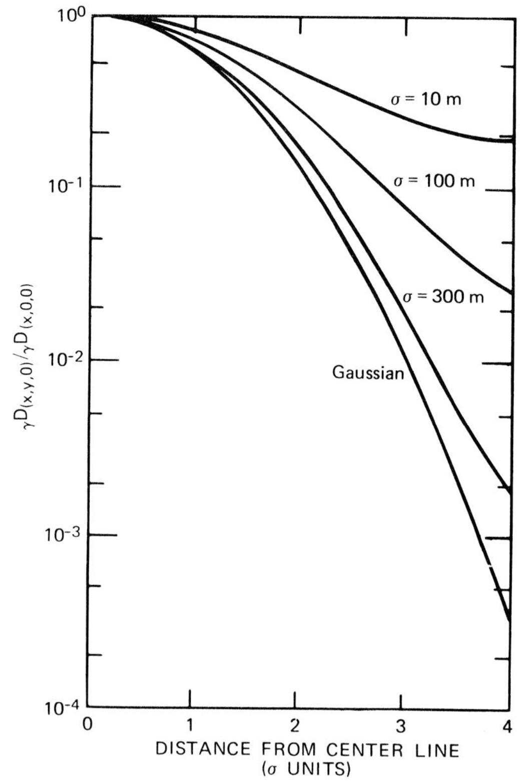

Fig. 16.11 Ratio of off-center-line to center-line doses from finite size clouds containing radionuclides emitting 0.7-MeV gammas compared to a Gaussian concentration distribution.

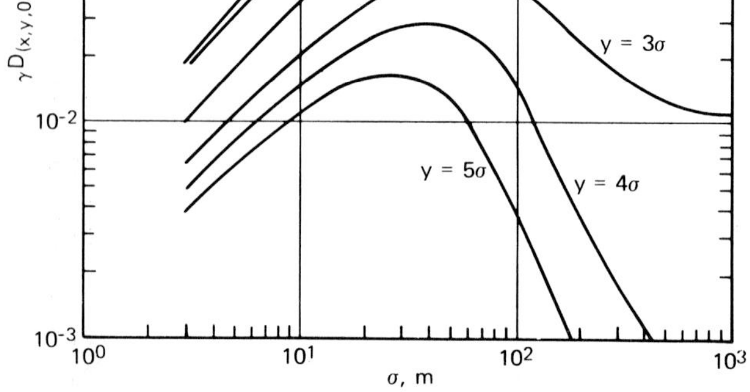

Fig. 16.10 Ratio of the gamma dose in a finite cloud to the gamma dose in an infinite cloud for 0.7-MeV gamma rays.

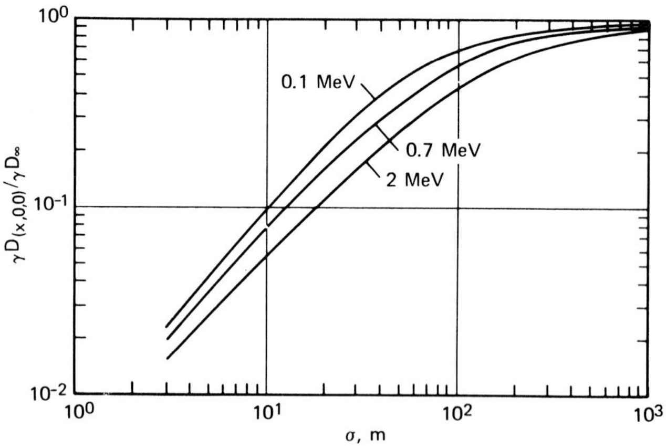

Fig. 16.12 Ratio of the gamma dose at center line in a finite cloud of gamma-emitting radionuclides to the dose in an infinite cloud with the same center-line concentrations.

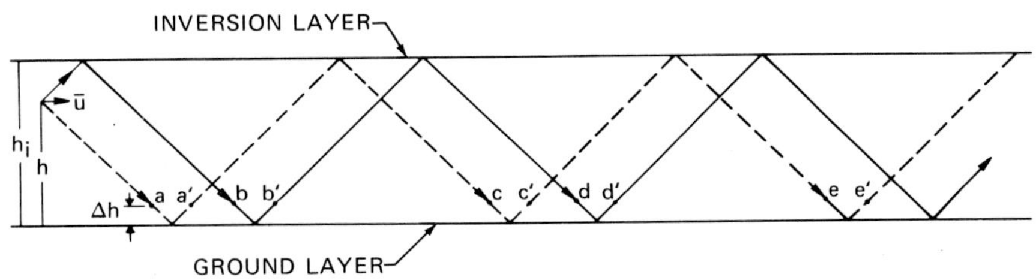

Fig. 16.13 Reference system used for calculating the cloud concentration under a capping inversion.

16-5.2.4 Gamma Dose from Releases Under Capping Inversions. It is not uncommon for a temperature inversion to exist above some modest altitude such that the vertical diffusion of the radioactive cloud will be restricted vertically. Values of the heights of capping inversions that might be expected have been presented by Holzworth (1964).

The concentration and concentration time integral of a cloud released in the layer between the ground and a capping inversion will be greater than similar values calculated with equations based on the uniform infinite atmosphere. The magnitude of this difference will depend on the relative magnitudes of the height of the inversion layer, the height of release of the source material, and the distance downwind from the release point. The following treatment of the problem, which is similar to that of Hewson, Gill, and Bierly, (1959), makes use of the reference system shown in Fig. 16.13.

Consider a source Q_0 released at a point h meters above the ground and below the base of the inversion at height h_i meters. As the cloud moves downwind, the source is depleted by ground deposition and radioactive decay, i.e., $Q_x = Q_0\ (Q_x/Q_0)$. An observer at the center of the cloud and moving with the cloud at the average wind speed $\bar{u}$ (m/s) would observe that the upper and lower cloud limits follow the paths indicated by the solid and dashed lines of Fig. 16.13, respectively. The path lengths computed with the reference points (a,a′,b,b′, . . .) are indicated in Table 16.7.

If an arbitrary index m = 1 is used for the first ground reflections, i.e., a,a′,b,b′, and m = 2 for the next pair of reflections, etc., the path lengths can be described in general terms:

Reference points	Distances
Lower Cloud	
(a,c,e, . . . m)	$(2m{-}1)h_i + h{-}\Delta h$
(a′,c′,e′, . . . m′)	$(2m{-}1)h_i + h{-}\Delta h$
Upper Cloud	
(b,d,f, . . . n)	$2mh_i - h - \Delta h$
(b′,d′,f′, . . . n′)	$2mh_i - h + \Delta h$

The concentration time integral at a point r meters from the center of an isotropic cloud can be described by

$$\psi_r = \psi_p \exp\left(-\frac{r^2}{2\sigma_r^2}\right) \tag{16.51}$$

where ψ_p is the concentration time integral of the cloud center line ($Ci{\cdot}s/m^3$).

With the path lengths developed for the capping inversion, the concentration time integral at a point x meters downwind directly under the cloud path and Δh meters above the ground can be described by

$$\psi_{(x,0,\Delta h)} = \psi_p \sum_{m=0}^{\infty} \exp\left\{-\left[\frac{(2m-1)h_i + h - \Delta h}{(2)^{1/2}\sigma_z}\right]^2 + \exp -\left[\frac{(2m-1)h_i + h + \Delta h}{(2)^{1/2}\sigma_z}\right]^2 + \exp -\left[\frac{(2m)h_i - h - \Delta h}{(2)^{1/2}\sigma_z}\right]^2 + \exp -\left[\frac{(2m)h_i - h + \Delta h}{(2)^{1/2}\sigma_z}\right]^2\right\} \tag{16.52}$$

For the special case of the receptor at ground level, Δh = 0, the first and second exponential terms are equal,

TABLE 16.7 Reflections from Capping Inversions (See Fig. 16.12)

Reference points	Direction	Distance
	Upper Cloud Limit	
b	↓	$2h_i - h - \Delta h$
b′	↑	$2h_i - h + \Delta h$
d	↓	$4h_i - h - \Delta h$
d′	↑	$4h_i - h + \Delta h$
•	•	•
•	•	•
n	↓	$nh_i - h - \Delta h$
n′	↑	$nh_i - h + \Delta h$
	Lower Cloud Limit	
a	↓	$h - \Delta h$
a′	↑	$h + \Delta h$
c	↓	$3h - \Delta h$
c′	↑	$3h + \Delta h$
•	•	•
•	•	•
m	↓	$mh - \Delta h$
m′	↑	$mh + \Delta h$

and the third and fourth terms are equal; therefore Eq. 16.52 becomes

$$\psi_{(x,0,0)} = 2\psi_p \sum_{m=0}^{\infty} \left\{ \exp\left[-\frac{(2m-1)h_i + h}{(2)^{1/2}\sigma_z} \right]^2 + \exp\left[-\frac{(2m)h_i - h}{(2)^{1/2}\sigma_z} \right]^2 \right\} \quad (16.52a)$$

For the special case of the inversion layer at a very-high altitude, only the ground reflection is important, and Eq. 16.51 becomes

$$\psi_{(x,0,0)} = 2\psi_p \exp\left(-\frac{h^2}{2\sigma_z^2} \right) \quad (16.52b)$$

The terms $f(\sigma_z, h, h_i)$, f_1, and f_2 are defined as

$$f(\sigma_z,h,h_i) = \sum_{m=0}^{\infty} \left\{ \exp\left[-\frac{(2m-1)h_i + h}{(2)^{1/2}\sigma_z} \right]^2 + \exp\left[-\frac{2mh_i - h}{(2)^{1/2}\sigma_z} \right]^2 \right\} \quad (16.53)$$

and

$$f_1(\sigma_z,h,h_i) = \frac{h}{(2)^{1/2}\sigma_z} f(\sigma_z,h,h_i)$$

$$f_2(\sigma_z,h,h_i) = \frac{h^2}{2\sigma_z^2} f(\sigma_z,h,h_i) \quad (16.54)$$

When $(2)^{1/2}\sigma_z > h_i$, where the vertical equilibrium has been established by multiple reflections of the cloud between the two planes, Eq. 16.54 becomes

$$f_1(\sigma_z,h,h_i) = \frac{(\pi)^{1/2}h}{2h_i}$$

and

$$f_2(\sigma_z,h,h_i) = \frac{(\pi)^{1/2}h^2}{(2)^{3/2}\sigma_z h_i} \quad (16.54a)$$

When $(2)^{1/2}\sigma_z < h_i$, Eq. 16.54 becomes

$$f_1(\sigma_z,h,h_i) = \frac{h \exp(-h^2/2\sigma_z^2)}{(2)^{1/2}\sigma_z}$$

and

$$f_2(\sigma_z,h,h_i) = \frac{h^2 \exp(-h^2/2\sigma_z^2)}{(2)\sigma_z^2} \quad (16.54b)$$

With these functions the equations for the cloud concentrations from an instantaneous point source released at a height h become

$$\chi(x,y,0,t) = \frac{Q_x\,\sigma_{zI}}{\pi^{3/2}(2)^{1/2}\sigma_{xI}\sigma_{yI}h^2} f_2(\sigma_{zI}h,h_i) \times \exp\left\{ -\left[\frac{(x-\bar{u}t)^2}{2\sigma_{xI}^2} + \frac{y^2}{2\sigma_{yI}^2} \right] \right\} \quad (16.55)$$

and, when $(2)^{1/2}\sigma_{zI} > h_i$,

$$\chi(x,y,0,t) = \frac{Q_x}{2\pi\sigma_{xI}\sigma_{yI}h_i} \times \exp\left\{ -\left[\frac{(x-\bar{u}t)^2}{2\sigma_{xI}^2} + \frac{y^2}{2\sigma_{yI}^2} \right] \right\} \quad (16.56)$$

16-5.2.5 Average Gamma Dose from Continuous Release over Long Time Periods. During routine release of radioactive effluents from stacks, the local gamma dose received over a long period of time is of interest for comparison with the radiation limits, which usually permit averaging of the dose over time intervals of up to 1 yr. Estimates of the dose permit the establishment of emission limits, which, in turn, indicate the plant features needed to meet these limits.

The average dose calculated over such a period of time must include the effects of shifting winds, as well as the more normal parameters for estimating turbulent diffusion. The frequency with which the wind blows in a given direction is given by the conventional wind rose, usually on an 8- or 16-point compass or to the nearest 10°. For an estimate of long-term average doses, the frequency of occurrence of each wind speed in each atmospheric stability class must be provided to augment the wind rose.

If it is assumed that a source of radioactive material is continuously emitted at a constant rate of Q_0' (Ci/s) at a height h (m) above the ground plane for a time period t_d (s), the equilibrium concentration in the plume at a point (x,y,z) is given by

$$\bar{\chi}(x,y,x) = \frac{\bar{Q}_x'}{2\pi\bar{u}\sigma_y\sigma_z} \exp\left(-\frac{y^2}{2\sigma_y^2} \right) \times \left\{ \exp\left[-\frac{(z-h)^2}{2\sigma_z^2} \right] + \exp\left[-\frac{(z+h)^2}{2\sigma_z^2} \right] \right\} \quad (16.57)$$

where $Q_x' = Q_0'(Q_x/Q_0)$ is the activity release rate for a specific radionuclide corrected for depletion en route to the receptor and the coordinate system is the same as that used previously. The last exponential term in the equation represents the reflected component (i.e., that portion of the plume that has diffused to the ground plane and diffused back into the plume). For calculation of the direct gamma dose rate from the plume, the term for the reflected component will be retained in this form which, essentially, disregards the presence of the ground plane as a barrier but correctly describes the location of the reflected component source contribution with respect to a ground-level receptor.

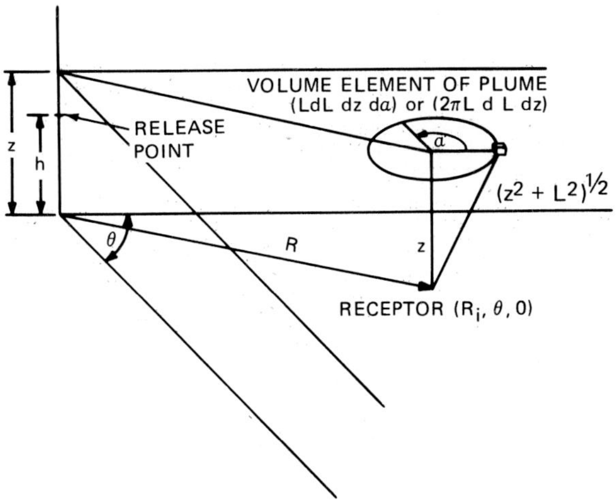

Fig. 16.14 Reference system for the long-term dose calculations.

The following treatment of the problem is based on the work of Gamertsfelder (1960). Figure 16.14 shows the reference system used in the following discussion. If the direction of the emission is continuously and uniformly shifting over a horizontal angle θ (radians) during the time period of concern, the concentration within the sector can be averaged in the horizontal direction across the sector. Integrating Eq. 16.57 in the y-direction from $-\infty$ to $+\infty$ and averaging the concentration over the width of the sector with R as the radial equivalent of x yields

$$\overline{\chi}(R,\theta,z) = \frac{Q'_R}{(2\pi)^{1/2}\sigma_z \bar{u}\, R\theta}\left\{\exp\left[-\frac{(z-h)^2}{2\sigma_z^2}\right] + \exp\left[-\frac{(z+h)^2}{2\sigma_z^2}\right]\right\} \qquad (16.58)$$

where $\overline{\chi}(R,\theta,z)$ is the average concentration (Ci/m^3) in the plume at a point R meters radially from the release point and z meters above the ground plane, θ is the angular width of the sector within which the plume is dispersed (radians), and Q'_R is the activity release rate corrected for depletion (Ci/s).

In cases where the distance between the source and the receptor is sufficiently great so that the change in concentration with distance is relatively small over the relaxation length of the gamma rays, the plume, averaged over the sector, can be considered to be comprised of a series of infinite horizontal plane sources. The plane sources vary in concentration according to a Gaussian distribution centered about the plume center plane, i.e., centered about the height of release. The gamma dose rate to tissue located a distance r (m) from a source increment of q curies was given by Eq. 16.35.

If an element of the source is located at a height z (m) above the ground plane and a horizontal distance L (m) from the receptor, the source is a distance $(L^2 + z^2)^{1/2}$ from the receptor. With the reference system shown in Fig. 16.14, an element of the source in the plane is $2\pi L$ dz. Substituting $(L^2 + z^2)^{1/2}$ for r in Eq. 16.35 and the expression for $\chi(R,\theta,z)$ in Eq. 16.58 for the incremental source q, the equation for the gamma dose rate from the plume can be written as

$$_\gamma D'(R,\theta,0) = \frac{0.0404\,\mu_a \bar{E}_\gamma Q'_x}{(2\pi)^{1/2}\bar{u}R\sigma_z\theta}\int_0^\infty\int_0^\infty \frac{1}{L^2+z^2} \times \left([1 + k\mu(L^2+z^2)^{1/2}]\left\{\exp\left[-\frac{(z-h)^2}{2\sigma_z^2}\right] + \exp\left[-\frac{(z+h)^2}{2\sigma_z^2}\right]\right\} \times \exp[-\mu(L^2+z^2)]\right) 2\pi L\, dL\, dz \qquad (16.59)$$

If $\bar{I}_1$ and $\bar{I}_2$ are defined as

$$\bar{I}_1 = \frac{1}{(2)^{3/2}\sigma_z}\int_0^\infty\int_0^\infty \frac{1}{L^2+z^2}\left\{\exp\left[-\frac{(z-h)^2}{2\sigma_z^2}\right] + \exp\left[-\frac{(z+h)^2}{2\sigma_z^2}\right]\right\}\exp\left[-\mu(L^2+z^2)^{1/2}\right] L\, dL\, dz$$

$$= \frac{1}{(2)^{3/2}\sigma_z}\int_0^\infty\left\{\exp\left[-\frac{(z-h)^2}{2\sigma_z^2}\right] + \exp\left[-\frac{(z+h)^2}{2\sigma_z^2}\right]\right\} \times \left\{\int_{\mu z}^\infty \frac{\exp[-\mu(L^2+z^2)^{1/2}]}{\mu(L^2+z^2)} \times d[\mu(L^2+z^2)^{1/2}]\right\} dz \qquad (16.59a)$$

$$\bar{I}_2 = \frac{1}{(2)^{3/2}\sigma_z}\int_0^\infty\int_0^\infty \frac{1}{(L^2+z^2)^{1/2}}\left(\left\{\exp\left[-\frac{(z-h)^2}{2\sigma_z^2}\right] + \exp\left[-\frac{(z+h)^2}{2\sigma_z^2}\right]\right\}\exp\left[-\mu(L^2+z^2)^{1/2}\right]\right)\mu L\, dL\, dz$$

$$= \frac{1}{(2)^{3/2}\sigma_z}\int_0^\infty\left\{\exp\left[-\frac{(z-h)^2}{2\sigma_z^2}\right] + \exp\left[-\frac{(z+h)^2}{2\sigma_z^2}\right]\right\} \times \exp(-\mu z)\, dz \qquad (16.59b)$$

Equation 16.59 can be written

$$_\gamma D'(R,\theta,0) = \frac{0.2865\,\mu_a Q'_x \bar{E}_\gamma}{\bar{u}R\theta}(\bar{I}_1 + k\bar{I}_2) \qquad (16.60)$$

Values of $\bar{I}_1$ and $\bar{I}_2$ are shown graphically in Figs. 16.15 and 16.16 as a function of $\mu\sigma_z$ for a variety of release heights. The dashed lines permit interpolating for release heights not given specifically. Equation 16.60 and others derived from the model will yield values that somewhat overestimate the gamma dose rate (or dose) if $\mu r\theta < 2$ but

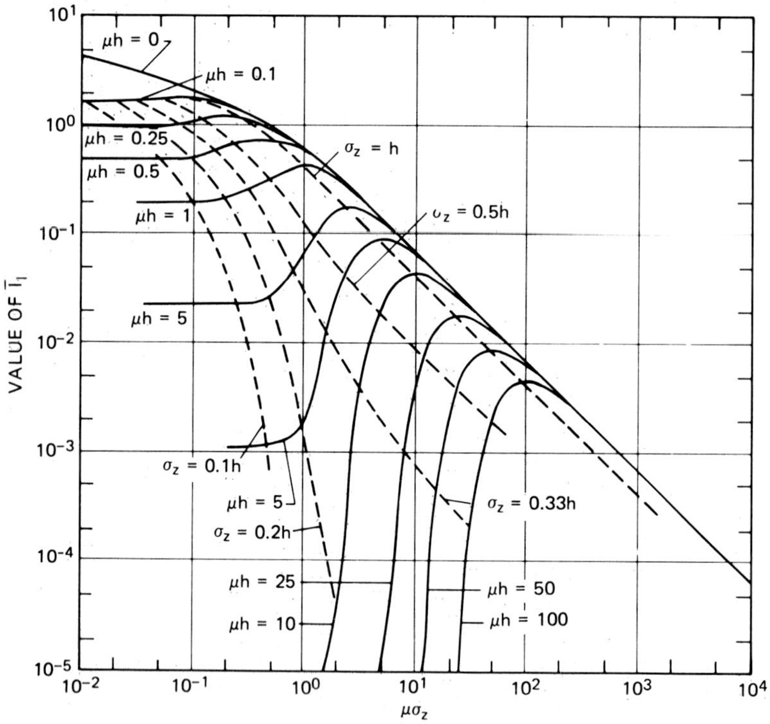

Fig. 16.15 Values of the $\overline{I}_1$ integral.

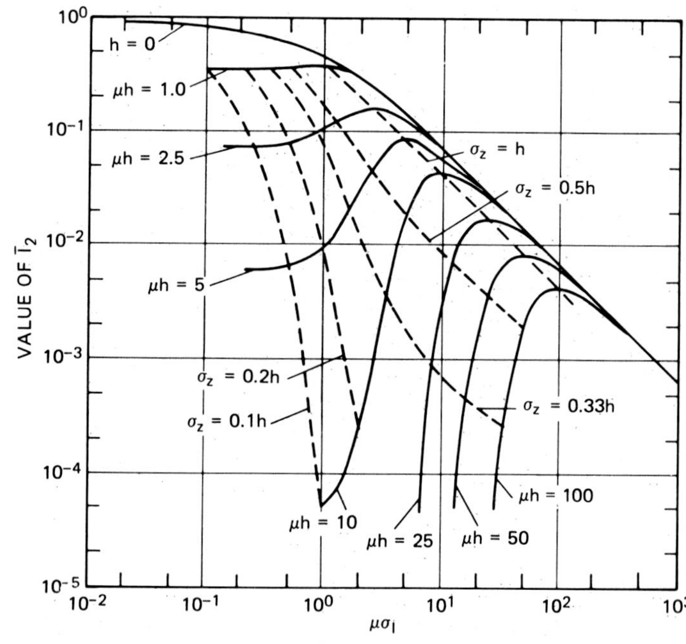

Fig. 16.16 Values of the $\overline{I}_2$ integral.

yield quite accurate values when $\mu r\theta \geqslant 2$. The gamma dose rate can be multiplied by the time period of discharge, t_d (s), to calculate the dose to the receptor. For a specific gamma energy, the term $[\overline{I}_1 + k\overline{I}_2]$ can be combined into a single term $\overline{I}_T$. Figure 16.17 gives the values of $\overline{I}_T$ for 0.7-MeV photons.

Equation 16.60 can be used to calculate the long-term direct gamma dose to receptors in the general vicinity of a point where a source of radioactive material is released continuously at a uniform rate over an extended period of time. The direct gamma dose from such a source to an individual located at $(R,r_1,0)$, a distance R (m) from the point of release in a sector of width θ (radians) at ground level, can be calculated by evaluating Eq. 16.61.

$$ {}_{\gamma}D_{(R,\theta_1,0)} = \frac{0.2865\,\Delta t}{\theta R} \left\{ \sum_{i=1}^{l} \sum_{j=1}^{m} \sum_{k=1}^{n} \frac{P_{\sigma_z(j,k,\theta_1)} Q'_{R(i,j,k)}}{\overline{u}_j} \times \mu_{a_i} \overline{E}_{\gamma_i} (\overline{I}_1 + k\overline{I}_2)_{(i,j,k)} \right\} \quad (16.61) $$

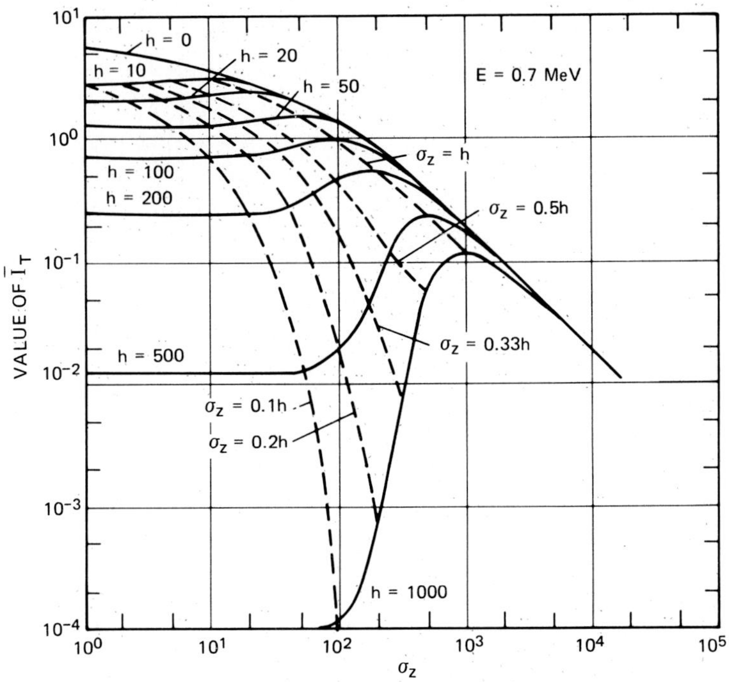

Fig. 16.17 Values of the $\bar{I}_T$ integral for 0.2-MeV gammas.

where $Q'_{R(i,j,k)}$ = source term (release rate) of nuclide type i corrected for depletion by ground deposition and radioactive decay en route to receptor location

$$= Q'_0 \left[\exp\left(-\lambda_i \frac{R}{\bar{u}_j}\right)\right](Q_x/Q_0)_{d(j,k)}\ \text{(Ci/s)}$$

$(Q_R/Q_0)_{d(j,k)}$ = depletion factor for ground deposition

$$= \exp\left[-\left(\frac{2}{\pi}\right)^{1/2} \frac{v_d}{\bar{u}_j}\right.$$

$$\left. \times \int_0^R \frac{\exp(-h^2/2\sigma_{z_k})}{\sigma_{z_k}}\right] dR$$

$(\bar{I}_1 + k\bar{I}_2)_{(i,j,k)}$ = function of μ_i, σ_{z_k}, and $\bar{u}_j$

$\bar{u}_j$ = average of j component of wind-speed range, m/s

σ_{z_k} = standard deviation of plume concentration in vertical plane for stability class k, m

$P_{\sigma_z(j,k,\theta_1)}$ = joint probability of wind speed and atmospheric stability into sector θ_1

Δt = exposure time interval, s

θ = angular width of sector θ_1, radians

R = distance from release point, m

At long distances from the point of release, σ_z becomes large and the cloud becomes uniform in concentration and can be regarded as an infinite cloud for which the gamma dose rate is given by Eq. 16.37, with the assumption that the value of χ used includes the reflected component of the cloud.

A receptor located on the line that defines the division between sectors would receive an intermediate dose between those doses calculated for the two sectors. In practice the doses for each of the sectors are calculated at several radial distances over the range of interest. The results are plotted (dose vs. distance), and the radial distances in each sector are determined for several selected dose values (usually differing by factors of 10^1 or $10^{0.5}$); these locations are plotted on circular coordinate paper for the center-line rays in each sector. Points representing the location of common dose values are then connected with a smooth curve to represent the dose isopleths. When the many approximations in developing the mathematical models and the difficulty in obtaining input data are considered, more exact methods are not warranted.

16-5.3 Gamma Dose from Radioactive Materials Deposited on the Ground

A passing cloud of radioactive particles or halogens will deposit radioactive isotopes on the ground by dry deposition, washout, or rainout. If gamma rays are emitted, the radioactive materials on the ground will result in a gamma-radiation dose for a period of time after the cloud has passed. The ground pattern of these materials can be estimated by the methods indicated in other chapters. The persistence of the dose is dependent on the radioactive decay of the materials involved and the amount of cleanup by natural forces or by decontamination efforts.

16-5.3.1 The Infinite Plane Source. When the contamination is uniformly spread over a wide area, the intensity, I, of the radiation for various distances above the ground is given by

$$I = \frac{S}{2} E_i(\mu r) + \frac{kS \exp(-\mu r)}{2} \tag{16.62}$$

where I = gamma-ray intensity photons/(m^2·s)
$E_i(\mu r)$ = exponential integral

$$\int_{\mu r}^{\infty} \exp -\frac{\mu r}{r}\, dr$$

S = source strength per unit area of the plane, photons/(m^2·s)

The exponential integral, $E_i(\mu r)$, occurs in many shielding problems and is given in tabular form by Jahnke, Emde, and Lösch (1960). Equation 16.62 can be converted to dose rate units to give

$$_\gamma D'_\infty = \frac{I\,\bar{E}_\gamma\,\mu_a\,(1.6 \times 10^{-6})(1.11)}{(1293)\,(100)} \tag{16.63}$$

$$_\gamma D'_\infty = 1.37 \times 10^{-11}\, I\bar{E}_\gamma \mu_a \tag{16.64}$$

If I is replaced by its equivalent, as given in Eq. 16.62, and one photon of average energy $\bar{E}_\gamma$ is assumed to be emitted per disintegration, and ω is the number of Ci/m^2 on the ground, the gamma dose rate is given by

$$_\gamma D'_\infty = 0.254\, \omega\, \bar{E}\lambda\, \mu_a \left[E_i(\mu b) + \frac{k \exp(-\mu b)}{2}\right] \tag{16.64a}$$

where $_\gamma D'_\infty$ is the gamma dose rate (rads/s) to a receptor located b meters above an infinite plane uniformly contaminated with ω Ci/m^2. The constant in the buildup factor has been divided by 2 to provide an approximate correction for the primary radiation that enters the ground and is scattered to the receptor with a much lower efficiency than that which enters the air. This equation considers all the radioactive material to be on the surface with no shielding by cover over the source or by uneven terrain. For this reason the results will be high in most practical situations. Figure 16.18 gives the dose rate at several heights above the ground for radiation with energies from 0.1 to 2 MeV as obtained from Eq. 16.64.

16-5.3.2 The Finite Plane Source. If radioactive materials are deposited from a plume, either by dry deposition or by rainout or washout, the deposition pattern will consist of a maximum at the center line, with the deposition decreasing with distance from the center line according to the Gaussian distribution of concentration in the cloud and for a given time of emission, t. This pattern can be described by

$$\omega_{(x,y,0,t)} = \omega_{(x,0,0,t)} \exp\left(-\frac{y^2}{2\sigma_y^2}\right) \tag{16.65}$$

where $\omega_{(x,0,0,t)}$ is the deposition (Ci/m^2) at distance x on the cloud center line and $\omega_{(x,y,0,t)}$ is the deposition at a distance y from the center line. Since the value of ω will vary as a function of time owing to radioactive decay or

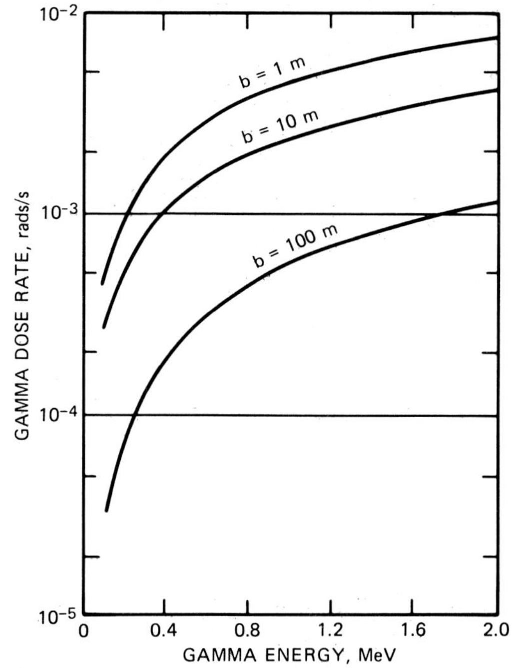

Fig. 16.18 Gamma dose rate at heights of 1, 10, and 100 m aboveground for an infinite plane source of 1 Ci/m^2.

decontamination efforts, the equation for ω must account for these parameters.

Gamertsfelder (1960) has derived an equation to give the gamma dose rate to tissue at a height b above the center line of the pattern. From the equation for a point source,

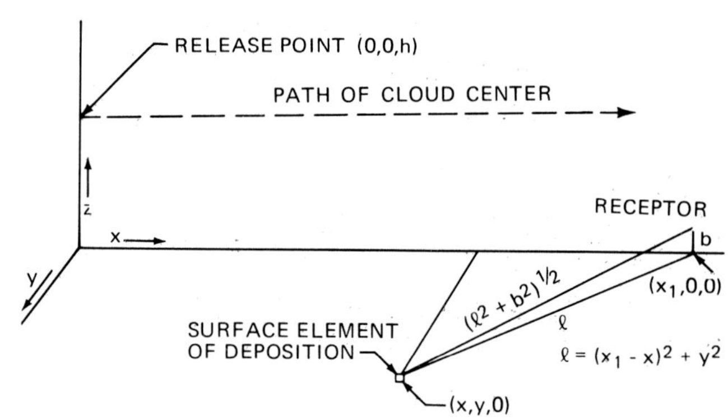

Fig. 16.19 Reference system for equations involving deposition patterns of finite size.

Eq. 16.35, and the coordinate system given in Fig. 16.19, the dose rate is given by

$$_\gamma D'_{(x_1,0,b,t)} = 0.0404 \mu_a \bar{E}_\gamma \omega_{(x_1,0,0,t)} \times \int_{-\infty}^{\infty}\int_0 \frac{[1 + (k/2)\,\mu(\ell^2 + b^2)^{1/2}]\exp[-\mu(\ell^2 + b^2)^{1/2}]}{\ell^2 + b^2} \times \exp\left(-\frac{y^2}{2\sigma_y^2}\right) dx\, dy \tag{16.66}$$

where $_\gamma D'_{(x_1,0,b,t)}$ is in rads per second at time t and (k/2) is used in the buildup factor to allow for loss of scattered radiation from that part of the primary radiation which starts in a downward direction. With G_1 and G_2 defined as

$$G_1 = \frac{1}{(2)^{3/2}\pi\mu\sigma_y} \times \int_{-\infty}^{\infty}\int_0^{\infty} \frac{\exp[-\mu(\ell^2+b^2)^{1/2}]\exp(-y^2/2\sigma_y^2)}{\ell^2+b^2} \times dx\,dy \tag{16.67}$$

$$G_2 = \frac{1}{(2)^{3/2}\pi\mu\sigma_y} \times \int_{-\infty}^{\infty}\int_0^{\infty} \frac{\mu\exp[-\mu(\ell^2+b^2)^{1/2}]\exp(-y^2/2\sigma_y^2)}{(\ell^2+b^2)^{1/2}} \times dx\,dy \tag{16.67a}$$

Then

$$_\gamma D'_{(x_1,0,b,t)} = 0.36\,\omega_{(x,0,0,t)}\,\sigma_y\mu\mu_a\bar{E}_\gamma \times \left(G_1 + \frac{k}{2}G_2\right) \tag{16.68}$$

For cases where the cloud is large with respect to the range of the gamma radiation ($\mu\sigma_y > 10$), the ground concentration is essentially uniform in all directions and the value of $\exp(-y^2/2\sigma_y^2)$ can be taken as unity for all values of y. Under these conditions the values of G_1 and G_2 can be written as

$$G_1 = \frac{1}{(2)^{1/2}\mu\sigma_y}\int_0 \frac{\exp[-\mu(\ell^2+b^2)^{1/2}]}{\ell^2+b^2}\,\ell\,d\ell = \frac{1}{(2)^{1/2}\mu\sigma_y}E_i[-\mu(\ell^2+b^2)^{1/2}] \tag{16.69}$$

$$G_2 = \frac{1}{(2)^{1/2}\mu\sigma_y}\int_0^{\infty} \frac{\mu\exp[-\mu(\ell^2+b^2)^{1/2}]}{(\ell^2+b^2)^{1/2}}\,\ell\,d\ell = \frac{1}{(2)^{1/2}\mu\sigma_y}\int_{\mu b}^{\infty}\frac{\exp(-\mu b)d(\mu b)}{(2)^{1/2}\mu\sigma_y} \tag{16.70}$$

The exponential integral, $E_i[-\mu(\ell^2+b^2)^{1/2}]$, is as described previously. These values for G_1 and G_2 provide the upper limits of the integrals. Values for smaller finite deposition patterns were obtained by Gamertsfelder by numerical evaluation of the integrals and are given in Figs. 16.20 and 16.21.

The ratio of the gamma dose rate above the center line of a deposition pattern smaller than that which can be considered as an infinite plane to the dose rate above an infinite plane can be obtained from Eqs. 16.64 and 16.68. Figure 16.22 indicates this ratio for 0.7-MeV gamma photons. For most deposition patterns of interest, the correction factor for cloud size is comparatively small. Calculations for other energies indicate that Fig. 16.21 gives reasonable estimates of the ratio for energies up to 2 MeV, but the difference may be 5 to 10% for the small pattern widths when the energy is as low as 0.1 MeV.

These equations provide an estimate of the gamma dose rate for a given amount of radioactive material on the ground at a specific time. If radioactive decay or weathering is of importance during the period of exposure, the total dose received must be corrected for the decrease in dose rate. In the simplest case, a release of a single nuclide and subsequent deposition, the decrease can be described by a single exponential, and the dose rate at any given time, t (s), after the cloud has passed the receptor can be written

$$_\gamma D'_{(x_1,0,b,t)} = 0.36\,\omega_{(x,0,0,0)}\,\sigma_y\mu\mu_a\bar{E}_\gamma \times \left(G_1 + \frac{k}{2}G_2\right)\exp(-\lambda_R t) \tag{16.71}$$

where $\omega_{(x,0,0,0)}$ is the center-line deposition based on the source at the time of release and λ_R is the decay constant ($0.693/T_{1/2}$, where $T_{1/2}$ is in seconds). Integrating from the cloud arrival time to the time at which the exposure estimate is desired gives the gamma-dose equation for the finite deposition pattern.

$$_\gamma D_{(x_1,0,b)} = 0.36\,\omega_{(x_1,0,0,t)}\,\sigma_y\mu\mu_a\bar{E}_\gamma \times \left(G_1 + \frac{k}{2}G_2\right)\left[\frac{1-\exp(-\lambda_R t)}{\lambda_R}\right] \tag{16.72}$$

In this equation, $_\gamma D_{(x_1,0,b)}$ is in rads, and t is the time interval in seconds after arrival over which the dose estimate is desired.

For more complex rates of removal, such as a power function, similar integrations over time can be carried out, but the resulting integrals will probably not be solved as easily and tables of integrals or numerical solutions may be needed.

16-5.4 Protection by Structures

The computational methods presented have assumed that the receptor is in the open, usually in the center of a large flat plain. In most cases, there will be some reduction of dose if the receptor is in a building. Although the integrals for estimating the dose received inside a building have not been evaluated exactly, some estimates of the protection factor expected can be obtained by assuming simple geometrical conditions for which analytical solutions can be found. *The following methods of estimating the protection factors afforded by structures are approximate and are given to supplement factors arrived at by more precise methods, either measurement or calculation.*

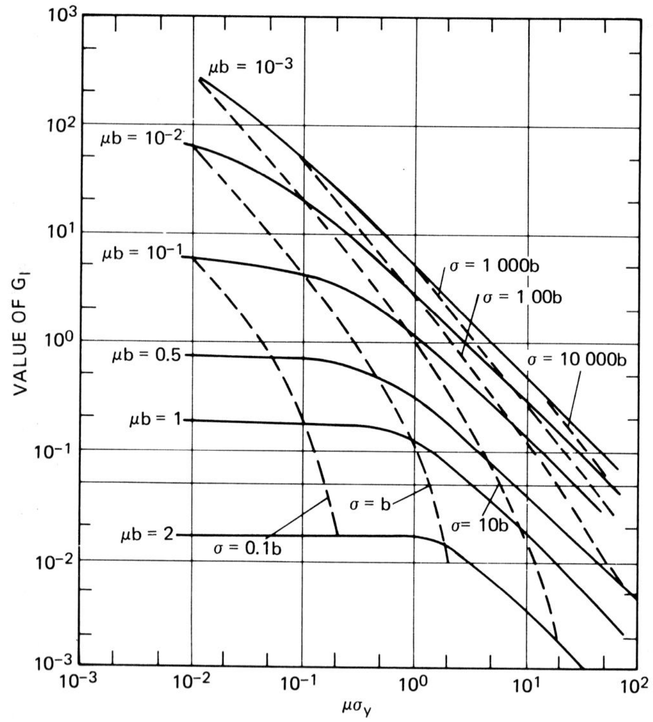

Fig. 16.20 Values of the G_1 integral.

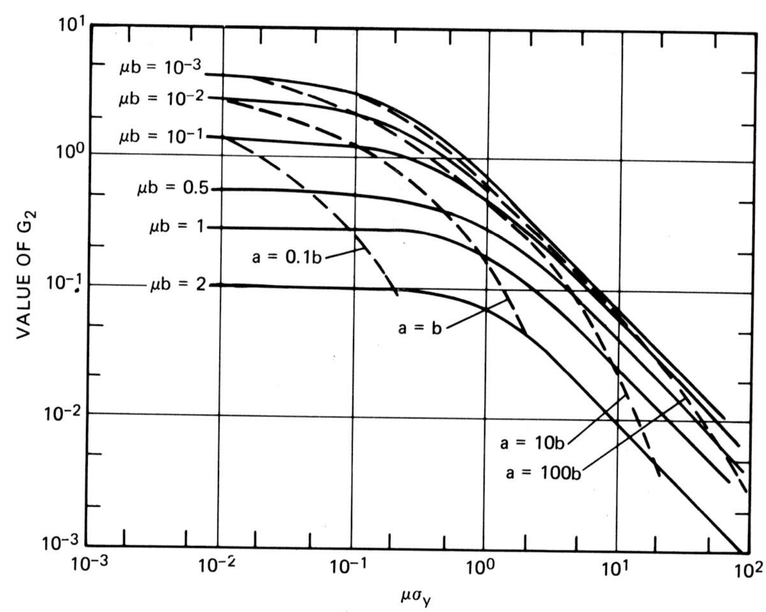

Fig. 16.21 Values of the G_2 integral.

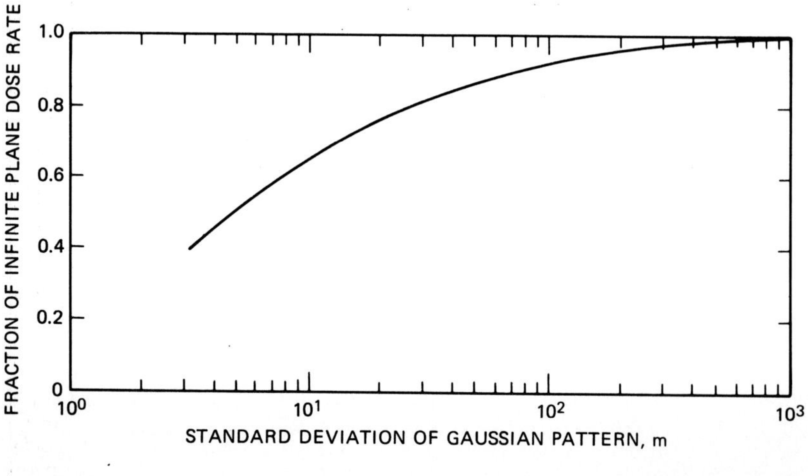

Fig. 16.22 Ratio of the gamma dose rate from a Gaussian pattern of finite size to that from an infinite plane pattern with deposition equal to the center line of the finite Gaussian pattern.

16-5.4.1 Cloud Gamma Dose. A structure can provide protection both by the geometrical effect of limiting the distance of approach of the cloud and by shielding afforded by the walls. A reasonable initial assumption for these calculations of geometrical protection is that the radioactive material does not enter the building.

A rough estimate of the protection factor afforded by various structures can be obtained by assuming the cloud to be hemispherical in shape and uniform in concentration throughout and by using a hemisphere of radius a to approximate the building. Under these conditions the gamma dose rate in the building will be given by

$$_\gamma D' = (\text{constant})\, B_x \exp(-\mu_x w) \times \int_a^{r_1} \frac{(1 + k\mu r) \exp(-\mu r)\, 2\pi r^2}{4\pi r^2}\, dr \qquad (16.73)$$

where w = thickness of shielding afforded by building walls, g/cm^2

μ_x = mass absorption coefficient for wall material, cm^2/g

r = uniform cloud radius, cm

a = building radius, cm

μ = absorption coefficient for air (Fig. 16.4), cm^2/g

B_x = dose buildup factor in the building material (dimensionless)

constant = dose-rate conversion factor

The dose buildup factor is the one appropriate for the building material and the energies and directions of radiation falling on the material.

Integrating Eq. 16.73 and substituting $B_x = (1 + k_x \mu_x w)$ gives

$$_\gamma D' = \frac{\text{constant}}{2\mu} (1 + k_x\mu_x w) \exp(-\mu_x w) \big[\exp(-\mu a) - \exp(-\mu r_1) - k(1 + \mu r_1) \exp(\mu r_1) + k(1 + \mu a) \exp(-\mu a)\big] \qquad (16.74)$$

If the building is absent, a and w equal zero, and Eq. 16.74 becomes

$$_\gamma D' = \frac{\text{constant}}{2\mu} [1 - \exp(-\mu r_1) - k(1 + \mu r_1) \times \exp(-\mu r_1) + k] \qquad (16.75)$$

The protection factor (P.F.) provided by the building is then the ratio of Eqs. 16.74 and 16.75

$$\text{P.F.} = (1 + k_x\mu_x w) \exp(-\mu_x w) \{[\exp(-\mu a) - \exp(-\mu r_1) - k(1 + \mu r_1) \exp(-\mu r_1) + k(1 + \mu a) \exp(-\mu a)] \times [1 - \exp(-\mu r_1) - k(1 + \mu r_1) \exp(-\mu r_1) + k]^{-1}\} \qquad (16.76)$$

The term in the braces expresses the protection factor from the geometrical considerations, whereas the term outside the braces expresses the shielding factor. These two factors have been evaluated separately.

Figure 16.23 presents the geometric protection factor for buildings up to 50 m in effective radius and for infinite clouds and clouds having a radius of 100 m. It is apparent that the geometrical protection is insignificant for most residential buildings in a very large cloud but becomes somewhat more important as the cloud becomes smaller.

The values given in Fig. 16.23 are for an individual in the center of the building. If he is against one wall, the reduction in dose would be approximately one-half of the reduction for a hemisphere twice as large. Thus, from Fig. 16.23, the reduction factor in the center of a 10-m-radius building for an infinite cloud is 0.96. If the receptor were against one wall, the reduction would be one-half of that for a 20-m building, or 1.0 − 0.075/2 = 0.963.

The shielding effectiveness was evaluated with the mass absorption coefficient for most construction materials considered to be about the same as that of air for the energies of interest. The mass absorption coefficients are given in Table 16.8 for materials and elements commonly used in construction (Wyckoff, 1959).

The attenuation through various thicknesses of material expressed as the dimensionless quantity μw was calculated from

$$\text{P.F.} = B_x \exp(-\mu_x w) \qquad (16.77)$$

where B_x is the buildup factor for radiation passing through a thickness w of a material with an absorption coefficient

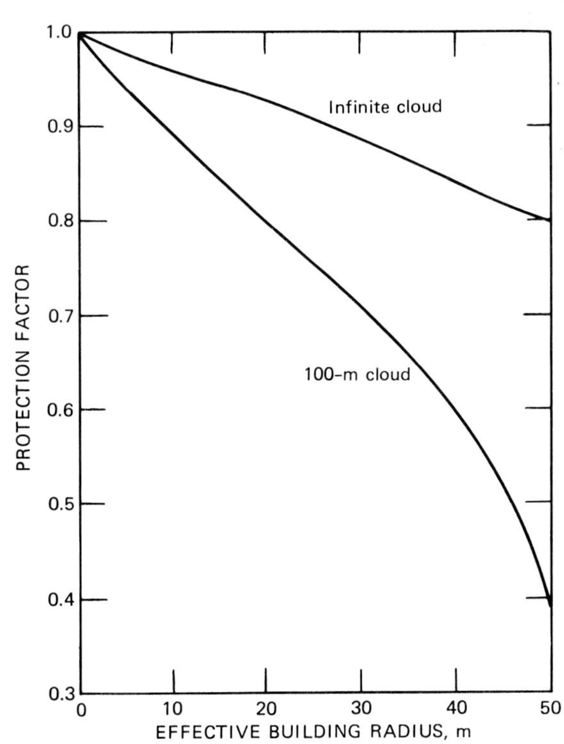

Fig. 16.23 Apparent dose reduction (protection factor) afforded by the geometrical configuration of buildings.

TABLE 16.8 Mass Absorption Coefficients for Air and Elements Found in Construction Materials

Energy, MeV	Mass absorption coefficients (μ_x), cm²/g						
	Air	Concrete	Iron	Calcium	Silicon	Aluminum	Oxygen
0.1	0.151	0.169	0.344	0.238	0.172	0.161	0.151
0.3	0.106	0.107	0.106	0.109	0.107	0.103	0.107
0.5	0.0868	0.0870	0.0828	0.0876	0.0869	0.0840	0.0870
0.6	0.0804	0.0804	0.0762	0.0809	0.0802	0.0777	0.0806
0.8	0.0706	0.0706	0.0664	0.0708	0.0706	0.0683	0.0708
1.0	0.0635	0.0635	0.0595	0.0634	0.0635	0.0614	0.0636
1.5	0.0517	0.0517	0.0485	0.0518	0.0517	0.0500	0.0518
2.0	0.0445	0.0445	0.0424	0.0451	0.0447	0.0423	0.0445

μ_x. The buildup factor varies with the energy of the radiation, the atomic number of the shielding material, and the direction of the radiation falling on the shield. Figures 16.24 and 16.25 illustrate the buildup factors for various thicknesses of water, aluminum, and iron for 0.5- and 2.0-MeV photons originating as a point source and as a plane monodirectional source. The value for a point source in air as computed by $(1 + k\mu w)$ is given for comparison. In this problem, one would expect the source to be between the point source and the monodirectional plane source with a fraction of the photons falling on the shield degraded in energy to a value below that of the initial energy.

In view of these factors and the simplifying assumptions already made, the buildup factor for air was used in

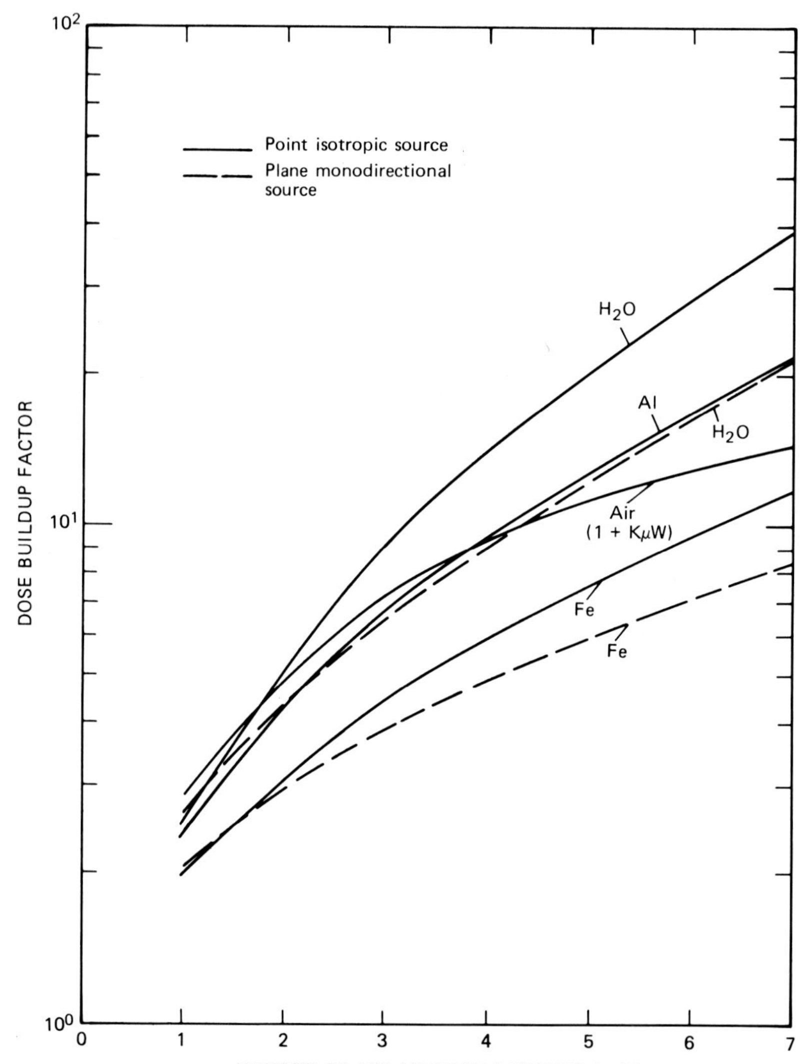

Fig. 16.24 Buildup factors for 0.5-MeV gamma radiation.

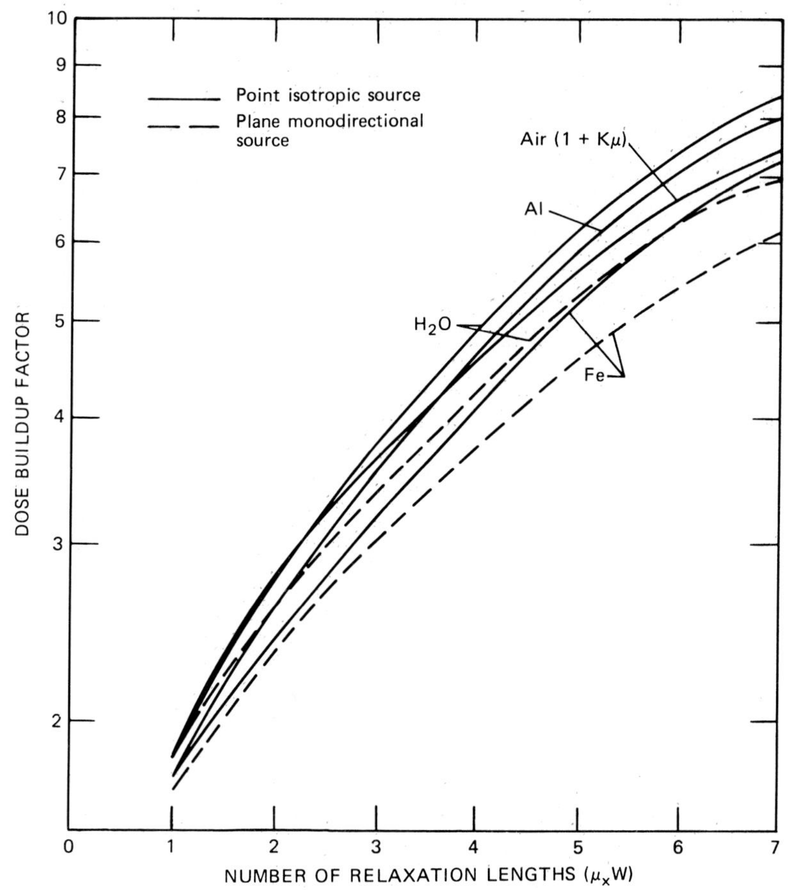

Fig. 16.25 Buildup factors for 2-MeV gamma radiation.

estimating the shielding protection factor. This buildup factor will underestimate the protection factor for low-shielding thickness of higher atomic numbers but may overestimate it for thicker structures of lower atomic numbers.

Figure 16.26 gives the protection factors for various thicknesses of shields for 0.3-MeV, 0.7-MeV, and 2-MeV gamma photons. For illustration, a building with 8-in.-thick brick walls and roof would have a shielding thickness of about 20 cm. For brick with a density of 1.9, the shielding mass thickness would be 38 g/cm^2. For a 0.7-MeV gamma photon with an absorption coefficient of 0.075 cm^2/g, the thickness in units of absorption coefficient times the mass thickness would be 2.85 or the estimated protection factor from shielding would be about 0.33. For buildings of lighter construction, such as most wooden residences, the shielding is negligible.

Most buildings have relatively thin roofs, and thus the shielding from material overhead is negligible. The overall protection factor can be estimated by considering the fraction of the solid angle subtended by the roof and the walls and applying the appropriate shielding and geometrical corrections. Figure 16.27 gives the fraction of a hemisphere that is subtended by the roof for buildings of the given ratio of effective diameter to height.

From these data the order of magnitude of dose reduction expected by structures can be estimated. For the usual wooden residence, about 20 by 20 by 10 m, only the geometric factor would be of any significance. From Fig. 16.23 this geometric factor would amount to 93 to 94% of the outside dose for an individual in the center of the ground floor. If the walls are of brick 8 in. (20 cm) thick, about 70% of the solid angle would be shielded and about 30% open (Fig. 16.27). Thus in an infinite cloud an individual in the center of the first floor would be exposed to 0.93 (0.7 × 0.33 + 0.3) = 0.5 of the cloud dose outside the building. If the individual were on the second floor halfway up the building (diameter/height ratio = 4), the angle subtended by the roof would be 0.55, and his protection factor would be 0.93 (0.45 × 0.33 + 0.55) = 0.65. Actually the geometric protection factor will not be as great as 0.93 in the latter case because the effective radius is smaller. If the individual were directly under the roof, the building would offer little or no protection. As a further illustration, the protection factor offered in the center of the ground floor of a large industrial building 100 by 100 by 20 m with 8-in. (20-cm)-thick brick walls for a large cloud emitting 0.7-MeV photons would be [a geometrical protection factor of 0.86 (35-m radius) and a roof fraction of 0.63 being assumed].

$$\text{P.F.} = 0.86\,(0.37 \times 0.33 + 0.63) = 0.65$$

Additional shielding may be available from floors or equipment; this would increase the protection factor.

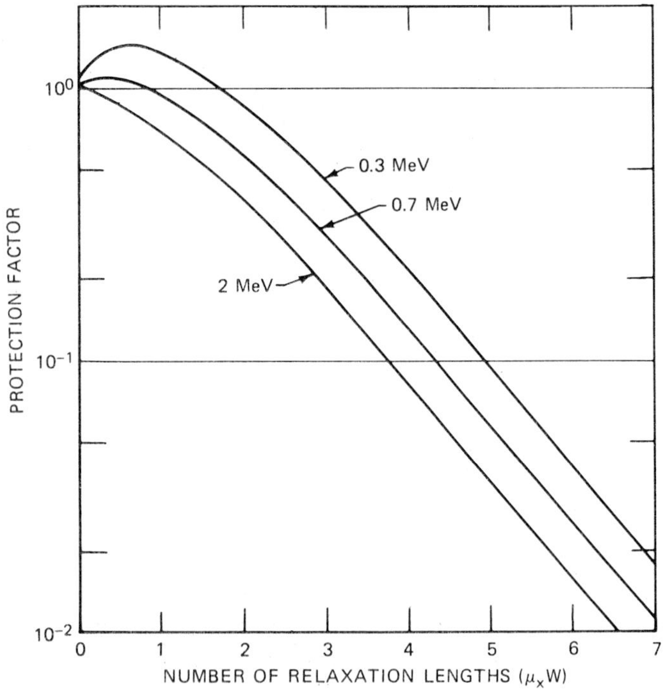

Fig. 16.26 Shielding protection factors.

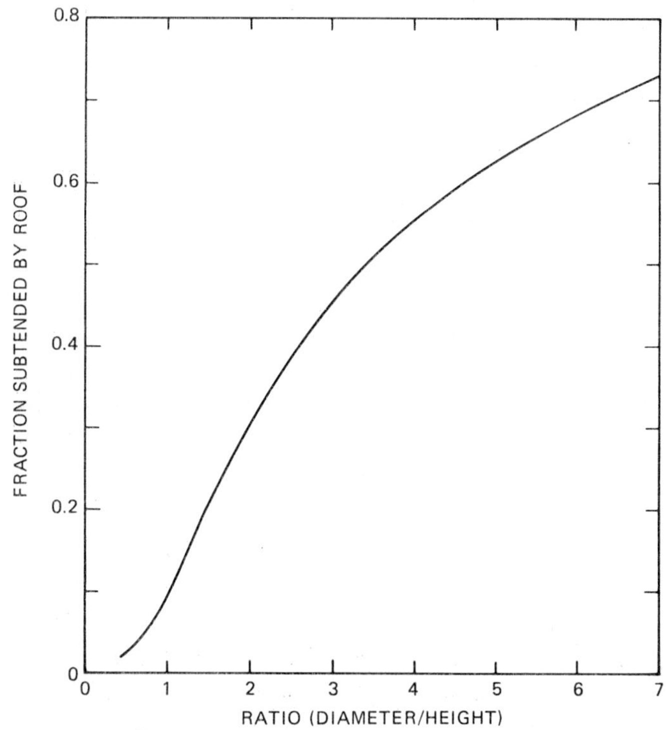

Fig. 16.27 Fraction of hemisphere subtended by roof of building with various ratios of diameter to height.

The foregoing methods lay no claim to accuracy, but they do illustrate the approximate protection against gamma radiation from a passing cloud. Except for small clouds, it appears that the ordinary residence provides little reduction in dose and that, for large clouds, only massive buildings with complete shielding are effective.

TABLE 16.9 Representative Protection Factors from Gamma Cloud Source*

Structure	Protection factor	Representative range
Wood frame house† (no basement)	0.9	
Basement of wood house	0.6	0.1 to 0.7‡
Masonry house (no basement)	0.6	0.4 to 0.7‡
Basement of masonry house	0.4	0.1 to 0.5‡
Large office or industrial building	0.2	0.1 to 0.3‡ §

*From U. S. Nuclear Regulatory Commission, Reactor Safety Study, An Assessment of Accident Risks in U. S. Commercial Nuclear Power Plants, NRC Report WASH-1400 (NUREG-75/014), 1975.

†A wood frame house with brick or stone veneer is approximately equivalent to a masonry house for shielding purposes.

‡This range is mainly due to different wall materials and different geometries.

§The reduction factor depends on where the personnel are located within the building (e.g., the basement or an inside room).

Burson and Profio (1975) have made more exact calculations for the gamma rays from a cloud of fission products produced in a major reactor accident. Their values were obtained along the same general principles given previously and are for a semiinfinite cloud surrounding the structure. Table 16.9 presents a summary of their results as given in the Rasmussen study (U. S. Nuclear Regulatory Commission, 1975).

16-5.4.2 Radiation from Deposited Materials. The radiation dose rate at a height b above a uniformly contaminated area is given by Eq. 16.64a. If the building provides a clean area around the receptor of radius a (m), the dose rate from the material outside the clean area is given in units of rads per second by

$$_\gamma D' = 0.254\,\omega \bar{E}_\gamma \mu_a \{E_i[\mu(a^2+b^2)^{1/2}] + (k/2)\exp[-\mu(a^2+b^2)^{1/2}]\} \quad (16.78)$$

The ratio of Eq. 16.78 to Eq. 16.64a then gives the geometrical protection factor

$$\text{P.F.} = \frac{E_i[\mu(a^2+b^2)^{1/2}] + (k/2)\exp[-\mu(a^2+b^2)^{1/2}]}{E_i(\mu b) + (k/2)\exp(-\mu b)} \quad (16.79)$$

Figure 16.28 gives the reduction in dose at 1 and 4 m above grade, compared to the dose 1 m above an infinite plane contaminated to a uniform level with radioactive materials emitting 0.7-MeV gamma rays. Note that the reduction in dose on the second floor of an unshielded building is negligible.

This treatment ignores any deposition on the roof and sides of the building and the resulting radiation dose. The dose from the roof can be approximated by considering the roof to be a disk source at a distance b_1 meters from the

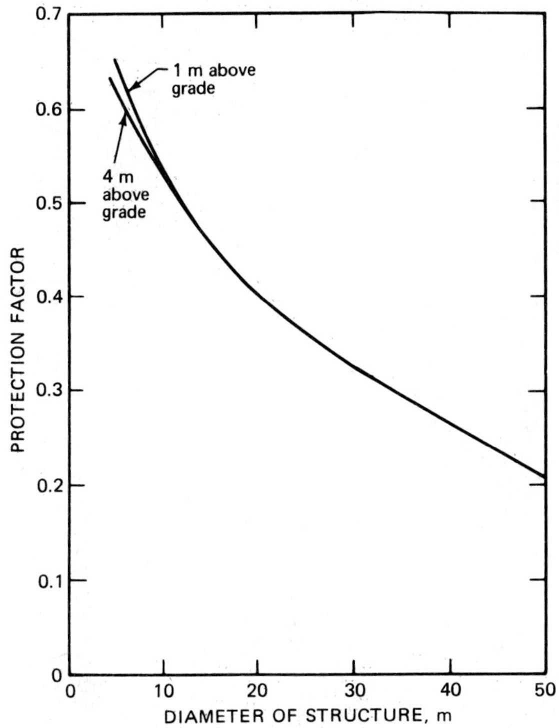

Fig. 16.28 Geometrical protection from ground deposition of radionuclides emitting 0.7-MeV gammas outside of a structure.

receptor. If the shielding of the roof and floor materials is ignored, this dose rate is given by

$$_{\gamma}D' = 0.254\,\omega \bar{E}_{\gamma}\mu_a \{E_i(\mu b_1) - E_1[\mu(b_1^2 + a^2)^{1/2}] + \frac{k}{2}\exp(-\mu b_1) - \frac{k}{2}\exp[-\mu(b_1^2 + a^2)^{1/2}]\} \quad (16.80)$$

where a is the diameter of the roof. Again, the ratio of Eq. 16.80 to Eq. 16.64a will give the geometrical protection factor.

$$\text{P.F.} = \{E_1(\mu b_1) - E_i[\mu(b_1^2 + a^2)^{1/2}] + (k/2\exp(-\mu b_1) - (k/2)\exp[-\mu(b_1^2 + a^2)^{1/2}]\} \times [E_i(\mu b) + (k/2)\exp(-\mu b)]^{-1} \quad (16.81)$$

This equation was solved by considering the ratio of the dose received at various distances from the roof to the dose received 1 m above a uniformly contaminated infinite plane emitting 0.7-MeV gamma rays. These results are given in Fig. 16.29.

To illustrate the protection afforded, consider a wood frame house 20 m square and 10 m high. From Fig. 16.28 the dose at the center of the ground floor will be about 40% of the infinite plane dose. At the ground floor the roof is 10 m away with a diameter of 20 m. From Fig. 16.29 the roof will contribute an additional 7%, for a total of 47%. On the second floor where the roof is 3 to 5 m away, the roof will contribute 20%, and the dose from the surrounding area is about the same. For this floor the protection factor drops to about 60% of the dose 1 m above an infinite plane. These calculations assume that the roof is contaminated to the same level as the surrounding area and that no radioactive material is adhering to the walls.

If the same house is made of brick 8-in. (31-cm)-thick, the radiation from the ground outside is reduced by a factor of 0.33 (Fig. 16.26) and the protection factor on the ground floor becomes about 0.2 on the ground floor and about 0.33 on the second floor.

Burson and Profio (1975) have evaluated the protection factor for individual structures using a point-kernel integration, including buildup from scatter. A summary of their results, obtained from the Rasmussen study (U. S. Nuclear Regulatory Commission, 1975), is given in Table 16.10.

16-5.4.3 Contamination Inside a Building. The interior of buildings in the path of a radioactive cloud will not remain free of contamination. Most buildings, unless specifically designed to be airtight, have changes of air from leaks in the structure resulting from pressure differences caused by wind speeds or by temperature differences in heated buildings. If windows or doors are open, interchange of air with the outside can be relatively rapid.

If a building is surrounded by a cloud of average concentration $\bar{\chi}$, the amount of activity, dq, entering is given by

$$dq = V\bar{\chi}\,dt - \frac{Vq}{s}\,dt \quad (16.82)$$

where V is the ventilation rate in cubic meters per time and s is the volume of the building (q/s is the concentration in the building). This equation assumes that the activity entering is uniformly mixed through the building on entry. Solving Eq. 16.82 and using the condition that q/s = 0 at zero time gives the concentration in the building as

$$\frac{q}{s} = \bar{\chi}\left[1 - \exp\left(-\frac{V}{s}t\right)\right] \quad (16.83)$$

If V/s is one change per hour (a reasonably tight building with light winds), the concentration inside will build to 10% of that outside in 6 min, to 50% in 40 min, and to 90% in 2.3 hr. If V/s = 6, the concentration inside will build to 10% of that outside in 1 min, to 50% in 7 min, and to 90% in 23 min.

Since the total inhalation depends on the integral of the concentration vs. time curve, it is interesting to compare the integral inside the building with that outside for a transient cloud. If the time of cloud passage is taken as t_1 (with the time of cloud arrival as t_0), then the integrated concentration is given by

$$\left(\frac{q}{s}\right)_f = \int_0^{t_1} \bar{\chi}\left[1 - \exp\left(-\frac{V}{s}t\right)\right]dt + \int_0^{\infty}\left(\frac{q}{s}\right)_{t_1} \times \exp\left(-\frac{nV}{s}t\right)dt \quad (16.84)$$

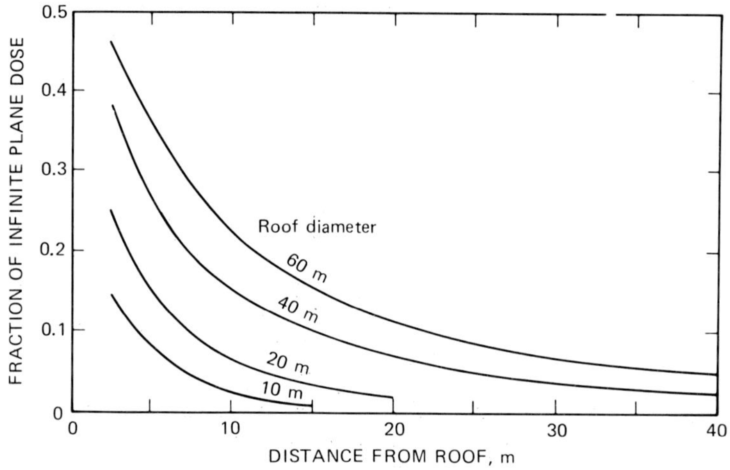

Fig. 16.29 Radiation received from deposit on roof given as a fraction of dose at 1 m above an infinite plane for 0.7-MeV gamma radiation.

TABLE 16.10 Representative Protection Factors for Surface Deposition

Structure	Representative protection factor	Representative range
One- and two-story wood frame house (no basement)	0.4	0.2–0.5
One- and two-story block-and-brick house (no basement)	0.2	0.04–0.40
House basement, one or two walls fully exposed	0.1	0.03–0.15
One story, less than 2 ft of basement, walls exposed	0.5	0.03–0.07
Two stories, less than 2 ft of basement, walls exposed	0.03	0.02–0.05
Three- or four-story structures, 5000 to 10 000 ft² per floor		
First and second floors	0.05	0.01–0.08
Basement	0.01	0.001–0.07
Multistory structures, >10 000 ft² per floor		
Upper floors	0.01	0.001–0.02
Basement	0.005	0.001–0.015

where $(q/s)_{t_1}$ is the concentration in the building after the cloud passage. The ventilation rate after cloud passage is taken as $\eta V/s$, where η is a factor describing possible changes in the ventilation rate due to changes in wind speed or to deliberate actions, such as opening windows or turning on ventilation.

Solving and substituting for q/s at t_1 from Eq. 16.83 gives

$$\left(\frac{q}{s}\right)_f = \bar{\chi}\left\{t_1 - \frac{s}{V}\left[1 - \exp\left(-\frac{V}{s}t_1\right)\right]\left(1 - \frac{1}{\eta}\right)\right\} \qquad (16.85)$$

If the ventilation rate remains the same, $\eta = 1$, the integral concentration inside becomes $\bar{\chi}t_1$, *which is the same as that outside.* If the ventilation rate decreases (as, for example, by diminishing wind speed), the integral concentration inside is *greater than the concentration time integral outside.* Equation 16.85 indicates that a defensive measure against a transient cloud would consist in closing windows and doors and sealing cracks during cloud passage and then increasing the ventilation rate as rapidly as possible after the cloud passage. This may be impractical since it depends on a knowledge of the exact time of cloud arrival and the exact time when the bulk of the cloud has passed.

In practice, the concentration inside may be lower than outside, depending on the physical and chemical characteristics of the contaminant, because the incoming air passes through narrow cracks or small holes (unless there are wide openings). During such passage, some materials may deposit due to inertial forces or chemical affinity and thus keep the material from entering the building. (In this case there will be localized areas of higher surface contamination.) Megaw (1962), for example, has measured the concentrations of ^{132}I as vapor inside and outside a hut with ventilation rates of 1.8 to 5.5 changes per hour and found integral concentrations inside ranging from about 20 to 70% of those outside. When the windows were open on one end of the hut, the ventilation rate increased from 1.8 to 5.5 to 5 to 8 changes per hour, and the integral concentrations inside were about 70% of those outside. Iodine vapor is known as a very strongly depositing material, and thus the influence of deposition mechanisms would be expected to be strong.

Megaw also studied the penetration of Aitken nuclei produced by burning paraffin-soaked sand in the same hut. For ventilation rates ranging from 0.8 to 2.5 changes per hour, the integral concentrations inside ranged from 34 to 78% of those outside.

Although the preceding treatment indicates that high concentrations inside the building are possible, the deposition inside buildings is expected to be considerably less than that outside, both due to lower turbulence levels inside and due to the restriction on the source by the limited rate of entry to the building. The experiments by Megaw again confirm this expectation since the deposition inside was only a few percent of that outside. The deposition velocity outside ranged from 0.3 to 3.1 cm/s and that inside ranged from 0.1 to 0.24 cm/s.

16-6 DOSE FROM INHALED MATERIAL

The inhalation of radioactive materials can result in radiation doses to (1) the lungs and upper respiratory tract from nuclides deposited in the passages, (2) the gastrointestinal (GI) tract from nuclides originally deposited in the respiratory tract and later removed by ciliary action and swallowed, and (3) other organs in which the nuclide may be deposited after they are absorbed into the bloodstream from either the lungs or GI tract.

The radiation dose to each organ will depend on the physical and chemical characteristics of the radionuclides in the cloud and the distribution and retention of these materials in the body. In general, the physical and chemical characteristics of the routine radioactive wastes normally emitted from a facility can be determined with reasonable accuracy, but the materials from an accidental release will not be well characterized in advance. Even with accurate characterization the metabolic parameters involved, such as retention in the lungs, solubility in body fluids, and translocation to other organs, are not known well enough to do more than provide an estimate of the radiation dose involved. The primary models for these calculations have been derived by the NCRP or the ICRP. The current maximum permissible concentrations (MPC's) for air and water depend on definitions of the organ size and metabolic models derived by the NCRP (National Committee on Radiation Protection and Measurements, 1959) and the ICRP (International Commission on Radiological Protection, 1960). These models have undergone changes with time. For example, a Task Group of the ICRP (International Commission on Radiological Protection, 1966) defined a lung model that describes the deposition in several regions of the lungs as a function of particle size along with clearance constants that depend on the chemical nature of the inhaled material. In addition, a GI tract model was developed by Eve (1966), a review of parameters for transuranium elements was proposed by the ICRP (International Commission on Radiological Protection, 1972), and a definition of the reference man was described by the ICRP (International Commission on Radiological Protection, 1975). These newer data have not been incorporated into the MPC's but should be used for calculations.

16-6.1 Intake, Retention, and Translocation of Inhaled Radioactive Material

The total amount of radioactive material inhaled by a person during cloud passage is determined by the product of the cloud concentration time integral ψ (Ci · s/m^3) during the time of exposure and the breathing rate B' (m^3/s) during cloud passage.

The breathing rate for the standard man is given as 10 m^3/8-h workday and 20 m^3/24-h day, or 3.48×10^{-4} and 2.31×10^{-4} m^3/s, respectively (International Commission on Radiological Protection, 1960). During times of extreme stress, the breathing rate can be greater, and a breathing rate of 5×10^{-4} m^3/s is frequently assumed for relatively short periods of time. The breathing rate selected depends on the details of the situation considered.

Two factors are of primary importance in assessing the dose to the lungs: (1) the fraction of the inhaled material that is deposited and (2) the time interval during which the deposited material is retained in the lungs. The deposition and site of deposition, whether in the nose or in the upper or lower pulmonary tract, will vary with particle size, shape, and density. Different fractions deposited and different patterns of deposition will occur, depending on whether an individual breathes through his nose or his mouth. The time of retention in the lungs will depend on the site of deposition and the rate of solution of the material in body fluids and transfer elsewhere in the body. Material deposited in the upper respiratory tract will be quickly cleared by ciliary action; that deposited deep in the lungs will be removed more slowly, with transfer of at least a portion to the bloodstream and the lymph nodes.

Two standard schemes for estimating the deposition, retentions, and transfers are available. One scheme is that used by the ICRP (International Commission on Radiological Protection, 1960) to evaluate the MPC's now in use. This is a simplified model intended only for situations where better data are not available. The MPC values for insoluble forms of the nuclides in most tabulations did not even use this simplified model because transfer to the body was not included. This lack of transfer makes such values suspect. A second and more elaborate scheme is the report of the ICRP Task Group on Lung Dynamics (International Commission on Radiological Protection, 1966). This model accounts for differences in deposition in three parts of the respiratory tract with particle size and specifies clearance rates for each part along with fractions transferred to the bloodstream and lymph nodes. This model, along with a revision specifically for transuranium nuclides (International Commission on Radiological Protection, 1972), is preferred for dose calculations in all situations except where gross simplifications are desired.

Although in many instances it is difficult to predict the biological fate of the source material released from a facility, radioiodine in most forms will be quickly absorbed from the lungs into the bloodstream and translocated to the thyroid. The noble gases do not enter into the body metabolism but may be dissolved to some extent in body fluids or fat.

16-6.2 Dose to a Body Organ from Internal Irradiation

The dose rate to any body organ from internal irradiation is given by

$$D' = (qf_2)\frac{(3.7 \times 10^{10})\,\bar{E}(1.6 \times 10^{-6})}{100\,M}$$

$$= 5.92 \times 10^2\,(qf_2)\frac{\bar{E}}{M} \qquad (16.86)$$

where D' = dose rate to the organ from internal irradiation, rads/s
M = mass of the organ, g
q = total quantity of radioactive material in the entire body, Ci
f_2 = fraction of the total-body radionuclide burden in the critical organ
(qf_2) = quantity of radioactive material in the critical organ, Ci
3.7×10^{10} = disintegration rate (dis/s) for 1.0 Ci
$\bar{E}$ = energy absorbed in the organ from one disintegration, MeV
1.6×10^{-6} = number of ergs per million electron volts
100 = number of ergs per gram per rad

The values of $\bar{E}$, the average energy absorbed in the organ per disintegration, can be taken as equal to the average energy released per disintegration for alpha and most beta emitters because the ranges of these particles are small compared to the organ dimensions, and nearly all the energy is released in the organ containing the radioactive material. For most gamma photons the range in tissue will be considerably greater than the organ dimensions, and thus only a portion of the energy released will be absorbed in the organ. This fraction can be estimated by assuming the organ to be approximately spherical. Under these conditions the fraction absorbed will be approximately

$${}_a\bar{E}_\gamma = {}_o\bar{E}_\gamma\,[1 - \exp(-\mu_a r_o)] \qquad (16.87)$$

where ${}_a\bar{E}_\gamma$ = average gamma energy absorbed per disintegration
${}_o\bar{E}_\gamma$ = average gamma energy emitted per disintegration
μ_a = energy absorption coefficient for the gamma photon energy
r_o = effective diameter of the organ, cm

Data on the average beta energy, including conversion electrons for a number of important nuclides, are given by the Medical Internal Radiation Dose Subcommittee (MIRD) of the Society of Nuclear Medicine (Dillman, 1969). MIRD has also tabulated absorbed fractions for gamma energies ranging from 0.01 to 4 MeV in a number of organs (Snyder et al., 1969).

For some nuclides that emit penetrating radiation, the presence of a quantity of this nuclide in one organ will produce a radiation dose to another organ. For example, a quantity of a gamma emitter in the liver will subject the kidneys to some dose. A computer code for calculating these doses has been developed by Dunning, Pleasant, and Killough (1977). In their report they tabulate such data for 22 source organs and 24 target organs with 52 radionuclides. The doses result from alpha particles, electrons, and photons as well as spontaneous fission. Such a methodology is useful in a detailed assessment of doses.

The dose delivered to an organ by internal irradiation depends on the retention of the radioactive material in the organ over the time period of concern. For the total dose the quantity of radioactive material in the organ must be integrated over the time period of concern, with proper allowance for accumulation, radioactive decay, and biological elimination. In the following discussion the intake is assumed to occur in a short time interval, such as for exposure to a relatively small cloud from a puff release. In such a case the organ dose is given as

$$D = 5.92 \times 10^2\,\frac{\bar{E}}{M}\int_{t_A}^{t_2} qf_2\,dt \qquad (16.88)$$

where D is the dose to the organ (rads), t_A is the time at which the irradiation starts (s), and t_2 is the time to which the total dose is to be calculated (s).

The dose commitment for short-lived radionuclides can be obtained by setting $t_2 = 0$; for long-lived nuclides the dose commitment from an intake can be obtained by setting t_2 equal to the time period desired. The value of (qf_2) at a given time after uptake depends on both the elimination by radioactive decay and the elimination by biological processes.

$$\int_{t_A}^{t_2}(qf_2)\,dt = (qf_2)_0\int_{t_A}^{t_2} f_R(t)f_B(t)\,dt \qquad (16.89)$$

where $(qf_2)_0$ is the quantity of radioactive material in the organ initially (Ci), $f_R(t)$ is the function chosen to describe the radioactive decay of the mixture in the organ, and $f_B(t)$ is the function chosen to describe the rate of biological elimination of the mixture from the organ.

The elimination from the body is usually expressed as a simple exponential function, implying that the eliminated material arises from a single compartment. This assumption is an approximation for many radionuclides since the actual elimination rate usually decreases with time after administration. Such a decrease in the elimination rate can be expressed as the sum of a series of exponentials, thereby describing a number of different compartments. In most cases the largest part of the dose originates from the compartment with the slowest elimination rate, and the fraction in this compartment and this rate is used when only a single exponential is used. An alternative method of describing elimination processes that are not exponentials is to use a power function of the elimination time (Healy, 1956). However, this model entails additional mathematical complication, and most workers prefer to use the series of exponentials.

Radioactive decay for a single nuclide can be described as an exponential function with the decay constant characteristic of the nuclide. For a number of nuclides, individual calculations permit description of the decay within the accuracy of the knowledge of the half-lives. For gross calculations of the fission-product dose where biological elimination of all nuclides can be considered to be the same, such as lung or GI dose from the inhalation of particles, an approximation of the dose can be obtained by using the Way–Wigner (1951) power-function relation of $t^{-1.2}$ decay for an instantaneous nuclear burst or $t^{-0.2}$ for a reactor operating for a long period of time.

In the following equation we will address only the exponential functions,

$$\int_{t_A}^{t_2} (qf_2)\,dt = -(qf_2)_0 \int_{t_A}^{t_2} [\exp(-\lambda_R t) \times \exp(-\lambda_B t)]\,dt \qquad (16.90)$$

where t_A = time of administration, s
t_2 = time to which dose is desired, s
λ_R = radioactive-decay constant and equals $0.693/T_R$, s^{-1}
λ_B = biological elimination constant and equals $0.693/T_B$, s^{-1}
T_R, T_B = radioactive and biological half-lives, s

Integration of Eq. 16.90 gives

$$\int_{t_A}^{t_2} (qf_2)\,dt = \frac{(qf_2)_0}{\lambda_B + \lambda_R}\{\exp[-(\lambda_B + \lambda_R)\,t_A] - \exp[-(\lambda_B + \lambda_R)\,t_2]\} \qquad (16.91)$$

If, compared to the human lifetime, the residence time in the body is short because of biological elimination or radioactive decay, then the total number of disintegrations in the organ (or dose instead of dose rate when Eq. 16.88 is involved) can be obtained by letting t_2 equal infinity,

$$\int_0^{\infty} (qf_2)\,dt = \frac{(qf_2)_0}{\lambda_B + \lambda_R} \qquad (16.92)$$

16-6.2.1 Dose to Organ Other Than Lung. A fraction of the radioactive material inhaled will be absorbed into the bloodstream from the lungs or the GI tract and transferred to other organs. For example, a significant fraction of radioiodine absorbed into the blood will be deposited in the thyroid and a significant fraction of radiostrontium will be deposited in the bone. Experiments on the inhalation of water-insoluble silver iodide and strontium sulfate have indicated that the translocation process for these materials is as efficient as if the iodine or strontium were present as water-soluble compounds (Willard and Bair, 1958).

For an individual exposed to a uniform concentration of a nuclide for a finite period of time, the change in the quantity of the nuclide in an organ during the period of exposure is given by

$$d(qf_2) = B'\bar{\chi} f_a\,dt - (qf_2)\,\lambda\,dt \qquad (16.93)$$

where B' is the breathing rate (m^3/s), f_a is the fraction of material inhaled that is translocated to the organ, and λ equals $\lambda_R + \lambda_B$ (s^{-1})

The cumulative dose in rads from the onset of plume inhalation until the exposure ceases is obtained by integrating the amount in the organ over the time of exposure and converting to dose, yielding

$$_1D = \frac{5.92 \times 10^2 \bar{E}}{M\lambda} f_a B'\bar{\chi}_{(x,y,0)} \left\{\tau - \frac{[1 - \exp(-\lambda\tau)]}{\lambda}\right\} \qquad (16.94)$$

where $_1D$ is the dose from onset of plume inhalation until exposure ceases (rads), τ is the time interval of exposure (s), and M is the mass of organ (g).

If the exposure to the plume ceases after the time interval τ (s), the irradiation of the organ will continue owing to the burden in the organ at that time. The dose received in the time interval τ_1 immediately following the exposure is obtained by substituting the body burden at the end of the exposure period into Eq. 16.88 and integrating.

$$_2D = \frac{5.92 \times 10^2 \bar{E}}{M\lambda^2} f_a B'\bar{\chi}_{(x,y,0)} \times [1 - \exp(-\lambda\tau)][1 - \exp(-\lambda\tau_1)] \qquad (16.95)$$

where $_2D$ is the dose received in time interval following exposure (rads) and τ_1 is the time interval following exposure (s). The total cumulative dose delivered from the time that exposure to the plume started until some time after the end of exposure, i.e., $T = (\tau + \tau_1)$ seconds, is the sum $_1D + {_2D}$ (rads),

$$D_{co} = \frac{5.92 \times 10^2 \bar{E}}{M\lambda^2} f_a B'\bar{\chi}_{(x,y,0)} \times \{\lambda\tau - \exp(-\lambda\tau_1) + \exp[-\lambda(\tau + \tau_1)]\} \qquad (16.96)$$

where D_{co} is the cumulative dose from the beginning of exposure to the plume until some time after end of exposure (rads).

There is one observation concerning this development that deserves special consideration. When the irradiation time τ_1 (time interval after cloud exposure) is long compared to the effective half-life of the radioactive material ($\lambda\tau_1 \to \infty$), the total dose to the critical organ is given by

$$_\infty D_{co} = \frac{5.92 \times 10^2 \bar{E}}{M\lambda} f_a B' \psi_{(x,y,o)} \qquad (16.97)$$

where $_\infty D_{co}$ is the total dose when the irradiation time is long compared to the half-life (rads) and $\psi_{(x,y,o)}$ equals $\bar{\chi}_{(x,y,o)}\tau$. The total dose (dose commitment) to the critical organ is dependent on the amount of radioactive material inhaled without regard to the time required to inhale the amount.

16-6.2.2 Lung Dose. The lung dose resulting from the inhalation of radioactive materials can be calculated in the manner described previously when the quantity deposited in the pulmonary region and its retention time can be established. The model used by the ICRP (International Commission on Radiological Protection, 1960) in establishing many of the present MPC's for air is crude and provides only rough approximations. As noted earlier the model was not used in deriving limits for insoluble materials in that the transfer to the body was ignored. Therefore, if these MPC's are used in a calculation by ratioing to the dose on which they are based, the insoluble values should be treated with great care.

The ICRP Task Group model (International Commission on Radiological Protection, 1966) for lung dynamics provides curves for the deposition in three regions: (1) the nasopharyngeal region, (2) the tracheobronchiolar region, and (3) the pulmonary region. Clearance from the nasopharyngeal and tracheobronchiolar regions is rapid (<1 d), and thus the dose to these regions is usually small. However, a fraction of that deposited in the pulmonary region can remain for some time, depending on the compound, and result in a dose to the lung. The ICRP has divided the compounds into three classes on the basis of their retention time in the lungs and their absorption into the body from all three regions. These are Class D (retention time in the pulmonary region in days), Class W (retention time in weeks), and Class Y (retention time in years).

A schematic diagram of the transfers to blood, GI tract, and lymphatic system by the ICRP Task Group (International Commission on Radiological Protection, 1966) is given in Fig. 16.30. Each of the fractions given symbolically refer to the transfer from a single region (i.e., the transfer factors from a given region of the pulmonary system sum to 1.0). The deposition fraction in each region for a particular particle size and the values of the transfer parameters, together with the half-lives associated with the transfer, are given by the ICRP (International Commission on Radiological Protection, 1966, 1972). The application of these concepts to calculate the dose to the lung and to other organs following continuous intake by inhalation will be

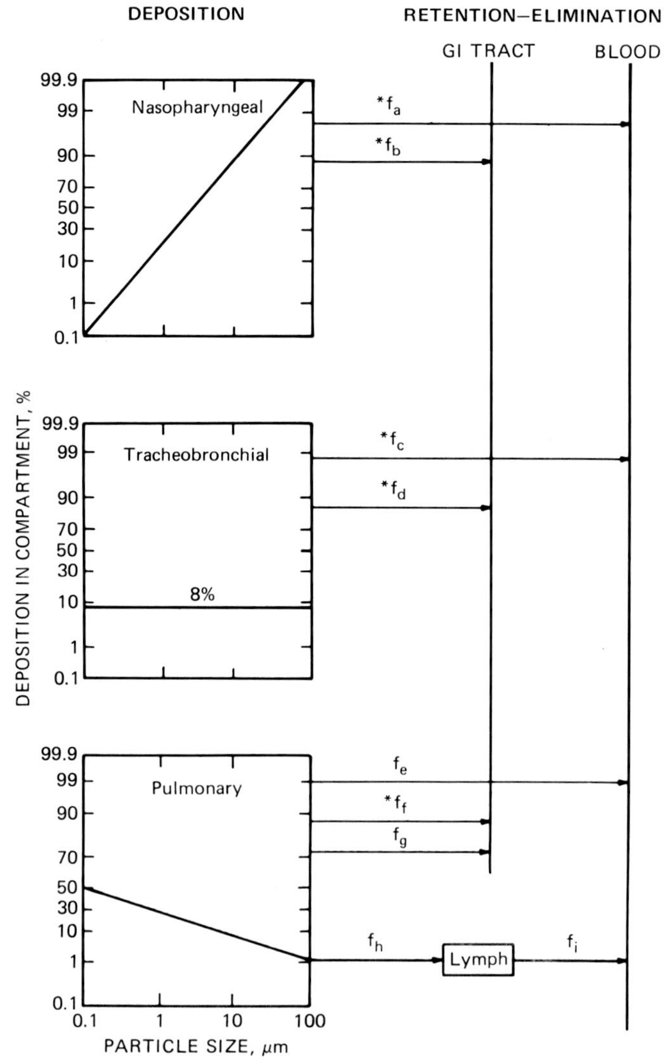

Fig. 16.30 The ICRP Task Group on Lung Dynamics Deposition and Transfer model for inhaled materials. Transfer coefficients marked with an asterisk have short half-lives. Values of the half-lives and fractional transfers are given by the ICRP (International Commission on Radiological Protection, 1966, 1972).

described here. For simplification the rapid transfers that occur with a half-life of 1 d or less will be considered to occur instantaneously.

The quantity in the lung following the intake, I (Ci/d), given by the product of the breathing rate and the average concentration, is

$$dL = ID_3(f_e + f_g + f_h)\,dt - L(\lambda_1 + \lambda_R)\,dt \qquad (16.98)$$

where L = quantity in the lungs, Ci

I = daily intake, Ci

D_3 = deposition in the lungs

f_e = fraction transferred to the blood with half-life λ_1

f_h = fraction transferred to the lymph with half-life λ_1

f_g = fraction transferred to the GI tract with half-life λ_1

Solving, with L = 0 when t = 0, gives

$$L = \frac{ID_3(f_e + f_g + f_h)}{\lambda_1 + \lambda_R} [1 - \exp - (\lambda_1 + \lambda_R)t] \qquad (16.99)$$

The rapid transport to blood and deposition in an organ (or organs) is

$$dO_1 = If_2'(D_1 f_a + D_2 f_c)\,dt + If_1 f_2'(D_1 f_b + D_2 f_d + D_3 f_f)\,dt - O_1(\lambda_2 + \lambda_R)\,dt \qquad (16.100)$$

where O_1 = organ content from rapid transfers, Ci
f_2' = transfer from the blood to an organ
D_1 = fractional deposition in the nasopharyngeal region
f_a = transfer to blood from the nasopharyngeal region
D_2 = deposition in the tracheobronchial region
f_c = transfer to blood from the nasopharyngeal region
f_1 = transfer from GI tract to blood
f_b = transfer to GI tract from the nasopharyngeal region
f_d = transfer to GI tract from the tracheobronchial region
f_f = transfer to GI tract from the pulmonary region
λ_2 = biological elimination constant for the organ, s^{-1}

Solving, again assuming that $O_1 = 0$ when t = 0,

$$O_1 = \frac{If_2'(D_1 f_a + D_2 f_c) + If_1 f_2'(D_1 f_b + D_2 f_d + D_3 f_f)}{\lambda_2 + \lambda_R} \times \{1 - \exp[-(\lambda_2 + \lambda_R)t]\} \qquad (16.101)$$

Additional delayed transfers to the organ occur from the fraction in the lung that is retained for a long time and the fraction in the lung that transfers through the lymph node. For the long-term transfer from the lung,

$$dO_2 = f_e f_2' D_3 \lambda_1 \frac{L}{f_e + f_g + f_h}\,dt - O_2(\lambda_2 + \lambda_R)\,dt \qquad (16.102)$$

where O_2 is the quantity in organ resulting from the long-term transfer from the pulmonary region (Ci), L is the quantity remaining in the lung as given by Eq. 16.99 (Ci), and f_e is the transfer of total deposition in the pulmonary region to the blood.

Substituting for L from Eq. 16.99 and using the condition that $O_2 = 0$ at t = 0 gives

$$O_2 = \frac{If_e f_2' D_3 \lambda_1}{\lambda_1 + \lambda_R} \left[\frac{1}{(\lambda_2 + \lambda_R)} - \frac{\exp[-(\lambda_1 + \lambda_R)t]}{\lambda_2 - \lambda_1} + \frac{(\lambda_1 + \lambda_R)\exp(-\lambda_2 t)}{(\lambda_2 + \lambda_R)(\lambda_2 - \lambda_1)} \right] \qquad (16.103)$$

For the transfer through the lymphatic system, it is first necessary to evaluate the fraction of the quantity in the lymph nodes that will eventually be transferred to the blood.

$$dN' = f_h f_i \lambda_1 \frac{L}{f_e + f_g + f_h}\,dt - N'(\lambda_3 + \lambda_R)\,dt \qquad (16.104)$$

where N' = quantity of material in the lymph nodes that will eventually be transferred to blood, Ci
f_h = transfer from lung to lymph
f_i = transfer from lymph to blood
λ_3 = biological half-life of nuclide in lymph nodes, s^{-1}

Substituting for L from Eq. 16.99 and solving with the condition that $N' = 0$ when t = 0 gives

$$N' = \frac{ID_3 f_h f_i \lambda_1}{\lambda_1 + \lambda_R} \left\{ \frac{1}{\lambda_3 + \lambda_R} - \frac{\exp[-(\lambda_1 + \lambda_R)t]}{\lambda_3 - \lambda_1} + \frac{(\lambda_1 + \lambda_R)\exp[-(\lambda_3 + \lambda_R)t]}{(\lambda_3 + \lambda_R)(\lambda_3 - \lambda_1)} \right\} \qquad (16.105)$$

It should be noted that N' is not the complete lymphatic burden for Class Y material, because 10% of that transferred to the lymph nodes remains there indefinitely.

For the transfer from the lymph nodes to the body,

$$dO_3 = N' f_2' \lambda_3\,dt - O_3(\lambda_2 + \lambda_3)\,dt \qquad (16.106)$$

where O_3 = organ burden from the transfer through the lymphatic system

$$O_3 = \frac{If_h f_i f_2' D_3 \lambda_1 \lambda_3}{\lambda_1 + \lambda_R} \left\{ \frac{1}{(\lambda_2 + \lambda_R)(\lambda_3 + \lambda_R)} - \frac{\exp[-(\lambda_1 + \lambda_R)t]}{(\lambda_1 - \lambda_3)(\lambda_1 - \lambda_2)} + \frac{(\lambda_1 + \lambda_R)\exp[-(\lambda_3 + \lambda_R)t]}{(\lambda_1 - \lambda_3)(\lambda_3 - \lambda_2)(\lambda_3 + \lambda_R)} - \exp[-(\lambda_2 + \lambda_R)t] \left[\frac{1}{(\lambda_2 + \lambda_R)(\lambda_3 + \lambda_R)} - \frac{1}{(\lambda_1 - \lambda_3)(\lambda_1 - \lambda_2)} + \frac{\lambda_1 + \lambda_R}{(\lambda_1 - \lambda_3)(\lambda_3 - \lambda_2)(\lambda_3 + \lambda_R)} \right] \right\} \qquad (16.107)$$

The total organ burden is then given by the sum of O_1, O_2, and O_3. The dose rate at the end of a given time period of exposure is then given by Eq. 16.86 where L or O is equal to (qf_2). The total dose to a given time, or for a given time interval can be obtained by substituting the organ burdens in Eq. 16.86 and integrating over the time interval desired. These integrations are relatively simple and are not performed here, but this procedure gives a better estimate of

TABLE 16.11 Characteristics of the GI Tract of the Standard Man

Portion of GI tract	Mass of contents, g	Time food remains, d	Flow rate, g/d		Effective radius, cm
			In*	Out†	
Stomach	250	$^{1}/_{24}$‡			10
Small intestine	400	$^{4}/_{24}$	7600	600	10
Upper large intestine	220	$^{13}/_{24}$	600	150	5
Lower large intestine	135	$^{24}/_{24}$	150	135	5

*V_1.
†V_2.
‡Read as $^{1}/_{24}$th of a day.

dose, particularly for the early periods of exposure, than does multiplying the dose rate at the end of an interval by the interval.

An approximate solution for long times of exposure can be obtained by considering all the transfers to occur instantaneously and solving for the organ burden after the long time. In this case a long time is defined as long with respect to the half-life in the lung of the class considered.

16-6.2.3 Dose to GI Tract from Inhalation of Radioactive Material. The majority of the radioactive material eliminated from the respiratory tract is moved to the throat where it is swallowed and can result in irradiation of the GI tract as it passes through to the feces. Eve (1966) points out that some portion of the GI tract is critical for the ICRP recommendations on MPC's (International Commission on Radiological Protection, 1960), with 40% of the values for soluble nuclides and 36% of the values for insoluble nuclides giving the highest dose to some portion of the GI tract following inhalation.

The dose to the GI tract arises from the admixture of the radionuclide with the food or food remains passing through the tract. The gut wall is irradiated from the contact between this material and the wall, with an additional gamma dose due to the transmission from one part of the gut to the next in the convoluted packing of these organs in the abdominal cavity. The GI tract is usually considered in four sections so that the changes in concentration of the radionuclide in the contents due to radioactive decay, absorption of the nuclide, and change in water content can be included. These four sections are the stomach, the small intestine, the upper large intestine, and the lower large intestine. An original model for these calculations was described by the ICRP (International Commission on Radiological Protection, 1960). A revision of this model was provided by Eve (1966) after a study of data available on the GI tract and, in so doing, new values of mass and retention time for use in GI dose calculations were proposed. These values are given in Table 16.11.

The methods of estimating dose resulting from the flow of radioactive material through the GI tract have been given by Dolphin and Eve (1966). Their procedures have served as the basis for this discussion. In all cases the dose to the wall of the GI tract is considered to be equal to the dose in contact with the contents or one-half of the dose within the contents (i.e., radiation to the wall is received only from one side). The gamma contribution can be estimated from Eq. 16.87 using the radius from Table 16.8. Dolphin and Eve (1966) point out that the gamma rays from one loop in the small intestine irradiating nearby loops is reasonably taken into account by the choice of 5 to 10 cm for the effective radius rather than the approximate physical dimension radius of the gut tube of about 1 cm.

The treatment of the GI dose selected for use here provides an estimate of the mean dose to the entire section of the GI tract of interest. This is in contrast to the method used by the ICRP (International Commission on Radiological Protection, 1960) where the dose to the small region of the organ receiving the maximum dose was used. This maximum region was determined by the balance between the decay of the radionuclide and the change in contents of the gut section to provide dilution. The mean organ dose corresponds more closely to the basis used for the dose estimate to other organs, and, as Dolphin and Eve (1966) point out, the amount of churning and mixing mean that such a region of maximum dose probably does not exist.

For the stomach, Dolphin and Eve (1966) assume an exponential emptying of the stomach at a rate λ_s. The dose to the wall of the stomach (surface of contents) from an intake, I (μCi), is thus given by

$$D_s = \frac{(3.7 \times 10^4)\,\bar{E}I\,(1.6 \times 10^{-6})}{m} \times \int_0^\infty \exp\,[(-\lambda_R + \lambda_s)t]\;dt \tag{16.108}$$

$$D_s = \frac{3 \times 10^{-4}\,\bar{E}I}{m(\lambda_R + \lambda_s)} \tag{16.109}$$

where D_s = dose to stomach wall, rads
3.7×10^4 = disintegrations per second per curie
1.6×10^{-6} = ergs per million electron volts
I = intake, μCi
m = mass of stomach contents, g
λ_R = radioactive decay constant, s^{-1}
λ_s = emptying constant for stomach, s^{-1}

A half-resident time of 45 min was used in the evaluation for λ_s, which corresponds to a mean life of about 65 min, compared to the emptying time of 60 min in the model by Eve (1966). Although Eq. 16.109 was derived for a single intake, it can be used for a series of intakes if I is chosen as the total intake over a period of time, such as 1 d.

The absorption of some of the gut contents occurs in all three of the remaining sections of the GI tract. However, the radioactive material is assumed to be absorbed only in the small intestine. The equations for the calculation of dose are similar for the small intestine, the upper large intestine, and the lower large intestine, with differences only in the content of the section, time of travel, and the absorption of the radionuclides in the small intestine.

As noted earlier the small intestine is considered to be unique in that those radioactive materials which will be absorbed are absorbed in this section. Thus an additional source of decrease of quantity of radionuclide exists for those nuclides. Again it is assumed that the decrease in the mass of the contents, as well as the decrease in the quantity of radionuclide, occurs exponentially with time of travel through the small intestine. Thus, for the movement of the gut contents, we can write

$$V = V_1 \exp[-\lambda_a(t_2 - t_1)] \tag{16.110}$$

where V, V_1 = flow rate in the gut as defined in Table 16.8
t_1 = time of entry into the small intestine
t_2 = time of leaving the small intestine
λ_a = decrease constant for the mass during its movement

The decrease in radioactive material due to absorption and radioactive decay at any point along the small intestine is given by

$$P/P_0 = \exp[-\lambda_R t - \lambda_b(t - t_1)] \tag{16.111}$$

In this case, P_0 is the amount of radioactive material taken into the GI tract, λ_b is the absorption constant in the small intestine, and P is the quantity of radioactive material remaining in the contents of the GI tract at a time t between t_1 and t_2. The difference in the times used for λ_R and λ_b is because the radioactive decay is effective as soon as the material enters the GI tract, and the absorption only starts when the radionuclide enters the small intestine.

The values to be used for λ_a and λ_b can be obtained from the assumed flow rates in the portion of the gut of interest and by the fraction absorbed from the gut during passage, f_1. The most complete tables of f_1 are in the 1960 Report of the ICRP (International Commission on Radiological Protection, 1960). The values for the elimination constant are

$$\lambda_a = \frac{1}{t_2 - t_1} \ln \frac{V_1}{V_2} \tag{16.112}$$

and

$$\lambda_b = \frac{1}{t_2 - t_1} \ln \frac{1}{1 - f_1} \tag{16.113}$$

The dose to a small element of the GI tract wall is then given by

$$\text{Dose} = \frac{k\bar{E}P_0 \exp[-\lambda_R t - \lambda_b(t - t_1)]}{V_1 \exp[-\lambda_a(t - t_1)]} \tag{16.114}$$

where t is a variable that indicates the distance along the GI tract which the mass has moved, $\bar{E}$ is the effective energy absorbed in the contents of the gut, and k is a constant one-half of that in Eq. 16.86 to allow for contact between the gut wall and the contents on only one side. If the amount arriving in the GI tract is in curies, the constant, k, will be 296 for the dose rate in rads per second.

The mean dose to the full small intestine is then

$$\text{Mean dose} = \frac{\dfrac{k\bar{E}P_0}{V_1}\dfrac{\exp(\lambda_b t_1)}{\exp(\lambda_a t_1)}\displaystyle\int_{t_1}^{t_2}\frac{\exp[-(\lambda_r + \lambda_b)t]}{\exp(-\lambda_a t)}}{\displaystyle\int_{t_1}^{t_2} dt} \tag{16.115}$$

or, after integrating,

$$\text{Mean dose} = \frac{k\bar{E}P_0 \exp(\lambda_b t_1)}{(\lambda_a - \lambda_r - \lambda_b)(t_2 - t_1)} \times \left\{\frac{\exp[-(\lambda_r + \lambda_b)t_2]}{V_2} - \frac{\exp[-(\lambda_R + \lambda_b)t_1]}{V_1}\right\} \tag{16.116}$$

If the nuclide is not absorbed in the GI tract, then $\lambda_b = 0$ and

$$\text{Mean dose} = \frac{k\bar{E}P_0}{(\lambda_a - \lambda_r)(t_2 - t_1)} \times \left[\frac{\exp(-\lambda_R t_2)}{V_2} - \frac{\exp(-\lambda_R t_1)}{V_2}\right] \tag{16.117}$$

The dose to the remaining sections of the GI tract can be calculated from Eq. 16.117 with, however, an additional multiplicative factor of $(1 - f_1)$ added to account for the radioactive material removed earlier in the GI tract.

The above doses are calculated for the surface of the GI wall. However, the critical cells are some distance beneath this wall. A U. S. Nuclear Regulatory Commission report on reactor accidents (U. S. Nuclear Regulatory Commission, 1975) indicates that, for a dog, the critical tissue is 0.5 to 0.75 mm below the mucosal surface. Such distances will result in considerable reduction in dose, particularly for low-energy beta particles; thus the dose calculated by the above equations will be conservative.

For alpha emitters the alpha-radiation energy is reduced by a factor of 100 because of the limited range of the alpha particles and the inability to reach the sensitive area (International Commission on Radiological Protection, 1960).

16-7 DOSE FROM INGESTED MATERIALS

The deposition of radioactive materials from the atmosphere can result in general contamination of an area so that, in addition to the direct radiation, radiation can be received from the later ingestion of water or of foodstuffs grown in the area. This type of exposure often provides the lowest limit for the release of wastes during normal operating conditions and requires countermeasures over the greatest area for accident conditions, for example, the confiscation of milk produced in the area around the reactor following the release of ^{131}I at Windscale (Chamberlain and Dunster, 1958).

Such contamination can result either by direct deposition of the radioactive materials on the foodstuff or by incorporation of the radioactive material into the nutrients of a food chain. Because of the many uncertainties in the deposition process, direct estimates of the potential contamination are difficult and uncertain. For this reason much of the information available on the deposition on foods relates to experience with individual radioisotopes in a given environment, and generalization of the experience is available in only a few cases. Thus only general consideration will be given to dose estimates from these chains.

16-7.1 Contamination of Vegetation

Consideration of the mechanisms of contamination of vegetation is of importance because the vegetation may serve directly as a source of diet for man or it may serve as an entry into a food chain for another source of diet, for example, the movement of iodine from contaminated pasture to the milk of cows.

16-7.1.1 Direct Deposition. The contamination of plants due to direct deposition on the foliage from a single cloud is a transitory problem in the sense that only a single crop is affected by this process, and the contamination is reduced by radioactive decay, weathering, plant growth (and consequent dilution by uncontaminated vegetation), and, eventually, harvesting or a new cycle of growth. Materials deposited in this fashion provide a direct path to man without the time delay inherent in the root uptake and metabolism in the plant. As a result, this deposition is important for relatively short-lived radionuclides that would be largely eliminated by radioactive decay before significant incorporation into the plant through the soil-uptake route. The importance of this mechanism for introducing radioactive materials into the diet will vary with the season of the year, that is, it will be of less importance in some areas during the winter when crops are not growing and when milk cows are usually fed with stored feed.

Certain chemical forms of some radionuclides can be absorbed through the plant foliage. About 4% of the strontium and 2% of the ruthenium applied to tomatoes is absorbed through the intact skin of the fruit (National Academy of Sciences—National Research Council, 1963). Gaseous ^{131}I will adsorb on plant leaves; about 35 to 40% penetrates the leaf epidermis, and about 2% moves to other parts of the plant. Stable iodine will not reduce the ^{131}I adsorption but does reduce the translocation (Selders and Hungate, 1956). The plant species is important in the absorption of elements. Wheat plants absorbed 85% of the applied strontium and 93% of the cesium, whereas cabbage absorbed only 5 to 6% of either of these elements.

The effectiveness of weathering as a mechanism for removing deposited materials has been estimated by the FRC (Federal Radiation Council, 1965): "There is a wide range of measured values for the half-time of radionuclides on grasses, most of which are close to 14 days." Unfortunately, since no references are given to the sources of these measured values and the types of contaminant are not indicated, it is difficult to assess the overall applicability of this half-time. However, Cataldo and Vaughan (1977) state that the early fallout work, resulting in retention half-times of 10 to 14 d, was done with large particles. They indicate that such short retention times are probably characteristic only of large particles and that smaller particles probably remain on the vegetation for much longer times. Thus, when the direct deposition pathway is being estimated, account must be made of the particle size of the contaminant. The NAS—NRC report (National Academy of Sciences—National Research Council, 1963) notes that the washing effect of rain can reduce the foliar intake of strontium from material sprayed on the leaves by a factor as large as 6, whereas cesium is reduced to a lesser extent. It is not indicated whether a similar reduction occurs in the total contamination including adsorbed isotopes. Over 50% of the iodine in foliar contamination is removed by washing. Washing can also remove ions from within the plant. Sodium, potassium, and manganese are readily leached while calcium, magnesium, sulfur, potassium, and strontium are moderately leached; but iron, zinc, phosphorous, and chlorine are leached with difficulty. Thus some reduction in the amount of radioactive materials on the vegetation may occur during food preparation from such processes as washing, peeling, husking, and cooking.

A possible mechanism for the uptake of materials not deposited directly on the foliage exists where there are root mats as in long-established pastures. Here the ^{90}Sr may be deposited directly on the roots and held long enough for absorption to occur.

If the deposition occurs as a result of a single release, as in an accident, the vegetation contamination occurs in a relatively short time. Under conditions where the vegetation serves directly as a source of food for man with continual harvesting (as in a truck farm or a home garden), the uptake in the body can be estimated by using the absorption from the GI tract and the deposition in specific organs as given by the ICRP (International Commission on Radiological Protection, 1960) for water. (Note that these factors may be different from that of water owing to the possible difference in chemical species of the radionuclide and the difference in chemical compounds associated with the radionuclide.) In this case the change in organ content can be written as

$$d(qf_2) = M_e f_w c_g \exp(-\lambda_v t)\, dt - (qf_2)(\lambda_R + \lambda_B)\, dt \qquad (16.118)$$

where qf_2 = organ burden, μCi
M_e = quantity eaten, g/d
f_w = fraction of nuclide ingested that is deposited in the critical organ
c_g = concentration of nuclide in foodstuff, μCi/g
λ_v = fraction of nuclide removed from the vegetation by decay, weathering, growth, etc., d^{-1}
λ_R = radioactive disintegration constant, d^{-1}
λ_B = fraction of isotope removed from the critical organ per day by biological processes, d^{-1}

Solving Eq. 16.118 with the condition that qf_2 equals zero at time zero (time is measured after ingestion starts) gives

$$qf_2 = \frac{M_e f_w c_g}{\lambda_R + \lambda_B - \lambda_v}\{\exp(-\lambda_v t) - \exp[-(\lambda_R + \lambda_B)\tau]\} \qquad (16.119)$$

where τ is the time interval after ingestion begins.

Equation 16.119 describes the quantity of radionuclide in the organ at any time after the start of ingestion. Substituting Eq. 16.119 for (qf_2) in Eq. 16.86 and integrating yields the equation for the total dose received over any time interval τ:

$$D = \frac{51\ \bar{E} M_e f_w c_g}{M(\lambda_R + \lambda_B - \lambda_v)}\left\{\frac{1 - \exp(-\lambda\tau)}{\lambda_v} - \frac{1 - \exp[-(\lambda_R + \lambda_B)\tau]}{\lambda_R + \lambda_B}\right\} \qquad (16.120)$$

where the values of λ_v, λ_R, and λ_B are in units of d^{-1}.

For the special case in which removal from both the organ and the vegetation is by radioactive decay only, i.e., $\lambda_v = \lambda_R$ and $\lambda_B = 0$,

$$qf_2 = M_e f_w c_g \tau \exp(-\lambda_R \tau) \qquad (16.121)$$

Substituting Eq. 16.121 for (qf_2) in Eq. 16.86 yields the dose received over the time period τ when removal is by radioactive decay only:

$$D = \frac{51\ \bar{E} M_e f_w c_g}{M\lambda_R^2}[1 - \exp(-\lambda_R\tau)(\lambda_R\tau - 1)] \qquad (16.122)$$

where λ_R is in units of d^{-1}.

The dose to the GI tract from ingestion of radioactive material can be estimated by the equations given in Sec. 16-6.2.3.

If the crop is harvested as a whole, the quantity remaining on the edible portions at the time of harvesting will depend on the rate of removal by weathering, growth, etc., and the time since the contamination event. This quantity will then remain constant except for radioactive decay. The quantity of nuclide accumulated in the critical organ of an individual eating M_e grams of this crop per day will then be

$$(qf_2) = \frac{M_e f_{w\,0} c_g \exp(-\lambda_v \tau_1)}{\lambda_R + \lambda_B} \times \{1 - \exp[(\lambda_R + \lambda_B)\tau_2]\} \qquad (16.123)$$

where ${}_0c_g$ is the initial deposition (μCi/g), τ_1 is the time interval between deposition and harvest, and τ_2 is the time over which the crop is eaten.

With a continuous release the rate of deposition when averaged over a long time period can be taken as essentially constant. Under these conditions the contamination levels in the foodstuffs will be given by

$$c_g = \frac{f v_d \bar{X}(x,y,0)}{M_g(\lambda_R + \lambda_v)}\{1 - \exp[-(\lambda_R + \lambda_v)t]\} \qquad (16.124)$$

where f = fraction of the deposited material which is on the edible portion
v_d = deposition velocity, m/s
M_g = mass of edible portion per unit area, g/m^2
t = time over which the deposited material is accumulated, d (i.e., effective time of formation of the edible portions)

The quantity in the critical organ of an individual eating the food can then be calculated for continuous intake over the contamination period or after harvesting as in the short-period release.

16-7.1.2 Uptake from the Soil. Soils can become contaminated from airborne materials by several processes:

—Direct deposition on the soil.

—Removal of material initially deposited on vegetation by weathering.

—Seasonal cycling of vegetation cover.

—Irrigation by water contaminated by direct fallout or by runoff from a contaminated area.

Once in the soils, the fate of the radioactive material depends on the physical nature of the material and the chemical behavior of the radionuclides. In general, soluble materials will be adsorbed to a variable degree by soil particles. This results in the gradual leaching of such materials as strontium or cesium leached through the soil. For example, ^{90}Sr from worldwide fallout was located primarily in the upper 2 in. of uncultivated soil during 1954 and 1955 (National Academy of Sciences—National Research Council, 1963).

The plant uptake will depend on the concentration of the radionuclide in the root environment and on the availability of the radionuclide in ionic form. Of the long-lived fission products, strontium and cesium are the most strongly absorbed. Since cesium is more tightly bound to the soils than strontium, it is usually regarded as less likely to be absorbed by plants. The uptake in a particular situation, however, can vary, depending on the local conditions. Cesium uptake, for instance, is increased by ammonia or potassium because these ions free the adsorbed cesium from the soil (U. S. Atomic Energy Commission, 1961). In addition, bean plants that were grown under conditions dry to the point of wilting contained five times the concentration of ^{137}Cs in stems when compared to those grown with the soil moisture kept at near field capacity.

Measurements of the uptake indicate that about 1% of the applied strontium and less than 0.1% of the other elements are taken up by single crops of plants; however, uptake of up to 4 to 8% has been noted in controlled pot experiments. When expressed as the ratio of plant-tissue concentration to soil concentration, factors for fallout constituents have been reported as 0.05 for the alkaline earth group and 0.009 for the rare earths in barley. The average maximum uptake of cesium in experiments appears to be about 0.1% of the applied dose. However, in a complex situation represented by long-continued worldwide fallout from nuclear debris, the ratio of ^{137}Cs to ^{90}Sr in Chicago foodstuffs was higher than in the contaminating mixture even for root vegetables and during periods when the rate of fallout was comparatively low (Hardy and Rivera, 1964).

Strontium tends to accumulate in the aboveground portions of plants. The greatest concentration is usually found in the older leaves. Cesium is distributed more uniformly through the plants. About 10% of the total cesium is in the grain of wheat and oats.

The root system of the plant plays an important role in uptake. As a result the depth of the contamination in the soil in relation to the root will also affect the uptake. Since grasses have a shallow root system, they will tend to absorb radionuclides occurring near the surface. In a grass–clover mixture, the strontium content was reduced 70% by plowing under the surface contamination. In other crops, cultivation tends to increase strontium uptake. Little variation in the availability of strontium has been observed over a period of years, and the uptake of strontium per unit of plant weight is fairly constant. The uptake of cesium can increase under prolonged cropping, possibly because of some potassium depletion in the soil.

It has been noted that plants growing in areas contaminated by a radionuclide that is poorly taken up through the roots will have significantly higher concentrations than would be predicted from laboratory studies of uptake. This phenomenon has been shown with measurements both at the Nevada Test Site (Romney, Wallace, and Gilbert, 1975) and at Savannah River (McLendon et al., 1975). The increased concentration is attributed to resuspension from the ground and direct accumulation of the contaminant bearing soil particles on the plant.

The organ burden of an individual eating such crops can be estimated from Eq. 16.123 and from the estimated contamination level of the crop. The dose to the organ can then be obtained from Eq. 16.86.

Special dose considerations are frequently used for ^{90}Sr because of its metabolic similarity to calcium. Both strontium and calcium are incorporated into new bone in the same ratio as they exist in blood. The dose to the mineral portion of the bone and that to the marrow surrounded by the mineralized bone then depend on the ^{90}Sr concentration in the calcified bone. The strontium/calcium ratio in the blood, however, is not the same as the strontium/calcium ratio in nature or even in the diet since there is some discrimination against the strontium. This discrimination is expressed as the Observed Ratio (OR) and is the ratio of strontium/calcium in bone to the strontium/calcium in diet. Values of the OR are quoted as about 0.1 for the fetus, 1 at birth, 0.5 at the age of 1 yr, and 0.25 for the adult (National Academy of Sciences–National Research Council, 1964). It is estimated that 10^{-3} μCi of ^{90}Sr per gram of calcium uniformly distributed in adult bone would result in a dose of about 2.7 rads/yr to the mineral bone and about 0.7 rad/yr to the bone marrow (Federal Radiation Council, 1965). The dose to mineral bone of the fetus and infant may be about one-half of the value for adults because of the lower calcification, but the dose to the bone marrow will be about the same as for the adult.

These concepts are applicable primarily to the ^{90}Sr uptake over a long period of time. For short-term exposures, about 50% of the deposited radiostrontium is localized in areas of active bone formation, and the remainder is diffusely distributed in the bone in lower concentrations.

16-7.2 Contamination of Animal Products

Animals living in a contaminated region will frequently ingest radioactive materials to a much larger extent than humans. A portion of some of these radionuclides will appear in the products used by man, such as meat, milk, and eggs. Chains of this nature that appear to be important for three of the radionuclides commonly found in fission are:

^{131}I → vegetation → cow → milk

Sr → vegetation → cow → milk (Sr → soil → vegetation)

^{137}Cs → vegetation → cow → milk, meat (^{137}Cs → soil → vegetation)

There is some evidence that eggs may be important in such chains, particularly for ^{131}I.

Information on these chains and their importance has developed both from experience and by experiment (Dunster, Howells, and Templeton, 1958; United Nations, 1972). Milk containing ^{131}I was the controlling factor in the environment following the Windscale release. Here a deposition of about 1 μCi/m^2 resulted in about 0.09 μCi/L of ^{131}I in milk.

If data are available on the uptake and retention (or excretion in milk) from a single administration of a radioisotope, the quantities in meat or milk can be estimated. For example, for a single contaminating event

TABLE 16.12 A Few Computer Codes for Radiation Dose Calculations

Code name	Dispersion	Deposition	Organ dose	External radiation	Ingestion	Inhalation	Food pathways	Language	Reference
INREM II			X		X	X		Fortran IV	Killough, Dunning, and Pleasant, 1978
AIRDOS	X		X	X	X	X	X	Fortran IV	Moore, 1975
AIREM	X	X	X	X		X		Fortran IV	Martin, Nelson, and Cury, 1974
DACRIN	X		X			X		Fortran IV	Strenge, 1975
KRONIC	X			X				Fortran IV	Strenge and Watson, 1973
SUBDOSA	X			X					Strenge, Watson, and Houston, 1975
COMRADEX-III	X		X	X		X		Fortran	Otter and Conners, 1975
ARRRG, CRITR, GRONK	X			X				Basic	Soldat, Robinson, and Baker, 1974

with ^{137}Cs followed by continued ingestion of the vegetation by cows, the average concentration in meat will be given by

$$\frac{dc_m}{dt} = \frac{M_e c_v f_w}{M_m} \exp(-\lambda_W t) - c_m \lambda_B \qquad (16.125)$$

where c_m = concentration of radioactive material in the animal's muscle, μCi/g
M_e = mass of food ingested by the animal per day, g/d
c_v = concentration of cesium in the animal's food, μCi/g
f_w = fraction of the radioactive material eaten that goes to the muscle
M_m = total mass of the animal muscle, g
λ_W = weathering constant for vegetation, d^{-1}
λ_B = elimination constant for the muscle, d^{-1}

Solving Eq. 16.125 gives

$$c_m = \frac{M_e c_v f_w}{M_m (\lambda_B - \lambda_W)} [\exp(-\lambda_W t) - \exp(-\lambda_B t)] \qquad (16.126)$$

16-7.3 Contamination of Water

Direct deposition into a stream or body of water is usually of minor importance compared with deposition on land because of the dilution and the relatively rapid removal of the contaminated portion by flushing. In special cases significant quantities of radioisotopes may deposit in sources of domestic water.

A special problem is the contamination of water supplies by runoff from land contaminated with fission products by direct deposition or by decontamination efforts following an accident. For example, if 1 mi^2 (2.59 km^2) of land were contaminated so that the gamma reading at 1 m above the ground was about 5 R/h, decontamination disposal to a stream with a flow of 5000 ft^3/s would result in an average concentration of 10^{-6} to 10^{-5} $\mu Ci/cm^3$ over a 1-month period.

Potential doses from the direct ingestion of water can be estimated by comparison with the maximum permissible concentration in water given by the NCRP (National Committee on Radiation Protection and Measurements, 1959), with appropriate allowances made for the exposure time compared to the half-life of the contaminating radioisotope.

Food chains are again important in many bodies of water because some elements are concentrated strongly in the lower organisms and in fish or wild fowl that feed on these organisms (Davis et al., 1958). This concentration in exposed organisms may make fish or wild fowl a source of significant radiation dose if they are eaten by people, even though the water may be within maximum permissible limits for drinking purposes. In addition, such accumulations provide a mechanism for retaining some contamination in a body of water even after the major contaminated portion of the stream has passed.

Detailed studies have been made of several radionuclides in particular environments (National Academy of Sciences–National Research Council, 1963; Dunster, 1958; White and Dunaway, 1977). Estimates of the potential effects can be made only after detailed study of the data available and their application to the environment in question.

16-8 COMPUTATIONAL METHODS

In this chapter we have attempted to provide some basis for calculating radiation doses that result from radioactive materials in the atmosphere. Emphasis has been placed on analytical solutions that are appropriate for hand calculations. However, as knowledge has been gained on the complexity of environmental–ecological processes and the need for estimates has been extended to large numbers of radionuclides, many computer programs have been written to permit (1) more detailed estimates of the dose chain, (2) use of different dispersion and deposition models, and (3) estimates of complex radionuclide mixtures. Although we feel that such programs should only be used by those who are thoroughly familiar with the calculational techniques and the coefficients used, a brief description of several computer programs is included for information. There has been no attempt to be comprehensive, and only a sampling of programs has been selected to illustrate the variety available. The programs listed in Table 16.12 were selected primarily from listings by Strenge, Watson, and Droppo (1976) and Miller (1978).

ACKNOWLEDGMENT

This work was performed under the auspices of the Department of Energy.

LIST OF SYMBOLS

Symbols used frequently in this chapter are listed here. The equation or section number indicates the first appearance of the symbols. The noted units for each symbol are those which are typically used in the calculations discussed in the text, although other units may be used in specific instances.

A	Atomic weight: relative weight of an atom based on a weight of 16 for oxygen, Eq. 16.14
A′	Activity (dis/s), Eq. 16.6a
a	Building radius (cm), Eq. 16.73
B′	Breathing rate (m^3/s), Eq. 16.86
B_y	Gamma-dose buildup factor, Eq. 16.73
b	Variable distance in ground gamma-dose problem (m), Eq. 16.66; also, height above an infinite plane for the beta dose (g/cm^2), Eq. 16.27
C	Parameter in the Bateman equations, Eq. 16.7; also, coulomb, Sec. 16-3.2.2

c — Parameter in the beta-dose equation, Eq. 16.26

c_g — Initial deposition on crops (μCi/g), Eq. 16.123

c_m — Concentration of radionuclide in animal muscle (μCi/g), Eq. 16.125

c_v — Concentration of radionuclide in animal food (μCi/g), Eq. 16.125

D_1 — Fractional deposition in the nasopharyngeal region, Eq. 16.100

D_2 — Fractional deposition in the tracheobronchial region, Eq. 16.100

D_3 — Fractional deposition in the pulmonary region, Eq. 16.98

D_s — Dose to the stomach wall (rads), Eq. 16.108

$_\beta D_d$ — Beta dose at a depth d (rads), Eq. 16.26

$_\beta D_\infty$ — Beta dose at the surface from an infinite cloud, Eq. 16.26

$_\beta D'_\infty$ — Beta dose rate from an infinite cloud (rads/s), Eq. 16.15

$_\gamma D$ — Gamma-ray dose (rads), Eq. 16.33

$_\gamma D_0$ — Gamma-ray dose without shielding (rads), Eq. 16.33

$_\gamma D'_\infty$ — Gamma dose above an infinite plane (rads/s), Eq. 16.64

d — Depth in the body (g/cm^2), Eq. 16.26

$\bar{E}$ — Energy absorbed in an organ from one disintegration (MeV), Eq. 16.86

E_0 — Maximum energy of the beta-ray spectrum (MeV), Eq. 16.26c

$E_i(\ )$ — Exponential integral, Eq. 16.62

$\bar{E}_\beta$ — Average beta energy per disintegration (MeV), Eq. 16.15

$\bar{E}_\gamma$ — Average gamma energy per disintegration (MeV), Eq. 16.34

F — Fission rate (fissions/s), Eq. 16.11

f_1 — Transfer of radionuclide from the gastrointestinal (GI) tract to the blood, Eq. 16.100

f_2 — Fraction of the total radionuclide in the body which is in an organ, Eq. 16.86

f'_2 — Transfer of the radionuclide passing from the blood to an organ, Eq. 16.100

f_a — Fraction of an inhaled material which is translocated to an organ, Eq. 16.93

f_b — Transfer to the GI tract from the nasopharyngeal region, Eq. 16.100

f_i — Transfer to the blood from the nasopharyngeal region, Eq. 16.100

f_d — Transfer to the GI tract from the tracheobronchial region, Eq. 16.100

f_e — Transfer from the long-term compartment in the pulmonary region to the blood, Eq. 16.98

f_f — Transfer to the GI tract from the pulmonary region, Eq. 16.100

f_g — Transfer from the lung to the GI tract, Eq. 16.98

f_h — Transfer from the lung to the lymph nodes, Eq. 16.98

f_i — Transfer from the lymph nodes to the blood, Eq. 16.104

f_w — Fraction of a nuclide ingested which deposits in an organ, Eq. 16.118

G_1, G_2 — Integrals used in the evaluation of the gamma dose from radioactive materials deposited on the ground, Eqs. 16.67 and 16.67c

h — Height of emission (m), Eq. 16.20; also, height of receptor in inversion calculation (m), Eq. 16.52

h_i — Height of inversion aboveground (m), Eq. 16.52

I — Daily intake of a nuclide (Ci), Eq. 16.98; also, gamma-ray intensity (photons/m$^2\cdot$s), Eq. 16.62

I_1, I_2, I_T — Integrals used in the evaluation of a gamma dose from a cloud, Eqs. 16.43a, 16.43b, and Fig. 16.7

k — Factor used in the buildup factor for gamma rays, where $k = (\mu - \mu_a)\mu_a$, Eq. 16.34

L — Quantity of a nuclide in the lungs (Ci), Eq. 16.98

ℓ — Variable distance in cloud gamma-dose problem (m), Eq. 16.66

M_e — Daily intake of food (g/d), Eq. 16.118

M_R — Mass of a radionuclide (g), Eq. 16.4

m — Mass of an organ or contents of the GI tract, Eq. 16.86; also, a distance used in deriving the gamma dose from a cloud (m), Eq. 16.39

N — Number of atoms, Eq. 16.1

N' — Quantity of radionuclide in the lymph nodes which will eventually be transferred to the blood (Ci), Eq. 16.104

N_0 — Avogadro's number

N_T — Number of target atoms, Eq. 16.8

nv — Neutron flux [neutrons/(cm$^2\cdot$s)], Eq. 16.8

O_1 — Organ content from rapid transfer from the pulmonary region (Ci), Eq. 16.100

O_2 — Organ content from long-term transfer from the pulmonary region (Ci), Eq. 16.102

O_3 — Organ content from transfer through the lymph nodes (Ci), Eq. 16.106

P — Quantity of radionuclide at any point along the GI tract (Ci), Eq. 16.111

P_0 — Quantity of radionuclide entering a section of the GI tract, Eq. 16.111

$P_{\beta 1}$ — Energy emission by beta emitters (MeV/s), Eq. 16.23

$P_{\beta 2}$ — Energy emission by beta emitters (ergs/s), Eq. 16.25

$P_{\gamma 1}$ — Energy emission by gamma emitters (MeV/s), Eq. 16.46

$P_{\gamma 2}$ — Energy emission by gamma emitters (W), Eq. 16.47

$P_{\sigma_z(j,k,\theta_1)}$ Joint probability of wind speed and atmospheric stability into sector θ_1, Eq. 16.61

Q_0 Quantity of radioactive material emitted to the atmosphere (Ci), Eq. 16.19a

Q_0' Release rate for radioactive material emitted to the atmosphere (Ci/s), Eq. 16.45

Q_x Quantity of radioactive material in a cloud at a receptor (Ci), Eq. 16.19a

q Quantity of radioactive material in a source (Ci), Eq. 16.14; also, quantity of radioactive material in the body (Ci), Eq. 16.86

(qf_2) Quantity of radioactivity in an organ (Ci), Eq. 16.86

R Distance from a release point (m), Eq. 16.61

r Distance from the center of an isotropic cloud (m), Eq. 16.51

r_o Radius of an organ (cm), Eq. 16.87

S Source strength per unit area of a plane source ($photons/m^2 \cdot s$), Eq. 16.62

s Volume of a building (m^3), Eq. 16.82

T_B Biological half-life of a nuclide (s), Eq. 16.90

T_R Radioactive half-life of a nuclide (s), Eq.16.3

t Time (s), Eq. 16.1

t_a Time of cloud arrival (s), Eq. 16.19

t_d Time since fission (d), Eq. 16.13; also, time interval of discharge (s), Eq. 16.29

$\bar{u}$ Mean wind speed (m/s), Eq. 16.19a

V Building ventilation rate (m^3/s), Eq. 16.82

V_1 Flow rate entering section of GI tract (g/d), Table 16.8

V_2 Flow rate leaving section of GI tract (g/d), Table 16.8

v_d Deposition velocity of a contaminant from the atmosphere (m/s), Eq. 16.20

w Thickness of building walls (g/cm), Eq. 16.73

x Downwind distance between point of release and receptor (m), Eq. 16.19a

y Crosswind distance from center of a Gaussian cloud (m), Eq. 16.28

α A parameter in the depth–dose equations for beta radiation, Eq. 16.26b

$\beta(t)$ Beta-particle emission rate (particles/s), Eq. 16.13

η Factor describing changes in building ventilation rate, Eq. 16.85

θ Sector width (radians), Eq. 16.61

λ $\lambda_R + \lambda_B$ (s^{-1}), Eq. 16.93

λ_1 Long biological half-life for a nuclide in the lungs (d^{-1}), Eq. 16.98

λ_2 Total elimination constant for an organ (d^{-1}), Eq. 16.100

λ_3 Total half-life of a nuclide in the lymph nodes (d^{-1}), Eq. 16.103

λ_a Rate of decrease in mass of contents of GI tract (s^{-1}), Eq. 16.110

λ_B Biological elimination constant (s^{-1}), Eq. 16.90

λ_b Absorption constant for radionuclides in the small intestine (s^{-1}), Eq. 16.113

λ_R Radioactive-decay constant (s^{-1}), Eq. 16.1

λ_s Emptying constant for the stomach (s^{-1}), Eq. 16.108

λ_v Removal constant for nuclides deposited on vegetation (d^{-1}), Eq. 16.118

μ Total gamma-ray absorption coefficient (m^{-1} or cm^2/g), Eq. 16.33

μ_a Energy-absorption coefficient (m^{-1} or cm^2/g), Eq. 16.34

μ_s Scatter absorption coefficient (m^{-1} or cm^2/g), Eq. 16.51

$\mu_x W$ Number of relaxation lengths

ν Apparent beta absorption coefficient (cm^2/g), Eq. 16.26c

ρ_0/ρ Correction factor for air at conditions other than standard, Eq. 16.36

σ Standard deviation of cloud concentration (m), Figs. 16.10 and 16.11

σ_a Activation cross section (barns), Eq. 16.8

σ_{av} $(\sigma_y, \sigma_z)^{1/2}$ (m), Eq. 16.48

σ_I Standard deviation of an isotropic cloud (m), Eq. 16.39

σ_y Standard deviation of concentration with crosswind distance (m), Eq. 16.28

σ_z Standard deviation of concentration with height (m), Eq. 16.20

χ Concentration in air (Ci/m^3), Eq. 16.18

ψ Concentration time integral ($Ci \cdot s/m^3$), Eq. 16.19

ω Surface activity (Ci/m^2), Eq. 16.27

REFERENCES

Bair, W. J., C. R. Richmond, and B. W. Wachholz, 1974, *A Radiobiological Assessment of the Spatial Distribution of Radiation Dose from Inhaled Plutonium,* USAEC Report WASH-1320, NTIS.

Bateman, H., 1910, The Solution of a System of Differential Equations Occurring in the Theory of Radioactive Transformation, *Proc. Cambridge Philos. Soc.,* 15: 423-427.

Braestrup, C. B., and H. O. Wyckoff, 1958, *Radiation Protection,* Charles C Thomas, Publisher, Springfield, Ill.

Briggs, G. A., I. Van der Hoven, R. J. Engelmann, and J. Halitsky, 1968, Processes Other Than Natural Turbulence Affecting Effluent Concentrations, in *Meteorology and Atomic Energy—1968,* David H. Slade (Ed.), USAEC Report TID-24190, NTIS.

Burson, Z. G., and A. E. Profio, 1975, *Structure Shielding from Cloud and Fallout Gamma Ray Sources for Assessing the Consequences of Reactor Accidents,* Report EGG-1183-1670, Edgerton, Germeshausen & Grier, Inc., NTIS.

Cataldo, D. A., and B. E. Vaughan, 1977, Retention, Absorption, and Translocation of Foliar Contaminants, in *Transuranics in Natural Environments,* M. G. White and P. B. Dunaway (Eds.), pp. 331-350, ERDA Report NVO-178, Nevada Operations Office, NTIS.

Cember, H., 1964, Radiogenic Lung Cancer, in *Prog. Exp. Tumor Res.,* 4: 251-305, available from Hafner Publishing Co., Inc., New York.

Chamberlain, A. C., and H. J. Dunster, 1958, Deposition of Radioactivity in Northwest England from the Accident at Windscale, *Nature*, 182: 629-630.

Couchman, J. C., 1960, *Determining External Dose from the Passage of a Radioactive Cloud,* Report FZM-2029, Convair, Fort Worth, Tex., NTIS.

Court-Brown, W. M., and R. Doll, 1957, *Leukemia and Aplastic Anemia in Patients Irradiated for Ankylosing Spondylitis,* Medical Research Council, Special Report 295, Her Majesty's Stationery Office, London.

Davis, J. J., R. W. Perkins, R. F. Palmer, W. C. Hanson, and J. F. Cline, 1958, Radioactive Materials in Aquatic and Terrestrial Organisms Exposed to Reactor Effluent Water, in *Proceedings of the Second United Nations International Conference on the Peaceful Uses of Atomic Energy, Geneva, 1958,* Vol. 18, pp. 423-428, United Nations, New York.

Dillman, L. T., 1969, Radionuclide Decay Schemes and Nuclear Parameters for Use in Radiation-Dose Estimation, *J. Nucl. Med.*, 10(Suppl. 2), Pamphlet 4.

Dolphin, G. W., and I. S. Eve, 1966, Dosimetry of the Gastrointestinal Tract, *Health Phys.*, 12: 162-163.

Dunning, D. E., Jr., J. C. Pleasant, and G. G. Killough, 1977, *SFACTOR: A Computer Code for Calculating Dose Equivalent to a Target Organ per Micro-Curie-Day Residence of a Radionuclide in a Source Organ,* Report ORNL/NUREG/TM-85, Oak Ridge National Laboratory, NTIS.

Dunster, H. J., 1958, The Disposal of Radioactive Liquid Wastes into Coastal Waters, in *Proceedings of the Second United Nations International Conference on the Peaceful Uses of Atomic Energy, Geneva, 1958,* Vol. 18, pp. 390-399, United Nations, New York.

——, H. Howells, and W. L. Templeton, 1958, District Surveys Following the Windscale Incident, in *Proceedings of the Second United Nations International Conference on the Peaceful Uses of Atomic Energy, Geneva, 1958,* Vol. 18, pp. 296-308, United Nations, New York.

Eisenbud, M., 1972, The Natural Versus the Unnatural, in *The Natural Radiation Environment II,* J. A. S. Adams, W. M. Lowder, and T. F. Gesell (Eds.), USAEC Report CONF-720805-P2, pp. 941-947, NTIS.

——, and H. G. Petrow, 1964, Radioactivity in the Atmospheric Effluents of Power Plants That Use Fossil Fuels, *Science,* 144: 288-289.

Environmental Protection Agency, 1976, *Radiological Quality of the Environment,* Report EPA-520/1-76-010, Office of Radiation Programs, NTIS.

Eve, I. S., 1966, A Review of the Physiology of the Gastrointestinal Tract in Relation to Radiation Dose from Radioactive Materials, *Health Phys.*, 12: 131-161.

Federal Radiation Council, 1960, *Background Material for the Development of Radiation Protection Standards,* NTIS.

——, 1965, *Background Material for the Development of Radiation Protection—Protective Action Guides for Strontium-89, Strontium-90, and Cesium-137,* Staff Report 7, NTIS.

Furth, J., and E. Lorenz, 1954, Carcinogenesis by Ionizing Radiation, in *Radiation Biology,* Vol. I, Part 2, pp. 1145-1201, McGraw-Hill Book Co., New York.

Gamertsfelder, G. G., 1960, *Hazards Calculation Notebook,* General Electric Co., Aircraft Nuclear Propulsion Department, unpublished.

Gifford, F. A., Jr., 1967, *The Rise of Strongly Radioactive Plumes,* Report AECL-2787, Atomic Energy of Canada Ltd.

Glasstone, S. (Ed.), 1962, *Effects of Nuclear Weapons,* GPO.

Goldstein, Herbert, and J. E. Wilkins, Jr., 1954, *Calculations of the Penetration of Gamma Rays,* Final Report, USAEC Report NYO-3075, Nuclear Development Associates, Inc., NTIS.

Gopal-Ayengar, A. R., K. Sundaram, K. B. Mistry, C. M. Sunta, K. S. V. Mambi, S. P. Kathuria, A. S. Basu, and M. David, 1972, Evaluation of the Long Term Effects of High Background Radiation on Selected Population Groups on the Kerala Coast, in *Proceedings of the Fourth International Conference on the Peaceful Uses of Atomic Energy, Geneva, 1971,* Vol. 11, pp. 31-51, United Nations, New York, 1972.

Hardy, E. P., Jr., and J. Rivera, 1964, *Fallout Program Quarterly Report, June 1—Sept. 1, 1964,* USAEC Report HASL-149, Health and Safety Laboratory, NTIS.

Healy, J. W., 1956, The Calculation of Maximum Permissible Concentrations for Long-Lived Radioisotopes, *Radiat. Res.*, 4: 367-372.

——, 1975, The Origin of Current Standards, *Health Phys.*, 29: 489-494.

——, C. R. Richmond, and E. C. Anderson, 1974, *A Review of the Natural Resources Defense Council Position Concerning Limits for Insoluble Alpha Emitters,* USAEC Report LA-5810-MS, Los Alamos Scientific Laboratory, NTIS.

Hewson, E. W., G. C. Gill, and E. Bierly, 1959, *Maximum Ground Concentrations in Relation to Stack Heights Under Various Meteorological Conditions at the Enrico Fermi Nuclear Plant near Monroe, Michigan,* Univ. of Michigan, Ann Arbor, Mich., unpublished.

Hine, G. J., and G. L. Brownell (Eds.), 1956, *Radiation Dosimetry,* Chap. 16, pp. 693-799, Academic Press, New York.

Holland, J. Z., 1958, Radiation from Clouds of Reactor Debris, in *Proceedings of the Second United Nations International Conference on the Peaceful Uses of Atomic Energy, Geneva, 1958,* Vol. 13, pp. 110-118, United Nations, New York.

Holzworth, G. C., 1964, Estimates of Mean Maximum Mixing Depth in the Contiguous United States, *Mon. Weather Rev.*, 92: 235-242.

International Commission on Radiological Protection, 1960, Report of ICRP Committee II on Permissible Dose for Internal Radiation (1959), with Bibliography for Biological, Mathematical, and Physical Data, *Health Phys.*, 3: 1-380.

——, 1965, Recommendations of the International Commission on Radiological Protection (Adopted September 17, 1965), ICRP Publication 9, Pergamon Press, New York.

——, 1966, Deposition and Retention Models for Internal Dosimetry of the Human Respiratory Tract, *Health Phys.*, 12: 173-207.

——, 1972, The Metabolism of Compounds of Plutonium and Other Actinides, ICRP Report 19, Pergamon Press, New York.

——, 1975, Report of the Task Group on Reference Man, A Report Prepared by a Task Group of Committee 2 of the International Commission on Radiological Protection, Pergamon Press, New York.

——, 1977, ICRP Publication 26, Recommendations of the International Commission on Radiological Protection, *Annals ICRP,* 1: 3.

International Commission on Radiological Units and Measurements, 1962, *Radiation Quantities and Units,* National Bureau of Standards, Handbook 84 (ICRU Report 10a), GPO.

——, 1971, Radiation Quantities and Units, ICRU Report 19.

Jahnke, E., and F. Emde, revised by F. Lösch, 1960, *Tables of Higher Functions,* 6th ed., McGraw-Hill Book Co., New York.

Killough, G. G., P. E. Dunning, Jr., and J. C. Pleasant, 1978, *INREM II: A Computer Implementation of Recent Models for Estimating the Dose Equivalent to Organs of Man from an Inhaled or Ingested Radionuclide,* NRC Report NUREG/CR-0114(ORNL/NUREG/TM-64), Oak Ridge National Laboratory, NTIS.

Kocher, D. C., 1977, *Nuclear Decay Data for Radionuclides Occurring in Routine Releases from Nuclear Fuel Cycle Facilities,* Report ORNL/NUREG/TM-102, Oak Ridge National Laboratory, NTIS.

Langham, W. H., and J. W. Healy, 1973, Maximum Permissible Body Burdens and Concentration of Plutonium: Biological Basis and History of Development, in *Handbuch der Experimentellen Pharmakologie,* H. C. Hodge, J. N. Stannard, and J. C. Hursh (Eds.), Chap. 12, pp. 569-572, Springer-Verlag, New York.

Lewis, E. B., 1957, Leukemia and Ionizing Radiation, *Science,* 125: 965-972.

Loevinger, R., E. M. Japha, and G. L. Brownell, 1956, Discrete Radioisotope Sources, in *Radiation Dosimetry,* G. J. Hine and G. L. Brownell (Eds.), Chap. 16, pp. 693-789, Academic Press, New York.

McBride, J. P., R. E. Moore, J. P. Witherspoon, and R. E. Blanco, 1978, Radiological Impact of Airborne Effluents of Coal and Nuclear Plants, *Science,* 202: 1045-1050.

McLendon, H. R., O. M. Stewart, A. L. Boni, J. C. Corey, K. W. McLeod, and J. E. Pinder, 1975, Relationships Among Plutonium Contents of Soil, Vegetation, and Animals Collected on and Adjacent to an Integrated Nuclear Complex in the Humid Southeastern United States, in *Transuranium Nuclides in the Environment,* Symposium Proceedings, San Francisco, pp. 347-363, STI/PUB/410, International Atomic Energy Agency, Vienna.

Magill, P. L., F. R. Holden, and C. Ackley (Eds.), 1956, *Air Pollution Handbook,* McGraw-Hill Book Co., New York.

March, H. C., 1950, Leukemia in Radiologists in a 20 Year Period, *Am. J. Med. Sci.,* 220: 282-286.

Martin, J. A., Jr., C. B. Nelson, and P. A. Cuny, 1974, *AIREM Program Manual: A Computer Code for Calculating Doses, Population Doses, and Atmospheric Emissions of Radionuclides,* Report EPA-5201/1-74-004, Environmental Protection Agency, NTIS.

Megaw, W. J., 1962, The Penetration of Iodine into Buildings, *Int. J. Air Water Pollut.,* 6: 121-128.

Merriam, G. R., and E. F. Focht, 1958, Radiation Dose to the Lens in Treatment of Tumors of the Eye and Adjacent Structures, Possibilities of Cataract Formation, *Radiology,* 71: 357-369.

Miller, C. W., 1978, *The Evaluation of Models Used for the Assessment of Radionuclide Releases to the Environment,* DOE Report ORNL-5382, Oak Ridge National Laboratory, NTIS.

Miller, K. M., and H. L. Beck, 1979, The Radiological Aspects of the Coal Fuel Cycle, *Health Phys.,* 37: 6 (Abstr. 802).

Moore, R. E., 1975, *AIRDOS: A Computer Code for Estimating Population and Individual Doses Resulting from Atmospheric Release of Radionuclides from Nuclear Facilities,* ERDA Report ORNL/TM-4687, Oak Ridge National Laboratory, NTIS.

National Academy of Sciences–National Research Council, 1963, *The Behavior of Radioactive Fallout in Soils and Plants,* Publication 1092, Committee on Effects of Atomic Radiation on Agriculture and Food Supplies.

——, 1964, Implications to Man of Irradiation by Internally Deposited Strontium-89, Strontium-90, and Cesium-137, Federal Radiation Council, Washington, D. C.

——, 1972, *The Effects on Populations of Exposure to Low Levels of Ionizing Radiation.*

——, 1976, *Health Effects of Alpha-Emitting Particles in the Respiratory Tract,* Environmental Protection Agency, Office of Radiation Programs, Report EPA-520/4-76-013, NTIS.

National Committee on Radiation Protection, 1958, *Permissible Dose from External Sources of Ionizing Radiation, Addendum: Maximum Permissible Radiation Exposures to Man,* National Bureau of Standards, Handbook 59, GPO.

National Committee on Radiation Protection and Measurements, 1959, *Maximum Permissible Body Burdens and Maximum Permissible Concentrations of Radionuclides in Air and in Water for Occupational Exposure,* National Bureau of Standards, Handbook 69, GPO.

——, 1962, *Exposure to Radiation in an Emergency,* Report 29, University of Chicago, Chicago, Ill.

National Council on Radiation Protection, 1971, *Basic Radiation Protection Criteria,* NCRP Publications, Washington, D. C.

——, 1974, *Radiological Factors Affecting Decision-Making in a Nuclear Attack,* NCRP Report 42, NCRP Publications, Washington, D. C.

——, 1975a, *Natural Background Radiation in the United States,* NCRP Report 45, NCRP Publications, Washington, D. C.

——, 1975b, *Alpha-Emitting Particles in Lungs,* NCRP Report 46, NCRP Publications, Washington, D. C.

——, 1976, *Environmental Radiation Measurements,* NCRP Report 50, NCRP Publications, Washington, D. C.

Nickson, J. J., and H. N. Bane, 1959, Physiological Effects of Radiation, in *Radiation Hygiene Handbook,* pp. 19-1–19-17, McGraw-Hill Book Co., New York.

Otter, J. M., and P. A. Conners, 1975, *Description of the COMRADEX-III Code,* Report TI-001-130-053, Atomics International, NTIS.

Patt, H. M., and A. M. Brues, 1954, The Pathological Physiology of Radiation Injury in the Mammal. I. Physical and Biological Factors in Radiation Action, in *Radiation Biology,* Chap. 14, pp. 919-958, A. Hollaender (Ed.), McGraw-Hill Book Co., New York.

Romney, E. M., A. Wallace, R. O. Gilbert, and Jean E. Kinnear, 1975, ^{239,240}Pu and ^{241}Am Contamination of Vegetation in Aged Fallout Areas, in *Transuranium Nuclides in the Environment,* Symposium Proceedings, San Francisco, pp. 479-491, STI/PUB/410, International Atomic Energy Agency, Vienna.

Selders, A. A., and F. P. Hungate, 1956, *The Foliar Sorption of Iodine by Plants,* USAEC Report HW-44890, Hanford Atomic Products Operation, NTIS.

Snyder, W. S., Mary G. Ford, G. G. Warner, and H. L. Fisher, Jr., 1969, Estimates of Absorbed Fractions for Monoenergetic Photon Sources Uniformly Distributed in Various Organs of a Heterogeneous Phantom, *J. Nucl. Med.,* 10(Suppl. 3), Pamphlet 5.

Soldat, J. K., N. M. Robinson, and D. A. Baker, 1974, *Models and Computer Codes for Evaluating Environmental Doses,* USAEC Report BNWL-1754, Battelle Pacific Northwest Laboratory, NTIS.

Strenge, D. L., 1975, *DACRIN: A Computer Program for Calculating Organ Dose from Acute or Chronic Radionuclide Inhalation; Modification for Gastrointestinal Tract Dose,* USAEC Report BNWL-B-389(Suppl.), Battelle Pacific Northwest Laboratory, NTIS.

——, and E. C. Watson, 1973, *KRONIC: A Computer Program for Calculating Annual Average External Doses from Chronic Atmospheric Releases of Radionuclides,* USAEC Report BNWL-B-264, Battelle Pacific Northwest Laboratory, NTIS.

——, E. C. Watson, and J. G. Droppo, 1976, *Review of Calculational Models and Computer Codes for Environmental Dose Assessment of Radioactive Releases,* Report BNWL-B-45, Battelle Pacific Northwest Laboratory, NTIS.

——, E. C. Watson, and J. R. Houston, 1975, *SUBDOSA: A Computer Program for Calculating External Doses from Accidental Atmospheric Releases of Radionuclides,* USAEC Report BNWL-B-351, Battelle Pacific Northwest Laboratory, NTIS.

United Nations Scientific Committee on the Effects of Atomic Radiation, 1972, Vol. II, Effects, United Nations, New York.

——, 1977, *Sources and Effects of Ionizing Radiation,* United Nations, New York.

U. S. Atomic Energy Commission, 1961, *Atomic Energy Research: Life and Physical Sciences, Reactor Development, Waste Management,* GPO.

——, 1975, *Code of Federal Regulations,* Title 10, Energy, Parts 0 to 199, *Federal Register (Special Edition),* GPO.

U. S. Nuclear Regulatory Commission, 1975, *Reactor Safety Study, An Assessment of Accident Risks in U. S. Commercial Nuclear Power Plants,* NRC Report WASH-1400 (NUREG-75/014), NTIS.

Wald, N., 1958, Leukemia in Hiroshima City Atomic Bomb Survivors, *Science,* 127(3300): 699-700.

Warren, S., 1956, Longevity and Causes of Death from Irradiation in Physicians, *J. Am. Med. Assoc.,* 162: 464-468.

Way, K., and E. P. Wigner, 1951, Rate of Decay of Fission Products, in *Radiochemical Studies: The Fission Products,* Division IV, Vol. 9, Book 1, Part IV, pp. 436-458, C. D. Corvell and Nathan

Sugarman (Eds.), National Nuclear Energy Series, McGraw-Hill Book Co., New York.

White, M. G., and P. B. Dunaway, 1977, *Transuranics in Natural Environments,* ERDA Report NVO-178, Nevada Operations Office, NTIS.

Willard, D. H., and W. J. Bair, 1958, *Behavior of I^{131} Following Its Inhalation as a Vapor and as a Particle,* USAEC Report HW-58221, Hanford Atomic Products Operation, NTIS.

World Health Organization, 1957, *Effect of Radiation on Human Heredity.*

Wyckoff, H. O., 1959, Radiation Attenuation Data, in *Radiation Hygiene Handbook,* pp. 8-1–8-69, H. Blatz (Ed.), McGraw-Hill Book Co., New York.

Chapter 16 covers research prior to January 1980.

17 TOXICOLOGICAL EFFECTS OF NONNUCLEAR POLLUTANTS

N. Irving Sax, Author and Consultant

17-1 INTRODUCTION

This chapter deals with the nonnuclear public health impact of fossil-fueled power plants (ffpp's), specifically the effects of emitted pollutants on populations. In this chapter atmospheric pollution is defined as the presence of contaminants in the ambient air of such kind, concentration, and duration as to cause a nuisance, discomfort, or actual or potential injury to human or animal life, vegetation, or property.

Since there can be no toxicological effects without contact between a toxicant and members of the population under study, we can use various meteorological methods to determine how much of a given toxicant is being emitted from a given source and then how much is made available to a population at a predetermined location. Listed in the following paragraphs are the components of stack effluents, such as fly ash, particulates, gases, vapors, and most of the elements or compounds thus included. Our mission in these paragraphs is to discuss ffpp contamination in a qualitative sense.

Thus we first discuss the principles of toxicology to acquaint you with the terminology of toxicology so that the reader can relate analytical data to threshold limit values (TLV's) or to mean lethal dose (LD_{50}) data on various materials, and when the amounts of contaminants become known, the toxicological hazards can be estimated.

Then, we list the most important elements and compounds known to be emitted by ffpp's, and we discuss the toxicology of each. References provide additional sources of information. Thus volatile substances are those that vaporize easily. Examples of some of the volatile elements known to appear in stack effluents are the halogens—fluorine, chlorine, iodine, and bromine. Examples of volatile compounds appearing in stack effluents are ammonia, ozone, and the oxides of sulfur, nitrogen, and carbon. These compounds and the halogens tend to remain completely in the gaseous phase after emission. By contrast, the following elements also volatilize but end up partly as components of fly ash:

Aluminum	Hafnium	Scandium
Barium	Potassium	Silicon
Calcium	Lanthanum	Samarium
Cerium	Magnesium	Strontium
Cobalt	Manganese	Tantalum
Europium	Rubidium	Thorium
Iron	Sodium	Titanium
Chromium	Nickel	Uranium
Cesium		Vanadium

To adequately describe the relationship between a polluted environment and the people, we will use the terminology of toxicology. Toxicology may be defined as the science of poisons, their effects, antidotes, and detection. With continual population exposure to many potentially harmful chemicals such as food additives, pesticides, fertilizers, cosmetics, therapeutic drugs, and other environmental pollutants, it is important to try to evaluate the added impact of ffpp emissions, and this assessment requires some knowledge of toxicology.

Basic to the science of toxicology is the fact that all materials vary in their impact on man and other living members of the environment. Although all things can be hazardous under the proper conditions, there is a classification of materials known as poisons of such high toxicity that they are hazardous to health even under normal conditions of use and encounter and even in small quantities (Sax, 1979). In addition, the toxicity, i.e., the dosage or amount of a chemical required to produce harm, is important, and the numbers used to express the toxicity constitute the quantitative aspect of toxicology.

17-2 TOXICITY RATINGS

One fast screening method for describing the toxicity of materials is the use of toxicity ratings as follows:

The rating *U = unknown* is assigned to chemicals for which insufficient toxicity data are available to enable a valid assessment of hazard to be made for one of the following reasons: (1) No toxicity information was found in the literature and none was known to the authors. (2) Limited information based on animal experiments was available, but in the opinion of the authors this information could not be applied to human exposures. In some cases the information is mentioned so that the reader may know what experimental work has been done. (3) The authors felt the published toxicity data to be of questionable validity.

Sax: 7 Royal Palm Way, Boca Raton, Florida.

The rating *0 = no toxicity rating* is assigned to materials which cause no harm under any conditions of normal use or which produce toxic effects on humans only under the most unusual conditions or because of overwhelming dosage, e.g., water can cause drowning.

All toxicity ratings are divided into four descriptive categories:

1. *Acute local.* The effects are noted at the points of contact on or in the body from single exposures to materials or from exposures lasting from seconds to hours.

2. *Acute systemic.* The effects apply anywhere in the body after materials are absorbed into the body following inhalation, ingestion, or skin absorption. The effects follow single exposures lasting seconds, minutes, or hours, or the ingestion of a single dose, regardless of the quantity absorbed or the extent of exposure.

3. *Chronic local.* The effects occur locally from continuous or repeated exposures to materials extending over periods of days, months, or years. The extent of exposure can be great or small.

4. *Chronic systemic.* The effects are from materials that are absorbed into the body (1) following inhalation, ingestion, or absorption through the skin and (2) following continuous or repeated exposures extending over days, months, or years. The extent of the exposure can be great or small, and the effects can show up anywhere in the body.

In general, the substances classified as having slight toxicity produce changes in the human body which are readily reversible and which will disappear after exposure is ended, either with or without medical treatment.

Substances classified as having moderate toxicity can produce irreversible or reversible changes in the human body, but not so severe as to threaten life or to produce serious physical impairment. The damage can occur anywhere in the body.

Substances classified as highly toxic cause injury of sufficient severity as to threaten life or to cause permanent physical impairment or disfigurement.

In the case of severe toxicity:

1. *Acute local.* Materials that on a single exposure lasting seconds or minutes cause injury to skin or mucous membranes of sufficient severity to threaten life or to cause permanent physical impairment or disfigurement.

2. *Acute systemic.* Materials which can be absorbed into the body by inhalation, by ingestion, or through the skin and which can cause injury of sufficient severity to threaten life following a single exposure lasting seconds, minutes, or hours or following ingestion of a single dose.

3. *Chronic local.* Materials that on continuous or repeated exposures extending over periods of days, months, or years can cause injury to skin or mucous membranes of sufficient severity to threaten life or cause permanent impairment, disfigurement, or irreversible change.

4. *Chronic systemic.* Materials which can be absorbed into the body by inhalation, by ingestion, or through the skin and which can cause death or serious physical impairment following continuous or repeated exposure to small amounts extending over periods of days, months, or years.

17-3 TYPES OF TOXICITY

Human toxicity is a unique and reproducible property of a material which causes it to produce injury once it touches or exposes a susceptible site in or on the human body.

The hazard potential of a material is the probability that exposure to it will occur and injury will ensue from its normal use. Thus, although hundreds of humans die annually from drowning in it, water is considered nonhazardous because in normal use, i.e., cooking, washing, and eating, it is in no way harmful. Thus hazard is variable and depends on use.

No damage is possible and therefore no hazard can be assigned to a chemical without exposure, but it is seldom possible to keep exposure to a material down to zero while working with it. Consequently, very toxic materials are automatically considered hazardous, and the two terms are often mistakenly interchanged.

Two often misunderstood terms referring to subdivisions of toxicity are "acute" and "chronic." Actually, they refer only to duration of exposure or effects therefrom.

Acute exposure means of short duration. As applied to materials that are inhaled, ingested, in contact with the skin, or absorbed through it, acute exposure refers to a single exposure of duration measured in seconds, minutes, or hours. As applied to effects, the term refers to the results of acute exposures. As applied to materials that are ingested, it refers generally to a single quantity or dose. The term "subacute" is also used, and it indicates a duration that is longer than strictly acute (up to 90 d) but shorter than chronic.

Chronic exposure, used in contrast to acute, means of long duration. For materials that are inhaled, ingested, or absorbed through the skin, chronic exposure refers to prolonged or repeated exposures of a duration measured in days, months, or years. As applied to effects, the term refers to the results of chronic exposures. As applied to materials that are ingested, it refers to repeated doses over a period of days, months, or years. The term "chronic" does not refer to severity of symptoms but carries the implication of exposures or doses that would be relatively harmless unless extended or repeated over long periods of time (days, months, or years).

17-4 ACUTE AND CHRONIC EFFECTS

It is important to explain the terms "acute effects" and "chronic effects." The expression chronic illness is used to indicate a long illness, i.e., an illness lasting months or years, such as cancer; whereas an acute illness or acute episode is an illness lasting hours or days, such as appendicitis or an episode of alcoholic intoxication. However in this chapter the terms acute effects and chronic effects will refer only to the effects produced by short or long exposures to toxic materials. Thus an acute exposure can produce a chronic illness. The chronic illness, however, would be an acute effect. The fact is that some compounds, including heavy metals and some carcinogens, have a strong tendency to produce chronic illness even though the exposure to it may be acute, i.e., only a single dose. Of course, their tendency to produce chronic sickness is accentuated by repeated doses. At the opposite extreme are compounds, such as cyanide, with which it is virtually impossible to produce chronic illness—even a single excessive dose will produce rapid death. Most compounds lie somewhere between the two extremes.

The words acute and chronic applied to illness indicate the duration and character of illness rather than the severity of the illness. The common cold, intoxication from social drinking, plague, and parathion poisoning are all acute illnesses. The first two are mild, and the last two are potentially fatal; all are brief with little tissue reaction. Pulmonary tuberculosis and lead poisoning are almost always chronic diseases. They are characterized by a prolonged course and by pathological changes in tissue that reflect continuing injury and perhaps ineffectual repair.

17-5 TOXIC EFFECTS

Toxic effects may be subdivided on the basis of site of action:

Local effect means that the action takes place at the point or area of contact. The site may be skin, mucous membranes of the eyes, nose, mouth, throat, or anywhere along the respiratory or gastrointestinal system. Absorption does not necessarily occur.

Systemic effect refers to a site of action other than the point of contact and presupposes that absorption has taken place. It is possible, however, for toxic agents to be absorbed through a channel (skin, lungs, or intestinal canal) and produce later manifestations on one of those channels which are not a result from the original direct contact. Thus some agents can produce harmful effects on a single organ or tissue as a result of both local and systemic action.

17-6 DOSAGE

The single most important factor in determining whether illness will occur as the result of exposure to a specific material is dosage or the amount of material in the exposure. The dosage concept leads to the conclusion that no material is completely safe nor so hazardous as to be unusable. The same idea was expressed by Paracelsus (1493–1541), who wrote "All things are poisons, for there is nothing without poisonous qualities. It is only the dose which makes a thing a poison."

Compounds vary tremendously in their toxicity. Usually nonhazardous and even essential substances, such as water and salt, can cause illness or death if consumed in sufficient amount. Even compounds recognized as poisons can differ in toxicity by a factor of at least 10 billion. The most toxic material presently known is botulinum toxin, which is more poisonous by orders of magnitude than any other material known. It is interesting to note that botulinum is a natural material of biological origin.

In comparing the toxicity of different compounds, it is convenient to use a standardized notation. The most commonly used notation is the median lethal dose, or LD_{50}, which is a statistical estimate of the dosage necessary to kill 50% of an infinite population of the test animals. The LD_{50} is usually expressed in terms of the weight of poison per unit of body weight, most often as milligram of poison per kilogram of animal (mg/kg). This measure also permits a meaningful comparison of the susceptibility of animals of different species, regardless of their size, to various poisons.

The LD_{50} is a special case of a more general measure of effect, the median effective dosage (ED_{50}). The ED_{50} is the dosage necessary to produce any specified effect in 50% of the test animals. The effect can be anything that can be observed, e.g., a specified degree of inhibition of an enzyme or the production of a tumor. These measures of effects for a particular compound have meaning only if experimental conditions are defined, including the species, age, and sex of the experimental animal, the number of doses, and the route of administration.

In addition to the dosage and the identity of the material involved, a number of less important factors play a role in determining whether illness can result from any particular exposure episode. Among these, the following factors should be considered:

1. Route of exposure. It is important to learn whether the toxicant was ingested, inhaled, or spilled on the skin or, as most often happens, some combination of these caused the exposure.
2. Type of formulation or state of dispersion of the toxicant.
3. Temperature.
4. Humidity.
5. Physiologic condition of the subject.
6. Interaction of the toxicant with other chemicals and drugs. Interaction of chemicals can occur either in the external environment (such as when exposure to a mixture of two industrial chemicals occurs), or inside the patient (such as occurs when previous dosage of a therapeutic drug may alter response to another drug). It is now recognized that the latter type of reaction occurs with some frequency as a result of stimulation of the drug-metabolizing enzyme system of the liver (microsomal enzymes) by various chemicals and drugs.

17-7 THRESHOLD LIMIT VALUES

Another concept for expressing and comparing toxic levels for different compounds is the use of TLV's, formerly known as maximum allowable concentrations, MAC's (Sax, 1979). It should be noted that the LD_{50} represents an experimentally derived value generally based on animal studies, whereas the TLV is arbitrarily set for human exposures on the basis of known experimental and other available data and a safety factor. The acute oral LD_{50} values are most useful if an investigator is comparing ingestion toxicities, while the TLV is meant to provide on-the-job safety for humans who work with toxic or hazardous materials. Thus TLV's make use of LD_{50} values but also consider many other factors.

In the United States, TLV's have received wide acceptance (Sax, 1979). The American Council of Government and Industrial Hygienists (ACGIH) has now set TLV's for about 800 chemical compounds. These values refer to airborne concentrations and represent conditions under which it is believed that most workers can be safely exposed for an 8-h day for 5 d a week. Since individual susceptibility varies widely, a few workers may experience some discomfort and even fewer may be affected more seriously by substances at concentrations at or near the TLV. Sensitivity testing should be carried out on new industrial workers being placed in jobs involving significant exposure to toxic chemicals so that persons who are hypersusceptible can be screened out. Also, since many more women are accepting factory and research positions, the problems of fetotoxicity and teratogenicity have also to be carefully considered in setting work standards.

With some few exceptions the TLV's are time-weighted average concentrations. Thus temporary excursions above the limit can be permitted occasionally provided they are compensated by equivalent excursions below the limit during a workday. In spite of the fact that no serious injury is expected to occur as a result of exposure to TLV concentrations, the best industrial hygiene practice is to maintain concentrations of all atmospheric contaminants at levels as low as practicable.

Closely related to TLV's are the so-called acceptable concentration standards promulgated by the American Standards Association (ASA). According to the ASA, these standards are designed to prevent (1) undesirable changes in body structure or biochemistry, (2) undesirable functional reactions that may have no discernible effects on health, and (3) irritation or other adverse sensory effects.

For gases and vapors the TLV is usually expressed in parts per million (ppm), i.e., parts of the gas or vapor per million parts of air. For fumes or mists and for some dusts, the TLV is usually listed as milligrams per cubic meter (mg/m^3) or per 10 m^3 of air. For asbestos the TLV is given as 0.2 to 2.0 fibers $> 5\ \mu m$ length/cm^3 of air. This TLV is, however, still under study. Under notice of intended changes in the 1981 TLV lists for ACGIH, asbestos is assigned a TLV = 0.2 to 2.0 fibers/cm^3 $> 5\ \mu m$ length and is labeled a human carcinogen. Intended is amosite = 0.5 fiber/cm^3; other forms = 2 fibers/cm^3. For some dusts, particularly those containing silica, the TLV is usually expressed as millions of particles per cubic foot of air (mppcf). However, for fibrous materials, i.e., natural diatomaceous earth = 1.5 mg/m^3 of respirable dust, for amor silica = 6 mg/m^3 of total dust and 3 mg/m^3 of respirable dust of $<5\ \mu m$, and for fibrous talc = 2 fibers/cm^3, $> 5\ \mu m$.

17-8 CLASSES OF TOXIC SUBSTANCES

Toxic or harmful substances encountered in industry or in some instances released into the environment (such as pesticides) can be physically classified in various ways. A simple and useful classification is given below, together with definitions adopted by the ASA.

Dusts. Solid particles generated by handling, crushing, grinding, rapid impact, detonation, and decrepitation of such organic or inorganic materials as rocks, ore, metal, coal, wood, and grain. Dusts do not tend to flocculate except under electrostatic forces; they do not diffuse in air but settle under the influence of gravity.

Fumes. Solid particles generated by condensation from the gaseous state, generally after volatilization from molten metals, etc., and often accompanied by a chemical reaction such as oxidation. Fumes flocculate and sometimes coalesce.

Mists. Suspended liquid droplets generated by condensation from the gaseous to the liquid state or by breaking up a liquid into a dispersed state, such as by splashing, foaming, and atomizing.

Vapors. The gaseous form of substances which are normally in the solid or liquid state and which can be changed to these states by either decreasing the pressure or increasing the temperature alone. Vapors diffuse.

Gases. Normally formless fluids which occupy the space of enclosure and which can be changed to the liquid or solid state only by the combined effect of increased pressure and decreased temperature. Gases diffuse.

This classification does not include the obvious categories of solids and liquids which may be harmful, nor does it encompass physical agents, i.e., radiation and noise. The latter, strictly speaking, cannot be considered "substances" that would appear in a comprehensive classification of industrial health hazards.

17-9 ROUTES OF ABSORPTION

In the physiological sense, a material is said to have been absorbed only when it has gained entry into the bloodstream and consequently is carried to all parts of the body. Something which is swallowed and which is later excreted more or less unchanged in the feces *has not been absorbed*, even though it may have remained within the gastrointestinal tract for hours or even days.

Although there are a number of ways by which an individual can be exposed to a toxic chemical, the three most important and most usually encountered routes are oral, dermal, and respiratory. In accidental poisoning, as is usually the case with children, ingestion or oral exposure is most common. In industrial situations, exposure is usually either respiratory or dermal or a combination of both.

17-9.1 Absorption Through the Skin

In general, materials are absorbed more slowly and less completely through the skin than through the gastrointestinal tract or the lungs, and thus toxic effects are less likely to occur when a toxicant is spilled on the skin than when it is ingested or inhaled. Fortunately, not many materials are absorbed to a significant degree by the skin. For example, botulinum toxin, the substance that produces botulism, is extremely toxic to man when taken orally but is relatively harmless when applied to the skin. The principal reason for this difference is that the botulinus toxin molecule is very large and is therefore not readily absorbed by the skin. However, when it enters the digestive tract, the toxin is immediately absorbed and transported to the susceptible sites of action. Thus the chemical and physical properties of the toxicant determine which routes of exposure will produce what toxic effects. There is greater magnitude of difference between materials with regard to skin absorption than for the other common exposure routes.

There are some compounds that are more toxic when spilled on the skin than when swallowed. Usually these are compounds rapidly detoxified by the liver. The detoxication site is important since materials absorbed from the skin or the lungs are distributed by the blood to the rest of the body at the same instant that a portion passes to the liver, and thus there is much less detoxification.

One of the problems with materials which present a skin-absorption hazard is that it is frequently difficult to convince a layman that a chemical can cause poisoning merely by spilling some on his skin.

Skin absorption attains its greatest importance in connection with organic solvents. It is generally recognized that significant quantities of these compounds may enter the body through the skin either as a result of direct accidental contamination or indirectly when the material has been spilled on the clothing. An additional source of exposure is found in the fairly common practice of using industrial solvents for removing grease and dirt from the hands and arms, i.e., for washing purposes. This procedure, incidentally, is a fruitful source of dermatitis.

It is recognized that skin absorption can be a significant factor in occupational mercury poisoning as well as in a number of other industrial diseases. In the case of metals other than mercury, however, entry through the skin is relatively unimportant except for some organometallic compounds such as lead tetraethyl.

A number of other important environmental and industrial chemicals are also absorbed to a significant degree through the skin. Some of the organic phosphorus pesticides, including parathion, are notable examples.

17-9.2 Gastrointestinal Absorption

In accidental poisoning of children, oral ingestion is the most common route of exposure; this is not true for adults. In the past it has been common practice to attribute certain cases of occupational (adult) poisoning to slovenly habits on the part of the victim, particularly on failure to wash his hands before eating. There is no doubt that some toxic materials used industrially can be absorbed through the intestinal tract, but it is now generally believed that with certain notable exceptions this portal of entry is of minor importance. One outstanding exception is the case of the radium-dial painters who followed the practice of "pointing" their brushes between their lips, thus ingesting lethal quantities of radioactive material. Accidental swallowing of harmful amounts of poisonous compounds in single large doses has also been known to occur. In general, it can be said that intestinal absorption of industrial poisons is of minor importance and that the "dirty hands" theory of poisoning has been pretty well discredited.

17-9.3 Absorption Through the Lungs

The inhalation of contaminated air is by far the most important means by which occupational poisons gain entry into the body. It seems safe to estimate that at least 90% of all industrial poisoning (exclusive of dermatitis) can be attributed to absorption of poisons through the lungs. Harmful substances may be suspended in the air in the form of dust, fume, mist, or vapor and then mixed with the respired air as the case of true gases. Since an individual under conditions of moderate exertion will breathe about 10 m^3 of air in the course of an ordinary 8-h working day, it is readily understood that any poisonous material present in the respired air and which may be trapped by the lungs offers a serious threat to health.

Fortunately, all foreign matter that is inhaled is not necessarily absorbed into the blood. A certain amount, particularly that in a very finely divided state, will be immediately exhaled. Another portion of respired particulate matter is trapped by the mucus which lines the air passages and is subsequently brought up in the sputum. Incidentally, some of the sputum may be conciously or unconsciously swallowed, thus affording an opportunity for intestinal absorption. Other particles are taken up by "scavenger cells" following which they may enter the blood stream or may be deposited in various tissues or organs. True gases will pass directly from the lungs into the blood in the same manner as the oxygen in inspired air. Because a great majority of the known industrial poisons may at some time be present as atmospheric contaminants and thus constitute a potential threat to health, programs directed toward the prevention of occupational poisoning generally place major emphasis on ventilation to reduce the hazard.

17-10 STORAGE AND EXCRETION

Some toxic substances can be retained or stored in the body for indefinite periods of time, being excreted but slowly over periods of months or years. Lead, for example, is stored primarily in the kidneys. Smaller amounts can be stored in other organs or tissues. Particulate matter, when inhaled, can be phagocytosed and remain in regional lymph nodes where it may have little effect, as in the case of coal dust, or it may produce pathological changes, as in the case of silica and beryllium.

The excretion of absorbed toxic agents takes place through the same channels as does absorption, namely, through the lungs, intestines, and skin, but the kidneys (urine) are the main excretory organs for many substances. Sweat, saliva, and other body fluids may participate to a small extent in the excretory process. Gases and volatile vapors are commonly excreted via the lungs and breath. This fact can sometimes be used as a measure of earlier absorption.

Many organic compounds are not excreted unchanged but pass through what is known as biotransformation. These processes are also known as "detoxication mechanism." The resulting new compounds, or metabolites, can subsequently be found in the urine and are used as evidence of absorption of the parent substance. Sometimes the metabolites are as, or more, toxic than the original materials.

17-11 INDIVIDUAL SUSCEPTIBILITY

The term "individual susceptibility" has long been used to express the well-known fact that under conditions of like exposure to potentially harmful substances there is usually a marked variability in the manner in which individuals respond. Some may show no evidence of intoxication whatsoever; others may show signs of mild poisoning, and still others may become severely or even fatally poisoned. Comparatively little is known about the factors that are responsible for this variability. It is believed that differences in the anatomical structure of noses may be concerned with different degrees of efficiency in filtering out harmful dusts in the inspired air. Previous infections of the lungs, particularly tuberculosis, is known to enhance susceptibility to silicosis. Most industrial toxicologists believe that obesity is an important predisposing factor among persons who are subject to occupational exposure to organic solvents and related compounds. Age and sex are also believed to play a part, and previous illnesses may be significant.

Other possible factors relating to individual susceptibility are even less understood than those just mentioned. It has been suggested that different rates of working speed, resulting in variations in respiratory rate, in depth of respiration, and in pulse rate, may play a part. The action of the cilia, those tiny hairs present in the cells which line the air passages, may have some importance. The permeability of the lungs may influence absorption and the efficiency of the kidneys may govern the rate at which toxic materials are excreted, but the underlying nature of these possible variations is not known. Since the liver plays a major role in the detoxification and excretion of harmful substances, subnormal functioning of that organ may lead to increased susceptibility also.

There is considerable literature purporting to show that nutritional factors may have something to do with individual susceptibility to occupational poisoning. Most of the published material is rather unscientific and unconvincing, but a few reports strongly suggest that there actually exists a relationship between the nature of the diet and susceptibility to poisoning from exposure. There is as yet no substantial evidence that the addition of vitamin concentrates, milk, or special foods has any protective value, but when diets are deficient in some of the essential nutritional elements, it appears that poisoning is more likely to occur. There is considerable evidence that indulgence in alcoholic beverages will significantly increase the possibility of occurrence of occupational poisoning, particularly from organic solvents.

17-12 ACUTE AND CHRONIC EXPOSURES

Industrial toxicology is generally concerned with the effects of low-grade sublethal exposures that are continued over a period of months or years. It is, of course, true that toxicological problems are frequently presented as a result of accidents which create sudden massive exposures to overwhelming concentrations of toxic compounds. The acute poisoning that results may cause unconsciousness, shock, collapse, severe inflammation of the lungs, or even sudden death. An understanding of the nature of the action of the offending agent may be of great value in the treatment of acute poisoning, but in some instances the only application of toxicological knowledge will be to establish the cause of death.

The detection of minute amounts of toxic agents in the atmosphere and in body fluids (blood and urine) and the recognition of the effects of exposure to small quantities of poisons are among the principal jobs of the industrial toxicologist. The manifestations of chronic poisoning are often so subtle that the keenest judgment is required to detect and interpret them. The most refined techniques of analytical chemistry and of clinical pathology are called into play, involving studies of the working environments and of exposed individuals.

To demonstrate that chronic industrial poisoning has taken place or is a possibility, we must show that an offending agent is present in significant concentrations, that it has been absorbed, and that it has produced in the exposed subject disturbances compatible with poisoning by the suspected substance. Significant concentrations are ordinarily expressed in terms of TLV's. Absorption of a substance may be proved by demonstrating its presence in the blood or urine in concentrations above those found in nonexposed persons or by finding certain metabolic products of the substance in the excreta. To prove that disturbances have occurred in an exposed subject may require the application of all the diagnostic procedures used in medicine, including a medical history, physical examination, blood counts, urinalysis, X-ray studies, and other measures.

A few of the more widely used industrial chemicals, notably lead and benzene, will produce changes in the blood in the very early stages of poisoning. Other chemicals, particularly chlorinated hydrocarbons, give no such early evidence of their presence. Heavy metals such as mercury and lead produce their chronic harmful effects through what is known as "cumulative action" which means that over a period of time the absorbed material is only partially excreted and that increasing amounts accumulate in the body; i.e., the toxicant is coming in faster than it is being excreted. Eventually the quantity in the body becomes great enough to cause physiological disturbances. Volatile compounds do not accumulate in the body but probably produce their chronic toxic effects by causing a series of small insults to one or more of the vital organs.

17-13 SITE OF ACTION OF POISONS

Brief mention has already been made of the fact that different poisons act on different parts of the body. Many substances can produce a local or direct action on the skin. The fumes and mists arising from such substances as strong acids, some of the war gases, and many other chemicals have a direct irritating effect on the eye, nose, and lower air passages. If they reach the lungs, they may set up a severe inflammatory reaction called chemical pneumonitis. These local effects are of greatest importance in connection with acute poisoning.

More important to the industrial toxicologist are the so-called *systemic effects* that occur when a toxic substance has been absorbed into the blood stream and has been distributed throughout the body. Some materials, such as arsenic, when absorbed in toxic amounts, may cause disturbances in several parts of the body: blood, nervous system, liver, kidneys, and skin. Benzene, on the other hand, appears to affect only one organ, namely, the blood-forming bone marrow. Carbon monoxide causes asphyxia by preventing the hemoglobin of the blood from carrying out its normal function of transporting oxygen from the lungs to the tissues of the body. Although oxygen starvation occurs equally in all parts of the body, brain tissue is most sensitive; consequently, the earliest manifestations are those due to effects on the brain. Among the more important functions of the industrial toxicologist are understanding what organ or organs can be damaged and understanding the nature and manifestations of the damage caused by various compounds.

At the cellular level, toxic agents may act on the cell surface or within the cell, depending on "receptors" or binding sites. A familiar example is the affinity of arsenic and mercury for sulfhydryl (SH) groups in biological material.

17-14 ABSORPTION AND POISONING

As mentioned previously, except for external irritants, for poisoning to occur, toxic substances generally must be absorbed into the body and distributed through the body by means of the blood stream. In other words, poisoning ordinarily does not occur without absorption. However, absorption does not necessarily or always result in poisoning. The human body has developed an elaborate system of protective mechanisms and is able to tolerate to an amazing degree the presence of many toxic materials. Some foreign materials are excreted unchanged through the urine and feces. Toxic gases, following absorption, may be excreted through the lungs. Some chemical compounds go through processes of metabolism and are excreted in an altered form. Some of these processes are known as detoxication (or detoxification) mechanisms. In some instances the biotransformation products in a detoxication process may be more toxic than the original substance, e.g., the conversion of parathion to the much more toxic paraoxon.

17-15 TOXICITY AND HAZARD

The toxicity of a material is a unique property of that material and can only be described by its effects on a living organism. As a general principle, toxic materials are toxic to all living things, whereas the individual susceptibility of each species to the toxic material varies, as the susceptibilities of the individual members within a species vary with age, sex, state of health, rate of dosage, diet, etc. Therefore, to be useful, toxicity data must include all of the preceding information and must be stated in terms of the specific test animal or organism used, the routes of administration of the toxic matter, and the time of exposure. Also, a sufficient number of test animals or organisms must have been tested to rule out some of the variables listed above and to make the data statistically significant. For purposes of this section, suffice it to state that all materials are more or less toxic. Those of any concern to this section are the ones which are or could be damaging to humans in amounts that ordinarily occur or might occur in the living environment or in the industrial, medical, or scientific environments. Here is where hazard enters the picture. For instance, it is conceivable for a material to have a high "toxicity" rating but a low "hazard" rating because the hazard rating is simply a measure of the *likelihood* of damage to humans working with or using the material in question.

17-16 ALLERGY AND HYPERSENSITIVITY

One of the most difficult toxic manifestations of chemical substances with which the toxicologist has to deal is *allergy* or *hypersensitivity.*

Allergy is estimated to be a health problem for 10% to 30% of the population. Generally, an allergic disease is not fatal; however, it may be disabling. In industry the problem of lost work time due to allergic response from on-the-job contact with chemical or other allergens is a perplexing one and is complicated by off-the-job contacts over which the industrial hygienist has little or no control, or even knowledge.

Any substance that produces an allergic reaction is called an *allergen.* Almost any substance can be allergenic to an allergy-predisposed individual. An allergen affects only the susceptible individual because only the allergic person exhibits an abnormal response to the particular allergen. Usually the substance is otherwise harmless; in fact, some usually wholesome foods, such as milk, eggs, or wheat, may be specific causes of allergic disease.

Allergy is related to a specific substance. Thus, if one is allergic to an animal, he may not be allergic to all animals, or to all furs, or to all chemicals. This specificity can usually permit a worker sensitized to one type of chemical to work safely with a substitute material. However, an allergy-prone person can become cross-sensitized, i.e., having become allergic to one chemical, he subsequently develops an allergy to related compounds.

Although the general belief is to the contrary, the allergic reaction is, in general, a quantitative one. This is because a very minute quantity of an allergen may precipitate a reaction and, in general, the amount of chemical required to produce an allergic response is less than that needed to bring about a systemic toxic response. Slight contact with an allergen produces light symptoms, whereas contact with a larger amount of the same material may produce a much more severe response. It is important to understand the quantitative nature of the allergic response to appreciate that simply decreasing the magnitude of an individual's exposure to an offending substance may eliminate the occurrence of allergic symptoms. Protective measures to guard personnel against allergens are the same as for toxic substances or for an individual to become sensitized to a material.

The nature of the allergic response depends to some extent on the route of exposure. Thus, since industrial exposure is primarily through the skin or respiratory tract, symptomatology most often involves the skin, sinuses, nasal or respiratory system, and eyes causing asthma, rhinitis, urticaria, angioedema, anaphylaxis, and possibly emphysema.

17-17 PARTICULATES

Pollution effects caused by fossil-fuel production of electric power are based on the calculation that about 95% of all the trace elements in coal goes into the slag, with only 5% committed to the atmosphere in the form of fly ash, particulates, and gases. The gases later react in the atmosphere to form more particulate matter, to become absorbed on particulates, or to be captured in precipitation processes and to appear subsequently in condensation nuclei, rain, mist, or fog, i.e., acid rain.

The particulate segment of power-plant air pollutants is composed of asbestos and fly ash (Linton, Loh, and Natusch, 1976). Fly ash is a complex material composed generally of hollow spheres (Fig. 17.1) of matrix material composed largely of aluminum, iron, and silicon and containing a variety of elements.

Knowledge of the characteristics and properties of small solid and liquid particles is required to understand their environmental effects. These characteristics and properties include particle size, size distribution, shape, specific gravity, and surface (interfacial) phenomena. Particle sizes of typical dispersions of environmental concern are summarized in Fig. 17.2. With a diameter of less than 100 μm and a specific gravity of 1 to 2, suspended particles have a terminal settling velocity, in air, less than about 50 cm^{-1}, slightly greater than large-scale vertical motion in the atmosphere. Consequently, particulates can be carried great distances by the wind before reaching ground level.

Other factors that affect particle behavior in the air include shape and density. Typical shapes encountered are near-spherical (droplets, fly ash cenospheres, ragweed pollen); rodlike (lint, hair, fibers; as in cellulosic, protein, or asbestos materials); cubical (sea salt); acicular, i.e., pointed; starlike [ZnO, $(NH_4)_2 \cdot SO_4$]; granular (mineral dusts); condensation flocs (C, metallurgical fumes as Fe_2O_3); platelike (mica, graphite); and general mixed, or unclassed. Extensive illustrations of particulate morphology are presented in the *Particle Atlas* (McCrone, Draftz, and Delly, 1970) in a series of color photomicrographs of specific materials of concern in atmospheric pollution source emissions. A given atmospheric sample will contain many materials, whereas a specific source emission may contain only a few. Particle density for most compounds of concern will range from about 0.8 to 20 g/cm^3. Natural coagulation (contact of two or more suspended particles while airborne) will produce an aggregate particle containing substantial void space, reducing the effective density to about 0.1 or less of the true density for the particular material.

Particulates may assume the form of dispersions of solids, such as dusts, fly ash, fumes, and smokes; of liquids, such as mists and fogs; and of combinations of the two, such as mists created by sprays or suspensions of dusts and paints. With rare exceptions, such as with organic phosphorus, insecticides, and, possibly, massive concentrations of the more highly toxic metals, particulate contaminants are not immediately dangerous to life.

The simultaneous effect of particle dimensions, particle shape, and density may result in aerodynamic behavior significantly different from that predicted on the basis of apparent size as measured microscopically. For example, an aggregate particle of Fe_2O_3 fume (true density 7 g/cm^3) consisting of long interconnected and extended chains, whose microscopic size appears to be 20 μm (across a major dimension), may behave in sedimentation as if it were very much smaller because of its reduced density caused by interstitial void spaces. The effect of shape on aerodynamic behavior is evident in fibrous materials such as asbestos, where the aerodynamic diameter† is approximately equivalent to the fiber diameter, without regard to fiber length. The practical consequences of this shape effect are such that a 0.25-μm-diameter asbestos fiber of the order of 10 μm long may penetrate deeply into the respiratory air passages. The major techniques commonly used for particle analyses are (1) microscopic analysis by light-optical systems for particles >0.5 μm combined with transmission or

†Diameter of a hypothetical spherical particle of density 1.0 g/cm^3 having the same terminal settling velocity as the actual particle under consideration.

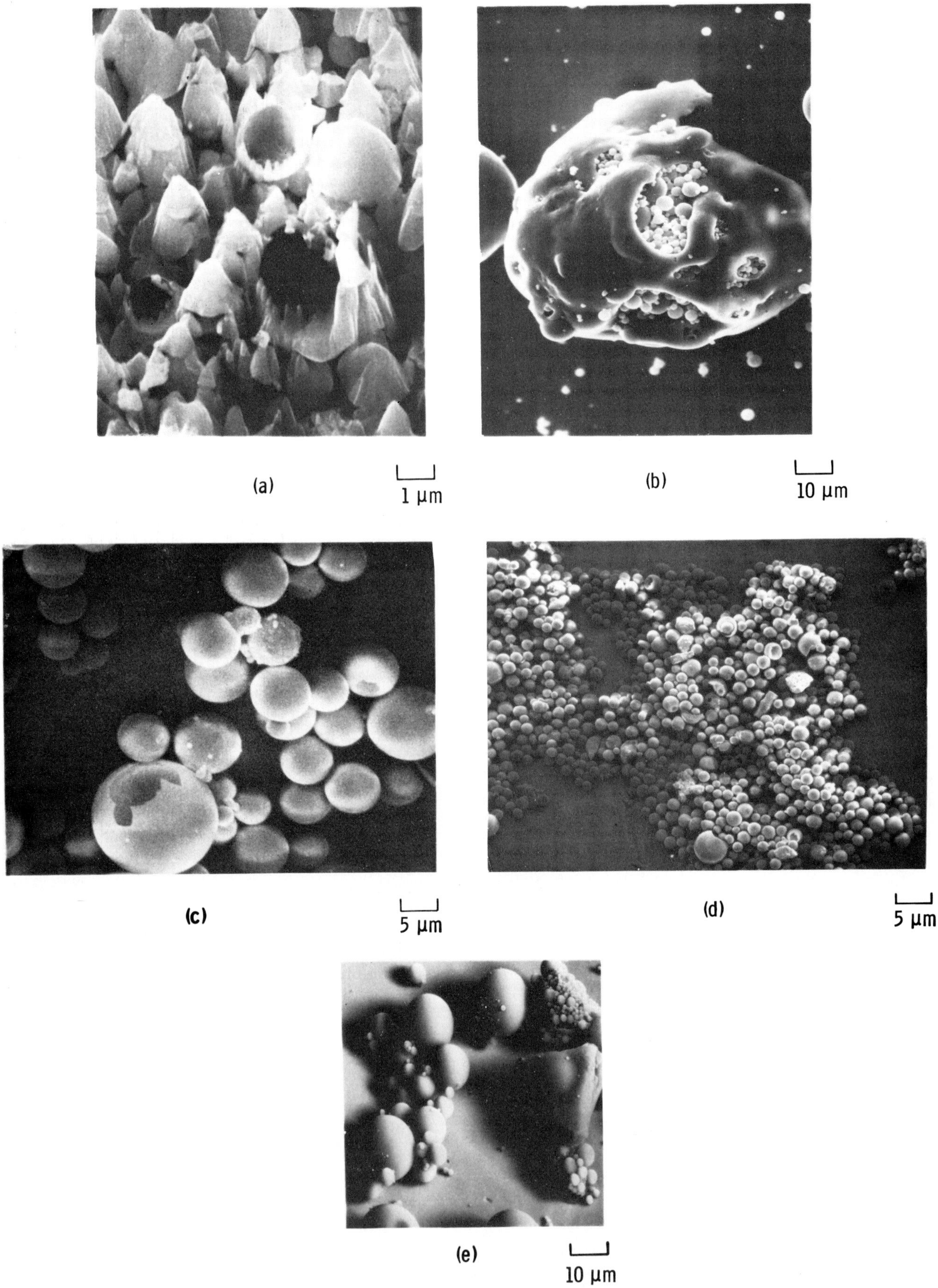

Fig. 17.1 Electron microscope slides of fly ash samples collected from electrostatic precipitators.

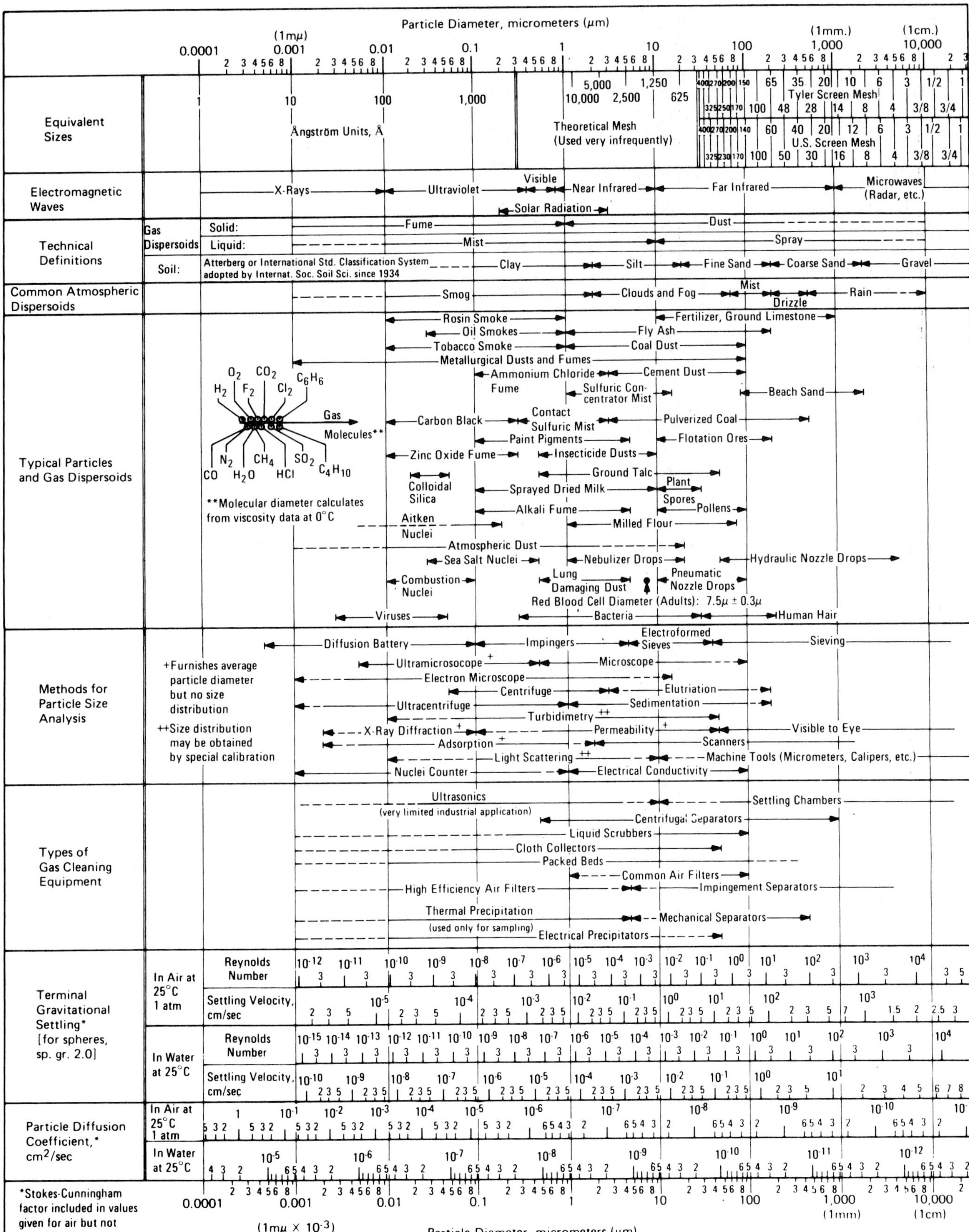

Fig. 17.2 Characteristics of particles and particle dispersoids.

scanning electron microscopy for particles down to 10^{-3} μm and (2) aerodynamic separators or classifiers such as sedimentation or centrifugal classifiers or cascade jet impactors for particles with aerodynamic diameters >0.1 μm. Many other techniques are available for specific investigations.

17-18 SPECTRAL DISTRIBUTION OF CHEMICAL CONSTITUENTS BY PARTICLE SIZE

The principal characteristic of particulate matter of concern in the evaluation of the hazard associated with environmental exposures lies in estimation of the fate of the individual particles with respect to transport, deposition, and interactions within the biological system. Hazard estimation for atmospheric pollutants considers the portal of entry to the biological system; for man, this includes primary inhalation, deposition site, clearance processes, and retention in organs of concentration. For other species, respiration, nutrition, and metabolism are of primary concern. The pathways have been discussed previously. In all instances the concentration of contaminant materials with respect to particle size becomes a question of significant concern.

Most federal regulations related to control of particulate contaminant materials contain a particle size concentration statement based on inhalation characteristics of the human respiratory system, if inhalation is the major portal of entry for the specific toxic material. Particle size concentration limits have not yet been adopted as part of regulations for air-pollution control. Ambient air-quality standards for total suspended particulate matter apply to all particles <100 μm, without regard to the respirable fraction, which is of the most significant effect in human health considerations. Recent emphasis on the fine respirable particles has prompted consideration of particle size limitations on emissions from certain sources.

Figure 17.2 summarizes a great deal of information about particles and particle dispersoids. The particle size distribution spectrum for particulate matter is presented in Chap. 11, Fig. 11.27.

The following is a terse toxicological summary of the known air pollutants emitted by the stacks of fossil-fuel-burning power plants. Well over 100 different entities have been detected in the atmosphere and have been traced to the combustion of fossil fuels. This discussion will include such categories as inorganic acids, aldehydes, alcohols, aluminum, ammonia, antimony, arsenic, asbestos particles, ketones, mercaptans, chlorinated hydrocarbons (both aliphatic and aromatic), organic acids, and some metallic elements and vapor, gases, and particulate matter. Categories of constituents are discussed rather than individual members of each category because the public health impact is much the same for any member of a given category. It must be kept in mind that ambient levels of contamination about ffpp's can be much lower than the levels we have data for.

Acids. In the respirable atmosphere, they are noted to be irritant to the skin, eyes, and mucous membranes, particularly of the lungs. Their presence simply adds to the irritant burden of our industrial atmosphere. Air containing tartaric or citric acid at 1 to 2 ppm has been known to cause etching of human teeth in about 6 months and eye irritation almost at once. Trace of nitric, sulfuric, and hydrochloric acids as air contaminants along with organic acids impart an irritant quality to the air. Usually, in industry, exposure to these materials is self-limiting in that the irritant qualities warn workers of their presence and something can be done about it. However, where the pollution is widespread, it imposes a hardship on populations that cannot escape it.

Alcohols. Methanol, ethanol, isopropanol, and propanol fumes are typical examples. They are irritants to the eyes and the mucous membranes. Exposure tends also to depress the central nervous system. Methanol liquid can produce blindness on ingestion. There are little data on long-term or chronic effects of inhalation or exposure to quite low levels of this or other alcohols. Consequently, this area merits more study since the use of fossil fuel is increasing. At concentrations of 10 to 100 ppm, they can have some acute effects. Below 10 ppm, they add to the irritant quality of industrial air, and the really long-term, chronic effects on populations are not yet well enough known.

Aldehydes. This category includes materials such as acrolein, crotonaldehyde, formaldehyde, and many others. Aldehydes are known irritants, even in the concentrations in which they occur in the environment surrounding ffpp's. They especially irritate the eyes, the skin, and the mucous membranes. One can get used to this type of irritant, although it is a hardship, particularly for the very young, the infirm, and the aged. Recent studies implicate some aldehydes as suspected carcinogens. Some can cause dermatitis and show allergenic activity as well. The absence of reliable chronic exposure data to low concentrations makes this category one to be studied because it may require regulation.

Aluminum is a widespread element usually found in very low concentrations in the general atmosphere and the environment surrounding power plants that burn fossil fuels. It is known that continual inhalation of high concentrations of aluminum powder can cause pulmonary changes and pneumoconioses. Long inhalation of aluminum powder has produced emphysema and nonnodular pulmonary fibrosis in man—in some cases leading to death. No data are available, however, on chronic exposure to subclinical concentrations of fly ash aluminum for large populations except that aluminum may be a factor in Alzheimer's disease.

Ammonia is a gaseous irritant and in high concentrations is toxic, caustic, and, if the concentration in the air is high enough, brief inhalations can cause death. It appears that in low concentrations it simply adds to the irritant load of industrially contaminated air. No data are available on chronic subclinical exposures to large populations or to the general environment.

Amines can be divided into aliphatic (so-called fatty amines) and aromatic or benzene derivatives type. The fatty amines are not very toxic. They are like ammonia, i.e., slightly to moderately irritant to the skin, eyes, and mucous membranes, but in general are more of a nuisance than a toxic hazard. However, the aromatic amines, so called because they usually have an aromatic odor, have already been implicated as experimental and even human carcinogens. The public health impact of continual exposure even to low concentrations of this material are unknown and need to be studied and controlled.

Antimony is a metallic element highly toxic to living things, particularly human beings. However, in the atmospheric concentrations expected from ffpp's, it would probably have only an irritating effect on the skin, mucous membranes, and eyes of humans. In the event of some unusual meteorological conditions, it could be concentrated enough to be moderately toxic on an acute basis. There are no data available on chronic effects from subclinical doses.

Arsenic is an extremely toxic and carcinogenic element. From ffpp's, only low concentrations are emitted from stacks, mainly in the vapor form. However, under the proper meteorological conditions, the concentration of arsenic, which is cumulative in the body, can rise to be of acute concern to exposed humans. Hueper (1974) has noted that chronic exposure to arsenic can produce a high percentage of cancers in populations so exposed. It would appear that more study of the potential cancer problem from arsenic emitted by ffpp's is in order. Chronic arsenic poisoning by ingestion or inhalation manifests itself in many ways. There may be disturbances of the digestive system such as loss of appetite, cramps, nausea, constipation, or diarrhea. Liver damage may occur, resulting in jaundice. Disturbances of the blood, kidneys, and nervous system are frequent. Arsenic can cause a variety of skin abnormalities, including itching, pigmentation, and cancerous changes. Characteristic of arsenic poisoning is the great variety of symptoms which can be produced. It is a carcinogen of the mouth, esophagus, larynx, bladder, and paranasal sinus. Its release to the atmosphere should be controlled.

Asbestos particulates. Recently Selikoff et al. determined that asbestos particles are responsible for cancers of the lining of the lungs, known as mesotheliomas, and of the intestines. This asbestos cancer usually requires 4 to 7 yr of exposure, and there is apparently less predisposing to tuberculosis than is the case with silicosis. Prolonged inhalation can cause cancer of the lung, pleura, and peritoneum and has experimentally produced cancers of the peritoneum and the intestines. Presently, it is not known whether there is a concentration of asbestos particles in air below which it is safe to inhale for 24 h/d. Ongoing studies, however, may clarify this situation. The use of fossil fuels is increasing and the amount of asbestos committed to the environment is also increasing (Sax, 1979). The ffpp release of asbestos to the environment should be controlled.

Barium compounds (soluble) are somewhat caustic and many are highly toxic. However, in the small amounts liberated by the combustion of fossil fuels, acute poisoning episodes are not to be expected. Little is known about the chronic exposure effects on populations to low concentrations of barium, either on contact with the eyes, skin, and mucous membranes, or when absorbed into the body.

Benzo-α-pyrene is a carcinogenic pyrolysis product of carbonaceous materials and along with fluoranthene, pyrene, coronene, and perylene, which are suspected carcinogens, are known as polynuclear hydrocarbons and are emitted from ffpp stacks (Billings, 1974; Sax, 1979). The quantities emitted are probably in the range of milligrams of carcinogen/10^6 Btu of heat input. Since there are no data on the exact relationship of air concentration of carcinogen, time of exposure, and the formation of cancers in humans, it behooves us to be extremely careful of the dissemination of this power plant stack effluent. It must be studied and controlled.

Beryllium is a toxic metallic element. In a finely divided oxide form, it is extremely toxic, causing berylliosis, and is considered responsible for a form of lung cancer. The symptoms of chronic beryllium intoxication can take as long as 25 yr to show. In industry, it has been noted that if the concentration of beryllium in the air is kept low enough the incidence of associated disease becomes very low. However, it is not known if any threshold exists below which it is entirely safe to inhale beryllium compounds. Therefore it would appear to be very important that the concentration of beryllium in the environment be studied very carefully, and any activity such as the burning of fossil fuels which contributes to environmental beryllium concentration should also be studied so that the release of beryllium to the environment can be controlled.

Benzol is very toxic to humans by inhalation, ingestion, or absorption through the intact skin. It is a recognized carcinogen (Sax, 1979) of the blood-forming organs. Not much is known about safe concentrations of benzene, but 100 ppm is known to be toxic while 1 ppm is considered safe to inhale for 40 h/week. As a leukemogen, its release to the environment must be controlled.

Bromine is a halogen very much like chlorine. Small concentrations of bromine or bromides in the air will probably result in no more than some skin and mucous membrane irritation to human beings. However, there is the possibility in the case of bromides that the long-term or chronic inhalation of even low concentrations may have untoward physiological effects.

Cadmium is a highly toxic metallic element by inhalation or ingestion (Nusbaum, 1965; Revelle, 1967; Sax, 1979; Linton, Loh, and Natusch, 1976). Ingestion of a toxic dose can result in sudden nausea, salivation, vomiting, diarrhea, and abdominal pain and discomfort. Symptoms begin almost immediately after ingestion. The oral LD_{50}(rat) = 14 mg/kg of body weight for the lactate. A yellow discoloration of teeth has been reported in workers exposed to it. Cadmium oxide fumes can cause metal fume fever resembling that caused by zinc oxide fumes. Cadmium compounds are carcinogens of the connective tissue, lung, and liver (NIOSH, 1979). Under these conditions, it would seem that almost any cadmium in the environment due to the combustion of fossil fuels for the production of energy should be minimized until careful epidemiologic studies have been made (Pihl and Parks, 1977).

Calcium in the form of its oxides or carbonates in the atmosphere in low concentrations causes no problem other than adding to the irritant nature of the atmosphere surrounding a source.

Carbon bisulfide is a highly toxic material by inhalation, ingestion, or absorption via the intact skin (Sax, 1979). However, in very low concentrations as might be encountered in the environment, it would give rise at most to chronic effects on the central nervous system. It is narcotic and anesthetic, but it mainly damages the central and peripheral nervous system. This damage may be permanent if the exposure is severe. Sensory symptoms usually precede motor involvement. A secondary anemia may be caused. These changes are accompanied by insomnia, loss of memory, and personality changes. Chronic fatigue is a very common complaint. Very little is known regarding long exposures to very low concentrations as may be expected in the atmosphere surrounding a ffpp.

Carbon dioxide. Truly enormous quantities are emitted from ffpp's. Carbon dioxide has no noticeable physiological effects on populations; it may have planetary effects. In large concentrations it can replace oxygen in the air and thus reduce oxygen availability and thereby become an asphyxiant. However, nothing like this is contemplated from the burning of fossil fuels. The planetary effects (greenhouse effect) are presently being intensively studied.

Carbon monoxide is a highly toxic gas since, when inhaled with air, it ties up that factor in the blood which transports oxygen so that an individual exposed to too much carbon monoxide can suffer and even die rapidly from asphyxiation although he is surrounded by oxygen. Not much is known about the overall effects of high concentrations of carbon monoxide on an environment because it is rapidly diluted and because there are large natural planetary sinks for it. Recently, Deichmann and Gerarde (1969) and Sax (1979) have reported on research which has pointed to possibly long-term or chronic effects from continuous inhalation of low concentrations of carbon monoxide. Apparently, it also crosses the human placenta to affect the fetus, and its fetotoxic effects must be studied.

Cerium or cerium compounds such as oxides or sulfates are present in fossil fuels. Consequently, some cerium finds its way into fly ash. This material is pharmacologically similar to aluminum. It is not highly toxic, and no problems are projected from the low concentrations in the environment due to ffpp's which will become available to populations.

Chlorinated hydrocarbons, aliphatic and aromatic, are somewhat narcotic by inhalation in high concentrations. They cause liver, heart, and kidney damage. The unsaturated aliphatic chlorinated hydrocarbons also are generally highly toxic in concentrations of a few hundred parts per million. In the case of the chlorinated aromatic hydrocarbons, it is difficult to predict acute toxicity. Most such compounds display an acute toxicity no greater than that of the corresponding aromatic hydrocarbon. However, both classes are suspected carcinogens of the liver, lung, skin, and blood-forming tissues and must be treated with extreme respect (Hueper, 1972, 1974; NIOSH, 1979; Sax, 1979). There is presently very little data on the effects of chronic exposures to these suspected or experimental carcinogens, and most toxicologists agree that there is no "safe" level for exposure to carcinogens. Therefore this situation must be studied at the earliest, particularly in the face of increased use of fossil fuels.

Cesium is a metallic element and very much resembles potassium in its chemical action. However, if it replaces potassium in the body, it can do great physiological damage, even cause death. Linton, Loh, and Natusch (1976), Rancitelli (1974), and others have noted that even in very low concentrations cesium oxide, or cesium hydroxide, is caustic and therefore irritating to the skin, eyes, and mucous membranes of humans who come in contact with it.

Chlorine is a gaseous element at normal temperatures and pressures and has been used in high concentrations as a military poison. It is a very common air contaminant and is highly toxic. Inhalation of perhaps 1000 ppm can quickly prove fatal. However, in ordinary environmental concentrations due to ffpp's, one can expect no more than irritation to the eyes, skin, and mucous membranes from it. Most of it would combine to form chlorides before finding its way to the lower atmosphere (Billings, 1974). One chloride formed is HCl, a constituent of acid rain.

Chloral, particularly in the form of hydrate, is a poisonous drug in the pure or nearly pure state. *The Merck Index* (1976) and Deichmann and Gerarde (1969) report the oral LD_{50} (rat) = 479 mg/kg. In the very small quantities normally produced from combustion of fossil fuels, no problem is expected although it is an experimental tumorigen (Sax, 1980) and therefore must be studied.

Chromium is a metallic element. Many chromium compounds are recognized carcinogens of the lungs, nasal cavity, and paranasal sinuses (also an experimental carcinogen of the stomach and larynx), and the hexavalent compounds are more acutely toxic than the trivalent. For chromous nitrate the acute oral LD_{50} (rats) = 3.25 g/kg. Eczematous dermatitis due to trivalent chromium compounds has been reported. Chromic acid and its salts have a corrosive action on the skin and mucous membranes creating lesions of the tissue. The lesions are confined to the exposed parts, affecting chiefly the skin of the hands and forearms and mucous membranes of the nasal septum. The characteristic lesion is a deep, penetrating ulcer, which for the most part does not tend to suppurate and which is slow to heal. The very least impact that can be expected from even low concentrations on exposed populations would be irritation of the skin, eyes, and mucous membranes and some increase in cancers. The problems caused by chromium released into the environment from the production of power needs epidemiologic study, particularly from the standpoint of chronic exposure to low concentrations. It should be controlled.

Cobalt is a metallic element. Sax (1979) has reported that its compounds are high to moderately toxic in an acute sense. However, it is an experimental NEO and ETA† (NIOSH, 1980). There are very little data on the effects of chronic exposure to low levels of cobalt in the atmosphere. Therefore its emission from power plant stacks should be controlled.

Copper, mainly as salts, is much used in industry and is not regarded as highly toxic to man. Copper causes a metal fume fever, and the dust has caused cirrhosis of the liver and pancreas or "bronze diabetes," as it is sometimes known. Exposure to copper salts can cause irritation of the skin and conjunctiva, which may be on an allergic basis. Cuprous oxide is irritating to the eyes and upper respiratory tract. Discoloration of the skin is often seen in persons handling copper, but this does not indicate any actual injury. The air pollution from ffpp's should be only a slight problem as far as copper is concerned because of the small amounts produced and the great dilution of the atmosphere.

o,m,p-Cresols are highly toxic and similar to phenol in action but somewhat less severe in effects. According to the Registry of Toxic Effects of Chemical Substances (NIOSH, 1979), the acute oral LD_{50} (rats) = 242 mg/kg, and they are corrosive to the skin, eyes, and mucous membranes and can produce severe chemical burns and dermatitis. However, usual environmental concentrations produce no more than some increased irritation on an acute basis and there are no data on chronic, subclinical exposures (Sax, 1979). Also, they are experimental neoplastigens via skin exposure.

Dimethylacetamide and dimethylformamide are moderately acutely toxic by inhalation, ingestion, or skin absorption. Acute oral LD_{50} (mice) = 1.1 g/kg for dimethylformamide and about 5 g/kg for dimethylacetamide. However, the concentrations of these materials normally found in even a polluted atmosphere are very small. The only effects we might expect from it would be that of irritation of the skin, eyes, and mucous membranes. There are insufficient data on the low-level chronic effects of these materials on man. Also, these materials are known experimental teratogens, and these effects must be studied.

Dimethylsulfate and its hydrolysis product monomethylsulfate have been found in coal fly ash and airborne particulate matter (Lee et al., 1980). Levels determined have been as high as 830 ppm. These materials are of high acute toxicity (Sax, 1979). A further problem concerns the carcinogenicity. This situation merits intensive study by public health agencies and may possibly have to be regulated.

Dimethylsulfide is an irritant and is moderately toxic. Its emission is not of great concern (Sax, 1979).

Diphenylsulfide is moderately toxic and is irritant to the skin, eyes, and mucous membranes (Sax, 1979). LD_{50} to rats is about 2 g/kg. The ffpp concentrations are expected to be very low, merely adding to the toxic load of the atmosphere.

Ethyl acrylate has an acrid, penetrating odor. It is highly to moderately toxic by inhalation, ingestion, and skin absorption and is a powerful irritant. Oral LD_{50} (rats) = 1020 mg/kg (NIOSH, 1979). Fairly high air concentrations cause great irritation to mucous membranes, eyes, nose, mouth, and lethargy, dyspnea, and convulsive movements in animals. Very low concentrations as might be expected from ffpp's merely would add to the toxic load of the atmosphere. Its carcinogenicity is indefinite and requires further study.

Europium as the chloride is moderately to highly toxic to mice, and the intraperitoneal LD_{50} is about 500 mg/kg. However, since this is a trace material in fossil fuels, concentrations in the air from the ffpp's must be very low, and therefore its impact on populations should be small (Sax, 1979; Hawley, 1981).

Fluorine or fluorides are powerful irritants and highly toxic by ingestion or inhalation on an acute basis. On a chronic basis, they are highly toxic by ingestion and inhalation and can cause osteoporosis, some enzyme system upset, and experimental teratogens. Oral LD_{50} (rats) = 180 mg/kg (NIOSH, 1979). Any fluorides or fluorine emitted into the environment will tend to render that atmosphere irritating and toxic to populations; therefore emission of this element must be regulated.

Fly ash is the very fine ash produced by the combustion of powdered coal with forced draft. It is considered the most abundant solid-waste product in America (Hulett et al., 1980). The ash is often carried off with the flue gases from such processes. It has been found that fly ash carries with it much of the metallic elements, some organics, and such materials that are the residue from the combustion of fossil fuels. The matrix elements of fly ash are mainly aluminum, iron, silicon, and magnesium. Some examples of minor elements accompanying fly ash are carbon, potassium, sodium, and titanium. There are also trace elements associated with fly ash such as beryllium, lithium, lead, vanadium, and zinc. The toxicity or pollution potential of fly ash depends on the elements that accompany it and their particle size. Since the introduction of electrostatic precipitators, only the finest particles of fly ash get away and are committed to the atmosphere (Hulett et al., 1980). These are precisely the particles which are respirable and therefore can be inhaled into the body and absorbed from there. Also fly ash has been shown to be mutagenic, experimentally, and therefore carcinogenic (Smith-Sonneborn et al., 1981; Chrisp, Fisher, and Lammert, 1978).

Gallium is a liquid metallic element, and neither it nor its compounds display much toxicity. Once the material gets into the tissues, it seems to behave more like bismuth or mercury. In laboratory animals, only 15 mg/kg of body weight has been injected intravenously and tolerated without harm. Larger doses produced hemorrhagic nephritis. Also, there are some indications it may yet prove to

†NEO = neoplastigen, a nonmetastasizing cancer; ETA = equivocal tumorigenic agent.

have a higher order of toxicity than present data show (Sollmann, 1957). As a trace contaminant of coal, very little will eventually escape into the environment.

Gold and its compounds manifest a moderate level of toxicity and are moderately powerful allergens. Salts of gold are used in medicine. However, toxic reactions to gold are not infrequent. Hives, itching, purpura, and skin rashes are common, and damage to the blood-forming organs resulting in aplastic anemia has been described. However, there are no data on chronic exposure to subclinical concentrations (Sollmann, 1957); ffpp would emit only traces of it.

Hafnium compounds display high toxicity via the intraperitoneal and chronic oral routes. The oral LD_{50} (mice) = 76 mg/kg; the intravenous LD_{50} (rats) = 112 mg/kg. Some liver damage from hafnium compounds has been noted by Sax (1979), Sollmann (1957), and Deichmann and Gerarde (1969). However, it is but a trace contaminant of coal, and very little is expected to be emitted.

Hydrogen chloride or hydrochloric acid is a combination of hydrogen and chlorine and is extremely irritating and in even moderate concentrations is very toxic and irritating to the skin, mucous membranes, and eyes. Taken internally, it could cause great damage to tissues and rapid death (Deichmann and Gerarde, 1969; *The Merck Index,* 1976; Sax, 1979). As an ffpp effluent component, it must be carefully regulated. It is a component of acid rain.

Hydrogen sulfide is more toxic than hydrogen cyanide, and some is emitted by ffpp's. In low concentrations its offensive (rotten egg) odor serves as a warning. However, as a continuous, pervasive atmosphere pollutant, many individuals are unable to escape it and, in fact, the sense of smell becomes accustomed to it and accepts it, and toxic effects may still ensue. Very little is known about long exposures to very low (subclinical) concentrations. Hydrogen sulfide is an irritant and an asphyxiant at levels of 20 to 150 ppm. Slightly higher levels cause upper respiratory irritation which, if prolonged, lead to pulmonary edema. Six hundred ppm inhaled for 30 min has been fatal to humans (Sax, 1979). The emission of this material should be controlled.

Iron dust can cause conjunctivitis, choroiditis, retinitis, and siderosis of tissues if the iron remains in those tissues. Iron ore dust can cause palpebral conjunctivitis, massive pulmonary fibrosis, and an increased incidence of lung cancer (Sax, 1979). Fresh iron oxide fumes can cause metal fume fever, and iron compounds are carcinogens of the lung, liver, connective tissue, and reticuloendothelial tissue (Hueper, 1974). Moderate quantities of iron are emitted in the combustion of fossil fuels for the production of power. Consequently, emissions of iron must be taken into consideration in the vicinity of present as well as proposed power plants.

Ketones generally display from low to moderate toxicity. The volatile ones have a somewhat anesthetic action and are central nervous system depressants. Recently, methyl-n-butyl ketone has been implicated as a neurotoxic causing a neuropathy or slowly developing weakness in the extremities (Sax, 1979). However, the concentration of this material, which might appear in the environment surrounding an ffpp, is orders of magnitude below that required for industrial intoxication (Sax, 1979). The acute toxicity is 1 g or more per kilogram to produce death in 14 d. No data are available on chronic exposure to subclinical quantities (Sollmann, 1957).

Lanthanum salts generally display moderate toxicity, although some lanthanons can cause delayed blood clotting, leading to hemorrhages. It has caused liver injury in experimental animals; however, as a trace material, the hazard from ffpp's is low. Acute oral LD_{50} (rats) = about 4 g/kg, whereas acute intraperitoneal LD_{50} (rats) = 275 to 450 mg/kg (Sax, 1979; *The Merck Index,* 1976; Sollmann, 1957). Thus it is highly toxic via the intraperitoneal route and moderately toxic via oral to rats, mice, guinea pigs, and frogs.

Lead is emitted in fly ash at very low concentrations from the stacks of ffpp's. Thus we must be concerned with lead as a problem to the populations surrounding power plants. Lead poisoning is also one of the commonest of occupational diseases. For example, the acute intraperitoneal LD_{50} (rats) = 200 mg/kg for lead acetate (*The Merck Index,* 1976; Browning, 1969; Gosselin et al., 1976). Lead must be in such form and so distributed, i.e., finely divided or as a soluble salt, as to gain entrance into the body by inhalation, on foods, or in water and in measurable quantity to do any harm (Sollmann, 1957). It is a suspected carcinogen of the lungs and kidney (Hueper, 1974). Lead emitted by stacks of fossil-fuel-burning power plants will be an additional toxic load on members of the population, the very young, aged, or infirm, who already are being exposed to lead from gasoline, canned foods, and many other sources. Consequently, each source (Pihl and Parks, 1977) must be treated as an environmental problem and is regulated (Billings, 1974; Sax, 1979; Chap. 20).

Lithium is an alkali metal and therefore in the oxide form very much resembles Na_2O, a powerful irritant. Small amounts in the environment can cause irritation of the eyes, skin, and mucous membranes, whereas large quantities ingested or inhaled are known to have central nervous system toxicity, cause dizziness, prostration, kidney damage, anorexia, nausea, apathy, coma, and even death; the chloride is a teratogen. Lithium oxide, hydroxide, carbonate, etc., are strong bases, and the water solutions are very caustic. However, ffpp's are not expected to limit more than traces to the environment.

Magnesium toxicity is systemically of a low order. However, if particles of the element penetrate the skin, they can cause an inflammation which, in extreme cases, causes a "chemical gas gangrene." Atmospheric pollution by magnesium from the combustion of fossil fuels to produce power is not considered a serious problem (Billings, 1974; Sax, 1979).

Manganese compounds are systemically highly toxic. Chronic manganese poisoning (Sax, 1979) usually begins with complaints of languor and sleepiness followed by weakness in the legs and the development of a stolid, masklike facies, and the patient speaks with a slow, monotonous voice. Then twitchings appear and a typical Parkinsonian slapping gait shows up. Chronic manganese poisoning is a clearly characterized disease that results from the inhalation of fumes of manganese. Exposure to heavy concentrations for as little as 3 months may produce the condition, but usually cases develop after 1 to 3 yr of exposure. The central nervous system is the chief site of damage, and if cases are removed from exposure shortly after the appearance of symptoms, some improvement in the patient's condition frequently occurs, although there may be some residual disturbances in gait and speech. However, when well established, the disease results in permanent disability. Moreover, individuals exposed to manganese fumes exhibit a much higher incidence of respiratory disease (i.e., pneumonia) than the general population. The manganese concentration in the atmosphere due to the combustion of fossil fuels for the production of power is normally not great enough to produce the type of exposure described above; however, there are no data on continual inhalation of very low concentrations in air.

Mercaptans are a class of sulfur materials that are highly odoriferous but not highly toxic. Their offensive odor is noticeable far below toxic concentrations. Excessive exposure can cause headache, nausea, and vertigo. Probably the worst problem from stack-emitted mercaptans is the offensive nuisance odor which is added to the atmosphere.

Methacrylates are low to moderately toxic and irritants in relatively high concentrations, i.e., oral LD_{50} (methyl methacrylate—rats) = 9 g/kg. In the concentrations that would be formed in the environment from the combustion of fossil fuels to produce power, these materials would be of very little or no concern. Some methacrylates are teratogens, however, and merit study on that account above.

Mercury is a liquid, metallic element also known as "quicksilver." It is emitted from the stacks of fossil-fuel-burning power plants in the form of a vapor. This vapor is reactive with oxygen and chlorine or can condense on cooling to form tiny droplets of mercury. Air can carry enough mercury vapor at ordinary conditions to cause intoxication. Air, saturated with mercury vapor, may contain nearly 20 mg/m^3. Poisoning has occurred from inhalation of 0.5 mg/m^3 of mercury in air for 16 h. Even much lower concentrations would soon prove toxic to human beings (Sollmann, 1957). Mercury vapor from the air can condense out on food, cigarettes, clothing, and the skin and thereby work its way into the body, where it acts as a protoplasmic poison. It circulates in the bloodstream and is stored in the liver, kidneys, spleen, and bone. It is eliminated in the urine, feces, sweat, saliva, and milk. The chief effect seems to be on the central nervous system and on the mouth and gums. Colitis has been reported frequently. The cardinal symptoms of industrial mercury poisoning (i.e., chronic poisoning) are stomatitis, tremors, and psychic disturbances. First, excessive salivation and pain on chewing are experienced. In severe cases, there may be gingivitis, with loosening of the teeth. Tremor and psychic disturbances are commonly seen in the slow, chronic form of the poisoning. The psychic disturbances, so-called "erethism," include such changes as loss of memory, insomnia, lack of confidence, irritability, vague fears, and depression. It is a suspected carcinogen (Hueper, 1974; NIOSH, 1979; Billings, 1974; Gosselin et al., 1976). Since fossil fuels containing traces of mercury are being used more for the production of power and since the concentration of people surrounding such installations is increasing, mercury is an air pollutant meriting further study in the very near future.

Molybdenum displays, at most, moderate toxicity. There are no reports of chronic or even acute intoxication in man. Deichmann and Gerarde (1969) note that in cattle a diet of 20 to 100 ppm of molybdenum can develop diseases, possibly because of interference with copper metabolism. Only minute traces are found in fly ash.

Nickel and most of its salts are highly toxic. Nickel is a recognized carcinogen of the nasal cavity, paranasal sinuses, and the lungs (Billings, 1974; Hueper, 1972, 1974). The most common acute effect resulting from exposure to airborne nickel compounds is the development of the "nickel itch," a form of dermatitis. Nickel is one of the elements that must be studied from the standpoint of rapidly rising use of fossil fuels to produce power and its effects on surrounding populations. We do not know at the present time what concentration of nickel in the environment produces how many cancers in people living there. Until this is known, we must be extremely careful of it. The fly ash content of nickel is low.

Niobium chloride and potassium niobate are highly toxic salts of niobium. For example, Deichmann and Gerarde (1969) report that the oral LD_{50} for $NbCl_5$ in mice is 61 mg/kg. Furthermore, direct contact with the material can cause severe localized damage to the skin, as well as ocular irritation, but with no permanent damage. Deichmann and Gerarde (1969) and Sax (1979) note that the $NbCl_5$ produces immediate cardiovascular collapse and respiratory paralysis to the cat at a dose of 5 mg/kg. It is not expected that the concentrations of niobium produced by the combustion of fossil fuels for the production of power will prove an environmental problem, although it is one that should be watched if the coal used should contain it.

Nitriles are organic cyanides, and butyronitrile, acrylonitrile, propionitrile, and some others resemble cyanides in toxicity. Other nitriles such as cyanamides and cyanates have no cyanide effect. Generally speaking, it requires a low body dose of nitriles in the 100 to 500 mg/kg for mice and rats to cause the acute toxic effects. Hawley (1981) and Sax (1979) note that 300 ppm in air is rapidly fatal for some and that the TLV for acetonitrile is 1 ppm in air. Unless care is taken, significant amounts of nitriles can be committed to the environment by ffpp's.

Nitrobenzene is highly toxic on a systemic basis. According to Sax (1979), nitrobenzene causes cyanosis due to formation of methemoglobin. It is a common air contaminant and the oral LD (rabbits) = 700 mg/kg (NIOSH, 1979; Gosselin et al., 1976). No available air data in the vicinity of ffpp's have been related to human effects.

Oxides of nitrogen (NO_x) consist of nitrogen dioxide (NO_2), nitric oxide (NO), and nitrous oxide (N_2O) as well as nitrogen trioxide (N_2O_3), nitrogen pentoxide (N_2O_5), and nitrogen tetroxide (N_2O_4). Nitrous oxide is a stable gas used extensively as an anesthetic. Inhaled in small amounts, it often produces a type of hysteria; hence its popular name "laughing gas." It is an experimental teratogen (Sax, 1979). Nitric oxide is colorless and highly endothermic and is the most stable of the nitrogen oxides at high temperatures. Nitrogen dioxide is a brownish-colored gas which is considered to be of greatest air-pollution interest because it is a major end product of combustion. Large quantities of nitrogen oxides are released to the atmosphere yearly from natural and anthropogenic sources. Oxides of nitrogen are somewhat soluble in water, reacting with it in the presence of oxygen to form nitric and nitrous acids. This same reaction takes place deep in the human respiratory system on inhalation. The acids formed are irritant, causing congestion of the throat and bronchi and edema of the lungs. The acids are neutralized by the alkaline tissues with the formation of nitrates and nitrites. The latter may cause some arterial dilation, fall in blood pressure, headache, and dizziness, and there may be some formation of methemoglobin. However, the acute nitrite effects are of secondary importance. Because of their relatively low solubility in water, the nitrogen oxides are only slightly irritating to the mucous membranes of the upper respiratory tract. Their warning power is therefore low, and dangerous amounts of the fumes may be breathed deeply before one notices any real discomfort. Higher concentrations, 60 to 100 ppm, cause immediate irritation of the nose and throat and cause coughing and burning in the throat and chest. These symptoms often clear up on breathing fresh air, and the victim may feel well for several hours. Some 6 to 12 h after exposure, however, he may develop a sensation of tightness, burning in the chest, shortness of breath, sleeplessness, and restlessness. Dyspnea and air hunger may increase rapidly with the development of cyanosis and loss of consciousness followed by death. In cases that recover from pulmonary edema, there is usually no permanent disability, but pneumonia may develop later. Concentrations of 100 to 150 ppm are dangerous for short exposures of 30 to 60 min. Concentrations of 200 to 700 ppm may be fatal after even very short exposures. Continued exposures to even low concentrations of the fumes, insufficient to cause pulmonary edema, is said to result in chronic irritation of the respiratory tract as cough, headache, loss of appetite, dyspepsia, erosion of the teeth, and gradual loss of strength. Exposure to these fumes is always potentially serious, and persons so exposed should be kept under observation for at least 48 h. The emission of NO_x in the combustion of fossil fuels in power production is so great a problem that if the use of fossil fuels is to continue or increase a great deal of study must be applied to it to increase and perfect the controls of the emission of this material (Browning, 1969; Sax, 1979; Mahoney, 1974; Deichmann and Gerarde, 1969).

Ozone (O_3) or the oxide of oxygen is not emitted directly from the stacks of fossil-fuel-burning power plants, but the emitted oxides SO_x, NO_x, and the presence of sunlight interact to form O_3 in the lower atmosphere. Too much O_3 in the lower atmosphere damages sensitive human lung tissues and decreases the yield of agricultural plants. Ozone is a strong eye irritant and an irritant of the upper respiratory system. Concentrations of 0.015 ppm produce a barely detectable odor. Concentrations of 1 ppm produce a disagreeable sulfur-like odor and may cause headache and irritation of the upper respiratory tract, which symptoms disappear after leaving the exposure. With concentrations higher than 1 ppm, dyspnea, cough, choking sensation, tachycardia, vertigo, lowering of blood pressure, and severe cramping chest pains may develop. Such symptoms may cause erroneous diagnoses as myocardial infarction. Pulmonary edema may develop, although the onset of the edema is usually delayed for an hour or more following exposure. It is possible that exposures for as little as 30 min, to 50 ppm, would prove fatal, while 2 ppm might be safe for several minutes, providing that no respiratory infection, latent or manifest, is present. It is therefore evident that O_3, which forms downwind of power plants, is of real concern to public health officers (*Science News*, 1974; Sax, 1979; Deichmann and Gerarde, 1969) and is regulated.

Potassium has very low toxicity. However, oxides, carbonates, peroxides, and hydroxides of potassium are very caustic, and even fine particles of these materials in contact with the moisture of the skin, eyes, or mucous membranes can be quite irritating (Billings, 1974; Sax, 1979).

Phenol is also known as carbolic acid. Some of it in the form of vapor is emitted from the stacks of fossil-fuel-burning power plants. Phenol is highly toxic and a cocarcinogen. Its oral LD_{50} for experimental animals runs about 44 mg/kg (Gosselin et al., 1976; Sollmann, 1957; Deichmann and Gerarde, 1969). On the skin the affected area from pure phenol becomes white, wrinkled, and softened and usually there is no immediate pain; later, intense burning is felt, followed by local anesthesia and, still later, by gangrene. Absorbed phenol is excreted partly by the kidneys and is partly oxidized. Part of the excreted portion is combined with sulfuric and glycuronic acids; the remainder is excreted unchanged. Death has resulted from the absorption of phenol through a skin area of 64 $in.^2$ Where death is delayed, damage to the kidneys, liver, pancreas, and spleen and edema of the lungs may result. The symptoms develop rapidly, frequently within 15 to 20 min following contact of phenol on the skin. Headache, dizziness, muscular weakness, dimness of vision, ringing in the ears, irregular and rapid breathing, weak pulse, and dyspnea may all develop and may be followed by loss of consciousness, collapse, and death. The situation in the vicinity of power plants is one of long-term or chronic effects from very low concentrations which still may do damage to the kidneys and liver and cause a dermatitis resulting from contact. This situation requires much more study.

Phosphorus is so reactive that it is always found in nature as a compound, usually the oxide. Phosphorus oxides are also the anhydrides of the various phosphorus and phosphoric acids. These materials are extremely irritating. The absorption of toxic quantities of phosphorus has an acute effect on the liver and is accompanied by vomiting and marked weakness. The long-continued absorption of small amounts of phosphorus can result in necrosis of the mandible or jawbone and is known as "phossy-jaw." Long continued absorption, particularly through the lungs and through the gastrointestinal tract, can cause a chronic poisoning, characterized by a generalized weakness, attended by anemia, loss of appetite, gastrointestinal weakness, and pallor. It can also cause changes in the long bones, and seriously affected bones may become brittle, leading to spontaneous fractures. It is especially hazardous to the eyes. It also has adverse effects on the teeth, and exposed people should have periodic dental examinations. The lethal dose of phosphorus for man is less than 1 mg/kg of body weight (Billings, 1974; Deichmann and Gerarde, 1969; *The Merck Index,* 1976; Sollmann, 1957; Gosselin et al., 1976). The problem of phosphorus contamination of the atmosphere from ffpp's merits further study.

Phosphine is a highly toxic gaseous compound of phosphorus and hydrogen. Chronic poisoning from it is characterized by anemia and bronchitis and gastrointestinal, visual, speech, and motor disturbances. Since phosphorus and phosphine are emitted from stacks of fossil-fuel-burning power plants, we must be concerned with the amount of it in the environment surrounding the populations that live near the power plants. If there are any long-term effects of very low concentrations, we do not know what they are. Since the toxicity of the material is so high, we must be concerned with even very low concentrations. Therefore more information about effects of various concentrations is needed (*The Merck Index,* 1976; Sollmann, 1957; Gosselin et al., 1976) and merits much more study.

Phosgene is carbonyl chloride or "choking gas" as it was known by the military during the first World War. Inhalation of phosgene may result in irritation of the membranes of the lower respiratory tract and pulmonary edema several hours after exposure. It is highly toxic. A brief exposure to 50 ppm may be fatal. Phosgene is formed when halogenated hydrocarbons are caused to decompose under oxidative conditions and heat (Billings, 1974; Sax, 1979; Deichmann and Gerarde, 1969; *The Merck Index,* 1976).

Polycyclic aromatic hydrocarbons (PAH's): These materials are a very possible contaminant of fly ash particulates. They can be both neurogenic and carcinogenic and as such may prove a problem in future expansion of the use of ffpp's. This situation merits further study.

Pyridine is a moderately toxic material. When the concentration of its vapor in the air reaches 10 ppm, it can have very disagreeable odor effects. It is irritating to the kidneys and liver, causes nausea, and may increase urinary frequency. It is absorbed through the intact skin. It also causes skin irritation. Emission from power plants must constitute an added burden of odor and irritation to populations so far as present-day research shows.

Rubidium is a metallic element much like potassium. The oxide, hydroxide, and carbonate are irritant to the skin, eyes, and mucous membranes. Therefore, if some of this element is committed to the atmosphere, it would simply increase the irritating qualities of the air.

Samarium, a rare earth, is a silvery metal classified as one of the lanthanons. All we know about it is that as a lanthanon it may, in high concentrations, interfere with the clotting ability of the blood (Billings, 1974; Sax, 1979). No problem is expected from this source.

Selenium toxicity is greater than that of tellurium and similar to that of arsenic, although traces appear to be essential to normal growth of some animals. Inhalation and ingestion of selenium compounds causes poisoning via rapid gut and lung absorption. It is most damaging to the liver, kidneys, spleen, bone marrow, and thyroid. It is slowly absorbed via intact skin; however, prolonged or repeated skin contact can cause inflammation. Selenium is a suspected carcinogen of the liver and thyroid. Acute LD_{50} (rats, mice) = 1 to 15 mg/kg from sodium selenate or sodium selenite. Chronic human selenosis may cause gastrointestinal problems, erythema, pallor, dermatitis, dizziness, metallic taste, and irritation of the eyes, nose, and throat. Nausea, lassitude, and garlicky breath have been described in exposed people. Airborne selenium is odoriferous, although olfactory fatigue can set in rapidly so that this warning property is soon lost. There have been studies of the selenium emission from ffpp's which indicate that much selenium can be found in fly ash, and of course a good deal will be emitted in the vapor form. In any case, it would be emitted in such form to be readily inhaled and possibly absorbed into the body (Gutenmann, 1976; Rancitelli et al., 1974; Andren, Klein, and Talmi, 1975; Robertus et al., 1975; Billings, 1974; Sax, 1979; Deichmann and Gerarde, 1969). This is a real problem and merits intensive study, particularly with the expected great increase in ffpp's.

Scandium is a rare element, and not very much of it is expected to get into the atmosphere via the combustion of fossil fuels. However, this material has some of the properties of boron, and since boron is toxic it may be wise, by analogy, to be careful of the scandium that is emitted into the atmosphere. There are very little data about the toxicological nature of this material, and since more and more of it is forthcoming, we must keep this in mind for the protection of populations.

Silver is present in fossil fuels, probably in trace quantities. Silver is concentrated and emitted via the fly ash and therefore finds itself in the atmosphere where it is deposited on foods and water and can be inhaled by individuals. Nevertheless, the concentration of silver in fossil fuels is very low, and no problem is expected to result from it at the present time (Billings, 1974; Sax, 1979; Deichmann and Gerarde, 1969).

Sodium is not acutely toxic, and its compounds display toxicity due to other parts of the molecule. As in the case of potassium, these compounds are irritating to the skin, eyes, and mucous membranes. Therefore, if it gets into the atmosphere, it could increase the irritative abilities of the atmosphere.

Styrene or phenyl ethylene or vinyl benzene or styrene monomer causes irritation, violent itching of the eyes, lachrymation, and severe human eye injuries. Its toxic effects are usually transient and result in irritation and possible narcosis. It is not considered a very toxic material because, under ordinary conditions, it does not vaporize sufficiently to reach a concentration sufficient to kill animals, such as rats and guinea pigs. Its emission would merely add to the irritant load of the atmosphere. Its oral LD_{50} (rats) = 5 g/kg (NIOSH, 1979). It is a mutagen, possible tumorigenic irritant. This material must be studied and possibly regulated.

Sulfur dichloride, in pure form, is a fuming, corrosive liquid with a penetrating odor that is very irritating to the eyes, lungs, and mucous membranes. It decomposes on contact with water and forms hydrochloric acid, thiosulfuric acid, and sulfur, which are highly irritating by-products so far as the skin and human mucous membranes are concerned. Its toxic effects are irritation of the upper respiratory tract, although the results of such intoxication are usually transitory in nature. However, hydrolysis is not complete in the upper respiratory tract, and so injury to the bronchioles and alveoli can result. Concentrations in air of 2 to 9 ppm have been found to be mildly irritating, whereas 150 ppm has been stated to be fatal to mice in 1 min. This material is worth study since we know that a great deal of sulfur in one form or another is emitted by the burning of fossil fuels and reasonable amounts of chlorine or chlorides are also emitted. Therefore irritant quantities of sulfur chloride may find their way into the atmosphere near ffpp's. Releases of this material may change with the species of fossil fuel being used (Deichmann and Gerarde, 1969; NIOSH, 1979; *The Merck Index*, 1976).

Sulfur oxides (SO_x) are emitted by fossil-fuel-burning power plants in the form of gases, or very finely divided particulates. It is therefore entirely possible that in areas surrounding power plant installations ground-level concentrations of SO_x such as sulfur dioxide can reach toxic levels. These materials are dangerous to the eyes; they cause irritation and inflammation of the conjunctivae. They have suffocating odors and are corrosive and poisonous. In moist air or fogs, they combine with water to form sulfurous acid and sulfuric acid. Sulfurous acid is very slowly oxidized to the sulfuric form. Concentrations of 6 to 12 ppm cause immediate irritation of the nose and throat, while 0.3 to 1 ppm can be detected by the average individual, possibly by taste more than the sense of smell. Three ppm has an easily noticeable odor, and 20 ppm is irritating to the eyes. However, 10 000 ppm is an irritant to the moist areas of the skin within a few minutes of exposure. These materials chiefly affect the upper respiratory tract and bronchi. They cause edema of the lungs or glottis and can produce respiratory paralysis. Concentrations of less than 1 ppm are believed to be injurious to plant foliage. These materials are so irritating that they provide their own warning of toxic concentrations. For example, 400 to 500 ppm is immediately dangerous to life, and 50 to 100 ppm is considered to be the maximum permissible concentration for exposure of 30 to 60 min. Excessive exposures to high enough concentrations can be fatal. The toxicity of SO_x is comparable to that of hydrogen chloride. Less than fatal concentrations can be borne for fair periods of time with no apparent permanent damage. It is used as a fumigant, insecticide, and fungicide and a chemical preservative food additive. Naturally, in a population of individuals living around power plants, the exposure might be for 24 h/d, and the very young, the aged, and the infirm, as well as those with respiratory problems, can suffer real damage (Billings, 1974; Sax, 1979; Mahoney, 1974; Deichmann and Gerarde, 1969; Sollmann, 1957). This has been shown to be true in areas that suffer from inversions and where sources of SO_x also exist, i.e., London, England, and Donora, Penn.

Tantalum. So far as we know, there are no acute toxic reactions, although it is carcinogenic to animals by implantation (Billings, 1974; Sax, 1979).

Terbium, a soft, malleable, metallic, silvery-gray lanthanon. Some of the lanthanons have the property of interfering with the coagulative properties of blood. However, it is a trace element, and the quantity that would be emitted from the stacks of fossil-fuel-burning power plants is very small and therefore no problem is expected from this source (Billings, 1974; Sax, 1979).

Tin. Elemental tin is not generally considered toxic. Some inorganic tin salts are irritants or can liberate toxic fumes on decomposition. The latter is particularly true of tin halogens. Alkyl tin compounds can be highly toxic and produce skin rashes. Dust of tin oxide has caused a pneumoconiosis which is relatively benign, and the symptoms of it are irritation of the mucous membranes of the throat, coughing, and retching. Organic tin compounds can be absorbed via the intact skin. Actually, we do not expect to find organic tin compounds in the environment surrounding power plants. We do expect to find a small amount of particulate compounds, such as the oxide or chloride, and from that we would expect no more than a slight increase in the irritation of the skin, eyes, and mucous membranes from the air (Billings, 1974; Sax, 1979).

Titanium. There is very little or no acute physiological toxicity from titanium to humans who inhale it or are exposed to it on the skin, eyes, or mucous membranes (Deichmann and Gerarde, 1969); however, it is a suspected implantation carcinogen presently under study (Sax, 1979).

Toluene (common solvent) is considered no more than moderately toxic. Up to 200 ppm, it causes few symptoms in most people. It is not expected that the small amounts emitted from the stacks of power plants will cause any specific public health problems.

Uranium is a recognized carcinogen due to its radioactivity and a highly toxic element on an acute basis. It would seem that even small quantities, if committed to the environment surrounding fossil-fuel-burning power plants, would have an impact on the populations around it. Many soluble uranium salts are highly toxic compared to the more insoluble salts. They are readily absorbed from the lungs, skin, and gastroenteric tract and they induce local irritation, inflammation of the capillaries, nephritis, hepatitis, acidosis, and edema. Following inhalation, the soluble compounds exert their primary effect on the kidneys, while the insoluble compounds act mainly on the lungs. Usually, there is a latent period of 1 to 4 d between exposure and the onset of acute symptoms. The toxic dose of a uranium compound induces a typical acute intoxication that results either in recovery within a month or early death (Deichmann and Gerarde, 1969; Sax, 1979).

Vanadium, chiefly in the form of trioxide, acts mainly as an irritant to the conjunctivae and respiratory tract. Prolonged exposures can lead to pulmonary involvement. There is controversy as to the effects of industrial exposure to vanadium on other systems of the body. Responses appear always to be acute. Early reports on vanadium poisoning in humans described widespread systemic effects, consisting of polycythemia, followed by red blood cell destruction and anemia, loss of appetite, pallor and emaciation, albuminuria and hematuria, gastrointestinal disorders, nervous complaints, and cough, sometimes severe enough to cause hemoptysis. Some fossil fuels contain significant concentrations of vanadium. Therefore it is in need of regulation.

Yttrium probably as the oxide, is emitted from the stacks of fossil-fuel-burning power plants. There is no known toxicology for it; however, as a lanthanon, it may interfere with the clotting qualities of the blood of exposed individuals. At this point it would not seem to be a potential hazard requiring study by public health agencies (Sax, 1979).

Zinc compounds such as the chloride are toxic, i.e., astringent, corrosive, and emetic. When ingested in toxic doses, they produce severe gastroenteric irritation, nausea, vomiting, diarrhea, and gastroenteric pain. While it is not expected that any large quantities of zinc will be emitted by a single power plant that burns fossil fuels, the material is irritating enough, even in low concentrations, so that it would certainly be a nuisance to the public and, probably worse than that, to the very young, the aged, or individuals with some form of respiratory problem. Many zinc compounds are experimental carcinogens. Therefore, the pollution of air by zinc should be controlled.

Zirconium is not an important industrial poison. Deaths in rabbits have been caused by intravenous injection of 150 mg/kg of body weight. Most zirconium compounds in common use are insoluble and are considered inert. Pulmonary granuloma in zirconium workers has been reported, and sodium zirconium lactate has been held responsible. One should avoid inhalation of zirconium-containing aerosols which can cause lung granulomas.

In the above paragraphs an effort has been made to discuss most of the important fly ash and ffpp emissions. We have missed some, but we hope that very few of the important ones have been overlooked. It must be kept in mind that ground-level concentrations of various emitted materials will depend to a great extent on the chemical composition of the coal (this can vary by ±100X), meteorological conditions, and, to estimate hazard, the population density. There are other variables also.

REFERENCES

Andren, A. W., D. H. Klein, and Y. Talmi, 1975, Selenium in Coal-Fired Steam-Plant Emissions, *Environ. Sci. Technol.,* 9: 856-858.

Billings, C. E., 1974, Technological Sources of Air Pollution, in *Industrial Pollution,* N. I. Sax (Ed.), Van Nostrand Reinhold Co., New York.

Browning, E., 1969, *Toxicology of Industrial Metals,* 2nd ed., Appleton-Century-Crofts, Inc., New York.

Chrisp. C. E., G. L. Fisher, and J. E. Lammert, 1978, Mutagenicity of Filtrates from Respirable Coal Fly Ash, *Science,* 199: 73-76.

Deichmann, W. B., and Horace W. Gerarde, 1969, *Toxicology of Drugs and Chemicals,* Academic Press, Inc., New York.

Fisher, G. L., D. P. Y. Chang, and M. Brummer, 1976, Fly Ash Collected from Electrostatic Precipitators: Microcrystalline Structures and the Mystery of the Spheres, *Science,* 192: 553-555.

Gosselin, R., Hodge, Smith, and Gleason, 1976, *Clinical Toxicology of Commercial Products,* 4th ed., Williams and Wilkins, Baltimore, Md.

Gutenmann, W. H., C. A. Bache, et al., 1976, Selenium in Fly Ash, *Science,* 191: 966-967.

Hall, H. J., 1975, False Threats of Mercury Emissions, *The Chemist.*

Hawley, G. G. (Ed.), 1977, *Condensed Chemical Dictionary,* 9th ed., Van Nostrand Reinhold Co., New York.

——, 1981, *Condensed Chemical Dictionary,* 10th ed., Van Nostrand Reinhold Co., New York.

——, and C. Hampel, 1973, *Encyclopedia of Chemistry,* 3rd ed., Van Nostrand Reinhold Co., New York, 1198 pp.

Hueper, W. C., 1972, Medico Legal Considerations of Occupational and Non-Occupational Environmental Cancers, in *Lawyers' Medical Cyclopedia.*

——, 1974, Cancer Risks in an Industrialized Economy, in *Industrial Pollution,* N. I. Sax (Ed.), Van Nostrand Reinhold Co., New York.

Hulett, L. D., Jr., A. J. Weinberger, K. J. Northcutt, and M. Ferguson, 1980, Chemical Species in Fly Ash from Coal-Burning Power Plants, *Science,* 210: 1356-1358.

Kaakinen, J. W., R. M. Jordan, M. H. Lawasani, and R. E. West, 1975, Trace Element Behaviour in Coal-Fired Power Plant, *Environ. Sci. Technol.,* 9: 862-869.

Korfmacher, W. F., D. F. S. Natusch, D. R. Taylor, G. Mamantov, and E. L. Wehry, 1980, Oxidative Transformations of Polycyclic Aromatic Hydrocarbons Adsorbed on Coal Fly Ash, *Science,* 207: 763-765.

Lee, M. L., D. W. Later, D. K. Rollins, D. J. Eatough, and L. D. Hansen, 1980, Dimethyl and Monomethyl Sulfate: Presence in Coal Fly Ash and Airborne Particulate Matter, *Science,* 207: 186-188.

Linton, R. W., A. Loh, and D. F. S. Natusch, 1976, Surface Predominance of Trace Elements in Airborne Particles, *Science,* 191: 852-854.

Mahoney, J. R., 1974, Meteorological Aspects of Air Pollution, in *Industrial Pollution,* N. I. Sax (Ed.), Van Nostrand Reinhold Co., New York.

McCrone, W. C., R. G. Draftz, and J. G. Delly, 1970, *Particle Atlas*, Ann Arbor Science Publishers, Inc., Ann Arbor, Mich.

National Coal Resources Data System from United States Geological Survey.

NIOSH (National Institute for Occupational Safety and Health), 1979, *Registry of Toxic Effects of Chemical Substances.*

——, 1980, *RTECS Quarterly Microfiche.*

Nusbaum, R. E., et al., 1965, Relation of Air Pollutants to Trace Metals in Bone, *Arch. Environ. Health,* Volume 10.

Pihl, R. O., and M. Parks, 1977, Hair Element Content in Learning Disabled Children, *Science,* 198: 204-206.

Rancitelli, L. A., et al., 1974, Elemental Content of Trace Elements in Vegetation Collected at Centralia, Washington, private communication.

Revelle, R., 1967, *Pollution and Cities,* Harvard University Center for Population Studies Publication.

Robertus, R. J., K. L. Nielsen, C. T. Crowe, and D. T. Pratt, 1975, An Attempt to Reduce NO_x Emissions from Pulverized Coal Furnaces, *Environ. Sci. Technol.,* 9(9): 859-862.

Sax, N. I., 1979, *Dangerous Properties of Industrial Materials,* 5th ed., Van Nostrand Reinhold Co., New York.

——, 1980, *Dangerous Properties of Industrial Materials Report.*

Science News, Power Plant Gases Source of Puzzling Rural Ozone Clouds, 1974, 106: 260.

Smith-Sonneborn, J., R. A. Palizzi, C. Herr, and G. L. Fisher, 1981, Mutagenicity of Fly Ash Particles in Paramecium, *Science,* 211: 180-182.

Sollmann, T., 1957, *A Manual of Pharmacology,* 8th ed., W. B. Saunders Company, Philadelphia.

The Merck Index, 1976, 9th ed., Merck and Co., Inc.

Chapter 17 covers research prior to May 1980.

METEOROLOGY AND ALTERNATIVE ENERGY RESOURCES

Joseph B. Knox, Paul H. Gudiksen, Donald M. Hardy, Frederick M. Luther, and Robert L. Zalkan, Lawrence Livermore National Laboratory

18

18-1 INTRODUCTION

During the past few years, an increasing interest in alternative energy resources has emerged in the United States as a result of the increasing cost of our fossil fuels. Consequently alternative energy sources are being sought. In fact, the U. S. Department of Energy (DOE) estimates that about 25% of the U. S. energy demand in the year 2020 will come from alternative energy sources (U. S. Energy Research and Development Administration, 1976). This trend is part of a movement away from our dependence on foreign energy resources to long-lasting domestic sources of energy. The alternative energy resources include geothermal energy, wind energy, several ways of extracting energy from the ocean, solar energy, and biomass conversion. The last four of these resources are normally called renewable in addition to alternative because the natural flows of energy that are tapped are resupplied by the sun. Most current studies state or imply that the environmental impacts of tapping renewable energy resources are either minimal or acceptable. Although this might be true, in-depth quantitative assessments of effects on weather and climate of tapping wind energy or the kinetic energy of ocean currents or thermal gradients are yet to be completed with modern methods. Topics normally excluded in discussions of alternative energy resources are fusion, the use of uranium extracted from the ocean, the replacement of uranium by thorium in reactors, controlled thermonuclear reactions (CTR), etc. Several of these options are also very long lasting and will be important in our energy future beyond the year 2000. We limit our present discussion to geothermal, wind energy, energy from the ocean, and solar—those popularly called alternative energy resources.

Man's use of alternative energy resources is of course very old; wind energy (dated to 400 B.C.), solar energy for the drying of crops, and the use of wood for heating (a form of biomass conversion) were important prior to the industrial revolution. Today, interest in alternative energy is rising in the less-developed countries of the world, with growing populations, rising energy costs, and more dispersed life-styles (Council on Environmental Quality, 1975). Since the oil embargo, the less-developed countries are harvesting forests more actively, town-size biomass conversion plants are being considered, and wind energy has been analyzed and found to be economical for these countries (Merriam, 1972). The environmental consequences of these energy options should be examined, and the basis for their implementation with minimal impact needs to be established in the next decade. Table 18.1 shows the estimated resource potential, the state of the technology, and the presently identified environmental concerns for each of the alternative energy resources considered here.

The role of the atmospheric scientist in planning, developing, and using alternative energy includes assessing the potential of the energy resource (wind and solar), screening selected sites for the most promising or identifying previously unrecognized sites (wind and solar), documenting regional energy resources through measurement and calculation, and assessing the environmental impact (all these activities are covered in Table 18.1). Modern techniques illustrating these various functions are discussed in the following sections since they are available from recent research. In some instances the assessment methods for treating some of the environmental concerns listed in Table 18.1 are still in development and research and thus cannot be discussed in depth. However, because of the timely nature of the subject of alternative energy resources, we will make the presentation insofar as it is possible at this time.

18-2 SOLAR ENERGY

18-2.1 Solar Energy Resource

As an alternative energy resource, solar energy has the advantage that it is a clean, safe, essentially inexhaustible source of energy. Its disadvantages are that it is noncontinuous and has a relatively low intensity (flux density), and thus enormous areas are required to collect substantial amounts of energy. Furthermore, some form of energy storage is required to overcome its diurnal nature.

The potential role of solar energy in the energy economy of the United States (Fig. 18.1) can be evaluated by looking at the solar energy budget. The solar radiation continuously intercepted by the outer

Knox, Gudiksen, Hardy, Luther, and Zalkan, Lawrence Livermore National Laboratory, University of California, Livermore, California.

TABLE 18.1 Alternative Energy Resources: Potential Technology and Environmental Concerns

Energy resource	Potential	State of technology	Potential environmental concern
		Solar Energy	
	3×10^9 MW·h annually by 2020	Solar-powered generating stations are in the developmental stage for possible implementation by 1985; the economics of solar heating and cooling units is being evaluated	Alteration of surface albedo with feedback on local or regional climate
		Wind Energy	
	2×10^6 MW	Modern technology is being developed to improve reliability and to lower capital investment	Alteration of regional airflow patterns, land use, and aesthetics
		Ocean Energy	
Ocean currents	1×10^4 MW	Low energy density requires bulky turbines, high capital costs, low efficiency, also must solve problem of shifting axis of current	Can only tap small percentage of available energy since potential climatic changes may occur
Salinity gradients	2×10^6 MW	Technical breakthrough in the development of large, efficient, and inexpensive semipermeable membranes is required	Could cause changes in salinity and temperature distribution important to estuarian marine life; may require rerouting of a percentage of river flow
Tidal	2×10^4 MW	Existing plants are in operation	Sedimentation, navigation, and sea life are affected by the existence of a tidal barrage
Surface waves	2×10^6 MW (2×10^7 MW if some regeneration is included)	Further testing is required to evaluate the current position of several promising designs	Effects will be seen on longshore currents, sediment transport, and beachlines, thus influencing Coastal Zone Management
Ocean thermal-energy conversion (OTEC)	4×10^6 MW	Basic technology is available; continual attention is being given to heat-exchanger refinements, biofouling research, and cold-water intake design	Change in sea surface temperature and albedo could result in climatic changes; change in salinity, temperature, and density distributions; use of antifouling agents; bringing to the euphotic zone large amounts of nutrients and CO_2
Wind (offshore)	10^{11} MW	Awaiting the development of shore-based wind energy	Since winds drive currents and cause waves, tapping wind energy over the ocean will affect these phenomena
		Geothermal Energy	
Vapor-dominated systems	3.5×10^4 MW(t)·centuries (500 MW is presently being generated at The Geysers, California)	Existing technology	H_2S emissions, cooling-tower drift, and noise
Hot water systems	1×10^6 MW(t)·centuries	Need to construct and operate small-scale demonstration plants using the multistage flash or binary-flow concepts	H_2S emissions, land subsidence, and induced seismicity
Hot igneous systems	3×10^7 MW·centuries	Demonstrate the feasibility of hydrofracturing the rock and extracting the energy by forcing cold water through the rock	H_2S emissions
Geopressured systems	1.5×10^7 MW(t)·centuries (does not include an additional 8×10^6 MW·centuries of methane gas)	Demonstrate the feasibility of using the flashed steam or binary systems for methane gas and thermal energy extraction	H_2S emissions, land subsidence

atmosphere of the earth is 1.73×10^{14} kW (which corresponds to 1.5×10^{18} kW·h annually). Solar radiation received at the ground consists of a direct beam component and a diffuse component that has been scattered, reflected, and absorbed by molecules and particulate matter in the atmosphere. Averaged globally, approximately one-half of the intercepted radiation reaches the earth's surface, and most of the energy is in the direct component. The amount of direct and diffuse solar radiation received varies with latitude, cloudiness, and atmospheric turbidity. The distribution of solar radiation received at the earth's surface on an annual average for the United States is shown in Fig. 18.2. The solar energy received annually in the continental United States is 9×10^{15} kW·h.

In 1972 the United States used 2.2×10^{13} kW·h of energy from all sources, of which 1.7×10^{12} kW·h was electric energy. If all the electric energy generated in 1972 had come from solar energy using photovoltaic conversion, it would have required 1.2×10^{10} m^2 of solar collectors (assuming a 10% efficiency for electrical conversion and an average daily insolation of 4 kW·h/m^2), or about 0.14% of the total land area of the United States (Kaplan, 1975; American Institute of Aeronautics and Astronautics, 1975). The total land area devoted to solar energy collection would have to be considerably greater to account for spacing between collectors. For perspective the total road and highway area in the United States today is 1.5% of the entire land area. Considering the anticipated growth in energy usage, a solar energy system capable of producing 1.7×10^{12} kW·h annually would provide only about 9% of the electric energy generation projected for the year 2020. In terms of total energy (electrical and thermal), DOE officials see solar conversion as supplying about 7% of the U. S. energy needs by the year 2000 and maybe 15% by 2020 (Josephson, 1976).

18-2.2 Solar Energy Conversion

Various technologies are being considered as the means of generating electric power in large quantities through solar energy. Included among these are photovoltaic, thermoelectric, thermionic, and solar thermal conversion (STC).

Photovoltaic cells, also referred to as solar cells or solar batteries, are thin wafers of semiconductor material. When exposed to light, electrons flow from one layer of the device to the other. A single cell produces an output in milliwatts with an electric potential of approximately 0.5 V. Individual cells are connected in series or parallel to raise the voltage or to yield more current. Although other materials, such as gallium arsenide and cadmium sulfide, can be used for photovoltaic conversion, silicon is particu-

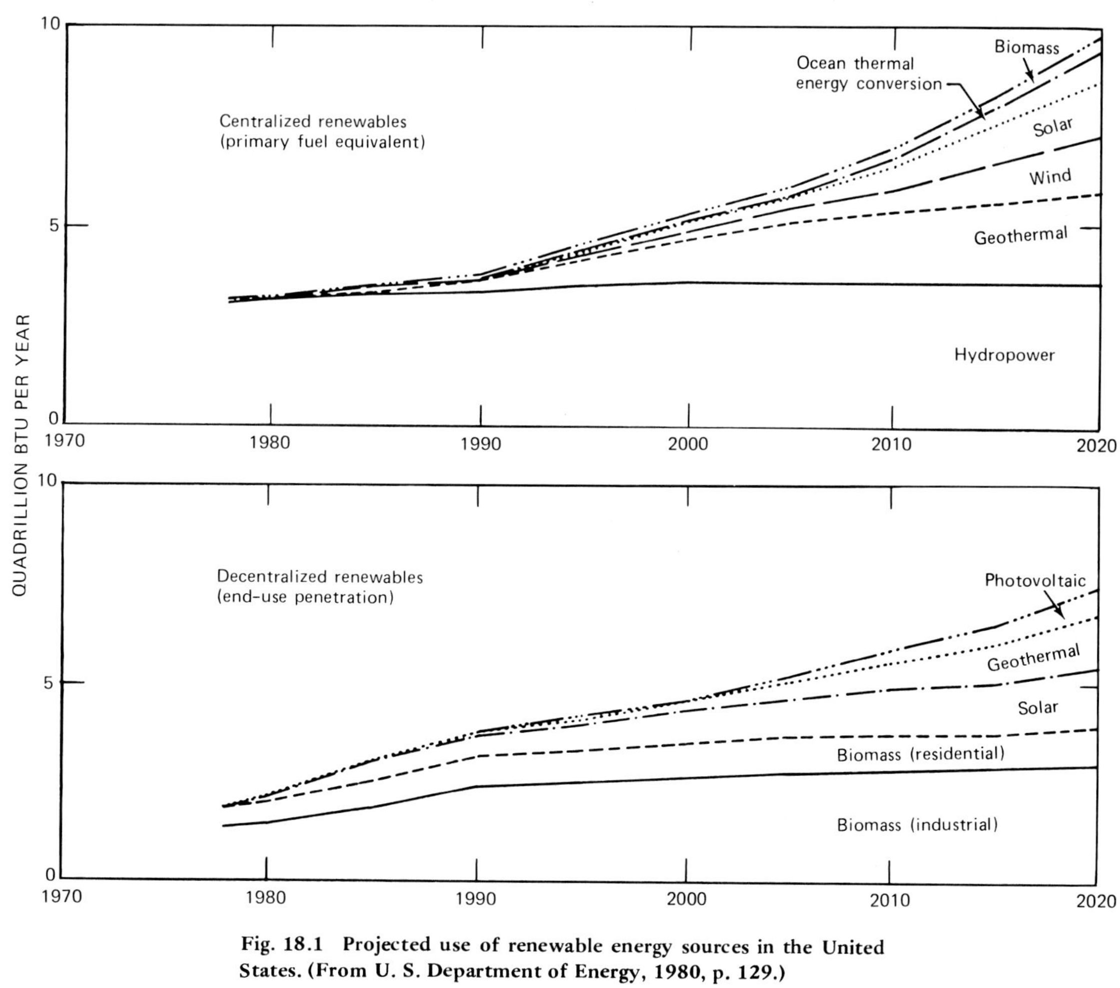

Fig. 18.1 Projected use of renewable energy sources in the United States. (From U. S. Department of Energy, 1980, p. 129.)

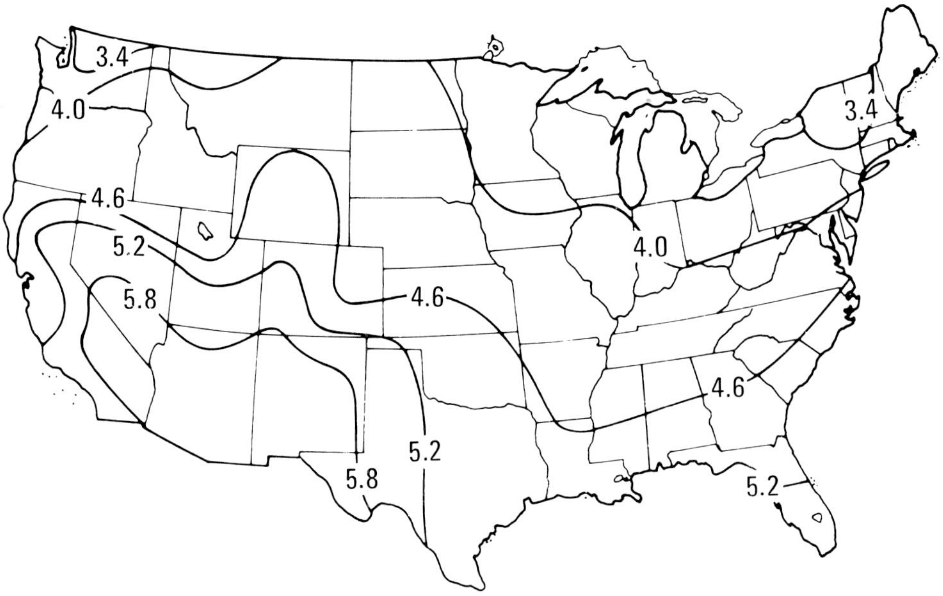

Fig. 18.2 Distribution of daily incident solar radiation (annual average). Units are kW·h/m^2.

larly attractive because of its great abundance, its high conversion efficiency (presently 10 to 15% with a theoretical upper limit of 26%), and the high level of development of silicon manufacturing technology (Lawrence Berkeley Laboratory, 1976). Most photovoltaic devices are responsive to a broad range of wavelengths, and the wavelength range can be modified to encompass the major part of the solar spectrum. The most important characteristic of a photovoltaic device is carrier (free charge) lifetime (Kaplan, 1975), which depends on the quality of the crystal. The purer the crystal, the longer the carrier lifetime and the higher the conversion efficiency. There is a trade-off, however, between purity requirements and manufacturing cost.

Research and fabrication advances are being reflected in lower prices, but the direct conversion of sunlight to electricity will still have to become far less expensive before these systems can come into widespread use. In 1977, single-crystal silicon flat-plate arrays cost \$11/W (1975 dollars at 20°C). It is possible that, by using concentrating systems to focus sunlight onto a smaller area of cell, the cost can be reduced to \$2/peak watt by the early 1980s (*Energy Today,* Feb. 10, 1978).

Thermoelectric conversion is based on the phenomenon known as the "Seebeck effect," whereby heat applied to the junction between two different metals produces a voltage. These devices, also called thermocouples, rely on solar concentrators to increase the temperature of the junction. Thermoelectric converters must dispose of waste heat, and a large temperature difference must be maintained between the hot and cold junctions of the thermocouple, thus making thermoelectric converters more complex than photovoltaic cells. With thermoelectric conversion the sunlight is focused by solar collectors onto thermocouples only a few centimeters in diameter, which results in a saving of expensive converter material. Such materials as lead telluride and zinc antimony produce power at 10% efficiency, and it is theoretically possible to achieve a significantly higher efficiency.

Thermionic conversion is based on the phenomenon that heat applied to a cathode causes the emission of electrons and the generation of an electric current. Present efficiencies of these devices, also called thermionic diodes, are about 15%, but efficiencies as high as 40% may be possible. The technology required for the thermionic converter is more demanding than either photovoltaic or thermoelectric converters.

The STC systems collect solar radiation and convert it into thermal energy in a working fluid for use either in a thermal–electric conversion system or in a total energy system that delivers both electrical and thermal energy (Fig. 18.3). There are two basic types of STC systems: the central receiver system and the distributed collector system. The central receiver system uses thousands of individually controlled mirrors to reflect solar radiation onto a central receiver on top of a tower where it heats an energy transport fluid. The fluid carries the thermal energy to the power plant where it drives a turbogenerator. In the distributed collector system, incident solar radiation is absorbed by a large number of collector units. The energy is then conveyed to a centrally located power plant by an extensive piping network.

The central receiver concepts proposed to date differ in heliostat (mirror) and receiver designs. The heliostats may be either individually mounted or "ganged" and receivers may have either a cavity configuration, which requires optical focusing of the energy from the reflectors, or an open-face configuration, which does not require focusing. The possibility of using fixed reflectors and a movable central receiver is also being considered. In general, the overall efficiencies are expected to be in the range of 18 to 25%. For a 20% net efficiency, 5 m^2 of mirror area would

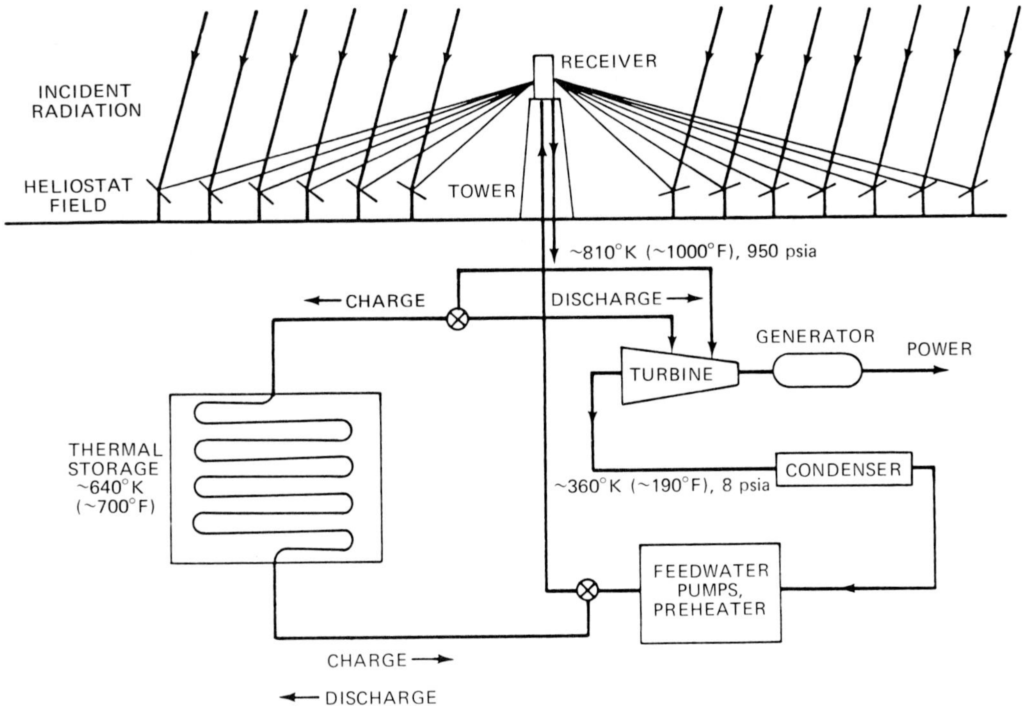

Fig. 18.3 Solar thermal power-plant schematic. (From U. S. Department of Energy, 1980, p. 302.)

be required for each kilowatt of electrical capacity. A mirror area of approximately 0.5 km^2 is needed for each 100 MW of electrical capacity. To minimize shading and blocking of adjacent heliostats, the land area will have to be considerably larger than the mirror area. Consequently a land area of 1.0 to 1.5 km^2 would be needed for a 100-MW plant.

Solar collector configurations for distributed systems may consist of flat plates, parabolic troughs, V-troughs, or paraboloidal dishes. Nonfocusing solar collectors use total insolation (direct plus diffuse), whereas focusing types use only direct insolation. Thus nonfocusing collectors work reasonably well on hazy or cloudy days. However, they can heat a working fluid only to 370 to 530°K (Gervais and Bos, 1975). These temperatures are relatively inefficient for power production compared to those in fossil-fueled power plants, where turbogenerators receive steam at close to 800°K. Temperatures greater than 800°K can be obtained using central receiver systems (Gervais and Bos, 1975). For electrical power generation, higher temperatures permit greater thermal–electric conversion efficiency, and thus less collector area is required for a given electrical output. A technical innovation that may enable temperatures in the range 600 to 800°K to be reached using flat-plate collectors is the use of selective surfaces that absorb most of the solar radiation and radiate away very little of it.

For elimination of problems associated with the diurnal nature of solar energy and the attenuation of the radiation transmitted through the earth's atmosphere, it has been proposed (Glaser, 1968) that large arrays of photovoltaic cells be put into space in a near-equatorial synchronous orbit where the sun would shine on them nearly 100% of the time. The direct-current power obtained from the arrays would be converted into microwave power and beamed to receiving stations on the earth. Because the rotation of the solar satellite would be synchronous with the earth, the microwave link would be fixed and operative at all times. The microwave beam density could be designed to meet international safety standards, and the thermal pollution in the reconversion of microwaves directly to electricity at the receiving antenna would be about one-quarter of that of conventional power plants. The receiving antenna would be substantially transparent to solar radiation and open enough for rain to reach the ground below it, thus providing opportunities for multiple-land use. The effects on birds and on aircraft flying through the beam have yet to be determined. Detailed feasibility and design studies are currently in progress at the National Aeronautics and Space Administration (NASA), the Jet Propulsion Laboratory, Boeing, Econ, Rockwell, and Arthur D. Little, Inc. (Glaser, 1977).

Technical problems exist in all solar energy schemes which will need solutions before commercial, large-scale applications can be carried out. Photovoltaic, thermoelectric, and thermionic conversion systems require technical breakthroughs to reduce cost and improve efficiency. In the solar thermal central-receiver concept, extensive developments will be required to reduce capital equipment and construction costs. Developmental problems of the distributed collector power plant are associated with the need to efficiently transfer collected thermal energy from a large area to a central point. Nevertheless, several pilot plants are planned or are already under construction.

A conceptual design has been chosen for the nation's first experimental solar central receiver system, a 10-MW electric-generation plant to be built by DOE near Barstow,

Calif., at a cost of about $100 million. In the concept chosen, a cylindrical configuration of boiler tubes sits on top of an 86-m-high tower and absorbs sunlight reflected from more than 1500 heliostats surrounding the tower. The heliostats will be multifaceted glass mirrors of approximately 60 m^2 supported by a metal structure, with each mirror mounted on a single concrete pile foundation. The field of heliostats will be controlled by a computer to continuously reflect the sun's energy onto the receiver. The pilot plant is scheduled for completion in 1980. The plant will be run for 1 yr, and the experience will be used to project the economic potential and engineering refinements of a larger demonstration plant. Similar plans call for the design and fabrication of a 10-MW(t) power plant using distributed collectors.

The various prototype boilers and heliostats proposed for the 10-MW pilot plant near Barstow, Calif., were tested at the Solar Thermal Test Facility (STTF) located on a 100-acre site at Kirtland Air Force Base outside of Albuquerque, N. Mex. The STTF, funded by DOE and constructed by Sandia Laboratories, will provide a crucial testing ground for techniques and components useful in solar-powered central-receiver electric-power plants. The STTF includes a 61-m-high concrete tower with a field of solar reflectors to the north. In addition to room at the top for boilers, the tower includes bays at 55-, 49-, and 43-m levels, which serve as test locations for small receivers and experimental equipment. The facility does not have a turbogenerator, but all other systems are presently producing roughly 5 MW of thermal energy. This potentially could be converted to 1.5 MW of electric power.

The interest in solar energy is international. In France a central-receiver pilot plant began operation in October 1976. The plant produces 70 to 100 kW of electricity, and this marked the first time that power from the sun had been fed into a utility grid. The $1 million pilot plant is located outside the village of Odeillo, a mile high in the French Pyrenees. Mirrors totaling 2900 m^2 located on a hillside reflect the sun's radiation to a huge parabolic reflector that concentrates and focuses the solar energy onto a boiler. The experience gained from operating the pilot plant will aid in the final design of a larger pilot plant. Construction is expected to begin in 1979 on a $30-million facility located in the south of France which will be capable of producing 10 MW of electrical power. This will be a central receiver system with the boiler mounted on a tower.

18-2.3 Potential Environmental Impact

For solar energy to contribute significantly to the total energy economy of the United States, large geographic areas will be required for solar collection, regardless of the solar conversion system used. To be cost effective, individual power plants will need many square kilometers of solar collectors. For example, it is estimated that central receiver systems will be the most economical for towers 250 to 500 m high, with solar collectors covering an area 1 to 2 km on a side (Vant-Hull and Easton, 1974). There is also a cost reduction from the clustering of towers together to a single power plant. Consequently commercial "solar farms" (using distributed collector systems) or central receiver systems can be expected to cover areas as large as a major city. The ideal location would be in generally flat, semidesert areas of the southwestern United States.

In estimating the environmental effect of such alterations to large land surfaces, we already know that cities alter both regional temperature patterns, through a "heat island" effect, and precipitation patterns (American Meteorological Society, 1977). Although the atmospheric perturbation is speculated to be acceptable (Williams, Kromer, and Weingart, 1977; Kaplan, 1975; U. S. Energy Research and Development Administration, 1977), no reliable, quantitative assessment has been made of the potential impact on regional temperature and precipitation patterns or ecosystems.

The interception and collection of solar radiation by large arrays of solar reflectors and absorbers will alter the regional radiation budget (solar and long wave) and change the moisture retention properties, heat capacity, and thermal conductivity of the earth's surface. Natural surfaces absorb 70 to 90% of the incident solar radiation; deserts absorb 70 to 75% and vegetated surfaces 80 to 90%. Although natural surfaces vary in their solar radiative properties, they have similar long-wave properties; the long-wave emissivity is about 90% (Sellers, 1969). Portions of the solar energy absorbed by the surface are convected away as sensible heat, used for evaporation, radiated away as long-wave radiation, and conducted into the ground. Variations in surface temperature depend strongly on the surface type and amount of moisture available for evaporation.

Arrays of photovoltaic cells can have a minimal effect on the regional radiation and moisture balances for desert locations, where there is little soil moisture. If the arrays were placed over soil with a significant moisture content, evaporation would be restricted, thus affecting both the moisture and radiation budgets. A method of counteracting the effect of large arrays of solar cells on regional temperature might be to group the arrays into smaller clusters, thereby increasing the ground area required to generate the energy. The albedo of the surface between clusters could be adjusted so as to compensate for the perturbation due to the arrays, but this process would increase the area over which the arrays must be distributed. Whether this method is economically feasible, or even necessary, has yet to be determined.

The STC systems present a greater problem in terms of waste-heat disposal. The large array of mirrors used with the central receiver system reduces the solar energy absorbed by the ground over a large area and concentrates that energy on a central receiver. About 90% of the energy incident on the receiver is absorbed (Gervais and Bos, 1975), and the rest is lost to the atmosphere. The energy that is absorbed powers a steam-turbine generator, which has an efficiency of up to 40%.

The absorbing panels used with the distributed collector system might significantly affect both the solar and long-wave radiation budgets. This would be most pronounced when selective absorbers are used to reduce the long-wave emission and increase the solar absorption. Reductions in surface albedo will reduce the amount of backscattered solar radiation incident on the ground, and changes in temperature and moisture patterns will affect convection and wind patterns, but these effects have not been assessed in detail.

Both the central and distributed collector systems use cooling towers (dry or wet) for the disposal of waste heat. Wet cooling towers would require large amounts of water. Since water is scarce in the deserts, one proposal calls for the use of seawater, which, when desalted, produces fresh water as a by-product. The release of waste heat and moisture from cooling towers and the disposal of brine from the desalting process could present major environmental problems.

Closely connected with the heat and water balance are the ecological impacts of solar energy systems. The magnitude of the effect will likely depend on the design of the solar collectors or heliostats. If the collectors or heliostats are raised or mounted on pedestals, then this would permit underlying vegetation and animal life. The shading and possible increase in moisture retention may provide an environment that is suitable for agriculture, perhaps as grazing land. The reduced surface winds and potential increase in vegetation could reduce wind-caused soil erosion and airborne dust.

Technology already exists for thermal and electric-power production from solar energy. However, large-scale use of solar energy is not economically attractive at the present time. Most of the technical problems associated with solar power relate to increasing efficiency and decreasing cost. Solar energy is not expected to become a significant fraction of the total energy economy of the United States until the end of this century. As solar energy use progresses toward larger scales, the environmental impact due to alteration of the radiation and moisture budgets of the earth's surface over areas of 10 to 100 km^2 has not been assessed. Atmospheric scientists will play a key role in these assessments. Because of the complex nature of the problem, a team effort will be required with expertise spanning many scientific disciplines. The research tasks will include the following experimental and theoretical efforts: assessing meteorological conditions at proposed sites; measuring radiative properties of materials used for solar collectors; monitoring meteorological data at pilot-plant facilities; numerical modeling of regional energy and moisture balances; and interpreting experimental and theoretical data. The atmospheric scientist will also be involved in providing initialization data for the numerical models and in model validation.

18-3 WIND ENERGY

18-3.1 Wind Energy Resource

The winds of the world are driven by the solar energy received over the globe. A fraction of the incident solar radiation is converted by the atmosphere into kinetic energy in the form of the winds. This natural conversion of the solar flux provides a source of mechanical energy which has long served mankind. The winds provided power for transportation at sea from the beginning of civilization and for agricultural purposes in Western Europe as early as the eleventh century. Estimates of available wind energy for modern uses vary significantly, but even the lowest represents a substantial energy potential.

For a realistic estimate of the wind resource, several factors must be considered. The mean kinetic energy of the atmosphere is a stored resource that greatly exceeds the natural rate of kinetic energy dissipation and replenishment in the atmosphere. The mean motion of the atmosphere represents, in effect, a large "flywheel" in which the equivalent of approximately 1 d of total incident solar flux is stored. Extraction of a large fraction of this stored energy, if possible, would obviously change the mean state of motion of the atmosphere with unknown environmental consequences. Thus it is the natural rate of kinetic energy replenishment that places an upper limit on wind power collection.

Because the atmosphere is in a state of global energy equilibrium, the replenishment rate can be estimated from the mean dissipation rate, which has been judged to be 2.5 W/m^2 (Gustavson, 1977) for the near-surface atmospheric flow. If the wind energy extraction average over the land area of the United States (8×10^{12} m^2) is limited to 10% of the normal dissipation rate, then the practical wind power potential of the United States is about 2 TW. Such a rough estimate, however, uses no information about the distribution of the resource in space. The development of the 3% of the United States with the currently documented highest annual wind speeds would yield a wind power resource of about 2.5 TW (Gustavson, 1977). The land area required is very large, being approximately equal to the size of the state of Colorado. Both of these estimates are large, being approximately equal to the mean rate of energy usage in the United States from all sources in 1972.

To estimate the wind power available to a wind turbine at a particular site, consider the mass of air meeting a wind turbine of area A normal to the wind in 1 s. The mass of air, m, moving with a velocity, V, has a kinetic energy of 0.5 mV^2. Since the mass of air is ρAV, where ρ is the density of air, the kinetic energy of air meeting a wind turbine of area A in 1 s is 0.5 ρAV^3. For example, a uniform wind of 10 m/s driving a wind turbine of 10 m^2 area would have an available power of 6 kW. The actual power extracted would depend on the efficiency of the wind turbine at this wind speed.

The distribution of wind energy varies markedly in space and in time over the nation. Continental scale analyses (Reed, 1975; Justus, Hargraves, and Yalcin, 1976) must necessarily rely on a large number of historical wind observations. Such wind observations are conducted primarily to support the agricultural, aviation, and marine industries. Many observations are available at airports that are not generally located in the more windy areas.

Traditional wind data can also be misleading in the mountainous, hilly, and coastal regions that characterize much of the nation. However, in geographic regions where individual wind measurements are representative of a broad surrounding area, available historical data can be directly used for wind energy estimates. A contour map based on Reed (1975) of wind power density over the United States is shown in Fig. 18.4.

High wind-speed areas with persistent winds of the order of 10 m/s within approximately 150 m of the surface are of primary interest in large-scale energy conversion systems. A neutral boundary layer and power law vertical profile can therefore be assumed in most cases. Terrain inhomogeneities that would cause severe turbulence or flow separation must be avoided in siting equipment. Unsteady situations, such as the cold air outflow from thunderstorms and associated gust fronts, represent potential hazards to wind turbine generators (WTGs) in the operation of a wind energy conversion system.

Hourly, monthly, and yearly variations of the average wind speed are also important in design studies. Statistical information, such as the cumulative probability function of hourly wind speeds, is used to determine the energy potential of specific areas. Maximum wind speeds expected are needed in designing equipment for a given lifetime of service. Determination of the maximum expected wind speed is difficult because systems will often be located in presently undeveloped areas where few wind data are available. The possibility of directional shear in the vertical wind profile must also be considered because of the resulting structural stresses that would be produced.

18-3.2 Energy Conversion Technology

Present concepts for capturing large amounts of wind energy emphasize the use of wind turbines to generate electricity. The generation of electricity has been of primary interest because of the many uses of electricity and the ability to transmit large amounts of energy through the existing electrical transmission system. Several wind turbine-generator designs are under development. Two representative examples of designs for large-scale power collection are a horizontal rotor-axis machine under development by NASA and DOE (Thomas et al., 1975) and a vertical rotor-axis machine being developed by Sandia Corp. (Blackwell, 1974; Blackwell et al., 1977) under a DOE funding arrangement.

The horizontal axis machine is a conventional propeller-type design scaled for power levels from 100 to 1500 kW. The machine designed for the 100- to 200-kW range has a rotor diameter of 37.5 m. The rotor and generator are mounted on a 30-m tower, with the rotor downwind of the tower. A schematic diagram of the rotor and tower is shown in Fig. 18.5. The experimental 100-kW prototype reaches full-power output in an 8-m/s wind.

The pitch of the turbine blades is controlled under varying wind speeds to maintain a constant rotation rate of 40 rpm and synchronous generation of 60-Hz three-phase power by an alternator. Under extreme wind conditions the blades must be feathered to prevent damage. The direction of the turbine is adjusted to match the wind direction by a yaw control, which rotates the rotor and generating equipment about the tower axis. Cut-in, rated, and cut-out

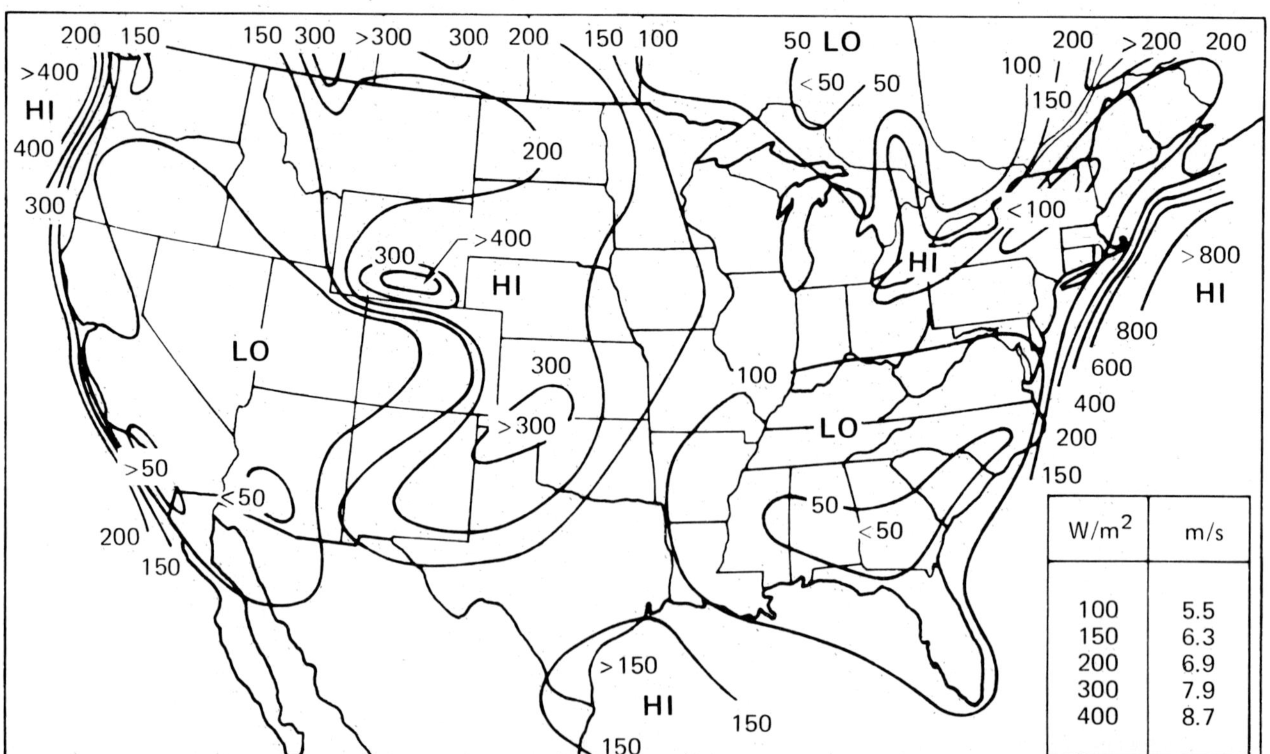

Fig. 18.4 Contour map showing the annual average available wind power. (Based on Reed, 1975.)

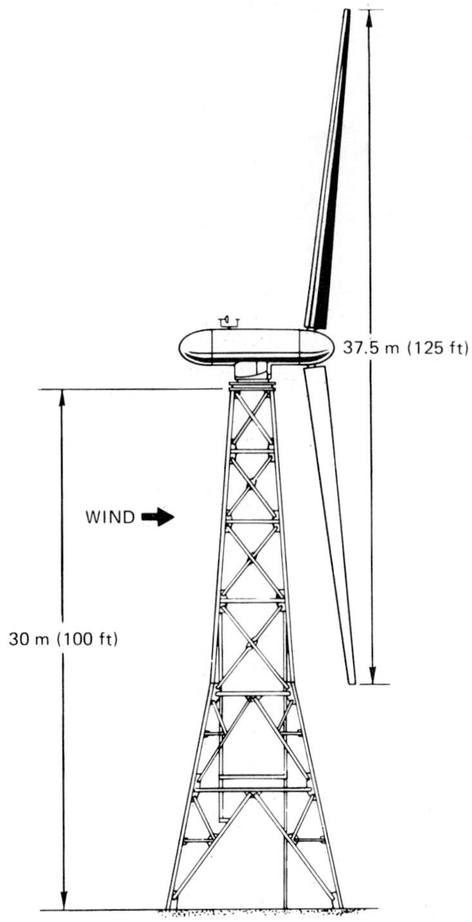

Fig. 18.5 A 100-kW experimental wind turbine generator.

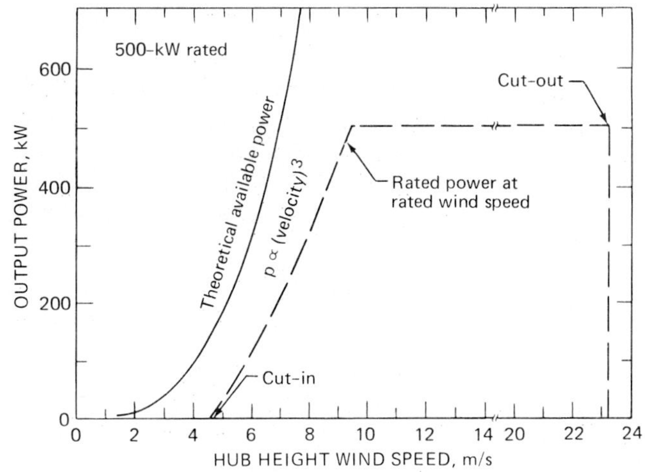

Fig. 18.6 Output power for a 500-kW rated power wind turbine.

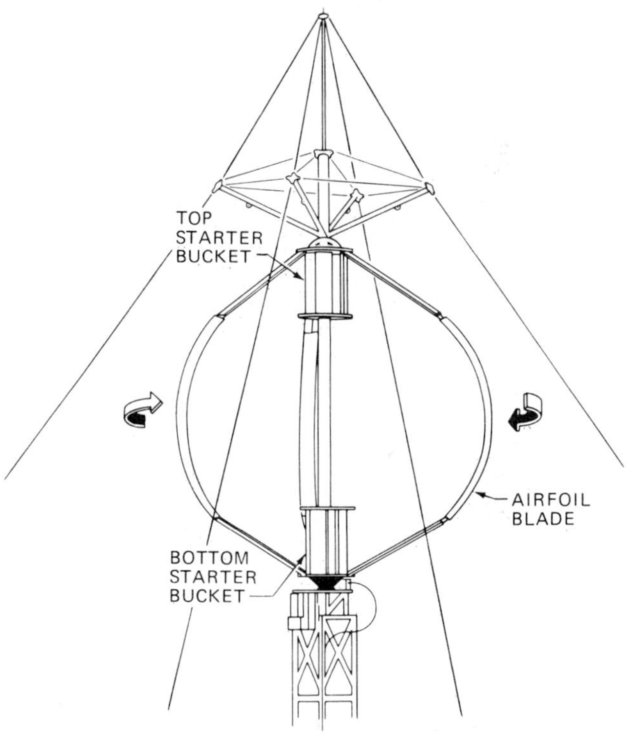

Fig. 18.7 Vertical axis wind turbine.

wind speeds and power output for a hypothetical 500-kW WTG are defined in Fig. 18.6.

Basic design features of the 100-kW WTG are being used to develop machines with rate capacities of 200 and 1500 kW. Machines will be designed for relatively low wind areas as well as for windy environments.

The vertical axis machine under development by Sandia is a Darrieus design, named for French inventor G. J. M. Darrieus who was issued a U. S. patent in 1931. A schematic diagram of this turbine is shown in Fig. 18.7. Machines of this type, with rotor diameters of 5 and 17 m, are being developed and tested by Sandia. The design uses two or three blades about the rotor axis. Each blade approximates a troposkien shape, which is the shape that an ideal cable would assume if it were spun about a vertical axis. The Darrieus turbine is not self-starting from zero rotational speed. Therefore Savonius rotors ("starter buckets") are also attached to the central axis to start the blades rotating at very low wind speeds.

The cross section of each curved blade has the shape of a symmetric airfoil. The velocity of the wind relative to the blade produces lift and drag forces on the airfoil. A positive driving torque is created as long as the projection of the lift force along the chord of the airfoil exceeds the drag force.

The wind velocity relative to the blade is the difference between the absolute wind velocity and the absolute blade velocity, which is the product of the blade radius and its angular velocity. The blade angle of attack with respect to the relative wind velocity varies during each revolution. At moderate speeds of rotation, when the absolute blade velocity is several times the absolute wind velocity, the blade is never stalled during a revolution.

At low rates of rotation, the blade is stalled over an appreciable fraction of a revolution, and power output is low. At very high rates of rotation, the angle of attack of the blade decreases, the projection of lift along the blade chord drops, and performance is lowered. Optimum performance thus occurs for a particular ratio of rotational speed to wind speed. The optimum efficiency of the vertical axis turbine is comparable to that of more conventional turbine designs.

In overall design the Darrieus vertical axis turbine is quite different from propeller-type wind turbines. One important feature is that it is nondirectional in regard to the horizontal wind. The turbine accepts winds from any direction and does not need to be oriented into the wind as the prevailing wind direction shifts. Power is produced along the vertical axis and can be transmitted directly to the base of the machine. Heavy electrical-generating equipment can therefore be mounted on the ground rather than on the tower. Also, the tower and turbine are in a sense integral and can be made of relatively light materials supported by guy wires.

18-3.3 Regional Wind Energy Assessment Methods

The collection of wind power at levels near 100 MW or more in a given locality represents a very significant undertaking. Large arrays of many turbines (each with power capacities of ~1 MW) are required. The overall system size must be planned, turbine sites within an array must be determined, and the array output variation over weeks, months, and years must be estimated. Existing weather stations can provide some but not all the necessary data. Improved techniques for locating areas with the most favorable wind conditions are needed.

Wind variations both in space and in time are important in evaluating the economic competitiveness of wind energy. The cost of land and of construction at different locations, the cost of energy transmission, and land-use restrictions must be considered in relation to the spatial distribution of wind energy. The varying energy demands of a locality by time of year and by time of day and the cost of energy storage are related to the temporal variations of wind energy. Good siting of expensive energy-conversion equipment is necessary for wind energy to be economically competitive.

Both wind measurements and modeling of atmospheric flow are needed in the development of the wind as an energy source. Numerical models that make use of wind observations and local terrain are important in siting and resource assessment. Mesoscale models can also include both thermal and terrain effects in areas where surface heating and topography are dynamically important. Wind-tunnel models (Meroney et al., 1976) can be used to simulate flow around hills or other specific terrain features. Site-specific measurements are an important part of evaluating potential WTG sites. In addition, certain site-specific data may be necessary to optimize designs, and on-site data will be required for performance evaluations at any location.

Mountainous regions experience complex flow variations, and rugged coastal areas are also affected by the land–sea influence on the large-scale weather systems. These complex flow situations require detailed measurement and modeling efforts to achieve optimum siting for energy conversion. Such an effort is justified because coastal hills and mountainous regions often have significant areas of enhanced wind speeds with a high wind power potential. Areas with linear dimensions ~100 km and subareas with dimensions ~10 km are of interest when energy collection at multimegawatt power levels is desired in complex terrain. Several subareas with a power potential in the range of 100 to 500 MW, comparable to a fossil-fuel facility, can exist in an area where terrain strongly influences the wind field.

General procedures for estimating the wind potential in rugged terrain are needed to study many parts of the nation, including the Pacific Coast, Rocky Mountain, and New England regions. Estimating the distribution of wind velocities in such areas is, however, very difficult. Terrain effects can substantially increase or decrease the prevailing wind speed at different locations.

Analysis of the complex interaction of terrain and the atmosphere is a challenging problem. Simulation modeling of the flow field and direct measurement of wind velocities under different meteorological conditions are valuable in assessing the wind resource (Hardy, 1977a). An example of the effect of terrain on the wind field is shown in Fig. 18.8 for the island of Oahu. A numerical objective-analysis model used the available wind observations to produce a three-dimensional wind field (Hardy, 1977b). Flow lines for the horizontal wind field 120 m above topography as computed by the model are shown in Fig. 18.8. Wind enhancement occurs at the northern and southern ends of the mountain ranges and at relative low spots along the ridge lines where wind data are not generally available. The lowest speeds occur east of the Koolau range, in the central valley, and at lower elevations west of the Waianae range. Calculations such as these were used in planning a field measurement program to instrument locations of expected high wind speeds, including sites near Kahuku Point and Kaena Point.

Remote measurements can provide information on the vertical velocity profile and map wind velocities in three

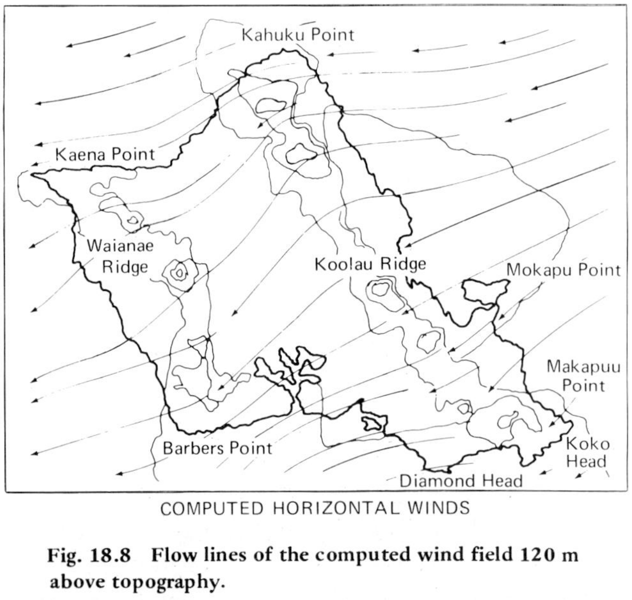

Fig. 18.8 Flow lines of the computed wind field 120 m above topography.

dimensions above the terrain and can provide direct measurement of the integrated wind speed over an extended area. Several methods of remote sensing are of interest in wind energy. In addition to multiple towers and conventional anemometry, radar, laser, and acoustic techniques of wind measurement are also available. However, the range, accuracy, resolution, and state of development of these instrument systems vary. The properties of several modern techniques are reviewed next.

Optical crosswind systems measure the path-average component of the wind perpendicular to a laser-beam path. Fluctuations in the optical refractive index in the atmosphere are advected by the mean wind. The time-delayed correlation between sensors at the end of the optical path is examined to determine the mean wind component perpendicular to the path. Beam paths of several kilometers are possible, and measurements can be made well aboveground by transmitting a beam from one hill to another.

The frequency shift in transmitted optical signals is used in Doppler laser systems to measure wind velocity. Atmospheric aerosols that are carried by the wind serve as the scatterers. Existing systems use an infrared (10.6-μm) CO_2 laser as the transmitter, with the larger aerosols in the diameter range of 2 to 5 μm as the most effective scatterers. Both pulsed and continuous-wave (CW) systems are in use (Huffaker, 1975). The pulsed system has a range of 10 km or more and can be used as an airborne measurement system. The CW measurement range is determined by focusing the transmitting and receiving optics and is limited to 1 km or less. High-quality measurements of the mean wind velocity and turbulent velocity fluctuations are possible. In most applications, only the wind velocity component along the laser beam is measured. However, a conical scan about a vertical axis can be used to infer three-dimensional vertical profiles.

Measurements over a large spatial scale can be made with pulsed Doppler radar systems (Gray et al., 1975; Baynton et al., 1976). Such scattering targets as clouds, precipitation, or chaff are needed. The velocity component along the transmitted beam can be measured for scattering volumes of approximately 75 m in length and 20 mrads in angle. Useful ranges extend to 60 km. Two coordinated radars can be used to simultaneously measure two components of the wind (Lhermitte, 1970 and 1975). The radars can be scanned through space, and the continuity equation can be used to determine a three-dimensional velocity field. Such techniques are well suited for measuring the large-scale wind field with respect to terrain.

A frequency-modulated (f-m) CW radar is sensitive to clear air refractive-index fluctuations and does not require precipitation or chaff (Strauch et al., 1975; Bean et al., 1971). The sample volume is similar to pulsed Doppler radar, but the range is restricted to 3 km or less. Only a single component of the wind is measured. However, a conical scan about a vertical axis can be used to infer all three velocity components in the volume sampled. The vertical profile can be measured by scanning at different levels.

Acoustic Doppler techniques are also useful for wind velocity measurements (Beran, 1971, 1975; Beran and Clifford, 1972). These systems are sensitive to the motion of thermal fluctuations that scatter the transmitted acoustic pulse. The returned signal is Doppler shifted by the motion of the temperature inhomogeneities. Separate receivers can be steered upward to scan a vertical line above the transmitter. Such devices can be used to obtain wind profiles to several hundred meters above a fixed site.

18-3.4 Potential Environmental Impact

As noted previously, possible adverse environmental effects of wind energy relate primarily to large-scale energy collection. Some early estimates of wind energy extraction exceed the rate of kinetic energy replenishment and thus would affect the mean state of motion of the atmosphere. It is unlikely that wind energy extraction at such levels will ever take place. A reasonable assumption is that some fraction of the normal rate of kinetic energy dissipation and replenishment represents an upper limit on global wind energy extraction. However, an exact upper limit cannot be set on the basis of current knowledge.

Local extraction rates might greatly exceed the global mean replenishment rate, with the risk of adverse effects depending strongly on the geographic area involved. The wind energy collected will be converted to mechanical, chemical, or electrical energy that can be transported to other locations for use. A fraction of the collected energy will then be released again in the near-surface environment. The ultimate scale of such energy redistribution and the locations involved will determine how seriously the impact of this process needs to be investigated.

There are several positive environmental effects of wind energy. The energy extracted from the wind will replace energy that would otherwise be produced from other sources. This replacement will reduce the consumption of fossil or nuclear fuel as well as the nuclear waste-disposal problem and will reduce the demands for cooling water at electric-generating facilities. Reduced fossil-fuel consumption will reduce air pollution, including CO_2 emissions that can have significant climatic impacts at massive fossil-fuel consumption rates. The environmental hazards of transporting fuels will be lessened by using wind energy or altered if the fuel is converted and transported as hydrogen. Land damage due to mining for conventional fuels will also be lessened, but this will have to be evaluated in comparison to the land-use requirements for wind energy collection.

Land use for wind energy is a concern. A very large amount of land, several percent of the U. S. surface area, is required for large-scale energy extraction. The aesthetic impact of many wind machines is difficult to judge. Scenic and recreational areas would be avoided. Other low population areas that are infrequently visited may present no problem. It is possible that some areas can be used for both wind energy and agricultural purposes with little interference between the two uses.

Development of wind energy involves the atmospheric scientist in many ways. Atmospheric scientists can contribute by answering important questions that must be resolved before and during wind energy development. Questions of importance include the amount of wind energy that is available in a practical sense and the meteorological perturbations that might result at different levels of wind energy extraction.

Areas of high wind energy need to be identified and their energy resource correlated with regional energy needs. Wind conditions that are unacceptable in terms of engineering and systems requirements must be identified and reliably avoided. Finally, procedures must be established to adequately document the available wind energy to support engineering and systems requirements.

Owing to the broad scope of the questions that must be addressed, atmospheric scientists involved in analytical and experimental studies ranging from climatology to micrometeorology can make important contributions. Steady wind conditions and unsteady meteorological situations, including extreme winds, will require documentation in wind energy development areas.

Improved methods of regional wind energy assessment should make use of atmospheric simulations, remote sensing, and new instrumentation developed to evaluate wind energy. Present anemometers and towers are too expensive for deployment in large numbers. New instruments to measure the limited range of wind speeds important for wind energy are needed. Brief high winds and the sequence and durations of calms are important parameters that might be measured directly by special instruments. Such wind energy measurements will be needed to fully establish the wind energy potential of various regions. Measurements at potential sites are needed to define the range of operational conditions for WTGs. Optimal selection of areas where large amounts of wind energy can be collected will minimize energy costs and maximize the usefulness of this intermittent energy resource.

18-4 OCEAN ENERGY CONVERSION

18-4.1 Ocean Energy Resource

Energy is found in the oceans in various forms: the kinetic energy of ocean currents, kinetic and potential energy of wave motion, the internal energy or heat stored by temperature gradients, the osmotic pressure of salinity gradients, and the potential and kinetic energy of the tides. In addition, wind energy and biomass conversion can be tapped at sea as well as on land. A summary of the state of technological development and environmental considerations of various methods of oceanic energy conversion is given in Table 18.1.

18-4.1.1 Energy from Ocean Currents. The kinetic energy of the ocean currents can be tapped as a usable source of power. For instance, the Florida Current has surface and near-surface velocities of 4 knots and carries 50 times the total flow of all the rivers of the world. The kinetic energy of the Florida Current is approximately 100 times the present electrical output of the United States. Energy-conversion techniques suggested include a system of parachute drogues (Steelman, 1974) and honeycomb arrays of turbines (Fisher, 1975). There appear to be no insurmountable engineering obstacles to tapping this energy source, and the costs may be comparable to those of other proposed energy systems for the 1980s (Soucie, 1975; Fisher, 1975). The low energy density of the current (equivalent to about a 10-in. head of water) requires large bulky turbines that are expensive both to build and to maintain. In addition, techniques must be developed to handle velocity fluctuations and the various east-west shifts of the current axis during the year.

The ecological aspects of the problem are complex and unevaluated at the present time. If a significant amount of the energy of the Florida Current were removed, then the total circulation of the North Atlantic Subtropical Gyre would be affected and might seriously disrupt climatic conditions to the north and east, perhaps making parts of Europe colder. For this reason an arbitrary limit of 2000 MW (the equivalent of two commercial nuclear power plants) has been suggested as a maximum amount for extraction from the Florida Current (Fisher, 1975).

Other oceanic currents with velocities high enough for consideration are the western boundary currents and the Cromwell Current. It appears that energy extraction from ocean currents will only be feasible at a limited number of locations; thus it will make only a small contribution as a future alternative energy source.

18-4.1.2 Energy from Salinity Gradients. Potential energy of the salinity differences produced by the separation of fresh water from salt water is represented by an osmotic pressure. This energy can be extracted by using the osmotic pressure to drive an "osmotic pump" and using the head established to drive a turbine (Levenspiel and deNevers, 1974). It is possible to develop an osmotic pressure of 2.3×10^6 Pa (23 atm) (equivalent to a 700-ft head of water). Although the device is thermodynamically feasible, its actual use also requires a technological breakthrough in the development of semipermeable membranes.

A reverse electrodialysis battery (power generated by alternate cells of fresh water and salt water separated by semipermeable membranes) has been developed, but it is only a small power device. Larger, more efficient, and less expensive semipermeable membranes must be developed before this source of energy is feasible on a large-scale basis (Wick and Isaacs, 1976).

Another approach was taken by the Lockheed Electronic Company, which has developed and successfully tested a battery using seawater as its electrolyte. Like the reverse electrodialysis battery, this one is also a very small power device (Charlier, 1969).

Although there is little hope for exploiting this energy source in the near future, within the distribution of salts in the oceans there clearly lies a vast resource awaiting the development of technology to unlock its potential.

18-4.1.3 Energy from the Tides. For ages, man has used energy from the tides to do his work. The principle of operation is simply to control the filling and emptying of a dammed coastal bay or estuary during tidal periods, causing the flow to drive a turbine. Various flow schemes have been devised to extract the energy and to allow the energy usage to be decoupled from the tidal cycles.

Two tidal power plants have been built and are in operation today: a 240-MW plant on the Rance River in France and a 0.4-MW experimental plant at Kislaya Guba in the Soviet Union. A tidal power plant has been proposed for construction at Passamaquoddy on the Bay of Fundy in the United States but has not received funding.

There are perhaps 100 sites around the world which are suitable for tidal power plants. One difficulty is that many of these sites are remote from users, causing transmission costs to be high. In fact, high capital cost has been the major factor against construction of more tidal plants.

The environmental impact of a tidal power plant is caused by a barrage across the entrance to a bay or estuary. The barrage acts as a barrier to migration by estuarine organisms. In addition, the controls placed on river inflow and outflow can drastically change sedimentation and sediment transport patterns.

18-4.1.4 Energy from Ocean Waves. Another possible source of energy in the ocean is the motion of ocean waves. There are problems involved in recovering the energy found in ocean waves because the waves are broadband in frequency, low in energy density, and come from a variety of directions.

Several proposals are being given serious consideration. The British Government is particularly interested in wave energy since it estimates that about 15% of the national electrical energy requirement can be met by wave energy by the year 2000 (Pierson, 1976). In particular, the British are funding the development of a linear array of rotating wave absorbers developed by Professor Salter of Edinburgh University (Salter, 1974; Richards, 1976) and a wave-contouring raft design by Wooley and Platts (1975). A point converter has been developed by Scripps Institution of Oceanography and is known as a "wave pump" (Isaacs, Wick, and Schmitt, 1976). The wave pump is detuned from the ocean wave spectra and can thus respond to all frequencies present, and, being a point converter, it can respond to waves from all directions.

Since wave energy density is increased near coastlines, these are logical places to locate wave energy converters. The environmental impacts that could occur involve the elimination or modification of wave-induced long-shore currents, changes in sediment transfer (particularly in wave shadow zones), and modifications to beaches, which are dependent on wave processes.

18-4.1.5 Ocean Thermal Energy Conversion Technology. Over much of the ocean, a layer of warm surface water heated by solar radiation overlays a layer of colder, more dense water which was formed near the poles and flows toward the equator underneath the surface layer. Over 70% of the surface of the world is covered by the oceans, and approximately 90% of the area between latitudes 25°N and 25°S is ocean. Thus the oceans serve as a large natural solar collector and energy storage system. Oceanic temperature gradients are very stable and can lead to temperature differences as large as 20°C near many populated areas. Our national oceanic energy effort has focused on the development of the capability to exploit this temperature difference. Charts showing the extent of the resource are shown in Fig. 18.9.

In 1881, Jacques d'Arsonval, a 30-yr-old French physicist, proposed that this temperature difference could be exploited as a means of producing energy (d'Arsonval, 1881). The operating principle was demonstrated by Georges Claude in 1928 when he drove a 50-kW turbine using the 20°C temperature difference between the outflow of cooling water from a blast furnace and the River Meuse. During 1929 and 1930, Claude was able to produce 22 kW from a shore-based demonstration plant near Matanzas, Cuba, where a 14°C temperature difference existed. In 1934 he operated a floating plant built aboard a freighter off the coast of Brazil. After World War II, the French government financed a 7-MW thermal plant at Abidjan on the Ivory Coast of Western Africa, but this plant was never placed into operation. There have been other attempts to harness thermal sea power at Brest, Toulon, and Dakar (Dugger, 1975; Griffin, 1975).

Claude's design was an open Rankine cycle using seawater as a working fluid. Later analysis found many problems with the design (Griffin, 1975; Heronemus and McGowan, 1975), such as:

1. The shore-based plant required a 2-km-long cold intake pipe that had to withstand the pounding of the surf. In addition, heat flow into the extremely long pipe reduced the temperature difference that was available for energy conversion.
2. Using seawater as a working fluid caused secondary-system components to be subjected to corrosion.
3. The use of a low-pressure water-vapor system in the region where the specific volume is high required a large turbine (8 m in diameter) that had low efficiency at low Reynolds numbers.
4. The high vacuum requirements of such a system meant that the dissolved gases had to be continually removed. An additional problem was experienced in maintaining large leaktight connections.

The obvious solution to the first problem is to locate the plant offshore on a floating structure where the length of the intake pipe is minimized (by making it vertical) and where the plant is removed from the destructive force of pounding surf. The other problems are solved by going to a closed Rankine cycle with a suitable working fluid (Fig. 18.10). Twelve such fluids with boiling points at temperatures and pressures which are obtainable within the ocean thermal energy conversion (OTEC) cycle have been identified. Propane and ammonia are the two working fluids that have received the most attention. The use of

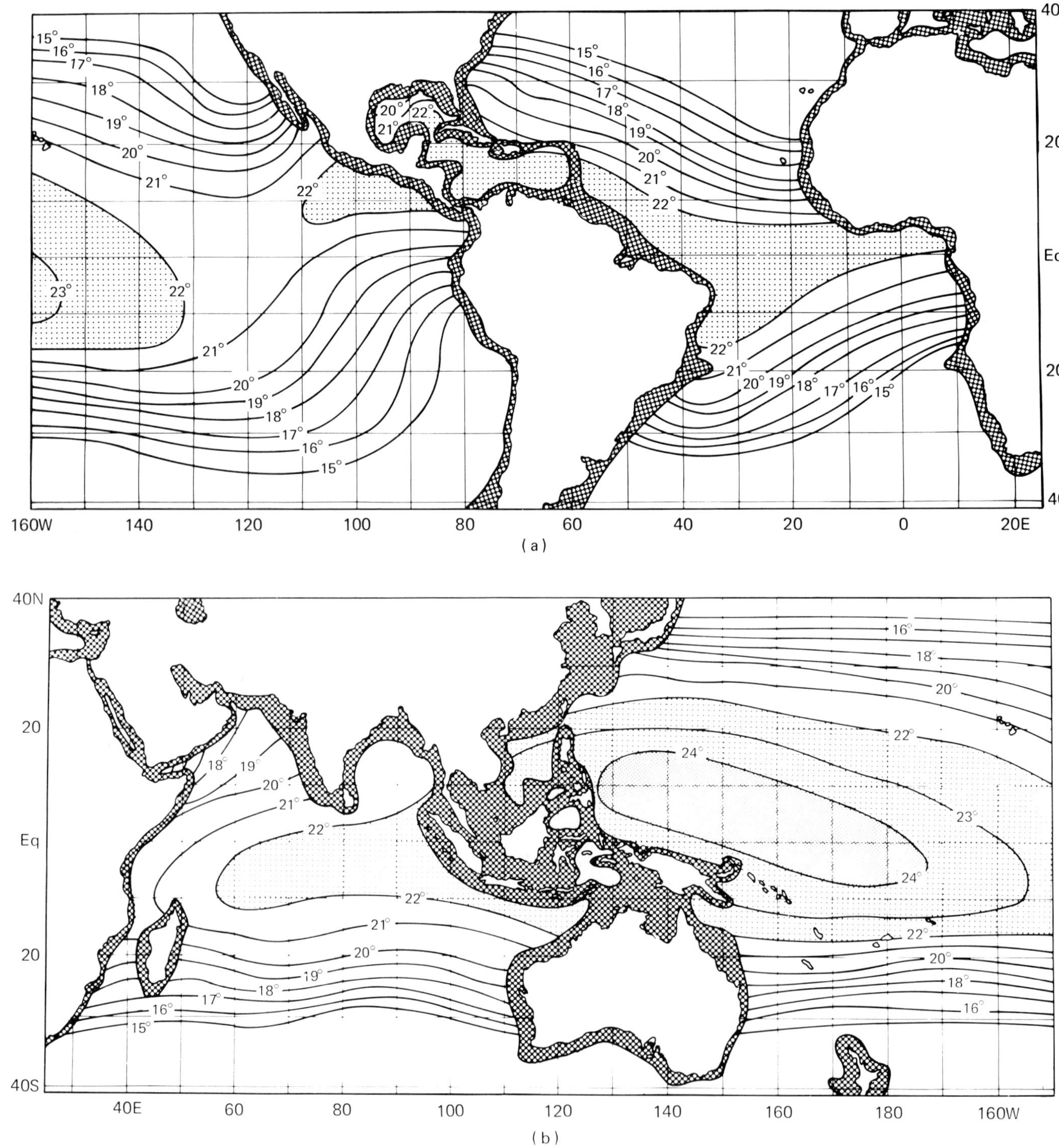

Fig. 18.9 Temperature difference (°K) between the ocean surface and a depth of 1000 m. Temperature differences greater than 22°K are desirable for OTEC plants. The dark shading indicates that the depth is less than 1000 m.

propane would reduce the turbine diameter from 7 m for water vapor down to 1.2 m for a 20-MW turbine. In addition, by locating the boiler at a depth of 85 m, the vapor pressure of propane on one side of the heat-exchanger wall would be balanced by the hydrostatic pressure on the other side and would thus permit the use of thinner walls, which in turn allows an increase in heat-transfer efficiency. A similar pressure balance would be achieved by locating the condenser at a depth of 47 m (Heronemus and McGowan, 1975).

Fossil-fueled plants can obtain efficiencies in the 30 to 40% range, but the small ΔT for sea thermal plants limits their efficiency to 3% or less. Such plants require 10 to 15 times as much heat as a fossil-fueled plant to obtain the same amount of power; thus large amounts of warm water must be circulated to extract the energy. However, because

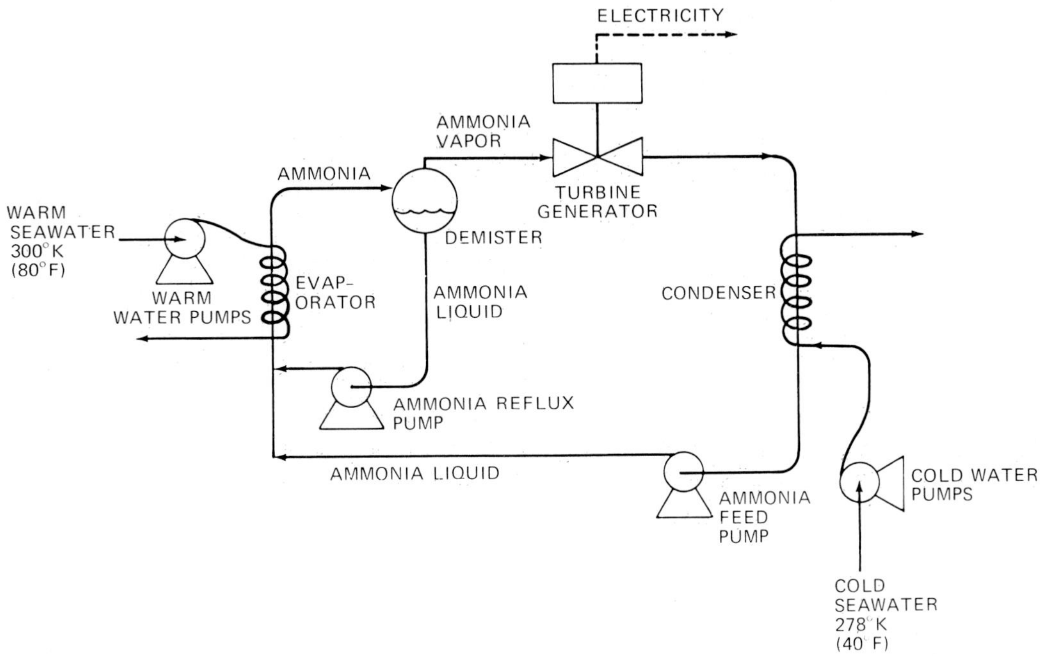

Fig. 18.10 Schematic diagram of an OTEC plant using a closed Rankine cycle. (From U. S. Department of Energy, 1980, p. 303.)

the fuel is free, high efficiency is not a prime requirement. What is important is the cost per unit power output (or perhaps the cost per kilowatt-hour), which in this case is largely determined by capital costs. Although the low efficiency does mean that large heat exchangers are required, they do not have to withstand the high pressure and temperature environment of a fossil-fueled plant, and cost trade-offs come into play.

Recent efforts have considered both the controlled flash evaporation open cycle and the closed Rankine cycle, which is favored by the preponderance of present thinking. Basic research has been accomplished at Sea Solar Power, Inc., Carnegie–Mellon University, and the University of Massachusetts. Technical and economic assessments were conducted by two industrial teams: (1) Lockheed Ocean Systems, Bechtel Corp., and T. Y. Lin International and (2) TRW, Global Marine Development, Inc., and United Engineers and Constructors (McGowan and Heronemus, 1975). The findings are that OTEC is feasible and that the basic technology to build such plants is currently available. Various designs of OTEC power plants have been proposed; see Smith (1974), Isaacs and Seymour (1973), Griffin (1975), and Zener (1976).

These findings do not mean that there are no problems left to solve, but only that the problems are of a design and development nature rather than of basic research needed to advance technology. In fact, it appears that more efficient heat-exchanger designs coupled with the ability for mass production are required before OTEC will be economically competitive with conventional power sources. Recent advances in heat-transfer surface design, including the use of less expensive materials, and the attention being given to eliminating biofouling and corrosion of the heat exchangers make the future look promising. The DOE plans include in situ tests of various heat-exchanger designs in the near future.

Estimates of the size of OTEC pilot plants have been based on the limitations caused by engineering constraints. The required efficiency for a closed-cycle plant would put the minimum size between 10 and 25 MW, although 400 MW is considered to be the maximum size feasible due to the size of the required heat exchangers and cold water supply tubes.

Evaluation of the economics of OTEC requires that additional consideration be given to various by-products that can be produced at competitive costs. These include freshwater production, recuperation of salt and chemicals (sodium sulfate, chlorine, hydrochloric acid, etc.), the harvesting of fish and plankton as a food supply and for the production of fertilizer, and the availability of cold water for local air conditioning and refrigeration. Ammonia can be produced using hydrogen from the hydrolysis of seawater and nitrogen from the air. If the ammonia were used in fertilizer production, the process would be cheaper than the processes currently in use, and it has the additional advantage of freeing the world from the dependence on petroleum-based fertilizers (Isaacs and Seymour, 1973).

Another benefit of OTEC development is that, as mass production lowers heat-exchanger costs, the concept can be used as a "bottom cycle" for conventional land-based plants. After energy is extracted from the working fluid of a conventional plant, waste heat must be removed in order to condense the fluid back to the liquid phase so that it can be efficiently pumped through the system again. The heat is transferred to cooling water, which then flows back into the body of water from which it came (e.g., river or bay), or it is cooled by means of a cooling-tower complex. The bottom-cycle concept is to run an OTEC-type system using the heated cooling water and intake cooling water. Thus "waste heat" would be removed from the cooling water and converted into electrical output. Such a system would have the advantages both of extracting more useful energy per kilowatt-hour of fuel and of lowering the thermal pollution in the waterways that supply plant cooling.

18-4.2 Potential Environmental Impact

There are environmental questions that must be answered before the deployment of large OTEC systems. Specifically the deployment of such a system could cause a slight lowering of the sea surface temperature. The climatic consequences of this alteration need quantification and assessment on both a regional and a global basis. For example, if one tapped the 20 million MW of thermal power estimated to be available in the Florida Current, would the climate of Western Europe be altered? Other environmental aspects that require investigation include the ecological aspects of changing salinity and temperature distributions coupled with the effects on density structure, the use of antifouling agents required in the heat exchangers, and the effects of any of the OTEC structural material on the environment (e.g., copper from Cu–Ni heat exchangers). In addition, the raising of large amounts of nutrient-rich deep water to the euphotic zone will effectively allow massive plankton blooms. Although these blooms could be used to develop food sources (as mentioned above) and/or biomass for ultimate energy conversion, the overall ecological effect of such a process must be evaluated. Considering that a massive bloom might alter the albedo of the sea surface and thus the ocean energy balance, we are led back to the requirement for an assessment of climatic effects. Recently the amount of CO_2 which could be released to the atmosphere by bringing large amounts of deep water near the air–sea interface has become an additional concern of atmospheric scientists.

On a more localized level, the choice of the working fluid could cause some environmental and/or occupational problems. Specifically let us consider a leak of ammonia to the sea. Since ammonia is biodegradable, rapidly taken up by many species of phytoplankton, and in fact a naturally occurring step in the nitrogen cycle, no long-range effects would be caused. On the other hand, consider the problems a leak of this toxic gas to the confined atmosphere of the crew manning the OTEC station would cause. Propane has just the opposite problems. Although not toxic in air, propane is not naturally degradable in seawater. These considerations must enter into the decision-making process when the selection of a system design is made.

The combined and coordinated efforts of researchers with backgrounds in many disciplines will be required in the future to answer these environmental questions. Oceanographers, atmospheric scientists, air–sea interaction specialists, biologists, marine engineers, and industrial engineers are all concerned with some aspect of OTEC's environmental impact. As future development occurs and in situ testing begins, information on these effects will be collected. This information will have to be used as input to model calculations in an effort to predict what the conditions would be if large numbers of stations were deployed. From these results, decisions can be made concerning the maximum density of OTEC stations and the total number of OTEC stations permitted as man strives for a balanced development of this alternative energy resource.

18-5 GEOTHERMAL ENERGY

18-5.1 Geothermal Resource Estimates and Uses

The earth is an enormous reservoir of energy in the form of heat. Most of this energy, however, is contained within the earth's core and mantle at depths too great to be accessible by present drilling technology. The energy stored near the surface (upper 10 km) is still vast but is generally too diffuse to be a potential energy resource. However, anomalies do occur near the surface in localized deposits where heat flow is significantly increased. These deposits are generally associated with tectonic boundaries, volcanic activity, and the deep parts of sedimentary basins. Thus many anomalies are found on the rim of the Pacific Ocean and along the mid-oceanic ridges. In the United States, the primary geothermal areas are found in the western states.

Geothermal resources can be classified into four types: vapor-dominated systems, hot water systems, hot igneous systems, and geopressured systems. The vapor-dominated systems that are of sufficient size to be developed economically exist in only a few places. Wells drilled into these systems produce saturated steam with no associated liquid. This steam can be used for direct conversion to electricity. At the present time a large fraction of the world's total geothermal power production is obtained from these rare vapor-dominated reservoirs (e.g., at The Geysers in California and at Larderello, Italy). The much more predominant hot water systems are characterized by liquid water, of varying temperature and salinity, as the convective working fluid. When the hot water in such a reservoir comes to the surface at the wellhead, the pressure is reduced, and some of the water flashes into steam. This steam can then be used for power production. There are numerous reservoirs of this type throughout the world. The principal ones in the United States are situated in the Imperial Valley, Long Valley, and Coso Hot Springs in California; Roosevelt Hot Springs, Utah; Raft River, Idaho; Valles Caldera, New Mexico; and Puna, Hawaii. Other reservoirs are located in New Zealand, Mexico, Japan, Iceland, Philippines, Turkey, Chile, El Salvador, and the USSR, to name a few. The hot igneous systems, consisting of magma and hot rock, contain vast quantities of stored heat. Recent experiments indicate that it is technologically feasible to extract this heat by fracturing large amounts of bedrock and forcing cold water down through it to carry the heat to the surface. Geopressured reservoirs are found principally in the north Gulf of Mexico basin. The reservoirs contain hot water at pressures exceeding normal hydrostatic pressures. In addition, considerable amounts of methane are dissolved in the hot water and may also be useful for energy production. At the present time no geopressured reservoir has been used for commercial power production.

Estimates of the U. S. geothermal resources vary widely due largely to an inadequate understanding of the nature and extent of the reservoirs (White, 1965; Grossling, 1972; Rex and Howell, 1973). A summary of estimates reported by White and Williams (1975) is given in Table 18.2 as a function of resource type.

TABLE 18.2 Estimated Heat Content of Geothermal Resource Base of the United States*

Type of resource	Estimated heat content, MW·centuries: Identified resources	Identified and undiscovered resources
Vapor-dominated systems	3.5×10^{10}	7×10^4
Hot water systems	1×10^6	4×10^6
Hot igneous systems (magma and hot rock)	3×10^7	1×10^8
Geopressured systems	1.5×10^7 †	6×10^7

*From White and Williams, 1975.

†Does not include the 8×10^6 MW·centuries derived from the dissolved methane.

TABLE 18.3 Total Atmospheric Emissions of Several Chemical Species Resulting from the Generation of 502 MW at The Geysers*

Chemical species	Emissions: g/kW·h	kg/d
H_2S	1.8	2.1×10^4
CO_2	26.0	3.1×10^5
NH_3	1.5	1.9×10^4
CH_4	1.5	1.9×10^4
H_2	0.44	5.4×10^3
N_2	0.41	5.1×10^3
As	1.9×10^{-6}	0.02
B	4.0×10^{-3}	48
Hg	1.2×10^{-4}†	0.4†
H_2O	6.7×10^3	8.1×10^7
Rn	7.5×10^{-13}	9.0×10^{-9}

*From Pacific Gas & Electric Co., 1975; and Anspaugh, 1976.

†Extrapolated from measurements on five generating units. [From D. E. Robertson et al., Mercury Emissions from Geothermal Power Plants, *Science*, 196: 1094-1097 (1977).]

For comparison, the total energy demand by the state of California during 1971 was approximately 2×10^3 MW·centuries (Stanford Research Institute, 1973). From the data in Table 18.3, it is evident that the vapor-dominated systems are much less abundant than the hot water systems. According to White and Williams (1975), only three vapor-dominated systems have been discovered in the United States: at The Geysers in California, Yellowstone National Park, and Mt. Lassen Volcanic National Park. The hot water systems, with fluids between 90 and 360°C, have been discovered at 287 sites throughout the western United States, Alaska, and Hawaii. The salinity of these systems ranges from very low to brines of extreme salinity. The most common range is from 0.1 to 1% dissolved salts; however, geothermal brines from California's Imperial Valley may contain as much as 20 to 30% dissolved salts. The estimated heat content of the igneous-related geothermal systems in Table 18.2 is highly speculative but is assumed to be enormous. Forty-eight systems of this type have been identified in the western United States, Alaska, and Hawaii. Vast quantities of heat are also stored in the geopressured zones in the northern part of the Gulf of Mexico basin. In addition to the geothermal energy, the geopressured systems can also provide energy from the dissolved methane and the abnormally high fluid pressures.

For increased geothermal energy production, it will be necessary to assess further the long-term potential of geothermal reservoirs, to improve or develop the energy extraction and conversion technology for the various types of resources, and to resolve the multitude of economic, institutional, and environmental issues. The primary goal set by Project Independence (Federal Energy Administration, 1974) is to encourage industry to augment the commercial production of electric power by 20 000 to 30 000 MW by 1985 and by 100 000 to 200 000 MW between 1990 and 2000. The goal set by DOE is to increase electric- and nonelectric-power production to 10 000 to 15 000 MW by 1985. However, more realistic forecasts set the goal at 2000 to 4000 MW by 1985 (Kruger, 1976).

Since vapor-dominated systems are the simplest to develop, the geothermal industry in the United States has primarily focused on this type of reservoir. At the present time, 502 MW is being generated at The Geysers, with a planned expansion to 1400 MW by the mid-1980s. Likewise the Italians are generating 365 MW from the vapor-dominated field at Larderello. Because of the relative abundance of hot water reservoirs, the industry is currently evaluating these reservoirs more extensively for temperature, pressure, and salinity. Investigations in the United States are under way to develop selected areas, such as Roosevelt Hot Springs, Utah, and Imperial Valley, Calif., for power generation. The high salinity brines encountered in the Imperial Valley pose the special engineering problems of scaling and corrosion. There are presently no power plants in the United States that use hot water reservoirs, but power is being generated at Wairakei, New Zealand (192 MW) and Cerro Prieto, Mexico (75 MW) from such systems.

Geothermal energy is also used for other purposes throughout the world. About 6000 MW of thermal power is used for space heating and process heat, primarily in the USSR (Anspaugh, 1976).

18-5.2 Technology

The technology for using dry superheated steam, characteristic of The Geysers and Larderello vapor-dominated fields, is well developed and analogous to that used in fossil-fueled power plants, where the steam is delivered directly to a turbine. After the geothermal steam leaves the turbine, it is condensed and fed to cooling towers or disposed of by reinjection. This is shown schematically in Fig. 18.11. Because of the characteristically lower temperatures and pressures of geothermal steam, the efficiency of geothermal power plants is generally about 10 to 15%, which is considerably lower than the near 40% efficiency for fossil-fueled plants. A detailed description of this type of plant is given by Finney (1973).

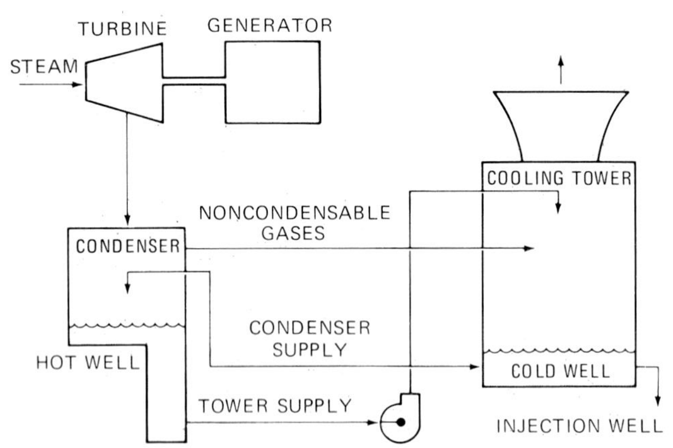

Fig. 18.11 A schematic flow diagram of a power plant at The Geysers. Steam is delivered directly to the turbine that drives the generator. After the steam leaves the turbine, it is condensed and fed to the cooling tower. Most of the fluid evaporates but any remaining collects in the bottom of the cooling tower where it can be recirculated to the condenser or reinjected. The noncondensable gases are sent directly to the cooling tower.

Power production from hot water reservoirs is often obtained by partially decompressing the water when it reaches the surface, causing part of it to flash into steam. The steam is then separated and used in a manner similar to that from vapor-dominated fields. The plants at Cerro Prieto, Mexico, and Wairakei, New Zealand, operate on this concept. Another technique for using the hot water reservoirs is the binary cycle (Anderson, 1973). This technique, illustrated in Fig. 18.12, involves passing the hot water through a heat exchanger to vaporize a low boiling secondary fluid, such as isobutane or Freon. The vapor is then allowed to expand through a turbine to drive the electrical generator. The geothermal water, after having lost most of its heat energy, can be disposed of by reinjection. This type of power plant has not been commercialized. However, a small unit (0.68 MW) of this design has operated in the USSR. A technique presently under development is the total flow concept for extracting the energy from hot water reservoirs, including high salinity brines. This process allows the total flow, water and steam, to be passed through a mixed-phase expander that converts the fluid's thermal energy to kinetic energy in the form of high-velocity jets driving a specially designed turbine. Theoretically this method produces about 60% more power than the other techniques (Austin, Higgins, and Howard, 1973).

Recent experiments at Valles Caldera, N. Mex., have been performed to extract energy from hot dry rock (igneous) reservoirs by fracturing large amounts of the rock by hydrofracturing. Heat can then be extracted from the fractured zone by forcing cold water down through it, thus creating steam that is piped to a turbine generator. The condensed water exhausted by the turbine can then be

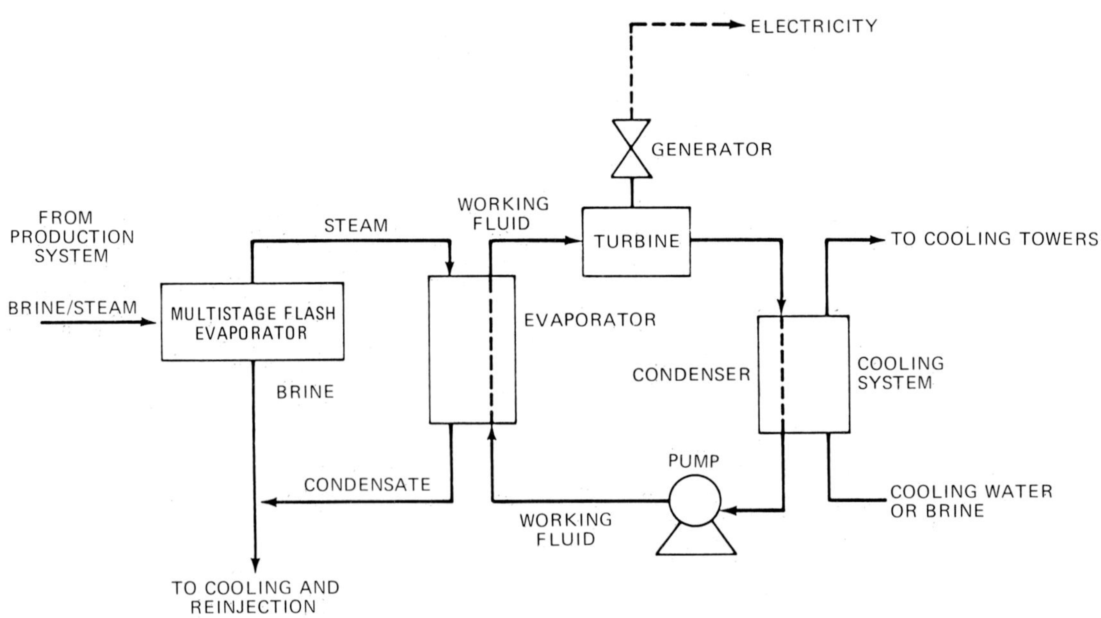

Fig. 18.12 A schematic flow diagram of a binary-cycle power plant. Geothermal fluid is pumped to the surface where it passes through a heat exchanger to vaporize a secondary fluid, such as isobutane. The cooled water is reinjected into a nearby well. The isobutane expands through a turbine that drives the generator. The exhaust gas is recondensed to the liquid state in a water-cooled condenser and returned to the heat exchanger. (From U. S. Department of Energy, 1980, p. 300.)

reinjected back into the fractured zone for additional heat extraction (Smith, 1973). Energy can be extracted from geopressured reservoirs by a combination of techniques (House, Johnson, and Towse, 1975). A significant fraction of the dissolved gases will be released by the pressure drop between the revervoir and the wellhead. Therefore the first step will be to separate the gases, primarily methane, for subsequent power or heat generation. The excess pressure of the fluid can be used by delivering the hot water directly to a hydraulic turbine linked to an electrical generator. The drop in pressure through the turbine would allow additional gas to be released. This gas can then be separated and combined with the first fraction. The thermal energy of the residual water can be extracted by the flashed steam, binary cycle, or total flow systems described previously. This type of extraction technique has not yet been demonstrated.

18-5.3 Potential Environmental Impact

One of the principal advantages of geothermally produced power over other methods of power production is that the environmental impacts are generally restricted to the immediate vicinity of the power plants because it is not economically feasible to pipe steam or hot water over distances of more than 2 to 5 km. Hence the geothermal "fuel cycle" is short when it is compared with other schemes since there is no need for mining, refining, enrichment, fuel reprocessing, etc. There are, however, other environmental considerations that must be examined before a geothermal resource area can be developed. The more important of these are:

1. Most geothermal development schemes will release noncondensable gases to the atmosphere with attendant potential for air-quality degradation in the region under development. Furthermore, significant quantities of water vapor and waste heat are also released to the atmosphere.
2. Land subsidence may occur where large amounts of water are withdrawn over long periods of time from geothermal reservoirs. Subsidence has occurred at Wairakei, New Zealand. Reinjection of the geothermal fluid into the reservoir can minimize or even circumvent this effect.
3. The steam and hot water from many reservoirs contain variable amounts of chemicals (for example, boron) which can cause environmental damage if they are emitted from cooling towers or discharged into streams. Again, reinjection of the fluid can minimize this impact.
4. Earthquakes may possibly be induced by high-pressure injection of geothermal fluids back into the reservoir, especially where seismically active faults are present.
5. The drilling of steam wells and the operation of power plants are inherently noisy and may affect the work force and nearby populations.

Items 2 to 5 will not be treated here because of the emphasis in this chapter on the potential impact of geothermal operations on air quality. The principal sources of pollutants emitted to the atmosphere from geothermal power plants are the noncondensable gases that are present in geothermal fluids. The fraction of noncondensable gas in the fluid is highly variable, from a minimum of less than 1% to as much as 30%. The main component of the noncondensable fraction is carbon dioxide, but lesser amounts of other gases are also present. As an example, Table 18.3 lists some of these gases and the quantities emitted by power plants at The Geysers. These gases are mostly released from the cooling towers. Hydrogen sulfide is the most objectionable pollutant because of its noxious odor and the sizable quantities released and hence presents the most serious problem. At The Geysers the state of California ambient air-quality standard of 30 ppbv for hydrogen sulfide is exceeded periodically. Thus emission rates from the existing plants, as well as those anticipated from future plants, will play a key role in the rate of expansion of the present generating capacity. Hydrogen sulfide and ammonia are eventually oxidized in the atmosphere at uncertain rates. Carbon dioxide and water vapor, of course, are the main components. Carbon dioxide is relatively harmless, and its main impact is to add to the global inventory of carbon dioxide. However, the contributions by geothermal emissions is essentially insignificant in comparison with other sources of carbon dioxide emissions. The release of large quantities of water vapor can, under certain meteorological conditions, produce fogging in the immediate vicinity of the plant. In addition, trace quantities of B, Na, Mg, Al, Si, Ca, Ti, Cr, Mn, Fe, Ni, Cu, Zn, As, Sr, Hg, and Pb have been detected in the drift from the cooling towers (Pacific Gas & Electric Co., 1975). Of these constituents, B and Hg appear to be of the greatest environmental concern.

Direct venting of the steam at the wellhead does occur in some operations, such as The Geysers, whenever a power plant is temporarily shut down. This causes a noticeable increase in the rate of discharge of the various chemical species normally released to the atmosphere. Such steam ventings, however, are rapidly being curtailed to reduce the environmental impact.

Several methods are currently under investigation for reducing the H_2S emissions from geothermal power plants. These involve oxidation by a variety of chemicals, such as SO_2, H_2O_2, O_3, $Fe_2(SO_4)_3$, and $CuSO_4$ (Stanford Research Institute, 1976). On the basis of laboratory studies, it appears that a reduction of about 90% in the H_2S emissions can be achieved by these means.

Since large-scale use of a geothermal resource can affect the air quality in the region to be developed, it is vital to make an assessment of this impact before development occurs. This assessment will permit one to distinguish between pollutants existing before development and those arising from geothermal operations. The scope of such an assessment should address problems specific to the area under development, for instance, its geographical size, extent and rate of development, nature of the geothermal reservoir, the amount and composition of the noncondensable gases in the geothermal fluids, types of power plants to be installed, abatement measures, and the complexity of terrain and wind flow patterns. In general, the assessment objectives should be to:

1. Perform baseline measurements of ambient air concentrations of selected air pollutants, such as H_2S and SO_2, throughout the region of interest before development.
2. Characterize the atmospheric transport on a regional basis by making appropriate meteorological measurements at specific sites.
3. Characterize the concentrations of specific pollutants in the vicinity of specialized sources connected with geothermal operations as well as other activities, either natural or man-made, pertinent to the development area.
4. Measure the concentrations of noncondensable gases in the geothermal fluids obtained from each distinct geothermal resource area for evaluating the potential rate of emissions from future geothermal power plants.
5. Use the above data in atmospheric transport models to predict the potential change in air quality due to the planned geothermal plants.

The baseline air-quality and meteorological measurements are usually obtained by installing a network of stationary monitoring stations throughout the area under development. This network is often supplemented by a mobile air-quality laboratory for measurements in the vicinity of existing pollutant sources or other areas of specific interest. The air-quality measurements should include H_2S and, preferably, SO_2, O_3, and total sulfur concentrations because of the atmospheric oxidation of H_2S to SO_2 and subsequently to sulfate. Additional ambient air measurements should include NH_3, Hg, and Rn because of their likely presence in the noncondensable gases. Particle concentrations as a function of size should be measured to evaluate ambient elemental and sulfate concentrations. The meteorological parameters of primary interest for estimating the extent to which pollutants are dispersed in the atmosphere are wind speed and direction, atmospheric stability, and mixing height. Temperature and relative humidity measurements should also be included.

Development of a predictive capability for estimating the impact of the geothermal development on air quality requires the use of modeling techniques. Numerous atmospheric transport models are available that range from the simple Gaussian solution of the transport equation to complex, three-dimensional, numerical advection-diffusion models. However, it is not the subject of this discussion to list the multitude of available models and their capabilities and limitations; the reader is referred to Chaps. 13 and 14 for this information. Selection of a particular model depends on a variety of factors, such as the complexity of the topography and wind flow patterns and the amount and type of meteorological data available. The meteorological data base should enable the atmospheric scientist to evaluate the flow patterns over the region that may be impacted by the geothermal development. Estimates of the noncondensable gas emissions from existing or future power plants as a function of height of release, temporal variability, and geographical location in conjunction with wind flow patterns can be used as input for the chosen transport model to predict the concentrations of the species of interest on a regional scale. The calculations should include the effects of diffusion, advection, surface deposition, chemical transformations (such as the oxidation of H_2S to SO_2 and, possibly, to sulfate), gravitational settling of particulates, and plume rise due to buoyancy. The results should have temporal resolutions consistent with appropriate air-quality standards ranging from 1 h to 1 yr. These calculated concentrations can then be related directly to the baseline air-quality measurements made before development to assess the potential change.

18-6 CONCLUSION

During the coming decade this nation will in all likelihood be implementing a national energy policy oriented toward reducing our dependence on imported oil, increasing the use of more indigenous energy resources, and transitioning from expendable energy resources to long-lasting or inexhaustible resources. The alternative energy resources discussed here are options that may well be implemented in the course of this important movement. Energy from the wind, solar energy flows, geothermal, and that obtained from the ocean by various means are frequently called benign technologies. A benign technology is presumably one for which the environmental impacts are believed to be small in comparison to those impacts associated with more conventional energy resources. However, as mentioned in this chapter, the tapping of the natural fluxes of energy flow in the planetary boundary layer for wind energy, at the earth's surface for solar, and from the ocean thermal structure or currents for ocean energy involves interference with some geophysical budgets whose regional responses must be evaluated before these so-called "benign energy technologies" can move ahead confidently to large-scale implementation and with the same environmental awareness and concern for impact acceptability now being focused on more conventional technologies.

REFERENCES

American Institute of Aeronautics and Astronautics, 1975, *Solar Energy for Earth*, New York, N. Y.

American Meteorological Society, 1977, *Proceedings of the Sixth Conference on Planned and Inadvertent Weather Modification*, Urbana, Ill., Oct. 10–13.

Anderson, J. H., 1973, The Vapor-Turbine Cycle for Geothermal Power Generation, in *Geothermal Energy*, pp. 163-175, P. Kruger and C. Otte (Eds.), Stanford University Press, Stanford.

Anspaugh, L. R. (Ed.), 1976, *Balanced Program Plan*, Vol. 9, Geothermal Energy: Analysis for Biomedical and Environmental Research, ERDA Report ERDA-76/116.

Austin, A. L., G. H. Higgins, and J. H. Howard, 1973, *The Total Flow Concept for Recovery of Energy from Geothermal Hot Brine Deposits*, USAEC Report UCRL-51366, Lawrence Livermore Laboratory, NTIS.

Baynton, H. W., C. L. Frush, R. J. Serafin, P. V. Hobbs, R. A. Houze, Jr. and J. E. Locatelli, 1976, Wind and Divergence

Measurements in Extratropical Cyclones from Doppler Radar, in *Proceedings of the Conference on Radar Meteorology, Seattle, Wash., Oct. 26–29, 1976,* American Meteorological Society, Boston, Mass.

Bean, B. R., R. E. McGavin, R. B. Chadwich, and B. D. Warner, 1971, Preliminary Results of Utilizing the High Resolution FM Radar as a Boundary-Layer Probe, *Boundary-Layer Meteorol.,* 1: 466-473.

Beran, D. W., 1971, Acoustics: A New Approach for Monitoring the Environment near Airports, *J. Aircr.,* 8: 934-936.

——, 1975, Wind Measurement with Acoustic Doppler, *Atmos. Technol.,* 6: 94-99.

——, and S. F. Clifford, 1972, Acoustic Doppler Measurements of the Total Wind Vector, in *Proceedings of the Second Symposium on Meteorological Observations and Instrumentation,* pp. 100-109, American Meteorological Society, Boston, Mass.

Blackwell, B. F., 1974, *The Vertical-Axis Wind Turbine "How It Works,"* USAEC Report SLA-74-0160, Sandia Laboratories, NTIS.

——, W. N. Sullivan, R. C. Reuter, and J. F. Banas, 1977, Engineering Development Status of the Darrieus Wind Turbine, *J. Energy,* 1: 50-64.

Brown, W. C., 1973, Satellite Power Stations: A New Source of Energy?, *IEEE Spectrum,* pp. 38-47 (March).

Charlier, R. H., 1969, Harnessing the Energy of the Oceans: A Review, *Mar. Technol. Soc. J.,* 3(4): 59-81.

Council on Environmental Quality, 1975, *Sixth Annual Report.*

d'Arsonval, J., 1881, Utilization des forces naturelles. Avenir de l'électricité, *La Rev. Sci.,* 17: 370-372.

Dugger, G. L. (Ed.), 1975, *Proceedings of the Third Workshop on Ocean Thermal Energy Conversion,* Houston, Tex., May 8–10, 1975, Applied Physics Laboratory, Johns Hopkins University, Report SR-75-2.

Federal Energy Administration, 1974, *Project Independence Blueprint, Final Task Force Report, Geothermal Energy.*

Finney, J. P., 1973, Design and Operation of The Geysers Power Plant, in *Geothermal Energy,* pp. 145-161, P. Kruger and C. Otte (Eds.), Stanford University Press, Stanford.

Fisher, A., 1975, Energy from the Sea . . . Waves, Tides, and Currents, *Pop. Sci.,* 124: 69-73.

Gervais, R. L., and P. B. Bos, 1975, Solar Thermal Electric Power, *Astronaut. Aeronaut.,* pp. 38-45 (November).

Glaser, P. E., 1968, Power from the Sun: Its Future, *Science,* 162: 857-861.

——, 1977, Solar Power from Satellites, *Phys. Today,* 30(2): 30-38.

Gray, G. R., R. J. Serafin, D. Atlas, R. E. Rinehart, and J. J. Boyajian, 1975, Real-Time Color Doppler Radar Display, *Bull. Am. Meteorol. Soc.,* 56: 580-588.

Griffin, O. M., 1975, The Ocean as a Renewable Source of Energy, *J. Eng. Ind.,* 97: 897-908 (February).

Grossling, B. F., 1972, An Appraisal of the Prospects of Geothermal Energy in the United States, *U. S. Energy Outlook,* Chap. 2, pp. 15-26, National Petroleum Council, Washington, D. C.

Gustavson, M. R., 1977, *Wind Energy Resource Parameters,* The MITRE Corp., Metrek Division, Report M77-29.

Hardy, D. M., 1977a, Wind Studies in Complex Terrain, in *Proceedings of the American Wind Energy Association Conference and Exposition,* May 11–14, 1977, available from the American Wind Energy Association, MucKwonago, Wis.

——, 1977b, Numerical and Measurement Methods of Wind Energy Assessment, in *Proceedings of the Third Biennial Conference on Wind Energy Conversion Systems,* Washington, D. C., GPO, September 1977.

Heronemus, W. E., and J. G. McGowan, 1975, Oceans Thermal Power and Wind Power Systems—Natural Solar Energy Conversion for Near-Term Impact on World Energy Markets, *Adv. Astronaut. Sci.,* 35(Suppl.): 491-506.

House, P. A., P. M. Johnson, and D. F. Towse, 1975, *Potential Power Generation and Gas Production from Gulf Coast Geopressure Reservoirs,* ERDA Report UCRL-51813, Lawrence Livermore Laboratory, NTIS.

Huffaker, R. M., 1975, CO_2 Laser Doppler Systems for the Measurement of Atmospheric Winds and Turbulence, *Atmos. Technol.,* 6: 71-76.

Isaacs, J. D., and R. J. Seymour, 1973, The Ocean as a Power Resource, *Int. J. Environ. Stud.,* 4: 201-205.

——, G. L. Wick, and W. R. Schmitt, 1976, Utilization of the Energy from Ocean Waves, in *Proceedings of the Wave and Salinity Gradient Energy Conversion Workshop,* Newark, Del., May 24–26, 1976, ERDA Report COO-2946-1, pp. F1-F36, NTIS.

Josephson, J., 1976, Energy: The U. S. at the Crossroads, *Environ. Sci. Technol.,* 10: 854-859.

Justus, C. G., W. R. Hargraves, and A. Yalcin, 1976, Nationwide Assessment of Potential Output from Wind-Powered Generators, *J. Appl. Meteorol.,* 15: 673-678.

Kaplan, G., 1975, For Solar Power: Sunny Days Ahead?, *IEEE Spectrum,* pp. 47-52 (December).

Kruger, P., 1976, Geothermal Energy Development, in *Proceedings of the First Workshop on Sampling Geothermal Effluents,* Report EPA-600/9-76-011, pp. 1-35, Environmental Protection Agency.

Lawrence Berkeley Laboratory, 1976, *Solar Energy Conversion,* input to the Balanced Program Plan of ERDA, Berkeley, Calif.

Lawrence, R. S., G. R. Ochs, and S. F. Clifford, 1972, Use of Scintillations to Measure Average Wind Across a Light Beam, *Appl. Opt.,* 11: 239-243.

Levenspiel, O., and N. deNevers, 1974, The Osmotic Pump, *Science,* 183: 157-160.

Lhermitte, R. M., 1970, Measurement of Wind and Wind Field by Microwave Doppler Radar Techniques, *Atmos. Environ.,* 5: 691-702.

——, 1975, Real-Time Monitoring of Convective Storm Processes by Dual-Doppler Radar, *Atmos. Technol.,* 6: 26-33

McGowan, J. G., and W. E. Heronemus, 1975, Ocean Thermal and Wind Power: Alternative Energy Sources Based on Natural Solar Collection, *Environ. Aff.,* 4: 629-660.

Meroney, R. N., V. A. Sandborn, R. J. B. Bouwmeester, and M. A. Rider, 1976, *Sites for Wind Power Installations: Wind Tunnel Simulation of the Influence of Two-Dimensional Ridges on Wind Speed and Turbulence, ERDA Report ERDA/NSF-00702/75/1,* NTIS.

Merriam, M. F., 1972, *Is There a Place for the Windmill in the Less Developed Countries,* Working Paper 20, East-West Technology and Development Institute, Honolulu, Hawaii.

Pacific Gas & Electric Co., 1975, *Amended Environmental Data Statement,* Geysers Unit 13, San Francisco, Calif.

Pierson, W. J., 1976, The Conference Session on the Resource Potential of Ocean Waves as a Source of Electrical Power, in *Proceedings of the Wave and Salinity Gradient Energy Conversion Workshop,* Newark, Del., May 24–26, 1976, ERDA Report COO-2946-1, pp. O1-O4, NTIS.

Reed, J. W., 1975, *Wind Power Climatology of the United States,* ERDA Report SAND-74-0348, Sandia Laboratories, NTIS.

Rex, R. W., and D. J. Howell, 1973, Assessment of U. S. Geothermal Resources, in *Geothermal Energy,* P. Kruger and C. Otte (Eds.), Stanford University Press, Stanford.

Richards, A. F., 1976, Extracting Energy from the Oceans: A Review, *Mar. Technol. Soc. J.,* 10(2): 5-24.

Robertson, D. E., E. A. Crecelius, J. S. Fruchter, and J. D. Ludwick, 1977, Mercury Emissions from Geothermal Power Plants, *Science,* 196: 1094-1097.

Salter, S. H., 1974, Wave Power, *Nature,* 249.

Sellers, W. D., 1969, *Physical Climatology,* University of Chicago Press, Chicago, p. 272.

Smith, F. G. W., 1974, Power from the Oceans, *Sea Front.,* 20(2): 87-99.

Smith, M. C., 1973, *Geothermal Energy,* USAEC Report LA-5289-MS, Los Alamos Scientific Laboratory, NTIS.

Soucie, G., 1975, Plugging in the Oceans for the Future Energy Needs, *Audubon*, pp. 115-122 (September).

Stanford Research Institute, 1976, *Environmental Analysis for Geothermal Energy Development in The Geysers Region*, Vol. 1, Project 5554.

——, 1973, *Meeting California's Energy Requirements, 1975–2000*, Project ECC-2355.

Steelman, G. E., 1974, An Invention Designed to Convert Ocean Currents into Usable Power, in *Proceedings of the MacArthur Workshop on the Feasibility of Extracting Usable Energy from the Florida Current*, Palm Beach Shores, Fla., Feb. 27–Mar. 1, 1974, M. B. Stewart, Jr. (Ed.), pp. 258-277, CONF-740252, NTIS.

Strauch, R. G., W. C. Campbell, R. B. Chadwick, and K. P. Moran, 1975, *Meteorological Applications of the FM Doppler Radar*, Report NOAA-TR-ERL 351-WPL-44, National Oceanic and Atmospheric Administration, NTIS.

Thomas, R., R. Puthoff, J. Savino, and W. Johnson, 1975, *Plans and Status of the NASA Lewis Research Center Wind Energy Project*, Report NASA-TM-71701, National Aeronautics and Space Administration, NTIS.

U. S. Department of Energy, 1980, *Annual Report to Congress*, Vol. 3, Forecasts, Report DOE/EIA-0173(80)/3, GPO.

U. S. Energy Research and Development Administration, 1976, *A National Plan for Energy Research and Development and Demonstration; Creative Energy Choices for the Future*, ERDA Report 76-1, NTIS.

——, 1977, *Solar Program Assessment: Environmental Factors, Solar Thermal Electric*, ERDA Report 77-47/4, NTIS.

Vant-Hull, L., and C. R. Easton, 1974, *Solar Thermal Power Systems Based on Optical Transmission*, Report NSF/RANN/SE/G1-39456/74/2, National Science Foundation.

White, D. E., 1965, *Geothermal Energy*, Circular No. 519, U. S. Geological Survey.

——, and D. L. Williams (Eds.), 1975, *Assessment of Geothermal Resources of the United States—1975*, Circular No. 726, U. S. Geological Survey.

Wick, G. L., and J. D. Isaacs, 1976, Utilization of the Energy from Salinity Gradients, in *Proceedings of the Wave and Salinity Gradient Energy Conversion Workshop*, Newark, Del., May 24–26, 1976, ERDA Report COO-2946-1, pp. A-14–A-34, NTIS.

Williams, J., G. Kromer, and J. Weingart (Eds.), 1977, *Proceedings of the International Workshop on Climate and Solar Energy Conversion, Dec. 8–10, 1976*, International Institute for Applied Systems Analysis, Laxenburg, Austria.

Wooley, M., and J. Platts, Jr., 1975, Energy on the Crest of a Wave, *New Sci.*, 66: 241-243.

Zener, C., 1976, Solar Sea Power, *Bull. At. Sci.*, 17-24.

Chapter 18 covers research prior to November 1979.

THE ENVIRONMENTAL IMPACT STATEMENT

19

Karl F. Zeller, Environmental Research and Technology, Inc.

THE ENVIRONMENTAL IMPACT STATEMENT

To declare a national policy which will encourage productive and enjoyable harmony between man and his environment; to promote efforts which will prevent or eliminate damage to the environment and biosphere and stimulate the health and welfare of man; to enrich the understanding of the ecological systems and natural resources important to the Nation, ...

—National Environmental Policy Act of 1969

...ill founded certainty is far worse than a realistic appreciation of the confusing issues ... .

—Schneider (1976), *The Genesis Strategy*

19-1 INTRODUCTION

The purpose of this chapter is to discuss the environmental impact statement (EIS) as it relates to the science of meteorology and air quality. This chapter does not deal specifically with air-quality permits as required by the Environmental Protection Agency (EPA) and state air-quality agencies. These permits are required to demonstrate project compliance with federal and state air-quality regulations and standards. At the present time both federal and state air-quality regulations are in a tremendous state of flux due to the new Clean Air Act Amendments of 1977 and subsequent court case (U. S. Court of Appeals for the District of Columbia, 1979). Typically, air-quality permits are secured prior to the preparation of an EIS. However, this is not necessarily the case, nor is there any regulation that specifies chronology except that the EIS should list all required federal permits and licenses. Since the EIS and air-quality permits deal with the same project and environment, they are not scientifically mutually exclusive. Therefore this chapter does contain a lot of information that is also useful for permit application. The sophisticated reader may find various parts of this chapter out of date; this is unfortunate but unavoidable since the laws and regulations are changing so fast. As an example, the Council on Environmental Quality EIS guidelines were revised and changed to regulations after the draft of this chapter had passed peer review.

19-1.1 Background

Since the President signed the National Environmental Policy Act of 1969 (NEPA) (Public Law 91-190) on Jan. 1, 1970, bringing birth to the EIS, many changes have taken place in the planning of environmental and private actions. Many managers have been perplexed on learning that their project required an EIS. Probably the biggest immediate cause of consternation, other than project delays, has been the lack of understanding and precedent outlining just what to do or how to do it. The content of the first EIS's prepared ran the gamut from a few sentences declaring little or no environmental effect to prepared volumes of data and analyses attempting to tell it all.† Now we have had a decade of experience in EIS preparation with respect to NEPA (there were 7334 EIS's filed as of December 1977). Furthermore, we have two decades of experience specific to meteorology and air quality due to environmental assessments required for licensing nuclear power plants by the Atomic Energy Act of 1954. A chapter dealing specifically with the EIS is appropriate since EIS requirements are a political reality and a main use for applying the information in the preceding chapters. The EIS is used to help us understand our environment and predict ahead of time the consequences of proposed actions of man.

19-1.2 General Comments

Although the EIS process has been in a continuous state of flux due to unspecific legal requirements, in addition to questions of who must prepare one and when, the air-quality and meteorological fundamentals of analysis in an EIS are related to physical laws that are inherently not dated or changed by man's laws. The approaches used in air-quality and meteorology data collection, preparation, and analysis are scientific and will remain pertinent no matter which man-enacted law applies or who is required to make an environmental assessment. Specific analysis techniques (various models, methodologies, etc.) may be improved, but the basic data collection requirements and parameters used for evaluations and calculations will remain usable, even though alterations in legal law or regulations may change types of comparisons and design thresholds.

This chapter initially discusses in general the development of the EIS and then focuses specifically on the air-quality and meteorology requirements. The chapter covers what an EIS is, discusses its evolution, relates the air-quality‡ and meteorology requirements to the overall EIS process, and presents useful

Zeller: Environmental Research and Technology, Inc., Ft. Collins, Colorado.

†The practice of filling volumes with extraneous data in an EIS is given the name "dandelion counts."

‡The term air quality, used as a noun throughout this chapter, quantifies the state of cleanliness/dirtiness of the air (e.g., The air quality is good. The air quality was degraded by SO_2, which averaged 40 mg/m^3 for 1976).

guidelines for the systematic preparation and analysis of the air-quality and meteorology portions of an EIS.

19-1.3 Purpose

A point to keep in mind when preparing an EIS is that the overall purposes of assessing potential environmental impacts, both positive and negative, are to give decision makers (industry, government, and the general public) information for balancing the environmental pros and cons of any specific project or action. Along these lines, remember that a portion of the users will not be scientists. There is a difference between mathematical models with their internal structures inherently rigorous and consistent and the required verbal explanations of the various symbols and calculated results which are not always intuitively understood and are sometimes ambiguous with respect to reality. Consequently, in the preparation of an EIS, it is important to distill the volumes of data and analyses "to make the EIS process more useful to decision makers and the public; and to reduce paperwork...in order to...focus on real environmental issues and alternatives."† In addition to recognizing the problem, Executive Order 11991 required the development of regulations that "require impact statements to be concise, clear, and to the point, and supported by evidence that agencies have made the necessary environmental analysis."† Such regulations have been proposed‡ and will be discussed below.

Oliver Houch, General Counsel for the National Wildlife Federation, made a few interesting comments on the EIS process§ at the Environmental Law Institute Conference, Nov. 8, 1977: An "ideal" impact statement would first state the action objective and would then analyze equally all reasonable alternatives to accomplish that objective. The preceding should be done before considering preferred alternatives, to which the initiating organization may be institutionally committed and compelled to justify. In practice, the responsibility for EIS analysis and preparation is usually "delegated downward" to the point where the perspective is one of justifying a specific approach rather than answering the overall objective. As an example, an alternative to a power-production facility to supply a given area with power may be to restrict growth in that area rather than immediately to fall into the questions considering the plant to be constructed. One environmental impact handbook (Cheremisinoff and Morresi, 1977) appropriately quotes Albert Einstein: "Concern for man himself and his fate must always form the chief interest of all technical endeavor. Never forget this in the midst of your diagrams and equations." Considered from a holistic approach to life, the introduction of the EIS into man's decision making process might be considered an orderly and positive factor in the evolution of man.

There are differences between environmental reports, environmental assessments, environmental statements, and the EIS. The semantics become confusing since each governmental agency and state has its own jargon. The official EIS required by NEPA should be a usable distillation of all the data collection and analysis that has gone into a larger and more complete environmental assessment study. The purpose of the NEPA EIS is to relay the essence of the impact of the proposed action to the decision makers, including the public.

†Executive Order 11991, May 24, 1977.

‡*Fed. Regist. (Washington, D. C.)*, 43: 25230, June 9, 1978.

§*Environmental Reporter*, "Current Developments," page 1037, 1977.

19-2 BASIC EIS LAWS, LEGAL INTENTS, AND EVOLUTION

There are several main documents or combination of documents that provide the rules, regulations, and guidelines for the meteorology/air-quality portion of the EIS process. They include The National Environmental Policy Act of 1969 (U. S. Congress, 1969); The Clean Air Act of 1963 as amended (U. S. Congress, 1963); Preparation of Environmental Impact Statements, 1973 guidelines issued by the Council on Environmental Quality [Council on Environmental Quality (CEQ), 1973]; and Executive Orders 11514 (U. S. President, 1970) and 11991 (U. S. President, 1977). As a result of Executive Order 11991, the CEQ published Regulations for Implementing the Procedural Provisions of the NEPA (CEQ, 1978c) which replaced the 1973 CEQ EIS guidelines. The Atomic Energy Act of 1954 has also been responsible for a lion's share of the development of meteorology and air-quality environmental assessment techniques particular to nuclear power production. In addition to the previously mentioned documents, numerous court cases have interpreted NEPA, CEQ guidelines, and the various regulations; each federal governmental agency has issued separate regulations and guidelines detailing its internal policies and procedures for implementing NEPA; and also, some states and cities have statutes, analogous to NEPA requirements, which require environmental impact statements ("little NEPA's") for private actions, such as shopping centers, refineries, etc. A discussion follows concerning the most important aspect of each of the preceding main documents.

19-2.1 Atomic Energy Act of 1954

This legislation is particularly important because, prior to NEPA, the Atomic Energy Act of 1954 was the only federal requirement for a systematic assessment of environmental impact. Specifically, the Act required the Atomic Energy Commission (AEC), whose programs are now administered by the Nuclear Regulatory Commission (NRC) and the Department of Energy (DOE),¶ to make a detailed environmental assessment before a nuclear construction permit and final operation license were issued. As a result of the decision from the case of Calvert Cliffs Coordinating Committee vs. the AEC in 1971 (U. S. Court of Appeals for the District of Columbia, 1971), the AEC was required to further revise its activities to systematically analyze environmental impacts in line with NEPA. Owing to the Atomic Energy Act and resultant NRC regulations, the methods of air-quality and meteorological assessment are somewhat more advanced than those of other disciplines. Valuable meteorological guidance can be attained from NRC guidelines and regulations. Van der Hoven (1975) has prepared a discussion and an excellent summary of NRC guidelines for preparation of EIS's. This list of guidelines has been updated and is listed below. The NRC guidelines can be categorized into those pertaining to environmental assessments which require realistic estimates and those pertaining to safety assessments which require very conservative evaluations:

¶In January 1975, the Atomic Energy Commission was divided between two newly formed organizations. The regulatory programs were assumed by the Nuclear Regulatory Commission, and the research and development programs became part of the Energy Research and Development Administration. In October 1977, the Energy Research and Development Administration was absorbed into the new Department of Energy.

For Environmental Assessments:

NRC Regulatory Guide 1.111, *Methods for Estimating Atmospheric Transport and Dispersion of Gaseous Effluents in Routine Releases from Light-Water-Cooled Reactors,* 1977

NRC Regulatory Guide 1.21, Rev. 2, *Measuring, Evaluating, and Reporting Radioactivity in Solid Wastes and Releases of Radioactive Materials in Liquid and Gaseous Effluents from Light-Water-Cooled Nuclear Power Plants,* 1974

NRC Regulatory Guide 1.23, *Onsite Meteorological Programs,* 1972

American National Standard Guide 2.5, *Standard for Obtaining Meteorological Information at Nuclear Power Sites,* 1977 (possible revision to NRC Regulatory Guide 1.23)

NRC Regulatory Guide 4.2, Rev. 2, *Preparation of Environmental Reports for Nuclear Power Plants,* 1976
- Section 2.8: Meteorology
- Section 5.3.3: Effects of Heat Dissipation Facilities
- Section 6.4: Environmental Measurements and Monitoring Programs: Meteorological

NRC Regulatory Guide 4.7, *General Site Suitability Criteria for Nuclear Power Stations,* 1974

For Safety Assessments:

NRC Standard Review Plan, 1975
- Section 2.3.1: Regional Climatology
- Section 2.3.2: Local Meteorology
- Section 2.3.3: Onsite Meteorological Measurements Programs
- Section 2.3.4: Short-Term Diffusion Estimates
- Section 2.3.5: Long-Term Diffusion Estimates

NRC Regulatory Guide 1.145, *Atmospheric Dispersion Models for Potential Accident Consequence Assessments at Nuclear Power Plants,* 1977

NRC Regulatory Guide 1.24, *Assumptions for Evaluating the Potential Radiological Consequences of a Pressurized Water Reactor Radioactive Gas Storage Tank Failure,* 1972

NRC Regulatory Guide 1.25, *Assumptions Used for Evaluating the Potential Radiological Consequences of a Fuel Handling Accident in the Fuel Handling and Storage Facility for Boiling and Pressurized Water Reactors,* 1972

NRC Regulatory Guide 1.27, Rev. 2, *Ultimate Heat Sink for Nuclear Power Plants,* 1976

NRC Regulatory Guide 1.3, Rev. 2, *Assumptions Used for Evaluating the Potential Radiological Consequences of a Loss of Coolant Accident for Boiling Water Reactors,* 1974

NRC Regulatory Guide 1.4, Rev. 2, *Assumptions Used for Evaluating the Potential Radiological Consequences of a Loss of Coolant Accident for Pressurized Water Reactors,* 1974. *Criteria for Preparation and Evaluation of Radiological Emergency Response Plans and Preparedness in Support of Nuclear Power Plants,* Report NUREG-0654, Nuclear Regulatory Commission, 1980

NRC Regulatory Guide 1.5, *Assumptions for Evaluating the Potential Radiological Consequences of a Steam Line Break Accident for Boiling Water Reactors,* 1971

NRC Regulatory Guide 1.70, Rev. 2, *Standard Format and Content of Safety Analysis Reports for Nuclear Power Plants,* 1975

NRC Regulatory Guide 1.76, *Design Basis Tornado for Nuclear Power Plants,* 1974

NRC Regulatory Guide 1.77, *Assumptions Used for Evaluating a Control Rod Ejection Accident for Pressurized Water Reactors,* 1974

NRC Regulatory Guide 1.78, *Assumptions for Evaluating the Habitability of a Nuclear Power Plant Control Room During a Postulated Hazardous Chemical Release,* 1974

NRC Regulatory Guide 1.98, *Assumptions Used for Evaluating the Potential Radiological Consequences of a Radioactive Offgas System Failure in a Boiling Water Reactor,* 1976

Federal Emergency Management Agency (FEMA), *Criteria for Preparation and Evaluation of Radiological Emergency Response Plans and Preparation in Support of Nuclear Power Plants,* Report NUREG-0654, Nuclear Regulatory Commission, 1980

These guidelines contain detailed information specific to requirements for impact analyses related to nuclear power generation. The NEPA EIS's do not generally require the exhaustive detail that nuclear related impact statements do.

19-2.2 National Environmental Policy Act of 1969

The NEPA is essentially Congress's attempt to require our country to consider the basics of ecology—that everything is connected to everything else and everything has to go somewhere—in all our nation's planning. The Act has two parts: Title I, a declaration of a national environmental policy, and Title II, which establishes the CEQ. Of particular importance to the subject of this chapter are the first three parts of Section 102 of Title I which specify that all agencies of the federal government shall:

> (A) utilize a systematic, interdisciplinary approach which will ensure the integrated use of the natural and social sciences and the environmental design arts in planning and in decision making which may have an impact on man's environment;
>
> (B) identify and develop methods and procedures, in consultation with the Council on Environmental Quality established by subchapter II of this chapter, which will insure that presently unqualified environmental amenities and values may be given appropriate consideration in decision making along with economic and technical considerations;
>
> (C) include in every recommendation or report on proposals for legislation and other major Federal actions significantly affecting the quality of the human environment, a detailed statement by the responsible official on:
>
> (i) the environmental impact of the proposed action,
>
> (ii) any adverse environmental effects which cannot be avoided should the proposal be implemented,
>
> (iii) alternatives to the proposed action,
>
> (iv) the relationship between local short-term uses of man's environment and the maintenance and enhancement of long-term productivity, and
>
> (v) any irreversible and irretrievable commitments of resources which would be involved in the proposed action should it be implemented.
>
> Prior to making any detailed statement, the responsible Federal official shall consult with and obtain the comments of any Federal agency which has jurisdiction by law or special expertise with respect to any environmental impact involved. Copies of such statement and comments and views of the appropriate Federal, State, and local agencies, which are authorized to develop and enforce environmental standards, shall be made available to the President, the Council on Environmental Quality and to the public as provided by Section 552 of Title 5, and shall accompany the proposal through the existing agency review processes.

Notice that it is Part C of Section 102 that requires the EIS and identifies five items to be addressed (items 3 to 7 of Table 19.1). The CEQ prepares a monthly report, *102 Monitor,*

TABLE 19.1 Content of Environmental Statements: CEQ Guidelines

(Revised as of Aug. 1, 1973)

1. A description of the proposed action, a statement of its purposes, and a description of the environment affected,...
2. The relationship of the proposed action to land-use plans, policies, and controls for the affected area
3. The probable impact (positive and negative) of the proposed action on the environment (reflects NEPA Sec. 102(c)(i) requirement)
4. Alternatives to the proposed action, including, where relevant, those not within the existing authority of the responsible agency (reflects NEPA Sec. 102(c)(iii) requirement)
5. Any probable adverse environmental effects that cannot be avoided should the proposal be implemented (reflects NEPA Sec. 102(c)(ii) requirement)
6. The relationship between local short-term uses of man's environment and maintenance and enhancement of long-term productivity (reflects NEPA Sec. 102(c)(iv) requirement)
7. Any irreversible and irretrievable commitments of resources that would be involved in the proposed action should it be implemented (reflects NEPA Sec. 102(c)(v) requirement)
8. An indication of what other interests and considerations of federal policy are thought to offset the adverse environmental effects of the proposed action identified in items 3 and 5 above
9. The final EIS must include a discussion of all important issues raised during the review of the draft EIS

available through the U. S. Government Printing Office, which is concerned with all current aspects of EIS's.

Title II of NEPA creates the CEQ as an Executive Office of the President. The CEQ has served as a central repository for final EIS's,† prepared guidelines applicable to all federal agencies for NEPA compliance, reviewed controversial draft EIS's, developed comparative analyses on the EIS process, and provided consultation to federal agencies on NEPA compliance and on the preparation of agency procedures and guidelines. As a result of Executive Order 11991, the CEQ's function has recently been expanded from providing EIS guidelines to issuing regulations to federal agencies concerned with EIS preparation.

19-2.3 Clean Air Act of 1963

The Clean Air Act of 1963 as amended by the Air Quality Act of 1967, the Clean Air Act Amendments of 1970, the Technical Amendments of 1973, and the Clean Air Act Amendments of 1977 (U. S. Congress, 1963) provide for, among other things: (1) National Ambient Air-Quality Standards (NAAQS), Table 19.2,‡ for eight pollutants: particulates, CO, NMHC, SO_2, NO_x, oxidants, lead, and H_2S; (2) emission standards for new and existing stationary sources: New Source Performance Standards (NSPS), Table 19.3, which limit the allowed emissions of NAAQS Pollutants; and (3) prevention of significant deterioration (PSD) air-quality increment allowances, Table 19.4, for three classes (Classes I, II, and III) of clean areas or "attainment" areas§ that are in compliance with NAAQS. The PSD class increment allowances are simply increments of further allowable air-quality deterioration beyond current background levels of January 1975. This increment must be shared by all potential pollution sources, i.e., once an increment has been consumed in a given area, no further sources would be allowed. The evaluation of incremental increases is based on modeling because the increase must be considered and a PSD permit must be granted before a facility is constructed. Possible deterioration of visibility is an "air-quality related value" that must also be considered in Class I areas as required by the 1977 amendments. Figure 19.1 shows the mandatory current federal Class I PSD areas as declared by Congress, and Figure 19.2 shows the potential federal Class I areas as recommended by Congress to be considered by the managing agency for reclassification to Class I from Class II.

For administrative reasons, EPA has proposed to ignore certain minuscule rates of emission or net increases in emissions. These rates are termed de minimis, a legal concept meaning that they are so negligible that they fall outside the strict requirements of the law. The de minimis exemptions apply either to emissions of pollutants from a new source or to "net increases" of emissions from modifications to existing sources. There are two de minimis tests. First, if the mass emissions are below a certain rate (Table 19.5), the source will not need a PSD permit, but it must notify EPA of the proposed construction. Second, even if the mass emissions are above the specified rate, if the ambient air-quality impact of those emissions is below the significant impact guideline (Table 19.6), the emissions do not need to be reviewed for air-quality impact (monitoring and modeling).

The preceding NAAQS and PSD increments, which may be made equal or more stringent by states in State Implementation Plans (SIPs) and state PSD class increments, and the de minimis levels provide specific concentrations to which to compare the results of an environmental analysis.

Under its Clean Air Act (Canada, 1970), the Canadian federal government is empowered to set three ranges of air-quality objectives (Wetstone, 1980): desirable, acceptable, and tolerable. The tolerable range is the equivalent of the U. S. primary standard and indicates the level at which there is a danger to public health. The acceptable range is comparable to the secondary standard in the United States and is intended to reflect the level at which "welfare" effects to vegetation, soil, water, or the general public comfort may occur. The desirable range represents a long-range pollution control goal. These objectives are only advisory, however, and have no legal effect unless they are incorporated into provincial or municipal legislation. Alberta, Manitoba, New Brunswick, Ontario, and Saskatchewan all have adopted ambient standards keyed to these objectives, whereas Labrador, Newfoundland, Nova Scotia, Prince Edward Island, and Quebec, in general the less important polluters, have not. The standards for Alberta, Canada, are included in Table 19.2 as a comparison to U. S. standards.

19-2.4 CEQ Guidelines

The CEQ first published guidelines for preparation of EIS's on Apr. 23, 1971 (CEQ, 1971). These were updated and

†Pursuant to the President's Reorganization Plan No. 1 (1978), the EPA is the official recipient for EIS's and is required to publish a list of EIS's received weekly.

‡Primary standards are based on protecting human health, and secondary standards are based on protecting the public welfare in general [reviews of these standards are required by the Clean Air Act Amendments of 1977, §109 (U. S. Congress, 1963)].

§Attainment areas have measured ambient air-quality concentrations below NAAQS. Nonattainment areas exceed NAAQS. The PSD regulations are replaced by an "offset" policy that requires a net decrease in emissions for nonattainment areas.

TABLE 19.2 National Ambient Air-Quality Standards†

Description	Adverse health effects observed at these concentrations	Air-quality standards: U. S. primary	U. S. secondary	Alberta, Canada
		Total Suspended Particulates		
Solid and liquid particles in the atmosphere including dust, smoke, mists, fumes, and spray from many sources	80 $\mu g/m^3$, annual mean	75 $\mu g/m^3$, annual geometric mean; 260 $\mu g/m^3$, maximum 24-h value may occur once each year	60 $\mu g/m^3$, annual geometric mean; 150 $\mu g/m^3$, maximum 24-h average	60 $\mu g/m^3$, annual geometric mean; 100 $\mu g/m^3$, 24-h average
		Sulfur Dioxide		
Heavy, pungent, colorless gas	115 $\mu g/m^3$, annual mean; 300 $\mu g/m^3$, 24-h average for 3 to 4 days	80 $\mu g/m^3$ (0.03 ppm), annual arithmetic mean; 365 $\mu g/m^3$ (0.14 ppm) maximum 24-h value may occur once each year	1300 $\mu g/m^3$, maximum 3-h average	30 $\mu g/m^3$ (0.01 ppm), annual arithmetic mean; 150 $\mu g/m^3$, 24-h concentration; 450 $\mu g/m^3$, 1-h concentration; 525 $\mu g/m^3$, 1½-h concentration
		Carbon Monoxide		
Invisible, odorless gas formed from combustion of gasoline, coal, and other; largest man-made fraction comes from motor vehicles	12 to 17 $\mu g/m^3$ for 8 h produces concentration of 2 to 2.5% carboxyhemoglobin; 35 $\mu g/m^3$ for 8 h produces concentration of 5% carboxyhemoglobin	10,000 $\mu g/m^3$ (9 ppm), maximum 8-h value may occur once each year; 40,000 $\mu g/m^3$ (35 ppm), maximum 1-h value may occur once each year	Same as primary	6000 $\mu g/m^3$ (5 ppm), 8-h concentration; 15000 $\mu g/m^3$, 1-h concentration
		Photochemical Oxidants (as O_1)		
Pungent, colorless toxic gas; one component of photochemical smog	130 $\mu g/m^3$ hourly average impaired performance of student athletes; 200 $\mu g/m^3$ instantaneous level increased eye irritation; 490 $\mu g/m^3$ peaks with 300 $\mu g/m^3$ hourly average increased asthma attacks	235 $\mu g/m^3$ (0.12 ppm), maximum 1-h value may occur once each year	Same as primary	None

(Table continues on the next page.)

TABLE 19.2 (Continued)

Description	Adverse health effects observed at these concentrations	Air-quality standards: U. S. primary	U. S. secondary	Alberta, Canada
		Nonmethane Hydrocarbons (NMHC)		
	With nonmethane hydrocarbon, 200 $\mu g/m^3$ in 3 h (6 to 9 a.m.) produced (2 to 4 h later) photochemical oxidant of up to 200 $\mu g/m^3$ that lasted 1 h; by extrapolation downward, concentration of HC of 100 $\mu g/m^3$ can produce lowest injurious level of photochemical oxidant	(Old: 160 $\mu g/m^3$ (0.05 ppm), maximum level that may occur 6 to 9 a.m. once each year) This standard is now considered a guide to achievement of other standards	Same as primary	None
		Nitrogen Dioxide		
Brown, toxic gas formed from fuel combustion; under certain conditions, it may be associated with ozone production	118 to 156 $\mu g/m^3$, 24-h mean over 6 months produced increase in acute bronchitis in infants and schoolchildren; this average associated with a 24-h maximum of 284 $\mu g/m^3$; 117 to 205 $\mu g/m^3$ 24-h mean over 6 months and mean suspended nitrate level of 3.8 $\mu g/m^3$ or more produced increased respiratory disease in family groups	(Old: 100 $\mu g/m^3$, annual arithmetic mean) Proposed: None proposed to date		60 $\mu g/m^3$ (0.03 ppm), annual arithmetic mean; 200 $\mu g/m^3$, 24-h concentration; 400 $\mu g/m^3$, 1-h concentration
		Lead		
		1.5 $\mu g/m^3$ per quarter	Same as primary	
		Hydrogen Sulfide		
			None	4 $\mu g/m^3$ (0.003 ppm), 24-h concentration; 14 $\mu g/m^3$, 1-h concentration; 17 $\mu g/m^3$, ½-h concentration

†Section 109 of the Clean Air Act Amendments of 1977 requires a review of NAAQS by Dec. 31, 1980, and subsequently at least every 5 yr.

TABLE 19.3 Industrial Source Categories Covered by EPA New Source Performance Standards†

(As of Sept. 1, 1978)

Fossil-fuel-fired steam generators
Incinerators
Portland cement plants
Nitric acid plants
Sulfuric acid plants

Petroleum refineries
Storage vessels for petroleum liquids
Secondary lead smelters
Secondary brass and bronze ingot production plants
Iron and steel plants

Sewage treatment plants
Primary copper smelters
Primary zinc smelters
Primary lead smelters
Primary aluminum reduction plants

Phosphate fertilizer industry
- Wet process phosphoric acid plants
- Superphosphoric acid plants
- Diammonium phosphate plants
- Triple superphosphate plants
- Granular triple superphosphate storage facilities

Coal preparation plants
Ferroalloy production facilities
Steel plants: electric arc furnaces
Kraft pulp mills
Lime plants
Grain elevators
Stationary gas turbines

†*Fed. Regist. (Washington, D. C.)*, 43: 38872 (Aug. 31, 1978) lists, in order of priority, 70 additional source categories that will have New Source Performance Standards imposed prior to August 1982.

TABLE 19.4 PSD Class Increments

For any Class I, II, or III area, the maximum allowable increase in concentrations of sulfur dioxide and particulate matter over the baseline concentration of such pollutants shall not exceed the amounts given below:

Pollutant	Maximum allowable increase, $\mu g/m^3$
Class I	
Particulate matter:	
Annual geometric mean	5
24-h maximum	10
Sulfur dioxide:	
Annual arithmetic mean	2
24-h maximum	5
3-h maximum	25
Class II	
Particulate matter:	
Annual geometric mean	19
24-h maximum	37
Sulfur dioxide:	
Annual arithmetic mean	20
24-h maximum	91
3-h maximum	512
Class III	
Particulate matter:	
Annual geometric mean	37
24-h maximum	75
Sulfur dioxide:	
Annual arithmetic mean	40
24-h maximum	182
3-h maximum	700

amended by a second set of guidelines published on Aug. 1, 1973 (CEQ, 1973). These guidelines added three new required EIS sections (Table 19.1) to the basic five required by NEPA: (1) a description of the proposed action as well as a description of the existing environment; (2) the relationship of the proposed action to existing land-use plans, policies, and controls in the affected areas; and (3) an indication of what other interests and considerations of federal policy are thought to offset any adverse environmental effects of the proposed action. In addition the final EIS should contain a discussion of the problems and objections brought to light during the draft EIS review process. Executive Order 11991 required that these guidelines be changed to regulation status under CEQ recommendations.

The CEQ guidelines contain a comprehensive list of areas of environmental concern. Table 19.7 is a partial list of federal agencies and federal–state agencies having jurisdiction by law, or special expertise, to comment on specific EIS subjects related to power production. The guidelines also contain a list of offices within federal agencies and federal–state agencies for information regarding the agencies' NEPA activities and for receiving and reviewing the EIS's of other agencies for which comments are requested.

19-2.5 Executive Order 11514

This Order (U. S. President, 1970) required agencies to implement NEPA and gave the CEQ the duty of issuing the guidelines discussed above. Each federal agency is required to have guidelines and regulations for the preparation of EIS's. In lieu of Executive Order 11991, there has been an attempt at systematizing these documents.

19-2.6 Executive Order 11991 and the CEQ EIS Regulations

A significant boost was given to the EIS program in 1977 when the President signed Executive Order 11991 (U. S. President, 1977) which required the CEQ to issue procedural EIS regulations to replace the guidelines. Major points of this Executive Order required CEQ regulations to:

1. "...make the environmental impact statement process more useful to decision makers and the public;..."
2. "...require impact statements to be concise, clear, and to the point,..."
3. provide for procedures (1) "...for the early preparations of EIS." and (2) "for the referral to the CEQ of conflicts between agencies concerning the implementation of the NEPA..."

Proposed NEPA regulations were issued for comment (CEQ, 1978b). Major provisions (CEQ, 1978a) of the new proposed EIS regulations are as follows:

1. The final EIS should not normally exceed 150 pages or, if unusually complex, 300 pages.

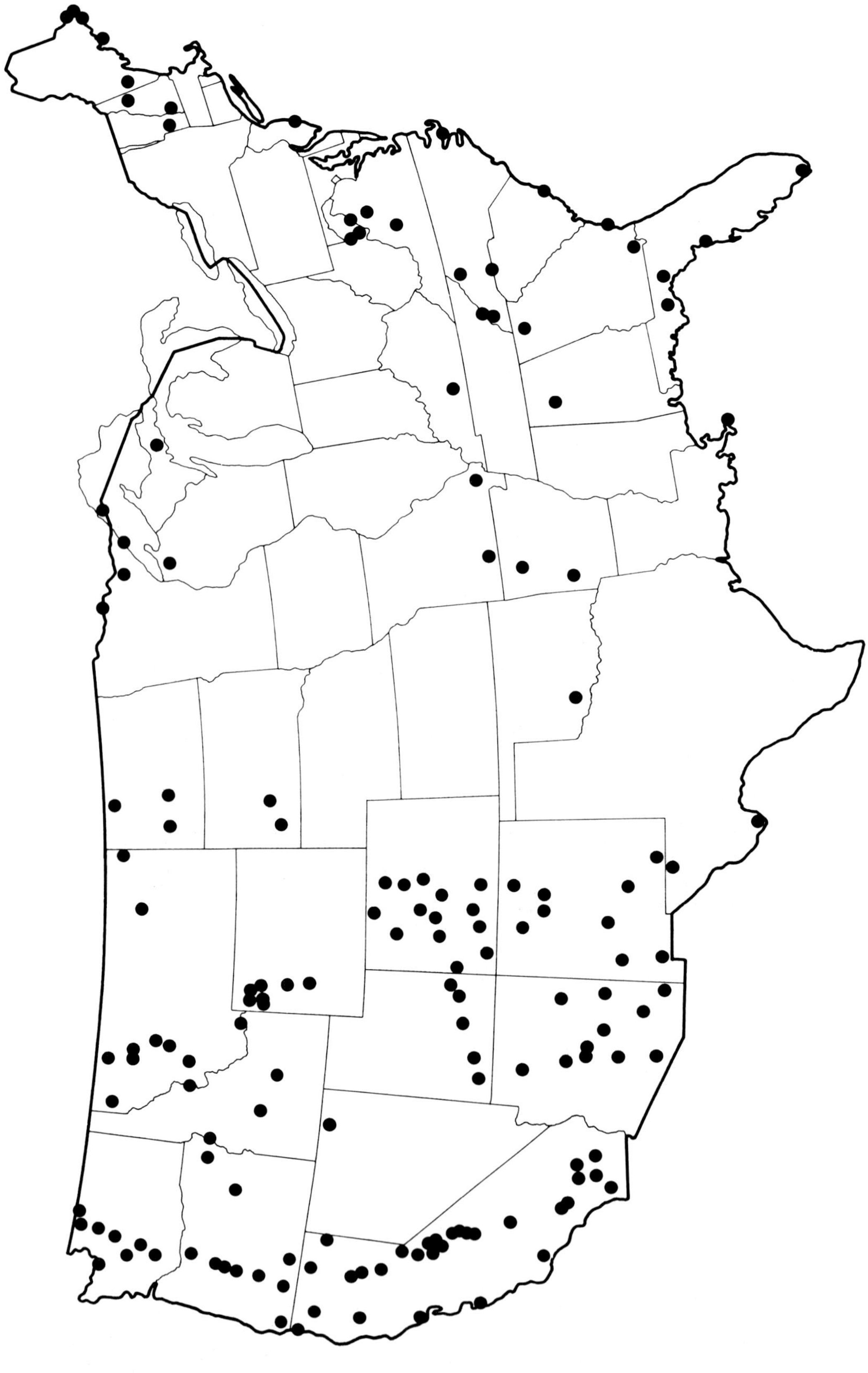

Fig. 19.1 Current federal Class I areas.

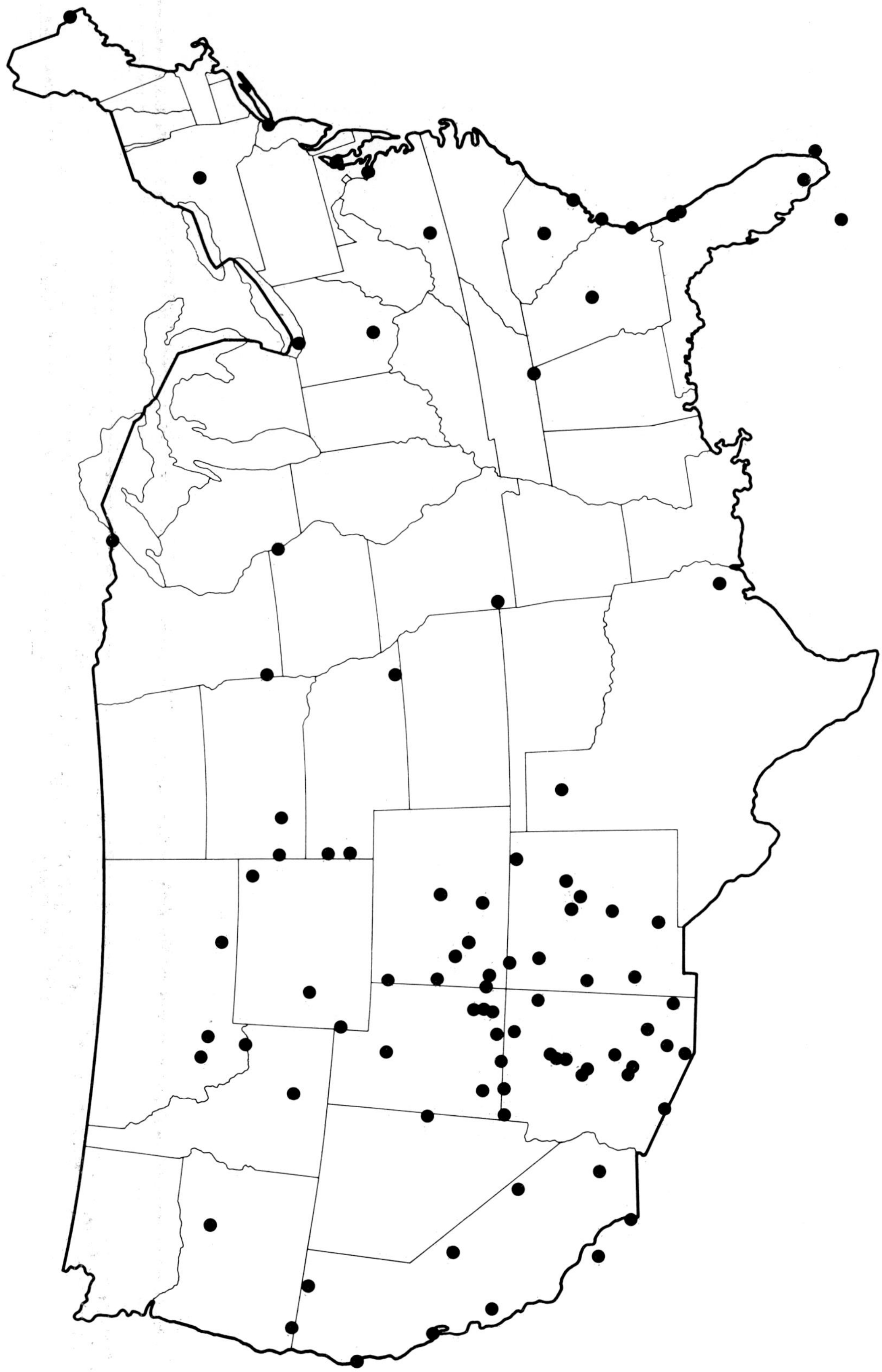

Fig. 19.2 Potential federal Class I areas.

TABLE 19.5 Guidelines for Significant Emission Rates

Pollutant	Emission rate, tons/yr
Carbon monoxide	100
Nitrogen dioxide	10
Total suspended particulates	10
Sulfur dioxide	10
Ozone	10 of volatile organic compounds
Lead	1
Mercury†	0.2
Beryllium†	0.004
Asbestos†	1
Fluoride†	0.02
Sulfuric acid mist†	1
Vinyl chloride†	1
Total reduced sulfur	
Hydrogen sulfide†	1
Methyl mercaptan†	1
Dimethyl sulfide†	1
Dimethyl disulfide†	1
Reduced sulfur compounds	
Hydrogen sulfide (see above)	10
Carbon disulfide†	10
Carbonyl sulfide†	10

†Noncriteria pollutants.

TABLE 19.6 Guidelines for Significant Ambient Air-Quality Impacts

Pollutant	Air-quality impact
Carbon monoxide	500 $\mu g/m^3$, 8-h average
Nitrogen dioxide	1 $\mu g/m^3$, annual
Total suspended particulates	5 $\mu g/m^3$, 24 h
Sulfur dioxide	5 $\mu g/m^3$, 24 h
Ozone	†
Lead	0.03 $\mu g/m^3$, 3 months
Mercury	0.1 $\mu g/m^3$, 30 d
Beryllium	0.005 $\mu g/m^3$, 24 h
Asbestos	1 $\mu g/m^3$, 1 h
Fluoride	0.01 $\mu g/m^3$, 24 h
Sulfuric acid mist	1 $\mu g/m^3$, 24 h
Vinyl chloride	1 $\mu g/m^3$, maximum value
Total reduced sulfur	
Hydrogen sulfide	1 $\mu g/m^3$, 1 h
Methyl mercaptan	0.5 $\mu g/m^3$, 1 h
Dimethyl sulfide	0.5 $\mu g/m^3$, 1 h
Dimethyl disulfide	2 $\mu g/m^3$, 1 h
Reduced sulfur compounds	
Hydrogen sulfide (see above)	
Carbon disulfide	200 $\mu g/m^3$, 1 h
Carbonyl disulfide	200 $\mu g/m^3$, 1 h

†No de minimis air-quality impact is proposed for ozone. However, any net increase of 100 tons per year of volatile organic compounds subject to PSD would be required to perform ambient impact analysis, including the gathering of ambient air-quality data.

2. The final EIS should reflect the desirability of emphasizing the real options among alternative courses of action, including the no action option.
3. The regulations spell out a standard EIS format for all EIS's.
4. The regulations call for the EIS to be written in "plain language."
5. The regulations attempt to improve accuracy by making the submitting agency responsible for the scientific integrity of the analysis that went into the final EIS.

These regulations were finalized and listed in the *Federal Register* on Nov. 29, 1978 (CEQ, 1978c). Their effect thus far is to produce short summary EIS's with more complex, backup technical reports on each discipline and/or impact attached as separate appendixes.

19-2.7 Court Cases and Federal and State Agency Guidelines

The NEPA, having been rushed through Congress late in 1969, is hampered by a great deal of ambiguity. As a result, the courts have had a major role in interpreting policy questions and formulating the NEPA process. It is not the intent of this chapter, however, to delve into specific cases. For further reading, two summaries that detail NEPA court highlights are found in *The Environmental Impact Statement Process* [Environmental Protection Agency (EPA), 1976b] and Chap. 2 of *Manual for Environmental Impact Evaluation* (Rosen, 1976).

As of May 1977, 25 states and 2 cities have their own requirements, "little NEPAs," for EIS's on state actions of one type or another (Table 19.8).

19-3 CHRONOLOGICAL PATH OF THE OVERALL EIS PROCESS

There are four major steps in the development of an EIS. It would be helpful to the air-quality engineer or to the meteorologist to be aware of these steps.

The first step is to decide whether an EIS would be required in a particular proposed action. Questions to be asked are: Is a federal action involved; is the action "major" and likely to "significantly" affect the environment; are federal permits or licenses required; are federal funds involved; is the activity directly and wholly undertaken by the federal government; do any state or local EIS laws apply? *The Environmental Impact Statement Process* (EPA, 1976b) is a reference on the subject of EIS determination. An "environmental assessment" is made in the beginning to determine if an EIS is necessary. If the responsible official at the agency determines that no EIS is required, the agency summarizes the results of the environmental assessment and makes a "negative declaration."†

The second step, assuming a negative declaration has not been filed, is to write a draft EIS (DEIS). Although NEPA does not specifically require a DEIS, the CEQ guidelines and subsequent regulations do.

†Negative declarations are also subject to review. For example, in a draft negative declaration for a Federal Highway Administration supported road in Wyoming (Cody–Greybull road), the Bureau of Land Management pointed out the proposal would disrupt a yearly rest-rotation grazing system and suggested mitigating measures which were adopted.

TABLE 19.7 Partial List of Power-Production-Related Areas of Environmental Impact and Federal Agencies with Jurisdiction by Law or Special Expertise to Comment Thereon

(Council on Environmental Quality, 1973)

AIR

Air Quality

Department of Agriculture, Forest Service (effects on vegetation)
Atomic Energy Commission (radioactive substances)
Department of Health, Education, and Welfare
Environmental Protection Agency
Department of the Interior:
  Bureau of Mines (fossil and gaseous fuel combustion)
  Bureau of Sport Fisheries and Wildlife (effect on wildlife)
  Bureau of Outdoor Recreation (effects on recreation)
  Bureau of Land Management (public lands)
  Bureau of Indian Affairs (Indian lands)
National Aeronautics and Space Administration (remote sensing, aircraft emissions)
Department of Transportation:
  Assistant Secretary for Systems Development and Technology (auto emissions)
  Coast Guard (vessel emissions)
  Federal Aviation Administration (aircraft emissions)

Weather Modification

Department of Agriculture, Forest Service
Department of Commerce, National Oceanic and Atmospheric Administration
Department of Defense, Department of the Air Force
Department of the Interior, Bureau of Reclamation

Natural Gas Development, Production, Transmission, and Use

Department of Housing and Urban Development (urban areas)
Department of the Interior:
  Office of Oil and Gas
  Geological Survey
  Bureau of Mines
  Bureau of Land Management (public lands)
  Bureau of Indian Affairs (Indian lands)
  Bureau of Sport Fisheries and Wildlife
  Bureau of Outdoor Recreation
  National Park Service
Department of Transportation (transport and safety)
Environmental Protection Agency
Federal Power Commission (production, transmission, and supply)
Interstate Commerce Commission

Coal and Minerals Development, Mining, Conversion, Processing, Transport, and Use

Appalachian Regional Commission
Department of Agriculture, Forest Service
Department of Commerce
Department of the Interior:
  Office of Coal Research
  Mining Enforcement and Safety Administration
  Bureau of Mines
  Geological Survey
  Bureau of Indian Affairs (Indian lands)
  Bureau of Land Management (public lands)
  Bureau of Sport Fisheries and Wildlife
  Bureau of Outdoor Recreation
  National Park Service
Department of Labor, Occupational Safety and Health Administration
Department of Transportation
Environmental Protection Agency
Interstate Commerce Commission
Tennessee Valley Authority

ENERGY SUPPLY AND NATURAL RESOURCES DEVELOPMENT

Electric Energy Development, Generation, Transmission, and Use

Atomic Energy Commission (nuclear)
Department of Agriculture, Rural Electrification Administration (rural areas)
Department of Defense, Army Corps of Engineers (hydro)
Department of Health, Education, and Welfare (radiation effects)
Department of Housing and Urban Development (urban areas)
Department of the Interior:
  Bureau of Indian Affairs (Indian lands)
  Bureau of Land Management (public lands)
  Bureau of Reclamation
  Power Marketing Administrations
  Geological Survey
  Bureau of Sport Fisheries and Wildlife
  Bureau of Outdoor Recreation
  National Park Service
Environmental Protection Agency
Federal Power Commission (hydro, transmission, and supply)
River Basin Commissions (as geographically appropriate)
Tennessee Valley Authority
Water Resources Council

Petroleum Development, Extraction, Refining, Transport, and Use

Department of the Interior:
  Office of Oil and Gas
  Bureau of Mines
  Geological Survey
  Bureau of Land Management (public lands and outer continental shelf)
  Bureau of Indian Affairs (Indian lands)
  Bureau of Sport Fisheries and Wildlife (effects on fish and wildlife)
  Bureau of Outdoor Recreation
  National Park Service
Department of Transportation (transport and pipeline safety)
Environmental Protection Agency
Interstate Commerce Commission

Energy and Natural Resource Conservation

Department of Agriculture:
  Forest Service
  Soil Conservation Service
Department of Commerce, National Bureau of Standards (energy efficiency)
Department of Housing and Urban Development, Federal Housing Administration (housing standards)
Department of the Interior:
  Office of Energy Conservation
  Bureau of Mines
  Bureau of Reclamation
  Geological Survey
  Power Marketing Administration
Department of Transportation
Environmental Protection Agency
Federal Power Commission
General Service Administration (design and operation of buildings)
Tennessee Valley Authority

Radiation

Atomic Energy Commission
Department of Commerce, National Bureau of Standards
Department of Health, Education, and Welfare

(Table continues on the next page.)

TABLE 19.7 (Continued)

ENERGY SUPPLY AND NATURAL RESOURCES DEVELOPMENT (continued)	HAZARDOUS SUBSTANCES (continued)
Radiation (continued)	**Pesticides**
Department of the Interior:	Department of Agriculture:
Bureau of Mines (uranium mines)	Agricultural Research Service (biological controls, food, and fiber production)
Mining Enforcement and Safety Administration (uranium mines)	Consumer and Marketing Service
Environmental Protection Agency	Forest Service
HAZARDOUS SUBSTANCES	Department of Commerce, National Oceanic and Atmospheric Administration
Toxic Materials	Department of Health, Education, and Welfare
Department of Agriculture, Consumer and Marketing Service (meat and poultry products)	Department of the Interior:
Department of Health, Education, and Welfare	Bureau of Sport Fisheries and Wildlife (fish and wildlife effects)
Environmental Protection Agency	Bureau of Land Management (public lands)
	Bureau of Indian Affairs (Indian lands)
	Bureau of Reclamation (irrigated lands)
	Environmental Protection Agency

TABLE 19.8 States and Cities with Environmental Protection Requirements†

States with Comprehensive Statutory Requirements

California, The Resources Agency
Connecticut, Department of Environmental Protection
Hawaii, Office of Environmental Quality Control
Indiana, Environmental Management Board
Maryland, Department of Natural Resources
Massachusetts, Executive Office of Environmental Affairs
Minnesota, Environmental Quality Council
Montana, Montana Environmental Quality Council
New York, New York State Department of Environmental Conservation
North Carolina, Department of Natural and Economic Resources
South Dakota, Department of Environmental Protection
Virginia, Council on the Environment
Washington, Department of Ecology
Wisconsin, Office of State Planning and Energy
Puerto Rico, Environmental Quality Board

States with Comprehensive Executive or Administrative Orders

Michigan, Department of Management and Budget
New Jersey, Department of Environmental Protection
Texas, Governor's Budget and Planning Office
Utah, Office of the Attorney General

States with Special or Limited EIS Requirements

Arizona, Wildlife Planning and Development Division
Delaware, Division of Environmental Control
Georgia, Department of Transportation
Mississippi, Mississippi Marine Resources Council
Nebraska, Office of Planning and Programming
Nevada, Department of Human Resources
New Jersey, Department of Environmental Protection

City NEPA's

Bowie, Maryland, Office of Planning and Community Development
New York City, New York City, Department of Environmental Protection

† Council on Environmental Quality (1977).

The third step is to circulate the DEIS to all groups interested in the proposed action in addition to the EPA and other federal agencies specified for review in the 1973 CEQ guidelines. After 45 to 60 d, a resonable time for comments on the DEIS to be prepared, the issuing agency revises the draft to reflect all major comments and prepares a final EIS.

The fourth and final step is for the agency to apply the final EIS to the decision-making process and then submit the final EIS to the CEQ for filing† if the action is to be undertaken. Of course, there is always the possibility that, on the basis of the final EIS, a suit will be brought against the agency by a group in order to gain a court decision on whether or not the proposed action should take place. It may be of interest to note that as of June 5, 1975, out of more than 6000 EIS's, only 654 cases involving NEPA had come to the courts. Of the 654, only 291 cases concerned the adequacy of the subject EIS's.

19-4 THE ROLE OF THE EPA

The EPA, the national environmental regulatory agency, is not the administrative agency for EIS's, nor does it accept or reject EIS's. It does, however, have the responsibility under §308 of the Clean Air Act "...to review and comment in writing on the environmental impact of any matter... ." Although other agencies are required by the CEQ guidelines to review EIS's in particular categories, EPA is required to review almost all government EIS's. In carrying out this requirement, EPA has issued an internal manual, *Review of Federal Actions Impacting the Environment* (EPA, 1975). Essentially, each DEIS is reviewed by EPA and is assigned a category 1, 2, or 3 (see Table 19.9) as to the adequacy of the EIS. If the category is a 1 or 2, EPA then comments on the environmental impact of the action (Table 19.9) with a rating of LO (lack of objections), ER (environmental reservations), or EU (environmentally unsatisfactory). It should be noted that CEQ relies heavily on EPA rating. If the DEIS received a 2 or 3 or an ER or EU, EPA will again review the final statement.

†The 1978 CEQ-proposed EIS regulations (CEQ, 1978b) require that EIS's be filed with EPA, Office of Federal Activities (A-104), 401 M Street SW, Washington, DC 20460.

TABLE 19.9 Definition of Ratings Used by EPA in Reviewing Federal Agency Actions and Environmental Impact Statements

Environmental Impact of the Action

LO—Lack of Objections
EPA has no objections to the proposed action as described in the draft impact statement or suggests only minor changes in the proposed action.

ER— Environmental Reservations
EPA has reservations concerning the environmental effects of certain aspects of the proposed action. EPA believes that further study of suggested alternatives or modifications is required and has asked the originating federal agency to reassess these aspects.

EU—Environmentally Unsatisfactory
EPA believes that the proposed action is unsatisfactory because of its potentially harmful effect on the environment. Furthermore, the agency believes that the potential safeguards that might be utilized may not adequately protect the environment from hazards arising from this action. The agency recommends that alternatives to the action be analyzed further (including the possibility of no action at all).

Adequacy of the Impact Statement

Category 1—Adequate
The draft impact statement adequately sets forth the environmental impact of the proposed project or action as well as alternatives reasonably available to the project or action.

Category 2—Insufficient information
EPA believes that the draft impact statement does not contain sufficient information to assess fully the environmental impact of the proposed project or action. However, from the information submitted, the agency is able to make a preliminary determination of the impact on the environment. EPA has requested that the originator provide the information that was not included in the draft statement.

Category 3—Inadequate
EPA believes that the draft impact statement does not adequately assess the environmental impact of the proposed project or action, or that the statement inadequately analyzes reasonably available alternatives. The agency has requested more information and analysis concerning the potential environmental hazards and has asked that substantial revision be made to the draft statement.

If a draft impact statement is assigned a Category 3, ordinarily no rating will be made of the project or action since a basis does not generally exist on which to make such determination.

19-5 PREPLANNING FOR EIS

The scope and detail of the meteorological and air-quality input and the analysis of any given EIS can vary from a short paragraph on the climate to a full blown all-inclusive meteorological/air-quality modeling effort. One certainty, however, is that the science of meteorology has a role in each and every EIS. There are several steps and considerations which should be treated before actually preparing the EIS. A certain amount of preplanning will save the EIS writer a great deal of distress and sometimes embarassment. There are several good handbooks available with detailed discussions of EIS preparation: Canter (1977), Burchell and Listokin (1975), and Cheremisinoff and Morresi (1977). Other references that give the reader a good EIS background are Rosen (1976), Heffer-man and Corwin (1975), Pantell (1976), and Black and Herrington (1974a,b). Persons preparing the meteorological portion of an EIS may find "Meteorological Content of Environmental Impact Assessments" (Mahoney and Spengler, 1975) and "Federal Government Requirements for Environmental Impact Assessment" (Van der Hoven, 1975) particularly useful. By no means does the author intend the preceding lists to be considered inclusive.

There is a basic first step that is sometimes overlooked and not usually suggested in available EIS checklists: namely, once you have been assigned the duty of preparing a section of an EIS, determine who will review it within and outside of your organization. Also, find out if there are any environmental groups that are concerned about a particular aspect of the action. The results of these inquiries may have an effect on how you proceed or gauge the level of effort of your EIS assignment. This author once recommended that photochemical, air-quality modeling not be included in an oil pipeline project because he felt the current state-of-the-art would not produce useful results compared to the modeling costs involved. An inventory of the photochemical constituents including insolation along with a general explanation of possible photochemical impacts should have been enough. Fortunately, the potential EPA EIS reviewer had a chance to comment on the contract being issued to prepare the EIS before its consummation. Owing to the local sensitive political nature of the seaport facility included in the pipeline project, the EPA reviewer expected photochemical modeling to be included, and three different photochemical modeling approaches at that, to emphatically ensure that an assessment of the photochemical impact was included. Hence political considerations very often lead to specific analyses not always directed by science.

Mahoney and Spengler (1975) offer good advice about preplanning for the meteorological portion of an EIS. The following considerations have been updated and should be given in any EIS.

19-5.1 Regulatory Requirements

Our environmental laws are changing continually. The trend is for stronger and stronger constraints on so-called environmental exploitation. In addition to federal laws and regulations, there are many state and some local environmental laws that must be considered. Most recently the federal government has significantly strengthened its PSD laws. The EIS must now consider PSD impact which involves even more detailed meteorological analysis than in the past. Some regulations require specific analysis to be made. It would behoove the

EIS preparer to take the time to investigate and become familiar with all the laws and regulations that may impact on the proposed action in question. In addition, nuclear power actions require specific regulations and standards that must be followed concerning meteorological analysis (see preceding discussion on the Atomic Energy Act of 1954).

19-5.2 Scope and Detail of EIS

There is a varying spectrum of efforts given on any EIS. The EIS preparer should assess the complexity of the action: volume of assessments (in air quality; number of sources or receptors); importance of impact (a small source in a large urban nonattainment area is handled differently from the same small source in a PSD Class I attainment area; also see de minimis levels Table 19.5); toxicity of emissions; and proximity to sensitive receptors (hospital, schools, etc.). Each of these factors may bring different focus and emphasis to the EIS. The new CEQ NEPA regulations (CEQ, 1978c, Part 1501.7) require a "scoping" meeting, or meetings, prior to EIS preparation. The meetings are to be held by the sponsoring agency with all participating or interested parties in attendance to decide which issues will be emphasized and which issues will be excluded in the EIS.

19-5.3 Adequacy of Available Data

The method and level of analysis that can be performed will depend greatly on the amount and adequacy of meteorological or air-quality data available at or near the location of the proposed action or whether the available data can be used to analyze the proposed action. Most large power projects now require at least 1 yr of on-site meteorological and background air-quality data. This requirement has been strengthened by the new PSD requirements of the 1977 amendments to the Clean Air Act. If new on-site data are required, considerations should be given to specific parameters needed for specific analyses (i.e., dispersion estimates, plume rise calculations, etc.). The EPA (1974) provides guidelines for air-quality monitoring and data analysis.

19-5.4 Standardization of Assessment

Where possible, an attempt should be made to keep the analysis standard. In air-quality works, specific data and calculation averaging times are usually evaluated in addition to worst case, second-worst case, expected exceedances, and normal ambient meteorological scenarios. Although it is also necessary to consider the uniqueness of each local site, evaluation of standard impacts such as yearly averages and averaging times specific to NAAQS and the location of maximum short-term impacts help to standardize the EIS review process. Overlooking a standard analysis or comparison may result in an EPA category 2 evaluation—insufficient information. The EPA (1977a) provides guidelines for evaluating air-quality impact of New Stationary Sources.

19-5.5 Assessment Method

The accuracy, completeness, and quantity of data available and the importance of a particular assessment will determine the type of atmospheric dispersion model used and the overall level of effort. A balance should be achieved between the required analysis, the accuracy of the data, and the sophistication of the evaluation methodology. The new PSD classifications address air-quality increments that, in some cases, are smaller than the capability of state-of-the-art air-quality monitors. Splitting hairs on these increment assessments and the technical justification for acceptance will rely greatly on the models chosen.

19-6 PREPARATION OF THE AIR QUALITY AND METEOROLOGICAL PORTIONS OF THE EIS

In the following discussion of an assessment framework outlining the meteorological and air-quality contents of an EIS, the author has tried to follow the order of NEPA requirements as specified by the CEQ guidelines (Table 19.1). This presentation is not meant to be inclusive or to suggest that these procedures must be followed exactly.

Essentially, the main portions germane to meteorology and air quality are the description of the existing and future environment, the assessment of impact, and the evaluation of alternatives. The alternatives are usually given and are assessed according to normal procedures; therefore there is no reason to cover them separately beyond mention here and to point out that the area covered or data gathered may be increased by alternatives. Keep in mind that the data gathered and analyzed may be voluminous, and for meaningful presentation the data must be consolidated. Usually a separate technical document is prepared that backs up a summary which is presented in the actual EIS.

19-6.1 Description of Existing and Future Environment

Before any action can be assessed, it is necessary to know the existing environment. The EIS should describe current baseline conditions of the environment. In addition to specific parameters that may be affected, the EIS preparer must consider the possible needs and related interests for all phases of the project. This includes the planning and design phase, the construction phase, and the operation phase. For example, heavy rains typical of a local climate during a certain season may significantly affect soil conditions and hamper any construction undertaken during that period. Such data are valuable during decision making and planning. Information demonstrating past actions that have altered the natural climate may be useful in the assessment of a no-action alternative. An impacted area can be divided into geographic areas for ease in describing the environment and for reporting on the environmental impact.

19-6.1.1 Existing Climatology. The eventual reviewers of the EIS are not always meteorologically oriented; therefore it is important to consider how to present the material effectively. For instance, a wind rose plot would be more effective in most cases than a presentation of wind frequency in table format. This is not to say that all the data should not be available in a separate appendix (or at the preparing office) for possible later use. Also it is important to predict specific climatological data that may be needed in the assessment stage (e.g., humidity data for the impact analysis of cooling towers). All the required data for later modeling should be assembled. The information gathered for the existing climate portion of the EIS may be used by other disciplines to evaluate such effects as impact on flora or fauna. Data can be presented as maps with overlays, tabulations, or narrative. Locations at which data were collected should always be presented on a map.

19-6.1.1.1 Data Not Available. It is important to decide whether existing data can be extrapolated to the site under consideration, whether area generalities will be sufficient for the given EIS, or whether data will have to be collected at the site.

Data gaps that will restrict the level of impact analysis should be identified in the EIS. Identification of such gaps is related to the level of the required impact assessment and to the need of these data for impact assessment. What are insufficient data in some statements may not be in others. Some projects will require on-site data collection (observations) regardless of other available data.

Mahoney and Spengler (1975) give some good advice for an on-site observation program:

1. Involve an experienced meteorologist in the selection of instruments and sites.
2. Review the measurement program with appropriate regulatory officers, consultants, or other persons directly involved with the impact assessment and request specific approval of the observation program.
3. Adopt standard procedures for operation, maintenance, and recalibration of equipment and for data analysis.
4. Consider the question of record length. Usually a 1-yr record is the bare minimum in any air-quality assessment.

The reader should be aware of an attempt by EPA to specify the type of on-site data to be collected to characterize dispersion data for impact prediction. In January 1980, an EPA workshop on "Meteorological Instrumentation for Characterization of Dispersion from Point Sources" was held with 30 prominent meteorologists in attendance. The results of this workshop are not published to date (Hoffnagle, 1980). However, one effect will probably be an increase in the number of parameters collected at a typical meteorological monitoring site in the future.

19-6.1.1.2 Data Used for Air-Quality Assessment. In the description of dispersion potential of a site, some meteorological parameters are handled in special ways. For example, the National Climatic Center Stability Array (STAR) is a widely accepted climatology for general site dispersion potential. The STAR data, which are frequency tables of Pasquill–Gifford stability classifications and other meteorological parameters, such as wind direction and speed, are generated from hourly National Weather Service observations. These observations should be used with care because they are obtained for aviation and general meteorological use rather than for air-quality modeling. Furthermore, the observations must be extrapolated to the proposed project location.

When upper air data are available, frequency of various temperature structures (inversions, stable layers, and mixing heights) are also very useful. The EPA publications by Holzworth (1972, 1979) are useful references on mixing heights.

When an on-site observation program is established, a climatology of eddy energies, Richardson numbers, or any other specific requirement can be calculated and recorded directly for later use in air-quality models. Some usual dispersion potential parameters are:

—Mixing heights (mechanical and/or thermal).
—Inversion frequency.
—Mean wind speeds.
—Potential episode (air stagnation) days.
—Historical record of pollution episodes in the area.
—Vertical and horizontal eddy energies.
—Horizontal and vertical wind-direction fluctuations (used to calculate σ_y and σ_z).
—Bulk Richardson number and/or delta temperature.

19-6.1.2 Existing Air Quality. A description of existing air quality involves data both on existing ambient air-quality levels and on emissions from existing sources of air pollution. It should be determined what specific pollutants will be emitted during the construction and operation phase of the proposed action. Discussion may be limited to these pollutants, but this should be verified. Existing air quality should be compared with NAAQS and typical background values found in nature (Tables 19.2 and 19.10). Of particular importance concerning the description of existing air quality is the consideration of what specific regulation will govern the potential increase in air-quality levels associated with your project. For instance, the following regulatory limitations should be considered:

NAAQS: National Ambient Air-Quality Standards (attainment or nonattainment)
SAAQS: State Ambient Air-Quality Standards
PSD Class I: Includes visibility degradation and a determination of Best Available Control Technology (BACT)
Class II
Class III
Nonattainment areas: Requires offsets and Lowest Achievable Emission Rate (LAER)
NSPS: New Source Performance Standards
Hazardous pollutants
De minimus: Emissions and air concentrations

All these factors will affect the assessment portion of the EIS. A list should be made of specific requirements of the preceding regulations, including numerical standards.

The description of the existing air-quality format must be compatible with the standards and the impact analysis. The existing air quality should be described by pollutant in terms of the maximum pollution concentrations observed for appropriate averaging times and record lengths. Some pollutants have specific time averages, e.g., 8 h for carbon monoxide or 3 months for lead. Next, the major sources of air pollution should be identified. Categorize the sources as to geometry (point, line, or area), frequency (continuous, intermittent, or instantaneous), and type (industrial, power plants, residential, highways, etc.). Climatology information causing specific air-pollution problems in the general area should then be discussed, e.g., a land–sea breeze situation that may cause local pollution recirculation of an industrial source.

A source of data on ambient air quality and emissions is provided by EPA's Aerometric and Emissions Reporting System (AEROS) (EPA, 1976a) which contains the subsystems: Storage and Retrieval of Aerometric Data (SAROAD) for air-quality concentrations and National Emissions Data System (NEDS) for emissions data. Other sources of data are state air-quality agencies, local air-pollution-control districts, private companies, and special studies or experiments continually being carried out by government and private nonprofit organizations. If it is necessary to compile an emission inventory from scratch, consult *Guide for Compiling a Comprehensive Emission Inventory* (EPA, 1973).

The new PSD requirements require that background air-quality levels be specified for any given area. Background levels can be established through extrapolation of existing data, modeling, or monitoring. If background monitoring is required, the comments given in the section on meteorological on-site observations apply.

19-6.1.3 Presentation of Data. In the preparation of the climatology inventory, flexibility is needed in cases where data are not available or in situations that may require addi-

TABLE 19.10 Summary of Sources, Annual Emissions, Background Concentrations, and Major Sinks of Atmospheric Gaseous Pollutants†

Pollutant	Major source		Estimated emission, 10^9 kg		Background concentration, $\mu g/m^3$	Major identified sinks
	Anthropogenic	Natural	Anthropogenic	Natural		
Sulfur						
SO_2	Combustion of coal and oil	Volcanoes	65	2	1–4	Scavenging; chemical reaction; soil and surface water absorption; dry deposition
H_2S	Chemical processes; sewage treatment	Volcanoes; biological decay	3	100	0.3	Oxidation to SO_2
Nitrogen						
N_2O	None	Biological decay	None	590	460–490	Photodissociation in stratosphere; surface water and soil absorption
NO	Combustion	Bacterial action in soil; photodissociation of N_2O and NO_2	53 combined with NO_2	768	0.25–2.5	Oxidation to NO_2
NO_2	Combustion	Bacterial action in soil; oxidation of NO			1.9–2.6	Photochemical reactions; oxidation to nitrate; scavenging
NH_3	Coal burning; fertilizer; waste treatment	Biological decay	4		4	Reaction with SO_2; oxidation to nitrate; scavenging
CO	Auto exhaust and other combustion processes	Oxidation of methane; photodissociation of CO_2; forest fires; oceans	360	3000	100	Soil absorption; chemical oxidation
O_3	None	Tropospheric reactions and transport from stratosphere		(?)	20–60	Photochemical reactions; absorption by land surfaces (soil and vegetation) and surface water
Nonreactive hydrocarbons	Auto exhaust; combustion of oil	Biological processes in swamps	70	300	$CH_4 = 1000$ non-$CH_4 < 1$	Biological action
Reactive hydrocarbons	Auto exhaust; combustion of oil	Biological processes in forests	27	175	<1	Photochemical oxidation

†Cheremisinoff and Morresi (1977).

tional climatological information not specified below. Comments should be incorporated specifying that data are not available in those cases. Data can be presented visually with overlays and isopleth maps, in tabular form, or by narrative. The following is an example of what may be included in the existing environment section of an EIS or in the backup technical document to the EIS. This does not imply that all the items below are required for an EIS; the list is presented for guidance only. In practice the existing environment section is usually less than 3 pages long.

A. Climate

1. Map Overlay
   Wherever distributions are displayed by zones or isopleths, the individual data points used for the analysis should be displayed. Where data are not sufficiently dense to prepare isopleths, it is better to document rather than to extrapolate.

   a. Show the mean annual precipitation pattern. Indicate by isohyets. If desirable, draw isohyets for precipitation by season or for growing season (i.e., April through October).
   b. Indicate the average yearly snow pack. Use inch isopleths in appropriate increments to display data.
   c. Plot isotherms of average maximum and minimum temperatures for January and July. Choice of isotherm increments would be chosen to best present data.
   d. Show the average length of frost-free period. Draw isopleths in 20-d increments. This parameter is particularly important in mountainous terrain where growing regions are terrain dominated.
   e. Plot wind roses showing frequency distribution of wind speed and wind direction for the surface and, if possible, for the upper air below 5000 ft. If there is significant diurnal or seasonal variation, plot separate wind roses according to time of day or season.

f. Locate weather stations within the area or adjacent to the area from which meteorological data are available. Indicate type of station and parameters measured (i.e., recording/nonrecording, temperature, wind speed, etc.).

2. Tabulations

Complete data tabulations are not always necessary. Wherever possible the National Climatic Center (NCC) Local Climatological Data (LCD) summaries should be used for the tabulation requirement. The NCC LCD summaries may contain more data than your EIS requires. Possible inclusions are the following tables or bar graphs:

a. Average monthly distribution of precipitation for representative stations within the study area. Show the average monthly precipitation, maximum monthly 24-h precipitation, and average monthly snowfall.
b. Average monthly frequency of thunderstorms. If hail and thunderstorm intensity data are available, list them also.
c. Average snow pack by month if the data are available. Indicate water content. These data are sometimes available through the Soil Conservation Service for mountain states. List the average first and, then, the last day of snowfall for each station if they are available.
d. Average maximum and minimum relative humidities for each month.
e. Average monthly mean, maximum, and minimum temperatures, and record high and low temperatures for each month.
f. Date of the first killing frost in the autumn, date of the last killing frost in the spring, and length of growing season for the last 10 yr. Indicate average date of first autumn freeze and last spring freeze.
g. Average monthly or seasonal wind direction/wind speed frequency distributions. (When wind averages are prepared from raw data, the vector mean direction and speed are customarily the mean values calculated and listed.
h. Average monthly evaporation rate.

3. Narrative

In the preparation of the narrative, keep in mind that the ultimate purpose of an EIS is to provide decision makers with pertinent information necessary for *good* decisions. It is recognized that the climate narrative uses are interdisciplinary. The narrative should not be limited to a recapitulation of overlay and tabulation data but should highlight climatic situations that may limit or aid particular aspects of the project.

a. Discuss the effect of precipitation amounts and seasonal precipitation distribution on the various parts of the proposed action. For example, what restrictions are imposed by precipitation, or the lack of it, at the construction site? Are there any major drought/flood occurrences? What are their duration and effects? What are the long-term trends in precipitation (amounts and distribution) and what influence do they have? Indicate occurrences of extreme precipitation episodes along with their effects. Discuss seasonal distribution of thunderstorms in terms of frequency and intensity. Areas repeatedly affected by severe thunderstorms with damaging winds and/or hail should be included.
b. Describe snow-pack areas by types of snow, depths, and water content. Discuss the average maximum and minimum snow pack, the mean duration, and the possible effects on the project.
c. Indicate average annual potential evaporation. Describe its seasonal distribution and compare with precipitation. Discuss its effects on vegetation and water resources that are affected by the proposed project. Describe the occurrence and effects of periods of extreme high or low humidity.
d. Discuss effects of temperature extremes, freezing, and thawing periods and the length of the average frost-free season.
e. Discuss any remaining climate-related topic that may be of specific importance to the proposed project.
f. Discuss effects of severe storms such as hurricanes and tornadoes. This is an especially important consideration in the safety assessment of nuclear power plants.

B. Air Quality

1. Map Overlay

a. Plot and identify significant existing pollution sources (point, line, and area sources) both within and immediately external to the area influenced by the proposed action.
b. Plot location and type of air-monitoring stations in and immediately adjacent to the proposed project.
c. Outline areas that are of particular importance to the air-quality evaluation: air-quality control region (AQCR) boundary lines, nonattainment areas, and areas where the topography can physically restrict the dispersion of pollutants.
d. Plot PSD Class I and III boundaries. Specific Class I boundaries can be obtained from the federal land manager who has jurisdiction over the land in question. To date there are no Class III areas; however, future Class III boundary areas will be delineated by the state having jurisdiction over the PSD regulations.
e. Show areas having the potential for reclassification under the PSD regulations.

2. Tabulations

Wherever possible, existing formats should be used [e.g., STAR data and EPA National Emissions Data System (NEDS) data]. Use evaluation formats referred to below as guidance when existing tables are not available:

a. Frequency of stagnation episodes as defined by mixing depths, wind speed, and frequency of occurrence and duration.
b. Inversions: type, time, seasonal variation, frequency of occurrence, duration, intensity, and depth.
c. Average mixing depth by season.

d. Average atmospheric stability conditions by season.
e. Existing atmospheric pollution concentrations compared to applicable standards.
f. Pollution sources: location, type of pollutant, height of emission, and amount emitted.
g. Average maximum and minimum visual range by season where available.

3. Narrative

a. Discuss the pollution dispersion potential in the area in terms of seasonal average mixing depths, wind speed and direction, stability conditions, and inversion frequency and intensities (both surface and subsidence).
b. Discuss existing visibility and pollution levels and other influencing pollution sources from both within and outside the project boundary. Relate visibility and pollution levels to the applicable short-term and annual air-quality standards. (The EPA is required to establish visibility regulations by Aug. 7, 1979.)
c. Discuss local trends in pollution levels. Relate to applicable air-quality standards and regulations.
d. Describe the current air-pollution effects on the environment. Relate items a, b, and c to potential future land uses and to the applicable air-quality standards and regulations.
e. Discuss the basis for potential change of PSD classification. Relate this to existing land uses, potential land uses, and level of air-quality degradation to be allowed. Discuss potential influence on adjacent areas, particularly Class I or potential Class I.

19-6.2 Future Environment

The discussion on the future environment in most EIS's is usually very brief. Only environmental changes that would be expected to occur without implementation of the proposal are covered. These changes would include developments in adjacent areas. The future time frame considered would depend on the projected economic life of the project. This discussion is not an impact assessment; it should only discuss in general what future climates or air-quality conditions may reasonably occur in the site area without the project.

19-6.3 Assessment of Impact

The impact due to an action is any change in the present environment that can be predicted in terms of magnitude, intensity, duration, and incidence. The impact assessment should answer: *how* the impact will be caused; *how much* in terms of magnitudes and potential ranges; *when* the impact will occur in terms of years, seasons, phase of project, and duration; *where* the impact will occur; the *significance* of the impact by comparing it to known standards; and the *methodology used* to identify and quantify the impact. Since it would be impossible to evaluate all possible impacts, professional judgment as to what to include as well as agreements from EIS scoping meetings will define specific assessments. It is also important to acknowledge situations where impact occurrence, magnitude, or duration is unknown.

19-6.3.1 Impact on Existing Climatology. The existing climatology or microclimatology, or both, is going to impact on the proposed action as well as possibly be impacted by it. The possibilities are very numerous and dependent on the specific climate and project. As an example of possible climatology effects on a project, Burchell and Listokin (1975) list such considerations as:

—Whether heavy precipitation will cause flooding.
—Likelihood of tornadoes and hurricanes.
—Need for project modifications due to heavy fog, areas of extremely strong winds, and extremes of temperature.

Examples of project effects on the microclimate include:

—Air pollutants filtering out sunlight.
—Local increases in cloudiness or fog.
—Topographic modifications or vegetation change affecting air circulation.
—Changes in surface heating and water runoff.

Generally the larger scale climate will not be affected,† possible exceptions being large-scale weather modification projects, power parks, large prescribed burning programs, or areas prone to major volcanic eruptions.

19-6.3.2 Impact on Existing Air Quality and Receptors. Assessment of the air-quality impact of a proposed action requires a description of the potential emissions along with specific source characteristics (geometry, emission duration, and effluent properties), the meteorology information required for the particular assessment methodology, and a decision on the receptor locations or areas or the circumstances to be investigated. It is also important to have topographical information and to choose the proper spatial and temporal resolution necessary to determine the required impacts.

19-6.3.2.1 Description of Emissions and Source Characteristics. It is best to start with a description of the emissions since this will logically lead to the choice of time and space-scale consideration which will then indicate possible affected receptors. Once these inputs are known, the time and space scales can be determined and a modeling approach can be chosen.

The emissions description serves two purposes: first, it provides the information necessary to compare with NSPS (Table 19.3), which are limiting factors that may require the inclusion of pollution control devices regardless of the ambient air quality impact. Second, the emissions description provides the necessary information for input into air-quality models.

To establish the potential emissions, it is necessary to have "emission factors" for the proposed process of interest. EPA provides and continually updates emission factors in *Compilation of Air Pollutant Emission Factors* (1976c). Usually, emission factors are given in terms of grams per time per weight of material burned, or grams per weight of material processed. Model inputs generally require emissions in g/s for point sources, $g \cdot s^{-1} \cdot m^{-1}$ for line sources, and $g \cdot s^{-1} \cdot m^{-2}$ for area sources. The sufficiency and accuracy of the air-quality calculation is dependent on and directly related to the completeness and accuracy of the emission data used.

19-6.3.2.2 Dispersion Potential and the Meteorological Scenario. The NAAQS are promulgated for various averaging periods (e.g., 1 h, 3 h, 24 h, and annually) in addition to specialized periods such as during peak traffic hours (8 h CO standard). It is difficult to determine what combination of

†It is recognized that there are long-term problems (increased CO_2 levels, desertification, and deforestation) related to large-scale climate change which go beyond the average EIS.

meteorological conditions and emissions should be considered in a model to effectively calculate the average and worst possible or second worst possible air-pollution situation. There are various methods of choosing a meteorological scenario of wind directions, wind speeds, dispersion probability, mixing heights, and duration and timing of sampling. These methods include choosing a specific scenario based on the modelers' experience with similar sources, terrain, and climate; sorting a long-term meteorological data set for worst case† situations; and calculating air-quality concentrations for each and every hour for an entire year ("brute force method"). The air-quality model used and the complexity of the required assessment determine the methods for choosing a meteorological scenario. For instance, EPA's CRSTER model (EPA, 1977b) uses an entire year of meteorological data as input.

19-6.3.2.3 Atmospheric Dispersion and Computer Modeling. The NEPA EIS requirements and the EPA air-quality regulations have encouraged a proliferation in the number of air-quality computer algorithms used to calculate air concentrations. The majority of the marketed models use the Gaussian approach, but there also are some that solve the dispersion equations by using finite differencing techniques. Many consulting firms encode the Gaussian model, incorporate in-house simplifications, add on emissions inventory, give the resulting "model" an appealing acronym, and market it as a priority, unique approach. Some of these models are more efficient and more soundly applied than others. In an attempt to provide greater consistency in modeling, EPA has published *Guideline on Air Quality Models* (EPA, 1978a). This guide makes specific recommendations concerning (1) air-quality models, (2) data bases, and (3) general requirements for concentration estimates. The National Commission on Air Quality (NCAQ) has also formed a special panel to review air-quality models.

Since May 1973, the EPA Meteorology Laboratory at Research Triangle Park, N. C., has made a number of models available to those wanting to make dispersion estimates. These models are collectively referred to as UNAMAP: User's Network for Applied Modeling of Air Pollution. The UNAMAP models are available through the National Technical Information Service, U. S. Department of Commerce, Springfield, Va. 22161.

The EPA also suggests a screening procedure that does not require a computer for evaluating whether a new source must undergo detailed modeling or not. These procedures are found in *Guidelines for Air Quality Maintenance Planning and Analysis, Vol. 10 (revised): Procedures for Evaluating Air Quality Impact of New Stationary Sources* (EPA, 1977a).

It is important for the EIS preparer to decide the modeling approach and depth of analysis that are appropriate for the specific evaluation in question. Considerations include magnitude, intensity, duration, incidence of occurrence, synergistic impacts, isopleth mapping, comparison of estimates with standards, model accuracy and confidence limits, micro–meso scale topography complications,‡ mitigating measures, and data reduction/presentation techniques. These considerations should be addressed during the "scoping" process. The two EPA publications cited and the information in this chapter should be adequate to guide the EIS preparer to the proper modeling approach.

19-6.3.2.4 Visibility Deterioration. Section 169A of the Clean Air Act as amended in 1977 (U. S. Congress, 1963) declares as a national goal the prevention of visibility impairment from man-made air pollution and the restoration of natural visibility in mandatory Class I federal areas. The EPA must complete a study recommending methods for (1) identifying, characterizing, determining, quantifying, and measuring visibility impairment, (2) modeling techniques for determining the extent of visual impairment, and (3) preventing and remedying visibility impairment by Feb. 7, 1979. The EPA was also required to promulgate visibility regulations by Aug. 7, 1979. This did not happen; however, EPA has now proposed partial regulations in the *Federal Register* of May 22, 1980, pp. 34762-34779.

Visibility is now an issue of environmental concern to be included in any environmental assessment. Because there is a time gap involving EPA visibility studies and the eventual promulgation of regulations, EIS preparers should pursue a reasonable approach to the problem of assessing visibility impacts in the interim. It is recognized that:

1. The identification and quantification of visibility parameters is a task currently underway.
2. Visibility assessment is an issue associated with social values and not strictly a problem in physical science.
3. There are no known existing reliable visibility data that can be used for completely quantifying the problem or for monitoring and documenting future deterioration.
4. Visibility models are generally not totally adequate to assess impacts properly (EPA, 1978b).

It is necessary, however, that we try to assess visibility deterioration with the tools on hand. In environmental assessments, visibility should be considered from both the social and the scientific point of view.

The following criteria, developed by the National Park Service in 1977 for assessing whether visibility was an important value in mandatory Class I areas (Fig. 19.1), provide good guidance for assessing social aspects of visibility reduction. The questions could be applied to each proposed action that potentially impacts areas where visibility may be an important value:

1. Does the area possess scenic values that are important to public enjoyment? Inventory and discuss social impact on each area. Discuss existing legislation and possible PSD area redesignations. Does the legislation for the potentially impacted areas indicate that scenic value was an important consideration for establishing the area?
2. Are any sweeping views of background features, panoramas, or views of middleground or background features present, as opposed to only views of foreground features less than 1-mile distance, e.g., streamside, trailsides? Inventory and discuss impact.
3. Do natural sources causing visibility restrictions seriously affect the ability of the public to appreciate visibility as an important value in any of the potentially affected areas? Inventory and discuss. For those areas in which natural sources of visibility impairment seriously affect public appreciation of scenic values, is the magnitude of the scenic value sufficient to warrant protection from man-caused sources?

†Worst case situations are not always associated with very stable slow wind speed conditions. In the case of a tall stack, the highest ground-level concentrations sometimes occur during unstable afternoons.

‡Section 165 of the Clean Air Act Amendments of 1977 (U. S. Congress, 1963) allows for different variance in the PSD increment values for high and low terrain.

4. Discuss the potential visual impacts to visitors in any potential Class I area (includes views from Class I area to sights beyond boundary of the Class I area).

5. Discuss any other potentially important impacts on the social aspects of visibility deterioration as a result of the proposed action.

Since there are no existing guidelines for assessing the scientific aspects of visibility deterioration, a certain degree of latitude will be necessary from project to project. The following discussion is offered as guidance:

1. National Weather Service (NWS) visibility data for NWS stations exist. However, the climatology of specific visibility distances is not directly usable owing to the data observing techniques and visibility thresholds used by the weather observers in recording visibility distances. The recorded weather associated with each NWS visibility record can be used to indicate visibility-reducing phenomena at that location, e.g., natural (blowing dust, sand, fog, etc.) or man-made (smoke, haze, smog, etc.) visibility-reducing phenomena. A summary of this weather information would be in order when it is available.

2. The size and nature of resulting aerosols predicted as a result of a proposed development will also affect visibility at different levels of degradation. An inventory of type and quantities of aerosol in relation to size distribution and chemical characteristics should be considered.

3. The color distribution of distant vistas are very important; therefore reduction in color contrasts to scenic views due to light-absorbing gases or particles (e.g., soot or nitrogen dioxide which absorb small wavelength light) should be addressed.

4. Loss of visibility (background contrast) because of scattering of light by sulfates, nitrates, fly ash, or any other aerosol resulting from pollution should be addressed and/or possibly modeled.

5. In some cases, air-pollution studies have been carried out in selected areas and visibility-related measurements (nephelometric, photographic, telephotometric, etc.) are sometimes available. These studies should be evaluated for use in a determination of the impact on visibility.

6. There are visibility models available that relate particulate concentrations to visual range and more sophisticated models that are wave length dependent and use chemistry and radiative transfer to calculate visual range and discoloring (EPA, 1978b; ERT, 1979). Although these models are recognized to be limited in the quantification of visual impacts, their use may be in order. Existing visibility models recognize the ability of the human eye to distinguish a 2% change in contrast.

This discussion of visibility is intended to provide guidance for a reasonable interim approach in assessing the social and physical impact of potential visibility reductions associated with proposed actions. Keep in mind that EPA visibility guidance and regulations and federal land manager guidance will be available in the future. The EPA (1979) has prepared a report to Congress as required in Sec. 169A on "Methods for Implementing the National Goal of Protecting Visibility" [Clean Air Act Amendments of 1977 (U. S. Congress, 1963)]. Also, the U. S. Forest Service has published proceedings from a speciality workshop on visibility values (Fox, Loomis, and Greene, 1979) which delves into complexities of assigning a deterioration value to visibility. The Air Pollution Control Association (1979) has also published *View on Visibility—Regulatory and Scientific,* proceedings of a speciality conference held on November 26–27, 1979.

19-7 CONCLUSION

Potential impacts on the environment, whether ecological, physical, chemical, aesthetic, or social, all require an input from the science of meteorology in order to make an impact evaluation. The guidance provided in this chapter can be used to put the EIS requirements and its meteorological input into perspective. Most important to a successful EIS is the "scoping" process since each EIS, like the meteorological snowflake, is always different. Whether that difference is in objective, size, time, geography, potential reviewers, or whatever, the EIS preparer should try to determine the concerns of all interested parties and should attempt to address them.

REFERENCES

Air Pollution Control Association, 1979, *View on Visibility—Regulatory and Scientific,* Symposium Proceedings, APCA, Pittsburg, Pa.

Black, P. E., and L. P. Herrington, 1974a, *Readings in Environmental Impact,* MSS Information Corp., New York.

——, and L. P. Herrington, 1974b, *Working with NEPA: Environmental Impact Analysis for the Resource Manager,* MSS Information Corp., New York.

Burchell, R. W., and D. Listokin, 1975, *The Environmental Impact Handbook,* p. 125, Center for Urban Policies Research, Rutgers—the State University, New Brunswick, N. J.

Bureau of National Affairs, 1977, Lawyers Discuss NEPA Issues, Environmental Impact Reform, in *Environmental Reporter Current Developments.*

Canada, 1970, Clean Air Act, Canada Statute C.47 (1970-71-72).

Canter, L. W., 1977, *Environmental Impact Assessment,* McGraw-Hill Book Co., New York.

Cheremisinoff, Paul H., and Angelo C. Morresi, 1977, *Environmental Assessment and Impact Statement Handbook,* Ann Arbor Science, Ann Arbor, Mich.

Council on Environmental Quality (CEQ), 1971, *Fed. Regist. (Washington, D. C.),* 36(79), Apr. 23, 1971.

——, 1973, Preparation of Environmental Impact Statements, *Fed. Regist. (Washington, D. C.),* 38(147), Aug. 1, 1973.

——, 1977, *Environmental Quality: The Eighth Annual Report of the Council on Environmental Quality,* pp. 130-135, GPO.

——, 1978a, *The Leaner Look of the New NEPA Regulations,* remarks by Charles Warren at the Environmental Law Conference, Feb. 10, 1978, Washington, D. C.

——, 1978b, National Environmental Policy Act Proposed Regulations for Implementory Procedural Provisions, *Fed. Regist. (Washington D. C.),* 43(112): 25230, June 9, 1978.

——, 1978c, Regulations for Implementing the Procedural Provisions of the NEPA, *Code of Federal Regulations,* Title 40, Parts 1500-1508; *Fed. Regist. (Washington, D. C.),* 43: 55978-56007, Nov. 29, 1978.

Environmental Protection Agency, 1973, *Guide for Compiling a Comprehensive Emissions Inventory,* Report APTD-1135.

——, 1974, *Guidelines for Air Quality Maintenance Planning and Analysis,* Vol. II: Air Quality Monitoring and Data Analysis, Report EPA-450/4-74-012.

——, 1975, *Review of Federal Actions Impacting the Environment,* transmitted to EPA Mar. 1, 1975 (internal manual).

——, 1976a, *Aeros Manual Series, Volume I: Aeros Overview,* Report EPA-450/2-76-001, EPA Office of Air Quality Planning and Standards guidelines.

——, 1976b, *The Environmental Impact Statement Process,* draft publication from the EPA Office of Federal Activities.

——, 1976c, *Compilation of Air Pollutant Emission Factors,* 2nd ed., Report AP-42, EPA Office of Air and Waste Management; available from Air Pollution Technical Information Center, Research Triangle Park, N. C.

———, 1977a, *Guidelines for Air Quality Maintenance Planning and Analysis,* Vol. 10 (revised): Procedures for Evaluating Air Quality Impact of New Stationary Sources, Report EPA-450/4-77-001, EPA Office of Air Quality Planning and Standards.

———, 1977b, *Users Manual for Single-Source (CRSTER) Model,* Report EPA-450/2-77-013.

———, 1978a, *OAQPS Guideline Series: Guideline on Air Quality Models,* Report EPA-450/2-78-027, EPA Office of Air Quality Planning and Standards.

———, 1978b, *The Development of Mathematical Models for the Prediction of Anthropogenic Visibility Impairment,* Report EPA-450/3-78-110 a,b,c, EPA Office of Air Quality Planning and Standards.

———, 1979, *Protecting Visibility—An EPA Report to Congress,* Report EPA-450/5-79-008, EPA Office of Air Quality Plannng and Standards.

ERT, Inc., 1979, *Assessment of the Principal Technical Issues Related to Visibility Protection Under the Clean Air Act,* prepared for U. S. Department of Energy, Contract No. EE-77-C-01-6036.

Fox, D., R. J. Loomis, and T. C. Greene, 1979, *Proceedings of the Workshop in Visibility Values, Ft. Collins, Colorado, January 28–February 1, 1979,* General Technical Report WO-18, U. S. Department of Agriculture, Forest Service, Washington, D. C.

Hefferman, P. H., and R. Corwin, 1975, *Environmental Impact Assessment,* Freeman Cooper and Company.

Hoffnagle, G., 1980, Workshop Leader for EPA Workshop on Meteorological Instrumentation for Characterization of Dispersion from Point Sources, ER&T, Inc., Houston, private communication.

Holzworth, G. C., 1972, *Mixing Heights, Wind Speeds, and Potential for Urban Air Pollution Throughout the Contiguous United States,* Report AP-101, Environmental Protection Agency, Office of Air Programs.

———, 1979, *Climatological Summaries of the Lower Few Kilometers of Rawinsonde Observations,* Report EPA-600/4-79-026, Environmental Protection Agency, Meteorology and Assessment Division.

Mahoney, J. R., and J. D. Spengler, 1975, *Meteorological Content of Environmental Impact Assessments,* Chap. 9, Lectures on Air Pollution and Environmental Impact Analysis, American Meteorological Society, Boston.

Pantell, R. H., 1976, *Techniques of Environmental System Analysis,* John Wiley & Sons, New York.

Rosen, Sherman J., 1976, *Manual for Environmental Impact Evaluation,* Chap. 2, Prentice-Hall, Englewood Cliffs, N. J.

Schneider, S. H., 1976, *The Genesis Strategy: Climate and Global Survival,* Plenum Press, New York.

U. S. Congress, 1954, Atomic Energy Act, *United States Code,* Title 42, §2011–2296.

———, 1963, Clean Air Act as Amended by Air Quality Act of 1967 (Public Law 90-148); The Clean Air Act Amendments of 1970 (Public Law 91-604); The Technical Amendments of 1973 (Public Law 92-157, Public Law 93-15, and Public Law 93-319); and The Clean Air Act Amendments of 1977 (Public Law 95-95), *United States Code,* Title 42, §1857 et. seq.

———, 1969, National Environmental Policy Act (Public Law 91-190), *United States Code,* Title 42, §4321-4327.

———, 1970, Environmental Quality Improvement Act, Title II of H. R. 4148.

U. S. Court of Appeals for the District of Columbia, 1971, Calvert Cliffs Coordinating Committee vs. U. S. Atomic Energy Commission, 449 F. 3d: 1109.

———, 1979, Alabama Power Company et al. vs. Douglas M. Costle, as Administrator, Environmental Protection Agency, et al., No. 78-1006, decided Dec. 14, 1979.

U. S. President, 1970, Executive Order 11514: Protection and Enhancement of Environmental Quality, *Fed. Regist. (Washington, D. C.),* 35: 4247, Mar. 7, 1970.

———, 1977, Executive Order 11991: Relating to Protection and Enhancement of Environmental Quality, May 24, 1977.

Van der Hoven, I., 1975, Federal Government Requirements for Environmental Impact Assessment, in *Lectures on Air Pollution and Environmental Impact Analysis,* American Meteorological Society, Boston.

Wetstone, G., 1980, Air Pollution Control Laws in North America and the Problem of Acid Rain and Snow, *Environmental Law Reporter,* 10(2): 50001.

Chapter 19 covers research prior to June 1980.

20 VISIBILITY AND AIR POLLUTION

Eric G. Walther, Visibility Research Center

20-1 INTRODUCTION

The combustion of coal, oil, natural gas, gasoline, wood, agricultural wastes, and similar fuels produces particulate matter, nitrogen dioxide, and other gases. These combustion products affect visibility, which has been explicitly protected by federal law since August 1977. This chapter reviews the federal visibility regulations, visibility theory, human perception of visibility, instruments for measuring visibility-related variables, and visibility modeling.

20-1.1 Federal Visibility Regulations

On Dec. 2, 1980, the Environmental Protection Agency (EPA) published regulations* to implement visibility protection provisions of the Clean Air Act (see Chap. 19). Key provisions of these regulations include:

1. Impairment. Visibility impairment is defined as "any humanly perceptible change in visibility (visual range, contrast, coloration) from that which would have existed under natural conditions."

2. Monitoring and Modeling Visibility Impairment. EPA has determined "that the present mathematical models and monitoring techniques show promise for being used in regulatory manner. However, these techniques must be further evaluated"

3. Phased Approach. The regulations recognize two distinct types of air pollution which impair visibility:

a. "smoke, dust, colored gas plumes or layered haze emitted from stacks which obscure the sky or horizon and are relatable to a single source or a small group of sources"

b. "widespread regionally homogeneous haze from a multitude of sources which impairs visibility in every direction over a large area."

Because of "certain scientific and technical limitations," EPA promulgated a phased approach to visibility protection regulations. Phase I of the program will require control of impairment that can be traced to a single existing stationary facility or small group of stationary facilities.

20-2 THEORY OF VISIBILITY

This discussion is limited to those aspects of visibility directly applicable to power production. More complete treatments can be found in Middleton (1952) and McCartney (1976). The discussion covers the optics of visibility and the contributing role of specific gaseous and particulate matter to modeling visibility.

20-2.1 Optics of Visibility

A person sees by the light reaching his eyes from objects. Light is electromagnetic radiation with wavelengths capable of stimulating the receptors in human eyes, covering the range of approximately 0.38 to 0.77 μm. Spectral radiance, N, is the amount of light energy per unit time received per unit area of detector, per unit solid angle of the field of view of the detector, per unit wavelength interval at a specific wavelength (Fig. 20.1). (See the List of Symbols at the end of this chapter.) Spectral radiance is

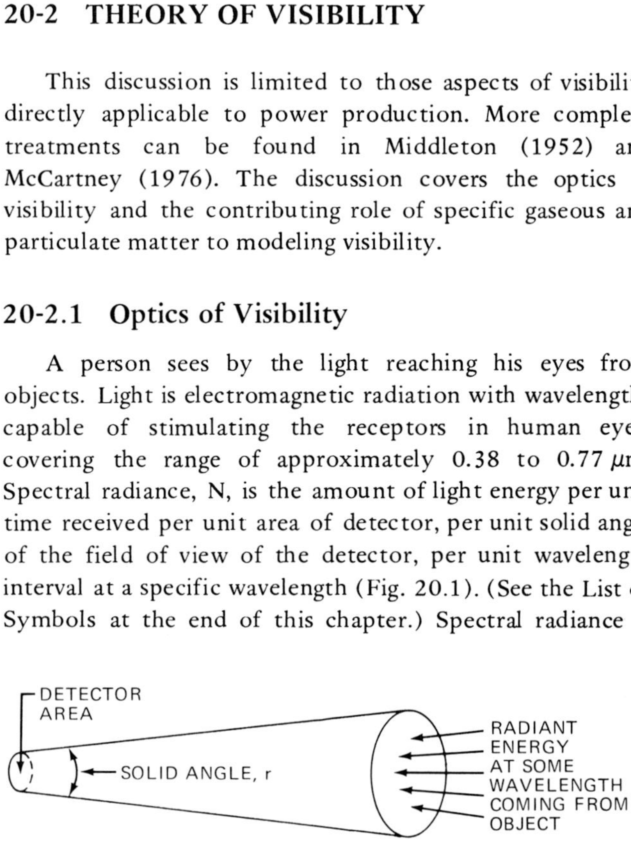

Fig. 20.1 Diagram of radiance received by detector.

called inherent radiance, N_0, if the detector is next to the object, and it is called apparent radiance, N_r, if the detector is located at distance, r, from the object. The contrast of a target against its background is the relative difference of light coming from the target and the adjacent background, usually the sky. As with radiance, contrast can be described as inherent or apparent, depending on the distance between the observer and the target. Inherent spectral contrast, C_0, is defined as

$$C_0 = \frac{{}_tN_0 - {}_sN_0}{{}_sN_0} \qquad (20.1)$$

Walther: Visibility Research Center, John Muir Institute, Department of Physics, University of Nevada, Las Vegas, Nevada.
Present address: Lockheed–EMSCO, P. O. Box 15027, Las Vegas, Nevada.

*_Fed. Regist. (Washington, D. C.)_, 45: 80084 (Dec. 2, 1980).

and apparent spectral contrast, C_r, is defined as

$$C_r = \frac{{}_tN_r - {}_sN_r}{{}_sN_r} \tag{20.2}$$

where ${}_tN_0$ = inherent spectral radiance of the target, t, at zero distance, $W/(m^2 \cdot sr \cdot \mu m)$

${}_sN_0$ = inherent spectral radiance of the sky, s, at the target, $W/(m^2 \cdot sr \cdot \mu m)$

${}_tN_r$ = apparent spectral radiance of the target, t, at distance r, $W/(m^2 \cdot sr \cdot \mu m)$

${}_sN_r$ = apparent spectral radiance of the sky, s, at distance r from the target, $W/(m^2 \cdot sr \cdot \mu m)$

Contrast is dimensionless because it is a ratio of radiances. The light coming from a target is attenuated by scattering and absorption (Figs. 20.2 and 20.3). Gas molecules and particulate matter scatter some of the inherent radiance out of the sight path and absorb another portion. Skylight and light reflected from the ground is scattered by particulate matter and gas molecules into the sight path toward the observer (Fig. 20.4). The result of these processes is illustrated in Fig. 20.5. A bright object loses radiance as the distance between it and the observer increases, approaching the limiting value of the adjacent horizon sky radiance. A perfectly black object has no inherent radiance. It acquires radiance as the path between it and the observer increases, again approaching the horizon sky radiance as the limiting value. A dark object is an intermediate case. The apparent radiance reaching the observer from a target has two parts: the attenuated inherent radiance and the path radiance added by scattering from the surrounding air. In equation form,

$$ {}_tN_r = {}_tN_0 T_r + N_r^* \tag{20.3}$$

where T_r is the transmittance of light from the target to the observer at distance r (dimensionless) and N_r^* is the spectral path radiance over distance r $[W/(m^2 \cdot sr \cdot \mu m)]$.

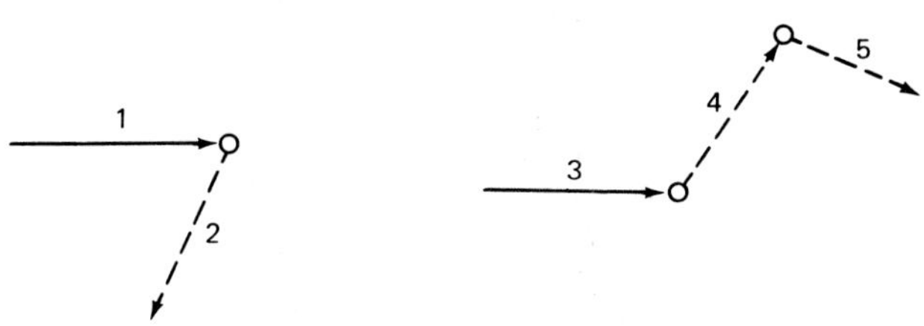

Fig. 20.2 Scattering: Photon on path 1 is backscattered along path 2. Photon on path 3 is forward scattered along paths 4 and 5.

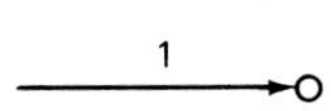

Fig. 20.3 Absorption: Photon on path 1 is absorbed by the gas molecule or particle.

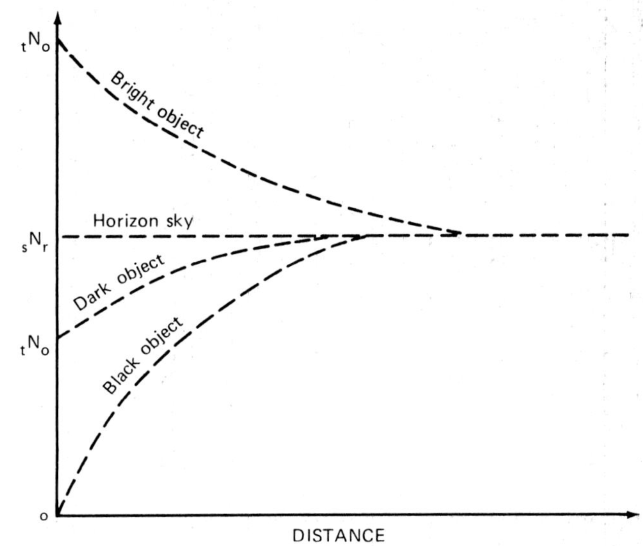

Fig. 20.5 Apparent radiance of bright, dark, or black objects as a function of distance and horizon sky radiance.

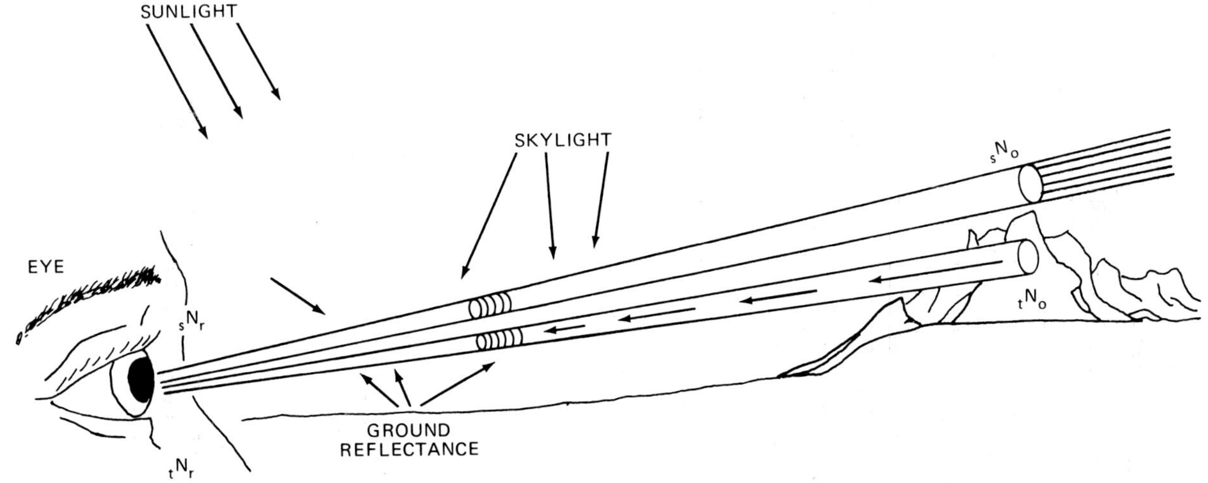

Fig. 20.4 Schematic representation of vision through the atmosphere.

Similarly, for the apparent background sky radiance,

$$_sN_r = {_sN_0}T_r + N_r^* \qquad (20.4)$$

If Eq. 20.4 is subtracted from Eq. 20.3, then

$$_tN_r - {_sN_r} = ({_tN_0} - {_sN_0})T_r \qquad (20.5)$$

Equation 20.5 expresses the fact that the difference in radiance between the target and sky is transmitted to the observer with the same attenuation as each image-forming ray of light. If we divide both sides of Eq. 20.5 by the background sky apparent radiance and multiply the right side of the equation by ${_sN_0}/{_sN_0}$, then

$$\frac{{_tN_r} - {_sN_r}}{{_sN_r}} = \frac{({_tN_0} - {_sN_0})}{{_sN_0}}\frac{{_sN_0}}{{_sN_r}}T_r \qquad (20.6)$$

Combining Eqs. 20.1, 20.2, and 20.6, we get the following relation for the apparent contrast of a target:

$$C_r = C_0 \frac{{_sN_0}}{{_sN_r}} T_r \qquad (20.7)$$

Apparent contrast depends on the inherent contrast of the target, which depends on the type and amount of vegetation on the target, the illumination of the target as a function of time of day, latitude, longitude, azimuth of the sight path, azimuth of the normal to the face of the target, and the slope angle of the target. The ratio of sky radiances at the target and at the observer is unity for the theoretical assumptions of a horizontal sight path through homogeneous air under uniform illumination on a flat earth.

If the atmosphere is homogeneous in composition along the entire sight path, then the transmission, T_r, can be expressed as a function of the extinction coefficient, b_{ext}:

$$T_r = e^{-rb_{ext}} \qquad (20.8)$$

Combining Eqs. 20.7 and 20.8 gives

$$C_r = C_0 e^{-r\bar{b}_{ext}} \frac{{_sN_0}}{{_sN_r}} \qquad (20.9)$$

If C_r and C_0 are measured with a teleradiometer and r is known, then Eq. 20.9 can be solved for the average extinction coefficient:

$$\bar{b}_{ext} = \frac{1}{r}\ln\left(\frac{C_0}{C_r}\frac{{_sN_0}}{{_sN_r}}\right)$$

Under these same conditions, visual range is defined as the distance at which the apparent contrast of a black target (CO = −1) is reduced to 2% (C_r = −0.02). Equation 20.9 can then be solved for visual range, VR, assuming ${_sNO} = {_sNR}$:

$$VR = \frac{3.912}{\bar{b}_{ext}} = \frac{3.912}{(1/r)\ln(C_0/C_r)} \qquad (20.10)$$

The choice of 2% as the threshold contrast is easily adjusted to values as high as 5%. Middleton (1952) discusses the experiments conducted by others to derive the threshold contrast. It is important not to interpret visual range too literally as the distance at which large black targets disappear. Hence the choice of threshold contrast is not critical but needs to be consistent for comparing different data sets.

The definition of visual range attempts to account for different distances between the observer and targets and different inherent contrasts. It does not account for targets viewed at different altitudes, for which the atmosphere has a different clean air (Rayleigh) extinction coefficient. Measurements of the apparent contrast of targets at different altitudes are standardized by correcting the total extinction for the Rayleigh component. This correction is made by subtracting the average Rayleigh extinction coefficient of the actual sight path from the measured total extinction coefficient and then adding the reference Rayleigh extinction coefficient. This reference is set at 0.01 km^{-1}, corresponding to an altitude of 1550 m (Elterman, 1968). Hence standard visual range, SVR, is defined by

$$SVR = \frac{3.912}{(1/r)\ln(C_0/C_r) - \bar{b}_{ext,R} + 0.01} \qquad (20.11)$$

where $\bar{b}_{ext,R}$ is the average Rayleigh extinction coefficient of the sight path.

The variables discussed thus far describe visibility without reference to what it would be in a Rayleigh atmosphere, completely unpolluted by natural or anthropogenic sources. The change in the apparent contrast of a target from its best possible value in a Rayleigh atmosphere is called delta contrast, ΔC, and is defined by

$$\Delta C = C_r - C_0 e^{-r\bar{b}_{ext,R}(z_m)}$$

where $\bar{b}_{ext,R}(z_m)$ is the average Rayleigh extinction coefficient as a function of the sight path midpoint z_m and $z_m = (z_0 + z_t)/2$. The second term is the apparent contrast of the target computed as if it were viewed through a Rayleigh atmosphere.

So far the discussion has focused on variables related to the clarity of the atmosphere and the ability to see distant objects. The discussion now shifts to the apparent color of objects viewed through the atmosphere. The color of a target is the sensation of our eye–brain system receiving and interpreting the various wavelengths of light from the target compared to the light from the surroundings (Land, 1977). Color and discoloration are important attributes of a vista, especially at national parks, such as the Grand Canyon. An "approved reference" way to describe color has not been specified by the EPA. This chapter describes several color variables.

Measurements of apparent spectral radiance, $N(\lambda_i)$, at selected wavelengths, λ_i, are used to compute the full wavelength dependence of radiance, N(λ). For example, a teleradiometer detecting four different wavelengths (i = 1,

2, 3, 4) can be used to develop four equations of the form $N(\lambda_i) = a_0 + a_1\lambda_i + a_2\lambda_i^2 + a_3\lambda_i^3$. The four polynomial coefficients are found by solving the four equations with the method of determinants. The resulting cubic polynomial equation for radiance,

$$N(\lambda) = a_0 + a_1\lambda + a_2\lambda^2 + a_3\lambda^3$$

is used to derive the tristimulus values, X, Y, and Z for the color:

$$X = \int_{\lambda_1}^{\lambda_2} \bar{x}_\lambda N_r(\lambda)\, d\lambda \qquad W/(m^2 \cdot sr)$$

$$Y = \int_{\lambda_1}^{\lambda_2} \bar{y}_\lambda N_r(\lambda)\, d\lambda \qquad W/(m^2 \cdot sr)$$

and

$$Z = \int_{\lambda_1}^{\lambda_2} \bar{z}_\lambda N_r(\lambda)\, d\lambda$$

where $\lambda_1 = 0.380\ \mu m$, $\lambda_2 = 0.770\ \mu m$, and $\bar{x}_\lambda$, $\bar{y}_\lambda$, and $\bar{z}_\lambda$ are the color mixture data for equi-energy spectrum (Middleton, 1952) or color matching functions for the 1931 International Commission on Illumination [Commission Internationale de l'Eclairage (CIE)] standard observer (Wyszecki and Stiles, 1967).

The tristimulus values of a color are the amounts of three reference lights that, when added together, give the color. The reference lights are specified by the color matching functions, which are the wavelength dependent tristimulus values of monochromatic lights of equal radiant energy for three given primary colors. The tristimulus values are used to compute the chromaticity coordinates of the color

$$x = \frac{X}{X + Y + Z} \qquad y = \frac{Y}{X + Y + Z} \qquad z = \frac{Z}{X + Y + Z}$$

The x and y coordinates can be plotted in a two-dimensional diagram as shown in Fig. 20.6. Only two of the coordinates need to be plotted because the third coordinate can be computed with the relation

$$x + y + z = 1$$

Chromaticity coordinates specify a color in terms of the amount of light in each wavelength reaching the eye from a target, regardless of the target surroundings.

To describe the color of plumes against a sky background, Latimer et al. (1978) defined a variable called the blue–red ratio as

$$R' = \frac{(N_p\lambda_1)/(N_h\lambda_1)}{(N_p\lambda_2)/(N_h\lambda_2)}$$

where p and h are subscripts for plume and horizon, and the numerical subscripts denote blue and red wavelengths, respectively. In terms of contrast of the plume against the horizon sky, the blue–red ratio is

$$R' = \frac{C_p(\lambda_1) + 1}{C_p(\lambda_2) + 1}$$

and this ratio can be used for targets other than plumes.

Chromaticity not only ignores the target surroundings but also does not differentiate color change caused by different brightness levels (Latimer et al., 1978), such as the change from yellow to brown. To quantify the brightness effect on color, a variable called value is defined as a function of the Y tristimulus value (Munsell Color Company, 1976).

According to Latimer et al. (1978) and Judd and Wyszecki (1975), the value of sky color, ${}_sL^*$, is

$${}_sL^* = 116\left(\frac{{}_sY}{Y_0}\right)^{1/3} - 16$$

where ${}_sY$ is the tristimulus value of sky and Y_0 is the tristimulus value of some reference color, such as diffuse reflectance of sunlight from a white card.

The value of a target color, ${}_tL^*$, is

$${}_tL^* = 116\left(\frac{{}_tY}{Y_0}\right)^{1/3} - 16$$

where ${}_tY$ is the tristimulus value of the target.

A color can be specified by L*, x, and y, but color differences are the key to describing discoloration by air pollution. Color differences should be specified by a variable such that it is constant for a perceptible color difference in any direction of color change. Delta E, ΔE, is such a variable and is defined in two ways by the CIE:

$$\Delta E_A = [({}_tL^* - {}_sL^*)^2 + ({}_tU^* - {}_sU^*)^2 + ({}_tV^* - {}_sV^*)^2]^{1/2}$$

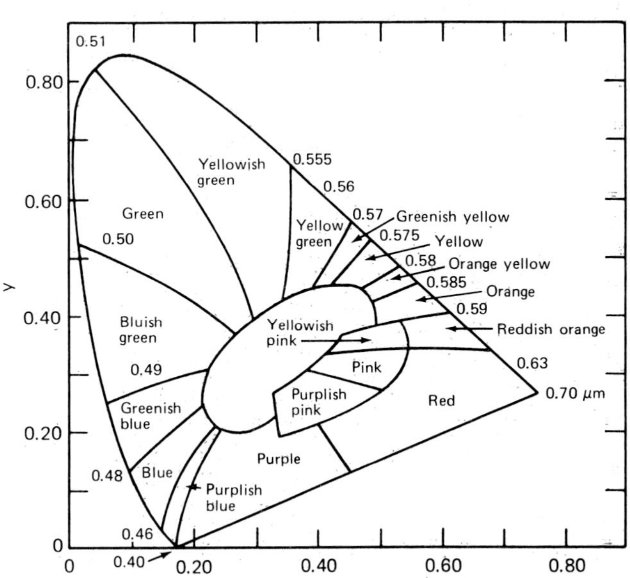

Fig. 20.6 Chromaticity diagram. (From Latimer et al., 1978.)

and

$$\Delta E_B = [({}_tL^* - {}_sL^*)^2 + ({}_ta^* - {}_sa^*)^2 + ({}_tb^* - {}_sb^*)^2]^{1/2}$$

where ΔE is relative to the sky,

$${}_tU^* = 13\,{}_tW^* ({}_tu - u_0)$$

$${}_tu = \frac{4\,{}_tX}{{}_tX + 15\,{}_tY + 3\,{}_tZ}$$

$$u_0 = \frac{4X_0}{X_0 + 15Y_0 + 3Z_0}$$

$${}_tV^* = 13\,{}_tW^* ({}_tv - v_0)$$

$${}_tv = \frac{6\,{}_tY}{{}_tX + 15\,{}_tY + 3\,{}_tZ}$$

$$v_0 = \frac{6Y_0}{X_0 + 15Y_0 + 3Z_0}$$

$${}_ta^* = 500 \left[\left(\frac{{}_tX}{X_0}\right)^{1/3} - \left(\frac{{}_tY}{Y_0}\right)^{1/3}\right]$$

$${}_tb^* = 200 \left[\left(\frac{{}_tY}{Y_0}\right)^{1/3} - \left(\frac{{}_tZ}{Z_0}\right)^{1/3}\right]$$

$${}_tW^* = 25\,{}_tY^{1/3} - 17$$

X_0, Y_0, and Z_0 are the three tristimulus values of the reference color, and ${}_tX$, ${}_tY$, and ${}_tZ$ are the three tristimulus values of the target color.

Similar equations can be written for the sky variables ${}_sU^*$, ${}_sV^*$, ${}_sW^*$, ${}_sa^*$, ${}_sb^*$, ${}_su$, and ${}_sv$. Although it takes several equations to define the terms in ΔE, there is no direct physical interpretation of each term. W^* is a lightness index, differing from value, L^*, by not being relative to a reference color. The first color difference, ΔE_A, is intended to be used with color differences on the order of magnitude of those shown in the *Munsell Book of Color,* whereas ΔE_B is intended for smaller color differences. ΔE has an advantage over chromaticity coordinates because it accounts for all three dimensions of a color (hue, saturation, and brightness).

Land (1977) has shown that we see colors almost independently of the amount of light coming from the target but that the surrounding colors are important to the perception of target color. In fact, we compare the light coming from a target to the lightest object in the scene. Unfortunately a computational scheme has not yet been demonstrated to quantitatively represent how we see colors under different lighting conditions. The best we can do is to describe one straightforward approach to compute the color contrast of a target. Let ${}_tN(\lambda)$ be the wavelength dependent radiance of the target and ${}_sN(\lambda)$ be the same for the sky, assumed here to be the brightest part of the scene. Then the target and sky luminances received are

$${}_tL = \int_\lambda p(\lambda)\,{}_tN(\lambda)\, d\lambda$$

and

$${}_sL = \int_\lambda p(\lambda)\,{}_sN(\lambda)\, d\lambda$$

where $\rho(\lambda)$ is the photopic response of the eyes. The apparent contrast of the target based on luminance is

$$C_L = \frac{{}_tL - {}_sL}{{}_sL}$$

Although our understanding of how we see color is incomplete, there is evidence that the perception of visual air quality is linearly related to color contrast. This relationship is discussed further in Sec. 20-3.

Now that several variables describing visibility have been covered, we are ready to discuss the physical processes by which particles and gas molecules affect the transfer of light through the atmosphere.

20-2.2 Relating Optics to Pollutants

Particulate matter and gases attenuate light by scattering and absorption as a function of the gaseous molecular structure, the size and composition of the particles, and the wavelength of light. Most absorption is caused by NO_2 and carbon particles, whereas most scattering is caused by particles. Any changes in source emissions and meteorology which cause higher concentrations of light scattering or absorbing materials will result in increased visibility impairment. The mathematical relationships of effluent concentration to source and meteorological factors are discussed in many of the preceding chapters. Scattering usually dominates absorption, especially in clean air, where 78 to 95% of the total attenuation is caused by scattering. Scattering is 55 to 63% of the total attenuation in highly polluted urban areas (Weiss et al., 1979). The proportional contribution of absorption to total extinction (attenuation) can be seen in Fig. 20.7.

20-2.2.1 Gaseous Scattering. Scattering by gases is treated separately from scattering by particles because of important differences. Gaseous scattering has an inverse fourth-power dependence on wavelength, which accounts for the blue color of skylight. In clean Rayleigh air, scattering is the dominant process because the nitrogen, oxygen, and other gases absorb a negligible amount of visible light. Rayleigh scattering by gases depends somewhat on the direction of observation, as shown in Fig. 20.8. Maximum forward and backward scattering is at observation angles of 0 and 180, and minimum scattering is at 90°.

More details on the scattering by an individual gas molecule are given by McCartney (1976). If the scattering by air molecules in a specific direction is summed over all possible directions, the total scattering can be found. Expressing total scattering as a coefficient, the effect of air (with a molecular density appropriate to sea level) is about

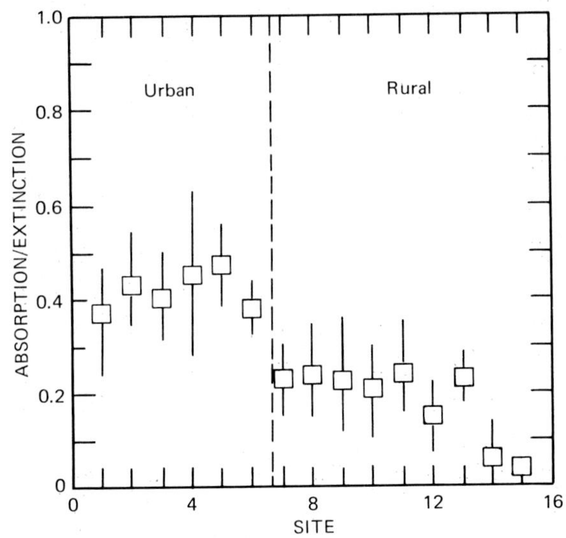

1. Industrial Seattle, Wash.
2. Downtown Portland, Oreg.
3. Industrial St. Louis, Mo.
4. Denver, Colo. (fairgrounds)
5. Denver, Colo. (trout farm)
6. Central Phoenix, Ariz.
7. Residential Seattle, Wash.
8. Residential St. Louis, Mo.
9. Tyson, Mo. (1973)
10. Tyson, Mo. (1975)
11. Milford, Mich.
12. Hall Mountain, Ark.
13. Puget Island, Wash.
14. Flagstaff, Ariz.
15. Mauna Loa Observatory, Hawaii

Fig. 20.7 Absorption/extinction measurements vs. site location. (From R. E. Weiss et al., 1979.)

10^{-2} km^{-1}. Hence Rayleigh scattering removes about 1% of the incident light per kilometer of horizontal path. When Eq. 20.10 is used, this Rayleigh scattering coefficient translates into a visual range of 391 km, assuming no absorption nor particulate scattering. Do not expect to see real objects over such distances. Mountains are not black objects, and they are not tall enough to be seen at such distances.

20-2.2.2 Particulate Scattering. Scattering by particles is more complex in its angular dependence, which itself depends on the size of the particle. Particulate scattering is often called Mie scattering, after the scientist who developed the first successful theory (Mie, 1908). As the particle size decreases, the Mie theory of scattering approaches the Rayleigh theory, appropriate to particles or gas molecules smaller than one-tenth of the wavelength of light (McCartney, 1976).

The Mie theory was developed for spherical particles of uniform composition and hence uniform index of refraction. However, ambient aerosol consists of spherical, irregular, plate-like, and rod-like particles. A compromise is

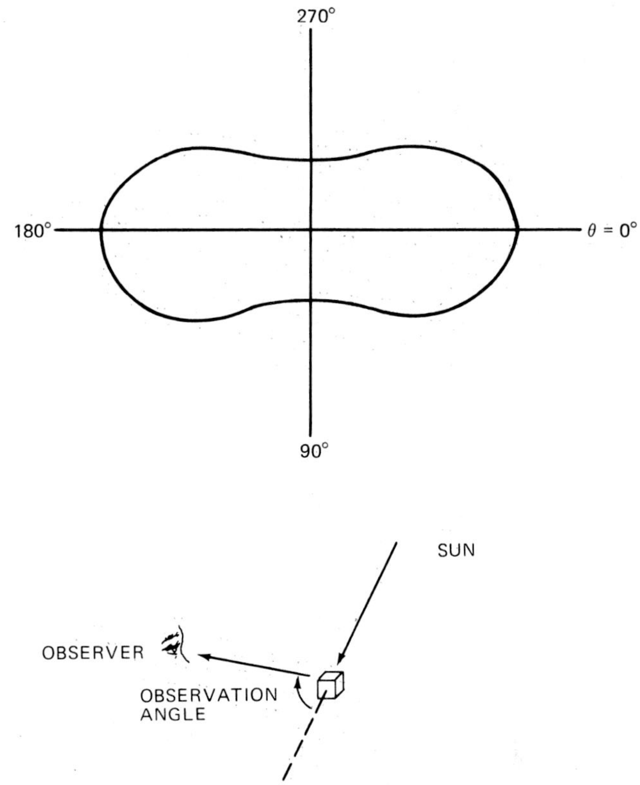

Fig. 20.8 Rayleigh scattering dependence on observation angle. (Adapted from Earl H. McCartney, *Optics of the Atmosphere*, John Wiley & Sons, New York, 1976.)

made to utilize Mie theory for nonspherical particles. Size distributions are measured by some instruments in terms of the aerodynamic behavior of the particles, from which an equivalent spherical diameter is computed. This diameter is then used in Mie theory to predict the approximate scattering of complex shaped aerosol. The angular dependence favors the forward direction, as shown in Fig. 20.9, with greater complexity as the particle size increases.

An important aspect of scattering is the efficiency with which particles of different size scatter incident light in all directions. The scattering efficiency factor is defined as the ratio of the total scattering cross section and the geometric cross section. For a spherical particle the geometric cross section is πr^2, where r is the radius of the particle. The total scattering cross section is "that cross section of an incident wave, acted on by the particle, having an area such that the power flowing across it is equal to the total power scattered in all directions" (McCartney, 1976). The dependence of the scattering efficiency factor on the size parameter, $\alpha = 2\pi r/\lambda$, is shown in Fig. 20.10. The relative size of the particle with respect to the wavelength of incident light, not its absolute size, is the important independent variable, as implied by the use of the size parameter, α. The curve with index of refraction m = 1.33 represents particles of water, whereas the m = 1.50 curve approximates silica or ammonium sulfate, two critical components related to soil and coal combustion sources, respectively. In Fig. 20.10 the scattering efficiency factor oscillates less and less around a

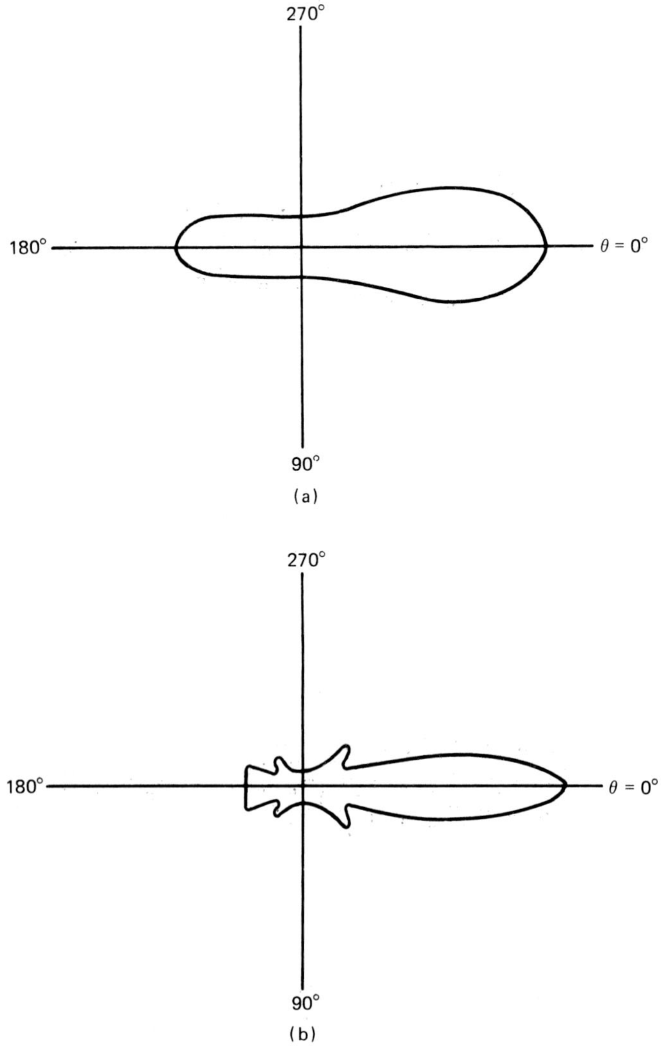

Fig. 20.9 Observation angle scattering dependence for particles. (a) Particle size ≃ λ/4; (b) λ less than particle size. (Adapted from Earl H. McCartney, *Optics of the Atmosphere*, John Wiley & Sons, New York, 1976.)

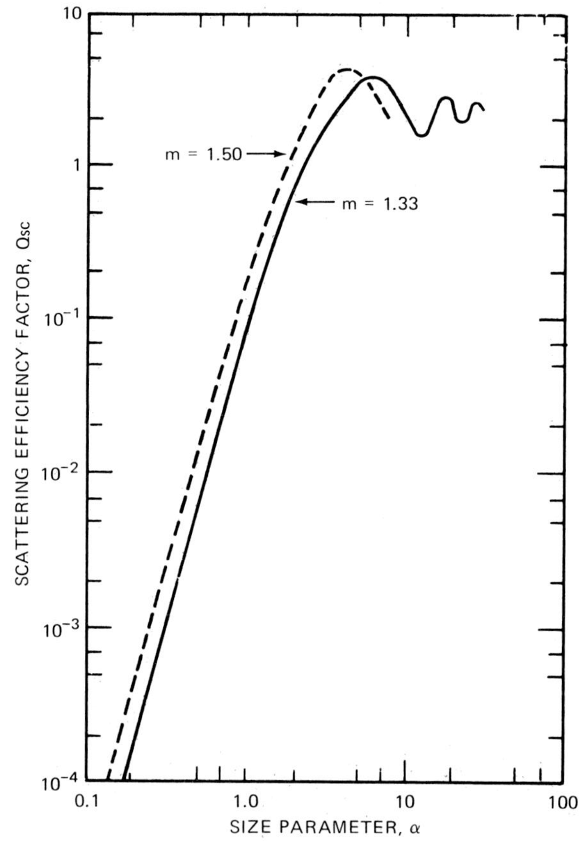

Fig. 20.10 Scattering efficiency factor as a function of size parameter and refractive index. (From Earl H. McCartney, *Optics of the Atmosphere*, John Wiley & Sons, New York, 1976.)

value of 2 as relative size becomes very large. At this value a particle refracts, reflects, and diffracts twice the radiant power incident on the geometric cross section (McCartney, 1976). White and Roberts (1977) found that sulfates and nitrates in the Los Angeles air basin scattered light more efficiently per unit mass concentration than other chemical fractions of the ambient aerosol. The high scattering efficiency of sulfates and the large stationary source emissions of sulfates led these authors to suggest that this source was comparable with the automobile in reducing visibility there.

The size of particles and the resulting light scattering is sensitive to the relative humidity of the air. When the relative humidity rises above 70%, water condenses on particles and enlarges them (Charlson, Waggoner, and Thielke, 1978). The composition of the particle affects the threshold relative humidity, above which water vapor condenses on the particle. The ratio of scattering coefficient at any relative humidity to that at 30% is plotted in Fig. 20.11.

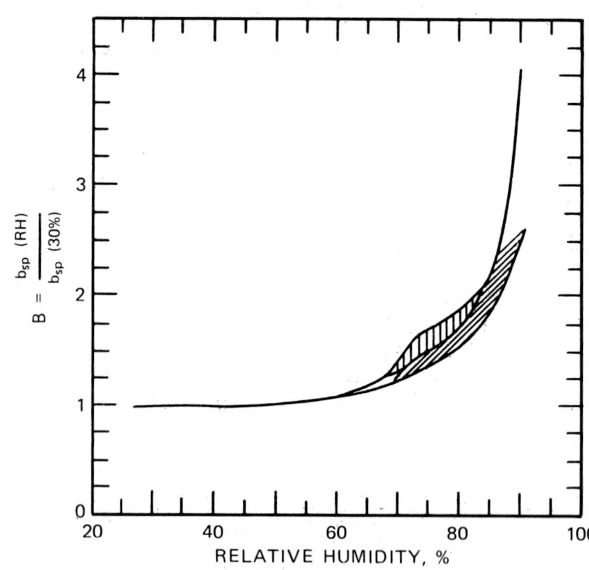

Fig. 20.11 Range of variability in humidogram data averaged by site; vertically hatched area includes strongly deliquescent aerosol at Point Reyes and Tyson. (Adapted from Charlson, Waggoner, and Thielke, 1980.)

*20-2.2.3 **Absorption.*** The absorption of light is the process by which the incident light at specific wavelengths is converted to internal energy of molecules (rotation, vibration, and electronic arrangement). A quantum of energy is absorbed for each discrete change in any of these forms of internal energy. The energy of a quantum is inversely proportional to the wavelength of the light.

The only absorption of enough consequence in the effluents released into the atmosphere from power production is that of nitrogen dioxide, NO_2. The absorptivity is the relative loss of incident light per unit length of absorbing path per unit concentration of pollutant. The absorptivity of NO_2 (and N_2O_4) as a function of wavelength of incident light is shown in Fig. 20.12, taken from Hall and Blacet (1952). The absorptivity is strongest in the blue. Also, there are many detailed absorption peaks in the overall curve, whose wavelengths correlate with specific changes in the internal energy of the NO_2 molecule. The strong NO_2 absorption of blue light causes plumes and urban hazes to appear brown. Simultaneous scattering of all wavelengths by particles and scattering of blue by air leads to a variety of brown, gray, and white colors, depending on the relative contribution of these processes.

Absorption of light by particles is attributed to their graphitic "soot" content (Rosen et al., 1979; Weiss et al., 1979). Faxvog and Roessler (1978) found that carbon particles were most effective in reducing visibility if their diameters were 15 to 50% of the wavelength of light. Roessler and Faxvog (1980) found that 85% of the acetylene smoke particle attenuation of 514 μm of light was caused by absorption and 15% was caused by scattering. Roessler and Faxvog (1981) found that absorbing aerosols increase the visual range computed for light objects viewed against the horizon sky and decrease the visual range for dark objects viewed against the horizon sky. The dependence of absorption on the wavelength of incident visible light is quite weak, as shown by the curves in Fig. 20.12 (Weiss et al., 1979). The effect of absorption in optical computations can be expressed in terms of an absorption coefficient, as plotted in Fig. 20.13.

The processes by which light is attenuated as it moves through air are the basic building blocks of visibility models and instruments. A discussion on how we construct

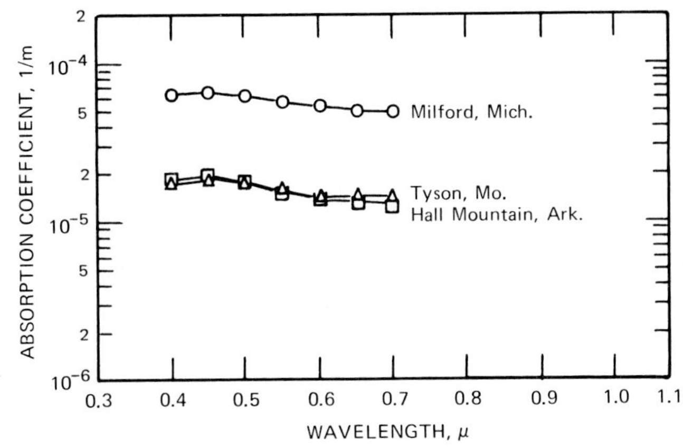

Fig. 20.13 Characteristic absorption wavelength dependence measurements. (From R. E. Weiss et al., 1979.)

complete models of visibility as a function of air pollution follows.

20-2.3 Modeling Visibility

Visibility models have been developed to predict the impact on visibility of various air-pollution sources (Latimer et al., 1978; Williams, Treiman, and Wecksung, 1980). These models "bootstrap" a visibility module on to some combination of modules describing emissions, transport, dispersion chemistry, and deposition (Fig. 20.14). These other modules will not be discussed here because they are covered in many of the preceding chapters.

The visibility module computes the NO_2 absorption and particulate scattering and absorption effects on light reaching the observer from targets, sky, and sight path elements. These radiance changes are then used to compute changes in apparent target spectral contrast, visual range, extinction coefficient, blue–red ratio, delta E, chromaticity coordinates, and color contrast. These radiance changes can also be combined with a photograph of the vista to create a new image of the vista (Williams, Treiman, and Wecksung, 1980).

A workbook that provides three levels of detail in modeling visibility has been prepared by Latimer and Ireson (1980). As one goes from the simplest to the most complex

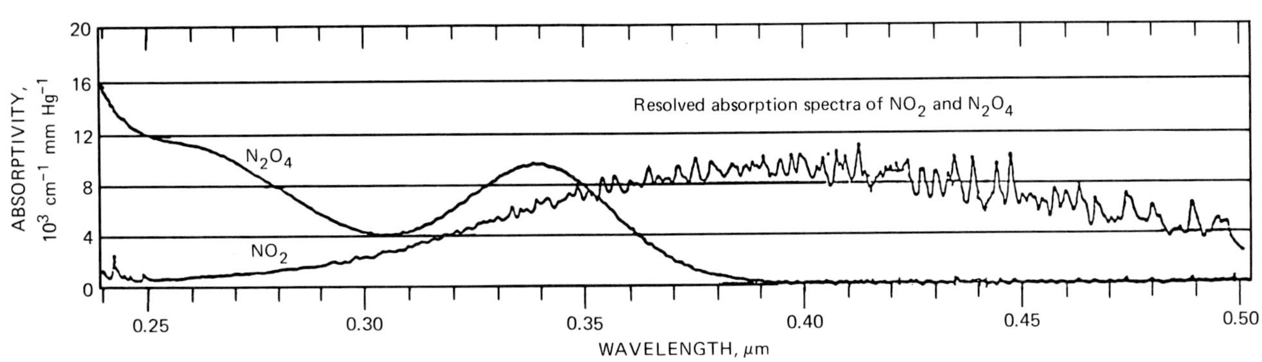

Fig. 20.12 Absorptivity of NO_2 and N_2O_4 vs. wavelength of incident light measured at 25°C. [From T. C. Hall, Jr., and F. E. Blacet, Separation of the Absorption Spectra of NO_2 and N_2O_4 in the Range of 2400–5000 Å, *J. Chem. Phys.*, 20(11): 1745-1749 (1952).]

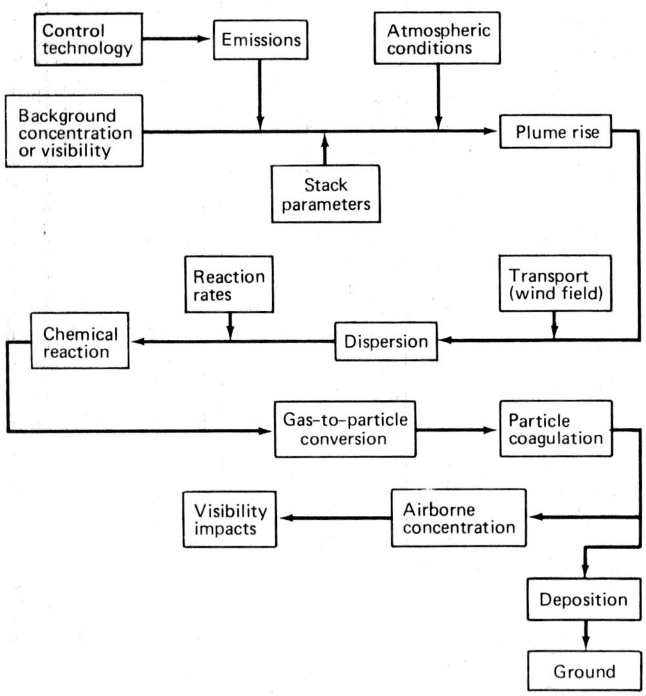

Fig. 20.14 Modules and their interrelationships.

level, more information is required on the size distribution and other properties of the particulate emissions, on the topography, and on the joint frequency distributions of wind speed, direction, stability, and other meteorological variables. At each level of analysis, tests are conducted to determine if there might be enough visibility impairment to warrant more refined analysis.

Following the approach of Latimer and Ireson (1980), the change in radiance, dN, in any increment, dr, of the sight path is given by

$$dN = -b_{ext}N\,dr + \frac{p(\theta)}{4\pi}\,b_{scat}F_s\,dr$$

where $p(\theta)$ is the scattering distribution for scattering angle, θ; F_s is the solar flux at the sight path increment [W/(m·μm)], and b_{scat} is the scattering coefficient caused by air molecules and particles (km^{-1}).

The first term on the right side of the equation is the loss of radiance by scattering and absorption within the sight path increment. The second term on the right is the radiance added by scattering of sunlight within the increment in the direction of the observer. This radiative transfer equation is used to derive the horizon sky radiance, ${}_sN$, in a homogeneous atmosphere,

$${}_sN = \frac{p(\theta)}{4\pi}\,\frac{b_{scat}}{b_{ext}}\,F_s$$

If a plume is located between an observer and the target, as shown in Fig. 20.15, then the radiance coming from the direction of the target is

$$\begin{aligned}{}_tN_r = {}_sN(1 - e^{-r_1 b_{ext}}) &+ \frac{p(\theta)}{4\pi}\,F_sW_p \\ &\times (1 - e^{-\tau_p})(e^{-r_1 b_{ext}}) \\ &+ {}_sN(1 - e^{-r_2 b_{ext}})(e^{-\tau_p})(e^{-r_1 b_{ext}}) \\ &+ {}_tN_0(e^{-r_2 b_{ext}})(e^{-\tau_p})(e^{-r_1 b_{ext}})\end{aligned}$$

The first term on the right is the radiance scattered toward the observer in path r_1. The second term is the light scattered toward the observer by the plume and then attenuated in path segment r_1. The third term is the

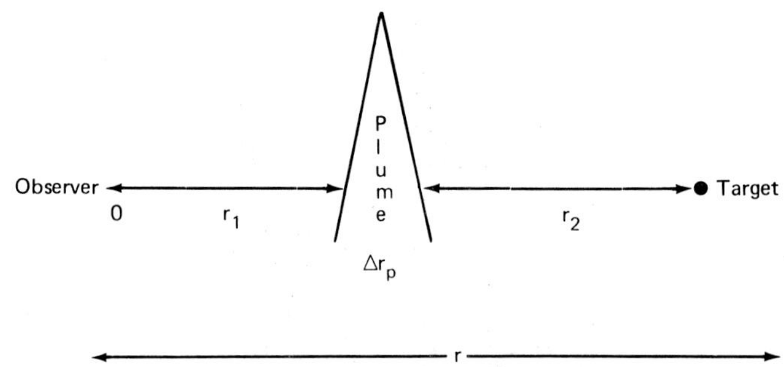

Fig. 20.15 Top-view geometry of plume and sight path.

radiance scattered toward the observer in segment r_2 and attenuated in the plume and in segment r_1. The fourth term is the inherent radiance of the target, attenuated by the plume and path segments r_1 and r_2. The plume albedo, W_p, is given by

$$W_p = \frac{\int_{plume} b_{scat}\,dr}{\int_{plume} b_{ext}\,dr}$$

and the plume optical thickness, τ_p, is given by

$$\tau_p = \int_{plume} b_{ext}\,dr$$

The contrast of a plume against the horizon sky is given by

$$C_p = \left\{\frac{[p(\theta)W]_p}{[p(\theta)W]_b} - 1\right\}(1 - e^{-\tau_p})(e^{-r_1 b_{ext}})$$

where $[p(\theta)W]_p$ is the product of scattering distribution and albedo of the plume and $[p(\theta)w]_b$ is the product of scattering distribution and albedo of the background air. The ratio of the products of scattering distribution and albedo determines whether the plume is brighter ($C_p > 0$) or darker ($C_p < 0$) than the horizon sky.

The change in contrast of a target caused by a plume is given by

$$\begin{aligned}\Delta C_r &= C_{rp} - C_r \\ &= -C_0(e^{-rb_{ext}})\left(1 - \frac{{}_sN}{{}_sN_p}\,e^{-\tau_p}\right)\end{aligned}$$

where C_{rp} is the target contrast through a plume; ${}_sN_p$ is the horizon sky radiance through a plume and is given by

$$
\begin{aligned}
{}_sN_p = {}_sN(1 - e^{-r_1 b_{ext}}) \\
+ \frac{p(\theta)}{4\pi} F_s W_p (1 - e^{-\tau_p})(e^{-r_1 b_{ext}}) \\
+ {}_sN(e^{-\tau_p})(e^{-r_1 b_{ext}})
\end{aligned}
$$

These equations for radiance, contrast, and contrast change are expressed in terms of plume optical thickness and atmospheric extinction coefficient. They can be expressed in terms of particle and NO_2 mass concentrations by choosing suitable particle scattering and NO_2 absorption coefficients per unit mass.

One test is applied to the change in apparent green ($\lambda = 0.550$ μm) contrast of a target. Latimer and Ireson (1980) suggest that, if this contrast change is less than 0.1, the visibility impairment is unlikely to be significant. The choice of 0.1 for this test is arbitrary. Under some conditions people can observe a contrast change of 0.02. The visibility regulations do not provide numerical values for visibility variables to be used in decision-making on new source review, but rely instead on the judgment of state officials and federal land managers.

Latimer and Ireson (1980) present a probabilistic approach to modeling visibility impairment which depends on such variables as wind speed, wind direction, atmospheric stability, background visibility, and ozone concentration. Ozone is important to NO to NO_2 conversion and to some gas-to-particle conversions.

Although it may have a low probability or frequency of occurrence, the worst-case conditions are usually used in at least one modeling run. The worst-case conditions for visibility impairment are not necessarily the same as for maximum ground-level concentrations and the associated potential health effects. A stable air mass with slow wind speeds can cause the appearance of the most obtrusive looking plume. It will affect the least field of view but will have the most psychological impact on an observer. In comparison, a moderate wind speed of 2 to 4 m/s throughout a 1-km-deep mixed layer (Chap. 5) will cause haze perceptible over a region extending 100 to 200 km from a source. On the other hand, maximum ground-level concentrations tend to occur under fumigation conditions or in a stagnant air mass.

Visibility models have a problem simulating the effects of clouds on radiative transfer. Further, the simple flat earth approximation in some radiative transfer models is inappropriate for clean atmospheres, although this assumption is acceptable for more optically thick, humid, hazy air.

In pristine air the modeling of visibility requires a spatial scale of hundreds of kilometers, and hence the transport and other modules must be appropriate to the same scale. On this scale the modules can ignore detailed topography but must account for the diurnal changes of mixing height and wind structure which are important in regional transport.

The modeling considered thus far covers the computation of physical variables. Malm et al. (1981a) proposed a model of human perception in which an incremental change, dP, is proportional to the perception, P, distance, r, and an incremental change in the extinction coefficient:

$$dP = -Prdb_{ext}$$

In Sec. 20-2.1 it was shown that

$$\bar{b}_{ext} = \frac{1}{r} \ln \left(\frac{C_0 \; {}_sN_0}{C_r \; {}_sN_r} \right)$$

For the perception of a target in a specific scene, r, C_0, and ${}_sN_0/{}_sN_r$ are constant. Therefore

$$\frac{db_{ext}}{dC_r} = \frac{-1}{rC_r}$$

The change in perception can be rewritten as

$$dP = \frac{PdC_r}{C_r}$$

Integrating this relation from an initial perception, P_0, and an initial (not inherent) contrast, C_{r0}, we get a linear relation for perception as a function of contrast,

$$P = \frac{P_0}{C_{r0}} C_r$$

This linear relation is discussed in the next section, which describes the field measurement of perception.

20-3 PERCEPTION OF VISIBILITY

People look into the distance and see mountains and other objects through an atmosphere with variable composition and lighting. Pollutants may be distributed uniformly enough so that a person cannot perceive boundaries, or the pollutants may be concentrated in a plume or ground-based layer. It is important to know the impact of air quality, sun angle, target distance, target color, and observer factors on the ability to see changes in visibility.

Studies have used a survey methodology of asking visitors to national parks to rate slides of selected vistas on a 1-to-10 scale (Malm, Leiker, and Molenar, 1980; Malm et al., 1981a; Malm, Kleine, and Kelley, 1980). Standard 2-in.-square slides were obtained from a regional visibility network photographic program (Malm et al., 1981b) operated by the Visibility Research Center of the John Muir Institute. The visitors participating in the surveys were read an introductory statement and shown selected slides to set up their 1-to-10 rating scales. They were shown slides of a few vistas with many different conditions of air quality, lighting, cloudiness, amount of snow on the target, and presence of elevated plumes and ground-level hazes. A control slide with some intermediate level of apparent contrast was shown at various times during the presentation of 48 or so test slides.

There is close agreement between the ratings of slides and the rating of the actual vista with the same visibility. This result was critical to making results from the survey technically credible. Stratifying the 1-to-10 ratings of the respondents revealed no effect of education, age, sex, or place of residence.

20-3.1 Perception of Regional Haze

Perception of regional or uniform haze depends on the vista composition and the amount of contrast in the features. Perception can be a linear relation with the apparent green contrast of a target if the vista is dominated by distant features. The view of distant mountains from a canyon rim with low terrain in between is an example of such vista composition. The linear relation can be seen in Fig. 20.16.

A vista dominated by large colorful foreground features will result in a nonlinear relation between perception and the apparent green contrast of a distant feature, as shown in Fig. 20.17. The amount of contrast in a vista depends on several factors, including sun angle, cloudiness, snow cover, and air quality. Contrast increases when air quality increases and when there are some cumulus clouds or snow in the scene. Contrast also increases when the illumination changes to add more color or texture to a scene. Figure 20.18 illustrates the increased perception that results from adding snow or illumination.

Malm, Leiker, and Molenar (1980) also showed the sensitivity of change in apparent green contrast as a function of target distance and fine particulate concentration (Fig. 20.19). For any specified fine particulate concentration, there is an optimum target distance to produce the maximum change in target contrast on the increase of fine particulate concentration by 2 $\mu g/m^3$.

20-3.2 Perception of Plumes and Layered Haze

Plumes and layered hazes have distinct borders. Malm, Kleine, and Kelley (1980) found that the perceived visual air quality drops like a step function as soon as the observer detects a plume viewed against the sky (Fig. 20.20). The slides of plumes were created by computer digital image processing, similar to that described by Williams, Treiman, and Wecksung (1980). This simulation technique used mathematical models of radiative transfer and dispersion to produce images of plumes of specified width, color, and intensity.

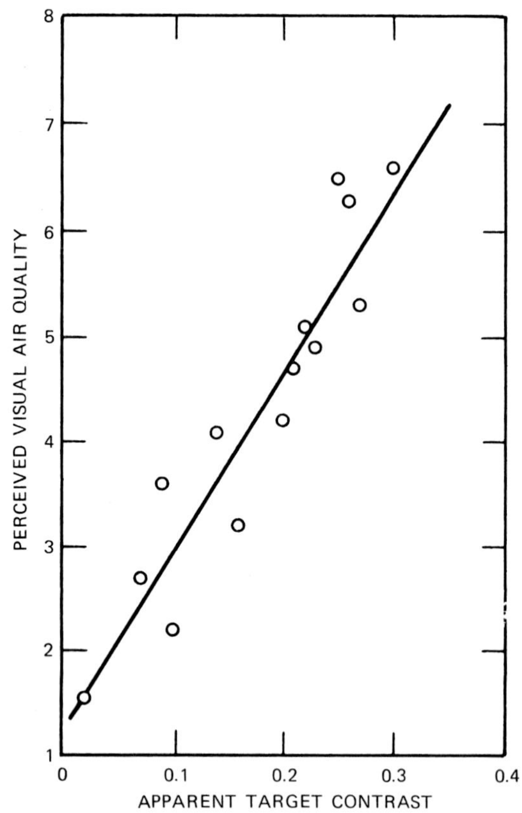

Fig. 20.16 Perceived visual air quality is plotted as a function of apparent contrast of Mount Trumbull as viewed from Hopi Point, Grand Canyon National Park (foreground features are absent). [From William Malm, Karen Kelley, John Molenar, and Terry Daniel, Human Perception of Visual Air Quality (Uniform Haze), *Atmos. Environ.*, 15(10/11): 1875-1890 (1981).]

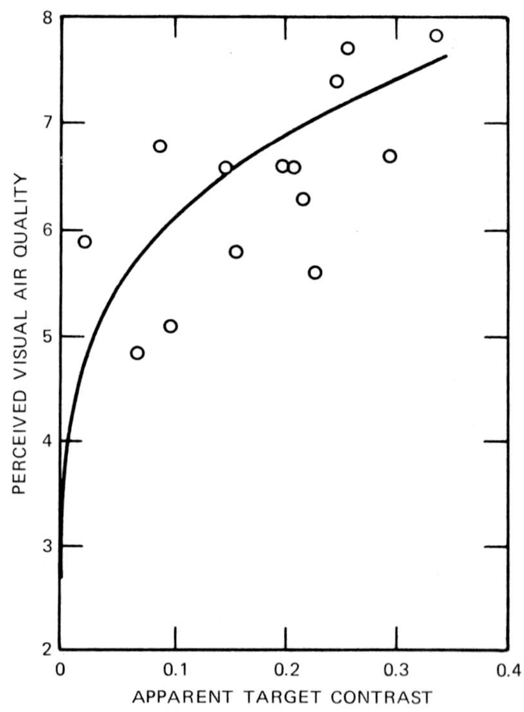

Fig. 20.17 Perceived visual air quality is plotted as a function of apparent contrast of Mount Trumbull (foreground features are present). [From William Malm, Karen Kelley, John Molenar, and Terry Daniel, Human Perception of Visual Air Quality (Uniform Haze), *Atmos. Environ.*, 15(10/11): 1875-1890 (1981).]

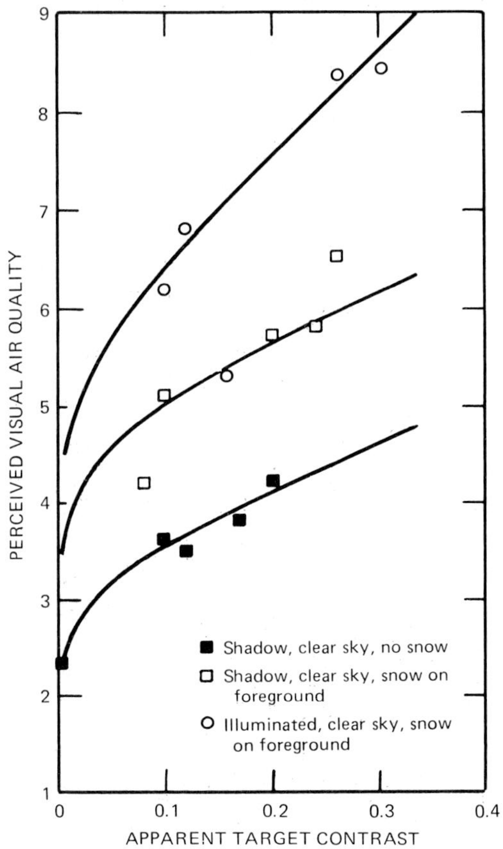

Fig. 20.18 Perceived visual air quality is plotted as a function of apparent contrast of Chuska Mountains as viewed from Mesa Verde National Park. [From William Malm, Karen Kelley, John Molenar, and Terry Daniel, Human Perception of Visual Air Quality (Uniform Haze), *Atmos. Environ.*, 15(10/11): 1875-1890 (1981).]

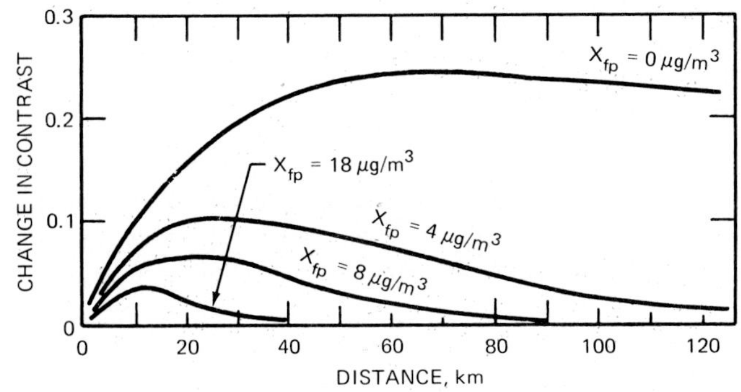

Fig. 20.19 A plot of contrast change resulting from an increase in fine particulate loading of 2 μg m^{-3} as a function of vista distance for initial fine particulate concentrations of 0, 4, 8, and 18 μg m^{-3}. [From William C. Malm, Karen K. Leiker, and John V. Molenar, Human Perception of Visual Air Quality, *J. Air Pollut. Control Assoc.*, 30(2): 122-131 (February 1980).]

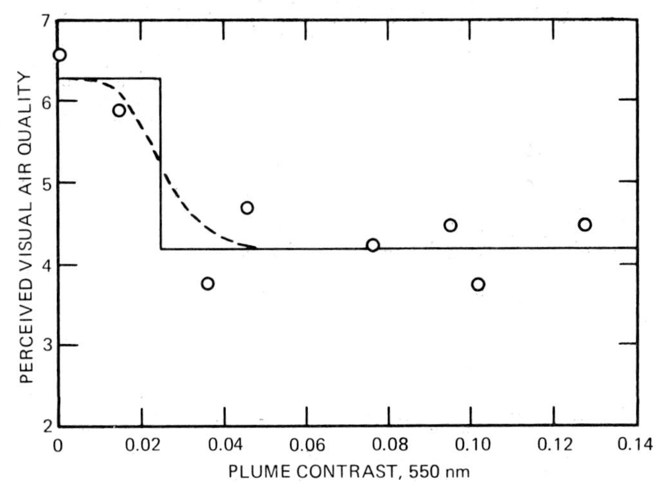

Fig. 20.20 Perceived visual air quality as a function of contrast for NO_2 plume. [From William Malm, Marvin Kleine, and Karen Kelley, Human Perception of Visual Air Quality (Layered Haze), in preparation (1982).]

20-4 INSTRUMENTATION FOR MEASURING VISIBILITY-RELATED VARIABLES

Visibility cannot be directly measured, but there are several classes of instruments that measure variables related to visibility. These variables were discussed in Sec. 20-2.1. Most of this discussion of instruments is taken from Malm and Walther (1980), which contains information on specific instruments developed since the 1800s.

There are several ways to classify visibility instrumentation; the classification here distinguishes between instruments that measure (1) the apparent spectral radiance and contrast of objects or their surroundings, (2) the light scattered from an air sample; (3) the light transmitted through an air sample; and (4) the physical properties of sampled aerosol.

Contrast measurements are further classified by the type of detector, including the unaided human eye, the eye aided by a telescope, a camera, and a photoelectric device used with a telescope. Scattering measurements are classified by the angle of scattering: integrated, backward, forward, and polar. The transmission measurements are classified by the source of light and by the type of detector.

Visibility depends on the properties of objects, the quality of the air along the sight path, the length and illumination of the path, the observation angle with respect to horizontal, and the properties of the instrument detector. Every visibility-related instrument has a detector, but not all instruments make measurements that include the effects of target properties, illumination of the path, and the observation angle.

Contrast-type instruments measure the amount of light reaching the detector from a target or its background. These instruments generally are called telephotometers and can directly measure the apparent spectral radiance needed to define color and its change. They should be called teleradiometers if they use narrow wavelength band-pass filters and called telephotometers if they use photopic filters that have a wavelength response like the human eye. The contrast of the target against its background can be calculated from the radiance measurements. Visual range can also be calculated after making a series of assumptions about the inherent contrast of the target, uniformity of the atmosphere along the sight path, and the angle of observation.

Teleradiometers measure light in a manner similar to the way it is detected by the human eye. Properties of the target, air quality, distance to the target, illumination of the sight path, and observation angle affect the measurement. A disadvantage of making such measurements with teleradiometers is that it is difficult to separate the different effects. This disadvantage is important if the goal is to isolate the effect of air quality on visibility.

Cameras can be classified as one subclass of teleradiometers, using photographic film as the detector. Technology is available to digitally measure the light in the picture in each of the three primary colors. Mathematical models of air pollution can be combined with the digitized image to create new images of the original scenes with uniformly reduced visual range or with the plume expected to result from a specified proposed source (Williams, Treiman, and Wecksung, 1980). This imaging technology can be tied to actual measurements of the original scene by using a teleradiometer on targets selected from the scene.

In addition to teleradiometers, human-eye observation is the other important contrast-type approach to measuring visibility. Human-eye observation from airport National Weather Service stations is the sole source of visibility data over many years. Unfortunately human-eye observations depend not only on all the effects previously mentioned but also on the effects of widely varying eyes and interpretation by different brains. Also, the natural targets available for viewing at different locations are not uniform in direction and distance from the observation sites, and thus a comparison of sites is made difficult. The selection of the most distant target sets the upper bound on the prevailing visibility reportable for each location.

The second major class of visibility-related instruments measures the light scattered from a small volume of air in specified directions. Scattering instruments measure a basic optical property of the air sample, the volume-scattering function, which is the proportion of incident light scattered in a specific direction per unit length of path per unit solid angle of scattered light. The volume-scattering function is independent of target properties, natural illumination of the atmosphere, and distance between the observer and the target. Many scattering instruments enclose the air sample and thus permit continuous day and night operation. Some unenclosed scattering instruments modulate the intensity of the light source to allow operation during daytime. Enclosed instruments can easily change the sampled air temperature, relative humidity, and the aerosol size distribution. Care must be taken that such change is not large enough to alter the measured scattering.

It is important to note that the transfer of light through the atmosphere depends on scattering and absorption. Most scattering instruments can measure only scattering. One exception is a polar nephelometer set up to measure the ellipticity of scattered light, which provides information on the complex part of the index of refraction and hence on absorption. Fortunately, scattering dominates absorption, especially in clean air. Although it is acceptable to neglect absorption in clean air, it is not acceptable in urban air or in the plumes of rurally located large-point sources like coal-fired power plants.

Scattering instruments measure the light scattered at various angles from the air sample. The choice of scattering angle allows the separation of instruments into integrating nephelometers, backscatter meters, forward-scatter meters, and polar nephelometers.

Integrating nephelometers measure the light scattered over a range of angles somewhat less than 0 to 180°; hence these instruments measure the sum of forward and backscattering. The instrument is calibrated to read out the total scattering coefficient of the air sample. This variable can be translated into the visual range of a black target only if the atmosphere is uniform over a distance as long as the visual range and if the target is viewed in a horizontal direction.

Backscatter meters measure the amount of light scattered at angles between 90 and 180°. The instruments that use a laser for the light source are usually called lidars. Other backscatter instruments use incandescent or spark lamps. Upward facing lidars have been used to measure the distance to horizontally stratified aerosol layers in the atmosphere. Except for the problem of eye safety, lidar could be used in the horizontal direction to measure the distance to plumes and other atmospheric nonuniformities. Lidar is capable of probing out to several tens of kilometers, but nonlaser backscatter instruments are usually limited to measuring visual ranges less than 20 km. No backscatter instrument is suitable for measuring the visual ranges of 100 to 300 km found in pristine areas.

Forward-scatter meters are similarly limited to measuring visual ranges less than 20 km, which is too low for use in pristine areas. These instruments measure the amount of light scattered from a collimated beam at angles between 0 and 90°. They are most commonly used for measuring visibility at airports and along stretches of highway where fog is a danger.

The polar nephelometer measures the light scattered from a collimated source in any specified direction. The volume-scattering function is usually measured at a number of scattering angles between 10 and 120°. The instrument can also measure the angle of polarization and ellipticity of scattered radiation, allowing the calculation of real and imaginary parts of the complex index of refraction. This information on polarization and ellipticity allows the calculation of light absorbed by the sampled air. The ability of the polar nephelometer to measure optical variables makes it a powerful research instrument, but it has not been automated nor offered inexpensively enough to be used widely.

Another type of scattering instrument measures sky radiation. Some of these instruments are used just like a teleradiometer, which measures the apparent spectral radiance reaching the detector from a small portion of the sky.

The measurement of sky radiation is used in calculating the visual range of targets viewed against a sky background.

Pyranometers are used to measure the downwelling irradiance, that is, the total amount of light coming from the sun and sky. The total sky irradiance can be measured without the sun irradiance if the instrument is used with an occulting disk to block the direct solar radiation. These total sky radiation instruments are not important to the study of visibility.

Transmission instruments measure the amount of light transmitted from a source to a receiver and thus permit the direct calculation of the average extinction coefficient of the air along the instrument path. The light lost along the path is scattered out of the path or absorbed by gas molecules and aerosols in the path. The path for transmission instruments is long in terms of the small volume measured by scattering instruments and short in comparison with the 40- to 100-km paths used with teleradiometers.

Transmission instruments do not include the effects of target properties and the natural illumination of the sight path. An exception is the pyrheliometer group that measures the apparent sun radiance. These measurements allow the calculation of the optical thickness of the total atmosphere along the path between the observer and the sun. Corrections for Rayleigh scattering and ozone absorption allow a calculation of aerosol scattering. Measurements at several different wavelengths provide some information on the vertical distribution of aerosol. Aerosol distribution, as a function of height above the earth's surface, is important to visibility and the proper interpretation of measurements made by teleradiometers.

Other transmission instruments called transmissometers use artificial light sources, including incandescent lamps, xenon spark lamps, and lasers. Transmissometers require the placement of either the receiver or reflectors at one end of a baseline and the transmitter at the other end. The fixed baseline does not allow the instrument to measure visibility-related variables in different directions easily. Laser transmissometers are sensitive to beam spreading by atmospheric turbulence, when they are used to measure transmission through clean air, and to beam steering. In addition, a laser transmissometer is limited to one wavelength. This wavelength is 0.633 μm for the commonly available helium–neon laser and is unrepresentative of the 550-μm peak sensitivity of the human eye. These instruments are not very portable and require considerable power for the light source.

Other methods and instruments are used to measure the size distribution, mass concentration, or number concentration of particulate matter. The Rayleigh theory of molecular scattering and the Mie theory of aerosol scattering allow a computation of scattering from measurements of the aerosol size distribution as a function of the atmospheric density. From these relationships, visual range and color change can be calculated, but not as precisely as with more direct measurements. No instrument measures visual range, but it can be directly calculated by measuring the target contrast, scattering coefficient, or extinction coefficient. These three variables are measured by teleradiometers, scattering instruments, and transmissometers, respectively. Only teleradiometers can measure target color.

Measurements of particle size distribution and composition help identify anthropogenic and natural sources of visibility-degrading particulate matter. These measurements do not account for other factors affecting visibility.

Some comparative advantages and disadvantages of multiwavelength teleradiometers, transmissometers, and integrating nephelometers are listed here.

Multiwavelength Teleradiometer

Measures: Sky or target radiance at various wavelengths.

Advantages:

1. Measurement of the spectral radiance of the sky or the target allows for the direct calculation of color variables; eliminates assumptions of ground reflectance, spatial variations of atmospheric aerosol, variations in volume scattering function, etc.
2. Is sensitive to spatial variation in aerosol scattering coefficients over hundreds of kilometers.
3. Measurement is sensitive to both atmospheric scattering and absorption.
4. Portable.

Disadvantages:

1. Measurement can be made only during daylight hours.

Transmissometer

Measures: Spatially averaged extinction coefficient.

Advantages:

1. Measurement includes absorption (by nitrogen dioxide, carbon, etc.) and scattering. scattering.
2. Is sensitive to spatial variations in extinction coefficient over approximately 10 km.
3. Measurement can be made day or night or under any cloud cover.

Disadvantages:

1. Is not sensitive to aerosol concentration gradients that might occur on a regional scale; for the clean atmospheres that exist in the western United States, a trans-

missometer would be closer to a "point" than to a "long path" measurement.
2. Not portable.

Additional comments: Should have capability to measure the extinction coefficient at more than one wavelength.

Integrating Nephelometer

Measures: Scattering coefficient at one point in space, typically ground level.

Advantages:
1. Measures atmospheric scattering coefficient day or night or under any cloud condition.
2. Semiportable.
3. Readily available.
4. Has been used in previous studies, and hence new data can be compared to an existing data base.

Disadvantages:
1. Does not measure absorption; consequently it will neglect effects of NO_2 and carbon.
2. Insensitive to spatial variations, both vertical and horizontal.
3. Measurement of the scattering coefficient at one point in the atmosphere cannot directly predict the impact air pollution has on visual range or color.

20-5 REGIONAL PATTERNS AND TRENDS

Trijonis and Yuan (1978a) used the observations from four urban and eight nonurban airports over the past 25 yr to show a significant deterioration of the generally good visibility in the Southwest. The non-Rayleigh extinction increased 20 to 70% and was correlated with the 10-yr sulfate concentration data from five National Air Surveillance Network sites. In a subsequent report, Trijonis and Yuan (1978b) use the same kind of human observations to show a deterioration of visibility in the Northeast, especially for nonurban locations and during the summer months. Trijonis (1980) found from the National Weather Service visibility data on California that there was general deterioration of visibility from 1950 to 1966 and general improvement from 1966 to 1975. Over the entire 1950–1975 period, the Los Angeles and San Francisco coastal areas improved slightly, whereas the San Joachim and Sacramento valleys deteriorated slightly.

Malm et al. (1981b) present measurements from summer 1978 to spring 1979 at 14 sites in southwestern national parks and monuments which show excellent visibility. Results from this network through spring 1980 (Walther and Newburn, 1981) show that winter has the highest visibility and spring the lowest.

20-6 CONCLUSIONS

Visibility in federally designated Class I areas must be protected under the Clean Air Act Amendments of 1977. The regulations require the prevention of future deterioration and the remedying of existing deterioration.

Visibility is an optical effect that is sensitive to small additions of particulate matter to clean air. Visibility also depends on the illumination of the sight path and target and on the composition of the target. Visibility can only be directly measured as a distance if there are targets located far enough away from the observer to result in a 2% contrast with the sky under the existing air quality.

Standard visual range is a convenient variable to describe visibility because it is expressed as a distance and attempts to account for the effects of target distance, target inherent contrast, and mean height of the sight path. There does not yet exist an approved reference method for computing target color.

Scattering of light by particulate matter usually dominates scattering by air molecules and absorption by NO_2 and carbon. Several kinds of instruments measure target contrast, scattering, or transmission, but no instruments can directly measure visibility or visual range.

Models describe visibility-related variables as functions of the emissions from specified sources, of the atmospheric transport, dispersion, and deposition of the emissions, and of the transfer of radiation from the sun through the atmosphere and along sight paths. Existing models do not account for the detailed effects of clouds on visual air quality, and most of the models are inaccurate when they are applied to very clean air.

One model predicts that human perception of a target is directly proportional to the target apparent contrast. Observations confirm this prediction if the target is in the background of a vista not dominated by foreground features.

The perceived visual air quality decreases rapidly as soon as a layered haze or plume is detected, after which it matters little whether the layered haze or plume is light or obvious in appearance.

Power production affects visibility by its emission of particles, sulfur dioxide, and nitrogen dioxide. Therefore this industry will have to continue its efforts to better understand and control its potential impact on visibility.

LIST OF SYMBOLS

a_0, a_1, a_2, a_3	Polynomial coefficients
b_{abs}	Absorption coefficient
b_{ext}	Extinction coefficient
$b_{ext,R}$	Rayleigh extinction coefficient
b_{scat}	Scattering coefficient

C_L	Apparent contrast of target, based on luminance
C_0	Inherent spectral contrast
C_p	Apparent spectral contrast of plume against horizon sky
C_r	Apparent spectral contrast
F_s	Solar flux at sight path increment
L^*	Value of color
${}_sL$	Apparent luminance of sky
${}_tL$	Apparent luminance of target
m	Refractive index of air at 10°C
N	Spectral radiance
N_r^*	Spectral path radiance over distance r
PVAQ	Perceived visual air quality
$p(\theta)$	Scattering distribution for scattering angle (θ)
r	(1) Distance between target and observation point (2) Particle radius
R'	Blue–red ratio
SVR	Standard visual range
N_h	Apparent spectral radiance of horizon
${}_sN_0$	Inherent spectral radiance of sky
N_p	Apparent spectral radiance of plume
${}_sN_r$	Apparent spectral radiance of sky
${}_tN_0$	Inherent spectral radiance of target
${}_tN_r$	Apparent spectral radiance of target
p	Photopic response function
P	Perception
T_r	Transmittance of light from the target to the observer
U^*,V^*,u,v,a^*,b^*,W^*	Functions of tristimulus values used in ΔE formulation
VR	Visual range
W^*	Lightness index
W_p	Albedo of plume
X,Y,Z	Tristimulus values
x,y,z	Chromaticity coordinates
$\bar{x}_\lambda,\bar{y}_\lambda,\bar{z}_\lambda$	Color mixture data for equienergy spectrum
Y_0	Tristimulus value of reference color, e.g., diffuse reflectance of sun from white card
${}_sY$	Tristimulus value of sky
${}_tY$	Tristimulus value of target
z_m	Altitude of sight path midpoint
z_0	Altitude of observation point
z_t	Altitude of target
α	Particle size parameter
ΔC	Delta contrast
ΔE	Color difference variable
λ	Wavelength
θ	Observation angle for scattering
τ_p	Optical thickness of plume
Ω	Solid angle

REFERENCES

Charlson, Robert J., Alan P. Waggoner, and John F. Thielke, 1978, *Visibility Protection for Class I Areas,* Report to Council on Environmental Quality.

Elterman, Louis, 1968, *UV, Visible, and Ir Attenuation for Altitudes to 50 km,* Report AD-671933, Air Force Cambridge Research Laboratories, NTIS.

Environmental Protection Agency, 1979, *Protecting Visibility: EPA Report to Congress,* Report EPA-450/5-79-008, NTIS, October.

Faxvog, Fred R., and David M. Roessler, 1978, Carbon Aerosol Visibility Versus Particle Size Distribution, *Appl. Opt.,* 17(16): 2612-2616 (August 15).

Hall, T. C., Jr., and F. E. Blacet, 1952, Separation of the Absorption Spectra of NO_2 and N_2O_4 in the Range of 2400–5000 A, *J. Chem. Phys.,* 20(11): 1745-1749.

Judd, Deane B., and Gunter Wyszecki, 1975, *Color in Business, Science, and Industry,* John Wiley & Sons, New York.

Land, Edwin H., 1977, The Retinex Theory of Color Vision, *Sci. Am.,* 237(6): 108-128 (December).

Latimer, Douglas A., Robert W. Bergstrom, Stanley R. Hayes, Nei-Kao Lui, John H. Seinfeld, Gary Z. Whitten, Michael A. Wojcik, and Martin J. Hillyer, 1978, *The Development of Mathematical Models for the Prediction of Anthropogenic Visibility Impairment,* Report EPA-450/3/78-110-a, Vol. I, Environmental Protection Agency, NTIS, September.

——, and Robert G. Ireson, 1980, *Workbook for Estimating Visibility Impairment,* Report EPA-450/4-80-031, Environmental Protection Agency, NTIS, November.

Malm, William, Karen Kelley, John Molenar, and Terry Daniel, 1981a, Human Perception of Visual Air Quality (Uniform Haze), *Atmos. Environ.,* 15(10/11): 1875-1890.

——, Marvin Kleine, and Karen Kelley, 1982, Human Perception of Visual Air Quality (Layered Haze), in preparation.

——, Karen K. Leiker, and John V. Molenar, 1980, Human Perception of Visual Air Quality, *J. Air Pollut. Control Assoc.,* 30(2): 122-131 (February).

——, and Eric G. Walther, 1980, A Review of Instruments Measuring Visibility-Related Variables, Report EPA-600/4-80-016, Environmental Protection Agency, NTIS, February.

——, Eric. G. Walther, Kenneth O'Dell, and Marvin Kleine, 1981b, Visibility in the Southwestern United States from Summer 1978 to Spring 1979, *Atmos. Environ.,* 15(10/11): 2031-2042.

McCartney, Earl H., 1976, *Optics of the Atmosphere,* John Wiley & Sons, New York.

Middleton, W. E. K., 1952, *Vision Through the Atmosphere,* University of Toronto Press.

Mie, G., 1908, A Contribution to the Optics of Turbid Media, Especially Colloidal Metallic Suspension (in German), *Ann. Phys.,* 25(4): 377-445.

Munsell Color Company, 1976, *Munsell Book of Color,* Baltimore, Md.

Roessler, David M., and Fred R. Faxvog, 1980, Photoacoustic Determination of Optical Absorption to Extinction Ratio in Aerosols, *Appl. Opt.,* 19(4): 578-581 (February 15).

——, 1981, Visibility in Absorbing Aerosols, *Atmos. Environ.,* 15: 151-155.

Rosen, N., D. D. A. Hansen, L. Gundel, and T. Novakov, 1979, *Identification of the Graphitic Carbon Component of Source and Ambient Particulates by Raman Spectroscopy and an Optical Attenuation Technique,* in Proceedings of Carbonaceous Particles in the Atmosphere, March 20–22, 1978, available from Lawrence Berkeley Laboratory, University of California.

Trijonis, John C., and Kung Yuan, 1978a, *Visibility in the Southwest,* Report EPA-600/3-79-039, Environmental Protection Agency, NTIS.

——, and Kung Yuan, 1978b, *Visibility in the Northeast,* Report EPA-600/3-78-075, Environmental Protection Agency, NTIS.

——, 1980, *Visibility in California,* Report TSC-PD-B612-3, Technology Service Corp.

Walther, Eric G., and Robert Mark Newburn, 1981, *Visibility Measured in the EPA/NPS Regional Network, June 1978 Through May 1980,* report to the Environmental Protection Agency from the Visibility Research Center of the John Muir Institute, University of Nevada, Las Vegas.

Weiss, R. E., A. P. Waggoner, R. J. Charlson, D. L. Thorsell, J. S. Hall, and L. A. Riley, 1979, *Studies of the Optical, Physical, and Chemical Properties of Light Absorbing Aerosols,* in Proceedings of Carbonaceous Particles in the Atmosphere, March 20–22, 1978, available from Lawrence Berkeley Laboratory, University of California.

White, Warren H., and P. T. Roberts, 1977, On the Nature and Origins of Visibility-Reducing Aerosols in the Los Angeles Air Basin, *Atmos. Environ.,* 11: 803-812.

Williams, Michael D., Evelyn Treiman, and Mona Wecksung, 1980, Plume Blight Visibility Modeling with a Simulated Photograph Technique, *J. Air Pollut. Control Assoc.,* 30(2): 131-134 (February).

Wyszecki, G., and W. S. Stiles, 1967, *Color Science,* John Wiley & Sons, New York.

Chapter 20 covers research prior to May 1981.

SUBJECT INDEX

NAME INDEX

☆U.S. GOVERNMENT PRINTING OFFICE: 1984 753 154 0